GENETICS
ANALYSIS & PRINCIPLES

Eighth Edition

ROBERT J. BROOKER

University of Minnesota

GENETICS: ANALYSIS & PRINCIPLES, EIGHTH EDITION

Published by McGraw Hill LLC, 1325 Avenue of the Americas, New York, NY 10019. Copyright ©2024 by McGraw Hill LLC. All rights reserved. Printed in the United States of America. Previous editions ©2021, 2018, and 2015. No part of this publication may be reproduced or distributed in any form or by any means, or stored in a database or retrieval system, without the prior written consent of McGraw Hill LLC, including, but not limited to, in any network or other electronic storage or transmission, or broadcast for distance learning.

Some ancillaries, including electronic and print components, may not be available to customers outside the United States.

This book is printed on acid-free paper.

1 2 3 4 5 6 7 8 9 LWI 28 27 26 25 24 23

ISBN 978-1-265-35079-6 (bound edition)
MHID 1-265-35079-5 (bound edition)
ISBN 978-1-266-82298-8 (loose-leaf edition)
MHID 1-266-82298-4 (loose-leaf edition)

Portfolio Manager: *Lora Neyens*
Product Developer: *Elizabeth M. Sievers, Joan Weber*
Marketing Manager: *Julie Ryer*
Content Project Managers: *Paula Patel, Brent Dela Cruz*
Manufacturing Project Manager: *Sandy Ludovissy*
Content Licensing Specialist: *Lori Hancock*
Cover Image: *Science Photo Library/Getty Images*
Compositor: *MPS Limited*

All credits appearing on page are considered to be an extension of the copyright page.

Library of Congress Cataloging-in-Publication Data

Names: Brooker, Robert J., author.
Title: Genetics : analysis & principles / Robert J. Brooker.
Description: Eighth edition. | New York, NY : McGraw Hill LLC, [2024] | Includes index.
Identifiers: LCCN 2023005503 (print) | LCCN 2023005504 (ebook) |
   ISBN 9781265350796 (hardcover ; alk. paper) | ISBN 1265350795 (hardcover ; alk. paper) |
   ISBN 9781266822988 (spiral bound ; alk. paper) | ISBN 1266822984 (spiral bound ; alk. paper) |
   ISBN 9781266823992 (ebook)
Subjects: LCSH: Genetics. | MESH: Genetic Phenomena | Genetic Techniques
Classification: LCC QH430 .B766 2024 (print) | LCC QH430 (ebook) |
   NLM QU450 | DDC 576.5—dc23/eng20230517
LC record available at https://lccn.loc.gov/2023005503
LC ebook record available at https://lccn.loc.gov/2023005504

The Internet addresses listed in the text were accurate at the time of publication. The inclusion of a website does not indicate an endorsement by the authors or McGraw Hill LLC, and McGraw Hill LLC does not guarantee the accuracy of the information presented at these sites.

BRIEF CONTENTS

TABLE OF CONTENTS

ABOUT THE AUTHOR

Robert J. Brooker is a professor in the Department of Genetics, Cell Biology, and Development and the Department of Biology Teaching and Learning at the University of Minnesota–Minneapolis. He received his B.A. in biology from Wittenberg University in 1978 and his Ph.D. in genetics from Yale University in 1983. At Harvard, he conducted postdoctoral studies on lactose permease, which is the product of the *lacY* gene of the *lac* operon. He continued to work on transporters at the University of Minnesota, with an emphasis on the structure, function, and regulation of iron transporters found in bacteria and *Caenorhabditis elegans*. At the University of Minnesota, he teaches undergraduate courses in biology and genetics.

©Robert Brooker

DEDICATION

To my wife, Deborah, and our children, Daniel, Nathan, and Sarah

PREFACE

In the 8th edition of *Genetics: Analysis & Principles*, the content has been updated to reflect current trends in the field. In addition, the presentation of the content has been improved in ways that foster active learning and inclusion. As an author, researcher, and teacher, I want a textbook that gets all students actively involved in learning genetics. To achieve this goal, I have worked with a talented team of editors, illustrators, and media specialists who have helped me to make the 8th edition of *Genetics: Analysis & Principles* a fun learning tool.

An effective textbook should strive to accomplish five goals:

1. The principles of diversity, inclusion, and equity should be followed.
2. The pedagogy should help students improve their critical-thinking skills. As described shortly, this goal is a key emphasis of this textbook.
3. Every effort must be made to make the content comprehensive, accurate, and up-to-date.
4. Students should be exposed to the techniques and methodologies they will need to become successful in their fields of work.
5. The pedagogy should inspire students.

The hard work that has gone into the 8th edition of *Genetics, Analysis & Principles* has been aimed at achieving these goals. Instead of being a collection of "facts and figures," *Genetics, Analysis & Principles* is intended to be an engaging and motivating textbook in which critical-thinking skills are strongly emphasized. We welcome your feedback so we can make future editions even better!

Diversity, Equity, and Inclusion

Diversity, equity, and inclusion (DEI) are goals that we all should strive to achieve. We need to change educational materials and practices that may not foster these goals. We all appreciate that understanding and improving DEI are complex tasks. However, developing an awareness of DEI concerns is not enough. We actually need to change things. The author is aware that textbook materials may conflict with issues of DEI. After revising the content of the material for the entire textbook, the author went through the entire textbook a second time with the sole goal of improving DEI. The following changes were made:

- A new section was added to Chapter 1, which overtly addresses the issue of Gender Identity (Section 1.5).
- In genetics, we sometimes speak of traits, such as red eyes in fruit flies, as being normal. The word "normal" has the potential to be exclusionary if its meaning connotes negatively. A new section in Chapter 1 (see Section 1.6) tackles this issue.
- When discussing researchers' work, the author has gone through the entire textbook and eliminated binary pronouns to make the discussion more gender-neutral.
- After reviewing the entire textbook, the author noted that a majority of references to researchers' nationalities were American or European. These references to nationalities have been eliminated from the 8th edition.
- The author has consciously added new examples to the textbook with an eye toward increasing diversity.

Improving DEI in this textbook is a work in progress. The author sincerely welcomes feedback regarding ways to alter the content in future editions or revisions to this edition in order to foster DEI.

Critical-thinking Skills

Critical thinking is the mental process of actively and skillfully conceptualizing, applying, analyzing, synthesizing, and/or evaluating information gathered from, or generated by, various experiences, including observation, experimentation, reflection, and/or communication. Critical thinking provides a guide to belief and action. A person who is skillful at critical thinking can draw reasonable conclusions from a collection of information, and can discriminate between useful and less useful details when solving problems and making decisions. If a primary goal for your students is to improve their critical-thinking skills, the 8th edition of *Genetics, Analysis & Principles* is an excellent choice. Some features that help develop critical thinking skills include:

1. **A feature called *Genetic TIPS* provides a consistent approach to help students solve problems in genetics.** Starting with Chapter 2, typically 2 to 3 of these TIPS are found in each chapter. They are directly aimed at helping students improve their critical-thinking skills.

GENETIC TIPS

*Problem solving is a skill that genetics students need to master. **Genetic TIPS** help students solve problems in genetics. This approach has three components: First, the student is made aware of the **T**opic at hand. Second, the question is evaluated with regard to the **I**nformation that is available to the student. Finally, the student is guided through a **P**roblem-Solving **S**trategy to tackle the question. **More Genetic TIPS** are presented at the end of the chapter, allowing for additional practice in strengthening problem-solving skills.*

GENETIC TIPS **THE QUESTION:** A cat is born with two X chromosomes and one Y chromosome. This is a rare event. One of the X chromosomes carries the black fur allele and the other carries the orange fur allele. Would you expect this cat to be a male or female? Would it be calico?

TOPIC: *What topics in genetics does this question address?* The topics are sex determination and X-chromosome inactivation.

INFORMATION: *What information do you know based on the question and your understanding of the topic?* From the question, you know the composition of sex chromosomes in a cat and the fur color alleles carried on the cat's X chromosomes. From your understanding of the topics, you may remember from Chapter 2 (refer back to Figure 2.15) that the Y chromosome determines maleness in mammals and that X-chromosome inactivation occurs and only one X chromosome remains active in somatic cells.

PROBLEM-SOLVING **S**TRATEGY: *Predict the outcome.* With regard to sex determination, you would predict that the cat is a male because the Y chromosome causes maleness. XCI occurs in a way that leaves only one X chromosome active in the cat's somatic cells. Because this cat has two X chromosomes, you

---

2. **In genetics and other fields of biology, a key component of critical thinking is the ability to follow the scientific method.** Many chapters in this textbook feature one or two experiments that are presented according to the scientific method. They begin with a hypothesis or goal and then describe the experiment in a series of steps. The data are then presented and discussed, so that students can understand how data analysis allows scientists to reach conclusions based on evidence.

EXPERIMENT 18A

The Genome of Tobacco Mosaic Virus Is Composed of RNA

We now know that bacteria, archaea, protists, fungi, plants, and animals all use DNA as their genetic material. In 1956, Alfred Gierer and Gerhard Schramm isolated RNA from tobacco mosaic virus (TMV), which infects plant cells. When this purified RNA was applied to plant tissue, the plants developed the same types of lesions that occurred when they were exposed to intact TMVs. Gierer and Schramm cor-

BACKGROUND OBSERVATIONS

Each experiment begins with a description of the information that led researchers to study a hypothesis-driven or discovery-based problem. Detailed information about the researchers and the experimental challenges they faced help students to understand actual research.

THE HYPOTHESIS OR THE GOAL

The student is given a possible explanation for the observed phenomenon that will be tested or the question researchers were hoping to answer. This section reinforces the scientific method and allows students to experience the process for themselves.

THE HYPOTHESIS

RNA is the genetic material of TMV.

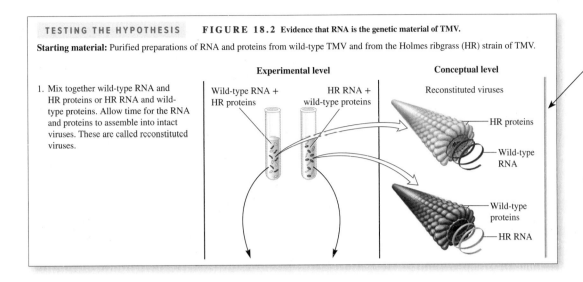

TESTING THE HYPOTHESIS   **FIGURE 18.2** Evidence that RNA is the genetic material of TMV.

Starting material: Purified preparations of RNA and proteins from wild-type TMV and from the Holmes ribgrass (HR) strain of TMV.

1. Mix together wild-type RNA and HR proteins or HR RNA and wild-type proteins. Allow time for the RNA and proteins to assemble into intact viruses. These are called reconstituted viruses.

TESTING THE HYPOTHESIS OR ACHIEVING THE GOAL

This section illustrates the experimental process, including the actual steps followed by scientists to test their hypothesis or study a question. Science comes alive for students with this detailed look at experimentation.

THE DATA

Actual data from the original research paper help students understand how real-life research results are reported. Each experiment's results are discussed in the context of the larger genetic principle to help students understand the implications and importance of the research.

THE DATA

Composition of reconstituted virus placed on tobacco leaves	Lesions on tobacco leaves	Amino acids found in newly made viral proteins following infection:	
		Methionine	Histidine
Wild-type RNA and HR protein	Like wild-type TMV	No	No
HR RNA and wild-type protein	Like HR TMV	Yes	Yes

Source: Fraenkel-Conrat, Heinz, and Singer, B., "Virus reconstitution: II. Combination of protein and nucleic acid from different strains," *Biochimica et Biophysica Acta*, vol. 24, January 17, 1957, 540–548.

INTERPRETING THE DATA

This discussion, which examines whether the experimental data supported or disproved the hypothesis or provided new information to propose a hypothesis, gives students an appreciation for scientific interpretation.

INTERPRETING THE DATA

As seen in the data, the outcome of infection depended on the RNA that was found in the reconstituted virus but not on the capsid protein. If wild-type RNA was used, the leaves developed lesions that were typical of wild-type TMV and the capsid proteins of newly made viruses lacked methionine or histidine. In contrast,

3. **"Genes → Traits"** These descriptions have been added to many figure legends to help students relate the concepts they have learned in molecular genetics with the traits that occur at the level of a whole organism their knowledge in new ways.

GENES → TRAITS

Because genetics is such a broad discipline ranging from the molecular level to populations, many students have trouble connecting the concepts they learn in molecular genetics with the traits that occur at the level of an organism. To make this connection more meaningful, certain figures have a "Genes→Traits" feature that reminds students that molecular and cellular phenomena ultimately lead to traits observed in organisms.

FIGURE 19.5 **The effects of a somatic mutation.** A mutation during embryonic development has caused this sheep to have a black spot on its side.

Genes→Traits This sheep has a patch of black hair on its side because a somatic mutation that occurred in a single cell during embryonic development caused pigmentation of the hair. This cell continued to divide, producing a patch of black hair.

©Robert Brooker

4. **Interactive Exercises allow students to make their own choices and predict outcomes.** Many of these exercises are focused on inheritance patterns and human genetic diseases (e.g., see Chapters 3 and 4 and 24). In addition, interactive exercises are found in the chapters in Units III and V, which largely focus on molecular genetics. These types of exercises engage students in the learning process. The interactive exercises are found in the eBook, and the corresponding material in the chapter is indicated with an Interactive Exercise icon.

5. **Our media specialists have created over 50 animations for a variety of genetic processes.** These animations were made specifically for this textbook and use the art from the textbook. The animations make many of the figures in the textbook literally "come to life." The animations are found in the eBook, and the corresponding material in the chapter is indicated with an Online Animation icon.

Flipping the Classroom

A trend in science education is the phenomenon that is sometimes called "flipping the classroom." This phrase refers to the idea that some of the activities that used to be done in class are now done out of class, and vice versa. For example, instead of spending the entire class time lecturing on textbook and other materials, some of the class time is spent engaging students in

various activities, such as problem solving, working through case studies, and designing experiments. This approach is also known as active learning.

For many instructors, the classroom has become more learner-centered rather than teacher-centered. A learner-centered classroom provides a rich environment in which students can interact with each other and with their instructors. Instructors and fellow students often provide formative assessment—immediate feedback that helps students understand if their learning is on the right track.

What are some advantages of active learning? Educational studies reveal that active learning usually promotes greater learning gains. In addition, active learning often focuses on skill development rather than the memorization of facts that are easily forgotten. Students become trained to "think like scientists" and to develop problem-solving skills that enable them to apply scientific reasoning. In other words, active learning fosters critical thinking. Another advantage is student motivation. Active-learning environments are inspiring and fun!

A common concern among instructors who are beginning to try out active learning is that they think they will be teaching their students less material. However, this may not be the case. Although students may be provided with online lectures, "flipping the classroom" typically gives students more responsibility for understanding the textbook material on their own.

Along these lines, the 8th edition of *Genetics, Analysis & Principles* is intended to provide students with a resource that can be effectively used out of the classroom. In particular, when students are expected to learn textbook material on their own, it is imperative that they are given formative assessment on a regular basis so they can gauge whether or not they are mastering the material. Formative assessment is a major feature of this textbook and is bolstered by McGraw Hill Connect®—a state-of-the art digital assignment and assessment platform. In this edition of *Genetics, Analysis & Principles,* formative assessment is provided in multiple ways:

1. Answers to students are provided for: (1) the multiple-choice questions throughout the chapter; (2) the Concept Check questions following the figure legends; and (3) the even-numbered end-of-chapter questions.
2. As described later, McGraw Hill **SmartBook 2.0** is an adaptive learning tool available in Connect that guides students through the textbook material. As they move through each chapter, students are asked questions to determine if they have mastered the material. If they get a question wrong, they can highlight the material that they need to review.
3. The set of learning outcomes at the beginning of each section provides a road map for students to appreciate what they should be learning. Likewise, the summaries at the end of each chapter reinforce the key concepts.

HOW WE EVALUATED YOUR NEEDS

Organization: Mendel First Versus Molecular First

In surveying many genetics instructors, it became apparent that most people fall into two camps: **Mendel first** versus **molecular first.** I have taught genetics both ways. As a teaching tool, this textbook has been written with these different teaching strategies in mind. The organization and content lend themselves to various teaching formats.

Chapters 2 through 8 are largely inheritance chapters, whereas Chapters 27 though 29 examine population, quantitative, and evolutionary genetics. The bulk of the molecular genetics is found in Chapters 9 through 19, although I have tried to weave a fair amount of molecular genetics into Chapters 2 through 8 as well. The information in Chapters 9 through 19 does not assume that a student has already covered Chapters 2 through 8. Actually, each chapter is written with the perspective that instructors may want to vary the order of their chapters to fit their students' needs.

For those who like to discuss inheritance patterns first, a common strategy would be to cover Chapters 1 through 8 first, and then possibly 27 through 29. (However, many instructors like to cover quantitative and population genetics at the end. Either way works fine.) The more molecular and technical aspects of genetics would then be covered using Chapters 9 through 19. Alternatively, if you like the "molecular first" approach, you would probably cover Chapter 1, then skip to Chapters 9 through 19, then return to Chapters 2 through 8, and then cover Chapters 27 through 29 at the end of the course. This textbook was written in such a way that either strategy works well.

Accuracy

Both the publisher and author acknowledge that inaccuracies can be a source of frustration for both the instructor and students. Therefore, throughout the writing and production of this textbook we have worked very hard to catch and correct errors during each phase of development and production.

Each chapter has been reviewed by faculty members who teach the course or conduct research in genetics or both. In addition, a developmental editor has gone through the material to check for accuracy in art and consistency between the text and art. When the problem sets were first developed, we had a team of students work through all of the problems and one developmental editor also checked them. The author personally checked every question and answer when the chapters were completed for this edition.

Writing Style

Motivation in learning often stems from enjoyment. If you enjoy what you're reading, you are more likely to spend longer amounts of time with it and focus your attention more sharply. The writing style of this book is meant to be interesting, down to earth, and easy to follow. Each section of every chapter begins with an overview of the contents of that section, often with a table or figure that summarizes the broad points. The section then examines how those broad points were discovered experimentally, and also explains many of the finer scientific details. Important terms appear in the text in a boldface font. These terms are also listed at the end of the chapter and defined in the glossary.

There are various ways to make a genetics book interesting and inspiring. The subject matter itself is pretty amazing, so it's not difficult to build on that. In addition to describing the concepts and experiments in ways that motivate students, it is important to draw on examples that bring the concepts to life. In a genetics book, many of these examples come from the medical realm. This textbook contains lots of examples of human diseases that convey some of the underlying principles of genetics. Students often say they remember certain genetic concepts because they remember how defects in certain genes can cause disease. For example, defects in DNA repair genes cause a higher predisposition to develop cancer. In addition, I have tried to be evenhanded in providing examples from the microbial and plant world. Finally, students are often interested in applications of genetics that affect their everyday lives. Because we frequently hear about genetics in the news, it's inspiring for students to learn the underlying basis for such technologies. Chapters 20 through 23 are devoted to genetic technologies, and applications of these and other technologies are found throughout this textbook. By the end of their genetics course, students should come away with a greater appreciation for the influence of genetics in their lives.

SIGNIFICANT CONTENT CHANGES IN THE 8TH EDITION

Due to the rapid pace of genetic discoveries and their relevance in human affairs, the 8th edition of this textbook has undergone some major changes. These include the following:

- *Impact of chromosome conformation capture methods on our understanding of eukaryotic chromosome structure and gene transcription.* Although many methods have contributed to our molecular understanding of chromosome structure and gene transcription, the method of chromosome conformation capture (3C method) has provided recent and unprecedented insight in these two areas. The method itself is described in Figure 10.22, and information discovered using this method has led to many new or revised figures presented in Chapters 10, 12, 15, and 16.
- *Recent knowledge regarding archaea.* Past editions of this textbook have largely focused on the domains Bacteria and Eukaryotes. In the 8th edition, new information that compares these two domains with the Archaea has been added to Chapters 10, 12, 13, and 15.
- *Genetic basis of medical technologies.* Two new sections that pertain to medical technologies have been added to the 8th edition. These are Section 21.2 on vaccines and Section 25.5 on cancer therapeutics.
- *Updates to genomic and bioinformatic methods.* Genomics and bioinformatics are perhaps the most rapidly changing areas in genetics. Much of Chapter 22 has been revised to reflect current approaches to genome sequencing, including a new section that provides an overview of genome sequencing (see Section 22.5). Furthermore, the topic of bioinformatics has been expanded into three sections in Chapter 23.
- *Expansion of coverage of population and evolutionary genetics.* The advent of molecular genetics continues to influence the fields of population and evolutionary genetics. In Chapter 27, the topic of natural selection has been expanded to two sections, and the coverage of the origin of species in Section 29.1 in the 7th edition has been expanded into three sections.

In addition to these major additions to the 8th edition, specific changes to each chapter are outlined next.

Examples of Specific Content Changes to Individual Chapters in the 8th Edition

- Chapter 1, **Overview of Genetics:** Two new sections have been added with the goal of fostering greater inclusivity in the discussion of genetics. These are Section 1.5 (Gender Identity Versus Sex) and Section 1.6 (The Meaning of Normal in Genetics).
- Chapter 2, **Reproduction and Chromosome Transmission:** Based on reviewer feedback, the order of Chapters 2 and 3

has been swapped. The chapter on reproduction and chromosome transmission now precedes the chapter on Mendelian inheritance. Figure 2.10 has been replaced with a new figure in which part (a) emphasizes the structure of the synaptonemal complex, and part (b) shows cohesion between sister chromatids in a bivalent after crossing over has occured. In Section 2.6, the topic of sex determination has been expanded in the areas of environmental effects on sex determination in animals and sex determination in dioecious plants. This section includes two new figures (see Figures 2.16 and 2.17).

- Chapter 3, **Mendelian Inheritance:** As mentioned, this chapter now follows the chapter on reproduction and chromosome transmission. The discussion of the chromosome theory of inheritance has been moved to this chapter and follows the sections that describe Mendel's laws of inheritance. Figure 3.15 regarding pedigree analysis has been revised to reflect only phenotype, which coincides with the pedigrees in the problem sets at the end of the chapter. The explanation of the chi square test has been changed from four steps to seven shorter steps to make it easier for students to understand how to calculate a chi square value and interpret its meaning.
- Chapter 4, **Extensions of Mendelian Inheritance:** The notation of alleles that follow a sex-influenced pattern of inheritance have been changed to a superscript notation to clarify that either allele can be dominant in a heterozygote, depending on the sex. A new subsection has been added that describes how gene modification plays a role in the color of parakeet feathers (see Table 4.4).
- Chapter 5, **Non-Mendelian Inheritance:** A new figure illustrates three cellular mechanisms that result in maternal inheritance (see Figure 5.16). A description of a new reproductive technology called three parent babies, which is used to avoid mitochondrial diseases, has been added.
- Chapter 6, **Genetic Linkage and Mapping in Eukaryotes:** The section on genetic mapping has been revised to emphasize that the pattern of allele linkage is deduced from the true-breeding P generation and that F_2 recombinants are produced by a crossover. Consistent with the revision in Chapter 3, an application of the chi square test to evaluate linkage has been changed from four steps to seven shorter steps to make it easier for students to understand how to calculate a chi square value and interpret its meaning.
- Chapter 7, **Genetic Transfer and Mapping in Bacteria:** The section on bacterial transduction has been simplified by eliminating the discussion of cotransduction, which is not commonly used.
- Chapter 8, **Variation in Chromosome Structure and Number:** In the discussion of gene families, a new table has been added that shows the oxygen affinity of the various forms of hemoglobin that are produced during different stages of development (see Table 8.1). A new subsection has been added that describes the microscopic and molecular methods that are

used to identify duplications and deletions (see Table 8.2). A new figure has been added showing an example of an allodiploid, a liger, which is a cross between a lion and a tiger (see Figure 8.25).

- Chapter 9, **Molecular Structure of DNA and RNA:** The data for the Hershey and Chase experiment has been added (see Figure 9.3).
- Chapter 10, **Molecular Structure of Chromosomes and Transposable Elements:** The material on bacterial chromosomes has been expanded to include chromosomes in archaea, including a new figure (see Figure 10.4). Section 10.6 (Structure of Eukaryotic Chromosomes in Nondividing Cells) has been largely reorganized. Four new figures have been added to this section that depict the loop extrusion model for how loops are formed in eukaryotic chromatin (see Figure 10.21); the general strategy of chromosome conformation capture (3C) methods (see Figure 10.22); the structure of topologically associating domains (TADS) (see Figure 10.23); and the four levels of chromosome organization and structure in a cell nucleus (see Figure 10.26). In Section 10.7, a new figure has been added that shows the structure of radial loop arrays in metaphase chromosomes (see Figure 10.29).
- Chapter 11, **DNA Replication:** The discussion regarding origins of replication in eukaryotes has been updated, including a new figure with an illustration of a G-quadruplex (see Figure 11.20). Also, new information has been added regarding the initiation of DNA replication in archaea.
- Chapter 12, **Gene Transcription and RNA Modification:** The coverage of the formation of the preinitiation complex in eukaryotes has been revised to better show how general transcription factors and mediator interact with RNA polymerase II (see Figure 12.14). A new figure describes the processing of tRNA molecules in eukaryotes (see Figure 12.17). New information has been added that compares the domain Archaea to both Bacteria and Eukaryotes with regard to gene transcription and RNA modification (see Table 12.5).
- Chapter 13, **Translation of mRNA:** A new subsection describes how changes in expression of translation initiation factors can lead to various types of cancer in humans (see Table 13.7). Analogous to Chapter 12, new information has been added that compares the domain Archaea to both Bacteria and Eukaryotes with regard to translation (see Table 13.9).
- Chapter 14, **Gene Regulation in Bacteria:** The topic of diauxic growth has been expanded, including a new graph (see Figure 14.8). The subsection on riboswitches has been updated to show that a riboswitch has an aptamer domain and an expression platform; Figures 14.17 and 14.18 have been revised.
- Chapter 15, **Gene Regulation in Eukaryotes I: General Features of Transcriptional Regulation:** This chapter has undergone a major revision in the 8th edition. Figures 15.2

and 15.10 have been revised to show that a loop must usually form so that an enhancer can be brought close to a core promoter. A new figure has been added that depicts the elongation phase of transcription in greater detail, including CTD phosphorylation, promoter escape, and proximal promoter pausing (see Figure 15.11). A new subsection describes how gene repression can occur. Also, a new section compares transcriptional regulation among bacteria, archaea, and eukaryotes (see Section 15.5).

- Chapter 16, **Gene Regulation in Eukaryotes II: Epigenetics:** The topic of liquid-liquid phase separation of heterochromatin is now described (see Figure 16.6). A new figure shows how heterochromatin structure is maintained after DNA replication (see Figure 16.9). A new subsection has been added that describes the function of pioneer factors; it includes a new figure (see Figure 16.12).
- Chapter 17, **Non-coding RNAs:** New discussion focuses on the importance of miRNAs in plants and includes two new figures and one new table that show the effects of miRNAs on flower phenotypes (see Figure 17.5), plant development (see Figure 17.8), and plant health (see Table 17.3). The topic of snoRNAs, which are involved with rRNA modifications, is now covered in its own section (see Section 17.4).
- Chapter 18, **Genetics of Viruses:** New information on the coronavirus has been added, including a description of different types of variants and a new micrograph (see Figure 18.5).
- Chapter 19, **Gene Mutation, DNA Repair, and Recombination:** Figure 19.5 provides a new example of a somatic mutation.
- Chapter 20, **Molecular Technologies:** Coverage of the method of Gibson assembly, which is used to link multiple DNA fragments, has been added (see Figure 20.5). The use of dead Cas9 to enhance transcription is described (see Figure 20.14). A new subsection discusses the ethical considerations of gene editing.
- Chapter 21, **Biotechnology:** Section 21.2 is a new section on the topic of vaccines, including a new figure on mRNA vaccines (see Figure 21.2). The information on transgenic crop usage has been updated (see Figure 21.13b).
- Chapter 22, **Genomics I: Analysis of DNA:** Discussion of the newer method of primer walking has been added. It is presented in Section 22.4, along with the traditional method of chromosome walking (see Figure 22.7). The topic of genome sequencing has been separated into two sections. Section 22.5 discusses de novo genome assembly and compares short- and long-read sequencing methods. This section also includes a new figure on Illumina sequencing and an updated table on popular DNA sequencing technologies (see Figure 22.9, Table 22.2). Section 22.6 focuses on the results of genome-sequencing projects.
- Chapter 23, **Genomics II: Functional Genomics, Proteomics, and Bioinformatics:** The section on bioinformatics has been separated into three sections

(Overview of Computer Analyses and Gene Prediction; Databases; and Homology). A new table has been added that describes databases that are focused on clinical aspects of genomics (see Table 23.5).

- Chapter 24, **Medical Genetics:** Noninvasive prenatal testing (NIPT) has been added to Section 24.3 (Genetic Testing and Screening). The discussion of human gene therapy has been updated, and the text covers various possible approaches, including CRISPR-Cas technology.
- Chapter 25, **Genetic Basis of Cancer:** Section 25.5 is a new section on cancer therapeutics, which includes a new table and a new figure (see Table 25.7 and Figure 25.10).
- Chapter 26, **Developmental Genetics:** The topic of Notch signaling has been added, and the coverage includes a new illustration (see Figure 26.3).
- Chapter 27, **Population Genetics:** A general selection model has been added to Section 27.3 (Overview of Natural Selection). The model is used to calculate changes in genotype and allele frequencies due to directional selection (see Tables 27.2, 27.4). The topic of natural selection is now divided into two sections: Section 27.3 (Overview of Natural Selection) and Section 27.4 (Patterns of Natural Selection). A new subsection has been added that describes how genome-wide selection scans can provide insight regarding the effects of natural selection (see Table 27.3).
- Chapter 28, **Complex and Quantitative Traits:** Rather than showing photographs, Figure 28.10 now presents a drawing that better illustrates the way in which the wild mustard plant was subjected to selective breaking to produce six popular agricultural strains.
- Chapter 29, **Evolutionary Genetics:** Section 29.1 in the 7th edition has been separated into three sections. The chapter contains a new subsection on hybrid zones, which includes three new figures (see Figures 29.4–29.6). The discussion of horizontal gene transfer has been updated, and Figure 29.13 has been revised.

Suggestions Welcome!

It seems very appropriate to use the word *evolution* to describe the continued development of this textbook. I welcome any and all comments. The refinement of any science textbook requires input from instructors and their students. These include comments regarding writing, illustrations, supplements, factual content, and topics that may need greater or less emphasis. You are invited to contact me at:

Dr. Rob Brooker
Dept. of Genetics, Cell Biology, and Development
Dept. of Biology Teaching and Learning
University of Minnesota
6-160 Jackson Hall
321 Church St.
Minneapolis, MN 55455
brook005@umn.edu
612-624-3053

ACKNOWLEDGMENTS

The production of a textbook is truly a collaborative effort, and I am deeply indebted to many people. All eight editions of this textbook went through multiple rounds of rigorous revision that involved the input of faculty, students, editors, and educational and media specialists. Their collective contributions are reflected in the final outcome.

Deborah Brooker (Freelance Developmental Editor) meticulously read the new material, analyzed every figure, and offered extensive feedback. Her attention to detail in this edition and previous editions has profoundly contributed to the accuracy and clarity of this textbook. I would also like to thank Jane Hoover (Freelance Copy Editor) for understanding the material and working extremely hard to improve the text's clarity. Her efforts are truly appreciated. She is at the top of the list of copy editors.

I would also like to acknowledge the many people at McGraw Hill whose skills and insights are amazing. My highest praise goes to Elizabeth Sievers (Senior Product Developer), who carefully checks all aspects of textbook development and makes sure that all of the pieces of the puzzle are in place. I am also grateful to Lora Neyens (Portfolio Manager) for overseeing this project. I would like to thank other people at McGraw Hill who have played key roles in producing the actual book and the supplements that go along with it. In particular, Paula Patel (Senior Content Project Manager) has done a superb job of managing the components that need to be assembled to produce a book. I would also like to thank Lori Hancock (Content Licensing Specialist), who acted as an interface between me and the photo company. Finally, I would like to thank Julie Ryer (Marketing Manager), whose major efforts begin when the eighth edition comes out!

I would also like to extend my thanks to everyone at MPS Limited who worked with great care in the paging of the book, making sure that the figures and relevant text are as close to each other as possible. Likewise, the people at Photo Affairs, Inc. have done a great job of locating many of the photographs that have been used in this edition.

Finally, I want to thank the many scientists who reviewed the chapters of this textbook. Their broad insights and constructive suggestions were an overriding factor that shaped its final content and organization. I am truly grateful for their time and compassion.

REVIEWERS

Amy Abdulovic-Cui, *Augusta University*
Peter Barrett, *Xavier University of Louisiana*
Anna L. Bass, *University of New England*
Grant Bledsoe, *University of Missouri– Kansas City*
Henry Chang, *Purdue University*

Sierra Colavito, *University of Wisconsin– La Crosse*
Kamal Dulai, *University of California, Merced*
Gerard Jozwiak, *Oakland University*
Todd Kelson, *Brigham Young University– Idaho*
Troy Larson, *Augustana College*
Audrey Majeske, *Oakland University*

Todd Osmundson, *University of Wisconsin–La Crosse*
Thomas R. Peavy, *California State University, Sacramento*
Janet Rinehart-Kim, *Old Dominion University*
Laura Rusche, *University at Buffalo, SUNY*
Sunitha Sukumaran, *Texas Tech University*
Yunqiu Wang, *University of Miami*

SUPPORT AT
every step

Students
Get Learning that Fits You

Effective tools for efficient studying

Connect is designed to help you be more productive with simple, flexible, intuitive tools that maximize your study time and meet your individual learning needs. Get learning that works for you with Connect.

Study anytime, anywhere

Download the free ReadAnywhere® app and access your online eBook, SmartBook® 2.0, or Adaptive Learning Assignments when it's convenient, even if you're offline. And since the app automatically syncs with your Connect account, all of your work is available every time you open it. Find out more at **mheducation.com/readanywhere**

"I really liked this app—it made it easy to study when you don't have your text-book in front of you."

- Jordan Cunningham, Eastern Washington University

iPhone: Getty Images

Everything you need in one place

Your Connect course has everything you need—whether reading your digital eBook or completing assignments for class—Connect makes it easy to get your work done.

Learning for everyone

McGraw Hill works directly with Accessibility Services Departments and faculty to meet the learning needs of all students. Please contact your Accessibility Services Office and ask them to email accessibility@mheducation.com, or visit **mheducation.com/about/accessibility** for more information.

Connect® Remote Proctoring & Browser-Locking Capabilities

Remote proctoring and browser-locking capabilities, hosted by Proctorio within Connect, provide control of the assessment environment by enabling security options and verifying the identity of the student.

Seamlessly integrated within Connect, these services allow instructors to control the assessment experience by verifying identification, restricting browser activity, and monitoring student actions.

Instant and detailed reporting gives instructors an at-a-glance view of potential academic integrity concerns, thereby avoiding personal bias and supporting evidence-based claims.

Writing Assignment

Available within McGraw Hill Connect® and McGraw Hill Connect® Master, the Writing Assignment tool delivers a learning experience to help students improve their written communication skills and conceptual understanding. As an instructor, you can assign, monitor, grade, and provide feedback on writing more efficiently and effectively.

Virtual **Labs**   *Virtual Labs*

While the biological sciences are hands-on disciplines, instructors are now often being asked to deliver some of their lab components online, as full online replacements, supplements to prepare for in-person labs, or make-up labs.

These simulations help each student learn the practical and conceptual skills needed, then check for understanding and provide feedback. With adaptive pre-lab assignment, found under Adaptive Learning Assignment, and post-lab assessment available under Coursewide Content, instructors can customize each assignment.

From the instructor's perspective, these simulations may be used in the lecture environment to help students visualize complex scientific processes, such as DNA technology or Gram staining, while at the same time providing a valuable connection between the lecture and lab environments.

SMARTBOOK® *SmartBook 2.0*

Connect's SmartBook 2.0 provides an adaptive learning experience that combines eBook reading for comprehension, as well as assessments that test understanding. Learning resources are also available at key points to further aid understanding. The reading experience and assessments adapt to individual student learning. This is an environment that develops self-awareness through meaningful, immediate feedback that improves student success.

ReadAnywhere

Read or study when it's convenient for you with McGraw Hill's free ReadAnywhere app. Available for iOS or Android smartphones or tablets, ReadAnywhere gives users access to McGraw Hill tools, including the eBook and SmartBook 2.0 or Adaptive Learning Assignments in Connect. Take notes, highlight, and complete assignments offline–all of your work will sync when you open the app with WiFi access. Log in with your McGraw Hill Connect username and password

CHAPTER OUTLINE

CC (for "carbon copy" or "copy cat"), the first cloned pet. In 2002, the cat shown here was produced by reproductive cloning, a procedure described in Chapter 21. Texas A&M University/FEMA/Handout/Getty Images

OVERVIEW OF GENETICS

Hardly a week goes by without a major news story announcing a genetic breakthrough. The increasing pace of genetic discoveries has become staggering. The Human Genome Project is a case in point. This project began in the United States in 1990, when the National Institutes of Health (NIH) and the Department of Energy (DOE) joined forces with international partners to decipher the massive amount of information contained in our **genome**—the DNA found within all of our chromosomes (**Figure 1.1**). Remarkably, in only a decade, the researchers working on this project determined the DNA sequence (the order of the bases A, T, G, and C) of over 90% of the human genome. The completed sequence, published in 2003, has an accuracy greater than 99.99%; less than 1 mistake was made in every 10,000 base pairs!

In 2008, a more massive undertaking, called the 1000 Genomes Project, was launched with the goal of attaining a detailed understanding of human genetic variation. In this international project, researchers set out to determine the DNA sequence of at least 1000 anonymous participants from around the globe. In 2015, the sequencing of over 2500 genomes was described in the journal *Nature*.

Studying the human genome allows us to explore fundamental details about ourselves at the molecular level. The results of human genome projects have shed considerable light on basic questions, like how many genes we have, how genes direct the activities of living cells, how species evolve, how single cells develop into complex tissues, and how defective genes cause disease. Furthermore, understanding our genome may lead to improvements in modern medicine by leading to better diagnoses of diseases and allowing the development of new treatments for them.

The quest to unravel the mysteries within our genes has involved the invention of many new technologies. For example, researchers have developed genetic techniques to produce medicines, such as human insulin, that would otherwise be difficult or impossible to make. Human insulin is synthesized in strains of *Escherichia coli* bacteria that have been genetically altered by the addition of genes that code the polypeptides that form this hormone. Grown in laboratories, these bacteria make large amounts of human insulin. As discussed in Chapter 21, the insulin is purified and administered to many people with insulin-dependent diabetes.

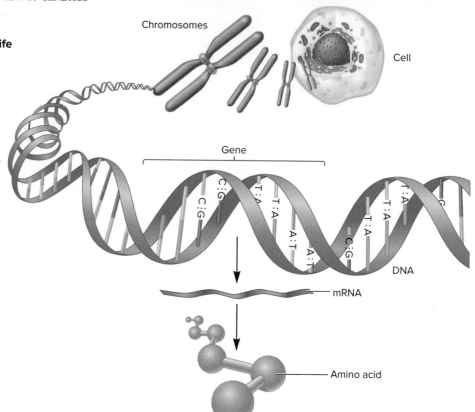

DNA, the molecule of life

The adult human body is composed of trillions of cells.

Most human cells contain the following:

• 46 human chromosomes, found in 23 pairs

• 2 meters of DNA

• Approximately 20,000 genes coding for proteins that perform most life functions

• Approximately 3 billion DNA base pairs per set of chromosomes, containing the bases A, T, G, and C

Chromosomes

Cell

Gene

DNA

mRNA

Amino acid

Protein (composed of amino acids)

FIGURE 1.1   The human genome. The human genome is a complete set of human chromosomes. People have two sets of chromosomes—one set from each parent—which are found in the cell nucleus. The Human Genome Project revealed that each set of chromosomes is composed of a DNA sequence that is approximately 3 billion base pairs long. As discussed later, most genes are first transcribed into mRNA and then the mRNA is used to make proteins. Estimates suggest that each set of chromosomes contains about 20,000 different protein-coding genes. This figure emphasizes the DNA found in the cell nucleus. Humans also have a small amount of DNA in their mitochondria, which has also been sequenced.

CONCEPT CHECK: How might a better understanding of our genes be used in the field of medicine?

New genetic technologies are often met with skepticism and sometimes even with disdain. An example is mammalian cloning. In 1997, Ian Wilmut and his colleagues created clones of sheep, using mammary cells from an adult animal (**Figure 1.2**). More recently, such cloning has been achieved in several mammalian species, including cows, mice, goats, pigs, and cats. In 2002, the first pet was cloned, a cat named CC (for "carbon copy" or "copy cat"; see the chapter-opening photo). The cloning of mammals provides the potential for many practical applications. With regard to livestock, cloning would enable farmers to use cells from their best individuals to create genetically homogeneous herds. This could be advantageous in terms of agricultural yield, although such a genetically homogeneous herd may be more susceptible to certain diseases. However, people have become greatly concerned about the possibility of human cloning. This prospect has raised serious ethical questions. Within the past few years, legislation that involves bans on human cloning has been introduced.

Finally, genetic technologies provide the means to modify the traits of animals and plants in ways that would have been unimaginable just a few decades ago. **Figure 1.3a** shows a striking example in which scientists introduced a gene from jellyfish into mice. Certain species of jellyfish emit a "green glow" produced by a bioluminescent protein called green fluorescent protein (GFP) coded by a

FIGURE 1.2   The cloning of a mammal. The lamb in the front is Dolly, the first mammal to be cloned. She was cloned from the cells of a Finn Dorset (a white-faced sheep). The sheep behind Dolly is her surrogate mother, a Blackface ewe. A description of how Dolly was produced is presented in Chapter 21.

R. Scott Horner KRT/Newscom

CONCEPT CHECK: What ethical issues may be associated with human cloning?

(a) GFP expressed in mice

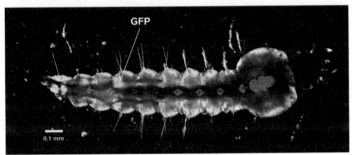

(b) GFP expressed in the gonads of a male mosquito

FIGURE 1.3 **Introduction of a jellyfish gene into laboratory mice and mosquitoes.** (a) A gene that naturally occurs in jellyfish codes a protein called green fluorescent protein (GFP). The *GFP* gene was cloned and introduced into mice. When these mice are exposed to UV light, GFP emits a bright green color. These mice glow green, just like the jellyfish! (b) The *GFP* gene was introduced next to a gene sequence that causes the expression of GFP only in the gonads of male mosquitoes. The resulting green glow allows researchers to identify and sort males from females.

(a): Eye of Science/Science Source; (b): Photo taken by Flaminia Catteruccia, Jason Benton and Andrea Crisanti, and assembled by www.luciariccidesign.com

CONCEPT CHECK: Why is it useful to sort male mosquitoes from females?

gene in the jellyfish genome. When exposed to blue or ultraviolet (UV) light, the protein emits a striking green-colored light. Scientists were able to clone the *GFP* gene from a sample of jellyfish cells and then introduce this gene into laboratory mice. The green fluorescent protein is made throughout the cells of their bodies. As a result, their skin, eyes, and organs give off an eerie green glow when exposed to UV light. Only their fur does not glow.

The expression of green fluorescent protein allows researchers to identify particular proteins in cells or specific body parts. For example, Andrea Crisanti and colleagues have altered mosquitoes to express GFP only in the gonads of males (**Figure 1.3b**). This enables the researchers to distinguish males from females and sort mosquitoes by sex. Why is this useful? Researchers can produce a population of mosquitoes and then sterilize the males. The ability to

distinguish males from females makes it possible to release the sterile males without the risk of releasing additional females. The release of sterile males may be an effective means of controlling mosquito populations because females mate only once before they die. Mating with a sterile male prevents a female from producing offspring. In 2008, Osamu Shimomura, Martin Chalfie, and Roger Tsien received the Nobel Prize in chemistry for the discovery and development of GFP, which has become a widely used tool in biology.

Overall, as we move forward in the twenty-first century, the excitement level in the field of genetics is high, perhaps higher than it has ever been. Nevertheless, new genetic knowledge and technologies will create many ethical and societal challenges. In this chapter, we begin with an overview of genetics and then explore the various fields of genetics and their experimental approaches. The meanings of the words *sex*, *gender*, and *normal*, as used in this textbook will also be discussed.

1.1 THE MOLECULAR EXPRESSION OF GENES

Learning Outcomes:

1. Describe the biochemical composition of cells.
2. Explain how proteins are largely responsible for cell structure and function.
3. Outline how DNA stores the information to make proteins.

Genetics is the branch of biology that focuses on heredity and variation. It stands as a unifying discipline in biology by allowing us to understand how life can exist at all levels of complexity, ranging from the molecular to the population level. Genetic variation is the root of the natural diversity that we observe among members of the same species and among different species.

Genetics is centered on the study of genes. A gene is classically defined as a unit of heredity, but such a vague definition does not do justice to the exciting characteristics of genes as intricate molecular units that manifest themselves as critical contributors to cell structure and function.

- At the molecular level, a **gene** is a segment of DNA that contains the information to produce a functional product. The functional product of most genes is a polypeptide, which is a linear sequence of amino acids that folds into a unit that constitutes a protein or part of a protein.
- At the organism level, genes are commonly described according to the way they affect **traits,** which are the characteristics of an organism. In humans, for example, we observe traits such as eye color, hair texture, and height. An ongoing theme of this textbook is the relationship between genes and traits. As an organism grows and develops, its collection of genes provides a blueprint that determines its traits.

In this section, we will examine the general features of life, beginning with the molecular level and ending with populations of organisms. As will become apparent, genetics is the common thread that explains the existence of life and its continuity from generation to generation. For most students, this chapter should serve as an overview of topics covered in other introductory courses such as general biology. Even so, it is usually helpful to see the "big picture" of genetics before delving into the finer details that are covered in Chapters 2 through 29.

Living Cells Are Composed of Biochemicals

To fully understand the relationship between genes and traits, we need to begin with an examination of the composition of living organisms. Every cell is constructed from intricately organized chemical substances. Small organic molecules such as glucose and amino acids are produced by the linkage of atoms via chemical bonds. The chemical properties of organic molecules are essential for cell vitality in two key ways.

- First, the breaking of chemical bonds during the degradation of small molecules provides energy to drive cellular processes.
- A second important function of these small organic molecules is their role as the building blocks for the synthesis of larger molecules. Four important categories of larger molecules are **nucleic acids** (i.e., DNA and RNA), **proteins, carbohydrates,** and **lipids.** Three of these—nucleic acids, proteins, and carbohydrates—exist as **macromolecules** that are composed of many repeating units of smaller building blocks. RNA, proteins, and some carbohydrates are made from hundreds or even thousands of repeating building blocks. DNA is the largest macromolecule found in living cells. A single DNA molecule can be composed of a linear sequence of hundreds of millions of building blocks called nucleotides!

The formation of cellular structures relies on the interactions of molecules and macromolecules. **Figure 1.4** illustrates this concept.

- Nucleotides are small organic molecules.
- Nucleotides are linked to each other and form the building blocks of DNA, which is a macromolecule.
- DNA is a component of chromosomes, which also contain proteins that contribute to chromosome structure.
- Within a eukaryotic cell, the chromosomes are contained in a compartment called the cell nucleus. The nucleus is bounded by a double membrane that is composed of lipids and proteins and shields the chromosomes from the rest of the cell. The nucleus is an example of an **organelle**—a membrane-bound compartment with a specialized function. The cell nucleus protects the chromosomes from mechanical damage and provides a single compartment for genetic activities such as gene transcription.
- Finally, cellular molecules, macromolecules, and organelles are organized to make a complete living cell.

Each Cell Contains Many Different Proteins That Determine Cell Structure and Function

To a great extent, the characteristics of a cell depend on the types of proteins that it makes. The entire collection of proteins that a cell makes at a given time is called its **proteome.** The range of functions among different types of proteins is truly remarkable. Some proteins

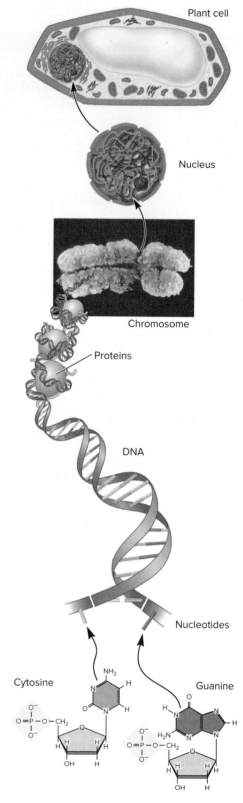

FIGURE 1.4 Molecular organization of a living cell. Cellular structures are constructed from smaller building blocks. In this example, DNA is formed from the linkage of nucleotides, producing a very long macromolecule. DNA associates with proteins to form a chromosome. The chromosomes are located within a membrane-bound organelle called the nucleus, which, along with many other different types of organelles, is found within a complete cell.

(inset) Biophoto Associates/Science Source

CONCEPT CHECK: Is DNA a small molecule, a macromolecule, or an organelle?

help determine the shape and structure of a given cell. For example, the protein known as tubulin assembles into large structures known as microtubules, which provide a cell with internal structure and organization. Other proteins are inserted into cell membranes and aid in the transport of ions and small molecules across the membrane. **Enzymes,** which accelerate chemical reactions, are a particularly important category of proteins. Some enzymes play a role in the breakdown of molecules or macromolecules into smaller units. Known as catabolic enzymes, these are important in the utilization of energy. Alternatively, anabolic enzymes and accessory proteins function in the synthesis of molecules and macromolecules throughout the cell. The construction of a cell greatly depends on its proteins that are involved in anabolism because these are required to synthesize all cellular macromolecules.

Molecular biologists have come to realize that the functions of proteins underlie the cellular characteristics of every organism. At the molecular level, proteins can be viewed as the active participants in the enterprise of life.

DNA Stores the Information for Protein Synthesis

The genetic material of living organisms is composed of a substance called **deoxyribonucleic acid,** abbreviated **DNA.** The DNA stores the information needed for the synthesis of all cellular proteins. In other words, the main function of the genetic blueprint is to code for the production of proteins in the correct cell, at the proper time, and in suitable amounts. This is an extremely complicated task because living cells make thousands of different proteins. Genetic analyses have shown that a typical bacterium can make a few thousand different proteins, and estimates for the numbers produced by complex eukaryotic species range in the tens of thousands.

DNA's ability to store information is based on its structure.

- DNA is composed of a linear sequence of **nucleotides.** Each nucleotide contains one of four nitrogen-containing bases: adenine (A), thymine (T), guanine (G), or cytosine (C).
- The linear order of these bases along a DNA molecule contains information similar to the way that groups of letters of the alphabet represent words. For example, the "meaning" of the sequence of bases ATGGGCCTTAGC differs from that of the sequence TTTAAGCTTGCC.
- DNA sequences within most genes contain the information to direct the order of amino acids within **polypeptides** according to the **genetic code.** In the code, a three-base sequence, called a **codon,** specifies one particular **amino acid** among the 20 possible choices (look ahead to Table 13.1).

DNA Sequence	Amino Acid Sequence
ATG GGC CTT AGC	Methionine Glycine Leucine Serine
TTT AAG CTT GCC	Phenylalanine Lysine Leucine Alanine

- The sequence of amino acids in a polypeptide causes it to fold into a particular structure. One or more polypeptides form a functional protein. In this way, the DNA stores the information to specify the proteins made by an organism.

In living cells, the DNA is found within large structures known as **chromosomes. Figure 1.5** is a micrograph of the 46 chromosomes

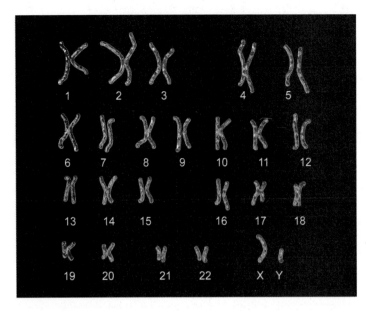

FIGURE 1.5   A micrograph of the 46 chromosomes found in a cell from a human male.
Kateryna Kon/Shutterstock

CONCEPT CHECK: Which types of macromolecules are found in chromosomes?

contained in a cell from a human male; this type of image is known as a **karyotype.** The DNA of an average human chromosome is an extraordinarily long, linear, double-stranded structure that contains well over 100 million nucleotides. Along the immense length of a chromosome, the genetic information is parceled into functional units known as genes. An average-sized human chromosome is expected to contain about 1000 different protein-coding genes.

The Information in DNA Is Accessed During the Process of Gene Expression

To synthesize its proteins, a cell must be able to access the information that is stored within its DNA. The process of using a gene sequence to affect the characteristics of cells and organisms is referred to as **gene expression.** At the molecular level, the information within genes is accessed in a stepwise process (**Figure 1.6**).

1. In the first step, known as **transcription,** the DNA sequence within a gene is copied into a nucleotide sequence of **ribonucleic acid (RNA). Protein-coding genes** carry the information for the amino acid sequence of a polypeptide. When a protein-coding gene is transcribed, the first product is an RNA molecule known as **messenger RNA (mRNA).**
2. During polypeptide synthesis—a process called **translation**—the sequence of nucleotides within the mRNA determines the sequence of amino acids in a polypeptide.
3. One or more polypeptides then fold and assemble into a functional protein. The synthesis of functional proteins largely determines an organism's traits. As discussed further in Chapter 12 (look ahead to Figure 12.1), the pathway of gene expression from DNA to RNA to protein is called the **central dogma of genetics** (also called the central dogma of molecular biology). It forms a cornerstone of our understanding of genetics at the molecular level.

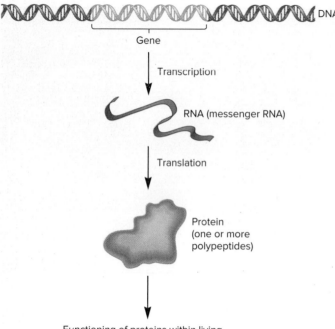

FIGURE 1.6 Gene expression at the molecular level. The expression of a gene is a multistep process. During transcription, one of the DNA strands is used as a template to make an RNA strand. During translation, the RNA strand is used to specify the sequence of amino acids within a polypeptide. One or more polypeptides form a functional protein, thereby influencing an organism's traits.

CONCEPT CHECK: Where is the information to make a polypeptide stored?

1.1 COMPREHENSION QUESTIONS

1. Which of the following is *not* a constituent of a cell's proteome?
   a. An enzyme
   b. A cytoskeletal protein
   c. A transport protein in the plasma membrane
   d. An mRNA

2. A gene is a segment of DNA that contains the information to produce a functional product. The functional product of most genes is
   a. DNA.   c. a polypeptide.
   b. mRNA.   d. none of the above.

3. The function of the genetic code is to
   a. promote transcription.
   b. specify the amino acids within a polypeptide.
   c. alter the sequence of DNA.
   d. do none of the above.

4. The direct result of the process of transcription is the synthesis of
   a. DNA.   c. a polypeptide.
   b. RNA.   d. all of the above.

1.2 THE RELATIONSHIP BETWEEN GENES AND TRAITS

Learning Outcomes:

1. Outline how the expression of genes leads to an organism's traits.
2. Define *genetic variation*.
3. Discuss the relationship between genes and traits.
4. Describe how genes are transmitted in sexually reproducing species.
5. Explain the process of evolution.

A trait is any characteristic that an organism displays. In genetics, we may place traits into different categories.

- **Morphological traits** affect the appearance, form, and structure of an organism. The color of a flower and the height of a pea plant are morphological traits. Geneticists frequently study these types of traits because they are easy to evaluate. For example, an experimenter can simply look at a plant and tell if it has red or white flowers.
- **Physiological traits** affect the ability of an organism to function. For example, the rate at which a bacterium metabolizes a sugar such as lactose is a physiological trait. Like morphological traits, physiological traits are controlled, in part, by the expression of genes.
- **Behavioral traits** affect the ways an organism responds to its environment. An example is the mating calls of bird species. In animals, the nervous system plays a key role in governing such traits.

In this section, we will examine the relationship between the expression of genes and an organism's traits.

The Molecular Expression of Genes Leads to an Organism's Traits

A complicated, yet very exciting, aspect of genetics is that the field's observations and theories span four levels of biological organization: molecules, cells, organisms, and populations. This broad scope can make it difficult to appreciate the relationship between genes and traits. To understand this connection, we need to relate the following phenomena:

1. Genes are expressed at the **molecular level.** In other words, gene transcription and translation lead to the production of a particular protein, which is a molecular process.
2. Proteins often function at the **cellular level.** The function of a protein within a cell affects the structure and workings of that cell.
3. An organism's traits are determined by the characteristics of its cells. We do not have microscopic vision, yet when we view morphological traits, we are really observing the properties of an individual's cells. For example, a red flower has its color because its cells make a red pigment. The trait of red flower color is an observation at the

organism level. Yet the trait is rooted in the molecular characteristics of the organism's cells.

4. A **species** is a group of organisms that maintains a distinctive set of attributes in nature. The occurrence of a trait within a species is an observation at the **population level.** Along with learning how a trait occurs, we also want to understand why a trait becomes prevalent in a particular species. In many cases, researchers discover that a trait predominates within a population because it promotes the reproductive success of the members of the population. This leads to the evolution of beneficial traits.

To illustrate the four levels of genetics with an example, **Figure 1.7** considers the trait of pigmentation in a species of butterflies. Some members of this species are dark colored and others are very light. Let's consider how we can explain this trait at the molecular, cellular, organism, and population levels.

At the molecular level, we need to understand the nature of the gene or genes that govern this trait. As shown in Figure 1.7a, a gene, which we will call the pigmentation gene, is responsible for the amount of pigment produced. The pigmentation gene exists in two different versions. Alternative versions of a specific gene are called **alleles.** In this example, one allele results in a dark pigmentation and the other causes a light pigmentation. Each of these alleles codes a protein that functions as a pigment-synthesizing enzyme. However, the DNA sequences of the two alleles differ slightly from each other. This difference in the DNA sequences leads to a variation in the structure and function of the respective pigment-synthesizing enzymes.

At the cellular level (Figure 1.7b), the functional differences between the two pigment-synthesizing enzymes affect the amount of pigment produced. The allele causing dark pigmentation, which is shown on the left, codes an enzyme that functions very well. Therefore, when this gene is expressed in the cells of the wings and other cells of the body, a large amount of pigment is made. By comparison, the allele causing light pigmentation codes an enzyme that functions poorly. Therefore, when this allele is the only pigmentation gene expressed, little pigment is made.

At the organism level (Figure 1.7c), the amount of pigment in the wing cells governs the color of the wings. If the pigment-synthesizing enzymes produce high amounts of pigment, the wings are dark colored. If the enzymes produce little pigment, the wings are light.

Finally, at the population level (Figure 1.7d), geneticists would like to know why a species of butterfly would contain some members with dark wings and other members with light wings. One possible explanation is differential predation. The butterflies with dark wings might better avoid being eaten by birds if they happened to live within a dimly lit forest. The dark wings would help to camouflage the butterfly if it were perched on a dark surface such as a tree trunk. In contrast, light-colored wings would be an advantage if the butterfly inhabited a brightly lit meadow. Under these conditions, a bird might be less likely to notice a light-colored butterfly that was perched on a sunlit surface. A population geneticist might study this species of butterfly and find that the dark-colored members usually live in forested areas and the light-colored members reside in unforested areas.

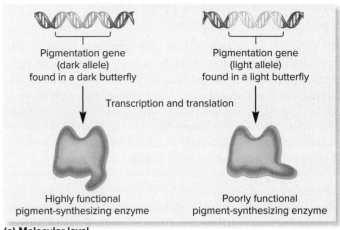

(a) Molecular level

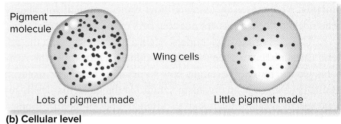

(b) Cellular level

(c) Organism level

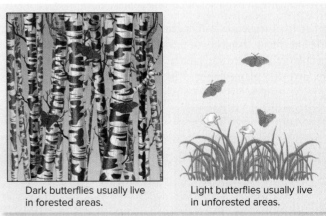

(d) Population level

FIGURE 1.7 The relationship between genes and traits at the **(a)** molecular, **(b)** cellular, **(c)** organism, and **(d)** population levels.

CONCEPT CHECK: Which butterfly has a more active pigment-synthesizing enzyme, the dark- or light-colored one?

Inherited Differences in Traits Are Due to Genetic Variation

In Figure 1.7, we considered how gene expression leads to variation in a trait of organisms, using the example of dark- versus light-colored wings in a species of butterflies. Variation in traits among members of the same species is very common. For example, some people have black hair and others have blond hair; some petunias have white flowers and others have purple flowers. These are examples of **genetic variation.** This term refers to the differences in inherited traits among individuals within a population.

In large populations that occupy a wide geographic range, genetic variation can be quite striking. Morphological differences have often led geneticists to misidentify two members of the same species as belonging to separate species. As an example, **Figure 1.8** shows two dyeing poison frogs that are members of the same species, *Dendrobates tinctorius.* They display dramatic differences in their markings. Such contrasting forms within a single species are termed **morphs.** You can easily imagine how someone might mistakenly conclude that these frogs are not members of the same species.

Changes in the nucleotide sequence of DNA underlie the genetic variation that we see among individuals. Throughout this textbook, we will routinely examine how variation in the genetic material results in changes in an organism's traits. At the molecular level, genetic variation can be attributed to different types of modifications.

- Small or large differences can occur within gene sequences. When such changes initially occur, they are called **gene mutations.** Mutations result in genetic variation in which a gene is found in two or more alleles, as previously described in Figure 1.7. In many cases, gene mutations alter the expression or function of a protein that a gene codes.
- Major alterations can also occur in the structure of a chromosome. A large segment of a chromosome can be lost, rearranged, or reattached to another chromosome.
- Variation may also occur in the total number of chromosomes. In some cases, an organism may inherit one too many or one too few chromosomes. In other cases, it may inherit an extra set of chromosomes.

Variations of sequences within genes are a common source of genetic variation among members of the same species. In humans, familiar examples of sequence variation involve genes for eye color, hair texture, and skin pigmentation. Chromosome variation—a change in

chromosome structure or number (or both)—is also found, but this type of change is often detrimental. Some human genetic disorders are the result of chromosomal alterations. The most common example is Down syndrome, which is due to the presence of an extra chromosome (**Figure 1.9a**). By comparison, chromosome variation in plants is common and often results in plants with superior characteristics, such as increased resistance to disease. Plant breeders have frequently exploited this observation. Cultivated varieties of wheat, for example, have six sets of chromosomes, whereas the wild species from which they were derived contain two sets (**Figure 1.9b**).

Traits Are Governed by Genes and by the Environment

In our discussion thus far, we have considered the role that genes play in determining an organism's traits. Another critical factor is the **environment**—the surroundings in which an organism exists. A variety of factors in an organism's environment profoundly affect its morphological and physiological features. For example, a person's diet greatly influences many traits, such as height, weight, and even intelligence. Likewise, the amount of sunlight a plant receives affects its growth rate and the color of its flowers.

An interesting example of the interplay between genes and the environment involves the human genetic disorder **phenylketonuria (PKU).** Humans have a gene that codes an enzyme known as phenylalanine hydroxylase. Most people have two functional copies of this gene. People with one or two functional copies of the gene can eat foods containing the amino acid phenylalanine and metabolize it properly.

A rare variation in the gene that codes phenylalanine hydroxylase results in a nonfunctional version of this enzyme. Individuals with two copies of this rare, inactive allele cannot metabolize phenylalanine properly. When given a standard diet containing phenylalanine, individuals with this disorder are unable to break down this amino acid.

FIGURE 1.8 **Two dyeing poison frogs (*Dendrobates tinctorius*) are examples of different morphs within a single species.**
(Left): Natalia Kuzmina/Shutterstock; (Right): Valt Ahyppo/Shutterstock

CONCEPT CHECK: Why do these two frogs look so different?

(a)   (b)

FIGURE 1.9 **Examples of chromosome variation.** **(a)** A person with Down syndrome. This individual has 47 chromosomes rather than the common number of 46, because of an extra copy of chromosome 21. **(b)** A wheat plant. Cultivated wheat has six sets of chromosomes.

(a): George Doyle/Stockbyte/Alamy Stock Photo; (b): Bryan Mullennix/Pixtal/age fotostock

CONCEPT CHECK: Do these examples constitute variation in chromosome structure or variation in chromosome number?

Phenylalanine accumulates and is converted into phenylketones, which are detected in the urine. Raised on a standard diet containing phenylalanine, PKU individuals manifest a variety of detrimental traits, including mental impairment, underdeveloped teeth, and foul-smelling urine. Fortunately, through routine screening of newborns, most affected babies in the United States are now diagnosed and treated early. Part of the treatment is a diet that restricts phenylalanine, which is present in high-protein foods such as eggs, meat, and dairy products. Restricting phenylalanine allows the affected child to develop normally. PKU provides a dramatic example of how the environment and an individual's genes can interact to influence the traits of the organism.

During Reproduction, Genes Are Passed from Parent to Offspring

Now that we have considered how genes and the environment govern the outcome of traits, we can turn to the topic of inheritance. How are traits passed from parents to offspring? The foundation for our understanding of inheritance came from Gregor Mendel's study of pea plants in the nineteenth century. Mendel's work revealed that the genetic determinants that govern traits, which we now call genes, are passed from parent to offspring as discrete units. We can predict the outcome of many genetic crosses based on Mendel's laws of inheritance.

The inheritance patterns identified by Mendel can be explained by the existence of chromosomes and their behavior during cell division.

- Like pea plants, sexually reproducing species are commonly **diploid.** This means that their cells contain two copies of each chromosome, one from each parent. The two copies are called **homologs** of each other.

- Because genes are located within chromosomes, diploid organisms have two copies of most genes. Humans, for example, have 46 chromosomes, which are found in homologous pairs (**Figure 1.10a**). With the exception of the sex chromosomes (X and Y), each chromosome in a homologous pair contains the same kinds of genes. For example, both copies of human chromosome 12 carry the gene that codes phenylalanine hydroxylase, the enzyme that is nonfunctional in people with PKU. Therefore, each individual has two copies of this gene, which may or may not be identical alleles.

- Most cells of the human body that are not directly involved in sexual reproduction contain 46 chromosomes. These cells are called **somatic cells.** In contrast, the **gametes**—sperm and egg cells—contain half that number (23) and are termed **haploid** (**Figure 1.10b**).

- The union of gametes during fertilization restores the diploid number of chromosomes. The primary advantage of sexual reproduction is that it enhances genetic variation. For example, a tall person with blue eyes and a short person with brown eyes may have short offspring with blue eyes or tall offspring with brown eyes. Therefore, sexual reproduction can result in new combinations of two or more traits that differ from those of either parent.

The Genetic Composition of a Species Evolves from Generation to Generation

As we have just seen, sexual reproduction has the potential to enhance genetic variation. This can be an advantage for a population of individuals as they struggle to survive and compete within their

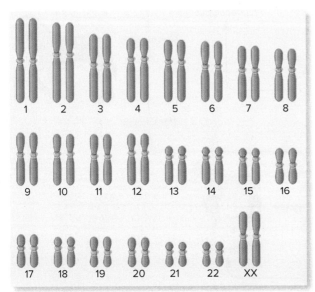
(a) Chromosomal composition found in human somatic cells of females (46 chromosomes)

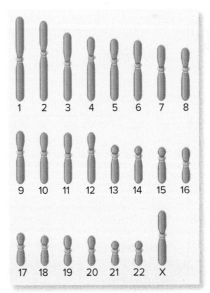
(b) Chromosomal composition found in a human gamete (23 chromosomes)

FIGURE 1.10   The complement of human chromosomes in somatic cells and gametes. (**a**) A schematic drawing of the 46 chromosomes of a human. With the exception of the sex chromosomes, these are always found in homologous pairs in somatic cells, such as skin or nerve cells. (**b**) The chromosomal composition of a gamete, which contains only 23 chromosomes, one from each pair. This gamete contains an X chromosome. Half of the gametes from human males contain a Y chromosome instead of an X chromosome.

CONCEPT CHECK: The leaf cells of a corn plant contain 20 chromosomes each. How many chromosomes are found in a gamete made by a corn plant?

natural environment. The term **biological evolution,** or simply, **evolution,** refers to the process of changes in the genetic makeup of a population from one generation to the next.

As suggested by Charles Darwin and Alfred Russel Wallace in the nineteenth century, the members of a species are in competition with one another for essential resources. Random genetic changes (i.e., mutations) occasionally occur within an individual's genes, and sometimes these changes lead to a modification of traits that promote reproductive success. For example, over the course of many generations, random gene mutations have lengthened the snout and extended the tongue of the anteater, enabling it to probe into the ground and feed on ants. When a mutation creates a new allele that is beneficial, the allele may become prevalent in future generations because the individuals carrying the allele are more likely to reproduce and pass the beneficial allele to their offspring. This process is known as **natural selection.** In this way, a species becomes better adapted to its environment.

Over a long period of time, the accumulation of many genetic changes may lead to rather striking modifications in a species' characteristics. As an example, **Figure 1.11** depicts the evolution of the modern-day horse. Over time, a variety of morphological changes occurred, including an increase in size, fewer toes, and modified jaw structure. The changes can be attributed to natural selection. Over North America, where much of horse evolution occurred, large areas of dense forests were replaced with grasslands. The increase in size and changes in foot structure enabled horses to escape predators more easily and travel greater distances in search of food. Natural selection favored the changes seen in horses' teeth, because such changes allowed them to eat grasses and other types of vegetation that are tougher and require more chewing.

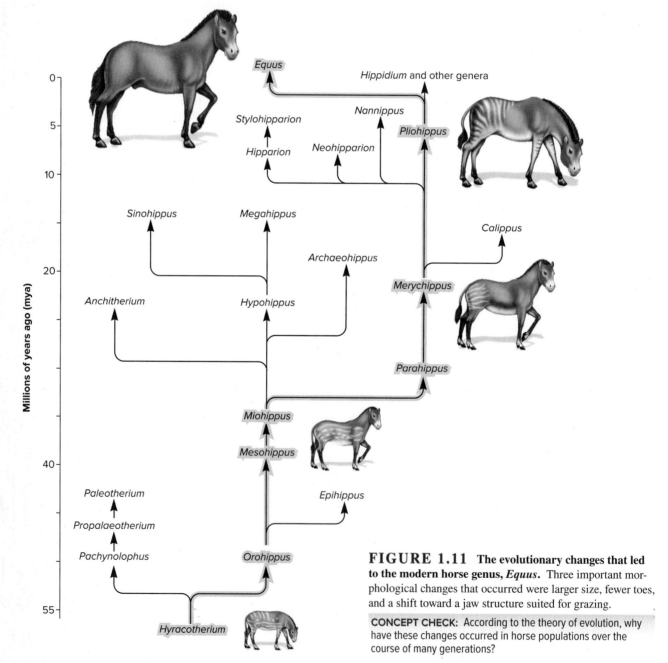

FIGURE 1.11 **The evolutionary changes that led to the modern horse genus, *Equus*.** Three important morphological changes that occurred were larger size, fewer toes, and a shift toward a jaw structure suited for grazing.

CONCEPT CHECK: According to the theory of evolution, why have these changes occurred in horse populations over the course of many generations?

1.2 COMPREHENSION QUESTIONS

1. At which of the following levels can gene expression be observed?
   a. Molecular and cellular levels
   b. Organism level
   c. Population level
   d. All of the above

2. Variation in the traits of organisms may be attributable to
   a. gene mutations.
   b. alterations in chromosome structure.
   c. variation in chromosome number.
   d. all of the above.

3. A human skin cell has 46 chromosomes. A human gamete (egg or sperm cell) has
   a. 23.
   b. 46.
   c. 92.
   d. None of the above is the number of chromosomes in a sperm cell.

4. Evolutionary change caused by natural selection results in species with
   a. greater complexity.
   b. less complexity.
   c. greater reproductive success in their environment.
   d. the ability to survive longer.

1.3 FIELDS OF GENETICS

Learning Outcomes:

1. Define *model organism*.
2. Compare and contrast the three major fields of genetics: transmission, molecular, and population genetics.

Genetics is a broad discipline encompassing molecular, cellular, organism, and population biology. Many scientists who are interested in genetics have been trained in supporting disciplines such as biochemistry, biophysics, cell biology, mathematics, microbiology, population biology, ecology, agriculture, and medicine. The study of genetics has been traditionally divided into three areas—transmission, molecular, and population genetics—although there is some overlap of these three fields. In this section, we will consider how some researchers study model organisms and compare these three fields of genetics.

Model Organisms Are Species That Are Studied by Many Different Researchers

Experimentally, geneticists often focus their efforts on **model organisms**—organisms studied by many different researchers so they can compare their results and determine scientific principles that apply more broadly to other species. **Figure 1.12** shows some examples of model organisms that are used by researchers around the globe: *Escherichia coli* (a bacterium), *Saccharomyces cerevisiae* (a yeast), *Drosophila melanogaster* (fruit fly), *Caenorhabditis*

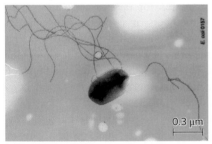
(a) *Escherichia coli*

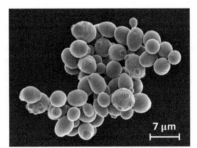
(b) *Saccharomyces cerevisiae*

(c) *Drosophila melanogaster*

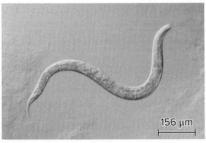
(d) *Caenorhabditis elegans*

(e) *Mus musculus*

(f) *Arabidopsis thaliana*

FIGURE 1.12 **Examples of model organisms studied by geneticists.** (a) *Escherichia coli* (a bacterium), (b) *Saccharomyces cerevisiae* (a yeast), (c) *Drosophila melanogaster* (fruit fly), (d) *Caenorhabditis elegans* (a nematode worm), (e) *Mus musculus* (mouse), and (f) *Arabidopsis thaliana* (a flowering plant). Note: The micrographs shown in parts (a), (b), and (d) are artificially colorized so the organisms can be seen more easily.

(a): Peggy S. Hayes & Elizabeth H. White, M.S./CDC; (b): Steve Gschmeissner/Science Photo Library/Alamy Stock Photo; (c): janeff/iStockphoto/Getty Images; (d): Sinclair Stammers/Science Source; (e): G.K. & Vikki Hart/Photodisc/Getty Images; (f): WILDLIFE GmbH/Alamy Stock Photo

CONCEPT CHECK: Can you think of another example of a model organism?

elegans (a nematode worm), *Mus musculus* (mouse), and *Arabidopsis thaliana* (a flowering plant). Model organisms offer experimental advantages over other species. For example, *E. coli* is a very simple organism that can be easily grown in the laboratory. By limiting their work to a few model organisms, researchers can more easily unravel the genetic mechanisms that govern the traits of a given species. Furthermore, the genes found in model organisms often function in a similar way to those found in humans.

Transmission Genetics Explores the Inheritance Patterns of Traits as They Are Passed from Parents to Offspring

A scientist working in the field of transmission genetics examines the relationship between the transmission of genes from parent to offspring and the outcome of the offspring's traits. For example, how can two brown-eyed parents produce a blue-eyed child? Or why do tall parents tend to produce tall children, but not always?

Our modern understanding of transmission genetics began with the studies of Gregor Mendel, whose work provided the conceptual framework for transmission genetics. In particular, Mendel originated the idea that factors, which we now call genes, are passed as discrete units from parents to offspring via sperm and egg cells. Since Mendel's pioneering studies in the 1860s, our knowledge of genetic transmission has skyrocketed. Many patterns of genetic transmission are more complex than the simple Mendelian patterns that are described in Chapter 3. The additional complexities of transmission genetics are examined in Chapters 4 through 8.

Experimentally, the fundamental technique used by a transmission geneticist is the **genetic cross**—the breeding of two selected individuals and then analyzing their offspring in an attempt to understand how traits are passed from parents to offspring. In the case of experimental organisms, the researcher chooses two parents with particular traits and then categorizes the offspring according to the traits they possess. In many cases, this analysis is quantitative in nature. For example, an experimenter may cross two tall pea plants and obtain 100 offspring that fall into two categories: 75 tall and 25 dwarf. As we will see in Chapter 3, the ratio of tall to dwarf offspring provides important information concerning the inheritance pattern of the height trait.

Molecular Genetics Focuses on a Biochemical Understanding of the Hereditary Material

The goal of molecular genetics, as the name of the field implies, is to understand how the genetic material works at the molecular level. In other words, molecular geneticists want to understand the molecular features of DNA and how these features underlie the expression of genes. The experiments of molecular geneticists are usually conducted within the confines of a laboratory. Their efforts frequently progress to a detailed analysis of DNA, RNA, and proteins, using a variety of techniques that are described throughout Parts III, IV, and V of this textbook.

Molecular geneticists often study mutant genes that have abnormal function. This is called a **genetic approach** to the

study of a research question. In many cases, researchers analyze the effects of a gene mutation that eliminates the function of a gene. This is called a **loss-of-function mutation,** and the resulting version of the gene is called a **loss-of-function allele.** Studying the effect of such a mutation often reveals the role of the functional, nonmutant gene. For example, let's suppose that a particular plant species produces purple flowers. If a loss-of-function mutation within a given gene causes a plant of that species to produce white flowers, you might suspect that the role of the functional gene involves the production of purple pigmentation.

Studies within molecular genetics interface with other disciplines such as biochemistry, biophysics, and cell biology. In addition, advances within molecular genetics have shed considerable light on the areas of transmission and population genetics. Our quest to understand molecular genetics has spawned a variety of modern molecular technologies and computer-based approaches. Furthermore, discoveries within molecular genetics have had widespread applications in agriculture, medicine, and biotechnology.

Population Genetics Is Concerned with Genetic Variation and Its Role in Evolution

The foundations of population genetics arose during the first few decades of the twentieth century. Although many scientists of this era did not accept the findings of Mendel or Darwin, the theories of population genetics provided a compelling way to connect the two viewpoints. Mendel's work and that of many succeeding geneticists gave insight into the nature of genes and how they are transmitted from parents to offspring. The work of Darwin provided a natural explanation for the variation in characteristics observed among the members of a species. To relate these two phenomena, population geneticists have developed mathematical theories to explain the prevalence of certain alleles within populations of individuals. This work helps us understand how processes such as natural selection have resulted in the prevalence of individuals that carry particular alleles.

Population geneticists are particularly interested in genetic variation and how that variation is related to an organism's environment. In this field, the frequencies of alleles within a population are of central importance.

1.3 COMPREHENSION QUESTIONS

1. Which of the following is *not* a model organism?
   a. *Mus musculus* (laboratory mouse)
   b. *Escherichia coli* (a bacterium)
   c. *Saccharomyces cerevisiae* (a yeast)
   d. *Sciurus carolinensis* (gray squirrel)
2. A person studying the rate of transcription of a particular gene is working in the field of
   a. molecular genetics.
   b. transmission genetics.
   c. population genetics.
   d. None of the above is correct.

1.4 THE SCIENCE OF GENETICS

Learning Outcomes:

1. Describe what makes genetics an experimental science.
2. Outline different strategies for solving problems in genetics.

Science is a way of knowing about our natural world. The science of genetics allows us to understand how the expression of our genes produces the traits that we possess. In this section, we will consider how scientists attempt to answer questions via experimentation. We will also consider general approaches for solving problems.

Genetics Is an Experimental Science

Researchers typically follow two general types of scientific approaches: hypothesis testing and discovery-based science. In **hypothesis testing,** also called the **scientific method,** scientists follow a series of steps to reach verifiable conclusions about the world. Although scientists arrive at their theories in different ways, the scientific method provides a way to validate (or invalidate) a particular hypothesis.

Alternatively, research may also involve the collection of data without a preconceived hypothesis. For example, researchers might analyze the genes found in cancer cells to identify those that have become mutant. In this case, the scientists may not have a hypothesis about which particular genes may be involved. The collection and analysis of data without the need for a preconceived hypothesis is called **discovery-based science** or, simply, discovery science.

In traditional science textbooks, the emphasis often lies on the product of science. That is, many textbooks are aimed primarily at teaching the student about the observations scientists have made and the hypotheses they have proposed to explain those observations. Along the way, the student is provided with many bits and pieces of experimental techniques and data. Likewise, this textbook also provides you with many observations and hypotheses. However, it attempts to go one step further. Most chapters contain one or two figures presenting experiments that have been "dissected" into five individual components to help you to understand the entire scientific process:

1. Background information is provided so that you can appreciate observations that were known prior to conducting the experiment.
2. Most experiments involve the testing of a hypothesis via the scientific method. In those cases, the figure presenting the experiment states the hypothesis the scientists were trying to test. In other words, what scientific question(s) were the researchers trying to answer?
3. Next, the figure follows the experimental steps the scientists took to test the hypothesis. The figure presents two parallel illustrations labeled "Experimental Level" and "Conceptual Level." The experimental level helps you to understand the techniques that were used. The conceptual level helps you to understand what is actually happening at each step in the procedure.

4. The raw data from the experiment are then presented.
5. Last, an interpretation of the data is offered within the text.

The rationale behind this approach is that it enables you to see the experimental process from beginning to end. As you read through the chapters, the experiments will help you to see the relationship between science and scientific theories.

As a student of genetics, you will be given the opportunity to involve your mind in the experimental process. As you are reading an experiment, you may find yourself thinking about different approaches and alternative hypotheses. Different people can view the same data and arrive at very different conclusions. As you advance through the experiments in this book, you will enjoy genetics far more if you try to improve your skills of formulating hypotheses, designing experiments, and interpreting data. Also, some of the questions in the problem sets are aimed at refining these skills.

Finally, it is worthwhile to point out that science is a social discipline. As you develop your skills at scrutinizing experiments, it is fun to discuss your ideas with other people, including fellow students and faculty members. Keep in mind that you do not need to "know all the answers" before you enter into a scientific discussion. Instead, it is more rewarding to view science as an ongoing and never-ending dialogue.

Genetic TIPS Will Help You to Improve Your Problem-Solving Skills

As you progress through this textbook, your learning will involve two general goals:

- You will gather foundational knowledge. In other words, you will be able to describe core concepts in genetics. For example, you will be able to explain how DNA replication occurs and describe the proteins that are involved in this process.
- You will develop problem-solving skills that allow you to apply that foundational knowledge in different ways. For example, you will learn how to use statistics to determine if a genetic hypothesis is consistent with experimental data.

The combination of foundational knowledge and problem-solving skills will enable you not only to understand genetics, but also to apply your knowledge in different situations. To help you develop these skills, Chapters 2 through 29 contain solved problems named Genetic TIPS, which stands for Topic, Information, and Problem-solving Strategy. These solved problems follow a consistent pattern.

GENETIC TIPS   THE QUESTION: All of the Genetic TIPS begin with a question. As an example, let's consider the following question:

The coding strand of DNA in a segment of a gene is as follows: ATG GGC CTT AGC. This strand carries the information to make a region of a polypeptide with the amino acid sequence, methionine-glycine-leucine-serine. What would be the consequences if a mutation changed the second cytosine (C) in this sequence to an adenine (A)?

T OPIC: *What topic in genetics does this question address?* The topic is gene expression. More specifically, the question is about the relationship between a gene sequence and the genetic code.

I NFORMATION: *What information do you know based on the question and your understanding of the topic?* In the question, you are given the base sequence of a short segment of a gene and told that one of the bases has been changed. From your understanding of the topic, you may remember that a polypeptide sequence is determined by reading the mRNA (transcribed from a gene) in groups of three bases called codons.

P ROBLEM-SOLVING **S** TRATEGY: *Compare and contrast.* One strategy to solve this problem is to compare the mRNA sequence (transcribed from this gene) before and after the mutation:

Original: AUG GGC **C**UU AGC

Mutant: AUG GGC **A**UU AGC

ANSWER: The mutation has changed the sequence of bases in the mRNA so the third codon has changed from CUU to AUU. Because codons specify amino acids, this alteration may change the third amino acid to something else. Note: If you look ahead to Chapter 13 (see Table 13.1), you will see that CUU specifies leucine, whereas AUU specifies isoleucine. Therefore, you would predict that the mutation would change the third amino acid from leucine to isoleucine.

Throughout Chapters 2 through 29, each chapter will contain several Genetic TIPS. Some of these will be within the chapter itself and some will precede the problem sets that are at the end of each chapter. Though there are many different problem-solving strategies, Genetic TIPS will focus on 10 strategies that will help you to solve problems. You will see these 10 strategies over and over again as you progress through the textbook:

1. *Define key terms.* In some cases, a question may be difficult to understand because you don't know the meaning of one or more key terms in the question. If so, you will need to begin your problem solving by defining such terms, either by looking them up in the glossary or by using the index to find the location in the text where the key terms are explained.
2. *Make a drawing.* Genetic problems are often difficult to solve in your head. Making a drawing may make a big difference in your ability to see the solution.
3. *Predict the outcome.* Geneticists may want to predict the outcome of an experiment. For example, in Chapters 3 through 6, you will learn about different ways to predict the outcome of genetic crosses. Becoming familiar with these methods will help you to predict the outcomes of particular experiments.
4. *Compare and contrast.* Making a direct comparison between two things, such as two RNA sequences, may help you to understand how they are similar and how they are different.
5. *Relate structure and function.* A recurring theme in biology and genetics is that structure determines function. This relationship holds true at many levels of biology, including the molecular, microscopic, and macroscopic levels. For some questions, you will need to understand how certain

structural features are related to their biological functions.
6. *Describe the steps.* At first, some questions may be difficult to understand because they may involve mechanisms that occur in a series of several steps. Sometimes, if you sort out the steps, you may identify the key step that you need to understand to solve the problem.
7. *Propose a hypothesis.* A hypothesis is an attempt to explain an observation or data. Hypotheses may be made in many forms, including statements, models, equations, and diagrams.
8. *Design an experiment.* Experimental design lies at the heart of science. In many cases, an experiment begins with some type of starting material, such as strains of organisms or purified molecules, and then the starting material is subjected to a series of steps. The experiments featured throughout the textbook will also help you refine the skill of designing experiments.
9. *Analyze data.* The results from an experiment produce data that can be analyzed in order to accept or reject a hypothesis. Many of the Genetic TIPS give you practice at analyzing data. For example, a variety of different statistical methods are used to analyze data and make conclusions about what the data mean.
10. *Make a calculation.* Genetics is a quantitative science. Researchers have devised mathematical relationships to understand and predict genetic phenomena. Becoming familiar with these mathematical relationships will help you to better understand genetic concepts and to make predictions.

For most problems throughout this textbook, one or more of these strategies may help you arrive at the correct solution. Genetic TIPS will provide you with practice in applying these 10 problem-solving strategies.

1.4 COMPREHENSION QUESTION

1. The scientific method involves which of the following?
   a. The collection of observations and the formulation of a hypothesis
   b. Experimentation
   c. Data analysis and interpretation
   d. All of the above

1.5 GENDER IDENTITY VERSUS SEX

Learning Outcomes:

1. Compare and contrast gender identity and sex.
2. Define *female* and *male,* as the terms will be used in this textbook.

Inclusion is the practice that strives to include all types of people, regardless of their differences, in many areas and environments, such as education and the workplace. This practice involves including and embracing people from various backgrounds and those who might otherwise be excluded or marginalized. Promoting inclusion is easier in theory than in actuality. Unfortunately, discrimination based on race, ethnicity, sex, gender, disability, and other characteristics is common.

Genetics, as an educational discipline, has the potential to marginalize people due to variation in gender identity, and it has done so historically. To achieve inclusion, recognition needs to be given to the entire spectrum of gender that people identify with. In this section, we will explore the meanings of gender identity and sex, and then discuss what the terms *male* and *female* mean in this textbook. Hopefully, this discussion will foster greater inclusion in the conversation of genetics.

Gender Identity Is a Person's Perception of Their Own Gender

Gender refers to the characteristics of women and men that are socially constructed, which may include specific social roles, norms, and expectations that differentiate women and men. These characteristics are subjective, and they differ among societies, social classes, and cultures. Gender is a social group's perception of what it means to be a woman or a man. In this regard, the words *woman* and *man* refer to a person's gender as do *boy* and *girl*. It is important to note that many cultures recognize more than two genders.

Gender identity is a person's deeply held inner feelings about whether they're a woman, man, both, or neither. Gender identity isn't seen by others. The research and medical community now accepts gender identity as being more complex than simply identifying as a woman or a man. Gender identity is a spectrum that includes transgender people and those whose identity is not solely that of a man or a woman (i.e., those who identify as nonbinary).

Sex Is Rooted in Anatomical and Cellular Characteristics

How can we promote gender inclusivity when discussing genetics? A core dilemma centers around the words we use and the clarity of their meanings. As described in Chapters 2 through 8, the field of transmission genetics focuses on the transmission of genes from parents to offspring. In the case of many species, including humans, this occurs via sexual reproduction, which involves the union of gametes. Is there a way to communicate sexual reproduction that acknowledges the spectrum of gender identity?

In recognition of the spectrum of gender identities among people, this textbook will clearly distinguish between sex and gender. In this textbook, the terms **female** and **male** refer to sex but not to gender. In the case of humans, **sex** has the following characteristics:

- Sex is a designation that is usually assigned at birth based on external body parts associated with reproduction. For example, females usually have a vulva and vagina, and males usually have a penis and scrotum. External structures that appear differently than these two common patterns are called **intersex structures.**

- Other characteristics typically associated with male or female sex include internal structures associated with reproduction, such as ovaries, testes, fallopian tubes, and seminal vesicles; levels of hormones including testosterone and estrogen; and secondary sex characteristics that appear after puberty. A wide variation in these sex characteristics is observed; intersex people have sex characteristics that do not fit into the two most common patterns seen in males and females.

- Sex determines the type of gamete that an individual has the potential to produce. Fertile females produce oocytes (eggs) and fertile males produce sperm.

- With regard to sex chromosome composition, females are usually XX and males are usually XY. However, other combinations of sex chromosomes, including XXX, XXY, or just a single X chromosome, may occur in humans.

Different Relationships May Occur Between Gender Identity and Sex

The prefixes *cis* and *trans* are used to describe the relationship between a person's sex and their gender identity. Let's consider a few examples:

- A cis-man, shorthand for cisgender man, has the gender identity of a man and the assigned sex at birth of a male.
- A trans-man, shorthand for transgender man, has the gender identity of a man and the assigned sex at birth that is not a male.
- A nonbinary person has a gender identity that is not solely that of a man or a woman and can have any set of sex characteristics.

Words such as *woman, man, mother, father, daughter,* and *son* are commonly used to denote gender. Likewise, pronouns such as *she, her, he,* and *him* can be used to refer to people based on gender. Because these words denote gender identity and because the author does not have knowledge regarding the gender identity of people mentioned in this textbook, these words will be avoided when referring to people. Instead, the terms female and male will be used, with the understanding that these terms are <u>not</u> indicative of gender identity, nor do they encompass the entirety of diversity in sex characteristics seen in humans. Female and male are solely defined by anatomical and cellular attributes. This textbook will also use the terms *maternal* and *paternal* to indicate inheritance derived from females and males, respectively. Like female and male, maternal and paternal are <u>not</u> meant to imply gender identity.

With regard to certain cellular and molecular processes, the gendered terms *mother, daughter,* and *sister* have been traditionally used to denote relationships. For example, mother cell and daughter cell are used when discussing cell division, and daughter chromosome and sister chromatids are used when discussing DNA replication. This convention will also be followed in this textbook with the understanding that cells and molecules do not have gender, nor are they part of families in the social sense that these words connote for humans. Like gender and sex, human family structures are very diverse, and can include both genetically related and unrelated individuals.

Sexual Orientation Is a Person's Attraction to Someone Else

Before leaving this topic, it is worthwhile to distinguish gender identity and sexual orientation. As mentioned, gender identity refers to one's deeply held inner feelings about whether they're a woman, man, both, or neither. It's about oneself. By comparison, **sexual orientation** (also called sexual preference or sexuality) is about the type of person someone is attracted to. It is an enduring pattern of romantic, emotional, and/or sexual attraction that can be oriented toward persons of a different sex or gender, the same sex or gender, more than one sex and/or gender, or no sex nor any gender. As with gender identity, sexual orientation falls along a spectrum.

1.5 COMPREHENSION QUESTIONS

1. Which of the following statements regarding gender identity is *false*?
   a. Gender identity is one's deeply held inner feelings about whether they're a woman, man, both, or neither.
   b. Gender identity can be the same or different from an individual's sex.
   c. Gender identity is determined by sex chromosomes; XX is a woman and XY is a man.
   d. Gender identity occurs along a spectrum.

2. In humans, sex
   a. is usually assigned at birth based on body parts associated with reproduction.
   b. determines the type of gametes that an individual has the potential to produce.
   c. is determined by sex chromosome composition; females are usually XX and males are usually XY.
   d. All of the above are ways to characterize sex.

1.6 THE MEANING OF NORMAL IN GENETICS

Learning Outcome:

1. With regard to genetics, define *normal*, *abnormal*, and *wild type*.

As mentioned in Section 1.5, inclusion is the practice that strives to include all types of people, regardless of their differences, in many areas and environments, such as education and the workplace. Unfortunately, a misunderstanding of the meaning of certain words in science can result in feelings that foster a lack of inclusion.

In Section 1.5, we considered the meaning of terms such as *sex*, *gender*, *gender identity*, *male* and *female*, and explained their meanings as they are used in this textbook. Another term that has the potential to be misunderstood is the word *normal*. If you look this word up in multiple online dictionaries, you will find a surprisingly large number of different definitions. Because of the many definitions of normal in everyday language, a person's perception of what this word means is often muddied. In this section, we will discuss the meaning of the term *normal* as it is used in this genetics textbook.

In Genetics, *Normal* Means Common

The reason why it is important to define *normal* in the field of genetics is because some definitions of that word in everyday language have a positive connotation, whereas *abnormal* may have a negative connotation. However, in this textbook, the term *normal* is meant to be a quantitative term that refers to the relative likelihood of one or more characteristics compared to other members of the same species.

- **Normal** means common. When an individual is said to be normal in this textbook, that individual displays one or more characteristics that are commonly observed among members of its species.
- **Abnormal** means rare. An abnormal individual has one or more characteristics that are not very common among members of its species.

In genetics, the term *wild type* means the same thing as *normal*. **Wild type** refers to any characteristic that is commonly found in the members of a given species in nature (i.e., in the wild). Often, a wild-type characteristic occurs in all or nearly all members of a population. For example, the wild-type color of the savanna elephant (*Loxodonta africana*) is gray. Quantitatively, geneticists usually define wild type as a characteristic that occurs in at least 1% of the population. In some cases, a species may display a characteristic that exhibits more than one wild type. As an example, let's consider a hypothetical species of flowering plants:

- 42.5% have red flowers.
- 47.3% have purple flowers.
- 10.1% have pink flowers.
- 0.1% have yellow flowers.

In this example, red, purple, and pink flowers are wild types. Another way of saying this is that having red, purple, or pink flower color is normal. Alternatively, the percentage of plants with yellow color is below the 1% threshold, and therefore that flower color would not be considered wild type. When a given characteristic is very rare in a population and that characteristic is caused by genetic variation, geneticists may use the term *mutant* to distinguish the rare characteristic from wild-type ones.

Throughout this textbook, we will consider genetics at various levels. Some chapters largely focus on characteristics at the individual level, such as red versus white eyes in *Drosophila* or purple versus white flowers in pea plants. Other chapters explore genetics at the cellular and molecular levels, focusing on the properties of genes and chromosomes. The terms *normal* and *wild-type* can pertain to these different levels. For example, we can speak of a wild-type eye color, which is at the individual level, or we can speak of a wild-type allele of an eye color gene, which is at the molecular level. Likewise, we may consider the normal number of chromosomes in a given species, and discuss the consequences when a rare individual has an abnormal number.

Rare Genetic Variation May Be Beneficial, Neutral, or Detrimental

In genetics, we often consider how genetic variation, such as variation in a particular gene, affects an individual's traits. Such variation may affect survival and reproductive success and thereby be acted upon by natural selection. From this perspective, traits can be beneficial, neutral, or detrimental.

The species on Earth are the result of over 3.5 billion years of evolution. Therefore, the genes in individuals that confer wild-type traits tend to work pretty well. Because of this long history, random mutations tend to be neutral or detrimental with regard to survival and reproductive success. However, on some occasions, a rare mutation may confer a trait that is beneficial. The normalcy of a beneficial allele may change due to natural selection.

As an example, let's consider a species of frogs in which the wild-type color is medium green. A mutation in a single individual might confer a dark green color and this dark green color may make the frog less susceptible to predation. Initially, when the mutation is rare in this frog species, we would describe dark green frogs as mutants. They would be considered abnormal. However, over the course of many generations, natural selection may result in an increase in the frequency of the trait and eventually dark green color may become wild type. Thus, a given trait may begin as a rare mutant (i.e., abnormal), but eventually become wild type and be viewed as normal. In this scenario, keep in mind that the trait itself, that is, dark green color, has not changed. What has changed is its relative frequency in a population over the course of many generations.

1.6 COMPREHENSION QUESTIONS

1. Which of the following terms have the same meaning?
   a. Wild type and mutant
   b. Wild type and normal
   c. Wild type and abnormal
   d. Mutant and normal
2. In genetics, normal and wild type refer to
   a. a characteristic that is common in a given species.
   b. a characteristic that is uncommon in a given species.
   c. a gene that all individuals of a given species have.
   d. a chromosome that all individuals of a given species have.

KEY TERMS

Introduction: genome
1.1: genetics, gene, traits, nucleic acids, proteins, carbohydrates, lipids, macromolecules, organelle, proteome, enzymes, deoxyribonucleic acid (DNA), nucleotides, polypeptides, genetic code, codon, amino acid, chromosomes, karyotype, gene expression, transcription, ribonucleic acid (RNA), protein-coding genes, messenger RNA (mRNA), translation, central dogma of genetics
1.2: morphological traits, physiological traits, behavioral traits, molecular level, cellular level, organism level, species, population level, alleles, genetic variation, morphs, gene mutations, environment, phenylketonuria (PKU), diploid, homologs, somatic cells, gametes, haploid, biological evolution (evolution), natural selection
1.3: model organisms, genetic cross, genetic approach, loss-of-function mutation, loss-of-function allele
1.4: hypothesis testing, scientific method, discovery-based science
1.5: inclusion, gender, gender identity, female, male, sexual orientation
1.6: normal, abnormal, wild type

CHAPTER SUMMARY

- The complete genetic composition of a cell is its genome. The genome codes all of the proteins a cell can make. Many key discoveries in genetics are related to the study of genes and genomes (see Figures 1.1, 1.2, 1.3).

1.1 The Molecular Expression of Genes

- Living cells are composed of nucleic acids (DNA and RNA), proteins, carbohydrates, and lipids (see Figure 1.4).
- The entire collection of proteins a cell makes at a given time is its proteome. The proteome largely determines the structure and function of a cell.
- DNA, which is found within chromosomes, stores the information needed to make cellular proteins (see Figure 1.5).

- Most genes code polypeptides, which are units that assemble into functional proteins. Gene expression at the molecular level involves transcription to produce mRNA and translation to produce a polypeptide (see Figure 1.6).

1.2 The Relationship Between Genes and Traits

- Genetics, which governs an organism's traits, is studied at the molecular, cellular, organism, and population levels (see Figure 1.7).
- Genetic variation underlies variation in traits. In addition, the environment plays a key role (see Figures 1.8, 1.9).
- During reproduction, genetic material is passed from parents to offspring. In many species, somatic cells are diploid and have

two sets of chromosomes, whereas gametes are haploid and have a single set (see Figure 1.10).

- Evolution refers to changes in the genetic composition of a population from one generation to the next (see Figure 1.11).

1.3 Fields of Genetics

- Model organisms are studied by many different researchers so they can compare their results and determine scientific principles that apply more broadly to other species (see Figure 1.12).
- Genetics is traditionally divided into transmission genetics, molecular genetics, and population genetics, though overlap occurs among these fields.

1.4 The Science of Genetics

- Researchers in genetics carry out hypothesis testing or discovery-based science.
- Genetic TIPS are aimed at improving your ability to solve problems.

1.5 Gender Identity Versus Sex

- Gender refers to the characteristics of women and men that are socially constructed. Gender identity is a person's deeply held inner feelings about whether they're a woman, man, both, or neither.
- Sex is a physical characteristic that is usually assigned at birth based on external body parts associated with reproduction. Sex determines the type of gametes that an individual has the potential to produce. With regard to sex chromosome composition, females are usually XX and males are usually XY.
- In this textbook, the terms *male* and *female* refer to sex but not to gender identity.

1.6 The Meaning of Normal in Genetics

- In genetics, *normal* is a quantitative term that means common. The term *wild type* also means common.

PROBLEM SETS & INSIGHTS

MORE GENETIC TIPS   1. A human gene called the *CFTR* gene (for <u>c</u>ystic <u>f</u>ibrosis <u>t</u>ransmembrane <u>r</u>egulator) codes a protein that functions in the transport of chloride ions across the cell membrane. Most people have two copies of a functional *CFTR* gene and do not have cystic fibrosis. However, a mutant version of the *CFTR* gene is found in some people. A person who has two mutant copies of the gene will develop the disease known as cystic fibrosis. Does each of the following descriptions relate to genetics at the molecular, cellular, organism, or population level?

A. People with cystic fibrosis have lung problems due to a buildup of thick mucus in their lungs.

B. The mutant *CFTR* gene codes a defective chloride transporter.

C. A defect in the chloride transporter causes a salt imbalance in lung cells.

D. Scientists have wondered why the mutant *CFTR* gene is relatively common. It is the most common mutant gene that causes a severe disease in people of Northern European descent. One possible explanation why cystic fibrosis is so common is that people who have one copy of the functional *CFTR* gene and one copy of the mutant gene may be more resistant to diarrheal diseases such as cholera. Therefore, even though individuals with two mutant copies have the disorder, people with one mutant copy and one functional copy might have a survival advantage over people with two functional copies of the gene.

TOPIC: *What topic in genetics does this question address?* The topic is the different levels at which genetics is studied, ranging from the molecular to the population level.

INFORMATION: *What information do you know based on the question and your understanding of the topic?* The question describes the disease called cystic fibrosis. Parts A through D give descriptions of various aspects of the disease. From your understanding of the topic, you may remember that genetics can be studied at the molecular, cellular, organism, and population level. This concept is described in Figure 1.7.

PROBLEM-SOLVING **S**TRATEGY: *Make a drawing. Compare and contrast.* One strategy for solving this problem is to make a drawing of what is described in each of parts A through D and decide if you are drawing something at the molecular, cellular, organism, or population level. For example, if you drew the description in part B, you would draw a protein, which is a molecule. If you drew the description in part C, you would draw a cell in which a salt imbalance is present. Another strategy to solve this problem is to compare and contrast parts A, B, C, and D with each other. For example, if you compared part A and part D, you might realize that part A is describing something in one person, whereas part D is describing the occurrence of the mutant gene in multiple people.

ANSWER:

A. Organism level. This is a description of a trait at the level of an entire individual.

B. Molecular level. This is a description of a gene and the protein it codes.

C. Cellular level. This is a description of how protein function affects the cell.

D. Population level. This is a possible explanation of why two alleles of the gene occur within a population.

---

2. Most genes code proteins. Explain how proteins produce an organism's traits. Provide examples.

T **OPIC:** *What topic in genetics does this question address?* The topic is the relationship between genes and traits. More specifically, the question is about how proteins, which are coded by genes, produce an organism's traits.

I **NFORMATION:** *What information do you know based on the question and your understanding of the topic?* In the question, you are reminded that most genes code proteins and that proteins play a role in producing an organism's traits. From your understanding of the topic, you may remember that proteins carry out a variety of functions that determine cell structure and function.

P **ROBLEM-SOLVING** **S** **TRATEGY:** *Relate structure and function.* One strategy for solving this problem is to consider the relationship between protein structure and function. Think about examples in which the structure and function of proteins govern the structure and function of living cells. Also, consider how the structures and functions of cells determine an organism's traits.

ANSWER: The structure and function of proteins govern the structure and function of living cells. For example, specific proteins help determine the shape and structure of a given cell. The protein known as tubulin can assemble into large structures known as microtubules, which provide the cell with internal structure and organization. The proteins that a cell makes are largely responsible for the cell's structure and function. For example, the proteins made by a nerve cell cause the cell to be very elongated and to be able to receive and transmit signals to and from other cells. The structure of a nerve cell provides animals with many traits, such as the ability to sense the temperature of their environment and the ability to send signals to their muscles to promote movement.

Conceptual Questions

C1. Pick any example of a genetic technology and describe how it has directly affected your life.

C2. At the molecular level, what is a gene? Where are genes located?

C3. Most genes code polypeptides, which are functional units of proteins. Explain how the structure and function of proteins produce an organism's traits.

C4. Briefly explain how gene expression occurs at the molecular level.

C5. A human gene called the β-globin gene codes a polypeptide that functions as a subunit of the protein known as hemoglobin. Hemoglobin is found within red blood cells; it carries oxygen. In human populations, the β-globin gene can be found as the common allele called the Hb^A allele, and it can also be found as the Hb^S allele. Individuals who have two copies of the Hb^S allele have the disease called sickle cell disease. Are the following descriptions examples of genetics at the molecular, cellular, organism, or population level?

   A. The Hb^S allele codes a polypeptide that functions slightly differently from the polypeptide coded by the Hb^A allele.

   B. If an individual has two copies of the Hb^S allele, that person's red blood cells take on a sickle shape.

   C. Individuals who have two copies of the Hb^A allele do not have sickle cell disease, but they are not resistant to malaria. People who have one Hb^A allele and one Hb^S allele do not have sickle cell disease, and they are resistant to malaria. People who have two copies of the Hb^S allele have sickle cell disease, and this disease may significantly shorten their lives.

   D. Individuals with sickle cell disease have anemia because their red blood cells are easily destroyed by the body.

C6. What is meant by the term *genetic variation*? Give two examples of genetic variation not discussed in this chapter. What causes genetic variation at the molecular level?

C7. What is the cause of Down syndrome?

C8. The text describes how the detrimental symptoms associated with the disease phenylketonuria (PKU) are caused by a faulty gene. However, a change in diet can prevent these symptoms. Pick a trait of your favorite plant species, and explain how genetics and the environment may play important roles in the outcome of that trait.

C9. What is meant by the term *diploid*? Which cells of the human body are diploid, and which cells are not?

C10. What is a DNA sequence?

C11. What is the genetic code?

C12. Explain the relationship between each of these pairs of genetic terms:

   A. Gene and trait

   B. Gene and chromosome

   C. Allele and gene

   D. DNA sequence and amino acid sequence

C13. With regard to biological evolution, which of the following statements is incorrect? Explain why.

   A. During its lifetime, an animal evolves to become better adapted to its environment.

   B. The process of biological evolution has produced species that are better adapted to their environments.

   C. When an animal is better adapted to its environment, the process of natural selection makes it more likely that the animal will reproduce.

C14. What are the primary interests of researchers working in the following fields of genetics?

   A. Transmission genetics

   B. Molecular genetics

   C. Population genetics

Experimental Questions

E1. What is a genetic cross?

E2. The technique known as DNA sequencing (described in Chapter 20) enables researchers to determine the DNA sequence of genes. Would this technique be used primarily by transmission geneticists, molecular geneticists, or population geneticists?

E3. Figure 1.5 shows a micrograph of the common number of chromosomes from a human cell. If you created this kind of image using a cell from a person with Down syndrome, what would you expect to see?

E4. Many organisms are studied by geneticists. Of the following species, do you think it is more likely that each of them would be studied by a transmission geneticist, a molecular geneticist, or a population geneticist? Explain your answer. Note: More than one answer may be possible for a given species.

A. Dogs

B. *E. coli*

C. Fruit flies

D. Leopards

E. Corn

E5. Pick any trait you like in any species of wild plant or animal. The trait must somehow vary among different members of the species (see Figure 1.7). Note: When picking a trait to answer this question, do not pick the trait of wing color in butterflies.

A. Summarize all of the background information that you already have (from personal observations) regarding this trait.

B. Propose a hypothesis that would explain the genetic variation within the species. For example, in the case of the butterflies, your hypothesis might be that the dark butterflies survive better in dark forests and the light butterflies survive better in sunlit fields.

C. Describe the experimental steps you would follow to test your hypothesis.

D. Describe the possible data you might collect.

E. Interpret your data.

Note: All answers are available for the instructor in Connect; the answers to the even-numbered questions and all of the Concept Check and Comprehension Questions are in Appendix B.

CHAPTER OUTLINE

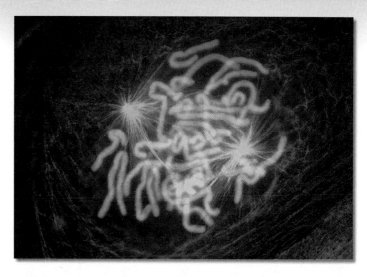

Chromosome sorting during cell division. *When eukaryotic cells divide, their chromosomes (shown in blue) are replicated and sorted, so that each cell receives the correct number.*

Photomicrograph by Dr. Conly L. Rieder, Wadsworth Center, Albany, New York 12201-0509

2 CHROMOSOME TRANSMISSION DURING CELL DIVISION AND SEXUAL REPRODUCTION

In this chapter, we will begin by considering the general features of chromosomes and how they are observed under the microscope. We then examine how bacterial and eukaryotic cells divide. During cell division, eukaryotic cells can sort their chromosomes in two different ways. One process, called mitosis, sorts chromosomes so that each daughter cell receives the same number and types of chromosomes as the original mother cell had. A second process, called meiosis, results in daughter cells with half the number of chromosomes that the mother cell had. Meiosis is necessary for sexual reproduction in eukaryotic species, such as animals and plants. During sexual reproduction, gametes with half the number of chromosomes unite at fertilization. We will also explore how chromosomes and environmental factors can determine whether individuals develop into males or females.

2.1 GENERAL FEATURES OF CHROMOSOMES

Learning Outcomes:

1. Define the term *chromosome*.
2. Outline the key differences between prokaryotic and eukaryotic cells.
3. Describe the procedure for making a karyotype.
4. Summarize the similarities and differences between homologous chromosomes.

The **chromosomes** are structures within living cells that contain the genetic material. The term *chromosome*—meaning "colored body"—refers to the microscopic observation of chromosomes after they have been stained with dyes. Genes are physically located within chromosomes. Biochemically, each chromosome

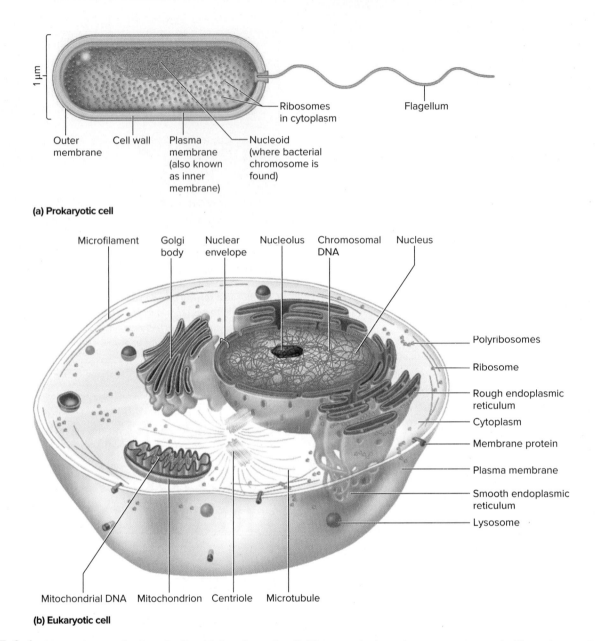

FIGURE 2.1 **The basic organization of cells.** (**a**) A prokaryotic cell. The example shown here represents a typical bacterium, such as *Escherichia coli*, which has an outer membrane. (**b**) A eukaryotic cell. The example shown here is a typical animal cell.

CONCEPT CHECK: Eukaryotic cells exhibit compartmentalization. Define *compartmentalization*.

contains a very long segment of DNA, which is the genetic material, and proteins, which are bound to the DNA and provide it with an organized structure. In eukaryotic cells, this complex between DNA and proteins is called **chromatin.** In this chapter, we will focus on the cellular mechanics of chromosome transmission during cell division to better understand the patterns of gene transmission that we will consider in Chapters 3 and 4. In particular, we will examine how chromosomes are copied and sorted into newly made cells.

Before we begin a description of chromosome transmission, we need to consider the distinctive cellular differences between prokaryotic and eukaryotic species. Bacteria and archaea are referred to as **prokaryotes,** from the Greek meaning

"prenucleus," because their chromosomes are not contained within a membrane-bound nucleus in the cell. Prokaryotes usually have a single type of circular chromosome in a region of the cytoplasm called the **nucleoid (Figure 2.1a)**. The cytoplasm is enclosed by a plasma membrane that regulates the uptake of nutrients and the excretion of waste products. Outside the plasma membrane is a rigid cell wall that protects the cell from breakage. Certain species of bacteria also have an outer membrane on the exterior side of the cell wall.

Eukaryotes, from the Greek meaning "true nucleus," include some simple species, such as single-celled protists and some fungi (such as yeast), and more complex multicellular species, such as plants, animals, and other fungi. The cells of eukaryotic species have internal membranes that enclose highly specialized compartments (**Figure 2.1b**). These compartments form membrane-bound **organelles** with specific functions. For example, the lysosomes play a role in the degradation of macromolecules. The endoplasmic reticulum and Golgi body play a role in protein modification and trafficking. A particularly conspicuous organelle is the **nucleus,** which is bounded by two membranes that constitute the nuclear envelope.

Most of the genetic material in a eukaryotic cell is found within chromosomes, which are located in the nucleus. In addition to the nucleus, certain organelles in eukaryotic cells contain a small amount of their own DNA. These include the mitochondrion, which functions in ATP synthesis, and the chloroplast, in plant and algal cells, which functions in photosynthesis. The DNA found in these organelles is referred to as extranuclear, or extrachromosomal, DNA to distinguish it from the DNA that is found in the cell nucleus. We will examine the role of mitochondrial and chloroplast DNA in Chapter 5.

In this section, we will focus on the composition of chromosomes found in the nucleus of eukaryotic cells. As you will learn, eukaryotic species contain genetic material that comes in sets of linear chromosomes.

Eukaryotic Chromosomes Are Examined Cytologically to Prepare a Karyotype

Insights into inheritance patterns have been gained by observing chromosomes under the microscope. **Cytogenetics** is the field of genetics that involves the microscopic examination of chromosomes. The most basic observation that a **cytogeneticist** can make is the examination of the chromosomal composition of a particular cell. For eukaryotic species, this is usually accomplished by observing the chromosomes as they are found in actively dividing cells. When a cell is preparing to divide, the chromosomes become more tightly coiled, which shortens them and increases their diameter. The consequence of this shortening is that distinctive shapes and numbers of chromosomes become visible with a light microscope.

Each species has a particular chromosome composition. For example, most human cells contain 23 pairs of chromosomes, for a total of 46. On rare occasions, some individuals may inherit an abnormal number of chromosomes or a chromosome with an abnormal structure. Such abnormalities can often be detected by a microscopic examination of the chromosomes within actively dividing cells. In addition, a cytogeneticist may examine chromosomes as a way to distinguish between two closely related species.

Figure 2.2a shows the general procedure for preparing human chromosomes to be viewed by microscopy. In this example, the cells were obtained from a sample of human blood; more specifically, the chromosomes within leukocytes (also called white blood cells) were examined. Blood cells are a type of **somatic cell.** This term refers to any cell of the body that is not a gamete or a precursor to a gamete. The **gametes** (sperm and egg cells or their precursors) are also called **germ cells.**

After the blood cells have been removed from the body, they are treated with one chemical that stimulates them to begin cell division and another chemical that halts cell division during mitosis, which is described later in this chapter. As shown in Figure 2.2a, these actively dividing cells are subjected to centrifugation to concentrate them. The concentrated preparation is then mixed with a hypotonic solution that makes the cells swell. This swelling causes the chromosomes to spread out within the cell, thereby making it easier to see each individual chromosome. Next, the cells are treated with a fixative that chemically freezes them so that the chromosomes will no longer move around. The cells are then treated with a chemical dye that binds to the chromosomes and stains them. As discussed in greater detail in Chapter 8, this gives chromosomes a distinctive banding pattern that greatly enhances the ability to visualize and to uniquely identify them (look ahead to Figure 8.1c, d). The cells are then placed on a slide and viewed with a light microscope.

In a cytogenetics laboratory, the microscopes are equipped with a camera that can photograph the chromosomes. In recent years, advances in technology have allowed cytogeneticists to view microscopic images on a computer screen (**Figure 2.2b**). On the screen, the chromosomes can be arranged in a standard way, usually from largest to smallest. As seen in **Figure 2.2c**, the human chromosomes are lined up, and a number is used to designate each type of chromosome. An exception is the sex chromosomes, which are designated with the letters X and Y. An organized representation of the chromosomes within a cell is called a **karyotype.** A karyotype reveals how many chromosomes are found within an actively dividing somatic cell.

Eukaryotic Chromosomes Are Inherited in Sets

Most eukaryotic species are **diploid** or have a diploid phase in their life cycle, which means that each type of chromosome is a member of a pair. A diploid cell has two sets of chromosomes. In humans, most somatic cells have 46 chromosomes—two sets of 23 each. Other diploid species, however, have different numbers of chromosomes in their somatic cells. For example, the dog has 39 chromosomes per set (78 total), the fruit fly has 4 chromosomes per set (8 total), and the tomato plant has 12 per set (24 total).

When a species is diploid, the members of a pair of chromosomes are called **homologs;** each type of chromosome is found in a homologous pair. As shown in Figure 2.2c, for example, a human somatic cell has two copies of chromosome 1, two copies of chromosome 2, and so forth. Within each pair, the chromosome on the left is a homolog to the one on the right, and vice versa. In each pair, one chromosome was inherited from the female parent, and its homolog was inherited from the male parent.

The two chromosomes in a homologous pair are usually identical in size, have the same banding pattern, and contain a similar composition of genetic material. If a particular gene is found on one copy of a chromosome, it is also found on the other homolog. However, the two homologs may carry different versions of a given gene, which are called **alleles.** As discussed in Chapter 3, some alleles are dominant, meaning that they mask the expression of recessive alleles. As an example, let's consider a gene in humans, called *Herc2*, which is one of a few different genes that affect eye color. The *Herc2* gene is located on

A sample of blood is collected and treated with chemicals that stimulate the cells to divide. Colchicine is added because it disrupts spindle formation and stops cells in mitosis when the chromosomes are highly compacted. The cells are then subjected to centrifugation.

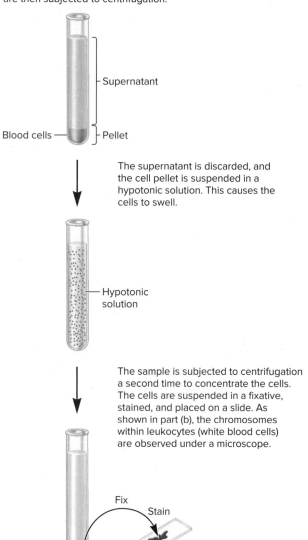

The supernatant is discarded, and the cell pellet is suspended in a hypotonic solution. This causes the cells to swell.

The sample is subjected to centrifugation a second time to concentrate the cells. The cells are suspended in a fixative, stained, and placed on a slide. As shown in part (b), the chromosomes within leukocytes (white blood cells) are observed under a microscope.

(a) Preparing cells for a karyotype

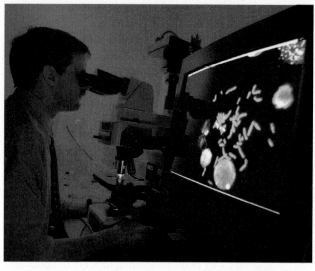

(b) The slide is viewed by a light microscope; the sample is seen on a computer screen. The chromosomes can be arranged electronically on the screen.

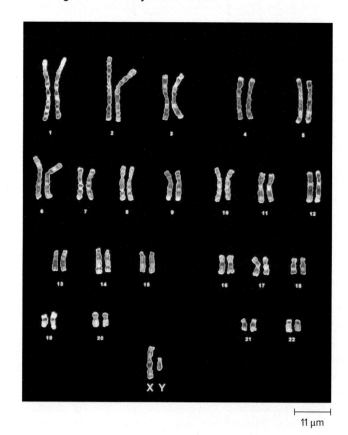

11 μm

(c) For a diploid human cell, two complete sets of chromosomes from that cell constitute a karyotype of the cell.

FIGURE 2.2   **The procedure for making a human karyotype.**

(b): David Parker/Science Source; (c): Leonard Lessin/Science Source

CONCEPT CHECK: How do you think the end results of karyotype preparation would be affected if the blood cells were not treated with a hypotonic solution?

chromosome 15 and comes in variants that result in brown or blue eyes. In a person with brown eyes, one copy of chromosome 15 may carry a dominant brown allele, whereas its homolog may carry a recessive blue allele.

At the molecular level, how similar are homologous chromosomes? The answer is that the sequence of bases of one homolog usually differs from the sequence of the other homolog by less than 1%. For example, the DNA sequence of chromosome 1 that you inherited from your female parent is more than 99% identical to the sequence of chromosome 1 that you inherited from your male parent.

It is worth emphasizing that the DNA sequences on homologous chromosomes are not completely identical. The slight differences in DNA sequences provide the allelic differences in genes. Again, if we use the eye color gene as an example, a slight difference in DNA sequence distinguishes the brown and blue alleles.

However, the striking similarities between homologous chromosomes do not apply to the pair of sex chromosomes—X and Y. These chromosomes differ in size and genetic composition. Certain genes that are found on the X chromosome are not found on the Y chromosome, and vice versa. The X and Y chromosomes are not considered homologous chromosomes even though they do have short regions of homology.

Figure 2.3 shows two homologous chromosomes with three different genes labeled. Dominant alleles are denoted with an uppercase letter, whereas recessive alleles are denoted with a lowercase letter. An individual having these two chromosomes is **homozygous** for the dominant allele of gene A, which means that both homologs carry the same allele. The individual is **heterozygous,** *Bb,* for the second gene, meaning that the homologs carry different alleles. For the third gene, the individual is homozygous for a recessive allele, *c.* The physical location of a gene is called its **locus (plural: loci).** As seen in Figure 2.3, for example, the locus of gene C is toward one end of this chromosome, whereas the locus of gene B is more in the middle.

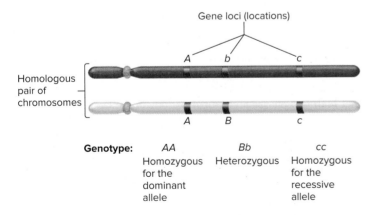

FIGURE 2.3 **A comparison of homologous chromosomes.** Each of the homologous chromosomes in a pair carries the same types of genes, but, as shown here, the alleles may or may not be different.

CONCEPT CHECK: How are homologs similar to each other, and how are they different?

2.1 COMPREHENSION QUESTIONS

1. Which of the following is *not* found in a prokaryotic cell?
   a. Plasma membrane
   b. Ribosome
   c. Cell nucleus
   d. Cytoplasm

2. When a karyotype is prepared, which of the following steps is carried out?
   a. Treat the cells with a chemical that causes them to begin cell division.
   b. Treat the cells with a hypotonic solution that causes them to swell.
   c. Expose the cells to chemical dyes that bind to the chromosomes and stain them.
   d. All of the above steps are carried out.

3. How many sets of chromosomes are found in a human somatic cell, and how many chromosomes are within one set?
   a. 2 sets, with 23 in each set
   b. 23 sets, with 2 in each set
   c. 1 set, with 23 in each set
   d. 23 sets, with 1 in each set

2.2 CELL DIVISION

Learning Outcomes:

1. Describe the process of binary fission in bacteria.
2. Outline the phases of the eukaryotic cell cycle.

Now that we have an appreciation for the chromosomal composition of living cells, we can consider how chromosomes are copied and transmitted when cells divide. One purpose of cell division is **asexual reproduction.** In this process, a preexisting cell divides to produce two new cells. By convention, the original cell is usually called the mother cell, and the two new cells are the daughter cells. In unicellular species, the mother cell is judged to be one organism, and the two daughter cells are two new separate organisms. Asexual reproduction is how bacterial cells proliferate. In addition, certain unicellular eukaryotes, such as the amoeba and baker's yeast (*Saccharomyces cerevisiae*), can reproduce asexually.

Another purpose of cell division is to achieve **multicellularity.** Species such as plants, animals, most fungi, and some protists are derived from a single cell that has undergone repeated cellular divisions. Humans, for example, begin as a single fertilized egg; repeated cell divisions produce an adult with trillions of cells. The precise transmission of chromosomes during every cell division is critical so that all cells of the body receive the correct amount of genetic material.

In this section, we will consider how the duplication, organization, and distribution of the chromosomes are critical to the process of cell division. In bacteria, which have a single circular chromosome, the division process is relatively simple. Prior to cell division, bacteria duplicate their circular chromosome; they then distribute a copy into each of the two daughter cells. This process, known as binary fission, is described first. Eukaryotes have multiple chromosomes that occur as sets. This added complexity in eukaryotic cells requires a more complicated sorting process, called mitosis, so that each newly made cell receives the correct number and types of chromosomes.

Bacteria Reproduce Asexually by Binary Fission

As discussed earlier in this chapter (see Figure 2.1a), bacterial species are typically unicellular, although individual bacteria may associate with each other to form pairs, chains, or clumps. Unlike eukaryotes, which have their chromosomes in a separate nucleus, the circular chromosome of a bacterium is in direct contact with the cytoplasm.

The capacity of bacteria to divide is really quite astounding. Some species, such as *Escherichia coli* (*E. coli*), a common species in the intestine, can divide every 20 to 30 minutes. Prior to cell division, bacterial cells copy, or replicate, their chromosomal DNA. This produces two identical copies of the genetic material, as shown at the top of **Figure 2.4**. Following DNA replication, a

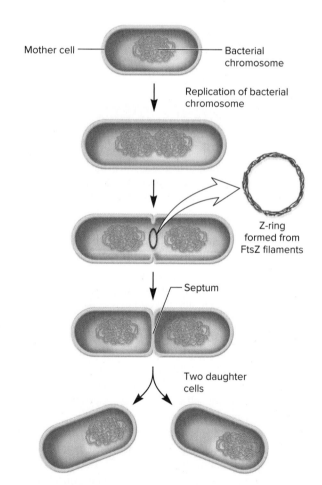

FIGURE 2.4 Binary fission: the process by which bacterial cells divide. Prior to division, the chromosome replicates to produce two identical copies. These two copies segregate from each other, with one copy going to each daughter cell.

CONCEPT CHECK: What is the function of the FtsZ protein during binary fission?

bacterial cell divides into two daughter cells by a process known as **binary fission.**

1. Prior to cell division, bacterial cells copy, or replicate, their chromosomal DNA. This produces two identical copies of the genetic material, as shown at the top of Figure 2.4.
2. A protein called FtsZ is the first protein to move to the division site that will separate the two daughter cells. Filaments composed of FtsZ protein assemble into a structure called a Z-ring (see inset in Figure 2.4).
3. FtsZ recruits other proteins to produce a septum, which is a new cell wall between the daughter cells. The filaments within the Z-ring are thought to pull on each other and tighten to promote the formation of a septum. FtsZ is evolutionarily related to a eukaryotic protein called tubulin. As discussed later in this chapter, tubulin is the main component of microtubules, which play a key role in chromosome sorting in eukaryotes. Both FtsZ and tubulin form structures that provide cells with organization and play key roles in cell division.
4. As a result of binary fission, a bacterial cell called the mother cell has divided into two daughter cells. Each daughter cell receives a copy of the chromosomal genetic material. Except when rare mutations occur, the daughter cells are usually genetically identical because they contain exact copies of the genetic material from the mother cell.

Binary fission is an asexual form of reproduction because it does not involve genetic contributions from two different gametes. On occasion, bacteria can transfer small pieces of genetic material to each other, which is described in Chapter 7.

Eukaryotic Cells Advance Through a Cell Cycle to Produce Genetically Identical Daughter Cells

The common outcome of eukaryotic cell division is the production of two daughter cells that have the same number and types of chromosomes as the original mother cell. This requires a replication and division process that is more complicated than simple binary fission. Eukaryotic cells that are destined to divide advance through a series of phases known as the **cell cycle** (**Figure 2.5**). These phases are named G for gap, S for synthesis (of the genetic material), and M for mitosis. There are two G phases: G_1 and G_2. The term *gap* originally described the gaps between S phase and mitosis in which it was not microscopically apparent that significant changes were occurring in the cell. However, we now know that both gap phases are critical periods in the cell cycle that involve many molecular changes. In actively dividing cells, the G_1, S, and G_2 phases are collectively known as **interphase.**

In addition, cells may remain permanently, or for long periods of time, in a phase of the cell cycle called G_0. A cell in the G_0 phase is either temporarily not advancing through the cell cycle or, in the case of terminally differentiated cells such as most nerve cells in an adult mammal, never dividing again. In other words, the G_0 phase is a nondividing stage.

Let's consider the key steps in these four phases.

1. During the **G_1 phase,** a cell may prepare to divide. Depending on the cell type and the conditions the cell

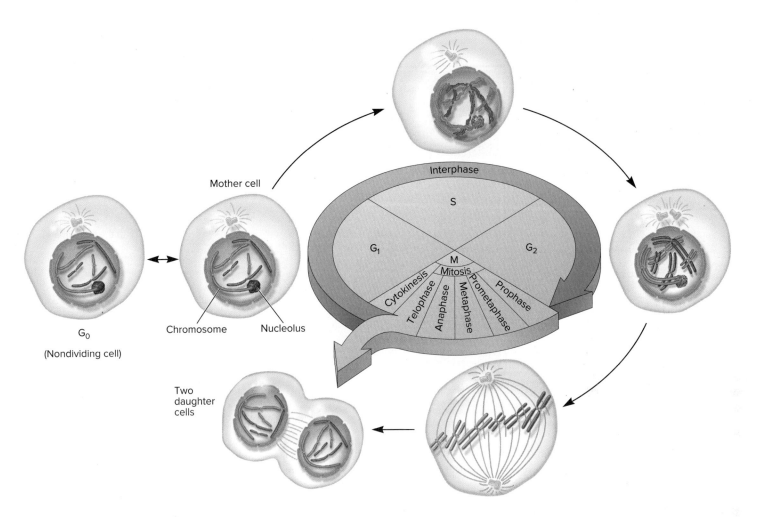

FIGURE 2.5 **The eukaryotic cell cycle.** Dividing cells advance through a series of phases, denoted G_1, S, G_2, and M. This diagram shows the advancement of a cell through mitosis to produce two daughter cells. The original diploid cell had three pairs of chromosomes, for a total of six individual chromosomes. During S phase, these have replicated to yield 12 chromatids found in six pairs of sister chromatids. After mitosis and cytokinesis are completed, each of the two daughter cells contains six individual chromosomes, just like the mother cell. Note: The chromosomes in G_0, G_1, S, and G_2 phases are not actually condensed as shown here. In this drawing, they are shown partially condensed so they can be easily counted.

CONCEPT CHECK: What is the difference between the G_0 and G_1 phases?

encounters, a cell in the G_1 phase may accumulate molecular changes (e.g., produce new proteins) that cause it to advance through the rest of the cell cycle. When this occurs, cell biologists say that a cell has reached a **restriction point** and is committed to a pathway that leads to cell division.

2. Once past the restriction point, the cell then advances to the **S phase,** during which the chromosomes are replicated. After replication, the two copies of a chromosome are called **chromatids.** They are joined to each other at a region of DNA called the **centromere** to form a unit known as a pair of **sister chromatids,** or a **dyad** (**Figure 2.6**). A single chromatid within a dyad is called a **monad.** An unreplicated chromosome can also be called a monad. The **kinetochore** is a group of proteins that are bound to the centromere. These proteins help to hold the sister chromatids together and also play a role in chromosome sorting, as discussed later in this chapter.

When S phase is completed, a cell has twice as many chromatids as it had chromosomes in the G_1 phase. For example, a human cell in the G_1 phase has 46 distinct

chromosomes, whereas in G_2, it has 46 pairs of sister chromatids, for a total of 92 chromatids. The term *chromosome* can be a bit confusing because it originally meant a distinct structure that is observable with the microscope. Therefore, *chromosome* can refer either to a pair of sister chromatids (a dyad) during G_2 and early stages of M phase or to a structure that is observed at the end of M phase and during G_1, which is a monad and contains the equivalent of one chromatid (refer back to Figure 2.5).

3. During the **G_2 phase,** the cell accumulates the materials necessary for nuclear and cell division.

4. Next, the cell advances into the **M phase** of the cell cycle, when **mitosis** occurs.

5. In most cases, two daughter cells are formed by a process called cytokinesis.

The primary purpose of mitosis is to distribute the replicated chromosomes, dividing one cell nucleus into two nuclei, so that each daughter cell receives the same complement of

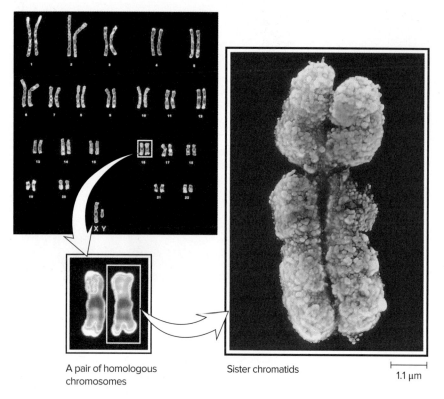

A pair of homologous
chromosomes

Sister chromatids

1.1 μm

(a) Homologous chromosomes and sister chromatids

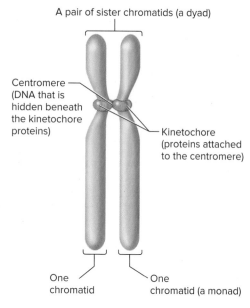

A pair of sister chromatids (a dyad)

Centromere
(DNA that is
hidden beneath
the kinetochore
proteins)

Kinetochore
(proteins attached
to the centromere)

One
chromatid

One
chromatid (a monad)

(b) Schematic drawing of sister chromatids

FIGURE 2.6 **Chromosomes following DNA replication.** **(a)** The photomicrograph on the upper left shows a human karyotype. The large pho-tomicrograph on the right shows a chromosome in the form called a dyad, or pair of sister chromatids. This chromosome is in the metaphase stage of mitosis, which is described later in this chapter. Note: Each of the 46 chromosomes that are viewed in a human karyotype (upper left) is actually a pair of sister chromatids. Look closely at the two insets. **(b)** A schematic drawing of sister chromatids. This structure has two chromatids that lie side by side. As seen here, each chromatid is a distinct unit, called a monad. The two chromatids are held together by kinetochore proteins that bind to each other and to the centromere of each chromatid.

(a): (top left & bottom inset) Leonard Lessin/Science Source; (right): Biophoto Associates/Science Source

CONCEPT CHECK: What is the difference between homologs and sister chromatids?

chromosomes. For example, a human cell in the G_2 phase has 92 chromatids, which are found in 46 pairs. During mitosis, these pairs of chromatids are separated and sorted in such a way that each daughter cell receives 46 chromosomes.

Mitosis was first observed microscopically in the 1870s by German biologist Walther Flemming, who coined the term *mitosis* (from the Greek *mitos*, meaning "thread"). He studied the dividing epithelial cells of salamander larvae and noticed that chromo-somes were constructed of two parallel "threads." These threads separated and moved apart, one going to each of the two daughter nuclei. We will examine the steps of mitosis in the next section.

2.2 COMPREHENSION QUESTIONS

1. Binary fission
   a. is a form of asexual reproduction.
   b. is a way for bacteria to reproduce.
   c. begins with a single mother cell and produces two geneti-cally identical daughter cells.
   d. All of the above are true of binary fission.

2. Which of the following is the correct order of phases of the eu-karyotic cell cycle?
   a. G_1, G_2, S, M
   b. G_1, S, G_2, M
   c. G_1, G_2, M, S
   d. G_1, S, M, G_2

3. What critical event occurs during the S phase of the eukaryotic cell cycle?
   a. The cell either prepares to divide or commits to not dividing.
   b. DNA replication produces pairs of sister chromatids.
   c. The chromosomes condense.
   d. The single nucleus is divided into two nuclei.

2.3   MITOSIS AND CYTOKINESIS

Learning Outcomes:

1. Describe the structure and function of the mitotic spindle apparatus.
2. List and describe the phases of mitosis.
3. Outline the key differences between animal and plant cells with regard to cytokinesis.

As we have seen, eukaryotic cell division involves a cell cycle in which the chromosomes are replicated and then sorted so that each daughter cell receives the same amount of genetic material. This process ensures genetic consistency from one generation of cells to the next. In this section, we will examine the stages of mitosis and cytokinesis in greater detail.

The Mitotic Spindle Apparatus Organizes and Sorts Eukaryotic Chromosomes

Before we discuss the events of mitosis, let's first consider the structure of the **mitotic spindle apparatus** (also known simply as the **spindle apparatus** or the **mitotic spindle**), which is involved in the organization and sorting of chromosomes (**Figure 2.7**). The spindle apparatus is formed from **microtubule-organizing centers (MTOCs)**, which are structures found in eukaryotic cells from which microtubules grow. Microtubules are produced from the rapid polymerization of tubulin proteins. In animal cells, the spindle apparatus is formed from two MTOCs called **centrosomes.** Each centrosome is located at a **spindle pole.** A pair of **centrioles** at right angles to each other is found within each centrosome. Centrosomes and centrioles are found in animal cells but not in all eukaryotic species. For example, plant cells do not have centrosomes. Instead, the nuclear envelope functions as an MTOC for the formation of the spindle apparatus in plant cells.

The spindle apparatus of a typical animal cell has three types of microtubules (see Figure 2.7).

- The **astral microtubules** emanate outward from the centrosome toward the plasma membrane. They are important for the positioning of the spindle apparatus within the cell.
- The **polar microtubules** project toward the region where the chromosomes will be found during mitosis—the region between the two spindle poles. Polar microtubules that overlap with each other play a role in the separation of the two poles. They help to "push" the poles away from each other.
- The **kinetochore microtubules** have attachments to **kinetochores,** which are protein complexes bound to the centromeres of individual chromosomes.

The spindle apparatus allows cells to organize and separate chromosomes so that each daughter cell receives the same complement of chromosomes. This sorting process, known as mitosis, is described next.

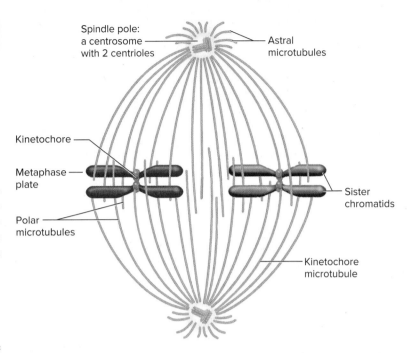

FIGURE 2.7   The structure of the mitotic spindle apparatus in a typical animal cell. During the cell cycle, a single centrosome duplicates in S phase and the two centrosomes separate at the beginning of M phase. The spindle apparatus is formed from microtubules that are rooted in the centrosomes. Each centrosome is located at a spindle pole. The astral microtubules emanate away from the region between the poles. They help position the spindle apparatus within the cell. However, the spindle apparatus formed in many species, such as plants, does not have astral microtubules. The polar microtubules project into the region between the two poles; they play a role in pole separation. The kinetochore microtubules are attached to the kinetochores of sister chromatids. Note: For simplicity, only a single pair of homologous chromosomes is shown; eukaryotic cells typically have many pairs of homologous chromosomes.

CONCEPT CHECK: Where are the two ends of a kinetochore microtubule?

The Transmission of Chromosomes During the Division of Eukaryotic Cells Requires a Process Known as Mitosis

The process of mitosis is shown for a diploid animal cell in **Figure 2.8**. In the simplified diagrams below the micrographs in the figure, the original mother cell contains six chromosomes; it is diploid ($2n$) and has three chromosomes per set ($n = 3$). One set is shown in blue, and the homologous set is shown in red. As discussed next, mitosis is subdivided into phases known as prophase, prometaphase, metaphase, anaphase, and telophase.

Prophase   Prior to mitosis, the cells are in interphase, during which the chromosomes are **decondensed**—less tightly compacted—and found in the nucleus (Figure 2.8a). At the start of mitosis, in **prophase,** the chromosomes have already replicated, resulting in 12 chromatids that are joined as six pairs of sister chromatids (Figure 2.8b). As prophase proceeds, the nuclear membrane begins to dissociate into small vesicles and the nucleolus becomes less visible. At the same time, the chromatids become

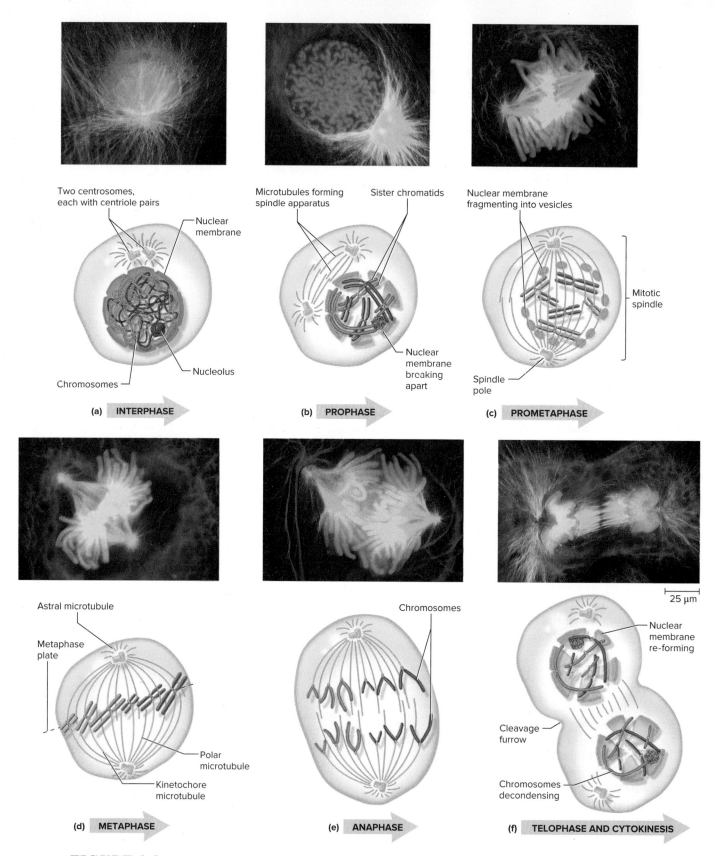

FIGURE 2.8 **The process of mitosis in an animal cell.** The two rows of micrographs illustrate cells of a fish embryo advancing through mitosis. The chromosomes are stained in blue and the microtubules are green. Below the micrographs are schematic diagrams that emphasize the sorting and separation of the chromosomes. In the diagrams, the original diploid cell is shown with six chromosomes (three in each set). At the start of mitosis, these have already replicated into 12 chromatids. The final result is two daughter cells, each containing six chromosomes.

ONLINE ANIMATION

(a–f): Photomicrographs by Dr. Conly L. Rieder, Wadsworth Center, Albany, New York 12201-0509

CONCEPT CHECK: During which phase of mitosis are sister chromatids separated and sent to opposite poles?

condensed into more compact structures that are readily visible by light microscopy. The two centrosomes move apart, and the spindle apparatus begins to form.

Prometaphase   As mitosis advances from prophase to prometaphase, the centrosomes move to opposite ends of the cell and establish two spindle poles, one within each of the future daughter cells. During **prometaphase,** the nuclear membrane is completely fragmented into vesicles, allowing the spindle fibers to interact with the sister chromatids (Figure 2.8c).

How do sister chromatids become attached to the spindle apparatus? Initially, microtubules form rapidly and can be seen growing out from the two poles. As a microtubule grows, if its end happens to make contact with a kinetochore, the end is said to be captured and remains firmly attached to the kinetochore. This random process is how sister chromatids become attached to kinetochore microtubules. Alternatively, if the end of a microtubule does not collide with a kinetochore, the microtubule eventually depolymerizes and retracts to the centrosome. As the end of prometaphase nears, the kinetochore on a pair of sister chromatids is attached to kinetochore microtubules from opposite poles. As these events are occurring, the sister chromatids undergo jerky movements as they are tugged, back and forth, between the two poles. By the end of prometaphase, the spindle apparatus is completely formed.

Metaphase   Eventually, the pairs of sister chromatids align themselves along a plane called the **metaphase plate.** As shown in Figure 2.8d, when this alignment is complete, the cell is in **metaphase** of mitosis. At this point, each pair of chromatids (each dyad) is attached to both poles by kinetochore microtubules. The pairs of sister chromatids have become organized into a single row along the metaphase plate. When this organizational process is finished, the chromatids can be equally distributed into two daughter cells.

Anaphase   At **anaphase,** the connection that is responsible for holding each pair of chromatids together is broken (Figure 2.8e). (We will examine the process of sister chromatid cohesion during prophase and separation during anaphase in more detail in Chapter 10; look ahead to Figure 10.30.) Each chromatid, or monad, now an individual chromosome, is linked to only one of the two poles. As anaphase proceeds, the chromosomes move toward the pole to which they are attached. This movement is due to the shortening of the kinetochore microtubules. In addition, the two poles themselves move farther apart due to the elongation of the polar microtubules, which slide in opposite directions as a result of the actions of motor proteins.

Telophase   During **telophase,** the chromosomes reach their respective poles and decondense. The nuclear membrane now reforms to produce two separate nuclei. In Figure 2.8f, this membrane re-formation has produced two nuclei that contain six chromosomes each. The nucleoli have also reappeared.

Cytokinesis   In most cases, mitosis is quickly followed by **cytokinesis,** in which the two nuclei are segregated into separate daughter cells. Likewise, cytokinesis also segregates cell organelles, such as mitochondria and chloroplasts, into daughter cells.

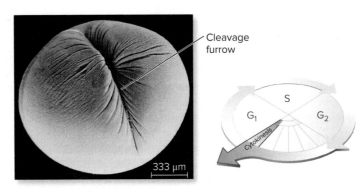

(a) Cleavage of an animal cell

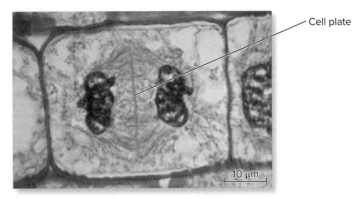

(b) Formation of a cell plate in a plant cell

 FIGURE 2.9   Cytokinesis in animal and plant cells. **(a)** In an animal cell, cytokinesis involves the formation of a cleavage furrow. **(b)** In a plant cell, cytokinesis occurs via the formation of a cell plate between the two daughter cells.
(a): Don W. Fawcett/Science Source; (b): Ed Reschke

CONCEPT CHECK: What causes the cleavage furrow in a dividing animal cell to ingress?

In animal cells, cytokinesis begins shortly after anaphase. A contractile ring, composed of myosin motor proteins and actin filaments, assembles at the cytoplasmic surface of the plasma membrane. Myosin hydrolyzes ATP, which shortens the ring, thereby constricting the plasma membrane to form a **cleavage furrow** that ingresses, or moves inward (**Figure 2.9a**). Ingression continues until a midbody structure is formed that physically pinches the single cell into two cells.

In plants, the two daughter cells are separated by the formation of a **cell plate** (**Figure 2.9b**). At the end of anaphase, Golgi-derived vesicles carrying cell wall materials are transported to the equator of a dividing cell. The fusion of these vesicles gives rise to the cell plate, which is a membrane-bound compartment. The cell plate begins in the middle of the cell and expands until it attaches to the mother cell's wall. Once this attachment has taken place, the cell plate undergoes a process of maturation and eventually separates the mother cell into two daughter cells.

Outcome of Mitotic Cell Division   Mitosis and cytokinesis ultimately produce two daughter cells having the same number of chromosomes as the mother cell. Barring rare mutations, the two daughter cells are genetically identical to each other and to the mother cell from which they were derived. The critical consequence of this sorting process is to ensure genetic consistency from one generation of somatic cells to the next. The development of multicellularity relies on the repeated process of mitosis and cytokinesis. In diploid organisms that are multicellular, most of the somatic cells are diploid and genetically identical to each other.

GENETIC **TIPS**   **THE QUESTION:** What are the functional

roles of the mitotic spindle apparatus in an animal cell? Explain how these functions are related to the three types of microtubules: astral, polar, and kinetochore microtubules.

TOPIC: *What topic in genetics does this question address?* The topic is mitosis. More specifically, the question is about the roles of the mitotic spindle apparatus.

INFORMATION: *What information do you know based on the question and your understanding of the topic?* From the question, you know there are three types of microtubules. From your understanding of the topic, you may remember the structure of the spindle apparatus, which is shown in Figure 2.7. Also, Figure 2.8 shows the roles that the spindle apparatus plays during mitosis.

PROBLEM-SOLVING **S**TRATEGY: *Define key terms. Describe the steps.* One strategy to begin solving this problem is to make sure you understand the key terms. In particular, you may want to look up the meaning of *mitotic spindle apparatus* and *microtubules*, if you don't already know what those terms mean. After you understand the key terms, another useful problem-solving strategy is to describe the steps of mitosis, and think about the roles of the types of microtubules in the various steps. These steps are shown in Figure 2.8. You may also want to refer back to Figure 2.7 to appreciate the structure of the spindle apparatus.

ANSWER: The spindle apparatus is involved in sorting the chromosomes and promoting the division of one cell into two daughter cells.

- The polar microtubules overlap with each other and push the poles apart during anaphase.
- The astral microtubules help to orient the spindle apparatus in the cell.
- The kinetochore microtubules attach to chromosomes and aid in their sorting. Their roles are to align the chromosomes at the metaphase plate and to pull the chromosomes to the poles during anaphase.

2.3 COMPREHENSION QUESTIONS

1. What is the function of the kinetochore during mitosis?
   a. It promotes the replication of a chromosome to form a dyad.
   b. It is a location where a kinetochore microtubule can attach to a chromosome.
   c. It promotes the condensation of chromosomes during prophase.
   d. Both a and b are correct.

2. Which phase of mitosis is depicted in the drawing below?

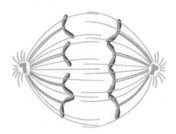

   a. Prophase
   b. Prometaphase
   c. Metaphase
   d. Anaphase
   e. Telophase

2.4   MEIOSIS

Learning Outcomes:

1. List and describe the phases of meiosis.
2. Outline the key differences between mitosis and meiosis.

In the previous section, we considered the process in which a eukaryotic cell can divide by mitosis and cytokinesis so that a mother cell produces two genetically identical daughter cells. Diploid eukaryotic cells may also divide by an alternative process called **meiosis** (from the Greek meaning "less"). During meiosis, **haploid** cells, which contain a single set of chromosomes, are produced from a cell that was originally diploid. For this to occur, the chromosomes must be correctly sorted and distributed in a way that reduces the chromosome number to half its original value. For example, in humans, haploid gametes (sperm or egg cells) are produced by meiosis. Each gamete must receive half the total number of 46 chromosomes, but not just any 23 chromosomes will do. A gamete must receive one chromosome from each of the 23 pairs. In this section, we will examine how the phases of meiosis lead to the formation of cells with a haploid complement of chromosomes.

Meiosis Produces Cells That Are Haploid

The process of meiosis bears striking similarities to mitosis. Like mitosis, meiosis begins after a cell has advanced through the G_1, S, and G_2 phases of the cell cycle. However, meiosis involves two successive divisions rather than one (as in mitosis). Prior to meiosis, the chromosomes are replicated in S phase to produce pairs of sister chromatids. This single replication event is then followed by two sequential cell divisions called meiosis I and II. Like mitosis, each of these divisions is subdivided into prophase, prometaphase, metaphase, anaphase, and telophase.

Prophase of Meiosis I   **Figure 2.10** emphasizes some of the important events that occur during prophase of meiosis I.

Synapsis. At the beginning of prophase, the replicated chromosomes begin to condense and become visible with a light microscope. The sister chromatids are connected to each other by proteins called cohesin. Next, a process called **synapsis** occurs in which the homologous chromosomes recognize each other and begin to align themselves along their entire length. In most species, this process involves the formation of a synaptonemal complex that physically connects the homologous chromosomes to each other (Figure. 10a). The associated chromatids now form a structure called a **bivalent,** which contains two pairs of sister chromatids, or a total of four chromatids. A bivalent is also called a **tetrad** (from the prefix *tetra-,* meaning "four") because it is composed of four chromatids—that is, four monads.

Chiasma Formation. The next key event is **chiasma formation,** also known as **crossing over.** Chiasma formation involves a physical exchange of chromosome pieces. At a chiasma, the DNA within a chromatid from each homolog breaks at the same location and then attaches to its homolog. The term **chiasma** (plural: **chiasmata**) was coined because a chiasma physically resembles the Greek letter chi, χ. In Figure 2.10b, two chiasmata have formed. We will consider the genetic consequences of crossing over in Chapter 6 and the molecular process of crossing over in Chapter 19.

Dissolution of the Synaptomenal Complex. After chiasmata formation, the synaptonemal complex dissociates from the chromatids. Even so, the homologs still remain attached to each other within a bivalent. What holds the homologs together? The answer is cohesin, which holds the sister chromatids together. In the distal region between the chiasmata and the ends of each chromosome, these cohesin links are responsible for keeping the homologs associated with each other and thereby maintain a bivalent structure. (See the regions within the dashed ovals in Figure 2.10b.)

Multiple Chiasmata. Depending on the size of the chromosome and the species, an average eukaryotic chromosome incurs from a couple to a couple of dozen crossovers. During spermatogenesis in humans, for example, an average chromosome undergoes slightly more than two crossovers, whereas chromosomes in certain plant species may undergo 20 or more crossovers. Figure 2.10c shows an electron micrograph of a bivalent with two chiasmata. Recent research has shown that crossing over is usually critical for the proper segregation of chromosomes. Abnormalities in chromosome segregation are often related to a defect in crossing over. In a high percentage of people with Down syndrome, in which an individual has three copies of chromosome 21 instead of two, research has shown that the presence of the extra chromosome is associated with a lack of crossing over between homologous chromosomes.

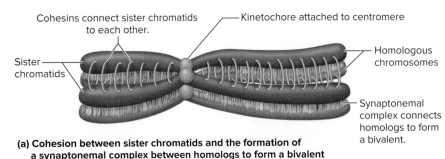

(a) Cohesion between sister chromatids and the formation of a synaptonemal complex between homologs to form a bivalent

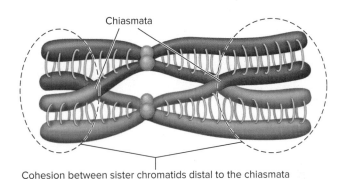

(b) Cohesion between sister chromatids that maintains the bivalent structure after the dissolution of the synaptonemal complex

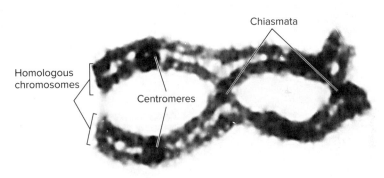

(c) Electron micrograph showing two chiasmata during prophase of meiosis

FIGURE 2.10 **Key events that occur during prophase of meiosis I.** **(a)** Prior to prophase, sister chromatids are held together via protein complexes called cohesins. (Note: Cohesins are described in greater detail in Chapter 10; see Section 10.7.) During early prophase, the formation of a synaptonemal complex causes homologous pairs of sister chromatids to associate with each other in a structure called a bivalent. **(b)** In the middle of prophase, chiasma formation (i.e., crossing over) involves a physical exchange of chromosome pieces. After chiasma formation, the synaptonemal complex dissociates. In the distal region between the chiasmata and the ends of each chromosome, the cohesin links keep the homologs associated with each other to maintain a bivalent structure. **(c)** A transmission electron micrograph showing two chiasmata.

(c): Stanley K. Sessions

CONCEPT CHECK: What is the end result of crossing over?

Prometaphase of Meiosis I Figure 2.10 emphasizes the pairing and crossing over that occur during prophase of meiosis I. In **Figure 2.11**, we turn our attention to the general events in meiosis. Prophase of meiosis I is followed by prometaphase, in which the

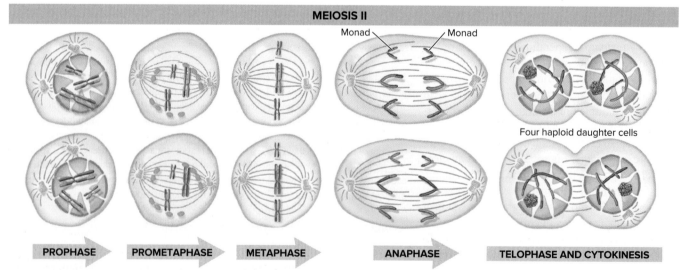

FIGURE 2.11 **The phases of meiosis in an animal cell.** See text for details.

CONCEPT CHECK: How do the four cells at the end of meiosis differ from the original mother cell?

ONLINE
ANIMATION

spindle apparatus is complete, and the chromatids are attached via kinetochore microtubules.

Metaphase of Meiosis I   At metaphase of meiosis I, the bivalents (tetrads) are organized along the metaphase plate. Before we consider the rest of meiosis I, a particularly critical feature for you to appreciate is how the bivalents are aligned along the metaphase plate. In particular, the pairs of sister chromatids are aligned in a double row rather than a single row, as occurs in mitosis (refer back to Figure 2.8d). Furthermore, the arrangement of sister chromatids (dyads) within this double row is random with regard to the blue and red homologs. In Figure 2.11, one of the blue homologs is above the metaphase plate and the other two are below, whereas one of the red homologs is below the metaphase plate and the other two are above.

In an organism that produces many gametes, meiosis can produce many different arrangements of homologs in various cells—for example, three blues above and none below, or none above and three below, and so on. As discussed in Chapter 3 (see Section 3.4), the random arrangement of homologs is consistent with Mendel's law of independent assortment.

Because most eukaryotic species have several chromosomes per set, the sister chromatids can be randomly aligned along the metaphase plate in many possible ways. Let's consider humans, who have 23 chromosomes per set. The possible number of different random alignments equals 2^n, where n is the number of chromosomes per set. Thus, in humans, there are 2^{23}, or over 8 million, possibilities! Because the homologs are genetically similar but not identical, we see from this calculation that the random alignment of homologous chromosomes provides a mechanism to promote a vast amount of genetic diversity.

In addition to the random arrangement of homologs within a double row, a second distinctive feature of metaphase of meiosis I is the attachment of kinetochore microtubules to the sister chromatids (**Figure 2.12**). One pair of sister chromatids is linked to one of the poles, and the homologous pair is linked to the opposite pole. This arrangement is quite different from the kinetochore attachment sites during mitosis, in which a pair of sister chromatids is linked to both poles (see Figure 2.8).

Anaphase of Meiosis I   At the start of anaphase of meiosis I, the cohesins along the arms of sister chromatids are released, thereby allowing the two pairs of sister chromatids within a bivalent to separate from each other (see Figure 2.11). However, the connection at the centromere remains intact so the sister chromatids remain attached to each other. Each joined pair of chromatids migrates to one pole, and the homologous pair of chromatids moves to the opposite pole. Another way of saying this is that the two dyads within a tetrad separate from each other and migrate to opposite poles.

Telophase of Meiosis I   Finally, at telophase of meiosis I, the sister chromatids have reached their respective poles, and decondensation occurs in most, but not all, species. In many species, the nuclear membrane re-forms to produce two separate nuclei. In the example of Figure 2.11, the end result of meiosis I is two cells,

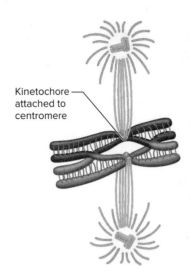

Kinetochore attached to centromere

FIGURE 2.12   Attachment of the kinetochore microtubules to replicated chromosomes at metaphase of meiosis I. The kinetochore microtubules from a given pole are attached to one pair of chromatids in a bivalent, but not both. Therefore, each pair of sister chromatids is attached to only one pole.

CONCEPT CHECK: How is the attachment of chromosomes to kinetochore microtubules during metaphase of meiosis I different from their attachment during metaphase of mitosis?

each with three pairs of sister chromatids. A reduction division has occurred. The original diploid cell had its chromosomes in homologous pairs, but the two cells produced at the end of meiosis I are considered to be haploid; they do not have pairs of homologous chromosomes. The reduction division occurs because the connection holding the sister chromatids together does not break during anaphase.

Meiosis II   The sorting events that occur during meiosis II are similar to those that occur during mitosis, but the starting point is different. For a diploid organism with six chromosomes, mitosis begins with 12 chromatids that are joined as six pairs of sister chromatids (refer back to Figure 2.8). In other words, mitosis begins with six dyads in this case. By comparison, in such a diploid organism, the two cells that begin meiosis II each have six chromatids that are joined as three pairs of sister chromatids; meiosis II begins with three dyads in this case. Otherwise, the steps that occur during prophase, prometaphase, metaphase, anaphase, and telophase of meiosis II are analogous to a mitotic division.

Meiosis Versus Mitosis   If we compare the outcome of meiosis (see Figure 2.11) to that of mitosis (see Figure 2.8), the results are quite different. In the examples we examined, mitosis produced two diploid daughter cells with six chromosomes each, whereas meiosis produced four haploid daughter cells with three chromosomes each. In other words, meiosis halved the number of chromosomes per cell. **Table 2.1** describes key differences between mitosis and meiosis that account for the different outcomes of the two processes.

TABLE 2.1
A Comparison of Mitosis, Meiosis I, and Meiosis II

Phase	Event	Mitosis	Meiosis I	Meiosis II
Prophase	Synapsis	No	Yes	No
Prophase	Crossing over	Rarely	Commonly	Rarely
Prometaphase	Attachment to the poles	A pair of sister chromatids to both poles	A pair of sister chromatids to one pole	A pair of sister chromatids to both poles
Metaphase	Alignment along the metaphase plate	Sister chromatids	Bivalents	Sister chromatids
Anaphase	Separation of:	Sister chromatids	Bivalents	Sister chromatids
End result		Two diploid cells		Four haploid cells

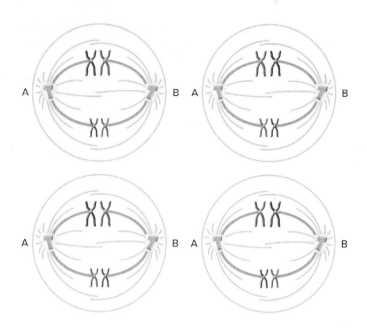

Another strategy is to make a calculation in which the number of different random alignments equals 2^n, where n is the number of chromosomes per set.

ANSWER: In the drawing, the number of random alignments is 4. From a calculation, the number of random alignments equals 2^n. So the possible number of arrangements in this case is 2^2, which equals 4.

With regard to alleles, the results of mitosis and meiosis are also different. The daughter cells produced by mitosis are genetically identical. However, the haploid cells produced by meiosis are not genetically identical to each other because they contain only one homologous chromosome from each pair. In Section 3.4 of Chapter 3, we will consider how the haploid cells may differ in the alleles that they carry on their homologous chromosomes (look ahead to Figures 3.12 and 3.13).

GENETIC TIPS   THE QUESTION: If a diploid cell
contains four chromosomes (i.e., two per set), how many possible random arrangements of homologs can occur during metaphase of meiosis I?

T OPIC: *What topic in genetics does this question address?* The topic is meiosis. More specifically, the question is about metaphase of meiosis I.

I NFORMATION: *What information do you know based on the question and your understanding of the topic?* From the question, you know that a cell that started with two pairs of homologous chromosomes has entered meiosis and is now in metaphase of meiosis I. From your understanding of the topic, you may remember that bivalents align along the metaphase plate (see Figure 2.11). The orientations of the homologs within the bivalents are random.

P ROBLEM-SOLVING S TRATEGY: *Make a drawing. Make a calculation.* One strategy to solve this problem is to make a drawing in which the homologs are different colors, such as red and blue. Note: The spindle poles are labeled A and B in the drawing in the upper right column. The alignment occurs relative to the spindle poles.

2.4 COMPREHENSION QUESTIONS

1. When does crossing over usually occur, and what is the end result?
   a. It occurs during prophase of meiosis I, and the end result is the exchange of pieces between homologous chromosomes.
   b. It occurs during prometaphase of meiosis I, and the end result is the exchange of pieces between homologous chromosomes.
   c. It occurs during prophase of meiosis I, and the end result is the separation of sister chromatids.
   d. It occurs during prometaphase of meiosis I, and the end result is the separation of sister chromatids.
2. Which phase of meiosis is depicted in the drawing to the right?
   a. Metaphase of meiosis I
   b. Metaphase of meiosis II
   c. Anaphase of meiosis I
   d. Anaphase of meiosis II

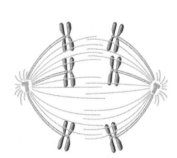

2.5 SEXUAL REPRODUCTION

Learning Outcomes:

1. Define *sexual reproduction*.
2. Describe how animals make sperm and egg cells.
3. Explain how plants alternate between haploid and diploid generations.

In the previous section, we considered how a diploid cell divides by meiosis to produce cells with half the genetic material of the original mother cell. This process is critical for sexual reproduction, which is a common way for eukaryotic organisms to produce offspring. During **sexual reproduction,** two gametes fuse with each other in the process of fertilization to begin the development of a new organism.

Gametes are highly specialized cells that are produced by a process called **gametogenesis.** As discussed previously, gametes are typically haploid, which means they contain half the number of chromosomes found in diploid cells. Haploid cells are represented by $1n$ and diploid cells by $2n$, where n refers to a set of chromosomes. A haploid gamete contains a single set of chromosomes, whereas a diploid cell has two sets. For example, a diploid human cell contains two sets of chromosomes, for a total of 46, but a human gamete (sperm or egg cell) contains only a single set of 23 chromosomes.

Some simple eukaryotic species are **isogamous,** which means that their gametes are morphologically similar. Examples of isogamous organisms include many species of fungi and algae. Most eukaryotic species, however, are **heterogamous**—they produce two morphologically different types of gametes. Male gametes, or **sperm cells,** are relatively small and usually travel relatively far distances to reach the female gamete—the **egg cell,** or **ovum.**

The mobility of the sperm is an important characteristic, making it likely that it will come in close proximity to the egg cell. The sperm of most animal species contain a single flagellum that enables them to swim. The sperm of ferns and nonvascular plants, such as bryophytes, may have multiple flagella. In flowering plants, however, the sperm are contained within pollen grains. A pollen grain is a small mobile structure that can be carried by the wind or on the feet or hairs of insects. In flowering plants, sperm are delivered to egg cells via pollen tubes. Compared with sperm cells, an egg cell is usually very large and nonmotile. In animal species, the egg stores a large amount of nutrients to nourish the growing embryo. In this section, we will examine how sperm and egg cells are made in animal and plant species.

In Animals, Spermatogenesis Produces Four Haploid Sperm Cells and Oogenesis Produces a Single Haploid Egg Cell

In male animals, **spermatogenesis,** the production of sperm cells, occurs within glands known as the testes. The testes contain spermatogonial cells that divide by mitosis to produce two cells. One of these remains a spermatogonial cell, and the other cell becomes a primary spermatocyte. As shown in **Figure 2.13a,** the spermatocyte advances through meiosis I and meiosis II to produce four

haploid cells, which are known as spermatids. These cells then mature into sperm cells. The structure of a sperm cell includes a long flagellum and a head. The head of the sperm contains little more than a haploid nucleus and an organelle known as an acrosome, at its tip. The acrosome contains digestive enzymes that are released when a sperm meets an egg cell. These enzymes enable the sperm to penetrate the outer protective layers of the egg cell and gain entry into the cell's cytosol. In animal species without a mating season, sperm production is a continuous process in mature males. A mature human male, for example, may produce several hundred million sperm each day.

In female animals, **oogenesis,** the production of egg cells, occurs within specialized diploid cells of the ovary known as oogonia. Quite early in the development of the ovary, the oogonia initiate meiosis to produce primary oocytes. For example, in human females, approximately 1 million primary oocytes per ovary are produced before birth. These primary oocytes are arrested—enter a dormant phase—at prophase of meiosis I, and they remain at this stage until the female becomes sexually mature. At maturity, primary oocytes are periodically activated to advance through meiosis I to an early stage of meiosis II. If fertilization occurs, the oocyte nucleus completes meiosis II.

During oocyte maturation, meiosis produces only one cell that is destined to become an egg, as opposed to the four gametes produced from each primary spermatocyte during spermatogenesis. How does oogenesis occur? As shown in **Figure 2.13b,** the first meiotic division is asymmetrical and produces a secondary oocyte and a much smaller cell, known as a polar body. Most of the cytoplasm is retained by the secondary oocyte and very little by the polar body, making the oocyte the larger cell with more stored nutrients. The secondary oocyte then begins meiosis II.

In mammals, the secondary oocyte is released from the ovary—an event called ovulation—and travels down the oviduct toward the uterus. During this journey, if a sperm cell penetrates the secondary oocyte, it is stimulated to complete meiosis II; in this case, the secondary oocyte produces a haploid egg and a second polar body. The haploid nuclei of the egg and sperm then unite to create the diploid nucleus of a new individual.

Plant Species Alternate Between Haploid (Gametophyte) and Diploid (Sporophyte) Generations

Most species of animals are diploid, and their haploid gametes are considered to be a specialized type of cell. By comparison, the life cycles of plant species alternate between haploid and diploid generations. The haploid generation is called the **gametophyte,** whereas the diploid generation is called the **sporophyte.** Certain cells in the sporophyte undergo meiosis and produce haploid cells called spores, which divide by mitosis to produce a gametophyte.

- In simpler plants, such as mosses, a haploid spore can produce a large multicellular gametophyte by repeated mitoses and cellular divisions.

- In flowering plants, spores develop into gametophytes that contain only a few cells. In this case, the organism that we think of as a plant is the sporophyte, whereas the gametophyte

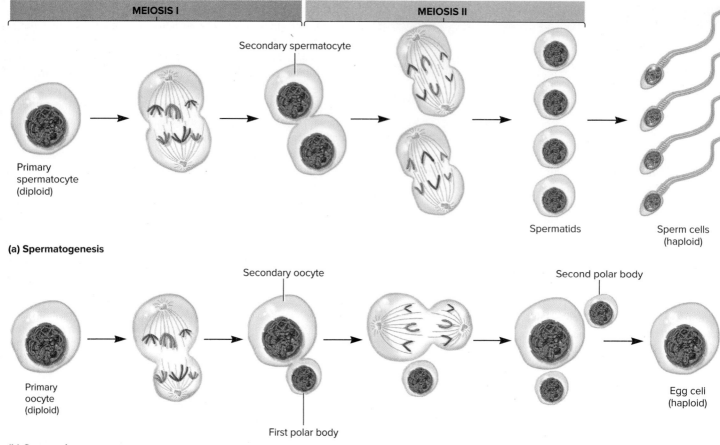

(a) Spermatogenesis

(b) Oogenesis

FIGURE 2.13 **Gametogenesis in animals.** **(a)** Spermatogenesis. A diploid spermatocyte undergoes meiosis to produce four haploid (1n) spermatids. These differentiate during spermatogenesis to become mature sperm. **(b)** Oogenesis. A diploid oocyte undergoes meiosis to produce one haploid egg cell and two or three polar bodies. In some species, the first polar body divides; in other species, it does not. Because of asymmetrical cytokinesis, the amount of cytoplasm the egg receives is maximized. The polar bodies degenerate. Note: In some species, such as humans, the oocyte completes meiosis II when fertilization takes place.

CONCEPT CHECK: What are polar bodies?

is very inconspicuous. The gametophytes of most plant species are small structures produced within the much larger sporophyte. Certain cells within the haploid gametophytes then become specialized as haploid gametes.

Figure 2.14 provides an overview of gametophyte development and gametogenesis in flowering plants, using as an example a flower from an angiosperm (a plant that produces seeds within an ovary). Meiosis occurs within cells found in two different structures of the sporophyte: the anthers and the ovaries, which produce male and female gametophytes, respectively.

In the anther, diploid cells called microsporocytes undergo meiosis to produce four haploid microspores. These separate into individual microspores. In many angiosperms, each microspore undergoes mitosis to produce a two-celled structure containing one tube cell and one generative cell, both of which are haploid. This structure differentiates into a **pollen grain,** which is the male gametophyte and has a thick cell wall. Later, the generative cell undergoes a mitotic cell division to produce two haploid sperm

cells. In most plant species, this division occurs only if the pollen grain germinates—if it lands on a stigma and forms a pollen tube (look ahead to Figure 3.2c).

By comparison, female gametophytes are produced within ovules found in the plant ovaries. A type of cell known as a megasporocyte undergoes meiosis to produce four haploid megaspores. Three of the four megaspores degenerate. The remaining haploid megaspore then undergoes three successive mitotic divisions accompanied by asymmetrical cytokinesis to produce seven individual cells—one egg, two synergids, three antipodals, and one central cell. This seven-celled structure, also known as the **embryo sac,** is the mature female gametophyte. Each embryo sac is contained within an ovule.

For fertilization to occur, specialized cells within the male and female gametophytes must meet. The steps of plant fertilization are described in Chapter 3.

1. To begin this process, a pollen grain lands on a stigma (look ahead to Figure 3.2c).

3. One of the sperm enters the central cell, which contains the two polar nuclei. This results in a cell that is triploid (3*n*). This cell divides mitotically to produce **endosperm,** which acts as a food-storing tissue. The other sperm enters the egg cell. The egg and sperm nuclei fuse to create a diploid cell, the zygote, which becomes a plant embryo. Therefore, fertilization in flowering plants is actually a double fertilization. The result is that the endosperm, whose production uses a large amount of plant resources, will develop only when an egg cell has been fertilized.

4. After fertilization is complete, the ovule develops into a seed, and the surrounding ovary develops into the fruit, which encloses one or more seeds.

When comparing animals and plants, it's interesting to consider how gametes are made. Animals produce gametes by meiosis. In contrast, plants produce gametes by mitosis. The gametophyte of plants is a haploid multicellular organism produced by mitotic cellular divisions of a haploid spore. Within the multicellular gametophyte, certain cells become specialized as gametes.

2.5 COMPREHENSION QUESTIONS

1. In animals, a key difference between spermatogenesis and oogenesis is that
   a. only oogenesis involves meiosis.
   b. only spermatogenesis involves meiosis.
   c. spermatogenesis produces four sperm, whereas oogenesis produces only one egg cell.
   d. None of the above describes a difference between the two processes.

2. Which of the following statements regarding plants is *false*?
   a. Meiosis within anthers produces spores that develop into pollen.
   b. Meiosis within ovules produces spores that develop into an embryo sac.
   c. The male gametophyte is a pollen grain, and the female gametophyte is an embryo sac.
   d. Meiosis directly produces sperm and egg cells in plants.

2.6 SEX CHROMOSOMES AND SEX DETERMINATION

Learning Outcomes:
1. Define *sex chromosomes* and *sex determination*.
2. Outline different mechanisms of sex determination.

In the preceding section, we considered how sexual reproduction involves the production and subsequent union of male and female gametes—sperm and egg. In most species of animals, sperm are produced by males, whereas eggs are produced by females. Males and females are said to be of different sexes. However, some animal

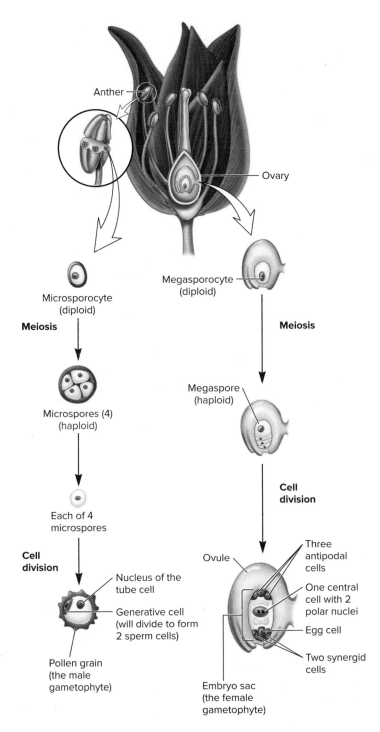

FIGURE 2.14 **The formation of male and female gametes by the gametophytes of angiosperms (flowering plants).**

CONCEPT CHECK: Are all cell nuclei in the embryo sac haploid or is just the egg cell haploid?

2. This stimulates the tube cell to sprout a pollen tube that grows through the style and eventually makes contact with an ovule. As this is occurring, the generative cell undergoes mitosis to produce two haploid sperm cells. The sperm cells migrate through the pollen tube and eventually reach the ovule.

species produce hermaphrodites in which the same individual can make both male and female gametes. In most species of flowering plants, a single individual (a sporophyte) produces both male gametophytes and female gametophytes. Such species are call **monoecious.** However, in other plant species, the sporophytes are divided into those that produce only male gametophytes and those that produce only female gametophytes. Such species are called **dioecious.**

In 1901, Clarence McClung, who studied grasshoppers, was the first to suggest that male and female sexes are due to the inheritance of particular chromosomes. The term **sex chromosomes** refers to chromosomes that differ between males and females. Sex chromosomes are found in most species of animals and a few species of dioecious plants. For example, in mammals and fruit flies, the sex chromosomes are designated X and Y. In this section, we will explore difference mechanisms that give rise to male or female individuals, a process call **sex determination.**

Sex Differences May Depend on the Presence of Sex Chromosomes or on the Number of Sets of Chromosomes

Since McClung's initial observations, we now know that a pair of chromosomes, called the sex chromosomes, determines sex in many different animal species. Some examples are described in **Figure 2.15**.

X-Y System. In the X-Y system of sex determination, which operates in mammals, the male has one X chromosome and one Y chromosome, whereas the female has two X chromosomes (Figure 2.15a). In this case, the male is called the **heterogametic sex.** Two types of sperm are produced: one type that carries only the X chromosome, and another type that carries the Y. The female is the **homogametic sex** because all eggs carry a single X chromosome.

The 46 chromosomes in humans consist of 1 pair of sex chromosomes and 22 pairs of **autosomes**—chromosomes that are not sex chromosomes. In the human male, each of the four sperm produced during gametogenesis contains 23 chromosomes. Two sperm contain an X chromosome, and the other two have a Y chromosome. The sex of the offspring is determined by whether the sperm that fertilizes the egg carries an X or a Y chromosome.

What causes an offspring to develop into a male or female? One possibility is that two X chromosomes are required for female development. A second possibility is that the Y chromosome promotes male development. In the case of mammals, the second possibility is correct. This is known from the analysis of rare individuals who carry chromosomal abnormalities. For example, mistakes that occasionally occur during meiosis may produce an individual who carries two X chromosomes and one Y chromosome. Such an individual develops into a male. As discussed in Chapter 26, the *SRY* gene on the Y chromosome plays a key role in causing the development of male characteristics.

X-0 System. The X-0 system of sex determination operates in many insects (Figure 2.15b). In such species, the male has one sex chromosome (the X) and is designated X0, whereas the female has a pair of sex chromosomes (two X's). In other insect species, such as *Drosophila melanogaster*, the male is XY. For both types of insect species (i.e., having X0 males and XX females or having XY males and XX females), the ratio between X chromosomes

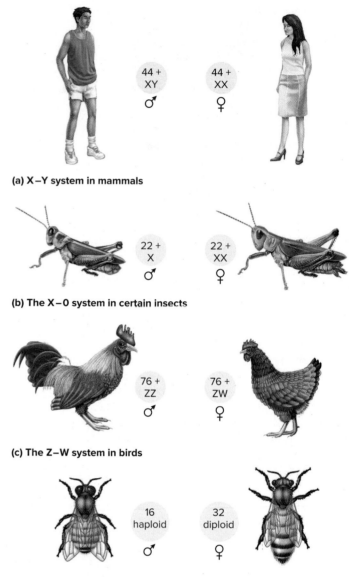

FIGURE 2.15 **Different mechanisms of sex determination in animals.** See the text for a description.

CONCEPT CHECK: What is the difference between the X-Y and X-0 systems of sex determination?

and the number of autosomal sets determines sex. If a fly has one X chromosome and is diploid for the autosomes (2*n*), the ratio is 1/2, or 0.5. This fly will become a male even if it does not receive a Y chromosome. In contrast to the X-Y system of mammals, the Y chromosome in the X-0 system does not determine maleness. If a fly receives two X chromosomes and is diploid, the ratio is 2/2, or 1.0, and the fly becomes a female.

Z-W System. In the Z-W system, which determines sex in birds and some fish, the male is ZZ and the female is ZW (Figure 2.15c). The letters Z and W are used to distinguish these types of sex chromosomes from those found in the X-Y pattern of sex determination of other species. In the Z-W system, the male is the homogametic sex, and the female is heterogametic.

Haplodiploid System. In addition to sex chromosomes, sex determination in other species is determined by the number of sets of chromosomes. For example, the haplodiploid system is found in bees (Figure 2.15d). The male bee, called the drone, is produced from unfertilized haploid eggs. Female bees, both worker bees and queen bees, are produced from fertilized eggs and therefore are diploid.

Sex Differences May Depend on the Environment

Although sex in many species is determined by chromosomes, other mechanisms are also known. In certain reptiles and fish, sex is controlled by environmental factors such as temperature. For example, in the American alligator (*Alligator mississippiensis*), temperature controls sex development (**Figure 2.16a**). The temperature-sensitive period is between 7 and 21 days after laying. Temperatures $\leqslant 30°C$ produce all female offspring, whereas those $\geqslant 34°C$ yield all male offspring. Temperatures between 30°C and 34°C result in batches of both sexes.

Recent research suggests that a thermosensitive protein called TRPV4 is present within the developing alligator gonad inside the egg shell. This protein is functional at warm temperatures, at or above 34°C, and can activate cell signaling by inducing calcium ion influx into cells. The results indicate that TRPV4 may play a key role in promoting the sex determination pathway that gives rise to males.

Another way that sex can be environmentally determined is via behavior. Clownfish of the genus *Amphiprion* are coral reef fish that live among anemones on the ocean floor (**Figure 2.16b**). One anemone typically harbors a harem of clownfish consisting of a large female, a medium-sized reproductive male, and small nonreproductive juveniles. Clownfish are **protandrous hermaphrodites**—they can switch from male to female! When the female of a harem dies, the reproductive male changes sex to become a female and the largest of the juveniles matures into a reproductive male. Unlike male and female humans, the sexes of clownfish are not determined by chromosome differences. Male and female clownfish have the same chromosomal composition.

How can a clownfish switch from female to male? A juvenile clownfish has both male and female immature sexual organs. Hormone levels, particularly those of an androgen called testosterone and an estrogen called estradiol, control the expression of particular genes. In nature, the first sexual change that usually happens is that a juvenile clownfish becomes a male. This occurs when the testosterone level becomes high, which promotes the expression of genes that code proteins that cause the male organs to mature. Later, when the female of the harem dies, the estradiol level in the reproductive male becomes high and testosterone is decreased. This alters gene expression in a way that leads to the synthesis of some new proteins and prevents the synthesis of others. When this occurs, the female organs grow and the male reproductive system degenerates. The male fish becomes female.

What factor determines the hormone levels in clownfish? A female seems to control the other clownfish in the harem through aggressive dominance, thereby preventing the formation of other females. This aggressive behavior suppresses an area of the brain in the other clownfish that is responsible for the production of certain hormones that are needed to promote female development.

(a) Sex determination via temperature: American alligator (*A. mississippiensis*)

(b) Sex determination via behavior: Clownfish (*Amphiprion ocellaris*)

FIGURE 2.16 **Sex determination caused by environmental factors.** **(a)** In the alligator, temperature determines whether an individual develops into a female or male. **(b)** In clownfish, males can change into females due to behavioral changes that occur when a dominant female dies.

(a) NASA; (b) Krzysztof Odziomek/iStock/Getty Images

CONCEPT CHECK: How might global warming affect alligator populations?

If a clownfish is left by itself in an aquarium, it will automatically develop into a female because this suppression does not occur.

Dioecious Plant Species Have Opposite Sexes

As mentioned, in most flowering plants, including pea plants discussed in Chapter 3, a single diploid individual (a sporophyte) produces both female and male gametophytes, which are haploid and produce egg or sperm cells, respectively (see Figure 2.14). However, some plant species are dioecious, which means that some individuals produce only male gametophytes, whereas others produce only female gametophytes. These species include hollies (**Figure 2.17a**), willows, and ginkgo trees.

(a) American holly (*I. opaca*)

     Female   **Male**

(b) Female and male flowers on separate individuals in white campion (*S. latifolia*)

FIGURE 2.17 **Examples of dioecious plants in which individuals produce only male gametophytes or only female gametophytes.** **(a)** American holly (*Ilex opaca*). The female sporophyte, which produces red berries, is shown here. **(b)** White campion (*Silene latifolia*), which is often studied by researchers.

(a) valentino cazzanti/Shutterstock; (b, c) Arco Images GmbH/Alamy Stock Photo

The genetics of sex determination in dioecious plant species is beginning to be understood. To study this process, many researchers have focused their attention on the white campion, *Silene latifolia*, which is a relatively small dioecious plant with a short generation time (**Figure 2.17b**). In this species, sex chromosomes, designated X and Y, are responsible for sex determination. The male plant has X and Y chromosomes, whereas the female plant is XX. Sex chromosomes are also found in other plant species such as papaya and spinach. However, in other dioecious species, cytological examination of the chromosomes does not reveal distinct types of sex chromosomes. Even so, in these plant species, the male plants usually appear to be the heterogametic sex.

2.6 COMPREHENSION QUESTIONS

1. Among different species, sex may be determined by
   a. differences in sex chromosomes.
   b. differences in the number of sets of chromosomes.
   c. environmental factors.
   d. all of the above.

2. In mammals, sex is determined by
   a. the *SRY* gene on the Y chromosome.
   b. having two copies of the X chromosome.
   c. having one copy of the X chromosome.
   d. both a and c.

3. An abnormal fruit fly has two sets of autosomes and is XXY. Such a fly is
   a. a male.
   b. a female.
   c. a hermaphrodite.
   d. none of the above.

KEY TERMS

2.1: chromosomes, chromatin, prokaryotes, nucleoid, eukaryotes, organelles, nucleus, cytogenetics, cytogeneticist, somatic cell, gametes, germ cells, karyotype, diploid, homologs, alleles, locus (loci)

2.2: asexual reproduction, multicellularity, binary fission, cell cycle, interphase, G_1 phase, restriction point, S phase, chromatids, centromere, sister chromatids, dyad, monad, kinetochore, G_2 phase, M phase, mitosis

2.3: mitotic spindle apparatus (spindle apparatus, mitotic spindle), microtubule-organizing centers (MTOCs), centrosomes, spindle pole, centrioles, astral microtubules, polar microtubules, kinetochore microtubules, kinetochore, decondensed, prophase, condensed, prometaphase, metaphase plate, metaphase, anaphase, telophase, cytokinesis, cleavage furrow, cell plate

2.4: meiosis, haploid, synapsis, bivalent, tetrad, chiasma formation (crossing over), chiasma (chiasmata)

2.5: sexual reproduction, gametogenesis, isogamous, heterogamous, sperm cell, egg cell, ovum, spermatogenesis, oogenesis, gametophyte, sporophyte, pollen grain, embryo sac, endosperm

2.6: monoecious, dioecious, sex chromosomes, sex determination, heterogametic sex, homogametic sex, autosomes, protandrous hermaphrodites

CHAPTER SUMMARY

2.1 General Features of Chromosomes

- Chromosomes are structures that contain the genetic material, which is DNA.
- Prokaryotic cells are simple and lack cell compartmentalization, whereas eukaryotic cells contain a cell nucleus and other compartments (see Figure 2.1).
- Chromosomes can be examined under the microscope. An organized representation of the chromosomes from a single cell is called a karyotype (see Figure 2.2).
- In eukaryotic species, the chromosomes are found in sets. Eukaryotic cells are often diploid, which means that each type of chromosome occurs in a homologous pair (see Figure 2.3).

2.2 Cell Division

- Bacteria divide by binary fission (see Figure 2.4).
- To divide, eukaryotic cells advance through a cell cycle (see Figure 2.5).
- Prior to cell division, eukaryotic chromosomes are replicated to form sister chromatids (see Figure 2.6).

2.3 Mitosis and Cytokinesis

- Chromosome sorting in eukaryotes is achieved via the mitotic spindle apparatus (see Figure 2.7).
- A common way for eukaryotic cells to divide is by mitosis and cytokinesis. Mitosis is divided into prophase, prometaphase, metaphase, anaphase, and telophase (see Figure 2.8).

- During cytokinesis, animal cells divide by forming a cleavage furrow and plant cells form a cell plate (Figure 2.9).

2.4 Meiosis

- Another way for eukaryotic cells to divide is via meiosis, which produces four haploid cells. During prophase of meiosis I, homologs synapse and crossing over may occur (see Figures 2.10, 2.11, 2.12, Table 2.1).

2.5 Sexual Reproduction

- Animals produce gametes via spermatogenesis and oogenesis (see Figure 2.13).
- Plants exhibit alternation of generations between a diploid sporophyte and a haploid gametophyte. The gametophyte produces gametes (see Figure 2.14).

2.6 Sex Chromosomes and Sex Determination

- Sex chromosomes are chromosomes that differ between males and females. Sex determination is the process that gives rise to male and female individuals.
- Mechanisms of sex determination may involve differences in chromosome composition or environmental factors (see Figures 2.15, 2.16, 2.17).

PROBLEM SETS & INSIGHTS

MORE GENETIC TIPS **1.** A diploid cell begins with eight chromosomes, four per set, and then proceeds through cell division. In the following diagram, in what phase of mitosis, meiosis I, or meiosis II is the cell?

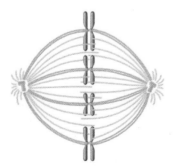

TOPIC: *What topic in genetics does this question address?* The topic is cell division. More specifically, the question is asking you to look at a diagram and discern which phase of cell division it shows.

INFORMATION: *What information do you know based on the question and your understanding of the topic?* In the question, you are given a diagram of a cell in a particular phase of the cell cycle. This cell is derived from a mother cell with four pairs of chromosomes. From your understanding of the topic, you may remember the various phases of mitosis, meiosis I, and meiosis II, which are described in Figures 2.8 and 2.11. If so, you may initially realize that the cell is in metaphase.

PROBLEM-SOLVING **S**TRATEGY: *Describe the steps.* To solve this problem, you may need to describe the steps, starting with a mother cell that has four pairs of chromosomes. Keep in mind that a mother cell with four pairs of chromosomes has eight chromosomes during G_1, which then replicate to form eight pairs of sister chromatids during S phase. Therefore, at the beginning of M phase, this mother cell will have eight pairs of sister chromatids. During metaphase of mitosis, the eight pairs of sister chromatids in the mother cell will align in a single row. During meiosis I, four bivalents will align along the metaphase plate in the mother cell. During meiosis II, four pairs of sister chromatids in the two daughter cells will align along the metaphase plate in a single row.

ANSWER: The cell is in metaphase of meiosis II. You can tell because the pairs of sister chromatids are lined up in a single row along the metaphase plate, and the cell has only four pairs of sister chromatids. If the diagram was showing mitosis, the cell would have eight pairs of sister chromatids in a single row. If it was showing meiosis I, four bivalents would be aligned along the metaphase plate.

2. What are the key differences in anaphase when comparing mitosis, meiosis I, and meiosis II?

T OPIC: *What topic in genetics does this question address?* The topic is cell division. More specifically, the question is about the events that occur during anaphase.

I NFORMATION: *What information do you know based on the question and your understanding of the topic?* From the question, you know you are supposed to distinguish the key differences among anaphase of mitosis, anaphase of meiosis I, and anaphase of meiosis II. From your understanding of the topic, you may remember that the separation of chromosomes is distinctly different in meiosis I compared to mitosis and meiosis II. Compare Figures 2.8 and 2.11.

P ROBLEM-SOLVING **S** TRATEGY: *Make a drawing. Compare and contrast.* One strategy to solve this problem is to make a drawing. If you make drawings of a cell in anaphase, like those shown in Figures 2.8 and 2.11, you may appreciate that sister chromatids (dyads) move to opposite poles during anaphase of meiosis I, whereas individual chromatids (monads) move to opposite poles during anaphase of mitosis and meiosis II. Another strategy is to compare and contrast what happens during meiosis I with the events of mitosis and meiosis II. During meiosis I, anaphase does not involve the splitting of centromeres, whereas centromeres split during mitosis and meioisis II, thereby separating sister chromatids.

ANSWER: During anaphase in mitosis and meiosis II, the centromeres split and individual chromatids move to their respective poles. In anaphase of meiosis I, the centromeres do not split. Instead, the bivalents separate and pairs of sister chromatids move to opposite poles.

3. Assuming that such a fly would be viable, what would be the sex of a fruit fly with each of the following chromosomal compositions?

   A. One X chromosome and two sets of autosomes

   B. Two X chromosomes, one Y chromosome, and two sets of autosomes

   C. Two X chromosomes and four sets of autosomes

   D. Four X chromosomes, two Y chromosomes, and four sets of autosomes

T OPIC: *What topic in genetics does this question address?* The topic is sex determination. More specifically, the question is about sex determination in fruit flies.

I NFORMATION: *What information do you know based on the question and your understanding of the topic?* In the question, you are given four examples that describe how many sex chromosomes and how many sets of autosomes a fruit fly has. From your understanding of the topic, you may remember that sex determination in fruit flies follows the X-0 system, in which the ratio of the number of X chromosomes to the number of sets of autosomes determines sex. If that ratio is 1, the fly becomes a female. If it is 0.5, the fly becomes a male. The presence of a Y chromosome does not determine sex in the X-0 system.

P ROBLEM-SOLVING **S** TRATEGY: *Make a calculation.* For each example, you divide the number of X chromosomes by the number of sets of autosomes.

   A. 1 divided by 2 equals 0.5.

   B. 2 divided by 2 equals 1.

   C. 2 divided by 4 equals 0.5.

   D. 4 divided by 4 equals 1.

ANSWER:

   A. Male

   B. Female

   C. Male

   D. Female

Conceptual Questions

C1. The process of binary fission begins with a single mother cell and ends with two daughter cells. Would you expect the mother and daughter cells to be genetically identical? Explain why or why not.

C2. What is a homolog? With regard to genes and alleles, how are homologs similar to and different from each other?

C3. What is a sister chromatid? Are sister chromatids genetically similar or identical? Explain.

C4. With regard to sister chromatids, which phase of mitosis is the organization phase, and which is the separation phase?

C5. A species is diploid and has three chromosomes per set. Make a drawing that shows what the chromosomes look like in the G_1 and G_2 phases of the cell cycle.

C6. How does the attachment of kinetochore microtubules to the kinetochore differ in metaphase of meiosis I compared to metaphase of mitosis? Discuss what you think would happen if a sister chromatid was not attached to a kinetochore microtubule.

C7. For the following events, specify whether each occurs during mitosis, meiosis I, or meiosis II:

   A. Separation of conjoined chromatids within a pair of sister chromatids

   B. Pairing of homologous chromosomes

   C. Alignment of chromatids along the metaphase plate

   D. Attachment of sister chromatids to both poles

C8. Identify the key events during meiosis that result in a 50% reduction in the amount of genetic material per cell.

C9. A cell is diploid and contains three chromosomes per set. Draw the arrangement of the chromosomes during metaphase of mitosis and during metaphase of meiosis I and II. In your drawings, make the sets of chromosomes different colors.

C10. The alignment of homologs along the metaphase plate during metaphase of meiosis I is random. In your own words, explain what this means.

C11. A eukaryotic cell is diploid and contains 10 chromosomes (5 in each set). In mitosis and meiosis, how many daughter cells will be produced, and how many chromosomes will each one contain?

C12. If a diploid cell contains six chromosomes (i.e., three per set), how many possible random arrangements of homologs can occur during metaphase of meiosis I?

C13. A cell has four pairs of chromosomes. Assuming that crossing over does not occur, what is the probability that a gamete will contain all of the paternal chromosomes? If n is the number of chromosomes in a set, which of the following expressions can be used to calculate the probability that a gamete will receive all of the paternal chromosomes: $(1/2)^n$, $(1/2)^{n-1}$, or $n^{1/2}$?

C14. With regard to question C13, how would the phenomenon of crossing over affect the results? In other words, would the probability of a gamete inheriting only paternal chromosomes be higher or lower? Explain your answer.

C15. Eukaryotic cells must sort their chromosomes during mitosis so that each daughter cell receives the correct number of chromosomes. Why don't bacteria need to sort their chromosomes?

C16. Why is it necessary for the chromosomes to condense during mitosis and meiosis? What do you think might happen if the chromosomes were not condensed?

C17. Nine-banded armadillos almost always give birth to four offspring that are genetically identical quadruplets. Explain how you think this happens.

C18. A diploid species has four chromosomes per set for a total of eight chromosomes in its somatic cells. Draw a diagram showing how such a cell would look in late prophase of meiosis II and in prophase of mitosis. Discuss how prophase of meiosis II and prophase of mitosis differ from each other, and explain how the difference originates.

C19. Explain why the products of meiosis may not be genetically identical, whereas the products of mitosis are.

C20. The period between meiosis I and meiosis II is called interphase II. Does DNA replication take place during interphase II?

C21. List several ways in which telophase appears to be the reverse of prophase and prometaphase.

C22. Corn has 10 chromosomes per set, and the sporophyte of the species is diploid. If you made karyotypes, what is the total number of chromosomes you would expect to see in each of the following types of corn cells?

A. A leaf cell

B. The sperm nucleus of a pollen grain

C. An endosperm cell after fertilization

D. A root cell

C23. The arctic fox has 50 chromosomes (25 per set), and the common red fox has 38 chromosomes (19 per set). These species can interbreed to produce viable but infertile offspring. How many chromosomes will the offspring have? What problems do you think may occur during meiosis that would explain the offspring's infertility?

C24. Let's suppose that a gene affecting pigmentation is found on the X chromosome (in mammals or insects) or the Z chromosome (in birds) but not on the Y or W chromosome. It is found on an autosome in bees. This gene exists in two alleles: D (dark) is dominant to d (light). What would be the phenotypic results of crosses between true-breeding dark females and true-breeding light males and of the reciprocal crosses involving true-breeding light females and true-breeding dark males for each of the following species? Refer to Figure 2.15 for the mechanism of sex determination in these species.

A. Birds   C. Bees

B. Fruit flies   D. Humans

C25. Describe the cellular differences between male and female gametes.

C26. At puberty in humans, the testes contain a finite number of cells but produce an enormous number of sperm cells during the life span of a male. Explain why testes do not run out of spermatogonial cells.

C27. Describe the timing of meiosis I and II during human oogenesis.

C28. Three genes (A, B, and C) are found on three different chromosomes. For the following diploid genotypes, list all of the possible gamete combinations.

A. *Aa Bb Cc*   C. *Aa BB Cc*

B. *AA Bb CC*   D. *Aa bb cc*

C29. A female with an abnormally long chromosome 13 (and a normal homolog of chromosome 13) has children with a male with an abnormally short chromosome 11 (and a normal homolog of chromosome 11). What is the probability of producing an offspring that will have both a long chromosome 13 and a short chromosome 11? If such a child is produced, what is the probability that this child will eventually pass both abnormal chromosomes to one of their offspring?

C30. *Anopheles gambiae* is the mosquito that transmits malaria to humans. Like *Drosophila*, the mosquitos carry X and Y sex chromosomes and follow the X-O mechanism of sex determination. Assuming that such a mosquito would be viable, what would be the sex of a mosquito with each of the following chromosomal compositions?

A. One X chromosome, one Y chromosome, and two sets of autosomes

B. Two X chromosomes, one Y chromosome, and two sets of autosomes

C. Two X chromosomes, four Y chromosomes, and four sets of autosomes

D. Four X chromosomes, two Y chromosomes, and four sets of autosomes

C31. What would be the sex of a human with each of the following sets of sex chromosomes?

A. XXX

B. X (also described as X0)

C. XYY

D. XXY

Experimental Questions

E1. When studying living cells in a laboratory, researchers sometimes use drugs to cause cells to remain in a particular phase of the cell cycle. For example, aphidicolin inhibits DNA synthesis in eukaryotic cells and causes them to remain in the G_1 phase because they cannot replicate their DNA. In what phase of the cell cycle—G_1, S, G_2, prophase, metaphase, anaphase, or telophase—would you expect somatic cells to stay if they were treated with each of following types of drug?

A. A drug that inhibits microtubule formation

B. A drug that allows microtubules to form but prevents them from shortening

C. A drug that inhibits cytokinesis

D. A drug that prevents chromosomal condensation

E2. How would you set up crosses to determine if a gene is located on the Y chromosome versus one that is located on the X chromosome?

E3. With regard to thickness and length, what do you think chromosomes would look like if you examined them microscopically during interphase? How would that compare with their appearance during metaphase?

E4. A rare form of very tall stature that also included hearing loss was found to run in a particular family. It is inherited as a dominant trait. It was discovered that an affected individual had one normal copy of chromosome 15 and one abnormal copy of chromosome 15 that was unusually long. How would you determine if the unusually long chromosome 15 was causing this disorder?

E5. Experimentally, how do you think researchers were able to determine that the Y chromosome causes maleness in mammals, whereas the ratio of X chromosomes to the sets of autosomes causes sex determination in fruit flies?

E6. The amount of DNA per cell can be measured experimentally. It is typically expressed in picograms (pg). Let's suppose a human cell in the G_1 phase contains 9 pg of DNA. How much DNA would you expect in a single cell in each of the following phases?

A. Metaphase of mitosis

B. Metaphase of meiosis I

C. Metaphase of meiosis II

D. After telophase of meiosis II and after cytokinesis

Questions for Student Discussion/Collaboration

1. Sex determination among different species is caused by a variety of mechanisms. With regard to evolution, discuss why you think this has happened. During the evolution of life on Earth over the past 4 billion years, do you think the phenomenon of opposite sexes is a relatively recent event?

2. A diploid eukaryotic cell has 10 chromosomes (5 per set). As a group, take turns having one student draw the cell as it would look during a phase of mitosis, meiosis I, or meiosis II; then have the other students guess which phase it is.

Note: All answers are available for the instructor in Connect; the answers to the even-numbered questions and all of the Concept Check and Comprehension Questions are in Appendix B.

The garden pea, studied by Mendel.
Zigzag Mountain Art/Alamy Stock Photo

3 MENDELIAN INHERITANCE

An appreciation for the concept of heredity can be traced far back in human history. Hippocrates, a physician, was the first person to provide an explanation for hereditary traits (around 400 B.C.E.). Hippocrates' incorrect explanation was that "seeds" are produced by all parts of the body and then collected and transmitted to the offspring at the time of conception. Furthermore, these seeds were hypothesized to cause certain traits of the offspring to resemble those of the parents. This idea, known as pangenesis, was the first attempt to explain the transmission of hereditary traits from generation to generation.

The first systematic studies of genetic crosses were carried out by Joseph Kölreuter from 1761 to 1766. In crosses between different varieties of tobacco plants, the offspring were usually intermediate in appearance between the two parents. This observation led Kölreuter to conclude that both parents make equal genetic contributions to their offspring. Furthermore, it was consistent with the blending hypothesis of inheritance. According to this idea, the factors that dictate hereditary traits could blend together from generation to generation. The blended traits would then be passed to the next generation. Before the 1860s, the popular view, which combined the notions of pangenesis and blending inheritance, was that hereditary traits were rather malleable and could change and blend over the course of one or two generations.

However, the pioneering work of Gregor Mendel would prove instrumental in refuting this viewpoint.

In this chapter, we will first examine Mendel's crosses of pea plants. We begin our inquiry into genetics here because the inheritance patterns observed in pea plants are fundamentally related to inheritance patterns found in other eukaryotic species, such as corn, fruit flies, mice, and humans. Mendel's experiments revealed some simple laws that govern the process of inheritance. We will then explore the chromosome theory of inheritance that explains the relationship between the transmission of chromosomes during sexual reproduction and the pattern of transmission of traits observed by Mendel. In addition, we will consider how researchers were able to confirm the chromosome theory of inheritance by showing that certain genes are located on sex chromosomes.

In the last section of this chapter, we will examine some general concepts in probability and statistics. How are statistical methods useful? First, probability calculations allow us to predict the outcomes of simple genetic crosses, as well as the outcomes of more complicated crosses described in later chapters. In addition, we will explore the use of statistics to test the validity of genetic hypotheses that attempt to explain the inheritance patterns of traits.

3.1 MENDEL'S STUDY OF PEA PLANTS

Learning Outcomes:

1. Describe the characteristics of pea plants that make them a suitable organism to study genetically.
2. Outline the steps that Mendel followed to make crosses between different strains of pea plants.
3. Define the following terms: *strain*, *character*, and *trait* (or *variant*).

Starting in 1856, Gregor Mendel grew and crossed thousands of pea plants in a small 23- by 115-foot garden (**Figure 3.1**). The pea plants differed with regard to various traits. For example, some had purple flowers and others had white flowers. Mendel gathered a large amount of quantitative data concerning the outcomes of crosses between plants with different traits.

The culmination of Mendel's work, entitled *Experiments in Plant Hybridization*, was published in 1866. This paper was largely ignored by scientists at that time, possibly because of its title, which did not reveal its key observations. Another reason this work went unrecognized could be a lack of understanding of chromosomes and their transmission, a topic discussed in Chapter 2. Mendel reflected, "My scientific work has brought me a great deal of satisfaction and I am convinced that it will be appreciated before long by the whole world."

In 1900, the studies of Mendel were independently rediscovered by three biologists with an interest in plant genetics: Hugo de Vries, Carl Correns, and Erich von Tschermak. Within a few years, the influence of these studies was felt around the world. In this section, we will begin by considering the process of sexual reproduction in pea plants. We will then examine key features of pea plants that were advantageous from an experimental point of view. In Sections 3.2 and 3.3, we will explore how Mendel's experiments revealed fundamental laws that govern the transmission of traits from parent to offspring.

Sexual Reproduction in Pea Plants Occurs Via Pollination and Fertilization

Before we delve into Mendel's experiments, let's first consider how flowering plants reproduce. The reproductive features of the garden pea, *Pisum sativum*, are shown in **Figure 3.2**.

- The term **gamete** is used to describe haploid reproductive cells that fuse to form a zygote.
- Male gametes (**sperm cells**) are produced within **pollen grains** that form in the **anthers.**
- Female gametes (**egg cells**) are produced within **ovules** that form in the **ovaries.**
- For fertilization to occur, a pollen grain first lands on the **stigma,** an event called pollination. This event then stimulates the growth of a pollen tube (Figure 3.2c). The pollen tube enables sperm cells to enter the stigma and migrate toward an ovule. Fertilization takes place when a sperm

FIGURE 3.1 Gregor Johann Mendel (1822–1884).
National Library of Medicine

enters the micropyle, an opening in the ovule wall, and fuses with an egg cell, resulting in the formation of a zygote.
- Via cell division, a zygote first develops into a plant embryo within a pea seed. After the seed is planted, the seed germinates, and the embryo develops into a seedling and eventually a mature pea plant.

Pea Plants Can Reproduce by Cross-Fertilization or Self-Fertilization

Mendel's study of genetics grew out of an interest in ornamental flowers. Many plant breeders had previously conducted experiments aimed at obtaining flowers with new colors. When two distinct individuals with different characteristics are bred to each other, the experiment is called a **cross,** or a **hybridization,** and the offspring are referred to as **hybrids.** For example, a hybridization experiment could involve a cross between a purple-flowered plant and a white-flowered plant.

Certain properties of the garden pea were particularly advantageous for studying plant hybridization. First, the flowers of the garden pea are relatively large, making it easy to manipulate the anthers where the pollen is produced. In some

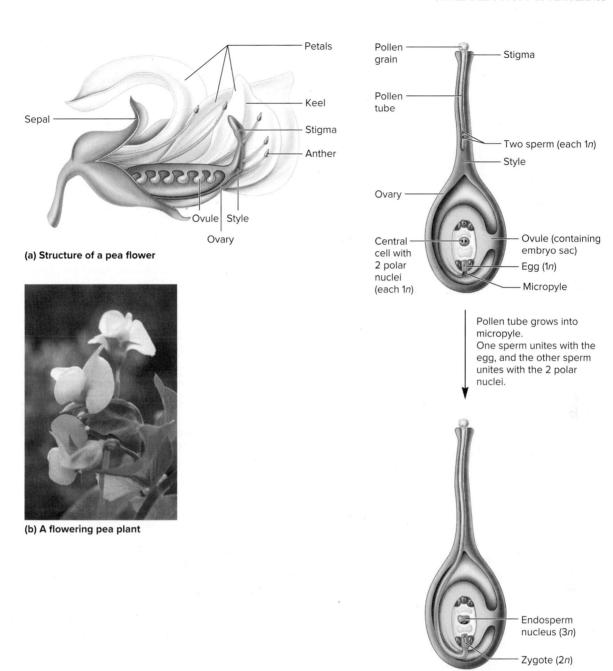

(a) Structure of a pea flower

(b) A flowering pea plant

(c) Pollination and fertilization

FIGURE 3.2 **Flower structure and pollination in pea plants.** (a) The pea flower produces both pollen and egg cells. The pollen grains are produced within the anthers, and the egg cells are produced within the ovules that are contained within the ovary. The keel is two modified petals that are fused and enclose the anthers and ovaries. In this drawing, some of the keel is not shown so that the internal reproductive structures of the flower can be seen. (b) Photograph of a flowering pea plant. (c) A pollen grain must first land on the stigma. After this occurs, the pollen grain sends out a long tube through which two sperm cells travel toward an ovule to reach an egg cell. The fusion between a sperm and an egg cell results in fertilization and creates a zygote. The second sperm fuses with a central cell containing two polar nuclei to create the endosperm. The endosperm, which is a major component of a pea seed, provides nutritive material for the developing embryo.

(b): np-e07/iStock/Getty Images

CONCEPT CHECK: Prior to fertilization, where are a pea plant's male gametes located?

experiments, Mendel wanted to make crosses between different plants. This process, known as **cross-fertilization,** requires that the pollen from one plant be placed on the stigma of another plant. The procedure is shown in **Figure 3.3**. Mendel was able to pry open immature flowers and remove the anthers before pollen was produced. Therefore, these flowers could not make their own pollen. Instead, pollen was obtained from another plant by gently touching its mature anthers with a

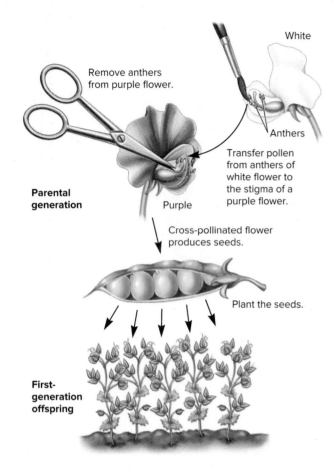

Remove anthers from purple flower.

White

Anthers

Parental generation

Purple

Transfer pollen from anthers of white flower to the stigma of a purple flower.

Cross-pollinated flower produces seeds.

Plant the seeds.

First-generation offspring

FIGURE 3.3   How Mendel cross-fertilized pea plants. This illustration depicts a cross between one plant with purple flowers and another with white flowers. The offspring from this cross are the result of pollination of the purple flower using pollen from a white flower.

CONCEPT CHECK: In this experiment, which plant, the white- or purple-flowered one, is providing the egg cells, and which is providing the sperm cells?

paintbrush. This pollen was applied to the stigma of the flower that already had its anthers removed. In this way, Mendel was able to cross-fertilize the pea plants, thereby obtaining hybrids.

In other experiments, Mendel allowed plants to reproduce by **self-fertilization,** which means that the pollen and eggs are derived from the same plant. In peas, two modified petals are fused to form a keel that encloses the reproductive structures of a flower. Because of this covering, pea plants naturally reproduce by self-fertilization. Usually, pollination occurs even before the flower opens. To achieve self-fertilization, Mendel simply let the pea plants reproduce naturally. In other words, the anthers were not removed.

Mendel Studied Seven Characteristics That Differed Among Strains of Pea Plants

A second experimental advantage of the garden pea is that it was available in different strains. Within agricultural species such as the garden pea, a **strain** is a genetically related group of individuals that display one or more differences compared to another group of individuals. In this case, several garden pea strains varied in height and in the appearance of their flowers, seeds, and pods. For example, one strain had purple flowers and another strain had white flowers. These color differences were due to variation in a gene that controls flower color.

- The general characteristics of an organism are called **characters.** For example, flower color is a character of peas.
- The term **trait,** or **variant,** is typically used to describe the specific properties of a character. For example, purple flower color is a trait (or variant) of peas.
- Different strains of pea plants differed with regard to their traits (or variants).

Over the course of 2 years, Mendel tested the pea strains to determine if their traits bred true. Breeding true means that a trait does not vary in appearance from generation to generation. For example, if the seeds from a pea plant were yellow, the next generation would also produce yellow seeds. Likewise, if these offspring were allowed to self-fertilize, all of their offspring would also produce yellow seeds, and so on. A strain that continues to produce the same trait after several generations of self-fertilization is called a **true-breeding strain.**

Mendel focused on the analysis of characters that were clearly distinguishable between different true-breeding strains. **Figure 3.4** illustrates the seven characters that Mendel eventually chose to follow in breeding experiments. All seven were found in two variants. For example, one character was height, which was found in two variants: tall and short plants.

3.1 COMPREHENSION QUESTIONS

1. Which of the following were experimental advantages of using pea plants?
   a. They came in several different varieties.
   b. They were capable of self-fertilization.
   c. They were easy to cross.
   d. All of the above were advantages.

2. With regard to Mendel's experiments, the term *cross* refers to an experiment in which
   a. the gametes come from different individuals.
   b. the gametes come from a single flower of the same individual.
   c. the gametes come from different flowers of the same individual.
   d. Both a and c are true.

3. To avoid self-fertilization in his pea plants, Mendel had to
   a. spray the plants with a chemical that damaged the pollen.
   b. remove the anthers from immature flowers.
   c. grow the plants in a greenhouse that did not contain pollinators (e.g., bees).
   d. do all of the above.

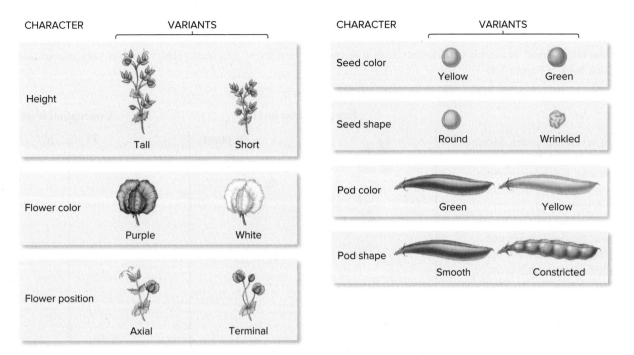

FIGURE 3.4 **An illustration of the seven characters that Mendel studied.** Each character was found as two variants that were decisively different from each other.

CONCEPT CHECK: What do we mean when we say a strain is true-breeding?

3.2 LAW OF SEGREGATION

Learning Outcomes:

1. Analyze Mendel's experiments involving single-factor crosses.
2. State Mendel's law of segregation, and explain how it is related to gamete formation and fertilization.
3. Predict the outcome of a single-factor cross or a self-fertilization experiment using a Punnett square.

In the previous section, we considered how cross-fertilization or self-fertilization experiments can be conducted on pea plants. In this section, we will examine how Mendel studied the inheritance of characters by crossing variants to each other. A cross in which an experimenter observes one character is called a **single-factor cross.** A cross between two parents with different variants for a given character produces single-character hybrids, also known as **monohybrids.** As you will learn, this type of experimental approach led Mendel to propose the law of segregation.

EXPERIMENT 3A

Mendel Followed the Outcome of a Single Character for Two Generations

Prior to conducting crosses in pea plants, Mendel did not have a hypothesis to explain the formation of hybrids. The goal of Mendel's experiments was to determine the quantitative relationships that govern the transmission of hereditary traits from parent to offspring. This rationale is called an **empirical approach.** Laws deduced from an empirical approach are known as empirical laws.

   Mendel's experimental procedure is shown in **Figure 3.5.** This type of experiment began with true-breeding plants that differed in a single character. These are termed the **parental generation,** or **P generation.** Crossing true-breeding parents to each other, called a P cross, produces offspring that constitute the

F_1 **generation,** or first filial generation (from the Latin *filius,* which refers to a male offspring). As seen in the data, all plants of the F_1 generation showed the trait of one parent but not the other. This prompted Mendel to follow the transmission of this character for one additional generation. To do so, the plants of the F_1 generation were allowed to self-fertilize to produce a second generation called the F_2 **generation,** or second filial generation.

THE GOAL (DISCOVERY-BASED SCIENCE)

Mendel speculated that the inheritance pattern for a single character may follow quantitative natural laws. The goal of this experiment was to uncover such laws.

Starting material: Mendel began his experiments with true-breeding strains of pea plants that varied in only one of seven different characters (look back at Figure 3.4).

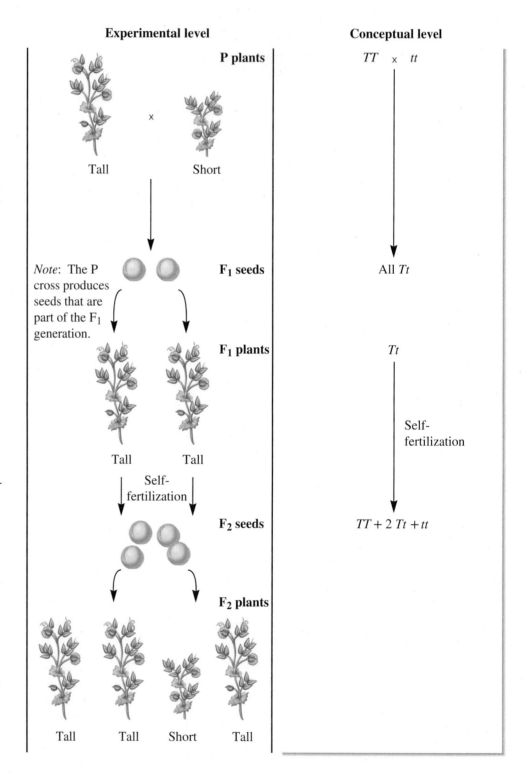

1. For each of seven characters, Mendel cross-fertilized two different true-breeding strains. Keep in mind that each cross involved two plants that differed in regard to only one of the seven characters studied. The illustration at the right shows one cross between a tall and short plant. This is called a P (parental) cross.

2. Collect the F_1 generation seeds. The following spring, plant the seeds and allow the plants to grow. These are the plants of the F_1 generation.

3. Allow the F_1 generation plants to self-fertilize. This produces seeds that are part of the F_2 generation.

4. Collect the F_2 generation seeds and plant them the following spring to obtain the F_2 generation plants.

5. Analyze the traits found in each generation.

Experimental level

P plants

Tall   x   Short

Note: The P cross produces seeds that are part of the F_1 generation.

F_1 seeds

F_1 plants

Tall   Tall

Self-fertilization

F_2 seeds

F_2 plants

Tall   Tall   Short   Tall

Conceptual level

TT x tt

All Tt

Tt

Self-fertilization

$TT + 2\,Tt + tt$

THE DATA

P cross	F_1 generation	F_2 generation	Ratio of traits in F_2 generation
Tall × short height	All tall	787 tall, 277 short	2.84:1
Purple × white flowers	All purple	705 purple, 224 white	3.15:1
Axial × terminal flowers	All axial	651 axial, 207 terminal	3.14:1
Yellow × green seeds	All yellow	6,022 yellow, 2,001 green	3.01:1
Round × wrinkled seeds	All round	5,474 round, 1,850 wrinkled	2.96:1
Green × yellow pods	All green	428 green, 152 yellow	2.82:1
Smooth × constricted pods	All smooth	882 smooth, 299 constricted	2.95:1
Total	All dominant	14,949 dominant, 5,010 recessive	2.98:1

Source: Mendel, Gregor, "Versuche über Plflanzenhybriden, Verhandlungen des naturforschenden Vereines in Brünn, Bd. IV für das Jahr 1865," *Abhandlungen*, 1866, 3–47.

INTERPRETING THE DATA

The data in the table are the results of producing an F_1 generation via cross-fertilization and an F_2 generation via self-fertilization of the F_1 plants. A quantitative analysis of these data allowed Mendel to propose three important ideas:

1. The data argued strongly against a blending mechanism of heredity. In all seven cases, the F_1 generation displayed traits that were distinctly like one of the two parents rather than traits that were intermediate in character. Mendel's first proposal was that one variant for a particular character is **dominant** to another variant. For example, the variant of green pods is dominant to that of yellow pods. The term **recessive** is used to describe a variant that is masked by the presence of a dominant trait but reappears in subsequent generations. Yellow pods and short height are examples of recessive variants. They can also be referred to as recessive traits.

2. When a true-breeding plant with a dominant trait was crossed to a true-breeding plant with a recessive trait, the dominant trait was always observed in the F_1 generation. In the F_2 generation, most offspring displayed the dominant trait, but some showed the recessive trait. How did Mendel explain this observation? Because the recessive trait appeared in the F_2 generation, a second proposal was made—the genetic determinants of traits are passed along as "unit factors" from generation to generation. The data were consistent with a **particulate theory of inheritance,** in which the factors that govern traits are inherited as discrete units that remain unchanged as they are passed from parent to offspring. Mendel referred to the genetic determinants as unit factors, but we now call them genes (from the Greek, *genesis*, meaning "birth," or, *genos*, meaning "origin").

3. When comparing the numbers of dominant and recessive traits in the F_2 generation, Mendel noticed a recurring pattern. Within experimental variation, a 3:1 ratio was observed between the dominant and the recessive trait. As described next, this quantitative approach allowed Mendel to make a third proposal—genes **segregate** from each other during the process that gives rise to gametes.

A 3:1 Phenotypic Ratio Is Consistent with the Law of Segregation

Mendel's research was aimed at understanding the laws that govern the inheritance of traits. At that time, scientists did not understand the molecular composition of the genetic material or its mode of transmission during gamete formation and fertilization. We now know that the genetic material is composed of deoxyribonucleic acid (DNA), a component of chromosomes. Each chromosome contains hundreds to thousands of shorter segments that function as genes—a term that was originally coined by Wilhelm Johannsen in 1909. A **gene** is defined as a unit of heredity that may influence the outcome of a trait in an organism. Each of the seven pea plant characters that Mendel studied is influenced by a different gene.

Most eukaryotic species, such as pea plants and humans, have their genetic material organized into pairs of chromosomes, as discussed in Chapter 2. For this reason, eukaryotes have two copies of most genes. These copies may be the same or they may differ. The term **allele** (from the Latin *alius* meaning "other")

refers to an alternative form of a particular gene. For example, the height gene in pea plants is found as a tall allele and a short allele. With this modern knowledge, the results shown in Figure 3.5 are consistent with the idea that each parent transmits only one copy of each gene (i.e., one allele) to each offspring. Using modern terminology, **Mendel's law of segregation** states the following:

The two copies of a gene segregate (or separate) from each other during the process that gives rise to gametes.

Therefore, only one copy of each gene is found in a gamete. At fertilization, two gametes combine randomly, potentially producing different allelic combinations.

Let's use Mendel's cross of tall and short pea plants to illustrate how alleles are passed from parents to offspring according to the law of segregation, and also to define some new genetic terms (**Figure 3.6**).

- The letters T and t are used to represent the alleles of the gene that determines plant height. By convention, the uppercase letter represents the dominant allele (T for tall

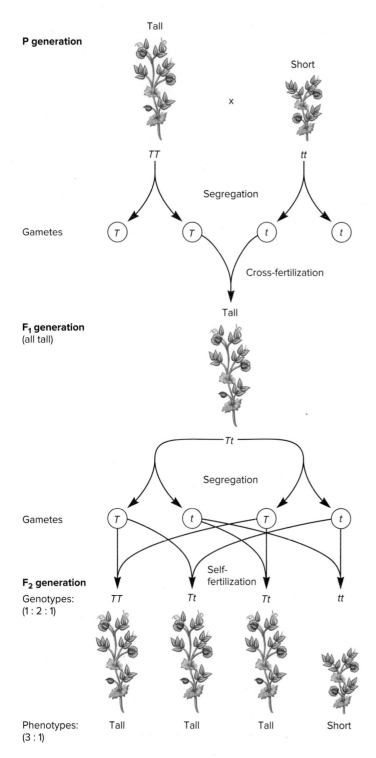

height, in this case), and the recessive allele is represented by the same letter in lowercase (*t*, for short height).

- For the P cross, both parents are true-breeding plants (see Figure 3.6). Therefore, each one has identical copies of the height gene. When an individual has two identical copies of a gene, the individual is said to be **homozygous** with respect to that gene. (The prefix *homo-* means "like," and the suffix *-zygo* means "pair.") In the P cross, the tall plant is homozygous (*TT*) for the tall allele, and the short plant is homozygous (*tt*) for the short allele. These alleles segregate during the process that gives rise to gametes.

- In contrast, the F_1 generation is **heterozygous**, *Tt*, because every individual carries one copy of the tall allele and one copy of the short allele. A heterozygous individual carries different alleles of a gene. (The prefix *hetero-* means "different.") The alleles in F_1 plants also segregate during the process that gives rise to gametes (see Figure 3.6). Their gametes can carry either a *T* allele or a *t* allele, but not both.

- The term **genotype** refers to the genetic composition of an individual. Following self-fertilization, *TT*, *Tt*, and *tt* are the possible genotypes of the F_2 generation. By randomly combining these alleles, the genotypes are produced in a 1 *TT* : 2 *Tt*: 1 *tt* ratio.

- The term **phenotype** refers to observable traits of an organism. Because *TT* and *Tt* both produce tall phenotypes, a 3 tall:1 short ratio of phenotypes is observed in the F_2 generation.

A Punnett Square Can Be Used to Predict the Outcome of Crosses and Self-Fertilization Experiments

An approach to predict the outcome of simple genetic crosses and self-fertilization experiments is to use a **Punnett square,** a method originally proposed by Reginald Punnett. To construct a Punnett square, you must know the genotypes of the parents. With this information, the Punnett square enables you to predict the types of offspring the parents are expected to produce and in what proportions.

Step 1. *Write down the genotypes of both parents. (In a self-fertilization experiment, a single parent provides the sperm and egg cells.)* Let's consider an example in which a heterozygous tall plant is crossed to another heterozygous tall plant. The plant providing the sperm (via pollen) is viewed as the male parent and the plant providing the eggs is the female parent.

Male parent: *Tt*

Female parent: *Tt*

Step 2. *Write down the possible gametes that each parent can make.* Remember that the law of segregation tells us that a gamete carries only one copy of each gene.

Male gametes: *T* or *t*

Female gametes: *T* or *t*

Step 3. *Create an empty Punnett square.* In the examples shown in this textbook, the number of columns equals the number of male gametes, and the number of rows equals

INTERACTIVE EXERCISE

FIGURE 3.6 **Mendel's law of segregation.** This illustration shows a cross between a true-breeding tall plant and a true-breeding short plant and the subsequent segregation of the tall (*T*) and short (*t*) alleles in the F_1 and F_2 generations.

CONCEPT CHECK: With regard to the *T* and *t* alleles of pea plants, explain what segregation means.

the number of female gametes. Our example has two rows and two columns. Place the male gametes across the top of the Punnett square and the female gametes along the side.

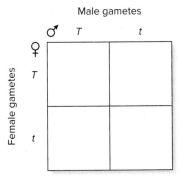

Male gametes

Step 4. *Fill in the possible genotypes of the offspring by combining the alleles of the gametes in the empty boxes.*

Male gametes

	♂ T	t
♀ T	TT	Tt
t	Tt	*tt*

Step 5. *Determine the relative proportions of genotypes and phenotypes of the offspring.* The genotypes are obtained directly from the Punnett square. These genotypes are contained within the boxes that have been filled in. In this example, the genotypes are *TT*, *Tt*, and *tt* in a 1:2:1 ratio. To determine the phenotypes, you must know the dominant/recessive relationship between the alleles. For plant height, *T* (tall) is dominant to *t* (short). The genotypes *TT* and *Tt* are tall, whereas the genotype *tt* is short. Therefore, our Punnett square shows us that the ratio of phenotypes is 3:1, or 3 tall plants : 1 short plant.

GENETIC TIPS   **THE QUESTION:** A pea plant that is heterozygous with regard to flower color (purple is dominant to white) is crossed to a pea plant with white flowers. What are the predicted outcomes of genotypes and phenotypes for the offspring?

TOPIC: *What topic in genetics does this question address?* The topic is Mendelian inheritance. More specifically, the question is about a single-factor cross.

INFORMATION: *What information do you know based on the question and your understanding of the topic?* From the question, you know that one plant is heterozygous for flower color. If *P* is the purple allele and *p* the white allele, the genotype of this plant is *Pp*. The other plant exhibits the recessive phenotype, so its

genotype must be *pp*. From your understanding of the topic, you may remember that alleles segregate during gamete formation and each parent passes one allele to their offspring; the two alleles combine at fertilization.

PROBLEM-SOLVING **S**TRATEGY: *Predict the outcome.* One strategy to solve this type of problem is to use a Punnett square to predict the outcome of the cross. The Punnett square is shown next.

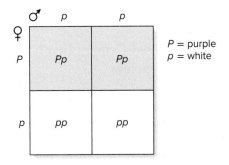

ANSWER: The ratio of offspring genotypes is 1 *Pp* : 1 *pp*. The ratio of the phenotypes is 1 purple : 1 white.

3.2 COMPREHENSION QUESTIONS

1. A pea plant is *Tt*. Which of the following statements is correct?
   a. Its genotype is *Tt*, and its phenotype is short.
   b. Its phenotype is *Tt*, and its genotype is short.
   c. Its genotype is *Tt*, and its phenotype is tall.
   d. Its phenotype is *Tt*, and its genotype is tall.
2. A *Tt* pea plant is crossed to a *tt* plant. What is the expected ratio of phenotypes for offspring from this cross?
   a. 3 tall : 1 short
   b. 1 tall : 1 short
   c. 1 tall : 3 short
   d. 2 tall : 1 short

3.3 LAW OF INDEPENDENT ASSORTMENT

Learning Outcomes:

1. Analyze Mendel's experiments involving two-factor crosses.
2. State Mendel's law of independent assortment.
3. Predict the outcome of a two-factor cross using a Punnett square.
4. Use the forked-line or multiplication method to predict the outcome of crosses involving three or more genes

Even though the experiments described in Figure 3.5 revealed important ideas regarding a hereditary law, Mendel realized that additional insights might be uncovered if more complicated experiments were carried out. In this section, we will examine crosses in which Mendel simultaneously investigated the pattern of inheritance for two different characters. Such

two-factor crosses followed the inheritance of two different characters within the same groups of individuals. These experiments led to the formulation of a second law—the law of independent assortment.

Mendel Also Analyzed Crosses Involving Two Different Characters

To illustrate Mendel's work, we will consider an experiment in which one of the characters involved was seed shape, found in round or wrinkled variants, and the second character was seed color, which existed as yellow and green variants. In this two-factor cross, Mendel followed the inheritance pattern for both characters simultaneously.

What results are possible from a two-factor cross? One possibility is that the genes for these two different characters are always linked to each other and inherited as a single unit (**Figure 3.7a**). If this were the case, the F_1 offspring could produce only two types of gametes, *RY* and *ry*. A second possibility is the genes are not linked and can assort themselves independently into gametes (**Figure 3.7b**). If independent assortment occurred, an F_1 offspring could produce four types of gametes: *RY*, *Ry*, *rY*, and *ry*. Keep in mind that the results of Figure 3.5 have already shown us that a gamete carries only one allele for each gene.

The experimental protocol for this two-factor cross is shown in **Figure 3.8**. Mendel began with two different strains of true-breeding pea plants that were different in seed shape and seed color. One plant was produced from seeds that were round and yellow; the other plant from seeds that were wrinkled and green. When these plants were crossed, the seeds, which contain the plant embryo, are considered part of the F_1 generation. As expected, the data revealed that the F_1 seeds displayed a phenotype of round and yellow. This phenotype was observed because round and yellow are dominant traits. It is the F_2 generation that supports the independent-assortment model and refutes the linked-assortment model.

THE HYPOTHESES

The inheritance pattern for two different characters follows one or more quantitative natural laws. Two possible hypotheses are described in Figure 3.7.

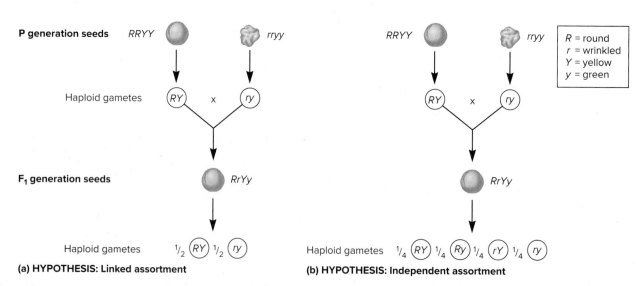

(a) HYPOTHESIS: Linked assortment

(b) HYPOTHESIS: Independent assortment

R = round
r = wrinkled
Y = yellow
y = green

FIGURE 3.7 **Two hypotheses to explain how two different genes assort during gamete formation.** **(a)** According to the hypothesis of linked assortment, the two genes always stay associated with each other. **(b)** In contrast, the independent-assortment hypothesis proposes that the two different genes randomly segregate into haploid cells.

CONCEPT CHECK: According to the hypothesis of linked assortment shown here, what is linked? Are two different genes linked, or are two different alleles of the same gene linked, or both?

TESTING THE HYPOTHESES   **FIGURE 3.8**   Mendel's analysis of a two-factor cross.

Starting material: In this experiment, Mendel began with two types of true-breeding strains of pea plants that were different with regard to two characters. One strain produced round, yellow seeds (*RRYY*); the other strain produced wrinkled, green seeds (*rryy*).

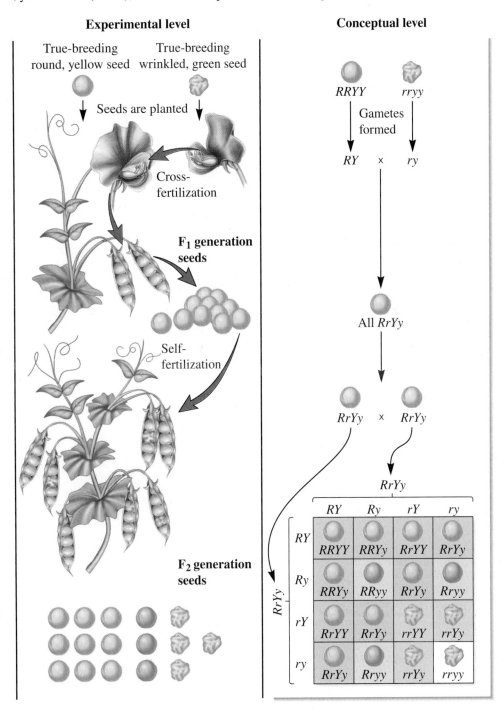

Experimental level

Conceptual level

1. Cross the two true-breeding strains to each other. This produces F₁ generation seeds.

2. Collect many seeds and record their phenotype.

3. F₁ seeds are planted and grown, and the F₁ plants are allowed to self-fertilize. This produces seeds that are part of the F₂ generation.

4. Analyze the characteristics found in the F₂ generation seeds.

THE DATA

P cross	F₁ generation	F₂ generation
Round, yellow seeds × wrinkled, green seeds	All round, yellow seeds	315 round, yellow seeds 108 round, green seeds 101 wrinkled, yellow seeds 32 wrinkled, green seeds

INTERPRETING THE DATA

As seen in the data, the F₂ generation had four categories of seeds. In addition to seeds that were like those of the parental generation, the F₂ generation also had seeds that were round and green and seeds that were wrinkled and yellow. These two categories of F₂ seeds are called **nonparental** because these combinations of traits were not found in the true-breeding plants of the parental generation.

The occurrence of nonparental variants contradicts the linked-assortment hypothesis (see Figure 3.7a). According to that model, the *R* and *Y* alleles should be linked together and so should the *r* and *y* alleles. If this were the case, the F₁ plants could only produce gametes that were *RY* or *ry*. These would combine to produce *RRYY* (round, yellow), *RrYy* (round, yellow), or *rryy* (wrinkled, green) seeds in a 1:2:1 ratio. Nonparental seeds could, therefore, not be produced. However, Mendel did not obtain this result. Instead, a phenotypic ratio of 9:3:3:1 was observed in the F₂ generation.

Mendel's Two-Factor Crosses Led to the Law of Independent Assortment

Mendel's results from many two-factor crosses rejected the linked-assortment hypothesis and, instead, supported the hypothesis that different characters assort themselves independently. Using modern terminology, **Mendel's law of independent assortment** states the following:

> *Two different genes will randomly assort their alleles during the process that gives rise to gametes.*

In other words, the allele for one gene will be found within a resulting gamete independently of whether the allele for a different gene is found in the same gamete. In the example given in Figure 3.8, the round and wrinkled alleles are assorted into haploid gametes independently of the yellow and green alleles. Therefore, a heterozygous *RrYy* parent can produce four different gametes—*RY*, *Ry*, *rY*, and *ry*—in equal proportions.

In an F₁ self-fertilization experiment, any two gametes can combine randomly during fertilization. This allows for 4², or 16, possible offspring, although some offspring will be genetically identical to each other. As shown in **Figure 3.9**, these 16 possible combinations result in seeds with the following phenotypes: 9 round, yellow; 3 round, green; 3 wrinkled, yellow; and 1 wrinkled, green. This 9:3:3:1 ratio is the expected outcome when a plant that is heterozygous for both genes is allowed to self-fertilize. Mendel was clever enough to realize that the data from the two-factor crosses were close to a 9:3:3:1 ratio.

As an example, let's consider the data in Figure 3.8. The F₁ generation produced F₂ seeds with the following phenotypes: 315 round, yellow; 108 round, green; 101 wrinkled, yellow; and 32 wrinkled, green. This yielded a total of 556 seeds. If you divide each phenotype by 556 (the total number of seeds) and multiply by 16 (the total number of possible genotypes), the phenotypic ratio of the F₂ generation is 9.1 : 3.1 : 2.9 : 0.9. Within experimental error, Mendel's data approximated the predicted 9:3:3:1 ratio for the F₂ generation.

The law of independent assortment held true for two-factor crosses involving all seven characters described earlier in Figure 3.4. However, in other cases, the inheritance pattern of two different genes is consistent with the linked-assortment hypothesis described in Figure 3.7a. In Chapter 6, we will examine the inheritance of genes that are linked because they are close to each other within the same chromosome. As we will see, linked genes do not assort independently.

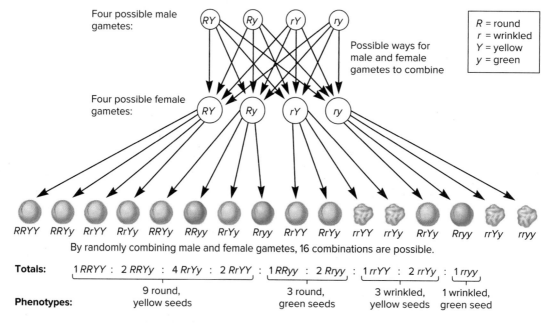

FIGURE 3.9 Mendel's law of independent assortment.

Genes→Traits This self-fertilization experiment involves a parent that is heterozygous for seed shape and seed color (*RrYy*). Four types of male gametes are possible: *RY*, *Ry*, *rY*, and *ry*. Likewise, four types of female gametes are possible: *RY*, *Ry*, *rY*, and *ry*. These four types of gametes are the result of the independent assortment of the seed shape and seed color alleles relative to each other. During fertilization, any one of the four types of male gametes can unite with any one of the four types of female gametes, resulting in 16 combinations.

CONCEPT CHECK: Why does independent assortment promote genetic variation?

Independent Assortment Promotes Genetic Diversity

An important consequence of the law of independent assortment is that a single individual can produce a vast array of genetically different gametes. As mentioned in Chapter 1, diploid species have pairs of homologous chromosomes, which may differ with respect to the alleles they carry. When an offspring receives a combination of alleles that differs from those in the parental generation, this phenomenon is termed **genetic recombination.** One mechanism that accounts for genetic recombination is independent assortment. A second mechanism, discussed in Chapter 6, is crossing over, which can reassort alleles that happen to be linked along the same chromosome.

The phenomenon of independent assortment is rooted in the random pattern in which the pairs of chromosomes assort themselves during the process of meiosis, a topic addressed in Section 3.4. When two different genes are found on different chromosomes, they randomly assort into haploid cells (look ahead to Figure 3.13). If a species contains a large number of chromosomes, this creates the potential for an enormous amount of genetic diversity.

Let's consider how independent assortment affects genetic diversity in humans. Human cells contain 23 pairs of chromosomes. These pairs separate and randomly assort into gametes during meiosis. The number of different gametes an individual can make equals 2^n, where n is the number of pairs of chromosomes. Therefore, a human can make 2^{23}, or over 8 million, possible gametes, due to independent assortment. The capacity to make so many genetically different gametes enables a species to produce a great diversity of individuals with different combinations of traits. This variety of phenotypes allows environmental factors to select for those combinations of traits that favor reproductive success.

A Punnett Square Can Be Used to Solve Independent Assortment Problems

As depicted in Figure 3.8, we can make a Punnett square to predict the outcome of experiments involving two or more genes that assort independently. Let's see how such a Punnett square is made by considering a cross between two plants that are heterozygous for height and seed color (**Figure 3.10**). This cross is $TtYy \times TtYy$. When we construct a Punnett square for this cross, we must keep in mind that each gamete has a single allele for each of two genes. In this example, the four possible gametes from each parent are

$$TY, Ty, tY, \text{ and } ty$$

In this two-factor cross, we need to make a Punnett square containing 16 boxes. The phenotypes of the resulting offspring are predicted to occur in a ratio of 9:3:3:1.

The Forked-Line and Multiplication Methods Can Also Be Used to Solve Independent Assortment Problems

In crosses involving three or more genes, the construction of a single large Punnett square becomes very unwieldy. For example, in a

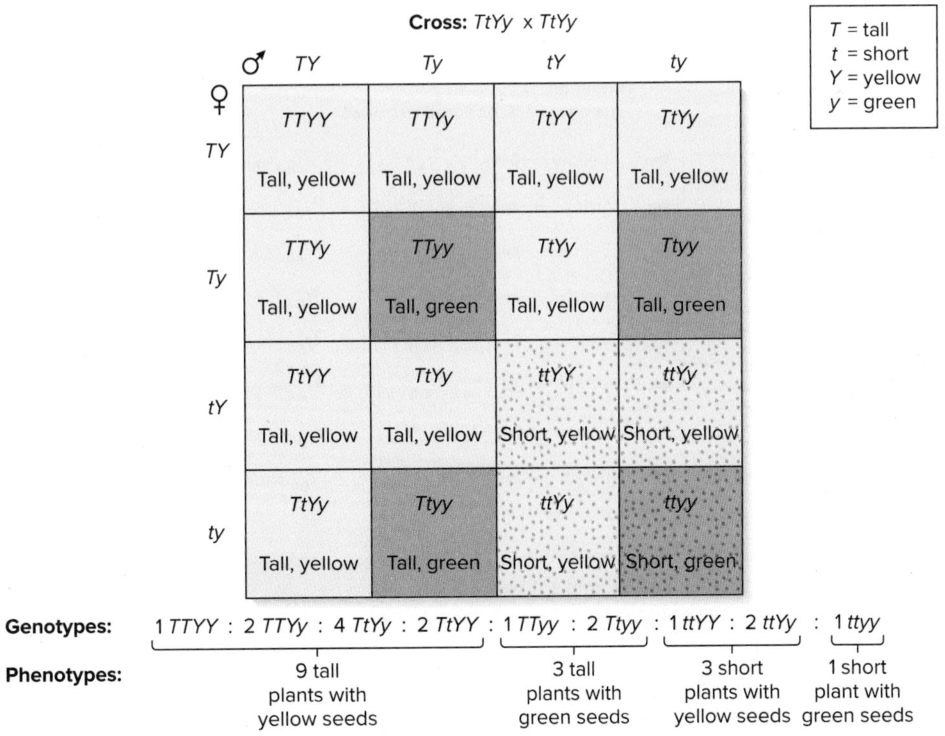

FIGURE 3.10 A Punnett square for a two-factor cross. The Punnett square shown here involves a cross between two pea plants that are heterozygous for height and seed color. The cross is $TtYy \times TtYy$.

CONCEPT CHECK: If a parent plant is $Ttyy$, how many different types of gametes can it make?

three-factor cross between two pea plants that are *Tt Rr Yy*, each parent can make 2^3, or 8, possible gametes. Therefore, the Punnett square must contain $8 \times 8 = 64$ boxes. As a more reasonable alternative, we can consider each gene separately and then algebraically combine them by multiplying together the expected outcomes for each gene. Two methods for doing this kind of analysis are termed the forked-line method and the multiplication method. To illustrate these methods, let's consider the following question:

> *Two pea plants are heterozygous for three genes (Tt Rr Yy), where T = tall, t = short, R = round seeds, r = wrinkled seeds, Y = yellow seeds, and y = green seeds. If these plants are crossed to each other, what are the predicted phenotypes of the offspring, and what fraction of the offspring will have each phenotype?*

You could solve this problem by constructing a large Punnett square and filling in the boxes. However, in this case, eight

different male gametes and eight different female gametes are possible: *TRY, TRy, TrY, tRY, trY, Try, tRy,* and *try*. It would become rather time-consuming to construct and fill in this Punnett square, which would contain 64 boxes. As an alternative, we can consider each gene separately and then algebraically combine them by multiplying together the expected phenotypic outcomes for each gene. In the cross *Tt Rr Yy × Tt Rr Yy*, a Punnett square can be made for each gene (**Figure 3.11a**).

In the **forked-line method,** a series of forked lines are created that connect the variants in this three-factor cross. The variants are tall or short, round or wrinkled, and yellow or green (**Figure 3.11b**). The genetic proportions of the 8 possible phenotype combinations are determined by multiplying the probabilities of each of the variants. For example, as shown at the top of Figure 3.11b, 3/4 tall × 3/4 round × 3/4 yellow = 27/64 tall, round, yellow.

An alternative to the forked-line method is the **multiplication method.** In this approach, you use the product rule (discussed

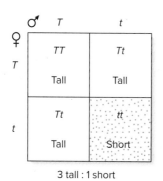 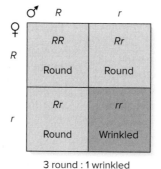 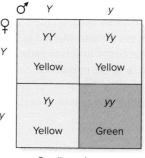

3 tall : 1 short   3 round : 1 wrinkled   3 yellow : 1 green

(a) Punnett squares for each gene

Tall or short	Round or wrinkled	Yellow or green	Observed product	Phenotype
3/4 tall	3/4 round	3/4 yellow	$(^3/_4)(^3/_4)(^3/_4)$ =	**27**/**64** tall, round, yellow
		1/4 green	$(^3/_4)(^3/_4)(^1/_4)$ =	**9**/**64** tall, round, green
	1/4 wrinkled	3/4 yellow	$(^3/_4)(^1/_4)(^3/_4)$ =	**9**/**64** tall, wrinkled, yellow
		1/4 green	$(^3/_4)(^1/_4)(^1/_4)$ =	**3**/**64** tall, wrinkled, green
1/4 short	3/4 round	3/4 yellow	$(^1/_4)(^3/_4)(^3/_4)$ =	**9**/**64** short, round, yellow
		1/4 green	$(^1/_4)(^3/_4)(^1/_4)$ =	**3**/**64** short, round, green
	1/4 wrinkled	3/4 yellow	$(^1/_4)(^1/_4)(^3/_4)$ =	**3**/**64** short, wrinkled, yellow
		1/4 green	$(^1/_4)(^1/_4)(^1/_4)$ =	**1**/**64** short, wrinkled, green

(b) Forked-line method

P = (3 tall + 1 short)(3 round + 1 wrinkled)(3 yellow + 1 green)

First, multiply (3 tall + 1 short) times (3 round + 1 wrinkled):

(3 tall + 1 short)(3 round + 1 wrinkled) = 9 tall, round + 3 tall, wrinkled + 3 short, round + 1 short, wrinkled

Next, multiply this product by (3 yellow + 1 green):

P = (9 tall, round + 3 tall, wrinkled + 3 short, round + 1 short, wrinkled) (3 yellow + 1 green) = 27 tall, round, yellow + 9 tall, round, green + 9 tall, wrinkled, yellow + 3 tall, wrinkled, green + 9 short, round, yellow + 3 short, round, green + 3 short, wrinkled, yellow + 1 short, wrinkled, green

(c) Multiplication method

FIGURE 3.11 **Two alternative ways to predict the outcome of a three-factor cross.** The cross is *Tt Rr Yy × Tt Rr Yy*. (**a**) For both methods, each gene is treated separately. The three Punnett squares predict the outcome for each gene. (**b**) Forked-line method. (**c**) Multiplication method.

later in Section 3.6) and multiply the phenotypic outcomes together in three sequential pairs: (3 tall + 1 short)(3 round + 1 wrinkled)(3 yellow + 1 green) (**Figure 3.11c**). Even though the multiplication steps are rather tedious, this approach is usually faster than making a Punnett square with 64 boxes, filling them in, deducing each phenotype, and then adding them up.

3.3 COMPREHENSION QUESTIONS

1. A pea plant has the genotype *rrYy*. How many different types of gametes can it make and in what proportions?
   a. 1 *rr* : 1 *Yy*
   b. 1 *rY* : 1 *ry*
   c. 3 *rY* : 1 *ry*
   d. 1 *RY* : 1 *rY* : 1 *Ry* : 1 *ry*

2. A cross is made between a pea plant that is *RrYy* and one that is *rrYy*. What is the predicted outcome of the seed phenotypes?
   a. 9 round, yellow : 3 round, green : 3 wrinkled, yellow : 1 wrinkled, green
   b. 3 round, yellow : 3 round, green : 1 wrinkled, yellow : 1 wrinkled, green
   c. 3 round, yellow : 1 round, green : 3 wrinkled, yellow : 1 wrinkled, green
   d. 1 round, yellow : 1 round, green : 1 wrinkled, yellow : 1 wrinkled, green

3. In a population of wild squirrels, most of them have gray fur, but an occasional squirrel is completely white. If we let *P* and *p* represent dominant and recessive alleles, respectively, of a gene that codes an enzyme necessary for pigment formation, which of the following statements do you think is most likely to be correct?
   a. The white squirrels are *pp*, and the *p* allele is a loss-of-function allele.
   b. The gray squirrels are *pp*, and the *p* allele is a loss-of-function allele.
   c. The white squirrels are *PP*, and the *P* allele is a loss-of-function allele.
   d. The gray squirrels are *PP*, and the *P* allele is a loss-of-function allele.

3.4 THE CHROMOSOME THEORY OF INHERITANCE

Learning Outcomes:
1. List the five principles of the chromosome theory of inheritance.
2. Explain the relationship between meiosis and Mendel's laws of inheritance.
3. Analyze the results of Morgan's experiment, which showed that a gene affecting eye color in fruit flies is located on the X chromosome.

The **chromosome theory of inheritance** explains the relationship between the transmission of chromosomes from parent to offspring and the transmission of traits from parent to offspring. This theory was a major breakthrough in the study of genetics because it established the framework for understanding how chromosomes carry and transmit the genetic determinants that govern the outcome of traits. In this section, we will examine the five principles of the chromosome theory of inheritance and consider how they relate to Mendel's laws of inheritance. We will also consider an experiment in fruit flies that provided evidence for this theory.

The Chromosome Theory of Inheritance Relates the Behavior of Chromosomes to the Mendelian Inheritance of Traits

This theory dramatically unfolded as a result of three lines of scientific inquiry.

- One of these concerned Mendel's breeding studies, in which the transmission of traits from parent to offspring were analyzed quantitatively.
- A second line of inquiry focused on the biochemical basis for heredity. Carl Nägeli and August Weismann championed the idea that a substance found in living cells is responsible for the transmission of traits from parent to offspring. Nägeli also suggested that both parents contribute equal amounts of this substance to their offspring. We now know that DNA within the chromosomes is the genetic material.
- The third line of inquiry involved the microscopic examination of the processes of meiosis and fertilization. When the work of Mendel was rediscovered, several scientists noted striking parallels between the segregation and assortment of traits noted by Mendel and the behavior of chromosomes during meiosis. Among them were Theodor Boveri and Walter Sutton, who independently proposed the chromosome theory of inheritance.

According to this theory, the inheritance patterns of traits can be explained by the transmission patterns of chromosomes during meiosis and fertilization. The chromosome theory of inheritance is based on a few fundamental principles:

1. Chromosomes contain the genetic material that is transmitted from parent to offspring and from cell to cell.
2. Chromosomes are replicated and passed along, generation after generation, from parent to offspring. They are also passed from cell to cell during the development of a multicellular organism. Each type of chromosome retains its individuality during cell division and gamete formation.
3. The nuclei of most eukaryotic cells contain chromosomes that are found in homologous pairs—the cells are diploid. One member of each pair is inherited from the female parent, the other from the male parent. At meiosis, one of the two members of each pair segregates into one daughter

cell, and its homolog segregates into the other daughter cell. Gametes contain one set of chromosomes—they are haploid.

4. During the formation of haploid cells, different types of (nonhomologous) chromosomes segregate independently of each other.

5. At fertilization, each parent contributes one set of chromosomes to its offspring. The maternal and paternal sets of homologous chromosomes are functionally equivalent; each set carries a full complement of genes.

The Law of Segregation Is Explained by the Separation of Homologs During Meiosis

The chromosome theory of inheritance allows us to see the relationship between Mendel's laws and chromosome transmission. As shown in **Figure 3.12**, Mendel's law of segregation can be explained by the homologous pairing and segregation of chromosomes during meiosis. This figure depicts the behavior of a pair of homologous chromosomes that carry a gene for seed color in pea plants. One of the chromosomes carries a dominant allele that confers yellow seed color, whereas the homologous chromosome carries a recessive allele that confers green color. A heterozygous individual passes only one of these alleles to each offspring. In other words, a gamete may contain the yellow allele or the green allele but not both. Because the homologs segregate during meiosis I and the sister chromatids separate during meiosis II, a gamete contains only one copy of each type of chromosome.

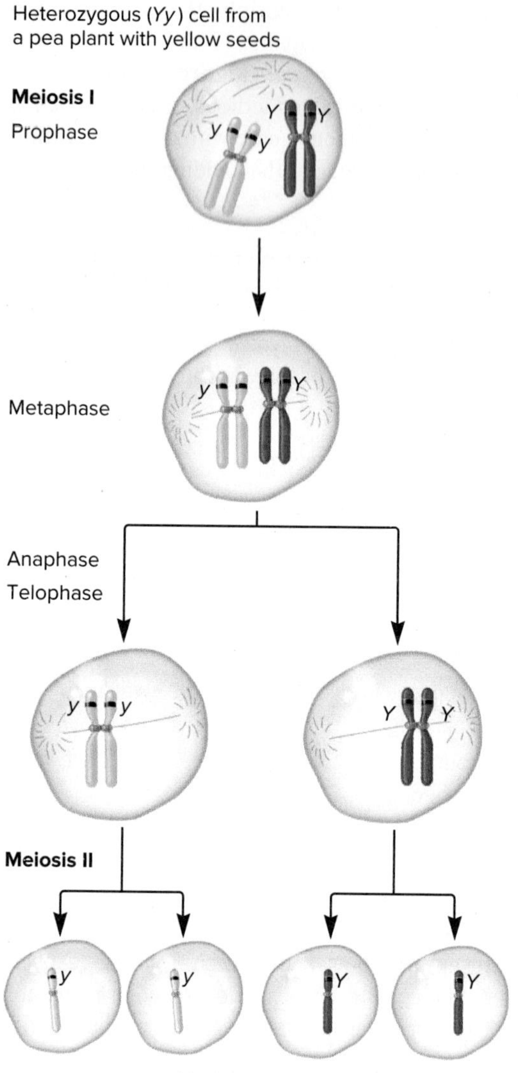

The Law of Independent Assortment Is Explained by the Random Alignment of Homologs During Meiosis

How is the law of independent assortment explained by the behavior of chromosomes? **Figure 3.13** considers the segregation of two types of chromosomes in a pea plant, each carrying a different gene.

- One pair of chromosomes carries the gene for seed color: The yellow (*Y*) allele is on one chromosome, and the green (*y*) allele is on the homolog.
- The other pair of (smaller) chromosomes carries the gene for seed shape: One copy has the round (*R*) allele, and the homolog carries the wrinkled (*r*) allele.
- At metaphase of meiosis I, the different types of chromosomes have randomly aligned along the metaphase plate. As shown in Figure 3.13, this can occur in more than one way. On the left, the *y* allele has sorted with the *R* allele, whereas the *Y* allele has sorted with the *r* allele. On the right, the opposite situation has occurred. Therefore, the random alignment of chromatid pairs during meiosis I can lead to an independent assortment of genes that are found on nonhomologous chromosomes.

FIGURE 3.12 Mendel's law of segregation can be explained by the segregation of homologs during meiosis. The two copies of a gene are located on homologous chromosomes. In this example using pea plant seed color, the two alleles are *Y* (yellow) and *y* (green). During meiosis, the homologous chromosomes segregate from each other, leading to segregation of the two alleles into separate gametes.

Genes→Traits The gene for seed color exists in two alleles, *Y* (yellow) and *y* (green). During meiosis, the homologous chromosomes that carry these alleles segregate from each other. The resulting cells receive either the *Y* or *y* allele, but not both. When two gametes unite during fertilization, the alleles they carry determine the traits of the resulting offspring. In this case, they affect seed color, producing yellow or green seeds.

CONCEPT CHECK: At which stage of meiosis do homologous chromosomes separate from each other?

As we will see in Chapter 6, this law is violated if two different genes are located close to one another on the same chromosome.

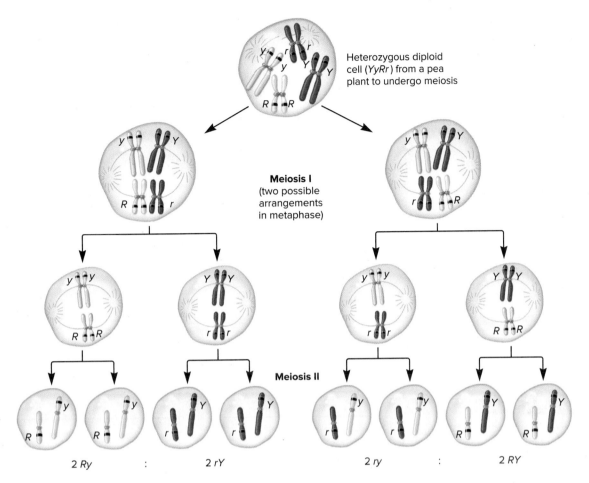

FIGURE 3.13 Mendel's law of independent assortment can be explained by the random alignment of bivalents during metaphase of meiosis I. This figure shows the assortment of two genes located on two different chromosomes in a pea plant, using seed color and shape as an example (*YyRr*). During metaphase of meiosis I, different possible arrangements of the homologs within bivalents can lead to different combinations of the alleles in the resulting gametes. For example, on the left, the recessive *y* allele has sorted with the dominant *R* allele; on the right, the dominant *Y* allele has sorted with the dominant *R* allele.

Genes→Traits Most species have several different chromosomes that carry many different genes. In this example, the gene for seed color exists in two alleles, *Y* (yellow) and *y* (green), and the gene for seed shape is found as *R* (round) and *r* (wrinkled) alleles. The two genes are located on different (nonhomologous) chromosomes. During meiosis, the homologous chromosomes that carry these alleles segregate from each other. In addition, the chromosomes carrying the *Y* or *y* alleles will independently assort from the chromosomes carrying the *R* or *r* alleles. As shown here, this sorting process can create haploid cells with different combinations of alleles. When two gametes unite during fertilization, the alleles they carry affect the traits of the resulting offspring.

CONCEPT CHECK: Let's suppose a pea plant is heterozygous for three genes, *Tt Yy Rr*, and each gene is on a different chromosome. How many different ways could the three pairs of homologous chromosomes line up during metaphase of meiosis I?

EXPERIMENT 3C

Morgan's Experiments Showed a Connection Between a Genetic Trait and the Inheritance of a Sex Chromosome in *Drosophila*

Even though an examination of meiosis provided compelling evidence that Mendel's laws could be explained by chromosome sorting, researchers still needed to correlate chromosome behavior with the inheritance of particular traits. In the early 1900s, American geneticist Thomas Hunt Morgan carried out a particularly influential study that confirmed the chromosome theory of inheritance by showing that a gene affecting eye color in fruit flies (*Drosophila melanogaster*) is located on the X chromosome.

Within a population of red-eyed flies, Morgan identified a male fruit fly with white eyes rather than the common (wild-type) red eyes. Because this had been a true-breeding strain of flies, this white-eyed male must have arisen from a new mutation that converted a red-eye allele (denoted w^+) into a white-eye allele (denoted w). Morgan is said to have carried this fly home with him in a jar, kept it on a bedside table at night, and then took it back to the laboratory during the day!

Morgan studied the inheritance of this white-eye trait by making crosses and quantitatively analyzing their outcome (**Figure 3.14**). First, the white-eyed male was crossed to a true-breeding red-eyed female. All of the F_1 offspring had red eyes, indicating that red is dominant to white. Next, the F_1 offspring were then mated to each other to obtain an F_2 generation.

THE GOAL (DISCOVERY-BASED SCIENCE)

This is an example of discovery-based science rather than hypothesis testing. In this case, a quantitative analysis of genetic crosses may reveal the inheritance pattern for the white-eye allele.

ACHIEVING THE GOAL   **FIGURE 3.14** **Inheritance pattern of an X-linked trait in fruit flies.**

Starting material: A true-breeding strain of red-eyed fruit flies plus one white-eyed male fly that was discovered in the population.

	Experimental level	Conceptual level

1. Cross the white-eyed male to a true-breeding red-eyed female.

P generation

$X^w Y$ × $X^{w^+} X^{w^+}$

2. Record the results of the F_1 generation. This involves noting the eye color and sex of many offspring.

F_1 generation

$X^{w^+}Y$ male offspring and $X^{w^+}X^w$ female offspring, both with red eyes

$X^{w^+}Y$ × $X^{w^+}X^w$

3. Cross F_1 offspring with each other to obtain F_2 offspring. Also record the eye color and sex of the F_2 offspring.

F_2 generation

$1 X^{w^+}Y$: $1 X^w Y$: $1 X^{w^+}X^w$: $1 X^{w^+}X^w$
1 red-eyed male : 1 white-eyed male : 2 red-eyed females

4. In a separate experiment, perform a testcross between a white-eyed male from the F_2 generation and a red-eyed female from the F_1 generation. Record the results.

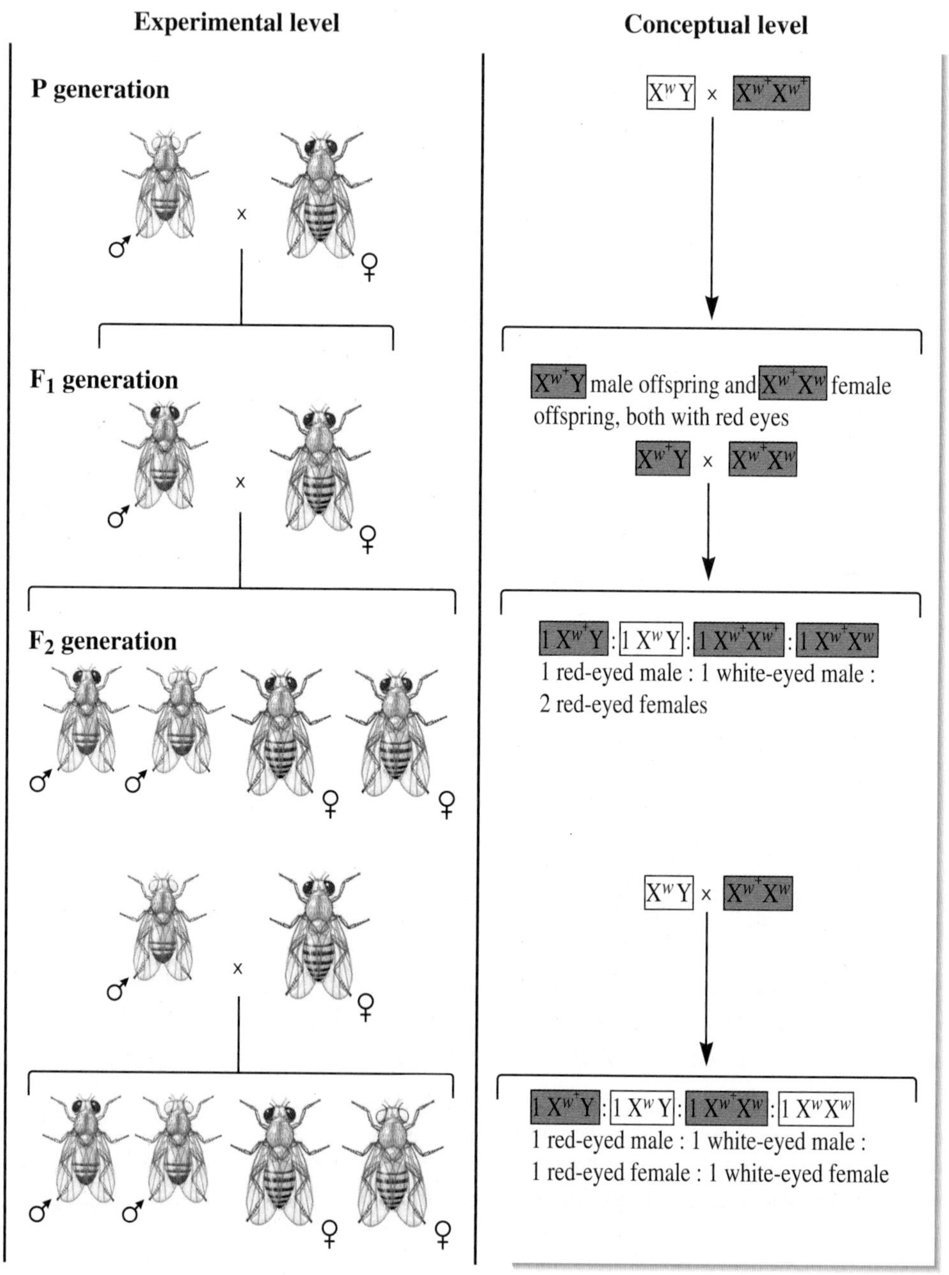

$X^w Y$ × $X^{w^+}X^w$

$1 X^{w^+}Y$: $1 X^w Y$: $1 X^{w^+}X^w$: $1 X^w X^w$
1 red-eyed male : 1 white-eyed male : 1 red-eyed female : 1 white-eyed female

THE DATA

Cross	Results	
Original white-eyed male to a true-breeding red-eyed female	F$_1$ generation:	All red-eyed flies
F$_1$ male to F$_1$ female	F$_2$ generation:	2459 red-eyed females 1011 red-eyed males 0 white-eyed females 782 white-eyed males
F$_2$ white-eyed male to F$_1$ female	Testcross:	129 red-eyed females 132 red-eyed males 88 white-eyed females 86 white-eyed males

Source: Morgan, Thomas H., "Sex limited inheritance in Drosophila," *Science*, vol. 32, no. 812, July 22, 1910, 120–122.

INTERPRETING THE DATA

As seen in the data table, the F$_2$ generation consisted of 2459 red-eyed females, 1011 red-eyed males, and 782 white-eyed males. Most notably, no white-eyed female offspring were observed in the F$_2$ generation. These results suggested that the pattern of transmission from parent to offspring depends on the sex of the offspring and on the alleles that they carry. As shown in the Punnett square below, the data are consistent with the idea that the eye color alleles are located on the X chromosome. This phenomenon is called **X-linked inheritance.**

F$_1$ male is X$^{w^+}$Y
F$_1$ female is X^{w}X^w

The Punnett square predicts that the F$_2$ generation will not have any white-eyed females. This prediction was confirmed experimentally. These results indicated that the eye color alleles are located on the X chromosome. Genes that are physically located within the X chromosome are called **X-linked genes,** or **X-linked alleles.**

However, it should also be pointed out that the experimental ratio of red eyes to white eyes in the F$_2$ generation is (2459 + 1011):782, which equals 4.4:1. This ratio deviates significantly from the predicted ratio of 3:1. How can this discrepancy be explained? Later work revealed that the lower-than-expected number of white-eyed flies is due to their decreased survival rate.

Morgan also conducted a **testcross** (see step 4, Figure 3.14) in which an individual with a dominant phenotype and unknown genotype is crossed to an individual with a recessive phenotype. In this case, an F$_1$ red-eyed female was mated to an F$_2$ white-eyed male. This cross produced red-eyed males and females in approximately equal numbers, and white-eyed males and females in approximately equal numbers. The testcross data are also consistent with an X-linked pattern of inheritance. As shown in the following Punnett square, a 1:1:1:1 ratio is predicted for this testcross:

Testcross:
Male is X^{w}Y
F$_1$ female is X$^{w^+}$X^w

The observed data are 129:132:88:86, which is a ratio of 1.5:1.5:1:1. Again, the lower-than-expected numbers of white-eyed males and females can be explained by a lower survival rate for white-eyed flies. Based on these data, Morgan concluded that red eye color and X (a sex factor that is present in two copies in the female) are combined and have never existed apart. In other words, this gene for eye color is on the X chromosome. Morgan was the first geneticist to receive a Nobel Prize.

3.4 COMPREHENSION QUESTIONS

1. Which of the following is *not* one of the principles of the chromosome theory of inheritance?
   a. Chromosomes contain the genetic material that is transmitted from parent to offspring and from cell to cell.
   b. Chromosomes are replicated and passed along, generation after generation, from parent to offspring.
   c. Chromosome replication occurs during S phase of the cell cycle.
   d. Each parent contributes one set of chromosomes to its offspring.

2. A pea plant has the genotype *TtRr*. The independent assortment of these two genes occurs at _____ because chromosomes carrying the _____ alleles line up independently of the chromosomes carrying the _____ alleles.
   a. metaphase of meiosis I, *T* and *t*, *R* and *r*
   b. metaphase of meiosis I, *T* and *R*, *t* and *r*
   c. metaphase of meiosis II, *T* and *t*, *R* and *r*
   d. metaphase of meiosis II, *T* and *R*, *t* and *r*

3.5 STUDYING INHERITANCE PATTERNS IN HUMANS

Learning Outcomes:

1. Describe the features of a pedigree.
2. Analyze a pedigree to determine if a trait or disease is dominant or recessive.

Before we end our discussion of simple Mendelian traits, let's address the question of how we can analyze inheritance patterns that occur in humans. As discussed in Sections 3.2 and 3.3, Mendel selectively made crosses and then analyzed a large number of offspring. When studying human traits, however, researchers cannot control parental crosses. Instead, they must rely on the information that is contained within family trees, or **pedigrees,** which are charts representing family relationships. This type of approach, known as **pedigree analysis,** is aimed at determining the type of inheritance pattern that a gene follows. Although this method may be less definitive than performing experiments like Mendel's, pedigree analyses often provide important clues concerning the pattern of inheritance of traits within human families. An expanded discussion of human pedigrees is found in Chapter 24, which concerns the inheritance patterns of many human diseases.

In order to discuss the applications of pedigree analysis, we need to understand the organization and symbols of a pedigree (**Figure 3.15**).

- The oldest generation is at the top of the pedigree, and the most recent generation is at the bottom.
- Vertical lines connect each succeeding generation.
- A male (square) and female (circle) who produce one or more offspring are directly connected by a horizontal line.
- A vertical line connects parents with their offspring.
- If parents produce two or more offspring, the group of siblings (male and/or female offspring) is denoted by two or more squares and/or circles projecting downward from the same horizontal line.

As discussed in Section 1.5 in Chapter 1, the terms *male* and *female* refer to an individuals' sex. These terms do not indicate gender identity.

The pedigree shown in Figure 3.15a involves the transmission of a human disorder called cystic fibrosis. Affected individuals are depicted by filled symbols (in this case, filled with

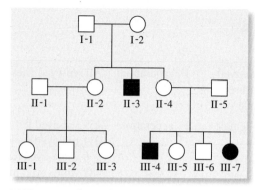

(a) Human pedigree showing cystic fibrosis

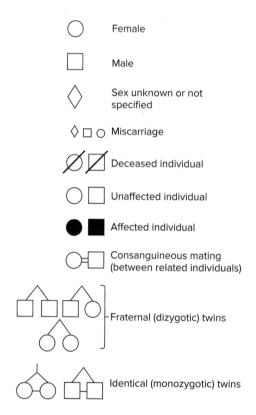

(b) Symbols used to denote phenotype in a human pedigree

FIGURE 3.15 Pedigree analysis. (a) A family pedigree in which some of the members are affected with cystic fibrosis. **(b)** The symbols used in a human pedigree. Note: The symbols shown here denote phenotypes, not genotypes.

CONCEPT CHECK: What are the two different meanings of horizontal lines in a pedigree?

black) that distinguish them from unaffected individuals. Each generation is given a roman numeral designation, and individuals within the same generation are numbered from left to right. A few examples of the genetic relationships in Figure 3.15a are described here:

- Individuals II-1 and II-2 are the parents of III-1, III-2, and III-3.
- Individuals I-1 and I-2 are the grandparents of III-1, III-2, III-3, III-4, III-5, III-6, and III-7.

- Individuals III-1, III-2, and III-3 are siblings.
- Individual III-4 is affected by a genetic disease.

The symbols in Figure 3.15 and those found in the problem sets at the end of the chapter indicate phenotypes. In such pedigrees, affected individuals are shown with filled symbols, and unaffected individuals, including those that might be heterozygous for a recessive disease, are depicted with unfilled symbols.

Pedigree analysis is commonly used to determine the inheritance pattern for human genetic diseases. Human geneticists are routinely interested in knowing whether a genetic disease is inherited as a recessive or dominant trait. One way to discern the dominant/recessive relationship between two alleles is by pedigree analysis. Genes that play a role in disease may exist as a common (wild-type) allele or a mutant allele that causes disease symptoms.

If the disease follows a simple Mendelian pattern of inheritance and is caused by a recessive allele, an individual must inherit two copies of the mutant allele to exhibit the disease. Therefore, a recessive pattern of inheritance makes two important predictions:

- Two heterozygous unaffected individuals will, on average, have 1/4 of their offspring affected.
- All offspring of two affected individuals will be affected.

Alternatively, with a dominant trait, affected individuals will have inherited the gene from at least one affected parent (unless a new mutation has occurred during gamete formation).

As mentioned, the pedigree in Figure 3.15a illustrates inheritance of a human genetic disease known as cystic fibrosis (CF). Among people of Northern European descent, approximately 3% of the population are heterozygous carriers of the recessive allele. In homozygotes, the disease symptoms include abnormalities of the pancreas, intestine, sweat glands, and lungs. These abnormalities are caused by an imbalance of ions across the plasma membranes of cells. In the lungs, this leads to a buildup of thick, sticky mucus. Respiratory problems may lead to early death, although modern treatments have greatly increased the life span of CF patients.

In the late 1980s, the gene for CF was identified. It codes a protein called the cystic fibrosis transmembrane conductance regulator (CFTR). This protein regulates ion balance across the cell membrane in tissues of the pancreas, intestine, sweat glands, and lungs. The mutant allele causing CF alters the CFTR protein in such a way that the altered protein is not correctly inserted into the plasma membrane, resulting in decreased function that causes the ion imbalance.

As seen in the pedigree, the pattern of affected and unaffected individuals is consistent with a recessive mode of inheritance. Two unaffected individuals can produce an affected offspring. Although not shown in this pedigree, a recessive mode of inheritance is also indicated by the observation that two affected individuals produce 100% affected offspring. However, for human genetic diseases that limit survival or fertility (or both), there may never be cases where two affected individuals produce offspring.

3.5 COMPREHENSION QUESTIONS

1. Which of the following would *not* be observed in a pedigree if a genetic disorder was inherited in a recessive manner?
   a. Two unaffected parents have an affected offspring.
   b. Two affected parents have an unaffected offspring.
   c. One affected and one unaffected parent have an unaffected offspring.
   d. All of the above are possible for a recessive disorder.
2. For the pedigree shown here, which pattern(s) of inheritance is/are possible? Affected individuals are shown with a filled symbol.

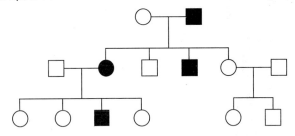

   a. Recessive
   b. Dominant
   c. Both recessive and dominant
   d. Neither recessive nor dominant

3.6 PROBABILITY AND STATISTICS

Learning Outcomes:
1. Define *probability*.
2. Predict the outcomes of crosses using the product rule or the binomial expansion equation.
3. Evaluate the validity of a hypothesis using a chi square test.

A powerful application of Mendel's work is that the laws of inheritance can be used to predict the outcomes of genetic crosses. In agriculture, for example, plant and animal breeders are concerned with the types of offspring resulting from their crosses. This information is used to produce commercially important crops and livestock. In addition, people are often interested in predicting the characteristics of the children they may have. This may be particularly important to individuals who carry alleles that cause inherited diseases. Of course, we cannot see into the future and definitively predict what will happen. Nevertheless, genetic counselors can help couples to predict the likelihood of having an affected child. This probability is one factor that may influence a couple's decision whether to have children.

In this section, we will see how probability calculations are used in genetic problems to predict the outcomes of crosses. We will consider two mathematical operations known as the product rule and the binomial expansion equation. These methods allow us to determine the probability that a cross between two individuals will produce a particular outcome. To apply these operations, we must have some knowledge regarding the genotypes of the parents and the pattern of inheritance of a given trait. The following operations are used to solve certain types of problems:

- The product rule is used for problems in which the outcomes are independent of each other.
- The binomial expansion equation is used for problems having an unordered combination of outcomes.

Probability calculations can also be used in hypothesis testing. In many situations, a researcher would like to discern the genotypes and patterns of inheritance for traits that are not yet understood. A traditional approach to this type of problem is to conduct crosses and then analyze their outcomes. The proportions of offspring may provide important clues that allow the experimenter to propose a hypothesis, based on the quantitative laws of inheritance, that explains the transmission of the trait from parents to offspring. Statistical methods, such as the chi square test, can then be used to evaluate how well the observed data from crosses fit the expected data. We will end this chapter with an example that applies the chi square test to a genetic cross.

Probability Is the Likelihood That an Outcome Will Occur

The chance that an outcome will occur in the future is called the outcome's **probability.** For example, if you flip a coin, the probability is 0.50, or 50%, that the head side will be showing when the coin lands. Probability depends on the number of possible outcomes. In this case, two possible outcomes (heads or tails) are equally likely. This allows us to predict a 50% chance that a coin flip will produce heads. The general formula for probability (P) is

$$\text{Probability} = \frac{\text{Number of times a particular outcome occurs}}{\text{Total number of possible outcomes}}$$

Thus, the probability of heads for a coin flip is

$$P_{\text{heads}} = 1 \text{ heads } / (1 \text{ heads } + 1 \text{ tails}) = 1/2, \text{ or } 50\%$$

In genetic problems, we are often interested in the probability that a particular type of offspring will be produced. Recall that when two heterozygous tall pea plants (Tt) are crossed, the phenotypic ratio of the offspring is 3 tall to 1 short. This information can be used to calculate the probability for either type of offspring:

$$\text{Probability} = \frac{\text{Number of individuals with a given phenotype}}{\text{Total number of individuals}}$$

$$P_{\text{tall}} = 3 \text{ tall } / (3 \text{ tall } + 1 \text{ short}) = 3/4, \text{ or } 75\%$$

$$P_{\text{short}} = 1 \text{ short } / (3 \text{ tall } + 1 \text{ short}) = 1/4, \text{ or } 25\%$$

The probability is 75% for offspring that are tall and 25% for offspring that are short. When we add together the probabilities of all possible outcomes (tall and short), we should get a sum of 100% (here, 75% + 25% = 100%).

A probability calculation allows us to predict the likelihood that an outcome will occur in the future. The accuracy of this prediction, however, depends to a great extent on the size of the sample. For example, if we toss a coin six times, our probability prediction suggests that 50% of the time we should get heads (i.e., three heads and three tails). With this small sample size, however, we would not be too surprised if we came up with four heads and two tails. Each time we toss a coin, there is a random chance that it will be heads or tails.

The deviation between the observed and expected outcomes is called the **random sampling error.** In a small sample of coin tosses, the error between the predicted percentage of heads and the actual percentage observed may be quite large. By comparison, if we flipped a coin 1000 times, the percentage of heads would be fairly close to the predicted 50% value. In a larger sample, we expect the random sampling error to be a much smaller percentage. For example, the fairly large data sets produced by Mendel had relatively small sampling errors (refer back to the data table following Figure 3.5).

The Product Rule Is Used to Predict the Probability of Independent Outcomes

We can use probability to make predictions regarding the likelihood of two or more independent outcomes from a genetic cross. When we say that outcomes are independent, we mean that the occurrence of one outcome does not affect the probability of another. As an example, let's consider a rare, recessive human trait known as congenital analgesia. Persons with this trait can distinguish between sharp and dull, and hot and cold, but do not perceive extremes of sensation as being painful. The first case of congenital analgesia, described in 1932, was a person who entertained the public as a "human pincushion."

For a phenotypically unaffected couple, in which each parent is heterozygous, Pp (where P is the common allele and p is the recessive allele causing congenital analgesia), we can ask, What is the probability that the couple's first three offspring will have congenital analgesia? To answer this question, the **product rule** is used. According to this rule,

The probability that two or more independent outcomes will occur is equal to the product of their individual probabilities.

A strategy for solving this type of problem is shown here.

The Cross: $Pp \times Pp$

The Question: What is the probability that this couple's first three offspring will have congenital analgesia?

Step 1. *Calculate the individual probability of this phenotype.* As described previously, this is accomplished using a Punnett square. The probability of an affected offspring is 1/4.

Step 2. *Multiply the individual probabilities.* In this case, we are asking about the first three offspring, and so we multiply 1/4 three times.

$$1/4 \times 1/4 \times 1/4 = 1/64 = 0.016, \text{ or } 1.6\%$$

In this case, the probability that the first three offspring will have this trait is 0.016. We predict that 1.6% of the time the first three offspring will all have congenital analgesia when both parents are heterozygotes. In this example, the phenotypes of the first, second, and third offspring are independent outcomes. The phenotype of the first offspring does not have an effect on the phenotype of the second or third offspring.

In the problem described here, we have used the product rule to determine the probability that the first three offspring will all have the same phenotype (congenital analgesia). We can also apply the rule to predict the probability of a sequence of outcomes that involves combinations of different offspring. For example, consider this question: What is the probability that the first offspring will be unaffected, the second offspring will have congenital analgesia, and the third offspring will be unaffected? Again, to solve this problem, we begin by calculating the individual probability of each phenotype.

Unaffected = 3/4

Congenital analgesia = 1/4

The probability that these three phenotypes will occur in this specified order is

$$3/4 \times 1/4 \times 3/4 = 9/64 = 0.14, \text{ or } 14\%$$

This sequence of outcomes is expected to occur 14% of the time.

The product rule can also be used to predict the outcome of a cross involving two or more genes. Let's suppose an individual with the genotype *Aa Bb CC* was crossed to an individual with the genotype *Aa bb Cc*. We could ask: What is the probability that an offspring will have the genotype *AA bb Cc*? If the three genes independently assort, the probability of inheriting alleles for each gene is independent of the probability for the two other genes. Therefore, we can separately calculate the probability of the desired outcome for each gene.

Cross: *Aa Bb CC* × *Aa bb Cc*

Probability that an offspring will be *AA* = 1/4

Probability that an offspring will be *bb* = 1/2

Probability that an offspring will be *Cc* = 1/2

We can use the product rule to determine the probability that an offspring will be *AA bb Cc*:

$$P = 1/4 \times 1/2 \times 1/2 = 1/16 = 0.0625, \text{ or } 6.25\%$$

The Binomial Expansion Equation Is Used to Predict the Probability of an Unordered Combination of Outcomes

With regard to probability, unordered combinations are outcomes in which the order of a set of events does not matter. For example, a group of children in a given family might be composed of two females and one male, but the order of their births does not matter. From first to last, the order could be: female, female, male; female, male, female; or male, female, female.

With regard to genetics, the probability of an unordered combination also depends on the relative probability of each phenotype. For example, we can calculate the relative probabilities of brown-eyed and blue-eyed children, if both parents are

heterozygotes. As seen in the following Punnett square, 3/4 of the offspring have brown eyes and 1/4 have blue eyes.

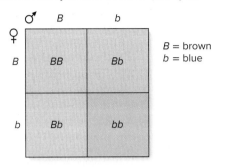

With the relative probabilities of producing children with each type of eye color, we can calculate the probability of an unordered combination of offspring. In this case, we are not concerned with the order in which the offspring are born. Instead, we are only concerned with the final numbers of blue-eyed and brown-eyed offspring. To solve this type of problem, the **binomial expansion equation** can be used. This equation represents all of the possibilities for a given set of two unordered outcomes.

$$P = \frac{n!}{x!(n-x)!}p^x q^{n-x}$$

where

P = the probability that the unordered outcome will occur

n = total number of outcomes

x = number of outcomes in one category (e.g., blue eyes)

p = individual probability of x

q = individual probability of the other category (e.g., brown eyes)

Note: In this case, $p + q = 1$.

The symbol ! denotes a factorial. The factorial $n!$ is the product of all integers from n down to 1. For example, $4! = 4 \times 3 \times 2 \times 1 = 24$. An exception is 0!, which equals 1.

The use of the binomial expansion equation is described next.

The Cross: *Bb* × *Bb*

The Question: What is the probability that two out of five offspring will have blue eyes?

Step 1. *Calculate the individual probabilities of the blue-eye and brown-eye phenotypes.* If we constructed a Punnett square, we would find the probability of blue eyes is 1/4 and the probability of brown eyes is 3/4:

$$p = 1/4$$
$$q = 3/4$$

Step 2. *Determine the number of outcomes in category x (in this case, blue eyes) versus the total number of outcomes.* In this example, the number of outcomes in category x is two blue-eyed children among a total number of five.

$$x = 2$$
$$n = 5$$

Step 3. *Substitute the values for p, q, x, and n in the binomial expansion equation.*

$$P = \frac{n!}{x!(n-x)!}p^x q^{n-x}$$

$$P = \frac{5!}{2!(5-2)!}(1/4)^2(3/4)^{5-2}$$

$$P = \frac{5 \times 4 \times 3 \times 2 \times 1}{(2 \times 1)(3 \times 2 \times 1)}(1/16)(27/64)$$

$$P = 0.26, \text{ or } 26\%$$

Thus, the probability is 0.26 that two out of five offspring will have blue eyes. In other words, 26% of the time we expect a $Bb \times Bb$ cross yielding five offspring to have two blue-eyed children and three brown-eyed children.

When more than two outcomes are possible, we use the **multinomial expansion equation** to solve a problem involving three or more types of unordered outcomes. A general expression for this equation is

$$P = \frac{n!}{a!b!c!\cdots}p^a q^b r^c \cdots$$

where

P = the probability that the unordered number of outcomes will occur

n = total number of outcomes

$a + b + c + \cdots = n$

$p + q + r + \cdots = 1$

(*p* is the likelihood of *a*, *q* is the likelihood of *b*, *r* is the likelihood of *c*, and so on)

The multinomial expansion equation can be useful for many genetic problems in which more than two combinations of offspring are possible. For example, this formula can be used to solve problems involving an unordered sequence of outcomes in a two-factor cross (see question 3 in More Genetic TIPS at the end of this chapter).

The Chi Square Test Is Used to Test the Validity of a Genetic Hypothesis

Let's now consider a different way to analyze genetic problems, namely **hypothesis testing.** Our goal here is to determine if the data from genetic crosses are consistent with a particular pattern of inheritance. For example, a geneticist may study the inheritance of body color and wing shape in fruit flies over the course of two generations. The following question may be asked about the F_2 generation: Do the observed numbers of offspring agree with the predicted numbers based on Mendel's laws of segregation and independent assortment? As we will see in Chapters 4 through 6, not all traits follow a Mendelian pattern of inheritance. Some genes do not segregate and independently assort themselves in the way that Mendel's seven characters did in pea plants.

To distinguish inheritance patterns that obey Mendel's laws from those that do not, a conventional strategy is to make crosses and then quantitatively analyze the offspring. Based on the observed outcome, an experimenter may make a tentative hypothesis. For example, the data may seem to obey Mendel's laws. Hypothesis testing provides an objective, statistical method to evaluate whether the observed data really agree with the hypothesis. In other words, we use statistical methods to determine whether the data that have been gathered from crosses are consistent with predictions based on quantitative laws of inheritance.

The rationale behind certain types of statistical approaches is to evaluate the **goodness of fit** between the observed data from experimentation and the expected data that are predicted from a hypothesis. This predictive hypothesis is also called a **null hypothesis** because it is the hypothesis that the experimenter may or may not nullify. The null hypothesis assumes that there is no real difference between the observed and expected data. Any actual differences that occur are presumed to be due to random sampling error. In some cases, statistical methods may reveal a poor fit between hypothesis and data. In other words, a high deviation may be found between the observed and expected values. If this occurs, the null hypothesis is rejected or nullified. Hopefully, the experimenter can subsequently propose an alternative hypothesis that has a better fit with the data.

In other cases, if the observed and expected data are very similar, we conclude that the hypothesis is consistent with the observed outcome. It is reasonable to accept the hypothesis. However, it should be emphasized that statistical methods can never prove that a hypothesis is correct. They can provide insight as to whether or not the observed data seem reasonably consistent with a hypothesis. Alternative hypotheses, perhaps even ones that the experimenter has failed to realize, may also be consistent with the data.

One commonly used statistical method to determine goodness of fit is the **chi square test** (the symbol for "chi square" is χ^2). We can use the chi square test to analyze population data in which the members of the population fall into different categories. We typically have this kind of data when we evaluate the outcomes of genetic crosses, because these usually produce a population of offspring that differ with regard to phenotypes. The general formula for the chi square test is

$$\chi^2 = \Sigma \frac{(O-E)^2}{E}$$

where

O = observed data in each category

E = expected data in each category based on the experimenter's hypothesis

The symbol Σ means that the data values for each category are added together. For example, if the population data fell into two categories, the chi square calculation would be

$$\chi^2 = \frac{(O_1-E_1)^2}{E_1} + \frac{(O_2-E_2)^2}{E_2}$$

We can use the chi square test to determine if a genetic hypothesis is consistent with the observed outcome of a genetic cross. The strategy described next provides a step-by-step outline for applying the chi square test. In this problem, the experimenter wants to determine if a two-factor cross obeys Mendel's laws. The experimental organism is *Drosophila melanogaster* (the common fruit fly), and the two characters involve wing shape and body color. Straight wing shape and curved wing shape are designated by c^+ and c, respectively; gray body color and ebony body color are designated by e^+ and e, respectively. Note: For certain species, such as *Drosophila melanogaster*, the convention is to designate the common (wild-type) allele with a plus sign. Recessive mutant alleles are designated with lowercase letters and dominant mutant alleles with capital letters.

The Cross: A true-breeding fly with straight wings and a gray body ($c^+c^+e^+e^+$) is crossed to a true-breeding fly with curved wings and an ebony body ($ccee$). The flies of the F_1 generation are then allowed to mate with each other to produce an F_2 generation.

The Outcome:

F_1 generation: All offspring have straight wings and gray bodies.

F_2 generation:
193 straight wings, gray bodies
69 straight wings, ebony bodies
64 curved wings, gray bodies
26 curved wings, ebony bodies

Total: 352

Step 1. *Propose a hypothesis that allows us to calculate the expected values based on Mendel's laws.* The phenotypes of the F_1 generation suggest that the trait of straight wings is dominant to curved wings and gray body color is dominant to ebony. Looking at the F_2 generation, it appears that the data show a 9:3:3:1 ratio. If so, this is consistent with an independent assortment of the two characters.

Based on these observations, the null hypothesis is the following:

Straight (c^+) is dominant to curved (c), and gray (e^+) is dominant to ebony (e). The two characters segregate and assort independently from generation to generation.

Step 2. *List the observed data, including the total number of observed offspring.*

193 straight wings, gray bodies
69 straight wings, ebony bodies
64 curved wings, gray bodies
26 curved wings, ebony bodies
352 Total

Step 3. *Based on the hypothesis and the total number of observed offspring, calculate the expected values of the four phenotypes.* We first need to calculate the individual probabilities of the four phenotypes. According to our hypothesis, the ratio of the F_2 generation should be 9:3:3:1. Therefore, the expected probabilities are as follows:

9/16 = straight wings, gray bodies
3/16 = straight wings, ebony bodies

3/16 = curved wings, gray bodies
1/16 = curved wings, ebony bodies

The observed F_2 generation contained a total of 352 individuals. Our next step is to calculate the expected numbers of each type of offspring when the total equals 352. This is accomplished by multiplying each individual probability by 352.

$9/16 \times 352 = 198$ (expected number with straight wings, gray bodies)
$3/16 \times 352 = 66$ (expected number with straight wings, ebony bodies)
$3/16 \times 352 = 66$ (expected number with curved wings, gray bodies)
$1/16 \times 352 = 22$ (expected number with curved wings, ebony bodies)

Step 4. *Apply the chi square formula, using the observed data listed in step 2 and the expected values that have been calculated in step 3.* In this case, the data include four categories, and thus the sum has four terms.

$$\chi^2 = \frac{(O_1 - E_1)^2}{E_1} + \frac{(O_2 - E_2)^2}{E_2} + \frac{(O_3 - E_3)^2}{E_3} + \frac{(O_4 - E_4)^2}{E_4}$$

$$\chi^2 = \frac{(193 - 198)^2}{198} + \frac{(69 - 66)^2}{66} + \frac{(64 - 66)^2}{66} + \frac{(26 - 22)^2}{22}$$

$$\chi^2 = 0.13 + 0.14 + 0.06 + 0.73 = 1.06$$

Step 5. *Determine the degrees of freedom.* Before we can determine the probability that this deviation occurred as a matter of random chance, we must first determine the degrees of freedom (df) in this experiment. The **degrees of freedom** indicate the number of categories that are independent of each other. When phenotype categories are derived from a Punnett square, the degrees of freedom are typically given by $n - 1$, where n equals the total number of categories. In our fruit fly problem, $n = 4$ (the categories are the phenotypes: straight wings and gray body; straight wings and ebony body; curved wings and gray body; and curved wings and ebony body); thus, the degrees of freedom equal 3.[*] We now have sufficient information to interpret our chi square value of 1.06.

Step 6. *Determine the P value for the chi square value calculated in step 4. This is done using a chi square table and the degrees of freedom determined in step 5.* We must use **Table 3.1** to determine a **P value,** which is the probability that the amount of variation indicated by a given chi square value is due to random chance alone, based on a particular hypothesis. The chi square value is 1.06 and the degrees of freedom equals 3. Begin in the left column at 3 degrees of freedom and slide to the right until you identify the range where 1.06 is located. In this case,

---

[*]If the hypothesis assumed that the law of segregation is obeyed, the degrees of freedom would be 1 (see Section 6.2).

TABLE 3.1
Chi Square Values and Probability

Degrees of Freedom	P = 0.99	0.95	0.80	0.50	0.20	Null Hypothesis Rejected	
						0.05	0.01
1	0.000157	0.00393	0.0642	0.455	1.642	3.841	6.635
2	0.020	0.103	0.446	1.386	3.219	5.991	9.210
3	0.115	0.352	1.005	2.366	4.642	7.815	11.345
4	0.297	0.711	1.649	3.357	5.989	9.488	13.277
5	0.554	1.145	2.343	4.351	7.289	11.070	15.086
6	0.872	1.635	3.070	5.348	8.558	12.592	16.812
7	1.239	2.167	3.822	6.346	9.803	14.067	18.475
8	1.646	2.733	4.594	7.344	11.030	15.507	20.090
9	2.088	3.325	5.380	8.343	12.242	16.919	21.666
10	2.558	3.940	6.179	9.342	13.442	18.307	23.209
15	5.229	7.261	10.307	14.339	19.311	24.996	30.578
20	8.260	10.851	14.578	19.337	25.038	31.410	37.566
25	11.524	14.611	18.940	24.337	30.675	37.652	44.314
30	14.953	18.493	23.364	29.336	36.250	43.773	50.892

Source: Fisher, Ronald A., and Yates, Frank, *Statistical Tables for Biological, Agricultural, and Medical Research.* London: Oliver and Boyd, 1943.

it is located between 1.005 and 2.366. If you then look up to the top of the table, 1.005 is a *P* value of 0.8 and 2.366 is 0.5. Our chi square value of 1.06 is pretty close to 1.005, so the P value is close to 0.8, or 80%.

Step 7. *Interpret the P value you have obtained.* As mentioned, a *P* value is the likelihood that the amount of variation indicated by a given chi square value is due to random chance alone, based on a particular hypothesis. In our example, the *P* value is approximately 0.8, or 80%. With 3 degrees of freedom, chi square values that are equal to or greater than 1.06 are expected to occur about 80% of the time when a hypothesis is correct. In other words, about 80 times out of 100 we expect that random chance alone will produce a deviation between the observed data and the expected data that is equal to or greater than 1.06. In general, such a low chi square value indicates a high probability that the observed deviations could be due to random chance alone. By comparison, if a high chi square value were obtained, an experimenter becomes suspicious that the high deviation has occurred because the hypothesis is incorrect.

A common convention is to reject the null hypothesis if the chi square value results in a probability that is less than 0.05 (less than 5%) or if the probability is less than 0.01 (less than 1%). These are called the 5% and 1% significance levels, respectively. Which level is better to choose? The choice is somewhat subjective. If you choose a 5% level rather than a 1% level, a disadvantage is that you are more likely to reject a null hypothesis that happens to be correct. Even so, choosing a 5% level

rather than a 1% level has the advantage that you are less likely to accept an incorrect null hypothesis.

For our problem involving flies with straight or curved wings and gray or ebony bodies, the calculated chi square value is 1.06, which gives a *P* value of approximately 0.80, or 80%. Because this *P* value is greater than 0.05, you accept the null hypothesis. To reject the null hypothesis at the 5% significance level, the chi square would have to be equal to or greater than 7.815 (see Table 3.1). Because it is actually far less than this value, you accept that the null hypothesis is correct. Keep in mind that the chi square test does not prove a hypothesis is correct! It is a statistical method for evaluating whether the data and hypothesis have a good fit. In other words, the chi square test provides an objective way to accept or reject a hypothesis based on the deviation between observed data and expected data.

3.6 COMPREHENSION QUESTIONS

1. A cross is made between *AA Bb Cc Dd* and *Aa Bb cc dd* individuals. Rather than making a very large Punnett square, which statistical operation could you use to solve this problem, and what would be the probability that the cross produces an offspring that is *AA bb Cc dd*?
   a. Product rule, 1/32
   b. Product rule, 1/4
   c. Binomial expansion, 1/32
   d. Binomial expansion, 1/4

2. In dogs, brown fur color (*B*) is dominant to white (*b*). A cross is made between two heterozygotes for fur color. If the litter consists of six puppies, what is the probability that half of them will be white?
   a. 0.066, or 6.6%
   b. 0.13, or 13%
   c. 0.25, or 25%
   d. 0.26, or 26%

3. Which of the following operations could be used for hypothesis testing?
   a. Product rule
   b. Binomial expansion equation
   c. Chi square test
   d. Any of the above operations could be used.

KEY TERMS

3.1: gamete, sperm cell, pollen grains, anthers, egg cell, ovules, ovaries, stigma, cross, hybridization, hybrids, cross-fertilization, self-fertilization, strain, characters, trait, variant, true-breeding strain

3.2: single-factor cross, monohybrids, empirical approach, parental generation (P generation), F_1 generation, F_2 generation, dominant, recessive, particulate theory of inheritance, segregate, gene, allele, Mendel's law of segregation, homozygous, heterozygous, genotype, phenotype, Punnett square

3.3: two-factor crosses, nonparental, Mendel's law of independent assortment, genetic recombination, forked-line method, multiplication method

3.4: chromosome theory of inheritance, X-linked inheritance, X-linked gene, X-linked allele, testcross

3.5: pedigrees, pedigree analysis

3.6: probability, random sampling error, product rule, binomial expansion equation, multinomial expansion equation, hypothesis testing, goodness of fit, null hypothesis, chi square test, degrees of freedom, *P* value

CHAPTER SUMMARY

- Early ideas regarding inheritance of traits included pangenesis and the blending hypothesis of inheritance. These ideas were later refuted by the work of Mendel.

3.1 Mendel's Study of Pea Plants

- Sexual reproduction in pea plants occurs via pollination followed by fertilization (see Figure 3.1).
- Mendel chose pea plants as his experimental organism because it was easy to carry out cross-fertilization or self-fertilization experiments with these plants, and because pea plants were available in several varieties in which a character existed in two distinct variants (see Figures 3.2, 3.3, 3.4).

3.2 Law of Segregation

- By conducting single-factor crosses, Mendel proposed three key ideas regarding inheritance: (1) Traits may be dominant or recessive. (2) Genes are passed unaltered from generation to generation. (3) The law of segregation states the following: The two copies of a gene segregate (or separate) from each other during transmission from parent to offspring (see Figures 3.5, 3.6).
- A Punnett square is used to predict the outcome of single-factor crosses and self-fertilization experiments.

3.3 Law of Independent Assortment

- By conducting two-factor crosses, Mendel formulated the law of independent assortment: Two different genes will randomly assort their alleles during the formation of haploid cells (see Figures 3.7, 3.8, 3.9).
- Independent assortment promotes genetic diversity.
- A Punnett square can be used to predict the outcomes of two-factor crosses (see Figure 3.10).
- The multiplication and forked-line methods are used to solve the outcomes of crosses involving three or more genes (see Figure 3.11).

3.4 The Chromosome Theory of Inheritance

- The chromosome theory of inheritance describes how the transmission of chromosomes can explain Mendel's laws. This theory is based on five fundamental principles.
- Mendel's law of segregation is explained by the separation of homologs during meiosis (see Figure 3.12).
- Mendel's law of independent assortment is explained by the random alignment of bivalents during metaphase of meiosis I (see Figure 3.13).
- Morgan's work provided strong evidence for the chromosome theory of inheritance by showing that a gene affecting eye color in fruit flies is located on the X chromosome (see Figure 3.14).

3.5 Studying Inheritance Patterns in Humans

- Human inheritance patterns are determined by analyzing charts that represent family relationships and are known as pedigrees (see Figure 3.15).

3.6 Probability and Statistics

- Probability is the number of times an outcome occurs divided by the total number of outcomes.
- According to the product rule, the probability that two or more independent outcomes will occur is equal to the product of their

individual probabilities. This rule can be used to predict the outcome of crosses involving two or more genes.
- The binomial expansion equation is used to predict the probability of a given set of two unordered outcomes.
- The chi square test is used to test the validity of a hypothesis (see Table 3.1).

PROBLEM SETS & INSIGHTS

MORE GENETIC TIPS
1. In dogs, black fur color is dominant to white. Two heterozygous black dogs are mated. What is the probability of the following combinations of offspring?

A. A litter of six pups, four with black fur and two with white fur

B. A first litter of six pups, four with black fur and two with white fur, and then a second litter of seven pups, five with black fur and two with white fur

TOPIC: *What topic in genetics does this question address?* The topic is Mendelian inheritance. More specifically, the question is about a single-factor cross involving fur color in dogs.

INFORMATION: *What information do you know based on the question and your understanding of the topic?* From the question, you know that black fur is dominant to white fur and that the parents are heterozygotes. If *B* is the black allele and *b* is the white allele, the genotype of each parent must be *Bb*. From your understanding of the topic, you may remember how to use a Punnett square to predict the outcome of a cross. You may also realize that each litter is an unordered combination of two different outcomes and, therefore, the binomial expansion can be used to calculate the probability of each litter.

PROBLEM-SOLVING **S**TRATEGIES: *Predict the outcome. Make a calculation.* To begin to solve this problem, you need to know the probabilities of producing black offspring and white offspring. These can be deduced from a Punnett square, which is shown next.

B = black
b = white

For part A of the question, you can derive probabilities for black and white fur from the Punnett square, and then use those values in the binomial expansion equation. For part B, you need two types of calculations. To determine the probability of each litter

occurring, you can use the binomial expansion equation. Because each litter is an independent outcome, you multiply the probability of the first litter times the probability of the second litter to get the probability of both litters occurring in this order.

ANSWER: From the Punnett square, you can deduce that the probability of black fur is 3/4, or 0.75, and the probability of white fur is 1/4, or 0.25.

A. Because this is an unordered combination of outcomes, you use the binomial expansion equation, where $n = 6$, $x = 4$, $p = 0.75$ (probability of black), and $q = 0.25$ (probability of white).

The answer is that such a litter will occur 0.297, or 29.7%, of the time.

B. The two litters occur in a row, so they are independent outcomes. Therefore, you use the product rule and multiply the probability of the first litter times the probability of the second litter. You need to use the binomial expansion equation for each litter: The result from this equation for the first litter is multiplied by the result for the second litter.

For the first litter, $n = 6$, $x = 4$, $p = 0.75$, $q = 0.25$. For the second litter, $n = 7$, $x = 5$, $p = 0.75$, $q = 0.25$.

The answer is that two such litters will occur in this order 0.092, or 9.2%, of the time.

2. A pea plant is heterozygous for three genes (*Tt Rr Yy*), where *T* = tall, *t* = short, *R* = round seeds, *r* = wrinkled seeds, and *Y* = yellow seeds, *y* = green seeds. Tall, round, and yellow are the dominant traits. What is the probability that an offspring from self-fertilization of this plant will be tall with wrinkled, yellow seeds?

TOPIC: *What topic in genetics does this question address?* The topic is Mendelian inheritance. More specifically, the question is about a three-factor self-fertilization experiment involving a pea plant.

INFORMATION: *What information do you know based on the question and your understanding of the topic?* In the question, you are given the genotype of a pea plant and told that it is self-fertilized. From your understanding of the topic, you may remember how to use a Punnett square to predict the outcome of a self-fertilization experiment. You may also realize that the outcome for each gene can be considered an independent outcome and the product rule can be used to solve this type of problem.

PROBLEM-SOLVING **S**TRATEGIES: *Predict the outcome. Make a calculation.* Instead of constructing a large, 64-box

Punnett square, you can make smaller Punnett squares to determine the probability of offspring inheriting each of the three genes and having particular phenotypes.

3 tall : 1 short

3 round : 1 wrinkled

3 yellow : 1 green

Next, you use the product rule. Multiply the specified combinations together. In this case, tall = 3/4, wrinkled = 1/4, and yellow = 3/4.

ANSWER: The probability of an offspring being tall with wrinkled, yellow seeds is

$$3/4 \times 1/4 \times 3/4 = 9/64 = 0.14, \text{ or } 14\%$$

3. A pea plant that is $(RrYy)$ is allowed to self-fertilize. Round seed (R) is dominant to wrinkled (r), and yellow seed (Y) is dominant to green (y). What is the probability of producing the following group of five seeds: two round, yellow; one round, green; one wrinkled, yellow; and one wrinkled, green?

T OPIC: *What topic in genetics does this question address?* The topic is Mendelian inheritance. More specifically, the question is about a two-factor self-fertilization experiment involving a pea plant.

I NFORMATION: *What information do you know based on the question and your understanding of the topic?* In the question, you are given the genotype of a pea plant and told that it is self-fertilizing. From your understanding of the topic, you may remember how to use a Punnett square to predict the outcome of a self-fertilization experiment. You may also recall how to use the multinomial expansion equation to predict the probability of an unordered combination of offspring having more than two phenotypes.

P ROBLEM-SOLVING S TRATEGIES: *Predict the outcome. Make a calculation.* To begin to solve this problem, you need to know the probability of producing the four types of offspring described in the question. As shown at the bottom of the Conceptual level in Figure 3.8, this can be determined from a Punnett square. The phenotypic ratio is: 9 round, yellow; 3 round, green; 3 wrinkled, yellow; and 1 wrinkled, green.

The probability of a round, yellow seed: $p = 9/16$

The probability of a round, green seed: $q = 3/16$

The probability of a wrinkled, yellow seed: $r = 3/16$

The probability of a wrinkled, green seed: $s = 1/16$

For the values in the multinomial expansion equation, you have

$$n = 5, a = 2, b = 1, c = 1, d = 1$$

ANSWER: Substitute the values in the multinomial expansion equation.

$$P = \frac{n!}{a!b!c!d!}p^a q^b r^c s^d$$
$$P = \frac{5!}{2!1!1!1!}(9/16)^2(3/16)^1(3/16)^1(1/16)^1$$
$$P = 0.04, \text{ or } 4\%$$

This means that 4% of the time you can expect to obtain five offspring with the phenotypes described in the question.

4. A cross was made between a hyacinth plant that has blue flowers and purple seeds and another hyacinth plant with white flowers and green seeds. The F_1 generation was then allowed to self-fertilize. The following data were obtained:

F_1 generation: All offspring have blue flowers with purple seeds.

F_2 generation: 208 offspring with blue flowers, purple seeds; 13 with blue flowers, green seeds; 19 with white flowers, purple seeds; and 60 with white flowers, green seeds. Total = 300 offspring.

Start with the hypothesis that blue flowers and purple seeds are dominant traits and that the two genes assort independently. Calculate a chi square value. What does this value mean with regard to your hypothesis? If you decide to reject the hypothesis, which aspect of the hypothesis do you think is incorrect (i.e., that blue flowers and purple seeds are dominant traits, or that the two genes assort independently)?

T OPIC: *What topic in genetics does this question address?* The topic is hypothesis testing. More specifically, the question is about evaluating the dominant and recessive relationships of two genes and determining if they are obeying the law of independent assortment.

INFORMATION: *What information do you know based on the question and your understanding of the topic?* From the question, you know the outcome of a two-factor cross. You are given a starting hypothesis. From your understanding of the topic, you may remember that this type of experiment should produce a 9:3:3:1 ratio of the four types of offspring in the F_2 generation, according to the law of independent assortment. Alternatively, you could set up a Punnett square and deduce the outcome for the F_2 generation. You may also remember that the chi square test can be used to evaluate the validity of a hypothesis.

PROBLEM-SOLVING **S**TRATEGY: *Analyze data.* One strategy is to analyze the data by carrying out a chi square test. According to the hypothesis, the F_2 generation should display a ratio of 9 offspring with blue flowers, purple seeds : 3 with blue flowers, green seeds : 3 with white flowers, purple seeds : 1 with white flowers, green seeds. Because a total of 300 offspring were produced, the expected numbers are as follows:

$9/16 \times 300 = 169$ blue flowers, purple seeds

$3/16 \times 300 = 56$ blue flowers, green seeds

$3/16 \times 300 = 56$ white flowers, purple seeds

$1/16 \times 300 = 19$ white flowers, green seeds

ANSWER: In this case, the data include four categories, and thus the sum for calculating the chi square value has four terms.

$$\chi^2 = \frac{(O_1 - E_1)^2}{E_1} + \frac{(O_2 - E_2)^2}{E_2} + \frac{(O_3 - E_3)^2}{E_3} + \frac{(O_4 - E_4)^2}{E_4}$$

$$\chi^2 = \frac{(208 - 169)^2}{169} + \frac{(13 - 56)^2}{56} + \frac{(19 - 56)^2}{56} + \frac{(60 - 19)^2}{19}$$

$$\chi^2 = 154.9$$

Looking up this value in the chi square table under 3 degrees of freedom, you see that it is much higher than would be expected 1% of the time by chance alone. Therefore, you reject the hypothesis. The idea that the two genes are assorting independently seems to be incorrect. The outcome for the F_1 generation supports the idea that blue flowers and purple seeds are dominant traits. Note: We will discuss why independent assortment may not occur in Chapter 6.

---

5. Calvin Bridges, who worked in the lab of Morgan, made crosses to study the inheritance of X-linked traits in fruit flies. One of Bridges' experiments concerned two different X-linked genes affecting eye color and wing length. For the eye color gene, the red-eye allele (w^+) is dominant to the white-eye allele (w). A second X-linked trait is wing length; the allele called *miniature* is recessive to the wild-type allele, which is called *long*. In this case, m represents the miniature allele and m^+ the long allele. A male fly carrying a miniature allele on its single X chromosome has small (miniature) wings. A female must be homozygous, mm, in order to have miniature wings.

Bridges made a cross between $X^{w,m^+}X^{w,m^+}$ female flies (white eyes and long wings) and $X^{w^+,m}Y$ male flies (red eyes and

miniature wings). Thousands of offspring were then examined with regard to their eye color, wing length, and sex. As expected, most of the offspring were females with red eyes and long wings or males with white eyes and long wings. On rare occasions (approximately 1 out of 1700 flies), however, female offspring with white eyes or males with red eyes were obtained. The wing length in these flies was also noted and then their chromosome composition was examined using a microscope. The following results were obtained:

Offspring	Eye Color	Wing Length	Sex Chromosomes
Expected females	Red	Long	XX
Expected males	White	Long	XY
Unexpected females (rare)	White	Long	XXY
Unexpected males (rare)	Red	Miniature	X0

Source: Bridges, Calvin B., "Non-disjunction as Proof of the Chromosome Theory of Heredity," *Genetics*, vol. 1, no. 2, March, 1916, 107–163.

Explain how the unexpected female and male offspring were produced.

TOPIC: *What topic in genetics does this question address?* The topic is X-linked inheritance—more specifically, the question is about X-linked inheritance and its relationship to abnormalities in chromosome number.

INFORMATION: *What information do you know based on the question and your understanding of the topic?* From the question, you know the outcome of a cross involving $X^{w,m^+}X^{w,m^+}$ female flies and $X^{w^+,m}Y$ male flies. From your understanding of the topic, you may remember that females transmit an X chromosome to both daughters and sons, whereas males transmit an X to their daughters and a Y to their sons. In fruit flies, sex is determined by the ratio of the number of X chromosomes to the number of sets of autosomes. It is not determined by the presence of the Y chromosome (see Figure 2.15).

PROBLEM-SOLVING **S**TRATEGY: *Compare and contrast.* One strategy to solve this problem is to compare and contrast the outcomes of this cross if the female transmits the wrong number of chromosomes or if the male transmits the wrong number.

The cross is $X^{w,m^+}X^{w,m^+} \times X^{w^+,m}Y$.

Female Transmits	Male Transmits	Offspring's Genotype	Offspring's Phenotype
$X^{w,m^+}X^{w,m^+}$	Y	$X^{w,m^+}X^{w,m^+}Y$	White eyes, long wings, female
$X^{w,m^+}X^{w,m^+}$	$X^{w^+,m}$	$X^{w,m^+}X^{w,m^+}$ $X^{w^+,m}$	Red eyes, long wings, female
No sex chromosomes	Y	Y	Inviable
No sex chromosomes	$X^{w^+,m}$	$X^{w^+,m}$	Red eyes, miniature wings, male
X^{w,m^+}	$X^{w^+,m}Y$	$X^{w,m^+}X^{w^+,m}Y$	Red eyes, long wings, female
X^{w,m^+}	No sex chromosomes	X^{w,m^+}	White eyes, long wings, male

ANSWER: The white-eyed female flies were due to the union between an abnormal XX female gamete and a normal Y male gamete. The unexpected male offspring had only one X chromosome and no Y. These male offspring were due to the union between an abnormal egg without any X chromosome and a normal sperm containing one X chromosome. The wing size of the unexpected males was a particularly significant result. The red-eyed males showed a miniature wing size. As noted by Bridges, this means that they inherited their X chromosome from their male parent rather than their female parent.

Conceptual Questions

C1. Why did Mendel's work refute the blending hypothesis of inheritance?

C2. What is the difference between cross-fertilization and self-fertilization?

C3. Explain the difference between genotype and phenotype. Give three examples. Is it possible for two individuals to have the same phenotype but different genotypes?

C4. With regard to genotypes, what is a true-breeding organism?

C5. How can you determine whether an organism is heterozygous or homozygous for a dominant trait?

C6. In your own words, describe Mendel's law of segregation. Do not use the word *segregation* in your answer.

C7. With regard to genes in pea plants that we have considered in this chapter, which statement(s) is/are *not* correct?

   A. The gene causing tall plants is an allele of the gene causing short plants.

   B. The gene causing tall plants is an allele of the gene causing purple flowers.

   C. The alleles causing tall plants and purple flowers are dominant.

C8. For a cross between a heterozygous tall pea plant and a short pea plant, predict the ratios of the offspring's genotypes and phenotypes.

C9. Do you know the genotype of an individual exhibiting a recessive trait or of an individual exhibiting a dominant trait? Explain your answer.

C10. A cross is made between a pea plant that has constricted pods (a recessive trait; smooth is dominant) and is heterozygous for seed color (yellow is dominant to green) and a pea plant that is heterozygous for both pod texture and seed color. Construct a Punnett square that depicts this cross. What are the predicted outcomes of genotypes and phenotypes of the offspring?

C11. A pea plant that is heterozygous with regard to seed color (yellow is dominant to green) is allowed to self-fertilize. What are the predicted outcomes of genotypes and phenotypes of the offspring?

C12. Describe the significance of nonparental combinations of traits with regard to the law of independent assortment. In other words, explain how the appearance of nonparental combinations refutes the hypothesis of linked assortment.

C13. From the following pedigrees, describe what is the likely inheritance pattern (dominant or recessive). Explain your reasoning. Filled (black) symbols indicate affected individuals.

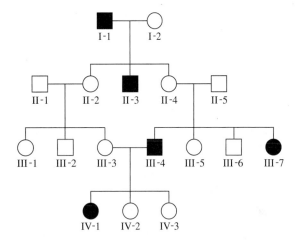

(a)

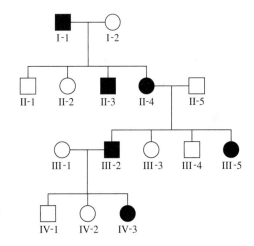

(b)

C14. Ectrodactyly is a recessive disorder in humans that affects the structure of the hands. If a phenotypically unaffected couple produces an offspring with ectrodactyly, what are the following probabilities?

   A. Both parents are heterozygotes.

   B. The next offspring is a heterozygote.

   C. The next three offspring will be phenotypically unaffected.

   D. Any two out of the next three offspring will be phenotypically unaffected.

C15. Identical twins are produced from the same sperm and egg. Sometime after the first mitotic division, the cells detach from each other and develop into distinct individuals. In contrast, fraternal twins are produced from separate sperm and separate egg cells. If two parents with brown eyes (a dominant trait) produce one twin child with blue eyes, what are the probabilities of the following?

A. The other twin, who is an identical twin, has blue eyes.

B. The other twin, who is a fraternal twin, has blue eyes.

C. If the other twin is fraternal, this twin will transmit the blue eye allele to an offspring.

D. The parents are both heterozygotes.

C16. In cocker spaniels, solid coat color is dominant over spotted coat color. If two heterozygous dogs were crossed to each other, what would be the probability of the following combinations of offspring?

A. A litter of five pups, four with solid fur and one with spotted fur

B. A first litter of six pups, four with solid fur and two with spotted fur, and then a second litter of five pups, all with solid fur

C. A first litter of five pups, the firstborn with solid fur, and then among the next four, three with solid fur and one with spotted fur, and then a second litter of seven pups in which the firstborn is spotted, the second born is spotted, and the remaining five are composed of four solid and one spotted animal

D. A litter of six pups, the firstborn with solid fur, the second born spotted, and among the remaining four pups, two with spotted fur and two with solid fur

C17. Crosses were made between a white male dog and two different black females. The first female gave birth to eight black pups, and the second female gave birth to four white and three black pups. What are the likely genotypes of the male parent and the two female parents? Explain whether you are uncertain about any of the genotypes.

C18. In humans, the allele for brown eye color (B) is dominant to that for blue eye color (b). If two parents who are heterozygous for eye color produce children, what are the following probabilities for those offspring?

A. The first two children have blue eyes.

B. Among a total of four children, two have blue eyes and the other two have brown eyes.

C. The first child has blue eyes, and the next two have brown eyes.

C19. Albinism, a condition characterized by a partial or total lack of skin pigment, is a recessive human trait. If a phenotypically unaffected couple produce a child with albinism, what is the probability that their next child will have albinism?

C20. A true-breeding tall plant was crossed to a short plant. Tallness is a dominant trait. The F_1 individuals were allowed to self-fertilize. What are the following probabilities for the F_2 generation?

A. The first plant is short.

B. The first plant is short or tall.

C. The first three plants are tall.

D. For any seven plants, three are tall and four are short.

E. The first plant is tall, and then among the next four, two are tall and the other two are short.

C21. For pea plants with the following genotypes, list the possible gametes that each plant can make.

A. $TT\ Yy\ Rr$ C. $Tt\ Yy\ Rr$

B. $Tt\ YY\ rr$ D. $tt\ Yy\ rr$

C22. An individual has the genotype $Aa\ Bb\ Cc$ and makes an abnormal gamete with the genotype $AaBc$. Does this gamete violate the law of independent assortment or the law of segregation (or both)? Explain your answer.

C23. In people with maple syrup urine disease, the body is unable to metabolize the amino acids leucine, isoleucine, and valine. One of the symptoms is that the urine smells like maple syrup. An unaffected couple produced six children in the following order: unaffected female, affected female, unaffected male, unaffected male, affected male, and unaffected male. The youngest unaffected male in this family and an unrelated unaffected female have three children in the following order: affected female, unaffected female, and unaffected male. Draw a pedigree that represents this family. What type of inheritance (dominant or recessive) would you propose to explain maple syrup urine disease?

C24. Marfan syndrome is a rare inherited human disorder characterized by unusually long limbs and digits plus defects in the heart (especially the aorta) and the eyes, among other symptoms. Following is a pedigree for this disorder in a certain family. Affected individuals are shown with filled (black) symbols. What type of inheritance pattern do you think is more likely?

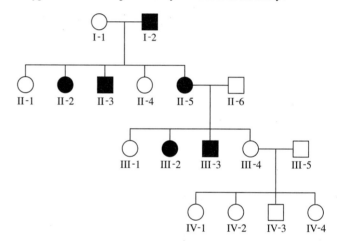

C25. A true-breeding pea plant with round and green seeds was crossed to a true-breeding pea plant with wrinkled and yellow seeds. Round and yellow seeds are the dominant traits. The F_1 plants were allowed to self-fertilize. What are the following probabilities for the F_2 generation?

A. An F_2 plant with wrinkled, yellow seeds

B. Three out of three F_2 plants with round, yellow seeds

C. Five F_2 plants in the following order: two have round, yellow seeds; one has round, green seeds; and two have wrinkled, green seeds

D. An F_2 plant with either round, green seeds, wrinkled, yellow seeds, or wrinkled, green seeds

C26. A true-breeding tall pea plant was crossed to a true-breeding short pea plant. What is the probability that an F_1 individual will be true-breeding? What is the probability that an F_1 individual will be a true-breeding tall plant?

C27. What are the expected phenotypic ratios from the following cross: $Tt\ Rr\ yy\ Aa \times Tt\ rr\ YY\ Aa$, where T = tall, t = short, R = round, r = wrinkled, Y = yellow, y = green, and A = axial, a = terminal; T, R, Y, and A are dominant alleles. Hint: Consider using the multiplication method in answering this problem.

C28. On rare occasions, an organism may have three copies of a chromosome and therefore three copies of the genes on that chromosome (instead of the usual number of two copies). The alleles for each gene usually segregate so that a gamete will contain one or two copies of the gene. Let's suppose that a rare pea plant has three copies of the chromosome that carries the height gene. Its genotype is *TTt*. The plant is also heterozygous for the seed color gene, *Yy*, which is found on a different chromosome. With regard to both genes, how many types of gametes can this plant make, and in what proportions? (Assume that it is equally likely that a gamete will contain one or two copies of the height gene.)

C29. Honeybees are unusual in that male bees (drones) have only one copy of each gene, but female bees have two copies of their genes. This difference arises because drones develop from eggs that have not been fertilized by sperm cells. In bees, the trait of long wings is dominant over short wings, and the trait of black eyes is dominant over white eyes. If a drone with short wings and black eyes was mated to a queen bee that is heterozygous for both genes, what are the predicted genotypes and phenotypes of male and female offspring? What are the phenotypic ratios if we assume an equal number of male and female offspring?

C30. A pea plant that is short with green, wrinkled seeds is crossed to a true-breeding pea plant that is tall with yellow, round seeds. The F_1 generation is then allowed to self-fertilize. What types of gametes, and in what proportions, will the F_1 generation make? What will be the ratios of genotypes and phenotypes of the F_2 generation?

C31. A true-breeding pea plant with round and green seeds is crossed to a true-breeding pea plant with wrinkled and yellow seeds. The F_1 plants are then allowed to self-fertilize. What is the probability of obtaining the following plants in the F_2 generation: two that have round, yellow seeds; one with round, green seeds; and two with wrinkled, green seeds?

C32. Wooly hair is a rare dominant trait found in people of Scandinavian descent; their hair resembles the wool of a sheep. A male with wooly hair, who has a female parent with straight hair, moves to an island that is inhabited by people who are not of Scandinavian descent. Assuming that no other Scandinavians immigrate to the island, what is the probability that a great-grandchild of this male will have wooly hair? (Hint: You may want to draw a pedigree to help you figure this out.) If this wooly-haired male has eight great-grandchildren, what is the probability that one out of eight will have wooly hair?

C33. Huntington disease is a rare dominant disorder that causes neurodegeneration later in life. A 34-year old person named Alonzo already has three children. One of Alonzo's parents has recently developed Huntington disease though the other parent is unaffected. If you assume that the unaffected parent does not carry the allele causing Huntington disease, what are the following probabilities?

A. Alonzo will develop Huntington disease.

B. Alonzo's first child will develop Huntington disease.

C. One out of three of Alonzo's children will develop Huntington disease.

C34. Brown eyes are caused by a dominant allele (*B*), whereas blue eyes are recessive (*b*). A person with brown eyes has seven children, all of whom have brown eyes. The other parent has blue eyes.

A. What is the probability of producing such a family if the brown-eyed parent is a heterozygote?

B. What is the probability that the brown-eyed parent is a heterozygote if an eighth child has blue eyes?

Experimental Questions

E1. List three advantages of using pea plants as an experimental organism.

E2. Explain the technical difference between a cross-fertilization experiment and a self-fertilization experiment.

E3. How long did it take Mendel to complete the experiment in Figure 3.5?

E4. For all seven characters described in the data table for Figure 3.5, Mendel allowed the F_2 plants to self-fertilize. When F_2 plants with recessive traits were allowed to self-fertilize, they always bred true. However, when F_2 plants with dominant traits were allowed to self-fertilize, some bred true but others did not. A summary of Mendel's results is shown in the following table.

F2 Parents	True-Breeding	Non-True-Breeding	Ratio
Round	193	372	1:1.9
Yellow	166	353	1:2.1
Gray	36	64	1:1.8
Smooth	29	71	1:2.4
Green	40	60	1:1.5
Axial	33	67	1:2.0
Tall	28	72	1:2.6
TOTAL:	525	1059	1:2.0

When considering the data in this table, keep in mind that they describe the F_3 generation that was produced from F_2 individuals with a dominant phenotype. These data were deduced by analyzing the outcome of the F_3 generation. Based on Mendel's laws, explain why the ratios are approximately 1:2.

E5. From the point of view of crosses and data collection, what are the experimental differences between a single-factor and a two-factor cross?

E6. As in many animals, albino coat color is a recessive trait in guinea pigs. Researchers removed the ovaries from an albino female guinea pig and then transplanted ovaries from a true-breeding black guinea pig. They then mated this albino female (with the transplanted ovaries) to an albino male. The albino female produced three offspring. What were their coat colors? Explain the results.

E7. The fungus *Melampsora lini* causes a disease known as flax rust. Different strains of *M. lini* cause varying degrees of the disease. Conversely, different strains of flax are resistant or sensitive to the various varieties of the fungus. The Bombay variety of flax is resistant to *M. lini* strain 22 but sensitive to *M. lini* strain 24. A strain of flax called 770B has the opposite characteristics; it is resistant to strain 24 but sensitive to strain 22. When 770B was crossed to Bombay, all F_1 individuals were resistant to both strain 22 and strain 24. When F_1 individuals were self-fertilized, the following data were obtained:

> 43 resistant to strain 22 but sensitive to strain 24
>
> 9 sensitive to strain 22 and strain 24
>
> 32 sensitive to strain 22 but resistant to strain 24
>
> 110 resistant to strain 22 and strain 24

Explain the inheritance pattern for flax resistance and sensitivity to *M. lini* strains.

E8. Using Mendel's data from the experiment in Figure 3.8, conduct a chi square test to determine if the data agree with Mendel's law of independent assortment.

E9. Would it be possible to deduce the law of independent assortment from a single-factor cross? Explain your answer.

E10. In fruit flies, curved wings are recessive to straight wings, and ebony body is recessive to gray body. A cross was made between true-breeding flies with curved wings and gray bodies and flies with straight wings and ebony bodies. The F_1 offspring were then mated to flies with curved wings and ebony bodies to produce an F_2 generation.

A. Describe the genotypes of this cross, starting with the parental generation and ending with the F_2 generation.

B. What is the predicted phenotypic ratio of the F_2 generation?

C. Let's suppose the following data were obtained for the F_2 generation:

> 114 curved wings, ebony body
>
> 105 curved wings, gray body
>
> 111 straight wings, gray body
>
> 114 straight wings, ebony body

Conduct a chi square test to determine if the experimental data are consistent with the expected outcome based on Mendel's laws.

E11. A recessive allele in mice results in an unusually long neck. Sometimes, during early embryonic development, the long neck causes the embryo to die. An experimenter began with a population of true-breeding mice with normal necks and true-breeding mice with long necks. Crosses were made between these two populations to produce an F_1 generation of mice with normal necks. The F_1 mice were then mated to each other to obtain an F_2 generation. For the mice that were born alive, the following data were obtained:

> 522 mice with normal necks
>
> 62 mice with long necks

What percentage of homozygous mice (that would have had long necks if they had survived) died during embryonic development?

E12. The data with Figure 3.5 show the results of the F_2 generation for seven of Mendel's experiments. Conduct a chi square test to determine if these data are consistent with the law of segregation.

E13. Let's suppose you conducted an experiment involving genetic crosses and calculated a chi square value of 1.005. There were four categories of offspring (i.e., the degrees of freedom equaled 3). Explain what the 1.005 value means. Your answer should include the phrase "80% of the time."

E14. In Morgan's experiments, which result do you think is the most convincing piece of evidence pointing to X-linkage of the eye color gene? Explain your answer.

E15. In the original studies presented in Figure 3.14, Morgan first suggested that the original white-eyed male had two copies of the white-eye allele. In this problem, let's assume that this means the fly was X^wY^w instead of X^wY. Are the data in Figure 3.14 consistent with this hypothesis? What crosses would need to be made to rule out the possibility that the Y chromosome carries a copy of the eye color gene?

E16. White-eyed flies have a lower survival rate than red-eyed flies. Based on the data in Figure 3.14, what percentage of white-eyed flies survived compared with red-eyed flies, assuming 100% survival of red-eyed flies?

E17. Discuss why crosses (i.e., the experiments of Mendel) and the microscopic observations of chromosomes during mitosis and meiosis were both needed to deduce the chromosome theory of inheritance.

Questions for Student Discussion/Collaboration

1. Consider this cross of pea plants: *Tt Rr yy Aa* × *Tt rr Yy Aa*, where *T* = tall, *t* = short, *R* = round, *r* = wrinkled, *Y* = yellow, *y* = green, and *A* = axial, *a* = terminal. What is the expected phenotypic outcome of this cross? One group of students should solve this problem by making one big Punnett square, and another group should solve it by making four single-gene Punnett squares and using the forked-line method. Time each other to see who gets done first.

2. A cross was made between two pea plants, *TtAa* and *Ttaa*, where *T* = tall, *t* = short, and *A* = axial, *a* = terminal. What is the probability that the first three offspring will be tall with axial flowers or short with terminal flowers and the fourth offspring will be tall with axial flowers? Discuss what operation(s) (e.g., product rule and/or binomial expansion equation) you used and in what order you used them.

3. Discuss the principles of the chromosome theory of inheritance. Which principles were deduced via light microscopy, and which were deduced from crosses? What modern techniques could be used to support the chromosome theory of inheritance?

Note: All answers are available for the instructor in Connect; the answers to the even-numbered questions and all of the Concept Check and Comprehension Questions are in Appendix B.

Inheritance patterns and alleles. *In petunia plants, multiple alleles result in flowers with several different colors, such as the ones shown here.*
dimitriosp/123RF

4 EXTENSIONS OF MENDELIAN INHERITANCE

Many traits in eukaryotic species follow a pattern known as **Mendelian inheritance.** Such traits obey two laws: the law of segregation and the law of independent assortment. Furthermore, the genes that influence such traits are not altered (except by rare mutations) as they are passed from parent to offspring. The traits that are displayed by the offspring depend on the alleles they inherit and also on environmental factors.

Until now, we have mainly considered traits that are affected by a single gene that is found in two different alleles. In these cases, one allele is dominant over the other. This type of inheritance is called **simple Mendelian inheritance** because the observed ratios of traits in the offspring clearly obey Mendel's laws. For example, when true-breeding tall and short pea plants are crossed and the F_1 generation is allowed to self-fertilize, the F_2 generation shows a 3:1 phenotypic ratio of tall to short offspring.

In this chapter, we will extend our understanding of Mendelian inheritance by first examining the transmission patterns for several traits that do not display a simple dominant/recessive relationship. Geneticists have discovered an amazing diversity of mechanisms by which alleles affect the outcome of traits. Many alleles don't produce the ratios of offspring that are expected with simple Mendelian inheritance. This does not mean that Mendel was wrong. Rather, the inheritance patterns of many traits are more complex and interesting than Mendel had realized. In this chapter, we will examine how the outcome of a trait may be influenced by a variety of factors such as the level of protein expression, the sex of the individual, the presence of multiple alleles of a given gene, and environmental effects. We will also explore how two different genes can contribute to the outcome of a single trait.

In Chapters 5 and 6, we will examine eukaryotic inheritance patterns that are not considered Mendelian. For example, some genes are located in mitochondrial DNA and do not obey the law of segregation. Others are closely linked along the same chromosome and violate the law of independent assortment. In addition, some genes are altered (by DNA methylation) during gamete formation, which affects their expression in the offspring that inherit them. This results in an epigenetic pattern of inheritance that we will consider in Chapter 5.

OVERVIEW OF MENDELIAN INHERITANCE PATTERNS

Learning Outcomes:

1. Compare and contrast the different types of Mendelian inheritance patterns.
2. Describe the molecular mechanisms that account for Mendelian inheritance patterns involving single genes.

Before we delve more deeply into inheritance patterns, let's first compare a variety of these patterns (**Table 4.1**). We have already discussed two of them—simple dominant/recessive inheritance in Chapter 2 and X-linked inheritance in Chapter 3. The various patterns occur because the outcome of a trait may be governed by two

or more alleles in several different ways. Geneticists want to understand Mendelian inheritance for two reasons:

- One goal is to predict the outcome of crosses. Many of the inheritance patterns described in Table 4.1 do not produce a 3:1 phenotypic ratio when two heterozygotes produce offspring.
- A second goal is to understand how the molecular expression of genes can account for an individual's phenotype. In other words, what is the underlying relationship between molecular genetics—the expression of genes to produce functional proteins—and the traits of individuals that inherit the genes?

The remaining sections of this chapter will explore several patterns of inheritance and consider their underlying molecular mechanisms.

TABLE 4.1

Mendelian Inheritance Patterns Involving Single Genes

Type	Description
Simple Mendelian inheritance	**Inheritance:** This term is commonly applied to the inheritance of alleles that obey Mendel's laws and follow a strict dominant/recessive relationship. In this chapter, we will see that some genes occur as three or more alleles, making the relationship more complex. **Molecular:** 50% of the protein, produced by a single copy of the dominant (functional) allele in the heterozygote, is sufficient to produce the dominant trait.
Incomplete penetrance	**Inheritance:** In the case of dominant traits, this pattern occurs when a dominant phenotype is not expressed even though an individual carries a dominant allele. An example is an individual who carries the polydactyly allele but has a normal number of fingers and toes. In the case of recessive traits, this pattern occurs when a homozygote carrying both recessive alleles does not exhibit the trait. **Molecular:** Even though a dominant allele is present or two recessive alleles are present, the protein coded by the gene may not exert its effects. This can be due to environmental influences or due to other genes that may code proteins that counteract the effects of the protein coded by the dominant allele.
Incomplete dominance	**Inheritance:** This pattern occurs when the heterozygote has a phenotype that is intermediate between either corresponding homozygote. For example, a cross between homozygous red-flowered and homozygous white-flowered parents produces heterozygous offspring with pink flowers. **Molecular:** 50% of the protein, produced by a single copy of the functional allele in the heterozygote, is not sufficient to produce the same trait as in a homozygote making 100% of that protein.
Heterozygote advantage	**Inheritance:** This pattern occurs when the heterozygote has a trait that confers a greater level of reproductive success than either homozygote has. **Molecular:** Three common ways that heterozygotes may gain benefits: (1) Their cells may have increased resistance to infection by microorganisms; (2) they may produce more forms of protein dimers with enhanced function; or (3) they may produce proteins that function under a wider range of conditions.
Codominance	**Inheritance:** This pattern occurs when the heterozygote expresses both alleles simultaneously without forming an intermediate phenotype. For example, with regard to human blood types, an individual carrying the A and B alleles will have an AB blood type. **Molecular:** The codominant alleles code proteins that function slightly differently from each other, and the function of each protein in the heterozygote affects the phenotype uniquely.
X-linked inheritance	**Inheritance:** This pattern involves the inheritance of genes that are located on the X chromosome. In mammals and fruit flies, males have one copy of X-linked genes, whereas females have two copies. **Molecular:** If a pair of X-linked alleles shows a simple dominant/recessive relationship, 50% of the protein, produced by a single copy of the dominant allele in a heterozygous female, is sufficient to produce the dominant trait. Males have only one copy of X-linked genes and therefore express the copy they carry.
Sex-influenced inheritance	**Inheritance:** This pattern refers to the effect of sex on the phenotype of the individual. Some alleles are recessive in males and dominant in females; others are dominant in males and recessive in females. **Molecular:** Sex hormones may regulate the molecular expression of genes. This regulation can influence the phenotypic effects of alleles.
Sex-limited inheritance	**Inheritance:** In this pattern, a trait occurs in only one sex. It may occur in males or females, but not both. An example is sperm production in male animals. **Molecular:** Sex hormones may regulate the molecular expression of genes. This regulation can influence the phenotypic effects of alleles. In this pattern of inheritance, sex hormones that are primarily produced in only one sex are essential for an individual to display a particular phenotype.
Lethal alleles	**Inheritance:** A lethal allele is one that has the potential of causing the death of an organism. **Molecular:** Lethal alleles are most commonly loss-of-function alleles that code proteins that are necessary for survival. In some cases, such an allele may be due to a mutation in a nonessential gene that changes a protein so that it functions with abnormal and detrimental consequences.

4.1 COMPREHENSION QUESTION

1. Which of the following statements is *true*?
   a. Not all inheritance patterns follow a strict dominant/recessive relationship.
   b. Geneticists want to understand both inheritance patterns and the underlying molecular mechanisms that cause them to happen.
   c. Different inheritance patterns are explained by a variety of different molecular mechanisms.
   d. All of the above are true.

4.2 DOMINANT AND RECESSIVE ALLELES

Learning Outcomes:

1. Define *wild-type allele* and *genetic polymorphism.*
2. Explain why loss-of-function alleles often follow a recessive pattern of inheritance.
3. Describe how traits can exhibit incomplete penetrance and vary in their expressivity.

In Chapter 2, we examined patterns of inheritance that showed a simple dominant/recessive relationship. This means that a heterozygote exhibits the dominant trait. In this section, we will take a closer look at why an allele may be dominant or recessive and discuss how dominant alleles may not always exert their effects.

Wild-Type Alleles Are Common in a Population

For any given gene, geneticists refer to prevalent alleles in a natural population as **wild-type alleles.** In large populations, more than one wild-type allele may occur—a phenomenon known as **genetic polymorphism.** For example, **Figure 4.1** illustrates a striking example of polymorphism in the elderflower orchid, *Dactylorhiza sambucina.* Throughout the range of this species in Europe, both yellow- and red-flowered individuals are common. Both colors are considered wild type. At the molecular level, a wild-type allele typically codes a protein that is made in the proper amount and functions in a particular way. As discussed in Chapter 27, wild-type alleles tend to promote the reproductive success of organisms in their native environments.

Recessive Mutant Alleles Often Cause a Reduction in the Amount or Function of the Coded Proteins

Random mutations occur in populations and alter preexisting alleles. Geneticists sometimes refer to these altered alleles as **mutant alleles** to distinguish them from the more common wild-type alleles. Because random mutations are more likely to disrupt gene function, mutant alleles are often defective in their ability to express a functional protein. These defective alleles are called **loss-of-function alleles.** Such mutant alleles tend to be rare in natural populations. They are typically, but not always, inherited in a recessive fashion.

FIGURE 4.1 **An example of genetic polymorphism.** Both yellow and red flowers are common in natural populations of the elderflower orchid, *Dactylorhiza sambucina*, and both are considered wild type.
Blickwinkel/Woike/Alamy Stock Photo

CONCEPT CHECK: Why are both of these colors considered to be wild type?

Among Mendel's seven characters discussed in Chapter 3, the wild-type alleles are those that produce tall plants, purple flowers, axial flowers, yellow seeds, round seeds, green pods, and smooth pods (refer back to Figure 3.4). The mutant alleles result in short plants, white flowers, terminal flowers, green seeds, wrinkled seeds, yellow pods, and constricted pods. You may have already noticed that the seven wild-type alleles are dominant over the seven mutant alleles. Likewise, red eyes and normal (long) wings are examples of traits produced by wild-type alleles in *Drosophila*, and white eyes and miniature wings are due to recessive mutant alleles.

The idea that recessive mutant alleles usually cause a substantial decrease in the expression of a functional protein is supported by the analysis of many human genetic diseases. Keep in mind that a genetic disease is usually caused by a mutant allele. **Table 4.2** lists several examples of human genetic diseases in which the recessive allele fails to produce a specific cellular protein in its active form. In many cases, molecular techniques have enabled researchers to clone these genes and determine the differences between the wild-type and mutant alleles. They have found that the recessive allele usually contains a mutation that causes a defect in the synthesis of a fully functional protein.

To understand why many defective mutant alleles are inherited recessively, we need to take a quantitative look at protein function. With the exception of sex-linked genes, diploid individuals usually have two copies of every gene. In a simple dominant/recessive relationship, the recessive allele does not affect the phenotype of the heterozygote. In other words, a single copy of the dominant allele is sufficient to mask the effects of the recessive allele. If the recessive allele cannot produce a functional protein, how do we explain the wild-type phenotype of the heterozygote? As shown in **Figure 4.2**, a common explanation is that 50% of the functional protein is adequate to provide the wild-type phenotype.

TABLE 4.2

Examples of Recessive Human Diseases

Disease	Protein That Is Produced by the Functional Gene*	Description
Phenylketonuria	Phenylalanine hydroxylase	Inability to metabolize phenylalanine. The effects of the disease can be prevented by following a phenylalanine-free diet. If the diet is not followed early in life, the result can be severe mental impairment and physical degeneration.
Albinism	Tyrosinase	Lack of pigmentation in the skin, eyes, and hair.
Tay-Sachs disease	Hexosaminidase A	Defect in lipid metabolism. Leads to paralysis, blindness, and early death.
Sandhoff disease	Hexosaminidase B	Defect in lipid metabolism. Muscle weakness in infancy, early blindness, and progressive mental and motor deterioration.
Cystic fibrosis	Chloride transporter	Inability to regulate ion balance across epithelial cells. Leads to production of thick mucus and results in chronic lung infections, poor weight gain, and organ malfunctions.
Lesch-Nyhan syndrome	Hypoxanthine-guanine phosphoribosyl transferase	Inability to metabolize purines, which are bases found in DNA and RNA. Leads to self-mutilation behavior, poor motor skills, and usually mental impairment and kidney failure.

*Individuals who exhibit the disease are either homozygous for a recessive allele or have only one copy of the gene in the case of X-linked genes in human males. The disease symptoms result from a defect in the amount or function of the normal protein.

Dominant (functional) allele: *P* (purple)
Recessive (defective) allele: *p* (white)

Genotype	*PP*	*Pp*	*pp*
Amount of functional protein P	100%	50%	0%
Phenotype	Purple	Purple	White
Simple dominant/ recessive relationship			

FIGURE 4.2 **A comparison of protein levels among homozygous (*PP* or *pp*) and heterozygous (*Pp*) genotypes.**

Genes→Traits In a simple dominant/recessive relationship, 50% of the protein coded by one copy of the dominant allele in the heterozygote is sufficient to produce the wild-type phenotype, in this case, purple flowers. A complete lack of the functional protein results in white flowers.

CONCEPT CHECK: Does a *PP* individual produce more of the protein coded by the *P* gene than is necessary for the purple color?

In this example, the *PP* homozygote and *Pp* heterozygote each make sufficient amounts of the functional protein to yield purple flowers.

A second possible explanation for other dominant alleles is that the heterozygote actually produces more than 50% of the functional protein. Due to gene regulation, the expression of the normal (functional) gene may be increased, or up-regulated, in the heterozygote to compensate for the lack of function of the defective allele. The topic of gene regulation is discussed in Chapters 14 and 15.

Dominant Mutant Alleles Usually Exert Their Effects in One of Three Ways

Though dominant mutant alleles are much less common than recessive mutant alleles, they do occur in natural populations. How can a mutant allele be dominant over a wild-type allele? One of three mechanisms accounts for most dominant mutant alleles: a gain-of-function mutation, a dominant-negative mutation, or haploinsufficiency.

- **Gain-of-function mutations** change the gene or the protein coded by a gene so that it gains a new or abnormal function. A mutant gene may be overexpressed or it may be expressed in the wrong cell type. For example, many forms of cancer are caused by gain-of-function mutations in genes that code proteins that promote cell division. Such mutations cause the cells to divide in an uncontrolled way.
- **Dominant-negative mutations** change a protein such that the mutant protein acts antagonistically to the normal protein. In a heterozygote, the mutant protein counteracts the effects of the normal protein, thereby altering the phenotype. An example is STAT-3 dominant-negative disease. STAT-3 codes a regulatory protein that is important for the function of the immune system in humans. In a heterozygote, the dominant-negative mutant protein inhibits the function of the wild-type protein. The inhibition is thought to occur because the mutant STAT-3 protein forms a dimer with the wild-type protein. The inhibition of the wild-type protein results in defects in immune system function.
- In **haploinsufficiency,** the dominant mutant allele is a loss-of-function allele. Haploinsufficiency is used to describe patterns of inheritance in which a heterozygote (with one functional allele and one inactive allele) exhibits an abnormal or disease phenotype. An example in humans is polydactyly, which is discussed next.

Traits May Skip a Generation Due to Incomplete Penetrance and Vary in Their Expressivity

As we have seen, dominant alleles are expected to influence the outcome of a trait when they are present in heterozygotes. Occasionally, however, this may not occur.

- The phenomenon called **incomplete penetrance** results in a pattern of inheritance in which an allele that is expected to cause a particular phenotype does not always do so.

Figure 4.3a presents a human pedigree for a dominant trait known as polydactyly. This trait causes the affected individual to have additional fingers or toes (or both) (**Figure 4.3b**). Polydactyly is due to an autosomal dominant allele—the allele is found in a gene located on an autosome (not a sex chromosome) and a single copy of this allele is sufficient to cause this condition.

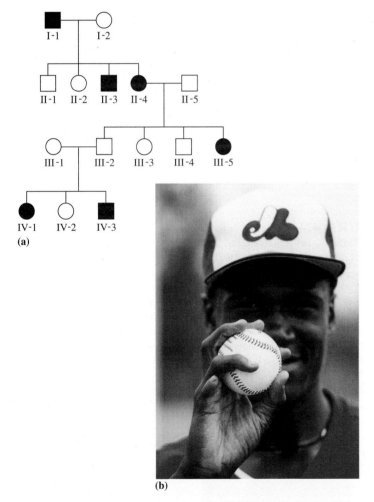

(a)

(b)

FIGURE 4.3 Polydactyly, a dominant trait that shows incomplete penetrance. **(a)** A family pedigree. Affected individuals are shown in black. Notice that offspring IV-1 and IV-3 have inherited the trait from a parent, III-2, who is heterozygous but does not exhibit polydactyly. **(b)** Antonio Alfonseca, a former baseball player with polydactyly.

(b): Bob Shanley/Palm Beach Post/ZUMAPRESS/Newscom

CONCEPT CHECK: Which individual(s) in this pedigree exhibit(s) the effect of incomplete penetrance?

Sometimes, however, individuals carry the dominant allele but do not exhibit the trait. In Figure 4.3a, individual III-2, who does not have polydactyly, has inherited the polydactyly allele from II-4 and passed the allele to a female and male offspring. These observations indicate that III-2 is an unaffected heterozygote. In the case of polydactyly, the dominant allele does not always "penetrate" into the phenotype of the individual. Alternatively, for recessive traits, incomplete penetrance occurs if a homozygote carrying the recessive allele does not exhibit the recessive trait.

The measure of penetrance is described at the population level. For example, if 60% of the heterozygotes carrying a dominant allele exhibit the trait, we say that this trait is 60% penetrant. At the individual level, the trait is either present or not.

Another term used to describe the outcome of traits is the degree to which the trait is expressed, or its **expressivity.** In the case of polydactyly, the number of extra digits can vary. For example, one individual may have an extra toe on only one foot, whereas a second individual may have extra digits on both the hands and feet. Using genetic terminology, a person with several extra digits would have high expressivity of this trait, whereas a person with a single extra digit would have low expressivity.

How do we explain incomplete penetrance and variable expressivity? Although the answer may not always be understood, the range of phenotypes is often due to two factors:

- The environment may affect the outcome of the phenotype.
- One or more modifier genes may also affect the phenotype. For example, a modifier gene may affect the expression of the gene associated with polydactly and thereby influence the number of fingers or toes.

We will consider the issue of the environment next. The effects of modifier genes will be discussed later in the chapter.

4.2 COMPREHENSION QUESTIONS

1. Which of the following phenotypes is *not* an example of a wild-type phenotype?
   a. Yellow-flowered elderflower orchid
   b. Red-flowered elderflower orchid
   c. A gray elephant
   d. An albino (white) elephant

2. Dominant alleles may result from a mutation that causes
   a. the overexpression of a gene or its protein product.
   b. production of a protein that inhibits the function of a normal protein.
   c. a protein to be inactive and 50% of the normal amount of the protein is insufficient for a normal phenotype.
   d. any of the above.

3. Polydactyly is a condition in which a person has extra fingers and/or toes. It is caused by a dominant allele. If a person carries

this allele but does not have any extra fingers or toes, this is an example of

a. haploinsufficiency.

b. a dominant-negative mutation.

c. incomplete penetrance.

d. a gain-of-function mutation.

4.3 ENVIRONMENTAL EFFECTS ON GENE EXPRESSION

Learning Outcomes:

1. Discuss the role of the environment with regard to an individual's traits.

2. Define *norm of reaction*.

Throughout this book, our study of genetics tends to focus on the roles of genes in the outcome of traits. In addition to genetic variation, environmental conditions have a great effect on the phenotype of the individual. An example is the coat color variation found in a Siamese cat (**Figure 4.4a**).

Dark coloration in the fur is seen at the extremities of the body, such as the ears, tail, and paws. How do we explain this phenotype? It is due to a mutation in the gene that codes tyrosinase, which is an enzyme involved in making the pigment melanin. The more melanin that there is, the darker the fur will be. The mutation changes one amino acid in tyrosinase and this causes the enzyme to work poorly in warmer parts of the body. However, in the cooler extremities, it is able to function and produce darker fur. This is an example of a **temperature-sensitive allele.** The phenotypic effects are dependent on the temperature. As you might expect, if Siamese cats are raised at a lower temperature, their fur tends to be darker than if they are raised at a higher one. For those cats that spend time outdoors in seasonal climates, their fur is lighter in the warm summer and darker in the cold winter.

A dramatic example of the relationship between environment and phenotype can be seen in the human genetic disease known as phenylketonuria (PKU). This autosomal recessive disease is caused by a loss-of-function mutation in a gene that codes the enzyme phenylalanine hydroxylase. Homozygous individuals with this defective allele are unable to metabolize the amino acid phenylalanine properly. When given a standard diet containing phenylalanine, which is found in most protein-rich foods, PKU individuals manifest a variety of detrimental traits including mental impairment, underdeveloped teeth, and foul-smelling urine. In contrast, when PKU is diagnosed early and patients follow a restricted diet low in phenylalanine, they develop properly (**Figure 4.4b**).

Since the 1960s, testing methods have been developed that can determine if an individual is lacking the phenylalanine hydroxylase enzyme. These tests permit the identification of infants who have PKU, and their diets can then be modified before the harmful effects of phenylalanine ingestion have occurred. As a result of government legislation, more than 90% of infants born in the United States are now tested for PKU. This test

(a) Siamese cat

(b) Healthy person with PKU

FIGURE 4.4 **Variation in the expression of traits due to environmental effects.** **(a)** A Siamese cat. This coat color pattern is also found in a breed of rabbit called a Himalayan. **(b)** A person with PKU who has followed a restricted diet and developed properly. **(c)** Norm of reaction. In this experiment, fertilized eggs from a population of genetically identical fruit flies (*Drosophila melanogaster*) were allowed to develop into adults at different environmental temperatures. The graph shows the relationship between temperature (an environmental factor) and facet number in the eyes of the resulting adult flies. The micrograph shows an eye of *D. melanogaster*.

(a) axelbueckert/iStock/Getty Images Plus; (b): ©Sally Haugen/Virginia Schuett, www.pkunews.org; (c) Tomatito/Shutterstock

CONCEPT CHECK: What are the two main factors that determine an organism's traits?

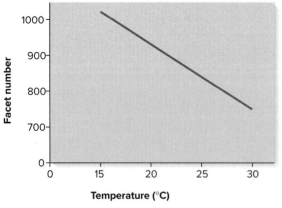

(c) Norm of reaction

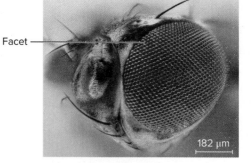

prevents a great deal of human suffering and is also cost-effective. In the United States, the annual cost of PKU testing is estimated to be a few million dollars, whereas the cost of treating severely affected individuals with the disease would be hundreds of millions of dollars.

The Siamese cat and individuals with PKU provide examples of the effects of different environmental conditions. When considering the environment, geneticists often examine a range of conditions, rather than simply observing phenotypes under two different conditions. The term **norm of reaction** refers to the effects of environmental variation on a phenotype. Specifically, it is the phenotypic range seen in individuals with a particular genotype. To evaluate the norm of reaction, researchers begin with true-breeding strains that have the same genotypes and subject them to different environmental conditions.

As an example, let's consider facet number in the eyes of fruit flies, *Drosophila melanogaster*. This species has compound eyes composed of many individual facets. **Figure 4.4c** shows the norm of reaction for facet number in genetically identical fruit flies that developed from fertilized eggs at different temperatures. As shown in the graph, the facet number varies with changes in temperature. At a lower temperature (15°C), the facet number is over 1000, whereas at a higher temperature (30°C), it is approximately 750.

4.3 COMPREHENSION QUESTION

1. The outcome of an individual's traits is controlled by
   a. genes.
   b. the environment.
   c. both genes and the environment.
   d. neither genes nor the environment.

4.4 INCOMPLETE DOMINANCE, HETEROZYGOTE ADVANTAGE, AND CODOMINANCE

Learning Outcomes:

1. Predict the outcome of crosses involving incomplete dominance, heterozygote advantage, and codominance.
2. Explain the underlying molecular mechanisms of incomplete dominance, heterozygote advantage, and codominance.

Thus far, we have considered inheritance patterns that follow a simple dominant/recessive inheritance pattern. In these cases, the heterozygote exhibits a phenotype that is the same as a homozygote that carries two copies of the dominant allele but different from the homozygote carrying two copies of the recessive allele. In this section, we will examine three different inheritance patterns in which the heterozygote shows a phenotype that is different from both types of homozygotes.

Incomplete Dominance Occurs When Two Alleles Produce an Intermediate Phenotype

Although many alleles display a simple dominant/recessive relationship, some do not.

- **Incomplete dominance** is a pattern of inheritance in which the phenotype of the heterozygote is intermediate between those of the corresponding homozygous individuals.

In 1905, Carl Correns first observed an example of incomplete dominance in the colors of the flowers of the four-o'clock plant (*Mirabilis jalapa*). **Figure 4.5** describes Correns' experiment, in which a homozygous red-flowered four-o'clock plant was crossed to a homozygous white-flowered plant. Because neither allele is dominant, each allele is designated with a superscript. The wild-type

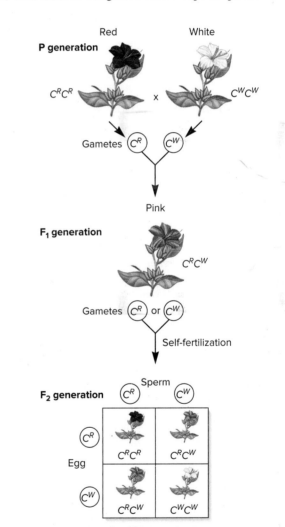

FIGURE 4.5 **Incomplete dominance in the four-o'clock plant, *Mirabilis jalapa*.**

INTERACTIVE EXERCISE

Genes→Traits When two different homozygotes ($C^R C^R$ and $C^W C^W$) of the four-o'clock plant are crossed, the resulting heterozygote, $C^R C^W$, has an intermediate phenotype of pink flowers. In the heterozygote, 50% of the functional protein coded by the C^R allele is not sufficient to produce a red phenotype.

CONCEPT CHECK: At the molecular level, what is the explanation for why the flowers of the heterozygous four-o'clock plant are pink instead of red?

allele for red flower color is designated C^R and the allele for white flower color is C^W. As shown in the figure, the offspring had pink flowers. When these F$_1$ offspring were allowed to self-fertilize, the F$_2$ generation consisted of 1/4 red-flowered plants, 1/2 pink-flowered plants, and 1/4 white-flowered plants. The pink plants in the F$_2$ generation were heterozygotes with an intermediate phenotype. As presented in the Punnett square in Figure 4.5, the F$_2$ generation displayed a 1:2:1 phenotypic ratio, which is different from the 3:1 ratio observed for simple Mendelian inheritance.

In Figure 4.5, incomplete dominance has occurred because a heterozygote has an intermediate phenotype. At the molecular level, the allele that causes a white phenotype is expected to result in a lack of a functional protein required for pigmentation. Depending on the effects of gene regulation, the heterozygotes may produce only 50% of the functional protein, but this amount is not sufficient to produce the same phenotype as the $C^R C^R$ homozygote, which may make twice as much of this protein. In this example, a reasonable explanation is that 50% of the functional protein cannot accomplish the same level of pigment synthesis that 100% of the protein can.

GENETIC TIPS

THE QUESTION: Two pink-flowered four-o'clocks were crossed to each other. What is the probability that a group of six offspring from this cross will be composed of one pink-, two white-, and three red-flowered plants?

T **OPIC:** *What topic in genetics does this question address?* The topic is Mendelian inheritance. More specifically, the question is about incomplete dominance in four-o'clock plants.

I **NFORMATION:** *What information do you know based on the question and your understanding of the topic?* From the question, you know that two pink-flowered four-o'clock plants are crossed to each other. From your understanding of the topic, you may remember that pink-flowered plants are heterozygous and show an intermediate phenotype. Also, from Chapter 3, you may recall that the multinomial expansion equation is used to solve problems involving three or more categories of offspring that are produced in an unordered fashion.

P **ROBLEM-SOLVING** **S** **TRATEGY:** *Predict the outcome. Make a calculation.* To begin to solve this problem, you need to know the probability of producing pink-, white-, and red-flowered offspring. This can be deduced from a Punnett square, which is shown below.

The cross is $C^R C^W \times C^R C^W$.

Next, you can use the probabilities derived from the Punnett square in the multinomial expansion equation.

ANSWER: From the Punnett square, the phenotypic ratio for the offspring is 1 red : 2 pink : 1 white. In other words, 1/4 are expected to be red, 1/2 pink, and 1/4 white.

$$P = \frac{n!}{a!b!c!}\, p^a q^b r^c$$

where

n = total number of offspring = 6

a = number of reds = 3

p = probability of red = 1/4

b = number of pinks = 1

q = probability of pink = 1/2

c = number of whites = 2

r = probability of white = 1/4

You substitute these values into the equation:

$$P = \frac{6!}{3!1!2!}\, (1/4)^3 (1/2)^1 (1/4)^2$$

$$P = 0.029, \text{ or } 2.9\%$$

This means that 2.9% of the time you expect to obtain six offspring plants of which three have red flowers, one has pink flowers, and two have white flowers.

Our Conclusions About Dominance May Depend on the Level of Examination

Our opinion of whether a trait is dominant or incompletely dominant may depend on how closely we examine the trait in individual organisms. The more closely we look, the more likely we are to discover that the heterozygote is not quite the same as the wild-type homozygote. For example, Mendel studied the characteristic of pea seed shape and visually concluded that the *RR* and *Rr* genotypes produced round seeds and the *rr* genotype produced wrinkled seeds. The peculiar morphology of the wrinkled seed is caused by a large decrease in the amount of starch deposition in the seed due to a defective *r* allele.

More recently, other scientists have dissected round and wrinkled seeds and examined their contents under a microscope. They have found that round seeds from heterozygotes actually contain an intermediate number of starch grains compared with seeds from the corresponding homozygotes (**Figure 4.6**). Within the seed, an intermediate amount of the functional protein is not enough to produce as many starch grains as in the homozygote carrying two copies of the *R* allele. Even so, at the level of our unaided eyes, heterozygotes produce seeds that appear to be round. With regard to phenotypes, the *R* allele is dominant to the *r* allele at the level of visual examination, but the *R* and *r* alleles show incomplete dominance at the level of starch biosynthesis.

Dominant (functional) allele: *R* (round)
Recessive (defective) allele: *r* (wrinkled)

Genotype	*RR*	*Rr*	*rr*
Amount of functional (starch-producing) protein	100%	50%	0%
Phenotype	Round	Round	Wrinkled
With unaided eye (simple dominant/ recessive relationship)			
With microscope (incomplete dominance)			

FIGURE 4.6   A comparison of phenotypes at the macroscopic and microscopic levels.

Genes→Traits This illustration shows the effects of a heterozygous pea plant having only 50% of the functional protein needed for starch production. The seed from the heterozygote appears to be as round as that from the homozygote carrying two copies of the *R* allele, but when examined microscopically, it has only half the amount of starch as is found in the homozygote's seed.

CONCEPT CHECK: At which level(s) is incomplete dominance more likely to be observed—the molecular, cellular, and/or organism level?

Heterozygote Advantage Occurs When Heterozygotes Have Greater Reproductive Success

As we have seen, the environment plays a key role in the outcome of traits. For certain genes, heterozygotes may display characteristics that are more beneficial for their survival in a particular environment. Such heterozygotes may be more likely to survive and reproduce. For example, a heterozygote may be larger, more disease-resistant, or better able to withstand harsh environmental conditions.

- The phenomenon in which a heterozygote has greater reproductive success compared with either of the corresponding homozygotes is called **heterozygote advantage.**

A well-documented example of heterozygote advantage involves a human allele that causes sickle cell disease in homozygous individuals. This disease is an autosomal recessive disorder in which the affected individual produces an altered form of the protein hemoglobin, which carries oxygen within red blood cells. Most people carry the Hb^A allele and make hemoglobin A. Individuals affected with sickle cell disease are homozygous for the Hb^S allele and produce only hemoglobin S. This causes their red blood cells to deform into a sickle shape (see **Figure 4.7a, b**) under conditions of low oxygen concentration as in the deoxygenated part of the cardiovascular system.

How does the sickling of red blood cells cause disease symptoms? The sickling phenomenon causes the life span of red blood cells to be greatly shortened to only a few weeks, compared

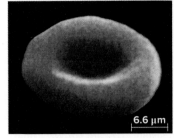

(a) Normal red blood cell   **(b) Sickled red blood cell**

$Hb^A Hb^S \times Hb^A Hb^S$

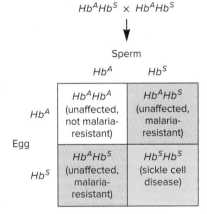

Sperm

(c) Example of sickle cell inheritance pattern

INTERACTIVE EXERCISE

FIGURE 4.7   Inheritance of sickle cell disease.
A comparison of **(a)** a normal red blood cell and **(b)** one from a person with sickle cell disease. **(c)** The outcome of a cross between two heterozygous individuals.

(a, b): Mary Martin/Science Source

CONCEPT CHECK: Why does the heterozygote have an advantage?

with a normal span of 4 months, and therefore, anemia results. In addition, sickled cells can become clogged in the capillaries throughout the body, leading to localized areas of oxygen depletion. Such an event causes pain and sometimes tissue and organ damage. For these reasons, the homozygous $Hb^S Hb^S$ individual usually has a shortened life span relative to an individual producing hemoglobin A.

In spite of the harmful consequences to homozygotes, the sickle cell allele is found at a fairly high frequency among human populations that are exposed to malaria. The protist genus that causes malaria, *Plasmodium*, spends part of its life cycle within the *Anopheles* mosquito and another part within the red blood cells of humans who have been bitten by an infected mosquito. When an $Hb^S Hb^S$ homozygote or an $Hb^A Hb^S$ heterozygote is infected with *Plasmodium*, the parasite's metabolic activity within red blood cells causes a lower oxygen concentration and thereby promotes sickling. The sickled cells are more likely to be phagocytized by the immune system, which diminishes the proliferation of the pathogen.

People who are heterozygotes have better resistance to malaria than do $Hb^A Hb^A$ homozygotes, and they do not suffer the ill effects of sickle cell disease. Therefore, even though the homozygous $Hb^S Hb^S$ condition is detrimental, the higher survival rate of heterozygotes has selected for the presence of the Hb^S allele within populations where malaria is prevalent. When viewing survival in such a region, heterozygote advantage explains the prevalence of the sickle cell allele. In Chapter 27, we will consider the role that natural selection plays in maintaining alleles that are beneficial to the heterozygote but harmful to the homozygote.

Figure 4.7c illustrates the predicted outcome when two heterozygotes have children. In this example, 1/4 of the offspring are $Hb^A Hb^A$ (unaffected, not malaria-resistant), 1/2 are $Hb^A Hb^S$ (unaffected, malaria-resistant), and 1/4 are $Hb^S Hb^S$ (have sickle cell disease). This 1:2:1 ratio deviates from a simple Mendelian 3:1 phenotypic ratio.

Heterozygote advantage is usually due to two alleles that produce proteins with slightly different amino acid sequences. How can we explain the observation that two protein variants in the $Hb^A Hb^S$ heterozygote produce a more favorable phenotype? Three common explanations are discussed next.

Disease Resistance. In the case of sickle cell disease, the phenotype is related to the infectivity of *Plasmodium* (**Figure 4.8a**). In the heterozygote, the infectious agent is less likely to propagate

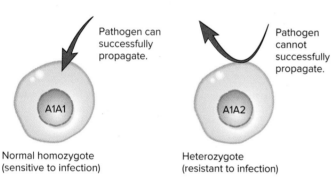

Normal homozygote
(sensitive to infection)

Heterozygote
(resistant to infection)

(a) Disease resistance

The homozygotes that are *A1A1* or *A2A2* will make homodimers that are A1A1 and A2A2, respectively. The *A1A2* heterozygote can make A1A1 and A2A2 and can also make A1A2 homodimers, which may have better functional activity.

(b) Homodimer formation

27°–32°C
(optimum temperature range)

30°–37°C
(optimum temperature range)

A heterozygote, *E1E2*, would produce both enzymes and have a broader temperature range (i.e., 27°–37°C) in which the enzyme would function.

(c) Variation in functional activity

FIGURE 4.8 **Three possible explanations for heterozygote advantage at the molecular level.**

CONCEPT CHECK: Which of these three scenarios explains heterozygote advantage with regard to the sickle cell allele?

within red blood cells. Interestingly, researchers have speculated that other alleles in humans may confer disease resistance in the heterozygous condition but are detrimental in the homozygous state. These include alleles involved in PKU, in which the heterozygous fetus may be resistant to miscarriage caused by a fungal toxin, and in Tay-Sachs disease, in which the heterozygote may be resistant to tuberculosis.

Subunit Composition of Proteins. A second way to explain heterozygote advantage is related to the subunit composition of proteins. In some cases, a protein functions as a complex of multiple subunits; each subunit is composed of one polypeptide. A protein composed of two subunits is called a dimer. When both subunits are coded by the same gene, the protein is a homodimer. The prefix *homo-* means that the subunits come from the same type of gene although the gene may exist in different alleles. **Figure 4.8b** considers a situation in which a gene exists in two alleles that code polypeptides designated A1 and A2. Homozygous individuals can produce only A1A1 or A2A2 homodimers, whereas a heterozygote can also produce an A1A2 homodimer. Thus, heterozygotes can produce three forms of the homodimer, homozygotes only one. For some proteins, A1A2 homodimers may have better functional activity because they are more stable or able to function under a wider range of conditions. The greater activity of the homodimer protein may be the underlying reason why a heterozygote has characteristics superior to either homozygote.

Differences in Protein Function. A third molecular explanation of heterozygote advantage is that the proteins coded by each allele exhibit differences in their functional activity. For example, suppose that a gene codes a metabolic enzyme that can be found in two forms (corresponding to the two alleles), one that functions better at a lower temperature and another that functions optimally at a higher temperature (**Figure 4.8c**). The heterozygote, which makes a mixture of both enzymes, may be at an advantage under a wider temperature range than either of the corresponding homozygotes.

Alleles of the ABO Blood Group Can Be Dominant, Recessive, or Codominant

Thus far, we have considered examples in which a gene exists in two different alleles. As researchers have probed genes at the molecular level within natural populations of organisms, they have discovered that most genes exist in **multiple alleles.** Within a population, genes are typically found in a few or even many alleles.

The ABO group of antigens, which determine blood type in humans, are produced in the human population under the control of multiple alleles; two of these alleles exhibit a relationship called codominance. To understand this concept, we first need to examine the molecular characteristics of human blood types. The plasma membranes of red blood cells have groups of interconnected sugars—oligosaccharides—that act as surface antigens (**Figure 4.9a**). Antigens are molecular structures that are recognized by antibodies produced by the immune system.

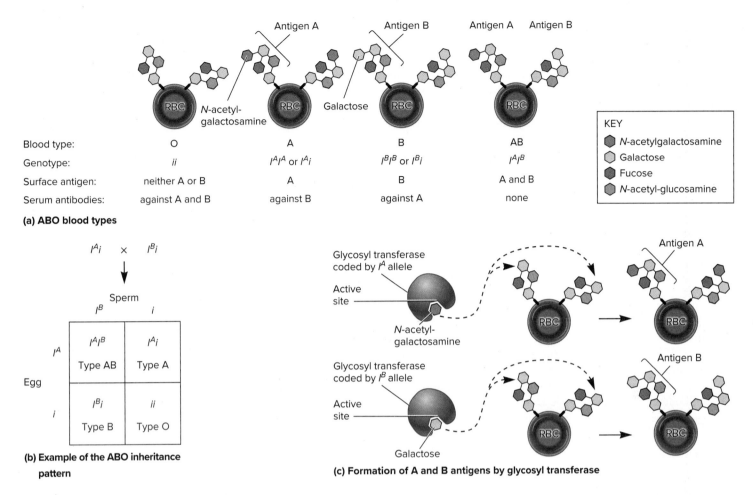

(a) ABO blood types

Blood type:	O	A	B	AB
Genotype:	ii	$I^A I^A$ or $I^A i$	$I^B I^B$ or $I^B i$	$I^A I^B$
Surface antigen:	neither A or B	A	B	A and B
Serum antibodies:	against A and B	against B	against A	none

KEY
- *N*-acetylgalactosamine
- Galactose
- Fucose
- *N*-acetyl-glucosamine

(b) Example of the ABO inheritance pattern

(c) Formation of A and B antigens by glycosyl transferase

INTERACTIVE EXERCISE

FIGURE 4.9 **ABO blood types.** (a) A schematic representation of human blood types at the cellular level. Note: This is not drawn to scale; a red blood cell is much larger than any oligosaccharide on its surface. (b) The predicted offspring from parents who are $I^A i$ and $I^B i$. (c) The glycosyl transferase coded by the I^A and I^B alleles recognizes different sugars due to changes in its active site. The i allele results in a nonfunctional enzyme. The antigen produced by type O individuals is called H antigen. Humans of all blood types do not usually produce antibodies against H antigen.

CONCEPT CHECK: Which allele is an example of a loss-of-function allele?

The synthesis of these surface antigens is controlled by three alleles, designated i, I^A, and I^B.

- The i allele is recessive to both I^A and I^B, which is why the i allele is designated with a lowercase letter. A person who is homozygous ii has type O blood and produces a relatively short oligosaccharide, which is called H antigen.
- A homozygous $I^A I^A$ or heterozygous $I^A i$ individual has type A blood. The red blood cells of this individual contain the surface antigen known as A.
- A homozygous $I^B I^B$ or heterozygous $I^B i$ individual produces surface antigen B. As Figure 4.9a indicates, surface antigens A and B have different molecular structures.
- A person who is $I^A I^B$ has the blood type AB and expresses both surface antigens A and B.
- The phenomenon in which two alleles are both expressed in the heterozygous individual is called **codominance.** In this case, the I^A and I^B alleles are codominant to each other.

Note: The allele designations for the I^A and I^B alleles are a capital I with a superscript (A or B) because both of them are dominant to the i allele, but they are not dominant to each other.

As an example of the inheritance of blood type, let's consider the possible offspring between two parents who are $I^A i$ and $I^B i$ (**Figure 4.9b**). The $I^A i$ parent makes I^A and i gametes, and the $I^B i$ parent makes I^B and i gametes. These combine to produce $I^A I^B$, $I^A i$, $I^B i$, and ii offspring in a 1:1:1:1 ratio. The resulting blood types of the offspring are AB, A, B, and O, respectively.

Biochemists have analyzed the oligosaccharides on the surfaces of cells of differing blood types. In type O, the oligosaccharide is smaller than in type A or type B because a sugar has not been attached to a specific site on the oligosaccharide. This idea is schematically shown in Figure 4.9a. How do we explain this difference at the molecular level? The gene that determines ABO blood type codes a type of enzyme called a glycosyl transferase that attaches a sugar to an oligosaccharide.

- The *i* allele carries a mutation that renders this enzyme inactive, which prevents the attachment of an additional sugar.
- The two types of glycosyl transferase coded by the I^A and I^B alleles have different structures in their active sites. The glycosyl transferase coded by the I^A allele has an active site that recognizes uridine diphosphate *N*-acetylgalactosamine and attaches *N*-acetylgalactosamine to the oligosaccharide (**Figure 4.9c**). This produces the molecular structure of surface antigen A.
- The glycosyl transferase coded by the I^B allele recognizes UDP-galactose and attaches galactose to the oligosaccharide. This produces the molecular structure of surface antigen B.
- A person with type AB blood makes both types of enzymes and thereby makes oligosaccharides with both types of sugar attached.

A small difference in the structure of the oligosaccharide, namely, a GalNAc in antigen A versus galactose in antigen B, explains why the two antigens are different from each other at the molecular level. These differences enable them to be recognized by different antibodies. A person who has blood type A makes antibodies to blood type B (refer back to Figure 4.9a). The antibodies against blood type B require a galactose on the oligosaccharide for their proper recognition. This person's antibodies will not recognize and destroy the person's own blood cells, but they will recognize and destroy the blood cells from a type B person.

With this in mind, let's consider why blood typing is essential for safe blood transfusions. The donor's blood must be an appropriate match with the recipient's blood. People with type O blood have the potential to produce antibodies against both A and B antigens if they receive type A, type B, or type AB blood. After the antibodies are produced in the recipient of the transfusion, they will react with the donated blood cells and cause them to agglutinate (clump together). This is a life-threatening situation that causes the blood vessels to clog. Other incompatible combinations include a type A person receiving type B or type AB blood and a type B person receiving type A or type AB blood. Because individuals with type AB blood do not produce antibodies to either A or B antigens, they can receive any type of blood and are known as universal recipients. By comparison, type O persons are universal donors because their blood can be given to recipients with type O, A, B, or AB.

4.4 COMPREHENSION QUESTIONS

1. A pink-flowered four-o'clock plant is crossed to a red-flowered plant. What is the expected outcome for the offspring's phenotypes?
   a. All pink
   b. All red
   c. 1 red : 2 pink : 1 white
   d. 1 red : 1 pink

2. A person with type AB blood has a child with a person with type O blood. What are the possible blood types of the child?
   a. A or B
   b. A, B, or O
   c. A, B, AB, or O
   d. O only

4.5 GENES ON SEX CHROMOSOMES

Learning Outcomes:

1. Predict the outcome of crosses for X-linked inheritance.
2. Explain pseudoautosomal inheritance.

The term **sex chromosomes** refers to chromosomes that differ between males and females. In mammals and fruit flies, the sex chromosomes are designated X and Y. In Chapter 3, we considered experiments with fruit flies that showed that an eye color gene is located on the X chromosome. In this section, we will further explore the inheritance of traits for which genes are located on sex chromosomes.

The Inheritance Pattern of X-Linked Genes Can Be Revealed by Reciprocal Crosses

As discussed in Chapter 2, many species have males and females that differ in their sex chromosome composition. In mammals, for example, females are XX and males are XY. In such species, certain traits are governed by genes that are located on a sex chromosome. For these traits, the outcome of crosses depends on the genotypes and sexes of the parents and offspring.

- When a gene is located on the X chromosome but not on the Y chromosome, it follows a pattern of transmission called **X-linked inheritance.**

The inheritance pattern of X-linked genes shows certain distinctive features. In mammals, males transmit X-linked genes only to their female offspring, and male offspring receive their X-linked genes only from their female parent. The term **hemizygous** is used to indicate that males have a single copy of an X-linked gene. A male mammal is said to be hemizygous for X-linked genes. Because males of certain species, such as humans, have a single X chromosome, another distinctive feature of X-linked inheritance is that males are more likely to be affected by rare, recessive X-linked disorders.

As an example, let's consider a human disease known as Duchenne muscular dystrophy (DMD), which was first described by Guillaume Duchenne in the 1860s. Affected individuals show signs of muscle weakness as early as age 3. The disease gradually weakens the skeletal muscles and eventually affects the heart and breathing muscles. Survival is rare beyond the early 30s. The gene for DMD, found on the X chromosome, codes a protein called dystrophin that is required inside muscle cells for structural

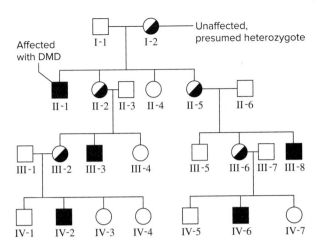

FIGURE 4.10 A human pedigree for Duchenne muscular dystrophy, an X-linked recessive trait. Affected individuals are shown with filled symbols. Females who are unaffected with the disease but have affected sons are presumed to be heterozygous carriers, shown with half-filled symbols.

CONCEPT CHECK: What features of this pedigree indicate that the allele for Duchenne muscular dystrophy is X-linked?

support. Dystrophin is thought to strengthen muscle cells by anchoring elements of the internal cytoskeleton to the plasma membrane. Without it, the plasma membrane becomes permeable and may rupture.

DMD follows an inheritance pattern called **X-linked recessive**—the allele causing the disease is recessive and located on the X chromosome. In the pedigree shown in **Figure 4.10**, several males are affected by DMD, as indicated by filled squares. The female parents of these males are presumed to be heterozygous for the X-linked recessive allele. This recessive disease is very rare among females because females would have to inherit a copy of the mutant allele from their female parent and a copy from an affected male parent.

X-linked muscular dystrophy has also been found in certain breeds of dogs such as golden retrievers (**Figure 4.11a**). As in humans, the mutation occurs in the dystrophin gene, and the symptoms include severe weakness and muscle atrophy that begin at about 6 to 8 weeks of age. Many dogs that inherit this disorder die within the first year of life, though some can live 3 to 5 years and reproduce.

Figure 4.11b (left side) considers a cross between an unaffected female dog with two copies of the wild-type gene and a male dog with muscular dystrophy that carries the mutant allele and has survived to reproductive age. When setting up a Punnett square involving X-linked traits, we must consider the alleles on the X chromosome as well as the observation that males may transmit a Y chromosome instead of the X chromosome. A male makes two types of gametes, one that carries the X chromosome and one that carries the Y. The Punnett square must also include the Y chromosome even though this chromosome does not carry any X-linked genes. The X chromosomes from the female and male are designated with their corresponding alleles. When the Punnett square is filled in, it predicts the X-linked genotypes and sexes of the offspring.

As seen on the left side of Figure 4.11b, none of the offspring from this cross are affected with the disorder, although all female offspring are carriers.

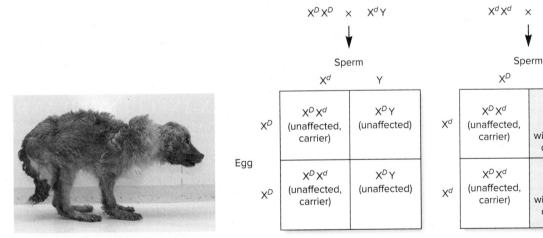

(a) Male golden retriever with X-linked muscular dystrophy

(b) Examples of X-linked muscular dystrophy inheritance patterns

INTERACTIVE EXERCISE

FIGURE 4.11 X-linked muscular dystrophy in dogs. (a) The male golden retriever shown here has the disease. **(b)** The Punnett square on the left shows a cross between an unaffected female and an affected male. The one on the right shows a reciprocal cross between an affected female and an unaffected male. *D* represents the common (non-disease-causing) allele for the dystrophin gene, and *d* is the mutant allele that causes a defect in dystrophin function.

CONCEPT CHECK: Explain why the reciprocal cross yields a different result from that obtained for the first cross.

Robriquet et al. (2015) Differential Gene Expression Profiling of Dystrophic Dog Muscle after MuStem Cell Transplantation. *PLoS ONE* 10(5): e0123336. doi:10.1371/journal.pone.0123336

The right side of Figure 4.11b shows a **reciprocal cross**—a second cross in which the sexes and phenotypes are reversed. In this case, an affected female dog is crossed to an unaffected male. This cross produces female offspring that are carriers and male offspring that are all affected with muscular dystrophy.

By comparing the two Punnett squares, we see that the outcome of the reciprocal cross yielded different results. This is expected with X-linked genes, because the male transmits the gene only to female offspring, but the female transmits an X chromosome to both male and female offspring. Male offspring receive a Y chromosome from their male parent. Therefore, a male parent does not contribute to X-linked phenotypes of male offspring. This explains why X-linked traits do not behave the same way in reciprocal crosses. Experimentally, the observation that reciprocal crosses do not yield the same results is an important clue that a trait may be X-linked.

Genes Located on Mammalian Sex Chromosomes Can Be Transmitted in an X-Linked, a Y-Linked, or a Pseudoautosomal Pattern

Our discussion of sex chromosomes has focused on genes that are located on the X chromosome but not on the Y chromosome.

- The term **sex-linked gene** refers to a gene that is found on one of the two types of sex chromosomes but not on both.
- **X-linked genes** are found only on the X chromosome.
- **Y-linked genes,** or **holandric genes,** are found only on the Y chromosome. An example of a Y-linked gene is the *SRY* gene found in mammals. As discussed in Chapter 26, its expression is necessary for proper male development. A Y-linked inheritance pattern is very distinctive—the gene is transmitted only from a male parent to a male offspring.

In humans, researchers estimate that the X chromosome carries between 800 and 900 protein-coding genes, whereas the Y chromosome has between 50 and 60.

In addition to sex-linked genes, the X and Y chromosomes contain short regions of homology where both chromosomes carry the same genes. Along with several smaller regions, the human sex chromosomes have three large homologous regions (**Figure 4.12**). These regions, which are evolutionarily related, promote the necessary pairing of the X and Y chromosomes that occurs during meiosis I of spermatogenesis. Relatively few genes are located in these homologous regions. One example is a human gene called *Mic2*, which codes a cell surface antigen. The *Mic2* gene is found on both the X and Y chromosomes. It follows a pattern of inheritance called **pseudoautosomal inheritance.** The term *pseudoautosomal* refers to the idea that the inheritance pattern of the *Mic2* gene is the same as the inheritance pattern of a gene located on an autosome even though the *Mic2* gene is actually located on the sex chromosomes. As in autosomal inheritance, males have two copies of pseudoautosomally inherited genes, and they can transmit the genes to both female and male offspring.

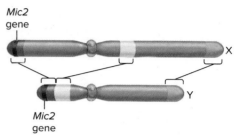

FIGURE 4.12 **A comparison of the homologous and nonhomologous regions of the X and Y chromosomes in humans.** The brackets show three regions of homology between the X and Y chromosomes. A few pseudoautosomal genes, such as *Mic2*, are found on both the X and Y chromosomes in these small regions of homology. Researchers estimate that the X chromosome has between 800 and 900 protein-coding genes and the Y chromosome has between 50 and 60.

CONCEPT CHECK: Why are the homologous regions of the X and Y chromosomes important during meiosis?

4.5 COMPREHENSION QUESTIONS

1. A cross is made between a white-eyed female fruit fly and a red-eyed male. What would be the reciprocal cross?
   a. Female is X^wX^w and male is X^wY.
   b. Female is $X^{w+}X^{w+}$ and male is $X^{w+}Y$.
   c. Female is X^wX^{w+} and male is X^wY.
   d. Female is X^wX^w and male is $X^{w+}Y$.

2. Hemophilia is a blood-clotting disorder in humans that follows an X-linked recessive pattern of inheritance. A male with hemophilia and a female without hemophilia have a female offspring with hemophilia. If you let H represent the common (non-disease-causing) allele and h the hemophilia-causing allele, what are the genotypes of the parents?
   a. Female parent is X^HX^h and male parent is X^hY.
   b. Female parent is X^hX^h and male parent is X^hY.
   c. Female parent is X^hX^h and male parent is X^HY.
   d. Female parent is X^HX^h and male parent is X^HY.

4.6   SEX-INFLUENCED AND SEX-LIMITED INHERITANCE

Learning Outcomes:

1. Compare and contrast sex-influenced inheritance and sex-limited inheritance.
2. Predict the outcome of crosses involving sex-influenced inheritance.

As we have just seen, the transmission pattern of sex-linked genes depends on whether the gene is on the X or Y chromosome and on the sex of the offspring. Sex can influence traits in other ways as well.

- The term **sex-influenced inheritance** refers to the phenomenon in which an allele is dominant in one sex but recessive in the other sex. Therefore, sex influence is a phenomenon of heterozygotes.

Sex-influenced inheritance should not be confused with sex-linked inheritance. The genes that govern sex-influenced traits are autosomal, not on the X or Y chromosome. Researchers once thought that human pattern baldness, which is characterized by hair loss on the front and top of the head but not on the sides, is an example of sex-influenced inheritance. However, recent research indicates that mutations in the androgen receptor gene, which is located on the X chromosome, often play a key role in pattern baldness. Therefore, pattern baldness often follows an X-linked pattern of inheritance. Even so, variation in other gene(s) located on chromosome 20 (an autosome) can be a contributing factor to baldness.

An example of sex-influenced inheritance is found in cattle. Certain breeds exhibit scurs, which are small hornlike growths on the frontal bone in the same locations where horns (in other breeds of cattle) would grow (**Figure 4.13**). This trait appears to be controlled by a single gene that exists in two alleles, Sc^P and Sc^A. The superscript P represents the allele in which scurs are present, whereas the superscript A represents the allele in which scurs are absent. The Sc^P allele is dominant in males and recessive in females, whereas the Sc^A allele is dominant in females and recessive in males:

| | Phenotype | |
Genotype	Males	Females
$Sc^P Sc^P$	Scurs	Scurs
$Sc^P Sc^A$	Scurs	No scurs
$Sc^A Sc^A$	No scurs	No scurs

INTERACTIVE EXERCISE

FIGURE 4.13 **Scurs in cattle, an example of a sex-influenced trait.**
Courtesy of Sheila M. Schmutz, Ph.D.

CONCEPT CHECK: What is the phenotype of a female cow that is heterozygous for the scurs alleles?

THE QUESTION: As we have seen, having scurs is an example of a sex-influenced trait in cattle that is dominant in males and recessive in females. A male and a female, neither of which has scurs, produce a male offspring with scurs. What are the genotypes of the parents?

TOPIC: *What topic in genetics does this question address?* The topic is Mendelian inheritance. More specifically, the question is about sex-influenced inheritance.

INFORMATION: *What information do you know based on the question and your understanding of the topic?* From the question, you know that a male and a female without scurs produced a male offspring with scurs. From your understanding of the topic, you may remember that the unique feature of sex-influenced inheritance is that the trait is dominant in one sex and recessive in the other.

PROBLEM-SOLVING **S**TRATEGY: *Predict the outcome.* One strategy to solve this type of problem is to use a Punnett square to predict the possible outcomes of this cross. Because the Sc^A allele is recessive in males, the male parent without scurs must be homozygous, $Sc^A Sc^A$. A female without scurs can be either $Sc^P Sc^A$ or $Sc^A Sc^A$. Therefore, two Punnett squares are possible, which are shown next.

♂	Sc^A	Sc^A
♀ Sc^P	$Sc^P Sc^A$	$Sc^P Sc^A$
Sc^A	$Sc^A Sc^A$	$Sc^A Sc^A$

♂	Sc^A	Sc^A
♀ Sc^A	$Sc^A Sc^A$	$Sc^A Sc^A$
Sc^A	$Sc^A Sc^A$	$Sc^A Sc^A$

ANSWER: From the question, you know that the male offspring has scurs. When comparing the two Punnett squares, only the one on the left can produce a male offspring with scurs (see the gray-shaded boxes). Therefore, the female parent must be $Sc^P Sc^A$ and the male parent is $Sc^A Sc^A$.

Some Traits Are Limited to One Sex

Another way in which sex affects an organism's phenotype involves traits that are found in males or females, but not both.

- In **sex-limited inheritance,** a trait occurs in only one sex.

Genes that produce such traits are controlled by sex hormones or by the pathway that leads to male and female development, which is described in Chapter 26. The genes that affect sex-limited traits may be autosomal or X-linked. In humans, examples of sex-limited traits are the presence of ovaries in females and the presence of testes in males. Due to these two sex-limited traits, fertile females only produce eggs, whereas fertile males only produce sperm.

(a) Hen   (b) Rooster

FIGURE 4.14 Differences in morphological features between (a) female and (b) male chickens, an example of sex-limited inheritance.

(a): Javier Larrea/Pixtal/age fotostock; (b): Image Source/PunchStock/Getty Images

CONCEPT CHECK: What is the molecular explanation for sex-limited inheritance?

Sex-limited traits are responsible for **sexual dimorphism,** in which the sexes in a particular species have different morphological features. This phenomenon is common among many animal species and is often striking in various species of birds in which the male has more ornate plumage than the female. As shown in **Figure 4.14,** roosters have a larger comb and wattles and longer neck and tail feathers than do hens. These features are limited to roosters.

4.6 COMPREHENSION QUESTION

1. A cow with scurs and a bull with no scurs have an offspring. This offspring could be
   a. a female with scurs or a male with scurs.
   b. a female with no scurs or a male with scurs.
   c. a female with scurs or a male with no scurs.
   d. a female with no scurs or a male with no scurs.

4.7   LETHAL ALLELES

Learning Outcomes:

1. Describe the different types of lethal alleles.
2. Predict how lethal alleles may affect the outcome of a cross.

Let's now turn our attention to alleles that have the most detrimental effect on phenotype—those that result in death. An allele that has the potential to cause the death of an organism is called a **lethal allele.** Such alleles are usually inherited in a recessive manner. When the absence of a specific protein results in a lethal phenotype, the gene that codes the protein is considered an **essential gene,** one that must be present for survival. Though the proportion varies by species, researchers estimate that approximately 1/3 of all genes are essential genes.

By comparison, **nonessential genes** are not absolutely required for survival, although they are likely to be beneficial to the

organism. A loss-of-function mutation in a nonessential gene will not usually cause death. On rare occasions, however, a nonessential gene may acquire a gain-of-function mutation that causes the gene product to be abnormally expressed in a way that may interfere with cell function and lead to a lethal phenotype. Therefore, not all lethal mutations occur in essential genes, although the great majority do.

Some lethal alleles may kill an organism only when certain environmental conditions prevail. Such **conditional lethal alleles** have been extensively studied in experimental organisms. For example, some conditional lethal alleles cause an organism to die only in a particular temperature range. These alleles, called **temperature-sensitive (ts) lethal alleles,** have been observed in many organisms, including *Drosophila.* A ts lethal allele may be fatal for a developing larva at a high temperature (30°C), but the larva survives if grown at a lower temperature (22°C). Temperature-sensitive lethal alleles are typically caused by mutations that alter the structure of the coded protein so that it does not function correctly at the nonpermissive temperature or becomes unfolded and is rapidly degraded.

Conditional lethal alleles may also be identified when an individual is exposed to a particular agent in the environment. For example, people with a defect in the gene that codes the enzyme glucose-6-phosphate dehydrogenase (G-6-PD) have a negative reaction to the ingestion of fava beans. This reaction can lead to an acute hemolytic syndrome with 10% mortality if not treated properly.

Finally, it is surprising that certain lethal alleles act only in some individuals. These are called **semilethal alleles.** Of course, any particular individual cannot be semidead. However, within a population, a semilethal allele will cause some individuals to die but not all of them. The reasons for semilethality are not always understood, but environmental conditions and the actions of other genes within the organism may help to prevent the detrimental effects of certain semilethal alleles. An example of a semilethal allele is the X-linked white-eyed allele in fruit flies, which is described in Chapter 3 (refer back to Figure 3.18). Depending on the growth conditions, approximately 1/4 to 1/3 of the flies that carry the white-eyed allele die during early stages of development.

In some cases, a lethal allele may produce ratios that seemingly deviate from Mendelian ratios. An example is an allele in a breed of cats known as Manx, which originated on the Isle of Man (**Figure 4.15a**). The Manx cat is a heterozygote that carries a dominant mutant allele affecting the spine. This allele shortens the tail, resulting in a range of tail lengths from normal to tailless. When two heterozygous Manx cats are crossed to each other, the ratio of offspring is 1 normal to 2 Manx. How do we explain the 1:2 ratio? The answer is that 1/4 of the offspring are homozygous for the dominant mutant allele, and they die during early embryonic development (**Figure 4.15b**). In this case, the Manx phenotype is dominant in heterozygotes, whereas the dominant mutant allele is lethal in the homozygous condition.

The time when a lethal allele exerts its effect can vary. Many lethal alleles disrupt proper cell division and thereby cause an

FIGURE 4.15
The Manx cat, which carries a lethal allele.
(a) Photo of a Manx cat, which typically has a shortened tail. **(b)** Outcome of a cross between two Manx cats. Animals that are homozygous for the dominant Manx allele (*M*) die during early embryonic development.

Juniors Bildarchiv/F215/Alamy Stock Photo

CONCEPT CHECK: Why do you think the *Mm* heterozygote offspring of two Manx cats survives with developmental abnormality, whereas the *MM* homozygote dies?

(a) A Manx cat

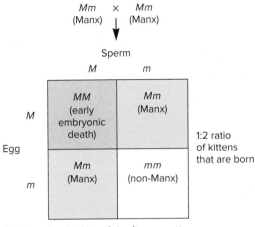

(b) Example of a Manx inheritance pattern

organism to die at a very early stage. Others may allow a short period of development before the organism dies. In the case of the Manx allele, the homozygote dies early in embryonic development. Certain inherited diseases in humans, such as Tay-Sachs (see Table 4.2), result in death during childhood. However, some lethal alleles may exert their effects later in life, or only under certain environmental conditions. For example, a human genetic disease known as Huntington disease is caused by a dominant allele. The disease is characterized by a progressive degeneration of the nervous system, dementia, and early death. The age when these symptoms appear, or the **age of onset,** is usually between 30 and 50.

4.7 COMPREHENSION QUESTION

1. The Manx phenotype in cats is caused by a dominant allele that is lethal in the homozygous state. A Manx cat is crossed to a normal (non-Manx) cat. What is the expected outcome for the surviving offspring?
   a. All Manx
   b. All normal
   c. 1 normal : 1 Manx
   d. 1 normal : 2 Manx

4.8 UNDERSTANDING COMPLEX PHENOTYPES CAUSED BY MUTATIONS IN SINGLE GENES

Learning Outcomes:
1. Explain the phenomenon of pleiotropy.
2. Describe how embryonic development determines certain coat patterns in animals.

Thus far, we have considered a variety of examples in which mutations in a single gene affect the outcome of a single trait. For some of these examples, such as ABO blood types, flower color in the four-o'clock plant, and sickle cell disease, the relationship between a mutation in a gene and its effect on a single trait is relatively easy to understand. However, for other genes, the phenotypic effects may be more complex, and researchers may have to dig deeper to understand how a mutation in a single gene can produce complex effects on phenotype. In this section, we will consider how the expression of a single gene can have multiple effects throughout the body and see how an understanding of embryonic development can explain certain complex phenotypes.

Pleiotropy Occurs When the Expression of a Single Gene Has Two or More Phenotypic Effects

Although we tend to discuss genes within the context of how they influence a single trait, most genes actually have multiple effects throughout a cell or throughout a multicellular organism. The multiple effects of a single gene on the phenotype of an organism is called **pleiotropy.** Pleiotropy occurs for several reasons, including the following:

- The expression of a single gene can affect cell function in more than one way. For example, a defect in a microtubule protein may affect cell division and cell movement.
- A gene may be expressed in different cell types in a multicellular organism.
- A gene may be expressed only at a specific stage of development.

In all or nearly all cases, the expression of a gene is pleiotropic with regard to the characteristics of an organism. The expression of any given gene influences the expression of many other genes in the genome, and vice versa. Pleiotropy is revealed when researchers study the effects of gene mutations. As an example of a pleiotropic mutation, let's consider cystic fibrosis, which is a recessive human disorder. In the late 1980s, the gene for cystic fibrosis was identified. It codes a protein called the cystic fibrosis transmembrane conductance regulator (CFTR), which regulates ionic balance by allowing the transport of chloride ions (Cl⁻) across epithelial cell membranes.

The mutation that causes cystic fibrosis diminishes the function of this Cl⁻ transporter, affecting several parts of the body in different ways:

- Because the movement of Cl⁻ affects water transport across membranes, the most severe symptom of cystic fibrosis is thick mucus in the lungs that occurs because of a water imbalance. This thickened mucus results in difficulty in breathing and frequent lung infections.
- Thick mucus can also block the tubes that carry digestive enzymes from the pancreas to the small intestine. Without these enzymes, certain nutrients are not properly absorbed into the body. As a result, persons with cystic fibrosis may show poor weight gain.
- Another effect is seen in the sweat glands. A functional Cl⁻ transporter is needed to recycle salt out of these glands and back into the skin before it can be lost to the outside world. Persons with cystic fibrosis have excessively salty sweat due to their inability to recycle salt back into their skin cells—a common test for cystic fibrosis is measurement of salt on the skin.

Taken together, these symptoms show that a defect in CFTR has multiple effects throughout the body.

Certain Coat-Color Patterns in Dogs Are Determined by Events During Embryonic Development

Many breeds of dogs and other mammals have a coat-color pattern, called white spotting, in which portions of an animal's fur lack pigmentation. The coat-color gene influencing this trait exists in multiple alleles that affect the amount of pigmentation of an animal's fur. The S^+ allele results in full pigmentation (no white spotting), whereas other alleles vary with regard to the amount of pigmentation produced; possible patterns include Irish spotting (s^I)

and extreme white spotting (s^w)(compare **Figure 4.16a,b**). In many breeds, the areas that are white include the legs, belly, neck, and the tip of the tail. For decades, researchers were baffled by this coat pattern in which some areas are pigmented and others are white.

Fur pigmentation is dependent on melanocytes, which are cells located in the bottom layer of the skin's epidermis and in the hair follicles. Melanocytes produce the pigment melanin, which is found in the skin, eyes, and hair. If melanin is not produced within hair follicles, a dog's fur remains white. An understanding of melanocyte development provides insight regarding the intriguing phenotype of white spotting.

During embryogenesis, melanocyte precursor cells, called melanoblasts, originate in the neural crest, which is a temporary group of cells that are associated with each other only during embryonic development. The neural crest is located dorsal to the neural tube, which gives rise to the brain and spinal cord. Melanoblasts originate only in the part of the neural crest that is located in the trunk region of the embryo (between the neck and the tail). From there, the melanoblasts migrate to other parts of the body (**Figure 4.16c**). As they migrate, the melanoblasts proliferate, which allows some of them to travel longer distances and reach more ventral regions of the embryo. Once the melanoblasts reach their final destination, they continue to proliferate and differentiate into pigment-producing melanocytes in places such as the epidermis and hair follicles.

Researchers speculate that the alleles conferring the white spotting phenotype cause a decrease in the number of melanocytes due to failure of melanoblast migration, proliferation, and/or survival during embryonic development. In 2007, a genome-wide association study determined that this coat-color gene codes a protein called microphthalmia-associated transcription factor (MITF). (Genome-wide association studies are described in Chapter 24.) In mice and humans, MITF expression is needed for proper migration, proliferation, and survival of melanoblasts. Reduced expression of MITF is expected to decrease the number of melanocytes in adult animals. Because the melanoblasts begin their journey from the neural crest in the trunk, this reduced expression of MITF

(a) Irish spotting

(b) Extreme white spotting

(c) Migration of melanoblasts during embryonic development

FIGURE 4.16 White spotting phenotype in dogs. (a) This animal shows a typical Irish spotting pattern seen in an $s^I s^I$ homozygote or an $s^I s^w$ heterozygote. (b) The extreme white spotting pattern of an $s^w s^w$ homozygote. (c) During embryonic development, melanoblasts migrate away from the trunk region of the neural crest. *Source (parts a and b):* Figure 1 in *Annu. Rev. Anim. Biosci.* 2013. *1*:125–156.

(a, b): Deanna Vout

causes the regions of the body in the adult that are farthest away from the spinal cord to contain fewer melanocytes and are therefore more likely to be white.

4.8 COMPREHENSION QUESTION

1. Which of the following is a possible explanation for pleiotropy?
   a. The expression of a single gene can affect cell function in more than one way.
   b. A gene may be expressed in different cell types in a multicellular organism.
   c. A gene may be expressed at different stages of development.
   d. All of the above are possible explanations.

4.9 GENE INTERACTIONS

Learning Outcomes:

1. Define *gene interaction*.
2. Predict the outcome of crosses involving epistasis, complementation, gene modifiers, and gene redundancy.
3. Describe examples that explain the molecular mechanisms of epistasis, complementation, gene modification, and gene redundancy.

Thus far, we have considered the effects of single genes on the outcome of traits. This approach helps us to understand the various ways that alleles influence traits. Researchers often examine the effects of a single gene on the outcome of a single trait as a way to simplify the genetic analysis. For example, Mendel studied one gene that affected the height of pea plants—a gene with tall and short alleles. Actually, many other genes in pea plants also affect height, but Mendel did not happen to study variants in those other height genes. How then did Mendel study the effects of a single gene? The answer lies in the genotypes of his strains. Although many genes affect the height of pea plants, Mendel chose true-breeding strains that differed with regard to only one of those genes. As a hypothetical example, let's suppose that pea plants have 10 genes affecting height, which we will call *K, L, M, N, O, P, Q, R, S,* and *T*. The genotypes of two hypothetical strains of pea plants may be

Tall strain: *KK LL MM NN OO PP QQ RR SS TT*
Short strain: *KK LL MM NN OO PP QQ RR SS tt*

In this example, the alleles affecting height differ at only a single gene. One strain is *TT* and the other is *tt*, and this accounts for the difference in their height. If we make crosses between these tall and short strains, the genotypes of the F₂ offspring may differ with regard to only one gene; the other nine genes will be identical in all of them. This approach allows a researcher to study the

TABLE 4.3

Types of Mendelian Inheritance Patterns Involving Two Genes

Type	Description
Epistasis	An inheritance pattern in which the alleles of one gene mask the phenotypic effects of the alleles of a different gene.
Complementation	A phenomenon in which two parents that express the same or similar recessive phenotypes produce offspring with a wild-type phenotype.
Gene modification	A phenomenon in which an allele of one gene modifies the phenotypic outcome of the alleles of a different gene.
Gene redundancy	A phenomenon in which the loss of function in a single gene has no phenotypic effect, but the loss of function of two genes has an effect. Functionality of only one of the two genes is necessary for a wild-type phenotype; the genes are functionally redundant.

effects of a single gene even though many genes may affect a single trait.

Researchers now appreciate that essentially all traits are affected by the contributions of many genes. Morphological features such as height, weight, growth rate, and pigmentation are all affected by the expression of many different genes in combination with environmental factors. In this section, we will further our understanding of genetics by considering how the allelic variants of two different genes affect a single trait. This phenomenon is known as **gene interaction. Table 4.3** describes several examples of inheritance patterns in which two different genes interact to influence the outcome of particular traits. In this section, we will examine these examples in greater detail.

Gene Interaction Can Exhibit Epistasis and Complementation

In the early 1900s, William Bateson and Reginald Punnett discovered an unexpected gene interaction when studying crosses involving the sweet pea, *Lathyrus odoratus.* The wild sweet pea has purple flowers. However, these researchers obtained several true-breeding mutant varieties with white flowers. Not surprisingly, when they crossed a true-breeding purple-flowered plant to a true-breeding white-flowered plant, the F₁ generation had all purple-flowered plants and the F₂ generation (produced by self-fertilization of the F₁ generation) consisted of purple- and white-flowered plants in a 3:1 ratio. These results indicated that white flowers are recessive to purple.

A surprising result came in an experiment where Bateson and Punnett crossed two different varieties of white-flowered plants (**Figure 4.17**). All of the F₁ generation plants had purple flowers! The researchers then allowed the F₁ offspring to self-fertilize. The F₂ generation resulted in purple- and white-flowered plants in a ratio of 9 purple to 7 white. From this result, Bateson and Punnett deduced that two different genes were involved, with the following relationship:

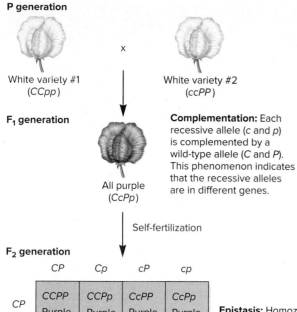

P generation

White variety #1
(*CCpp*)

x

White variety #2
(*ccPP*)

F₁ generation

All purple
(*CcPp*)

Complementation: Each recessive allele (*c* and *p*) is complemented by a wild-type allele (*C* and *P*). This phenomenon indicates that the recessive alleles are in different genes.

Self-fertilization

F₂ generation

	CP	Cp	cP	cp
CP	CCPP Purple	CCPp Purple	CcPP Purple	CcPp Purple
Cp	CCPp Purple	CCpp White	CcPp Purple	Ccpp White
cP	CcPP Purple	CcPp Purple	ccPP White	ccPp White
cp	CcPp Purple	Ccpp White	ccPp White	ccpp White

Epistasis: Homozygosity for the recessive allele of either gene results in a white phenotype, thereby masking the purple (wild-type) phenotype. Both gene products coded by the wild-type alleles (*C* and *P*) are needed for a purple phenotype.

INTERACTIVE EXERCISE

FIGURE 4.17 **A cross between two different white varieties of the sweet pea.**

Genes→Traits The color of the sweet pea flower is controlled by two genes, which are epistatic to each other and show complementation. Each gene is necessary for the production of an enzyme required for pigment synthesis. The recessive allele of either gene codes a defective enzyme. If an individual is homozygous for the recessive allele of either of the two genes, the purple pigment cannot be synthesized. The result is a white phenotype.

CONCEPT CHECK: What do the terms *epistasis* and *complementation* mean?

- *C* (one purple-color-producing) allele is dominant to *c* (white).
- *P* (another purple-color-producing) allele is dominant to *p* (white).
- *cc* or *pp* masks the *P* or *C* allele, producing white color.

When the alleles of one gene mask the phenotypic effects of the alleles of another gene, the phenomenon is called **epistasis.** Geneticists describe epistasis relative to a particular phenotype. If possible, geneticists use the wild-type phenotype as their reference phenotype when describing an epistatic interaction. In the case of sweet peas, purple flowers are wild type. Homozygosity for the

white allele of one gene masks the expression of the purple-producing allele of another gene. In other words, the *cc* genotype is epistatic to a purple phenotype, and the *pp* genotype is also epistatic to a purple phenotype. At the level of genotypes, *cc* is epistatic to *PP* or *Pp*, and *pp* is epistatic to *CC* or *Cc*. This is an example of **recessive epistasis.** As seen in Figure 4.17, this epistatic interaction produces only two phenotypes—purple or white flowers—in a 9:7 ratio.

Epistasis often occurs because two (or more) different proteins participate in a common function. For example, two (or more) proteins may be part of an enzymatic pathway leading to the formation of a single product. To illustrate this idea, let's consider the formation of a purple pigment in the sweet pea flower:

$$\text{colorless precursor} \xrightarrow{\text{Enzyme C}} \text{colorless intermediate} \xrightarrow{\text{Enzyme P}} \textbf{purple pigment}$$

In this example, a colorless precursor molecule must be acted on by two different enzymes to produce the purple pigment. Gene *C* codes a functional protein called enzyme C, which converts the colorless precursor into a colorless intermediate. Gene *P* codes a functional enzyme P, which converts the colorless intermediate into the purple pigment. If a plant is homozygous for either recessive allele (*cc* or *pp*), it will not make any functional enzyme C or enzyme P, respectively. When one of these enzymes is missing, purple pigment cannot be made, and the flowers remain white.

The parental cross shown in Figure 4.17 illustrates another genetic phenomenon called **complementation.** This term refers to the production of offspring with a wild-type phenotype by parents that both display the same or similar recessive phenotype. In the example shown in the figure, purple-flowered F₁ offspring were obtained from two white-flowered parents. Complementation typically occurs because the recessive phenotype in the parents is due to homozygosity of two different genes. In our sweet pea example, one parent is *CCpp* and the other is *ccPP*. In the F₁ offspring, the *C* and *P* alleles, which are wild-type and dominant, complement the *c* and *p* alleles, which are recessive. The offspring must have one wild-type allele of both genes to display the wild-type phenotype. Why is complementation an important experimental observation? When geneticists observe complementation in a genetic cross, this outcome suggests that the recessive phenotype in the two parent strains is caused by mutant alleles in two different genes.

Feather Coloration in Parakeets Provides an Example of Gene Modification

The Australian parakeet (*Melopsittacus undulatus*), also called a budgerigar or a "budgie," is a small, slender parrot that is native to Australia. Since the late 1800s, it has become a common house pet worldwide. The wild-type parakeet has a yellow face and a green underside (see upper left side of **Table 4.4**). These colors are caused by psittacofulvin and eumelanin—two types of pigments. In parakeets, psittacofulvin is a yellow pigment and eumelanin is black.

TABLE 4.4
Feather Coloration in Parakeets

Phenotype:				
Genotype:	*YYAA* (wild type)	*yyAA*	*YYaa*	*yyaa*
Pigment in the feathers on the face:	psittacofulvin	none	psittacofulvin	none
Face color:	yellow	white	yellow	white
Pigment(s) in the feathers on the underside:	psittacofulvin and eumelanin	eumelanin	psittacofulvin	none
Underside color:	green	blue	yellow	white

left: Dlb08d/Getty Images; (second from left) Devonyu/Getty Images; (second from right) suppixz/Getty Images; right: Cheryl Bridges/Alamy Stock Photo

Psittacofulvin and eumelanin are distributed in the feathers in varying patterns across the bird's body. With regard to feather coloration on the face and underside, wild-type parakeets have a base body color that is yellow due to psittacofulvin pigmentation. In addition, a secondary coloration occurs on the underside of the bird, but not on the face. The secondary coloration is due to eumelanin synthesis. Although eumelanin is a black pigment, tiny air pockets and eumelanin crystals reflect only blue light and absorb the other wavelengths. The reflected blue color combined with the yellow color from psittacofulvin results in a green color.

This green feather coloration provides an example of **gene modification**—the phenomenon in which an allele of one gene modifies the phenotypic outcome of the alleles of a different gene. In this case, the blue color modifies yellow to green on the undersides of the bird.

How does allelic variation in these two genes affect the phenotypes of parakeets (Table 4.4)? The gene for psittacofulvin synthesis is designated *Y* for yellow. The dominant allele (*Y*) results in psittacofulvin synthesis, whereas a loss-of-function recessive allele (*y*) prevents synthesis of that pigment. The eumelanin gene is typically designated *A* for albino, because its loss of function in many different species results in an albino phenotype. The dominant allele (*A*) results in eumelanin synthesis and the loss-of-function recessive allele (*a*) prevents that synthesis. As seen in Table 4.4, in places where only psittacofulvin is synthesized, the feathers are yellow. If only eumelanin is made, the feathers are blue. If both pigments are made, they are green. If neither pigment is made, the feathers are white.

Parakeets have been bred in captivity for over a century. During that time, over 30 new alleles in several different genes have been identified that have a gene modifying effect on feather color. Some of these alleles have a direct effect on pigment synthesis. Alternatively, other alleles may affect feather structure in a way that causes light refraction; this refraction alters the wavelength of light and thereby changes the feather color.

The alleles that affect feather color can be dominant or recessive or exhibit incomplete dominance. An interesting example of incomplete dominance occurs with an allele called dark. Let's designate the wild-type allele for this gene as *D* and the dark allele as *D'* (*D* prime). The gene-modifying effect of the dark allele on the underside color depends on whether it is present in one or two copies, and also on whether psittacofulvin synthesis occurs.

- *YYAADD*: green underside (see wild-type parakeet at the upper left in Table 4.4)
- *YYAADD'*: dark green underside
- *YYAAD'D'*: olive underside
- *yyAADD*: blue underside (as in the image that is second from the left in Table 4.4)
- *yyAADD'*: cobalt underside
- *yyAAD'D'*: mauve underside

Due to Gene Redundancy, Loss-of-Function Alleles May Have No Effect on Phenotype

During the past several decades, researchers have discovered new kinds of gene interactions by studying model organisms such as *Escherichia coli* (a bacterium), *Saccharomyces cerevisiae* (baker's yeast), *Arabidopsis thaliana* (a flowering plant), *Drosophila melanogaster* (fruit fly), *Caenorhabditis elegans* (a nematode worm), and *Mus musculus* (the laboratory mouse). The isolation of mutant alleles that alter the phenotypes of these organisms has become a powerful tool for investigating gene function and has provided ways for researchers to identify new kinds of gene interactions. With the advent of modern molecular techniques (described in Chapters 20, 22, and 23), a common approach for investigating gene function is to intentionally produce loss-of-function alleles in a gene of interest. When a geneticist abolishes gene function by creating an organism that is homozygous for a loss-of-function allele, the resulting organism is said to have undergone a **gene knockout.**

Why are gene knockouts useful? The primary reason for making a gene knockout is to understand how that gene affects the structure and function of cells or the phenotypes of organisms. For example, if a researcher knocked out a particular gene in a mouse and the resulting animal was unable to hear, the researcher would suspect that the role of the functional gene is to promote the formation of ear structures that are vital for hearing.

Interestingly, by studying many gene knockouts in a variety of experimental organisms, geneticists have discovered that many knockouts have no obvious effect on phenotype at the cellular level or the level of discernible traits. To explore gene function further, researchers may make two or more gene knockouts in the same organism. In some cases, gene knockouts in two different genes produce a phenotypic change even though the single knockouts

have no effect (**Figure 4.18**). Geneticists may attribute this change to **gene redundancy**—the phenomenon in which one gene compensates for the loss of function of another gene.

Gene redundancy may be due to different underlying causes:

- One common reason is gene duplication. Certain genes have been duplicated during evolution, so a species may have two or more copies of similar genes. These copies, which are not identical due to the accumulation of random changes during evolution, are called **paralogs** (look ahead to Figures 8.6 and 8.7). When one gene is missing, a paralog may be able to carry out the missing function. For example, genes *A* and *B* in Figure 4.18 could be paralogs of each other.
- Alternatively, gene redundancy may involve proteins that are involved in a common cellular function. When one of the proteins is missing due to a gene knockout, the function of another protein may be increased to compensate for the missing protein and thereby overcome the defect.

Let's explore the consequences of gene redundancy in a genetic cross. George Shull conducted one of the first studies that illustrated the phenomenon of gene redundancy. This work involved a weed known as shepherd's purse, a member of the mustard family. The trait Shull followed was the shape of the seed capsule, which is commonly triangular (**Figure 4.19**). Strains producing smaller ovate capsules are homozygous for loss-of-function alleles in two different genes (*ttvv*). The ovate strain is an example of a double gene knockout.

When Shull crossed a true-breeding plant with triangular capsules to a plant having ovate capsules, the F₁ generation all had triangular capsules. When the F₁ plants were self-fertilized,

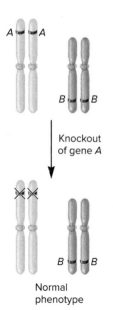

Knockout of gene *A*

Normal phenotype

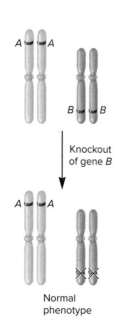

Knockout of gene *B*

Normal phenotype

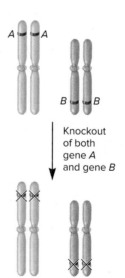

Knockout of both gene *A* and gene *B*

Altered phenotype— genes *A* and *B* are redundant.

FIGURE 4.18 A molecular explanation for gene redundancy. To have a normal phenotype, an organism must have a functional copy of either gene *A* or gene *B*. If both gene *A* and gene *B* are knocked out, an altered phenotype occurs.

CONCEPT CHECK: Explain why a single gene knockout does not always have an effect on phenotype.

P generation

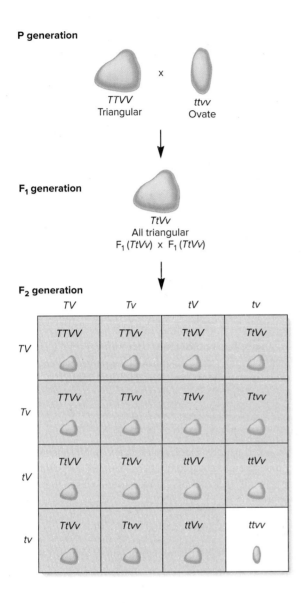

TTVV
Triangular

×

ttvv
Ovate

F₁ generation

TtVv
All triangular
F₁ (*TtVv*) × F₁ (*TtVv*)

F₂ generation

	TV	Tv	tV	tv
TV	TTVV	TTVv	TtVV	TtVv
Tv	TTVv	TTvv	TtVv	Ttvv
tV	TtVV	TtVv	ttVV	ttVv
tv	TtVv	Ttvv	ttVv	ttvv

FIGURE 4.19 Inheritance of seed capsule shape in shepherd's purse, an example of gene redundancy. In this case, triangular seed capsule shape requires a dominant allele in one of two genes, but not both. The *T* and *V* alleles are redundant.

CONCEPT CHECK: At the molecular level (with regard to loss-of-function alleles), explain why the *ttvv* homozygote has an ovate seed capsule.

a surprising result came in the F₂ generation. Shull observed a 15:1 ratio of plants having triangular capsules to ovate capsules. The result can be explained by gene redundancy. Having at least one functional copy of either gene (*T* or *V*) is sufficient to produce the triangular phenotype. *T* and *V* are functional alleles of redundant genes. Only one of them is necessary for a triangular shape. When the functions of both genes are knocked out, as in the *ttvv* homozygote, the capsule becomes smaller and ovate.

4.9 COMPREHENSION QUESTIONS

1. Two different strains of sweet peas are true-breeding and have white flowers. When plants of these two strains are crossed, the F₁ offspring all have purple flowers. The outcome of purple flowers in the F₁ generation is due to
   a. epistasis.
   b. complementation.
   c. incomplete dominance.
   d. incomplete penetrance.

2. If the F₁ offspring from question 1 are allowed to self-fertilize, what is the expected outcome for the F₂ offspring?
   a. All white
   b. All purple
   c. 3 purple : 1 white
   d. 9 purple : 7 white

KEY TERMS

Introduction: Mendelian inheritance, simple Mendelian inheritance
4.2: wild-type alleles, genetic polymorphism, mutant alleles, loss-of-function alleles, gain-of-function mutations, dominant-negative mutations, haploinsufficiency, incomplete penetrance, expressivity
4.3: temperature-sensitive allele, norm of reaction
4.4: incomplete dominance, heterozygote advantage, multiple alleles, codominance
4.5: sex chromosomes, X-linked inheritance, hemizygous, X-linked recessive, reciprocal cross, sex-linked gene, X-linked

genes, Y-linked genes (holandric genes), pseudoautosomal inheritance
4.6: sex-influenced inheritance, sex-limited inheritance, sexual dimorphism
4.7: lethal allele, essential gene, nonessential genes, conditional lethal alleles, temperature-sensitive (ts) lethal alleles, semi-lethal alleles, age of onset
4.8: pleiotropy
4.9: gene interaction, epistasis, recessive epistasis, complementation, gene modification, gene knockout, gene redundancy, paralogs

CHAPTER SUMMARY

- Mendelian inheritance patterns obey Mendel's laws.

4.1 Overview of Mendelian Inheritance Patterns

- Several inheritance patterns involving single genes differ from those observed by Mendel (see Table 4.1).

4.2 Dominant and Recessive Alleles

- Wild-type alleles are those that are prevalent in a natural population. When a gene exists in two or more wild-type alleles in a population, the phenomenon is called genetic polymorphism (see Figure 4.1).
- Recessive alleles are often due to mutations that result in a reduction or loss of function of the coded protein (see Figure 4.2 and Table 4.2).
- Dominant mutant alleles are most commonly caused by gain-of-function mutations, dominant-negative mutations, or haploinsufficiency.
- Incomplete penetrance occurs when an allele that is expected to be expressed is not expressed (see Figure 4.3).
- Traits may vary in their expressivity.

4.3 Environmental Effects on Gene Expression

- The outcome of traits is influenced by the environment (see Figure 4.4).

4.4 Incomplete Dominance, Heterozygote Advantage, and Codominance

- Incomplete dominance is an inheritance pattern in which the heterozygote has an intermediate phenotype (see Figure 4.5).
- Whether we consider an allele to be dominant or incompletely dominant may depend on how closely we examine the phenotype (see Figure 4.6).
- Heterozygote advantage is an inheritance pattern in which the heterozygote has greater reproductive success than either homozygote (see Figures 4.7, 4.8).
- Most genes exist in multiple alleles in a population. Some alleles, such as those that produce A and B blood antigens, are codominant (see Figure 4.9).

4.5 Genes on Sex Chromosomes

- X-linked inheritance patterns show differences between males and females and are revealed in reciprocal crosses (see Figures 4.10, 4.11).
- The X and Y chromosomes carry different sets of genes, but they do have short regions of homology that can lead to pseudoautosomal inheritance (see Figure 4.12).

4.6 Sex-Influenced and Sex-Limited Inheritance

- For sex-influenced traits, heterozygous males and females have different phenotypes (see Figure 4.13).
- Sex-limited traits are expressed in only one sex, thereby resulting in sexual dimorphism (see Figure 4.14).

4.7 Lethal Alleles

- Lethal alleles most commonly occur in essential genes.
- Lethal alleles may result in inheritance patterns that yield unexpected ratios of phenotypes (see Figure 4.15).

4.8 Understanding Complex Phenotypes Caused by Mutations in Single Genes

- Single genes usually exhibit pleiotropy, which means that they exert multiple phenotypic effects.
- Some phenotypes are best understood within the context of development (see Figure 4.16).

4.9 Gene Interactions

- A gene interaction is the phenomenon in which two or more genes affect a single phenotype (see Table 4.3).
- Epistasis occurs when the alleles of one gene mask the phenotypic expression of the alleles of a different gene. Complementation is the phenomenon in which two individuals with similar recessive phenotypes produce offspring with a wild-type phenotype (see Figure 4.17).
- Feather coloration in parakeets is an example of gene modification (see Table 4.4).
- Two different genes may have redundant functions, which is revealed by a double gene knockout (see Figures 4.18, 4.19).

PROBLEM SETS & INSIGHTS

MORE GENETIC TIPS 1. In Ayrshire cattle, the coat can be either red and white or mahogany and white. Both phenotypes are viable and have equal reproductive success. The mahogany and white phenotype is caused by the allele S^M. The red and white phenotype is controlled by the allele S^R. The following table shows the relationship between genotype and phenotype for females and males:

Genotype	Phenotype Females	Males
$S^M S^M$	Mahogany and white	Mahogany and white
$S^M S^R$	Red and white	Mahogany and white
$S^R S^R$	Red and white	Red and white

Explain the pattern of inheritance.

T **OPIC:** *What topic in genetics does this question address?* The topic concerns patterns of Mendelian inheritance. More specifically, the question asks you to determine which of several patterns (described in Table 4.1) is demonstrated by the coat coloration in Ayrshire cattle.

I **NFORMATION:** *What information do you know based on the question and your understanding of the topic?* From the question, you know the relationship between genotype, phenotype, and sex with regard to coat colors in Ayrshire cattle. From your understanding of the topic, you may remember that certain inheritance patterns result in differences between males and females.

P **ROBLEM-SOLVING** **S** **TRATEGY:** *Compare and contrast.* One strategy to solve this problem is to compare and contrast these results with the inheritance patterns described in Table 4.1. This allows you to rule out certain patterns. The information displayed in the table is not consistent with simple Mendelian inheritance because male and female heterozygotes differ in phenotypes. Because you are not given a pedigree, you don't have evidence for incomplete penetrance. The pattern is not incomplete dominance or heterozygote advantage, because the heterozygote does not have an intermediate phenotype or greater reproductive success, respectively. The pattern does not exhibit codominance, because the heterozygote is not expressing two phenotypes uniquely. It can't be X-linked because males carry two copies of the gene. It is not sex-limited inheritance because neither phenotype is unique to a particular sex. Finally, the alleles are not lethal.

ANSWER: The inheritance pattern for this trait is sex-influenced inheritance. The S^M allele is dominant in males but recessive in females, whereas the S^R allele is dominant in females but recessive in males.

---

2. The following pedigree represents a family in which a single gene causes an inherited disease. (Affected individuals are shown as filled symbols.) Assuming that incomplete penetrance is *not* occurring, indicate which of the following inheritance patterns is/are *not* possible in this case, and explain why.

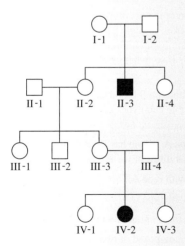

A. Recessive

B. Dominant

C. X-linked recessive

D. Sex-influenced, dominant in females

E. Sex-limited, recessive in females

T **OPIC:** *What topic in genetics does this question address?* The topic is different patterns of Mendelian inheritance. More specifically, the question asks you to determine which of five patterns is/are *not* possible.

I **NFORMATION:** *What information do you know based on the question and your understanding of the topic?* In the question, you are given a pedigree involving a human genetic disorder. From your understanding of the topic, you may remember how the five patterns of inheritance differ from each other.

P **ROBLEM-SOLVING** **S** **TRATEGIES:** *Analyze data. Predict the outcome.* To solve this problem, you need to analyze the pedigree and determine if the offspring produced by each set of parents are consistent with any of the five patterns of inheritance. In other words, you need to predict what types of offspring each set of parents could produce for the five inheritance patterns. This allows you to rule out certain patterns.

ANSWER:

A. It could be recessive.

B. It cannot be dominant (and completely penetrant) because both affected offspring have two unaffected parents.

C. It cannot be X-linked recessive because IV-2 is an affected female. An affected female would have to inherit the disease-causing allele from both parents. Because males are hemizygous for X-linked traits, this means that III-4 would also be affected. However, III-4 is unaffected.

D. Note: If a sex-influenced allele is dominant in females, it must be recessive in males. This allele cannot be sex-influenced and dominant in females because individual II-3 is an affected male and would have to be homozygous for the disease-causing allele if it was recessive in males. If so, I-1 would have to carry at least one copy of the disease-causing allele, and therefore, I-1, who is a female, would be affected if the allele was dominant in females. However, I-1 is not affected with the disease.

E. It cannot be sex-limited because individual II-3 is an affected male and IV-2 is an affected female.

---

3. As shown in Figure 4.9, a gene in humans that occurs as the *i*, I^A, and I^B alleles codes a glycosyl transferase that is involved in attaching galactose or *N*-acetylgalactosamine to an oligosaccharide on the surface of red blood cells. In addition, another gene, called the *H* gene, codes a different glycosyl transferase that is needed to attach the sugar fucose onto the oligosaccharide and thereby make H antigen, which is the antigen found in people with type O blood.

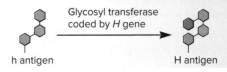

The *H* gene exists as the common allele, *H*, and as a very rare, recessive, loss-of-function allele, *h*. An individual who is *hh* is unable to attach fucose, and the resulting oligosaccharide, which is smaller, is called h antigen. A female with blood type O has an offspring with type B blood. The offspring's genotype is $I^B i$. Surprisingly, the male parent of this offspring does not have type B or type AB blood. The offspring's I^B allele is not due to a new mutation. How is this possible? In your answer, identify the possible genotypes of the parents.

TOPIC: *What topic in genetics does this question address?* The topic is Mendelian patterns of inheritance. More specifically, the question asks you to form a hypothesis to explain an unexpected genetic outcome.

INFORMATION: *What information do you know based on the question and your understanding of the topic?* In the question, you have learned about a gene that codes a glycosyl transferase that attaches fucose to the oligosaccharide on the surface of red blood cells. You also are given information about the blood types of the parents and their offspring, and you are given the genotype of the offspring. From your understanding of the topic, you may remember that another gene is involved with blood type (see Figure 4.9). You know that *i* is recessive to I^B. Even though the male parent does not have type B or AB blood, you know that the offspring must have received the I^B allele from this male parent. This seems mysterious.

PROBLEM-SOLVING **S**TRATEGIES: *Relate structure and function. Propose a hypothesis. Predict the outcome.* It

appears that the offspring inherited the I^B allele from the male parent. Your task is to understand why the male parent is not expressing the I^B allele. Because the expression of the dominant I^B allele is being masked, this is a case of epistasis.

One strategy to begin to solve this problem is to think about the relationship between the structure of the oligosaccharide and the functions of the two types of glycosyl transferases. The glycosyl transferase coded by the I^B allele recognizes H antigen and attaches a galactose to the oligosaccharide. The structure of H antigen is important because it allows the glycosyl transferase coded by the I^B allele to recognize the oligosaccharide and attach an additional galactose. If the oligosaccharide is smaller because it is missing fucose, it will not be recognized by the glycosyl transferase coded by the I^B allele. To predict the outcome of this cross, you could assume that the female parent is *HHii* and the male parent is *hh* and carries at least one copy of the I^B allele. If that were the case, the offspring would be heterozygous, *Hh*, and would make H antigen.

ANSWER: One hypothesis to explain these results is that the male parent is *hh* and that the glycosyl transferase that attaches galactose is unable to recognize the smaller oligosaccharide on the surface of red blood cells. Another way of saying this is that the *hh* genotype is epistatic to I^B. To account for the offspring receiving the I^B allele, the male parent's genotype could be $hhI^B i$, $hhI^B I^B$, or $hhI^A I^B$. The female parent's genotype is *HHii*, and the offspring's genotype is $HhI^B i$. Note: The *hh* genotype, which is very rare, is also called the Bombay genotype because the first case was reported in Mumbai (formerly Bombay).

Conceptual Questions

C1. Describe the differences among dominance, incomplete dominance, codominance, and heterozygote advantage.

C2. Discuss the differences among sex-influenced, sex-limited, and sex-linked inheritance. Give examples.

C3. What is meant by a gene interaction? How can a gene interaction be explained at the molecular level?

C4. Let's suppose a recessive allele codes a completely defective protein. If the functional allele is dominant, what does that tell you about the amount of the functional protein that is sufficient to cause the phenotype? What if the allele shows incomplete dominance?

C5. Over the course of a few months, the fur on an adult Siamese cat became noticeably lighter in color. During that time, was the tyrosinase enzyme functioning better or worse than it had been prior to the color change? Why? Provide two or more ideas for what the cat was experiencing to explain the change in fur color.

C6. An allele in *Drosophila* produces a star-eye trait in the heterozygous individual. However, the star-eye allele is lethal in homozygotes. What would be the phenotypic ratio of surviving offspring if star-eyed flies were crossed to each other?

C7. A seed dealer wants to sell four-o'clock seeds that will produce only a single color of flowers (red, white, or pink). Explain how this should be done.

C8. The blood serum from one individual (individual 1) is known to agglutinate the red blood cells from a second individual (individual 2). List the pairwise combinations of possible genotypes that individuals 1 and 2 could have. If individual 1 is the parent of individual 2, what are their possible genotypes?

C9. Which blood type phenotype(s) (A, B, AB, and/or O) can be produced only from a single genotype? Is it possible for a couple to produce a family of children in which all four blood types are represented? If so, what would the genotypes of the parents have to be?

C10. A female with type B blood has a child with type O blood. What are the possible genotypes and blood types of the male parent?

C11. A female with type A blood has parents who are type O and type A. If this type A female has children with a type AB male, what are the probabilities of the following offspring?

  A. A child with type AB

  B. A child with type O

  C. The first three children with type AB

  D. A family composed of two children with type B blood and one child with type AB

C12. In Shorthorn cattle, coat color is controlled by a single gene that can exist as a red allele (H^R) or a white allele (H^W). The heterozygotes ($H^R H^W$) have a color called roan that looks less red than the color of the $H^R H^R$ homozygotes. However, when examined carefully, the roan phenotype is actually due to a mixture of completely red hairs and completely white hairs. Should this effect be called incomplete dominance, codominance, or something else? Explain your reasoning.

C13. In chickens, the Leghorn variety has white feathers due to an autosomal dominant allele. Silkies have white feathers due to a recessive allele in a second (different) gene. If a true-breeding white Leghorn is crossed to a true-breeding white Silkie, what is the expected phenotype of the F₁ generation? If members of the F₁ generation are mated to each other, what is the expected phenotypic ratio of the F₂ generation? Assume that the chickens in the parental generation are homozygous for the white allele of one gene and homozygous for the brown allele of the other gene. In subsequent generations, nonwhite birds will be brown.

C14. Propose the most likely mode of inheritance (autosomal dominant, autosomal recessive, or X-linked recessive) for the disease represented in the following pedigrees. Affected individuals are shown as filled (black) symbols.

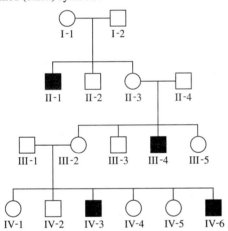

(A)

(B)

C15. A human disease known as vitamin D–resistant rickets is inherited as an X-linked dominant trait. If a male with the disease has children with a female who does not have the disease, what is the expected ratio of affected and unaffected offspring?

C16. Hemophilia is an X-linked recessive disorder in humans. If a heterozygous female has children with an unaffected male, what is the probability of each of the following combinations of offspring?

A. An affected male offspring

B. Four unaffected offspring in a row

C. An unaffected female or male offspring

D. Two affected offspring out of five

C17. Incontinentia pigmenti, a rare, X-linked dominant disorder in humans, is characterized by swirls of pigment in the skin. If an affected female, who had an unaffected father, has children with an unaffected male, what are the predicted ratios of affected and unaffected female and male offspring?

C18. Scurs in cattle is a sex-influenced trait. A cow with no scurs whose female parent had scurs has offspring with a bull with scurs whose male parent did not have scurs. What is the probability of each of the following for their offspring?

A. The first offspring will not have scurs.

B. The first offspring will be a male with no scurs.

C. The first three offspring will be females with no scurs.

C19. In rabbits, the color of body fat is controlled by a single gene with two alleles, designated Y and y. The outcome of this trait is affected by the diet of the rabbit. When raised on a standard vegetarian diet, the dominant Y allele confers white body fat, and the y allele confers yellow body fat. However, when raised on a xanthophyll-free diet, a homozygote yy rabbit has white body fat. If a heterozygous rabbit is crossed to a rabbit with yellow body fat and the offspring are raised on a standard vegetarian diet, what are the proportions of offspring with white and yellow body fat? How do the proportions change if the offspring are raised on a xanthophyll-free diet?

C20. In cats, a temperature-sensitive allele produces the Siamese phenotype, in which the cooler extremities are dark and the warmer trunk area is lighter. A Siamese cat that spends most of its time outside was accidentally injured in a trap and required several stitches in its right front paw. The veterinarian had to shave the fur from the paw and leg, which originally had rather dark fur. Later, when the fur grew back, it was much lighter than the fur on the other three legs. Do you think the injury occurred in the hot summer or cold winter? Explain your answer.

C21. The trait of feathering in fowls is a sex-limited trait controlled by a single gene. Females always exhibit hen-feathering, as do HH and Hh males. Only hh males show cock-feathering. Starting with a female and a male that are both heterozygous, explain how you could obtain a true-breeding line in which the male offspring always had cock feathers.

C22. Based on the following pedigree for a trait determined by a single gene (affected individuals are shown as filled symbols), state whether it would be possible for the trait to be inherited in each of the following ways:

A. Recessive

B. X-linked recessive

C. Dominant, complete penetrance

D. Sex-influenced, dominant in males

E. Sex-limited

F. Dominant, incomplete penetrance

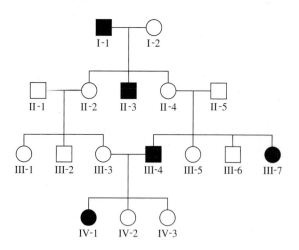

C23. The pedigree shown next involves a trait determined by a single gene (affected individuals are shown as filled symbols). Which of the following patterns of inheritance are possible for this trait?

A. Recessive

B. X-linked recessive

C. Dominant

D. Sex-influenced, recessive in males

E. Sex-limited

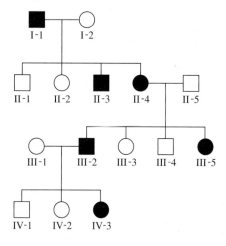

C24. Let's suppose you have pedigree data from thousands of different families concerning a particular genetic disease. How would you decide whether the disease is inherited as a recessive trait as opposed to being dominant with incomplete penetrance?

C25. Compare phenotypes at the molecular, cellular, and organism levels for individuals who are homozygous for the hemoglobin A allele, $Hb^A Hb^A$, and the sickle cell allele, $Hb^S Hb^S$.

C26. In humans, a very rare dominant allele that causes the little finger to be crooked has a penetrance of 80%. In other words, 80% of heterozygotes carrying the allele will have a crooked little finger. If a homozygous unaffected person has children with a heterozygote carrying this mutant allele, what is the probability that an offspring will have a little finger that is crooked?

C27. A sex-influenced trait in humans affects the length of the index finger. A short allele is dominant in males and recessive in females. Heterozygous males have an index finger that is significantly shorter than the ring finger. In contrast, the long allele is dominant in females and recessive in males. The gene affecting index finger length is located on an autosome. A female with short index fingers has five children with a male who has long index fingers. Their order of birth is: female, male, male, female, male. The oldest female offspring and a male with long index fingers have one male offspring. The youngest male among the five children and a female with short index fingers have two male offspring. Draw the pedigree for this family. Indicate the phenotypes of every individual (filled symbols for individuals with short index fingers and open symbols for individuals with long index fingers).

C28. Three coat-color patterns that occur in some breeds of horses are termed cremello (beige), chestnut (brown), and palomino (golden with light mane and tail). If two palomino horses are mated, they produce offspring in a ratio of about 1/4 cremello : 1/4 chestnut : 1/2 palomino. In contrast, cremello horses and chestnut horses breed true. (In other words, two cremello horses will produce only cremello offspring, and two chestnut horses will produce only chestnut offspring.) Explain this pattern of inheritance.

Experimental Questions

E1. Mexican hairless dogs have little hair and few teeth. When a Mexican hairless is mated to another breed of dog, about half of the puppies are hairless. When two Mexican hairless dogs are mated to each other, about 1/3 of the surviving puppies have hair and about 2/3 of the surviving puppies are hairless. However, about two out of eight puppies from this type of cross are born grossly deformed and do not survive. Explain this pattern of inheritance.

E2. In chickens, some varieties have feathered shanks (legs), but others do not. In a cross between a Black Langshan (feathered shanks) and a Buff Rock (unfeathered shanks), the shanks of the F_1 generation are all feathered. When chickens from the F_1 generation are crossed to each other, the F_2 generation exhibits a 15:1 ratio of

chickens with feathered shanks to those with unfeathered shanks. Suggest an explanation for this result.

E3. In sheep, the formation of horns is a sex-influenced trait; the allele that results in horns is dominant in males and recessive in females. Females must be homozygous for the horned allele to have horns. A horned ram was crossed to a polled (unhorned) ewe, and the first offspring was a horned ewe. What are the genotypes of the parents?

E4. A particular breed of dog can have long hair or short hair. When true-breeding long-haired animals were crossed to true-breeding short-haired animals, the offspring all had long hair. The F_2 generation showed a 3:1 ratio of long- to short-haired offspring. A second gene affects the texture of the hair. The two variants are wiry hair and straight hair. F_1 offspring from a cross of these two varieties all had wiry hair, and F_2 offspring showed a 3:1 ratio of wiry-haired to straight-haired puppies. Recently, a breeder of the short- and wiry-haired dogs found a female puppy that was albino. Similarly, another breeder of the long- and straight-haired dogs found a male puppy that was albino. Because the albino trait is always due to a recessive allele, the two breeders got together and mated the two dogs. Surprisingly, all of the puppies in the litter had black hair. How can you explain this result?

E5. In the clover butterfly, males are always yellow, but females can be yellow or white. In females, white is a dominant allele. Two yellow butterflies were crossed to yield an F_1 generation consisting of 50% yellow males, 25% yellow females, and 25% white females. Describe how this trait is inherited, and identify the genotypes of the parents.

E6. The *Mic2* gene in humans is present on both the X and Y chromosomes. Let's suppose the *Mic2* gene exists in a dominant *Mic2* allele, which results in normal surface antigen production, and a recessive *mic2* allele, which results in defective surface antigen production. Using molecular techniques, it is possible to distinguish homozygous and heterozygous individuals. By following the transmission of the *Mic2* and *mic2* alleles in a large human pedigree, would it be possible to distinguish between pseudoautosomal inheritance and autosomal inheritance? Explain your answer.

E7. Duroc Jersey pigs are typically red, but a sandy phenotype also occurs. When two different strains of true-breeding sandy pigs were crossed to each other, they produced F_1 offspring that were red. When these F_1 offspring were crossed to each other, they produced red, sandy, and white pigs in a 9:6:1 ratio. Explain this pattern of inheritance.

E8. As shown in Table 4.4, feather color in parakeets depends on two pigments, coded by the *Y* and *A* genes. If a *YyAa* male is crossed to a *yyAa* female, what is the predicted ratio of offspring phenotypes?

E9. Summer squash exist in long (i.e., cylindrical), round, or disk shapes. When a true-breeding long-shaped strain was crossed to a true-breeding disk-shaped strain, all the F_1 offspring were disk-shaped. When the F_1 offspring were allowed to self-fertilize, the F_2 generation consisted of a ratio of 9 disk-shaped to 6 round-shaped to 1 long-shaped. Assuming that the shape of summer squash is governed by two different genes, with each gene existing in two alleles, propose a mechanism to account for this 9:6:1 ratio.

E10. In a species of plant, two genes control flower color. The red allele (*R*) is dominant to the white allele (*r*); the color-producing allele (*C*) is dominant to the non-color-producing allele (*c*). You suspect that either an *rr* homozygote or a *cc* homozygote will produce white flowers. In other words, *rr* is epistatic to *C*, and *cc* is

epistatic to *R*. To test your hypothesis, you allow heterozygous plants (*RrCc*) to self-fertilize and count the phenotypes of the offspring. You obtain the following data: 201 plants with red flowers and 144 with white flowers. Conduct a chi square test to see if your observed data are consistent with your hypothesis.

E11. Red eyes is the wild-type phenotype in *Drosophila*, and several different genes (with each gene existing in two or more alleles) affect eye color. One allele causes purple eyes, and a different allele causes vermilion eyes. Both of these alleles are recessive to the red eye color allele. Two types of crosses provided the following results:

Cross 1: Males with vermilion eyes × females with purple eyes

  354 offspring, all with red eyes

Cross 2: Males with purple eyes × females with vermilion eyes

  212 male offspring, all with vermilion eyes

  221 female offspring, all with red eyes

Explain the pattern of inheritance based on these results. What additional crosses might you make to confirm your hypothesis?

E12. As mentioned in question E11, red eyes is the wild-type phenotype in *Drosophila*. Several different genes (with each gene existing in two or more alleles) are known to affect eye color. One allele causes purple eyes, and a different allele causes sepia eyes. Both of these alleles are recessive to the red eye color allele. When flies with purple eyes were crossed to flies with sepia eyes, all of the F_1 offspring had red eyes. When the F_1 offspring were allowed to mate with each other, the following results were obtained:

  146 purple eyes

  151 sepia eyes

  50 purplish sepia eyes

  444 red eyes

Explain this pattern of inheritance. Conduct a chi square test to see if the experimental data fit your hypothesis.

E13. Let's suppose you were looking at a bottle containing fruit flies in your laboratory and noticed a male fly with pink eyes. What crosses would you make to determine if the pink allele is on the X chromosome? What crosses would you make to determine if the pink allele is an allele of the same X-linked gene that occurs as a white allele?

E14. When examining a human pedigree, what features do you look for to distinguish between X-linked recessive inheritance and autosomal recessive inheritance? How would you distinguish X-linked dominant inheritance from autosomal dominant inheritance in a human pedigree?

Questions for Student Discussion/Collaboration

1. Let's suppose a gene exists as a functional wild-type allele and a nonfunctional mutant allele. At the organism level (i.e., at the level of visible traits), the wild-type allele is dominant. In a heterozygote, discuss whether dominance occurs at the cellular or molecular level. Discuss examples in which the existence of dominance depends on the level of examination.

2. In oats, the color of the chaff is determined by a two-gene interaction. When a true-breeding black chaff plant was crossed to a true-breeding white chaff plant, the F_1 generation was composed of all

black chaff plants. When the F_1 offspring were crossed to each other, the phenotypic ratio of the F_2 generation was 12 black : 3 gray : 1 white. First, construct a Punnett square that accounts for this pattern of inheritance. Which genotypes produce the gray chaff phenotype? Second, at the level of protein function, how would you explain this type of inheritance?

Note: All answers are available for the instructor in Connect; the answers to the even-numbered questions and all of the Concept Check and Comprehension Questions are in Appendix B.

Shell coiling in the water snail, Lymnaea peregra. *In this species of snail, some shells coil to the left, and others coil to the right. These different phenotypes are due to an inheritance pattern called the maternal effect.*
©John Mendenhall Institute for Cellular and Molecular Biology, University of Texas at Austin

5 NON-MENDELIAN INHERITANCE

As discussed in Chapter 4, Mendelian inheritance patterns involve genes that directly influence the outcome of an offspring's traits and obey Mendel's laws. Most genes in eukaryotic species follow a Mendelian pattern of inheritance. To predict phenotypes, we must consider several factors. These include the dominant/recessive relationship of alleles, gene interactions that may affect the expression of a single trait, and the roles that sex and the environment play in influencing the individual's phenotype. Once these factors are understood, we can predict the phenotypes of offspring from their genotypes. Genes that follow a Mendelian inheritance pattern conform to four rules:

1. The expression of the genes in the offspring directly influences their traits.
2. Except in the case of rare mutations, the genes are passed unaltered from generation to generation.
3. The genes obey Mendel's law of segregation.
4. For crosses involving two or more genes, the genes obey Mendel's law of independent assortment.

In this and the following chapter, we will examine several additional types of inheritance patterns that deviate from a Mendelian pattern because one of these four rules is broken.

- We begin this chapter by analyzing a non-Mendelian pattern called maternal effect in which rule 1 is broken. Traits that follow a maternal effect inheritance pattern are controlled by gene products that are transported into egg cells, rather than by the genes of the offspring.
- We then turn our attention to epigenetic inheritance, which breaks rule 2, because the genes are altered in the offspring. As we will see, in certain types of epigenetic inheritance, genes are methylated, which alters their expression.
- Finally, in the last section of this chapter, we will examine inheritance patterns that arise because some genetic material is not located in the cell nucleus. Certain organelles, such as mitochondria and chloroplasts, contain their own genetic material. We will consider examples in which traits are determined by genes within these organelles. These traits do not conform to rule 3; they do not obey the law of segregation.
- In Chapter 6, we will examine inheritance patterns that do not conform to rule 4; they do not obey the law of independent assortment.

5.1 MATERNAL EFFECT

Learning Outcomes:

1. Define *maternal effect.*
2. Predict the outcome of crosses for genes that exhibit a maternal effect pattern of inheritance.
3. Explain the mechanism of maternal effect at the molecular and cellular levels.

We begin by considering genes that exhibit a **maternal effect.** For these genes, the genotype of the female parent directly determines the phenotype of the offspring. Surprisingly, for maternal effect genes, the genotypes of the male parent and the offspring themselves do not affect the phenotype of the offspring. Therefore, you cannot use a Punnett square to predict the phenotype of the offspring. We will see that maternal effect inheritance is explained by the accumulation of gene products that the female parent provides to developing oocytes (immature eggs).

The Genotype of the Female Parent Determines the Phenotype of the Offspring for Maternal Effect Genes

The first example of a maternal effect gene was studied in the 1920s by Arthur Boycott, and it affected morphological features of the water snail, *Lymnaea peregra*. In this species, the shell and internal organs can be arranged in either a right-handed (dextral) or left-handed (sinistral) direction (see the chapter-opening photo). The dextral orientation is more common and is dominant to the sinistral orientation.

Figure 5.1 describes the results of a genetic analysis of shell coiling in the water snail. In this experiment, Boycott began with two different true-breeding strains of snails, one with dextral morphology and the other with sinistral morphology. Many combinations of crosses produced results that could not be explained by a Mendelian pattern of inheritance.

F₁ Generation. When a dextral female (*DD*) was crossed to a sinistral male (*dd*), all F₁ offspring were dextral. However, in the **reciprocal cross,** where a sinistral female (*dd*) was crossed to a dextral male (*DD*), all F₁ offspring were sinistral. Taken together, these results contradict a Mendelian pattern of inheritance.

How can we explain the unusual results obtained in the experiment in Figure 5.1? Alfred Sturtevant proposed the idea that shell coiling is due to a maternal effect gene that exists as a dextral (*D*) or sinistral (*d*) allele. This proposal was based on the inheritance patterns of the F₂ and F₃ generations.

F₂ Generation. The genotype of the F₁ generation is expected to be heterozygous (*Dd*). When these F₁ individuals were crossed to each other, the predicted genotypic ratio for the F₂ generation was 1 *DD* : 2 *Dd* : 1 *dd*. Because the *D* allele is dominant to the *d* allele, a 3:1 phenotypic ratio of dextral to sinistral snails should be observed, according to a Mendelian pattern of inheritance. Instead of this predicted phenotypic ratio, however, the F₂ generation was composed of all dextral snails. This incongruity with Mendelian inheritance is

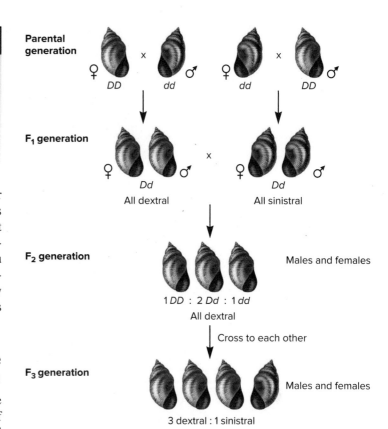

FIGURE 5.1 Experiment showing the inheritance pattern of shell coiling. In this species of snail, *D* (dextral) is dominant to *d* (sinistral). The genotype of the female parent determines the phenotype of the offspring. This phenomenon is known as the maternal effect. In this case, a *DD* or *Dd* female produces dextral offspring, and a *dd* female produces sinistral offspring. The genotypes of the male parent and the offspring do not affect the offspring's phenotype.

CONCEPT CHECK: Explain why all offspring in the F₂ generation are dextral even though some of them are *dd*.

due to the maternal effect. The phenotype of the offspring depended solely on the genotype of the female parent. The F₁ females were *Dd*. The *D* allele is dominant to the *d* allele and caused the offspring to be dextral, even when the offspring's genotype was *dd*!

F₃ Generation. When the members of the F₂ generation were crossed, the F₃ generation exhibited a 3:1 ratio of dextral to sinistral snails. This ratio corresponds to the genotypes of the F₂ females, which were the female parents of the F₃ generation. The ratio of F₂ females was 1 *DD* : 2 *Dd* : 1 *dd*. The *DD* and *Dd* females produced dextral offspring, whereas the *dd* females produced sinistral offspring. This explains the 3:1 ratio of dextral and sinistral offspring in the F₃ generation.

Oocytes Receive Gene Products That Affect Early Developmental Stages of the Embryo

At the molecular and cellular levels, the non-Mendelian inheritance pattern of maternal effect genes can be explained by the process of oogenesis in female animals (**Figure 5.2a**). As an

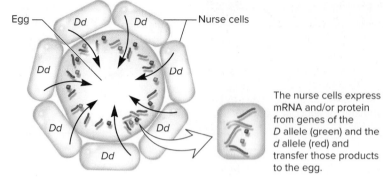

(a) Transfer of gene products from nurse cells to egg

The nurse cells express mRNA and/or protein from genes of the *D* allele (green) and the *d* allele (red) and transfer those products to the egg.

Female is *DD*.
Egg is *D*.

All offspring are dextral because the egg received the gene product of the *D* allele.

Female is *Dd*.
Egg can be *D* or *d*.

All offspring are dextral because the egg received the gene product of the dominant *D* allele.

Female is *dd*.
Egg is *d*.

All offspring are sinistral because the egg only received the gene product of the *d* allele.

(b) Maternal effect in shell coiling

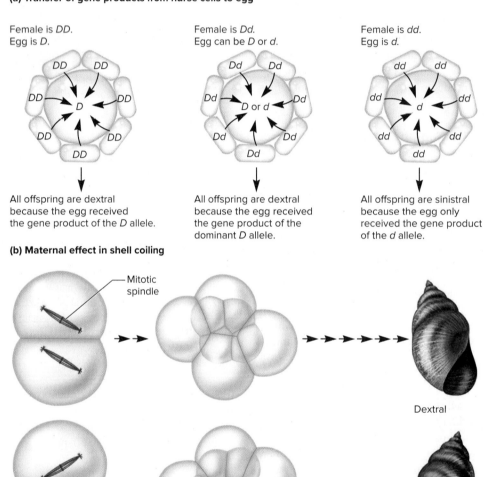

Mitotic spindle

Dextral

Sinistral

(c) An explanation of coiling direction at the cellular level

FIGURE 5.2 **The mechanism of maternal effect in shell coiling.** (a) Transfer of gene products from nurse cells to an oocyte. In this example, a female is heterozygous (*Dd*), which means the nurse cells are also heterozygous (*Dd*). Both the *D* and *d* alleles are expressed in the nurse cells to produce *D* and *d* gene products (mRNA or proteins, or both). These products are transported into the cytoplasm of the oocyte, where they accumulate to significant amounts. (b) Explanation of the maternal effect in shell coiling. (c) The direction of shell coiling is determined by differences in the cleavage planes during early embryonic development.

Genes→Traits If the nurse cells are *DD* or *Dd*, they will transfer the *D* gene product to the oocyte and thereby cause the resulting offspring to be dextral. If the nurse cells are *dd*, only the *d* gene product will be transferred to the oocyte, so the resulting offspring will be sinistral.

CONCEPT CHECK: If a female snail is heterozygous, *Dd*, which gene products will the oocyte receive?

animal oocyte matures, many surrounding maternal cells called nurse cells provide the oocyte with nutrients and other materials. In Figure 5.2a, a female is heterozygous for the shell-coiling maternal effect gene, with the alleles designated *D* and *d*. Depending on the outcome of meiosis, the haploid oocyte may receive the *D* allele or the *d* allele, but not both. The surrounding nurse cells, however, produce both *D* and *d* gene products (mRNA and proteins). These gene products are then transported into the oocyte. As shown in Figure 5.2a, the oocyte has received both the *D* gene product and the *d* gene product. These gene products persist for a significant time after the egg has been fertilized and embryonic development has begun. In this way, the gene products of the nurse cells, which reflect the genotype of the female parent, influence the early developmental stages of the embryo.

Now that you have an understanding of the relationship between oogenesis and maternal effect genes, let's reconsider the topic of snail shell coiling.

- As shown in **Figure 5.2b**, a female snail that is *DD* transmits only the *D* gene product to the oocyte via nurse cells. During the early stages of embryonic development, this gene product causes the embryo cleavage to occur in a way that produces a right-handed body plan.
- A heterozygous female transmits both *D* and *d* gene products. Because the *D* allele is dominant, the maternal effect also causes a right-handed body plan.
- A *dd* female contributes only the *d* gene product that produces a left-handed body plan, even if the egg is fertilized by a sperm carrying a *D* allele. The sperm's genotype is irrelevant, because the expression of the sperm's gene will occur too late.

The origin of dextral and sinistral coiling can be traced to the orientation of the mitotic spindle at the two- to four-cell stage of embryonic development. The dextral and sinistral snails develop as mirror images of each other (**Figure 5.2c**).

Since these initial studies, researchers have found that maternal effect genes code proteins that are important in the early steps of embryogenesis. The accumulation of maternal gene products in the oocyte allows embryogenesis to proceed quickly after fertilization. Maternal effect genes often play a role in cell division, cleavage pattern, and body axis orientation. Therefore, defective alleles in maternal effect genes tend to have a dramatic effect on the phenotype of the offspring, altering major features of morphology, often with dire consequences.

Our understanding of maternal effect genes has been greatly aided by their identification in experimental organisms such as *Drosophila melanogaster*. In such organisms with a short generation time, geneticists have successfully searched for mutant alleles that alter the normal process of embryonic development. In *Drosophila*, geneticists have identified several maternal effect genes with profound effects on the early stages of development. The pattern of development of a *Drosophila* embryo occurs along axes, such as the anteroposterior axis and the dorsoventral axis. The proper development along each axis requires a distinct set of maternal gene products.

For example, the maternal effect gene called *bicoid* produces a gene product that accumulates in a region of the oocyte that will become the anterior end of the embryo. After fertilization, the bicoid protein promotes the development of anterior structures in the developing embryo. A female that is homozygous for a defective *bicoid* allele will produce embryos that have two posterior ends; this is a lethal condition. More recently, several maternal effect genes that are required for proper embryonic development have been identified in mice and humans. Chapter 26 examines the relationships among the actions of several maternal effect genes during embryonic development.

GENETIC TIPS
THE QUESTION: A female snail has offspring whose shells all coil to the right. What are the possible genotypes of this female snail?

T OPIC: *What topic in genetics does this question address?* The topic is non-Mendelian inheritance. More specifically, the question is about maternal effect inheritance.

I NFORMATION: *What information do you know based on the question and your understanding of the topic?* From the question, you know a female snail has offspring whose shells all coil to the right. From your understanding of the topic, you may remember that this trait shows a maternal effect pattern of inheritance and that the female parent's genotype determines the offspring's phenotype.

P ROBLEM-SOLVING **S** TRATEGY: *Predict the outcome.* A strategy to solve this problem is to predict the outcome for each possible genotype of the female parent. For maternal effect genes, you cannot use a Punnett square to predict the phenotype of the offspring. Instead, to predict their phenotype, you have to know the female parent's genotype. In this question, you already know the offspring's phenotype. With this information, you can deduce the possible genotype(s) of the female parent. Because the offspring's shells coil to the right, you may realize that the female parent must have at least one *D* allele. If the female was *DD*, all of the offspring's shells would coil to the right. If the female was *Dd*, all of the offspring's shells would also coil to the right, because *D* is dominant. If the female was *dd*, all of the offspring's shells would coil to the left.

ANSWER: The female snail's genotype could be either *DD* or *Dd*.

5.1 COMPREHENSION QUESTIONS

1. A snail that we will call female A produces a female offspring that we will call female B. The shell of female B coils to the left. Female B has many offspring and all of their shells coil to the right. What are the genotypes of female A and female B, respectively?
   a. *DD, dd*
   b. *Dd, Dd*
   c. *Dd, dd*
   d. *dd, Dd*

2. What is the explanation for maternal effect inheritance at the molecular and cellular levels?
   a. The gene from the male parent is silenced at fertilization.
   b. During oogenesis, nurse cells transfer gene products to the oocyte.
   c. The gene products from the nurse cells exert effects on the embryo during very early stages of development.
   d. Both b and c are correct.

5.2 EPIGENETICS: DOSAGE COMPENSATION

Learning Outcomes:

1. Define *epigenetics*.
2. Compare and contrast the mechanisms of dosage compensation in different animal species.
3. Describe the process of X-chromosome inactivation in mammals.
4. Explain how X-chromosome inactivation may affect the phenotype of female mammals.

Epigenetics is the study of modifications that occur to a gene or chromosome and alter gene expression, but are not permanent over the course of many generations. As we will see, epigenetic effects can be the result of DNA and chromosomal modifications that occur during oogenesis, spermatogenesis, or early stages of embryogenesis. In this chapter, we will consider the general ways that epigenetic changes occur and how they affect traits. In Chapter 16, we will examine several molecular mechanisms of epigenetic changes in greater detail.

Once they are initiated, epigenetic changes alter the expression of particular genes in a way that may persist during an individual's lifetime. Therefore, epigenetic changes can permanently affect the phenotype of the individual. However, epigenetic modifications are not permanent over the course of many generations, and they do not change the actual DNA sequence. For example, a gene may undergo an epigenetic change that inactivates it for the lifetime of an individual. However, when this individual makes gametes, the gene may become activated and remain active during the lifetime of an offspring who inherits the gene.

In this section, we will examine an epigenetic change called dosage compensation, which has the effect of offsetting differences in the number of sex chromosomes. One of the sex chromosomes is altered, with the result that males and females have similar levels of gene expression even though they do not possess the same complement of sex chromosomes. We will largely focus on mammals in which dosage compensation is initiated during the early stages of embryonic development.

Dosage Compensation Results in Similar Levels of Gene Expression Between the Sexes

Dosage compensation refers to the phenomenon in which the level of expression of many genes on the sex chromosomes (such as the X chromosome) is similar in females and males even though the two sexes have a different complement of sex chromosomes. For example, female mammals are XX, whereas males are XY. The term *dosage compensation* was coined in 1932 by Hermann Muller to explain the effects of eye color mutations in *Drosophila*. Muller observed that female flies homozygous for certain X-linked eye color alleles had a phenotype similar to that of hemizygous males, which have only one copy of those alleles.

As a specific example, Muller noted that an X-linked gene conferring an apricot eye color produces a very similar phenotype in homozygous females and hemizygous males. In contrast, a female that has one copy of the apricot allele and a deletion of the apricot gene on the other X chromosome has eyes of paler color. Therefore, one copy of the allele in the female is not equivalent to

TABLE 5.1

Mechanisms of Dosage Compensation Among Different Species

Species	Sex Chromosomes in: Females	Males	Mechanism of Compensation
Placental mammals	XX	XY	One of the X chromosomes in the somatic cells of females is inactivated. In certain species, the X chromosome from the male parent is inactivated, and in other species, such as humans, either of the two X chromosomes is randomly inactivated throughout the somatic cells of females.
Marsupial mammals	XX	XY	The X chromosome from the male parent is inactivated in the somatic cells of females.
Drosophila melanogaster	XX	XY	The level of expression of genes on the X chromosome in males is doubled.
Caenorhabditis elegans	XX*	X0	The level of expression of genes on each X chromosome in hermaphrodites is decreased to 50% of the level occurring on the X chromosome in males.

*In *C. elegans*, an XX individual is a hermaphrodite, not a female.

one copy of the allele in the male. Instead, two copies of the allele in the female produce a phenotype that is similar to that produced by one copy in the male. In other words, the difference in gene dosage—two copies in females versus one copy in males—is being compensated for at the level of gene expression.

Since these initial studies, dosage compensation has been studied extensively in mammals, *Drosophila*, and *Caenorhabditis elegans* (a nematode). Depending on the species, dosage compensation occurs via different mechanisms (**Table 5.1**):

- Female mammals equalize the expression of X-linked genes by turning off one of their two X chromosomes. This process is known as **X-chromosome inactivation (XCI).**
- In *Drosophila*, the male accomplishes dosage compensation by doubling the expression of most X-linked genes.
- In *C. elegans*, the XX animal is a hermaphrodite that produces both sperm and egg cells, and an animal carrying a single X chromosome is a male that produces only sperm. The XX hermaphrodite diminishes the expression of X-linked genes on both X chromosomes to approximately 50% of that in the male.
- In birds, the Z chromosome is a large chromosome, usually the fourth or fifth largest, which contains almost all of the known sex-linked genes. The W chromosome is generally a much smaller chromosome containing a high proportion of repeat sequence DNA, which is discussed in Chapter 10, that does not code genes. Male birds are ZZ, and females are ZW. The expression of hundreds of Z-linked genes has been examined in chickens. The results suggest that chickens lack a general mechanism of dosage compensation that controls the expression of most Z-linked genes. Even so, the pattern of gene expression between males and females was found to vary a great deal for certain Z-linked genes. Overall, the results suggest that some Z-linked genes may be dosage-compensated, but many of them are not.

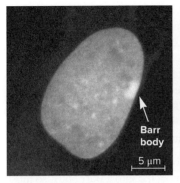
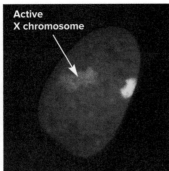

(a) Nucleus with a Barr body

(b) A calico cat

FIGURE 5.3 X-chromosome inactivation in female mammals. (a) The left micrograph shows the Barr body on the periphery of a human nucleus after staining with a DNA-specific dye. Because it is compact, the Barr body is more brightly visible than other chromosomes in the nucleus. The right micrograph shows the same nucleus using a yellow fluorescent probe that recognizes the X chromosome. The Barr body is more compact than the active X chromosome, which is to the left of the Barr body. **(b)** The fur pattern of a calico cat.

Genes→Traits The pattern of black and orange fur on this cat is due to random X-chromosome inactivation during embryonic development. The orange patches of fur are due to the inactivation of the X chromosome that carries a black allele; the black patches are due to the inactivation of the X chromosome that carries the orange allele. In general, only heterozygous female cats can be calico. A rare exception would be a male cat that has an abnormal composition of sex chromosomes (XXY).

(a, both): Irine Solovei, University of Munich (LMU); (b): cgbaldauf/Getty Images

CONCEPT CHECK: Why is the Barr body more brightly visible after staining by a DNA-specific dye than other chromosomes in the nucleus?

Dosage Compensation Occurs in Female Mammals by the Inactivation of One X Chromosome

In 1961, Mary Lyon proposed that dosage compensation in mammals occurs by the inactivation of a single X chromosome in females. Liane Russell proposed the same idea around the same time. This proposal brought together evidence from two lines of study. The first type of evidence came from cytological studies. In 1949, Murray Barr and Ewart Bertram identified a highly condensed structure in the interphase nuclei of somatic cells in female cats that was not found in male cats. This structure became known as the **Barr body** (**Figure 5.3a**). In 1960, Susumu Ohno

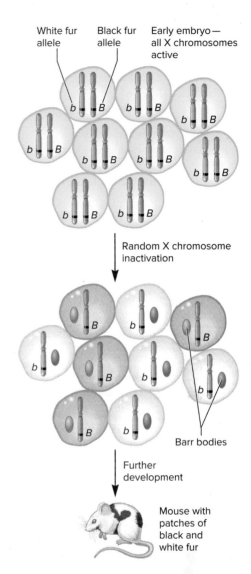

FIGURE 5.4 **The mechanism of X-chromosome inactivation.**

Genes→Traits The top of this figure represents a mass of several cells that compose the early embryo. Initially, both X chromosomes in each cell are active. At an early stage of embryonic development, random inactivation of one X chromosome occurs in each somatic cell. This inactivation pattern is maintained as the embryo matures into an adult.

CONCEPT CHECK: At which stage of development does XCI initially occur?

correctly proposed that the Barr body is a highly condensed X chromosome.

In addition to this cytological evidence, Lyon was also aware of examples in which the coat color of mammals has a variegated pattern. **Figure 5.3b** is a photo of a calico cat, which is a female that is heterozygous for an X-linked gene that can occur as an orange or a black allele. (The cat's white underside is due to a dominant allele in a different gene.) The orange and black patches are randomly distributed in different female individuals. The calico pattern does not occur in male cats, but similar kinds of mosaic patterns have been identified in the female mouse. Lyon suggested that both the Barr body and the calico pattern are the result of XCI in the cells of female mammals.

The proposed mechanism of XCI, known as the **Lyon hypothesis,** is schematically illustrated in **Figure 5.4**. This example involves a white and black variegated coat found in certain strains of mice. As shown here, a female mouse has inherited an X chromosome from its female parent that carries an allele conferring white coat color (X^b). The X chromosome from its male parent carries a black coat color allele (X^B).

How can XCI explain a variegated coat pattern?

1. Initially, both X chromosomes are active.
2. At an early stage of embryonic development, one of the two X chromosomes is randomly inactivated in each somatic cell and becomes a Barr body. For example, one embryonic cell may have the X^B chromosome inactivated.
3. As the embryo continues to grow and mature, this embryonic cell will divide and may eventually give rise to billions of cells in the adult animal. The epithelial (skin) cells that are derived from this embryonic cell will produce a patch of white fur because the X^b chromosome has been permanently inactivated. Alternatively, another embryonic cell may have the X^b chromosome inactivated. The epithelial cells derived from this embryonic cell will produce a patch of black fur.
4. Because the primary event of XCI is a random process that occurs at an early stage of development, the result is an animal with some patches of white fur and other patches of black fur. The basis of this variegated phenotype is the random inactivation of one X chromosome in each embryonic cell.

During XCI, the chromosomal DNA of the inactivated X chromosome becomes highly compacted into a Barr body, so most genes on that chromosome cannot be expressed. When cell division occurs and the inactivated X chromosome is replicated, both copies remain highly compacted and inactive. Likewise, during subsequent cell divisions, the inactivated X chromosome is passed along to all future somatic cells.

EXPERIMENT 5A

In Adult Female Mammals, One X Chromosome Has Been Permanently Inactivated

According to the Lyon hypothesis, each somatic cell of female mammals expresses the genes on one of the X chromosomes, but not both. If an adult female is heterozygous for an X-linked gene, only one of two alleles will be expressed in any given cell. In 1963, Ronald Davidson, Harold Nitowsky, and Barton Childs set out to test the Lyon hypothesis at the cellular level. To do so, they analyzed the expression of a human X-linked gene that codes an enzyme involved with sugar metabolism known as glucose-6-phosphate dehydrogenase (G-6-PD).

Prior to the formulation of the Lyon hypothesis, biochemists had found that individuals vary with regard to the G-6-PD enzyme, which is composed of a single polypeptide. This variation can be detected when the enzyme is subjected to gel electrophoresis (see Appendix A for a description of gel electrophoresis). One *G-6-PD* allele codes a G-6-PD enzyme that migrates very quickly during gel electrophoresis (the fast enzyme), whereas another *G-6-PD* allele produces an enzyme that migrates more slowly (the slow enzyme).

As shown in **Figure 5.5**, when multiple skin samples are collected from a heterozygous adult female and mixed together, both types of enzymes are observed on a gel. In contrast, a hemizygous male produces either the fast or slow type, but not both. The difference in migration between the fast and slow G-6-PD enzymes is due to minor differences in the amino acid sequences of these enzymes. These minor differences do not significantly affect G-6-PD function, but they do enable geneticists to distinguish the proteins coded by the two X-linked alleles.

As shown in **Figure 5.6**, Davidson, Nitowsky, and Childs tested the Lyon hypothesis using cell culturing techniques. They removed several small samples of epithelial cells from a heterozygous female and grew them in the laboratory. When combined, these samples contained a mixture of both types of enzymes because the adult cells were derived from many different embryonic cells, some that had the slow allele inactivated and some that had the fast allele inactivated. In the experiment of Figure 5.6, these cells were sparsely plated onto solid growth media. After several days, each cell grew and divided to produce a **clone** of cells. All cells within a clone were derived from a single cell. The researchers reasoned that all cells within a single clone would express only one of the two *G-6-PD* alleles if the Lyon hypothesis was correct. Nine clones were grown in liquid cultures, and then the cells were lysed to release the G-6-PD enzymes inside of them. The enzymes were then subjected to sodium dodecyl sulfate (SDS) gel electrophoresis.

THE HYPOTHESIS

According to the Lyon hypothesis, an adult female who is heterozygous for the fast and slow *G-6-PD* alleles should express only one of the two alleles in any particular somatic cell and its descendants, but not both.

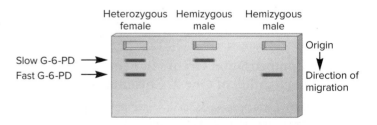

FIGURE 5.5 Mobility of G-6-PD enzymes on a gel. *G-6-PD* exists as an allele that codes a fast enzyme that migrates more quickly to the bottom of the gel and an allele that codes a slow enzyme that migrates more slowly. The fast enzyme is closer to the bottom of the gel.

CONCEPT CHECK: Why do these two forms of G-6-PD migrate differently?

TESTING THE HYPOTHESIS

FIGURE 5.6 **Evidence that adult female mammals contain one X chromosome that has been permanently inactivated.**
Lutz Slomianka

Starting material: Several small skin samples taken from a human female who was heterozygous for the fast and slow alleles of *G-6-PD*.

1. Mince the tissue to separate the individual cells.

2. Grow the cells in a liquid growth medium and then plate (sparsely) onto solid growth medium. Each cell divides to form a clone of many cells.

3. Take nine isolated clones and grow in liquid cultures. (Only three are shown here.)

4. Take cells from the liquid cultures, lyse cells to obtain enzymes, and subject to gel electrophoresis. (This technique is described in Appendix A.)

Note: As a control, lyse cells from step 1, and subject the enzymes to gel electrophoresis. The control is derived from several small skin samples from a heterozygous female.

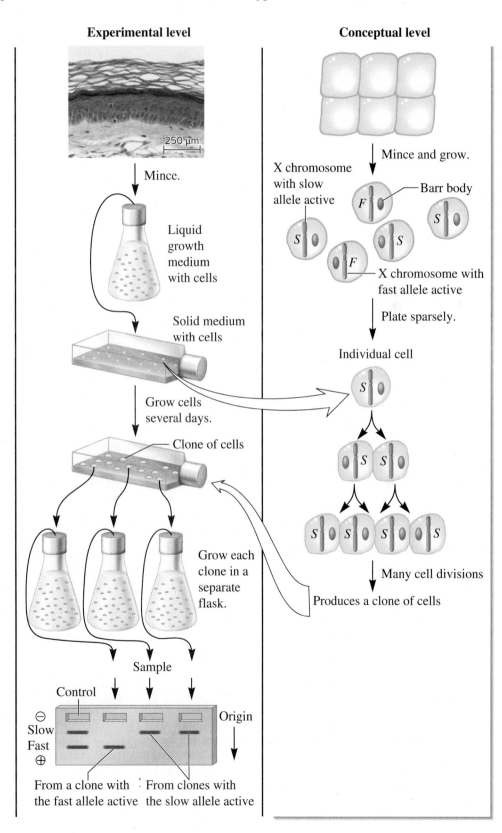

Experimental level

Conceptual level

250 μm

Mince.

Liquid growth medium with cells

Solid medium with cells

Grow cells several days.

Clone of cells

Grow each clone in a separate flask.

Sample

Control

⊖
Slow
Fast
⊕

Origin

From a clone with the fast allele active   From clones with the slow allele active

Mince and grow.

X chromosome with slow allele active

Barr body

X chromosome with fast allele active

Plate sparsely.

Individual cell

Many cell divisions

Produces a clone of cells

THE DATA

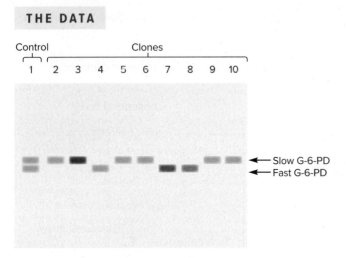

Control   Clones
1   2 3 4 5 6 7 8 9 10

← Slow G-6-PD
← Fast G-6-PD

Adapted from Ronald G. Davidson, Harold M. Nitowsky, and Barton Childs (1963), Demonstration of two populations of cells in the human female heterozygous for glucose-6-phosphate dehydrogenase variants, *PNAS*, 50(3): 481–485, Fig. 2.

INTERPRETING THE DATA

In the data, the control (lane 1) was an enzyme sample obtained from a mixture of epithelial cells from a heterozygous female who produced both types of G-6-PD enzymes. Bands corresponding to the fast and slow enzymes were observed in this lane. As described in steps 2 to 4 of Figure 5.6, this mixture of epithelial cells was also used to generate nine clones. The enzymes obtained from these clones are shown in lanes 2 through 10. Each clone was a population of cells independently derived from a single epithelial cell.

Because the epithelial cells were obtained from an adult female, the Lyon hypothesis predicts that each epithelial cell would already have one of its X chromosomes permanently inactivated and would pass this trait to its daughter cells. For example, suppose that an epithelial cell had inactivated the X chromosome that codes the fast G-6-PD. If this cell was allowed to form a clone of cells on a plate, all cells in this clonal population would be expected to have the same X chromosome inactivated—the one coding the fast G-6-PD. Therefore, this clone of cells should express only the slow G-6-PD. As shown in the data, all nine clones expressed either the fast or slow G-6-PD protein, but not both. These results are consistent with the hypothesis that X-chromosome inactivation has already occurred in any given epithelial cell and that this pattern of inactivation is passed to all of its daughter cells.

Mammals Maintain One Active X Chromosome in Their Somatic Cells

Since the Lyon hypothesis was confirmed, the genetic control of XCI has been investigated further by several laboratories. Research has shown that mammalian somatic cells have the ability to count the X chromosomes they contain and allow only one of them to remain active. How was this determined? A key observation came from comparisons of the chromosome composition of people who were born with different combinations of sex chromosomes.

Sex	Chromosome Composition	Number of X Chromosomes	Number of Barr Bodies	Number of Active X Chromosomes
Female	XX	2	1	1
Male	XY	1	0	1
Female (Turner syndrome)	X0	1	0	1
Female (Triple X syndrome)	XXX	3	2	1
Male (Klinefelter syndrome)	XXY	2	1	1

Most human females are XX; two X chromosomes are counted and one is inactivated. Most human males are XY; one X chromosome is counted and it is not inactivated. On rare occasions, a female may inherit a single X chromosome. (X0 means X zero, indicating that a second sex chromosome is not present.) This female has Turner syndrome; the single X chromosome is not inactivated. Another rare occurrence is when a female inherits three X chromosomes. In triple X syndrome, three X chromosomes are counted and two are converted to Barr bodies. Finally, a third rare syndrome involves males that carry two X chromosomes and one Y chromosome (Klinefelter syndrome). In this case, two X chromosomes are counted and one is inactivated. As shown in the column on the right, in all five cases, XCI allows only one X chromosome to remain active.

X-Chromosome Inactivation in Mammals Depends on the X-Inactivation Center and Occurs in Three Phases

Although the genetic control of XCI is not entirely understood at the molecular level, a short region on the X chromosome called the **X-inactivation center (Xic)** is known to play a critical role. Eeva Therman and Klaus Patau discovered that if one of the two X chromosomes in a female is missing its Xic due to a chromosome mutation, a cell counts only one Xic and X-chromosome inactivation does not occur. Having two active X chromosomes is a lethal condition for a human female embryo.

The process of XCI can be divided into three phases: nucleation, spreading, and maintenance (**Figure 5.7**). Note: The molecular details of XCI are described in Chapter 16 (look ahead to Figure 16.11).

- During embryonic development, the number of Xics is counted and one of the X chromosomes is chosen to be inactivated. The *Xist* gene on the chosen X chromosome is expressed, and some *Xist* RNA binds to the Xic to create a nucleation site.

- During the spreading phase, the chosen X chromosome is inactivated. This spreading requires the further expression

Choosing an X chromosome for inactivation:
Occurs during embryonic development.
The number of X-inactivation centers (Xics) is
counted and one of the X chromosomes remains
active and the other is targeted for inactivation.

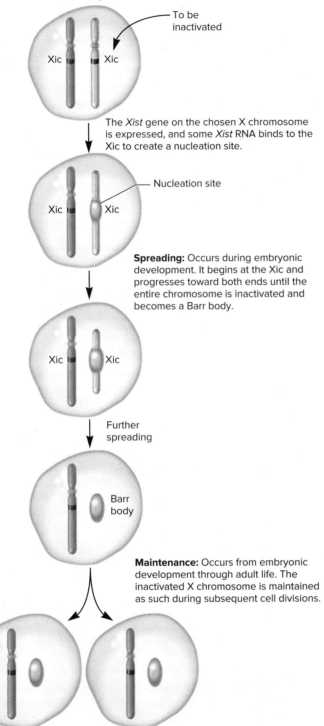

To be inactivated

Xic   Xic

The *Xist* gene on the chosen X chromosome
is expressed, and some *Xist* RNA binds to the
Xic to create a nucleation site.

Nucleation site

Xic   Xic

Spreading: Occurs during embryonic
development. It begins at the Xic and
progresses toward both ends until the
entire chromosome is inactivated and
becomes a Barr body.

Xic   Xic

Further
spreading

Barr
body

Maintenance: Occurs from embryonic
development through adult life. The
inactivated X chromosome is maintained
as such during subsequent cell divisions.

ONLINE
ANIMATION

FIGURE 5.7 **The function of the Xic during
X-chromosome inactivation.**

CONCEPT CHECK: Which of the phases of XCI occurs in an
adult female?

of the *Xist* gene. The *Xist* RNA coats the inactivated
X chromosome and recruits proteins that promote compaction. The spreading phase is so named because inactivation
begins at the nucleation site and spreads in both directions
along the X chromosome.

- Once the nucleation and spreading phases occur for a
given X chromosome, the inactivated X chromosome is
maintained as a Barr body during future cell divisions.
When a cell divides, the Barr body is replicated, and both
copies remain compacted. This maintenance phase continues from the embryonic stage through adulthood.

Some genes on the inactivated X chromosome are expressed
in the somatic cells of adult female mammals. These genes are
said to escape the effects of XCI. In humans, up to a quarter of
X-linked genes may escape inactivation to some degree. Many of
these genes occur in clusters. Among these are the pseudoautosomal genes found on the X and Y chromosomes in the regions of
homology described in Chapter 4. Dosage compensation is not
necessary for X-linked pseudoautosomal genes because they are
located on both the X and Y chromosomes. How are genes on the
Barr body expressed? Although the mechanism is not understood,
these genes may be found in localized regions where the chromatin is less tightly packed and able to be transcribed.

GENETIC TIPS   THE QUESTION: A cat is born with
two X chromosomes and one Y chromosome. This is a rare event.
One of the X chromosomes carries the black fur allele and the other
carries the orange fur allele. Would you expect this cat to be a male
or female? Would it be calico?

TOPIC: *What topics in genetics does this question
address?* The topics are sex determination and X-chromosome
inactivation.

INFORMATION: *What information do you know based on the
question and your understanding of the topic?* From the question,
you know the composition of sex chromosomes in a cat and the fur
color alleles carried on the cat's X chromosomes. From your
understanding of the topics, you may remember from Chapter 2 (refer
back to Figure 2.15) that the Y chromosome determines maleness in
mammals and that X-chromosome inactivation occurs and only one
X chromosome remains active in somatic cells.

PROBLEM-SOLVING **S**TRATEGY: *Predict the outcome.* With
regard to sex determination, you would predict that the cat is a
male because the Y chromosome causes maleness. XCI occurs in
a way that leaves only one X chromosome active in the cat's
somatic cells. Because this cat has two X chromosomes, you
would predict that one out of the two will be randomly inactivated
throughout the cat's body. The cat is heterozygous for the orange
and black fur alleles, resulting in some patches of orange fur and
some patches of black.

ANSWER: It is a male cat with a calico coat.

5.2 COMPREHENSION QUESTIONS

1. In fruit flies, dosage compensation is achieved by
   a. X-chromosome inactivation.
   b. doubling the expression of genes on the single X chromosome in the male.
   c. halving the expression of genes on each X chromosome in the female.
   d. all of the above.

2. According to the Lyon hypothesis,
   a. one of the X chromosomes is converted to a Barr body in somatic cells of female mammals.
   b. one of the X chromosomes is converted to a Barr body in all cells of female mammals.
   c. both of the X chromosomes are converted to Barr bodies in somatic cells of female mammals.
   d. both of the X chromosomes are converted to Barr bodies in all cells of female mammals.

3. Which of the following is *not* a phase of XCI?
   a. Nucleation   c. Maintenance
   b. Spreading   d. Erasure

5.3 EPIGENETICS: GENOMIC IMPRINTING

Learning Outcomes:

1. Define *genomic imprinting*.
2. Predict the outcome of crosses involving imprinted genes.
3. Explain the molecular mechanism of imprinting.

As we have just seen, dosage compensation changes the level of expression of many genes located on the X chromosome. We now turn to another epigenetic phenomenon known as genomic imprinting, or simply imprinting. The term *imprinting* implies a type of marking process that has a memory. For example, newly hatched birds identify marks on their parents, which allows them to distinguish their parents from other individuals. The term **genomic imprinting,** or simply **imprinting,** refers to an analogous situation in which a segment of DNA is marked and that mark is retained and recognized throughout the life of the organism inheriting the marked DNA. Genomic imprinting happens prior to fertilization; it involves a change in a single gene or chromosome during gamete formation. Depending on whether the modification occurs during spermatogenesis or oogenesis, imprinting governs whether an offspring expresses a gene that has been inherited from its female parent or male parent, but not both.

The Expression of an Imprinted Gene Depends on the Sex of the Parent from Which the Gene Was Inherited

The phenotypes that result from the expression of imprinted genes follow a non-Mendelian pattern of inheritance because the marking process causes the offspring to distinguish between the alleles received from the male and female parents. Depending on how the genes are marked, each offspring expresses only one of the two alleles. This phenomenon is termed **monoallelic expression.**

To understand genomic imprinting, let's consider a specific example. In the mouse, a gene designated *Igf2* codes a protein called insulin-like growth factor 2.

- Imprinting occurs in a way that results in the expression of the *Igf2* allele received from the male parent but not the allele from the female parent.
- The allele inherited from the male parent is transcribed into RNA, but the allele from the female parent is transcriptionally silent.

With regard to phenotype, a functional *Igf2* gene is necessary for normal size. A loss-of-function allele of this gene, designated *Igf2⁻*, is defective in the synthesis of a functional Igf2 protein. This may cause a mouse to be smaller than normal, but the occurrence of small size depends on whether the mutant allele is inherited from the male or female parent, as shown in **Figure 5.8**.

- On the left side, an offspring has inherited the *Igf2* allele from its male parent and the *Igf2⁻* allele from its female parent. Due to imprinting, only the *Igf2* allele is expressed in the offspring. Therefore, this mouse grows to a normal size.
- In the reciprocal cross on the right side, an individual has inherited the *Igf2⁻* allele from its male parent and the *Igf2* allele from its female parent. In this case, the *Igf2* allele is not expressed. In this mouse, the *Igf2⁻* allele would be transcribed into mRNA, but the mutation renders the Igf2 protein defective. Therefore, the offspring on the right grows to a small size.
- Both offspring shown in Figure 5.8 have the same genotype; they are heterozygous for the *Igf2* alleles (i.e., *Igf2 Igf2⁻*). They are phenotypically different, however, because only the allele received from the male parent is expressed.

For imprinted genes, you cannot use a Punnett square to predict an offspring's phenotype. Instead, you need two pieces of information:

- You need to know if the offspring expresses the allele that is inherited from the female or male parent.
- You need to know which allele was inherited from the female parent and which allele was inherited from the male parent.

With this information, you can predict an offspring's phenotype. As an example, let's consider the *Igf2* gene. We know that the allele from the male parent is expressed. If the offspring inherits the *Igf2* allele from the male parent, it will be normal size. Alternatively, if the offspring inherits the *Igf2⁻* allele from the male parent, it will be small size.

The Imprint Is Established During Gametogenesis

At the cellular level, imprinting is an epigenetic process that can be divided into three stages: (1) the establishment of the imprint during gametogenesis, (2) the maintenance of the imprint during embryogenesis and in adult somatic cells, and (3) the erasure and reestablishment of the imprint in the germ cells. These stages are described in **Figure 5.9**, which shows the imprinting of the *Igf2* gene.

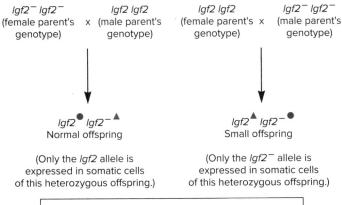

▲ Denotes an allele that is silent in the offspring

● Denotes an allele that is expressed in the offspring

FIGURE 5.8 **An example of genomic imprinting in the mouse.** In the cross on the left, a homozygous female carrying a defective allele, designated *Igf2⁻*, is crossed to a homozygous male with the functional *Igf2* allele. The offspring is heterozygous and normal size because the allele inherited from the male parent is active. In the reciprocal cross on the right, a female that is homozygous for the functional *Igf2* allele is crossed to a homozygous male carrying the defective allele. In this case, the offspring is heterozygous and small size. The offspring is small size because the allele from the male parent is defective due to mutation and the allele from the female parent is not expressed. The photograph shows normal-size (left) and small-size littermates (right) derived from a cross between a wild-type female and a heterozygous male (*Igf2 Igf2⁻*) carrying one copy of a loss-of-function allele. The loss-of-function allele was created using gene knockout methods.

Dr. Argiris Efstratiadis

CONCEPT CHECK: What would be the outcome of a cross between a heterozygous female and a male that carries two functional copies of the *Igf2* gene?

1. The two mice shown here have inherited the *Igf2* allele from their male parent and the *Igf2⁻* allele from their female parent. Imprinting was established during gametogenesis in the parents of these mice.

2. Due to imprinting, both mice express the *Igf2* allele in their somatic cells, and the pattern of imprinting is maintained in the somatic cells throughout development.

3. In the germ-line cells that give rise to gametes (i.e., sperm or eggs), the imprint is erased; it will be reestablished according to the sex of the animal. The female mouse on the left in Figure 5.9 will transmit only transcriptionally inactive alleles to offspring. The male mouse on the right will transmit transcriptionally active alleles. However, this male is a heterozygote and

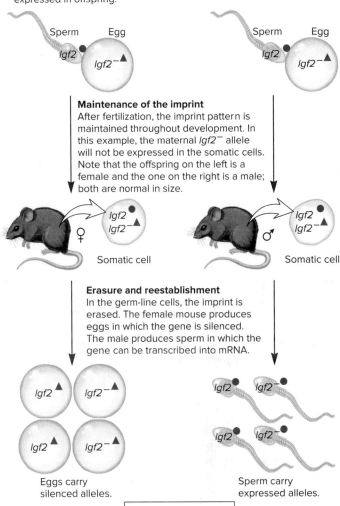

Establishment of the imprint
In this example, imprinting of the *Igf2* gene occurs during gametogenesis. The sperm carries the *Igf2* allele and the egg carries the *Igf2⁻* allele. Only the paternal allele will be expressed in offspring.

Maintenance of the imprint
After fertilization, the imprint pattern is maintained throughout development. In this example, the maternal *Igf2⁻* allele will not be expressed in the somatic cells. Note that the offspring on the left is a female and the one on the right is a male; both are normal in size.

Erasure and reestablishment
In the germ-line cells, the imprint is erased. The female mouse produces eggs in which the gene is silenced. The male produces sperm in which the gene can be transcribed into mRNA.

Eggs carry silenced alleles.

Sperm carry expressed alleles.

▲ Silenced allele
● Transcribed allele

FIGURE 5.9 **Genomic imprinting during gametogenesis.** This example involves a mouse gene, *Igf2*, which is found in two alleles designated *Igf2* and *Igf2⁻*. The left side shows a female mouse that was produced from a sperm carrying the *Igf2* allele and an egg carrying the *Igf2⁻* allele. In the somatic cells of this female animal, the *Igf2* allele is active. However, when this female produces eggs, both alleles are transcriptionally inactive when they are transmitted to offspring. The right side of this figure shows a male mouse that was also produced from a sperm carrying the *Igf2* allele and an egg carrying the *Igf2⁻* allele. In the somatic cells of this male animal, the *Igf2* allele is active. The sperm from this male may carry either a functionally active *Igf2* allele or a functionally defective *Igf2⁻* allele. The allele inherited from the male is expressed, resulting in normal-size or small-size offspring, respectively.

CONCEPT CHECK: Look ahead to Figure 5.10. Explain why the erasure stage of imprinting is necessary in eggs.

will transmit either a functionally active *Igf2* allele or a functionally defective mutant allele (*Igf2⁻*). If this heterozygous male transmits the *Igf2* allele to an offspring, the offspring will be normal size. In contrast, if this male transmits an *Igf2⁻* allele, the offspring will be small size. Although an *Igf2⁻* allele inherited from a male mouse can be transcribed into mRNA, it will not produce a functional Igf2 protein, which is the reason for the small-size phenotype.

As seen in Figure 5.9, genomic imprinting is permanent in the somatic cells of an animal, but the marking of alleles can be altered from generation to generation. For example, the female mouse on the left has an active copy of the *Igf2* allele, but any allele this female transmits to its offspring will be transcriptionally inactive.

Genomic imprinting occurs in several species, including numerous insects, mammals, and flowering plants. Imprinting may involve a single gene, a part of a chromosome, an entire chromosome, or even all of the chromosomes from one parent. Helen Crouse discovered the first example of imprinting, which involved an entire chromosome in the housefly, *Sciara coprophila*. In this species, an offspring normally inherits three sex chromosomes, rather than two as in most other species. One X chromosome is inherited from the female, and two are inherited from the male. In male offspring, both X chromosomes from the male parent are lost from somatic cells during embryogenesis. In female offspring, only one of those two X chromosomes is lost. In both sexes, the X chromosome from the female parent is never lost. These results indicate that either the X chromosome from the female parent is marked to promote its retention or the X chromosomes from the male parent are marked to promote their loss.

Genomic imprinting can also be involved in the process of X-chromosome inactivation, described in the preceding section. In certain species, imprinting plays a role in the choice of the X chromosome that will be inactivated. For example, in marsupials, the X chromosome inherited from the male parent is marked so that it is the X chromosome that is always inactivated in the somatic cells of females. In marsupials, XCI is not random; the X chromosome from the female parent is always active.

The Imprinting of Genes Is a Molecular Marking Process That Involves DNA Methylation

As we have seen, genomic imprinting must involve a marking process. A particular gene or chromosome must be marked differently during spermatogenesis compared to oogenesis. After fertilization takes place, this differential marking affects the expression of particular genes. What is the molecular explanation for genomic imprinting? As discussed in Chapter 15, **DNA methylation**—the attachment of a methyl group onto a cytosine base—is a common way that eukaryotic genes may be regulated. Research indicates that genomic imprinting involves an **imprinting control region (ICR)** that is located near the imprinted gene (look ahead to Figure 16.10). Depending on the particular gene, the ICR is methylated in the egg or the sperm, but not both. The ICR contains binding sites for one or more proteins that regulate the transcription of the imprinted gene.

Let's now consider the methylation process from one generation to the next. In the example shown in **Figure 5.10**, the allele

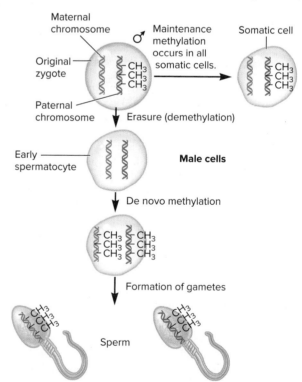

FIGURE 5.10   The pattern of methylation from one generation to the next. In this example, a female and a male offspring have inherited a nonmethylated gene and a methylated gene from their female parent and male parent, respectively. Maintenance methylation retains the imprinting in somatic cells during embryogenesis and in adulthood. Demethylation occurs in cells that are destined to become gametes. In this example, de novo methylation occurs only in cells that are destined to become sperm (see right side). Haploid male gametes transmit a methylated gene, whereas haploid female gametes transmit an unmethylated gene.

CONCEPT CHECK: What is the difference between maintenance methylation and de novo methylation? In what cell type (somatic cells or germ-line cells) does each process occur?

for a particular gene that was inherited from the male parent is methylated but the allele from the female parent is not. A female (left side) and male (right side) have inherited an unmethylated gene from their female parent and a methylated gene from their male parent. This pattern of imprinting is maintained in the somatic cells of both individuals. However, when the female makes gametes, the imprinting is erased during early oogenesis, so the female will pass an unmethylated gene to offspring. In the male, the imprinting is also erased during early spermatogenesis, but then de novo (new) methylation occurs in both genes. Therefore, the male will transmit a methylated gene to offspring.

Imprinting Plays a Role in the Inheritance of Certain Human Genetic Diseases

About 100 genes have been shown to be imprinted in humans (**Table 5.2**). Some human diseases, such as Prader-Willi syndrome (PWS) and Angelman syndrome (AS), are influenced by imprinting. PWS is characterized by reduced motor function, obesity, and small hands and feet. Individuals with AS are thin and hyperactive, have unusual seizures and repetitive symmetrical muscle movements, and exhibit mental deficiencies. Most commonly, both PWS and AS arise from a small deletion in human chromosome 15. If this deletion is inherited from the female parent, it leads to Angelman syndrome; if inherited from the male parent, it leads to Prader-Willi syndrome (**Figure 5.11**).

Researchers have discovered that this region of chromosome 15 contains closely linked but distinct genes that are maternally or paternally imprinted. AS results from the lack of expression of a single gene (*UBE3A*) that codes a protein that regulates protein degradation. If the allele from the female parent is deleted, as in the left side of Figure 5.11, the individual will not have an active copy of the *UBE3A* gene and therefore will develop AS.

The symptoms of PWS appear to be caused by the loss of function (e.g., deletion) of several genes in this region of chromosome 15. These include the *SNRPN* gene, the *NDN* gene, and a

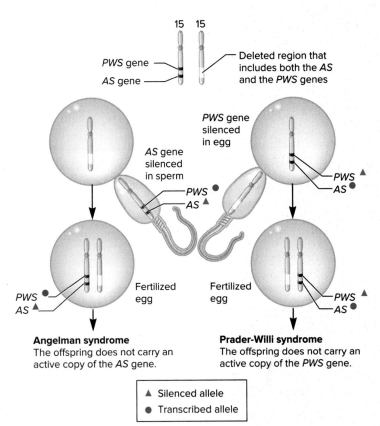

FIGURE 5.11 **The role of imprinting in the development of Angelman or Prader-Willi syndrome.**

Genes→Traits A small region on chromosome 15 contains two different loci designated the *AS* gene and *PWS* gene in this figure. (The PWS gene is actually a group of several genes in this region.) The *AS* gene is silenced during sperm formation, and the *PWS* gene is silenced during egg formation. If a deletion of this small region of chromosome 15 is inherited from the female parent, Angelman syndrome occurs because the offspring does not inherit an active copy of the *AS* gene (left). Alternatively, if the chromosome 15 deletion is inherited from the male parent, Prader-Willi syndrome occurs because the offspring does not inherit an active copy of the *PWS* gene (right).

CONCEPT CHECK: Explain why the offspring on the left side of this figure does not have Prader-Willi syndrome but does have Angelman syndrome.

cluster of genes that code RNAs called snoRNAs. The *SNRPN* gene codes a small nuclear ribonucleoprotein polypeptide N, which is part of a complex that controls RNA splicing. The *NDN* gene codes a protein called necdin that functions as a growth suppressor for neurons in the brain. The function of snoRNAs is described in Chapter 17.

TABLE 5.2

Examples of Human Genes That Are Imprinted*

Gene	Allele Expressed	Function
WT1	Maternal	Wilms tumor-suppressor gene; suppresses cell growth
INS	Paternal	Insulin; hormone involved in cell growth and metabolism
Igf2	Paternal	Insulin-like growth factor 2; similar to insulin
Igf2R	Maternal	Receptor for insulin-like growth factor 2
UBE3A	Maternal	Regulates protein degradation; involved in Angelman syndrome
SNRPN	Paternal	Splicing factor; involved in Prader-Willi syndrome
Gabrb	Maternal	Neurotransmitter receptor

*Researchers estimate that approximately 1–2% of human genes are subject to genomic imprinting, but only about 100 have actually been demonstrated to be imprinted.

5.3 COMPREHENSION QUESTIONS

1. In mice, the copy of the *Igf2* gene that is inherited from the female parent is never expressed in the offspring. This happens because the *Igf2* gene from the female parent
   a. always undergoes a mutation that inactivates its function.
   b. is deleted during oogenesis.
   c. is deleted during embryonic development.
   d. is not transcribed in the somatic cells of the offspring.

2. A female mouse that is *Igf2 Igf2⁻* is crossed to a male that is also *Igf2 Igf2⁻*. The expected outcome for the phenotypes of the offspring for this cross is
   a. all normal size.
   b. all small size.
   c. 1 normal size : 1 small size.
   d. 3 normal size : 1 small size.

3. The marking process for genomic imprinting initially occurs during
   a. gametogenesis.
   b. fertilization.
   c. embryonic development.
   d. adulthood.

4. A female born with Angelman syndrome carries a deletion in the *AS* gene (i.e., the *UBE3A* gene). Which parent transmitted the deletion?
   a. Male parent
   b. Female parent
   c. Could be either the female or male parent

5.4   EXTRANUCLEAR INHERITANCE

Learning Outcomes:

1. Describe the general features of the mitochondrial and chloroplast genomes.
2. Predict the outcome of crosses involving extranuclear inheritance.
3. Describe three cellular mechanisms of maternal inheritance in animals.
4. Explain how mutations in mitochondrial genes cause human diseases.
5. Evaluate the endosymbiosis theory.

Thus far, we have considered several types of non-Mendelian inheritance patterns: maternal effect, dosage compensation, and genomic imprinting. All of these inheritance patterns involve **nuclear genes**—genes located on chromosomes that are in the cell nucleus.

Another cause of non-Mendelian inheritance patterns involves genes that are not located in the cell nucleus. In eukaryotic species, the most biologically important example of extranuclear inheritance is due to genetic material in cellular organelles. In addition to the cell nucleus, the mitochondria and chloroplasts contain their own genetic material. Because these organelles are found within the cytoplasm of cells, the inheritance of organellar genetic material is called **extranuclear inheritance** (the prefix *extra*- means "outside of"), or **cytoplasmic inheritance.** In this section, we will examine the genetic composition of mitochondria and chloroplasts and explore the pattern of transmission of these organelles from parent to offspring.

Mitochondria and Chloroplasts Contain Circular Chromosomes with Many Genes

In 1951, Yukako Chiba was the first to suggest that chloroplasts contain their own DNA. This conclusion was based on the staining properties of a DNA-specific dye known as Feulgen. Researchers later developed techniques to purify organellar DNA. In addition, electron microscopy studies provided interesting insights into the organization and composition of mitochondrial and chloroplast chromosomes. More recently, the advent of molecular genetic techniques in the 1970s and 1980s has allowed researchers to determine the genome sequences of organellar DNAs. From these types of studies, researchers found that the chromosomes of mitochondria and chloroplasts resemble smaller versions of bacterial chromosomes.

The genetic material of a mitochondrion or chloroplast is located inside the organelle in a region known as the **nucleoid** (**Figure 5.12**). The genome is a single circular chromosome, although a nucleoid contains several copies of this chromosome.

FIGURE 5.12 **Nucleoids within (a) a mitochondrion and (b) a chloroplast.** The mitochondrial and chloroplast chromosomes are found within the nucleoid of the organelle.

(a): From: Prachar J., "Mouse and human mitochondrial nucleoid–detailed structure in relation to function," *Gen Physiol Biophys*. 2010 Jun, 29(2):160–174. Fig 3A; (b): Biophoto Associates/Science Source

CONCEPT CHECK: How is a nucleoid different from a cell nucleus?

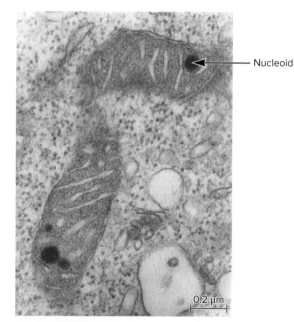

Nucleoid

0.2 μm

(a) Mitochondrion

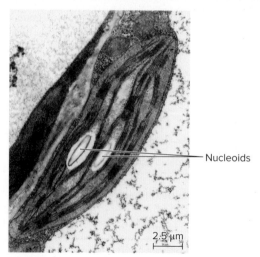

Nucleoids

2.5 μm

(b) Chloroplast

TABLE 5.3

Genetic Composition of Mitochondria and Chloroplasts

Organism(s)	Organelle	Nucleoids per Organelle	Number of Chromosomes per Nucleoid
Tetrahymena	Mitochondrion	1	6–8
Mouse	Mitochondrion	1–3	2–6
Chlamydomonas	Chloroplast	5–6	∽15
Euglena	Chloroplast	20–34	10–15
Flowering plants	Chloroplast	12–25	3–5

Source: Gillham, Nicholas W., *Organelle Genes and Genomes*. New York, NY: Oxford University Press, 1994.

In addition, a mitochondrion or chloroplast often has more than one nucleoid. In mice, for example, each mitochondrion has one to three nucleoids, with each nucleoid containing two to six copies of the circular mitochondrial genome. However, this number varies depending on the type of cell and the stage of development. In comparison, the chloroplasts of algae and plants tend to have more nucleoids per organelle. **Table 5.3** describes the genetic composition of mitochondria and chloroplasts for a few selected organisms.

The sizes of mitochondrial and chloroplast genomes also vary greatly among different species. For example, a 400-fold variation is found in the sizes of mitochondrial chromosomes. In general, the mitochondrial genomes of animal species tend to be fairly small; those of fungi and protists are intermediate in size; and those of plant cells tend to be fairly large. Among algae and plants, substantial variation is also found in the sizes of chloroplast chromosomes.

Figure 5.13 shows a map of human **mitochondrial DNA (mtDNA)**. Each copy of the mitochondrial chromosome consists of a circular DNA molecule that contains only 17,000 base pairs (bp). This size is less than 1% of a typical bacterial chromosome. The human mtDNA carries relatively few genes. Thirteen genes code proteins that function within the mitochondrion. In addition, the mtDNA carries genes that code ribosomal RNA and transfer RNA. These rRNAs and tRNAs are necessary for the synthesis of the 13 polypeptides that are coded by the mtDNA. The primary role of mitochondria is to provide cells with the bulk of their adenosine triphosphate (ATP), which is used as an energy source to drive cellular reactions. The 13 polypeptides synthesized in mitochondria are subunits of proteins that function in a process known as oxidative phosphorylation, in which mitochondria use oxygen and synthesize ATP.

Mitochondria require many additional proteins to carry out oxidative phosphorylation and other mitochondrial functions. Most mitochondrial proteins are coded by genes within the cell nucleus. When these nuclear genes are expressed, the mitochondrial polypeptides are first synthesized outside the mitochondria in the cytosol of the cell. They are then transported into the mitochondria where they may associate with other polypeptides and become functional proteins.

Chloroplast genomes tend to be larger than mitochondrial genomes, and they have a correspondingly greater number of genes. A typical chloroplast genome has approximately 100,000 to 200,000 bp, which is about 10 times larger than the mitochondrial genome of mammalian cells. For example, the **chloroplast DNA (cpDNA)** of the tobacco plant is a circular DNA molecule that contains 156,000 bp and carries between 110 and 120 different genes. These genes code ribosomal RNAs, transfer RNAs, and many proteins required for photosynthesis. As with mitochondria, many chloroplast proteins are coded by genes found in the plant cell nucleus. These proteins contain chloroplast-targeting signals that direct them into the chloroplasts.

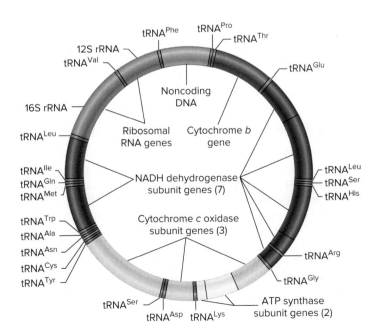

	Ribosomal RNA genes
	Transfer RNA genes
	NADH dehydrogenase subunit genes (7)
	Cytochrome *b* gene
	Cytochrome *c* oxidase subunit genes (3)
	ATP synthase subunit genes (2)
	Noncoding DNA

FIGURE 5.13 **A genetic map of human mitochondrial DNA (mtDNA).** This diagram illustrates the locations of many genes along the circular mitochondrial chromosome. The genes that code ribosomal RNA are shown in light brown. The genes shown in red code transfer RNAs. For example, tRNA^{Arg} codes a tRNA that carries arginine. The other genes code proteins that function within the mitochondrion. In addition, the mitochondrial genome has a noncoding region, shown in blue.

CONCEPT CHECK: Why do mitochondria need genes that code rRNAs and tRNAs?

Extranuclear Inheritance Produces Non-Mendelian Results in Reciprocal Crosses

In diploid eukaryotic species, most genes within the nucleus obey a Mendelian pattern of inheritance because the homologous pairs of chromosomes segregate and independently assort during meiosis. Except for genes that determine sex-linked traits, offspring inherit one copy of each gene from both the female and male parent. The sorting of chromosomes during meiosis explains the inheritance patterns of nuclear genes. By comparison, the inheritance of extranuclear genetic material does not display a Mendelian pattern. Mitochondria and chloroplasts are not sorted by the spindle apparatus during meiosis and therefore do not segregate into gametes in the same way as nuclear chromosomes do.

In 1909, Carl Correns discovered a trait that showed a non-Mendelian pattern of inheritance for pigmentation in *Mirabilis jalapa* (the four-o'clock plant). Leaves can be green, white, or variegated (with both green and white sectors). Correns demonstrated that the pigmentation of the offspring depended solely on the female parent that provides the egg cells (**Figure 5.14**).

- If the female parent had white leaves, all offspring had white leaves.
- If the female was green, all offspring were green.
- When the female was variegated, the offspring could be green, white, or variegated.

The pattern of inheritance observed by Correns is a type of extranuclear inheritance called **maternal inheritance** (not to be confused with maternal effect). Chloroplasts are a type of plastid that makes chlorophyll, a green photosynthetic pigment. Maternal inheritance occurs because the chloroplasts are inherited only through the cytoplasm of the egg. The pollen grains of *M. jalapa* do not transmit chloroplasts to the offspring.

The phenotypes of leaves can be explained by the types of chloroplasts within the leaf cells. The green phenotype, which is the wild-type condition, is due to the presence of normal chloroplasts that make green pigment. By comparison, the white phenotype is due to a mutation in a gene within the chloroplast DNA that diminishes the synthesis of green pigment.

How does a variegated phenotype occur? **Figure 5.15** considers the leaf of a plant that began from a fertilized egg that contained both types of chloroplasts, a condition known as **heteroplasmy**. A leaf cell containing both types of chloroplasts is green because the chloroplasts produce a sizeable amount green pigment. As a plant grows, the two types of chloroplasts are irregularly distributed to daughter cells. On occasion, a cell may receive only the chloroplasts that have a defect in making green pigment. Such a cell continues to divide and produces a sector of the plant that is entirely white. In this way, the variegated phenotype is produced. Similarly, if we consider the results of Figure 5.14, a variegated female parent may transmit green, white, or a mixture of these types of chloroplasts to the egg cell, thereby producing green, white, or variegated offspring, respectively.

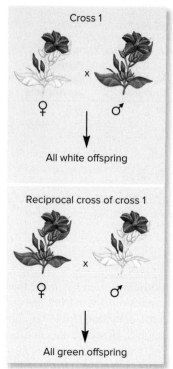
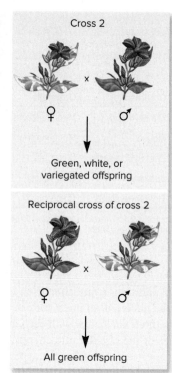

FIGURE 5.14 Maternal inheritance in the four-o'clock plant, *Mirabilis jalapa.* The reciprocal crosses of four-o'clock plants by Correns consisted of a pair of crosses between white-leaved and green-leaved plants, and a second pair of crosses between variegated-leaved and green-leaved plants.

Genes→Traits In this example, the green phenotype is due to the synthesis of chlorophyll by wild-type chloroplasts, whereas the white phenotype is due to chloroplasts that carry a mutant allele that diminishes green pigmentation. The variegated phenotype is explained by a mixture of chloroplasts (see Figure 5.15). In the crosses shown here, the parent providing the eggs determines the phenotypes of the offspring. This outcome is due to maternal inheritance. The egg contains the chloroplasts that are inherited by the offspring. (Note: The defective chloroplasts that give rise to white sectors are not completely defective in chlorophyll synthesis. Therefore, plants with entirely white leaves can survive, though they are smaller than green or variegated plants.)

CONCEPT CHECK: What is a reciprocal cross?

The Pattern of Inheritance of Mitochondria and Chloroplasts Varies Among Different Species

The inheritance of traits via genetic material within mitochondria and chloroplasts is now a well-established phenomenon that geneticists have investigated in many different species. In **heterogamous** species, two kinds of gametes are made. The female gamete tends to be large and provides most of the cytoplasm to the zygote, whereas the male gamete is small and often provides little more than a nucleus. Therefore, mitochondria and chloroplasts are most often inherited from the female parent. However, this is not always the case. **Table 5.4** describes the inheritance patterns of mitochondria and chloroplasts in several selected species.

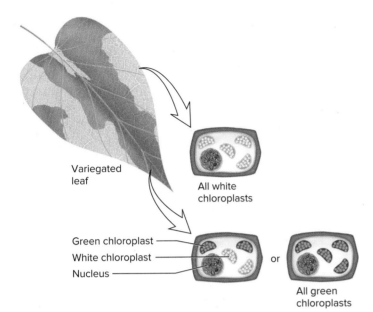

FIGURE 5.15 A cellular explanation of the variegated phenotype in *Mirabilis jalapa*. This plant inherited two types of chloroplasts—those that produce green pigment and those that are defective. As the plant grows, the two types of chloroplasts are irregularly distributed to daughter cells. On occasion, a leaf cell may receive only the chloroplasts that are defective at making green pigment. Such a cell continues to divide and produces a sector of the leaf that is entirely white. Cells that contain both types of chloroplasts or cells that contain only green chloroplasts produce green tissue, which may be adjacent to a sector of white tissue. This is the basis for the variegated phenotype of the leaves.

CONCEPT CHECK: During leaf growth, can a patch of tissue with a white phenotype give rise to a patch with a green phenotype? Explain.

TABLE 5.4

Transmission of Organelles Among Different Organisms

Organism	Organelle(s)	Transmission
S. cerevisiae (Yeast)	Mitochondria	Biparental inheritance
Molds	Mitochondria	Usually maternal inheritance; paternal inheritance has been found in the genus *Allomyces*
Chlamydomonas (Alga)	Mitochondria	*Chlamydomonas* exists in two mating types (*mt⁺* and *mt⁻*). Inherited from the parent with the *mt⁻* mating type
Chlamydomonas	Chloroplasts	Inherited from the parent with the *mt⁺* mating type
Angiosperms (Plants)	Mitochondria and chloroplasts	Often maternal inheritance, although biparental inheritance is found among some species
Gymnosperms (Plants)	Mitochondria and chloroplasts	Usually paternal inheritance
Nearly all animals	Mitochondria	Maternal inheritance

For extranuclear genes, you cannot use a Punnett square to predict an offspring's phenotype. Instead, you need to have two pieces of information:

- You need to know if the offspring inherits the gene from the female parent, the male parent, or both.
- You need to know if heteroplasmy is present in the parent(s) who transmit(s) the extranuclear gene.

With this information, you can predict an offspring's phenotype. For example, let's consider a gene that affects leaf pigmentation. If the gene is inherited from a female parent with white leaves, you can predict that the offspring will receive the white allele and all of them will have white leaves. Alternatively, the female parent may exhibit heteroplasmy and have variegated leaves. In this case, the offspring could have green, white, or variegated leaves.

In Most Animals, Mitochondria Are Inherited Via the Egg

Although a few exceptions are known, maternal inheritance is observed in nearly all animal species. Interestingly, the mechanism that is responsible for maternal inheritance varies greatly among different animal species. Most of these mechanisms fall into three general categories:

Lack of Entry of Sperm Mitochondria. In some species, the sperm mitochondria do not enter the cytoplasm of the egg (**Figure 5.16a**). For example, in Chinese hamsters (*Cricetulus griseus*), which have unusually large sperm, the contents of the sperm midpiece including the paternal mitochondria remain outside the egg and do not enter the egg cytoplasm after fertilization. Therefore, the fertilized egg only contains maternal mitochondria.

Destruction of Sperm Mitochondrial DNA Prior to Fertilization. In some animal species, such as *Drosophila*, the process of sperm maturation involves the destruction of mitochondrial DNA (**Figure 5.16b**). An endonuclease cleaves the mitochondrial DNA into many small pieces. The destruction of mitochondrial DNA prevents the paternal mitochondria from proliferating.

Destruction of Sperm Mitochondria After Fertilization. In many animal species, the sperm mitochondria and their mtDNA do enter the oocyte cytoplasm upon fertilization (**Figure 5.16c**). How are these paternal mitochondria eliminated? In most mammals including humans, the paternal mitochondria that enter the egg are modified by the attachment of ubiquitin, which is a small regulatory protein that targets molecules and organelles for destruction via the proteasome or via fusion with lysosomes. All paternal mitochondria disappear by the 4- to 8-cell stage of embryonic development.

Many Human Diseases Are Caused by Mitochondrial Mutations

Researchers have identified many diseases that are due to mutations in the mitochondrial DNA. Such diseases can occur in two ways:

1. Mitochondrial mutations may be transmitted from the female parent to offspring. As mentioned, human mtDNA is maternally inherited via the cytoplasm of the egg.

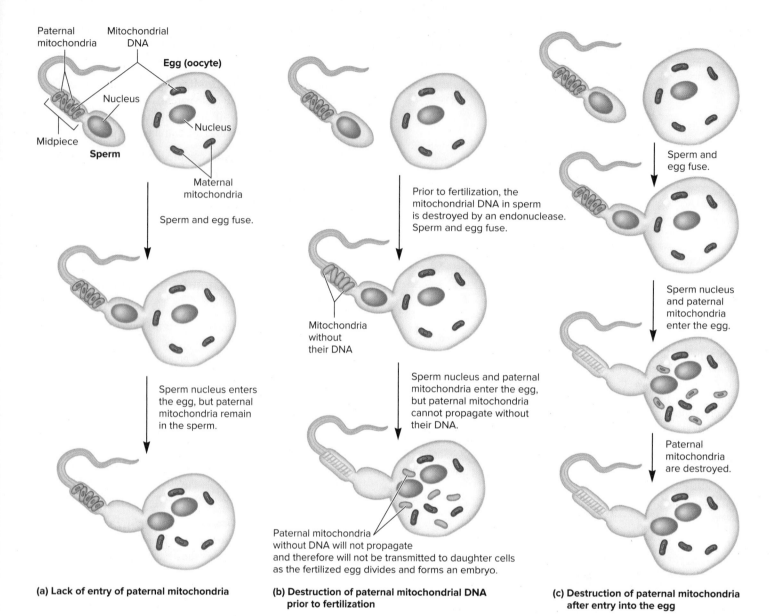

(a) Lack of entry of paternal mitochondria

(b) Destruction of paternal mitochondrial DNA prior to fertilization

(c) Destruction of paternal mitochondria after entry into the egg

FIGURE 5.16 Cellular mechanisms of maternal inheritance of mitochondria in animals. In animals, the mitochondria in sperm are typically found in the midpiece of the sperm's tail and supply the energy for sperm to swim toward the egg. The mitochondria are often arranged in a spiral configuration in the midpiece. Note: The sperm and egg shown here are not drawn to scale. The egg is much larger than the sperm. In part (c), the paternal mitochondria are destroyed, but all of them may not be destroyed until the 4- to 8-cell stage of embryonic development.

Therefore, the transmission of inherited human mitochondrial diseases follows a maternal inheritance pattern.

2. Mitochondrial mutations may occur in somatic cells and accumulate as a person ages. Researchers have discovered that mitochondria are particularly susceptible to DNA damage. When more oxygen is consumed than is actually used to make ATP, mitochondria tend to produce free radicals that damage DNA. Unlike nuclear DNA, which can be repaired by several different repair systems, mitochondria have very limited DNA repair abilities and almost no protective ability against free radical damage.

Table 5.5 describes a few mitochondrial diseases that have been discovered in humans and are caused by mutations in mitochondrial genes. Over 200 diseases associated with defective

TABLE 5.5

Examples of Human Mitochondrial Diseases

Disease	Mitochondrial Gene Mutated
Leber hereditary optic neuropathy (LHON)	A mutation in one of several mitochondrial genes that code the respiratory chain proteins ND1, ND2, CO1, ND4, ND5, ND6, and cytb; tends to affect males more than females
Neurogenic muscle weakness	A mutation in the *ATPase6* gene that codes a subunit of the mitochondrial ATP-synthase, which is required for ATP synthesis
Mitochondrial myopathy	A mutation in a gene that codes a tRNA for leucine
Maternal myopathy and cardiomyopathy	A mutation in a gene that codes a tRNA for leucine

mitochondria have been identified. These are usually chronic degenerative disorders that affect cells requiring a high level of ATP, such as nerve and muscle cells. For example, Leber hereditary optic neuropathy (LHON) affects the optic nerve and may lead to the progressive loss of vision in one or both eyes. LHON can be caused by a defective mutation in one of several different mitochondrial genes. Researchers are still investigating how a defect in these mitochondrial genes produces the symptoms of this disease.

An important factor in mitochondrial diseases is heteroplasmy, a condition in which a cell contains a mixed population of mitochondria. Within a single cell, some mitochondria may carry a disease-causing mutation but others may not. As cells divide, mutant and normal mitochondria randomly segregate into the resulting daughter cells. Some daughter cells may receive a high ratio of mutant to normal mitochondria, whereas others may receive a low ratio. The ratio of mutant to normal mitochondria must exceed a certain threshold value before disease symptoms are observed. Because of the random nature of heteroplasmy, the symptoms of mitochondrial diseases often vary widely within a given family. Some members that carry the mutant mitochondria may be asymptomatic, whereas others exhibit mild to severe symptoms.

The occurrence of mitochondrial diseases in the human population has spawned a new reproductive technology called **three parent babies,** which was first carried out by John Zhang and colleagues in 2016. In this case, a female wanting to have more children had a mutation in a mitochondrial gene that can cause Leigh syndrome, a disorder that affects the developing nervous system. Approximately 25% of this female's mitochondria carried the disease-causing mutation, but this percentage was not high enough to cause disease symptoms. Unfortunately, the first two offspring of this female with heteroplasmy died, presumably because the percentage of disease-causing mitochondria in these offspring exceeded 25% in the nervous system.

Using the three parent baby method, a baby was produced by combining genetic material from three sources: the nuclear DNA from this female with heteroplasmy, the nuclear DNA from a male partner, and the mitochrondrial DNA from a donor female. How was this accomplished? A donor female provided an egg in which the nuclear DNA had been removed, but the egg still carried normal (nonmutant) mitochondria. The nucleus was removed from an egg provided by the female with heteroplasmy, which was then inserted into this donor egg. The resulting egg—with nuclear DNA from the female with heteroplasmy and mitochondrial DNA from a female donor—was then fertilized with the male partner's sperm. Thus, the genetic contributions of three individuals contributed to the resulting baby.

Extranuclear Genomes of Chloroplasts and Mitochondria Evolved from an Endosymbiotic Relationship

The idea that the nucleus, chloroplasts, and mitochondria contain their own separate genetic material may at first seem puzzling. Wouldn't it be simpler to have all of the genetic material in one place in the cell? The underlying reason for the distinct genomes of chloroplasts and mitochondria can be traced back to their evolutionary origin, which is thought to have involved a symbiotic association.

A symbiotic relationship occurs when two different species live together in a close association. The symbiont is the smaller of the two species; the host is the larger. The term **endosymbiosis** describes a symbiotic relationship in which the symbiont actually lives inside (*endo*- means "inside") the host. In 1883, Andreas Schimper proposed that chloroplasts were descended from an endosymbiotic relationship between cyanobacteria and eukaryotic cells. This idea, now known as the **endosymbiosis theory,** proposes that the ancient origin of chloroplasts was initiated when a cyanobacterium took up residence within a primordial eukaryotic cell (**Figure 5.17**). Over the course of evolution, the characteristics of the intracellular bacterial cell gradually changed to those of a chloroplast. In 1922, Ivan Wallin proposed an endosymbiotic origin for mitochondria also.

In spite of these hypotheses, the question of whether endosymbiosis had occurred was largely ignored until researchers in the 1950s discovered that chloroplasts and mitochondria contain their own genetic material. The issue of endosymbiosis was hotly debated after Lynn Margulis published a book entitled *Origin of Eukaryotic Cells* (1970). During the 1970s and 1980s, the advent of molecular genetic techniques allowed researchers to analyze genes from chloroplasts, mitochondria, bacteria, and eukaryotic nuclear genomes. They found that genes in chloroplasts and

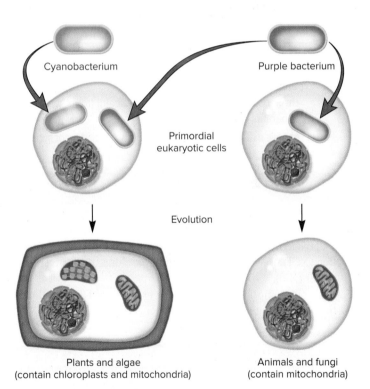

FIGURE 5.17   The endosymbiotic origin of chloroplasts and mitochondria. According to the endosymbiosis theory, chloroplasts descended from an endosymbiotic relationship between cyanobacteria and eukaryotic cells. This relationship arose when a bacterium took up residence within a primordial eukaryotic cell. Over the course of evolution, the intracellular bacterial cell gradually changed its characteristics, eventually becoming a chloroplast. Similarly, mitochondria are derived from an endosymbiotic relationship between purple bacteria and eukaryotic cells.

CONCEPT CHECK: How have chloroplast and mitochondrial genomes changed since the initial endosymbiosis events, which occurred hundreds of millions of years ago?

mitochondria are very similar to bacterial genes but not as similar to those found within the nuclei of eukaryotic cells. This observation provided strong support for the endosymbiotic origin of chloroplasts and mitochondria, which is now widely accepted.

The endosymbiosis theory proposes that the relationship provided eukaryotic cells with useful cellular characteristics. Chloroplasts were derived from cyanobacteria, a bacterial species that is capable of photosynthesis. The ability to carry out photosynthesis enabled algal and plant cells to use the energy from sunlight. By comparison, mitochondria are thought to have been derived from a different type of bacteria known as gram-negative nonsulfur purple bacteria. In this case, the endosymbiotic relationship enabled eukaryotic cells to synthesize greater amounts of ATP. It is less clear how the relationship would have been beneficial to cyanobacteria or purple bacteria, though the cytosol of a eukaryotic cell may have provided a stable environment with an adequate supply of nutrients.

During the evolution of eukaryotic species, most genes that were originally found in the genome of the primordial cyanobacteria and purple bacteria have been lost or transferred from the organelles to the nucleus. Such changes have occurred many times throughout evolution, so modern chloroplasts and mitochondria have lost most of the genes that are still found in present-day cyanobacteria and purple bacteria.

5.4 COMPREHENSION QUESTIONS

1. Extranuclear inheritance occurs due to
   a. chromosomes that have become detached from the spindle apparatus during meiosis.
   b. genetic material that is found in chloroplasts and mitochondria.
   c. mutations that disrupt the integrity of the nuclear membrane.
   d. none of the above.
2. A cross is made between a green-leaved four-o'clock plant and one with variegated leaves. If the variegated plant provides the pollen, the expected outcome of the phenotypes of the offspring will be
   a. all plants with green leaves.
   b. 3 plants with green leaves to 1 plant with variegated leaves.
   c. 3 plants with green leaves to 1 plant with white leaves.
   d. some plants with green leaves, some with variegated leaves, and some with white leaves.
3. Some human diseases are caused by mutations in mitochondrial genes. Which of the following statements is *false*?
   a. Human mitochondrial diseases follow a maternal inheritance pattern.
   b. Mutations associated with mitochondrial diseases often affect cells with a high demand for ATP.
   c. The symptoms associated with mitochondrial diseases tend to improve with age.
   d. Heteroplasmy plays a key role in the severity of mitochondrial disease symptoms.
4. Chloroplasts and mitochondria evolved from an endosymbiotic relationship involving
   a. purple bacteria and cyanobacteria, respectively.
   b. cyanobacteria and purple bacteria, respectively.
   c. cyanobacteria.
   d. purple bacteria.

KEY TERMS

5.1: maternal effect, reciprocal cross
5.2: epigenetics, dosage compensation, X-chromosome inactivation (XCI), Barr body, Lyon hypothesis, clone, X-inactivation center (Xic)
5.3: genomic imprinting, monoallelic expression, DNA methylation, imprinting control region (ICR)

5.4: nuclear genes, extranuclear inheritance (cytoplasmic inheritance), nucleoid, mitochondrial DNA (mtDNA), chloroplast DNA (cpDNA), maternal inheritance, heteroplasmy, heterogamous, three parent babies, endosymbiosis, endosymbiosis theory

CHAPTER SUMMARY

- Non-Mendelian inheritance refers to inheritance patterns that do not conform to (at least) one of the four rules of Mendelian inheritance.

5.1 Maternal Effect

- Maternal effect is an inheritance pattern in which the genotype of the female parent determines the phenotype of the offspring. It occurs because gene products of maternal effect genes are transferred from nurse cells to the oocyte. These gene products affect early stages of development (see Figures 5.1, 5.2).

5.2 Epigenetics: Dosage Compensation

- Epigenetic effects occur when a gene or chromosome is modified and gene expression is altered, but the modification is not permanent over the course of many generations.
- Dosage compensation often occurs in species in which males and females differ in their sex chromosomes (see Table 5.1).

- In mammals, the process of X-chromosome inactivation in females compensates for the single X chromosome found in males. The inactivated X chromosome is called a Barr body. The process can lead to a variegated phenotype, such as a calico cat (see Figures 5.3, 5.4).
- After it occurs during embryonic development, X-chromosome inactivation (XCI) is maintained when somatic cells divide (see Figures 5.5, 5.6).
- X-chromosome inactivation is controlled by the X-inactivation center (Xic). XCI occurs via nucleation, spreading, and maintenance phases (see Figure 5.7).

5.3 Epigenetics: Genomic Imprinting

- Genomic imprinting refers to a marking process in which an offspring expresses a gene that is inherited from one parent but not both (see Figures 5.8, 5.9).
- DNA methylation at an imprinting control region is the marking process that causes imprinting (see Figure 5.10).
- Human diseases such as Prader-Willi syndrome and Angelman syndrome are associated with genomic imprinting (see Figure 5.11, Table 5.2).

5.4 Extranuclear Inheritance

- Extranuclear inheritance is the inheritance of genetic material that is not found within the cell nucleus. It occurs via genetic material in mitochondria or chloroplasts.
- Mitochondria and chloroplasts have circular chromosomes in one or more nucleoids. These circular chromosomes carry relatively few genes compared to the number in the cell nucleus (see Figures 5.12, 5.13, Table 5.3).
- Maternal inheritance occurs when organelles, such as mitochondria or chloroplasts, are transmitted via the egg (see Figure 5.14).
- Heteroplasmy of chloroplasts can result in a variegated phenotype (see Figure 5.15).
- The transmission patterns of mitochondria and chloroplasts vary among different organisms (see Table 5.4).
- Three different cellular mechanisms can result in maternal inheritance (see Figure 5.16).
- Many diseases are caused by mutations in mitochondrial DNA (see Table 5.5).
- Chloroplasts and mitochondria were derived from ancient endosymbiotic relationships (see Figure 5.17).

PROBLEM SETS & INSIGHTS

MORE GENETIC TIPS 1. One strain of periwinkle plants has green leaves and another strain has white leaves. Both strains are true-breeding. You do not know if the phenotypic difference is due to alleles of a nuclear gene or an organellar gene. The two strains were analyzed using reciprocal crosses, and the following results were obtained:

A plant with green leaves is pollinated by a plant with white leaves: All offspring have green leaves.

A plant with white leaves is pollinated by a plant with green leaves: All offspring have white leaves.

Is this pattern of inheritance consistent with simple Mendelian inheritance, where green is dominant to white, and/or consistent with maternal inheritance?

TOPIC: *What topic in genetics does this question address?* The topic is inheritance patterns. More specifically, the question is about distinguishing nuclear and extranuclear inheritance patterns.

INFORMATION: *What information do you know based on the question and your understanding of the topic?* From the question, you know there are two strains of periwinkles, green- and white-leaved. From your understanding of the topic, you may remember that some genes are in the nucleus, whereas others are found in organelles, such as chloroplasts and mitochondria. Because nuclear genes segregate differently from organellar genes, one way to distinguish these inheritance patterns is to analyze the results of reciprocal crosses.

PROBLEM-SOLVING **S**TRATEGY: *Analyze data. Predict the outcome. Compare and contrast.* The data from reciprocal crosses may yield different results depending on the mode of inheritance. For example, if the gene is a nuclear gene and the green allele is dominant, you would predict that all of the offspring will be green-leaved for both crosses. This result was not obtained. On the other hand, if the gene is in the chloroplasts and follows maternal inheritance, the phenotype of the offspring will depend on which plant contributed the egg.

ANSWER: The data are consistent with maternal inheritance, because the phenotype of the offspring correlates with inheriting the gene from the plant contributing the egg cells.

2. A human male named Phillip has an X chromosome that is missing its Xic. Is this caused by a new mutation (one that occurred during gametogenesis), or could this mutation have occurred in an earlier generation and be found in the somatic cells of one of Phillip's parents? Explain your answer. How would this mutation affect Phillip's ability to produce viable offspring?

TOPIC: *What topic in genetics does this question address?* The topic is X-chromosome inactivation. More specifically, the question is about the inheritance of a mutation that removes the Xic from an X chromosome.

INFORMATION: *What information do you know based on the question and your understanding of the topic?* From the question, you know that Phillip has an X chromosome that is missing its Xic. From your understanding of the topic, you may remember that an X chromosome that is missing its Xic will not be inactivated. This is a lethal condition in females.

PROBLEM-SOLVING **S**TRATEGY: *Predict the outcome. Compare and contrast.* One strategy to solve this problem is to compare the outcomes if the missing Xic occurred in a previous generation (denoted with an asterisk) or if it occurred as a new mutation during oogenesis or spermatogenesis in one of Phillip's parents. Let's compare the following crosses, in which X* indicates an X chromosome that is missing its Xic:

X*X × XY: Female parent would have been inviable, so this cross is not possible.

XX × X*Y: Male parent could not pass X* to Phillip, who is a male.

XX × XY (new mutation during oogenesis): A male offspring with a missing Xic could be produced.

XX × XY (new mutation during spermatogenesis): A male parent could not pass X* to Phillip, who is a male.

ANSWER: The missing Xic must be due to a new mutation that occurred during oogenesis in Phillip's female parent. Phillip will pass this mutant X chromosome to female offspring and a Y chromosome to male offspring. Therefore, Phillip cannot produce living female offspring, because a missing Xic is lethal in females. However, Phillip can produce living male offspring.

---

3. A maternal effect gene in *Drosophila*, called *torso*, is found as a functional allele (*torso$^+$*) and a nonfunctional, recessive allele (*torso$^-$*) that prevents the correct development of anterior- and posterior-most structures. A wild-type male (*torso$^+$ torso$^+$*) is crossed to a female of unknown genotype. This mating produces larvae of which 100% are missing their anterior- and posterior-most structures and therefore die during early development. What are the genotype and the phenotype of the female fly in this cross? What are the genotypes and phenotypes of the female fly's parents?

TOPIC: *What topic in genetics does this question address?* The topic is non-Mendelian inheritance. More specifically, the question is about a maternal effect gene.

INFORMATION: *What information do you know based on the question and your understanding of the topic?* From the question, you know that a gene called *torso* can exist as a recessive allele that prevents the proper development of anterior- and posterior-most structures. You are also given the results of a cross in which all of the offspring are missing their anterior- and posterior-most structures, and die at the larval stage. From your understanding of the topic, you may remember that when a trait shows a maternal effect pattern of inheritance, the female parent's genotype determines the offspring's phenotype.

PROBLEM-SOLVING **S**TRATEGY: *Predict the outcome.* A strategy to solve this problem is to predict the outcome depending on the female parent's genotype. For maternal effect genes, you cannot use a Punnett square to predict the phenotype of the offspring. Instead, to predict their phenotype, you have to know the female parent's genotype. In this question, you already know the offspring's phenotype. With this information, you can deduce the possible genotype(s) of their female parent. Because the offspring are missing their anterior- and posterior-most structures,

you may realize that the female parent must be homozygous for the recessive *torso$^-$* allele because the female parent's genotype determines the offspring's phenotype.

ANSWER: Because the cross produced 100% abnormal offspring that were missing their anterior- and posterior-most structures, the female parent of the abnormal offspring must have been homozygous, *torso$^-$ torso$^-$*. Even so, this female must be phenotypically normal in order to reproduce. As shown below, the female of the abnormal offspring had a female parent that was heterozygous for the *torso* alleles and a male parent that was either heterozygous or homozygous for the *torso$^-$* allele.

$$torso^+\ torso^- \qquad \times \qquad torso^+\ torso^-\ \text{or}\ torso^-\ torso^-$$
$$\text{(female grandparent)} \qquad \downarrow \qquad \text{(male grandparent)}$$
$$torso^-\ torso^-$$
$$\text{(female parent of 100\%}$$
$$\text{abnormal offspring)}$$

The female parent of the abnormal offspring is phenotypically normal because the female grandparent was heterozygous and provided the gene product of the *torso$^+$* allele from the nurse cells. However, this homozygous female, whose genotype is *torso$^-$ torso$^-$*, will produce only abnormal offspring because those offspring did not receive the functional *torso$^+$* gene product.

---

4. An individual named Pat with Prader-Willi syndrome (PWS) produced an offspring named Lee with Angelman syndrome (AS). The other parent of Lee does not have either syndrome. How might this occur? What are the sexes of Pat and Lee?

TOPIC: *What topic in genetics does this question address?* The topic is inheritance of genetic diseases. More specifically, the question is about PWS and AS, which are associated with deletions and genomic imprinting.

INFORMATION: *What information do you know based on the question and your understanding of the topic?* From the question, you know an individual with Prader-Willi syndrome produced an offspring with Angelman syndrome. From your understanding of the topic, you may remember that the *PWS* gene is silenced during oogenesis and the *AS* gene is silenced during spermatogenesis. The *PWS* and *AS* genes are closely linked along the same chromosome and a small deletion may remove both of them.

PROBLEM-SOLVING **S**TRATEGY: *Predict the outcome. Compare and contrast.* You cannot use a Punnett square to solve a problem involving genomic imprinting. An offspring with one of these syndromes does not have an active copy of the gene associated with the syndrome. In this case, Lee has AS and therefore does not have an active copy of the *AS* gene. You would assume that Lee inherited the deletion from Pat, because the other parent does not have either syndrome. Because the male parent naturally silences the *AS* gene, the deletion was inherited from the female parent (see Figure 5.11). Therefore, Lee does not have an active copy of the *AS* gene and has Angelman syndrome.

ANSWER: Pat is a female, who carried a deletion that encompassed both the *PWS* and *AS* genes. If a female transmits the deletion to either a female or male offspring, that offspring will have AS. Therefore, Lee could be either a female or male.

Conceptual Questions

C1. Define the term *epigenetics*, and describe two examples of epigenetic effects.

C2. Describe the inheritance pattern of maternal effect genes. Explain how the maternal effect occurs at the cellular level. What are the expected functional roles of the proteins that are coded by maternal effect genes?

C3. A maternal effect gene exists in a dominant *N* (functional) allele and a recessive *n* (nonfunctional) allele. The *n* allele causes a smaller body. What are the ratios of genotypes and phenotypes for the offspring of the following crosses?

  A. *nn* female × *NN* male

  B. *NN* female × *nn* male

  C. *Nn* female × *Nn* male

C4. A *Drosophila* embryo dies during early embryogenesis due to a recessive allele of a maternal effect gene called *bicoid⁻*. The wild-type allele is designated *bicoid⁺*. What are the genotypes and phenotypes of the embryo's female parent and maternal grandparents? For clarity, let's refer to the embryo's female parent as female P, and the maternal grandparents as female GP and male GP, where P stands for parent and GP stands for grandparent.

C5. For Mendelian inheritance, the nuclear genotype (i.e., the alleles found on chromosomes in the cell nucleus) directly influences an offspring's traits. In contrast, for non-Mendelian inheritance patterns, the offspring's phenotype cannot be reliably predicted solely from its genotype. For each of the following traits, what do you need to know to predict the phenotypic outcome?

  A. Small size due to a mutant *Igf2* allele

  B. Direction of snail shell coiling

  C. Leber hereditary optic neuropathy (LHON)

C6. Suppose a maternal effect gene exists as a functional dominant allele and a nonfunctional recessive allele that causes a developmental disorder. A female with the disorder produces all offspring without the disorder. Identify the genotype of the female parent, and explain your answer.

C7. Suppose that a maternal effect gene affects the anterior morphology in house flies. The gene exists in a dominant (functional) allele, *H*, and a recessive (nonfunctional) allele, *h*, which causes a small head. A female fly with a normal-size head is mated to a true-breeding male with a small head. All of the offspring have small heads. What are the genotypes of the female parent and offspring? Explain your answer.

C8. Explain why maternal effect genes exert their effects during the early stages of development.

C9. As described in Chapter 21, researchers have been able to clone mammals by fusing a cell having a diploid nucleus (i.e., a somatic cell) with an egg that has had its nucleus removed.

  A. With regard to maternal effect genes, would the phenotype of such a cloned animal be determined by the animal that donated the egg or by the animal that donated the somatic cell? Explain.

  B. Would the cloned animal inherit extranuclear traits from the animal that donated the egg or from the animal that donated the somatic cell? Explain.

  C. In what ways would you expect this cloned animal to be similar to or different from the animal that donated the somatic cell? Is it fair to call such an animal a clone of the animal that donated the diploid nucleus?

C10. With regard to the numbers of sex chromosomes, explain why dosage compensation is necessary.

C11. What is a Barr body? How is its structure different from that of other chromosomes in the cell? How does the structure of a Barr body affect the level of X-linked gene expression?

C12. Describe three distinct mechanisms for accomplishing dosage compensation in different species.

C13. Describe when X-chromosome inactivation occurs and how this process leads to phenotypic outcomes at the organism level. In your answer, you should explain why XCI causes results such as variegated coat patterns in mammals. Why do two different calico cats have their patches of orange and black fur in different places? Explain whether or not a variegated coat pattern due to XCI could occur in marsupials.

C14. Describe the molecular process of X-chromosome inactivation. This description should include the three phases of inactivation and the role of the Xic. Explain what happens to the X chromosomes during embryogenesis, in adult somatic cells, and during oogenesis.

C15. On rare occasions, a human male is born who is somewhat feminized compared with other males. Microscopic examination of the cells of one such individual revealed that this male has a single Barr body in each cell. What is the chromosomal composition of this individual?

C16. How many Barr bodies would you expect to find in humans with the following abnormal compositions of sex chromosomes?

  A. XXY

  B. XYY

  C. XXX

  D. X0 (a person with just a single X chromosome)

C17. Certain forms of human color blindness are inherited as X-linked recessive traits. Hemizygous males are color blind, but heterozygous females are not. However, heterozygous females sometimes have partial color blindness.

  A. Discuss why heterozygous females may have partial color blindness.

  B. Doctors identified an unusual case in which a heterozygous female was color blind in the right eye but had normal color vision in the left eye. Explain how this might have occurred.

C18. A black female cat (X^BX^B) and an orange male cat (X^OY) were mated to each other and produced a male cat that was calico. Which sex chromosomes did this male offspring inherit from its female parent and male parent? Remember that the presence of the Y chromosome determines maleness in mammals.

C19. What is the spreading phase of X-chromosome inactivation? Why do you think it is called a spreading phase?

C20. When does the erasure and reestablishment phase of genomic imprinting occur? Explain why it is necessary to erase an imprint and then reestablish it in order to maintain imprinting from the parent of the same sex.

C21. In what types of cells would you expect de novo methylation to occur? In what cell types would it not occur?

C22. On rare occasions, people are born with a condition known as uniparental disomy. Such an individual inherits both copies of a chromosome from one parent and no copies from the other parent. This occurs when two abnormal gametes happen to complement each other to produce a diploid zygote. For example, an abnormal sperm that lacks chromosome 15 could fertilize an egg that contains two copies of chromosome 15. In this situation, the individual has maternal uniparental disomy 15 because both copies of chromosome 15 were inherited from the female parent. Alternatively, an abnormal sperm with two copies of chromosome 15 could fertilize an egg with no copies. This is known as paternal uniparental disomy 15. If a female is born with paternal uniparental disomy 15, would you expect this female to have Angelman syndrome (AS), have Prader-Willi syndrome (PWS), or be unaffected? Explain. Would you expect this female to produce unaffected offspring or offspring affected with AS or PWS?

C23. Genes that cause Prader-Willi syndrome and Angelman syndrome are closely linked along chromosome 15. Although people with these syndromes do not usually reproduce, let's suppose that a couple produces two children with Angelman syndrome. The oldest child (named Pat) grows up and has two children with Prader-Willi syndrome. The second child (named Robin) grows up and has one child with Angelman syndrome.

A. Are Pat and Robin's parents both phenotypically unaffected or is one of them affected with Angelman or Prader-Willi syndrome? If one of them does have one of these syndromes, state whether it is the female parent or the male parent, and explain how you know.

B. What are the sexes of Pat and Robin? Explain.

C24. How is the process of X-chromosome inactivation similar to genomic imprinting? How is it different?

C25. What is extranuclear inheritance? Describe two examples.

C26. What is a reciprocal cross? Suppose that a gene in fruit flies occurs as a wild-type allele and a recessive mutant (nonfunctional) allele. The recessive allele results in an inability to fly. If the wild-type allele is represented by F, and the recessive allele by f, describe how you would set up a cross and a reciprocal cross involving a true-breeding individual carrying the F allele and a true-breeding individual carrying the mutant allele, f? How would the results differ if the gene was autosomally inherited versus maternally inherited?

C27. Among different species, does extranuclear inheritance always follow a maternal inheritance pattern?

C28. Describe three cellular mechanisms that can result in maternal inheritance in animals. Give one specific example for each of the three types.

C29. Discuss the structure and organization of the mitochondrial and chloroplast genomes. How large are they, how many genes do they contain, and how many copies of the genome are found in each organelle?

C30. Explain the likely evolutionary origin of mitochondrial and chloroplast genomes. How have the sizes of the mitochondrial and chloroplast genomes changed since their origin? How did this occur?

C31. Is each of the following traits or diseases determined by nuclear genes?

A. Direction of snail shell coiling

B. Prader-Willi syndrome

C. Leber hereditary optic neuropathy (LHON)

C32. Acute murine leukemia virus (AMLV) causes leukemia in mice. This virus is easily passed from female parent to offspring through the milk. (Note: Even though newborn offspring acquire the virus, they may not develop leukemia until much later in life. Testing can determine if an animal carries the virus.) Describe how the development of leukemia due to AMLV resembles a maternal inheritance pattern. How could you show experimentally that this form of leukemia is not caused by extranuclear inheritance?

C33. Describe how a biparental pattern of extranuclear inheritance could resemble a Mendelian pattern of inheritance for a particular gene. How would the patterns differ?

Experimental Questions

E1. Figure 5.1 illustrates an example of a maternal effect gene. Explain how Sturtevant deduced a maternal effect gene based on the inheritance patterns of the F_2 and F_3 generations.

E2. Discuss the types of experimental observations that Lyon brought together to propose a hypothesis for X-chromosome inactivation. In your own words, explain how these observations were consistent with Lyon's hypothesis.

E3. Chapter 20 describes two blotting methods, Northern blotting and Western blotting, which are used to detect gene products. Northern blotting detects RNA and Western blotting detects proteins. Suppose that a female fruit fly is heterozygous for a maternal effect gene, gene B. The female is Bb. The dominant allele, B, codes a functional mRNA that is 550 nucleotides long. The recessive allele, b, codes a shorter mRNA that is 375 nucleotides long. (Allele b is due to a deletion within this gene.) How could you use one or both of the blotting methods to show that nurse cells transfer gene products from gene B to developing oocytes? You may assume that you can dissect the ovaries of fruit flies and isolate oocytes separately from nurse cells. In your answer, describe your expected results.

E4. Suppose that a trait in some mice is an unusually long tail. You initially have a true-breeding strain with normal-length tails and a true-breeding strain with long tails. You then make the following types of crosses:

Cross 1: When true-breeding females with normal tails are crossed to true-breeding males with long tails, all F_1 offspring have long tails.

Cross 2: When true-breeding females with long tails are crossed to true-breeding males with normal tails, all F_1 offspring have normal tails.

Cross 3: When F_1 females from cross 1 are crossed to true-breeding males with normal tails, all offspring have normal tails.

Cross 4: When F_1 males from cross 1 are crossed to true-breeding females with long tails, half of the offspring have normal tails and half have long tails.

Explain the pattern of inheritance of this trait.

E5. You have a female snail with a shell that coils to the right, but you do not know its genotype. You may assume that right coiling (*D*) is dominant to left coiling (*d*). You also have male snails of known genotype. How would you determine the genotype of this female snail? In your answer, describe your expected results depending on whether the female is *DD*, *Dd*, or *dd*.

E6. On a camping trip on an uninhabited island, you find one male snail with a shell that coils to the right. However, in this same area, you find several shells (not containing living snails) that coil to the left. Therefore, you conclude that you are not certain of the genotype of this male snail. On a different island, you find a large colony of snails of the same species. All of these snails' shells coil to the right, and every snail shell that you find on this second island coils to the right. With regard to the maternal effect gene that determines coiling pattern, how would you determine the genotype of the male snail that you found on the uninhabited island? In your answer, describe your expected results.

E7. Figure 5.6 describes certain results of X-chromosome inactivation in mammals. If fast and slow alleles that code glucose-6-phosphate dehydrogenase (G-6-PD) exist in other species, what would be the expected results of gel electrophoresis for a heterozygous female of each of the following species?

A. Marsupial

B. *Drosophila melanogaster*

C. *Caenorhabditis elegans* (Note: We are considering the hermaphrodite in *C. elegans* to be equivalent to a female.)

E8. Two male mice, male A and male B, are both normal size. Male A was from a litter that contained half normal-size mice and half small-size mice. The female parent of male A was known to be homozygous for the functional *Igf2* allele. Male B was from a litter of eight mice that were all normal size. The parents of male B were a normal-size male and a small-size female. Male A and male B were put into a cage with two female mice, female A and female B. Female A is small size, and female B is normal size. The parents of these two females were unknown, although it was known that they were from the same litter. The mice were allowed to mate with each other, and the following data were obtained:

Female A gave birth to three small-size babies and four normal-size babies.

Female B gave birth to four normal-size babies and two small-size babies.

Which male(s) mated with female A and female B? Explain.

E9. In the experiment of Figure 5.6, why does a clone of cells produce only one type of G-6-PD enzyme? What would you expect to happen if a clone was derived from an early embryonic cell? Why does the initial sample of tissue produce both forms of G-6-PD?

E10. Chapter 20 describes a blotting method known as Northern blotting that is used to determine the amount of mRNA produced by a particular gene. In this method, the amount of a specific mRNA produced by cells is detected as a band on a gel. If one type of cell produces twice as much of a particular mRNA as another cell, the band will appear twice as thick. Also, sometimes mutations affect the length of mRNA that is transcribed from a gene. For example, a small deletion within a gene may shorten an mRNA. Northern blotting also can discern the sizes of mRNAs.

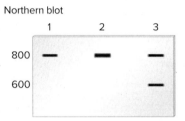

Northern blot

Lane 1 is a Northern blot of mRNA from cell type A that is 800 nucleotides long.

Lane 2 is a Northern blot of the same mRNA from cell type B. (Cell type B produces twice as much of this RNA as cell type A.)

Lane 3 shows a heterozygote in which one of the two genes has a deletion, which shortens the mRNA by 200 nucleotides.

Here is the question: Suppose an X-linked gene exists as two alleles: *B* and *b*. Allele *B* codes an mRNA that is 750 nucleotides long, and allele *b* codes a shorter mRNA that is 675 nucleotides long. Draw the expected results of a Northern blot using mRNA isolated from the same type of somatic cells taken from the following individuals:

A. First lane is mRNA from an $X^b Y$ male fruit fly.

Second lane is mRNA from an $X^b X^b$ female fruit fly.

Third lane is mRNA from an $X^B X^b$ female fruit fly.

B. First lane is mRNA from an $X^B Y$ male mouse.

Second lane is mRNA from an $X^B X^b$ female mouse.*

Third lane is mRNA from an $X^B X^B$ female mouse.*

C. First lane is mRNA from an $X^B 0$ male *C. elegans*.

Second lane is mRNA from an $X^B X^b$ hermaphrodite *C. elegans*.

Third lane is mRNA from an $X^B X^B$ hermaphrodite *C. elegans*.

*The sample is taken from an adult female mouse. It is not a clone of cells. It is a tissue sample, like the one described in the experiment of Figure 5.6.

Questions for Student Discussion/Collaboration

1. Let's suppose a recessive maternal effect allele is lethal in fruit flies. How would you identify heterozygous individuals that are carrying the recessive maternal effect allele? How would you maintain this strain of flies in a laboratory over many generations?

2. According to the endosymbiosis theory, mitochondria and chloroplasts are derived from bacteria that took up residence within eukaryotic cells. At one time, prior to being taken up by eukaryotic

cells, these bacteria were free-living organisms. However, we cannot take a mitochondrion or chloroplast out of a living eukaryotic cell and get it to survive and replicate on its own. Discuss why not.

Note: All answers are available for the instructor in Connect; the answers to the even-numbered questions and all of the Concept Check and Comprehension Questions are in Appendix B.

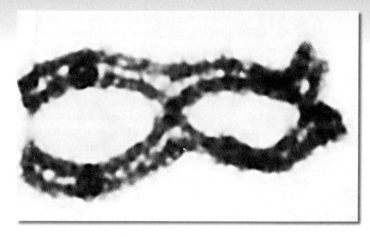

Crossing over during meiosis. This event provides a way to reassort the alleles of genes that are located on the same chromosome.
Stanley K. Sessions

6
GENETIC LINKAGE AND MAPPING IN EUKARYOTES

In Chapter 3, we focused on Mendel's laws of inheritance. According to these principles, we expect that two different genes will segregate and independently assort themselves during the process that creates gametes. After Mendel's work was rediscovered at the turn of the twentieth century, chromosomes were identified as the cellular structures that carry genes. The chromosome theory of inheritance explained how the transmission of chromosomes is responsible for the passage of genes from parents to offspring.

When geneticists first realized that chromosomes contain the genetic material, they began to suspect that discrepancies might sometimes occur between the outcomes predicted by the law of independent assortment of genes and the actual behavior of chromosomes during meiosis. In particular, geneticists assumed that each species must contain thousands of different genes, yet cytological studies revealed that most species have at most a few dozen chromosomes. Therefore, it seemed likely, and turned out to be true, that each chromosome carries many hundreds or even thousands of different genes. The transmission of genes located close to each other on the same chromosome violates the law of independent assortment.

In this chapter, we will consider patterns of inheritance that occur when different genes are situated on the same chromosome.

In addition, we will briefly explore how the data from genetic crosses are used to construct a **genetic map**—a diagram that shows the order of genes along a chromosome. Newer methods of gene mapping are examined in Chapter 22. However, an understanding of traditional mapping studies, as described in this chapter, will strengthen your appreciation for these more recently developed molecular approaches. More importantly, traditional mapping studies further illustrate how the location of two or more genes on the same chromosome can affect the patterns of gene transmission from parents to offspring.

6.1 OVERVIEW OF LINKAGE

Learning Outcomes:

1. Define *genetic linkage*.
2. Explain how linkage affects the outcome of crosses.

In eukaryotic species, each linear chromosome contains a very long segment of DNA along with many different proteins. A chromosome contains many individual functional units—called genes—that influence an organism's traits. A typical chromosome is expected to contain many hundreds to perhaps a few thousand different genes.

- The term **synteny** refers to the situation in which two or more genes are located on the same chromosome. Genes that are syntenic are physically linked to each other, because each eukaryotic chromosome contains a single, continuous, linear molecule of DNA.
- **Genetic linkage** is the phenomenon in which genes that are close together on the same chromosome tend to be transmitted as a unit. Therefore, genetic linkage has an influence on inheritance patterns.

Chromosomes are sometimes called **linkage groups,** because a chromosome contains a group of genes that are physically linked together. In species that have been characterized genetically, the number of linkage groups equals the number of chromosome types. For example, human somatic cells have 46 chromosomes, which are composed of 22 types of autosomes that come in pairs plus one pair of sex chromosomes, X and Y. Therefore, humans have 22 autosomal linkage groups and an X chromosome linkage group, and human males also have a Y chromosome linkage group. In addition, the human mitochondrial genome is another linkage group.

Geneticists are often interested in the transmission of two or more characters in a genetic cross. An experiment that follows the variants of two different characters in a cross is called a **two-factor cross;** one that follows three characters is a **three-factor cross;** and so on. The outcome of a two-factor or three-factor cross depends on whether or not the genes are linked to each other on the same chromosome. Next, we examine how linkage affects the transmission patterns of two characters. Later sections will examine crosses involving three characters.

Bateson and Punnett Discovered Two Genes That Did Not Assort Independently

An early study indicating that some genes may not assort independently was carried out by William Bateson and Reginald Punnett in 1905. According to Mendel's law of independent assortment, a two-factor cross between two individuals that are heterozygous for two genes should yield a 9:3:3:1 phenotypic ratio among the offspring. However, a surprising result occurred when Bateson and Punnett conducted a cross of sweet peas involving two different characters: flower color and pollen shape.

As seen in **Figure 6.1**, they began by crossing a true-breeding strain with purple flowers (*PP*) and long pollen (*LL*) to a true-breeding strain with red flowers (*pp*) and round pollen (*ll*). This cross yielded an F_1 generation of plants that all had purple flowers and long pollen (*PpLl*). An unexpected result came from the F_2 generation. Even though the F_2 generation had four different phenotypic categories, the observed numbers of offspring with the various phenotypes did not conform to a 9:3:3:1 ratio. Bateson and Punnett found that the F_2 generation had a much greater proportion of the two phenotypes found in the P generation—purple flowers with long pollen and red flowers with round pollen. Therefore, they suggested that the transmission of these two characters from the P generation to the F_2 generation was somehow coupled; that is, the alleles were not assorted in an independent manner. However, Bateson and

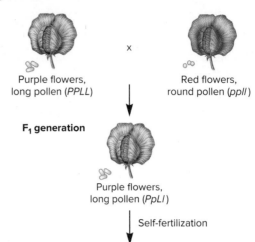

P generation

Purple flowers, long pollen (*PPLL*)   x   Red flowers, round pollen (*ppll*)

F_1 generation

Purple flowers, long pollen (*PpLl*)

Self-fertilization

F_2 generation	Observed number	Ratio	Expected number	Ratio
Purple flowers, long pollen	296	15.6	240	9
Purple flowers, round pollen	19	1.0	80	3
Red flowers, long pollen	27	1.4	80	3
Red flowers, round pollen	85	4.5	27	1

FIGURE 6.1   An experiment of Bateson and Punnett with sweet peas, showing that independent assortment does not always occur. Note: The expected numbers are rounded to the nearest whole number.

CONCEPT CHECK: Which types of offspring in the F_2 generation of this experiment are found in excess of what Mendel's law of independent assortment predicts?

Punnett did not realize that this coupling was due to the linkage of the flower color gene and the pollen shape gene on the same chromosome.

6.1 COMPREHENSION QUESTIONS

1. Genetic linkage occurs because
   a. genes that are on the same chromosome may affect the same character.
   b. genes that are close together on the same chromosome tend to be transmitted together to offspring.
   c. genes that are on different chromosomes are independently assorted.
   d. None of the above explains why linkage occurs.

2. In the experiment by Bateson and Punnett, which of the following observations suggested genetic linkage in the sweet pea?
   a. A 9:3:3:1 ratio was observed in the F_2 offspring.
   b. A 9:3:3:1 ratio was not observed in the F_2 offspring.
   c. An unusually high number of F_2 offspring had the phenotypes of the P generation.
   d. Both b and c suggested genetic linkage.

6.2 RELATIONSHIP BETWEEN LINKAGE AND CROSSING OVER

Learning Outcomes:

1. Describe how crossing over can change the arrangements of alleles along a chromosome.
2. Explain how the distance between linked genes affects the proportions of recombinant and nonrecombinant offspring.
3. Apply a chi square test to distinguish between linkage and independent assortment.
4. Analyze the experiment of Stern, and explain how it indicated that recombinant offspring carry chromosomes that are the result of crossing over.

Even when the alleles for different genes are linked on the same chromosome, the linkage can be altered during meiosis. In diploid eukaryotic species, homologous chromosomes can exchange pieces with each other, a phenomenon called **crossing over.** This event most commonly occurs during prophase of meiosis I. As discussed in Chapter 2, the replicated chromosomes, known as sister chromatids, associate with the homologous sister chromatids to form a structure known as a **bivalent.** A bivalent is composed of two pairs of sister chromatids. In prophase of meiosis I, a sister chromatid of one pair may cross over with a sister chromatid from the homologous pair (refer back to Figure 2.10). In this section, we will consider how crossing over affects the pattern of inheritance for genes that are linked on the same chromosome.

Crossing Over May Produce Recombinant Genotypes

Figure 6.2 considers meiosis when two genes are linked on the same chromosome. Prior to meiosis, one of the chromosomes carries the *B* and *A* alleles, while the homolog carries the *b* and *a* alleles. In Figure 6.2a, no crossing over has occurred. Therefore, the resulting haploid cells contain the same combination of alleles as the original chromosomes. In this case, two haploid cells carry the dominant *B* and *A* alleles, and the other two carry the recessive *b* and *a* alleles. The arrangement of linked alleles has not been altered.

In contrast, Figure 6.2b illustrates what can happen when crossing over occurs. The two lowermost haploid cells contain combinations of alleles, namely, *B* and *a* in one or *b* and *A* in the other, which differ from those in the original chromosomes. In these two cells, the grouping of linked alleles has changed. The haploid cells carrying the *B* and *a* alleles or the *b* and *A* alleles are called **recombinant** cells. If such haploid cells are gametes that participate in fertilization, the resulting offspring are called recombinant offspring. These offspring can display combinations of traits that are different from those of either parent. In contrast, offspring that have inherited chromosomes carrying the same combinations of alleles found in the chromosomes of their parents are known as **nonrecombinant** offspring.

When offspring inherit a combination of two or more alleles or traits that are different from either of their parents, this event is known as **genetic recombination.** It commonly occurs in two ways:

1. When two or more genes are linked on the same chromosome, crossing over during meiosis can result in genetic recombination.

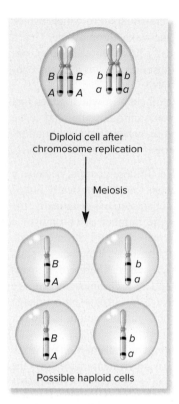

(a) Without crossing over, linked alleles segregate together.

(b) Crossing over can reassort linked alleles.

FIGURE 6.2 Consequences of crossing over during meiosis. **(a)** In the absence of crossing over, the *B* and *A* alleles and the *b* and *a* alleles are maintained in the same arrangement. **(b)** Crossing over has occurred in the region between the two genes. The lowermost two cells are recombinant haploid cells, each with a new combination of alleles.

CONCEPT CHECK: If a crossover occurred in the short region between gene *A* and the tip of the chromosome, would this event affect the arrangement of the *B* and *A* alleles?

2. When two or more genes are on different chromosomes, the independent assortment of those chromosomes during meiosis can result in genetic recombination.

In this chapter, our definition of **recombinant offspring** is the following:

Recombinant offspring are produced by crossing over between two homologous chromosomes during meiosis in a parent, leading to a different combination of alleles along a chromosome compared to that of either parent.

In other words, recombinant offspring carry one or more chromosomes that are the product of crossing over.

Morgan Provided Evidence for the Linkage of X-Linked Genes and Proposed That Crossing Over Between X Chromosomes Can Occur

The first direct evidence that different genes are physically located on the same chromosome came from the studies of Thomas Hunt Morgan in 1911, who investigated the inheritance of different characters that had been shown to follow an X-linked pattern of

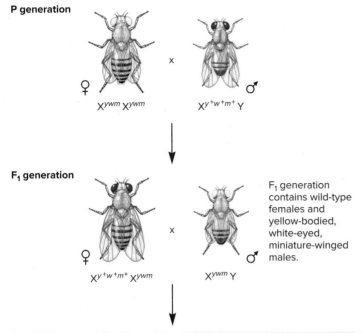

P generation

$X^{ywm} X^{ywm}$ ♀   × $X^{y^+w^+m^+} Y$ ♂

F₁ generation

$X^{y^+w^+m^+} X^{ywm}$ ♀   × $X^{ywm} Y$ ♂

F₁ generation contains wild-type females and yellow-bodied, white-eyed, miniature-winged males.

F₂ generation	Females	Males	Total
Gray body, red eyes, long wings	439	319	758
Gray body, red eyes, miniature wings	208	193	401
Gray body, white eyes, long wings	1	0	1
Gray body, white eyes, miniature wings	5	11	16
Yellow body, red eyes, long wings	7	5	12
Yellow body, red eyes, miniature wings	0	0	0
Yellow body, white eyes, long wings	178	139	317
Yellow body, white eyes, miniature wings	365	335	700

FIGURE 6.3 **Morgan's three-factor crosses involving three X-linked traits in *Drosophila*.**

Genes→Traits Three genes that govern body color, eye color, and wing length are all found on the X chromosome of fruit flies. Therefore, the offspring tend to inherit a nonrecombinant pattern of alleles ($y^+w^+m^+$ or ywm). Figure 6.5 explains how single and double crossovers can create a recombinant pattern of alleles.

CONCEPT CHECK: Of the eight possible phenotypic combinations in the F₂ generation, which ones are the product of a single crossover?

inheritance. **Figure 6.3** illustrates an experiment involving three characters that Morgan studied. The P-generation crosses were between female fruit flies that had yellow bodies (yy), white eyes (ww), and miniature wings (mm) and wild-type males. The wild-type alleles for these three genes are designated y^+ (gray body), w^+ (red eyes), and m^+ (long wings). As expected, the phenotypes of the F₁ generation were wild-type females and males with yellow bodies, white eyes, and miniature wings. The linkage of these genes was revealed when the F₁ flies were mated to each other and the F₂ generation examined.

Instead of equal proportions of the eight possible phenotypes, Morgan observed much higher proportions of the combinations of traits found in the P generation.

- There were 758 flies with gray bodies, red eyes, and long wings, and 700 flies with yellow bodies, white eyes, and miniature wings. These combinations of traits were found in the P generation.

- Morgan's explanation for this higher proportion of nonrecombinant patterns of traits was that all three genes are located on the X chromosome and, therefore, tend to be transmitted together as a unit.

However, to fully account for the data shown in Figure 6.3, Morgan needed to explain why a significant proportion of the F₂ generation had recombinant arrangements of traits. Along with the two nonrecombinant phenotypes, five other phenotypic combinations appeared that were not found in the P generation. To explain these data, Morgan considered studies conducted in 1909 by cytologist Frans Janssens, who observed chiasmata under the microscope and proposed that crossing over involves a physical exchange between homologous chromosomes.

- Morgan hypothesized that the genes for body color, eye color, and wing length are all located on the same chromosome, namely, the X chromosome. Therefore, the alleles for all three genes are most likely to be inherited together.
- Due to crossing over, Morgan also proposed that the homologous X chromosomes (in the female) can exchange pieces of chromosomes and produce new (recombinant) patterns of alleles and recombinant patterns of traits in the F₂ generation.
- Note: It was assumed that crossing over did not occur between the X chromosome and Y chromosome and that these three genes are not found on the Y chromosome.

To appreciate Morgan's proposals, let's simplify the data and consider only two of the three genes: those that affect body color and eye color. If we use the data from Figure 6.3, the following totals are obtained:

Gray body, red eyes	1159	
Yellow body, white eyes	1017	
Gray body, white eyes	17	} Recombinant
Yellow body, red eyes	12	} offspring
Total	2205	

Figure 6.4 considers how Morgan's proposals could account for these data. The nonrecombinant offspring with gray bodies and red eyes or yellow bodies and white eyes were produced when no crossing over had occurred between the two genes (Figure 6.4a). This was the more common situation. By comparison, crossing over could alter the pattern of alleles along each chromosome and account for the recombinant offspring (Figure 6.4b). Why were there relatively few recombinant offspring? These two genes are very close together on the same chromosome, which makes it unlikely that a crossover would be initiated between them. As described next, the distance between two genes is an important factor in determining the relative proportions of recombinant offspring.

The Likelihood of Crossing Over Between Two Genes Depends on the Distance Between Them

In the experiment of Figure 6.3, Morgan also noticed a quantitative difference between recombinant offspring involving body color and eye color versus those involving eye color and wing length. This quantitative difference is revealed by reorganizing the data of Figure 6.3 by pairs of genes.

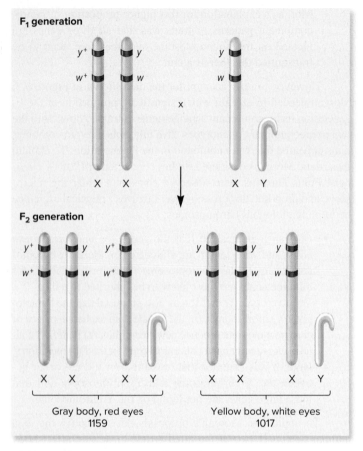

(a) No crossing over, nonrecombinant offspring

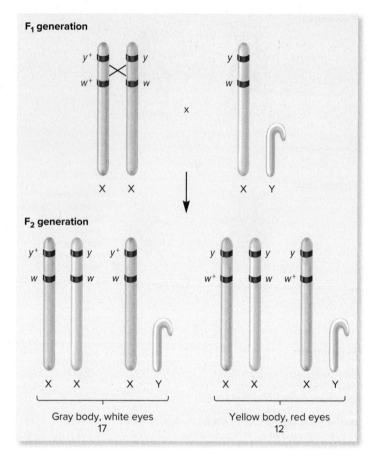

(b) Crossing over, recombinant offspring

FIGURE 6.4 **Morgan's explanation for nonrecombinant and recombinant offspring.** As described in Chapter 2, crossing over occurs when chromosomes form bivalents, that is, when homologous pairs of sister chromosome align with each other. For simplicity, this figure shows only two X chromosomes (one from each homolog) rather than four chromatids, which would be present when bivalents form during meiosis. Also note that this figure shows only a portion of the X chromosome. A map of the entire X chromosome is shown in Figure 6.7.

CONCEPT CHECK: Why are the nonrecombinant offspring more common than the recombinant offspring in this case?

Gray body, red eyes	1159	
Yellow body, white eyes	1017	
Gray body, white eyes	17	Recombinant
Yellow body, red eyes	12	offspring
Total	2205	
Red eyes, long wings	770	
White eyes, miniature wings	716	
Red eyes, miniature wings	401	Recombinant
White eyes, long wings	318	offspring
Total	2205	

Morgan found a substantial difference between the numbers of recombinant offspring when pairs of genes were considered separately. Recombinant patterns involving only eye color and wing length were fairly common—401 + 318 recombinant offspring. In sharp contrast, recombinant patterns for body color and eye color were quite rare—17 + 12 recombinant offspring.

To explain these data, Morgan proposed the following:

• The likelihood of crossing over depends on the distance between two genes.
• If two genes are far apart from each other, crossing over is more likely to occur in the region between them than it is in the shorter region between two genes that are close together.

Figure 6.5 illustrates the possible events that occurred in the F_1 female flies of Morgan's experiment. One of the X chromosomes carried all three dominant alleles; the other had all three recessive alleles. During oogenesis in the F_1 female flies, crossing over may or may not have occurred in this region of the X chromosome.

• If no crossing over occurred, the nonrecombinant phenotypes were produced in the F_2 offspring (Figure 6.5a).
• A crossover sometimes occurred between the eye color gene and the wing length gene to produce recombinant offspring with gray bodies, red eyes, and miniature wings or with yellow bodies, white eyes, and long wings (Figure 6.5b). According to Morgan's proposal, such an

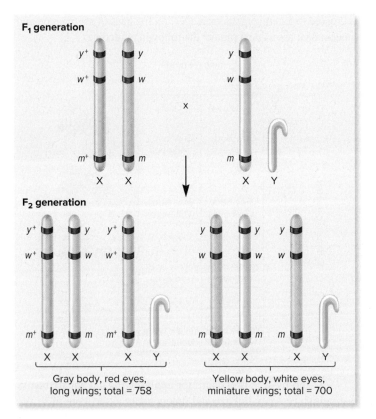

(a) No crossing over in this region, very common

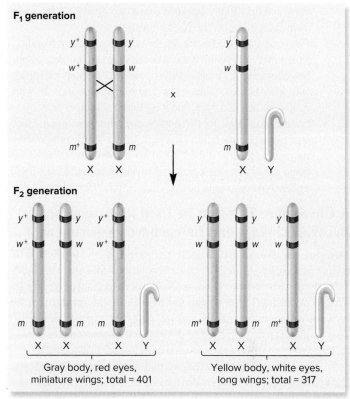

(b) Crossover between eye color and wing length genes, fairly common

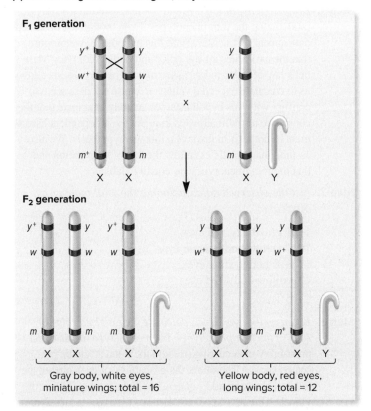

(c) Crossover between body color and eye color genes, uncommon

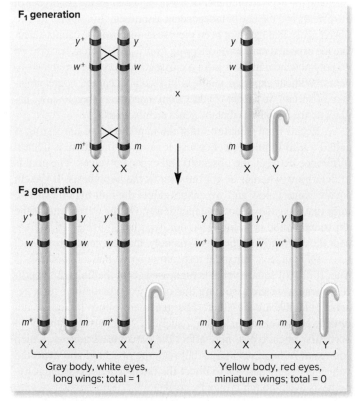

(d) Double crossover, very uncommon

FIGURE 6.5 Morgan's explanation for different proportions of recombinant offspring. Crossing over is more likely to occur between two genes that are relatively far apart than between two genes that are very close together. A double crossover is particularly uncommon.

INTERACTIVE
EXERCISE

CONCEPT CHECK: Why are the types of F₂ offspring shown in part (b) of Figure 6.5 more numerous than those shown in part (c)?

event is fairly likely because these two genes are far apart from each other on the X chromosome.

- The body color and eye color genes are very close together, which makes crossing over between them an unlikely event. Nevertheless, it occasionally occurred, yielding offspring with gray bodies, white eyes, and miniature wings or with yellow bodies, red eyes, and long wings (Figure 6.5c).
- It was also possible for two homologous chromosomes to cross over twice (Figure 6.5d). However, such a double crossover is very unlikely. Among the 2205 offspring Morgan examined, he found only 1 fly with a gray body, white eyes, and long wings, a phenotype that could be explained by this phenomenon.

A Chi Square Test Can Be Used to Distinguish Between Linkage and Independent Assortment

Now that you have an appreciation for linkage and the production of recombinant offspring, let's consider how an experimenter can objectively decide whether two genes are linked or assort independently. In Chapter 3, we used a chi square test to evaluate the goodness of fit between a genetic hypothesis and the observed experimental data. This method can be employed to determine if the outcome of a two-factor cross is consistent with linkage or independent assortment.

To conduct a chi square test, we must first propose a hypothesis. For a two-factor cross, the standard hypothesis is that the two genes are not linked. This hypothesis is chosen even if the observed data suggest linkage, because an independent assortment hypothesis allows us to calculate the expected number of offspring based on the genotypes of the parents and the law of independent assortment. In contrast, for two linked genes that have not been previously mapped, we cannot calculate the expected number of offspring from a genetic cross because we do not know how likely it is for a crossover to occur between the two genes. Without expected numbers of recombinant and nonrecombinant offspring, we cannot conduct a chi square test. Therefore, we begin with the hypothesis that the genes are not linked.

Recall from Chapter 3 that the hypothesis we are testing is called a **null hypothesis,** because it assumes that there is no real difference between the observed and expected values. The goal is to determine whether or not the data fit the hypothesis. If the chi square value is low and we cannot reject the null hypothesis, we infer that the genes assort independently. On the other hand, if the chi square value is so high that our hypothesis is rejected, we accept the alternative hypothesis, namely, that the genes are linked.

Of course, a statistical test cannot prove that a hypothesis is true. If the chi square value is high, we accept the linkage hypothesis because we are assuming that only two explanations for a genetic outcome are possible: The genes are either linked or not linked. However, other factors, such as a decreased viability of particular phenotypes, may affect the outcome of a cross, which can result in large deviations between the observed and expected values and can cause us to reject the independent assortment hypothesis even though it may be correct.

To carry out a chi square test, let's reconsider Morgan's data concerning body color and eye color (see Figure 6.4). The cross produced the following offspring: 1159 with gray body, red eyes; 1017 with yellow body, white eyes; 17 with gray body, white eyes; and 12 with yellow body, red eyes. When a heterozygous female ($X^{y^+w^+}X^{yw}$)

is crossed to a hemizygous male ($X^{yw}Y$), the laws of segregation and independent assortment predict the following outcome:

A step-by-step procedure for applying the chi square test to distinguish between linkage and independent assortment in a two-factor cross is as follows:

Step 1. *Propose a hypothesis.* Mendel's laws predict a 1:1:1:1 ratio among the four phenotypes. Even though the observed data appear inconsistent with this hypothesis, we propose that the two genes for eye color and body color obey Mendel's law of independent assortment. This hypothesis allows us to calculate expected values. Because the data seem to conflict with this hypothesis, we actually anticipate that the chi square test will allow us to reject the independent assortment hypothesis in favor of a linkage hypothesis. We also assume that the alleles follow the law of segregation and that the four phenotypes are equally viable.

Step 2. *List the observed data, including the total number of offspring.*

1159 gray body, red eyes
1017 yellow body, white eyes
17 gray body, white eyes
12 yellow body, red eyes
2205 Total

Step 3. *Based on the hypothesis, calculate the expected value of each of the four phenotypes.* Each phenotype has an equal probability of occurring (refer to the Punnett square presented above). Therefore, the probability of each phenotype is 1/4. The observed F_2 generation had a total of 2205 individuals. Our next step is to calculate the expected number of offspring with each phenotype when the total equals 2205; 1/4 of the offspring should have each of the four phenotypes:

$1/4 \times 2205 = 551$ (expected number of each phenotype, rounded to the nearest whole number)

Step 4. *Apply the chi square formula, using the observed values (O) in step 2 and the expected values (E) that have been calculated in step 3.* In this case, the data consist of four phenotypes.

$$\chi^2 = \frac{(O_1 - E_1)^2}{E_1} + \frac{(O_2 - E_2)^2}{E_2} + \frac{(O_3 - E_3)^2}{E_3} + \frac{(O_4 - E_4)^2}{E_4}$$

$$\chi^2 = \frac{(1159 - 551)^2}{551} + \frac{(17 - 551)^2}{551} +$$

$$\frac{(12 - 551)^2}{551} + \frac{(1017 - 551)^2}{551}$$

$$\chi^2 = 670.9 + 517.5 + 527.3 + 394.1 = 2109.8$$

Step 5. *Determine the degrees of freedom.* The four phenotypes are based on the law of segregation and the law of independent assortment. By itself, the law of independent assortment predicts only two categories: recombinant and nonrecombinant. Therefore, based on a hypothesis of independent assortment alone, the degrees of freedom equal $n - 1$, which is $2 - 1$, or 1.

Step 6. *Determine the P value for the chi square value calculated in step 4. This is done using a chi square table and the degrees of freedom determined in step 5.* The calculated chi square value of 2109.8 is enormous! This means that the deviation between observed and expected values is very large. With 1 degree of freedom, such a large deviation is expected to occur by chance alone less than 1% of the time (refer back to Table 3.1).

Step 7. *Interpret the calculated chi square value.* Because the *P* value is less than 0.01 (or 1%), we reject the hypothesis that the two genes assort independently. As an alternative, we accept the hypothesis that the genes are linked. Even so, it should be emphasized that rejecting the null hypothesis does not prove that the linkage hypothesis is correct. For example, some of the non-Mendelian patterns described in Chapter 5 can produce results that do not conform to those expected with independent assortment.

Research Studies Provided Evidence That Recombinant Offspring Have Inherited a Chromosome That Is the Product of a Crossover

As we have seen, Morgan's studies were consistent with the hypothesis that crossing over occurs between homologous chromosomes to produce new combinations of alleles. In the 1930s, experimentation provided direct evidence that recombinant offspring have inherited a chromosome that is the product of a crossover. One of these studies involved the analysis of corn kernels and was carried out by Harriet Creighton and Barbara McClintock. Another study by Curt Stern involved crosses in *Drosophila*. In both studies, the researchers first made crosses involving two linked genes to produce nonrecombinant and recombinant offspring. Second, they used a microscope to view the structures of the chromosomes in the parents and in the offspring. Because the chromosomes had some unusual structural features, the researchers could microscopically distinguish between the two homologous chromosomes within a pair. Therefore, they were able to correlate the occurrence of

recombinant offspring with microscopically observable exchanges in segments of homologous chromosomes.

Let's consider the experiments of Stern, which involved strains of flies with microscopically detectable abnormalities in the X chromosome. In one strain, the X chromosome was shorter than normal due to a deletion at one end (this is the completely blue chromosome on the left side of **Figure 6.6a**). In another strain, the X chromosome was longer than normal because a piece of the Y chromosome was attached at its other end, where the centromere is located (this is the completely purple chromosome in Figure 6.6a).

On the short X chromosome two mutant alleles were located: a dominant allele (*B*) that causes bar-shaped eyes and a recessive allele (*car*) that results in carnation-colored eyes, which are less red than wild-type eyes. On the long X chromosome were located the wild-type alleles for these two genes (designated B^+ and car^+), which confer round eyes and red eyes, respectively.

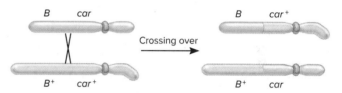

(a) X chromosomes found in Stern's flies and the result of them crossing over

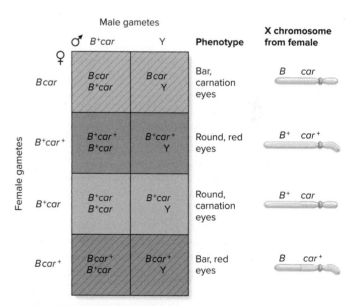

(b) The results of Stern's crosses

FIGURE 6.6 **Stern's experiments showing a correlation between recombinant offspring and crossing over.** **(a)** The left side shows the original X chromosomes found in Stern's female flies. On the top left (completely blue chromosome), the X chromosome has a deletion on the left end. On the bottom left (completely purple chromosome), a segment of the Y chromosome has been attached to the right end of the X chromosome. The right side shows the results of a crossover. **(b)** The results obtained by Stern. The female parent carried the two X chromosomes shown on the left side in part (a). The male parent had a normal X chromosome carrying B^+ and *car* and a Y chromosome.

Stern realized that a crossover between the two X chromosomes in such female flies would result in recombinant chromosomes that would be cytologically distinguishable from the starting chromosomes. If a crossover occurred between the *B* and *car* genes on the X chromosome, it would be expected to produce the following two chromosomes (see the right side of Figure 6.6a):

- An abnormal X chromosome with a deletion at one end and an extra piece of the Y chromosome at the other end; the genotype would be *B car*$^+$.
- A normal-sized X chromosome; the genotype would be *B*$^+$ *car*.

Stern crossed female flies carrying *B car* (see the completely blue chromosome in Figure 6.6a) and *B*$^+$*car*$^+$ (see the completely purple chromosome in Figure 6.6a) to male flies that had a normal-sized X chromosome with both recessive alleles (*B*$^+$*car*). Using a microscope, it was possible to discriminate morphologically between the starting chromosomes—those in the original parental flies—and the recombinant chromosomes that might occur in the offspring.

As shown in **Figure 6.6b**, Stern's experiment showed a correlation between recombinant phenotypes and the inheritance of chromosomes that were the product of a crossover. The arrangement of alleles on the two X chromosomes in the female flies was known, making it possible to predict the phenotypes of nonrecombinant and recombinant offspring. The male flies could contribute the *B*$^+$ and *car* alleles (on a normal-sized X chromosome) or contribute a Y chromosome. In the absence of crossing over, the female flies could contribute a short X chromosome with the *B* and *car* alleles or a long X chromosome with the *B*$^+$ and *car*$^+$ alleles. If crossing over occurred in the region between these two genes, the female flies would contribute recombinant X chromosomes. One possible recombinant X chromosome would be normal-sized and carry the *B*$^+$ and *car* alleles, and the other recombinant X chromosome would have a deletion at one end with a piece of the Y chromosome attached to the other end and would carry the *B* and *car*$^+$ alleles. When combined with an X or Y chromosome from the males, the nonrecombinant offspring would have bar, carnation eyes or wild-type eyes; the recombinant offspring would have round, carnation eyes or bar, red eyes.

The results shown in Figure 6.6b are the actual results that Stern observed. The interpretation was that crossing over between homologous chromosomes—in this case, X chromosomes—accounts for the formation of offspring with recombinant phenotypes.

6.2 COMPREHENSION QUESTIONS

1. With regard to linked genes on the same chromosome, which of the following statements is *false*?
   a. Crossing over is needed to produce nonrecombinant offspring.
   b. Crossing over is needed to produce recombinant offspring.
   c. Crossing over is more likely to separate alleles if they are far apart on the same chromosome.
   d. Crossing over that separates linked alleles occurs during prophase of meiosis I.

2. Morgan observed a higher number of recombinant offspring involving eye color and wing length (401 + 318) than those involving body color and eye color (17 + 12). These results occurred because
   a. the genes affecting eye color and wing length are farther apart on the X chromosome than are the genes affecting body color and eye color.
   b. the genes affecting eye color and wing length are closer together on the X chromosome than are the genes affecting body color and eye color.
   c. the gene affecting wing length is not on the X chromosome.
   d. the gene affecting body color is not on the X chromosome.

3. For a chi square test involving genes that may be linked, which of the following statements is correct?
   a. An independent assortment hypothesis is not proposed because the data usually suggest linkage.
   b. An independent assortment hypothesis is proposed because it allows the expected numbers of offspring to be calculated.
   c. A large chi square value suggests that the observed and expected data are in good agreement.
   d. The null hypothesis is rejected when the chi square value is very low.

6.3 GENETIC MAPPING IN PLANTS AND ANIMALS

Learning Outcomes:

1. Describe why genetic mapping is useful.
2. Calculate the map distance between linked genes using data from a testcross.
3. Explain how positive interference affects the number of double crossovers.

The purpose of **genetic mapping,** also known as *gene mapping* or *chromosome mapping*, is to determine the linear order and distance of separation among genes that are linked to each other on the same chromosome. Shown in **Figure 6.7** is a depiction of the linear arrangement of genes, known as a **genetic map,** or a **genetic linkage map.** This diagram is a simplified genetic map of *Drosophila melanogaster*, depicting the locations of many different genes along the individual chromosomes. As seen here, each gene has its own unique **locus**—the site where the gene is found within a particular chromosome. For example, the gene designated *brown eyes* (*bw*), which affects eye color, is located near one end of chromosome 2. The gene designated *black body* (*b*), which affects body color, is found near the middle of the same chromosome.

Why is genetic mapping useful?

- It allows geneticists to understand the overall complexity and organization of the genome of a particular species. The genetic map of a species reveals the underlying basis for the inherited traits an organism of that species displays.
- The known locus of a gene within a genetic map can help molecular geneticists to clone that gene and thereby obtain greater information about its molecular features.

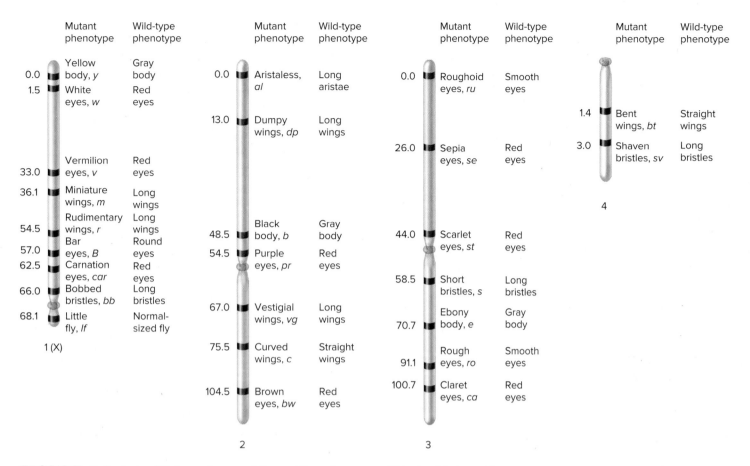

FIGURE 6.7 **A simplified genetic map of *Drosophila melanogaster.*** This simplified map illustrates a few of the many thousands of genes that have been identified in this organism.

CONCEPT CHECK: List five reasons why genetic maps are useful.

- Genetic mapping is useful from an evolutionary point of view. A comparison of the genetic maps for different species can improve our understanding of the evolutionary relationships among them.
- Many human genes that play a role in diseases have been genetically mapped. This information can be used to diagnose and perhaps someday treat inherited human diseases. It can also help genetic counselors predict the likelihood that a couple will produce children with certain inherited diseases.
- Genetic mapping is gaining increasing importance in agriculture. A genetic map can provide plant and animal breeders with helpful information for improving agriculturally important strains through selective breeding programs.

In this section, we will examine traditional genetic mapping techniques that involve an analysis of crosses of individuals that are heterozygous for two or more genes. Determining the number of recombinant offspring due to crossing over provides a way to deduce the linear order of genes along a chromosome. Such genetic maps have been constructed for several plant species and certain species of animals, including *Drosophila.* Mapping has also been carried out on fungi, as described in Section 6.4. For many organisms, however, traditional mapping approaches are unsuitable due to long generation times or the inability to carry out experimental crosses (for example, with humans). Fortunately,

many alternative methods of gene mapping have been developed to replace the need to carry out crosses. As described in Chapter 22, molecular approaches are also used to map genes.

A Testcross Is Conducted to Produce a Genetic Linkage Map

Genetic mapping relies on the estimation of the relative distances between linked genes, based on how frequently a crossover occurs between them. If two genes are very close together on the same chromosome, a crossover is unlikely to begin in the region between them. However, if two genes are very far apart, a crossover is more likely to be initiated in the region between them, and thereby recombine the alleles of the two genes. Experimentally, the basis for genetic mapping is that the percentage of recombinant offspring is correlated with the distance between two genes. If two genes are far apart, many recombinant offspring will be produced. However, if two genes are close together, very few recombinant offspring will be observed.

To interpret a genetic mapping experiment, the experimenter must know if the characteristics of an offspring are due to crossing over during meiosis in a parent. This is accomplished by conducting a **testcross.** Most testcrosses are between an individual that is heterozygous for two or more genes and an individual that is homozygous for the recessive alleles of the same genes. In this type of testcross, the goal is to determine if recombination has occurred during meiosis in the heterozygous parent. Therefore, genetic mapping is based on the level of recombination that occurs in just one parent—the heterozygote. In the testcross, new combinations of alleles cannot occur in the gametes of the other parent, the one that is homozygous for these genes.

Figure 6.8 illustrates the strategy for conducting a testcross to determine the distance between two genes based on the relative numbers of recombinant and nonrecombinant offspring. The experiment begins with a true-breeding P generation. This cross focuses on two linked genes affecting bristle length and body color

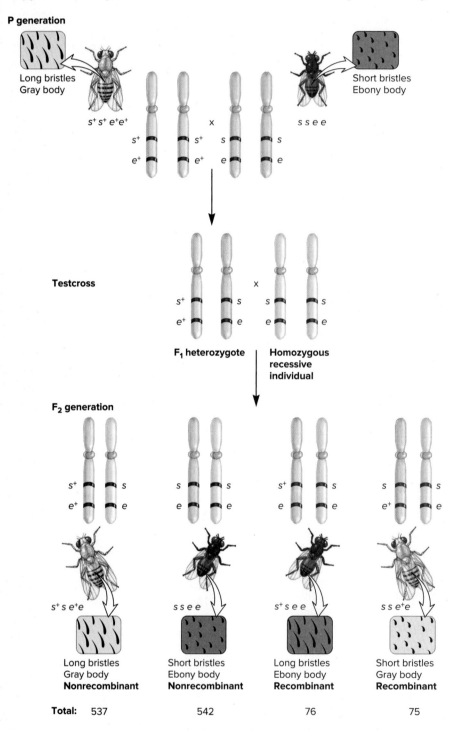

FIGURE 6.8 Use of a testcross to distinguish between recombinant and nonrecombinant offspring. The P generation consists of flies from two different true-breeding strains, and the cross produces F_1 heterozygotes. In the testcross, F_1 females that are heterozygous for both genes ($s^+s\ e^+e$) are crossed to males that are homozygous recessive for short bristles (ss) and ebony body (ee). The F_2 recombinant offspring carry a chromosome that is the product of a crossover. (Note: Crossing over does not occur during sperm formation in *Drosophila*, which is unusual among eukaryotes. Therefore, the heterozygotes in a testcross involving *Drosophila* must be females.)

CONCEPT CHECK: When and in which fly or flies did crossing over occur in order to produce the F_2 recombinant offspring?

in fruit flies. The dominant (wild-type) alleles are s^+ (long bristles) and e^+ (gray body), and the recessive alleles are s (short bristles) and e (ebony body). Because the P generation is true-breeding, the experimenter knows the arrangement of linked alleles. In the female of the P generation, s^+ is linked to e^+, and in the male, s is linked to e. Therefore, in the F_1 offspring, we know that the s^+ and e^+ alleles are located on one chromosome and the corresponding s and e alleles are located on the homologous chromosome. In the testcross, the F_1 heterozygote is crossed to an individual that is homozygous for the recessive alleles of the two genes (*ssee*).

Now let's take a look at the four possible types of F_2 offspring. Their phenotypes are long bristles, gray body; short bristles, ebony body; long bristles, ebony body; and short bristles, gray body. All four types of F_2 offspring have inherited a chromosome carrying the s and e alleles from their homozygous parent, which is the blue chromosome shown on the right in each pair of chromosomes. Now, consider the other chromosome in each pair. The offspring with long bristles and gray bodies have inherited a chromosome carrying the s^+ and e^+ alleles from the heterozygous parent. This chromosome is not the product of a crossover. The offspring with short bristles and ebony bodies have inherited a chromosome carrying the s and e alleles from the heterozygous parent. Again, this chromosome is not the product of a crossover.

The two types of recombinant F_2 offspring, however, can be produced only if crossing over occurred in the region between these two genes. The offspring with long bristles and ebony bodies and those with short bristles and gray bodies have inherited a chromosome that is the product of a crossover during oogenesis in the F_1 female. A key point for you to observe is that the recombinant offspring of the F_2 generation must carry a chromosome that is the product of a crossover. As noted in Figure 6.8, the recombinant offspring are fewer in number than are the nonrecombinant offspring.

The relative number of recombinant offspring can be used as an estimate of the physical distance between two genes on the same chromosome. The **map distance,** also called the **recombination frequency,** is defined as the number of recombinant offspring divided by the total number of offspring, multiplied by 100. We can calculate the map distance between the bristle length gene and the body color gene using this formula:

$$\text{Map distance} = \frac{\text{Number of recombinant offspring}}{\text{Total number of offspring}} \times 100$$
$$= \frac{76 + 75}{537 + 542 + 76 + 75} \times 100$$
$$= 12.3 \text{ map units}$$

The units of map distance are called **map units (mu),** or sometimes **centiMorgans (cM)** in honor of Thomas Hunt Morgan. One map unit is equivalent to 1% recombinant offspring in a testcross. In this example, we would conclude that the bristle length gene and the body color gene are 12.3 mu apart from each other on the same chromosome.

The likelihood of multiple crossovers sets a quantitative limit on the relationship between map distance and the percentage of recombinant offspring. Even though two different genes can be on the same chromosome and more than 50 mu apart, a testcross is

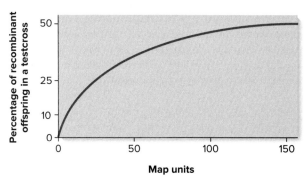

FIGURE 6.9 **Relationship between the percentage of recombinant offspring observed in a testcross and the actual map distance between genes.** The y-axis depicts the percentage of recombinant offspring that would be observed in a two-factor testcross. The actual map distance, shown on the x-axis, is calculated by analyzing the percentages of recombinant offspring from a series of many two-factor crosses involving closely linked genes. Even though two genes may be more than 50 mu apart, the percentage of recombinant offspring will not exceed 50%.

CONCEPT CHECK: What phenomenon explains why the maximum percentage of recombinant offspring does not exceed 50%?

expected to yield a maximum of only 50% recombinant offspring (**Figure 6.9**). What accounts for this 50% limit? The answer lies in the pattern of multiple crossovers. As shown in the figure with question 4 of More Genetic TIPS at the end of the chapter, a double crossover between two genes could involve four, three, or two chromatids, which would yield 100%, 50%, or 0% recombinants, respectively. Because all of these double crossovers are equally likely, we take the average of them to determine the maximum value. This average equals 50%. Therefore, when two different genes are more than 50 mu apart, they follow the law of independent assortment in a testcross and only 50% recombinant offspring are observed.

GENETIC TIPS   **THE QUESTION:** In the testcross example in Figure 6.8, the dominant alleles were on one chromosome and the recessive alleles were on the homolog. Let's consider a two-factor cross in which the dominant allele for one gene is on one chromosome, but the dominant allele for a second gene is on the homolog. A cross is made between *AAbb* and *aaBB* parents. The F_1 offspring are *AaBb*. A testcross of these F_1 heterozygotes with *aabb* individuals is then carried out. Which F_2 offspring are recombinant?

TOPIC: *What topic in genetics does this question address?* The topic is linkage and genetic mapping. More specifically, the question is about identifying recombinant offspring in a two-factor testcross.

INFORMATION: *What information do you know based on the question and your understanding of the topic?* From the question, you know that a cross involves a P generation that is *AAbb* and *aaBB*, and then a testcross involves the F_1 heterozygotes and *aabb* individuals. From your understanding of the topic, you may remember that F_2 recombinant offspring are produced by crossing over in the F_1 heterozygotes.

PROBLEM-SOLVING **S**TRATEGIES: *Make a drawing. Predict the outcome.* In solving genetic linkage problems, it can be very helpful to draw the chromosomes in the F₁ heterozygote and thereby deduce the possible haploid cells that the heterozygote can produce. Such a drawing is shown below.

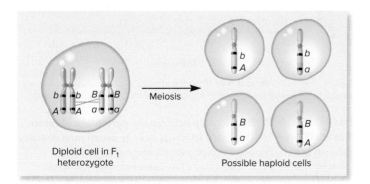

Allele *A* is initially linked to *b*, and *a* is linked to *B*. If crossing over occurs in the region between these two genes, it will produce chromosomes in which *A* is linked to *B* and *a* is linked to *b* (as shown in the above drawing). All F₂ offspring will inherit *ab* from the *aabb* parent. The F₂ nonrecombinant offspring, which are not produced by crossing over in the F₁ heterozygote, will inherit *Ab* or *aB* from the F₁ heterozygous parent. Their genotypes will be *Aabb* and *aaBb*. The F₂ recombinant offspring, which are the result of crossing over, will inherit *AB* or *ab* from the F₁ parent. Their genotypes will be *AaBb* and *aabb*.

ANSWER: The F₂ recombinant offspring are those with the genotypes *AaBb* and *aabb*. Note: You might be surprised that the recombinant offspring in this example have genotypes that are the same as those of their parents that were involved in the testcross. Even so, the arrangement of alleles in the recombinant offspring is different from their heterozygous parent. In the heterozygous (*AaBb*) parent, *A* is linked to *b*, and *a* is linked to *B*. In the recombinant *AaBb* offspring, *A* is linked *B* and *a* is linked to *b*. In the recombinant *aabb* offspring, *a* is linked to *b*. It is important to remember that in this chapter we define recombinant offspring as offspring that have inherited a chromosome that is the product of a crossover.

Three-Factor Crosses Can Be Used to Determine the Order and Distance Between Linked Genes

Thus far, we have considered the construction of genetic maps using two-factor testcrosses to compute map distance. The data from three-factor crosses can yield additional information about map distance and gene order. In a three-factor cross, the experimenter crosses two individuals that differ in three characters. The following experiment outlines a common strategy for using three-factor crosses to map genes. In this experiment, the P generation consists of fruit flies that differ in body color, eye color, and wing shape. We begin with true-breeding strains, so we know which alleles are initially linked to each other on the same chromosome.

Step 1. *Cross two true-breeding strains that differ with regard to three alleles.* In this example, we will cross a fly that has a black body (*bb*), purple eyes (*prpr*), and vestigial wings (*vgvg*) to a homozygous wild-type fly with a gray body (*b⁺b⁺*), red eyes (*pr⁺pr⁺*), and long wings (*vg⁺vg⁺*):

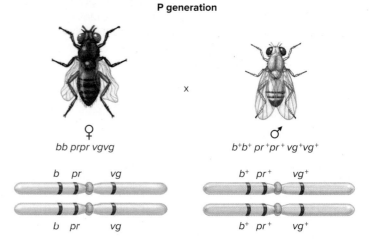

P generation

The goal in this step is to obtain F₁ individuals that are heterozygous for all three genes. In the F₁ heterozygotes, all dominant alleles are located on one chromosome, and all recessive alleles are on the other homologous chromosome.

Step 2. *Perform a testcross by mating female F₁ heterozygotes to male flies that are homozygous recessive for all three alleles (bb prpr vgvg).*

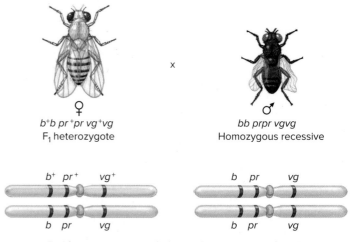

During gametogenesis in the heterozygous female F₁ flies, crossovers may produce new combinations of the three alleles.

Step 3. *Collect data for the F₂ generation.* As shown in **Table 6.1**, eight phenotypic combinations are possible. An analysis of the F₂ generation flies allows us to map these three genes. Because each of the three genes exists as two alleles, we have 2^3, or 8, possible combinations of phenotypes in the offspring. If these alleles assorted independently, all eight combinations would occur in equal proportions. However, we see that the proportions of the eight phenotypes are far from equal.

TABLE 6.1

Data from a Three-Factor Cross (see step 2)

Phenotype	Number of Observed Offspring (Males and Females)	Chromosome Inherited from F₁ Female
Gray body, red eyes, long wings	411	b^+ pr^+ vg^+
Gray body, red eyes, vestigial wings	61	b^+ pr^+ vg
Gray body, purple eyes, long wings	2	b^+ pr vg^+
Gray body, purple eyes, vestigial wings	30	b^+ pr vg
Black body, red eyes, long wings	28	b pr^+ vg^+
Black body, red eyes, vestigial wings	1	b pr^+ vg
Black body, purple eyes, long wings	60	b pr vg^+
Black body, purple eyes, vestigial wings	412	b pr vg
Total	1005	

The genotypes of the P generation correspond to two phenotypes: gray body, red eyes, long wings; and black body, purple eyes, vestigial wings. In crosses involving linked genes, the nonrecombinant phenotypes occur most frequently in the offspring. The remaining six phenotypes in the F₂ generation, which are due to crossing over, are recombinants.

A double crossover is always expected to cause the least frequent types of offspring. Two of the phenotypes—gray body, purple eyes, long wings; and black body, red eyes, vestigial wings—arose from a double crossover between two pairs of genes. Also, the combination of traits involved in the double crossover tells us which gene is in the middle along the chromosome. When a chromatid undergoes a double crossover, the gene in the middle becomes separated from the other two genes at either end.

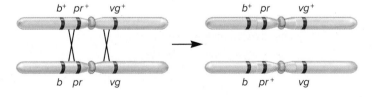

In the types of offspring resulting from a double cross-over, the recessive purple eye allele is separated from the other two recessive alleles. In the testcross, this yields flies with either gray bodies, purple eyes, and long wings

or black bodies, red eyes, and vestigial wings. This observation indicates that the gene for eye color lies between the genes for body color and wing shape.

Step 4. *Calculate the map distance between pairs of genes.* To do this, we need to understand which allele combinations are recombinant and which are nonrecombinant. The recombinant offspring are due to crossing over in the heterozygous female parent. If you look back at step 2, you can see the arrangement of alleles in the heterozygous female parent in the absence of crossing over. Let's consider this arrangement with regard to gene pairs:

b^+ is linked to pr^+, and b is linked to pr

pr^+ is linked to vg^+, and pr is linked to vg

b^+ is linked to vg^+, and b is linked to vg

With regard to body color and eye color, the recombinant offspring have gray bodies and purple eyes (2 + 30) or black bodies and red eyes (28 + 1). As shown along the right side of Table 6.1, these offspring were produced by crossovers in the female parents. The total number of these recombinant offspring is 61. The map distance between the body color and eye color genes is

$$\text{Map distance} = \frac{61}{944 + 61} \times 100 = 6.1 \, \text{mu}$$

With regard to eye color and wing shape, the recombinant offspring have red eyes and vestigial wings (61 + 1) or purple eyes and long wings (2 + 60). The total number is 124. The map distance between the eye color and wing shape genes is

$$\text{Map distance} = \frac{124}{881 + 124} \times 100 = 12.3 \, \text{mu}$$

With regard to body color and wing shape, the recombinant offspring have gray bodies and vestigial wings (61 + 30) or black bodies and long wings (28 + 60). The total number is 179. The map distance between the body color and wing shape genes is

$$\text{Map distance} = \frac{179}{826 + 179} \times 100 = 17.8 \, \text{mu}$$

Step 5. *Construct the map.* Based on the map unit calculation, the body color (*b*) and wing shape (*vg*) genes are farthest apart. The eye color gene (*pr*) must lie in the middle. As mentioned earlier, this order of genes is also confirmed by the pattern of traits found in the double crossovers. To construct the map, we use the distances between the genes that are closest together.

In our example, we have placed the body color gene first and the wing shape gene last. The data also are consistent with a map in which the wing shape gene comes first and the body color gene comes last. In detailed genetic maps, the locations of genes are mapped relative to the centromere.

You may have noticed that our calculations underestimate the distance between the body color and wing shape genes. We obtained a value of 17.8 mu even though the distance seems to be 18.4 mu when we add together the distance between body color and eye color genes (6.1 mu) and the distance between eye color and wing shape genes (12.3 mu).

What accounts for this discrepancy? The answer is double crossovers. If you look at the data in Table 6.1, the offspring with gray bodies, purple eyes, long wings and those with black bodies, red eyes, vestigial wings are due to a double crossover. From a phenotypic perspective, these offspring are nonrecombinant with regard to the body color and wing shape alleles. Even so, we know they arose from a double crossover between these two genes. Therefore, we should consider these crossovers when calculating the distance between the body color and wing shape genes. In this case, three offspring (2 + 1) were due to double crossovers. The number of double crossovers (2 + 1) is multiplied by 2 and we add this number to our previous value of recombinant offspring:

$$\text{Map distance} = \frac{179 + 2(2+1)}{826 + 179} \times 100 = 18.4\,\text{mu}$$

Interference Can Influence the Number of Double Crossovers That Occur in a Short Region

In Chapter 3, we used the product rule to determine the probability that two independent events will both occur. The product rule allows us to predict the expected likelihood of a double crossover, provided we know the individual probabilities of each single crossover. Let's reconsider the data for the three-factor testcross just described to see if the frequency of double crossovers is what we would expect based on the product rule. The map distance between b and pr is 6.1 mu and the distance between pr and vg is 12.3 mu. Because map distances are calculated as the percentage of recombinant offspring, we divide the map distances by 100 to compute the crossover frequency between each pair of genes. In this case, the crossover frequency between b and pr is 0.061 and that between pr and vg is 0.123. The product rule predicts

Expected likelihood of a double crossover = $0.061 \times 0.123 = 0.0075$, or 0.75%

Expected number of offspring due to a double crossover, based on a total of 1005 offspring produced = $1005 \times 0.0075 = 7.5$

In other words, we would expect about 7 or 8 offspring to be produced as a result of a double crossover. The observed number of offspring was only 3 (namely, 2 with gray bodies, purple eyes, and long wings, and 1 with a black body, red eyes, and vestigial wings). What accounts for the lower number? This lower-than-expected value is probably not due to random sampling error. Instead, the likely cause is a common genetic phenomenon known as **positive interference,** in which the occurrence of a crossover in one region of a chromosome decreases the probability that a second crossover will occur nearby. In other words, the first crossover interferes

with the ability to form a second crossover in the immediate vicinity. To provide interference with a quantitative value, we first calculate the coefficient of coincidence (C), which is the ratio of the observed number of double crossovers to the expected number.

$$C = \frac{\text{Observed number of double crossovers}}{\text{Expected number of double crossovers}}$$

Interference (I) is expressed as

$$I = 1 - C$$

For the data from the three-factor testcross, the observed number of crossovers is 3 and the expected number is 7.5, so the coefficient of coincidence equals $3/7.5 = 0.40$. In other words, only 40% of the expected number of double crossovers was actually observed. The value for interference equals $1 - 0.4 = 0.60$, or 60%. This means that 60% of the expected number of crossovers did not occur. Because I has a positive value here, this is called positive interference. Although the molecular mechanisms that cause interference are not entirely understood, the number of crossovers in most organisms is regulated so that very few occur per chromosome.

6.3 COMPREHENSION QUESTIONS

Answer the multiple-choice questions based on the following crosses:

P generation: True-breeding fruit flies with red eyes and long wings were crossed to flies with white eyes and miniature wings. All F_1 offspring had red eyes and long wings.

The F_1 female flies were then crossed to males with white eyes and miniature wings. The following results were obtained for the F_2 generation:

129 red eyes, long wings

133 white eyes, miniature wings

71 red eyes, miniature wings

67 white eyes, long wings

1. What is/are the phenotype(s) of the recombinant offspring of the F_2 generation?
   a. red eyes, long wings
   b. white eyes, miniature wings
   c. red eyes, long wings; and white eyes, miniature wings
   d. red eyes, miniature wings; and white eyes, long wings

2. The recombinant offspring of the F_2 generation were due to crossing over that occurred
   a. during spermatogenesis in the P generation males.
   b. during oogenesis in the P generation females.
   c. during spermatogenesis in the F_1 males.
   d. during oogenesis in the F_1 females.

3. What is the map distance between the two genes for eye color and wing length?
   a. 32.3 mu
   b. 34.5 mu
   c. 16.2 mu
   d. 17.3 mu

6.4 GENETIC MAPPING IN HAPLOID EUKARYOTES

Learning Outcomes:

1. Explain the experimental advantage of genetic mapping of fungi.
2. Calculate the map distance between genes in fungi using tetrad analysis.

Before ending our discussion of genetic mapping, let's consider how species of simple eukaryotes, which spend most of their life cycle in the haploid state, have also been the subjects in genetic mapping studies. The sac fungi, called ascomycetes, have been particularly useful to geneticists because of their unique style of sexual reproduction.

Fungi may be unicellular or multicellular organisms. Fungal cells are typically haploid (1*n*) and can reproduce asexually. In addition, fungi can also reproduce sexually by the fusion of two haploid cells to create a diploid zygote (2*n*) (**Figure 6.10**). The diploid zygote then proceeds through meiosis to produce four haploid cells, which are called **spores.** This group of four spores is known as a **tetrad** (not to be confused with a tetrad of four sister chromatids). In some species, meiosis is followed by a mitotic cell division to produce an octad.

In ascomycete fungi, the cells of a tetrad or octad are contained within a sac, which is called an **ascus** (plural: **asci**). In other words, the products of a single meiotic division are contained within one sac. This mode of reproduction does not occur in other eukaryotic groups.

An experimenter can conduct a two-factor cross, remove the spores from each ascus, and determine the phenotypes of the spores. This analysis can determine if two genes are linked or assort independently. If two genes are linked, a tetrad analysis can also be used to compute map distance.

Figure 6.11 illustrates the possible outcomes starting with two haploid strains of yeast (*Saccharomyces cerevisiae*). One strain carries the wild-type alleles *ura+* and *arg+*, which are required for uracil and arginine biosynthesis, respectively. The other strain has defective alleles *ura-2* and *arg-3*; these result in yeast strains that require uracil and arginine in the growth medium. A diploid zygote with the genotype *ura+ura-2 arg+arg-3* was produced by the fusion of haploid cells from these two strains. The diploid cell then proceeded through meiosis to produce four haploid cells. Upon completion of meiosis, three distinct types of tetrads can result in this case:

- One possibility is that the ascus will contain four haploid cells (spores) with nonrecombinant combinations of alleles. This ascus is said to have the **parental ditype (PD).**
- Alternatively, an ascus may have two nonrecombinant cells and two recombinant cells, which is called a **tetratype (T).**
- Finally, an ascus with a **nonparental ditype (NPD)** contains four cells with recombinant genotypes.

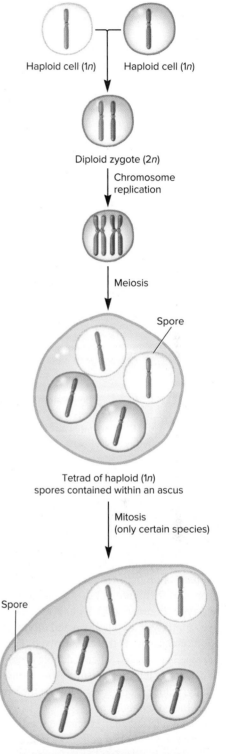

FIGURE 6.10 Sexual reproduction in ascomycetes. For simplicity, this diagram shows each haploid cell as having only one chromosome per haploid set. However, fungal species actually have several chromosomes per haploid set. In some species of fungi, meiosis is followed by a mitotic division to produce eight spores.

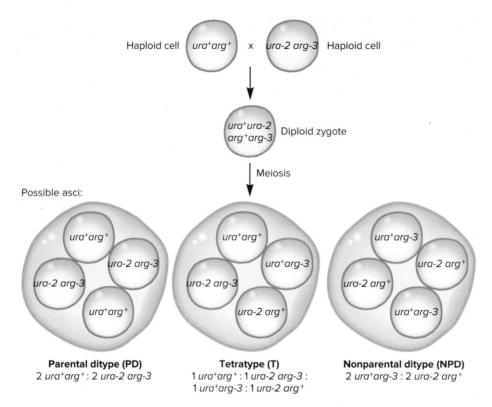

Possible asci:

Parental ditype (PD)
2 *ura⁺arg⁺* : 2 *ura-2 arg-3*

Tetratype (T)
1 *ura⁺arg⁺* : 1 *ura-2 arg-3* :
1 *ura⁺arg-3* : 1 *ura-2 arg⁺*

Nonparental ditype (NPD)
2 *ura⁺arg-3* : 2 *ura-2 arg⁺*

FIGURE 6.11 **The possible assortment of two genes in a tetrad.** If the tetrad consists of 100% nonrecombinant cells, the ascus has the parental ditype (PD). If the tetrad has 50% nonrecombinant and 50% recombinant cells, the ascus is a tetratype (T). Finally, an ascus with 100% recombinant cells is called a nonparental ditype (NPD). This figure does not illustrate the chromosomal locations of the alleles. In this type of experiment, the goal is to determine whether the two genes are linked on the same chromosome and, if they are linked, how far apart they are.

INTERACTIVE EXERCISE

When two genes assort independently, the number of asci having a parental ditype is expected to equal the number having a nonparental ditype, thus yielding 50% recombinant spores. For linked genes, **Figure 6.12** illustrates the relationship between crossing over and the type of ascus that will result. If no crossing over occurs in the region between the two genes, the parental ditype will be produced (Figure 6.12a). A single crossover event produces a tetratype (Figure 6.12b). Double crossovers can yield a parental ditype, a tetratype, or a nonparental ditype, depending on the combination of chromatids that are involved (Figure 6.12c). A nonparental ditype is produced when a double crossover involves all four chromatids. A tetratype results from a three-chromatid crossover. Finally, a double crossover between the same two chromatids produces the parental ditype.

The data from a tetrad analysis can be used to calculate the map distance between two linked genes. As in conventional mapping, the map distance is calculated as the percentage of offspring that carry recombinant chromosomes. As mentioned, a tetratype contains 50% recombinant chromosomes, and a nonparental ditype, 100%. Therefore, the map distance is computed as

$$\text{Map distance} = \frac{\text{NPD} + (1/2)(\text{T})}{\text{total number of asci}} \times 100$$

Over short map distances, this calculation provides a fairly reliable measure of the distance. However, it does not adequately account for double crossovers. When two genes are far apart on

the same chromosome, the map distance calculated using this equation underestimates the actual map distance, because the occurrence of double crossovers is not taken into account. Fortunately, a particular strength of tetrad analysis is that we can derive another equation that accounts for double crossovers, thereby providing a more accurate value for map distance. To begin this derivation, let's consider a more precise way to calculate map distance:

$$\text{Map distance} = \frac{\begin{array}{c}\text{Single crossover tetrads} + \\ (2)\,(\text{Double crossover tetrads})\end{array}}{\text{Total number of asci}} \times 0.5 \times 100$$

This equation includes the numbers of tetrads resulting from single and double crossovers in the computation of map distance. The total number of crossovers equals the number of single crossovers plus two times the number of double crossovers. Overall, the tetrads that are derived from single and double crossovers contain 50% nonrecombinant chromosomes. To calculate map distance, therefore, we divide the total number of crossovers by the total number of asci and multiply by 0.5 and 100.

To be useful, we need to relate this equation to the number of parental ditypes, nonparental ditypes, and tetratypes that are obtained by experimentation. To derive this relationship, we must consider the types of tetrads produced from no crossing over, a single crossover, and double crossovers. To do so, let's take

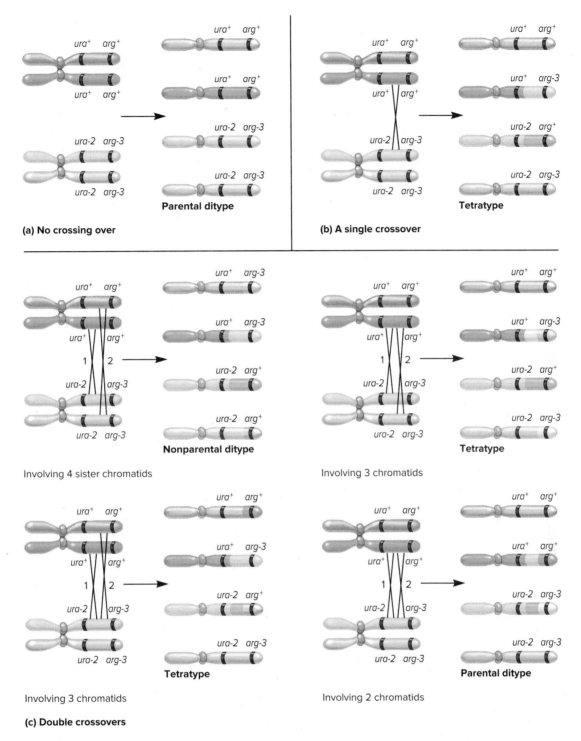

FIGURE 6.12 **Relationship between crossing over and the production of parental ditype, tetratype, or nonparental ditype for two linked genes.** Note: For part (c), the outcome assumes that the crossover on the left (labeled 1) occurred before the crossover on the right (labeled 2).

another look at Figure 6.12. As shown there, the parental ditype and tetratype are ambiguous. The parental ditype can be derived from no crossovers or a double crossover; the tetratype can be derived from a single crossover or a double crossover. However, the nonparental ditype is unambiguous, because it can be produced only from a double crossover. We can use this observation as a way to determine the actual number of single and double

crossovers. As seen in Figure 6.12, 1/4 of all the double crossovers are nonparental ditypes. Therefore, the total number of double crossovers equals four times the number of nonparental ditypes (4NPD).

Next, we need to know the number of single crossovers. A single crossover yields a tetratype, but double crossovers can also yield a tetratype. Therefore, the total number of tetratypes overestimates the true number of single crossovers. Fortunately, we can compensate for this overestimation. Because two types of tetratypes are due to a double crossover, the actual number of tetratypes arising from a double crossover should equal 2NPD. Therefore, the true number of single crossovers is calculated as T − 2NPD.

Now we have accurate measures of both single and double crossovers. The number of single crossovers equals T − 2NPD, and the number of double crossovers equals 4NPD. We can substitute these values into our previous equation.

$$\text{Map distance} = \frac{(T-2NPD) + (2)(4NPD)}{\text{Total number of asci}} \times 0.5 \times 100$$

$$= \frac{T + 6NPD}{\text{Total number of asci}} \times 0.5 \times 100$$

This equation provides a more accurate measure of map distance because it takes into account both single and double crossovers.

6.4 COMPREHENSION QUESTIONS

1. A tetrad of spores in an ascus is the product of
   a. one meiotic division.
   b. two meiotic divisions.
   c. one meiotic division followed by one mitotic division.
   d. one mitotic division followed by one meiotic division.

2. One yeast strain carries the alleles *lys*⁺ and *arg*⁺, whereas another strain has *lys-3* and *arg-2*. The two strains were crossed to each other, and an ascus obtained from this cross has four spores with the following genotypes: *lys*⁺ *arg*⁺, *lys*⁺ *arg-2*, *lys-3 arg*⁺, and *lys-3 arg 2*. This ascus has
   a. a parental ditype.
   b. a tetratype.
   c. a nonparental ditype.
   d. either a tetratype or a nonparental ditype.

6.5   MITOTIC RECOMBINATION

Learning Outcome:

1. Describe the process of mitotic recombination, and explain how it can produce a twin spot.

Thus far, we have considered how the arrangement of linked alleles along a chromosome can be rearranged by crossing over. This event produces cells and offspring with a recombinant pattern of traits. In such cases, crossing over occurs during meiosis, when the homologous chromosomes replicate and form bivalents.

In multicellular organisms, the union of egg and sperm is followed by many cellular divisions, which occur in conjunction with mitotic divisions of the cell nuclei. As discussed in Chapter 2, mitosis normally does not involve the homologous pairing of

chromosomes to form bivalents. Therefore, crossing over during mitosis is expected to occur much less frequently than crossing over during meiosis. Nevertheless, it does happen on rare occasions. Mitotic crossing over may produce a pair of recombinant chromosomes that have a new combination of alleles, an event known as **mitotic recombination.** If this recombination occurs during an early stage of embryonic development, the daughter cells containing one recombinant chromosome and one nonrecombinant chromosome will continue to divide many times to produce a patch of tissue in the adult. This may result in a portion of tissue with characteristics different from those of the surrounding tissue in the organism.

In 1936, Curt Stern identified unusual patches on the bodies of flies from certain *Drosophila* strains. This work involved strains carrying X-linked alleles affecting body color and bristle morphology (**Figure 6.13**). A recessive allele confers yellow body color (*y*), and another recessive allele causes shorter body bristles that look singed (*sn*). The corresponding wild-type alleles result in gray body color (*y*⁺) and long bristles (*sn*⁺). Females with the genotype *y*⁺*y sn*⁺*sn* are expected to have gray bodies and long bristles. This was generally the case. However, when looking at the bodies of these female flies under a low-power microscope, Stern occasionally noticed places in which two adjacent regions were phenotypically different from each other and also different from the rest of the body. Such a place is called a *twin spot*. Stern concluded that twin spotting was too frequent to be explained by the random positioning of two independent single spots that happened to occur close together. How was the phenomenon of twin spotting explained? Stern proposed that a twin spot is due to a single mitotic recombination within one cell during embryonic development.

As shown in Figure 6.13, the X chromosomes of the fertilized egg were *y*⁺ *sn* and *y sn*⁺. During development, a rare crossover occurred during mitosis to produce two adjacent daughter cells that were *y*⁺*y*⁺ *snsn* and *yy sn*⁺*sn*⁺. As embryonic development proceeded, the cell on the left continued to divide to produce many cells, eventually producing a patch on the body that had gray color with singed bristles. The daughter cell next to it produced a patch of body tissue that was yellow with long bristles. These two adjacent patches—a twin spot—were surrounded by cells that were *y*⁺*y sn*⁺*sn* and had gray color and long bristles. Twin spots provide evidence that mitotic recombination occasionally occurs.

6.5 COMPREHENSION QUESTION

1. The process of mitotic recombination involves the
   a. exchange of chromosomal regions between homologs during gamete formation.
   b. exchange of chromosomal regions between homologs during the division of somatic cells.
   c. reassortment of alleles that occurs at fertilization.
   d. reassortment of alleles that occurs during gamete formation.

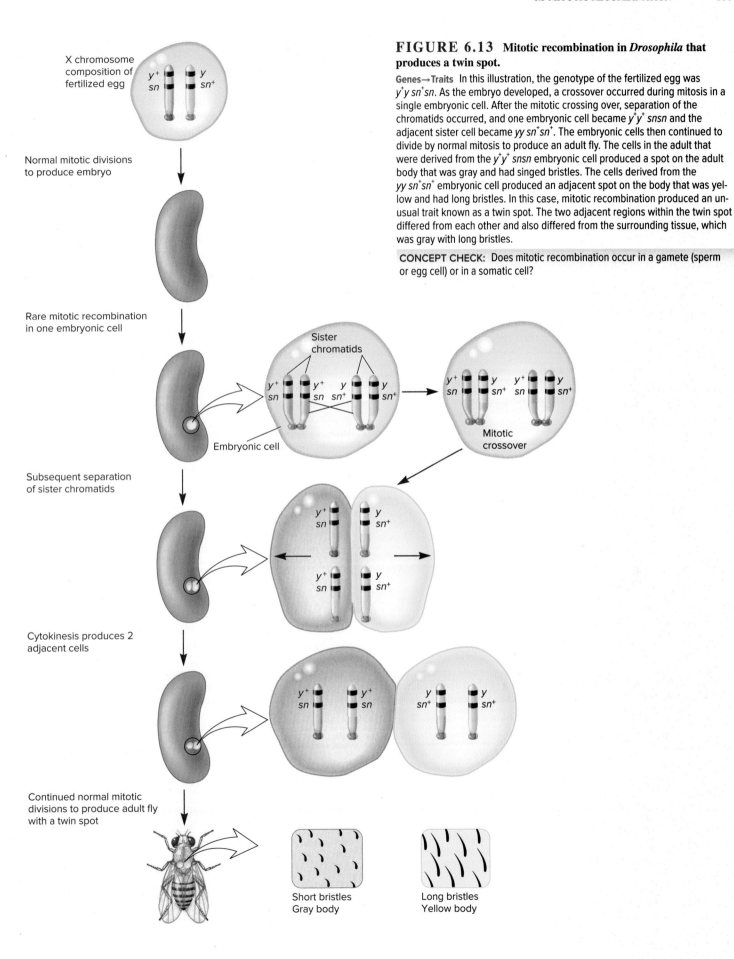

FIGURE 6.13 **Mitotic recombination in *Drosophila* that produces a twin spot.**

Genes→Traits In this illustration, the genotype of the fertilized egg was $y^+y\ sn^+sn$. As the embryo developed, a crossover occurred during mitosis in a single embryonic cell. After the mitotic crossing over, separation of the chromatids occurred, and one embryonic cell became $y^+y^+\ snsn$ and the adjacent sister cell became $yy\ sn^+sn^+$. The embryonic cells then continued to divide by normal mitosis to produce an adult fly. The cells in the adult that were derived from the $y^+y^+\ snsn$ embryonic cell produced a spot on the adult body that was gray and had singed bristles. The cells derived from the $yy\ sn^+sn^+$ embryonic cell produced an adjacent spot on the body that was yellow and had long bristles. In this case, mitotic recombination produced an unusual trait known as a twin spot. The two adjacent regions within the twin spot differed from each other and also differed from the surrounding tissue, which was gray with long bristles.

CONCEPT CHECK: Does mitotic recombination occur in a gamete (sperm or egg cell) or in a somatic cell?

X chromosome composition of fertilized egg

y^+ sn y sn^+

Normal mitotic divisions to produce embryo

Rare mitotic recombination in one embryonic cell

Sister chromatids

Embryonic cell

y^+ sn y^+ sn y sn^+ y sn^+

y^+ sn y sn^+ y^+ sn y sn^+

Mitotic crossover

Subsequent separation of sister chromatids

y^+ sn y sn^+

y^+ sn y sn^+

Cytokinesis produces 2 adjacent cells

y^+ sn y^+ sn y sn^+ y sn^+

Continued normal mitotic divisions to produce adult fly with a twin spot

Short bristles
Gray body

Long bristles
Yellow body

KEY TERMS

Introduction: genetic map
6.1: synteny, genetic linkage, linkage groups, two-factor cross, three-factor cross
6.2: crossing over, bivalent, recombinant, nonrecombinant, genetic recombination, recombinant offspring, null hypothesis

6.3: genetic mapping, genetic map (genetic linkage map), locus, testcross, map distance (recombination frequency), map unit (mu), centiMorgan (cM), positive interference
6.4: spores, tetrad, ascus (plural: asci), parental ditype (PD), tetratype (T), nonparental ditype (NPD)
6.5: mitotic recombination

CHAPTER SUMMARY

6.1 Overview of Linkage

- Synteny refers to a situation where two or more genes are located on the same chromosome. Linkage means that the alleles of two or more genes tend to be transmitted as a unit because they are relatively close to each other on the same chromosome.
- Bateson and Punnett discovered the first example of genetic linkage in sweet peas (see Figure 6.1).

6.2 Relationship Between Linkage and Crossing Over

- Crossing over during meiosis can alter the pattern of linked alleles along a chromosome (see Figure 6.2).
- Morgan discovered genetic linkage in *Drosophila* and proposed that recombinant offspring are produced when crossing over occurs during meiosis (see Figures 6.3, 6.4).
- When genes are linked, the relative proportions of recombinant offspring depend on the distance between the genes (see Figure 6.5).
- A chi square test can be used to determine whether two genes are linked or assort independently.
- The studies of Creighton and McClintock and those of Stern were able to correlate the formation of recombinant offspring with the presence of chromosomes that had exchanged pieces due to crossing over (see Figure 6.6).

6.3 Genetic Mapping in Plants and Animals

- A genetic map is a diagram that shows the order and relative locations of genes along one or more chromosomes (see Figure 6.7).
- A testcross can be performed to map the distance between two or more linked genes (see Figure 6.8).
- Due to the effects of multiple crossovers, the map distance between two genes obtained from a testcross cannot exceed 50% (see Figure 6.9).
- The data from a three-factor cross can be used to map three genes on a chromosome (see Table 6.1).
- Positive interference is the phenomenon in which a crossover in a given region of a chromosome decreases the probability that another crossover will occur nearby.

6.4 Genetic Mapping in Haploid Eukaryotes

- Haploid eukaryotes have been the subjects of genetic mapping studies. Ascomycetes have haploid cells that are the product of a single meiosis and are contained with an ascus (see Figure 6.10).
- The analysis of tetrads in yeast asci is used to map the distance between two linked genes (see Figures 6.11, 6.12).

6.5 Mitotic Recombination

- Mitotic recombination occurs on rare occasions and may produce twin spots (see Figure 6.13).

PROBLEM SETS & INSIGHTS

MORE GENETIC TIPS 1. In the garden pea, orange pods (*orp*) are recessive to green pods (*Orp*), and sensitivity to pea mosaic virus (*mo*) is recessive to resistance to the virus (*Mo*). A plant with orange pods and sensitivity to the virus was crossed to a true-breeding plant with green pods and resistance to the virus. The F_1 plants were then testcrossed to plants with orange pods and sensitivity to the virus. The following results were obtained for the F_2 offspring:

160 orange pods, virus-sensitive

165 green pods, virus-resistant

36 orange pods, virus-resistant

39 green pods, virus-sensitive

_____

400 total

A. Conduct a chi square test to see if these two genes are linked.

B. If they are linked, calculate the map distance between the two genes.

T OPIC: *What topic in genetics does this question address?* The topic is linkage. More specifically, the question asks you to determine if the outcome of a testcross is consistent with linkage or independent assortment.

I NFORMATION: *What information do you know based on the question and your understanding of the topic?* From the question, you know the outcome of a two-factor cross and a subsequent testcross. From your understanding of the topic, you may remember that linked genes do not independently assort.

A chi square test can be used to evaluate whether two genes are likely to be linked. You may also recall the equation for calculating map distance if two genes are linked.

PROBLEM-SOLVING **S**TRATEGIES: *Propose a hypothesis. Predict the outcome. Analyze data. Make a calculation.* To answer part A of this question, one strategy is to follow the seven steps for conducting a chi square test, which are described in Section 6.2.

Step 1. *Propose a hypothesis.* In this case, your hypothesis is that the genes are not linked. This hypothesis allows you to predict the outcome of the testcross based on independent assortment.

Step 2. *List the observed data, including the total number of offspring.* See the list in the question.

Step 3. *Based on the hypothesis, calculate the expected value of each of the four phenotypes.* The testcross is

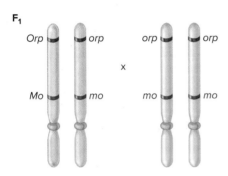

F$_1$

Orp   orp   orp   orp

×

Mo   mo   mo   mo

For this testcross, a 1:1:1:1 ratio of the four phenotypes is predicted if the genes are not linked. In other words, the phenotypes of the offspring should occur in these proportions: 1/4 orange pods, virus-sensitive; 1/4 green pods, virus-resistant; 1/4 orange pods, virus-resistant; and 1/4 green pods, virus-sensitive. Because a total of 400 offspring were produced, your hypothesis predicts 100 offspring in each phenotypic category.

Step 4. *Apply the chi square formula, using the data for the observed values (O) and the expected values (E) that have been calculated in step 3.* In this case, the expected values are 100, as calculated in step 3, and the observed values consist of the numbers of offspring with the four phenotypes, given in the question.

$$\chi^2 = \frac{(O_1-E_1)^2}{E_1} + \frac{(O_2-E_2)^2}{E_2} + \frac{(O_3-E_3)^2}{E_3} + \frac{(O_4-E_4)^2}{E_4}$$

$$\chi^2 = \frac{(160-100)^2}{100} + \frac{(165-100)^2}{100} + \frac{(36-100)^2}{100} + \frac{(39-100)^2}{100}$$

$$\chi^2 = 36 + 42.3 + 41 + 37.2 = 156.5$$

Step 5. *Determine the degrees of freedom.* As discussed in Section 6.2, based on a hypothesis of independent assortment, the degrees of freedom equal $n - 1$, which is $2 - 1$, or 1.

Step 6. *Determine the P value for the chi square value calculated in step 4. This is done using a chi square table and the degrees of freedom determined in step 5.* The calculated chi square value is 156.5. This indicates that the deviation between observed and expected values is very high. For 1 degree of freedom, Table 3.1 indicates that such a large deviation is expected to occur by chance alone less than 1% of the time.

Step 7. *Interpret the calculated chi square value.* You reject the hypothesis that the genes assort independently. As an alternative, you may infer that the two genes are linked.

Use the equation given in Section 6.3 to calculate the map distance between the two genes:

$$\text{Map distance} = \frac{(\text{Number of recombinant offspring})}{\text{Total number of offspring}} \times 100$$

$$= \frac{36 + 39}{36 + 39 + 160 + 165} \times 100$$

$$= 18.8 \, \text{mu}$$

ANSWER:

A. You reject the hypothesis that the genes are not linked.

B. The genes are approximately 18.8 mu apart.

2. Two recessive traits in mice—droopy ears and flaky tail—are caused by genes that are located 6 mu apart on the same chromosome. A true-breeding mouse with normal ears (*De*) and a flaky tail (*ft*) was crossed to a true-breeding mouse with droopy ears (*de*) and a normal tail (*Ft*). The F$_1$ offspring were then crossed to mice with droopy ears and flaky tails. If this testcross produced 100 offspring, what is the expected phenotypic outcome?

TOPIC: *What topic in genetics does this question address?* The topic is linkage. More specifically, the question asks you to predict the outcome of a testcross when you know the map distance between two linked genes.

INFORMATION: *What information do you know based on the question and your understanding of the topic?* From the question, you know that two genes are 6 mu apart on the same chromosome. You also know the phenotypes of the F$_1$ offspring. From your understanding of the topic, you may remember that the map distance is a measure of the percentage of recombinant offspring.

PROBLEM-SOLVING **S**TRATEGIES: *Make a drawing. Predict the outcome. Make a calculation.* In solving linkage problems, you may find it helpful to make a drawing of the chromosomes to help you distinguish the recombinant and nonrecombinant offspring. The recombinant offspring are produced by a crossover in the F$_1$ parent. The testcross is shown below, with the F$_1$ parent on the left side.

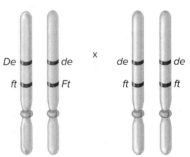

De   *de*   ×   *de*   *de*

ft   *Ft*   *ft*   *ft*

If the F$_1$ parent transmits a chromosome that is not the product of a crossover, the nonrecombinant offspring are as follows:

Dede ftft   normal ears, flaky tail
dede Ftft   droopy ears, normal tail

Alternatively, if a crossover occurs in the region between these two genes, the recombinant offspring are as follows:

dede ftft   droopy ears, flaky tail
Dede Ftft   normal ears, normal tail

Note: In this testcross, both dominant alleles were not on the same chromosome in the F$_1$ parent. This situation is different from that in many of the previous problems. Crossing over produces chromosomes in which one chromosome carries both dominant alleles and the other carries both recessive alleles. Because the two genes are located 6 mu apart on the same chromosome, 6% of the offspring will be recombinants.

ANSWER: The expected phenotypic outcome for 100 offspring is
  3 droopy ears, flaky tail
  3 normal ears, normal tail
  47 normal ears, flaky tail
  47 droopy ears, normal tail

3. The following X-linked recessive traits are found in fruit flies: vermilion eyes are recessive to red eyes, miniature wings are recessive to long wings, and sable body is recessive to gray body. A cross was made between wild-type males with red eyes, long wings, and gray bodies and females with vermilion eyes, miniature wings, and sable bodies. The F_1 heterozygous female offspring from this cross, which had red eyes, long wings, and gray bodies, were then crossed to males with vermilion eyes, miniature wings, and sable bodies. The following data were obtained for the F_2 generation (including both males and females):

  1320 vermilion eyes, miniature wings, sable body
  1346 red eyes, long wings, gray body
  102 vermilion eyes, miniature wings, gray body
  90 red eyes, long wings, sable body
  42 vermilion eyes, long wings, gray body
  48 red eyes, miniature wings, sable body
  2 vermilion eyes, long wings, sable body
  1 red eyes, miniature wings, gray body

A. Calculate the map distances separating the three genes.

B. Is positive interference occurring?

T OPIC: *What topic in genetics does this question address?* The topic is linkage. More specifically, you are asked to calculate the map distances separating three linked genes and to determine if positive interference is occurring.

I NFORMATION: *What information do you know based on the question and your understanding of the topic?* From the question, you know that the three different genes are on the X chromosome. You also know the results of a three-factor testcross. The F_1 offspring have all of the dominant alleles on one chromosome and all of the recessive alleles on another chromosome. (Make a drawing of the chromosomes involved in these crosses if this isn't obvious.) From your understanding of the topic, you may remember that the map distance is a measure of the percentage of recombinant offspring. In a three-factor cross, double crossovers result in the rarest category of offspring, in which the gene in the middle becomes separated from the genes at the ends.

P ROBLEM-SOLVING **S** TRATEGIES: *Analyze data. Make a drawing. Make a calculation.* The first step is to analyze the data and make a drawing that depicts the order of the three genes. You can do this by evaluating the pattern of inheritance for the double crossovers. The double crossovers occur with the lowest frequency. Thus, the double crossovers produced vermilion eyes, long wings, sable body and red eyes, miniature wings, gray body. Comparing these recombinant patterns of traits with the nonrecombinant patterns (vermilion eyes, miniature wings, sable body and red eyes, long wings, gray body) shows that the gene for wing length has been reassorted. Two flies have

long wings associated with vermilion eyes and sable body, and one fly has miniature wings associated with red eyes and gray body. Taken together, these results indicate that the wing length gene is found in between the eye color and body color genes.

$$v \qquad\qquad m \qquad\qquad s$$

You can now calculate the distance between the genes for eye color and wing length and between wing length and body color. To do this, you can consider the data on numbers of offspring according to gene pairs, first for eye color and wing length:

  vermilion eyes, miniature wings = 1320 + 102 = 1422
  red eyes, long wings = 1346 + 90 = 1436
  vermilion eyes, long wings = 42 + 2 = 44
  red eyes, miniature wings = 48 + 1 = 49

The recombinants are the offspring with vermilion eyes, long wings and red eyes, miniature wings. The map distance between these two genes is

$$(44 + 49)/(1422 + 1436 + 44 + 49) \times 100 = 3.2 \text{ mu}$$

Likewise, for the other pair of genes for wing length and body color, the numbers of offspring are

  miniature wings, sable body = 1320 + 48 = 1368
  long wings, gray body = 1346 + 42 = 1388
  miniature wings, gray body = 102 + 1 = 103
  long wings, sable body = 90 + 2 = 92

The recombinants are the offspring with miniature wings, gray body and long wings, sable body. The map distance between these two genes is

$$(103 + 92)/(1368 + 1388 + 103 + 92) \times 100 = 6.6 \text{ mu}$$

The genetic map is shown in the answer.

To calculate the interference value, you must first calculate the coefficient of coincidence:

$$C = \frac{\text{Observed number of double crossovers}}{\text{Expected number of double crossovers}}$$

Based on your calculated map distances, the percentages of single crossovers equal 3.2% (0.032) and 6.6% (0.066). The expected number of double crossovers equals 0.032 × 0.066, which is 0.002, or 0.2%. A total of 2951 offspring were produced. If you multiply 2951 × 0.002, you get 5.9, which is the expected number of double crossovers. The observed number was 3. Therefore,

$$C = 3/5.9 = 0.51$$
$$I = 1 - C = 1 - 0.51 = 0.49, \text{ or } 49\%$$

ANSWER:

A. The genetic map shown below is consistent with these data:

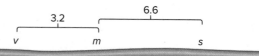

B. Approximately 49% of the expected double crossovers did not occur due to positive interference.

4. As illustrated in Figure 6.9, a limit exists for the relationship between map distance and the percentage of recombinant offspring. Even though two genes on the same chromosome may be much more than 50 mu apart, a testcross is not expected to yield more than 50% recombinant offspring. You may be wondering why this is so. The answer lies in the pattern of multiple crossovers. At prophase of meiosis I, a single crossover in the region between two genes produces only 50% recombinant chromosomes (see Figure 6.2b). Therefore, to exceed a 50% recombinant level, it would seem that multiple crossovers must occur within a bivalent.

Let's suppose that two genes are far apart on the same chromosome. A testcross is made between a heterozygous individual, *AaBb*, and a homozygous individual, *aabb*. In the heterozygous individual, the dominant alleles (*A* and *B*) are linked on the same chromosome, and the recessive alleles (*a* and *b*) are linked on the same chromosome. What are all of the possible double crossovers (between two, three, or four chromatids)? What is the average number of recombinant offspring, assuming an equal probability of occurrence for all of the double crossovers?

TOPIC: *What topic in genetics does this question address?* The topic is linkage. More specifically, the question asks you to determine how double crossovers affect the maximum map distance in a testcross.

INFORMATION: *What information do you know based on the question and your understanding of the topic?* In the question, you are reminded that the maximum percentage of recombinant offspring in a testcross is 50%. From your understanding of the topic, you may remember that the map distance is a measure of the percentage of recombinant offspring, which are produced by crossing over.

PROBLEM-SOLVING **S**TRATEGY: *Make a drawing. Predict the outcome. Make a calculation.* One strategy to solve this problem is to make a drawing that shows how the sister chromatids within a bivalent could cross over. A double crossover affecting the two genes could involve two chromatids, three chromatids, or four chromatids. From the drawing, you should be able to predict the outcome of the different types of double crossovers, and then calculate the average percentage.

ANSWER: A double crossover between the two genes could involve two chromatids, three chromatids, or four chromatids. The possibilities for all types of double crossovers are shown in the drawing in the right column.

This drawing considers two crossovers that occur in the region between the two genes. Because a bivalent is composed of two pairs of homologs, a double crossover between homologs could occur in several possible ways. In each part of the drawing, the crossover on the right has occurred first. Because all of these double crossovers are equally probable, you take the average of them, which is 100% + 50% + 50% + 0% divided by 4, to determine the maximum number of recombinants. This average equals 50%.

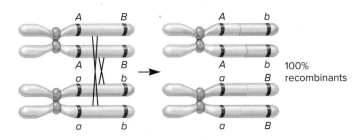

Double crossover (involving 4 chromatids)

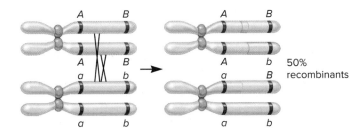

Double crossover (involving 3 chromatids)

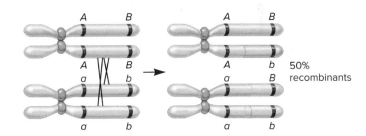

Double crossover (involving 3 chromatids)

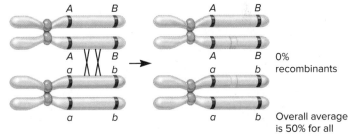

Double crossover (involving 2 chromatids)

Overall average is 50% for all 4 possibilities.

Conceptual Questions

C1. What is the difference in meaning between the terms *genetic recombination* and *crossing over*?

C2. When a chi square test is applied to solve a linkage problem, explain why an independent assortment hypothesis is proposed.

C3. A crossover has occurred in the bivalent shown here.

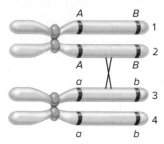

If a second crossover occurs in the same region between these two genes, which two chromatids would be involved to produce the following outcomes?

A. 100% recombinants

B. 0% recombinants

C. 50% recombinants

C4. A crossover has occurred in the bivalent shown here.

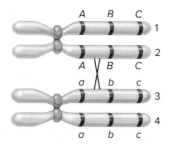

What is the outcome of this single crossover? If a second crossover occurs somewhere between *A* and *C*, explain which two chromatids would be involved and where the crossover would occur (i.e., between which two genes) to produce each of the following arrangements of alleles along the chromosomes:

A. *ABC, AbC, aBc,* and *abc*

B. *Abc, Abc, aBC,* and *aBC*

C. *ABc, Abc, aBC,* and *abC*

D. *ABC, ABC, abc,* and *abc*

C5. A diploid organism has a total of 14 chromosomes and about 20,000 genes per haploid genome. Approximately how many genes are in each linkage group?

C6. If you try to throw a basketball into a basket, the likelihood of succeeding depends on the size of the basket. It is more likely that you will get the ball into the basket if the basket is bigger. In your own words, explain how this analogy applies to the idea that the likelihood of crossing over is greater when two genes are far apart than when they are close together.

C7. By conducting testcrosses, researchers have found that the sweet pea plant has seven linkage groups. How many chromosomes would you expect to find in leaf cells of sweet pea plants?

C8. In humans, a rare dominant disorder known as nail-patella syndrome causes abnormalities in the fingernails, toenails, and kneecaps. Researchers have examined pedigrees of families in which this disorder occurs and have also determined the blood types of individuals within each pedigree. (A description of blood genotypes is found in Chapter 4.) In the following pedigree, individuals affected with nail-patella syndrome are shown as filled symbols. The genotype

of each individual with regard to ABO blood type is also shown. Does this pedigree suggest any linkage between the gene that causes nail-patella syndrome and the gene that determines blood type?

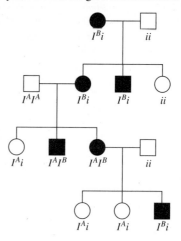

C9. When true-breeding mice with brown fur and short tails (*BBtt*) were crossed to true-breeding mice with white fur and long tails (*bbTT*), all of the F_1 offspring had brown fur and long tails. The F_1 offspring were crossed to mice with white fur and short tails. What are the possible phenotypes of the F_2 offspring? Which F_2 offspring are recombinant, and which are nonrecombinant? What is the ratio of phenotypes of the F_2 offspring if independent assortment is taking place? How is the ratio affected by linkage?

C10. Though we often think of genes in terms of the phenotypes they produce (e.g., curly leaves, flaky tail, brown eyes), the molecular function of most genes is to code proteins. Many cellular proteins function as enzymes. The following table gives the map distances separating pairs of genes for six different genes that code six different enzymes: *Ada*, adenosine deaminase; *Hao-1*, hydroxyacid oxidase-1; *Hdc*, histidine decarboxylase; *Odc-2*, ornithine decarboxylase-2; *Sdh-1*, sorbitol dehydrogenase-1; and *Ass-1*, arginosuccinate synthetase-1.

Map distances between pairs of these genes are as follows:

	Ada	*Hao-1*	*Hdc*	*Odc-2*	*Sdh-1*	*Ass-1*
Ada		14		8	28	
Hao-1	14		9		14	
Hdc		9		15	5	
Odc-2	8		15			63
Sdh-1	28	14	5			43
Ass-1				63	43	

Construct a genetic map that shows the locations of all six genes.

C11. If the likelihood of a single crossover in a particular chromosomal region is 10%, what is the theoretical likelihood of a double or triple crossover in that same region? How would positive interference affect these theoretical values?

C12. In most two-factor crosses involving linked genes, you cannot tell if a double crossover between the two genes has occurred because the offspring will inherit the nonrecombinant pattern of alleles. How does the inability to detect double crossovers affect the calculation of map distance? Is map distance underestimated or overestimated because of this inability to detect double crossovers? Explain your answer.

C13. Researchers have discovered that some regions of chromosomes are much more likely than others to cross over. We might call such a region a "hot spot" for crossing over. Let's suppose that two genes, gene *A* and gene *B*, are 5,000,000 bp apart on the same chromosome. Genes *A* and *B* are in a hot spot for crossing over. Two other genes, let's call them gene *C* and gene *D*, are also 5,000,000 bp apart but are not in a hot spot for recombination. If you conducted two-factor crosses to compute the map distance between genes *A* and *B* and other two-factor crosses to compute the map distance between genes *C* and *D*, would the calculated map distance between *A* and *B* be the same as that between *C* and *D*? Explain.

C14. Describe the unique features of ascomycetes that facilitate genetic analysis of these fungi.

C15. What is mitotic recombination? A heterozygous individual (*Bb*) with brown eyes has one eye with a small patch of blue. Provide two or more explanations for how the blue patch may have occurred.

C16. Mitotic recombination can occasionally produce a twin spot. Let's suppose a mouse is heterozygous for two genes that govern fur color and length: One gene affects pigmentation, with dark pigmentation (*A*) dominant to albino (*a*); the other gene affects hair length, with long hair (*L*) dominant to short hair (*l*). The two genes are linked on the same chromosome: *A* is linked to *l*, and *a* is linked to *L*. Draw the chromosomes carrying these alleles, and explain how mitotic recombination could produce a twin spot, with one spot having albino pigmentation and long fur and the other having dark pigmentation and short fur.

Experimental Questions (Including Many Mapping Questions)

E1. Figure 6.1 shows the first experimental results that indicated linkage between two different genes. Conduct a chi square test to confirm that the genes are really linked (i.e., the data could not be explained by independent assortment).

E2. In the experiment in Figure 6.6, Stern studied the inheritance pattern of crosses in which female fruit flies carried two abnormal X chromosomes, in order to find a relationship between genetic recombination and the physical exchange of chromosome pieces. Is it necessary to use a strain carrying two abnormal chromosomes, or could you use a strain in which females carried one normal X chromosome and one abnormal X chromosome with a deletion at one end and an extra piece of the Y chromosome at the other end?

E3 Explain the rationale behind a testcross. Is it necessary for one of the parents to be homozygous recessive for the genes of interest? In the heterozygous parent in a testcross, must all dominant alleles be linked on the same chromosome and all of the recessive alleles be linked on the homolog?

E4. In your own words, explain why a testcross cannot produce more than 50% recombinant offspring. When a testcross does produce 50% recombinant offspring, what does this result mean?

E5. Explain why the percentage of recombinant offspring from a testcross is a more accurate measure of map distance when two genes are close together. When two genes are far apart, is the percentage of recombinant offspring an underestimate or overestimate of the actual map distance?

E6. If two genes are more than 50 mu apart, how would you ever be able to show experimentally that they are located on the same chromosome?

E7. From the three-factor crosses of Figure 6.3, Morgan realized that crossing over was more frequent between the eye color and wing length genes than between the body color and eye color genes. Explain how Morgan determined this.

E8. Two genes are located on the same chromosome and are known to be 12 mu apart. An *AABB* individual was crossed to an *aabb* individual to produce *AaBb* offspring. The *AaBb* offspring were then testcrossed to *aabb* individuals.

   A. If the testcross produces 1000 offspring, what are the predicted numbers of offspring with each of these four genotypes: *AaBb*, *Aabb*, *aaBb*, and *aabb*?

   B. What would be the predicted numbers of offspring with the four genotypes if the P generation had been *AAbb* and *aaBB* instead of *AABB* and *aabb*?

E9. Two genes, designated *A* and *B*, are located 10 mu from each other. A third gene, designated *C*, is located 15 mu from *B* and 5 mu from *A*. The P generation consisted of *AA bb CC* and *aa BB cc* individuals that were crossed to each other. The F_1 heterozygotes were then testcrossed to *aa bb cc* individuals. Assuming that no double crossovers occur, what percentage of F_2 offspring would you expect with each of the following genotypes?

   A. *Aa Bb Cc*

   B. *aa Bb Cc*

   C. *Aa bb cc*

E10. Two genes in tomato plants are 61 mu apart; normal fruit (*F*) is dominant to fasciated (flattened) fruit (*f*), and normal number of leaves (*Lf*) is dominant to leafy (*lf*). A true-breeding plant with a normal number of leaves and normal fruit was crossed to a leafy plant with fasciated fruit. The F_1 offspring were then crossed to leafy plants with fasciated fruit. If this cross produced 600 offspring, what are the expected numbers of plants in each of the four possible categories: normal number of leaves, normal fruit; normal number of leaves, fasciated fruit; leafy, normal fruit; and leafy, fasciated fruit?

E11. In tomato plants, three genes are linked on the same chromosome. Tall is dominant to short, skin that is smooth is dominant to skin that is peachy (fuzzy), and fruit with a (wild-type) round tomato shape is dominant to fruit with an oblate (flattened) shape. A plant that is true-breeding for the dominant traits was crossed to a short plant with peachy skin and oblate fruit. The F_1 plants were then testcrossed to short plants with peachy skin and oblate fruit. The following results were obtained for the F_2 offspring:

   151 tall, smooth, round

   33 tall, smooth, oblate

   11 tall, peachy, oblate

   2 tall, peachy, round

   155 short, peachy, oblate

   29 short, peachy, round

   12 short, smooth, round

   0 short, smooth, oblate

Construct a genetic map that shows the order of these three genes and the map distances between them.

E12. A trait in garden peas involves the curling of leaves. A two-factor cross was made by crossing a plant with yellow pods and curling leaves to a wild-type plant with green pods and normal leaves. All of the F_1 offspring had green pods and normal leaves. The F_1 plants were then crossed to plants with yellow pods and curling leaves. The following results were obtained for the F_2 offspring:

117 green pods, normal leaves

115 yellow pods, curling leaves

78 green pods, curling leaves

80 yellow pods, normal leaves

A. Conduct a chi square test to determine if these two genes are linked.

B. If they are linked, calculate the map distance between the two genes. How accurate do you think this calculated distance is?

E13. In mice, the gene that codes the enzyme inosine triphosphatase is 12 mu from the gene that codes the enzyme ornithine decarboxylase. Suppose you have identified a strain of mice homozygous for a defective inosine triphosphatase gene that does not produce any of this enzyme and also homozygous for a defective ornithine decarboxylase gene. In other words, this strain of mice cannot make either enzyme. You cross this homozygous recessive strain to a normal strain of mice to produce F_1 heterozygotes.

A. Let's suppose that F_1 heterozygotes are then crossed to the strain that cannot produce either enzyme. What is the probability of obtaining a mouse that cannot make either enzyme?

B. Alternatively, if two F_1 heterozygotes are crossed to each other, what is the probability of obtaining a mouse that cannot make either enzyme?

E14. In the garden pea, several different genes affect pod characteristics. A gene affecting pod color (green is dominant to yellow) is approximately 7 mu away from a gene affecting pod width (wide is dominant to narrow). Both genes are located on chromosome 5. A third gene, located on chromosome 4, affects pod length (long is dominant to short). A true-breeding wild-type plant (green, wide, long pods) was crossed to a plant with yellow, narrow, short pods. The F_1 offspring were then testcrossed to plants with yellow, narrow, short pods. If the testcross produced 800 offspring, what are the expected numbers of the eight possible phenotypes?

E15. A sex-influenced trait of a certain species of animal is dominant in males and causes bushy tails. The same trait is recessive in females. Fur color is not sex-influenced. Yellow fur is dominant to white fur. A true-breeding female with a bushy tail and yellow fur was crossed to a white male without a bushy tail (i.e., a normal tail). The F_1 females were then crossed to white males without bushy tails. The following results were obtained for the F_2 offspring:

Males	Females
28 normal tails, yellow fur	102 normal tails, yellow fur
72 normal tails, white fur	96 normal tails, white fur
68 bushy tails, yellow fur	0 bushy tails, yellow fur
29 bushy tails, white fur	0 bushy tails, white fur

A. Conduct a chi square test to determine if these two genes are linked.

B. If the genes are linked, calculate the map distance between them. Explain which data you used in your calculation.

E16. Three recessive traits in garden pea plants are as follows: yellow pods are recessive to green pods; bluish green seedlings are recessive to green seedlings; creeper (a plant that cannot stand up) is recessive to normal. A true-breeding normal plant with green pods and green seedlings was crossed to a creeper with yellow pods and bluish green seedlings. The F_1 plants were then crossed to creepers with yellow pods and bluish green seedlings. The following results were obtained for the F_2 offspring:

2059 green pods, green seedlings, normal

151 green pods, green seedlings, creeper

281 green pods, bluish green seedlings, normal

15 green pods, bluish green seedlings, creeper

2041 yellow pods, bluish green seedlings, creeper

157 yellow pods, bluish green seedlings, normal

282 yellow pods, green seedlings, creeper

11 yellow pods, green seedlings, normal

Construct a genetic map that shows the map distances between these three genes.

E17. In mice, a trait called snubnose is recessive to a normal nose, a trait called pintail is dominant to a normal tail, and a trait called jerker (a defect in motor skills) is recessive to a normal gait. Jerker mice with a snubnose and a pintail were crossed to normal mice, and then the F_1 mice were crossed to jerker mice that have a snubnose and a normal tail. The outcome of this cross was as follows:

560 jerker, snubnose, pintail

548 normal gait, normal nose, normal tail

102 jerker, snubnose, normal tail

104 normal gait, normal nose, pintail

77 jerker, normal nose, normal tail

71 normal gait, snubnose, pintail

11 jerker, normal nose, pintail

9 normal gait, snubnose, normal tail

Construct a genetic map that shows the order of these genes and the map distances between them.

E18. In *Drosophila*, an allele causing vestigial wings is 12.5 mu away from another allele that causes purple eyes. A third gene that affects body color has an allele that causes black body color. This third gene is 18.5 mu away from the vestigial wings allele and 6 mu away from the allele causing purple eyes. The alleles causing vestigial wings, purple eyes, and black body are all recessive. The dominant (wild-type) traits are long wings, red eyes, and gray body. A researcher crossed wild-type flies to flies with vestigial wings, purple eyes, and black bodies. All F_1 flies were wild-type. F_1 female flies were then crossed to male flies with vestigial wings, purple eyes, and black bodies. If 1000 offspring were observed, what are the expected numbers of the following phenotypes?

long wings, red eyes, gray body

long wings, purple eyes, gray body

long wings, red eyes, black body

long wings, purple eyes, black body

short wings, red eyes, gray body

short wings, purple eyes, gray body

short wings, red eyes, black body

short wings, purple eyes, black body

Which of these phenotypes can be produced only by a double crossover event?

E19. Three autosomal genes are linked along the same chromosome. The distance between gene *A* and gene *B* is 7 mu, the distance between *B* and *C* is 11 mu, and the distance between *A* and *C* is 4 mu. An individual that is *AA bb CC* was crossed to an individual that is *aa BB cc* to produce heterozygous F_1 offspring. The F_1 offspring were then crossed to homozygous *aa bb cc* individuals to produce F_2 offspring.

   A. Draw the arrangement of the alleles on the chromosomes in the parents and in the F_1 offspring.

   B. Where would a crossover have to occur to produce an F_2 offspring that was heterozygous for all three genes?

   C. If you assume that no double crossovers occur, what percentage of F_2 offspring is likely to be homozygous for all three genes?

E20. Let's suppose that two different X-linked genes, designated by the letters *N* and *L*, are found in mice. Gene *N* exists in a dominant, normal allele and in a recessive allele, *n*, that is lethal. Similarly, gene *L* exists in a dominant, normal allele and in a recessive allele, *l*, that is lethal. Heterozygous females are normal, but males that carry either recessive allele are born dead. Explain whether or not it would be possible to map the distance between these two genes by making crosses and analyzing the number of living and dead offspring. You may assume that you have strains of mice in which females are heterozygous for one or both genes.

E21. The alleles *his-5* and *lys-1*, found in baker's yeast, result in cells that require histidine and lysine for growth, respectively. A cross was made between two haploid yeast strains that are *his-5 lys-1* and *his+ lys+*. From the analysis of 818 tetrads, the following results were obtained:

   2 spores with *his-5 lys+* + 2 spores with *his+ lys-1* = 4

   2 spores with *his-5 lys-1* + 2 spores with *his+ lys+* = 502

   1 spore with *his-5 lys-1* + 1 spore with *his-5 lys+* + 1 spore with *his+ lys-1* + 1 spore with *his+ lys+* = 312

   A. Compute the map distance between these two genes using first the method of calculation that considers double crossovers and then the one that does not. Which method gives a higher value? Explain why.

   B. What is the frequency of single crossovers between these two genes?

   C. Based on your answer to part B, how many NPDs are expected from this cross? Explain your answer. Is positive interference occurring?

E22. The alleles *trp-3* and *his-2*, found in baker's yeast, result in cells that require tryptophan and histidine for growth, respectively. These two genes are closely linked. A cross was made between two haploid yeast strains that are *trp-3 his+* and *trp+ his-2*. Tetrads were obtained with the following genotypes:

   2 spores with *trp-3 his+* + 2 spores with *trp+ his-2*

   2 spores with *trp-3 his-2* + 2 spores with *trp+ his+*

   1 spore with *trp-3 lys-1* + 1 spore with *trp-3 his+* + 1 spore with *trp+ lys-1* + 1 spore with *trp+ his+*

   A. Which of these types of tetrads is a parental ditype, a tetratype, or a nonparental ditype?

   B. Which of the three types of tetrads contain spores that are due to crossing over?

Questions for Student Discussion/Collaboration

1. In mice, a dominant allele that causes a short tail is located on chromosome 2. On chromosome 3, a recessive allele causing droopy ears is 6 mu away from another recessive allele that causes a flaky tail. A recessive allele that causes a jerker (uncoordinated) phenotype is located on chromosome 4. A jerker mouse with droopy ears and a short, flaky tail was crossed to a normal mouse. All of the mice in the F_1 generation were phenotypically normal, except they had short tails. These F_1 mice were then testcrossed to jerker mice with droopy ears and long, flaky tails. If this testcross produces 400 offspring, what are the expected numbers of the 16 possible phenotypic categories?

2. In Chapter 4, we discussed the observation that the X and Y chromosomes have a small number of genes in common. These genes are inherited in a pseudoautosomal pattern. With this phenomenon in mind, discuss whether or not the X and Y chromosomes are really distinct linkage groups.

3. Mendel studied seven traits in pea plants, and the garden pea happens to have seven different chromosomes. It has been pointed out that Mendel was very lucky not to have conducted crosses involving two traits governed by genes that are closely linked on the same chromosome because the results would have confounded the law of independent assortment. It has even been suggested that Mendel may not have published data involving traits that were linked! An article by Stig Blixt ("Why Didn't Gregor Mendel Find Linkage?" *Nature* 256:206, 1975) considers this issue. Read this article, and discuss why Mendel did not find linkage.

Note: All answers are available for the instructor in Connect; the answers to the even-numbered questions and all of the Concept Check and Comprehension Questions are in Appendix B.

CHAPTER OUTLINE

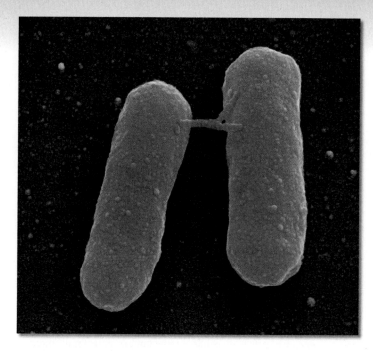

Conjugating bacteria. *The bacteria shown here are transferring genetic material by a process called conjugation.*
Eye of Science/Science Source

7

GENETIC TRANSFER AND MAPPING IN BACTERIA

Thus far, our attention in Part II of this textbook has focused on genetic analyses of eukaryotic species such as plants and animals. As we have seen, these organisms are amenable to genetic studies for two reasons. First, characteristics, such as tall versus short height in pea plants and red versus white eyes in *Drosophila*, provide readily discernible traits for distinguishing individuals. Second, because most eukaryotic species reproduce sexually, crosses can be made, and the pattern of transmission of traits from parent to offspring can be analyzed. The ability to follow allelic differences in a genetic cross is a basic tool in the genetic examination of eukaryotic species.

In this chapter, we turn our attention to the genetic analysis of bacteria. Like their eukaryotic counterparts, bacteria often possess allelic differences that affect their cellular traits. Common allelic variations among bacteria involve traits such as sensitivity to antibiotics and differences in nutrient requirements for growth.

In these cases, the allelic differences are between different strains of bacteria, because any given bacterium is usually haploid with regard to a particular gene. Throughout this chapter, we will consider interesting experiments that examine bacterial strains with allelic differences.

Compared with eukaryotes, a striking difference among bacterial species is their mode of reproduction. Because bacteria reproduce asexually, researchers do not use crosses in genetic analyses of bacterial species. Instead, they rely on a similar mechanism, called genetic transfer, in which a segment of bacterial DNA is transferred from one bacterium to another. In this chapter, we will explore the different routes of genetic transfer in bacteria and see how researchers have used this process to map the locations of genes along the chromosome of a few bacterial species. We will also consider the medical relevance of bacterial genetic transfer.

7.1 OVERVIEW OF GENETIC TRANSFER IN BACTERIA

Learning Outcome:

1. Compare and contrast the three mechanisms of genetic transfer in bacteria.

Genetic transfer is a process by which one bacterium transfers genetic material to another bacterium. Why is genetic transfer an advantage? Like sexual reproduction in eukaryotes, genetic transfer in bacteria is thought to enhance the genetic diversity of bacterial species. For example, a bacterial cell carrying a gene that provides antibiotic resistance may transfer this gene to another bacterial cell, enabling that cell to survive exposure to the antibiotic.

Bacteria can transfer genetic material naturally via three mechanisms (**Table 7.1**).

- **Conjugation** involves a direct physical interaction between two bacterial cells. One bacterium acts as a donor and transfers genetic material to a recipient cell.
- During **transduction,** a virus infects a bacterium and then transfers bacterial genetic material from that bacterium to another.

TABLE 7.1

Three Mechanisms of Genetic Transfer Between Bacteria

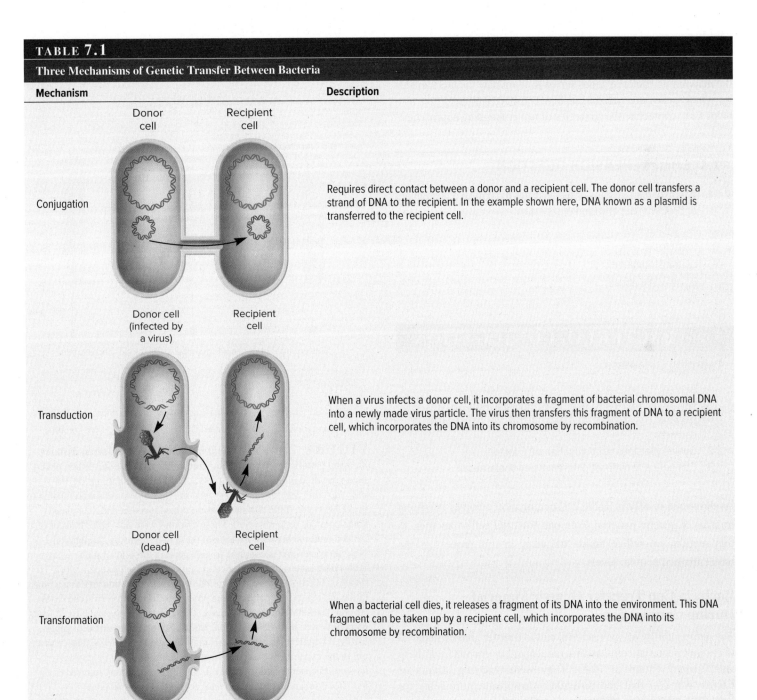

Mechanism	Description
Conjugation	Requires direct contact between a donor and a recipient cell. The donor cell transfers a strand of DNA to the recipient. In the example shown here, DNA known as a plasmid is transferred to the recipient cell.
Transduction	When a virus infects a donor cell, it incorporates a fragment of bacterial chromosomal DNA into a newly made virus particle. The virus then transfers this fragment of DNA to a recipient cell, which incorporates the DNA into its chromosome by recombination.
Transformation	When a bacterial cell dies, it releases a fragment of its DNA into the environment. This DNA fragment can be taken up by a recipient cell, which incorporates the DNA into its chromosome by recombination.

- **Transformation** is a process in which genetic material is released into the environment when a bacterial cell dies. This material then binds to a living bacterial cell, which can take it up.

These three mechanisms of genetic transfer have been extensively investigated in research laboratories, and their molecular mechanisms continue to be studied with great interest. In later sections of this chapter, we will examine these three mechanisms in greater detail. Although this chapter focuses on bacteria, it is worth noting that prokaryotes include two domains: Bacteria and Archaea. Gene transfer has also been observed in archaea, and the molecular mechanisms are beginning to be identified.

We will also examine how genetic transfer between bacterial cells has provided unique ways to accurately map bacterial genes. The mapping methods described in this chapter have been largely replaced by molecular approaches described in Chapter 22. Even so, the mapping of bacterial genes serves to illuminate the mechanisms by which genes are transferred between bacterial cells and also helps us to appreciate the strategies of newer mapping approaches.

7.1 COMPREHENSION QUESTION

1. A form of genetic transfer that involves the uptake of a fragment of DNA from the environment is called
   a. conjugation.
   b. transduction.
   c. transformation.
   d. all of the above.

7.2   BACTERIAL CONJUGATION

Learning Outcomes:

1. Analyze the work of Lederberg and Tatum and that of Davis, and explain how the data indicated that some strains of bacteria can transfer genetic material via direct physical contact.
2. Outline the steps of conjugation via F factors.
3. Compare and contrast different types of plasmids.

As described briefly in Table 7.1, conjugation involves the direct transfer of genetic material from one bacterial cell to another. In this section, we will examine the steps in this process at the molecular and cellular levels.

Bacteria Can Transfer Genetic Material During Conjugation

The natural ability of one bacterial cell to transfer genetic material to another bacterial cell was first recognized by Joshua Lederberg and Edward Tatum in 1946. They were studying strains of *Escherichia coli* that had different nutritional requirements for growth. A **minimal medium** is a growth medium that contains the essential nutrients for a wild-type (nonmutant) bacterial species to

grow. Researchers often study bacterial strains that harbor mutations and cannot grow on a minimal medium.

- A strain that cannot synthesize a particular nutrient and needs that nutrient to be added to its growth medium is called an **auxotroph.** For example, a strain that cannot make the amino acid methionine would not grow on a minimal medium because the minimal medium does not contain methionine. Such a strain would need to have methionine added to its growth medium and would be called a methionine auxotroph.
- By comparison, a strain that could make this amino acid would be termed a methionine prototroph. A **prototroph** does not need a particular nutrient included in its growth medium.

The experiment in **Figure 7.1** considers one strain, designated $met^-\ bio^-\ thr^+\ leu^+\ thi^+$, which required the addition to its

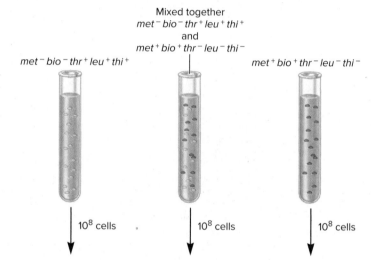

Mixed together
$met^-\ bio^-\ thr^+\ leu^+\ thi^+$
and
$met^+\ bio^+\ thr^-\ leu^-\ thi^-$

$met^-\ bio^-\ thr^+\ leu^+\ thi^+$ $met^+\ bio^+\ thr^-\ leu^-\ thi^-$

10^8 cells 10^8 cells 10^8 cells

Nutrient agar plates lacking amino acids, biotin, and thiamine

No colonies   Bacterial colonies   No colonies

FIGURE 7.1 Experiment of Lederberg and Tatum demonstrating genetic transfer during conjugation in *E. coli*. When plated on a growth medium lacking amino acids, biotin, and thiamine, the $met^-\ bio^-\ thr^+\ leu^+\ thi^+$ and $met^+\ bio^+\ thr^-\ leu^-\ thi^-$ strains were unable to grow. However, if the two strains were mixed together and then plated, some colonies were observed. These colonies were due to the transfer of genetic material between bacteria of these two strains by conjugation. Note: In bacteria, it is common to give genes a three-letter name (shown in italics) that is related to the function of the gene. A plus superscript (+) indicates a functional gene, and a minus superscript (−) indicates that a mutation has caused the gene or gene product to be inactive. In some cases, several genes have related functions. These may have the same three-letter name followed by different capital letters. For example, different genes involved with leucine biosynthesis may be called *leuA*, *leuB*, *leuC*, and so on. In the experiment described here, the genes involved in leucine biosynthesis were not distinguished, so the gene is simply referred to as leu^+ (for a functional gene) and leu^- (for a nonfunctional gene).

CONCEPT CHECK: Describe how genetic transfer can explain the growth of colonies on the middle plate.

growth medium of one amino acid, methionine (met), and one vitamin, biotin (bio). This strain did not require the amino acids threonine (thr) and leucine (leu) or the vitamin thiamine (thi) to be added to its growth medium. Another strain, designated *met$^+$ bio$^+$ thr$^-$ leu$^-$ thi$^-$*, had just the opposite requirements. It was an auxotroph for threonine, leucine, and thiamine, but a prototroph for methionine and biotin. These differences in nutritional requirements correspond to variations in the genetic material of the two strains. The first strain had two defective genes that would, if functional, code enzymes necessary for methionine and biotin synthesis. The second strain had three defective genes that would, if functional, allow synthesis of threonine, leucine, and thiamine.

Figure 7.1 compares the results when the two strains were mixed together and when they were not mixed.

- Without mixing, about 100 million (10^8) *met$^-$ bio$^-$ thr$^+$ leu$^+$ thi$^+$* cells were applied to plates on a growth medium lacking amino acids, biotin, and thiamine; no colonies were observed to grow. This result is expected because the medium did not contain methionine or biotin.
- Likewise, when 10^8 *met$^+$ bio$^+$ thr$^-$ leu$^-$ thi$^-$* cells were plated, no colonies were observed because threonine, leucine, and thiamine were missing from this growth medium.
- When the two strains were mixed together and then 10^8 cells plated, approximately 10 bacterial colonies formed. Because growth occurred, the genotype of the cells within these colonies must have been *met$^+$ bio$^+$ thr$^+$ leu$^+$ thi$^+$*.

How could the *met$^+$ bio$^+$ thr$^+$ leu$^+$ thi$^+$* genotype occur? Because no colonies were observed on either plate in which the two strains were not mixed, Lederberg and Tatum concluded that this genotype was not due to mutations that converted *met$^-$ bio$^-$* to *met$^+$ bio$^+$* or to mutations that converted *thr$^-$ leu$^-$ thi$^-$* to *thr$^+$ leu$^+$ thi$^+$*. Instead, they hypothesized that some genetic material was transferred between the two strains.

- One possibility is that the genetic material providing the ability to synthesize methionine and biotin (*met$^+$ bio$^+$*) was transferred to the *met$^-$ bio$^-$ thr$^+$ leu$^+$ thi$^+$* strain.
- Alternatively, genetic material providing the ability to synthesize threonine, leucine, and thiamine (*thr$^+$ leu$^+$ thi$^+$*) may have been transferred to the *met$^+$ bio$^+$ thr$^-$ leu$^-$ thi$^-$* cells.

The results of this experiment did not distinguish between these two possibilities.

Conjugation Requires Direct Physical Contact

In 1950, Bernard Davis conducted experiments showing that two strains of bacteria must make physical contact with each other to transfer genetic material. The apparatus he used, known as a U-tube, is shown in **Figure 7.2**. At the bottom of the U-tube is a filter with pores small enough to allow the passage of genetic material (i.e., DNA molecules) but too small to permit the passage of bacterial cells.

1. On one side of the filter, Davis added a bacterial strain with a certain combination of nutritional requirements

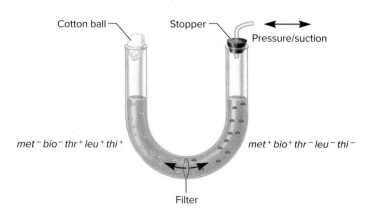

Cotton ball | Stopper | Pressure/suction

met$^-$ bio$^-$ thr$^+$ leu$^+$ thi$^+$ *met$^+$ bio$^+$ thr$^-$ leu$^-$ thi$^-$*

Filter

FIGURE 7.2 **A U-tube apparatus like that used by Davis.** The fluid in the tube is forced through the filter by alternating suction and pressure. However, the pores in the filter are too small for the passage of bacteria.

CONCEPT CHECK: With regard to studying the mechanism of conjugation, what is the purpose of using a U-tube?

(the *met$^-$ bio$^-$ thr$^+$ leu$^+$ thi$^+$* strain). On the other side, he added a different bacterial strain (the *met$^+$ bio$^+$ thr$^-$ leu$^-$ thi$^-$* strain).

2. The application of alternating pressure and suction promoted the movement of liquid through the filter. Because the bacteria were too large to pass through the pores, the movement of liquid did not allow the two types of bacterial strains to mix with each other. However, any genetic material that was released from a bacterium could pass through the filter.

3. After incubation in a U-tube, bacteria from either side of the tube were placed on a medium that could select for the growth of cells that were *met$^+$ bio$^+$ thr$^+$ leu$^+$ thi$^+$*. This minimal medium lacked methionine, biotin, threonine, leucine, and thiamine, but contained all other nutrients essential for growth.

4. In this case, no bacterial colonies grew on the plates. The experiment showed that, without physical contact, the two bacterial strains did not transfer genetic material to one another.

The term *conjugation* is now used to describe the natural process of genetic transfer between bacterial cells that requires direct cell-to-cell contact. Many, but not all, species of bacteria can conjugate.

An F$^+$ Strain Transfers an F factor to an F$^-$ Strain During Conjugation

We now know that certain donor strains of *E. coli* contain a small circular segment of genetic material known as an **F factor** (for fertility factor) in addition to their circular chromosome. Strains of *E. coli* that contain an F factor are designated F$^+$, and strains without an F factor are designated F$^-$. F factors carry several genes that are required for conjugation to occur. For example, **Figure 7.3** shows the arrangement of genes on the F factor found in certain strains of *E. coli*. The proteins coded by these genes

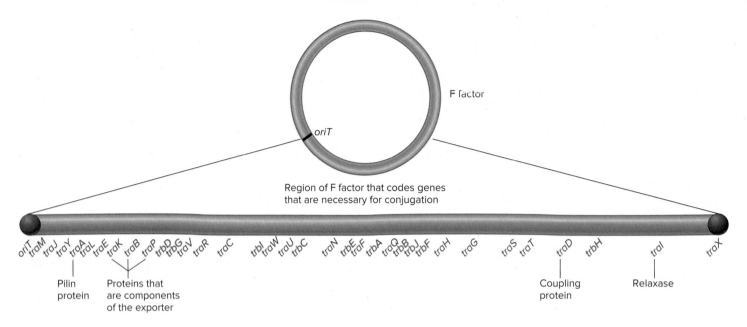

FIGURE 7.3   Genes on the F factor that play a role during conjugation.   A region of the F factor carries genes that play a role in the conjugation process. Because they play a role in the transfer of DNA from donor to recipient cell, the genes are designated with the three-letter names of *tra* or *trb*, followed by a capital letter. (Note: The *tr-* prefix stands for "transfer.") The *tra* genes are shown in red, and the *trb* genes are shown in blue. The functions of a few examples are indicated. The origin of transfer is designated *oriT*.

CONCEPT CHECK: Would this circular DNA molecule be found in an F⁺ or F⁻ strain?

are needed to transfer a strand of DNA from the donor cell to a recipient cell.

Figure 7.4a shows the molecular events that occur during conjugation in *E. coli*.

1. Contact between donor and recipient cells is a key step that initiates the conjugation process. **Sex pili** (singular: **pilus**) are produced by F⁺ strains (**Figure 7.4b**). The gene coding the pilin protein (*traA*) is located on the F factor. The pili act as attachment sites that promote the binding of bacteria to each other. In this way, an F⁺ strain makes physical contact with an F⁻ strain. In certain species, such as *E. coli*, long pili project from F⁺ cells and attempt to make contact with nearby F⁻ cells. Once contact is made, the pili shorten, thereby drawing the donor and recipient cells closer together. A **conjugation bridge** is later formed between the two cells, which provides a passageway for DNA transfer.

2. The successful contact between donor and recipient cells stimulates the donor cell to begin the transfer process. Genes within the F factor code a protein complex called the **relaxosome.** This complex first recognizes a DNA sequence in the F factor known as the **origin of transfer** (see Figure 7.4a). Upon recognition, the relaxosome cuts one DNA strand at that site in the F factor.

3. The relaxosome also catalyzes the separation of the DNA strands, and only the cut DNA strand, called the **T DNA,** is transferred to the recipient cell. As the DNA strands separate, most of the proteins within the relaxosome are

released, but one protein, called relaxase, remains bound to the end of the cut DNA strand. The complex between the single-stranded DNA and relaxase is called a **nucleoprotein** because it contains both nucleic acid (DNA) and a protein (relaxase).

4. The next phase of conjugation involves the export of the nucleoprotein complex from the donor cell to the recipient cell. To begin this process, the T DNA/relaxase complex is recognized by a coupling factor that promotes the entry of the nucleoprotein into the exporter, a complex of proteins that spans both inner and outer membranes of the donor cell. In various bacterial species, this complex is composed of 10 to 15 different proteins that are coded by genes within the F factor.

5. Once the DNA/relaxase complex is pumped out of the donor cell, it travels through the conjugation bridge and then into the recipient cell. As shown in Figure 7.4a, the other strand of the F-factor DNA remains in the donor cell, where DNA replication restores this DNA to its original double-stranded condition. After the recipient cell receives a single strand of the F-factor DNA, relaxase catalyzes the joining of the ends of this linear DNA molecule to form a circular molecule. This single-stranded DNA is replicated in the recipient cell to become double-stranded.

The result of conjugation is that the recipient cell has acquired an F factor, converting it from an F⁻ to an F⁺ cell. The genetic composition of the donor cell has not changed.

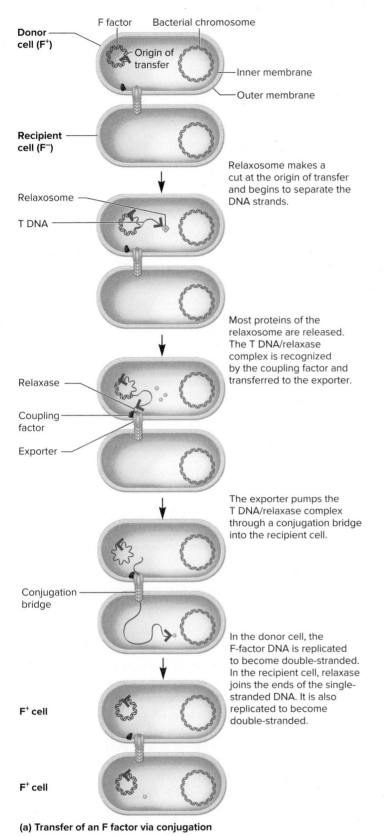

Donor cell (F⁺)

F factor   Bacterial chromosome

Origin of transfer

Inner membrane

Outer membrane

Recipient cell (F⁻)

Relaxosome makes a cut at the origin of transfer and begins to separate the DNA strands.

Relaxosome

T DNA

Most proteins of the relaxosome are released. The T DNA/relaxase complex is recognized by the coupling factor and transferred to the exporter.

Relaxase

Coupling factor

Exporter

The exporter pumps the T DNA/relaxase complex through a conjugation bridge into the recipient cell.

Conjugation bridge

In the donor cell, the F-factor DNA is replicated to become double-stranded. In the recipient cell, relaxase joins the ends of the single-stranded DNA. It is also replicated to become double-stranded.

F⁺ cell

F⁺ cell

(a) Transfer of an F factor via conjugation

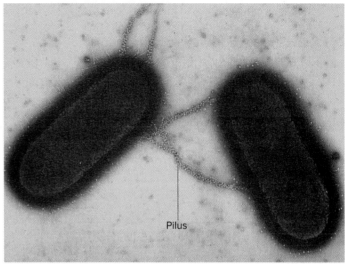

Pilus

(b) Conjugating _E. coli_

ONLINE ANIMATION

FIGURE 7.4 **The transfer of an F factor during bacterial conjugation.** **(a)** The mechanism of transfer. The end result is that both cells have an F factor. **(b)** Two _E. coli_ cells in the act of conjugation. The cell on the left is F⁺, and the one on the right is F⁻. The two cells make contact with each other via sex pili that are produced by the F⁺ cell.

(b): Dr. L. Caro/SPL/Science Source

CONCEPT CHECK: What are the functions of relaxase, coupling factor, and the exporter in the process of conjugation?

Bacteria May Contain Different Types of Plasmids

Thus far, we have considered F factors, which are one type of DNA that can exist independently of the chromosomal DNA. The more general term for this type of DNA molecule is a **plasmid.** Most known plasmids are circular, although some are linear. Plasmids occur naturally in many strains of bacteria and in a few types of eukaryotic cells such as yeast. The smallest plasmids consist of just a few thousand base pairs (bp) and carry only a gene or two; the largest are in the range of 100,000 to 500,000 bp and carry several dozen or even hundreds of genes. Some plasmids, such as F factors, can integrate into a chromosome. These plasmids are also called **episomes.**

A plasmid has its own origin of replication that allows it to be replicated independently of the bacterial chromosome. The DNA sequence of the origin of replication influences how many copies of the plasmid are found within a cell. Some origins are said to be very strong because they result in many copies of the plasmid, perhaps as many as 100 per cell. Other origins of replication have sequences that are described as much weaker, because the number of copies created is relatively low, such as one or two per cell.

Why do bacteria have plasmids? Plasmids are not usually necessary for bacterial survival. However, in many cases, certain genes within a plasmid provide some type of growth advantage to the cell. By studying plasmids in many different species, researchers have discovered that most plasmids fall into five different categories:

1. Fertility plasmids, also known as F factors, allow bacteria to conjugate with each other.

2. Resistance plasmids, also known as R factors, contain genes that confer resistance
against antibiotics and other types of toxins.

3. Degradative plasmids carry genes that enable
the bacterium to digest and utilize an unusual
substance. For example, a degradative plasmid may carry genes that allow a bacterium
to digest an organic solvent such as toluene.

4. Col-plasmids contain genes that code colicins,
which are proteins that kill other bacteria.

5. Virulence plasmids carry genes that turn a
bacterium into a pathogenic strain.

7.2 COMPREHENSION QUESTIONS

1. A bacterial cell with an F factor conjugates with
   an F⁻ cell. Following conjugation, the two cells
   will be
   a. F⁺.
   b. F⁻.
   c. one F⁺ and one F⁻.
   d. none of the above.

2. Which of the following is a type of plasmid?
   a. F factor (fertility factor)
   b. R factor (resistance plasmid)
   c. Virulence plasmids
   d. All of the above are types of plasmids.

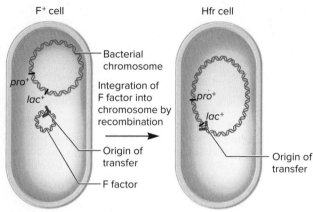

(a) When an F factor integrates into the chromosome, it creates an Hfr cell.

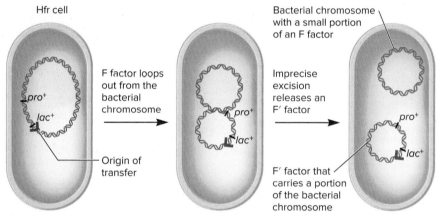

(b) When an F factor excises imprecisely, an F′ factor is created.

FIGURE 7.5 **Integration of an F factor to form an Hfr cell and its subsequent excision to form an F′ factor.** **(a)** An Hfr cell is created when an F factor integrates into the bacterial chromosome. **(b)** When an F factor is imprecisely excised, an F′ factor is created, and it carries a portion of the bacterial chromosome.

CONCEPT CHECK: How is an F′ factor different from an F factor?

<table>
<tr><td>

7.3

</td><td>

CONJUGATION AND MAPPING VIA HFR STRAINS

</td></tr>
</table>

Learning Outcomes:

1. Explain how an Hfr strain is produced.
2. Describe how bacterial cells of an Hfr strain can transfer genes to recipient cells.
3. Construct a genetic map using data from conjugation experiments.

Thus far, we have considered how conjugation may involve the transfer of an F factor from a donor to a recipient cell. In addition to this
form of conjugation, certain donor strains of *E. coli*, called Hfr strains,
are capable of conjugation and can transfer a portion of chromosomal
DNA. In this section, we will examine how Hfr strains are formed and
how they transfer genes to recipient cells. We will also explore the use
of Hfr strains to map genes along the *E. coli* chromosome.

Hfr Strains Have an F Factor Integrated into the Bacterial Chromosome

Luca Cavalli-Sforza discovered a strain of *E. coli* that was very
efficient at transferring many chromosomal genes to recipient F⁻
strains. Cavalli-Sforza designated this bacterial strain an

Hfr strain (for "high frequency of recombination"). How is an
Hfr strain formed? As shown in **Figure 7.5a**, an F factor may align
with a similar region found in the bacterial chromosome. Due to
recombination, which is described in Chapter 19, the F factor may
integrate into the bacterial chromosome. In the example shown
in Figure 7.5a, the F factor has integrated next to a *lac⁺* gene.
F factors can integrate into several different sites that are scattered
around the *E. coli* chromosome.

Occasionally, the integrated F factor in an Hfr strain is
excised from the bacterial chromosome. This process involves the
looping out of the F-factor DNA from the chromosome, which is
followed by recombination that releases the F factor from the
chromosome (Figure 7.5b). In the example shown in Figure 7.5b,
the excision is imprecise. This produces an F factor that carries a
portion of the bacterial chromosome and leaves behind some of
the F-factor DNA in the bacterial chromosome. F factors that

FIGURE 7.6 **Transfer of bacterial genes from an Hfr cell to an F⁻ cell.** The transfer of the bacterial chromosome begins at the origin of transfer and then proceeds around the circular chromosome. After a segment of chromosome has been transferred to the F⁻ recipient cell, it recombines with that cell's chromosome.

Genes→Traits The F⁻ recipient cell was originally *lac⁻* (unable to metabolize lactose) and *pro⁻* (unable to synthesize proline). If conjugation proceeds for a short period of time, the recipient cell acquires *lac⁺*, allowing it to metabolize lactose. If conjugation proceeds for a longer period of time, the recipient cell also acquires *pro⁺*, enabling it to synthesize proline.

CONCEPT CHECK: With regard to the timing of conjugation, explain why the recipient cell at the top right in this figure is *pro⁻*, whereas the recipient cell at the bottom right is *pro⁺*.

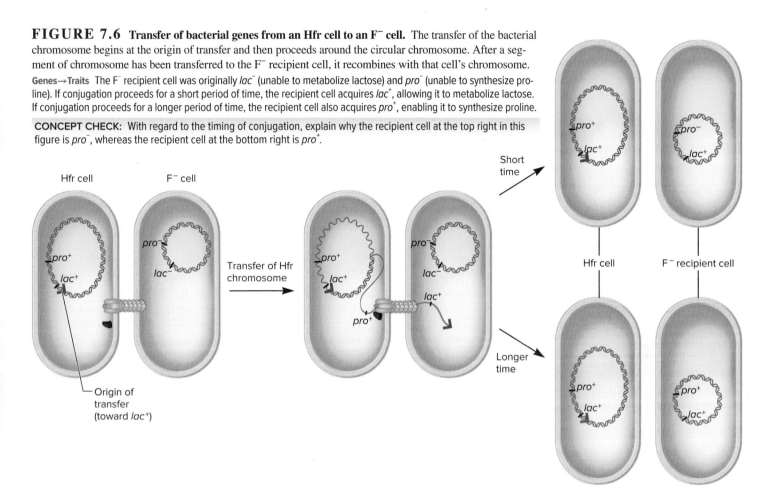

Hfr Strains Can Transfer a Portion of the Bacterial Chromosome to Recipient Cells Via Conjugation

William Hayes determined that conjugation between an Hfr strain and an F⁻ strain involves the transfer of a portion of the bacterial chromosome from an Hfr cell to an F⁻ cell (**Figure 7.6**). The origin of transfer within the integrated F factor determines the starting point and direction of this transfer process.

1. One of the DNA strands is cut at the origin of transfer. This cut, or nicked, site is the starting point from which the Hfr chromosome starts to enter the F⁻ recipient cell.
2. From this starting point, one strand of DNA from the Hfr chromosome begins to enter the F⁻ cell in a linear manner. The transfer process occurs in conjunction with chromosomal replication, so the Hfr cell retains its original chromosomal composition. About 1.5 to 2 hours is required for the entire Hfr chromosome to pass into the F⁻ cell.

carry a portion of the bacterial chromosome are called **F′ factors** (read "F prime factors"). We will consider F′ factors further in Chapter 14 when we discuss mechanisms of bacterial gene regulation.

Because most conjugations do not proceed for that long, usually only a portion of the Hfr chromosome is transmitted to the F⁻ cell.

3. Once inside the F⁻ cell, the chromosomal material from the Hfr cell can swap, or recombine, with the homologous region of the recipient cell's chromosome. (Chapter 19 describes the process of homologous recombination.)

How does this process affect the recipient cell? As illustrated in Figure 7.6, this recombination may provide the recipient cell with a new combination of alleles. In this example, the recipient strain was originally *lac⁻* (unable to metabolize lactose) and *pro⁻* (unable to synthesize proline).

- If conjugation proceeds for a short time, the recipient cell will receive a short segment of chromosomal DNA from the donor. In this case, the recipient cell becomes *lac⁺* but remains *pro⁻*.
- If the conjugation is prolonged, the recipient cell will receive a longer segment of chromosomal DNA from the donor. After a longer conjugation, the recipient becomes *lac⁺* and *pro⁺*.

As shown in Figure 7.6, an important feature of Hfr conjugation is that the bacterial chromosome is transferred linearly to the recipient strain. In this example, *lac*⁺ is always transferred first, and *pro*⁺ is transferred later.

In any particular Hfr strain, the origin of transfer has a specific orientation that promotes either a counterclockwise or clockwise transfer of genes. Among different Hfr strains, the origin of transfer

may be located in different regions of the chromosome. Therefore, the order in which the bacterial genes are transferred depends on the location and orientation of the origin of transfer. For example, an Hfr strain different from that in Figure 7.6 could have its origin of transfer next to *pro*⁺ and thus transfer *pro*⁺ first and *lac*⁺ later.

EXPERIMENT 7A

Conjugation Experiments Can Map Genes Along the *E. coli* Chromosome

The first genetic mapping experiments in bacteria were carried out by Élie Wollman and François Jacob in the 1950s. These researchers were aware of previous microbiological studies involving bacteriophages—viruses that bind to bacterial cells and subsequently infect them. Those studies showed that bacteriophages can be sheared from the surface of *E. coli* cells if the cells are spun in a blender. In this treatment, the bacteriophages are detached from the surface of the bacterial cells, but the bacteria themselves remain healthy and viable. Wollman and Jacob reasoned that a blender treatment could be used to separate bacterial cells that were in the act of conjugation without killing them. This technique is known as **interrupted mating.**

The rationale behind Wollman and Jacob's mapping strategy is that the time it takes for genes to enter a recipient cell is directly related to their order along the bacterial chromosome. They hypothesized that the chromosome of the donor strain in an Hfr conjugation is transferred in a linear manner to the recipient strain. If so, the order of genes along the chromosome can be deduced by determining the time it takes various genes to enter the recipient strain. Assuming that the Hfr chromosome is transferred linearly, Wollman and Jacob reasoned that interruptions of conjugation at different times would lead to various lengths of the Hfr chromosome being transferred to the F⁻ recipient cell. If two bacterial cells had conjugated for a short period of time, only a small segment of the Hfr chromosome would be transferred to the recipient bacterium. However, if the bacterial cells were allowed to conjugate for a longer period before being interrupted, a longer segment of the Hfr chromosome could be transferred (see Figure 7.6). By determining which genes were transferred during short conjugations and which required longer times, Wollman and Jacob were able to deduce the order of certain genes along the *E. coli* chromosome.

As shown in the experiment in **Figure 7.7**, Wollman and Jacob began with two *E. coli* strains. The donor (Hfr) strain had the following genetic composition:

thr⁺:   able to synthesize threonine, an essential amino acid for growth

leu⁺:   able to synthesize leucine, an essential amino acid for growth

*azi*ˢ: sensitive (*s* stands for "sensitive") to killing by azide (a toxic chemical)

*ton*ˢ: sensitive to infection by bacteriophage T1

lac⁺:   able to metabolize lactose and use it for growth

gal⁺:   able to metabolize galactose and use it for growth

*str*ˢ: sensitive to killing by streptomycin (an antibiotic)

The recipient (F⁻) strain had the opposite genotype: *thr*⁻ *leu*⁻ *azi*ʳ *ton*ʳ *lac*⁻ *gal*⁻ *str*ʳ (*r* stands for "resistant"). Before the experiment, Wollman and Jacob already knew the *thr*⁺ gene was transferred first, followed by the *leu*⁺ gene, and both were transferred relatively soon (5–10 minutes) after conjugation began. Their main goal in this experiment was to determine the times at which the other genes (*azi*ˢ, *ton*ˢ, *lac*⁺, and *gal*⁺) were transferred to the recipient strain. The transfer of the *str*ˢ gene was not examined because streptomycin was used to kill the donor strain following conjugation.

Before discussing the conclusions of this experiment, let's consider how Wollman and Jacob monitored genetic transfer. To determine if particular genes had been transferred after conjugation, the conjugated bacterial cells were first plated on a growth medium that lacked threonine (thr) and leucine (leu) but contained streptomycin (str). On these plates, the original donor and recipient strains could not grow because the donor strain was streptomycin-sensitive and the recipient strain required threonine and leucine. However, recipient cells into which the donor strain had transferred chromosomal DNA carrying the *thr*⁺ and *leu*⁺ genes were able to grow.

To determine the order of genetic transfer of the *azi*ˢ, *ton*ˢ, *lac*⁺, and *gal*⁺ genes, Wollman and Jacob picked colonies from the first plates and restreaked them on a medium that contained azide or bacteriophage T1 or on a medium that contained lactose or galactose as the sole source of energy for growth. The plates were incubated overnight to observe the formation of visible bacterial growth. Whether or not the bacteria could grow depended on their genotypes. For example, a cell that is *azi*ˢ cannot grow on a medium containing azide, and a cell that is *lac*⁻ cannot grow on a medium containing lactose as the carbon source for growth. By comparison, a cell that is *azi*ʳ and *lac*⁺ can grow on both types of media.

THE GOAL (DISCOVERY-BASED SCIENCE)

The chromosome of the donor strain in an Hfr conjugation is transferred in a linear manner to the recipient strain. The order of genes along the chromosome can be deduced by determining the time various genes take to enter the recipient strain.

ACHIEVING THE GOAL   **FIGURE 7.7**   The use of conjugation to map the order of genes along the *E. coli* chromosome.

Starting materials: The two *E. coli* strains already described, one Hfr strain (*thr⁺ leu⁺ azi^s ton^s lac⁺ gal⁺ str^s*) and one F⁻ strain (*thr⁻ leu⁻ azi^r ton^r lac⁻ gal⁻ str^r*).

Experimental level	Conceptual level

1. Mix together a large number of Hfr donor and F⁻ recipient cells.

2. After different periods of time, take a sample of cells and interrupt conjugation in a blender.

3. Plate the cells on solid growth medium lacking threonine and leucine but containing streptomycin. Note: The general methods for growing bacteria in a laboratory are described in Appendix A.

4. Pick each surviving colony, which would have to be *thr⁺ leu⁺ str^r*, and test to see if it is sensitive to killing by azide, sensitive to infection by T1 bacteriophage, and able to metabolize lactose or galactose.

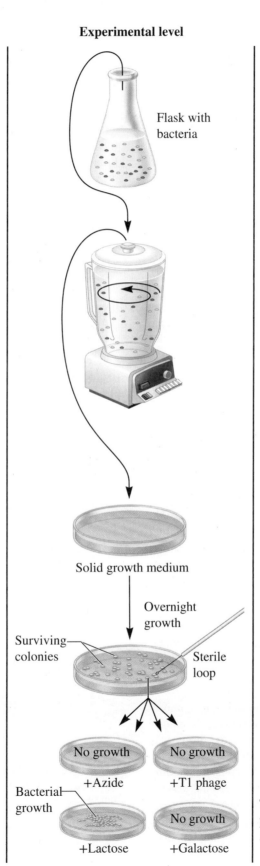

Flask with bacteria

Solid growth medium

Overnight growth

Surviving colonies

Sterile loop

Bacterial growth

No growth +Azide

No growth +T1 phage

+Lactose

No growth +Galactose

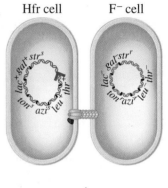

Hfr cell   F⁻ cell

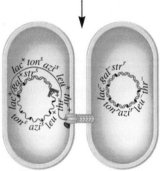

Separate by blending; donor DNA recombines with recipient cell chromosome.

In this conceptual example, the cells have been incubated about 20 minutes.

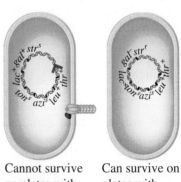

Cannot survive on plates with streptomycin

Can survive on plates with streptomycin

Additional tests

The conclusion is that the colony that was picked contained cells with a genotype of *thr⁺ leu⁺ azi^s ton^s lac⁺ gal⁻ str^r*.

THE DATA

Number of minutes that bacterial cells were allowed to conjugate before blender treatment

Number of minutes that bacterial cells were allowed to conjugate before blender treatment	Percentage of surviving bacterial colonies with each of the following genotypes:				
	thr⁺ leu⁺	aziˢ	tonˢ	lac⁺	gal⁺
5*	—	—	—	—	—
10	100	12	3	0	0
15	100	70	31	0	0
20	100	88	71	12	0
25	100	92	80	28	0.6
30	100	90	75	36	5
40	100	90	75	38	20
50	100	91	78	42	27
60	100	91	78	42	27

*There were no surviving colonies within the first 5 minutes of conjugation.

Source: Wollman, Élie L., and Jacob, François, *Sexuality and the Genetics of Bacteria*. New York, NY: Academic Press, 1961.

INTERPRETING THE DATA

Now let's discuss the data shown in the table. After the first plating, all colonies were composed of cells in which the *thr⁺* and *leu⁺* alleles had been transferred to the F⁻ recipient strain, which was already streptomycin-resistant. As seen in the data, 5 minutes was not sufficient time to transfer the *thr⁺* and *leu⁺* alleles because no surviving colonies were observed. After 10 minutes or longer, however, surviving bacterial colonies with the *thr⁺ leu⁺* genotype were obtained.

To determine the order of the remaining genes (*aziˢ*, *tonˢ*, *lac⁺*, and *gal⁺*), each surviving colony was tested to see if it was sensitive to killing by azide, sensitive to infection by T1 bacteriophage, able to use lactose for growth, or able to use galactose for growth. The likelihood of colonies surviving depended on whether the *aziˢ*, *tonˢ*, *lac⁺*, and *gal⁺* genes were close to the origin of transfer or farther away. For example, when cells were allowed to conjugate for 25 minutes, 80% carried the *tonˢ* gene, whereas only 0.6% carried the *gal⁺* gene. These results indicate that the *tonˢ* gene is closer to the origin of transfer than the *gal⁺* gene is.

When Wollman and Jacob reviewed all of the data, a consistent pattern emerged. The gene that conferred sensitivity to azide (*aziˢ*) was transferred first, followed by *tonˢ*, *lac⁺*, and finally, *gal⁺*. From these data, as well as results from other experiments, Wollman and Jacob constructed a genetic map that depicted the order of these genes along the *E. coli* chromosome.

| thr | leu | azi | ton | lac | gal |

This work provided the first method for bacterial geneticists to map the order of genes along a bacterial chromosome. Throughout the course of their studies, Wollman and Jacob identified several different Hfr strains in which the origin of transfer had been integrated at different places along the bacterial chromosome. When they compared the order of genes among different Hfr strains, their results were consistent with the idea that the *E. coli* chromosome is circular (see question 1 in More Genetic TIPS at the end of the chapter).

A Genetic Map of the *E. coli* Chromosome Has Been Obtained from Many Conjugation Studies

Conjugation experiments have been used to map more than 1000 genes along the circular *E. coli* chromosome. A map of that chromosome is shown in **Figure 7.8**. This simplified map shows the locations of only a few dozen genes. Because the chromosome is circular, a starting point on the map must be arbitrarily assigned; in this case, it is the gene *thrA*. Researchers scale genetic maps from bacterial conjugation studies in units of **minutes.** This unit refers to the relative time it takes for genes to first enter an F⁻ recipient strain during a conjugation experiment. The *E. coli* genetic map shown in Figure 7.8 is 100 minutes long, which is approximately the time that it takes to transfer the complete chromosome during an Hfr conjugation.

The distance between two genes is determined by comparing their times of entry into the recipient strain during a conjugation experiment. As shown in **Figure 7.9**, the time of entry is found by conducting conjugation experiments that proceed for different time intervals before interruption. We compute the time of entry by extrapolating the data back to the horizontal axis. In this experiment, the time of entry of the *lacZ* gene was approximately 16 minutes, and that of the *galE* gene was 25 minutes. Therefore, these two genes are approximately 9 minutes apart from each other along the *E. coli* chromosome.

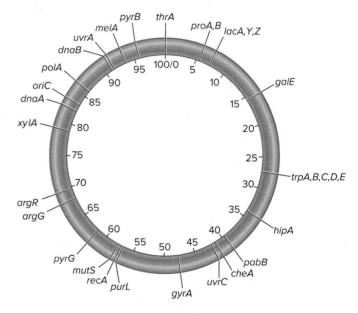

FIGURE 7.8 A simplified genetic map of the *E. coli* chromosome indicating the positions of several genes. *E. coli* has a circular chromosome with about 4300 different genes. This map shows the locations of some of them. The map is scaled in units of minutes, and the scale runs in the clockwise direction. The starting point on the map is the gene *thrA*.

CONCEPT CHECK: Why is the scale of this map in minutes?

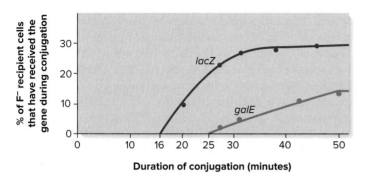

FIGURE 7.9 **Time course of an interrupted *E. coli* conjugation experiment.** By extrapolating the data back to the horizontal axis, the approximate time of entry of the *lacZ* gene is found to be 16 minutes; that of the *galE* gene is 25 minutes. Therefore, the distance between these two genes is 9 minutes.

CONCEPT CHECK: Which of the two genes, *lacZ* or *galE*, is closer to the origin of transfer?

7.3 COMPREHENSION QUESTIONS

1. With regard to conjugation, a key difference between F$^+$ and Hfr cells is that an Hfr cell
   a. is unable to conjugate.
   b. transfers a plasmid to the recipient cell.
   c. transfers a portion of the bacterial chromosome to the recipient cell.
   d. becomes an F$^-$ cell after conjugation.

2. In mapping experiments, _____ strains are conjugated to F$^-$ strains. The distance between two genes is determined by comparing their _____ during a conjugation experiment.
   a. F$^+$, times of entry
   b. Hfr, times of entry
   c. F$^+$, expression levels
   d. Hfr, expression levels

7.4   BACTERIAL TRANSDUCTION

Learning Outcome:

1. Outline the steps of bacterial transduction.

We now turn to a second method of genetic transfer, one that involves **bacteriophages** (also known as *phages*), which are viruses that infect bacterial cells. Following infection of the cell, new viral particles are made, which are then released from the cell in an event called lysis. In this section, we will examine how mistakes in the phage reproductive cycle can lead to the transfer of genetic material from one bacterial cell to another.

Bacteriophages Transfer Genetic Material from One Bacterial Cell to Another via Transduction

Before we discuss the ability of bacteriophages to transfer genetic material between bacterial cells, let's consider some general features of a phage's reproductive cycle. Bacteriophages are composed of genetic material that is surrounded by a protein coat. As described in Chapter 18, certain bacteriophages bind to the surface of a bacterium and inject their genetic material into the bacterial cytoplasm. Depending on the specific type of bacteriophage and its growth conditions, a phage may follow a lytic cycle or a lysogenic cycle. During the **lytic cycle,** the phage directs the synthesis of many copies of its own genetic material and coat proteins (look ahead to Figure 18.4, left side). These components then assemble to make new phages. When the synthesis and assembly of phages are completed, the bacterial host cell is lysed (broken apart), releasing the newly made phages into the environment.

When bacteriophages follow the lytic cycle, bacterial genes may on rare occasions be transferred from one bacterial cell to another. As noted earlier in this chapter, this process is called transduction. Examples of bacteriophages that can transfer bacterial chromosomal DNA from one bacterium to another are the P22 and P1 bacteriophages, which infect the bacterial species *Salmonella typhimurium* and *E. coli*, respectively.

How does a bacteriophage transfer bacterial chromosomal genes from one cell to another? As shown in **Figure 7.10**, when a bacteriophage infects a bacterial cell and follows the lytic cycle, the bacterial chromosome is digested into fragments of DNA. The bacteriophage DNA directs the synthesis of more phage DNA and proteins, which then assemble to make new phages. Occasionally, a mistake can happen in which a fragment of bacterial DNA assembles with bacteriophage proteins. This creates a phage that contains bacterial chromosomal DNA. When phage synthesis is completed, the bacterial cell is lysed and releases the newly made phages into the environment. Following release, the abnormal phage can bind to a living bacterial cell and inject its genetic material into that bacterium. The DNA fragment, which was derived from the chromosomal DNA of the first bacterium, can then recombine with the recipient cell's bacterial chromosome. In this case, the recipient bacterium has been changed from a cell that was *his*$^-$ (unable to synthesize histidine) to a cell that is *his*$^+$ (able to synthesize histidine).

7.4 COMPREHENSION QUESTION

1. During transduction involving a P1 phage,
   a. any small fragment of the bacterial chromosome may be transferred to another bacterium by a new phage.
   b. only a specific fragment of DNA may be transferred to another bacterium by a new phage.
   c. any small fragment of the bacterial chromosome may be transferred during conjugation.
   d. only a specific fragment of DNA may be transferred during conjugation.

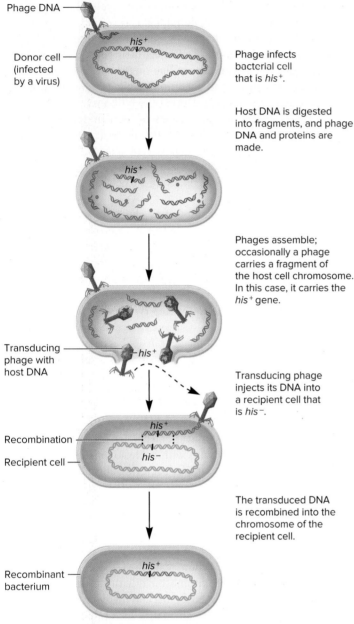

Phage DNA

Donor cell
(infected
by a virus)

his⁺

Phage infects
bacterial cell
that is *his*⁺.

Host DNA is digested
into fragments, and phage
DNA and proteins are
made.

his⁺

Phages assemble;
occasionally a phage
carries a fragment of
the host cell chromosome.
In this case, it carries the
his⁺ gene.

Transducing
phage with
host DNA

his⁺

Transducing phage
injects its DNA into
a recipient cell that
is *his*⁻.

Recombination

Recipient cell

his⁺

his⁻

The transduced DNA
is recombined into the
chromosome of the
recipient cell.

Recombinant
bacterium

his⁺

The recombinant bacterium's genotype has
changed from *his*⁻ to *his*⁺.

FIGURE 7.10   **Transduction in bacteria.**

Genes→Traits During transduction, a phage carries a segment
of bacterial DNA from a donor to a recipient cell. In this case, the
phage carried a segment of DNA with the *his*⁺ gene and trans-
ferred this gene to a recipient cell that was originally *his*⁻ (unable to synthesize
histidine). Following transduction, the recipient cell became *his*⁺, and thus able
to synthesize histidine.

CONCEPT CHECK: Transduction is sometimes described as a mistake in the
bacteriophage reproductive cycle. Explain how it can be viewed as a mistake.

7.5   BACTERIAL TRANSFORMATION

Learning Outcomes:

1. Outline the steps of bacterial transformation.
2. Explain how certain bacterial species preferentially take up
   DNA from their own species.

A third mechanism for the transfer of genetic material from one
bacterium to another is transformation. This process was first
discovered by Frederick Griffith in 1928 while he was working
with strains of *Streptococcus pneumoniae* (see Chapter 9). During
transformation, a living bacterial cell takes up DNA that was
released from a dead bacterium. This acquired DNA may then
recombine into the living bacterium's chromosome, producing a
bacterium with genetic material that it has received from the dead
bacterium. In this section, we will examine the steps in the
transformation process.

Transformation Involves the Uptake of DNA Molecules into Bacterial Cells

Transformation occurs as a natural process that has evolved in certain
species of bacteria, in which case it is called **natural transformation.**
This form of genetic transfer has been reported in a wide variety of
bacterial species. Bacterial cells that are able to take up DNA from the
environment are known as **competent cells.** Those that can take up
DNA naturally carry genes that code proteins called **competence
factors.** These proteins facilitate the binding of DNA fragments to the
cell surface, the uptake of the DNA into the cytoplasm, and its
subsequent incorporation into the bacterial chromosome. Temperature,
ionic conditions, and nutrient availability can affect whether or not a
bacterium is competent to take up genetic material from its
environment. These conditions influence the expression of the genes
that code proteins that function as competence factors.

In recent years, geneticists have unraveled some of the steps
that occur when competent bacterial cells are transformed by
genetic material in their environment. **Figure 7.11** shows the steps
of transformation.

1. A large fragment of genetic material binds to the surface of
   the bacterial cell. Competent cells express DNA receptors
   that promote such binding.
2. Before entering the cell, however, this large piece of chromosomal DNA must be cut into smaller fragments. This
   cutting is accomplished by an extracellular bacterial enzyme known as an endonuclease, which makes occasional
   random cuts in the long piece of DNA. At this stage, the
   DNA fragments are composed of double-stranded DNA.
3. In the next step, the DNA fragment begins its entry into the
   bacterial cytoplasm. For this to occur, the double-stranded
   DNA interacts with proteins in the bacterial membrane.
   One of the DNA strands is degraded, and the other strand
   enters the bacterial cytoplasm via an uptake system, which
   is structurally similar to the exporter involved in conjugation (shown in Figure 7.4a) but is involved with DNA
   uptake rather than export.
4. To be stably inherited, the DNA strand must be incorporated into the bacterial chromosome. If the DNA strand has
   a sequence that is similar to a region of DNA in the bacterial chromosome, the DNA may be incorporated into the
   chromosome by a process known as **homologous recombination,** discussed in detail in Chapter 19. For this to occur,
   the single-stranded DNA aligns itself with a homologous
   location on the bacterial chromosome. In the example

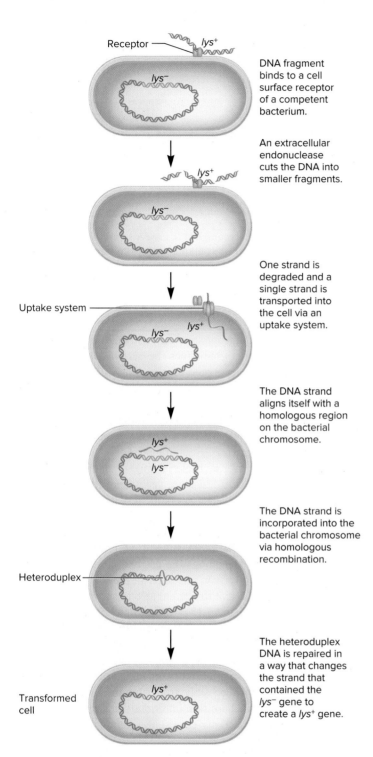

Receptor — *lys*⁺

lys⁻

DNA fragment binds to a cell surface receptor of a competent bacterium.

lys⁺

lys⁻

An extracellular endonuclease cuts the DNA into smaller fragments.

Uptake system

lys⁻   *lys*⁺

One strand is degraded and a single strand is transported into the cell via an uptake system.

lys⁺

lys⁻

The DNA strand aligns itself with a homologous region on the bacterial chromosome.

Heteroduplex

The DNA strand is incorporated into the bacterial chromosome via homologous recombination.

Transformed cell

lys⁺

The heteroduplex DNA is repaired in a way that changes the strand that contained the *lys*⁻ gene to create a *lys*⁺ gene.

ONLINE ANIMATION

FIGURE 7.11 The steps of bacterial transformation. In this example, a fragment of DNA carrying a *lys*⁺ gene enters the competent cell and recombines with the chromosome, transforming the bacterium from *lys*⁻ to *lys*⁺.

Genes→Traits Bacterial transformation can lead to new traits for the recipient cell. The recipient cell was *lys*⁻ (unable to synthesize the amino acid lysine). Following transformation, it became *lys*⁺. This result transforms the recipient bacterial cell into a cell that can synthesize lysine and grow on a medium that lacks this amino acid. Before transformation, the recipient *lys*⁻ cell would not have been able to grow on a medium lacking lysine.

CONCEPT CHECK: If the recipient cell did not already have a *lys*⁻ gene, could the *lys*⁺ DNA become incorporated into the bacterial chromosome?

shown in Figure 7.11, the foreign DNA carries a functional *lys*⁺ gene that aligns itself with a nonfunctional (mutant) *lys*⁻ gene already present within the bacterial chromosome.

5. The foreign DNA then recombines with one of the strands in the bacterial chromosome of the competent cell. In other words, the foreign DNA replaces one of the chromosomal strands of DNA, which is subsequently degraded. During homologous recombination, alignment of the *lys*⁻ and the *lys*⁺ alleles results in a region of double-stranded DNA called a **heteroduplex,** which contains one or more base mismatches.

6. The heteroduplex exists only temporarily. DNA repair enzymes in the recipient cell recognize the heteroduplex and repair it. In this example, the heteroduplex has been repaired by eliminating the mutation that caused the *lys*⁻ genotype, thereby creating a *lys*⁺ gene. Therefore, the recipient cell has been transformed from a *lys*⁻ strain to a *lys*⁺ strain. Alternatively, a DNA fragment that has entered a cell may not be homologous to any genes that are already found in the bacterial chromosome. In this case, the DNA strand may be incorporated at a random site in the chromosome. This process is known as **nonhomologous recombination.**

Transformation is also commonly used as a laboratory method to introduce plasmid DNA into cells, such as *E. coli*. This process is called **artificial transformation** to distinguish it from the natural process. Different approaches can be used to achieve transformation. One method is to treat the cells with calcium chloride, followed by a brief period of high temperature (a heat shock). These conditions make the cells permeable to small DNA molecules. Another method, called electroporation, makes cells permeable to DNA by using an externally applied electric field. In Chapters 20 and 21, we will explore how plasmids are used in a variety of genetic research methods and biotechnological applications.

Some Bacterial Species Have Evolved Ways to Take Up DNA from Their Own Species

Some bacteria preferentially take up DNA fragments from other bacteria of the same species or closely related species. How does this occur? Recent research has shown that the mechanism can vary among different species. In *S. pneumoniae*, the cells secrete a short peptide called the **competence-stimulating peptide (CSP).** When many *S. pneumoniae* cells are in the vicinity of one another, the concentration of CSP becomes high, which stimulates the cells to express the competence proteins needed for the uptake of DNA and its incorporation in the chromosome. Because competence requires a high external concentration of CSP, *S. pneuomoniae* cells are more likely to take up DNA from nearby *S. pneumoniae* cells that have died and released their DNA into the environment.

Other bacterial species promote the uptake of DNA among members of their own species via **DNA uptake signal sequences.** In the human pathogens *Neisseria meningitidis* (a causative agent of meningitis), *N. gonorrhoeae* (a causative agent of gonorrhea), and *Haemophilus influenzae* (a causative agent of ear, sinus, and respiratory infections), such sequences are found at many locations within the genomes. For example, the genome of

H. influenzae has approximately 1500 copies of the sequence 5′-AAGTGCGGT-3′, and that of *N. meningitidis* has about 1900 copies of the sequence 5′-GCCGTCTGAA-3′. DNA fragments that contain the same uptake signal sequence are preferentially taken up by these species instead of other DNA fragments. For example, *H. influenzae* is much more likely to take up a DNA fragment with the sequence 5′-AAGTGCGGT-3′. For this reason, transformation is more likely to involve DNA uptake between members of the same species.

7.5 COMPREHENSION QUESTIONS

1. What is the correct order for the steps of transformation given in the following list?
   1. Recombination with the bacterial chromosome
   2. Binding of a large DNA fragment to the surface of a bacterial cell
   3. Cutting a large DNA fragment into smaller pieces
   4. Uptake of DNA into the cytoplasm
   5. Degradation of one of the DNA strands
      a. 1, 2, 3, 4, 5
      b. 2, 3, 5, 4, 1
      c. 2, 3, 4, 5, 1
      d. 2, 5, 4, 3, 1

2. Some bacterial species preferentially take up DNA fragments from members of their own species. This uptake can be promoted by
   a. competence-stimulating peptide (CSP).
   b. DNA uptake signal sequences.
   c. either a or b.
   d. none of the above.

7.6 MEDICAL RELEVANCE OF HORIZONTAL GENE TRANSFER

Learning Outcomes:

1. Define *horizontal gene transfer.*
2. Explain the relevance of bacterial horizontal gene transfer in medicine.

The term **horizontal gene transfer,** also called *lateral gene transfer*, refers to a process in which an organism incorporates genetic material from another organism and is not the offspring of that organism. Conjugation, transformation, and transduction are examples of horizontal gene transfer. This process can occur between members of the same species or members of different species.

A key reason why horizontal gene transfer is important is its medical relevance. One area of great concern is the phenomenon of antibiotic resistance. Antibiotics are widely prescribed to treat bacterial infections in humans. They are also used in agriculture to control bacterial diseases in livestock. Unfortunately, the widespread use of antibiotics has increased the prevalence of antibiotic-resistant strains of bacteria, strains that have a selective

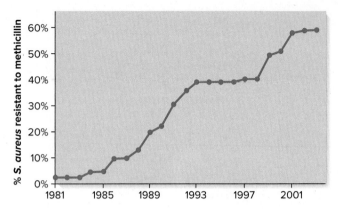

FIGURE 7.12 **Percentage of *S. aureus* strains that were found to be resistant to the antibiotic, methicillin, over a 20-year period.** The bacteria were isolated from patients who had been admitted into a particular hospital. From 1981 to 2001, the percentage of resistant strains rose from a very low level to nearly 60%.

advantage over those that are susceptible to antibiotics. Resistant strains carry genes that counteract the action of antibiotics in various ways. A resistance gene may code a protein that breaks down the antibiotic, pumps it out of the cell, or prevents it from inhibiting cellular processes.

The term **acquired antibiotic resistance** refers to the common phenomenon in which a previously susceptible strain of bacteria becomes resistant to a specific antibiotic. This change may result from genetic alterations in the bacterial genome, but it is often due to the horizontal transfer of resistance genes from a resistant strain. As reported in the news media, antibiotic resistance has increased dramatically worldwide over the past few decades, with resistant strains emerging in almost all pathogenic strains of bacteria.

Some *Staphylococcus aureus* strains have developed resistance to methicillin and all penicillins. The use and, in some cases, the overuse of these antibiotics have increased the prevalence of resistant strains, because the resistant strains are able to survive in the presence of the antibiotic. **Figure 7.12** shows the dramatic increase in the percentage of methicillin-resistant *S. aureus* strains over a 20-year period in the United States. Evidence suggests that these methicillin-resistant strains of *Staphlococcus aureus* (MRSA, pronounced "mersa") acquired the methicillin-resistance gene by horizontal gene transfer, possibly from a strain of *Enterococcus faecalis*. MRSA strains cause skin infections that are more difficult to treat than staph infections caused by nonresistant strains of *S. aureus*.

7.6 COMPREHENSION QUESTION

1. Which of the following is an example of horizontal gene transfer?
   a. The transfer of a gene from one strain of *E. coli* to a different strain via conjugation
   b. The transfer of a gene from one strain of *E. coli* to a different strain via transduction
   c. The transfer of an antibiotic resistance gene from *E. coli* to *Salmonella typhimurium* via transformation
   d. All of the above are examples of horizontal gene transfer.

KEY TERMS

7.1: genetic transfer, conjugation, transduction, transformation
7.2: minimal medium, auxotroph, prototroph, F factor, sex pili (pilus), conjugation bridge, relaxosome, origin of transfer, T DNA, nucleoprotein, plasmid, episomes
7.3: Hfr strain, F′ factor, interrupted mating, minute
7.4: bacteriophage (phage), lytic cycle

7.5: natural transformation, competent cells, competence factors, homologous recombination, heteroduplex, nonhomologous recombination, artificial transformation, competence-stimulating peptide (CSP), DNA uptake signal sequences
7.6: horizontal gene transfer, acquired antibiotic resistance

CHAPTER SUMMARY

7.1 Overview of Genetic Transfer in Bacteria

• Three general mechanisms for genetic transfer observed in various species of bacteria are conjugation, transduction, and transformation (see Table 7.1).

7.2 Bacterial Conjugation

• Lederberg and Tatum discovered conjugation in *E. coli* by analyzing auxotrophic strains (see Figure 7.1).
• Using a U-tube apparatus, Davis showed that conjugation requires cell-to-cell contact (see Figure 7.2).
• Certain strains of bacteria have F factors that they can transfer to other bacteria via conjugation in a series of steps (see Figures 7.3, 7.4).
• Many bacterial strains carry plasmids, which are circular pieces of DNA that replicate independently of the chromosomal DNA.

7.3 Conjugation and Mapping via Hfr Strains

• Hfr strains are formed when an F factor integrates into the bacterial chromosome. An imprecise excision can produce an F′ factor that carries a portion of the bacterial chromosome (see Figure 7.5).
• Hfr strains transfer a portion of the bacterial chromosome to a recipient cell during conjugation (see Figure 7.6).
• Wollman and Jacob showed that conjugation can be used to map the locations of genes along the bacterial chromosome, thereby creating a genetic map (see Figures 7.7, 7.8, 7.9).

7.4 Bacterial Transduction

• During transduction, a portion of a bacterial chromosome from a dead bacterium is transferred to a living recipient cell by a bacteriophage (see Figure 7.10).

7.5 Bacterial Transformation

• During transformation, a segment of DNA from a dead bacterium is taken up by a living bacterial cell and then incorporated into the bacterial chromosome (see Figure 7.11).
• Some bacteria have evolved ways to take up DNA from their own species.

7.6 Medical Relevance of Horizontal Gene Transfer

• Horizontal gene transfer is a process in which an organism incorporates genetic material from another organism and is not the offspring of that organism. Conjugation, transformation, and transduction are examples of horizontal gene transfer.
• Horizontal gene transfer has occurred with genes that confer antibiotic resistance. When pathogenic strains acquire such genes, they become more difficult to treat with commonly prescribed antibiotics (see Figure 7.12).

PROBLEM SETS & INSIGHTS

MORE GENETIC TIPS 1. By conducting conjugation experiments between Hfr and recipient strains, Wollman and Jacob mapped the order of many bacterial genes. Throughout the course of their studies, they identified several different Hfr strains in which the F-factor DNA had been integrated at different places along the bacterial chromosome. A sample of their experimental results is shown in the following table:

Hfr Strain	First								Last
H	thr	leu	azi	ton	pro	lac	gal	str	met
1	leu	thr	met	str	gal	lac	pro	ton	azi
2	pro	ton	azi	leu	thr	met	str	gal	lac
3	lac	pro	ton	azi	leu	thr	met	str	gal
4	met	str	gal	lac	pro	ton	azi	leu	thr
5	met	thr	leu	azi	ton	pro	lac	gal	str
6	met	thr	leu	azi	ton	pro	lac	gal	str
7	ton	azi	leu	thr	met	str	gal	lac	pro

A. Explain how these results are consistent with the idea that the bacterial chromosome is circular.

B. Draw a map of the bacterial chromosome that shows the order of these genes and the locations of the origins of transfer among these different Hfr strains.

T OPIC: *What topic in genetics does this question address?* The topic is genetic mapping. More specifically, the question is about using data from conjugation experiments to construct a genetic map.

I NFORMATION: *What information do you know based on the question and your understanding of the topic?* From the question, you know the order of genetic transfer from several conjugation experiments. From your understanding of the topic, you may remember that genes are transferred linearly, from donor to recipient cell, starting at the origin of transfer.

P ROBLEM-SOLVING S TRATEGY: *Analyze data. Compare and contrast. Make a drawing.* One strategy to solve this problem is to analyze the data by comparing and contrasting the orders of transfer in these conjugation experiments, which involve eight strains and nine different genes.

ANSWER:

A. In the data for the different Hfr strains, the order of the nine genes is always the same or is the reverse of that order. For example, HfrH and Hfr4 transfer the same genes but their orders are reversed relative to each other. In addition, the Hfr strains showed an overlapping pattern of transfer with regard to the origin. For example, Hfr1 and Hfr2 had the same order of genes, but Hfr1 began with *leu* and ended with *azi*, whereas Hfr2 began with *pro* and ended with *lac*. From these findings, Wollman and Jacob concluded that the segment of DNA that was the origin of transfer had been integrated at different points within the circular *E. coli* chromosome in different Hfr strains. A circular chromosome accounts for the overlapping pattern of transfer. The researchers also concluded that the origin of transfer can be oriented in either direction, so the direction of genetic transfer can be clockwise or counterclockwise around the circular bacterial chromosome.

B. A genetic map consistent with these results is shown here.

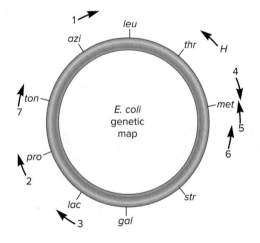

2. An Hfr strain that is *leuA*⁺ and *thiL*⁺ was mixed with a strain that is *leuA*⁻ and *thiL*⁻. The data points in the following graph were obtained when conjugation was interrupted at different time points and the percentage of recombinants for each gene was determined by streaking on a medium that lacked either leucine or thiamine.

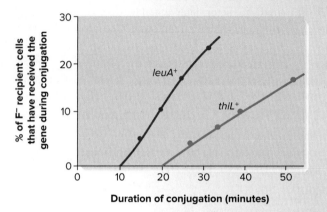

What is the distance (in minutes) between these two genes on the bacterial chromosome?

T OPIC: *What topic in genetics does this question address?* The topic is bacterial genetic mapping via a conjugation experiment. More specifically, the question asks you to use the experimental results to find the distance between two genes on the bacterial chromosome.

I NFORMATION: *What information do you know based on the question and your understanding of the topic?* In the question, you are given the data from a conjugation experiment involving two genes. From your understanding of the topic, you may remember that genes are transferred linearly during conjugation, so the order of transfer is determined by the locations of the genes relative to the origin of transfer.

P ROBLEM-SOLVING S TRATEGY: *Make a calculation.* One strategy to solve this problem is by extrapolating the data points to the horizontal axis to determine the time of entry. For *leuA*⁺, they extrapolate to 10 minutes. For *thiL*⁺, they extrapolate to 20 minutes.

ANSWER: The distance between the two genes is approximately 10 minutes.

Conceptual Questions

C1. The terms *conjugation*, *transduction*, and *transformation* are used to describe three different natural forms of genetic transfer between bacterial cells. Briefly discuss the similarities and differences among these processes.

C2. Conjugation is sometimes called bacterial mating. Is it a form of sexual reproduction? Explain.

C3. If you mix together an equal number of F⁺ and F⁻ cells, how would you expect the proportions to change over time? In other words, do you expect an increase in the relative proportion of F$^+$ or of F$^-$ cells? Explain your answer.

C4. What is the difference between an F$^+$ and an Hfr strain? Which type of strain can transfer many bacterial genes to recipient cells?

C5. What is the role of the origin of transfer during conjugation involving F$^+$ and Hfr strains? What is the significance of the direction of transfer in Hfr conjugation?

C6. What is the role of sex pili during conjugation?

C7. Think about the structure and transmission of F factors, and discuss how you think F factors may have originated.

C8. Each species of bacteria has its own distinctive cell surface. The characteristics of the cell surface play an important role in processes such as conjugation and transduction. For example, certain strains of *E. coli* have pili on their cell surface. These pili enable *E. coli* to conjugate with other *E. coli* and also enable certain bacteriophages (such as M13) to bind to the surface of the *E. coli* and gain entry into the cytoplasm. With these observations in mind, explain which form(s) of genetic transfer (i.e., conjugation, transduction, and/or transformation) is/are more likely to occur between different species of bacteria. Discuss some of the potential consequences of interspecies genetic transfer.

C9. Describe the steps that occur during bacterial transformation. What is a competent cell? What factors may determine whether a cell will be competent?

C10. Which mechanism of bacterial genetic transfer does not require recombination with the bacterial chromosome?

C11. Researchers who study the molecular mechanism of transformation have identified many proteins in bacteria that function in the uptake of DNA from the environment and its recombination into the host cell's chromosome. This means that bacteria have evolved molecular mechanisms for the purpose of transformation by extracellular DNA. What advantage(s) does a bacterium gain from importing DNA from the environment and/or incorporating it into its chromosome?

C12. Antibiotics such as tetracycline, streptomycin, and bacitracin are small organic molecules that are synthesized by particular species of bacteria. Microbiologists have hypothesized that the reason why certain bacteria make antibiotics is to kill other species that occupy the same environment. Bacteria that produce an antibiotic may be able to kill competing species. Eliminating competitors provides more resources for the antibiotic-producing bacteria. In addition, bacteria that have the genes necessary for antibiotic biosynthesis contain genes that confer resistance to the same antibiotic. For example, tetracycline is made by the soil bacterium *Streptomyces aureofaciens*. Besides the genes that are needed to make tetracycline, *S. aureofaciens* also has genes that confer tetracycline resistance; otherwise, it would kill itself when producing tetracycline. In recent years, however, many other species of bacteria that do not synthesize tetracycline have acquired the genes that confer tetracycline resistance. For example, certain strains of *E. coli* carry tetracycline-resistance genes, even though *E. coli* does not synthesize tetracycline. When these genes were analyzed at the molecular level, it was found that they are evolutionarily related to the genes in *S. aureofaciens*. This observation indicates that the genes from *S. aureofaciens* have been transferred to *E. coli*.

A. What form of genetic transfer (i.e., conjugation, transduction, or transformation) is the most likely mechanism of interspecies gene transfer?

B. Because *S. aureofaciens* is a nonpathogenic soil bacterium and *E. coli* is a bacterium found in the intestinal tract, do you think the genetic transfer was direct, or do you think it may have occurred in multiple steps (i.e., from *S. aureofaciens* to other bacterial species and then to *E. coli*)?

C. How could the widespread use of antibiotics to treat diseases have contributed to the proliferation of many bacterial species that are resistant to antibiotics?

Experimental Questions

E1. In the experiment of Figure 7.1, a *met$^-$ bio$^-$ thr$^+$ leu$^+$ thi$^+$* cell could become *met$^+$ bio$^+$ thr$^+$ leu$^+$ thi$^+$* by a (rare) double mutation that converts the *met$^-$ bio$^-$* genes into *met$^+$ bio$^+$*. Likewise, a *met$^+$ bio$^+$ thr$^-$ leu$^-$ thi$^-$* cell could become *met$^+$ bio$^+$ thr$^+$ leu$^+$ thi$^+$* by three mutations that convert the *thr$^-$ leu$^-$ thi$^-$* genes into *thr$^+$ leu$^+$ thi$^+$*. From the results of Figure 7.1, how do you know that the occurrence of 10 *met$^+$ bio$^+$ thr$^+$ leu$^+$ thi$^+$* colonies is not due to these rare double or triple mutations?

E2. In the experiment of Figure 7.1, Lederberg and Tatum could not discern whether *met$^+$ bio$^+$* genetic material was transferred to the *met$^-$ bio$^-$ thr$^+$ leu$^+$ thi$^+$* strain or if *thr$^+$ leu$^+$ thi$^+$* genetic material was transferred to the *met$^+$ bio$^+$ thr$^-$ leu$^-$ thi$^-$* strain. Let's suppose that one strain is streptomycin-resistant (say, *met$^+$ bio$^+$ thr$^-$ leu$^-$ thi$^-$*) and the other strain is sensitive to streptomycin. Describe an experiment that could determine whether the *met$^+$ bio$^+$* genetic material was transferred to the *met$^-$ bio$^-$ thr$^+$ leu$^+$ thi$^+$* strain or the *thr$^+$ leu$^+$ thi$^+$* genetic material was transferred to the *met$^+$ bio$^+$ thr$^-$ leu$^-$ thi$^-$* strain.

E3. Explain how a U-tube apparatus can be used to distinguish between genetic transfer involving conjugation and genetic transfer involving transduction. Do you think a U-tube could be used to distinguish between transduction and transformation?

E4. What is an interrupted mating experiment? What type of experimental information can be obtained from this type of study? Why is it necessary to interrupt conjugation?

E5. In a conjugation experiment, what is meant by the time of entry? How is the time of entry determined experimentally?

E6. In your laboratory, you have an F$^-$ strain of *E. coli* that is resistant to streptomycin and is unable to metabolize lactose, but it can metabolize glucose. Therefore, this strain can grow on a medium that contains glucose and streptomycin, but it cannot grow on a medium containing only lactose. A researcher has sent you two *E. coli* strains in two separate tubes. One strain, let's call it strain *A*, has an F$'$ factor that carries the genes that are required for lactose metabolism. On its chromosome, this strain also has the genes that are required for glucose metabolism. However, it is sensitive to streptomycin. This strain can grow on a medium containing lactose or glucose, but it cannot grow if streptomycin is added to the medium. The second strain, let's call it strain *B*, is an F$^-$ strain. On its chromosome, it has the genes that are required for lactose and glucose metabolism. Strain *B* is also sensitive to streptomycin. Unfortunately, when strains *A* and *B* were sent to you, the labels had fallen off the tubes. Describe how you could determine which tubes contain strain *A* and strain *B*.

E7.   As mentioned in question 1 of More Genetic TIPS, origins of transfer can be located in many different places on a bacterial chromosome, and the direction of transfer can be clockwise or counterclockwise. Let's suppose a researcher conjugated six different Hfr strains that were *thr*+ *leu*+ *ton*^{s} *str*^{r} *azi*^{s} *lac*+ *gal*+ *pro*+ *met*+ to an F⁻ strain that was *thr*⁻ *leu*⁻ *ton*^{r} *str*^{s} *azi*^{r} *lac*⁻ *gal*⁻ *pro*⁻ *met*⁻, and obtained the following results:

Strain	Order of Gene Transfer
1	*ton*^{s} *azi*^{s} *leu*+ *thr*+ *met*+ *str*^{r} *gal*+ *lac*+ *pro*+
2	*leu*+ *azi*^{s} *ton*^{s} *pro*+ *lac*+ *gal*+ *str*^{r} *met*+ *thr*+
3	*lac*+ *gal*+ *str*^{r} *met*+ *thr*+ *leu*+ *azi*^{s} *ton*^{s} *pro*+
4	*leu*+ *thr*+ *met*+ *str*^{r} *gal*+ *lac*+ *pro*+ *ton*^{s} *azi*^{s}
5	*ton*^{s} *pro*+ *lac*+ *gal*+ *str*^{r} *met*+ *thr*+ *leu*+ *azi*^{s}
6	*met*+ *str*^{r} *gal*+ *lac*+ *pro*+ *ton*^{s} *azi*^{s} *leu*+ *thr*+

Draw a map of the circular *E. coli* chromosome that shows the locations and orientations of the origins of transfer in these six Hfr strains.

E8.   An Hfr strain that is *hisE*+ and *pheA*+ was mixed with a strain that is *hisE*⁻ and *pheA*⁻. The conjugation was interrupted, and the percentage of recombinants carrying each gene was determined by streaking on a medium that lacked either histidine or phenylalanine. The following results were obtained:

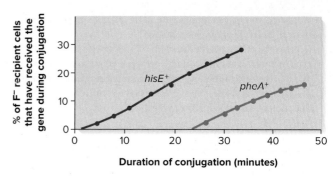

Duration of conjugation (minutes)

A.   Determine the distance (in minutes) between these two genes.

B.   In a previous experiment, it was found that *hisE* is 4 minutes away from *pabB* and that *pheA* is 17 minutes from *pabB*. Draw a genetic map showing the locations of all three genes.

E9.   Acridine orange is a chemical that inhibits the replication of F-factor DNA but does not affect the replication of chromosomal DNA, even if the chromosomal DNA is that of an Hfr strain. Let's suppose that you have an *E. coli* strain that is unable to metabolize lactose and has an F factor that carries a streptomycin-resistance gene. You also have an F⁻ strain of *E. coli* that is sensitive to streptomycin and has the genes that allow the bacterium to metabolize lactose. This second strain can grow on a lactose-containing medium. How would you generate an Hfr strain that is resistant to streptomycin and can metabolize lactose? (Hint: F factors occasionally integrate into the chromosome to yield Hfr strains, and occasionally Hfr strains excise their DNA from the chromosome and become F⁺ strains that carry an F′ factor.)

Questions for Student Discussion/Collaboration

1.   Discuss the advantages of the genetic analysis of bacteria. Make a list of the types of allelic differences among bacteria that are suitable for genetic analyses.

2.   Of the three types of bacterial genetic transfer, discuss which one(s) is/are more likely to occur between members of different species.

Discuss some of the potential consequences of interspecies genetic transfer.

Note: All answers are available for the instructor in Connect; the answers to the even-numbered questions and all of the Concept Check and Comprehension Questions are in Appendix B.

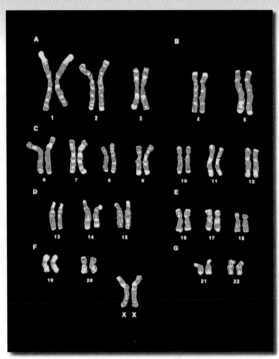

The chromosome composition of humans. *Somatic cells in humans contain 46 chromosomes, which come in 23 pairs.*
Leonard Lessin/Science Source

VARIATION IN CHROMOSOME STRUCTURE AND NUMBER

The term **genome** refers to all of the chromosomes and DNA sequences that an organism or species can possess. When comparing genomes, **genetic variation** pertains to genetic differences among members of the same species or among different species. Throughout Chapters 2–7, we focused primarily on variation in specific genes, which is called **allelic variation.** In this chapter, our emphasis will shift to broader genetic changes that affect the structure or number of eukaryotic chromosomes. These larger alterations may affect the expression of many genes and thereby influence phenotypes. Variations in chromosome structure and number are of great importance in the field of genetics because they are critical in the evolution of new species and have widespread medical relevance. In addition, agricultural geneticists have discovered that such variation can lead to the development of new crops, which may be hardier and more productive.

We begin this chapter by exploring how the structure of a eukaryotic chromosome can be modified, either by altering the total amount of genetic material or by rearranging the order of genes along the chromosome. The rest of the chapter is concerned with changes in the total number of chromosomes. We will explore how variation in chromosome number occurs and consider examples in which it has significant phenotypic consequences. We will also examine how changes in chromosome number can be induced through experimental treatments and how these approaches have applications in research and agriculture.

8.1 MICROSCOPIC EXAMINATION OF EUKARYOTIC CHROMOSOMES

Learning Outcome:

1. Describe the characteristics that are used to classify and identify chromosomes.

To identify changes in chromosome structure and number, researchers need to start with a reference point: the chromosomal composition of most members of a given species. For example,

183

most people have two sets of chromosomes with 23 specific chromosomes in each set, resulting in a total of 46 chromosomes. On relatively rare occasions, however, a person may have a chromosomal composition different from that of most other people. Such a person may have a chromosome that has an unusual structure or may have too few or too many chromosomes. **Cytogeneticists**—scientists who study chromosomes microscopically—examine the chromosomes from many members of a given species to determine the common chromosomal composition and to identify rare individuals that show variation in chromosome structure and/or number. In addition, cytogeneticists may be interested in analyzing the chromosomal compositions of two or more species to see how they are similar and how they are different.

Because chromosomes are more compact and microscopically visible during cell division, cytogeneticists usually analyze chromosomes in actively dividing cells. **Figure 8.1a** shows micrographs of chromosomes from individuals of three species: a human, a fruit fly, and a corn plant. As seen here, a cell from a human has 46 chromosomes (23 pairs), one from a fruit fly has 8 chromosomes (4 pairs), and one from a corn plant has 20 chromosomes (10 pairs). Except for the sex chromosomes, which differ between males and females in some species, most members of the same species have very similar chromosomes. By comparison, the chromosomal compositions of distantly related species, such as humans and fruit flies, may be very different. A total of 46 chromosomes is the usual number for humans, whereas 8 chromosomes is the norm for fruit flies.

Cytogeneticists have various ways to classify and identify chromosomes. The three most commonly used features are location of the centromere, size, and banding patterns that are revealed when the chromosomes are treated with stains. As shown in **Figure 8.1b**, chromosomes are classified with regard to centromere location as follows:

- **metacentric** (in which the centromere is near the middle);
- **submetacentric** (in which the centromere is slightly off center);
- **acrocentric** (in which the centromere is significantly off center but not at the end);
- **telocentric** (in which the centromere is at one end).

Because the centromere is never exactly in the center of a chromosome, each chromosome has a short arm and a long arm. For human chromosomes, the short arm is designated with the letter p (for the French word *petit*, meaning "little"), and the long arm is designated with the next letter in the alphabet, q. In the case of telocentric chromosomes, the short arm may be nearly nonexistent.

A **karyotype** is a photographic (microscopic) representation in which all of the chromosomes within a single cell have been arranged in a standard fashion. **Figure 8.1c** shows a human karyotype. The procedure for making a karyotype is described in Chapter 2 (refer back to Figure 2.2). As seen in Figure 8.1c, the chromosomes are aligned with the short arms on top and the long arms on the bottom. By convention, the chromosomes are numbered according to their size, with the largest chromosomes having the smallest numbers. For example, human chromosomes 1, 2, and 3 are relatively large, whereas 21 and 22 are the two smallest.

Not included in the numbering system are the sex chromosomes, which are designated with letters (for humans, X and Y).

Because different chromosomes often have similar sizes and centromere locations (e.g., compare human chromosomes 8, 9, and 10), geneticists must use additional methods to accurately identify each type of chromosome within a karyotype. For detailed identification, chromosomes are treated with stains to produce characteristic banding patterns. Several different staining procedures are used by cytogeneticists to identify specific chromosomes.

An example is the procedure that produces **G bands,** as shown in Figure 8.1c. In this procedure, chromosomes are treated with mild heat or with proteolytic enzymes that partially digest chromosomal proteins. When exposed to the stain called Giemsa (named after its inventor Gustav Giemsa), some chromosomal regions bind the stain molecules heavily, resulting in dark bands. In other regions, the stain hardly binds at all and light bands result. Though the mechanism of staining is not completely understood, the dark bands are thought to represent regions that are more tightly compacted. As shown in Figure 8.1c and 8.1d, the alternating pattern of G bands is a unique feature for each chromosome.

In the case of human chromosomes, approximately 300 G bands can usually be distinguished during metaphase. A larger number of G bands (in the range of 800) can be observed in prometaphase because the chromosomes are less compacted compared to metaphase. **Figure 8.1d** shows the conventional numbering system that is used to designate G bands along a set of human chromosomes. The left chromatid in each pair of sister chromatids shows the expected banding pattern during metaphase, and the right chromatid shows the banding pattern as it appears during prometaphase.

Why is the banding pattern of eukaryotic chromosomes useful?

- When stained, individual chromosomes can be distinguished from each other, even if they have similar sizes and centromere locations. For example, compare the differences in banding patterns between human chromosomes 8 and 9 (see Figure 8.1d). These differences permit us to distinguish these two chromosomes even though their sizes and centromere locations are very similar.
- Banding patterns are used to detect changes in chromosome structure. Chromosomal rearrangements or changes in the total amount of genetic material are more easily detected in banded chromosomes.
- Chromosome banding is used to assess evolutionary relationships between species. Research studies have shown that the similarity of chromosome banding patterns is a good measure of genetic relatedness.

8.1 COMPREHENSION QUESTIONS

1. A chromosome that is metacentric has its centromere
   a. at the very tip.
   b. near one end, but not at the very tip.
   c. near the middle.
   d. at two distinct locations.

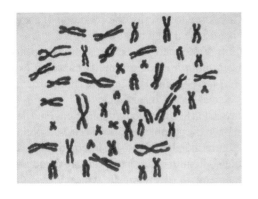
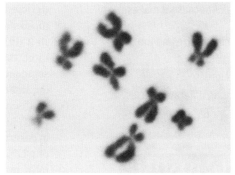

Human   Fruit fly   Corn

(a) Micrographs of metaphase chromosomes

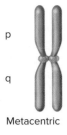
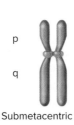
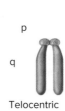

Metacentric   Submetacentric   Acrocentric   Telocentric

(b) A comparison of centromere locations

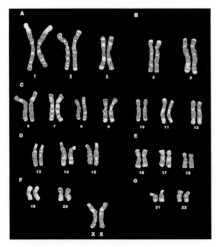
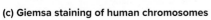

(c) Giemsa staining of human chromosomes

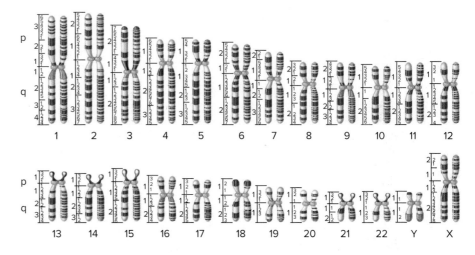

(d) Conventional numbering system of G bands in human chromosomes

FIGURE 8.1 **Features of chromosomes under the microscope. (a)** Micrographs of chromosomes from a human, a fruit fly, and a corn plant. **(b)** A comparison of centromere locations. Chromosomes can be classified by centromere location as metacentric, submetacentric, acrocentric (near one end), or telocentric (at the end). **(c)** Human chromosomes that have been treated with Giemsa stain. This micrograph has been colorized so that the banding patterns can be more easily discerned. **(d)** The conventional numbering of bands in Giemsa-stained human chromosomes. Each chromosome is divided into broad regions, which then are subdivided into smaller regions. The numbers increase as regions get farther away from the centromere. For example, in the left chromatid of chromosome 1, the uppermost dark band is at a location designated p35. The banding patterns of chromatids change as the chromatids condense. The left chromatid of each pair of sister chromatids shows the banding pattern of a chromatid in metaphase, and the right chromatid shows the banding pattern as it would appear in prometaphase. Note: In prometaphase, the chromatids are less compacted than in metaphase.

(a): (left): Scott Camazine/Science Source; (middle): Michael Abbey/Science Source; (right): Carlos R. Carvalho; (c): Leonard Lessin/Science Source

CONCEPT CHECK: Why is it useful to stain chromosomes?

2. Staining eukaryotic chromosomes is useful because it makes it possible to
    a. distinguish chromosomes that are similar in size and the location of the centromere.
    b. identify changes in chromosome structure.
    c. explore evolutionary relationships among different species.
    d. do all of the above.

8.2 CHANGES IN CHROMOSOME STRUCTURE: AN OVERVIEW

Learning Outcome:

1. Compare and contrast the four types of changes in chromosome structure.

Now that you understand that chromosomes typically come in a variety of shapes and sizes, let's consider how the structures of normal chromosomes can be modified by mutation. In some cases, the total amount of genetic material within a single chromosome can be decreased or increased significantly. Alternatively, the genetic material in one or more chromosomes may be rearranged without affecting the total amount of material. As shown in **Figure 8.2**, these mutations are categorized as deletions, duplications, inversions, and translocations.

Deletions and duplications are changes in the total amount of genetic material within a single chromosome. In Figure 8.2, human chromosomes are labeled according to their normal G-banding patterns.

- When a **deletion** occurs, a segment of chromosomal material is missing. In other words, the affected chromosome is deficient in a significant amount of genetic material. The term

deficiency is also used to describe the condition of a chromosome that is missing a region.

- In a **duplication,** a section of a chromosome is repeated more than once within a chromosome. Because an extra segment of DNA has been inserted into the chromosome, a duplication is also called an **insertion.**
- As discussed in Section 8.3, insertions (duplications) and deletions are common in the genomes of many species, especially in eukaryotes. The term **indel** refers to the insertion or deletion of a DNA segment. In most cases, indels that are found naturally in genomes are relatively small, typically less than 10,000 bp. Many of them are one or only a few bp in length.

Inversions and translocations are chromosomal rearrangements.

- An **inversion** involves a change in the direction of the genetic material along a single chromosome. For example, in Figure 8.2c a segment of one chromosome has been inverted, so the order of four G bands is opposite to what it was originally.
- A **translocation** occurs when one segment of a chromosome becomes attached to a different chromosome or to a different part of the same chromosome. A **simple translocation** occurs when a single piece of chromosome is attached to another chromosome. In a **reciprocal translocation,** two different chromosomes exchange pieces, thereby altering both of them.

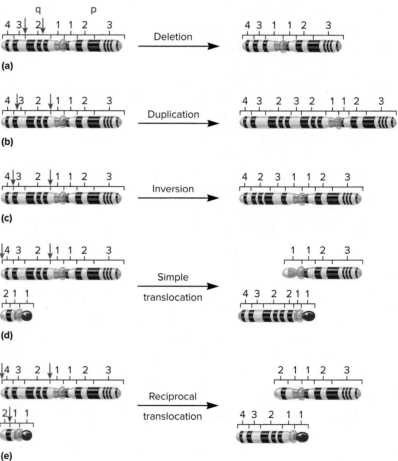

FIGURE 8.2   Types of changes in chromosome structure. The large chromosome shown throughout is human chromosome 1. The smaller chromosome seen in (d) and (e) is human chromosome 21. The red arrows indicate the ends of the affected portion. (**a**) A deletion removes a large portion of the q2 region, indicated by the red arrows. (**b**) A duplication doubles the q2–q3 region. (**c**) An inversion inverts the q2–q3 region. (**d**) The q2–q4 region of chromosome 1 is translocated to chromosome 21. A region of a chromosome cannot be attached directly to the tip of another chromosome because telomeres at the tips of chromosomes prevent such an event. In this example, a small piece at the end of chromosome 21 must be removed for the q2–q4 region of chromosome 1 to be attached to chromosome 21. (**e**) The q2–q4 region of chromosome 1 is exchanged with the q1–q2 region of chromosome 21.

CONCEPT CHECK: Which of these changes in chromosome structure alter the total amount of genetic material?

8.2 COMPREHENSION QUESTION

1. A change in chromosome structure that does *not* involve a change in the total amount of genetic material is
   a. a deletion.
   b. a duplication.
   c. an inversion.
   d. none of the above.

8.3 DELETIONS AND DUPLICATIONS

Learning Outcomes:

1. Explain how deletions and duplications occur.
2. Describe how deletions and duplications may affect the phenotype of an organism.
3. Define *copy number variation*.
4. List some different experimental methods used to detect deletions and duplications.
5. Interpret the results of an experiment that uses the technique of comparative genomic hybridization.

As we have seen, deletions and duplications alter the total amount of genetic material within a chromosome. In this section, we will examine how these changes occur, how they are detected experimentally, and how they affect the phenotypes of the individuals that inherit them.

Deletions Can Happen in Different Ways and Tend to Be Detrimental to an Organism's Phenotype

Different mechanisms may produce a chromosomal deletion. One way is by chromosomal breakage. A deletion may occur when a chromosome breaks in one or more places and a fragment of the chromosome is lost.

- In **Figure 8.3a**, a normal chromosome has broken into two separate pieces. The piece without the centromere will be eventually lost from future daughter cells because it usually does not find its way into the nucleus following mitosis, and is then degraded in the cytosol. The other piece is a chromosome with a **terminal deletion.**
- In **Figure 8.3b**, a chromosome has broken in two places to produce three chromosomal fragments. The central fragment is lost, and the two outer pieces reattach to each other. This process creates a chromosome with an **interstitial deletion.**

Deletions can also happen when recombination takes place at incorrect locations between two homologous chromosomes. The products of this type of aberrant recombination event are one chromosome with a deletion and another chromosome with a duplication. This process is examined later in this section (look ahead to Figure 8.5). As described in Section 8.4, deletions can also occur

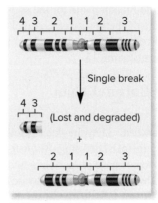

(a) Terminal deletion

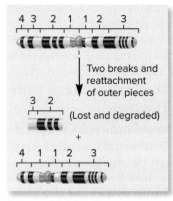

(b) Interstitial deletion

FIGURE 8.3 **Production of terminal and interstitial deletions.** This illustration shows the production of deletions in human chromosome 1.

CONCEPT CHECK: Why is a chromosomal fragment without a centromere eventually degraded?

when a chromosome carrying an inversion undergoes crossing over with a chromosome that lacks an inversion.

The phenotypic consequences of a chromosomal deletion depend on the size of the deletion and whether it includes genes or portions of genes that are vital to the development of the organism. When a deletion has a phenotypic effect, it is usually detrimental. Larger deletions tend to be more harmful because more genes are missing.

Many examples are known in which deletions affect phenotype. For example, a human genetic disease known as cri-du-chat syndrome is caused by a deletion of a segment in the short arm of human chromosome 5 (**Figure 8.4a**). Individuals who carry a single copy of this abnormal chromosome along with a normal chromosome 5 display an array of abnormalities, including mental deficiencies, unique facial anomalies, and, in infancy, an unusual

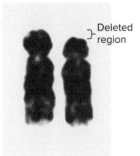

Deleted region

(a) Chromosome 5   **(b) A child with cri-du-chat syndrome**

FIGURE 8.4 **Cri-du-chat syndrome.** (a) Chromosome 5 from the karyotype of an individual with this disorder. A section of the short arm of one copy of chromosome 5 is missing. (b) An affected individual.

Genes→Traits Compared with an individual who has two copies of each gene on chromosome 5, an individual with cri-du-chat syndrome has only one copy of the genes that are located within the missing segment. This genetic imbalance (one versus two copies of many genes on chromosome 5) causes the phenotypic characteristics of this disorder, which include a catlike cry in infancy, short stature, characteristic facial anomalies (e.g., a triangular face, almond-shaped eyes, broad nasal bridge, and low-set ears), and microencephaly (a smaller than normal brain).

(a): Biophoto Associates/Science Source; (b): Ryan Garza/TNS/Newscom

catlike cry, which is the meaning of the French name for the syndrome (**Figure 8.4b**). Two other human genetic diseases, Angelman syndrome and Prader-Willi syndrome, which are described in Chapter 5, are due to a deletion in chromosome 15.

Duplications Tend to Be Less Harmful Than Deletions

Duplications result in extra genetic material. They may be caused by abnormal crossover events. During meiosis, crossing over usually occurs after homologous chromosomes have properly aligned with each other. On rare occasions, however, a crossover may occur at misaligned sites on homologs (**Figure 8.5**). What causes the misalignment? In some cases, a chromosome may carry two or more homologous segments of DNA that have identical or similar sequences. These are called **repetitive sequences** because they occur multiple times. An example of repetitive sequences is transposable elements, which are described in Chapter 10.

In Figure 8.5, the repetitive sequence on the right (in the upper homolog) has lined up with the repetitive sequence on the left (in the lower homolog). A misaligned crossover then occurs. This is called **nonallelic homologous recombination** because it has occurred at

homologous sites (such as repetitive sequences), but the sites are not alleles of the same gene. The result is that one chromatid has a duplication and another chromatid has a deletion. In Figure 8.5, the resulting chromosome with the extra genetic material carries a **gene duplication,** because the number of copies of gene C has been increased from one to two. In most cases, gene duplications happen as rare, sporadic events during the evolution of species.

As with deletions, the phenotypic consequences of duplications tend to be correlated with size. Duplications are more likely to have phenotypic effects if they involve a large piece of a chromosome. In general, small duplications are less likely to have harmful effects than are deletions of comparable size. This observation suggests that having only one copy of a gene is more harmful than having three copies. In humans, relatively few well-defined syndromes are caused by small chromosomal duplications. An example is Charcot-Marie-Tooth disease (type 1A), a peripheral neuropathy characterized by numbness in the hands and feet that is caused by a small duplication on the short arm of chromosome 17.

Duplications Provide Additional Material for Gene Evolution, Sometimes Leading to the Formation of Gene Families

In contrast to the gene duplication that causes Charcot-Marie-Tooth disease, the majority of small chromosomal duplications have no phenotypic effect. Nevertheless, they are vitally important because they provide raw material for the addition of more genes into a species' chromosomes. Over the course of many generations, this can lead to the formation of a **gene family** consisting of two or more genes in a particular species that are similar to each other. As shown in **Figure 8.6**, the members of a gene family are derived from the same ancestral gene. Over time, two copies of an ancestral gene can accumulate different mutations. Therefore,

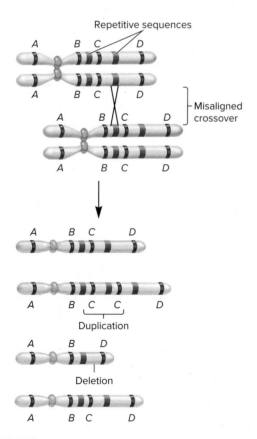

FIGURE 8.5 Nonallelic homologous recombination, leading to a duplication and a deletion. Repetitive sequences, shown in red, have promoted the misalignment of homologous chromosomes. A crossover has occurred at sites between genes C and D in one chromatid and between genes B and C in another chromatid. After meiosis is completed, one chromosome contains a duplication, and another contains a deletion.

CONCEPT CHECK: In this example, what is the underlying cause of nonallelic homologous recombination?

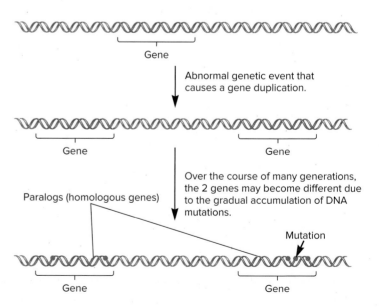

ONLINE
ANIMATION

FIGURE 8.6 Gene duplication and the evolution of paralogs. An abnormal crossover event like the one shown in Figure 8.5 leads to a gene duplication. Over time, each gene accumulates different mutations.

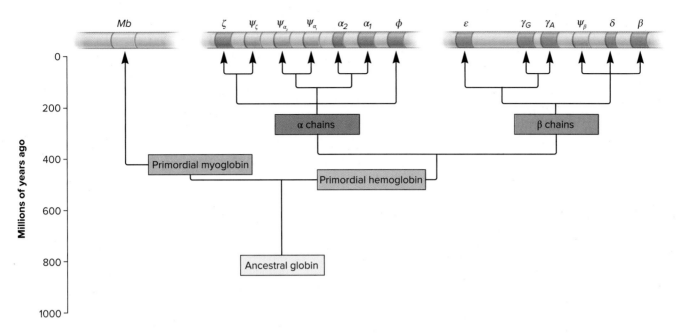

FIGURE 8.7 **The evolution of the globin gene family in humans.** The globin gene family evolved from a single ancestral globin gene. The first gene duplication produced two genes, one that codes myoglobin (on chromosome 22) and a primordial hemoglobin gene that duplicated several times to produce several α-chain and β-chain genes, which are found on chromosomes 16 and 11, respectively. The four genes shown in gray are non-functional pseudogenes.

after many generations, the two genes will be similar but not identical. During evolution, this type of event can occur several times, creating a family of many similar genes.

When two or more genes are derived from a single ancestral gene, the genes are said to be **homologous.** Homologous genes within a single species are called **paralogs** and constitute a gene family. A well-studied example of a gene family is shown in **Figure 8.7**, which illustrates the evolution of the globin gene family found in humans. The globin genes code polypeptides that are subunits of proteins that function in oxygen binding. One such protein is hemoglobin found in red blood cells. The globin gene family is composed of 14 paralogs that were originally derived from a single ancestral globin gene. According to an evolutionary analysis, the ancestral globin gene first duplicated about 500 mya (million years ago) and became separate genes coding myoglobin and the hemoglobin group of genes. The primordial hemoglobin gene duplicated into an α-chain gene and a β-chain gene, which subsequently duplicated to produce several genes located on chromosomes 16 and 11, respectively. Currently, 14 globin genes are found on three different human chromosomes.

Why is it advantageous to have a family of globin genes? Although all globin polypeptides are subunits of proteins that play

a role in oxygen binding, the accumulation of different mutations in the various family members has produced globins that are more specialized in their function. For example, myoglobin is better at binding and storing oxygen in muscle cells, and the hemoglobins are better at binding and transporting oxygen via the red blood cells. Also, different globin genes are expressed during different stages of human development.

- The ζ-globin and ε-globin genes are expressed very early in embryonic life.
- The α-globin and γ-globin genes are expressed during the second and third trimesters of gestation.
- Following birth, the α-globin genes remain turned on, but the γ-globin genes are turned off and the β-globin gene is turned on.

These differences in the expression of the globin genes reflect the differences in the oxygen transport needs of humans during the embryonic, fetal, and postpartum stages of life. As described in **Table 8.1**, the embryonic and fetal forms of hemoglobin have higher affinities for oxygen compared to the adult form. This allows the embryo and fetus to capture oxygen from the hemoglobin in the female parent's bloodstream.

TABLE 8.1			
Globin Gene Expression During Human Development			
Stage of Development	**Globin Genes Expressed**	**Hemoglobin Composition**	**Oxygen Affinity (P50)***
Embryo	ε-globin and ζ-globin	Two ε-globin and two ζ-globin subunits	5–13.5 mmHg
Fetus	γ-globin and α-globin	Two γ-globin and two α-globin subunits	19.5 mmHg
Birth to adult	β-globin and α-globin	Two β-globin and two α-globin subunits	26.5 mmHg

*P50 values represent the partial pressure of oxygen required to half-saturate hemoglobin. A lower P50 indicates a higher affinity of hemoglobin for oxygen.

Copy Number Variation Is Relatively Common Among Members of the Same Species

As we have seen, deletions and duplication can alter the number of copies of a given gene. In recent years, geneticists have analyzed the occurrence of deletions and duplications within modern populations. The term **copy number variation (CNV)** refers to a type of structural variation in which a segment of DNA, which is 1000 bp or more in length, commonly exhibits copy number differences among members of the same species. In other words, copy number variation is a phenomenon that occurs at the population level.

One possibility is that some members of a species may carry a chromosome that is missing a particular gene or part of a gene. Alternatively, a CNV may be due to a duplication. For example, some members of a diploid species may have one copy of gene A on both homologs of a chromosome, and thus have two copies of the gene (**Figure 8.8**). By comparison, other members of the same species might have one copy of gene A on a particular chromosome and two copies on its homolog for a total of three copies. The homolog with two copies of gene A is said to have undergone a **segmental duplication.**

In the past few decades, researchers have discovered that copy number variation is relatively common in animal and plant species. Though the analysis of CNV is a relatively new area of investigation, researchers estimate that 0.1% to 10% of the genome of a typical species of animal or plant may show CNV. In humans, approximately 0.4% of the genomes of two unrelated individuals typically differ with respect to copy number. Researchers estimate that 5% to 10% of all human genes may exhibit CNV, though CNV in many of those genes is relatively rare.

Most CNV is inherited and happened in the distant past, but CNV may also be caused by new mutations. A variety of mechanisms may bring about CNV. One common cause is nonallelic homologous recombination, which was illustrated in Figure 8.5. This type of event can produce a chromosome with a duplication or deletion, thereby altering the copy number of genes. Researchers also speculate that the proliferation of transposable elements may increase the copy number of DNA segments. A third mechanism that underlies CNV may involve errors in DNA replication, which is described in Chapter 11.

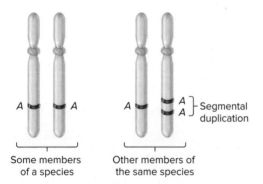

FIGURE 8.8 An example of copy number variation. Among members of the same species, some individuals carry two copies of gene A (left side), whereas others carry three copies (right side).

What are the phenotypic consequences of CNV? In many cases, CNV has no obvious phenotypic consequences. However, in some cases, CNV may confer a survival or reproductive advantage:

- An example of such an effect involves the human *CCL3* gene. This gene codes a chemokine protein, which is a signaling protein that plays a role in immunity. In human populations, the copy number of this gene varies from one to six. In people infected with HIV (human immunodeficiency virus), copy number variation of *CCL3* may affect the progression of AIDS (acquired immune deficiency syndrome). Individuals with a higher copy number of *CCL3* produce more chemokine protein and often show a slower advancement of AIDS.

- A second example involves the gene in humans that codes amylase, an enzyme responsible for starch digestion. Starch digestion begins in the mouth and continues in the small intestine. *Amy1*, which is the gene that codes salivary amylase, shows extensive CNV, with a reported number of gene copies ranging from 2 to 18. Research indicates that CNV for *Amy1* is significantly different among human populations. A higher average copy number is associated with populations that traditionally have subsisted on a high-starch diet.

Recent medical research has also revealed that some CNV is associated with specific human diseases. These include the following:

- schizophrenia
- autism
- certain forms of learning disabilities
- certain types of congenital heart defects
- many types of cancer

Microscopy and Molecular Techniques Are Used to Identify Deletions and Duplications

How do researchers and clinicians identify duplications and deletions, also called indels, in an individual's genome? Most experimental approaches can be placed into two different categories: microscopic methods and molecular techniques (see **Table 8.2**). When choosing a technique, an important issue is sensitivity. Certain methods can only detect relatively large indels. For example, conventional light microscopy, which is used for making a karyotype (see Section 8.1), can detect indels that are greater than 5 million bp in length. By comparison, molecular methods, such as DNA sequencing, can detect changes at the level of a single base pair!

Another consideration in choosing a method is the primary objective of intel identification. Some methods are aimed at indel detection involving a specific gene or a group of genes. These methods are commonly used in human disease diagnosis. Clinicians may want to know if a particular disease is caused by a deletion or duplication involving one particular gene or a group of genes that are already known to play a role in causing a disease. Other methods are broader in their ability to detect indels. For example, whole-genome sequencing and comparative genomic hybridization can be used to identify indels at the level of a whole genome.

TABLE 8.2
Methods to Identify Duplications and Deletions

Method	Description
Microscopy	
Karyotyping	Described in Section 8.1. Can be used to detect indels on a genome-wide level.
Fluorescence in situ hybridization	Described in Chapter 22 (look ahead to Figure 22.2). One or more fluorescently labeled probes are hybridized to intact chromosomes. Can be used to detect indels involving one gene or multiple genes.
Comparative genomic hybridization (CGH)	Described in Figure 8.9. Fluorescently labeled probes from two different cells types, such as noncancerous cells and cancer cells, are hybridized to an intact set of chromosomes. Can be used to detect indels on a genome-wide level.
Molecular Techniques	
DNA sequencing	Described in Chapter 20 (look ahead to Figure 20.12). The base sequences of DNA fragments are determined. Can be used to detect indels in one gene or a group of genes. In addition, as described in Chapter 22, whole-genome sequencing can be performed to identify all indels in an individual's genome.
RNA sequencing	Described in Chapter 23 (look ahead to Figure 23.2). RNA is isolated from a sample of cells, and used as a template to create DNA for DNA sequencing. Can be used to detect indels on a genome-wide level.
Polymerase chain reaction (PCR)	Described in Chapter 20 (look ahead to Figure 20.7). Oligonucleotide primers are used to amplify a specific segment of DNA in a genome. Can be used to detect indels involving one gene or multiple genes.
Array comparative genomic hybridization (aCGH)	DNA microarrays are described in Chapter 23 (look ahead to Figure 23.1). Fluorescently labeled probes from different cells types are hybridized to a DNA microarray. Can be used to identify indels on a genome-wide level.

EXPERIMENT 8A

Comparative Genomic Hybridization Can Be Used to Detect Indels in Cancer patients

As we have seen, chromosome deletions and duplications (indels) may influence the phenotypes of individuals who inherit them. One very important reason why researchers have become interested in these types of chromosomal changes is that some are related to cancer. As discussed in Chapter 25, chromosomal deletions and duplications have been associated with many types of human cancers. Though such changes may be detectable by traditional chromosomal staining and karyotyping methods, small deletions and duplications may be difficult to detect in this manner. Fortunately, researchers have been able to develop more sensitive methods for identifying changes in chromosome structure.

In 1992, Anne Kallioniemi, Dan Pinkel, and colleagues devised a method called **comparative genomic hybridization (CGH).** This technique can be used to determine if cancer cells have changes in chromosome structure, such as deletions or duplications. To begin this procedure, DNA is isolated from a test sample, which in this case was a sample of breast cancer cells, and also from cells of noncancerous breast tissue (**Figure 8.9**). The DNA from the breast cancer cells was used as a template to make green fluorescent DNA, and the DNA from noncancerous cells was used to make red fluorescent DNA. These green or red DNA molecules averaged 800 bp in length and were made from sites that were scattered all along each chromosome. The green and red DNA molecules were then denatured by heat treatment.

Equal amounts of the two fluorescently labeled DNA samples were mixed together and applied to metaphase chromosomes that were known to carry no deletions or duplications. These metaphase chromosomes were obtained from white blood cells and the DNA within these chromosomes was also denatured. The fluorescently labeled DNA strands can bind to complementary regions on the metaphase chromosomes. This process is called **hybridization** because the DNA from one sample (a green or red DNA strand) forms a double-stranded region with a DNA strand from another sample (an unlabeled metaphase chromosome). Following hybridization, the metaphase chromosomes were visualized using a fluorescence microscope, and the images were analyzed by a computer that determines the relative intensities of green and red fluorescence.

What are the expected results? If a chromosomal region is present in both breast cancer cells and noncancerous cells in the same amount, the ratio between green and red fluorescence should be 1. If a chromosomal region is deleted in the breast cancer cell line, the ratio will be less than 1, or if a region is duplicated, it will be greater than 1.

THE GOAL (DISCOVERY-BASED SCIENCE)

Deletions or duplications in cancer cells can be detected by comparing the ability of fluorescently labeled DNA from cancer cells and noncancerous cells to bind (hybridize) to metaphase chromosomes.

ACHIEVING THE GOAL   **FIGURE 8.9**   The use of comparative genomic hybridization (CGH) to detect chromosomal deletions and duplications in cancer cells.

Starting materials: Breast cancer cells and noncancerous breast cells.

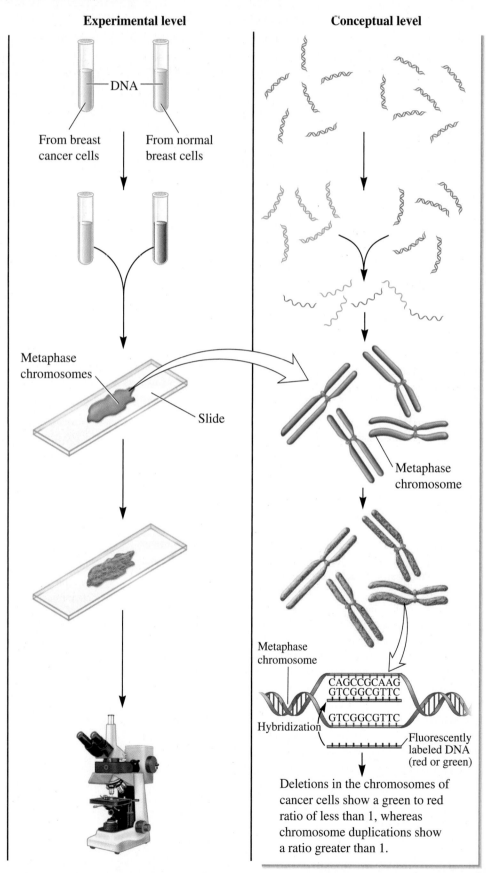

Experimental level   **Conceptual level**

1. Isolate DNA from human breast cancer cells and noncancerous breast cells. This involves breaking open the cells and isolating DNA fragments by chromatography. (See Appendix A for a description of chromatography.)

DNA

From breast cancer cells

From normal breast cells

2. Label the breast cancer DNA with a green fluorescent molecule and the DNA from noncancerous cells with a red fluorescent molecule. This is done by using the DNA from step 1 as a template, and incorporating fluorescently labeled nucleotides into newly made DNA strands.

3. The DNA strands are then denatured by heat treatment. Mix together equal amounts of fluorescently labeled DNA and add it to a preparation of metaphase chromosomes from white blood cells. The procedure for preparing metaphase chromosomes is described in Figure 3.2. The metaphase chromosomes are partially denatured.

Metaphase chromosomes

Slide

Metaphase chromosome

4. Allow the fluorescently labeled DNA to hybridize to the metaphase chromosomes.

Metaphase chromosome

CAGCCGCAAG
GTCGGCGTTC

GTCGGCGTTC

Hybridization

5. Visualize the chromosomes with a fluorescence microscope. Analyze the amount of green and red fluorescence along each chromosome with a computer.

Fluorescently labeled DNA (red or green)

Deletions in the chromosomes of cancer cells show a green to red ratio of less than 1, whereas chromosome duplications show a ratio greater than 1.

THE DATA

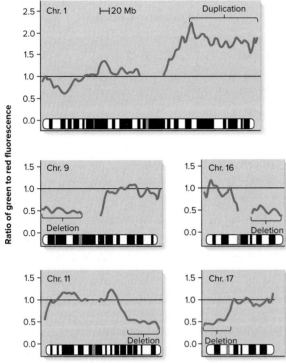

Note: Unlabeled repetitive DNA was also included in this experiment to decrease the level of nonspecific, background labeling. This repetitive DNA also prevents labeling near the centromere. In the graphs presenting the data, regions in the chromosomes where the curves are missing are due to the presence of highly repetitive sequences near the centromere.

Source: Kallioniemi, Anne, Kallioniemi, Olli-P., Rutovitz, Denis, Gray, Joe W., Waldman, Fred, and Pinkel, Dan, "Comparative Genomic Hybridization for Molecular Cytogenetic Analysis of Solid Tumors," *Science*, vol. 258, no. 5083, October 30, 1992, 818–821.

INTERPRETING THE DATA

The data show the ratio of green (DNA from cancer cells) to red (DNA from noncancerous cells) fluorescence along five different metaphase chromosomes. Chromosome 1 shows a large duplication, as indicated by the ratio of 2. One interpretation of this observation is that both copies of chromosome 1 carry a duplication. In comparison, chromosomes 9, 11, 16, and 17 have regions with a value of 0.5. This value indicates that one of the two chromosomes of these four types in the cancer cells carries a deletion, but the other chromosome does not. (A value of 0 would indicate both copies of a chromosome had deletions of the same region.) Overall, these results illustrate how this technique can be used to map chromosomal duplications and deletions in cancer cells.

This method is named comparative genomic hybridization because a comparison is made between the ability of two DNA samples (from cancer cells and from noncancerous cells) to hybridize to an entire genome. In this case, the entire genome is in the form of metaphase chromosomes.

As indicated in Table 8.2, the fluorescently labeled DNAs can also be hybridized to a DNA microarray instead of to metaphase chromosomes. This newer method, called array comparative genomic hybridization (aCGH), is gaining widespread use in the analysis of cancer cells. In this method, the two cellular samples, control (noncancerous cells) and experimental (cancer cells) are labeled with different fluorescent colors and then hybridized to complementary sequences on a DNA microarray. The relative amounts of differently colored DNA can then be measured:

- Higher levels of control DNA color indicate a deletion in the cancer cell sample.
- Higher levels of experimental DNA color indicate a duplication in the cancer cell sample.

GENETIC TIPS   THE QUESTION: A female parent and a

male offspring both have an inherited disorder that affects the nervous system. How would you determine if the disorder is caused by a change in chromosome structure, such as a deletion or duplication?

TOPIC: *What topic in genetics does this question address?*
The topic is how to determine if a genetic disorder is caused by a change in chromosome structure.

INFORMATION: *What information do you know based on the question and your understanding of the topic?* From the question, you know that a female parent and a male offspring have an inherited disorder. From your understanding of the topic, you may remember that some disorders are caused by variation in chromosome structure, but some are caused by single gene mutations. You may also recall that karyotyping and comparative genomic hybridization are techniques that can be used to detect deletions and duplications.

PROBLEM-SOLVING **S**TRATEGIES: *Design an experiment. Compare and contrast.* To solve this problem, you need to design an experiment in which you analyze the chromosomes of the affected individuals. You also need to examine the chromosomes of unaffected individuals as a control.

ANSWER:

1. Obtain a sample of cells, such as leukocytes, from the two affected individuals. As a control, obtain cells from the unaffected male parent and any unaffected siblings.

2. Subject the samples to one or more of the methods described in Table 8.2.

3. Compare the results of affected and unaffected individuals.

Expected results: If the disorder is caused by a change in chromosome structure, you would expect to find a change, such as a deletion or duplication, in both the female parent and male offspring, but you would not see the change in the unaffected male parent or any unaffected siblings.

8.3 COMPREHENSION QUESTIONS

1. Which of the following statements is correct?
   a. If a deletion and a duplication are the same size, the deletion is more likely to be harmful.
   b. If a deletion and a duplication are the same size, the duplication is more likely to be harmful.
   c. If a deletion and a duplication are the same size, the likelihood of causing harm is about the same.
   d. A deletion is always harmful, whereas a duplication is always beneficial.

2. With regard to gene duplications, which of the following statement(s) is/are correct?
   a. Gene duplications may be caused by nonallelic homologous recombination.
   b. Large gene duplications are more likely to be harmful than smaller ones.
   c. Gene duplications are responsible for creating gene families that code proteins with similar and specialized functions.
   d. All of the above statements are correct.

8.4 INVERSIONS AND TRANSLOCATIONS

Learning Outcomes:

1. Define *pericentric inversion* and *paracentric inversion*.
2. Diagram the production of abnormal chromosomes as the result of crossing over in inversion heterozygotes.
3. Explain two mechanisms that result in reciprocal translocations.
4. Describe how reciprocal translocations align during meiosis and how they segregate.
5. Explain how inversions and translocations may have phenotypic consequences and affect fertility.

As discussed earlier in this chapter, inversions and translocations are types of chromosomal rearrangements. In this section, we will explore how they occur and how they may affect an individual's phenotype and fertility.

Inversions Often Occur Without Phenotypic Consequences

A chromosome with an inversion contains a segment that has been flipped so that it runs in the opposite direction. This can occur when a chromosome breaks at two sites and an internal segment flips around and reconnects in the opposite direction (see Figure 8.2c). Geneticists classify inversions according to the location of the centromere:

- If the centromere lies within the inverted region of the chromosome, the inverted region is known as a **pericentric inversion (Figure 8.10b)**.
- Alternatively, if the centromere is found outside the inverted region, the inverted region is called a **paracentric inversion (Figure 8.10c)**.

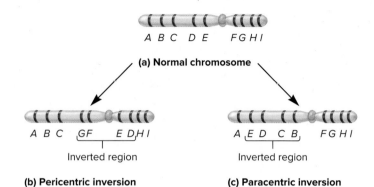

(a) Normal chromosome

A B C GF E D H I   A E D C B F G H I
Inverted region   Inverted region

(b) Pericentric inversion   **(c) Paracentric inversion**

FIGURE 8.10 Types of inversions. (a) A normal chromosome with the genes ordered from *A* through *I*. A pericentric inversion **(b)** includes the centromere, whereas a paracentric inversion **(c)** does not.

When a chromosome contains an inversion, the total amount of genetic material remains the same as in a normal chromosome. Therefore, the great majority of inversions do not have any phenotypic consequences. In rare cases, however, an inversion can alter the phenotype of an individual. Whether or not this occurs is related to the boundaries of the inverted segment. When an inversion occurs, the chromosome is broken in two places, and the center piece flips around to produce the inversion.

- If either breakpoint occurs within a vital gene, the function of the gene is expected to be disrupted, possibly producing a phenotypic effect. For example, some people with hemophilia (type A) have inherited an X-linked inversion in which a breakpoint has inactivated the gene for factor VIII—a blood-clotting protein.
- In other cases, an inversion (or translocation) may reposition a gene on a chromosome in a way that alters its normal level of expression. This is a type of **position effect**—a change in phenotype that occurs when the location of a gene changes from one chromosomal site to a different one. This topic is also discussed in Chapter 19 (see Figures 19.2 and 19.3).

Because inversions seem like an unusual genetic phenomenon, it is perhaps surprising that they are found in human populations in significant numbers. About 2% of the human population carries inversions that are detectable with a light microscope. In most cases, such individuals show no phenotypic effect and live their lives without knowing they carry an inversion. In a few cases, however, an individual with a chromosomal inversion may produce offspring with phenotypic abnormalities, as discussed next. This event may prompt a physician to request a microscopic examination of the individual's chromosomes. In this way, phenotypically unaffected individuals may discover they have a chromosome with an inversion.

Inversion Heterozygotes May Produce Abnormal Chromosomes Due to Crossing Over

An individual carrying one normal copy of a chromosome and one copy of the chromosome with an inversion is known as an **inversion heterozygote.** Such an individual, though possibly phenotypically normal, may have a high probability of producing haploid cells that are abnormal in their total genetic content.

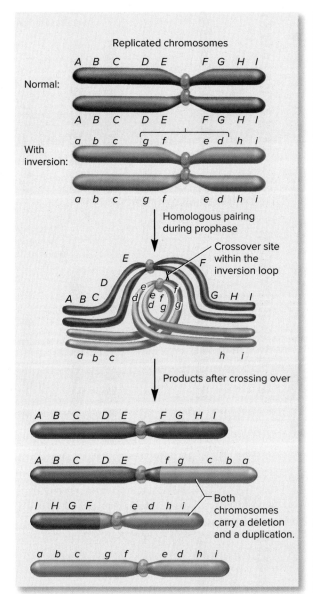

(a) Pericentric inversion

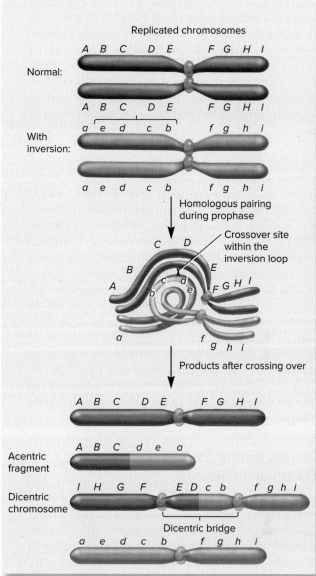

(b) Paracentric inversion

FIGURE 8.11 **The consequences of crossing over within the inversion loop.** (a) Crossover within a pericentric inversion. (b) Crossover within a paracentric inversion. Note: Following the completion of meiosis, just one of the four chromosomes found at the bottom of parts (a) and (b) would be transmitted to each of the resulting haploid cells (e.g., gametes). If a gamete receives a chromosome with a large deletion and duplication, the gamete itself may be inviable or it may produce an inviable or phenotypically affected offspring following fertilization.

CONCEPT CHECK: Explain why the homologous chromosomes of an inversion heterozygote can synapse only if an inversion loop forms.

The underlying cause of gamete abnormality is the phenomenon of crossing over within the inverted region. During meiosis I, pairs of homologous sister chromatids synapse with each other. **Figure 8.11** illustrates how this occurs in an inversion heterozygote. For the normal chromosome and inverted chromosome to synapse properly, an **inversion loop** must form so that the homologous genes on both chromosomes can align next to each other, despite the inverted sequence. If a crossover occurs within the inversion loop, highly abnormal chromosomes are produced. A crossover is more likely to occur in this region if the inversion is

large. Therefore, individuals carrying large inversions are more likely to produce abnormal gametes.

The consequences of this type of crossover depend on whether the inversion is pericentric or paracentric. Figure 8.11a shows a crossover within the inversion loop when one of the homologs has a pericentric inversion, in which the centromere lies within the inverted region of the chromosome. This event consists of a single crossover that involves only two of the four sister chromatids. Following the completion of meiosis, this single crossover yields two chromosomes that have a segment that is deleted and a different segment that is duplicated. In this example, one of the chromosomes is missing genes *H* and *I* and has an extra copy of genes *A*, *B*, and *C*. The other chromosome has the opposite situation; it is missing genes *A*, *B*, and *C* and has an extra copy of genes *H* and *I*. These abnormal chromosomes may result in gametes that are inviable. Alternatively, if these abnormal chromosomes are passed to offspring, they are likely to produce phenotypic abnormalities, depending on the amount and nature of the duplicated and deleted genetic material. A large deletion is likely to be lethal.

Figure 8.11b shows the outcome of a crossover involving a paracentric inversion, in which the centromere lies outside the inverted region. This single crossover event produces a very strange outcome. One product is a piece of chromosome without any centromere—an **acentric fragment,** which is lost and degraded in subsequent cell divisions. The other product is a **dicentric** chromosome that contains two centromeres. The region connecting the two centromeres in such a chromosome is a **dicentric bridge.** The dicentric chromosome is a temporary structure. If the two centromeres try to move toward opposite poles during anaphase, the dicentric bridge will be forced to break at some random location. Therefore, the net result of this crossover is to produce one normal chromosome, one chromosome with an inversion, and two chromosomes that contain deletions. The two chromosomes with deletions result from the breakage of the dicentric chromosome. They are missing the genes that were located on the acentric fragment.

Translocations Involve Exchanges Between Different Chromosomes

Another type of chromosomal rearrangement is a translocation in which a piece from one chromosome is attached to another chromosome. Eukaryotic chromosomes have telomeres, which tend to prevent translocations from occurring. As described in Chapters 10 and 11, **telomeres**—specialized repeated sequences of DNA—are found at the ends of normal chromosomes. Telomeres allow cells to identify where a chromosome ends and prevent the attachment of chromosomal DNA to the natural ends of a chromosome.

If cells are exposed to agents that cause chromosomes to break, the broken ends lack telomeres and are said to be reactive—a reactive end readily binds to another reactive end. If a single chromosome break occurs, DNA repair enzymes will usually recognize the two reactive ends and join them back together; the chromosome is repaired properly. However, if multiple chromosomes are broken, the reactive ends may be joined incorrectly to produce a reciprocal translocation (**Figure 8.12a**). Alternatively, **Figure 8.12b** shows another way that a reciprocal translocation can be produced: On rare occasions, nonhomologous chromosomes may crossover and thereby exchange pieces.

The reciprocal translocations we have considered thus far are also called **balanced translocations** because the total amount of genetic material is not altered. Like inversions, balanced translocations usually occur without any phenotypic consequences because the individual has a normal amount of genetic material. In a few cases, balanced translocations can result in position effects

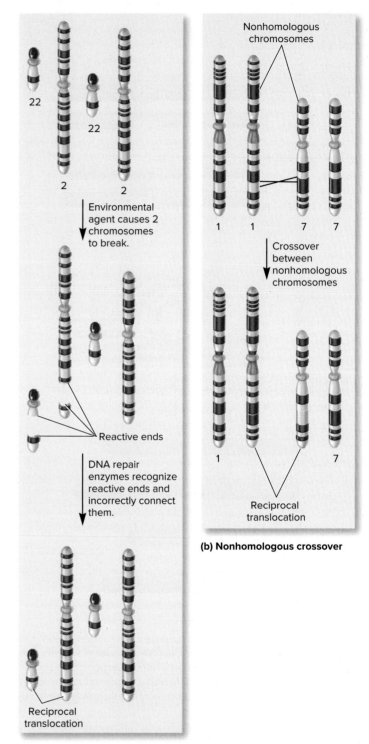

(b) Nonhomologous crossover

(a) Chromosomal breakage and DNA repair

FIGURE 8.12   Two mechanisms that produce a reciprocal translocation. **(a)** When two different chromosomes break, the reactive ends are recognized by DNA repair enzymes, which attempt to reattach them. If two different chromosomes are broken at the same time, the broken pieces may be reattached incorrectly. **(b)** A nonhomologous crossover has occurred between chromosome 1 and chromosome 7. This crossover yields two chromosomes that carry translocations.

CONCEPT CHECK: Which of these two mechanisms might be promoted by the presence of the same repetitive sequence, such as a transposable element, in many places in a species' genome?

similar to those that can occur with inversions. In addition, carriers of a balanced translocation are at risk of having offspring with an **unbalanced translocation,** in which a significant portion of genetic material is duplicated and/or deleted. Unbalanced translocations are generally associated with phenotypic abnormalities or may even be lethal.

Let's consider how a person with a balanced translocation may produce gametes and offspring with an unbalanced translocation. An inherited human syndrome known as familial Down syndrome provides an example. Due to a translocation, a balanced carrier may have one copy of chromosome 14, one copy of chromosome 21, and one copy of a chromosome that is a fusion between chromosome 14 and 21 (**Figure 8.13a**). This individual has a normal phenotype because the total amount of genetic material is similar to that of a person with the normal number of 46 chromosomes except that the short arms of chromosomes 14 and 21 are missing; these short arms do not carry vital genetic material.

During meiosis, the three chromosomes of the balanced carrier replicate and segregate from each other. However, because the three chromosomes cannot segregate evenly, six possible types of gametes may be produced.

- One gamete is normal.
- One carries a balanced translocated chromosome.
- The four gametes to the right, however, are unbalanced, either containing too much or too little material from chromosome 14 or 21.

If fertilization with a normal gamete occurs, the unbalanced gamete that carries chromosome 21 and the fused chromosome results in an offspring with familial Down syndrome (also see the schematic karyotype in **Figure 8.13b**). When this type of unbalanced gamete combines with a normal gamete, the resulting offspring has three copies of the genes that are found on the long arm of chromosome 21 and therefore has familial Down syndrome. In contrast, the other three types of unbalanced gametes may be inviable, or they may combine with a normal gamete. The three possible offspring along the bottom right in Figure 8.13b will not survive.

Figure 8.13c shows a person with familial Down syndrome, who has characteristics similar to those of an individual with the more prevalent form of Down syndrome that is due to having three complete copies of chromosome 21. We will examine this common form of Down syndrome later in this chapter.

The abnormal chromosome that is involved in familial Down syndrome is an example of a **Robertsonian translocation,** named after William Robertson, who first described this type of rearrangement in grasshoppers. A Robertsonian translocation arises from breaks near the centromeres of two nonhomologous acrocentric chromosomes. In the example shown in Figure 8.13, the long arms of chromosomes 14 and 21 have fused, creating one large single chromosome; the two short arms are lost. This type of translocation between two nonhomologous acrocentric chromosomes is the most common type of chromosome rearrangement in humans, occurring at a frequency of approximately 1 in 900 live births. In humans, Robertsonian translocations involve only the acrocentric chromosomes 13, 14, 15, 21, and 22.

Individuals with Reciprocal Translocations May Produce Abnormal Gametes Due to the Segregation of Chromosomes

As we have seen, individuals who carry balanced translocations have a greater risk of producing gametes with unbalanced combinations of chromosomes. Whether or not this occurs depends on the segregation pattern during meiosis I (**Figure 8.14**). In the example in the figure, the parent carries a reciprocal translocation and is likely to be unaffected phenotypically. During meiosis, the homologous chromosomes attempt to synapse with each other. Because of the translocations, the pairing of homologous regions leads to the formation of an unusual structure that contains four pairs of sister chromatids (i.e., eight chromatids), termed a **translocation cross.**

The chromosomes within the translocation cross can be segregated in three ways, as shown in Figure 8.14:

- One possibility is alternate segregation. As shown in Figure 8.14a, the chromosomes diagonal to each other within the translocation cross sort into the same cell. One daughter cell receives two normal chromosomes, and the other cell gets two translocated chromosomes. Following meiosis II, four haploid cells are produced: Two have normal chromosomes, and two have reciprocal (balanced) translocations.
- Another possible segregation pattern is called adjacent-1 segregation (Figure 8.14b). In this case, adjacent chromosomes (one with each type of centromere) sort into the same cell. Each daughter cell receives one normal chromosome and one translocated chromosome. After meiosis II is completed, four haploid cells are produced, all of which are genetically unbalanced because part of one chromosome is missing and part of another is duplicated. If these haploid cells give rise to gametes that unite with a normal gamete, the zygote is expected to be abnormal genetically and possibly phenotypically.
- On very rare occasions, adjacent-2 segregation can occur (Figure 8.14c). In this case, the centromeres do not segregate as they should. One daughter cell has received both copies of the centromere on chromosome 1; the other, both copies of the centromere on chromosome 2. This rare segregation pattern also yields four abnormal haploid cells that contain an unbalanced combination of chromosomes.

For these translocated chromosomes, the expected segregation pattern is governed by the centromeres. Each haploid cell should receive one centromere located on chromosome 1 and one centromere located on chromosome 2. Therefore, alternate and adjacent-1 segregation patterns are the likely outcomes when an individual carries a reciprocal translocation. Depending on the sizes of the translocated segments, both types may be equally likely to occur. In many cases, the haploid cells from adjacent-1 segregation are not viable, thereby lowering the fertility of the individual. This condition is called **semisterility.**

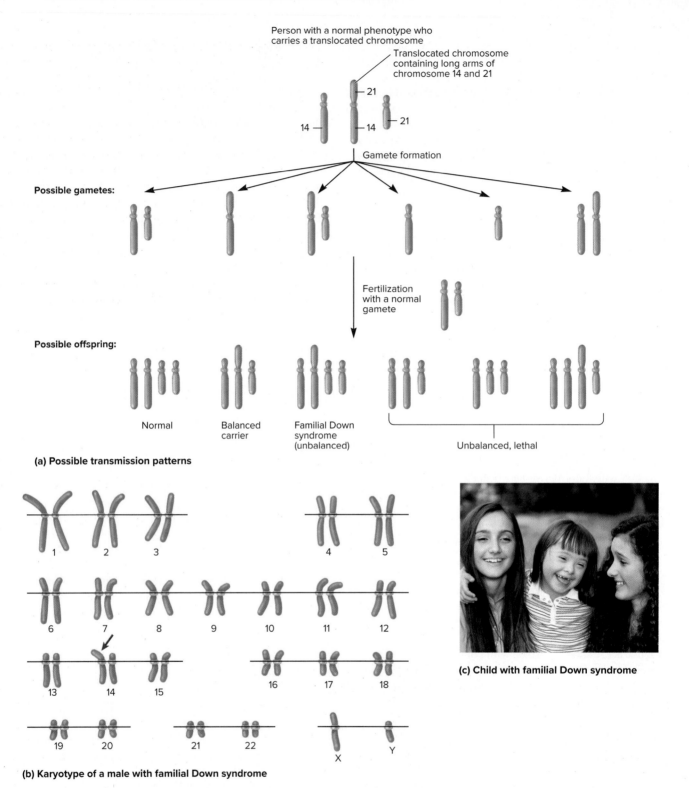

(a) Possible transmission patterns

(b) Karyotype of a male with familial Down syndrome

(c) Child with familial Down syndrome

FIGURE 8.13 **Transmission of familial Down syndrome.** **(a)** Potential transmission of familial Down syndrome. The individual with the chromosome composition shown at the top may produce a gamete carrying chromosome 21 and a fused chromosome consisting of the long arms of chromosomes 14 and 21. Such a gamete can give rise to an offspring with familial Down syndrome. **(b)** A schematic karyotype of an individual with familial Down syndrome. This karyotype shows that the long arm of chromosome 21 has been translocated to chromosome 14 (see red arrow). In addition, the individual also carries two normal copies of chromosome 21. **(c)** An individual (in the center) with familial Down syndrome.

(c): Denys Kuvaiev/Alamy Stock Photo

Genes→Traits As shown in part (a) of this figure, different offspring are possible depending on how the three chromosomes at the top are segregated into gametes. Two types of offspring (labeled Normal or Balanced carrier) are expected to have a normal (unaffected) phenotype. An offspring with familial Down syndrome has three copies of most of the genes that are located on chromosome 21 and therefore exhibits characteristics of Down syndrome.

CONCEPT CHECK: If the segregation patterns in part (a) are equally likely, what is the probability that a gamete produced by the individual who carries the translocated chromosome will result in a viable offspring with a normal phenotype?

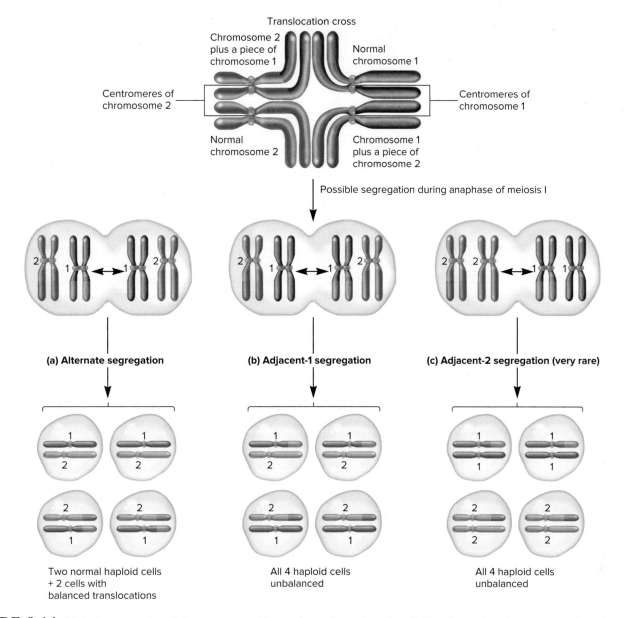

FIGURE 8.14 **Meiotic segregation of chromosomes with a reciprocal translocation.** Follow the numbered centromeres through each process. **(a)** Alternate segregation gives rise to balanced haploid cells, whereas **(b)** adjacent-1 segregation and **(c)** adjacent-2 segregation produce haploid cells with an unbalanced amount of genetic material.

CONCEPT CHECK: Explain why the chromosomes that have balanced translocations and the normal chromosomes form a translocation cross during prophase of meiosis I.

8.4 COMPREHENSION QUESTIONS

1. A paracentric inversion
   a. includes the centromere within the inverted region.
   b. does not include the centromere within the inverted region.
   c. has two adjacent inverted regions.
   d. is an inverted region at the very end of a chromosome.

2. Due to crossing over within an inversion loop, a heterozygote with a pericentric inversion may produce gametes that carry
   a. a deletion.
   b. a duplication.
   c. a translocation.
   d. both a deletion and a duplication.

3. A mechanism that can produce a translocation is
   a. the joining of reactive ends when two different chromosomes break.
   b. crossing over between nonhomologous chromosomes.
   c. crossing over between homologous chromosomes.
   d. either a or b.

8.5 CHANGES IN CHROMOSOME NUMBER: AN OVERVIEW

Learning Outcomes:

1. Define *euploid* and *aneuploid*.
2. Compare and contrast polyploidy and aneuploidy.

As we saw in previous sections of this chapter, chromosome structure can be altered in a variety of ways. Furthermore, the total number of chromosomes can vary. Eukaryotic species typically have several chromosomes that are inherited as one or more sets. Variations in chromosome number can be categorized in two ways: variation in the number of sets of chromosomes and variation in the number of particular chromosomes within a set.

Organisms that are **euploid** have a chromosome number that is an exact multiple of a chromosome set. In other words, euploid organisms have the same number of each type of chromosome (with the possible exception of the sex chromosomes). A diploid organism has two copies of chromosome 1, two copies of chromosome 2, two copies of chromosome 3, and so on.

Figure 8.15 shows euploid variation in *Drosophila melanogaster.* In this species, a single set is composed of 4 different chromosomes. The species is diploid, having two sets of 4 chromosomes

each (Figure 8.15a). A normal (wild-type) fruit fly is euploid because 8 chromosomes divided by 4 chromosomes per set equals exactly 2 sets. On rare occasions, an abnormal fruit fly can have 12 chromosomes—that is, 3 sets of 4 chromosomes each. This fly with 12 chromosomes is **triploid.** Such a fly is also euploid because it has exactly three sets of chromosomes. A **tetraploid** fly, one with four sets of chromosomes, is also euploid. Organisms with three or more sets of chromosomes are also called **polyploid** (Figure 8.15b). Note that diploid, triploid, and tetraploid individuals are all considered euploid. Geneticists use the letter *n* to represent a set of chromosomes. A diploid organism is referred to as $2n$, a triploid organism as $3n$, a tetraploid organism as $4n$, and so on.

A second way in which chromosome number can vary is when an organism is **aneuploid.** Such variation involves an alteration in the number of particular chromosomes, so the total number of chromosomes is not an exact multiple of a set. For example, an abnormal fruit fly could have 9 chromosomes instead of 8 because it has three copies of chromosome 2 instead of the usual two copies (Figure 8.15c). Such an animal is said to have trisomy 2 or to be **trisomic.** Instead of being perfectly diploid ($2n$), a trisomic animal is $2n + 1$. By comparison, a fruit fly could be lacking a single chromosome, such as chromosome 1, and thus have a total of 7 chromosomes ($2n - 1$). This animal is **monosomic** and said to have monosomy 1.

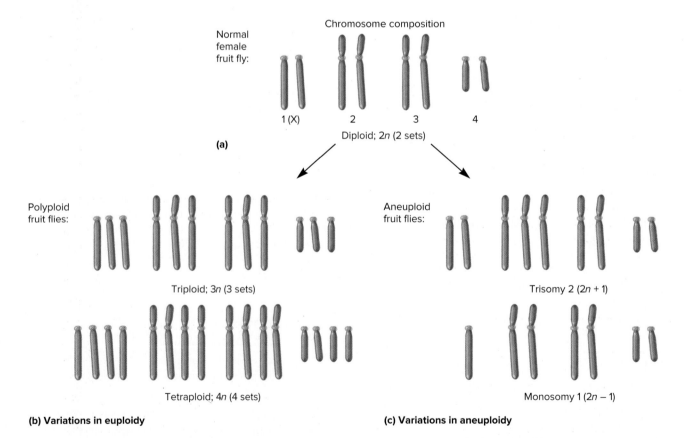

(a) Normal female fruit fly: Chromosome composition — 1 (X) 2 3 4 — Diploid; $2n$ (2 sets)

Polyploid fruit flies: Triploid; $3n$ (3 sets) — Tetraploid; $4n$ (4 sets)

(b) Variations in euploidy

Aneuploid fruit flies: Trisomy 2 ($2n + 1$) — Monosomy 1 ($2n - 1$)

(c) Variations in aneuploidy

FIGURE 8.15 Types of variation in chromosome number. (a) Depicts the normal diploid number of chromosomes in *Drosophila.* **(b)** Examples of polyploidy. **(c)** Examples of aneuploidy.

CONCEPT CHECK: What terms can be used to describe a fruit fly that has a total of 7 chromosomes because it is missing one copy of chromosome 3?

Previously in this chapter, we discussed how deletions and duplications can alter the total amount of genetic material. Deletions are also called **partial monosomy** and duplications are called **partial trisomy.** The word partial indicates that the individual is monosomic or trisomic for just a part of chromosome. For example, a person with cri-du-chat syndrome has a partial monosomy for a region of chromosome 5 (see Figure 8.4).

8.5 COMPREHENSION QUESTION

1. Humans have 23 chromosomes per set. A person with 45 chromosomes can be described as being
   a. euploid.   c. monoploid.
   b. aneuploid.   d. trisomic.

8.6 VARIATION IN THE NUMBER OF CHROMOSOMES WITHIN A SET: ANEUPLOIDY

Learning Outcomes:

1. Explain why aneuploidy usually has a detrimental effect on phenotype.
2. Describe examples of aneuploidy in humans.

In this section, we will consider several examples of aneuploidy. As you will learn, this is regarded as an abnormal condition that usually has a negative effect on phenotype.

Aneuploidy Causes an Imbalance in Gene Expression That Is Often Detrimental to the Phenotype of the Individual

The phenotype of every eukaryotic species is influenced by thousands of different genes. In humans, for example, a single set of chromosomes has approximately 20,000 different protein-coding genes. To produce a phenotypically normal individual, the expression of thousands of genes requires intricate coordination. In the case of humans and other diploid species, evolution has resulted in a developmental process that works correctly when somatic cells have two copies of each chromosome. In other words, when a human is diploid, the balance of gene expression among many different genes usually produces a person with a normal phenotype.

Aneuploidy commonly causes a detrimental phenotype. To understand why, let's consider the relationship between gene expression and chromosome number in a species that has three pairs of chromosomes (**Figure 8.16**). The level of gene expression is influenced by the number of genes per cell. Compared with a diploid cell, if a cell has a chromosome that is present in three copies instead of two, more of the product of a gene on that chromosome is typically made. For example, a gene present in three copies instead of two may produce 150% of the gene product, though that number may vary due to effects of gene regulation. Alternatively,

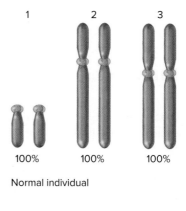
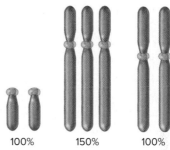
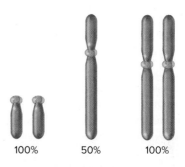

FIGURE 8.16 Imbalance of gene products in trisomic and monosomic individuals. Aneuploidy of chromosome 2 (i.e., trisomy or monosomy) leads to an imbalance in the amount of gene products from chromosome 2 compared with the amounts from chromosomes 1 and 3.

CONCEPT CHECK: Describe the imbalance in gene products that occurs in an individual with monosomy 2.

if only one copy of a gene is present, due to a missing chromosome, less of the gene product is usually made, perhaps only 50% as much. Therefore, in trisomic and monosomic individuals, an imbalance occurs between the level of gene expression for the chromosomes found in pairs and the level of gene expression for the ones with extra or missing copies.

At first glance, the difference in gene expression between euploid and aneuploid individuals may not seem terribly dramatic. Keep in mind, however, that a eukaryotic chromosome carries hundreds or even thousands of different genes. Therefore, when an organism is trisomic or monosomic, many gene products occur in excessive or deficient amounts. This imbalance among many genes appears to underlie the phenotypic abnormalities that aneuploidy frequently causes. In most cases, these effects are detrimental and produce an individual that is less likely to survive than a euploid individual.

Aneuploidy in Humans Causes Detrimental Phenotypes

A key reason why geneticists are so interested in aneuploidy is its relationship to certain inherited disorders in humans. Even though most people are born with 46 chromosomes, alterations in chromosome number occur fairly frequently during gamete formation. About 5%–10% of all fertilized human eggs result in an embryo with an abnormality in chromosome number. In most cases, such an embryo does not develop properly, and a spontaneous abortion occurs very early in pregnancy. Approximately 50% of all spontaneous abortions are due to abnormalities in chromosome number.

In some cases, an abnormality in chromosome number produces an offspring that survives to birth or longer. Several human disorders involve abnormalities in chromosome number. The most common are trisomies of chromosome 13, 18, or 21 and abnormalities in the number of the sex chromosomes (**Table 8.3**). Most of the known trisomies involve chromosomes that are relatively small—chromosome 13, 18, or 21—and carry fewer genes than the larger chromosomes. Trisomies of the other human autosomes and monosomies of all autosomes are presumed to produce a lethal phenotype, and many have been found in spontaneously aborted embryos and fetuses. For example, all possible human trisomies have been found in spontaneously aborted embryos except trisomy 1. It is believed that trisomy 1 either causes gametes to be inviable or it is lethal at such an early stage of development that it prevents the successful implantation of the embryo.

Variation in the number of X chromosomes, unlike that of other large chromosomes, is often nonlethal. The survival of

TABLE 8.3
Aneuploid Conditions in Humans

Condition	Frequency	Syndrome	Characteristics
Autosomal			
Trisomy 13	1/15,000	Patau	Mental and physical deficiencies, wide variety of defects in organs, large triangular nose, early death
Trisomy 18	1/6000	Edward	Mental and physical deficiencies, facial abnormalities, extreme muscle tone, early death
Trisomy 21	1/800	Down	Mental deficiencies, abnormal pattern of palm creases, slanted eyes, flattened face, short stature
Sex Chromosomal			
XXY	1/1000	Klinefelter	Sexual immaturity (no sperm [males]), breast swelling
XYY	1/1000	Jacobs	Tall and thin (males)
XXX	1/1500	Triple X	Tall and thin, menstrual irregularity (females)
X0	1/5000	Turner	Short stature, webbed neck, sexually undeveloped (females)

trisomy X individuals may be explained by X-chromosome inactivation, which is described in Chapter 5. In an individual with more than one X chromosome, all additional X chromosomes are converted to Barr bodies in the somatic cells of adult tissues. In an individual with trisomy X, for example, two out of three X chromosomes are converted to inactive Barr bodies. Unlike the level of expression for autosomal genes, the normal level of expression for most X-linked genes is from a single X chromosome. In other words, the correct level of mammalian gene expression results from two copies of each autosomal gene and one copy of each X-linked gene. This explains how the expression of X-linked genes in males (XY) can be maintained at the same levels as in females (XX). It may also explain why trisomy X is not a lethal condition.

The phenotypic effects noted in Table 8.3 involving X chromosome abnormalities may be due to the expression of X-linked genes prior to embryonic X-chromosome inactivation or to the expression of genes on the inactivated X chromosome. As described in Chapter 5, pseudoautosomal genes and some other genes on the inactivated X chromosome are expressed in humans. Having three copies or one copy of the X chromosome results in overexpression or underexpression of these X-linked genes, respectively.

Human abnormalities in chromosome number are influenced by the age of the parents. Older parents are more likely to produce children with abnormalities in chromosome number. Down syndrome provides an example. The common form of this disorder is caused by the inheritance of three copies of chromosome 21. The incidence of Down syndrome rises with the age of either parent. In males, however, the rise occurs relatively late in life, usually past the age when most males have children. By comparison, the likelihood of having a child with Down syndrome rises dramatically with age during a female's reproductive years (**Figure 8.17**). This syndrome was first described by John Langdon Down in 1866. The association between maternal age and Down syndrome was later discovered by L. S. Penrose in 1933, even before the chromosomal basis for the disorder was identified by Jérôme Lejeune in 1959. Down syndrome is most commonly caused by **nondisjunction,** which means that the chromosomes do not segregate properly. (Nondisjunction is discussed later in this chapter.) In this case, nondisjunction of chromosome 21 most commonly occurs during meiosis I in the oocyte.

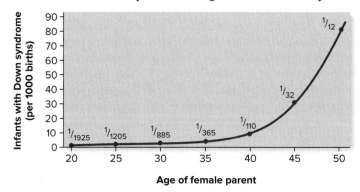

FIGURE 8.17 The incidence of Down syndrome births according to the age of the female parent. The y-axis shows the number of infants born with Down syndrome per 1000 live births, and the x-axis plots the age of the female parent at the time of birth. The data points indicate the fraction of live offspring born with Down syndrome.

Different hypotheses have been proposed to explain the relationship between maternal age and Down syndrome. One popular idea suggests that it may be due to the age of the oocytes. Human primary oocytes are produced within the ovaries of the female fetus prior to birth, are arrested at prophase of meiosis I, and remain in this stage until the time of ovulation. Therefore, as a human female ages, the primary oocytes have been in prophase I for a progressively longer period of time. This added length of time may contribute to an increased frequency of nondisjunction. About 5% of the time, Down syndrome is due to an extra paternal chromosome.

8.6 COMPREHENSION QUESTIONS

1. In a person with trisomy 21 (Down syndrome), a genetic imbalance occurs because
   a. genes on chromosome 21 are overexpressed.
   b. genes on chromosome 21 are underexpressed.
   c. genes on the other chromosomes are overexpressed.
   d. genes on the other chromosomes are underexpressed.

2. Humans with aneuploidy who survive usually have an incorrect number of chromosome 13, 18, or 21 or of one of the sex chromosomes. A possible explanation why these abnormalities permit survival is because
   a. the chromosomes have clusters of genes that aid in embryonic growth.
   b. the chromosomes are small and carry relatively few genes.
   c. X-chromosome inactivation allows only one X chromosome to remain active.
   d. Both b and c are possible explanations.

8.7 VARIATION IN THE NUMBER OF SETS OF CHROMOSOMES

Learning Outcomes:

1. Describe examples in animals that involve variations in euploidy.
2. Define *endopolyploidy*.
3. Outline the process of polytene chromosome formation.
4. Discuss the effects of polyploidy in plant species and its impact in agriculture.

We now turn our attention to changes in the number of sets of chromosomes, referred to as variations in euploidy. In some cases, such changes are detrimental. However, in many species, particularly plants, additional sets of chromosomes are very common and are often beneficial with regard to an individual's phenotype.

Variations in Euploidy Occur Naturally in a Few Animal Species

Most species of animals are diploid. In some cases, changes in euploidy are not well tolerated. For example, polyploidy in mammals is generally a lethal condition. However, many examples of

naturally occurring variations in euploidy occur. In species that are **haplodiploid,** which include many species of bees, wasps, and ants, one of the sexes is haploid, usually the male, and the other is diploid. For example, male bees, which are called drones, contain a single set of chromosomes. They are produced from unfertilized eggs. By comparison, female bees are produced from fertilized eggs and are diploid.

Examples of vertebrate polyploid animals have been discovered. Interestingly, in several cases, animals that are morphologically very similar are members of a diploid species or a separate polyploid species. This situation occurs among certain amphibians and reptiles. **Figure 8.18** shows photographs of a diploid frog and a tetraploid (4n) frog. As you can see, in terms of outward appearance, they are nearly indistinguishable from each other. Their difference can be revealed only by an examination of the chromosome

(a) Hyla chrysoscelis

(b) Hyla versicolor

FIGURE 8.18 **Differences in euploidy in two closely related frog species.** The frog in (**a**) is diploid, whereas the frog in (**b**) is tetraploid.

Genes→Traits Though similar in appearance, these two species differ in their number of chromosome sets. At the level of gene expression, this observation suggests that the number of copies of each gene (two versus four) does not critically affect the phenotype of these two species.

(a and b): A.B. Sheldon

number in the somatic cells of the animals and by their mating calls—*Hyla chrysoscelis* has a faster trill rate than *Hyla versicolor*.

Variations in Euploidy Can Occur in Certain Tissues Within an Animal

Thus far, we have considered variations in chromosome number that occur at fertilization, so all somatic cells of an individual contain this variation. Euploidy may also change after fertilization. In many animals, certain tissues of the body display normal variations in the number of sets of chromosomes. For example, the cells of the human liver are typically polyploid. Liver cells are often tetraploid, and may even be octaploid ($8n$). The occurrence of polyploid cells in organisms that are otherwise diploid is known as **endopolyploidy.** What is the biological significance of endopolyploidy? One possibility is that the increase in chromosome number in certain cells may enhance their ability to produce specific gene products that are needed in great abundance.

An interesting example of natural variation in the euploidy of somatic cells occurs in *Drosophila* and some other insects.

Drosophila cells contain eight chromosomes (two sets of four chromosomes each; see Figure 8.15a). Within certain tissues, such as the salivary glands, the homologous chromosomes synapse with each other and undergo repeated rounds of chromosome replication without separating from each other. For example, in the salivary gland cells of *Drosophila*, the homologous chromosomes double approximately nine times ($2^9 = 512$).

As shown in **Figure 8.19a**, repeated rounds of chromosomal replication produce a bundle of chromosomes that lie together in a parallel fashion. During this process, the four types of *Drosophila* chromosomes aggregate to form an enormous **polytene chromosome** with several arms (**Figure 8.19b**). The central point where the chromosomes aggregate is known as the **chromocenter** (**Figure 8.19c**). Each of the four types of chromosome is attached to the chromocenter near its centromere. The X and Y chromosomes and chromosome 4 are telocentric, and chromosomes 2 and 3 are metacentric. Therefore, the X chromosome and/or Y chromosome and chromosome 4 form a single arm projecting from the chromocenter, whereas chromosomes 2 and 3 form two arms.

How does endopolyploidy occur? Polyploid cells within certain tissues of diploid animals undergo an altered version of the cell

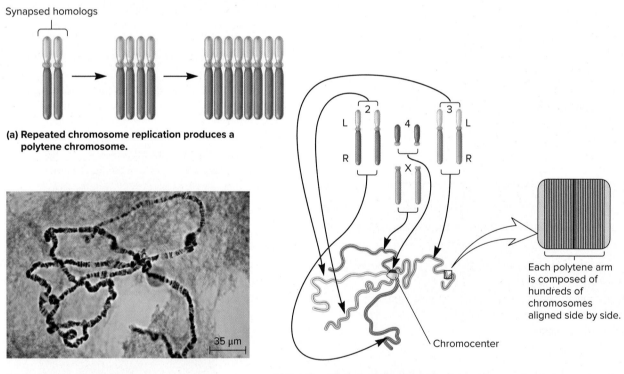

(a) Repeated chromosome replication produces a polytene chromosome.

35 μm

(b) A polytene chromosome

(c) Relationship between a polytene chromosome and regular *Drosophila* chromosomes

Each polytene arm is composed of hundreds of chromosomes aligned side by side.

Chromocenter

FIGURE 8.19 Polytene chromosomes in *Drosophila*. **(a)** A schematic illustration of the formation of a polytene chromosome. Homologous chromosomes synapse and undergo several rounds of replication without separating from each other. This results in a bundle of chromosomes that lie parallel to each other. Note: This replication does not occur in highly condensed, heterochromatic DNA near the centromere. **(b)** A micrograph of a polytene chromosome. **(c)** This drawing shows the relationship between the four types of homologous chromosomes and the formation of a polytene chromosome in a salivary gland cell of *Drosophila*. The heterochromatic regions of the chromosomes aggregate at the chromocenter, and the arms of the chromosomes project outward. In chromosomes with two arms, the short arm is labeled L and the long arm is labeled R.

(b): David M. Phillips/Science Source

CONCEPT CHECK: Approximately how many copies of chromosome 2 are found in a polytene chromosome of *Drosophila*?

cycle, which is called an endocycle, in which S phases alternate with gap phases, but there is no cell division. (Note: the eukaryotic cell cycle is described in Chapter 2.) In the case of polytene chromosomes, M phase is bypassed completely. Interestingly, mammalian liver cells exhibit different types of polyploid cells. Some liver cells contain polyploid nuclei because the nucleus fails to undergo nuclear division during the endocycle. Alternatively, other liver cells undergo M phase and the nucleus divides, but cytokinesis does not occur. This mechanism gives rise to multinucleate liver cells.

Variations in Euploidy Are Common in Plants

We now turn our attention to variations of euploidy that occur in plants. Compared with animals, plants more commonly exhibit polyploidy. Among ferns and seed plants, at least 30%–35% of species are polyploid. Polyploidy is also important in agriculture. Many of the fruits and grains we eat are produced from polyploid plants. For example, the species of wheat that we use to make bread, *Triticum aestivum*, is a hexaploid (6*n*) that arose from the union between three diploid species (**Figure 8.20**). Different

Cultivated wheat, a hexaploid species

FIGURE 8.20 Example of a polyploid plant. Cultivated wheat, *Triticum aestivum*, is a hexaploid. It was derived from three different species of diploid wheat that originally were found in the Middle East and were cultivated by ancient farmers in that region.

Genes→Traits An increase in chromosome number from diploid to tetraploid or hexaploid affects the phenotype of the individual. In the case of many plant species, a polyploid individual is larger and more robust than its diploid counterpart. This suggests that, for plants, having additional copies of each gene is somewhat better than having two copies of each gene. The situation is rather different in animals. Tetraploidy in animals may have little effect (as shown in Figure 8.18b), but it is also common for polyploidy in animals to be detrimental.

Jim Steinberg/Science Source

CONCEPT CHECK: What are some common advantages of polyploidy in plants?

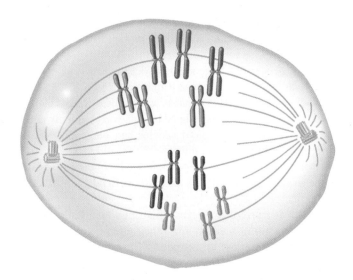

FIGURE 8.21 Schematic representation of anaphase of meiosis I in a triploid organism containing three sets of four chromosomes. In this example, the homologous chromosomes (three each) do not evenly separate during anaphase. Each cell receives one copy of some chromosomes and two copies of other chromosomes. This produces aneuploid gametes.

CONCEPT CHECK: Explain why a triploid individual is usually infertile.

species of strawberries are diploid, tetraploid, hexaploid, and even octaploid!

In many instances, polyploid strains of plants display outstanding agricultural characteristics. They are often larger in size and more robust. These traits are clearly advantageous in the production of food. In addition, polyploid plants tend to exhibit a greater adaptability, which allows them to withstand harsher environmental conditions. Also, polyploid ornamental plants often produce larger flowers than their diploid counterparts.

Polyploid plants having an odd number of chromosome sets, such as triploids (3*n*) or pentaploids (5*n*), usually cannot reproduce. Why are they sterile? The sterility arises because these plants produce highly aneuploid gametes. During prophase of meiosis I, homologous pairs of sister chromatids form bivalents. However, organisms with an odd number of chromosome sets, such as three, display an unequal separation of homologous chromosomes during anaphase of meiosis I (**Figure 8.21**). An odd number cannot be divided equally between two daughter cells. For each type of chromosome, a daughter cell randomly gets either one or two copies. For example, one daughter cell might receive one copy of chromosome 1, two copies of chromosome 2, two copies of chromosome 3, one copy of chromosome 4, and so forth. For a triploid species containing many different chromosomes in a set, meiosis is very unlikely to produce a daughter cell that is euploid.

If we assume that a daughter cell receives either one copy or two copies of each kind of chromosome, the probability that meiosis will produce a cell that is perfectly haploid or diploid is $(1/2)^{n-1}$, where n is the number of chromosomes in a set. As an example, in a triploid organism containing 20 chromosomes per set, the probability of producing a haploid or diploid cell is 0.000001907, or 1 in 524,288. Thus, meiosis is almost certain to produce cells that contain one copy of some chromosomes and two copies of the others. This high probability of aneuploidy underlies the reason for triploid sterility.

Though sterility is generally a detrimental trait, it can be desirable agriculturally because it may result in a seedless fruit. For example, domesticated bananas and seedless watermelons are triploid varieties. The domesticated banana was originally derived from a seed-producing diploid species and has been asexually propagated by humans via cuttings. The small black spots in the center of such a banana are degenerate seeds. In the case of flowers, the seedless phenotype can also be beneficial. Seed producers such as Burpee have developed triploid varieties of flowering plants such as marigolds. Because the triploid marigolds are sterile and unable to set seed, more of their energy goes into flower production. According to Burpee, "They bloom and bloom, unweakened by seed bearing."

8.7 COMPREHENSION QUESTIONS

1. The term *endopolyploidy* refers to the phenomenon of having
   a. too many chromosomes.
   b. extra chromosomes inside the cell nucleus.
   c. extra sets of chromosomes in certain cells of the body.
   d. extra sets of chromosomes in gametes.
2. In agriculture, an advantage of triploidy in plants is that the plants are
   a. more fertile.
   b. often seedless.
   c. always disease-resistant.
   d. all of the above.

8.8 NATURAL AND EXPERIMENTAL MECHANISMS THAT PRODUCE VARIATION IN CHROMOSOME NUMBER

Learning Outcomes:

1. Describe the mechanisms of meiotic and mitotic nondisjunction and their possible phenotypic consequences.
2. Compare and contrast autopolyploidy, alloploidy, and allopolyploidy.
3. Describe how colchicine is used to produce polyploid species.

As we have seen, variations in chromosome number are fairly widespread and usually have a significant effect on the phenotypes of plants and animals. For these reasons, researchers have wanted to understand the cellular mechanisms that cause variations in chromosome number.

In some cases, a change in chromosome number is the result of nondisjunction, which refers to an event in which the

chromosomes do not segregate properly. **Meiotic nondisjunction,** which is an improper separation of chromosomes during meiosis, produces haploid cells that have too many or too few chromosomes. If such a cell gives rise to a gamete that fuses with a normal gamete during fertilization, the resulting offspring will have an abnormal chromosome number in all of its cells. A nondisjunction also may occur after fertilization in one of the somatic cells of the body. This event is known as **mitotic nondisjunction.** When this occurs during embryonic stages of development, it may lead to a patch of tissue in the organism that has an altered chromosome number.

Another common way in which the chromosome number of an organism can vary is by interspecies crosses. An **alloploid** organism contains sets of chromosomes from two or more different species. This term refers to the occurrence of chromosome sets (ploidy) from the genomes of different (allo) species. In this section, we will examine these three natural mechanisms in greater detail, and also explore how chromosome number can be changed experimentally.

Meiotic Nondisjunction Can Produce Aneuploidy or Polyploidy

The process of meiosis is described in Chapter 2. Nondisjunction can occur during anaphase of meiosis I or meiosis II. If it happens during meiosis I, an entire bivalent migrates to one pole (**Figure 8.22a**). Following the completion of meiosis, the four resulting haploid cells are abnormal. If nondisjunction occurs during anaphase of meiosis II (**Figure 8.22b**), the net result is two abnormal and two normal haploid cells. If a gamete carrying an extra chromosome unites with a normal gamete, the offspring will be trisomic. Alternatively, if a gamete that is missing a chromosome is viable and participates in fertilization, the resulting offspring is monosomic for the missing chromosomes.

In rare cases, all of the chromosomes may undergo nondisjunction and migrate to one of the daughter cells, an event called **complete nondisjunction.** The daughter cell that does not receive any chromosomes is inviable. In contrast, the daughter cell receiving all of the chromosomes may complete meiosis and form two diploid cells, which give rise to diploid gametes. If a diploid gamete participates in fertilization with a normal haploid gamete, a triploid individual is produced. Therefore, complete nondisjunction can result in individuals that are polyploid.

Mitotic Nondisjunction or Chromosome Loss Can Produce a Patch of Tissue with an Altered Chromosome Number

Abnormalities in chromosome number occasionally occur after fertilization takes place. In this case, the abnormal event happens during mitosis rather than meiosis. One possibility is that the sister chromatids separate improperly, so one daughter cell receives three copies of a chromosome, whereas the other daughter cell gets only one (**Figure 8.23a**). Alternatively, the sister chromatids can separate during anaphase of mitosis, but one of the chromosomes is improperly attached to the mitotic spindle apparatus and so does not migrate to a pole (**Figure 8.23b**). A chromosome will

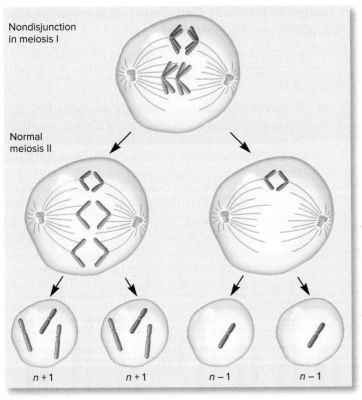

(a) Nondisjunction in meiosis I

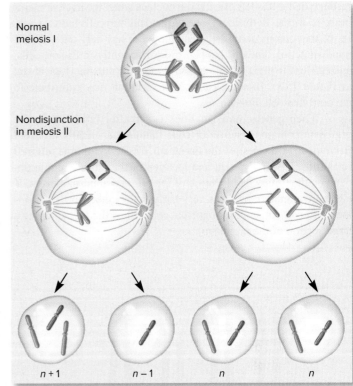

(b) Nondisjunction in meiosis II

FIGURE 8.22 **Nondisjunction during meiosis I or II.** The chromosomes shown in purple are behaving properly during meiosis I and II, so each haploid cell receives one copy of the purple chromosome. The chromosomes shown in blue are not segregating correctly. In **(a)**, nondisjunction occurs in meiosis I, so the resulting four cells receive either two copies of a blue chromosome or zero copies. In **(b)**, nondisjunction occurs during meiosis II, so one cell has two blue chromosomes and another cell has zero. The remaining two cells are normal.

CONCEPT CHECK: Explain what the word *nondisjunction* means.

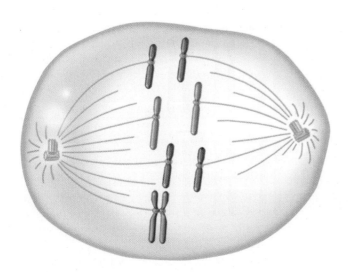

(a) **Mitotic nondisjunction**

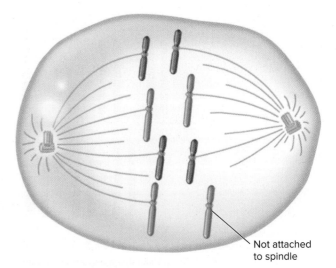

Not attached to spindle

(b) **Chromosome loss**

FIGURE 8.23 **Nondisjunction and chromosome loss during mitosis in somatic cells.** **(a)** Mitotic nondisjunction produces trisomic and monosomic daughter cells. **(b)** Chromosome loss produces normal and monosomic daughter cells.

be degraded if it is left outside the nucleus when the nuclear membrane re-forms. In this case, one of the daughter cells has two copies of that chromosome, whereas the other has only one. Mitotic nondisjunction tends to happen more frequently in cancer cells, and can lead to great variation in chromosome number (look ahead to Figure 25.8). In some cases, this variation may contribute to uncontrolled cell division.

When genetic abnormalities occur after fertilization, the organism contains a subset of cells that are genetically different from those of the rest of the organism. This condition is referred to as **mosaicism.** The size and location of the mosaic region depend on the timing and location of the original abnormal event. If a genetic alteration happens very early in the embryonic development of an organism, the abnormal cell will be the precursor for a large section of the organism.

Changes in Euploidy Can Result in Autopolyploidy, Alloploidy, or Allopolyploidy

Different mechanisms account for changes in the number of chromosome sets among natural populations of plants and animals (**Figure 8.24**).

- As previously mentioned, complete nondisjunction, due to a general defect in the spindle apparatus, can produce an individual with one or more extra sets of chromosomes. This individual is known as an **autopolyploid** (Figure 8.24a). The prefix *auto-* (meaning "self") and term *polyploid* (meaning "many sets of chromosomes") refer to an increase in the number of chromosome sets within a single species.

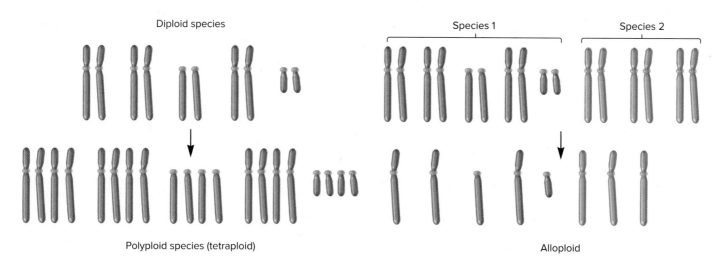

(a) Autopolyploidy (tetraploid)

(b) Alloploidy (allodiploid)

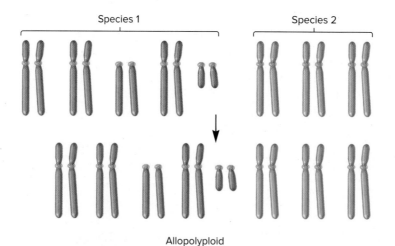

(c) Allopolyploidy (allotetraploid)

FIGURE 8.24 A comparison of autopolyploidy, alloploidy, and allopolyploidy.

CONCEPT CHECK: What is the key difference between autopolyploidy and allopolyploidy?

FIGURE 8.25 A liger. This animal is an allodiploid that is produced from a mating between a male lion and female tiger.
old3310/Getty Images

- An alloploid is a result of interspecies crosses (Figure 8.24b). An alloploid that has one set of chromosomes from each of two different species is called an **allodiploid.** An interspecies cross is most likely to occur between species that are close evolutionary relatives. For example, closely related species of grasses may interbreed to produce allodiploids. Another example is the liger, which is an allodiploid offspring that is produced from a mating between a male lion (*Panthera leo*) and a female tiger (*Panthera tigris*) (**Figure 8.25**).
- As shown in Figure 8.24c, an **allopolyploid** contains two (or more) sets of chromosomes from two (or more) species. In the example shown, the **allotetraploid** contains two complete sets of chromosomes from two different species, for a total of four sets.

In nature, allotetraploids usually arise from allodiploids. This can occur when a somatic cell in an allodiploid undergoes complete nondisjunction to create an allotetraploid cell. In plants, such a cell can continue to grow and produce a section of the plant that is allotetraploid. If this part of the plant produces seeds by self-pollination, the seeds give rise to allotetraploid offspring. Cultivated wheat (refer back to Figure 8.20) is derived from a series of events in which two diploid species produced an allotetraploid, and then a third diploid species interbred with the allotetraploid to produce an allohexaploid.

Experimental Treatments Can Promote Polyploidy

Because autopolyploid and allopolyploid plants often exhibit desirable traits, the development of polyploids is of considerable interest among plant breeders. Experimental studies on the ability of environmental agents to promote polyploidy began in the early twentieth century. Since that time, various treatments have been shown to promote nondisjunction, thereby leading to polyploidy. These treatments include abrupt temperature changes during the initial stages of seedling growth and exposure of plants to chemical agents that interfere with the formation of the spindle apparatus.

The drug colchicine is one commonly used agent for promoting polyploidy. Once inside the cell, colchicine binds to tubulin (a protein found in the spindle apparatus) and thereby interferes with normal chromosome segregation during mitosis or meiosis. In 1937, Alfred Blakeslee and Amos Avery applied colchicine to plant tissue and found that high doses of the drug caused complete mitotic nondisjunction and produced polyploidy in plant cells.

Colchicine can be applied to seeds, young embryos, or rapidly growing regions of a plant (**Figure 8.26**). This application may produce aneuploidy, which is usually an undesirable outcome, but it often produces polyploid cells, which may grow faster than the surrounding diploid tissue. In a diploid plant, colchicine may cause complete mitotic nondisjunction, yielding tetraploid ($4n$) cells.

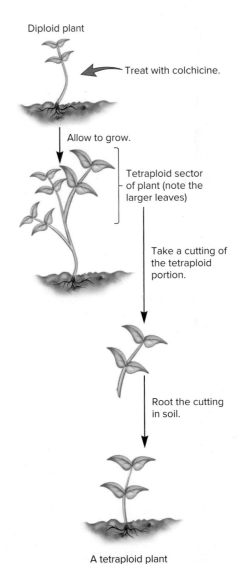

Diploid plant

Treat with colchicine.

Allow to grow.

Tetraploid sector of plant (note the larger leaves)

Take a cutting of the tetraploid portion.

Root the cutting in soil.

A tetraploid plant

FIGURE 8.26 Use of colchicine to promote polyploidy in plants. Colchicine interferes with the mitotic spindle apparatus and promotes mitotic nondisjunction. If complete nondisjunction occurs in a diploid cell, a tetraploid daughter cell will be formed. Such a tetraploid cell may continue to divide and produce a segment of the plant with more robust characteristics. This segment may be cut from the rest of the plant and rooted. In this way, a tetraploid plant can be propagated.

As the tetraploid cells continue to divide, they generate a sector that is often morphologically distinguishable from the rest of the plant. For example, a tetraploid stem may have a larger diameter and produce larger leaves and flowers.

Because individual plants can be propagated asexually from pieces of plant tissue (i.e., cuttings), the polyploid portion of the plant can be removed, treated with the proper growth hormones, and grown as a separate plant. Alternatively, the tetraploid region of a plant may have flowers that produce seeds by self-pollination. For many plant species, a tetraploid flower produces diploid pollen and eggs, which can combine to produce tetraploid offspring. In this way, colchicine provides a straightforward method of producing polyploid strains of plants.

8.8 COMPREHENSION QUESTIONS

1. In a diploid species, complete nondisjunction during meiosis I may produce a viable cell that is
   a. trisomic.   c. diploid.
   b. haploid.   d. triploid.

2. The somatic cells of an allotetraploid contain
   a. one set of chromosomes from four different species.
   b. two sets of chromosomes from two different species.
   c. four sets of chromosomes from one species.
   d. one set of chromosomes from two different species.

KEY TERMS

Introduction: genome, genetic variation, allelic variation

8.1: cytogeneticist, metacentric, submetacentric, acrocentric, telocentric, karyotype, G bands

8.2: deletion, deficiency, duplication, insertion, indel, inversion, translocation, simple translocation, reciprocal translocation

8.3: terminal deletion, interstitial deletion, repetitive sequences, nonallelic homologous recombination, gene duplication, gene family, homologous, paralogs, copy number variation (CNV), segmental duplication, comparative genomic hybridization (CGH), hybridization

8.4: pericentric inversion, paracentric inversion, position effect, inversion heterozygote, inversion loop, acentric fragment,

dicentric, dicentric bridge, telomeres, balanced translocation, unbalanced translocation, Robertsonian translocation, translocation cross, semisterility

8.5: euploid, triploid, tetraploid, polyploid, aneuploid, trisomic, monosomic, partial monosomy, partial trisomy

8.6: nondisjunction

8.7: haplodiploid, endopolyploidy, polytene chromosome, chromocenter

8.8: meiotic nondisjunction, mitotic nondisjunction, alloploid, complete nondisjunction, mosaicism, autopolyploid, allodiploid, allopolyploid, allotetraploid

CHAPTER SUMMARY

8.1 Microscopic Examination of Eukaryotic Chromosomes

- Among different species, natural variation exists with regard to chromosome structure and number. Three features of chromosomes that aid in their identification are location of the centromere, size, and banding pattern (see Figure 8.1).

8.2 Changes in Chromosome Structure: An Overview

- Within a species, variations in chromosome structure include deletions, duplications, inversions, and translocations (see Figure 8.2).

8.3 Deletions and Duplications

- Chromosome breaks can create terminal or interstitial deletions. Some deletions are associated with human genetic disorders such as cri-du-chat syndrome (see Figures 8.3, 8.4).
- Nonallelic homologous recombination produces gene duplications and deletions. Over time, gene duplications can lead to the formation of gene families, such as the globin gene family (see Figures 8.5, 8.6, 8.7, Table 8.1).

- Copy number variation (CNV) among members of a species is fairly common. In humans, CNV is associated with certain diseases (see Figure 8.8).
- Microscopy and molecular techniques are used to identify deletions and duplications. One example is comparative genomic hybridization (CGH), which is used in the analysis of cancer cells (see Table 8.2, Figure 8.9).

8.4 Inversions and Translocations

- Inversions can be pericentric or paracentric. In an inversion heterozygote, crossing over within an inversion loop creates deletions and duplications in the resulting chromosomes (see Figures 8.10, 8.11).
- Two mechanisms can produce reciprocal translocations: (1) chromosome breakage with subsequent repair, and (2) nonhomologous crossing over (see Figure 8.12).
- Familial Down syndrome is due to a translocation between chromosomes 14 and 21 (see Figure 8.13).
- Individuals that carry a balanced translocation have a greater risk of producing unbalanced gametes due to the formation of a translocation cross (see Figure 8.14).

8.5 Changes in Chromosome Number: An Overview

- Variation in chromosome number may involve changes in the number of sets (euploidy) or changes in the number of particular chromosomes within a set (aneuploidy) (see Figure 8.15).

8.6 Variation in the Number of Chromosomes Within a Set: Aneuploidy

- Aneuploidy is often detrimental because it results in an imbalance in gene expression. Down syndrome, an example of aneuploidy in humans, increases in frequency with maternal age (see Figures 8.16, 8.17, Table 8.3).

8.7 Variation in the Number of Sets of Chromosomes

- Among animals, variations in euploidy are relatively rare, though they do occur. Some tissues within a diploid animal may exhibit polyploidy. An example is the polytene chromosomes found in salivary gland cells of *Drosophila* (see Figures 8.18, 8.19).
- Polyploidy in plants is relatively common and has many advantages for agriculture. Triploid plants are usually seedless because their chromosomes cannot segregate properly during meiosis (see Figures 8.20, 8.21).

8.8 Natural and Experimental Mechanisms That Produce Variation in Chromosome Number

- Meiotic nondisjunction, mitotic nondisjunction, and chromosome loss result in changes in chromosome number (see Figures 8.22, 8.23).
- Autopolyploidy is an increased number of sets of chromosomes within a single species. Interspecies matings result in alloploids. Alloploids with multiple sets of chromosomes from two (or more) species are allopolyploids (see Figures 8.24, 8.25).
- The drug colchicine promotes nondisjunction and is used to produce polyploid plants (see Figure 8.26).

PROBLEM SETS & INSIGHTS

MORE GENETIC TIPS   **1.** Describe how a gene family is produced. Discuss the common and unique features of the members of the globin gene family.

TOPIC: *What topic in genetics does this question address?* The topic is gene families. More specifically, the question is about the globin gene family in humans.

INFORMATION: *What information do you know based on the question and your understanding of the topic?* In the question, you are reminded that humans carry a globin gene family in their genome. From your understanding of the topic, you may remember that gene families are produced by gene duplication events, and you may remember some of the special features of the globin gene family.

PROBLEM-SOLVING **S**TRATEGIES: *Make a drawing. Compare and contrast.* To begin to solve this problem, you may want to make a drawing that shows how gene duplications may occur (see right column). You also need to compare and contrast the general features of the members of the globin gene family.

ANSWER: A gene family is produced when a single gene is copied one or more times by a duplication event. This duplication may occur because of a misaligned crossover, which produces a chromosome with a deletion and another chromosome with a duplication.

Over time, this type of duplication may occur several times to produce many copies of a particular gene. In addition, translocations may

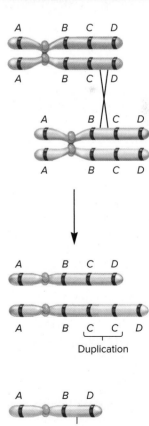

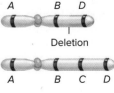

move the duplicated genes to other chromosomes, so the members of the gene family may be dispersed among two or more different chromosomes. Eventually, each member of a gene family will accumulate mutations, which may subtly alter its function.

All members of the globin gene family have gene products that bind oxygen. Myoglobin tends to bind oxygen more tightly; therefore, it is good at storing oxygen. Hemoglobin binds oxygen more loosely, so it can transport oxygen throughout the body (via red blood cells) and release it to the tissues that need it. The polypeptides that form hemoglobins are expressed by globin genes in red blood cells, whereas the myoglobin gene is expressed in many different cell types. The expression pattern of the globin genes changes during different stages of development. The ε-globin and ζ-globin genes are expressed in the early embryo. They are turned off near the end of the first trimester, and then the α-globin and γ-globin genes are expressed. Following birth, the γ-globin genes are silenced, and the β-globin gene is expressed for the rest of a person's life. These differences in the expression of the globin genes reflect the differences in the oxygen transport needs of humans during the different stages of life. Overall, the evolution of gene families has resulted in gene products that are better suited to particular tissues or stages of development.

2. An inversion heterozygote has the following inverted chromosome:

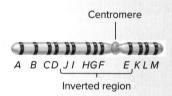

What is the result if crossing over occurs between genes *F* and *G* on an inverted and a normal chromosome in this individual?

T **OPIC: *What topic in genetics does this question address?*** The topic is changes in chromosome structure. More specifically, the question is about the consequences of crossing over between chromosomes when one of the chromosomes carries an inversion.

I **NFORMATION: *What information do you know based on the question and your understanding of the topic?*** From the question, you know the features of a chromosome carrying an inversion and the site of a crossover between the inverted chromosome and a normal chromosome. From your understanding of the topic, you may remember that the inverted chromosome has a pericentric inversion. For crossing over to occur, an inversion loop must form so that the inverted and normal chromosomes can pair up.

P **ROBLEM-SOLVING** **S** **TRATEGIES: *Make a drawing. Predict the outcome.*** One strategy to begin to solve this problem is to make a drawing that shows the alignment of the inverted and normal chromosomes and includes the crossover site.

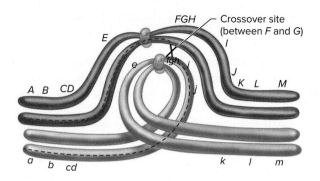

To determine the products of the crossover, start at one end of the normal (purple) chromosome (at gene *A*) and trace along the chromosome with a pencil, as shown by the red dashed line. At the crossover site, shift your pencil to the inverted (blue) chromosome. Keep your pencil going in the same direction as you go through the crossover site. (Just before the crossover site, your pencil will be going away from the centromere. Keep it going away from the centromere when you shift to the blue chromosome.) When your pencil gets to the end of the blue chromosome, it will have traced one of the chromosomes that is produced, which has a duplication and a deletion. Next, start at the other end of the normal chromosome and do the same thing. This will yield the other chromosome, which has a different duplication and a deletion.

ANSWER: The resulting products are four chromosomes. One chromosome is normal, one has an inversion, and two have a duplication and a deletion. The two chromosomes with duplicated and deleted parts are shown here:

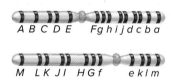

3. In humans, the number of chromosomes per set equals 23. Even though the following conditions are lethal, what would be the total number of chromosomes for an individual with each condition?

A. Trisomy 22
B. Monosomy 11
C. Triploidy

T **OPIC: *What topic in genetics does this question address?*** The topic is variation in chromosome number. More specifically, the question is about predicting the number of chromosomes in individuals with different types of chromosomal abnormalities.

I **NFORMATION: *What information do you know based on the question and your understanding of the topic?*** From the question, you know that humans have 23 chromosomes per set, and you are given the chromosomal composition for three conditions that are due to variation in chromosome number. From your understanding of the topic, you may remember that humans are diploid and you may recall the consequences of trisomy, monosomy, and triploidy.

PROBLEM-SOLVING **S**TRATEGIES: *Define key terms. Make a calculation.* One strategy to begin to solve this problem is to define trisomy, monosomy, and triploidy. If *n* represents one set of chromosomes, trisomy is $2n + 1$, monosomy is $2n - 1$, and triploidy is $3n$. To calculate the numbers of chromosomes in three individuals with these conditions, you substitute 23 for *n*. For example, $2n + 1$ is $2(23) + 1$, which equals 47.

ANSWER:

A. 47 (the diploid number, 46, plus 1)

B. 45 (the diploid number, 46, minus 1)

C. 69 (3 times 23)

---

4. A diploid species with 44 chromosomes (i.e., 22 per set) is crossed to another diploid species with 38 chromosomes (i.e., 19 per set). How many chromosomes are found in an allodiploid or allotetraploid offspring from this cross? Would you expect the offspring to be sterile or fertile?

TOPIC: *What topic in genetics does this question address?* The topic is variation in chromosome number. More specifically, the question is about calculating the number of chromosomes in different types of alloploids, and then predicting if those alloploids would be sterile or fertile.

INFORMATION: *What information do you know based on the question and your understanding of the topic?* From the question, you know that both species are diploid; one has 22 chromosomes per set and the other has 19. From your understanding of the topic, you may remember that allodiploids have one set of chromosomes from each species and that allotetraploids have two sets from each species. Also, you may recall that individuals with even numbers of homologous chromosomes are usually fertile.

PROBLEM-SOLVING **S**TRATEGIES: *Define key terms. Make a calculation. Predict the outcome.* One strategy to begin to solve this problem is to define *allodiploid* and *allotetraploid*. As noted in Section 8.8, an allodiploid has one set of chromosomes from each species and an allotetraploid has two sets from each species. If n_1 represents a set of chromosomes from one species and n_2 represents a set of chromosomes from another species, an allodiploid will be $n_1 + n_2$, whereas an allotetraploid will be $2n_1 + 2n_2$. The allotetraploid will have its chromosomes in homologous pairs.

ANSWER: The allodiploid will have $22 + 19 = 41$ chromosomes. During meiosis in the allodiploid, not all of the chromosomes will have similar partners to pair with. This unequal pairing results in aneuploidy, which usually produces inviable gametes and causes sterility. An allotetraploid will have $44 + 38 = 82$ chromosomes. Because each chromosome has a homologous partner, the allotetraploid is likely to be fertile.

Conceptual Questions

C1. Which changes in chromosome structure cause a change in the total amount of genetic material, and which do not?

C2. Explain why small deletions and duplications are less likely than large ones to have a detrimental effect on an individual's phenotype. If a small deletion within a single chromosome happens to have a phenotypic effect, what would you conclude about the genes in the region affected by the deletion?

C3. How does a chromosomal duplication occur?

C4. What is a gene family? How are gene families produced over time? With regard to gene function, what is the biological significance of a gene family?

C5. Following a gene duplication, the two genes will accumulate different mutations, causing them to have slightly different sequences. In Figure 8.7, which pair of genes would you expect to have more similar sequences: α_1 and α_2 or ψ_{α_1} and α_2? Explain your answer.

C6. Two chromosomes have the following orders of their genes:

Normal: *A B C* centromere *D E F G H I*
Abnormal: *A B G F E D* centromere *C H I*

Does the abnormal chromosome have a pericentric or a paracentric inversion? Draw a sketch showing how these two chromosomes would align with each other during prophase of meiosis I.

C7. An inversion heterozygote has the following inverted chromosome:

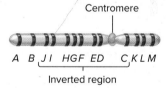

What would be the products if a crossover occurred between genes *H* and *I* on the inverted chromosome and a normal chromosome?

C8. An inversion heterozygote has the following inverted chromosome:

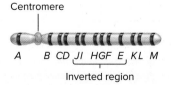

What would be the products if a crossover occurred between genes *H* and *I* on the inverted chromosome and a normal chromosome?

C9. Explain why inversions and reciprocal translocations do not usually cause a phenotypic effect. Then explain how they can do so in certain cases.

C10. An individual has the following reciprocal translocation:

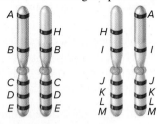

What would be the outcome of alternate segregation and of adjacent-1 segregation?

C11. A phenotypically normal individual has the following combinations of normal and abnormal chromosomes:

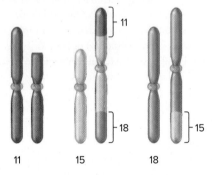

The normal chromosome is shown on the left in each pair. Suggest a series of events (breaks, translocations, crossovers, etc.) that may have produced these combinations of chromosomes.

C12. Two phenotypically normal parents produce a phenotypically abnormal child in which chromosome 5 is missing part of its long arm but has a piece of chromosome 7 attached to it. The child also has one normal copy of chromosome 5 and two normal copies of chromosome 7. With regard to chromosomes 5 and 7, what do you think are the chromosomal compositions of the parents?

C13. With regard to the proper segregation of centromeres, why is adjacent-2 segregation less frequent than alternate or adjacent-1 segregation?

C14. Which of the following types of chromosomal changes would you expect to have phenotypic consequences? Explain your choices.

   A. Pericentric inversion

   B. Reciprocal translocation

   C. Deletion

   D. Unbalanced translocation

C15. Explain why a translocation cross occurs during metaphase of meiosis I when a cell contains chromosomes with a reciprocal translocation.

C16. An individual affected with a genetic disorder has a phenotypically unaffected male parent with an inversion on one copy of chromosome 7 and a phenotypically unaffected female parent without any changes in chromosome structure. The orders of genes along the two copies of chromosome 7 in the male parent are as follows:

   $R\ T\ D\ M$ centromere $P\ U\ X\ Z\ C$ (normal chromosome 7)

   $R\ T\ D\ U\ P$ centromere $M\ X\ Z\ C$ (inverted chromosome 7)

The phenotypically affected offspring has one copy of chromosome 7 with the following order of genes:

$R\ T\ D\ M$ centromere $P\ U\ D\ T\ R$

Using a sketch, explain how this chromosome was formed. In your answer, explain where the crossover occurred (i.e., between which two genes).

C17. A diploid fruit fly has 8 chromosomes. How many chromosomes would be found in a fly with each of the following chromosomal compositions?

   A. Tetraploid

   B. Trisomy 2

   C. Monosomy 3

   D. $3n$

   E. $4n + 1$

C18. A person is born with one X chromosome, no Y chromosome, trisomy 21, and two copies of the other chromosomes. How many chromosomes does this person have altogether? Explain whether this person is euploid or aneuploid.

C19. Two phenotypically unaffected parents produce two children with familial Down syndrome. With regard to chromosomes 14 and 21, what are the chromosomal compositions of the parents?

C20. Aneuploidy is typically detrimental, whereas polyploidy is sometimes beneficial, particularly in plants. Discuss why you think this is the case.

C21. Explain how aneuploidy, deletions, and duplications cause genetic imbalances. Why do you think that deletions and monosomies are more detrimental than duplications and trisomies?

C22. Female fruit flies homozygous for the X-linked white-eye allele are crossed to males with red eyes. On very rare occasions, an offspring of such a cross is a male with red eyes. Assuming that these rare offspring are not due to a new mutation in one of the female parent's X chromosomes that converted the white-eye allele into a red-eye allele, explain how a red-eyed male arises.

C23. A cytogeneticist has collected tissue samples from members of a certain butterfly species. Some of the butterflies were located in Canada, and others were found in Mexico. Through karyotyping, the cytogeneticist discovered that chromosome 5 of the Canadian butterflies had a large inversion compared with chromosome 5 of the Mexican butterflies. The Canadian butterflies were inversion homozygotes, whereas the Mexican butterflies had two normal copies of chromosome 5.

   A. Explain whether a cross between Canadian and Mexican butterflies would produce phenotypically normal offspring.

   B. Explain whether the offspring of a cross between Canadian and Mexican butterflies would be fertile.

C24. Why do you think that humans with trisomy 13, 18, or 21 can survive but other trisomies are lethal? Even though X chromosomes are large, aneuploidy of this chromosome is also tolerated. Explain why.

C25. A zookeeper has obtained a male and a female lizard that look like they belong to the same species. They mate with each other and produce phenotypically normal offspring. However, the offspring are sterile. Suggest one or more explanations for their sterility.

C26. What is endopolyploidy? What is its biological significance?

C27. What is mosaicism? How is it produced?

C28. Explain how polytene chromosomes of *Drosophila* are produced, and describe the six-armed structure they form.

C29. Describe some of the advantages of polyploid plants. What are the consequences of having an odd number of chromosome sets?

C30. While conducting field studies on a chain of islands, you decide to karyotype turtles from two phenotypically identical groups, which are found on different islands. The turtles on one island have 24 chromosomes, but those on another island have 48 chromosomes. How would you explain this observation? How do you think the turtles with 48 chromosomes came into being? If you crossed the two types of turtles, would you expect the offspring to look phenotypically normal? Would you expect them to be fertile? Explain.

C31. A diploid fruit fly has 8 chromosomes. Which of the following terms should *not* be used to describe a fruit fly with four sets of chromosomes?

   A. Polyploid

   B. Aneuploid

   C. Euploid

   D. Tetraploid

   E. $4n$

C32. Which of the following terms should *not* be used to describe a human with three copies of chromosome 12?

   A. Polyploid

   B. Triploid

   C. Aneuploid

   D. Euploid

   E. $2n + 1$

   F. Trisomy 12

C33. The kidney bean plant, *Phaseolus vulgaris*, is a diploid species containing a total of 22 chromosomes in somatic cells. How many possible types of trisomic individuals could be produced in this species?

C34. The karyotype of a young child who is affected with familial Down syndrome revealed a total of 46 chromosomes. An older sibling, however, who is phenotypically unaffected, has 45 chromosomes. Explain how this could happen. What would you expect to be the numbers of chromosomes in the parents of these two children?

C35. A triploid plant has 18 chromosomes (i.e., 6 chromosomes per set). If you assume that a gamete has an equal probability of receiving one or two copies of each of the 6 types of chromosome, what are the odds of this plant producing a haploid or a diploid gamete? If the plant is allowed to self-fertilize, what are the odds of producing a euploid offspring?

C36. Describe three naturally occurring ways that chromosome number can change.

C37. Meiotic nondisjunction is much more likely than mitotic nondisjunction. Based on this observation, would you conclude that meiotic nondisjunction is usually due to nondisjunction during meiosis I or meiosis II? Explain your reasoning.

C38. A person who is heterozygous, *Bb*, has brown eyes; *B* (brown) is the dominant allele, and *b* (blue) is recessive. One eye of this person, however, has a patch of blue color. Give three different explanations for how this might have occurred.

C39. What is an allodiploid? What factor determines the fertility of an allodiploid? Why are allotetraploids more likely than allodiploids to be fertile?

C40. Meiotic nondisjunction usually occurs during meiosis I. What is not separating properly: bivalents or sister chromatids? What is not separating properly during mitotic nondisjunction?

C41. Table 8.3 shows that Turner syndrome occurs when an individual inherits one X chromosome but lacks a second sex chromosome. Can Turner syndrome be due to nondisjunction during oogenesis, spermatogenesis, or both? If a phenotypically normal couple has a color-blind child (due to a recessive X-linked allele) with Turner syndrome, did nondisjunction occur during oogenesis or spermatogenesis in this child's parents? Explain your answer.

C42. Male honeybees, which are haploid, produce sperm by meiosis. Explain what unusual event (compared with events of sperm production in other animals) must occur during spermatogenesis in honeybees to produce sperm. Does this unusual event occur during meiosis I or meiosis II?

Experimental Questions

E1. What is the main goal of comparative genomic hybridization? Explain how the ratio of green to red fluorescence observed along a chromosome provides information about chromosome structure.

E2. Let's suppose a researcher conducted comparative genomic hybridization (see Figure 8.9) and accidentally added twice as much DNA from noncancerous cells (labeled with red fluorescence) relative to DNA from cancer cells. What ratio of green-to-red fluorescence would you expect in a region on a chromosome from cancer cells that carried a duplication on both copies of that chromosome? What ratio would be observed in a region that was deleted on just one of the copies of that chromosome from cancer cells?

E3. With regard to the analysis of chromosome structure, explain the experimental advantage that polytene chromosomes offer. Discuss why changes in chromosome structure are more easily detected in polytene chromosomes than in ordinary (nonpolytene) chromosomes.

E4. Describe how colchicine can be used to alter chromosome number.

E5. Describe the steps you would take to produce a tetraploid plant from a diploid plant.

E6. It is an exciting time to be a plant breeder because so many options are available for the development of new types of agriculturally useful plants. Let's suppose you wish to develop a seedless tomato that can grow in a very hot climate and is resistant to a viral pathogen that commonly infects tomato plants. At your disposal, you have a seed-bearing tomato strain that is heat-resistant and produces great-tasting tomatoes. You also have a wild strain of tomato plants (which have lousy-tasting tomatoes) that are resistant to the viral pathogen. Suggest a series of steps you might follow to produce a great-tasting, seedless tomato that is resistant to heat and the viral pathogen.

E7.   What are G bands? Discuss how G bands are useful in the analysis of chromosome structure.

E8.   A female fruit fly has one normal X chromosome and one X chromosome with a deletion. The deletion occurred in the middle of the X chromosome and removed about 10% of the entire length of the X chromosome. Suppose you stained and observed the chromosomes in salivary gland cells of this female fruit fly. Draw the arm of the polytene chromosome that contains the X chromosome. Explain your drawing.

Questions for Student Discussion/Collaboration

1.   A chromosome that was involved in a reciprocal translocation also has an inversion. In addition, the cell contains two normal chromosomes.

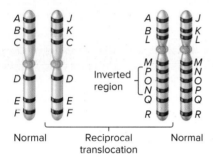

Make a drawing that shows how these chromosomes will pair during metaphase of meiosis I.

2.   Besides the ones mentioned in this textbook, look for other examples of variations in euploidy. Perhaps you might look in more advanced textbooks on population genetics, ecology, or other related fields. Discuss the phenotypic consequences of these changes.

3.   Cell biology textbooks often discuss proteins coded by genes that are members of a gene family. Examples of such proteins include myosins and glucose transporters. Look through a cell biology textbook, and identify some proteins coded by members of gene families. Discuss the importance of gene families at the cellular level.

4.   Discuss how variation in chromosome number has been useful in agriculture.

Note: All answers are available for the instructor in Connect; the answers to the even-numbered questions and all of the Concept Check and Comprehension Questions are in Appendix B.

UNIT III. MOLECULAR STRUCTURE AND REPLICATION OF THE GENETIC MATERIAL

CHAPTER OUTLINE

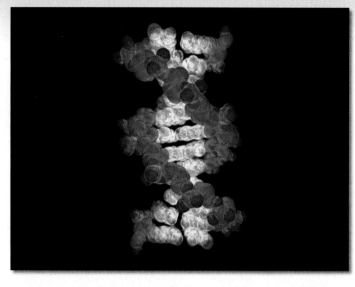

A molecular model showing the structure of the DNA double helix.
scott camazine/123RF

9

MOLECULAR STRUCTURE OF DNA AND RNA

In Chapters 2 through 8, we focused on the relationship between the inheritance of genes and chromosomes and the outcome of an organism's traits. In this chapter, we will shift our attention to **molecular genetics**—the study of DNA structure and function at the molecular level. An exciting goal of molecular genetics is to use our knowledge of DNA structure to understand how DNA functions as the genetic material. Using molecular techniques, researchers have determined the organization of many genes. This information, in turn, has helped us understand how the expression of genes governs the outcome of an individual's inherited traits.

The past several decades have seen dramatic advances in techniques and approaches used to investigate and even to alter the genetic material. These advances have greatly expanded our understanding of molecular genetics and also have provided key insights into the mechanisms underlying transmission and population genetics. Molecular genetics technology is also widely used in supporting disciplines such as biochemistry, cell biology, and microbiology.

To a large extent, our understanding of genetics comes from our knowledge of the molecular structure of **deoxyribonucleic acid (DNA)** and **ribonucleic acid (RNA).** In this chapter, we will begin by considering classic experiments that showed that DNA is the genetic material. We will then survey the molecular features of DNA and RNA that underlie their function.

9.1 IDENTIFICATION OF DNA AS THE GENETIC MATERIAL

Learning Outcomes:

1. List the four criteria that the genetic material must meet.
2. Analyze the results of (1) Griffith, (2) Avery, MacLeod, and McCarty, and (3) Hershey and Chase, and explain how they indicate that DNA is the genetic material.

To fulfill its role, the genetic material must meet four criteria:

1. *Information.* The genetic material must contain the information necessary to construct an entire organism. In other words, it must provide the blueprint for determining the inherited traits of an organism.
2. *Transmission.* During reproduction, the genetic material must be passed from parents to offspring.
3. *Replication.* Because the genetic material is passed from parents to offspring, and from mother cell to daughter cells during cell division, it must be copied.
4. *Variation.* Within any species, a significant amount of phenotypic variability occurs. For example, Mendel studied several characteristics in pea plants that varied among different strains. These included height (tall versus short) and seed color (yellow versus green). Therefore, the genetic material must also vary in ways that can account for the known phenotypic differences within each species.

In the nineteenth century, the data of many geneticists, including Mendel, were consistent with these four properties of genetic material. However, the experimental study of genetic crosses cannot, by itself, identify the chemical nature of the genetic material.

In the 1880s, August Weismann and Carl Nägeli championed the idea that a chemical substance within living cells is responsible for the transmission of traits from parents to offspring. The chromosome theory of inheritance was developed, and experimentation demonstrated that the chromosomes are the carriers of the genetic material (see Chapter 3). Nevertheless, the story was not complete because chromosomes contain both DNA and proteins. Also, RNA is associated with chromosomes. Therefore, further research was needed to precisely identify the genetic material. In this section, we will examine the first experimental attempts to achieve this goal.

Griffith's Experiments Indicated That Genetic Material Can Transform *Streptococcus*

Some early work in microbiology was important in developing an experimental strategy for identifying the genetic material. Frederick Griffith studied a type of bacterium known then as pneumococci and now classified as *Streptococcus pneumoniae*.

Certain strains of *S. pneumoniae* secrete a polysaccharide capsule, whereas other strains do not. When streaked onto petri plates containing a solid growth medium, capsule-secreting strains produce colonies with a smooth morphology, whereas those strains unable to secrete a capsule produce colonies with a rough appearance.

The different forms of *S. pneumoniae* also vary in their pathogenicity, or ability to cause disease. When smooth strains of *S. pneumoniae* infect a mouse, the capsule allows the bacteria to escape attack by the mouse's immune system. As a result, the bacteria can grow and eventually kill the mouse. In contrast, the non-encapsulated (rough) bacteria are destroyed by the animal's immune system. In 1928, Griffith conducted experiments that involved injecting live and/or heat-killed bacteria into mice, and then observing whether or not the bacteria caused a lethal infection.

Griffith was working with two strains of *S. pneumoniae*, a type S (for "smooth") and a type R (for "rough").

- When injected into a live mouse, the type S bacteria proliferated within the mouse's bloodstream and ultimately killed the mouse (**Figure 9.1a**). Following the death of the mouse, Griffith found many type S bacteria in the mouse's blood.

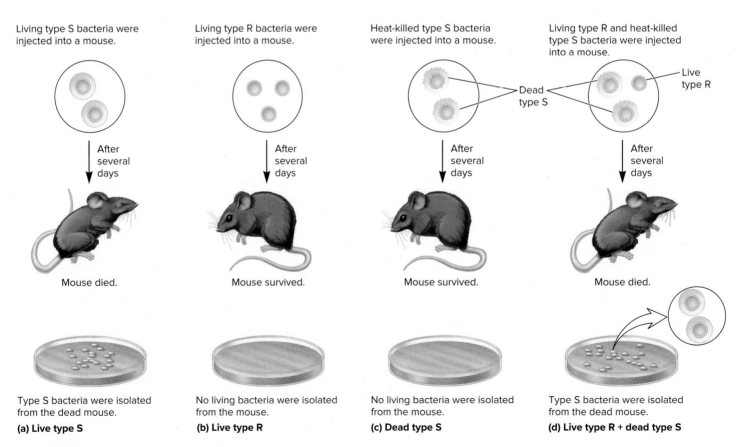

FIGURE 9.1 Griffith's experiments on genetic transformation in *S. pneumoniae*.

CONCEPT CHECK: Explain why the mouse in part (d) died.

- In contrast, when type R bacteria were injected into a mouse, the mouse lived (**Figure 9.1b**).
- To verify that the proliferation of the smooth bacteria was causing the death of the mouse, Griffith killed the smooth bacteria with heat treatment before injecting them into a mouse. In this case, the mouse also survived (**Figure 9.1c**).
- The critical and unexpected result was obtained in the experiment shown in **Figure 9.1d**. In this experiment, live type R bacteria were mixed with heat-killed type S bacteria. Several days after receiving this mixture, the mouse died. Furthermore, extracts from tissues of the dead mouse were found to contain living type S bacteria!

What can account for these results? Because living type R bacteria alone could not proliferate and kill the mouse (Figure 9.1b), the interpretation of the result in Figure 9.1d is that something from the dead type S bacteria was transforming the type R bacteria into type S bacteria. Griffith called this process **transformation,** and the unidentified substance causing this to occur was termed the transforming principle. The steps of bacterial transformation are described in Chapter 7 (refer back to Figure 7.11).

At this point, let's look at what Griffith's observations mean in genetic terms. The transformed bacteria acquired the *information* to make a capsule. Among different strains of *S. pneumoniae, variation* exists in the ability to create a capsule and to cause mortality in mice. The genetic material that is necessary to create a capsule must be *replicated* so it can be *transmitted* from mother to daughter cells during cell division. Taken together, these observations are consistent with the idea that the formation of a capsule is governed by the bacteria's genetic material, meeting the four criteria described previously. These experiments showed that some genetic material from the dead bacteria had been transferred to the living bacteria and provided them with a new trait. However, Griffith did not know what the transforming substance was.

Avery, MacLeod, and McCarty Showed That DNA Is the Substance That Transforms Bacteria

Important scientific discoveries often take place when researchers recognize that someone else's experimental observations can be used to address a particular scientific question. Oswald Avery, Colin MacLeod, and Maclyn McCarty realized that Griffith's observations could be used as part of an experimental strategy to identify the genetic material. They asked: What substance is being transferred from the dead type S bacteria to the live type R?

At the time of these experiments in the 1940s, researchers already knew that DNA, RNA, proteins, and carbohydrates are major constituents of living cells. To separate these components and to determine if any of them was the genetic material, Avery, MacLeod, and McCarty used established biochemical purification procedures and prepared extracts from type S bacterial strains that contained each type of these molecules. After many repeated attempts with the different types of extracts, they discovered that only one of the extracts, namely, the one that contained purified DNA from type S bacteria, was able to convert type R bacteria into type S. As shown in **Figure 9.2**, when this extract was mixed

with type R bacteria, some of the bacteria were converted to type S. However, if no DNA extract was added, no type S bacterial colonies were observed on the petri plates.

A biochemist might point out that a DNA extract may not be 100% pure. In fact, any purified extract might contain small traces of some other substances. Therefore, one can argue that a small amount of contaminating material in the DNA extract might actually be the genetic material. The most likely contaminating substance in this case would be RNA or protein. To further verify that the DNA in the extract was responsible for the transformation, Avery, MacLeod, and McCarty treated samples of the DNA extract with enzymes that digest DNA (called **DNase**), RNA (**RNase**), or protein (**protease**) (see Figure 9.2).

- When the DNA extracts were treated with RNase or protease, they still converted type R bacteria into type S. These results indicated that any RNA or protein in the extract was not acting as the genetic material.
- However, when the extract was treated with DNase, it lost its ability to convert type R into type S bacteria.

These results indicated that the degradation of the DNA in the extract by DNase prevented conversion of type R to type S. This interpretation is consistent with the hypothesis that DNA is the genetic material. A more elegant way of saying this is that DNA is the transforming principle.

Hershey and Chase Provided Evidence That DNA Is the Genetic Material of T2 Phage

A second set of experimental results indicating that DNA is the genetic material came from the studies of Alfred Hershey and Martha Chase in 1952. Their research centered on the study of a virus known as T2. This virus infects *Escherichia coli* bacterial cells and is therefore known as a **bacteriophage,** or simply, a **phage.** The structure of the T2 phage consists of genetic material that is packaged inside a phage coat. From a molecular perspective, this phage is rather simple, because it is composed of only two types of macromolecules: DNA and proteins. During infection, the phage coat remains attached on the outside of the bacterium and does not enter the cell. Only the genetic material of the phage enters the bacterial cell (look ahead to Figure 18.3a).

Hershey and Chase asked: What is the biochemical composition of the genetic material that enters the bacterial cell during infection? They used radioisotopes to distinguish proteins from DNA. Sulfur atoms are found in proteins but not in DNA, whereas phosphorus atoms are found in DNA but not in phage proteins. Therefore, ^{35}S (a radioisotope of sulfur) and ^{32}P (a radioisotope of phosphorus) were used to specifically label phage proteins and DNA, respectively. After phages were given sufficient time to infect bacterial cells, the researchers separated the phage coats from the bacterial cells. As shown in **Figure 9.3**, most of the ^{32}P had entered the bacterial cells, whereas most of the ^{35}S remained outside the cells. These results were consistent with the idea that the genetic material of bacteriophages is DNA, not proteins.

We now know that bacteria, archaea, protists, fungi, plants, and animals all use DNA as their genetic material. Many viruses,

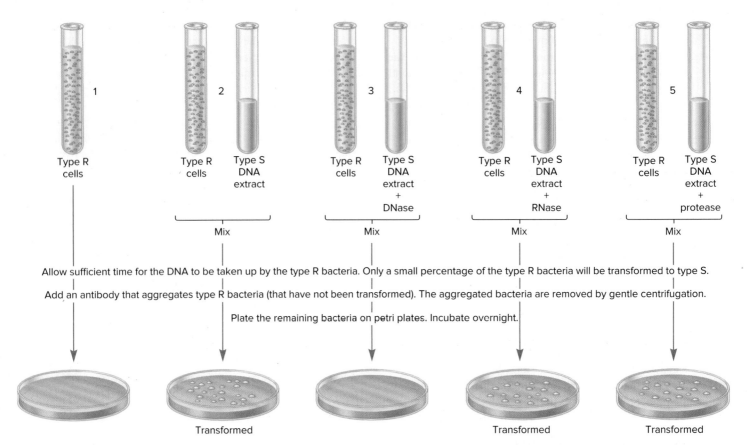

FIGURE 9.2 **Experiments of Avery, MacLeod, and McCarty to identify the transforming principle.** Samples of *S. pneumoniae* cells were either not exposed to a type S DNA extract (experiment 1, left side) or exposed to a type S DNA extract (experiments 2–5). Extracts used in experiments 3, 4, and 5 also contained DNase, RNase, or protease, respectively. After incubation, the cells were exposed to antibodies, which are molecules that can specifically recognize the molecular structure of macromolecules. In this experiment, the antibodies recognized the cell surface of type R bacteria and caused the bacteria to clump together. The clumped bacteria were removed by a gentle centrifugation step. Only the bacteria that were not recognized by the antibody (namely, the type S bacteria) remained in the supernatant. The cells in the supernatant were plated on solid growth media. After overnight incubation, visible colonies may be observed.

CONCEPT CHECK: What was the purpose of adding RNase or protease to a DNA extract?

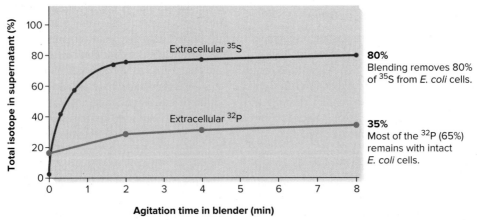

Data from A. D. Hershey and Martha Chase (1952) Independent Functions of Viral Protein and Nucleic Acid in Growth of Bacteriophage. *Journal of General Physiology 36*, 39–56.

ONLINE
ANIMATION

FIGURE 9.3   **Results from the Hershey and Chase experiment.**

such as T2, also use DNA as their genetic material. However, as discussed in Chapter 18, the genetic material of some viruses is RNA, rather than DNA.

9.1 COMPREHENSION QUESTION

1. In the experiment of Avery, McLeod, and McCarty, the addition of RNase or protease to a DNA extract
   a. prevented the conversion of type S bacteria into type R bacteria.
   b. allowed the conversion of type S bacteria into type R bacteria.
   c. prevented the conversion of type R bacteria into type S bacteria.
   d. allowed the conversion of type R bacteria into type S bacteria.

9.2 OVERVIEW OF DNA AND RNA STRUCTURE

Learning Outcomes:

1. Define *nucleic acid.*
2. Describe the four levels of complexity of DNA.

DNA and its molecular cousin, RNA, are known as **nucleic acids.** This term is derived from the discovery of DNA by Friedrich Miescher in 1869, who identified a previously unknown phosphorus-containing substance that was isolated from the nuclei of white blood cells found in waste surgical bandages. Miescher named this substance nuclein. As the structure of DNA and RNA became better understood, it was determined that they are acidic molecules, which means they release hydrogen ions (H^+) in solution and have a net negative charge at neutral pH. Thus, the name nucleic acid was coined.

Geneticists, biochemists, and biophysicists have been interested in the molecular structure of nucleic acids for several decades. Both DNA and RNA are macromolecules composed of smaller building blocks. To fully appreciate their structures, we need to consider four levels of complexity (**Figure 9.4**):

1. **Nucleotides** form the repeating structural unit of nucleic acids.
2. Nucleotides are linked together in a linear manner to form a **strand** of DNA or RNA.
3. Two strands of DNA (or sometimes strands of RNA) interact with each other to form a **double helix.**
4. The three-dimensional structure of DNA results from the folding and bending of the double helix. Within living cells, DNA is associated with a wide variety of proteins that influence its structure. Chapter 10 examines the roles of these proteins in creating the three-dimensional structure of DNA found within chromosomes.

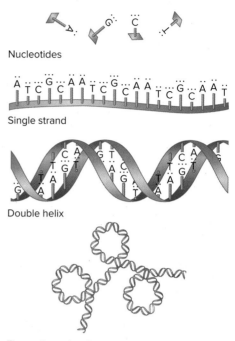

Nucleotides

Single strand

Double helix

Three-dimensional structure

FIGURE 9.4 **Levels of nucleic acid structure.**

9.2 COMPREHENSION QUESTION

1. Going from simple to complex, which of the following is the proper order for the structure of DNA?
   a. Nucleotide, double helix, DNA strand, chromosome
   b. Nucleotide, chromosome, double helix, DNA strand
   c. Nucleotide, DNA strand, double helix, chromosome
   d. Chromosome, nucleotide, DNA strand, double helix

9.3 NUCLEOTIDE STRUCTURE

Learning Outcomes:

1. Describe the structure of a nucleotide.
2. Compare and contrast the structures of nucleotides found in DNA and in RNA.

The nucleotide is the repeating structural unit of both DNA and RNA. A nucleotide has three components: a pentose sugar, a nitrogenous (nitrogen-containing) base, and at least one phosphate group. As shown in **Figure 9.5**, nucleotides vary with regard to the sugar and the nitrogenous base.

- The two types of sugars are **deoxyribose** and **ribose,** which are found in DNA and RNA, respectively.
- The five different bases are subdivided into two categories: the **purines** and the **pyrimidines.**
- The purine bases, **adenine (A)** and **guanine (G),** contain a double-ring structure; the pyrimidine bases, **thymine (T), cytosine (C),** and **uracil (U),** contain a single-ring structure.

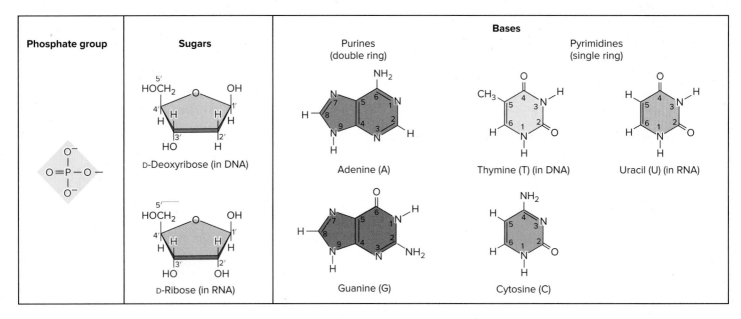

FIGURE 9.5 **The components of nucleotides.** The three building blocks of a nucleotide are a sugar, a base, and one, two, or three phosphate groups. The bases are categorized as purines (adenine and guanine) and pyrimidines (thymine, cytosine, and uracil). Note: The location of a double bond in a phosphate group is not fixed. Because the sharing of electrons between phosphorus and the oxygen atoms is delocalized, phosphate exists as multiple resonance structures.

CONCEPT CHECK: Which of these components of nucleotides are not found in DNA?

- The base thymine is found in DNA, whereas uracil is found in RNA instead of thymine. Adenine, guanine, and cytosine occur in both DNA and RNA.

As shown in Figure 9.5, the bases and sugars have a standard numbering system. The nitrogen and carbon atoms found in the ring structures of the bases are given numbers 1 through 9 for the purines and 1 through 6 for the pyrimidines. In comparison, the numbers given to the five carbons found in the sugars have primes, as in 1′, to distinguish them from the numbers used for the bases.

Figure 9.6 shows the repeating units of nucleotides found in DNA and RNA. The locations of the attachment sites of the base and phosphate to the sugar molecule are important to the nucleotide's function. In the sugar ring, carbon atoms are numbered in a clockwise direction, beginning with a carbon atom adjacent to the ring's oxygen atom. The fifth carbon is outside the ring. In a single nucleotide, the base is always attached to the 1′ carbon atom, and one or more phosphate groups are attached at the 5′ position. As discussed later, the —OH group attached to the 3′ carbon is important in allowing nucleotides to form covalent linkages with each other.

The terminology used to describe nucleic acid units is based on three structural features: the type of sugar, the type of base, and the number of phosphate groups. When a sugar is attached to only a base, this pair is a **nucleoside.** If ribose is attached to adenine, this nucleoside is called adenosine (**Figure 9.7**). Nucleosides composed of ribose and guanine, cytosine, or uracil are named guanosine, cytidine, and uridine, respectively. Nucleosides made of deoxyribose and adenine, guanine, thymine, or cytosine are called deoxyadenosine, deoxyguanosine, deoxythymidine, and deoxycytidine, respectively.

The covalent attachment of one or more phosphate molecules to a nucleoside creates a nucleotide. One or more phosphate groups are attached to a sugar via an ester bond. If a nucleotide contains ribose, adenine, and one phosphate, it is named adenosine monophosphate, abbreviated AMP. A nucleotide composed of ribose, adenine, and three phosphate groups is called adenosine triphosphate, or ATP. By comparison, a nucleotide made of deoxyribose, adenine, and three phosphate groups is referred to as deoxyadenosine triphosphate (dATP).

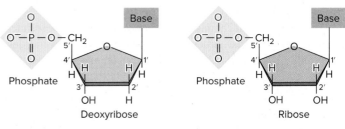

(a) Repeating unit of deoxyribonucleic acid (DNA)

(b) Repeating unit of ribonucleic acid (RNA)

FIGURE 9.6 **The structure of nucleotides found in (a) DNA and (b) RNA.** DNA contains deoxyribose as its sugar and the bases A, T, G, and C. RNA contains ribose as its sugar and the bases A, U, G, and C. In a DNA or RNA strand, the oxygen on the 3′ carbon of the sugar is linked to the phosphorus atom of the phosphate in the adjacent nucleotide. The two atoms (O and H) shown in red are found within individual nucleotides but not when nucleotides are joined together to make strands of DNA and RNA.

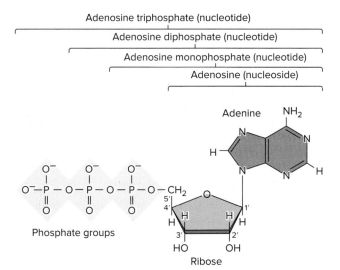

Adenosine triphosphate (nucleotide)

Adenosine diphosphate (nucleotide)

Adenosine monophosphate (nucleotide)

Adenosine (nucleoside)

Adenine

Phosphate groups

Ribose

FIGURE 9.7 **A comparison of the structures of an adenine-containing nucleoside and nucleotides.**

INTERACTIVE EXERCISE

GENETIC TIPS **THE QUESTION:** A molecule contains adenine, deoxyribose, and one phosphate. Is it a nucleoside or a nucleotide? Would it be found in DNA or RNA?

T OPIC: *What topic in genetics does this question address?* The topic is the structure of nucleosides and nucleotides, and whether they are found in DNA or RNA.

I NFORMATION: *What information do you know based on the question and your understanding of the topic?* In the question, you are given the components of a molecule. From your understanding of the topic, you may remember that a nucleoside and a nucleotide differ with regard to their phosphate content, whereas the nucleotides found in DNA and RNA differ with regard to their sugars and the presence of thymine or uracil.

P ROBLEM-SOLVING S TRATEGIES: *Define key terms. Compare and contrast.* A nucleoside is composed of a sugar and a base, whereas a nucleotide has a sugar, a base, and one or more phosphate groups. Compare the components found in DNA and RNA.

ANSWER: The molecule is a nucleotide. It is composed of a sugar, a base, and a phosphate. Because it contains deoxyribose, it could be found in DNA, but not in RNA.

9.3 COMPREHENSION QUESTIONS

1. Which of the following could be the components of a single nucleotide found in DNA?
   a. Deoxyribose, adenine, and thymine
   b. Ribose, phosphate, and cytosine
   c. Deoxyribose, phosphate, and thymine
   d. Ribose, phosphate, and uracil

2. A key difference between the nucleotides found in DNA and those in RNA is that
   a. those in DNA have phosphate, and those in RNA do not.
   b. those in DNA have deoxyribose, and those in RNA have ribose.
   c. those in DNA have thymine, and those in RNA have uracil.
   d. Both b and c are correct.

9.4 STRUCTURE OF A DNA STRAND

Learning Outcome:

1. Describe the structural features of a DNA strand.

A strand of DNA (or RNA) has nucleotides that are linked to each other in a linear fashion. **Figure 9.8** depicts a short strand

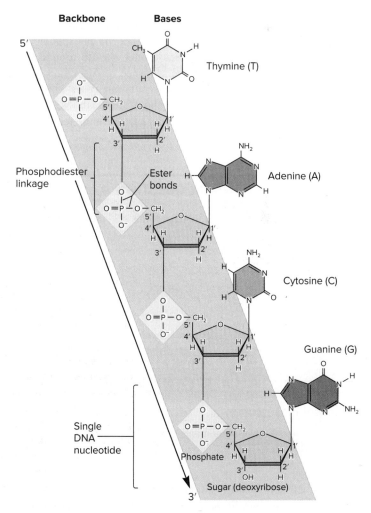

FIGURE 9.8 **A short strand of DNA containing four nucleotides.** Nucleotides are covalently linked together to form a strand of DNA.

CONCEPT CHECK: Which components of nucleotides form the backbone of a DNA strand?

of DNA with four nucleotides. A few structural features are worth noting.

- A phosphate group connects two sugar molecules via two ester bonds. For this reason, the linkage in DNA (or RNA) strands is called a **phosphodiester linkage.**
- The phosphates and sugar molecules form the **backbone** of the strand. The bases project from the backbone. The backbone is negatively charged due to the negative charge on each phosphate.
- A phosphodiester linkage involves attachment of a phosphate to the 3′ carbon in one nucleotide and to the 5′ carbon in the next nucleotide. In a strand, all sugar molecules are oriented in the same direction. Therefore, a strand has a **directionality** because all sugar molecules have the same orientation. In Figure 9.8, the direction of the strand is 5′ to 3′ proceeding from top to bottom.

A critical aspect regarding DNA and RNA structure is that a strand contains a specific sequence of bases. In Figure 9.8, the sequence of bases is thymine–adenine–cytosine–guanine, abbreviated TACG. Furthermore, to show the directionality, the abbreviation for the sequence is written 5′–TACG–3′. The nucleotides within a strand are covalently attached to each other via phosphodiester linkages, so the sequence of bases cannot shuffle around and become rearranged. Therefore, the sequence of bases in a DNA strand remains the same over time, except in rare cases when mutations occur. As we will see throughout this textbook, the sequence of bases within DNA and RNA is the defining feature that allows them to carry information.

9.4 COMPREHENSION QUESTION

1. In a DNA strand, a phosphate connects a 3′ carbon atom in one deoxyribose to
   a. a 5′ carbon in an adjacent deoxyribose.
   b. a 3′ carbon in an adjacent deoxyribose.
   c. a base in an adjacent nucleotide.
   d. none of the above.

9.5 DISCOVERY OF THE DOUBLE HELIX

Learning Outcome:

1. Outline the key experiments that led to the discovery of the DNA double helix.

A major discovery in molecular genetics was made in 1953 by James Watson and Francis Crick. At that time, DNA was already known to be composed of nucleotides. However, it was not understood how the nucleotides are bonded together to form the structure of DNA. Watson and Crick committed

themselves to determining the structure of DNA because they felt this knowledge was needed to understand the functioning of genes. Other researchers, such as Rosalind Franklin and Maurice Wilkins, shared this view. Before we examine the characteristics of the double helix, let's consider the events that provided the scientific framework for Watson and Crick's breakthrough.

A Few Key Events Led to the Discovery of the Double-Helix Structure

One method that proved important in the discovery of the structure of the DNA double helix was model building. In the early 1950s, Linus Pauling proposed that regions of proteins can fold into a secondary structure known as an α helix (**Figure 9.9a**). To elucidate this structure, Pauling built large models by linking together simple ball-and-stick units (**Figure 9.9b**). By carefully scaling the objects in these models, it was possible to visualize whether atoms fit together properly in a complex three-dimensional structure. As we will see, Watson and Crick also used ball-and-stick modeling to solve the structure of the DNA double helix. Interestingly, they were well aware that Pauling might figure out the structure of DNA before they did. This

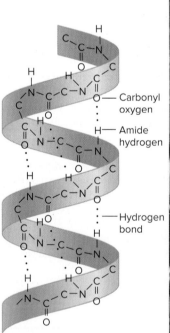

Carbonyl oxygen

Amide hydrogen

Hydrogen bond

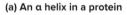

(a) An α helix in a protein   **(b) Linus Pauling**

FIGURE 9.9 **Linus Pauling and the α-helix protein structure.** **(a)** An α helix is a secondary structure found in proteins. The polypeptide backbone is shown as a tan ribbon. Hydrogen bonding between hydrogen and oxygen atoms stabilizes the helical conformation. **(b)** Pauling with a ball-and-stick model.

(b): Tom Hollyman/Science Source

stimulated a rivalry between the researchers. It is worth noting that the use of models is still an important tool in understanding structural features of biomolecules. However, modern-day molecular geneticists usually construct their three-dimensional models on computers.

A second development that led to the elucidation of the double helix was improvement in the use of X-ray diffraction. When a purified substance, such as DNA, is subjected to X-rays, it produces a well-defined diffraction pattern if the molecular structure has a repeating pattern. An interpretation of the diffraction pattern (using mathematical theory) can ultimately provide information concerning the structure of the molecule. Rosalind Franklin (**Figure 9.10a**), working in the same laboratory as Maurice Wilkins, used X-ray diffraction to study wet DNA fibers. Franklin made marked advances in X-ray diffraction techniques while working with DNA.

- The equipment was modified to produce an extremely fine beam of X-rays.
- Finer DNA fibers were extracted than ever before, and these were arranged in parallel bundles.
- The fibers' reactions to humid conditions were also studied.

A diffraction pattern of Franklin's DNA fibers is shown in **Figure 9.10b**. It suggested several key features of DNA:

- The pattern was consistent with a helical structure.
- The diameter of the helical structure was too wide to be only a single-stranded helix.
- The diffraction pattern indicated that the helix contains about 10 base pairs (bp) per complete turn.

These observations were instrumental in solving the structure of DNA.

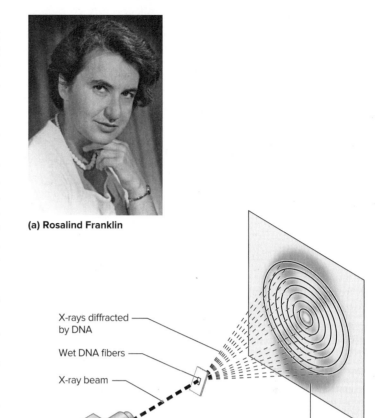

(a) Rosalind Franklin

X-rays diffracted by DNA

Wet DNA fibers

X-ray beam

The pattern represents the atomic array in wet fibers.

(b) X-ray diffraction of wet DNA fibers

FIGURE 9.10 **X-ray diffraction of DNA fibers.**
(a): World History Archive/Alamy Stock Photo

EXPERIMENT 9A

Chargaff Found That DNA Has a Biochemical Composition in Which the Amounts of A and T Are Equal, As Are the Amounts of G and C

Another piece of information that led to the discovery of the double-helix structure came from the studies of Erwin Chargaff. In the 1940s and 1950s, Chargaff pioneered many of the biochemical techniques for the isolation, purification, and measurement of nucleic acids from living cells. This was no trivial undertaking, because the biochemical composition of living cells is complex. At the time of Chargaff's work, researchers already knew that the building blocks of DNA are nucleotides containing the bases adenine, thymine, guanine, or cytosine. Chargaff analyzed the base composition of DNA from many different species, hoping that the results might provide important clues concerning the structure of DNA.

Chargaff's experimental protocol is described in **Figure 9.11**. As the starting material, samples of cells were obtained from various species. The chromosomes were extracted from cells and then treated with protease to separate the DNA from chromosomal proteins. The DNA was then treated with a strong acid, which cleaved the bonds between the sugars and bases, thereby releasing the individual bases from the DNA strands. This mixture of bases was then subjected to paper chromatography to separate the four types. The amounts of the four bases were determined spectroscopically.

THE GOAL (DISCOVERY-BASED SCIENCE)

An analysis of the base composition of DNA in different organisms may reveal important features about the structure of DNA.

FIGURE 9.11 An analysis of base composition among different DNA samples.

Starting material: The following types of cells were obtained: *Escherichia coli*, *Streptococcus pneumoniae*, yeast, turtle red blood cells, salmon sperm cells, chicken red blood cells, and human liver cells.

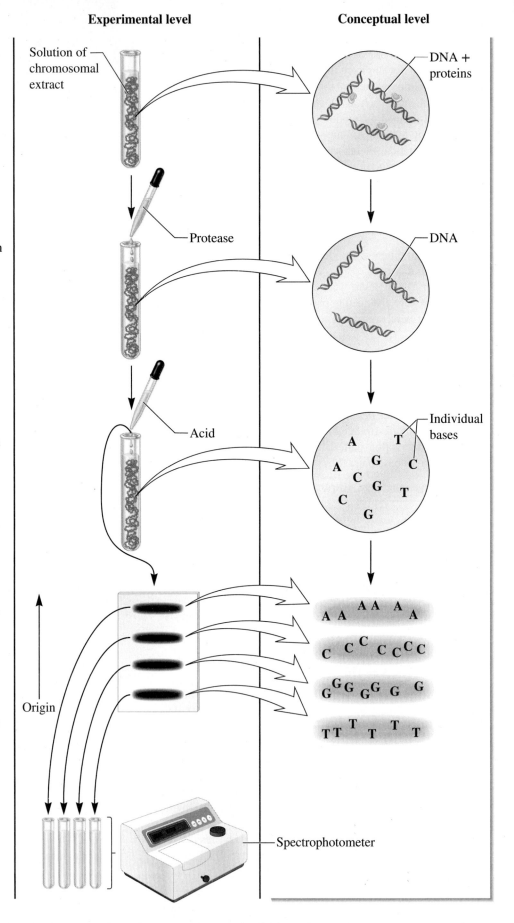

Experimental level

Conceptual level

1. For each type of cell, extract the chromosomal material. This can be done in a variety of ways, including treatment with concentrated salt solution, detergent, or mild alkali. Note: The chromosomes contain both DNA and protein.

Solution of chromosomal extract

DNA + proteins

2. Remove the protein. This can be done in several ways, including treatment with protease.

Protease

DNA

3. Hydrolyze the DNA to release the bases from the DNA strands. A common way to do this is by strong acid treatment.

Acid

Individual bases

4. Separate the bases by chromatography. Paper chromatography provides an easy way to separate the four types of bases. (The technique of chromatography is described in Appendix A.)

Origin

5. Extract bands from paper into solutions and determine the amounts of each base by spectroscopy. Each base will absorb light at a particular wavelength. By examining the absorption profile of a sample of base, it is then possible to calculate the amount of the base. (Spectroscopy is described in Appendix A.)

Spectrophotometer

6. Compare the base content in the DNA from different organisms.

THE DATA

Base Content in the DNA from a Variety of Organisms*

Organism	Percentage of Bases (Based on Molarity)			
	Adenine	Thymine	Guanine	Cytosine
E. coli	26.0	23.9	24.9	25.2
S. pneumoniae	29.8	31.6	20.5	18.0
Yeast	31.7	32.6	18.3	17.4
Turtle red blood cells	28.7	27.9	22.0	21.3
Salmon sperm	29.7	29.1	20.8	20.4
Chicken red blood cells	28.0	28.4	22.0	21.6
Human liver cells	30.3	30.3	19.5	19.9

*When the base compositions from different tissues within the same species were measured, similar results were obtained. These data were compiled from several sources. See E. Chargaff and J. Davidson, Eds. (1995) *The Nucleic Acids.* Academic Press, New York.

Source: Chargaff, Erwin, and Davidson, Norman, The Nucleic Acids. New York, NY: Academic Press, 1960.

INTERPRETING THE DATA

The data shown in the table are only a small sampling of Chargaff's results. During the late 1940s and early 1950s, Chargaff published many papers concerned with the chemical composition of DNA from biological sources. Hundreds of measurements were made. The compelling observation was that the amount of adenine was similar to that of thymine, and the amount of guanine was similar to that of cytosine. The observation that the amount of A in DNA equals the amount of T, and the amount of G equals that of C, is known as **Chargaff's rule.**

These results were not sufficient to propose a model for the structure of DNA. However, they provided the important clue that DNA is structured so that each molecule of adenine interacts with a thymine, and each molecule of guanine interacts with a cytosine. A DNA structure in which A binds to T and G binds to C would explain the equal amounts of A and T and of G and C observed in Chargaff's experiments. As we will see, this observation became crucial evidence that Watson and Crick used to elucidate the structure of the double helix.

Watson and Crick Deduced the Double-Helix Structure of DNA

Thus far, we have examined key pieces of information used to determine the structure of DNA. In particular, the X-ray diffraction work of Franklin suggested a helical structure composed of two or more strands with 10 bases per turn. In addition, Chargaff's work indicated that the amounts of A and T are equal and so are the amounts of G and C. Furthermore, Watson and Crick were familiar with Pauling's success in using ball-and-stick models to deduce the secondary structure of proteins. With these key observations, they set out to solve the structure of DNA.

Watson and Crick assumed that DNA is composed of nucleotides linked together in a linear fashion. They also assumed that the chemical linkage between two nucleotides is always the same. With these ideas in mind, they tried to build ball-and-stick models that incorporated the known experimental observations. During this time, Franklin had produced even clearer X-ray diffraction patterns, which provided greater detail concerning the relative locations of the bases and backbone of DNA. This major breakthrough suggested a two-strand interaction that was helical.

In their model building, Watson and Crick's emphasis shifted to models containing the two backbones on the outside of the model, with the bases projecting toward each other. At first, a structure was considered in which the bases form hydrogen bonds with the identical base in the opposite strand (A to A, T to T, G to G, and C to C). However, the model building revealed that the bases could not fit together this way. The final hurdle was overcome when it was realized that the hydrogen bonding of adenine to thymine was structurally similar to that of guanine to cytosine. With interactions between A and T and between G and C, the ball-and-stick model showed that

the two strands fit together properly. This ball-and-stick model, shown in **Figure 9.12**, was consistent with all of the known data regarding DNA structure.

FIGURE 9.12 Watson and Crick and their model of the DNA double helix. James Watson is on the left and Francis Crick on the right, next to the molecular model they originally proposed for the double helix. Each strand contains a sugar-phosphate backbone. Between the opposite strands, A hydrogen bonds with T, and G hydrogen bonds with C.

A. Barrington Brown/Science Source

For their work, Watson, Crick, and Wilkins were awarded the 1962 Nobel Prize in physiology or medicine. The contribution of Franklin to the discovery of the double helix was also critical and has been acknowledged in several books and articles. Franklin was independently trying to solve the structure of DNA. However, Wilkins, who worked in the same laboratory, shared Franklin's X-ray diffraction data with Watson and Crick, presumably without Franklin's knowledge. This important information helped them solve the structure of DNA, which was published in the journal *Nature* in April 1953. Franklin was not given credit in the original publication of the double-helix structure. However, the key contribution of the X-ray diffraction data became known in later years. Unfortunately, Franklin died in 1958, and so could not share in the Nobel Prize because it is not awarded posthumously.

9.5 COMPREHENSION QUESTIONS

1. Evidence that led to the discovery of the DNA double helix included
   a. the determination of structures using ball-and-stick models.
   b. the X-ray diffraction data of Franklin.
   c. the base composition data of Chargaff.
   d. all of the above.
2. Chargaff's analysis of the base composition of DNA is consistent with base pairing between
   a. A and G, and T and C.
   b. A and A, G and G, T and T, and C and C.
   c. A and T, and G and C.
   d. A and C, and T and G.

9.6 STRUCTURE OF THE DNA DOUBLE HELIX

Learning Outcomes:
1. Outline the key structural features of the DNA double helix.
2. Compare and contrast B DNA and Z DNA.

As we have seen, the discovery of the DNA double helix required different kinds of evidence. In this section, we will examine the double-helix structure in greater detail.

The Molecular Structure of the DNA Double Helix Has Several Key Features

The general structural features of the double helix are shown in **Figure 9.13**.

- In a DNA double helix, two DNA strands are twisted together around a common axis to form a structure that resembles a spiral staircase.
- This double-stranded structure is stabilized by **base pairs (bp)**—pairs of bases in opposite strands that are hydrogen

bonded to each other. If you count the bases along one strand, once you reach 10, you will have gone 360° around the axis of the helix. The linear distance along a strand through such a complete turn is 3.4 nm; each base pair traverses 0.34 nm.
- A distinguishing feature of the hydrogen bonding between base pairs is its specificity. An adenine in one strand hydrogen bonds with a thymine in the opposite strand, or a guanine hydrogen bonds with a cytosine. This **AT/GC rule** indicates that purines (A and G) always bond with pyrimidines (T and C). This pairing of bases keeps the width of the double helix relatively constant.
- Three hydrogen bonds occur between G and C but only two between A and T. For this reason, DNA sequences with a high proportion of G and C tend to form more stable double-stranded structures.
- The AT/GC rule allows us to predict the sequence in one DNA strand if the sequence in the opposite strand is known. For example, let's consider a DNA strand with the sequence 5'–ATGGCGGATTT–3'. The opposite strand has to be 3'–TACCGCCTAAA–5'. Using genetic terms, we say that these two sequences are **complementary** to each other.
- The direction of DNA strands is depicted in the inset to **Figure 9.13**. Going from the top of this figure to the bottom, one strand is running in the 5' to 3' direction, and the other strand in the 3' to 5' direction. This opposite orientation of the two DNA strands is referred to as an **antiparallel** arrangement.

Figure 9.14a is a schematic model that emphasizes certain molecular features of DNA structure. The bases in this model are depicted as flat rectangular structures that hydrogen bond to form base pairs. (The hydrogen bonds are represented by the dotted lines.) Although the bases are not actually rectangular, they do have flattened planar structures. Within DNA, the base pairs are oriented so that the flat sides of the bases are facing each other, an arrangement referred to as **base stacking.** In other words, if you think of the bases as flat plates, these plates are stacked on top of each other in the double-stranded DNA structure. Along with hydrogen bonding, base stacking is a structural feature that stabilizes the double helix by excluding water molecules, which are polar. The stability of the helical structure of the DNA backbone depends on the hydrogen bonding between bases in the base pairs and also on base stacking.

By convention, the direction of the DNA double helix shown in Figure 9.14a spirals in a direction that is called right-handed. To understand this terminology, imagine that a double helix is laid on your desk; one end of the helix is close to you, and the other end is at the opposite side of the desk. As it spirals away from you, a right-handed helix turns in a clockwise direction. By comparison, a left-handed helix spirals in a counterclockwise direction. Both strands in Figure 9.14a spiral in a right-handed direction.

Figure 9.14b is a space-filling model of DNA in which the atoms are represented by spheres. This model emphasizes the surface features of DNA. Note that the backbone—composed of

Key Features

- Two strands of DNA form a right-handed double helix.

- The bases in opposite strands hydrogen bond according to the AT/GC rule.

- The 2 strands are antiparallel with regard to their 5′ to 3′ directionality.

- There are ~10.0 nucleotides in each strand per complete 360° turn of the helix.

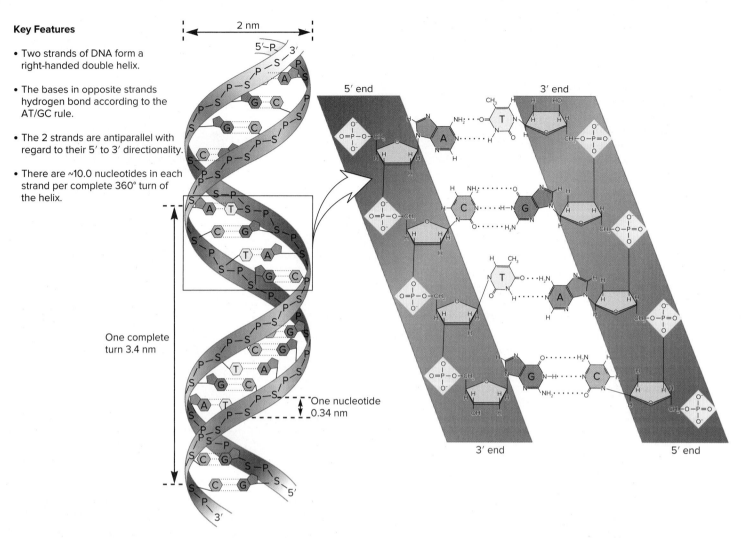

FIGURE 9.13 **Key features of the structure of the double helix.** Note: In the drawing on the left and in the inset, the planes of the bases and sugars are shown parallel to the central axis of the double helix in order to show the hydrogen bonding between the bases. In an actual DNA molecule, the bases are rotated about 90°, so their planes are perpendicular to the helix's axis, as shown in Figure 9.14a.

CONCEPT CHECK: What holds the DNA strands together?

sugar and phosphate groups—is on the outermost surface. In a living cell, the backbone has the most direct contact with water. In contrast, the bases are more internally located within the double-stranded structure. Biochemists use the term **grooves** to describe the indentations where atoms of the bases are in contact with the water in the surrounding cellular fluid. The DNA helix has two grooves winding around its outer surface: a narrow **minor groove** and a wider **major groove.**

As we will discuss in later chapters, proteins can bind to DNA and affect its conformation and function. For example, some proteins hydrogen bond to the bases within the major groove. This hydrogen bonding can be very precise, allowing a protein to interact with a particular sequence of bases. In this way, a protein can recognize a specific gene and then affect its ability to be transcribed. We will consider such proteins in Chapters 12, 14, and 15. Alternatively, other proteins bind to the DNA backbone. For example, histone proteins, which are discussed in

Chapter 10, form ionic interactions with the negatively charged phosphates in the DNA backbone. The histones are important for the proper compaction of DNA in eukaryotic cells and also play a role in gene transcription.

DNA Forms Alternative Types of Double Helices

The DNA double helix can form different types of structures. **Figure 9.15** compares the structures of **B DNA** and **Z DNA.** The highly detailed structures shown here were deduced by X-ray crystallography on short segments of DNA.

- B DNA is the predominant form of DNA in living cells, though some of that DNA is found in the Z DNA conformation.
- B DNA is a right-handed helix, whereas Z DNA is left-handed. In addition, the helical backbone in Z DNA appears to zigzag slightly as it winds itself around the double-helix structure.

FIGURE 9.14
Two models of the double helix.
(a) Ball-and-stick model of the double helix. The deoxyribose-phosphate backbone is shown in detail, whereas the bases are depicted as flattened rectangles. **(b)** Space-filling model of the double helix.

(b): Laguna Design/Science Source

CONCEPT CHECK: Describe the major and minor grooves.

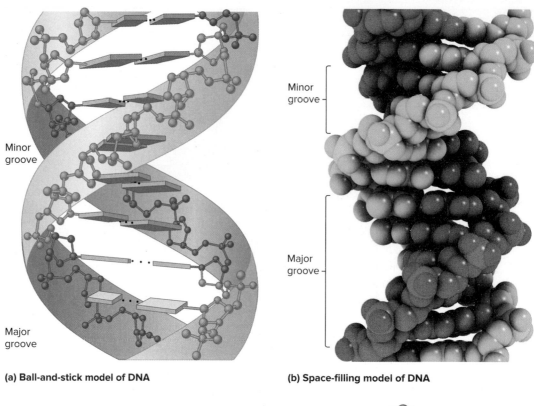

(a) Ball-and-stick model of DNA

(b) Space-filling model of DNA

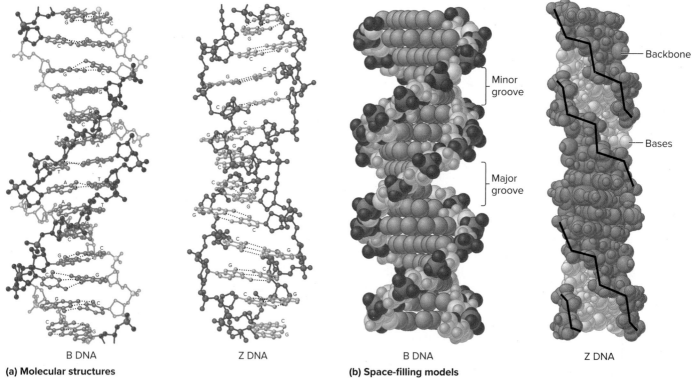

B DNA   Z DNA   B DNA   Z DNA

(a) Molecular structures   **(b) Space-filling models**

FIGURE 9.15 A comparison of the structures of B DNA and Z DNA. **(a)** The highly detailed structures shown here were deduced by X-ray crystallography performed on short segments of DNA. In contrast to the less detailed structures obtained from DNA wet fibers, the diffraction pattern obtained from the crystallization of short segments of DNA provides much greater detail concerning the exact placement of atoms within a double-helical structure. Alexander Rich, Richard Dickerson, and their colleagues were the first researchers to crystallize a short piece of DNA. **(b)** Space-filling models of the B DNA and Z DNA structures. In the case of Z DNA, the black lines connect the phosphate groups in the DNA backbone. As seen here, they travel along the backbone in a zigzag pattern.

llustration, Irving Geis. Image from Irving Geiss Collection/Howard Hughes Medical Institute. Rights owned by HHMI. Not to be reproduced without permission.

CONCEPT CHECK: What are the structural differences between B DNA and Z DNA?

- The numbers of base pairs per 360° turn are 10.0 in B DNA and 12.0 in Z DNA.
- In B DNA, the bases tend to be centrally located, and the hydrogen bonds between base pairs are oriented perpendicular to the central axis. In contrast, the bases in Z DNA are substantially tilted relative to the central axis.

Interestingly, other forms of DNA, such as A DNA, have been discovered when DNA is extracted from cells and analyzed by X-ray crystallography. The A DNA form is also a right-handed helix, but its structure (e.g., number of bases per turn) is slightly different from B. However, A DNA is not known to exist in living cells.

The ability of the predominant B DNA to adopt a Z-DNA conformation depends on various factors. At high ionic strength (i.e., high salt concentration), formation of a Z-DNA conformation is favored by a sequence of bases that alternates between purines and pyrimidines. One such sequence is

$$5'-GCGCGCGCG-3'$$
$$3'-CGCGCGCGC-5'$$

At lower ionic strength, the methylation of cytosine bases favors Z-DNA formation. Cytosine **methylation** occurs when a cellular enzyme attaches a methyl group ($-CH_3$) to the cytosine base. In addition, negative supercoiling (discussed in Chapter 10) favors the Z-DNA conformation.

What is the biological significance of Z DNA? Accumulating evidence suggests a possible role for Z DNA in the process of transcription. Research has identified cellular proteins that specifically recognize Z DNA. In 2005, Alexander Rich and colleagues reported that the Z-DNA-binding region of one such protein played a role in regulating the transcription of particular genes. In addition, other research has suggested that Z DNA may play a role in determining chromosome structure by affecting the level of compaction.

GENETIC TIPS   THE QUESTION: A double-stranded molecule of B DNA contains 340 nucleotides. How many complete turns occur in this double helix?

T OPIC: *What topic in genetics does this question address?* The topic is DNA structure. More specifically, the question is about determining the number of turns in a DNA molecule.

I NFORMATION: *What information do you know based on the question and your understanding of the topic?* From the question, you know that a DNA molecule contains 340 nucleotides. From your understanding of the topic, you may remember that B DNA is double-stranded and has 10 bp per turn.

P ROBLEM-SOLVING **S** TRATEGY: *Make a calculation.* To solve this problem, the first thing you need to do is to determine the number of base pairs. A molecule that has 340 nucleotides will contain 340/2, or 170 bp. To calculate the number of turns, you divide 170 bp by 10.

ANSWER: This DNA molecule has 17 complete turns.

9.6 COMPREHENSION QUESTIONS

1. Which of the following is *not* a feature of the DNA double helix?
   a. It obeys the AT/GC rule.
   b. The DNA strands are antiparallel.
   c. The structure is stabilized by base stacking.
   d. All of the above are features of the DNA double helix.

2. A groove in a DNA double helix refers to
   a. the indentations where the bases are in contact with the surrounding water.
   b. the interactions between bases in the DNA.
   c. the spiral structure of the DNA.
   d. all of the above.

3. A key difference between B DNA and Z DNA is that
   a. B DNA is right-handed, whereas Z DNA is left-handed.
   b. B DNA obeys the AT/GC rule, whereas Z DNA does not.
   c. Z DNA allows ribose in its structure, whereas B DNA uses deoxyribose.
   d. Z DNA allows uracil in its structure, whereas B DNA uses thymine.

9.7   RNA STRUCTURE

Learning Outcome:

1. Outline the key structural features of RNA.

Let's now turn our attention to RNA structure, which has many similarities with DNA structure. The structure of an RNA strand is much like that of a DNA strand (**Figure 9.16**). Strands of RNA are usually a few hundred to several thousand nucleotides in length—much shorter than chromosomal DNA, which is typically millions of base pairs long. When RNA is made during transcription, the DNA is used as a template. In most cases, only one of the two DNA strands is used as a template for RNA synthesis. Therefore, only one complementary strand of RNA is usually made.

Base pairing between A and U and between G and C may occur within one RNA molecule or between two different RNA molecules. This base pairing causes short segments of RNA to form a double-stranded region that is helical. As shown in **Figure 9.17**, different arrangements of base pairing are possible, which result in structures called a bulge loop, an internal loop, a multibranched junction, and a stem-loop (also called a hairpin). These structures contain regions of complementarity punctuated by regions of noncomplementarity. As shown in Figure 9.17, the complementary regions are held together by hydrogen bonds between base pairs, whereas the noncomplementary regions have their bases projecting away from the double-stranded region.

Many factors contribute to the structure of RNA molecules.

- After folding, some parts of RNA molecules may become double-stranded because the base sequences are complementary. The complementary regions are held together by

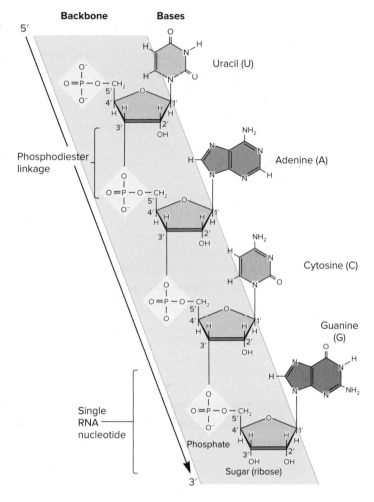

FIGURE 9.16   A strand of RNA. This structure is very similar to a DNA strand (see Figure 9.8), except that the sugar is ribose instead of deoxyribose, and uracil is substituted for thymine.

CONCEPT CHECK: What types of bonds hold nucleotides together in an RNA strand?

connecting hydrogen bonds. Other regions of RNA molecules remain single-stranded.

• The structure of RNA is stabilized by hydrogen bonding between bases in base pairs, stacking between bases, and hydrogen bonding between bases and backbone regions.

• In addition, interactions with ions, small molecules, and large proteins may influence RNA structure.

Figure 9.18 depicts the structure of a transfer RNA molecule known as tRNA^{Phe}, which is a tRNA molecule that carries the amino acid phenylalanine (Phe). It was the first naturally occurring RNA to have its structure elucidated. This RNA molecule has several double-stranded and single-stranded regions. RNA double helices are antiparallel and right-handed, with 11–12 bp per turn. In a living cell, various regions of an RNA molecule fold and interact with each other to produce the three-dimensional structure.

The folding of RNA into a three-dimensional structure is important for its function. For example, as discussed in Chapter 13, a tRNA molecule has two key functional sites—an anticodon and an acceptor site at the 3′ end—that play important roles in translation. In a folded tRNA molecule, these sites are exposed on the surface of the molecule so they can perform their roles (see Figure 9.18a). Many other examples are known in which RNA folding is key to the molecule's structure and function. These include the folding of ribosomal RNAs (rRNAs), which are important components of ribosomes, and ribozymes, which are RNA molecules with catalytic function.

9.7 COMPREHENSION QUESTION

1. A double-stranded region of RNA
   a. forms a helical structure.
   b. obeys the AU/GC rule.
   c. may result in the formation of a structure such as a bulge loop or a stem-loop.
   d. does all of the above.

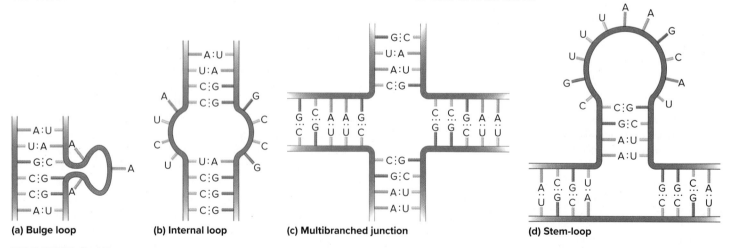

(a) Bulge loop   (b) Internal loop   (c) Multibranched junction   (d) Stem-loop

FIGURE 9.17   Possible structures of RNA molecules. The double-stranded regions are formed when two hydrogen bonds between A and U or three hydrogen bonds between G and C (represented by two or three dots, respectively) connect complementary bases. Double-stranded regions can form within a single RNA molecule or between two different RNA molecules. Note: Though not shown here, double-stranded regions are in a helical conformation.

CONCEPT CHECK: What are the base-pairing rules for RNA?

FIGURE 9.18 The structure of tRNA^Phe, a transfer RNA molecule that carries phenylalanine. (a) The double-stranded regions of the molecule are shown as antiparallel ribbons. (b) A space-filling model of tRNA^Phe.

(b): Alfred Pasieka/Science Source

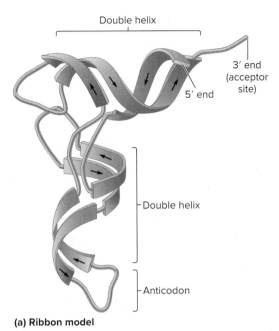
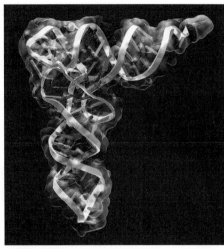

(a) Ribbon model

(b) Space-filling model

KEY TERMS

Introduction: molecular genetics, (deoxyribonucleic acid) DNA, (ribonucleic acid) RNA
9.1: transformation, DNase, RNase, protease, bacteriophage (phage)
9.2: nucleic acids, nucleotides, strand, double helix
9.3: deoxyribose, ribose, purine, pyrimidine, adenine (A), guanine (G), thymine (T), cytosine (C), uracil (U), nucleoside

9.4: phosphodiester linkage, backbone, directionality
9.5: Chargaff's rule
9.6: base pairs (bp), AT/GC rule, complementary, antiparallel, base stacking, grooves, minor groove, major groove, B DNA, Z DNA, methylation

---

CHAPTER SUMMARY

- Molecular genetics is the study of DNA structure and function at the molecular level.

9.1 Identification of DNA as the Genetic Material

- To fulfill its role, genetic material must meet four criteria: information, transmission, replication, and variation.
- Griffith showed that the genetic material from type S bacteria could transform type R bacteria into type S (see Figure 9.1).
- Avery, MacLeod, and McCarty discovered that the transforming substance is DNA (see Figure 9.2).
- Hershey and Chase determined that the genetic material of T2 phage is DNA (see Figure 9.3).

9.2 Overview of DNA and RNA Structure

- DNA and RNA are types of nucleic acids.
- In DNA, nucleotides are linked together to form strands, which then form a double helix that is found within chromosomes (see Figure 9.4).

9.3 Nucleotide Structure

- A nucleotide is composed of one or more phosphates, a sugar, and a base. The purine bases are adenine and guanine, and the pyrimidine bases are thymine (DNA only), cytosine, and uracil (RNA only) (see Figures 9.5, 9.6, 9.7).

9.4 Structure of a DNA Strand

- In a DNA strand, nucleotides are covalently attached to one another via phosphodiester linkages (see Figure 9.8).

9.5 Discovery of the Double Helix

- Pauling used ball-and-stick models to deduce the structure of an α helix in a protein (see Figure 9.9).
- Franklin performed X-ray diffraction studies that helped to determine the structure of DNA (see Figure 9.10).
- Chargaff determined that, in DNA, the amounts of A and T are equal, and so are the amounts of G and C (see Figure 9.11).

- Watson and Crick deduced the structure of DNA by building a model that was consistent with the available data (see Figure 9.12).

9.6 Structure of the DNA Double Helix

- DNA is a right-handed double helix in which adenine (A) hydrogen bonds with thymine (T) and guanine (G) hydrogen bonds with cytosine (C). The two strands are antiparallel and contain about 10 bp per turn (see Figure 9.13).
- Base stacking stabilizes the DNA double helix by excluding water. The double helix of DNA has a major groove and a minor groove (see Figure 9.14).

- B DNA is the major form of DNA found in living cells. Z DNA is an alternative conformation of DNA (see Figure 9.15).

9.7 RNA Structure

- RNA is composed of a strand of nucleotides in which the sugar is ribose and uracil is substituted for thymine (see Figure 9.16).
- RNA can form double-stranded helical regions and fold into a three-dimensional structure (see Figures 9.17, 9.18).

PROBLEM SETS & INSIGHTS

MORE GENETIC TIPS

1. A hypothetical base sequence of an RNA molecule is

5′–AUUUGCCCUAGCAAACGUAGCAAACG–3′

Using two of the three underlined parts of this sequence, draw two possible stem-loop structures that might form in this RNA.

T OPIC: *What topic in genetics does this question address?* The topic is RNA structure. More specifically, the question is about the ability of RNA to form stem-loop structures.

I NFORMATION: *What information do you know based on the question and your understanding of the topic?* From the question, you know the base sequence of part of an RNA molecule and are given a choice regarding parts of that sequence that could interact to form stem-loops. From your understanding of the topic, you may remember that RNA molecules can form double-stranded regions that are antiparallel and complementary.

P ROBLEM-SOLVING **S** TRATEGY: *Make a drawing.* One strategy to solve this problem is to make a drawing that shows how parts of the given base sequence can bind to each other because they are antiparallel and complementary.

ANSWER:

2. Within living cells, many different proteins play important functional roles by binding to DNA. Some proteins bind to DNA but not in a sequence-specific manner. For example, histones are proteins important in the formation of chromosome structure. The positively charged histone proteins bind to the negatively charged phosphate groups in DNA. In addition, several other proteins interact with DNA but do not require a specific nucleotide sequence to carry out their function. For example, DNA polymerase, which catalyzes the synthesis of new DNA strands, does not bind to DNA in a sequence-dependent way. By comparison, many other proteins do interact with nucleic acids in a sequence-dependent fashion. This means that a specific sequence of bases can provide a structure that is recognized by a particular protein. Throughout this textbook, the functions of many of these proteins will be described. Some examples include transcription factors that affect the rate of gene transcription and proteins that bind to origins of replication in bacteria. With regard to the three-dimensional structure of DNA, where would you expect DNA-binding proteins to bind if they recognize a specific base sequence? What about DNA-binding proteins that do not recognize a base sequence?

T OPIC: *What topic in genetics does this question address?* The topic is DNA-binding proteins. More specifically, the question is about deciding where on a DNA molecule a protein will bind if it recognizes a specific base sequence, or if it does not.

I NFORMATION: *What information do you know based on the question and your understanding of the topic?* From the question, you know that proteins can bind to specific places on a DNA molecule, such as the backbone or a particular sequence of bases. From your understanding of the topic, you may remember that the bases are accessible in the major and minor grooves.

P ROBLEM-SOLVING **S** TRATEGY: *Relate structure and function.* One strategy to solve this problem is to relate the function of a protein to the structure of DNA. Some proteins function by binding to a base sequence in the DNA. Others bind to the DNA but do not recognize a specific sequence of bases.

ANSWER: DNA-binding proteins that recognize a base sequence must bind to it in the major or minor groove of the DNA, which is where the bases are accessible to a DNA-binding protein. Most DNA-binding proteins, which recognize a base sequence, fit into the major groove. By comparison, other DNA-binding proteins, such as histones, which do not recognize a base sequence, typically bind to the DNA backbone.

3. As described in Experiment 9A (see Figure 9.11), Chargaff determined the base composition of DNA from a variety of different sources. Explain how the data are consistent with the AT/GC rule.

T OPIC: *What topic in genetics does this question address?* The topic is Chargaff's experiments and how they relate to the AT/GC rule.

I NFORMATION: *What information do you know based on the question and your understanding of the topic?* In the question, you are reminded that Chargaff determined the base composition using DNA from a variety of sources. The data are shown with Figure 9.11. From your understanding of the topic, you may remember that, in DNA, A binds to T and G binds to C. This observation was the basis of the AT/GC rule.

P ROBLEM-SOLVING S TRATEGIES: *Analyze data. Compare and contrast.* To solve this problem, you need to take a look at the data with Figure 9.11. If you compare the numbers in each row and contrast them with each other, you will notice that the amount of A is roughly the same as the amount of T and the amount of G is roughly the same as the amount of C.

ANSWER: Because the amount of A equals that of T and the amount of G equals that of C, the data are consistent with a DNA structure in which A binds to T and G binds to C.

Conceptual Questions

C1. What is the meaning of the term *genetic material*?

C2. After type R bacteria are exposed to DNA from type S bacteria, list all of the steps that you think must occur for the type R bacteria to start making a capsule and thereby become type S.

C3. Look up the meaning of the word *transformation* in a dictionary, and explain whether it is an appropriate word to describe the transfer of genetic material from one organism to another.

C4. What are the building blocks of a nucleotide? With regard to the 5′ and 3′ positions on a sugar molecule, how are nucleotides linked together to form a strand of DNA?

C5. Draw the structures of guanine, guanosine, and deoxyguanosine triphosphate.

C6. Draw the structure of a phosphodiester linkage.

C7. Describe how bases interact with each other in the double helix. This description should include the concepts of complementarity, hydrogen bonding, and base stacking.

C8. If one DNA strand is 5′–GGCATTACACTAGGCCT–3′, what is the sequence of the complementary strand?

C9. What is meant by the term *DNA sequence*?

C10. Make a side-by-side drawing of two DNA helices: one with 10 bp per 360° turn and the other with 15 bp per 360° turn.

C11. Discuss the differences in the structural features of B DNA and Z DNA.

C12. What part(s) of a nucleotide (namely, phosphate, sugar, and/or base) is/are accessible in the major and minor grooves of double-stranded DNA, and what part(s) is/are found in the DNA backbone? If a DNA-binding protein does not recognize a specific nucleotide sequence, do you expect that it binds in the major groove, in the minor groove, or to the DNA backbone? Explain.

C13. List the structural differences between DNA and RNA.

C14. Draw the structure of deoxyribose and number the carbon atoms. Describe the numbering of the carbon atoms in deoxyribose with regard to the directionality of a DNA strand. With regard to a DNA double helix, what does the term *antiparallel* mean?

C15. Write a base sequence for an RNA molecule that could form a stem-loop with 24 nucleotides in the stem and 16 nucleotides in the loop.

C16. Compare the structural features of a double-stranded RNA molecule with those of a DNA double helix.

C17. Which of the following DNA double helices would be more difficult to separate into single-stranded molecules by treatment with heat, which breaks hydrogen bonds?

A. GGCGTACCAGCGCAT
   CCGCATGGTCGCGTA

B. ATACGATTTACGAGA
   TATGCTAAATGCTCT

Explain your choice.

C18. What structural feature allows DNA to store information?

C19. Discuss the structural significance of complementarity in DNA and in RNA.

C20. The total amount of G plus C in an organism's DNA is 64% of the total base content of that DNA. What are the percentages of A, T, G, and C in the DNA?

C21. A DNA-binding protein recognizes the following double-stranded sequence:

5′–GCCCGGGC–3′
3′–CGGGCCCG–5′

This type of double-stranded structure could also occur within the stem region of an RNA stem-loop. Discuss the structural differences between RNA and DNA that might prevent the DNA-binding protein from recognizing a double-stranded RNA molecule.

C22. Within a protein, certain amino acids are positively charged (e.g., lysine and arginine), some are negatively charged (e.g., glutamate and aspartate), some are polar but uncharged, and some are nonpolar. If you knew that a DNA-binding protein was recognizing the DNA backbone rather than a base sequence, which amino acids in the protein would be good candidates for interacting with the DNA?

C23. In what ways are the structures of an α helix in a protein and the double helix of DNA similar, and in what ways are they different?

C24. A double-stranded DNA molecule contains 560 nucleotides. How many complete turns occur in this double helix?

C25. As the minor and major grooves wind around a DNA double helix, do they ever intersect each other, or do they always run parallel to each other?

C26. What chemical group (phosphate group, hydroxyl group, or a nitrogenous base) is found at the 3′ end of a DNA strand? What group is found at the 5′ end?

C27. The base composition of an RNA virus was analyzed and found to be 14.1% A, 14.0% U, 36.2% G, and 35.7% C. Would you conclude that this viral genetic material is single-stranded RNA or double-stranded RNA?

C28. The genetic material found within some viruses is single-stranded DNA. Would this genetic material contain equal amounts of A and T and equal amounts of G and C?

C29. A medium-sized human chromosome contains about 100 million bp. If the DNA of this chromosome was stretched out in a linear manner, how long would it be?

C30. A double-stranded DNA molecule is 1 cm long, and the percentage of adenine it contains is 15%. How many cytosines does this DNA molecule contain?

C31. Could single-stranded DNA form a stem-loop structure? Why or why not?

C32. As described in Chapter 15, the methylation of cytosine bases can have an important effect on gene expression. For example, such

methylation may inhibit the transcription of genes. A methylated cytosine base has the following structure:

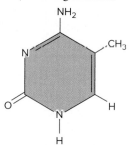

Would you expect the methylation of cytosine to affect the hydrogen bonding between cytosine and guanine in a DNA double helix? Why or why not? (Hint: See Figure 9.13 for help.) Look back at question 2 in More Genetic TIPS and speculate as to how methylation could affect gene expression.

C33. An RNA molecule has the following sequence:

Region 1	Region 2	Region 3

5′–CAUCC<u>AUCCAUUCCCC</u>AUCCGAUAA<u>GGGGA</u>AUGG<u>A</u>UCC<u>GAAUGGAU</u>AAC–3′

Parts of region 1 can form a stem-loop with region 2 and with region 3. Can region 1 form a stem-loop with region 2 and region 3 at the same time? Why or why not? Which stem-loop would you predict to be more stable: one due to a region 1/region 2 interaction or one due to a region 1/region 3 interaction? Explain your choice.

Experimental Questions

E1. Genetic material acts as a blueprint for an organism's traits. Explain how Griffith's experiments indicated that genetic material was being transferred to the type R bacteria.

E2. With regard to the experiment shown in Figure 9.2, answer the following:

A. List several possible reasons why only a small percentage of the type R bacteria were converted to type S.

B. Explain why an antibody was used to remove the bacteria that were not transformed. What would the results look like, in all five cases, if the antibody and centrifugation steps had not been included in the experimental procedure?

C. The DNA extract was treated with DNase, RNase, or protease. Why was this done? In other words, what were the researchers trying to determine?

E3. An interesting trait that some bacteria exhibit is resistance to being killed by antibiotics. For example, certain strains of bacteria are resistant to tetracycline, whereas other strains are sensitive to tetracycline. Describe an experiment you would carry out to

demonstrate that tetracycline resistance is an inherited trait coded by the DNA of the resistant strain.

E4. The type of model building used by Pauling and by Watson and Crick involved the use of ball-and-stick units. Model building can now be done with computer software. Even though you may not be familiar with this approach, discuss potential advantages of using computers in molecular model building.

E5. With regard to Chargaff's experiment described in Figure 9.11, answer the following:

A. What is the purpose of using paper chromatography?

B. Explain why it is necessary to remove the bases in order to determine the base composition of DNA.

C. Would Chargaff's experiments have been convincing if they had been done on DNA from only one species? Discuss.

E6. Alfred Gierer and Gerhard Schramm exposed plant tissue to purified RNA from tobacco mosaic virus, and the plants developed the same types of lesions as if they had been exposed to the virus itself. What would be the results if the RNA was treated with DNase, RNase, or protease prior to its exposure to the plant tissue?

Questions for Student Discussion/Collaboration

1. Try to propose structures for a genetic material that are substantially different from the double helix. Remember that the genetic material must have a way to store information and a way to be faithfully replicated.

2. How might you provide evidence that DNA is the genetic material in mice?

Note: All answers are available for the instructor in Connect; the answers to the even-numbered questions and all of the Concept Check and Comprehension Questions are in Appendix B.

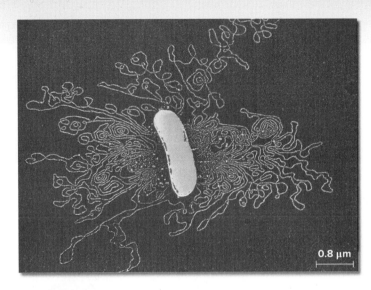

Structure of a bacterial chromosome. Electron micrograph of a bacterial chromosome, which has been released from a bacterial cell.
Dr. Gopal Murti/Science Source

10 MOLECULAR STRUCTURE OF CHROMOSOMES AND TRANSPOSABLE ELEMENTS

Chromosomes are the structures within living cells that contain the genetic material. The term **genome** refers to the entire complement of genetic material in an organism or species. For prokaryotes, the genome is typically a single circular chromosome. For eukaryotes, genetic material is found in different cellular compartments. The nuclear genome in humans includes 22 autosomes, the X chromosome, and (in males) the Y chromosome. Eukaryotes also have a mitochondrial genome, and plants have a chloroplast genome.

The primary function of the genetic material is to store the information needed to produce the characteristics of an organism. As we saw in Chapter 9, the sequence of bases in a DNA molecule stores information. To fulfill their role at the molecular level, chromosomal sequences facilitate four important processes:

- the synthesis of RNA and cellular proteins
- the replication of chromosomes
- the proper segregation of chromosomes
- the compaction of chromosomes so that they fit within living cells

In this chapter, we will examine key features of prokaryotic and eukaryotic chromosomes. First, we will consider the general organization of functional sites along a chromosome. Second, we will examine the molecular structures of chromosomes. Third, we will explore a process called transposition, in which short segments of DNA, called **transposable elements (TEs),** are able to move to multiple sites within chromosomes and accumulate in large numbers. Finally, we will examine the molecular mechanisms that make chromosomes more compact.

10.1 ORGANIZATION OF FUNCTIONAL SITES ALONG PROKARYOTIC CHROMOSOMES

Learning Outcome:

1. Describe the general organization of functional sites along bacterial and archaeal chromosomes.

Both bacteria and archaea are considered prokaryotes. They are mostly unicellular species that lack the complex cellular compartmentalization found in eukaryotes. A hallmark characteristic

FIGURE 10.1 **Organization of sequences in prokaryotic chromosomal DNA.**

CONCEPT CHECK: What types of sequences constitute most of a prokaryotic genome?

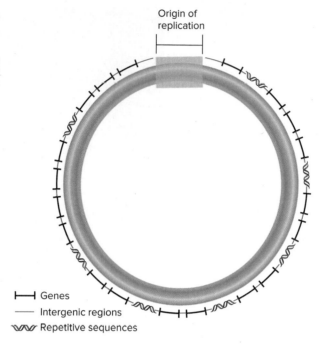

Origin of replication

⊢⊣ Genes
— Intergenic regions
ᔜᔜ Repetitive sequences

Key features:

• Most, but not all, prokaryotic species contain circular chromosomal DNA.

• Most prokaryotic species contain a single type of chromosome, but it may be present in multiple copies.

• A typical chromosome is a few million base pairs in length.

• Several thousand different genes are interspersed throughout the chromosome. The short regions between adjacent genes are called intergenic regions.

• At least one origin of replication is required to initiate DNA replication.

• Repetitive sequences may be interspersed throughout the chromosome.

of prokaryotic cells is that they lack a cell nucleus that is bounded by a double membrane.

Figure 10.1 shows the general features of a prokaryotic chromosome.

• In most species, the chromosomal DNA is a circular molecule, though some species have linear chromosomes (Figure 10.1). Although bacteria and archaea usually contain a single type of chromosome, more than one copy of that chromosome may be found within one cell.

• A typical prokaryotic chromosome is a few million base pairs (bp) in length. For example, the chromosome of *Escherichia coli* has approximately 4.6 million bp, and that of *Haemophilus influenzae* has roughly 1.8 million bp. A prokaryotic chromosome commonly has a few thousand different genes, which are interspersed throughout the entire chromosome. **Protein-coding genes** account for the majority of the chromosomal DNA. The nontranscribed regions of DNA located between adjacent genes are termed **intergenic regions.**

• Bacterial chromosomes usually have one **origin of replication** as do some archaeal species (see Figure 10.1), though other archaea have multiple origins. The origin is a DNA sequence that is a few hundred base pairs in length. This nucleotide sequence functions as an initiation site for the assembly of several proteins required for DNA replication. However, the mechanism for how DNA replication is initiated in bacteria is different from how it starts in archaea (see Chapter 11).

• A wide variety of **repetitive sequences** have been identified in prokaryotic species. These sequences are found in multiple copies and are usually interspersed within the intergenic regions throughout the chromosome. Repetitive

sequences sometimes play a role in a variety of genetic processes, including DNA folding, gene regulation, and genetic recombination. As discussed in Section 10.5, some repetitive sequences are transposable elements that can move throughout the genome.

10.1 COMPREHENSION QUESTION

1. A bacterial chromosome typically contains
   a. a few thousand genes.
   b. one origin of replication.
   c. some repetitive sequences.
   d. all of the above.

10.2 STRUCTURE OF PROKARYOTIC CHROMOSOMES

Learning Outcomes:

1. Outline the processes that make a prokaryotic chromosome more compact.
2. Describe how DNA gyrase causes DNA supercoiling.

Inside a prokaryotic cell, a chromosome is highly compacted and found within a region of the cell known as a **nucleoid.** Depending on the growth conditions and phase of the cell cycle, prokaryotes typically have one to four identical chromosomes per cell. In addition, the number of copies varies depending on the species. Having

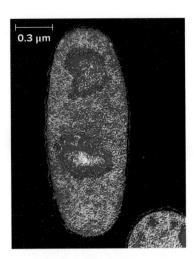

FIGURE 10.2 **The localization of nucleoids within the bacterium *Bacillus subtilis*.** The nucleoids are fluorescently labeled, so they appear as bright blue regions within the bacterial cytoplasm. Note that two or more nucleoids may be found within a cell.

M. Wurtz/Biozentrum, University of Basel/Science Source

CONCEPT CHECK: How many nucleoids are in this bacterial cell?

multiple nucleoids may enhance a cell's ability to synthesize more proteins.

Figure 10.2 is a micrograph in which the nucleoids are fluorescently labeled in blue. Each chromosome is found within its own distinct nucleoid in the cell. Unlike the eukaryotic nucleus, the nucleoid is not a separate cellular compartment surrounded by a membrane. Rather, the DNA in a nucleoid is in direct contact with the cytoplasm of the cell. In this section, we will explore the structure of prokaryotic chromosomes and the processes by which they are compacted to fit within a nucleoid.

The Formation of Chromosomal Loops Helps Make the Bacterial Chromosome More Compact

To fit within the average-sized prokaryotic cell, the chromosomal DNA must be compacted about 1000-fold. The mechanism of bacterial chromosome compaction is not entirely understood, and it may vary among different species. **Figure 10.3** shows a schematic drawing of a chromosome that has been removed from an *E. coli* cell. As the drawing shows, the chromosome has a central core with many loops emanating from the core.

- Based on microscopy studies, the loops that emanate from the core, which are called **microdomains,** are typically 10,000 base pairs (10 kbp) in length. An *E. coli* chromosome is expected to have about 400 to 500 microdomains. The lengths and boundaries of these microdomains are thought to be dynamic, changing in response to environmental conditions.
- In *E. coli*, many adjacent microdomains are further organized into **macrodomains** that are about 800 to 1000 kbp in length; each macrodomain contains about 80 to 100 microdomains. One proposed model suggests that the *E. coli* chromosome has 4 macrodomains and two nonstructured

FIGURE 10.3 **Core and microdomains of a bacterial chromosome.** This is a schematic drawing of an *E. coli* chromosome that has been extracted from a cell and viewed by electron microscopy. The core is in the center with many loops (microdomains) emanating from it. Not all bacterial species have their chromosomes organized into microdomains and macrodomains.

Source: Wang, Xindan, Llopis, Paula Montero, and Rudner, David Z., "Organization and Segregation of Bacterial Chromosomes," *Nature Reviews Genetics*, vol. 14, no. 3, March, 2013, 191–203.

regions. The existence of macrodomains was first determined by measuring the frequency of recombination (crossing over) between particular sites scattered throughout the *E. coli* chromosome. Recombination is much more frequent between sites within a macrodomain than between sites in different macrodomains. Therefore, macrodomains are also called chromosomally interacting domains (CIDs). Because the identification of macrodomains is not based on electron microscopy, they are not evident in Figure 10.3.

To form microdomains and macrodomains, bacteria use a set of DNA-binding proteins called **nucleoid-associated proteins (NAPs)** that facilitate chromosome compaction and organization. These proteins either bend the DNA or act as bridges that cause different regions of DNA to bind to each other. NAPs also facilitate chromosome segregation and play a role in gene regulation. Examples of NAPs include <u>h</u>istone-like <u>n</u>ucleoid <u>s</u>tructuring (H-NS) proteins and <u>s</u>tructural <u>m</u>aintenance of <u>c</u>hromosomes (SMC) proteins. SMCs are also found in eukaryotes, and later in this chapter, we will examine how they tether segments of DNA to each other (look ahead to Figure 10.21).

GENETIC TIPS

THE QUESTION: As noted in Chapter 9, 1 bp of DNA is approximately 0.34 nm in length. A bacterial chromosome is about 4 million bp in length. The dimensions of the cytoplasm of a bacterium, such as *E. coli*, are roughly 0.5 μm wide and 1.0 μm long.

A. A microdomain is a loop of DNA that contains about 10,000 bp. If it was stretched out linearly, how long (in micrometers) would a microdomain be?

B. If a bacterial microdomain was circular, what would be its diameter? (Note: Circumference = πD, where D is the diameter of the circle.)

C. Is the diameter of the circular microdomain calculated in part B small enough to fit inside a bacterium?

T OPIC: *What topic in genetics does this question address?* The topic is the dimensions of a bacterial chromosome. More specifically, the question asks you to calculate the dimensions of a microdomain.

I NFORMATION: *What information do you know based on the question and your understanding of the topic?* In the question, you are reminded that the length of 1 bp of DNA is about 0.34 nm and that a bacterial chromosome is about 4 million bp in length. One microdomain is a loop with about 10,000 bp. You are also told that the bacterial cytoplasm is about 0.5 μm wide and 1.0 μm long and given the equation for calculating the circumference of a circle.

P ROBLEM-SOLVING **S** TRATEGIES: *Make a calculation. Compare and contrast.* For part A, you simply multiply 10,000 by 0.34 nm, which is the length of 1 bp. For part B, you use the equation that is given. The circumference is the linear length of the DNA. For part C, you compare the answer to part B to the dimensions of the bacterial cytoplasm.

ANSWER:

A. One microdomain is 10,000 bp. One base pair is 0.34 nm, which equals 0.00034 μm. You multiply these two numbers:

(10,000) (0.00034 μm) = 3.4 μm

B. Circumference = πD

3.4 μm = πD

D = 1.1 μm

C. The diameter is a little too big to fit inside a bacterium such as *E. coli*. NAPs and supercoiling make the microdomains much more compact so that a single chromosome can occupy a nucleoid within the bacterial cell.

Some Archaeal Chromosomes Contain Histone Proteins that Form Nucleosomes

The structure of the archaeal chromosome varies among different species depending on the DNA-binding proteins they express. Some archaeal species produce bacterial-like nucleoid-associated proteins. These species organize their chromosomes in a way that appears to be similar to bacterial species. (see Figure 10.3).

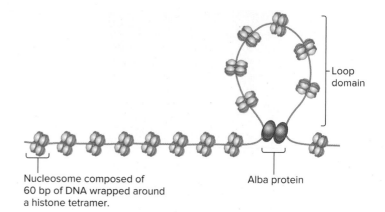

FIGURE 10.4 **Chromosome structure in some archaea.** This illustration shows a short region of an archaeal chromosome. In this example, 60 bp of DNA is wrapped around a histone tetramer to form a nucleosome. The repeating occurrence of nucleosomes resembles beads on a string. Loop domains are created by the Alba protein, which forms a bridge between two different segments of DNA. Though not shown in this figure, the DNA is also supercoiled to make the chromosome more compact.

By contrast, other archaeal species produce eukaryotic-like histone proteins. (Histones are described in Section 10.6.) In these species, the DNA is wrapped around histone proteins to form **nucleosomes** and also organized into **loop domains** by a protein called Alba (**Figure 10.4**). The number of histone proteins within archaeal nucleosomes varies among different species. In some archaea, 60 bp of DNA is wrapped around a histone tetramer to form a "beads-on-a-string" structure (see Figure 10.19a). This structure is similar to eukaryotic chromatin except that eukaryotic nucleosomes contain histone octamers (look ahead to Figure 10.17).

DNA Supercoiling Further Compacts a Prokaryotic Chromosome

Because DNA is a long thin molecule, twisting forces can dramatically change its conformation. This effect is similar to what happens when you twist a rubber band. If twisted in one direction, a rubber band eventually coils itself into a compact structure as it absorbs the energy applied by the twisting motion. Because the two strands within DNA already coil around each other, the formation of additional coils due to twisting forces is referred to as **DNA supercoiling.** The DNA within microdomains and loop domains is further compacted because of DNA supercoiling.

How do twisting forces affect DNA structure? **Figure 10.5** illustrates four possibilities. In Figure 10.5a, a double-stranded DNA molecule with five complete turns is anchored between two plates. In this hypothetical example, the ends of the DNA molecule cannot rotate freely. Both underwinding and overwinding of the DNA double helix can cause supercoiling of the helix. As described in Chapter 9, the predominant form of DNA, called B DNA, is a right-handed helix. Therefore, underwinding is a left-handed twisting motion, and overwinding is a right-handed twisting motion. Along the left side of Figure 10.5, one of the plates has been given a left-handed twist that tends to

FIGURE 10.5 **Schematic representation of DNA supercoiling.** In this example, the DNA in **(a)** is anchored between two plates and given a twist as noted by the arrows. A left-handed twist (underwinding) can produce either **(b)** fewer turns or **(c)** a negative supercoil. A right-handed twist (overwinding) can produce **(d)** more turns or **(e)** a positive supercoil. The structures shown in (b) and (d) are unstable.

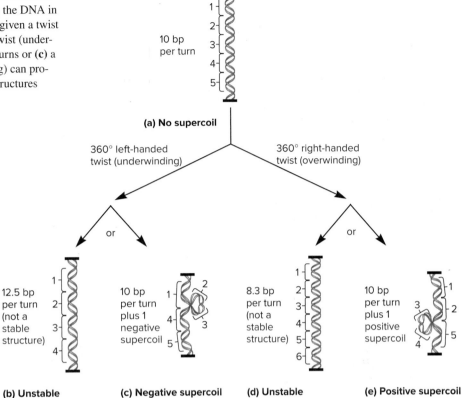

unwind the helix. As the helix absorbs this force, the underwinding can cause:

- fewer turns (Figure 10.5b) or
- the formation of a negative supercoil (Figure 10.5c).

On the right side of Figure 10.5, one of the plates has been given a right-handed twist, which overwinds the double helix. Overwinding can cause:

- more turns (Figure 10.5d) or
- the formation of a positive supercoil (Figure 10.5e).

The DNA conformations shown in Figure 10.5a, c, and e differ only with regard to supercoiling. These three DNA conformations are referred to as **topoisomers** of each other. The DNA conformations shown in Figure 10.5b and d are not structurally stable and do not occur in living cells.

Chromosome Function Is Influenced by DNA Supercoiling

The chromosomal DNA in living bacteria and archaea is negatively supercoiled. For example, in the chromosome of *E. coli*, about one negative supercoil occurs per 40 turns of the double helix. Negative supercoiling has important consequences. As already mentioned, the supercoiling of chromosomal DNA makes it much more compact. Therefore, supercoiling helps to greatly decrease the size of a prokaryotic chromosome.

In addition, negative supercoiling also affects DNA function. To understand how it does so, remember that negative

supercoiling is due to an underwinding force on the DNA. Therefore, negative supercoiling creates tension on the DNA strands that may be released by their separation (**Figure 10.6**). Although most of the chromosomal DNA is negatively supercoiled and compact, the force of negative supercoiling may promote DNA strand separation in small regions. This enhances genetic activities such as replication and transcription that require the DNA strands to be separated.

How does DNA become supercoiled? In 1976, Martin Gellert and colleagues discovered the enzyme **DNA gyrase,** also known as topoisomerase II. This enzyme, which contains four subunits (two A and two B subunits), introduces negative supercoils (or relaxes positive supercoils) using energy from

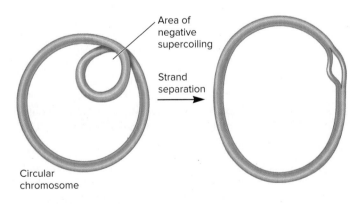

FIGURE 10.6 **Negative supercoiling promotes strand separation.**
CONCEPT CHECK: Why is strand separation beneficial?

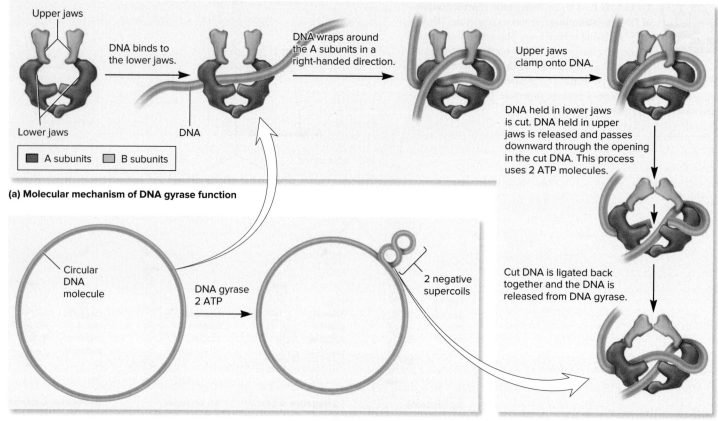

(a) Molecular mechanism of DNA gyrase function

(b) Overview of DNA gyrase function

FIGURE 10.7 **The action of DNA gyrase.** (a) DNA gyrase, also known as topoisomerase II, is composed of two A and two B subunits. The lower jaws (shown in red) first bind to the DNA, and then the DNA forms a loop and a second site in the DNA binds to the upper jaws (shown in green). The DNA in the lower jaws is then cut (i.e., a double-stranded break is made). The unbroken segment of DNA is released from the upper jaws and passes through the break. The break is repaired. The B subunits capture the energy from the hydrolysis of two ATP molecules to catalyze this process. (b) The result is that two negative turns have been introduced into the DNA molecule.

CONCEPT CHECK: In your own words, describe the step that requires the use of ATP.

ATP (**Figure 10.7a**). To alter supercoiling, DNA gyrase has two sets of jaws that allow it to grab onto two regions of DNA.

1. To begin the process, a DNA region is grabbed by the lower jaws.
2. The DNA is then wrapped in a right-handed direction around the two A subunits.
3. The upper jaws then clamp onto another region of DNA.
4. The DNA in the lower jaws is cut in both strands, and the other region of DNA is then released from the upper jaws and passed through this double-stranded break.
5. To complete the process, the double-stranded break is ligated back together. The net result is that two negative supercoils have been introduced into the DNA molecule (**Figure 10.7b**).

In addition, DNA gyrase can untangle DNA molecules. For example, as discussed in Chapter 11, circular DNA molecules are sometimes intertwined following DNA replication (see Figure 11.14). Such interlocked molecules can be separated by DNA gyrase.

A second type of enzyme, **topoisomerase I**, relaxes negative supercoils. This enzyme can bind to a negatively supercoiled

region and introduce a break in one of the DNA strands. After one DNA strand has been broken, the DNA molecule rotates to relieve the tension that is caused by negative supercoiling. This rotation relaxes negative supercoiling. The broken strand is then repaired. The competing actions of DNA gyrase and topoisomerase I govern the overall supercoiling of the bacterial DNA.

The ability of DNA gyrase to introduce negative supercoils into DNA is critical for bacterial (and archaeal) cell survival. Therefore, much research has been aimed at identifying drugs that specifically block this enzyme's function as a way to cure or alleviate diseases caused by bacteria. Two main classes—quinolones and coumarins—inhibit gyrase and other bacterial topoisomerases, thereby blocking bacterial cell growth. These drugs do not inhibit eukaryotic topoisomerases, which are structurally different from their bacterial counterparts. This finding has been the basis for the production of many drugs with important antibacterial applications. An example is ciprofloxacin (known also by the brand name Cipro), which is used to treat a wide spectrum of bacterial diseases, including anthrax.

10.2 COMPREHENSION QUESTIONS

1. Mechanisms that make the bacterial chromosome more compact include
   a. the formation of microdomains and macrodomains.
   b. DNA supercoiling.
   c. crossing over.
   d. both a and b.

2. Negative supercoiling can enhance RNA transcription and DNA replication because it
   a. allows the binding of proteins in the major groove.
   b. promotes DNA strand separation.
   c. makes the DNA more compact.
   d. causes all of the above.

3. DNA gyrase
   a. promotes negative supercoiling.
   b. relaxes positive supercoils.
   c. cuts DNA strands as part of its function.
   d. does all of the above.

10.3 ORGANIZATION OF FUNCTIONAL SITES ALONG EUKARYOTIC CHROMOSOMES

Learning Outcome:

1. Describe the organization of functional sites along a eukaryotic chromosome.

Figure 10.8 shows the general features of a eukaryotic chromosome and the functional sites along it.

- Each eukaryotic chromosome contains a long, linear DNA molecule.
- Eukaryotic species have one or more sets of chromosomes in the cell nucleus; each set is composed of several different linear chromosomes (refer back to Figure 8.1). Humans, for example, have two sets of 23 chromosomes each, for a total of 46.
- A typical eukaryotic chromosome is typically tens of millions to hundreds of millions of base pairs in length.
- A single chromosome usually carries hundreds to several thousand different genes. A typical eukaryotic gene is several thousand to tens of thousands of base pairs in length. In less complex eukaryotes such as yeast, genes are relatively small, often several hundred to a few thousand base pairs long. In more complex eukaryotes such as mammals and flowering plants, protein-coding genes tend to be much longer due to the presence of **introns**—noncoding intervening sequences. The size of introns ranges from less than 100 bp to more than 10,000 bp. Therefore, the presence of large introns can greatly increase the lengths of eukaryotic genes.

- Origins of replication are chromosomal sites that are necessary to initiate DNA replication. Unlike most bacterial chromosomes, which contain only one origin of replication, eukaryotic chromosomes contain many origins, interspersed approximately every 100,000 bp. The function of origins of replication is discussed in greater detail in Chapter 11.
- **Centromeres** are regions that play a role in the proper segregation of chromosomes during mitosis and meiosis. In most eukaryotic species, each chromosome contains a single centromere, which usually appears as a constricted region of a mitotic chromosome. A centromere functions as a site for the formation of a kinetochore, which

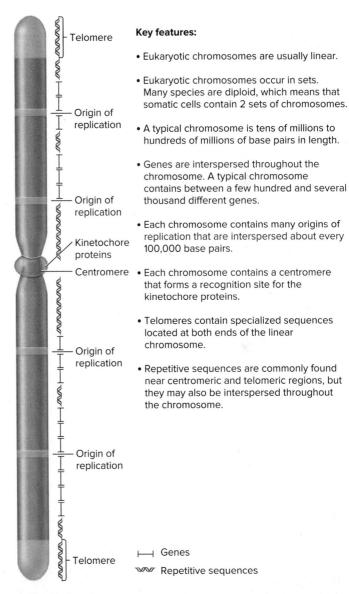

Key features:

- Eukaryotic chromosomes are usually linear.

- Eukaryotic chromosomes occur in sets. Many species are diploid, which means that somatic cells contain 2 sets of chromosomes.

- A typical chromosome is tens of millions to hundreds of millions of base pairs in length.

- Genes are interspersed throughout the chromosome. A typical chromosome contains between a few hundred and several thousand different genes.

- Each chromosome contains many origins of replication that are interspersed about every 100,000 base pairs.

- Each chromosome contains a centromere that forms a recognition site for the kinetochore proteins.

- Telomeres contain specialized sequences located at both ends of the linear chromosome.

- Repetitive sequences are commonly found near centromeric and telomeric regions, but they may also be interspersed throughout the chromosome.

FIGURE 10.8 General features of a eukaryotic chromosome.
Note: This is meant to be a schematic representation that depicts a metaphase chromosome. The genetic elements are not drawn to scale. The numbers of origins of replication, genes, and repetitive sequences are much higher than shown here.

CONCEPT CHECK: What are some differences between the types of sequences found in eukaryotic chromosomes and those in prokaryotic chromosomes?

assembles just before and during the very early stages of mitosis and meiosis.

In certain yeast species, such as *Saccharomyces cerevisiae*, the centromere has a defined DNA sequence that is about 125 bp in length. This type of centromere is called a point centromere. By comparison, the centromeres found in more complex eukaryotes are much larger and contain tandem arrays of short repetitive DNA sequences. (Tandem arrays are described in Section 10.4.) These are called regional centromeres. They can range in length from several thousand to more than a million base pairs. By themselves, the repeated DNA sequences within regional centromeres are not necessary or sufficient to form a functional centromere with a kinetochore. Other features must be present in a functional centromere. For example, a distinctive feature of all eukaryotic centromeres is that histone protein H3 is replaced with a histone variant called CENP-A. (Histone variants are described in Chapter 15.) However, researchers are still trying to identify all of the biochemical properties that distinguish regional centromeres and understand how these properties are transmitted during cell division.

- The **kinetochore** is composed of a group of proteins that link the centromere to the spindle apparatus during mitosis and meiosis, ensuring the proper segregation of the chromosomes to each daughter cell.

- The ends of linear eukaryotic chromosomes have specialized regions known as **telomeres.** Telomeres serve several important functions in the replication and stability of the chromosome. As discussed in Chapter 8, telomeres prevent chromosomal rearrangements such as translocations. In addition, they prevent chromosome shortening in two ways. First, the telomeres protect chromosomes from digestion via enzymes called exonucleases that recognize the ends of DNA. Second, a specialized form of DNA replication occurs at the telomeres so that eukaryotic chromosomes do not become shortened with each round of DNA replication (see Chapter 11). However, shortening does occur in adult somatic cells as a part of the aging process.

10.3 COMPREHENSION QUESTIONS

1. The chromosomes of eukaryotes typically contain
   a. a few hundred to several thousand different genes.
   b. multiple origins of replication.
   c. a centromere.
   d. telomeres at their ends.
   e. all of the above.

2. The kinetochore is attached to _____ and its function is to _____.
   a. a gene, promote transcription
   b. the centromere, promote chromosome segregation during mitosis and meiosis
   c. a telomere, prevent chromosome shortening
   d. the centromere, promote chromosome replication

10.4 SIZES OF EUKARYOTIC GENOMES AND REPETITIVE SEQUENCES

Learning Outcomes:
1. Describe the variation in size of eukaryotic genomes.
2. Define *repetitive sequence*, and explain how this type of sequence affects genome sizes.

The total amount of DNA in cells of eukaryotic species is usually much greater than the amount in prokaryotic cells. In addition, eukaryotic genomes contain many more genes than their prokaryotic counterparts. In this section, we will examine the sizes of eukaryotic genomes and consider how repetitive sequences may have a significant effect on those sizes.

The Sizes of Eukaryotic Genomes Vary Substantially

Different eukaryotic species vary dramatically in the size of their genomes (**Figure 10.9a**; note that the graph uses a log scale). In many cases, this variation is not related to the complexity of the species. For example, two closely related species of salamander, *Plethodon richmondi* and *Plethodon larselli*, differ considerably in genome size (**Figure 10.9b, c**). The genome of *P. larselli* is more than twice as large as the genome of *P. richmondi*. However, the genome of *P. larselli* probably doesn't contain more genes. How do we explain the difference in genome size? The additional DNA in *P. larselli* is due to the accumulation of many copies of repetitive DNA sequences. In some species, the amounts of these repetitive sequences have reached enormous levels. Such repetitive sequences do not code proteins, and their function remains a matter of controversy and great interest. The structure and significance of repetitive DNA sequences are discussed next.

The Genomes of Eukaryotes Contain Sequences That Are Unique, Moderately Repetitive, or Highly Repetitive

The term **sequence complexity** refers to the number of times a particular base sequence appears throughout the genome of a species. Unique, or nonrepetitive, sequences are those found once or a few times within a genome. Protein-coding genes are typically unique sequences of DNA. The vast majority of proteins in eukaryotic cells are coded by genes present in one or a few copies. In the case of humans, unique sequences make up roughly 41% of the entire genome (**Figure 10.10**). These unique sequences include the protein-coding regions of genes (2%), introns (24%), and unique regions that are not found within genes (15%).

Moderately repetitive sequences are found a few hundred to several thousand times in a genome. In a few cases, moderately repetitive sequences are multiple copies of the same gene. For example, the genes that code ribosomal RNA (rRNA) are found in many copies. Ribosomal RNA is necessary for the functioning of ribosomes. Cells need a large amount of rRNA for making

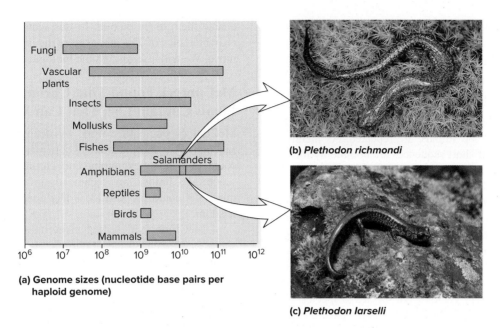

(b) *Plethodon richmondi*

(c) *Plethodon larselli*

(a) **Genome sizes (nucleotide base pairs per haploid genome)**

FIGURE 10.9 **Haploid genome sizes among groups of eukaryotic species.** **(a)** Ranges of genome sizes among different groups of eukaryotes. **(b)** A species of salamander, *Plethodon richmondi*, and **(c)** a close relative, *Plethodon larselli*. The genome of *P. larselli* is more than twice as large as that of *P. richmondi*.

Genes→Traits The two species of salamander shown here have very similar traits, even though the genome of *P. larselli* is more than twice as large as that of *P. richmondi*. However, the genome of *P. larselli* is not likely to contain twice as many genes. Rather, the additional DNA is due to the accumulation of short repetitive DNA sequences that do not contain functional genes and are present in many copies.

(a): Source: Gregory, T. Ryan, "Eukaryotic Genome Size Databases," *Nucleic Acids Research*, vol. 35, January, 2007, D332–D338.; (b): Ann & Rob Simpson; (c): Gary Nafis

CONCEPT CHECK: What are two reasons for the wide variation in genome sizes among eukaryotic species?

ribosomes, and producing such an amount is facilitated by having multiple copies of the genes that code rRNA. Likewise, the genes that code histone proteins are also found in multiple copies because a large number of histone proteins are needed for the structure of chromosomes.

In addition, other types of functionally important sequences are moderately repetitive. For example, moderately repetitive

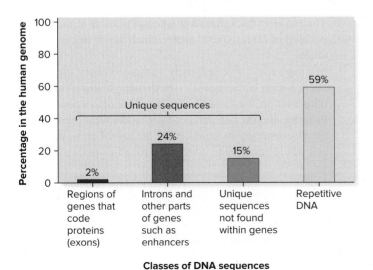

Classes of DNA sequences

FIGURE 10.10 **Relative amounts of unique and repetitive DNA sequences in the human genome.**

sequences may play a role in the regulation of gene transcription and translation. By comparison, some moderately repetitive sequences do not play a functional role and are derived from transposable elements (TEs)—short segments of DNA that have the ability to move within a genome. This category of repetitive sequences is discussed in greater detail in Section 10.5.

Highly repetitive sequences are found tens of thousands or even millions of times throughout a genome. Each copy of a highly repetitive sequence is relatively short, ranging from a few nucleotides to several hundred in length. A widely studied example is the *Alu* family of sequences found in humans and other primates. The *Alu* sequence is approximately 300 bp long. This sequence derives its name from the observation that it contains a site for cleavage by a restriction enzyme known as *Alu*I. (The function of restriction enzymes is described in Chapter 20.) The *Alu* sequence represents about 10% of the total human DNA and occurs approximately every 5000–6000 bp! Evolutionary studies suggest that the *Alu* sequence arose 65 mya from a section of a single ancestral gene known as *7SL RNA*. Since that time, this gene has become a type of TE called a retrotransposon, which is transcribed into RNA, copied into DNA, and inserted into the genome (see Section 10.5). Over the past 65 million years, the *Alu* sequence has been copied and inserted into the human genome many times and is now present in about 1,000,000 copies.

Repetitive sequences, like those of the *Alu* family, are interspersed throughout the genome. However, some moderately and highly repetitive sequences are clustered together in a **tandem array,** also known as a tandem repeat. In a tandem array, a very short nucleotide sequence is repeated many times in a row. In *Drosophila*, for example, 19% of the chromosomal DNA consists of highly repetitive sequences found in tandem arrays. An example is shown here.

AATATAATATAATATAATATAATATATAATAT
TTATATTATATTATATTATATTATATATTATA

In this particular tandem array, two related sequences, AATAT and AATATAT (in the top strand), are repeated. As mentioned earlier, tandem arrays of short sequences are commonly found in centromeric regions of chromosomes and can be quite long, sometimes more than 1,000,000 bp in length!

What is the functional significance of highly repetitive sequences? Whether they have any significant function is controversial. Some experiments in *Drosophila* indicate that highly repetitive sequences may be important in the proper segregation of chromosomes during meiosis. It is not yet clear if highly repetitive DNA plays the same role in other species. The sequences within highly repetitive DNA vary greatly from species to species. Likewise, the amount of highly repetitive DNA can vary a great deal even among closely related species (as noted in Figure 10.9).

10.4 COMPREHENSION QUESTION

1. Which of the following is/are moderately repetitive sequences?
   a. Genes that code rRNA
   b. Most protein-coding genes
   c. Both a and b
   d. None of the above

10.5 TRANSPOSITION

Learning Outcomes:

1. Summarize the studies of McClintock, and explain how they revealed the existence of transposable elements.
2. Describe the organization of sequences within different types of transposable elements.
3. Explain how transposons and retrotransposons move to new locations in a genome.
4. Discuss the effects of transposable elements on gene function.

As we have seen, sizeable portions of many species' genomes are composed of repetitive sequences. In many cases, the repetitive sequences are due to **transposition**, the process in which a DNA segment is inserted into a new location in the genome. The DNA segments that transpose themselves are known as transposable elements (TEs). TEs have sometimes been referred to as "jumping genes" because they are inherently mobile.

Transposable elements were first identified by Barbara McClintock in the early 1950s during classic studies with corn plants. Since that time, geneticists have discovered many different types of TEs in organisms as diverse as bacteria, archaea, fungi, plants, and animals. The advent of molecular techniques has allowed scientists to better understand the characteristics of TEs that

enable them to be mobile. In this section, we will examine the characteristics of TEs and explore the mechanisms by which they move. We will also discuss the biological significance of TEs.

McClintock Found That Chromosomes of Corn Plants Contain Loci That Can Move

McClintock's scientific work was focused on the structure and function of the chromosomes of corn plants. This research involved countless hours of examining corn chromosomes under the microscope. McClintock was technically gifted and had a theoretical mind that could propose ideas that conflicted with conventional wisdom.

McClintock identified many unusual features of chromosomes in different strains of corn. In one strain, a particular site in chromosome 9 had the strange characteristic of showing a fairly high rate of breakage. McClintock termed this a **mutable site,** or *mutable locus*. The mutable locus was named *Ds* (for dissociation), because chromosomal breakage occurred frequently there.

McClintock identified strains of corn in which the *Ds* locus was found in different places within the corn genome. In one case, *Ds* was located in the middle of a gene affecting kernel color. The *C* allele provides dark red color, whereas *c* is a recessive allele of the same gene and causes a colorless kernel. The endosperm of corn kernels is triploid. The drawing below shows the genotype of chromosome 9 in the endosperm of one of McClintock's strains.

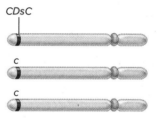

This strain had an interesting phenotype. Most of the corn kernel was colorless, but it also contained some red sectors. McClintock explained this phenotype in the following way:

1. The colorless background of a kernel was due to the transposition of *Ds* into the *C* allele, which would inactivate that allele.
2. In a few cells, *Ds* occasionally transposed out of the *C* allele during kernel growth (see drawing below). During transposition, *Ds* moved out of the *C* allele to a new location, and the two parts of the *C* allele were rejoined, thereby restoring its function. As the kernel grew, such a cell would continue to divide, resulting in a red sector.

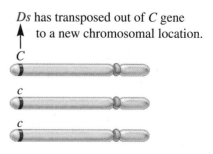

Ds has transposed out of *C* gene to a new chromosomal location.

On rare occasions, when McClintock crossed a strain carrying *Ds* in the middle of the *C* allele to a strain carrying the recessive *c* allele, the cross produced a kernel that was completely red. In this case, *Ds* had transposed out of the *C* allele prior to kernel growth, probably during gamete formation. In offspring that grew from a solid red kernel, McClintock determined that the *Ds* locus had moved out of the *C* allele to a new location. In addition, the restored *C* allele behaved normally. In other words, the *C* allele was no longer mutable; the kernels did not show a sectoring phenotype. Taken together, the results were consistent with the hypothesis that the *Ds* locus can move around the corn genome by transposition.

When McClintock published these results in 1951, they were met with great skepticism. Some geneticists of that time were unable to accept the idea that the genetic material was susceptible to frequent rearrangement. Instead, they believed that the genetic material was very stable and permanent in its structure. Over the next several decades, however, the scientific community came to realize that TEs are a widespread phenomenon. McClintock was awarded the Nobel Prize in Physiology or Medicine in 1983, more than 30 years after the original discovery of transposable elements.

Transposable Elements Move by Different Transposition Pathways

Since McClintock's pioneering studies, many different TEs have been found in bacteria, archaea, fungi, plants, and animals. Two main types of transposition mechanisms have been identified: simple transposition and retrotransposition.

Simple Transposition. In **simple transposition,** the TE is removed from its original site and transferred to a new target site (**Figure 10.11a**). This mechanism is called a cut-and-paste mechanism because the element is cut out of its original site and pasted into a new one. Transposable elements that move via simple transposition are widely found in bacterial and eukaryotic species. Such TEs are also called **transposons.**

Retrotransposition. Another type of transposable element moves via an RNA intermediate. This form of transposition, termed **retrotransposition,** is found only in eukaryotic species, where it is very common (**Figure 10.11b**). Transposable elements that move via retrotransposition are known as **retrotransposons,** or **retroelements.** In retrotransposition, the element is transcribed into RNA. An enzyme called reverse transcriptase uses the RNA as a template to synthesize a DNA molecule that is integrated into a new region of the genome. Retrotransposons increase in number during retrotransposition.

Each Type of Transposable Element Has a Characteristic Pattern of DNA Sequences

Research on TEs from many species has established that the DNA sequences within them are organized in several different ways. **Figure 10.12** presents a few of those ways, although many variations are possible. All TEs are flanked by **direct repeats (DRs),**

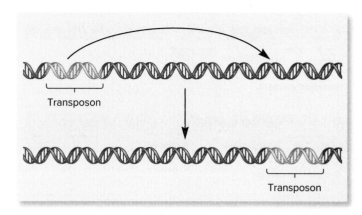

(a) Simple transposition

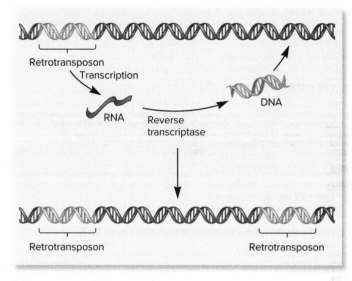

(b) Retrotransposition

FIGURE 10.11 Different mechanisms of transposition.

CONCEPT CHECK: Which of these mechanisms causes the TE to increase in number?

also called **target-site duplications,** which are identical base sequences that are oriented in the same direction and repeated. Direct repeats are adjacent to both ends of any TE.

Insertion Elements. The simplest TE is known as an **insertion element (IS element).** As shown in Figure 10.12a, an IS element has two important characteristics. First, both ends of the element contain **inverted repeats (IRs).** Inverted repeats are DNA sequences that are identical (or very similar) but run in opposite directions, such as the following:

5′–CTGACTCTT–3′   and   5′–AAGAGTCAG–3′
3′–GACTGAGAA–5′   3′–TTCTCAGTC–5′

Depending on the particular IS element, the inverted repeats range from 9 to 40 bp in length. In addition, IS elements may contain a central region that codes the enzyme **transposase,** which catalyzes the transposition event.

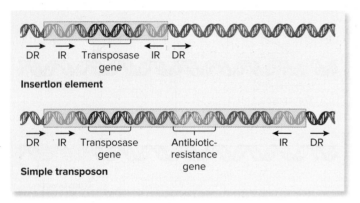

(a) Elements that move by simple transposition

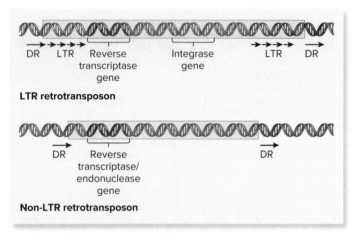

(b) Elements that move by retrotransposition (via an RNA intermediate)

FIGURE 10.12 **Common organization of DNA sequences in transposable elements.** Direct repeats (DRs) are identical sequences found on both sides of all TEs. Inverted repeats (IRs) are at the ends of some transposable elements. Long terminal repeats (LTRs) are regions containing a large number of tandem repeats.

Simple Transposons   By comparison, a **simple transposon** carries one or more genes that are not required for transposition to occur. For example, the simple transposon shown in Figure 10.12a carries an antibiotic-resistance gene.

Retrotransposons   The organization of retrotransposons varies greatly. They are categorized based on their evolutionary relationship to retroviruses. As described in Chapter 18, retroviruses are RNA viruses that make a DNA copy that integrates into the host's genome.

   LTR retrotransposons are evolutionarily related to retroviruses. These TEs have retained the ability to move around the genome, though, in most cases, they do not produce mature viral particles. LTR retrotransposons are so named because they contain **long terminal repeats (LTRs)** at both ends (Figure 10.12b). The LTRs are typically a few hundred base pairs in length. Like their viral counterparts, LTR retrotransposons may code virally related proteins, such as reverse transcriptase and integrase, that are needed for the retrotransposition process.

By comparison, **non-LTR retrotransposons** do not resemble retroviruses in having LTRs. They may contain a gene that codes a protein that functions as both a reverse transcriptase and an endonuclease (see Figure 10.12b). As discussed later, these functions are needed for retrotransposition. Some non-LTR retrotransposons are evolutionarily derived from normal eukaryotic genes. For example, the *Alu* family of repetitive sequences found in humans is derived from a single ancestral gene known as *7SL RNA* (that codes a component of the complex called signal recognition particle, which targets newly made proteins to the endoplasmic reticulum). This gene sequence has been copied via retrotransposition many times, and the current number of copies in the human genome is approximately 1 million.

Transposable elements are considered to be complete elements, or **autonomous elements,** when they contain all of the information necessary for transposition or retrotransposition to take place. However, TEs are often incomplete, or nonautonomous. A **nonautonomous element** typically lacks a gene such as one that codes transposase or reverse transcriptase, which is necessary for transposition to occur.

   The *Ds* locus, which is the mutable site in corn discussed previously, is a nonautonomous element, because it lacks a transposase gene. An element that is similar to *Ds* but contains a functional transposase gene is called the *Ac* element, which stands for <u>ac</u>tivator element. An *Ac* element provides a transposase gene that enables *Ds* to transpose. Therefore, nonautonomous TEs such as *Ds* can transpose only when an *Ac* element is present at another region in the genome. The *Ac* element was present in McClintock's strains.

Transposase Catalyzes the Excision and Insertion of Transposons

Now that we have considered the typical organization of TEs, let's examine the steps of the transposition process. The enzyme **transposase** catalyzes the removal of a transposon from its original site in the chromosome and its subsequent insertion at another location. A general scheme for simple transposition is shown in **Figure 10.13a**.

1. Transposase monomers first bind to the inverted repeat sequences at the ends of the TE.
2. The monomers then dimerize, which brings the inverted repeats close together.
3. The DNA is cleaved between the inverted and direct repeats, which excises the TE from its original site within the chromosome.
4. Transposase carries the TE to a new site and cleaves the target DNA sequence at staggered recognition sites. The TE is then inserted into the target DNA and ligated to it.

As shown in **Figure 10.13b**, the ligation of the transposable element into its new site initially leaves short gaps in the target DNA. Notice that the DNA sequences in these gaps are complementary to each other (in this case, ATGCT and TACGA). Therefore, when they are filled in by DNA gap repair synthesis, the repair produces direct repeats that flank both ends of the TE. These direct repeats are common features found adjacent to all TEs (see Figure 10.12).

   Although the transposition process depicted in Figure 10.13 does not directly alter the number of TEs, simple transposition is known to increase their numbers in genomes, in some cases to

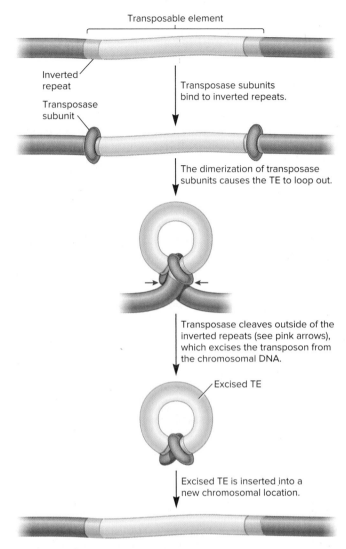

(a) Movement of a transposon via transposase

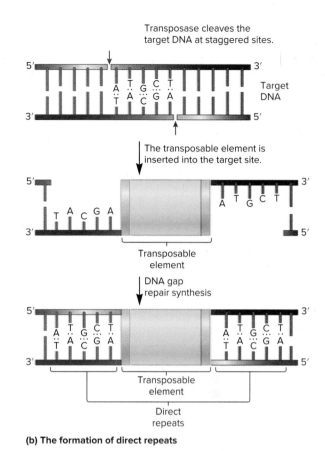

(b) The formation of direct repeats

FIGURE 10.13 **Simple transposition.** **(a)** Transposase removes the TE from its original site and inserts it into a new site. **(b)** A closer look at how the insertion process creates direct repeats.

ONLINE ANIMATION

fairly high levels. How can this happen? The answer is that transposition often occurs around the time of DNA replication (**Figure 10.14**). After a replication fork has passed a region containing a TE, two TEs will be found behind the fork—one in each of the replicated regions. One of these TEs could then transpose from its original location into a region ahead of the replication fork. After the replication fork has passed this second region and DNA replication is completed, two TEs will be found in one of the chromosomes and one TE in the other chromosome. In this way, simple transposition can lead to an increase in TE number. We will discuss the biological significance of transposon proliferation later in this section.

Retrotransposons Use Reverse Transcriptase for Retrotransposition

Thus far, we have considered how transposons can move throughout a genome. By comparison, retrotransposons use an RNA intermediate in their transposition mechanism. Let's begin with LTR retrotransposons. As shown in **Figure 10.15**, the movement of LTR retrotransposons requires two key enzymes: reverse

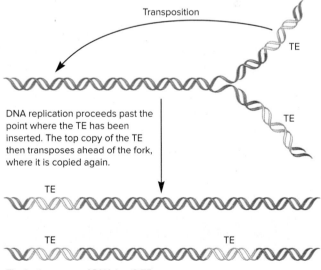

The bottom copy of DNA has 2 TEs.

FIGURE 10.14 **Increase in the number of copies of a transposable element (TE) via simple transposition.** In this example, a TE that has already been replicated transposes to a new site that has not yet replicated. Following the completion of DNA replication, the TE has increased in number.

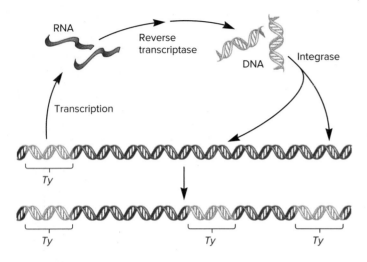

FIGURE 10.15 **Retrotransposition of an LTR retrotransposon.** *Ty* is a retrotransposon found in yeast.

CONCEPT CHECK: What is the function of reverse transcriptase?

transcriptase and integrase. In this example, the cell already contains a retrotransposon known as *Ty* within its genome.

1. This retrotransposon is transcribed into RNA.
2. In a series of steps, **reverse transcriptase** uses this RNA as a template to synthesize a double-stranded DNA molecule.
3. The LTRs at the ends of the double-stranded DNA are then recognized by **integrase,** which makes staggered cuts at a target site in the host chromosome and catalyzes the insertion of the TE into the site.

The integration of a retrotransposon can occur at many locations within the genome. Figure 10.15 shows the integration occurring at two locations. Furthermore, because a single retrotransposon can be copied into many RNA transcripts, retrotransposons may accumulate rapidly within a genome.

Though the mechanism of non-LTR retrotransposition is not entirely understood, one popular model for the replication and integration of non-LTR retrotransposons is called **target-site primed reverse transcription (TPRT).** A simplified version of this model is shown in **Figure 10.16**.

1. The retrotransposon is first transcribed into RNA with a polyA tail at the 3′ end.
2. The target DNA site is recognized by an endonuclease, which may be coded by the retrotransposon. This endonuclease recognizes a consensus sequence of 5′-TTTTA-3′, and initially cuts just one of the DNA strands.
3. The polyA tail of the retrotransposon RNA binds to this nicked site due to A-T base pairing.
4. Reverse transcriptase then uses the target DNA as a primer and makes a DNA copy of the RNA, which is why the process is named target-site primed reverse transcription.
5. An endonuclease makes a second cut in the other DNA strand usually about 7–20 nucleotides away from the first cut.

6. The retrotransposon DNA is then integrated into the target site.
7. The gaps in the DNA are filled in, perhaps by DNA gap repair synthesis, described in Chapter 19 (see Section 19.6).

Transposable Elements May Have Important Influences on Mutation and Evolution

Over the past few decades, researchers have found that TEs probably occur in the genomes of all species. **Table 10.1** describes a few TEs that have been studied in great detail. As discussed earlier in this chapter, the genomes of eukaryotic species typically contain moderately and highly repetitive sequences. In some cases, these repetitive sequences are due to the proliferation of TEs. In the genomes of mammals, for example, **LINEs** are long interspersed elements that are usually 1000 to 10,000 bp in length and occur in 20,000 to 1,000,000 copies per genome. In humans, a particular family of related LINEs called LINE-1, or L1, is found in about 500,000 copies and represents about 17% of the total human DNA! By comparison, **SINEs** are short interspersed elements that are less than 500 bp in length. A specific example of a SINE is the *Alu* sequence, present in about 1 million copies in the human genome. About 10% of the human genome is composed of this particular TE.

LINEs and SINEs continue to proliferate in the human genome, but at a fairly low rate. In about 1 live birth in 100, an *Alu* or an L1 (or both) sequence has been inserted into a new site in the human genome. On rare occasions, a new insertion can disrupt a gene and cause phenotypic abnormalities. For example, new insertions of L1 or *Alu* sequences into particular genes have been shown, on occasion, to be associated with diseases such as hemophilia, muscular dystrophy, and breast and colon cancer.

The relative abundance of TEs varies widely among different species. As shown in **Table 10.2,** TEs can be quite prevalent in amphibians, mammals, and flowering plants, but tend to be less abundant in simpler organisms such as bacteria and yeast. The biological significance of TEs in the evolution of prokaryotic and eukaryotic species remains a matter of debate. According to the **selfish DNA hypothesis,** TEs exist because they have characteristics that allow them to multiply within the chromosomal DNA of living cells. In other words, they resemble parasites in the sense that they inhabit a cell without offering any selective advantage to the organism. They can proliferate as long as they do not harm the organism to the extent that they significantly disrupt survival.

Alternatively, other geneticists have argued that transpositional events are often deleterious. Therefore, TEs would be eliminated from the genome by natural selection if they did not also offer a compensating advantage. Several potential advantages have been suggested. For example, TEs may cause greater genetic variation by promoting recombination. In addition, bacterial TEs often carry an antibiotic-resistance gene that provides the organism with a survival advantage. Researchers have also suggested that transposition may cause the insertion of exons from one gene into another gene, thereby producing a new gene with novel function(s). This phenomenon, called **exon shuffling,** is described in Chapter 27.

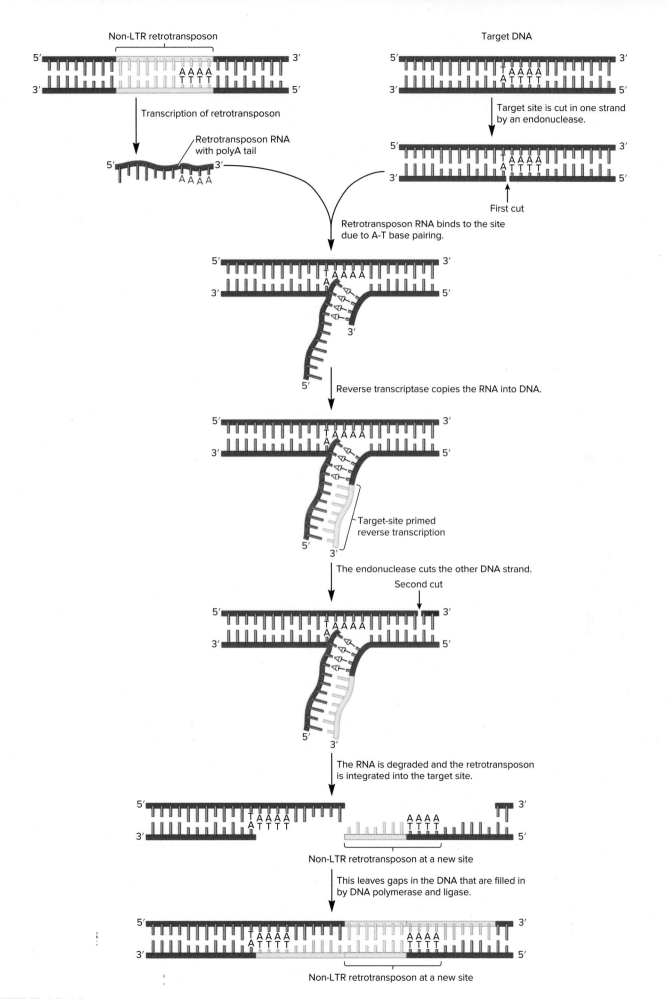

FIGURE 10.16 A simplified model for retrotransposition of a non-LTR retrotransposon.

TABLE 10.1

Examples of Transposable Elements

Element	Type	Approximate Length (bp)	Description
Bacterial			
IS1	Transposon	768	An insertion element that is commonly found in five to eight copies in *E. coli*.
Tn10	Transposon	9300	One of many different bacterial transposons that carry an antibiotic-resistance gene.
Tn951	Transposon	16,600	A transposon that provides bacteria with genes that allow them to metabolize lactose.
Yeast			
Ty element	Retrotransposon	6300	Found in *S. cerevisiae* in about 35 copies per genome.
Fruit Fly			
P element	Transposon	500–3000	A transposon that may be found in 30–50 copies in P strains of *Drosophila*. It is absent from M strains.
Copia-like element	Retrotransposon	5000–8000	One of a family of TEs found in *Drosophila*, which vary slightly in their lengths and sequences. Typically, each family member is found in about 5–100 copies per genome.
Humans			
Alu sequence	Retrotransposon	300	A SINE found in about 1,000,000 copies in the human genome.
L1	Retrotransposon	6500	A LINE found in about 500,000 copies in the human genome.
Plants			
Ac or *Ds*	Transposon	4500	*Ac* is an autonomous transposon found in corn and other plant species. It carries a transposase gene. *Ds* is a nonautonomous version that lacks a functional transposase gene.
Opie	Retrotransposon	9000	A retrotransposon found in plants that is related to the *copia*-like elements found in animals.

This controversy remains unresolved, but studies have shown that TEs can rapidly enter the genome of an organism and proliferate quickly. In *Drosophila melanogaster*, for example, a TE known as a P element was probably introduced into this species in the 1950s. Laboratory stocks of *Drosophila* collected prior to this time do not contain P elements. Remarkably, in the last 60 years, the P element has expanded throughout *Drosophila* populations worldwide. The only strains without the P element are laboratory strains collected prior to the 1950s. This observation underscores the surprising ability of TEs to infiltrate a population of organisms.

Transposable elements have a variety of effects on chromosome structure and gene expression (**Table 10.3**). Many of these outcomes are likely to be harmful. Usually, transposition is a relatively rare event that occurs only in a few individuals under certain conditions, such as exposure to radiation. As described in Chapter 17, prokaryotes and eukaryotes have mechanisms that greatly decrease the movement of TEs.

When transposition occurs at a high rate, it is likely to be detrimental. In *Drosophila*, M strain females lack P elements and lack the ability to inhibit P element transposition. If these M strain females are crossed with males that contain numerous P elements (P strain males), the egg cells allow the P elements to transpose at a high rate. The resulting hybrid offspring exhibit a variety of abnormalities, which include a high rate of sterility, mutation, and

TABLE 10.2

Abundance of Transposable Elements in the Genomes of Selected Species

Species	Percentage of the Total Genome Composed of TEs*
Frog (*Xenopus laevis*)	77
Corn (*Zea mays*)	60
Human (*Homo sapiens*)	45
Mouse (*Mus musculus*)	40
Fruit fly (*Drosophila melanogaster*)	20
Nematode (*Caenorhabditis elegans*)	12
Yeast (*Saccharomyces cerevisiae*)	4
Bacterium (*Escherichia coli*)	0.3

*In some cases, the abundance of TEs may vary somewhat among different strains of the same species. The values reported here are typical values.

TABLE 10.3

Possible Consequences of Transposition

Consequence	Cause
Chromosome Structure	
Chromosome breakage	Excision of a TE.
Chromosomal rearrangements	Homologous recombination between TEs located at different positions in the genome.
Gene Expression	
Mutation	Incorrect excision of TEs.
Gene inactivation	Insertion of a TE into a gene.
Alteration in gene regulation	Transposition of a gene next to a regulatory sequence or the transposition of a regulatory sequence next to a gene.
Alteration in the exon content of a gene	Insertion of exons into the coding sequence of a gene via TEs. This phenomenon is called exon shuffling.

chromosome breakage. This deleterious outcome, which is called **hybrid dysgenesis,** occurs because the P elements were able to insert into a variety of locations in the genome.

10.5 COMPREHENSION QUESTIONS

1. Which of the following types of transposable elements rely on an RNA intermediate for transposition?
   a. Insertion elements
   b. Simple transposons
   c. Retrotransposons
   d. All of the above
2. The function of transposase is
   a. to recognize inverted repeats.
   b. to remove a TE from its original site.
   c. to insert a TE into a new site.
   d. all of the above.
3. According to the selfish DNA hypothesis, TEs exist because
   a. they offer the host a selective advantage.
   b. they have characteristics that allow them to multiply within the chromosomal DNA of living cells.
   c. they promote the expression of certain beneficial genes.
   d. all of the above.

10.6 STRUCTURE OF EUKARYOTIC CHROMOSOMES IN NONDIVIDING CELLS

Learning Outcomes:

1. Define *chromatin.*
2. Describe the structures of (1) nucleosomes, (2) nucleosome interactions according to the zigzag model, and (3) loop domains.
3. Analyze Noll's results, and explain how they support the beads-on-a-string model.
4. Describe how loop domains are formed according to the loop extrusion model.
5. Describe the key features of a topologically interacting domain.
6. Explain the general strategy of chromosome conformation capture methods.
7. Compare and contrast euchromatin and heterochromatin, and distinguish between constitutive heterochromatin and facultative heterochromatin.
8. Explain the meaning of *chromosome territory.*
9. List the four levels of chromosome structure in a nondividing cell.

A distinguishing feature of eukaryotic cells is that their chromosomes are located within a cellular compartment known as the **nucleus.** The DNA within a typical eukaryotic chromosome is a single, linear, double-stranded molecule that may be hundreds of millions of base pairs in length. If the DNA from a single set of human chromosomes were stretched from end to end, the length would be over 1 meter! By comparison, most eukaryotic cells are only 10–100 μm in diameter, and the cell nucleus is only about 2–4 μm in

diameter. Therefore, the DNA in a eukaryotic cell must be folded and compacted to a staggering extent to fit inside the nucleus. In eukaryotic chromosomes, as in prokaryotic chromosomes, this is accomplished by the binding of many different proteins to the DNA.

The DNA-protein complex found within eukaryotic chromosomes is termed **chromatin.** In recent years, it has become increasingly clear that the proteins bound to chromosomal DNA are subject to change over the life of the cell. These changes in protein composition, in turn, affect the degree of compaction of the chromatin. As discussed in Chapter 17, non-coding RNA molecules also play a role in chromatin structure. In this section, we will consider the structures of chromosomes during interphase—the period of the cell cycle that includes the G_1, S, and G_2 phases. In Section 10.7, we will examine the additional compaction that is necessary to produce the highly condensed chromosomes present during M phase.

Linear DNA Wraps Around Histone Proteins to Form Nucleosomes, the Repeating Structural Unit of Chromatin

As in some archaea, the repeating structural unit within eukaryotic chromatin is the nucleosome. In eukaryotes, a nucleosome is a double-stranded segment of DNA wrapped around an octamer of **histone proteins,** or **histones (Figure 10.17a).**

- Each octamer contains two copies each of four different histone proteins: H2A, H2B, H3, and H4. These are called the core histones.
- Each of the histone proteins consists of a globular domain and a flexible, charged amino terminus called an amino-terminal tail.
- The DNA is negatively supercoiled over the surface of an octamer; it makes 1.65 negative superhelical turns around the octamer. Positively charged amino acids lysine and arginine in the histone proteins play a major role in binding to the negatively charged phosphate groups along the DNA backbone.
- The amount of DNA required to wrap around the histone octamer is 146 or 147 bp.
- At its widest point, a single nucleosome is about 11 nm in diameter.
- In 1997, Timothy Richmond and colleagues determined the structure of a nucleosome by X-ray crystallography **(Figure 10.17b).**

The chromatin of eukaryotic cells displays a repeating pattern in which the nucleosomes are connected by linker regions of DNA that vary in length from 20 to 100 bp, depending on the species and cell type. It has been suggested that the overall structure of connected nucleosomes resembles beads on a string. This structure shortens the length of the DNA molecule about sevenfold.

In addition to the core histones, another histone protein, H1, is found in most eukaryotic cells and is called the linker histone. It binds to the DNA in the linker region between nucleosomes and helps to organize adjacent nucleosomes **(Figure 10.17c).** The linker histone is less tightly bound to the DNA than are the core histones. In addition, nonhistone proteins bound to the linker region play a role in the organization and compaction of chromosomes, and their presence may affect the expression of nearby genes.

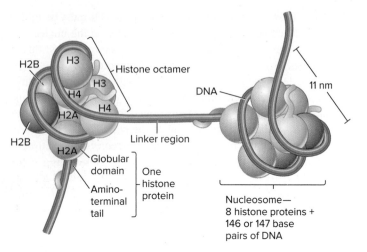

(a) Nucleosomes showing core histone proteins

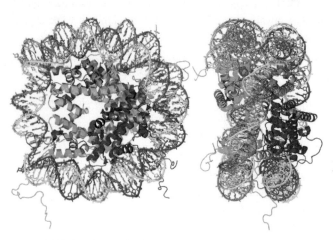

(b) Molecular model for nucleosome structure

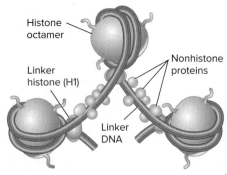

(c) Nucleosomes showing linker histones and nonhistone proteins

FIGURE 10.17 **Nucleosome structure.** **(a)** A nucleosome consists of 146 or 147 bp of DNA wrapped around an octamer of core histone proteins. **(b)** A model for the structure of a nucleosome as determined by X-ray crystallography. The drawing shows two views of a nucleosome that are at right angles to each other. **(c)** The linker region of DNA connects adjacent nucleosomes. The linker histone H1 and the nonhistone proteins also bind to this linker region. Note: In part (a), the structures of the histone proteins are shown as spheres, which are meant to be schematic. H2A-H2B and H3-H4 actually form intertwined dimers, as shown in part (b).

(b): Laguna Design/SPL/Science Source

CONCEPT CHECK: What is the diameter of a nucleosome?

EXPERIMENT 10A

The Repeating Nucleosome Structure Is Revealed by Digestion of the Linker Region

The model of eukaryotic nucleosome structure was originally proposed by Roger Kornberg in 1974. This proposal was based on several observations. Biochemical experiments had shown that chromatin contains a ratio of one molecule of each of the four core histone proteins (namely, H2A, H2B, H3, and H4) per 100 bp of DNA. Approximately one H1 protein was found per 200 bp of DNA. In addition, purified core histone proteins were observed to bind to each other via specific pairwise interactions. Subsequent X-ray diffraction studies showed that chromatin is composed of a repeating pattern of smaller units. Finally, electron microscopy of chromatin fibers revealed a diameter of approximately 11 nm. Taken together, these observations led Kornberg to propose a model in which the DNA double helix is wrapped around an octamer of core histone proteins. Including the linker region, this structure involves about 200 bp of DNA.

Markus Noll decided to test Kornberg's model by digesting chromatin with DNase I, an enzyme that cuts the DNA backbone. He then accurately measured the molecular mass of the DNA fragments by gel electrophoresis. Noll assumed that the linker region of DNA is more accessible to DNase I and, therefore, that enzyme is more likely to make cuts in the linker region than in the 146-bp region that is tightly bound to the core histones. If this was correct, incubation with DNase I was expected to make cuts in the linker region, thereby producing DNA pieces approximately 200 bp in length. The size of the DNA fragments might vary somewhat because the linker region is not of constant length and because the cut within the linker region may occur at different sites.

Figure 10.18 describes Noll's experimental protocol. He began with nuclei from rat liver cells and incubated them with low, medium, or high concentrations of DNase I. The DNA was extracted into an aqueous phase and then loaded onto an agarose gel that separated the fragments according to their molecular mass. The DNA fragments within the gel were stained with a UV-sensitive dye, ethidium bromide, which made it possible to view the DNA fragments under UV illumination.

THE HYPOTHESIS

The aim of this experiment was to test the beads-on-a-string model for chromatin structure. According to this model, DNase I should preferentially cut the DNA in the linker region, thereby producing DNA pieces that are about 200 bp in length.

TESTING THE HYPOTHESIS   **FIGURE 10.18** DNase I cuts chromatin into repeating units containing about 200 bp of DNA.

Starting material: Nuclei from rat liver cells.

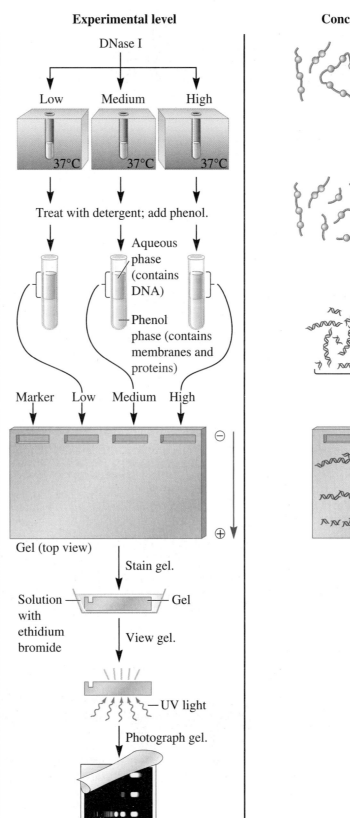
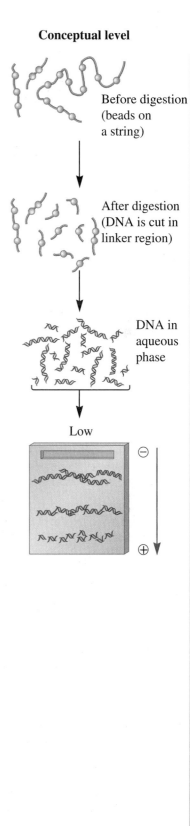

Experimental level

Conceptual level

1. Incubate the nuclei with low, medium, and high concentrations of DNase I. The conceptual level illustrates a low DNase I concentration.

2. Isolate the DNA. This involves dissolving the nuclear membrane with detergent and treating the sample with the organic solvent phenol.

3. Load the DNA into a well of an agarose gel and run the gel to separate the DNA pieces according to size. On this gel, also load DNA fragments of known molecular mass (marker lane).

4. Visualize the DNA fragments by staining the DNA with ethidium bromide, a dye that binds to DNA and is fluorescent when excited by UV light.

DNase I

Low   Medium   High

37°C   37°C   37°C

Treat with detergent; add phenol.

Aqueous phase (contains DNA)

Phenol phase (contains membranes and proteins)

Marker   Low   Medium   High

⊖

⊕

Gel (top view)

Stain gel.

Solution with ethidium bromide — Gel

View gel.

UV light

Photograph gel.

Before digestion (beads on a string)

After digestion (DNA is cut in linker region)

DNA in aqueous phase

Low

⊖

⊕

THE DATA

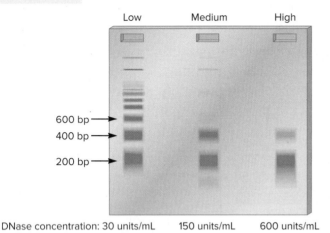

DNase concentration: 30 units/mL   150 units/mL   600 units/mL

Note: The marker lane is omitted from this schematic drawing.

Source: Adapted from Noll, Markus, "Subunit Structure of Chromatin," *Nature*, vol. 251, no. 5472, September 20, 1974, 249–251.

INTERPRETING THE DATA

As shown in the data, at a high DNase I concentration, the entire sample of chromosomal DNA was digested into fragments of approximately 200 bp in length. This result is predicted by the beads-on-a-string model. Furthermore, at a low or medium DNase I concentration, longer pieces were observed, and these were in multiples of 200 bp (400, 600, etc.). How do we explain these longer pieces? They occurred because occasional linker regions remained uncut at a low or medium DNase I concentration. For example, if one linker region was not cut, a DNA piece would contain two nucleosomes and be about 400 bp in length. If two consecutive linker regions were not cut, this would produce a piece with three nucleosomes containing about 600 bp of DNA. Taken together, these results strongly supported the nucleosome model for chromatin structure.

Nucleosomes Associate with Each Other in a Zigzag Manner

In eukaryotic chromatin, nucleosomes associate with each other to form a more compact structure. Evidence for the packaging of nucleosomes was obtained in microscopy studies by Fritz Thoma in 1977. Chromatin samples were exposed to a solid resin that could bind to histone H1 and remove it from the DNA. However, the removal of H1 depended on the NaCl concentration. A buffered solution with a moderate salt concentration (100 mM NaCl) removed H1, but a solution with no added NaCl did not remove H1. Both types of samples were then observed with an electron microscope.

- At moderate salt concentrations (**Figure 10.19a**), the chromatin exhibited the classic beads-on-a-string morphology.
- Without added NaCl (when H1 is expected to remain bound to the DNA), these "beads" associated with each other in a more compact conformation (**Figure 10.19b**).

These results suggest that the nucleosomes are packaged into a more compact unit and that H1 has a role in the organization and compaction of nucleosomes.

The experiment of Figure 10.18 and other experiments have established that nucleosome units can associate with each into a more compact structure. A model for how nucleosomes interact with each other, advocated by Rachel Horowitz, Christopher Woodcock, and others is known as the **zigzag model.** This model is based on research using techniques such as cryoelectron microscopy (electron microscopy at low temperature). According to this model, nucleosomes zigzag back and forth connected by straight linker regions (**Figure 10.20**). As seen in the figure, each nucleosome preferentially interacts with a nucleosome that is two nucleosomes away from it.

In 2005, Timothy Richmond and colleagues were the first to solve the crystal structure of a segment of DNA containing multiple nucleosomes, in this case four. The structure with four

nucleosomes revealed that the linker DNA zigzags back and forth between nucleosomes, a feature consistent with the zigzag model. More recently, results from a cryoelectron microscopy study by Feng Song, Guohong Li, and colleagues were also consistent with the zigzag model.

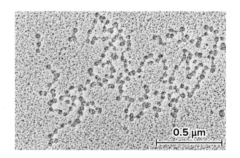

(a) H1 histone not bound—beads on a string

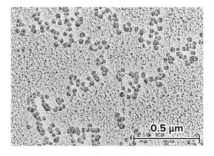

(b) H1 histone bound to linker region—nucleosomes interactions occur

FIGURE 10.19 **The nucleosome structure of eukaryotic chromatin as viewed by electron microscopy.** The chromatin in (a) has been treated with a moderately concentrated NaCl solution to remove the linker histone, H1. It exhibits the classic beads-on-a-string morphology. The chromatin in (b) has been incubated without added NaCl and shows nucleosome interactions.

(a and b): ©Fritz Thoma, ETH Zurich

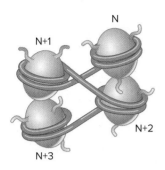

FIGURE 10.20 Zigzag model. In the zigzag model, the linker DNA is relatively straight and the nucleosomes form a zigzag arrangement. The nucleosomes (N) are labeled according to their arrangement if they were stretched out linearly: N, N+1, N+2, and N+3. The zigzag model is consistent with more recent data regarding chromatin conformation.

For several decades, many researchers proposed that nucleosomes associate over long stretches to form a secondary structure with a 30-nm diameter, called the 30-nm fiber. This proposal was largely based on the ability of chromatin to form a 30-nm fiber in vitro when chromatin was extracted from nuclei. However, the ability to observe a 30-nm fiber in living cells remained elusive. Based on more recent studies, there is little evidence to support that such long stretches of secondary structure exist in vivo to a great extent. Instead, these newer studies indicate that short

tri- and tetranucleosomes are arranged in a zigzag manner and prevalent in vivo. However, they do not appear to form longer, fiber-like structures.

Chromatin Is Further Compacted by the Formation of Loop Domains

Thus far, we have examined two mechanisms that compact eukaryotic DNA: the wrapping of DNA within nucleosomes and the interaction of nucleosomes to form a zigzag arrangement. To further compact chromatin, segments of nucleosomes are folded into loops, which are called loop domains. The formation of loop domains is a key mechanism that contributes to chromatin compaction.

How do loops form? According to the **loop extrusion model,** two different proteins play key roles in loop formation. An **SMC protein** forms a dimer that can wrap itself around DNA and initially form a small loop **Figure 10.21**). (SMC stands for structural maintenance of chromosomes.) The chromatin is forced through the SMC protein as the loop grows. The term *extrusion* refers to forcing or pushing chromatin through a small opening in order to form a loop.

The other key player is a protein named the **CCCTC binding factor (CTCF),** which binds to three regularly spaced repeats of the core sequence CCCTC. As shown in Figure 10.21, two different CTCFs are bound to the DNA outside of the region where the loop is forming. They get closer to each other as the loop grows. Eventually, these two CTCFs bind to each other. This stops the further growth of the loop and stabilizes the bottom of the loop.

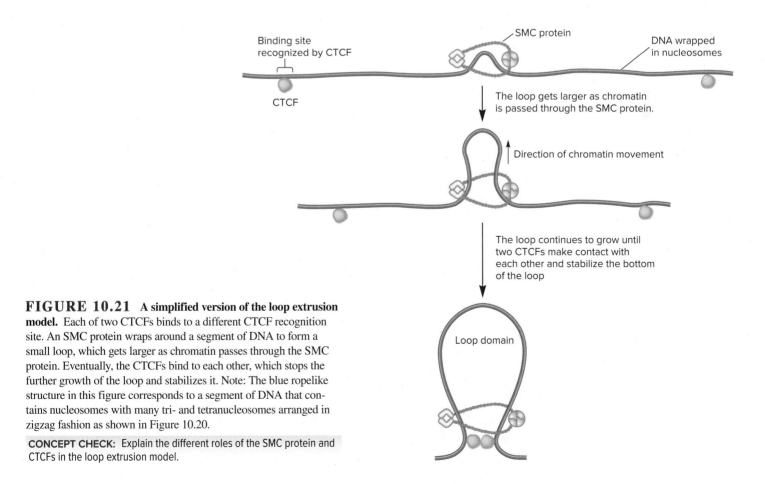

FIGURE 10.21 A simplified version of the loop extrusion model. Each of two CTCFs binds to a different CTCF recognition site. An SMC protein wraps around a segment of DNA to form a small loop, which gets larger as chromatin passes through the SMC protein. Eventually, the CTCFs bind to each other, which stops the further growth of the loop and stabilizes it. Note: The blue ropelike structure in this figure corresponds to a segment of DNA that contains nucleosomes with many tri- and tetranucleosomes arranged in zigzag fashion as shown in Figure 10.20.

CONCEPT CHECK: Explain the different roles of the SMC protein and CTCFs in the loop extrusion model.

The loop extrusion model involving SMC proteins and CTCFs is a primary mechanism for loop formation in eukaryotic chromatin. However, chromatin loops can form in other ways. For example, loops can form when proteins bound to regulatory sequences interact with gene promoters. This topic is discussed in Chapter 15 (look ahead to Figure 15.2a)

Additional loop formation is required during interphase so that chromosomes can fit inside the cell nucleus. In Section 10.7, we will take a closer look at how SMC proteins are also involved in chromosome condensation that occurs during mitosis.

Topologically Associated Domains Are Revealed by Chromosome Conformation Capture Methods

<u>C</u>hromosome <u>c</u>onformation <u>c</u>apture methods (abbreviated 3C methods) are a set of related techniques used to analyze the spatial organization of chromatin in a living cell. They were pioneered by Job Dekker and colleagues in 2002. These methods identify interactions between sites in the genome that are near one another in three-dimensional space, but may be separated by many base pairs in the linear DNA sequence. For example, Figure 10.21 shows how the bottom of a loop domain is held together by CTCFs. A 3C method can determine that the DNA segments bound by the CTCFs at the bottom of this loop are close to each other in three-dimensional space.

How do 3C methods work? Although the methods vary in their details, a general strategy for 3C methods is shown in **Figure 10.22**.

1. The chromatin in living cells is subjected to a crosslinking reagent that catalyzes covalent crosslinks between proteins and between DNA and proteins.
2. The DNA is then digested into small fragments with an endonuclease.
3. DNA pieces that are being held close to each other due to crosslinking are ligated to each other.
4. The crosslinking is undone.
5. The resulting DNA fragments are subjected to DNA sequencing.
6. As shown at the bottom of Figure 10.22, DNA sequencing will reveal areas where the DNA sequence switches from one region to another, revealing that these two regions are close to each other in three-dimensional space.

3C methods have provided insight regarding the organization of chromatin in nondividing cells. A key observation is that chromatin is organized into regions called **topologically associating domains (TADs).** The segments of DNA within a TAD are more likely to interact with each other than they are with segments in other neighboring TADs. Each TAD is typically 100 kb to 1 Mb in length and may include 1 to 5 genes.

As shown in **Figure 10.23**, TAD boundaries are determined by SMC proteins and CTCFs. Within each TAD, additional loops may be formed due to interactions between various types of proteins, such as interactions between regulatory transcription factors and general transcription factors (see Chapter 15). Organization of chromatin into TADs is thought to

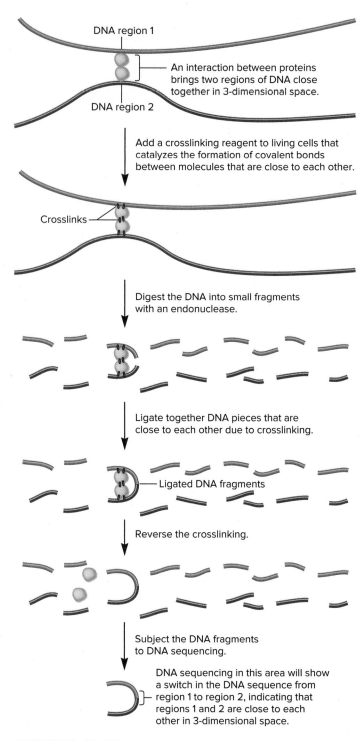

FIGURE 10.22 **The general strategy used in chromosome conformation capture methods.** In this illustration, two segments of DNA are shown in different colors (blue and red). For simplicity, the DNA is not shown in nucleosomes.

facilitate interactions between regulatory elements, regulatory transcription factor proteins, and their target promoters. Even so, changes in TAD boundaries sometimes do not have dramatic effects on gene regulation, suggesting that other factors are also important for gene regulation.

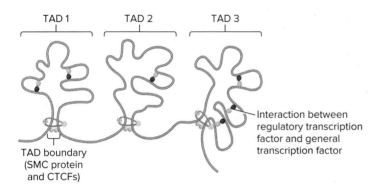

FIGURE 10.23 **Topologically associating domains.** Three TADs are shown here. Each TAD has a boundary formed by an SMC protein and two CTCFs. With each TAD, smaller loops can form due to protein-protein interactions such as interactions between regulatory transcription and general transcription factors. Note: The blue ropelike structure in this figure corresponds to a segment of DNA that contains nucleosomes with many tri- and tetranucleosomes arranged in zigzag fashion as shown in Figure 10.20.

CONCEPT CHECK: What are two ways that loops are formed in a TAD?

In addition to promoting interactions within a TAD, another function of TAD boundaries is to prevent interactions between different TADs. For example, a regulatory element in one TAD will not affect the transcription of a gene in a different TAD. In this regard, the TAD boundaries are acting as **insulators,** which are structures that limit genetic interactions to a specific region. A gene within a given TAD is insulated from the effects of regulatory elements in other TADs.

Some Chromatin Is Highly Compacted During Interphase

During interphase, when chromosomes are in the cell nucleus, the compaction level of each chromosome varies. This variability can be seen with a light microscope and was first observed by Emil Heitz in 1928.

- The term **heterochromatin** describes the tightly compacted regions of chromosomes in nondividing cells. These regions of a chromosome are usually transcriptionally inactive.
- Less compacted regions, known as **euchromatin,** are capable of gene transcription.

In euchromatin, the TADs are less compacted and thereby accessible to factors that are needed for gene transcription. In heterochromatin, however, clusters of many nucleosomes are very close to each other. The function, structure, and formation of heterochromatin are described in greater detail in Chapter 16 (look ahead to Section 16.2).

Figure 10.24 is a schematic drawing that shows the distribution of heterochromatin and euchromatin in a typical eukaryotic chromosome during interphase. The chromosome contains regions of both heterochromatin and euchromatin. Heterochromatin is most abundant in the centromeric regions of the chromosome and, to a lesser extent, in the telomeric regions. The term **constitutive heterochromatin** refers to chromosomal regions that are heterochromatic in all cell types of a multicellular organism. Constitutive heterochromatin usually contains highly repetitive DNA sequences, such as tandem repeats, rather than gene sequences. It is typically found in the centromeric and telomeric regions.

Facultative heterochromatin refers to heterochromatin that differs among different cells of the body. For example, the locations of facultative heterochromatin in a muscle cell is different from the locations in a nerve cell. Facultative heterochromatin usually occurs in small regions in between the centromere and the telomeres, where many genes are located. In Chapter 16, we will consider how chromatin can interconvert between euchromatin and facultative heterochromatin as a way to regulate the expression of genes.

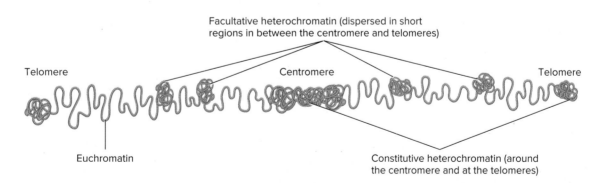

FIGURE 10.24 **Chromatin structure during interphase.** Constitutive heterochromatic regions are more highly condensed and tend to be localized in centromeric and telomeric regions. Facultative heterochromatic regions occur as shorter regions in between the centromere and telomeres. Note: To show the euchromatic and heterochromatic regions, this chromosome is depicted in a linear manner. In an actual cell, it would be found in a more oval-shaped chromosome territory within the cell nucleus, as shown in Figure 10.25b. Also, the heterochromatic regions are found close to the inner nuclear membrane as described in Chapter 16.

CONCEPT CHECK: Would you expect to find active genes in regions of heterochromatin or euchromatin?

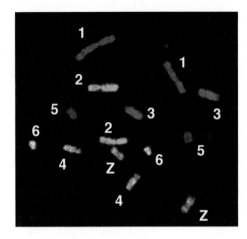
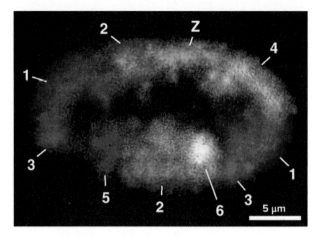

(a) Metaphase chromosomes   (b) Chromosomes in the cell nucleus during interphase

FIGURE 10.25 **Chromosome territories in the cell nucleus.** **(a)** Several metaphase chromosomes from chicken cells were labeled with chromosome-specific fluorescent molecules. Each of seven types of chicken chromosomes (i.e., 1, 2, 3, 4, 5, 6, and Z) is labeled a different color. **(b)** The same fluorescent molecules were used to label interphase chromosomes in the cell nucleus. Each of these chromosomes occupies its own distinct, nonoverlapping territory within the cell nucleus. (Note: Chicken cells are diploid, with two copies of each chromosome.)

(a and b): Courtesy of Felix Habermann and Irina Solovei, University of Munich (LMU, Biocenter)

CONCEPT CHECK: What is a chromosome territory?

In the Cell Nucleus, Each Chromosome Occupies Its Own Distinct Territory

Each chromosome in the cell nucleus is located in a discrete **chromosome territory.** As shown in studies by Thomas Cremer, Christoph Cremer, and others, these territories can be viewed when interphase cells are exposed to multiple fluorescent molecules that recognize specific sequences on particular chromosomes. **Figure 10.25** illustrates an experiment in which chicken cells were exposed to multiple fluorescent molecules that recognize specific sites within the larger chromosomes found in this species (*Gallus gallus*). Figure 10.25a shows the chromosomes in metaphase. The fluorescent molecules label each type of metaphase chromosome with a different color. Figure 10.25b shows the chromosomes labeled with the same fluorescent molecules during interphase, when the chromosomes are less condensed and found in the cell nucleus. As seen here, each chromosome occupies its own distinct territory.

Chromosome Organization and Structure in the Cell Nucleus has a Four-Level Hierarchy

How are chromosomes organized in nondividing cells? Based on a variety of methods including microscopy and crosslinking studies, a few organizing principles have emerged. The arrangement and folding of chromosomes in three-dimensional space is driven by at least four major processes. Going from a larger to a smaller scale, these are as follows (**Figure 10.26**):

Level 1.   At the scale of an entire nucleus, chromosomes occupy distinct territories (as shown in Figure 10.25b). The arrangement of the chromosomes within the nucleus is determined by a variety of factors. Chromosomal arrangements in the nucleus are partially driven by specific interchromosomal interactions. For example, at the interface between two adjacent territories, the telomere of one

chromosome may specifically interact with the telomere of another chromosome. The arrangement of chromosomes in the nucleus is also determined by chromosomal interactions with other nuclear structures. In particular, chromosomes bind to the proteins in the nuclear lamina that lines the inner nuclear membrane. This phenomenon is described in Chapter 16 (look ahead to Figure 16.5).

Level 2.   Within each territory, chromosomes are next organized on a megabase-scale into compartments in which active genes associate with other active genes while repressed genes cluster together. Euchromatin contains genes that are able to be transcribed into RNA, whereas heterochromatin largely contains repetitive sequences and genes that are transcriptionally silent. Heterochromatic regions tend to cluster next to each other and bind to the nuclear lamina. By contrast, euchromatic regions, which are less tightly compacted, tend to be found more toward the interior of the nucleus. The euchromatic area is called compartment A, and the heterochromatic area is called compartment B.

Level 3.   At a scale of 100 kb to 1 Mb, which typically includes 1 to 5 genes, chromosomes are organized into structural domains known as TADs. Organization of chromosomes into TADs is thought to be largely driven by loop extrusion, as described earlier in Figure 10.21.

Level 4.   Finally, chromosome organization occurs at a scale of 1 to 10 nucleosomes. This organization is largely determined by the affinity of nucleosomes for one another. Short tri- or tetranucleosomes, which are arranged in zigzag manner, are prevalent. As discussed in Chapter 15, short regions of DNA that lack nucleosomes are occasionally found (look ahead to Figure 15.9). In heterochromatin, certain proteins such as HP1, also cause nucleosomes to form close associations (look ahead to Figure 16.5).

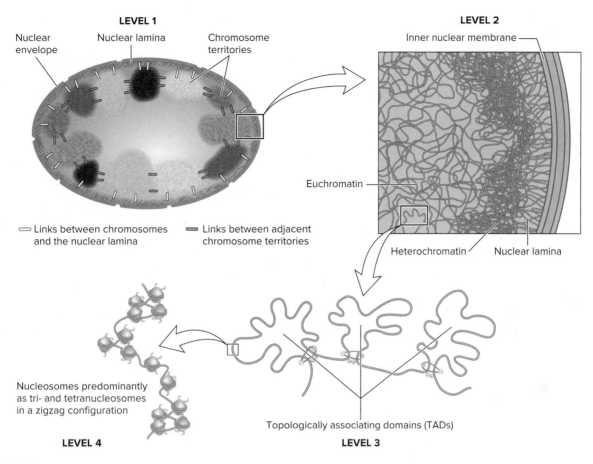

FIGURE 10.26 **Levels of chromosome organization in nondividing eukaryotic cells.** In level 2, the compaction of heterochromatin is much greater than that of euchromatin. Level 3 represents topologically associating domains found in euchromatin. Likewise, level 4 represents nucleosome arrangements in euchromatin. The structure of heterochromatin is described in Chapter 16 (look ahead to Figure 16.5).

10.6 COMPREHENSION QUESTIONS

1. What are the components of a single nucleosome?
   a. About 146 bp of DNA and four core histone proteins
   b. About 146 bp of DNA and eight core histone proteins
   c. About 200 bp of DNA and four core histone proteins
   d. About 200 bp of DNA and eight core histone proteins

2. In Noll's experiment to test the beads-on-a-string model, exposure of nuclei from rat liver cells to a low concentration of DNase I resulted in
   a. a single band of DNA with a size of approximately 200 bp.
   b. several bands of DNA in multiples of 200 bp.
   c. a single band of DNA with a size of 100 bp.
   d. several bands of DNA in multiples of 100 bp.

3. A chromosome territory is a region
   a. along a chromosome where many genes are clustered.
   b. along a chromosome where the nucleosomes are close together.
   c. in a cell nucleus where a single chromosome is located.
   d. in a cell nucleus where multiple chromosomes are located.

10.7 STRUCTURE OF EUKARYOTIC CHROMOSOMES DURING CELL DIVISION

Learning Outcomes:

1. Describe the levels of compaction that lead to a metaphase chromosome.
2. Explain the functions of condensin and cohesin.

As described in Chapter 2, when eukaryotic cells prepare to divide, the chromosomes become very condensed, or compacted. This compaction aids in their proper sorting during M phase. The highly compacted chromosomes that are viewed during metaphase of mitosis are aptly called **metaphase chromosomes.** In this section, we will examine the compaction level of metaphase chromosomes and discuss certain proteins that are involved in the process.

Metaphase Chromosomes Are Highly Compacted

As first described by Walther Flemming in 1882, when eukaryotic cells prepare to divide, their chromosomes become compacted

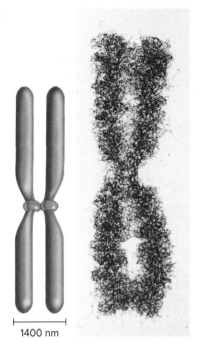

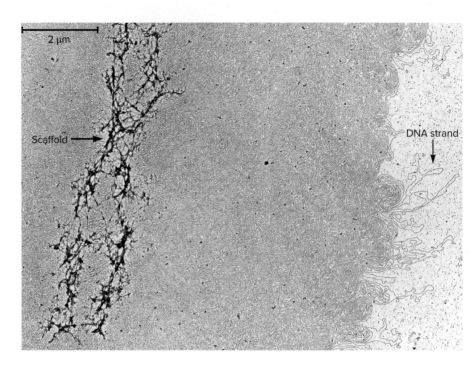

(a) Metaphase chromosome   (b) Metaphase chromosome treated with a highly concentrated salt solution to remove histone proteins

FIGURE 10.27   **The importance of histone proteins and scaffolding proteins in the compaction of eukaryotic chromosomes.**   (a) A metaphase chromosome. **(b)** A metaphase chromosome following treatment with a highly concentrated salt solution to remove the histone proteins. The arrow on the right points to an elongated strand of DNA. The arrow on the left points to the scaffold (composed of nonhistone proteins), which anchors the bases of the loops.

(a): Department of Virology, Haartman Institute/Peter Engelhardt; (b): Don W. Fawcett/Science Source

even further and typically form a characteristic X-shaped structure at metaphase (**Figure 10.27a**). Such metaphase chromosomes are highly condensed, which means they are highly compacted. This additional level of compaction greatly shortens the overall length of a chromosome and produces a diameter of approximately 700 nm. Two parallel chromatids have a diameter of approximately 1400 nm and a much shorter length than interphase chromosomes. The overall size of a metaphase chromosome is much smaller than a chromosome territory found in the cell nucleus during interphase (see Figure 10.25b).

The chromatin within metaphase chromosomes is organized along a protein scaffold. The chromosome scaffold was originally identified as a proteinaceous structure that could be derived from metaphase chromosomes after nuclease digestion or the removal of histone proteins. **Figure 10.27b** shows a metaphase chromosome that was treated with a highly concentrated salt solution to remove both the core and linker histones. As shown in this image, the highly compact configuration is lost, but the bottoms of the elongated loops remain attached to the darkly staining scaffold, which is composed of nonhistone proteins. A black arrow points to an elongated DNA strand emanating from the darkly stained protein scaffold. Remarkably, the scaffold retains the shape of the original metaphase chromosome even though the DNA strands have become greatly elongated.

Early biochemical analyses revealed that the scaffold contains a large amount of SMC proteins, suggesting that the association of

SMC proteins within the scaffold may help to organize chromatin loops. This conclusion was confirmed using biochemical, microscopic, and 3C methods, as described next.

Condensin I and II Promote the Condensation of Metaphase Chromosomes into Radial Loop Arrays

Researchers discovered that two multiprotein complexes, called **condensin I** and **II,** play a critical role in chromosomal condensation. Both of these complexes contain SMC proteins, which were described earlier in this chapter (look back at Figure 10.21).

Let's take a closer look at SMC proteins. They use energy from ATP to catalyze changes in chromosome structure. Together with topoisomerases and other subunits within condensins, SMC proteins have been shown to promote major changes in DNA structure. An emerging theme is that SMC proteins actively fold, tether, and manipulate DNA strands. These proteins generally have a dimeric structure, with the monomers connected at a hinge region and having a long coiled arm and a head region that binds ATP (**Figure 10.28**). The length of each monomer is about 50 nm, which is equivalent to approximately 150 bp of DNA.

The function of condensin is to facilitate the condensation of chromatin during mitosis and meiosis. Many eukaryotic species possess two different types of condensin complexes, known as condensin I and condensin II. Condensin II is able to enter the

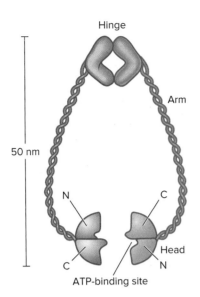

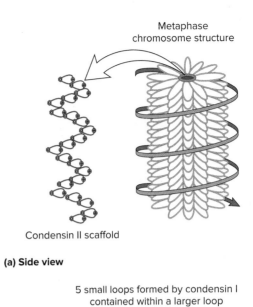

FIGURE 10.28 The structure of SMC proteins. This figure shows the generalized structure of SMC proteins, which are dimers that have hinge, arm, and head regions. The head regions bind and hydrolyze ATP. Condensins have additional protein subunits not shown here. Note: N indicates the amino terminus; C indicates the carboxyl terminus.

nucleus and plays a role in chromosome condensation during early prophase. Condensin I can bind to chromatin only after the nuclear envelope breaks apart. The coating of individual chromatids with both types of condensins is associated with the compaction of chromosomes into the characteristic (X-shaped) form they have at metaphase.

How do condensin I and II promote the formation of metaphase chromosomes? 3C methods conducted by Job Dekker and colleagues in 2018 have shed insight regarding the pathway for metaphase chromosome formation. Upon entry into prophase, interphase features such as compartment A (euchromatin), compartment B (heterochromatin), and TADs are lost within minutes. This loss is dependent on the presence of condensin II.

After the interphase organization is lost, the chromosomes become reorganized into **radial loop arrays (Figure 10.29)**. This reorganization is largely attributed to the concerted actions of condensin II and condensin I.

- Condensin II plays two key roles. First, condensin II complexes interact with each other to form a protein scaffold that has a helical shape (Figure 10.29a). Second, condensin II promotes the formation of loops that are anchored to this scaffold. These loops continue to grow during prophase and prometaphase, ultimately reaching a size of about 400 kb.
- Condensin I functions during prometaphase. It creates smaller loops within the loops generated by condensin II. The end result is ~80-kb loops that are nested within the ~400-kb loops produced by condensin II.

Thus, chromosome condensation is achieved by the creation of two types of structures. First, loops of chromatin become anchored

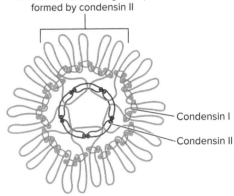

FIGURE 10.29 Structure of metaphase chromosomes. Because the loops are organized in a circular manner around a condensin II scaffold, the arrangement of chromatin loops is called a radial loop array. Note: Condensin I and II are multiprotein complexes. For simplicity, only the SMC proteins within these condensins are shown. The blue ropelike structures in this figure correspond to a segment of DNA that contains nucleosomes with many tri- and tetranucleosomes arranged in zigzag fashion as shown in Figure 10.20. Note: condensin I and II are similar in size, but condensin II is shown larger here in order to illustrate how condensin II proteins make contact with each other to form a scaffold.

CONCEPT CHECK: Describe two key functions of condensin II.

to a spiral scaffold via condensin II. Further compaction occurs because condensin I creates smaller loops within the large loops created by condensin II. Because the loops are organized in a circular and spiral manner around a condensin II scaffold, this arrangement of chromatin loops in metaphase chromosomes is called a radial loop array.

Cohesin Promotes the Binding of Sister Chromatids to Each Other

Cohesin is a multiprotein complex that also contains an SMC protein. Its function is to promote binding (i.e., cohesion) between

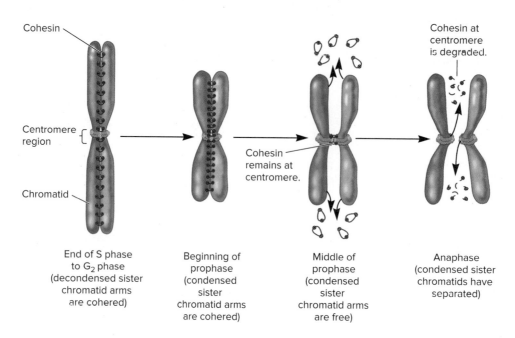

FIGURE 10.30   The alignment of sister chromatids via cohesin during mitosis. After S phase is completed, many cohesins bind along each chromatid, thereby facilitating attachment of the sister chromatids to each other. During the middle of prophase, cohesin is released from the chromosome arms, but some cohesin remains in the centromere region. At anaphase, the remaining cohesins are rapidly degraded by separase, which promotes separation of the sister chromatids. Note: during meiosis I, the cohesin proteins along the arms of sister chromatid are not released until the start of anaphase, which allows the separation of bivalents. During meiosis II, the cohesins at the centromere are degraded at the start of anaphase.

CONCEPT CHECK: Describe what happens to cohesin from the beginning of prophase through anaphase during mitosis.

sister chromatids during mitosis and meiosis. After S phase and until the middle of prophase, sister chromatids remain attached to each other along their length. As shown in **Figure 10.30**, this attachment is promoted by cohesin, which is found along the entire length of each chromatid. Cohesin dimers form rings around DNA in adjacent sister chromatids.

In certain species, such as mammals, cohesins located along the chromosome arms are released during prophase, which allows the arms to separate. However, some cohesins remain attached, primarily to the centromere region, leaving that region as the main linkage before anaphase. At anaphase, the cohesins bound to the centromere are rapidly degraded by a protease aptly named separase, thereby allowing separation of the sister chromatids.

10.7 COMPREHENSION QUESTIONS

1. The compaction leading to a metaphase chromosome involves which of the following?
   a. The anchoring of chromatin loops to a scaffold
   b. The formation of large loops via condensin II
   c. The formation of smaller loops via condensin I
   d. All of the above
2. The role of cohesin is to
   a. make chromosomes more compact.
   b. allow for the replication of chromosomes.
   c. hold sister chromatids together.
   d. promote the separation of sister chromatids.

KEY TERMS

Introduction: chromosomes, genome, transposable element (TE)

10.1: protein-coding genes, intergenic region, origin of replication, repetitive sequences

10.2: nucleoid, microdomain, macrodomain, nucleoid-associated proteins (NAPs), nucleosome, loop domain, DNA supercoiling, topoisomers, DNA gyrase, topoisomerase I

10.3: intron, exon, centromere, kinetochore, telomeres

10.4: sequence complexity, moderately repetitive sequences, highly repetitive sequences, tandem array

10.5: transposition, mutable site, simple transposition, transposon, retrotransposition, retrotransposon (retroelement), direct repeats (DRs), target-site duplications, insertion element (IS element), inverted repeats (IRs), transposase, simple transposon, LTR retrotransposon, long terminal repeats (LTRs), non-LTR retrotransposon, autonomous element, nonautonomous element, transposase, reverse transcriptase, integrase, target-site primed reverse transcription (TPRT), LINEs, SINEs, selfish DNA hypothesis, exon shuffling, hybrid dysgenesis

10.6: nucleus, chromatin, histone proteins (histones), zigzag model, loop extrusion model, SMC protein, CCCTC-binding factor (CTCF), chromosome conformation capture (3C) methods, topologically associating domains, insulators, heterochromatin, euchromatin, constitutive heterochromatin, facultative heterochromatin, chromosome territory

10.7: metaphase chromosome, condensin I and II, radial loop arrays, cohesin

CHAPTER SUMMARY

- Chromosomes contain the genetic material, which is DNA. A genome refers to a complete set of genetic material that a cell or species possesses.

10.1 Organization of Functional Sites Along Prokaryotic Chromosomes

- Bacterial and archaeal chromosomes are typically circular, have one or more origins of replication, and carry a few thousand genes (see Figure 10.1).

10.2 Structure of Prokaryotic Chromosomes

- A prokaryotic chromosome is found in a nucleoid of the cell (see Figure 10.2).
- Bacterial chromosomes are made more compact by the formation of microdomains and macrodomains (see Figure 10.3).
- The chromosome of some archaeal species is made more compact by the formation of nucleosomes, which contain histone proteins (see Figure 10.4).
- DNA supercoiling also makes prokaryotic chromosomes more compact (see Figure 10.5).
- Negative DNA supercoiling promotes DNA strand separation (see Figure 10.6).
- DNA gyrase (topoisomerase II) is an enzyme that introduces negative supercoils. Topoisomerase I relaxes negative supercoils (see Figure 10.7).

10.3 Organization of Functional Sites Along Eukaryotic Chromosomes

- Eukaryotic chromosomes are usually linear and contain a centromere, telomeres, multiple origins of replication, and many genes (see Figure 10.8).

10.4 Sizes of Eukaryotic Genomes and Repetitive Sequences

- The genome sizes of eukaryotes vary greatly. Some of this variation is due to the accumulation of repetitive sequences (see Figure 10.9).
- The human genome contains about 41% unique sequences and 59% repetitive sequences (see Figure 10.10).

10.5 Transposition

- McClintock discovered the phenomenon of transposition, in which a segment of DNA called a transposable element can move to multiple sites in a genome.

- Transposable elements can move via simple transposition (a cut-and-paste mechanism) or retrotransposition (see Figure 10.11).
- Each type of transposable element has a characteristic pattern of DNA sequences, which is always flanked by direct repeats (see Figure 10.12).
- Transposase catalyzes the excision and insertion of transposons at new sites via simple transposition (see Figure 10.13).
- Simple transposition can increase the number of copies of a transposon if it occurs just after a transposon has been replicated (see Figure 10.14).
- Retrotransposition of LTR retrotransposons requires reverse transcriptase and integrase, whereas retrotransposition of non-LTR retrotransposons occurs via target-site primed reverse transcription (see Figures 10.15, 10.16).
- Many different transposable elements are found among living organisms. Their abundance varies among different species (see Tables 10.1, 10.2).
- Transposition can have a variety of effects on chromosome structure and gene expression (see Table 10.3).

10.6 Structure of Eukaryotic Chromosomes in Nondividing Cells

- Chromatin is the DNA-protein complex found within eukaryotic chromosomes.
- Eukaryotic DNA wraps around an octamer of core histone proteins to form a nucleosome (see Figure 10.17).
- Noll tested Kornberg's nucleosome model by digesting eukaryotic chromatin with DNase I in varying concentrations (see Figure 10.18).
- The linker histone, H1, plays a role in nucleosome compaction (see Figure 10.19).
- A zigzag model has been proposed that describes the arrangement of nucleosomes (see Figure 10.20).
- SMC proteins and CTCFs form loop domains according to the loop extrusion model (see Figure 10.21).
- Chromosome conformation capture (3C) methods are used to analyze the spatial organization of chromatin in a living cell (see Figure 10.22).
- Eukaryotic chromatin is organized into regions called topologically associating domains (TADs) in which segments of DNA within a TAD are more likely to interact with each other than they are with segments in other neighboring TADs. Each TAD is insulated from genetic regulatory elements in other TADs (see Figure 10.23).

- In nondividing cells, each chromosome has highly compacted regions called heterochromatin and less compacted regions called euchromatin (see Figure 10.24).
- Within the cell nucleus, each eukaryotic chromosome occupies its own unique chromosome territory (see Figure 10.25).
- Chromosome organization and structure in the cell nucleus has a four-level hierarchy (see Figure 10.26).

10.7 Structure of Eukaryotic Chromosomes During Cell Division

- Metaphase chromosomes are highly condensed. The chromatin within metaphase chromosomes is organized along a protein scaffold (see Figure 10.27).
- Condensin I and condensin II contain SMC proteins and cause chromosomes to form compact structures organized into radial loop arrays (see Figures 10.28, 10.29).
- Cohesin promotes sister chromatid cohesion (see Figure 10.30).

PROBLEM SETS & INSIGHTS

MORE GENETIC TIPS

1. Suppose that a bacterial DNA molecule is given a left-handed twist. How does this affect the structure and function of the DNA?

TOPIC: *What topic in genetics does this question address?* The topic is DNA supercoiling. More specifically, the question is about the effects of giving a DNA molecule a left-handed twist.

INFORMATION: *What information do you know based on the question and your understanding of the topic?* From the question, you know that a DNA molecule has been given a left-handed twist. From your understanding of the topic, you may remember that DNA forms a right-handed double helix.

PROBLEM-SOLVING **S**TRATEGY: *Relate structure and function.* One strategy to solve this problem is to begin by considering DNA structure. Because DNA is right-handed, a left-handed twist could have either of two potential effects. It could add a negative supercoil, or it could promote DNA strand separation. Negative supercoiling makes the DNA more compact, and strand separation makes the DNA strands more accessible.

ANSWER: Negative supercoiling makes the bacterial chromosome more compact, so it fits within the cell. Alternatively, negative supercoiling promotes strand separation and thereby enhances DNA functions such as replication and transcription.

2. Describe the differences between unique and highly repetitive sequences in DNA.

TOPIC: *What topic in genetics does this question address?* The topic is the complexity of DNA sequences. More specifically, the question is about the differences between unique and highly repetitive sequences.

INFORMATION: *What information do you know based on the question and your understanding of the topic?* From the question, you know that some sequences in DNA are unique and some are highly repetitive. From your understanding of the topic, you may remember that unique DNA occurs once per haploid genome, whereas highly repetitive DNA occurs multiple times.

PROBLEM-SOLVING **S**TRATEGY: *Compare and contrast.* To answer this question, you can compare and contrast the features of unique and highly repetitive sequences.

ANSWER: Unique DNA occurs once per haploid genome. Many genes in a genome are unique. By comparison, a highly repetitive sequence, as its name suggests, is repeated many times, from tens of thousands to millions of times throughout a genome. It can be interspersed in the genome or clustered in a tandem array, in which a short nucleotide sequence is repeated many times in a row.

3. To hold bacterial DNA in a more compact configuration, specific proteins must bind to the DNA and stabilize its conformation. Several different proteins are involved in this process. Some of these proteins, such as H-NS, have been referred to as histone-like, because of their functional similarity to the histone proteins found in eukaryotes. Based on your knowledge of eukaryotic histone proteins, what biochemical properties would you expect bacterial histone-like proteins to have?

TOPIC: *What topic in genetics does this question address?* The topic is DNA compaction. More specifically, the question is asking you to predict the properties of the bacterial proteins that make a bacterial chromosome more compact.

INFORMATION: *What information do you know based on the question and your understanding of the topic?* From the question, you know that bacterial chromosomes have histone-like proteins. From your understanding of eukaryotic chromosomes, you may recall that the negatively charged DNA backbone wraps around positively charged core histone proteins.

PROBLEM-SOLVING **S**TRATEGY: *Relate structure and function.* One strategy to solve this problem is to consider the structure of nucleosomes, in which eukaryotic DNA is wrapped around core histone proteins.

ANSWER: The histone-like proteins have the properties expected for proteins involved in DNA folding. They are all small proteins found in relative abundance within the bacterial cell. In some cases, the histone-like proteins are biochemically similar to eukaryotic histones. For example, they tend to be basic (positively charged) and bind to the negatively charged DNA backbone in a non-sequence-dependent fashion.

4. If you assume that the average length of a DNA linker region is 50 bp, approximately how many nucleosomes can be found in the haploid human genome, which contains 3 billion bp?

T OPIC: *What topic in genetics does this question address?* The topic is chromosome structure. More specifically, the topic is the number of nucleosomes in the haploid human genome.

I NFORMATION: *What information do you know based on the question and your understanding of the topic?* From the question, you know that the haploid human genome has about 3 billion bp, and you are asked to assume that the average length of a DNA linker is 50 bp. From your understanding of the topic, you may remember that approximately 146 bp are found within one nucleosome.

P ROBLEM-SOLVING S TRATEGY: *Make a calculation.* The repeating unit is a nucleosome with 146 bp plus a linker region with 50 bp, which equals 196 bp. To determine the maximum number of nucleosomes, you divide 3 billion by 196.

ANSWER: $3,000,000,000/196 = 15,306,122$, or about 15.3 million. Note: This is the maximum number. The actual number is somewhat less because a small percentage of eukaryotic DNA occurs in nucleosome-free regions, a topic discussed in Chapter 15.

Conceptual Questions

C1. What is a nucleoid? With regard to cellular membranes, what is the difference between a prokaryotic nucleoid and a eukaryotic nucleus?

C2. In Part II of this text, we considered inheritance patterns for diploid eukaryotic species. Prokaryotic cells frequently contain two or more nucleoids. With regard to genes and alleles, how is a bacterium that contains two nucleoids similar to a diploid eukaryotic cell, and how is it different?

C3. Describe the mechanisms by which bacterial DNA becomes compacted.

C4. Why is DNA supercoiling called supercoiling rather than just coiling? Why is positive supercoiling called overwinding and negative supercoiling called underwinding? How would you define the terms *positive* and *negative supercoiling* for Z DNA (described in Chapter 9)?

C5. Coumarins and quinolones are two classes of drugs that inhibit the growth of prokaryotes by directly inhibiting DNA gyrase. Discuss two reasons why inhibiting DNA gyrase also inhibits prokaryotic cell growth.

C6. Take two pieces of string that are approximately 10 inches long and create a double helix by wrapping them around each other to make 10 complete turns. Tape one end of the strings to a table, and then twist the strings three times (360° each time) in a right-handed direction. Note: As you are looking down at the strings from above, a right-handed twist is in the clockwise direction.

A. Did the three twists create more or fewer turns in your double helix? How many turns does your double helix have after you twisted it?

B. Is your double helix right-handed or left-handed? Explain your answer.

C. Did the three twists create any supercoils?

D. If you had coated your double helix with rubber cement and allowed the cement to dry before making the three additional right-handed twists, would the rubber cement make it more or less likely for the twists to create supercoiling? Would a pair of cemented strings be more or less like a real DNA double helix than an uncemented pair of strings? Explain your answer.

C7. Try to explain the function of DNA gyrase with a drawing.

C8. How are two topoisomers different from each other? How are they the same?

C9. On rare occasions, a chromosome can suffer a small deletion that removes the centromere. When this occurs, the chromosome usually is not found within subsequent daughter cells. Explain why a chromosome without a centromere is not transmitted very efficiently from mother to daughter cells. (Note: If a chromosome is located outside the nucleus after telophase, it is degraded.)

C10. What is the function of a centromere? At what stage of the cell cycle would you expect the centromere to be the most important?

C11. Describe the characteristics of highly repetitive DNA sequences.

C12. If you were examining a sequence of chromosomal DNA, what characteristics would cause you to believe that the sequence contained a transposable element?

C13. For insertion elements and simple transposons, what is the function of the inverted repeat sequences during transposition?

C14. Why does transposition always produce direct repeats in the chromosomal DNA?

C15. Which types of TEs have the greatest potential for proliferation: insertion elements, simple transposons, or retrotransposons? Explain your choice.

C16. Do you consider TEs to be mutagens? In other words, do TEs cause mutations? Explain.

C17. Let's suppose that a species of mosquito has two different types of simple transposons that we will call X elements and Z elements. The X elements appear quite stable. In a population of 100 mosquitoes, it is found that every mosquito has 6 X elements, and they are always located in the same chromosomal locations among different individuals. In contrast, the Z elements seem to move around quite a bit. Within the same 100 mosquitoes, the number of Z elements ranges from 2 to 14, and the locations of the Z elements tend to vary considerably among different individuals. Explain how one simple transposon can be stable and another simple transposon can be mobile, within the same group of individuals.

C18. This chapter describes different types of TEs, including insertion elements, simple transposons, LTR retrotransposons, and non-LTR retrotransposons. Which of these four types of TEs have the following features?

   A. Require reverse transcriptase to transpose

   B. Require transposase to transpose

   C. Are flanked by direct repeats

   D. Have inverted repeats

C19. What features distinguish a transposon from a retrotransposon? How are their DNA sequences different, and how are their mechanisms of transposition different?

C20. The occurrence of multiple transposable elements (e.g., transposons) within the genome of organisms has been suggested as a possible cause of chromosomal rearrangements such as deletions, translocations, and inversions. How could the occurrence of transposons promote these types of structural rearrangements?

C21. What is the difference between an autonomous element and a nonautonomous element? Is it possible for nonautonomous elements to move? If yes, explain how.

C22. Describe the structure of a eukaryotic nucleosome, and explain how these nucleosomes are arranged according to the zigzag model.

C23. Let's assume the linker region of DNA averages 54 bp in length. How many molecules of H2A would you expect to find in a DNA sample that is 46,000 bp in length?

C24. Explain how loop domains are formed in some archaea via Alba and in eukaryotes.

C25. What is a topologically associating domain (TAD)? Describe two different mechanisms for loop formation in a TAD.

C26. Compare heterochromatin and euchromatin. What are the differences between them? Discuss the difference between constitutive and facultative heterochromatin.

C27. What is a chromosome territory? Describe two types of interactions that facilitate the arrangement of chromosome territories in the cell nucleus.

C28. Discuss the differences between the compaction level of metaphase chromosomes and that of interphase chromosomes. When would you expect gene transcription and DNA replication to take place, during M phase or interphase? Explain why.

C29. During the formation of radial loop arrays in metaphase chromosomes, compare and contrast the roles of condensin I and II.

C30. With regard to metaphase chromosomes, what is the function of the protein scaffold? What is the main protein that constitutes the scaffold?

C31. What are the roles of the core histones and of histone H1 in the compaction of eukaryotic DNA?

C32. Which of the following proteins or protein complexes contain an SMC protein?

   A. Alba protein

   B. Condensin I and II

   C. Cohesin

C33. Describe the structure and function of SMC proteins.

C34. What is the function of cohesin during cell division?

Experimental Questions

E1. Two circular DNA molecules, molecule A and molecule B, are topoisomers of each other. When viewed under an electron microscope, molecule A appears more compact than molecule B. The level of gene transcription is much lower for molecule A. Which of the following three possibilities could account for these observations?

   First possibility: Molecule A has three positive supercoils, and molecule B has three negative supercoils.

   Second possibility: Molecule A has four positive supercoils, and molecule B has one negative supercoil.

   Third possibility: Molecule A has zero supercoils, and molecule B has three negative supercoils.

E2. Let's suppose you have isolated DNA from a cell and viewed it under a microscope. It looks supercoiled. What experiment would you perform to determine if it is positively or negatively supercoiled? In your answer, describe your expected results. You may assume that you have purified topoisomerases at your disposal.

E3. We seem to know more about the structure of eukaryotic chromosomal DNA than that of bacterial DNA. Discuss why you think this is so, and list several experimental procedures that have yielded important information concerning the compaction of eukaryotic chromatin.

E4. In Noll's experiment of Figure 10.18, explain where DNase I cuts the DNA. Why were the bands on the gel in multiples of 200 bp at lower DNase I concentrations?

E5. When chromatin is treated with a NaCl solution of moderate concentration, the linker histone H1 is removed (see Figure 10.19a). A higher salt concentration removes the rest of the histone proteins (see Figure 10.27b). If the experiment presented in Figure 10.18 was carried out after the DNA was treated with a moderately or highly concentrated salt solution, what would be the expected results?

E6. Let's suppose you have isolated chromatin from some bizarre eukaryote that has a DNA linker region that is usually 300–350 bp in length. The nucleosome structure is the same as in other eukaryotes. If you digested this eukaryotic organism's chromatin with a high concentration of DNase I, what would be your expected results?

E7. If you were given a sample of chromosomal DNA and asked to determine if it is bacterial or eukaryotic, what experiment would you perform, and what would be your expected results?

E8. Consider how histone proteins bind to DNA and explain why a salt solution with a high concentration can remove them from DNA (as shown in Figure 10.27b).

E9. In your own words, describe the experimental strategy of chromosome conformation capture (3C) methods. Let's suppose a researcher made a mistake and did the ligation step after the step in which the crosslinks are reversed, rather than before. How would this mistake affect the results?

E10. In Chapter 22, the technique of fluorescence in situ hybridization (FISH) is described. This is another method for examining sequence complexity within a genome. With this method, a DNA sequence, such as a particular gene sequence, can be detected within an intact chromosome by using a fluorescently labeled DNA molecule that is complementary to the sequence. For example, let's consider the β-globin gene, which is found on human chromosome 11. A labeled DNA molecule that is complementary to the β-globin gene binds to that gene and shows up as a brightly colored spot on human chromosome 11. In this way, researchers can detect where the β-globin gene is located within a set of chromosomes. Because the β-globin gene is unique and because human cells are diploid (i.e., have two copies of each chromosome), a FISH analysis shows two bright spots per cell; the labeled molecule binds to each copy of chromosome 11. What would you expect to see if you used the following types of labeled DNA molecules?

A. A labeled DNA molecule complementary to the *Alu* sequence

B. A labeled DNA molecule complementary to a tandem array near the centromere of the X chromosome

Questions for Student Discussion/Collaboration

1. Prokaryotic and eukaryotic chromosomes are very compact. Discuss the advantages and disadvantages of a compact chromosomal structure.

2. The prevalence of highly repetitive sequences seems rather strange to many geneticists. Does this phenomenon seem strange to you? Why or why not? Discuss whether or not you think highly repetitive sequences have an important function.

3. Discuss and make a list of the similarities and differences between prokaryotic and eukaryotic chromosomes.

Note: All answers are available for the instructor in Connect; the answers to the even-numbered questions and all of the Concept Check and Comprehension Questions are in Appendix B.

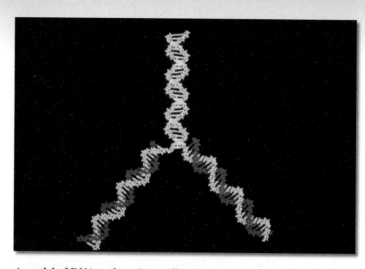

A model of DNA undergoing replication. This molecular model shows a DNA replication fork, the site where new DNA strands are made. In this model, the original DNA strands are yellow and blue. The newly made strands are purple.

Clive Freeman/The Royal Institution/Science Source

11

DNA REPLICATION

As discussed throughout Chapters 2–8, genetic material is transmitted from parent to offspring and from cell to cell. For transmission to occur, the genetic material must be copied. During this process, known as **DNA replication,** the original DNA strands are used as templates for the synthesis of new DNA strands. We will begin this chapter with a consideration of the structural features of the double helix that underlie the replication process. Then we will examine how chromosomes are replicated within living cells, addressing the following questions: Where does DNA replication begin, how does it proceed, and where does it end? We will first consider bacterial DNA replication and examine how DNA replication occurs within living cells, and we will then turn our attention to the unique features of the replication of eukaryotic DNA. At the molecular level, it is rather remarkable that the replication of chromosomal DNA occurs very quickly, very accurately, and at the appropriate time in the life of the cell. For this to happen, many cellular proteins play vital roles. In this chapter, we will examine the functions of several proteins involved in the process of DNA replication.

11.1 STRUCTURAL OVERVIEW OF DNA REPLICATION

Learning Outcomes:

1. Describe the structural features of DNA that enable it to be replicated.
2. Analyze the experiment of Meselson and Stahl, and explain how the results were consistent with the semiconservative model of DNA replication.

Because they bear directly on the replication process, let's begin by recalling a few important structural features of the double helix from Chapter 9. The double helix is composed of two DNA strands, and the individual building blocks of each strand are nucleotides. Each nucleotide contains one of four bases: adenine, thymine, guanine, or cytosine. The double-stranded structure is held together by base stacking and by hydrogen bonding between the bases in opposite strands. A critical feature of the double-helix structure is that adenine forms two hydrogen bonds with thymine, and guanine forms three hydrogen bonds with cytosine. This structural feature gave rise to the AT/GC rule, which describes the complementarity of the base sequences in double-stranded DNA.

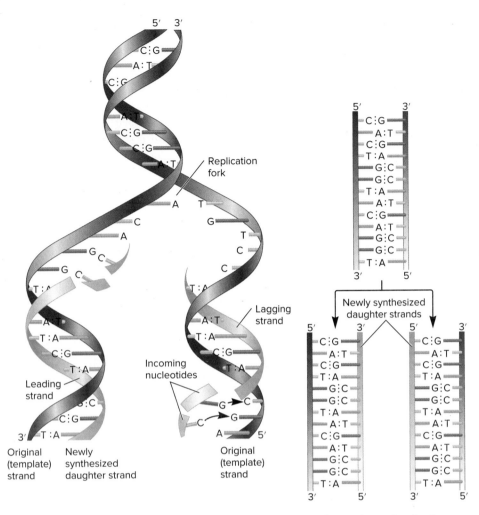

(a) The mechanism of DNA replication

(b) The products of replication

ONLINE ANIMATION

FIGURE 11.1 **The structural basis for DNA replication.** **(a)** The mechanism of DNA replication as originally proposed by Watson and Crick. As discussed later, the synthesis of one newly made strand (the leading strand) occurs in the direction toward the replication fork, whereas the synthesis of the other newly made strand (the lagging strand) occurs in small segments away from the replication fork (look ahead to Figure 11.10). **(b)** DNA replication produces two copies of DNA with the same base sequence as the original DNA molecule.

CONCEPT CHECK: What features of the structure of DNA enable it to be replicated?

Another feature worth noting is that the strands within a double helix have an antiparallel alignment. This directionality is determined by the orientation of sugar molecules within the sugar-phosphate backbone. If one strand is running in the 5′ to 3′ direction, the complementary strand is running in the 3′ to 5′ direction. The concept of directionality will be important when we consider the function of the enzymes that synthesize new DNA strands. In this section, we will consider how the structure of the DNA double helix provides the basis for DNA replication.

Existing DNA Strands Act as Templates for the Synthesis of New Strands

As shown in **Figure 11.1a**, DNA replication relies on the complementarity of DNA strands, based on the AT/GC rule.

- During the replication process, the two complementary strands of DNA come apart and serve as **template strands,** or **parental strands,** for the synthesis of two new strands of DNA.
- After the double helix has separated, individual nucleotides have access to the template strands. Hydrogen bonding between individual nucleotides and the template strands must obey the AT/GC rule.
- To complete the replication process, a covalent bond is formed between a phosphate of one nucleotide and the sugar of the previous nucleotide. The two newly made strands are referred to as the **daughter strands.**
- The base sequences are identical in both double-stranded molecules after replication (**Figure 11.1b**). That is, DNA is replicated in such a way that both copies retain the same information—the same base sequence—as in the original molecule.

Three Different Models Were Proposed to Describe the End Result of DNA Replication

Scientists in the late 1950s considered three different mechanisms to explain the end result of DNA replication. These mechanisms are shown in **Figure 11.2**.

- In the **conservative model,** both parental strands of DNA remain together following DNA replication. In this model, the original arrangement of parental strands is completely conserved, and the two newly made daughter strands also remain together following replication.
- In the **semiconservative model,** the double-stranded DNA is half conserved following the replication process. In other words, the newly made double-stranded DNA contains one parental strand and one daughter strand.
- In the **dispersive model,** segments of parental DNA and newly made DNA are interspersed in both strands following the replication process.

Only the semiconservative model shown in Figure 11.2b is correct. How did scientists determine this? In 1958, Matthew Meselson and Franklin Stahl devised a method to experimentally distinguish newly made daughter strands from the original parental strands. Their technique involved labeling DNA with a heavy isotope of nitrogen. Nitrogen, which is found in the bases of DNA, occurs in both a heavy (^{15}N) and a light (^{14}N) form. Prior to their experiment, Meselson and Stahl grew *Escherichia coli* cells in the presence of ^{15}N for many generations. This produced a population of cells in which all of the DNA was heavy labeled.

At the start of their experiment, shown in **Figure 11.3** (generation 0), they switched the bacteria to a medium that contained only ^{14}N and then collected samples of cells after various time points. Under the growth conditions they employed, the cells replicated their DNA and divided into daughter cells every 30 minutes. After each doubling, the new daughter cells were viewed as part of a new generation. Because the bacteria were doubling in a medium that contained only ^{14}N, all of the newly made DNA strands were labeled with light nitrogen, but the original strands remained in the heavy form.

Meselson and Stahl then analyzed the density of the DNA by centrifugation, using a cesium chloride (CsCl) gradient. (The procedure of gradient centrifugation is described in Appendix A.) If both DNA strands contained ^{14}N, the DNA had a light density and remained near the top of the tube. If one strand contained ^{14}N and the other strand contained ^{15}N, the DNA was half-heavy and had an intermediate density. Finally, if both strands contained ^{15}N, the DNA was heavy and moved closer to the bottom of the centrifuge tube.

THE HYPOTHESIS

Based on Watson's and Crick's ideas, the hypothesis was that DNA replication is semiconservative. Figure 11.2 also shows the two alternative models.

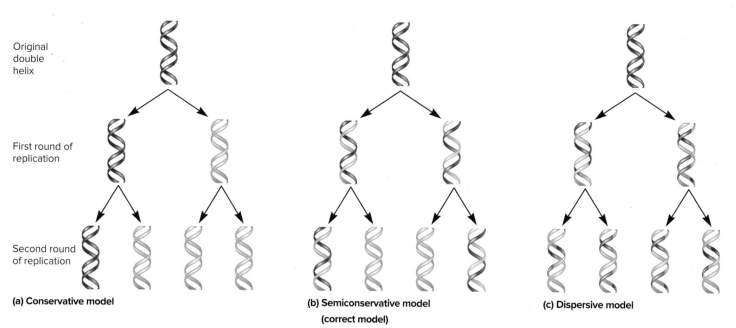

(a) Conservative model

(b) Semiconservative model
(correct model)

(c) Dispersive model

FIGURE 11.2 **Three possible models for DNA replication.** The two original parental strands of DNA are shown in purple, and the newly made strands after one and two generations are shown in light blue.

TESTING THE HYPOTHESIS   **FIGURE 11.3** Evidence that DNA replication is semiconservative.

Starting material: A strain of *E. coli* that has been grown for many generations in the presence of ^{15}N. All of the bases in the bacterial DNA are initially labeled with ^{15}N.

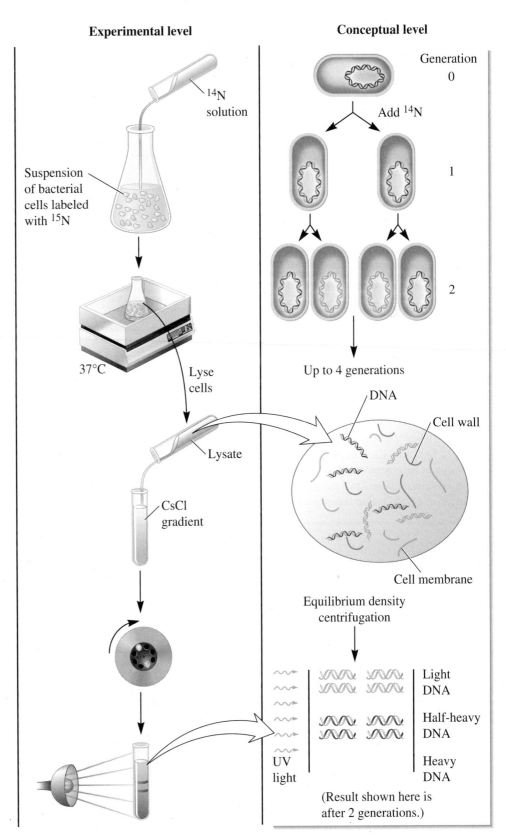

Experimental level   **Conceptual level**

1. Add an excess of ^{14}N-containing compounds to the growth medium so all of the newly made DNA will contain ^{14}N.

^{14}N solution

Generation 0

Suspension of bacterial cells labeled with ^{15}N

Add ^{14}N

1

2. Incubate the cells for various lengths of time. Note: The ^{15}N-labeled DNA is shown in purple, and the ^{14}N-labeled DNA is shown in blue.

2

3. Lyse the cells by adding lysozyme and detergent, which disrupt the bacterial cell wall and cell membrane, respectively.

37°C   Lyse cells

Up to 4 generations

DNA

Cell wall

Lysate

4. Load a sample of the lysate onto a CsCl gradient. Note: The average density of DNA is around 1.7 g/cm^3, which is sufficiently different from other cellular macromolecules.

CsCl gradient

Cell membrane

Equilibrium density centrifugation

5. Centrifuge the gradients until the DNA molecules reach their equilibrium densities.

Light DNA

Half-heavy DNA

6. DNA within the gradient can be observed under a UV light.

UV light

Heavy DNA

(Result shown here is after 2 generations.)

THE DATA

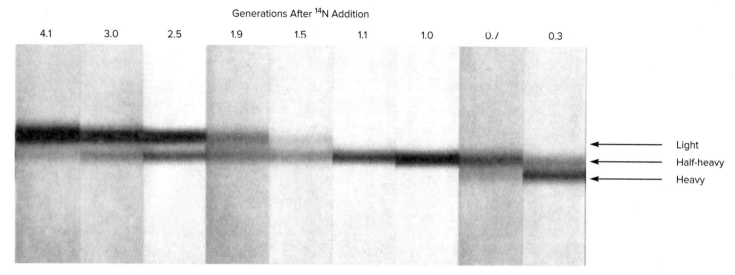

Data from M. Meselson & F. Stahl (1958), "The Replication of DNA in *Escherichia coli*," *PNAS*, 44(7): 671–682, Fig. 4A. Courtesy of M. Meselson.

INTERPRETING THE DATA

As seen in the data for Figure 11.3, after one round of DNA replication (i.e., after one generation), all of the DNA sedimented at a density that was half-heavy. Which of the three models is consistent with this result? Both the semiconservative and dispersive models are consistent. In contrast, the conservative model predicts two separate DNA types: a light type and a heavy type. Because all of the DNA had sedimented as a single band, this model was disproved. According to the semiconservative model, the replicated DNA would contain one original strand (a heavy strand) and a newly made daughter strand (a light strand). Likewise, in a dispersive model, all of the DNA should have been half-heavy after one generation as well. To determine which of these two remaining models is correct, therefore, Meselson and Stahl had to investigate future generations.

After approximately two rounds of DNA replication (i.e., after 1.9 generations), a mixture of light DNA and half-heavy DNA was observed. This result was consistent with the semiconservative model of DNA replication, because some DNA molecules should contain all light DNA, and other molecules should be half-heavy (see Figure 11.2b). The dispersive model predicts that after two generations, the heavy nitrogen would be evenly dispersed among four strands, each strand containing 1/4 heavy nitrogen and 3/4 light nitrogen (see Figure 11.2c). However, this result was not obtained. Instead, the results of the Meselson and Stahl experiment provided compelling evidence in favor of only the semiconservative model for DNA replication.

11.1 COMPREHENSION QUESTIONS

1. The complementarity of DNA strands is based on
   a. the chemical properties of a phosphodiester linkage.
   b. the binding of proteins to the DNA.
   c. the AT/GC rule.
   d. none of the above.

2. To make a new DNA strand, which of the following is/are necessary?
   a. A template strand   c. Heavy nitrogen
   b. Nucleotides   d. Both a and b

3. The model that correctly describes the process of DNA replication is
   a. the conservative model.
   b. the semiconservative model.
   c. the dispersive model.
   d. All of the above models describe the process correctly.

11.2 BACTERIAL DNA REPLICATION: THE FORMATION OF TWO REPLICATION FORKS AT THE ORIGIN OF REPLICATION

Learning Outcomes:

1. Describe the key features of a bacterial origin of replication.
2. Explain how DnaA proteins initiate DNA replication.

Thus far, we have considered how a complementary, double-stranded structure underlies the ability of DNA to be copied. In addition, the experiment of Meselson and Stahl showed that DNA replication results in two double helices, each one containing an original parental strand and a newly made daughter strand. We now turn our attention to how DNA replication occurs within

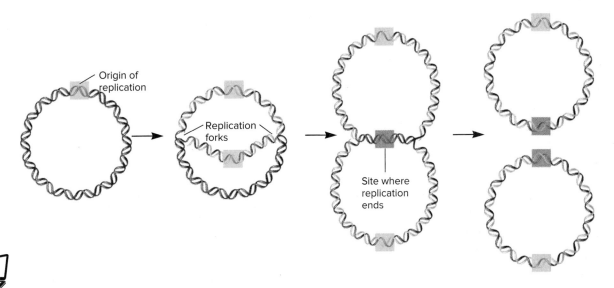

ONLINE ANIMATION

FIGURE 11.4 The process of bacterial DNA replication.

living cells. Much research has focused on understanding the process in the bacterium *E. coli*. In this section, we will examine how bacterial DNA replication begins.

Bacterial Chromosomes Contain a Single Origin of Replication

Figure 11.4 presents an overview of the process of bacterial chromosomal replication. The site on the bacterial chromosome where DNA synthesis begins is known as the **origin of replication.** Bacterial chromosomes have a single origin of replication. The synthesis of new daughter strands is initiated within the origin and proceeds in both directions, or **bidirectionally,** around the

bacterial chromosome. A **replication fork** is the region where the parental DNA strands have separated and new daughter strands are being made. Two replication forks move in opposite directions outward from the origin. Eventually, the replication forks meet each other on the opposite side of the bacterial chromosome to complete the replication process.

Replication Is Initiated by the Binding of DnaA Proteins to the Origin of Replication

Considerable research has focused on the origin of replication in *E. coli*. This origin is named *oriC*, for <u>ori</u>gin of <u>c</u>hromosomal replication (**Figure 11.5**). Three types of DNA sequences are found

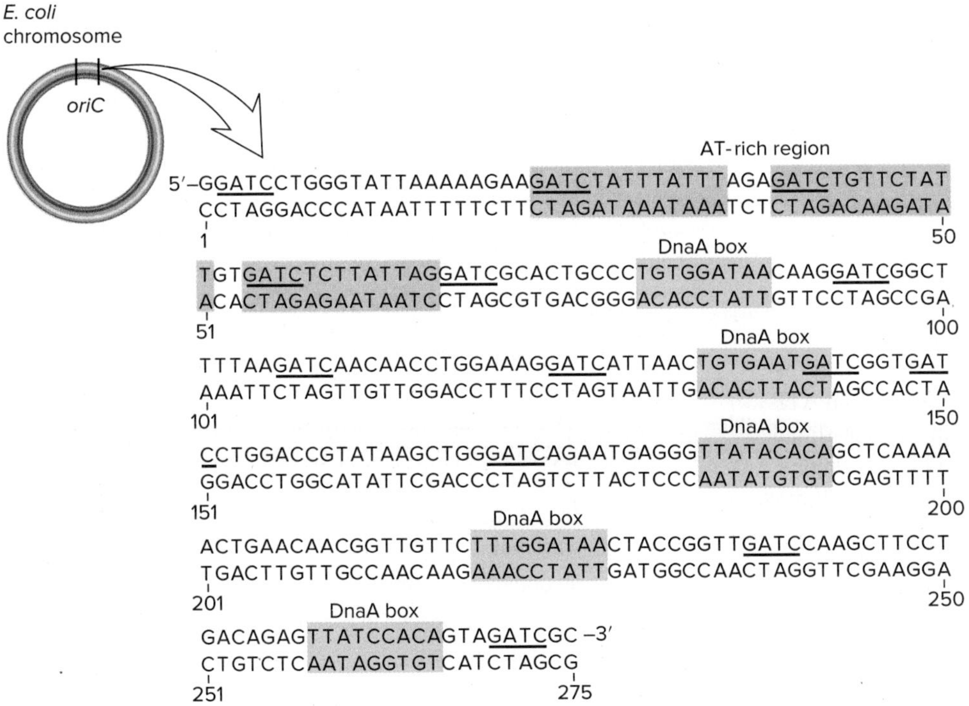

FIGURE 11.5 The sequence of *oriC* in *E. coli.* The AT-rich region is composed of three similar sequences that are 13 bp long and highlighted in blue. The five DnaA boxes are highlighted in gold. The GATC methylation sites are underlined.

CONCEPT CHECK: What are the functions of the AT-rich region and DnaA boxes?

E. coli chromosome

oriC

AT-rich region

```
5′–GGATCCTGGGTATTAAAAAGAAGATCTATTTATTTAGAGATCTGTTCTAT
   CCTAGGACCCATAATTTTTCTTCTAGATAAATAAATCTCTAGACAAGATA
   |                                                  |
   1                                      DnaA box   50

   TGTGATCTCTTATTAGGATCGCACTGCCCTGTGGATAACAAGGATCGGCT
   ACACTAGAGAATAATCCTAGCGTGACGGGACACCTATTGTTCCTAGCCGA
   |                                                  |
   51                                     DnaA box  100

   TTTAAGATCAACAACCTGGAAAGGGATCATTAACTGTGAATGATCGGTGAT
   AAATTCTAGTTGTTGGACCTTTCCTAGTAATTGACACTTACTAGCCACTA
   |                                                  |
   101                                    DnaA box  150

   CCTGGACCGTATAAGCTGGGGATCAGAATGAGGGTTATACACAGCTCAAAA
   GGACCTGGCATATTCGACCCTAGTCTTACTCCCAATATGTGTCGAGTTTT
   |                                                  |
   151                   DnaA box                    200

   ACTGAACAACGGTTGTTCTTTGGATAACTACCGGTTGATCCAAGCTTCCT
   TGACTTGTTGCCAACAAGAAACCTATTGATGGCCAACTAGGTTCGAAGGA
   |                                                  |
   201            DnaA box                           250

   GACAGAGTTATCCACAGTAGATCGC –3′
   CTGTCTCAATAGGTGTCATCTAGCG
   |                      |
   251                   275
```

FIGURE 11.6 **The events that occur at *oriC* to initiate the DNA replication process.** To initiate DNA replication, DnaA proteins bind to sequences in the five DnaA boxes, which causes the DNA strands to separate at the AT-rich region. The DnaA protein has a DNA-binding domain, shown in red, which binds to the DnaA boxes. It also has a domain called an oligomerization domain, shown in green, which promotes the binding of DnaA proteins to each other. DnaA and DnaC proteins (not shown) recruit DNA helicase (DnaB) to this region. Each DNA helicase is composed of six subunits, which form a ring around one DNA strand and travel in the 5′ to 3′ direction. As shown here, the movement of two DNA helicases separates the DNA strands beyond the *oriC* region.

CONCEPT CHECK: How many replication forks are formed at the origin of replication?

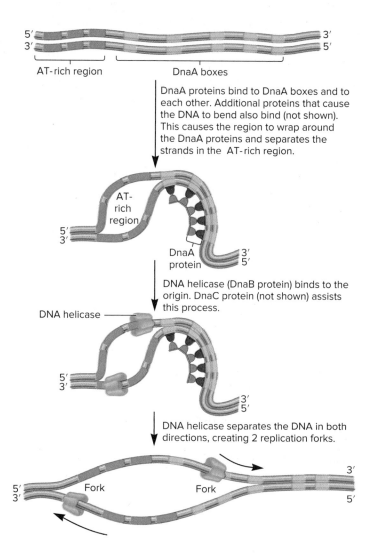

within *oriC:* an AT-rich region, DnaA box sequences, and GATC methylation sites.

- The **AT-rich region** is where the DNA strands will initially separate.
- The **DnaA boxes** are sequences that provide binding sites for DnaA proteins, which initiate the formation of a replication fork.
- As discussed later, GATC methylation sites provide a mechanism for regulating DNA replication.

Figure 11.6 describes the steps that occur to form two DNA replication forks.

1. DNA replication begins with the binding of **DnaA proteins** to DnaA boxes within the origin of replication. When DnaA proteins are in their ATP-bound form, they bind to the five DnaA boxes in *oriC* to initiate DNA replication. DnaA proteins also bind to each other to form a complex.
2. Other DNA-binding proteins not shown in Figure 11.6, such as HU and IHF, cause the DNA to bend around the complex of DnaA proteins, which results in the separation of the strands at the AT-rich region. Because only two hydrogen bonds form between the bases A and T, whereas three hydrogen bonds occur between G and C, the DNA strands are more easily separated at an AT-rich region.
3. Following separation at the AT-rich region, the DnaA proteins, with the help of DnaC proteins, recruit **DNA helicase** to this site. When DNA helicase encounters a double-stranded region, it breaks the hydrogen bonds between the two strands, thereby generating two single strands.
4. Two DNA helicases begin strand separation within the *oriC* region and continue to separate the DNA strands beyond the origin. These proteins use the energy from ATP hydrolysis to catalyze the separation of the double-stranded parental DNA. In *E. coli,* DNA helicases bind to single-stranded DNA and travel along the DNA in a 5′ to 3′ direction to keep the replication fork moving. As shown in Figure 11.6, the action of DNA helicases promotes the movement of two replication forks outward from *oriC* in opposite directions. This initiates the replication of the bacterial chromosome in both directions, a phenomenon termed **bidirectional replication.**

The GATC methylation sites within *oriC* are involved with regulating DNA replication. These sites are recognized and methylated by an enzyme known as <u>D</u>NA <u>a</u>denine <u>m</u>ethyltransferase (Dam). Prior to DNA replication, the GATC sites are methylated in both strands. This full methylation facilitates the initiation of DNA replication at the origin. Following DNA replication, the newly made strands are not methylated, because adenine rather than methyladenine is incorporated into the daughter strands. The initiation of DNA replication at the origin does not readily occur until after the GATC sites are fully methylated. Because it takes several minutes for Dam to methylate the GATC sites within this region, DNA replication does not occur again until after that methylation is completed.

11.2 COMPREHENSION QUESTIONS

1. A site in a chromosome where DNA replication begins is
   a. a promoter.
   c. an operator.
   b. an origin of replication.
   d. a replication fork.
2. The origin of replication in *E. coli* contains
   a. an AT-rich region.
   c. GATC methylation sites.
   b. DnaA boxes.
   d. all of the above.

11.3 BACTERIAL DNA REPLICATION: SYNTHESIS OF NEW DNA STRANDS

Learning Outcomes:

1. Describe how DNA helicase, DNA gyrase, and single-strand binding protein are important for creating a functional replication fork.
2. Explain how primase, DNA polymerase, and DNA ligase function in synthesizing new strands of DNA at the replication fork.
3. Compare and contrast how the leading and lagging strands are made.
4. List the components of the replisome.
5. Describe how DNA replication is terminated.
6. Explain how the isolation of mutants was instrumental in our understanding of DNA replication.

As we have seen, bacterial DNA replication begins at the origin of replication. The synthesis of new DNA strands is a stepwise process in which many cellular proteins participate. In this section, we will examine how DNA strands are made at a replication fork.

Several Proteins Are Required for DNA Replication at the Replication Fork

Figure 11.7 provides an overview of the molecular events that occur as one of the two replication forks moves around the bacterial chromosome, and **Table 11.1** summarizes the functions of the major proteins involved in *E. coli* DNA replication.

Unwinding of the Helix. Let's begin with strand separation. To act as the templates for DNA replication, the strands of a double helix must separate.

- As mentioned previously, the function of DNA helicase is to break the hydrogen bonds between base pairs and thereby unwind the strands. This action generates positive supercoiling ahead of each replication fork.
- As shown in Figure 11.7, an enzyme known as **DNA gyrase** (also called **topoisomerase II**) travels in front of DNA helicase and relaxes positive supercoiling.
- After the two parental DNA strands have been separated and the supercoiling relaxed, the strands must be kept that way until the complementary daughter strands have been made. What prevents the DNA strands from coming back together? DNA replication requires **single-strand binding proteins,** which bind to the strands of parental DNA and prevent them from re-forming a double helix. In this way, the bases within the parental strands are kept in an exposed condition that allows them to hydrogen bond with individual nucleotides.

Synthesis of RNA Primers via Primase. The next event in DNA replication is the synthesis of short strands of RNA (rather than DNA) called **RNA primers.** These strands of RNA are synthesized by the linkage of ribonucleotides via an enzyme known as

Functions of key proteins involved with bacterial DNA replication

- DNA helicase breaks the hydrogen bonds between the DNA strands.
- DNA gyrase alleviates positive supercoiling.
- Single-strand binding proteins keep the parental strands apart.
- Primase synthesizes an RNA primer.
- DNA polymerase III synthesizes a daughter strand of DNA.
- DNA polymerase I excises the RNA primers and fills in with DNA (not shown).
- DNA ligase covalently links the Okazaki fragments together.

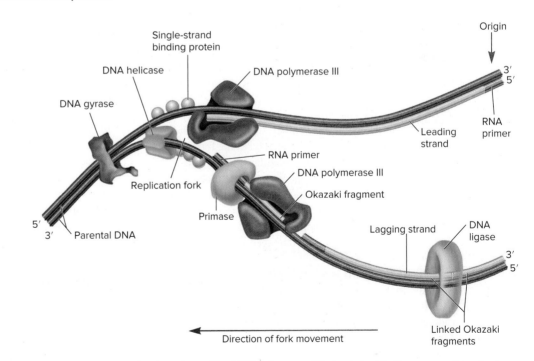

FIGURE 11.7 **The proteins involved in bacterial DNA replication.** Note: The DNA polymerase III shown in this diagram consists of only the catalytic subunit that synthesizes DNA.

CONCEPT CHECK: Look ahead to Figure 11.9. Why is primase needed for DNA replication?

TABLE 11.1

Proteins Involved in *E. coli* DNA Replication

Common Name	Function
DnaA proteins	Bind to DnaA box sequences within the origin of replication to initiate DNA replication
DnaC proteins	Aid DnaA in the recruitment of DNA helicase to the origin
DNA helicase (DnaB)	Separates double-stranded DNA
DNA gyrase (topoisomerase II)	Removes positive supercoiling ahead of the replication fork
Single-strand binding proteins	Bind to single-stranded DNA and prevent it from re-forming a double-stranded structure
Primase	Synthesizes short RNA primers
DNA polymerase III	Synthesizes DNA in the leading and lagging strands
DNA polymerase I	Removes RNA primers, fills in gaps with DNA
DNA ligase	Covalently attaches adjacent Okazaki fragments
Tus	Binds to ter sequences and prevents the advancement of the replication fork

TABLE 11.2

Subunit Composition of DNA Polymerase III Holoenzyme from *E. coli*

Subunit(s)	Function
α	Synthesizes DNA
ε	Contains the 3′ to 5′ exonuclease site that removes mismatched nucleotides (proofreading), which is discussed in Section 11.4
θ	Accessory protein that stimulates the proofreading function
β	Clamp protein, which allows DNA polymerase to slide along the DNA without falling off
τ, γ, δ, δ', ψ, and χ	Clamp loader complex, involved with helping the clamp protein bind to the DNA

Though the various DNA polymerases in *E. coli* and other bacterial species vary in their subunit composition, several common structural features have emerged. The catalytic subunit of all DNA polymerases has a structure that resembles a human hand. As shown in **Figure 11.8**, the template DNA is threaded through the palm of the hand; the thumb and fingers are wrapped around the DNA. The incoming deoxyribonucleoside triphosphates (dNTPs) enter the catalytic site, bind to the template strand according to the AT/GC rule, and then are covalently attached to the 3′ end of the growing strand. DNA polymerase also contains a 3′ exonuclease site that removes mismatched bases, as described later.

DNA Polymerase Requires an RNA Primer or an Existing DNA Strand and It Synthesizes DNA Only in the 5′ to 3′ Direction

As researchers began to unravel the function of DNA polymerase, two features seemed unusual (**Figure 11.9**).

- DNA polymerase cannot begin DNA synthesis by linking together the first two individual nucleotides. In other words, it cannot begin to make DNA using a bare template strand. Rather, DNA polymerase can only elongate a DNA strand starting with an RNA primer on the existing DNA strand (Figure 11.9a).
- A second unusual feature is the directionality of strand synthesis. DNA polymerase can attach nucleotides only in the 5′ to 3′ direction, not in the 3′ to 5′ direction (Figure 11.9b).

The Synthesis of the Leading and Lagging Strands Is Distinctly Different

Due to the two unusual features described in Figure 11.9, the synthesis of the leading and lagging strands shows distinctive differences (**Figure 11.10**). The synthesis of RNA primers by primase allows DNA polymerase III to begin the synthesis of complementary daughter strands of DNA. DNA polymerase III catalyzes the attachment of nucleotides to the 3′ end of each primer, in a 5′ to 3′ direction.

primase. This enzyme synthesizes short strands of RNA, typically 10–12 nucleotides in length. These short RNA strands start, or prime, the process of DNA replication.

- In the **leading strand,** a single primer is made at the origin of replication.
- In the **lagging strand,** multiple primers are made.

As discussed later, the RNA primers are eventually removed.

Synthesis of DNA via DNA Polymerase.   An enzyme known as **DNA polymerase** is responsible for synthesizing the DNA in the leading and lagging strands. This enzyme catalyzes the formation of covalent bonds between adjacent nucleotides, thereby producing the new daughter strands. In *E. coli,* five distinct proteins function as DNA polymerases and are designated I, II, III, IV, and V. DNA polymerases I and III are involved in DNA replication during cell division, whereas DNA polymerases II, IV, and V play a role in DNA repair and the replication of damaged DNA.

DNA polymerase III is responsible for most of the DNA replication during cell division. It is a large enzyme consisting of 10 different subunits that play various roles in the DNA replication process (**Table 11.2**). The α subunit catalyzes the bond formation between adjacent nucleotides, and the remaining nine subunits fulfill other functions. The complex of all 10 subunits together is called DNA polymerase III holoenzyme. By comparison, DNA polymerase I is composed of a single subunit. Its role during DNA replication is to remove the RNA primers and fill in the vacant regions with DNA.

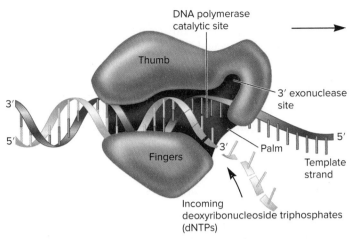

(a) Schematic side view of DNA polymerase bound to DNA.

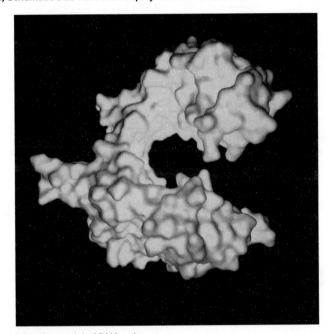

(b) Molecular model of DNA polymerase

FIGURE 11.8 **The structure and function of DNA polymerase.** (**a**) DNA polymerase slides along the template strand as it synthesizes a new strand by connecting deoxyribonucleoside triphosphates (dNTPs) in the 5′ to 3′ direction. The catalytic subunit of DNA polymerase resembles a hand that is wrapped around the template strand. Thus, the movement of DNA polymerase along the template strand is similar to the sliding of a hand along a rope. (**b**) Molecular model of the structure of the catalytic portion of DNA polymerase.

(b): Luk Cox/Alamy Stock Photo

CONCEPT CHECK: Is the template strand read in the 5′ to 3′ or the 3′ to 5′ direction?

Leading Strand.

In the leading strand, one RNA primer is made at the origin, and then DNA polymerase III attaches nucleotides in a 5′ to 3′ direction as it slides toward the opening of the replication fork. The synthesis of the leading strand is continuous.

Lagging Strand.

In the lagging strand, the synthesis of DNA also proceeds in a 5′ to 3′ direction. However, as it is drawn in

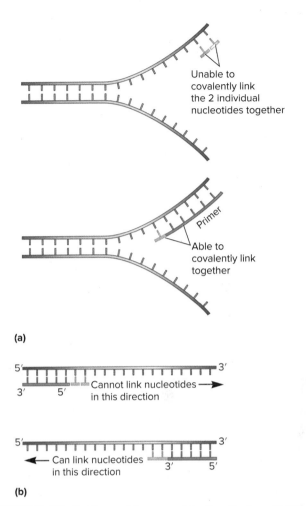

FIGURE 11.9 **Unusual features of the functioning of DNA polymerase.** (**a**) DNA polymerase can only elongate a strand starting with an RNA primer or an existing DNA strand. (**b**) DNA polymerase can attach nucleotides only in a 5′ to 3′ direction. Note that the template strand is read in the opposite, 3′ to 5′, direction.

Figure 11.10, the direction of synthesis is away from the replication fork. In the lagging strand, RNA primers repeatedly initiate the synthesis of short segments of DNA; the synthesis is discontinuous. The length of these fragments in bacteria is typically 1000–2000 nucleotides. In eukaryotes, the fragments are shorter: 100–200 nucleotides. Each fragment contains a short RNA primer at the 5′ end, which is made by primase. The remainder of the fragment is a strand of DNA made by DNA polymerase. The DNA fragments made in this manner are known as **Okazaki fragments,** after Reiji and Tsuneko Okazaki, who initially discovered them in the late 1960s.

To complete the synthesis of Okazaki fragments for the lagging strand, three additional events must occur: removal of the RNA primers, synthesis of DNA in the area where the primers have been removed, and the covalent attachment of adjacent fragments of DNA (see Figure 11.10 and refer back to Figure 11.7). In *E. coli*, the RNA primers are removed by the action of DNA polymerase I. This enzyme has a 5′ to 3′ exonuclease activity, which means that DNA polymerase I digests away the RNA primers in a 5′ to 3′ direction, leaving a vacant area. DNA polymerase I then

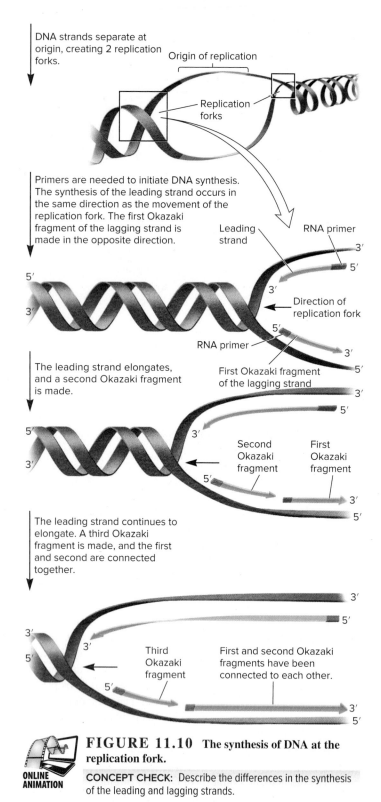

DNA strands separate at origin, creating 2 replication forks.

Origin of replication

Replication forks

Primers are needed to initiate DNA synthesis. The synthesis of the leading strand occurs in the same direction as the movement of the replication fork. The first Okazaki fragment of the lagging strand is made in the opposite direction.

Leading strand

RNA primer

Direction of replication fork

RNA primer

First Okazaki fragment of the lagging strand

The leading strand elongates, and a second Okazaki fragment is made.

Second Okazaki fragment

First Okazaki fragment

The leading strand continues to elongate. A third Okazaki fragment is made, and the first and second are connected together.

Third Okazaki fragment

First and second Okazaki fragments have been connected to each other.

FIGURE 11.10 **The synthesis of DNA at the replication fork.**

ONLINE ANIMATION

CONCEPT CHECK: Describe the differences in the synthesis of the leading and lagging strands.

synthesizes DNA to fill in this region. It uses the 3′ end of an adjacent Okazaki fragment as a primer. For example, in Figure 11.10, DNA polymerase I will remove the RNA primer from the first Okazaki fragment and then synthesize DNA in the vacant region by attaching nucleotides to the 3′ end of the second Okazaki fragment.

After the gap has been completely filled in, a covalent bond is still missing between the last nucleotide added by DNA polymerase I (on the second Okazaki fragment) and the adjacent DNA strand of the first Okazaki fragment (see Figure 11.10). An enzyme known as **DNA ligase** catalyzes the formation of a covalent bond between adjacent Okazaki fragments to complete the replication process in the lagging strand (refer back to Figure 11.7). In *E. coli*, DNA ligase requires NAD^+ to carry out this reaction, whereas the DNA ligases found in archaea and eukaryotes require ATP.

Bidirectional DNA Replication. Now that we have seen how the leading and lagging strands are made, **Figure 11.11** shows how new strands are constructed from a single origin of replication. As mentioned, this process, called bidirectional replication, involves two replication forks. To the left of the origin, the top strand is made continuously, whereas to the right of the origin, it is made in Okazaki fragments. By comparison, the synthesis of the bottom strand is just the opposite. To the left of the origin, this strand is made in Okazaki fragments; to the right of the origin, the synthesis is continuous.

GENETIC TIPS THE QUESTION: What are the similarities and differences in the synthesis of the leading and lagging strands of DNA in *E. coli*?

T OPIC: *What topic in genetics does this question address?* The topic is DNA replication. More specifically, the question is about comparing the synthesis of the leading and lagging strands.

I NFORMATION: *What information do you know based on the question and your understanding of the topic?* From the question, you know that DNA replication occurs with both lagging and leading strands. From your understanding of the topic, you may remember that the leading strand is made continuously, in the direction of the replication fork, whereas the lagging strand is made as Okazaki fragments in the direction away from the fork.

P ROBLEM-SOLVING **S** TRATEGIES: *Compare and contrast. Describe the steps.* One strategy to solve this problem is to compare your knowledge of DNA replication along the leading and lagging strands. It may be helpful to break down the process into individual steps.

ANSWER: The leading strand is primed once at the origin, and then DNA polymerase III synthesizes DNA continuously in the direction of the replication fork. The DNA is made in the 5′ to 3′ direction. In the lagging strand, many short pieces of DNA (Okazaki fragments) are made. This requires many RNA primers. The primers are removed by DNA polymerase I, which then fills in the gaps with DNA. Like the leading strand, the lagging strand is made in the 5′ to 3′ direction. DNA ligase then covalently connects the Okazaki fragments together.

FIGURE 11.11 The synthesis of leading and lagging strands outward from a single origin of replication.

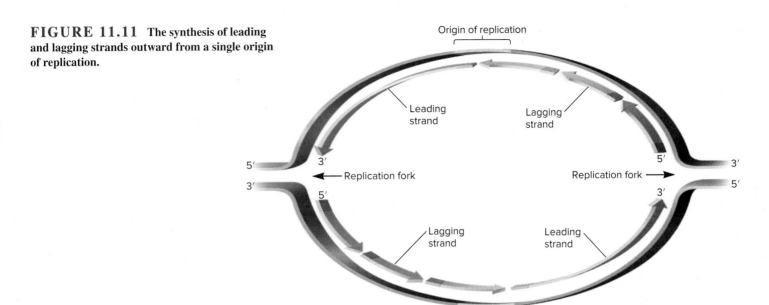

Certain Enzymes Involved in DNA Replication Bind to Each Other to Form a Complex

Figure 11.12 provides a more three-dimensional view of the DNA replication process. DNA helicase and primase are physically bound to each other to form a protein complex known as a **primosome.** This complex leads the way at the replication fork. The primosome tracks along the DNA, separating the parental strands and synthesizing RNA primers at regular intervals along the lagging strand. Being associated together within a complex allows the actions of DNA helicase and primase to be better coordinated.

The primosome is physically associated with two DNA polymerase holoenzymes to form a **replisome.** As shown in Figure 11.12, two DNA polymerase III proteins act in concert to replicate the leading and lagging strands. The term **dimeric DNA polymerase** is used to describe a complex of two DNA polymerase holoenzymes that move as a unit during DNA replication. To facilitate this movement, the lagging strand is looped out with respect to the DNA polymerase that synthesizes the lagging strand. This loop allows the lagging-strand polymerase to make DNA in a 5′ to 3′ direction yet move as a unit with the leading-strand polymerase. Interestingly, when the lagging-strand polymerase reaches the end of an Okazaki fragment, it must be released from the template DNA and "hop" to the RNA primer that is closest to the fork. The clamp loader complex (see Table 11.2), which is part of DNA polymerase holoenzyme, then reloads the holoenzyme at the site where the next RNA primer has been made. Similarly, after

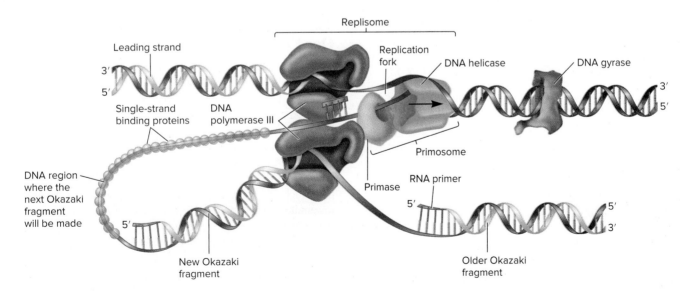

FIGURE 11.12 **A three-dimensional view of DNA replication.** DNA helicase and primase are bound together to form a primosome. The primosome associates with two DNA polymerase holoenzymes to form a replisome.

CONCEPT CHECK: What is the advantage of having the replication enzymes associated with one another in a complex?

primase synthesizes an RNA primer in the 5′ to 3′ direction, it must hop over the primer and synthesize the next primer closer to the replication fork.

Replication Is Terminated When the Replication Forks Meet at the Termination Sequences

On the opposite side of the *E. coli* chromosome from *oriC* is a pair of **termination sequences,** known as *ter sequences*. A protein known as the termination utilization substance (Tus) binds to the ter sequences and stops the movement of the replication forks. As shown in **Figure 11.13**, one of the ter sequences, designated T1, prevents the advancement of the fork that is moving left to right, but allows the movement of the other fork (see the inset in Figure 11.13). Alternatively, the ter sequence T2 prevents the advancement of the fork that is moving right to left, but allows the advancement of the other fork. In any given cell, only one ter sequence is required to stop the advancement of one replication fork, and then the other fork ends its synthesis of DNA when it reaches the halted fork. In other words, DNA replication ends when oppositely advancing forks meet, usually at T1 or T2. Finally, DNA ligase covalently links the two daughter strands, creating two circular, double-stranded molecules.

After DNA replication is completed, one problem may exist. DNA replication of prokaryotic chromosomes often results in two intertwined circular DNA molecules known as **catenanes** (**Figure 11.14**). Fortunately, catenanes are only transient structures in DNA replication. In *E. coli*, DNA gyrase introduces a temporary break in the DNA strands and then rejoins them after the strands have become unlocked. This allows the catenanes to be separated into individual circular molecules.

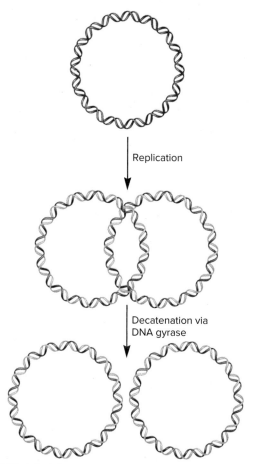

FIGURE 11.14   Separation of catenanes. DNA replication can result in two intertwined circular DNA molecules called catenanes. These catenanes are separated by the action of DNA gyrase.

CONCEPT CHECK: Is DNA strand breakage necessary for catenane separation?

The Isolation of Mutants Has Been Instrumental to Our Understanding of DNA Replication

Thus far, we have considered how a variety of proteins play a role in the replication of bacterial DNA. An important experimental approach that led to the identification of most of these proteins involved the isolation of mutants in *E. coli* that have abnormalities in DNA replication. The first such mutant, which was identified by Paula DeLucia and John Cairns in 1969, was in the gene that codes DNA polymerase I.

Because DNA replication is vital for cell division, mutations that block DNA replication prevent cell growth. For this reason, if researchers want to identify loss-of-function mutations in vital genes, they must screen for **conditional mutants.** One type of conditional mutant is a **temperature-sensitive (ts) mutant.** For example, a ts mutant might grow at 30°C (a permissive temperature) but fail to grow at a higher temperature, such as 42°C. The higher temperature at which the mutant strain fails to grow is called a nonpermissive temperature. The mutant strain fails to grow because the higher temperature inactivates the function of the protein coded by the mutant gene.

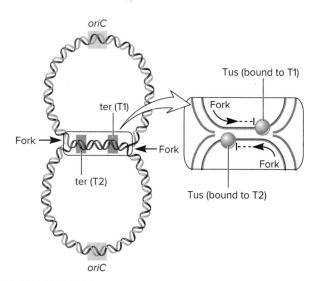

FIGURE 11.13   The termination of DNA replication. Two sites in the bacterial chromosome, highlighted with rose-colored rectangles, are ter sequences designated T1 and T2. The T1 sequence prevents the further advancement of the fork moving from left to right, and the T2 sequence prevents the advancement of the fork moving from right to left. As shown in the inset, the binding of Tus prevents the replication forks from proceeding past the ter sequences in either direction.

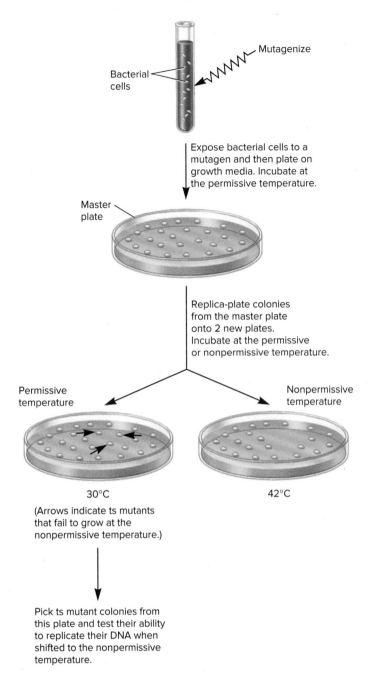

FIGURE 11.15 A strategy to identify ts mutants. In this approach, bacteria are mutagenized, which increases the likelihood of mutation, and then grown at the permissive temperature. Colonies are then replica-plated and grown at both the permissive and nonpermissive temperatures. (Note: The procedure of replica plating is shown in Chapter 19; see Figure 19.6.) The ts mutants fail to grow at the nonpermissive temperature. The appropriate colonies are picked from the master plate, grown at the permissive temperature, and analyzed to see if DNA replication is altered at the nonpermissive temperature.

Figure 11.15 shows a general strategy for the isolation of ts mutants. Researchers expose bacterial cells to a mutagen that increases the likelihood of mutations. The mutagenized cells are plated on growth media and incubated at the permissive temperature. The colonies are then replica-plated onto two plates: one

incubated at the permissive temperature and one at the nonpermissive temperature. (The technique of replica plating is described in Chapter 19; see Figure 19.6.) As seen in Figure 11.15, this method enables researchers to identify ts mutants that are unable to grow at the nonpermissive temperature.

While studying DNA replication, researchers analyzed a large number of ts mutants to discover if any of them had a defect in DNA replication. For example, one approach involved exposing a ts mutant strain to radiolabeled thymine (a base that is incorporated into DNA), shifting to the nonpermissive temperature, and determining if the mutant strain could make radiolabeled DNA. Because *E. coli* has many vital genes not involved with DNA replication, only a small subset of ts mutants would be expected to have mutations in genes that code proteins that are critical to the replication process. Therefore, researchers had to screen many thousands of ts mutants to identify the few involved in DNA replication. This approach is sometimes called a "brute force" genetic screen.

Table 11.3 summarizes some of the genes that were identified using this type of strategy. The genes were originally designated with names beginning with *dna*, followed by a capital letter that generally referred to the order in which they were discovered. When shifted to a nonpermissive temperature, certain mutants showed a rapid arrest of DNA synthesis. These so-called rapid-stop mutants had mutations in genes that code proteins needed for DNA synthesis. By comparison, other mutants were able to complete their current round of replication but could not start another round. These slow-stop mutants involved genes that code proteins needed for the initiation of replication at the origin. In later studies, the proteins coded by these genes were purified, and their functions were studied in vitro.

TABLE 11.3

Examples of ts Mutants Involved in DNA Replication in *E. coli*

Gene Name	Protein Function
Rapid-Stop Mutants	
dnaE	α subunit of DNA polymerase III; synthesizes DNA
dnaX	τ subunit of DNA polymerase III; part of the clamp loader complex and also promotes the dimerization of two DNA polymerase III proteins at the replication fork
dnaN	β subunit of DNA polymerase III; functions as a clamp protein that makes DNA polymerase a processive enzyme
dnaZ	γ subunit of DNA polymerase III; helps the β subunit bind to the DNA
dnaG	Primase; needed to make RNA primers
dnaB	Helicase; needed to unwind the DNA strands during replication
Slow-Stop Mutants	
dnaA	DnaA protein that recognizes the DnaA boxes at the origin
dnaC	DnaC protein that recruits DNA helicase to the origin

11.3 COMPREHENSION QUESTIONS

1. The enzyme known as _____ uses _____ and separates the DNA strands at the replication fork.
   a. DNA helicase, ATP   c. DNA gyrase, ATP
   b. DNA helicase, GTP   d. DNA gyrase, GTP

2. In the lagging strand, DNA is made in the direction _____ the replication fork and is made as _____.
   a. toward, one continuous strand
   b. away from, one continuous strand
   c. toward, Okazaki fragments
   d. away from, Okazaki fragments

11.4 BACTERIAL DNA REPLICATION: CHEMISTRY AND ACCURACY

Learning Outcomes:

1. Describe how nucleotides are connected to a growing DNA strand.
2. Define *processivity*.
3. Explain the proofreading function of DNA polymerase.

In Sections 11.2 and 11.3, we examined the origin of replication and considered how DNA strands are synthesized. In this section, we will take a closer look at the process in which nucleotides are attached to a growing DNA strand and discuss the amazing accuracy of DNA replication.

DNA Polymerase III Is a Processive Enzyme That Uses Deoxyribonucleoside Triphosphates

Let's now turn our attention to the chemistry of DNA replication, in which DNA polymerase catalyzes the covalent attachment between a phosphate in one nucleotide and the sugar in the previous nucleotide (**Figure 11.16**). The formation of this covalent bond requires an input of energy.

1. Prior to bond formation, the nucleotide about to be attached to the growing strand is a deoxyribonucleoside triphosphate (dNTP). It contains three phosphate groups attached at the 5′ carbon (C) atom of deoxyribose. The dNTP first enters the catalytic site of DNA polymerase and binds to the template strand according to the AT/GC rule.

2. The 3′ hydroxyl (—OH) group in the nucleotide on the end of the growing strand reacts with the phosphate group (PO_4^{2-}) attached to the sugar on the incoming nucleotide. This reaction is highly exergonic and provides the energy to form a covalent bond between the sugar at the 3′ end of the DNA strand and the PO_4^{2-} of the incoming nucleotide.

3. The formation of this covalent bond causes the newly made strand to grow in the 5′ to 3′ direction. As shown in Figure 11.16, pyrophosphate (PP_i) is released. It is later broken down into two phosphates.

As noted in Chapter 9 (refer back to Figure 9.8), a phosphodiester linkage is the linkage between a phosphate and two sugar molecules. As its name implies, a phosphodiester linkage involves two ester bonds. In comparison, as a DNA strand grows, a single covalent (ester) bond is formed between adjacent nucleotides (see Figure 11.16). The other ester bond in the phosphodiester linkage—the bond between the 5′-oxygen and phosphorus—is already present in the incoming nucleotide.

DNA polymerase catalyzes the covalent attachment of nucleotides with great speed. In *E. coli*, DNA polymerase III attaches approximately 750 nucleotides per second! DNA polymerase III can catalyze the synthesis of the daughter strands so quickly because it is a **processive enzyme.** This means it does not dissociate from the template strand after it has catalyzed the covalent joining of two nucleotides in the growing daughter strand. Rather, as depicted in Figure 11.8a, it remains clamped to the DNA template strand and slides along that strand as it catalyzes the synthesis of the daughter strand. The β subunit of the holoenzyme, also known as the clamp protein, promotes the association of the holoenzyme with the DNA as it glides along the template strand (refer back to Table 11.2). The β subunit forms a dimer in the shape of a ring. The hole of the ring is large enough to accommodate a double-stranded DNA molecule, and its width is equal to about one turn of DNA. A complex of several subunits functions as a clamp loader that allows the DNA polymerase holoenzyme to initially clamp onto the DNA.

The effects of processivity are really quite remarkable. In the absence of the β subunit, DNA polymerase can synthesize DNA at a rate of only about 20 nucleotides per second and typically falls off the DNA template after approximately 10 nucleotides have been linked together in the daughter strand. By comparison, when the β subunit is present, as it is in the holoenzyme, the synthesis rate is approximately 750 nucleotides per second. In the leading strand, DNA polymerase III has been estimated to synthesize a segment of DNA that is over 500,000 nucleotides in length before it falls off.

The High Fidelity of DNA Replication Is Due to Three Mechanisms

With replication occurring so rapidly, you might imagine that mistakes could happen in which the wrong nucleotide is incorporated into the growing daughter strand. Although mistakes do happen during DNA replication, they are extraordinarily rare. In the case of DNA synthesis via DNA polymerase III, only 1 mistake is made per 100 million nucleotides. Therefore, DNA synthesis occurs with a high degree of accuracy, or **fidelity.** Three factors account for this remarkable accuracy.

Stability of Base Pairing. The hydrogen bonding between G and C or between A and T is much more stable than that between mismatched pairs of bases. Due to this stability alone, only 1 mistake per 1000 nucleotides would be made.

Structure of the Active Site of DNA Polymerase. The active site of DNA polymerase preferentially catalyzes the attachment of nucleotides when the correct bases are located in opposite strands. Helix distortions caused by mispairing usually prevent an incorrect

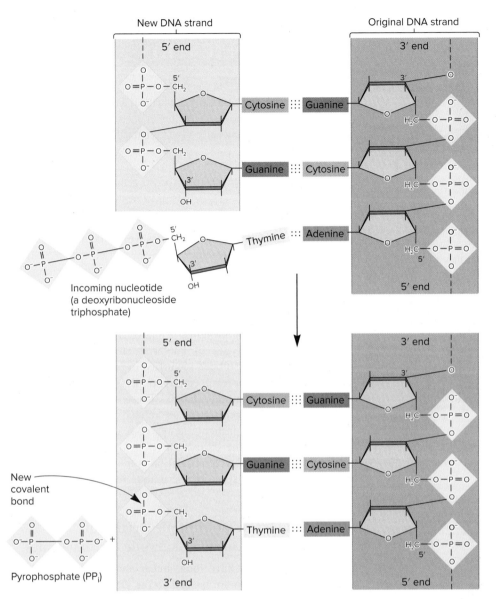

FIGURE 11.16 The **enzymatic action of DNA polymerase.** An incoming deoxyribonucleoside triphosphate (dNTP) is cleaved to form a nucleoside monophosphate and pyrophosphate (PP$_i$). The energy released from this exergonic reaction allows the nucleoside monophosphate to form a covalent bond at the 3′ end of the growing strand. This reaction is catalyzed by DNA polymerase. PP$_i$ is released.

CONCEPT CHECK: Does the oxygen in the newly made ester bond come from the phosphate or from the sugar?

nucleotide from properly occupying the active site of DNA polymerase. By comparison, the correct nucleotide occupies the active site with precision and promotes induced fit, which is a conformational change in the enzyme that is necessary for catalysis. The inability of incorrect nucleotides to promote induced fit decreases the error rate to a range from 1 in 100,000 to 1 in 1 million.

Proofreading. DNA polymerase decreases the error rate even further by the enzymatic removal of mismatched nucleotides. As shown in **Figure 11.17**, DNA polymerase can identify a mismatched nucleotide and remove it from the daughter strand. This occurs by exonuclease cleavage of the bonds between adjacent nucleotides at the 3′ end of the newly made strand. The ability to remove mismatched bases by this mechanism is called the **proofreading function** of DNA polymerase. Proofreading occurs by the removal of nucleotides in the 3′ to 5′ direction at the 3′ exonuclease site. After the mismatched nucleotide is removed, DNA polymerase resumes DNA synthesis in the 5′ to 3′ direction.

11.4 COMPREHENSION QUESTIONS

1. DNA polymerase III is a processive enzyme, which means that
   a. it does not dissociate from the template strand after it has attached a nucleotide to the 3′ end of the daughter strand.
   b. it makes a new strand very quickly.
   c. it proceeds toward the opening of the replication fork.
   d. it copies DNA with relatively few errors.
2. The proofreading function of DNA polymerase involves the recognition of a _____ and the removal of a short segment of DNA in the _____ direction.
   a. missing base, 5′ to 3′
   b. base pair mismatch, 5′ to 3′
   c. missing base, 3′ to 5′
   d. base pair mismatch, 3′ to 5′

Mismatch causes DNA polymerase to pause,
leaving mismatched nucleotide near the 3′ end.

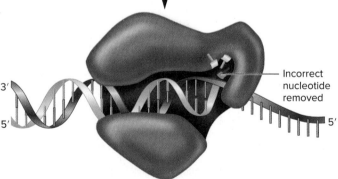

3′ exonuclease
site

Template
strand

5′

3′

5′

Base pair
mismatch
near the
3′ end

The 3′ end enters the
exonuclease site.

3′

5′

5′

At the 3′ exonuclease site,
the strand is digested in
the 3′ to 5′ direction until the
incorrect nucleotide is
removed.

3′

5′

5′

Incorrect
nucleotide
removed

**ONLINE
ANIMATION**

**FIGURE 11.17   The proofreading function of
DNA polymerase.**   When a base pair mismatch is found,
the end of the newly made strand is shifted into the 3′ exonuclease site. The DNA is digested in the 3′ to 5′ direction
to release the incorrect nucleotide, and replication is resumed in the 5′ to
3′ direction.

11.5   EUKARYOTIC DNA REPLICATION

Learning Outcomes:

1. Compare and contrast the origins of replication in bacteria and eukaryotes.
2. Outline the functions of different DNA polymerases in eukaryotes.
3. Describe how RNA primers are removed in eukaryotes.
4. Explain how DNA replication occurs at the ends of eukaryotic chromosomes.

Eukaryotic DNA replication is not as well understood as bacterial replication. Much research has been carried out on a variety of experimental organisms, particularly yeast and mammalian cells. Many of these studies have found extensive similarities between the general features of DNA replication in prokaryotes and eukaryotes. For example, DNA helicase, topoisomerase II, single-strand binding proteins, primase, DNA polymerase, and DNA ligase—the types of bacterial enzymes described in Table 11.1—have also been identified in eukaryotes. Nevertheless, at the molecular level, eukaryotic DNA replication appears to be substantially more complex. These additional intricacies of eukaryotic DNA replication are related to several features of eukaryotic cells. In particular, eukaryotic cells have larger, linear chromosomes, the chromatin is tightly packed within nucleosomes, and cell-cycle regulation is more complicated. This section emphasizes some of the unique features of eukaryotic DNA replication.

Initiation Occurs at Multiple Origins of Replication on Linear Eukaryotic Chromosomes

Because eukaryotes have long, linear chromosomes, multiple origins of replication are needed so the DNA can be replicated within a reasonable length of time. In 1968, Joel Huberman and Arthur Riggs provided evidence for multiple origins of replication.

1. They added a radiolabeled nucleoside (^{3}H-deoxythymidine) to a culture of actively dividing cells. For a brief period, the radiolabeled deoxythymidine was taken up by the cells and incorporated into their newly made DNA strands.
2. The cells were then exposed to a high concentration of unlabeled deoxythymidine, which prevented the further incorporation of radiolabeled deoxythymidine.
3. The chromosomes were then isolated from the cells and subjected to autoradiography.
4. As seen in **Figure 11.18**, radiolabeled segments were interspersed among nonlabeled segments. This result is consistent with the hypothesis that eukaryotic chromosomes contain multiple origins of replication.

As shown schematically in **Figure 11.19**, eukaryotic DNA replication proceeds bidirectionally from many origins of replication. As discussed in Chapter 2, this occurs during S phase of the

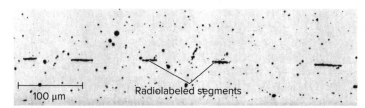

FIGURE 11.18 **Evidence for multiple origins of replication in eukaryotic chromosomes.** In this experiment, cells were first exposed to ³H-deoxythymidine and then exposed to a high amount of unlabeled deoxythymidine, a procedure called a pulse-chase. The chromosomes were isolated and subjected to autoradiography. In this micrograph, radiolabeled segments were interspersed among nonlabeled segments, indicating that eukaryotic chromosomes contain multiple origins of replication.
Courtesy of Dr. Joel A. Huberman

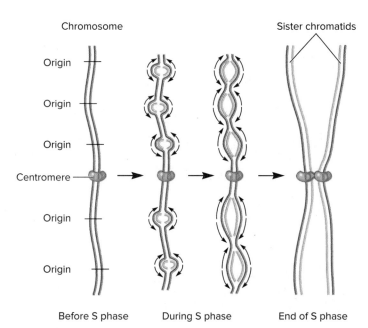

FIGURE 11.19 **The replication of eukaryotic chromosomes.** Replication begins from multiple origins of replication, and the replication forks move bidirectionally to replicate the DNA. Eventually, all of the replication forks will merge. The net result is two sister chromatids attached to each other.

CONCEPT CHECK: Why do eukaryotes need multiple origins of replication in their chromosomes?

cell cycle (refer back to Figure 2.5). The multiple replication forks eventually make contact with each other to complete the replication process.

Origins of Replication Differ Among Budding Yeast and Other Eukaryotes

The molecular features of eukaryotic origins of replication vary among different species. At the molecular level, eukaryotic origins of replication have been extensively studied in the budding yeast *Saccharomyces cerevisiae*. In this organism, many replication origins have been identified and sequenced. They have been

named **ARS elements** (for <u>a</u>utonomously <u>r</u>eplicating <u>s</u>equence) (**Figure 11.20a**). They have the following properties:

- ARS elements are about 100–200 bp in length and are necessary for the initiation of DNA replication.
- They contain a high percentage of A and T bases.
- ARS elements have a copy of the ARS consensus sequence (ACS), which is ATTTAT(A or G)TTTA, along with additional sites, such as B1 and B2, that also enhance the function of the origin of replication. Note: This arrangement in

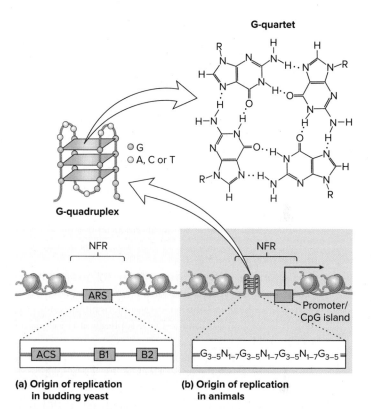

FIGURE 11.20 **General features of origins of replication in eukaryotes.** (a) In budding yeast, an origin has an element called an autonomously replicating sequence (ARS), which is composed of the ARS consensus sequence (ACS) and B1 and B2 sequences. (b) In animals, most origins have a G4 motif that can form a G-quadruplex. The term G4 refers to series of <u>4</u> groups of 3 to 5 guanines (<u>G</u>) that are separated by other bases (N) as seen at the bottom of part (b). In this notation, G_{3-5} indicates a sequence of 3 to 5 guanine-containing nucleotides in a row and N_{1-7} indicates a sequence of 1 to 7 nucleotides that do not contain guanines. The inset shows how a DNA strand that contains a G4 motif can form a four-stranded structure (a G-quadruplex) that is stabilized by hydrogen bonding among four guanines, a structure called a G-quartet. The hydrogen bonds are depicted as two dots. In this example, the G-quadruplex is stabilized by the formation of three G-quartets. The letter R indicates where each of the four guanines is attached to the DNA backbone. In many origins found in animals, a promoter and CpG island are located next to the G4 motif. In both (a) and (b), the origin contains a nucleosome-free region (NFR) that is flanked by well-positioned nucleosomes. As discussed in the text, the covalent modification of histones within these nucleosomes plays a role in DNA replication.

CONCEPT CHECK: What type of bonding interactions promote the formation of a G-quadruplex?

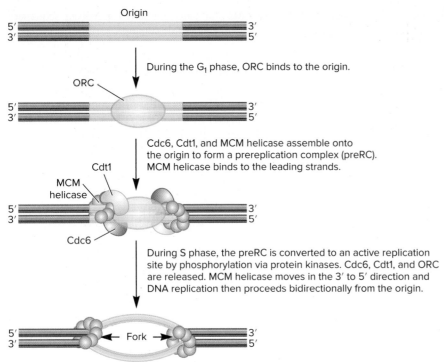

FIGURE 11.21 **The formation of a prereplication complex in eukaryotes.** This is a simplified model; more proteins are involved in this process than are shown here.

budding yeast is similar to that in bacterial origins of replication, which also have an AT-rich region and specific elements, such as DnaA boxes.

- The separation of DNA occurs within B2, induced by attachment of a protein called ARS binding factor. The ACS and B1 sequences bind to origin recognition complex, which is required to initiate DNA replication (look ahead to Figure 11.21).
- Functional ARS elements are found within a nucleosome-free region (NFR) that is flanked by well-positioned nucleosomes.
- Histone modifications within the nucleosomes that promote a more open chromatin conformation play a role in facilitating DNA replication. These include acetylation and methylation. Note: Histone modifications are discussed in Chapter 15.

Unlike the origins of replication in budding yeast, those in other eukaryotes do not contain a consensus sequence analogous to the ARS consensus sequence. **Figure 11.20b** shows some key features that are commonly found in origins of replication in animals.

- Most of these origins contain G-rich sequences called G4 motifs. These motifs form a four-stranded helical structure termed a **G-quadraplex** (Figure 11.20b). Some evidence suggests that origin recognition complex binds preferentially to DNA that contains an G-quadraplex.
- Like ARS elements in yeast, the G4 motif in animals must be found in a nucleosome-free region.
- The flanking histones need to be modified in a way that favors an open conformation. For example, the methylation of a particular lysine in histone H4 promotes the binding of origin recognition complex.

- Promoters and CpG islands, which are discussed in Chapter 15 (look ahead to Section 15.1), are frequently found in the NFR.

In more complex eukaryotes such as animals, origins of replication tend be more dynamic, meaning that the same origins are not always used. Replication origins are in excess relative to their use in each cell cycle. They fall into three main classes:

Constitutive: Origins that are used all the time.

Flexible: Those used in an apparently stochastic (random) manner. These are the most common type.

Dormant: Those used only at a specific stage of development or during cell differentiation.

The Origin Recognition Complex Initiates the Process of DNA Replication

DNA replication in eukaryotes requires the assembly of a **prereplication complex (preRC)** during the G_1 phase of the cell cycle (**Figure 11.21**). Part of the preRC is a group of five or six proteins called the **origin recognition complex (ORC)** that acts as the first initiator of preRC assembly. ORC promotes the binding of Cdc6, Cdt1, and a group of six proteins called **MCM helicase.** The binding of MCM helicase to the leading strands completes a process called **DNA replication licensing.** Those origins of replication with MCM helicases are able to begin the process of DNA synthesis.

As S phase approaches, the preRC is converted to an active replication site by phosphorylation via protein kinases. This phosphorylation promotes the release of Cdc6, Cdt1, and ORC and the assembly of additional replication factors and DNA polymerases. (These additional replication factors and DNA polymerases are not shown in Figure 11.21.) The MCM helicases move in the 3' to 5' direction, and DNA replication then proceeds bidirectionally from the origin.

The initiation of DNA replication in archaea appears to be a simpler process than that seen in eukaryotes. The archaeal origin of replication is usually AT-rich and flanked by several conserved repeated motifs known as origin recognition boxes. Archaea also make Orc1, which is one of the proteins that is found in eukaryotic ORCs, along with Cdc6 and MCM helicase. The initiation of DNA replication involves the binding of Orc1 and Cdc6 to an origin and the subsequent binding of MCM helicase.

Eukaryotes Have Many Different DNA Polymerases

Eukaryotes have many types of DNA polymerases. For example, mammalian cells have well over a dozen different DNA polymerases (**Table 11.4**). Four of these, designated α (alpha), ε (epsilon), δ (delta), and γ (gamma), have the primary function of replicating DNA. DNA polymerase γ functions in the mitochondria to replicate mitochondrial DNA, whereas α, ε, and δ are involved with DNA replication in the cell nucleus.

- DNA polymerase α is the only eukaryotic polymerase that associates with primase. The functional role of the DNA polymerase α/primase complex is to synthesize a short RNA-DNA primer of approximately 10 RNA nucleotides followed by 20–30 DNA nucleotides. The DNA polymerase α/primase complex dissociates from the replication fork and is replaced by DNA polymerase ε or δ. This exchange is called a **polymerase switch.**
- DNA polymerase ε is involved with leading-strand synthesis.
- DNA polymerase δ is responsible for lagging-strand synthesis.

Both DNA polymerases ε and δ are processive enzymes, whereas DNA polymerase α is not processive. For this reason,

the polymerase switch is necessary to allow DNA replication to occur at a fast rate.

What are the functions of the other DNA polymerases? Several of them play an important role in DNA repair, a topic that will be examined in Chapter 19. DNA polymerase β, which is not involved in the replication of normal DNA, plays an important role in removing incorrect bases from damaged DNA. More recently, several additional DNA polymerases have been identified. Although their precise roles have not been elucidated, many are in a category called **translesion-replicating polymerases.** When DNA polymerases α, δ, and ε encounter abnormalities in DNA structure, such as abnormal bases or crosslinks, they may be unable to replicate over the aberration. When this occurs, a translesion-replicating polymerase is attracted to the damaged DNA, and its special properties enable it to synthesize a complementary strand over the abnormal region. The various types of translesion-replicating polymerases are able to replicate over different kinds of DNA damage. For example, polymerase κ can replicate over DNA lesions caused by benzo[α]pyrene, an agent found in cigarette smoke, whereas polymerase η can replicate over thymine dimers, which are caused by UV light.

Flap Endonuclease Removes RNA Primers During Eukaryotic DNA Replication

Another key difference between bacterial and eukaryotic DNA replication is the way in which RNA primers are removed. As discussed earlier in this chapter, bacterial RNA primers are removed by DNA polymerase I. By comparison, a DNA polymerase does not play this role in eukaryotes. Instead, an enzyme called flap endonuclease is primarily responsible for RNA primer removal.

Flap endonuclease gets its name because it removes small RNA flaps that are generated by the action of DNA polymerase δ. In the diagram shown in **Figure 11.22**, DNA polymerase δ elongates the left Okazaki fragment until it runs into the RNA primer of the adjacent Okazaki fragment on the right. This causes a portion of the RNA primer to form a short flap, which is removed by flap endonuclease. As DNA polymerase δ continues to elongate the DNA, short flaps continue to be generated, which are sequentially removed by flap endonuclease. Eventually, all of the RNA primer is removed, and DNA ligase seals the DNA fragments together.

Though flap endonuclease is thought to be the primary agent for RNA primer removal in eukaryotes, it cannot remove a flap that is too long. In such cases, the long flap is cleaved by an enzyme called Dna2 nuclease/helicase. This enzyme can cut a long flap, thereby generating a short flap. The short flap is then removed by flap endonuclease.

TABLE 11.4
Eukaryotic DNA Polymerases

Polymerase Type(s)*	Function
α	Initiates DNA replication in conjunction with primase
ε	Replication of the leading strand
δ	Replication of the lagging strand
γ	Replication of mitochondrial DNA
η, κ, ι, ξ (translesion-replicating polymerases)	Replication of damaged DNA
α, β, δ, ε, σ, λ, μ, φ, θ, η	DNA repair or other functions†

*The designations are those of mammalian enzymes.
†Many DNA polymerases have dual functions. For example, DNA polymerases α, δ, and ε are involved in the replication of normal DNA and also play a role in DNA repair. In cells of the immune system, certain genes that code antibodies (i.e., immunoglobulin genes) undergo a phenomenon known as hypermutation. This increases the variation in the kinds of antibodies the cells can make. Certain polymerases in this list, such as η, may play a role in hypermutation of immunoglobulin genes.

The Ends of Eukaryotic Chromosomes Are Replicated by Telomerase

Linear eukaryotic chromosomes have telomeres at both ends. The term **telomere** refers to the telomeric repeat sequences within the

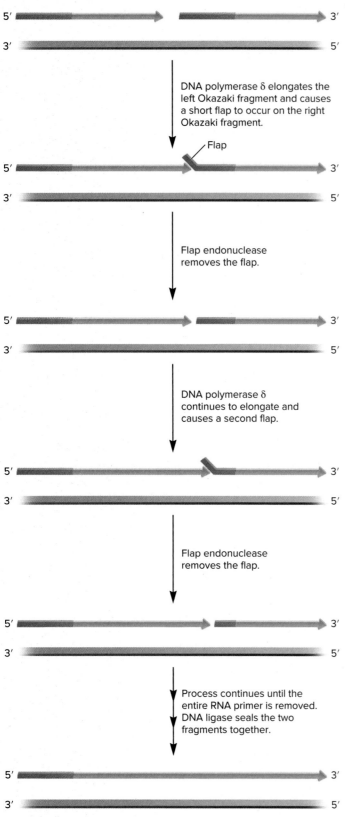

DNA polymerase δ elongates the left Okazaki fragment and causes a short flap to occur on the right Okazaki fragment.

Flap

Flap endonuclease removes the flap.

DNA polymerase δ continues to elongate and causes a second flap.

Flap endonuclease removes the flap.

Process continues until the entire RNA primer is removed. DNA ligase seals the two fragments together.

FIGURE 11.22 **Removal of an RNA primer by flap endonuclease.**

DNA and the specific proteins that are bound to those sequences. Telomeres consist of a tandemly repeated sequence and a 3′ overhang region that is 12–16 nucleotides in length (**Figure 11.23**).

The repeat sequences that occur within telomeres have been studied in a wide variety of eukaryotic organisms. A common feature is that the telomeric sequence contains several guanine nucleotides and often many thymine nucleotides (**Table 11.5**). Depending on the species and the cell type, this sequence can be tandemly repeated up to several hundred times in the telomeric region.

One reason telomeric repeat sequences are needed is that DNA polymerase is unable to replicate the 3′ ends of DNA strands. Why is DNA polymerase unable to replicate this region? The answer lies in the two unusual functional features of this enzyme. As discussed previously, DNA polymerase synthesizes DNA only in a 5′ to 3′ direction, and it cannot link together the first two individual nucleotides; it can elongate only preexisting strands. These two functional features of DNA polymerase pose a problem at the 3′ ends of linear chromosomes. As shown in **Figure 11.24**, the 3′ end of a DNA strand cannot be replicated by DNA polymerase because a primer cannot be made upstream from this point. Therefore, if this problem were not solved, the chromosome would become progressively shorter with each round of DNA replication.

To prevent the loss of genetic information due to chromosome shortening, additional DNA sequences are attached to the ends of telomeres. In 1984, Carol Greider and Elizabeth Blackburn discovered an enzyme called **telomerase** that catalyzes a specialized form of DNA replication that lengthens the telomere. It recognizes the telomeric sequences at the ends of eukaryotic chromosomes and synthesizes additional repeats of those sequences. This process prevents chromosome shortening so that vital genetic information, such as important genes, is not lost. Greider and Blackburn received the 2009 Nobel Prize in Physiology or Medicine for their discovery.

In animals that advance through developmental stages from a fertilized egg to an embryo and ultimately to an adult, the level of telomerase function varies with age and among different cell types. In germ-line cells that give rise to gametes, telomerase function tends to be very high, thereby preventing the passage of shortened chromosomes from parents to offspring. In actively dividing somatic cells, however, telomerase function tends to decrease with age, thereby resulting in shorter telomeres in older individuals. For example, the average length of telomeres of the DNA in human blood cells is about 8000 bp at birth, but can be as short as 1500 bp in an elderly person.

Figure 11.25 shows the interesting mechanism by which telomerase works. Telomere lengthening occurs in three phases: binding, polymerization, and translocation.

Binding of Telomerase. Telomerase contains both protein subunits and RNA. The RNA part of telomerase, known as **telomerase RNA component (TERC),** contains a sequence complementary to that found in the telomeric repeat sequence. This allows telomerase to bind to the 3′ overhang region of the telomere.

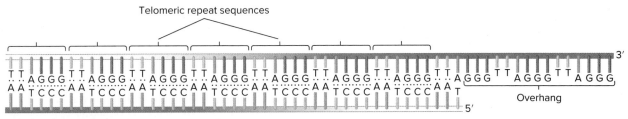

Telomeric repeat sequences

FIGURE 11.23 **General structure of telomeric sequences.** The DNA in telomeres consists of a tandemly repeated sequence and an overhang consisting of 12–16 nucleotides.

TABLE 11.5		
Telomeric Repeat Sequences Within Selected Organisms		
Group	**Example**	**Telomeric Repeat Sequence**
Mammals	Humans	TTAGGG
Slime molds	*Physarum, Didymium*	TTAGGG
	Dictyostelium	$AG_{(1-8)}$
Filamentous fungi	*Neurospora*	TTAGGG
Budding yeast	*Saccharomyces cerevisiae*	$TG_{(1-3)}$
Ciliates	*Tetrahymena*	TTGGGG
	Paramecium	TTGGG(T/G)
	Euplotes	TTTTGGGG
Flowering plants	*Arabidopsis*	TTTAGGG

Polymerization. Following binding, the RNA sequence beyond the binding site functions as a template for the synthesis of a six-nucleotide sequence at the end of the DNA strand. This synthesis is called polymerization, because it is analogous to the process carried out by DNA polymerase. Telomere lengthening is catalyzed by two identical protein subunits of telomerase called **telomerase reverse transcriptase (TERT).** TERT's name indicates that it catalyzes the reverse of transcription; it uses an RNA template to synthesize DNA.

Translocation Following polymerization, telomerase then moves—a process called translocation—to the new end of the DNA strand and attaches another six nucleotides to the end. This binding-polymerization-translocation cycle occurs many times in a row, thereby greatly lengthening the 3′ end of the DNA strand in the telomeric region. The complementary strand is synthesized by primase, DNA polymerase, and DNA ligase, as described earlier in this chapter. The RNA primer is later removed, which leaves a 3′ overhang.

Telomere Length May Play a Role in Aging and Cancer

When telomeres are too short, the cells become **senescent,** which means they lose their ability to divide. Researchers are investigating whether the shortening of telomeres is simply a sign of aging, like gray hair, or if the shortening is a programmed process that contributes to aging. In 1998, Andrea Bodnar and her colleagues inserted a gene that codes a highly active telomerase into human cells grown in the laboratory. The expression of telomerase prevented telomere shortening and senescence!

If telomeres tend to shorten with each cell division, how do cancer cells keep dividing? Though some types of cancer cells have very short telomeres, it is common for cancer cells to carry mutations that increase the activity of telomerase, thereby preventing telomere shortening. This increase in telomerase function is thought to prevent senescence. In the laboratory, when cancer cells are treated with drugs that inhibit telomerase, they often stop dividing. For this reason, researchers are interested in whether such drugs could be used to combat cancer. However, additional research is needed before these drugs can be used. A major concern is how they could affect normal cells of the body, possibly accelerating senescence.

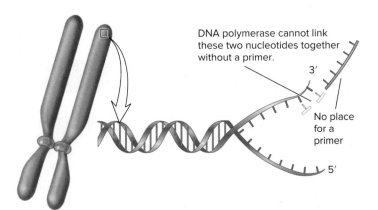

DNA polymerase cannot link these two nucleotides together without a primer.

3′

No place for a primer

5′

FIGURE 11.24 **The replication problem at the ends of linear chromosomes.** DNA polymerase cannot synthesize a DNA strand that is complementary to the 3′ end because a primer cannot be made upstream from this site.

11.5 COMPREHENSION QUESTIONS

1. In eukaryotes, DNA replication is initiated at an origin of replication by
   a. DnaA proteins.
   b. the origin recognition complex.
   c. DNA polymerase δ.
   d. MCM helicase.

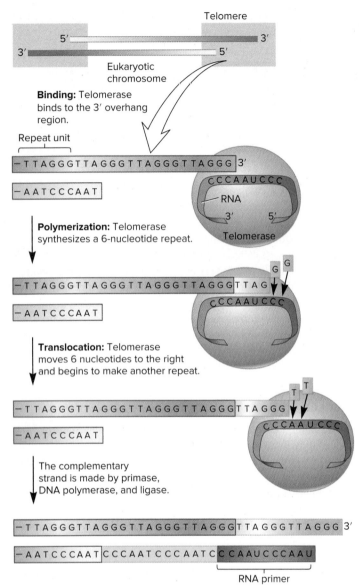

Telomere

Binding: Telomerase binds to the 3' overhang region.

Repeat unit

Polymerization: Telomerase synthesizes a 6-nucleotide repeat.

Translocation: Telomerase moves 6 nucleotides to the right and begins to make another repeat.

The complementary strand is made by primase, DNA polymerase, and ligase.

RNA primer

FIGURE 11.25 The enzymatic action of telomerase. A short, three-nucleotide segment of RNA within telomerase allows it to bind to the 3' overhang. The adjacent part of the RNA is used as a template to make a short, six-nucleotide repeat of the end of the DNA strand. After the repeat is made, telomerase moves six nucleotides to the right and then synthesizes another repeat. This process is repeated many times to lengthen the top strand (as shown in this figure). The bottom strand is made by DNA polymerase, using an RNA primer at the end of the chromosome that is complementary to the telomeric repeat sequence in the top strand. DNA polymerase fills in the region, which is sealed by ligase. Note: Though not shown in the figure, the RNA primer is eventually removed, which leaves a 3' overhang.

CONCEPT CHECK: How many times would telomerase have to bind to a different site in the telomere to make a segment of DNA that is 36 nucleotides in length?

2. Which of the following statements regarding DNA polymerases in eukaryotes is *not* correct?
   a. DNA polymerase α synthesizes a short RNA-DNA primer.
   b. DNA polymerases ε and δ synthesize most of the leading and lagging strands, respectively.
   c. Translesion-replicating DNA polymerases can replicate over damaged DNA.
   d. All of the above statements are correct.

3. In eukaryotes, RNA primers are primarily removed by
   a. DNA polymerase I.
   b. DNA polymerase α.
   c. flap endonuclease.
   d. helicase.

4. To synthesize DNA, what does telomerase use as a template?
   a. It uses the DNA in the 3' overhang region.
   b. It uses RNA that is a component of telomerase.
   c. No template is used.
   d. Both a and b are correct.

KEY TERMS

Introduction: DNA replication

11.1: template strand (parental strand), daughter strands, conservative model, semiconservative model, dispersive model

11.2: origin of replication, bidirectionally, replication fork, AT-rich region, DnaA box, DnaA protein, DNA helicase, bidirectional replication

11.3: DNA gyrase (topoisomerase II), single-strand binding protein, RNA primer, primase, leading strand, lagging strand, DNA polymerase, Okazaki fragments, DNA ligase, primosome, replisome, dimeric DNA polymerase, termination (ter)

sequences, catenanes, conditional mutant, temperature-sensitive (ts) mutant

11.4: processive enzyme, fidelity, proofreading function

11.5: ARS elements, G-quadraplex, prereplication complex (preRC), origin recognition complex (ORC), MCM helicase, DNA replication licensing, polymerase switch, translesion-replicating polymerase, flap endonuclease, telomeres, telomerase, telemerase RNA component (TERC), telomerase reverse transcriptase (TERT), senescent

CHAPTER SUMMARY

- DNA replication is the process in which existing DNA strands are used as templates to make new DNA strands.

11.1 Structural Overview of DNA Replication

- DNA replication occurs when the strands of DNA unwind and each strand is used as a template to make a new strand according to the AT/GC rule. The resulting DNA molecules have the same base sequence as the original DNA (see Figure 11.1).
- By labeling DNA with heavy and light isotopes of nitrogen and using centrifugation, Meselson and Stahl showed that DNA replication is semiconservative (see Figures 11.2, 11.3).

11.2 Bacterial DNA Replication: The Formation of Two Replication Forks at the Origin of Replication

- Bacterial DNA replication begins at a single origin of replication and proceeds bidirectionally around the circular chromosome (see Figure 11.4).
- In *E. coli*, DNA replication is initiated when DnaA proteins bind to sequences in five DnaA boxes at the origin of replication. This causes the AT-rich region to unwind. DNA helicases then promote the movement of two forks (see Figures 11.5, 11.6).

11.3 Bacterial DNA Replication: Synthesis of New DNA Strands

- At each replication fork, DNA helicase unwinds the DNA, and DNA gyrase relaxes positive supercoiling. Single-strand binding proteins coat the DNA to prevent the strands from coming back together. Primase synthesizes RNA primers, and DNA polymerase synthesizes complementary strands of DNA. DNA ligase joins adjacent Okazaki fragments by catalyzing the formation of a covalent bond in the DNA backbone (see Figure 11.7, Table 11.1).
- DNA polymerase III is an enzyme in *E. coli* with several subunits. The catalytic subunit wraps around the DNA like a hand (see Figure 11.8, Table 11.2).
- DNA polymerase needs an RNA primer or an existing DNA strand to synthesize DNA and make a new DNA strand in a 5′ to 3′ direction (see Figure 11.9).
- During DNA synthesis, the leading strand is made continuously in the direction toward the replication fork, whereas the lagging strand is made as Okazaki fragments in the direction away from the fork (see Figures 11.10, 11.11).
- A primosome is a complex between DNA helicase and primase. A replisome is a complex between a primosome and two DNA polymerase holoenzymes (see Figure 11.12).

- In *E. coli*, DNA replication is terminated at ter sequences (see Figure 11.13).
- Following DNA replication, intertwined catenanes sometimes need to be separated by the action of DNA gyrase (see Figure 11.14).
- The isolation and characterization of temperature-sensitive mutants was a useful strategy for identifying genes involved with DNA replication (see Figure 11.15, Table 11.3).

11.4 Bacterial DNA Replication: Chemistry and Accuracy

- DNA polymerase III is a processive enzyme that uses deoxynucleoside triphosphates to make new DNA strands (see Figure 11.16).
- The high fidelity of DNA replication is a result of (1) the stability of hydrogen bonding between the correct pairs of bases, (2) the phenomenon of induced fit, and (3) the proofreading function of DNA polymerase (see Figure 11.17).

11.5 Eukaryotic DNA Replication

- Eukaryotic chromosomes contain multiple origins of replication (see Figures 11.18, 11.19).
- Eukaryotic origins of replication may contain an ARS element (yeast) or a G-quadraplex (animals) within a nucleosome-free region that is flanked by nucleosomes in which the histones are covalently modified. Those in animals may also contain promoters and CpG islands (Figure 11.20).
- Part of the prereplication complex (preRC) is a group of six proteins called the origin recognition complex (ORC). The binding of MCM helicase to the leading strands completes a process called DNA replication licensing (see Figure 11.21).
- Eukaryotes have many different DNA polymerases with specialized roles. Different types of DNA polymerases switch with each other during the process of DNA replication (see Table 11.4).
- Flap endonuclease is an enzyme that removes flaps from RNA primers on Okazaki fragments (see Figure 11.22).
- The ends of eukaryotic chromosomes have telomeres, which are composed of tandemly repeated sequences and proteins (see Figure 11.23, Table 11.5).
- DNA polymerase is unable to replicate the 3′ ends of a eukaryotic chromosome (see Figure 11.24).
- Telomerase uses a short RNA molecule as a template to add repeat sequences onto telomeres (see Figure 11.25).
- Telomere shortening leads to cell senescence, whereas a lack of telomere shortening is associated with some forms of cancer.

PROBLEM SETS & INSIGHTS

MORE GENETIC TIPS

1. Describe three factors that account for the high fidelity of DNA replication. Discuss the quantitative contribution of each of the three.

TOPIC: *What topic in genetics does this question address?* The topic is the high fidelity of DNA replication. More specifically, the question asks you to describe and quantify the factors that account for that high fidelity.

INFORMATION: *What information do you know based on the question and your understanding of the topic?* From the question, you know there are three factors that account for the high fidelity of DNA replication. From your understanding of the topic, you may recall that these are proper base pairing, induced fit, and proofreading.

PROBLEM-SOLVING **S**TRATEGY: *Relate structure and function.* One strategy to solve this problem is to consider how the structures of DNA and DNA polymerase are related to the fidelity of DNA replication.

ANSWER:

First: A-T and G-C pairs are much more likely to form than are other types of base pairs. This base pairing limits mistakes to around 1 mistake per 1000 nucleotides added.

Second: Induced fit by DNA polymerase prevents covalent bond formation unless the proper nucleotides are in place. This increases fidelity another 100-fold to 1000-fold, to a range from 1 error in 100,000 to 1 error in 1 million.

Third: Exonuclease proofreading by DNA polymerase increases the fidelity another 100-fold to 1000-fold, to about 1 error per 100 million nucleotides added.

2. Summarize the process of chromosomal DNA replication in *E. coli.*

TOPIC: *What topic in genetics does this question address?* The topic is DNA replication in bacteria. More specifically, the question asks you to summarize the replication process in *E. coli.*

INFORMATION: *What information do you know based on the question and your understanding of the topic?* From the question, you know that DNA replication occurs in *E. coli.* From your understanding of the topic, you may recall that DNA replication in bacteria begins at an origin of replication and then two replication forks proceed bidirectionally around the circular chromosome.

PROBLEM-SOLVING **S**TRATEGY: *Describe the steps.* DNA replication is a complicated process. One strategy to solve this problem is to break it down into its individual steps.

ANSWER:

Step 1. DnaA proteins bind to the origin of replication, resulting in the separation of the AT-rich region.

Step 2. DNA helicase breaks the hydrogen bonds between the DNA strands, DNA gyrase relaxes positive supercoiling, and single-strand binding proteins keep the parental strands apart. Two replication forks move bidirectionally outward from the origin.

Step 3. At each of the two forks, primase synthesizes one RNA primer in the leading strand and many RNA primers in the lagging strand. DNA polymerase III then synthesizes the daughter strands of DNA. In the lagging strand, many short segments of DNA (Okazaki fragments) are made. DNA polymerase I removes the RNA primers and fills in with DNA, and DNA ligase covalently links the Okazaki fragments together.

Step 4. The processes described in steps 2 and 3 continue until the two replication forks meet each other on the other side of the circular bacterial chromosome at a ter sequence (T1 or T2).

Step 5. In some cases, the chromosomes are intertwined as catenanes. These are separated via DNA gyrase.

3. The ability of DNA polymerase to digest a DNA strand from one end is called its exonuclease activity. Exonuclease activity is involved in digesting RNA primers and also in removing mismatched nucleotides from a newly made DNA strand as part of the proofreading function of DNA polymerase. Note: DNA polymerase I does not change direction while it is removing an RNA primer and synthesizing new DNA. It does change direction during proofreading.

    A. In which direction, 5′ to 3′ or 3′ to 5′, is the exonuclease activity occurring during the removal of RNA primers?

    B. Figure 11.17 shows the location of the 3′ exonuclease site. Do you think this site would be used by DNA polymerase I to remove RNA primers? Why or why not?

TOPIC: *What topic in genetics does this question address?* The topic is about how DNA replication occurs in bacteria. More specifically, the question is about the exonuclease activity of DNA polymerase I.

INFORMATION: *What information do you know based on the question and your understanding of the topic?* From the question, you know that DNA polymerase I uses its exonuclease activity to remove RNA primers. It removes primers in the same direction that it synthesizes DNA. From your understanding of the topic, you may recall that DNA replication occurs in the 5′ to 3′ direction. You may also remember that proofreading by DNA polymerase III starts at the 3′ end of a strand and moves toward the 5′ end.

PROBLEM-SOLVING **S**TRATEGIES: *Make a drawing. Relate structure and function.* One strategy to solve this problem is make a drawing and think about the directionality of these processes while you are looking at the drawing. For part A, you could make a drawing similar to Figure 11.7. That will help you see the direction in which DNA polymerase I must move to remove the RNA primer and synthesize DNA without reversing direction. For part B, you could also see from your drawing that the proofreading function moves in a 3' to 5' direction.

ANSWER:

A. The removal of RNA primers occurs in the 5' to 3' direction, which is the same direction as DNA synthesis. Therefore, DNA polymerase I does not have to reverse direction.

B. No. The removal of RNA primers begins at the 5' end of the strand and moves in the 3' direction, whereas proofreading begins at the 3' end and moves in the 5' direction.

Conceptual Questions

C1. What key structural features of the DNA molecule underlie its ability to be accurately replicated?

C2. With regard to DNA replication, define *bidirectional replication*.

C3. Which of the following statements is *not* true? Explain why.

A. A DNA strand can serve as a template strand on many occasions.

B. Following semiconservative DNA replication, one strand of a DNA double helix is a newly made daughter strand and the other strand is a parental strand.

C. A DNA double helix may contain two strands of DNA that were made at the same time.

D. A DNA double helix obeys the AT/GC rule.

E. A DNA double helix could contain one strand that is 10 generations older than its complementary strand.

C4. The compound known as nitrous acid is a reactive chemical that replaces amino groups (–NH_2) with keto groups (=O). When nitrous acid reacts with the bases in DNA, it can change cytosine to uracil and change adenine to hypoxanthine. A segment of a DNA double helix has the following sequence:

TTGGATGCTGG
AACCTACGACC

A. What would be the sequence of this segment of the double helix immediately after reaction with nitrous acid? Let the letter H represent hypoxanthine and U represent uracil.

B. Let's suppose this DNA was treated with nitrous acid. The nitrous acid was then removed, and the DNA was replicated for two generations. What would be the sequences of the DNA products after the DNA had replicated twice? Your answer should contain the sequences of four segments of double helices. Note: During DNA replication, uracil hydrogen bonds with adenine, and hypoxanthine hydrogen bonds with cytosine.

C5. One way that bacterial cells regulate DNA replication is through GATC methylation sites within the origin of replication. Would this mechanism work if the DNA was conservatively (rather than semiconservatively) replicated?

C6. The chromosome of *E. coli* contains about 4.6 million bp. How long will it take to replicate its DNA? Assuming that DNA polymerase III is the primary enzyme involved and that it can actively proofread during DNA synthesis, how many base pair mistakes will be made in one round of DNA replication in a bacterial population containing 1000 bacteria?

C7. Here are two strands of DNA:

————————————————DNA polymerase→

————————————————

The one on the bottom is a template strand, and the one on the top is being synthesized by DNA polymerase in the direction shown by the arrow. Label the 5' and 3' ends of the top and bottom strands.

C8. A DNA strand includes the following sequence:

5'–GATCCCGATCCGCATACATTTACCAGATCACCACC–3'

In which direction would DNA polymerase slide along this strand (from left to right or from right to left)? If this strand was used as a template by DNA polymerase, what would be the sequence of the newly made strand? Indicate the 5' and 3' ends of the newly made strand.

C9. List and briefly describe the three types of functionally important sequences within bacterial origins of replication.

C10. As shown in Figure 11.5, five DnaA boxes are found within the origin of replication in *E. coli*. Take a look at these five sequences carefully.

A. Are the sequences of the five DnaA boxes very similar to each other? (Hint: Remember that DNA is double-stranded; think about these sequences in the forward and reverse directions.)

B. What is the most common sequence for a DnaA box? In other words, what is the most common base in the first position, second position, and so on until the ninth position? The most common sequence is called the consensus sequence.

C. The *E. coli* chromosome is about 4.6 million bp long. Based on random chance, is it likely that the consensus sequence for a DnaA box occurs elsewhere in the *E. coli* chromosome? If so, why doesn't the *E. coli* chromosome have multiple origins of replication?

C11. Obtain two strings of different colors (e.g., black and white) that are the same length. A length of 20 inches is sufficient. Tie a knot at one end of the black string and another knot at one end of the white string. Each knot designates the 5' end of a string. Make a double helix with your two strings. Now tape one end of the double helix to a table so that the tape is covering the knot on the black string.

A. Pretend your hand is DNA helicase and use your hand to unravel the double helix, beginning at the end that is not taped to the table. Should your hand be sliding along the white string or the black string?

B. As shown in Figure 11.12, imagine that your two hands together form a dimeric DNA polymerase. Unravel your two strings halfway to create a replication fork. Grasp the black string with your left hand and the white string with your right hand. Your thumbs should point toward the 5' end of each string. You need to loop one of the strings so that one of the DNA polymerases can synthesize the lagging strand. With such a loop, dimeric DNA polymerase can move toward the replication fork and synthesize both DNA strands in the 5′ to 3′ direction. In other words, with such a loop, your two hands can touch each other with both of your thumbs pointing toward the fork. Should the black string be looped, or should the white string be looped?

C12. Sometimes DNA polymerase makes a mistake, and the wrong nucleotide is added to the growing DNA strand. With regard to pyrimidines and purines, two general types of mistakes are possible. The addition of an incorrect pyrimidine instead of the correct pyrimidine (e.g., adding cytosine where thymine should be added) is called a transition. If a pyrimidine is incorrectly added to the growing strand instead of purine (e.g., adding cytosine where an adenine should be added), this type of mistake is called a transversion. If a transition or transversion is not detected by DNA polymerase, a mutation is created that permanently changes the DNA sequence. Though both types of mutations are rare, transition mutations are more frequent than transversion mutations. Based on your understanding of DNA replication and DNA polymerase, offer three explanations why transition mutations are more common.

C13. A short DNA sequence, which may be recognized by primase, is repeated many times throughout the *E. coli* chromosome. Researchers have hypothesized that primase may recognize this sequence as a site to begin the synthesis of an RNA primer for DNA replication. The *E. coli* chromosome is roughly 4.6 million bp in length. How many copies of the primase recognition sequence would be necessary to replicate the entire *E. coli* chromosome?

C14. Single-strand binding proteins keep the two parental strands of DNA separated from each other until DNA polymerase has an opportunity to replicate the strands. Suggest how single-strand binding proteins keep the strands separated and yet do not impede the ability of DNA polymerase to replicate the strands.

C15. In the following drawing, the top strand of the DNA is the template strand, and the bottom strand shows the lagging strand prior to the action of DNA polymerase I. The lagging strand contains three Okazaki fragments. The RNA primers have not yet been removed.

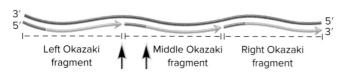

Left Okazaki fragment   Middle Okazaki fragment   Right Okazaki fragment

A. Which Okazaki fragment was made first, the one on the left or the one on the right?

B. Which RNA primer will be the first one to be removed by DNA polymerase I, the primer on the left or the primer on the right? For this primer to be removed by DNA polymerase I and for the gap to be filled in, is it necessary for the Okazaki fragment in the middle to have already been synthesized? Explain.

C. Let's consider how DNA ligase connects the left Okazaki fragment with the middle Okazaki fragment. After DNA polymerase I removes the middle RNA primer and fills in the gap with DNA, where does DNA ligase function? See the arrows on either side of the middle RNA primer. Is ligase needed at the left arrow, at the right arrow, or both?

D. When connecting two Okazaki fragments, DNA ligase uses NAD⁺ or ATP as a source of energy to catalyze this reaction.

Explain why DNA ligase needs another source of energy to connect two nucleotides, but DNA polymerase needs nothing more than the incoming nucleotide and the existing DNA strand. Note: You may want to refer to Figure 11.16 to answer this question.

C16. Describe the three important functions of DnaA proteins.

C17. Make a drawing that illustrates how DNA helicase works.

C18. What is an Okazaki fragment? In which strand of replicating DNA are Okazaki fragments found? Based on the properties of DNA polymerase, why is it necessary to make these fragments?

C19. Discuss the similarities and differences in the synthesis of DNA in the lagging and leading strands. What is the advantage of having a primosome and a replisome as opposed to having all replication enzymes functioning independently of each other?

C20. Explain the proofreading function of DNA polymerase.

C21. What is a processive enzyme? Explain why processivity is an important feature of DNA polymerase.

C22. What enzymatic features of DNA polymerase prevent it from replicating one of the DNA strands at the ends of linear chromosomes? Compared with DNA polymerase, how is telomerase different in its ability to synthesize a DNA strand? What does telomerase use as its template for the synthesis of a DNA strand? How does the use of this template result in a telomere sequence that is tandemly repetitive?

C23. As shown in Figure 11.25, telomerase adds additional DNA, six nucleotides at a time, to the ends of eukaryotic chromosomes. However, it works on only one DNA strand. Describe how the opposite strand is replicated.

C24. If a eukaryotic chromosome has 25 origins of replication, how many replication forks does it have at the beginning of DNA replication?

C25. In eukaryotes, what is meant by the term *DNA replication licensing*? How does the process occur?

C26. A diagram of a linear chromosome is shown here:

5′–A————————————————————————B–3′

3′–C————————————————————————D–5′

The ends of the two strands are labeled with A, B, C, or D. Which ends could *not* be replicated by DNA polymerase? Why not?

C27. As discussed in Chapter 18, some viruses contain RNA as their genetic material. Certain RNA viruses can exist as a provirus in which the viral genetic material has been inserted into the chromosomal DNA of the host cell. For this to happen, the viral RNA must be copied into a strand of DNA. An enzyme called reverse transcriptase, coded by the viral genome, copies the viral RNA into a complementary strand of DNA. The strand of DNA is then used as a template to make a double-stranded DNA molecule. This double-stranded DNA molecule is then inserted into the chromosomal DNA, where it may exist as a provirus for a long period of time.

A. How is the function of reverse transcriptase similar to the function of telomerase?

B. Unlike DNA polymerase, reverse transcriptase does not have a proofreading function. How might this affect the proliferation of the virus?

C28. Telomeres contain a 3′ overhang region, as shown in Figure 11.23. Does telomerase require a 3′ overhang to replicate the telomere region? Explain.

Experimental Questions

E1. Answer the following questions pertaining to the experiment of Figure 11.3.

A. What would be the expected results if the Meselson and Stahl experiment was carried out for four or five generations?

B. What would be the expected results of the Meselson and Stahl experiment after three generations if the mechanism of DNA replication was dispersive?

C. Considering the data from the experiment, explain why three different bands (i.e., light, half-heavy, and heavy) can be observed in the CsCl gradient.

E2. An absentminded researcher follows the steps of Figure 11.3, and, when viewing the gradient under UV light, does not see any bands at all. Which of the following mistakes could account for this observation? Explain how.

A. The researcher forgot to add ^{14}N-containing compounds.

B. The researcher forgot to add lysozyme.

C. The researcher forgot to turn on the UV lamp.

E3. Figure 11.4 shows an illustration of a replicating bacterial chromosome. If you analyzed many replicating chromosomes, what types of information could you learn about the mechanism of DNA replication?

E4. Table 11.3 lists rapid-stop and slow-stop mutants that affect DNA replication. What is the difference between a rapid-stop and a slow-stop mutant? What are different roles of the proteins that are defective in rapid-stop and slow-stop mutants?

E5. The technique of dideoxy sequencing of DNA is described in Chapter 20. The technique relies on the use of dideoxyribonucleotides (shown in Figures 20.11 and 20.12). A dideoxyribonucleotide has a hydrogen atom attached to the 3′carbon atom instead of a hydroxyl (—OH) group. When a dideoxyribonucleotide is incorporated into a newly made strand, the strand cannot grow any longer. Explain why.

E6. Another technique described in Chapter 20 is polymerase chain reaction (PCR) (see Figures 20.6 and 20.7), which is based on our understanding of DNA replication. In this method, a small amount of double-stranded template DNA is mixed with a high concentration of primers. Nucleotides and DNA polymerase are also added. The template DNA strands are separated by heat treatment, and when the temperature is lowered, the primers bind to the single-stranded DNA, and then DNA polymerase replicates the DNA. This increases the amount of DNA made from the primers. This cycle of steps (i.e., heat treatment, lower temperature, and allowing DNA replication to occur) is repeated again and again. Because the cycle is repeated many times, this method is called a chain reaction. It is called polymerase chain reaction because DNA polymerase is the enzyme needed to increase the amount of DNA with each cycle. In a PCR experiment, the template DNA is placed in a tube, and the primers, nucleotides, and DNA polymerase are added to the tube. The tube is then placed in a machine called a thermocycler, which raises and lowers the temperature. During one cycle, the temperature is raised (e.g., to 95°C) for a brief period and then lowered (e.g., to 60°C) to allow the primers to bind. The sample is then incubated at a slightly higher temperature for a few minutes to allow DNA replication to proceed. In a typical PCR experiment, the tube may be left in the thermocycler for 25–30 cycles. The total time for a PCR experiment is a few hours.

A. Why is DNA helicase not needed in a PCR experiment?

B. How is the sequence of each primer important in a PCR experiment? Do the two primers recognize the same strand or opposite strands?

C. The DNA polymerase used in PCR experiments is isolated from thermophilic bacteria that grow at high temperatures. Why is this kind of polymerase used?

D. If a tube initially contained 10 copies of double-stranded DNA, how many copies of double-stranded DNA (in the region flanked by the two primers) would be in the tube after 27 cycles?

Questions for Student Discussion/Collaboration

1. The complementarity of its two strands is the underlying reason that DNA can be accurately copied. Propose alternative chemical structures that could be accurately copied.

2. Compare and contrast DNA replication in bacteria and eukaryotes.

3. DNA replication is fast, virtually error-free, and coordinated with cell division. Discuss which of these three features you think is the most important.

Note: All answers are available for the instructor in Connect; the answers to the even-numbered questions and all of the Concept Check and Comprehension Questions are in Appendix B.

CHAPTER OUTLINE

A molecular model showing the enzyme RNA polymerase (gray and blue) in the act of sliding along a DNA molecule (yellow and orange) and synthesizing an RNA molecule (red), using DNA as a template.
Ramon Andrade/Science Source

12

GENE TRANSCRIPTION AND RNA MODIFICATION

The primary function of the genetic material, which is DNA, is to store the information necessary to create a living organism. The information is contained within units called genes. At the molecular level, a **gene** is defined as a segment of DNA that contains the information to make a functional product, either an RNA molecule or a polypeptide.

How is the information within a gene accessed? The first step in this process is called **transcription,** which literally means the act or process of making a copy. In genetics, this term refers to the process of synthesizing RNA from a DNA template (**Figure 12.1**). The structure of DNA is not altered as a result of transcription. Rather, the DNA base sequence has only been accessed to make a copy in the form of RNA. Therefore, the same DNA can continue to store information. DNA replication, which was discussed in Chapter 11, provides a mechanism for copying that stored information so that it can be transmitted to new daughter cells and from parent to offspring.

Protein-coding genes carry the information for the amino acid sequence of a polypeptide. When a protein-coding gene is transcribed, the first product is an RNA molecule known as **messenger RNA (mRNA).** During polypeptide synthesis—a process called **translation**—the sequence of nucleotides within the mRNA determines the sequence of amino acids in a polypeptide. One or more polypeptides then assemble into a functional protein. The structures and functions of proteins largely determine an organism's traits.

The model depicted in Figure 12.1, which is called the **central dogma of genetics** (also called the central dogma of molecular biology), was first enunciated by Francis Crick in 1958. The model forms a cornerstone of our understanding of genetics at the molecular level. The flow of genetic information occurs from DNA to mRNA to polypeptide.

In this chapter, we begin to study the molecular steps in gene expression, with an emphasis on transcription and the modifications that may occur to an RNA transcript after it has been made. Chapter 13 will examine the process of translation.

12.1 OVERVIEW OF TRANSCRIPTION

Learning Outcomes:

1. Describe the organization of a protein-coding gene and its mRNA transcript.
2. Outline the three stages of transcription.

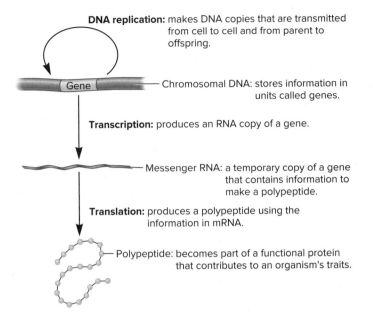

DNA replication: makes DNA copies that are transmitted from cell to cell and from parent to offspring.

Gene

Chromosomal DNA: stores information in units called genes.

Transcription: produces an RNA copy of a gene.

Messenger RNA: a temporary copy of a gene that contains information to make a polypeptide.

Translation: produces a polypeptide using the information in mRNA.

Polypeptide: becomes part of a functional protein that contributes to an organism's traits.

ONLINE
ANIMATION

FIGURE 12.1 **The central dogma of genetics.** The usual flow of genetic information is from DNA to mRNA to polypeptide. Note: The direction of informational flow shown in this figure is the most common direction found in living organisms, but exceptions do occur. For example, RNA viruses use an enzyme called reverse transcriptase to make a copy of DNA from RNA.

One key phenomenon regarding transcription is that specific base sequences define the beginning and ending of a gene and also play a role in regulating the level of RNA synthesis. In this section, we begin by examining the sequences that determine where transcription starts and ends, and then we will also briefly consider DNA sequences, called regulatory elements, that influence whether a gene is turned on or off. The functions of regulatory elements will be examined in greater detail in Chapters 14, 15, and 16. A second important aspect of transcription is the role that proteins play. DNA sequences, in and of themselves, just exist. For genes to be actively transcribed, proteins must recognize particular DNA sequences and act on them in a way that affects the transcription process. Later in this section, we will consider how proteins participate in the general steps of transcription.

Gene Expression Requires Base Sequences That Perform Different Functional Roles

At the molecular level, **gene expression** is the overall process by which the information within a gene is used to produce a functional product, such as a polypeptide. Along with environmental factors, the molecular expression of genes determines an organism's traits. **Figure 12.2** shows a common organization of base sequences that make up a protein-coding gene that functions in a bacterium such as *E. coli*.

The promoter and terminator are base sequences used during gene transcription.

- The **promoter** provides a site for beginning transcription.
- The **terminator** specifies the end of transcription.

These two sequences cause RNA synthesis to occur within a defined location. As shown in Figure 12.2, the DNA is transcribed into RNA from the promoter to the terminator.

As described later, the base sequence in the RNA transcript is complementary to the **template strand** of DNA. The opposite

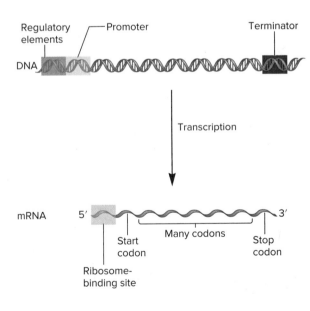

Regulatory elements ⌐ Promoter   Terminator

DNA

Transcription

mRNA   5′   3′

Start codon   Many codons   Stop codon

Ribosome-binding site

DNA:

- **Regulatory element:** site for the binding of regulatory proteins; the role of regulatory proteins is to influence the rate of transcription. Regulatory elements can be found in a variety of locations.

- **Promoter:** site for RNA polymerase binding; signals the beginning of transcription.

- **Terminator:** signals the end of transcription.

mRNA:

- **Ribosome-binding site:** site for ribosome binding; translation begins near this site in the mRNA. In eukaryotes, the ribosome scans the mRNA for a start codon.

- **Start codon:** specifies the first amino acid in a polypeptide sequence, usually a formylmethionine (in bacteria) or a methionine (in eukaryotes).

- **Codons:** 3-nucleotide sequences within the mRNA that specify particular amino acids. The sequence of codons within mRNA determines the sequence of amino acids within a polypeptide.

- **Stop codon:** specifies the end of polypeptide synthesis.

- Bacterial mRNA may be polycistronic, which means it codes two or more polypeptides.

FIGURE 12.2 **Organization of sequences of a bacterial gene and its mRNA transcript.** This figure depicts the general organization of sequences that are needed to form a functional gene that codes an mRNA.

CONCEPT CHECK: If a mutation changed the start codon into a stop codon, would the mutation affect the length of the mRNA? Explain.

strand of DNA is the **nontemplate strand.** For protein-coding genes, the nontemplate strand is also called the **coding strand,** or sense strand, because its sequence is the same as the transcribed mRNA that codes a polypeptide, except that the DNA has thymine (T) in places where the mRNA contains uracil (U). By comparison, the template strand is also called the noncoding strand, or antisense strand.

Transcription factors are a category of proteins that bind to genes and control the rate of transcription. Some transcription factors bind directly to the promoter and facilitate transcription. Other transcription factors recognize **regulatory elements** (also called **regulatory sequences** or **response elements**). A regulatory element is a short segment of DNA involved in the regulation of transcription. Certain transcription factors bind to regulatory elements and increase the rate of transcription, whereas others inhibit transcription.

As noted at the bottom of Figure 12.2, base sequences within an mRNA are used during the translation process.

- In bacteria, a short sequence within the mRNA, the **ribosome-binding site** (also known as the **Shine-Dalgarno sequence**), provides a location for a ribosome to bind and begin translation. The bacterial ribosome recognizes this site because it is complementary to a sequence in ribosomal RNA.
- Each mRNA contains a series of **codons,** read as groups of three nucleotides, which contain the information for the amino acid sequence of a polypeptide.
- The first codon, which is very close to the ribosome-binding site, is the **start codon.**
- The start codon is followed by many more codons that dictate the sequence of amino acids within the synthesized polypeptide.
- A **stop codon** signals the end of translation.

Chapter 13 will examine the process of translation in greater detail.

The Three Stages of Transcription Are Initiation, Elongation, and Termination

Transcription occurs in three stages: **initiation; elongation,** or synthesis of the RNA transcript; and **termination (Figure 12.3).** These steps involve protein-DNA interactions in which proteins such as **RNA polymerase,** the enzyme that synthesizes RNA, interact with DNA sequences. What causes transcription to begin? The initiation

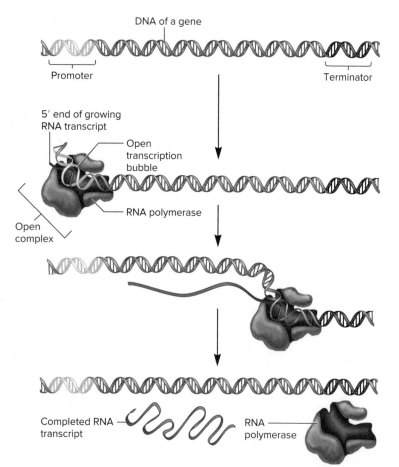

Initiation: The promoter functions as a recognition site for transcription factors (not shown). The transcription factors enable RNA polymerase to bind to the promoter. Following binding, the DNA is denatured to form an open complex.

Elongation/synthesis of the RNA transcript: RNA polymerase slides along the DNA in an open complex to synthesize RNA.

Termination: A terminator is reached that causes RNA polymerase and the RNA transcript to dissociate from the DNA.

ONLINE ANIMATION

FIGURE 12.3 **Stages of transcription.**

Genes→Traits The ability of genes to produce an organism's traits relies on the molecular process of gene expression. Transcription is the first step in gene expression. During transcription, the gene's sequence within the DNA is used as a template to make a complementary molecule of RNA. In Chapter 13, we will examine how the sequence in mRNA is translated into a polypeptide. After polypeptides are made within a living cell, they fold into functional proteins that govern an organism's traits.

stage in the transcription process is a recognition step. The sequence of bases within the promoter is recognized by one or more transcription factors. The specific binding of transcription factors to the promoter identifies the starting site for transcription.

The transcription factor(s) and RNA polymerase first bind to the promoter when the DNA is in the form of a double helix. For transcription to occur, the DNA strands must be separated into a bubble-like structure called the open transcription bubble. This allows one of the two strands to be used as a template for the synthesis of a complementary strand of RNA. The complex between RNA polymerase and an open transcription bubble is called the **open complex.** The open complex slides along the DNA, making an RNA transcript as it goes. Eventually, RNA polymerase reaches a terminator, which causes both RNA polymerase and the newly made RNA transcript to dissociate from the DNA.

12.1 COMPREHENSION QUESTIONS

1. Which of the following base sequences is used during transcription?
   a. Promoter and terminator
   b. Start and stop codons
   c. Ribosome-binding site
   d. Both a and b

2. The three stages of transcription are
   a. initiation, ribosome binding, and termination.
   b. elongation, ribosome binding, and termination.
   c. initiation, elongation, and termination.
   d. initiation, regulation, and termination.

12.2 TRANSCRIPTION IN BACTERIA

Learning Outcomes:

1. Describe the characteristics of a bacterial promoter.
2. Explain how RNA polymerase transcribes a bacterial gene.
3. Compare and contrast two mechanisms for transcriptional termination in bacteria.

Our understanding of gene transcription at the molecular level initially came from studies involving bacteria and bacteriophages. Several early investigations focused on the production of viral RNA after bacteriophage infection. The first suggestion that RNA is derived from the transcription of DNA was made by Elliot Volkin and Lazarus Astrachan in 1956. When these researchers exposed E. coli cells to T2 bacteriophages, they observed that RNA made immediately after infection had a base composition substantially different from the base composition of RNA made prior to infection. Furthermore, the base composition of the RNA after infection was very similar to the base composition of the T2 DNA, except that the RNA contained uracil instead of thymine. These results were consistent with the idea that bacteriophage DNA is used as a template for the synthesis of bacteriophage RNA.

In 1960, Matthew Meselson and François Jacob found that proteins are synthesized on ribosomes. One year later, Jacob and colleague Jacques Monod proposed that a certain type of RNA acts as a genetic messenger (from the DNA to the ribosome) to provide the information for protein synthesis. Jacob and Monod hypothesized that this RNA, which they called messenger RNA (mRNA), is transcribed from the base sequence within DNA and then directs the synthesis of particular polypeptides. This proposal was a remarkable insight, considering that it was made before the actual isolation and characterization of mRNA molecules in vitro. In 1961, the hypothesis was confirmed by Sydney Brenner, in collaboration with Jacob and Meselson. These researchers found that when a virus infects a bacterial cell, a virus-specific RNA is made and rapidly associates with preexisting ribosomes in the cell.

Since these pioneering studies, a great deal has been learned about the molecular features of bacterial gene transcription. Much of our knowledge comes from studies using E. coli. In this section, we will examine the three steps in the gene transcription process as they occur in bacteria.

A Promoter Is a Short Sequence of DNA That Is Necessary to Initiate Transcription

The type of DNA sequence known as a promoter gets its name from the idea that it "promotes" gene expression. More precisely, this sequence of bases directs the exact location for the initiation of transcription. Most of the promoter is located just ahead of, or upstream from, the site where transcription of a gene actually begins. By convention, the bases in a promoter sequence are numbered in relation to the **transcriptional start site (Figure 12.4).** This site is the first nucleotide used as a template for transcription and is denoted +1. The bases preceding this site are numbered in a negative direction. No base is numbered zero. Therefore, most of the promoter is labeled with negative numbers that identify the number of bases preceding the beginning of transcription.

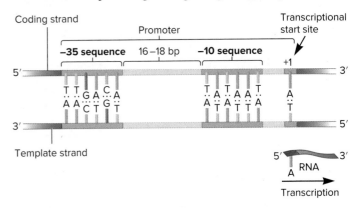

FIGURE 12.4 The conventional numbering system of promoters. The first nucleotide that acts as a template for transcription is designated +1. The numbering of nucleotides to the left of this spot is in a negative direction, whereas the numbering to the right is in a positive direction. For example, the nucleotide that is immediately to the left of the +1 nucleotide is numbered −1, and the nucleotide to the right of the +1 nucleotide is numbered +2. There is no zero nucleotide in this numbering system. In many bacterial promoters, the −35 and −10 sequences play a key role in promoting transcription.

Although the promoter may encompass a region several dozen nucleotides in length, short sequences are particularly critical for promoter recognition. By comparing the sequence of DNA bases within many promoters, researchers have learned that certain sequences of bases are necessary for a promoter to be functional. In many promoters found in *E. coli* and similar species, two sequences, which are located at approximately the −35 and −10 sites in the promoter, are particularly important (see Figure 12.4). The sequence at the −35 site in the top strand in Figure 12.4 is 5′–TTGACA–3′, and the one at the −10 site is 5′–TATAAT–3′. The TATAAT sequence is called the **Pribnow box** after David Pribnow, who initially discovered it in 1975.

Sequences within DNA, such as those found in promoters or regulatory elements, vary among different genes. The set of the most commonly occurring bases within a specific type of sequence is called the **consensus sequence.** As an example, let's consider how sequences may vary at the −35 and −10 sites among different genes. **Figure 12.5** illustrates the sequences found in several different *E. coli* promoters. The consensus sequence for this group of promoters is shown at the bottom. This sequence is efficiently recognized by proteins that initiate transcription. For many bacterial genes, a strong correlation is found between the maximal rate of transcription and the degree to which the −35 and −10 sequences agree with their consensus sequences. Mutations in the −35 or −10 sequences that lessen their similarity to the consensus sequences typically slow down the rate of transcription.

Bacterial Transcription Is Initiated When RNA Polymerase Holoenzyme Binds at a Promoter

Thus far, we have considered the DNA sequences that constitute a functional promoter. Let's now turn our attention to the proteins that recognize those sequences and carry out the transcription process. The enzyme that catalyzes the synthesis of RNA is RNA polymerase.

- In *E. coli*, the **core enzyme** is composed of five subunits (the subscript 2 indicates that there are two α subunits): $\alpha_2\beta\beta'\omega$.
- The association of a sixth subunit, **sigma (σ) factor,** with the core enzyme creates what is referred to as RNA polymerase **holoenzyme** (look ahead to Figure 12.7).
- The two α subunits are important in the proper assembly of the holoenzyme and in the process of binding to DNA.
- The β and β′ subunits are also needed for binding to the DNA, and they carry out the catalytic synthesis of RNA.
- The ω (omega) subunit is important for the proper assembly of the core enzyme.
- The holoenzyme is required to initiate transcription; the primary role of σ factor is to recognize the promoter. Proteins such as σ factor that influence the function of RNA polymerase are types of transcription factors.

After the six subunits assemble with each other, RNA polymerase holoenzyme binds loosely to the DNA and then slides along the DNA, much like a train rolls along tracks. How is a promoter identified? When the holoenzyme encounters a promoter, σ factor recognizes both the −35 and −10 sequences. The σ-factor protein contains a structure called a **helix-turn-helix motif** that can bind tightly to these sequences. Alpha (α) helices within the protein fit into the major groove of the DNA double helix and form hydrogen bonds with the bases. This phenomenon of molecular recognition is shown in **Figure 12.6**. Hydrogen bonding occurs between nucleotides in the −35 and −10 sequences of the promoter and amino acid side chains in the helix-turn-helix motif of σ factor.

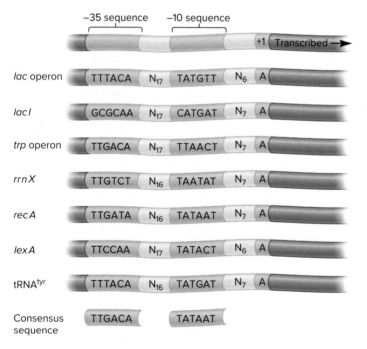

FIGURE 12.5 Examples of −35 and −10 sequences within several *E. coli* promoters. This figure shows the −35 and −10 sequences within one DNA strand in seven different *E. coli* promoters. The consensus sequence is shown at the bottom. The spacer regions are labeled to show the number of nucleotides between the −35 and −10 sequences or between the −10 sequence and the transcriptional start site. For example, N_{17} means there are 17 nucleotides between the end of the −35 sequence and the beginning of the −10 sequence.

CONCEPT CHECK: What does the term *consensus sequence* mean?

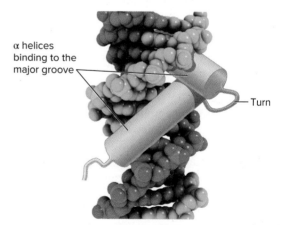

FIGURE 12.6 The binding of two α helices in σ-factor protein to the DNA of a promoter. The σ-factor protein contains two α helices connected by a turn, forming a structure termed a helix-turn-helix motif. Two α helices of the protein fit within the major groove of the DNA. Amino acids within the α helices form hydrogen bonds with the bases in the DNA. The DNA strands are shown in red and green.
Laguna Design/Science Source

CONCEPT CHECK: Why is it necessary for portions of σ-factor protein to fit into the major groove?

As shown in **Figure 12.7**, the process of transcription is initiated when σ factor within the holoenzyme binds to the promoter to form a **closed complex,** so named because the DNA strands at the promoter have not yet separated from each other. For transcription to begin, the double-stranded DNA must be unwound to form an open complex. This unwinding begins at the TATAAT sequence in the −10 site, which contains only A-T base pairs, as shown in Figure 12.4. A-T base pairs form only two hydrogen bonds, whereas G-C pairs form three. Therefore, DNA in an AT-rich region is more easily separated because fewer hydrogen bonds must be broken. A short strand of RNA is made within the open complex, and then σ factor is released from the core enzyme. The release of σ factor marks the transition to the elongation phase of transcription. The core enzyme may now slide along the DNA to synthesize a strand of RNA.

The RNA Transcript Is Synthesized During the Elongation Stage

After the initiation stage of transcription is completed, the RNA transcript is made during the elongation stage.

- During the synthesis of the RNA transcript, RNA polymerase moves along the DNA, causing it to unwind (**Figure 12.8**).
- The DNA strand used as a template for RNA synthesis is called the template strand.
- As RNA polymerase moves along the DNA, it creates an open complex that is approximately 17 bp long.
- On average, the rate of RNA synthesis is about 43 nucleotides per second!
- Behind the open complex, the DNA rewinds back into a double helix.

FIGURE 12.7 The initiation stage of transcription in bacteria. The σ-factor subunit of RNA polymerase holoenzyme recognizes the −35 and −10 sequences of the promoter. The DNA unwinds at the −10 sequence to form an open complex, and a short RNA is made. Then σ factor dissociates from the holoenzyme, and RNA polymerase core enzyme proceeds along the DNA, synthesizing RNA and forming an open complex as it goes.

CONCEPT CHECK: What feature of the −10 sequence makes the DNA at that location easier to unwind?

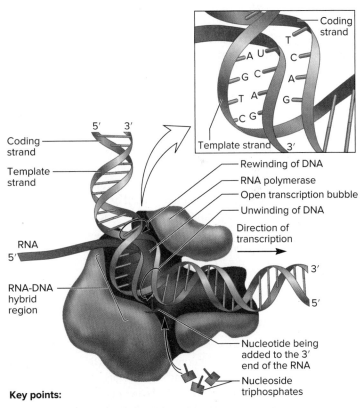

Key points:

- RNA polymerase slides along the DNA, creating an open complex as it moves.

- The DNA strand known as the template strand is used to make a complementary copy of RNA, resulting in an RNA-DNA hybrid.

- RNA polymerase moves along the template strand in a 3′ to 5′ direction, and RNA is synthesized in a 5′ to 3′ direction using nucleoside triphosphates as precursors. Pyrophosphate is released (not shown).

- The complementarity rule is the similar to the AT/GC rule except that U is substituted for T in the RNA.

FIGURE 12.8 Synthesis of the RNA transcript.

FIGURE 12.9 **The tran-scription of three different genes found in the same chromosome.** RNA polymerase synthesizes each RNA transcript in a 5′ to 3′ direction, sliding along a DNA template strand in a 3′ to 5′ direction. However, which strand is used as the template strand varies from gene to gene. For example, genes *A* and *B* are transcribed from the bottom strand, but gene *C* is transcribed from the top strand.

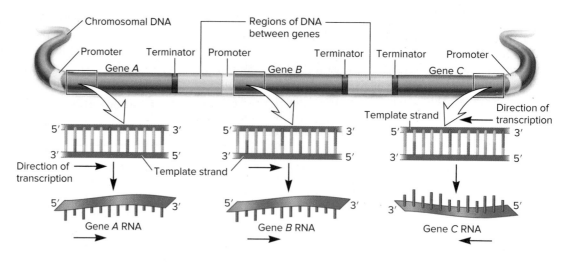

As summarized in Figure 12.8, the chemistry of transcription by RNA polymerase is similar to the replication of DNA via DNA polymerase, which is discussed in Chapter 11. RNA polymerase always connects nucleotides in the 5′ to 3′ direction. During this process, RNA polymerase catalyzes the formation of a bond between the 5′ PO_4^{2-} group on one nucleotide and the 3′ —OH group on the previous nucleotide. The complementarity rule is similar to the AT/GC rule, except that uracil substitutes for thymine in the RNA. In other words, RNA synthesis obeys an $A_{DNA}U_{RNA}/T_{DNA}A_{RNA}/G_{DNA}C_{RNA}/C_{DNA}G_{RNA}$ rule.

For transcription of multiple genes within a chromosome, the direction of transcription and the DNA strand that is used as a template vary among different genes. **Figure 12.9** shows three genes adjacent to each other within a chromosome. Genes *A* and *B* are transcribed from left to right, and the bottom DNA strand is used as a template. By comparison, gene *C* is transcribed from right to left, and the top DNA strand is used as a template. Note that in all three cases, the template strand is read in the 3′ to 5′ direction, and the synthesis of the RNA transcript occurs in a 5′ to 3′ direction. Within a given gene, only the template strand is used for RNA synthesis.

Transcription Is Terminated by Either an RNA-Binding Protein or an Intrinsic Terminator

The end of RNA synthesis is referred to as termination. Prior to termination, the hydrogen bonding between the DNA and RNA within the open complex is of central importance in preventing dissociation of RNA polymerase from the template strand. Termination occurs when this short RNA-DNA hybrid region is forced to separate, thereby releasing RNA polymerase as well as the newly made RNA transcript. In *E. coli*, two different mechanisms for termination have been identified. For certain bacterial genes, an RNA-binding protein known as **rho (ρ) protein** is responsible for terminating transcription, in a mechanism called ρ-dependent termination. For other genes, termination does not require the involvement of ρ protein—and in these cases, it is referred to as ρ-independent termination.

ρ-Dependent Termination.   In **ρ-dependent termination,** the termination process requires two nucleotide sequences. First, a rho recognition site upstream from the terminator, called the *rut* site for rho utilization site, acts as a recognition site for the binding of rho (ρ) protein (**Figure 12.10**). How does ρ protein facilitate termination? It functions as a helicase, a protein that can separate RNA-DNA hybrid regions. After the *rut* site is synthesized in the RNA, ρ protein binds to the RNA and moves in the direction of RNA polymerase.

The second nucleotide sequence of ρ-dependent termination is near the site where termination actually takes place. At this site, the DNA codes an RNA sequence containing several G-C base pairs that form a stem-loop structure. RNA synthesis terminates several nucleotides beyond this stem-loop. As discussed in Chapter 9, a stem-loop, also called a hairpin, can form due to complementary sequences within the RNA (refer back to Figure 9.17). This stem-loop forms almost immediately after the RNA sequence is synthesized and quickly binds to RNA polymerase. This binding results in a conformational change that causes RNA polymerase to pause in its synthesis of RNA. The pause allows ρ protein to catch up to the stem-loop, pass through it, and break the hydrogen bonds between the DNA and RNA within the open complex. When this occurs, the completed RNA strand is separated from the DNA along with RNA polymerase.

ρ-Independent Termination.   The process of **ρ-independent termination** does not require ρ protein. In this case, the terminator includes two adjacent nucleotide sequences (**Figure 12.11**).

- One sequence promotes the formation of a stem-loop.
- The second sequence, which is downstream from the stem-loop, is a uracil-rich sequence located at the 3′ end of the RNA.

As shown in Figure 12.11, the formation of the stem-loop causes RNA polymerase to pause in its synthesis of RNA. This pausing is stabilized by other proteins that bind to RNA polymerase. For example, a protein called NusA binds to RNA polymerase and promotes pausing at stem-loop sequences. At the precise time that RNA polymerase pauses, the uracil-rich sequence in the RNA transcript is bound to the DNA template strand. As previously

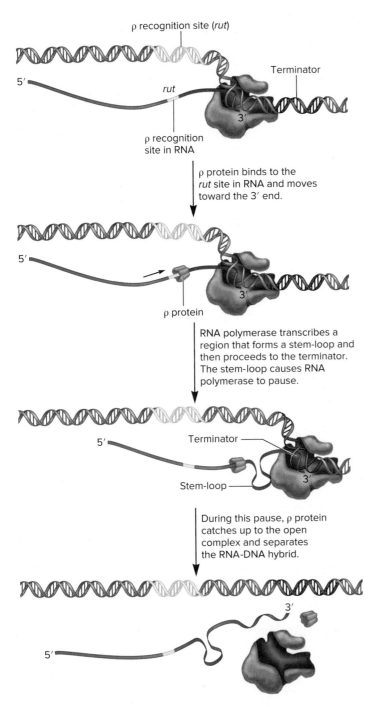

FIGURE 12.10 ρ-dependent termination.

CONCEPT CHECK: What would be the consequences if a mutation removed the *rut* site from the RNA molecule?

mentioned, the hydrogen bonding of RNA to DNA keeps RNA polymerase clamped onto the DNA. However, the binding of this uracil-rich sequence to the DNA template strand is relatively weak, causing the RNA transcript to spontaneously dissociate from the DNA and stop further transcription. Because ρ-independent termination does not require the ρ protein to physically remove the RNA transcript from the DNA, it is also referred to as **intrinsic termination.** In *E. coli*, about half of the genes show intrinsic termination, and the other half are terminated by ρ protein.

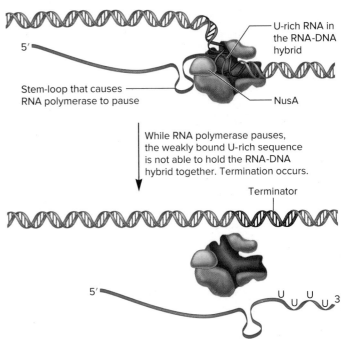

FIGURE 12.11 ρ-independent (or intrinsic) termination. When RNA polymerase reaches the end of the terminator, it transcribes a uracil-rich sequence. As this uracil-rich sequence is transcribed, a stem-loop forms just upstream from the open complex. The formation of this stem-loop causes RNA polymerase to pause in its synthesis of the transcript. This pausing is stabilized by the protein NusA, which binds near the region where RNA exits the open complex. At the point where RNA polymerase pauses, the RNA segment in the RNA–DNA hybrid is a uracil-rich sequence. Because only two hydrogen bonds form between U and A, the binding between RNA and DNA is relatively weak in this region, and the transcript and RNA polymerase dissociate from the DNA.

CONCEPT CHECK: Why is NusA important for this termination process?

GENETIC TIPS THE QUESTION: The technique of Northern blotting, which is described in Chapter 20, is used to determine how much RNA is transcribed from any particular gene. Figure 12.5 shows the sequences of promoters from several bacterial genes. Let's focus on the *lac* operon promoter. The transcription of the *lac* operon is induced when *E. coli* cells are exposed to lactose. Let's suppose you can use the technique of gene editing, which is described in Chapter 20, to change the *lac* operon promoter sequence in any way you like. How would you determine if the similarity of a gene's promoter sequence to the consensus sequence is an important factor affecting the level of gene transcription?

T OPIC: *What topic in genetics does this question address?* The topic is transcription. More specifically, the question is about the importance of the similarity of the promoter sequence to the consensus sequence.

I NFORMATION: *What information do you know based on the question and your understanding of the topic?* From the question, you know that Northern blotting can be used to measure the amounts of RNA transcribed from a gene, and gene editing can be used to alter a gene's sequence. From your understanding of the topic, you may remember that the consensus sequence is efficiently recognized by proteins that initiate transcription.

PROBLEM-SOLVING STRATEGY: *Design an experiment.*
One strategy to solve this problem is to design an experiment that compares the functioning of *lac* operon promoters that differ in their similarity to the consensus sequence.

ANSWER: Starting material is a (wild-type) strain of *E. coli* that carries a normal *lac* operon.

1. Use gene editing to create different *E. coli* strains in which the *lac* operon promoter sequence is altered to become either more similar to the consensus sequence or less similar to the consensus sequence.

2. Grow the wild-type and mutant strains and induce transcription by adding lactose.

3. Determine the resulting amounts of *lac* operon RNA using Northern blotting.

Expected results: Mutations that make the promoter sequence more like the consensus sequence will result in higher amounts of *lac* operon RNA compared to the amount produced by the wild-type strain. The bands on the gel would appear darker. Mutations that make the promoter sequence less like the consensus sequence will result in lower amounts of that RNA.

12.2 COMPREHENSION QUESTIONS

1. Within a promoter, a transcriptional start site is
   a. located at the −35 sequence and is recognized by σ factor.
   b. located at the −35 sequence and is where the first base is used as a template for transcription.
   c. located at the +1 site and is recognized by σ factor.
   d. located at the +1 site and is where the first base is used as a template for transcription.

2. For the following five sequences, what is the consensus sequence?
   5′–GGGAGCG–3′
   5′–GAGAGCG–3′
   5′–GAGTGCG–3′
   5′–GAGAACG–3′
   5′–GAGAGCA–3′
   a. 5′–GGGAGCG–3′
   b. 5′–GAGAGCG–3′
   c. 5′–GAGTGCG–3′
   d. 5′–GAGAACG–3′

3. Sigma (σ) factor is needed during which stage(s) of transcription?
   a. Initiation   c. Termination
   b. Elongation   d. All of these

4. A uracil-rich sequence occurs at the end of the RNA in
   a. ρ-dependent termination.   c. both a and b.
   b. ρ-independent termination.   d. none of these.

12.3 TRANSCRIPTION IN EUKARYOTES

Learning Outcomes:
1. List the functions of the three types of RNA polymerases in eukaryotes.
2. Describe the characteristics of a eukaryotic promoter for a protein-coding gene.
3. Explain how general transcription factors and RNA polymerase assemble at the promoter and form an open complex.
4. Compare and contrast two possible mechanisms for transcriptional termination in eukaryotes.

Many of the basic features of gene transcription are very similar in bacterial and eukaryotic species. Much of our understanding of transcription has come from studies of *Saccharomyces cerevisiae* (baker's yeast) and other eukaryotic species, including mammals. In general, gene transcription in eukaryotes is more complex than it is in bacteria. Eukaryotic cells are larger than bacterial cells and contain a variety of compartments known as organelles. This added level of cellular complexity requires that eukaryotes make many more proteins, and consequently they have many more protein-coding genes. In addition, most eukaryotic species are multicellular, that is, composed of many different cell types. Multicellularity adds a requirement that genes must be transcribed in the correct type of cell and during the proper stage of development. Therefore, in any given eukaryotic species, the transcription of the thousands of different genes that an organism possesses requires appropriate timing and coordination.

In this section, we will examine the basic features of gene transcription in eukaryotes. We will focus on the proteins that are needed to make an RNA transcript. In addition, an important factor that affects eukaryotic gene transcription is chromatin structure. Eukaryotic gene transcription requires changes in the positions and structures of nucleosomes. However, because these changes are important for regulating transcription, they are described in Chapter 15 rather than in this chapter.

Eukaryotes Have Multiple RNA Polymerases That Are Structurally Similar to the Bacterial Enzyme

The genetic material within the nucleus of a eukaryotic cell is transcribed by three different RNA polymerase enzymes, designated RNA polymerase I, II, and III. What are the roles of these enzymes? Each of the three RNA polymerases transcribes different categories of genes.

- RNA polymerase I: transcribes all of the genes for ribosomal RNAs (rRNAs) except for 5S rRNA.
- RNA polymerase II: transcribes all protein-coding genes. Therefore, it is responsible for the synthesis of all mRNAs. It also transcribes the genes for most snRNAs, which are needed for RNA splicing (discussed later in this chapter). In addition, it transcribes several types of genes that produce other noncoding RNAs (described in Chapter 17), including most long non-coding RNAs, microRNAs, and snoRNAs.
- RNA polymerase III: transcribes all tRNA genes and the 5S rRNA gene. To a much lesser extent than RNA polymerase II, it also transcribes a few genes that produce other

non-coding RNAs, such as snRNAs, long non-coding RNAs, microRNAs, and snoRNAs.

All three eukaryotic RNA polymerases are structurally very similar to each other and are composed of many subunits. They contain two large catalytic subunits similar to the β and β′ subunits of bacterial RNA polymerase. The structures of RNA polymerase from a few different species have been determined by X-ray crystallography. A remarkable similarity also exists between the bacterial enzyme and its eukaryotic counterparts. **Figure 12.12a** compares the structures of a bacterial RNA polymerase with RNA polymerase II from yeast. As you can see, the two enzymes have similar structures.

Interestingly, the structure of RNA polymerase provides a way to envision how the transcription process works. As seen in **Figure 12.12b**, DNA enters the enzyme through the jaw and lies on a surface within RNA polymerase termed the bridge. The part of the enzyme called the clamp is thought to control the movement of the DNA through RNA polymerase. A wall in the enzyme forces the RNA-DNA hybrid to make a right-angle turn. This bend facilitates the ability of nucleotides to bind to the template strand. Mg^{2+} is located at the catalytic site, which is precisely at the 3′ end of the growing RNA strand. Nucleoside triphosphates (NTPs) enter the catalytic site via a pore. The correct nucleotide binds to the template DNA and is covalently attached to the 3′ end. As RNA polymerase slides along the template strand, a rudder, which is about 9 bp away from the 3′ end of the RNA, forces the RNA-DNA hybrid apart. The DNA and the single-stranded RNA then exit under a small lid.

Eukaryotic Protein-Coding Genes Have a Core Promoter and Regulatory Elements

To ensure that transcription occurs at an appropriate rate, protein-coding genes have two features: a core promoter and regulatory elements. **Figure 12.13** shows a common pattern of sequences found in protein-coding genes. The **core promoter** is a relatively short DNA sequence that is necessary for transcription to take place. It typically consists of a TATAAA sequence called the **TATA box;** the transcriptional start site, where transcription begins; and downstream promoter elements (DPEs). The TATA box, which is usually about 25 bp upstream from the transcriptional start site, is important in determining the precise starting point for transcription. If it is missing from the core promoter, the transcription start site becomes undefined, and transcription may start at a variety of different locations. The core promoter, by itself, produces a very low level of transcription. This level is termed **basal transcription.**

In eukaryotes, transcription is influenced by **enhancers,** which are DNA segments, usually 50–1000 bp in length, that contain one or more regulatory elements. Enhancers are recognized by **regulatory transcription factors (RTFs)**—proteins that affect the ability of RNA polymerase to recognize the core promoter and begin the process of transcription. RTFs fall into two broad categories:

- **Activators** are proteins that recognize enhancers and stimulate the rate of transcription. Without such stimulation, most eukaryotic genes have very low levels of basal transcription.
- Under certain conditions, it may be necessary to prevent transcription of a given gene. This occurs via **repressors,** which also bind to enhancers but inhibit the rate of transcription.

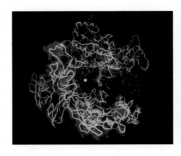 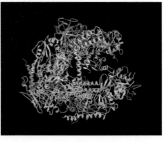

(a) Structure of a bacterial RNA polymerase   **Structure of a eukaryotic RNA polymerase II (yeast)**

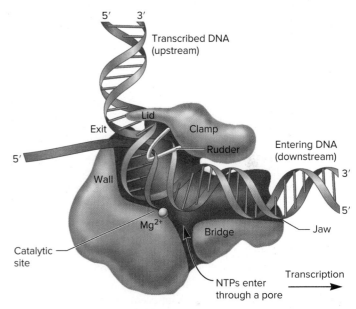

(b) Schematic structure of RNA polymerase

FIGURE 12.12 Structure and molecular function of RNA polymerase. **(a)** A comparison of the crystal structures of a bacterial RNA polymerase (left) and a eukaryotic RNA polymerase II (right). The bacterial enzyme is from *Thermus aquaticus.* The eukaryotic enzyme is from *Saccharomyces cerevisiae.* **(b)** A mechanism for transcription based on the enzyme's structure. In this diagram, the direction of transcription is from left to right. The double-stranded DNA enters the polymerase along a bridge surface that is between the jaw and clamp. At a region termed the wall, the RNA-DNA hybrid is forced to make a right-angle turn, which enables nucleotides to bind to the template strand of the DNA. Mg^{2+} is located at the catalytic site. Nucleoside triphosphates (NTPs) enter the catalytic site via a pore and bind to the template strand. At the catalytic site, the nucleotides are covalently attached to the 3′ end of the RNA. As RNA polymerase slides along the template strand, a small region of the protein termed the rudder separates the RNA-DNA hybrid. The DNA and single-stranded RNA then exit under a small lid. Note: RNA polymerase II in eukaryotes also has a carboxyl terminal domain as shown later in Figure 12.14.

(a, left): Courtesy of Dr. Seth Darst; (a, right): Image of PDB ID 1K83 (Bushnell, D.A., Cramer, P., Kornberg, R.D. (2002) Structural basis of transcription: alpha-amanitin-RNA polymerase II cocrystal at 2.8 Å resolution. *PNAS* 99: (1218)) created using ProteinWorkshop, a product of the RCSB PDB, and built using the Molecular Biology toolkit Developed by John Moreland and Apostol Gramada. The MBT is financed by grant GM63208.

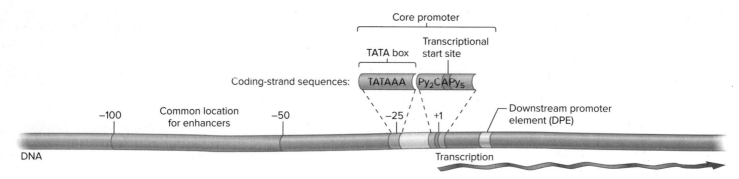

FIGURE 12.13 **A common pattern found within the promoter of protein-coding genes recognized by RNA polymerase II.** The start site usually occurs at adenine (A); two pyrimidines (Py: cytosine or thymine) and a cytosine (C) are to the left of this adenine, and five pyrimidines (Py) are to the right. A TATA box is approximately 25 bp upstream from the start site. However, the sequences that constitute eukaryotic promoters are quite diverse, and not all protein-coding genes have a TATA box. Enhancers, which are discussed in Chapter 15, vary in their locations but some are found in the −50 to −100 region. Downstream regulatory elements (DPEs) are found after the transcriptional start site. The core promoters for RNA polymerase I and III are quite different from this one. A single upstream regulatory element is involved in the binding of RNA polymerase I to its promoter, whereas two regulatory elements, called A and B boxes, facilitate the binding of RNA polymerase III.

CONCEPT CHECK: What is the functional role of the TATA box?

The mechanisms whereby activators and repressors bind to enhancers and exert their effects are described in Chapter 15. Note: In bacteria, segments of DNA that are recognized by activators or repressors are called operators rather than enhancers (see Chapter 14).

As seen in Figure 12.13, a possible location for regulatory elements is the −50 to −100 region. However, the locations of regulatory elements vary considerably among different eukaryotic genes. These elements can be far away from the core promoter yet strongly influence the ability of RNA polymerase to initiate transcription.

DNA sequences such as the TATA box and enhancers exert their effects only on a particular gene. They are called ***cis*-acting elements.** The term *cis* comes from chemistry nomenclature and means "next to." *Cis*-acting elements, though possibly located far away from the core promoter, are always found within the same chromosome as the genes they regulate. By comparison, the regulatory transcription factors that bind to such elements are called ***trans*-acting factors** (the term *trans* means "across from"). The transcription factors that control the expression of a gene are themselves coded by genes; regulatory genes that code transcription factors may be far away from the genes they control, even on a different chromosome. When a gene coding a *trans*-acting factor is expressed, the protein product can diffuse into the cell nucleus and bind to its appropriate *cis*-acting element. Let's now turn our attention to the function of such proteins.

Transcription of Eukaryotic Protein-Coding Genes Is Initiated When RNA Polymerase II and General Transcription Factors Bind to a Promoter Sequence

Thus far, we have considered the DNA sequences that play a role in the core promoter of eukaryotic protein-coding genes. By studying transcription in a variety of eukaryotic species, researchers have discovered that three categories of proteins are needed for basal transcription at the core promoter: RNA polymerase II, general transcription factors, and a complex called mediator (**Table 12.1**).

TABLE 12.1

Proteins Needed for Transcription via the Core Promoter of Eukaryotic Protein-Coding Genes

RNA polymerase II: The enzyme that catalyzes the linkage of nucleotides in the 5′ to 3′ direction, using DNA as a template. Most eukaryotic RNA polymerase II enzymes are composed of 12 subunits. The two largest subunits are structurally similar to the β and β′ subunits found in *E. coli* RNA polymerase.

General transcription factors:

TFIID: Composed of TATA-binding protein (TBP) and other TBP-associated factors (TAFs). Recognizes the TATA box of the core promoter of eukaryotic protein-coding genes.

TFIIA: Binds to TFIID and promotes its binding to the TATA box.

TFIIB: Binds to TFIID and then enables RNA polymerase II to bind to the core promoter. Also promotes TFIIF binding.

TFIIF: Binds to RNA polymerase II and plays a role in its ability to bind to TFIIB and the core promoter. Also plays a role in the ability of TFIIE and TFIIH to bind to RNA polymerase II.

TFIIE: Plays a role in the formation or the maintenance (or both) of the open complex. It may exert its effects by facilitating the binding of TFIIH to RNA polymerase II and regulating the activity of TFIIH.

TFIIH: A multisubunit protein that has multiple roles. First, TFIIH has a subunit that functions as a DNA translocase, tracking along DNA and leaving unwound DNA in its wake. This subunit separates the DNA strands to convert the preinitiation complex to an open complex. Other subunits phosphorylate the carboxyl-terminal domain (CTD) of RNA polymerase II, which releases it from interacting with TFIIB, thereby allowing RNA polymerase II to proceed to the elongation phase.

Mediator: A multisubunit complex that mediates the effects of regulatory transcription factors on the function of RNA polymerase II. Though mediator typically has certain core subunits, many of its subunits vary, depending on the cell type and environmental conditions. The ability of mediator to affect the function of RNA polymerase II occurs via the carboxyl-terminal domain (CTD) of RNA polymerase II. Mediator can influence the ability of TFIIH to phosphorylate CTD, and subunits within mediator itself have the ability to phosphorylate CTD. Because CTD phosphorylation is required to release RNA polymerase II from TFIIB, mediator plays a key role in the ability of RNA polymerase II to switch from the initiation to the elongation stage of transcription.

Six different **general transcription factors (GTFs)** and a protein complex called mediator are needed for RNA polymerase II to initiate transcription of protein-coding genes. **Figure 12.14** shows the assembly of GTFs, mediator, and RNA polymerase II at the TATA box. As shown in the figure, a series of interactions leads to the formation of the open complex.

1. Transcription factor IID (TFIID) is a very large protein complex that first binds to the TATA box and to one or more DPEs and thereby plays a critical role in the recognition of the core promoter. The binding of TFIIA to TFIID enhances TFIID's ability to bind to the TATA box. TFIID is composed of 13 to 14 subunits, including TATA-binding protein (TBP), which directly binds to the TATA box, and several other proteins called TBP-associated factors (TAFs). TFIID is larger than RNA polymerase II.

2. After TFIID binds to the TATA box, it associates with TFIIB.

3. TFIIB promotes the binding of RNA polymerase II and TFIIF.

4. Lastly, TFIIE, TFIIH, and mediator bind to the complex. **Mediator** derives its name from the observation that it mediates interactions between RNA polymerase II and regulatory transcription factors that bind to enhancers. The function of mediator during eukaryotic gene regulation is explored in greater detail in Chapter 15. Overall, the binding of TFIIE, TFIIH, and mediator completes the assembly of proteins to form a closed complex. For protein-coding genes in eukaryotes, the closed complex is more commonly known as the **preinitiation complex.**

5. TFIIH and mediator play a major role in the formation of the open complex. TFIIH has several subunits that perform different functions. Certain subunits act as helicases, which break the hydrogen bonds between the two strands of the DNA and are needed to initially form an open complex. Another subunit hydrolyzes ATP and phosphorylates a domain in RNA polymerase II known as the carboxyl-terminal domain (CTD). Similarly, mediator and other cellular proteins are able to phosphorylate the CTD.

6. Phosphorylation of the CTD breaks the contact between RNA polymerase II and TFIIB. Next, TFIIB, TFIIE, TFIIH, and mediator dissociate, and RNA polymerase II is free to proceed to the elongation stage of transcription. Note: The TFIID/TFIIA complex may remain bound to the TATA box and thereby initiate another round of transcription at a later time.

In vitro, when researchers mix together TFIID, TFIIB, TFIIF, TFIIE, TFIIH, RNA polymerase II, and a DNA sequence containing a TATA box and transcriptional start site, the DNA is transcribed into RNA, but at a very low rate. Therefore, these components are referred to as the **basal transcription apparatus.** In a living cell, however, additional components, such as TFIIA, mediator, enhancers, activators, and coactivators, regulate transcription and allow it to proceed at a reasonable rate. The functions of these components are explored in greater detail in Chapter 15.

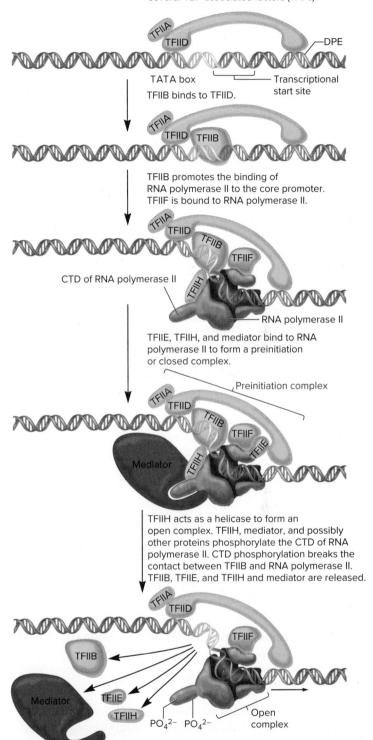

FIGURE 12.14 **Steps leading to the formation of the open complex.** The strands of the DNA are initially separated at the TATA box to form an open complex. In this diagram, the open complex has moved to the transcriptional start site, which is usually about 25 bp away from the TATA box. Note: The shapes of the GTFs and mediator are depicted schematically in order to show how they interact with RNA polymerase.

CONCEPT CHECK: Why is the phosphorylation of the carboxyl-terminal domain (CTD) functionally important?

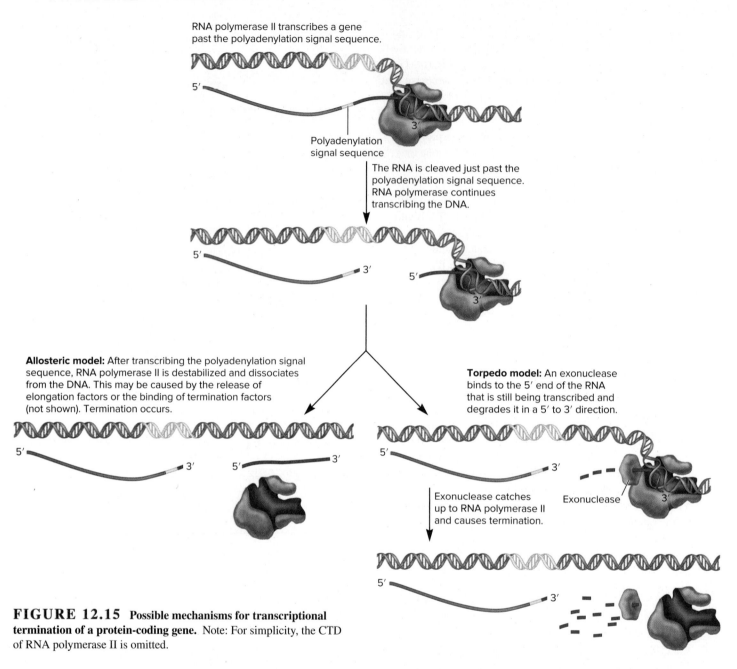

FIGURE 12.15 **Possible mechanisms for transcriptional termination of a protein-coding gene.** Note: For simplicity, the CTD of RNA polymerase II is omitted.

Transcriptional Termination Occurs After the 3′ End of the Transcript Is Cleaved Near the Polyadenylation Signal Sequence

As discussed later in this chapter, eukaryotic mRNAs are modified by cleavage near their 3′ end and the subsequent attachment of a string of adenine nucleotides (look ahead to Figure 12.24). This process, which is called polyadenylation, requires the transcription of a polyadenylation signal sequence that directs the cleavage of the mRNA. Transcription via RNA polymerase II typically terminates about 500–2000 nucleotides downstream from the polyadenylation signal sequence.

Figure 12.15 shows a simplified scheme for the transcriptional termination of a protein-coding gene. After RNA polymerase II has transcribed the polyadenylation signal sequence, the RNA is cleaved just downstream from this sequence. This

cleavage occurs before transcriptional termination. Two models have been proposed for transcriptional termination:

- According to the allosteric model, RNA polymerase II becomes destabilized after it has transcribed the polyadenylation signal sequence, and it eventually dissociates from the DNA. This destabilization may be caused by the release of proteins that function as elongation factors or by the binding of proteins that function as termination factors.
- A second model, called the torpedo model, suggests that RNA polymerase II is physically removed from the DNA. According to this model, the region of RNA that is still being transcribed and is downstream from the polyadenylation signal sequence is cleaved by an exonuclease that degrades the transcript in the 5′ to 3′ direction. When the

exonuclease catches up to RNA polymerase II, this causes RNA polymerase II to dissociate from the DNA.

Which of these two models is correct? Additional research is needed, but the results of studies over the past few years have provided evidence that the two models are not mutually exclusive. Therefore, both mechanisms may play a role in transcriptional termination.

12.3 COMPREHENSION QUESTIONS

1. Which RNA polymerase in eukaryotes is responsible for the transcription of genes that code proteins?
   a. RNA polymerase I
   b. RNA polymerase II
   c. RNA polymerase III
   d. All of the above transcribe protein-coding genes.
2. An enhancer is a _____ that regulates the _____.
   a. *trans*-acting factor, rate of transcription
   b. *trans*-acting factor, termination of transcription
   c. *cis*-acting element, rate of transcription
   d. *cis*-acting element, termination of transcription
3. The basal transcription apparatus is composed of
   a. five general transcription factors.
   b. RNA polymerase II.
   c. a DNA sequence containing a TATA box and transcriptional start site.
   d. all of the above.
4. With regard to transcriptional termination in eukaryotes, which model(s) suggest(s) that RNA polymerase is physically removed from the DNA?
   a. Allosteric model
   b. Torpedo model
   c. Both models
   d. Neither model

12.4 RNA MODIFICATION

Learning Outcomes:

1. List the different types of RNA modifications.
2. Describe the processing of rRNAs and tRNAs.
3. Compare and contrast different mechanisms of RNA splicing.
4. Outline how alternative splicing occurs, and describe its benefits.
5. Explain how eukaryotic mRNAs are modified so that they have a cap and a tail.
6. Describe the process of RNA editing.

During the 1960s and 1970s, studies of bacteria established the physical structure of the gene. The analysis of bacterial genes showed that the sequence of DNA within the coding strand corresponds to the sequence of nucleotides in the mRNA, except that T is replaced with U. During translation, the sequence of codons in the mRNA is read, providing the instructions for the correct amino acid sequence in a polypeptide. The correspondence between the sequence of codons in

the DNA coding strand and the amino acid sequence of the polypeptide has been termed the **colinearity** of gene expression.

The situation changed dramatically in the late 1970s, when the tools became available to study eukaryotic genes at the molecular level. The scientific community was astonished by the discovery that eukaryotic protein-coding genes are not always colinear with their functional mRNAs. Instead, the coding sequences within many eukaryotic genes are separated by DNA sequences that are not translated into proteins. The coding sequences are found within **exons,** which are regions that are contained within functional mRNA. By comparison, the sequences that are found between the exons are called **intervening sequences,** or **introns.** During transcription, a **pre-mRNA** is made, and it corresponds to the entire gene sequence that was transcribed. To produce a functional, or mature, mRNA, the sequences in the pre-mRNA that correspond to the introns are removed and the exons are connected, or spliced, together. This process is called **RNA splicing (Table 12.2).** Since the 1970s, research has revealed that splicing is a common genetic phenomenon in eukaryotic species. Splicing occurs occasionally in bacteria as well.

Aside from splicing, research has also shown that RNA transcripts are modified in several other ways. Table 12.2 describes several types of RNA modifications. For example, rRNAs and tRNAs are synthesized as long transcripts that are processed into smaller functional pieces. In addition, most mature mRNAs in eukaryotes have a cap attached to their 5′ end and a tail attached at their 3′ end. In this section, we will examine the molecular mechanisms that account for these types of RNA modifications and consider why they are functionally important.

rRNAs and tRNAs Are Made as Longer Transcripts That Are Cleaved into Smaller Functional Transcripts

For some non-protein-coding genes, the RNA transcript initially made during gene transcription is processed or cleaved into smaller pieces. As an example, **Figure 12.16** shows the processing of mammalian ribosomal RNA. The ribosomal RNA gene is transcribed by RNA polymerase I, resulting in a long primary transcript known as

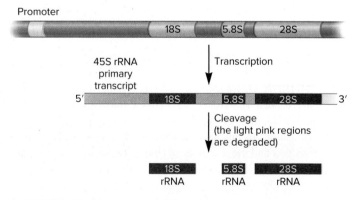

FIGURE 12.16 The processing of ribosomal RNA in eukaryotes. The ribosomal RNA gene is transcribed into a long 45S rRNA primary transcript. This transcript is cleaved to produce 18S, 5.8S, and 28S rRNA molecules, which become associated with protein subunits in the ribosome. This processing occurs within the nucleolus in the cell nucleus.

TABLE 12.2

Modifications That May Occur to RNAs

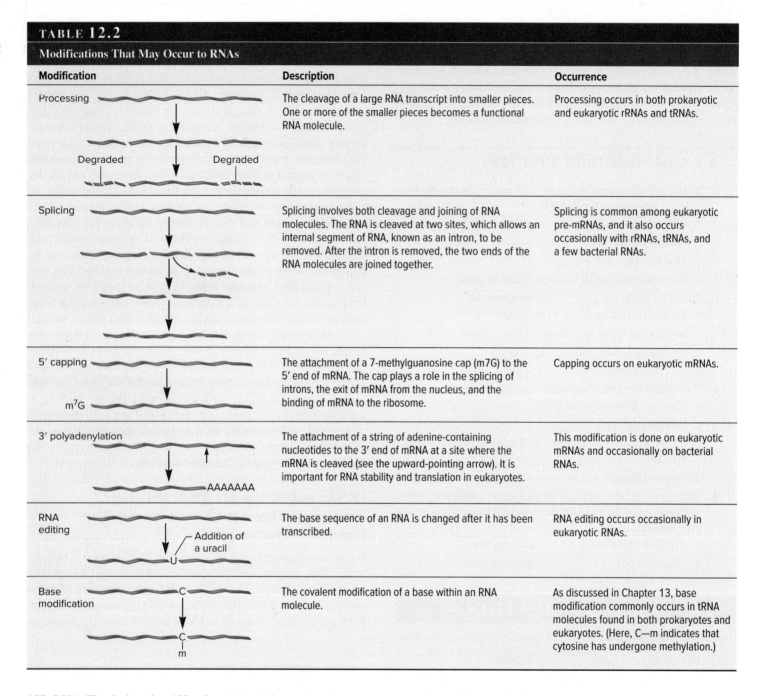

Modification	Description	Occurrence
Processing	The cleavage of a large RNA transcript into smaller pieces. One or more of the smaller pieces becomes a functional RNA molecule.	Processing occurs in both prokaryotic and eukaryotic rRNAs and tRNAs.
Splicing	Splicing involves both cleavage and joining of RNA molecules. The RNA is cleaved at two sites, which allows an internal segment of RNA, known as an intron, to be removed. After the intron is removed, the two ends of the RNA molecules are joined together.	Splicing is common among eukaryotic pre-mRNAs, and it also occurs occasionally with rRNAs, tRNAs, and a few bacterial RNAs.
5' capping	The attachment of a 7-methylguanosine cap (m7G) to the 5' end of mRNA. The cap plays a role in the splicing of introns, the exit of mRNA from the nucleus, and the binding of mRNA to the ribosome.	Capping occurs on eukaryotic mRNAs.
3' polyadenylation	The attachment of a string of adenine-containing nucleotides to the 3' end of mRNA at a site where the mRNA is cleaved (see the upward-pointing arrow). It is important for RNA stability and translation in eukaryotes.	This modification is done on eukaryotic mRNAs and occasionally on bacterial RNAs.
RNA editing	The base sequence of an RNA is changed after it has been transcribed.	RNA editing occurs occasionally in eukaryotic RNAs.
Base modification	The covalent modification of a base within an RNA molecule.	As discussed in Chapter 13, base modification commonly occurs in tRNA molecules found in both prokaryotes and eukaryotes. (Here, C—m indicates that cytosine has undergone methylation.)

45S rRNA. The designation 45S refers to the sedimentation characteristics of this transcript in Svedberg units (S). Following synthesis of the 45S rRNA, cleavage occurs at several points to produce three fragments, termed 18S rRNA, 5.8S rRNA, and 28S rRNA. These functional rRNA molecules play a key role in forming the structure of the ribosome. In eukaryotes, the cleavage of 45S rRNA into smaller rRNAs and the assembly of ribosomal subunits occur in an organelle within the cell nucleus known as the **nucleolus.**

The production of tRNA molecules requires processing via endonucleases and sometimes exonucleases. An **endonuclease** cleaves the bond between two adjacent nucleotides within a strand. By comparison, an **exonuclease** cleaves a covalent bond between two nucleotides at one end of a strand. Starting at one end, an exonuclease digests a strand, removing one nucleotide at a time.

Some exonucleases begin this digestion only from the 3' end, traveling in the 3' to 5' direction, whereas others begin only at the 5' end and digest in the 5' to 3' direction.

Like rRNAs, tRNAs are synthesized as longer precursor molecules that must be cleaved to produce mature, functional tRNAs that carry amino acids during translation. The details of tRNA processing vary among different tRNAs and among different species. **Figure 12.17** shows a common series of steps that occur in the processing of tRNA molecules in eukaryotes.

- Near the 5' end, the precursor tRNA is recognized by RNaseP, which is an endonuclease that cuts the precursor tRNA. The action of RNaseP produces the correct 5' end of the mature tRNA in all species.

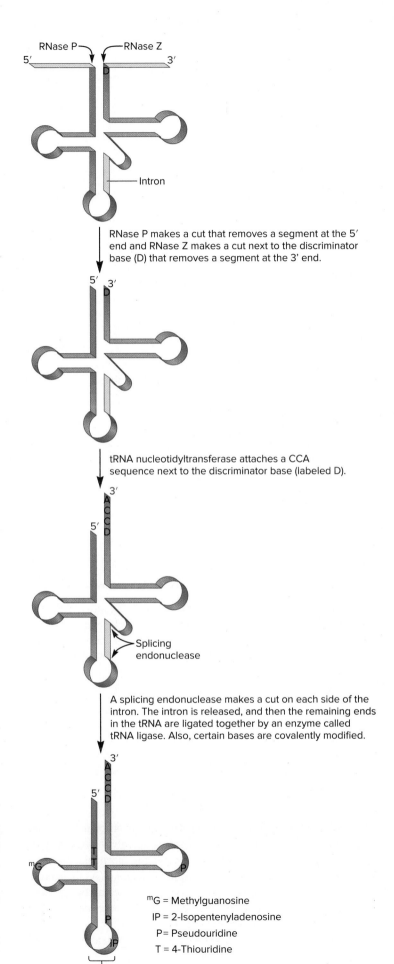

RNase P makes a cut that removes a segment at the 5′ end and RNase Z makes a cut next to the discriminator base (D) that removes a segment at the 3′ end.

tRNA nucleotidyltransferase attaches a CCA sequence next to the discriminator base (labeled D).

A splicing endonuclease makes a cut on each side of the intron. The intron is released, and then the remaining ends in the tRNA are ligated together by an enzyme called tRNA ligase. Also, certain bases are covalently modified.

^mG = Methylguanosine
IP = 2-Isopentenyladenosine
P = Pseudouridine
T = 4-Thiouridine

Anticodon

FIGURE 12.17 **Common steps in the processing of a precursor tRNA molecule in eukaryotes.** RNaseP is an endonuclease that makes a cut that creates the 5′ end of the mature tRNA. At the 3′ end, another endonuclease, RNase Z, makes a cut, which removes a segment at the 3′ end. Next, the intron is spliced out and some bases are modified to other bases (methylguanosine, 2-isopentenyladenosine, pseudouridine, and 4-thiourine), as indicated. Note: RNase Z, splicing endonuclease, and tRNA ligase are proteins. RNase P is a complex between an RNA molecule and a protein.

CONCEPT CHECK: In the mechanism shown here, in how many places has the precursor tRNA molecule been cut?

- RNase Z, also called 3′-tRNase, cleaves next to a nucleotide that contains the discriminator base, which is labeled D in Figure 12.17. Alternatively, in some bacterial species, one or more exonucleases may be involved in removing the 3′ end.
- All tRNAs have a CCA sequence at their 3′ end. In eukaryotes, the CCA sequence is added next to the nucleotide with the discriminator base by an enzyme called tRNA nucleotidyltransferase. However, in some bacterial tRNAs, the CCA sequence is coded in the tRNA gene.
- Some pre-tRNAs contain introns. The intron is removed by a splicing endonuclease that makes a cut on each side of the intron. The intron is released, and then the remaining ends in the tRNA are ligated together by an enzyme called tRNA ligase.
- Certain bases in tRNA molecules are covalently modified to alter their structure. The functional importance of modified bases in tRNAs is discussed in Chapter 13.

As researchers studied tRNA processing, they discovered certain features that were very unusual and exciting, changing the way biologists view the actions of catalysts. RNaseP is a catalyst that contains both RNA and protein subunits. In 1983, Sidney Altman and colleagues made the surprising discovery that the RNA portion of RNaseP, not the protein subunit, contains the catalytic ability to cleave the precursor tRNA. RNaseP is an example of a **ribozyme,** an RNA molecule with catalytic activity. Prior to the study of RNaseP and the identification of self-splicing RNAs (discussed next), biochemists had staunchly believed that only protein molecules could function as biological catalysts.

Different Splicing Mechanisms Remove Introns

As mentioned, splicing is the removal of intron RNA and the covalent connection of the exon RNA. Eukaryotic introns were first detected by comparing the base sequence of viral genes and their mRNA transcripts during viral infection of mammalian cells by adenovirus. This research was carried out in 1977 by two groups headed by Philip Sharp and Richard Roberts. Several other research groups, including those of Pierre Chambon, Bert O'Malley, and Philip Leder, later identified introns in eukaryotic protein-coding genes.

In Figure 12.17, we saw how an intron is removed from a precursor tRNA molecule by a splicing endonuclease and then the ends of the resulting tRNA are connected to each other

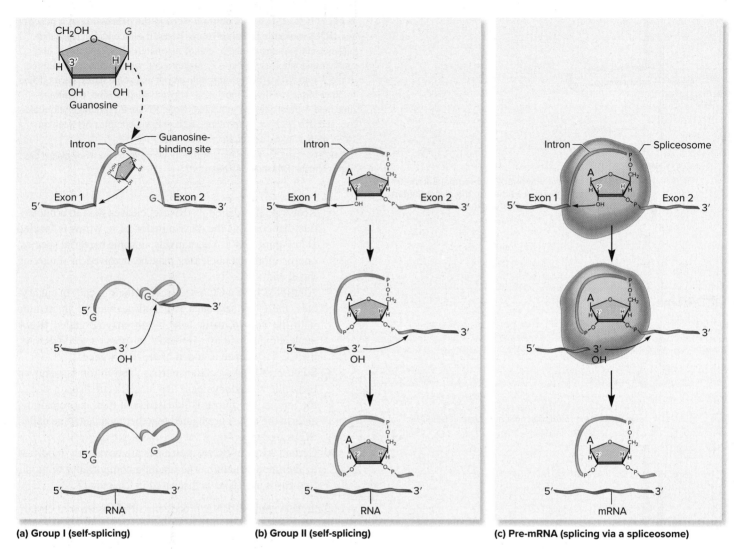

(a) Group I (self-splicing)   **(b) Group II (self-splicing)**   **(c) Pre-mRNA (splicing via a spliceosome)**

FIGURE 12.18 **Mechanisms of RNA splicing.** Group I and II introns are self-splicing. **(a)** The splicing of group I introns involves the binding of a guanosine to a site within the intron, leading to the cleavage of RNA at the 3′ end of exon 1. The bond between a different nucleotide in the intron strand (in this case, guanine) and the 5′ end of exon 2 is cleaved. The 3′ end of exon 1 then forms a covalent bond with the 5′ end of exon 2. **(b)** In group II introns, self-splicing occurs by a similar mechanism, except that the 2′ —OH group on an adenine nucleotide (already within the intron) begins the catalytic process. **(c)** Pre-mRNA splicing requires the aid of a complex known as a spliceosome.

CONCEPT CHECK: Which of these three mechanisms is very common in eukaryotes?

(i.e., spliced together) by tRNA ligase. In addition, researchers have discovered a few other splicing mechanisms in bacteria and eukaryotes (**Figure 12.18**).

The splicing of **group I** and **group II introns** occurs via **self-splicing**—splicing that does not require the aid of any proteins or other RNAs. Instead, a self-splicing RNA functions as its own ribozyme. Introns of groups I and II differ in the ways in which they are removed and the exons are connected. Group I introns that occur within the rRNA of *Tetrahymena* (a protist) have been studied extensively by Thomas Cech and colleagues. In this organism, the splicing process involves the binding of a single guanosine to a guanosine-binding site within the intron (Figure 12.18a).

1. The guanosine breaks the bond between the first exon and the intron. The guanosine becomes attached to the 5′ end of the intron.

2. The 3′ —OH group of exon 1 then breaks the bond next to a different nucleotide (in this example, a guanine, G) that lies at the boundary between the end of the intron and exon 2; exon 1 forms a covalent bond with the 5′ end of exon 2.

3. The intron RNA is subsequently degraded.

In this example, the RNA molecule functions as its own ribozyme, because it splices itself without the aid of a catalytic protein.

With group II introns, a similar splicing mechanism occurs, except the 2′ —OH group on ribose in an adenine (A) nucleotide already within the intron strand begins the catalytic process (Figure 12.18b). Experimentally, self-splicing of group I and II introns can occur in vitro without the addition of any proteins. However, in a living cell, proteins known as **maturases** often enhance the rate of this self-splicing.

Type of Intron	Mechanism of Removal	Occurrence
Group I	Self-splicing	Found in rRNA genes within the nucleus of *Tetrahymena* and other simple eukaryotes. Found in a few protein-coding, tRNA, and rRNA genes within mitochondrial DNA (in fungi and plants) and in chloroplast DNA. Found very rarely in tRNA genes within bacteria.
Group II	Self-splicing	Found in a few protein-coding, tRNA, and rRNA genes within mitochondrial DNA (in fungi and plants) and in chloroplast DNA. Also found rarely in bacterial genes.
Pre-mRNA	Spliceosome	Very commonly found in protein-coding genes within the nucleus of eukaryotic cells.

TABLE 12.3 Occurrence of Introns

In eukaryotes, the transcription of protein-coding genes produces a long transcript known as pre-mRNA, which is made in the nucleus. This pre-mRNA is usually altered by splicing and other modifications before it exits the nucleus. Unlike group I and II introns, which may undergo self-splicing, pre-mRNA splicing requires the aid of a complex known as a **spliceosome.** As discussed shortly, the spliceosome recognizes the boundaries of the intron so that it can be removed properly.

Table 12.3 summarizes the occurrence of introns among the genes of different groups of organisms. The splicing of group I and II introns is relatively uncommon. By comparison, pre-mRNA splicing is a widespread phenomenon among complex eukaryotes. In mammals and flowering plants, most protein-coding genes have at least one intron that can be located anywhere within the gene. An average human gene has about eight introns. In some cases, a single gene can have a large number of introns. As a striking example, the human dystrophin gene, which, when mutated, causes Duchenne muscular dystrophy, has 79 exons punctuated by 78 introns.

Pre-mRNA Splicing Occurs by the Action of a Spliceosome

As noted previously, the spliceosome is a large complex that splices pre-mRNA in eukaryotes. It is composed of five subunits (U1, U2, U4, U5, and U6) known as **snRNPs** (pronounced

"snurps"). Each snRNP consists of small nuclear RNA and a set of proteins. During splicing, the subunits of a spliceosome carry out several functions. First, spliceosome subunits bind to an intron sequence and precisely recognize the intron-exon boundaries. In addition, the spliceosome must hold the pre-mRNA in the correct configuration to ensure the splicing together of the exons. And finally, the spliceosome catalyzes the chemical reactions that cause the intron to be removed and the exons to be covalently linked.

Intron RNA is defined by base sequences located within the intron and at the intron-exon boundaries. The consensus sequences for the splicing of mammalian pre-mRNA are shown in **Figure 12.19**. The bases most commonly found at these sites—those that are highly conserved evolutionarily—are shown in bold. The 5′ and 3′ splice sites occur at the ends of the intron, whereas the branch site is somewhere in the middle. These sites are recognized by subunits of the spliceosome.

The molecular mechanism of pre-mRNA splicing is depicted in **Figure 12.20**. The snRNP designated U1 binds to the 5′ splice site, and U2 binds to the branch site. This is followed by the binding of a trimer of three snRNPs: a U4/U6 dimer plus U5. The intron loops outward, and the two exons are brought closer together. The 5′ splice site is then cut, and the 5′ end of the intron becomes covalently attached to the 2′ —OH group of a specific adenine nucleotide in the branch site. U1 and U4 are released. In the final step, the 3′ splice site is cut, and then the exons are covalently attached to each other. The three snRNPs—U2, U5, and U6—remain attached to the intron, which is in a lariat configuration. Eventually, the intron is degraded, and the snRNPs are used again to splice other pre-mRNAs.

An snRNA molecule within the U6 subunit of the spliceosome plays a catalytic role in the removal of introns and the connection of exons. Two catalytic metal ions (Mg^{2+}) are specifically bound to this snRNA and form an active site that catalyzes the cleavage and ligation reactions shown in Figure 12.20. In other words, the U6 subunit functions as a ribozyme that cleaves the RNA at the exon-intron boundaries and connects the remaining exons. More specifically, this subunit is called a **metalloribozyme** due to the role played by Mg^{2+} ions in its catalytic action. Interestingly, the active site in the spliceosome has a very similar structure to the active site of RNAs with group II introns, which are self-splicing. This observation suggests that the active sites of the spliceosome and the self-splicing RNAs are evolutionarily related to each other.

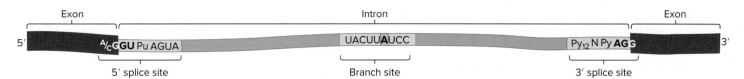

FIGURE 12.19 **Consensus sequences for pre-mRNA splicing in complex eukaryotes.** Consensus sequences occur at the intron-exon boundaries and at a branch site found within the intron itself. The adenine nucleotide shown in blue in this figure corresponds to the adenine nucleotide at the branch site in Figure 12.20. The nucleotides shown in bold are highly conserved evolutionarily. Designations: A/C = A or C, Pu = purine, Py = pyrimidine, N = any of the four bases.

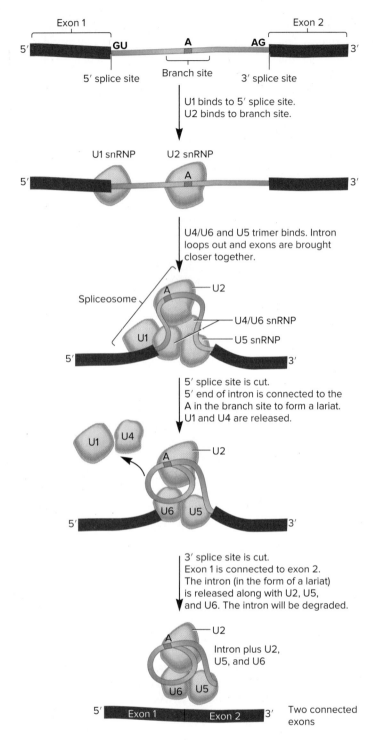

FIGURE 12.20 **A simplified model for splicing of pre-mRNA via a spliceosome.**

CONCEPT CHECK: Describe the roles of snRNPs in the splicing process.

Alternative Splicing Regulates Which Exons Occur in a Mature mRNA, Allowing Different Polypeptides to Be Made from the Same Gene

When it was first discovered, the phenomenon of splicing seemed like a wasteful process. During transcription, energy is used to synthesize intron sequences, which are subsequently removed by spliceosomes. Making and then removing introns uses a significant amount of energy. This observation intrigued many geneticists, because natural selection tends to eliminate wasteful processes. Therefore, researchers expected to find that pre-mRNA splicing has one or more important biological roles. In recent years, one very important biological advantage of pre-mRNA splicing has become apparent. This advantage is due to **alternative splicing,** which refers to the phenomenon that a pre-mRNA can be spliced in more than one way.

Why is alternative splicing advantageous? To understand the biological effects of alternative splicing, remember that the sequence of amino acids within a polypeptide determines the structure and function of a protein. Alternative splicing produces two or more polypeptides from the same gene that have differences in their amino acid sequences, leading to possible changes in their functions. In most cases, the alternative versions of the protein have similar functions, because most of their amino acid sequences are identical to each other. Nevertheless, alternative splicing produces differences in amino acid sequences that provide each polypeptide with its own unique characteristics. Because alternative splicing allows two or more different polypeptide sequences to be derived from a single gene, some geneticists have speculated that an important advantage of this process is that it allows an organism to carry fewer genes in its genome.

The degree of splicing and alternative splicing varies greatly among different species. Baker's yeast (*S. cerevisiae*), for example, contains about 6300 genes and approximately 300 (i.e., approximately 5%) of those code pre-mRNAs that are spliced. Of these pre-mRNAs, only a few have been shown to be alternatively spliced. Therefore, in this unicellular eukaryote, alternative splicing is not a major mechanism for generating protein diversity. In comparison, complex multicellular organisms rely much more heavily on alternative splicing. Humans have approximately 20,000 different protein-coding genes, and most of these contain one or more introns. Recent estimates suggest that about 70% of all human pre-mRNAs are alternatively spliced. Furthermore, certain pre-mRNAs are alternatively spliced to an extraordinary extent. Some pre-mRNAs can be alternatively spliced to produce dozens of different mRNAs. This level of alternative splicing provides a much greater potential for human cells to create protein diversity.

Figure 12.21 considers an example of alternative splicing for a mammalian gene that codes a protein known as α-tropomyosin, which functions in the regulation of cell contraction. It is located along the thin filaments found in smooth muscle cells, such as those in the uterus and small intestine, and in striated muscle cells that are found in cardiac and skeletal muscle. The protein α-tropomyosin is also synthesized in many types of nonmuscle cells but in lower amounts. Within a multicellular organism, different types of cells must regulate their contractibility in subtly different ways. One way this variation in function may be accomplished is by the production of different forms of α-tropomyosin.

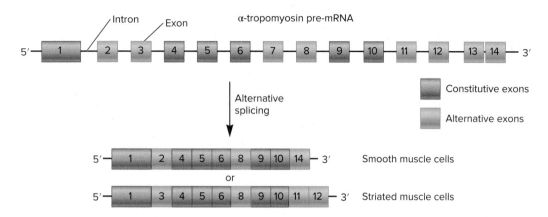

FIGURE 12.21 Alternative ways that the rat α-tropomyosin pre-mRNA can be spliced. The top part of this figure depicts the structure of the rat α-tropomyosin pre-mRNA. Exons are shown as colored rectangles, and introns as connecting black lines. The lower part of the figure depicts the mature mRNAs of smooth and striated muscle cells. The constitutive exons (red) are found in the mature α-tropomyosin mRNAs in all cell types. Alternative exons (green) are also found in mRNAs, but they vary from one cell type to another. Though exon 8 in this figure is found in both mature mRNAs, it is considered an alternative exon because it is not found in the mature mRNAs for α-tropomyosin in some other cell types.

Genes→Traits The protein α-tropomyosin functions in the regulation of cell contraction in muscle and nonmuscle cells. Alternative splicing of the pre-mRNA provides a way to vary contractibility in different types of cells by modifying the function of α-tropomyosin. As shown here, the alternatively spliced versions of the pre-mRNA produce α-tropomyosin proteins that differ from each other in their structure (i.e., amino acid sequence). These alternatively spliced versions vary in function to meet the needs of the cell type in which they are found. For example, the sequence of exons 1–2–4–5–6–8–9–10–14 produces an α-tropomyosin protein that functions suitably in smooth muscle cells. Overall, alternative splicing affects the traits of an organism by allowing a single gene to code several versions of a protein, each optimally suited to the cell type in which it is made.

The intron-exon structure of the rat α-tropomyosin pre-mRNA and two alternative ways that the pre-mRNA can be spliced are presented in Figure 12.21.

- The pre-mRNA contains 14 exons, 6 of which are **constitutive exons** (shown in red), which are always found in the mature mRNAs from all cell types. Presumably, constitutive exons code polypeptide segments of the α-tropomyosin protein that are necessary for its general structure and function.

- The mature mRNA also contains other exons, called **alternative exons** (shown in green), which are not always found in the mRNA after splicing of the pre-mRNA has occurred. The amino acid sequences coded by alternative exons may subtly change the function of α-tropomyosin to meet the needs of the cell type in which it is found. For example, Figure 12.21 shows the predominant splicing products found in smooth muscle cells and striated muscle cells. Exon 2 codes a segment of the α-tropomyosin protein that alters its function to make it suitable for smooth muscle cells. By comparison, the α-tropomyosin mRNA found in striated muscle cells does not include exon 2. Instead, this mRNA contains exon 3, which is more suitable for that cell type.

Alternative splicing is not a random event. Rather, the specific pattern of splicing is regulated in any given cell. The molecular mechanism for the regulation of alternative splicing involves proteins known as **splicing factors,** which play a key role in the choice of particular splice sites. **SR proteins** are one type of splicing factor. They contain a domain at their carboxyl-terminal end that is rich in the amino acids serine (S) and arginine (R) and is involved in protein-protein recognition. They also contain an RNA-binding domain at their amino-terminal end.

As shown previously in Figure 12.20, the components of a spliceosome recognize the 5′ and 3′ splice sites and then remove the intervening intron. The function of splicing factors is to modulate the ability of a spliceosome to choose 5′ and 3′ splice sites. This can occur in two ways:

- Some splicing factors act as repressors that inhibit the ability of the spliceosome to recognize a splice site. In **Figure 12.22a**, a splicing repressor binds to a 3′ splice site and prevents the spliceosome from recognizing the site. Instead, the spliceosome binds to the next available 3′ splice site. The splicing repressor causes exon 2 to be spliced out of the pre-mRNA and not included in the mature mRNA, an event called **exon skipping.**

- Alternatively, other splicing factors enhance the ability of the spliceosome to recognize particular splice sites. In **Figure 12.22b**, splicing enhancers bind to the 3′ and 5′ splice sites that flank exon 3, which results in the inclusion of exon 3 in the mature mRNA.

Alternative splicing occurs because each cell type has its own characteristic concentration of many kinds of splicing factors. Furthermore, splicing factors may be regulated by the binding of small effector molecules, protein-protein interactions, and covalent modifications. Overall, the differences in the composition of splicing factors and the regulation of their activities form the basis for alternative splicing outcomes.

The Ends of Eukaryotic Pre-mRNAs Have a 5′ Cap and a 3′ Tail

In addition to splicing, pre-mRNAs in eukaryotes are also subjected to modifications at their 5′ and 3′ ends. At their 5′ end, most mature mRNAs have a 7-methylguanosine covalently attached—an event known as **capping.** Capping occurs while the pre-mRNA is being made by RNA polymerase II, usually when the transcript

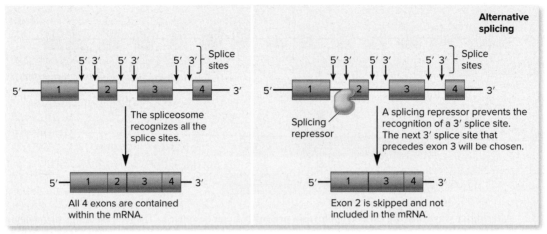

(a) Splicing repressors

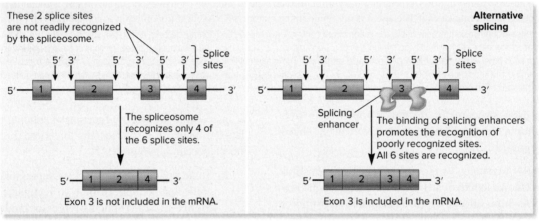

(b) Splicing enhancers

FIGURE 12.22 **The roles of splicing factors during alternative splicing.** **(a)** Splicing factors can act as repressors to prevent the recognition of splice sites. In this example, the presence of a splicing repressor causes exon 2 to be skipped and thus not included in the mRNA. **(b)** Other splicing factors can enhance the recognition of splice sites. In this example, splicing enhancers promote the recognition of sites that flank exon 3, thereby causing its inclusion in the mRNA.

CONCEPT CHECK: A pre-mRNA with 7 exons and 6 introns is recognized by just one splicing repressor that binds to the 3′ end of the third intron. The third intron is located between exon 3 and exon 4. After splicing in the presence of the splicing repressor is complete, would you expect the mRNA to include exon 3 and/or exon 4?

is only 20–25 nucleotides in length. As shown in **Figure 12.23,** capping is a three-step process:

1. The nucleotide at the 5′ end of the transcript initially has three phosphate groups. An enzyme called RNA 5′-triphosphatase removes one of the phosphates.
2. A second enzyme, guanylyltransferase, hydrolyzes guanosine triphosphate (GTP) and attaches a guanosine monophosphate (GMP) to the 5′ end.
3. Finally, a methyltransferase attaches a methyl group to the nitrogen at position 7 in the base guanine.

What are the functions of the 7-methylguanosine cap? The cap structure is recognized by cap-binding proteins, which perform various roles:

- Cap-binding proteins are required for the proper exit of most mRNAs from the nucleus.

- The cap structure is recognized by an initiation factor that causes the mRNA to bind to a ribosome during the initiation stage of translation.
- The cap structure may be important in the efficient splicing of introns, particularly the intron that is closest to the 5′ end.

Let's now turn our attention to the 3′ end of the mRNA molecule. At that end, most mature mRNAs have a string of adenine nucleotides, referred to as a **polyA tail,** which is important for mRNA stability, the exit of mRNA from the nucleus, and in the synthesis of polypeptides. The polyA tail is not coded in the gene sequence. Instead, it is added enzymatically after the pre-mRNA has been completely transcribed—a process termed **polyadenylation.**

To allow the addition of a polyA tail, the pre-mRNA contains a polyadenylation signal sequence near its 3′ end. In mammals, the consensus sequence is AAUAAA. This sequence is

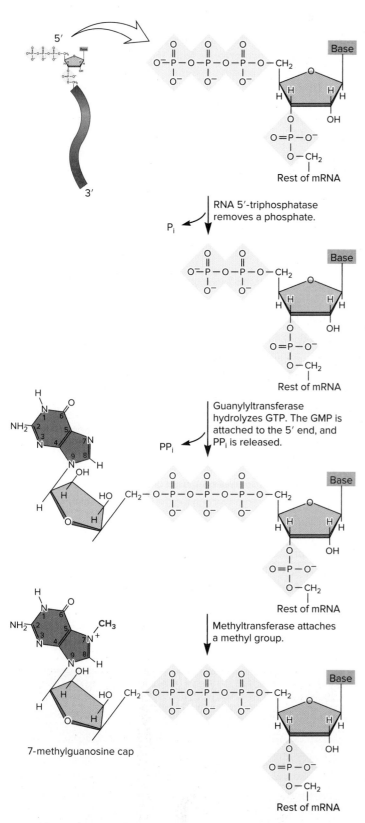

RNA 5'-triphosphatase
removes a phosphate.

Guanylyltransferase
hydrolyzes GTP. The GMP is
attached to the 5' end, and
PP_i is released.

Methyltransferase attaches
a methyl group.

7-methylguanosine cap

FIGURE 12.23 **Attachment of a 7-methylguanosine cap to the 5'
end of an mRNA.** When the transcript is about 20–25 nucleotides in
length, RNA 5'-triphosphatase removes one of the three phosphates, and
then a second enzyme, guanylyltransferase, attaches GMP to the 5' end.
Finally, methyltransferase attaches a methyl group to the base guanine.

CONCEPT CHECK: What are three functional roles of the
7-methylguanosine cap?

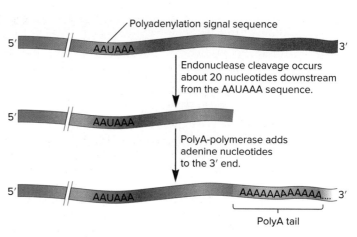

FIGURE 12.24 **Attachment of a polyA tail.** First, an endonuclease cuts the RNA at a location that is about 20 nucleotides downstream from the AAUAAA polyadenylation signal sequence, making the
RNA shorter at its 3' end. Adenine-containing nucleotides are then attached, one at a time, to the 3' end by the enzyme polyA-polymerase.

downstream (toward the 3' end) from the stop codon in the pre-
mRNA. The steps required to synthesize a polyA tail are shown in
Figure 12.24.

1. An endonuclease recognizes the signal sequence and
cleaves the pre-mRNA at a location that is about 20 nucleotides beyond the 3' end of the AAUAAA sequence. The
fragment beyond the 3' cut is degraded.
2. Next, an enzyme known as polyA-polymerase attaches
many adenine-containing nucleotides.

The length of the polyA tail varies among different mRNAs; the
maximum length is typically around 250 nucleotides. A long
polyA tail facilitates mRNA export from the nucleus, stability of
mRNA in the cytosol, and translation.

Some bacterial RNAs are also polyadenylated. However, in
bacteria, polyadenylation targets the RNA to a structure called the
degradosome, where it is degraded.

The Base Sequence of RNA Can Be Modified
by RNA Editing

The term **RNA editing** refers to the process of making a change in
the base sequence of an RNA molecule that involves additions or
deletions of particular bases or conversion of one type of base to
another, such as a cytosine to a uracil. In the case of mRNAs, editing can have various effects, such as generating start codons, generating stop codons, and changing the coding sequence for a
polypeptide.

The phenomenon of RNA editing was first discovered in trypanosomes, the protists that cause sleeping sickness. As with the
discovery of RNA splicing, the initial finding of RNA editing was
met with great skepticism. Since that time, however, RNA editing
has been shown to occur in various organisms and in a variety of
ways, although its functional significance is only slowly emerging
(**Table 12.4**). In the specific case of trypanosomes, the editing process involves the addition or deletion of one or more uracils.

TABLE 12.4
Examples of RNA Editing

Organism	Type of Editing	Found In
Trypanosomes (protists)	Primarily addition but occasionally deletion of one or more uracils	Many mitochondrial mRNAs
Slime mold	Addition of cytosines	Many mitochondrial mRNAs
Plants	Cytosine-to-uracil conversion	Many mitochondrial and chloroplast mRNAs, tRNAs, and rRNAs
Mammals	Cytosine-to-uracil conversion	Apoliproprotein B mRNA and NFI mRNA, which codes a tumor-suppressor protein
	Adenine-to-hypoxanthine conversion	Glutamate receptor mRNA and many tRNAs
Drosophila	Adenine-to-hypoxanthine conversion	mRNA for calcium and sodium channel proteins

A more widespread type of RNA editing involves changes of one type of base to another. In this form of editing, a base in the RNA is deaminated, meaning an amino group is removed from the base. When cytosine is deaminated, uracil is formed, and when adenine is deaminated, hypoxanthine (H) is formed (**Figure 12.25**). Hypoxanthine is recognized as guanine during translation.

An example of RNA editing that occurs in mammals involves an mRNA that codes a protein called apolipoprotein B. In the liver, translation of an unedited mRNA produces apolipoprotein B-100, a protein that is essential for the transport of cholesterol in the blood. In intestinal cells, the mRNA may be edited so that a single cytosine (C) is changed to a uracil (U). What is the significance of this base substitution? This change converts a glutamine codon (CAA) to a stop codon (UAA), resulting in a shorter apolipoprotein. In this case, RNA editing produces an

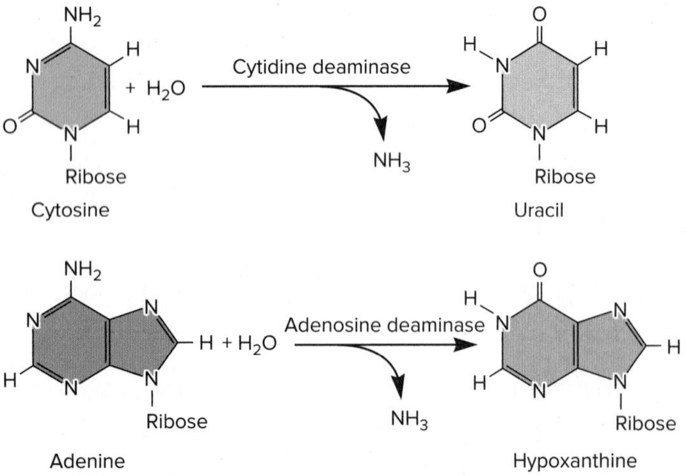

FIGURE 12.25 RNA editing by deamination. Cytidine deaminase removes an amino group from cytosine, thereby creating uracil. Adenosine deaminase removes an amino group from adenine to make hypoxanthine.

CONCEPT CHECK: What is a functional consequence of RNA editing?

apolipoprotein B with an altered structure. Therefore, RNA editing can produce two proteins from the same gene, much like the phenomenon of alternative splicing.

How widespread is RNA editing that involves the conversion of cytosine to uracil or of adenine to hypoxantine? In invertebrates such as *Drosophila*, researchers estimate that 50–100 pre-mRNAs are edited in a way that changes the RNA coding sequence. In mammals, the pre-mRNAs from fewer than 25 genes are currently known to be edited.

12.4 COMPREHENSION QUESTIONS

1. Which of the following are examples of RNA modifications?
   a. Splicing
   b. Capping with 7-methylguanosine
   c. Adding a polyA tail
   d. All of the above are examples of RNA modifications.
2. A ribozyme is
   a. a complex between RNA and a protein.
   b. an RNA that codes a protein that functions as an enzyme.
   c. an RNA molecule with catalytic function.
   d. a protein that degrades RNA molecules.
3. Which of the following statements about a spliceosome is false?
   a. A spliceosome splices pre-mRNA molecules.
   b. A spliceosome only removes exons from RNA molecules.
   c. A spliceosome is composed of snRNPs.
   d. A spliceosome recognizes the exon-intron boundaries and the branch site.
4. Which of the following is a function of the 7-methylguanosine cap?
   a. Exit of mRNA from the nucleus
   b. Efficient splicing of pre-mRNAs
   c. Initiation of translation
   d. All of the above are functions of the cap.

12.5 A COMPARISON OF TRANSCRIPTION AND RNA MODIFICATION IN BACTERIA, ARCHAEA, AND EUKARYOTES

Learning Outcome:

1. Compare and contrast the processes of transcription and RNA modification among bacteria, archaea, and eukaryotes.

Throughout this chapter, we have considered the processes of transcription and RNA modification. The focus has been on bacteria and eukaryotes. In this section, we will also consider the third domain of life, the Archaea. This domain comprises a group of single-celled organisms with distinct molecular characteristics separating them from the domain Bacteria.

As mentioned in Chapter 10, bacteria and archaea are considered prokaryotes. From an evolutionary perspective, prokaryotes arose about 3.5 billion years ago, whereas the first eukaryotes emerged significantly later, about 1.8 billion years ago. The closest modern relative to eukaryotes is thought to be a superphylum of archaea called Asgard archaea. In recent years, researchers have discovered that many molecular processes in eukaryotes are more similar to those in archaea than to those in bacteria. At the molecular level, certain processes in archaea, such as transcription and RNA modification, tend to be more similar to the processes in eukaryotes than to those in bacteria. However, there are some exceptions. For example, in both bacteria and archaea, a 7-methylguanosine cap is not attached to the 5′ end of mRNAs. **Table 12.5** summarizes many of the key similarities and differences among the three domains of life with regard to transcription and RNA modification.

12.5 COMPREHENSION QUESTION

1. Which of the following is not a key difference between bacteria and eukaryotes?
   a. Initiation of transcription requires more proteins in eukaryotes.
   b. A 7-methylguanosine cap is added only to eukaryotic pre-mRNAs.
   c. Splicing is common in complex eukaryotes but not in bacteria and archaea.
   d. All of the above are key differences.

TABLE 12.5

Key Similarities and Differences Among Bacteria, Archaea, and Eukaryotes with Regard to Transcription and RNA Modification*

Component	Bacteria	Archaea	Eukaryotes
Promoter	Consists of −35 and −10 sequences	For protein-coding genes, the core promoter consists of a TATA box and a transcriptional start site. It also contains an upstream element called a B recognition element (BRE).	For protein-coding genes, the core promoter often consists of a TATA box, a transcriptional start site, and downstream promoter elements.
RNA polymerase	A single RNA polymerase	A single RNA polymerase. Its structure is more similar to eukaryotic RNA polymerase II than to bacterial RNA polymerase, but it does not contain a carboxyl-terminal domain (CTD).	Three types of RNA polymerases; RNA polymerase II transcribes protein-coding genes.
Initiation	Sigma (σ) factor is needed for promoter recognition.	The process is similar to but simpler than eukaryotic initiation. Archaea have homologs to eukaryotic TBP (TATA-binding protein) and to subunits within TFIIB and TFIIE.	Six general transcription factors and mediator assemble at the core promoter.
Elongation	Requires the release of σ factor	The archaeal protein called TFE, which is homologous to a subunit in eukaryotic TFIIE, is replaced with proteins called Sp4 and Sp5.	Mediator controls the switch to the elongation phase via phosphorylation of the CTD domain.
Termination	ρ-dependent or ρ-independent	Not well understood. For some genes, a U-rich region at the 3′ end of the transcribed RNA can cause intrinsic termination, but the 3′ end does not require an RNA hairpin structure for termination, which is necessary for ρ-independent termination in bacteria.	According to the allosteric or torpedo model
Splicing	Rare; self-splicing	Rare; the splicing of all introns occurs by the same mechanism that removes introns from tRNAs in eukaryotes.	Commonly occurs in protein-coding pre-mRNAs in complex eukaryotes via a spliceosome; self-splicing occurs rarely. The removal of introns in tRNAs is catalyzed by a splice endonuclease and tRNA ligase.
7-methyl-guanosine cap	Does not occur	Does not occur	Occurs on nearly all mRNAs
PolyA tail	Sometimes added to the 3′ end; promotes degradation	Sometimes added to the 3′ end; promotes degradation	Almost always added to the 3′ end of mRNAs; promotes stability
RNA editing	Not known to occur	Occurs rarely; conversion of cytosine to uracil	Occurs occasionally

*Note: This table does not include the process of gene regulation, which is described in Chapters 14, 15, and 16.

KEY TERMS

Introduction: gene, transcription, protein-coding genes, messenger RNA (mRNA), translation, central dogma of genetics
12.1: gene expression, promoter, terminator, template strand, nontemplate strand, coding strand, transcription factors, regulatory element (regulatory sequence or response element), ribosome-binding site (Shine-Dalgarno sequence), codon, start codon, stop codon, initiation, elongation, termination, RNA polymerase, open complex
12.2: transcriptional start site, Pribnow box, consensus sequence, core enzyme, sigma (σ) factor, holoenzyme, helix-turn-helix motif, closed complex, rho (ρ) protein, ρ-dependent termination, ρ-independent termination, intrinsic termination

12.3: core promoter, TATA box, basal transcription, enhancer, regulatory transcription factors (RTFs), activators, repressors, *cis*-acting element, *trans*-acting factor, general transcription factor (GTFs), mediator, preinitiation complex, basal transcription apparatus
12.4: colinearity, exon, intervening sequence (introns), pre-mRNA, RNA splicing, nucleolus, endonuclease, exonuclease, ribozyme, group I intron, group II intron, self-splicing, maturase, spliceosome, snRNPs, metalloribozyme, alternative splicing, constitutive exon, alternative exon, splicing factor, SR protein, exon skipping, capping, polyA tail, polyadenylation, RNA editing

CHAPTER SUMMARY

- According to the central dogma of genetics, DNA is transcribed into mRNA, and mRNA is translated into a polypeptide. DNA replication allows the DNA to be passed from cell to cell and from parent to offspring (see Figure 12.1).

12.1 Overview of Transcription

- Gene expression is the process of producing a functional product from the information within a gene. A gene is an organization of DNA sequences. A promoter signals the start of transcription, and a terminator signals the end. Regulatory elements control the rate of transcription. For genes that code polypeptides, the gene sequence also includes a start codon, a stop codon, and many codons in between. Bacterial genes also specify a ribosome-binding site (see Figure 12.2).
- Transcription occurs in three phases called initiation, elongation, and termination (see Figure 12.3).

12.2 Transcription in Bacteria

- Many bacterial promoters have sequence elements at the −35 and −10 sites. The transcriptional start site is at +1 (see Figures 12.4, 12.5).
- During the initiation phase of transcription in *E. coli*, sigma (σ) factor, which is bound to RNA polymerase, binds into the major groove of DNA and recognizes sequence elements in the promoter. This process forms a closed complex. Following the formation of an open complex, σ factor is released (see Figures 12.6, 12.7).
- During the elongation phase of transcription, RNA polymerase slides along the DNA and maintains an open complex as it goes. RNA is made in the 5′ to 3′ direction, and base pairing follows a complementarity rule similar to the AT/GC rule except that uracil (U) replaces thymine in RNA (see Figure 12.8).
- In a given chromosome, the strand that is used as the template strand varies from gene to gene (see Figure 12.9).

- Transcriptional termination in *E. coli* occurs by a ρ-dependent or ρ-independent mechanism (see Figures 12.10, 12.11).

12.3 Transcription in Eukaryotes

- Eukaryotes use RNA polymerases I, II, and III to transcribe different categories of genes. Prokaryotic and eukaryotic RNA polymerases have similar structures (see Figure 12.12).
- Eukaryotic protein-coding genes have a core promoter and enhancers (see Figure 12.13).
- Transcription of protein-coding genes in eukaryotes requires RNA polymerase II, six general transcription factors, and mediator. The six general transcription factors and RNA polymerase assemble together to form an open complex (see Table 12.1, Figure 12.14).
- Transcriptional termination of a protein-coding gene may proceed according to an allosteric model or a torpedo model (see Figure 12.15).

12.4 RNA Modification

- RNA transcripts can be modified in a variety of ways, which include cleavage (processing), splicing, capping at the 5′ end, addition of a polyA tail at the 3′ end, RNA editing, and base modification (see Table 12.2).
- Certain RNA molecules such as ribosomal RNAs and precursor tRNAs are processed via cleavage and sometimes by splicing to yield smaller, functional molecules (see Figures 12.16, 12.17).
- Group I and group II introns are removed by self-splicing. Pre-mRNA introns are removed via a spliceosome (see Table 12.3, Figure 12.18).
- The spliceosome is a multisubunit structure that recognizes intron sequences and removes them from pre-mRNA (see Figures 12.19, 12.20).

- During alternative splicing, proteins called splicing factors regulate which exons are included in a mature mRNA (see Figures 12.21, 12.22).
- In eukaryotes, mRNAs have a 7-methylguanosine cap at the 5′ end and a polyA tail at the 3′ end (see Figures 12.23, 12.24).
- RNA editing changes the base sequence of an RNA molecule after it has been synthesized (see Table 12.4, Figure 12.25).

12.5 A Comparison of Transcription and RNA Modification in Bacteria, Archaea, and Eukaryotes

- Differences and similarities have been found between transcription and RNA modification among bacteria, archaea, and eukaryotes. Archaea tend to be more similar to eukaryotes than bacteria are (see Table 12.5).

PROBLEM SETS & INSIGHTS

MORE GENETIC TIPS

1. Describe the important events that occur during gene transcription in bacteria. What proteins play critical roles in the three stages?

T **OPIC:** *What topic in genetics does this question address?* The topic is gene transcription in bacteria. More specifically, the question is about the critical roles played by proteins in that process.

I **NFORMATION:** *What information do you know based on the question and your understanding of the topic?* In the question, you are reminded that gene transcription in bacteria has three stages. From your understanding of the topic, you may remember that the stages of transcription are initiation, elongation, and termination.

P **ROBLEM-SOLVING** **S** **TRATEGY:** *Describe the steps.* One strategy to solve this problem is to break down transcription into its three stages and describe each one separately.

ANSWER: The three stages are initiation, elongation, and termination.

Initiation: RNA polymerase holoenzyme slides along the DNA until σ factor recognizes a promoter. Next, σ factor binds tightly to this sequence, forming a closed complex. The DNA strands are then separated to form an open complex.

Elongation: After σ factor is released from the RNA polymerase holoenzyme, the core enzyme slides along the DNA, synthesizing RNA as it goes. The α subunits of RNA polymerase keep the enzyme bound to the DNA, while the β subunits are responsible for binding and for the catalytic synthesis of RNA. The ω (omega) subunit is also important for the proper assembly of the core enzyme. During elongation, RNA is made according to the AU/GC rule, with nucleotides added in the 5′ to 3′ direction.

Termination: RNA polymerase eventually reaches a sequence at the end of the gene that signals the end of transcription. In ρ-independent termination, the properties of the termination sequences in the DNA are sufficient to cause termination. In ρ-dependent termination, ρ protein recognizes a sequence within the RNA, binds there, and travels toward RNA polymerase. When the formation of a stem-loop causes RNA polymerase to pause, ρ protein catches up and separates the RNA-DNA hybrid, releasing RNA polymerase.

2. The consensus sequence for the −35 sequence of a bacterial promoter is 5′–TTGACA–3′. The −35 sequence of a particular bacterial gene is 5′–TTAACA–3′. A mutation changes the fifth base from a C to a G. Would you expect this mutation to increase or decrease the rate of transcription?

T **OPIC:** *What topic in genetics does this question address?* The topic is transcription in bacteria. More specifically, the question is about the effects of a mutation at the promoter.

I **NFORMATION:** *What information do you know based on the question and your understanding of the topic?* In the question, you are reminded of the consensus sequence for the −35 sequence of bacterial promoters and given information regarding a particular mutation in that sequence of a bacterial gene. From your understanding of the topic, you may recall that the consensus sequence is the most efficiently recognized sequence for initiating transcription.

P **ROBLEM-SOLVING** **S** **TRATEGIES:** *Compare and contrast. Predict the outcome.* One way to solve this problem is to compare the sequences of the nonmutant and mutant promoters and contrast them with the consensus sequence. If the mutation makes the sequence more like the consensus sequence, it will increase transcription, whereas a mutation that makes the sequence less like the consensus sequence will inhibit transcription.

Consensus: 5′–TTGACA–3′

Nonmutant promoter: 5′–TTAACA–3′

Mutant promoter: 5′–TTAAGA–3′

The bases that are different from those in the consensus sequence are highlighted in red.

ANSWER: The mutation is predicted to decrease transcription. The mutant promoter has two bases that deviate from the consensus sequence, whereas the nonmutant promoter has only one.

3. When RNA polymerase transcribes DNA, only one of the two DNA strands is used as a template. Referring to Figure 12.4, explain how RNA polymerase determines which DNA strand is the template strand.

T **OPIC:** *What topic in genetics does this question address?* The topic is transcription. More specifically, the question is about how RNA polymerase identifies the template strand.

I **NFORMATION:** *What information do you know based on the question and your understanding of the topic?* From the question, you know that only one DNA strand is used as a template. From your understanding of the topic, you may remember that specific base sequences form a promoter that determines the starting point for transcription.

P **ROBLEM-SOLVING** **S** **TRATEGIES:** *Make a drawing. Relate structure and function.* One strategy to begin solving this problem is to make a drawing of a bacterial promoter that is based on Figure 12.4. You want to relate the structure of the promoter sequence to the function of RNA polymerase in choosing the correct strand as the template strand.

ANSWER: The binding of σ factor and RNA polymerase depends on the sequence of the promoter. RNA polymerase binds to the promoter in such a way that the −35 sequence TTGACA and the −10 sequence TATAAT are within the coding strand, whereas the −35 sequence AACTGT and the −10 sequence ATATTA are within the template strand.

---

4. As shown in the following diagram, a pre-mRNA contains seven exons, which are numbered in black and found within red rectangles, and six introns, which are numbered in green. A splicing repressor binds at the 3′ splice site at the end of intron 4, which is just before exon 5. What exons will be included in the mature mRNA?

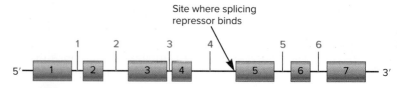

Site where splicing repressor binds

T **OPIC:** *What topic in genetics does this question address?* The topic is RNA splicing. More specifically, the question asks you to predict the effects of a splicing repressor.

I **NFORMATION:** *What information do you know based on the question and your understanding of the topic?* From the question, you know that a splicing repressor binds at the end of intron 4. From your understanding of the topic, you may remember that such a repressor will prevent the spliceosome from recognizing that 3′ splice site.

P **ROBLEM-SOLVING** **S** **TRATEGY:** *Compare and contrast.* One strategy to solve this problem is to look at the diagram provided with the question, and compare how splicing will occur in the presence and absence of the repressor. In the absence of the repressor, all six introns will be removed and all seven exons will be included in the mature mRNA. In the presence of the splicing repressor, the splicesome will not recognize the 3′ splice site at the end of intron 4. Instead, it will chose the next available 3′ splice site, which is the one that is at the 3′ end of intron 5. Therefore, in the presence of the repressor, the spliceosome will choose the 5′ splice site at the beginning of intron 4 and the 3′ splice site at the end of intron 5. The region encompassing intron 4, exon 5, and intron 5 will be spliced out as a single piece.

ANSWER: The splicing repressor will cause exon skipping. In this case, exon 5 will be skipped, so the mature mRNA will include exons 1, 2, 3, 4, 6, and 7.

Conceptual Questions

C1. Explain the central dogma of genetics at the molecular level.

C2. In bacteria, what event marks the end of the initiation stage of transcription?

C3. What is the meaning of the term *consensus sequence*? Give an example. Describe the locations of consensus sequences within bacterial promoters. What are the functions of those sequences?

C4. What is the consensus sequence for the following six DNA sequences?

    GGCATTGACT

    GCCATTGTCA

    CGCATAGTCA

    GGAAATGGGA

    GGCTTTGTCA

    GGCATAGTCA

C5. Mutations in bacterial promoters may increase or decrease the rate of gene transcription. Promoter mutations that increase the transcription rate are termed up-promoter mutations, and those that decrease the transcription rate are termed down-promoter mutations. As shown in Figure 12.5, the −10 sequence of the promoter for the *lac* operon is

TATGTT. Would you expect each of the following mutations to be an up-promoter or a down-promoter mutation?

A. TATGTT to TATATT

B. TATGTT to TTTGTT

C. TATGTT to TATGAT

C6. According to the examples shown in Figure 12.5, which positions of the −35 sequence (i.e., first, second, third, fourth, fifth, or sixth) are more tolerant of changes? Do you think these positions play a more important or less important role in the binding of σ factor? Explain why.

C7. In Chapter 9, we considered the dimensions of the double helix (see Figure 9.13). In an α helix of a protein, there are 3.6 amino acids per complete turn. Each amino acid adds 0.15 nm to the length of the α helix; a complete turn of the α helix is 0.54 nm in length. As shown in Figure 12.6, two α helices of a transcription factor occupy the major groove of the DNA. Referring to Figure 12.6, estimate the number of amino acids that bind to this region. How many complete turns of the α helices occupy the major groove of DNA?

C8. A mutation within the DNA sequence of a certain gene changes the start codon to a stop codon. How will this mutation affect the transcription of this gene?

C9. What is the subunit composition of bacterial RNA polymerase holoenzyme? What are the functional roles of the different subunits?

C10. At the molecular level, describe how σ factor recognizes a bacterial promoter. Be specific about the structure of σ factor and the type of chemical bonding.

C11. Let's suppose a DNA mutation changes the consensus sequence at the −35 site in a way that inhibits binding of σ factor to the DNA. Explain how a mutation could inhibit that binding. Look at Figure 12.5, and identify two specific base substitutions you think would inhibit the binding of σ factor. Explain why your base substitutions would have this effect.

C12. What is the complementarity rule that governs the synthesis of an RNA molecule during transcription? An RNA transcript has the following sequence:

5′–GGCAUGCAUUACGGCAUCACACUAGGGAUC–3′

What are the sequences of the template and coding strands of the DNA that codes this RNA? Toward which end (5′ or 3′) of the template strand is the promoter located?

C13. Describe the movement of the open complex along the DNA.

C14. Describe what happens to the chemical bonding interactions when transcriptional termination occurs. Be specific about the type of chemical bonding.

C15. Discuss the differences between ρ-dependent and ρ-independent termination.

C16. In Chapter 11, we discussed the function of DNA helicase, which is involved in DNA replication. Discuss how the functions of ρ-protein and DNA helicase are similar and how they are different.

C17. Discuss the similarities and differences between RNA polymerase (described in this chapter) and DNA polymerase (described in Chapter 11).

C18. Mutations that occur at the end of a gene may alter the sequence of the gene and prevent transcriptional termination.

   A. What types of mutations would prevent ρ-independent termination?

   B. What types of mutations would prevent ρ-dependent termination?

   C. If a mutation prevented transcriptional termination at the end of a gene, where would gene transcription end? Or would it end?

C19. If each of the following RNA polymerases was missing from a eukaryotic cell, what types of genes would not be transcribed?

   A. RNA polymerase I

   B. RNA polymerase II

   C. RNA polymerase III

C20. What sequence elements are found within the core promoter of protein-coding genes in eukaryotes? Describe their locations and specific functions.

C21. For each of the following transcription factors, explain how eukaryotic transcriptional initiation would be affected if it were missing.

   A. TFIIB

   B. TFIID

   C. TFIIH

C22. Describe the allosteric and torpedo models for transcriptional termination of RNA polymerase II. Which model is more similar to ρ-dependent termination in bacteria, and which model is more similar to ρ-independent termination?

C23. Which eukaryotic transcription factor(s) shown in Figure 12.14 play(s) a role that is equivalent to that of σ factor in bacterial cells?

C24. The initiation phase of eukaryotic transcription via RNA polymerase II is considered an assembly-and-disassembly process. Which types of biochemical interactions—hydrogen bonding, ionic bonding, covalent bonding, and/or hydrophobic interactions—would you expect to drive the assembly-and-disassembly process? How would temperature and salt concentration affect this process?

C25. A eukaryotic protein-coding gene contains two introns and three exons: exon 1–intron 1–exon 2–intron 2–exon 3. The 5′ splice site at the boundary between exon 2 and intron 2 has been eliminated by a small deletion in the gene. Describe how the pre-mRNA coded by this mutant gene will be spliced. Indicate which introns and exons will be found in the mRNA after splicing occurs.

C26. Describe the processing events that occur during the production of tRNA in eukaryotes.

C27. Describe the structure and function of a spliceosome. Speculate why the spliceosome subunits contain snRNA. In other words, what do you think is/are the functional role(s) of snRNA during splicing?

C28. What is the unique functional feature of ribozymes? Give two examples described in this chapter.

C29. What does it mean to say that gene expression is colinear?

C30. What is meant by the term *self-splicing*? What types of introns are self-splicing?

C31. In eukaryotes, what types of modifications occur to pre-mRNAs?

C32. What is alternative splicing? What is its biological significance?

C33. What is the function of a splicing factor? Explain how splicing factors can regulate the cell-specific splicing of mRNAs.

C34. Figure 12.21 shows the products of alternative splicing for the α-tropomyosin pre-mRNA. Let's suppose that smooth muscle cells produce splicing factors that are not produced in other cell types. Explain where you think such splicing factors bind and how they influence the splicing of the α-tropomyosin pre-mRNA.

C35. The processing of ribosomal RNA in eukaryotes is shown in Figure 12.16. Why is this called cleavage or processing but not splicing?

C36. In the splicing of a group I intron shown in Figure 12.18a, does the 5′ end of the intron have a phosphate group? Explain.

C37. According to the mechanism shown in Figure 12.20, several snRNPs play different roles in the splicing of pre-mRNA. Identify the snRNP that recognizes each of the following sites:

   A. 5′ splice site

   B. 3′ splice site

   C. Branch site

C38. After the intron (which is in a lariat configuration) is released during pre-mRNA splicing, a brief time passes before the two exons are connected to each other. Which snRNP(s) hold(s) the exons in place so they can be covalently connected to each other?

Experimental Questions

E1. A research group has sequenced the cDNA and genomic DNA for a particular gene. The cDNA is derived from mRNA, so it does not contain introns. Here are the DNA sequences.

cDNA:

5'-ATTGCATCCAGCGTATACTATCTCGGGCCCAATTAATGCCA–
GCGGCCAGACTATCACCCAACTCGGTTACCTACTAGTATATC–
CCATATACTAGCATATATTTTACCCATAATTTGTGTGTGGGTATA–
CAGTATAATCATATA-3'

Genomic DNA (contains one intron):

5'-ATTGCATCCAGCGTATACTATCTCGGGCCCAATTAATGCCAG
CGGCCAGACTATCACCCAACTCGGCCCACCCCCCAGGTTTA–
CACAGTCATACCATACATACAAAAATCGCAGTTACTTATCCCA–
AAAAAACCTAGATACCCCACATACTATTAACTCTTTCTTTCTAG–
GTTACCTACTAGTATATCCCATATACTAGCATATATTTTAC–
CCATAATTTGTGTGTGGGTATACAGTATAATCATATA-3'

Indicate where the intron is located. Does the intron contain the normal consensus sequences for splicing, based on those shown in Figure 12.19? Underline the splice site sequences, and indicate whether or not they agree with the consensus sequences.

E2. Chapter 20 describes a technique known as Northern blotting that is used to detect RNA transcribed from a particular gene. In this method, a specific type of RNA is detected using a short segment of cloned DNA that is labeled. The labeled DNA is complementary to the RNA that the researcher wishes to detect. After the labeled DNA binds to the RNA, the RNA is run on a gel and then visualized as a labeled (dark) band. As shown below, Northern blotting can be used to determine the amount of a particular RNA transcribed in a given cell type. If one type of cell produces twice as much of a particular mRNA as another type of cell does, the band will appear twice as intense. Also, the method can distinguish whether alternative RNA splicing has occurred to produce an RNA that has a different molecular mass.

Northern blot

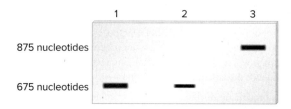

Lane 1 is a sample of RNA isolated from nerve cells.

Lane 2 is a sample of RNA isolated from kidney cells. Nerve cells produce twice as much of this RNA as do kidney cells.

Lane 3 is a sample of RNA isolated from spleen cells. Spleen cells produce an alternatively spliced version of this RNA that is about 200 nucleotides longer than the RNA produced in nerve and kidney cells.

Let's suppose a researcher is interested in the effects of mutations on the expression of a particular protein-coding gene in eukaryotes. The gene has one intron that is 450 nucleotides long. After this intron is removed from the pre-mRNA, the mRNA transcript is 1100 nucleotides in length. Diploid somatic cells have two copies of this gene. Make a drawing that shows the expected results of a Northern blot using mRNA from the cytosol of somatic cells, which were obtained from the following individuals:

Lane 1: A normal individual

Lane 2: A homozygote for a deletion that removed the −50 to −100 region of the gene that codes this mRNA

Lane 3: A heterozygote in which one gene is normal and the other gene has a deletion that removed the −50 to −100 region

Lane 4: A homozygote for a mutation that introduced an early stop codon into the middle of the coding sequence of the gene

Lane 5: A homozygote for a two-nucleotide deletion that removed the sequence AG at the 3' splice site

E3. An electrophoretic mobility shift assay (EMSA) can be used to study the binding of proteins to a segment of DNA. This method is described in Chapter 20. When a protein binds to a segment of DNA, it slows the movement of the DNA through a gel, so the DNA appears at a higher point in the gel, as shown in the following example:

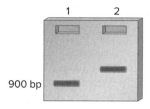

Lane 1: 900-bp fragment alone

Lane 2: 900-bp fragment plus a protein that binds to the 900-bp fragment

In this example, the segment of DNA is 900 bp in length, and the binding of a protein causes the DNA to appear at a higher point in the gel. Assuming that this 900-bp fragment of DNA contains a core promoter of a eukaryotic protein-coding gene, draw a gel that shows the relative locations of the 900-bp fragment under the following conditions:

Lane 1: 900 bp plus TFIID and TFIIA

Lane 2: 900 bp plus TFIIB

Lane 3: 900 bp plus TFIID, TFIIA, and TFIIB

Lane 4: 900 bp plus TFIIB and RNA polymerase II

Lane 5: 900 bp plus TFIID, TFIIA, TFIIB, and RNA polymerase II/TFIIF

E4. As described in Chapter 20 and in question E3, an electrophoretic mobility shift assay (EMSA) can be used to determine if a protein binds to DNA. This method can also determine whether a protein binds to RNA. For each of the following combinations, would you expect the movement of the RNA through the gel to be slowed down due to the binding of the protein?

A. mRNA from a gene that is terminated in a ρ-independent manner plus ρ protein

B. mRNA from a gene that is terminated in a ρ-dependent manner plus ρ protein

C. Pre-mRNA from a protein-coding gene that contains two introns plus the snRNP called U1

D. Mature mRNA from a protein-coding gene that contains two introns plus the snRNP called U1

E5. The technique of DNase I footprinting is described in Chapter 20. If a protein binds over a region of DNA, it will protect the DNA in that region from digestion by DNase I. To carry out a DNase I footprinting experiment, a researcher starts with a sample of a cloned DNA fragment. The fragments are exposed to DNase I in the presence and absence of a DNA-binding protein. Regions of the DNA fragment not covered by the DNA-binding protein will be digested by DNase I, producing a series of bands on a gel. Regions of the DNA fragment not digested by DNase I (because a DNA-binding protein is preventing DNase I from gaining access to the DNA) will be revealed, because a region of the gel will not contain any bands.

In the gel from a DNase I footprinting experiment shown next, a researcher began with a sample of cloned DNA 300 bp in length. This DNA contained a eukaryotic promoter for RNA polymerase II. For the sample loaded in lane 1, no proteins were added. For the sample loaded in lane 2, the 300-bp fragment was mixed with RNA polymerase II plus TFIID, TFIIA, and TFIIB.

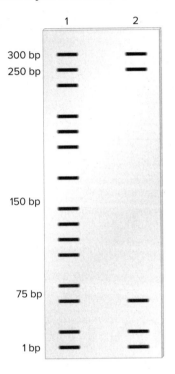

A. How long is the region of DNA that is covered up by the binding of RNA polymerase II and the transcription factors?

B. Describe how this binding would occur if the DNA was within a nucleosome. (Note: The structure of a nucleosome is described in Chapter 10.) Do you think that the DNA was in a nucleosome when RNA polymerase and transcription factors were bound to the promoter? Explain why or why not.

E6. Many researchers are interested in the transcription of protein-coding genes in eukaryotes. Such researchers want to study mRNA. One method that is used to isolate mRNA is column chromatography. (Note: See Appendix A for a general description of chromatography.) Researchers can covalently attach short pieces of DNA that contain stretches of thymines (i.e., TTTTTTTTTTTTT) to the column matrix, creating what is called a poly-dT column. When a cell extract is poured over this column, mRNA binds to the column, but other types of RNA do not.

A. Explain how you would use a poly-dT column to obtain a purified preparation of mRNA from eukaryotic cells. In your description, explain why mRNA binds to this column, and describe what you would do to release the mRNA from the column.

B. Can you think of ways to purify other types of RNA, such as tRNA or rRNA?

Questions for Student Discussion/Collaboration

1. Based on your knowledge of introns and pre-mRNA splicing, discuss whether or not you think alternative splicing fully explains the existence of introns. Can you think of other possible reasons to explain their existence?

2. Discuss the types of RNA transcripts and the functional roles they play. Why do you think some RNAs form complexes with protein subunits?

Note: All answers are available for the instructor in Connect; the answers to the even-numbered questions and all of the Concept Check and Comprehension Questions are in Appendix B.

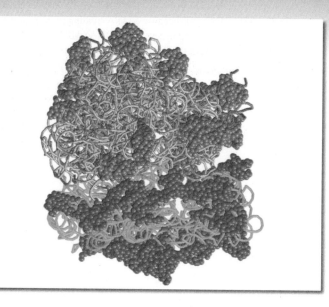

A molecular model for the structure of a ribosome. This model of ribosome structure is based on X-ray diffraction studies. Ribosomes synthesize polypeptides, using mRNAs as a source of information. A detailed look at this model is provided in Figure 13.13b.
Tom Pantages

13

TRANSLATION OF mRNA

Translation is the process in which the sequence of codons within mRNA provides the information to synthesize the sequence of amino acids that constitutes a polypeptide. One or more polypeptides then fold and assemble to create a functional protein. In this chapter, we will explore the current state of knowledge regarding translation, focusing on the specific molecular interactions responsible for this process. During the past few decades, the combined efforts of geneticists, cell biologists, and biochemists have advanced our understanding of translation in profound ways. Even so, many questions remain unanswered, and this process continues to be an exciting area of investigation.

We will begin this chapter by considering classic experiments that revealed that some genes code proteins that function as enzymes. Next, we examine how the genetic code is used to decipher the information within mRNA to produce a polypeptide with a specific amino acid sequence. The rest of the chapter is devoted to understanding translation at the molecular level as it occurs in living cells. To accomplish this, we will need to examine the cellular components—including many different proteins, RNAs, and small molecules—that are involved in the translation process.

We will consider the structure and function of tRNA molecules, which act as the translators of the genetic information within mRNA, and then examine the composition of ribosomes. Finally, we will explore the stages of translation and identify differences in the translation process between bacterial cells and eukaryotic cells.

13.1 THE GENETIC BASIS FOR PROTEIN SYNTHESIS

Learning Outcomes:

1. Explain how the work of Garrod indicated that some genes code enzymes.
2. Analyze the experiments of Beadle and Tatum, and explain how their results were consistent with the idea that certain genes code a single enzyme.

Proteins, which are composed of one or more polypeptides, are critically important as active participants in cell structure and function. The primary role of DNA is to store the information needed for the synthesis of all the proteins that an organism makes. As discussed in Chapter 12, genes that code the amino acid sequence of polypeptides are known as **protein-coding genes.** The RNA transcribed from a protein-coding gene is called **messenger RNA (mRNA).**

The main function of the genetic material is to code the production of proteins in the correct cell, at the proper time, and in suitable amounts. This is an extremely complicated task because living cells make thousands of different proteins. Genetic analyses have shown that a typical bacterium can make a few thousand different proteins, and estimates for eukaryotes range from several thousand in simple eukaryotic organisms, such as yeast, to tens of thousands in plants and animals. In this section, we will consider early experiments showing that the role of some genes is to code proteins that function as enzymes.

Garrod Proposed That Some Genes Code for the Production of Enzymes

The idea that a relationship exists between genes and the production of proteins was first suggested at the beginning of the twentieth century by physician Archibald Garrod. Prior to Garrod's studies, biochemists had studied many metabolic pathways within living cells. These pathways consist of a series of metabolic conversions of one molecule to another, each step catalyzed by a specific enzyme. Each enzyme is a distinctly different protein that catalyzes a particular chemical reaction.

Figure 13.1 illustrates part of the metabolic pathway for the degradation of phenylalanine, an amino acid commonly found in human diets. The enzyme phenylalanine hydroxylase catalyzes the conversion of phenylalanine to tyrosine, and a different enzyme (tyrosine aminotransferase) converts tyrosine into *p*-hydroxyphenylpyruvic acid, and so on. In all of the steps shown in Figure 13.1, a specific enzyme catalyzes a single type of chemical reaction.

Garrod studied patients who had defects in their ability to metabolize certain compounds. In **alkaptonuria,** which is inherited, the patient's body accumulates abnormal levels of homogentisic acid (also called alkapton), which is excreted in the urine, causing it to appear black on exposure to air. In addition, this disorder is characterized by bluish-black discoloration of cartilage and skin (ochronosis). Garrod proposed that the accumulation of homogentisic acid in these patients is due to a defective enzyme, namely, homogentisic acid oxidase (see Figure 13.1).

Garrod already knew that alkaptonuria is an inherited trait that follows an autosomal recessive pattern of inheritance. Therefore, an individual with alkaptonuria must have inherited the mutant (defective) gene that causes this disorder from both parents. From these observations, Garrod proposed that a relationship exists between the inheritance of the trait and the inheritance of a defective enzyme. Namely, if someone inherited the mutant gene (which causes a loss of enzyme function) from both parents, such a person would not produce any normal enzyme and would be unable to metabolize homogentisic acid. Garrod described alkaptonuria as an **inborn error of metabolism.** This hypothesis was the first suggestion that a connection exists between the function

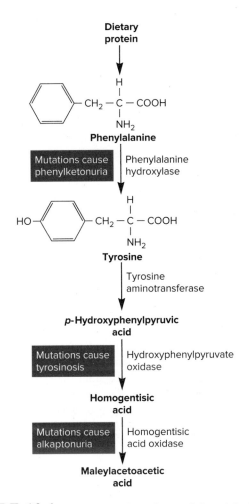

FIGURE 13.1   The metabolic pathway of phenylalanine breakdown. This diagram shows part of the pathway of phenylalanine metabolism, which consists of specific enzymes (shown to the right of the arrows) that successively convert one molecule to another. Certain human genetic disorders (noted in red boxes) are caused when enzymes in this pathway are missing or defective.

Genes→Traits Someone who inherits two defective copies of the gene that codes homogentisic acid oxidase cannot convert homogentisic acid into maleylacetoacetic acid. Such a person accumulates large amounts of homogentisic acid in the body and has other symptoms of the genetic disorder known as alkaptonuria. Similarly, someone with two defective copies of the gene coding phenylalanine hydroxylase is unable to synthesize the enzyme phenylalanine hydroxylase and has the disorder called phenylketonuria (PKU).

CONCEPT CHECK: Which genetic disorder occurs when homogentisic acid oxidase is defective?

of genes and the production of enzymes. At the time of Garrod's studies, this idea was particularly insightful, because the structure and function of the genetic material were completely unknown.

Beadle and Tatum's Experiments with *Neurospora* Led Them to Propose the One-Gene/One-Enzyme Hypothesis

In the early 1940s, George Beadle and Edward Tatum were also interested in the relationship among genes, enzymes, and traits. They developed an experimental system for investigating the connection between genes and the production of particular enzymes.

Consistent with Garrod's hypothesis, the underlying assumption behind their approach was that a relationship exists between genes and the production of enzymes. However, the quantitative nature of this relationship was unclear. In particular, Beadle and Tatum asked, "Does one gene control the production of one enzyme, or does one gene control the synthesis of many enzymes involved in a complex biochemical pathway?"

At the time of their research, many geneticists were trying to understand the nature of the gene by studying morphological traits. However, Beadle and Tatum realized that morphological traits are likely to be based on systems of biochemical reactions so complex as to make analysis exceedingly difficult. Therefore, they turned their genetic studies to the analysis of simple nutritional requirements in *Neurospora crassa*, a common bread mold. *Neurospora* can be easily grown in the laboratory and has few nutritional requirements: a carbon source (sugar), inorganic salts, and the vitamin biotin. Normal *Neurospora* cells produce many different enzymes that can synthesize the organic molecules, such as amino acids and other vitamins, that are essential for growth.

Beadle and Tatum wanted to understand how enzymes are controlled by genes. They reasoned that a mutation in a gene, causing a defect in an enzyme needed for the synthesis of an essential molecule, would prevent that mutant strain from growing on a minimal medium, which contains only a carbon source, inorganic salts, and biotin. In the study described in **Figure 13.2**, they isolated several different mutant strains that required methionine for growth. They hypothesized that each mutant strain might be blocked at only a single step in the consecutive series of reactions that lead to methionine synthesis.

To test Beadle and Tatum's hypothesis, the strains were examined for their ability to grow in the presence of *O*-acetylhomoserine, cystathionine, homocysteine, or methionine. *O*-acetylhomoserine, cystathionine, and homocysteine are intermediates in the synthesis of methionine from homoserine (Figure 13.2a). The wild-type strain could grow on a minimal medium that contained the minimum set of nutrients that is required for growth. The minimal medium did not contain *O*-acetylhomoserine, cystathionine, homocysteine, or methionine.

Based on their growth properties, the mutant strains that had been originally identified as requiring methionine for growth could be placed into four groups designated strains 1, 2, 3, and 4 in Figure 13.2.

- A strain 1 mutant was missing enzyme 1, needed for the conversion of homoserine into *O*-acetylhomoserine. The cells could grow only if *O*-acetylhomoserine, cystathionine, homocysteine, or methionine was added to the growth medium.
- A strain 2 mutant was missing the second enzyme in this pathway, which is needed for the conversion of *O*-acetylhomoserine into cystathionine.

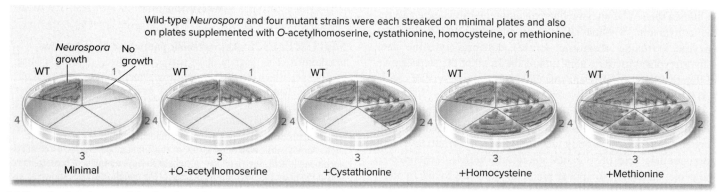

Wild-type *Neurospora* and four mutant strains were each streaked on minimal plates and also on plates supplemented with *O*-acetylhomoserine, cystathionine, homocysteine, or methionine.

Minimal   +O-acetylhomoserine   +Cystathionine   +Homocysteine   +Methionine

(a) Growth of strains on minimal and supplemented growth media

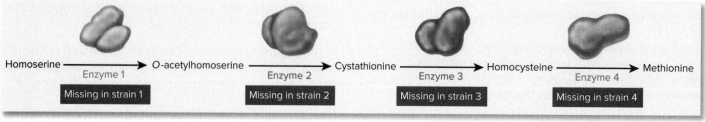

Homoserine → Enzyme 1 → O-acetylhomoserine → Enzyme 2 → Cystathionine → Enzyme 3 → Homocysteine → Enzyme 4 → Methionine

Missing in strain 1   Missing in strain 2   Missing in strain 3   Missing in strain 4

(b) Simplified pathway for methionine biosynthesis

FIGURE 13.2   An example of an experiment that supported Beadle and Tatum's one-gene/one-enzyme hypothesis. (a) Growth of wild-type (WT) and mutant strains of *Neurospora crassa* on a minimal medium or in the presence of *O*-acetylhomoserine, cystathionine, homocysteine, or methionine. **(b)** A simplified pathway for methionine biosynthesis. Note: The first compound in this pathway, homoserine, is made by *Neurospora* via enzymes and precursor molecules not discussed in this experiment.

CONCEPT CHECK: What enzymatic function is missing in the strain 2 mutants?

- A strain 3 mutant was unable to convert cystathionine into homocysteine.
- A strain 4 mutant could not make methionine from homocysteine.

Based on these results, the researchers could order the enzymes into a biochemical pathway, depicted in Figure 13.2b. The analysis of the four types of mutants allowed Beadle and Tatum to conclude that a single gene controlled the synthesis of a single enzyme. This was referred to as the **one-gene/one-enzyme hypothesis.**

In later decades, this hypothesis had to be modified in four ways:

1. Enzymes are only one category of proteins. All proteins are coded by genes, and many of them do not function as enzymes.
2. Some proteins are composed of two or more different polypeptides. Therefore, it is more accurate to say that a protein-coding gene codes a polypeptide. The term **polypeptide** refers to a structure; it is a linear sequence of amino acids. By comparison, the term **protein** denotes function. Some proteins are composed of one polypeptide. In such cases, a single gene does code a single protein. In other cases, however, a functional protein is composed of two or more different polypeptides. An example is hemoglobin, which is composed of two α-globin and two β-globin polypeptides. In this case, the expression of two genes—the α-globin and β-globin genes—is needed to create one functional protein.
3. Many genes do not code polypeptides. As discussed in Chapter 17, several types of genes specify functional RNA molecules that do not code polypeptides.
4. As discussed in Chapter 12, one gene can code multiple polypeptides due to alternative splicing and RNA editing.

13.1 COMPREHENSION QUESTIONS

1. An inborn error of metabolism is caused by
   a. a mutation in a gene that causes an enzyme to be inactive.
   b. a mutation in a gene that occurs in somatic cells.
   c. the consumption of foods that disrupt metabolic processes.
   d. any of the above.
2. The reason why Beadle and Tatum observed four different categories of mutants that could not grow on media without methionine is because
   a. the enzyme involved in methionine biosynthesis is composed of four different subunits.
   b. the enzyme involved in methionine biosynthesis is present in four copies in the *Neurospora* genome.
   c. four different enzymes are involved in a pathway for methionine biosynthesis.
   d. a lack of methionine biosynthesis can inhibit *Neurospora* growth in four different ways.

13.2 THE RELATIONSHIP BETWEEN THE GENETIC CODE AND PROTEIN SYNTHESIS

Learning Outcomes:

1. Outline how the information within DNA is used to make an mRNA and a polypeptide.
2. Explain the function of the genetic code.
3. List a few exceptions to the genetic code.
4. Describe the four levels of protein structure.
5. Compare and contrast the functions of several types of proteins.

Thus far, we have considered experiments that led to the conclusion that some genes code enzymes. The sequence of a protein-coding gene provides a template for the synthesis of mRNA. In turn, the mRNA contains the information to synthesize a polypeptide. In this section, we will examine the general features of the genetic code—the sequence of bases in a codon that specifies an amino acid or the end of translation. In addition, we will look at the biochemistry of polypeptide synthesis and explore the structure and function of proteins. Ultimately, proteins are largely responsible for determining the characteristics of living cells and an organism's traits.

During Translation, the Codons in mRNA Provide the Information to Make a Polypeptide with a Specific Amino Acid Sequence

Why have researchers given the name *translation* to the process of polypeptide synthesis? At the molecular level, translation involves an interpretation of one language—the language of mRNA, a nucleotide sequence—into the language of proteins—an amino acid sequence. **Figure 13.3** emphasizes how a gene stores information to make a polypeptide. As discussed in Chapter 12, the first step to access this information, in which mRNA is made, is called transcription. During the second step, called translation, the information within mRNA is used to make a polypeptide.

The ability of mRNA to be translated into a specific sequence of amino acids relies on the **genetic code.** The sequence of bases within an mRNA molecule provides coded information that is read in groups of three nucleotides known as codons (see Figure 13.3).

- The sequence of three bases in most codons specifies a particular amino acid. These codons are termed **sense codons.** For example, the codon AGC specifies the amino acid serine.
- The codon AUG, which specifies methionine, functions as a **start codon;** it is usually the first codon that begins a polypeptide sequence. The AUG codon can also be used to specify additional methionines later in the coding sequence.
- Three codons, UAA, UAG, and UGA, which are known as **stop codons,** are signals that end the process of translation.

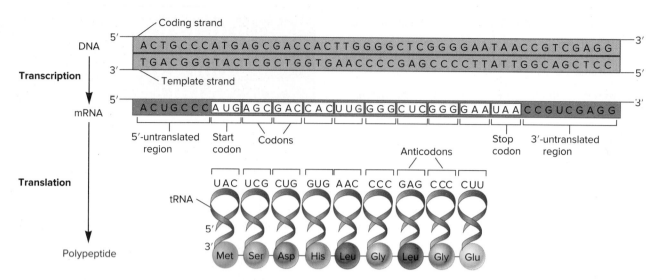

FIGURE 13.3 **The relationships among the DNA coding sequence, mRNA codons, tRNA anticodons, and amino acids in a polypeptide.** The sequence of nucleotides within DNA is transcribed to make a complementary sequence of nucleotides within mRNA. This sequence of nucleotides in mRNA is translated into a sequence of amino acids in a polypeptide. tRNA molecules act as intermediates in this translation process. Note: As discussed later, the tRNAs are detached from the polypeptide, one at a time, as it is being synthesized. Also, this gene does not contain any introns.

CONCEPT CHECK: Describe the role of DNA in the synthesis of a polypeptide.

TABLE 13.1
The Genetic Code

		Second base					
First base		U	C	A	G		Third base
U		UUU UUC Phenylalanine (Phe)	UCU UCC Serine (Ser)	UAU UAC Tyrosine (Tyr)	UGU UGC Cysteine (Cys)	U C	
		UUA UUG Leucine (Leu)	UCA UCG	UAA Stop codon UAG Stop codon	UGA Stop codon UGG Tryptophan (Trp)	A G	
C		CUU CUC CUA CUG Leucine (Leu)	CCU CCC CCA CCG Proline (Pro)	CAU CAC Histidine (His) CAA CAG Glutamine (Gln)	CGU CGC CGA CGG Arginine (Arg)	U C A G	
A		AUU AUC Isoleucine (Ile) AUA AUG Methionine (Met); start codon	ACU ACC ACA ACG Threonine (Thr)	AAU AAC Asparagine (Asn) AAA AAG Lysine (Lys)	AGU AGC Serine (Ser) AGA AGG Arginine (Arg)	U C A G	
G		GUU GUC GUA GUG Valine (Val)	GCU GCC GCA GCG Alanine (Ala)	GAU GAC Aspartic acid (Asp) GAA GAG Glutamic acid (Glu)	GGU GGC GGA GGG Glycine (Gly)	U C A G	

Stop codons are also known as **termination codons,** or **nonsense codons.**

- An mRNA molecule also has regions that precede the start codon and follow the stop codon. Because these regions do not code a polypeptide, they are called the **5′-untranslated region** and **3′-untranslated region,** respectively.

The codons in mRNA are recognized by the anticodons in **transfer RNA (tRNA)** molecules (see Figure 13.3). **Anticodons** are three-nucleotide sequences that are complementary to codons in mRNA. The tRNA molecules carry the amino acids that are specified by the codons in the mRNA. In this way, the order of codons in mRNA dictates the order of amino acids within a polypeptide.

The genetic code is composed of 64 different codons, as shown in **Table 13.1.** Because polypeptides are composed of 20 different kinds of amino acids, a minimum of 20 codons is needed to specify all of the amino acids. With four types of bases in mRNA (A, U, G, and C), a genetic code containing two bases in a codon would not be sufficient because it would specify only 4^2, or 16, possible types. By comparison, a three-base codon system can form 4^3, or 64, different codons. Because the number of possible codons exceeds 20—which is the number of different types of amino acids—the genetic code is said to possess **degeneracy**. This means that more than one codon can specify the same amino acid. For example, the codons GGU, GGC, GGA, and GGG all specify the amino acid glycine. Such codons are termed **synonymous codons.**

The start codon (AUG) defines the **reading frame** of an mRNA—a series of codons determined by reading the bases in groups of three, beginning with the start codon as a frame of reference. This concept is best understood with a few examples. The mRNA sequence shown below codes a short polypeptide with seven amino acids:

5′–AUGCCCGGAGGCACCGUCCAAU–3′
Met–Pro–Gly–Gly–Thr–Val–Gln

If we remove one base (C) adjacent to the start codon, this changes the reading frame to produce a polypeptide with a different amino acid sequence:

5′–AUGCCGGAGGCACCGUCCAAU–3′
Met–Pro–Glu–Ala–Pro–Ser–Asn

Alternatively, if we remove three bases (CCC) next to the start codon, the reading frame of the mRNA is the same as the first example, but one amino acid (Pro, proline) has been deleted from the amino acid sequence:

5′–AUGGGAGGCACCGUCCAAU–3′
Met–Gly–Gly–Thr–Val–Gln

GENETIC **TIPS**   THE QUESTION: An mRNA has the following base sequence:

5′–CAGGCGGCGAUGGACAAUAAAGCGGGCCUGUAAGC–3′

Identify the start codon, and determine the complete amino acid sequence that will be translated from this mRNA.

T OPIC: *What topic in genetics does this question address?* The topic is translation. More specifically, the question is about predicting a polypeptide sequence based on an mRNA sequence.

I NFORMATION: *What information do you know based on the question and your understanding of the topic?* From the question, you know the base sequence of an mRNA. From your understanding of the topic, you may remember that the genetic code determines the amino acid sequence of a polypeptide. Furthermore, certain codons function as start and stop codons.

P ROBLEM-SOLVING **S** TRATEGY: *Predict the outcome.* One strategy to solve this problem is to first identify the start codon (AUG) and then determine the adjacent codons. You need to use Table 13.1 to solve this problem.

ANSWER: The start codon is AUG (shown in red). The amino acid sequence is shown below the mRNA sequence:

5′–CAGGCGGCGAUGGACAAUAAAGCGGGCCAUUAAGC–3′
Met Asp Asn Lys Ala Gly Leu STOP

Exceptions to the Genetic Code Include the Incorporation of Selenocysteine and Pyrrolysine into Polypeptides

From the analysis of many different species, including bacteria, archaea, protists, fungi, plants, and animals, researchers have found that the genetic code is nearly universal. However, a few exceptions to the genetic code have been noted (**Table 13.2**). The eukaryotic organelles known as mitochondria have their own DNA, which includes a few protein-coding genes. In vertebrates, the mitochondrial genetic code differs from the nuclear genetic code in certain ways. For example, AUA codes for methionine and UGA codes for tryptophan. Also, in mitochondria and certain ciliated protists, AGA and AGG function as stop codons instead of specifying arginine.

Selenocysteine (Sec) and **pyrrolysine** (Pyl) are sometimes called the twenty-first and twenty-second amino acids found in polypeptides. Their structures are shown later in Figure 13.5f. Selenocysteine is found in several enzymes involved in oxidation-reduction reactions in bacteria, archaea, and eukaryotes. Pyrrolysine is found in a few enzymes of methane-producing archaea. Selenocysteine and pyrrolysine are coded by the codons UGA and UAG, respectively, which usually function as stop codons. Like the standard 20 amino acids, selenocysteine and pyrrolysine are bound to tRNAs that specifically carry them to the ribosomes for their incorporation into polypeptides. The anticodon of the tRNA that carries selenocysteine is complementary to a UGA codon, and the

TABLE 13.2

Examples of Exceptions to the Genetic Code

Codon	Universal Meaning	Exception*
AUA	Isoleucine	Methionine in yeast and in vertebrate mitochondria
UGA	Stop	Tryptophan in vertebrate mitochondria
CUU, CUC, CUA, CUG	Leucine	Threonine in yeast mitochondria
AGA, AGG	Arginine	Stop codon in ciliated protozoa and in yeast and vertebrate mitochondria
UAA, UAG	Stop	Glutamine in ciliated protozoa
UGA	Stop	Selenocysteine in certain genes found in bacteria, archaea, and eukaryotes
UAG	Stop	Pyrrolysine in certain genes found in methane-producing archaea

*Several other exceptions, found sporadically in various species, are also known.

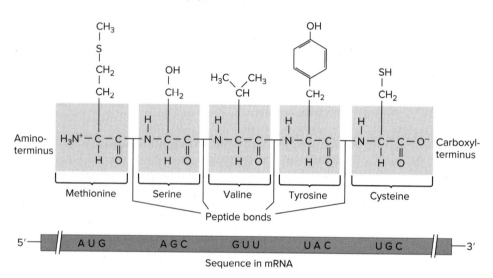

FIGURE 13.4 The directionality of polypeptide synthesis.
(a) An amino acid is connected to a growing polypeptide via a condensation reaction that releases a water molecule. The letter R is a general designation for an amino acid side chain. **(b)** The first amino acid in a polypeptide (usually methionine) is located at the amino-terminus, and the last amino acid is at the carboxyl-terminus. Thus, the directionality of amino acids in a polypeptide is from the amino-terminus to the carboxyl-terminus, which corresponds to the 5′ to 3′ orientation of codons in mRNA.

Last peptide bond formed in the growing chain of amino acids

(a) Attachment of an amino acid to a peptide chain

(b) Directionality in a polypeptide and mRNA

between the carboxyl group in the last amino acid of the polypeptide and the amino group in the amino acid being added. As shown in **Figure 13.4a**, this occurs via a condensation reaction that releases a water molecule. The newest amino acid added to a growing polypeptide always has a free carboxyl group. **Figure 13.4b** compares the sequence of a very short polypeptide with the mRNA that codes it.

- The first amino acid is said to be at the **amino-terminus,** or **N-terminus,** of the polypeptide. An amino group (NH_3^+) is found at this site. The term *N-terminus* refers to the presence of a nitrogen atom (N) at this end. The first amino acid is specified by a codon that is near the 5′ end of the mRNA.
- The last amino acid in a completed polypeptide is located at the **carboxyl-terminus,** or **C-terminus.** A carboxyl group (COO^-) is always found at this site in the polypeptide. This last amino acid is specified by a codon that is closer to the 3′ end of the mRNA.

tRNA that carries pyrrolysine has an anticodon that is complementary to UAG.

How does a UGA or UAG codon occasionally specify the incorporation of selenocysteine or pyrrolysine, respectively? In the case of a codon that specifies selenocysteine, the UGA codon is followed by a sequence called the selenocysteine insertion sequence (SECIS), which forms a stem-loop. In bacteria, a SECIS may be located immediately following the UGA codon, whereas in archaea and eukaryotes, the SECIS may be further downstream in the 3′-untranslated region of the mRNA. The SECIS is recognized by proteins that favor the binding of a UGA codon to a tRNA carrying selenocysteine instead of the binding of release factors that are needed for polypeptide termination. Similarly, pyrrolysine incorporation may involve sequences downstream from a UAG codon that form a stem-loop.

A Polypeptide Has Directionality from Its Amino-Terminus to Its Carboxyl-Terminus

Let's now turn our attention to polypeptide biochemistry. Polypeptide synthesis has a directionality that parallels the order of codons in the mRNA. As a polypeptide is made, a **peptide bond** is formed

The Amino Acid Sequences of Polypeptides Determine the Structure and Function of Proteins

Now that we have examined how mRNAs code polypeptides, let's consider the structure and function of the gene products, namely, polypeptides. **Figure 13.5** shows the 20 different amino acids that are most commonly found within polypeptides. Each amino acid contains a unique **side chain,** or **R group,** that has its own particular chemical properties. For example, aliphatic and aromatic amino acids are relatively nonpolar, which means that they are less likely to associate with water. These hydrophobic (meaning "water-fearing") amino acids are often buried within the interior of a folded protein. In contrast, the polar amino acids are hydrophilic ("water-loving") and are more likely to be on the surface of a protein, where they can favorably interact with the surrounding water in cells or extracellular fluids. The chemical properties of the amino acids and their sequences in a polypeptide are critical factors that determine the unique structure of that polypeptide.

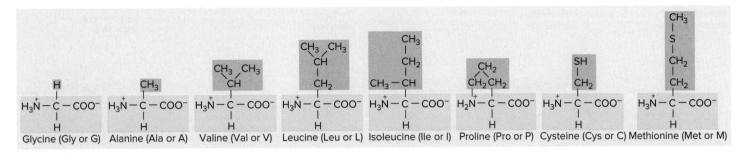

(a) Nonpolar, aliphatic amino acids

Glycine (Gly or G) Alanine (Ala or A) Valine (Val or V) Leucine (Leu or L) Isoleucine (Ile or I) Proline (Pro or P) Cysteine (Cys or C) Methionine (Met or M)

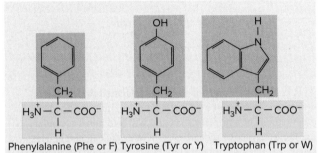

(b) Aromatic amino acids

Phenylalanine (Phe or F) Tyrosine (Tyr or Y) Tryptophan (Trp or W)

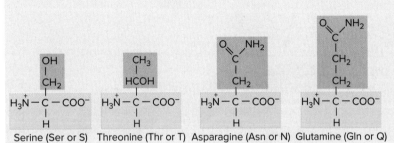

(c) Polar, neutral amino acids

Serine (Ser or S) Threonine (Thr or T) Asparagine (Asn or N) Glutamine (Gln or Q)

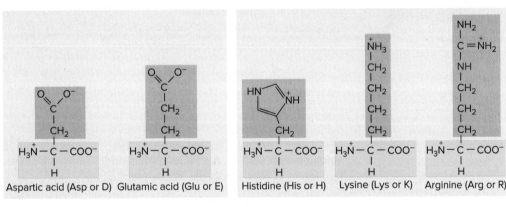

(d) Polar, acidic amino acids

Aspartic acid (Asp or D) Glutamic acid (Glu or E)

(e) Polar, basic amino acids

Histidine (His or H) Lysine (Lys or K) Arginine (Arg or R)

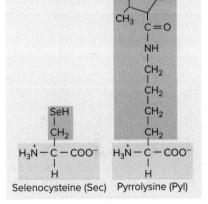

(f) Nonstandard amino acids

Selenocysteine (Sec) Pyrrolysine (Pyl)

FIGURE 13.5 **The amino acids that are incorporated into polypeptides during translation.** Parts (a) through (e) show the 20 standard amino acids, and part (f) shows two amino acids that are occasionally incorporated into polypeptides through the use of stop codons (see Table 13.2). Each amino acid has a three-letter abbreviation, given in parentheses, and each standard amino acid also has a one-letter abbreviation. The structures of amino acid side chains can also be covalently modified after a polypeptide is made, a phenomenon called posttranslational modification.

CONCEPT CHECK: Which two amino acids do you think are the least soluble in water?

Primary Structure. Following gene transcription and mRNA translation, the end result is a polypeptide with a defined amino acid sequence. This sequence is the **primary structure** of a polypeptide (**Figure 13.6a**). The primary structure of a typical polypeptide may be a few hundred or even a couple of thousand amino acids in length. Within a living cell, a newly made polypeptide is not usually found in a long linear state for a significant length of time. Rather, to become a functional unit, most

polypeptides quickly adopt a compact three-dimensional structure. The folding process begins while the polypeptide is still being translated. The progression from the primary structure of a polypeptide to the three-dimensional structure of a protein is dictated by the amino acid sequence of the polypeptide. In particular, the chemical properties of the amino acid side chains play a central role in determining the folding pattern of a protein. In addition, the folding of some polypeptides is aided by

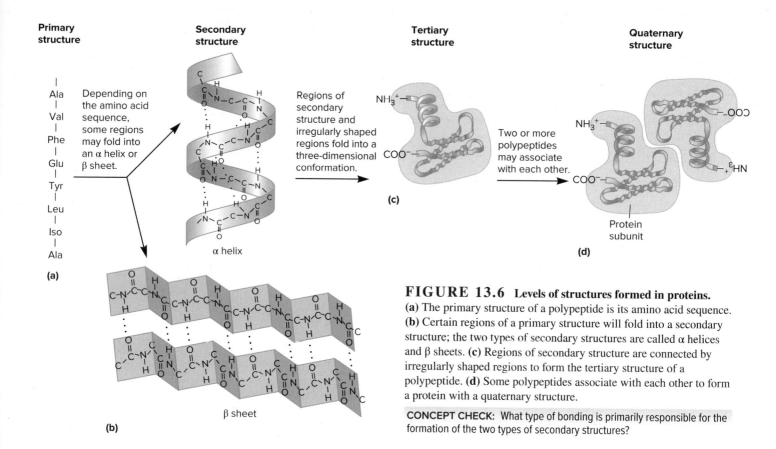

FIGURE 13.6 Levels of structures formed in proteins.
(a) The primary structure of a polypeptide is its amino acid sequence.
(b) Certain regions of a primary structure will fold into a secondary structure; the two types of secondary structures are called α helices and β sheets. (c) Regions of secondary structure are connected by irregularly shaped regions to form the tertiary structure of a polypeptide. (d) Some polypeptides associate with each other to form a protein with a quaternary structure.

CONCEPT CHECK: What type of bonding is primarily responsible for the formation of the two types of secondary structures?

chaperones—proteins that bind to polypeptides and facilitate their proper folding.

Secondary Structure. The folding of polypeptides is governed by their primary structure and occurs in multiple stages. The first stage involves the formation of a regular, repeating shape known as a **secondary structure.** The two main types of secondary structures are the **α helix** and the **β sheet** (**Figure 13.6b**). A single polypeptide may have some regions that fold into an α helix and other regions that fold into a β sheet. Secondary structures within polypeptides are primarily stabilized by the formation of hydrogen bonds between atoms that are located in the polypeptide backbone. In addition, some regions do not form a repeating secondary structure. Such regions have shapes that look very irregular in their structure because they do not follow a repeating folding pattern.

Tertiary Structure. The short regions of secondary structure within a polypeptide are folded relative to each other to make the **tertiary structure** of the polypeptide. As shown in **Figure 13.6c**, α-helical regions and β-sheet regions are connected by irregularly shaped segments to determine this tertiary structure. The folding of a polypeptide into its secondary and then tertiary conformation is thermodynamically favorable. The structure is determined by various interactions, including the tendency of hydrophobic amino acids to avoid water, ionic

interactions among charged amino acids, hydrogen bonding among amino acids in the folded polypeptide, and weak attractive forces known as van der Waals interactions.

Quaternary Structure. A protein is a functional unit that can be composed of one or more polypeptides. Some proteins are composed of a single polypeptide. Many proteins, however, are composed of two or more polypeptides that associate with each other to form a **quaternary structure** (**Figure 13.6d**). The individual polypeptides are called **subunits** of the resulting functional protein, and each of them has its own tertiary structure.

Proteins Are Primarily Responsible for the Characteristics of Living Cells and an Organism's Traits

Why is the genetic material largely devoted to storing the information to make proteins? To a great extent, the characteristics of a cell depend on the types of proteins that it makes. In turn, the traits of multicellular organisms are determined by the properties of their cells. Proteins perform a variety of roles that are critical to the life of cells and to the morphology and function of organisms. **Table 13.3** describes several examples of how proteins function.

TABLE 13.3
Functions of Selected Cellular Proteins

Function	Examples
Cell shape and organization	Tubulin: forms cytoskeletal structures known as microtubules
Transport	Sodium channels: transport sodium ions across the nerve cell membranes
	Hemoglobin: transports oxygen in red blood cells
Movement	Myosin: involved in muscle cell contraction
Cell signaling	Insulin: a hormone that influences cell metabolism and growth
	Insulin receptor: recognizes insulin and initiates a cell response
Cell surface recognition	Integrins: bind to large extracellular proteins
Enzymes	Hexokinase: phosphorylates glucose during the first step in glycolysis
	β-Galactosidase: cleaves lactose into glucose and galactose
	Glycogen synthetase: uses glucose molecules as building blocks to synthesize a large carbohydrate known as glycogen
	RNA polymerase: uses ribonucleotides as building blocks to synthesize RNA
	DNA polymerase: uses deoxyribonucleotides as building blocks to synthesize DNA

13.2 COMPREHENSION QUESTIONS

1. What is the genetic code?
   a. The relationship between a three-base codon sequence and an amino acid or the end of translation
   b. The entire base sequence of an mRNA molecule
   c. The entire sequence from the promoter to the terminator of a gene
   d. The binding of tRNA to mRNA

2. The reading frame begins at a _____ and is read
   _____.
   a. promoter, one base at a time
   b. promoter, in groups of three bases
   c. start codon, one base at a time
   d. start codon, in groups of three bases

3. The fourth codon in an mRNA is GGG, which specifies glycine. Assuming that no amino acids are removed from the polypeptide, which of the following statements is correct?
   a. The third amino acid from the N-terminus is glycine.
   b. The fourth amino acid from the N-terminus is glycine.
   c. The third amino acid from the C-terminus is glycine.
   d. The fourth amino acid from the C-terminus is glycine.

4. A type of secondary structure found in proteins is
   a. an α helix.
   b. a β sheet.
   c. Both a and b are secondary structures in proteins.
   d. Neither a nor b is a secondary structure in proteins.

13.3 EXPERIMENTAL DETERMINATION OF THE GENETIC CODE

Learning Outcome:

1. Compare and contrast the experiments of (1) Nirenberg and Matthaei, (2) Khorana, and (3) Nirenberg and Leder that were instrumental in deciphering the genetic code.

In the previous section, we examined how the genetic code determines the amino acid sequence of a polypeptide. In this section, we will consider the experimental approaches that deduced the genetic code.

EXPERIMENT 13A

Synthetic RNA Helped to Decipher the Genetic Code

How did scientists determine the functions of the 64 codons of the genetic code? During the early 1960s, three research groups headed by Marshall Nirenberg, Severo Ochoa, and H. Gobind Khorana set out to decipher the genetic code. Though they used different methods, all of these groups used synthetic mRNA in their experimental approaches to "crack the code." We first consider the work of Nirenberg and his colleagues. Prior to their studies, several laboratories had already determined that extracts from bacterial cells, containing a mixture of components including ribosomes, tRNAs, and other factors required for translation, are able to synthesize polypeptides if mRNA and amino acids are

added. This mixture is termed an in vitro translation system, or a **cell-free translation system.** If radiolabeled amino acids are added to a cell-free translation system, the synthesized polypeptides are radiolabeled and easy to detect.

To decipher the genetic code, Nirenberg and colleagues needed to gather information regarding the relationship between mRNA composition and polypeptide composition. To accomplish this goal, they made mRNA molecules of a known base composition, added them to a cell-free translation system, and then analyzed the amino acid composition of the resultant polypeptides. For example, if an mRNA molecule consisted of a string of adenine-containing nucleotides (e.g., 5′–AAAAAAAAAAAAAAAA–3′), researchers could add this polyA mRNA to a cell-free translation system in order to answer the question "Which amino acid is specified by a codon that contains only adenine nucleotides?" (As Table 13.1 shows, it is lysine.)

Before discussing the details of this type of experiment, let's consider how the synthetic mRNA molecules were made. To synthesize mRNA, an enzyme known as polynucleotide phosphorylase was used. In the presence of excess ribonucleoside diphosphates, also called nucleoside diphosphates (NDPs), this enzyme catalyzes the covalent linkage of nucleotides to make a polymer of RNA. Because it does not use a DNA template, the order of the nucleotides is random. For example, if nucleotides containing two different bases, such as uracil and guanine, are added, then polynucleotide phosphorylase makes random polymers containing both nucleotides (e.g., 5′–GGGUGUGUG-GUGGGUG–3′). An experimenter can control the amounts of the nucleotides that are added. For example, if 70% G and 30% U are mixed together with polynucleotide phosphorylase, the predicted amounts of the codons within the random polymer are as follows:

Codon possibilities	*Percentage in the random polymer*
GGG	$0.7 \times 0.7 \times 0.7 = 0.34 =$ 34%
GGU	$0.7 \times 0.7 \times 0.3 = 0.15 =$ 15%
GUU	$0.7 \times 0.3 \times 0.3 = 0.06 =$ 6%
UUU	$0.3 \times 0.3 \times 0.3 = 0.03 =$ 3%
UUG	$0.3 \times 0.3 \times 0.7 = 0.06 =$ 6%
UGG	$0.3 \times 0.7 \times 0.7 = 0.15 =$ 15%
UGU	$0.3 \times 0.7 \times 0.3 = 0.06 =$ 6%
GUG	$0.7 \times 0.3 \times 0.7 = 0.15 =$ 15%
	100%

Thus, if the amounts of the NDPs in the phosphorylase reaction are controlled, the relative amounts of the possible codons can be predicted.

The first experiment that demonstrated the ability to synthesize polypeptides from synthetic mRNA was performed by Marshall Nirenberg and J. Heinrich Matthaei in 1961. As shown in **Figure 13.7**, a cell-free translation system was placed into 20 different tubes. An mRNA template made using polynucleotide phosphorylase was then added to each tube. In this example, the mRNA was made from 70% G and 30% U. Next, the 20 amino acids were added to each tube, but each tube differed with regard to which of the amino acids was radiolabeled. For example, radiolabeled glycine would be found in only 1 of the 20 tubes.

The tubes were incubated for a sufficient length of time to allow translation to occur. The newly made polypeptides were then precipitated by treatment with trichloroacetic acid (TCA). This treatment precipitates polypeptides but not individual amino acids. The contents of each tube were then subjected to filtration. The precipitated polypeptides were captured on a filter, whereas amino acids that had not been incorporated into polypeptides remained in solution and passed through the filter. Finally, the amount of radioactivity captured on the filter was determined using a scintillation counter.

THE GOAL (DISCOVERY-BASED SCIENCE)

The researchers assumed that the sequence of bases in mRNA determines the incorporation of specific amino acids into a polypeptide. The goal of this experiment was to provide information that would help to decipher the relationship between base composition and particular amino acids.

ACHIEVING THE GOAL   **FIGURE 13.7**   Elucidation of the genetic code.

Starting material: A cell-free translation system that can synthesize polypeptides if mRNA and amino acids are added. The translation system contained ribosomes, tRNAs, and other factors required for translation. These other factors included enzymes that attach amino acids to tRNA molecules.

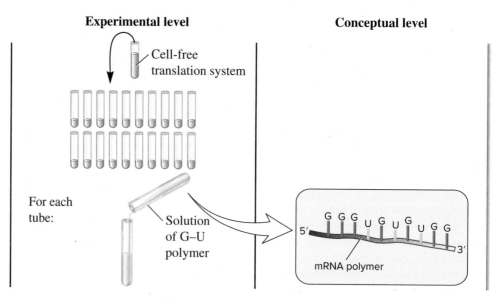

Experimental level

Conceptual level

1. Place the cell-free translation system into 20 tubes.

Cell-free translation system

2. To each tube, add random mRNA polymers of G and U made via polynucleotide phosphorylase using 70% G and 30% U.

For each tube:

Solution of G–U polymer

5′ G G G U G U G U G G 3′

mRNA polymer

3. Add a different radiolabeled amino acid to each tube, and add the other 19 nonradiolabeled amino acids. In this example, glycine was radiolabeled.

4. Incubate for 60 minutes to allow translation to occur.

5. Add 15% trichloroacetic acid (TCA), which precipitates polypeptides but not amino acids.

6. Capture the precipitated polypeptides on a filter. Note: Amino acids that were not incorporated into polypeptides pass through the filter.

7. Count the radioactivity on the filter in a scintillation counter (see Appendix A for a description).

8. Calculate the amount of radiolabeled amino acids in the precipitated polypeptides.

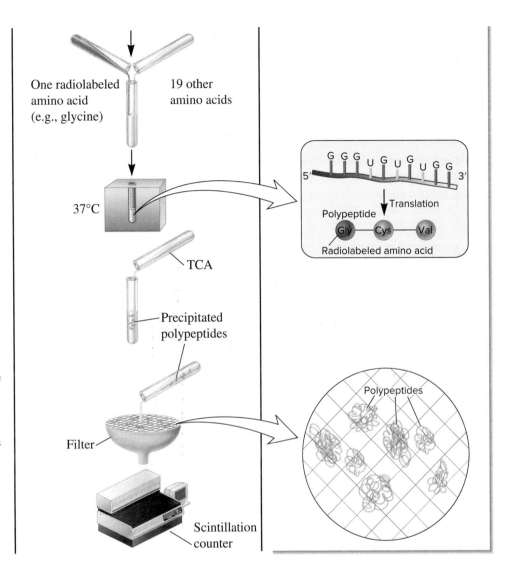

THE DATA

Radiolabeled Amino Acid Added	Relative Amount of Radiolabeled Amino Acid Incorporated into Translated Polypeptides (% of total)	Radiolabeled Amino Acid Added	Relative Amount of Radiolabeled Amino Acid Incorporated into Translated Polypeptides (% of total)
Alanine	0	Leucine	6
Arginine	0	Lysine	0
Asparagine	0	Methionine	0
Aspartic acid	0	Phenylalanine	3
Cysteine	6	Proline	0
Glutamic acid	0	Serine	0
Glutamine	0	Threonine	0
Glycine	49	Tryptophan	15
Histidine	0	Tyrosine	0
Isoleucine	0	Valine	21

Source: Nirenberg, Marshall W., and Matthaei, J. Heinrich, "The Dependence of Cell-Free Protein Synthesis in *E. coli* upon Naturally Occurring or Synthetic Polyribonucleotides," *Proceedings of the National Academy of Sciences*, vol. 47, no. 10, October 1, 1961, 1588–1602.

According to the calculation previously described, codons should occur in the following percentages: 34% GGG, 15% GGU, 6% GUU, 3% UUU, 6% UUG, 15% UGG, 6% UGU, and 15% GUG. We now know that the value of 49% for glycine is due to two codons: GGG (34%) and GGU (15%). The 6% cysteine is due to UGU, and so on. Keep in mind that the genetic code was not deciphered in a single experiment such as the one described here.

Furthermore, this kind of experiment yields information regarding only the nucleotide content of codons, not the specific order of bases within a single codon. For example, this experiment indicates that a cysteine codon contains two U's and one G. However, it does not tell us that a cysteine codon is UGU. Based on these data alone, a cysteine codon could be UUG, GUU, or UGU. However, by comparing many different RNA polymers, the laboratories of Nirenberg and Ochoa established patterns between the specific base sequences of codons and the amino acids they code.

In their first experiments, Nirenberg and Matthaei showed that a random polymer containing only uracil produced a polypeptide containing only phenylalanine. From this result, they inferred that UUU specifies phenylalanine. As shown in the data, this idea is also consistent with the results involving a random 70% G and 30% U polymer. In this case, 3% of the codons will be UUU. Likewise, 3% of the amino acids within the polypeptides were found to be phenylalanine.

The Use of RNA Copolymers and the Triplet-Binding Assay Also Helped to Crack the Genetic Code

In the 1960s, H. Gobind Khorana and colleagues developed a novel method for synthesizing RNA. First, they developed a chemical method to make DNA molecules that had repeating two-, three-, and four-nucleotide sequences. This type of molecule is called a copolymer, because it is made from the linkage of smaller di-, tri-, or tetra-nucleotide sequences. These synthetic DNA molecules were then used as templates to make RNA molecules with repeating sequences. For example, RNA molecules with the repeating trinucleotide sequence 5′–AUC–3′ were made:

5′–AUCAUCAUCAUCAUCAUCAUCAUCAUC–3′

Depending on the reading frame, such an RNA copolymer contains three different codons: AUC, UCA, and CAU. In a cell-free translation system like the one used in the experiment by Nirenberg and Matthaei, this copolymer produced polypeptides containing isoleucine, serine, and histidine. **Table 13.4** summarizes some of the copolymers that were made using this approach and the amino acids that were incorporated into polypeptides.

Finally, another method that helped to decipher the genetic code also involved the chemical synthesis of short RNA molecules. In 1964, Marshall Nirenberg and Philip Leder discovered that an RNA molecule composed of three nucleotides—a triplet—can cause a tRNA to bind to a ribosome. In other words, an RNA triplet acts like a codon. An example is shown in **Figure 13.8**.

1. In this experiment, the researchers began with a sample of ribosomes that were mixed with 5′–CCC–3′ triplets.
2. Portions of this sample were then added to 20 different tubes that had tRNAs with different radiolabeled amino acids. For example, one tube contained radiolabeled histidine, a second tube had radiolabeled proline, a third tube contained radiolabeled glycine, and so on. Only one radiolabeled amino acid was added to each tube.
3. After allowing sufficient time for tRNAs to bind to the ribosomes, the samples were filtered; only the large ribosomes and anything bound to them were trapped on the filter. Unbound tRNAs passed through the filter.
4. Finally, the researchers determined the amount of radioactivity trapped on each filter. If the filter contained a large amount of radioactivity, the results indicated that the added triplet coded the amino acid that was radiolabeled.

Using the triplet-binding assay, Nirenberg and Leder were able to establish relationships between particular RNA triplet sequences and the binding of tRNAs carrying specific (radiolabeled) amino acids. In the case of the 5′–CCC–3′ triplet, they determined that tRNAs carrying radiolabeled proline were bound to the ribosomes. Unfortunately, in some cases, a triplet could not promote sufficient tRNA binding to yield unambiguous results. Nevertheless, the triplet-binding assay was an important tool in the identification of the majority of codons.

TABLE 13.4

Examples of Copolymers That Were Analyzed by Khorana and Colleagues

Synthetic RNA*	Codon Possibilities	Amino Acids Incorporated into Polypeptides
UC	UCU, CUC	Serine, leucine
AG	AGA, GAG	Arginine, glutamic acid
UG	UGU, GUG	Cysteine, valine
AC	ACA, CAC	Threonine, histidine
UUC	UUC, UCU, CUU	Phenylalanine, serine, leucine
AAG	AAG, AGA, GAA	Lysine, arginine, glutamic acid
UUG	UUG, UGU, GUU	Leucine, cysteine, valine
CAA	CAA, AAC, ACA	Glutamine, asparagine, threonine
UAUC	UAU, AUC, UCU, CUA	Tyrosine, isoleucine, serine, leucine
UUAC	UUA, UAC, ACU, CUU	Leucine, tyrosine, threonine

*The synthetic RNAs were made using DNA templates that were composed of copolymers.

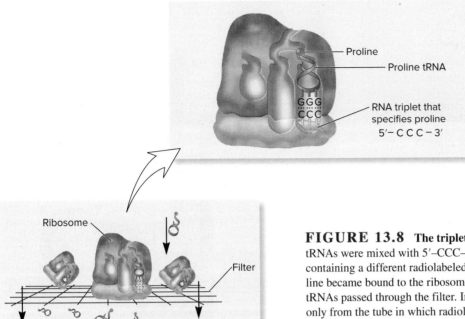

FIGURE 13.8 **The triplet-binding assay.** In this experiment, ribosomes and tRNAs were mixed with 5′–CCC–3′ RNA triplets in 20 separate tubes, with each tube containing a different radiolabeled amino acid (not shown). Only tRNAs carrying proline became bound to the ribosomes, which were trapped on the filter. Unbound tRNAs passed through the filter. In this case, radioactivity was trapped on the filter only from the tube in which radiolabeled proline was added.

CONCEPT CHECK: Explain how the use of radiolabeled amino acids in the experiment by Nierenberg and Leder helped to reveal the genetic code.

13.3 COMPREHENSION QUESTIONS

1. Let's suppose a researcher mixed together nucleotides with the following percentages of bases: 30% G, 30% C, and 40% A. If RNA was made via polynucleotide phosphorylase, what percentage of the codons would be 5′-GGC-3′?
   a. 30%
   b. 9%
   c. 2.7%
   d. 0%

2. In the triplet-binding assay carried out by Nirenberg and Leder, an RNA triplet composed of three bases was able to cause the
   a. translation of a polypeptide.
   b. binding of a tRNA carrying the appropriate amino acid.
   c. termination of translation.
   d. release of the amino acid from the tRNA.

13.4 STRUCTURE AND FUNCTION OF tRNA

Learning Outcomes:

1. Describe the specificity between the amino acid carried by a tRNA and a codon in mRNA.
2. Describe the key structural features of a tRNA molecule.
3. Explain how an amino acid is attached to a tRNA via aminoacyl-tRNA synthetase.
4. Outline the wobble rules.

Thus far, we have considered the general features of translation and surveyed the structure and functional significance of proteins. The rest of this chapter is devoted to a molecular understanding of translation as it occurs in living cells. Biochemical studies of protein synthesis began in the 1950s. As work progressed toward an understanding of translation, research revealed that certain types of RNA molecules are involved in the incorporation of amino acids into growing polypeptides.

Francis Crick proposed the **adaptor hypothesis.** According to this idea, the position of an amino acid within a polypeptide is determined by the binding between the mRNA and an adaptor molecule carrying a specific amino acid. At the time, Crick did not know what that adaptor molecule was. Later, work by Paul Zamecnik and Mahlon Hoagland showed that the adaptor molecule is tRNA. During translation, a tRNA has two functions: (1) It recognizes a three-base codon sequence in mRNA, and (2) it carries an amino acid specific for that codon. In this section, we will examine the general function of tRNA molecules. We begin by exploring the important structural features that underlie tRNA function.

The Function of a tRNA Depends on the Specificity Between the Amino Acid It Carries and Its Anticodon

The adaptor hypothesis proposes that tRNA molecules recognize the codons within mRNA and carry the correct amino acids to the site of polypeptide synthesis. During mRNA-tRNA recognition, the anticodon in a tRNA molecule binds to a codon in mRNA in an antiparallel manner and according to the AU/GC rule (**Figure 13.9**). For example, if the anticodon in the tRNA is 3′–AAG–5′, it will bind to a 5′–UUC–3′ codon. Importantly, the anticodon in the tRNA corresponds to the amino acid that it carries. According to the genetic code, described earlier in this chapter, the UUC codon specifies phenylalanine. Therefore, a tRNA with a 3′–AAG–5′ anticodon must carry a phenylalanine. As another example, if the tRNA has a 3′–GGC–5′ anticodon, it will bind to a 5′–CCG–3′ codon that

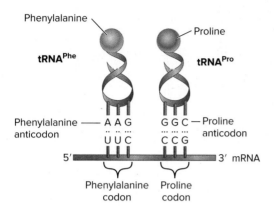

FIGURE 13.9 Recognition between tRNAs and mRNA. The anticodon in the tRNA binds to a complementary sequence in the mRNA. At its 3′ end, the tRNA carries the amino acid that corresponds to the codon in the mRNA according to the genetic code.

specifies proline. This tRNA must carry proline. A tRNA molecule is named according to the type of amino acid it carries. For example, a tRNA that carries phenylalanine is referred to as tRNA^{Phe}, whereas a tRNA that carries proline is tRNA^{Pro}.

Recall that the genetic code has 64 codons. Of these, 61 are sense codons that specify the 20 amino acids. Therefore, to synthesize proteins, a cell must produce many different tRNA molecules having specific anticodon sequences. To accomplish this, the chromosomal DNA contains many distinct tRNA genes that code tRNA molecules with different base sequences.

Common Structural Features Are Shared by All tRNAs

To understand how tRNAs are able to carry the correct amino acids during translation, researchers have examined the structural characteristics of these molecules in great detail. Though a cell makes many different tRNAs, all tRNAs share common structural features (**Figure 13.10**).

- As originally proposed by Robert Holley in 1965, the secondary structure of tRNAs resembles a cloverleaf pattern because it has three stem-loops. The anticodon is located in the second loop region.
- The acceptor stem is where an amino acid becomes attached to a tRNA (see the inset in the figure). All tRNA molecules have the sequence CCA at their 3′ ends. These three nucleotides are usually added enzymatically by the enzyme tRNA nucleotidyltransferase after the tRNA is made.
- Among different types of tRNA molecules, the variable sites (shown in blue in the figure) can differ in the number of nucleotides they contain.

The three-dimensional, or tertiary, structure of tRNA molecules involves additional folding of the secondary structure. In the tertiary structure of tRNA, the stem-loops are folded into a much more compact molecule. The ability of RNA molecules to form stem-loops and the tertiary folding of tRNA molecules are described in Chapter 9 (see Figures 9.17 and 9.18). In addition to the standard A, U, G, and C bases, tRNA molecules commonly contain modified bases. For example, Figure 13.10 illustrates a tRNA that

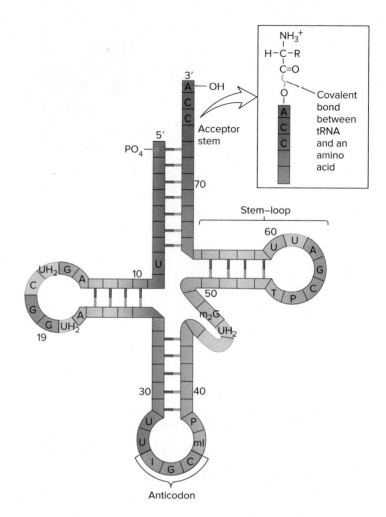

FIGURE 13.10 Secondary structure of tRNA. The conventional numbering of nucleotides within a tRNA begins at the 5′ end and proceeds toward the 3′ end. The bicolored pegs between bases within a stem represent hydrogen bonds between complementary bases. The figure also shows the locations of a few modified bases specifically found in a yeast tRNA that carries alanine. The modified bases are as follows: UH$_2$ = dihydrouridine, I = inosine, mI = methylinosine, P = pseudouridine, m$_2$G = dimethylguanosine, and T = ribothymidine. The inset shows an amino acid covalently attached to the 3′ end of a tRNA. Note: The 5′ to 3′ orientation of the anticodon in this drawing (left to right) is opposite to that of tRNAs shown in other drawings, such as Figure 13.9.

CONCEPT CHECK: What are the two key functional sites of a tRNA molecule?

contains several modified bases. Among many different species, researchers have found that more than 80 different base modifications can occur in tRNA molecules. We will explore the significance of modified bases in codon recognition later in this section.

Aminoacyl-tRNA Synthetases Charge tRNAs by Attaching the Appropriate Amino Acid

To function correctly, each type of tRNA must have the appropriate amino acid attached to its 3′ end. How does an amino acid get attached to the correct tRNA? Enzymes in the cell known as **aminoacyl-tRNA synthetases** catalyze the attachment of amino acids to tRNA molecules. Cells produce 20 different aminoacyl-tRNA synthetase enzymes, one for each of the 20 distinct amino acids. Each aminoacyl-tRNA synthetase is named for the specific amino acid it attaches to tRNA. For example, alanyl-tRNA synthetase recognizes a tRNA with an alanine anticodon—tRNA^{Ala}—and attaches an alanine to it.

Aminoacyl-tRNA synthetases catalyze a chemical reaction involving three different molecules: an amino acid, a tRNA molecule, and ATP.

1. In the first step of the reaction, a synthetase binds to a specific amino acid and also ATP (**Figure 13.11**).
2. The ATP is hydrolyzed, and AMP becomes attached to the amino acid; pyrophosphate is released.
3. The correct tRNA then binds to the synthetase. The amino acid becomes covalently attached to the 3′ end of the tRNA molecule at the acceptor stem, and AMP is released.
4. Finally, the tRNA with its attached amino acid is released from the enzyme.

At this stage, the tRNA is called a **charged tRNA,** or an **aminoacyl-tRNA.** In a charged tRNA molecule, the amino acid is attached to the 3′ end of the tRNA by a covalent bond (refer back to the inset in Figure 13.10).

The ability of aminoacyl-tRNA synthetases to recognize tRNAs has sometimes been called the "second genetic code." This recognition process is necessary to maintain the fidelity of genetic information. The frequency of error for aminoacyl-tRNA synthetases is less than 10^{-4}. In other words, the wrong amino acid is attached to a tRNA less than once in 10,000 times! As you might expect, the anticodon region of the tRNA is usually important for precise recognition by the correct aminoacyl-tRNA synthetase. In studies of *E. coli* synthetases, 17 of the 20 types of aminoacyl-tRNA synthetases recognize the anticodon region of the tRNA. However, other regions of the tRNA are also important recognition sites. These include the acceptor stem and bases in the stem-loops. Also, most aminoacyl-tRNA synthetases have a proofreading ability. They can enzymatically remove the wrong amino acid from the 3′ end.

As mentioned previously, tRNA molecules frequently contain bases within their structure that have been chemically modified. These modified bases have important effects on tRNA function. For example, modified bases within tRNA molecules affect the rate of translation and the recognition of tRNAs by aminoacyl-tRNA synthetases. Positions 34 and 37 contain the largest variety of modified nucleotides; position 34 is the first base in the anticodon that matches the third base in the codon of mRNA. As discussed next, a modified base at position 34 can have important effects on codon-anticodon recognition.

The Wobble Rules Describe Mismatches That Are Allowed at the Third Position in Codon-Anticodon Pairing

Having considered the structure and function of tRNA molecules, let's reexamine some subtle features of the genetic code. As discussed earlier, the genetic code possesses degeneracy, which means that more than one codon can specify the same amino acid. Degeneracy usually occurs at the third position in a codon. For example, valine is specified by GUU, GUC, GUA, and GUG. In all four

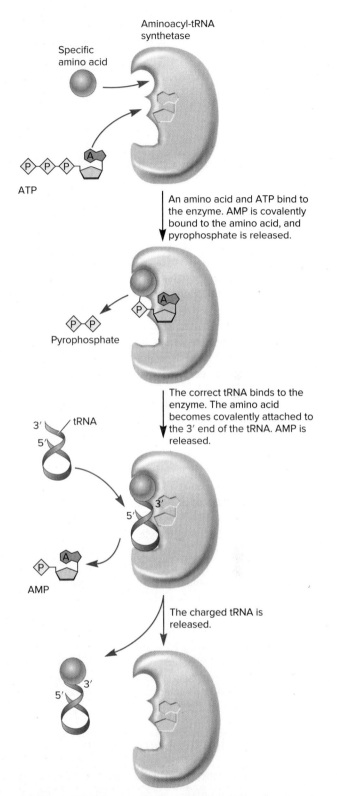

FIGURE 13.11 **Catalytic function of aminoacyl-tRNA synthetase.** Aminoacyl-tRNA synthetase has binding sites for a specific amino acid, ATP, and a particular tRNA. In the first step, the enzyme catalyzes the covalent attachment of AMP to an amino acid, yielding an activated amino acid. In the second step, the activated amino acid is attached to the appropriate tRNA. The charged tRNA is then released.

CONCEPT CHECK: What is the difference between a charged tRNA and an uncharged tRNA?

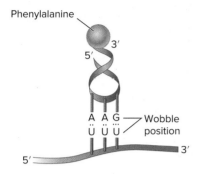

(a) Location of wobble position

Third base of mRNA codon:	Base in anticodon can be:
A	U, I, xo⁵U
U	A, G, U, I, xo⁵U
G	C, A, U, xo⁵U
C	G, A, I

(b) Revised wobble rules

FIGURE 13.12   Wobble position and base-pairing rules. (a) The wobble position occurs at the third base in the mRNA codon and the corresponding base in the anticodon. **(b)** The revised wobble rules are slightly different from those originally proposed by Crick. The standard bases found in RNA are G, C, A, and U. In addition, the structures of bases in tRNAs may be modified. Two examples of modified bases that may occur in the wobble position in tRNA are: I = inosine and xo⁵U = 5-hydroxyuridine.

CONCEPT CHECK: How do the wobble rules affect the total number of different tRNAs that are needed to carry out translation?

cases, the first two bases are always G and U. The third base, however, can be U, C, A, or G. To explain this pattern of degeneracy, Francis Crick proposed in 1966 that it is due to "wobble" at the third position in the codon-anticodon recognition process. According to the **wobble rules,** the first two positions pair strictly according to the AU/GC rule. However, the third position can tolerate certain types of mismatchcs (**Figurc 13.12**). This proposal suggested that the base at the third position in the codon does not have to hydrogen bond as precisely with the corresponding base in the anticodon.

Because of the wobble rules, some flexibility is observed in the recognition between a codon and anticodon during the process of translation. When two or more tRNAs that differ at the wobble position are able to recognize the same codon, they are termed **isoacceptor tRNAs.** As an example, tRNAs with an anticodon of 3′–CCA–5′ or 3′–CCG–5′, which both carry glycine, can recognize a codon with the sequence 5′–GGU–3′. In addition, the wobble rules enable a single type of tRNA to recognize more than one codon. For example, a tRNA with an anticodon sequence of 3′–AAG–5′, which carries phenylalanine, can recognize a 5′–UUC–3′ and a 5′–UUU–3′ codon. The 5′–UUC–3′ codon is a perfect match with this tRNA. The 5′–UUU–3′ codon is mismatched according to the standard RNA-RNA hybridization rules (namely, G in the anticodon is mismatched to U in the codon), but the G can fit with the U according to the wobble rules presented in Figure 13.12. The ability of a single tRNA to recognize more than one codon makes it unnecessary for a cell to make 61 different tRNA molecules with anticodons that are complementary to the 61 possible sense codons. *E. coli* cells, for example, make a population of tRNA molecules that have just 40 different anticodon sequences.

13.4   COMPREHENSION QUESTIONS

1. If a tRNA has an anticodon with the sequence 3′-GAC-5′, which amino acid does it carry?
   a. Aspartic acid   c. Leucine
   b. Valine   d. Glutamine
2. The anticodon of a tRNA is located in the
   a. 3′ single-stranded region of the acceptor stem.
   b. loop of the first stem-loop.
   c. loop of the second stem-loop.
   d. loop of the third stem-loop.

3. An enzyme known as _____ attaches an amino acid to the _____ of a tRNA, thereby producing _____.
   a. aminoacyl-tRNA synthetase, anticodon, a charged tRNA
   b. aminoacyl-tRNA synthetase, 3′ single-stranded region of the acceptor stem, a charged tRNA
   c. polynucleotide phosphorylase, anticodon, a charged tRNA
   d. polynucleotide phosphorylase, anticodon, an aminoacyl-tRNA

13.5   RIBOSOME STRUCTURE AND ASSEMBLY

Learning Outcome:

1. Outline the structural features of ribosomes.

In Section 13.4, we examined the important role of tRNA molecules in translation. According to the adaptor hypothesis, tRNAs bind to mRNAs due to complementarity between the anticodons and codons. Concurrently, the tRNA molecules have the correct amino acid attached to their 3′ ends.

To synthesize a polypeptide, additional events must occur. In particular, the bond between the 3′ end of the tRNA and the amino acid must be broken, and a peptide bond must be formed between the adjacent amino acids. To facilitate these events, translation occurs on a macromolecular complex known as a **ribosome.** The ribosome can be thought of as the macromolecular arena where translation takes place.

In this section, we will begin by outlining the biochemical compositions of ribosomes in bacterial and eukaryotic cells. We will then examine the key functional sites on ribosomes where the translation process is carried out.

Bacterial and Eukaryotic Ribosomes Are Assembled from rRNA and Proteins

Bacterial cells have one type of ribosome that is found within the cytoplasm. Eukaryotic cells contain biochemically distinct ribosomes in different cellular locations. The most abundant type of eukaryotic ribosome functions in the cytosol, which is the region of the cell that is inside the plasma membrane but outside the membrane-bound organelles. In addition to the cytosolic ribosomes, all eukaryotic cells have ribosomes within the mitochondria. Plant and algal cells also have ribosomes in their chloroplasts. The compositions of mitochondrial and chloroplast ribosomes are quite different from that of the cytosolic ribosomes. Unless otherwise noted, the term eukaryotic ribosome refers to a ribosome in the cytosol, not to those found within organelles. Likewise, the description of eukaryotic translation refers to translation via cytosolic ribosomes.

Each ribosome is composed of structures called the large and small subunits. This term is perhaps misleading because each ribosomal subunit itself is an assembly of many different proteins and one or more RNA molecules called **ribosomal RNA (rRNA).** In bacterial ribosomes, the 30S subunit is formed from 21 different ribosomal proteins and a 16S rRNA molecule; the 50S subunit contains 34 different proteins and 5S and 23S rRNA molecules (**Table 13.5**). With regard to the ribosomal subunits, the designations 30S and 50S refer to the rate at which these subunits sediment when subjected to a centrifugal force. This rate is given as a sedimentation coefficient in Svedberg units (S), in honor of Theodor Svedberg, who invented the ultracentrifuge. Together, the 30S and 50S subunits form a 70S ribosome. (Note: Svedberg units do not add up linearly.) In bacteria, the ribosomal proteins and rRNA molecules are synthesized in the cytoplasm, and the ribosomal subunits are assembled there.

The synthesis of eukaryotic rRNAs occurs within the nucleus, and the ribosomal proteins are made in the cytosol, where translation takes place. The 40S subunit is composed of 33 proteins and an 18S rRNA; the 60S subunit is made of 49 proteins and 5S, 5.8S, and 28S rRNAs (see Table 13.5).

The assembly of the rRNAs and ribosomal proteins to make the 40S and 60S subunits occurs within the **nucleolus.** The nucleolus is an example of a **droplet organelle** in which aggregated solutes, such as proteins and RNA molecules, separate from the bulk solvent and form a droplet via a process called liquid-liquid phase separation. This separation is much like the formation of an oil droplet in water. The droplet has a spherical shape with a measurable surface tension and viscosity. Molecules can diffuse within the droplet and occasionally leave it and pass into the surrounding liquid phase. The 40S and 60S subunits assemble within the nucleolous and are then exported into the cytosol, where they associate to form an 80S ribosome during translation.

Components of Ribosomal Subunits Form Functional Sites for Translation

In recent years, many advances have been made toward understanding ribosomes on a molecular level. Microscopic and biophysical methods have been used to study ribosome structure. An electron micrograph of bacterial ribosomes in the act of translation is shown in **Figure 13.13a**. In this example, many ribosomes are translating a single mRNA. The term **polyribosome,** or **polysome,** is used to describe an mRNA transcript that has many bound ribosomes in the act of translation.

More recently, a few research groups have succeeded in crystallizing ribosomal subunits, and even intact ribosomes. This is an amazing technical feat, because it is difficult to find the right conditions under which large macromolecules will form highly ordered crystals. **Figure 13.13b** is a schematic drawing of the crystal structure of bacterial ribosomal subunits. The overall shape of each subunit is largely determined by the structure of the rRNAs, which constitute most of the mass of the ribosome. The interface between the 30S and 50S subunits is primarily composed of rRNA. Ribosomal proteins cluster on the outer surface of the ribosome and on the periphery of the interface.

During bacterial translation, the mRNA lies on the surface of the 30S subunit within a space between the 30S and 50S subunits. As the polypeptide is being synthesized, it exits through a channel within the 50S subunit (**Figure 13.13c**). Ribosomes contain discrete

---

TABLE 13.5

Composition of Bacterial and Eukaryotic Ribosomes

	Small subunit	Large subunit	Assembled ribosome
Bacterial			
Sedimentation coefficient	30S	50S	70S
Number of proteins	21	34	55
rRNA molecule(s)	16S rRNA	5S rRNA, 23S rRNA	16S rRNA, 5S rRNA, 23S rRNA
Eukaryotic			
Sedimentation coefficient	40S	60S	80S
Number of proteins	33	49	82
rRNA molecule(s)	18S rRNA	5S rRNA, 5.8S rRNA, 28S rRNA	18S rRNA, 5S rRNA, 5.8S rRNA, 28S rRNA

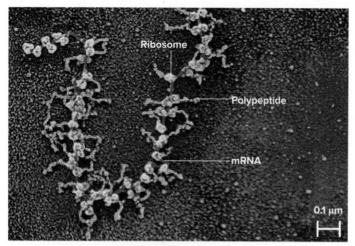

(a) Ribosomes in the act of translation

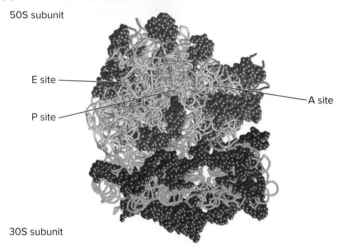

(b) Bacterial ribosome model based on X-ray diffraction studies

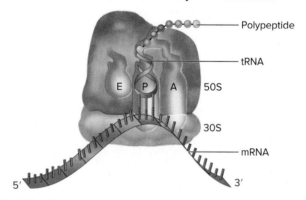

(c) Model for ribosome structure

FIGURE 13.13   Ribosomal structure. **(a)** A colorized transmission electron micrograph of ribosomes in the act of translation. Ribosomes are blue, mRNA is red, and polypeptides are green. **(b)** Crystal structure of the 50S and 30S subunits in bacterial ribosomes. The rRNA is shown as gray strands (50S subunit) and turquoise strands (30S subunit), and proteins are magenta (50S subunit) and navy blue (30S subunit). **(c)** A model depicting the sites where tRNA and mRNA bind to an intact ribosome. The mRNA lies on the surface of the 30S subunit. The E, P, and A sites are formed at the interface between the large and small subunits. The growing polypeptide exits through a hole in the 50S subunit.

(a): Dr. Elena Kiseleva/SPL/Science Source; (b): Tom Pantages

sites where tRNAs bind and the polypeptide is synthesized. In 1964, James Watson was the first to propose a two-site model for tRNA binding to a ribosome. These sites are known as the **peptidyl site (P site)** and the **aminoacyl site (A site).** In 1981, Knud Nierhaus, Hans Sternbach, and Hans-Jörg Rheinberger proposed a three-site model. This model incorporated the observation that uncharged tRNA molecules bind to a site on the ribosome that is distinct from the P and A sites. This third site is now known as the **exit site (E site).** The locations of the E, P, and A sites are shown in Figure 13.13c. In the next section, we will examine the roles of these sites during the three stages of translation.

13.5 COMPREHENSION QUESTIONS

1. Each ribosomal subunit is composed of
   a. multiple proteins.   c. tRNA
   b. rRNA.   d. both a and b.

2. The site(s) on a ribosome where tRNA molecules may be located include
   a. the A site.   c. the E site.
   b. the P site.   d. all of these.

13.6   STAGES OF TRANSLATION

Learning Outcomes:

1. Outline the three stages of translation.
2. Describe the steps that occur during the initiation, elongation, and termination stages of translation.

Like transcription, the process of translation can be viewed as occurring in three stages: initiation, elongation, and termination. **Figure 13.14** presents an overview of these stages, which are similar between bacteria and eukaryotes. During **initiation,** the ribosomal subunits, mRNA, and the first tRNA assemble to form a complex. After the initiation complex is formed, the ribosome slides along the mRNA in the 5′ to 3′ direction, moving over the codons. This is the **elongation** stage of translation. As the ribosome moves, tRNA molecules sequentially bind to the mRNA at the A site in the ribosome, bringing with them the appropriate amino acids. Therefore, amino acids are linked in the order dictated by the codon sequence in the mRNA. Finally, a stop codon is reached, signaling the **termination** of translation. At this point, disassembly occurs, and the newly made polypeptide is released. In this section, we will examine the components required for the translation process and consider their functional roles during the three stages of translation.

The Initiation Stage Involves the Binding of mRNA and the Initiator tRNA to the Ribosomal Subunits

During initiation, an mRNA and the first tRNA bind to the ribosomal subunits. A specific tRNA functions as the **initiator tRNA,** which recognizes the start codon in the mRNA. In bacteria, the initiator tRNA, which is designated tRNA^{fMet}, carries a methionine that has

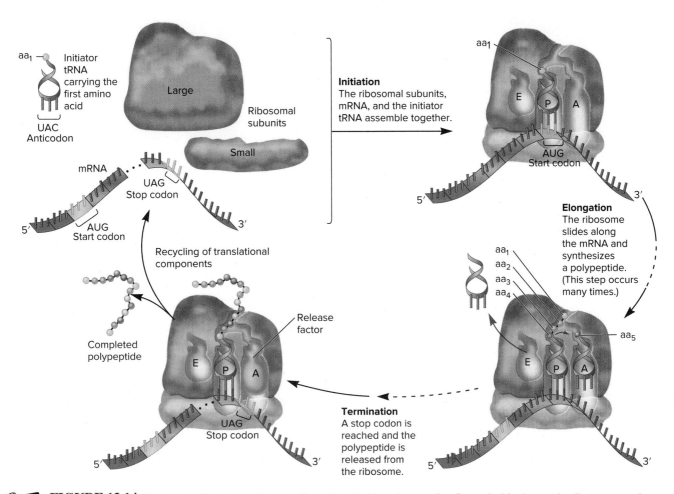

FIGURE 13.14 Overview of the stages of translation. Note: In this and succeeding figures in this chapter, the ribosomes are drawn schematically to emphasize different aspects of the translation process. A more detailed illustration of ribosome structure is shown in Figure 13.13b.

Genes→Traits The ability of genes to produce an organism's traits relies on the molecular process of gene expression. During translation, the codon sequence within mRNA (which is derived from a gene sequence during transcription) is translated into a polypeptide sequence. After polypeptides are made within a living cell, they function as proteins or components of proteins to govern an organism's traits. For example, once the β-globin polypeptide is made, it functions within the hemoglobin protein and provides red blood cells with the ability to carry oxygen, a vital trait for survival. Translation allows functional proteins to be made within living cells.

CONCEPT CHECK: Describe the roles that mRNA plays in the three stages of translation.

been covalently modified to *N*-formylmethionine. In this modification, a formyl group (—CHO) is attached to the nitrogen atom in methionine after the methionine has been attached to the tRNA.

Figure 13.15 shows the initiation stage of translation in bacteria, during which the mRNA, tRNA^{fMet}, and ribosomal subunits associate with each other to form an initiation complex. The formation of this complex requires the participation of three initiation factors: IF1, IF2, and IF3.

1. First, IF1 and IF3 bind to the 30S ribosomal subunit, which prevents the association of the 50S subunit.
2. Next, the mRNA binds to the 30S subunit. This binding is facilitated by a nine-nucleotide sequence within the bacterial mRNA called the **Shine-Dalgarno sequence.** The location of this sequence is shown in Figure 13.15 and in more detail in **Figure 13.16**. How does the Shine-Dalgarno sequence facilitate the binding of mRNA to the ribosome? The Shine-Dalgarno sequence is complementary to a short

sequence within the 16S rRNA, which promotes hydrogen bonding of the mRNA to the 30S subunit.

3. The initiator tRNA (tRNA^{fMet}) binds to the mRNA already bound to the 30S subunit (see Figure 13.15). This step requires the function of IF2, which has GTP bound to it. IF2 promotes the binding of the tRNA^{fMet} to the start codon, which is typically a few nucleotides downstream from the Shine-Dalgarno sequence. The start codon is usually AUG, but in some cases it can be GUG or UUG. Even when the start codon is GUG (which normally codes valine) or UUG (which normally codes leucine), the first amino acid in the polypeptide is still an *N*-formylmethionine because only a tRNA^{fMet} can initiate translation. During or after translation of the entire polypeptide, the formyl group or the entire *N*-formylmethionine may be removed. Therefore, some polypeptides may not have *N*-formylmethionine as their first amino acid. As noted in Figure 13.15, the tRNA^{fMet}

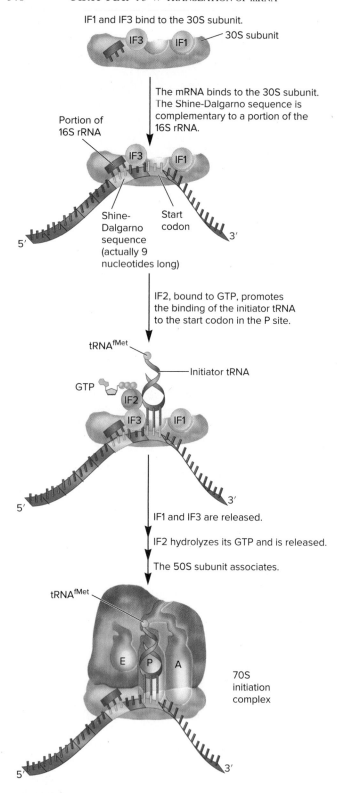

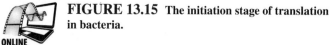

FIGURE 13.15 The initiation stage of translation in bacteria.

TABLE **13.6**

A Simplified Comparison of Translational Protein Factors in Bacteria and Eukaryotes

Bacterial Factors	Eukaryotic Factors*	Function
Initiation Factors		
	eIF4	Recognizes the 7-methylguanosine cap at the 5′ end of mRNA and facilitates the binding of the mRNA to the small ribosomal subunit
IF1, IF3	eIF1, eIF3, eIF6	Prevent the association between the small and large ribosomal subunits and favor their dissociation
IF3	eIF1	Participate in start codon selection
IF2	eIF2	Promote the binding of the initiator tRNA to the small ribosomal subunit
	eIF5	Helps to dissociate the initiation factors, which allows the two ribosomal subunits to assemble
Elongation Factors		
EF-Tu	eEF1α	Involved in the binding of tRNAs to the A site
EF-Ts	eEF1βγ	Nucleotide exchange factors required for the functioning of EF-Tu and eEF1α, respectively
EF-G	eEF2	Required for translocation
Release Factors		
RF1, RF2	eRF1	Recognize a stop codon and trigger the cleavage of the polypeptide from the tRNA
RF3	eRF3	GTPases that are also involved in termination

*Eukaryotic translation factors are typically composed of multiple proteins.

4. After the mRNA and tRNA^{fMet} have become bound to the 30S subunit, IF1 and IF3 are released, and then IF2 hydrolyzes its GTP and is also released. This allows the 50S ribosomal subunit to associate with the 30S subunit. After translation is completed, IF1 binding is necessary to dissociate the 50S and 30S ribosomal subunits so that the 30S subunit can reinitiate with another mRNA molecule.

In eukaryotes, the assembly of the initiation complex has similarities to what occurs in bacteria. However, as shown in **Table 13.6**, additional factors are required for the eukaryotic initiation process. Note that the initiation factors are designated eIF (for eukaryotic initiation factor) to distinguish them from bacterial initiation factors. The initiator tRNA in eukaryotes carries methionine rather than *N*-formylmethionine, as in bacteria. A eukaryotic initiation factor, eIF2, binds directly to tRNA^{Met} to recruit it to the 40S

binds to the P site on the ribosome. IF1 is thought to occupy a portion of the A site, thereby preventing the binding of tRNA^{fMet} to the A site during initiation. By comparison, during the elongation stage, discussed later in this section, all of the other tRNAs initially bind to the A site.

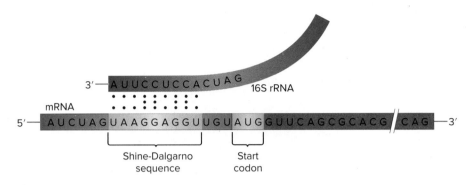

FIGURE 13.16   The locations of the Shine-Dalgarno sequence and the start codon in bacterial mRNA. The Shine-Dalgarno sequence is complementary to a sequence in the 16S rRNA. It hydrogen bonds with the 16S rRNA to promote initiation. The start codon is typically a few nucleotides downstream (toward the 3′ end) from the Shine-Dalgarno sequence.

CONCEPT CHECK: Why does bacterial mRNA bind specifically to the small ribosomal subunit?

subunit. Eukaryotic mRNAs do not have a Shine-Dalgarno sequence. How then are eukaryotic mRNAs recognized by the ribosome? The mRNA is recognized by eIF4, which is a multiprotein complex that recognizes the 7-methylguanosine cap at the 5′ end of the mRNA. eIF4 then facilitates the binding of the 5′ end of the mRNA to the 40S ribosomal subunit.

The identification of the correct AUG start codon in eukaryotes occurs in a different way than it does in bacteria. After the initial binding of mRNA to the ribosome, the next step is locating an AUG start codon that is somewhere downstream from the 7-methylguanosine cap. In 1986, Marilyn Kozak proposed that the ribosome begins at the 5′ end and then scans along the mRNA in the 3′ direction in search of an AUG start codon. In many, but not all, cases, the ribosome uses the first AUG codon that it encounters as a start codon. When a start codon is identified, eIF5 causes the release of the other initiation factors, which enables the 60S subunit to associate with the 40S subunit.

By analyzing the sequences of many eukaryotic mRNAs, researchers have found that not all AUG codons near the 5′ end of mRNA can function as start codons. In some cases, the scanning ribosome passes over the first AUG codon and chooses an AUG farther along within the mRNA. The sequence of bases around the AUG codon plays an important role in determining whether or not it is selected as the start codon by a scanning ribosome. The consensus sequence for optimal start codon recognition in complex eukaryotes, such as vertebrates and vascular plants, is shown in the upper right column.

						Start codon			
G	C	C	(A/G)	C	C	A	U	G	G
−6	−5	−4	−3	−2	−1	+1	+2	+3	+4

Aside from an AUG codon itself, a guanine at the +4 position and a purine, preferably an adenine, at the −3 position are the most important sites for start codon selection. These conditions for optimal initiation of translation are called **Kozak's rules.** How does scanning occur at the molecular level? A subunit within eIF1 called eIF1A binds to the A site. Its structure contains N- and C-terminal extensions that are necessary for scanning along the mRNA.

Changes in the Expression of eIFs Are Associated with Different Forms of Human Cancers

One benefit of studying translation at the molecular level is that it may provide insight into the underlying causes of human diseases. An area of research that has received great attention is the relationship between translation and various forms of cancer. Many studies have shown that changes in the expression of translation initiation factors (eIFs) can result in an increase in the translation of mRNAs coding proteins involved in tumorigenesis, metastasis, or resistance to anticancer drugs. These effects can promote cancer or lessen the effectiveness of anticancer drugs. **Table 13.7** describes several examples in which changes in eIF expression are associated with different types of cancer.

TABLE 13.7

Changes in Expression of eIFs That Are Associated with Human Cancers

Translation Initiation Factor*	Change in Expression	Types of Cancer
eIF2a	Overexpression	Non-Hodgkin lymphoma, melanocytic neoplasm, gastrointestinal cancer, and brain cancer
eIF3a	Overexpression	Brain cancer, cervical cancer, lung cancer, stomach cancer, and colorectal cancer
eIF3e	Underexpression	Breast cancer and prostate cancer
eIF3f	Underexpression	Melanocytic neoplasm, pancreatic cancer, breast cancer, and ovarian cancer
eIF4E	Overexpression	Breast cancer, lung cancer, prostate cancer, colorectal cancer, skin cancer, leukemia, and cervical cancer
eIF5A	Overexpression	Cervical cancer and colorectal cancer
eIF6	Overexpression	Colorectal cancer and malignant mesothelioma

*The upper- or lowercase letters following each number indicate a particular subunit within a eukaryotic initiation factor. eIF6 is composed of just one type of protein, so it doesn't have this type of designation.

FIGURE 13.17   The elongation stage of translation in bacteria. This process begins with the binding of an incoming charged tRNA. The hydrolysis of GTP by the elongation factor EF-Tu provides the energy for the binding of the tRNA to the A site. A peptide bond is then formed between the amino acid at the A site and the last amino acid in the growing polypeptide. This moves the polypeptide to the A site. The ribosome then translocates in the 3′ direction so that the two tRNAs are moved to the E and P sites. The tRNA carrying the polypeptide is now back in the P site. This translocation requires the hydrolysis of GTP by the elongation factor EF-G. The uncharged tRNA in the E site is released from the ribosome. Now the process is ready to begin again. Each cycle of elongation causes the polypeptide to grow by one amino acid.

CONCEPT CHECK: What is the role of peptidyl transferase during the elongation stage?

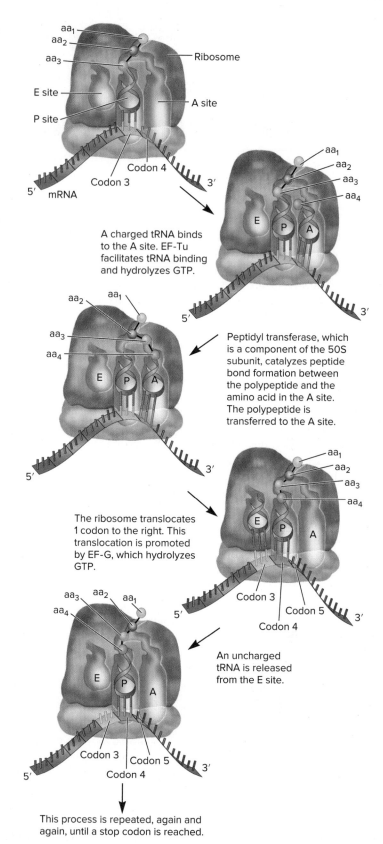

The discovery that changes in the expression of eIFs are associated with a wide variety of cancer types has sparked great interest in developing anticancer drugs that target eIFs.

- One example is ribavirin, which is an eIF4E inhibitor. It is already approved to treat certain types of viral infections. In clinical studies, ribavirin showed benefits for patients with acute myeloid leukemia.
- Another avenue in the development of anticancer drugs focuses on antisense oligonucleotides. These are short synthetic RNA molecules that bind to the mRNAs that code eIFs and thereby inhibit their synthesis.

Polypeptide Synthesis Occurs During the Elongation Stage

During the elongation stage of translation, amino acids are added, one at a time, to a growing polypeptide (**Figure 13.17**). Even though this process is rather complex, it occurs with remarkable speed. Under normal cellular conditions, a polypeptide can elongate at a rate of 15–20 amino acids per second in bacteria or 2–6 amino acids per second in eukaryotes!

Binding of a Charged tRNA to the A Site. To begin elongation, a charged tRNA brings a new amino acid to the ribosome so that it can be attached to the end of the growing polypeptide. At the top of Figure 13.17, which describes bacterial translation, a short polypeptide is attached to the tRNA located at the P site of the ribosome. A charged tRNA carrying a single amino acid binds to the A site. This binding occurs because the anticodon in the tRNA is complementary to the codon in the mRNA. The hydrolysis of GTP by the elongation factor EF-Tu provides energy for the binding of a tRNA to the A site. In addition, the 16S rRNA, which is a component of the small 30S ribosomal subunit, plays a key role by ensuring the proper recognition between the mRNA and the correct tRNA. The 16S rRNA can detect when an incorrect tRNA is bound at the A site and will prevent elongation until the mispaired tRNA is released from the A site. This phenomenon, termed the **decoding function** of the ribosome, is important in maintaining high fidelity of mRNA translation. An incorrect amino acid is incorporated into a growing polypeptide at a rate of approximately 1 mistake per 10,000 amino acids, or 10^{-4}.

Peptidyl Transfer Reaction. The next step of elongation is a reaction called **peptidyl transfer**—the polypeptide is removed from the tRNA in the P site and transferred to the amino acid at the A site. This transfer is accompanied by the formation of a peptide bond between the amino acid at the A site and the polypeptide, lengthening the polypeptide by one amino acid. The peptidyl transfer reaction is catalyzed by a component of the 50S subunit known as **peptidyl transferase,** which is composed of several proteins and rRNA. Interestingly, based on the crystal structure of the 50S subunit, Thomas Steitz, Peter Moore, and their colleagues concluded that 23S rRNA (a component of peptidyl transferase)—not a ribosomal protein—catalyzes the bond formation between adjacent amino acids that occurs during peptidyl transfer. In other words, the ribosome is a ribozyme!

Translocation. After the peptidyl transfer reaction is complete, the ribosome moves, or translocates, to the next codon in the mRNA. This moves the tRNAs at the P and A sites to the E and P sites, respectively. Finally, the uncharged tRNA exits the E site. You should notice that the next codon in the mRNA is now exposed in the unoccupied A site. At this point, a charged tRNA can enter the empty A site, and the same series of steps adds the next amino acid to the growing polypeptide.

As you may have realized, the A, P, and E sites are named for the role of the tRNA that is usually found there. The A site binds an aminoacyl-tRNA (also called a charged tRNA), the P site usually contains the peptidyl-tRNA (a tRNA with an attached peptide), and the E site is where the uncharged tRNA exits.

Termination Occurs When a Stop Codon Is Reached in the mRNA

The final stage of translation, known as termination, occurs when a stop codon is reached in the mRNA. In most species, the three stop codons are UAA, UAG, and UGA. The stop codons are not recognized by a tRNA with a complementary sequence. Instead, they are recognized by proteins known as **release factors** (see Table 13.6). Interestingly, the release factor proteins are "molecular mimics" in that their three-dimensional structures resemble those of tRNAs. Release factors can specifically bind to a stop codon sequence. In bacteria, RF1 recognizes UAA and UAG, and RF2 recognizes UAA and UGA. A third release factor, RF3, is also required. In eukaryotes, a single release factor, eRF1, recognizes all three stop codons, and eRF3 is also required for termination.

Figure 13.18 illustrates the termination stage of translation in bacteria. At the top of this figure, the completed polypeptide is attached to a tRNA in the P site. A stop codon is located at the A site.

1. In the first step, RF1 or RF2 binds to the stop codon at the A site and RF3 (not shown) binds at a different location on the ribosome.
2. After RF1 (or RF2) and RF3 have bound, the bond between the polypeptide and the tRNA is hydrolyzed. The polypeptide and tRNA are then released from the ribosome.
3. The final step in the termination stage is the disassembly of ribosomal subunits, mRNA, and the release factors.

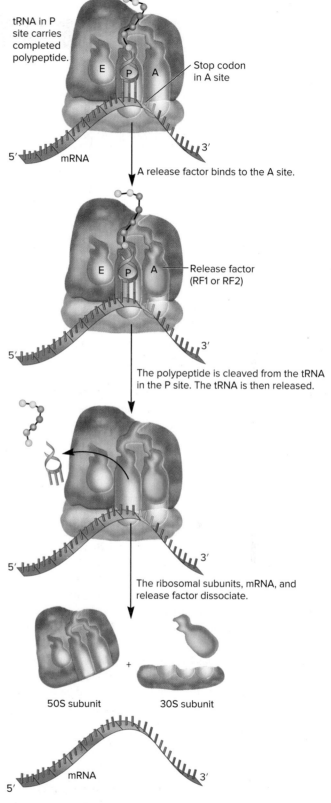

 FIGURE 13.18 The termination stage of translation in bacteria. When a stop codon is reached, a release factor (either RF1 or RF2) binds to the A site. (RF3 binds elsewhere and uses GTP to facilitate the termination process.) The polypeptide is cleaved from the tRNA in the P site and released. The tRNA is released, and the rest of the components disassemble.

CONCEPT CHECK: Explain why release factors are called "molecular mimics."

TABLE 13.8

Mechanisms of Inhibition of Bacterial Translation by Selected Antibiotics

Antibiotic	Description
Chloramphenicol	Blocks elongation by acting as competitive inhibitor of peptidyl transferase.
Erythromycin	Binds to the 23S RNA and blocks elongation by interfering with the translocation step.
Puromycin	Binds to the A site and causes premature release of the polypeptide. This early termination of translation results in polypeptides that are shorter than normal.
Tetracycline	Blocks elongation by inhibiting the binding of aminoacyl-tRNAs to the ribosome.
Streptomycin	Interferes with normal pairing between aminoacyl-tRNAs and codons. This causes misreading, thereby producing abnormal polypeptides.

Bacterial Translation Can Begin Before Transcription Is Completed

Microscopic, biochemical, and genetic studies have shown that the translation of a bacterial protein-coding gene begins before the mRNA transcript is completed. In other words, as soon as an mRNA strand is long enough, a ribosome attaches to the 5′ end and begins translation, even before RNA polymerase has reached the transcriptional termination site within the gene. This phenomenon in bacterial cells is termed coupling between transcription and translation. Note that coupling of these processes does not usually occur in eukaryotes, because transcription takes place in the nucleus of eukaryotic cells, whereas translation occurs in the cytosol.

Antibiotics That Inhibit Bacterial Translation Are Used to Treat Bacterial Diseases

Many different diseases that affect people and domesticated animals are caused by pathogenic bacteria. An **antibiotic** is any substance produced by a microorganism that inhibits the growth of other microorganisms, such as pathogenic bacteria. Most antibiotics are small organic molecules, with masses less than 2000 daltons (Da). In some cases, antibiotics exert their effect because they inhibit or interfere with bacterial translation. Because the components of translation differ somewhat between bacteria and eukaryotes, some antibiotics inhibit bacterial translation without affecting eukaryotic translation. Therefore, they can be used to treat bacterial infections in humans, pets, and livestock. **Table 13.8** describes a few examples.

13.6 COMPREHENSION QUESTIONS

1. During the initiation stage of translation in bacteria, which of the following events occur(s)?
   a. IF1 and IF3 bind to the 30S subunit.
   b. The mRNA binds to the 30S subunit, and tRNA^fMet binds to the start codon in the mRNA.
   c. IF2 hydrolyzes its GTP and is released; the 50S subunit binds to the 30S subunit.
   d. All of the above events occur.

2. Kozak's rules determine
   a. the choice of the start codon in complex eukaryotes.
   b. the choice of the start codon in bacteria.
   c. the site in the mRNA where translation ends.
   d. how fast the mRNA is translated.

3. During the peptidyl transfer reaction, the polypeptide, which is attached to a tRNA in the _____, becomes bound via _____ to an amino acid attached to a tRNA in the _____.
   a. A site, several hydrogen bonds, P site
   b. A site, a peptide bond, P site
   c. P site, a peptide bond, A site
   d. P site, several hydrogen bonds, A site

4. A release factor is referred to as a "molecular mimic" because its structure is similar to that of
   a. a ribosome.
   b. an mRNA.
   c. a tRNA.
   d. an elongation factor.

13.7 REGULATION OF TRANSLATION

Learning Outcome:

1. Explain how the translation and degradation of mRNAs may be controlled by RNA-binding proteins in mammals.

Thus far, we have focused on the components of translation and how translation occurs in living cells. After an mRNA is completely made, its expression can be regulated in a variety of ways. In Chapter 14, we will consider how bacteria may regulate mRNAs via antisense RNA. In Chapter 17, we will see how mRNAs are also regulated by small non-coding RNA molecules (look ahead to Section 17.3). In this section, we will focus on how the process of mRNA translation may be regulated by RNA-binding proteins that prevent ribosomes from initiating the translation process or affect mRNA degradation.

An Iron Response Element Regulates Translation and mRNA Degradation of Particular mRNAs

Specific mRNAs are sometimes regulated by RNA-binding proteins that bind to elements within the noncoding region of the mRNA and directly affect translation of the mRNA or its degradation. The regulation of iron assimilation in mammals provides a well-studied example in which both of these phenomena occur. Before discussing this form of control, let's consider the biology of iron uptake.

Iron is an essential element for the survival of living organisms because it is required for the function of many different proteins. The pathway by which mammalian cells take up iron is depicted in **Figure 13.19**.

1. Iron (Fe^{3+}) ingested by an animal becomes bound to transferrin, a protein that carries iron in the blood. The transferrin-Fe^{3+} complex is recognized by a transferrin receptor on the surface of cells.
2. The complex binds to the receptor and then is transported into the cytosol by endocytosis, forming an endocytic vesicle.
3. The iron within the vesicle is then released from transferrin and transported out of the vesicle. At this stage, Fe^{3+} may bind to enzymes that require iron for their

activity. Alternatively, if too much iron is present, the excess iron is stored within a hollow, spherical protein known as ferritin. The storage of excess iron within ferritin helps to prevent the toxic buildup of too much iron within the cell.

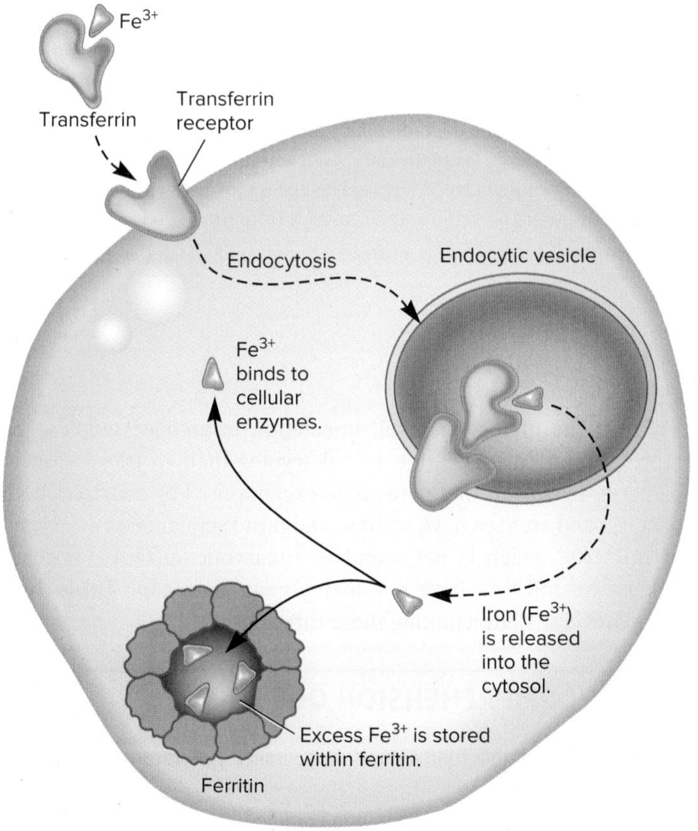

Iron is a vital yet potentially toxic substance, and animals have evolved an interesting way to regulate its assimilation. The two mRNAs that code ferritin and the transferrin receptor are both influenced by an RNA-binding protein known as the **iron regulatory protein (IRP).** How does IRP exert its effects? This protein binds to a regulatory element within these two mRNAs known as the **iron response element (IRE).**

Ferritin mRNA. The ferritin mRNA has an IRE in its 5′-untranslated region (5′-UTR), which is the region between the 5′ end and the start codon. When iron levels in the cell are low, IRP binds to this IRE, and the IRE/IRP complex prevents the translation of the ferritin mRNA (**Figure 13.20a**, left). In other words, the IRE/IRP complex acts as a blocker of translation. However, when iron is abundant in the cytosol, the iron binds directly to the IRP and causes a conformational change that prevents IRP from binding to the IRE. Under these conditions, the ferritin mRNA is translated to make more ferritin protein (Figure 13.20a, right), which stores the excess iron and thereby prevents the toxic buildup of iron within the cytosol.

Transferrin Receptor mRNA. The transferrin receptor mRNA also contains an iron response element, but it is located in the 3′-untranslated region (3′-UTR), which is the region between the stop codon and the 3′ end. When IRP binds to this IRE, it does not inhibit translation. Instead, the binding of IRP prevents the degradation of the mRNA by blocking the action of endonucleases that degrade the RNA. Therefore, when the cytosolic levels of iron are very low, this mRNA can be translated to make more transferrin receptor proteins (**Figure 13.20b**, left). This promotes the uptake of iron into the cell. In contrast, when iron is abundant within the cytosol, the iron binds to IRP, which causes IRP to dissociate from transferrin receptor mRNA. After this occurs, the mRNA is rapidly degraded because the removal of IRP exposes sites that are recognized by endonucleases (Figure 13.20b, right). The degradation of mRNA leads to a decrease in the amount of transferrin receptor, thereby helping to prevent the uptake of too much iron into the cell.

FIGURE 13.19 **The uptake of iron (Fe^{3+}) into mammalian cells.**

13.7 COMPREHENSION QUESTION

1. The binding of iron regulatory protein (IRP) to the iron response element (IRE) in the 5′-UTR of the ferritin mRNA causes
   a. the inhibition of translation of the ferritin mRNA.
   b. the stimulation of translation of the ferritin mRNA.
   c. the degradation of the ferritin mRNA.
   d. both a and c.

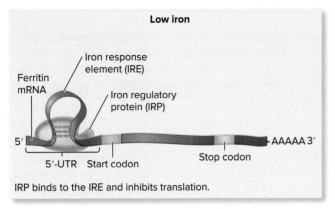
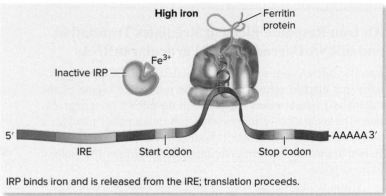

(a) Regulation of translation: ferritin mRNA

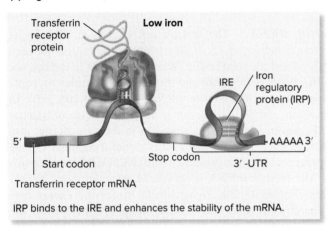
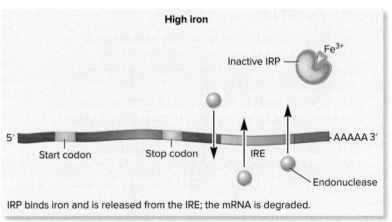

(b) Regulation of RNA stability: transferrin receptor mRNA

FIGURE 13.20   The regulation of iron assimilation genes by IRP and IRE. (a) When the Fe^{3+} concentration is low, the binding of iron regulatory protein (IRP) to the iron response element (IRE) in the 5'-UTR of ferritin mRNA inhibits translation (left). When the Fe^{3+} concentration is high, Fe^{3+} binds to IRP and causes a conformation change that prevents IRP from binding to ferritin mRNA. Under these conditions, translation can proceed (right). (b) The binding of IRP to the IRE in the 3'-UTR of the transferrin receptor mRNA prevents the degradation of that mRNA. Therefore, the mRNA is available for the synthesis of transferrin receptor proteins when the Fe^{3+} concentration is low (left). When the Fe^{3+} concentration is high and iron binds to IRP, that protein dissociates from the IRE, and the transferrin receptor mRNA is cleaved by endonucleases. This cleavage removes the polyA tail, and then the mRNA is further degraded by an exonuclease (not shown) that degrades the rest of the transferrin receptor mRNA from the 3' end.

CONCEPT CHECK: If a mutation prevented IRP from binding to the IRE in the ferritin mRNA, how would that mutation affect the regulation of ferritin synthesis? Do you think there would be too much or too little ferritin in the cell?

13.8   A COMPARISON OF TRANSLATION IN BACTERIA, ARCHAEA, AND EUKARYOTES

Learning Outcome:

1. Compare and contrast the process of translation among bacteria, archaea, and eukaryotes.

Throughout this chapter, we have considered the processes of translation with a focus on bacteria and eukaryotes. In this section, we will also consider the third domain of life, the Archaea. As mentioned in Chapter 12 (see Section 12.5), this domain comprises a group of single-celled organisms with distinct molecular characteristics separating them from the domain, Bacteria.

At the molecular level, translation in archaea tends to be more similar to translation in eukaryotes than to translation in bacteria. However, there are some exceptions. For example, both bacteria and archaea have a Shine-Dalgarno sequence at the 5' end of mRNAs, which is not found in eukaryotic mRNAs. Instead, eukaryotic mRNAs have a 7-methylguanosine cap. **Table 13.9** compares translation among these three domains.

13.8   COMPREHENSION QUESTION

1. Which of the following is not similar between archaea and eukaryotes?
   a. Formation of the initiation complex
   b. Initial binding of mRNA to the ribosome
   c. Termination of translation
   d. All of the above are similar between archaea and eukaryotes.

TABLE 13.9
A Comparison of Bacterial, Archaeal and Eukaryotic Translation

	Bacterial	Archaeal	Eukaryotic
Ribosome composition:	70S ribosomes: 30S subunit—21 proteins + 1 rRNA 50S subunit—34 proteins + 2 rRNAs	70S ribosomes: 30S subunit—25 proteins + 1 rRNA 50S subunit—39 proteins + 2 rRNAs Note: The size of an archaeal ribosome is more similar to that of a bacterial ribosome, but the sequences of archaeal rRNAs are more similar to those of eukaryotic rRNAs and their ribosomal proteins are more similar to those of eukaryotes.	80S ribosomes: 40S subunit—33 proteins + 1 rRNA 60S subunit—49 proteins + 3 rRNAs
Initiator tRNA:	tRNA^{fMet}	tRNA^{Met}	tRNA^{Met}
Formation of the initiation complex:	Requires IF1, IF2, and IF3	Requires more initiation factors; many of them are homologous to eIFs	Requires more initiation factors than in bacteria
Initial binding of mRNA to the ribosome:	Requires a Shine-Dalgarno sequence	Requires a Shine-Dalgarno sequence or has a very short 5′-UTR	Requires a 7-methylguanosine cap
Selection of a start codon:	AUG, GUG, or UUG located just downstream from the Shine-Dalgarno sequence	Depends on an initiation factor, aIF1, which is homologous to a eukaryotic initiation factor, and binds to the A site on the ribosome	According to Kozak's rules
Elongation rate:	Typically 10–20 amino acids per second	Not well established	Typically 2–6 amino acids per second
Termination:	Requires RF1, RF2, and RF3	Similar to eukaryotic termination; requires two proteins that are similar in structure and function to eRF1 and ERF3	Requires eRF1 and eRF3
Location of translation:	Cytoplasm	Cytoplasm	Cytosol
Coupled to transcription:	Yes	Yes	No

KEY TERMS

Introduction: translation

13.1: protein-coding genes, messenger RNA (mRNA), alkaptonuria, inborn error of metabolism, one-gene/one-enzyme hypothesis, polypeptide, protein

13.2: genetic code, sense codon, start codon, stop codon (termination codon or nonsense codon), 5′-untranslated region, 3′-untranslated region, transfer RNA (tRNA), anticodon, degeneracy, synonymous codons, reading frame, selenocysteine, pyrrolysine, peptide bond, amino-terminus (N-terminus), carboxyl-terminus (C-terminus), side chain, R group, primary structure, chaperone, secondary structure, α helix, β sheet, tertiary structure, quaternary structure, subunit

13.3: cell-free translation system

13.4: adaptor hypothesis, aminoacyl-tRNA synthetase, charged tRNA (aminoacyl-tRNA), wobble rules, isoacceptor tRNAs

13.5: ribosome, ribosomal rRNA (rRNA), nucleolus, droplet organelle, polyribosome (polysome), peptidyl site (P site), aminoacyl site (A site), exit site (E site)

13.6: initiation, elongation, termination, initiator tRNA, Shine-Dalgarno sequence, Kozak's rules, decoding function, peptidyl transfer, peptidyl transferase, release factor, antibiotic

13.7: iron regulatory protein (IRP), iron response element (IRE)

CHAPTER SUMMARY

- Cellular proteins are made via the translation of mRNAs.

13.1 The Genetic Basis for Protein Synthesis

- Garrod studied the genetic disorder called alkaptonuria and suggested that some genes code enzymes (see Figure 13.1).

- Beadle and Tatum studied *Neurospora* mutants that were altered in their nutritional requirements and hypothesized that one gene codes one enzyme. This one-gene/one-enzyme hypothesis was later modified because (1) some proteins are not enzymes; (2) some proteins are composed of two or more different polypeptides; (3) some genes code RNAs that are not translated into polypeptides; and (4) some mRNAs are alternatively spliced or edited (see Figure 13.2).

13.2 The Relationship Between the Genetic Code and Protein Synthesis

- During translation, the codons in mRNA are recognized by tRNA molecules to make a polypeptide with a specific amino acid sequence (see Figure 13.3).
- The genetic code refers to the relationship between the three-base codons in the mRNA and the amino acids that are incorporated into a polypeptide. It is composed of 64 codons. One codon (AUG) is a start codon, which determines the reading frame of the mRNA. Three codons (UAA, UAG, and UGA) function as stop codons (see Table 13.1).
- The genetic code is largely universal, but some exceptions are known to occur (see Table 13.2).
- A polypeptide is made by the formation of peptide bonds between adjacent amino acids. Each polypeptide has a directionality from its amino-terminus to its carboxyl-terminus, which parallels the arrangement of codons in mRNA in the 5′ to 3′ direction (see Figure 13.4).
- Amino acids have unique side chains (see Figure 13.5).
- Protein structure can be viewed at different levels: primary structure (sequence of amino acids), secondary structure (repeating folding patterns such as the α helix and the β sheet), tertiary structure (additional folding), and quaternary structure (the association of two or more polypeptide subunits with each other) (see Figure 13.6).
- Proteins carry out a variety of functions. The structures and functions of proteins are largely responsible for an organism's traits (see Table 13.3).

13.3 Experimental Determination of the Genetic Code

- Nirenberg and colleagues used synthetic RNA and a cell-free translation system to decipher the genetic code (see Figure 13.7).
- Other methods used to decipher the genetic code included the synthesis of RNA copolymers by Khorana and colleagues and the triplet-binding assays conducted by Nirenberg and Leder (see Table 13.4, Figure 13.8).

13.4 Structure and Function of tRNA

- The anticodon in a tRNA binds to a codon in mRNA in an antiparallel manner that obeys the AU/GC rule. The tRNA carries a specific amino acid that corresponds to the codon in the mRNA according to the genetic code (see Figure 13.9).
- The secondary structure of tRNA resembles a cloverleaf. The anticodon is in the second loop and the amino acid is attached to the 3′ end (see Figure 13.10).
- An aminoacyl-tRNA synthetase is one of a group of enzymes that attach the correct amino acid to a tRNA. The resulting tRNA is called a charged tRNA, or an aminoacyl-tRNA (see Figure 13.11).

- Mismatches are allowed between the pairing of tRNAs and mRNA according to the wobble rules (see Figure 13.12).

13.5 Ribosome Structure and Assembly

- Ribosomes are the site of polypeptide synthesis. The small and large subunits of ribosomes are composed of rRNAs and multiple proteins (see Table 13.5).
- A ribosome has A (aminoacyl), P (peptidyl), and E (exit) sites, which are occupied by tRNA molecules (see Figure 13.13).

13.6 Stages of Translation

- The three stages of translation are initiation, elongation, and termination (see Figure 13.14).
- During the initiation stage of translation, the mRNA, initiator tRNA, and ribosomal subunits assemble. Initiation factors are involved in the process. In bacteria, the Shine-Dalgarno sequence promotes the binding of the mRNA to the small ribosomal subunit (see Figures 13.15, 13.16, Table 13.6).
- Selection of a start codon in complex eukaryotes follows Kozak's rules.
- Changes in the expression of eIFs are associated with different forms of human cancers (see Table 13.7).
- During elongation, tRNAs bring amino acids to the A site, and a series of peptidyl transfers creates a polypeptide. At each step, the polypeptide is transferred from the P site to the A site. The uncharged tRNAs are released from the E site. Elongation factors are involved in this process (see Figure 13.17).
- During termination, a release factor binds to a stop codon in the A site. This promotes the cleavage of the polypeptide from the tRNA and the subsequent disassembly of the ribosomal subunits, mRNA, and release factor (see Figure 13.18).
- Bacterial translation can begin before transcription is completed.
- Antibiotics that inhibit translation are used to treat bacterial diseases (see Table 13.8).

13.7 Regulation of Translation

- The regulation of iron uptake and storage is needed to prevent the toxic buildup of iron in cells. When iron levels are low, the binding of iron regulatory protein (IRP) to an iron response element (IRE) found in the 5′-UTR of ferritin mRNA inhibits translation. By comparison, the binding of IRP to an IRE in the 3′-UTR of the transferrin receptor mRNA promotes stability of that mRNA when iron levels are low (see Figures 13.19, 13.20).

13.8 A Comparision of Translation in Bacteria, Archaea, and Eukaryotes

- Bacterial, archaeal, and eukaryotic translation have many similarities and differences; archaeal translation is more similar to eukaryotic translation than it is to bacterial translation (see Table 13.9).

PROBLEM SETS & INSIGHTS

MORE GENETIC TIPS

1. The first amino acid in a purified bacterial polypeptide is methionine. The start codon in the mRNA is GUG, which codes for valine. Why isn't the first amino acid *N*-formylmethionine or valine?

T **OPIC:** *What topic in genetics does this question address?* The topic is translation. More specifically, the question is about what determines the first amino acid in a polypeptide.

I **NFORMATION:** *What information do you know based on the question and your understanding of the topic?* From the question, you know that the start codon in an mRNA is GUG, but the first amino acid in the resulting polypeptide is methionine. From your understanding of the topic, you may remember that the initiator tRNA carries the first amino acid in a polypeptide.

P **ROBLEM-SOLVING** **S** **TRATEGY:** *Describe the steps.* One strategy to solve this problem is to describe the steps of translation. This may help you to see how the first amino acid gets incorporated into a polypeptide.

ANSWER: During polypeptide synthesis, the first amino acid is carried by the initiator tRNA. This initiator tRNA always carries *N*-formylmethionine even when the start codon is GUG (valine) or UUG (leucine). The formyl group can be removed later to leave methionine as the first amino acid.

2. A tRNA has the anticodon sequence 3′–CAG–5′. What amino acid does it carry?

T **OPIC:** *What topic in genetics does this question address?* The topic is translation. More specifically, the question asks you to determine the amino acid that a tRNA carries.

I **NFORMATION:** *What information do you know based on the question and your understanding of the topic?* From the question, you know that a tRNA has an anticodon that is 3′–CAG–5′. From your understanding of the topic, you may remember that the anticodon and codon are complementary and antiparallel.

P **ROBLEM-SOLVING** **S** **TRATEGY:** *Make a drawing.* One strategy to solve this problem is to make a drawing showing how the anticodon in a tRNA binds to a codon in an mRNA.

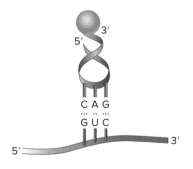

ANSWER: An anticodon that is 3′–CAG–5′ is complementary to a codon with the sequence 5′–GUC–3′. According to the genetic code, this codon specifies the amino acid valine. Therefore, this tRNA must carry valine at its acceptor stem.

3. An antibiotic is a drug that kills or inhibits the growth of microorganisms. The use of antibiotics has been of great importance in the battle against many infectious diseases caused by microorganisms. The mode of action for many antibiotics is to inhibit the translation process within bacterial cells. Certain antibiotics selectively bind to bacterial (70S) ribosomes but do not inhibit eukaryotic (80S) ribosomes. Why would an antibiotic bind to a bacterial ribosome but not to a eukaryotic ribosome? Why does this binding inhibit growth?

T **OPIC:** *What topic in genetics does this question address?* The topic is translation. More specifically, the question is about how an antibiotic may inhibit translation in bacteria, but not in eukaryotes.

I **NFORMATION:** *What information do you know based on the question and your understanding of the topic?* From the question, you know that some antibiotics bind to bacterial ribosomes but not to eukaryotic ribosomes. From your understanding of the topic, you may remember that ribosomes must function correctly for translation to occur.

P **ROBLEM-SOLVING** **S** **TRATEGY:** *Relate structure and function.* One strategy to solve this problem is to consider how the structure of bacterial ribosomes differs from that of eukaryotic ribosomes and how such structural differences may affect the ability of antibiotics to bind to ribosomal components. The binding of an antibiotic may inhibit a key step in translation (see Table 13.8).

ANSWER: Bacterial ribosomes are composed of proteins and rRNAs whose structures differ from those of eukaryotic ribosomes. Certain antibiotics can recognize these components and bind specifically to bacterial ribosomes, thereby interfering with the process of translation. In other words, the surface of a bacterial ribosome must be somewhat different from the surface of a eukaryotic ribosome so that antibiotic molecules are able to bind only to the surfaces of bacterial ribosomes. If a bacterial cell is exposed to certain antibiotics, it cannot synthesize new polypeptides because the antibiotic inhibits ribosome function. Because polypeptides form functional proteins needed for processes such as cell division, the bacterium is unable to grow and proliferate.

Conceptual Questions

C1. An mRNA has the following sequence:

5′–GGCGAUGGGCAAUAAACCGGGCCAGUAAGC–3′

Identify the start codon, and determine the complete amino acid sequence that will be translated from this mRNA.

C2. What does it mean when we say that the genetic code possesses degeneracy? Discuss the universality of the genetic code.

C3. According to the adaptor hypothesis, is each of the following statements true or false?

A. The sequence of anticodons in tRNA directly recognizes codon sequences in mRNA, with some allowance for wobble.

B. The amino acid attached to the tRNA directly recognizes codon sequences in mRNA.

C. The amino acid attached to the tRNA affects the binding of the tRNA to a codon sequence in mRNA.

C4. Researchers have isolated strains of bacteria that carry mutations within genes that code tRNAs. These mutations can change the sequence of the anticodon of a tRNA. For example, a normal tRNA^{Trp} gene codes a tRNA with the anticodon 3′–ACC–5′. A mutation can change this sequence to 3′–CCC–5′. When this mutation occurs, the tRNA still carries a tryptophan at its 3′ acceptor stem, even though the anticodon sequence has been altered.

A. How would this mutation affect the synthesis of polypeptides within the bacterium?

B. What does this mutation tell you about the recognition between tryptophanyl-tRNA synthetase and tRNA^{Trp}? Does the enzyme primarily recognize the anticodon or not?

C5. The covalent attachment of an amino acid to a tRNA is an endergonic reaction. In other words, an input of energy is required for the reaction to proceed. Where does the energy needed to attach amino acids to tRNA molecules come from?

C6. The wobble rules for tRNA-mRNA pairing are shown in Figure 13.12. If you assume that the tRNAs do not contain modified bases, what is the minimum number of tRNAs needed to recognize the codons for each of the following amino acids?

A. Leucine

B. Methionine

C. Serine

C7. How many different mRNA codon sequences could code a peptide with the sequence proline-glycine-methionine-serine?

C8. If a tRNA molecule carries a glutamic acid, what are the two possible anticodon sequences that it could contain? Be specific about the 5′ and 3′ ends.

C9. A tRNA anticodon has the sequence 3′–GGU–5′. What amino acid does it carry?

C10. If a tRNA anticodon is 3′–CCI–5′, what codon(s) can it recognize?

C11. Specify the anticodon of a single tRNA that could recognize the codons 5′–AAC–3′ and 5′–AAU–3′. What type(s) of base modification to this tRNA would allow it to also recognize 5′–AAA–3′?

C12. Describe the structural features that all tRNA molecules have in common.

C13. In the tertiary structure of tRNA, where is the anticodon located relative to the attachment site for the amino acid? Are these locations adjacent to each other?

C14. What is the role of aminoacyl-tRNA synthetase? The ability of aminoacyl-tRNA synthetases to recognize tRNAs has sometimes been called the "second genetic code." Why has the function of this type of enzyme been described this way?

C15. What is an activated amino acid?

C16. Discuss the significance of modified bases within tRNA molecules.

C17. How and when does *N*-formylmethionine become attached to the initiator tRNA in bacteria?

C18. Is it necessary for a cell to make 61 different tRNA molecules, corresponding to the 61 codons for amino acids? Explain your answer.

C19. List the components required for translation. Describe the relative sizes of these different components. In other words, which components are small molecules, macromolecules, or assemblies of macromolecules?

C20. Describe the components of eukaryotic ribosomal subunits and the location where the assembly of the subunits occurs within living cells.

C21. The term *subunit* can be used in a variety of ways. What is the difference between a protein subunit and a ribosomal subunit?

C22. Do the following events during bacterial translation occur primarily within the 30S subunit, within the 50S subunit, or at the interface between these two ribosomal subunits?

A. mRNA-tRNA recognition

B Peptidyl transfer reaction

C. Exit of the polypeptide from the ribosome

D. Binding of initiation factors IF1, IF2, and IF3

C23. What are the three stages of translation? Discuss the main events that occur during each of these three stages.

C24. Describe the sequence in bacterial mRNA that promotes recognition by the 30S subunit.

C25. For each of the following initiation factors, how would eukaryotic initiation of translation be affected if it were missing?

A. eIF2

B. eIF4

C. eIF5

C26. How does a eukaryotic ribosome select its start codon? Describe the sequences in eukaryotic mRNAs that optimize start codon recognition.

C27. Rank the following sequences in order (from best to worst) in terms of their suitability to initiate translation according to Kozak's rules.

GACGCCAUGG

GCCUCCAUGC

GCCAUCAAGG

GCCACCAUGG

C28. Explain the functional roles of the A, P, and E sites during translation.

C29. An mRNA has the following sequence:

5′–AUG UAC UAU GGG GCG UAA–3′

Describe the amino acid sequence of the polypeptide that would be coded by this mRNA. Specify the amino-terminus and carboxyl-terminus.

C30. Which steps during the translation of bacterial mRNA involve an interaction between complementary strands of RNA?

C31. What is the function of the nucleolus?

C32. In which of the ribosomal sites, A site, P site, and/or E site, could the following be found?

A. A tRNA without an amino acid attached

B. A tRNA with a polypeptide attached

C. A tRNA with a single amino acid attached

C33. What is a polysome?

C34. Referring to Figure 13.17, explain why the ribosome translocates along the mRNA in the 5′ to 3′ direction rather than the 3′ to 5′ direction.

C35. Lactose permease, a protein produced in *E. coli*, is composed of a single polypeptide that is 417 amino acids long. By convention, the amino acids within a polypeptide are numbered from the amino-terminus to the carboxyl-terminus. Are the following questions about lactose permease true or false?

A. Because the 64th amino acid is glycine and the 68th amino acid is aspartic acid, the codon for glycine, codon 64, is closer to the 3′ end of the mRNA than the codon for aspartic acid, codon 68.

B. The mRNA that codes lactose permease must have more than 1241 nucleotides.

C36. An mRNA codes a polypeptide that is 312 amino acids long. The 53rd codon in this polypeptide is a tryptophan codon. A mutation in the gene that codes this polypeptide changes this tryptophan codon into a stop codon. How many amino acids would be in the resulting polypeptide: 52, 53, 259, or 260?

C37. Explain what is meant by the coupling of transcription and translation in bacteria. Does coupling occur in bacterial and/or eukaryotic cells? Explain.

C38. Describe how the binding of IRP to an IRE affects the mRNAs for ferritin and the transferrin receptor. How does iron (Fe^{3+}) influence this process?

Experimental Questions

E1. In the experiment of Figure 13.7, what would be the predicted amounts of amino acids incorporated into the polypeptides if the RNA was a random polymer containing 50% C and 50% G?

E2. Polypeptides can be translated in vitro. Would a bacterial mRNA be translated in vitro by eukaryotic ribosomes? Would a eukaryotic mRNA be translated in vitro by bacterial ribosomes? Why or why not?

E3. Discuss how the elucidation of the structure of the ribosome can help us to understand its function.

E4. Describe the structure of a polysome, which is depicted in Figure 13.13a.

E5. Chapter 20 describes a blotting method known as Western blotting that can be used to detect the production of a polypeptide that is translated from a particular mRNA. In this method, a protein is detected using an antibody that specifically recognizes and binds to an amino acid sequence in the protein. The antibody acts as a probe that can label and thereby detect the protein. In a Western blotting experiment, gel electrophoresis is used to separate a mixture of proteins according to their molecular masses. After the antibody has bound to the protein of interest within a blot of a gel, the protein is visualized as a dark band. For example, an antibody that recognizes the β-globin polypeptide could be used to specifically detect the β-globin polypeptide in a Western blot. As shown here, the method of Western blotting can be used to determine the amount and relative size of a particular protein that is produced in a given cell type.

Western blot

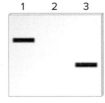

Lane 1 is a sample of proteins isolated from normal red blood cells.

Lane 2 is a sample of proteins isolated from kidney cells. Kidney cells do not produce the β-globin polypeptide.

Lane 3 is a sample of proteins isolated from red blood cells from a patient with β-thalassemia. This patient is homozygous for a mutation that results in the shortening of the β-globin polypeptide.

Here is the question. A protein called troponin contains 334 amino acids. The molecular mass of this protein is about 40,000 Da, or 40 kDa. Troponin functions in muscle cells, and it is not expressed in nerve cells. Draw the expected results of a Western blot for the following samples:

Lane 1: Proteins isolated from muscle cells

Lane 2: Proteins isolated from nerve cells

Lane 3: Proteins isolated from the muscle cells of an individual who is homozygous for a mutation that introduces a stop codon at codon 177 in the gene that codes troponin

E6. The technique of Western blotting is described in Chapter 20 and also in question E5. Let's suppose a researcher is interested in the effects of mutations on the expression of a protein-coding gene that codes a protein we will call protein X. This protein is expressed in skin cells and contains 572 amino acids. Its molecular mass is approximately 68,600 Da, or 68.6 kDa. Make a drawing that shows the expected results of a Western blot using proteins isolated from the skin cells obtained from the following individuals:

Lane 1: A normal individual

Lane 2: An individual who is homozygous for a deletion that removes the promoter for this gene

Lane 3: An individual who is heterozygous such that one gene is normal and the other gene has a mutation that introduces an early stop codon at codon 421

Lane 4: An individual who is homozygous for a mutation that introduces an early stop codon at codon 421

Lane 5: An individual who is homozygous for a mutation that changes codon 198 from a valine codon to a leucine codon

E7. The protein known as tyrosinase is needed to make certain types of pigments. Tyrosinase is composed of a single polypeptide with 511 amino acids. The molecular mass of this protein is approximately 61,300 Da, or 61.3 kDa. People who carry two defective copies of the tyrosinase gene have the condition known as albinism. They are unable to make pigment in the skin, eyes, and hair. Western blotting is used to detect proteins that are translated from a particular mRNA. This method is described in Chapter 20 and also in question E5. Skin samples were collected from a pigmented individual (lane 1) and from three unrelated albino individuals (lanes 2, 3, and 4) and subjected to a Western blot analysis using an antibody that recognizes tyrosinase. Explain the possible cause of albinism in the three albino individuals.

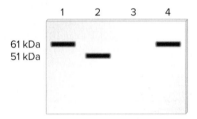

E8. Although 61 codons specify the 20 amino acids, most species display a codon bias, which means that certain codons are used much more frequently than other codons. For example, UUA, UUG, CUU, CUC, CUA, and CUG all specify leucine. In yeast, however, the UUG codon is used to specify leucine approximately 80% of the time.

A. The experiment of Figure 13.7 shows the use of an in vitro, or cell-free, translation system. In this experiment, the RNA that was used for translation was chemically synthesized. Instead of using a chemically synthesized RNA, researchers can isolate mRNA from living cells and then add the mRNA to the cell-free translation system. If a researcher isolated mRNA from kangaroo cells and then added it to a cell-free translation

system that came from yeast cells, how might the phenomenon of codon bias affect the production of proteins?

B. Discuss potential advantages and disadvantages of codon bias for translation in a given species.

E9. Chapter 20 describes a blotting method known as Northern blotting, in which a short segment of cloned DNA is used as a probe to detect RNA that is transcribed from a particular gene. The DNA probe, which is labeled, is complementary to the RNA that the researcher wishes to detect. After the labeled probe DNA binds to the RNA within a blot of a gel, the RNA is visualized as a dark band. The method of Northern blotting can be used to determine the amount of a particular RNA transcribed in a given cell type. If one type of cell produces twice as much of a particular mRNA compared to another cell, the band appears twice as thick.

Suppose that a researcher has a segment of labeled DNA complementary to the ferritin mRNA. This labeled DNA can be used to specifically detect the amount of ferritin mRNA on a gel. The researcher began with two flasks containing human skin cells. One flask had a very low concentration of iron, and the other flask had a high concentration of iron. The mRNA was isolated from these cells and then subjected to Northern blotting, using a segment of labeled DNA that is complementary to the ferritin mRNA. The sample loaded in lane 1 was from the cells exposed to a low concentration of iron, and the sample in lane 2 was from the cells exposed to a high concentration of iron. Three Northern blots are shown next, but only one of them is correct. Based on your understanding of ferritin mRNA regulation, which blot (a, b, or c) would be the expected result? Explain. Which blot (a, b, or c) would be the expected result if the segment of labeled DNA was complementary to the transferrin receptor mRNA?

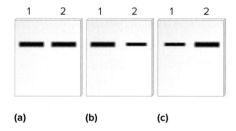

(a)   (b)   (c)

Questions for Student Discussion/Collaboration

1. Discuss why you think ribosomes need to contain so many proteins and rRNA molecules. Does it seem like a waste of cellular energy to make such a large structure so that translation can occur?

2. Discuss and make a list of the similarities and differences in the events that occur during the initiation, elongation, and termination stages of transcription (see Chapter 12) and those of translation (discussed in this chapter).

3. Which events during translation involve molecular recognition between base sequences within different RNAs? Which events involve recognition between different protein molecules?

Note: All answers are available for the instructor in Connect; the answers to the even-numbered questions and all of the Concept Check and Comprehension Questions are in Appendix B.

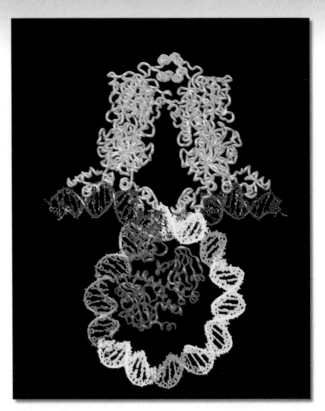

A model showing the binding of a genetic regulatory protein to DNA, which results in a DNA loop. This model illustrates the lac repressor protein found in E. coli binding to the operator site in the lac operon.
SPL/Science Source

14

GENE REGULATION IN BACTERIA

Chromosomes of bacteria, such as *Escherichia coli*, contain a few thousand different genes. **Gene regulation** is the phenomenon in which the level of gene expression can vary under different conditions. In comparison, unregulated genes have essentially constant levels of expression in all conditions over time. Unregulated genes are also called **constitutive genes.** Frequently, constitutive genes code proteins that are continuously needed for the survival of the bacterium. In contrast, the majority of genes are regulated so that the proteins they code can be produced at the proper times and in the proper amounts.

A key benefit of gene regulation is that the coded proteins are produced only when they are required. Therefore, the cell avoids wasting valuable energy making proteins it does not need. From the viewpoint of natural selection, this enables a bacterium to compete as efficiently as possible for limited resources. Gene regulation is particularly important because bacteria exist in an environment that is frequently changing with regard to temperature, nutrients, and many other factors. The following are a few common processes regulated at the genetic level:

1. ***Metabolism.*** Some proteins function in the metabolism of small molecules. For example, certain enzymes are needed for a bacterium to metabolize particular sugars. These enzymes are required only when the sugars are present in the bacterium's environment.
2. ***Response to environmental stress.*** Certain proteins help a bacterium to survive an environmental stress such as osmotic shock or heat shock. These proteins are required only when the bacterium is confronted with the stress.
3. ***Cell division.*** Some proteins are needed for cell division. These are necessary only when the bacterial cell is getting ready to divide.

The expression of protein-coding genes, which code polypeptides, ultimately leads to the production of functional proteins. As we saw in Chapters 12 and 13, gene expression is a multistep process that proceeds from transcription to translation, and it may involve posttranslational effects on protein structure and function. As shown in **Figure 14.1**, gene regulation can occur at any of these steps in the pathway of gene expression. In this chapter, we will examine the molecular mechanisms of these types of gene regulation.

REGULATION OF
GENE EXPRESSION

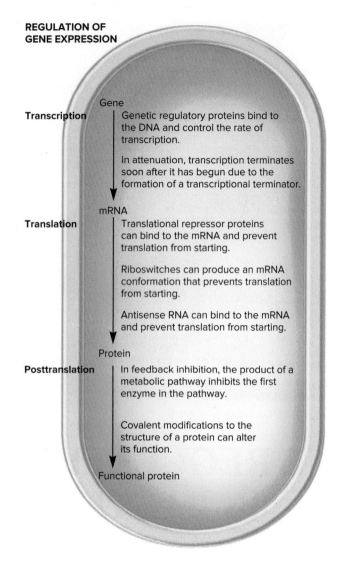

Transcription

Gene

Genetic regulatory proteins bind to the DNA and control the rate of transcription.

In attenuation, transcription terminates soon after it has begun due to the formation of a transcriptional terminator.

mRNA

Translation

Translational repressor proteins can bind to the mRNA and prevent translation from starting.

Riboswitches can produce an mRNA conformation that prevents translation from starting.

Antisense RNA can bind to the mRNA and prevent translation from starting.

Protein

Posttranslation

In feedback inhibition, the product of a metabolic pathway inhibits the first enzyme in the pathway.

Covalent modifications to the structure of a protein can alter its function.

Functional protein

FIGURE 14.1 Common points where regulation of gene expression occurs in bacteria.

CONCEPT CHECK: What is an advantage of gene regulation?

14.1 OVERVIEW OF TRANSCRIPTIONAL REGULATION

Learning Outcomes:

1. Describe the functions of activators and repressors.
2. Explain how small effector molecules affect the functions of activators and repressors.

In bacteria, the most common way to regulate gene expression is by influencing the rate at which transcription is initiated. Although we frequently refer to genes as being "turned on or off," it is more accurate to say that the level of gene expression is increased or decreased. At the level of transcription, this means that the rate of RNA synthesis can be increased or decreased.

In most cases, transcriptional regulation involves the actions of proteins called **regulatory transcription factors (RTFs),** which can bind to regulatory elements in the DNA and affect the rate of transcription of one or more nearby genes. In bacteria, two types of RTFs are common:

- A **repressor** is an RTF that binds to a regulatory element in the DNA and inhibits the rate of transcription.
- An **activator** is an RTF that binds to a regulatory element in the DNA and increases the rate of transcription.
- Transcriptional regulation by a repressor is termed **negative control.**
- Regulation by an activator is considered to be **positive control.**

In conjunction with RTFs, small effector molecules often play a critical role in transcriptional regulation. However, small effector molecules do not bind directly to the DNA to alter transcription. Rather, an effector molecule exerts its effects by binding to a repressor or an activator. The binding of the effector molecule causes a conformational change in the RTF and thereby influences whether or not the protein can bind to a regulatory element in the DNA. RTFs that respond to small effector molecules typically have two binding sites. One site is where the protein binds to a regulatory element; the other is the binding site for the effector molecule.

RTFs are given names describing how they affect transcription when they are bound to the DNA (repressor or activator). In contrast, small effector molecules are given names that describe how they affect transcription when they are present in the cell at a sufficient concentration to exert their effect (**Figure 14.2**).

- An **inducer** is a small effector molecule that causes the rate of transcription to increase. An inducer may accomplish this in two ways: It can bind to a repressor protein and prevent it from binding to the DNA, or it can bind to an activator protein and cause it to bind to the DNA. In either case, the transcription rate is increased because the initiation of transcription is faster. Genes that are regulated in this manner are called **inducible genes.**

Alternatively, the presence of a small effector molecule may inhibit transcription. This can occur in two ways:

- A **corepressor** is a small effector molecule that binds to a repressor protein, thereby causing the protein to bind to the DNA.
- An **inhibitor** binds to an activator protein and prevents it from binding to the DNA.

Both corepressors and inhibitors act to reduce the rate of transcription by inhibiting the initiation of transcription. Therefore, the genes they regulate are termed **repressible genes.** Unfortunately, this terminology can be confusing because a repressible gene could be controlled by an activator protein, or an inducible gene could be controlled by a repressor protein.

In the absence of the inducer, this repressor protein blocks transcription. The presence of the inducer causes a conformational change that inhibits the ability of the repressor protein to bind to the DNA. Transcription proceeds.

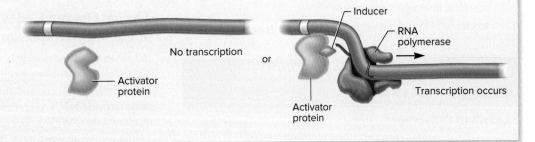

(a) Repressor protein, inducer molecule, inducible gene

This activator protein cannot bind to the DNA unless an inducer is present. When the inducer is bound to the activator protein, this enables the activator protein to bind to the DNA and activate transcription.

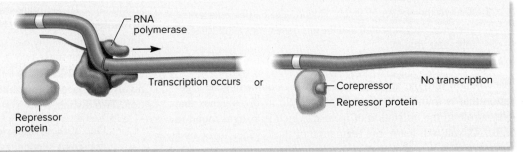

(b) Activator protein, inducer molecule, inducible gene

In the absence of a corepressor, this repressor protein will not bind to the DNA. Therefore, transcription can occur. When the corepressor is bound to the repressor protein, a conformational change occurs that allows the repressor to bind to the DNA and inhibit transcription.

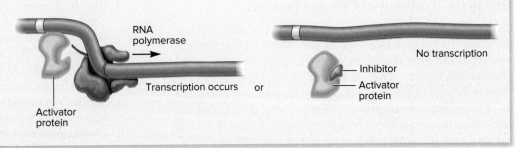

(c) Repressor protein, corepressor molecule, repressible gene

This activator protein will bind to the DNA without the aid of an effector molecule. The presence of an inhibitor causes a conformational change that inhibits the ability of the activator protein to bind to the DNA. This inhibits transcription.

(d) Activator protein, inhibitor molecule, repressible gene

FIGURE 14.2 **Binding sites on regulatory transcription factors.** In these examples, each RTF (repressor or activator) has two binding sites: one for DNA and one for a small effector molecule. The binding of the small effector molecule changes the conformation of the RTF, which alters the structure of its DNA-binding site, thereby influencing whether the protein can bind to the DNA.

14.1 COMPREHENSION QUESTIONS

1. A repressor is a _____ that _____ transcription.
   a. small effector molecule, inhibits
   b. small effector molecule, enhances
   c. regulatory transcription factor, inhibits
   d. regulatory transcription factor, enhances

2. Which of the following combinations will cause the rate of transcription to increase?
   a. A repressor plus an inducer
   b. A repressor plus a corepressor
   c. An activator plus an inhibitor
   d. None of the above will increase the rate of transcription.

14.2 REGULATION OF THE *lac* OPERON

Learning Outcomes:

1. Describe the organization of the *lac* operon.
2. Explain how the *lac* operon is regulated by lac repressor and by catabolite activator protein.
3. Analyze the results of Jacob, Monod, and Pardee, and explain how they indicated that the *lacI* gene codes a diffusible repressor protein.

We will now turn our attention to a specific example of gene regulation that is found in *E. coli*. This example involves genes that play a role in the utilization of lactose, which is a sugar found in milk. As you will learn, the regulation of these genes involves a repressor protein and also an activator protein.

The Phenomenon of Enzyme Adaptation Is Due to the Synthesis of Cellular Proteins

Our initial understanding of gene regulation can be traced back to the studies of François Jacob and Jacques Monod. Their research into genes and gene regulation stemmed from an interest in the phenomenon known as **enzyme adaptation,** which had been identified at the turn of the twentieth century. Enzymes are composed of proteins. Enzyme adaptation refers to the observation that a particular enzyme appears within a living cell only after the cell has been exposed to the substrate for that enzyme. When a bacterium is not exposed to a particular substance, it does not make the enzyme(s) needed to metabolize that substance.

To investigate this phenomenon, Jacob and Monod focused their attention on lactose metabolism in *E. coli*. Several key experimental observations led to an understanding of this genetic system:

1. The exposure of bacterial cells to lactose increased the levels of lactose-utilizing enzymes by 1000- to 10,000-fold.
2. Antibody and labeling techniques revealed that the increase in the activity of these enzymes was due to the increased synthesis of the proteins that form the enzymes.
3. The removal of lactose from the environment caused an abrupt termination in the synthesis of the enzymes.

4. The analysis of mutations in the *lac* operon revealed that each protein involved with lactose utilization is coded by a separate gene.

These critical observations indicated to Jacob and Monod that enzyme adaptation is due to the synthesis of specific proteins in response to lactose in the environment. Next, we will examine how Jacob and Monod discovered that this phenomenon is due to the interactions between regulatory transcription factors and small effector molecules. In other words, we will see that enzyme adaptation is due to the transcriptional regulation of genes.

The *lac* Operon Codes Proteins Involved in Lactose Metabolism

In bacteria and archaea, it is common for genes to be arranged together in an **operon**—a group of two or more genes that are under the transcriptional control of a single promoter. An operon codes a **polycistronic mRNA,** an RNA that contains the sequences of two or more genes. Why do operons occur in bacteria? One biological advantage of an operon is that it allows a bacterium to coordinately regulate a group of two or more genes that are involved with a common functional goal; the expression of the genes occurs as a single unit. To facilitate transcription, an operon is flanked by a **promoter** that signals the beginning of transcription and a **terminator** that signals the end of transcription. Two or more genes are found between these two sequences.

Figure 14.3a shows the organization of the genes in *E. coli* involved in lactose utilization and in their transcriptional regulation. Two distinct transcriptional units are present.

- One of these units, known as the *lac* operon, contains several DNA sequences: a CAP site; a *lac* promoter (*lacP*); an operator site (*lacO*); three protein-coding genes, *lacZ*, *lacY*, and *lacA*; and a terminator.
- The *lacZ* gene codes β-galactosidase, an enzyme that cleaves lactose into galactose and glucose. As a side reaction, β-galactosidase also converts a small percentage of lactose into allolactose, a structurally similar sugar (**Figure 14.3b**). As we will see later, allolactose acts as a small effector molecule for regulating the *lac* operon.
- The *lacY* gene codes lactose permease, a membrane protein required for the active transport of lactose into the cytoplasm of the bacterium.
- The *lacA* gene codes galactoside transacetylase, an enzyme that covalently modifies lactose and lactose analogs by the attachment of hydrophobic acetyl groups. The acetylation of nonmetabolizable lactose analogs prevents their toxic buildup within the bacterial cytoplasm by allowing them to diffuse out of the cell.

The CAP site and the operator site are **regulatory elements**—short segments of DNA that are recognized by regulatory transcription factors (RTFs). The binding of the RTF affects the rate of transcription.

- The **CAP site** is a DNA sequence recognized by an activator protein called **catabolite activator protein (CAP).**

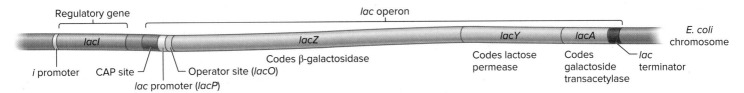

(a) **Organization of DNA sequences in the *lac* region of the *E. coli* chromosome**

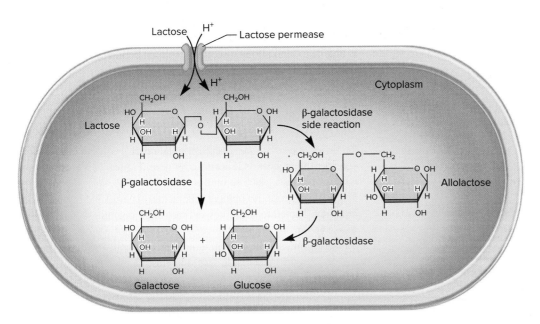

(b) **Functions of lactose permease and β-galactosidase**

FIGURE 14.3 **Organization of the *lac* operon and other genes involved with lactose metabolism in *E. coli*.** (a) The CAP site is the binding site for the catabolite activator protein (CAP). The operator site is a binding site for lac repressor. The promoter (*lacP*) is responsible for the transcription of the *lacZ*, *lacY*, and *lacA* genes as a single unit, which ends at the *lac* terminator. The *i* promoter is responsible for the transcription of the *lacI* gene. (b) Lactose permease allows the uptake of lactose into the bacterial cytoplasm. It cotransports lactose with H^+. Because bacteria maintain an H^+ gradient across their cytoplasmic membrane, this cotransport permits the active accumulation of lactose against a gradient. β-Galactosidase is a cytoplasmic enzyme that cleaves lactose and related compounds into galactose and glucose. As a minor side reaction, β-galactosidase also converts lactose into allolactose. Allolactose can also be broken down into galactose and glucose.

CONCEPT CHECK: Which genes are under the control of the *lac* promoter?

- The **operator site** (also referred to simply as the **operator**) is a sequence of bases that provides a binding site for a repressor protein called lac repressor.

A second transcriptional unit is the *lacI* gene (see Figure 14.3a), which is not part of the *lac* operon. The *lacI* gene has its own promoter called the *i* promoter and is constitutively expressed at fairly low levels. The *lacI* gene codes **lac repressor,** a protein that regulates the *lac* operon by binding to the operator site and inhibiting transcription. This repressor is a homotetramer, a protein composed of four identical subunits. Only a small amount of lac repressor is needed to repress the *lac* operon. The mechanism of action of lac repressor is described next.

The *lac* Operon Is Regulated by a Repressor Protein

The *lac* operon can be transcriptionally regulated in more than one way. The first mechanism that we will examine is inducible and under negative control. As shown in **Figure 14.4**, this form of regulation involves lac repressor, which is a protein that binds to the sequence of nucleotides found within the *lac* operator site.

Once bound, lac repressor prevents RNA polymerase from transcribing the *lacZ*, *lacY*, and *lacA* genes (Figure 14.4a). The binding of the repressor to the operator site is a reversible process. In the absence of allolactose, lac repressor is bound to the operator site most of the time.

The ability of lac repressor to bind to the operator site depends on whether or not allolactose is bound to it. Each of the repressor protein's four subunits has a single binding site for allolactose, the inducer. How does a small molecule like allolactose exert its effects? When four molecules of allolactose bind to lac repressor, a conformational change occurs in the repressor that prevents it from binding to the operator site. Under these conditions, RNA polymerase is free to transcribe the operon (Figure 14.4b).

In genetic terms, we say that the operon has been **induced.** The action of a small effector molecule, such as allolactose, is called **allosteric regulation.** The functioning of allosteric proteins, such as lac repressor, is controlled by effector molecules that bind to the proteins' **allosteric sites.** In the case of lac repressor, the binding of allolactose alters the function of lac repressor by preventing it from binding to the DNA.

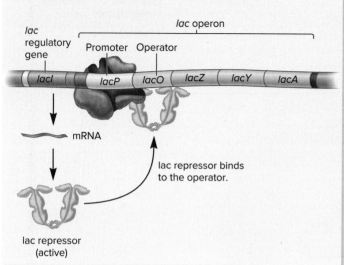

In the absence of the inducer allolactose, the repressor protein is tightly bound to the operator site, thereby inhibiting the ability of RNA polymerase to transcribe the operon.

(a) No lactose in the environment

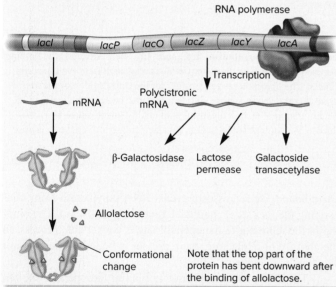

When allolactose is available, it binds to the repressor. This alters the conformation of the repressor protein, which prevents it from binding to the operator site. Therefore, RNA polymerase can transcribe the operon.

(b) Lactose present

ONLINE ANIMATION

FIGURE 14.4 **Mechanism of induction of the *lac* operon.** Note: The CAP site is not labeled in these diagrams, but it is shown in purple.

Rare mutations in the *lacI* gene that alter the regulation of the *lac* operon reveal that each subunit of lac repressor has a region that binds to the DNA and another region that contains the allolactose-binding site. As shown later in this section, in Figure 14.7, researchers have identified *lacI⁻* mutations that result in the constitutive expression of the *lac* operon, which means that it is expressed in both the presence and the absence of lactose. Such

mutations may result in an inability to synthesize any repressor protein, or they may produce a repressor protein that is unable to bind to the DNA at the *lac* operator site. If lac repressor is unable to bind to the DNA, the *lac* operon cannot be repressed.

By comparison, *lacIˢ* mutations have the opposite effect: The *lac* operon cannot be induced even in the presence of lactose. These mutations, which are called super-repressor mutations, alter the region of lac repressor that binds allolactose. The mutation usually results in a lac repressor protein that cannot bind allolactose. If lac repressor is unable to bind allolactose, it will remain bound to the *lac* operator site and therefore induction cannot occur.

The Regulation of the *lac* Operon Allows a Bacterium to Respond to Environmental Change

To better appreciate *lac* operon regulation at the cellular level, let's consider the process as it occurs over time. **Figure 14.5** illustrates the effects of external lactose on the regulation of the *lac* operon. In the absence of lactose, no inducer (allolactose) is available to bind to lac repressor. Therefore, lac repressor binds to the operator site and inhibits transcription. In reality, the repressor does not completely inhibit transcription, so very small amounts of β-galactosidase, lactose permease, and galactoside transacetylase are made. However, the levels are far too low to enable the bacterium to readily use lactose.

When the bacterium is exposed to lactose, a small amount can be transported into the cytoplasm via lactose permease, and β-galactosidase converts some of that lactose to allolactose. As this occurs, the cytoplasmic level of allolactose gradually rises; eventually, allolactose binds to lac repressor. The binding of allolactose promotes a conformational change that prevents the repressor from binding to the *lac* operator site, thereby allowing transcription of the *lacZ*, *lacY*, and *lacA* genes to occur. Translation of the coded polypeptides produces the proteins needed for lactose uptake and metabolism.

To understand how the induction process is shut off when lactose in the environment has been depleted, let's consider the interaction between allolactose and lac repressor. The repressor protein has a measurable affinity for allolactose. The binding of allolactose to lac repressor is reversible. The likelihood that allolactose will bind to the repressor depends on the allolactose concentration. During induction of the operon, the concentration of allolactose (the inducer) rises and approaches the affinity for the repressor protein. This makes it likely that allolactose will bind to lac repressor, thereby preventing the repressor from binding to the operator site.

The intracellular concentration of allolactose remains high as long as lactose is available in the environment. However, when lactose is depleted from the environment, the concentration of allolactose also becomes lower due to the action of metabolic enzymes. Eventually, the concentration of allolactose drops below its affinity for the repressor. At this point, allolactose is unlikely to be bound to lac repressor. When allolactose is released, lac repressor returns to the conformation that binds to the operator site. In this way, the binding of the repressor shuts down the *lac* operon when lactose is depleted from the environment. After repression occurs, mRNA molecules and proteins coded by the *lac* operon are eventually degraded (see Figure 14.5).

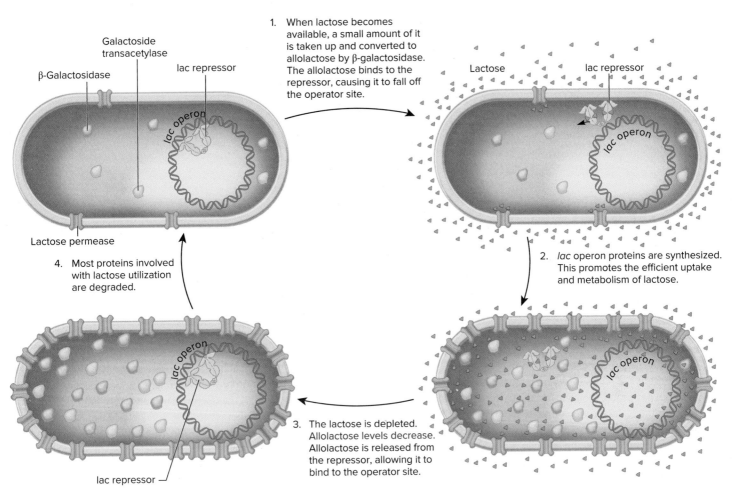

1. When lactose becomes available, a small amount of it is taken up and converted to allolactose by β-galactosidase. The allolactose binds to the repressor, causing it to fall off the operator site.

2. *lac* operon proteins are synthesized. This promotes the efficient uptake and metabolism of lactose.

3. The lactose is depleted. Allolactose levels decrease. Allolactose is released from the repressor, allowing it to bind to the operator site.

4. Most proteins involved with lactose utilization are degraded.

β-Galactosidase

Galactoside transacetylase

lac repressor

Lactose permease

Lactose

lac repressor

lac operon

lac repressor

FIGURE 14.5 **The cycle of *lac* operon induction and repression.**

Genes→Traits The genes of the *lac* operon give a bacterium the ability to metabolize lactose when it is present in the bacterium's environment. When lactose is present, the genes of the *lac* operon are induced, and the proteins needed for the efficient uptake and metabolism of lactose are synthesized. When lactose is absent, these genes are repressed so the bacterium does not waste energy expressing them. Note: The proteins involved with lactose utilization are fairly stable, but they will eventually be degraded.

CONCEPT CHECK: Under what condition is lac repressor bound to the *lac* operon?

EXPERIMENT 14A

The *lacI* Gene Codes a Diffusible Repressor Protein

Now that we have examined the *lac* operon, let's consider one of the experimental approaches that was used to elucidate its regulation. In the 1950s, Jacob, Monod, and their colleague Arthur Pardee identified a few rare mutant strains of bacteria that had abnormal lactose adaptation. As mentioned earlier, one type of mutant, designated *lacI⁻*, showed constitutive expression of the *lac* operon even in the absence of lactose. As shown in **Figure 14.6a**, the correct explanation is that a loss-of-function mutation in the *lacI* gene prevented lac repressor from binding to the *lac* operator site and inhibiting transcription. At the time of this work, however, the function of lac repressor was not yet known. Instead, the researchers incorrectly hypothesized that the *lacI⁻* mutation resulted in the synthesis of an internal activator, making it unnecessary for cells to be exposed to lactose for the expression of the *lac* operon (**Figure 14.6b**).

To explore the nature of this mutation, Jacob, Monod, and Pardee applied a genetic approach. In order to understand their approach, let's briefly consider the process of bacterial conjugation (described in Chapter 7). The earliest studies of Jacob, Monod, and Pardee in 1959 involved conjugations between recipient cells, termed F⁻, and donor cells, which were Hfr strains that transferred a portion of the bacterial chromosome. Later experiments in 1961 involved the transfer of circular segments of DNA known as F factors. We consider the latter type of experiment here. Sometimes an F factor also carries genes that were originally within the bacterial chromosome. These types of F factors are called F′ factors ("F prime" factors). In their studies, Jacob, Monod, and Pardee identified F′ factors that carried the *lacI* gene and the *lac* operon. These F′ factors can be transferred from one cell to another by bacterial conjugation. A strain of bacteria containing F′ factor genes is called a **merozygote,** or partial diploid.

The production of merozygotes was instrumental in allowing Jacob, Monod, and Pardee to determine the function of the *lacI* gene. This experimental approach was based on two key points. First, the

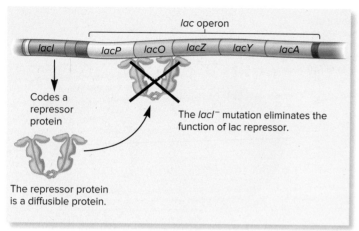

(a) Correct explanation

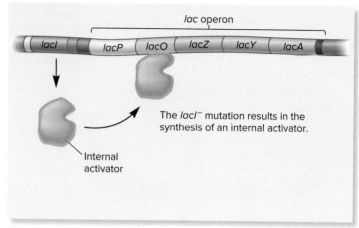

(b) Internal activator hypothesis

FIGURE 14.6 **Alternative hypotheses to explain how a *lacI⁻* mutation could cause the constitutive expression of the *lac* operon.** (a) The correct explanation in which the *lacI⁻* mutation eliminates the function of lac repressor, which prevents it from repressing the *lac* operon. (b) The hypothesis of Jacob, Monod, and Pardee. In this case, the *lacI⁻* mutation results in the synthesis of an internal activator that turns on the *lac* operon.

two *lacI* genes in a merozygote may be different alleles. For example, the *lacI* gene on the chromosome may be a *lacI⁻* allele that causes constitutive expression, whereas the *lacI* gene on the F′ factor may be normal. Second, the genes on the bacterial chromosome and the genes on the F′ factor are not physically adjacent to each other. As we now know, the expression of a normal *lacI* gene on an F′ factor produces repressor proteins that can diffuse within the cell and eventually bind to the operator site of the *lac* operon located on the chromosome and also to the operator site on an F′ factor.

Figure 14.7 shows one experiment of Jacob, Monod, and Pardee in which they analyzed a *lacI⁻* mutant strain that was already known to constitutively express the *lac* operon and compared it with the corresponding merozygote. The merozygote had a *lacI⁻* mutant gene on the chromosome and a normal *lacI* gene on an F′ factor.

Each strain was grown, and then the bacteria were divided into two tubes. In one tube of each pair, lactose was omitted; the other tube contained lactose in order to determine if lactose was needed to induce the expression of the operon. The cells were subjected to sonication, which caused them to release β-galactosidase. Next, a lactose analog, β-*o*-nitrophenylgalactoside (β-ONPG), was added. This molecule is colorless, but β-galactosidase cleaves it to yield a product that has a yellow color. Therefore, the amount of yellow color produced in a given amount of time is a measure of the amount of β-galactosidase being expressed from the *lac* operon.

THE HYPOTHESIS

The *lacI⁻* mutation results in the synthesis of an internal activator.

TESTING THE HYPOTHESIS

FIGURE 14.7 **Evidence that the *lacI* gene codes a diffusible repressor protein.**

Starting material: The genotype of the mutant strain was *lacI⁻ lacZ⁺ lacY⁺ lacA⁺*. The merozygote strain had an F′ factor that was *lacI⁺ lacZ⁺ lacY⁺ lacA⁺*. The F′ factor had been introduced into the mutant strain via conjugation.

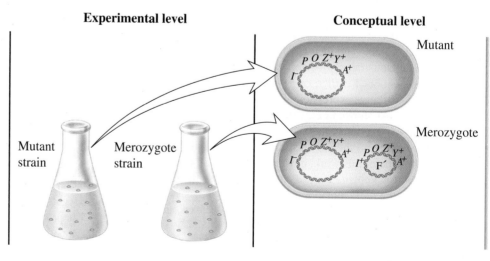

1. Grow mutant strain and merozygote strain separately.

2. Divide each strain into two tubes.

3. To one of the two tubes, add lactose.

4. Incubate the cells long enough to allow *lac* operon induction.

5. Lyse the cells with a sonicator. This allows β-galactosidase to escape from the cells.

6. Add β-*o*-nitrophenylgalactoside (β-ONPG). This is a colorless compound. If β-galactosidase is present, it will cleave the compound to produce galactose and *o*-nitrophenol (O-NP). *o*-Nitrophenol has a yellow color. The deeper the yellow color, the more β-galactosidase was produced.

7. Incubate the sonicated cells to allow β-galactosidase time to cleave β-*o*-nitrophenylgalactoside.

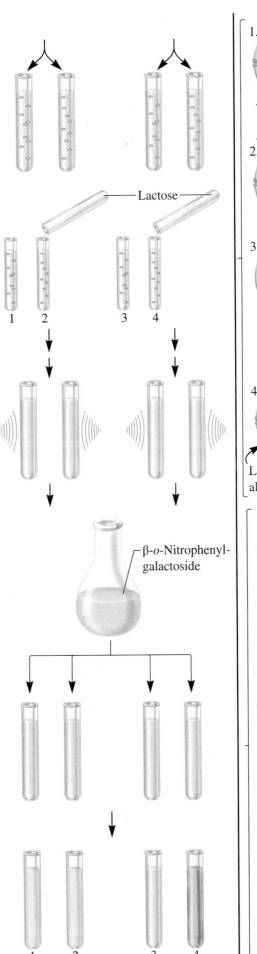

Lactose

1 2   3 4

β-*o*-Nitrophenyl-galactoside

1 2   3 4

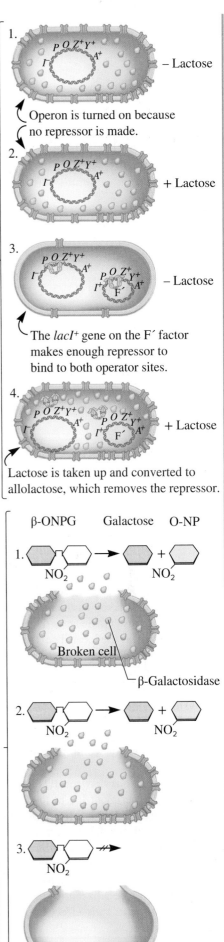

8. Measure the yellow color produced
with a spectrophotometer. (See
Appendix A for a description
of spectrophotometry.)

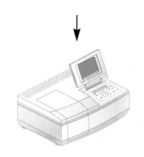

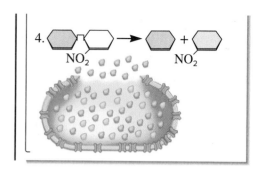

THE DATA

Strain	Addition of Lactose	Amount of β-Galactosidase (%)
Mutant	No	100
Mutant	Yes	100
Merozygote	No	<1
Merozygote	Yes	220

Source: Jacob, François, and Monod, Jacques, "Genetic Regulatory Mechanisms
in the Synthesis of Proteins," *Journal of Molecular Biology,* vol. 3, no. 3, June,
1961, 318–356.

INTERPRETING THE DATA

As seen in the data, the amount of β-galactosidase produced by the
original mutant strain was the same in either the absence or the presence of lactose. This result is expected because β-galactosidase was
known to be constitutively expressed in the *lacI⁻* mutant strain. In
other words, the presence of lactose was not needed to induce the
operon, due to a defective *lacI* gene.

With the merozygote strain, however, a different result was
obtained. In the absence of lactose, the *lac* operons were repressed—even the operon on the bacterial chromosome. How do
we explain these results? Because the normal *lacI* gene on the F′
factor was not physically located next to the chromosomal *lac* operon, this result is consistent with the idea that the *lacI* gene codes
for a repressor protein that can diffuse throughout the cell and bind
to any *lac* operon. The hypothesis that the *lacI⁻* mutation resulted
in the synthesis of an internal activator was rejected. If that hypothesis had been correct, the merozygote strain would have still
made an internal inducer, and the *lac* operons in the merozygote
would have been expressed in the absence of lactose. This result
was not obtained.

The interactions between RTFs and DNA sequences observed in this experiment led to the definition of two genetic
terms. A ***trans*-effect** is a form of genetic regulation that can
occur even though two DNA segments are not physically adjacent. The action of lac repressor on the *lac* operon is a *trans*-
effect. An RTF, such as lac repressor, is called a ***trans*-acting
factor.** In bacteria (and archaea), a ***cis*-acting element** is a DNA
segment that must be adjacent to the gene(s) that it regulates,
and it is said to have a ***cis*-effect** on gene expression. The *lac*

operator site is an example of a *cis*-acting element. A *trans*-
effect is mediated by genes that code RTFs, whereas a *cis*-effect
is mediated by DNA sequences (that is, regulatory elements)
that are binding sites for RTFs.

Jacob and Monod also isolated constitutive mutants that affected the operator site, *lacO*. **Table 14.1** summarizes the effects
of mutations based on their location in the *lacI* gene or in *lacO* and
their presence in merozygotes. As seen here, a loss-of-function
mutation in a gene coding a repressor protein has the same effect
as a mutation in an operator site that cannot bind a repressor protein. In both cases, the genes of the *lac* operon are constitutively
expressed.

With a merozygote, however, the results are quite different.
When a *lacI⁺* gene and a normal *lac* operon carried on an F′ factor
are introduced into a cell harboring a defective (mutant) *lacI* gene
on the bacterial chromosome, the *lacI⁺* gene on the F′ factor can
express enough lac repressor to regulate both operons. In contrast,
when a *lacI⁺* gene and a normal *lac* operon carried on an F′ factor
are introduced into a cell with a defective (mutant) operator site on
the bacterial chromosome, the *lac* operon on the chromosome continues to be expressed without lactose present. This occurs because
the repressor cannot bind to the defective operator site on the chromosome. Overall, a mutation in a *trans*-acting factor can be complemented by the introduction of a second gene with normal function.
However, a mutation in a *cis*-acting element is not affected by the
introduction of a normal *cis*-acting element into the cell.

TABLE 14.1

**A Comparison of Loss-of-Function Mutations in the *lacI* Gene or
in the Operator Site**

Chromosome	F′ factor	Expression of the *lac* Operon (%)	
		With Lactose	**Without Lactose**
Wild type	None	100	<1
lacI⁻	None	100	100
lacO⁻	None	100	100
lacI⁻	*lacI⁺* and a normal *lac* operon	200	<1
lacO⁻	*lacI⁺* and a normal *lac* operon	200	100

The *lac* Operon Is Also Regulated by an Activator Protein

The *lac* operon is transcriptionally regulated in a second way, known as **catabolite repression**. This form of transcriptional regulation is influenced by the presence of glucose, which is a catabolite—a substance that is broken down inside the cell. The presence of glucose ultimately leads to repression of the *lac* operon. When exposed to both glucose and lactose, *E. coli* cells first use glucose, and catabolite repression prevents the use of lactose. Why is this an advantage? The explanation is efficiency. The bacterium does not have to express all of the genes necessary for both glucose and lactose metabolism. If the glucose is used up, catabolite repression is alleviated, and the bacterium then expresses the *lac* operon.

The sequential use of two sugars by a population of dividing bacterial cells, known as **diauxic growth,** is a common phenomenon among many bacterial species. In the experiment whose results are shown in **Figure 14.8**, *E. coli* cells were given both glucose and lactose at time zero. Over a period of 10 hours, the number of *E. coli* cells and the concentrations of extracellular glucose and extracellular lactose were monitored over the course of 10 hours. The cells first used glucose to increase in number. The presence of glucose inhibits the expression of the *lac* operon. After glucose was used up, a brief lag phase occurred during which the cells were switching to their ability to use lactose. In other words, the expression of the *lac* operon increased after glucose was used up. This lag phase was followed by a second increase in cell number until lactose was eventually depleted and growth leveled off.

At the molecular level, how do we explain the phenomenon of diauxic growth? In other words, how does the presence of glucose prevent the cells from expressing the *lac* operon and using lactose? Diauxic growth is a form of regulation that involves a small effector molecule, **cyclic-AMP (cAMP),** which is produced from ATP via an enzyme known as adenylyl cyclase. When a bacterium is exposed to glucose, the transport of glucose into the cell stimulates a signaling pathway that causes the intracellular concentration of cAMP to decrease because the pathway inhibits adenylyl cyclase, the enzyme needed for cAMP synthesis. The effect of cAMP on the *lac* operon is mediated by an activator protein called catabolite activator protein (CAP). CAP is composed of two subunits, each of which binds one molecule of cAMP.

Figure 14.9 shows how the interplay between lac repressor and CAP determines whether the *lac* operon is expressed in the presence or absence of lactose and/or glucose.

- When only lactose is present, allolactose and cAMP levels are high (Figure 14.9a). Allolactose binds to lac repressor and prevents it from binding to the DNA. The effector molecule cAMP binds to CAP, and then CAP binds to the CAP site. A domain in CAP interacts with RNA polymerase, which facilitates the binding of RNA polymerase to the promoter. Under these conditions, transcription proceeds at a high rate.
- In the absence of both lactose and glucose, cAMP levels are also high, so CAP is bound to the DNA (Figure 14.9b). However, the binding of lac repressor is a very strong inhibitor of transcription. For this reason, the transcription rate is very low.
- Figure 14.9c shows the situation in which both sugars are present. The presence of lactose causes lac repressor to be inactive, which prevents it from binding to the operator site. Even so, the presence of glucose decreases cAMP levels, and so cAMP is released from CAP, which prevents CAP from binding to the CAP site. The lack of CAP binding does not completely inhibit transcription. Therefore, the transcription of the *lac* operon is low in the presence of both sugars.
- Figure 14.9d illustrates what happens when only glucose is present. The transcription of the *lac* operon is very low because lac repressor is bound to the operator site and strongly inhibits transcription.

The effect of glucose, called catabolite repression, may seem like a puzzling way to describe this process because this regulation involves the action of an inducer (cAMP) and an activator protein (CAP), not a repressor. The term was coined before the action of the cAMP-CAP complex was understood. At that time, the primary observation was that glucose (a catabolite) inhibited (repressed) lactose metabolism.

CAP can activate transcription at more that 100 operons in *E. coli*. Some of these operons code genes that are needed for the breakdown of sugars. These include the *lac* operon, as well as other operons that code genes involved in the breakdown of different sugars, such as maltose, arabinose, and melibiose. Therefore, when glucose levels are high, these operons are inhibited. This promotes diauxic growth.

Further Studies Have Revealed That the *lac* Operon Has Three Operator Sites for Lac Repressor

The traditional view of the regulation of the *lac* operon has been modified as we have gained a greater understanding of the molecular process. In particular, detailed genetic and crystallographic studies have shown that the binding of lac repressor is more

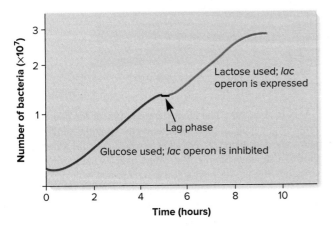

FIGURE 14.8 Diauxic growth of *E. coli* when given both glucose and lactose. At time zero, both glucose and lactose were added to a culture of *E. coli* cells. The division of cells to make more cells was then monitored for 10 hours.

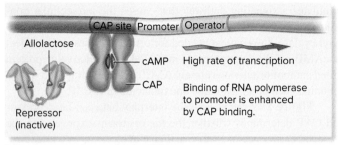

(a) Lactose, no glucose (high cAMP)

- Allolactose
- Repressor (inactive)
- cAMP
- CAP
- High rate of transcription
- Binding of RNA polymerase to promoter is enhanced by CAP binding.

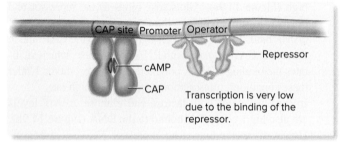

(b) No lactose or glucose (high cAMP)

- Repressor
- cAMP
- CAP
- Transcription is very low due to the binding of the repressor.

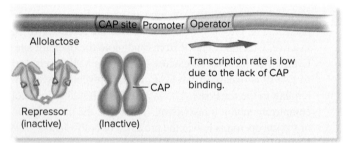

(c) Lactose and glucose (low cAMP)

- Allolactose
- Repressor (inactive)
- (Inactive)
- CAP
- Transcription rate is low due to the lack of CAP binding.

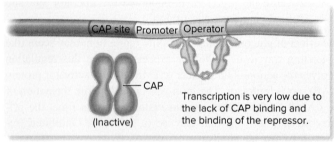

(d) Glucose, no lactose (low cAMP)

- CAP
- (Inactive)
- Transcription is very low due to the lack of CAP binding and the binding of the repressor.

ONLINE ANIMATION   INTERACTIVE EXERCISE

FIGURE 14.9 The roles of lac repressor and catabolite activator protein (CAP) in the regulation of the *lac* operon. This figure illustrates how the *lac* operon is regulated depending on the presence or absence of lactose and/or glucose.

Genes→Traits The mechanism of catabolite repression provides the bacterium with the ability to choose between two sugars. When exposed to both glucose and lactose, the bacterium metabolizes glucose first. After the glucose is used up, the bacterium expresses the genes necessary for lactose metabolism. This trait allows the bacterium to more efficiently use sugars from its environment.

CONCEPT CHECK: Why is it beneficial for the bacterium to regulate the *lac* operon with both a repressor protein and an activator protein?

complex than originally realized. The site in the *lac* operon commonly called the operator site was first identified by mutations that prevented lac repressor from binding. These mutations, called *lacO⁻* or *lacOᶜ* mutations, resulted in the constitutive expression of the *lac* operon even in strains that make a normal lac repressor protein. The *lacOᶜ* mutations were localized in the *lac* operator site, which is now known as O_1. This led to the view that binding of lac repressor to a single operator site inhibits transcription, as shown in Figure 14.4.

In the late 1970s and 1980s, two additional operator sites were identified. As shown at the top of **Figure 14.10**, these sites are called O_2 and O_3.

- O_1 is the operator site slightly downstream from the promoter.
- O_2 is located farther downstream in the *lacZ* coding sequence.
- O_3 is located slightly upstream from the promoter.

The O_2 and O_3 operator sites were initially called pseudo-operators, because substantial repression occurred in the absence of either one of them. However, studies by Benno Müller-Hill and his colleagues revealed a surprising result. As shown in Figure 14.10 (fourth example from the top), if both O_2 and O_3 are missing, repression is dramatically reduced even when O_1 is present.

How were these results interpreted? The data of Figure 14.10 supported a hypothesis that lac repressor must

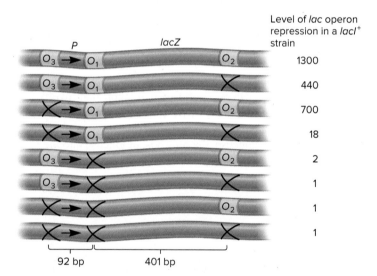

FIGURE 14.10 The identification of three *lac* operator sites. The drawing at the top of this figure shows the locations of three *lac* operator sites, designated O_1, O_2, and O_3, where O_1 is the *lac* operator site shown in previous figures. The arrows indicate the starting site for transcription. Missing operator sites are represented with an X. When all three operator sites are present, the repression of the *lac* operon is 1300-fold; that is, there is 1/1300 of the level of expression that occurs with an induced *lac* operon. This figure also shows the amount of repression when one or more operator sites are removed. A repression value of 1 indicates that no repression is occurring.

CONCEPT CHECK: Which data provide the strongest evidence that O_1 is not the only operator site?

bind to two operator sites to repress the *lac* operon. The data indicate that lac repressor can bind to O_1 and O_2, or to O_1 and O_3, but not to O_2 and O_3. If either O_2 or O_3 is missing, maximal repression is not achieved because it is less likely that the repressor will bind when only two operator sites are present. When O_1 is missing, even in the presence of the other operator sites, repression is nearly abolished because lac repressor cannot bind to O_2 and O_3. Look at Figure 14.10 and you will notice that the operator sites are a fair distance from each other. For this reason, Müller-Hill and colleagues proposed that the binding of lac repressor to two operator sites requires the DNA to form a loop. A loop in the DNA brings the operator sites closer together, thereby facilitating the binding of the repressor protein (**Figure 14.11a**).

In 1996, the proposal that lac repressor binds to two operator sites was confirmed by studies in which lac repressor was crystallized by Mitchell Lewis and colleagues. The crystal structure of lac repressor provided exciting insights into its mechanism of action. As mentioned earlier in this chapter, lac repressor is a tetramer of four identical subunits. The crystal structure revealed that each dimer within the tetramer recognizes one operator site.

Figure 14.11b is a molecular model illustrating the binding of lac repressor to the O_1 and O_3 sites. The side chains of specific amino acids in the protein interact directly with bases in the major groove of the DNA double helix. This is how RTFs recognize specific DNA sequences. Because each dimer within the tetramer recognizes a single operator site, the association of two dimers to form a tetramer requires that the two operator sites be close to each other. For this to occur, a loop must form in the DNA. The formation of this loop dramatically inhibits the ability of RNA polymerase to slide past the O_1 site and transcribe the operon.

Figure 14.11b also shows the binding of the cAMP-CAP complex to the CAP site (see the dark blue protein within the loop). A particularly striking observation is that the binding of the cAMP-CAP complex to the DNA causes a bend in the DNA structure. When the repressor is active—not bound to allolactose—the cAMP-CAP complex facilitates the formation of a loop in which lac repressor binds to the O_1 and O_3 sites. When the repressor is inactive, the bending of the DNA also appears to be important in the ability of RNA polymerase to initiate transcription slightly downstream from the bend.

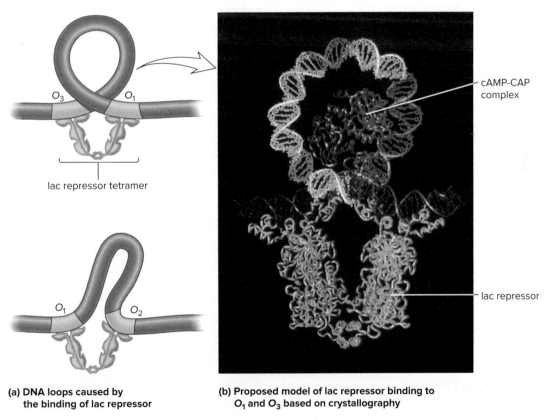

(a) **DNA loops caused by the binding of lac repressor**

(b) **Proposed model of lac repressor binding to O_1 and O_3 based on crystallography**

FIGURE 14.11 **The binding of lac repressor to two operator sites.** (a) The binding of lac repressor protein to the O_1 and O_3 or to the O_1 and O_2 operator sites. Because the two sites are far apart in each case, a loop must form in the DNA. (b) A molecular model for the binding of lac repressor to O_1 and O_3. Each repressor dimer binds to one operator site, so the repressor tetramer brings the two operator sites together. This causes the formation of a DNA loop in the intervening region. Note that the DNA loop contains the −35 and −10 sequences (shown in green) (refer back to Figure 12.4). These sequences are recognized by σ factor of RNA polymerase. This loop also contains the binding site for the cAMP-CAP complex, which is the blue protein within the loop.

GENETIC TIPS   **THE QUESTION:** Let's suppose you have isolated a mutant strain of *E. coli* in which the *lac* operon is constitutively expressed. In other words, the operon is turned on in the presence or absence of lactose. One possibility is that the mutation may block the transcription of the *lacI* gene, thereby preventing the synthesis of lac repressor. A second possibility is that the mutation could alter the sequence of the *lac* operon in a way that prevents lac repressor from binding to the operator. How would you distinguish between these two possibilities?

T OPIC: *What topic in genetics does this question address?*
The topic is gene regulation. More specifically, the question is about how a mutation could potentially alter the expression of the *lac* operon.

I NFORMATION: *What information do you know based on the question and your understanding of the topic?* From the question, you know that a mutation may either inhibit the expression of *lacI* or alter *lacO* in a way that prevents lac repressor from binding. From your understanding of the topic, you may remember that *lacI* exerts a *trans*-effect because it codes a diffusible protein, whereas *lacO* exerts a *cis*-effect.

P ROBLEM-SOLVING S TRATEGY: *Design an experiment.*
One strategy to solve this problem is to design an experiment that can distinguish between a mutation that has a *cis*-effect versus one that has a *trans*-effect. The use of merozygotes is one way to accomplish that goal.

ANSWER: Starting materials: The constitutive strain and a merozygote strain that carries a normal *lac* operon and a normal *lacI* gene on an F′ factor (see Figure 14.7).

1. Place strains into separate tubes with or without lactose.
2. Allow induction to occur.
3. Burst the cells.
4. Add β-ONPG, and measure the intensity of yellow color produced.

Expected results: If the mutation is in *lacI*, the repressor coded on the F′ factor will inhibit the expression of the *lac* operon on the chromosome and the one on the F′ factor. There will be very little yellow color in the absence of lactose in the merozygote strain. (This was the result obtained in Figure 14.7.) Alternatively, if the mutation is in *lacO*, the *lac* operon on the chromosome will still be turned on even in the absence of lactose (also see Table 14.1). The merozygote will produce a high level of yellow color in the absence of lactose.

14.2 COMPREHENSION QUESTIONS

1. What is an operon?
   a. A site in the DNA where a regulatory transcription factor binds
   b. A group of genes under the control of a single promoter
   c. An mRNA that codes several genes
   d. All of the above are true of an operon.

2. The binding of _____ to lac repressor causes lac repressor to _____ to the operator site, thereby _____ transcription.
   a. glucose, bind, inhibiting
   b. allolactose, bind, inhibiting
   c. glucose, not bind, increasing
   d. allolactose, not bind, increasing

3. On its chromosome, an *E. coli* cell has the genotype *lacI⁻ lacZ⁺ lacY⁺ lacA⁺*. It has an F′ factor with the genotype *lacI⁺ lacZ⁺ lacY⁺ lacA⁺*. What is the expected level of expression of the *lac* operon genes (*lacZ⁺ lacY⁺ lacA⁺*) in the absence of lactose?
   a. Both *lac* operons will be expressed.
   b. Neither *lac* operon will be expressed.
   c. Only the chromosomal *lac* operon will be expressed.
   d. Only the *lac* operon on the F′ factor will be expressed.

4. How does exposing an *E. coli* cell to glucose affect the regulation of the *lac* operon via CAP?
   a. cAMP binds to CAP and transcription is increased.
   b. cAMP binds to CAP and transcription is decreased.
   c. cAMP does not bind to CAP and transcription is increased.
   d. cAMP does not bind to CAP and transcription is decreased.

14.3   REGULATION OF THE *trp* OPERON

Learning Outcomes:
1. Describe the organization of the *trp* operon.
2. Explain how the *trp* operon is regulated by trp repressor and by attenuation.

We now turn our attention to a second operon in *E. coli*, called the *trp* operon (pronounced "trip"), which codes enzymes involved with the synthesis of the amino acid tryptophan. Like the *lac* operon, the *trp* operon is regulated by a repressor protein. In addition, the *trp* operon is regulated by another mechanism called **attenuation,** in which transcription begins but is stopped prematurely, or attenuated, before most of the *trp* operon is transcribed.

The *trp* Operon Is Regulated by a Repressor Protein

The *trp* operon contains five genes, designated *trpE*, *trpD*, *trpC*, *trpB*, and *trpA*, which code enzymes involved in tryptophan biosynthesis (**Figure 14.12a**). In addition, two genes, *trpR* and *trpL*, play a role in the regulation of the *trp* operon. The *trpL* gene is part of the *trp* operon, whereas the *trpR* gene has its own promoter and is not part of the *trp* operon.

Let's first consider how the *trp* operon is regulated by **trp repressor,** which is a protein coded by the *trpR* gene.

- When tryptophan levels within the cell are very low, trp repressor cannot bind to the operator site. Under these conditions, RNA polymerase transcribes the *trp* operon, and the cell expresses the genes required for the synthesis of tryptophan (see Figure 14.12a).

When tryptophan levels are low, tryptophan does not bind to trp repressor, which prevents the repressor protein from binding to the operator site. Under these conditions, RNA polymerase can transcribe the operon, which leads to the expression of the *trpE, trpD, trpC, trpB,* and *trpA* genes. These genes code enzymes involved in tryptophan biosynthesis.

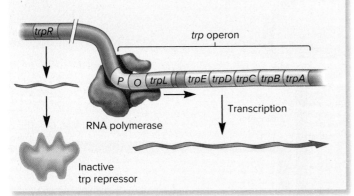

(a) Low tryptophan levels, transcription of the entire *trp* operon occurs

When tryptophan levels are high, tryptophan acts as a corepressor that binds to trp repressor. The tryptophan-trp repressor complex then binds to the operator site to inhibit transcription.

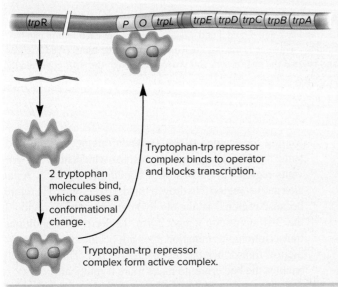

(b) High tryptophan levels, repression occurs

Another mechanism of regulation is attenuation. When attenuation occurs, the RNA is transcribed only to the attenuator sequence, and then transcription is terminated.

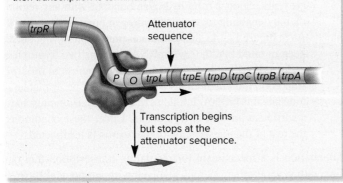

(c) High tryptophan levels, attenuation occurs

• When the tryptophan levels within the cell become high, tryptophan acts as a corepressor that binds to trp repressor. This causes a conformational change in trp repressor that allows it to bind to the *trp* operator site (**Figure 14.12b**). This binding inhibits the ability of RNA polymerase to transcribe the operon. Therefore, when a high level of tryptophan is present within the cell—when the cell does not need to make more tryptophan—the *trp* operon is turned off.

The *trp* Operon Is Also Regulated by Attenuation

In the 1970s, Charles Yanofsky and coworkers discovered a second regulatory mechanism for the *trp* operon, called attenuation, which is mediated by the region that includes the *trpL* gene (**Figure 14.12c**). During attenuation, transcription actually begins, but it is terminated before the entire mRNA is made. A segment of DNA, termed the **attenuator sequence,** is important in facilitating this termination. When attenuation occurs, the mRNA from the *trp* operon is made as a short piece that terminates at the attenuator sequence, which is just downstream from the *trpL* gene (see Figure 14.12c). Because this short mRNA has been terminated before RNA polymerase has transcribed the *trpE, trpD, trpC, trpB,* and *trpA* genes, it will not code the proteins required for tryptophan biosynthesis. In this way, attenuation inhibits the further production of tryptophan in the cell.

The segment of the *trp* operon immediately downstream from the operator site plays a critical role during attenuation. The first gene in the *trp* operon is the *trpL* gene. The mRNA made from the *trpL* gene contains codons for 14 amino acids that form the trp leader peptide (**Figure 14.13**). One key feature of attenuation is that two tryptophan (Trp) codons are found within the mRNA. As we will see later, these two codons provide a way for the bacterium to sense whether or not it has sufficient tryptophan to synthesize its proteins.

A second key feature that underlies attenuation is that the mRNA made from the *trpL* gene has four regions that are complementary to each other, which causes the mRNA to form stem-loops. Different combinations of stem-loops are possible due to interactions among these four regions (see the color key in Figure 14.13).

• Region 2 is complementary to region 1 and also to region 3.
• Region 3 is complementary to region 2 as well as to region 4.

Three stem-loops are possible: 1–2, 2–3, and 3–4. However, a particular segment of RNA can participate in the formation of only one stem-loop. For example, if region 2 forms a stem-loop with region 1, it cannot (at the same time) form a stem-loop with region 3. Alternatively, if region 2 forms a stem-loop with region 3, then region 3 cannot form a stem-loop with region 4.

ONLINE ANIMATION

FIGURE 14.12 Organization of the *trp* operon and regulation via trp repressor and attenuation.

CONCEPT CHECK: How does tryptophan affect the function of trp repressor?

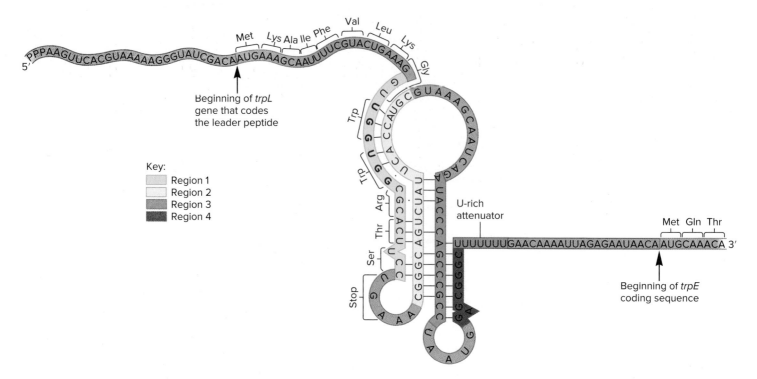

FIGURE 14.13 **Sequence of the mRNA produced at the beginning of the *trp* operon.** As shown here, this mRNA has several regions that are complementary to each other. The possible hydrogen bonding between regions 1 and 2, 2 and 3, and 3 and 4 is also shown, but each region can hydrogen bond to only one other region at a given time. The last U in the purple attenuator sequence is the last nucleotide that will be transcribed if attenuation occurs. At very low tryptophan concentrations, however, transcription occurs beyond the end of *trpL* and proceeds through the *trpE* gene and the rest of the *trp* operon.

CONCEPT CHECK: What type of bonding interaction causes stem-loops to form?

Though three stem-loops are possible, the 3–4 stem-loop is functionally unique. The 3–4 stem-loop in combination with the U-rich attenuator sequence results in intrinsic termination, also called ρ-independent termination, as described in Chapter 12. Therefore, the formation of the 3–4 stem-loop causes RNA polymerase to pause, and the U-rich sequence dissociates from the DNA. This terminates transcription at the U-rich attenuator. In comparison, if region 3 forms a stem-loop with region 2, transcription will not be terminated because a 3–4 stem-loop cannot form.

Conditions that favor the formation of the 3–4 stem-loop rely on the translation of the *trpL* gene and on the amount of tryptophan in the cell. As shown in **Figure 14.14**, three scenarios are possible:

- On some occasions, translation is not coupled with transcription (Figure 14.14a). As the *trpL* gene is being transcribed, region 1 rapidly hydrogen bonds to region 2, and region 3 is left to hydrogen bond to region 4. Therefore, the terminator stem-loop forms, and transcription is terminated just past the *trpL* gene at the U-rich attenuator.
- In bacteria, transcription and translation are often coupled, which means that translation begins as transcription is occurring. As Figure 14.14b indicates, this coupling happens under conditions in which the tryptophan concentration is low and the cell cannot make a sufficient amount of charged tRNA^{Trp}. As shown in Figure 14.14b, the ribosome pauses at the Trp codons in the *trpL* mRNA because it is

waiting for charged tRNA^{Trp}. When this occurs, the ribosome shields region 1 of the mRNA, which sterically prevents region 1 from hydrogen bonding to region 2. As an alternative, region 2 hydrogen bonds to region 3. Therefore, because region 3 is already hydrogen bonded to region 2, the 3–4 stem-loop cannot form. Under these conditions, transcriptional termination does not occur, and RNA polymerase transcribes the rest of the operon. This ultimately enables the bacterium to make more tryptophan.

- As illustrated in Figure 14.14c, coupled transcription and translation occur under conditions in which a sufficient amount of tryptophan is present in the cell. In this case, translation of the *trpL* mRNA progresses to its stop codon, where the ribosome pauses. The pausing at the stop codon prevents region 2 from hydrogen bonding with any region, thereby enabling region 3 to hydrogen bond with region 4. As in Figure 14.14a, this terminates transcription. Of course, keep in mind that the *trpL* mRNA contains two tryptophan codons. For the ribosome to smoothly progress to the *trpL* stop codon, enough charged tRNA^{Trp} must be available to translate this mRNA. It follows that the bacterium must have a sufficient amount of tryptophan. Under these conditions, the rest of the transcription of the operon is terminated.

Attenuation is a mechanism for regulating transcription that occurs with several other operons involved with amino acid biosynthesis. In all cases, the mRNAs that code leader peptides have

When translation is not coupled with transcription, region 1 hydrogen bonds to region 2 and region 3 hydrogen bonds to region 4. Because a 3–4 terminator stem-loop forms, transcription will be terminated at the U-rich attenuator.

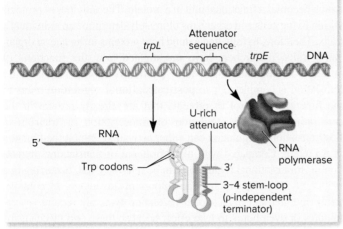

(a) No translation, 1–2 and 3–4 stem-loops form

Coupled transcription and translation occur under conditions in which the tryptophan concentration is very low. The ribosome pauses at the Trp codons in the *trpL* gene because insufficient amounts of charged tRNA^{Trp} are present. This pause blocks region 1 of the mRNA, so region 2 can hydrogen bond only with region 3. When this happens, the 3–4 stem-loop structure cannot form. Transcriptional termination does not occur, and RNA polymerase transcribes the rest of the operon.

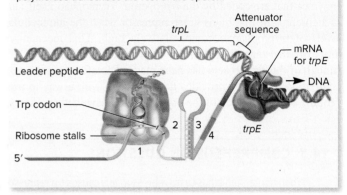

(b) Low tryptophan levels, 2–3 stem-loop forms

Coupled transcription and translation occur under conditions in which a sufficient amount of tryptophan is present in the cell. Translation of the *trpL* gene progresses to its stop codon, where the ribosome pauses. This blocks region 2 from hydrogen bonding with any region and thereby enables region 3 to hydrogen bond with region 4. This terminates transcription at the U-rich attenuator.

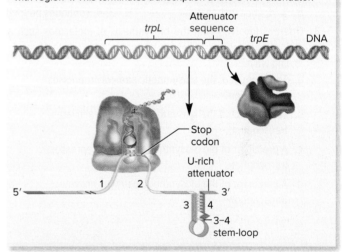

(c) High tryptophan levels, 3–4 stem-loop forms

ONLINE ANIMATION

FIGURE 14.14 **Possible stem-loops formed in *trpL* mRNA under different conditions of translation.** Attenuation occurs in parts **(a)** and **(c)** due to the formation of a 3–4 stem-loop.

CONCEPT CHECK: Explain how the presence of tryptophan favors the formation of the 3–4 stem-loop.

codons for the amino acid that is synthesized by the enzymes coded by the operon. For example, the mRNA for the leader peptide of the histidine operon has seven histidine codons, and the mRNA for the leader peptide of the leucine operon has four leucine codons. Like the *trp* operon, these other operons have alternative stem-loops, one of which is a transcriptional terminator.

Inducible Operons Usually Code Catabolic Enzymes, and Repressible Operons Commonly Code Anabolic Enzymes

Thus far, we have seen that bacterial genes can be transcriptionally regulated in a positive or negative way—and sometimes both. The *lac* operon is an inducible operon that is regulated by sugar molecules.

By comparison, the *trp* operon is a repressible operon regulated by tryptophan, a corepressor that binds to the repressor and turns the operon off. In addition, an abundance of charged tRNA^{Trp} in the cytoplasm can turn the *trp* operon off via attenuation.

By studying the genetic regulation of many operons, geneticists have discovered a general trend concerning inducible versus repressible regulation. When the genes in an operon code proteins that function in the catabolism, or breakdown, of a substance, they are usually regulated in an inducible manner. The substance to be broken down or a related compound often acts as the inducer. For example, allolactose acts as an inducer of the *lac* operon. An inducible form of regulation allows the bacterium to express the appropriate genes only when they are needed to break down the sugars or other substances.

In contrast, other enzymes are important for the anabolism, or synthesis, of small molecules. The genes that code these anabolic enzymes tend to be regulated by a repressible mechanism. The corepressor or inhibitor is commonly the small molecule that is the product of the enzymes' biosynthetic activities. For example, tryptophan is produced by the sequential action of several enzymes that are coded by the *trp* operon. Tryptophan itself acts as a corepressor that binds to trp repressor when the intracellular levels of tryptophan become relatively high. This mechanism turns off the genes required for tryptophan biosynthesis when enough of this amino acid has been made. Therefore, genetic regulation via repression provides the bacterium with a way to prevent the overproduction of the product of a biosynthetic pathway.

14.3 COMPREHENSION QUESTIONS

1. When tryptophan binds to trp repressor, this causes trp repressor to _____ to the *trp* operator and _____ transcription.
   a. bind, inhibit
   c. bind, activate
   b. not bind, inhibit
   d. not bind, activate

2. During attenuation, when tryptophan levels are high, the _____ stem-loop forms and transcription _____ the *trpL* gene.
   a. 1–2, ends just past
   c. 1–2, continues beyond
   b. 3–4, ends just past
   d. 3–4, continues beyond

3. Operons involved with the biosynthesis of molecules such as amino acids are most likely to be regulated in which of the following ways?
   a. The product of the biosynthetic pathway represses transcription.
   b. The product of the biosynthetic pathway activates transcription.
   c. A precursor of the biosynthetic pathway represses transcription.
   d. A precursor of the biosynthetic pathway activates transcription.

14.4 TRANSLATIONAL AND POSTTRANSLATIONAL REGULATION

Learning Outcomes:
1. Explain how translational regulatory proteins and antisense RNAs regulate translation.
2. Summarize how feedback inhibition and posttranslational modifications regulate protein function.

Though genetic regulation in bacteria occurs predominantly at transcription, many examples are known in which regulation takes place at a later stage in gene expression. In some cases, specialized mechanisms have evolved to regulate the translation of certain mRNAs. Recall that the translation of mRNA occurs in three stages: initiation,

elongation, and termination. Genetic regulation of translation is usually aimed at preventing the initiation step. In addition, as described in Section 14.5, translation can be regulated by riboswitches.

The net result of translation is the synthesis of a polypeptide, which becomes a functional unit in a protein. The activities of proteins within living cells and organisms ultimately determine an individual's traits. Therefore, to fully understand how proteins influence an organism's traits, researchers have investigated how the functions of proteins are regulated. The term **posttranslational** means "after translation is completed"; so **posttranslational regulation** refers to the functional control of proteins that are already present in the cell rather than the regulation of transcription or translation. Posttranslational regulation can either activate or inhibit the function of a protein. Compared with transcriptional or translational regulation, posttranslational regulation can be relatively fast, occurring in a matter of seconds, which is an important advantage. In contrast, transcriptional and translational regulation typically require several minutes or even hours to take effect because these two mechanisms involve the synthesis and turnover of mRNAs and polypeptides. In this section, we will examine some of the ways that bacteria can regulate the initiation of translation, as well as ways that protein function can be regulated posttranslationally.

Repressor Proteins and Antisense RNA Can Inhibit Translation

For some bacterial genes, the translation of mRNA is regulated by the binding of proteins or other RNA molecules that influence the ability of ribosomes to translate the mRNA into a polypeptide. A **translational regulatory protein** recognizes sequences within the mRNA, similar to the way that transcription factors recognize DNA sequences. In most cases, translational regulatory proteins act to inhibit translation. These are known as **translational repressors.**

When a translational repressor protein binds to the mRNA, it can inhibit translational initiation in one of two ways:

- One possibility is that it can bind in the vicinity of the Shine-Dalgarno sequence and/or the start codon. As discussed in Chapter 13, the Shine-Dalgarno sequence is needed for the mRNA to bind to a ribosome, and the start codon is the site where translation begins. A translational repressor can bind to either of these two sites, thereby sterically blocking the ribosome's ability to initiate translation in this region.
- Alternatively, a repressor protein may bind to the mRNA at a location that is not near the Shine-Dalgarno sequence or the start codon, but the binding stabilizes an mRNA secondary structure that prevents initiation. For example, a translational repressor could stabilize an mRNA secondary structure in which the Shine-Dalgarno sequence is inaccessible. This would prevent the mRNA from binding to the ribosome, thereby blocking translation.

Translational repression is also a form of genetic regulation found in eukaryotic species, and a specific example was described in Chapter 13 (refer back to Figure 13.20a).

A second way that bacteria regulate translation is via the synthesis of **antisense RNA,** an RNA strand that is complementary

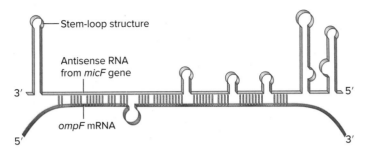

FIGURE 14.15 **The double-stranded RNA structure formed between the *micF* antisense RNA and the *ompF* mRNA.** Because these two RNA molecules have regions that are complementary to each other, the *micF* antisense RNA hydrogen bonds with the *ompF* mRNA to form a double-stranded structure that prevents the *ompF* mRNA from being translated.

CONCEPT CHECK: How does *micF* antisense RNA affect the translation of *ompF* mRNA?

to a strand of mRNA. To understand this form of genetic regulation, let's consider a trait known as osmoregulation, which is essential for the survival of most bacteria. Osmoregulation refers to the ability of a cell to control the amount of water inside it. Because the solute concentrations in the external environment may rapidly change between hypotonic and hypertonic conditions, bacteria must have an osmoregulation mechanism to maintain their internal cell volume. Otherwise, bacterial cells would be susceptible to the harmful effects of swelling or shrinking.

In *E. coli*, an outer membrane protein coded by the *ompF* gene is important in osmoregulation. At low osmolarity, the ompF protein is preferentially produced, whereas at high osmolarity, its synthesis is decreased. The expression of another gene, known as *micF*, is responsible for inhibiting the expression of the *ompF* gene at high osmolarity. As shown in **Figure 14.15**, the inhibition occurs because the RNA transcribed from the *micF* gene is complementary to the *ompF* mRNA; it is an antisense strand of RNA. When the *micF* gene is transcribed, its RNA product binds to the *ompF* mRNA via hydrogen bonding between their complementary regions. The binding of the *micF* RNA to the *ompF* mRNA prevents the *ompF* mRNA from being translated. The RNA transcribed from the *micF* gene is called antisense RNA because it is complementary to the *ompF* mRNA, which is a sense strand of mRNA that codes a polypeptide. The *micF* RNA does not code a polypeptide.

Posttranslational Regulation Can Occur via Feedback Inhibition or Covalent Modifications

Let's now turn our attention to ways that protein function is regulated posttranslationally. A common mechanism for regulating the activity of metabolic enzymes is **feedback inhibition.** The synthesis of many cellular molecules such as amino acids, vitamins, and nucleotides occurs via the action of a series of enzymes that convert precursor molecules to particular products. The final product in a metabolic pathway then inhibits an enzyme that acts early in the pathway.

Figure 14.16 depicts feedback inhibition in a metabolic pathway. Enzyme 1 is an example of an **allosteric enzyme,** an enzyme

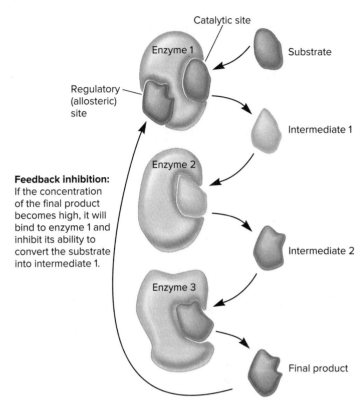

Feedback inhibition: If the concentration of the final product becomes high, it will bind to enzyme 1 and inhibit its ability to convert the substrate into intermediate 1.

FIGURE 14.16 **Feedback inhibition in a metabolic pathway.** The substrate is converted to a product by the sequential action of three different enzymes. Enzyme 1 has a catalytic site that recognizes the substrate, and it also has a regulatory site (also called an allosteric site) that recognizes the final product. When the final product binds to the regulatory site, it inhibits enzyme 1.

CONCEPT CHECK: Why is feedback inhibition an advantage to a bacterium?

that contains two different binding sites. The catalytic site is responsible for the binding of the substrate and its conversion to intermediate 1. The second site is a regulatory, or allosteric, site. This site binds the final product of the metabolic pathway. When bound to the regulatory site, the final product inhibits the catalytic ability of enzyme 1.

To appreciate feedback inhibition at the cellular level, we can consider the relationship between the product concentration and the regulatory site on enzyme 1. As the final product is made within the cell, its concentration gradually increases. Once the final product concentration has reached a level that is similar to the product's affinity for enzyme 1, the product is likely to bind to the regulatory site on enzyme 1 and inhibit its function. In this way, the accumulation of the final product of a metabolic pathway inhibits the further synthesis of more product. Under these conditions, the concentration of the final product has reached a level sufficient for the cell's needs.

A second strategy to control the function of proteins is the covalent modification of their structure, a process called **posttranslational covalent modification** (look ahead to Figure 23.3b). Certain types of modifications, such as phosphorylation (addition of $—PO_4$), acetylation (addition of $—COCH_3$), and methylation (addition of $—CH_3$), are often reversible modifications that transiently affect the function of a protein.

14.4 COMPREHENSION QUESTIONS

1. Translation can be regulated by
   a. translational repressors.
   b. antisense RNA.
   c. attenuation.
   d. both a and b.

2. An example of a posttranslational covalent modification that may affect protein function is
   a. phosphorylation.
   b. acetylation.
   c. methylation.
   d. Any of the above can affect protein function.

14.5   RIBOSWITCHES

Learning Outcome:

1. Explain how riboswitches can regulate transcription and translation.

In 2001 and 2002, researchers in a few different laboratories discovered a mechanism of gene regulation that involves a **riboswitch**—an element within an mRNA, usually at the 5′ end, that can bind a metabolite or ion as a ligand and switch the mRNA to a different secondary conformation. In this form of regulation, an RNA molecule can exist in two different secondary conformations. The conversion from one conformation to the other is due to the binding of a small molecule or ion. As described in **Table 14.2**, a riboswitch can regulate transcription, translation, RNA stability, or splicing.

Riboswitches are fairly common in bacteria. Researchers estimate that 3–5% of all bacterial genes may be regulated by riboswitches. The bacterial genes that are subject to this form of regulation are associated with the biosynthesis of purines, amino acids, vitamins, and other essential molecules. Riboswitches are

TABLE 14.2

Types of Riboswitches

Type of Regulation	Description
Transcription	The 5′ region of an mRNA may exist in one conformation that forms a terminator stem-loop, which causes attenuation of transcription. The other conformation does not form such a terminator and is completely transcribed.
Translation	The 5′ region of an mRNA may exist in one conformation in which the Shine-Dalgarno sequence cannot be recognized by the ribosome, whereas the other conformation has an accessible Shine-Dalgarno sequence that allows the mRNA to be translated.
RNA stability	One mRNA conformation may be stable, whereas the other conformation acts as ribozyme that causes self-degradation.
Splicing	In eukaryotes, one pre-mRNA conformation may be spliced in one way, whereas another conformation is spliced in a different way.

also found in archaea, algae, fungi, and plants. In this section, we will examine two examples of riboswitches in bacteria. The first example shows how a riboswitch can regulate transcription, and the second example involves translational regulation.

A Riboswitch Can Regulate Transcription

Thiamin, also called vitamin B$_1$, is an important organic molecule for bacteria, archaea, and eukaryotes. The active form of this vitamin is thiamin pyrophosphate (TPP). TPP is an essential coenzyme for the functioning of a variety of enzymes, such as certain enzymes in the citric acid cycle. In bacteria, TPP is made via biosynthetic enzymes that are coded in the bacterial genome. For example, in *Bacillus subtilis*, the majority of genes involved in TPP synthesis are found within the *thi* operon, which contains seven genes.

The *thi* operon in *B. subtilis* is regulated by a riboswitch that controls transcription (**Figure 14.17**). As the polycistronic mRNA

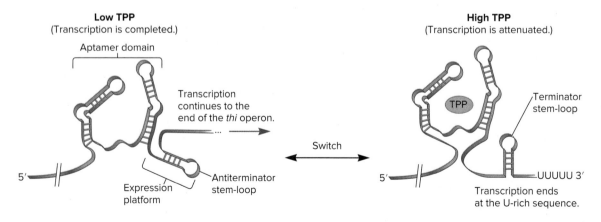

FIGURE 14.17   A TPP riboswitch in *B. subtilis* that regulates transcription of the *thi* operon.

CONCEPT CHECK: Which RNA conformation favors transcription—the form with the antiterminator stem-loop or the form with the terminator stem-loop?

for the *thi* operon is being made, the 5′ end quickly folds into a secondary structure. The riboswitch at the 5′ end contains two domains, an aptamer domain, which forms a ligand binding pocket for TPP, and a second domain, called an expression platform, that affects mRNA transcription.

- When TPP levels are low, TPP is not bound to the aptamer domain and the secondary structure of the expression platform has a stem-loop called an **antiterminator,** which prevents the formation of the terminator stem-loop. Therefore, under these conditions, transcription of the entire *thi* operon occurs. In this way, the bacterium is able to make more TPP, which is in short supply.
- When TPP levels are high, TPP binds to the aptamer domain and causes a switch in the secondary structure of the expression platform. As shown on the right side of Figure 14.17, the expression platform contains a terminator stem-loop instead of the antiterminator stem-loop. This terminator stem-loop causes ρ-independent termination, which is described in Chapter 12 (look back at Figure 12.11). Because the formation of this terminator stem-loop abruptly stops transcription, it inhibits the production of the enzymes that are needed to make more TPP. This is an example of attenuation. (Compare Figures 14.14a and the right side of Figure 14.17.)

A Riboswitch Can Regulate Translation

As we have just seen, the regulation of TPP biosynthetic enzymes in *B. subtilis* occurs via a riboswitch that controls transcription. However, in gram-negative bacteria, such as *E. coli*, the regulation of TPP biosynthetic enzymes occurs via a riboswitch that controls translation. In *E. coli*, the *thiMD* operon codes two enzymes involved with TPP biosynthesis. **Figure 14.18** shows how a riboswitch can regulate translation.

- When TPP levels are low, TPP does not bind to the aptamer domain at the 5′ end of the mRNA and the expression platform adopts a secondary structure that contains a stem-loop called the Shine-Dalgarno antisequestor. When this stem-loop forms, the Shine-Dalgarno sequence is accessible to ribosome binding. Therefore, the mRNA is translated when TPP is in short supply.
- When TPP levels are high, TPP binds to the aptamer domain and causes a switch in the secondary structure of the expression platform. As shown on the right side of Figure 14.18, the expression platform has a stem-loop that contains the Shine-Dalgarno sequence and the start codon. The formation of this stem-loop sequesters the Shine-Dalgarno sequence, thereby preventing ribosomal binding. This blocks the translation of the enzymes that are needed to make more TPP.

When comparing Figures 14.17 and 14.18, it is worth noting that the 5′ end of the *thiMD* mRNA of *E. coli* can exist in two different secondary structures that greatly resemble those that occur at the 5′ end of the *thi* operon of *B. subtilis*. However, the effects on regulation are quite different. The TPP riboswitch in *E. coli* controls translation, whereas the TPP riboswitch in *B. subtilis* controls transcription.

14.5 COMPREHENSION QUESTION

1. For a riboswitch that controls transcription, the binding of a small molecule such as TPP controls whether the mRNA
   a. has an antiterminator or terminator stem-loop.
   b. has a Shine-Dalgarno antisequestor or the Shine-Dalgarno sequence within a stem-loop.
   c. is degraded from its 5′ end.
   d. has both a and b.

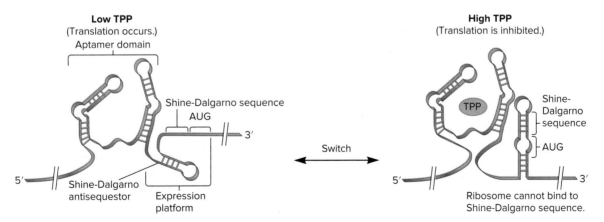

FIGURE 14.18 A TPP riboswitch in *E. coli* that regulates translation of the *thiMD* operon. Note: In both cases, the mRNA for the *thiMD* operon is completely transcribed. The double slashes near the 5′ and 3′ ends of the RNA indicate that this figure is not showing a large portion of the mRNA from this operon.

CONCEPT CHECK: Which RNA conformation favors translation—the form with the Shine-Dalgarno antisequestor or the form in which the Shine-Dalgarno sequence is within a stem-loop?

CHAPTER SUMMARY

- Gene regulation is the phenomenon in which the level of gene expression can vary under different conditions. By comparison, unregulated, or constitutive, genes are expressed at relatively constant levels. Gene regulation can occur during transcription, during translation, or posttranslationally (see Figure 14.1).

14.1 Overview of Transcriptional Regulation

- Repressors and activators are regulatory transcription factors (RTFs) that exert negative control and positive control, respectively, on transcription. Small effector molecules that control the functioning of repressors and activators include inducers, corepressors, and inhibitors (see Figure 14.2).

14.2 Regulation of the *lac* Operon

- Enzyme adaptation is the phenomenon in which an enzyme appears within a living cell only after the cell has been exposed to the substrate for that enzyme.
- The *lac* operon is transcribed into a polycistronic mRNA that codes proteins involved with the uptake and metabolism of lactose (see Figure 14.3).
- Lac repressor binds to the *lac* operator and inhibits transcription. Allolactose binds to the repressor and causes a conformational change that prevents the repressor from binding to the operator site. This event induces transcription (see Figure 14.4).
- The regulation of the *lac* operon enables the bacterium *E. coli* to respond to changes in the level of lactose in its environment (see Figure 14.5).
- Jacob, Monod, and Pardee constructed merozygotes to show that the *lacI* gene codes a diffusible repressor protein (see Figures 14.6, 14.7).
- Mutations in the *lacI* gene and the *lac* operator site may have different effects in a merozygote (see Table 14.1).

- CAP is an activator protein for the *lac* operon. It binds to the CAP site when cAMP levels are high. The presence of glucose causes cAMP levels to decrease, which results in diauxic growth by inhibiting the *lac* operon (see Figures 14.8 14.9).
- Lac repressor, which is a tetramer, binds to two operator sites: either O_1 and O_2, or O_1 and O_3 (see Figures 14.10, 14.11).

14.3 Regulation of the *trp* Operon

- The *trp* operon is regulated by trp repressor, which binds to the *trp* operator when tryptophan, a corepressor, is present (see Figure 14.12).
- A second way that the *trp* operon is regulated is via attenuation, in which the formation of a terminator stem-loop causes early termination of transcription (see Figures 14.13, 14.14).
- Inducible operons typically code catabolic enzymes, whereas repressible operons often code anabolic enzymes.

14.4 Translational and Posttranslational Regulation

- Translation can be regulated by translational regulatory proteins and by antisense RNA (see Figure 14.15).
- Posttranslational control of protein function may involve feedback inhibition or posttranslational covalent modifications (see Figure 14.16).

14.5 Riboswitches

- A riboswitch is an element within an mRNA, usually at the 5′ end, that can bind a metabolite or ion as a ligand and switch the mRNA to a different secondary conformation.
- A riboswitch can regulate transcription, translation, RNA stability, or splicing (see Table 14.2, Figures 14.17, 14.18).

PROBLEM SETS & INSIGHTS

MORE GENETIC TIPS

1. Researchers have identified mutations in the promoter region of the *lacI* gene that make it more difficult for the *lac* operon to be induced. These are called *lacI^Q* mutants, because their effect is that a greater quantity of lac repressor is made. Explain why increased transcription of the *lacI* gene makes it more difficult to induce the *lac* operon.

T **OPIC:** *What topic in genetics does this question address?* The topic is gene regulation. More specifically, the question is about how the amount of a repressor protein will affect transcription.

I **NFORMATION:** *What information do you know based on the question and your understanding of the topic?* From the question, you know that certain mutations result in an overproduction of lac repressor. From your understanding of the topic, you may remember that the binding of lac repressor to the *lac* operator inhibits transcription.

P **ROBLEM-SOLVING** **S** **TRATEGIES:** *Relate structure and function. Predict the outcome.* One strategy to solve this problem is to consider the structure and function of lac repressor. It forms a tetramer and then binds to two operators in the absence of allolactose. When allolactose binds to lac repressor, the result is a conformational change that causes the repressor to not bind to the operators.

ANSWER: An increase in the amount of lac repressor makes it easier for tetramers to form and then repress the *lac* operon. When the cell is exposed to lactose, allolactose levels slowly rise. Some of the allolactose binds to lac repressor and prevents it from binding to the operator sites. If many lac repressor proteins accumulate within the cell, more allolactose is needed to bind to those proteins to prevent them from repressing the *lac* operon.

2. Explain how the pausing of the ribosome in the presence or absence of tryptophan affects the formation of a terminator (3–4) stem-loop, and describe how this affects transcription.

T **OPIC:** *What topic in genetics does this question address?* The topic is gene regulation. More specifically, the question is about a form of gene regulation called attenuation.

I **NFORMATION:** *What information do you know based on the question and your understanding of the topic?* In the question, you are reminded that tryptophan affects the formation of a terminator stem-loop. From your understanding of the topic, you may remember that the location where a ribosome pauses during translation of the *trpL* gene can influence which type(s) of stem-loop(s) can form. The three possible types are 1–2, 2–3, and 3–4. The formation of a 2–3 stem-loop prevents the formation of a 3–4 stem-loop.

P **ROBLEM-SOLVING** **S** **TRATEGIES:** *Make a drawing. Compare and contrast.* One strategy to solve this problem is to make two drawings similar to those in Figure 14.14, parts (b) and (c).

In part (b), the ribosome is shielding region 1, and in part (c), it is shielding region 2. Compare the two drawings and think about which stem-loops are able to form.

ANSWER: The key factor is the location where the ribosome pauses. When tryptophan levels are low, it pauses over the Trp codons in the *trpL* mRNA. Pausing at this site shields region 1 in the mRNA. Because region 1 is unavailable to hydrogen bond with region 2, region 2 hydrogen bonds with region 3. Therefore, region 3 cannot form a terminator stem-loop with region 4. Alternatively, if tryptophan levels in the cell are sufficiently high, the ribosome pauses over the stop codon in the *trpL* mRNA. In this case, the ribosome shields region 2. Therefore, regions 3 and 4 hydrogen bond with each other to form a terminator stem-loop, which abruptly halts the continued transcription of the *trp* operon.

3. The 5′ region of the TPP riboswitch in *B. subtilis* is very similar to the TPP riboswitch in *E. coli*. However, the riboswitch in *B. subtilis* regulates transcription, whereas the one in *E. coli* regulates translation. What is the role of the 5′ region in both riboswitches? How can one riboswitch regulate transcription while the other regulates translation?

T **OPIC:** *What topic in genetics does this question address?* The topic is riboswitches. More specifically, the question is about TPP riboswitches found in two different species of bacteria.

I **NFORMATION:** *What information do you know based on the question and your understanding of the topic?* From the question, you know that *B. subtilis* and *E. coli* have TPP riboswitches with similar 5′ regions, yet one riboswitch regulates transcription and the other regulates translation. From your understanding of the topic, you may remember that some riboswitches function by forming stem-loops that may affect transcription or translation.

P **ROBLEM-SOLVING** **S** **TRATEGIES:** *Relate structure and function. Compare and contrast.* One strategy to solve this problem is to first consider the structure and function of the 5′ region and then compare the effects of conformational changes on the two types of riboswitches.

ANSWER: With regard to RNA structure, the 5′ region of both riboswitches contains an aptamer domain, which is a binding site for TPP. The binding of TPP alters the structure of the 5′ region in a way that transmits a subsequent alteration in RNA structure to the expression platform. In the case of the *B. subtilis* TPP riboswitch, the subsequent alteration in RNA structure causes a terminator stem-loop to form, which ends transcription. In this way, the *B. subtilis* riboswitch controls transcription. By comparison, the binding of TPP to the *E. coli* TPP riboswitch causes the Shine-Dalgarno sequence and AUG start codon to be sequestered in a stem-loop, which inhibits translation. Therefore, the *E. coli* riboswitch controls translation.

Conceptual Questions

C1. What is the difference between a constitutive gene and a regulated gene?

C2. In general, why is it important to regulate genes? Give examples of situations in which it would be advantageous for a bacterial cell to regulate genes.

C3. If a gene is repressible and under positive control, what kind of effector molecule and RTF are involved in its regulation? Explain how the binding of the effector molecule affects the RTF.

C4. Transcriptional regulation often involves an RTF that binds to a segment of DNA and a small effector molecule that binds to the RTF. Is each of the following an RTF, a segment of DNA, or a small effector molecule?

   A. Repressor

   B. Inducer

   C. Operator site

   D. Corepressor

   E. Activator

   F. Attenuator

   G. Inhibitor

C5. A certain operon is repressible—a small effector molecule turns off its transcription. Which combination(s) of small effector molecule and RTF could be involved in this process?

   A. An inducer plus a repressor

   B. A corepressor plus a repressor

   C. An inhibitor plus an activator

   D. An inducer plus an activator

C6. Some mutations have a *cis*-effect, whereas others have a *trans*-effect. Explain the molecular differences between these two types of mutations. Which type of mutation can be complemented in a merozygote experiment?

C7. What is enzyme adaptation? From a genetic point of view, how does it occur?

C8. How would the expression of genes in the *lac* operon be affected if each one of the following sequences was missing?

   A. *lac* operon promoter

   B. Operator site

   C. *lacA* gene

C9. If an abnormal repressor protein could still bind allolactose but the binding of allolactose did not alter the conformation of lac repressor, how would the expression of the *lac* operon be affected?

C10. What is diauxic growth? Explain the roles of cAMP and CAP in this process.

C11. Mutations may have an effect on the expression of the *lac* operon and the *trp* operon. Would each of the following mutations have a *cis*- or *trans*-effect on the expression of the protein-coding genes in the operon?

   A. A mutation in the operator site that prevents lac repressor from binding to it

   B. A mutation in the *lacI* gene that prevents lac repressor from binding to DNA

   C. A mutation in the *trpL* gene that prevents attenuation

C12. Would a mutation that inactivated lac repressor and prevented it from binding to the *lac* operator site result in the constitutive expression of the *lac* operon under all conditions? Explain. What is the disadvantage to the bacterium of having a constitutive *lac* operon?

C13. What is meant by the term *attenuation*? Is it an example of gene regulation during transcription or translation? Explain your answer.

C14. As shown in Figure 14.13, four regions within the *trpL* mRNA can form stem-loops. Let's suppose that mutations have been previously identified that prevent the ability of a particular region to form a stem-loop with a complementary region. For example, a region 1 mutant cannot form a 1–2 stem-loop, but it can still form a 2–3 or 3–4 stem-loop. Likewise, a region 4 mutant can form a 1–2 or 2–3 stem-loop but not a 3–4 stem-loop. Under each of the following sets of conditions, would attenuation occur?

   A. Region 1 is mutant, tryptophan is high, and translation is not occurring.

   B. Region 2 is mutant, tryptophan is low, and translation is occurring.

   C. Region 3 is mutant, tryptophan is high, and translation is not occurring.

   D. Region 4 is mutant, tryptophan is low, and translation is not occurring.

C15. As described in Chapter 13, enzymes known as aminoacyl-tRNA synthetases are responsible for attaching amino acids to tRNAs. Let's suppose that in a mutant bacterium tryptophanyl-tRNA synthetase has a reduced ability to attach tryptophan to tRNA; its activity is only 10% of that found in a normal bacterium. How would attenuation of the *trp* operon be affected? Would the operon be more or less likely to be attenuated? Explain your answer.

C16. The combination of a 3–4 stem-loop and a U-rich attenuator sequence in the *trp* operon (see Figure 14.13) is an example of a ρ-independent terminator. The function of ρ-independent terminators is described in Chapter 12. Would you expect attenuation to occur if the tryptophan levels were high and mutations changed the attenuator sequence from UUUUUUUU to UGGUUGUC? Explain why or why not.

C17. Mutations in genes that code tRNAs can create tRNAs that recognize stop codons. Because stop codons are sometimes called nonsense codons, these types of mutations that affect tRNAs are called nonsense suppressors. For example, a normal tRNA^{Gly} has an anticodon sequence CCU that recognizes a glycine codon in mRNA (GGA) and inserts a glycine during translation. However, a mutation in the gene that codes tRNA^{Gly} could change the anticodon to ACU. This mutant tRNA^{Gly} would still carry glycine, but it would recognize the stop codon UGA. Would this mutation affect attenuation of the *trp* operon? Explain why or why not. Note: To answer this question, you need to look carefully at Figure 14.13 and see if you can identify any stop codons that may exist beyond the UGA stop codon that is located after region 1.

C18. Translational control is usually aimed at preventing the initiation of translation. With regard to cellular efficiency, why do you think this is the case?

C19. What is antisense RNA? How does it affect the translation of a complementary mRNA?

C20. A species of bacteria can synthesize the amino acid histidine, so these bacteria do not require histidine in their growth medium. A key enzyme, which we will call histidine synthetase, is necessary for histidine biosynthesis. When these bacteria are given histidine in their growth medium, they stop synthesizing histidine intracellularly. Based on this observation alone, propose three different regulatory mechanisms to explain why histidine biosynthesis ceases when histidine is in the growth medium. To explore this phenomenon further, you measure the amount of intracellular histidine synthetase when bacterial cells are grown in the presence and absence of histidine. In both conditions, the amount of this protein is identical. Which mechanism of regulation is consistent with this observation?

C21. Using three examples, describe how allosteric sites are important in the function of genetic regulatory proteins.

C22. How are the actions of lac repressor and trp repressor similar and how are they different with regard to their binding to operator sites, their effects on transcription, and the influences of small effector molecules?

C23. Transcriptional repressor proteins (e.g., lac repressor), antisense RNA, and feedback inhibition are three different mechanisms that inhibit gene expression and gene products. Which of these three mechanisms will be effective in each of the following situations?

A. Shutting down the synthesis of a polypeptide

B. Shutting down the synthesis of mRNA

C. Shutting off the function of a protein

For your answers to parts A–C that list more than one mechanism, which mechanism will be the fastest or the most efficient?

Experimental Questions

E1. Answer the following questions regarding the experiment of Figure 14.7.

A. Why was β-ONPG used? Why was no yellow color observed in one of the four tubes? Can you propose alternative methods to measure the level of expression of the *lac* operon?

B. In the presence of lactose, the yellow color as measured by spectrophotometry was twice as intense for the merozygote strain as for the mutant strain. Why was this result obtained?

E2. Chapter 20 describes a blotting method known as Northern blotting, which can be used to detect RNA transcribed from a particular gene or a particular operon. In this method, a specific RNA is detected by using a short segment of labeled DNA, which is complementary to the target RNA. After the labeled DNA binds to the RNA within a blot of a gel, the RNA is visualized as a dark band. For example, a segment of labeled DNA complementary to the mRNA of the lac operon could be used to specifically detect the lac operon mRNA on a gel blot. As shown here, the method of Northern blotting can be used to determine the amount of a particular RNA transcribed under different types of growth conditions. In this Northern blot, bacteria containing a normal *lac* operon were grown under different types of conditions, and then the mRNA was isolated from the cells and subjected to Northern blotting, using a segment of labeled DNA that is complementary to the mRNA of the *lac* operon.

Lane 1. Growth in media containing glucose
Lane 2. Growth in media containing lactose
Lane 3. Growth in media containing glucose and lactose
Lane 4. Growth in media that doesn't contain glucose or lactose

Based on your understanding of the regulation of the *lac* operon, explain these results. Which is more effective at shutting down the *lac* operon, the binding of lac repressor or the removal of CAP? Explain your answer based on the results shown in the Northern blot.

E3. As described in question E2 and also in Chapter 20, the technique of Northern blotting can be used to detect the level of transcription of a specific RNA. Draw the results you would expect from a Northern blot if bacteria were grown in media that contained lactose (and no glucose) but had the following mutations:

Lane 1. Normal strain

Lane 2. Strain with a mutation that inactivates lac repressor

Lane 3. Strain with a mutation that prevents allolactose from binding to lac repressor

Lane 4. Strain with a mutation that inactivates CAP

How would your results differ if these bacterial strains were grown in media that did not contain lactose or glucose?

E4. An absentminded researcher follows the protocol described in Figure 14.7 and (at the end of the experiment) does not observe any yellow color in any of the tubes. Yikes! Which of the following mistakes could account for this observation?

A. Forgot to sonicate the cells

B. Forgot to add lactose to two of the tubes

C. Forgot to add β-ONPG to the four tubes

E5. Explain how the data shown in Figure 14.10 indicate that two operator sites are necessary for repression of the *lac* operon. What would the results have been if all three operator sites were required for the binding of lac repressor?

E6. A mutant strain has a defective *lac* operator site that results in the constitutive expression of the *lac* operon. Outline an experiment you would carry out to demonstrate that the operator site must be physically adjacent to the genes that it influences. Based on your knowledge of the *lac* operon, describe the results you would expect.

E7. Suppose that you have isolated a mutant strain of *E. coli* in which the *lac* operon is constitutively expressed. To understand the nature of this defect, you create a merozygote in which the mutant strain contains an F′ factor with a normal *lac* operon and a normal *lacI* gene. You then compare the mutant strain and the merozygote with regard to the amount of β-galactosidase produced in the presence and absence of lactose. You obtain the following results:

	Addition of Lactose	Amount of β-Galactosidase (% of mutant strain in the presence of lactose)
Mutant	No	100
Mutant	Yes	100
Merozygote	No	100
Merozygote	Yes	200

Explain the nature of the defect in the mutant strain.

E8. What is diauxic growth? With regard to *lac* operon expression, explain what is happening at the molecular level in the three phases (i.e., first growth phase, lag phase, and second growth phase) when bacteria are exposed to both glucose and lactose.

Questions for Student Discussion/Collaboration

1. Discuss the advantages and disadvantages of genetic regulation at the different points identified in Figure 14.1.

2. Looking at Figure 14.11, discuss possible "molecular" ways that the cAMP-CAP complex and lac repressor may influence RNA polymerase function. In other words, try to explain how the bending and looping in DNA may affect the ability of RNA polymerase to initiate transcription.

Note: All answers are available for the instructor in Connect; the answers to the even-numbered questions and all of the Concept Check and Comprehension Questions are in Appendix B.

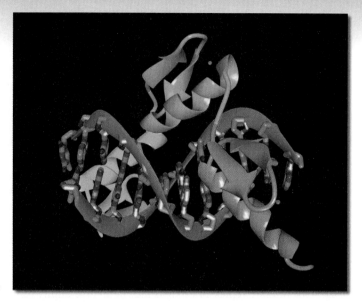

Binding of a regulatory transcription factor to DNA. Certain proteins, known as regulatory transcription factors, have the ability to bind into the major groove of DNA and regulate gene transcription.

Song Tan, Penn State University, www.bmb.psu.edu/faculty/tan/lab

15 GENE REGULATION IN EUKARYOTES I: GENERAL FEATURES OF TRANSCRIPTIONAL REGULATION

Gene expression refers to the process by which the information within a gene is accessed, first to synthesize RNA and polypeptides, and eventually to affect the properties of cells and the phenotype of multicellular organisms. For cells and organisms to develop and function properly, the level of expression of most genes must be carefully controlled so it is not too high or not too low. **Gene regulation** refers to the phenomenon in which the level of gene expression can vary under different conditions—the level can be increased or decreased. From a molecular perspective, gene expression in eukaryotes can be regulated at many different steps in the process that gives rise to polypeptides and functional proteins. These steps include the following:

Transcription (discussed in this chapter and Chapter 16)

1. Regulatory transcription factors can activate or inhibit transcription.
2. The arrangements and composition of nucleosomes influence transcription.
3. DNA methylation (usually) inhibits transcription.

RNA modification (discussed in Chapter 12)

1. Alternative splicing alters exon choices.
2. RNA editing alters the base sequence of RNAs.

Translation (discussed in Chapters 13 and 17)

1. Proteins that bind to the 5′ of mRNA regulate translation and those that bind to the 3′ end may influence mRNA degradation.
2. Small RNAs, called miRNAs and siRNAs, silence the translation of mRNA, a process called RNA interference.

Posttranslation (discussed in Chapters 14 and 23)

Feedback inhibition and covalent modifications regulate protein function.

The ability to regulate gene expression provides many benefits to eukaryotic organisms—a category that includes protists, fungi, plants, and animals.

- Like their prokaryotic counterparts, eukaryotic cells need to adapt to changes in their environment. For example, eukaryotic cells respond to changes in nutrient availability through enzyme adaptation, much as prokaryotic cells do.
- Among plants and animals, multicellularity and a more complex cell structure demand a much greater level of gene regulation than occurs in bacteria. Gene regulation is necessary to establish and maintain the differences in structure and function among distinct cell types. It is amazing that the various cells within a multicellular organism usually contain the same genetic material, yet phenotypically may

be quite different. For example, the appearance of a human nerve cell seems about as similar to that of a muscle cell as an amoeba is to a paramecium. In spite of these phenotypic differences, a human nerve cell and muscle cell actually contain the same complement of human chromosomes. Nerve and muscle cells look strikingly different because of gene regulation rather than differences in DNA content.

In this chapter, our focus is on the regulation of gene transcription. Researchers have discovered that most eukaryotic genes, particularly those found in multicellular species, are transcriptionally regulated by many factors. This phenomenon is called **combinatorial control** because the combination of many factors determines the expression of any given gene. At the level of transcription, the following factors commonly contribute to combinatorial control:

1. One or more activator proteins may stimulate the ability of RNA polymerase to initiate transcription.
2. One or more repressor proteins may inhibit the ability of RNA polymerase to initiate transcription.
3. The function of activators and repressors may be modulated in a variety of ways, including the binding of small effector molecules, protein-protein interactions, and covalent modifications.
4. Regulatory proteins may alter the composition or arrangement of nucleosomes in the vicinity of a promoter, thereby affecting transcription. Also, histone proteins may be covalently modified.
5. DNA methylation may inhibit transcription, either by preventing the binding of an activator protein or by recruiting proteins that change the structure of chromatin in a way that inhibits transcription.
6. As discussed in Chapter 16, the formation of heterochromatin may inhibit gene expression in localized regions of a chromosome.

All six of these factors can contribute to the regulation of a single gene, or possibly only three or four will play a role. In most cases, transcriptional regulation is aimed at controlling the initiation of transcription at the promoter or at an early point in the elongation phase. In Section 15.1, we will survey the first three factors that may contribute to combinatorial control of transcription. In the next two sections, we will consider the fourth and fifth factors; the sixth factor is considered in Chapter 16. The process of gene activation as it occurs in living cells is described in Section 15.4, which also includes specific examples. Finally, the last section compares transcriptional regulation in bacteria, archaea, and eukaryotes.

15.1 REGULATORY TRANSCRIPTION FACTORS AND ENHANCERS

Learning Outcomes:

1. Distinguish between general and regulatory transcription factors.
2. Describe how a regulatory transcription factor binds to an enhancer.
3. Describe three ways that the function of a regulatory transcription factor can be modulated.

The term **transcription factor** is broadly used to describe a category of proteins that influence the ability of RNA polymerase to transcribe DNA into RNA. We will focus much of our attention on transcription factors that affect the ability of RNA polymerase to begin the transcription process. Such transcription factors regulate the binding of the preinitiation complex to the core promoter and/or control the switch from the initiation to the elongation stage of transcription.

Two types of transcription factors play a key role in these processes. In Chapter 12, we considered **general transcription factors (GTFs),** which are required for the binding of RNA polymerase to the core promoter and for progression to the elongation stage. General transcription factors are necessary for any transcription to occur. In addition, eukaryotic cells possess a diverse array of **regulatory transcription factors (RTFs)** that serve to regulate the rate of transcription of target genes.

Structural Features of Regulatory Transcription Factors Allow Them to Bind to DNA

Genes that code general and regulatory transcription factors have been identified and sequenced from a wide variety of eukaryotic species, including yeast, plants, and animals. Several different families of evolutionarily related transcription factors have been discovered. In recent years, the molecular structures of transcription factor proteins have become an area of intense research. These proteins contain regions, called **domains,** that have specific functions. For example, one domain of a transcription factor may have a DNA-binding function, and another may provide a binding site for a small effector molecule. When a domain or a portion of a domain has a very similar structure in many different proteins, the structurally similar region is called a **motif.**

Figure 15.1 depicts several different motifs found in transcription factor proteins. The protein secondary structure known as an α helix occurs frequently in transcription factors. Why is the α helix common in such proteins? The explanation is that the α helix is the proper width to bind into the major groove of the DNA double helix. In helix-turn-helix and helix-loop-helix motifs, an α helix called the recognition helix makes contact with and recognizes a base sequence along the major groove of the DNA (Figure 15.1a, b). Such motifs are able to recognize a particular base sequence because they have a surface that is biochemically complementary to that of the DNA, allowing for a series of favorable electrostatic and van der Waals interactions between the protein and the base pairs. Also, such proteins make a large number of contacts with the DNA backbone, including salt bridges between negatively charged phosphates and positively charged amino acid side chains and hydrogen bonds with uncharged amino acids.

Recall that the major groove is a region of the DNA double helix where the nucleotide bases are in contact with the water in cellular fluid. Hydrogen bonding between the amino acid side chains in an α helix and the nucleotide bases in the DNA is one way that a transcription factor binds to a specific DNA sequence. In addition, the recognition helix often contains many positively charged amino acids (e.g., arginine and lysine) that favorably interact with the DNA backbone, which is negatively charged.

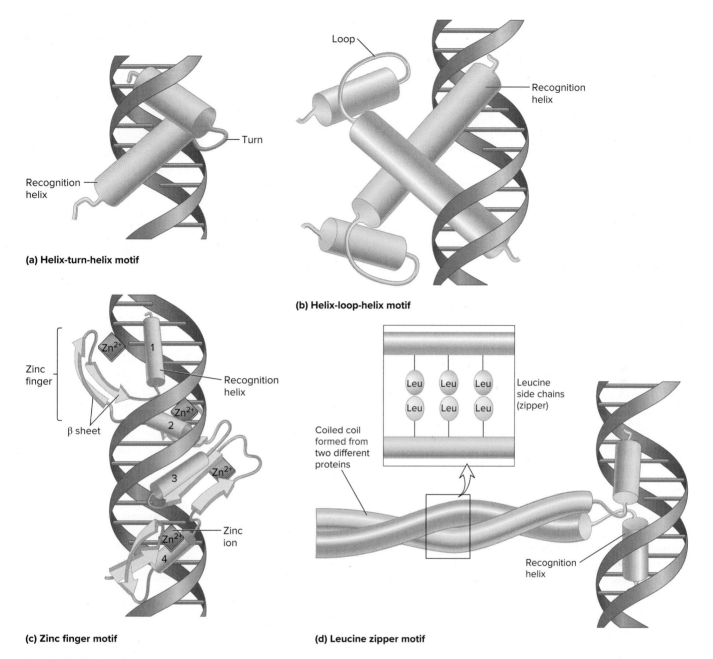

(a) Helix-turn-helix motif

(b) Helix-loop-helix motif

(c) Zinc finger motif

(d) Leucine zipper motif

FIGURE 15.1 **Structural motifs found in transcription factor proteins.** Certain types of protein secondary structures are found in many different transcription factors. In this figure, α helices are shown as cylinders and β sheets as flattened arrows. **(a)** Helix-turn-helix motif: Two α helices are connected by a turn. The α helices bind to the DNA within the major groove. **(b)** Helix-loop-helix motif: A short α helix is connected to a longer α helix by a loop. In this illustration, a dimer is formed from the interactions of two helix-loop-helix motifs, and the longer helices are binding to the DNA. **(c)** Zinc finger motif: Each zinc finger is composed of one α helix and two antiparallel β sheets. A zinc ion (Zn^{2+}), shaded red, holds the zinc finger together. This illustration shows four zinc fingers in a row. **(d)** Leucine zipper motif: The leucine zipper promotes the dimerization of two transcription factor proteins. Two α helices (termed a coiled coil) are intertwined due to interactions between their leucines (see inset).

CONCEPT CHECK: Explain how an α helix in a transcription factor protein is able to function as a recognition helix.

A zinc finger motif is composed of one α helix and two β sheets that are held together by a zinc (Zn^{2+}) metal ion (Figure 15.1c). The zinc finger can also recognize DNA sequences within the major groove.

A second interesting feature of certain motifs is that they promote protein dimerization. The leucine zipper (Figure 15.1d) and helix-loop-helix motif (see Figure 15.1b) mediate protein dimerization. For example, Figure 15.1d depicts the

dimerization and DNA binding of two proteins that have several leucine amino acids (a zipper). Leucines in both proteins interact ("zip up"), resulting in protein dimerization. Two identical transcription factors may come together to form a **homodimer,** or two different transcription factors can form a **heterodimer.** As discussed later, the dimerization of transcription factors can be an important way to modulate their function.

Regulatory Transcription Factors Recognize Regulatory Elements Within Enhancers

Some RTFs exert their effects by influencing the ability of RNA polymerase to begin transcription of a particular gene. They typically recognize *cis*-acting regulatory sequences that are located in the same chromosome as the gene they regulate; such elements may be close to or far way from the core promoter. These DNA sequences are analogous to the operator sites found near bacterial promoters. Such sequences are generally known as **regulatory elements** (also called **regulatory sequences** or **response elements**). As described in Sections 15.2 and 15.4, changes in chromatin structure are often needed for RTFs to bind to regulatory elements.

An **enhancer** is a DNA region, usually 50–1000 bp in length, that contains one or more regulatory elements. If the binding of an RTF to an enhancer increases the rate of transcription, such a regulatory transcription factor is termed an **activator.** In **Figure 15.2a**, the enhancer is not adjacent to the core promoter so a bend in the DNA must occur to bring the enhancer and core promoter close to each other. DNA-bending proteins facilitate this process.

Alternatively, regulatory transcription factors may act as **repressors** by binding to enhancers and preventing transcription from occurring (**Figure 15.2b**). Repressors also bind to enhancers but they do not promote the binding of GTFs, mediator, and RNA polymerase to the core promoter or they do not allow RNA polymerase to proceed to the elongation phase of transcription. The mechanisms by which an activator stimulates transcription or a repressor inhibits transcription are discussed in Section 15.4.

When an activator binds to an enhancer, such binding can stimulate transcription 10- to 1000-fold, a phenomenon known as **up regulation.** Alternatively, the binding of repressors that inhibit transcription result in **down regulation.**

Regulatory elements and the enhancers that contain them are **orientation-independent,** or **bidirectional.** This means that a regulatory element can function in either the forward or the reverse direction. For example, let's consider an element with a forward orientation as shown below:

5′–GATA–3′
3′–CTAT–5′

This element is also bound by a regulatory transcription factor and enhances transcription even when it is rotated 180° and oriented in the reverse direction:

5′–TATC–3′
3′–ATAG–5′

Striking variation is also observed in the locations of enhancers relative to a gene's promoter. Enhancers are sometimes located in a region within 200 bp upstream from the core promoter. More commonly, they are fairly distant from the core promoter, even 100,000 bp away, yet exert strong effects on the ability of RNA polymerase to initiate transcription at the core promoter!

Enhancers were first discovered by Susumu Tonegawa and coworkers in the 1980s. While studying genes that play a

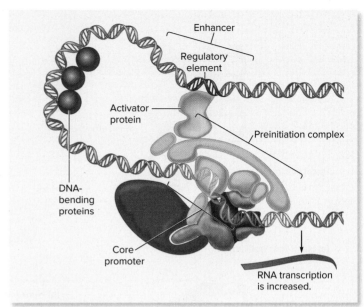

(a) Gene activation

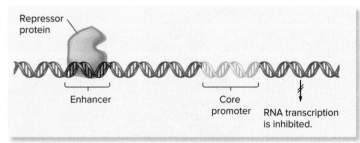

(b) Gene repression

FIGURE 15.2   Overview of transcriptional regulation by regulatory transcription factors.   A regulatory transcription factor can act as either an activator to increase the rate of transcription or a repressor to decrease the rate of transcription. **(a)** The preinitiation complex is bound to the core promoter; the DNA of the core promoter is shown in tan. The activator binds to a regulatory element within a distant enhancer. A loop in the DNA allows the activator to be close to the core promoter and preinitiation complex. Some activators, such as the one shown here, may interact with the preinitiation complex in a way that increases the rate of transcription. **(b)** When a repressor binds to an enhancer, it may inhibit the formation of the preinitiation complex, as shown here, or it may prevent the preinitiation complex from proceeding to the elongation phase of transcription.

role in immunity, these researchers identified a region that is far away from the core promoter, but is needed for high levels of transcription to take place. In some cases, enhancers are located downstream from the promoter site and may even be found within introns, the noncoding parts of genes. As you may imagine, the variation in the orientation and location of enhancers profoundly complicates the efforts of geneticists to identify those regulatory elements that affect the expression of any given gene.

The Functions of RTFs Can Be Modulated in Three Ways

Thus far in this section, we have considered the structures of RTFs and their abilities to bind to DNA and influence transcription. The functions of the RTFs themselves must also be modulated. Why is this necessary? The answer is that the genes they control must be turned on at the proper time, in the correct cell type, and under the appropriate environmental conditions. Therefore, eukaryotes have evolved different ways to modulate the functions of these proteins.

The functions of RTFs are controlled in three common ways:

- The binding of a small effector molecule
- Protein-protein interactions
- Covalent modifications

Figure 15.3 depicts these three mechanisms for modulating the functions of RTFs. Usually, one or more of these modulating effects are important in determining whether a transcription factor binds to the DNA or influences transcription by RNA polymerase. For example, a small effector molecule may bind to an RTF and promote its binding to DNA (Figure 15.3a). In Section 15.4, we will see that steroid hormones function in this manner. Another important mechanism of modulation is via protein-protein interactions (Figure 15.3b). The formation of homodimers and heterodimers is a fairly common means of controlling transcription. Finally, the function of an RTF can be affected by covalent modifications, such as the attachment of a phosphate group (Figure 15.3c). As discussed in Section 15.4, the phosphorylation of activators can control their ability to stimulate transcription.

15.1 COMPREHENSION QUESTIONS

1. As described in the chapter introduction, combinatorial control refers to the phenomenon that
   a. transcription factors always combine with each other when regulating genes.
   b. the combination of many factors determines the expression of any given gene.
   c. small effector molecules and regulatory transcription factors are found in many different combinations.
   d. genes and regulatory transcription factors must combine with each other during gene regulation.

2. An RTF that binds to a regulatory element typically contains _____ that binds to the _____ of the DNA.
   a. an α helix, backbone
   b. an α helix, major groove
   c. a β sheet, backbone
   d. a β sheet, major groove

3. A bidirectional regulatory element has the following sequence:
   5′–GTCA–3′
   3′–CAGT–5′
   Which of the following sequences would also be a functional regulatory element?
   a. 5′–ACTG–3′   c. 3′–GTCA–5′
      3′–TGAC–5′   5′–CAGT–3′
   b. 5′–TGAC–3′   d. 3′–TGAC–5′
      3′–ACTG–5′   5′–ACTG–3′

4. RTFs can be modulated by
   a. the binding of small effector molecules.
   b. protein-protein interactions.
   c. covalent modifications.
   d. any of the above.

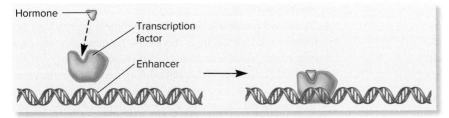

(a) Binding of a small effector molecule such as a hormone

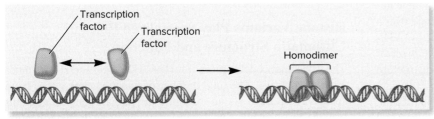

(b) Protein-protein interaction

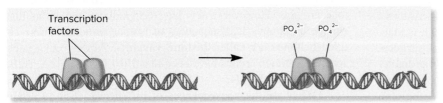

(c) Covalent modification by phosphorylation

FIGURE 15.3 Common ways to modulate the function of regulatory transcription factors. (a) The binding of an effector molecule such as a hormone may influence the ability of an RTF to bind to the DNA. (b) Protein-protein interactions among transcription factor proteins may influence their functions. (c) Covalent modifications, such as phosphorylation, may alter transcription factor function.

15.2 CHROMATIN REMODELING, HISTONE VARIANTS, AND HISTONE MODIFICATIONS

Learning Outcomes:

1. Describe how chromatin-remodeling complexes alter nucleosomes.
2. Define *histone variant*, and explain why histone variants are functionally important.
3. Explain how histone modifications affect transcription.

In Section 15.1, we considered how RTFs can bind to regulatory elements within enhancers and thereby influence transcription. In eukaryotes, certain features of chromatin structure may act as obstacles for the binding of RTFs to enhancers. If chromatin is in a **closed conformation,** transcription may be difficult or impossible. By comparison, chromatin that is in an **open conformation** is more easily accessed by transcription factors and RNA polymerase, allowing transcription to occur. As discussed in Chapter 16, a closed conformation may occur via the conversion of euchromatin, which is more loosely compacted, into tightly compacted heterochromatin. In addition, researchers have determined that the following factors can affect gene transcription:

- Precise positioning of nucleosomes at or near promoters
- Presence of histone variants
- Covalent modification of histones

In this section, we will examine molecular mechanisms that explain how changes in these three factors can occur.

Chromatin-Remodeling Complexes Alter the Positions and Compositions of Nucleosomes

The term **ATP-dependent chromatin remodeling,** or simply **chromatin remodeling,** refers to dynamic changes in the structure of chromatin that occur during the life of a cell. These changes range from local alterations in the positioning of one or a few nucleosomes to larger changes that affect chromatin structure over a longer distance. Chromatin remodeling is carried out by ATP-dependent chromatin-remodeling complexes, which are a set of diverse multiprotein machines that reposition and restructure nucleosomes.

In recent years, geneticists have been trying to identify the steps that promote the interconversion between the closed and open conformations of chromatin. Nucleosomes have been shown to have different positions in cells that normally express a particular gene compared with cells in which the gene is inactive. For example, in reticulocytes that express the β-globin gene, an alteration in nucleosome positioning occurs in the promoter region from nucleotide −500 to nucleotide +200. This alteration is thought to be an important step in expressing the β-globin gene. Based on the analysis of many genes, researchers have discovered that a key role of some transcriptional activators is to

orchestrate a change in chromatin from closed to open conformation by altering nucleosomes.

One way to change chromatin structure is through ATP-dependent chromatin remodeling. In this process, the energy of ATP hydrolysis is used to drive changes in the positions and/or compositions of nucleosomes, thereby making the DNA more or less amenable to transcription. Therefore, chromatin remodeling is important for both the activation and repression of transcription.

The remodeling process is carried out by a protein complex that recognizes nucleosomes and uses ATP to alter their configuration. All chromatin-remodeling complexes have a catalytic ATPase subunit called **DNA translocase;** this ATPase subunit, similar to what is found in motor proteins, moves along the DNA. Eukaryotes have multiple families of chromatin remodeling complexes. Though their names may differ depending on the species, common families of chromatin-remodeling complexes include the SWI/SNF-family, the ISWI-family, the INO80-family, and the Mi-2-family. The names of these complexes sometimes refer to the effects of mutations in genes that code them. For example, the abbreviations SWI and SNF refer to the effects that occur in yeast when these remodeling complexes are defective. SWI mutants are defective in mating-type s̲witching, and SNF mutations create a s̲ucrose n̲onfermenting phenotype.

How do chromatin remodeling complexes change chromatin structure? Three effects are possible:

- One result of ATP-dependent chromatin remodeling is a change in the positions of nucleosomes (**Figure 15.4a**). This may involve shifts in nucleosomes to new locations or changes in the relative spacing of nucleosomes over a long stretch of DNA.
- A second possible effect of remodeling is that histone octamers are evicted from the DNA, thereby creating gaps where nucleosomes are not found (**Figure 15.4b**).
- A third possibility is that remodeling may change the composition of nucleosomes by removing standard histones and replacing them with histone variants (**Figure 15.4c**). The functions of histone variants are described next.

Histone Variants Play Specialized Roles in Chromatin Structure and Function

The genes that code histones H1, H2A, H2B, H3, and H4 are moderately repetitive. The total number of histone genes varies from species to species. As an example, the human genome contains over 70 histone genes that have been produced by gene duplication events during evolution. Most of these genes code standard histone proteins. However, a few have accumulated mutations that change the amino acid sequences of histone proteins. These altered histones are called **histone variants.** Among eukaryotic species, histone variants have been identified for H1, H2A, H2B, and H3, but not for H4.

What are the consequences of histone variation? Certain histone variants play specialized roles in chromatin structure and function. In all eukaryotes, histone variants are incorporated

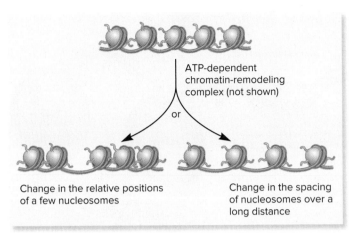

(a) Change in nucleosome location

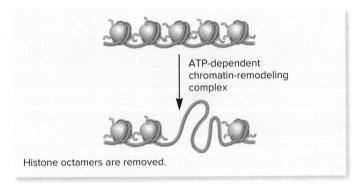

(b) Eviction of histone octamers

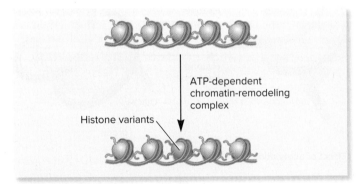

(c) Replacement with histone variants

FIGURE 15.4 ATP-dependent chromatin remodeling. The top part of each illustration shows five nucleosomes. Chromatin-remodeling complexes may **(a)** change the locations of nucleosomes, **(b)** remove histone octamers from the DNA, or **(c)** replace core histones with histone variants. The chromatin-remodeling complex is not shown in this figure; only its effects are shown.

CONCEPT CHECK: How might eviction of histone octamers affect transcription?

TABLE 15.1			
Standard Human Histones and Examples of Histone Variants			
Histone	**Type**	**Number of Genes in Humans**	**Function**
H1	Standard	11	Standard linker histone
H1⁰	Variant	1	Linker histone associated with chromatin compaction and gene repression
H2A	Standard	15	Standard core histone
MacroH2A	Variant	1	Core histone that is abundant on the inactivated X chromosome in female mammals. Plays a role in chromatin compaction.
H2A.Z	Variant	1	Core histone that is usually found in nucleosomes that flank the transcriptional start site of promoters. Plays a role in gene transcription.
H2A.Bbd	Variant	1	Core histone that promotes open chromatin. Plays a role in gene activation.
H2A.X	Variant	1	Plays a role in DNA repair
H2B	Standard	17	Standard core histone
spH2B	Variant	1	Core histone found in the telomeres of sperm cells
H3	Standard	10	Standard core histone
cenH3	Variant	1	Core histone found at centromeres. Involved with the binding of kinetochore proteins.
H3.3	Variant	2	Core histone that promotes open chromatin. Plays a role in gene activation.
H4	Standard	14	Standard core histone

Table 15.1 describes the standard histones and a few histone variants that are found in humans. A key role of many histone variants is to regulate the structure of chromatin, thereby influencing gene transcription. Such variants can have opposing effects. The incorporation of histone H2A.Bbd into a chromosomal region where a particular gene is found favors gene activation. In contrast, the incorporation of histone H1⁰ represses gene expression.

Although our focus in this chapter is on gene regulation, it is worth noting that histone variants play other important roles. For example, histone cenH3 (also called CENP-A), which is a variant of histone H3, is found at the centromere of each chromosome and functions in the binding of kinetochore proteins. Histone cenH3 is required for the proper segregation of eukaryotic chromosomes. Other histone variants are primarily found at specialized sites in certain cells. Histone macroH2A is found on the inactivated X chromosome in female mammals, whereas spH2B is found in the telomeres in sperm cells. Finally, certain histone variants appear to play a role in DNA repair. For example, histone H2A.X becomes

into a subset of nucleosomes to create functionally specialized regions of chromatin. In most cases, the standard histones are incorporated into the nucleosomes while new DNA is synthesized during S phase of the cell cycle. Later, some of the standard histones are replaced by histone variants via chromatin-remodeling complexes.

phosphorylated where a double-stranded DNA break occurs. This phosphorylation is thought to be important for the proper repair of that break.

The Histone Code Also Controls Gene Transcription

As described in Chapter 10, each of the core histone proteins consists of a globular domain and a flexible, charged amino-terminus called an amino-terminal tail (refer back to Figure 10.17a). The DNA wraps around the globular domains, and the amino-terminal tails protrude from the chromatin. Particular amino acids in the amino-terminal tails of both standard histones and histone variants are subject to several types of covalent modifications, including acetylation, methylation, and phosphorylation. Over 50 different enzymes have been identified in mammals that selectively modify the amino-terminal tails of histones. **Figure 15.5a** shows examples of sites in the tails of H2A, H2B, H3, and H4 that can be modified.

How do histone modifications affect the level of transcription? First, they may directly influence interactions within nucleosomes. For example, positively charged lysines within the core histone proteins can be acetylated by **histone acetyltransferases (HATs).** The attachment of the acetyl group ($-COCH_3$) eliminates the positive charge on the lysine side chain, thereby disrupting the electrostatic attraction between the histone protein and the negatively charged DNA backbone and favoring the open conformation (**Figure 15.5b**). However, histone acetylation is highly reversible. **Histone deacetylases (HDACs)** remove acetyl groups from acetylated histones and thereby favor a tighter contact between histones and the DNA.

In addition, histone modifications occur in patterns that are recognized by proteins. According to the **histone code hypothesis,** proposed by Brian Strahl, C. David Allis, and Bryan Turner in 2000, the pattern of histone modification acts much like a language or code in specifying alterations in chromatin structure. For example, one pattern might specify phosphorylation of the serine at the first position in H2A and acetylation of the fifth and eighth amino acids in H4, which are lysines. A different pattern could invoke acetylation of the fifth amino acid, a lysine, in H2B and methylation of the third amino acid in H4, which is an arginine.

The pattern of covalent modifications to the amino-terminal tails provides binding sites for proteins that subsequently affect the degree of transcription. One pattern of histone modification may attract proteins that inhibit transcription, which would silence the transcription of genes in the region. A different combination of histone modifications may attract proteins, such as chromatin-remodeling complexes, that would alter the positions of nucleosomes in a way that promotes gene transcription. For example, the acetylation of histones attracts certain chromatin-remodeling complexes that shift nucleosomes or evict histone octamers, thereby aiding in the transcription of genes.

Overall, the histone code is known to play an important role in determining whether the information within the genomes of eukaryotic species is accessed. Researchers are trying to decipher

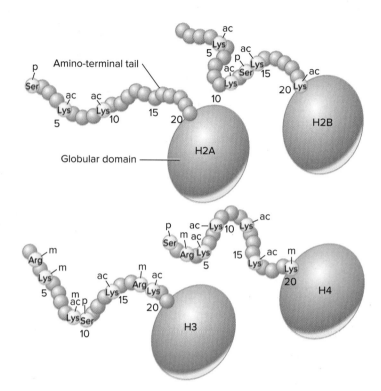

(a) Examples of possible histone modifications

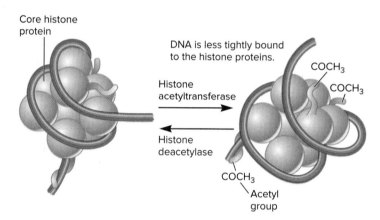

(b) Effect of acetylation

FIGURE 15.5 **Histone modifications and their effects on nucleosome structure.** **(a)** Examples of histone modifications that may occur on the amino-terminal tail of each of the four core histone proteins. The abbreviations are: p, phosphate; ac, acetyl group; and m, methyl group. **(b)** Effect of acetylation. When the core histones are acetylated via histone acetyltransferase, the DNA becomes less tightly bound to the histones. Histone deacetylase removes the acetyl groups. Note: The structures of the histone proteins in part (a) are meant to be schematic. H2A and H2B as well as H3 and H4 actually form intertwined dimers.

CONCEPT CHECK: Describe two different ways that histone modifications may alter chromatin structure.

the effects of the covalent modifications that make up the histone code. In Chapter 16, we will consider specific histone modifications that promote heterochromatin formation and thereby inhibit gene transcription.

15.2 COMPREHENSION QUESTIONS

1. A chromatin-remodeling complex may
   a. change the locations of nucleosomes.
   b. evict histones from DNA.
   c. replace standard histones with histone variants.
   d. do any of the above.
2. According to the histone code hypothesis, the pattern of histone modifications acts like a language that
   a. influences chromatin structure.
   b. promotes transcriptional termination.
   c. inhibits the elongation of RNA polymerase.
   d. does all of the above.

15.3 DNA METHYLATION

Learning Outcomes:

1. Define *DNA methylation*, and explain how it affects transcription.
2. Explain how DNA methylation is heritable.

We now turn our attention to a regulatory mechanism that involves a direct change in DNA structure. A methyl group (CH_3) can be attached to the cytosine base in DNA, a process called **DNA methylation.** The methylation of cytosines is common in some, but not all, eukaryotic species. For example, yeast and *Drosophila* have little or no detectable methylation of their DNA, whereas DNA methylation in vertebrates and plants is relatively abundant. In mammals, approximately 2–7% of the DNA is methylated. In this section, we will examine how DNA methylation occurs and how it controls gene expression.

DNA Methylation Occurs on the Cytosine Base and Usually Inhibits Gene Transcription

As shown in **Figure 15.6,** eukaryotic DNA methylation occurs via an enzyme called **DNA methyltransferase,** which attaches a methyl group to the carbon at the number 5 position of the cytosine base, forming 5-methylcytosine. The sequence that is usually methylated is shown here:

$$
\begin{array}{c}
CH_3 \\
| \\
5'-CG-3' \\
3'-GC-5' \\
| \\
CH_3
\end{array}
$$

Note that this sequence contains cytosines in both strands. Methylation of the cytosine in both strands is termed full methylation, whereas methylation of the cytosine in only one strand is called hemimethylation.

DNA methylation typically inhibits the transcription of eukaryotic genes, particularly when it occurs in the vicinity of the promoter. In vertebrates and plants, **CpG islands** occur near many promoters of genes. (Note: CpG refers to a dinucleotide of C and G in DNA that is

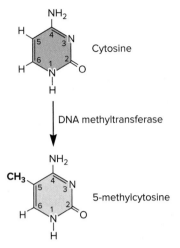

(a) The methylation of cytosine

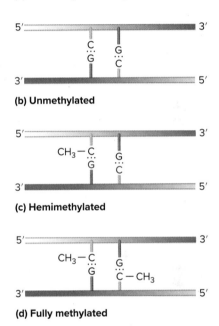

(b) Unmethylated

(c) Hemimethylated

(d) Fully methylated

FIGURE 15.6 DNA methylation on cytosine bases.
(a) Methylation occurs via an enzyme known as DNA methyltransferase, which attaches a methyl group to the number 5 carbon of cytosine. The CG sequence can be (b) unmethylated, (c) hemimethylated, or (d) fully methylated. In (b) and (c), the C in CH_3 refers to carbon, whereas the C in the DNA refers to cytosine.

connected by a phosphodiester linkage.) These CpG islands are commonly 1000–2000 bp in length and contain a large number of CpG sites. In the case of **housekeeping genes**—genes that code proteins required in most cells of a multicellular organism—the cytosine bases in the CpG islands are unmethylated. Therefore, housekeeping genes tend to be expressed in most cell types.

By comparison, other genes are highly regulated and may be expressed only in a particular cell type. These are **tissue-specific genes.** In some cases, it has been found that the expression of such genes may be silenced by the methylation of CpG islands. In general, unmethylated CpG islands are correlated with active genes, whereas suppressed genes contain methylated CpG islands. Thus, DNA methylation can play an important role in the silencing

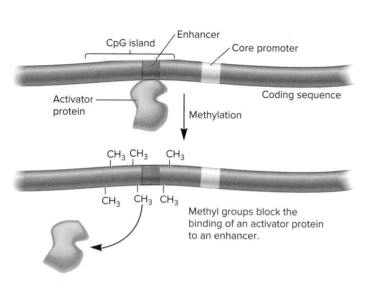

(a) **Methylation inhibits the binding of an activator protein.**

FIGURE 15.7 **Transcriptional silencing via methylation.**
(a) The methylation of a CpG island may inhibit the binding of a transcriptional activator protein to the promoter region. (b) The binding of a methyl-CpG-binding protein to a CpG island may lead to the recruitment of other proteins, such as histone deacetylase, that convert chromatin to a closed conformation and thus suppress transcription.

CONCEPT CHECK: Explain why the events shown in part (a) inhibit transcription.

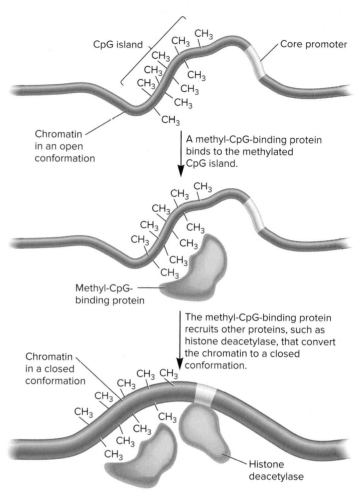

(b) **Methyl-CpG-binding protein recruits other proteins that change the chromatin to a closed conformation.**

of tissue-specific genes and thereby prevents them from being expressed in the wrong tissue. Methylation can affect transcription in two general ways, as described next.

Alteration in the Binding of Regulatory Transcription Factors.

Because the methyl group on cytosine protrudes into the major groove of the DNA, methylation can affect the binding of proteins into that groove, and thereby affect transcription. For example, methylation of CpG islands may prevent or enhance the binding of regulatory transcription factors to the promoter region. In some cases, methylated CpG islands prevent the binding of an activator protein to an enhancer (**Figure 15.7a**). The inability of an activator protein to bind to the DNA inhibits the initiation of transcription. However, CG methylation does not slow down the movement of RNA polymerase along a gene. In vertebrates and plants, coding regions downstream from the core promoter usually contain methylated cytosines, but these do not hinder the elongation phase of transcription. This observation suggests that methylation must occur in the vicinity of the promoter to have an effect on transcription.

Binding of Methyl-CpG-Binding Proteins.

A second way that methylation affects transcription is via proteins known as **methyl-CpG-binding proteins,** which bind to methylated CpG islands (**Figure 15.7b**). These proteins contain a domain called the methyl-binding domain that specifically recognizes a methylated

CpG island. Once bound to the DNA, the methyl-CpG-binding protein recruits to that region other proteins that inhibit transcription. For example, methyl-CpG-binding proteins may recruit histone deacetylase to a methylated CpG island near a promoter. Histone deacetylation removes acetyl groups from the histone proteins, which makes it more difficult for nucleosomes to be removed from the DNA. In this way, deacetylation tends to inhibit transcription.

DNA Methylation Is Heritable

Methylated DNA sequences are inherited during cell division. Experimentally, if fully methylated DNA is introduced into a plant or vertebrate cell, the DNA will remain fully methylated even in subsequently produced daughter cells. However, if the same sequence of nonmethylated DNA is introduced into a cell, it will remain nonmethylated in the daughter cells. These observations indicate that the pattern of methylation is retained following DNA replication and, therefore, is inherited in future daughter cells.

How can methylation be inherited from cell to cell? **Figure 15.8** illustrates a molecular model that explains this process, which was originally proposed by Arthur Riggs, Robin Holliday, and J. E. Pugh.

1. The DNA in a particular cell may become methylated by **de novo methylation**—the methylation of DNA that was previously unmethylated.

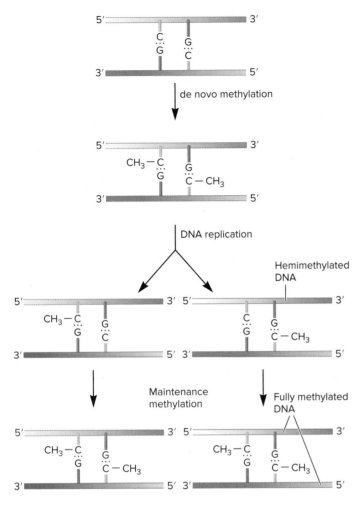

FIGURE 15.8 **A molecular model for the inheritance of DNA methylation.** The DNA initially undergoes de novo methylation, which is a rare, highly regulated event. Once this occurs, DNA replication produces hemimethylated DNA molecules, which are then fully methylated by DNA methyltransferase. This process, called maintenance methylation, is a routine event that is expected to occur for all hemimethylated DNA.

CONCEPT CHECK: What is the difference between de novo methylation and maintenance methylation?

2. When a fully methylated segment of DNA replicates in preparation for cell division, the newly made daughter strands contain unmethylated cytosines. Such DNA is said to be hemimethylated.

3. This hemimethylated DNA is efficiently recognized by DNA methyltransferase, which makes it fully methylated. This process is called **maintenance methylation,** because it preserves the methylated condition in future cells. However, maintenance methylation does not act on unmethylated DNA.

Overall, maintenance methylation appears to be an efficient process that routinely occurs within vertebrate and plant cells. By comparison, de novo methylation and demethylation are infrequent and highly regulated events. According to this view, the initial methylation or demethylation of a given gene can be regulated so that it occurs in a specific cell type or stage of development.

Once methylation has occurred, it can then be transmitted from mother to daughter cells via maintenance methylation.

The methylation mechanism shown in Figure 15.8 can explain the phenomenon of genomic imprinting, which is described in Chapters 5 and 16. In this case, specific genes are methylated during oogenesis or spermatogenesis, but not during both of these processes. Following fertilization, the pattern of methylation is maintained in the offspring. For example, if a gene is methylated only during spermatogenesis, the allele that is inherited from the father will be methylated in the somatic cells of the offspring, but the maternal allele will remain unmethylated. Along these lines, geneticists are eager to determine how variations in DNA methylation patterns may be important for cell differentiation. Methylation may be a key way to silence genes in different cell types. However, additional research is necessary to fully understand how specific genes may be targeted for de novo methylation or demethylation in specific cell types or during different stages of development.

15.3 COMPREHENSION QUESTIONS

1. How can methylation affect transcription?
   a. It may prevent the binding of regulatory transcription factors.
   b. It may enhance the binding of regulatory transcription factors.
   c. It may promote the binding of methyl-CpG-binding proteins, which inhibit transcription, to a methylated sequence.
   d. All of the above are possible ways for methylation to affect transcription.

2. The process in which completely unmethylated DNA becomes methylated is called
   a. maintenance methylation.
   b. de novo methylation.
   c. primary methylation.
   d. demethylation.

15.4 GENE ACTIVATION AND GENE REPRESSION

Learning Outcomes:

1. Describe the organization of nucleosome-free regions for an active gene.
2. Explain how activators bind to enhancers and facilitate the binding of the preinitiation complex.
3. Describe the key events that occur during the elongation phase of transcription.
4. Outline the steps of gene regulation via glucocorticoid receptors and CREB proteins.
5. Distinguish between short- and long-term repression.
6. Describe the method of chromatin immunoprecipitation sequencing, and explain what it is used for.

Gene activation is a series of events that enable a gene to be transcribed to produce an RNA molecule. For eukaryotic genes that are controlled by regulatory transcription factors, this process can be viewed as occurring in sequential phases:

1. One or more regulatory transcription factors (activators) bind to an enhancer.
2. The activators recruit coactivators such as chromatin remodeling complexes and histone-modifying enzymes.
3. RNA polymerase binds to the core promoter to form a preinitiation complex.
4. RNA polymerase proceeds to the elongation phase and makes an RNA transcript.

In previous sections, we have considered some structural features of RTFs, and also explored various types of changes that may alter chromatin and DNA structure. These changes include the following:

- Chromatin remodeling
- The presence of histone variants
- Histone modifications
- DNA methylation

These types of changes can favor transcription or inhibit it. For example, chromatin remodeling may favor an open conformation when histones are evicted, whereas DNA methylation typically inhibits transcription. In the case of histone modifications, some of them enhance transcription (e.g., acetylation) and others inhibit transcription (e.g., deacetylation). For a gene to be transcribed, it must undergo changes in chromatin structure that favor transcription, and also not be subjected to inhibitory modifications such as DNA methylation. In this section, we will first examine the steps that occur when RTFs activate the transcription of a eukaryotic gene, and consider some specific examples. After that, we will consider mechanisms of gene repression.

Eukaryotic Genes Are Flanked by Nucleosome-Free Regions and Well-Positioned Nucleosomes

Studies in yeast, *Drosophila*, humans, and other eukaryotic species have revealed that many eukaryotic genes show a common pattern of nucleosome organization (**Figure 15.9**). For constitutive genes or genes that are able to be activated, the transcriptional start site at the core promoter is found in a **nucleosome-free region (NFR),** which is a region of DNA where nucleosomes are not found. An NFR is typically 150 bp in length but it can be much longer. Although the NFR may be required for transcription, it is not, by itself, sufficient for gene activation. At any given time in the life of a cell, many genes that contain an NFR are not being actively transcribed.

The NFR at the transcriptional start site (TSS) is flanked by two well-positioned nucleosomes that are termed the −1 and +1 nucleosomes. In yeast, the TSS is usually at the boundary between the NFR and the +1 nucleosome. However, in animals, the TSS is about 60 bp farther upstream and within the NFR. The +1 nucleosome typically contains histone variants H2A.Z and H3.3. Depending on the species and the gene, these variants may also be found in the −1 nucleosome and in some of the nucleosomes that immediately follow the +1 nucleosome in the transcribed region. For example, the +2 nucleosome is likely to contain H2A.Z, but not as likely to do so as the +1 nucleosome is. Similarly, the +3 nucleosome is likely to contain H2A.Z, but not as likely as the +2 nucleosome is.

The nucleosomes downstream from the +1 nucleosome tend to be evenly spaced near the beginning of a eukaryotic gene, but their spacing becomes less regular farther downstream. The ends of many eukaryotic genes appear to have a well-positioned nucleosome that is followed by an NFR. This arrangement at the ends of genes may be important for transcriptional termination.

How are NFRs formed at particular sites in a cell's genome?

- Certain DNA sequences favor NFRs. For example, NFRs often have stretches that are AT-rich. However, such sequences, by themselves, are not sufficient to create an NFR.
- Certain transcription factors can bind to particular DNA sequences and recruit chromatin-remodeling complexes that evict histones to create an NFR. For example, a protein called Reb1 in yeast binds to a 22-bp sequence and recruits to that site chromatin-remodeling complexes that evict histones. The eviction of histones is facilitated by AT-rich sequences.

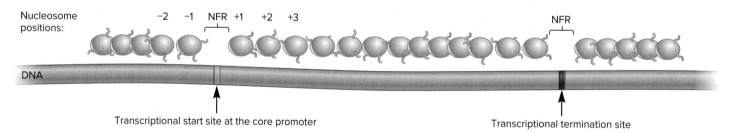

A nucleosome-free region (NFR) is found at the beginning and end of many genes. Nucleosomes tend to be precisely positioned near the beginning and end of a gene, but are less regularly distributed elsewhere.

FIGURE 15.9 Nucleosome arrangements and composition in the vicinity of a protein-coding gene.

CONCEPT CHECK: Why is an NFR needed at the core promoter for transcription to occur?

- Pioneer factors can bind to nucleosomes and recruit chromatin-remodeling complexes and histone modifying enzymes that carry out epigenetic changes such as histone eviction and covalent modifications (look ahead to Figure 16.12). Because pioneer factors play a role in tissue-specific gene expression, they are discussed in Chapter 16.

Another role of NFRs is to act as barriers to heterochromatin formation. As discussed in Chapter 16, the formation of heterochromatin involves a spreading phase (look ahead to Figure 16.7). By acting as barriers to heterochromatin spreading, NFRs allow genes to remain in a euchromatic (more open) conformation that is accessible to transcription.

The Formation of a PreInitiation Complex Involves the Binding of Activator Proteins That Promote Changes in Nucleosome Position, Changes in Nucleosome Composition, and Histone Modifications

As we have seen, genes in eukaryotes that are able to be transcribed are flanked by NFRs. Additional chromatin changes are needed for RNA polymerase to bind to the core promoter and to proceed to the elongation phase of transcription. A key role of RTFs is to recruit chromatin-remodeling complexes and histone-modifying enzymes to the promoter region. Though the order of recruitment may differ among specific transcriptional activators, this appears to be critical for transcriptional initiation and elongation. **Figure 15.10** presents a general scheme for

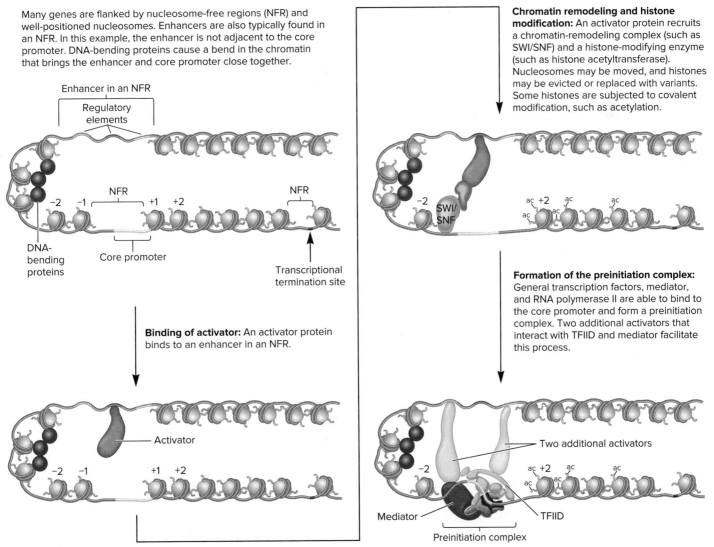

FIGURE 15.10 A simplified model for the formation of a preinitiation complex at a eukaryotic gene promoter. Note: For simplicity, the first activator and then the second and third activators are shown to bind sequentially. Commonly, multiple activators and coactivators bind simultaneously to form a large complex to initiate transcription. The assembly of the preinitiation complex is described in Chapter 12 (refer back to Figure 12.14).

CONCEPT CHECK: Explain how chromatin-remodeling complexes facilitate the binding of the preinitiation complex.

how transcriptional activators may facilitate the formation of a preinitiation complex.

1. ***Binding of activators to an enhancer:*** In the example in Figure 15.10, three activators bind to an enhancer. This enhancer is likely to be in an NFR and is typically located at a site that is distant from the core promoter. For activators to exert their effect, a bend must form in the chromatin so that the enhancer and core promoter come close to each other. This bend is facilitated by the binding of DNA-bending proteins, such as the high mobility group (HMG) proteins.

2. ***Recruitment of coactivators:*** The activators, which bind directly to the DNA of regulatory elements within the enhancer, then recruit **coactivators**—proteins that increase the rate of transcription but do not bind directly to the DNA. Coactivators can enhance transcription in a variety of ways, including chromatin remodeling, histone modification, recruitment or stimulation of the preinitiation complex, and stimulation of transcriptional elongation.

3. ***Chromatin remodeling and histone modifications:*** In the example of Figure 15.10, a chromatin-remodeling complex and a histone-modifying enzyme act as coactivators to modify chromatin structure. The chromatin-remodeling complexes may shift nucleosomes or temporarily evict histone octamers from the promoter region. Nucleosomes containing the histone variant H2A.Z, which are typically found at the +1 nucleosome, are thought to be more easily removed from the DNA than those containing the standard histone H2A. Histone-modifying enzymes, such as histone acetyltransferase, covalently modify histone proteins and may affect the contact between nucleosomes and the DNA.

4. ***Formation of the preinitiation complex:*** The actions of chromatin-remodeling complexes and histone-modifying enzymes facilitate the binding of general transcription factors, mediator, and RNA polymerase II to the core promoter, thereby allowing the formation of a **preinitiation complex** (see Figure 15.10). Activators may also be involved with recruitment or stimulation of the preinitiation complex, typically by interacting with mediator or general transcription factors such as TFIID.

As mentioned in Section 15.1, an enhancer and a promoter can be quite far away from each other. As shown in Figure 15.10, they can be brought close together by the formation of a loop. Because loops can bring enhancers and promoters close together, you may be wondering how enhancers are prevented from regulating lots of different genes that they are not supposed to regulate. One way is via the formation of topologically associating domains (TADs), which are described in Chapter 10 (refer back to Figure 10.23). The segments within a TAD are more likely to interact with each other than they are with segments in other TADs. In this regard, each TAD boundary acts as an **insulator,** which is a structure that limits genetic interactions to a specific region. A gene within a given TAD is insulated from the effects of enhancers in other TADs.

Formation of an Open Complex, Promoter Escape, Proximal Promoter Pausing, and Histone Eviction Are Key Events During the Elongation Phase of Transcription

After the preinitiation complex has formed, the next phase of transcription is elongation, or synthesis of an RNA transcript. As shown in **Figure 15.11,** different events are required for elongation to occur.

1. ***Formation of an open complex:*** In the preinitiation complex, also called the **closed complex,** the DNA strands are still bound to each other. For transcription to occur, the strands must be separated into an **open complex** so that one of them can act as a template for RNA synthesis. TFIIH, which a component of the preinitiation complex (refer back to Figure 12.14), has a subunit that functions as a DNA translocase, tracking along the DNA and leaving unwound DNA in its wake. This subunit separates the DNA strands to convert the preinitiation complex to an open complex.

2. ***Promoter escape:*** At the core promoter, GTFs and mediator bind to RNA polymerase II and prevent it from traveling along the template strand. For elongation to occur, RNA polymerase II must be released from this binding, an event called **promoter escape.** The CTD of RNA polymerase II contains many serines that can be phosphorylated by a variety of protein kinases, which are enzymes that attach phosphate to amino acid side chains such as serine. In particular, protein kinase subunits with TFIIH and mediator play key roles in phosphorylating CTD. Transcriptional activators and/or coactivators stimulate the ability of mediator to cause the phosphorylation of the carboxyl-terminal domain, thereby facilitating the switch between the initiation and elongation stages. The phosphorylation of CTD allows promoter escape.

3. ***Proximal promoter pausing:*** In complex eukaryotes, such as *Drosophila* and mammals, many or most genes are transcriptionally regulated by **proximal promoter pausing,** in which RNA polymerase II pauses in RNA synthesis while it is still fairly close to the transcriptional start site, usually within 60 bp.

    Proximal promoter pausing involves the binding of two factors, DRB sensitivity-inducing factor (DSIF) and negative elongation factor (NELF). To alleviate the pausing, a subunit within a protein complex called positive transcriptional elongation factor b (P-TEFb) phosphorylates both DSIF and NELF. This phosphorylation causes the release of NELF and also causes DSIF to facilitate elongation. RNA polymerase II is now able to transcribe the rest of the gene.

    Why does proximal promoter pausing happen? Several possible reasons have been proposed:

    - For many genes, the alleviation of pausing is the rate-limiting step for transcription. Therefore, pausing can be a key point to regulate the level of transcription. Various proteins can affect the rate of transcription by controlling the ability of P-TEFb to phosphorylate DSIF and NELF.
    - Pausing helps to maintain the nucleosome-free region by blocking nucleosome assembly over the core promoter. Therefore, it keeps the promoter region accessible for the binding of activators and coactivators.

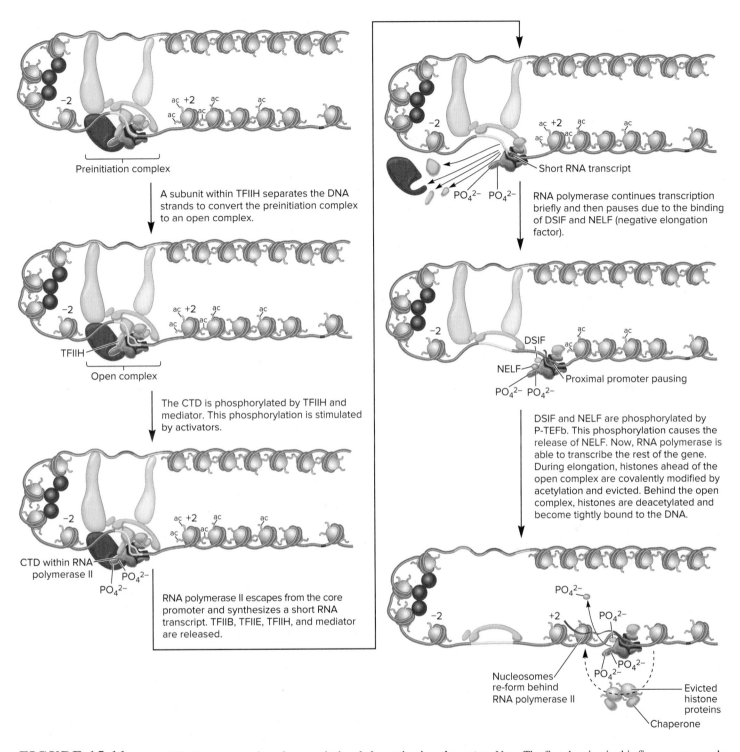

FIGURE 15.11 A simplified representation of transcriptional elongation in eukaryotes. Note: The first drawing in this figure corresponds to the last drawing in Figure 15.10. The names of the proteins in the preinitiation complex are described in Figure 12.14.

CONCEPT CHECK: When an activator interacts with mediator, how does this affect the function of RNA polymerase II?

- Pausing may provide time for the recruitment of factors that play a role in RNA modifications such as 5′ capping and splicing.
- Pausing may provide time for the binding of transcriptional elongation factors that provide stability and facilitate the transcription process.

4. **Histone modifications:** Histone-modifying enzymes also play a key role in histone removal and replacement during the elongation phase of transcription. Histone-modifying enzymes have been found to travel with RNA polymerase II during the elongation phase of transcription. These include enzymes that carry out histone acetylation, H3 methylation,

and H2B ubiquitination. These modifications facilitate histone removal ahead of the traveling RNA polymerase II. Behind RNA polymerase II, histone deacetylase removes the acetyl groups, thereby favoring the binding of histones to the DNA to form nucleosomes. The re-formation of nucleosomes behind RNA polymerase II is thought to be critical to maintaining the fidelity of transcription. If nucleosome re-formation did not occur properly, transcriptional initiation might occur at multiple points in a gene, thereby producing faulty transcripts.

Changes in the relative amounts of histone variants also occur within actively transcribed genes. For example, histone variant H3.3 is often found in the transcribed regions of genes, but is less common in silent genes. H3.3 may facilitate the eviction and replacement of nucleosomes ahead and behind of RNA polymerase II, respectively. Also, genes with very high levels of transcription may be largely devoid of nucleosomes, because multiple RNA polymerases are transcribing them at the same time.

GENETIC TIPS   THE QUESTION: Discuss the roles of
histones in eukaryotic transcription. How might histones inhibit transcription, and how are histones modified or moved to allow transcription to occur?

T OPIC: *What topic in genetics does this question address?*
The topic is the roles of histones in eukaryotic transcription.

I NFORMATION: *What information do you know based on the question and your understanding of the topic?* From the question, you know that histones may inhibit transcription. From your understanding of the topic, you may remember that histones may be moved, evicted, or replaced with histone variants by chromatin-remodeling complexes. You also may recall that histones can undergo covalent modifications that alter their functional properties.

P ROBLEM-SOLVING **S** TRATEGY: *Describe the steps.*
Eukaryotic transcription is a complex process. One strategy to solve this problem is to look at each step in the process, which is shown in Figures 15.10 and 15.11, and consider how histones could affect each step.

ANSWER:

1. Histones are removed at the core promoter of a gene. Activators also bind to enhancers in a nucleosome-free region.

2. Chromatin-remodeling complexes remove more histones to allow the preinitiation complex to form.

3. Histone variants are placed near the transcriptional start site; such variants are thought to be more easily removed during transcription.

4. Histones are covalently modified. Some modifications, such as acetylation, make it easier for histones to be removed so that transcription can take place. Other modifications, such as deacetylation, allow evicted histones to rebind to the DNA after RNA polymerase has passed.

5. A nucleosome-free region is also found at the end of a eukaryotic gene, which may facilitate transcriptional termination.

Steroid Hormones Exert Their Effects by Binding to a Regulatory Transcription Factor

Now that you have a general understanding of transcriptional activation, let's turn our attention to specific examples that illustrate how activators carry out their roles within living cells. Our first example is a category that responds to steroid hormones. This type of regulatory transcription factor is known as a **steroid receptor,** because the steroid hormone binds directly to the protein.

The ultimate action of a steroid hormone is to affect gene transcription, usually by increasing the rate of transcription. In animals, steroid hormones act as signaling molecules and are synthesized by endocrine glands and secreted into the bloodstream. The hormones are then taken up by cells that respond to these substances in different ways. For example, glucocorticoid hormones influence nutrient metabolism in most body cells. Other steroid hormones, such as estrogen and testosterone, are called gonadocorticoids because they influence the growth and function of the gonads (i.e., ovaries and testes).

Figure 15.12 shows the stepwise action of glucocorticoid hormones, which are produced in mammals.

1. In the example shown in the figure, the hormone enters the cytosol of a cell by diffusing through the plasma membrane.
2. Once inside, the hormone specifically binds to a **glucocorticoid receptor.** Prior to this binding, the glucocorticoid receptor is complexed with proteins known as heat shock proteins (HSPs), one example of which is HSP90.
3. After the hormone binds to the glucocorticoid receptor, HSP90 is released, thereby exposing a nuclear localization signal (NLS)—a sequence of amino acids within the protein that directs the protein into the nucleus.
4. Two glucocorticoid receptors form a homodimer and then travel through a nuclear pore into the nucleus.
5. In the nucleus, the glucocorticoid receptor homodimer binds to a pair of regulatory elements with the following consensus sequence:

$$5'-AGRACA-3'$$
$$3'-TCYTGT-5'$$

where R is a purine and Y is a pyrimidine. A glucocorticoid response element (GRE) contains two of these sequences running in opposite directions (Figure 15.12). This GRE is found within an enhancer that is next to many different genes. The binding of a homodimer to a GRE has two important effects. Compared to a monomer, a homodimer has a greater binding affinity for the DNA. Second, it promotes specificity. As a matter of random chance, a $5'-AGRACA-3'$ sequence could occur at many sites in a genome. However, two of them must be close together to create a GRE.

6. The binding of the glucocorticoid receptor homodimer to a GRE activates the transcription of the adjacent gene, eventually leading to the synthesis of the coded protein. Though not shown in Figure 15.12, the glucocorticoid receptor interacts with various types of coactivators, including those called steroid receptor coactivators (SRCs). An SRC functions as a histone acetyltransferase, and thereby promotes the formation of the preinitiation complex (see Figure 15.10).

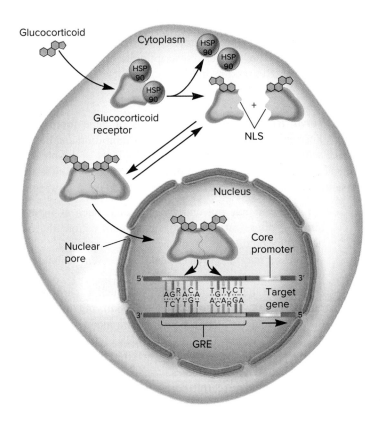

Glucocorticoid

Cytoplasm

HSP 90

HSP 90

HSP 90

Glucocorticoid receptor

NLS

+

Nuclear pore

Nucleus

Core promoter

Target gene

5′   3′

AG·R·A·C·A   T·G·T·Y·C·T
T·C·T·T·G·T   A·C·A·R·G·A

3′   5′

GRE

ONLINE ANIMATION

FIGURE 15.12 **The action of glucocorticoid hormones.** Once inside the cell, the glucocorticoid hormone binds to the glucocorticoid receptor, which then releases a heat shock protein known as HSP90. This exposes a nuclear localization signal (NLS). Two glucocorticoid receptors then form a homodimer and travel into the nucleus, where the dimer binds to a glucocorticoid response element (GRE) that is next to a particular gene. This binding of the glucocorticoid receptors to the GRE activates the transcription of the adjacent target gene. Note: This GRE could be in an enhancer that is relatively far away from the target gene, as in Figure 15.10.

Genes→Traits Glucocorticoid hormones are produced by the endocrine glands in response to fasting and activity. They enable the body to regulate its metabolism properly. When glucocorticoids are produced, they are taken into cells and bind to glucocorticoid receptors. This binding eventually leads to the activation of genes that code proteins involved in the synthesis of glucose, the breakdown of proteins, and the mobilization of fats.

CONCEPT CHECK: Explain why the glucocorticoid receptor binds next to the core promoter of some genes, but not next to the core promoter of most genes.

Mammalian cells usually have a large number of glucocorticoid receptors within the cytoplasm. Because GREs are located near dozens of different genes, the uptake of many hormone molecules can activate many glucocorticoid receptors, thereby stimulating the transcription of many different genes. For this reason, a cell can respond to the presence of the hormone in a very complex way. Glucocorticoid hormones stimulate many genes that code proteins involved in several different cellular processes, including the synthesis of glucose, the breakdown of proteins, and the mobilization of fats. Although the genes are not physically adjacent to each other, the regulation of multiple genes via glucocorticoid hormones is much like the simultaneous control of the expression of several genes by bacterial operons.

The CREB Protein Is an Example of an Activator That Is Modulated by Covalent Modification

As we have just seen, steroid hormones function as signaling molecules that bind directly to regulatory transcription factors to alter their function. This modulation of transcription factors enables a cell to respond to a hormone by up-regulating a particular set of genes. Most extracellular signaling molecules, however, do not enter the cell or bind directly to transcription factors. Instead, most signaling molecules must bind to receptors in the plasma membrane. This binding activates the receptors and may lead to the synthesis of an intracellular signal that causes a cellular response. One type of cellular response is an effect on the transcription of particular genes within the cell.

As our second example of the functioning of regulatory transcription factors within living cells, we will examine the

cAMP response element-binding protein (CREB protein). The CREB protein is a regulatory transcription factor that becomes activated in response to cell-signaling molecules that cause an increase in the cytoplasmic concentration of the molecule cyclic adenosine monophosphate (cAMP). The CREB protein, which is composed of two identical subunits, recognizes a DNA site that has two adjacent copies of the following consensus sequence:

$$5′\text{–TGACGTCA–}3′$$
$$3′\text{–ACTGCAGT–}5′$$

This response element, which is found near many different genes, is called the cAMP response element (CRE).

Figure 15.13 shows the steps leading to the activation of the CREB protein. A wide variety of hormones, growth factors, neurotransmitters, and other signaling molecules bind to plasma membrane receptors to initiate an intracellular response. In this case, the response involves the production of a second messenger, cAMP. The extracellular signaling molecule itself is considered the primary messenger.

1. When the signaling molecule binds to the receptor, it activates a G protein that subsequently activates the enzyme adenylyl cyclase.
2. The activated adenylyl cyclase catalyzes the synthesis of cAMP.
3. The cAMP molecule then binds to a second enzyme, protein kinase A, and activates it.
4. Protein kinase A travels into the nucleus and phosphorylates several different cellular proteins, including CREB protein.
5. This phosphorylation activates CREB protein by allowing it to bind to CREB-binding protein (CBP), which is a coactivator. CBP functions as a histone acetyltransferase that acetylates histones at the core promoter and thereby promotes the formation of the preinitiation complex. In contrast, unphosphorylated CREB protein can still bind to a CRE, but the binding does not lead to the activation of RNA polymerase because CREB cannot bind to the coactivator.

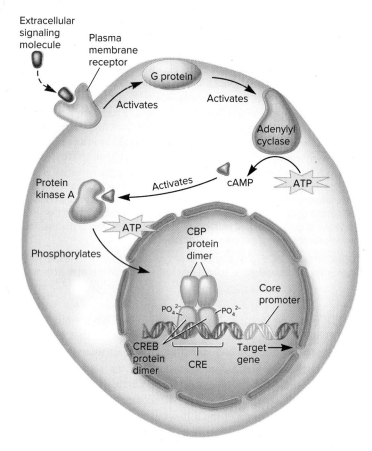

FIGURE 15.13 **Pathway that activates the CREB protein, and thereby leads to the transcriptional activation of a target gene.** An extracellular signaling molecule binds to a receptor in the plasma membrane, thereby activating a G protein, which then activates adenylyl cyclase, leading to the synthesis of cAMP. Next, cAMP binds to protein kinase A, which activates that enzyme. Protein kinase A then travels into the nucleus and phosphorylates the CREB protein. Once phosphorylated, CREB protein acts as a transcriptional activator by promoting the binding of CBP, which is a coactivator. Note: The CRE could be in an enhancer that is relatively far away from the target, as in Figure 15.10.

Different Mechanisms Can Result in Gene Repression

Thus far, we have been considering regulatory mechanisms that result in gene activation, a process that enhances the rate of transcription. We will now turn our attention to **gene repression,** which refers to any mechanism that inhibits the transcription of a gene and thereby results in a lower level of RNA synthesis from that gene. A repressor is a protein that binds directly to a DNA sequence, such as a regulatory element within an enhancer, and inhibits transcription. In eukaryotes, repressors may exert their effects by interacting with other proteins or protein complexes called **corepressors.**

Recall from Chapter 12 that basal transcription is transcription that occurs from a core promoter without the aid of enhancers, activators, and coactivators. Basal transcription from

a core promoter in eukaryotes results in a very low level of RNA synthesis. Therefore, gene repression is aimed at preventing gene activation and thereby causes transcription to occur at or below the basal level. Repression may be a short-term or long-term event.

Short-Term Repression. Some forms of gene repression cause transcription to be inhibited, but that inhibition can be reversed within a short time. These mechanisms are typically aimed at preventing one or more of the activation steps that are presented in Figures 15.10-15.13. These include the following:

- An activator may not undergo a change, such as ligand binding, dimerization, or covalent modification, that is needed for binding to an enhancer.
- A repressor may bind to an enhancer and prevent the binding of an activator.
- A repressor may recruit corepressors, such as chromatin-remodeling complexes or histone-modifying enzymes, that favor a closed conformation of chromatin. For example, a repressor may recruit histone deacetylase, which removes acetyl groups, thereby making the DNA more tightly bound to nucleosomes.
- A repressor may recruit corepressors that interact with TFIID and/or mediator in a way that prevents promoter escape.
- Various cellular factors may prevent P-TEFb from phosphorylating DSIF and NELF, and thereby maintain proximal promoter pausing.

Long-Term Repression. In long-term repression, more commonly called **gene silencing,** a gene is subjected to changes in chromatin structure that are more permanent in nature. These changes may be passed from mother cell to daughter cells during cell division, a phenomenon called epigenetics. Epigenetic changes are responsible for the establishment and long-term maintenance of gene repression. In multicellular organisms, such changes are important in silencing genes in a tissue-specific manner. For example, during embryonic development, certain genes are silenced in cells that are destined to become muscle cells. When muscle cells are formed later in development, the silenced genes will not be expressed.

How does gene silencing occur? In multicellular species, a common way for gene silencing to occur is via the conversion of euchromatin, which is more loosely packed, into heterochromatin, which is more tightly packed. This topic is discussed in Chapter 16 (see Section 16.2).

Chromatin Immunoprecipitation Sequencing Allows Researchers to Determine the Locations Where Specific Proteins Bind Throughout Entire Genomes

We have just considered how the binding of proteins to DNA elicits a series of events that may result in gene activation or gene repression. To better understand these events, researchers have developed methods that enable them to identify the DNA site(s) in a genome where a particular protein binds. One widely used method is called **chromatin immunoprecipitation sequencing (ChIP-Seq).**

As shown in **Figure 15.14**, ChIP-Seq begins with living cells.

1. The cells are treated with formaldehyde, which covalently links proteins to the DNA. Such a linkage is called a crosslink.

2. Following crosslinking, the cells are lysed, and the DNA is broken by sonication (or endonuclease digestion) into pieces approximately 200–1000 bp long. This yields a collection of millions of DNA fragments.

3. An antibody that is specific for the protein of interest is added. For example, if a researcher wanted to know the locations where CREB protein binds in a genome, the researcher would add an antibody that specifically recognizes CREB protein. Antibodies are made by the immune systems of mammals, and they specifically bind to antigens. In this case, the antigen is the protein of interest that is thought to be a DNA-binding protein. In Figure 15.14, the antibodies are attached to heavy beads. The antibodies bind to DNA-protein complexes and cause the complexes to form a pellet following centrifugation. Because an antibody is made by the immune system of an animal, this step is called **immunoprecipitation.**

4. In the next step, the proteins are removed by treatment with chemicals that break the covalent crosslinks.

5. Linkers are attached to the DNA fragments. These linkers are complementary to DNA primers that are used to subject the fragments to DNA sequencing.

6. How are the DNA sequences analyzed? The ChIP-Seq method is used on samples of cells from species in which the entire genome has already been sequenced, such as yeast, *Drosophila*, humans, and *Arabidopsis*. A genome map describes the locations of DNA sequences along each chromosome in a genome. Using computer software, the sequences obtained via ChIP-Seq can be matched to identical sequences on a genome map. This allows a researcher to identify where particular proteins bind.

15.4 COMPREHENSION QUESTIONS

1. A coactivator may
   a. covalently modify histones.
   b. evict histones from DNA.
   c. replace standard histones with histone variants.
   d. do any of the above.

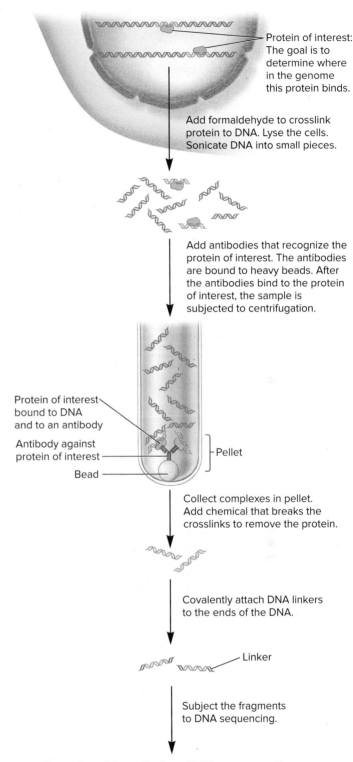

Protein of interest: The goal is to determine where in the genome this protein binds.

Add formaldehyde to crosslink protein to DNA. Lyse the cells. Sonicate DNA into small pieces.

Add antibodies that recognize the protein of interest. The antibodies are bound to heavy beads. After the antibodies bind to the protein of interest, the sample is subjected to centrifugation.

Protein of interest bound to DNA and to an antibody

Antibody against protein of interest

Bead

Pellet

Collect complexes in pellet. Add chemical that breaks the crosslinks to remove the protein.

Covalently attach DNA linkers to the ends of the DNA.

Linker

Subject the fragments to DNA sequencing.

The end result is a collection of DNA sequences. These sequences can be matched up with sequences in the genome to determine where the protein of interest binds in the genome.

FIGURE 15.14 **Chromatin immunoprecipitation sequencing (ChIP-Seq).**

CONCEPT CHECK: Why is an antibody used in this experiment?

2. A key event that allows promoter escape to occur is
   a. phosphorylation of the CTD.
   b. the formation of the open complex.
   c. the deacetylation of histones.
   d. All of the above facilitate promoter escape.

3. Which of the following characteristics is typical of a eukaryotic gene that can be transcribed?
   a. The core promoter is wrapped around a nucleosome.
   b. The core promoter is found in a nucleosome-free region.
   c. The terminator is wrapped around a nucleosome.
   d. None of the above characteristics is typical of a eukaryotic gene.

4. Transcriptional activation of eukaryotic genes involves which of the following events?
   a. Changes in chromatin structure
   b. Promoter escape
   c. Alleviation of proximal promoter pausing
   d. All of the above

5. The method of ChIP-Seq is used to
   a. measure the level of transcription of a particular gene.
   b. determine the locations in the genome where particular proteins can bind.
   c. map the locations of genes on a genome-wide level.
   d. determine the sequence of gene promoters on a genome-wide level.

15.5 A COMPARISON OF TRANSCRIPTIONAL REGULATION IN BACTERIA, ARCHAEA, AND EUKARYOTES

Learning Outcome:

1. Compare and contrast the process of transcriptional regulation among bacteria, archaea, and eukaryotes.

Throughout this chapter, we have considered mechanisms of transcriptional regulation with a focus on eukaryotes. In Chapter 14, we examined transcriptional regulation in bacteria. In this section, we will also include the third domain of life, the Archaea. As mentioned in Chapter 12 (see Section 12.5), this domain comprises a group of single-celled organisms with distinct molecular characteristics that distinguish them from the domain Bacteria. **Table 15.2** compares some general features of transcriptional regulation among these three domains of life.

15.5 COMPREHENSION QUESTION

1. Which of the following features is not similar in archaea and eukaryotes?
   a. Common location of regulatory elements
   b. Presence of mediator
   c. Gene organization
   d. All of the above features are not similar in archaea and eukaryotes.

TABLE 15.2

A Comparison of Transcriptional Regulation Among Bacterial, Archaeal, and Eukaryotic Cells

	Bacterial	Archaeal	Eukaryotic
Gene organization	Often in operons	Often in operons	Usually in single genes; though operons are observed in some species such as *Caenorhabditis elegans* (nemotode worm)
Location of regulatory elements	Operators are typically located close to the promoter.	Operators are typically located close to the promoter.	Enhancers are typically located at more distant sites from the core promoter, and a loop must form in the DNA to bring the enhancer and core promoter close together.
GTFs	Sigma (σ) factor is needed for promoter recognition.	Archaea have homologs to eukaryotic TBP (TATA-binding protein) and to subunits within TFIIB and TFIIE.	Six general transcription factors and mediator assemble at the core promoter.
Mediator	Absent	Absent	Present
Activators and Repressors	Bind to an operator site. Some of them interact directly with RNA polymerase, while others make local changes in DNA structure, like promoting the formation of a loop or a bend. Repressors may sterically inhibit RNA polymerase.	Bind to an operator site. Some of them interact directly with RNA polymerase, while others make local changes in DNA structure, like promoting the formation of a loop or a bend. Repressors may sterically inhibit RNA polymerase.	Bind to an enhancer and recruit coactivator complexes. The coactivators may alter chromatin structure and interact with GTFs and mediator to influence promoter escape and promoter pausing.
DNA methylation	Yes	Yes	Yes
Riboswitches	Yes	Yes	Yes (but not in animals)

KEY TERMS

Introduction: gene expression, gene regulation, combinatorial control

15.1: transcription factor, general transcription factor (GTF), regulatory transcription factor (RTF), domain, motif, homodimer, heterodimer, regulatory element (regulatory sequence or response element), enhancer, activator, repressor, up regulation, down regulation, orientation-independent (bidirectional)

15.2: closed conformation, open conformation, ATP-dependent chromatin remodeling (chromatin remodeling), DNA translocase, histone variants, histone acetyltransferase (HAT), histone deacetylase (HDAC), histone code hypothesis

15.3: DNA methylation, DNA methyltransferase, CpG island, housekeeping gene, tissue-specific gene, methyl-CpG-binding protein, de novo methylation, maintenance methylation

15.4: gene activation, nucleosome-free region (NFR), coactivator, preinitiation complex (closed complex), insulator, open complex, promoter escape, proximal promoter pausing, steroid receptor, glucocorticoid receptor, cAMP response-element-binding (CREB) protein, gene repression, corepressor, gene silencing, chromatin immunoprecipitation sequencing (ChIP-Seq), immunoprecipitation

CHAPTER SUMMARY

- Gene regulation refers to the phenomenon in which the level of gene expression can be controlled so that genes can be expressed at high or low levels. Gene regulation can occur at many points during gene expression.
- Combinatorial control means that the expression of a gene is regulated by a combination of many factors.

15.1 Regulatory Transcription Factors and Enhancers

- Regulatory transcription factors can be activators that bind to enhancers and increase transcription, or they can be repressors that inhibit transcription (see Figure 15.1).
- Transcription factor proteins contain specific domains that may be involved in a variety of processes, such as DNA binding and protein dimerization (see Figure 15.2).
- Regulatory elements are usually orientation-independent.
- The functions of regulatory transcription factors can be modulated by the binding of small effector molecules, protein-protein interactions, and covalent modifications (see Figure 15.3).

15.2 Chromatin Remodeling, Histone Variants, and Histone Modifications

- Chromatin remodeling occurs via ATP-dependent chromatin-remodeling complexes that alter the positions and compositions of nucleosomes (see Figure 15.4).
- Histone variants, which have amino acid sequences that differ slightly from those of the standard histones, play specialized roles in chromatin structure and function (see Table 15.1).
- The amino-terminal tails of histones are subject to covalent modifications that act as a histone code for the binding of proteins that affect chromatin structure and gene expression (see Figure 15.5).

15.3 DNA Methylation

- DNA methylation in eukaryotes is the attachment of a methyl group to a cytosine base via DNA methyltransferase (see Figure 15.6).

- The methylation of CpG islands near promoters usually silences transcription. Methylation may affect the binding of regulatory transcription factors, or it may inhibit transcription via methyl-CpG-binding proteins (see Figure 15.7).
- Maintenance methylation preserves the methylated condition of DNA following cell division (see Figure 15.8).

15.4 Gene Activation and Gene Repression

- Many eukaryotic genes are flanked by nucleosome-free regions (NFRs) and well-positioned nucleosomes (see Figure 15.9).
- During the initial stages of gene activation, the binding of activators to an enhancer recruits coactivators that facilitate the formation of a preinitiation complex (see Figure 15.10).
- During elongation, three key events are phosphorylation of CTD to elicit promoter escape, proximal promoter pausing, and histone eviction (see Figure 15.11).
- Steroid hormones, such as glucocorticoids, bind to receptors that function as transcriptional activators (see Figure 15.12).
- The CREB protein activates transcription after it has been phosphorylated by protein kinase A (see Figure 15.13).
- Gene repression can be short-term, usually by directly inhibiting one or more steps that are needed for gene activation, or it can be more permanent, as in gene silencing via the formation of heterochromatin.
- Chromatin immunoprecipitation sequencing (ChIP-Seq) is a method that is used to determine where particular proteins bind within an entire genome (see Figure 15.14).

15.5 A Comparison of Transcriptional Regulation in Bacteria, Archaea, and Eukaryotes

- With regard to transcriptional regulation, bacteria, archaea, and eukaryotes show similarities and differences (Table 15.2).

PROBLEM SETS & INSIGHTS

MORE GENETIC TIPS

1. The glucocorticoid response element (GRE) has two copies of the sequence 5′-AGRACA-3′, where R is a purine. Given that the size of the human genome is about 3 billion bp, how many times would you expect this sequence to occur, due to random chance? Would the glucocorticoid receptor bind to all of those sites? Why or why not?

T **OPIC: *What topic in genetics does this question address?*** The topic is the frequency of a sequence in the human genome and whether the glucocorticoid receptor would bind to all of the instances of the sequence.

I **NFORMATION: *What information do you know based on the question and your understanding of the topic?*** From the question, you know that a GRE has two copies of a certain sequence, and you are reminded of the size of the human genome. From your understanding of the topic, you may remember that the glucocorticoid receptor forms a dimer and binds to two adjacent copies of this sequence.

P **ROBLEM-SOLVING** **S** **TRATEGY: *Make a calculation.*** One strategy to solve this problem is to first consider the likelihood of the given six-base sequence occurring in the human genome, based on random chance. You can use the product rule, which is discussed in Chapter 3. The likelihood of any particular base in a given position is 1/4 (because there are 4 types of bases) and the likelihood of a purine is 1/2. If you assume random base arrangements, the likelihood of the six-base sequence is (1/4)(1/4)(1/2)(1/4)(1/4)(1/4), which equals 1/2048. In other words, this sequence would occur every 2048 bp by random chance alone.

ANSWER: If you divide 3 billion by 2048, you find that this sequence should occur about 1,464,844 times in the human genome! However, for the glucocorticoid receptor to bind tightly, two copies of this sequence must be close together. So, it would not bind to most of them.

---

2. A drug called garcinol, isolated from *Garcinia indica* (a fruit-bearing tree commonly known as kokum), is a potent inhibitor of histone acetyltransferase. Would you expect this drug to enhance or inhibit transcriptional initiation and elongation?

T **OPIC: *What topic in genetics does this question address?*** The topic is the role of histone modifications in eukaryotic transcription. More specifically, the question is about how the attachment of acetyl groups affects transcriptional initiation and elongation.

I **NFORMATION: *What information do you know based on the question and your understanding of the topic?*** From the question, you know that garcinol inhibits histone acetyltransferase. From your understanding of the topic, you may remember that acetylated histones do not bind as tightly to DNA and are more easily removed than are nonacetylated histones.

P **ROBLEM-SOLVING** **S** **TRATEGIES: *Relate structure and function. Predict the outcome.*** One strategy to solve this problem is to first consider how the drug will affect the ability of histones to bind to the DNA. You can then relate that effect to what is involved in the processes of transcriptional initiation and elongation.

ANSWER: Garcinol will inhibit the acetylation of histones. Histones that are not acetylated bind more tightly to DNA. This tighter binding will inhibit transcriptional initiation and elongation, which both require the removal of histones.

---

3. An approach to identifying genetic sequences that play a role in the transcriptional regulation of a gene is the strategy sometimes called promoter bashing. This approach requires gene cloning methods, which are described in Chapter 20. A clone is obtained that has the coding region for a protein-coding gene as well as a long region that is upstream from the core promoter. This upstream region may contain regulatory elements within enhancers. The diagram below depicts a cloned DNA region that contains the upstream region (shown in green), the core promoter (shown in tan), and the coding sequence for a protein that is expressed in human liver cells. The upstream region may be several thousand base pairs in length.

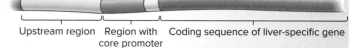

Upstream region   Region with core promoter   Coding sequence of liver-specific gene

To determine if promoter bashing has an effect on transcription, it is helpful to have an easy way to measure the level of gene expression. One way to accomplish this is to swap the coding sequence of the gene of interest with the coding sequence of another gene. For example, the coding sequence of the *lacZ* gene, which codes β-galactosidase, is frequently swapped because it is easy to measure the activity of β-galactosidase by performing an assay for its enzymatic activity. The *lacZ* gene is called a reporter gene because its activity is easy to measure. As shown here, the coding sequence of the *lacZ* gene has been swapped with the coding sequence of the liver-specific gene. In this new genetic construct, the expression and transcriptional regulation of the *lacZ* gene are under the control of the core promoter and an upstream region of the liver-specific gene.

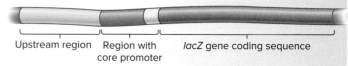

Upstream region   Region with core promoter   *lacZ* gene coding sequence

Now comes the "bashing" part of the experiment. Different segments of the upstream region are deleted, and then the DNA is transformed into living cells. In this case, the researcher would probably transform the DNA into liver cells, because those are the cells where the gene is normally expressed. The last step is to measure the β-galactosidase activity in the transformed liver cells.

In the following diagram, the upstream region and the region containing the core promoter (shown in tan) have been divided into five regions, labeled A–E.

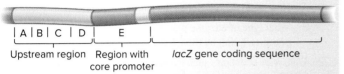

| A | B | C | D | E |

Upstream region   Region with core promoter   *lacZ* gene coding sequence

One of these regions was deleted (i.e., bashed out), and the rest of the DNA segment was transformed into liver cells. The data shown next are the results from this experiment.

Region deleted	Percentage of β-galactosidase activity*
None	100
A	100
B	330
C	100
D	5
E	<1

*The amount of β-galactosidase activity in the cells carrying an undeleted upstream region and core promoter was assigned a value of 100%. The amounts of activity in the cells carrying a deletion were expressed relative to this 100% value.

Explain what these results mean.

TOPICS: *What topic in genetics does this question address?* The topic is the regulation of gene transcription. More specifically, the question is about the effects of deletions on transcription.

INFORMATION: *What information do you know based on the question and your understanding of the topic?* In the question, you learned that researchers can connect a reporter gene to the promoter for a protein-coding gene. The region that is upstream from the promoter is likely to contain one or more enhancers. From your understanding of the topic, you may remember that activators bind to enhancers and increase the rate of transcription, whereas repressors bind to enhancers and decrease the rate of transcription. The core promoter is needed for a basal level of transcription.

PROBLEM-SOLVING **S**TRATEGIES: *Analyze data. Compare and contrast.* One way to solve this problem is to analyze the effects of the deletions by comparing the resulting levels of β-galactosidase activity with the level observed when there are no deletions.

ANSWER: The amount of β-galactosidase activity found in liver cells that do not carry a deletion reflects the amount of expression under normal circumstances. If the region with the core promoter (region E) is deleted, very little expression is observed. This is expected because a core promoter is needed for transcription. If enhancers that bind to activators are deleted, the activity should be less than 100%. It appears that one or more of this type of enhancer is/are located in region D. If enhancers that bind to repressors are deleted, the activity should be above 100%. From the data shown, it appears that one or more enhancers of this type are located in region B. Finally, if a deletion has no effect, the region may not contain any enhancers. This was observed for regions A and C.

Note: The deletion of an enhancer has an effect on β-galactosidase activity only if the cell is expressing one or more regulatory transcription factors that bind to the enhancer and activate or repress transcription. In the data given in this problem, the liver cells must be expressing the activator(s) and repressor(s) that recognize the enhancers found in regions D and B, respectively.

Conceptual Questions

C1. Discuss the common points of control in eukaryotic gene expression.

C2. With regard to eukaryotic gene regulation, what is meant by *combinatorial control*? Briefly describe six common types of such control.

C3. At the level of protein secondary structure, describe how regulatory transcription factors bind to DNA and are able to recognize a sequence of bases.

C4. Discuss the structure and function of regulatory elements. Where are they located relative to the core promoter?

C5. List and describe three general ways that the functions of transcription factors can be modulated.

C6. What are the functions of activator proteins and repressor proteins in transcription? Explain how these proteins work at the molecular level.

C7. Is each of the following statements true or false?

A. An enhancer contains one or more regulatory elements.

B. A core promoter is a type of regulatory element.

C. Regulatory transcription factors bind to regulatory elements.

D. The binding of a repressor to an enhancer may cause the down regulation of transcription.

C8. Transcription factors usually contain one or more motifs that play key roles in their function. What is the function of each of the following motifs?

A. Helix-turn-helix

B. Zinc finger

C. Leucine zipper

C9. The binding of small effector molecules, protein-protein interactions, and covalent modifications are three common ways to modulate the activities of transcription factors. Which of these three mechanisms are used by steroid receptors and by the CREB protein?

C10. Briefly describe three ways that ATP-dependent chromatin-remodeling complexes may change chromatin structure.

C11. What is a histone variant?

C12. Explain how the acetylation of core histones may reduce chromatin compaction.

C.13. What is meant by the term *histone code*? With regard to gene regulation, what is the proposed role of the histone code?

C14. What is DNA methylation? When we say that DNA methylation is heritable, what do we mean? How is it passed from a mother cell to a daughter cell?

C15. Let's suppose that a vertebrate organism carries a mutation that causes some cells that normally differentiate into nerve cells to differentiate into muscle cells. A molecular analysis reveals that this mutation is in a gene that codes a DNA methyltransferase. Explain how an alteration in a DNA methyltransferase could produce this phenotype.

C16. What is a CpG island? Where do you expect one to be located? How does the methylation of CpG islands affect gene expression?

C17. What is a nucleosome-free region? Where are such regions typically found in a genome? What is believed to be the functional importance of nucleosome-free regions?

C18. Beginning with the binding of an activator protein at an enhancer, describe the steps that occur to promote the formation of the preinitiation complex.

C19. Explain why DNA-bending proteins are needed for the transcriptional activation of most genes.

C20. What is the key difference between a preinitiation complex and an open complex?

C21. Define *promoter escape*. What key event facilitates promoter escape, and how is this event regulated?

C22. Histones are thought to be displaced as RNA polymerase II is transcribing a gene. What would be the potentially harmful consequence if histones were not put back onto a gene after RNA polymerase II had passed?

C23. Define *proximal promoter pausing*. How is the pausing alleviated? List possible reasons why it occurs.

C24. Describe the steps that need to occur for the glucocorticoid receptor to bind to a GRE.

C25. Suppose that a mutation in the glucocorticoid receptor does not prevent the binding of the glucocorticoid hormone to the protein but prevents the ability of the receptor to activate transcription. Make a list of all the possible defects that may explain why transcription cannot be activated.

C26. Explain how phosphorylation affects the function of the CREB protein.

C27. A particular drug inhibits protein kinase A, which is responsible for phosphorylating the CREB protein. How would this drug affect the following events?

A. The ability of the CREB protein to bind to a CRE

B. The ability of extracellular hormones to enhance cAMP levels

C. The ability of the CREB protein to stimulate transcription

D. The ability of the CREB protein to dimerize

C28. The glucocorticoid receptor and the CREB protein are two examples of transcriptional activators. These proteins bind to response elements and activate transcription. (Note: The answers to this question are not directly described in this chapter. You have to rely on your understanding of the functioning of other proteins that are modulated by the binding of effector molecules, such as lac repressor.)

A. How could the function of the glucocorticoid receptor be shut off?

B. What type of enzyme would be needed to block the activation of transcription by the CREB protein?

C29. Transcription factors such as the glucocorticoid receptor and the CREB protein form homodimers and activate transcription. Other transcription factors form heterodimers. For example, a transcription factor known as myogenic bHLH forms a heterodimer with a protein called the E protein. This heterodimer activates the transcription of genes that promote muscle cell differentiation. However, when myogenic bHLH forms a heterodimer with a protein called the Id protein, transcriptional activation does not occur. (Note: Id stands for "inhibitor of differentiation.") Which of the following possibilities best explains this observation? Only one possibility is correct.

	Myogenic bHLH	E protein	Id protein
Possibility 1			
DNA-binding domain:	Yes	No	No
Leucine zipper:	Yes	No	Yes
Possibility 2			
DNA-binding domain:	Yes	Yes	No
Leucine zipper:	Yes	Yes	Yes
Possibility 3			
DNA-binding domain:	Yes	No	Yes
Leucine zipper:	Yes	No	No

C30. A regulatory element, located upstream from a gene, has the following sequence:

$$5'-GTAG-3'$$
$$3'-CATC-5'$$

This element is orientation-independent. Which of the following sequences also works as a regulatory element?

A. 5'–CTAC–3'
   3'–GATG–5'

B. 5'–GATG–3'
   3'–CTAC–5'

C. 5'–CATC–3'
   3'–GTAG–5'

C31. The DNA-binding domain of each CREB protein subunit recognizes the sequence 5'–TGACGTCA–3'. Due to random chance, how often would you expect this sequence to occur in the human genome, which contains approximately 3 billion base pairs? Actually, only a few dozen genes are activated by the CREB protein. Does the value of a few dozen agree with the number of random occurrences expected in the human genome? If the number of random occurrences of the sequence in the human genome is much higher than a few dozen, provide at least one explanation why the CREB protein is not activating more than a few dozen genes.

C32. The gene that codes the enzyme called tyrosine hydroxylase is known to be activated by the CREB protein. Tyrosine hydroxylase is expressed in nerve cells and is involved in the synthesis of catecholamine, a neurotransmitter. The exposure of cells to adrenaline normally up-regulates the transcription of the tyrosine hydroxylase gene. A mutant cell was identified in which the tyrosine hydroxylase gene was not up-regulated when exposed to adrenaline. List all the possible mutations that could explain this defect. How would you explain the defect if only the tyrosine hydroxylase gene was not up-regulated by the CREB protein,

whereas other genes having CREs were properly up-regulated in response to adrenaline in the mutant cell?

C33. Explain the difference between short-term and long-term gene repression. List three different ways that short-term repression may occur.

Experimental Questions

E1. Briefly describe the method of chromatin immunoprecipitation sequencing (ChIP-Seq). How is it used to determine where regulatory transcription factors bind within a genome?

E2. Researchers can isolate a sample of cells, such as skin fibroblasts, and grow them in the laboratory. This procedure is called a cell culture. A cell culture can be exposed to a sample of DNA. If the cells are treated with agents that make their membranes permeable to DNA, the cells may take up the DNA and incorporate it into their chromosomes. This process is called transformation, or transfection. Scientists have transformed human skin fibroblasts with methylated DNA and then allowed the fibroblasts to divide for several cellular generations. The DNA in the daughter cells was then isolated, and the segment that corresponded to the transformed DNA was examined. This DNA segment in the daughter cells was also found to be methylated. However, if the original skin fibroblasts were transformed with unmethylated DNA, the DNA found in the daughter cells was also unmethylated. With regard to the transformed DNA, do fibroblasts perform de novo methylation, maintenance methylation, or both? Explain your answer.

E3. Restriction enzymes, discussed in Chapter 20, are enzymes that recognize a particular DNA sequence and cleave the DNA (along the DNA backbone) at that site. The restriction enzyme known as *Not*I recognizes the sequence

5′–GCGGCCGC–3′
3′–CGCCGGCG–5′

However, if the cytosines in this sequence have been methylated, *Not*I will not cleave the DNA at this site. For this reason, *Not*I is commonly used to investigate the methylation state of CpG islands.

A researcher has studied a gene, which we will call gene *T*, that is found in corn. This gene codes a transporter involved in the uptake of phosphate from the soil. A CpG island is located near the core promoter of gene *T*. The CpG island has a single *Not*I site. The arrangement of gene *T* is shown here:

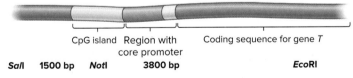

A *Sal*I restriction site is located upstream from the CpG island, and an *Eco*RI restriction site is located near the end of the coding sequence for gene *T*. The distance between the *Sal*I and *Not*I sites is 1500 bp, and the distance between the *Not*I and *Eco*RI sites is 3800 bp. No other sites for *Sal*I, *Not*I, or *Eco*RI are found in this region.

Here is the question. Suppose a researcher has isolated DNA samples from four different tissues in a corn plant: the leaf, the tassels,

a section of stem, and a section of root. The DNA was then digested with all three restriction enzymes, separated by gel electrophoresis, and then probed with a labeled DNA fragment complementary to the gene *T* coding sequence. The results are shown here.

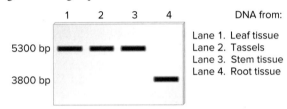

In which type(s) of tissue is the CpG island methylated? Does this make sense based on the function of the protein coded by gene *T*?

E4. You need to understand the approach described in question 3 in More Genetic TIPS before answering this question. A muscle-specific gene was cloned and then subjected to promoter bashing. As shown here, six regions, labeled A–F, were deleted, and then the DNA was transformed into muscle cells.

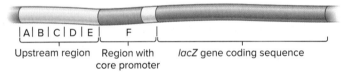

The following data show the results from this experiment.

Region deleted	Percentage of β-galactosidase activity
None	100
A	20
B	330
C	100
D	5
E	15
F	<1

Explain these results.

E5. You need to understand the approach described in question 3 in More Genetic TIPS before answering this question. A gene that is normally expressed in pancreatic cells was cloned and then subjected to promoter bashing. As shown here, four regions, labeled A–D, were individually deleted, and then the DNA was transformed into pancreatic cells or into kidney cells.

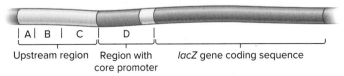

The data in the following table are the results from this experiment.

Region deleted	Cell type transformed	Percentage of β-galactosidase activity
None	Pancreatic	100
A	Pancreatic	5
B	Pancreatic	100
C	Pancreatic	100
D	Pancreatic	<1
None	Kidney	<1
A	Kidney	<1
B	Kidney	100
C	Kidney	<1
D	Kidney	<1

Assume that the upstream region has one enhancer that is recognized by one or more activators and another enhancer that is recognized by one or more repressors, and answer the following questions:

A. Where are these two enhancers located?

B. In pancreatic cells, why don't we detect the presence of the enhancer that is recognized by one or more represssors?

C. Why isn't this gene normally expressed in kidney cells?

E6. As described in Chapter 20, an electrophoretic mobility shift assay (EMSA) can be used to determine if a protein binds to a segment of DNA. When a segment of DNA is bound by a protein, its mobility through a gel will be slowed down, and the DNA band will be shifted to a placc that is higher in the gel. An EMSA is conducted with a cloned gene fragment that is 750 bp in length and contains a regulatory element that is recognized by a transcription factor called protein X. The results are shown below. Previous experiments showed that the presence of hormone X results in transcriptional activation by protein X.

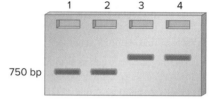

Lane 1. No additions
Lane 2. Plus hormone X
Lane 3. Plus protein X
Lane 4. Plus protein X and
          hormone X

Explain the action of hormone X.

Questions for Student Discussion/Collaboration

1. Explain how DNA methylation could be used to regulate gene expression in a tissue-specific way. When and where would de novo methylation occur, and when would demethylation occur? What would occur in the cells that give rise to eggs and sperm?

2. Enhancers can occur almost anywhere in DNA and affect the transcription of a gene. Let's suppose you are studying a gene that is within a DNA fragment that is 50,000 bp in length. Using molecular methods described in Chapter 20, you can cut out short segments from this 50,000-bp fragment and then reintroduce the smaller fragments into a cell that can express the gene. You would like to know if any enhancers are within the 50,000-bp region that may affect the expression of the gene. Discuss the most efficient strategy you can think of to trim your 50,000-bp fragment, thereby locating enhancers. You can assume that the coding sequence of the gene is in the center of the 50,000-bp fragment and that you can trim pieces of any size from the 50,000-bp fragment using molecular techniques described in Chapter 20.

Note: All answers are available for the instructor in Connect; the answers to the even-numbered questions and all of the Concept Check and Comprehension Questions are in Appendix B.

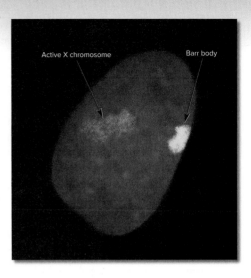

X-chromosome inactivation in female mammals. *This micrograph shows the nucleus from a human female cell in which a yellow fluorescent probe has been used to highlight the X chromosomes. The Barr body is more compact than the active X chromosome, which is to the left of the Barr body. The compaction of the Barr body is due to epigenetic modifications.*

Courtesy of I. Solovei, University of Munich (LMU)

16

GENE REGULATION IN EUKARYOTES II: EPIGENETICS

In Chapter 15, we began our discussion of gene regulation in eukaryotes with an examination of regulatory transcription factors. We also explored mechanisms that alter chromatin or DNA structure and thereby affect gene expression. These mechanisms include chromatin remodeling, histone variants, the covalent modification of histones, and DNA methylation. In this chapter, we will continue our discussion of eukaryotic gene regulation by examining epigenetics at the molecular level, currently one of the hottest topics in molecular genetics.

Epigenetics is a field of genetics that explores changes in gene expression that may be permanent over the course of an individual's life but are not permanent over the course of multiple generations. Epigenetic changes are responsible for the establishment and maintenance of gene activation or repression. Such changes enable cells to "remember" past events, such as developmental alterations in embryonic cells or prior exposure to environmental agents. In Chapter 5, we considered X-chromosome inactivation and genomic imprinting—both of which are explained by epigenetic gene regulation.

In this chapter, we will begin with an overview of the types of epigenetic modifications that affect gene expression. We will then examine how some epigenetic modifications are programmed to occur during development and others are the result of environmental factors.

16.1 OVERVIEW OF EPIGENETICS

Learning Outcomes:
1. Define *epigenetics* and *epigenetic inheritance.*
2. Outline the types of molecular changes that underlie epigenetic gene regulation.
3. Distinguish between *cis-* and *trans-*epigenetic mechanisms that maintain epigenetic changes.
4. Compare and contrast epigenetic changes that are programmed during development versus those that are caused by environmental agents.

The term *epigenetics* was coined by Conrad Waddington in the 1940s. The prefix *epi-*, which means "over," suggests that some types of changes in gene expression are at a level that goes beyond changes in DNA sequences. How do geneticists distinguish epigenetic effects from other types of gene regulation, such as those described in Chapters 14 and 15? An epigenetic effect begins with an initial event that causes a change in gene expression. For example, DNA methylation may inhibit transcription. However, for this to be an epigenetic effect, the change must be passed from cell to cell and must not involve a change in the base sequence of DNA.

Thus, a key feature of an epigenetic effect is the long-term maintenance of a change in gene expression. As an example, let's consider liver cells in humans. Some genes in the human genome should not be expressed in liver cells. During embryonic development, these genes in embryonic cells that are destined to become liver cells are inhibited by epigenetic changes such as DNA methylation. As the embryo grows and eventually becomes an adult, the epigenetic changes are passed from cell to cell so that adult liver cells do not express these inhibited genes.

Some epigenetic changes, such as those involving the silencing of genes in liver cells, are relatively permanent during the life of a single individual. Alternatively, other epigenetic changes may be reversible during the life of an individual, or they may be reversible from one generation to the next. For example, a gene that is silenced in one individual may be active in the offspring of that individual.

Although researchers are still debating the proper definition, one way to define epigenetics is the following.

- **Epigenetics** is the study of mechanisms that lead to changes in gene expression that can be passed from cell to cell and are reversible, but do not involve a change in the base sequence of DNA. This type of change may also be called an **epimutation**—a heritable change in gene expression that does not alter the sequence of DNA.

In multicellular species that reproduce via gametes (i.e., sperm and egg cells), when an epigenetic change is passed from parent to offspring, the phenomenon is called **epigenetic inheritance,** or **transgenerational epigenetic inheritance.** For example, as discussed in Chapter 5, genomic imprinting is an epigenetic change in which a gene is modified differently during oogenesis and spermatogenesis; an offspring expresses the gene from one parent but not both. However, not all epigenetic changes are passed to offspring via gametes. For example, an organism may be exposed to an environmental agent in cigarette smoke that causes an epigenetic change in a lung cell that promotes lung cancer and is subsequently transmitted from cell to cell. Such an epigenetic change in lung cells would not be transmitted to offspring.

In this section, we will begin with an examination of the molecular changes that have an epigenetic effect on gene expression. We will then consider how such changes may be programmed into an organism's development or caused by environmental agents.

Different Types of Molecular Changes Underlie Epigenetic Gene Regulation

The molecular mechanisms that promote epigenetic gene regulation are the subject of a large amount of recent research. The most common types of molecular changes that underlie epigenetic control are **DNA methylation, chromatin remodeling, covalent histone modification,** the localization of **histone variants,** and **feedforward loops (Table 16.1).** These types of changes can also be involved in transient (nonepigenetic) gene regulation. The details of the first four mechanisms were examined in Chapter 15. In Sections 16.3 and 16.5, we will explore specific examples in which epigenetic gene regulation occurs by the first four mechanisms. In some cases, epigenetic changes stimulate the transcription of a given gene, and in other cases they repress gene transcription.

TABLE 16.1
Molecular Mechanisms That Underlie Epigenetic Gene Regulation

Type of Modification	Description
DNA methylation	Methyl groups may be attached to cytosine bases in DNA. When methylation occurs near promoters, transcription is usually inhibited.
Chromatin remodeling	Nucleosomes may be moved to new positions or evicted. When such changes occur in the vicinity of promoters, the level of transcription may be altered. Also, larger-scale changes in chromatin structure may occur, such as those that happen during X-chromosome inactivation in female mammals.
Covalent histone modification	Specific amino acid side chains found in the amino-terminal tails of histones can be covalently modified. For example, they can be acetylated or phosphorylated. Such modifications may enhance or inhibit transcription.
Localization of histone variants	Histone variants may become localized to specific positions, such as near the promoters of genes, and affect transcription.
Feedforward loop	The activation of a gene that codes a transcription factor may result in a feedforward loop in which that transcription factor continues to stimulate its own expression. This mechanism is more common in microorganisms.

Epigenetic Changes May Be Targeted to Specific Genes by Transcription Factors or Non-coding RNAs

How are specific genes or chromosomes targeted for the types of epigenetic changes described in Table 16.1? In some cases, transcription factors may bind to a specific gene and initiate a series of events that leads to an epigenetic modification. For example, particular transcription factors in embryonic cells initiate epigenetic modifications that cause the cells to follow a specific pathway of development. For this to occur, the transcription factors recognize specific sites in the genome and recruit proteins, such as histone-modifying enzymes and DNA methyltransferase, to those sites. This recruitment leads to epigenetic changes, such as changes in chromatin structure and DNA methylation. A simplified illustration of this process is presented in **Figure 16.1a.**

In other cases, **non-coding RNAs (ncRNAs)**—RNAs that do not code polypeptides—are involved in establishing an epigenetic modification. Chapter 17 is devoted to the topic of ncRNAs and explores several examples in which ncRNAs facilitate epigenetic changes. Later in this chapter, we will consider how X-chromosome inactivation is mediated by an ncRNA. In some cases, ncRNAs act as bridges between specific sites in the DNA and proteins that alter chromatin or DNA structure, such as histone-modifying enzymes and DNA methyltransferase (**Figure 16.1b**).

Epigenetic Changes May Be Maintained by *cis*- or *trans*-Epigenetic Mechanisms

By studying epigenetics at the molecular and cellular levels, researchers have discovered that the types of epigenetic changes described in Table 16.1 can be maintained in two general ways,

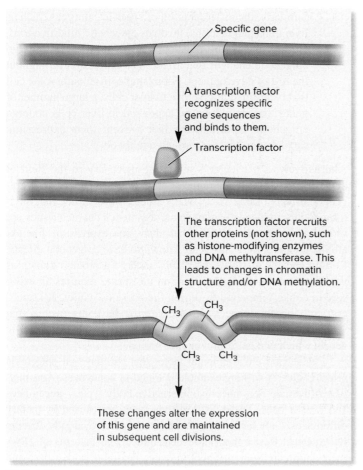

(a) Targeting of a gene for epigenetic modification by a transcription factor

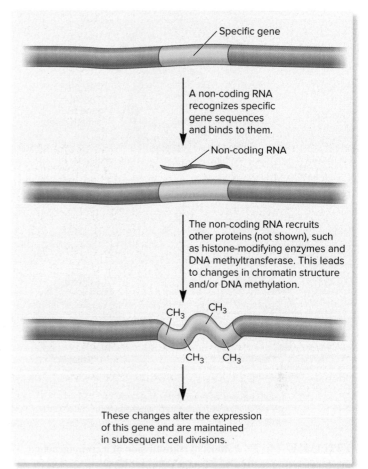

(b) Targeting of a gene for epigenetic modification by a non-coding RNA

FIGURE 16.1 **Establishing epigenetic modifications.** Two common ways are **(a)** via proteins such as transcription factors and **(b)** via non-coding RNAs.

called *cis-* and *trans-***epigenetic mechanisms.** These two mechanisms are distinguished by studying the epigenetic modification of a specific gene that occurs in two copies within a given cell (**Figure 16.2**).

With a *cis*-epigenetic mechanism, the epigenetic change at a given site is maintained only at that site; it does not affect the expression of the same gene located elsewhere in the cell nucleus. For example, if one copy of gene *B* is modified by DNA methylation and the other copy is not, a *cis*-epigenetic mechanism will maintain this pattern from one cell division to the next (see Figure 16.2). Genomic imprinting and X-chromosome inactivation, which are discussed in Chapter 5 and Section 16.3, are examples of *cis*-epigenetic mechanisms.

Alternatively, other epigenetic phenomena are explained by *trans*-epigenetic mechanisms that occur via diffusible molecules, such as proteins or ncRNAs. One example of a *trans*-epigenetic mechanism is paramutation, which is discussed in Section 16.4. Another example is a feedforward loop. In this mechanism, an epigenetic change is established by activating a gene that codes a transcription factor. After the transcription factor is initially made, it stimulates its own expression. Furthermore, if the gene that codes this transcription factor is present in

two copies in the cell, both copies will be activated because the transcription factor is a diffusible protein and many of these proteins are made when the gene is expressed. This pattern in which both copies of a gene are activated will be maintained following cell division. *Trans*-epigenetic mechanisms involving feedforward loops are more commonly found in prokaryotes and single-celled eukaryotes.

Epigenetic Gene Regulation May Occur as a Programmed Developmental Change or Be Caused by Environmental Agents

Many epigenetic modifications that regulate gene expression are programmed changes that occur at specific stages of development (**Table 16.2**).

- In Chapter 5, we examined genomic imprinting and X-chromosome inactivation. Genomic imprinting of the *Igf2* gene occurs during gametogenesis—the maternal allele is silenced, whereas the paternal allele remains active.
- By comparison, X-chromosome inactivation occurs during embryogenesis in female mammals. In early embryonic

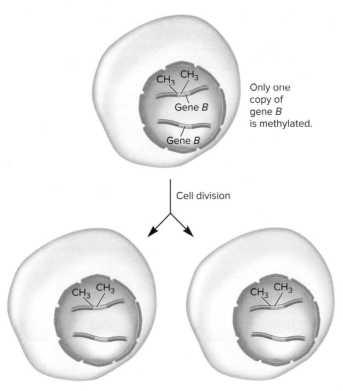

Only one copy of gene B is methylated.

Cell division

In subsequent cell divisions, one copy of gene B is always methylated, whereas the other copy is not.

FIGURE 16.2   **Pattern of transmission of a *cis*-epigenetic mechanism that maintains an epigenetic modification.**

CONCEPT CHECK: Explain how DNA methylation could be maintained by a *cis*-epigenetic mechanism.

cells, one of the X chromosomes of a female is inactivated and forms a Barr body, whereas the other remains active. This pattern is maintained as the cells divide and eventually form an adult organism.

- Similarly, the differentiation of specific cell types, such as liver cells and neurons, involves epigenetic modifications. During embryonic development, certain genes undergo epigenetic changes that affect their expression throughout the rest of development. For example, in an embryonic cell that is destined to give rise to liver cells, a large number of genes that should not be expressed in liver cells undergo epigenetic modifications that prevent their expression; such changes persist through adulthood.

Alternatively, a particularly surprising discovery in the field of epigenetics is that a wide range of environmental agents have epigenetic effects (Table 16.2). Many studies have shown that environmentally induced changes in an organism's characteristics are rooted in epigenetic changes that alter gene expression. For example, temperature changes are known to have epigenetic effects. Certain species of flowering plants undergo a process known as vernalization, in which flowering in the spring requires an exposure to colder temperatures during the previous winter. Researchers studying this process in *Arabidopsis* have discovered that vernalization involves covalent histone modifications of specific genes, which persist from winter to spring.

A second environmental factor that can have an epigenetic effect is diet. A striking example is found in honeybees (*Apis mellifora*). Female bees have two alternative body types—queen bees and worker bees. These distinct body types are caused by dietary differences. Only larvae that are persistently fed royal jelly develop into queens. Researchers have determined that patterns of DNA methylation are quite different in queen bees and worker bees and that the pattern of methylation affects the expression of many genes.

A third environmental factor that is of great interest to many geneticists is environmental toxins that can cause epigenetic changes. In humans, exposure to tobacco smoke has been shown to alter DNA methylation and covalent histone modifications of specific genes in lung cells. As discussed in Chapter 25, such changes may play a role in the development of cancer.

TABLE **16.2**	
Factors That Promote Epigenetic Changes	
Factor	**Examples**
Programmed Changes During Development	
Genomic imprinting	Certain genes, such as the *Igf2* gene discussed in Chapter 5, undergo different patterns of DNA methylation during oogenesis and spermatogenesis. Such patterns affect whether the maternal or paternal allele is expressed in offspring.
X-chromosome inactivation	As described in Chapter 5 and later in this chapter, X-chromosome inactivation occurs during embryogenesis in female mammals.
Cell differentiation	The differentiation of cells into particular cell types involves epigenetic changes such as DNA methylation and covalent histone modification.
Environmental Agents	
Temperature	In some species of flowering plants, cold winter temperatures cause specific types of covalent histone modifications that affect the expression of specific genes the following spring. This process may be necessary for seed germination or flowering in the spring.
Diet	The different diets of queen and worker bees alter DNA methylation patterns, which affects the expression of many genes. Such effects underlie the different body types of queen and worker bees.
Toxins	Cigarette smoke contains a variety of toxins that affect DNA methylation and covalent histone modifications in lung cells. These epigenetic changes may play a role in the development of lung cancer. In addition, metals, such as cadmium and nickel, and certain chemicals found in pesticides and herbicides, cause epigenetic changes that can affect gene expression.

16.1 COMPREHENSION QUESTIONS

1. Which of the following are examples of molecular changes that can have an epigenetic effect on gene expression?
   a. Chromatin remodeling
   b. Covalent histone modification
   c. Localization of histone variants
   d. DNA methylation
   e. Feedforward loops
   f. All of the above can have an epigenetic effect on gene expression.

2. An epigenetic modification to a specific gene may initially be established by
   a. a transcription factor.   c. either a or b.
   b. a non-coding RNA.   d. none of these.

3. In one cell, one allele of gene *C* is expressed, whereas the other allele is silenced due to an epigenetic modification. Following cell division, both of the daughter cells show the same pattern of expression: One copy of gene *C* is active and the other is silenced. This observation could be explained by
   a. a *trans*-epigenetic mechanism.
   b. a *cis*-epigenetic mechanism.
   c. neither a nor b.
   d. both a and b.

4. Epigenetic changes may
   a. be programmed during development.
   b. be caused by environmental factors.
   c. involve changes in the DNA sequence of a gene.
   d. Both a and b are true of epigenetic changes.

16.2 HETEROCHROMATIN: FUNCTION, STRUCTURE, FORMATION, AND MAINTENANCE

Learning Outcomes:

1. Explain the key differences between euchromatin and heterochromatin.
2. Describe the functional roles of heterochromatin.
3. Distinguish between constitutive heterochromatin and facultative heterochromatin, and describe their similarities and differences.
4. Explain how the pattern of heterochromatin is retained during cell division.
5. List the molecular changes that promote heterochromatin formation, and explain how they silence genes, lead to higher-order structures, and are maintained during cell division.
6. Describe examples of human inherited diseases that are caused by abnormalities in heterochromatin formation.

In eukaryotic cells, chromosomal DNA is packaged within the cell nucleus by the formation of chromatin, in which nucleosomes are the basic repeating units (refer back to Figure 10.17). In a nucleosome,

about 147 bp of DNA is wrapped around an octamer of histone proteins (H2A, H2B, H3, and H4). Originally, chromatin was thought to be composed only of DNA and proteins. However, we now know that non-coding RNAs, which are discussed in Chapter 17, are a critical component of chromatin.

As described in Chapter 10, chromatin occurs in two general forms: **euchromatin** and **heterochromatin.** The term *heterochromatin* was coined in 1928 by Emil Heitz. Using staining techniques and microscopy, Heitz found that chromosomes are composed of regions that are not stained during interphase (euchromatin) and those that remain stained throughout the entire cell cycle (heterochromatin). Heterochromatin is characterized by its greater level of compaction and by its inhibitory effect on gene expression. Heterochromatin and euchromatin typically occupy different nuclear regions (**Figure 16.3**). Heterochromatin is mostly localized along the periphery of the cell nucleus and is attached to the nuclear lamina that lines the inner nuclear membrane. By contrast, euchromatin occupies a more interior position.

In this section, we will begin with a brief discussion of the general functions of heterochromatin. We then discuss its key structural features, how it forms, and how it is passed from cell to cell during cell division. Finally, we will consider how abnormal heterochromatin formation may play a role in certain human diseases.

Heterochromatin Formation Plays Different Functional Roles in Eukaryotic Cells

The formation of heterochromatin plays several key functional roles in eukaryotic cells.

Gene Silencing. As mentioned earlier, heterochromatin formation is associated with the inhibition of transcription. Because of the more compact structure of heterochromatin, one mechanism by which heterochromatin formation silences the expression of

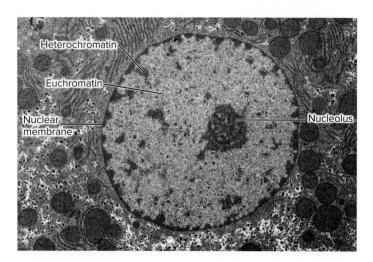

FIGURE 16.3 **Distribution of heterochromatin and euchromatin in the cell nucleus.** This transmission electron micrograph of a human liver cell shows darkly stained heterochromatin along the inner nuclear membrane. The lightly stained euchromatin is located more toward the interior of the nucleus.

Photo Researchers/Science History Images/Alamy Stock Photo

genes is by limiting the access of DNA-binding proteins, such as activators, to the DNA. Even so, some studies have shown that some activators can bind to enhancers that are found in heterochromatic regions. These studies suggest that, in some cases, heterochromatin formation may inhibit other aspects of the transcription process, such as the recruitment of general transcription factors or coactivators.

Prevention of Transposable Element Movement.
In Chapter 10, we examined the properties of transposable elements (TEs), which are mobile segments of DNA that can insert themselves into multiple sites within a genome. The random insertion of a TE into a gene is likely to inactivate that gene's function. Therefore, transposition can have a negative effect on gene expression, which may be detrimental with regard to cell structure and function. Eukaryotic species have evolved ways to prevent the movement of TEs and thereby minimize the detrimental consequences of TE movement.

How is TE movement inhibited? The sites where TEs are located are converted to heterochromatin, which silences genes that are needed for the transposition process. Because the mechanism of TE silencing involves non-coding RNAs, it is discussed in Chapter 17 (look ahead to Figure 17.12).

Prevention of Viral Proliferation.
Another role of heterochromatin formation is to prevent the proliferation of viruses. As a part of their reproductive cycle, some viruses integrate their DNA into the host genome in the form of a provirus. One way to prevent the provirus from becoming active and producing new viruses is to convert the region containing proviral DNA into heterochromatin, thereby inhibiting the expression of viral genes that are needed to produce new viruses.

Constitutive Heterochromatin and Facultative Heterochromatin Differ in Their Chromosomal Locations and Molecular Features

Another discovery made by Heitz is that some regions of heterochromatin are stained only in certain cells; these regions were later termed **facultative heterochromatin** to distinguish them from **constitutive heterochromatin,** which is heterochromatic in the same locations in all cell types. By comparison, the locations of facultative heterochromatin vary among different cell types. For example, a segment of DNA in a liver cell may occur within facultative heterochromatin and contain several genes whose expression is silenced. However, in a neuron, that same DNA segment may be packaged in euchromatin and the genes within the segment are not silenced. In this way, interconversions between euchromatin and heterochromatin are thought to play a key role in the silencing of genes in a cell-specific manner.

Although the structure of heterochromatin shows wide variation, recurring similarities and differences are found when comparing constitutive and facultative heterochromatin. They have been characterized with regard to their chromosomal locations, occurrence of repetitive sequences, DNA methylation, histone modifications, and histone variants.

Constitutive Heterochromatin.
Constitutive heterochromatin has some consistent structural features with regard to DNA sequences, DNA modifications, and histone modifications. Much of our understanding of constitutive heterochromatin comes from studies of yeast and mammalian cells.

- *Chromosomal locations:* Constitutive heterochromatin tends to be located close to a centromere, in an area called the pericentric region, and also at telomeres. These regions carry relatively few genes.
- *Repeat sequences:* The DNA within constitutive heterochromatin is largely composed of numerous, short, tandemly repeated sequences. These repeated DNA sequences are able to fold on themselves and may have an important role in the formation of the highly compact structure of constitutive heterochromatin. The specific base sequences found within tandem repeats in the pericentric regions of chromosomes vary among different species.
- *DNA methylation:* The DNA of constitutive heterochromatin is highly methylated on cytosines in vertebrates and plants.
- *Histone modifications:* As discussed in Chapter 10, the amino-terminal tails of histone proteins are subject to a variety of posttranslational modifications (PTMs). The trimethylation of a lysine at the ninth position in histone H3 (this modification is abbreviated H3K9me3) is a hallmark characteristic of constitutive heterochromatin in yeast and animal cells. In plants, such as *Arabidopsis thaliana*, a similar modification is found, except that it is a dimethylation (H3K9me2) instead of a trimethylation.

What are the consequences of PTMs, such as H3K9me3? Specific proteins have domains known as **reader domains,** which bind to particular nucleosome positions that have undergone PTMs (**Figure 16.4**). Reader domains can be found in the same proteins that also contain domains that modify chromatin. These chromatin-modifying domains are called **writer domains** if they catalyze the addition of PTMs (as in Figure 16.4a) or **eraser domains** if they remove PTMs. Alternatively, proteins with reader domains may bind to chromatin-modifying enzymes or chromatin-remodeling complexes and thereby recruit them to a nucleosome (Figure 16.4b). This recruitment leads to PTMs and chromatin remodeling that contribute to the formation of heterochromatin.

Some histone PTMs play a structural role. For example, a common modification found in constitutive heterochromatin is that many of the lysines in the histone tails are not acetylated, a phenomenon called hypoacetylation. As discussed in Chapter 15, the acetylation of lysines loosens the contact between histones and DNA, because it removes the positive charge on each lysine. By contrast, hypoacetylated histones bind DNA more tightly, which tends to inhibit transcription.

Facultative Heterochromatin.
Unlike constitutive heterochromatin, which tends to be the same in all cell types, the formation of facultative heterochromatin is variable among different cells types and is sometimes reversible. A heterochromatic state

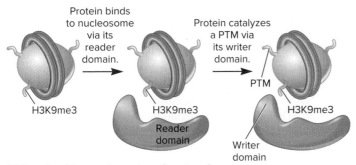

(a) Protein with a reader and a writer domain

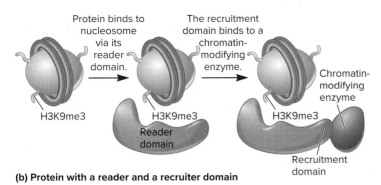

(b) Protein with a reader and a recruiter domain

FIGURE 16.4 **Possible functions of proteins with reader domains.** **(a)** In this example, a protein first binds to the H3K9me3 modification of a nucleosome via its reader domain. The protein also has a writer domain that catalyzes a PTM. **(b)** This protein also has a reader domain that binds to the H3K9me3 modification. Its recruitment domain binds to a chromatin-modifying enzyme.

CONCEPT CHECK: What is the function of a reader domain?

depends on the stage of development or the cell type examined. The formation of facultative heterochromatin plays a role in epigenetic processes such as genomic imprinting, gene silencing during development, and X-chromosome inactivation, which are described in Section 16.3. Facultative heterochromatin has some distinct characteristics.

- *Chromosomal locations:* Facultative heterochromatin may occur at multiple discrete sites that are located between the centromere and telomeres. This part of a eukaryotic chromosome contains many genes.
- *Repeat sequences:* In animals, the DNA of facultative heterochromatin is characterized by the presence of LINE-type repeated sequences, such as L1 in humans (see Chapter 10, Table 10.1). These sequences, which are dispersed throughout the genome, may initiate the propagation of a condensed chromatin structure.
- *DNA methylation:* The methylation of DNA in facultative heterochromatin is more discrete than in constitutive heterochromatin. Methylation often occurs in CpG islands, which are located in regulatory regions of genes. Such methylation plays a role in gene silencing.
- *Histone modifications:* Although the H3K9me3 modification has been mostly studied in constitutive heterochromatin, this PTM is also found in facultative heterochromatin.

In addition, facultative heterochromatin in animals often shows trimethylation of the 27th lysine of histone H3 (a PTM that is abbreviated H3K27me3). As discussed in Section 16.3, this PTM is associated with heterochromatin formation that silences genes in a cell-specific manner.

A Series of Molecular Events Results in Gene Silencing and Produces Heterochromatin with a Higher-Order Structure

In Chapter 10, we considered chromatin structure in nondividing cells and during cell division. Within regard to chromatin, the term **higher-order structure** is used to describe any assemblage of nucleosomes that assumes a reproducible conformation in three-dimensional space. For example, radial loop arrays are a higher-order structure found in metaphase chromosomes (refer back to Figure 10.29). Researchers are trying to characterize higher-order structures that are specific to heterochromatin. However, this quest has been challenging, in part, because heterochromatin is thought to be a very dynamic structure. Although much remains to be learned, heterochromatin formation is currently thought to involve the following molecular events:

- Posttranslational modifications of histones
- Binding of proteins to nucleosomes
- Chromatin remodeling
- DNA methylation (in vertebrates and plants)
- Binding of non-coding RNAs

What are the consequences of these molecular events? First, some of them silence gene expression. For example, DNA methylation of CpG islands and the removal of acetyl groups from histone proteins are known to inhibit transcription.

Another consequence of these molecular events is that they may promote the formation of higher-order structures in which heterochromatin has the following characteristics:

- Closer, more stable contacts of nucleosomes with each other
- Binding of heterochromatin to the nuclear lamina
- Liquid-liquid phase separation

Let's consider these three higher-order structures in more detail.

Closer, More Stable Contacts of Nucleosomes. As mentioned, the histone modification H3K9me3, which occurs in some species of yeast and in animals, is a hallmark of constitutive heterochromatin and is also found in regions of facultative heterochromatin. This histone modification is recognized by a reader domain in heterochromatin protein 1 (HP1). In 2018, Matthias Wolf, Hitoshi Kurumizaka, and colleagues used the technique of cryogenic (low-temperature) electron microscopy to visualize the interaction between HP1 and nucleosomes (**Figure 16.5**). HP1 forms a dimer and bridges two nucleosomes by binding to the H3K9me3 modification in each of them. In this way, it causes a closer association of nucleosomes and thereby makes the chromatin more compact.

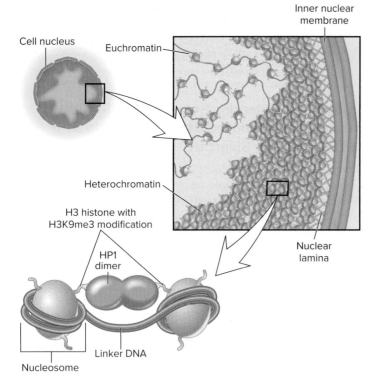

FIGURE 16.5 Bridging of nucleosomes via HP1. The HP1 protein forms a dimer that binds to two nucleosomes carrying the H3K9me3 modification, thereby holding the two nucleosomes in close association.

CONCEPT CHECK: How many H3K9me3 modifications can occur within a single nucleosome?

As seen in Figure 16.5, the HP1 dimer is localized in an accessible position between the nucleosomes. Though not shown in Figure 16.5, a recruitment domain within HP1 recognizes chromatin-modifying enzymes and chromatin-remodeling complexes and thereby localizes them to the site. The recruitment of these additional components leads to further changes in heterochromatin structure.

Binding of Heterochromatin to the Nuclear Lamina.

In most eukaryotic cells, the inner nuclear membrane of the cell nucleus is lined by the **nuclear lamina (NL),** which is composed of a fibrous layer of proteins (see Figure 16.5). In animal cells, these proteins are primarily intermediate filaments named lamins. In the 1960s, microscopy studies revealed the tight apposition of heterochromatin to the NL. More recently, molecular methods, such as ChIP-Seq (described in Chapter 15), have identified chromosomal regions that are in close contact with the NL, termed **lamina-associated domains (LADs).** LADs are thought to play two important roles:

- NL-anchoring via LADs may help to organize interphase chromosomes into chromosome territories, which are described in Chapter 10.
- Most of the thousands of genes in LADs are expressed at very low levels, suggesting a role for LADs in gene repression.

Researchers are beginning to identify the locations and biochemical characteristics of LADs.

- They are found in both constitutive and facultative heterochromatin.
- The average LAD contains about 2500 nucleosomes, and an average chromosome contains a few dozen LADs.
- LADs have CTCF recognition sites at their periphery. As discussed in Chapter 10 (refer back to Figure 10.21), two CTCFs bind to each other at the bottom of a loop domain.
- LADs show high levels of histone modification H3K9me3 (or H3K9me2 in plants), which is typical of heterochromatin. The H3K27me3 modification, which is specific to facultative heterochromatin, is also present in high levels at LAD boundaries in some cell types.

Researchers are trying to identify proteins that anchor LADs to the NL. Though other proteins may be involved with LAD formation, a mammalian protein, named PRR14, tethers heterochromatin to the nuclear lamina. PRR14 contains both an NL-binding domain and a separate domain that binds to HP1.

Liquid-Liquid Phase Separation.

Phase separation refers to the process in which a mixture of components in solution become unmixed (separate) into two or more distinct phases with different physical and chemical properties, like oil droplets in water. In living cells, liquid-like compartments are formed by macromolecules that become concentrated in a given location and come out of solution, a phenomenon called **liquid-liquid phase separation (LLPS).** As discussed in Chapter 13, the nucleolus, which is inside the cell nucleus, is an example of an organelle that is formed by LLPS.

Recent research suggests that heterochromatin may undergo LLPS. HP1 has regions that can form liquid droplets in vitro and liquid-like droplets in vivo. In addition, other components in heterochromatin may contribute to LLPS. The importance of LLPS in heterochromatin formation is an area under active research investigation.

How would LLPS affect chromatin structure and gene transcription? The environments within phase-separated droplets may favor the import of certain substances and the export of others. As shown in **Figure 16.6**, the environment within a heterochromatic droplet may favor the export of components that are involved with gene transcription and the import of components that are needed for heterochromatin formation. According to this scenario, the chromatin within the droplet would be transcriptionally inactive and would maintain a heterochromatic structure.

The Formation of Heterochromatin Involves Nucleation, Spreading, and Barriers

Thus far, we have considered the key molecular properties of constitutive and facultative heterochromatin and seen how those molecular properties inhibit gene expression and promote the formation of higher-order structures. Let's now look at how these two types of heterochromatin form at the chromosomal level, a process that has been most extensively studied in yeast and animal cells. The initial formation of constitutive and facultative

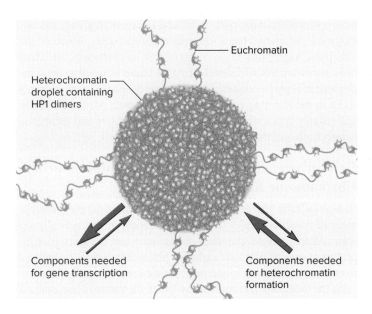

FIGURE 16.6 **Liquid-liquid phase separation (LLPS) of heterochromatin.** The phenomenon of LLPS causes regions of heterochromatin to form droplets.

CONCEPT CHECK: How does LLPS occur?

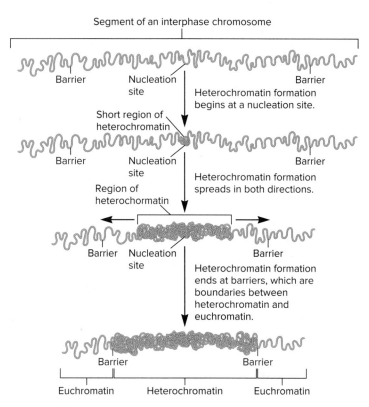

FIGURE 16.7 **Phases of heterochromatin formation.** Heterochromatin formation begins at a nucleation site and spreads in both directions until it reaches barriers. These barriers are the boundaries between adjacent heterochromatin and euchromatin.

heterochromatin occurs in three phases, as shown schematically in **Figure 16.7**.

During nucleation, a short chromosomal site attracts chromatin-modifying enzymes and chromatin-remodeling complexes that convert the site from euchromatin to heterochromatin. This nucleation site then spreads, usually in both directions, by converting adjacent euchromatin into heterochromatin. In interphase chromosomes, the spreading is stopped when it reaches a site called a barrier.

Let's consider each phase in more detail.

Nucleation. The term *nucleation* refers to a site where a process begins. Heterochromatin formation is initiated at a nucleation site by either sequence-specific DNA-binding proteins or by noncoding RNAs, both of which recruit histone deacetylases and histone methyltransferases. This recruitment typically results in the hypoacetylation of histones and the hypermethylation of H3K9 and/or H3K27 at the nucleation site.

Let's consider two examples. In *Drosophila*, nucleation in facultatitve heterochromatin begins with the binding of a protein to a specific DNA sequence called a polycomb response element (look ahead to Figure 16.13). This binding leads to the recruitment of protein complexes that result in the formation of heterochromatin. As a second example, other nucleation sites use components of the RNA interference (RNAi) machinery, which is described in Chapter 17. This second type of mechanism is involved in silencing TEs via heterochromatin formation (look ahead to Figure 17.12).

Spreading. How does heterochromatin formation spread? After chromatin-modifying enzymes are recruited to a nucleation site via HP1, they modify adjacent nucleosomes. For example,

methyltransferase promotes H3K9me3 modifications on adjacent nucleosomes, which leads to the binding of HP1 to those newly methylated sites. HP1 bound to the adjacent, newly methylated sites then recruits even more methyltransferases, thereby establishing a self-propagating mechanism to methylate adjacent nucleosomes and promote other changes in chromatin structure. Taken together, these combined actions lead to the bidirectional spreading of heterochromatin along a large region of chromatin in a DNA sequence-independent manner.

Barriers. Researchers have wanted to understand how the spreading phase is stopped so that a euchromatic region can be protected from adjacent heterochromatin. Barriers to heterochromatin spreading are thought to have different forms.

- One type of barrier is a nucleosome-free region. As discussed in Chapter 15, a nucleosome-free region is devoid of histone proteins. Therefore, PTMs of histones that are required for heterochromatin formation cannot occur in such a region. In this way, a nucleosome-free region can act as a boundary between heterochromatin and euchromatin.
- Alternatively, a barrier may be a site containing many anti-silencing proteins that counteract the effects of proteins required for heterochromatin formation. For example, a barrier may have histone-modifying enzymes that catalyze

the demethylation and acetylation of histones. Such enzymes favor the formation of euchromatin and inhibit the formation of heterochromatin.

- A third type of barrier is the formation of a loop domain. (A loop domain is not shown in Figure 16.7.) As discussed in Chapter 10 (refer back to Figure 10.21), the bottom of loop domains are held together by SMC proteins and interactions between CTCFs. These interactions can create a barrier that prevents the spread of heterochromatin into a loop domain. Alternatively, CTCFs could prevent the spread of heterochromatin out of a loop domain.

In Chapter 10, we discussed how insulators can prevent enhancers from regulating genes that are in different topologically associating domains. The barriers described for heterochromatin formation can also be categorized as insulators. In this case, an adjacent region of chromatin is insulated from the spreading of heterochromatin.

The Pattern of Heterochromatin in a Given Chromosome Is Passed from Cell to Cell

A defining feature of an epigenetic modification is that it can be passed from cell to cell during cell division. In other words, the epigenetic changes are maintained. Let's consider heterochromatin structure over the course of one cell division (**Figure 16.8**). (For simplicity, this drawing shows only one chromosome, whereas eukaryotic cells typically have two sets of several chromosomes.) At the top of the figure is a schematic illustration of a eukaryotic chromosome during interphase. This chromosome exhibits constitutive

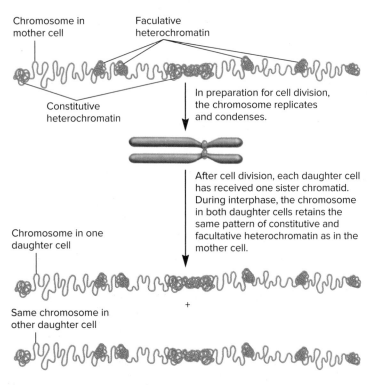

FIGURE 16.8 Pattern of heterochromatin before and after cell division. The pattern of constitutive and facultative heterochromatin is the same in the daughter cells as it was in the mother cell.

heterochromatin in the pericentric and telomeric regions, and also has some facultative heterochromatin at other sites. Much of this interphase chromosome is composed of euchromatin, in which most genes are not silenced. During M phase of mitosis, these euchromatic regions condense. Following M phase, each chromosome in the resulting two daughter cells, which are in interphase, will usually retain the same pattern of constitutive and facultative heterochromatin that was found in the mother cell.

Heterochromatin Structure Is Maintained During Chromosome Replication

In multicellular species, such as plants and animals, the establishment of constitutive and facultative heterochromatin typically occurs in embryonic development. As we have seen in Figure 16.8, the pattern of constitutive and facultative heterochromatin is maintained from one cell division to the next. Researchers are trying to determine the molecular mechanisms by which the pattern is maintained.

Prior to DNA replication, a region that is heterochromatic contains sites with fully methylated DNA and also contains histone modifications, such as H3K9me3, which promotes the binding of HP1 dimers and histone-modifying enzymes (**Figure 16.9**). After a DNA replication fork passes, each daughter chromatid is hemimethylated rather than fully methylated. Also, the original histones, which carry PTMs that promote heterochromatin formation, are randomly incorporated into both daughter chromatids behind the replication fork; about half of the original histones are distributed to each daughter chromatid and the other half of the histones in the daughter chromatids are newly synthesized. Therefore, the nucleosomes associated with newly synthesized DNA consist of both original and new histones. The new histones do not carry the histone modifications that promote heterochromatin formation.

Maintaining the epigenetic changes that promote heterochromatin formation requires mechanisms to quickly and reliably reestablish those modifications after DNA replication (Figure 16.9). Although the mechanisms are not entirely understood, they are thought to include the following:

- **DNA methylation:** As described in Chapter 15, hemimethylated DNA in eukaryotic cells becomes fully methylated by a process known as maintenance methylation, which occurs routinely in many types of eukaryotic cells.
- **Histone modifications:** The original histones that carry heterochromatic markers have previously recruited chromatin-modifying enzymes and chromatin-remodeling complexes to the heterochromatic region. For example, the original histones with the H3K9me3 modification have recruited histone methyltransferases that trimethylate lysine on histone H3. These enzymes will catalyzes the formation of H3K9me3 modifications on the newly made histones, which, in turn, will promote the binding of new HP1 proteins to form dimers that link nucleosomes together (Figure 16.9).
- **Role of DNA polymerase:** Components of DNA polymerase recruit chromatin-modifying enzymes that help to reestablish heterochromatin structure. These components, which are not shown in Figure 16.9, have been directly linked to heterochromatin formation in yeast, plants, and mammals.

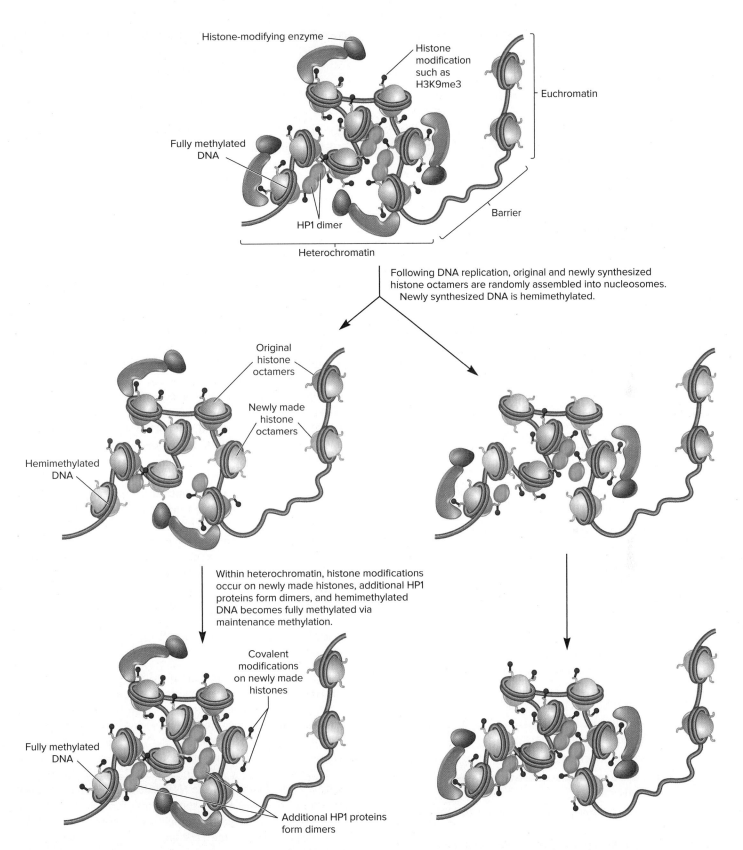

FIGURE 16.9 **Maintaining the epigenetic modifications that promote heterochromatin formation during DNA replication.** The events seen here occur during S phase of the cell cycle. Note: This figure emphasizes epigenetic changes in heterochromatin. Though not shown in this figure, other types of epigenetic changes would also maintain structural features of euchromatin.

CONCEPT CHECK: Describe how original and newly synthesized histone octamers are distributed after DNA replication occurs.

Abnormalities in Heterochromatin Formation Underlie Certain Inherited Human Diseases

In recent years, researchers have determined that certain inherited human diseases are caused by abnormalities in heterochromatin formation. An example is ICF syndrome, so named because it is associated with immunodeficiency, centromere instability, and facial anomalies. In some patients, this autosomal recessive disease is due to a loss-of-function mutation in the *Dnmt3b* gene, which codes a DNA methyltransferase. At the molecular level, certain regions that normally form constitutive heterochromatin are undermethylated. This hypomethylation causes the abnormal segregation of sister chromatids, chromosomal deletions, and other chromosomal abnormalities.

A second example of an inherited disease related to abnormal heterochromatin formation is Roberts syndrome, which is characterized by prenatal growth inhibition, craniofacial abnormalities, and limb malformations. This autosomal recessive disease is caused by loss-of-function mutations in the *Esco2* gene, which codes an acetyltransferase. This enzyme plays a key role in the establishment of sister chromatid cohesion during S phase. *Esco2* mutations result in delayed cell division, increased cell death, and impaired cell proliferation. The decrease in cell numbers during embryogenesis is thought to be responsible for the developmental defects observed in Roberts syndrome.

16.2 COMPREHENSION QUESTIONS

1. Heterochromatin is characterized by which of the following?
   a. Higher level of compaction
   b. Silencing of gene transcription
   c. Localization at the periphery of the cell nucleus
   d. All of the above are characteristics of heterochromatin.

2. Which of the following is *not* an important function of heterochromatin formation?
   a. Gene silencing
   b. Prevention of viral proliferation
   c. Splicing of pre-mRNA
   d. Prevention of movement of transposable elements

3. A form of chromatin that silences gene expression and varies in location among different types of cells is
   a. euchromatin.
   b. constitutive heterochromatin.
   c. facultative heterochromain.
   d. Both b and c are forms of chromatin with these characteristics.

4. Which of the following statements is/are true regarding a barrier?
   a. A barrier stops the spreading of heterochromatin.
   b. A barrier forms a boundary between euchromatin and heterochromatin.
   c. A barrier prevents DNA replication.
   d. Both a and b are true.

16.3   EPIGENETICS AND DEVELOPMENT

Learning Outcomes:

1. Describe the mechanism of genomic imprinting of the *Igf2* gene in mammals.
2. Outline the process of X-chromosome inactivation.
3. Describe the function of pioneer factors.
4. Explain how epigenetic modifications are involved in developmental changes that lead to the formation of specific cell types.

Beginning with gametes—sperm and egg cells—**development** in multicellular species involves a series of genetically programmed stages in which a fertilized egg becomes an embryo and eventually develops into an adult. These stages are discussed in Chapter 26. Over the past few decades, researchers have determined that epigenetic changes play key roles in the process of development in animals and plants.

At the molecular level, cells in the adult are able to "remember" events that happened much earlier in development. For example, in Chapter 5, we examined genomic imprinting. For the *Igf2* gene in mammals, an offspring expresses the copy that was inherited from the male parent, but not the copy that was inherited from the female parent. From an epigenetic perspective, the cells in the offspring are able to "remember" an event that occurred during gamete formation in their parents.

Likewise, during embryonic development, cells become destined to embark on pathways that lead to particular cell types. For example, an embryonic cell may give rise to a lineage of daughter cells that become a group of liver cells. The liver cells in the adult "remember" an event that occurred during embryonic development. In this section, we will explore the epigenetic mechanisms that explain how cells can remember events that occurred during specific stages of development.

Genomic Imprinting Occurs During Gamete Formation

In Chapter 5, we considered the imprinting of the *Igf2* gene in mice as an example of epigenetic inheritance (refer back to Figure 5.8). This gene codes a protein, insulin-like growth factor 2, that is required for proper growth.

The molecular mechanism of *Igf2* imprinting is due to different patterns of methylation during oogenesis and spermatogenesis (**Figure 16.10**). The *Igf2* gene is located next to another gene called *H19*. The function of the *H19* gene is not well understood, but it appears to play a role in some forms of cancer. Methylation may occur at a site called the **imprinting control region (ICR)** that is located between the *H19* and *Igf2* genes. A second site called a **differentially methylated region (DMR)** may also be methylated.

Oogenesis. During oogenesis (look at the upper right in Figure 16.10), methylation does not occur at either site. A protein

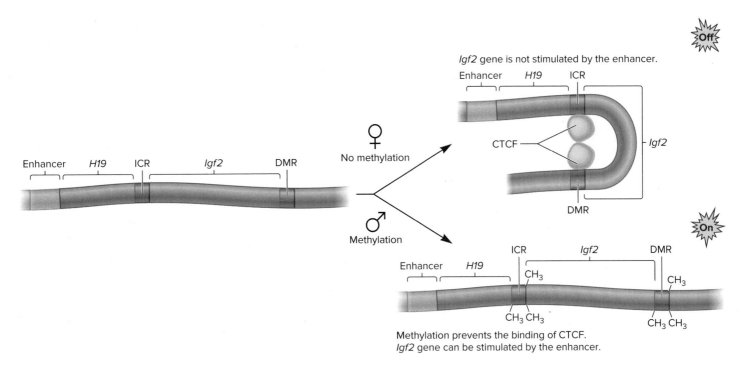

FIGURE 16.10 **A simplified molecular mechanism of *Igf2* imprinting.**   During oogenesis, the lack of methylation allows CTCFs to stabilize a loop, which inhibits the expression of *Igf2*. During spermatogenesis, methylation prevents CTCFs from binding, thereby preventing the stabilization of this loop. When this loop is not present, the enhancer can activate the expression of *Igf2*. Note: The formation and stabilization of loops are described in Chapter 10 (refer back to Figure 10.21).

Genes→Traits The phenomenon of genomic imprinting causes offspring to express the allele of one parent, but not both. In the case of the *Igf2* gene, offspring will exhibit the trait that is determined by the paternal allele.

called the CC<u>CTC</u>-binding factor (CTCF) binds to copies of a DNA sequence containing CCCTC (cytosine-cytosine-cytosine-thymine-cytosine) that are found in both the ICR and the DMR. The CTCFs bound to these sites may bind to each other to stabilize a loop in the DNA. This stabilization is needed to maintain the existence of the loop.

How can this loop affect the expression of *Igf2*? To understand how, we need to consider the effects of an enhancer that is located next to the *H19* gene. Even though it is fairly far away, this enhancer can stimulate transcription of the *Igf2* gene. However, when a loop is stabilized due to the interactions between two CTCFs, the change in chromatin structure prevents the enhancer from stimulating *Igf2*. Under these conditions, the maternal allele of *Igf2* is turned off.

Spermatogenesis.   Alternatively, if the ICR and the DMR are methylated, which occurs during sperm formation, CTCFs are unable to bind to these sites (look at the bottom right in Figure 16.10). This prevents loop stabilization via the two CTCFs, which allows the enhancer to stimulate the *Igf2* gene. Therefore, the paternally inherited *Igf2* allele is transcriptionally activated. The methylation that occurs during sperm formation is **de novo methylation,** which is the methylation of a completely unmethylated site.

Offspring.   After fertilization occurs, the somatic cells of the offspring maintain the methylated and nonmethylated states of the *Igf2* gene that occurred during gamete formation. The maternal allele stays nonmethylated, whereas the paternal allele is methylated. This is an example of *cis*-epigenetic mechanism (refer back to Figure 16.2). How does the paternal allele stay methylated in the somatic cells of the offspring? The persistence of methylation of the paternal allele is due to **maintenance methylation,** which is the methylation of hemimethylated sites (refer back to Figure 15.8).

Effects of Methylation.   As we have just seen, DNA methylation causes the *Igf2* gene to be transcriptionally active. However, it is worth noting that DNA methylation more commonly has the opposite effect—it inhibits transcription. How do we explain this difference? The answer is that it depends on the type of DNA sequence that is methylated and the events that occur after methylation:

- For nonimprinted genes, DNA methylation at CpG islands in the vicinity of promoters may have two effects (see Chapter 15). It may inhibit the binding of activator proteins or it may attract CpG-binding proteins to the promoter, which then convert the chromatin to a closed conformation. In both cases, the result is an inhibition of transcription.
- For imprinted genes such as *Igf2*, DNA methylation at ICRs and DMRs affects the binding of CTCFs. DNA methylation blocks CTCF binding, which prevents loop stabilization. The result is an activation of transcription.

GENETIC TIPS   **1.** Let's suppose a small deletion removes the DMR that is found after the *Igf2* gene, but it does not delete any part of the *Igf2* gene.

A. How would this deletion affect the expression of an allele inherited from a male parent?

B. How would it affect the expression of an allele inherited from a female parent?

T OPIC: *What topic in genetics does this question address?* The topic is genomic imprinting.

I NFORMATION: *What information do you know based on the question and your understanding of the topic?* In the question, you are given a description of a deletion that removes the DMR that is found after the *Igf2* gene. From your understanding of the topic, you may recall that methylation of the ICR and the DMR only occurs during sperm formation. Methylation inhibits CTCF binding, which prevents loop stabilization and allows the paternal allele to be expressed in offspring.

P ROBLEM-SOLVING **S** TRATEGY: *Make a drawing.* One strategy to solve this problem is to make a drawing and consider how the deletion will affect CTCF binding and whether a loop will be present during spermatogenesis and oogenesis.

Spermatogenesis:

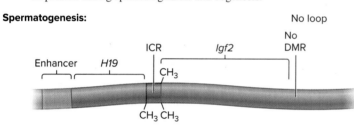

Oogenesis:

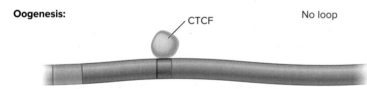

ANSWER:

A. During spermatogenesis, the deletion of the DMR doesn't really matter. Methylation of the ICR (and of the DMR if it were present) prevents loop stabilization. Because the loop cannot be stabilized, the paternal allele will be expressed.

B. During oogenesis, the deletion of the DMR does matter. To stabilize a loop, a CTCF must bind to both the ICR and the DMR, and then the two CTCFs bind to each other. However, this loop stabilization will not happen if the DMR is deleted. Because the loop is not present, the maternal allele will also be expressed.

X-Chromosome Inactivation in Mammals Occurs During Embryogenesis

As described in Chapter 5, **X-chromosome inactivation (XCI)** occurs in female mammals. One of the two X chromosomes in somatic cells is inactivated and becomes a condensed Barr body.

Because females are XX and males are XY, the process of XCI achieves dosage compensation—both females and males express a single copy of most X-linked genes. In addition, XCI is responsible for traits such as the calico coat pattern observed in certain female cats (refer back to Figure 5.3).

XCI is an epigenetic phenomenon. During early embryonic development in female eutherian (placental) mammals, one of the X chromosomes in each somatic cell is randomly chosen for inactivation. After this occurs, the same X chromosome is maintained in an inactivated state during subsequent cell divisions (refer back to Figure 5.4). A region of the X chromosome called the X-inactivation center (Xic) plays a key role in this process (**Figure 16.11a**). Within the Xic are two genes called *Xist*, for X̲ i̲nactive-̲specific t̲ranscript, and *Tsix*. *Xist* is highly expressed from the inactivated X chromosome, whereas *Tsix* is expressed from the active X chromosome. The two genes are transcribed in opposite directions. (Note: *Tsix* is *Xist* spelled backward; this naming reflects the opposite direction of transcription of the two genes.)

The mechanism of XCI is not completely understood, and researchers are still investigating key aspects of the process. **Figure 16.11b** shows a simplified model of how XCI occurs at the molecular level.

1. *Before X inactivation:* The expression of the *Tsix* gene from both X chromosomes inhibits the expression of the *Xist* gene. Because the genes are overlapping, *Tsix* transcription may directly inhibit *Xist* transcription, and also the *Tsix* RNA associates with DNA methylation machinery to silence the *Xist* promoter. At this very early embryonic stage, both X chromosomes are active (denoted Xa).

2. *Choosing the inactive X chromosome:* The next step is a random process in which one of the X chromosomes is chosen to be inactivated. This chromosome expresses the *Xist* gene, whereas the active X chromosome continues to express the *Tsix* gene. What governs this choice? Several proteins are known to bind to this region and influence the transcription of *Tsix* and/or *Xist*. Fluctuations in the concentrations of such factors may asymmetrically affect the *Xist* promoter on the two X chromosomes. For example, at the onset of XCI, a protein called RIF1 is lost from one of the active X chromosomes but remains associated with the

Portion of the X chromosome

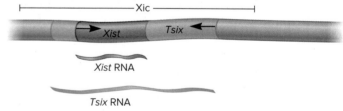

(a) The X-inactivation center (XIC)

ONLINE ANIMATION

FIGURE 16.11   The process of X-chromosome inactivation in eutherian mammals. Note: This is a simplified description of XCI. Several more proteins and ncRNAs that are not shown in this figure are involved in the process.

CONCEPT CHECK: In X-chromosome inactivation, when is the choice made as to which X chromosome is inactivated? Does this choice occur in embryonic cells, in adult somatic cells, or both?

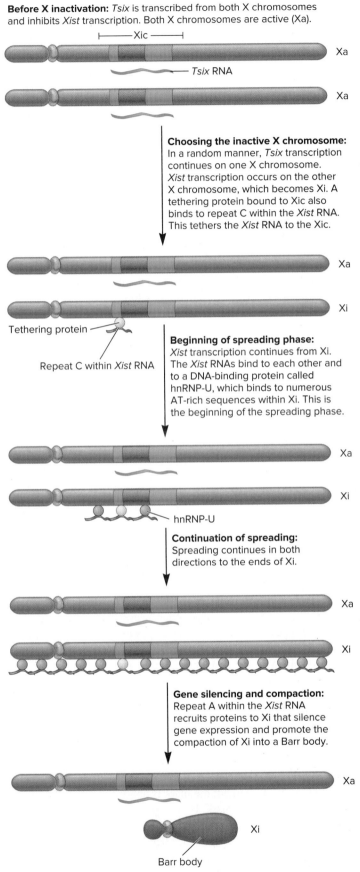

Before X inactivation: *Tsix* is transcribed from both X chromosomes and inhibits *Xist* transcription. Both X chromosomes are active (Xa).

Xic

Xa

Tsix RNA

Xa

Choosing the inactive X chromosome: In a random manner, *Tsix* transcription continues on one X chromosome. *Xist* transcription occurs on the other X chromosome, which becomes Xi. A tethering protein bound to Xic also binds to repeat C within the *Xist* RNA. This tethers the *Xist* RNA to the Xic.

Xa

Xi

Tethering protein

Repeat C within *Xist* RNA

Beginning of spreading phase: *Xist* transcription continues from Xi. The *Xist* RNAs bind to each other and to a DNA-binding protein called hnRNP-U, which binds to numerous AT-rich sequences within Xi. This is the beginning of the spreading phase.

Xa

Xi

hnRNP-U

Continuation of spreading: Spreading continues in both directions to the ends of Xi.

Xa

Xi

Gene silencing and compaction: Repeat A within the *Xist* RNA recruits proteins to Xi that silence gene expression and promote the compaction of Xi into a Barr body.

Xa

Xi

Barr body

(b) Mechanism of X-chromosome inactivation during embryonic development

FIGURE 16.11 Continued.

other one. The X chromosome with bound RIF1 expresses the *Xist* gene and is chosen to be inactivated.

Xist RNA is a 17-kb non-coding RNA (in humans) that is transcribed from Xi. This RNA consists of six repetitive sequences (called repeats A–F). Repeat A is located at the 5′ end of the *Xist* RNA and is necessary for later events that lead to the formation of a Barr body. Repeat C is necessary for the binding of *Xist* RNA to Xic via a tethering protein.

3. *Spreading:* The next phase of X-chromosome inactivation, called the spreading phase, involves the coating of Xi with *Xist* RNA. The transcription of the *Xist* gene produces many *Xist* RNA molecules. These *Xist* RNAs bind to each other and to a protein called hnRNP-U, which binds to numerous AT-rich sites along Xi.

4. *Gene compaction and silencing:* The first repeat sequence (repeat A) within the *Xist* RNA recruits protein complexes to Xi, thereby leading to epigenetic changes, which include the following:

   - *Xist* recruits a transcriptional repressor called SPEN to enhancers and promoters to trigger their silencing.
   - Covalent histone modifications, such as H3K27me3, occur and promote the formation of heterochromatin.
   - A histone variant, called macroH2A, is incorporated into nucleosomes at many sites along Xi.
   - DNA methyltransferases are recruited to Xi, which leads to DNA methylation.

Collectively, transcriptional repressors, covalent histone modifications, the incorporation of macroH2A into nucleosomes, and the methylation of many CpG islands are thought to play key roles in the silencing of genes on Xi and its compaction into a Barr body. These epigenetic changes in Xi are then maintained in subsequent cell divisions.

Pioneer Factors Bind Directly to Nucleosomes and Promote Changes in Cell Fate During Embryonic Development

As discussed in Chapter 15, many regulatory transcription factors (RTFs) bind to DNA sites that are found within nucleosome-free regions (NFRs). For example, RTFs bind to enhancers that are typically contained within NFRs (refer back to Figure 15.10). By comparison, a category of transcription factors called **pioneer factors** have the unique ability to recognize and bind to specific DNA sequences that are exposed on the surface of a nucleosome. Some of them bind directly to the center of a nucleosome, while others bind on the edge.

After binding to a nucleosome, pioneer factors subsequently recruit chromatin-remodeling complexes and histone-modifying enzymes that carry out epigenetic changes such as histone eviction and covalent modifications (**Figure 16.12**). These events may influence the ability of other transcription factors to bind to enhancer sequences. Pioneer factors can also decrease the level of DNA methylation by binding to CpG islands and thereby blocking access by DNA methyltransferases. At the genome level, pioneer factors are involved in the activation of some genes and the silencing of others.

During embryonic development in animals and plants, coordinated changes in gene expression allow a fertilized egg to develop

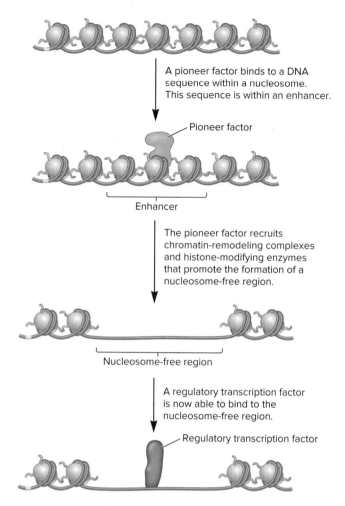

FIGURE 16.12 The action of pioneer factors. The pioneer factor binds directly to a DNA sequence within a nucleosome. After the pioneer factor binds, it recruits chromatin-remodeling complexes and histone-modifying enzymes that promote the formation of a nucleosome-free region (NFR). The formation of an NFR enables the binding of a regulatory transcription factor.

into a complex multicellular organism with many different cell types. For example, humans have liver cells, skin cells, and many others. The strikingly different characteristics of various cells types is due to gene regulation. Certain genes that are expressed in one cell type, such as a liver cell, are not expressed or are expressed at lower levels in another cell type, such as a skin cell. These differences in expression are largely controlled by the ability of transcription factors to gain access to and transcribe particular genes but not others.

Following fertilization and during early stages of embryonic development in animals, pioneer factors have been shown to play a key role in changing chromatin structure, which can have positive or negative effects on transcription. In the zygote, pioneer factors drive a reprogramming of the genome during the initial steps of development. The gene expression patterns of the egg and sperm are lost, and initially, the genome is reprogrammed to be pluripotent—the daughter cells in the early embryo are capable of differentiating into one of many cell types. As embryogenesis continues, the cells within the embryo adopt a cell fate; a cell and its subsequent daughter cells are destined to become a particular cell type or a few types.

Many different types of pioneer and nonpioneer transcription factors are known, and their levels of expression vary during different stages of embryonic development and among different cell types. For example, two pioneer factors, called FoxA and GATA4, bind to particular liver-specific enhancers in early embryonic cells that are destined to become liver cells. The binding of FoxA and GATA4 occurs long before gene expression is activated. These observations indicate that these pioneer factors prime certain genes for later expression, which occurs when the cells actually differentiate into liver cells. The emergent picture is that pioneer factors in conjunction with non-pioneer transcription factors enable some genes to be activated and others to be repressed. In each embryonic cell lineage, the functioning of pioneer factors and nonpioneer transcription factors ultimately causes cells to differentiate into particular cell types.

Pioneer factors are also important in differentiated cells in adults. For example, the glucocorticoid receptor, which is discussed in Chapter 15 (see Section 15.4), is a pioneer factor that orchestrates changes in chromatin structure.

Trithorax and Polycomb Group Complexes Are Key Regulators of Epigenetic Changes During Development

As we have seen, pioneer factors and nonpioneer transcription factors play important roles during development. In addition to these transcription factors, researchers have discovered that two competing groups of protein complexes—the **trithorax group (TrxG)** and the **polycomb group (PcG)**—are key regulators of epigenetic changes that are programmed during development. TrxG complexes are involved with gene activation, whereas PcG complexes cause gene repression. Both types of complexes are found in multicellular species, such as animals and plants, where they are required for proper development. TrxG and PcG complexes were discovered in genetic studies of *Drosophila*. The terms *trithorax* and *polycomb* refer to altered body parts associated with mutations in genes that code TrxG and PcG proteins, respectively.

TrxG and PcG complexes regulate many different genes, particularly those that code transcription factors that control developmental changes and cell differentiation. For example, PcG complexes regulate *Hox* genes, which are involved in specifying the structures that form along the anteroposterior axis in animals. The functions of *Hox* genes are described in Chapter 26 (look ahead to Figure 26.18).

At the molecular level, a key function of specific proteins within the TrxG and PcG complexes is the covalent modification of histones. For example, a component of TrxG recognizes histone H3 and attaches three methyl groups to a lysine at position 4—a process called **trimethylation.** (Note: This is methylation of a histone protein, not methylation of DNA.) The methylated histone is said to have been marked; the mark is abbreviated H3K4me3 and is called an activating mark. By comparison, a component of certain PcG complexes recognizes histone H3 and trimethylates a lysine at position 27. This mark (H3K27me3) is a repressive mark. Multiple marks are made by TrxG and PcG proteins. As discussed in Section 16.2, the H3K27me3 mark is associated with the formation of facultative heterochromatin.

Figure 16.13 is a simplified molecular model of how a gene or a region containing multiple genes may be targeted for

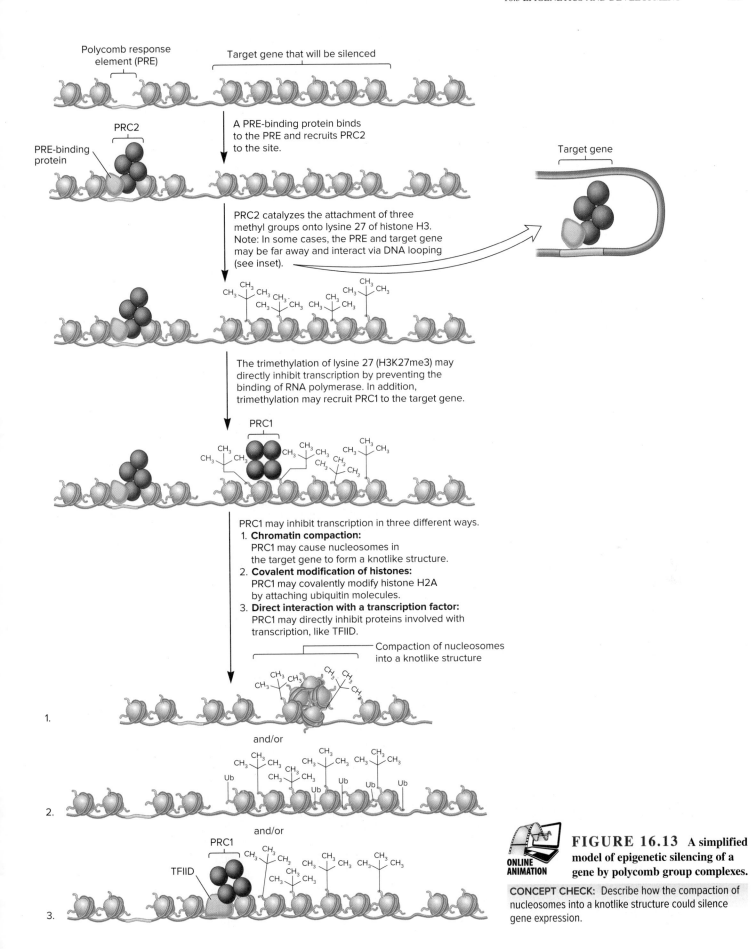

FIGURE 16.13 A simplified model of epigenetic silencing of a gene by polycomb group complexes.

ONLINE ANIMATION

CONCEPT CHECK: Describe how the compaction of nucleosomes into a knotlike structure could silence gene expression.

epigenetic silencing by PcG complexes. Keep in mind that some of these steps are not entirely understood and that the details of silencing may not be the same for different genes. This model considers the actions of two different types of PcG complexes, which are named polycomb repressive complex 1 and 2 (PRC1 and PRC2).

1. The first step involves the binding of PRC2 to a chromosomal site that is near a gene controlled by PcG complexes. In *Drosophila*, PRC2 binds to a DNA element called a **polycomb response element (PRE)**. The response element is initially recognized by a PRE-binding protein, which then recruits PRC2 to the site. In some cases, the PRE and the target gene are far apart and are brought close together by the formation of a DNA loop (see the inset in Figure 16.13). In mammals, the binding of PRC2 to specific genes may occur at PREs or CpG islands, or non-coding RNAs may recruit PRC2 to specific genes.

2. After PRC2 binds to a chromosomal site, a protein within that PcG complex catalyzes the covalent modification of histone H3 (see Figure 16.13). As mentioned earlier, certain PcG complexes recognize histone H3 and trimethylate a lysine at position 27.

3. The effect of trimethylation may have different effects. Primarily, trimethylation promotes the binding of PRC1. However, H3K27me3 modification may not always be required for PRC1 binding.

4. PRC1 complexes then silence gene expression. Three mechanisms of silencing have been proposed, which are not mutually exclusive:

   • The first mechanism involves chromatin compaction. PRC1 can catalyze the aggregation of nucleosomes into a more compact knotlike structure, which would silence gene expression (see Figure 16.13).

   • A second mechanism involves another covalent modification—the attachment of molecules of ubiquitin (Ub) to histone H2A. In some cases, this event is associated with the recruitment of PRC2.

   • A third possibility is a direct interaction with a transcription factor. In the model shown in Figure 16.13, PRC1 interacts with TFIID (described in Chapter 15), thereby inhibiting transcription.

After chromosomal regions have undergone epigenetic changes due to the actions of PcG complexes, these changes are maintained during subsequent cell divisions. In this way, epigenetic changes that occur during embryonic development can be transmitted to a population of cells that gives rise to a particular type of tissue, such as muscle tissue. The molecular mechanism (or mechanisms) by which such PcG-mediated changes are maintained during subsequent cell divisions is not well understood. One possibility is that PcG complexes may remain bound to their chromosomal sites during the process of DNA replication, thereby facilitating covalent histone modification and changes in chromatin structure after replication has occurred.

16.3 COMPREHENSION QUESTIONS

1. For the *Igf2* gene, where do de novo methylation and maintenance methylation occur?
   a. De novo methylation occurs in sperm, and maintenance methylation occurs in egg cells.
   b. De novo methylation occurs in egg cells, and maintenance methylation occurs in sperm.
   c. De novo methylation occurs in sperm, and maintenance methylation occurs in the somatic cells of offspring.
   d. De novo methylation occurs in egg cells, and maintenance methylation occurs in the somatic cells of offspring.

2. During the process of XCI, which chromosome expresses the *Xist* gene and which one expresses the *Tsix* gene?
   a. *Xist* is expressed only by Xa, and *Tsix* is expressed only by Xi.
   b. *Xist* is expressed only by Xi, and *Tsix* is expressed only by Xa.
   c. *Xist* is expressed by Xa, and *Tsix* is also expressed by Xa.
   d. *Xist* is expressed by Xi, and *Tsix* is also expressed by Xi.

3. Which of the following mechanisms may be involved when PRC1 complexes silence gene expression?
   a. The compaction of nucleosomes
   b. The attachment of ubiquitin to histone proteins
   c. The direct inhibition of transcription factors, such as TFIID
   d. Any of the above may be involved in silencing of gene expression by PRC1 complexes.

16.4 PARAMUTATION

Learning Outcomes:
1. Define *paramutation*.
2. Explain how paramutations can persist from generation to generation.
3. Describe the epigenetic changes that underlie paramutations.

A **paramutation** is an interaction between two alleles of a given gene in which one allele induces a heritable change in the other allele without changing that allele's DNA sequence. This phenomenon was first discovered in maize (corn, or *Zea mays*) by Alex Brink in the 1950s. Brink studied a gene called *red1*, which confers red color to corn kernels. Some alleles of this gene are weakly expressed, resulting in less red color. Brink discovered that certain weakly expressed alleles, designated *red1'* (read as "red one prime") can alter the expression of more strongly expressed *red1* alleles, changing them to the weak expression state. In future generations, the weak expression state adopted by a changed allele is heritable and can also act to reduce the expression of other strongly expressed alleles.

The *red1'* allele is said to be **paramutagenic,** because it can change the expression of another allele. The allele that can be altered by paramutation, such as the strongly expressed *red1* allele, is termed **paramutable.** After the paramutable allele is altered, its expression is the same or similar to the paramutagenic allele. Other alleles of this gene are not subject to paramutation, and these are called neutral alleles. In this section, we will examine the inheritance patterns that are caused by paramutation. We will also explore how epigenetic modifications are the underlying cause of this phenomenon.

Paramutations Occur When One Allele Alters the Expression of Another Allele

The phenomenon of paramutation has been most extensively studied in plants, particularly maize. Thus far, relatively few genes are known to undergo paramutation. However, paramutation has not been easy to detect unless it results in a readily discernible phenotype. For example, in maize, the majority of research on paramutation has centered on genes that code regulators of pigment synthesis. Alleles with reduced expression produce visible phenotypes, such as a change in stalk or kernel color. With the advent of newer molecular techniques to detect epigenetic changes, researchers may discover more genes that undergo paramutation. **Table 16.3** describes examples of paramutation in a plant, a fungus, and an animal.

A well-studied example of paramutation involves a gene in maize designated *b1*. This gene codes a transcription factor that regulates the synthesis of a pigment called anthocyanin. A dominant allele of this gene, designated *B-I*, is expressed at very high levels, which confers purple color to corn stalks and husks. (Note: The letter *I* in *B-I* refers to the intensity of the purple color.) Researchers discovered that the *B-I* allele can undergo paramutation, which causes a great reduction in its expression. The allele that has undergone paramutation is designated *B'*.

The *B'* allele has two striking properties:

- The paramutation changes the stalk and husk color from purple to green. The letter *I* is dropped from the allele's designation because the stalks and husks do not have an intense purple color.

TABLE 16.3
Examples of Paramutation

Species	Gene	Phenotype Conferred by Paramutagenic Allele*	Phenotype Conferred by Paramutable Allele
Maize (*Zea mays*)	*b1*	Green stalks and husks	Purple stalks and husks
Maize (*Zea mays*)	*red1*	Light pink to red kernels	Dark red kernels
Maize (*Zea mays*)	*pl1*	Light anthers	Red anthers
Maize (*Zea mays*)	*p1*	Light floral bracts	Red floral bracts
Fungus (*Ascobolus immersus*)	*b2*	White spores	Dark-brown spores
Mouse (*Mus musculus*)	*kit*	White feet and white spots on tail, especially the tip	Brown feet and brown tail

*The phenotypes listed are examples. Other phenotypic differences may be conferred by paramutagenic and paramutable alleles.

- The *B'* allele causes another *B-I* allele to undergo paramutation and become a *B'* allele. In other words, the *B'* allele is paramutagenic. This is an example of a *trans*-epigenetic mechanism (refer back to Section 16.1).

Figure 16.14 shows a cross between a strain that is homozygous for the weak *B'* allele and a strain that is homozygous for the strong *B-I* allele. This cross results in offspring that are all homozygous for the weak allele. How do we explain these results for the F_1 generation? They indicate that a paramutation occurred after fertilization that converted the *B-I* allele to the weaker *B'* allele.

In future generations, such as the F_2 generation in Figure 16.14, the reduced expression exhibited by the changed allele is heritable and can change the expression of other strongly expressed alleles, a process termed secondary paramutation. In this way, the influence of a paramutation can persist for many generations.

Depending on the particular gene, the effects of paramutation may vary in two ways: (1) the likelihood that the paramutagenic allele will alter the paramutable allele; and (2) the stability of the paramutagenic allele over the course of several generations.

1. Some paramutagenic alleles, such as *B'*, are highly paramutagenic. In crosses, such as the ones shown in Figure 16.14, *B'* always causes the paramutation of the *B-I* allele. For other genes, however, the rate of paramutation may be lower. For example, a gene designated *spt*, which confers antibiotic resistance in tobacco, is paramutated about 60% of the time. Among several different genes that can undergo paramutation, the reported frequency of conversion of the paramutable allele varies between 9% and 100%.
2. In some cases, a paramutagenic allele may be very stable, meaning that it remains paramutagenic every time it is passed from parent to offspring via meiosis. An example is the *B'* allele, featured in Figure 16.14. Other paramutagenic alleles are less stable and sometimes revert back to stronger alleles that are not paramutagenic. For example, in the fungus *Ascobolus immersus*, a gene designated *b2* exists as a paramutagenic allele that confers white spores and a paramutable allele that confers dark-brown spores. During meiosis, about 10–40% of the paramutagenic alleles revert back to paramutable alleles, which are not paramutagenic.

Paramutations Are Due to Epigenetic Changes That Are Transmitted from Parent to Offspring

As mentioned, paramutations do not alter the DNA sequence. Instead, they affect gene expression by promoting epigenetic changes. In other words, the reduced expression of a paramutagenic allele is due to epigenetic changes that decrease or silence its transcription. These epigenetic changes can be transferred to a paramutable allele, which also decreases its expression and transforms it into a paramutagenic allele.

The molecular mechanism by which paramutation occurs is not well understood. However, some research observations suggest that the silencing of paramutagenic alleles may occur by a mechanism similar to those described in Chapter 17, which

Strain A: *B′ B′*   (This strain is homozygous for the weak allele. The ′ symbol indicates that the *B-I* allele has been altered by a paramutation, which decreases its expression and causes the stalks and husk to be more green. B′ is also paramutagenic—it can convert a strong allele to a weak allele.)

Strain B: *B-I B-I*   (This strain is homozygous for the strong allele, resulting in purple stalks and husks. Note: The alleles of strain B are underlined so they can be distinguished from the alleles of strain A in the crosses below.)

Cross strain *A* to strain *B*   *B′ B′*   ×   *B-I B-I*

 Green stalks and husks   **Purple** stalks and husks

All F₁ offspring inherit a *B′* allele from strain *A*, and a *B-I* allele from strain *B*. The *B-I* allele is converted to *B′* by a paramutation.

F₁ offspring   *B′ B′*

 Green stalks and husks

Cross F₁ offspring to strain *B*:   *B′ B′*   ×   *B-I B-I*

 Green stalks and husks   **Purple** stalks and husks

F₂ offspring inherit either a *B′* or *B′* allele from the F₁ parent on the left, and a *B-I* allele from the strain *B* parent on the right. In both cases, a paramutation converts the *B-I* allele to *B′*.

F₂ offspring: *B′ B′*   **or**   *B′ B′*

 Green stalks and husks   **Green** stalks and husks

FIGURE 16.14   **The phenomenon of paramutation in the *b1* gene in maize.**

(photos) Vicki Chandler, University of Arizona

Genes→Traits   The *B-I* allele causes the stalks and husks to be very dark purple. The *B′* allele causes paramutation of the *B-I* allele. This results in stalks and husks that are mostly green, though they do have some purple pigment.

CONCEPT CHECK: In the F₁ offspring, what happened to the *B-I* allele that was inherited from the parent at the top right?

involve the production of short ncRNA molecules, such as siRNAs and piRNAs. These short RNAs are incorporated into RNA-induced silencing complexes and direct the complexes to particular genes (if the short RNAs are siRNAs) or to transposable elements (if they are piRNAs). The complexes inhibit transcription by promoting epigenetic changes, such as DNA methylation (look ahead to the left side of Figure 17.12).

What observations suggest that a similar mechanism may underlie paramutation?

- Multiple tandem repeat sequences are located close to the coding sequences of paramutagenic and paramutable alleles. For example, near the *b1* gene in maize, multiple tandem repeats are found about 100 kb upstream from the coding region. These repeats are required for paramutation to occur and may be used to make siRNAs, similar to the piRNAs shown in Figure 17.12. Neutral alleles of the *b1* gene do not have such tandem repeats.

- A functional *mop1* gene (mediator of paramutation 1 gene) is required for paramutation to occur. Studies of *Arabidopsis* reveal that this gene codes an RNA-dependent RNA polymerase. This enzyme is involved in producing double-stranded RNA (siRNA); one strand of the siRNA becomes incorporated into an RNA-induced silencing complex (as in Figure 17.12).

Future research will be needed to determine if paramutation is mediated by a mechanism that is similar to the one shown in Figure 17.12.

16.4 COMPREHENSION QUESTIONS

1. Which of the following statements about paramutation is *false*?
   a. A paramutagenic allele can alter the expression of a paramutable allele.
   b. A paramutagenic allele has a lower level of expression than does a paramutable allele.
   c. The paramutation alters the DNA sequence of the paramutagenic allele.
   d. A paramutation can persist for many generations.
2. The effects of paramutation may vary with regard to
   a. the likelihood that the paramutagenic allele will alter the paramutable allele.
   b. the stability of the paramutagenic allele over the course of several generations.
   c. the ability of the paramutagenic allele to alter the DNA sequence of the paramutable allele.
   d. both a and b.

16.5 EPIGENETICS AND ENVIRONMENTAL AGENTS

Learning Outcomes:
1. Explain how coat color in mice is epigenetically modified by dietary factors.
2. Describe the evidence that the development of honeybee queens is due to exposure to royal jelly.
3. Explain how flowering in certain species of plants is controlled by cold temperatures.

One of the most active fields in genetics is the study of how certain environmental agents cause epigenetic changes that affect gene expression. Two areas that have received a great deal of attention are the effects of diet on epigenetic modifications and the potential effects of toxic agents, such as carcinogens (cancer-causing agents). In this section, we will consider examples of both types of effects.

Exposure to Environmental Agents at Early Stages of Development May Cause Epigenetic Changes That Affect Phenotype

A striking example of how the environment can promote epigenetic changes is illustrated by studies of the *Agouti* gene (also designated *A*) found in mice. This gene codes the agouti signaling peptide that controls the deposition of yellow pigment in developing hairs. In wild-type mice (*AA*), the expression of this gene promotes the synthesis of pheomelanin, a yellow pigment. During the growth of a hair, melanocytes (pigment-producing cells) within a hair follicle initially make eumelanin, which is black. The transient expression of the *Agouti* gene causes the cells to express pheomelanin. The melanocytes then revert back to making black pigment. The result is a band of yellow pigment sandwiched between layers of black pigment, which gives a brown color. The yellow pigment is not synthesized near the tip of the hair, so the hairs of wild-type mice are brown with black tips.

Researchers have identified many mutations that affect the expression of the *Agouti* gene. For example, mice that are homozygous for a loss-of-function mutation (*aa*) have black fur because pheomelanin is not made. Alternatively, a gain-of-function mutation that causes the *Agouti* gene to be overexpressed results in a mouse with yellow fur. One such mutation is designated A^{vy} (*A* for *Agouti*, *v* for *viable*, and *y* for *yellow*; the letter *v* was used because some mutations of the *Agouti* gene are lethal). By characterizing the A^{vy} allele at the molecular level, researchers determined that it is created by the insertion of a transposable element (TE) upstream from the normal promoter of the *Agouti* gene (**Figure 16.15a**). (TEs are described in Chapter 10.) This TE carries an active promoter that causes the overexpression of the *Agouti* gene.

An intriguing observation about mice carrying the A^{vy} allele is that they exhibit a wide phenotypic variation, ranging from yellow to mottled to pseudo-agouti (**Figure 16.15b**). Why should mice with the same genotype show such a wide range of phenotypic variation? Although the answer is not entirely understood, researchers have speculated that TEs are particularly sensitive to epigenetic modifications. In the case of the A^{vy} allele, epigenetic modifications may affect the function of the promoter within the TE that is responsible for overexpressing the *Agouti* gene. For example, DNA methylation could inhibit this promoter. Furthermore, a variety of environmental factors may cause such epigenetic changes to occur. The sensitivity of TEs to epigenetic modifications together with variation in environmental factors may explain the phenotypic variation seen in these mice.

One environmental factor that may affect epigenetic modification is diet. With regard to the A^{vy} allele, the exposure of pregnant female mice to different types of diets can have a significant effect on the phenotypes of the resulting offspring. For example, in 2003, Robert Waterland and Randy Jirtle conducted a study in which they investigated the effects of certain dietary supplements. Their goal was to determine if nutrients that are known to affect DNA methylation would alter the expression of the *Agouti* gene and thereby affect coat color. When DNA is methylated, DNA methyltransferase removes a methyl group from *S*-adenosyl methionine and transfers it to a cytosine base in DNA. A variety of nutrients can increase the synthesis of *S*-adenosyl methionine in cells. These include folic acid, vitamin B_{12}, betaine, and choline chloride.

Waterland and Jirtle divided female mice into a control group that was fed a normal diet and an experimental group that was fed a diet supplemented with folic acid, vitamin B_{12}, betaine, and choline chloride. Both groups were fed their respective diets before and during pregnancy and up to the stage of weaning. Offspring carrying the A^{vy} allele were then analyzed with regard to their coat color and levels of DNA methylation. As expected, a range of coat colors was observed among the offspring

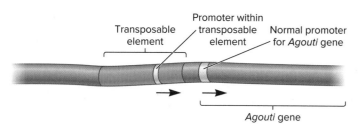

(a) The insertion of a transposable element to create the *A^{vy}* allele

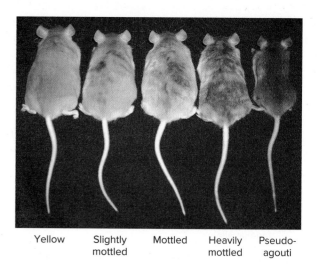

(b) Range in coat-color phenotypes in *A^{vy}a* mice due to epigenetic changes

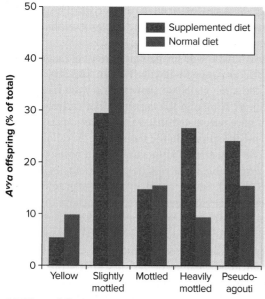

(c) Effect of diet on coat color

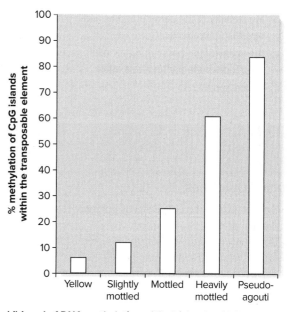

(d) Level of DNA methylation of CpG islands within the TE among mice with different coat colors

FIGURE 16.15 **Dietary effects on coat color in mice.** (**a**) A mutation in the *Agouti* gene, designated A^{vy}, is caused by the insertion of a TE upstream from the normal *Agouti* promoter. The TE promoter is very active and causes the overexpression of the *Agouti* gene. (**b**) Mice carrying the A^{vy} allele exhibit a range of phenotypes. The mice shown here are heterozygotes, $A^{vy}a$; they carry the mutant A^{vy} allele and a loss-of-function allele, *a*. (**c**) Effects of diet supplementation on coat color. Blue bars represent offspring from females given a normal diet, and red bars represent offspring from females given a supplemented diet. (**d**) DNA methylation patterns among mice with different coat colors. The samples to determine DNA methylation were obtained from cells in the tail.

(a, c, d): Source: Waterland, Robert A., and Jirtle, Randy L., "Transposable Elements: Targets for Early Nutritional Effects on Epigenetic Gene Regulation," *Molecular and Cellular Microbiology*, vol. 23, no.15, August 2003, 5293–5300. (b): Source: D.C. Dolinoy et al., "Maternal genistein alters coat color and protects A^{vy} mouse offspring from obesity by modifying the fetal epigenome," *Environ. Health Perspect.* 2006 Apr, 114(4): 567–572.

Genes→Traits The mice shown in part (b) are genetically identical. Their differences in coat color are due to epigenetic modifications that occur during early stages of development.

(**Figure 16.15c**). However, the offspring of females that had been fed a supplemented diet tended to have darker coats. For example, over 25% of the offspring with heavily mottled coats had mothers that were fed a supplemented diet (red bars), whereas less than 10% had mothers that were given a normal diet (blue bars).

The coat colors of the offspring largely correlated with the degree of methylation that occurred at CpG islands in the TE—offspring with darker coats had greater levels of DNA methylation (**Figure 16.15d**). How do we explain these results? In the mice that are more yellow, the TE has undergone very little methylation.

Therefore, the promoter remains active, leading to the transcription of the *Agouti* gene and the overproduction of yellow pigment. By contrast, the TE in the darker mice has undergone extensive methylation. Such methylation is expected to inhibit the overexpression of the *Agouti* gene, thereby preventing the overproduction of yellow pigment and resulting in a darker coat.

The Different Body Types of Queen and Worker Honeybees Depend on Factors in Royal Jelly That Promote Epigenetic Changes

Evidence that diet may affect DNA methylation also comes from the study of honeybees (*Apis mellifera*). Female honeybees are of two types: queen bees and worker bees (**Figure 16.16**). Queens are larger, live for years, and produce up to 2000 eggs each day. By comparison, the smaller worker bees are sterile, typically live only for weeks, and engage in specialized types of work, which include the cleaning and constructing of comb cells, nurturing larvae, guarding the hive entrance, and foraging for pollen and nectar.

The striking differences between queens and worker bees are largely caused by differences in their diets. Certain worker bees, called nurse bees, produce a secretion called royal jelly from glands in their mouths. All female larvae are initially fed royal jelly, but those that are bathed in royal jelly throughout their entire larval development and fed it into adulthood become queens. In contrast, female larvae that are weaned at an early stage of development and switched to a diet of pollen and nectar become worker bees.

In 2008, a study conducted by Ryszard Maleszka and colleagues indicated that DNA methylation may play a role in

FIGURE 16.16 **Dietary effects on honeybee development.**

Thiriet/Andia/Alamy Stock Photo

Genes→Traits Female honeybees that are fed royal jelly throughout the entire larval stage and into adulthood develop into queen bees. The larger queen bee is shown with a blue disk labeled 68. By comparison, those larvae that do not receive this diet become smaller worker bees. These differences in development are caused by epigenetic modifications.

CONCEPT CHECK: Are queens and worker bees genetically different from each other?

controlling developmental pathways that result in differing morphologies for queens and worker bees. Bee larvae were fed a diet that should produce worker bees. These larvae were injected with a substance that inhibits DNA methyltransferase. The result was that most of them became queen bees with fully developed ovaries! While other factors may contribute to the development of queens, these results are consistent with the hypothesis that royal jelly may contain a substance that inhibits DNA methylation. Such inhibition is thought to allow the expression of genes that contribute to the development of traits that are observed in queen bees.

Flowering in Certain Species of Plants Is Dependent on Vernalization

Vernalization refers to the phenomenon that certain species of plants must be exposed to the cold before they can flower. For example, many plant species must endure a cold winter to flower the following spring. After vernalization, plants do not necessarily initiate flowering, but they acquire the ability to flower when they are later exposed to favorable growth conditions.

The mechanism of vernalization has been extensively studied in *Arabidopsis thaliana,* which is an annual flowering plant. Strains of this species exist as summer-annual and winter-annual types. The summer-annuals grow from spring to fall, whereas the winter-annuals grow from fall to spring. The summer-annual type of *Arabidopsis* does not require vernalization to flower, whereas the winter-annual type does. By comparing the genetic differences between the summer- and winter-annual types of *Arabidopsis* and by identifying mutants that are altered in their vernalization requirements, researchers have been able to unravel how vernalization occurs at the molecular level. A key aspect of the molecular mechanism involves epigenetic changes that are induced by cold temperatures.

Figure 16.17 describes a simplified pathway for vernalization in the winter-annual type of *Arabidopsis*, but the mechanism of vernalization is known to vary among different species of flowering plants. For flowering to occur in the winter-annual type of *Arabidopsis*, two genes, *FT* and *SOC1*, must be expressed (Figure 16.17a). *FT* expression is activated by various factors, such as favorable temperatures, sufficient hours of sunlight, and plant hormones. The protein coded by the *FT* gene is a transcription factor that activates the *SOC1* gene, which also codes a transcription factor. Together, the FT and SOC1 proteins activate several genes that lead to flower development.

A key player in the vernalization pathway is FLC, which is a repressor protein that inhibits the *FT* and *SOC1* genes (upper right in Figure 16.17a). This repression prevents flowering even if the plants have been exposed to favorable temperatures, sufficient hours of sunlight, and the proper hormones. For flowering to occur, the FLC protein must be eliminated so that the *FT* and *SOC1* genes can be expressed.

Figure 16.17b shows how the expression of *FLC* gene is inhibited by cold weather. The exposure of plants to prolonged periods of cold induces the expression of the *VIN3* and *COLDAIR* genes. The VIN3 protein forms a complex with PRC2 (shown

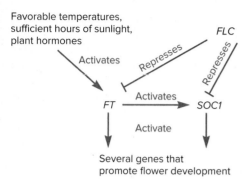

Favorable temperatures, sufficient hours of sunlight, plant hormones

Activates

Represses

Represses

FLC

Activates

FT → SOC1

Activate

Several genes that promote flower development

(a) Roles of *FT, SOC1,* and *FLC* in *Arabidopsis* flowering

Prolonged cold temperatures activate the *VIN3* and *COLDAIR* genes. The *VIN3* mRNA is translated into *VIN3* protein.

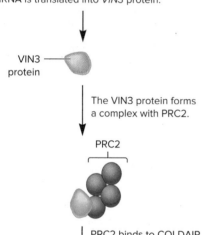

VIN3 protein

The VIN3 protein forms a complex with PRC2.

PRC2

PRC2 binds to COLDAIR RNA as it is being transcribed, thereby guiding the VIN3/PRC2 complex to the *FLC* gene.

COLDAIR RNA

COLDAIR gene

FLC gene

The VIN3/PRC2 complex causes chromatin modifications, such as the trimethylation of lysine 27 on histone H3 (H3K27). This modification causes repression of the *FLC* gene.

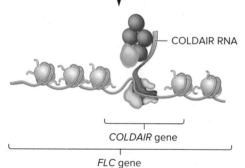

CH_3 CH_3 CH_3 CH_3 CH_3 CH_3

(b) Vernalization: repression of the *FLC* gene via VIN3, PRC2, and COLDAIR

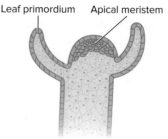

Leaf primordium   Apical meristem

Exposure of cold temperatures activates *VIN3* and *COLDAIR* genes within cells of the apical meristem. The VIN3/PRC2 complex represses *FLC*, which allows early flower buds to form.

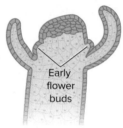

Early flower buds

Later, VRN1 and VRN2 form complexes with PRC2. These complexes maintain repression of *FLC*, which is passed to daughter cells, as the flower buds continue to grow when growth conditions are favorable.

(c) Initiation and maintenance of flower bud growth via VIN3, COLDAIR, VNR1, VRN2, and PRC2

(d) An *Arabidopsis* flower

FIGURE 16.17 **Vernalization in *Arabidopsis*.** (a) The abbreviations are as follows: *FT* (<u>f</u>lowering locus <u>T</u>); *SOC1* (<u>s</u>uppressor <u>o</u>f <u>C</u>onstans <u>1</u>); and *FLC* (<u>f</u>lowering locus <u>C</u>). Note: Though not discussed in this chapter, the *Constans* gene plays a role in activating the *FT* gene. (b) The abbreviations are as follows: *VIN3* (<u>v</u>ernalization <u>in</u>sensitive <u>3</u>) and PRC2 (<u>p</u>olycomb <u>r</u>epressive <u>c</u>omplex <u>2</u>). Note: The function of PRC2 is outlined in Figure 16.13. (c) The abbreviations are as follows: VRN1 (<u>v</u>ernalization <u>1</u>) and VRN2 (<u>v</u>ernalization <u>2</u>). A leaf primordium is an immature leaf. (d) Photograph of a flowering *Arabidopsis* plant.
(d): Peggy Greb/USDA

Genes→Traits The process of vernalization is controlled by several genes that provide the winter-annual type of *Arabidopsis* with the trait of flowering only after exposure to cold weather.

earlier in Figure 16.13), and this complex causes chromatin modifications. For example, the complex recognizes histone H3 and trimethylates a lysine at position 27. This modification occurs at multiple locations within the *FLC* gene and causes gene repression (refer back to Figure 16.13).

How does the VIN3/PRC2 complex get targeted to the *FLC* gene? While the mechanism is not entirely understood, recent evidence indicates that *COLDAIR* may play an important role. Interestingly, the *COLDAIR* gene is found within an intron of the *FLC* gene; it is a gene within a gene! The COLDAIR RNA has a binding site for PRC2. When the transcription of the *COLDAIR* gene is induced by cold temperatures, the VIN3/PRC2 complex binds to the *COLDAIR* transcript as it is being made. Because the *FLC* and *COLDAIR* genes overlap each other, this guides the VIN3/PRC2 complex to the *FLC* gene as well (see Figure 16.17b).

The primary location where vernalization takes place is at the growing tips of plants, which are called apical meristems (Figure 16.17c). When exposed to cold temperatures, the *VIN3* and *COLDAIR* genes within apical meristem cells are expressed for a brief period. They play a key role in initiating the repression of the *FLC* gene, and thereby allow the formation of early flower buds. Later on, other proteins, such as VRN1 and VRN2, form complexes with PRC2, which are needed to maintain the repression of the *FLC* gene. Such repression is transmitted from cell to cell as the flower buds continue to grow when conditions are favorable. In other words, the growing flower bud "remembers" it has been vernalized. This allows winter-annual *Arabidopsis* plants to produce flowers in the spring (Figure 16.17d).

16.5 COMPREHENSION QUESTIONS

1. When mice carrying the A^{vy} allele exhibit a darker coat, this phenotype is thought to be caused by dietary factors that result in
   a. a higher level of DNA methylation and a decrease in the expression of the *Agouti* gene.
   b. a lower level of DNA methylation and a decrease in the expression of the *Agouti* gene.
   c. a higher level of DNA methylation and overexpression of the *Agouti* gene.
   d. a lower level of DNA methylation and overexpression of the *Agouti* gene.

2. If the *VIN3* gene had a loss-of-function mutation, how do you think that would affect the phenotypes of summer-annual and winter-annual *Arabidopsis* plants?
   a. Neither type would flower.
   b. Both types would flower at the proper time.
   c. The summer type would flower, but the winter type would not.
   d. The winter type would flower, but the summer type would not.

KEY TERMS

16.1: epigenetics, epimutation, epigenetic inheritance, transgenerational epigenetic inheritance, DNA methylation, chromatin remodeling, covalent histone modification, histone variants, feedforward loops, non-coding RNAs (ncRNAs), *cis*-epigenetic mechanism, *trans*-epigenetic mechanism

16.2: euchromatin, heterochromatin, facultative heterochromatin, constitutive heterochromatin, reader domains, writer domains, eraser domains, higher-order structure, nuclear lamina (NL), lamina-associated domains (LADs), liquid-liquid phase separation (LLPS)

16.3: development, imprinting control region (ICR), differentially methylated region (DMR), de novo methylation, maintenance methylation, X-chromosome inactivation (XCI), pioneer factors, trithorax group (TrxG), polycomb group (PcG), trimethylation, polycomb response element (PRE)

16.4: paramutation, paramutagenic, paramutable

16.5: vernalization

CHAPTER SUMMARY

16.1 Overview of Epigenetics

- Epigenetics is the study of mechanisms that lead to changes in gene expression that are passed from cell to cell and are reversible, but do not involve a change in the sequence of DNA. The transmission of epigenetic changes from one generation to the next is referred to as epigenetic inheritance.
- The most common types of molecular changes that underlie epigenetic control are DNA methylation, chromatin remodeling, covalent histone modification, localization of histone variants, and feedforward loops (see Table 16.1).
- Epigenetic changes can be established by transcription factors or non-coding RNAs (see Figure 16.1).
- Epigenetic changes may be maintained by *cis-* or *trans*-epigenetic mechanisms (see Figure 16.2).
- Some epigenetic changes are programmed during development, and others are caused by environmental agents (see Table 16.2).

16.2 Heterochromatin: Function, Structure, Formation, and Maintenance

- Heterochromatin is more compact than euchromatin, has an inhibitory effect on gene transcription, and is located at the periphery of the cell nucleus (see Figure 16.3).
- Key functions of heterochromatin include gene silencing, the prevention of the movement of transposable elements, and the prevention of viral proliferation.
- Constitutive and facultative heterochromatin differ in their chromosomal locations and molecular features (see Figure 16.4).
- Molecular events result in gene silencing and produce heterochromatin with higher-order structures (see Figures 16.5 and 16.6).
- The formation of heterochromatin begins at a nucleation site and spreads outward from there until it reaches a barrier (see Figure 16.7).
- The pattern of heterochromatin in a given chromosome is passed from cell to cell and is maintained following cell division (see Figures 16.8 and 16.9).

- Abnormalities in heterochromatin formation underlic ccrtain inherited human diseases.

16.3 Epigenetics and Development

- The imprinting of the *Igf2* gene in mammals occurs during gametogenesis and involves DNA methylation that happens during spermatogenesis but not during oogenesis (see Figure 16.10).
- X-chromosome inactivation in female mammals involves epigenetic changes that are initiated during embryogenesis and are maintained throughout the rest of development (see Figure 16.11).
- Pioneer factors bind to DNA sequences that are exposed on the surface of a nucleosome and recruit chromatin-remodeling complexes and histone-modifying enzymes (see Figure 16.12).
- The trithorax and polycomb groups of protein complexes promote epigenetic changes that are important in development and in the formation of specific cell types (see Figure 16.13).

16.4 Paramutation

- A paramutation is an interaction between two alleles of a given gene in which one allele induces a heritable change in the other allele without changing that allele's DNA sequence.
- Paramutations are due to epigenetic changes that are transmitted from parent to offspring (see Table 16.3, Figure 16.14).

16.5 Epigenetics and Environmental Agents

- Dietary factors during early stages of development can cause epigenetic changes that affect phenotype (see Figures 16.15, 16.16).
- Vernalization refers to the phenomenon that certain species of plants must be exposed to the cold before they can flower. The exposure to cold results in epigenetic changes that are needed for flowering to occur (see Figure 16.17).

PROBLEM SETS & INSIGHTS

MORE GENETIC TIPS **1.** Is each of the following events best explained by mutation or an epigenetic modification?

 A. Imprinting of the *Igf2* gene
 B. Variation in coat color in mice carrying the A^{vy} allele
 C. Formation of cancer cells
 D. Variation in flower color between different strains of pea plants, such as purple versus white
 E. X-chromosome inactivation

T OPIC: *What topic in genetics does this question address?* The topic is how mutation and epimutation differ.

I NFORMATION: *What information do you know based on the question and your understanding of the topic?* In the question, you are given a list of events that involve variation in gene or chromosome expression. From your understanding of the topic, you may recall that a mutation involves a heritable change that alters the DNA (e.g., the nucleotide sequence in a gene), whereas an epimutation is a heritable change in gene expression that does not alter the DNA sequence.

PROBLEM-SOLVING **S**TRATEGY: *Compare and contrast.* One strategy to solve this problem is to consider each event and determine if the phenomenon involves a change in the DNA sequence (mutation) and/or an epigenetic modification.

ANSWER:

A. Epigenetic modification: due to differences in DNA methylation

B. Epigenetic modification: due to differences in DNA methylation

C. Usually both mutation and epigenetic modification

D. Mutation: due to variation in DNA sequences

E. Epigenetic modification: due to changes in the chromatin structure of the inactivated X chromosome

---

2. Explain how a non-coding RNA could play a role in establishing an epigenetic modification at a specific site in a chromosome.

TOPIC: *What topic in genetics does this question address?* The topic is epigenetic modification. More specifically, the question asks how an ncRNA can contribute to an epigenetic modification.

INFORMATION: *What information do you know based on the question and your understanding of the topic?* In the question, you are reminded that an ncRNA can play a role in establishing an epigenetic modification. From your understanding of the topic, you may recall that a ncRNA may bind to: (1) DNA, (2) a protein that is already bound to DNA, or (3) a chromatin-modifying protein.

PROBLEM-SOLVING **S**TRATEGY: *Relate structure and function.* One strategy to solve this problem is to first consider the structure of an ncRNA and then figure out how its structure could initiate a chromatin modification at a specific site in a chromosome.

ANSWER: A non-coding RNA can bind to a specific site in a chromosome due to complementary base pairing, or it may bind to a protein that is already attached to a specific chromosomal site. After binding, the ncRNA can act as a bridge by binding to one or more proteins that cause epigenetic modifications, such as DNA methylation and covalent histone modification.

---

3. Let's suppose a mutation in an early embryonic cell inhibits the promoter for the *Xist* gene on one of the X chromosomes, but has no effect on the promoter for the *Tsix* gene. How would this mutation affect the process of X-chromosome inactivation?

TOPIC: *What topic in genetics does this question address?* The topic is X-chromosome inactivation.

INFORMATION: *What information do you know based on the question and your understanding of the topic?* In the question, you are given information about a mutation that inhibits the promoter for the *Xist* gene on one of the X chromosomes. From your understanding of the topic, you may recall that the inactive X chromosome expresses the *Xist* gene.

PROBLEM-SOLVING **S**TRATEGY: *Describe the steps.* One strategy to solve this problem is to sort out the steps of X-chromosome inactivation and determine how the expression of the *Xist* gene is important for XCI (see Figure 16.11).

ANSWER: The choosing of the X chromosome to inactivate occurs when the *Xist* gene is expressed on that chromosome; the coating of *Xist* RNA causes the formation of a Barr body. If an X chromosome carried a mutation that inhibited *Xist* gene expression, it could not become inactivated. Therefore, it is likely that the other X chromosome, which does not carry the mutation, would become inactivated.

Conceptual Questions

C1. Define *epigenetics*. Are all epigenetic changes passed from parent to offspring? Explain.

C2. List and briefly describe five types of molecular mechanisms that may underlie epigenetic gene regulation.

C3. Explain how epigenetic changes may be targeted to specific genes.

C4. What is the key difference between *cis*- and *trans*-epigenetic mechanisms for maintaining an epigenetic modification? In Chapter 5, we considered genomic imprinting of the *Igf2* gene, in which offspring express the copy of the gene they inherit from their male parent, but not the copy they inherit from their female parent. Is this due to a *cis*- or a *trans*-epigenetic mechanism?

C5. What are some key functions of heterochromatin?

C6. With regard to chromosomal locations, how do constitutive and facultative heterochromatin differ?

C7. Which of the following types of drugs would you expect to inhibit heterochromatin formation?

A. A drug that inhibits DNA methyltransferase

B. A drug that inhibits histone methyltransferase

C. A drug that inhibits histone acetyltransferase

D. A drug that inhibits histone deacetylase

C8. Briefly describe three higher-order structures that occur in heterochromatin.

C9. What causes HP1 to bind to nucleosomes? How does HP1 promote the formation of higher-order structures in heterochromatin?

C10. What is liquid-liquid phase separation? How does it influence heterochromatin maintenance and gene transcription within heterochromatin?

C11. List and briefly describe the three phases of heterochromatin formation at the chromosome level.

C12. At the molecular level, what events promote the maintenance of heterochromatin formation during DNA replication and cell division?

C13. Explain how DNA methylation and the formation of a DNA loop control the expression of the *Igf2* gene in mammals. How is this gene imprinted so that only the paternal copy is expressed in offspring?

C14. Let's suppose a mutation removes the ICR next to the *Igf2* gene. If this mutation is inherited from the female parent, will the *Igf2* gene (from the female parent) be silenced or expressed? Explain.

C15. Outline the molecular steps in the process of X-chromosome inactivation (XCI). Which step plays a key role in choosing which of the X chromosomes will remain active and which will be inactivated?

C16. Following X-chromosome inactivation, most genes on the inactivated X chromosome are silenced. Explain how. Name one gene that is not silenced.

C17. Briefly explain the key features of the X-inactivation center (Xic).

C18. With regard to chromatin binding, what is a unique feature of a pioneer factor compared with other regulatory transcription factors. Describe how pioneer factors may promote gene activation.

C19. In general, explain how epigenetic modifications provide an important mechanism for developmental changes that lead to specialized body parts and cell types. How do the protein complexes called the trithorax and polycomb groups participate in this process?

C20. What are the contrasting roles of trithorax and polycomb groups during development in animals and plants?

C21. Describe the molecular steps by which polycomb group complexes cause epigenetic gene silencing.

C22. With regard to development, what dire consequences would result if polycomb group complexes did not function properly?

C23. Using coat color in mice and the development of female honeybees as examples, explain how dietary factors can cause epigenetic modifications, leading to phenotypic effects.

C24. Researchers have determined that environmental agents that do not cause gene mutations can contribute to cancer. Explain how. Would these epigenetic changes be passed to offspring?

C25. Is paramutation a *cis*- or a *trans*-epigenetic mechanism?

C26. If a winter-annual strain of *Arabidopsis* is grown in a greenhouse and not exposed to cold temperatures, its ability to flower is inhibited. Which gene is responsible for this inhibition?

C27. Explain how the VIN3/PRC2 complex specifically binds to the *FLC* gene in *Arabidopsis*.

Experimental Questions

E1. Chromosome conformation capture methods (3C methods) are described in Chapter 10 (refer back to Figure 10.22). Which of the following higher-order structures of heterochromatin could be revealed by 3C methods?

A. HP1 dimers bringing nucleosomes close together

B. Attachment of heterochromatin to the nuclear lamina

C. Formation of heterochromatin droplets due to LLPS

E2. Referring back to Figure 16.15, explain the relationship between the coat color of mice and DNA methylation. How is coat color related to the diet of the female parent?

E3. 5-Azocytidine is an inhibitor of DNA methyltransferase. If this drug was fed to female mice during pregnancy, explain how you think it would affect the coat color of offspring carrying the A^{vy} allele.

E4. Look back at Figure 16.14. If you crossed an F_2 offspring to a homozygous *B-I B-I* plant, what phenotypic results would you expect for the F_3 offspring?

E5. Researchers can introduce loss-of-function mutations into genes using CRISPR/Cas technology, described in Chapter 20. If you used this technology to produce a homozygous loss-of-function mutation in each of the following genes in a winter-annual strain of *Arabidopsis*, describe how each mutation would affect flowering and the requirement for vernalization.

A. *FLC*

B. *VIN3*

C. *COLDAIR*

Questions for Student Discussion/Collaboration

1. Go to the PubMed website and search the words *heterochromatin* and *disease*. Scan through the journal articles you retrieve and make a list of heterochromatin abnormalities that may contribute to human diseases.

2. Discuss the similarities and differences of phenotypic variations that are caused by epigenetic gene regulation versus variation in gene sequences (epigenetics versus genetics).

Note: All answers are available for the instructor in Connect; the answers to the even-numbered questions and all of the Concept Check and Comprehension Questions are in Appendix B.

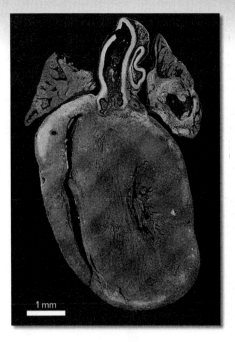

Non-coding RNAs and heart repair.
The muscles of the mammalian heart have poor regenerating abilities. Researchers have identified several non-coding RNAs that stimulate cardiac muscle regeneration. This heart, from a mouse that was treated with a type of non-coding RNA called a microRNA, showed a significant increase in proliferating cells. This discovery may lead to new therapies to help heart attack victims regenerate new cardiac muscle.

Mauro Giacca, Ana Eulalio, Miguel Mano

17

NON-CODING RNAs

In Chapters 12 and 13, we focused our attention on gene expression at the molecular level. The emphasis was on protein-coding genes, which are transcribed into mRNAs. During translation, the information within mRNAs is used to make polypeptides, which then assemble into functional proteins. The human genome has about 20,000 protein-coding genes. In contrast, other genes are transcribed into **non-coding RNAs (ncRNAs)**, which are RNA molecules that do not code polypeptides. In humans, the number of genes that specify non-coding RNAs is still difficult to measure and remains a matter of controversy. Estimates range from several thousand to tens of thousands.

In the past, educators have tended to emphasize DNA and proteins in the teaching of genetics at the molecular level. Although DNA, RNA, and proteins are all key molecular players in living cells, the emphasis on RNA has been much less. With a few exceptions, education about RNA has been limited to explaining its role in making proteins (as discussed in Chapter 13).

Why such a big time gap in developing our understanding of protein-coding genes versus other types of genes? It's all about tools. The experimental tools to study the structure and function of proteins and to identify protein-coding genes have been around for a long time. By contrast, the tools to study RNA structure and function and to identify genes that do not specify proteins are much newer and are under rapid development. We are witnessing a revolution in molecular biology that is uncovering an unprecedented number of functions for RNA molecules.

New molecular tools have enabled researchers to discover that ncRNAs perform a spectacular array of cellular functions in bacteria, archaea, and eukaryotes, including important roles in a variety of processes, such as DNA replication, chromatin modification, transcription, translation, and genome defense. In most cell types, ncRNAs are more abundant than mRNAs. For example, in a typical human cell, only about 20% of transcription involves the production of mRNAs, whereas 80% of it is associated with making ncRNAs! This observation underscores the importance of ncRNAs in the enterprise of life, and indicates why they deserve greater recognition. Furthermore, abnormalities in ncRNAs are associated with a wide range of human diseases, including cancer, neurological disorders, and cardiovascular diseases.

In this chapter, we will begin with an overview of the general properties of ncRNAs and then examine specific examples of how they perform their functions. We will end the chapter by considering the role of ncRNAs in different human diseases and in plant health.

17.1 OVERVIEW OF NON-CODING RNAs

Learning Outcomes:

1. Describe the ability of ncRNAs to bind to other molecules and macromolecules.
2. Outline the general functions of ncRNAs.
3. Define *ribozyme*.
4. List several examples of ncRNAs, and describe their functions.
5. Describe the RNA world, and explain its significance for modern cells.

The study of ncRNAs is rapidly expanding. The functions of many ncRNAs are not yet known, and researchers speculate that many others have yet to be discovered. Also, due to the relatively young age of this field, not all researchers agree on the names of certain ncRNAs or their primary functions. Even so, some broad themes are beginning to emerge. In this section, we will survey the general features of ncRNAs. In later sections of the chapter, we will discuss specific examples in greater detail.

ncRNAs Can Bind to Different Types of Molecules

The ability of ncRNAs to carry out an amazing array of functions is largely related to their ability to bind to different types of molecules. **Figure 17.1a** shows four common types of molecules that are recognized by ncRNAs. Some ncRNAs bind to DNA or to another RNA through complementary base pairing. This binding allows the ncRNAs to affect processes such as transcription and translation. In addition, ncRNAs can bind to proteins or small molecules.

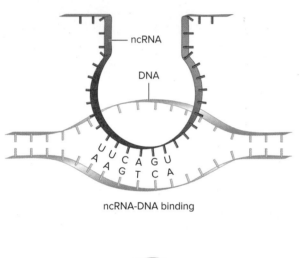

ncRNA-DNA binding

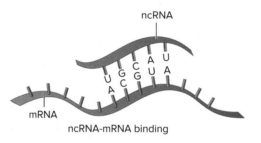

ncRNA-mRNA binding

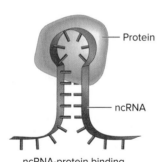

ncRNA-protein binding

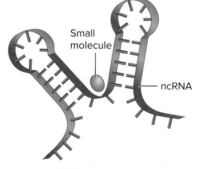

ncRNA-small molecule binding

(a) Common binding interactions between ncRNA and other molecules

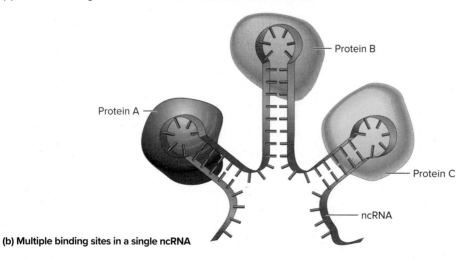

(b) Multiple binding sites in a single ncRNA

FIGURE 17.1 **Ability of ncRNAs to bind to other molecules.** (a) ncRNA molecules can bind to DNA, mRNA, proteins, or small molecules. (b) Some ncRNAs have multiple binding sites that can interact with several different molecules, such as proteins.

CONCEPT CHECK: What types of molecules can bind to a non-coding RNA?

As described in Chapter 9, RNA molecules can form stem-loops, which may bind to pockets on the surface of proteins. Furthermore, multiple stem-loops may form a binding site for a small molecule. In some cases, a single ncRNA may contain multiple binding sites. The presence of multiple binding sites in an ncRNA facilitates the formation of a large structure composed of multiple molecules, such as the one consisting of an ncRNA and three different proteins shown in **Figure 17.1b**.

ncRNAs Can Perform a Diverse Set of Functions

In recent decades, researchers have uncovered many examples in which ncRNAs play a critical role in different biological processes. Let's first consider six general functions that ncRNAs can perform.

Scaffold. Some ncRNAs contain binding sites for multiple components, such as a group of different proteins. In this way, an ncRNA can act as a scaffold for the formation of a complex, as shown in Figure 17.1b.

Guide. Another function related to having multiple binding sites is that some ncRNAs can guide one molecule to a specific location in a cell (**Figure 17.2**). For example, an ncRNA may bind to a protein and guide it to a specific site in the cell's DNA that is part of a particular gene. This specificity is due to the structure of the ncRNA: It has a binding site for the protein and another binding site for the DNA.

Alteration of Protein Function or Stability. When an ncRNA binds to a protein, it can alter that protein's structure, which in turn, can have a variety of effects. The binding of the ncRNA may affect:

- the ability of the protein to act as a catalyst;
- the ability of the protein to bind to another molecule, such as another protein, DNA, or RNA;
- the stability of the protein.

Ribozyme. Another interesting feature of some ncRNAs is that they function as **ribozymes,** which are RNA molecules with catalytic activity. For example, in Chapter 12, we considered RNaseP (refer back to Figure 12.17). RNaseP recognizes tRNA molecules. The RNA component of RNaseP functions as an endonuclease that cleaves a tRNA, reducing its size.

Blocker. An ncRNA may physically prevent or block a cellular process from happening. For example, as described in Chapter 14, an antisense RNA is a type of ncRNA that is complementary to an mRNA. When an antisense RNA binds to an mRNA in the vicinity of the start codon, it blocks the ability of the mRNA to be translated (refer back to Figure 14.15).

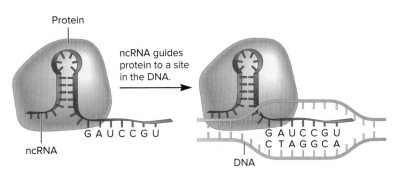

FIGURE 17.2 Ability of an ncRNA to act as a guide.

Decoy. Some ncRNAs recognize other ncRNAs and sequester them, thereby preventing them from functioning (**Figure 17.3**). For example, a decoy ncRNA may bind to a different ncRNA called a microRNA (miRNA), which is described later in this chapter. The function of an miRNA is to inhibit the translation of a particular mRNA. However, as shown below, if a decoy RNA binds to an miRNA, that miRNA is unable to carry out its function. When multiple decoy ncRNAs are found in a cell, they collectively act like a sponge by binding to the miRNAs and preventing them from functioning.

In these examples, the key difference between a blocker and a decoy is what they bind to. A blocker binds to a molecule that is not an ncRNA, and this binding sterically prevents that molecule from interacting with something else. In contrast, a decoy binds to an ncRNA molecule and prevents that ncRNA from carrying out its function.

The Functions of Some ncRNAs Are Understood

Table 17.1 describes several examples of ncRNAs that have been well characterized. Some of these are discussed in other chapters. In the remaining sections of this chapter, we will focus on the functions of ncRNAs that have not been discussed elsewhere in the text.

As shown in Table 17.1, ncRNAs are broadly categorized according to their length. **Long non-coding RNAs (lncRNAs)** are ncRNA molecules that are longer than 200 nucleotides. This arbitrary limit distinguishes lncRNAs from **small regulatory RNAs** (also called short ncRNAs), which are shorter than 200 nucleotides. An example is a microRNA, which is usually 20–25 nucleotides in length.

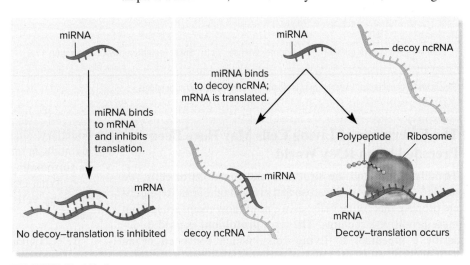

FIGURE 17.3 Ability of an ncRNA to act as a decoy.

TABLE **17.1**

Examples of Non-coding RNA Molecules

Type of ncRNA	Plays a role in:	lncRNA or Small Regulatory RNA?	Discussed in:	Description
Telomerase RNA component (TERC)	DNA replication	lncRNA	Chapter 11	TERC facilitates the binding of telomerase to the telomere and acts as a template for DNA replication.
X inactive-specific transcript (*Xist* RNA)	Chromatin structure, transcription	lncRNA	Chapter 16	*Xist* RNA coats one of the X chromosomes in female mammals and plays a role in its compaction and resulting inactivation.
Hox transcript antisense intergenic RNA (HOTAIR)	Chromatin structure, transcription	lncRNA	Section 17.2	HOTAIR alters chromatin structure and thereby represses transcription by guiding histone-modifying enzymes to target genes.
COLDAIR	Chromatin structure, transcription	lncRNA	Chapter 16	COLDAIR alters chromatin structure and thereby represses transcription by recognizing histone-modifying complexes, which modify the *FLC* gene in *Arabidopsis*. This modification allows flowering to occur.
RNaseP RNA	Processing of tRNA molecules	lncRNA	Chapter 12	RNaseP is involved in the processing of tRNA molecules. The RNaseP RNA is the catalytic component.
Small nuclear RNA (snRNA)	Splicing	Small regulatory RNA	Chapter 12	snRNAs associate with proteins to form subunits of the spliceosome, which splices pre-mRNAs in eukaryotes.
Transfer RNA (tRNA)	Translation	Small regulatory RNA	Chapter 13	tRNA molecules recognize mRNA codons during translation and carry the appropriate amino acid.
Ribosomal RNA (rRNA)	Translation	Variable*	Chapter 13	rRNAs are components of ribosomes, which are the site of polypeptide synthesis.
Antisense RNA	Translation	Variable	Chapter 14	An antisense RNA is complementary to an mRNA. The binding of the antisense RNA to the mRNA blocks translation.
MicroRNA (miRNA), small-interfering RNA (siRNA)	Translation and RNA degradation	Small regulatory RNAs	Section 17.3	miRNAs and siRNAs regulate the expression and degradation of mRNAs.
Small nucleolar RNA (snoRNA)	RNA modification	Variable	Section 17.4	A snoRNA facilitates covalent modifications to rRNAs. These modifications include ribose methylation and pseudouridylation.
RNA component of signal recognition particle (SRP RNA)	Protein targeting and secretion	lncRNA	Section 17.5	An SRP is composed of one ncRNA and one or more proteins. In prokaryotes, SRP directs certain polypeptides synthesized in the cell to the plasma membrane. In eukaryotes, it directs them to the endoplasmic reticulum.
CRISPR RNA (crRNA)	Genome defense	Small regulatory RNA	Section 17.6	crRNA, found in prokaryotes, guides an endonuclease to foreign DNA, such as the DNA of a bacteriophage. Some CRISPR-Cas systems can target RNA.
PIWI-interacting RNA (piRNA)	Genome defense	Small regulatory RNA	Section 17.6	piRNA associates with PIWI proteins and prevents the movement of transposable elements.

*The ncRNAs identified as "Variable" may be shorter or longer than 200 nucleotides.

The Emergence of Living Cells May Have Been Preceded by an RNA World

To fully understand the structures and functions of cells, we need to consider how life arose and evolved on Earth. Researchers propose that living cells, as we now know them, arose from more primitive structures. The term **protobiont** is used to describe a precursor to living cells, which consisted of an aggregate of molecules and macromolecules that acquired a boundary, such as a lipid bilayer. A protobiont was able to maintain an internal chemical environment distinct from that of its surroundings.

What characteristics make protobionts possible precursors of living cells? Scientists envision the existence of four key features:

1. A boundary, such as a membrane, separated the internal contents of the protobiont from the external environment.

2. Polymers inside the protobiont contained information.

3. Polymers inside the protobiont had catalytic functions.

4. The protobionts eventually developed the capability of self-replication.

According to this scenario, metabolic pathways became more complex, and the ability of protobionts to self-replicate became more refined over time. Eventually, protobionts acquired the complex characteristics that we attribute to living cells.

The majority of scientists favor the idea that RNA was the first macromolecule found in protobionts. The term **RNA world** refers to a period on Earth in which RNA molecules, but not DNA or proteins, were found within protobionts. If this were the case, researchers speculate that RNA would have carried out three key functions:

1. RNA was capable of storing information in its nucleotide base sequence.

2. RNA had the capacity for self-replication. An RNA molecule could function as a ribozyme and use an RNA molecule as a template to make a complementary RNA molecule according to the AU/GC rule.

3. Ribozymes carried out a variety of catalytic functions, such as the synthesis of polypeptides and other kinds of organic molecules.

By comparison, DNA and proteins are not as versatile as RNA. Research has shown that DNA has very limited catalytic activity, and proteins are not known to undergo self-replication. RNA can perform functions that are characteristic of proteins and, at the same time, can serve as genetic material with replicative and informational functions.

Assuming that an RNA world was the origin of life, researchers have asked, "Why and how did the RNA world evolve into the DNA/RNA/protein world we see today?" The RNA world may have been superseded by a DNA/RNA world or by an RNA/protein world before the emergence of the modern DNA/RNA/protein world. Let's consider the advantages of a DNA/RNA/protein world as opposed to the simpler RNA world and explore how this modern biological world might have come into being.

Information Storage.   RNA can store information in its base sequence. Because that is the case, why did DNA take over the information-storing function in modern cells? During the RNA world, RNA had to perform two roles: the long-term storage of information and the catalysis of chemical reactions. Scientists have speculated that the incorporation of DNA into cells would have relieved RNA of its long-term storage role, thereby allowing RNA to perform a variety of other functions. For example, if DNA stored the information for the synthesis of RNA molecules, such RNA molecules could bind cofactors, have modified bases, or bind peptides that might enhance their catalytic function. Cells with both DNA and RNA would have had an advantage over those with just RNA. Another advantage of DNA is its stability. Compared with RNA, DNA strands are less likely to spontaneously break.

How did DNA come into existence in an RNA world? Scientists have proposed that an ancestral RNA molecule had the ability to make DNA using RNA as a template. This function, known as reverse transcription, is described in Chapter 18 in the discussion of viral reproductive cycles. Interestingly, modern eukaryotic cells can use RNA as a template to make DNA. For example, an RNA sequence in the enzyme telomerase acts as a template to synthesize the ends of chromosomes, thus preventing the progressive shortening of the chromosomes (refer back to Figure 11.25).

Metabolism and Other Cellular Functions.   Now let's consider the origin of proteins. The emergence of proteins as catalysts would have been a great benefit to early cells. Due to the different chemical properties of the 20 amino acids, proteins have vastly greater catalytic ability than do RNA molecules, again providing a major advantage to cells that had both RNA and proteins. In modern cells, proteins have taken over most, but not all, catalytic functions. In addition, proteins can perform other important tasks. For example, cytoskeletal proteins carry out structural roles, and certain membrane proteins are responsible for the uptake of substances into living cells.

How did proteins come into existence in an RNA world? Although the answer to this question is not entirely clear, RNA is known to play a central role in protein synthesis in modern cells. First, mRNA provides the information for a polypeptide sequence. Second, tRNA molecules act as adaptors for the formation of a polypeptide. And finally, ribosomes containing rRNA provide a site for polypeptide synthesis. Furthermore, a particular rRNA acts as a ribozyme to catalyze peptide bond formation. Taken together, the analysis of translation in modern cells is consistent with an evolutionary history in which RNA molecules were instrumental in the emergence and formation of proteins.

17.1 COMPREHENSION QUESTIONS

1. Which of the following can bind to ncRNAs?
   a. DNA
   b. RNA
   c. Proteins
   d. Small molecules
   e. All of the above

2. When an ncRNA functions as a decoy, it
   a. contains binding sites for many different proteins, thereby promoting the formation of a large complex.
   b. recognizes other ncRNAs and sequesters them, thereby preventing them from functioning.
   c. can alter its conformation, allowing it to switch between active and inactive conformations.
   d. may physically prevent or block a cellular process from happening.

3. Scientists propose that the first macromolecules in protobionts were
   a. DNA molecules.
   b. RNA molecules.
   c. proteins.
   d. all of the above.

17.2 NON-CODING RNAs: EFFECTS ON CHROMATIN STRUCTURE AND TRANSCRIPTION

Learning Outcome:

1. Explain how the ncRNA known as HOTAIR plays a role in gene repression.

<u>Ho</u>x transcript <u>a</u>ntisense <u>i</u>ntergenic <u>R</u>NA, known as HOTAIR, is a recently discovered ncRNA that alters chromatin structure in a way that represses gene transcription. The gene that codes HOTAIR is located on human chromosome 12 within a cluster of genes called the *HoxC* genes. (*Hox* genes are discussed in Chapter 26.) HOTAIR is a 2.2-kb-long ncRNA that is transcribed from the opposite (antisense) strand relative to the strand that is used as a template to transcribe the *HoxC* genes. Though the name HOTAIR refers to the antisense direction of its transcription, this ncRNA does not function like the antisense RNAs discussed in Chapter 14, which bind directly to mRNAs (refer back to Figure 14.15). Instead, researchers have discovered that HOTAIR acts as a scaffold that guides two histone-modifying enzymes to target genes.

Figure 17.4 shows a simplified version of the mechanism by which HOTAIR represses gene transcription. HOTAIR acts as a scaffold for the binding of two protein complexes known as <u>p</u>olycomb <u>r</u>epressive <u>c</u>omplex <u>2</u> (PRC2) and <u>l</u>ysine-<u>s</u>pecific <u>d</u>emethylase <u>1</u> (LSD1). PRC2 binds to the 5′ end of HOTAIR, and LSD1 binds to the 3′ end. HOTAIR then guides PRC2 and LSD1 to a target gene by binding to a region near the gene that contains many purines, which is called a GA-rich region. For example, HOTAIR binds to a GA-rich region that is next to a specific *HoxD* gene on human chromosome 2. A portion of HOTAIR is complementary to this GA-rich region.

The next event involves histone modifications. PRC2 and LSD1 contain different types of histone-modifying enzymes. As described in Chapter 16, PRC2 functions as a histone methyltransferase that trimethylates lysine 27 on histone H3. LSD1 demethylates mono- and dimethylated lysines. In particular, it removes methyl groups from lysine 4 on histone H3. Although the mechanism of gene repression is not well understood, these histone modifications (H3K27 trimethylation and H3K4 demethylation) may inhibit transcription in two ways:

- The modifications may directly inhibit the ability of RNA polymerase to transcribe the target gene. For example, these histone modifications may prevent RNA polymerase from forming a preinitiation complex.

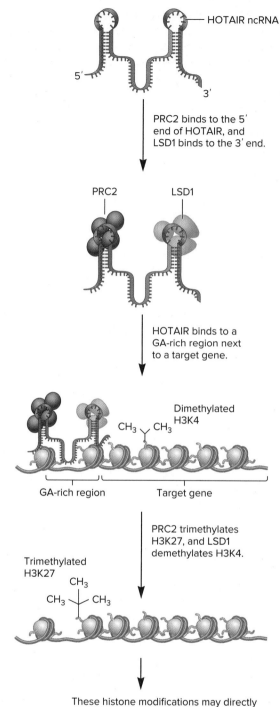

FIGURE 17.4 Simplified mechanism for repression of transcription by HOTAIR. This is just one proposed role of HOTAIR. This ncRNA is known to interact with other proteins as well. Note: For simplicity, this drawing shows just one trimethylation event and one demethylation event. PRC2 and LSD1 would trimethylate and demethylate many lysine residues, respectively. Also, the structure of HOTAIR is schematic. The actual structure is more complicated than that shown here.

CONCEPT CHECK: Explain why HOTAIR binds next to the target gene.

- Rather than directly affecting transcription, the histone modifications carried out by PRC2 and LSD1 may attract other chromatin-modifying enzymes to the target gene. For example, they may attract a histone deacetylase to the target gene, which would promote a closed chromatin conformation.

Of great interest to researchers investigating HOTAIR is its role in human disease. As discussed later in this chapter, certain types of cancer, such as breast cancer, may develop when HOTAIR is not functioning properly.

17.2 COMPREHENSION QUESTION

1. Which of the following functions does HOTAIR perform?
   a. Decoy
   b. Scaffold
   c. Guide
   d. Both b and c

17.3 NON-CODING RNAs: EFFECTS ON TRANSLATION AND mRNA DEGRADATION

Learning Outcomes:

1. Explain how additional copies of a pigmentation gene can cause less pigmentation in petunia flowers.
2. Analyze the experimental evidence that double-stranded RNA is more potent than antisense RNA at inhibiting mRNA.
3. Outline the steps of RNA interference.
4. Describe the roles of miRNAs during plant development.

In the previous section, we considered how an ncRNA can affect the process of transcription. In recent years, researchers have discovered that ncRNAs often exert their effects on RNA molecules that are already made. In this section, we will explore how ncRNAs inhibit the ability of mRNAs to be translated or cause them to be degraded.

RNA Silencing Was Revealed by Effects on Flower Color in Petunias

In 1990, researchers from the laboratories of Richard Jorgensen and Antoine Stuitje were attempting to produce strains of petunias with deeper flower colors. Their approach involved inserting into the petunia genome one or more copies of cloned genes that coded chalcone synthase or dihydroflavono-l4-reductase, which are enzymes involved in the synthesis of flower pigments. The cloned genes had been modified to contain a very strong promoter. Unexpectedly, flower pigmentation in some cases did not deepen, but instead showed variegation, ranging from a nearly complete loss of color to streaks of pigment of varying widths (**Figure 17.5a**).

The researchers concluded that additional copies of a gene can sometimes suppress the expression of both itself and its endogenous counterpart. For example, the level of chalcone synthase mRNA isolated from white flowers was reduced 50-fold compared to control petals from plants that did not carry additional copies of this cloned pigmentation gene.

How do we explain the reduction in levels of mRNAs made from the pigmentation genes? A few years later, researchers discovered that the production of double-stranded RNA is involved in the lowering of mRNA levels (via mechanisms shown

(a) Flower-color variation after introduction of a pigmentation gene

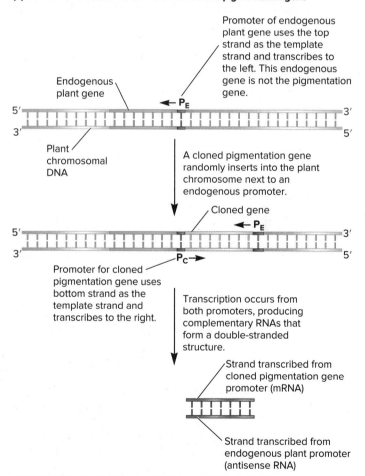

(b) Possible mechanism for double-stranded RNA production

FIGURE 17.5 An example of RNA silencing in petunias. **(a)** Flower-color variation in some strains after a cloned pigmentation gene was introduced into the petunia genome. **(b)** Explanation for the production of double-stranded RNA.

(a): Jen-Chih Chen

CONCEPT CHECK: Explain why the white flower color is unexpected.

in Figures 17.6 and 17.7). A possible mechanism for the production of double-stranded RNA for the pigmentation gene is shown in **Figure 17.5b**. As shown here, the cloned pigmentation gene contained its own active promoter, which caused the synthesis of mRNA. If this cloned gene inserted next to a promoter for some other endogenous gene in the petunia genome, the orientation of that endogenous promoter may result in the synthesis of the opposite strand, called antisense RNA. The mRNA and antisense RNA are complementary to each other. Therefore, when both of them are made, they can bind to each other to form double-stranded RNA.

EXPERIMENT 17A

Fire and Mello Showed That Double-Stranded RNA Is More Potent than Antisense RNA at Silencing mRNA

In addition to the work with petunias, other studies revealed cases in which mRNA levels can be suppressed. In the nematode worm *Caenorhabditis elegans*, researchers had often introduced antisense RNA into cells as a way to inhibit mRNA translation. Because antisense RNA is complementary to mRNA, it was assumed that the antisense RNA binds to the mRNA, thereby preventing translation. Oddly, in other experiments, researchers introduced sense RNA (RNA with the same sequence as mRNA) into cells instead, and, unexpectedly, this also inhibited mRNA expression. Another curious observation was that the effects of antisense RNA often persisted for a very long time, much longer than would have been predicted by the relatively short half-lives of most RNA molecules in the cell. These two unusual observations caused Andrew Fire, Craig Mello, and colleagues to investigate how the introduction of RNA into cells inhibits mRNA.

C. elegans was used as the experimental organism because it is relatively easy to inject with RNA and the expression of many of its genes had already been established. In this study, published in 1998, Fire and Mello investigated the effects of both antisense and sense RNA on the expression of specific mRNAs in *C. elegans*. **Figure 17.6** focuses on one of their experiments involving an mRNA coded by a gene called *mex-3*. This mRNA had already been shown to be made in high amounts in early embryos of *C. elegans*.

Prior to this work, the *mex-3* gene had been identified and inserted into a plasmid. The process of inserting genes into plasmids is described in Chapter 20 (look ahead to Figure 20.2). Let's first look at the plasmid shown at the top of the Conceptual level in step 1 of Figure 17.6. When RNA polymerase, nucleotides, and this plasmid are mixed together in a test tube, *mex-3* mRNA is made; this mRNA is called the sense strand. In living cells, the sense strand is used to make the mex-3 protein. Fire and Mello also switched the location of the promoter to the other end of the gene and transcribed the opposite strand, which is called the antisense strand. The sense and antisense strands are complementary to each other.

Next, Fire and Mello injected these RNAs into the gonads of *C. elegans*. Into some worms, they injected antisense RNA alone. Alternatively, they mixed sense and antisense RNA, which formed double-stranded RNA, and injected this double-stranded RNA into the gonads of other worms. They also used uninjected worms as controls. After injection, the RNA was taken up by eggs, which later developed into embryos. To determine the amount of *mex-3* mRNA present, Fire and Mello incubated the embryos with a segment of labeled DNA that was complementary to this mRNA. The labeled DNA bound to *mex-3* mRNA could be observed under the microscope. After this incubation step, any labeled DNA that was not bound to this mRNA was washed away.

THE GOAL (DISCOVERY-BASED SCIENCE)

The goal was to further understand how the experimental injection of RNA was responsible for the silencing of particular mRNAs.

ACHIEVING THE GOAL   **FIGURE 17.6** Injection of antisense and double-stranded RNAs into *C. elegans* to compare their effects on mRNA silencing.

Starting material: The researchers used *C. elegans* as their model organism. They also had the cloned *mex-3* gene, which had been previously shown to be highly expressed in the early embryos of this worm.

Experimental level   **Conceptual level**

1. Make sense and antisense *mex-3* RNA in vitro using cloned genes for *mex-3* with promoters on either side of the gene. RNA polymerase and nucleotides are added to synthesize the RNAs.

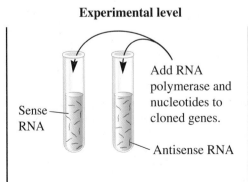

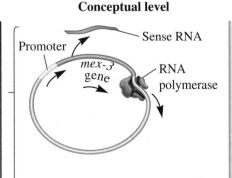

2. Inject either *mex-3* antisense RNA or a mixture of *mex-3* sense and antisense RNA into the gonads of *C. elegans*. This RNA is taken up by the eggs and early embryos. As a control, do not inject any RNA.

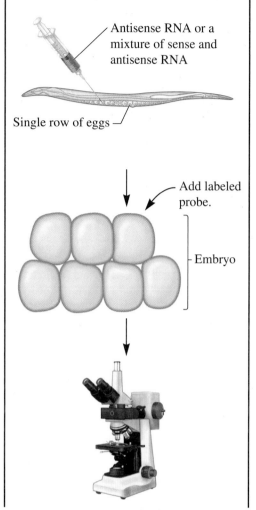
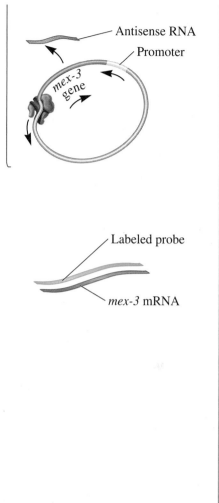

3. Incubate and then subject early embryos to *in situ* hybridization. In this method, a labeled segment of DNA is added that is complementary to the *mex-3* mRNA. If cells express *mex-3*, the mRNA in the cells will bind to the DNA and become labeled. After incubation with the labeled DNA, the cells are washed to remove unbound DNA.

4. Observe embryos under the microscope.

THE DATA

Control

Injected with *mex-3* antisense RNA

Injected with double-stranded RNA (both *mex-3* sense and antisense RNAs)

Source: Fire, Andrew, Xu, SiQun, Montgomery, Mary K., Kostas, Steven A., Driver, Samuel E., and Mello, Craig C., "Potent and specific genetic interference by double-stranded RNA in *Caenorhabditis elegans*," *Nature*, vol. 391, February 19, 1998, 806–811.

INTERPRETING THE DATA

As seen in the schematic data, the control embryos were very darkly labeled, as shown by the green color. These results indicated that the control embryos contained a high amount of *mex-3* mRNA, which was known from previous research. In the embryos that had received antisense RNA, *mex-3* mRNA levels were decreased, but detectable, as shown by the light-green color. Remarkably, in embryos that received double-stranded RNA, no

mex-3 mRNA was detected! These results indicated that double-stranded RNA is more potent at silencing mRNA than is antisense RNA. In this case, the double-stranded RNA caused the *mex-3* mRNA to be degraded.

Fire and Mello used the term **RNA interference (RNAi)** to describe the phenomenon in which double-stranded RNA causes the silencing of mRNA. This surprising observation led these researchers to investigate the underlying molecular mechanism that accounts for this phenomenon, as described next.

RNA Interference Is Mediated by MicroRNAs or Small Interfering RNAs via an RNA-Induced Silencing Complex

RNA interference is found in most eukaryotic species. The RNAs that promote interference are of two types: microRNAs and small interfering RNAs. **MicroRNAs (miRNAs)** are ncRNAs that are transcribed from endogenous eukaryotic genes—genes that are normally found in the genome. They play key roles in regulating gene expression, particularly during embryonic development in animals and plants. Most commonly, a single type of miRNA inhibits the translation of several different mRNAs. An miRNA and an mRNA bind to each other because they have sequences that are partially complementary. In humans, nearly 2000 genes code miRNAs, though that number may be an underestimate. Research suggests that 60% of human protein-coding genes are regulated by miRNAs.

By comparison, **small interfering RNAs (siRNAs)** are ncRNAs that usually originate from sources that are exogenous, which means they are not normally made by cells. The source of siRNAs can be viruses that infect a cell, or researchers can make siRNAs to study gene function experimentally, as Fire and Mello did in their experiment in Figure 17.6. In most cases, an siRNA is a perfect match to a single type of mRNA. The functioning of siRNAs is thought to play a key role in preventing certain types of viral infections. In addition, siRNAs have become important experimental tools in molecular biology.

How do miRNAs and siRNAs cause the silencing of specific mRNAs? **Figure 17.7** shows how an miRNA or siRNA achieves RNA interference. An miRNA is first synthesized as a pri-miRNA (for primary-miRNA) in the nucleus. Due to complementary base pairing, the pri-miRNA forms a hairpin structure (a stem-loop) with long single-stranded 5′ and 3′ ends. The pri-miRNA is recognized in the nucleus by two proteins, Drosha and DGCR8, and is cleaved at both ends. In animal cells, this typically results in a 70-nucleotide RNA molecule that is called a pre-miRNA (for precursor-miRNA; not to be confused with pri-miRNA). The pre-miRNA is then exported from the nucleus with the aid of a protein called exportin 5.

As shown in Figure 17.7, siRNAs do not go through the processing events that occur in the nucleus. Instead, precursor-siRNAs (pre-siRNAs) are usually derived from viral RNAs or may be made by researchers and taken up by cells. For example, in the experiment of Fire and Mello presented in Figure 17.6, the double-stranded *mex-3* RNA is an example of a pre-siRNA. In Figure 17.7, the pre-siRNA is formed from two complementary RNA molecules that base-pair with each other.

In the cytosol, both pre-miRNAs and pre-siRNAs are cut by an endonuclease called dicer (see Figure 17.7), releasing a double-stranded RNA molecule that is 20–25 bp long. This double-stranded RNA associates with proteins to form a complex called the **RNA-induced silencing complex (RISC).** One of the RNA strands is degraded. The remaining single-stranded miRNA or siRNA is complementary to specific mRNAs that will be silenced. The miRNA or siRNA acts as a guide that causes RISC to recognize and bind to such mRNA molecules.

After RISC binds to an mRNA, any of the following three outcomes may result:

- RISC may inhibit translation without degrading the mRNA. This outcome is more common for miRNAs, which often are only partially complementary to their target mRNAs.
- The RISC-mRNA complex may remain in a cellular structure called a **processing body (P-body),** where it can be stored until it is reused or degraded. In some cases, the inhibition of translation by an miRNA may be only temporary.
- RISC may direct the degradation of the mRNA. One of the proteins in RISC, which is called Argonaute, can cleave the mRNA. This outcome usually occurs with siRNAs, which are typically a perfect match to their target mRNAs.

The effect is termed RNA interference because the miRNA or siRNA interferes with the proper expression of an mRNA. In 2006, Fire and Mello received the Nobel Prize in Physiology or Medicine for their discovery of this mechanism.

RNA interference has at least two benefits:

- RNAi provides a defense against viruses. This mechanism is widely used by plants to prevent viral infections.
- RNAi is also an important means of gene regulation. When genes coding pri-miRNAs are turned on, the production of miRNAs silences the expression of specific mRNAs. This phenomenon is particularly important during early stages of development in plants and animals, in which miRNAs silence the expression of mRNAs that code transcription factors that are needed for developmental changes. For example, an miRNA called miR-156 is produced in plant seedlings, and it silences the expression of mRNAs that code transcription factors that are needed to transition into adult vegetative growth (**Figure 17.8**). When seedlings are exposed to the proper growth conditions, the level of miR-156 drops due to gene repression caused by PRC1 and PRC2, which are discussed in Chapter 16. The lowering of the miR-156 level allows the expression of the mRNA for a transcription factor called SPL. The synthesis of the SPL protein plays a key role in the transition to adult vegetative growth.

In addition to mRNA degradation, siRNAs can inhibit transcription by causing chromatin modifications. This effect is also seen with another type of small regulatory RNA, called PIWI-interacting RNA (piRNA). The ability of piRNAs to inhibit transcription is described later in this chapter (look ahead to Figure 17.12).

17.3 COMPREHENSION QUESTIONS

1. The process of RNA interference may lead to
   a. the degradation of an mRNA.
   b. the inhibition of translation of an mRNA.
   c. the synthesis of an mRNA.
   d. both a and b.

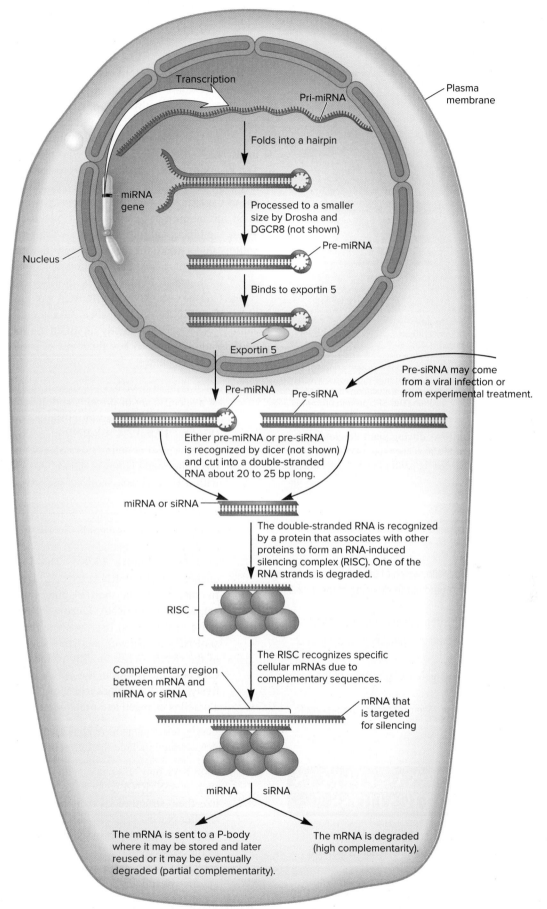

Transcription

Pri-miRNA

Plasma membrane

Folds into a hairpin

miRNA gene

Nucleus

Processed to a smaller size by Drosha and DGCR8 (not shown)

Pre-miRNA

Binds to exportin 5

Exportin 5

Pre-siRNA may come from a viral infection or from experimental treatment.

Pre-miRNA   Pre-siRNA

Either pre-miRNA or pre-siRNA is recognized by dicer (not shown) and cut into a double-stranded RNA about 20 to 25 bp long.

miRNA or siRNA

The double-stranded RNA is recognized by a protein that associates with other proteins to form an RNA-induced silencing complex (RISC). One of the RNA strands is degraded.

RISC

The RISC recognizes specific cellular mRNAs due to complementary sequences.

Complementary region between mRNA and miRNA or siRNA

mRNA that is targeted for silencing

miRNA   siRNA

The mRNA is sent to a P-body where it may be stored and later reused or it may be eventually degraded (partial complementarity).

The mRNA is degraded (high complementarity).

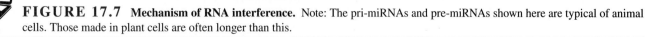

ONLINE ANIMATION

FIGURE 17.7 **Mechanism of RNA interference.** Note: The pri-miRNAs and pre-miRNAs shown here are typical of animal cells. Those made in plant cells are often longer than this.

CONCEPT CHECK: Explain why RISC binds to a specific mRNA. What type of bonding occurs?

451

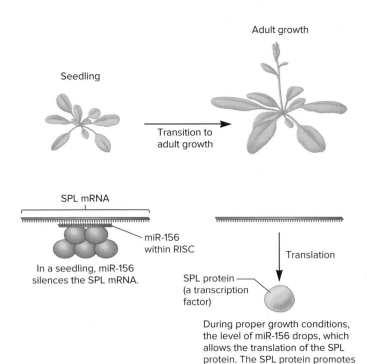

Adult growth

Seedling

SPL mRNA

miR-156 within RISC

In a seedling, miR-156 silences the SPL mRNA.

Translation

SPL protein (a transcription factor)

During proper growth conditions, the level of miR-156 drops, which allows the translation of the SPL protein. The SPL protein promotes the transition to adult growth.

FIGURE 17.8 **Role of an miRNA during plant development.** In this example, an miRNA called miR-156 silences an mRNA that codes a transcription factor that is needed for adult growth. Under the proper growth conditions, miR-156 is no longer made, which permits the translation of the SPL mRNA, and thereby favors the transition to adult growth.

CONCEPT CHECK: What molecule prevents the transition to adult growth?

2. In the experiment of Fire and Mello, which of the following statements is most accurate with regard to the silencing of the *mex-3* mRNA?

a. Antisense RNA blocked the translation of the *mex-3* RNA.

b. Double-stranded RNA blocked the translation of the *mex-3* RNA.

c. Antisense RNA caused the greatest degradation of the *mex-3* RNA.

d. Double-stranded RNA caused the greatest degradation of the *mex-3* RNA.

17.4 NON-CODING RNAs: EFFECTS ON RNA MODIFICATIONS

Learning Outcome:

1. Explain how snoRNAs direct covalent modifications of rRNAs.

Small nucleolar RNAs (snoRNAs) are so named because they are found in high amounts in the nucleolus, which is a darkly staining region located in the nucleus of eukaryotic cells. As described in Chapter 13, the nucleolus plays a role in the synthesis of ribosomal RNAs (rRNAs) and the assembly of ribosomal subunits.

A key function of snoRNAs is to covalently modify rRNAs. Two common covalent modifications of rRNAs that are facilitated by snoRNAs are the methylation of ribose on the 2′ hydroxyl group and the conversion of uracil to pseudouracil (**Figure 17.9a**).

The two types of modifications in Figure 17.9a are common in rRNAs, but they do not occur with most other types of RNA. The locations of these modifications are concentrated in functionally important regions of the rRNAs. For example, they are present in the rRNA of peptidyl transferase, which catalyzes peptide bond formation during translation (refer back to Figure 13.17). While the functional roles of these covalent modifications are not entirely understood, the current view is that they fine-tune rRNA structure so that rRNAs can function in an optimal manner.

As shown in **Figure 17.9b**, snoRNAs exist in two general types: C/D box snoRNA and H/ACA box snoRNA. These two types of snoRNAs differ in their structures. How do snoRNAs cause rRNAs to be covalently modified via methylation or pseudouridylation? The two key functions of snoRNAs are to act as scaffolds and as guides.

- Each type of snoRNA provides a scaffold for the binding of a specific set of proteins and thereby forms a complex called a **small nucleolar ribonucleoprotein (snoRNP)** (**Figure 17.9c**). A snoRNP that contains a C/D box snoRNA also contains two copies of a protein that catalyzes the methylation of ribose. Alternatively, a snoRNP with a H/ACA box snoRNA has two copies of a protein that catalyzes the conversion of uracil to pseudouracil.

- After a snoRNP complex has formed, the second role of a snoRNA is to act as a guide. To accomplish this, snoRNAs have antisense sequences that are complementary to sites in rRNAs (**Figure 17.9d**). These complementary sequences enable snoRNAs to recognize and bind to rRNAs. In other words, the snoRNA guides the snoRNP to its target, which is an rRNA. Once an rRNA binds to the antisense sequences in the snoRNA, the proteins within the snoRNP (not shown) catalyze the chemical modification of the rRNA. When an rRNA binds to a snoRNA with a C/D box, the ribose in the rRNA is methylated. Alternatively, if an rRNA binds to a snoRNA with an H/ACA box, a uracil is changed to a pseudouracil in two places.

More recently, researchers have identified other possible functions of snoRNAs, which include the following:

- snoRNAs may associate with long primary transcripts, such as 45S rRNA (refer back to Figure 12.16), and stabilize their structure so that they are properly cleaved to smaller pieces.

- snoRNAs may be processed to a smaller size and function as miRNAs.

- snoRNAs can become part of snRNPs and play a role in alternative splicing.

Before leaving this topic, it is worth noting that snoRNAs are structurally related to scaRNAs (small Cajal body-specific RNAs). Cajal bodies are located in the cell nucleus and function in the assembly and maturation of spliceosome subunits, which are

FIGURE 17.9 Functions of small nucleolar RNAs.
(a) snoRNAs facilitate the methylation of ribose and the conversion of uracil to pseudouracil in rRNAs. **(b)** The two general types of snoRNAs. The snoRNA on the left has short sequences called a C box and a D box, which are critical for the function of this type of RNA. It has similar sequences called C′ and D′ boxes. The snoRNA on the right has H box and ACA box sequences. Note: R stands for a purine (A or G), and N stands for any base (A, U, G, or C). **(c)** snoRNAs provide a scaffold for the binding of several proteins to form a complex called a snoRNP. The complex on the left contains a C/D box snoRNA and also has two copies of a protein, shown in green, that methylates ribose. The complex on the right has an H/ACA box snoRNA and includes two copies of a protein, shown in purple, that catalyzes pseudouridylation. Both complexes have additional proteins, shown in tan, that are important for snoRNP structure and function. **(d)** The snoRNAs have antisense sequences that are complementary to sequences in an rRNA. As shown on the left side, the C/D box snoRNA has two antisense sequences that are complementary to rRNAs. After two different rRNAs bind to the snoRNP, they are modified by ribose methylation. On the right side, a single rRNA binds to multiple antisense sequences and a uracil is changed to pseudouracil in two places. Pseudouracil is represented by the Greek letter psi (ψ). Note: The covalent modifications are carried out by snoRNPs. For simplicity, the proteins have been omitted from part (d), but they are shown in part (c).

CONCEPT CHECK: Which type of snoRNA causes an rRNA to be methylated?

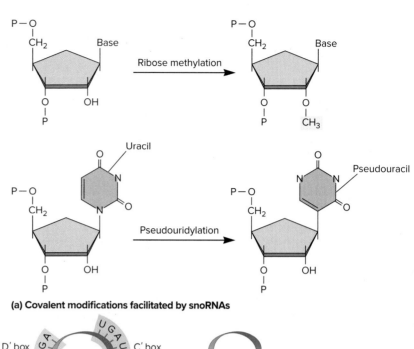

(a) Covalent modifications facilitated by snoRNAs

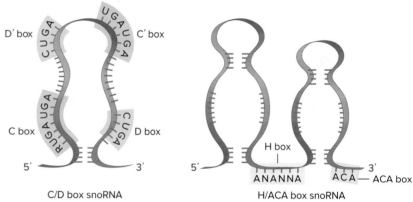

C/D box snoRNA

H/ACA box snoRNA

(b) General types of snoRNAs

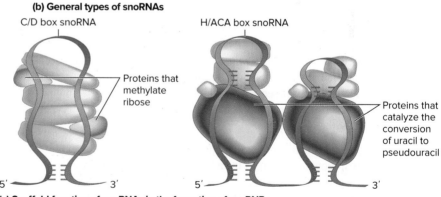

(c) Scaffold function of snoRNAs in the formation of snoRNPs

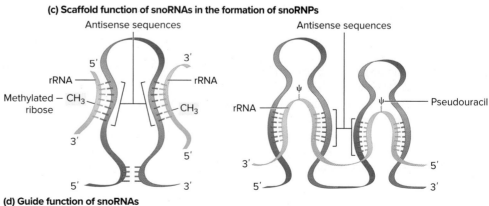

(d) Guide function of snoRNAs

described in Chapter 12 (refer back to Figure 12.20). scaRNAs direct the methylation and pseudouridylation of RNAs that are found within the spliceosome subunits.

17.4 COMPREHENSION QUESTION

1. When facilitating the methylation or pseudouridylation of an rRNA, a snoRNA functions as
   a. a decoy.
   b. a scaffold.
   c. a guide.
   d. both b and c.

17.5 NON-CODING RNAs AND PROTEIN TARGETING

Learning Outcome:

1. Describe the function of SRP, and explain the roles of SRP RNA with regard to SRP function.

To carry out their functions, proteins need to be targeted to a particular location. For example, some proteins function extracellularly and need to be secreted from the cell. For such proteins to be secreted, they must first be targeted to the plasma membrane in bacteria or archaea or to the endoplasmic reticulum (ER) membrane in eukaryotes. This targeting process is facilitated by an RNA-protein complex called **signal recognition particle (SRP).** In bacteria, SRP is composed of one ncRNA and one protein. In eukaryotes, SRP is composed of one ncRNA and six different proteins.

Figure 17.10 shows how SRP works in eukaryotes. To be directed to the ER membrane, a polypeptide must contain a sorting signal called an **ER signal sequence,** which is a sequence of about 6–12 amino acids that are predominantly hydrophobic and usually located near the N-terminus.

1. As the ribosome is making the polypeptide in the cytosol, the ER signal sequence emerges from the ribosome and is recognized by a protein in SRP.
2. The binding of SRP pauses translation.
3. SRP then binds to an SRP receptor in the ER membrane, which docks the ribosome over a channel. For this binding to occur, proteins within SRP and the SRP receptor must also be bound by guanosine triphosphate (GTP).
4. Next, these GTP-binding proteins hydrolyze the bound GTP, which causes the release of SRP from the SRP receptor and the polypeptide.
5. Once SRP is released, translation resumes, and the growing polypeptide is threaded through a channel to cross the ER membrane. If the newly made polypeptide is to be secreted, it travels through the Golgi apparatus and then to the plasma membrane, where it is released outside of the cell.

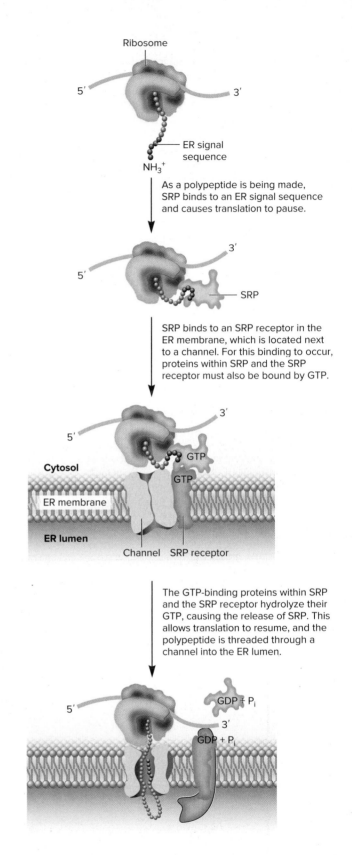

FIGURE 17.10 Targeting of a polypeptide to the endoplasmic reticulum membrane via SRP. In eukaryotes, several categories of proteins are first targeted to the ER membrane via SRP. These include secreted proteins as well as proteins that are destined to stay in the ER, Golgi apparatus, lysosomes, or vacuoles.

CONCEPT CHECK: Why is GTP necessary for this process?

Researchers have identified at least two key roles for SRP RNA:

- SRP RNA provides a scaffold for the binding of SRP proteins.
- After SRP binds to the SRP receptor in the ER membrane, the SRP RNA stimulates proteins within both SRP and the SRP receptor to hydrolyze GTP. In other words, SRP RNA alters the structures of these proteins to enhance their GTPase activities. This stimulation is essential for the release of SRP.

17.5 COMPREHENSION QUESTION

1. Which of the following is a function of SRP?
   a. Pausing translation of a polypeptide via an ER signal sequence
   b. Binding to an SRP receptor in the ER membrane
   c. Docking the ribosome over a channel
   d. All of the above are functions of SRP.

17.6 NON-CODING RNAs AND GENOME DEFENSE

Learning Outcomes:

1. Explain how the CRISPR-Cas system defends bacteria against bacteriophages.
2. Describe the two mechanisms by which piRNAs and PIWI proteins inhibit transposable elements.

Much like the immune system found in vertebrates, some prokaryotes have a system, called the **CRISPR-Cas system,** that defends against foreign invaders. In many prokaryotes, CRISPR-Cas systems provide an effective defense against bacteriophages (discussed in Chapter 18), plasmids (discussed in Chapter 10), and transposons (discussed in Chapter 10). ncRNAs play a key role in these systems. About half of all bacterial species and most archaeal species have a CRISPR-Cas system. Three general types are known. In this section, we will focus on the type II system and its role in providing bacteria with defense against bacteriophages.

We will also consider another defense system found in animals that involves a type of ncRNA called a PIWI-interacting RNA. As you will learn, this ncRNA interacts with PIWI proteins and inhibits the movement of transposable elements.

The CRISPR-Cas System Provides Bacteria with Defense Against Bacteriophages

In 1993, Francisco Mojica and colleagues were the first to recognize that different species of prokaryotes have a site in their chromosome, now called the CRISPR locus, which contains a series of repeated sequences. In 2005, by analyzing the DNA sequences of the CRISPR locus in a variety of prokaryotes, Mojica, Gilles Vergnaud, and Alexander Bolotin independently proposed that it

provides protection against bacteriophage infection. This hypothesis was based on the observation that the CRISPR locus contains segments that are derived from bacteriophage DNA. Their hypothesis was confirmed in 2007 by Philippe Horvath and colleagues, who showed experimentally that the CRISPR-Cas system provides defense against bacterophage infection.

Figure 17.11a shows a common organization of the CRISPR-Cas system (also called the CRISPR locus) in a bacterial chromosome. In this example, the system has five genes: *tracr, Cas9, Cas1, Cas2,* and *Crispr.* A key feature of the *Crispr* gene is a group of clustered, regularly interspaced, short, palindromic repeats, hence the name CRISPR. The repeats are interspersed with short, unique sequences, which are called spacers. The CRISPR-Cas type II system also employs a gene that codes an ncRNA called tracrRNA and a few protein-coding *Crispr*-associated genes (*Cas* genes), which are usually adjacent to the *Crispr* gene. These genes are needed to mediate the defense against bacteriophages.

The CRISPR-Cas system is considered an adaptive defense system because a bacterial cell or one of its ancestors must first be exposed to an agent, such as a bacteriophage, to elicit a response. The defense mechanism has three phases: adaptation, expression, and interference.

Adaptation. The process of adaptation (also called spacer acquisition) occurs after a bacterial cell has been exposed to a bacteriophage. The proteins coded by the *Cas1* and *Cas2* genes form a complex that recognizes the bacteriophage DNA as being foreign and cleaves it into small pieces. As shown in **Figure 17.11b**, a piece of bacteriophage DNA, usually between 20 and 50 bp in length, is inserted into the *Crispr* gene. The newly inserted piece of bacteriophage DNA is called a spacer, because it acts as a space between adjacent repeats. The different spacers found in the *Crispr* gene of modern bacterial species are derived from past bacteriophage infections. Each spacer provides a bacterium with defense against a particular bacteriophage. Once a bacterial cell has become adapted to a particular bacteriophage, it will pass this trait on to its daughter cells.

Cleavage of the bacteriophage DNA into pieces during the adaptation phase can protect a bacterial cell, because this action inactivates the phage. In addition, another way of destroying phages is provided by the expression and interference phases of this defense system.

Expression. If a bacterial cell has already been adapted to a bacteriophage, a subsequent bacteriophage infection will result in the expression phase, in which the system gets ready for action by expressing the *Crispr, tracr,* and *Cas9* genes (**Figure 17.11c**). The *Crispr* gene is transcribed from a single promoter and produces a long ncRNA called pre-crRNA. The gene coding the tracrRNA is also transcribed, which produces many molecules of tracrRNA, another kind of ncRNA. A region of the tracrRNA is complementary to the repeat sequences of the pre-crRNA. Several molecules of tracrRNA base-pair with the pre-crRNA. The pre-crRNA is then cleaved into many small molecules, now called crRNAs. Each crRNA is attached to a tracrRNA. A region of the tracrRNA is recognized by the Cas9 protein. The tracrRNA acts as a guide that causes the tracrRNA-crRNA complex to bind to a Cas9 protein.

Interference. After the tracrRNA-crRNA-Cas9 complex is formed, the bacterial cell is ready to destroy the bacteriophage DNA. This phase is called interference, because it resembles the process of RNA interference described earlier in this chapter (refer back to Figure 17.7). Each spacer within a crRNA is complementary to one of the strands of the bacteriophage DNA. Therefore, the crRNA acts as a guide that causes the tracrRNA-crRNA-Cas9 complex to bind to that strand (**Figure 17.11d**). After binding, the Cas9 protein functions as an endonuclease that makes double-strand breaks in the bacteriophage DNA. This cleavage inactivates the phage, and thereby prevents its proliferation.

Since the discovery of the CRISPR-Cas system in prokaryotes, researchers have been able to modify certain components of this system and use them to alter genes in living cells. We will consider this technology, which is called gene editing, in Chapter 20.

PIWI-Interacting RNAs Silence Transposable Elements in Animals

As described in Chapter 10, **transposable elements (TEs)** are segments of DNA that can become integrated into chromosomes. Various types of TEs are found in all species. Transposition is the process in which a TE is inserted into a new site in the genome. If a TE is inserted into a gene, this event is likely to inactivate the gene, which is an undesirable outcome.

All species of animals, from sponges to mammals, have evolved a defense mechanism using ncRNAs as way to inhibit the integration of TEs into new sites. Such ncRNAs are called **PIWI-interacting RNAs (piRNAs)** because they associate with a class of proteins called PIWI proteins. The name PIWI is derived from a phenotype in *Drosophila* in which a gene that codes a PIWI protein is mutated. Male fruit flies carrying a mutant version of this gene allow a transposable element called a P-element to proliferate at a high rate; they also have smaller testes. PIWI is an acronym for P-element induced wimpy testes. PIWI proteins are

FIGURE 17.11 The CRISPR-Cas system of genome defense in prokaryotes. The system shown here is a type II system, which is found in certain bacterial species but not in archaeal species. **(a)** Organization of the CRISPR-Cas system in a bacterial chromosome. This drawing shows a typical organization, but it can vary among different species. **(b–d)** A simplified mechanism for the functioning of the CRISPR-Cas system. The defense occurs in three phases called adaptation, expression, and interference.

CONCEPT CHECK: Which component of the CRISPR-Cas system directly recognizes the bacteriophage DNA?

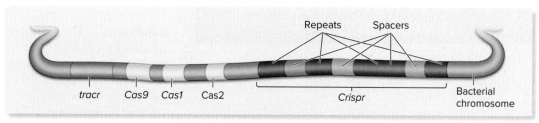

(a) A simplified diagram of the organization of the CRISPR-Cas system in the bacterial chromosome

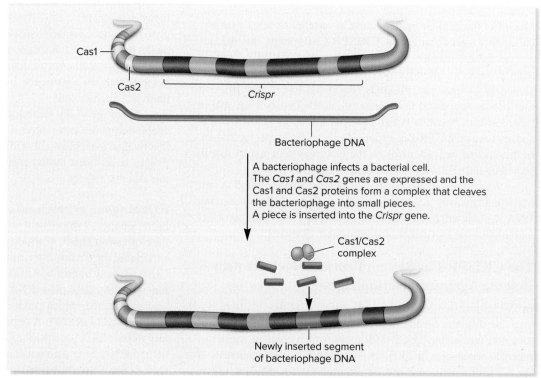

A bacteriophage infects a bacterial cell. The *Cas1* and *Cas2* genes are expressed and the Cas1 and Cas2 proteins form a complex that cleaves the bacteriophage into small pieces. A piece is inserted into the *Crispr* gene.

Cas1/Cas2 complex

Newly inserted segment of bacteriophage DNA

(b) Adaptation

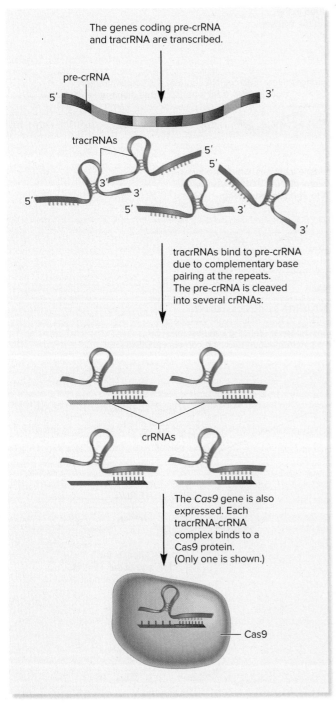

The genes coding pre-crRNA and tracrRNA are transcribed.

pre-crRNA

tracrRNAs

tracrRNAs bind to pre-crRNA due to complementary base pairing at the repeats. The pre-crRNA is cleaved into several crRNAs.

crRNAs

The *Cas9* gene is also expressed. Each tracrRNA-crRNA complex binds to a Cas9 protein. (Only one is shown.)

Cas9

(c) Expression

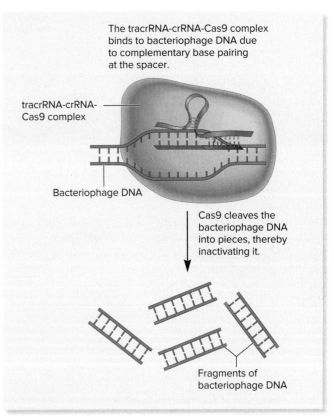

The tracrRNA-crRNA-Cas9 complex binds to bacteriophage DNA due to complementary base pairing at the spacer.

tracrRNA-crRNA-Cas9 complex

Bacteriophage DNA

Cas9 cleaves the bacteriophage DNA into pieces, thereby inactivating it.

Fragments of bacteriophage DNA

(d) Interference

FIGURE 17.11 *continued.*

to the RNA that is transcribed from a particular TE. These piRNA sequences are separated by sequences (shown in red) that are not complementary to the TE RNA. After the pre-piRNA is made, it is processed into one or more piRNAs, which are usually 24–31 nucleotides in length. These piRNAs associate with PIWI proteins to form one of two types of complexes: a **piRNA-induced transcriptional silencing complex (piRITS)** or a **piRNA-induced silencing complex (piRISC)**. These two types of complexes differ in their protein composition and in the way they silence TEs.

As shown on the lower left side of Figure 17.12, a piRITS enters the nucleus. This piRITS binds to an RNA molecule that is being transcribed from a TE. This binding occurs because the piRNA and the TE RNA are complementary. Following the binding, the silencing complex directs the methylation of DNA and the trimethylation of lysine 9 on histone H3 (producing H3K9me3). As discussed in Chapter 16, these modifications recruit proteins that convert euchromatin into more densely packed heterochromatin. The conversion of euchromatin to heterochromatin inhibits the transcription of the TE and thereby prevents its ability to move to a new site.

A piRISC exerts its effect by directly degrading an RNA made from a TE. The mechanism of this inhibition is very similar to the mechanism of mRNA silencing via siRNAs, described earlier in Figure 17.7. However, the processing of the pre-piRNAs that bind to piRISC does not involve dicer. As shown on the lower right side of Figure 17.12, the piRISC binds to the TE RNA in the cytosol, and a protein within piRISC called an Argonaute protein (not shown) cuts the TE RNA into pieces, thereby inactivating it.

primarily expressed in germ-line cells—cells that give rise to sperm or egg cells. Because they inhibit the movement of transposable elements and are expressed in germ-line cells, PIWI proteins prevent undesirable TE insertions from being passed from parent to offspring.

Figure 17.12 presents a simplified mechanism of how piRNAs and PIWI proteins exert their effects. The genes that code piRNAs are usually organized into clusters in which a single pre-piRNA transcript contains several different piRNA sequences (shown in different colors). Each piRNA sequence is complementary

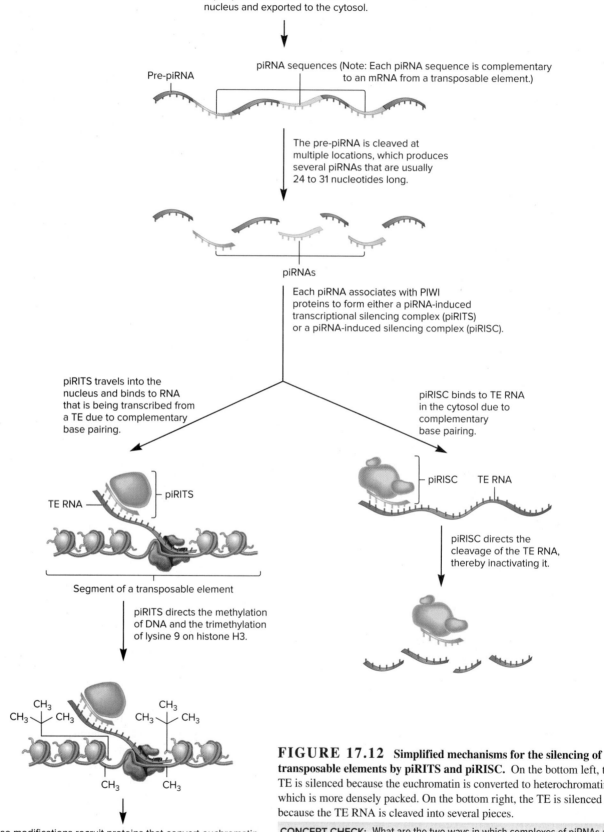

A pre-piRNA is transcribed in the nucleus and exported to the cytosol.

Pre-piRNA

piRNA sequences (Note: Each piRNA sequence is complementary to an mRNA from a transposable element.)

The pre-piRNA is cleaved at multiple locations, which produces several piRNAs that are usually 24 to 31 nucleotides long.

piRNAs

Each piRNA associates with PIWI proteins to form either a piRNA-induced transcriptional silencing complex (piRITS) or a piRNA-induced silencing complex (piRISC).

piRITS travels into the nucleus and binds to RNA that is being transcribed from a TE due to complementary base pairing.

piRISC binds to TE RNA in the cytosol due to complementary base pairing.

TE RNA — piRITS

piRISC   TE RNA

Segment of a transposable element

piRISC directs the cleavage of the TE RNA, thereby inactivating it.

piRITS directs the methylation of DNA and the trimethylation of lysine 9 on histone H3.

CH_3 CH_3
CH_3 CH_3
CH_3 CH_3
CH_3 CH_3

CH_3 CH_3

FIGURE 17.12 Simplified mechanisms for the silencing of transposable elements by piRITS and piRISC. On the bottom left, the TE is silenced because the euchromatin is converted to heterochromatin, which is more densely packed. On the bottom right, the TE is silenced because the TE RNA is cleaved into several pieces.

These modifications recruit proteins that convert euchromatin to heterochromatin, which stops the transcription of the TE.

CONCEPT CHECK: What are the two ways in which complexes of piRNAs and PIWI proteins silence transposable elements?

In recent years, researchers have discovered that piRNAs are an abundant class of ncRNAs. They are thought to be more diverse than any other known class of cellular RNAs and to constitute the largest class of ncRNAs. Although their main function is to prevent the movement and integration of TEs, piRNAs are also known to regulate a few protein-coding genes.

17.6 COMPREHENSION QUESTIONS

1. Which of the following components are needed for the adaptation phase of the CRISPR-Cas system?
   a. crRNA and Cas1 protein
   b. crRNA and Cas2 protein
   c. crRNA and Cas9 protein
   d. Cas1 and Cas2 proteins

2. In the CRISPR-Cas system, what does tracrRNA bind to?
   a. crRNA and Cas1 protein
   b. crRNA and Cas2 protein
   c. crRNA and Cas9 protein
   d. Cas1 and Cas2 proteins

3. Which of the following function(s) is/are carried out by piRITS or piRISC?
   a. Inhibits transcription of TEs
   b. Causes the degradation of TE RNA
   c. Causes chromosome breakage
   d. Both a and b are functions of piRITS or piRISC.

17.7 ROLE OF NON-CODING RNAs IN HUMAN DISEASES AND PLANT HEALTH

Learning Outcomes:

1. List examples of ncRNAs that are associated with human diseases.
2. Describe the role of ncRNAs in plant health.

During the past two decades, researchers have discovered that abnormalities in the expression of ncRNAs are associated with a wide range of human diseases. In 2001, the gene coding an RNA molecule that is a component of RNaseMRP was the first ncRNA gene shown to be associated with a genetic disease. RNaseMRP is an RNA-protein complex involved in processing some rRNAs, mRNAs, and mitochondrial RNAs. The RNA component is a ribozyme. Mutations in this gene cause a disorder called cartilage-hair hypoplasia (CHH). CHH patients have a short stature, underdeveloped hair, and short limbs. In addition, they have a predisposition to develop lymphomas and other cancers, and they suffer from defects in T-cell immunity.

Since the identification of the first CHH-associated mutations in 2001, many ncRNAs have been shown to play a key role in human diseases. Researchers speculate that they have seen only the "tip of the iceberg" with regard to identifying the roles of ncRNAs in human pathology. In this section, we will consider examples in which ncRNAs are associated with human diseases. We will also examine their roles in plant health.

ncRNAs Play a Role in Many Forms of Cancer and Other Human Diseases

As we have seen throughout this chapter, ncRNAs play important roles in regulating chromatin modification, gene transcription, mRNA translation, and protein function. When certain ncRNAs are expressed abnormally, that is, at too high or too low a level, disease conditions are known to occur. Such abnormal expression levels can be caused by mutations in specific genes or by epigenetic changes that alter the expression of the genes that code ncRNAs. Several examples of human diseases associated with the abnormal expression of ncRNAs are listed in **Table 17.2**. We will focus on the roles of ncRNAs in the development of cancer, neurological disorders, and cardiovascular diseases.

ncRNAs and Cancer. The roles of ncRNAs in cancer have been most thoroughly studied with respect to miRNAs. In nearly all forms of human cancer, levels of expression of particular miRNAs differ between normal and cancer cells. In some cases, the genes that code certain miRNAs act as oncogenes; their overexpression promotes cancer. In other cases, the genes behave like tumor-suppressor genes, because a lower level of expression of particular miRNAs allows tumor growth.

TABLE **17.2**	
Examples of Non-coding RNAs Associated with Human Diseases	
ncRNA(s) or Processing Protein(s) Associated with Disease(s)	**Disease(s)***
Group of miRNAs called the miR-200 family	Several types of cancer, including bladder cancer, melanoma, stomach cancer, and colorectal cancer
HOTAIR	Several types of cancer, including breast cancer, lung cancer, and colorectal cancer
piRNAs/PIWI proteins	Certain types of testicular cancer
Drosha	Familial amyotrophic lateral sclerosis
Many miRNAs	Alzheimer's disease
Many miRNAs	Multiple sclerosis
An miRNA called miR-1	Heart arrhythmias
Many different miRNAs	Damage to heart tissue due to heart failure
Several different miRNAs, including miR-10a, miR-145, and miR-143	Formation of arterial plaques
snoRNAs	Lung cancer

*The diseases listed here show an association with abnormal levels of the particular ncRNAs or processing proteins. In many cases, it is not yet clear if the disease symptoms are caused, in part, by the abnormal levels of ncRNAs or proteins or if the abnormal levels are a consequence of the disease symptoms.

A well-studied example of the role of miRNAs in cancer involves a group of several different miRNAs called the miR-200 family. These miRNAs are often involved in cancers that are derived from epithelial cells, such as skin or intestinal cells. Low levels of expression of miR-200 members have been associated with many types of cancer, including bladder cancer, melanoma, stomach cancer, and colorectal cancer.

The miR-200 family plays an essential role in tumor suppression by inhibiting an event called the epithelial-mesenchymal transition (EMT), which is the initiating step of metastasis. (The process of metastasis is described in Chapter 25.) The EMT occurs as part of normal embryonic development, and it shares many similarities with cancer progression. During the EMT, cells lose their adhesion to neighboring cells. This loss of adhesion is associated with a decrease in expression of E-cadherin, which is a membrane protein that adheres adjacent cells to each other. When E-cadherin levels are low, cells can more easily move to new sites in the body. In an adult, when the miR-200 family of miRNAs is expressed at normal levels, the miRNAs inhibit the EMT. This inhibition maintains a normal level of E-cadherin and thereby prevents metastasis.

Though they have been less well studied, other ncRNAs are also associated with particular types of human cancers. HOTAIR, which was discussed in Section 17.2, is an lncRNA that is highly expressed in a variety of cancers, including breast cancer, lung cancer, and colorectal cancer. In this regard, HOTAIR behaves like an oncogene. High levels of HOTAIR expression in primary breast tumors are a significant predictor of metastasis and death. HOTAIR is known to interact with a variety of cellular components. Although the mechanism by which it promotes cancer is not well understood, it seems to be related to its ability to advance the cell cycle.

ncRNAs and Neurological Disorders. Many miRNAs are essential for the proper development and functioning of the nervous system. Approximately 70% of all human miRNAs are expressed in the brain, and many of them are specific to neurons. miRNAs are involved in neuron growth and the overall development of the nervous system. Abnormal levels of expression of miRNAs have been associated with nearly all neurological disorders in which they have been investigated!

Table 17.2 lists some examples in which the altered expression of miRNAs has been associated with neurological disorders. The changes involve two categories of genes: genes that code miRNAs and genes that code proteins that are involved in processing miRNAs.

- Specific miRNAs have been linked to particular neurological diseases. For example, in Alzheimer's disease, abnormally expressed miRNAs are thought to be involved in down-regulating the expression of the enzyme β-secretase, which leads to the overproduction of certain β-amyloid peptides. miRNAs are also known to control the inflammatory process that leads to the development of multiple sclerosis.
- Mutations or epigenetic changes may alter the expression of genes that code miRNA-processing proteins, such as Drosha, dicer, and others (refer back to Figure 17.7). For example, mutations in components of Drosha cause up to

50% of all cases of familial amyotrophic lateral sclerosis (ALS) (sometimes called Lou Gehrig disease). These mutations result in a generalized decrease in the processing of many different miRNAs.

ncRNAs and Cardiovascular Diseases. Abnormalities in miRNA levels have been linked to several cardiovascular diseases.

- A particular miRNA called miR-1 is associated with the development of heart arrhythmias—irregularities in the rate or rhythm of the heartbeat. This miRNA regulates the expression of genes that code ion channel proteins, which are important for proper signaling between cardiac muscle cells.
- Cardiac tissue from patients with heart failure has a distinctly different expression pattern of many different miRNAs compared to healthy cardiac tissue.
- Particular miRNAs appear to play a role in cardiovascular disease. The formation of arterial plaques is associated with abnormal expression levels of several miRNAs, including miR-10a, miR-145, and miR 143.

miRNAs Play Key Roles in Plant Health

Plant researchers have also determined that levels of particular ncRNAs, especially miRNAs, can dramatically change in response to factors that have an impact on plant health. These include limitation of certain nutrients, such as phosphate and sulfate, exposure to pathogens, and environmental stress (**Table 17.3**). In some cases, the level of an ncRNA increases, whereas in others, it decreases.

17.7 COMPREHENSION QUESTIONS

1. Abnormalities in the expression of ncRNAs are associated with
   a. many forms of cancer.
   b. neurological disorders.
   c. cardiovascular diseases.
   d. all of the above.
2. With regard to plant health, changes in the level of particular miRNAs are known to be associated with
   a. drought stress.   c. heat stress.
   b. salt stress.   d. all of these.

TABLE 17.3

Examples of Non-coding RNAs Associated with Plant Health*

ncRNA(s)	Expression	During
HlLinc1 (a long ncRNA)	Increased	Heat stress
miR-395	Increased	Sulfate starvation, heat stress
miR-482	Decreased	Pathogen exposure
miR-851, miR-397b	Increased	Drought stress
miR-778	Increased	Salt stress

*The examples shown here are from *Arabidopsis*. In some cases, such as miR-482, the microRNAs are part of a family of related microRNAs.

Introduction: non-coding RNAs (ncRNAs)
17.1: ribozyme, long non-coding RNA (lncRNA), small regulatory RNA, protobiont, RNA world
17.3: RNA interference (RNAi), microRNAs (miRNAs), small interfering RNAs (siRNAs), RNA-induced silencing complex (RISC), processing body (P-body)
17.4: small nucleolar RNAs (snoRNAs), small nucleolar ribonucleoprotein (snoRNP)

17.5: signal recognition particle (SRP), ER signal sequence
17.6: CRISPR-Cas system, transposable element (TE), PIWI-interacting RNA (piRNA), piRNA-induced transcriptional silencing complex (piRITS), piRNA-induced silencing complex (piRISC)

CHAPTER SUMMARY

- Non-coding RNAs (ncRNAs) are RNA molecules that do not code polypeptides.

17.1 Overview of Non-coding RNAs

- ncRNAs bind to different types of molecules, including DNA, other RNAs, proteins, and small molecules (see Figure 17.1).
- An ncRNA can perform any of several functions: provide a scaffold, act as a guide, alter protein function or stability, function as a ribozyme, function as a blocker, and/or act as a decoy (see Figures 17.2 and 17.3).
- With regard to cell structure and function, ncRNAs play a role in DNA replication, chromatin structure, transcription, processing of tRNAs, splicing, translation, RNA degradation, RNA modification, protein targeting and secretion, and genome defense (see Table 17.1).
- The emergence of living cells may have been preceded by an RNA world in which RNA molecules, but not DNA or proteins, were the first macromolecules found within protobionts.

17.2 Non-coding RNAs: Effects on Chromatin Structure and Transcription

- HOTAIR is a long ncRNA found in humans that regulates transcription by guiding PRC2 and LSD1 to particular genes. These complexes produce histone modifications that silence the genes (see Figure 17.4).

17.3 Non-coding RNAs: Effects on Translation and mRNA Degradation

- Early studies revealed that the addition of a cloned gene into a plant's genome can silence endogenous genes (see Figure 17.5).
- Fire and Mello showed that double-stranded RNA is more potent at silencing mRNA than is antisense RNA (see Figure 17.6).
- RNA interference is a mechanism of RNA silencing in which miRNA or siRNA becomes part of an RNA-induced silencing complex (RISC) that inhibits the translation of a specific mRNA or causes its degradation (see Figure 17.7).
- RNA interference plays an important role in the development of plants and animals (see Figure 17.8).

17.4 Non-coding RNAs: Effects on RNA Modifications

- snoRNAs become part of complexes called snoRNPs that direct the methylation or pseudouridylation of rRNAs (see Figure 17.9).

17.5 Non-coding RNAs and Protein Targeting

- Signal recognition particle (SRP), which is composed of one or more proteins and an ncRNA, plays a role in the targeting of proteins to the plasma membrane of prokaryotic cells or to the ER membrane of eukaryotic cells (see Figure 17.10).

17.6 Non-coding RNAs and Genome Defense

- The CRISPR-Cas system found in bacteria and archaea provides defense against bacteriophages, plasmids, and transposons. The defense occurs in three phases: adaptation, expression, and interference (see Figure 17.11).
- PIWI-interacting RNAs (piRNAs) silence transposable elements in animals (see Figure 17.12).

17.7 Role of Non-coding RNAs in Human Diseases and Plant Health

- Abnormalities in the expression of ncRNAs have been associated with many diseases, including cancer, neurological disorders, and cardiovascular diseases (see Table 17.2).
- miRNAs play key roles in plant health (see Table 17.3).

MORE GENETIC TIPS   1. An ncRNA may have the
following functions: scaffold, guide, alterer of protein function or
stability, ribozyme, blocker, and/or decoy. Which of these functions
are carried out by each type of ncRNA listed next? Note: A single
ncRNA may have more than one function.

A. tRNA

B. rRNA

C. SRP RNA

D. piRNA

T OPIC: *What topic in genetics does this question address?*
The topic is the function(s) of specific types of ncRNAs.

I NFORMATION: *What information do you know based on the
question and your understanding of the topic?* In the question,
you are given a list of six general functions of ncRNAs and four
specific examples of ncRNAs. From your understanding of the
topic, you may recall the functions of each of these four examples.

P ROBLEM-SOLVING S TRATEGY: *Compare and contrast.*
To solve this problem, you need to compare the six general
functions of ncRNAs with your understanding of the functions of
each of the four specific ncRNAs.

ANSWER:

A. Guide: A tRNA binds to an mRNA and carries the correct
amino acid that is the next one to be added to a polypeptide.

B. Scaffold, guide, and ribozyme: rRNA functions as a scaffold for
the binding of ribosomal proteins; in bacteria, a segment of
rRNA acts as a guide for the binding of the Shine-Dalgarno sequence in mRNA; and the rRNA in peptidyl transferase functions as a ribozyme that catalyzes peptide bond formation.

C. Scaffold and alterer of protein function: SRP RNA acts as a scaffold for the binding of SRP proteins; it also affects the GTPase
activity of certain proteins in SRP and in the SRP receptor.

D. Guide: piRNA guides PIWI proteins to genes or to mRNAs.

2. An rRNA binds to a snoRNP that contains a C/D box snoRNA.
What type of covalent modification would you expect this rRNA to
undergo? Would it be methylated or pseudouridylated, or both?

T OPIC: *What topic in genetics does this question address?*
The topic is the covalent modifications carried out by snoRNPs.

I NFORMATION: *What information do you know based on the
question and your understanding of the topic?* In the question,
you are reminded that snoRNPs recognize rRNAs and then

methylate or pseudouridylate them. From your understanding of
the topic, you may remember there are two types of snoRNAs.
One type forms a complex containing a protein that methylates
RNA, whereas the other type forms a complex with a protein that
catalyzes pseudouridylation.

P ROBLEM-SOLVING S TRATEGY: *Relate structure and
function.* To solve this problem, consider the structures of the two
types of snoRNPs and their corresponding functions. The C/D box
snoRNA is part of an snoRNP that has a protein that methylates
the target RNA.

ANSWER: The rRNA would be methylated.

3. With regard to the CRISPR-Cas system that defends bacteria
against bacteriophage attack, what happens during the adaptation,
expression, and interference phases? When a bacterium is exposed to
a particular bacteriophage, is the adaptation phase always necessary?

T OPIC: *What topic in genetics does this question address?*
The topic is the CRISPR-Cas system that provides bacteria with
defense against bacteriophages. More specifically, the question
asks you to sort out what happens during each phase and decide
whether or not the first phase is always needed.

I NFORMATION: *What information do you know based on the
question and your understanding of the topic?* In the question,
you are reminded that the CRISPR-Cas system defends bacteria
against bacteriophages and that the defense process occurs in three
phases. From your understanding of the topic, you may recall what
happens during each phase.

P ROBLEM-SOLVING S TRATEGY: *Describe the steps.*
To solve this problem, one strategy is to sort out the steps of this
genome defense process.

ANSWER: During adaptation, a portion of bacteriophage DNA is
inserted into the *Crispr* gene. This phase requires the help of the proteins Cas1 and Cas2. During the expression phase, tracrRNA, crRNA,
and Cas9 are produced. Finally, during the interference phase,
tracrRNA, crRNA, and Cas9 come together and cleave the bacteriophage DNA, thereby inactivating it.

If the bacterium or one of its ancestors was already exposed to the
bacteriophage that is currently infecting it, the adaptation phase is
not necessary. Prior exposure to a bacteriophage likely resulted in
the insertion of a portion of the bacteriophage DNA into the *Crispr*
gene. This alteration would be passed on to daughter cells. Therefore, if a bacterium already had a portion of the bacteriophage DNA
in its *Crispr* gene, it would not have to go through the adaptation
phase; it would already be adapted to defend itself against that
bacteriophage.

Conceptual Questions

C1. List and briefly describe four types of molecules that can bind to an ncRNA.

C2. An ncRNA may have the following functions: scaffold, guide, alterer of protein function or stability, ribozyme, blocker, and/or decoy. Which of those functions is/are exhibited by each of the ncRNAs listed next? (Note: A single ncRNA may have more than one function.)

A. HOTAIR

B. RNA of RNaseP

C. microRNA

D. crRNA

C3. What is meant by the term *RNA world*? Describe observations and evidence that support this hypothesized period of life on Earth. From the perspective of living cells, what are the advantages of having had the RNA world be superseded by a DNA/RNA/protein world?

C4. Explain how HOTAIR plays a role in the transcriptional regulation of particular genes.

C5. What is the phenomenon of RNA interference (RNAi)? During RNAi, explain how the double-stranded RNA is processed and how that processing leads to the silencing of a complementary mRNA.

C6. With regard to RNAi, what are three possible sources for the double-stranded RNA?

C7. What is the difference between an miRNA and an siRNA? How do these ncRNAs affect mRNAs?

C8. Together with a specific set of proteins, snoRNAs direct the methylation or pseudouridylation of rRNAs. Does the snoRNA function as a scaffold, guide, ribozyme, blocker, decoy, and/or alterer of protein function or stability?

C9. Describe the structure of SRP in eukaryotes, and outline its role in targeting proteins to the ER membrane.

C10. Look at Figure 17.10 and predict what would happen if the SRP RNA was unable to stimulate the GTPase activities of the GTP-binding proteins within SRP and the SRP receptor.

C11. Compare and contrast the roles of crRNA and tracrRNA in the defense against bacteriophages provided by the CRISPR-Cas system.

C12. In the CRISPR-Cas system, does the tracrRNA act as a scaffold, guide, ribozyme, blocker, decoy, and/or alterer of protein function or stability?

C13. What are the roles of Cas1, Cas2, and Cas9 proteins in bacterial genome defense?

C14. Outline the steps that occur when piRITS or piRISC silences transposable elements by repressing transcription and by directly inhibiting TE RNAs, respectively. What is the role of piRNAs in this process?

C15. List five types of cancer in which ncRNAs can be involved.

C16. Explain how the miR-200 family of miRNAs behave as tumor-suppressor genes. What happens when their expression is blocked or decreased?

Experimental Questions

E1. A protein called trypsin, which plays a role in digestion in vertebrates, is made by pancreatic cells and secreted from those cells. Starting with a sample of pancreatic cells, a researcher modified the gene that codes trypsin by mutating the ER signal sequence, so it was no longer recognized by SRP. How would this mutation affect the targeting of trypsin?

E2. In the experiment presented in Figure 17.6, were Fire and Mello injecting pre-miRNA or pre-siRNA? Explain.

E3. Explain how the data of Fire and Mello suggested that double-stranded RNA is responsible for the silencing of *mex-3* mRNA.

E4. As described in Chapter 20, the CRISPR-Cas system has been modified so that it can be used as a gene editing tool (look ahead to Figure 20.13). Describe how that tool works, and explain how the natural CRISPR-Cas system is altered to produce it.

Questions for Student Discussion/Collaboration

1. Review the concept of an RNA world described in Section 17.1. Discuss which ncRNAs described in Table 17.1 may have arisen during the RNA world and which probably arose after the modern DNA/RNA/protein world came into existence.

2. Go to the PubMed website and do a search using the words *non-coding RNA* and *disease*. Scan through the journal articles you retrieve and make a list of the roles that ncRNAs may play in human diseases.

Note: All answers are available for the instructor in Connect; the answers to the even-numbered questions and all of the Concept Check and Comprehension Questions are in Appendix B.

A plant infected with tobacco mosaic virus. As this photo shows, the infection causes a mosaic pattern on the leaves, in which the normal green color is interspersed with yellow patches caused by damage due to the virus.
Nigel Cattlin/Science Source

18

GENETICS OF VIRUSES

Viruses are nonliving, infectious particles with nucleic acid genomes, either DNA or RNA. Viruses are considered nonliving because they do not exhibit all of the properties associated with living organisms. For example, viruses are not composed of cells, and by themselves, they do not carry out metabolism, use energy, maintain homeostasis, or even reproduce. A virus or its genetic material must be taken up by a living cell to replicate.

The first virus to be discovered was tobacco mosaic virus (TMV), which infects many species of plants. TMV causes mosaic-like patterns on the leaves, in which normal-colored patches are interspersed with light green or yellow patches (see the chapter-opening photo). TMV damages leaves, flowers, and fruit, but almost never kills the plant. In 1883, Adolf Mayer determined that this disease could be spread by spraying the sap from one plant onto another. By subjecting this sap to filtration, Dmitri Ivanovski demonstrated that the disease-causing agent was not a bacterium. Sap that had been passed through filters with pores small enough to trap bacterial cells was still able to spread the disease.

At first, some researchers suggested the agent was a chemical toxin. However, Martinus Beijerinck ruled out this possibility by showing that sap could continue to transmit the disease after many plant generations. A toxin would have been diluted after many generations, whereas Beijerinck's results indicated the disease agent was multiplying in the plant. Around the same time, animal viruses were discovered in connection with a livestock infection called foot-and-mouth disease. In 1900, the first human virus—the virus that causes yellow fever—was identified. Since that time, researchers have identified hundreds of different viruses that infect humans!

For many decades, microbiologists, geneticists, and molecular biologists have taken great interest in the structure, genetic composition, and replication of viruses. All organisms are susceptible to infection by viruses. Once a cell is infected, the genetic material of a virus orchestrates a series of events that ultimately leads to the production of new virus particles. In this chapter, we will first examine the structure and genetic composition of viruses and explore the general features of viral reproductive cycles. We will then take a closer look at the reproductive cycles of a virus called phage λ (lambda), which infects *E. coli* cells, and examine the reproductive cycle of human immunodeficiency virus (HIV).

18.1 VIRUS STRUCTURE AND GENETIC COMPOSITION

Learning Outcomes:

1. Compare and contrast viruses with regard to host range, structure, and genome composition.
2. Analyze the results of an experiment indicating that the genome of tobacco mosaic virus is RNA.

Researchers have identified and studied thousands of different viruses. In this section, we will survey the basic features of viruses and consider an early experiment showing that some viruses use RNA as their genetic material.

Viruses Differ in Their Host Range, Structure, and Genome Composition

Although all viruses share some similarities, such as small size and the reliance on a living cell for replication, they vary greatly in their characteristics, including their host range, structure, and genome composition. Some of the major differences are described next, and characteristics of selected viruses are presented in **Table 18.1**.

Host Range. A cell that is infected by a virus is called a **host cell,** and a species that can be infected by a specific virus is called a host species for that virus. Viruses differ greatly in their **host range**—the number of species they can infect. Table 18.1 lists a few examples of viruses with widely different host ranges. Tobacco mosaic virus has a broad host range, being known to infect hundreds of different plant species. By comparison, other viruses have a narrow host range, with some infecting only a single species. Furthermore, a virus may infect only certain cell types in a host species. For example, HIV specifically infects a type of lymphocyte called a helper T cell.

Structure. Although the existence of viruses was postulated in the 1890s, viruses were not observed until the 1930s, when the electron microscope was invented. Viruses cannot be resolved by even the best light microscope. Most of them are smaller than the wavelength of visible light. Viruses range in diameter from about 20 nm to 400 nm (1 nanometer = 10^{-9} meter). For comparison, a typical bacterium is 1000 nm in diameter, and the diameter of

most eukaryotic cells is 10 to 1000 times that of a bacterium. Adenoviruses, which cause infections of the respiratory and gastrointestinal tracts, have an average diameter of 75 nm. Over 50 million adenoviruses could fit into an average-sized human cell.

What are the common structural features of all viruses? As shown in **Figure 18.1**, all viruses have a protein coat called a **capsid,** which encloses a genome consisting of DNA or RNA. Capsids are composed of one or more types of protein subunits called capsomers. Capsids have a variety of shapes, including helical and polyhedral. Figure 18.1a shows the structure of TMV, which has a helical capsid made of identical capsomers. Figure 18.1b shows an adenovirus, which has a polyhedral capsid. Protein fibers with a terminal knob are located at the corners of the polyhedral capsid. Many viruses that infect animal cells, such as the influenza virus shown in Figure 18.1c, have a **viral envelope** enclosing the capsid. The envelope consists of a lipid bilayer that is derived from the plasma membrane of the host cell and is embedded with virally coded spike glycoproteins, also called spikes or peplomers.

In addition to encasing and protecting the genetic material, the capsid and envelope enable viruses to infect their hosts. The capsids or envelopes have specialized proteins, including protein fibers with a knob (Figure 18.1b) or spike glycoproteins (Figure 18.1c), that help them bind to the surface of a host cell. The spike glycoproteins in certain viruses, such as coronaviruses, are particularly conspicuous (look ahead to Figure 18.5). In electron micrographs, the spike glyoproteins on the surface of a coronavirus form very prominent projections.

Viruses that infect bacteria, called **bacteriophages,** or **phages,** may have more complex capsids, with accessory structures that are used for anchoring the virus to a host cell and injecting the viral nucleic acid (Figure 18.1d). As discussed later, the tail fibers of a bacteriophage attach the virus to the bacterial cell wall.

TABLE 18.1

Hosts and Characteristics of Selected Viruses

Virus or Group of Viruses	Host	Effect on Host	Nucleic Acid*	Genome Size[†]	Number of Genes[‡]
Phage fd	E. coli	Slows growth	ssDNA	6.4	10
Phage λ (lambda)	E. coli	Can exist harmlessly in the host cell or cause lysis	dsDNA	48.5	36
Phage T4	E. coli	Causes lysis	dsDNA	169	288
Phage Qß	E. coli	Slows growth	ssRNA	4.2	4
Tobacco mosaic virus (TMV)	Many plants	Causes mottling and necrosis of leaves and other plant parts	ssRNA	6.4	6
Baculoviruses	Insects	Usually kill the insect	dsDNA	133.9	154
Parvovirus	Mammals	Causes respiratory, flulike symptoms	ssDNA	5.0	5
Influenza virus	Mammals	Causes classic flu symptoms—fever, cough, sore throat, and headache	ssRNA	13.5	11
Coronavirus	Humans	Causes COVID-19—fever, cough, and respiratory symptoms	ssRNA	29.8	12
Epstein-Barr virus	Humans	Causes mononucleosis, with fever, sore throat, and fatigue	dsDNA	172	80
Adenovirus	Humans	Causes respiratory symptoms and diarrhea	dsDNA	34	35
Herpes simplex type II	Humans	Causes blistering sores around the genital region	dsDNA	158.4	77
HIV	Humans	Causes AIDS, an immunodeficiency syndrome that can lead to death	ssRNA	9.7	9

*ss stands for "single-stranded" and ds stands for "double-stranded." Though examples are not shown in this table, some viruses are known to contain double-stranded RNA.
[†]Given in thousands of nucleotides (bases) or thousands of nucleotide (base) pairs.
[‡]This number refers to the number of protein-coding units. In some cases, two or more proteins are made from a single gene as a result of events such as protein processing.

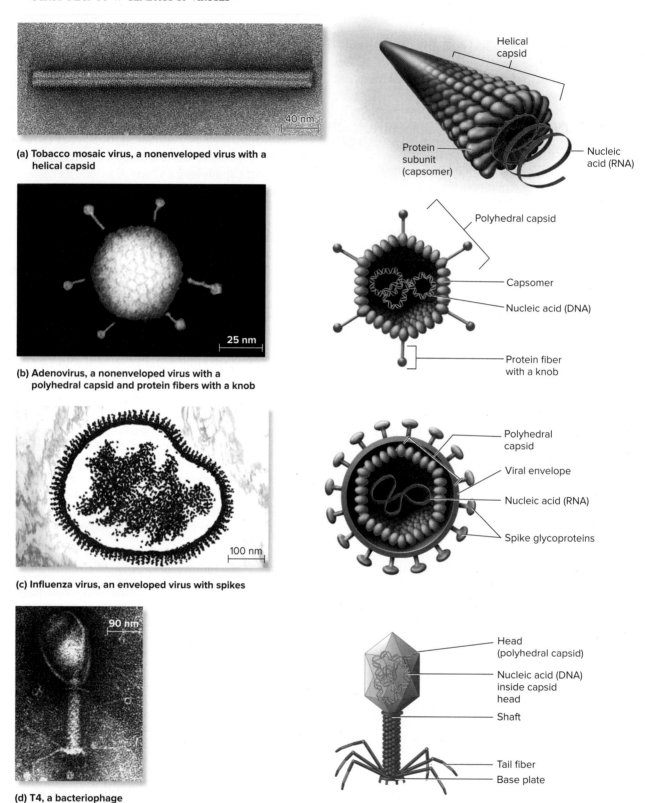

(a) Tobacco mosaic virus, a nonenveloped virus with a helical capsid

Helical capsid

Protein subunit (capsomer)

Nucleic acid (RNA)

(b) Adenovirus, a nonenveloped virus with a polyhedral capsid and protein fibers with a knob

Polyhedral capsid

Capsomer

Nucleic acid (DNA)

Protein fiber with a knob

(c) Influenza virus, an enveloped virus with spikes

Polyhedral capsid

Viral envelope

Nucleic acid (RNA)

Spike glycoproteins

(d) T4, a bacteriophage

Head (polyhedral capsid)

Nucleic acid (DNA) inside capsid head

Shaft

Tail fiber

Base plate

FIGURE 18.1   **Variations in the structure of viruses, as observed by transmission electron microscopy.**   All viruses contain nucleic acid (DNA or RNA) surrounded by a protein capsid. They may or may not have an outer envelope surrounding the capsid. **(a)** Tobacco mosaic virus (TMV) has a helical capsid made of 2130 identical protein subunits (capsomers), arranged around a strand of RNA. **(b)** Adenovirus has a polyhedral capsid containing protein fibers with a knob. **(c)** Many animal viruses, including influenza virus, have an envelope composed of a lipid bilayer and spike glycoproteins. The lipid bilayer is obtained from the host cell when the virus buds from the plasma membrane. **(d)** Some bacteriophages, such as T4, have capsids with accessory structures, such as tail fibers and base plates, that facilitate invasion of a bacterial cell.

(a): Omikron/Science Source; (b): Dr. Linda M. Stannard, University of Cape Town/SPL/Science Source; (c): Chris Bjornberg/Science Source; (d): Omikron/Science Source

CONCEPT CHECK: What features vary among different types of viruses?

Genome Composition.   The genetic material of a virus is called a **viral genome.** The composition of viral genomes varies markedly among different types of viruses, as suggested by the examples in Table 18.1. The nucleic acid in some viruses is DNA, whereas in others it is RNA; these are referred to as DNA viruses and RNA viruses, respectively. It is striking that some viruses use RNA for their genome, whereas all living organisms use DNA. In some viruses, the nucleic acid is single-stranded, whereas in others, it is double-stranded. The genome can be linear or circular, depending on the type of virus. Some kinds of viruses have more than one copy of the genome.

Viral genomes also vary considerably in size, ranging from a few thousand to more than a hundred thousand nucleotides or nucleotide pairs in length (see Table 18.1). For example, the genomes of some simple viruses, such as phage Qβ, are only a few thousand nucleotides in length and contain only a few genes. Other viruses, particularly those with a complex structure, such as phage T4, contain many more genes. These extra genes code many different proteins that are involved in the formation of the elaborate structure of the phage, as shown in Figure 18.1d.

EXPERIMENT 18A

The Genome of Tobacco Mosaic Virus Is Composed of RNA

We now know that bacteria, archaea, protists, fungi, plants, and animals all use DNA as their genetic material. In 1956, Alfred Gierer and Gerhard Schramm isolated RNA from tobacco mosaic virus (TMV), which infects plant cells. When this purified RNA was applied to plant tissue, the plants developed the same types of lesions that occurred when they were exposed to intact TMVs. Gierer and Schramm correctly concluded that the viral genome of TMV is composed of RNA.

To further confirm that TMV uses RNA as its genetic material, Heinz Fraenkel-Conrat and Beatrice Singer conducted additional research that involved different strains of TMV. They focused their efforts on the wild-type strain and a mutant strain called the Holmes ribgrass (HR) strain. The two strains differ in two ways:

- They cause significantly different lesions when they infect plants. In particular, the wild-type strain produces a mottled area with yellow and green irregularly shaped lesions on infected leaves (see the chapter-opening photo), whereas the HR strain often produces streaks along the veins and ringlike markings on other parts of the leaves.
- The capsid protein in the HR strain has two amino acids (histidine and methionine) that are not found in the capsid protein of the wild-type strain.

Previous experiments had shown that when purified capsid proteins and purified RNA molecules from TMVs are mixed together, they will self-assemble into intact viruses. Such a procedure is referred to as a reconstitution experiment because intact viruses are made from their individual parts. In the experiment described in **Figure 18.2,** Fraenkel-Conrat and Singer mixed wild-type RNA with HR proteins or HR RNA with wild-type proteins and then placed the reconstituted viruses onto tobacco leaves. Following infection, they then observed the symptoms caused by the viruses and analyzed the amino acid composition of the proteins of viruses produced after the infection.

THE HYPOTHESIS

RNA is the genetic material of TMV.

TESTING THE HYPOTHESIS   **FIGURE 18.2** **Evidence that RNA is the genetic material of TMV.**

Starting material: Purified preparations of RNA and proteins from wild-type TMV and from the Holmes ribgrass (HR) strain of TMV.

1. Mix together wild-type RNA and HR proteins or HR RNA and wild-type proteins. Allow time for the RNA and proteins to assemble into intact viruses. These are called reconstituted viruses.

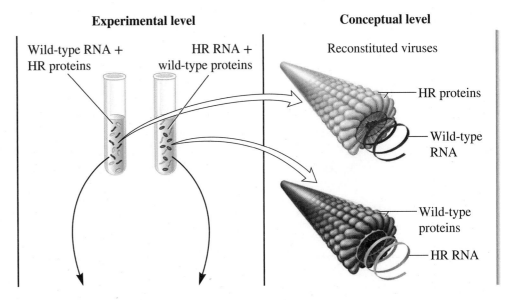

Experimental level

Wild-type RNA + HR proteins

HR RNA + wild-type proteins

Conceptual level

Reconstituted viruses

HR proteins

Wild-type RNA

Wild-type proteins

HR RNA

2. Inoculate a small amount of reconstituted viruses onto healthy tobacco leaves. Allow time for infection to occur.

3. Observe the types of lesions that form on the leaves.

4. Take plant tissue containing viral lesions and isolate newly made viral proteins. This is done by extracting the protein with mild alkali.

5. Determine the amino acid composition of the newly made viral proteins. This involves hydrolyzing the proteins into individual amino acids and then separating the amino acids by chromatography.

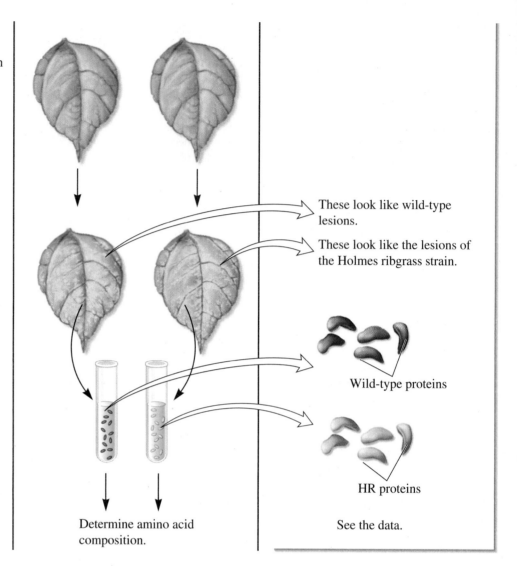

These look like wild-type lesions.

These look like the lesions of the Holmes ribgrass strain.

Wild-type proteins

HR proteins

Determine amino acid composition.

See the data.

THE DATA

Composition of reconstituted virus placed on tobacco leaves	Lesions on tobacco leaves	Amino acids found in newly made viral proteins following infection:	
		Methionine	*Histidine*
Wild-type RNA and HR protein	Like wild-type TMV	No	No
HR RNA and wild-type protein	Like HR TMV	Yes	Yes

Source: Fraenkel-Conrat, Heinz, and Singer, B., "Virus reconstitution: II. Combination of protein and nucleic acid from different strains," *Biochimica et Biophysica Acta*, vol. 24, January 17, 1957, 540–548.

INTERPRETING THE DATA

As seen in the data, the outcome of infection depended on the RNA that was found in the reconstituted virus but not on the capsid protein. If wild-type RNA was used, the leaves developed lesions that were typical of wild-type TMV and the capsid proteins of newly made viruses lacked methionine or histidine. In contrast, if the reconstituted viruses had HR RNA, the lesions were those of the HR TMV strain and the newly made capsid proteins contained both methionine and histidine. Taken together, these results are consistent with the hypothesis that the RNA component of TMV is its genetic material.

18.1 COMPREHENSION QUESTIONS

1. What is a common feature of all viruses?
   a. An envelope
   b. DNA
   c. Nucleic acid surrounded by a protein capsid
   d. All of the above are common features.
2. Viral genomes can be
   a. DNA or RNA.
   b. single-stranded or double-stranded.
   c. linear or circular.
   d. All of the above can characterize a viral genome.

18.2 OVERVIEW OF VIRAL REPRODUCTIVE CYCLES

Learning Outcomes:

1. Compare and contrast the reproductive cycles of phage λ and HIV.
2. Define *latency*, and explain how it occurs for phage λ and HIV.
3. Describe the properties of emerging viruses.

When a virus infects a host cell, the expression of the viral genome leads to a series of steps, called a **viral reproductive cycle,** that results in the production of new viruses. The details of the steps differ greatly among various types of viruses, and even the same virus may have the capacity to follow alternative cycles. Even so, by studying the reproductive cycles of hundreds of different viruses, researchers have determined that the viral reproductive cycle consists of five or six basic steps. In this section, we will examine these basic steps for two different viruses and consider how new viruses can quickly spread through a population.

Viruses Follow a Reproductive Cycle That Leads to the Synthesis of New Viruses

To illustrate the general features of viral reproductive cycles, **Figure 18.3** considers the basic steps for two viruses. Figure 18.3a shows the cycle of **phage λ** (lambda), a bacteriophage with double-stranded DNA as its genome, and Figure 18.3b depicts the cycle of **human immunodeficiency virus (HIV),** an enveloped animal virus that contains two copies of single-stranded RNA and infects humans. The descriptions that follow compare the reproductive cycles of these two very different viruses.

Step 1: Attachment.   In the first step of a viral reproductive cycle, the virus must attach to the surface of a host cell. This attachment is usually specific for one or just a few types of cells because proteins in the virus recognize and bind to specific molecules on the cell surface. In the case of phage λ, the tail fibers bind to proteins in the outer cell membrane of the bacterium *E. coli.* In the case of HIV, spike glycoproteins in the viral envelope bind to receptors in the plasma membrane of human blood cells called helper T cells.

Step 2: Entry.   After attachment, the viral genome enters the host cell. Attachment of phage λ stimulates a conformational change in the coat proteins of the phage, so the shaft contracts, and the phage injects its DNA into the bacterial cytoplasm. In contrast, the envelope of HIV fuses with the plasma membrane of the host cell, so both the capsid and its contents are released into the cytosol. Some of the HIV capsid proteins are then removed by host-cell enzymes, a process called uncoating. This releases two copies of the viral RNA, as well as molecules of enzymes called reverse transcriptase and integrase, into the cytosol. The functions of these enzymes are described later in this section.

Once a viral genome has entered the cell, specific viral genes may be expressed immediately due to the action of host-cell enzymes and ribosomes. Expression of these key genes leads quickly to either step 3 or step 4 of the reproductive cycle, depending on the type of virus. The genome of some viruses, including both phage λ and HIV, can integrate into a chromosome of the host cell. For such viruses, the cycle may proceed from step 2 to step 3 as described next, delaying the production of new viruses. Alternatively, the cycle for phage λ and some other viruses may proceed directly from step 2 to step 4 and quickly lead to the production of new viruses.

Step 3: Integration.   Viruses capable of integration carry a gene that codes an enzyme called **integrase.** For some viruses, such as phage λ, the gene within the phage genome that codes integrase is expressed soon after entry, which leads to the synthesis of the integrase protein. For other viruses, such as HIV, integrase proteins are packaged into newly made viruses, which are released during the uncoating process. The function of integrase is to cut the host's chromosomal DNA and integrate the viral genome into the chromosome. In the case of phage λ, the double-stranded DNA that entered the cell can be directly integrated into the double-stranded DNA of the bacterial chromosome. Once integrated, the phage DNA in a bacterium is called a **prophage.** When the phage DNA exists as a prophage, the viral reproductive cycle is called the **lysogenic cycle.** As discussed later, new phages are not made during the lysogenic cycle, and the host cell is not destroyed. The lysogenic cycle is favored when nutrients are in short supply. Under these conditions, the bacterium may lack the resources to make new phages.

How can an RNA virus, such as HIV, integrate its genome into the host-cell DNA? For this to occur, the viral genome must be copied into DNA. HIV accomplishes this by means of a viral enzyme called **reverse transcriptase,** which is carried within the capsid and released into the host cell along with the viral RNA. Reverse transcriptase uses the viral RNA strand to make a complementary strand of DNA, and it then uses that DNA strand as a template to make double-stranded viral DNA. This process is called reverse transcription because it is the reverse of the usual transcription process, in which a DNA strand is used to make a complementary strand of RNA. The viral double-stranded DNA enters the host-cell nucleus and is integrated into a host chromosome via integrase. Once integrated, the viral DNA in a eukaryotic cell is called a **provirus.** Viruses that follow this mechanism are called retroviruses.

Step 4: Synthesis of Viral Components.   The production of new viruses by a host cell involves the replication of the viral

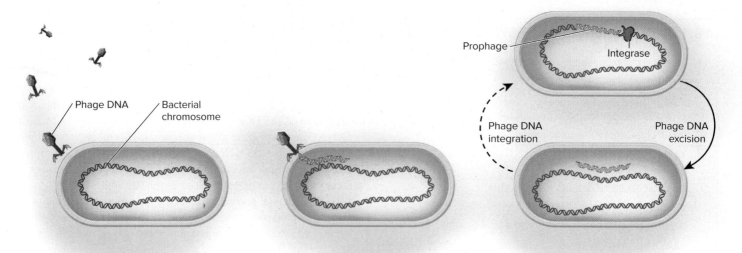

1 Attachment:
The phage binds specifically to proteins in the outer bacterial cell membrane.

(a) Reproductive cycle of phage λ

2 Entry:
The phage injects its DNA into the bacterial cytoplasm.

3 Integration:
Phage DNA may integrate into the bacterial chromosome via integrase. The host cell carrying a prophage may then undergo repeated divisions, which is called the lysogenic cycle. To end the lysogenic cycle and switch to the lytic cycle, the phage DNA is excised. Alternatively, the reproductive cycle may completely skip the lysogenic cycle and proceed directly to step 4.

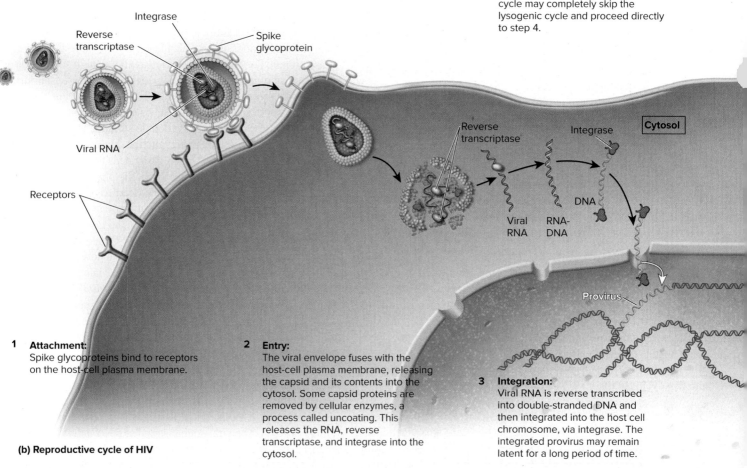

1 Attachment:
Spike glycoproteins bind to receptors on the host-cell plasma membrane.

(b) Reproductive cycle of HIV

2 Entry:
The viral envelope fuses with the host-cell plasma membrane, releasing the capsid and its contents into the cytosol. Some capsid proteins are removed by cellular enzymes, a process called uncoating. This releases the RNA, reverse transcriptase, and integrase into the cytosol.

3 Integration:
Viral RNA is reverse transcribed into double-stranded DNA and then integrated into the host cell chromosome, via integrase. The integrated provirus may remain latent for a long period of time.

ONLINE ANIMATION

FIGURE 18.3 **Comparison of the steps of two viral reproductive cycles.** (a) The reproductive cycles of phage λ, a bacteriophage with a double-stranded DNA genome. Note: As described later in this chapter, the phage DNA is in a circular form when it integrates into the bacterial chromosome. (b) The reproductive cycle of HIV, an enveloped virus with a single-stranded RNA genome that infects humans.

CONCEPT CHECK: During which step of the reproductive cycle can a virus remain latent?

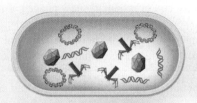

4 Synthesis of viral components:
In the lytic cycle, phage DNA directs the synthesis of viral components. During this process, the phage DNA circularizes, and the host chromosomal DNA is digested into fragments.

5 Viral assembly:
Phage components are assembled with the help of noncapsid proteins to make many new phages. The phage DNA becomes linear inside the capsid.

6 Release:
The viral enzyme called lysozyme causes cell lysis, and new phages are released from the broken cell.

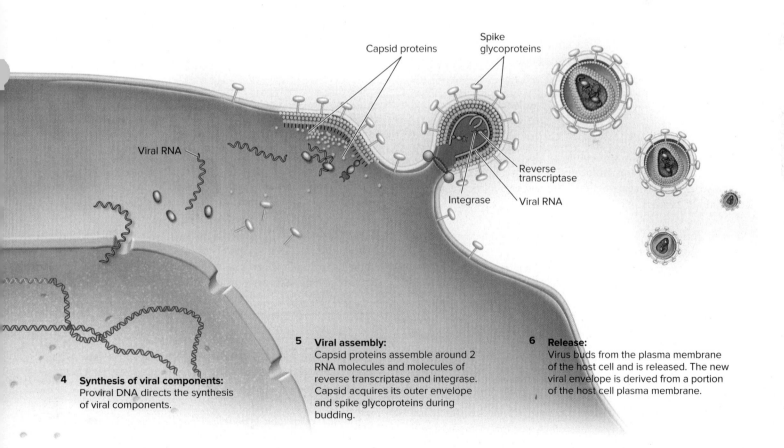

Capsid proteins

Spike glycoproteins

Viral RNA

Reverse transcriptase

Integrase

Viral RNA

4 Synthesis of viral components:
Proviral DNA directs the synthesis of viral components.

5 Viral assembly:
Capsid proteins assemble around 2 RNA molecules and molecules of reverse transcriptase and integrase. Capsid acquires its outer envelope and spike glycoproteins during budding.

6 Release:
Virus buds from the plasma membrane of the host cell and is released. The new viral envelope is derived from a portion of the host cell plasma membrane.

genome and the synthesis of viral proteins that make up the capsid. When phage λ has been integrated into the host chromosome, the prophage must be excised before synthesis of new viral components can occur. A viral enzyme called excisionase plays a role in the excision process. Following excision, many copies of the phage DNA are made via DNA polymerase from the host cell, and the genes within these copies are transcribed into mRNA using RNA polymerase from the host cell. This viral mRNA is used to synthesize viral proteins via host-cell ribosomes. The expression of phage genes also leads to the degradation of the host chromosomal DNA.

In the case of HIV, the DNA of the provirus is not excised from the host chromosome. Instead, it is transcribed in the nucleus to produce many copies of viral RNA. These viral RNA molecules enter the cytosol, where they are used to make viral proteins and serve as the genome for new viral particles.

Step 5: Viral Assembly.

After all of the necessary components have been synthesized, they must be assembled into new viruses. Some viruses with a simple structure self-assemble, meaning that viral components spontaneously bind to each other to form a complete virus particle. An example of a self-assembling virus is TMV, which we examined earlier (see Figure 18.1a). The capsid proteins assemble around the RNA genome, which lies inside the hollow capsid (see question 1 in More Genetic TIPS at the end of the chapter). This assembly process can occur in vitro if purified capsid proteins and RNA are mixed together.

Other viruses, including the two featured in Figure 18.3, do not self-assemble. The correct assembly of phage λ requires the help of noncapsid proteins not found in the completed phage particle. Some of these noncapsid proteins function as enzymes that modify capsid proteins, and others serve as scaffolding for the assembly of the capsid. The assembly of HIV components occurs at the plasma membrane. Capsid proteins assemble around two molecules of viral RNA and molecules of reverse transcriptase and integrase. As this is occurring, the virus acquires its viral envelope in a budding process.

Step 6: Release.

The last step of a viral reproductive cycle is the release of new viruses from the host cell. The release of bacteriophages is a dramatic event. Because bacterial cells are surrounded by a rigid cell wall, the phages must cause their host cells to burst, or lyse, in order to escape. After the phages have been assembled, a phage-coded enzyme called lysozyme digests the bacterial cell wall, causing the cell to burst. Lysis releases many new phages into the environment, where they can infect other bacteria and begin the cycle again. Collectively, steps 1, 2, 4, 5, and 6 are called the **lytic cycle** because they lead to cell lysis.

The release of enveloped viruses from an animal cell is far less dramatic. This type of virus escapes by a mechanism called budding, which does not lyse the cell. A portion of the host-cell plasma membrane enfolds the viral capsid and eventually buds from the cell surface.

Latency in Bacteriophages.

As we saw in step 3, viruses can integrate their genomes into a host chromosome. In some cases, the prophage or provirus may remain inactive, or **latent,** for a long time. Most of the viral genes are silent during latency, and the viral reproductive cycle does not progress to step 4. Latency in bacteriophages is also called **lysogeny.** When latency occurs, both the prophage and its host cell are said to be lysogenic. When a lysogenic bacterium prepares to divide, it copies the prophage DNA along with its own DNA, so each daughter cell inherits a copy of the prophage. The prophage DNA can be replicated repeatedly in this way without killing the host cell or producing new phage particles. As mentioned earlier, this process is called the lysogenic cycle.

Many bacteriophages can alternate between lysogenic and lytic cycles (**Figure 18.4**). A bacteriophage that can spend some of its time in a lysogenic cycle is called a **temperate phage.** Phage λ is an example of a temperate phage. Upon infection, it can either enter the lysogenic cycle or proceed directly to the lytic cycle. Other phages, called **virulent phages,** can only follow a lytic cycle. The genome of a virulent phage is not capable of integration into a host chromosome.

Latency in Human Viruses.

Latency among human viruses can occur in two different ways. For HIV, latency occurs because the virus has integrated into the host genome and may remain dormant for long periods of time. Alternatively, a viral genome can exist as an **episome**—a genetic element, such as a plasmid, that can replicate independently of the chromosomal DNA but can also occasionally integrate into the chromosomal DNA. Examples of viral genomes that exist as episomes include different types of herpes viruses that cause cold sores (herpes simplex type I), genital herpes (herpes simplex type II), and chickenpox (varicella-zoster virus). A person infected with a given type of herpes virus may have periodic outbreaks of disease symptoms when the virus switches from the latent, episomal form to the active form that produces new virus particles.

As an example, let's consider the herpes virus called varicella-zoster virus. The initial infection by this virus causes chickenpox, after which the virus may remain latent for many years as an episome. The disease called shingles usually occurs decades later, when varicella-zoster virus switches from the latent state and starts making new virus particles. It usually occurs in adults over the age of 60 or in people with weakened immune systems. Shingles begins as a painful rash that eventually erupts into blisters. The blisters follow the path of the nerve cells that carry the latent varicella-zoster virus, and often form a ring around the back of the patient's body. The term *shingles* is derived from a Latin word meaning "girdle," referring to the observation that the blisters girdle a part of the body.

GENETIC TIPS   THE QUESTION: From the perspective of a bacteriophage, what is the advantage of being able to follow either a lytic or a lysogenic cycle?

TOPIC: *What topic in genetics does this question address?* The topic is the lytic and lysogenic cycles of bacteriophages. More specifically, the question asks you to identify the advantage to a phage of being able to follow either type of cycle.

INFORMATION: *What information do you know based on the question and your understanding of the topic?* From the question, you know that some viruses can follow either a lytic or a

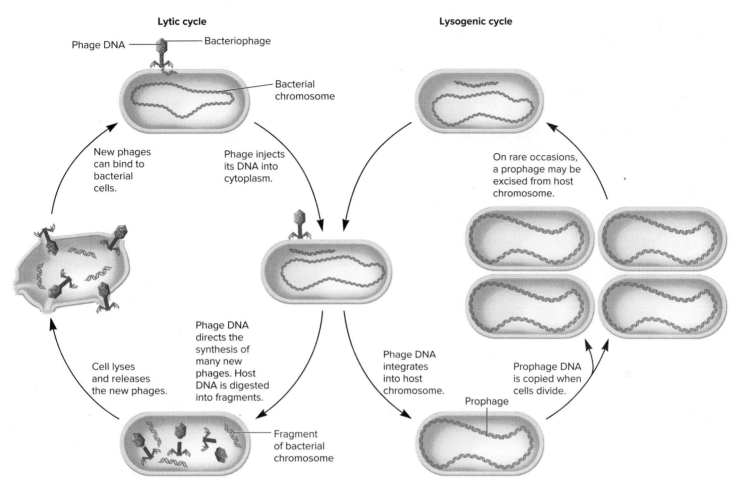

Lytic cycle

Phage DNA — Bacteriophage

— Bacterial chromosome

New phages can bind to bacterial cells.

Phage injects its DNA into cytoplasm.

Phage DNA directs the synthesis of many new phages. Host DNA is digested into fragments.

Cell lyses and releases the new phages.

Fragment of bacterial chromosome

Lysogenic cycle

On rare occasions, a prophage may be excised from host chromosome.

Phage DNA integrates into host chromosome.

Prophage DNA is copied when cells divide.

Prophage

ONLINE ANIMATION

FIGURE 18.4 **The lytic and lysogenic reproductive cycles of certain bacteriophages.** Some bacteriophages, known as virulent phages, can only follow a lytic cycle. Other phages, called temperate phages, can follow either a lysogenic or lytic cycle.

CONCEPT CHECK: Which reproductive cycle produces new phage particles?

lysogenic cycle. From your understanding of the topic, you may remember that the lytic cycle involves the synthesis of new viruses, whereas the lysogenic cycle involves the integration of the viral DNA into the host chromosome. The choice between the two cycles may be driven by the availability of nutrients to the host cell.

P **ROBLEM-SOLVING** **S** **TRATEGY:** *Compare and contrast.* One strategy to solve this problem is to compare the ability of the host cell to make new viruses when nutrients are abundant versus when nutrients are limiting.

ANSWER: For phages such as phage λ that can follow either type of reproductive cycle, environmental conditions influence whether the lytic or lysogenic cycle prevails. If nutrients are readily available, phage λ usually proceeds directly to the lytic cycle after its DNA enters the cell. This allows the phage to rapidly proliferate. Alternatively, if nutrients are in short supply, the lysogenic cycle is usually favored because sufficient nutrients to make new viruses are not available. If more nutrients become available later, the prophage DNA may be excised, and the viral reproductive cycle will switch to the lytic cycle. Therefore, under more favorable conditions, new viruses are made and released.

Emerging Viruses, such as HIV and Coronavirus, Have Arisen Recently and Are More Likely Than Previous Strains to Cause Infection

A primary reason researchers have been interested in viral reproductive cycles is the ability of many viruses to cause diseases in humans and other hosts. Some examples of viruses that infect humans were presented earlier in Table 18.1. **Emerging viruses** are viruses that have arisen recently and are more likely than previous strains to cause infection. Such viruses may lead to a significant loss of human life and often cause public alarm.

Cornonavirus. A very recent example of an emerging virus is a coronavirus named severe acute respiratory coronavirus 2 (SARS-CoV-2), which causes a respiratory infection called coronavirus disease 2019 (COVID-19). This coronavirus is an enveloped virus with an RNA genome (see Table 18.1). The name coronavirus is derived from the viral spike glycoproteins, colored red in **Figure 18.5**, which resemble the ring of gases surrounding the Sun called the solar corona. The COVID-19 outbreak is thought to have originated in Wuhan, China, and, within a matter of a few months, spread across the world. On March 11, 2020, the

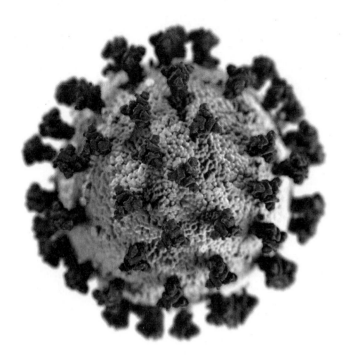

FIGURE 18.5 An electron micrograph of SARS-CoV-2, the type of coronavirus that causes COVID-19. The viral spike proteins, which are colored in red, create the look of a corona surrounding the virus.

Alissa Eckert, MS; Dan Higgins, MAMS/CDC

CONCEPT CHECK: What is an emerging virus?

World Health Organization characterized the COVID-19 outbreak as a **pandemic**—a disease that occurs over a wide geographic area and usually affects a high proportion of the population.

The coronavirus that causes COVID-19 is related to other human coronaviruses, such as those causing Middle East respiratory syndrome (MERS-CoV) and severe acute respiratory syndrome (SARS-CoV). Molecular studies have indicated that SARS-CoV-2 is derived from a coronavirus that infects bats.

Common symptoms of COVID-19 infection include fever, a dry cough, shortness of breath, and other respiratory symptoms. In more severe cases, which are more likely to occur in the elderly or those with weakened immune systems, pneumonia and even death may occur. In the period from March 2020 to the end of 2022, over 1 million people died from the disease in the United States, and over 6.6 million people died worldwide.

As SARS-CoV-2 spread across the world, mutations occurred that created new strains of SARS-CoV-2, which are termed *variants*. Many different variants have arisen due to mutations in the SARS-CoV-2 genome. Let's compare the original SARS-CoV-2 virus with two of the most common variants, delta and omicron.

- *Original SARS-CoV-2 virus:* fairly transmissible; the symptoms range from mild to severe. It was first detected in late 2019 in China.
- *Delta variant:* more transmissible than the original SARS-CoV-2 virus with symptoms similar to those caused by the original virus. The delta variant was first detected in India in late 2020. Compared to the original virus, the delta

variant carries at least 13 mutations, many of which alter the structure of the spike glycoprotein. Because the spike glycoprotein plays a key role in binding to host cells, at least some of these mutations play a role in the higher transmissibility of the delta variant compared to the original virus.

- *Omicron variant:* more transmissible than the delta variant with symptoms typically less severe than those of the original virus and the delta variant, though many deaths have still been reported. The omicron variant was first detected in South Africa in late 2021. This variant has more than 50 mutations compared with the original SARS-CoV-2 virus. About 30 of these mutations change amino acids in the spike glycoprotein. The omicron variant did not evolve from the delta variant. How do we explain the large number of mutations in the omicron variant? Some researchers have proposed that the omicron variant developed in an immunocomprised and chronically infected COVID-19 patient, such as someone whose immune response was impaired by another illness or a drug.

Influenza Virus. New strains of influenza virus arise fairly regularly due to new mutations. An example is the strain H1N1, which causes what is called swine flu. It was called swine flu because laboratory testing revealed that the H1N1 virus carries two genes that are normally found in flu viruses that infect pigs in Europe and Asia. In the United States, over 30,000 people die annually from this disease.

The Centers for Disease Control and Prevention (CDC) recommends vaccination as the first and most important step in preventing infection or lessening disease symptoms by influenza virus and the corona virus that causes COVID-19. The topic of vaccines is discussed in Chapter 21 (look ahead to Section 21.2).

Human Immunodeficiency Virus. During the past few decades, another dramatic example of an emerging virus has been human immunodeficiency virus (HIV), the causative agent of acquired immune deficiency syndrome (AIDS). Research studies indicate that HIV is a mutant form of a virus found in chimpanzees in West Africa, which is called simian immunodeficiency virus (SIV). SIV was most likely transmitted to humans and mutated into HIV when humans hunted chimpanzees for meat and came into contact with their infected blood. Scientists speculate that humans were first infected in the early 1900s.

HIV is primarily spread by sexual contact with an infected individual, but it can also be spread by the transfusion of HIV-infected blood, by the sharing of needles among drug users, and from an infected female to an unborn child. The total number of AIDS deaths between 1981 and the end of 2021 was about 36 million; more than 700,000 of these deaths occurred in the United States. Worldwide, approximately 38 million people are living with HIV; about 1.5 million new infections occurred in 2021.

The symptoms of AIDS result from viral destruction of helper T cells, a type of white blood cell that plays an essential role in the immune system of mammals. **Figure 18.6** shows HIV particles invading a helper T cell, which normally interacts with other cells of the immune system to facilitate the production of antibodies and other molecules that target and kill foreign invaders

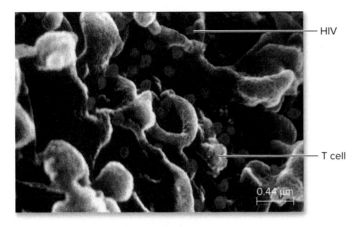

HIV

T cell

0.44 μm

FIGURE 18.6 Micrograph of HIV invading a human helper T cell. In this colorized scanning electron micrograph, the surface of the helper T cell is purple, and HIV particles are red.

Richard J. Green/Science Source

of the body. When large numbers of helper T cells are destroyed by HIV, the function of the immune system is seriously compromised, and the individual becomes susceptible to opportunistic infections—diseases that would not normally occur in a healthy person. For example, *Pneumocystis jiroveci*, a fungus that causes pneumonia, is easily destroyed by a healthy immune system. However, in people with AIDS, infection by this fungus can be fatal.

An insidious feature of HIV replication, which is outlined in Figure 18.3b, is that reverse transcriptase, the enzyme that copies the RNA genome into DNA, lacks a proofreading function. Because reverse transcriptase lacks this function, it makes more errors and thereby tends to create many mutant strains of HIV. This undermines the ability of the body to combat HIV because mutant strains may not be destroyed by the body's defenses.

In addition, mutant strains of HIV may be resistant to antiviral drugs. For this reason, current HIV treatments typically contain three different drugs, making it difficult for any mutation to promote resistance to all three of them. Such HIV treatments can be effective at preventing the proliferation of the virus, but they do not completely eliminate the virus from the body.

18.2 COMPREHENSION QUESTIONS

1. The viral reproductive cycle consists of six steps. Which of these steps is skipped in the phage λ lytic cycle?
   a. Entry
   b. Attachment
   c. Integration
   d. Synthesis of viral components
   e. Viral assembly
   f. Release of newly made viruses

2. An example of an emerging virus is
   a. phage λ.
   b. HIV, which causes AIDS.
   c. a strain of influenza virus called H1N1 that causes the flu.
   d. Both b and c are emerging viruses.

18.3 BACTERIOPHAGE λ REPRODUCTIVE CYCLES

Learning Outcomes:

1. Compare and contrast the lysogenic and lytic cycles of phage λ.
2. Explain how gene regulation determines the choice between the lysogenic and lytic cycles.

During the past several decades, the reproduction of viruses has presented an interesting and challenging problem for geneticists to investigate. The study of bacteriophages has greatly advanced our basic knowledge of how regulatory proteins work. In this section, we will largely focus on the function of bacteriophage genes that code the regulatory proteins that control the choice between the lysogenic and lytic cycles of phage λ. The general features of this phage's reproductive cycles were presented earlier, in Figure 18.3a. Since phage λ was discovered in 1951 by microbiologist Esther Lederberg, it has been investigated extensively and has provided geneticists with a model on which to base our understanding of viral proliferation.

Phage λ Can Follow a Lysogenic or Lytic Cycle

Phage λ binds to the surface of a bacterium and injects its genetic material into the bacterial cytoplasm. Inside the virus particle, phage λ DNA is linear. After injection into the bacterium, the two ends of the DNA become covalently attached to each other to form a circular piece of DNA. **Figure 18.7** shows the genome of phage λ. The organization of the genes reflects the two alternative cycles of this virus—the lysogenic and lytic cycles—which were described earlier in this chapter (refer back to Figure 18.4). The genes shaded orange are transcribed very soon after infection. As we will see, the pattern of expression of these early genes determines the choice between the lysogenic or lytic cycle.

After injection, the phage proceeds along one of two alternative cycles. If the lysogenic cycle prevails, the integrase (*int*) gene is turned on. The integrase gene codes an enzyme that integrates the phage λ DNA into the bacterial chromosome, where it becomes a prophage. Integration of the λ DNA into the *E. coli* chromosome requires sequences known as **attachment sites.** As shown at the top of **Figure 18.8**, a common sequence within the attachment sites is identical in the phage λ DNA and the *E. coli* chromosome. The attP sequence is in the λ DNA, and the attB sequence is in the *E. coli* chromosome. An enzyme known as integrase is coded by a gene in the λ DNA. Several integrase proteins recognize the attP and attB sequences and bring them close together. Integrase then makes staggered cuts in both the λ and *E. coli* attachment sites. The strands are then exchanged, and the ends are ligated together. In this way, the phage DNA is integrated into the host cell chromosome.

As a prophage, phage λ may remain latent for many generations. Certain environmental conditions, such as exposure to UV light, may alter gene expression in a way that causes the excision of the prophage from the host DNA. Excision also requires

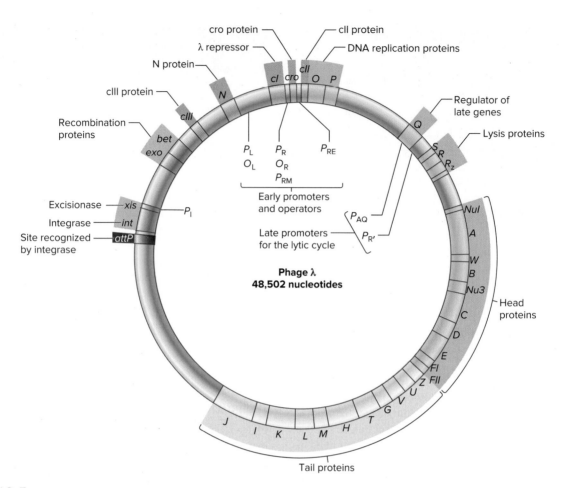

FIGURE 18.7 **The genome of phage λ.** The genes shaded orange code regulatory proteins that determine whether the lysogenic or lytic cycle prevails. Genes shaded blue code proteins necessary for the lysogenic cycle. The attP site, shaded red, is recognized by integrase. The genes shaded dark and light tan code proteins required for the lytic cycle.

integrase, which catalyzes the reverse reaction, as well as a second protein known as excisionase.

In contrast to the lysogenic cycle, the lytic cycle involves the synthesis of many copies of the phage genetic material and coat proteins that are then assembled to make new phages. When synthesis and assembly are completed, the bacterial host cell is lysed, and the newly made phages are released into the environment. If the lytic cycle is chosen, the genes on the right side and bottom of the genome in Figure 18.7 are transcribed. These genes are necessary for the synthesis of new phages. They code replication proteins, coat proteins (that form the phage head, shaft, and tail fibers), proteins involved in coat assembly, proteins involved in packaging the DNA into the phage head, and enzymes that cause the bacterium to lyse.

The Choice Between the Lysogenic or Lytic Cycle Depends on the Relative Levels of the cII and cro Proteins

Now that we have discussed the reproductive cycles and genome organization of phage λ, let's examine how the lysogenic or lytic cycle is chosen. This choice depends on the actions of regulatory proteins and their effects on transcription. **Table 18.2**

summarizes the key regulatory elements and proteins involved in this process.

Soon after λ DNA enters the bacterial cell, two promoters—designated P_L and P_R—are used for transcription. This initiates a competition between the lysogenic and lytic cycles (**Figure 18.9**). Initially, transcription from P_L and P_R results in the synthesis of two short RNA transcripts that code two proteins called the N protein and the cro protein, both of which are regulatory proteins. We will examine the function of cro protein, which is involved in the lytic cycle, later in this section.

The N protein is a regulatory protein with an interesting function that we have not yet considered. This function, known as **antitermination,** is to prevent transcriptional termination. The N protein inhibits termination at three sites, designated t_L, t_{R1}, and t_{R2}. The N protein binds to RNA polymerase and prevents transcriptional termination when these sites are being transcribed.

- When the N protein prevents termination at t_{R1} and t_{R2}, the transcript made from P_R is extended to include the cII, O, P, and Q genes. The cII gene codes an activator protein, the O and P genes code enzymes needed for the initiation of λ DNA synthesis, and the Q gene codes another antiterminator that is required for the lytic cycle.

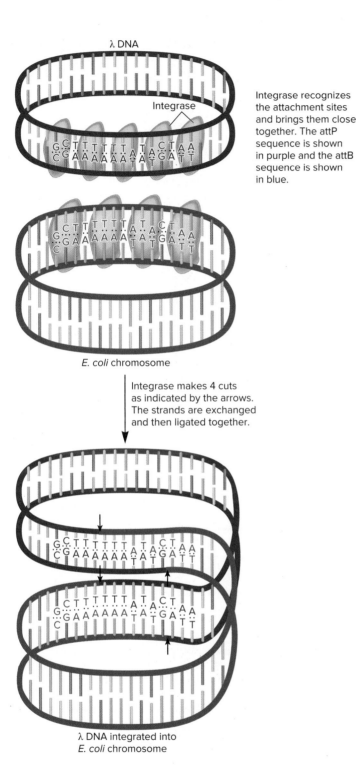

λ DNA

Integrase

Integrase recognizes the attachment sites and brings them close together. The attP sequence is shown in purple and the attB sequence is shown in blue.

E. coli chromosome

Integrase makes 4 cuts as indicated by the arrows. The strands are exchanged and then ligated together.

λ DNA integrated into E. coli chromosome

FIGURE 18.8 **The integration of λ DNA into the _E. coli_ chromosome.** The attP sequence in the λ DNA attaches to the attB sequence in the _E. coli_ chromosome. As noted here, the attP and attB sequences are identical and thereby provide recognition sites for integrase. Note: The λ DNA and _E. coli_ chromosome are not drawn to scale. The _E. coli_ chromosome is much larger.

- When the N protein prevents termination at t_L, the transcript made from P_L is extended to include the _int_, _xis_, and _cIII_ genes. The _int_ gene codes integrase, which is involved with integrating λ DNA into the _E. coli_ chromosome, and the _xis_ gene codes excisionase, which plays a role in excising the λ DNA if a switch is made from the lysogenic to the lytic cycle. The _cIII_ gene codes the cIII protein, which inhibits a cellular protease and thereby makes cII protein less vulnerable to protease digestion.

As shown at the bottom of Figure 18.9, if the cII protein accumulates to sufficient levels, the lysogenic cycle is favored. (We will explore this protein's function in greater detail later in the section.) Alternatively, if the level of the cro protein becomes high, the lytic cycle occurs. What environmental factors determine whether the lysogenic or lytic cycle prevails? A critical issue is that the cII protein is easily degraded by a cellular protease that is produced by _E. coli_. Whether or not this protease is made at high levels depends on the environmental conditions. If the growth conditions are very favorable, such as a rich growth medium, the intracellular protease levels are relatively high, and the cII protein tends to be degraded. The cro protein is not sensitive to such protease degradation and slowly accumulates to high levels. Therefore, environmental conditions that are favorable for growth promote the lytic cycle. This makes sense, because a sufficient supply of nutrients is necessary to synthesize new bacteriophages.

Alternatively, starvation conditions favor the lysogenic cycle. When nutrients are limited, the protease level in the cell is relatively low. Under these conditions, the cII protein builds up much more quickly than the cro protein. This situation favors the lysogenic cycle. From the perspective of the bacteriophage, lysogeny may be the better choice under starvation conditions because nutrients may be insufficient for the production of new λ phages.

The λ Repressor and Integrase Control the Lysogenic Cycle

Let's now consider how the lysogenic cycle is chosen (**Figure 18.10a**). If the level of cII becomes high, the cII protein activates two different promoters in the λ genome, P_{RE} and P_I. When the cII protein binds to P_{RE}, it turns on the transcription of _cI_, a gene that codes the λ repressor. The cII protein also activates the _int_ gene by binding to P_I. The λ repressor and integrase proteins promote the lysogenic cycle. When the λ repressor is made in sufficient quantities, it binds to operator sites (O_L and O_R) that are adjacent to P_L and P_R. When the λ repressor is bound to O_R, it inhibits the expression of the genes required for the lytic cycle.

Notice in Figure 18.10a that the binding of the λ repressor to O_R inhibits the expression of _cII_. This may seem counterintuitive, because the cII protein was initially required to bind to P_{RE} and activate the _cI_ gene that codes the λ repressor. You may be thinking that the inhibition of the _cII_ gene would eventually prevent the expression of the _cI_ gene and ultimately stop the synthesis of the λ repressor protein. What prevents the inhibition of _cI_ gene expression? The explanation is that the _cI_ gene has two promoters: P_{RE}, which is activated by the cII protein, and a second promoter called P_{RM}.

TABLE 18.2

Genetic Regulatory Elements and Proteins of Phage λ

Type	Description
Promoters	
P_L	Promoter for the *N* and *cIII* genes, which code N protein and cIII protein, respectively
P_R	Promoter for the *cro, cII, O,* and *P* genes; the *cro* and *cII* genes code cro protein and cII protein, respectively, whereas the *O* and *P* genes code proteins involved with the replication of λ DNA.
P_{RE}	Promoter for the *cI* gene, which codes λ repressor; this promoter is used to establish the lysogenic cycle.
P_{RM}	Promoter for the *cI* gene, which codes λ repressor; this promoter is used to maintain the lysogenic cycle.
P_I	Promoter for the *int* and *xis* genes, which code integrase and excisionase, respectively
$P_{R'}$	Promoter for a large operon that codes many of the proteins necessary for the lytic cycle
Operators	
O_L	Operator that controls P_L
O_R	Operator that controls P_R and P_{RM}
Regulatory Transcription Factors	
N protein	Promotes antitermination by binding to RNA polymerase and allowing transcription past t_L, t_{R1}, and t_{R2}; leads to the transcription of the *cIII, cII, O,* and *P* genes
cII protein	Favors the lysogenic cycle; binds to P_{RE} and P_I and activates their transcription
λ repressor	Establishes and maintains the lysogenic cycle; binds to O_L and inhibits transcription from P_L; binds to O_R and inhibits transcription from P_R and activates transcription from P_{RM}
cro protein	Favors the lytic cycle; binds to O_L and inhibits transcription from P_L; binds to O_R and inhibits transcription from P_{RM} and later from P_R
Q protein	Promotes antitermination by binding to RNA polymerase and allowing the transcription from $P_{R'}$ that is needed for the lytic cycle

FIGURE 18.9 **The events that lead to the beginning of the lysogenic or lytic cycle of phage λ.** This figure shows the region of the phage λ genome that regulates the choice between the lysogenic and lytic cycles. DNA is shown in green. The names of genes that code proteins are shown above the DNA. Promoters, operators, and terminators are shown below the DNA. In this figure and Figure 18.10, RNA transcripts are shown in red. The key regulatory proteins are represented as spheres. Immediately after infection, P_L and P_R are used to make two short mRNAs. These mRNAs code two proteins, designated N and cro. The N protein is an antiterminator that prevents transcriptional termination at three sites in the RNA (t_L, t_{R1}, and t_{R2}). This allows the transcription of several genes, which include *int, xis, cIII, cII, O, P,* and *Q*. The expression of the *O, P,* and *Q* genes is necessary only for the lytic cycle. If the lysogenic cycle is chosen, transcription of these genes is abruptly inhibited.

CONCEPT CHECK: Suppose a certain drug inhibits the function of the N protein. Would such a drug favor the lysogenic cycle, favor the lytic cycle, or prevent both cycles from occurring?

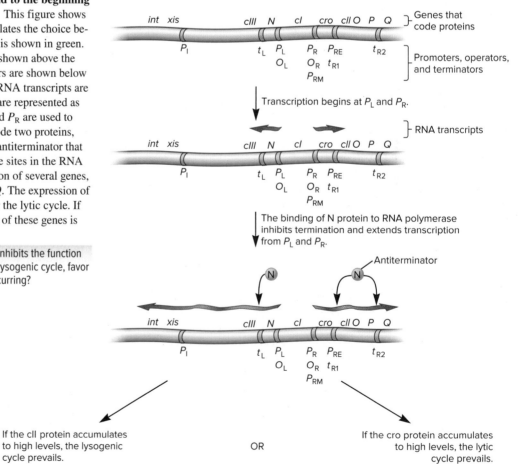

Regulatory region of λ genome

Transcription begins at P_L and P_R.

The binding of N protein to RNA polymerase inhibits termination and extends transcription from P_L and P_R.

If the cII protein accumulates to high levels, the lysogenic cycle prevails.

OR

If the cro protein accumulates to high levels, the lytic cycle prevails.

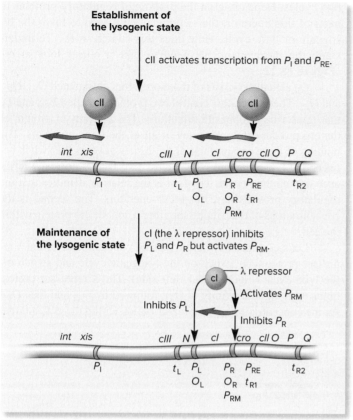

(a) Lysogenic cycle

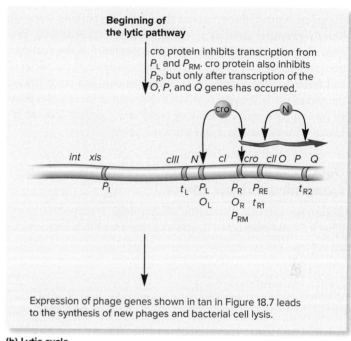

Expression of phage genes shown in tan in Figure 18.7 leads to the synthesis of new phages and bacterial cell lysis.

(b) Lytic cycle

FIGURE 18.10 **Phage λ gene regulation during the lysogenic and lytic cycles.** (a) Gene regulation pattern that leads to the lysogenic cycle. (b) Steps that lead to the lytic cycle.

Genes→Traits The ability to choose between two alternative reproductive cycles can be viewed as a trait of this bacteriophage. As described here, the choice between the two cycles depends on the pattern of gene regulation.

CONCEPT CHECK: What is the function of P_{RM}?

Transcription from P_{RE} occurs at the beginning of the lysogenic cycle. P_{RE} gets its name because this promoter is needed for the expression of the λ repressor during the establishment of the lysogenic cycle. The transcript made from P_{RE} is very stable and quickly leads to a buildup of the λ repressor protein. This causes an abrupt inhibition of the lytic cycle because the binding of the λ repressor protein to O_R blocks the P_R promoter. Later in the lysogenic cycle, large amounts of the λ repressor are no longer necessary. At this point, the P_{RM} promoter makes enough repressor protein to maintain the lysogenic cycle. Interestingly, the P_{RM} promoter is activated by the λ repressor protein. The λ repressor was named when it was understood that it repressed the lytic cycle. Later studies revealed that it also activates its own transcription from P_{RM}.

After the lysogenic cycle is established, certain environmental conditions favor induction of the lytic cycle. For example, exposure to UV light promotes induction. In this case, a protein known as recA detects the DNA damage from UV light and is activated to become a mediator of protein cleavage. RecA protein mediates the cleavage of the λ repressor, thereby inactivating it. This allows transcription from P_R and eventually leads to the

accumulation of the cro protein, which favors the lytic cycle. Under these conditions, it may be advantageous for the λ DNA to direct the synthesis of new phages and lyse the cell, because the exposure to UV light may have already damaged the bacterium to the point where further growth and division are prevented.

The Lytic Cycle Depends on the Action of the cro Protein

If the activity of the cro protein exceeds that of the cII protein, the lytic cycle prevails (**Figure 18.10b**). As mentioned, an early step in the expression of phage λ genes is transcription extending from P_R to produce the cro protein. If the concentration of the cro protein builds to sufficient levels, it will bind to two operators: O_L and O_R. The binding of the cro protein to O_L inhibits transcription from P_L; the binding of this protein to O_R has two effects. First, the binding of the cro protein to O_R inhibits transcription from P_{RM} in the leftward direction. This inhibition prevents the expression of the *cI* gene, which codes the λ repressor; the λ repressor is needed to maintain the lysogenic state. Therefore, the λ repressor cannot successfully shut down transcription from P_R.

Second, the binding of the cro protein to O_R inhibits transcription extending from P_R in the rightward direction. However, this inhibition occurs after the transcription of the O, P, and Q genes. The O and P proteins are necessary for the replication of the phage DNA. The Q protein is an antiterminator protein that permits transcription through another promoter, designated $P_{R'}$ (note the prime symbol; this promoter should not be confused with P_R). The $P_{R'}$ promoter controls a very large operon that codes the proteins necessary for the phage coat, the assembly of the coat proteins, the packaging of the λ DNA, and the lysis of the bacterial cell (refer back to Figure 18.7). These proteins are made toward the end of the lytic cycle. The late expression of these genes leads to the synthesis and assembly of many new λ phages that are released from the bacterial cell when it lyses.

O_R Acts as a Genetic Switch Between the Lysogenic and Lytic Cycles

Before we end this discussion of the phage λ reproductive cycles, let's consider how O_R acts as a genetic switch between the two cycles. Depending on the binding of regulatory proteins to parts of this operator, the switch can be turned to favor the lysogenic or lytic cycle. How does this switch work? To understand the mechanism, we need to take a closer look at O_R (**Figure 18.11**).

O_R actually consists of three operators, designated O_{R3}, O_{R2}, and O_{R1}. These operators control two promoters called P_{RM} and P_R that transcribe in opposite directions. The λ repressor protein or the cro protein can bind to any or all of the three operators. The binding of these two proteins at these sites governs the switch between the lysogenic and lytic cycles. Two factors critically influence this binding event. The first is the relative affinities that the regulatory proteins have for these operators. The second is the concentrations of the λ repressor protein and the cro protein within the cell.

Let's consider how an increasing concentration of the λ repressor protein can switch on the lysogenic cycle and switch off the lytic cycle (Figure 18.11, left side). The λ repressor protein binds with highest affinity to O_{R1}, followed by O_{R2} and then O_{R3}. As the concentration of λ repressor builds within the cell, a dimer

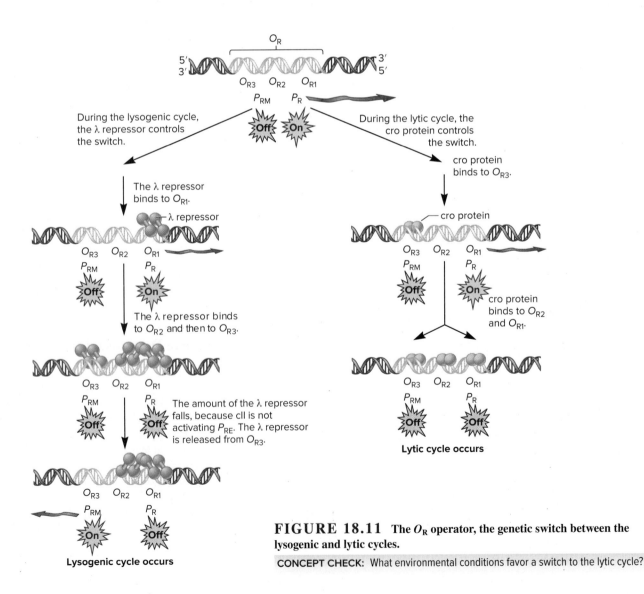

FIGURE 18.11 The O_R operator, the genetic switch between the lysogenic and lytic cycles.

CONCEPT CHECK: What environmental conditions favor a switch to the lytic cycle?

of this protein first binds to O_{R1} because it has the highest affinity for this operator. Next, a second λ repressor dimer binds to O_{R2}. This pair of events occurs very rapidly, because the binding of the first dimer to O_{R1} favors the binding of a second dimer to O_{R2}, in what is called a cooperative interaction. The binding of the λ repressor to O_{R1} and O_{R2} inhibits transcription from P_R and switches off the lytic cycle. Early in the lysogenic cycle, the λ repressor protein concentration may become so high that the protein occupies O_{R3}.

Eventually, the λ repressor concentration begins to drop, because the binding of the λ repressor to O_{R1} and O_{R2} inhibits P_R and thereby decreases the synthesis of the cII protein. Recall that cII initially activates the λ repressor gene from P_{RE}. As the λ repressor concentration gradually falls, this protein is removed first from O_{R3}. This allows transcription from P_{RM}. As mentioned earlier, the binding of λ repressor at only O_{R1} and O_{R2} acts as an activator of P_{RM}. The ability of the λ repressor to activate its own transcription allows the lysogenic cycle to be maintained.

In the lytic cycle (Figure 18.11, right side), the binding of the cro protein controls the switch. The cro protein has its highest affinity for O_{R3} and has lower and similar affinities for O_{R2} and O_{R1}. Under conditions that favor the lytic cycle, the cro protein accumulates, and a cro dimer binds first to O_{R3}. This blocks transcription from P_{RM}, thereby switching off the lysogenic cycle. Later in the lytic cycle, the cro protein concentration continues to rise, so eventually it binds to O_{R2} and O_{R1}. This inhibits transcription from P_R that is not needed in the later stages of the lytic cycle.

Genetic switches, like the one just described for phage λ, represent an important form of genetic regulation. As we have seen, a genetic switch can be used to control two alternative reproductive cycles of a bacteriophage. In addition, genetic switches are also important in the developmental pathways of bacteria and eukaryotes. As we will see in Chapter 26, they play key roles in the initiation of cell differentiation during development. Studies of the phage λ life cycle have provided fundamental information that is applicable to studies of how these other switches can operate at the molecular level.

18.3 COMPREHENSION QUESTIONS

1. A mutation in phage λ results in a 10-fold increase of transcription of the *cII* gene. How do you think this mutation would affect the reproductive cycle of the phage?
   a. The lytic cycle would be favored.
   b. The lysogenic cycle would be favored.
   c. Neither cycle could occur.
   d. None of the above would happen.

2. The *cI* gene that codes the λ repressor has two promoters called P_{RE} and P_{RM}. Which of the following statements is *false*?
   a. P_{RE} is activated by the cII protein.
   b. P_{RE} is activated by the λ repressor.
   c. P_{RM} is turned on after P_{RE}.
   d. P_{RM} is needed to maintain the lysogenic cycle.

18.4 HIV REPRODUCTIVE CYCLE

Learning Outcomes:

1. Outline the organization of the HIV genome.
2. Explain how HIV is reverse transcribed and integrated into the DNA of the host cell.
3. Describe the steps that lead to the formation of new HIV particles.

In Section 18.3, we focused on the reproductive cycles of phage λ, which infects bacteria. In this section, we will focus on HIV, which infects helper T cells in humans. Helper T cells are a type of lymphocyte that play a key role in cell-mediated immunity. They are distinguished from other lymphocytes by the presence of T-cell receptors in their plasma membrane. T-cell receptors are responsible for recognizing antigens—molecules that elicit an immune response. The activation of helper T cells to fight infection is initiated by the activation of T-cell receptors. Over time, the destruction of helper T cells by HIV compromises immune system function, thereby leading to an inability to fight infections or kill cancer cells.

An understanding of the reproductive cycle of HIV has been critical in the development of drugs to combat the virus. For example, azidothymidine (AZT), which is used to prevent the proliferation of HIV, was developed using knowledge of HIV reproduction. In this section, we will explore how HIV infects helper T cells and how new HIV particles are made.

The HIV-1 Genome Has Nine Genes

Researchers have identified different strains of HIV. We will focus on HIV-1, which is much more common than the other strains. The organization of the HIV genome is shown in **Figure 18.12a**. The HIV-1 genome found within the virus particles is composed of single-stranded RNA that is 9749 nucleotides long; each end has sequences called long terminal repeats (LTRs), which are the same at both ends. The HIV genome contains nine genes, but some of these genes code more than one protein. Most of the viral proteins are found within mature HIV particles (**Figure 18.12b**).

A great deal of research has been aimed at determining the functions of HIV proteins. Many of the HIV proteins have multiple functions during the viral reproductive cycle. The following is an overview of HIV genes and some of the best-characterized functions of the proteins they code.

- *gag:* **proteins used for viral assembly and capsid formation.** The *gag* gene codes the Gag polyprotein, which is cleaved into four different proteins, called matrix protein (p17), capsid protein (p24), nucleocapsid protein (p7), and p6 protein. The p17 protein acts as a matrix to anchor the capsid to the viral envelope. The p24 protein is the major capsid protein. The p7 protein is a nucleocapsid protein that binds to the viral RNA and protects it from cellular digestion by nucleases. The p6 protein facilitates the incorporation of certain proteins, such as Vpr (a regulatory protein described shortly), into HIV particles.

Key

■ Long terminal repeats

□ Viral assembly and capsid formation

▨ Enzymes for viral replication and assembly

▨ Infectivity and budding

□ Regulatory functions

■ Viral envelope proteins

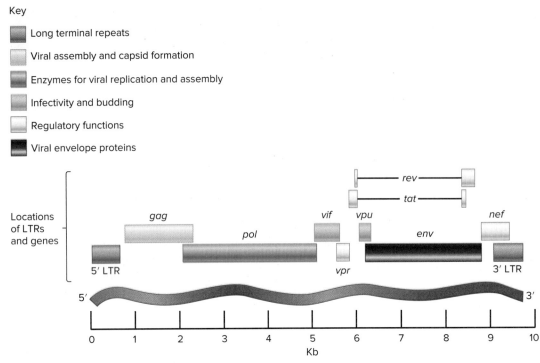

Locations of LTRs and genes

(a) Map of the RNA genome of HIV

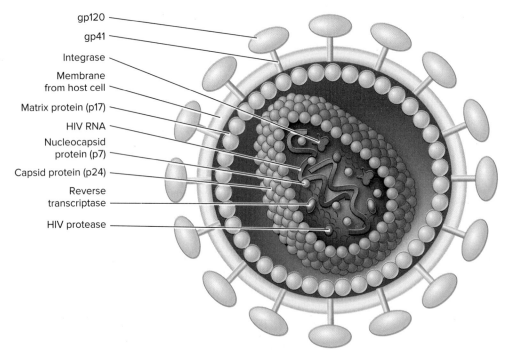

(b) A simplified model of HIV structure

FIGURE 18.12 Structure of HIV.
(a) Organization of the RNA genome of HIV. The HIV RNA has a long terminal repeat at each end and carries nine genes. Note: The *rev* and *tat* genes have two exons that become connected via splicing. **(b)** A simplified model of the structure of HIV, also called an HIV particle. Most HIV proteins are thought to be contained within mature virus particles, but their relative amounts and locations within a particle are not completely understood. For simplicity, this figure does not show all of the proteins found in an HIV particle. In addition, several host-cell proteins are incorporated into a mature virus.

- *pol:* **enzymes needed for viral replication and viral assembly.** The *pol* gene codes a polyprotein that is cleaved into three enzymes: HIV protease, reverse transcriptase, and integrase. HIV protease processes proteins made from the HIV genome so they can assemble into mature HIV particles. Reverse transcriptase is required for the production of DNA from the RNA genome of HIV. A portion of reverse transcriptase called RNase H digests RNA during

reverse transcription. Integrase is used to incorporate HIV DNA into the host genome.

- *vif, vpu:* **proteins that promote infectivity and budding.** The Vif protein is incorporated into mature HIV particles and promotes infectivity by interacting with host-cell proteins. The Vpu protein promotes budding.

- *vpr, rev, tat, nef:* **proteins with regulatory functions.** The Vpr protein has several functions including the transport of the HIV genome into the cell nucleus. The Rev and Tat proteins are involved in the expression of HIV genes. The Nef protein affects the expression of several host-cell genes.

- *env:* **proteins that are part of the viral envelope.** The *env* gene codes a protein that is cleaved into two membrane glycoproteins, gp41 and gp120, which are components of the viral envelope.

HIV RNA Is Reverse Transcribed into Double-Stranded DNA

As briefly summarized in Figure 18.3, the reproductive cycle of HIV involves reverse transcription and integration of viral DNA into the host-cell's genome. The synthesis of double-stranded DNA from viral RNA occurs via reverse transcriptase. This viral enzyme is found within the HIV particle and is released during the uncoating process. We will now take a closer look at how viral RNA is used to make double-stranded DNA.

At the top left in **Figure 18.13** is a representation of HIV RNA that emphasizes the regions that are involved in the synthesis of double-stranded HIV DNA. Let's first focus on the ends of the RNA. Identical repeat sequences (LTRs) are found at both ends. Throughout Figure 18.13, these are labeled simply "Repeats." Next to the 5′ repeat sequence is a unique sequence (U5) and then a site called the primer binding site (PBS), which is a site for the binding of a primer that will initiate the synthesis of HIV DNA. Next to the 3′ repeat sequence is a unique sequence (U3) and a site called a polypurine tract (PPT), which also plays a role in viral DNA synthesis. This diagram of HIV RNA also shows the locations of the *gag, pol,* and *env* genes, but leaves out the other six HIV genes for simplicity.

Let's now consider the steps in the reverse transcription of HIV RNA:

1. Once HIV enters a host cell and uncoating occurs, the viral RNA is recognized by a host tRNA that binds to PBS.
2. The 3′ end of the tRNA acts as a primer for DNA synthesis. Reverse transcriptase first synthesizes the 5′ repeats and U5 into DNA.
3. After this occurs, the region of the RNA strand encompassing the repeats and U5 is degraded by RNase H, which is a component of reverse transcriptase. The tRNA with an attached DNA strand containing the repeats and U5 is released. Because the 5′ and 3′ repeats are the same, the newly made DNA with the repeats and U5 (shown in blue) can bind to the 3′ repeats in the RNA (shown in red) due to complementary base pairing.
4. Reverse transcriptase then synthesizes the rest of the viral genome, except for the 5′ repeats and U5, which were previously degraded.
5. Next, the remaining viral RNA except for the PPT site is degraded by RNase H.
6. The RNA at the PPT site acts as a primer to synthesize the region encompassing U3, the repeats, U5, and PBS. This synthesis occurs via reverse transcriptase, which also can use a DNA template to make a complementary DNA strand.

At this stage, all of the RNA is degraded, which includes the RNA at the PPT site and the tRNA that initiated the synthesis of HIV DNA.

7. The short purple DNA strand and the longer blue DNA strand bind to each other at their complementary PBS regions. Though not shown in Figure 18.13, this results in a circular structure (to see this, look at the online animation). Reverse transcriptase then uses the 3′ ends of both DNA strands to complete the synthesis of viral DNA.

At the bottom right in Figure 18.13 is the double-stranded DNA of HIV after DNA synthesis is completed. Note that the ends of the double-stranded DNA are not identical to the ends of the single-stranded RNA. In particular, U3 and U5 are found at both ends in the DNA but not in the RNA.

Double-Stranded HIV DNA Is Integrated into the Host Genome

The enzyme called integrase is involved in the insertion of the double-stranded HIV DNA into the genome of the host cell. Like reverse transcriptase, integrase is found within HIV particles and is released during the uncoating process.

A simplified description of HIV integration is presented in **Figure 18.14**.

1. For integration to occur, a dimer of integrase first binds to specific sequences in the repeats at each end of the HIV DNA. This causes the HIV DNA to form a circular structure (see the first inset).
2. Integrase makes cuts that remove two nucleotides at both 3′ ends.
3. Additional proteins associate with the circular HIV DNA to form a **preintegration complex.** One such protein is Vpr, which facilitates the transport of the preintegration complex into the nucleus of the host cell through the nuclear pores.
4. The preintegration complex then binds to a site in a chromosome of the host cell.
5. At this site of integration, integrase makes staggered cuts in the host-cell DNA and then joins the 3′ ends of the HIV DNA strands to the 5′ ends of the host-cell DNA strands.
6. The two nucleotides at the 5′ ends of the HIV DNA, which do not base-pair with host-cell DNA, are removed.
7. Due to the staggered cuts and the removal of the two nucleotides at the 5′ ends of the HIV DNA, small gaps occur between the 5′ ends of the viral DNA strands and the 3′ ends of the host DNA. These gaps are filled in by DNA gap repair synthesis (shown in gray), which is described in Chapter 19.

At this stage, the double-stranded HIV DNA is fully integrated into a site in the host-cell genome. In this condition, it is referred to as a provirus.

The Production of New HIV Particles Begins with the Synthesis of HIV RNA and Viral Proteins

After the double-stranded HIV DNA has been integrated into the host-cell genome, it may remain there in a dormant state for a very

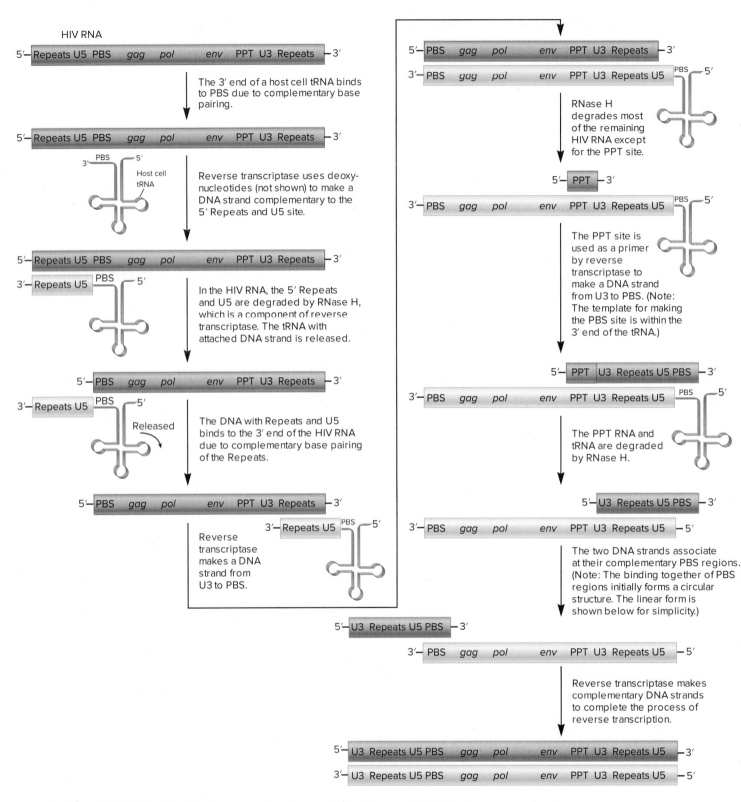

FIGURE 18.13 Synthesis of double-stranded DNA from HIV RNA via reverse transcriptase.

CONCEPT CHECK: What are the two enzymatic functions of reverse transcriptase?

ONLINE
ANIMATION

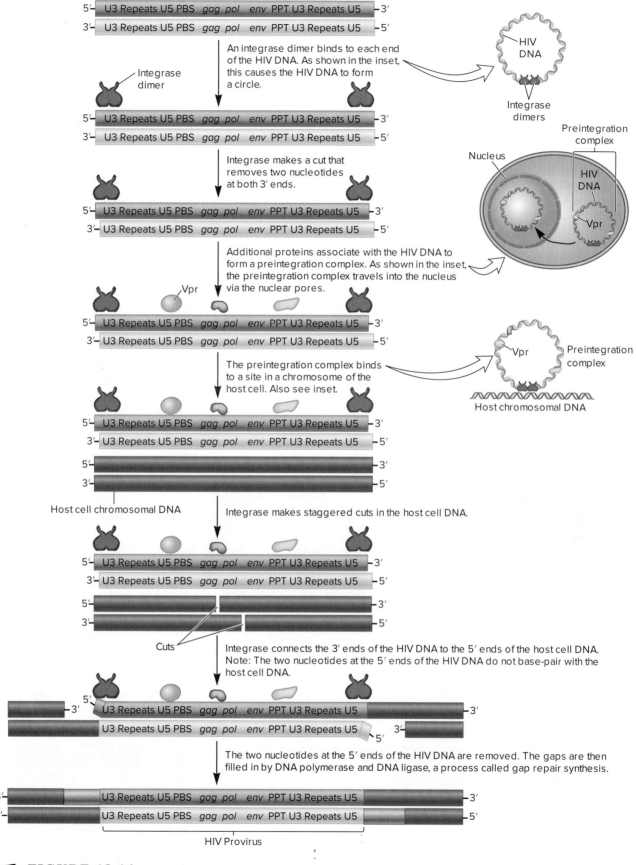

FIGURE 18.14 **Integration of the double-stranded HIV DNA into the host-cell genome.** Note: This is a simplified version of HIV integration that focuses on the role of integrase. Other proteins are necessary for integration.

long time—several years or more. As mentioned previously, this condition is called latency. To end the latent state and begin production of new virus particles, certain transcription factors must transcribe the HIV provirus. In particular, a transcription factor called NF-κB plays a key role in activating the transcription of the HIV provirus. NF-κB is usually present in an inactive state in the cytosol of T cells.

How does NF-κB become active? If an antigen binds to a T-cell receptor, this initiates a signal cascade that causes the activation of NF-κB and its movement into the nucleus. Once in the nucleus, NF-κB binds to the HIV proviral DNA and stimulates transcription. In this way, those T cells that are currently fighting infection are most likely to produce new HIV particles and to be killed.

Figure 18.15 shows a simplified version of the process of synthesizing new HIV components. The 5′ LTR of HIV contains a single promoter and the binding sites for several transcription factors, including NF-κB. These transcription factors stimulate the transcription of HIV RNA from the promoter. The HIV RNA contains several 5′ and 3′ splicing sites, so it can be spliced in more than 30 different ways, or it may not be spliced at all. Though the splicing process is complex, we can view splicing as occurring in three general ways, described next.

Fully Spliced HIV RNA. Early in the process of synthesizing new HIV components, the HIV RNA is fully spliced, which means that two introns have been removed. This form of HIV RNA is able to exit the nucleus and is used for the translation of

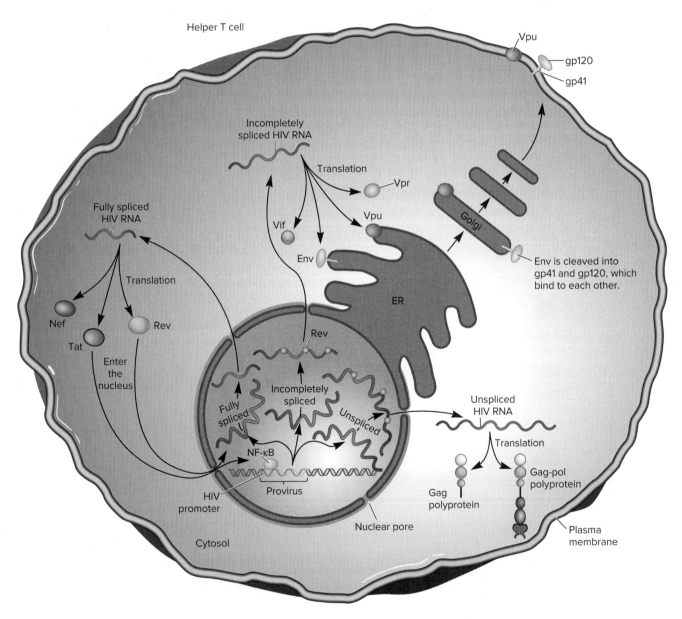

FIGURE 18.15 **Synthesis of new HIV components: HIV RNA and proteins.**

ONLINE
ANIMATION

CONCEPT CHECK: Which form of HIV RNA, fully spliced, incompletely spiced, or unspliced, is needed during the early stage of the synthesis of HIV components?

the Nef, Tat, and Rev proteins—the Nef, Tat, and Rev proteins are early gene products. The Nef protein maintains T-cell activation and inhibits the production of host-cell proteins that are involved with cellular defense against viruses. The Tat and Rev proteins enter the nucleus, where the Tat protein greatly increases the level of transcription of HIV RNA. The function of the Rev protein is described next.

Incompletely Spliced HIV RNA.   Incompletely spliced RNA is also produced during the transcription of HIV RNA, but eukaryotic cells possess mechanisms that prevent RNAs containing introns from exiting the cell nucleus. After the Rev protein is translated from the fully spliced HIV RNA, it enters the nucleus. Multiple copies of the Rev protein bind to incompletely spliced versions of the HIV RNA. This allows these RNAs to exit the nucleus. Incompletely spliced versions of HIV RNA are needed for the translation of the Vif, Env, Vpu, and Vpr proteins. The Env and Vpu proteins are integral membrane proteins that are initially inserted into the endoplasmic reticulum (ER) membrane and then move to the Golgi apparatus and finally to the plasma membrane via membrane vesicles. During its transport, the Env protein is cleaved into two components called gp41 and gp120. The gp41 protein is a transmembrane protein that anchors gp120 to the membrane (see Figure 18.12b). Vpu is needed for the release of HIV particles from the host cell.

Unspliced HIV RNA.   Rev proteins also bind to unspliced HIV RNA to allow its export from the nucleus. Unspliced HIV RNA has two key functions. First, unspliced HIV RNA is packaged into HIV particles. Second, it is used for the translation of Gag and Gag-pol polyproteins. The Gag-pol polyprotein is a larger molecule in which the Gag polyprotein is connected to the Pol polyprotein. How is the Gag-pol polyprotein made? On rare occasions the stop codon at the end of *gag* is not recognized. The resulting read-through past the stop codon results in the synthesis of Gag-pol polyprotein. However, because the read-through is a rare event, the amount of Gag-pol polyprotein produced is much less than the amount of Gag polyprotein produced.

HIV Components Are Assembled and Bud from the Host-Cell Plasma Membrane

The final event in the HIV reproductive cycle involves the formation of HIV particles that are released from the host cell, which is a helper T cell. This process occurs in three overlapping stages: assembly, budding, and maturation.

Assembly.   As we have seen, the expression of HIV proviral DNA leads to the synthesis of viral components, including unspliced HIV RNA and viral proteins. The assembly of HIV particles occurs at the host-cell plasma membrane and is mediated by the Gag polyprotein. **Figure 18.16** shows a simplified model of HIV assembly. The Gag polyprotein has four domains called MA, CA, NC, and P6, which play different roles in the assembly process. (Note: These four domains correspond to the matrix protein, capsid protein, nucleocapsid protein, and p6 protein, respectively.)

These four domains are released later as individual proteins following the cleavage of Gag polyprotein during viral maturation.

- The MA domain (shown in yellow) binds to the plasma membrane. It also binds to gp41, which is an integral membrane protein.
- The CA domain (shown in green) facilitates protein-protein interactions required for assembly of immature virus particles.
- The NC domain (shown in gray) captures the HIV RNA genome.
- The P6 domain contains binding sites for several other proteins that will be contained within virus particles, such as Vpr. The P6 domain also recognizes Vpu and cellular proteins that are required for budding, but are not contained within mature HIV particles.

Budding.   As seen in Figure 18.16, as assembly occurs, the interactions of many Gag polyproteins result in the formation of a spherical structure that protrudes from the plasma membrane. The release of this bud requires Vpu (an HIV protein) and host-cell proteins, including ones that are components of the ESCRT (<u>e</u>ndo-somal <u>s</u>orting <u>c</u>omplex <u>r</u>equired for <u>t</u>ransport) pathway. The P6 domain of the Gag polyprotein recognizes certain proteins of the ESCRT pathway. This initiates a series of events that leads to membrane fusion at the neck of the bud, thereby releasing the bud from the surface of the cell. The released bud is an immature HIV particle.

The HIV protein called Vpu is also important for budding because it enhances the release of immature HIV particles from the cell surface. Vpu exerts its effects by inhibiting a protein known as tetherin (not shown in Figure 18.16). In the absence of Vpu, tetherin binds to the membrane of the viral envelope and tethers it to the cell surface. When Vpu inhibits tetherin, immature HIV particles can be released.

Maturation.   The final stage of the HIV reproductive cycle is the maturation of immature HIV particles into mature virus particles. As a polyprotein, Gag promotes the assembly of the components that constitute an HIV particle. However, for a capsid to be formed and surround the HIV RNA, the Gag polyprotein must be cleaved into its individual components. HIV protease plays a key role in this maturation process. HIV protease cleaves the Gag polyprotein at multiple sites, which releases MA, CA, NC, and P6 as individual proteins. The matrix protein lines the inside of the HIV envelope, helping to anchor the gp41/gp120 spikes to the envelope. Once released from the Gag polyprotein, the capsid protein is able to form a capsid structure that encloses two molecules of HIV RNA along with several different proteins (see Figure 18.12b). The nucleocapsid and p6 proteins are found inside the capsid. They protect the HIV RNA from nuclease digestion and have binding sites that promote the incorporation of other proteins into the capsid. For example, p6 has binding sites for Vpr.

As mentioned earlier, another polyprotein called the Gag-pol polyprotein plays a role in the maturation process, but is made in a much smaller amount than is the Gag polyprotein. As shown

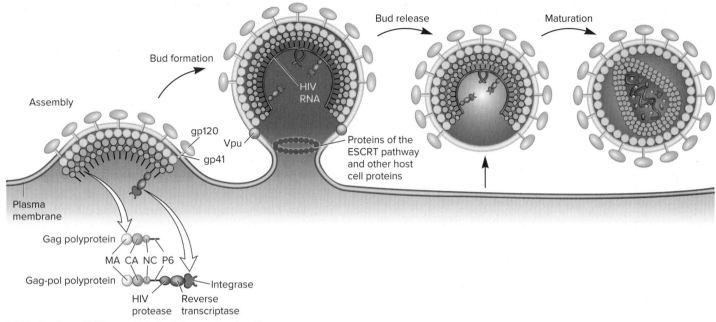

(a) Mechanism of HIV assembly, budding, and maturation

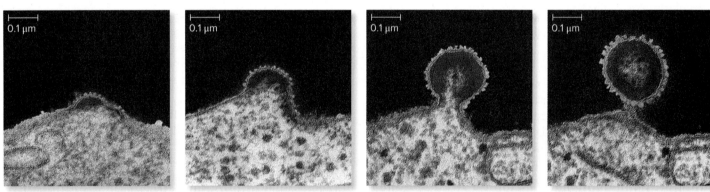

(b) Transmission electron micrograph of HIV assembly, budding, and maturation

FIGURE 18.16 **Assembly, budding, and maturation of new HIV particles.** Part **(a)** is a schematic drawing, and part **(b)** shows colorized transmission electron micrographs of the process. Note: Certain colors in parts (a) and (b) don't match each other. In part (a), the matrix protein is yellow but in part (b), it is red. The capsid protein is green in both (a) and (b).

(b): Eye of Science/Science Source

in Figure 18.16, a few copies of the Gag-pol polyprotein are incorporated into immature virus particles. When the Gag-pol polyprotein is cleaved by HIV protease, this cleavage releases more HIV protease, along with reverse transcriptase and integrase. These proteins are captured within the capsid. As shown in Figures 18.13 and 18.14, they are necessary for the reverse transcription and integration of the HIV genome in a newly infected host cell.

18.4 COMPREHENSION QUESTIONS

1. A viral protein that is needed to make HIV DNA is
   a. integrase.
   b. reverse transcriptase.
   c. Vpr.
   d. Gag polyprotein.

2. Which form of HIV RNA is packaged into HIV particles?
   a. Fully spliced RNA
   b. Incompletely spliced RNA
   c. Unspliced RNA
   d. All three forms of RNA are packaged into HIV particles.

3. After HIV components are made, during which phase are HIV particles released from the host cell?
   a. Synthesis of viral components
   b. Assembly
   c. Budding
   d. Maturation

KEY TERMS

KEY TERMS

Introduction: virus
18.1: host cell, host range, capsid, viral envelope, bacteriophage (phage), viral genome
18.2: viral reproductive cycle, phage λ, human immunodeficiency virus (HIV), integrase, prophage, lysogenic cycle,

reverse transcriptase, provirus, lytic cycle, latent, lysogeny, temperate phage, virulent phage, episome, emerging virus, pandemic
18.3: attachment sites, antitermination
18.4: preintegration complex

CHAPTER SUMMARY

18.1 Virus Structure and Genetic Composition

- Tobacco mosaic virus (TMV) was the first virus to be discovered. It infects many species of plants.
- Viruses vary with regard to their host range, structure, and genome composition (see Table 18.1, Figure 18.1).
- The study of reconstituted viruses confirmed that the genome of TMV is composed of RNA (see Figure 18.2).

18.2 Overview of Viral Reproductive Cycles

- A viral reproductive cycle consists of five or six basic steps: attachment, entry, integration, synthesis of viral components, viral assembly, and release (see Figure 18.3).
- Some bacteriophages can alternate between two reproductive cycles: the lysogenic and lytic cycles (see Figure 18.4).
- Emerging viruses are those that have arisen recently and are more likely than previous strains to cause infection, such as the coronavirus that causes COVID-19 (see Figure 18.5).
- The disease AIDS is caused by human immunodeficiency virus (HIV). This virus is a retrovirus whose reproductive cycle involves the integration of the viral genome into a chromosome in a helper T cell (see Figure 18.6).

18.3 Bacteriophage λ Reproductive Cycles

- Phage λ has different sets of genes that allow it to follow either a lysogenic or lytic cycle during reproduction (see Figure 18.7).
- The genome of phage λ is inserted into the *E. coli* host chromosome via integrase (see Figure 18.8).
- The expression of the cII protein favors the lysogenic cycle, whereas the expression of the cro protein favors the lytic cycle (see Figures 18.9, 18.10, Table 18.2).
- O_R consists of three operators that are binding sites for the λ repressor and cro proteins. O_R acts as a genetic switch between the lysogenic and lytic cycles (see Figure 18.11).

18.4 HIV Reproductive Cycle

- The HIV genome is a single-stranded RNA molecule that contains nine genes, which are needed to make HIV particles (Figure 18.12).
- Following infection, reverse transcriptase catalyzes the synthesis of HIV DNA, and integrase is involved in the integration of the HIV DNA into the host-cell genome (Figures 18.13, 18.14).
- The transcription of HIV DNA leads to the synthesis of HIV RNA and viral proteins (Figure 18.15).
- HIV components are assembled at the host-cell plasma membrane. Buds are released that mature into HIV particles (Figure 18.16).

PROBLEM SETS & INSIGHTS

MORE GENETIC TIPS **1.** Discuss how the components of viruses assemble to make new virus particles. What is the difference between a virus that can self-assemble and one that cannot? Give examples.

🅣 **OPIC:** *What topic in genetics does this question address?* The topic is the synthesis of new viruses. More specifically, the question is about the assembly of components into viral particles.

🅘 **NFORMATION:** *What information do you know based on the question and your understanding of the topic?* From the question, you know that viruses are composed of various components that must assemble to make a virus particle. From your understanding of the topic, you may remember that viral assembly occurs in a series of steps.

🅟 **ROBLEM-SOLVING** 🅢 **TRATEGY:** *Describe the steps.* One strategy to solve this problem is to break down the assembly of viruses into a series of steps.

ANSWER: Some viruses are composed of only nucleic acid (DNA or RNA) surrounded by a capsid of proteins. First, the viral nucleic acid and viral proteins are made by the host cells. Next, the proteins bind to each other to form a capsid that surrounds the viral nucleic acid. For viruses that also have an envelope, such as HIV, the third step is the budding of the viral particle from the cell, during which it acquires its envelope.

Many viruses with a simple structure are able to self-assemble, meaning that viral components spontaneously bind to each other to form a complete virus particle. An example of a self-assembling virus is the tobacco mosaic virus (TMV). The capsid proteins assemble around the RNA genome, which lies inside the hollow capsid.

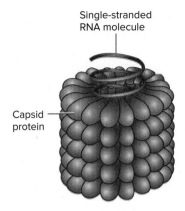

Single-stranded
RNA molecule

Capsid
protein

Other viruses do not self-assemble. For example, the assembly of phage λ requires the help of noncapsid proteins not found in the completed viral particle. Some of these noncapsid proteins function as enzymes that modify capsid proteins, whereas others provide scaffolding for the assembly of the capsid. The synthesis of HIV particles occurs in three stages. First, HIV RNA and viral proteins assemble at the plasma membrane. Next, the forming virus acquires its outer envelope in a budding process. After the bud is released from the cell, it matures into a virus particle.

---

2. Viruses may be latent for a long period of time. For example, HIV may be latent for many years, during which new viruses are not made. What are three different mechanisms of viral latency?

TOPIC: *What topic in genetics does this question address?* The topic is latency in viruses. More specifically, the question asks you to explain how latency occurs.

INFORMATION: *What information do you know based on the question and your understanding of the topic?* From the question, you know that viruses may become latent, during which time they do not produce new viruses. From your understanding of the topic, you may remember that latency can occur in different ways, depending on the type of virus.

PROBLEM-SOLVING **S**TRATEGY: *Compare and contrast.* One strategy to solve this problem is to consider how different types of viruses become latent and contrast the different mechanisms.

ANSWER: Some bacteriophages enter the lysogenic cycle, in which the phage DNA becomes incorporated into the bacterial chromosome as a prophage. In this case, the phage DNA is directly incorporated into the host-cell chromosome. Eukaryotic viruses can become latent in other ways. Retroviruses, which have an RNA genome, are reverse

transcribed and the viral DNA is incorporated into a host-cell chromosome as a provirus. Alternatively, herpes viruses can exist as episomes that remain latent for long periods of time.

---

3. What is an emerging virus? Give two examples. Propose an experiment to explain how an emerging virus could arise.

TOPIC: *What topic in genetics does this question address?* The topic is emerging viruses. More specifically, the question asks you to describe what an emerging virus is and propose an experiment to explain how it might come into existence.

INFORMATION: *What information do you know based on the question and your understanding of the topic?* From your understanding of the topic, you may remember that emerging viruses have arisen recently and have the potential to cause widespread infection and disease. You may also remember that emerging viruses arise via mutations in existing viruses.

PROBLEM-SOLVING **S**TRATEGIES: *Define key terms. Design an experiment.* To begin to solve this problem, you need to define a key term: emerging virus. To design an experiment that explains how such viruses arise, you may want to consider that new forms of viruses generally come from the genetic alteration of preexisting viruses. Another aspect of viruses to consider is that the base sequences of many viruses have already been determined.

ANSWER: An emerging virus is a virus that has arisen recently and is more likely than previous strains to cause infection. Examples include the coronavirus that causes COVID-19 and HIV, which causes AIDS. The rationale behind designing an experiment is based on the premise that emerging viruses arise from genetic alterations in preexisting viruses.

1. Determine the base sequence of the emerging virus that you are interested in. (Note: The technique of DNA sequencing is described in Chapter 20.)

2. Compare the base sequence with the known sequences of other viruses; the expectation is that your emerging virus will be closely related to some other virus that is already known.

3. Analyze the differences in base sequences between your emerging virus and its closest relative.

RESULTS: You may identify changes in particular viral genes that may have altered the infectivity of the virus. For example, a mutation could occur that alters a viral protein in a way that allows the virus to bind more easily to host cells and enter them.

Conceptual Questions

C1. Explain why viruses are considered nonliving.

C2. What structural features are common to all viruses? Which features are found only in certain types of viruses?

C3. What are the similarities and differences among viral genomes?

C4. What is a viral envelope? Describe how it is made.

C5. What do the terms *host cell* and *host range* mean?

C6. Describe why the attachment step in a viral reproductive cycle is usually specific for one or just a few cell types.

C7. Compare and contrast the entry step of the viral reproductive cycles of phage λ and HIV.

C8. What is the role of reverse transcriptase in the reproductive cycle of HIV and other retroviruses?

C9. Describe how lytic bacteriophages are released from their host cells.

C10. What is the difference between a temperate phage and a virulent phage?

C11. What are a prophage, a provirus, and an episome? What is their common role in a viral reproductive cycle?

C12. What key features distinguish the lytic from the lysogenic cycle?

C13. Describe the role that integrase plays during the insertion of phage λ DNA into the host-cell chromosome.

C14. With regard to promoting the lytic or lysogenic cycle, what would happen if the following genes were missing from the phage λ genome?

A. *cro*

B. *cI*

C. *cII*

D. *int*

E. *cII* and *cro*

C15. How do the λ repressor and the cro protein affect the transcription from P_R and P_{RM}? Explain where these proteins are binding to cause their effects.

C16. In your own words, explain why it is necessary for the *cI* gene to have two promoters. What would happen if it had only P_{RE}?

C17. Figure 18.11 shows a genetic switch that controls the choice between the lytic and lysogenic cycles of phage λ. What is a genetic switch? Compare the roles of a genetic switch and a simple operator site (like the one found in the *lac* operon) in gene regulation.

C18. Describe the process of reverse transcription of HIV RNA.

C19. Why is a host-cell tRNA needed for reverse transcription?

C20. Explain the role of RNase H (a component of reverse transcriptase) during the synthesis of HIV DNA.

C21. Describe how HIV DNA is integrated into a chromosome of the host cell.

C22. What is the role of the Vpr protein during the process of HIV DNA integration?

C23. Why is gap repair synthesis needed during HIV DNA integration?

C24. Compare and contrast the roles of fully spliced, incompletely spliced, and unspliced HIV RNA. Which type is needed in the early stages of HIV proliferation, and which is needed in later stages?

C25. Describe the role of the Gag polyprotein during the assembly of HIV components at the host-cell plasma membrane.

C26. How does an HIV particle acquire its envelope?

C27. Explain the role of HIV protease during the process of HIV maturation.

Experimental Questions

E1. Discuss how researchers determined that TMV is a virus that causes damage to plants.

E2. What technique must be used to visualize a virus?

E3. What is a reconstituted virus?

E4. Following the infection of healthy tobacco leaves by reconstituted viruses, what two characteristics did Fraenkel-Conrat and Singer analyze? Explain how their results were consistent with the idea that the RNA of TMV is responsible for the symptoms of viral infection and the properties of viral proteins.

E5. Certain drugs to combat human viral diseases affect spike glycoproteins in the viral envelope. Discuss how you think such drugs may prevent viral infection.

E6. Some drugs that inhibit HIV proliferation are inhibitors of HIV protease. Explain how these drugs would help to stop the spread of HIV.

E7. A researcher identified a mutation in P_R of phage λ that causes its transcription rate to be increased 10-fold. Do you think this mutation would favor the lytic or lysogenic cycle? Explain your answer.

E8. Experimentally, when an *E. coli* bacterium already has a λ prophage integrated into its chromosome, another λ phage cannot usually infect the cell and establish the lysogenic or lytic cycle. Based on your understanding of the genetic regulation of the phage λ reproductive cycles, why do you think the second phage would be unsuccessful?

E9. A bacterium is exposed to a drug that inhibits the N protein. What would you expect to happen if the bacterium was later infected by phage λ? Would phage λ follow the lytic cycle, the lysogenic cycle, or neither? Explain your answer.

E10. This question combines your knowledge of bacterial conjugation (described in Chapter 7) and the genetic regulation that directs the phage λ reproductive cycles. When researchers mix donor Hfr strains with recipient F⁻ bacteria that are lysogenic for phage λ, the conjugated cells survive normally. However, if donor Hfr strains that are lysogenic for phage λ conjugate with recipient F⁻ bacteria that do not contain any phage λ, the recipient cells often lyse, due to the induction of λ into the lytic cycle. Based on your knowledge of the regulation of the two reproductive cycles of phage λ, explain this experimental observation.

Questions for Student Discussion/Collaboration

1. Discuss the properties of emerging viruses. What are the challenges associated with combating them?

2. Certain environmental conditions such as exposure to UV light are known to activate lysogenic λ prophages and cause them to progress into the lytic cycle. UV light initially causes the repressor protein to be proteolytically degraded. Make a flow diagram showing the events following this degradation that lead to the lytic cycle. (Note: The *xis* gene codes for an enzyme that is necessary to excise the λ prophage from the *E. coli* chromosome. The enzyme integrase is also necessary for this excision.)

3. Browse the Internet to determine the drugs that are used to treat people with AIDS. Which proteins do these drugs affect? Discuss how an understanding of the HIV reproductive cycle has been helpful in developing treatments for AIDS.

Note: All answers are available for the instructor in Connect; the answers to the even-numbered questions and all of the Concept Check and Comprehension Questions are in Appendix B.

A child with xeroderma pigmentosum. *This child carries a mutation that decreases the ability to repair DNA, which makes the skin highly susceptible to the harmful effects of UV light.*
ZUMA Press, Inc./Alamy Stock Photo

19

GENE MUTATION, DNA REPAIR, AND RECOMBINATION

The primary function of DNA is to store information for the synthesis of proteins. A key aspect of the process of gene expression is that the DNA itself does not normally change. This stability allows DNA to function as a permanent storage unit. However, on relatively rare occasions, a mutation can occur. The term **mutation** refers to a heritable change in the genetic material. When a mutation occurs, the structure of DNA is changed permanently, and this alteration can be passed from mother to daughter cells during cell division. If a mutation occurs in reproductive cells, it may also be passed from parent to offspring.

The phenomenon of mutation is centrally important in all fields of genetics, including molecular genetics, Mendelian inheritance, and population genetics. Mutations provide the allelic variation that we have discussed throughout this textbook. For example, phenotypic differences, such as tall versus short pea plants, are due to mutations that alter the expression of particular genes.

With regard to their phenotypic effects, mutations can be beneficial, neutral, or detrimental. On the positive side, mutations are essential to the continuity of life. They provide the

variation that enables species to adapt to their environment via natural selection. Mutations are the foundation for evolutionary change. On the negative side, however, new mutations are much more likely to be harmful rather than beneficial to the individual. The genes within each species have evolved to work properly. They have functional promoters, coding sequences, terminators, and so on, that allow them to be expressed. Random mutations are more likely to disrupt these sequences than to improve their function. For example, many inherited human diseases result from mutated genes. In addition, diseases such as skin and lung cancer can be caused by environmental agents that are known to cause DNA mutations. For these and many other reasons, understanding the molecular nature of mutations is a deeply compelling area of research. In this chapter, we will consider the nature of mutations and their consequences for gene expression at the molecular level.

Species have evolved several ways to repair damaged DNA and thereby decrease the rate of mutations. DNA repair systems reverse DNA damage before it results in a mutation that could potentially have negative consequences. DNA repair systems have been studied extensively in many organisms, particularly *Escherichia coli*, yeast, mammals, and plants. A variety of systems repair different types of DNA lesions. We will examine how several of these DNA repair systems operate.

In the last section of this chapter, we will consider molecular processes in which segments of chromosomal DNA become rearranged. **Homologous recombination** is the process whereby identical or similar DNA segments are exchanged between homologous chromosomes. Homologous recombination occurs when chromosomes cross over during meiosis (see Chapter 2). Not only does homologous recombination enhance genetic diversity, it also helps to repair DNA and ensures the proper segregation of chromosomes.

19.1 EFFECTS OF MUTATIONS ON GENE STRUCTURE AND FUNCTION

Learning Outcomes:

1. Define *point mutation*.
2. Describe how a mutation within the coding sequence of a gene may alter a polypeptide's structure.
3. Explain how a mutation within a non-coding sequence may alter gene function.
4. Compare and contrast intragenic and intergenic suppressors.
5. Explain how changes in chromosome structure may affect gene expression.
6. Distinguish between germ-line and somatic mutations.

How do mutations affect phenotype? To answer this question, we must appreciate how changes in DNA structure can ultimately affect gene function. Much of our understanding of mutations has come from the study of experimental organisms, such as bacteria, yeast, and *Drosophila*. Researchers can expose these organisms to environmental agents that cause mutations and then study the consequences of the induced mutations. In addition, because these organisms have a short generation time, researchers can investigate the effects of mutations when they are passed from cell to cell and from parent to offspring.

As discussed in Chapter 8, changes in chromosome structure and number are important occurrences within natural populations of eukaryotic organisms. These types of changes are considered to be mutations because the genetic material has been altered in a way that can be inherited. Changes in chromosome structure and number usually affect the expression of many genes. By comparison, a mutation in a single gene is a relatively small change in DNA structure. In this section, we will be primarily concerned with the ways that mutations affect the molecular and phenotypic expression of single genes. We will also consider why the timing of mutations during an organism's development has important consequences.

Gene Mutations Are Molecular Changes in the DNA Sequence of a Gene

A gene mutation occurs when the sequence of DNA within a gene is altered in a permanent way. It can involve a base substitution or a removal or addition of one or more base pairs.

A **point mutation** is a change in a single base pair within the DNA. For example, the DNA sequence shown here has been altered by a **base substitution,** in which one base is substituted for another:

$$5'-\text{AACGCTAGATC}-3' \rightarrow \overset{\downarrow}{5'-\text{AACGCGAGATC}-3'}$$
$$3'-\text{TTGCGATCTAG}-5' \quad\;\; 3'-\text{TTGCGCTCTAG}-5'$$

A change of a pyrimidine to another pyrimidine, such as C to T, or a purine to another purine, such as A to G, is called a **transition.** By contrast, in a **transversion,** a purine and a pyrimidine are interchanged. The example just shown is a transversion (a change from T to G in the top strand and from A to C in the bottom strand). During DNA replication, an interchange between a purine and a pyrimidine causes a base-pair mismatch that results in a substantial distortion of the DNA double helix. This distortion tends to be readily recognized by DNA polymerase, and the incorrect base is removed by its proofreading function (refer back to Figure 11.17). For this reason, transversions are much less frequent than transitions.

Another way a gene mutation may occur is if a short sequence of base pairs is deleted from or added to chromosomal DNA:

$$5'-\text{AACGCTAGATC}-3' \rightarrow \overset{\downarrow}{5'-\text{AACGCTC}-3'}$$
$$3'-\text{TTGCGATCTAG}-5' \quad\;\; 3'-\text{TTGCGAG}-5'$$
(Deletion of 4 bp)

$$5'-\text{AACGCTAGATC}-3' \rightarrow \overset{-\downarrow-}{5'-\text{AACAGTCGCTAGATC}-3'}$$
$$3'-\text{TTGCGATCTAG}-5' \quad\;\; 3'-\text{TTGTCAGCGATCTAG}-5'$$
(Addition of 4 bp)

As we will see next, small deletions or additions to the sequence of a gene can significantly affect the function of the coded protein.

Gene Mutations Can Alter the Coding Sequence Within a Gene

How might a mutation within the coding sequence of a protein-coding gene affect the amino acid sequence of the polypeptide that is coded by the gene? **Table 19.1** describes the possible effects of point mutations.

- **Silent mutations** are those that do not alter the amino acid sequence of the polypeptide even though the base sequence has changed. Because the genetic code is degenerate, silent mutations can occur at certain bases within a codon, usually the third base, and the specific amino acid is not changed.
- **Missense mutations** are base substitutions for which an amino acid change does result. An example of a missense mutation is the one that causes the human disease known

TABLE 19.1

Consequences of Point Mutations Within a Coding Sequence

Type of Change	Mutation in the DNA	Example*	Amino Acids Altered	Likely Effect on Protein Function
None	None	5′–A–T–G–A–C–C–G–A–C–C–C–G–A–A–A–G–G–G–A–C–C–3′ Met – Thr – Asp – Pro – Lys – Gly – Thr –	None	None
Silent	Base substitution	↓ 5′–A–T–G–A–C–C–G–A–C–C–C–C–A–A–A–G–G–G–A–C–C–3′ Met – Thr – Asp – Pro – Lys – Gly – Thr –	None	None
Missense	Base substitution	↓ 5′–A–T–G–C–C–C–G–A–C–C–C–G–A–A–A–G–G–G–A–C–C–3′ Met – Pro – Asp – Pro – Lys – Gly – Thr –	One	Neutral or inhibitory
Nonsense	Base substitution	↓ 5′–A–T–G–A–C–C–G–A–C–C–C–G–T–A–A–G–G–G–A–C–C–3′ Met – Thr – Asp – Pro – STOP!	Many	Negative
Frameshift	Addition/deletion	↓ 5′–A–T–G–A–C–C–G–A–C–G–C–C–G–A–A–A–G–G–G–A–C–C–3′ Met – Thr – Asp – Ala – Glu – Arg – Asp –	Many	Negative

*DNA sequence in the coding strand. Note that this sequence is the same as the mRNA sequence except that the RNA contains uracil (U) instead of thymine (T). For the nonmutant sequence, the complementary sequence of the other strand would be: 3′–T–A–C–T–G–G–C–T–G–G–G–C–T–T–T–C–C–C–T–G–G–3′. The three-base codons are shown in alternating black and red. Mutations are shown in green. Changes in the amino acid sequence are shown in blue.

as sickle cell disease. This disease involves a mutation in the β-globin gene, which alters the polypeptide sequence such that the sixth amino acid is changed from glutamic acid to valine. This single amino acid substitution alters the structure and function of the hemoglobin protein. One consequence of this alteration is that under conditions of low oxygen, the red blood cells assume a sickle shape (**Figure 19.1**). In this case, a single amino acid substitution has a profound effect on the phenotype of cells and causes a serious disease.

- **Nonsense mutations** involve a change from a codon that specifies an amino acid to a stop codon. This change terminates the translation of the polypeptide earlier than normal, producing a truncated polypeptide (see Table 19.1).
- **Frameshift mutations** involve the addition or deletion of a number of nucleotides that is not divisible by 3. Because the codons are read in multiples of 3, this type of mutation shifts the reading frame. The translation of the mRNA then results in a completely different amino acid sequence downstream from the mutation.

Except for silent mutations, new mutations are more likely to produce polypeptides that have reduced rather than enhanced function. For example, when nonsense mutations produce polypeptides that are substantially shorter, those shorter polypeptides are unlikely to function properly. Likewise, frameshift mutations dramatically alter the amino acid sequence of polypeptides and are thereby likely to disrupt function. Missense mutations are less likely to alter function because they involve a change of a single amino acid within polypeptides that typically contain hundreds of amino acids. When a missense mutation has no detectable effect on protein function, it is referred to as a **neutral mutation.** A missense

mutation that substitutes an amino acid with a chemistry similar to that of the original amino acid is likely to be neutral or nearly neutral. For example, a missense mutation that substitutes a glutamic acid for an aspartic acid may be neutral because both amino acids are negatively charged and have similar side chain structures. Silent mutations are also considered a type of neutral mutation.

A mutation can occasionally produce a polypeptide with an enhanced ability to function. Although such a favorable mutation is relatively rare, it may result in an organism with a greater likelihood of surviving and reproducing. If this is the case, natural selection may cause the favorable mutation to increase in frequency within a population. This topic will be discussed later in this chapter and also in Chapter 27.

Gene Mutations Can Occur Outside of a Coding Sequence and Influence Gene Expression

Thus far, we have focused our attention on mutations in the coding regions of genes and their effects on polypeptide structure and protein function. In previous chapters, we learned how various sequences outside of coding sequences play important roles during the process of gene expression. A mutation can occur within a noncoding sequence, thereby affecting gene expression (**Table 19.2**).

- Promoter mutations that increase the rate of transcription are termed **up promoter mutations.** Mutations that make a sequence more like the consensus sequence are likely to be up promoter mutations.
- In contrast, a **down promoter mutation** causes the promoter to become less like the consensus sequence, decreasing its affinity for transcription factors and decreasing the transcription rate.

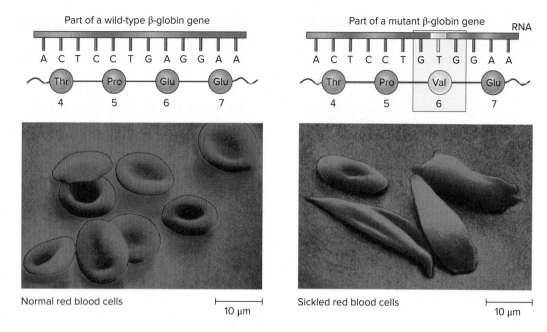

FIGURE 19.1 **Missense mutation in sickle cell disease.** Colorized electron micrographs of normal red blood cells (left) and sickled red blood cells (right). Above the micrographs is a comparison of the amino acid sequence of the normal β-globin polypeptide and the polypeptide coded by the sickle cell allele. This comparison shows only a portion of the polypeptide sequence, from amino acid 4 to amino acid 7. As seen here, a missense mutation changes the sixth amino acid from a glutamic acid to a valine.

Genes→Traits A missense mutation alters the structure of β globin, which is a subunit of hemoglobin, the oxygen-carrying protein in the red blood cells. When an individual is homozygous for the sickle cell allele, this missense mutation causes the red blood cells to sickle under conditions of low oxygen concentration. The sickling phenomenon is a description of the trait at the cellular level. At the organism level, the sickled cells can clog the capillaries, causing painful episodes that can result in organ damage. The shortened life span of the red blood cells leads to symptoms of anemia.

(a-b): Photo Researchers/Science History Images/Alamy Stock Photo

In Chapter 14, we considered how mutations can affect regulatory elements. For example, mutations in the *lac* operator site, called *lacO*^{C} mutations, prevent the binding of the protein called lac repressor. This causes the *lac* operon to be constitutively expressed even in the absence of lactose. Strains of *E. coli* with *lacO*^{C} mutations are at a selective disadvantage compared with wild-type *E. coli* because they waste their energy expressing the *lac* operon even when the proteins required for lactose metabolism are not needed.

As noted in Table 19.2, mutations can also occur in other non-coding regions of a gene and alter gene expression in a way that may affect phenotype. For example, in Chapter 13, we discussed how an iron response element (IRE) plays a role in regulating translation or causes the mRNA produced by translation to be degraded (refer back to Figure 13.20). Mutations in the 5′-UTR of the ferritin mRNA may alter the sequence of the IRE, thereby affecting the translation of the mRNA. In addition, mutations in eukaryotic genes can alter splice recognition sequences, thereby affecting the order or number of exons contained within an mRNA.

Gene Mutations Are Also Given Names That Describe How They Affect the Wild-Type Genotype and Phenotype

Thus far, several genetic terms have been introduced that describe the molecular effects of mutations. Gene mutations are also given names that describe their effects relative to a wild-type genotype or phenotype.

- In a natural population, a **wild type** is a relatively prevalent genotype. Many or most genes exist as multiple alleles, so a population may have two or more wild-type alleles.
- A mutation may change a wild-type genotype by altering the DNA sequence of a gene. When such a mutation is rare in a population, the result is generally referred to as a **mutant allele.**
- A reverse mutation, more commonly called a **reversion,** changes a mutant allele back to a wild-type allele.

TABLE 19.2

Possible Consequences of Gene Mutations Outside of a Coding Sequence

Sequence	Effect of Mutation
Promoter	May increase or decrease the rate of transcription
Enhancer/operator site	May disrupt the ability of the gene to be properly regulated.
5′-UTR/3′-UTR	May alter the ability of mRNA to be translated; may alter mRNA stability
Splice recognition sequence	May alter the ability of pre-mRNA to be properly spliced

Another way to describe a mutation is based on its influence on the wild-type phenotype. Mutants are often characterized by their differential ability to survive.

- A neutral mutation does not alter protein function, so it does not affect survival or reproductive success.
- A **deleterious mutation,** however, decreases the chances of survival and reproduction.
- The extreme example of a deleterious mutation is a **lethal mutation,** which results in the death of a cell or organism.
- On the other hand, a **beneficial mutation** enhances the survival or reproductive success of an organism.

In some cases, a mutated allele may be either deleterious or beneficial, depending on the genotype and/or the environmental conditions. An example is the sickle cell allele. In the homozygous state, the sickle cell allele lessens the chances of survival. However, an individual who is heterozygous for the sickle cell and wild-type alleles may have an increased chance of survival due to resistance to malarial infection.

Finally, some mutations result in **conditional mutants,** in which the phenotype is affected only under a defined set of conditions. Geneticists often study conditional mutants of microorganisms; a common example is a temperature-sensitive (*ts*) mutant. A bacterium that has a *ts* mutation grows normally in one temperature range—the permissive temperature range—but exhibits defective growth at a different temperature range—the nonpermissive range. For example, an *E. coli* strain carrying a *ts* mutation may be able to grow in the range 33°–38°C but not in the range 40°–42°C, whereas a wild-type strain can grow in either temperature range.

Suppressor Mutations Reverse the Phenotypic Effects of Another Mutation

A second mutation sometimes affects the phenotypic expression of a first mutation. As an example, let's consider a mutation that causes an organism to grow very slowly. A second mutation at another site in the organism's DNA may restore the normal growth rate, converting the mutant back to the wild-type phenotype. Geneticists call these second-site mutations **suppressors,** or **suppressor mutations.** The name indicates that this type of mutation acts to suppress the phenotypic effects of another mutation. A suppressor mutation differs from a reversion, because it occurs at a different site in the DNA from the first mutation. Suppressor mutations are classified according to their relative locations with regard to the mutation they suppress. **Table 19.3** provides a few examples.

Intragenic Suppressors.   When the second mutation site is within the same gene as the first, the mutation is termed an **intragenic suppressor.** This type of suppressor often produces a change in protein structure that compensates for an abnormality in protein structure caused by the first mutation.

Researchers often isolate suppressor mutations by setting up an experiment in which cells or organisms carrying the suppressor mutation are easy to identify. For example, Robert Brooker and colleagues have isolated many intragenic suppressors in the *lacY* gene of *E. coli,* which codes lactose permease, as described in Chapter 14. These researchers began with single mutations that altered amino acids and prevented lactose transport. When cells harboring these single mutations were placed on media in which lactose was the sole carbon source for growth, such mutants failed to grow. However, cells with rare suppressor mutations were easy to identify because they grew and formed visible colonies. By isolating many such cells and analyzing the locations of their suppressor mutations, the researchers determined that certain regions in the protein are critical because they undergo conformational changes that are required for lactose transport.

Intergenic Suppressors.   Alternatively, a suppressor mutation can occur in a different gene than the one containing the first mutation—in which case, it is called an **intergenic suppressor.** How do intergenic suppressors work? These suppressor mutations usually involve a change in the expression of one gene that compensates for a loss-of-function mutation affecting another gene (see Table 19.3).

- *Redundant function.* A first mutation may cause one protein to be partially or completely defective. An intergenic suppressor in a different protein-coding gene might overcome this defect by altering the structure of a second protein so that it can take over the functional role of the defective protein.
- *Common pathway.* Intergenic suppressors may affect proteins that participate in a common cellular pathway. When a first mutation affects the activity of a protein, a suppressor mutation could alter the function of a second protein involved in this pathway, thereby compensating for the defect in the first protein.
- *Protein-protein interactions.* In some cases, intergenic suppressors have effects on protein-protein interactions, in which one protein or protein subunit coded by one gene interacts with a protein or protein subunit coded by a different gene. A mutation in one gene that inhibits function may be compensated by a mutation in the other gene. In research studies, the identification of these types of suppressor mutations often reveals protein-protein interactions that were previously unknown.
- *Transcription factors.* Another type of intergenic suppressor involves mutations in genetic regulatory proteins such as transcription factors. When a first mutation causes a protein to be defective, a suppressor mutation may occur in a gene that codes a transcription factor. The mutant transcription factor transcriptionally activates another gene that can compensate for the loss-of-function mutation in the first gene.

Changes in Chromosome Structure Can Affect the Expression of a Gene

Thus far, we have considered small changes in the DNA sequences of particular genes. A change in chromosome structure can also be

TABLE 19.3

Examples of Suppressor Mutations

Type	No Mutation	First Mutation (x)	Second Mutation (x)	Description
Intragenic	Transport can occur	Transport inhibited	Transport can occur	A first mutation disrupts normal protein function, and a suppressor mutation affecting the same protein restores function. In this example, the first mutation inhibits transport function, and the second mutation restores it.
Intergenic Redundant function	Functional enzyme	Nonfunctional enzyme	Gain of a new functional enzyme	A first mutation inhibits the function of a protein, and a second mutation alters a different protein to carry out that function. In this example, the proteins function as enzymes.
Common pathway	Precursor, Fast → Intermediate, Slow → Product	Precursor, Slow → Intermediate, Slow → Little product	Precursor, Slow → Intermediate, Fast → Product	Two or more different proteins may function as enzymes in a common pathway. A mutation that causes a defect in one enzyme may be compensated for by a mutation that increases the function of a different enzyme in the same pathway.
Protein-protein interactions	Active	Inactive	Active	A mutation in a gene coding one protein or protein subunit that inhibits function may be suppressed by a mutation in a gene that codes a different protein or protein subunit. The function of the protein(s) in the double-mutant is restored.
Transcription factor	Normal function	Loss of function		A first mutation causes loss of function of a particular protein. A second mutation may alter a transcription factor and cause it to activate the expression of another gene. This other gene codes a protein that can compensate for the loss of function caused by the first mutation.

Mutant transcription factor turns on a gene that compensates for the loss of function.

Transcription factor

Compensates for inactive protein, so function is restored.

Causes expression of this protein.

associated with an alteration in the expression of single genes. Quite commonly, an inversion or translocation has no obvious phenotypic consequence. However, in 1925, Alfred Sturtevant was the first to recognize that chromosomal rearrangements in *Drosophila* can influence phenotypic expression (namely, eye morphology). In some cases, a chromosomal rearrangement may affect a gene because a chromosomal **breakpoint**—a region where two chromosome pieces break apart and rejoin with other chromosome pieces—occurs within a gene. A breakpoint in the middle of a gene is very likely to inhibit gene function because it separates the gene into two pieces.

In other cases, a gene may be left intact, but its expression may be altered when it is moved to a new chromosomal location. When this occurs, the change in gene location is said to have a **position effect.** How do position effects alter gene expression? Researchers have discovered two common explanations. **Figure 19.2** depicts a schematic example in which a piece of one chromosome has been inverted or translocated to a different chromosome.

- One possibility is that a gene may be moved next to regulatory sequences for a different gene, such as an enhancer, and the enhancer then influences the expression of the relocated gene (Figure 19.2a).
- Alternatively, a chromosomal rearrangement may reposition a gene from a less condensed, or euchromatic region of a chromosome, where it is active, to a very highly condensed, or heterochromatic region of a chromosome. When the gene is moved to a heterochromatic region, its expression may be turned off (Figure 19.2b). This second type of position effect may produce a variegated phenotype in which the expression of the gene is variable. For genes that affect pigmentation, for example, the phenotypic outcome may be a variegated appearance rather than an even color.

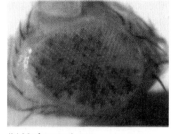

(a) Normal eye   (b) Variegated eye

FIGURE 19.3 **A position effect that alters eye color in *Drosophila*.** (a) A normal red eye. (b) An eye in which an eye color gene has been relocated to a heterochromatic region of a chromosome. This inactivates the gene in some cells and produces a variegated phenotype.
Genes→Traits Variegated eye color occurs because the degree of heterochromatin formation varies throughout different regions of the eye, which is composed of hundreds of facets. In some facets, heterochromatin formation occurs and turns off the eye color gene, thereby leading to the white phenotype. In other facets, the region containing the eye color allele remains euchromatic, yielding a red phenotype.
(a-b): ©Dr. Jack R. Girton

CONCEPT CHECK: Has the DNA sequence of the eye color gene been changed in part (b) compared with part (a)? How can the phenotypic difference be explained?

Figure 19.3 shows a position effect that alters eye color in *Drosophila*. Figure 19.3a shows a normal red eye of a fruit fly, and Figure 19.3b shows the eye of a mutant fly that has inherited a chromosomal rearrangement in which a gene affecting eye color has been relocated to a heterochromatic region of a chromosome. The variegated appearance of the eye occurs because the degree of heterochromatin formation varies across different regions of the eye. In cells where heterochromatin formation has turned off the eye color gene, a white phenotype occurs, but in

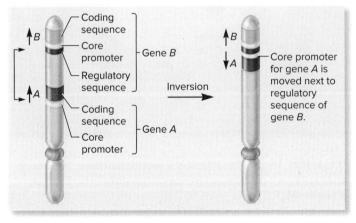

(a) Position effect due to regulatory sequences

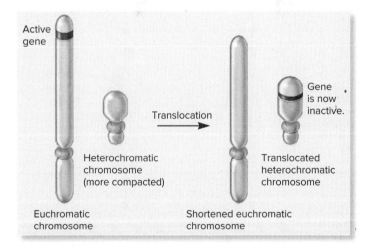

(b) Position effect due to translocation to a heterochromatic chromosome

FIGURE 19.2 **Causes of position effects.** (a) A chromosomal inversion has repositioned the core promoter of gene *A* next to the regulatory sequences for gene *B*. Because regulatory sequences are often bidirectional, the regulatory sequences for gene *B* may regulate the transcription of gene *A*. (b) A translocation has moved a gene from a euchromatic to a heterochromatic chromosome. This type of position effect inhibits the expression of the relocated gene.

CONCEPT CHECK: Explain what happens when a position effect occurs.

other cells, this region remains euchromatic and produces a red phenotype.

Mutations Can Occur in Germ-Line or Somatic Cells

In this section, we have considered many different ways that mutations affect gene expression. For multicellular organisms, the timing of mutations also plays an important role. A mutation can occur very early in life, such as in a gamete or a fertilized egg, or it may occur later in life, such as in the embryonic or adult stages. The exact time when a mutation occurs can be important with regard to the severity of the genetic effect and whether the mutation is passed from parent to offspring. Geneticists classify the cells of animals into two types: germ-line and somatic cells.

Germ-Line Mutations.
The term **germ line** refers to the lineage of cells that gives rise to gametes, such as eggs and sperm. A **germ-line mutation** can occur directly in a sperm or egg cell, or it can occur in a precursor cell that produces the gametes. If a mutant gamete participates in fertilization, all cells of the resulting offspring will contain the mutation (**Figure 19.4a**). Likewise,

when an individual with a germ-line mutation produces gametes, the mutation may be passed along to future generations of offspring.

Somatic Mutations.
The **somatic cells** comprise all cells of the body excluding germ-line cells and gametes. Examples include liver cells, neurons, and skin cells. Mutations can also happen within somatic cells at early or late stages of development. **Figure 19.4b** illustrates the consequences of a mutation that took place during the embryonic stage. In this example, a **somatic mutation** has occurred within a single embryonic cell. As the embryo grows, this single cell is the precursor for many cells of the adult organism. Therefore, in the adult, a portion of the body contains the mutation. The size of the affected region depends on the timing of the mutation. In general, the earlier the mutation occurs during development, the larger the affected region. An individual that has somatic regions that differ genotypically from each other is called a **genetic mosaic**. A somatic mutation cannot be passed from parent to offspring via sperm or egg cells.

Figure 19.5 is a photo of a sheep with a somatic mutation that occurred during an early stage of development. Presumably, a single mutation initially occurred in this individual in an embryonic cell that ultimately gave rise to a patch of epidermis that produced the black hair. Usually, the hair on the side of the body is entirely white in this breed of sheep.

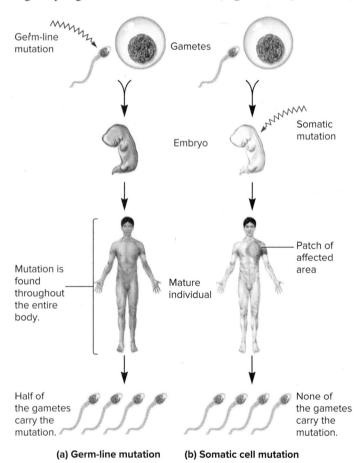

FIGURE 19.4 The effects of germ-line versus somatic mutations.

FIGURE 19.5 **The effects of a somatic mutation.** A mutation during embryonic development has caused this sheep to have a black spot on its side.

Genes→Traits This sheep has a patch of black hair on its side because a somatic mutation that occurred in a single cell during embryonic development caused pigmentation of the hair. This cell continued to divide, producing a patch of black hair.

©Robert Brooker

CONCEPT CHECK: Can this trait be passed to offspring?

Although a patch of black hair is not a harmful phenotypic effect, mutations during early stages of life can be quite detrimental, especially if they disrupt essential developmental processes. Therefore, even though it is smart to avoid environmental agents that cause mutations during all stages of life, the possibility of somatic mutations is a rather compelling reason to avoid them during the very early stages of life such as embryonic development, infancy, and early childhood. For example, the possibility of somatic mutations in an embryo is a reason why it is good to avoid exposure to X-rays during pregnancy.

19.1 COMPREHENSION QUESTIONS

1. A mutation changes a codon that specifies tyrosine into a stop codon. This type of mutation is a
   a. missense mutation.
   b. nonsense mutation.
   c. frameshift mutation.
   d. neutral mutation.

2. A down promoter mutation causes the promoter of a gene to be _____ like the consensus sequence and _____ transcription.
   a. less, stimulates
   b. more, stimulates
   c. less, inhibits
   d. more, inhibits

3. A mutation in one gene that reverses the phenotypic effects of a mutation in a different gene is
   a. an intergenic suppressor.
   b. an intragenic suppressor.
   c. an up promoter mutation.
   d. a position effect.

4. Which of the following is an example of a somatic mutation?
   a. A mutation in an embryonic skin cell
   b. A mutation in a sperm cell
   c. A mutation in an adult liver cell
   d. Both a and c are examples of somatic mutations.

19.2 RANDOM NATURE OF MUTATIONS

Learning Outcome:

1. Analyze the results obtained by the Lederbergs, and explain how they are consistent with the random mutation theory.

For a couple of centuries, biologists had questioned whether heritable changes occur as a result of behavior or exposure to particular environmental conditions or if they are events that happen randomly.

In the nineteenth century, naturalist Jean-Baptiste Lamarck proposed that physiological events—such as the use or disuse of muscles—determine whether traits are passed along to offspring. For example, Lamarck's hypothesis suggested that an individual who practiced and became adept at a physical activity and developed muscular legs would pass that characteristic on to the next generation.

The alternative point of view is that genetic variation exists in a population as a matter of random chance, and natural selection results in the differential reproductive success of organisms that are better adapted to their environments. Those individuals who, by chance, happen to have mutations that turn out to be beneficial will be more likely to survive and pass these genes to their offspring.

These opposing ideas of the nineteenth century—one termed *physiological adaptation* and the other termed *random mutation*—were tested in bacterial studies in the 1940s and 1950s. One of these studies is described next.

EXPERIMENT 19A

Replica-Plating Experiments of the Lederbergs Confirmed the Random Mutation Theory

To distinguish between the physiological adaptation and random mutation hypotheses, Joshua and Esther Lederberg developed a technique known as **replica plating** in the 1950s. As shown in **Figure 19.6**, they plated a large number of bacteria onto a master plate that did not contain any selective agent (namely, no T1 phage). The cells then divided many times to form bacterial colonies. Next, a sterile piece of velvet cloth was later lightly touched to this plate in order to pick up many bacterial cells from each colony. These replicas were then transferred to two secondary plates that contained an agent that selected for the growth of bacterial cells with a particular mutation. In the example shown in Figure 19.6, the secondary plates contained T1 bacteriophages, which lyse the wild-type cells.

THE HYPOTHESIS

Mutations are random events.

TESTING THE HYPOTHESIS   **FIGURE 19.6** **Evidence that mutations are random events.**

Starting material: The Lederbergs began with a wild-type strain of *E. coli*. They also had a preparation of T1 bacteriophage.

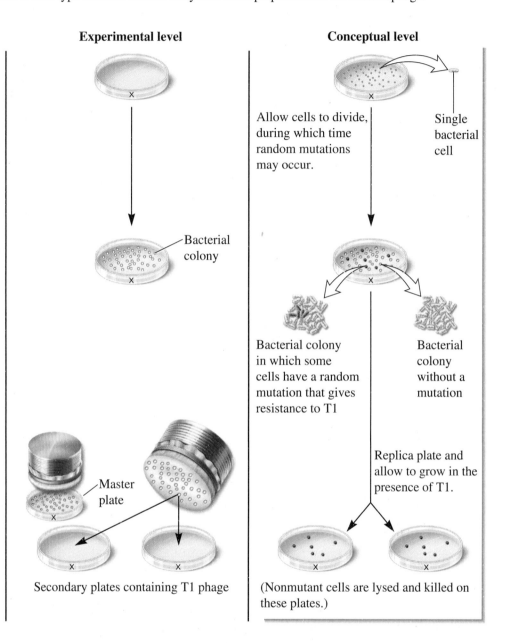

Experimental level

Conceptual level

1. Place individual bacterial cells onto growth media.

Allow cells to divide, during which time random mutations may occur.

Single bacterial cell

2. Incubate overnight to allow the formation of bacterial colonies. This is called the master plate.

Bacterial colony

Bacterial colony in which some cells have a random mutation that gives resistance to T1

Bacterial colony without a mutation

3. Press a velvet cloth (wrapped over a cylinder) onto the master plate, and then lift gently to obtain a replica of each bacterial colony. Press the replica onto 2 secondary plates that contain T1 phage. Incubate overnight to allow growth of mutant cells.

Master plate

Replica plate and allow to grow in the presence of T1.

Secondary plates containing T1 phage

(Nonmutant cells are lysed and killed on these plates.)

THE DATA

Colonies on each plate are in the same locations.

Note: The black × indicates the alignment of the velvet cloth with the plates.
Source: Lederberg, Joshua, and Lederberg, Esther M., "Replica Plating and Indirect Selection of Bacterial Mutants," *Journal of Bacteriology*, vol. 63, no. 3, March 1952, 399–406.

INTERPRETING THE DATA

On the secondary plates, only those mutant cells that are resistant to lysis by T1 phage (called *ton^r* mutants) could grow. A few colonies were observed. Strikingly, they occupied the same location on each plate. These results indicated that the *ton^r* mutations occurred randomly while the cells were growing on the nonselective master plate. The presence of the T1 phage in the secondary plates simply selected for the growth of previously occurring *ton^r* mutants. These results supported the random mutation hypothesis. In contrast, the physiological adaptation hypothesis would have predicted that *ton^r* mutants would occur after the exposure to the selective agent on the secondary plates. If that had been the case, the colonies would be expected to arise in different patterns on the two secondary plates.

The results of the Lederbergs supported the random mutation hypothesis, now known as the **random mutation theory.** According to this theory, mutation is a random process—it can occur in any gene and does not require the exposure of an organism to an environmental condition that causes specific types of mutations to happen. For example, exposure to T1 phage does not cause *ton^r* mutations. In some cases, a random mutation may provide a mutant organism with an advantage, such as resistance to T1 phage. Although such mutations occur as a matter of random chance, growth conditions may select for organisms that happen to carry them.

As researchers have learned more about mutation at the molecular level, the view that mutations are a totally random process has required some modification. Within the same individual, some genes mutate at a much higher rate than other genes. Why does this happen? Some genes are larger than others, which provides a greater chance for mutation. Also, the relative locations of genes within a chromosome may cause some genes to be more susceptible to mutation than others. Even within a single gene, **hot spots** are usually found—certain regions of a gene that are more likely to mutate than other regions.

19.2 COMPREHENSION QUESTION

1. In the replica-plating experiments of the Lederbergs, bacterial colonies appeared at the same locations on each of two secondary plates because
   a. T1 phage caused the mutations to happen.
   b. the mutations occurred on the master plate prior to T1 exposure and prior to replica plating.
   c. Both a and b are true.
   d. Neither a nor b is true.

19.3 SPONTANEOUS MUTATIONS

Learning Outcomes:

1. Distinguish between spontaneous and induced mutations.
2. List examples of spontaneous mutations.
3. Outline how mutations arise by depurination, deamination, and tautomeric shifts.
4. Describe how reactive oxygen species alter DNA structure and cause mutations.
5. Explain the mechanism of trinucleotide repeat expansion.

Mutations can have a wide variety of effects on the phenotypic expression of genes. For this reason, geneticists have expended a great deal of effort identifying the causes of mutations. This task has been truly challenging, because a staggering number of agents can alter the structure of DNA, thereby causing mutation. Geneticists categorize the causes of mutations in one of two ways: **Spontaneous mutations** are changes in DNA structure that result from natural biological or chemical processes, whereas **induced mutations** are caused by environmental agents (**Table 19.4**). In this section, we will focus on spontaneous mutations; the following section will consider induced mutations.

Many causes of spontaneous mutations are examined in other chapters throughout this textbook.

- As discussed in Chapter 8, abnormalities in crossing over can produce mutations such as deletions, duplications, translocations, and inversions. In addition, abnormal

TABLE 19.4

Causes of Mutations

Common Causes of Mutations	Description
Spontaneous	
Aberrant recombination	Abnormal crossing over may cause deletions, duplications, translocations, and inversions (see Chapter 8).
Aberrant segregation	Abnormal chromosomal segregation may cause aneuploidy or polyploidy (see Chapter 8).
Errors in DNA replication	A mistake by DNA polymerase may cause a base-pair mismatch (see Chapter 11).
Toxic metabolic products	The products of normal metabolic processes, such as reactive oxygen species, may be chemically reactive agents that can alter the structure of DNA.
Transposable elements	Transposable elements can insert themselves into the sequence of a gene (see Chapter 10).
Depurination	On rare occasions, the linkage between a purine (i.e., adenine or guanine) and deoxyribose can spontaneously break. If not repaired, this can lead to mutation.
Deamination	Cytosine or 5-methylcytosine can spontaneously deaminate to create uracil or thymine.
Tautomeric shifts	Spontaneous changes in base structure can cause mutations if they occur immediately prior to DNA replication.
Induced	
Chemical agents	Chemical substances may cause changes in the structure of DNA.
Physical agents	Physical phenomena such as UV light and X-rays can damage DNA.

chromosomal segregation during meiosis can cause changes in chromosome number.

- As discussed in Chapter 10, transposable elements can alter gene sequences by inserting themselves into genes.
- In Chapter 11, we examined how DNA polymerase can make a mistake during DNA replication and put the wrong base in a newly synthesized daughter strand. Errors in DNA replication are usually infrequent except in certain viruses, such as HIV, that have relatively high rates of spontaneous mutations.

In this section, we will explore additional mechanisms, such as depurination, deamination, and tautomeric shifts, by which some spontaneous mutations occur. We will also consider how oxidative stress and trinucleotide repeat expansions cause mutations.

Spontaneous Mutations Can Arise by Depurination, Deamination, or Tautomeric Shifts

How can molecular changes in DNA structure cause mutation? Our first examples concern changes that can occur spontaneously, albeit at a low rate.

Depurination. The most common type of chemical change that occurs naturally in DNA is **depurination,** which involves the removal of a purine (adenine or guanine). The covalent bond between deoxyribose and a purine base is somewhat unstable and occasionally undergoes a spontaneous reaction with water that releases the base from the sugar, thereby creating an **apurinic site** (**Figure 19.7a**). In a typical mammalian cell, approximately 10,000 purines are lost from the DNA in a 24-hour period at 37°C. The rate of loss is higher if the DNA is exposed to agents that cause certain types of base modifications such as the attachment of alkyl groups (methyl or ethyl groups). Fortunately, as discussed later in this chapter, apurinic sites are recognized by DNA repair enzymes, which can fix the problem. If the repair

system fails, however, a mutation may result during subsequent rounds of DNA replication. What happens at an apurinic site during DNA replication? Because a complementary base is not present to specify the incoming base for the new strand, any of the four bases are added to the new strand in the region that is opposite the apurinic site (**Figure 19.7b**). This may produce a new mutation.

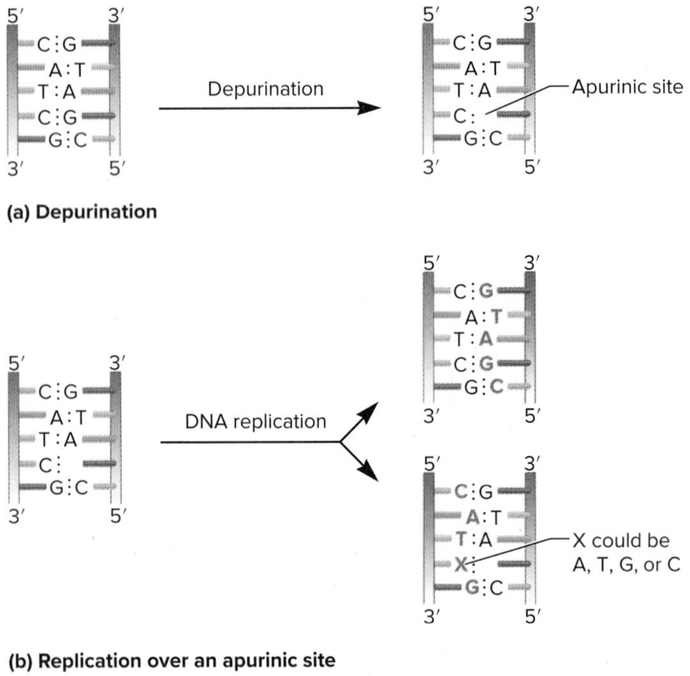

Deamination. A second spontaneous chemical change that may occur in DNA is the **deamination** of a cytosine base. The other bases are not readily deaminated. As shown in **Figure 19.8a**, deamination removes an amino group from the cytosine base. This produces uracil. As discussed later, DNA repair enzymes can recognize uracil as an inappropriate base within DNA and subsequently remove it. However, if such a repair does not take place, a mutation may result because uracil hydrogen bonds with adenine during DNA replication. Therefore, if a DNA template strand has uracil instead of cytosine, a newly made strand will incorporate adenine instead of guanine.

Figure 19.8b shows the deamination of 5-methylcytosine. As discussed in Chapter 15, the methylation of cytosine occurs in many eukaryotic species, forming 5-methylcytosine. This methylation also occurs in prokaryotes. If 5-methylcytosine is deaminated, the resulting base is thymine, which is a normal constituent of DNA. This result poses a problem for DNA repair because DNA repair systems cannot distinguish which is the incorrect base—the thymine that was produced by deamination or the guanine in the opposite strand that originally base-paired with the methylated cytosine. For this reason, methylated cytosine bases tend to produce hotspots for mutation. As an example,

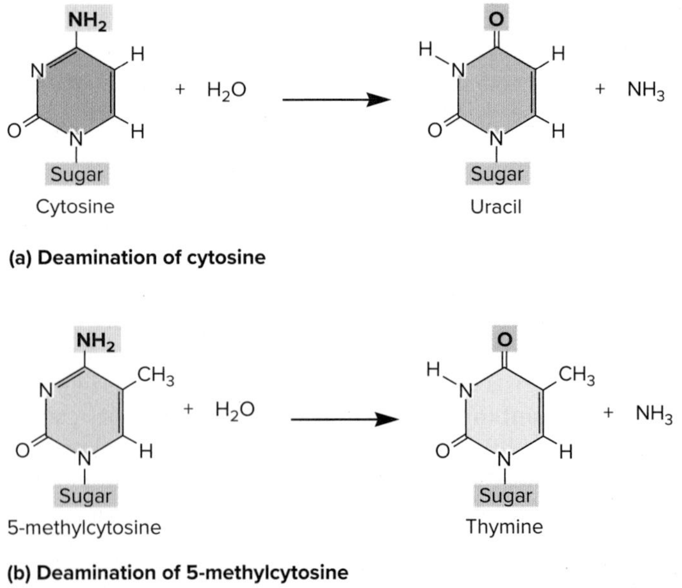

(a) Depurination

(b) Replication over an apurinic site

FIGURE 19.7 Spontaneous depurination. (a) The bond between guanine and deoxyribose is broken, thereby releasing the base. This leaves an apurinic site in the DNA. (b) If an apurinic site remains in the DNA as it is being replicated, any of the four nucleotides can be added to the newly made strand.

CONCEPT CHECK: When DNA that has an apurinic site is replicated, what is the probability that a mutation will occur?

(a) Deamination of cytosine

(b) Deamination of 5-methylcytosine

FIGURE 19.8 Spontaneous deamination of cytosine and 5-methylcytosine. (a) The deamination of cytosine produces uracil. (b) The deamination of 5-methylcytosine produces thymine.

CONCEPT CHECK: Which of these two changes is more difficult for DNA repair enzymes to fix correctly? Explain why.

researchers analyzed 55 spontaneous mutations that occurred within the *lacI* gene of *E. coli* and determined that 44 of them involved changes at sites that were originally occupied by a methylated cytosine base.

Tautomeric Shifts. A third way that mutations may arise spontaneously involves a temporary change in base structure called a **tautomeric shift.** In this case, the **tautomers** are bases, which exist in keto and enol forms or amino and imino forms. The two forms in each of these pairs can interconvert by a chemical reaction that involves the migration of a hydrogen atom and a switch of a single bond and an adjacent double bond. The common, stable form of guanine and thymine is the keto form; the common form of adenine and cytosine is the amino form (**Figure 19.9a**). At a low rate, G or T can convert to an enol form, and A or C can change to an imino form. Though the amounts of the enol and imino forms of these bases are relatively small, their presence can cause a mutation because these rare forms do not conform to the AT/GC rule of base pairing. Instead, if one of the bases is in the enol or imino form, hydrogen bonding will promote T-G and C-A base pairs, as shown in **Figure 19.9b.**

How does a tautomeric shift cause a mutation? The answer is that the shift must occur immediately prior to DNA replication. When DNA is double-stranded, the base pairing usually holds the bases in their more stable forms. After the strands unwind, however, a tautomeric shift may occur. In the example shown in **Figure 19.9c**, a thymine base in the template strand has undergone a tautomeric shift just prior to the replication of the complementary daughter strand. During replication, the daughter strand incorporates a guanine opposite this thymine, creating a base mismatch. This mismatch could be repaired via the proofreading function of DNA polymerase or via a mismatch repair system (discussed later in this chapter). However, if these repair mechanisms do not occur, the next round of DNA replication will produce a double helix with a C-G base pair, whereas the correct base pair should be T-A. As shown in the right side of Figure 19.9c, one of four daughter cells inherits this C-G mutation.

Oxidative Stress May Also Lead to DNA Damage and Mutation

Aerobic organisms use oxygen as a terminal acceptor in their electron transport chains. **Reactive oxygen species (ROS),** such as hydrogen peroxide, superoxide, and hydroxyl radical, are products of oxygen metabolism in all aerobic organisms. In eukaryotes, ROS are naturally produced as unwanted by-products of energy production in mitochondria. They may also be produced during certain types of immune responses and by a variety of detoxification reactions in the cell. If ROS accumulate, they can damage cellular molecules, including DNA, proteins, and lipids. To prevent this from happening, cells use a variety of enzymes, such as superoxide dismutase and catalase, to prevent the buildup of ROS.

Small molecules, such as vitamin C, may act as antioxidants. Also, certain foods contain other chemicals that act as antioxidants. Colorful fruits and vegetables, including grapes, blueberries, cranberries, citrus fruits, spinach, broccoli, beets,

beans, red peppers, carrots, and strawberries, are usually high in antioxidants.

In humans, the overaccumulation of ROS has been implicated in a wide variety of medical conditions, including cardiovascular disease, Alzheimer's disease, chronic fatigue syndrome, and aging. However, the production of ROS is not always harmful. ROS are produced by the immune system as a means of killing pathogens. In addition, some ROS are used in cell signaling.

Oxidative stress refers to an imbalance between the production of ROS and an organism's ability to break them down. If ROS overaccumulate, one particularly harmful consequence is **oxidative DNA damage,** which refers to changes in DNA structure that are caused by ROS. DNA bases are very susceptible to oxidation. Guanine bases are particularly vulnerable to oxidation, which can lead to several different oxidized products. The most thoroughly studied guanine oxidation product is 7,8-dihydro-8-oxoguanine, which is commonly known as 8-oxoguanine (8-oxoG) (**Figure 19.10**). Researchers often measure the amount of 8-oxoG in a sample of DNA to determine the extent of oxidative stress. Why are oxidized bases harmful? In the case of 8-oxoG, it base-pairs with adenine during DNA replication, causing mutations in which a G-C base pair becomes a T-A base pair. This is a transversion mutation.

Although oxidative DNA damage can occur spontaneously, it also results from environmental agents, such as ultraviolet light, X-rays, and many chemicals, including those found in cigarette smoke. Later in this chapter, we will discuss such environment agents and their abilities to cause mutation.

DNA Sequences Known as Trinucleotide Repeats Are Hot Spots for Mutation

Researchers have discovered that several human genetic diseases are caused by an unusual form of mutation known as **trinucleotide repeat expansion (TNRE).** The term refers to the phenomenon in which a repeated sequence of three nucleotides can readily increase in number from one generation to the next. In humans and other species, certain genes and chromosomal locations contain regions where trinucleotide sequences are repeated in tandem. These sequences are usually transmitted normally from parent to offspring without mutation. However, in persons with TNRE disorders, the length of a trinucleotide repeat has increased above a critical size, thus causing disease symptoms.

Table 19.5 describes several human diseases that involve these types of expansions: spinal and bulbar muscular atrophy (SBMA), Huntington disease (HD), spinocerebellar ataxia (SCA1), fragile X syndromes (FRAXA and FRAXE), and myotonic muscular dystrophy (also called dystrophia myotonica, DM). In some cases, the expansion is within the coding sequence of a gene. Typically, such an expansion involves CAG repeats. Because CAG codes glutamine, these repeats cause the coded proteins to contain long tracts of glutamine. The presence of glutamine tracts causes the proteins to aggregate. This aggregation of proteins or protein fragments carrying glutamine repeats is correlated with the progression of the disease. However, recent evidence suggests that the aggregated proteins may not directly cause the disease symptoms.

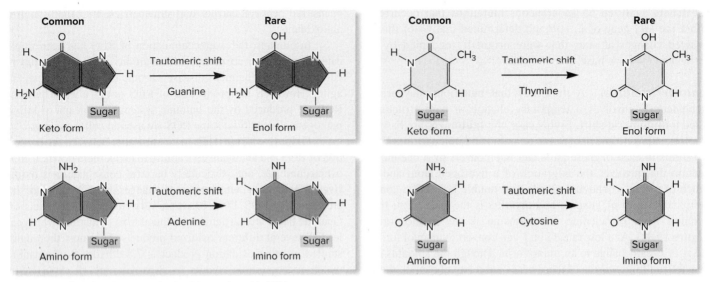

(a) Tautomeric shifts that occur in the 4 bases found in DNA

(b) Mistakes in base pairing due to tautomeric shifts

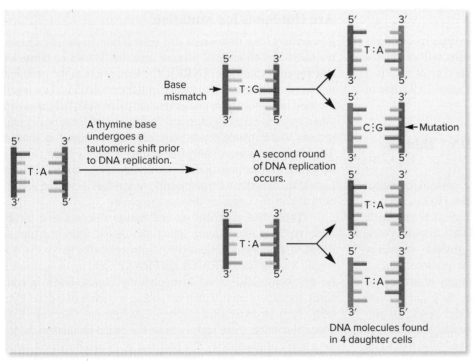

(c) A tautomeric shift just prior to DNA replication, producing a mutation

FIGURE 19.9 **Tautomeric shifts and their ability to cause mutation.** **(a)** The common forms of the bases are shown on the left, and the rare forms produced by tautomeric shifts are shown on the right. **(b)** On the left, the rare enol form of thymine pairs with the common keto form of guanine (instead of pairing with adenine); on the right, the rare imino form of cytosine pairs with the common amino form of adenine (instead of guanine). **(c)** A tautomeric shift occurred in a thymine base just prior to replication, causing the formation of a T-G base pair. If this mismatch is not repaired, a second round of replication will lead to the formation of a permanent C-G mutation. Note: A tautomeric shift is a very temporary condition. During the second round of replication, the thymine base that shifted prior to the first round of DNA replication is likely to have shifted back to its normal form. Therefore, during the second round of replication, an adenine base is found opposite this thymine.

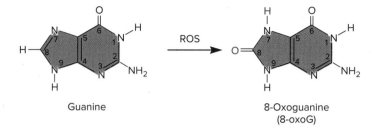

Guanine

8-Oxoguanine
(8-oxoG)

FIGURE 19.10 **Oxidation of guanine to 8-oxoguanine by a reactive oxygen species (ROS).**

CONCEPT CHECK: What is a reactive oxygen species?

In other TNRE disorders, the expansions are located in the non-coding regions of genes. In the case of the two fragile X syndromes, the expansions produce CpG islands that become methylated. As discussed in Chapter 15, DNA methylation can silence gene transcription. For DM, it has been hypothesized that the trinucleotide repeat expansions cause abnormal changes in RNA structure, which produce disease symptoms.

Some TNRE disorders have the unusual feature of worsening in severity in subsequent generations—a phenomenon called **anticipation.** In the example depicted here, the repeat of the triplet CAG has expanded from 11 tandem copies to 18, from one generation to the next:

CAGCAGCAGCAGCAGCAGCAGCAGCAGCAGCAG $n = 11$

to

CAGCAGCAGCAGCAGCAGCAGCAGCAGCAGCAGCAG
CAGCAGCAGCAGCAGCAG $n = 18$

However, anticipation does not frequently occur with all TNRE disorders and usually depends on whether the disease is inherited from the mother or father. In the case of HD, anticipation is likely to occur if the mutant gene is inherited from the male parent. In contrast, DM is more likely to get worse if the gene is inherited from the female parent. These results suggest that TNRE happens more frequently during oogenesis or spermatogenesis, depending on the particular gene involved. The phenomenon of anticipation makes it particularly difficult for genetic counselors to advise couples about the likely severity of these diseases if they are passed to their children.

How does TNRE occur? Though it may occur in more than one way, researchers have determined that a key aspect of TNRE is that the triplet repeat can form a hairpin, also called a stem-loop. A consistent feature of the triplet sequences associated with TNRE is they contain at least one C and one G (see Table 19.5). As shown in **Figure 19.11a,** such a sequence can form a hairpin due to the formation of C-G base pairs.

The formation of a hairpin during DNA replication can lead to an increase in the length of a DNA region if it occurs in the newly made daughter strand (**Figure 19.11b**).

1. DNA replication proceeds just past the TNRE.
2. When the hairpin forms, DNA polymerase temporarily slips off the template strand.
3. Next, DNA polymerase essentially backs up and hops back onto the template strand and resumes DNA replication from the end of the hairpin. When this occurs, DNA polymerase is synthesizing most of the hairpin region twice.
4. Hairpins within DNA are short-lived. The hairpin can spread out, which leaves a gap in the opposite strand.
5. This gap is later filled in by DNA polymerase and ligase as shown by the black portion of the DNA strand in Figure 19.11b. The end result is that the TNRE has become longer.

When the trinucleotide repeat sequence is abnormally long, such expansions may frequently occur during gamete formation, and therefore offspring in successive generations may have trinucleotide repeat sequences that are even longer than those in their parents.

TABLE **19.5**						
TNRE Disorders						
Disease	Spinal and bulbar muscular atrophy	Huntington disease	Spino-cerebellar ataxia	Fragile X syndrome-Type A	Fragile X syndrome-Type E	Myotonic muscular dystrophy
Repeated Triplet	CAG	CAG	CAG	CGG	GCC	CTG
Location of Repeat	Coding sequence	Coding sequence	Coding sequence	5′-UTR	5′-UTR	3′-UTR
Number of Repeats in Unaffected Individuals	11–33	6–37	6–44	6–53	6–35	5–37
Number of Repeats in Affected Individuals	36–62	27–121	43–81	>200	>200	>200
Pattern of Inheritance	X-linked	Autosomal dominant	Autosomal dominant	X-linked	X-linked	Autosomal dominant
Disease Symptoms	Neuro-degenerative	Neuro-degenerative	Neuro-degenerative	Mental impairment	Mental impairment	Muscle disease
Anticipation[†]	None	Male	Male	Female	None	Female

†Indicates the sex of the parent from whom the disease is usually inherited when anticipation occurs.

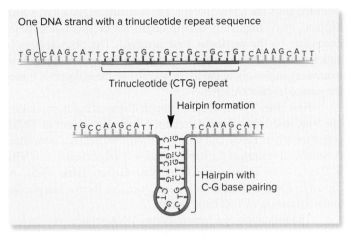

(a) Formation of a hairpin with a trinucleotide (CTG) repeat sequence

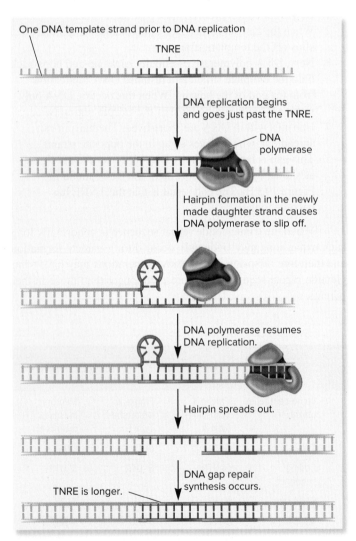

(b) Mechanism of trinucleotide repeat expansion

FIGURE 19.11   Proposed mechanism of trinucleotide repeat expansion (TNRE). **(a)** Trinucleotide repeats can form hairpin structures due to C-G base pairing. **(b)** Formation of a trinucleotide repeat expansion.

19.3 COMPREHENSION QUESTIONS

1. Which of the following is *not* an example of a spontaneous mutation?
   a. A mutation caused by an error in DNA replication
   b. A mutation caused by a tautomeric shift
   c. A mutation caused by UV light
   d. All of the above are spontaneous mutations.
2. A point mutation could be caused by
   a. depurination.
   b. deamination.
   c. tautomeric shift.
   d. any of the above.
3. One way that TNRE may occur involves the formation of _____ that disrupts _____.
   a. a double-strand break, chromosome segregation
   b. an apurinic site, DNA replication
   c. a hairpin, DNA replication
   d. a free radical, DNA structure

19.4   INDUCED MUTATIONS

Learning Outcomes:

1. Define *mutagen.*
2. Distinguish between chemical and physical mutagens, and provide examples.
3. Define *mutation rate.*
4. Analyze the results of an Ames test.

As we have seen, spontaneous mutations occur in a wide variety of ways. They result from natural biological and chemical processes. In this section, we turn our attention to induced mutations, which are caused by environmental agents. These agents can be either chemical or physical. They enter cells and lead to changes in DNA structure. Agents known to alter the structure of DNA and thereby cause mutations are called **mutagens.** In this section, we will explore several mechanisms by which mutagens alter the structure of DNA. We will also consider laboratory tests that can identify potential mutagens.

Mutagens Alter DNA Structure in Different Ways

Over the past few decades, researchers have found that an enormous array of agents act as mutagens that permanently alter the structure of DNA. We often hear in the news media that we should avoid these agents in our food and living environment. For example, we use sunscreens to help us avoid the mutagenic effects of ultraviolet (UV) light. The public is concerned about mutagens for two important reasons. First, mutagenic agents are often involved in the development of human cancers. In addition, because new mutations may be harmful, people want to avoid mutagens to prevent gene mutations that may have detrimental effects on their future offspring.

Mutagenic agents are usually classified as chemical or physical mutagens. Examples of both types of agents are listed in **Table 19.6.**

TABLE 19.6
Examples of Mutagens

Mutagen	Effect(s) on DNA Structure
Chemical	
Nitrous acid	Deaminates bases
Nitrogen mustard	Alkylates bases
Ethyl methanesulfonate	Alkylates bases
Proflavin	Intercalates within DNA helix
5-Bromouracil	Functions as a base analog
2-Aminopurine	Functions as a base analog
Physical	
X-rays, γ rays	Cause base deletions, single-stranded and double-strand breaks in the DNA backbone, crosslinking, and the oxidation of bases; can penetrate deeply into biological materials.
UV light	Promotes the formation of pyrimidine dimers, such as thymine dimers, and causes the oxidation of bases; usually affects the surface of biological materials.
β particles	Cause base deletions, single-stranded and double-strand breaks in the DNA backbone, crosslinking, and the oxidation of bases; usually affects the surface of biological materials.

In some cases, chemicals that are not mutagenic can be altered to a mutagenically active form after being ingested. Cellular enzymes such as oxidases have been shown to activate some mutagens.

Base Modifications. Some mutagens act by covalently modifying the structure of bases. For example, **nitrous acid** (HNO_2) replaces amino groups with keto groups (changing $=NH_2$ to $=O$), a process called deamination. Deamination changes cytosine to uracil and adenine to hypoxanthine. When the altered DNA replicates, mutations can occur because these modified bases pair differently than the original bases. For example, uracil pairs with adenine, and hypoxanthine pairs with cytosine (**Figure 19.12**).

Other chemical mutagens also disrupt the appropriate pairing between nucleotides by alkylating bases within the DNA. During alkylation, methyl or ethyl groups are covalently attached to the bases. Examples of alkylating agents include **nitrogen mustard** (a type of mustard gas) and **ethyl methanesulfonate (EMS).** Mustard gas was used as a chemical weapon during World War I. Such agents severely damage the skin, eyes, mucous membranes, lungs, and blood-forming organs.

Intercalating Agents. Some mutagens exert their effects by directly interfering with the DNA replication process. For example, **acridine dyes,** such as **proflavin,** contain flat structures that intercalate, or insert, themselves between adjacent base pairs, thereby distorting the helical structure. When DNA containing these mutagens is replicated, single-nucleotide additions and/or deletions can occur in the newly made daughter strands, causing frameshift mutations.

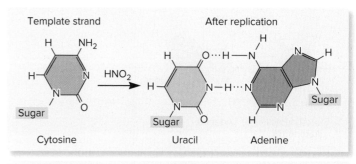

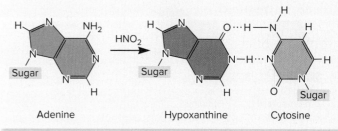

FIGURE 19.12 **Base pairing of modified bases that have been deaminated by nitrous acid.** Nitrous acid converts cytosine to uracil, and adenine to hypoxanthine. During DNA replication, uracil pairs with adenine, and hypoxanthine pairs with cytosine. This incorrect pairing creates mutations in the newly replicated strand during DNA replication.

Base Analogs. **5-Bromouracil (5BU)** and **2-aminopurine** are base analogs that become incorporated into daughter strands during DNA replication. 5BU is a thymine analog that can be incorporated into DNA instead of thymine. Like thymine, 5BU base-pairs with adenine. However, at a relatively high rate, 5BU undergoes a tautomeric shift and base-pairs with guanine (**Figure 19.13a**). When this occurs during DNA replication, 5BU causes a mutation in which an A-T base pair is changed to a G-5BU base pair (**Figure 19.13b**). This mutation is a transition, because the adenine has been changed to a guanine, and both of these bases are purines. During the next round of DNA replication, the template strand containing the guanine base produces a G-C base pair. In this way, 5BU promotes a change of an A-T base pair into a G-C base pair.

Ionizing Radiation. DNA molecules are also sensitive to physical agents such as radiation. In particular, radiation of short wavelength and high energy, known as ionizing radiation, can alter DNA structure. This type of radiation includes X-rays and γ (gamma) rays. Ionizing radiation can penetrate deeply into biological tissues, where it produces chemically reactive molecules known as free radicals. These molecules alter the structure of DNA in a variety of ways. Exposure to high doses of ionizing radiation results in base deletions, single-strand and double-strand breaks in the DNA backbone, crosslinking, and the oxidation of bases.

Non-ionizing Radiation. Non-ionizing radiation, such as UV light, has less energy, and so it penetrates only the surface of an organism, such as the skin. Nevertheless, UV light is known to cause DNA mutations. As shown in **Figure 19.14**, the energy in

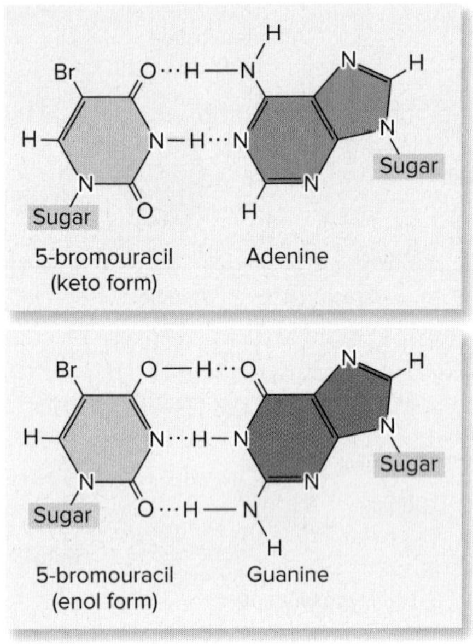

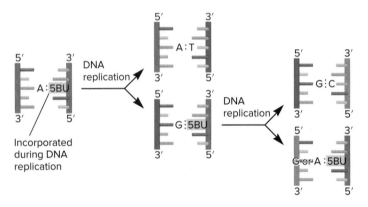

(a) Base pairing of 5BU (a thymine analog) with adenine or guanine

(b) How 5BU causes a mutation in a base pair during DNA replication

FIGURE 19.13 **Base pairing of 5-bromouracil and its ability to cause mutation.** **(a)** 5BU is a thymine analog. In its keto form, 5BU bonds with adenine; in its enol form, it bonds with guanine. **(b)** During DNA replication, guanine may be incorporated into a daughter strand by pairing with 5BU. After a second round of replication, the DNA contains a G-C base pair instead of the original A-T base pair.

CONCEPT CHECK: Does 5-bromouracil cause a transition or a transversion?

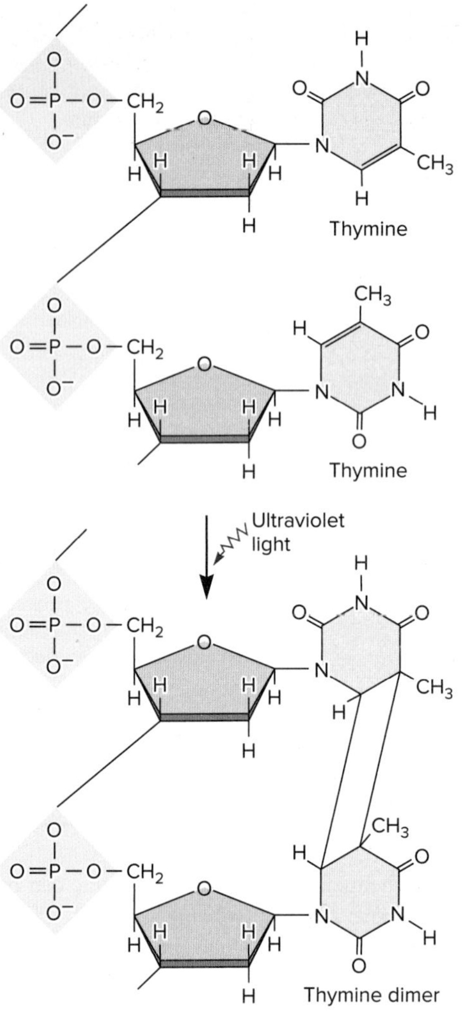

UV light causes the formation of **thymine dimers,** which are adjacent thymine bases that have become covalently linked.

Thymine dimers do not base-pair properly during DNA replication, and therefore can produce a mutation when the DNA strand is replicated. Plants, in particular, must have effective ways of preventing UV damage because they are exposed to sunlight throughout the day. Tanning greatly increases a person's exposure to UV light, raising the potential for thymine dimers and mutation. This explains the higher incidence of skin cancer among people who have been exposed to large amounts of sunlight during their lifetime. Because of the known link between skin cancer and sun exposure, people now apply sunscreen to their skin to prevent the harmful effects of UV light. Most sunscreens contain organic

FIGURE 19.14 **Formation and structure of a thymine dimer.**

CONCEPT CHECK: What is a common cause of thymine dimer formation in people, and in what cell type(s) is it most likely to occur?

compounds, such as oxybenzone, which absorb UV light, and/or opaque ingredients, such as zinc oxide, that reflect UV light.

The Mutation Rate Is the Likelihood of a New Mutation

Because mutations occur spontaneously and may be induced by environmental agents, geneticists are greatly interested in learning how prevalent they are. The **mutation rate** is the likelihood that a gene will be altered by a new mutation. This rate is commonly expressed as the number of new mutations in a given gene per cell generation. The spontaneous mutation rate for a particular gene is typically in the range from 1 in 100,000 to 1 in 1 billion, or 10^{-5}–10^{-9} per cell generation. In addition, mutations can occur at other sites in a genome, not only in genes. For example, people usually carry about 100–200 new mutations in their entire genome that were not present in their parents. Most of these are single-nucleotide changes that do not occur within the coding sequences of genes. Given the human genome size of approximately 3,200,000,000 bp, these numbers tell us that a mutation is a relatively infrequent event.

However, the mutation rate is not a constant number. The presence of certain environmental agents, such as X-rays, can increase the rate of induced mutations to a much higher value than the spontaneous mutation rate. In addition, mutation rates vary substantially from species to species and even within different strains of the same species. One explanation for this variation is that there are many different causes of mutations (refer back to Table 19.4).

The rate of new mutation is different from the concept of mutation frequency. The **mutation frequency** for a gene is the number of mutant genes divided by the total number of copies of that gene within a population. If 1 million bacteria were plated and 100 were found to carry a mutation in a particular gene, the mutation frequency for that gene would be 1 in 10,000, or 10^{-4}. The mutation frequency is an important genetic concept, particularly in the field of population genetics. As we will see in Chapter 27, the mutation frequency may rise above the mutation rate due to evolutionary factors such as natural selection and genetic drift.

Testing Methods Can Determine If an Agent Is a Mutagen

To determine if an agent is mutagenic, researchers use testing methods that monitor whether or not the agent increases the mutation rate. One common method is the **Ames test,** which was developed by Bruce Ames. This test uses strains of a bacterium, *Salmonella typhimurium,* that cannot synthesize the amino acid histidine. These strains contain a point mutation within a gene that codes an enzyme required for histidine biosynthesis. The mutation renders the enzyme inactive. Therefore, the bacteria cannot grow on petri plates unless histidine has been added to the growth medium. However, a second mutation—a reversion—may occur that restores the ability to synthesize histidine. In other words, a second mutation can cause a reversion back to the wild-type condition. The Ames test monitors the rate at which this second mutation occurs, thereby indicating whether an agent increases the mutation rate above the spontaneous rate.

Figure 19.15 outlines the steps in the Ames test.

1. The suspected mutagen is mixed with a rat liver extract and a strain of *S. typhimurium* that cannot synthesize histidine. A mutagen may require activation by cellular enzymes, which are provided by the rat liver extract. This step improves the ability of the test to identify agents that may cause mutation in mammals.
2. After the incubation period, a large number of bacteria are then plated on a growth medium that does not contain histidine.
3. The bacterial cells of this *Salmonella* strain are not expected to grow on these plates. However, if a mutation has occurred that allows a bacterial cell to synthesize histidine, such a cell can divide many times and form a visible bacterial colony composed of millions of cells.

To estimate the mutation rate, the colonies that grow on the media are counted and compared with the total number of bacterial cells that were originally streaked on the plate. For example, if 10,000,000 bacterial cells were originally added to each test tube and 10 growing colonies were observed after the cells were placed on the

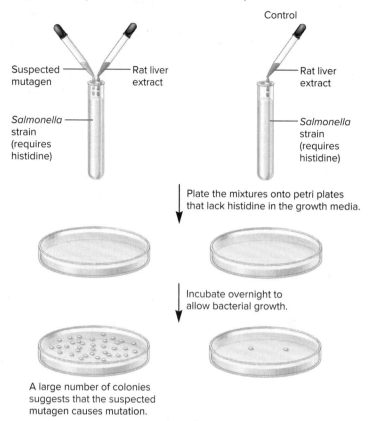

Mix together the suspected mutagen, a rat liver extract, and many bacterial cells from a *Salmonella* strain that cannot synthesize histidine. The suspected mutagen is omitted from the control sample.

Plate the mixtures onto petri plates that lack histidine in the growth media.

Incubate overnight to allow bacterial growth.

A large number of colonies suggests that the suspected mutagen causes mutation.

FIGURE 19.15 The Ames test for mutagenicity.

CONCEPT CHECK: What is the purpose of the rat liver extract in this procedure?

growth media without histidine, the rate of mutation would be 10 out of 10,000,000. This equals 1 in 10^6, or simply 10^{-6}. As a control, bacteria that have not been exposed to the mutagen are also tested, because a low level of spontaneous mutation is expected to occur.

How do we judge if an agent is a mutagen? Researchers compare the mutation rate in the presence and absence of the suspected mutagen. The test shown in Figure 19.15 is conducted several times. If statistic analysis reveals that the mutation rate in the experimental and control samples are significantly different, researchers may tentatively conclude that the agent may be a mutagen. Many studies have been conducted in which researchers used the Ames test to compare the urine from cigarette smokers to that from nonsmokers. This research has shown that the urine from smokers contains much higher levels of mutagens.

Many different mutagen tests have been developed, each with its own advantages and disadvantages. An advantage of the Ames test is its relative simplicity. However, researchers have pointed out particular limitations. For example, the Ames test uses mutant strains of bacteria, which are prokaryotic cells, and therefore not a perfect model for eukaryotic mammalian cells. Also, a rat liver extract is used in conjunction with the Ames test. Because differences exist in metabolic pathways between rats and humans, the mutagenic effect of any given substance may be significantly different in humans.

GENETIC TIPS   THE QUESTION: A researcher studied the effects of a suspected mutagen, called mutagen X, using the procedure for an Ames test shown in Figure 19.15. The following data were obtained after placing 2 million cells on each plate:

Trial	Control (number of colonies)	Plus mutagen (number of colonies)
1	3	62
2	2	77
3	5	46
4	2	55

Calculate the average mutation rate in the presence and absence of mutagen X. Conduct a *t*-test to determine if suspected mutagen X is significantly affecting the mutation rate.

T OPIC: *What topic in genetics does this question address?* The topic is testing for mutagens. More specifically, the question is about the Ames test.

I NFORMATION: *What information do you know based on the question and your understanding of the topic?* In the question, you are given data regarding the outcome of four trials using the Ames test. From your understanding of the topic, you may remember that a higher number of colonies on the experimental plates may indicate that a substance is a mutagen.

P ROBLEM-SOLVING S TRATEGIES: *Make a calculation. Analyze data.* To begin to solve this problem, you first need to calculate the mutation rate. To do so, you take the average of the four trials and then divide the average number of mutant colonies by the total number of cells applied to each plate (in this case, 2 million). You also need to conduct a *t*-test to determine if the control and experimental data are significantly different. A description of a *t*-test can be found in various statistics textbooks and a Statistics Primer is available in Connect®.

ANSWER: For the control data, the mutation rate is 1.5 in 1 million, or 1.5×10^{-6}. In the presence of the suspected mutagen, the mutation rate is 30 in a million, or 30×10^{-6}. If you conduct a *t*-test on these data, you will find that $P < 0.01$, so you can reject the null hypothesis that the control and experimental data are not different from each other. Therefore, you can accept the hypothesis that the suspected mutagen is causing a higher mutation rate. Keep in mind that this hypothesis is not proven; you simply are able to accept it based on this statistical outcome.

19.4 COMPREHENSION QUESTIONS

1. Nitrous acid replaces amino groups with keto groups, a process called
   a. alkylation.
   b. deamination.
   c. depurination.
   d. crosslinking.

2. A mutagen that is a base analog is
   a. ethyl methanesulfonate (EMS).
   b. 5-bromouracil.
   c. UV light.
   d. proflavin.

3. In an Ames test, a _____ number of colonies is observed if a substance _____ a mutagen, compared with the number of colonies for a control sample that is not exposed to the suspected mutagen.
   a. significantly higher, is
   b. significantly higher, is not
   c. significantly lower, is
   d. significantly lower, is not

19.5   DNA REPAIR

Learning Outcomes:

1. Compare and contrast the different types of DNA repair systems.
2. Describe how specialized DNA polymerases are able to synthesize DNA over a damaged region.

Because mutations are often harmful, DNA repair systems are vital to the survival of all organisms. If DNA repair systems did not exist, spontaneous and environmentally induced mutations would be so prevalent that few species, if any, would survive. The necessity of DNA repair systems becomes evident when they are missing. Bacteria contain several different DNA repair systems. Yet, when even a single system is absent, bacteria have a much higher rate of mutation. In fact, the rate of mutation is so high that these bacterial strains are sometimes called mutator strains. Likewise, in humans, an individual who is defective in only a single DNA repair system may manifest various disease symptoms, including a higher risk of skin cancer. This increased risk is due to the inability to repair UV-induced mutations.

Living cells have several DNA repair systems, which can fix different types of DNA alterations (**Table 19.7**). Each repair system is composed of one or more proteins that play specific roles in the repair mechanism. In most cases, DNA repair is a multistep process. First, one or more proteins in the DNA repair system detect an irregularity in DNA structure. Next, the abnormality is removed by the action of DNA repair enzymes. Finally, normal DNA is synthesized via DNA replication enzymes. In this section, we will examine several different repair systems that have been characterized in bacteria, yeast, mammals, and plants. Their diverse ways of repairing DNA underscore the necessity of proper maintenance of the structure of DNA.

Damaged Bases Can Be Directly Repaired

In a few cases, the covalent modification of nucleotides by mutagens can be reversed by specific DNA repair enzymes. As discussed earlier in this chapter, UV light causes the formation of thymine dimers. Bacteria, fungi, most plants, and some animals produce an enzyme called **photolyase** that recognizes thymine dimers and splits them, which returns the DNA to its original condition (**Figure 19.16a**). Photolyase contains two light-sensitive cofactors. The repair mechanism itself requires light and is known as **photoreactivation.**

ethyl group from the base to a cysteine side chain within itself (**Figure 19.16b**). Surprisingly, this action permanently inactivates alkyltransferase, which means it can be used only once!

Base Excision Repair Removes a Damaged Base

A second type of repair system, called **base excision repair (BER)**, is primarily responsible for eliminating non-helix-distorting changes that affect the structure of individual bases. BER involves the function of a category of enzymes known as **DNA N-glycosylases.** This type of enzyme can precisely recognize a site in the DNA where an abnormal base is located. Living organisms produce multiple types of DNA N-glycosylases, each recognizing particular types of abnormal base structures. Depending on the DNA N-glycosylase involved, this repair system can eliminate abnormal bases such as uracil, 3-methyladenine, and 7-methylguanine. BER is particularly important for the repair of oxidative DNA damage.

Figure 19.17 illustrates the general steps involved in DNA repair via N-glycosylase.

1. In the example featured in the figure, the DNA contains a uracil in its sequence. This could have happened spontaneously or by the action of a chemical mutagen.
2. After N-glycosylase recognizes a uracil within the DNA, the enzyme flips the altered base out of the double helix and then cleaves the bond between the base and the sugar in the DNA backbone. This cleavage releases the uracil base and leaves behind an apyrimidinic site.
3. This abnormality is recognized by a second enzyme, **AP endonuclease,** which makes a cut (a nick) in the DNA backbone on the 5′ side of the site.
4. Following this cut by AP endonuclease, one of three things can happen: In some species such as *E. coli*, DNA polymerase I,

TABLE 19.7
Common Types of DNA Repair Systems

System	Description
Direct repair	An enzyme recognizes an incorrect alteration in DNA structure and directly converts the structure back to the correct form.
Base excision repair and nucleotide excision repair	An abnormal base or nucleotide is first recognized and removed from the DNA, and a segment of DNA in this region is excised. Then the complementary DNA strand is used as a template to synthesize a normal DNA strand.
Mismatch repair	Similar to excision repair except that the DNA defect is a base-pair mismatch in the DNA, not an abnormal nucleotide. The mismatch is recognized, and a segment of DNA in this region is removed. The parental strand is used to synthesize a normal daughter strand of DNA.
Homologous recombination	Occurs at double-strand breaks or when DNA repair damage causes a gap in synthesis during DNA replication. The strands of a normal sister chromatid are used to repair a damaged sister chromatid.
Nonhomologous end joining	Occurs at double-strand breaks. The broken ends are recognized by proteins that keep the ends together; the broken ends are eventually rejoined.

This process directly restores the structure of DNA. Because plants are exposed to sunlight throughout the day, photolyase is a critical DNA repair enzyme for many plant species.

An enzyme known as **alkyltransferase** can remove methyl or ethyl groups from guanine bases that have been mutagenized by alkylating agents such as nitrogen mustard and EMS. This protein is called alkyltransferase because it transfers the methyl or

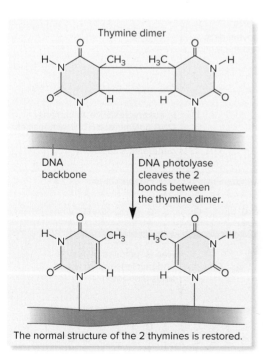

(a) Direct repair of a thymine dimer

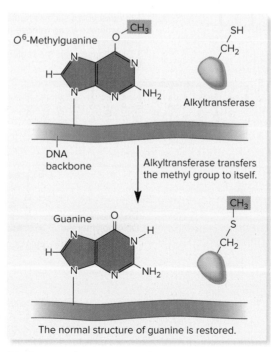

(b) Direct repair of a methylated base

FIGURE 19.16
Direct repair of damaged bases in DNA.
(a) The repair of a thymine dimer by photolyase.
(b) The repair of methylguanine by the transfer of the methyl group to alkyltransferase.

CONCEPT CHECK: Which of these repair systems is particularly valuable to plants?

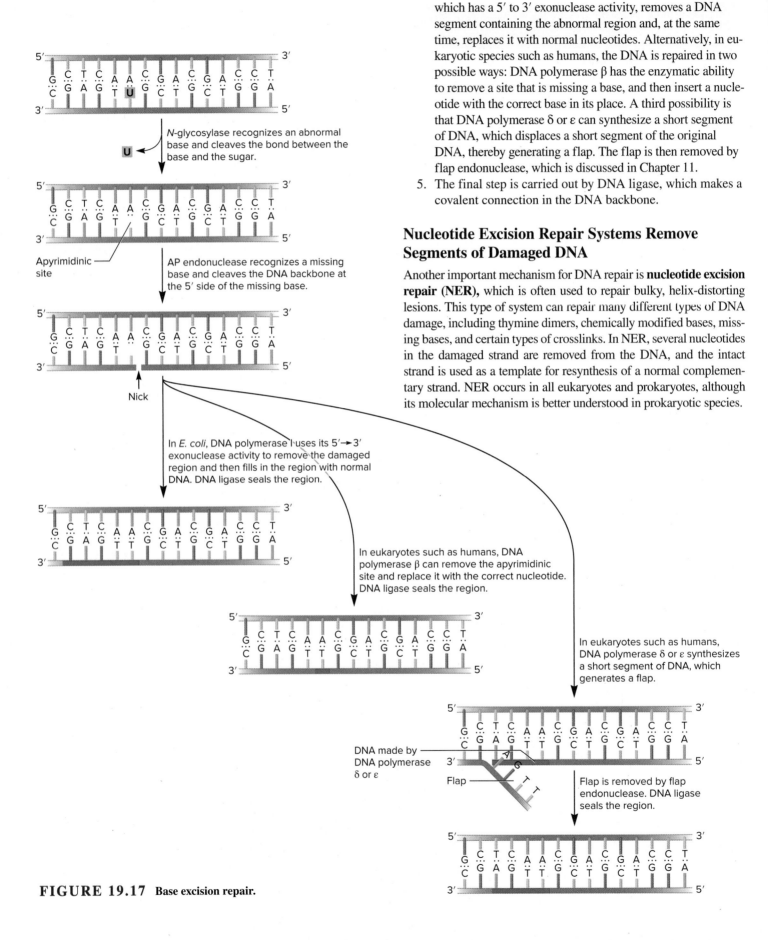

FIGURE 19.17   Base excision repair.

which has a 5′ to 3′ exonuclease activity, removes a DNA segment containing the abnormal region and, at the same time, replaces it with normal nucleotides. Alternatively, in eukaryotic species such as humans, the DNA is repaired in two possible ways: DNA polymerase β has the enzymatic ability to remove a site that is missing a base, and then insert a nucleotide with the correct base in its place. A third possibility is that DNA polymerase δ or ε can synthesize a short segment of DNA, which displaces a short segment of the original DNA, thereby generating a flap. The flap is then removed by flap endonuclease, which is discussed in Chapter 11.

5. The final step is carried out by DNA ligase, which makes a covalent connection in the DNA backbone.

Nucleotide Excision Repair Systems Remove Segments of Damaged DNA

Another important mechanism for DNA repair is **nucleotide excision repair (NER),** which is often used to repair bulky, helix-distorting lesions. This type of system can repair many different types of DNA damage, including thymine dimers, chemically modified bases, missing bases, and certain types of crosslinks. In NER, several nucleotides in the damaged strand are removed from the DNA, and the intact strand is used as a template for resynthesis of a normal complementary strand. NER occurs in all eukaryotes and prokaryotes, although its molecular mechanism is better understood in prokaryotic species.

In *E. coli*, the NER system requires four key proteins, designated UvrA, UvrB, UvrC, and UvrD, plus DNA polymerase and DNA ligase. The four Uvr proteins recognize and remove a short segment of a damaged DNA strand. These proteins are named Uvr because they are involved in <u>u</u>ltra<u>v</u>iolet light <u>r</u>epair of thymine dimers, although they are also important in repairing chemically damaged DNA.

Figure 19.18 outlines the steps involved in the *E. coli* NER system.

1. A protein complex consisting of two UvrA molecules and one UvrB molecule tracks along the DNA in search of damaged DNA. Such DNA has a distorted double helix, which is detected by the UvrA/UvrB complex.
2. When a damaged segment is identified, the two UvrA proteins are released, and UvrC binds to the site.
3. The UvrC protein makes cuts in the damaged strand on both sides of the damaged site. Typically, the damaged strand is cut 8 nucleotides from the 5′ end of the damaged site and 4–5 nucleotides away from the 3′ end.
4. After this process, UvrD, which is a helicase, recognizes the region and separates the two strands of DNA. This releases a short DNA segment that contains the damaged region, and UvrB and UvrC are also released.
5. Following the excision of the damaged DNA, DNA polymerase fills in the gap, using the undamaged strand as a template. Finally, DNA ligase makes the covalent connection between the newly made DNA and the original DNA strand.

NER Systems and Disease. In eukaryotes, NER systems are thought to operate similarly to those in bacteria, though more proteins are involved. Several human diseases are due to inherited defects in genes that code the proteins involved in NER. These include xeroderma pigmentosum (XP) and Cockayne syndrome (CS). A common symptom in both syndromes is an increased sensitivity to sunlight because of an inability to repair UV-induced lesions. The chapter opening photo shows a child with XP. Such individuals have pigmentation abnormalities and many premalignant lesions, and they are highly predisposed to developing skin cancer. They may also develop early degeneration of the nervous system.

Genetic analyses of patients with XP and CS have revealed that these syndromes result from defects in a variety of different genes that code NER proteins. For example, XP can be caused by defects in any of seven different NER genes. In all cases, individuals have a defective NER mechanism. In recent years, several human NER genes have been successfully cloned and sequenced. Although more research is needed to completely understand this mechanism of DNA repair, the identification of NER genes has helped unravel the complexities of NER in human cells.

Mismatch Repair Systems Recognize and Correct a Base-Pair Mismatch

Thus far, we have considered several DNA repair systems that recognize abnormal nucleotide structures within DNA, including thymine dimers, alkylated bases, and the presence of uracil in the DNA. Another type of abnormality that should not occur in DNA is a **base-pair mismatch.** The structure of the DNA double helix

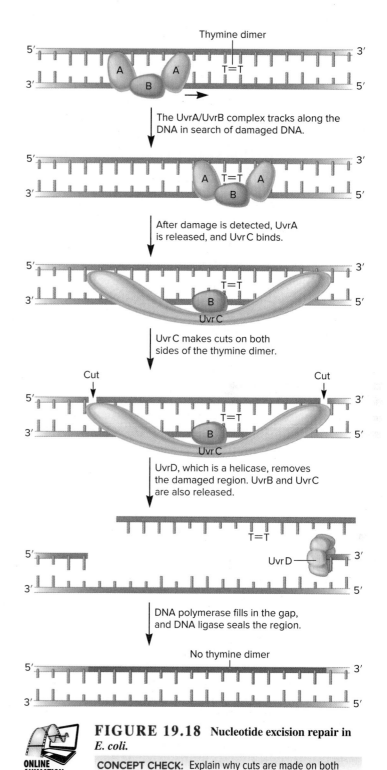

FIGURE 19.18 Nucleotide excision repair in *E. coli.*

ONLINE ANIMATION

CONCEPT CHECK: Explain why cuts are made on both sides of the damaged region of the DNA.

obeys the AT/GC rule of base pairing. During the normal course of DNA replication, however, an incorrect nucleotide may be added to the growing daughter strand by mistake. This produces a mismatch between a base in the parental strand and one in the daughter strand. Various DNA repair mechanisms can recognize and remove this mismatch. For example, as described in Chapter 11, DNA polymerase has a 3′ to 5′ proofreading ability that detects mismatches and removes them. However, if this proofreading ability fails, cells

have additional DNA repair systems that detect base-pair mismatches and fix them. An interesting DNA repair system that exists
in all species is the **mismatch repair system.**

In the case of a base-pair mismatch, how does a DNA repair
system determine which base to remove? If the mismatch is due to an
error in DNA replication, the newly made daughter strand contains
the incorrect base, whereas the parental strand is normal. Therefore,
an important aspect of mismatch repair is that it specifically repairs
the newly made strand rather than the parental template strand. Prior
to DNA replication, the parental DNA has already been methylated.
Immediately after DNA replication, some time must pass before a
newly made strand is methylated. Therefore, newly replicated DNA
is hemimethylated—only the parental DNA strand is methylated.
Hemimethylation provides a way for a DNA repair system to distinguish between the parental DNA strand and the daughter strand.

The molecular mechanism of mismatch repair has been stud-
ied extensively in *E. coli.* As shown in **Figure 19.19**, three proteins,
designated MutS, MutL, and MutH, detect the mismatch and direct
the removal of the mismatched base from the newly made strand.
These proteins are named Mut because their absence leads to a
much higher <u>mut</u>ation rate than occurs in normal strains of *E. coli.*

1. The role of MutS is to locate mismatches. Once a mismatch
   is detected, MutS forms a complex with MutL. MutL acts
   as a linker that binds to MutH, forming a loop in the DNA.
2. The role of MutH is to identify the methylated strand of
   DNA, which is the nonmutated parental strand. The formation of the MutS/MutL/MutH complex stimulates MutH,
   which is already bound to a hemimethylated site, to make a
   cut in the newly made, nonmethylated DNA strand. After
   the strand is cut, MutU, which functions as a helicase, separates the strands, and an exonuclease then digests the nonmethylated DNA strand in the direction of the mismatch
   and proceeds beyond the site where MutS is located.
3. Step 2 leaves a gap in the daughter strand that is repaired
   by DNA polymerase and DNA ligase.

The net result is that the mismatch has been corrected by removing the incorrect region in the daughter strand and then resynthe-
sizing the correct sequence using the parental DNA as a template.

Eukaryotic species have homologs to MutS and MutL, along
with many other proteins that are needed for mismatch repair. Eukaryotes do not have a MutH homolog. Instead, the eukaryotic MutL homolog has the ability to make a cut in the nonmethylated DNA strand.

As with defects in nucleotide excision repair systems, mutations that affect the human mismatch repair system are associated
with particular types of cancer. For example, mutations in two
human mismatch repair genes, *hMSH2* and *hMLH1*, play a role in
the development of a type of colon cancer known as hereditary
nonpolyposis colorectal cancer.

Double-Strand Breaks Can Be Repaired by Homologous Recombination Repair or by Nonhomologous End Joining

Of the many types of DNA damage that can occur within living
cells, the breakage of chromosomes—called a DNA double-strand
break (DSB)—is perhaps the most dangerous. DSBs can be caused

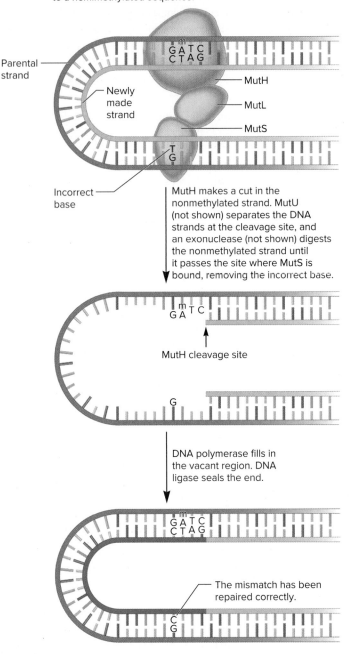

The MutS protein slides along the DNA and finds a mismatch.
The MutS/MutL complex binds to MutH, which is already bound
to a hemimethylated sequence.

MutH makes a cut in the nonmethylated strand. MutU
(not shown) separates the DNA strands at the cleavage site, and
an exonuclease (not shown) digests the nonmethylated strand until
it passes the site where MutS is bound, removing the incorrect base.

MutH cleavage site

DNA polymerase fills in the vacant region. DNA ligase seals the end.

The mismatch has been repaired correctly.

FIGURE 19.19 Mismatch repair in *E. coli.* In
this example, the methylated adenine is designated with
an m.

CONCEPT CHECK: Which of the three Mut proteins is
responsible for ensuring that the mismatched base in the
newly made daughter strand is the one that is removed?

by ionizing radiation (X-rays or gamma rays), chemical mutagens,
and certain drugs used for chemotherapy. In addition, reactive
oxygen species that are the by-products of aerobic metabolism can
cause double-strand breaks. Surprisingly, researchers estimate that
naturally occurring double-strand breaks in a typical human cell
occur at a rate of 10 to 100 breaks per cell per day!

Such breaks are harmful in a variety of ways. First, DSBs
can result in chromosomal rearrangements such as inversions and

translocations (refer back to Figure 8.2). In addition, DSBs can lead to terminal or interstitial deletions (refer back to Figure 8.3). Such genetic changes have the potential to result in detrimental phenotypic effects.

How are DSBs repaired? The two main mechanisms are **homologous recombination repair (HRR)** (also called **homology-directed repair**) and **nonhomologous end joining (NHEJ)**.

Homologous Recombination Repair. HRR occurs when homologous DNA strands, usually from a sister chromatid, are used to repair a DSB in the other sister chromatid (**Figure 19.20**).

1. First, a double-strand break occurs in the DNA.
2. This DSB is processed by the digestion of short segments of the DNA strands at the break site.
3. This processing event is followed by the exchange of DNA strands between the broken and unbroken sister chromatids.
4. The unbroken strands are then used as templates to synthesize DNA in the region where the break occurred.
5. Finally, the crisscrossed strands are resolved, which means they are broken and then rejoined in a way that produces separate chromatids.

Because sister chromatids are genetically identical, an advantage is that homologous recombination repair can be an error-free mechanism for repairing a DSB. A disadvantage is that sister chromatids are available only during the S and G_2 phases of the cell cycle in eukaryotes or following DNA replication in bacteria. Although sister chromatids are strongly preferred, HRR may also occur between homologous regions that are not identical. Therefore, HRR may occasionally happen when sister chromatids are unavailable. The proteins involved in homologous recombination repair are described later in this chapter (see Table 19.8).

Nonhomologous End Joining. During NHEJ, the two broken ends of DNA are simply pieced back together (**Figure 19.21**). This mechanism requires the participation of several proteins that play key roles in the process.

1. First, a double-strand break occurs in the DNA.
2. The DSB is recognized by end-binding proteins. These proteins are then recognized by additional proteins that form a crossbridge that prevents the two ends from drifting apart.
3. Next, additional proteins are recruited to the region, and they may process the ends of the broken chromosome by digesting particular DNA strands. This processing may result in the deletion of a small amount of genetic material from the region.
4. Finally, any gaps are filled in by DNA polymerase, and the DNA ends are ligated together.

One advantage of NHEJ is that it doesn't involve a sister chromatid, so it can occur at any stage of the cell cycle. However, a disadvantage is that NHEJ may result in a small deletion in the region that has been repaired.

Damaged DNA May Be Replicated by Translesion DNA Polymerases

Despite the efficient action of numerous repair systems that remove lesions in DNA in an error-free manner, it is inevitable that

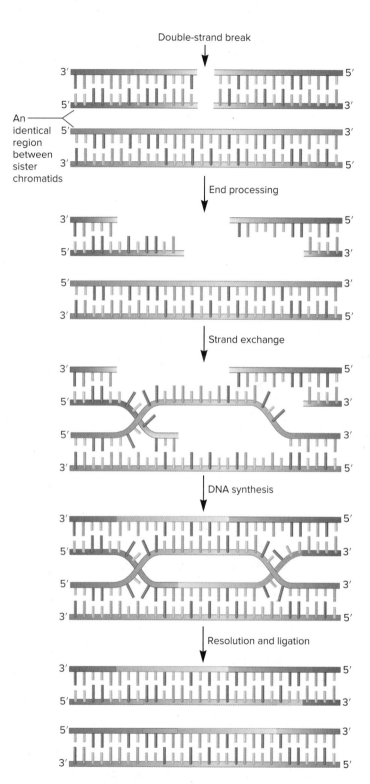

FIGURE 19.20 **Repair of a double-strand break in DNA via homologous recombination repair.**

some lesions may not be removed by these mechanisms. Such lesions may be present when DNA is being replicated. If so, replicative DNA polymerases, such as pol III in *E. coli*, which are highly sensitive to geometric distortions in DNA, are unable to replicate through the DNA lesions.

During the past decade, researchers have discovered that cells are equipped with specialized DNA polymerases that assist

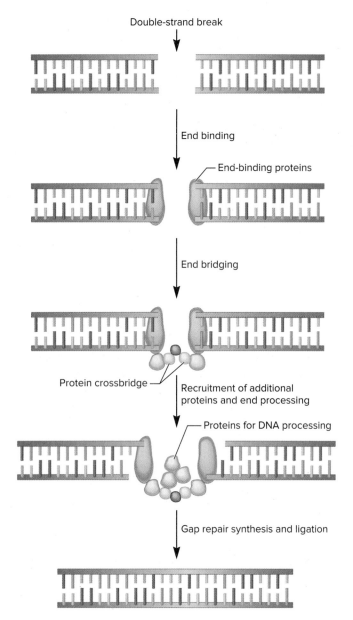

FIGURE 19.21 **Repair of a double-strand break in DNA via nonhomologous end joining.**

CONCEPT CHECK: What are an advantage and a disadvantage of this repair system?

the replicative DNA polymerases during the process of **translesion synthesis (TLS)**—the synthesis of DNA over a template strand that harbors some type of DNA damage. These translesion-replicating polymerases, which are also described in Chapter 11 (refer back to Table 11.4), contain an active site with a loose, flexible pocket that can accommodate aberrant structures in the template strand. When a replicative DNA polymerase encounters a damaged region, it is swapped with a translesion-replicating polymerase.

A negative consequence of translesion synthesis is low fidelity. Due to their flexible active site, translesion-replicating polymerases are much more likely to incorporate the wrong

base into a newly made daughter strand. The mutation rate is typically in the range from 10^{-2} to 10^{-3}. Therefore, translesion synthesis is referred to as **error-prone replication.** By comparison, replicative DNA polymerases are highly intolerant of the geometric distortions imposed on DNA by the incorporation of incorrect nucleotides, and consequently, they incorporate wrong bases with a very low frequency of approximately 10^{-8}. In other words, they copy DNA with a high degree of fidelity.

19.5 COMPREHENSION QUESTIONS

1. The function of photolyase is to repair
   a. double-strand breaks.
   b. apurinic sites.
   c. thymine dimers.
   d. all of these.

2. Which of the following DNA repair systems may involve the removal of a segment of a DNA strand?
   a. Base excision repair
   b. Nucleotide excision repair
   c. Mismatch repair
   d. All of the above

3. In nucleotide excision repair in *E. coli*, the function of the UvrA/UvrB complex is to
   a. detect DNA damage.
   b. make cuts on both sides of the damage.
   c. remove the damaged piece of DNA.
   d. replace the damaged DNA with undamaged DNA.

4. Double-strand breaks can be repaired by
   a. homologous recombination repair (HRR).
   b. nonhomologous end joining (NHEJ).
   c. nucleotide excision repair (NER).
   d. either a or b.

5. An advantage of translesion-replicating polymerases is that they can replicate _____, but a disadvantage is that they _____.
   a. very quickly, have low fidelity
   b. over damaged DNA, have low fidelity
   c. when resources are limited, are very slow
   d. over damaged DNA, are very slow

19.6 HOMOLOGOUS RECOMBINATION

Learning Outcomes:

1. Describe the Holliday model and the double-strand break model for homologous recombination.
2. Explain how gene conversion can occur via mismatch repair and DNA gap repair synthesis.

Homologous recombination involves an exchange of DNA segments that are similar or identical in their DNA sequences.

Eukaryotic chromosomes that have similar or identical sequences frequently participate in crossing over during meiosis I and occasionally during mitosis. Crossing over involves the alignment of a pair of homologous chromosomes, followed by the breakage of two chromatids at analogous locations, and the subsequent exchange of the corresponding segments (refer to Figure 2.10).

Figure 19.22 shows two types of homologous recombination that may occur between replicated chromosomes in a diploid species. When such recombination takes place between sister chromatids, the process is called **sister chromatid exchange (SCE).** Because sister chromatids are genetically identical, SCE does not produce a new combination of alleles (Figure 19.22a). By comparison, crossing over may occur between homologous chromosomes during meiosis. As shown in Figure 19.22b, this form of homologous recombination may produce new combinations of alleles in the resulting chromosomes. In this second case, homologous recombination has resulted in **genetic recombination,** which refers to the shuffling of genetic material to create a new combination of alleles that differs from the original. Homologous recombination is an important mechanism for fostering genetic recombination.

Bacteria are usually haploid. They do not have pairs of homologous chromosomes. Even so, bacteria can undergo homologous recombination. How can the exchange of DNA segments occur in a haploid organism? First, bacteria may have more than one copy of a chromosome per cell, though the copies are usually identical. These copies can exchange genetic material via homologous recombination. Second, during DNA replication, the replicated regions may also undergo homologous recombination. In bacteria, homologous recombination is particularly important in the repair of DNA segments that have been damaged. In this section, we will focus our attention on the molecular mechanisms that underlie homologous recombination.

The Holliday Model Describes a Molecular Mechanism for the Recombination Process

We now turn our attention to genetic exchange that occurs between homologous chromosomes. Perhaps it is surprising that the first molecular model of homologous recombination did not come from a biochemical analysis of DNA or from electron microscopy studies. Instead, it was deduced from the outcomes of genetic crosses in fungi.

As discussed in Chapter 6, geneticists have learned a great deal from the analysis of fungal asci, because an ascus is a sac that contains the products of a single meiosis. When two haploid fungi that differ at a single gene are crossed to each other, the ascus is expected to contain an equal proportion of each genotype. For example, if a pigmented strain of *Neurospora* producing orange spores is crossed to an albino strain producing white spores, the resulting group of eight cells, called an octad, should contain four orange spores and four white spores. In 1934, H. Zickler noticed that unequal proportions of the spores sometimes occurred within asci from such crosses. He occasionally observed octads with six orange spores and two white spores, or six white spores and two orange spores.

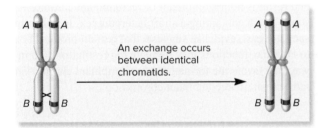

(a) Sister chromatid exchange

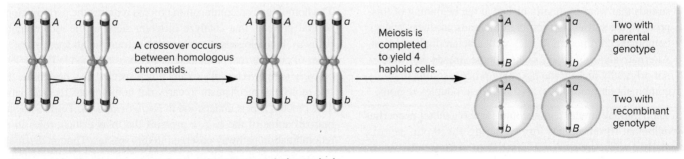

(b) Recombination between homologous chromosomes during meiosis

FIGURE 19.22   Two types of homologous recombination in eukaryotes. (a) Homologous recombination between sister chromatids. which are identical, does not produce a new combination of alleles. (b) Homologous recombination between homologous chromatids. This second form of homologous recombination may lead to a new combination of alleles, which is called a recombinant (or nonparental) genotype.

Genes→Traits Homologous recombination is particularly important with regard to the relationships between multiple genes and multiple traits. For example, if the X chromosome in a female fruit fly carries alleles for red eyes and gray body and its homolog carries alleles for white eyes and yellow body, homologous recombination could produce recombinant chromosomes that carry alleles for red eyes and yellow body or carry alleles for white eyes and gray body. Therefore, new combinations of two or more alleles can arise when homologous recombination takes place.

CONCEPT CHECK: What is the advantage of genetic recombination, which occurs in part (b)?

Zickler used the term **gene conversion** to describe the phenomenon in which one allele is converted to the allele on the homologous chromosome. Subsequent studies by several researchers confirmed this phenomenon in yeast and *Neurospora*. Gene conversion occurred at too high a rate to be explained by new mutations. In addition, research showed that gene conversion often occurs in a chromosomal region where a crossover has taken place.

Based on studies involving gene conversion, Robin Holliday proposed a model in 1964 to explain the molecular steps that occur during homologous recombination. We will first consider the steps in the Holliday model and then consider a more recent model. Later, we will examine how the Holliday model can explain the phenomenon of gene conversion.

The **Holliday model** is shown in **Figure 19.23a**.

1. At the beginning of the process, two homologous chromatids are aligned with each other. According to the model, a break or nick occurs at identical sites in one strand of each of the two homologous chromatids.
2. The strands then invade the opposite helices and base-pair with the complementary strands. This event is followed by a covalent linkage to create a **Holliday junction.**
3. The Holliday junction can migrate in a lateral direction. As it does so, a DNA strand in one helix is swapped for a DNA strand in the other helix. This process is called **branch migration** because the junction connecting the two double helices migrates laterally. Since the DNA sequences in the homologous chromosomes are similar but may not be identical, the swapping of the DNA strands during branch migration may produce a **heteroduplex,** a region in the double-stranded DNA that contains one or more base-pair mismatches.
4. The last event in the recombination process is called **resolution** because it involves the breakage and rejoining of two DNA strands to create two separate chromosomes. In other words, the entangled DNA strands become resolved into two separate structures. The bottom left side of Figure 19.23a shows the Holliday junction viewed in two different ways. If breakage occurs in the same two DNA strands that were originally nicked at the beginning of this process, the subsequent joining of strands produces nonrecombinant chromosomes, each with a heteroduplex region. Alternatively, if breakage occurs in the strands that were not originally nicked, the rejoining process results in recombinant chromosomes, also with heteroduplex regions.

The Holliday model can account for the general properties of recombinant chromosomes formed during eukaryotic meiosis. Particularly convincing evidence came from electron microscopy studies in which recombination structures could be visualized. **Figure 19.23b** shows a schematic drawing of two DNA fragments in the process of recombination. This structure has been called a chi form because its shape is similar to the Greek letter χ (chi).

The Double-Strand Break Model Has Refined the Molecular Steps of Homologous Recombination

As more detailed studies of homologous recombination have become available, certain steps in the Holliday model have been reconsidered. In particular, a more recent model has modified the initiation phase of recombination. Researchers now propose that homologous recombination is not likely to involve nicks at identical sites in one strand in each of two homologous chromatids. Instead, it is more likely for a DNA helix to incur a break in both strands of one chromatid.

In 1975, Jack Szostak, Terry Orr-Weaver, Rodney Rothstein, and Franklin Stahl proposed a model in which such a double-strand break initiates the recombination process; therefore, this model is called the **double-strand break model.** Though recombination may occur via more than one mechanism, research suggests that double-strand breaks commonly promote homologous recombination during meiosis and during DNA repair.

Figure 19.24 shows the general steps in the double-strand break model.

1. The top chromatid has experienced a double-strand break.
2. A small region near the break is degraded.
3. Strand degradation generates a single-stranded DNA segment that can invade the intact bottom chromatid. The strand displaced by the invading segment forms a structure called a displacement loop (D-loop).
4. After the D-loop is formed, two regions have a gap in the DNA. How is the problem fixed? DNA synthesis occurs in the relatively short gaps where a DNA strand is missing. This DNA synthesis is called **DNA gap repair synthesis.**
5. After this synthesis is completed, two Holliday junctions are produced. Depending on the way these are resolved, the end result is nonrecombinant or recombinant chromosomes, each containing a short heteroduplex. In eukaryotes such as yeast, evidence suggests that certain proteins bound to Holliday junctions may regulate the resolution step in a way that favors the formation of recombinant chromosomes rather than nonrecombinant chromosomes.

Several Proteins Facilitate Homologous Recombination

The homologous recombination process requires the participation of several proteins that catalyze different steps in the recombination pathway. Homologous recombination is found in all species, and the types of proteins that participate in the steps outlined in Figure 19.24 are very similar. The cells of any given species may have more than one molecular mechanism to carry out homologous recombination. This process is best understood in *Escherichia coli*. **Table 19.8** summarizes some of the *E. coli* proteins that play critical roles in one recombination pathway observed in this species. Though discussing them is beyond the scope of this textbook, *E. coli* has other pathways to carry out homologous recombination.

Before ending this discussion of the molecular mechanism of homologous recombination, let's consider crossing over during meiosis in eukaryotic cells. As described in Chapter 2, crossing over between homologous chromosomes is an important event during prophase of meiosis I. An intriguing question is this: How are crossover sites between two homologous chromosomes chosen?

Molecular studies of two different yeast species, *Saccharomyces cerevisiae* and *Schizosaccharomyces pombe*, suggest that

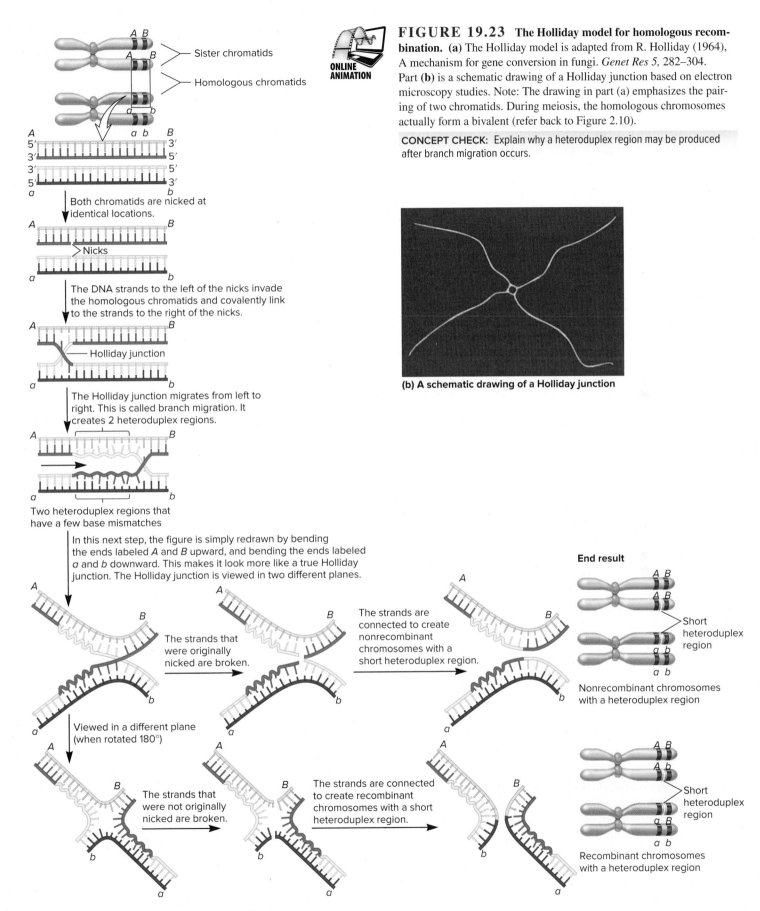

FIGURE 19.23 **The Holliday model for homologous recombination.** (a) The Holliday model is adapted from R. Holliday (1964), A mechanism for gene conversion in fungi. *Genet Res 5*, 282–304. Part (b) is a schematic drawing of a Holliday junction based on electron microscopy studies. Note: The drawing in part (a) emphasizes the pairing of two chromatids. During meiosis, the homologous chromosomes actually form a bivalent (refer back to Figure 2.10).

CONCEPT CHECK: Explain why a heteroduplex region may be produced after branch migration occurs.

(b) A schematic drawing of a Holliday junction

(a) The Holliday model for homologous recombination

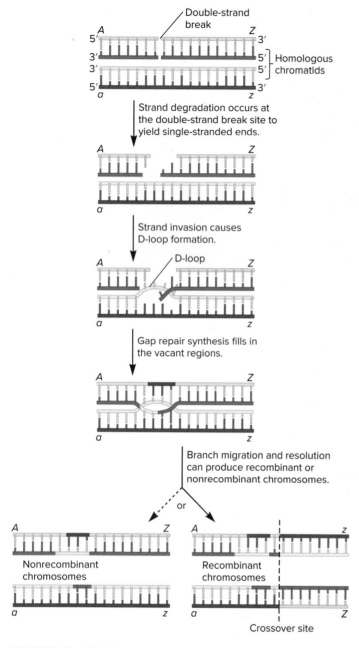

FIGURE 19.24 **A simplified version of the double-strand break model.** For simplicity, this illustration does not include the formation of heteroduplexes. The dashed arrow indicates that the pathway to the left may be less favored.

CONCEPT CHECK: Describe the structure and location of a D-loop.

double-strand breaks initiate the homologous recombination that occurs during meiosis. In other words, double-strand breaks create sites where a crossover will occur. In *S. cerevisiae*, the formation of DNA double-strand breaks that initiate meiotic recombination requires at least 10 different proteins. One particular protein, termed Spo11, is thought to be instrumental in cleaving the DNA, thereby creating a double-strand break. Once a double-strand break is made, homologous recombination can then occur according to the model described in Figure 19.24.

TABLE 19.8	
E. coli Proteins That Play a Role in Homologous Recombination	
Protein	**Description**
RecBCD	A complex of three proteins that tracks along the DNA and recognizes double-strand breaks. The complex partially degrades the double-stranded regions to generate single-stranded regions that can participate in strand invasion. RecBCD is also involved in loading RecA onto single-stranded DNA. In addition, RecBCD can create single-strand breaks that are used to initiate homologous recombination.
Single-strand binding protein	Coats broken ends of chromosomes and prevents excessive strand degradation.
RecA	Binds to single-stranded DNA and promotes strand invasion, which enables homologous strands to find each other. It also promotes the displacement of the complementary strand to generate a D-loop.
RuvABC	Protein complex that binds to Holliday junctions. RuvAB promotes branch migration. RuvC is an endonuclease that cuts the crossed or uncrossed strands to resolve Holliday junctions into separate chromosomes.
RecG	Protein that can also promote branch migration of Holliday junctions.

Gene Conversion May Result from DNA Mismatch Repair or DNA Gap Repair Synthesis

As mentioned earlier, homologous recombination can lead to gene conversion, in which one allele is converted to the allele on the homologous chromosome. The original Holliday model was based on this phenomenon. How can homologous recombination account for gene conversion? Researchers have identified two possible ways this can occur: DNA mismatch repair and DNA gap repair synthesis.

DNA Mismatch Repair. One mechanism for gene conversion involves DNA mismatch repair, which was discussed earlier in this chapter. To understand how this works, let's take a closer look at the heteroduplexes formed during branch migration of a Holliday junction (see Figure 19.23a). A heteroduplex contains a DNA strand from each of the two original parental chromosomes. The two parental chromosomes may contain an allelic difference within this region. In other words, this short region may contain DNA sequence differences. If this is the case, the heteroduplex formed after branch migration will contain one or more sites of base-pair mismatch. Gene conversion occurs when recombinant chromosomes are repaired and result in two copies of the same allele.

As shown in **Figure 19.25**, mismatch repair of a heteroduplex may result in gene conversion. In this example, the two chromosomes had different alleles due to a single base-pair difference in their DNA sequences, as shown at the top of the figure. During recombination, branch migration has occurred across this region, thereby creating two heteroduplexes with base-pair mismatches. Such mismatches will be recognized by DNA repair systems and repaired to yield a double helix that obeys the AT/GC rule. These two mismatches can be repaired in four possible ways. As shown here, two possibilities produce no gene conversion, whereas the other two lead to gene conversion.

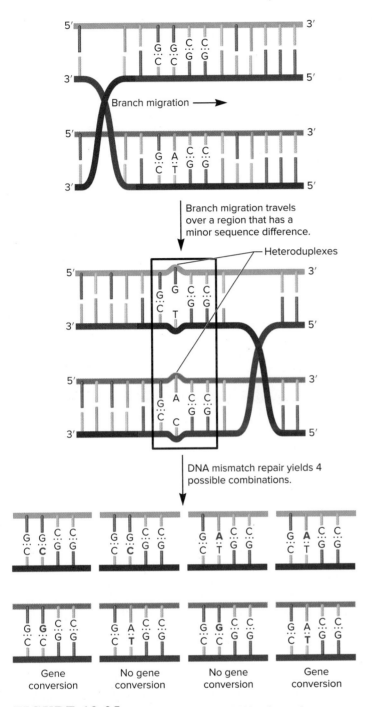

FIGURE 19.25 Gene conversion by DNA mismatch repair. A junction migrates past a homologous region that contains slightly different DNA sequences. This produces two heteroduplexes: DNA double helices with base pair mismatches. The mismatches can be repaired in four possible ways by the mismatch repair system described earlier in this chapter. Two of these ways result in gene conversion. The repaired base is shown in red.

DNA Gap Repair Synthesis. A second mechanism that can result in gene conversion is DNA gap repair synthesis. This mechanism is thought to be a more common explanation for gene conversion, particularly when two different alleles have base sequence differences at multiple sites. **Figure 19.26** illustrates how gap

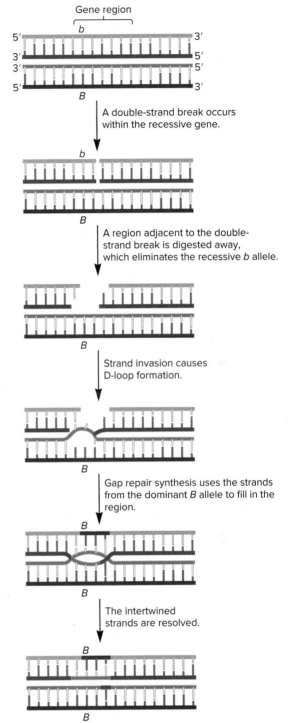

FIGURE 19.26 Gene conversion by DNA gap repair synthesis according to the double-strand break model. A gene occurs as two alleles, designated *b* and *B*. A double-strand break occurs in the DNA coding the *b* allele. Both of these DNA strands are digested away, thereby eliminating the genetic variation that created the *b* allele. A complementary DNA strand that codes the *B* allele migrates to this region and provides the template to synthesize a new double-stranded region. Following resolution, both DNA double helices carry the *B* allele.

CONCEPT CHECK: Explain what happened to the *b* allele that allowed gene conversion to occur.

repair synthesis can lead to gene conversion according to the double-strand break model.

1. The top chromatid, which carries the recessive *b* allele, has incurred a double-strand break in this gene.
2. A gap is created by the digestion of the DNA in the double helix. This digestion eliminates a segment of the *b* allele.
3. The two template strands used in gap repair synthesis are from the chromatid carrying the dominant *B* allele.
4. After gap repair synthesis takes place, the top chromatid carries the *B* allele, as does the bottom one. Gene conversion has changed the recessive *b* allele to a dominant *B* allele.

19.6 COMPREHENSION QUESTIONS

1. Homologous recombination refers to the exchange of DNA segments that are
   a. similar or identical in their DNA sequences.
   b. in close proximity to one another.
   c. broken due to ionizing radiation.
   d. misaligned along a chromosome.

2. During the molecular process of homologous recombination between homologous chromosomes,
   a. a Holliday junction forms.
   b. branch migration occurs.
   c. a heteroduplex region forms.
   d. all of the above occur.

3. A key difference between the original Holliday model and the double-strand break model is the way that
   a. the DNA strands are initially broken.
   b. branch migration occurs.
   c. a heteroduplex is formed.
   d. resolution occurs.

4. Which of the following mechanisms can result in gene conversion?
   a. DNA mismatch repair
   b. DNA gap repair
   c. Resolution of a Holliday junction
   d. Both a and b can result in gene conversion.

KEY TERMS

Introduction: mutation, homologous recombination
19.1: point mutation, base substitution, transition, transversion, silent mutation, missense mutation, nonsense mutation, frameshift mutation, neutral mutation, up promoter mutation, down promoter mutation, wild type, mutant allele, reversion, deleterious mutation, lethal mutation, beneficial mutation, conditional mutant, suppressor (suppressor mutation), intragenic suppressor, intergenic suppressor, breakpoint, position effect, germ line, germ-line mutation, somatic cell, somatic mutation, genetic mosaic
19.2: replica plating, random mutation theory, hot spots
19.3: spontaneous mutation, induced mutation, depurination, apurinic site, deamination, tautomeric shift, tautomers, reactive oxygen species (ROS), oxidative stress, oxidative DNA damage, trinucletoide repeat expansion (TNRE), anticipation

19.4: mutagen, nitrous acid, nitrogen mustard, ethyl methanesulfonate (EMS), acridine dye, proflavin, 5-bromouracil (5BU), 2-aminopurine, thymine dimer, mutation rate, mutation frequency, Ames test
19.5: photolyase, photoreactivation, alkyltransferase, base excision repair (BER), DNA *N*-glycosylase, AP endonuclease, nucleotide excision repair (NER), base pair mismatch, mismatch repair system, homologous recombination repair (HRR) (homology-directed repair), nonhomologous end joining (NHEJ), translesion synthesis (TLS), error-prone replication
19.6: sister chromatid exchange (SCE), genetic recombination, gene conversion, Holliday model, Holliday junction, branch migration, heteroduplex, resolution, double-strand break model, DNA gap repair synthesis

CHAPTER SUMMARY

- A mutation is a heritable change in the genetic material.
- Homologous recombination is the process whereby identical or similar DNA segments are exchanged between homologous chromosomes.

19.1 Effects of Mutations on Gene Structure and Function

- A point mutation is a change in a single base pair in the DNA. Such a mutation can be a transition or a transversion.
- Silent, missense, nonsense, and frameshift mutations may occur within the coding sequence of a gene (see Table 19.1, Figure 19.1).

- Mutations may also occur within a non-coding sequence of a gene and affect gene expression (see Table 19.2).
- Gene mutations are given names that describe how they affect the wild-type genotype and phenotype.
- Suppressor mutations reverse the phenotypic effects of another mutation. They can be intragenic or intergenic (see Table 19.3).
- Changes in chromosome structure can have a position effect that alters gene expression (see Figures 19.2, 19.3).
- If a mutation occurs early in development, a larger region of the individual is likely to carry the mutation. Also, mutations can occur in germ-line cells, which can be passed to offspring, or they can occur in somatic cells, which cannot be passed to offspring (see Figures 19.4, 19.5).

19.2 Random Nature of Mutations

- Results of experiments performed by Lederberg and Lederberg were consistent with the random mutation theory (see Figure 19.6).

19.3 Spontaneous Mutations

- Spontaneous mutations result from natural biological or chemical processes, whereas induced mutations are caused by environmental agents (see Table 19.4).
- Three common ways that mutations can arise spontaneously is by depurination, deamination, and tautomeric shifts (see Figures 19.7, 19.8, 19.9).
- Reactive oxygen species (ROS) can cause spontaneous mutations by oxidizing bases in DNA (see Figure 19.10).
- In individuals with a trinucleotide repeat expansion (TNRE), the number of repeats of a trinucleotide sequence has increased above a critical level and become prone to frequent expansion. This type of mutation is responsible for certain types of human diseases. The repeats can expand due to hairpin formation during DNA replication (see Table 19.5, Figure 19.11).

19.4 Induced Mutations

- A mutagen is an agent that can cause a mutation. Researchers have identified many different mutagens that change DNA structure in a variety of ways (see Table 19.6, Figures 19.12, 19.13, 19.14).
- Mutation rate is the likelihood that a new mutation will occur. Mutation frequency is the number of mutant genes divided by the total number of copies of that gene in a population.
- The Ames test is used to determine if an agent is a mutagen (see Figure 19.15).

19.5 DNA Repair

- All species have a variety of DNA repair systems to avoid the harmful effects of mutations (see Table 19.7).
- Photolyase and alkyltransferase can directly repair certain types of damaged bases (see Figure 19.16).
- Base excision repair (BER) recognizes and removes a damaged base (see Figure 19.17).
- Nucleotide excision repair (NER) removes damaged bases and damaged segments of DNA. Several inherited human diseases involve defects in nucleotide excision repair (see Figure 19.18).
- The mismatch repair system recognizes a base-pair mismatch and removes the segment of the DNA strand containing the incorrect base (see Figure 19.19).
- Double-strand breaks can be repaired by homologous recombination repair (HRR) or by nonhomologous end joining (NHEJ) (see Figures 19.20, 19.21).
- Damaged DNA may be replicated by translesion-replicating polymerases but the replication process is error-prone.

19.6 Homologous Recombination

- Homologous recombination can occur between sister chromatids or between homologous chromosomes (see Figure 19.22).
- The Holliday model describes the molecular steps that occur during homologous recombination (see Figure 19.23).
- The initiation of homologous recombination usually occurs with a double-strand break (see Figure 19.24).
- Several different proteins are involved in homologous recombination (see Table 19.8).
- Two different mechanisms, DNA mismatch repair and gap repair synthesis, can result in gene conversion during homologous recombination (see Figures 19.25, 19.26).

PROBLEM SETS & INSIGHTS

MORE GENETIC TIPS

1. If the mutation rate is 10^{-5} per gene per cell generation, how many new mutations per gene would you expect in a population of 1 million bacteria?

T OPIC: *What topic in genetics does this question address?* The topic is the mutation rate in a population of bacteria.

I NFORMATION: *What information do you know based on the question and your understanding of the topic?* From the question, you know the mutation rate and are asked to predict the number of new mutations per gene in a population composed of 1 million bacteria. From your understanding of the topic, you may remember that the mutation rate is the number of new mutations per gene per cell generation.

P ROBLEM-SOLVING S TRATEGY: *Make a calculation.* To solve this problem, you multiply the mutation rate times the number of bacteria.

ANSWER: The mutation rate times the number of bacteria is $10^{-5} \times 10^{6}$, which equals 10. Therefore, you would expect about 10 new mutations in a particular gene per 1 million bacteria.

2. A reversion is a mutation that restores a mutant codon back to a codon that gives a wild-type phenotype. At the DNA level, this type of mutation can be an exact reversion or an equivalent reversion:

GAG (glutamic acid)	First mutation	→	GTG (valine)	Exact reversion	→	GAG (glutamic acid)
GAG (glutamic acid)	First mutation	→	GTG (valine)	Equivalent reversion	→	GAA (glutamic acid)
GAG (glutamic acid)	First mutation	→	GTG (valine)	Equivalent reversion	→	GAT (aspartic acid)

An equivalent reversion produces a protein that is equivalent to the wild-type protein in structure and function. This outcome can occur in two ways. In some cases, the reversion produces the wild-type amino acid (in this case, glutamic acid), but it is expressed by a different codon than that in the wild-type gene. Alternatively, an equivalent reversion may substitute an amino acid structurally similar to the wild-type amino acid. In our example, an equivalent reversion has changed valine to an aspartic acid. Because aspartic and glutamic acids are structurally similar—they are acidic amino acids—this type of reversion can restore wild-type structure and function.

Here is the question. The template strand within the coding sequence of a gene has the following sequence:

3′–TACCCCTTCGACCCCGGA–5′

This template produces the following mRNA:

5′–AUGGGGAAGCUGGGGCCU–3′

The mRNA codes a polypeptide with the following sequence: methionine–glycine–lysine–leucine–glycine–proline.

A mutation changes the template strand to this sequence:

3′–TACCCCTACGACCCCGGA–5′

After the first mutation, another mutation occurs to change this sequence again. Is each of the following second mutations an exact reversion, an equivalent reversion, or neither?

A. 3′–TACCCCTCCGACCCCGGA–5′

B. 3′–TACCCCTTCGACCCCGGA–5′

C. 3′–TACCCCGACGACCCCGGA–5′

T OPIC: *What topic in genetics does this question address?* The topic is mutations called reversions. More specifically, the question is about the effects of such mutations on the coding sequence of a gene.

I NFORMATION: *What information do you know based on the question and your understanding of the topic?* From the question, you know the difference between an exact reversion and an equivalent reversion. From your understanding of the topic, you may remember that the template strand is used to make a complementary strand of mRNA. You can look at Table 13.1 to find out which amino acid is specified by any codon and determine whether a particular mutation will alter the amino acid sequence. You may also recall that AUG is the start codon.

P ROBLEM-SOLVING **S** TRATEGIES: *Relate structure and function. Predict the outcome.* To begin solving this problem, you first need to write out the sequence of the mRNA that will be produced from the wild-type, mutant, and reversion sequences. Note: Every other codon is shown in red, beginning with the AUG start codon:

Wild-type:	5′–AUGGGGAAGCUGGGGCCU–3′
Mutant:	5′–AUGGGG**AUG**CUGGGGCCU–3′
Reversion A:	5′–AUGGGG**AGG**CUGGGGCCU–3′
Reversion B:	5′–AUGGGGAAGCUGGGGCCU–3′
Reversion C:	5′–AUGGGG**CUG**CUGGGGCCU–3′

Changes to the wild-type sequence are shown in bold. The mutations occur in the third codon. You need to look up the amino acids that correspond to these codons in Table 13.1. Finally, you need to decide if a change in amino acid sequence is likely to affect protein function. In general, when an amino acid is substituted with a closely related amino acid, the change is less likely to inhibit protein function.

ANSWER:

A. This is probably an equivalent reversion. The third codon, which codes a lysine in the wild-type gene, is now an arginine codon. Arginine and lysine are both basic amino acids, so the polypeptide would probably function normally.

B. This is an exact reversion.

C. The third codon, which is a lysine codon in the wild-type gene, has been changed to a leucine codon. It is difficult to say if this would be an equivalent reversion or not. Lysine is a basic amino acid, and leucine is a nonpolar, aliphatic amino acid. The protein may still function normally with a leucine at the third codon, or it may function abnormally. You would need to test the function of the protein to determine if this was an equivalent reversion or not.

3. In the following schematic drawing of a Holliday junction, one chromatid is shown in red, and the homologous chromatid is shown in blue. The red chromatid carries a dominant allele labeled *A* and a recessive allele labeled *b*, whereas the blue chromatid carries a recessive allele labeled *a* and a dominant allele labeled *B*.

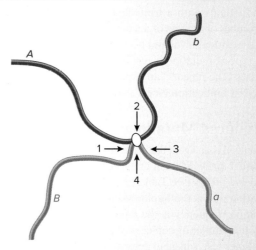

Where would the DNA strands have to be cut to produce recombinant chromosomes? Would they be cut at sites 1 and 3, or at sites 2 and 4? What would be the genotypes of the two recombinant chromosomes?

T OPIC: *What topic in genetics does this question address?* The topic is homologous recombination. More specifically, the question is about the resolution steps of this process.

I NFORMATION: *What information do you know based on the question and your understanding of the topic?* In the question, you are given a drawing of a Holliday junction. From your understanding of the topic, you may remember that resolution can occur in two ways: Either the strands that were originally nicked are cut again, or the strands that were not nicked are cut. The second type of event results in recombinant chromosomes with heteroduplex regions.

P ROBLEM-SOLVING **S** TRATEGIES: *Describe the steps. Predict the outcome.* One strategy to solve this problem is to consider the two ways that the DNA strands can be cut. Compare the drawing in this question with Figure 19.23.

ANSWER: The cuts would have to occur at the sites labeled 2 and 4. This would leave the *A* allele and the *B* allele on the same chromosome. The *a* allele in the other homolog would be on the same chromosome as the *b* allele. In other words, one recombinant chromosome would have the genotype *AB*, and the homolog would have *ab*.

Conceptual Questions

C1. Is each of the following mutations a transition, transversion, addition, or deletion? The original DNA strand is

5′-GGACTAGATAC-3′

(Note: Only the coding strand is shown.)

A. 5′-GAACTAGATAC-3′

B. 5′-GGACTAGAGAC-3′

C. 5′-GGACTAGTAC-3′

D. 5′-GGAGTAGATAC-3′

C2. A gene mutation changes a base pair from A-T to G-C. This change causes a gene to code a truncated protein that is nonfunctional. An organism that carries this mutation cannot survive at high temperatures. Make a list of all the genetic terms that could be used to describe this type of mutation.

C3. What does a suppressor mutation suppress? What is the difference between an intragenic and an intergenic suppressor?

C4. How would each of the following types of mutations affect protein function or the amount of functional protein that is expressed from a gene?

A. Nonsense mutation

B. Missense mutation

C. Up promoter mutation

D. Mutation that affects splicing

C5. X-rays strike a chromosome in a living cell and ultimately cause the cell to die. Did the X-rays produce a mutation? Explain why or why not.

C6. Lactose permease is coded by the *lacY* gene of the *lac* operon. Suppose a mutation that occurred at codon 64 of this gene changed the normal glycine codon to a valine codon. The mutant lactose permease is unable to function. However, a second mutation, which changes codon 50 from an alanine codon to a threonine codon, is able to restore function. Is each of the following terms appropriate or inappropriate to describe this second mutation?

A. Reversion

B. Intragenic suppressor

C. Intergenic suppressor

D. Missense mutation

C7. Is each of the following mutations a silent, missense, nonsense, or frameshift mutation? The sequence in the original DNA strand is 5′-ATGGGACTAGATACC-3′. (Note: Only the coding strand is shown; the first codon corresponds to methionine.)

A. 5′-ATGGGTCTAGATACC-3′

B. 5′-ATGCGACTAGATACC-3′

C. 5′-ATGGGACTAGTTACC-3′

D. 5′-ATGGGACTAAGATACC-3′

C8. A point mutation occurs in the middle of the coding sequence of a gene. Which types of mutation—silent, missense, nonsense, and frameshift—would be most likely to disrupt protein function and which would be least likely?

C9. In Chapters 12 through 16, we discussed many sequences that are outside of a gene's coding sequence but are important for gene expression. Look up two of these sequences and write them down. Explain how a mutation could change these sequences, thereby altering gene expression.

C10. Explain two ways that a chromosomal rearrangement can cause a position effect.

C11. Is a random mutation more likely to be beneficial or harmful? Explain your answer.

C12. Which of the following alterations of DNA could be appropriately described as having a position effect?

A. A point mutation at the −10 position in the promoter region prevents transcription.

B. A translocation places the coding sequence of a muscle-specific gene next to an enhancer that is turned on in nerve cells.

C. An inversion flips a gene from the long arm of chromosome 17 (which is euchromatic) to the short arm (which is heterochromatic).

C13. As discussed in Chapter 25, most forms of cancer are caused by environmental agents that produce mutations in somatic cells. Is an individual with cancer considered a genetic mosaic? Explain why or why not.

C14. Discuss the consequences of a germ-line mutation versus a somatic mutation.

C15. Make a drawing that shows how alkylating agents alter the structure of DNA, and explain the process.

C16. Explain how a mutagen can interfere with DNA replication and cause a mutation. Give two examples.

C17. What type of mutation (transition, transversion, or frameshift) would you expect each of the following mutagens to cause?

A. Nitrous acid

B. 5-Bromouracil

C. Proflavin

C18. Explain what happens to the sequence of DNA during trinucleotide repeat expansion (TNRE). If someone was mildly affected with a disease caused by TNRE, what issues would be important when considering possible effects in future offspring?

C19. Distinguish between spontaneous and induced mutations. Which are more harmful? Which are avoidable?

C20. Are mutations random events? Explain your answer.

C21. Give an example of a mutagen that can change cytosine to uracil. Which DNA repair system(s) would be able to repair this defect?

C22. If a mutagen causes bases to be removed from nucleotides within DNA, what repair system could fix this damage?

C23. Trinucleotide repeat expansions (TNREs) are associated with several different inherited human diseases. Certain types of TNREs produce a long tract of the amino acid glutamine within the coded protein. If a TNRE produces a glutamine tract, is each of the following statements true or false?

A. The TNRE is within the coding sequence of the gene.

B. The TNRE prevents RNA polymerase from transcribing the gene properly.

C. The trinucleotide sequence is CAG.

D. The trinucleotide sequence is CCG.

C24. With regard to TNRE, what is meant by the term *anticipation*?

C25. What is the difference between the mutation rate and the mutation frequency?

C26. Achondroplasia is a rare disorder that results in short stature. It is caused by an autosomal dominant mutation within a single gene. Among 1,422,000 live births, the number of babies born with achondroplasia was 31. Among those 31 babies, 18 of them had one parent with achondroplasia. The remaining babies had two unaffected parents. What is the mutation frequency for this disorder among these 1,422,000 babies? What is the mutation rate for achondroplasia?

C27. A segment of DNA has the following sequence:

TTGGATGCTG
AACCTACGAC

A. What would the sequence be immediately after reaction with nitrous acid? Let H represent hypoxanthine and U represent uracil.

B. Suppose this DNA was treated with nitrous acid. The nitrous acid was then removed, and the DNA replicated for two generations. What would be the sequences of the DNA products after the DNA replicated two times? (Note: Hypoxanthine pairs with cytosine.) Your answer should contain the sequences of four double strands.

C28. In the treatment of cancer, the basis for many types of chemotherapy and radiation therapy is that mutagens are more effective at killing dividing cells than nondividing cells. Explain why. What are possible harmful side effects of chemotherapy and radiation therapy?

C29. An individual carries a somatic mutation that changes a lysine codon to a glutamic acid codon. Prior to acquiring this mutation, the individual had been exposed to UV light, proflavin, and 5-bromouracil. Which of these three agents would be the most likely to have caused this somatic mutation? Explain your answer.

C30. Which of the following examples is likely to be caused by a somatic mutation?

A. A purple flower has a small patch of white tissue.

B. One child, in a family of seven, has albinism.

C. One apple tree, in a very large orchard, produces its apples 2 weeks earlier than any of the other trees.

D. A 60-year-old smoker develops lung cancer.

C31. How would nucleotide excision repair be affected if each one of the following proteins was missing? Describe the resulting condition of the DNA if the repair was attempted in the absence of the protein.

A. UvrA

B. UvrC

C. UvrD

D. DNA polymerase

C32. During mismatch repair, why is it necessary to distinguish between the template strand and the newly made daughter strand? How is this accomplished?

C33. What are the two main mechanisms by which cells repair double-strand breaks? Briefly describe each one.

C34. With regard to the repair of double-strand breaks, what are the advantages and disadvantages of homologous recombination repair versus nonhomologous end joining?

C35. When DNA *N*-glycosylase recognizes a thymine dimer, it detects only the thymine located on the 5′ side of the dimer as being abnormal. Make a drawing, and explain the steps whereby a thymine dimer is repaired by the consecutive actions of DNA *N*-glycosylase, AP endonuclease, and DNA polymerase.

C36. What is the underlying genetic defect that causes xeroderma pigmentosum? How can the symptoms of this disease be explained by the genetic defect?

C37. Three common ways to repair changes in DNA structure are nucleotide excision repair, mismatch repair, and homologous recombination repair. Which of these three mechanisms would be used to fix the following types of DNA changes?

A. A change in the structure of a base caused by a mutagen in a nondividing eukaryotic cell

B. A change in DNA sequence caused by a mistake made by DNA polymerase

C. A thymine dimer in the DNA of an actively dividing bacterial cell

C38. Discuss the similarities and differences between nucleotide excision repair and the mismatch repair system.

C39. In *E. coli*, a methyltransferase enzyme coded by the gene *dam* recognizes the sequence 5′–GATC–3′ and attaches a methyl group to the nitrogen at position 6 of adenine. *E. coli* strains that have the *dam* gene deleted are known to have a higher spontaneous mutation rate than normal strains. Explain why.

C40. Describe the similarities and differences between homologous recombination involving sister chromatid exchange (SCE) and that involving homologs. Would you expect the same types of proteins to be involved in both processes? Explain.

C41. The molecular mechanism of SCE is similar to homologous recombination between homologs except that the two segments of DNA are sister chromatids instead of homologous chromatids. If branch migration occurs during SCE, will a heteroduplex be formed? Explain why or why not. Can gene conversion occur during sister chromatid exchange?

C42. Which step(s) in the double-strand break model for recombination would be inhibited if each of the following proteins was missing?

A. RecBCD

B. RecA

C. RecG

D. RuvABC

C43. What two molecular mechanisms can result in gene conversion? Do both occur in the double-strand break model?

C44. Is homologous recombination an example of mutation? Explain.

C45. What are recombinant chromosomes? How do they differ from the original parental chromosomes from which they are derived?

C46. In the Holliday model for homologous recombination (see Figure 19.23), the resolution steps can produce recombinant or nonrecombinant chromosomes. Explain how this can occur.

C47. What is gene conversion?

C48. Make a list of the differences between the Holliday model and the double-strand break model.

C49. In recombinant chromosomes, where is gene conversion likely to take place: near the breakpoint or far away from the breakpoint? Explain.

C50. What events does the RecA protein facilitate?

C51. According to the double-strand break model, does gene conversion necessarily involve DNA mismatch repair? Explain.

C52. What type of DNA structure is recognized by RecG and RuvABC? Do you think these proteins recognize DNA sequences? Be specific about what type(s) of molecular recognition these proteins can perform.

Experimental Questions

E1. Explain how the use of the technique of replica plating supported the random mutation theory and not the physiological adaptation hypothesis.

E2. Outline how you would use the technique of replica plating to show that antibiotic resistance is due to random mutations.

E3. From an experimental point of view, is it better to use haploid or diploid organisms for mutagen testing? Consider the Ames test when preparing your answer.

E4. How would you modify the Ames test to evaluate physical mutagens? Would it be necessary to add the rat liver extract? Explain why or why not.

E5. During an Ames test, bacteria were exposed to a potential mutagen. Also, as a control, another sample of bacteria was not exposed to the mutagen. In both cases, 10 million bacteria were plated and the following results were obtained:

No mutagen: 17 colonies

With mutagen: 2017 colonies

Calculate the mutation rate in the presence and absence of the mutagen. How much does the mutagen increase the rate of mutation?

E6. Richard Boyce and Paul Howard-Flanders conducted an experiment that provided biochemical evidence that thymine dimers are removed from DNA by a DNA repair system. In their studies, bacterial DNA was radiolabeled in such a way that the amount of radioactivity given off by the DNA reflected the amount of thymine dimers it contained. The DNA was then subjected to UV light, causing the formation of thymine dimers. When radioactivity was observed in the soluble fraction, thymine dimers had been excised from the DNA by a DNA repair system.

But when the radioactivity was observed in the insoluble fraction, the thymine dimers had been retained within the DNA. The following table provides some of the experimental results involving a normal strain of *E. coli* and a mutant strain for which exposure to UV light tended to be lethal:

Strain	Treatment	Radioactivity in the insoluble fraction (cpm*)	Radioactivity in the soluble fraction (cpm)
Normal	No UV	<100	<40
Normal	UV-treated, incubated 2 hours at 37°C	357	940
Mutant	No UV	<100	<40
Mutant	UV-treated, incubated 2 hours at 37°C	890	<40

Source: Boyce, Richard P., and Howard-Flanders, Paul, "Release of Ultraviolet Light–Induced Thymine Dimers from DNA in E. Coli K-12," *Proceedings of the National Academy of Sciences of the United States of America*, vol. 51, no. 2, February 1964, 293–300.
*The abbreviation cpm stands for "counts per minute," which is a measure of the number of radioactive emissions from the sample.

Explain the results presented in this table. Why is the mutant strain sensitive to UV light?

E7. A gene exists in two alleles, *B* and *b*. The gene is 1123 bp in length, and the *B* and *b* alleles exhibit single base-pair differences at six different sites that are relatively close together. If you determined experimentally that gene conversion changed the *b* allele to the *B* allele, do you think that it is more likely to have occurred via DNA mismatch repair or gap repair synthesis? Explain your choice.

Questions for Student Discussion/Collaboration

1. In *E. coli*, a variety of mutator strains have been identified in which the spontaneous rate of mutation is much higher than in normal strains. Make a list of the types of abnormalities that could cause a strain of bacteria to become a mutator strain. Which abnormalities do you think would give the highest rate of spontaneous mutation?

2. Discuss the times in a person's life when it is most important to avoid mutagens. Which parts of a person's body should be the most highly protected from mutagens?

3. A large amount of research is aimed at studying mutations. However, there is not an infinite amount of research money.

Where do you think money for mutation research should be directed?

A. Testing of potential mutagens

B. Investigating molecular effects of mutagens

C. Investigating DNA repair mechanisms

D. Some other topic

Note: All answers are available for the instructor in Connect; the answers to the even-numbered questions and all of the Concept Check and Comprehension Questions are in Appendix B.

UNIT V. GENETIC TECHNOLOGIES

CHAPTER OUTLINE

Detection of DNA bands on a gel. DNA can be cut into fragments that can be separated via gel electrophoresis and then stained with a dye called ethidium bromide, which fluoresces under UV light. The cutting and pasting of DNA fragments allows researchers to clone genes.
Eurelios/Science Source

MOLECULAR TECHNOLOGIES

In this chapter, we will focus on methods that are used for manipulating and analyzing DNA and gene products at the molecular level. In some cases, these technologies involve the cutting and pasting of DNA segments to produce new arrangements. **Recombinant DNA technology** is the use of in vitro molecular techniques to isolate and manipulate fragments of DNA to produce new arrangements.

In the early 1970s, the first successes in making recombinant DNA molecules were accomplished independently by two groups at Stanford University: David Jackson, Robert Symons, and Paul Berg; and Peter Lobban and A. Dale Kaiser. Both groups were able to isolate and purify pieces of DNA in a test tube and then covalently link DNA fragments from two different sources. In other words, they constructed molecules called **recombinant DNA molecules.** Shortly thereafter, researchers were able to introduce such recombinant DNA molecules into living cells. Once inside a host cell, the recombinant molecules are replicated to produce many identical copies of a gene—a process called **gene cloning.**

DNA technologies, such as gene cloning, have enabled geneticists to investigate relationships between gene sequences and phenotypic consequences, and these technologies have been widely used to increase our understanding of gene structure and function. Researchers in molecular genetics employ different methods to study gene structure and gene expression. Also, many practical applications of DNA technologies have been developed, including exciting advances such as gene therapy, screening for human diseases, recombinant vaccines, and the production of agriculturally useful transgenic plants and animals, by transferring a cloned gene from one species to some other species. Transgenic organisms have also been important in basic research.

In this chapter, we will focus primarily on the use of technologies as a way to further our understanding of gene structure and function. We will look at the materials and molecular techniques used in gene cloning and explore a technique called polymerase chain reaction (PCR), which can make many copies of DNA within a defined region. We will then discuss how scientists analyze and alter DNA sequences through the techniques of DNA sequencing (a method that enables researchers to determine the base sequence of a DNA strand) and gene editing (a procedure that allows researchers to alter the sequence within genes). Finally, we will examine techniques for identifying gene products, as well as methods for detecting the binding of proteins to DNA or RNA sequences.

In Chapter 21, we will consider many of the practical applications that have arisen as a result of these technologies. Chapters 22 and 23 are devoted to genomics, the molecular analysis of genes and even the entire genome of a species.

20.1 GENE CLONING USING VECTORS

Learning Outcomes:

1. Outline the procedure for cloning a gene into a vector.
2. Describe how cDNA is made.
3. Compare and contrast a genomic library and a cDNA library.
4. Explain how the method called Gibson assembly can be used to link multiple DNA fragments together.

Molecular biologists want to understand how the molecules within living cells contribute to cell structure and function. Because proteins are the workhorses of cells and because they are the products of genes, many molecular biologists focus their attention on the structure and function of proteins or the genes that code them. Researchers may focus their efforts on the study of just one or perhaps a few different genes or proteins. At the molecular level, this poses a daunting task. In all species, any given cell expresses hundreds or thousands of different proteins, making the study of any single gene or protein akin to a "needle-in-a-haystack" exploration.

To overcome this formidable obstacle, researchers frequently take the approach of cloning the genes that code their proteins of interest. Gene cloning is the process of making many copies of a gene. The laboratory methods to clone a gene were devised in the 1970s and 1980s. Since then, many technical advances have enabled gene cloning to become a widely used procedure among scientists, including geneticists, cell biologists, biochemists, plant biologists, microbiologists, evolutionary biologists, clinicians, and biotechnologists.

Table 20.1 summarizes some of the common uses of gene cloning. In modern molecular biology, the diversity of uses for gene cloning is remarkable. For this reason, gene cloning has provided the foundation for critical technical advances in a variety of disciplines, including molecular biology, genetics, cell biology, biochemistry, and medicine. In this and the following section, we will examine two general strategies used to make copies of a gene: the insertion of a gene into a vector that is then propagated in living cells, and cloning via polymerase chain reaction (PCR). Later sections in this chapter, and the other chapters in this unit of this textbook, will consider the uses of gene cloning that are described in Table 20.1.

Cloning Experiments May Involve Two Kinds of DNA Molecules: Chromosomal DNA and Vector DNA

If a scientist wants to clone a particular gene, a common source of the gene is the chromosomal DNA (also called genomic DNA) of the species that carries the gene. For example, if the goal is to clone the rat β-globin gene, this gene is found within the chromosomal DNA of rat cells. In this case, the rat's chromosomal DNA is one type of DNA needed in a cloning experiment. To prepare chromosomal DNA, an experimenter first obtains cellular tissue from the organism of interest. The preparation of chromosomal DNA then involves the breaking open of cells and the extraction and purification of the DNA using biochemical techniques such as chromatography and centrifugation (see Appendix A for a description of these techniques).

Let's begin our discussion of gene cloning by considering a DNA technology in which a gene is removed from its native site within a chromosome and inserted into a smaller segment of DNA

TABLE 20.1
Some Uses of Gene Cloning

Technique	Description
DNA sequencing	A cloned gene provides enough DNA to allow the gene to be analyzed via DNA sequencing (described in Section 20.3). The sequence of the gene can reveal the gene's promoter, regulatory sequences, and coding sequence. DNA sequencing is also important in the identification of alleles that cause cancer and inherited human diseases.
DNA probes	Labeled DNA strands from a cloned gene can be used as probes for identifying RNA. This method of analysis, known as Northern blotting, is described in Section 20.5. Fluorescently labeled DNA molecules are also used to localize genes within intact chromosomes, a method known as fluorescence in situ hybridization (FISH) (see Chapter 22).
Expression of cloned genes	Cloned genes can be introduced into a different cell type or a different species. The expression of cloned genes has many uses: *Research* 1. The expression of a cloned gene can help to elucidate its cellular function. 2. The coding sequence of a gene can be placed next to an active promoter and then introduced into cells that express a large amount of the protein. This allows production of large amounts of purified protein that may be needed for biochemical or biophysical studies. *Biotechnology* 1. Cloned genes can be introduced into bacteria to make pharmaceutical products such as insulin (see Chapter 21). 2. Cloned genes can be introduced into plants and animals to make transgenic species with desirable traits (see Chapter 21). *Clinical trials* 1. Cloned genes have been used in clinical trials involving gene therapy (see Chapter 24).

known as a **vector**—a small DNA molecule that replicates independently of the chromosomal DNA and produces many identical copies of an inserted gene. The purpose of vector DNA is to act as a carrier of the DNA segment to be cloned. In cloning experiments, a vector may carry a small segment of chromosomal DNA, perhaps only a single gene. By comparison, a chromosome carries many more genes, perhaps a few hundred or a few thousand. Like a chromosome, a vector is replicated within a living cell; a cell that harbors a vector is called a **host cell.** When a vector is replicated within a host cell, the DNA that it carries is also replicated.

The vectors commonly used in gene-cloning experiments were derived originally from two natural sources: plasmids or viruses.

Plasmids.   Most vectors are **plasmids,** which are small pieces of DNA that are usually circular. As discussed in Chapter 7, plasmids are found naturally in many strains of bacteria and occasionally in eukaryotic cells.

- Plasmids contain a DNA sequence, known as an **origin of replication,** which is recognized by the replication enzymes of the host cell and allows the plasmid to be replicated. The sequence of the origin of replication determines whether or not the vector can replicate in a particular type of host cell. In cloning experiments, researchers must choose a vector that replicates in the appropriate cell type(s) for their experiments. For example, if researchers want a cloned gene to be propagated in *Escherichia coli,* the vector they employ must have an origin of replication that is recognized by this species of bacterium. The origin of replication also determines the copy number of a plasmid. Some plasmids are said to have strong origins because they achieve a high copy number—perhaps 100–200 copies of the plasmid per cell. Others have weaker origins, so only one or two copies are produced per cell.
- Commercially available plasmids have been genetically engineered for effective use in cloning experiments. They contain unique sites where geneticists insert pieces of DNA.
- Most plasmids used in gene cloning carry one or more genes that confer resistance to antibiotics or other toxic substances. Such a gene is called a **selectable marker** because expression of the gene selects for the growth of the host cells. For example, the gene *amp^R* codes an enzyme known as β-lactamase. This enzyme degrades ampicillin, an antibiotic that normally kills bacteria. Bacteria carrying the *amp^R* gene can grow and form visible colonies on media containing ampicillin, because they can degrade it. In a cloning experiment in which the *amp^R* gene is found within the plasmid, the growth of cells in the presence of ampicillin identifies bacteria that carry the plasmid. In contrast, those cells that do not have the plasmid are ampicillin-sensitive and do not grow.

Viral Vectors.   An alternative type of vector used in cloning experiments is a viral vector. As discussed in Chapter 18, viruses infect living cells and propagate themselves by taking control of

the host cell's metabolic machinery. When a chromosomal gene is inserted into a viral genome, the gene is replicated when the viral DNA is replicated. Therefore, viruses can be used as vectors to carry other pieces of DNA.

Molecular biologists may choose from hundreds of different vectors to use in their cloning experiments. **Table 20.2** provides a general description of several types of vectors that are used to clone DNA segments of various sizes. Vectors designed to introduce genes into plants and animals are discussed in Chapter 21.

TABLE 20.2
Some Vectors Used in Cloning Experiments

Example	Type	Description
pBluescript	Plasmid	A type of vector like the one shown in Figure 20.2. It is used to clone small segments of DNA (e.g., a few thousand base pairs in length) and propagate them in *E. coli.*
YEp24	Plasmid	This plasmid is an example of a **shuttle vector,** which can replicate in two different host species, in this case, *E. coli* and *Saccharomyces cerevisiae.* It carries origins of replication for both species.
λgt11	Viral	This vector is derived from bacteriophage λ, which is described in Chapter 18. It also contains a promoter from the *lac* operon. When fragments of DNA are inserted next to this promoter, the DNA is expressed in *E. coli.* This is an example of an **expression vector.** An expression vector is designed to clone the coding sequences of genes so that they are transcribed and translated correctly.
c2RB	Cosmid	Cosmid vectors are hybrids between plasmids and phage λ vectors. They are designed to clone fragments of DNA that are tens of thousands of bp in length; the DNA can be isolated as a plasmid or from viral particles.
SV40	Viral	This virus naturally infects mammalian cells. Genetically altered derivatives of the SV40 viral DNA are used as vectors for the cloning and expression of genes in mammalian cells that are grown in the laboratory.
Baculovirus	Viral	This virus naturally infects insect cells. Unlike many other types of eukaryotic cells, insect cells grown in liquid media often express large amounts of proteins coded by cloned genes. To make a large amount of a protein, researchers can clone the gene that codes the protein into baculovirus and then purify the protein from insect cells.
YACs, BACs	Artificial chromosome	These vectors behave like chromosomes when inside living cells. A **yeast artificial chromosome (YAC)** can propagate in yeast. An insert within a YAC can be from several hundred thousand base pairs to perhaps 2 million bp in length. **Bacterial artificial chromosomes (BACs)** were developed from bacterial F factors, which are described in Chapter 7. BACs typically can contain inserts with lengths up to 300,000 bp and sometimes larger.

Restriction Enzymes Cut DNA into Pieces and DNA Ligase Joins the Pieces Together

A key step in a cloning experiment is the insertion of chromosomal DNA into a plasmid or a viral vector. One approach to achieve this goal involves the cutting and pasting of DNA fragments. To cut DNA, researchers can use enzymes known as **restriction enzymes,** or **restriction endonucleases.** The restriction enzymes used in cloning experiments bind to a specific base sequence and then cleave the DNA backbone at two defined locations, one in each strand (see **Figure 20.1**). Proposed

by Werner Arber in the 1960s and discovered by Hamilton Smith and Daniel Nathans in the 1970s, restriction enzymes are made naturally by many species of bacteria and protect bacterial cells from invasion by foreign DNA, particularly that of bacteriophages.

Figure 20.1 shows the role of a restriction enzyme, called *Eco*RI, in producing a recombinant DNA molecule.

1. Certain types of restriction enzymes are useful in cloning because they cut the DNA into fragments with "sticky ends." As shown in Figure 20.1, the sticky ends are

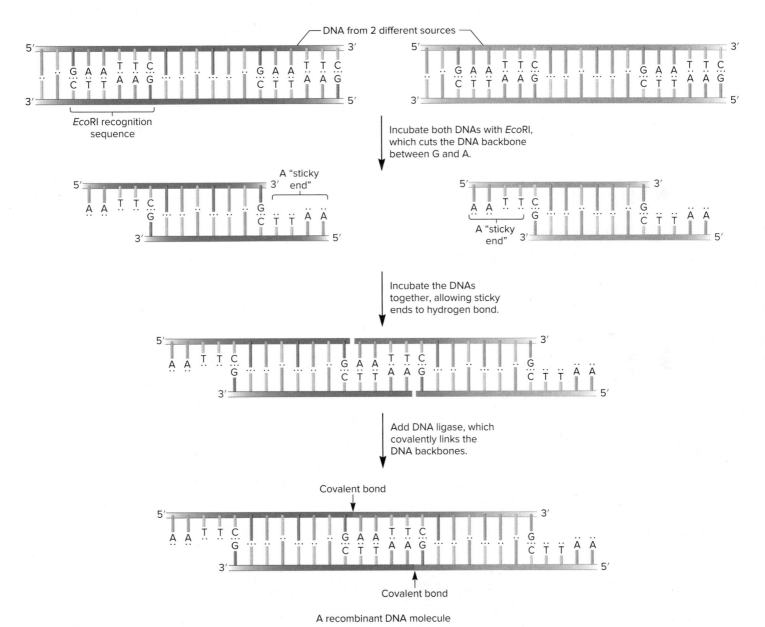

FIGURE 20.1   **The action of a restriction enzyme and the production of recombinant DNA.** The restriction enzyme *Eco*RI binds to a specific sequence, in this case 5′–GAATTC–3′, in both strands. It cleaves the DNA backbone between G and A, producing DNA fragments. The single-stranded ends of different DNA fragments hydrogen bond with each other, because they have complementary sequences. DNA ligase catalyzes the formation of covalent bonds in the DNA backbones of the fragments.

CONCEPT CHECK: Prior to the action of DNA ligase, how many hydrogen bonds are holding these two DNA fragments together?

single-stranded regions of DNA that hydrogen bond to a complementary sequence of DNA from a different source.

2. The ends of two different DNA pieces hydrogen bond to each other because of their complementary sticky ends.

3. The hydrogen bonding between the sticky ends of DNA fragments promotes a temporary interaction between the two fragments. However, this interaction is not stable because it involves only a few hydrogen bonds between complementary bases. How can this interaction be made more permanent? The answer is that the sugar-phosphate backbones of the DNA strands must be covalently linked. Experimentally, this linkage is catalyzed by the addition of **DNA ligase,** which catalyzes covalent bond formation in the sugar-phosphate backbones of both DNA strands after the sticky ends have hydrogen-bonded with each other.

Hundreds of different restriction enzymes from many bacterial species have been identified and are available commercially to molecular biologists. **Table 20.3** gives a few examples. Restriction enzymes usually recognize sequences that are **palindromic;** that is, the sequence in one strand is the same as that in the complementary strand when read in the opposite direction. For example, the sequence recognized by *Eco*RI is 5′–GAATTC–3′ in the top strand. Read in the opposite direction in the bottom strand, this sequence is also 5′–GAATTC–3′.

TABLE 20.3		
Some Restriction Enzymes Used in Gene Cloning		
Restriction Enzyme*	**Bacterial Source**	**DNA Sequence Recognized†**
*Bam*HI	*Bacillus amyloliquefaciens* H	↓ 5′–GGATCC–3′ 3′–CCTAGG–5′ ↑
*Sau*3AI	*Staphylococcus aureus* 3A	↓ 5′–GATC–3′ 3′–CTAG–5′ ↑
*Eco*RI	*Escherichia coli* RY13	↓ 5′–GAATTC–3′ 3′–CTTAAG–5′ ↑
*Nae*I	*Nocardia aerocolonigenes*	↓ 5′–GCCGGC–3′ 3′–CGGCCG–5′ ↑
*Pst*I	*Providencia stuartii*	↓ 5′–CTGCAG–3′ 3′–GACGTC–5′ ↑

*Restriction enzymes are named according to the species in which they are found. The first three letters are italicized because they indicate the genus and species names. Because a species may produce more than one restriction enzyme, the enzymes are designated I, II, III, and so on, to indicate the order in which they were discovered in a given species. Some restriction enzymes, like *Eco*RI, produce a sticky end with a 5′ overhang (see Figure 20.1), whereas others, such as *Pst*I, produce a sticky end with a 3′ overhang. However, not all restriction enzymes cut DNA to produce sticky ends. For example, the enzyme *Nae*I cuts DNA to produce blunt ends.
†The arrows show the locations in the upper and lower DNA strands where the restriction enzyme cleaves the DNA backbone.

Gene Cloning Involves the Insertion of DNA Fragments into Vectors, Which Are Propagated in Host Cells

Now that you are familiar with the materials, let's outline a general strategy for a cloning experiment. In the procedure shown in **Figure 20.2,** the goal is to clone a gene of interest into a plasmid vector that already carries the amp^R gene.

1. To begin this experiment, the chromosomal DNA is isolated and digested with a restriction enzyme. This enzyme cuts the chromosomes into many small fragments. Many copies of the plasmid DNA are also cut with the same restriction enzyme. However, the plasmid has only one unique site for the restriction enzyme.

2. After cutting, the plasmid has two ends that are complementary to the sticky ends of the chromosomal DNA fragments. The digested chromosomal DNA and plasmid DNA are mixed together and incubated under conditions that promote the binding of these complementary sticky ends.

3. DNA ligase is added to catalyze the covalent linkage within each DNA backbone. In some cases, the two ends of the vector simply ligate back together, restoring the vector to its original structure. This is called a recircularized vector. In other cases, a fragment of chromosomal DNA may become ligated to both ends of the vector. In this way, a segment of chromosomal DNA has been inserted into the vector. The vector containing a piece of chromosomal DNA is a **recombinant vector.**

4. Following ligation, the DNA is introduced into host cells treated with agents that render them permeable to DNA molecules. Cells that can take up DNA from the extracellular medium are called **competent cells.** This step in the procedure is called **transformation.** Only a very small percentage of bacterial cells actually take up a plasmid.

5. How can an experimenter distinguish between bacterial cells that have taken up a plasmid and those that have not? In the experiment shown in Figure 20.2, a plasmid is introduced into bacterial cells that were originally sensitive to ampicillin. The bacteria are then streaked onto plates containing bacterial growth media and ampicillin. A bacterium that has taken up a plasmid carrying the amp^R gene continues to divide and forms a bacterial colony containing tens of millions of cells. Because each cell within a single colony is derived from the same original cell, all cells within a colony contain the same type of plasmid DNA.

In the experiment shown in Figure 20.2, how can the experimenter distinguish between colonies of bacteria that have a recircularized vector and those that have a recombinant vector carrying a piece of chromosomal DNA? As shown in the figure, the chromosomal DNA has been inserted into a region of the vector that contains the *lacZ* gene, which codes the enzyme β-galactosidase (see Chapter 14). The insertion of chromosomal DNA into the vector

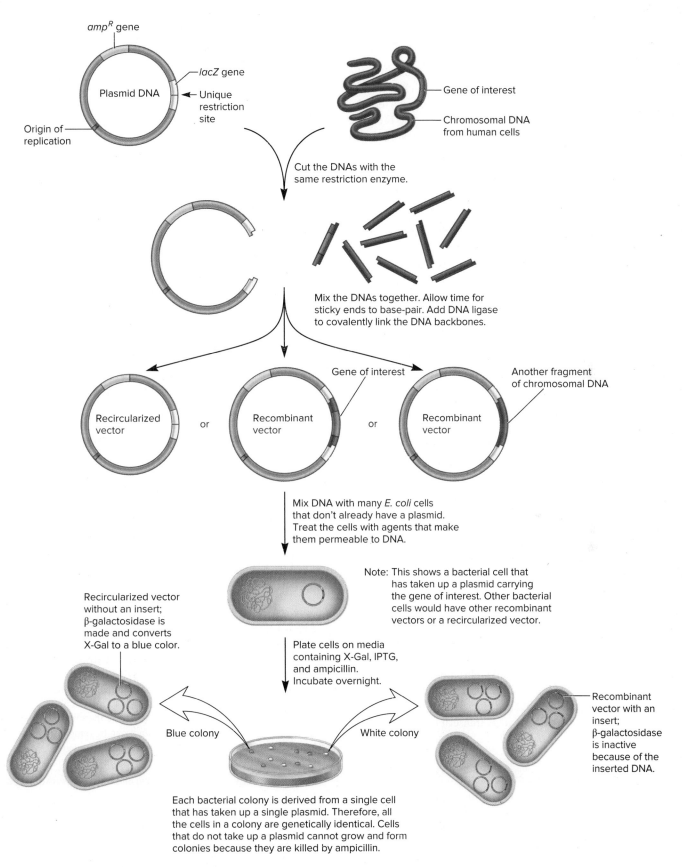

FIGURE 20.2 **The steps in gene cloning.** Note: X-Gal refers to the colorless compound 5-bromo-4-chloro-3-indolyl-β-ᴅ-galactoside. It is converted by β-galactosidase into a blue dye. IPTG is an abbreviation for isopropyl-β-ᴅ-thiogalactopyranoside, which is a nonmetabolizable lactose analog that induces the *lac* promoter.

ONLINE
ANIMATION

CONCEPT CHECK: Explain the role of the gene that is the selectable marker in this experiment.

disrupts the *lacZ* gene, so it no longer produces a functional enzyme. By comparison, a recircularized vector has a functional *lacZ* gene.

The functionality of *lacZ* is determined by including in the growth medium a colorless compound, X-Gal (5-bromo-4-chloro-3-indolyl-β-D-galactoside), which is converted by β-galactosidase into a blue dye. Bacteria grown in the presence of X-Gal and IPTG (isopropyl-β-D-thiogalactopyranoside, an inducer of the *lacZ* gene) form blue colonies if they have a functional *lacZ* gene and white colonies if they do not. In this experiment, therefore, bacterial colonies containing recircularized vectors are blue, whereas colonies containing recombinant vectors are white.

In the example of Figure 20.2, one of the white colonies contains cells with a recombinant vector that carries a human gene of interest; the segment containing the human gene is shown in red. The goal of gene cloning is to produce an enormous number of copies of a recombinant vector that carry the gene of interest. During transformation, a single bacterial cell usually takes up a single copy of a recombinant vector. However, two subsequent events lead to the amplification of the cloned gene. First, because the vector has an origin of replication, the bacterial host cell replicates the recombinant vector to produce many identical copies per cell. Second, the bacterial cells divide approximately every 20 minutes. After overnight growth, a bacterial colony may be composed of 10 million cells, with each cell containing 50 copies of the recombinant vector. Therefore, this bacterial colony would contain 500 million copies of the cloned gene!

cDNA Can Be Made from mRNA via Reverse Transcriptase

In the example of gene cloning in Figure 20.2, chromosomal DNA and plasmid DNA were used as the materials to clone a gene. Alternatively, instead of using chromosomal DNA, a sample of RNA can provide a starting point for cloning. As described in Chapter 18, the enzyme **reverse transcriptase** uses RNA as a template to make a single-stranded or double-stranded DNA molecule. This enzyme is coded in the genome of retroviruses and provides a way for retroviruses to copy their RNA genome into DNA molecules that then integrate into the host cell's chromosomes. Likewise, reverse transcriptase is coded in some retrotransposons and functions in the retrotransposition of such elements.

Researchers use purified reverse transcriptase in a strategy for cloning genes, with mRNA as the starting material (**Figure 20.3**).

1. To begin this experiment, mRNAs, which naturally contain a polyA tail at their 3′ ends, are purified from a sample of eukaryotic cells. The mRNAs are mixed with a primer composed of a string of thymine-containing nucleotides. This short strand of DNA is called a poly-dT primer. The poly-dT primer is complementary to the 3′ end of the mRNAs.
2. Reverse transcriptase and deoxyribonucleotides (dNTPs) are then added to make a DNA strand that is complementary to the mRNA.

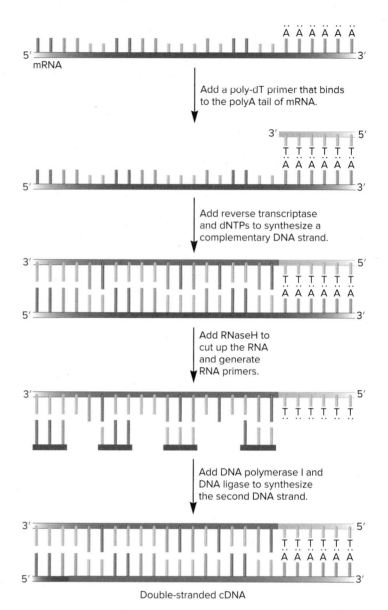

Double-stranded cDNA

ONLINE ANIMATION

FIGURE 20.3 Synthesis of cDNA. A poly-dT primer binds to the 3′ end of eukaryotic mRNAs. Reverse transcriptase catalyzes the synthesis of a complementary DNA strand (cDNA). RNaseH digests the mRNA into short pieces that are used as primers by DNA polymerase I to synthesize the second DNA strand. The 5′ to 3′ exonuclease function of DNA polymerase I removes all of the RNA primers except the one at the 5′ end (because there is no primer upstream from this site). This RNA primer can be removed by the subsequent addition of an RNase. After the double-stranded cDNA is made, it can then be inserted into a vector, as shown in Figure 20.4b.

CONCEPT CHECK: Explain the meaning of the name *reverse transcriptase*.

3. One way to make the other DNA strand is to use RNaseH, which partially digests the RNA, generating short RNAs that are used as primers.
4. DNA polymerase I uses these primers to make segments of DNA that are complementary to the strand made by

reverse transcriptase. During this process, it removes the RNA primers, except the one at the 5′ end. Finally, DNA ligase connects the DNA segments made by DNA polymerase I and thereby produces a continuous DNA strand, as shown at the bottom of Figure 20.3.

When DNA is made using an RNA template, the DNA is called **complementary DNA (cDNA).** The term originally referred to the single strand of DNA that is complementary to the RNA template. However, cDNA now refers to any DNA, whether it is single- or double-stranded, that is made using RNA as the starting material.

Why is cDNA cloning useful? From a research perspective, an important advantage of cDNA is that it lacks introns, which are often found in eukaryotic genes. Because introns can be quite large and numerous, it is simpler to insert cDNAs into vectors if researchers want to focus their attention on the coding sequence of a gene. For example, if the primary goal is to determine the coding sequence of a protein-coding gene, a researcher inserts cDNA into a vector and then determines the DNA sequence of the insert, as described later in this chapter. Similarly, if a scientist wants to express a protein of interest in a cell that does not splice out the introns properly (e.g., in a bacterial cell), it is necessary to make cDNA clones of the gene that codes for that protein.

A DNA Library May Be Constructed Using Genomic DNA or cDNA

In a typical cloning experiment that involves the use of vectors (see Figure 20.2), the treatment of the chromosomal DNA with restriction enzymes yields tens of thousands of different DNA fragments. Therefore, after the DNA fragments are ligated individually to vectors, the researcher has a collection of recombinant vectors, with each vector containing a particular fragment of chromosomal DNA. A collection of recombinant vectors is known as a **DNA library.** When the starting material is chromosomal DNA, the library is called a **genomic library** (**Figure 20.4a**).

Researchers may also make a **cDNA library** that contains recombinant vectors with cDNA inserts. **Figure 20.4b** illustrates how cDNAs are made and inserted into vectors. As shown in Figure 20.3, cDNA is first made via reverse transcriptase. To insert the cDNAs into vectors, short oligonucleotides called linkers are attached to the cDNAs via DNA ligase. The linkers contain DNA sequences with a unique site for a restriction enzyme. After the linkers are attached to the cDNAs, the cDNAs and the vectors are cut with restriction enzymes and then ligated to each other. They now comprise a cDNA library.

After a genomic or cDNA library has been made, researchers can screen that library to identify colonies that carry a particular gene. This screening typically involves the use of labeled probes. For example, a short segment of DNA that is complementary to a specific gene (e.g., gene X) can be labeled, such as with a fluorescent dye. Bacterial colonies are then exposed to the labeled DNA and those that carry gene X will bind the labeled probe very tightly. After any unbound probe is

washed away, only those colonies containing gene X will remain labeled.

Gibson Assembly Is a Cloning Method for Joining Three or More DNA Fragments

Thus far, we have considered how DNA can be digested with a restriction enzyme to produce DNA fragments with sticky ends and how such fragments can be inserted into a vector. Figures 20.1, 20.2, and 20.4 illustrate two key events:

- DNA fragments are made that contain complementary (sticky) ends that can hydrogen bond with each other.
- After hydrogen bonding, the backbones of such DNA fragments are covalently linked to each other via DNA ligase to produce recombinant DNA molecules.

Although the use of restriction enzymes for the construction of recombinant DNA molecules has been widely used, the strategy of employing restriction enzymes to produce complementary ends between different DNA fragments has its drawbacks. These include the following:

- When DNA fragments are generated from digestion by a single restriction enzyme, a researcher cannot control the directionality of the hydrogen-bonding process. For example, in Figure 20.1, the right side of the blue DNA fragment formed hydrogen bonds with the left side of the purple DNA fragment. If you rotate the blue DNA fragment 180 degrees, it is also possible for the left side of the blue DNA fragment to hydrogen bond with the left side of the purple DNA fragment.
- When connecting three or more DNA fragments, it is difficult to control the order in which those fragments will be connected to each other.
- When subjecting chromosomal DNA to restriction enzyme digestion, it is necessary to isolate and purify the desired DNA fragment(s).

To overcome these drawbacks, researchers have devised alternative cloning strategies. One commonly used method is called **Gibson assembly,** which was invented by Daniel Gibson and colleagues in 2009. **Figure 20.5** illustrates this method. The goal here is to insert two different DNA fragments into a vector in a particular order. Although the vector is circular, it is shown in a linear form for simplicity. At the start of this experiment, the two DNA fragments could have already been isolated and purified, or they could still be contained within chromosomal DNA.

1. The first step is to design a set of PCR primers that have complementary ends with other DNA fragments. The procedure of PCR is described in Section 20.2. Look closely at the three PCR primers in Figure 20.5 that are oriented in the 3′ to 5′ direction. Each of them has two distinct regions that are shown in different colors.

2. The PCR primers are used in a PCR reaction to create a new set of three DNA molecules that have short attachments at

FIGURE 20.4 **The construction of a DNA library.** **(a)** To make a genomic library, chromosomal DNA is first digested with a restriction enzyme to produce many fragments. The fragment containing the gene of interest is highlighted in red. Following ligation, each vector carries a different piece of chromosomal DNA. The library shown here is composed of four colonies. An actual genomic library would contain thousands of different bacterial colonies, each one carrying a different piece of chromosomal DNA. **(b)** To make a cDNA library, oligonucleotide linkers that contain a restriction site are attached to the cDNAs so they can be inserted into vectors.

CONCEPT CHECK: What is an advantage of making a cDNA library rather than a genomic library?

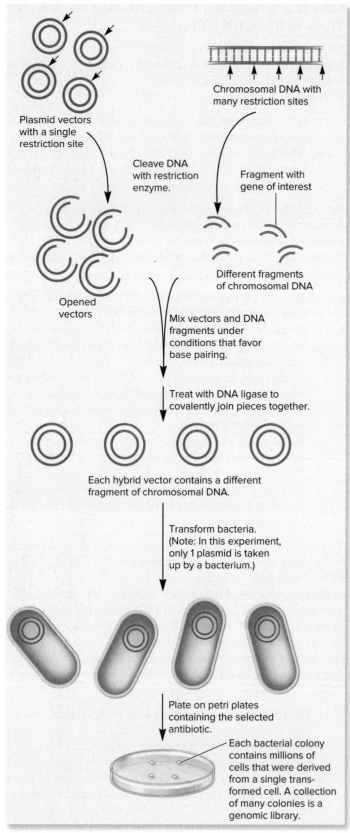

(a) Making a genomic library

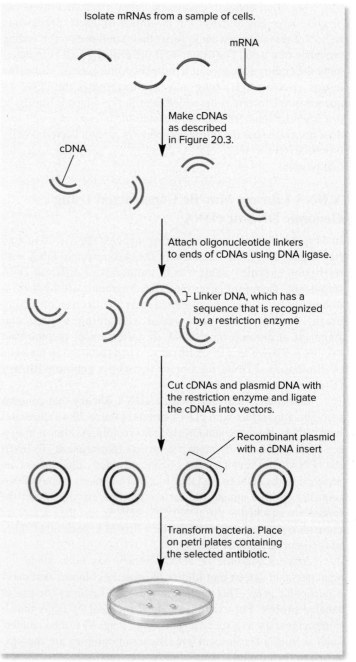

(b) Making a cDNA library

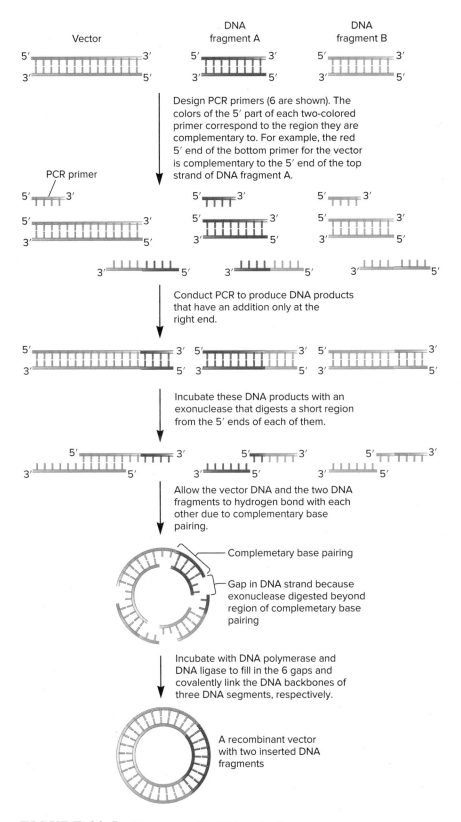

FIGURE 20.5   Gibson assembly. This method is used to connect two or more DNA fragments to each other in a specified way.

CONCEPT CHECK: What is the purpose of using an exonuclease in this procedure?

their right ends. However, at this point, the ends of the DNA fragments are not single-stranded. Therefore they cannot hydrogen bond to other DNA molecules.

3. To produce DNA molecules with single-stranded ends, the DNA molecules are incubated with an exonuclease that digests DNA strands in the 5′ to 3′ direction. This generates DNA molecules with 3′ single-stranded regions.

4. The single-stranded regions are then allowed to hydrogen bond with complementary sites on other DNA molecules. This occurs in a very specific way. The right end of the vector hydrogen bonds with the left end of fragment A. The right end of fragment A hydrogen bonds with the left end of fragment B. And the right end of fragment B hydrogen bonds with the left end of the vector, thereby forming a circular molecule.

5. Because the exonuclease digests away a region that goes beyond the site of complementary binding, short gaps occur after the hydrogen bonding takes place. These gaps are filled in by DNA polymerase, and then the DNA backbones of the three DNA molecules are covalently connected to each other via DNA ligase.

Figure 20.5 shows how three DNA molecules can be connected to each other in a specified way. Depending on the fragment size, Gibson assembly can be used to connect 6 to 15 fragments together in a single experiment!

20.1 COMPREHENSION QUESTIONS

1. Which of the following may be used as a vector in a gene-cloning experiment?
   a. mRNA
   c. Virus
   b. Plasmid
   d. Either b or c

2. The restriction enzymes used in gene cloning experiments _____, which generates sticky ends that can _____.
   a. cut the DNA, enter bacterial cells
   b. cut the DNA, hydrogen bond with complementary sticky ends
   c. methylate DNA, enter bacterial cells
   d. methylate DNA, hydrogen bond with complementary sticky ends

3. Which is the proper order of the following steps in a gene-cloning experiment involving vectors?
   1. Add DNA ligase.
   2. Incubate the chromosomal DNA and the vector DNA with a restriction enzyme.

3. Introduce the DNA into living cells.
4. Mix the digested chromosomal DNA and vector DNA together.

   a. 1, 2, 3, 4   c. 2, 4, 1, 3

   b. 2, 3, 1, 4   d. 1, 2, 4, 3

4. The function of reverse transcriptase is to
  a. use RNA as a template to make DNA.
  b. use DNA as a template to make RNA.
  c. translate RNA into protein.
  d. translate DNA into protein.

5. An advantage of Gibson assembly is that
  a. multiple DNA molecules can be connected in a specified order.
  b. a vector is not needed to achieve gene cloning.
  c. DNA ligase is not needed to connect DNA molecules to each other.
  d. all of the above are advantages.

20.2 POLYMERASE CHAIN REACTION

Learning Outcomes:

1. Describe the three steps of a PCR cycle.
2. Explain how reverse transcriptase PCR is carried out.
3. Outline the method of quantitative PCR, and discuss why it is used.

In the method of gene cloning discussed in Section 20.1, the DNA of interest is inserted into a vector, which is then introduced into a host cell. The replication of the vector within the host cell, followed by the proliferation of that cell, results in the production of many copies of the DNA. Another way to copy DNA, without the aid of vectors and host cells, is to use a technique called **polymerase chain reaction (PCR),** which was developed by Kary Mullis in 1985. He was awarded the Nobel Prize in Chemistry in 1993 for inventing this technique. In this section, we will begin with a general description of PCR and then examine how it can be used to quantitate the amount of DNA or RNA in a biological sample.

Each Cycle of PCR Involves Three Steps: Denaturation, Primer Annealing, and Primer Extension

The PCR method is used to make large amounts of DNA that is located in a defined region flanked by two **primers.** The primers are oligonucleotides, which are short segments of DNA, usually about 15–20 nucleotides in length. As shown in **Figure 20.6a,** the starting material of a PCR experiment can be a complex mixture of DNA. The two primers bind to specific sites in the DNA because their bases are complementary at these sites. The end result of PCR is that the region that is flanked by the primers, which contains the gene of interest, is amplified. The term *amplification*

means the production of many copies of the region between the two primers. In other words, the region has been cloned.

The primers are key reagents in a PCR experiment. They are made chemically, typically not in research or clinical laboratories, but instead at university or industrial facilities. Researchers or clinicians simply order primers that have specified sequences. How does someone choose the primer sequences? In most cases, PCR is conducted on DNA samples in which a scientist already knows the DNA sequences that flank the region of interest. Because the DNA sequence of the entire genome has already been determined for many species, sequence information can come from the genome database of the species of interest. For example, if you wanted to clone a human gene or a portion of a human gene, such as the β-globin gene, you would look up that gene sequence in the human genome database and decide where you want your two primers to bind. You would then order primers with base sequences that are complementary to those sites.

Several other reagents are also needed for a PCR experiment. In the experiment of Figure 20.6a, a sample of chromosomal DNA that contains a gene of interest, which is called **template DNA,** is mixed with the primers. In addition, deoxyribonucleoside triphosphates (dNTPs) are added, as is a thermostable form of DNA polymerase such as *Taq* **polymerase,** isolated from the bacterium *Thermus aquaticus*. A thermostable form of DNA polymerase is necessary because PCR includes heating steps that inactivate most other natural forms of DNA polymerase (which are thermolabile, or readily denatured by heat).

As outlined in **Figure 20.6b**, PCR involves three steps: denaturation, primer annealing, and primer extension.

1. To make copies of the DNA, the template DNA is first denatured by heat treatment, causing the strands to separate (look at the online animation).
2. As the temperature is lowered, oligonucleotide primers hydrogen bond to complementary sequences in the DNA, which is a process called **primer annealing.**
3. Once the primers have annealed, the temperature is raised slightly, and *Taq* polymerase catalyzes the synthesis of complementary DNA strands in the 5′ to 3′ direction, starting at the primers. This process, which is called **primer extension,** doubles the amount of the template DNA.

The three steps shown in Figure 20.6b constitute one cycle of a PCR reaction. The three-step cycle of denaturation, primer annealing, and primer extension is repeated many times in a row, doubling the starting amount of template DNA each time. This process is called a chain reaction because the products of each previous reaction (the newly made DNA strands) are used as reactants (as the template strands) in subsequent reactions.

Figure 20.7 follows a PCR experiment through four cycles. Because the region of interest in the template DNA is doubled with each cycle, the end result of four cycles is 2^4, or 16, copies of that region. Because the starting material contains long strands of chromosomal DNA, some of the products of a PCR experiment contain the region of interest plus some additional DNA at either end. However, after many cycles, the products that contain only the region of interest greatly predominate in the mixture.

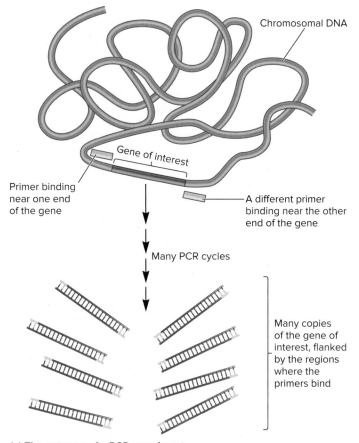

Chromosomal DNA

Gene of interest

Primer binding near one end of the gene

A different primer binding near the other end of the gene

Many PCR cycles

Many copies of the gene of interest, flanked by the regions where the primers bind

(a) The outcome of a PCR experiment

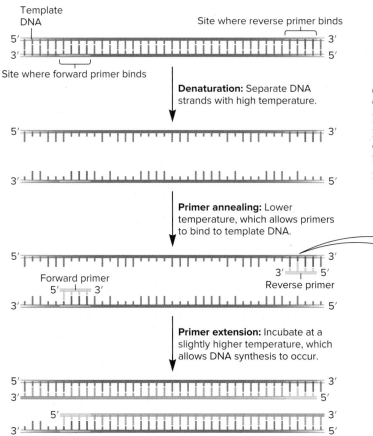

Template DNA

Site where reverse primer binds

5′   3′
3′   5′

Site where forward primer binds

Denaturation: Separate DNA strands with high temperature.

5′   3′

3′   5′

Primer annealing: Lower temperature, which allows primers to bind to template DNA.

5′   3′
                     3′   5′

Forward primer   Reverse primer

5′ 3′

3′   5′

Primer extension: Incubate at a slightly higher temperature, which allows DNA synthesis to occur.

5′   3′
3′   5′

5′   3′
3′   5′

(b) The three steps of a PCR cycle

PCR is carried out in a machine, known as a **thermocycler,** that automates the timing of the temperature changes in each cycle. The experimenter mixes the DNA sample, dNTPs, *Taq* polymerase, and an excess amount of primers together in a single tube. The tube is placed in a thermocycler, and the experimenter sets the machine to operate within a defined temperature range and number of cycles. During each cycle, the thermocycler increases the temperature to denature the DNA strands and then lowers the temperature to allow annealing and extension to take place. Usually, each cycle lasts 2–3 minutes and is then repeated. A typical PCR run is likely to involve 20–30 cycles of replication and takes a couple of hours to complete. The PCR technique can amplify the amount of DNA by a staggering amount. Assuming 100% efficiency, the number of copies of the intervening region between the two primers increases 2^{20}-fold after 20 cycles. This is approximately a million-fold!

An important advantage of PCR is that it can amplify a particular region of DNA from a very complex mixture of template DNA. For example, if a researcher uses two primers that anneal to the human β-globin gene, PCR can amplify just the β-globin gene from a DNA sample that contains all of the human chromosomes!

Alternatively, PCR can be used to amplify a sample of chromosomal DNA nonspecifically. A nonspecific approach uses a mixture of short PCR primers with many different random sequences. These primers anneal randomly throughout the genome and amplify most of the chromosomal DNA. Nonspecific DNA amplification is used to increase the total amount of DNA in very small samples, such as blood stains found at crime scenes.

ONLINE ANIMATION

FIGURE 20.6 **The technique of polymerase chain reaction (PCR).** **(a)** Because the two primers anneal (hydrogen bond) to specific sites, PCR amplifies a defined region of DNA that is located between the two primers. **(b)** During each cycle of PCR, three steps occur: denaturation, primer annealing, and primer extension. The primers used in actual PCR experiments are usually about 15–20 nucleotides in length. The region between the two primers is typically hundreds of nucleotides in length, not just several nucleotides, as shown here.

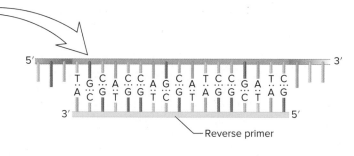

5′   3′
       T G C A C C A G C A T C C G A T C
       A C G T G G T C G T A G G C T A G
3′   5′

Reverse primer

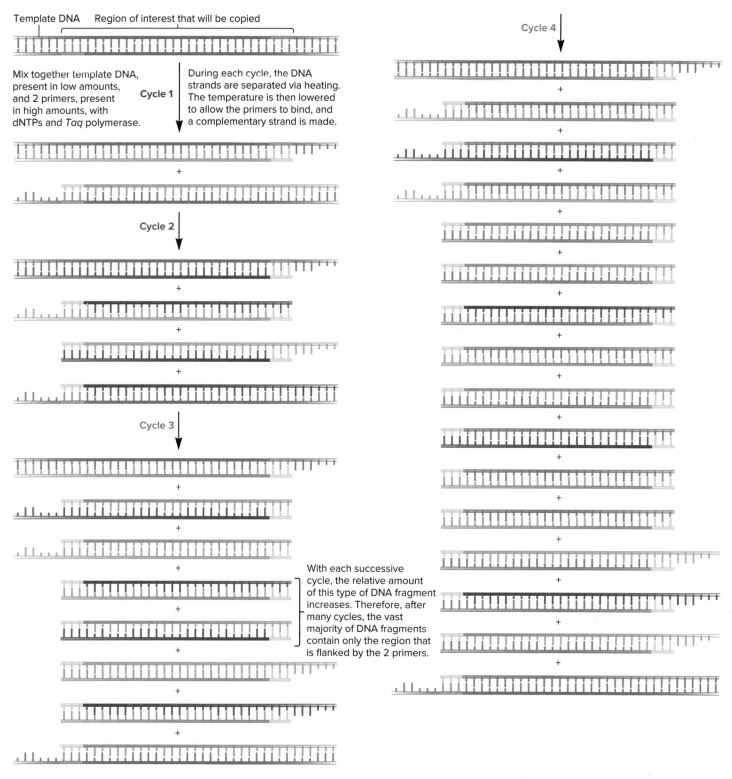

Template DNA   Region of interest that will be copied

Mix together template DNA, present in low amounts, and 2 primers, present in high amounts, with dNTPs and *Taq* polymerase.

Cycle 1

During each cycle, the DNA strands are separated via heating. The temperature is then lowered to allow the primers to bind, and a complementary strand is made.

Cycle 2

Cycle 3

Cycle 4

With each successive cycle, the relative amount of this type of DNA fragment increases. Therefore, after many cycles, the vast majority of DNA fragments contain only the region that is flanked by the 2 primers.

ONLINE ANIMATION

FIGURE 20.7   The technique of PCR carried out for four cycles. During each cycle, oligonucleotides (green) that are complementary to the ends of the targeted DNA sequence bind to the DNA and act as primers for the synthesis of this DNA region. Note: The original DNA strands are purple, and those strands made during the first, second, third, and fourth cycles are blue, red, gray, and orange, respectively.

CONCEPT CHECK: After four cycles of PCR, which type of product predominates? Explain why.

Reverse Transcriptase PCR Is Used to Amplify RNA

PCR is also used to detect and quantitate the amount of specific RNAs in living cells. To accomplish this goal, RNA is isolated from a sample and mixed with reverse transcriptase, a primer that binds near the 3′ end of the RNA of interest, and deoxyribonucleotides (**Figure 20.8**). The product is a single-stranded cDNA, which then can be used as the template DNA in a conventional PCR reaction. The end result is that the RNA has been amplified to produce many copies of DNA.

This method, called **reverse transcriptase PCR (RT-PCR),** is extraordinarily sensitive. RT-PCR can detect the expression of small amounts of RNA from a single cell! As discussed next, certain modifications to PCR allow researchers to observe the accumulation of PCR products and to quantitate the amount of DNA or RNA in a biological sample.

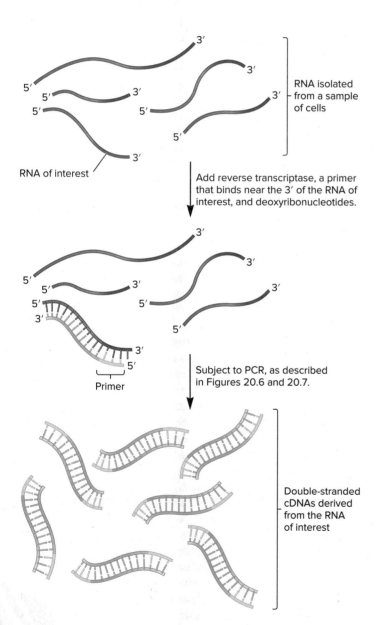

FIGURE 20.8 The technique of reverse transcriptase PCR.

Quantitative PCR Is Used to Quantitate the Amount of a Specific Gene or mRNA in a Sample

In some applications of PCR, the goal is to obtain large amounts of a DNA region. To determine if PCR is successful, a researcher typically runs a sample of DNA on a gel, stains the DNA with ethidium bromide (EtBr), and then observes the gel under UV light, which causes EtBr to fluoresce. If a band of the correct size is seen, the experiment is likely to have been successful. This PCR approach is sometimes called endpoint analysis, because the success of the experiment is judged after PCR is completed.

By comparison, the method of **quantitative PCR (qPCR)** allows a researcher to follow the amount of a specific PCR product in real time as PCR is taking place in a thermocycler. Because the PCR products are ultimately derived from the template DNA that was initially added to the reaction, this approach allows researchers to determine how much DNA, such as the DNA that codes a specific gene, was originally in the sample before PCR was conducted. Similarly, if the starting material is mRNA that is reverse transcribed into DNA, quantitative PCR can be used to determine how much mRNA from a specific gene was in the sample. This approach, called RT-qPCR, provides a way to quantitatively measure gene expression.

How do researchers determine the amount of a PCR product during qPCR or RT-qPCR? The procedure is carried out in a thermocycler that has the capacity to measure changes in the level of fluorescence that is emitted from probes that are added to the PCR mixture. The fluorescence given off by the probes depends on the amount of the PCR product. Several probes have been developed. We will consider one type called TaqMan.

The TaqMan probe is an oligonucleotide that has a reporter molecule at one end and a quencher molecule at the other (**Figure 20.9a**). The oligonucleotide is complementary to a site within the PCR product of interest. The reporter molecule emits fluorescence at a certain wavelength, but that fluorescence is largely absorbed by the nearby quencher. Therefore, the close proximity of the reporter molecule to the quencher molecule prevents the detection of fluorescence from the reporter molecule.

As shown in **Figure 20.9b**, a primer and the TaqMan probe both anneal to the template DNA during the primer annealing step. During the primer extension step, the 5′ to 3′ exonuclease activity of *Taq* polymerase cleaves the oligonucleotide in the TaqMan probe into individual nucleotides, thereby separating the reporter from the quencher. This allows the reporter to emit (unquenched) fluorescence that can be measured within the thermocycler. As PCR products accumulate, more and more of the TaqMan probes are digested, and therefore the level of fluorescence increases.

Stages of qPCR. **Figure 20.10a** considers a qPCR experiment as it occurs over the course of many cycles, such as 20 to 40. qPCR goes through three main phases. Initially, when the amount of PCR products is small and reagents are not limiting, the synthesis of the PCR products occurs with close to 100% efficiency—the amount produced nearly doubles with each cycle. This exponential accumulation is difficult to detect in the earliest cycles because the amounts of PCR products are small. Therefore, the amount of TaqMan probe that is cleaved is also small. In the second phase, as PCR products

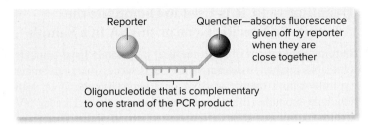

(a) TaqMan probe

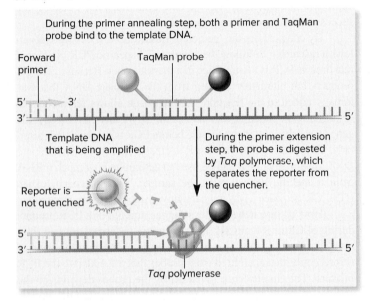

(b) Use of a TaqMan probe in qPCR

FIGURE 20.9 An example of the use of a probe, called Taq-Man, in a quantitative PCR experiment.

CONCEPT CHECK: What must happen to prevent the fluorescence emitted by the reporter molecule from being quenched?

continue to accumulate, their accumulation shows linear growth, but the reaction efficiency falls as reagents become limiting. Finally, in the third phase, the accumulation of PCR products reaches a plateau as one or more reagents are used up.

Cycle Threshold Method for Measuring the Initial Template Concentration.   During the exponential phase of qPCR, the amount of PCR product is proportional to the amount of starting template that was initially added. Therefore, the exponential phase of qPCR is analyzed to quantitate the initial template concentration. The commonly used method for this analysis is called the **cycle threshold method (C_t method).** The cycle threshold (C_t) is reached when the accumulation of fluorescence is significantly greater than the background level. During the early cycles, the fluorescence signal due to the background level of fluorescence is greater than that derived from the amplification of the PCR product. Once the C_t value is exceeded, the exponential accumulation of product can be measured. **Figure 20.10b** considers three PCR runs in which the template DNA was initially added at a high, medium, or low concentration. When the initial concentration of the template DNA is higher, the C_t is reached at an earlier cycle number.

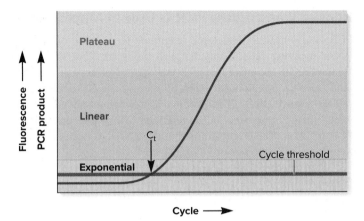

(a) Phases of PCR

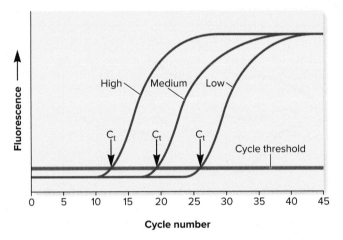

(b) qPCR at high, medium, and low starting concentrations of the template DNA

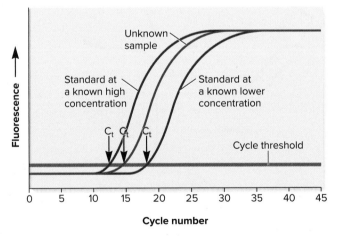

(c) A comparison between an unknown sample and standards of known concentrations

FIGURE 20.10 Examples of data that are obtained from a qPCR experiment. **(a)** The three phases that occur in a typical PCR experiment. **(b)** PCR carried out at three different starting concentrations of the template DNA. **(c)** A comparison between a sample with an unknown DNA concentration and a standard at a high and low concentration.

Using Standards to Determine the Initial Concentration of Template DNA. To determine the amount of starting template DNA, the sample of interest that has an unknown amount of starting template DNA is compared with some type of standard. qPCR involves the coamplification of two templates: the sample of interest and the standard. The two types of PCR products can be monitored simultaneously using molecules that fluoresce in different colors.

What types of standards are commonly used? One possibility is a standard of known concentration that is added to the PCR mixture. For example, plasmid DNA carrying a specific gene can be added to the PCR mixture in known amounts, and the amplification of this plasmid-coded gene provides a standard. By comparing the C_t values for the standard and the unknown sample, researchers can determine the concentration of the unknown sample. This is shown schematically in Figure 20.10c, but is actually accomplished with computer software. Alternatively, researchers may use an internal standard in which another gene that is already present in the sample is also amplified. For example, a gene that codes the cytoskeletal protein called actin may be used. This relative quantitation method is somewhat simpler. The amount of unknown template DNA of interest is expressed relative to the internal standard.

20.2 COMPREHENSION QUESTIONS

1. In one PCR cycle, the correct order of steps is
   a. primer annealing, primer extension, denaturation.
   b. primer annealing, denaturation, primer extension.
   c. denaturation, primer annealing, primer extension.
   d. denaturation, primer extension, primer annealing.

2. In reverse transcriptase PCR, the starting biological material is
   a. chromosomal DNA.
   b. mRNA.
   c. proteins.
   d. all of the above.

3. During qPCR, the synthesis of PCR products is analyzed
   a. at the very end of the reaction by gel electrophoresis.
   b. at the very end of the reaction by fluorescence that is emitted within the thermocycler.
   c. during the PCR cycles by gel electrophoresis.
   d. during the PCR cycles by fluorescence that is emitted within the thermocycler.

20.3 DNA SEQUENCING

Learning Outcome:

1. Outline the steps in automated DNA sequencing via the dideoxy method.

As we have seen throughout this text, our knowledge of genetics can be largely attributed to an understanding of DNA structure and function. The feature that underlies all aspects of inherited traits is the DNA sequence. For this reason, analyzing DNA sequences is a powerful approach for understanding genetics. A technique called **DNA sequencing** enables researchers to determine the base sequence of DNA found in genes and other chromosomal regions. It is one of the most important tools for exploring genetics at the molecular level. Molecular geneticists often want to determine DNA base sequences as a first step toward understanding the function and expression of genes. For example, the investigation of genetic sequences has been vital to our understanding of promoters, regulatory elements, and the genetic code itself. Furthermore, an examination of sequences has facilitated our understanding of origins of replication, centromeres, telomeres, and transposable elements.

During the 1970s, two methods for DNA sequencing were devised. One method, developed by Allan Maxam and Walter Gilbert, involved the base-specific chemical cleavage of DNA. Another method, developed by Frederick Sanger and colleagues, is known as **dideoxy sequencing.** Gilbert and Sanger won the 1980 Nobel Prize in Chemistry for their contributions to DNA sequencing. Because it became the more popular method of DNA sequencing, we will consider the dideoxy method here. In addition, Chapter 22 considers some newer methods of sequencing DNA that are not based on the Sanger dideoxy method. These newer methods are commonly used in projects aimed at determining the DNA sequence of an entire genome.

The dideoxy method of DNA sequencing is based on our knowledge of DNA replication but includes a clever twist. As described in Chapter 11, DNA polymerase connects adjacent deoxyribonucleotides by catalyzing formation of a covalent bond between the 5′ phosphate on one nucleotide and the 3′—OH group on the previous nucleotide (refer back to Figure 11.16). However, chemists can synthesize deoxyribonucleotides that are missing the —OH group at the 3′ position (**Figure 20.11**). These synthetic nucleotides are called **dideoxyribonucleotides (ddNTPs).** [Note: The prefix *dideoxy-* indicates that two (di) oxygens (oxy) have been removed (de) from the nucleotide's sugar; in comparison, ribose has —OH groups at both the 2′ and 3′ positions.] Sanger reasoned that if a dideoxyribonucleotide is added to a growing DNA strand, the strand can no longer grow because the dideoxyribonucleotide is missing the 3′ —OH group. The incorporation of a dideoxyribonucleotide into a growing DNA strand is therefore referred to as **chain termination.**

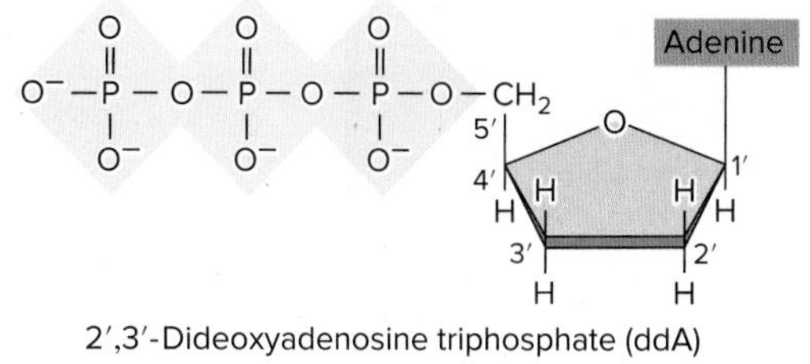

2′,3′-Dideoxyadenosine triphosphate (ddA)

FIGURE 20.11 **The structure of a dideoxyribonucleotide.** Note that the 3′ group is a hydrogen rather than an —OH group. For this reason, another nucleotide cannot be attached at the 3′ position.

To detect the incorporation of dideoxyribonucleotides during DNA replication, the newly made DNA strands must be labeled in some way. When dideoxy sequencing was first invented, researchers labeled the nucleotides with radioisotopes, which allowed the newly made strands to be detected via autoradiography. This traditional DNA sequencing method was modified by labelling each type of dideoxyribonucleotide with a different-colored fluorescent dye. For example, a common way to fluorescently label ddNTPs is the

following: ddA is green, ddT is red, ddG is yellow, and ddC is blue. This newer method is called **automated DNA sequencing** because the sequence of colored nucleotides is read by a machine.

For automated DNA sequencing, the segment of DNA to be sequenced is obtained in large amounts. This can be accomplished using gene cloning, which was described earlier in this chapter. In the example in **Figure 20.12**, the segment of DNA to be sequenced, which we will call the target DNA, was cloned into a vector at a

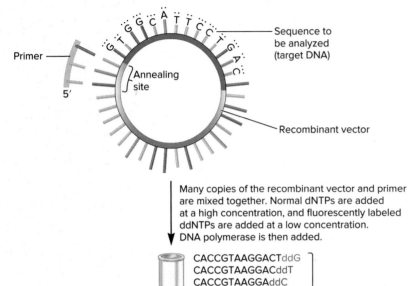

FIGURE 20.12 The protocol for DNA sequencing by the dideoxy method. (a) The method begins with a recombinant vector into which the target DNA has been inserted. For the primer to bind, the recombinant vector must be denatured into single-stranded DNA at the beginning of the experiment. Only the strand needed for DNA sequencing is shown here. This diagram schematically depicts a series of bands on a gel; the four colors of the bands occur because each type of dideoxyribonucleotide is labeled with a different-colored fluorescent dye. As each band passes a laser, the fluorescent dye is excited by the laser beam, and the fluorescence emission is recorded by a machine with a fluorescence detector. The detector reads the level of fluorescence at four wavelengths corresponding to the colors of the four dyes. (b) As shown in the printout, the peaks of fluorescence correspond to the DNA sequence that is complementary to the target DNA. Note that ddG is usually labeled with a yellow dye, but black ink is used instead of yellow on the printout shown in Figure 20.12b for ease of reading.

CONCEPT CHECK: What are two key ways that the DNA segments forming the bands on the gel differ from each other?

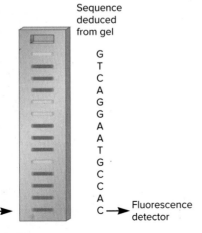

(a) **Automated DNA sequencing**

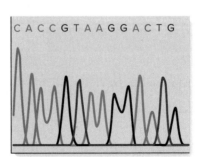

(b) **Output from automated sequencing**

defined location. The target DNA was inserted next to a site in the vector where a primer will bind, which is called the annealing site. The aim of the experiment is to determine the base sequence of the target DNA next to the annealing site. In the experiment shown in Figure 20.12, the recombinant vector DNA has been previously denatured into single strands, usually via heat treatment. Only the strand needed for DNA sequencing is shown here.

Let's now consider the steps that are involved in DNA sequencing by the dideoxy method (Figure 20.12a).

1. A sample containing many copies of the single-stranded recombinant vector is mixed with many primers that will bind to the annealing site. The primer binds to the DNA because it is complementary to the annealing site.

2. All four types of deoxyribonucleotides are added at a high concentration, and all four dideoxyribonucleotides (ddA, ddT, ddG, and ddC), which are fluorescently labeled, are added at a low concentration.

3. DNA polymerase is then added, which causes the synthesis of strands that are complementary to the target DNA sequence. DNA synthesis continues until a dideoxyribonucleotide is incorporated into a growing strand. For example, chain termination can occasionally occur at the sixth or thirteenth position of the newly synthesized DNA strand if a ddT becomes incorporated at either of these sites. Note that the base A (adenine), the base that is complementary to T (thymine), is found at the sixth and thirteenth position in the target DNA. Therefore, we expect to obtain DNA strands that terminate at the sixth or thirteenth positions and have a ddT at their ends. Because these DNA strands contain a ddT, they are fluorescently labeled in red. Alternatively, ddA causes chain termination at the second, seventh, eighth, or eleventh position because the complementary base T base is found at the corresponding position in the target strand. Strands that are terminated with ddA are fluorescently labeled in green.

4. After the samples have been incubated for several minutes, mixtures of DNA strands of different lengths are made, depending on the number of nucleotides attached to the primer. These DNA strands are separated according to their lengths by running them on a slab gel or more commonly by running them through a gel-filled capillary tube. The shorter strands move to the bottom of the gel more quickly than the longer strands.

As mentioned, each of the four types of bases in a dideoxynucleotide is labeled with a different color. Because the incorporation of a dideoxynucleotide stops the further growth of a DNA strand, only the last nucleotide in a strand is labeled. For example, if a ddT was incorporated at the end of a DNA strand, the strand would be red, but if a ddC was incorporated, the strand would be blue. The DNA sequence that is complementary to the target DNA is deduced by determining the sequence of colors in a series of DNA strands that differ in their lengths. Reading the base sequence, from bottom to top, is much like climbing a ladder of colored bands. For this reason, the series of bands obtained by this method is referred to as a **sequencing ladder** (see Figure 20.12a).

Theoretically, it is possible to read this base sequence directly from the gel. From a practical perspective, however, it is faster and more efficient to automate the procedure using a machine with a laser and a fluorescence detector. As the gel is running, each band passes the laser and the laser beam excites the fluorescent dye. The fluorescence detector records the amount of fluorescence emission from the excited dye. The detector reads the level of fluorescence at four wavelengths, corresponding to the four different colors of the dyes. An example of the printout from the fluorescence detector is shown in Figure 20.12b. As can be seen there, the peaks of fluorescence correspond to the DNA sequence that is complementary to the target DNA. Though improvements in automated sequencing continue to be made, a typical sequencing run can determine a DNA sequence that is approximately 700–900 bases long, or perhaps even longer.

20.3 COMPREHENSION QUESTION

1. When a dideoxyribonucleotide is incorporated into a growing DNA strand,
   a. the strand elongates faster.
   b. the strand cannot elongate.
   c. the strand becomes more susceptible to DNase I cleavage.
   d. none of the above events occurs.

20.4 GENE EDITING VIA CRISPR-CAS TECHNOLOGY

Learning Outcomes:

1. Describe the method of CRISPR-Cas technology, and give examples of its uses.
2. Explain how dead Cas9 can be used to increase the expression of genes.
3. Discuss some ethical considerations concerning human gene editing.

To understand how the genetic material functions, researchers often analyze mutations that alter normal DNA sequences, thereby affecting the expression of genes and the outcome of traits. For example, geneticists have discovered that many inherited human diseases, such as sickle cell disease and hemophilia, involve mutations within specific genes. These mutations provide insight into the function of the genes in unaffected individuals. Hemophilia, for example, is caused by mutations in genes that code blood-clotting factors.

Because the analysis of mutations can provide important information about normal genetic processes, researchers often wish to produce mutant organisms. As discussed in Chapter 19, mutations can arise spontaneously or can be induced by environmental agents. Mendel's pea plants are a classic example of allelic strains with different phenotypes that arose from spontaneous mutations. In addition, experimental organisms can be treated with mutagens that increase the rate of mutations.

In this section, we will consider a relatively new technology that allows researchers to make mutations within genes in living cells. Altering the sequence of a gene experimentally is called **gene editing.** We will also explore how gene-editing components can also be used to alter gene expression without changing the sequence of the gene.

Genes in Living Cells Can Be Edited Using CRISPR-Cas Technology

In Chapter 17, we considered how the CRISPR-Cas system provides bacteria with a defense against invasion by bacteriophages (refer back to Figure 17.11). Researchers realized that the components of this system can be used to edit genes in living cells. This application is called **CRISPR-Cas technology.** In 2020, Emmanuelle Charpentier and Jennifer Doudna received the Nobel Prize in Chemistry for their contributions to the development of this technology.

In the natural (type II) system found in bacteria, two different non-coding RNAs (tracrRNA and crRNA) play key roles. The tracrRNA binds to the Cas9 protein and also to crRNA. The crRNA binds to a target DNA, such as a DNA segment within a bacteriophage. These binding interactions guide the Cas9 protein to the bacteriophage DNA, and then Cas9 makes a double-strand break.

Researchers have made a modification to this system to make it efficient for gene editing. They create a single RNA in which the tracrRNA and crRNA are linked to each other (**Figure 20.13a**). This molecule is called a single guide RNA (sgRNA). The spacer region of the sgRNA is designed to be complementary to one of the strands of the gene that is to be edited. The sgRNA binds to Cas9 and guides it to the gene of interest. Cas9 then makes a double-strand break in this gene.

Following this break, two different repair events are possible. If the break is repaired by nonhomologous end joining (NHEJ), the region may incur a small deletion (see the lower left side of **Figure 20.13b**). This deletion may inactivate the gene, particularly if it causes a frameshift mutation in the coding sequence. Alternatively, the double-strand break can be repaired by homologous recombination repair (HRR). Both of these repair mechanisms are described in Chapter 19. For HRR to proceed, the researcher needs to include a segment of DNA, called the donor DNA, that is homologous to the region where the break occurs (see the lower right side of Figure 20.13b). This homologous DNA can be designed to carry a particular mutation, such as a point mutation, that the researcher wants to make. The HRR system swaps in the donor DNA by a double crossover.

The procedure presented in Figure 20.13b can be performed on different cell types and even on whole organisms. For example, this type of experiment could be carried out on a fertilized mouse oocyte. In this case, a researcher would inject a segment of DNA that codes an sgRNA and the Cas9 protein into the oocyte. After these two genes are expressed, the sgRNA-Cas9 complex would either carry out gene inactivation or produce a point mutation if the researcher also injected donor DNA. A major advantage of CRISPR-Cas technology is that it can be directly conducted on living cells, such as mouse oocytes. It has also been applied to mouse embryos, cells within adult mice, human cell lines, roundworm cells, and variety of cells from different species of plants.

During the past decade or so, CRISPR-Cas technology has been used to edit thousands of genes in many different species including animals, plants, and microorganisms. **Table 20.4** describes a few examples.

Gene Expression Can Be Altered Using Dead Cas9

In the CRISPR-Cas technology we have just considered, Cas9 makes a double-strand break in the target gene, which subsequently results in nonhomologous end joining or homologous recombination repair. Researchers have also devised a way to alter gene expression using a mutant version of Cas9 in which its endonuclease function has been inactivated. This mutant version of Cas9 is called dead Cas9 (dCas9), because it cannot make a double-strand break in the target gene.

Figure 20.14 shows how CRISPR-dCas9 technology works. The dCas9 is linked to a transcriptional activator protein. When the spacer region of the sgRNA binds upstream of the promoter for the target gene, the activator is brought close to the gene's promoter and thereby activates its expression. This results in a higher level of transcription of the target gene. Note: This technology is not an example of gene editing; the sequence of the target gene has not been altered. However, the expression of the gene has been altered.

Gene Editing Has Great Potential But Raises Ethical Issues

The ethics of gene editing is a complex topic. While researchers appreciate the potential of CRISPR-Cas technology to modify genes in people afflicted with serious genetic diseases, such as cystic fibrosis and muscular dystrophy, many feel that gene modification is "tampering with nature" and should be banned.

Discussions regarding the ethics of gene manipulation have been around since the 1970s, when gene cloning was invented. However, the meteoric rise of CRISPR-Cas technology has imparted to these discussions a greater sense of urgency because it can be applied to living cells, including oocytes, zygotes, and somatic cells in adult mammals. Many of the recent ethical discussions have centered on the possibility of gene editing in humans, which places this technology in a biomedical context.

In biomedical ethics, four tenets are typically emphasized:

(1) Respect for autonomy: Autonomy requires medical personnel to consult people and obtain their agreement before medical procedures are done to them. Medical confidentiality is another facet of respecting people's autonomy.

(2) Beneficence and (3) non-maleficence: A moral obligation of medicine is to provide a net medical benefit to patients with minimal harm—that is, beneficence with little or no maleficence. The obligation to provide a net benefit to patients requires clear communication about potential harm and benefit because these can differ depending on various factors, including the patient's age and financial situation.

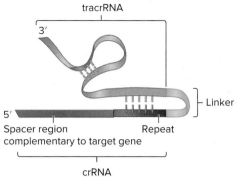

tracrRNA

3'

Linker

5'

Spacer region
complementary to target gene

Repeat

crRNA

(a) Structure of an sgRNA

FIGURE 20.13 **The use of CRISPR-Cas technology for gene editing.** **(a)** Structure of an sgRNA, which is composed of a crRNA that is connected to a tracrRNA via a linker. **(b)** Use of CRISPR-Cas technology. On the left side, the target gene has been repaired by NHEJ and has incurred a small deletion. This may result in gene inactivation. On the right side, the target gene has been repaired by HRR and now carries a point mutation. In this example, the donor DNA is double-stranded, but single-stranded DNA can also be used.

CONCEPT CHECK: How is sgRNA different from certain components of the bacterial defense system described in Chapter 17 (look back at Figure 17.11)?

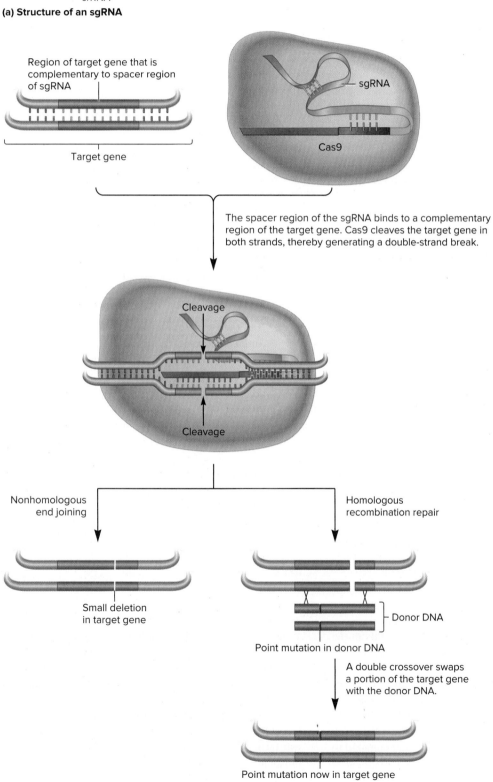

Region of target gene that is
complementary to spacer region
of sgRNA

sgRNA

Target gene

Cas9

The spacer region of the sgRNA binds to a complementary region of the target gene. Cas9 cleaves the target gene in both strands, thereby generating a double-strand break.

Cleavage

Cleavage

Nonhomologous
end joining

Homologous
recombination repair

Small deletion
in target gene

Donor DNA

Point mutation in donor DNA

A double crossover swaps
a portion of the target gene
with the donor DNA.

Point mutation now in target gene

(b) Use of CRISPR-Cas technology to inactivate a gene or create a point mutation

---

TABLE 20.4

Examples of Gene Editing Using CRISPR-Cas Technology and Their Potential Applications

Edited gene(s)	Effects of gene editing
β-globin gene in humans	Gene therapy: Sickle cell disease, which is described in Chapter 27, is caused by a missense mutation in the β-globin gene. Clinical trials are underway to determine if CRISPR-Cas technology can be used to treat this disease by correcting the mutation.
Receptor genes in human T cells	Cancer treatment: In CAR-T therapy, a patient's own immune cells (T cells) are extracted, and the receptors on the extracted cells are altered via CRISPR-Cas technology so they express the chimeric antigen receptor (CAR) that is known to attack cancer cells. The modified cells are reintroduced to the patient where CAR-T cells find and kill the cancer cells. In 2017, the FDA approved two new CAR-T therapies for specific cancers.
Cytokinin oxidase gene in wheat	Agriculture: Inactivation of this gene using CRISPR-Cas technology has been shown to increase yield in wheat.
Genes involved with lipid production	Biofuels: Much effort is underway in the development of photosynthetic algae to produce biofuels, which are fuels derived from living organisms. CRISPR-Cas technology has been used to modify algal genes involved with lipid production, thereby enhancing the production of biofuels.

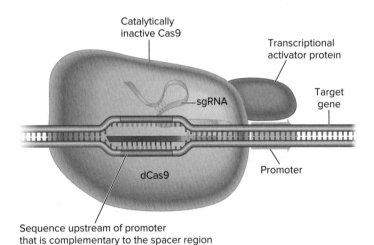

FIGURE 20.14 Gene activation by CRISPR-dCas9 technology. The sgRNA binds upstream of the promoter for the target gene. An activator protein that is linked to dCas9 is then able to increase the transcription of the target gene.

(4) Justice: The fourth ethical principle is justice, which is often regarded as being synonymous with fairness. It commonly includes a fair distribution of scarce resources, respect for people's rights, and respect for morally acceptable laws.

For three decades, these four principles have formed the ethical basis for most decision making in both research settings and clinical practice.

Let's consider some of the key areas of debate with regard to the possible use of CRISPR-Cas technology on humans.

Safety. Due to the possibility of off-target effects (edits in the wrong gene) and mosaicism (when some cells carry the gene edit but others do not), safety is of primary concern. Within the past several years, the level of safety has been evaluated in several clinical trials with varying degrees of success. Though safety is always a primary concern, it also must be evaluated in the context of risk versus benefit. In medical practice, a greater safety risk may be tolerated for treating a severe or fatal disease compared to one that has less serious symptoms. For example, the treatment of terminal cancer would likely tolerate a greater safety risk than the treatment of color blindness.

Somatic Cell Versus Germ-Line Editing. The components of CRISPR-Cas technology can be packaged into viruses and then those viruses can be used to deliver the CRISPR components to various cell types, including oocytes, zygotes, and somatic cells in children or adults.

- Gene editing in somatic cells cannot be passed from parent to offspring. When used to treat a disease, gene editing of somatic cells is considered by many to be less controversial than gene editing of germ-line cells because it has a limited scope and affects only an individual.
- Germ-line editing alters genes in oocytes, zygotes, or early embryos. It has a direct effect on offspring, as well as the potential to be passed to future generations. Germ-line editing is usually considered more controversial because it is a technology that is directly involved with human reproduction.

Fixing Diseases versus Enhancements. CRISPR-Cas technology has considerable promise to treat a wide variety of diseases. Around the world, several dozen clinical trials have been or are currently being conducted, including those focused on many types of cancer, HIV, β-thalassemia, and sickle cell disease. Hundreds of patients have had their genes altered as part of clinical trials. Although debate still continues, the medical community is continuing to pursue CRISPR-Cas technology as a treatment option for a variety of diseases. Many predict that this technology will become an accepted practice in the future for somatic cell gene editing in children and adults.

One common concern about gene editing is that it will be used as a tool for changing human traits and not merely to treat or prevent disease. As we learn more about the effects of human genetic variation on traits, gene editing could be directed at **enhancements**—changing traits that are not involved with causing a disease. For example, certain kinds of genetic variations result in people who have traits that are deemed desirable, such as those affecting appearance, athletic skills, etc. Theoretically, parents could choose germ-line gene editing as a way to produce offspring that harbor such traits. Many consider this concept, called designer babies, a sobering possibility.

Some bioethicists have argued that allowing any type of gene editing places humankind on a "slippery slope" in which society initially accepts gene editing for disease treatment and gradually moves toward the acceptance of enhancements. To stay off the slippery slope, they argue that all forms of gene editing should be banned. The counterargument is that we have a moral obligation to help people who are afflicted with diseases that could be alleviated by CRISRP-Cas technology, based on the principles of beneficence and non-maleficence. The debate continues.

20.4 COMPREHENSION QUESTION

1. The purpose of CRISPR-Cas technology is to
   a. determine if a protein binds to a DNA segment.
   b. determine the sequence of a segment of DNA.
   c. alter the sequence of a segment of DNA.
   d. determine if a gene is expressed.

20.5 BLOTTING METHODS TO DETECT GENE PRODUCTS

Learning Outcome:

1. Explain how Northern and Western blotting are used to detect specific RNAs and proteins, respectively.

Thus far, this chapter has largely focused on the analysis and manipulation of DNA. Scientists also need techniques for the identification of gene products, such as the RNA that is transcribed from a particular gene or the protein that is coded by an mRNA. In this section, we will consider methodologies for the detection of RNAs and proteins.

Northern Blotting Is Used to Detect a Specific RNA

As discussed earlier in this chapter, RT-PCR and RT-qPCR are used to detect and quantitate the amount of RNA in a biological sample, respectively. Another approach, known as **Northern blotting,**[*] is also used to identify a specific RNA within a mixture of many RNA molecules. Researchers employ Northern blotting to investigate the transcription of genes at the molecular level.

- This method can determine if a specific gene is transcribed in a particular cell type, such as nerve or muscle cells, or at a particular stage of development, such as in fetal or adult cells.

*A technique called Southern blotting was named after Edwin Southern. The name *Northern blotting* arose due to the method's technical similarity to Southern blotting, which can detect DNA fragments but is no longer widely used.

- Also, Northern blotting can reveal if a pre-mRNA transcript is alternatively spliced into two or more mRNAs of different sizes.

To conduct a Northern blotting experiment, RNA is extracted and purified from living cells. This RNA can be isolated from a particular cell type under a given set of conditions or during a particular stage of development. Any particular cell produces thousands of different types of RNA molecules, because cells express many genes at any given time. After the RNA is extracted from cells and purified, it is loaded onto a gel that separates the RNA transcripts according to their size. (The technique of gel electrophoresis is described in Appendix A.) The RNAs within the gel are then blotted onto a nylon membrane and probed with a labeled fragment of DNA from a cloned gene. RNAs that are complementary to the DNA probe are detected as labeled bands (shown in purple in **Figure 20.15**).

In the experiment whose results are presented in Figure 20.15, the probe was complementary to an mRNA that codes a protein called tropomyosin. In lane 1, the mRNA was isolated from smooth muscle cells; in lane 2, it was from striated muscle cells; and in lane 3, from brain cells. As seen here, smooth and striated muscle cells contain a large amount of this mRNA. This result is expected because tropomyosin plays a role in the regulation of cell contraction. By comparison, brain cells have much less of this mRNA.

In addition, we see from the locations of the bands that the molecular masses of the three mRNAs differ among the three cell types (shorter mRNAs with less mass move farther down on the gel). This observation indicates that the pre-mRNA is alternatively spliced to contain different combinations of exons.

Western Blotting Is Used to Detect a Specific Protein

For protein-coding genes, the end result of gene expression is the synthesis of proteins. A particular protein within a mixture

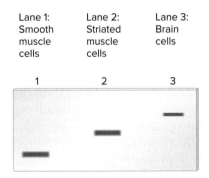

FIGURE 20.15 **The results of Northern blotting.** mRNA is isolated from a sample of cells and then separated by gel electrophoresis. The separated bands are blotted onto a nylon membrane and then placed in a solution containing a labeled DNA probe that is complementary to tropomyosin mRNA. mRNA molecules that are complementary to the probe appear as labeled bands (shown in purple).

of many different protein molecules can be identified by **Western blotting.** This method can determine if a specific protein is made in a particular cell type or at a particular stage of development.

In a Western blotting experiment, proteins are extracted from living cells. As with RNAs, a given cell produces many different proteins at any time, because it is expressing many different genes. After the proteins have been extracted from cells, they are loaded onto a gel that separates them by molecular mass. To perform the separation step, the proteins are first dissolved in <u>s</u>odium <u>d</u>odecyl <u>s</u>ulfate (SDS), a detergent that denatures proteins and coats them with negative charges. The negatively charged proteins are then separated in a gel made of polyacrylamide. This method of separating proteins is called SDS-PAGE (<u>p</u>oly<u>a</u>crylamide <u>g</u>el <u>e</u>lectrophoresis).

Following SDS-PAGE, the proteins within the gel are blotted onto a nylon membrane. The next step is to use a probe that recognizes a specific protein of interest. An important difference between Western blotting and Northern blotting is that Western blotting uses an **antibody** as a probe, rather than a labeled DNA strand. Antibodies bind to sites known as **epitopes.** An epitope has a structure that is recognized by an antibody. The term **antigen** refers to any molecule that is recognized by an antibody. An antigen contains one or more epitopes. In the case of proteins, an epitope is a short sequence of amino acids. Because the amino acid sequence is a unique feature of each protein, any given antibody specifically recognizes a particular protein. In Western blotting, this antibody is called the primary antibody for that protein (**Figure 20.16a**).

After the primary antibody has been given sufficient time to recognize the protein of interest, any unbound primary antibody is washed away, and a secondary antibody is added. A secondary antibody is an antibody that specifically recognizes and binds to a region in the primary antibody. Secondary antibodies, which may be labeled or attached to an enzyme, are used for convenience, because secondary antibodies are available commercially. In general, it is easier for researchers to obtain these antibodies from commercial sources rather than labeling their own primary antibodies.

In a Western blotting experiment, the secondary antibody provides a way to detect the protein of interest in a gel blot. For example, it is common for the secondary antibody to be linked to the enzyme alkaline phosphatase. When the colorless compound XP (5-bromo-4-chloro-3-indolyl phosphate) is added to the blotting solution, alkaline phosphatase converts XP to a dark purple dye (Figure 20.16a). Because the secondary antibody binds to the primary antibody, a protein band that is recognized by the primary antibody becomes dark purple (nearly black).

In the Western blot shown in **Figure 20.16b**, the primary antibody recognized the β-globin polypeptide. The proteins loaded into lane 1 were isolated from mouse red blood cells. The labeled band indicates that β globin is made in these cells. By comparison, the proteins loaded into lanes 2 and 3 were from brain cells and intestinal cells, respectively. The absence

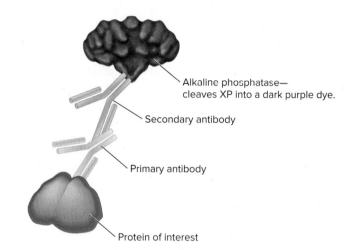

(a) **Interactions between the protein of interest and antibodies**

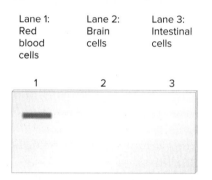

(b) **Results from a Western blotting experiment**

FIGURE 20.16   Western blotting. (a) After blotting, a primary antibody is added that binds to the protein of interest. Then a secondary antibody is added that binds to the primary antibody. In this example, the secondary antibody is also attached to an enzyme called alkaline phosphatase. When the colorless compound XP (5-bromo-4-chloro-3-indolyl phosphate) is added, alkaline phosphatase converts it to a dark purple dye. **(b)** The dark purple band indicates where the primary antibody has recognized the protein of interest, in this case, the β-globin polypeptide. In lane 1, proteins were isolated from mouse red blood cells. As seen here, the β-globin polypeptide is made in these cells. By comparison, lanes 2 and 3 were samples of proteins from brain cells and intestinal cells, respectively, which do not synthesize β globin.

CONCEPT CHECK: What is the purpose of using a secondary antibody in Western blotting?

of any bands indicates that these cell types do not synthesize β globin.

GENETIC TIPS   **THE QUESTION:** In the Western blot shown next, proteins were extracted from red blood cells obtained from tissue samples at different stages of human development. An equal amount of all cellular proteins made from a sample of cells was loaded into each lane. The primary antibody recognizes the β-globin polypeptide that is found in the hemoglobin protein. Explain these results.

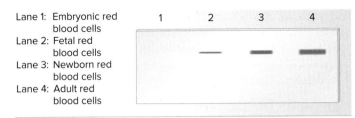

Lane 1: Embryonic red blood cells
Lane 2: Fetal red blood cells
Lane 3: Newborn red blood cells
Lane 4: Adult red blood cells

TOPIC: *What topic in genetics does this question address?* The topic is the use of Western blotting to study gene expression in different cell types.

INFORMATION: *What information do you know based on the question and your understanding of the topic?* From the question, you know the results of a Western blotting experiment. From your understanding of the topic, you may remember that the thickness of the band reflects the amount of the protein of interest, in this case, the β-globin polypeptide.

PROBLEM-SOLVING **S**TRATEGIES: *Analyze the results. Compare and contrast.* One strategy to solve this problem is to look at the results, and compare and contrast the bands that are found in different lanes on the Western blot.

ANSWER: As the blot shows, the amount of β globin increases during development. No detectable β globin is produced during embryonic development. The amount increases significantly during fetal development and becomes maximal in the adult. These results indicate that the β-globin gene is "turned on" in later stages of development, resulting in the synthesis of the β-globin polypeptide. This experiment illustrates how a Western blot can provide information concerning the relative amount of a specific protein within living cells.

20.5 COMPREHENSION QUESTIONS

1. Which of the following methods use(s) a labeled nucleic acid probe, such as a labeled fragment of DNA?
   a. CRISPR-Cas technology
   b. Northern blotting
   c. Western blotting
   d. Both a and b

2. Which of the following methods is used to detect a specific RNA within a mixture of many different RNAs?
   a. CRISPR-Cas technology
   b. Northern blotting
   c. Western blotting
   d. None of the above

3. During Western blotting, the primary antibody recognizes and binds to
   a. the secondary antibody.
   b. the protein of interest.
   c. an mRNA of interest.
   d. a specific fragment of chromosomal DNA.

20.6 METHODS FOR ANALYZING DNA- AND RNA-BINDING PROTEINS

Learning Outcomes:

1. Describe how an electrophoretic mobility shift assay is used to determine if a protein binds to DNA or RNA.
2. Outline the steps in DNase I footprinting, and analyze the results.

Researchers often want to study the binding of proteins to specific sites on a DNA or RNA molecule. For example, the molecular investigation of transcription factors requires methods for identifying interactions between transcription factor proteins and specific DNA sequences. In Chapter 15, we considered the technique of chromatin immunoprecipitation sequencing (ChIP-Seq), which can be used to identify where particular proteins, such as regulatory transcription factors, bind to the DNA (refer back to Figure 15.14). In this section, we will consider two additional methods, called the electrophoretic mobility shift assay and DNase I footprinting, which are used to study protein-DNA interactions. The electrophoretic mobility shift assay is also used to study protein-RNA interactions.

The Electrophoretic Mobility Shift Assay Is Used to Determine If a Protein Binds to a Specific DNA Fragment or RNA Molecule

A technically simple, widely used method for identifying DNA- or RNA-binding proteins is the **electrophoretic mobility shift assay (EMSA),** also known as the **gel retardation assay.** This technique was used originally to study interactions between specific proteins and rRNA molecules and quickly became popular after its success in studying protein-DNA interactions in the *lac* operon. Now it is commonly used as a technique for detecting interactions between RNA-binding proteins and mRNAs and between transcription factors and DNA regulatory elements.

In the case of DNA-binding proteins, the technical basis for an EMSA is that the binding of a protein to a DNA fragment slows down the fragment's ability to move through a polyacrylamide or agarose gel. During electrophoresis, DNA fragments are pulled through the gel matrix toward the bottom of the gel by a voltage gradient. Smaller molecules migrate more quickly through the gel matrix than do larger ones. The binding of a protein to a DNA fragment slows down the DNA's rate of movement through the gel matrix, because the protein-DNA complex has a higher mass than the DNA alone. Compared to the band on the gel for a DNA fragment alone, the band for a protein-DNA complex is shifted to a higher location, because the complex has a slower mobility (**Figure 20.17**). The bands are visualized by staining the DNA with a dye such as EtBr. To increase the sensitivity of the electrophoretic mobility shift assay, the DNA can also be labeled with a fluorescent molecule.

An EMSA must be carried out under nondenaturing conditions. This means that the buffers and gel cannot cause the

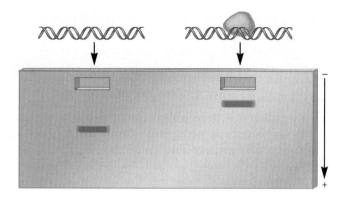

FIGURE 20.17   The results of an electrophoretic mobility shift assay (EMSA). The binding of a protein to a labeled fragment of DNA slows its rate of movement through a gel. For the results shown in the lane on the right, if the concentration of the DNA fragment was higher than the concentration of the protein, there would be two bands: one band with protein bound (at a higher molecular mass) and one band without protein bound (corresponding to the band seen in the left lane).

unfolding of proteins or the separation of the DNA double helix. Avoiding denaturation is necessary so that the proteins and DNA retain their proper structures and are thereby able to bind to each other. The nondenaturing conditions of EMSA differ from those of the more common SDS-PAGE technique, a type of gel electrophoresis in which the proteins are denatured by the detergent SDS.

DNase I Footprinting Shows Detailed Interactions Between a Protein and DNA

Another method for studying protein-DNA interactions is **DNase I footprinting,** a technique described by David Galas and Albert Schmitz in 1978. A DNase I footprinting experiment attempts to identify one or more regions of DNA that interact with a DNA-binding protein. Compared with an electrophoretic mobility shift assay, DNase I footprinting provides more detailed information about the interactions between a protein and DNA. One drawback is that DNase I footprinting is a more complicated technique than EMSA.

To understand the basis of a DNase I footprinting experiment, we need to consider the interactions among three types of molecules: a fragment of DNA, DNA-binding proteins, and reagents that alter DNA structure. As an example, let's examine the binding of RNA polymerase to a bacterial promoter, a topic discussed in Chapter 12. When RNA polymerase holoenzyme binds to the promoter to form a closed complex, it binds tightly to the −35 and −10 sites, but the protein covers up an even larger region of the DNA. Therefore, the holoenzyme bound at the promoter prevents other molecules from gaining access to this region of the DNA. The enzyme DNase I, which cleaves covalent bonds in the DNA backbone, is used as a reagent for determining if a DNA region has a protein bound to it. Galas and Schmitz reasoned that DNase I cannot cleave the DNA at locations where a protein is bound. In this example, it is expected that RNA polymerase holoenzyme will bind to a promoter and protect this DNA region from DNase I cleavage.

Figure 20.18 shows the results of a DNase I footprinting experiment. In this experiment, a sample of many identical DNA

fragments, all of which are 150 bp in length, were labeled at only one end. The sample of fragments was then divided into two tubes: tube A, which did not contain any holoenzyme, and tube B, which contained RNA polymerase holoenzyme. DNase I was then added to both tubes. The tubes were incubated long enough for DNase I

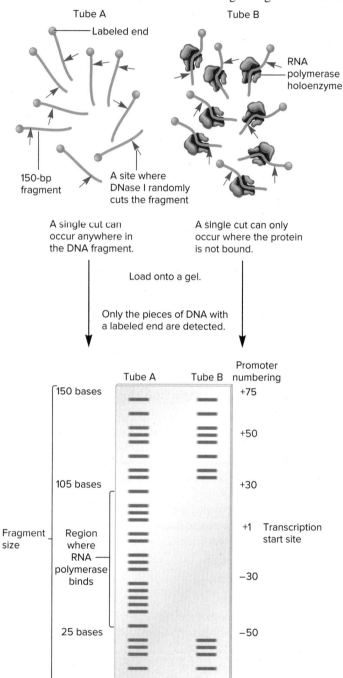

FIGURE 20.18   A DNase I footprinting experiment. Both tubes contained 150-bp fragments of DNA that were incubated with DNase I. Tube B also contained RNA polymerase holoenzyme. The binding of RNA polymerase holoenzyme protected a region with a length of about 80 bp (namely, the −50 region to the +30 region) from DNase I digestion. (Note: The promoter numbering convention shown here is the same as that presented in Figure 12.4.)

CONCEPT CHECK: How does the binding of a protein to DNA influence the ability of DNase I to cleave the DNA?

to cleave the DNA at a single site in each DNA fragment. Each tube contained many 150-bp DNA fragments, and the cleaving of any DNA strand by DNase I occurred randomly. Therefore, the DNase I treatment should produce a mixture of many smaller DNA fragments. A key point, however, is that DNase I cannot cleave the DNA in a region where RNA polymerase holoenzyme is bound. After DNase I treatment, the DNA fragments within the two tubes were separated by gel electrophoresis, producing a series of labeled bands. Note: The traditional use of gel electrophoresis shown in Figure 20.18 has been superseded by the more rapid capillary electrophoresis, which involves electrophoresis through small (capillary) tubes.

In the absence of RNA polymerase holoenzyme (as in tube A), DNase I cleaves each 150-bp fragment at a single random location. Therefore, a continuous range of fragment sizes occurs (see Figure 20.18). However, if you look at the gel lane from tube B, no bands are observed in the size range from 25 to 105 nucleotides. Why are these bands missing? The answer is that DNase I cannot cleave the DNA within the region where the holoenzyme is bound. The middle portion of the 150-bp fragment contains a promoter sequence that binds to the RNA polymerase holoenzyme. Along the right side of the gel, the bases are numbered according to their position within the gene. (The site labeled +1 is where transcription begins.) As seen here, RNA polymerase covers up a fairly large region (its "footprint") of about 80 bp, from the −50 region to the +30 region.

As illustrated in this experiment, DNase I footprinting can identify the DNA region that interacts with a DNA-binding protein. Along with RNA polymerase-promoter binding, DNase I footprinting has been used to identify the binding sites for many other types of DNA-binding proteins, such as eukaryotic transcription factors. This technique has greatly facilitated our understanding of protein-DNA interactions.

20.6 COMPREHENSION QUESTIONS

1. In an EMSA, the binding of a protein to DNA
   a. prevents the DNA from being digested with a restriction enzyme.
   b. causes the DNA to migrate more slowly through a gel.
   c. causes the DNA to migrate more quickly through a gel.
   d. inhibits the expression of any genes within the DNA.
2. The basis for DNase I footprinting is that the binding of a protein to DNA
   a. prevents the DNA from being digested with a restriction enzyme.
   b. enhances the ability of the DNA to be digested with a restriction enzyme.
   c. prevents the DNA from being cleaved by DNase I in that region.
   d. enhances the ability of the DNA to be cleaved by DNase I.

KEY TERMS

Introduction: recombinant DNA technology, recombinant DNA molecules, gene cloning

20.1: vector, host cell, plasmid, origin of replication, selectable marker, shuttle vector, expression vector, yeast artificial chromosome (YAC), bacterial artificial chromosome (BAC), restriction enzyme (restriction endonuclease), DNA ligase, palindromic, recombinant vector, competent cells, transformation, reverse transcriptase, complementary DNA (cDNA), DNA library, genomic library, cDNA library, Gibson assembly

20.2: polymerase chain reaction (PCR), primer, template DNA, *Taq* polymerase, primer annealing, primer extension, thermocycler, reverse transcriptase PCR (RT-PCR), quantitative PCR (qPCR), cycle threshold method (C_t method)

20.3: DNA sequencing, dideoxy sequencing, dideoxyribonucleotide (ddNTP), chain termination, automated DNA sequencing, sequencing ladder

20.4: gene editing, CRISPR-Cas technology, enhancement

20.5: Northern blotting, Western blotting, antibody, epitope, antigen

20.6: electrophoretic mobility shift assay (EMSA) (gel retardation assay), DNase I footprinting

CHAPTER SUMMARY

- Recombinant DNA technology is the use of in vitro molecular techniques to manipulate fragments of DNA to produce new arrangements.

20.1 Gene Cloning Using Vectors

- Gene cloning has many uses, including DNA sequencing, the use of genes as probes, and the expression of cloned genes (see Table 20.1).
- Cloning vectors are derived from plasmids or viruses (see Table 20.2).

- Restriction enzymes, also called restriction endonucleases, cut chromosomal DNA and vector DNA to produce sticky ends that will hydrogen bond with each other. DNA ligase then covalently links the DNA backbones (see Figure 20.1, Table 20.3).
- Gene cloning using vectors involves the insertion of a DNA fragment into a vector, followed by its propagation in a living cell such as *E. coli* (see Figure 20.2).
- Complementary DNA (cDNA) is made when reverse transcriptase uses mRNA as a template to make DNA (see Figure 20.3).

- A DNA library is a collection of recombinant vectors. The inserts can be chromosomal DNA or cDNA (see Figure 20.4).
- Gibson assembly is a cloning method for joining three or more DNA fragments (see Figure 20.5).

20.2 Polymerase Chain Reaction

- Polymerase chain reaction (PCR) uses oligonucleotide primers to copy a specific region of DNA. Each cycle of PCR involves three steps: denaturation, primer annealing, and primer extension (see Figures 20.6, 20.7).
- Reverse transcriptase PCR (RT-PCR) begins with mRNA and is used to study gene expression (see Figure 20.8).
- Quantitative PCR (qPCR) monitors PCR as it occurs in a thermocycler. It is used to quantitate the starting amount of DNA or mRNA in a sample. qPCR uses fluorescent molecules to follow the PCR reaction. A sample that has an unknown amount of DNA is compared with a standard (see Figures 20.9, 20.10).

20.3 DNA Sequencing

- A method of DNA sequencing, called dideoxy sequencing, uses fluorescently labeled dideoxyribonucleotides that cause chain termination (see Figures 20.11, 20.12).

20.4 Gene Editing via CRISPR-Cas Technology

- In CRISPR-Cas technology, components of a bacterial defense system are used to edit genes in living cells (see Figure 20.13, Table 20.4).
- Dead Cas9 can be used to activate the expression of a target gene (see Figure 20.14).
- Gene editing raises several ethical issues.

20.5 Blotting Methods to Detect Gene Products

- Northern blotting uses a labeled DNA probe to detect a specific RNA within a mixture of many different RNAs (see Figure 20.15).
- Western blotting uses antibodies to detect a specific protein within a mixture of many different proteins (see Figure 20.16).

20.6 Methods for Analyzing DNA- and RNA-Binding Proteins

- An electrophoretic mobility shift assay (EMSA) can determine if a protein binds to a specific DNA fragment or RNA molecule because the binding of the protein slows down the movement of the DNA or RNA through a gel (see Figure 20.17).
- DNase I footprinting can identify the regions of a DNA molecule that interact with a DNA-binding protein (see Figure 20.18).

PROBLEM SETS & INSIGHTS

MORE GENETIC TIPS   **1.** RNA was isolated from four different cell types and probed with labeled DNA strands from a cloned gene that is called gene *X*. The results are shown here.

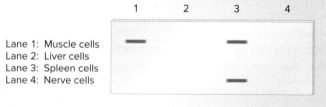

Lane 1: Muscle cells
Lane 2: Liver cells
Lane 3: Spleen cells
Lane 4: Nerve cells

Explain the results of this experiment.

T OPIC: *What topic in genetics does this question address?* The topic is the use of Northern blotting to study gene expression in different cell types.

I NFORMATION: *What information do you know based on the question and your understanding of the topic?* From the question, you know the results of a Northern blotting experiment. From your understanding of the topic, you may remember that the thickness of a band reflects the amount of RNA that is made from a specific gene. Also, if the locations of the bands indicate that the RNAs have different molecular masses, alternative splicing has occurred.

P ROBLEM-SOLVING **S** TRATEGIES: *Analyze data. Compare and contrast.* One strategy to solve this problem is to look at the results of the experiment, and compare and contrast the relative thicknesses and locations of the bands that are found in different lanes on the gel.

ANSWER: In this Northern blot, bands appear in those lanes where RNA was isolated from muscle and spleen cells but not from liver and nerve cells. These results indicate that the muscle and spleen cells express a significant amount of RNA from gene *X*, but the liver and nerve cells do not. The muscle cells show a single band, whereas the spleen cells show this band plus a second band of lower molecular mass. An interpretation of these results is that the spleen cells can alternatively splice the pre-mRNA to produce a second RNA containing fewer or shorter exons.

2. The sequence of a region of interest in a DNA template strand is 3′–ATACGACTAGTCGGGACCATATC–5′. If the primer in a dideoxy sequencing experiment anneals just to the left of this sequence, draw the sequencing ladder that will be obtained.

T OPIC: *What topic in genetics does this question address?* The topic is DNA sequencing. More specifically, the question is about predicting the banding pattern that will appear on a gel in a DNA sequencing experiment.

I NFORMATION: *What information do you know based on the question and your understanding of the topic?* From the question, you know the sequence of a region of a template strand that is adjacent to the annealing site. From your understanding of the topic, you may remember that each dideoxyribonucleotide is labeled with a differently colored fluorescent dye and that the new strand of DNA has a sequence that is complementary to the template strand. The bases that are closer to the annealing site will be contained in smaller DNA fragments and will move more quickly to the bottom of the gel.

PROBLEM-SOLVING **S**TRATEGY: *Make a drawing.* A strategy to solve this problem is to write out the complementary sequence: 5′–TATGCTGATCAGCCCTGGTATAG–3′. The first T (at the 5′ end) will be near the bottom of the gel, the next A will be slightly higher in the gel, the third base, a T, will be slightly higher, and so on.

ANSWER:

G = Yellow
A = Green
T = Red
C = Blue

3. Gene editing is used to explore the structure and function of proteins. For example, changes can be made to the coding sequence of a gene to determine how alterations in the amino acid sequence affect the function of a protein. Let's suppose that you are interested in the functional importance of an asparagine (an amino acid) at a particular location within a protein you are studying. Using CRISPR-Cas technology, you make mutant proteins in which this asparagine has been changed to other amino acids. You then test the coded mutant proteins for functionality. The results are shown next:

	Functionality (%)
Wild-type (normal) protein	100
Mutant proteins containing	
Leucine	7
Phenylalanine	3
Glutamine	98
Proline	4

From these results, what would you conclude about the functional significance of this asparagine within the protein?

TOPIC: *What topic in genetics does this question address?* The topic is gene editing. More specifically, the question is about the effects of gene editing on protein structure and function.

INFORMATION: *What information do you know based on the question and your understanding of the topic?* From the question, you are reminded that gene editing can be used to study protein structure and function. You are also given data on the results of a gene editing experiment. From your understanding of the topic, you may recall that if an amino acid is important for protein structure and function, then conservative substitutions (i.e., of amino acids with similar side chains) are more likely to retain function compared to nonconservative ones.

PROBLEM-SOLVING **S**TRATEGIES: *Analyze data. Compare and contrast.* An approach to solving this problem is to look at the results of this experiment and compare and contrast the level of function that the mutant proteins have compared to the wild-type protein.

ANSWER: These results suggest that the asparagine is important for this protein's function. When this asparagine is replaced with glutamine, which has a very similar structure, the protein retains most of its functionality. However, if it is replaced with other amino acids, most of the functionality is lost.

Conceptual Questions

C1. Discuss three important advances that have resulted from gene cloning.

C2. What structure does a restriction enzyme recognize? What type of chemical bond does it cleave? Be as specific as possible.

C3. Write a double-stranded DNA sequence that is 20 bp long and is palindromic.

C4. What is cDNA? How does cDNA differ from genomic DNA in eukaryotes?

C5. Draw the structural feature of a dideoxyribonucleotide that causes chain termination. Explain how it does this.

Experimental Questions

E1. What is the functional significance of sticky ends in a gene cloning experiment? What type of bonding makes the ends sticky?

E2. Table 20.3 shows the cleavage sites of five different restriction enzymes. After these restriction enzymes have cleaved the DNA, four of them produce sticky ends that can hydrogen bond with complementary sticky ends, as shown in Figure 20.1. The efficiency with which sticky ends bind together depends on the number of hydrogen bonds; more hydrogen bonds make the ends "stickier" and more likely to stay attached. Rank the five restriction enzymes from Table 20.3 (from best to worst) with regard to the efficiency of binding of the sticky ends.

E3. Describe the important features of cloning vectors. Explain the purpose of selectable markers in cloning experiments.

E4. How does gene cloning produce many copies of a gene?

E5. In your own words, describe the series of steps necessary to clone a gene.

E6. What is a recombinant vector? How is a recombinant vector constructed? Explain how X-Gal is used in a method of identifying recombinant vectors that contain segments of chromosomal DNA.

E7. What is a DNA library? Do you think this name is appropriate?

E8. Some vectors used in cloning experiments contain bacterial promoters that are adjacent to unique cloning sites. This makes it possible to insert a gene sequence next to the bacterial promoter and express the gene in bacterial cells. These vectors are called expression vectors. If you wanted to express a eukaryotic protein in bacterial cells, would you clone genomic DNA or cDNA into the expression vector? Explain your choice.

E9. In the experiment shown in Figure 20.5, fragment A is attached to the right end of the vector and fragment B is attached to the left end of the vector. How could you redesign the primers so that fragment B was attached to the right end of the vector and fragment A was attached to the left end of the vector?

E10. Discuss the advantages of Gibson assembly.

E11. Why is a thermostable form of DNA polymerase (e.g., *Taq* polymerase) used in PCR? Is it necessary to use a thermostable form of DNA polymerase in dideoxy sequencing?

E12. Starting with a sample of RNA that contains the mRNA from the β-globin gene, explain how you could create many copies of β-globin cDNA using reverse transcriptase PCR.

E13. What type of probe is used for qPCR? Explain how the level of fluorescence correlates with the amount of PCR product.

E14. What phase of qPCR (exponential, linear, or plateau) is analyzed to quantitate the amount of DNA or RNA in a sample? Explain why this phase is chosen.

E15. DNA sequencing can help researchers identify mutations within genes. The following data were obtained from an experiment in which a normal gene and a mutant gene have been sequenced:

G = Yellow
A = Green
T = Red
C = Blue

Normal   Mutant

Locate and describe the mutation.

E16. A sample of DNA was subjected to automated DNA sequencing and the output is shown here.

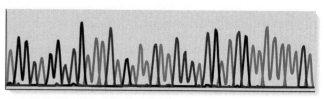

G = Black   T = Red
A = Green   C = Blue

What is the sequence of this DNA segment?

E17. A portion of the coding sequence of a gene is shown here:

5′–GCCCCCGATCTACATCATTACGGCGAT–3′
3′–CGGGGGCTAGATGTAGTAATGCCGCTA–5′

This portion of the gene codes a polypeptide with the amino acid sequence alanine–proline–aspartic acid–leucine–histidine–histidine–tyrosine–glycine–aspartic acid. Using CRISPR-Cas technology, a researcher wants to change the leucine codon to an arginine codon. What is the sequence of the donor DNA that should be used?

E18. Suppose you want to use gene editing to investigate a DNA sequence that functions as a response element for hormone-binding. From previous work, you have narrowed down the response element to a region of DNA that is 20 bp in length with the following sequence:

5′–GGACTGACTTATCCATCGGT–3′
3′–CCTGACTGAATAGGTAGCCA–5′

As a strategy to pinpoint the actual response element sequence, you decide to make 10 different mutants and then analyze their effects using EMSA. What mutations would you make? What results would you expect to obtain?

E19. Gene editing is often used to explore the structure and function of proteins. For example, changes can be made to the coding sequence of a gene to determine how alterations in the amino acid sequence affect the function of a protein. Let's suppose that you are interested in the functional importance of a particular glutamic acid (an amino acid) within a protein you are studying. By gene editing, you make mutant proteins in which the glutamic acid has been changed to other amino acids. You then test the mutant proteins for functionality. The results are as follows:

	Functionality (%)
Normal protein	100
Mutant proteins containing	
Tyrosine	5
Phenylalanine	3
Aspartic acid	94
Glycine	4

From these results, what would you conclude about the functional significance of this glutamic acid within the protein?

E20. Northern blotting depends on the binding of a labeled DNA segment to an mRNA. Explain why this binding occurs.

E21. In Northern and Western blotting, what is the purpose of gel electrophoresis?

E22. What is the purpose of a Northern blotting experiment? What types of information about the transcription of a gene can the results provide?

E23. Suppose an X-linked gene in mice exists as two alleles: *B* and *b*. X-chromosome inactivation, a process in which one X chromosome is turned off, occurs in the somatic cells of female mammals (see Chapter 5). Allele *B* codes an mRNA that is 900 nucleotides long, whereas allele *b* contains a small deletion that shortens the mRNA to a length of 825 nucleotides. Draw the expected Northern blot that will be obtained using mRNA isolated from somatic tissue from the following mice:

Lane 1. mRNA from an X^bY male mouse

Lane 2. mRNA from an X^bX^b female mouse

Lane 3. mRNA from an X^BX^b female mouse

Note: The samples taken from the female mice are not from a clone of cells. They are from a tissue sample, like that shown at the beginning of the experiment in Figure 5.6.

E24. The method of Northern blotting is used to determine the amount and size of a particular RNA transcribed in a given cell type. Alternative splicing (discussed in Chapter 12) produces mRNAs of different lengths from the same gene. The Northern blot shown here was made using a DNA probe that is complementary to the mRNA coded by a particular gene. The mRNA in lanes 1 through 4 was isolated from different cell types, and equal amounts of total cellular mRNA were loaded onto all four lanes.

Lane 1: mRNA isolated from nerve cells
Lane 2: mRNA isolated from kidney cells
Lane 3: mRNA isolated from spleen cells
Lane 4: mRNA isolated from muscle cells

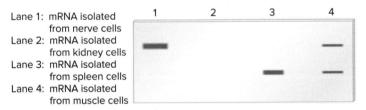

Explain these results.

E25. In the Western blot shown here, proteins were isolated from red blood cells and muscle cells from two different individuals. One individual was unaffected, and the other suffered from a disease known as thalassemia, which involves a defect in hemoglobin. The blot was exposed to an antibody that recognizes β globin, which is one of the polypeptides that constitute hemoglobin. Equal amounts of total cellular proteins were loaded onto all four lanes.

Lane 1: Proteins isolated from normal red blood cells
Lane 2: Proteins isolated from the red blood cells of a thalassemia patient
Lane 3: Proteins isolated from normal muscle cells
Lane 4: Proteins isolated from the muscle cells of a thalassemia patient

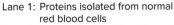
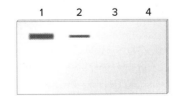

Explain these results.

E26. If you wanted to know if a protein was made during a particular stage of development, what technique would you choose?

E27. Let's suppose a researcher was interested in the effects of mutations on the expression of a protein-coding gene that codes a polypeptide that is 472 amino acids in length. This polypeptide is expressed in leaf cells of *Arabidopsis thaliana*. Because the average molecular mass of an amino acid is 120 daltons, this polypeptide has a molecular mass of approximately 56,640 daltons. Make a drawing that shows the expected results of a Western blot using polypeptides isolated from leaf cells that were obtained from the following plants:

Lane 1. A normal plant

Lane 2. A plant that is homozygous for a deletion that removes the promoter for this gene

Lane 3. A plant that is heterozygous, with one gene that is normal and the other gene with a mutation that introduces an early stop codon at codon 112

Lane 4. A plant that is homozygous for a mutation that introduces an early stop codon at codon 112

Lane 5. A plant that is homozygous for a mutation that changes codon 108 from a phenylalanine codon to a leucine codon

E28. Explain the basis for using an antibody as a probe in a Western blotting experiment.

E29. A cloned gene fragment contains a regulatory element that is recognized by a regulatory transcription factor. Previous experiments have shown that the presence of a hormone results in transcriptional activation by this transcription factor. To study this effect, you conduct a electrophoretic mobility shift assay and obtain the following results:

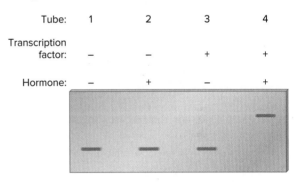

Explain the action of the hormone.

E30. Describe the rationale underlying the technique of an electrophoretic mobility shift assay.

E31. Certain hormones, such as epinephrine, can increase the levels of cAMP within cells. Suppose you pretreat one sample of cells with epinephrine and do not expose a second sample of cells to this hormone. For each cell sample, you prepare a cell extract that contains the CREB protein (see Chapter 15 for a description of this protein). You then use an electrophoretic mobility shift assay to analyze the ability of the CREB protein to bind to a DNA fragment containing a cAMP response element (CRE). Describe what the expected results will be.

E32. An electrophoretic mobility shift assay can be used to study the binding of proteins to a segment of DNA. In the results shown here, an EMSA was used to examine the requirements for the binding of RNA polymerase II (from eukaryotic cells) to the promoter of a protein-coding gene. The assembly of general transcription factors and RNA polymerase II at the core promoter is described in Chapter 12 (Figure 12.14). In this experiment, the segment of DNA containing a promoter sequence was 1100 bp in length.

The fragment was mixed with various combinations of proteins and then subjected to an EMSA.

Lane 1: No proteins added
Lane 2: TFIIA/TFIID
Lane 3: TFIIB
Lane 4: RNA polymerase II
Lane 5: TFIIA/TFIID + TFIIB
Lane 6: TFIIA/TFIID + RNA polymerase II
Lane 7: TFIIA/TFIID + TFIIB + RNA polymerase II

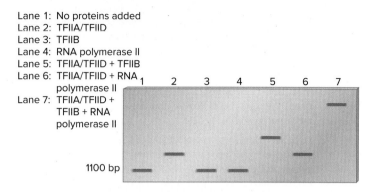

Which proteins (TFIIA/TFIID, TFIIB, and/or RNA polymerase II) are able to bind to this DNA fragment by themselves? Which transcription factor(s) (i.e., TFIIA/TFIID and/or TFIIB) is/are needed for the binding of RNA polymerase II?

E33. As described in Chapter 15 (Figures 15.12 and 15.13), certain regulatory transcription factors bind to DNA and activate RNA polymerase II. When glucocorticoid binds to the glucocorticoid receptor (a regulatory transcription factor), this changes the conformation of the receptor and allows it ultimately to bind to DNA. The glucocorticoid receptor binds to a DNA sequence called a glucocorticoid response element (GRE). In contrast, other regulatory transcription factors, such as the CREB protein, do not require hormone binding in order to bind to DNA. The CREB protein can bind to DNA in the absence of any hormone, but it does not activate RNA polymerase II unless the CREB protein is phosphorylated. (Phosphorylation is stimulated by certain hormones.) The CREB protein binds to a DNA sequence called a cAMP response element (CRE). With these observations in mind, draw the expected results of an EMSA conducted on the following samples:

Lane 1. A 600-bp fragment containing a GRE, plus the glucocorticoid receptor

Lane 2. A 600-bp fragment containing a GRE, plus the glucocorticoid receptor, plus glucocorticoid hormone

Lane 3. A 600-bp fragment containing a GRE, plus the CREB protein

Lane 4. A 700-bp fragment containing a CRE, plus the CREB protein

Lane 5. A 700-bp fragment containing a CRE, plus the CREB protein, plus a hormone (such as epinephrine) that causes the phosphorylation of the CREB protein

Lane 6. A 700-bp fragment containing a CRE, plus the glucocorticoid receptor, plus glucocorticoid hormone

E34. In the technique of DNase I footprinting, the binding of a protein to a region of DNA protects that region from cleavage by DNase I by blocking the ability of DNase I to gain access to the DNA. In the DNase I footprinting experiment shown here, a researcher began with a sample of cloned DNA 400 bp in length. This DNA contained a eukaryotic promoter for RNA polymerase II. The assembly of general transcription factors and RNA polymerase II at the core promoter is described in Chapter 12 (see Figure 12.14). For the sample loaded in lane 1, no proteins were added. For the sample loaded in lane 2, the 400-bp fragment was mixed with RNA polymerase II plus TFIIA/TFIID and TFIIB.

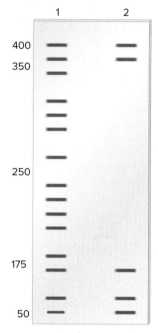

Which region of this 400-bp fragment of DNA is bound by RNA polymerase II and TFIIA/TFIID and TFIIB?

E35. Explain the rationale underlying a DNase I footprinting experiment.

Questions for Student Discussion/Collaboration

1. Discuss and make a list of some of the reasons why determining the amount of a particular gene product can be useful to a geneticist. Use specific examples of known genes (e.g., β-globin gene and others) when making your list.

2. Make a list of possible research questions that could be answered using CRISPR-Cas technology.

Note: All answers are available for the instructor in Connect; the answers to the even-numbered questions and all of the Concept Check and Comprehension Questions are in Appendix B.

*A **cloned animal.*** *The young sheep on the left, named Dolly, was cloned using genetic material from a somatic cell of another sheep.*
R. Scott Horner KRT/Newscom

21

BIOTECHNOLOGY

Biotechnology is broadly defined as the use of living organisms or materials they produce in the development of products or processes that are beneficial to humans. Biotechnology is not a new phenomenon. It began several thousand years ago when humans began to domesticate animals and plants for the production of food. Since that time, many species of microorganisms, animals, and plants have become routinely utilized by people.

More recently, the term *biotechnology* has become associated with molecular genetics. Since the 1970s, molecular genetic tools have provided novel ways to make use of living organisms. As discussed in Chapter 20, recombinant DNA techniques can be used to genetically engineer microorganisms. Recombinant methods also enable the introduction of genetic material into animals and plants, resulting in **genetically modified organisms (GMOs).** An organism that has received genetic material from a different species is called a **transgenic organism.** A gene from one species that is introduced into another species is called a **transgene.**

In the 1980s, court rulings made it possible to patent recombinant organisms such as transgenic microorganisms, animals, and plants. This was one factor that contributed to the growth of many biotechnology industries. In this chapter, we will examine how molecular techniques have expanded our knowledge of the genetic characteristics of commercially important species. We will also discuss examples in which recombinant microorganisms and transgenic microorganisms, animals, and plants have been given characteristics that are useful in the treatment of disease or in agricultural production. These examples include recombinant bacteria that make human insulin, transgenic livestock that produce human proteins in their milk, and transgenic corn that is resistant to insects. In addition, the topics of vaccine development, mammalian cloning, and stem cell research are examined from a technical point of view. In our discussions, we will occasionally touch upon some of the ethical issues associated with these technologies.

21.1   USES OF MICROORGANISMS IN BIOTECHNOLOGY

Learning Outcomes:

1. Explain how bacteria are genetically engineered to produce human insulin.
2. Define *biological control* and *bioremediation,* and describe how microorganisms may play a role in these two processes.

Microorganisms are used to benefit humans in various ways (**Table 21.1**). In this section, we will examine how molecular genetic tools have become increasingly important for improving our use of microorganisms. Such tools can produce recombinant microorganisms with genes that have been manipulated in vitro.

TABLE 21.1
Common Uses of Microorganisms

Application	Examples
Production of medicines	Antibiotics, vitamins
	Synthesis of human insulin by recombinant *E. coli*
Food and beverage production via fermentation	Production of cheese, yogurt, vinegar, wine, and beer
Biological control	Control of plant diseases, insect pests, and weeds
	Symbiotic nitrogen fixation
Bioremediation	Cleanup of environmental pollutants such as petroleum hydrocarbons and synthetics that are difficult to degrade

TABLE 21.2
Examples of Medical Agents Produced by Recombinant Microorganisms

Drug	Action	Condition Treated
Insulin	A hormone that promotes glucose uptake	Diabetes
Tissue plasminogen activator (TPA)	Dissolves blood clots	Heart attacks and other vascular occlusions
Superoxide dismutase	Antioxidant	Heart attacks and tissue damage
Factor VIII	Blood-clotting factor	Certain types of hemophilias
Renin inhibitor	Lowers blood pressure	Hypertension
Erythropoietin	Stimulates the production of red blood cells	Anemia

Why are recombinant organisms useful? Recombinant techniques can improve strains of microorganisms and have even yielded strains that make products not normally produced by microorganisms. For example, genes have been introduced into bacteria to produce medically important products such as human insulin and human growth hormone. As discussed in this section, several recombinant strains are in widespread use.

However, in some areas of biotechnology and in some parts of the world, the commercialization of recombinant strains has proceeded very slowly. This is particularly true of applications in which recombinant microorganisms are used to produce food products or where they are released into the environment. In such cases, safety and environmental concerns, along with negative public perceptions, have slowed or even halted the commercial use of recombinant microorganisms. Nevertheless, molecular genetic research continues, and many biotechnologists expect the use of recombinant microbes to expand in the future.

Many Important Medicines Are Produced by Recombinant Microorganisms

During the 1970s, geneticists became aware of the great potential of recombinant DNA technology to produce therapeutic agents for treating certain human diseases. Healthy individuals express many different genes that code peptide and polypeptide hormones. Diseases can result when people are unable to produce these hormones in sufficient amounts.

In 1976, Robert Swanson and Herbert Boyer formed Genentech Inc. The aspiration of this company was to engineer bacteria to synthesize useful products, particularly peptide and polypeptide hormones. The company's first contract was with researchers Keiichi Itakura and Arthur Riggs, who were able to engineer a bacterial strain that produced somatostatin, a human hormone that inhibits the secretion of a number of other hormones, including growth hormone, insulin, and glucagon. Somatostatin was not chosen for its commercial potential. Instead, it was chosen because the researchers thought it would be technically less difficult to produce than other hormones. Somatostatin is very small (only 14 amino acids long), so it requires a short coding sequence; in

addition, it can be detected easily. Since this pioneering work, recombinant DNA technology has been used to develop bacterial strains that synthesize several other medical agents, a few of which are described in **Table 21.2**.

In 1982, the U.S. Food and Drug Administration approved the sale of the first genetically engineered drug, human insulin, which was produced by Genentech Inc., and marketed by Eli Lilly. In non-diabetic individuals, insulin is produced by the β cells of the pancreas. Insulin functions to regulate several physiological processes, particularly the uptake of glucose into fat and muscle cells. Persons with insulin-dependent diabetes cannot synthesize an adequate amount of insulin due to a loss of β cells. Prior to 1982, insulin was isolated from pancreases removed from cattle and pigs. Unfortunately, in some cases, diabetic individuals became allergic to such insulin and had to use expensive combinations of insulin from other animals and human cadavers. Today, people with diabetes can use genetically engineered human insulin to treat their disease.

Insulin is a hormone composed of two different polypeptide chains, called the A and B chains. To make this hormone using bacteria, the coding sequence of either the A or B chain is placed next to the coding sequence of a native *E. coli* protein, β-galactosidase (**Figure 21.1**). This combined sequence creates a fusion protein comprising β-galactosidase and either the A or B chain. This step is necessary because the A and B chains are rapidly degraded when expressed in bacterial cells by themselves. The fusion proteins, however, are not. How are the two fusion proteins used to make human insulin? After the fusion proteins are expressed in bacteria, they can be purified and then treated with cyanogen bromide (CNBr), which cleaves each fusion protein after a methionine that is found at the junction between β-galactosidase and the A or B chain. This cleavage step separates β-galactosidase from the A or B chain. The A and B chains are then purified and mixed together under conditions in which they refold and form disulfide bonds with each other to make active insulin hormone, which has the same structure as the human-made hormone.

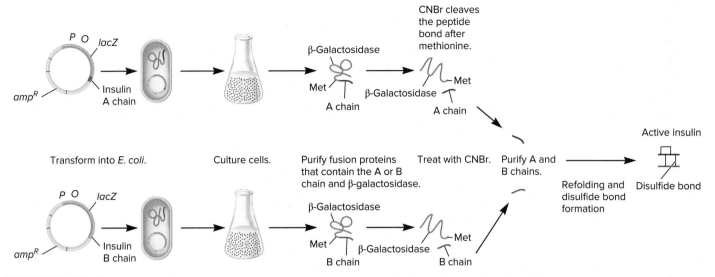

FIGURE 21.1 The use of bacteria to make human insulin. In recent forms of manufactured insulin, slight changes have been made to the insulin amino acid sequence. These changes prevent insulin molecules from clumping together, and thereby improve the manufactured insulin's biological properties.

Genes→Traits The synthesis of human insulin is not a trait that bacteria normally possess. However, genetic engineers can introduce the genetic sequences that code the A and B chains of human insulin via recombinant DNA technology, yielding bacteria that make these polypeptides in the form of fusion proteins that also contain β-galactosidase.

CONCEPT CHECK: What is the function of CNBr in the process of synthesizing human insulin?

Bacteria Are Used as Biological Control Agents

The term **biological control** refers to the control of a pest by the introduction of a natural enemy, predator, or a biological product. During the past 40 years, much research effort has been focused on the biological control of plant diseases and insect pests as an alternative to chemical pesticides. Biological control agents can prevent disease in several ways. In some cases, nonpathogenic microorganisms are used to compete effectively against pathogenic strains for nutrients or space. In other cases, microorganisms may produce a toxin that inhibits other pathogenic microorganisms or insects without harming the plant.

Biological control may involve the use of naturally occurring microorganisms. A successful example is the use of *Agrobacterium radiobacter* to prevent crown gall disease caused by *Agrobacterium tumefaciens*. The disease gets its name from the large swellings (galls) produced by a plant in response to *A. tumefaciens*. The microorganism *A. radiobacter* acts as a biological control agent by producing agrocin 84, an antibiotic that kills *A. tumefaciens*. Molecular geneticists have determined that *A. radiobacter* contains a plasmid with genes that code enzymes responsible for the synthesis of agrocin 84. The plasmid also carries genes that confer resistance to this antibiotic. Unfortunately, this plasmid is occasionally transferred from *A. radiobacter* to *A. tumefaciens* during interspecies conjugation. When this occurs, *A. tumefaciens* gains resistance to agrocin 84. Researchers have identified *A. radiobacter* strains in which this plasmid has been altered genetically to prevent its transfer during conjugation. This conjugation-deficient strain is now used commercially worldwide to prevent crown gall disease.

Another biological control agent is *Bacillus thuringiensis*, usually referred to as Bt (pronounced "bee-tee"). This naturally occurring bacterium produces toxins that are lethal to many caterpillars and beetles that feed on a wide variety of food crops and ornamental plants. Bt is generally harmless to plants and other animals, such as humans, and does not usually harm beneficial insects that act as pollinators. Therefore, it is viewed as an environmentally friendly pesticide. Commercially, Bt is sold in a powder form that is used as a dust or mixed with water as a foliage spray and applied to plants that are under attack by caterpillars or beetles. The pests ingest the bacteria as they eat the leaves, flowers, or fruits. The toxins produced by Bt bring about paralysis of the insect's digestive tract, causing it to stop feeding within hours and die within a few days. Geneticists have cloned the genes that code Bt toxins, which are proteins. As discussed in Section 21.5, such genes have been introduced into crops, such as corn, to produce transgenic plants resistant to insect attack.

Microorganisms Can Reduce Environmental Pollutants

The term **bioremediation** refers to the use of living organisms or their products to decrease pollutants in the environment. As its name suggests, this approach is a biological remedy for pollution. During bioremediation via microorganisms, enzymes produced by a microorganism modify a toxic pollutant by altering or transforming its structure. This event is called **biotransformation.** In many cases, biotransformation results in **biodegradation,** in which the toxic pollutant is degraded, yielding less complex, nontoxic metabolites. Alternatively, biotransformation without biodegradation can also occur. For example, toxic heavy metals can be rendered less toxic by oxidation or reduction reactions carried out by microorganisms. Another way to alter the toxicity of organic pollutants is by promoting polymerization. In many cases, polymerized toxic compounds are less likely to leach from the soil and, therefore, are less environmentally toxic than their parent compounds.

Since the early 1900s, microorganisms have been used in the treatment and degradation of sewage. More recently, the field of bioremediation has expanded into the treatment of hazardous and refractory chemical wastes—chemicals that are difficult to degrade and usually associated with industrial activity. These pollutants include petroleum hydrocarbons, halogenated organic compounds, pesticides, herbicides, and organic solvents. Many new applications that use microorganisms to degrade these pollutants are being tested. The field of bioremediation has been advanced, to a large extent, by better knowledge of how pollutants are degraded by microorganisms, by the identification of new and useful strains of microbes, and by the ability to enhance the bioremediation capabilities of microbes through genetic engineering.

Molecular genetic technology is key in identifying genes that code enzymes involved in bioremediation. The characterization of the relevant genes greatly enhances our understanding of how microbes can modify toxic pollutants. In addition, recombinant strains created in the laboratory can be more efficient at degrading certain types of pollutants.

In 1980, in a landmark case (*Diamond v. Chakrabarty*), the U.S. Supreme Court ruled that a live, recombinant microorganism is patentable as a "manufacture or composition of matter." The first recombinant microorganism to be patented was an oil-eating bacterium that contained a laboratory-constructed plasmid. This strain can oxidize the hydrocarbons commonly found in petroleum. It grew faster on crude oil than did any of the naturally occurring strains tested. However, this recombinant strain has not been a commercial success because it metabolizes only a limited number of toxic compounds, a fraction of the more than 3000 such compounds present in crude oil. Unfortunately, the recombinant strain did not degrade many higher-molecular-weight compounds, which tend to persist in the environment.

Thus far, most bioremediation has involved the use of natural microorganisms rather than recombinant ones. Currently, bioremediation should be considered a developing industry. Many studies are currently underway aimed at elucidating the mechanisms whereby microorganisms degrade toxic pollutants. In the future, recombinant microorganisms may provide an effective way to decrease the levels of toxic chemicals in the environment. However, this approach will require careful studies to demonstrate that recombinant organisms are effective at reducing pollutants and are safe and able to survive when released into the environment.

21.1 COMPREHENSION QUESTIONS

1. Which of the following uses of microorganisms is/are important to humans?
   a. Production of medicines
   b. Food fermentation
   c. Biological control
   d. All of the above are important to humans.

2. What is the key reason why the A and B chains of human insulin are made as fusion proteins with β-galactosidase?
   a. To make purification easier
   b. To prevent their degradation
   c. To be secreted from the cell
   d. All of the above are reasons for making the chains as fusion proteins.

3. Which of the following was the first living organism to be patented?
   a. A strain of *E. coli* that makes somatostatin
   b. A strain of *E. coli* that makes insulin
   c. An oil-eating bacterium
   d. A strain of *B. thuringiensis* that makes an insecticide

21.2   VACCINES

Learning Outcomes:
1. Explain what a vaccine is.
2. Describe the general composition of four different categories of vaccines.
3. List and describe the three types of vaccines against COVID-19.

A **vaccine** is a biological preparation that provides active acquired immunity to a particular infectious disease or to a disease such as cancer. The practice of administering a vaccine is called **vaccination.** The term *vaccination* is from the Latin word *vaccina* meaning "derived from a cow" because development of the first vaccine involved a cowpox virus.

A vaccine typically contains one or more components from a disease-causing agent (e.g., a virus or bacterium) or from a disease-causing cell (e.g., a cancer cell). These components contain antigens that stimulate the body's immune system to produce antibodies that recognize the component. After antibodies are made, they can destroy pathogenic viruses, bacteria, or cancer cells. Furthermore, the immune system will generate memory cells that retain the ability to fight future infections.

Most vaccines are administered to people prior to acquiring a disease, often by an injection into the arm. Such vaccines are prophylactic—they prevent infections or lessen disease symptoms. However, some vaccines are therapeutic. They are used to fight a disease that has already occurred, such as cancer. In this section, we will begin by discussing the discovery of the first human vaccine for smallpox. We will then survey the four general types of vaccines in use today, and explore recent advances in the development of vaccines against COVID-19.

The First Human Vaccine Was Against Smallpox

A physician, Edward Jenner, became intrigued by the observation that dairy cattle workers who were exposed to the cowpox virus were not susceptible to human smallpox, which is a more serious and often fatal disease. In 1796, Jenner inoculated an 8-year-old child by taking pus from cowpox lesions on a milkmaid's hands and introducing the fluid into a cut made in the child's arm. The child developed mild disease symptoms. Six weeks later, when Jenner exposed the child to smallpox, the child did not develop the infection, and remained resistant to infection on 20 subsequent exposures.

Soon after Jenner's work, exposure to cowpox became a widely practiced method to vaccinate people against smallpox. It saved millions of lives. Due to a worldwide vaccination effort, smallpox was eradicated; the last known case occurred in 1978. In spite of the tremendous positive impact of smallpox vaccination, many clinicians and ethicists have raised ethical concerns regarding Jenner's work, particularly because it involved experimentation on a child. In modern times, new disease treatments are done under carefully monitored clinical trials.

As mentioned, cowpox causes mild symptoms in humans, but provides immunity against human smallpox, which is a more serious and often fatal disease. You may be wondering how exposure to cowpox confers immunity to smallpox? The explanation is that the cowpox and smallpox viruses are closely related and produce viral proteins that are very similar in their molecular structures. Therefore, antibodies that are made against the cowpox virus can also recognize certain proteins from the smallpox virus and thereby prevent a smallpox infection. This phenomenon is called **cross immunity.**

Vaccines Are Diverse in the Types of Components They Contain

Vaccination has been widely practiced for over 200 years. During that time, researchers have developed many different types of vaccines, which vary in their compositions.

Whole-Pathogen Vaccines. Whole-pathogen vaccines consist of entire pathogens that have been completely inactivated or weakened so they cannot cause serious disease symptoms. They were the first types of vaccines that were developed and administered to people. Whole-pathogen vaccines include inactivated and attenuated vaccines:

Inactivated Vaccines. Inactivated vaccines contain a pathogen that has been treated in such a way that it cannot cause an infection. Such vaccines typically provide significant short-term immunity, but may not be as strong as attenuated vaccines (described next). Examples of inactivated vaccines include those against influenza, hepatitis A, and rabies.

Attenuated Vaccines. An attenuated vaccine is created by reducing the virulence of a pathogen, but keeping it viable. For example, the infectious agent used in the vaccine might contain mutations that render it less virulent while still retaining its ability to induce protective immunity. In comparison to inactivated vaccines, attenuated vaccines typically promote a strong and effective immune response that is long-lasting. Examples include vaccinations against measles, mumps, and rubella (MMR combined vaccine), chickenpox, and yellow fever.

Though whole-pathogen vaccines are still widely used, the advent of molecular methods described in Chapter 20 has allowed the development of vaccines that contain only a portion of a disease-causing agent. One advantage of these newer vaccines is that they are easier to produce on a large scale. These include viral vector vaccines, subunit vaccines, and nucleic acid vaccines, which are described next.

Viral Vector Vaccines. Viral vector vaccines use a modified version of a virus that is different from the virus that the vaccine is directed against. For example, the viral vector could be an adenovirus (a cause of the common cold), but the vaccine could be targeted against something else, such as SARS-CoV-2, which causes COVID-19 (see Chapter 18).

How do viral vector vaccines work? The viral vector is stripped of any disease-causing genes and sometimes also genes that can enable the viral vector to replicate in human cells. The gene(s) coding for one or more protein antigens from the target pathogen, such as SARS-CoV-2, are then inserted into the viral vector's genome. When the viral vector infects a person's cells, it delivers these genetic instructions to synthesize one or more proteins from the target pathogen, thereby eliciting an immune response. Thus far, viral vector vaccines have been approved for treatment against Ebola virus and SARS-CoV-2.

Subunit Vaccines. Instead of using any type of viral vector, subunit vaccines contain only certain components, or antigens, that best stimulate the immune system. The term *subunit* means that the vaccine contains only a portion of a disease-causing agent. For example, a subunit vaccine might contain only a single viral glycoprotein that is found in the envelop of a particular virus. This formulation can make a vaccine safer and easier to produce.

Subunit vaccines often require the addition of adjuvants, which are substances that accelerate, enhance, and/or prolong immune responses. Vaccine adjuvants are often needed because the antigens alone are not sufficient to induce adequate short-term or long-term immunity. Another limitation of subunit vaccines is that you may need booster shots to get ongoing protection against diseases.

Subunit vaccines vary in the types of subunits they contain, which include the following:

Proteins. As mentioned, a subunit vaccine against a virus may contain a viral spike glycoprotein. In some therapeutic cancer vaccines, the vaccine contains proteins that are produced by cancer cells. Such proteins can be made by using recombinant DNA technology, as discussed in Chapter 20. Exposure to these proteins can bolster the immune system to kill the cancer cells.

Polysaccharides. Polysaccharides are complex carbohydrates composed of sugar units that are linked to each other. They are found in the protective capsule that surrounds certain types of pathogenic bacteria. For example, pneumococcal polysaccharide (PPS) vaccine contains a capsular polysaccharide that results in the production of antibodies against 23 species of pathogenic bacteria. These include *Streptococcus pneumoniae*, which can cause both pneumonia and meningitis (infection of the tissue covering the brain and spinal cord). The antibodies that are produced following PPS vaccination allow immune system cells to phagocytize (engulf) bacteria and thereby destroy them.

Nucleic Acid Vaccines. A more recent approach to vaccine production involves introducing genetic material coding the protein antigen or antigens against which an immune response is sought. The body's own cells then use this genetic material to produce the antigen(s). In humans, a nucleic acid vaccine is typically injected into the upper arm and muscle cells take up the nucleic acid and express the antigen. Potential advantages of this approach include the relative ease of large-scale manufacturing of the vaccine and the ability to modify the vaccine when new mutant variants arise.

Nucleic acid vaccines comprise two main types:

DNA Plasmid Vaccines. These vaccines contain a small circular piece of DNA called a plasmid that carries genes coding proteins from a specific pathogen. As of 2022, a few veterinary DNA plasmid vaccines have been approved, including a vaccine against West Nile virus infection in horses.

mRNA Vaccines. Vaccines based on messenger RNA (mRNA) usually work by introducing an mRNA that codes a viral spike glycoprotein, which is normally found on the virus's outer membrane. The mRNA is encapsulated in a lipid nanoparticle. After injection into the arm, muscle cells take up the encapsulated mRNA (**Figure 21.2**). The mRNA is then translated into the spike glycoprotein, which is transported to the plasma membrane of muscle cells.

Because it has an amino acid sequence that is different from any human protein, the spike glycoprotein is recognized as a foreign antigen. This activates the immune system, which produces antibodies against the spike glycoprotein. Antibody-producing cells are derived from immune system cells that are called B cells. Those B cells that happen to recognize the spike glycoprotein will greatly increase in number when a person is vaccinated against or infected with SARS-CoV-2. Most of these B cells will differentiate into plasma cells that will produce antibodies against the spike glycoprotein. Such antibodies have the ability to kill virally infected cells and neutralize virus that is circulating in the body.

Some of the B cells that are produced in response to vaccination or viral infection become memory cells that can persist for many months and possibly years. The memory cells will rapidly proliferate if an individual has a subsequent exposure to the virus, which leads to a quick immune response that typically prevents or diminishes the severity of an infection.

Several Different Vaccines Have Been Developed Against COVID-19

As was noted in Chapter 18, the COVID-19 outbreak is thought to have originated in Wuhan, China; it spread across the world within a matter of months. On March 11, 2020, the World Health Organization characterized the COVID-19 outbreak as a **pandemic**—a disease that occurs over a wide geographic area and usually affects a high proportion of the population. The pandemic sparked a great sense of urgency regarding vaccine development, which occurred at an unprecedented pace.

To be approved for widespread use, vaccine development must go through a series of tests and trials.

Preclinical Testing. Researchers test a new vaccine on cells grown in the laboratory and on experimental animals to determine if it elicits an immune response.

Phase 1 Safety Trials. The vaccine is administered to a small number of people to test its safety and dosage, and to determine if it stimulates an immune response.

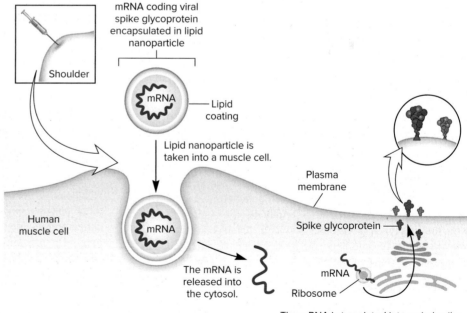

mRNA coding viral spike glycoprotein encapsulated in lipid nanoparticle

Shoulder

mRNA — Lipid coating

Lipid nanoparticle is taken into a muscle cell.

Human muscle cell

mRNA

The mRNA is released into the cytosol.

Plasma membrane

Spike glycoprotein

mRNA

Ribosome

The mRNA is translated into a viral spike glycoprotein, which travels through the endoplasmic reticulum and Golgi apparatus to reach the plasma membrane.

FIGURE 21.2 **Mechanism of an mRNA vaccine.** In this example, the mRNA codes a viral spike glycoprotein. After the spike glycoprotein is made, it elicits an immune response because it is a foreign antigen.

TABLE 21.3

Examples of COVID-19 Vaccines That Have Been Commonly Used Worldwide

Producer of the Vaccine	Vaccine Type	Description
Pfizer-BioNTech	Nucleic acid vaccine	An mRNA codes the viral spike glycoprotein.
Moderna	Nucleic acid vaccine	An mRNA codes the viral spike glycoprotein.
Johnson & Johnson/Janssen	Viral vector	The gene coding the viral spike glycoprotein was inserted into an adenovirus.
AstraZeneca/Oxford	Viral vector	The gene coding the viral spike glycoprotein was inserted into an adenovirus.
Novovax	Subunit vaccine	Contains purified viral spike glycoproteins

Phase 2 Expanded Trials. The vaccine is administered to hundreds of people split into various types of groups, such as different age groups, to see if the vaccine acts differently in them. These trials further test the vaccine's safety.

Phase 3 Efficacy Trials. The vaccine is given to thousands of people. The goal is to wait and see how many become infected, compared with volunteers who are given a placebo. Phase 3 trials are usually large enough to reveal evidence of relatively rare side effects. Phase 3 trials are aimed at determining an efficacy rate. If a vaccine achieves an efficacy of 90% in an efficacy trial, this means that there was a 90% reduction in disease cases in the vaccinated group compared to the unvaccinated (or placebo) group.

Phase 1 safety trials for a COVID-19 vaccine began in the spring of 2020. The first phase 3 efficacy trials for vaccines developed by Pfizer-BioNTech and Moderna began in the summer of 2020. Both companies reported efficacy rates of greater than 90% in adults. In the United States, the FDA granted emergency authorization for these two vaccines in the fall of 2020. **Table 21.3** describes a few COVID-19 vaccines that have been commonly used worldwide.

21.2 COMPREHENSION QUESTIONS

1. Which of the following is *not* an example of a whole-pathogen vaccine?
   a. Inactivated vaccine
   b. Attenuated vaccine
   c. Viral vector vaccine
   d. All of the above are whole-pathogen vaccines.

2. A vaccine that contains a purified component from a pathogen, such as a protein or a polysaccharide, is a _____ vaccine.
   a. whole-pathogen
   b. viral vector
   c. subunit
   d. nucleic acid

3. Which of the following is a type of vaccine used against COVID-19?
   a. A nucleic acid vaccine
   b. A viral vector vaccine
   c. A subunit vaccine
   d. All of the above are types of vaccines used to combat COVID-19

21.3 GENETICALLY MODIFIED ANIMALS

Learning Outcomes:

1. Distinguish between gene replacement and gene addition.
2. Explain how gene knockins and knockouts are produced in mice, and list some of their important uses.
3. Outline how transgenic livestock can produce human medicines in their milk.

As mentioned at the beginning of this chapter, transgenic organisms contain recombinant DNA from another species that has been integrated into their genome. A dramatic example of such a genetically modified organism (GMO) is shown in **Figure 21.3**. The larger Atlantic salmon (*Salmo salar*) in the background carries a growth hormone-regulating gene from Pacific Chinook salmon (*Oncorhynchus tshawytscha*) with a promoter from ocean pout (*Zoarces americanus*). This genetic modification enables the Atlantic salmon to grow year-round instead of only during spring and summer.

The production of genetically modified animals is a relatively new achievement of biotechnology. In recent years, a few transgenic species have reached the stage of commercialization. Many researchers believe that the production of GMOs holds great promise for innovations in biotechnology. However, the degree to

FIGURE 21.3 **A comparison of a normal salmon and a genetically modified salmon that overexpresses a growth hormone-regulating gene.** The two Atlantic salmon are the same age and same species. The larger GMO is in the back.

Genes→Traits The transgenic salmon (in the back) carries a growth hormone-regulating gene that was modified from another fish species. The introduction of this gene into the fish's genome has caused it to grow faster and larger.

2014 AquaBounty Technologies, Inc.

which this potential may be realized depends, in part, on the public's concern about the production and consumption of transgenic species. In this section, we will explore technologies for modifying or adding genes to animal cells and examine the potential uses of such genetic modifications.

The Genomes of Animals Can Be Altered by Gene Modification or Gene Addition

Researchers have developed a variety of methods to alter the genomes of animals. Such methods often have the goal of (1) altering specific genes that are already present in an animal's genome or (2) introducing a cloned gene into the genome.

- **Gene editing** alters the DNA sequence of a gene experimentally. Different approaches can be followed to modify genes. For example, in Chapter 20, we considered how CRISPR-Cas technology can inactivate a gene by introducing a deletion in it (refer back to Figure 20.13). Such technology can be applied to mice. If a mouse gene has been rendered inactive, researchers can study how the loss of normal gene function affects the phenotype of the mouse. For diploid species, when both copies of a gene are rendered inactive, the organism is said to carry a **gene knockout.** Also, technologies such as CRISPR-Cas can alter a gene sequence by introducing a specific mutation, such as a missense mutation, into a gene. As discussed later, researchers may want to produce missense mutations in mice that mimic disease-causing mutations in humans. In this way, they can study the effects of the disease in mice.
- **Gene addition** involves the insertion of a cloned gene into a genome, such as the genome of a mouse. In some cases, researchers may introduce additional copies of a gene that is already present in the genome. Alternatively, they may introduce a gene that is not already present in the genome.

To accomplish gene addition in mice, researchers can produce a **gene knockin,** in which a gene of interest has been inserted into a particular site in the mouse genome (**Figure 21.4**). In this example, a cloned gene is injected into a fertilized oocyte. Because the cloned gene is flanked by DNA pieces that are homologous to a noncritical site in the mouse genome, the cloned gene can be inserted into the noncritical site by homologous recombination. The noncritical site does not contain any mouse genes, so this insertion will not disrupt any critical genes. Such knockins tend to result in a more consistent level of expression of the gene of interest than gene additions that may occur randomly in another place in the genome. Also, because the insertion does not interfere with any critical mouse genes, a researcher can be more certain that any resulting phenotypic effect is due to the expression of the gene of interest. After the cloned gene has been inserted, the fertilized oocyte is implanted into the uterus of a female mouse, where it develops into a baby mouse that carries the inserted gene.

An interesting example of gene addition involves the production of aquarium fish that glow, the aptly named GloFish. **GloFish** are

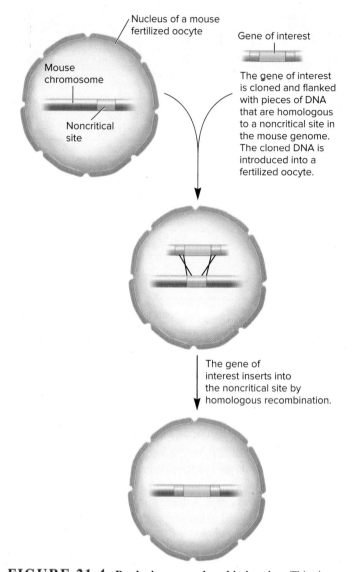

FIGURE 21.4 Producing a gene knockin in mice. This simplified diagram illustrates the technique of producing a gene knockin.

CONCEPT CHECK: What is the difference between a gene knockout and a gene knockin?

a strain of transgenic zebrafish (*Danio rerio*) that glow with bright green, red, or yellow fluorescent color (**Figure 21.5**). How were these fish produced? In 1999, Zhiyuan Gong and colleagues started with a gene from jellyfish that codes a green fluorescent protein (GFP) and inserted it into the zebrafish genome by gene addition, causing the zebrafish to glow green. By placing the gene next to a gene promoter that turns the gene on in the presence of certain environmental toxins, their long-term goal was to develop a fish that could be used to detect water pollution. However, as a first step in this process, they initially placed the *GFP* gene next to a promoter that was continually expressed. The researchers subsequently collaborated with a company to market the fish for aquarium use. They also developed a red fluorescent zebrafish by adding a gene from a sea coral, and a yellow fluorescent zebrafish by adding a variant of the jellyfish *GFP* gene. In 2003, the

FIGURE 21.5 **The use of gene addition to produce fish that glow.** The aquarium fish shown here, which are named GloFish, are transgenic organisms that have received a gene from jellyfish or sea coral that codes a fluorescent protein, causing them to glow green, red, or yellow. CB2/ZOB/WENN/Newscom

GloFish became the first genetically modified organism to be sold as a pet. GloFish have been successfully marketed in several countries, including the United States, although the sale of GloFish is banned in certain states.

Gene Knockouts and Knockins Are Produced in Mice to Understand Gene Function and Human Disease

As already mentioned, researchers can inactivate a normal mouse gene using a method such as CRISPR-Cas technology. Remember that mice are diploid and have two copies of most genes. Initially, only one copy of the gene of interest may be inactivated. In other words, the resulting mouse may be heterozygous: One copy is normal, and the other copy is inactivated. When such heterozygous mice are crossed to each other, one-fourth of the offspring will be homozygous for the inactivated gene. This type of mouse is said to carry a gene knockout.

By creating gene knockouts, researchers can study how the loss of normal gene function affects the organism. Gene knockouts frequently have specific effects on the phenotype of a mouse, which helps researchers to determine that the function of a gene is critical within a particular organ or during a specific stage of development. For example, if a gene knockout produced a phenotype in which a mouse had an enlarged heart, researchers would speculate that the normal gene plays a role in the proper development of the heart.

In many cases, however, a gene knockout produces no obvious phenotypic effect. One explanation is that a single gene may make such a small contribution to an organism's phenotype that its loss may be difficult to detect. Alternatively, another possible explanation for a lack of observable phenotypic change in a knockout mouse is the phenomenon of **gene redundancy.** If gene redundancy is present, when one type of gene is inactivated, another gene with a similar function may be able to compensate for the inactive gene. A third explanation for why a gene knockout has no obvious effect is that the effects of the knockout may be observed only under certain types of environmental conditions.

A particularly exciting avenue of gene knockout research is its application in the study and treatment of human diseases. How is a gene knockout useful in this kind of application? Knocking out the function of a gene may provide clues about what that gene normally does. Because humans share many genes with mice, observing the characteristics of knockout mice gives researchers information that can be used to better understand how a similar gene mutation may cause or contribute to a disease in humans. Examples of research areas in which knockout mice have been useful include the study of cancer, obesity, heart disease, diabetes, and many inherited disorders.

To study human diseases, researchers have produced strains of transgenic mice that harbor both gene knockouts and gene knockins. A strain of mice engineered to carry a mutation that is analogous to a disease-causing mutation of a human gene is termed a **mouse model.** As an example, let's consider sickle cell disease (refer back to Figure 4.7), which is due to a mutation in the human β-globin gene. This gene codes a polypeptide called β globin; adult hemoglobin is composed of both α-globin and β-globin polypeptides. When researchers produced a gene knockin by introducing the mutant human β-globin gene into mice, the resulting mice showed only mild symptoms of the disease. However, Chris Pászty and Edward Rubin produced a mouse model with multiple gene knockins and gene knockouts. In particular, the mice have gene knockins for the normal human α-globin gene and the mutant β-globin gene from patients with sickle cell disease. They also have gene knockouts of the mouse α-globin gene and β-globin gene. In summary, this mouse model has four alterations to its genome:

- Normal human α-globin gene knockin
- Mutant human β-globin gene knockin
- Mouse α-globin gene knockout
- Mouse β-globin gene knockout

Therefore, these mice make adult hemoglobin just like that produced by people with sickle cell disease, but they do not produce any adult mouse hemoglobin. This mouse model exhibits the major features of sickle cell disease—sickled red blood cells, anemia, and multiple organ pathology. Such mice have been useful as a model for studying the disease and testing potential therapies.

Another area of biotechnology research involving knockouts is investigation of their application in **xenotransplantation**—the transplantation of cells, tissues, or organs from one animal species to another. For example, work is currently under way to produce genetically modified pigs whose organs are resistant to the rejection mechanisms that occur following transplantation. Strains of pigs have been made in which the gene that codes 1,3-α-galactosyltransferase has been knocked out. This enzyme attaches sugars to cell surface proteins. When it is knocked out, the cell surface is much less immunogenic and therefore pig organs from this strain are less likely to be rejected when transplanted into humans. Although further genetic modifications may be necessary before pig organs can be transplanted into humans, some researchers are predicting that clinical trials for the transplantation of pig organs into humans may begin within a few years.

GENETIC TIPS   THE QUESTION: Researchers have
identified a gene in humans that (when mutated) causes tremors and unstable walking due to neurological problems. This disorder is inherited in an autosomal recessive manner, and the mutant allele is known to result from a loss-of-function mutation. The same gene has been found in mice, although a mutant mouse version has not been discovered. To develop an effective drug therapy to treat this disorder in humans, it would be experimentally useful to have a mouse model. In other words, it would be desirable to develop a strain of mice that carry the mutant allele in the homozygous condition. How would you develop such a strain?

T OPIC: *What topic in genetics does this question address?*
The topic is mouse models. More specifically, the question is about developing a mouse model to study a neurological disorder that affects humans.

I NFORMATION: *What information do you know based on the question and your understanding of the topic?* From the question, you know that a recessive disorder in humans has symptoms of tremors and unstable walking. From your understanding of the topic, you may remember how to produce gene knockouts in mice.

P ROBLEM-SOLVING **S** TRATEGY: *Design an experiment.* One strategy to solve this problem is to design an experiment in which the mouse gene is knocked out.

ANSWER: There is more than one way to inactivate a gene. One method is CRISPR-Cas technology, which is described in Chapter 20 (refer back to Figure 20.13). You would begin with a fertilized mouse oocyte and inactivate the gene using this technology. Initially, just one copy of the gene may be inactivated, although both copies may be inactivated a small percentage of the time. If only one copy was inactivated, you would have to mate heterozygotes to each other to obtain homozygotes that carry two copies of the inactivated gene. The knockout mice would then be analyzed to see if they exhibit symptoms similar to those experienced by humans: tremors and an unstable walk. If so, they could be used as a mouse model to study the human disorder.

Using Biotechnology in the Production of Transgenic Livestock Holds Promise

The technology for creating transgenic mice has been extended to other animals, and much research is under way to develop transgenic species of livestock, including fish, sheep, pigs, goats, and cattle. A growing area of biotechnology research focuses on the production of medically important proteins in the mammary glands of livestock, which is sometimes called **molecular pharming.** (The term is also used to describe the production of medical products by agricultural plants.)

As shown in **Table 21.4**, several human proteins have been successfully produced in the milk of livestock. Compared with the production of proteins in bacteria, one advantage of molecular pharming is that certain proteins are more likely to function properly when expressed in mammals. This may be due to covalent modifications, such as the attachment of carbohydrate groups, which occur in eukaryotes but not in bacteria. In addition, certain proteins may be degraded rapidly or folded improperly when expressed in bacteria. Furthermore, the yield of recombinant proteins in milk can be quite large. Dairy cows, for example, produce about 10,000 L of milk per year per cow. In some cases, a transgenic cow can produce approximately 1 g of the medically useful protein per liter of milk.

To introduce a human gene into an animal so that the coded protein will be secreted into the animal's milk, the gene is inserted next to a milk-specific promoter. Eukaryotic genes often are expressed in a tissue-specific fashion. In mammals, certain genes are expressed specifically within the mammary gland, so their protein product is secreted into the milk. Examples of milk-specific genes include genes that code milk proteins such as β-lactoglobulin, casein, and whey acidic protein.

To express a human gene that codes a specific protein, such as a hormone, into a domesticated animal's milk, the promoter for a milk-specific gene is linked to the coding sequence for the human gene (**Figure 21.6**). The DNA is then injected into an oocyte, where it is integrated into the genome. The fertilized oocyte is then implanted into the uterus of a female animal, which later

TABLE 21.4		
Proteins That Are Produced in the Milk of Domesticated Animals		
Protein	**Animal Source**	**Use**
Lactoferrin	Cattle	Used as an iron supplement in infant formula
Tissue plasminogen activator (TPA)	Goat	Dissolves blood clots
Antibodies	Cattle	Used to combat specific infectious diseases
α₁-Antitrypsin	Sheep	Treatment of emphysema
Factor IX	Sheep	Treatment of certain inherited forms of hemophilia
Insulin-like growth factor	Cattle	Treatment of diabetes

Human
hormone
gene

Using recombinant DNA technology (described in Chapter 20), insert the coding sequence of a human hormone gene next to a sheep β-lactoglobulin promoter. This promoter is functional only in mammary cells, so the protein product is secreted into the milk.

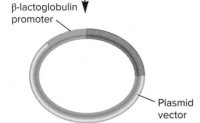

β-lactoglobulin
promoter

Plasmid
vector

Inject this DNA into a sheep oocyte. The plasmid DNA will integrate into the chromosomal DNA, resulting in the addition of the human hormone gene into the sheep's genome.

Oocyte

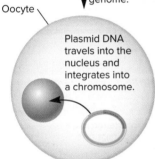

Plasmid DNA travels into the nucleus and integrates into a chromosome.

Implant the fertilized oocyte into a female sheep, which then gives birth to a transgenic sheep offspring.

Transgenic
sheep

Obtain milk from female transgenic sheep. The milk contains a human hormone.

Purify the hormone from the milk.

FIGURE 21.6 **Strategy for expressing human genes in a domesticated animal's milk.** The β-lactoglobulin gene is normally expressed in mammary cells, whereas the human hormone gene is not. To express the human hormone gene in milk, the promoter from the milk-specific gene in a sheep is linked to the coding sequence of the human hormone gene. In addition to the promoter, a short signal sequence may also be required so that the protein is secreted from the mammary cells and into the milk.

Genes→Traits By using genetic engineering, researchers can give sheep the trait of producing a human hormone in their milk. This hormone can be purified from the milk and used to treat humans.

CONCEPT CHECK: Why is the β-lactoglobulin promoter necessary in this process?

gives birth to a transgenic offspring. If this offspring is a female, the protein hormone coded by the human gene is expressed within the mammary gland and secreted into the milk. The milk can then be obtained from the animal, and the human hormone is isolated and purified.

21.3 COMPREHENSION QUESTIONS

1. When a cloned gene is inserted into a noncritical site in the mouse genome by homologous recombination, the result is
   a. gene addition.
   b. gene modification.
   c. gene knockout.
   d. both a and b.
2. One strategy for producing a protein in the milk of a cow is to place the coding sequence of the gene of interest next to a _____ and then inject the gene into a _____.
   a. *lac* operon promoter, cow oocyte
   b. β-lactoglobulin promoter, cow oocyte
   c. *lac* operon promoter, cow mammary cell
   d. β-lactoglobulin promoter, cow mammary cell

21.4 REPRODUCTIVE CLONING AND STEM CELLS

Learning Outcomes:
1. Outline the steps of reproductive cloning in mammals.
2. Define *stem cells,* and describe their two key properties.
3. Distinguish between totipotent, pluripotent, multipotent, and unipotent stem cells.
4. List examples of potential uses of stem cells to treat human diseases.

The previous section focused on the use of biotechnology for gene modification and gene addition in animals. Another area of biotechnology concerns the cloning of whole organisms or the manipulation of stem cells. In this section, we will consider mammalian cloning and stem cell research. These topics have received enormous public attention due to the complex ethical issues they raise.

Researchers Have Succeeded in Cloning Mammals from Somatic Cells

The term *cloning* has more than one meaning. In Chapter 20, we discussed gene cloning, which involves methods that produce many copies of a gene. The cloning of an entire organism is a different matter. **Reproductive cloning** refers to biotechnology methods that produce two or more genetically identical individuals. Such individuals occasionally occur in nature; identical twins are genetic clones that began from the same fertilized egg. Similarly, researchers can take mammalian embryos at an early stage of development (e.g., the two-cell to eight-cell stage), separate the cells, implant them into the uterus of a female, and obtain multiple births of genetically identical individuals.

In the case of plants, cloning is an easier undertaking, which we will explore in Section 21.5. Plants can be cloned from somatic cells. In most cases, it is relatively easy to take a cutting from a plant, expose it to growth hormones, and obtain a separate plant that is genetically identical to the original. However, this approach has not been possible with mammals. For several decades, scientists believed that chromosomes within the somatic cells of mammals had incurred irreversible genetic changes that render them unsuitable for cloning. In 1997, this hypothesis was proven to be incorrect. Ian Wilmut and his colleagues announced that a sheep, named Dolly, had been cloned using the genetic material from somatic cells.

How was Dolly produced? As shown in **Figure 21.7**, the researchers removed mammary cells from an adult female sheep and grew them in the laboratory. The researchers then extracted the nucleus from an egg cell of a different sheep and used electrical pulses to fuse the diploid mammary cell with the enucleated egg cell. After fusion, the zygote began embryonic development, and the resulting embryo was implanted into the uterus of a surrogate female sheep. One hundred and forty-eight days later, Dolly was born.

Though the production of Dolly via reproductive cloning was a technical breakthrough, one disturbing finding was the length of Dolly's chromosomes. As mammals age, chromosomes in somatic cells tend to shorten at the telomeres—the ends of eukaryotic chromosomes. Therefore, older individuals have shorter chromosomes in their somatic cells than younger ones do. This shortening does not seem to occur in the cells of the germ line, however. When researchers analyzed the chromosomes in Dolly's somatic cells at 3 years old, the lengths of those chromosomes were consistent with a sheep that was significantly older, say, 9–10 years old.

How do we explain these findings? The sheep that donated the mammary cell that produced Dolly was 6 years old, and the mammary cells had been grown in culture for several cell doublings before a mammary cell was fused with an oocyte. This led researchers to postulate that Dolly's shorter chromosomes were a result of telomere shortening in the somatic cells (i.e., mammary cells) of the sheep that donated the nucleus. In 2003, the Roslin Institute announced the decision to euthanize 6-year-old Dolly after an examination showed progressive lung disease. This diagnosis raised concerns among experts that the techniques used to produce Dolly could have caused premature aging.

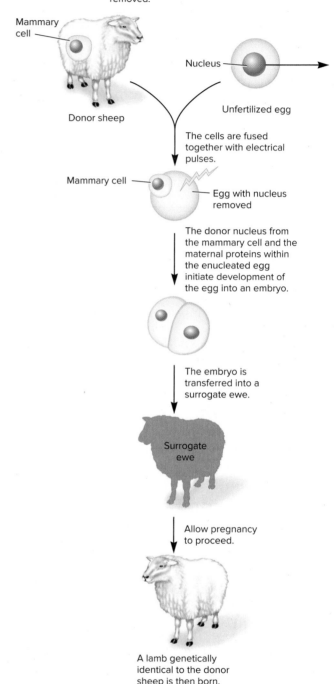

FIGURE 21.7 Protocol for the successful cloning of a sheep.

Genes→Traits Dolly was genetically identical to the sheep that donated a mammary cell. Dolly and that donor sheep were genetically identical in the same way that identical twins are; they carried the same set of genes and looked remarkably similar. However, they may have had minor genetic differences due to possible variation in their mitochondrial DNA and may have exhibited some phenotypic differences due to maternal effect genes or genomic imprinting.

CONCEPT CHECK: In the cloning protocol, why is the nucleus of the oocyte removed?

With regard to telomere length, research in mice and cattle has shown different results; the telomeres of cloned mice appear to be the correct length for their age. For example, cloning was conducted on mice using the method described in Figure 21.7 for six consecutive generations. The cloned mice of the sixth generation had appropriate telomere lengths. Further research is necessary to determine if cloning via somatic cells has an effect on the length of telomeres in subsequent generations. However, other studies in mice point to various types of genetic flaws in cloned animals. For example, Rudolf Jaenisch and colleagues used DNA microarray technology (described in Chapter 23) to analyze the transcription patterns of over 10,000 genes in cloned mice. As many as 4% of those genes were not expressed normally. Furthermore, research has shown that cloned mice die at a younger age than their naturally bred counterparts.

Mammalian cloning is still at an early stage of development. Nevertheless, the successful creation of Dolly showed that it is technically possible. In recent years, cloning from somatic cells has been achieved in several mammalian species, including sheep, cattle, mice, goats, and pigs. In 2002, the first pet was cloned. This cat was named Carbon Copy, but also called Copy Cat (**Figure 21.8**). Mammalian cloning may potentially have many practical applications. Cloning livestock would enable farmers to use the somatic cells from their best individuals to create genetically homogeneous herds, which could increase agricultural yield. A possible disadvantage, however, could be that animals in a genetically homogeneous herd may be more susceptible to rare diseases.

FIGURE 21.8 Carbon Copy, the first cloned pet. The animal shown here was produced using a procedure similar to the one shown in Figure 21.7.

Texas A&M University/FEMA/Handout/Getty Images

CONCEPT CHECK: Is Carbon Copy a transgenic animal?

Though some people are concerned about the use of cloning with agricultural species, a majority have become very concerned about the possibility of human cloning. This prospect has raised a host of serious ethical questions. For example, some people feel that cloning humans is morally wrong and threatens the basic fabric of parenthood and family. Others feel that it is a technology that could offer a new avenue for reproduction, one that could be offered to infertile couples, for example. In the public sector, the sentiment toward human cloning has been generally negative. Many countries have issued an all-out ban on human cloning, but others permit limited research in this area. Because the technology for cloning exists, our society will continue to wrestle with the legal and ethical aspects of cloning, not only of animals but also of people.

Stem Cells Have the Ability to Divide and Differentiate into Different Cell Types

Stem cells supply the various kinds of cells required to construct a multicellular organism, such as an animal or plant, starting from a fertilized egg. In adults, stem cells also replenish worn-out or damaged cells. Stem cells have two common characteristics that allow them to carry out these functions:

- They have the capacity to divide.
- They can differentiate into one or more specialized cell types.

As shown in **Figure 21.9**, the two daughter cells produced from the division of a stem cell can have different fates. One of the cells may remain an undifferentiated stem cell, while the other daughter

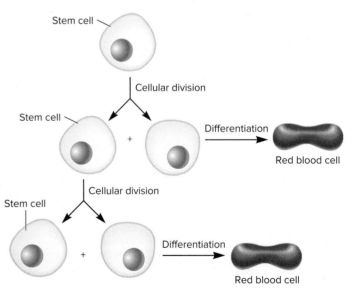

FIGURE 21.9 Growth pattern of stem cells. The two main traits that stem cells exhibit are an ability to divide and an ability to differentiate. When a stem cell divides, one of the two daughter cells remains a stem cell, and the other daughter cell differentiates into a specialized cell type.

CONCEPT CHECK: Explain why stem cells are not depleted during the life of an organism.

cell can differentiate into a specialized cell type. With this type of asymmetrical division/differentiation pattern, the population of stem cells remains relatively constant, yet the stem cells provide a population of specialized cells. In an adult, this type of mechanism is needed to replenish cells that have a finite life span, such as skin epithelial cells and red blood cells.

In mammals, stem cells are commonly categorized according to their developmental stage and their ability to differentiate (**Figure 21.10**).

- A fertilized egg is considered **totipotent,** because it can give rise to all cell types in the adult organism.
- The early mammalian embryo contains **embryonic stem cells (ES cells),** which are found in the inner cell mass of

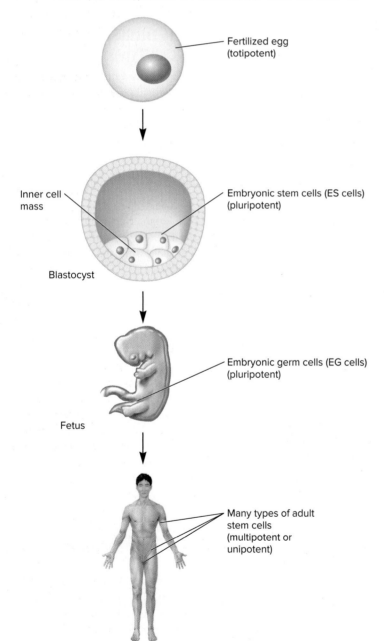

FIGURE 21.10 Occurrence of stem cells at different stages of human development.

the blastocyst. The blastocyst is the stage of embryonic development prior to uterine implantation—the preimplantation embryo. ES cells are **pluripotent,** which means they can differentiate into any or almost any cell type of the body. However, a single ES cell has lost the ability to produce an entire, intact individual. During the early fetal stage of development, the germ-line cells found in the gonads also are pluripotent. These cells are called **embryonic germ cells (EG cells).**

- As mentioned, adults also contain stem cells, but these are thought to be multipotent or unipotent. A **multipotent** stem cell can differentiate into several cell types but far fewer than an ES cell. For example, hematopoietic stem cells (HSCs) found in the bone marrow supply cells that populate two different tissues, namely, blood and lymphoid tissue (**Figure 21.11**). Furthermore, each of these tissues contains several cell types. Multipotent HSCs follow a pathway in which cell division produces a myeloid progenitor cell, which then differentiates into a red blood cell, megakaryocyte, basophil, monocyte, eosinophil, neutrophil, or dendritic cell. Alternatively, an HSC follows a path in which it becomes a lymphoid progenitor cell, which then differentiates into a T cell, B cell, natural killer cell, or dendritic cell.
- Other stem cells found in adults seem to be **unipotent.** For example, primordial germ cells in the testis can differentiate only into a single cell type, the sperm.

Stem Cells Have the Potential to Treat a Variety of Diseases

Interest in stem cells centers on two main areas. Because stem cells have the capacity to differentiate into multiple cell types, their study may help us to understand basic genetic mechanisms that underlie the process of development, the details of which are described in Chapter 26.

A second compelling reason why people have become interested in stem cells is their potential to treat human diseases or injuries that cause cell and tissue damage. This application has already become a reality in certain cases. For example, bone marrow transplantation is used to treat patients with certain forms of cancer. Such patients may be given radiation treatments that destroy their immune systems. When such a patient is injected with bone marrow from a healthy person, the stem cells within the transplanted marrow have the ability to proliferate and differentiate within the patient's body and provide a functioning immune system.

Renewed interest in the potential use of stem cells in the treatment of many other diseases was fostered in 1998 by two separate studies, directed by James Thomson and John Gearhart, showing that embryonic cells, either ES or EG cells, can be successfully propagated in the laboratory. As mentioned, ES and EG cells are pluripotent and therefore have the capacity to produce many different kinds of tissue. As shown in **Table 21.5,** embryonic cells could potentially be used to treat a wide variety of diseases associated with cell and tissue damage. By comparison, it would be difficult, based on our current knowledge, to treat these

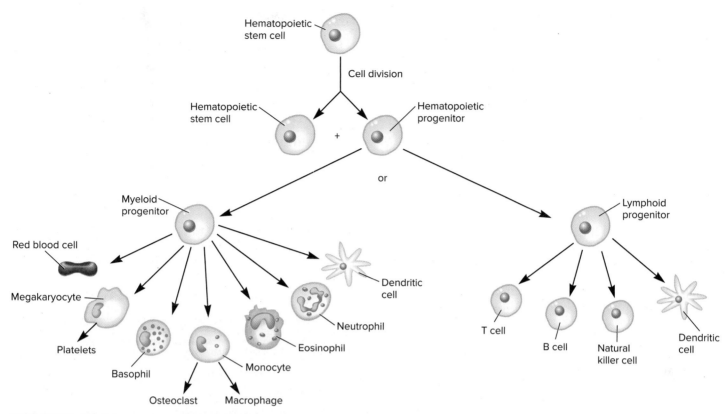

FIGURE 21.11 Fates of hematopoietic stem cells.

CONCEPT CHECK: Are hematopoietic stem cells unipotent, multipotent, or pluripotent?

diseases with adult stem cells because of the inability to locate most types of adult stem cells within the body and successfully grow them in the laboratory.

Even HSCs are elusive. In the bone marrow, about 1 cell in 10,000 is a stem cell, yet that is enough to populate all of the blood and lymphoid cells of the body. The stem cells of most other adult tissues are equally difficult to locate, if not more so. In addition, with the exception of stem cells in the blood, other types of stem cells in the adult body are difficult to remove in sufficient numbers

TABLE 21.5	
Potential Uses of Stem Cells to Treat Diseases	
Cell/Tissue Type	**Disease Treatment**
Neural	Implantation of cells into the brain to treat Parkinson disease
	Treatment of injuries such as those to the spinal cord
Skin	Treatment of burns and other types of skin disorders
Cardiac	Repair of heart damage associated with heart attacks
Cartilage	Repair of joints damaged by injury or arthritis
Bone	Repair of damaged bone or replacement with new bone
Liver	Repair or replacement of liver tissue that has been damaged by injury or disease
Skeletal muscle	Repair or replacement of damaged muscle

for transplantation. By comparison, ES and EG cells are easy to identify and have the great advantage of rapid growth in the laboratory. For these reasons, ES and EG cells offer a greater potential for transplantation, based on our current knowledge of stem cell biology.

For ES or EG cells to be used in transplantation, researchers need to derive methods that cause the cells to differentiate into the appropriate type of tissue. For example, if the goal was to repair a spinal cord injury, ES or EG cells would need the appropriate cues to cause them to differentiate into neural tissue. Research has shown that a complex array of factors determine the developmental fates of stem cells. These include internal factors within the stem cells themselves, as well as external factors such as the properties of neighboring cells and the presence of hormones and growth factors. In vitro laboratory studies have shown great promise in identifying the factors that cause stem cells to differentiate into particular cell types. However, achieving a similar effect in vivo remains challenging.

From an ethical perspective, the primary issue that prompts debate is the source of the stem cells for research and potential treatments. Most ES cells have been derived from human embryos that were produced from in vitro fertilization and were subsequently not used. Most EG cells are obtained from aborted fetuses. Some feel that it is morally wrong to use such tissue in research and/or the treatment of disease, or they fear that such use could lead to intentional abortions for the sole purpose of obtaining fetal tissues for transplantation.

Alternatively, others feel that the embryos and fetuses that provide the ES and EG cells are not going to become living individuals, and therefore, it is beneficial to study these cells and use them in a positive way to treat human diseases and injury. It is not clear whether these two opposing viewpoints can reach a common ground.

If stem cells could be obtained from adult cells and propagated in the laboratory, an ethical dilemma may be avoided because most people do not have serious moral objections to current procedures such as bone marrow transplantation. In 2006, work by Shinya Yamanaka and colleagues showed that adult mouse fibroblasts (a type of connective tissue cell) could become pluripotent via the insertion into the genome of four different genes that code transcription factors. In 2007, Yamanaka's laboratory and two other research groups showed that such induced pluripotent stem cells (iPS cells) can differentiate into all cell types when injected into mouse blastocysts and grown into baby mice. Though further research is still needed, these recent results indicate that adult cells can be reprogrammed to become embryonic stem cells.

21.4 COMPREHENSION QUESTIONS

1. During mammalian reproductive cloning, _____ is fused with _____.
   a. a somatic cell, a stem cell
   b. a somatic cell, an egg cell
   c. a somatic cell, an enucleated egg cell
   d. an enucleated somatic cell, an egg cell
2. Which of the following is a key feature of stem cells?
   a. They have the ability to divide.
   b. They have the ability to differentiate.
   c. They are always pluripotent.
   d. Both a and b are true of stem cells.

21.5 GENETICALLY MODIFIED PLANTS

Learning Outcomes:
1. List examples of transgenic plants that are useful to people.
2. Outline the steps in making transgenic plants using *Agrobacterium tumefaciens*.

For centuries, agriculture has relied on selective breeding programs to produce plants and animals with desirable characteristics. For agriculturally important species, selective breeding is often aimed at the production of strains that are larger, are more disease-resistant, and yield high-quality food. Agricultural scientists can now complement traditional breeding strategies with modern molecular genetic approaches. In the mid-1990s, genetically modified crops first became commercialized. Since that

time, their use has progressively increased. In 2022, over 40% of all agricultural crops were transgenic. Worldwide, about 190 million hectares (470 million acres) of transgenic crops were planted. In this section, we will discuss some current and potential uses of transgenic plants in agriculture, and examine the methods that scientists use to make them.

Transgenic Plants May Have Characteristics That Are Agriculturally Useful

Various traits can be modified in transgenic plants (**Table 21.6**). Most commonly, researchers have sought to produce transgenic plant strains resistant to herbicides, disease, and insects. For example, the Monsanto Company has produced transgenic plant strains that are tolerant of glyphosate, the active agent in the herbicide Roundup. The herbicide remains effective against weeds, but the herbicide-resistant crop is spared (**Figure 21.12**).

TABLE 21.6

Traits That Have Been Modified in Transgenic Plants

Trait	Example(s)
Plant Protection	
Resistance to herbicides	Transgenic plants can express proteins that render them resistant to particular herbicides (see Figure 21.12).
Resistance to viral, bacterial, and fungal pathogens	Transgenic plants that express the pokeweed antiviral protein are resistant to a variety of viral pathogens.
Resistance to insects	Transgenic plants that express the CryIA protein (a toxin) from *Bacillus thuringiensis* are resistant to a variety of insects (see Figure 21.13).
Plant Quality	
Improvement in storage life	Transgenic plants can express antisense RNA that silences a gene involved in fruit softening.
Change in plant composition	Transgenic strains of canola have been altered with regard to oil composition; the seeds of the Brazil nut have been rendered methionine-rich via transgenic technology.
New Products	
Biodegradable plastics	Transgenic plants have been made that can synthesize polyhydroxyalkanoates, which are used as biodegradable plastics.
Vaccines	Transgenic plants have been modified so their leaves produce vaccines against many human and animal diseases, including hepatitis B, cholera, and malaria.
Pharmaceuticals	Transgenic plants have been made that produce a variety of medicines, including human interferon-α (to fight viral diseases and cancer), human epidermal growth factor (for wound repair), and human aprotinin (for reducing blood loss during transplantation surgery).
Antibodies	Human antibodies have been made in transgenic plants to battle various diseases such as non-Hodgkin lymphoma.

FIGURE 21.12 Transgenic plants that are resistant to glyphosate.

Genes→Traits This field of corn plants has been treated with glyphosate. The larger, healthy plants in the background have been genetically engineered to carry a herbicide-resistance gene. They are resistant to killing by glyphosate. By comparison, the stunted plants in the front row do not carry this gene.

Claudius Thiriet/Biosphoto/ardea.com

(a) A field of Bt cotton

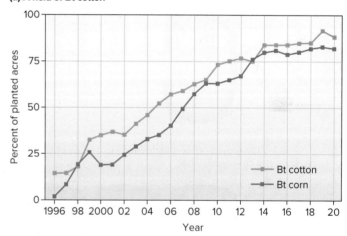

(b) Bt corn and Bt cotton usage in the United States since 1996

FIGURE 21.13 The production of Bt corn and cotton.
(a) A field of Bt cotton. These cotton plants carry an endotoxin gene from *Bacillus thuringiensis* that provides them with resistance to certain insects. (b) A graph showing the increase in usage of Bt corn and Bt cotton in the United States since their commercial introduction in 1996.
(a): Design Pics Inc/Alamy Stock Photo

Another important approach is to make plant strains that are disease-resistant. In many cases, virus-resistant plants have been developed by introducing a gene that codes a viral coat protein. When the plant cells express the viral coat protein, they become resistant to infection by that pathogenic virus.

A very successful example of the use of transgenic plants has involved the introduction of genes from *Bacillus thuringiensis* (Bt). As discussed earlier in this chapter, this bacterium produces toxins that kill certain types of caterpillars and beetles and has been widely used as a biological control agent for several decades. The toxins are proteins coded in the genome of *B. thuringiensis*. Researchers have succeeded in cloning toxin genes from *B. thuringiensis* and transferring those genes into plants. Such Bt varieties of plants produce the toxins themselves and therefore

are resistant to many types of caterpillars and beetles. Examples of commercial crops include Bt cotton (**Figure 21.13a**) and Bt corn. Since their introduction in 1996, the commercial use of these two Bt crops has steadily increased in the United States (**Figure 21.13b**).

The introduction of transgenic plants into agriculture has been strongly opposed by some people. What are the perceived risks? One potential risk is that transgenes in commercial crops could endanger native species. For example, Bt crops may kill pollinators of native species. Another worry is that the planting of transgenic crops could potentially lead to the proliferation of resistant insects. To prevent this from happening, researchers are producing transgenic strains that carry more than one toxin gene, which makes it more difficult for insect resistance to arise. Despite these and other concerns, many farmers are embracing transgenic crops, and their use continues to rise.

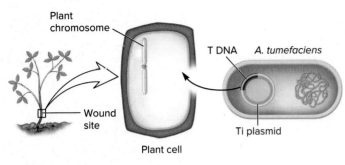

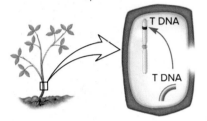

Agrobacterium tumefaciens is found within the soil. A wound on the plant enables the bacterium to infect the plant cells.

During infection, the T DNA within the Ti plasmid is transferred to the plant cell. The T DNA becomes integrated into the plant cell's DNA. Genes within the T DNA promote uncontrolled plant cell growth.

The growth of the recombinant plant cells produces a crown gall tumor.

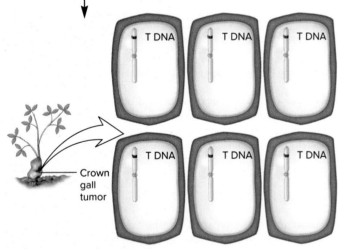

(a) The production of a crown gall tumor by *A. tumefaciens* infection

FIGURE 21.14 *Agrobacterium tumefaciens* infecting a plant and causing a crown gall tumor.

(b): Nigel Cattlin/Alamy Stock Photo

(b) A crown gall tumor on a chrysanthemum plant

Transformation of *Agrobacterium tumefaciens* and Other Methods Are Used to Make Transgenic Plants

As we have seen, the introduction of cloned genes into embryonic cells can produce transgenic animals. The production of transgenic plants is somewhat easier, because some plant somatic cells are totipotent, which means they are capable of developing into an entire organism. Therefore, a transgenic plant can be made by the introduction of cloned genes into somatic tissue, such as the tissue of a leaf. After the cells of a leaf have become transgenic, an entire plant can be regenerated by the treatment of the leaf with plant growth hormones, which cause it to form roots and shoots.

Molecular biologists can use the bacterium *Agrobacterium tumefaciens*, which naturally infects plant cells, to produce transgenic plants. A plasmid from the bacterium, known as the Ti plasmid (tumor-inducing plasmid), carries genes that cause tumor formation after a plant has been infected (**Figure 21.14a**). In particular, a segment of the plasmid DNA, known as **T DNA** (for transferred DNA), is naturally transferred from the bacterium to the infected plant cells. The T DNA from the Ti plasmid is integrated into the chromosomal DNA of the plant cell by recombination. After this occurs, genes within the T DNA that code plant growth hormones cause uncontrolled plant cell growth. This produces a cancerous plant growth known as a crown gall tumor (**Figure 21.14b**).

Because *A. tumefaciens* inserts its T DNA into the chromosomal DNA of plant cells, it can be used as a vector to introduce cloned genes into plants. Molecular geneticists have been able to modify the Ti plasmid to make this an efficient process. Such vectors are called **T-DNA vectors.** The T-DNA genes that cause the development of a gall have been identified. Fortunately for genetic engineers, when these genes are removed, the T DNA is still taken up into plant cells and integrated into the plant chromosomal DNA. However, a gall tumor does not form. In addition, geneticists have inserted genes that are selectable markers into the T DNA to allow selection of plant cells that have taken up the T DNA. A gene (kan^R) that provides resistance to the antibiotic kanamycin is a commonly used selectable marker. The T-DNA vectors used in cloning experiments are also modified to contain unique restriction sites for the convenient insertion of any gene.

Figure 21.15 shows the general strategy for producing transgenic plants via T-DNA-mediated gene transfer.

1. A gene of interest is inserted into a genetically engineered T-DNA vector.
2. The vector is then transformed into *A. tumefaciens.*
3. Plant cells are exposed to the transformed *A. tumefaciens.* After allowing time for infection, the plant cells are exposed to the antibiotics carbenicillin and kanamycin. Carbenicillin kills *A. tumefaciens,* and kanamycin kills any plant cells that have not taken up the T DNA with the antibiotic resistance gene. Therefore, the only surviving cells are those plant cells that have integrated the T DNA into their genome. Because the T DNA also contains the cloned gene of interest, the selected plant cells are expected to have received this cloned gene as well.
4. The cells are then transferred to a medium that contains the growth hormones necessary for the regeneration of entire plants. These plants can then be analyzed to verify they are transgenic plants containing the cloned gene.

A. tumefaciens infects a wide range of plant species, including most dicots, most gymnosperms, and some monocots. However, not all plant species are infected by this bacterium. Fortunately, other methods can be used to introduce genes into plant cells.

- One way to produce transgenic plants is an approach known as **biolistic gene transfer.** In this method, plant cells are bombarded with high-velocity microprojectiles coated with DNA. When fired from a "gene gun," the microprojectiles penetrate the cell wall and membrane, thereby entering the plant cell. The cells that take up the DNA are identified with a selectable marker and regenerated into new plants.
- DNA can also enter plant cells via **microinjection,** which involves the use of microscopic-sized needles.
- **Electroporation** is a technique that uses electric current to create temporary pores in the plasma membrane, thereby allowing the entry of DNA into a cell.
- Because the rigid plant cell wall presents a strong barrier to DNA entry, other approaches involve the use of protoplasts, which are plant cells that have had their cell walls removed. DNA can be introduced into protoplasts using a variety of methods, including treatment with polyethylene glycol and calcium phosphate.

21.5 COMPREHENSION QUESTIONS

1. When *A. tumefaciens* is used to make a transgenic plant, a gene of interest is inserted into a _____, which is first transformed into _____. The _____ is then transferred to a plant.
   a. viral vector, *E. coli*, gene of interest
   b. T-DNA vector, *A. tumefaciens*, T DNA carrying the gene of interest
   c. T-DNA vector, *A. tumefaciens*, gene of interest alone
   d. T-DNA vector, *E. coli*, gene of interest alone

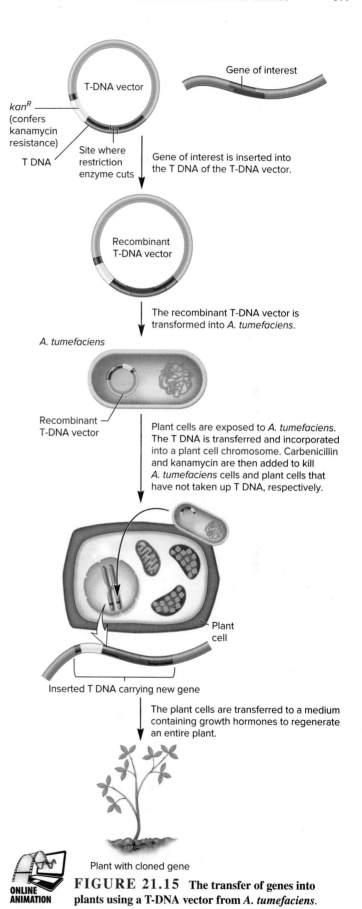

Gene of interest is inserted into the T DNA of the T-DNA vector.

The recombinant T-DNA vector is transformed into *A. tumefaciens.*

Plant cells are exposed to *A. tumefaciens.* The T DNA is transferred and incorporated into a plant cell chromosome. Carbenicillin and kanamycin are then added to kill *A. tumefaciens* cells and plant cells that have not taken up T DNA, respectively.

Inserted T DNA carrying new gene

The plant cells are transferred to a medium containing growth hormones to regenerate an entire plant.

Plant with cloned gene

FIGURE 21.15 **The transfer of genes into plants using a T-DNA vector from *A. tumefaciens.***

CONCEPT CHECK: Which portion of a T-DNA vector is transferred to a plant?

2. In addition to the use of T-DNA vectors, other methods to produce transgenic plants include
   a. biolistic gene transfer.
   b. microinjection.
   c. electroporation.
   d. all of these.

Introduction: biotechnology, genetically modified organism (GMO), transgenic organism, transgene

21.1: biological control, bioremediation, biotransformation, biodegradation

21.2: vaccine, vaccination, cross immunity, pandemic

21.3: gene editing, gene knockout, gene addition, gene knockin, GloFish, gene redundancy, mouse model, xenotransplantation, molecular pharming

21.4: reproductive cloning, stem cell, totipotent, embryonic stem cell (ES cell), pluripotent, embryonic germ cell (EG cell), multipotent, unipotent

21.5: T DNA, T-DNA vector, biolistic gene transfer, microinjection, electroporation

- Biotechnology is broadly defined as the use of living organisms or materials they produce in the development of products or processes that are beneficial to humans. Genetically modified organisms (GMOs) have received genetic material via recombinant DNA technology.

21.1 Uses of Microorganisms in Biotechnology

- Microorganisms have many uses that are beneficial to humans (see Table 21.1).
- Recombinant microorganisms are used to make human hormones, such as insulin (see Table 21.2, Figure 21.1).
- Microorganisms may be used as biological control agents.
- Microorganisms may be used in bioremediation, which is the use of living organisms or their products to decrease pollutants in the environment.

21.2 Vaccines

- A vaccine is a biological preparation that provides active acquired immunity to a particular infectious disease or to a disease such as cancer.
- The first human vaccine was against smallpox.
- The four general types of vaccines are whole-pathogen vaccines, viral vector vaccines, subunit vaccines, and nucleic acid vaccines (see Figure 21.2).
- Different types of vaccines have been developed to combat COVID-19 (see Table 21.3)

21.3 Genetically Modified Animals

- A transgenic organism carries a gene from another species (see Figure 21.3).

- The genetic manipulation of animals may involve gene modification or gene addition. To achieve gene addition in mice, researchers may insert a gene into a noncritical site in the mouse genome (see Figures 21.4, 21.5).
- Gene knockouts and knockins in mice are used to study gene function and to produce mouse models for studying human diseases.
- Transgenic livestock may produce human proteins, such as hormones, in their milk (see Table 21.4, Figure 21.6).

21.4 Reproductive Cloning and Stem Cells

- Mammalian reproductive cloning can be achieved using somatic cells and oocytes with their nuclei removed (see Figures 21.7, 21.8).
- Stem cells have the ability to divide and to differentiate. Stem cells may be totipotent, pluripotent, multipotent, or unipotent (see Figures 21.9, 21.10, 21.11).
- Stem cells have the potential to treat a variety of human diseases (see Table 21.5).

21.5 Genetically Modified Plants

- Researchers have made many transgenic plants that have traits that increase the plants' usefulness to humans, including herbicide and pesticide resistance (see Table 21.6, Figures 21.12, 21.13).
- *Agrobacterium tumefaciens* transfers T DNA into plant cells. Researchers have used T-DNA vectors to make genetically modified plants (see Figures 21.14, 21.15).

MORE GENETIC TIPS **1.** Which of the following can appropriately be described as a transgenic organism?

A. The sheep Dolly produced by reproductive cloning

B. A sheep that produces human α_1-antitrypsin in its milk

C. Bt corn

D. A hybrid strain of corn produced from crossing two inbred strains of corn (which were not transgenic)

TOPIC: *What topic in genetics does this question address?* The topic is transgenic organisms.

INFORMATION: *What information do you know based on the question and your understanding of the topic?* In the question, you are given a list of organisms and asked to decide whether each one is transgenic. From your understanding of the topic, you may remember that a transgenic organism is an organism that has received genetic material from a different species via recombinant DNA technology.

PROBLEM-SOLVING **S**TRATEGIES: *Define key terms. Compare and contrast.* One approach to solve this problem is to define what a transgenic organism is and then compare that definition to the characteristics of each of the organisms listed in parts A–D.

ANSWER:

A. No, Dolly does not carry genetic material from a different species.

B. Yes

C. Yes

D. No, the hybrid strain contains chromosomal genes from two different parental strains, which are of the same species.

2. What experimental approach would you follow to produce a human hormone in the milk of livestock?

TOPIC: *What topic in genetics does this question address?* The topic is molecular pharming. More specifically, the question is about producing a human hormone in the milk of livestock.

INFORMATION: *What information do you know based on the question and your understanding of the topic?* In the question, you are asked to propose a strategy for producing a human hormone in the milk of livestock. From your understanding of the topic, you may remember that this involves the insertion of the hormone-coding gene next to a milk-specific promoter and then introducing the gene into the oocyte of a sheep or cow.

PROBLEM-SOLVING **S**TRATEGIES: *Design an experiment. Describe the steps.* One approach to solve this problem is to outline the experimental steps involved in producing hormones in the milk of livestock.

ANSWER: Milk proteins are coded by genes that have promoters and regulatory sequences that direct the expression of the genes within the cells of a mammary gland.

1. To express a human hormone in an animal's mammary gland, the strategy is to use recombinant DNA techniques to link the promoter and regulatory sequences from a milk-specific gene to the coding sequence of the gene that codes the human hormone.

2. The gene is introduced into an oocyte, which is fertilized and implanted into a sheep or cow.

3. The sheep or cow gives birth to a transgenic offspring. If it is a female, the milk it produces may contain the hormone of interest.

4. Purify the hormone from the milk.

Conceptual Questions

C1. What is a recombinant microorganism? Discuss examples.

C2. A conjugation-deficient strain of *A. radiobacter* is used to combat crown gall disease. Explain how this bacterium prevents the disease, and describe the advantage of using a conjugation-deficient strain.

C3. What is bioremediation? What is the difference between biotransformation and biodegradation?

C4. What is a biological control agent? Briefly describe two examples.

C5. As shown in Table 21.2, several medical agents are now commercially produced by genetically engineered microorganisms. Discuss the advantages and disadvantages of making these agents this way.

C6. List and briefly describe the four general types of vaccines. Give examples of diseases that each type of vaccine is used against.

C7. Explain how the mRNA COVID-19 vaccine made by Pfizer-BioNTech and Moderna elicits an immune response.

C8. What is a mouse model for a human disease?

C9. What is a transgenic organism? Give three examples.

C10. What part of the *A. tumefaciens* DNA gets transferred to the genome of a plant cell during infection?

C11. Explain the difference between gene modification and gene addition. Is each of the following an example of gene modification or gene addition?

A. Creation of a mouse model to study cystic fibrosis

B. Introduction of a pesticide-resistance gene into corn using the T-DNA vector of *A. tumefaciens*

C12. As described in Chapter 5, not all inherited traits are determined by nuclear genes (i.e., genes located in the cell nucleus) that are expressed during the life of an individual. In particular, maternal effect genes and mitochondrial DNA are notable exceptions. With these ideas in mind, let's consider the cloning of a sheep (e.g., Dolly).

A. With regard to maternal effect genes, is the phenotype of such a cloned animal determined by the animal that donated the enucleated egg or by the animal that donated the somatic cell nucleus? Explain.

B. Does the cloned animal inherit extranuclear traits from the animal that donated the egg or from the animal that donated the somatic cell? Explain.

C. In what ways would you expect the cloned animal to be similar to or different from the animal that donated the somatic cell? Is it accurate to call such an animal a "clone" of the animal that donated the nucleus?

C13. Discuss some of the worthwhile traits that can be modified in transgenic plants.

C14. Discuss the concerns that some people have with regard to the uses of genetically modified organisms.

Experimental Questions

E1. Recombinant bacteria can produce hormones that are normally produced in humans. Briefly describe how this is accomplished.

E2. *Bacillus thuringiensis* makes toxins that kill insects. These toxins must be applied several times during the growth season to prevent insect damage. As an alternative to these repeated applications, one strategy is to apply bacteria directly to leaves. However, *B. thuringiensis* does not survive very long in the field. Other bacteria, such as *Pseudomonas syringae*, do survive. Propose a way to alter *P. syringae* so it could be used as an insecticide. Discuss advantages and disadvantages of this approach compared with the repeated applications of the toxins from *B. thuringiensis*.

E3. In the procedure in Figure 21.1, why was it necessary to link the coding sequence for the A or B chain to the sequence for β-galactosidase? How were the A or B chains separated from β-galactosidase after the fusion protein was synthesized in *E. coli*?

E4. To produce transgenic plants, plant tissue is exposed to *A. tumefaciens* and then grown in media containing kanamycin, carbenicillin, and plant growth hormones. Explain the purpose of using each of these three agents. What would happen if kanamycin was omitted from the media?

E5. List and briefly describe five methods for the introduction of cloned genes into plants.

E6. What is a gene knockout? Is an animal or plant with a gene knockout a heterozygote or a homozygote? What might you conclude if a gene knockout does not have a phenotypic effect?

E7. In the study of plants and animals, it is relatively common for researchers to identify a gene using molecular techniques without knowing the function of the gene. In the case of mice, the function of the gene can be investigated by creating a gene knockout. A knockout that causes a phenotypic change in the mouse may provide an important clue regarding the function of a gene. For example, a gene knockout that produced an albino mouse would indicate that the knocked-out gene probably plays a role in pigment formation. The experimental strategy of first identifying a gene based on its molecular properties and then investigating its function by making a knockout is called reverse genetics. Explain how this approach is opposite to (or the reverse of) the conventional way that geneticists study the function of genes.

E8. Evidence [see P. G. Shiels, A. J. Kind, K. H. Campbell, et al. (1999), Analysis of telomere lengths in cloned sheep, *Nature 399*, 316–317] suggests that Dolly may have been genetically older than her actual age. As mammals age, the chromosomes in somatic cells tend to shorten at the telomeres. Therefore, older individuals have shorter chromosomes in their somatic cells than do younger ones. When researchers analyzed the chromosomes in the somatic cells of Dolly at 3 years old, the lengths of Dolly's chromosomes were consistent with those of a sheep that was significantly older, say, 9–10 years old. (Note: As described in the chapter, the sheep that donated the somatic cell that produced Dolly was 6 years old, and the mammary cells had been grown in culture for several cell doublings before one of the cells was fused with an oocyte.)

A. Suggest an explanation as to why Dolly's chromosomes appeared older than they should have been.

B. Let's suppose that a female sheep (like Dolly), which was produced via reproductive cloning, was mated at age 11 to a normal male sheep and then gave birth to a lamb named Molly. When Molly was 8 years old, a sample of somatic cells was analyzed. How old would you expect Molly's chromosomes to appear, based on the phenomenon of telomere shortening? Explain your answer.

C. Discuss how the observation of chromosome shortening, which was observed in Dolly, might affect the popularity of reproductive cloning.

E9. What is molecular pharming? Compared with the production of proteins by bacteria, why might it be advantageous?

E10. What is reproductive cloning? Are human identical twins considered to be clones? With regard to agricultural species, what are some potential advantages to reproductive cloning?

E11. Researchers have identified a gene in humans that (when mutant) causes severe anemia and mental impairment. This disorder is inherited in an autosomal recessive manner, and the mutant allele is known to be a loss-of-function mutation. The same gene has been found in mice, although a mutant version of the mouse gene has not been discovered. To develop drugs and an effective therapy to treat this disorder in humans, it would be experimentally useful to have a mouse model. In other words, it would be desirable to develop a strain of mice that carry the mutant allele in the homozygous condition. Experimentally, how would you develop such a strain?

Questions for Student Discussion/Collaboration

1. A commercially available strain of *P. syringae* marketed as Frostban B is used to combat frost damage. This naturally occurring strain carries a loss-of-function mutation in a gene that codes a protein that is expressed on the surface of the bacterium and nucleates frost formation. In addition, researchers have used recombinant DNA techniques to eliminate the function of this gene. Discuss the advantages and disadvantages of using the nonrecombinant strain compared with a recombinant version.

2. Make a list of the types of traits you would like to see altered in transgenic plants and animals. Suggest techniques to accomplish these alterations.

Note: All answers are available for the instructor in Connect; the answers to the even-numbered questions and all of the Concept Check and Comprehension Questions are in Appendix B.

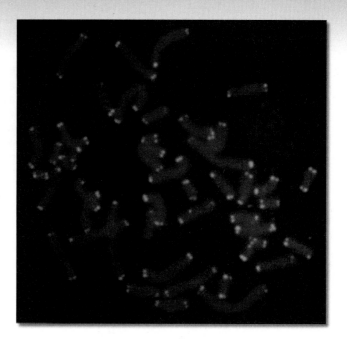

Labeling the ends of chromosomes. In this micrograph, the telomeric sequences at the ends of chromosomes are labeled with an orange fluorescent probe, and the rest of the chromosomes are labeled red. This method, called fluorescence in situ hybridization, allows geneticists to identify particular sequences within intact chromosomes.

Los Alamos National Laboratory/ Getty Images

22

GENOMICS I: ANALYSIS OF DNA

The term **genome** refers to the total genetic composition of a cell, an organism, or a species. It can also refer to a particular type of genome within a species. For example, the nuclear genome of humans is composed of 22 different autosomes and X and (in males) Y chromosomes. In addition, human cells have a mitochondrial genome composed of a single circular chromosome.

As genetic technology has progressed over the past few decades, researchers have gained an increasing ability to analyze the composition of genomes as whole units. The term **genomics** refers to the molecular analysis of the entire genome of a species. Genome analysis is a molecular dissection process applied to a complete set of chromosomes. Segments of chromosomes are analyzed in progressively smaller regions, whose locations on the intact chromosomes are known. The physical mapping of the genome ultimately advances to the determination of the complete DNA sequence for all of a species' chromosomes.

In 1995, a team of researchers headed by J. Craig Venter and Hamilton Smith completed the first entire DNA sequence of a genome, from the bacterium *Haemophilus influenzae*. As discussed later in this chapter, this genome is composed of a single circular chromosome that is composed of 1.83 million base pairs (bp) of DNA and carries approximately 1743 genes. In 1996, the first entire DNA sequence of a eukaryote, *Saccharomyces cerevisiae* (baker's yeast), was completed. The yeast genome contains 16 linear chromosomes, which have a combined length of about 12.1 million bp and contain approximately 6300 genes. Since that time, complete genome sequences from many prokaryotes and eukaryotes have been determined.

In this chapter, we will focus on methods used for analyzing a species' genome. We will consider three mapping strategies—cytogenetic mapping, linkage mapping, and physical mapping—and the techniques used to carry them out. Then we will explore genome-sequencing projects—research endeavors that have the ultimate goal of determining the sequence of DNA bases of the entire genome of a given species. We will examine the methods, goals, and results of these large undertakings, which include the Human Genome Project.

Once a genome sequence is known, researchers can examine how the numerous genes that make up a genome interact to produce the traits of an organism. This research area is called **functional genomics.** Scientists are also interested in determining the functions of all of the proteins that a given species can make. Techniques aimed at understanding the functions of many different proteins at once form the basis for the area of study called **proteomics.** As discussed in Chapter 17, researchers are interested in the functions of non-coding RNAs as well. In this chapter, we will explore genomics at the level of DNA segments and sequences. In Chapter 23, we will consider some recent advances in functional genomics, proteomics, and bioinformatics.

22.1 OVERVIEW OF CHROMOSOME MAPPING

Learning Outcomes:

1. Define *mapping*.
2. Distinguish among cytogenetic, linkage, and physical mapping.

In genetics, the term **mapping** refers to the experimental process of determining the relative locations of genes or other segments of DNA on individual chromosomes. Researchers may follow any of three general approaches to map a chromosome: cytogenetic, linkage, or physical mapping.

- **Cytogenetic mapping** (also called **cytological mapping**) is aimed at determining the locations of specific sequences, such as gene sequences, within chromosomes that are viewed microscopically. This type of mapping is primarily applied to eukaryotic species. When stained, each chromosome of a given eukaryotic species has a characteristic banding pattern, and genes are mapped cytogenetically relative to a band location using probes that recognize specific genes, as discussed in Section 22.2.
- By comparison, in Chapter 6, we considered the use of genetic crosses to map the relative locations of genes within a chromosome. This approach, known as **linkage mapping,** uses the frequency of genetic recombination between different genes to determine their relative spacing and order along a chromosome. In eukaryotes, linkage mapping involves crosses among organisms that are heterozygous for two or more genes. The number of recombinant offspring provides a relative measure of the distance between genes, which is expressed in map units (mu).
- A third approach is **physical mapping,** in which DNA-cloning or DNA-sequencing techniques are used to determine the location of and distance between genes and other DNA regions. In a physical map, the distances are computed as the number of base pairs between genes.

A **genetic map,** or **chromosome map,** is a diagram that shows the relative locations of genes or other DNA segments on a chromosome. The term **locus** (plural, loci) refers to the site within a genetic map where a specific gene or other DNA segment is found. **Figure 22.1** compares genetic maps that show the loci for two X-linked genes, *sc* (for "scute," a gene affecting bristle morphology) and *w* (a gene affecting eye color), in *Drosophila melanogaster.*

- In the cytogenetic map (at the top of the figure), the *sc* gene is located at band 1A8, and the *w* gene is located at band 3B6.
- For the linkage map (in the middle), genetic crosses indicated that the two genes are approximately 1.5 map units (mu) apart.
- The physical map (at the bottom) shows that the two genes are approximately 2.4×10^6 bp apart on the X chromosome.

Correlations among cytogenetic, linkage, and physical maps often vary from species to species and from one region of the chromosome to another. For example, a distance of 1 mu may correspond to 1–2 million bp in one region of a chromosome, but other regions

Cytogenetic map:

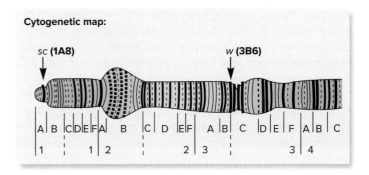

Linkage map:

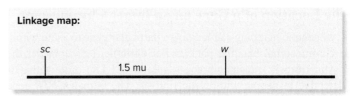

Physical map:

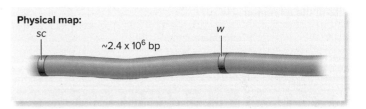

FIGURE 22.1 **A comparison of cytogenetic, linkage, and physical maps.** Each of these maps shows the distance between the *sc* and *w* genes on the X chromosome in *Drosophila melanogaster.* The cytogenetic map is that of the polytene chromosome.

CONCEPT CHECK: What is a genetic map?

may recombine at a much lower rate, so a distance of 1 mu in such a region may correspond to a much longer segment of DNA.

22.1 COMPREHENSION QUESTION

1. What type of chromosome mapping relies on microscopy?
   a. Cytogenetic mapping
   b. Linkage mapping
   c. Physical mapping
   d. All of the above rely on microscopy.

22.2 CYTOGENETIC MAPPING VIA MICROSCOPY

Learning Outcomes:

1. Outline the method of fluorescence in situ hybridization.
2. Describe the technique of chromosome painting, and explain why it is used.

As mentioned in Section 22.1, cytogenetic mapping determines the locations of specific sequences within chromosomes that are viewed microscopically. This type of mapping is commonly used in genetic studies of eukaryotes, which have very large chromosomes relative to those of bacteria. Microscopically, eukaryotic chromosomes can be distinguished from one another by their size, centromere location, and banding patterns (refer to Figure 8.1). Treating chromosomes with particular dyes produces discrete banding patterns. Cytogeneticists use a chromosome's banding pattern as a way to identify specific regions along its length. In this section, we will explore techniques that are aimed at producing cytogenetic maps.

A Goal of Cytogenetic Mapping Is to Determine the Location of a Gene on an Intact Chromosome

Cytogenetic mapping can localize a particular gene to a site within a chromosomal banding pattern. For example, let's consider the cystic fibrosis transmembrane regulator (CFTR) protein that is defective in people with cystic fibrosis. The human *CFTR* gene that codes this protein is located on chromosome 7, at a specific site in the q3 region.

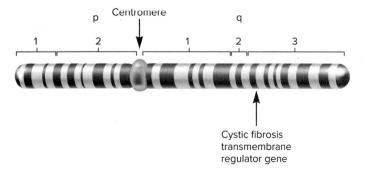

Cystic fibrosis transmembrane regulator gene

Cytogenetic mapping may be used as a first step in the localization of genes in plants and animals. However, because this mapping relies on light microscopy, it has a fairly crude limit of resolution. In most species, cytogenetic mapping is accurate only within limits of approximately 5 million bp along a chromosome. In species that have large polytene chromosomes, such as *Drosophila*, the resolution is much better.

In Situ Hybridization Localizes Genes Along Particular Chromosomes

The technique of **in situ hybridization** is used to cytogenetically map the locations of genes or other DNA sequences within large eukaryotic chromosomes. The term *in situ* (Latin for "in place") indicates that the procedure is conducted on chromosomes that are being held in place—adhered to a surface. The term **hybridization** indicates that a labeled strand of DNA base-pairs with and thereby forms a hybrid with an intact chromosome.

To map a gene via in situ hybridization, researchers use a labeled probe to detect the location of the gene within a set of chromosomes. If the gene of interest has been cloned previously, as described in Chapter 20, the DNA strand of the cloned gene can be used as a probe. Because a DNA strand will bind to, or hybridize, only to its complementary sequence on a particular chromosome, this technique provides a way of localizing the gene of interest. For example, let's consider the gene that causes the white-eye phenotype in

Drosophila when it carries a loss-of-function mutation. This gene has already been cloned. If a single-stranded piece of this cloned DNA is mixed with *Drosophila* chromosomes in which the DNA has been denatured, the cloned piece will bind only to the X chromosome at the location corresponding to the site of the eye color gene.

The most common method of in situ hybridization uses fluorescently labeled DNA probes and is referred to as **fluorescence in situ hybridization (FISH)**. **Figure 22.2** shows the steps of the FISH procedure.

1. The cells, which are prepared using a technique that keeps the chromosomes intact, are treated with agents that cause them to swell, and their contents are fixed to the slide.
2. The chromosomal DNA is then denatured.
3. A DNA probe is added. For example, the added DNA probe might be a cloned piece of single-stranded DNA that is complementary to a specific gene. In this case, the goal of the FISH experiment is to determine the location of the gene within a set of chromosomes. The probe binds to a site in the chromosomes where the gene is located because the probe and chromosomal gene line up and hydrogen bond with each other.
4. To detect where the probe has bound to a chromosome, the probe is subsequently tagged with a fluorescent molecule. Tagging may be accomplished by first incorporating biotin-labeled nucleotides into the probe. Biotin, a small, nonfluorescent molecule, has a very high affinity for a protein called avidin. When fluorescently labeled avidin is then added, it binds tightly to the biotin, thereby labeling the probe.
5. How is the fluorescently labeled probe detected? A fluorescent molecule absorbs light at a particular wavelength and then emits light at a longer wavelength. To detect the light emitted by a fluorescently labeled probe, a fluorescence microscope is used. Such a microscope contains filters that only allow the passage of light whose wavelength is within a defined range. The sample is illuminated at the wavelength of light that is absorbed by the fluorescent molecule. The fluorescent molecule then emits light at a longer wavelength, and the fluorescence microscope allows the transmission of the emitted light. Because filters prevent the transmission of light of other wavelengths, only the emitted light is viewed and the background of the sample is dark. Therefore, the fluorescence is seen as a brightly glowing color on a dark background.

For many FISH experiments, chromosomes are counterstained by a fluorescent dye that is specific for DNA. A commonly used dye is DAPI (4′,6-diamidino-2-phenylindol), which is excited by UV light. This dye gives all of the DNA a blue background color. The results of a FISH experiment are then compared with a sample of chromosomes that have been treated with Giemsa stain to produce banding. This comparison allows the location of a probe to be mapped relative to the chromosomal banding pattern.

Figure 22.3 illustrates the results of a FISH experiment involving six different DNA probes. The six probes were strands of DNA corresponding to six different DNA segments located on human chromosome 5. In this experiment, each probe was labeled with a different fluorescent molecule. This enabled researchers to distinguish the probes when they became bound to their corresponding locations on

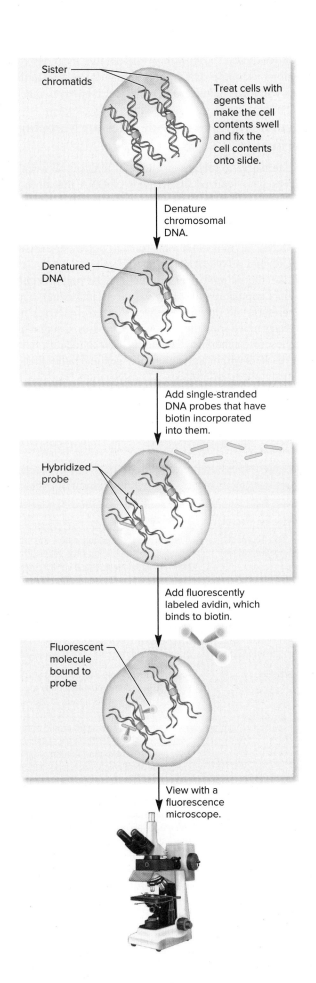

Sister chromatids

Treat cells with agents that make the cell contents swell and fix the cell contents onto slide.

Denature chromosomal DNA.

Denatured DNA

Add single-stranded DNA probes that have biotin incorporated into them.

Hybridized probe

Add fluorescently labeled avidin, which binds to biotin.

Fluorescent molecule bound to probe

View with a fluorescence microscope.

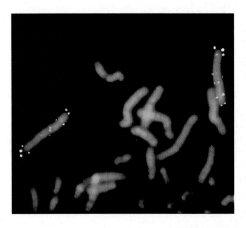

FIGURE 22.3 Chromosome painting via fluorescence in situ hybridization. In this experiment, six different probes were used to locate six different sites along human chromosome 5. The colors are due to computer imaging of the fluorescence emission; they are not the actual colors of the fluorescent labels. Two spots are usually seen at each site because the probe binds to both sister chromatids.

From: T. Ried, A. Baldini, T.C. Rand, & D.C. Ward (2002), "Simultaneous visualization of seven different DNA probes by in situ hybridization using combinatorial fluorescence and digital imaging microscopy," *PNAS*, *89*(4): 1388–1392. Courtesy Thomas Ried.

chromosome 5. In this experiment, computer-imaging methods were used to assign each fluorescently labeled probe a different color. This method, called **chromosome painting,** allowed the researchers to visualize the sites along chromosome 5 corresponding to the six different probes as different-colored spots. FISH was used here to determine the order and relative distances between six specific sites along a single chromosome. FISH is commonly used in genetics and cell biology research, and its use has become more widespread in clinical applications. For example, clinicians may use FISH to detect changes in chromosome structure, such as deletions, duplications, and translocations, which may occur in patients with genetic disorders.

GENETIC TIPS   **THE QUESTION:** Phenylketonuria (PKU) is a recessive disorder in humans that is due to a loss-of-function mutation involving the gene that codes the enzyme phenylalanine hydroxylase. Some people with the disorder carry a point mutation that causes the loss of function, whereas other individuals have a deletion of the entire gene. Explain how you could use fluorescence in situ hybridization to distinguish a point mutation from a deletion. Describe your expected results.

ONLINE ANIMATION

FIGURE 22.2 The technique of fluorescence in situ hybridization (FISH). The probe hybridizes to the denatured chromosomal DNA only at a specific complementary site. Note that the chromosomes are highly condensed metaphase chromosomes that have already replicated. These are sister chromatids. Therefore, each X-shaped chromosome actually contains two copies of a particular gene. Because the sister chromatids are identical, a probe that recognizes a site on one sister chromatid will also bind to the same site on the other.

CONCEPT CHECK: Why does the probe bind to a specific site on a chromosome?

TOPIC: *What topic in genetics does this question address?* The topic is the use of FISH to distinguish a point mutation from a deletion.

INFORMATION: *What information do you know based on the question and your understanding of the topic?* From the question, you know that some people with PKU have a point mutation in the gene that codes phenylalanine hydroxylase, whereas others have experienced a deletion of that gene. From your understanding of the topic, you may remember that FISH is used to detect a gene's location within an intact set of chromosomes.

PROBLEM-SOLVING **S**TRATEGY: *Design an experiment.* One strategy to solve this problem is to design an experiment using FISH.

ANSWER: You basically want to follow the procedure illustrated in Figure 22.2.

1. Obtain a blood sample from the affected individual. As a control, obtain a blood sample from an unaffected and unrelated individual. Treat the cells with agents that cause them to swell, and fix their contents to a slide.

2. Denature the chromosomal DNA.

3. Add a fluorescently labeled probe. In this case, the probe would be a single-stranded DNA segment that is complementary to the phenylalanine hydroxylase gene.

4. View under a fluorescence microscope.

Expected results: In the control, you would see bright spots where the phenylalanine hydroxylase gene is located. If the affected individual had a point mutation, you would still see the spots. If there had been a deletion of both copies of the gene, you would not see any fluorescently labeled spots.

22.2 COMPREHENSION QUESTION

1. The technique of fluorescence in situ hybridization involves the use of a _____ that hybridizes to a _____.
   a. radiolabeled probe, band on a gel
   b. radiolabeled probe, specific site on an intact chromosome
   c. fluorescent probe, band on a gel
   d. fluorescent probe, specific site on an intact chromosome

22.3 LINKAGE MAPPING VIA CROSSES

Learning Outcomes:

1. Define *molecular marker.*
2. Explain the use of molecular markers in linkage mapping.

Let's now turn to linkage mapping, which relies on the frequency of recombinant offspring to determine the distance between sites located along the same chromosome. We already considered linkage mapping methods in Chapter 6, where allelic differences

between genes were used to map the relative locations of those genes along a chromosome by conducting testcrosses. In this section, we will focus on the use of molecular markers to map genes.

Molecular Markers Provide Sites for Mapping Experiments

As an alternative to relying on allelic differences between genes, geneticists have realized that regions of DNA that do not code genes can be used as markers along a chromosome. A **molecular marker** is a segment of DNA that is found at a specific site along a chromosome and has properties that enable it to be uniquely recognized using molecular tools, such as polymerase chain reaction (PCR) and gel electrophoresis.

Like alleles, molecular markers may be **polymorphic;** that is, within a population, they may vary from individual to individual. Therefore, the distances between linked molecular markers can be determined from the outcomes of crosses. Using molecular techniques, researchers have found it easier to identify many molecular markers within a given species' genome rather than identifying many allelic differences that affect traits. For this reason, geneticists have increasingly turned to molecular markers as points of reference along genetic maps. As **Table 22.1** indicates, many different kinds of molecular markers are used by geneticists.

TABLE 22.1	
Common Types of Molecular Markers	
Marker	**Description**
Restriction fragment length polymorphism (RFLP)	A region in a chromosome where the distance between two restriction sites varies among different individuals. These variable regions are identified by digesting chromosomal DNA with restriction enzymes and observing DNA fragments that differ in length.
Amplified restriction fragment length polymorphism (AFLP)	This type of marker is similar to an RFLP; the DNA sequence of interest is amplified via PCR instead of isolating the chromosomal DNA.
Microsatellite	A site in the genome that contains short sequences that are repeated many times in a row. The total length is usually in the range of 10–200 bp, and the length of a given microsatellite may be polymorphic within a population. Microsatellites are isolated via PCR. They are also called short tandem repeats (STRs) and simple sequence repeats (SSRs).
Single-nucleotide polymorphism (SNP)	A site in a genome where a single nucleotide is polymorphic among different individuals. These sites occur commonly in all genomes and are widely used in the mapping of disease-causing alleles and genes that contribute to quantitative traits that are valuable in agriculture (see Chapter 28).
Sequence-tagged site (STS)	When a pair of PCR primers copies a single stretch of DNA within a set of chromosomes, the entire PCR product or a portion of it can be sequenced to produce a genetic marker called a sequence-tagged site (STS). An STS can include a microsatellite or a SNP.

Researchers have constructed detailed genetic maps in which a series of many molecular markers have been identified on each chromosome of certain species. These species include humans, model organisms, agricultural species, and many others. Why are molecular markers useful? One key reason is that molecular markers can be used to determine the approximate location of an unknown gene that causes a human disease. Clinical geneticists sometimes follow the transmission patterns of polymorphic molecular markers in family pedigrees to locate genes that, when they are mutated, cause human disease. The identification of a particular marker in those who have the disease can indicate that the marker is close to the disease-causing allele (look ahead to Figures 24.7 and 24.8). This may help researchers identify the gene by cloning methods, such as chromosome walking or primer walking, which we will examine in Section 22.4.

In addition, molecular markers may help researchers identify the locations of genes involved in quantitative traits, such as fruit yield and meat weight, that are valuable in agriculture. The use of molecular markers to identify such genes is described in Chapter 28 (see Figure 28.5). Large sets of genetic markers are also used by evolutionary biologists to determine patterns of genetic variation within a species and the evolutionary relatedness of different species.

As an example of a map using molecular markers, **Figure 22.4** shows a simplified map of two chromosomes found in the plant *Arabidopsis thaliana*, which is one of the favorite model organisms of plant molecular geneticists. Many RFLPs, which are described in Table 22.1, have been mapped to different locations on the *Arabidopsis* chromosomes.

Linkage Mapping Uses Molecular Markers Such as Microsatellites

To make a highly refined map of a genome, many different polymorphic sites must be identified and their transmission followed from parent to offspring over many generations. RFLPs were among the first molecular markers studied by geneticists. More recently, other molecular markers have been used because they are easier to amplify via PCR. As an example, let's consider **microsatellites,** which are short repetitive sequences that are interspersed throughout a species' genome and tend to vary in length among different individuals.

Microsatellites usually contain di-, tri-, tetra-, or pentanucleotide sequences that are repeated many times in a row. For example, the most common microsatellite encountered in humans is a dinucleotide sequence $(CA)_n$, where n ranges from 5 to around 50. In other words, this dinucleotide sequence can be tandemly repeated 5–50 or more times. The $(CA)_n$ microsatellite is found, on average, about every 10,000 bases in the human genome. Researchers have identified thousands of different DNA segments that contain $(CA)_n$ microsatellites, located at many distinct sites within the human genome.

How do researchers identify a specific microsatellite within a chromosome? The procedure is shown in **Figure 22.5**.

1. The starting material is a sample that contains all of the chromosomes.
2. Using primers complementary to the unique DNA sequences that flank a specific microsatellite allows that microsatellite to be amplified by PCR. In other words, the PCR

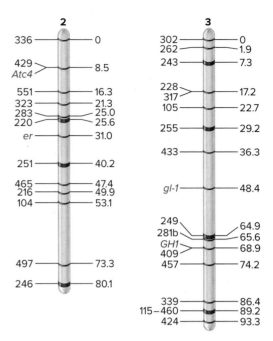

FIGURE 22.4 **A linkage map of RFLPs on two chromosomes in *Arabidopsis thaliana*.** This plant has five different chromosomes, but only the maps of chromosomes 2 and 3 are shown here. The maps show the locations of many RFLP markers. The numbers along the left side of each chromosome designate the locations of those markers. The numbers along the right side of each chromosome are the distances in map units (mu). For example, the RFLPs at map positions 16.3 and 40.2 on chromosome 2 are designated 551 and 251, respectively, and are 23.9 mu apart. The marker at the upper end of each chromosome was arbitrarily assigned as the starting point (zero) for that chromosome. In addition, the map shows the locations of a few known genes (shown in red): *Atc4* codes the protein actin, *er* codes erecta, *gl-1* codes glabra-1, and *GH1* codes acetolactate synthase.

primers copy only a particular microsatellite, but not the thousands of others that are interspersed throughout the genome. If a pair of PCR primers copies a single site within a set of chromosomes, the amplified region is called a **sequence-tagged site (STS).**

3. The PCR products are then run on a gel to identify their size(s). When DNA is collected from a haploid cell, an STS produces only a single band on a gel. In a diploid species, an individual has two copies of a given STS. When an STS contains a microsatellite, the two PCR products may be identical and will produce a single band on a gel if the region is the same length in both copies (i.e., if the individual is homozygous for the microsatellite). However, if an individual has two copies that differ in the number of repeats in the microsatellite sequence (i.e., if the individual is heterozygous for the microsatellite), the two PCR products obtained will be different in length, as is the case for the products in Figure 22.5. The DNA fragments found in the two bands on the gel in this figure were made via PCR, using primers that flank a particular microsatellite on chromosome 2. The DNA fragment in the higher band is longer, because it has more repeat sequences than does the lower band.

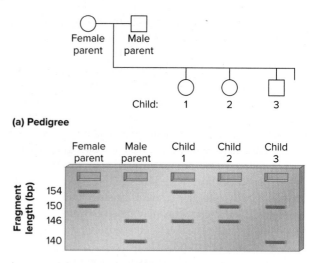

(a) Pedigree

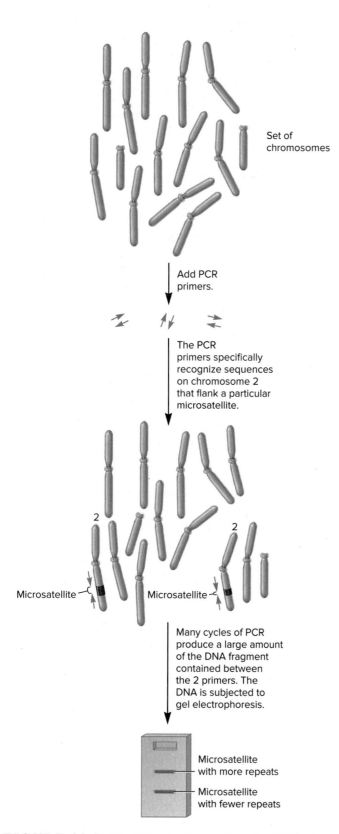

(b) High-resolution gel of PCR products for a polymorphic microsatellite found in the family in (a).

FIGURE 22.6 Inheritance pattern of a polymorphic microsatellite in a human pedigree. This microsatellite is autosomal.

When microsatellites have length polymorphisms, researchers can follow their transmission from parent to offspring. PCR amplification of particular microsatellites provides a strategy in the genetic analysis of human pedigrees, as shown in **Figure 22.6**. Prior to this analysis, a unique segment of DNA containing a microsatellite had been identified. Using PCR primers complementary to this microsatellite's unique flanking segments, two parents and their three offspring were tested for the inheritance of this microsatellite. A small sample of cells was obtained from each individual and subjected to PCR amplification, as shown in Figure 22.5. The amplified PCR products were then analyzed by high-resolution gel electrophoresis, which detects small differences in the lengths of DNA fragments.

- The female parent's PCR products were 154 and 150 bp in length.
- The male parents's were 146 and 140 bp.
- Their first offspring inherited the 154-bp microsatellite from the female parent and the 146-bp one from the male parent.
- The second inherited the 150-bp product from the female parent and the 146-bp one from the male parent.
- The third inherited the 150-bp product from the female parent and the 140-bp one from the male parent.

As shown in the figure, the transmission of polymorphic microsatellites is relatively easy to follow from generation to generation.

The simple pedigree analysis shown in Figure 22.6 illustrates the general method used to follow the transmission of a single microsatellite that is polymorphic for length. In linkage studies, the goal is to follow the transmission of many different microsatellites to determine which ones are linked along the same chromosome and which ones are not. Those that are not linked will independently assort from generation to generation. Those

FIGURE 22.5 Identifying a microsatellite using PCR primers.

CONCEPT CHECK: Explain how a microsatellite can be polymorphic?

that are linked tend to be transmitted together to the same offspring.

In a large pedigree, it is possible to identify cases in which linked microsatellites have segregated due to crossing over. The frequency of crossing over provides a measure of the map distance, in this case, between the different microsatellites. This approach can help researchers obtain a finely detailed linkage map of the human chromosomes without having to depend on alleles of closely linked genes that affect phenotype.

Pedigree analysis involving STSs, such as polymorphic microsatellites, enables researchers to identify the locations of disease-causing alleles. The assumption behind this approach is that a disease-causing allele had its origin in a single individual known as a **founder**, who lived many generations ago. Since that time, the allele has spread throughout portions of the human population.

A second assumption is that the founder is likely to have had a polymorphic molecular marker that lies somewhere near the mutant allele. This is a reasonable assumption, because all people carry many polymorphic markers throughout their genomes. If a polymorphic marker lies very close to a disease-causing allele, a crossover is unlikely to occur in the intervening region. Therefore, such a polymorphic marker may be linked to the disease-causing allele for many generations. By following the transmission of many polymorphic markers within large family pedigrees, it may be possible to determine that particular markers are found in people who carry specific disease-causing alleles. After the identification of a closely linked marker, a disease-causing allele may be identified (look ahead to Figure 24.7).

22.3 COMPREHENSION QUESTIONS

1. A molecular marker is a _____ that is found at a specific site on a chromosome and has properties that allow it to be _____.
   a. colored dye, visualized via microscopy
   b. colored dye, visualized on a gel
   c. segment of DNA, uniquely identified using molecular tools
   d. segment of DNA, visualized via microscopy

2. Which of the following is an example of a molecular marker?
   a. RFLP
   b. Microsatellite
   c. Single-nucleotide polymorphism
   d. All of the above are types of molecular markers.

3. To map the distance between molecular markers via testcrosses, the markers must be
   a. polymorphic.
   b. monomorphic.
   c. fluorescently labeled.
   d. on different chromosomes.

22.4 CHROMOSOME WALKING AND PRIMER WALKING

Learning Outcome:

1. Describe the techniques of chromosome walking and primer walking, and explain why they are used.

An important application of genomics is the identification of genetic variation that contributes to human disease. Various approaches are used to identify disease-causing alleles in human populations. One strategy, called a genome-wide association study (GWAS), relies on analysis of genetic variation in a large population (look ahead to Figure 24.8). A different approach that stems from linkage mapping is **chromosome walking.** This method has been successful in the identification of many genes that may exist as disease-causing alleles, including those that cause cystic fibrosis, Huntington disease, and Duchenne muscular dystrophy.

To conduct chromosome walking, a disease-causing allele's position relative to another gene or to a molecular marker must be known from previous studies. For example, when analyzing large pedigrees, linkage mapping studies may reveal a strong correlation between the presence of a particular molecular marker and a disease-causing allele. In other words, the disease-causing allele and the molecular marker are closely linked along a chromosome. As discussed in Section 22.3, this close linkage is based on the phenomenon that a disease-causing allele originated in a single founder. A closely linked marker provides a starting point from which to molecularly "walk" toward the gene of interest.

Figure 22.7a considers a chromosome walk in which the goal is to locate a gene that we will call gene *A*. In this example, linkage mapping studies had previously revealed that gene *A* is relatively close to a particular marker. Those mapping studies deduced that the marker and gene *A* are approximately 1 mu apart.

1. To begin this chromosome walk, a DNA fragment that contains the marker and its flanking sequences is inserted into a cloning vector.
2. A small piece of DNA from the starting clone is inserted into another vector. This procedure is called **subcloning.** Note that the subcloned piece is toward the end of the insert, away from the marker.
3. The subcloned DNA is labeled and used as a probe. The labeled probe is exposed to the members of a genomic library; the production of a genomic library is described in Chapter 20 (refer back to Figure 20.4a).
4. When the labeled probe binds to a member of the library, this enables the researcher to identify a second clone that extends into the region that is closer to gene *A*.
5. A subclone from this second clone is then used to screen the library a second time. This allows the researchers to identify a clone that is even closer to gene *A*.
6. This repeated pattern of subcloning and library screening is used to reach gene *A*.

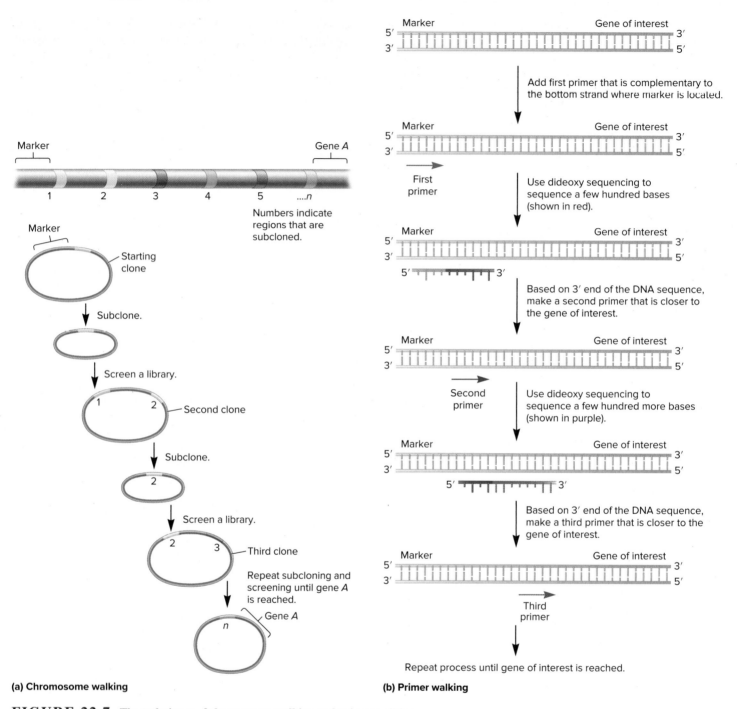

(a) Chromosome walking

(b) Primer walking

FIGURE 22.7 The techniques of chromosome walking and primer walking.

CONCEPT CHECK: For what purpose is the technique of chromosome walking used?

The term *chromosome walking* is used because each clone advances the analysis a step closer to the gene of interest. When starting at the marker in Figure 22.7a, researchers will also have identified different markers on the other side of marker in the starting clone to ensure their clones are not walking in the wrong direction.

The number of steps required to reach the gene of interest depends on the distance between the starting and ending points and on the sizes of the DNA inserts in the library. If the two points are 1 mu apart, they are expected to be approximately 1 million bp apart,

although the correlation between map units and physical distances can vary greatly. In a typical chromosome walking experiment, each clone might have an average insert size of 50,000 bp. Therefore, it takes about 20 steps to reach the gene of interest. Researchers conducting a chromosome walking experiment want to locate a starting point that is as close as possible to the gene they wish to identify.

How do researchers know when they have reached a gene of interest? In the case of a gene that causes a disease when mutant, researchers conduct their walking steps on DNA from both an unaffected and an affected individual. Each set of clones is subjected to DNA sequencing, and those DNA sequences are compared with each other. When the researchers reach a spot where the DNA sequences differ between the unaffected and the affected individual, such a site may be within the gene of interest. However, this has to be confirmed by sequencing the region from several unaffected and affected individuals to be certain the change in DNA sequence is correlated with the disease.

Chromosome walking was invented at a time when inserting DNA fragments into a cloning vector was needed to carry out dideoxy sequencing of DNA. With the advent of new DNA sequencing methods that do not require cloning vectors, a modification to chromosome walking is a newer method called **primer walking.** This method is also called directed DNA sequencing because the first primer is designed from a known region of DNA and it guides the sequencing in a specific direction.

Figure 22.7b shows the steps in the process.

1. A first primer is made that is complementary to one of the DNA strands where the marker is found.
2. Dideoxy sequencing is conducted in the 3′ direction to obtain a sequence that is several hundred bases long.
3. The 3′ end of this sequence is used to make a second primer.
4. This second primer is used to sequence another stretch of DNA, and the 3′ end of this second sequence is used to make a third primer.
5. This process is repeated over and over until the gene of interest has been reached.

22.4 COMPREHENSION QUESTION

1. Chromosome walking is a technique in which a researcher begins at a specific site on a chromosome and analyzes _____ until the gene of interest is reached.
   a. bands on gel
   b. a series of subclones
   c. the ability of primers to bind to sites along a chromosome
   d. both a and c

22.5 OVERVIEW OF GENOME SEQUENCING

Learning Outcomes:

1. Describe how shotgun sequencing and de novo genome assembly are used to obtain a complete DNA sequence of a species' genome.
2. Compare and contrast short-read sequencing and long-read sequencing.
3. Outline different methods of DNA sequencing, and explain how Illumina sequencing technology works.

As described in Section 22.1, the distances between genes in a physical map are computed as the number of base pairs. A key goal of physical mapping is to obtain a complete DNA sequence of a chromosome or a genome. In a physical map, the DNA sequence of a chromosome is annotated to show the locations of genes, molecular markers, and other sequences of interest.

Due to their small sizes, the earliest genomes to be sequenced were those of viruses, such as phage λ (discussed in Chapter 18). As we will see in Section 22.6, the first genome sequence of a bacterium was obtained in 1995. Since then, the genomes of hundreds of thousands of species have been sequenced.

Historically, genome sequencing in the 1990s and early twenty-first century began with the insertion of chromosomal DNA fragments into cloning vectors. The inserts within cloning vectors were then subjected to DNA sequencing via the Sanger dideoxy sequencing method. (Methods of vector cloning and dideoxy sequencing are discussed in Chapter 20.) However, as methods of DNA sequencing have advanced over the last few decades, the need to insert chromosomal DNA fragments into vectors has become obsolete. In this section, we will focus on recent advances in genome sequencing methods that do not rely on vector cloning.

Genome Sequencing Involves the Sequencing of Many Fragments of Chromosomal DNA

When sequencing an entire genome, researchers must consider factors such as genome size, the efficiency of the methods used to sequence DNA, and the costs of the project. Since genome-sequencing projects began in the 1990s, researchers have learned that the most efficient and inexpensive way to sequence genomes is via an approach called **shotgun sequencing.** This sequencing strategy involves breaking the genome into a collection of many DNA fragments that are sequenced individually. The production of a genome sequence involves two main methods:

1. Determine the base sequence of many DNA fragments that are derived from chromosomal (genomic) DNA.
2. Determine the order of the DNA fragments as they would occur along the chromosome(s) in a given species' genome. This order is deduced from overlaps of the DNA sequences among different DNA fragments, a process called **de novo genome assembly.**

As a matter of chance, some DNA fragments will have overlapping regions, as shown schematically in **Figure 22.8**. The DNA sequences in the two different fragments are identical in the overlapping region. This allows researchers to order those fragments as they would occur in the intact chromosome.

De novo genome assembly is done with the aid of computer software. The data from the sequencing of many DNA fragments are directly entered into sequence files. These files are analyzed using software that identifies overlapping regions. Based on these overlapping regions, the software will generate a contiguous DNA sequence that occurs along a chromosome. However, certain problems may be encountered during the de novo genome assembly process:

- Some regions of chromosomes, such as telomeres and centromeres in eukaryotic species, may contain long stretches of highly repetitive sequences (refer back to Chapter 10). The lengths of these regions are often difficult to

Overlapping region

```
TTACGGTACCAGTTACAAATTCCAGACCTAGTACC
AATGCCATGGTCAATGTTTAAGGTCTGGATCATGG
                  GACCTAGTACCGGACTTATTCGATCCCCAATTTTGCAT
                  CTGGATCATGGCCTGAATAAGCTAGGGGTTAAAACGTA
```

FIGURE 22.8 **A comparison of two DNA fragments that contain an overlapping region.**

determine due to an inability to accurately identify overlaps within them.

- Even though researchers may sequence many DNA fragments, an occasional chromosomal region may not be included as a matter of chance. This leaves one or more gaps in the physical map. Various methods can be used to close gaps, such as using long-read DNA sequencing methods, which are described next.

Genome Sequences Can Be Produced Using Short-Read and Long-Read DNA Sequencing Methods

DNA sequencing technologies can be broadly categorized as short-read sequencing or long-read sequencing depending on the length of the base sequences that are obtained from each fragment of chromosomal DNA. **Short-read sequencing (SRS)** produces base sequences up to a few hundred bases in length, whereas **long-read sequencing (LRS)** typically generates sequences with lengths in the range from 10,000 to 30,000 bases, and sometimes longer. Specific examples of short-read and long-read sequencing methods are described later in Table 22.2.

What are the advantages and disadvantages of short-read versus long-read methods?

Cost. SRS is usually less expensive than LRS, but the cost of LRS is coming down.

Accuracy. SRS is currently more accurate than LRS, but the accuracy of LRS is improving.

De Novo Genome Assembly. LRS is better because the assembly process involves the alignment of far fewer DNA sequences, due to the greater lengths of the fragments. LRS is also better at assembling regions that contain repetitive sequences.

Detection of Genetic Variation. Both SRS and LRS are good at detecting genetic variation among different individuals, though SRS is considered better due to its higher accuracy. Even so, an advantage of LRS is that it can more easily determine if genetic variation in an offspring is inherited from the female or male parent. LRS is also better at detecting structural variations, such as an inversion.

Some challenges in genome sequencing can be resolved by using a combination of both short-read and long-read technologies. LRS can be used for an initial de novo assembly of an entire genome, and then SRS can be used to improve the accuracy of the sequencing. The use of SRS to improve the accuracy of LRS is sometimes called "polishing" the DNA sequence.

Innovations in DNA Sequencing Have Made It Less Expensive and Faster

Since DNA sequencing was invented in the 1970s, technological advances have been aimed at making it less expensive and faster.

- As discussed in Section 22.6, the Human Genome Project, which began in the early 1990s, was originally estimated to cost about $3 billion to sequence a single genome. However, cost reductions due to innovations in DNA-sequencing technology drove the actual cost down to about $300 million. By the end of the project, researchers estimated that if they were starting again, they could have sequenced the genome for less than $50 million. The project took about 13 years to complete the sequencing of a single human genome.
- In 2007, researchers undertook the sequencing of James Watson's genome, which cost less than $1 million. It took 2 months to complete.
- By 2011, the cost of sequencing a human genome had been reduced to about $5000 and could be completed in a matter of weeks.
- Depending on the method used, the sequencing of a single human genome currently costs $600 or less, and it typically takes one to two days! Such innovation makes it feasible to sequence an individual's genome as a routine diagnostic procedure.

The ability to rapidly sequence large amounts of genomic DNA is referred to as **high-throughput sequencing.** Different types of technological advances have made this possible.

- First, different aspects of DNA sequencing have become automated, so samples can now be processed rapidly in a machine. For example, in Chapter 20, we considered how fluorescent labeling of nucleotides can automate the reading of a DNA sequence using a machine with a laser and a fluorescence detector.
- High-throughput sequencers are able to use samples that contain mixtures of DNA fragments that have not been subjected to vector-based cloning. By comparison, the earliest methods of genome sequencing, such as the one featured later in Figure 22.10, involved the insertion of DNA fragments into cloning vectors. At that time, dideoxy sequencing required cloned DNA. The elimination of such cloning steps in newer methods saves a great deal of time and money.
- Another advance in sequencing technologies involves parallel sequencing, which allows multiple samples to be processed at once. The first parallel sequencing machines could simultaneously perform many sequencing runs via

TABLE 22.2

Examples of Next-Generation Sequencing Technologies

Technology*	Short or Long Read	PCR Amplified DNA or Single Molecule	Description
Illumina	Short read: a few hundred bases	PCR amplified	Adaptors are attached to the ends of short DNA fragments followed by a bridge amplification step. The sequences are determined by sequencing by synthesis, one nucleotide at a time, using fluorescently tagged dNTPs. See Figure 22.9.
Ion-torrent (Thermo Fisher)	Short read: several hundred bases	PCR amplified	Uses microwells on a semiconductor chip to measure changes in pH during the sequencing cycles.
DNA nanoball	Short read: <100 bases	PCR amplified	Linkers are attached to small fragments of DNA that are amplified by rolling circle replication into DNA nanoballs. The nanoballs are then bound to a flow cell surface and the DNA sequence is determined using fluorescently labeled nucleotides.
PacBio SMRT	Long read: tens of thousands of bases	Single molecule	Also known as Single Molecule Real-Time (SMRT) sequencing. It is based on monitoring the activity of DNA polymerase molecules attached to the bottom surface of nano-sized sequencing units called zero-mode-waveguides (ZMWs) using fluorescently labeled nucleotides.
Oxford Nanopore	Long read: depends on sample preparation; can be over 1 million bases	Single molecule	Uses nanopores inserted into an artificial lipid bilayer, placed in individual microwells tens of micrometers wide, and arrayed on a sensor chip. As each single DNA strand travels through a channel, it disrupts the current running through the pore, and the change is measured by a semiconductor sensor.

*Most include the company name associated with the technology.

multiple gel-filled capillary tubes. Those in use today can process thousands or even millions of sequence reads in parallel. As shown later in Figure 22.9, one way that this massive parallel processing is made possible is by attaching many DNA strands to a flow cell.

- Some sequencing technologies use PCR to amplify the DNA, whereas others actually read single DNA molecules.

Although Sanger dideoxy sequencing is still in use for smaller DNA-sequencing projects, high-throughput sequencing methods are used for the sequencing of whole genomes. **Table 22.2** describes a few of these methods, which are often referred to as **next-generation sequencing technologies** because they have superseded the dideoxy method for large sequencing projects. Those that sequence single DNA molecules are sometimes called **third-generation sequencing technologies.** They are used for long-read sequencing.

The newer sequencing technologies rely on a complex interplay of enzymology, chemistry, high-resolution optics, electronics, and new approaches to processing the data. These instruments allow for easy sample preparation steps prior to DNA sequencing. We will consider the details of Illumina sequencing technology next.

Illumina Sequencing Technology Provides a Short-Read Method for DNA Sequencing

A very popular method for the sequencing of DNA, particularly the sequencing of whole genomes, is Illumina sequencing technology. It is called **sequence by synthesis (SBS)** because the base sequence is determined as a DNA strand is being made. The general steps in Illumina sequencing are shown in **Figure 22.9.**

1. Chromosomal DNA is isolated from a sample of cells and broken into small fragments about 200–600 bp in length.
2. Short sequences of DNA called adaptors are attached to the ends of the DNA fragments.
3. The DNA fragments with attached adaptors are denatured into single strands.
4. The DNA strands are then exposed to the surface of a flow cell. Because it has a complementary base sequence, each adaptor binds to a primer that is permanently attached to the surface of the flow cell. The two types of primers, which are complementary to the two types of adaptors, are arranged in rows along the flow cell.
5. The single-stranded DNA molecules form bridges in which the adaptors on both ends are bound to primers in nearby rows. Double-stranded DNA is produced from these molecules via PCR.
6. The double-stranded DNA is then denatured into single-stranded DNA.
7. The steps of bridge formation, PCR amplification, and denaturation are then repeated many times to form clusters of identical single-stranded DNA molecules. Note: The reason for generating clusters is that a cluster of identical DNA molecules will emit a stronger fluorescent signal than a single DNA molecule would.

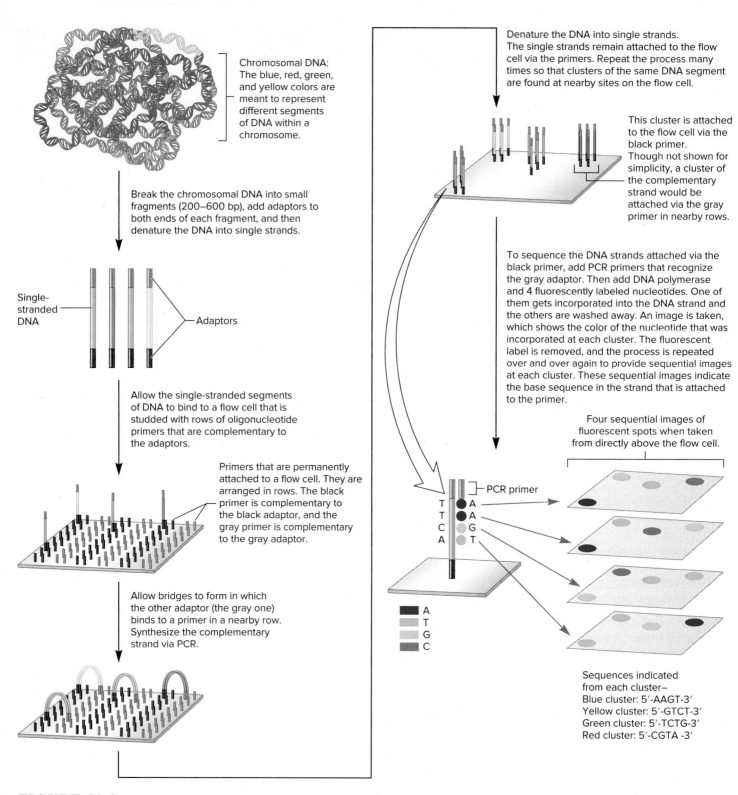

FIGURE 22.9 Illumina sequencing technology, an example of next-generation sequencing.

CONCEPT CHECK: Illumina sequencing technology is a sequence by synthesis method. Explain what that term means.

8. Primers that recognize one of the adaptors are added. A solution containing DNA polymerase and all four fluorescently labeled nucleotides is passed across the flow cell. The base in each of the four types of nucleotides is labeled in a different color. (In Figure 22.9, the colors used are

purple for A, green for T, yellow for G, and red for C.) After a single nucleotide has been added, another nucleotide cannot be added because the fluorescent molecule attached to the base temporarily prevents further strand growth. The nucleotides that were not added to the growing

DNA strand are washed away. A laser then activates the fluorescent label on the nucleotide that has just been added to the DNA strand. This fluorescence is detected by a camera and recorded on a computer. Note: The laser activates all of the clusters on the flow cell; each cluster produces a colored spot, and the color depends on which base (A, T, G, or C) was just added to the DNA strand.

9. The fluorescent molecule attached to the most recently added nucleotide is then removed, which permits further strand growth. Next, another solution containing DNA polymerase and the four fluorescently labeled nucleotides is passed across the flow cell, and the process is repeated many times in a row. This results in a series of images that reveal the base sequence in each cluster by analyzing the order of the colored spots for each cluster. For example, if the spots on sequential images from a cluster are purple, purple, yellow, and then green, the sequence is 5′-AAGT-3′. A typical Illumina sequencing run produces a series of a few hundred images, thereby revealing DNA sequences that are a few hundred bases in length.

22.5 COMPREHENSION QUESTIONS

1. Following the sequencing of many DNA fragments, de novo genome assembly is possible by analyzing
   a. the relative lengths of different DNA fragments.
   b. overlaps of the DNA sequences among different DNA fragments.
   c. the relative numbers of A-T base pairs compared to G-C base pairs.
   d. both b and c.

2. Which of the following is not an advantage of long-read DNA sequencing?
   a. Greater ease of de novo genome assembly
   b. Greater ease of determining if genetic variation was inherited from the female or male parent
   c. Greater ease of detecting structural variations, such as an inversion
   d. Greater accuracy

3. During Illumina sequencing, the two types of primers are arranged in nearby rows on a flow cell. This arrangement facilitates
   a. the incorporation of fluorescently labeled nucleotides.
   b. bridge formation.
   c. the ability of primers to bind to multiple sites along a DNA strand.
   d. both a and c

22.6 GENOME-SEQUENCING PROJECTS

Learning Outcomes:

1. Describe the approach that was used to sequence the first bacterial genome.
2. List the goals of the Human Genome Project.
3. Compare and contrast the sizes of different species' genomes.

Genome-sequencing projects are research endeavors with the ultimate goal of determining the sequence of DNA bases of the entire genome of a given species. Such projects involve scientists who isolate DNA and perform DNA-sequencing reactions, as well as theoreticians who gather the DNA sequence information and assemble it into a complete DNA sequence for each chromosome. For bacteria and archaea, which usually have just one chromosome, the genome sequence is that of a single chromosome. For eukaryotes, each chromosome must be sequenced. Thus, for humans, the genome sequence includes sequences of 22 autosomes, 2 sex chromosomes, and the mitochondrial genome.

In just a couple of decades, our ability to map and sequence genomes has improved dramatically. The complete genome sequences have been obtained for hundreds of thousands of different species. Considering that the first genome sequence was generated in 1995, the progress of genome-sequencing projects since then has been truly remarkable! In this section, we will examine the approaches that researchers follow when tackling such large projects. We will also survey some of the general goals of the Human Genome Project, the largest of its kind, and compare the results from the genome sequencing of various species.

EXPERIMENT 22A

Venter, Smith, and Colleagues Sequenced the First Genome in 1995

The first genome to be entirely sequenced was that of the bacterium *Haemophilus influenzae*. This bacterium causes a variety of diseases in humans, including respiratory illnesses and bacterial meningitis. *H. influenzae* has a relatively small genome of approximately 1.8 Mb (i.e., 1.8 million bp) in a single circular chromosome.

To obtain a complete sequence of a genome with the shotgun approach, how do researchers decide how many fragments to sequence? We can calculate the probability that a base will not be sequenced using this approach with the following equation:

$$P = e^{-m}$$

where

P is the probability that a base will be left unsequenced;

e is the base of the natural logarithm, $e \approx 2.72$;

m is the number of bases sequenced divided by the total genome size.

For example, in the case of *H. influenzae* with a genome size of 1.8 Mb, if researchers sequence 9 Mb, $m = 5$ (i.e., 9.0 Mb divided by 1.8 Mb), the calculated probability is

$$P = e^{-m} = e^{-5} = 0.0067, \text{ or } 0.67\%$$

This result means that if researchers randomly sequence 9.0 Mb, which is five times the length of this genome, they are likely to miss only 0.67% of the genome. Given a genome size of 1.8 Mb, they would miss about 12,000 bases out of approximately 1,800,000. Such missed sequences are typically located on small DNA fragments that—as a matter of random chance—did not happen to be sequenced. The missing links in the genome can be sequenced later via PCR using primers that flank the gaps, or as discussed in Section 22.5, using long-read sequencing methods.

The general protocol followed by J. Craig Venter, Hamilton Smith, and colleagues for sequencing the *H. influenzae* genome is presented in **Figure 22.10**. This is a shotgun DNA-sequencing approach using dideoxy sequencing, which is short-read sequencing. The researchers isolated chromosomal DNA from *H. influenzae* and used sound waves to break the DNA into small fragments of approximately 2000 bp in length. These fragments were randomly cloned into vectors, allowing the DNA to be propagated in *E. coli*. Each *E. coli* clone carried a vector with a different piece of DNA from *H. influenzae*. The researchers then subjected many of these pieces of DNA to dideoxy sequencing. They sequenced a total of approximately 10.8 Mb of DNA.

THE GOAL (DISCOVERY-BASED SCIENCE)

The goal was to obtain the entire genome sequence of *H. influenzae*. This information reveals the genome's size as well as the genes that compose it.

ACHIEVING THE GOAL   **FIGURE 22.10**   The use of shotgun DNA sequencing to determine the first genome sequence of a bacterial species.

Starting materials: A strain of *H. influenzae*.

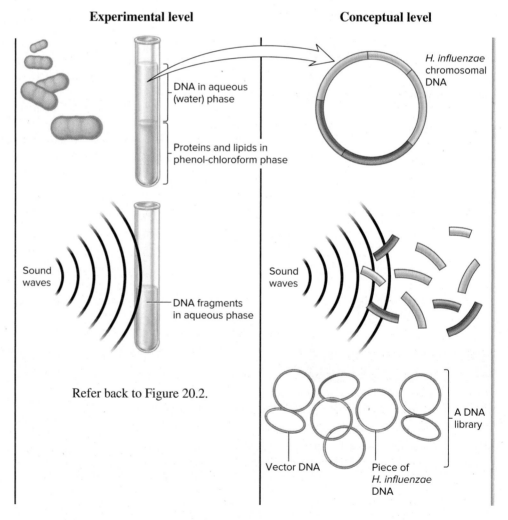

Experimental level   Conceptual level

1. Purify DNA from a strain of *H. influenzae*. This involves breaking the cells open by adding phenol and chloroform. Most protein and lipid components go into the phenol-chloroform phase. DNA remains in the aqueous (water) phase, which is removed and used in step 2.

DNA in aqueous (water) phase

Proteins and lipids in phenol-chloroform phase

H. influenzae chromosomal DNA

2. Sonicate the DNA to break it into small fragments about 2000 bp long.

Sound waves

DNA fragments in aqueous phase

Sound waves

3. Clone the DNA fragments into vectors. The procedures for cloning are described in Chapter 20. This produces a DNA library.

Refer back to Figure 20.2.

Vector DNA

Piece of *H. influenzae* DNA

A DNA library

4. Subject many clones to the procedure of dideoxy DNA sequencing, also described in Chapter 20. A total of 10.8 Mb was sequenced.

5. Use tools of bioinformatics, described in Chapter 23, to identify various types of genes in the genome.

Refer back to Figure 20.11.

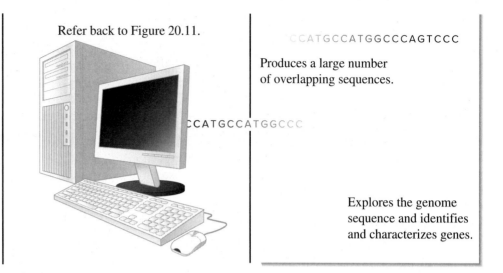

CCATGCCATGGCCCAGTCCC

Produces a large number of overlapping sequences.

CCATGCCATGGCCC

Explores the genome sequence and identifies and characterizes genes.

THE DATA

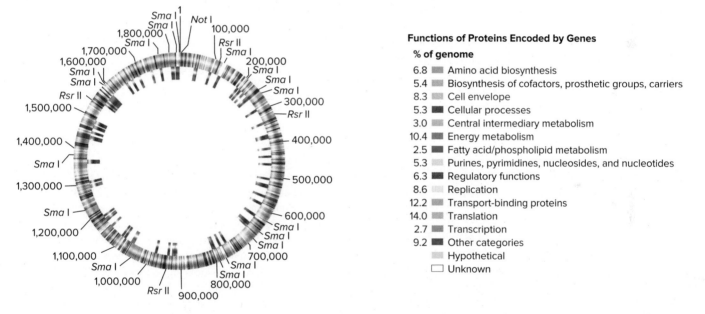

Functions of Proteins Encoded by Genes

% of genome

6.8	Amino acid biosynthesis
5.4	Biosynthesis of cofactors, prosthetic groups, carriers
8.3	Cell envelope
5.3	Cellular processes
3.0	Central intermediary metabolism
10.4	Energy metabolism
2.5	Fatty acid/phospholipid metabolism
5.3	Purines, pyrimidines, nucleosides, and nucleotides
6.3	Regulatory functions
8.6	Replication
12.2	Transport-binding proteins
14.0	Translation
2.7	Transcription
9.2	Other categories
	Hypothetical
	Unknown

Source: Fleischmann, R. D., and Adams, M. D., "Whole-Genome Random Sequencing and Assembly of *Haemophilus influenzae* Rd," *Science*, vol. 269, no. 5223, 1995, 496–512.

INTERPRETING THE DATA

The outcome of this genome-sequencing project was a very long DNA sequence. In 1995, Venter, Smith, and colleagues published the entire DNA sequence of *H. influenzae*. The researchers then analyzed the genome sequence using computer software to obtain information about the properties of the genome. They asked, "How many genes does the genome contain, and what are the likely functions of those genes?" In Chapter 23, we will learn how scientists can answer such questions with the use of computers.

The data shown in the chromosome diagram summarize the results that the researchers obtained. The *H. influenzae* genome is composed of 1,830,137 bp of DNA. The computer analysis predicted 1743 genes. Based on their similarities to known genes in other species, the researchers also predicted the functions of nearly two-thirds of these genes. The diagram displaying the data by Venter, Smith, and colleagues places genes into various categories based on the predicted functions of their coded proteins. These results gave the first comprehensive genome picture of a living organism!

The Human Genome Project Was the Largest Genome-Sequencing Project in History

Due to the large size of the human genome, sequencing it for the first time was an enormous undertaking. Scientists had been discussing how to tackle this project since the mid-1980s. In 1988, the National Institutes of Health established an Office of Human Genome Research, with James Watson as its first director. The **Human Genome Project,** which officially began on October 1, 1990, was a 13-year effort coordinated and funded by the U.S. Department of Energy and the National Institutes of Health. The human DNA that was used in the Human Genome Project was obtained from several volunteers; their identity was not revealed to protect their privacy. From its outset, the Human Genome Project had the following goals:

1. ***To obtain a linkage map of the human genome.*** This involved the identification of millions of genetic markers and their locations on the autosomes and sex chromosomes.
2. ***To obtain a physical map of the human genome.*** This required the cloning of many segments of chromosomal DNA into vectors.
3. ***To obtain the DNA sequence of the entire human genome.*** The first (nearly complete) sequence was published in February 2001. This was considered a first draft. A second draft was published in 2003, and the completed maps and sequences for all of the human chromosomes were published by 2006. The entire genome is approximately 3 billion bp in length. If the entire base sequence of the human genome were typed in a textbook like this, the book would be nearly 1 million pages long!
4. ***To develop technology for the management of human genome information.*** The amount of information obtained from this project is staggering, to say the least. The Human Genome Project developed user-friendly tools to provide scientists easy access to up-to-date information obtained from the project. The Human Genome Project also developed analytical tools for interpreting genome information.
5. ***To analyze the genomes of model organisms.*** These include bacterial species (e.g., *Escherichia coli* and *Bacillus subtilis*), *Drosophila melanogaster* (fruit fly), *Caenorhabditis elegans* (a nematode), *Arabidopsis thaliana* (a flowering plant), and *Mus musculus* (mouse).
6. ***To develop programs focused on understanding and addressing the ethical, legal, and social implications of the results obtained from the Human Genome Project.*** The Human Genome Project sought to identify the major genetic issues that will affect members of our society and to develop policies to address these issues. For example, what is an individual's right to privacy regarding genetic information? Some people are worried that their medical insurance company may discriminate against them if it is found that they carry a disease-causing or otherwise deleterious gene.
7. ***To develop technological advances in genetic methodologies.*** Some of the efforts of the Human Genome Project involved improvements in molecular genetic technology such as gene cloning and DNA sequencing. The project also developed computer technology for data processing, storage, and the analysis of sequence information (see Chapter 23).

A great benefit from the characterization of the human genome is the ability to identify and study the sequences of our genes. Genetic variation in many different genes is known to be correlated with human diseases, which include cancer, heart disease, and many other abnormalities. The identification of mutant genes that cause inherited diseases was a strong motivation for the Human Genome Project. A detailed genetic and physical map has made it considerably easier for researchers to locate such genes. Furthermore, a complete DNA sequence of the human genome provides researchers with insight into the types of proteins coded by these genes. The identification of disease-causing alleles is expected to play an increasingly important role in the diagnosis and treatment of disease.

In 2008, a more massive undertaking, called the 1000 Genomes Project, was launched to establish a detailed understanding of human genetic variation. In this international project, researchers set out to determine the DNA sequence of at least 1000 anonymous participants from around the globe. In 2012, the sequencing of 1092 genomes was announced in a publication in the journal *Nature.* Since then, many more human genomes have been sequenced. Scientists are on a path to sequence 1 million or more human genomes and use this enormous data set to unlock genetic secrets, such as finding correlations between genetic variation and many types of genetic diseases.

Many Genome Sequences Have Been Determined

The amazing advances in DNA-sequencing technology have enabled researchers to determine the complete genome sequences of hundreds of thousands of species. Though estimates vary, complete or draft genome sequences for more than 200,000 bacteria and archaea species have been uploaded to public databases as a result of the development of next-generation sequencing technology. The current number of eukaryotic genomes is in the range of tens of thousands. In 2017, the Vertebrate Genomes Project (VGP) was launched. This international project involves over 50 research institutions working to generate complete genome sequences of all living vertebrate species, which total approximately 66,000. **Table 22.3** describes the results of several genome-sequencing projects that have been completed.

How do researchers decide which genomes to sequence? Motivation behind genome-sequencing projects comes from a variety of sources:

- Basic research scientists can greatly benefit from a genome sequence. It allows them to know which genes a given species has, and it aids in the cloning and characterization of such genes. This has been the impetus for genome projects involving model organisms such as *Escherichia coli, Saccharomyces cerevisiae, Arabidopsis thaliana, Caenorhabditis elegans, Drosophila melanogaster,* and the mouse.
- A second impetus for genome sequencing is the treatment or prevention human disease. As noted previously, researchers expect that the sequencing of the human genome will aid in the identification of genes that, when mutant, play a role in diseases. Likewise, the decision to sequence many bacterial, protist, and fungal genomes has been related to the roles of these species in infectious diseases.

TABLE 22.3

Examples of Genomes That Have Been Sequenced

Species	Genome Size (bp)*	Approximate Number of Genes†	Description
Prokaryotic genomes			
Bacteria			
Mycoplasma genitalium	580,000	521	Bacterial inhabitant of the human genital tract that has a very small genome
Helicobacter pylori	1,668,000	1590	Bacterial inhabitant of the stomach that may cause gastritis, peptic ulcer, and gastric cancer
Mycobacterium tuberculosis	4,412,000	4294	Bacterial species that causes tuberculosis
Escherichia coli	4,639,000	4377	Widespread bacterial inhabitant of the guts of animals; also a model organism for research
Archaea			
Thermoplasma volcanium	1,580,000	1494	Archaeon with an optimal growth temperature of 60°C and optimal growth pH of <2.0
Pyrococcus abyssi	1,760,000	1765	Archaeon that was originally isolated from samples taken close to a hot spring situated 3500 m deep in the southeast Pacific; has optimal growth conditions of 103°C and 200 atm
Sulfolobus solfataricus	2,990,000	2977	Archaeon found in terrestrial volcanic hot springs, with optimal growth occurring at a temperature of 75–80°C and pH of 2–3
Eukaryotic genomes			
Protists			
Plasmodium falciparum	22,900,000	5268	Parasitic protist that causes malaria in humans
Entamoeba histolytica	23,800,000	9938	Amoeba that causes dysentery in humans
Fungi			
Saccharomyces cerevisiae	12,100,000	6464	Baker's yeast, a structurally simple eukaryotic species that has been extensively studied by researchers to understand eukaryotic genetics and cell biology
Neurospora crassa	40,000,000	10,082	Common bread mold, also a structurally simple eukaryotic species that has been extensively studied by researchers
Plants			
Arabidopsis thaliana	142,000,000	26,000	A small flowering plant of the mustard family used as a model organism by plant biologists
Oryza sativa	440,000,000	40,000	Rice, a cereal grain that has a relatively small genome and is very important worldwide as a food crop
Populus trichocarpa	550,000,000	45,555	Balsam poplar, a tree with a relatively small genome
Animals			
Caenorhabditis elegans	97,000,000	19,000	Nematode worm that has been used as a model organism to study animal development
Drosophila melanogaster	175,000,000	14,000	Fruit fly, a model organism used to study many genetic phenomena, including development
Anopheles gambiae	278,000,000	13,683	Mosquito that carries the malaria parasite, *Plasmodium falciparum*
Canis lupus familiaris	2,400,000,000	20,000	Dog, a common house pet
Mus musculus	2,500,000,000	20,000	House mouse, a rodent that is a model organism studied by researchers
Pan troglodytes	3,100,000,000	20,000	Chimpanzee, a primate and the closest living relative to humans
Homo sapiens	3,200,000,000	20,000	Human

*In some cases, the values indicate the estimated amount of DNA. For eukaryotic genomes, DNA sequencing is considered completed in the euchromatic regions. The DNA in certain heterochromatic regions containing highly repetitive sequences cannot be accurately sequenced, and the total amount is difficult to estimate.
†The numbers of genes were predicted using computer methods described in Chapter 23. These numbers should be considered as estimates of the total gene number. For plants and animals, the numbers listed are for protein-coding genes.

Thousands of microbial genomes have been sequenced. Many of them are from species that are pathogenic in humans. The sequencing of such genomes may help us understand which genes are involved in the infection process.

- In addition, the genomes of agriculturally important species have been the subject of genome sequencing. An understanding of these species' genomes has aided in the development of new strains of livestock or crops that have improved traits from an agricultural perspective.
- Finally, genome-sequencing projects help us to better understand the evolutionary relationships among living species. This approach, called **comparative genomics,** uses information from genome projects to understand the genetic variation among different species. We will explore this topic in Chapter 29.

Newly completed genome sequences are emerging rapidly, particularly those of microbial species (Table 22.3). As we obtain more genome sequences, it becomes progressively more interesting to compare them to each other as a way to understand the process of evolution. As we will explore in Chapter 23, the field of functional genomics enables researchers to study the roles of many genes as they interact to generate the phenotypic traits of a species.

22.6 COMPREHENSION QUESTIONS

1. Shotgun sequencing was used to sequence the *H. influenzae* genome. It is a method of DNA sequencing in which
   a. the DNA fragments to be sequenced are randomly generated from larger DNA fragments.
   b. the sequencing reactions are carried out in rapid succession.
   c. the samples to be sequenced are rapidly generated by PCR.
   d. all of the above occur.

2. Which of the following was *not* a goal of the Human Genome Project?
   a. To obtain the DNA sequence of the entire human genome
   b. To successfully clone a mammal
   c. To develop technology for the management of human genome information
   d. To analyze the genomes of model organisms

3. A prokaryotic genome is about 4 million bp in length. About how many genes would you expect it to contain?
   a. 400   c. 40,000
   b. 4000   d. 400,000

22.7 METAGENOMICS

Learning Outcomes:
1. Define *metagenomics.*
2. Describe the general strategy of metagenomics, and outline its uses.

Most microorganisms that exist in soil, water, and the human intestinal tract have not been successfully grown in the laboratory

because researchers do not fully understand their growth requirements and because some of those microbes require the presence of a complex microbial community to survive. In the past, such unculturable microbes have been very difficult to study.

In the 1980s and early 1990s, Norman Pace and colleagues showed that genes that code 16S rRNA, which is described in Chapter 13, could be analyzed from samples of different microbes using PCR. Many of the 16S rRNA genes that were sequenced from environmental samples were found to be different from 16S rRNA genes of bacteria that had been grown in the laboratory. Therefore, this work revealed that environmental samples contain an abundance of unculturable microbes. Though such studies were exciting, a limitation of PCR is that it amplifies specific genes, leaving the rest of the genomes of unculturable microbes unknown.

This limitation was overcome by the development of **metagenomics,** which is the study of a complex mixture of genetic material obtained from an environmental sample. The term **metagenome** refers to a collection of genes from a particular environmental sample. Such a sample can be analyzed in a way analogous to that used for a single genome.

Figure 22.11 outlines a general strategy for a metagenomic study.

1. First, researchers obtain a sample from the environment. This may be a soil sample, a water sample from the ocean, or a fecal sample from a person.
2. The sample is filtered, the cells within the sample are lysed, and the DNA is extracted and purified. During this procedure, the purified DNA is sheared, which breaks the DNA into small fragments.
3. Usually, the main goal is to obtain DNA sequence data. This is accomplished using next-generation sequencing technology (see the left side below the bifurcated arrow).
4. In some cases, researchers may wish to obtain clones of genes from the metagenome (see the right side below the bifurcated arrow). For example, researchers may be studying the metagenome collected from contaminated soil samples in hopes of identifying and cloning microbial genes that play a role in bioremediation. (The topic of bioremediation is discussed in Chapter 21.) To accomplish this goal, DNA fragments from the metagenome are randomly inserted into cloning vectors and transformed into host cells. The result is a DNA library in which each host cell carries a DNA fragment from the metagenome. This is called a metagenomic library. The cells of the library are then subjected to DNA sequencing to identify the genes they may contain.

What are the uses of metagenomics? The following are some of the common applications of metagenomics:

1. ***Human medicine.*** Various places in the body, such as the mouth and intestines, support a complex array of microorganisms. Metagenomics is being used to characterize these populations of organisms and to study changes in the relative compositions of microorganisms that may be associated with different diseases.

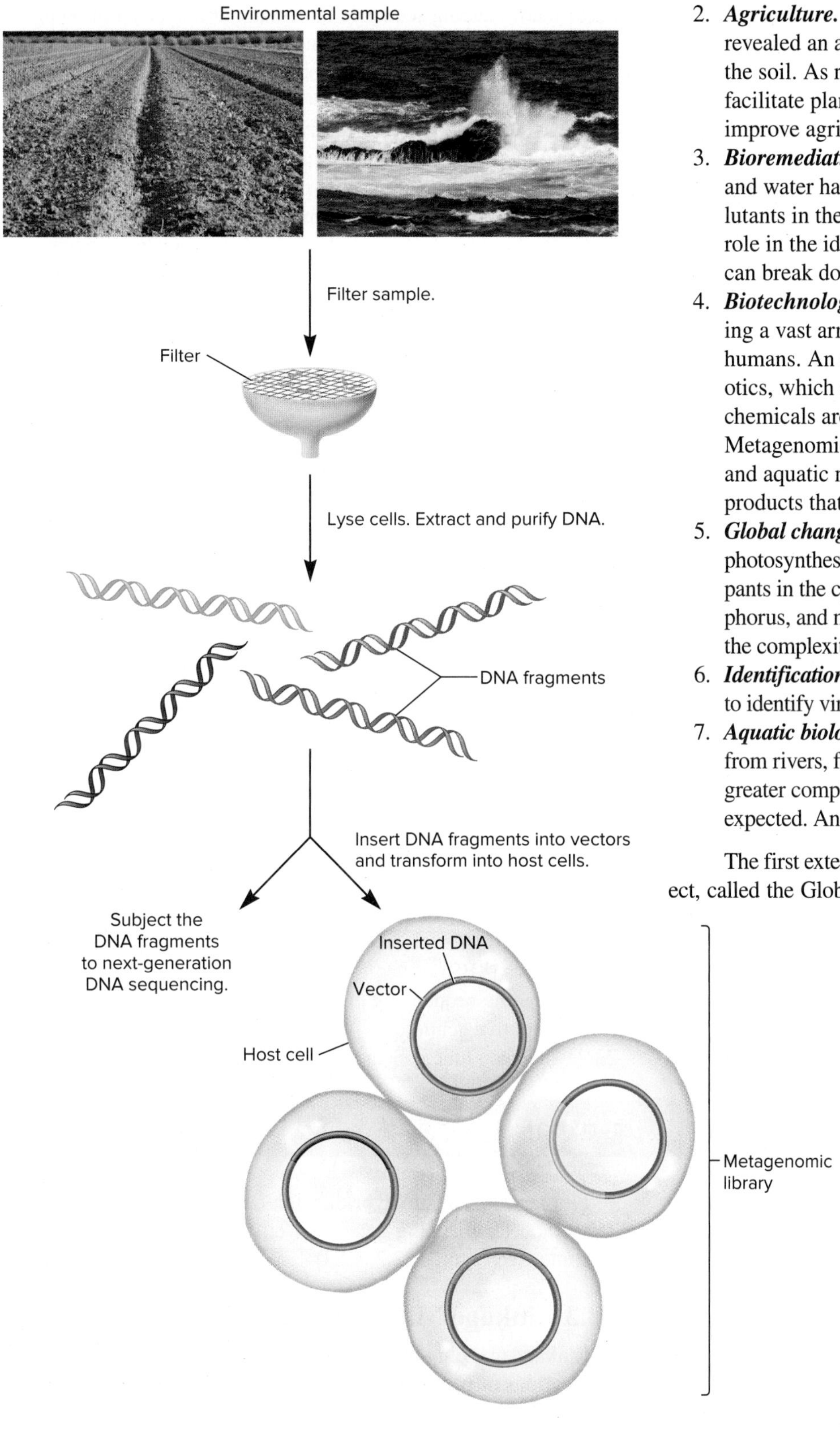

Environmental sample

Filter sample.

Filter

Lyse cells. Extract and purify DNA.

DNA fragments

Insert DNA fragments into vectors and transform into host cells.

Subject the DNA fragments to next-generation DNA sequencing.

Inserted DNA

Vector

Host cell

Metagenomic library

FIGURE 22.11 The general strategy of metagenomics.

(left): Neil Guegan/Image Source; (right): Joseph Gareri/Getty Images

CONCEPT CHECK: What type of data are produced from a metagenomics study?

2. **Agriculture.** The metagenomic analysis of soil samples has revealed an astonishing complexity of microorganisms in the soil. As researchers learn more about which microbes facilitate plant growth, such knowledge may be used to improve agricultural yields.

3. **Bioremediation.** The types of microorganisms found in soil and water have a great effect on the decomposition of pollutants in the environment. Metagenomics may play a key role in the identification and use of microorganisms that can break down specific types of pollutants.

4. **Biotechnology.** Microorganisms are capable of synthesizing a vast array of chemicals, some of which are useful to humans. An example is the category of drugs called antibiotics, which are used to treat bacterial infections. Such chemicals are made by enzymes coded by microbial genes. Metagenomics is being used to identify such genes in soil and aquatic microorganisms as a way to discover possible products that the microorganisms can make.

5. **Global change.** Microorganisms carry out about half of the photosynthesis that takes place on Earth and are key participants in the cycling of various elements such as carbon, phosphorus, and nitrogen. Metagenomics is helping us understand the complexity of microbial communities in these processes.

6. **Identification of viruses.** Environmental samples are analyzed to identify viruses that infect humans and other organisms.

7. **Aquatic biology.** The metagenomic analysis of water samples from rivers, freshwater lakes, and oceans has revealed a much greater complexity of microbial communities than had been expected. An example of such a study is described next.

The first extensive large-scale environmental sequencing project, called the Global Ocean Sampling Expedition, was carried out by J. Craig Venter and colleagues. It began in 2003 as a pilot project to sample the microbial population of the nutrient-limited Sargasso Sea, a region of the Atlantic Ocean close to Bermuda. This site was chosen because it was thought to contain a relatively simple population of microbes compared with other aquatic locations. The samples from the Sargasso Sea revealed several hundred previously undiscovered genes for the light-harvesting protein proteorhodopsin. This discovery may help us understand the role of this protein in energy metabolism under low nutrient conditions. In addition, over 1800 new species of microorganisms were identified.

Venter announced the full expedition in 2004. In the spirit of Darwin's travels on the *Beagle*, Venter's 95-foot sailboat (the *Sorcerer II*) was outfitted as a research vessel. Using a sample size of 200 liters, the team traveled over 32,000 miles, sampling every 200 miles; they collected about 40 different samples (**Figure 22.12**). After being frozen and shipped to a land-based research center, the samples were sequenced using shotgun techniques. A total of 7.7 million sequencing runs were

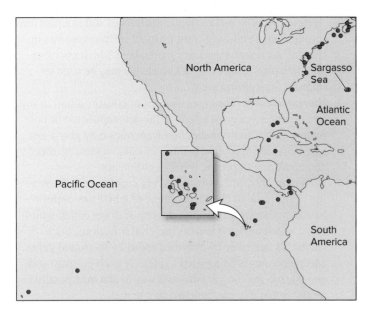

FIGURE 22.12 **Sampling sites of the Global Ocean Sampling Expedition.** The sampling sites are shown as red dots. The first two sites were in the Sargasso Sea. The next year the expedition started in Nova Scotia and proceeded southwest, ending in the Pacific Ocean.

Source: Rusch, D. B., Halpern, A. L., Sutton, G., Heidelberg, K. B., and Williamson, S., "The *Sorcerer II* Global Ocean Sampling Expedition: Northwest Atlantic Through Eastern Tropical Pacific." *PLOS Biology*, vol. 5, no. 3, March 13, 2007, 77.

performed, yielding sequence information on over 6 billion bp of DNA. Most of the DNA sequences were unique, reflecting the incredible diversity in naturally occurring microbial populations.

Though the expedition yielded many exciting results, one of the key findings that emerged involved the identification of species variants, called subtypes, that exist alongside each other. As communities adapt and change, previous research had suggested that certain subtypes tend to dominate and outcompete others. In contrast, this expedition found many closely related species and subtypes that were living in the same environment. Further research will be needed to explain this paradox.

22.7 COMPREHENSION QUESTION

1. Metagenomics is aimed at
   a. determining the complete genome sequence of newly identified microorganisms.
   b. mapping the genes on chromosomes of newly identified microorganisms.
   c. determining the sequences of DNA fragments from environmental samples.
   d. determining the functions of all of the genes in a given species' genome.

KEY TERMS

Introduction: genome, genomics, functional genomics, proteomics
22.1: mapping, cytogenetic mapping, linkage mapping, physical mapping, genetic map (chromosome map), locus (pl. loci)
22.2: in situ hybridization, hybridization, fluorescence in situ hybridization (FISH), chromosome painting
22.3: molecular marker, polymorphic, microsatellite, sequence-tagged site (STS), founder
22.4: chromosome walking, subcloning, primer walking

22.5: shotgun sequencing, de novo genome assembly, short-read sequencing (SRS), long-read sequencing (LRS), high-throughput sequencing, next-generation sequencing technologies, third-generation sequencing technologies, sequence by synthesis (SBS)
22.6: genome-sequencing projects, Human Genome Project, comparative genomics
22.7: metagenomics, metagenome

CHAPTER SUMMARY

- A genome is the total genetic composition of a cell, an organism, or a species. Genomics refers to the molecular analysis of the entire genome of a species. Functional genomics is aimed at studying the expression of genes that make up a genome, whereas proteomics is focused on the functions of proteins.

22.1 Overview of Chromosome Mapping

- Mapping is the experimental process of determining the relative locations of genes or other DNA segments on a chromosome, thereby producing a diagram called a genetic map. The process may involve the use of cytogenetic, linkage, or physical mapping techniques (see Figure 22.1).

22.2 Cytogenetic Mapping via Microscopy

- Cytogenetic mapping attempts to determine the locations of particular genes relative to a banding pattern of a chromosome.

- Fluorescence in situ hybridization (FISH) is a commonly used method for mapping genes and other segments on a chromosome (see Figures 22.2, 22.3).

22.3 Linkage Mapping via Crosses

- Linkage mapping often relies on molecular markers to map the locations of genes on chromosomes (see Table 22.1).
- Restriction fragment length polymorphisms (RFLPs), microsatellites, and single-nucleotide polymorphisms (SNPs) are types of molecular markers (see Figures 22.4, 22.5, 22.6).

22.4 Chromosome Walking and Primer Walking

- Chromosome walking and primer walking are techniques for identifying a disease-causing allele that is close to a known marker (see Figure 22.7).

22.5 Overview of Genome Sequencing

- The first step in the production of a physical map of the chromosome(s) within a genome involves the determination the base sequences of many DNA segments that are derived from chromosomal (genomic) DNA. This can be achieved via short-read sequencing (SRS) or long-read sequencing (LRS), each of which has its advantages and disadvantages.
- The second step in the production of a physical map is de novo genome assembly, in which overlapping regions allow researchers to align the sequences of DNA segments to form a complete DNA sequence as it would occur along a chromosome (see Figure 22.8).
- Newer methods of DNA sequencing can process many samples of DNA simultaneously and are superseding the dideoxy method. An example of these next-generation sequencing technologies is Illumina sequencing (see Table 22.2, Figure 22.9).

22.6 Genome-Sequencing Projects

- Shotgun sequencing has been commonly used to sequence the DNA of many species. Overlapping regions allow researchers to determine a complete DNA sequence (see Figure 20.10).
- The Human Genome Project resulted in the sequencing of the entire human genome.
- Since 1995, the genomes of hundreds of thousands of species have been sequenced (see Table 22.3).

22.7 Metagenomics

- Metagenomics is the study of a complex mixture of genetic material obtained from an environmental sample (see Figures 22.11, 22.12).

PROBLEM SETS & INSIGHTS

MORE GENETIC TIPS **1.** An RFLP marker is located 1 million bp away from a gene of interest. Your goal is to start at this RFLP and walk to the gene. The average insert size in the genomic library is 55,000 bp, and the average overlap at each end of an insert is 5000 bp. Approximately how many steps will it take to get there?

T **OPIC:** *What topic in genetics does this question address?* The topic is physical mapping. More specifically, the question is about the number of steps required in a chromosome walk.

I **NFORMATION:** *What information do you know based on the question and your understanding of the topic?* From the question, you know that a gene is 1 million bp from an RFLP, and that each step gets 55,000 bp closer to the gene. From your understanding of the topic, you may remember that chromosome walking involves a series of subcloning steps.

P **ROBLEM-SOLVING** **S** **TRATEGY:** *Make a calculation.* One strategy to solve this problem is to consider the length of the individual steps in the chromosome walk and calculate the number of steps needed to complete the walk.

ANSWER: Each step is 50,000 bp (i.e., 55,000 minus 5000), because you have to subtract the overlap between adjacent inserts, which is 5000 bp, from the average insert size. Therefore, dividing 1 million bp by 5000 bp gives about 20 steps as the number needed.

---

2. Does a molecular marker have to be polymorphic to be useful in linkage mapping (i.e., involving family pedigree studies or genetic crosses)? Explain why or why not.

T **OPIC:** *What topic in genetics does this question address?* The topic is molecular markers. More specifically, the question is about whether molecular markers need to be polymorphic to be used in linkage mapping.

I **NFORMATION:** *What information do you know based on the question and your understanding of the topic?* The question asks you to consider polymorphism in molecular markers. From your understanding of the topic, you may remember that linkage mapping involves crosses.

P **ROBLEM-SOLVING** **S** **TRATEGY:** *Describe the steps.* One strategy to solve this problem is to sort out the steps in linkage mapping. A key step is that you must conduct crosses in which one or both parents are heterozygous for a marker.

ANSWER: In linkage mapping studies, a marker must be polymorphic to be useful. Polymorphic molecular markers can be RFLPs, microsatellites, or SNPs. To compute map distances through linkage mapping, researchers must study individuals that are heterozygous for two or more markers (or genes). An individual could not be heterozygous for a marker if it was monomorphic. With experimental organisms, heterozygotes are testcrossed to homozygotes, and then the numbers of recombinant offspring and nonrecombinant offspring are determined. For markers that do not assort independently (i.e., linked markers), the map distance is computed as the number of recombinant offspring divided by the total number of offspring, times 100.

---

3. The distance between two molecular markers that are linked on the same chromosome can be determined by analyzing the outcomes of crosses. This can be done in humans by analyzing a family's pedigree. However, the accuracy of linkage mapping using human pedigrees is fairly limited because the number of people in most families is relatively small. As an alternative, researchers can analyze a population of sperm, produced from a single male, and compute a map distance in this manner.

As an example, let's suppose a male is heterozygous for two polymorphic STSs. STS-1 exists in two sizes: 234 bp and 198 bp. STS-2 also exists in two sizes: 423 bp and 322 bp. A sample of sperm was collected from this male, and individual sperm were placed into 40 separate tubes. In other words, there was one sperm in each tube. Believe it or not, PCR is sensitive enough to allow analysis of DNA in a single sperm! Into each of the 40 tubes were added the primers that amplify STS-1 and STS-2, and then the samples were subjected to PCR and gel electrophoresis. The following results were obtained.

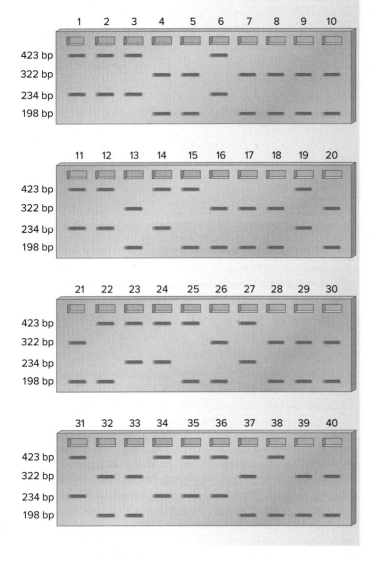

A. What is the arrangement of these two STSs in this male?

B. What is the map distance between STS-1 and STS-2?

TOPIC: *What topic in genetics does this question address?* The topic is linkage mapping. More specifically, the question is about using DNA in sperm to compute the map distance between two molecular markers.

INFORMATION: *What information do you know based on the question and your understanding of the topic?* From the question, you know the patterns of STS-1 and STS-2 in a population of 40 sperm. From your understanding of the topic, you may remember that linkage mapping involves the identification of recombinants, which are produced by crossovers. The map distance is the percentage of recombinants.

PROBLEM-SOLVING **S**TRATEGY: *Compare and contrast. Make a calculation.* One strategy to solve this problem is to compare and contrast the lanes on the gel. When markers are linked, the nonrecombinant pattern of bands will be more common than the recombinant pattern. Map distance is computed as the number of recombinants divided by the total number of sperm analyzed, times 100.

ANSWER: Keep in mind that mature sperm are haploid, so they have only one copy of STS-1 and one copy of STS-2.

A. If you look at the 40 lanes, most of them (i.e., 36) have either the 234-bp and 423-bp STSs or the 198-bp and 322-bp ones. This is the arrangement of STSs in this male. One chromosome has STS-1 that is 234 bp and STS-2 that is 423 bp, and the homologous chromosome has STS-1 that is 198 bp and STS-2 that is 322 bp.

B. There are four recombinant sperm, shown in lanes 15, 22, 25, and 38.

$$\text{Map distance} = \frac{4}{40} \times 100$$

$$= 10.0 \text{ mu}$$

(Note: This is a relatively easy experiment compared with a pedigree analysis, which would involve contacting lots of relatives and collecting samples from each of them.)

Conceptual Questions

C1. A sample of chromosomes from a person with a rare genetic disease is subjected to fluorescence in situ hybridization using a probe that is known to recognize band p11 on chromosome 7. Even though the chromosomes look cytologically normal, the probe does not bind to them. How would you explain these results? How would you use this information to positionally clone the gene that is related to this disease?

C2. For each of the following, decide if it could be appropriately described as a genome:

A. The *E. coli* chromosome

B. Human chromosome 11

C. A complete set of 10 chromosomes in corn

D. A copy of the single-stranded RNA packaged into human immunodeficiency virus (HIV)

C3. Which of the following statements about molecular markers are true?

A. All molecular markers are segments of DNA that carry specific genes.

B. A molecular marker is a segment of DNA that is found at a specific location in a genome.

C. We can follow the transmission of a molecular marker by analyzing the phenotype (i.e., the physical characteristics) of offspring.

D. We can follow the transmission of molecular markers using molecular techniques such as gel electrophoresis.

E. An STS is a molecular marker.

Experimental Questions

E1. Is each of the following a method used in linkage, cytogenetic, or physical mapping?

   A. Fluorescence in situ hybridization (FISH)

   B. Conducting two-factor crosses to compute map distances

   C. Chromosome walking

   D. Examination of polytene chromosomes in *Drosophila*

   E. Use of RFLPs in crosses

E2. In a fluorescence in situ hybridization (FISH) experiment, what is the relationship between the base sequence of the probe DNA and the site on the chromosomal DNA where the probe binds?

E3. Describe the technique of fluorescence in situ hybridization. Explain how it can be used to map genes.

E4. The cells from a person's malignant tumor were subjected to fluorescence in situ hybridization using a probe that recognizes a unique sequence on chromosome 14. The probe was detected only once in each of the cells. Explain this result, and speculate on its significance with regard to the malignant characteristics of these cells.

E5. Figure 22.2 illustrates the technique of FISH. Why is it necessary to fix the cell contents, including the chromosomes, to the slides? What does it mean to fix them? Why is it necessary to denature the chromosomal DNA?

E6. Explain how DNA probes with different fluorescence emission wavelengths can be used in a single FISH experiment to map the locations of two or more genes. This method is called chromosome painting. Explain why this is an appropriate term.

E7. A researcher is interested in a gene found on human chromosome 21. Describe the expected results of a FISH experiment using a probe that is complementary to this gene. How many spots would you see if the probe was used on a sample from an individual with 46 chromosomes or if it was used on a sample from an individual with Down syndrome?

E8. What are the two main steps in the physical mapping of a species' genome?

E9. Compare and contrast the advantages and disadvantages of short-read sequencing and long-read sequencing.

E10. What is the overall aim of both chromosome walking and primer walking? What are the key differences between these two methods?

E11. A female had five children with two different males. This group of seven individuals is analyzed with regard to three different STSs: STS-1 is 146 bp and 122 bp; STS-2 is 102 bp and 88 bp; and STS-3 is 188 bp and 204 bp. The female parent is homozygous for all three STSs: STS-1 = 122, STS-2 = 88, and STS-3 = 188. Male parent 1 is homozygous for STS-1 = 122 and STS-2 = 102, and heterozygous for STS-3 = 188 and 204. Male parent 2 is heterozygous for STS-1 = 122 and 146, STS-2 = 88 and 102, and homozygous for STS-3 = 204.

The five children show the following results:

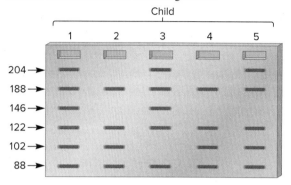

Which children can you definitely assign to male parent 1 or male parent 2?

E12. What is meant by shotgun sequencing?

E13. In the Human Genome Project, researchers collected linkage mapping data from many crosses in which the male was heterozygous for molecular markers and many crosses in which the female was heterozygous for the same markers. The distance between these two markers, computed in map units, differs between males and females. In other words, the linkage maps for human males and females are not the same. Propose an explanation for this discrepancy. Do you think the sizes of chromosomes (excluding the Y chromosome) in human males and females are different? How could physical mapping resolve this discrepancy?

E14. Look back at question 3 in More Genetic TIPS. Let's suppose a male is heterozygous for two polymorphic sequence-tagged sites. STS-1 exists in two sizes: 211 bp and 289 bp. STS-2 also exists in two sizes: 115 bp and 422 bp. A sample of sperm was collected from this male, and individual sperm were placed into 30 separate tubes. Into each of the 30 tubes were added the primers that amplify STS-1 and STS-2, and then the samples were subjected to PCR and gel electrophoresis. The following results were obtained:

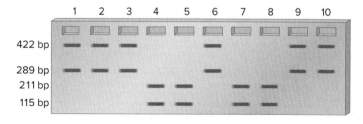

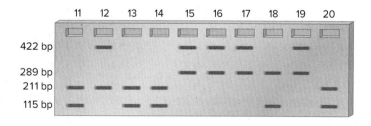

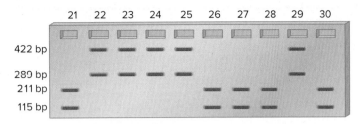

A. What is the arrangement of these STSs in this individual?

B. What is the map distance between STS-1 and STS-2?

C. Could this approach to analyzing a population of sperm be applied using RFLPs?

E15. What is an STS? How are STSs generated experimentally? What are the uses of STSs? Explain how a microsatellite can be a polymorphic STS.

E16. A human gene, which we will call gene *X*, is located on chromosome 11 and is found as a normal allele and a recessive disease-causing allele. The location of gene *X* has been approximated on the map shown next (upper right column), which also shows the locations of four STSs, labeled STS-1, STS-2, STS-3, and STS-4.

STS-1	STS-2	STS-3	Gene *X*	STS-4

A. Explain the general strategy of chromosome walking.

B. If you applied the approach of chromosome walking to clone gene *X*, where would you begin? As you progressed in your cloning efforts, how would you know if you were walking toward or away from gene *X*?

C. How would you know if you had reached gene *X*? (Keep in mind that gene *X* exists as a normal allele and a disease-causing allele.)

E17. Describe how you would clone a gene by chromosome walking.

E18. A bacterium has a genome size of 4.4 Mb. If a researcher carries out shotgun sequencing and sequences a total of 19 Mb of the bacterial DNA, what is the probability that a base will be left unsequenced? What percentage of the total genome will be left unsequenced?

E19. Discuss the advantages of next-generation sequencing technologies.

E20. What is meant by *sequence by synthesis*?

E21. Outline the general strategy used in metagenomics.

Questions for Student Discussion/Collaboration

1. What is a molecular marker? Give two examples. Discuss why it is generally easier to locate and map molecular markers rather than functional genes.

2. Which goals of the Human Genome Project do you think are the most important? Why? Discuss the types of ethical problems that might arise as a result of identifying all of our genes.

Note: All answers are available for the instructor in Connect; the answers to the even-numbered questions and all of the Concept Check and Comprehension Questions are in Appendix B.

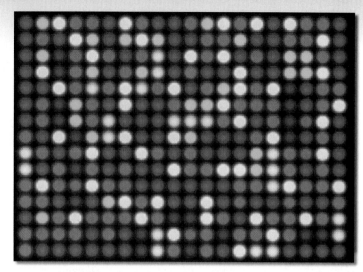

A DNA microarray for measuring gene expression at the genome level. Each spot on the array corresponds to a specific gene. The color of each spot, which is produced via computer imaging techniques, indicates the amount of RNA transcribed from that gene. A DNA microarray allows researchers to simultaneously analyze the expression of many genes.
Alfred Pasieka/Science Source

23

GENOMICS II: FUNCTIONAL GENOMICS, PROTEOMICS, AND BIOINFORMATICS

Chapter 22 discussed how genomics involves the mapping of an entire genome and the determination of a species' complete DNA sequence. The amount of information contained within a species' genome is enormous. The goal of **functional genomics** is to understand the roles of genetic sequences—DNA and RNA sequences—in a given species. In most cases, functional genomics is aimed at understanding gene function. At the genomic level, researchers can study genes as large groups. For example, the information gained from a genome-sequencing project can help researchers study entire metabolic pathways. Such research provides a description of the ways in which gene products interact to carry out cellular processes.

Because most genes code proteins, a goal of many molecular biologists is to understand the functional roles of all of the proteins a species produces. The entire collection of proteins that a given cell or organism makes is called its **proteome,** and the study of the functions and interactions of these proteins is termed **proteomics.** An objective of researchers in the field of proteomics is to understand the interplay among many proteins as they

function to create and maintain cells and, ultimately, to determine the traits of a given species.

From a research perspective, functional genomics and proteomics can be broadly categorized in two ways: experimental and computational. The experimental approach involves the study of groups of genes or proteins using molecular techniques in the laboratory. In the first two sections in this chapter, we will focus on these techniques. In the last section, we will consider bioinformatics. As a very general definition, **bioinformatics** is the use of computers, mathematical tools, and statistical techniques to record, store, and analyze biological information. We often think of bioinformatics in the context of examining genetic research data, such as DNA sequences. In addition, bioinformatics can be applied to information from other sources, such as clinical data. This rapidly developing branch of biology is highly interdisciplinary, incorporating principles from mathematics, statistics, information science, chemistry, and physics. We will see how the field of bioinformatics has provided great insights in our understanding of functional genomics and proteomics.

23.1 FUNCTIONAL GENOMICS

Learning Outcomes:

1. Describe the composition of a DNA microarray, and explain how it is used.
2. Outline the method of RNA sequencing (RNA-Seq).
3. Define *gene knockout,* and explain why gene knockouts are useful.

Though the rapid sequencing of genomes, particularly the human genome, has generated great excitement among geneticists, many would argue that an understanding of genome function is fundamentally more interesting. In the past, our ability to study genes involved many of the techniques described in Chapter 20, such as gene cloning, Northern blotting, and gene editing. These approaches continue to provide a solid foundation for our understanding of gene function. More recently, genome-sequencing projects have enabled researchers to consider gene function at a more complex level. It is now possible to analyze groups of many genes simultaneously to determine how they work as integrated units to produce the characteristics of cells and the traits of multicellular organisms.

In this section, we will examine two methods, DNA microarrays and RNA sequencing, that enable researchers to monitor the expression of thousands of genes simultaneously. We will also consider why researchers are producing gene knockout collections in which each gene of a given species is separately inactivated in order to understand gene function.

A DNA Microarray Can Quantify Gene Transcription at the Whole Genome Level

In the 1990s, researchers developed a technology, called a **DNA microarray** (also called a **gene chip**), that makes it possible to quantify the expression of thousands of genes simultaneously. A DNA microarray is a small silica, glass, or plastic slide that is dotted with many different DNA sequences, each corresponding to a short sequence within a known gene. For example, one spot in a microarray may correspond to a sequence within the β-globin gene, whereas another could correspond to a gene that codes actin, which is a cytoskeletal protein. A single slide may contain tens of thousands of different spots in an area the size of a postage stamp. The relative location of each gene represented in the array is known.

How are microarrays made?

- Some are produced by spotting different samples of DNA onto a slide, much like the way an inkjet printer works. For many species, researchers know the entire genome sequence. With this information, they can make primers that flank any given gene and use PCR to synthesize the DNA from a specific gene. The DNA segments from many different gene sequences are then individually spotted onto the slide. Such DNA segments are typically 500–5000 nucleotides in length, and a few thousand to tens of thousands are spotted to make a single array.
- Alternatively, other microarrays contain shorter DNA segments—oligonucleotides—that are directly synthesized on the surface of the slide. In this case, the DNA sequence at a given spot is produced by selectively controlling the growth of an oligonucleotide using narrow beams of light. Such oligonucleotides are typically 25–30 nucleotides in length. Hundreds of thousands of different spots can be found on a single array.

Overall, the technology of making DNA microarrays is pretty amazing.

A DNA microarray is used as a hybridization tool, as shown in **Figure 23.1**.

1. To begin this experiment, mRNAs were isolated from a sample of cells and then used to make fluorescently labeled cDNA. In this simplified example, the cells made three different mRNAs, from genes A, D, and F.
2. The mRNAs were mixed with fluorescently labeled nucleotides and reverse transcriptase to make fluorescently labeled cDNA.
3. The labeled cDNAs were then applied onto a DNA microarray. The cDNAs will be complementary to some of the DNA spots in the microarray. The cDNAs bind to the DNA in these spots—that is, they hybridize—thereby becoming bound to the microarray.
4. The array is then washed with a buffer to remove any unbound cDNAs and placed in a microscopy device called a laser scanner, which produces higher-resolution images than a conventional optical microscope. The device scans each pixel—the smallest element in a visual image—and after correction for local background, the final fluorescence intensity for each spot is obtained by averaging across the pixels in each spot. This results in a group of fluorescent spots at defined locations in the microarray.

High intensity of fluorescence in a particular spot means that a large amount of the cDNAs in the sample hybridized to the DNA at that location. Because the DNA sequence of each spot is already known, a fluorescent spot identifies cDNAs that are complementary to each DNA sequence. Furthermore, because the cDNAs were generated from mRNAs, this technique identifies RNAs that have been made in a particular cell type under a given set of conditions.

The technology of DNA microarrays has found many important uses (**Table 23.1**). Its most common use is for studying gene expression patterns. Such studies help us understand how genes are regulated in a cell-specific manner and how environmental conditions can induce or repress the transcription of genes. In some cases, microarrays can even help identify which genes code the proteins that participate in a complicated metabolic pathway. Microarrays can also be used as identification tools. For example, gene expression patterns can aid in the categorization of various tumors. Such identification is used to determine the best course of clinical treatment for a patient.

Instead of using labeled cDNAs, researchers can also hybridize labeled genomic DNA to a microarray. This technique can be used to identify mutant alleles in a population and to detect deletions and duplications. In addition, it is proving useful in correctly identifying closely related bacterial species and subspecies.

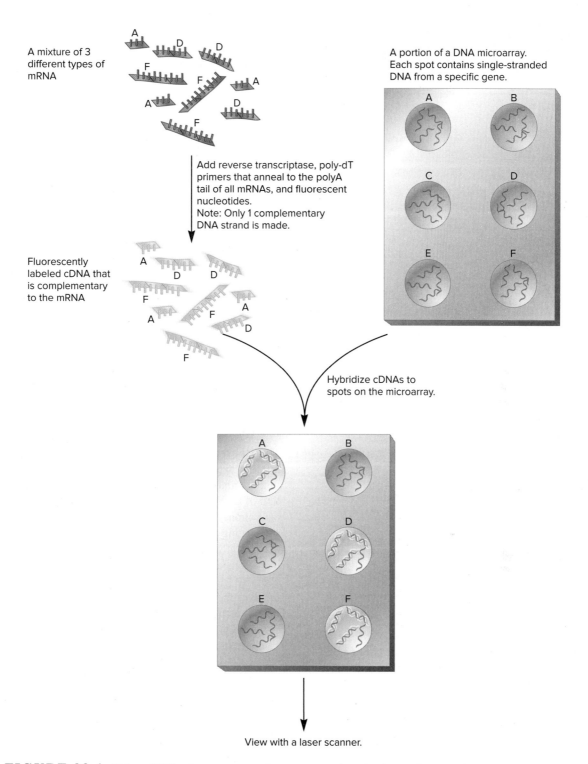

A mixture of 3 different types of mRNA

A portion of a DNA microarray. Each spot contains single-stranded DNA from a specific gene.

Add reverse transcriptase, poly-dT primers that anneal to the polyA tail of all mRNAs, and fluorescent nucleotides.
Note: Only 1 complementary DNA strand is made.

Fluorescently labeled cDNA that is complementary to the mRNA

Hybridize cDNAs to spots on the microarray.

View with a laser scanner.

ONLINE ANIMATION

FIGURE 23.1 Using a DNA microarray to study gene expression. A mixture of mRNAs isolated from a sample of cells is used to create cDNAs that are fluorescently labeled. The cDNAs are applied to the microarray. In this simplified example, three cDNAs specifically hybridize to spots on the microarray. In an actual experiment, the sample produces hundreds or thousands of different cDNAs and the array contains tens of thousands of different spots. After hybridization, some spots become fluorescent and can be visualized using a laser scanner.

CONCEPT CHECK: Explain how this experiment provides information regarding the expression of genes.

TABLE 23.1

Applications of DNA Microarrays

Application	Description
Cell-specific gene expression	A comparison of microarray data using cDNAs derived from RNAs of different cell types can identify genes that are expressed in a cell-specific manner.
Gene regulation	Environmental conditions play an important role in gene regulation. A comparison of microarray data may reveal genes that are induced under one set of conditions and repressed under another.
Elucidation of metabolic pathways	Genes that code proteins that participate in a common metabolic pathway are often expressed in a parallel manner. This application overlaps with the study of gene regulation via microarrays.
Tumor profiling	Different types of cancer cells exhibit striking differences in their profiles of gene expression. Such a profile can be revealed by a DNA microarray analysis. This approach can be used as a method to classify tumors that are sometimes morphologically indistinguishable, which may provide information that can improve a patient's clinical treatment.
Genetic variation	A mutant allele may not hybridize to a spot on a microarray as well as a wild-type allele. Therefore, microarrays have been used as a tool for detecting genetic variation. For example, they are used to identify disease-causing alleles in humans and mutations that contribute to quantitative traits in plants and other species.
Microbial strain identification	Microarrays can distinguish between closely related bacterial species and subspecies.
DNA-protein binding	Chromatin immunoprecipitation, which is described in Chapter 15 (refer back to Figure 15.14), can be used with DNA microarrays to determine where in the genome a particular protein binds to the DNA.

GENETIC TIPS
THE QUESTION: Samples of liver cells were collected from a healthy donor and from an individual with liver cancer. mRNAs were isolated from both samples of cells and subjected to DNA microarray analysis. In the results from the two samples, 77 spots on the microarray for the cells from the cancer patient were much brighter compared to those for the cells from the healthy donor. How would you interpret these results? Explain their meaning with regard to the growth of the cancer cells. (Note: Assume that each spot corresponds to a different gene.)

T OPIC: *What topic in genetics does this question address?*
The topic is DNA microarray analysis. More specifically, the question is about applying the technique to compare healthy cells with cancer cells.

I NFORMATION: *What information do you know based on the question and your understanding of the topic?* From the question, you know the results from a DNA microarray analysis that compares healthy cells and cancer cells. From your understanding of the topic, you may remember that the brightness of a spot on a microarray indicates how much of the cDNA, which was reverse transcribed from the cells' mRNA, hybridized to the known DNA sequence at that location.

P ROBLEM-SOLVING **S** TRATEGY: *Analyze data. Compare and contrast.* One strategy to solve this problem is compare the results from the cancer cells and the healthy cells and relate them to the transcription levels in the cells.

ANSWER: The cancer cells are overexpressing 77 genes that the normal cells are not overexpressing. The overexpression of some of these genes is likely to be contributing to the cancerous growth.

RNA Sequencing (RNA-Seq) Is a Newer Method for Identifying Expressed Genes

The **transcriptome** is the set of all RNA molecules, including mRNAs and non-coding RNAs, that are transcribed in one cell or in a population of cells. Researchers may focus on the identification of each type of RNA molecule and also on the relative concentrations of the different types. The invention of next-generation sequencing technologies, described in Chapter 22, has changed the way in which transcriptomes are studied. A particularly exciting advance is **RNA sequencing (RNA-Seq),** which was developed by Michael Snyder and colleagues in 2008. This method involves the sequencing of complementary DNAs (derived from RNAs) using next-generation DNA-sequencing methods.

RNA-Seq has several important applications. It is used to compare transcriptomes in the following ways:

- In different cell types
- In healthy versus diseased cells
- At different stages of development
- In response to different environmental agents, such as exposure to a hormone or to toxic chemicals

Figure 23.2 outlines a general strategy for RNA-Seq, but some steps may vary depending on the types of RNA molecules that are being studied and the method of next-generation sequencing that is used. The procedure always begins with the isolation of RNA molecules from a sample of one or more types of cells. Researchers or clinicians may want to analyze the entire population of RNAs, or they may want to analyze a subpopulation. For example, if researchers wanted to focus on eukaryotic mRNAs, they could "pull out" mRNAs by using heavy beads that are attached to polyT oligonucleotides. PolyT oligonucleotides bind to the polyA tails of eukaryotic mRNAs. Because the polyT oligonucleotides are

Isolate RNA from a sample of cells. In some cases, a researcher may want to focus on a subpopulation of RNAs, such as mRNAs or short non-coding RNAs. The illustration below shows three different types of RNAs in different colors. In an actual experiment, there would be hundreds or thousands of different RNAs. Note: The green RNA is highly expressed, and the red RNA is expressed at a low level.

attached to heavy beads, the mRNAs can be separated from the rest of the RNAs by centrifugation.

After the desired population of RNAs has been obtained, the next step is to produce cDNAs. First, the RNAs are fragmented into small pieces. One way to make cDNAs is to attach short segments of DNA, called linkers, to the 5′ and 3′ ends of the RNA fragments. After the attachment of linkers, primers are added that are complementary to the linkers, and cDNAs are made via reverse transcriptase PCR, which is described in Chapter 20. The population of cDNAs is then subjected to next-generation DNA sequencing (see Chapter 22). This DNA sequencing produces a diverse collection of cDNA sequences.

The next phase of RNA-Seq is to compare the collection of cDNA sequences with the already known genome sequence of the organism from which the RNA was isolated. In other words, the cDNA sequences are aligned with the genomic DNA sequence (as shown at the bottom of Figure 23.3). This phase is accomplished with computer technology. When a cDNA sequence aligns with a gene sequence within the genome, this result means that the gene was expressed, because each cDNA sequence is derived from an RNA molecule. Also, as discussed in Chapter 12, eukaryotic pre-mRNAs often undergo splicing and alternative splicing (refer back to Figures 12.20 and 12.21). The alignment of cDNAs with the genomic DNA allows researchers to determine the pattern of RNA splicing that is found in a particular cell type under a given set of conditions.

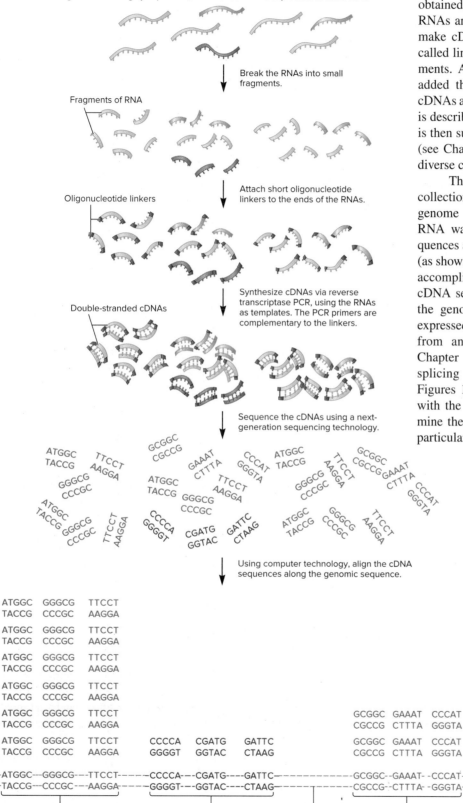

FIGURE 23.2 The technique of RNA sequencing (RNA-Seq). This simplified example involves a population of three different RNA molecules, each shown in a different color. Protein-coding genes in complex eukaryotes typically contain introns, which are spliced out of the pre-mRNAs. The alignment at the bottom of the figure corresponds to the sequences of the mature mRNAs. The gaps between the cDNA sequences in the alignment indicate the locations of introns. (Note: In actual RNA-Seq experiments, the RNA fragments are much longer than the ones shown here. Though the optimal length varies depending on the method of DNA sequencing used, a common length for the RNA fragments and their corresponding cDNA sequences is 100–300 nucleotides.)

Compared to the use of microarrays, RNA-Seq has several advantages:

- RNA-Seq is more accurate at quantifying the amount of each RNA transcript; the number of times that a particular cDNA sequence aligns with a gene is a measurement of the gene's expression level.
- It is superior at detecting RNA transcripts that are in low abundance.
- It identifies the precise boundaries between exons and introns and allows researchers to discover new splice variants.
- It identifies the 5′ and 3′ ends of RNA transcripts and aids in the identification of transcriptional start sites.

Gene Knockout Collections Allow Researchers to Study Gene Function at the Genomic Level

One broad goal of functional genomics is to determine the functions of all of the genes in a species' genome. Because each species has thousands of different genes, this is a very complicated task. One approach to achieving this goal is to produce collections of organisms of the same species in which each strain has one of its genes knocked out. For example, in *E. coli*, which has 4377 different genes, a complete knockout collection would be composed of 4377 different strains, with a different gene knocked out in each strain. A **gene knockout** is the alteration of a gene in a way that inactivates its function.

Why are knockout collections useful? Consider, for example, the phenotype produced by a particular gene knockout, which causes deafness in mice. Such a result suggests that the function of the normal gene is critical for hearing. Geneticists may also produce knockouts involving two or more genes to understand how the protein products of genes participate in a particular cellular pathway or contribute to a complex trait. In addition, gene knockouts in mice are used in the study of inherited human diseases. For example, as discussed in Chapter 21, gene knockouts have been used to study sickle cell disease.

Knockout collections are made in different ways. One way is via transposable elements. When a transposable element is inserted into a gene, it often inactivates the gene's function. Another way to produce knockouts is via CRISPR-Cas technology, which was described in Chapter 20. In 2006, the National Institutes of Health (NIH) launched the Knockout Mouse Project. The goal of this program is to build a comprehensive and publicly available resource comprising a collection of mouse embryonic stem cells (ES cells) containing a loss-of-function mutation in every gene in the mouse genome.

The NIH Knockout Mouse Project collaborates with other large-scale efforts to produce mouse knockouts, including one in Canada, called the North American Conditional Mouse Mutagenesis Project (NorCOMM), and one in Europe, called the European Conditional Mouse Mutagenesis Program (EUCOMM). The collective goal of these programs is to create at least one loss-of-function mutation in each of the approximately 20,000 protein-coding genes in the mouse genome. In addition, knockout collections are currently available for other model organisms, including *E. coli*, *S. cerevisiae*, and *C. elegans*.

23.1 COMPREHENSION QUESTIONS

1. A DNA microarray is a slide that is dotted with
   a. mRNAs from a sample of cells.
   b. fluorescently labeled cDNAs.
   c. known sequences of DNA.
   d. known cellular proteins.
2. For the method of RNA sequencing (RNA-Seq), which of the following is the correct order of steps?
   a. Isolate RNAs, synthesize cDNAs, fragment RNAs, sequence cDNAs, align cDNA sequences
   b. Synthesize cDNAs, sequence cDNAs, isolate RNAs, fragment RNAs, align cDNA sequences
   c. Isolate RNAs, fragment RNAs, synthesize cDNAs, sequence cDNAs, align cDNA sequences
   d. Synthesize cDNAs, isolate RNAs, fragment RNAs, sequence cDNAs, align cDNA sequences
3. A gene knockout results in a gene
   a. whose function has been inactivated.
   b. that has been transferred to a different species.
   c. that has been moved to a new location in the genome.
   d. that has been eliminated from a species during evolution.

23.2 PROTEOMICS

Learning Outcomes:

1. List reasons why the proteome is larger than the genome.
2. Outline the techniques of two-dimensional gel electrophoresis and tandem mass spectroscopy, and explain why they are used.
3. Describe two different types of protein microarrays, and discuss their uses.

Thus far, we have considered ways to characterize the genome of a given species and study its function. Because most genes code proteins, a logical next step is to examine the functional roles of the proteins that a cell or a species can make. As noted earlier in this chapter, this field is called proteomics, and the entire collection of a cell's or organism's proteins is its proteome.

Genomics represents only the first step in our comprehensive understanding of protein structure and function. Researchers often use genomic information to initiate proteomic studies, but such information must be followed up with research that involves the direct analysis of proteins. For example, as discussed in Section 23.1, the use of DNA microarrays and RNA-Seq can provide insights regarding the level of transcription of particular genes. However, an mRNA level may not provide an accurate measure of the abundance of a protein coded by a given gene. Protein levels are greatly affected, not only by the levels of mRNAs, but also by the rate of mRNA translation and by the turnover rate for a given protein. Therefore, data from DNA microarrays and RNA-Seq must be corroborated using other methods, such as Western blotting (discussed in Chapter 20), which directly determine the abundance of a protein in a given cell type.

As we move forward in the twenty-first century, the study of proteomes represents a key challenge facing molecular biologists. Much like genomics, proteomics will require the collective contributions of many research scientists, as well as improvements in technologies that are aimed at unraveling the complexities of the proteome. In this section, we will discuss the phenomena that increase protein diversity beyond genetic diversity. In addition, we will see how the techniques of two-dimensional gel electrophoresis and mass spectrometry can be used to isolate and identify cellular proteins and how protein microarrays are used to study protein expression and function.

The Proteome Is Much Larger Than the Genome

From the sequencing and analysis of an entire genome, researchers can identify all or nearly all of the genes for a given species. The entire proteome of a species, however, is larger than the genome, and its actual size is somewhat more difficult to determine.

What phenomena account for the larger size of the proteome? First, changes in pre-mRNAs may ultimately affect the resulting amino acid sequence of a protein.

- The most important alteration to pre-mRNAs that occurs commonly in eukaryotic species is **alternative splicing** (**Figure 23.3a**). A single pre-mRNA coded by a particular gene is frequently spliced into more than one mRNA. The splicing is often cell-specific, or it may be related to environmental conditions. As discussed in Chapter 12, alternative splicing is widespread, particularly among complex multicellular organisms. It can lead to the production of several or perhaps dozens of different polypeptide sequences from the same pre-mRNA, which greatly increases the number of potential proteins in the proteome.
- Similarly, **RNA editing,** the process of making a change in the nucleotide sequence of RNA after it has been transcribed (see Chapter 12), can lead to changes in the coding sequence of an mRNA. However, RNA editing is much less common than alternative splicing.

Another process that greatly diversifies the composition of a proteome is the phenomenon of **posttranslational covalent modification** (**Figure 23.3b**).

- Certain types of modifications can occur during the assembly and construction of a functional protein. These alterations include proteolytic processing, disulfide bond formation, and the covalent attachment of molecules, such as sugars or lipids. These are typically irreversible changes that are necessary to produce a functional protein.
- Other types of alterations, such as phosphorylation, acetylation, and methylation, are often reversible modifications that transiently affect the function of a protein.

A protein may be subject to several different types of posttranslational covalent modification, which can greatly increase the number of different forms of the protein found in a cell at any given time.

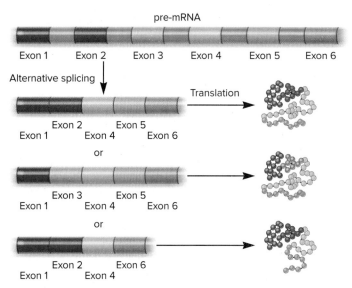

(a) Alternative splicing

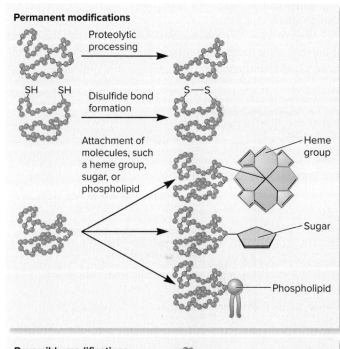

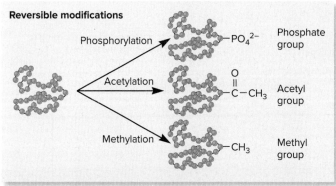

(b) Posttranslational covalent modification

FIGURE 23.3 Cellular mechanisms that increase protein diversity. (**a**) Due to alternative splicing, the patterns of exons that remain in mature mRNAs can be different, creating multiple types of transcripts from the same gene. (**b**) After a protein is made, it can be modified in a variety of ways, some of which are permanent and some reversible. These changes are called posttranslational covalent modifications.

CONCEPT CHECK: Explain how these mechanisms affect protein diversity.

Protein Purification Methods Are Often Used in Proteomics

As we have just discussed, a species' proteome is usually much larger than its genome. However, any given cell within a complex multicellular organism produces only a subset of the proteins found in the proteome of that species. For example, the human genome has approximately 20,000 different protein-coding genes, yet a muscle cell expresses only a subset of those genes at significant levels, perhaps 12,000 or so. The proteins a cell makes depend primarily on the type of cell, the stage of development, and the environmental conditions. An objective of researchers in the field of proteomics is the identification and functional characterization of all the proteins a particular type of cell can make. Because cells produce thousands of different proteins, this is a daunting task. Nevertheless, as in genomics, the past decade has seen important advances in researchers' ability to isolate and identify specific proteins.

The first step in protein identification is to purify the protein of interest. Most methods of protein purification rely on the technique of **chromatography,** which separates proteins based on

their chemical and/or physical properties. A sample containing a mixture of many different proteins is dissolved in a liquid solvent and exposed to some type of matrix, such as a gel, a column containing beads, or a strip of paper. Appendix A describes the general principles of chromatography (see Section A-3).

One type of chromatography that can be used to separate and purify proteins is **two-dimensional (2D) gel electrophoresis.** This technique can separate hundreds or even thousands of different proteins within a cell extract. The steps in this procedure are shown in **Figure 23.4.** As its name suggests, the technique involves two sequential gel electrophoresis procedures.

1. A sample of cells is lysed, and the proteins are loaded onto the top of a tube gel that separates them according to their net charge at a given pH.
2. A protein migrates to the point in the gel where its net charge is zero. This step is termed **isoelectric focusing.**
3. After the tube gel has been run, it is laid horizontally on top of a polyacrylamide slab gel, which is a flat, platelike gel that contains sodium dodecyl sulfate (SDS). The SDS coats the proteins with negative charges and denatures them.

Lyse a sample of cells and load the resulting mixture of proteins onto an isoelectric focusing gel.

pH 4.0

Proteins migrate until they reach the pH where their net charge is 0. At this point, a single band could contain 2 or more different proteins.

pH 10.0

Lay the tube gel onto an SDS gel and separate proteins according to their molecular mass.

SDS gel

pH 4.0   pH 10.0

200 kDa

10 kDa

(a) The technique of two-dimensional gel electrophoresis

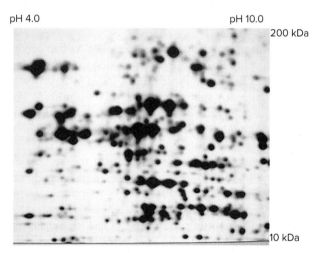

pH 4.0   pH 10.0

200 kDa

10 kDa

(b) An SDS slab gel that has been stained to visualize proteins following 2D gel electrophoresis

FIGURE 23.4 **Using two-dimensional gel electrophoresis to separate a mixture of cellular proteins.** **(a)** The technique involves two electrophoresis procedures. First, a mixture of proteins is separated on an isoelectric focusing gel that has the shape of a tube. Proteins migrate to the pH in the gel where their net charge is zero. This tube gel is placed into a long well on top of a sodium dodecyl sulfate (SDS) polyacrylamide gel. This SDS slab gel separates the proteins according to their molecular mass. In this diagram, only a few spots are shown, but an actual experiment involves a mixture of hundreds or thousands of different proteins. **(b)** A photograph of an SDS slab gel that has been stained to visualize proteins following 2D gel electrophoresis. Each spot represents a unique cellular protein.

SPL/Science Source

4. As proteins move through the SDS slab gel, they are separated according to their molecular mass. Smaller proteins move toward the bottom of the gel more quickly than larger ones.

5. After the SDS slab gel has run, the proteins within the gel are stained with a dye. As seen in Figure 23.4, the end result is a collection of spots, each of which corresponds to a unique cellular protein, with the spots for proteins having greater molecular masses remaining higher in the gel.

The resolving power of 2D gel electrophoresis is extraordinary. Proteins that differ by a single charged amino acid can be resolved as two distinct spots using this method.

Various approaches can be followed to identify spots on a 2D gel that may be of interest to researchers.

- One possibility is that the gel for a given cell type may show a few very large spots that are not found when proteins from other cell types are analyzed. The relative abundance of those spots may indicate that a particular protein is important to that cell's structure or function.

- Secondly, certain spots on a gel may be seen only under a given set of conditions. For example, a researcher may be interested in the effects of a hormone on the function of a particular cell type. Two-dimensional gel electrophoresis could be conducted on a sample in which the cells had not been exposed to the hormone and on another sample in which they had. Comparison of the results may reveal particular spots that are present only when the cells are exposed to the given hormone.

- Abnormal cells, such as cancer cells, often express proteins that are not found in normal cells. A researcher may compare normal and cancer cells via two-dimensional gel electrophoresis to identify proteins expressed only in the cancer cells.

Mass Spectrometry Is Used to Identify Proteins

Two-dimensional gel electrophoresis may be used as the first step in the separation of cellular proteins. The next goal is to correlate a given spot on a 2D gel with a particular protein. To accomplish this goal, a spot on a 2D gel can be cut out of the gel to obtain a tiny amount of the protein within the spot. In essence, the two-dimensional gel electrophoresis purifies a small amount of the cellular protein of interest. The next step is to identify that protein. This can be accomplished via **mass spectrometry,** a technique that accurately measures the mass of a molecule, such as a peptide that is a fragment of a protein.

Figure 23.5 shows how mass spectrometry can be used to determine the amino acid sequence of a protein. The procedure shown here has two mass spectrometry steps and therefore is called **tandem mass spectrometry.** It begins with a purified protein that is digested into small peptides. These peptides are subjected to the first mass spectrometry step. Figure 23.5 does not show the steps in mass spectrometry, but they are listed here:

1. The peptides are mixed with an organic acid, and the mixture is dried on a metal slide.

2. The sample is then subjected to a laser beam. This causes the peptides to be ejected from the slide in the form of an ionized gas in which each peptide contains one or more positive charges.

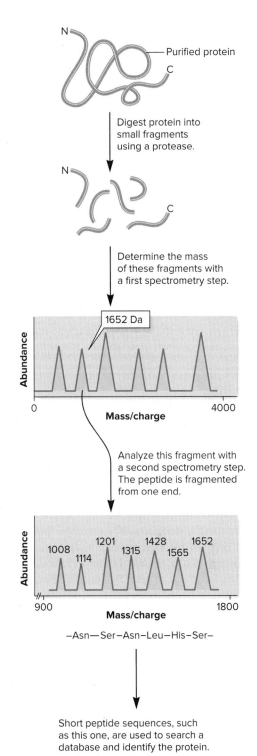

FIGURE 23.5 **The use of tandem mass spectrometry to determine the amino acid sequence of a peptide and identify a protein.**

CONCEPT CHECK: What is the purpose of tandem mass spectrometry?

3. The charged peptides are then accelerated via an electric field and fly toward a detector. The time it takes for them to reach the detector is directly related to their mass/charge ratio, which provides an extremely accurate way of determining the mass of a peptide.

In Figure 23.5, the first mass spectrometry step determines the masses of six different peptides; we will focus on the second peptide, which has a mass of 1652 daltons (Da). This 1652-Da peptide is then broken down into many smaller fragments. The mixture of smaller fragments is analyzed by a second mass spectrometry step.

The second step reveals the amino acid sequence of the peptide, because the masses of all 20 amino acids are known.

- As shown at the bottom of Figure 23.5, the starting peptide had a mass of 1652 Da.
- When one amino acid at the end was removed, the resulting smaller peptide had a mass that was 87 Da less (i.e., 1565 Da); this indicated that a serine is at one end of the peptide, because the mass of serine within a polypeptide chain is 87 Da.
- When two amino acids were removed at one end, the mass was 224 Da less; this corresponds to the removal of one serine (87 Da) and one histidine (137 Da).
- When three amino acids were removed, the mass was decreased by 337 Da; this corresponds to the removal of one serine (87 Da), one histidine (137 Da), and one leucine (113 Da).

Therefore, from these measurements, we conclude that the amino acid sequence at one end of the 1652-Da peptide is serine-histidine-leucine.

How does this information lead to the identification of a specific protein? As discussed in Chapter 22, the genome sequences of many species have already been determined. This information has allowed researchers to predict the amino acid sequences of most proteins that such species make and enter those sequences into a database. With computer software described in Section 23.5, the amino acid sequences obtained by tandem mass spectrometry can be used as query sequences to search a large database that contains a collection of protein sequences. The computer program may locate a match between the experimental sequences and a specific protein within a particular species. In this way, tandem mass spectrometry makes it possible to identify a purified protein.

Tandem mass spectrometry can also identify covalent modifications that have been made to proteins. For example, if an amino acid within a peptide was phosphorylated, the mass of the peptide is increased by the mass of a phosphate group. This increase in mass can be detected via tandem mass spectrometry.

Protein Microarrays Are Used to Study Protein Expression and Function

Earlier in this chapter, we considered DNA microarrays, which have been widely used to study gene expression at the RNA level. The technology for making DNA microarrays is also being applied to make **protein microarrays.** In this type of technology, proteins, rather than DNA molecules, are spotted onto a small silica, glass, or plastic slide.

The production of protein microarrays is more challenging than the production of DNA microarrays because proteins are much more easily damaged by the manipulations that occur during microarray formation. For example, the three-dimensional structure of a protein may be severely damaged by drying, which usually occurs during preparation of a microarray. This tendency for damage to occur has created additional challenges for researchers who are developing the

TABLE 23.2	
Some Applications of Protein Microarrays	
Application	**Description**
Protein expression	An antibody microarray can measure protein expression because each antibody in a given spot recognizes a specific amino acid sequence. Such a microarray can be used to study the expression of proteins in a cell-specific manner. It can also be used to determine how environmental conditions affect the levels of particular proteins.
Protein function	The substrate specificity and enzymatic activities of groups of proteins can be analyzed by exposing a functional protein microarray to a variety of substrates.
Protein-protein interactions	The ability of proteins to interact with each other can be determined by exposing a functional protein microarray to fluorescently labeled proteins.
Pharmacology	The ability of drugs to bind to cellular proteins can be determined by exposing a functional protein microarray to different kinds of labeled drugs. This type of experiment can help to identify the proteins within a cell to which a given drug may bind.

technology of protein microarrays. In addition, the synthesis and purification of proteins tend to be more time-consuming than the production of DNA, which can be amplified by PCR or directly synthesized on the microarray itself. In spite of these technical difficulties, the last few years have seen progress in the production and uses of protein microarrays (**Table 23.2**).

The two common types of protein microarrays are antibody microarrays and functional protein microarrays. The purpose of an **antibody microarray** is to quantify the amounts of particular proteins that are made by cells. Antibodies are proteins that are made by the immune systems of mammals and recognize antigens. One type of antigen that an antibody can recognize is a short peptide sequence found within another protein. Therefore, an antibody can specifically recognize a cellular protein. Researchers and commercial laboratories can produce different antibodies; each antibody recognizes a specific peptide sequence. The antibodies can be spotted onto a microarray. Cellular proteins can then be isolated, fluorescently labeled, and exposed to the antibody microarray. When a given protein is recognized by an antibody on the microarray, it is captured by the antibody and remains bound to that spot. The level of fluorescence at a given spot indicates the amount of the protein that is recognized by the particular antibody.

The other type of array is a **functional protein microarray.** To make this type of array, researchers must purify cellular proteins and then spot them onto a microarray. The microarray can then be analyzed with regard to specific kinds of protein function. In 2000, for example, Heng Zhu, Michael Snyder, and colleagues purified 119 proteins from yeast that were known to function as protein kinases. These kinds of proteins attach phosphate groups to other cellular proteins. A microarray was made consisting of different possible proteins that may or may not be phosphorylated by these 119 kinases, and then the array was exposed to each of the kinases in the presence of radiolabeled ATP. By following the incorporation of phosphate into the proteins on the array, the researchers determined the protein specificity of each kinase.

On a much larger scale, the same group of researchers purified 5800 different yeast proteins and spotted them onto a microarray. The array was then exposed to fluorescently labeled calmodulin, which is a regulatory protein that binds calcium ions. Several proteins in the microarray were found to bind calmodulin. Although some of these were already known to do so, others that had not been previously known to bind calmodulin were identified.

23.2 COMPREHENSION QUESTIONS

1. Which of the following is a reason why the proteome of a eukaryotic cell is usually much larger than its genome?
   a. Alternative splicing
   b. RNA editing
   c. Posttranslational covalent modifications
   d. All of the above are reasons for the larger size of the proteome of a eukaryotic cell.

2. During two-dimensional gel electrophoresis, proteins are separated based on
   a. their net charge at a given pH.
   b. their mass.
   c. their ability to bind to a specific resin.
   d. both a and b.

3. The technique of tandem mass spectrometry is used to determine
   a. the amino acid sequence of a peptide fragment.
   b. the nucleotide sequence of a segment of RNA.
   c. the nucleotide sequence of a segment of DNA.
   d. the number of genes in a species' genome.

4. Which of the following can be analyzed using a protein microarray?
   a. The amounts of particular proteins made by a sample of cells
   b. Protein function
   c. Protein-protein interactions
   d. All of the above

23.3 BIOINFORMATICS I: OVERVIEW OF COMPUTER ANALYSES AND GENE PREDICTION

Learning Outcomes:

1. Describe how sequence files may be analyzed by computer programs.
2. Outline different strategies for identifying gene sequences.

Geneticists use computers to collect, store, manipulate, and analyze data. Molecular genetic data are particularly amenable to computer analysis because they come in the form of sequences, such as those of DNA, RNA, or proteins. The ability of computers to analyze data at a rate of millions or even billions of operations per second has made it possible to solve problems concerning genetic information that were thought intractable a few decades ago.

In recent years, the marriage between genetics and computational tools has yielded an important branch of science known as bioinformatics. As mentioned earlier, bioinformatics is the use of computers, mathematical tools, and statistical techniques to record, store, and analyze biological information. In this section, we first consider the fundamental concepts that underlie the analysis of genetic sequences. We then explore how these methods are used to provide insights into functional genomics and proteomics. Chapter 29 will describe applications of bioinformatics in the area of evolutionary biology.

In addition to reading this section, you may wish to actually run computer programs, which are widely available at university and government websites (e.g., see www.ncbi.nlm.nih.gov/Tools). This type of hands-on learning will help you to see how valuable the computer has become as a tool for analysis of genetic data.

Sequence Files Are Analyzed by Computer Programs

Most people are familiar with **computer programs,** which consist of a series of operations that can manipulate and analyze data in a desired way. Simple computer programs that work on mobile devices are referred to as apps (short for "applications"). In functional genomics, many different types of computer programs have been designed. For example, a computer program can begin with a DNA sequence within a protein-coding gene and translate it into an amino acid sequence.

A first step in the computer analysis of genetic data is the creation of a **computer data file** to store the data. Such a file is simply a collection of information in a form suitable for storage and manipulation on a computer. In genetic studies, a computer data file might contain an experimentally obtained DNA, RNA, or amino acid sequence. For example, a file could contain the DNA sequence of one strand of the *lacY* gene from *Escherichia coli* (**Figure 23.6**). The numbers to the left represent the position in the sequence file of the first base in each row.

```
   1 ATGTACTATT TAAAAAACAC AAACTTTTGG ATGTTCGGTT TATTCTTTTT
  51 CTTTTACTTT TTTATCATGG GAGCCTACTT CCCGTTTTTC CCGATTTGGC
 101 TACATGACAT CAACCATATC AGCAAAGTG ATACGGGTAT TATTTTGCC
 151 GCTATTTCTC TGTTCTCGCT ATTATTCCAA CCGCTGTTTG GTCTGCTTC
 201 TGACAAACTC GGGCTGCGCA AATACCTGCT GTGGATTATT ACGGCATGT
 251 TAGTCATGTT TGCGCCGTTC TTTATTTTTA TCTTCGGGCC ACTGTTACAA
 301 TACAACATTT TAGTAGGATC GATTGTTGGT GGTATTTATC TAGGCTTTTG
 351 TTTTAACGCC GGTGCGCCAG CAGTAGAGGC ATTTATTGAG AAAGTCAGCC
 401 GTCGCAGTAA TTTCGAATTT GGTCGCGCGC GGATGTTTGG CTGTGTTGGC
 451 TGGGCGCTGT GTGCCTGAT TGTCGGCATC ATGTTCACCA TCAATAATCA
 501 GTTTGTTTTC TGGCTGGGCT CTGGCTGTGC ACTCATCCTC GCCGTTTTAC
 551 TCTTTTTCGC CAAAACGGAT GCGCCCTCTT CTGCCACGGT TGCCAATGCG
 601 GTAGGTGCCA ACCATTCGGC ATTTAGCCTT AAGTGGCAC TGGAACTGTT
 651 CAGACAGCCA AAACTGTGGT TTTGTCTACT GTATGTTATT GGCGTTTCCT
 701 GCACCTACGA TGTTTTTGAC CAACAGTTTG CTAATTTCTT TACTTCGTTC
 751 TTTGCTACCG GTGAACAGGG TACGCGGGTA TTTGGCTACG TAACGACAAT
 801 GGGCGCAATTA CTTAACGCCT CGATTATCTT CTTGCGCCA CTGATCATTA
 851 ATCGCATCGG TGGGAAAAAC GCCCTGCCTGC TGGCTGGCAC TATTATCTCT
 901 CTACGTATTA TTGGCTCATC GTTCGCCACC TCAGCGCTGG AAGTGGTTAT
 951 TCTGAAAACG CTGCATATGT TTGAAGTACC GTTCCTGCTG GTGGGCTGCT
1,001 TTAAATATAT TACCAGCCAG TTTGAAGTGC GTTTTTCAGC GACGATTTAT
1,051 CTGGTCTGTT TCTGCTTCTT TAAGCAACTG GCATGATTT TTATGTCTGT
1,101 ACTGGCGGGC AATATGTATG AAAGCATCGG TTTCCAGGGC GCTTATCTGG
1,151 TGCTGGGTCT GGTGGCGCTG GGCTTCACCT TAATTTCCGT GTTCACGCTT
1,201 AGCGGCCCCG GCCCGCTTTC CCTGCTGCGT CGTCAGGTGA ATGAAGTCGC
1,251 TTAA
```

FIGURE 23.6   A file of the DNA sequence of the *lacY* gene from *E. coli.*

To store data in a computer data file, a scientist creates the file via laboratory instruments such as densitometers and fluorescence detectors. These instruments are able to read data, such as a series of colored spots from Illumina sequencing images (see Figure 22.9), and enter the DNA sequence information directly into a computer file.

The purpose of making a computer file that contains a genetic sequence is to take advantage of the swift speed with which computers can analyze this information. Genetic sequence data in a computer file can be investigated in many different ways, corresponding to the many questions a researcher might ask about the sequence and its functional significance. These include the following:

1. Does a sequence contain a gene?
2. Where are functional sequences such as promoters, regulatory sites, and splice sites located within a particular gene?
3. Does a sequence code a polypeptide? If so, what is the amino acid sequence of the polypeptide?
4. Is a sequence homologous to any other known sequences?
5. What is the evolutionary relationship between two or more genetic sequences?

To answer these and many other questions, different computer programs have been written to analyze genetic sequences in particular ways. As an example, let's consider a computer program that can translate a DNA sequence into an amino acid sequence and consider how it might work in practice. The geneticist—the user—has a DNA sequence file and wants to have the DNA sequence translated into an amino acid sequence.

1. The user is sitting at a computer and opens a program that can translate a DNA sequence into an amino acid sequence.
2. The user provides the program with a DNA sequence file to be translated. In this case, the file is the DNA sequence of the *lacY* gene (see Figure 23.7).
3. The user also specifies which portion of the sequence should be translated. In this case, the user decides to begin the translation at the first nucleotide in the sequence file and end the translation at nucleotide number 1254.
4. With regard to output, the user specifies that the program should translate the sequence in all three forward reading frames, but only show the longest reading frame—the longest amino acid sequence that is uninterrupted by a stop codon.
5. This translated sequence is saved in a file whose contents are shown below. In this file, which was created by a computer program, each of the 20 amino acids is represented by a single-letter abbreviation (see Figure 13.5).

```
  1 MYYLKNTNFW MFGLFFFFYF FIMGAYFPFF PIWLHDINHI SKSDTGIIFA
 51 AISLFSLLFQ PLFGLLSDKL GLRKYLLWII TGMLVMFAPF FIFIFGPLLQ
101 YNILVGSIVG GIYLGFCFNA GAPAVEAFIE KVSRRSNFEF GRARMFGCVG
151 WALCASIVGI MFTINNQFVF WLGSGCALIL AVLLFFAKTD APSSATVANA
201 VGANHSAFSL KLALELFRQP KLWFLSLYVI GVSCTYDVFD QQFANFFTSF
251 FATGEQGTRV FGYVTTMGEL LNASIMFFAP LIINRIGGKN ALLLAGTIMS
301 VRIIGSSFAT SALEVVIKLT LHMFEVPFLL VGCFKYITSQ FEVRFSATIY
351 LVCFCFFKQL AMIFMSVLAG NMYESIGFQG AYLVLGLVAL GFTLISVFTL
401 SGPGPLSLLR RQVNEVA
```

Why is such a program useful? The advantages of running this program are speed and accuracy. It can translate a relatively long genetic sequence within milliseconds. By comparison, it would

probably take you a few hours to look up each codon in a genetic code table and write out the sequence in the correct order.

If you visit a website and actually run a program like this, you will discover that such a program can translate a genetic sequence into six reading frames—three forward and three reverse. This capability is useful if a researcher does not know where the start codon is located and/or does not know the direction of the coding sequence.

In genetic research, large software packages typically contain many computer programs that can analyze genetic sequences in different ways. For example, one program can translate a DNA sequence into an amino acid sequence, and another program can locate introns within genes. These software packages are found at universities, government facilities, hospitals, and businesses. At such locations, a central computer with substantial memory and high-speed computational abilities runs the software, and individuals can connect to this central computer via their personal computer (e.g., a laptop). Many such programs are freely available on the Internet. For example, as mentioned earlier in this section, www.ncbi.nlm.nih.gov/Tools has a variety of useful programs.

Different Computational Strategies Can Identify Functional Genetic Sequences

At the molecular level, the function of the genetic material is based largely on specific genetic sequences that play distinct roles. For example, codons are three-base sequences that specify particular amino acids, and promoters are sequences that provide a binding site for RNA polymerase to initiate transcription. Computer programs can be designed to scan very long sequences, such as those obtained from genome-sequencing projects, and locate meaningful features within them. To illustrate this concept, let's first consider the following sequence file, which contains 54 letters:

Sequence file:

```
GJTRLLAMAQLHEOGYLTOBWENTMNMTORXXXTGOODNTHEQ
ALLRTLSTORE
```

Now let's compare how three different computer programs might analyze this sequence to identify meaningful features. Suppose the first program is able to locate all of the English words within this sequence. If we ran this program, we would obtain the following result:

```
GJTRLLAMAQLHEOGYLTOBWENTMNMTORXXXTGOODNTHEQ
ALLRTLSTORE
```

In this case, a computer program has identified locations where the sequence of letters forms a word. Several words (which are in red and underlined) have been located within the sequence file.

A second computer program might be aimed at locating a series of words that are organized in the correct order to form a grammatically logical English sentence. If we used our sequence file and ran this program, we would obtain the following result:

```
GJTRLLAMAQLHEOGYLTOBWENTMNMTORXXXTGOODNTHEQ
ALLRTLSTORE
```

The second program has identified five words that form a logical sentence.

Finally, a computer program might identify patterns of letters, rather than words. For example, a computer program could locate a series of five letters that occurs in both the forward and reverse directions. If we applied this program to our sequence file, we would obtain the following:

GJ<u>TRLLA</u>MAQLHEOGYLTOBWENTMNMTORXXX<u>TGOODN</u>THEQ
<u>ALLRTL</u>STORE

In this case, the program has identified a pattern in which five letters occur in both the forward and reverse directions.

In these three examples, we can distinguish between **sequence recognition** (as in the first example) and **pattern recognition** (as in the third example). In sequence recognition, the program has the information that a specific sequence of symbols has a specialized meaning. This information must be supplied to the computer program. For example, the first program has access to the information from a dictionary with all known English words. With this information, the first program can identify sequences of letters that make words. By comparison, the third program does not rely on specialized sequence information. Rather, it is looking for a pattern of symbols that can occur within any group of symbol arrangements.

Overall, the simple programs we have considered illustrate three general types of identification strategies that computer programs can employ:

1. *Locate specialized sequences within a very long sequence.* A specialized sequence with a particular meaning or function is called a **sequence element,** or a **motif.** As in the first example, a computer program is supplied with a list of predefined sequence elements, in this case words, and can identify such elements within a sequence of interest. When applied in genetics, sequence elements, or motifs, include sequences of nucleotides, such as a promoter sequence, or sequences of amino acids that form functional domains in proteins.
2. *Locate an organization of sequences.* As illustrated by the second example, a program could recognize an organization of sequence elements or an organization of a pattern of symbols.
3. *Locate a pattern of symbols.* The third example is a program that locates a pattern of symbols.

The great power of computer analysis is that these types of operations can be performed with great speed and accuracy on sequences that may be enormously long.

As discussed throughout this textbook, many short nucleotide sequences play specialized roles in the structure or function of genetic material. A geneticist may want to locate a short sequence element within a longer nucleotide sequence in a data file. For example, a sequence of chromosomal DNA might be tens of thousands of nucleotides in length, and a geneticist may want to know whether a sequence element within many eukaryotic gene promoters, such as a TATA box, is found at one or more sites within the chromosomal DNA. To do so, a researcher could visually examine the long chromosomal DNA sequence in search of a TATA box. Of course, this process would be tedious and prone to human error.

TABLE 23.3

Short Sequence Elements That Can Be Identified by Computer Analysis

Type of Sequence	Examples*
Promoter	Many *E. coli* promoters contain TTGACA (–35 site) and TATAAT (–10 site). Eukaryotic core promoters may contain a TATA box, which is TATAAA.
Response elements	Glucocorticoid response element (AGRACA), cAMP response element (GTGACGTRA)
Start codon	ATG
Stop codons	TAA, TAG, TGA
Splice site	GTRAGT———YNYTRAC(Y)^{n}AG
Polyadenylation signal	AATAAA
Highly repetitive sequences	Relatively short sequences that are repeated many times throughout a genome
Transposable elements	Often characterized by a pattern in which direct repeats flank inverted repeats

*The sequences shown in this table are found in DNA. For gene sequences, only the coding strand is shown. R = purine (A or G); Y = pyrimidine (T or C); N = A, T, G, or C; U in RNA = T in DNA.

By comparison, the appropriate computer program can locate a sequence element in seconds and is thus very useful for this type of application. **Table 23.3** lists some examples of sequence elements that can be identified by computer analysis.

Computer-Based Approaches Can Identify Genes Within a Long Genomic Sequence

Gene prediction refers to the process of identifying regions of genomic DNA that code genes. These include protein-coding genes and genes for non-coding RNAs. One way to identify a gene is based on its ability to be transcribed into RNA. For example, the method of RNA-Seq described earlier in this chapter can be used to identify genes. However, at any given time, a cell transcribes only a subset of its genes. Therefore, a method such as RNA-Seq is not able to identify all of the genes that are found within a species' genome. As an additional tool, geneticists may use computer programs that are aimed at identifying genes in long genomic DNA sequences.

How do computer programs identify a protein-coding gene within a long DNA sequence? Such programs may employ different strategies.

Search by Signal. In a **search by signal** approach, the signals are known sequences, such as those found in promoters, the start codon, splice sites, etc. The program also contains information regarding the proper order of those signals for a gene. With this information, the program searches for the signals and tries to locate a region that contains a promoter sequence, followed by a start codon, a coding sequence, a stop codon, and a transcriptional terminator, in that order. When such a region is found, it has the correct properties to be a gene.

Search by Content. A second strategy for identifying genes is called **search by content.** The goal here is to identify sequences whose base content differs significantly from a random distribution. Within protein-coding genes, such a difference occurs primarily as a result of preferential codon usage. Although there are 64 codons, most species display a **codon bias** within the coding regions of genes. This means that certain codons are used much more frequently than others. For example, UUA, UUG, CUU, CUC, CUA, and CUG all specify leucine. In yeast (*Saccharomyces cerevisiae*), however, 80% of the leucine codons are UUG. Codon bias allows organisms to more efficiently rely on a smaller population of tRNA molecules. A search-by-content strategy, therefore, attempts to locate coding sequences by identifying regions where the nucleotide content displays a known codon bias.

Search for a Long Open Reading Frame. A third way to locate protein-coding genes within a DNA sequence is to examine translational reading frames. Recall that the reading frame is a sequence of codons determined by reading bases in groups of three. When analyzing a new DNA sequence, researchers must consider that the reading of codons (in groups of three nucleotides) could begin with the first nucleotide (reading frame 1), the second nucleotide (reading frame 2), or the third nucleotide (reading frame 3).

An **open reading frame (ORF)** is a region of a genetic sequence that does not contain any stop codons. Because most proteins are several hundred amino acids in length, a relatively long reading frame is required to code them. In bacteria, such long ORFs are contained within protein-coding genes. In **Figure 23.7**, a bacterial DNA sequence has been translated using all three reading frames. Only one of the three reading frames (frame 3) has a very long ORF with just one stop codon near the end, suggesting that this DNA sequence codes a protein. In eukaryotic genes, the approach of identifying a long ORF may not always be successful,

DNA sequence

Reading frame

1	S S	S S S	S	S		S	

2	S	S S S	S	S	S S	S

3						S

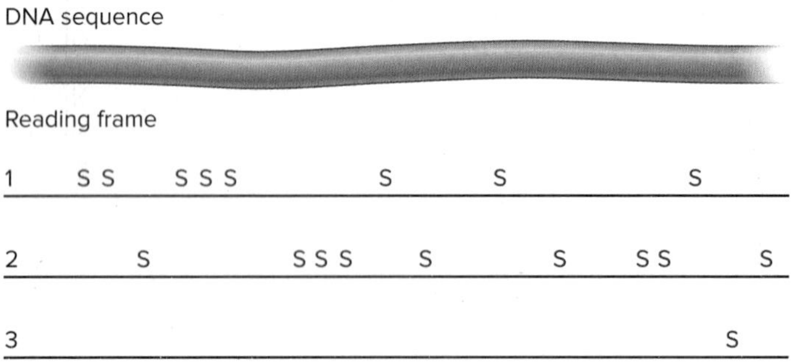

FIGURE 23.7 **Translation of a bacterial DNA sequence in all three reading frames.** The three lines represent the translation of a gene sequence according to each of three forward reading frames; the reading frames proceed from left to right. The letter S indicates the location of a stop codon. Reading frame 3 has a very long open reading frame (ORF), suggesting that the sequence may be a protein-coding gene. [Note: When analyzing genomic sequences, researchers may not know the direction of transcription for a given gene. In such cases, six reading frames (i.e., three forward and three reverse) are evaluated. Only the three forward frames are shown here.]

CONCEPT CHECK: Explain why the first two reading frames are unlikely to be the correct one.

because ORFs may be interrupted by multiple introns, which makes the ORFs relatively short.

Even though computer programs are valuable tools, they do not always accurately predict gene sequences. In particular, programs may not predict the correct start codon or the precise intron-exon boundaries. In some cases, computer programs may even suggest that a region codes a gene when it does not. Therefore, although a bioinformatics approach is a relatively easy way to identify potential genes, it should not be viewed as a definitive method. The confirmation that a DNA region contains an actual gene requires laboratory experimentation to show that the sequence is transcribed into RNA.

23.3 COMPREHENSION QUESTIONS

1. The identification of a stop codon for a particular gene is an example of
   a. sequence recognition.
   b. pattern recognition.
   c. both a and b.
   d. none of the above.

2. Which gene prediction strategy is aimed at locating a region in which the base content differs significantly from a random distribution due to codon bias?
   a. Search by signal
   b. Search by content
   c. Search for a long open reading frame
   d. Both a and b

23.4 BIOINFORMATICS II: DATABASES

Learning Outcome:

1. Define *database*, and list some examples.

One of the outcomes of the genetic technologies discussed in Chapters 20 and 22 is the generation of an enormous amount of molecular data. Much of these data are in the form of sequences, particularly those found in DNA, RNA, and proteins. From a research and clinical perspective, the ability to readily share this information is extremely useful. Therefore, groups of researchers and governmental agencies have taken the initiative to create websites that permit the sharing of molecular data.

A large number of computer data files collected and stored in a single location is called a **database.** A database can be accessed electronically at a website. In addition to genetic sequences, the files within databases contain **annotations,** or **metadata,** which consist of additional information such as a concise description of a sequence, the name of the organism from which it was obtained, and the function of the coded protein, if known. The annotations in a file may also describe other features of significance and cite a published scientific journal reference that describes the sequence.

TABLE 23.4

Examples of Major Computer Databases

Type	Description
Nucleotide sequence	DNA sequence data are collected in three internationally collaborating databases: GenBank (a U.S. database), EMBL (European Molecular Biology Laboratory Nucleotide Sequence Database), and DDBJ (DNA Databank of Japan). These databases receive sequence and sequence annotation data from genome projects, sequencing centers, individual scientists, and patent offices. These databases are accessed via the Internet.
Amino acid sequence	Amino acid sequence data are collected in a few international databases including Swissprot (Swiss protein database), PIR (Protein Information Resource), Genpept (translated peptide sequences from the GenBank database), and TrEMBL (translated sequences from the EMBL database).
Three-dimensional structure	PDB (Protein Data Bank) collects the three-dimensional structures of biological macromolecules with an emphasis on protein structure. These are primarily structures that have been determined by X-ray crystallography and nuclear magnetic resonance (NMR). These structures are stored in files that can be viewed on a computer with the appropriate software.
Protein motifs	Prosite is a database containing a collection of amino acid sequence motifs that are characteristic of a protein family, domain structure, or certain posttranslational modifications. Pfam is a database of protein families with multiple amino acid sequence alignments.
Gene expression data	Gene Expression Omnibus (GEO) contains data regarding the expression patterns of genes within a data set, such as the data obtained from microarrays or ChIP-chip assays.

TABLE 23.5

Examples of Clinical Databases Related to Genetics

Type	Description
Online Mendelian Inheritance in Man (OMIM)	OMIM is an online catalog of human genes and genetic disorders. The purpose of this database is to support human genetics research and education and to facilitate the practice of clinical genetics.
ClinVar	ClinVar is a database containing reports that are focused on the relationships among human genetic variation and phenotypes. The emphasis is on the clinical significance of such variation.
Clinical Genomic Database (CGD)	CGD is database of conditions with known genetic causes, focusing on medically significant genetic data with available interventions.
Catalogue of Somatic Mutations in Cancer (Cosmic)	Cosmic houses a very large database that focuses on the impact of somatic mutations in human cancer. It contains a catalogue of genes with mutations that are causally implicated in cancer.
The Cancer Genome Atlas (TCGA)	TCGA has collected, characterized, and analyzed cancer samples from more than 11,000 patients over more than a decade.

The scientific community has collected genetic information from thousands of research labs and created several large databases. **Table 23.4** describes some of the major databases in use worldwide. These databases enable genetic researchers to access and compare sequences obtained by many laboratories. In Section 23.5, we will consider how researchers use databases to analyze genetic sequences.

Another application of databases is in the area of human disease. Several different databases have been created that provide comprehensive information regarding the role of genetic variation in human health. Examples are described in **Table 23.5**. Some of these databases primarily focus on cancer.

23.4 COMPREHENSION QUESTION

1. Which of the following is/are important features of a database?
   a. Collection and storage of many files
   b. Annotation of files
   c. Electronic accessibility
   d. All of the above are important features.

23.5 BIOINFORMATICS III: HOMOLOGY

Learning Outcomes:

1. Define *homology*, and explain why the BLAST program is used.
2. Explain how a multiple sequence alignment can identify functional sites in a genetic sequence.

The term **homology** refers to similarities among various species that occur because the species are derived from a common ancestor. Attributes that are the result of homology are said to be **homologous.**

Homology can occur at the organism level and at the molecular level. At the organism level, the wing of a bat, the flipper of a dolphin, and the arm of a human are homologous, because they were evolutionarily derived from a forelimb in a mammalian ancestor. In this section, we will focus on homology at the molecular level. Homologous genes are derived from a common ancestral gene. Using bioinformatic tools, the study of homology has been a

```
              151                                           200
E. coli       TCTTTTTCTT TTACTTTTTT ATCATGGGAG CCTACTTCCC GTTTTTCCCG
K. pneumoniae TCTTTTTCTT TTACTATTTC ATTATGTCAG CCTACTTTCC TTTTTTTCCG

              201                                           250
              ATTTGGCTAC ATGACATCAA CCATATCAGC AAAAGTGATA CGGGTATTAT
              GTGTGGCTGG CGGAAGTTAA CCATTTAACC AAAACCGAGA CGGGTATTAT
```

(a) A comparison of DNA sequences in a portion of the *lacY* gene from *E. coli* and *K. pneumoniae*

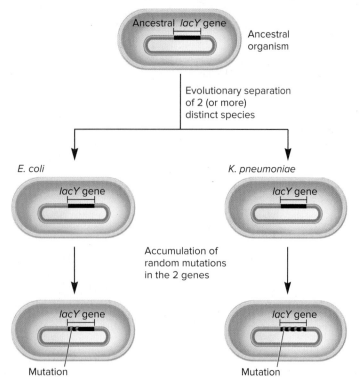

(b) The formation of homologous genes, known as orthologs, in two species of bacteria

FIGURE 23.8 The origin of homologous *lacY* genes in *Escherichia coli* and *Klebsiella pneumoniae.* **(a)** A comparison of the DNA sequences of a short region of the *lacY* genes from *E. coli* and *K. pneumoniae.* Bases that differ between the two genes are shown in red. **(b)** This figure emphasizes a single gene within an ancestral organism. During evolution, the ancestral organism diverged into many different species, including *E. coli* and *K. pneumoniae.* After this divergence, the *lacY* gene in the two species accumulated mutations, yielding *lacY* genes with somewhat different sequences.

Genes→Traits After two species diverge evolutionarily, their genes accumulate different random mutations. This example concerns the *lacY* gene, which codes lactose permease. In both species, the function of lactose permease is to transport lactose into the cell. The *lacY* gene in these two species has accumulated different mutations that alter the amino acid sequence of the protein. Researchers have determined that these two species transport lactose at significantly different rates. Therefore, the changes in gene sequences have affected the ability of these two species to transport lactose.

CONCEPT CHECK: Explain why the sequences of these two genes are similar to each other but not identical.

particularly powerful tool in elucidating the functions of genes and the proteins they code.

The Sequences of Homologous Genes Are Similar to Each Other

When comparing genetic sequences, researchers frequently find two or more similar sequences. For example, the sequence of the *lacY* gene that codes lactose permease in *E. coli* is similar to that of the *lacY* gene that codes lactose permease in another bacterium, *Klebsiella pneumoniae.* When segments of the two *lacY* genes are lined up, approximately 78% of their bases are a perfect match (**Figure 23.8a**).

In this case, the two sequences are similar because the genes are homologous, meaning they have been derived from the same ancestral gene. This idea is shown schematically in **Figure 23.8b**. An ancestral *lacY* gene was located in a bacterium that preceded the evolutionary divergence of *E. coli* and *K. pneumoniae.* After these two species diverged from each other, their *lacY* genes accumulated distinct mutations that produced

somewhat different base sequences for the gene. Therefore, in these two species of bacteria, the *lacY* genes are similar but not identical. When two homologous genes are found in different species and continue to serve the same function, they are termed **orthologs.**

Two or more homologous genes can also be found within a single species. As discussed in Chapter 8, abnormal gene duplications may happen several times during evolution, which results in multiple copies of a gene. These multiple copies of a gene within a single species are called **paralogs.** A **gene family** consists of two or more paralogs within the genome of a single species. During evolution, the functions of paralogs may change. For example, the globin genes have become more specialized to function at different stages of mammalian development (see Figure 8.7). A comparison of the amino acid sequences of globin orthologs and paralogs is found in Chapter 29 (look ahead to Figure 29.14).

When a gene family is present in a genome, the concept of orthologs becomes more complex. Again, let's consider the globin gene family found in mammals. Researchers would say that the β-globin gene in humans is an ortholog to the β-globin gene found in mice. Likewise, α-globin genes found in both species are considered orthologs. However, the α-globin gene in humans is not called an ortholog of the β-globin gene in mice, though they could be called homologous. The most closely related genes in two different species are considered orthologs.

A Database Can Be Searched to Identify Homologous Sequences

Homologous genes usually code proteins that carry out similar or identical functions. The members of the globin gene family code proteins involved with binding and transporting oxygen. Likewise, the *lacY* genes in *E. coli* and *K. pneumoniae* both code lactose permease, a protein that transports lactose across the bacterial cell membrane.

A very strong correlation is typically found between homology and function. How is this relationship useful with regard to bioinformatics? In many cases, the first indication of the function of a newly determined sequence is through homology to known sequences in a database. An example is the *CFTR* gene that is altered in people with cystic fibrosis. After the *CFTR* gene was identified in humans, a database search revealed that it is homologous to several genes found in other species. Moreover, a few of the homologous genes were already known to code proteins that function in the transport of ions and small molecules across the plasma membrane. This observation provided an important clue that cystic fibrosis involves a defect in ion transport.

The ability of computer programs to identify homology between genetic sequences provides a powerful tool for predicting the function of genetic sequences. In 1990, Stephen Altschul, David Lipman, and colleagues developed a program called **BLAST** (**b**asic **l**ocal **a**lignment **s**earch **t**ool). BLAST has become a very important bioinformatic tool that is used by many molecular biologists. This computer program starts with a particular genetic sequence and then locates similar sequences within a large database.

To see how the BLAST program works, let's consider the human enzyme phenylalanine hydroxylase, which functions in the metabolism of phenylalanine, an amino acid. Recessive mutations in the gene that codes this enzyme are responsible for the disease called phenylketonuria (PKU). The computational experiment whose results are shown in **Table 23.6** started with the amino acid

sequence of this protein and used the BLAST program to search the Swissprot database, which contains millions of different protein sequences. The BLAST program determines which sequences in this database are the closest matches to the amino acid sequence of human phenylalanine hydroxylase.

Table 23.6 shows a portion of the results—10 selected matches to human phenylalanine hydroxylase that were identified by the program. Because this enzyme is found in nearly all eukaryotic species, the program identified phenylalanine hydroxylase from many different species. The column to the right of the match number shows the percentage of amino acids that are identical between the enzyme in the species indicated and the human enzyme. Because the human phenylalanine hydroxylase sequence is already in the Swissprot database, the closest match of human phenylalanine hydroxylase is to itself (100%). The next nine sequences are in order of similarity. The most similar sequence is from the orangutan (99%), a close relative of humans. This is followed by two mammals, the mouse and rat, and then four vertebrates that are not mammals. The ninth and tenth best matches are from *Drosophila* and *C. elegans*, which are invertebrates.

As shown in the right column of Table 23.6, the relationship between the query sequence and each matching sequence is given an E-value (short for "Expect value"). The **E-value** represents the number of times that the match or a better one would be expected to occur purely by random chance in a search of the entire database. An E-value that is very small indicates that the similarity between the query sequence and the matching sequence is unlikely to have occurred by random chance. Instead, researchers would accept the hypothesis that the two sequences are homologous, which means they are derived from the same ancestral sequence.

E-values depend on several parameters, such as the length of the query sequence and the database size. As a general rule, if the E-value is less than 1×10^{-50}, the matching sequence is very similar to the query sequence, and the sequences are likely to be

TABLE 23.6
Results from a BLAST Program Comparing Human Phenylalanine Hydroxylase with Database Sequences

Match*	Percentage of Identical Amino Acids†	Species	Function of Sequence‡	E-value
1	100	Human (*Homo sapiens*)	Phenylalanine hydroxylase	0
2	99	Orangutan (*Pongo pygmaeus*)	Phenylalanine hydroxylase	0
3	92	Mouse (*Mus musculus*)	Phenylalanine hydroxylase	0
4	92	Rat (*Rattus norvegicus*)	Phenylalanine hydroxylase	0
5	83	Chicken (*Gallus gallus*)	Phenylalanine hydroxylase	0
6	78	Western clawed frog (*Xenopus tropicalis*)	Phenylalanine hydroxylase	0
7	75	Zebrafish (*Danio rerio*)	Phenylalanine hydroxylase	0
8	72	Japanese pufferfish (*Takifugu rubripes*)	Phenylalanine hydroxylase	0
9	62	Fruit fly (*Drosophila melanogaster*)	Phenylalanine hydroxylase	10^{-154}
10	57	Nematode (*Caenorhabditis elegans*)	Phenylalanine hydroxylase	10^{-141}

*The 10 examples shown here were randomly chosen from the results of running a BLAST program using human phenylalanine hydroxylase as the starting sequence.
†The number indicates the percentage of amino acids that are identical to the amino acid sequence of human phenylalanine hydroxylase.
‡In some cases, the function of the sequence was determined by biochemical assay. In other cases, the function was inferred due to the high degree of sequence similarity with other species.

homologous. (Values that are much less than 10^{-100} are reported as zero by the BLAST program.) If the value lies between 1×10^{-50} and 1×10^{-10}, the matching sequence, or part of it, is likely to be homologous to the query sequence. If the value is between 1×10^{-10} and 1×10^{-2}, the matching sequence has a significant chance of being related to the query sequence, whereas values between 1 and 1×10^{-2} indicate a relatively low probability of homology. Matching sequences with values above 1 are usually not evolutionarily related to the query sequence. As seen in Table 23.6, all of the E-values are below 1×10^{-140}, which suggests that all of these matching sequences are homologous to the query sequence.

The BLAST program searches for sequences that are similar to the query sequence, with the best matches at the top. Because the DNA sequences of closely related species tend to be more similar to each other than they are to distantly related ones, the order of the matches typically follows the evolutionary relatedness of the various species to the species from which the query sequence came. Among the species listed in Table 23.6, the human sequence is most similar to that of the orangutan, a closely related primate. The next most similar sequences are found in other mammals, followed by other vertebrates, and finally invertebrates. However, because the BLAST program is not designed to infer evolutionary relatedness, the best match(es) may not always be the most closely related species.

Homologous Genetic Sequences Can Be Aligned to Identify Conserved Sites That Are Likely to Be Functionally Important

After researchers identify homologous genes in a database, they may also take a closer look at them to identify particular sites that are functionally important, such as a short sequence of DNA within a gene. Other types of functionally important sites of interest include a short amino acid sequence within a polypeptide, or even a single amino acid at a particular site in a polypeptide.

A **conserved site** is a particular base within a gene or a particular amino acid within a polypeptide that is identical or similar across multiple species. Conserved sites are more likely to be functionally important compared to nonconserved sites. With regard to protein-coding genes, the maintenance of proper protein function is often critical for the survival and reproductive success of any given individual. Therefore, natural selection tends to eliminate mutations from a population if they inhibit protein function. For this reason, sites that are critical for function tend to be conserved—that is, stay the same—during the course of evolution.

One way to identify conserved sites is via **multiple-sequence alignment,** an approach in which a computer program aligns two or more homologous sequences and puts in gaps where the sequences do not match up. This approach was originated in 1970 by Saul Needleman and Christian Wunsch, who demonstrated that whale myoglobin and human β hemoglobin have similar sequences. To illustrate the usefulness of a multiple-sequence alignment, let's use the general methods of Needleman and Wunsch and apply them to the globin gene family. Hemoglobin, a protein found in red blood cells, is responsible for carrying oxygen through the bloodstream.

In humans, nine paralogous globin genes are expressed. (There are also four pseudogenes that are not expressed and one myoglobin gene.) The nine globin genes fall into two categories: those coding α chains and those coding β chains. The α-chain genes are α_1, α_2, θ, and ξ; the β-chain genes are β, δ, γ_A, γ_G, and ε. Each hemoglobin protein is composed of two α chains and two β chains. Because the globin genes are expressed at different stages of human development, the composition of hemoglobin changes during the course of growth. For example, the ξ and ε genes are expressed during early embryonic development, whereas the α_1, α_2, and β genes are expressed in the adult.

Insights into the structure and function of the hemoglobin polypeptides can be gained by comparing their sequences. In **Figure 23.9**, the amino acid sequences of the human globin polypeptides are compared in a multiple-sequence alignment. An inspection of a multiple-sequence alignment may reveal important features concerning the similarities and differences within a gene family. In this alignment, dots are shown where it is necessary to create gaps to keep the amino acid sequences aligned. As we can see, the sequence similarity is very high among α_1, α_2, θ, and ζ. In fact, the amino acid sequences coded by the α_1 and α_2 genes are identical. This suggests that the four types of α chain likely carry out very similar functions. Likewise, the β chains coded by the β, δ, γ_A, γ_G, and ε genes are very similar to each other. In the globin gene family, the α chains are much more similar to each other than they are to the β chains, and vice versa.

As mentioned, amino acids that are highly conserved within a gene family are more likely to be important functionally. The arrows in the multiple-sequence alignment point to histidine amino acids that are conserved in all nine members of the globin gene family. These histidines, which are highlighted in red, are involved in the necessary function of binding the heme molecule to the globin polypeptides.

Overall, the multiple-sequence alignment shown in Figure 23.9 illustrates the type of information that can be derived using this approach. In this case, multiple-sequence alignment has shown that a group of nine genes fall into two closely related subgroups. The alignment has also identified particular amino acids within the polypeptide sequences that are highly conserved. This conservation is consistent with these amino acids having an important role in this protein's function.

23.5 COMPREHENSION QUESTIONS

1. Homologous genes
   a. are derived from the same ancestral gene.
   b. are likely to carry out the same or similar functions.
   c. have similar DNA sequences.
   d. exhibit all of the above features.

2. The BLAST program begins with a particular genetic sequence and
   a. translates it into an amino acid sequence.
   b. determines if it contains one or more genes.
   c. identifies similar sequences within a database.
   d. does all of the above.

FIGURE 23.9 A multiple-sequence alignment for selected human globin polypeptides.

CONCEPT CHECK: Explain why functionally important sites are more likely to be conserved during evolution.

```
                    Gap
                    1                                                    50
α₁ alpha-1   VLSP.ADKTN VKAAWGKVGA HAGEGAEAL ERMFLSFPTT KTYFPHF.DL
α₂ alpha-2   VLSP.ADKTN VKAAWGKVGA HAGEGAEAL ERMFLSFPTT KTYFPHF.DL
θ theta      ALSA.EDRAL VRALWKKLGS NVGVYTTEAL ERTFLAFPAT KTYFSHL.DL
ζ zeta       SLTK.TERTI IVSMWAKIST QADTIGTETL ERLFLSHPQT KTYFPHF.DL
β beta       VHLTPEEKSA VTALWGKV.. NVDEVGGEAL GRLLVVYPWT QRFFESFGDL
δ delta      VHLTPEEKSA VNALWGKV.. NVDAVGGEAL GRLLVVYPWT QRFFESFGDL
γA gamma-A   GHFTEEDKAT ITSLWGKV.. NVEDAGGEAL GRLLVVYPWT QRFFESFGDL
γG gamma-G   GHFTEEDKAT ITSLWGKV.. NVEDAGGEAL GRLLVVYPWT QRFFESFGDL
ε epsilon    VHFTAEEKAA VTSLWSKM.. NVEEAGGEAL GRLLVVYPWT QRFFESFGDL

                      Gap           Gap
                    51                                                   100
α₁ alpha-1   SHGSA..... QVKGHGKKVA DALTNAVAHV DDMPNALSAL SDLHAHKLRV
α₂ alpha-2   SHGSA..... QVKGHGKKVA DALTNAVAHV DDMPNALSAL SDLHAHKLRV
θ theta      SPGSS..... QVKAHGQKVA DALSLAVERL DDLPHALSAL SHLHACQLRV
ζ zeta       HPGSA..... QLRAHGSKVV AAVGDAVKSI DDIGGALSKL SELHAYQLRV
β beta       STPDAVMGNP KVKAHGKKVL GAFSDGLAHL DNLKGTFATL SELHCDKLHV
δ delta      SSPDAVMGNP KVKAHGKKVL GAFSDGLAHL DNLKGTFSQL SELHCDKLHV
γA gamma-A   SSASAIMGNP KVKAHGKKVL TSLGDAIKHL DDLKGTFAQL SELHCDKLHV
γG gamma-G   SSASAIMGNP KVKAHGKKVL TSLGDAIKHL DDLKGTFAQL SELHCDKLHV
ε epsilon    SSPSAILGNP KVKAHGKKVL TSFGDAIKNM DNLKPAFAKL SELHCDKLHV

                    101                                                  148
α₁ alpha-1   DPVNFKLLSH CLLVTLAAHL PAEFTPAVHA SLDKFLASVS TVLTSKYR
α₂ alpha-2   DPVNFKLLSH CLLVTLAAHL PAEFTPAVHA SLDKFLASVS TVLTSKYR
θ theta      DPASFQLLGH CLLVTLARHL PGDFSPALQA SLDKFLSHSVI SALVSEYR
ζ zeta       DPVNFKLLSH CLLVTLAARF PADFTAEAHA AWDKFLSVVS SVLTEKYR
β beta       DPENFRLLGN VLVCVLAHHF GKEFTPPVQA AYQKVVAGVA NALAHKYH
δ delta      DPENFRLLGN VLVCVLARNF GKEFTPQMQA AYQKVVAGVA NALAHKYH
γA gamma-A   DPENFRLLGN VLVCVLAIHF GKEFTPEVQA SWQKMVTAVA SALSSRYH
γG gamma-G   DPENFRLLGN VLVCVLAIHF GKEFTPEVQA SWQKMVTAVA SALSSRYH
ε epsilon    DPENFRLLGN VMVIILATHF GKEFTPEVQA AWQKLVSAVA IALAHKYH
```

KEY TERMS

Introduction: functional genomics, proteome, proteomics, bioinformatics

23.1: DNA microarray (gene chip), RNA sequencing (RNA-Seq), gene knockout

23.2: alternative splicing, RNA editing, posttranslational covalent modification, two-dimensional (2D) gel electrophoresis, isoelectric focusing, mass spectrometry, tandem mass spectrometry, protein microarray, antibody microarray, functional protein microarray

23.3: computer program, computer data file, sequence recognition, pattern recognition, sequence element (motif), gene prediction, search by signal, search by content, codon bias, open reading frame (ORF)

23.4: database, annotation

23.5: homology, homologous, orthologs, paralogs, gene family, BLAST (basic local alignment search tool), E-value, conserved site, multiple-sequence alignment

CHAPTER SUMMARY

- The goal of functional genomics is to understand the roles of genetic sequences in a given species.

23.1 Functional Genomics

- A DNA microarray is a slide dotted with samples of many different DNA sequences. It is used to study the expression of many genes simultaneously, among other uses (see Figure 23.1, Table 23.1).

- In the method known as RNA sequencing (RNA-Seq), RNA is isolated from cells, converted to cDNA, and then sequenced using a next-generation sequencing technology. The cDNA sequences are aligned with the genomic sequence (Figure 23.2). This method has uses similar to those of DNA microarrays.

- Researchers have produced gene knockout collections for certain species, such as mice, to determine the functions of genes at the genomic level.

23.2 Proteomics

- The proteome is the entire collection of proteins that a cell or organism makes. The study of the functions of such a collection of proteins is called proteomics.
- The proteome is much larger than the genome due to alternative splicing, RNA editing, and posttranslational covalent modifications (see Figure 23.3).
- Two-dimensional gel electrophoresis separates a complex mixture of cellular proteins (see Figure 23.4).
- Tandem mass spectrometry identifies short amino acid sequences within a purified protein. These short sequences can be used to identify the protein (see Figure 23.5).
- Protein microarrays are used to study protein expression, protein function, protein-protein interactions, and protein-drug interactions (see Table 23.2).

23.3 Bioinformatics I: Overview of Computer Analyses and Gene Prediction

- Bioinformatics is the use of computers, mathematical tools, and statistical techniques to record, store, and analyze biological information, such as DNA sequences.
- Sequence files are analyzed by computer programs (see Figure 23.6).

- Different computational strategies, such as sequence recognition and pattern recognition, are applied to identify functional genetic sequences. Sequence recognition identifies sequence elements (see Table 23.3).
- Genes may be identified using computational strategies such as search by signal and search by content. A search for a long open reading frame (ORF) may also identify a gene (see Figure 23.7).

23.4 Bioinformatics II: Databases

- Researchers have collected many files of genetic sequences and other genetic information and compiled them in large databases (see Tables 23.4, 23.5).

23.5 Bioinformatics III: Homology

- Homologous genes are derived from the same ancestral gene. They can be orthologs (genes in different species) or paralogs (genes in the same species) (see Figure 23.8).
- The BLAST program starts with a particular genetic sequence and identifies homologous sequences within a large database (see Table 23.6).
- Researchers may use multiple-sequence alignment to compare the sequences of several homologous genes and identify conserved sites that may be functionally important (see Figure 23.9).

PROBLEM SETS & INSIGHTS

MORE GENETIC TIPS **1.** Which of the following statements uses the term *homologous* correctly?

A. The two X chromosomes in female mammalian cells are homologous to each other.
B. The α-tubulin gene in *Saccharomyces cerevisiae* is homologous to the α-tubulin gene in *Arabidopsis thaliana*.
C. The promoter of the *lac* operon is homologous to the promoter of the *trp* operon.
D. The *lacY* genes of *E. coli* and *K. pneumoniae* are approximately 78% homologous.

TOPIC: *What topic in genetics does this question address?* The topic is the meaning of *homologous*.

INFORMATION: *What information do you know based on the question and your understanding of the topic?* In the question, you are given four different comparisons that use the word *homologous*. From your understanding of the topic, you may remember that homology is similarity that is due to descent from a common ancestor.

PROBLEM-SOLVING STRATEGY: *Define key terms.* To solve this problem, you need to define *homology*, which means similarity due to descent from a common ancestor, and then decide if each comparison fits that definition.

ANSWER:
A. Correct
B. Correct

C. Incorrect; the promoters are short sequences that are similar to each other.
D. Incorrect; the genes are simply homologous. Two genes are either homologous (i.e., derived from a common ancestral gene) or not. However, you could say that the genetic sequences of the two species are 78% identical.

2. A number of computer programs identify sequence elements within a long segment of DNA. What is a sequence element? Give two examples. How is the specific sequence of a sequence element determined? In other words, is it determined by the computer program or by experimentation? Explain.

TOPIC: *What topic in genetics does this question address?* The topic is the use of computer programs to identify sequence elements.

INFORMATION: *What information do you know based on the question and your understanding of the topic?* From the question, you know that some computer programs can identify sequence elements. From your understanding of the topic, you may remember what a sequence element is.

PROBLEM-SOLVING STRATEGIES: *Define key terms. Compare and contrast.* One strategy to begin to solve this problem is to define *sequence element*, which is a specialized sequence (i.e., a base sequence or amino acid sequence) that has a particular meaning or function in DNA, RNA, or a polypeptide.

ANSWER: A sequence element is a particular sequence that has some type of known functional role. One example is a promoter that is needed for transcription. Another example is a start codon that is needed to begin the process of translation. The sequence of a sequence element is determined by experimentation.

---

3. To answer this question, you need to look back at the evolution of the globin gene family in humans, which is shown in Figure 8.7. Throughout the evolution of this gene family, mutations have occurred that have resulted in globin polypeptides with similar but significantly different amino acid sequences. If you look at the multiple-sequence alignment in Figure 23.9, you can make logical guesses regarding the timing of mutations, based on a comparison of the amino acid sequences of the globin polypeptides. What is/are the most probable time(s) that mutations occurred to produce the following amino acid differences?

  A. Val-111 and Cys-111
  B. Met-112 and Leu-112
  C. Ser-141, Asn-141, Ile-141, and Thr-141

T OPIC: *What topic in genetics does this question address?*
The topic is the analysis of results of a multiple-sequence alignment to infer the timing of evolutionary changes.

I NFORMATION: *What information do you know based on the question and your understanding of the topic?* In the question, you are asked to refer to Figure 8.7 and Figure 23.9. From your understanding of evolution, you may recall that if a particular amino acid is different in one gene family member compared to others, the amino acid change is likely to have occurred after that gene was formed by a gene duplication event and diverged. For example, if one family member has a methionine at position 112 while all of the other family members have a leucine at that position, your best guess would be that a mutation changed the leucine to a methionine after that one gene family member was formed and diverged. If the change had occurred prior to divergence, methionine would be found at that position in other family members.

P ROBLEM-SOLVING S TRATEGIES: *Analyze data. Compare and contrast.* Results of a multiple-sequence alignment can be viewed as a form of data. You need to compare the alignment in Figure 23.9 with the evolutionary time scale in Figure 8.7 to answer this question.

ANSWER:

  A. You do not know if the original globin gene coded a cysteine or a valine at codon 111. The mutation could have changed cysteine to valine or valine to cysteine. The mutation probably occurred after the duplication that produced the α-globin family and β-globin family (about 300 mya) but before the gene duplications that occurred in the last 200 million years to produce the multiple copies of the globin genes on chromosome 11 and chromosome 16. Therefore, all of the globin genes on chromosome 11 have a valine at codon 111, and all of the globin genes on chromosome 16 have a cysteine.

  B. Met-112 occurs only in the ε-globin polypeptide; all of the other globin polypeptides contain a leucine at position 112. Therefore, the ancestral globin gene probably contained a leucine codon at position 112. After the gene duplication that produced the ε-globin gene, a mutation occurred that changed this leucine codon into a methionine codon. This would have occurred since the evolution of primates (i.e., within the last 10–20 million years).

  C. If you look at the amino acids found at position 141 (i.e., Ser-141, Asn-141, Ile-141, and Thr-141), you will notice that a serine is found there in θ globin, ξ globin, and γ globin. Because the θ- and ξ-globin genes are found on chromosome 16 and the γ-globin genes are found on chromosome 11, it is probable that the ancestral codon coded serine and that the other codons (for asparagine, isoleucine, and threonine) arose later by mutation of the serine codon. If this is correct, the Thr-141 mutation arose before the gene duplication that produced the α-globin genes. The Asn-141 and Ile-141 mutations arose after the gene duplications that produced the γ-globin genes. Therefore, the Thr-141, Asn-141, and Ile-141 mutations occurred since the evolution of primates (i.e., within the last 10–20 million years).

Conceptual Questions

C1. Define the following terms: *genomics, functional genomics,* and *proteomics.*

C2. Discuss the reasons why the proteome of a given species is larger than the genome.

C3. What is a database? What types of information are stored within a database? Where does the information come from?

C4. Besides the examples listed in Table 23.3, list five types of short sequence elements that a geneticist might want to locate within a DNA sequence.

C5. Discuss the distinction between sequence recognition and pattern recognition.

C6. A multiple-sequence alignment of five homologous proteins is shown next:

```
  1                                                           50
1 MLAFLNQVRK PTLDLPLEVR RKMWFKPFM. QSYLVVFIGY LTMYLIRKNF
2 MLAFLNQVRK PTLDLALDVR RKMWFKPFM. QSYLVVFIGY LTMYLIRKNF
3 MLPFLKAPAD APL.MTDKYE IDARYRYWRR HILLTIWLGY ALFYFTRKSF
4 MLSFLKAPAN APL.ITDKHE VDARYRYWRR HILITIWLGY ALFYFTRKSF
5 MLSIFKPAPH KAR.LPAA.E IDPTYRRLRW QIFLGIFFGY AAYYLVRKNF

  51                                                          100
1 NIAQNDMIST YGLSMTQLGM IGLGFSITYG VGKTLVSYYA DGKNTKQFLP
2 NIAQNDMIST YGLSMTELGM IGLGFSITYG VGKTLVSYYA DGKNTKQFLP
3 NAAVPEILAN GVLSRSDIGL LATLFYITYG VSKFVSGIVS DRSNARYFMG
4 NAAAPEILAS GILTRSDIGL LATLFYITYG VSKFVSGIVS DRSNARYFMG
5 ALAMPYLVEQ .GFSRGDLGF ALSGISIAYG FSKFIMGSVS DRSNPRVFLP
```

Discuss some of the interesting features that this alignment reveals.

C7. What is the difference between similarity and homology?

C8. When comparing (i.e., aligning) two or more genetic sequences, it is sometimes necessary to put in gaps. Explain why. Discuss two changes (i.e., two types of mutations) that could happen during the evolution of homologous genes that would explain the occurrence of gaps in a multiple-sequence alignment.

Experimental Questions

E1. With regard to DNA microarrays, answer the following questions:

A. What is attached to the slide? Be specific about the number of spots, the lengths of DNA fragments, and the origin of the DNA fragments.

B. What is hybridized to the microarray?

C. How is hybridization detected?

E2. In the procedure called RNA sequencing (RNA-Seq), what type of molecule is actually sequenced?

E3. Can two-dimensional gel electrophoresis be used as a purification technique? Explain.

E4. Explain how tandem mass spectroscopy is used to determine the sequence of a peptide. Once a peptide sequence is known, how is this information used to determine the sequence of the entire protein?

E5. Describe the two general types of protein microarrays. What are their possible applications?

E6. Describe three bioinformatics approaches that can be used to identify a protein-coding gene.

E7. What is a motif? Why is it useful for computer programs to identify functional motifs within amino acid sequences?

E8. Discuss why it is useful to search a database to identify sequences that are homologous to a newly determined sequence.

E9. In this chapter, we considered a computer program that can translate a DNA sequence into a polypeptide sequence. A researcher has a sequence file that contains the amino acid sequence of a polypeptide and can run a program called BACKTRANSLATE that does the opposite of the program described in this chapter: It takes an amino acid sequence file and determines the sequence of DNA that would code such a polypeptide. How does this program work? In other words, what genetic principles underlie this program? What type of sequence file would this program generate: a nucleotide sequence or an amino acid sequence? Would the BACKTRANSLATE program produce only a single sequence file? Explain why or why not.

E10. In this chapter, we considered a computer program that translates a DNA sequence into a polypeptide sequence. Instead of running this program, a researcher could simply look the codons up in a genetic code table and determine the sequence by hand. What are the advantages of running the program rather than doing the translation the old-fashioned way, by hand?

E11. To answer each of the following types of genetic questions, would a computer program use sequence recognition, pattern recognition, or both?

A. Whether a segment of *Drosophila* DNA contains a P element (which is a specific type of transposable element)

B. Whether a segment of DNA contains a stop codon

C. Whether there is an inversion in a segment of DNA on a chromosome compared to a homologous chromosome from a different individual

D. Whether a long segment of bacterial DNA contains one or more genes

E12. The goal of many computer programs is to identify sequence elements within a long segment of DNA. What is a sequence element? Give two examples. How is the specific sequence of a sequence element determined? In other words, is it determined by a computer program or by genetic studies? Explain.

E13. Look again at the multiple-sequence alignment of the globin polypeptides in Figure 23.9, focusing on amino acids 101 to 148.

A. Which of these amino acids are likely to be most important for globin structure and function? Explain why.

B. Which are likely to be least important?

E14. Refer to question 3 in More Genetic TIPS before answering this question. Based on the multiple-sequence alignment in Figure 23.9, what is/are the most probable time(s) that mutations occurred in the human globin gene family to produce the following amino acid differences?

A. His-119 and Arg-119

B. Gly-121 and Pro-121

C. Glu-103, Val-103, and Ala-103

E15. Below is a short nucleotide sequence from a gene. Use a computer program from the Internet (e.g., www.ncbi.nlm.nih.gov/Tools) to determine what gene this sequence is from. Also, determine the species in which this gene sequence is found.

5′–GGGCGCAATTACTTAACGCCTCGATTATCTTCTTGCGCCACTGATCATTA–3′

E16. Refer back to question 3 in More Genetic TIPS and the codon table in Chapter 13 (Table 13.1). Assuming that a mutation causing a single base change is more likely than one causing a double base change, propose how the codons for Asn-141, Ile-141, and Thr-141 arose. In your answer, state which of the six possible serine codons is most likely to be the ancestral serine codon of the globin gene family, and explain how that codon changed to produce Asn-141, Ile-141, and Thr-141.

Questions for Student Discussion/Collaboration

1. Let's suppose you are in charge of organizing a database for the mouse genome. Make a list of innovative strategies you would initiate to make the mouse genome database useful and effective.

2. If you have access to the necessary computer software, make a sequence file and analyze it in the following ways: What is the translated sequence in all three forward reading frames? What is the longest open reading frame? Is the sequence homologous to any known sequences? If so, does this provide any clues about the function of the sequence?

Note: All answers are available for the instructor in Connect; the answers to the even-numbered questions and all of the Concept Check and Comprehension Questions are in Appendix B.

CHAPTER OUTLINE

An African girl with albinism. *This condition results in very light skin and hair color.*
Radu Sigheti/REUTERS/Alamy Stock Photo

24 MEDICAL GENETICS

Insight from genetic studies is expected to bring about revolutionary changes in medical practices. In fact, changes are already under way. Currently, several hundred genetic tests are in clinical use, with many more under development. Most of these tests detect mutations associated with rare genetic disorders that follow Mendelian inheritance patterns. These include Duchenne muscular dystrophy, cystic fibrosis, sickle cell disease, and Huntington disease. In addition, other genetic tests can detect a predisposition to develop certain forms of cancer.

As noted in Chapter 22, DNA sequencing technologies have advanced to the point where the sequencing of a person's entire genome is inexpensive enough to be done as a routine diagnostic procedure. Many people in the medical field expect that this advance will usher in the era of **personalized medicine**—the use of information about a patient's genotype and other clinical data in order to select a medication, therapy, or preventative measure that is specifically suited to that patient. We will explore this topic at the end of this chapter.

Thousands of genetic diseases are known to afflict people. Most of the genetic disorders discussed in the first part of this chapter are the direct result of a mutation in one gene. However, many diseases, including common medical disorders such as diabetes, asthma, and mental illness, have a complex pattern of inheritance involving several genes. In these cases, a single mutant gene does not determine whether a person has the disease. Instead, a number of genes may each make a subtle contribution to a person's susceptibility to the disease. Unraveling these complexities will be a challenge for some

time to come. The availability of the DNA sequence of the human genome, discussed in Chapter 22, will be of great help.

In this chapter, we will focus our attention on ways that mutant genes contribute to human disease. In the first part of the chapter, we will explore the molecular basis of several genetic disorders and their patterns of inheritance. We then consider how genetic variation can be used to identify disease-causing alleles in human populations. We will also examine how genetic testing can determine if an individual carries a disease-causing allele, and discuss the current and potential uses of human gene therapy—the introduction of cloned genes into living cells in the treatment of a disease. (Chapter 25 will consider an important genetic disease that affects many people, namely, cancer.) Finally, we will explore the topic of personalized medicine.

24.1 INHERITANCE PATTERNS OF GENETIC DISEASES

Learning Outcomes:

1. List seven observations that suggest a disease may have a genetic component.
2. Analyze human pedigrees, and distinguish autosomal recessive, autosomal dominant, X-linked recessive, and X-linked dominant patterns.
3. Define *locus heterogeneity,* and explain how it can confound pedigree analysis.

Human genetics is a topic that is hard to resist. Almost everyone who looks at a newborn is tempted to speculate whether the baby resembles one of the parents, a sibling, or perhaps a more distant relative. In this section, we will focus primarily on the inheritance of human genetic diseases rather than common traits found in the general population.

Even so, the study of human genetic diseases often provides insights about such common traits. The disease hemophilia illustrates this point. Hemophilia is a condition in which the blood does not clot properly. By analyzing people with this disorder, researchers have identified genes that participate in the process of blood clotting. The study of hemophilia has helped to identify a clotting pathway involving several different proteins. Therefore, as with the study of mutations in model organisms such as *Drosophila*, mice, and yeast, when we study the inheritance of genetic diseases, we often learn a great deal about the genetic basis for normal physiological processes as well.

Because thousands of human diseases have an underlying genetic basis, human genetic analysis is of great medical importance. In this section, we will examine the causes and inheritance patterns of human genetic diseases that result from defects in single genes. As you will learn, mutant genes that cause diseases often follow simple Mendelian inheritance patterns.

A Genetic Basis for a Human Disease May Be Suggested by a Variety of Observations

When we view the characteristics of people, we usually think that some traits are inherited, whereas others are caused by environmental factors. For example, when the facial features of two related individuals look strikingly similar, we think that this similarity has a genetic basis. The striking resemblance of identical twins is an obvious example. By comparison, other traits have environmental causes. If we see someone with purple hair, we likely suspect that the person has used hair dye as opposed to showing an unusual genetic trait.

For human diseases, geneticists would like to know the relative contributions from genetics and the environment. Is a disease caused by a pathogenic microorganism, a toxic agent in the environment, or a mutant gene? Unlike the case with experimental organisms, we cannot conduct human crosses to determine the genetic basis for diseases. Instead, we must rely on analyzing the occurrence of a disease in families that already exist. As described in the following list, several observations are consistent with the idea that a disease is caused, at least in part, by the inheritance of mutant genes. When the occurrence of a disease correlates with several of these observations, a geneticist becomes increasingly confident that the disease has a genetic basis.

1. *An individual who exhibits a disease is more likely to have genetic relatives with the disorder than is someone in the general population.* For example, someone with cystic fibrosis is more likely to have relatives with this disease than would a randomly chosen member of the general population.
2. *Identical twins share the disease more often than nonidentical twins.* Identical twins, also called **monozygotic (MZ) twins,** are genetically identical to each

other, because they were formed from the same sperm and egg. By comparison, nonidentical twins, also called fraternal twins, or **dizygotic (DZ) twins,** are formed from separate pairs of sperm and egg cells. Fraternal twins share, on average, 50% of their genetic material. When a disorder has a genetic component, both identical twins are more likely to exhibit the disorder than are fraternal twins.

Geneticists evaluate a disorder's **concordance,** the degree to which it is inherited, by calculating the percentage of twin pairs in which both twins exhibit the disorder relative to pairs where only one twin shows the disorder. Theoretically, for diseases caused by a single gene, concordance among identical twins should be 100%. For fraternal twins, concordance for dominant disorders is expected to be 50%, assuming only one parent is heterozygous for the disease. For recessive diseases, concordance among fraternal twins is expected to be 25% if we assume both parents are heterozygous carriers. However, the actual concordance values observed for most single-gene disorders are usually less than such theoretical values for a variety of reasons. For example, some disorders exhibit **incomplete penetrance,** meaning that the symptoms associated with the disorder are not always produced (refer back to Table 4.1). Also, one twin may have a disorder due to a new mutation that occurred after fertilization; it would be very unlikely for the other twin to have the same mutation.

3. *The disease does not spread to individuals sharing similar environmental situations.* Inherited disorders cannot spread from person to person. The only way genetic diseases can be transmitted is from parent to offspring.
4. *Different populations tend to have different frequencies of the disease.* Because mutations are rare events, they may arise in one population but not another. Also, each population is exposed to its own unique set of environmental conditions that may influence the prevalence of a given allele. Therefore, the frequencies of genetic diseases due to mutant alleles usually vary among different populations of humans. For example, the frequency of sickle cell disease is highest among certain African and Asian populations and relatively low in other parts of the world (look ahead to Figure 27.11).
5. *The disease tends to develop at a characteristic age.* Many genetic disorders exhibit a characteristic **age of onset** at which the symptoms of the disease first appear. Some mutant genes exert their effects during embryonic and fetal development, so their effects are apparent at birth. Other genetic disorders tend to develop much later in life.
6. *The human disorder may resemble a disorder that is already known to have a genetic basis in an animal.* In animals, on which we can conduct experiments, various traits are known to be governed by genes. For example, the albino phenotype is found in many animals as well as in humans (**Figure 24.1**).
7. *A correlation is observed between a disease and a mutant human gene or a chromosomal alteration.* A particularly convincing piece of evidence that a disease may have a

FIGURE 24.1 **The albino phenotype in a human and a wildebeest.**

Genes→Traits Certain enzymes (coded by genes) are necessary for the production of pigments. A homozygote with two defective alleles in one of these pigmentation genes exhibits an albino phenotype. This phenotype can occur in humans and other animals.

(top): Friedrich Stark/Alamy Stock Photo; (bottom): Mitch Reardon/Science Source

genetic basis is the identification of altered genes or chromosomes that occur only in people exhibiting the disorder. When comparing two individuals, one with a disease and one without, we expect to see differences in their genetic material if the disorder has a genetic component. Alterations in gene sequences are determined by DNA-sequencing techniques (see Chapter 20). Also, changes in chromosome structure and number can be detected by the microscopic examination of chromosomes (see Chapter 8).

Simple Mendelian Inheritance Patterns of Human Diseases May Be Determined via Pedigree Analysis

When a human disorder is caused by a mutation in a single gene, the pattern of inheritance is called **simple Mendelian inheritance,** and it can be deduced by analyzing human pedigrees. How is this accomplished? A geneticist must obtain data from many large pedigrees containing several individuals who exhibit the disorder and then follow its pattern of inheritance from generation to generation. To appreciate the basic features of pedigree analysis, we will examine a few pedigrees that involve diseases inherited in different ways. You may wish to look back at the discussion in Chapter 3 (see Figure 3.15) on the organization and symbols of pedigrees.

Autosomal Recessive Inheritance. The pedigree shown in **Figure 24.2** focuses on a genetic disorder called Tay-Sachs disease (TSD), first described by Warren Tay and Bernard Sachs in the 1880s. Affected individuals appear healthy at birth but then develop neurodegenerative symptoms at 4–6 months of age. The primary characteristics are cerebral degeneration, blindness, and loss of motor function. Individuals with TSD typically die in the third or fourth year of life. This disease is particularly prevalent in Ashkenazi (eastern European) Jewish populations, in which it has

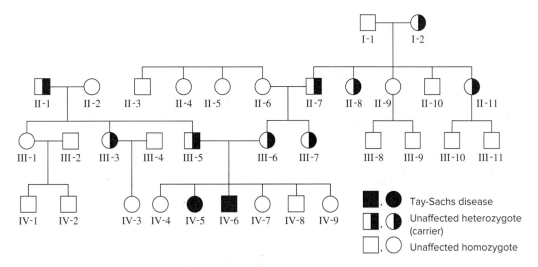

Tay-Sachs disease

Unaffected heterozygote (carrier)

Unaffected homozygote

INTERACTIVE EXERCISE

FIGURE 24.2 **A family pedigree for Tay-Sachs disease, indicating autosomal recessive inheritance.** Heterozygous carriers were determined by genetic testing.

CONCEPT CHECK: What feature(s) of this pedigree indicate(s) autosomal recessive inheritance?

TABLE 24.1

Examples of Human Disorders Inherited in an Autosomal Recessive Manner

Disorder	Chromosomal Location of Gene	Gene Product	Effect of Disease-Causing Allele
Albinism (type I)	11q	Tyrosinase	Inability to synthesize melanin, resulting in light-colored skin, hair, etc.
Cystic fibrosis (CF)	7q	Cystic fibrosis transmembrane conductance regulator	Water imbalance in tissues of the pancreas, intestine, sweat glands, and lungs due to impaired ion transport; leads to lung damage
Phenylketonuria (PKU)	12q	Phenylalanine hydroxylase	Foul-smelling urine, neurological abnormalities, mental impairment; may be remedied by diet modification starting at birth
Sickle cell disease	11p	β Globin	Anemia, blockages in blood circulation
Tay-Sachs disease (TSD)	15q	Hexosaminidase A (HexA)	Progressive neurodegeneration

a frequency of about 1 in 3600 births, which is over 100 times more frequent than in most other human populations.

At the molecular level, the mutation that causes TSD is in a gene that codes the enzyme hexosaminidase A (HexA). HexA is responsible for the breakdown of a category of lipids called G_{M2}-gangliosides, which are prevalent in the cells of the central nervous system. A defect in the ability to break down these lipids leads to their excessive accumulation in neurons and eventually causes the neurodegenerative symptoms characteristic of TSD. This defect in lipid breakdown was recognized long before the *hexA* gene was identified as being defective in these patients.

As illustrated in Figure 24.2, Tay-Sachs disease is inherited in an autosomal recessive manner. The term **autosomal** means that the gene associated with the disease is on an autosome, not on a sex chromosome. As discussed in Chapter 3, the term **recessive** refers to traits that are expressed in a homozygote but are masked in a heterozygote carrying a dominant allele. Four common features of autosomal recessive inheritance are as follows:

1. *Frequently, an affected offspring has two unaffected parents.* For rare recessive traits, the parents are usually unaffected, meaning they do not exhibit the disease. For deleterious alleles that cause early death or infertility, the two parents must be unaffected. This is always the case for TSD.
2. *When two unaffected heterozygotes have children, the percentage of affected children is (on average) 25%.*
3. *Two affected individuals have 100% affected children.* This observation can be made only when a recessive trait produces fertile, viable individuals. In the case of TSD, the affected individual dies in early childhood, and so it is not possible to observe crosses between two affected people.
4. *The trait occurs with the same frequency in both sexes.*

Autosomal recessive inheritance is a common mode of transmission for genetic disorders, particularly those that involve defective enzymes. Human recessive alleles are often caused by mutations that result in a loss of function in the coded enzyme. In the case of Tay-Sachs disease, a heterozygous carrier has approximately 50% of the functional enzyme, which is sufficient for a normal phenotype. However, for other genetic diseases, the level of functional protein may vary due to the effects of gene

regulation. Thousands of human genetic diseases are inherited in a recessive manner, and in many cases, the mutant genes have been identified. A few of these diseases are described in **Table 24.1**.

Autosomal Dominant Inheritance. Now let's examine a human pedigree involving an autosomal dominant disease in which a single copy of the dominant allele causes disease symptoms; most affected individuals are heterozygotes (**Figure 24.3**). In this example, the affected individuals have a disorder called Huntington disease. The major symptoms of this disease, which usually develops during middle age, are due to the degeneration of certain types of neurons in the brain, leading to personality changes, dementia, and early death.

In 1993, the gene involved in Huntington disease was identified and sequenced. It codes a protein called huntingtin that is expressed in neurons but is also found in some cells not affected in Huntington disease. In persons with this disorder, a mutation called a trinucleotide repeat expansion (TNRE) produces a polyglutamine tract—many glutamines in a row—within the huntingtin protein (refer back to Table 19.5). This causes an aggregation

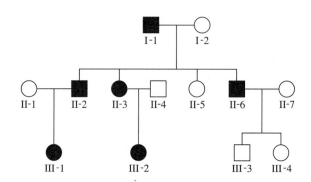

FIGURE 24.3 A family pedigree for Huntington disease, indicating autosomal dominant inheritance. Because the dominant allele is rare, most affected individuals are heterozygotes. Rare cases of people who are homozygous for the disease-causing allele have been reported. Such individuals tend to have more severe symptoms.

CONCEPT CHECK: What feature(s) of this pedigree indicate(s) autosomal dominant inheritance?

of the protein in neurons. However, additional research is needed to understand the molecular relationship between the abnormality in the huntingtin protein and the disease symptoms.

Five common features of autosomal dominant inheritance are as follows:

1. *An affected offspring usually has one or two affected parents.* However, this is not always the case. Some dominant traits show incomplete penetrance (see Chapter 4), so a heterozygote may not exhibit the trait even though it may be passed to offspring who do exhibit the trait. Also, a mutation that produces the dominant allele may occur during gametogenesis, so two unaffected parents may produce an affected offspring.
2. *An affected individual with only one affected parent is expected to produce 50% affected offspring (on average).*
3. *Two affected, heterozygous individuals have (on average) 25% unaffected offspring.*
4. *The trait occurs with the same frequency in both sexes.*
5. *For most dominant, disease-causing alleles, the homozygote is more severely affected with the disorder than the heterozygote. In some cases, a dominant allele may be lethal in the homozygous condition.*

Numerous autosomal dominant diseases have been identified in humans (**Table 24.2**). The three common explanations for dominant disorders are haploinsufficiency, a gain-of-function mutation, or a dominant-negative mutation. Let's consider examples of all three types.

- **Haploinsufficiency** is the phenomenon in which a person has only a single functional copy of a gene, and that single functional copy does not produce a normal phenotype. In these disorders, 50% of the functional protein is not sufficient to produce a normal phenotype. Haploinsufficiency shows a dominant pattern of inheritance because a heterozygote (with one functional allele and one inactive allele) has the disease. An example is aniridia, which is a rare disorder that results in an absence of the iris of the eye. Aniridia leads to visual impairment and blindness in severe cases.
- **Gain-of-function mutations** change the gene product so that it gains a new or abnormal function. An example of such a disorder is achondroplasia, which is characterized by abnormal bone growth that results in short stature with relatively short arms and legs. This disorder is caused by a mutation that occurs in the gene that codes fibroblast growth factor receptor-3. In achondroplasia, the mutant form of the receptor is overactive. This overactivity disrupts a signaling pathway and leads to severely shortened bones.
- **Dominant-negative mutations** alter the gene product in a way that acts antagonistically to the normal gene product. In humans, Marfan syndrome, which is due to a mutation in the *fibrillin-1* gene, is an example. The *fibrillin-1* gene codes a glycoprotein that is a structural component of the extracellular matrix that provides structure and elasticity to tissues. The mutant gene codes a glycoprotein that opposes the effects of the normal protein, thereby weakening the elasticity of certain body parts. For example, the walls of the major arteries such as the aorta, the large artery that leaves the heart, are often affected.

X-Linked Recessive Inheritance. Let's now turn to another inheritance pattern common in humans, called X-linked recessive inheritance. In this pattern, the mutant gene causing a human disorder is located only on the X chromosome (**Table 24.3**). X-linked recessive inheritance of diseases poses a special problem for males. Why are males more likely to be affected? X-linked genes lack a counterpart on the Y chromosome. Males are hemizygous—have a single copy—for these genes. Therefore, a female who is heterozygous for an X-linked recessive allele will not have the disorder but will pass this trait to 50% of male offspring, as shown in the following Punnett square for hemophilia. In this case,

TABLE 24.2

Examples of Human Disorders Inherited in an Autosomal Dominant Manner

Disorder	Chromosomal Location of Gene	Gene Product	Effects of Disease-Causing Allele
Aniridia	11p	Pax6 transcription factor	An absence of the iris of the eye, leading to visual impairment and sometimes blindness
Achondroplasia	4p	Fibroblast growth factor receptor-3	A disorder in which an overactive receptor causes a defect in the growth of long bones
Marfan syndrome	15q	Fibrillin-1	Tall and thin individuals with abnormalities in the skeletal, ocular, and cardiovascular systems due to a weakening in the elasticity of certain body tissues
Familial hypercholesterolemia	19p	LDL receptor	Very high serum levels of low-density lipoprotein (LDL), a predisposing factor in heart disease
Huntington disease	4p	Huntingtin	Neurodegeneration that occurs relatively late in life, usually in middle age

TABLE 24.3

Examples of Human Disorders Inherited in an X-Linked Recessive Manner

Disorder	Gene Product	Effects of Disease-Causing Allele
Duchenne muscular dystrophy	Dystrophin	Progressive degeneration of muscles that begins in early childhood
Hemophilia A	Clotting factor VIII	Defect in blood clotting
Hemophilia B	Clotting factor IX	Defect in blood clotting
Androgen insensitivity syndrome	Androgen receptor	Missing male steroid hormone receptor; XY individuals have external features that are female but internally have undescended testes and no uterus

X^H is an X chromosome that carries the wild-type allele, whereas X^{h-A} carries a recessive mutant allele causing hemophilia.

As mentioned previously, hemophilia is a disorder in which the blood cannot clot properly after an injury. For individuals with this trait, a minor cut may bleed for a very long time, and small injuries can lead to serious internal bleeding because internal broken capillaries may leak blood profusely before they are repaired. Hemophilia A, also called classic hemophilia, is caused by a loss-of-function mutation in an X-linked gene that codes a protein called clotting factor VIII. This disease has also been called the "royal disease," because it affected many members of European royal families. The pedigree shown in **Figure 24.4** illustrates the prevalence of hemophilia A among the descendants of Queen Victoria of England.

X-linked recessive inheritance is revealed by the following observations:

1. *Males are much more likely to exhibit the trait.*
2. *A female parent of an affected male offspring often has at least one male sibling or a male parent who is also affected.*
3. *Female offspring of affected males produce, on average, 50% affected male offspring.*
4. *Females give an X chromosome to both female and male offspring, whereas males give an X chromosome only to their female offspring. So, a female can transmit a recessive X-linked allele to male offspring (who would be affected) and to female offspring (who would be carriers if they are heterozygous).*

X-Linked Dominant Inheritance.   Relatively few genetic disorders in humans follow an X-linked dominant inheritance pattern. In most cases, males are more severely affected than females, probably because one of the X chromosomes carried by females usually has a normal copy of the gene in question. With most of the X-linked dominant disorders listed in **Table 24.4**, male embryos die at an early stage of development, so most individuals exhibiting the disorder are females. Also, due to their dominant nature and severity, persons with some of the disorders listed in Table 24.4 do not reproduce. Therefore, these dominant disorders, which include Rett syndrome and Aicardi syndrome, are not passed from parent to offspring. Instead, they are caused by new mutations that occur during gamete formation or early embryogenesis. For those X-linked dominant disorders in which the offspring can reproduce, the following pattern is often observed:

1. *Only females exhibit the trait when it is lethal to males.*
2. *Affected females have a 50% chance of passing the trait to female offspring.* Note: Affected females also have a 50% chance of passing the trait to male offspring, but for many of these disorders, affected males are not observed because of lethality.

Many Genetic Disorders Exhibit Locus Heterogeneity

Hemophilia, which we considered earlier in this section, illustrates another concept in genetics called **locus heterogeneity.**

TABLE 24.4

Examples of Human Disorders Inherited in an X-Linked Dominant Manner

Disorder	Gene Product	Effects of Disease-Causing Allele
Vitamin D–resistant rickets	Metallopeptidase	Defects in bone mineralization at the sites of bone growth or remodeling, leading to bone deformity and stunted growth in children
Rett syndrome	Methyl-CpG-binding protein-2	A neurodevelopmental disorder that includes a deceleration of head growth and small hands and feet; fatal in males
Aicardi syndrome	Unknown	Characterized by the partial or complete absence of a key structure in the brain called the corpus callosum, and the presence of retinal abnormalities; fatal in males
Incontinentia pigmenti	NF-κB essential modulator	Characterized by morphological and pigmentation abnormalities in the skin, hair, teeth, and nails; fatal in males

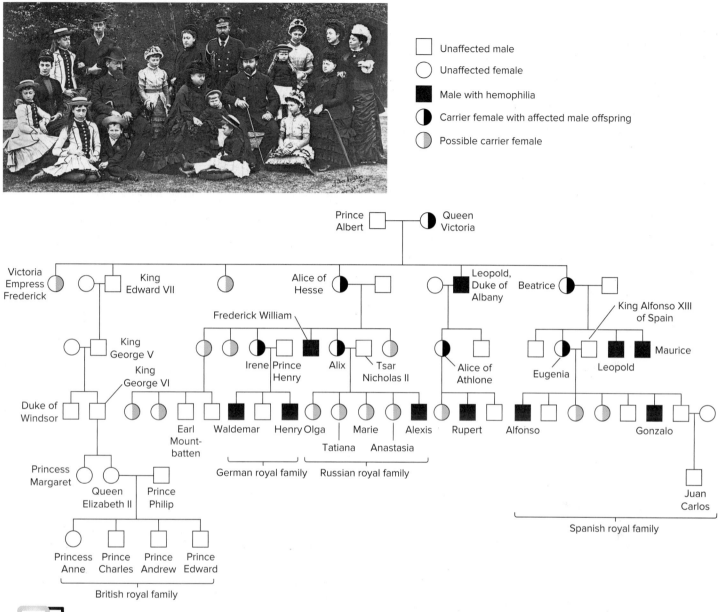

FIGURE 24.4 **A family tree for hemophilia A in the royal families of Europe, indicating an X-linked recessive inheritance.** Pictured are Queen Victoria and Prince Albert of Great Britain with some of their descendants.
Hulton-Deutsch Collection/Corbis/Getty Images

INTERACTIVE EXERCISE

CONCEPT CHECK: What feature(s) of this pedigree indicate(s) X-linked recessive inheritance?

This term refers to the phenomenon in which a particular type of disorder or trait may be caused by mutations in two or more different genes. For example, blood clotting involves the participation of several different proteins that take part in a cellular cascade that leads to the formation of a clot. Hemophilia is usually caused by a defect in one of three different clotting factors. In hemophilia A, a protein called factor VIII is missing. Hemophilia B is a deficiency in a different clotting factor, called factor IX. Factors VIII and IX are coded by different genes on the X chromosome. These two types of hemophilia show an X-linked recessive pattern of inheritance. By comparison, hemophilia C is due to a factor XI deficiency. The gene coding factor XI is found on chromosome 4, and this form of hemophilia follows an autosomal recessive pattern of inheritance.

Unfortunately, locus heterogeneity may greatly confound pedigree analysis. For example, a human pedigree might contain individuals with X-linked hemophilia and other individuals with hemophilia C. A geneticist who assumed all affected individuals had defects in the same gene would be unable to explain the resulting pattern of inheritance. For disorders such as hemophilia, pedigree analysis is not a major problem because the biochemical basis for this disease is well understood. However, for rare diseases that are poorly understood at the molecular level, locus heterogeneity may obscure the pattern of inheritance.

24.1 COMPREHENSION QUESTIONS

1. Which of the following would *not* be consistent with the idea that a disorder has a genetic component?
   a. The disorder is more likely to occur among an affected person's relatives than in the general population.
   b. The disorder can spread to individuals sharing similar environments.
   c. The disorder tends to develop at a characteristic age.
   d. A correlation is observed between the disorder and a mutant gene.

2. Assuming complete penetrance, which type of inheritance pattern is consistent with the pedigree shown here? Affected individuals are shown with filled (black) symbols.

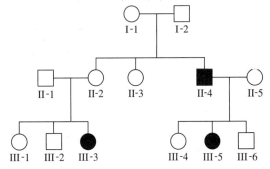

   a. Autosomal recessive
   b. Autosomal dominant
   c. X-linked recessive
   d. X-linked dominant

3. Which of the following is *not* a common explanation for a dominant disorder?
   a. Haploinsufficiency
   b. A change in chromosome number
   c. A gain-of-function mutation
   d. A dominant-negative mutation

4. Locus heterogeneity means that a genetic disorder
   a. has a heterogeneous phenotype.
   b. is caused by mutations in two or more different genes.
   c. involves a structural change in multiple chromosomes.
   d. is inherited from both parents.

24.2 DETECTION OF DISEASE-CAUSING ALLELES VIA HAPLOTYPES AND VIA GENOME-WIDE ASSOCIATION STUDIES

Learning Outcomes:

1. Define *haplotype*.
2. Explain how haplotypes are analyzed to identify disease-causing alleles in humans.
3. Describe the purpose of genome-wide association studies.

Because mutant genes are known to play a role in thousands of diseases, researchers have devoted great effort to identifying alleles associated with genetic diseases. In this section, we will explore various approaches used to identify mutant alleles that cause disease.

A Haplotype Is a Series of Linked Genetic Variations on a Chromosome

To identify disease-causing alleles, researchers often rely on the known chromosomal locations of genes and molecular markers that have been characterized in human populations. A disease-causing allele may be identified due to its proximity to another known gene or its proximity to known molecular markers.

As discussed in Chapter 22, researchers can characterize chromosomes at the molecular level and determine the precise locations of genes and molecular markers on each chromosome. During the course of evolution, new mutations arise that alter the DNA sequences of genes and molecular markers. For this reason, homologous chromosomes exhibit gene differences (i.e., allelic variation) and show variation in their molecular markers.

As an example, **Figure 24.5** compares a pair of homologous chromosomes from two different individuals and focuses on four sites (called 1, 2, 3, and 4) that occur at particular locations on those chromosomes. These sites could be within particular genes, or they could be molecular markers used in mapping studies. In this drawing, each site is also given a letter designation (A, B, or C) depending on the variation in the DNA sequence at the site.

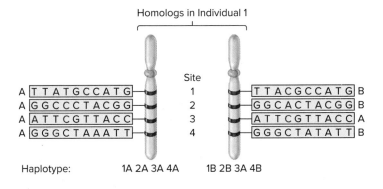

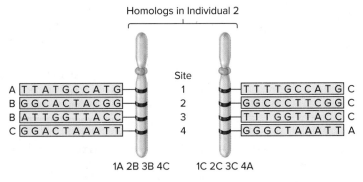

FIGURE 24.5 A schematic representation of haplotypes on human chromosomes. This example considers four sites on pairs of homologous chromosomes that may exist in different versions, designated A, B, or C. Bases that differ are shown in red. (Note: At each site, only one of the two DNA strands is shown.)

CONCEPT CHECK: What is a haplotype?

In individual 1, sites 1, 2, and 4 differ at one base pair between the homologs. In individual 2, all four sites differ at one or two base pairs. Also note that individuals 1 and 2 differ with regard to some of these sites.

The term **haplotype,** which is a contraction of <u>hapl</u>oid gen<u>otype,</u> refers to the linkage of particular alleles and/or molecular markers on a single chromosome. In Figure 24.5, the haplotypes for the four sites are shown at the bottom of each chromosome. For example, let's consider the left chromosome in individual 2. The designation 1A 2B 3B 4C means the allele or molecular marker at site 1 is represented by A; the allele or molecular marker at site 2 and 3 are represented by B; and the allele or molecular marker at site 4 is represented by C. Therefore, the haplotype of this homolog in individual 2 is 1A 2B 3B 4C.

Because mutations are rare events, haplotypes do not dramatically change from one generation to the next due to new mutations. By comparison, haplotypes are more likely to change over the course of a few generations due to crossing over. The likelihood of changing a haplotype via crossing over depends on the distance between the alleles or molecular markers. If two sites are far apart, a crossover is more likely to alter their pattern than if they are close together. If sites 1, 2, 3, and 4 were very close together on this chromosome, the haplotypes shown in this figure would be likely to stay the same after a few generations. For example, a great-great-great grandchild of individual 2 might inherit the haplotype 1A 2B 3B 4C or 1C 2C 3C 4A. In contrast, the inheritance of either haplotype would be much less likely if the sites were far apart and could frequently recombine by crossing over.

Haplotype Association Studies Are Conducted to Identify Disease-Causing Alleles

How do geneticists identify genes that cause disease when they are mutant? Although a variety of approaches may be followed, the hunt often begins with family pedigrees. The goal is to localize a disease-causing allele to a small region on a chromosome that is distinguished by its haplotype. This approach is based on two assumptions:

1. The disease-causing allele had its origin in a single individual known as a **founder,** who lived many generations ago. Since that time, the allele has spread throughout portions of the human population.
2. When the disease-causing allele originated in the founder, it occurred in a region of a chromosome with a particular haplotype. The haplotype is not likely to have changed over the course of several generations if the disease-causing allele and markers in this region are very close together.

By comparing the transmission patterns of many molecular markers with the occurrence of an inherited disease, researchers can pinpoint particular markers that are closely linked to the disease-causing mutant allele.

To clarify this concept, **Figure 24.6** shows a situation in which an individual—a founder—has incurred a new mutation that results in a disease-causing allele. The molecular markers designated 1A, 2C, 3B, and 4B are close to the location of this mutant allele. In succeeding generations, the disease-causing allele would be more likely to be present in individuals with

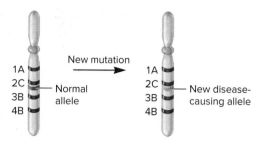

FIGURE 24.6   The occurrence of a new mutation in a founder. In this example, the new mutation occurred in a region with a specific haplotype: 1A 2C 3B 4B. Due to the close linkage between the 2C marker and the mutant allele, the same haplotype is likely to be found in individuals of succeeding generations who have inherited the mutant allele.

haplotype 1A 2C 3B 4B than in those without this haplotype. Furthermore, people who have inherited the disease-causing allele would be particularly likely to inherit the 2C marker because it is the closest to the disease-causing allele. Due to their close proximity, it would be unlikely that a crossover would separate the 2C marker and the disease-causing allele from each other and create a different haplotype.

Researchers can analyze the occurrence of specific molecular markers in many people who carry a particular disease-causing allele. When an allele and one or more molecular markers are associated with each other at a frequency that is significantly higher than expected by random chance, they are said to exhibit **linkage disequilibrium.** The phenomenon of linkage disequilibrium is common when a disease-causing allele arose in a founder and the allele is closely linked to specific markers on the same chromosome.

GENETIC TIPS   THE QUESTION: Figure 24.6 shows the location of a disease-causing allele that arose in a founder. The haplotypes of five grandchildren of this founder are as follows:

Grandchild 1: 1A 2B 3C 4B/1A 2B 3A 4C

Grandchild 2: 1A 2C 3B 4B/1B 2A 3B 4A

Grandchild 3: 1B 2A 3A 4B/1C 2B 3A 4A

Grandchild 4: 1B 2C 3B 4B/1A 2B 3A 4C

Grandchild 5: 1A 2B 3A 4C/1B 2A 3B 4A

What are the relative likelihoods that these grandchildren have inherited the disease-causing allele, based on their haplotypes?

T OPIC: *What topic in genetics does this question address?* The topic is the use of haplotypes to predict the likelihood that an individual may be carrying a disease-causing allele.

I NFORMATION: *What information do you know based on the question and your understanding of the topic?* From the question, you know the location of a disease-causing allele in a founder and the haplotypes of five grandchildren of that individual. From your understanding of the topic, you may remember that markers in the founder that were closely linked to the disease-causing allele are likely to be found in individuals who have inherited the disease-causing allele.

Haplotype Association Studies Identified the Mutant Gene That Causes Huntington Disease

As an example of haplotype association studies, let's consider Huntington disease. Nancy Wexler has studied Huntington disease among a population of related individuals in Venezuela. Over 15,000 individuals have been analyzed. **Figure 24.7** shows a simplified version of a pedigree for a large family from this Venezuelan population in which many individuals were affected with Huntington disease. A specific molecular marker, called G8, is found in four different versions, named A, B, C, or D. In this Venezuelan population, pedigree analysis revealed that the G8-C

marker, which is located near the tip of the short arm of chromosome 4, is almost always associated with the mutant gene causing Huntington disease. In other words, the G8-C marker is closely linked to the allele that causes Huntington disease. The gene was named the *HTT* gene and the coded protein is called huntingtin, which is expressed in neurons.

Once a disease-causing mutant gene has been localized to a short chromosomal region, the next step is to determine which gene in the region is responsible for the disease. Modern haplotype association studies typically localize a gene to a chromosome region that is about 1 Mb (1 million base pairs) in length. One way to identify a disease-causing allele is chromosome walking (refer back to Figure 22.7). That is how the *HTT* gene was identified.

However, the technique of chromosome walking is now less widely used because the entire human genome has been sequenced and most genes have been identified. Therefore, researchers can analyze the 1-Mb region to which a gene has been mapped to determine if a mutant gene in this region is responsible for a disease. In the human genome, a 1-Mb region usually contains about 5–10 different genes, though the number can vary greatly. Therefore, mapping does not definitively tell researchers which gene may play a role in human disease, but it usually narrows down the list to a few candidate genes.

How does a researcher determine which of the candidate genes is the correct one? To further reduce the list, researchers may also consider biological function. As the scientific community explores the functions of genes experimentally, the data are published in the research literature and placed into databases. In some cases,

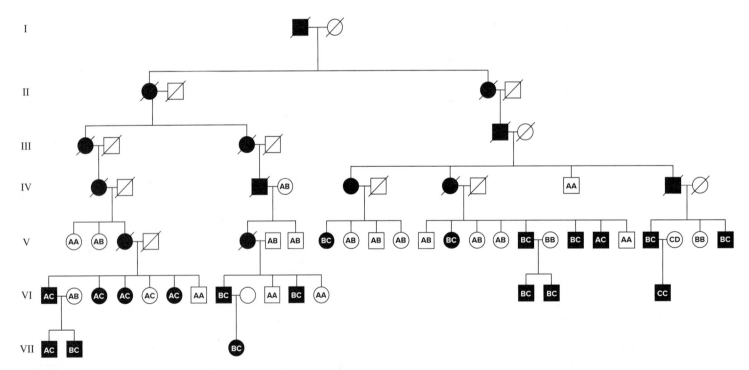

FIGURE 24.7 **The transmission pattern of a molecular marker for Huntington disease.** Affected individuals are shown with black symbols. The letters within each symbol indicate the forms of the G8 marker (A, B, C, or D) the individual carries. Affected individuals always carry the C version of this marker. In rare cases, an unaffected individual may also carry the G8-C marker. Symbols with slashes indicate deceased individuals.

Source: Gusella, James F., Wexler, Nancy S., and Conneally, Michael P., "A Polymorphic DNA Marker Genetically Linked to Huntington's Disease," *Nature*, vol. 306, November 17, 1983, 234–238.

CONCEPT CHECK: Explain the connection between the founder in whom the allele causing Huntington disease first appeared and the G8-C marker in this pedigree.

this information may help to narrow down the list of candidate genes. For example, if the disease of interest is neurological, researchers may discover that only certain genes in the mapped region are expressed in neurons. Also, researchers may compare data from other organisms. If a mutant gene in a mouse causes neurological problems and a human homolog of the mouse gene is found in the mapped region, this human gene would be a good candidate for being responsible for the human disease symptoms.

After researchers have narrowed down the number of candidate genes to as short a list as possible, the next phase is to sequence the candidate gene(s) from many affected and unaffected individuals, using the DNA-sequencing methods described in Chapter 22. The goal is to identify a gene in which affected individuals always carry a mutation. Identifying a gene that has a genetic change in affected individuals is strong evidence that the candidate gene causes the disease symptoms.

Why is it useful to identify disease-causing alleles? In many cases, the identification of these alleles helps us to understand how genes contribute to pathogenesis and may even aid in developing strategies aimed at the treatment of a disease. As described in Section 24.3, the identification of such genes may also result in the development of genetic tests that can enable people to determine if they or their unborn offspring are carrying disease-causing alleles.

The International HapMap Project Is a Worldwide Effort to Identify Haplotypes in Human Populations

The genetic sequences of different people are usually very similar. When the chromosomes of two different humans are compared, their DNA sequences differ at about one in every 1200 bases. As illustrated earlier in Figure 24.5, differences in individual bases are by far the most common type of genetic variation. Such differences, which are known as single-nucleotide polymorphisms (SNPs), were discussed in Chapter 22. The **International HapMap Project** is a worldwide effort to identify SNPs and other types of human genetic variations. Researchers estimate that approximately 10 million SNPs are commonly found in the human genome.

This project is producing **HapMap**—an extensive catalog of common genetic variants that occur in human beings. It describes what these variants are, where they are located in the human genome, and how they are distributed among human populations throughout the world. The International HapMap Project is not using the information in HapMap to understand connections between particular genetic variants and diseases. Instead, the goal of the project is to provide HapMap so that other researchers can find links between genetic variants and the risk of developing specific diseases. Such links are expected to lead to new methods of diagnosing and treating illnesses.

Genome-Wide Association Studies Can Identify Disease-Causing Alleles in Human Populations

As we have seen, haplotypes can be analyzed to identify disease-causing alleles via linkage to molecular markers. Instead of depending on linkage, a newer approach is to directly identify genetic variation that causes or contributes to human diseases by analyzing human genome sequences. With the advent of next-generation sequencing technologies described in Chapter 22, the DNA sequencing of a person's genome is relatively inexpensive (less than $1000) and can be completed in a day or two. Therefore, researchers now have access to thousands of genome sequences and can compare them to each other using computer software. These comparisons can directly identify disease-causing alleles on a broad scale, because genetic variation can be analyzed at the whole genome level. This type of work is done within the context of preserving the privacy of the people whose genomes are being analyzed.

A **genome-wide association study (GWAS)** is an examination of a genome-wide set of genetic variants among many different individuals to see if any variant is associated with a disease or other type of trait. In other words, a GWAS compares the genome sequences from many different people; some of the people have one or more diseases with an underlying genetic basis. A GWAS usually tries to find an association between one or more single-nucleotide polymorphisms (SNPs) and a disease or other trait. (SNPs are described in Chapter 22.) Many genome-wide association studies have focused on finding associations between SNPs and human diseases, but they can also be directed at other types of genetic variation (e.g., deletions or duplications) and at other species.

How is a GWAS carried out? One common approach is to compare the SNPs of two large groups of individuals—a control group that does not have a disease and another group in which all of the individuals are affected with the same disease, such as type 1 diabetes. Although the number of SNPs that are analyzed varies from study to study, researchers commonly compare a million or more. The goal is to identify SNPs for which the frequency is significantly different between the control group and the group of affected individuals.

To identify such SNPs, researchers determine the odds ratio, which is the odds of having a disease for individuals carrying a specific SNP versus the odds of having the disease for individuals who do not carry that same SNP. A P value for the significance of the odds ratio is typically calculated using a test that is similar to a chi square test, which is described in Chapter 3. In a standard chi square test, when $P < 0.05$, the results are considered significant. However, in a GWAS, researchers often analyze millions of SNPs, so using such a criterion for the P value would result in many false positives. It is common in GWA studies to set the level of significance at 0.05/(1,000,000 SNPs). In other words, a value of $P < 5 \times 10^{-8}$ is viewed as statistically significant.

After odds ratios and P values have been calculated for all SNPs, the data are used to produce a Manhattan plot. This name was chosen because the plot looks somewhat like the skyline of Manhattan in New York City. The data are plotted as the negative logarithm to the base 10 of the P value for a given SNP as a function of its location in the genome. **Figure 24.8** shows a Manhattan plot for a GWAS that looked for an association between SNPs and type 1 diabetes. In this plot, the SNPs with significant associations are indicated by stacks of green dots. These results indicate that SNPs on certain chromosomes are associated with type 1 diabetes.

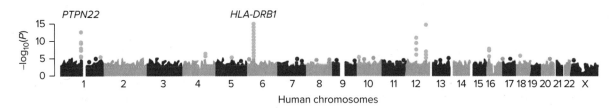

FIGURE 24.8 **Data from a GWAS of type 1 diabetes in humans.** The *y*-axis indicates the *P* values of SNPs, and the *x*-axis shows their locations on the human chromosomes. Those SNPs that are not statistically significant are shown in dark and light blue, whereas those that are statistically significant are green. Many statistically significant SNPs are found in the *PTPN22* and *HLA-DRB1* genes. *PTPN22* codes a protein called tyrosine phosphatase that affects the responsiveness of T-cell and B-cell receptors; mutations in this gene are associated with increases or decreases in risks of autoimmune diseases. The *HLA-DRB1* gene is part of a family of genes that code proteins that are present on the surface of certain immune system cells.

Source: The Wellcome Trust Case Control Consortium, "Genome-Wide Association Study of 14,000 Cases of Seven Common Diseases and 3,000 Shared Controls," *Nature*, vol. 447, no. 7145, June 7, 2007, 661–678.

How do we interpret the results of a GWAS? As discussed in Chapter 25 (see Section 25.4), an association does not necessarily imply a cause-and-effect relationship. Therefore, further work is usually needed to understand why a SNP is associated with a human disease. In some cases, a SNP may not be involved in a cause-and-effect relationship but is closely linked to a particular disease-causing allele due to the founder effect, described earlier in this section (see Figure 24.6). In other cases where a cause-and-effect relationship does exist, a SNP may have a variety of effects, including a change in any of the following:

- The coding sequence of a gene
- The promoter
- A regulatory sequence, such as an enhancer

Such changes can either alter gene expression or change the function of the coded protein and thereby contribute to disease symptoms. Thus far, genome-wide association studies have identified over 4000 SNPs that are associated with common human diseases or traits. In addition to type 1 diabetes, featured in Figure 24.8, these include type 2 diabetes, Alzheimer disease, Parkinson disease, and Crohn disease.

24.2 COMPREHENSION QUESTIONS

1. What is a haplotype?
   a. A species with one set of chromosomes
   b. A cell with one set of chromosomes
   c. The linkage of alleles or molecular markers on a chromosome
   d. All of the above describe a haplotype.

2. In order to understand human diseases that have a genetic basis, genome-wide association studies are usually aimed at the identification of a particular _____ that is found only in _____ individuals.
   a. deletion, affected
   b. deletion, unaffected
   c. SNP, affected
   d. SNP, unaffected

24.3 GENETIC TESTING AND SCREENING

Learning Outcomes:

1. Compare and contrast genetic testing and genetic screening.
2. List different methods used to test for genetic abnormalities.
3. Describe how genetic testing can be conducted before birth.

Because genetic abnormalities occur in the human population at a significant level, researchers have sought ways to determine whether individuals carry disease-causing alleles or other types of genetic abnormalities. The term **genetic testing** refers to the use of testing methods to determine if an individual carries a genetic abnormality. By comparison, the term **genetic screening** refers to population-wide genetic testing. In this section, we will examine both approaches.

Genetic Testing Is Used to Identify Many Inherited Human Diseases

Table 24.5 describes several different genetic testing methods. In many cases, single-gene mutations that affect the function of proteins can be examined at the protein level. If a gene codes an enzyme, biochemical assays to measure that enzyme's activity may be available. As mentioned earlier, Tay-Sachs disease is due to a defect in the enzyme hexosaminidase A (HexA). Enzymatic assays for this enzyme involve the use of an artificial substrate in which 4-methylumbelliferone (MU) is covalently linked to *N*-acetylglucosamine (GlcNAc). HexA cleaves this covalent bond and releases MU, which is fluorescent:

$$\text{MU–GlcNAc} \xrightarrow{\text{HexA}} \text{MU} + \text{GlcNAc}$$
$$\text{(nonfluorescent)} \qquad\qquad \text{(fluorescent)}$$

To perform the assay for the HexA enzyme, a small sample of cells is collected and incubated with MU–GlcNAc, and the fluorescence is measured. Samples from individuals affected with Tay-Sachs, who do not produce HexA, show little or no fluorescence, whereas samples from individuals who are homozygous for

TABLE 24.5

Testing Methods for Genetic Abnormalities

Method	Description
Protein Level	
Biochemical	As mentioned earlier concerning Tay-Sachs disease, the enzymatic activity of a protein can be assayed in vitro.
Immunological	The presence of a protein can be detected using antibodies that specifically recognize that protein. Western blotting is an example of a technique used for this type of testing (see Chapter 20).
DNA or Chromosomal Level	
DNA sequencing	If a gene associated with a disease has already been identified and sequenced, that gene can be amplified from a sample of cells using PCR and then subjected to DNA sequencing, as described in Chapter 20.
In situ hybridization	A DNA probe that hybridizes to a particular gene or gene segment can be used to determine if the gene is present, absent, or altered in an individual. The technique of fluorescence in situ hybridization (FISH) is described in Chapter 22 (see Figure 22.2).
Karyotyping	The chromosomes from a sample of cells can be stained and then analyzed microscopically for abnormalities in chromosome structure and number (see Figure 2.2).

the functional *hexA* allele exhibit a high level of fluorescence. Samples from heterozygotes, who have 50% HexA activity, produce intermediate levels of fluorescence.

An alternative and more common approach is to detect single-gene mutations at the DNA level. To apply this testing strategy, researchers must have previously identified the mutant gene using molecular techniques. The identification of many human genes, such as those involved in Duchenne muscular dystrophy, cystic fibrosis, and Huntington disease, has made it possible to test for affected individuals or those who may be heterozygous for the disease-causing allele (see Table 24.5).

Many human genetic abnormalities involve changes in chromosome number and/or structure. Changes in chromosome number are a common type of human genetic abnormality. Most of these changes result in spontaneous abortions. However, approximately 1 in 200 live births are aneuploid—have an abnormal number of chromosomes (see Table 8.1). About 5% of infant and childhood deaths are related to such genetic abnormalities. Changes in chromosome number and many changes in chromosome structure can be detected by karyotyping the chromosomes using a light microscope.

Genetic Screening Identifies Genetic Abnormalities at the Population Level

In the United States, genetic screening for certain disorders has become common medical practice. For example, pregnant females older than 35 often have tests conducted to see if their fetuses are carrying

chromosomal abnormalities. As discussed in Chapter 8, these tests are indicated because the rate of such defects increases with the age of the female parent. Another example is the widespread screening for phenylketonuria (PKU). An inexpensive test can determine if newborns have this disease. Those who test positive can then be given a low-phenylalanine diet to avoid PKU's devastating effects.

Genetic screening has also been conducted on specific populations in which a genetic disease is prevalent. For example, in 1971, community-based screening for heterozygous carriers of Tay-Sachs disease was begun among specific Ashkenazi Jewish populations. With the use of this screening, over the course of one generation, the incidence of births of children with Tay-Sachs disease in these populations was reduced by 90%. For most rare genetic abnormalities, however, genetic screening is not routine practice. Rather, genetic testing is performed only when a family history reveals a strong likelihood that a couple may produce an affected child. Typically, such a couple already has an affected child or has other relatives with a genetic disease.

Genetic testing and genetic screening are medical practices with many social and ethical dimensions. People must decide whether or not they want to make use of available tests, particularly when the disease in question has no cure. For example, Huntington disease typically does not affect people until their 40s and can last 20 years. People who learn that they are carriers of a genetic disease such as Huntington disease can be devastated by the news. Some argue that people have a right to know about their genetic makeup; others assert that it does more harm than good.

Another issue is privacy. Who should have access to personal genetic information, and how could it be used? Could routine genetic testing lead to discrimination by employers or medical insurance companies? In the coming years, we will become increasingly aware of our genetic makeup and the underlying causes of genetic diseases. As a society, establishing guidelines for the uses of genetic testing will be a necessary, yet very difficult, task.

Genetic Testing Can Be Performed Prior to Birth

Genetic testing can be performed during pregnancy, which may affect the decision to terminate a pregnancy. The three common ways of obtaining genetic material from an embryo or fetus for the purpose of genetic testing are noninvasive prenatal testing, amniocentesis, and chorionic villus sampling. In addition, preimplantation genetic diagnosis can be performed on embryos before they are implanted into the uterus.

Noninvasive Prenatal Testing (NIPT). **Noninvasive prenatal testing (NIPT)** is done by collecting a blood sample from a pregnant female and analyzing cell-free DNA (cfDNA). This DNA is released from worn-out cells that have been degraded. During pregnancy, the bloodstream contains a mix of cfDNA that comes from the pregnant female's own cells and also from placental cells that are shed into the female's bloodstream throughout pregnancy. The DNA in placental cells is usually identical to the DNA of the fetus. Analyzing cfDNA provides a noninvasive way to detect genetic abnormalities in the fetus. NIPT can be conducted as early as week 10 of pregnancy. It is usually used to test for abnormalities in chromosome number.

Amniocentesis.   In **amniocentesis,** amniotic fluid containing fetal cells is removed using a needle that is passed through the abdominal wall (**Figure 24.9**). Because relatively few fetal cells are found in the amniotic fluid, the cells are grown in the laboratory for 1–2 weeks and then karyotyped to determine the number of chromosomes per cell and whether changes in chromosome structure have occurred. If indicated, other genetic tests may be conducted, such as DNA sequencing to determine if particular genes carry disease-causing mutations.

Chorionic Villus Sampling.   In **chorionic villus sampling,** a small piece of the chorion (the fetal part of the placenta) is removed, and a karyotype is prepared directly from the collected cells (Figure 24.9). Chorionic villus sampling can be performed earlier during pregnancy than amniocentesis, usually around week 10 to week 12, compared to week 14 to week 20 for amniocentesis,

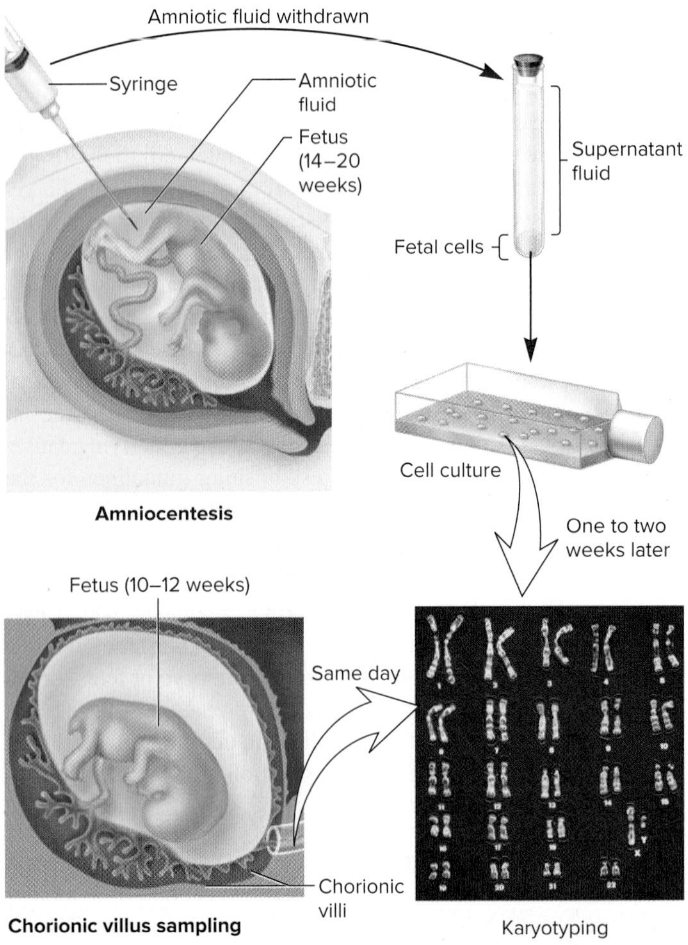

and results from chorionic villus sampling are available sooner. Weighed against these advantages, however, is that this procedure may pose a slightly greater risk of causing a miscarriage.

Preimplantation Genetic Diagnosis (PGD).   Another method of genetic testing performed prior to birth is called **preimplantation genetic diagnosis (PGD).** This approach, which is conducted before pregnancy even occurs, involves the genetic testing of embryos that have been produced by **in vitro fertilization (IVF)**—a process in which sperm and egg are mixed together in a laboratory and then fertilization occurs there. PGD can also determine if an embryo has the correct number of chromosomes (called aneuploidy screening).

PGD is done by removing one or two cells usually at about the eight-cell stage, which occurs 3 days after fertilization. This process is called embryo biopsy, or blastomere biopsy. Genetic tests described in Table 24.5 are then conducted on the removed cell(s) to either check for a particular genetic disease or determine the chromosome composition. The testing can usually be completed in a day or so. Depending on the outcome of the results, a decision can be made whether or not to transfer the embryo into the uterus of the prospective female parent in hopes of implantation and the eventual birth of a baby. In most cases, only embryos that do not carry genetic abnormalities are used. As with the genetic screening of adults, the screening of embryos and fetuses raises many ethical questions.

Amniocentesis

Fetus (10–12 weeks)

Chorionic villus sampling

Karyotyping

FIGURE 24.9 **Techniques for determining genetic abnormalities during pregnancy.** In amniocentesis, amniotic fluid is withdrawn, and fetal cells are collected by centrifugation. The cells are then grown in a laboratory culture medium for 1–2 weeks prior to karyotyping. In chorionic villus sampling, a small piece of the chorion is removed. These cells can be prepared directly for karyotyping. In addition to karyotyping, fetal samples obtained from amniocentesis or chorionic villus sampling can be subjected to other genetic tests, such as screening for gene mutations.
CNRI/Science Source

24.3 COMPREHENSION QUESTIONS

1. Which of the following is *not* a method used in genetic testing?
   a. Chromosome walking
   b. DNA sequencing
   c. In situ hybridization
   d. Karyotyping

2. Which of the following prenatal genetic testing methods is done in conjunction with in vitro fertilization?
   a. Amniocentesis
   b. Chorionic villus sampling
   c. Preimplantation genetic diagnosis
   d. All of the above are usually performed in conjunction with in vitro fertilization.

24.4 PRIONS

Learning Outcomes:
1. Define *prion*.
2. Explain how prions cause disease.

We now turn to an unusual mechanism by which agents known as prions cause disease. As shown in **Table 24.6**, prions cause several types of neurodegenerative diseases that affect humans and livestock, including mad cow disease. In the 1960s, Tikvah Alper and John Stanley Griffith discovered that preparations from animals with certain neurodegenerative diseases remained infectious

TABLE 24.6

Neurodegenerative Diseases Caused by Prions*

Disease	Description
Infectious Diseases	
Kuru	A human disease that was once common in New Guinea. It begins with a loss of coordination, usually followed by dementia.
Scrapie	A disease of sheep and pigs characterized by intense itching in which the animals tend to scrape themselves against trees, followed by neurodegeneration
Mad cow disease	Begins with changes in posture and temperament, followed by loss of coordination and neurodegeneration
Human Inherited Diseases	
Creutzfeldt-Jakob disease	Characterized by loss of coordination and dementia
Gerstmann-Straüssler-Scheinker disease	Characterized by loss of coordination and dementia
Familial fatal insomnia	Begins with sleeping and autonomic nervous system disturbances, followed by insomnia and dementia

*All of these diseases are eventually fatal.

even after exposure to radiation that would destroy any DNA or RNA. They suggested that the infectious agent was a protein. Furthermore, Alper and Griffith speculated that the protein usually preferred one folding pattern but could sometimes misfold and then catalyze other proteins to do the same.

In the early 1970s, Stanley Prusiner, moved by the death of a patient from a neurodegenerative disease, began to search for the causative agent. In 1982, Prusiner isolated a disease-causing agent composed entirely of protein, which was named a **prion.** The term emphasizes the prion's unusual character as a proteinaceous infectious agent. Before the discovery of prions, all known infectious agents such as viruses and bacteria contained their own genetic material (either DNA or RNA).

Prion-related diseases arise from the ability of a prion protein to exist in two conformational states: a normal form, PrP^C, which does not cause disease, and an abnormal form, PrP^{Sc}, which does. (Note: The superscript C refers to the normal conformation, and the superscript Sc refers to the abnormal conformation, such as the one found in the disease called scrapie.) The gene coding a prion protein (*PrP*) is found in humans and other mammals, and the protein is expressed at low levels in certain types of cells such as neurons.

The abnormal conformation of the prion protein can come from two sources. An individual may eat products from an animal that had the disease. Alternatively, some people carry rare alleles of the *PrP* gene that cause their prion protein to convert spontaneously to the abnormal conformation at a very low rate. These individuals have an inherited predisposition to develop a prion-related disease. An example of an inherited prion disease is familial fatal insomnia (see Table 24.6).

What is the molecular mechanism through which prions cause disease? As shown in **Figure 24.10**, the abnormal conformation, PrP^{Sc}, acts as a catalyst to convert normal prion proteins

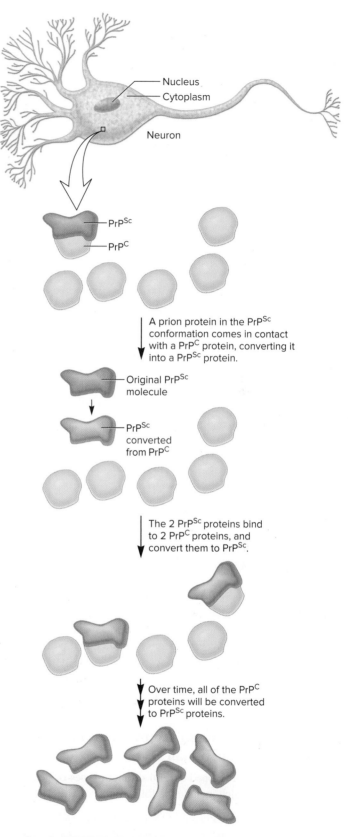

FIGURE 24.10 A proposed molecular mechanism of prion diseases. The PrP^{Sc} protein catalyzes the conversion of a PrP^C protein to PrP^{Sc}. Over time, the PrP^{Sc} conformation accumulates to high levels, leading to symptoms of the prion disease.

 ONLINE ANIMATION

CONCEPT CHECK: Where does the PrP^{Sc} protein come from?

within the cell to the misfolded conformation. As a prion disease progresses, the PrP^{Sc} proteins form dense aggregates in the cells of the brain and peripheral nervous tissues. This deposition is correlated with the disease symptoms affecting the nervous system. Some of the abnormal prion proteins are also excreted from infected cells and travel through the bloodstream to infect other cells.

24.4 COMPREHENSION QUESTION

1. A prion is a disease-causing agent composed of
   a. cells.
   b. nucleic acid with a protein coat.
   c. protein alone.
   d. nucleic acid alone.

24.5   HUMAN GENE THERAPY

Learning Outcomes:

1. Define *gene therapy*.
2. List different types of approaches for implementing gene therapy.
3. Analyze the results of the first gene therapy study involving adenosine deaminase deficiency.

Gene therapy is a method that involves the introduction of a gene into a patient or altering a gene already inside a patient's cells in an effort to treat or cure a disease. Gene therapy may be aimed at replacing or editing a mutant disease-causing gene, inactivating, or "knocking out," a mutant gene that is functioning improperly, or introducing a new gene into the body to help fight a disease.

Many current research efforts in gene therapy are aimed at alleviating inherited human diseases. Over 10,000 human genetic diseases are known to involve a single gene abnormality. Familiar examples include cystic fibrosis, sickle cell disease, and hemophilia. In addition, gene therapies have also been aimed at treating diseases such as cancer and cardiovascular disease, which may occur later in life. Some scientists are even pursuing research that will use gene therapy to combat infectious diseases such as AIDS. In this section, we will begin by surveying possible gene therapy approaches and consider some potential diseases they may be used to treat. We will then examine the first attempt at gene therapy in humans.

Different Gene Therapy Approaches Have the Potential to Treat Human Diseases

Researchers have developed several different types of gene therapy approaches, which include the following:

- *Plasmid DNA.* Plasmid DNA is a circular DNA molecule that can be genetically engineered to carry genes into human cells.

- *Viruses.* Certain viruses have a natural ability to infect human cells. Once viruses have been modified to remove their ability to cause infectious disease, they can be genetically engineered to carry genes into human cells.

- *Bacteria.* Bacteria can also be modified to prevent them from causing infectious disease and then used to carry genes into human cells.

- *Human gene editing technology.* Methods such as CRISPR-Cas technology (see Figure 20.13) can be used to edit genes that are already present in human cells.

- *Patient-derived cellular gene therapy.* Cells are removed from the patient, genetically modified, and then returned to the patient. This approach is described later in Experiment 24A.

Many current research efforts in gene therapy are aimed at alleviating human diseases caused by mutant genes. Examples include cystic fibrosis, sickle cell disease, phenylketonuria, hemophilia, and severe combined immunodeficiency disease (SCID). **Table 24.7** describes several types of diseases that are being investigated as potential targets of gene therapy.

Unfortunately, experimental gene therapy in humans has been associated with adverse reactions that have raised questions about certain safety issues. In 1999, a patient in a clinical trial died from an immune reaction to gene therapy treatment. In a French gene-therapy study begun in 2000, 4 out of the 10 treated children developed leukemia—a form of cancer involving the proliferation of white blood cells. In these cases, the disease occurred because the integration of a retroviral vector into the patients' genome caused a normal gene to become an oncogene, a gene that promotes cancer. The development of leukemia in these patients has raised concerns regarding the safety of gene therapy involving the use of viral vectors. Thus, it has prompted researchers to develop alternative methods of introducing genes into human cells.

Even though a large amount of research has already been conducted, success has been limited, and relatively few patients have been treated with gene therapy. However, some results have been promising. In 2017, the FDA approved two different gene therapies that target immune system cells. The first helps children and young adults battle a form of leukemia called B-cell acute lymphoblastic leukemia. A second is a gene therapy for adults with non-Hodgkin lymphoma.

TABLE 24.7	
Future Prospects in Gene Therapy	
Type of Disease	**Example(s) That May Be Treated by Gene Therapy**
Blood	Sickle cell disease, hemophilia, severe combined immunodeficiency disease (SCID), lymphoblastic leukemia, non-Hodgkin lymphoma
Metabolic	Glycogen storage diseases, lysosomal storage diseases, phenylketonuria
Muscular	Duchenne muscular dystrophy, dystrophia myotonica (myotonic muscular dystrophy)
Lung	Cystic fibrosis
Cancer	Brain tumors, breast cancer, colorectal cancer, malignant melanoma, ovarian cancer, several other types of malignancies
Cardiovascular	Atherosclerosis, essential hypertension
Infectious	AIDS, possibly other viral diseases that involve latent infections

EXPERIMENT 24A

Adenosine Deaminase Deficiency Was the First Inherited Disease Treated with Gene Therapy

Adenosine deaminase (ADA) is an enzyme involved in purine metabolism. If both copies of the *ADA* gene are defective, deoxyadenosine accumulates within the cells of the individual. At high concentrations, deoxyadenosine is particularly toxic to lymphocytes in the immune system, namely, T cells and B cells. In affected individuals, the destruction of T and B cells leads to a form of severe combined immunodeficiency disease (SCID). If left untreated, SCID is typically fatal at an early age (generally, 1–2 years old), because the immune system of these individuals is severely compromised and cannot fight infections.

Three approaches can be used to treat ADA deficiency. In some cases, a patient may receive a bone marrow transplant from a compatible donor. A second method is to treat SCID patients with purified ADA that is coupled to polyethylene glycol (PEG). The PEG-ADA complex is taken up by lymphocytes and can correct the ADA deficiency. Unfortunately, these two approaches are not always available and/or successful. A third, more recently developed approach is to treat ADA patients with gene therapy.

On September 14, 1990, the first human gene therapy was approved for a young child with ADA deficiency. This work was carried out by a large team of researchers composed of R. Michael Blaese, Kenneth Culver, W. French Anderson, and colleagues. Prior to this clinical trial, the normal gene for ADA had been cloned into a retroviral vector that could infect lymphocytes. The general aim of this therapy was to obtain a sample of lymphocytes from the blood of the child with SCID, introduce the normal *ADA* gene into those cells, and then return them to the child's bloodstream.

Figure 24.11 outlines the protocol for the experimental treatment. Lymphocytes (i.e., T cells) were removed from the patient and cultured in a laboratory. The lymphocytes were then transfected with a nonpathogenic retrovirus that had been genetically engineered to contain the normal *ADA* gene. During the life cycle of a retrovirus, the retroviral genetic material is inserted into the host cell's DNA. Therefore, because this retrovirus contained the normal *ADA* gene, this gene also was inserted into the chromosomal DNA of the child's lymphocytes. After this occurred in the laboratory, the cells were reintroduced back into the patient. This approach is called an **ex vivo approach** because the genetic manipulations occur outside the body, and the products are reintroduced into the body.

THE HYPOTHESIS

Infecting lymphocytes with a retrovirus containing the normal *ADA* gene will correct the inherited deficiency of the mutant *ADA* gene in patients with ADA deficiency.

TESTING THE HYPOTHESIS

FIGURE 24.11 **The first human gene therapy for adenosine deaminase deficiency carried out by Blaese and colleagues.**

Starting material: A retrovirus carrying the normal *ADA* gene.

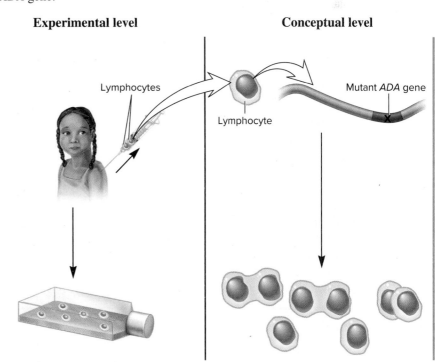

Experimental level   **Conceptual level**

1. Remove ADA-deficient lymphocytes from the patient with severe combined immunodeficiency disease (SCID).

   Lymphocytes

   Lymphocyte

   Mutant *ADA* gene

2. Culture the cells in a laboratory.

3. Transfect the cells with a retrovirus that contains the normal *ADA* gene. Retroviruses insert their DNA into the host cell chromosome as part of their reproductive cycle.

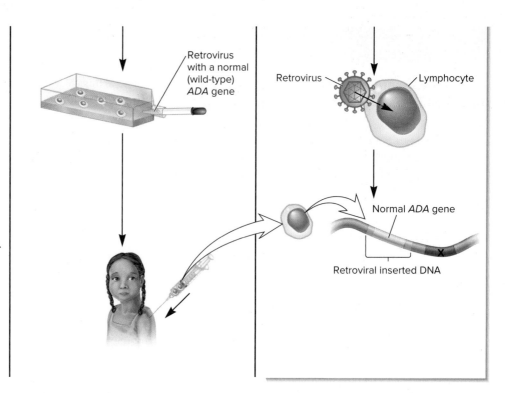

4. Infuse the *ADA*-gene-corrected lymphocytes back into the SCID patient.

THE DATA

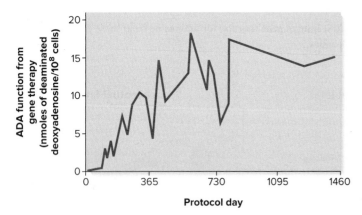

Source: Blaese, R. Michael, Culver, Kenneth W., and Miller, A. Dusty, "T Lymphocyte-Directed Gene Therapy for ADA–SCID: Initial Trial Results After 4 Years," *Science*, vol. 270, no. 5235, October 20, 1995, 475–480.

INTERPRETING THE DATA

In this clinical trial, two patients were enrolled, and a third patient was later treated in Japan. Was the treatment a success? The data show the level of ADA function for one patient involved in this study. The patient began with negligible ADA function. Even after several years, significantly higher levels of ADA were observed. However, because the patients also received a low dose of PEG-ADA, researchers could not determine whether or not gene transfer into T cells by itself was of significant clinical benefit.

Another form of SCID, termed SCID-X1, is inherited as an X-linked trait. SCID-X1 is characterized by a block in T-cell growth and differentiation. This block is caused by mutations in the gene that codes the γ_c cytokine receptor, which plays a key role in the recognition of signals that are needed to promote the growth, survival, and differentiation of T cells. A gene therapy trial for SCID-X1 similar to the trial shown in Figure 24.12 was initiated in 2000.

In this SCID-X1 trial, a gene coding the normal γ_c cytokine receptor was cloned into a retroviral vector and then introduced into lymphocytes from patients with SCID-X1. The lymphocytes were then reintroduced back into the patients' bodies. At a 10-month follow-up, T cells expressing the normal γ_c cytokine receptor were detected in two patients. Most importantly, the T-cell counts in these two patients had risen to levels that were comparable to those in normal individuals. This clinical trial was the first clear demonstration that gene therapy can offer clinical benefit, providing in these cases what seemed to be a complete correction of the disease phenotype.

24.5 COMPREHENSION QUESTIONS

1. Which of the following is/are a possible gene therapy approach?
   a. Delivery of normal genes via plasmids, viruses, or bacteria
   b. Gene editing via CRISPR-Cas technology
   c. Patient-derived cellular gene therapy
   d. All of the above are possible gene therapy approaches.

2. Which of the following best describes the approach that was used in the first gene therapy trial for treating SCID?
   a. The normal *ADA* gene was introduced by injecting liposomes directly into the patients' bodies.
   b. Lymphocytes were removed from a SCID patient, the normal *ADA* gene was transferred into the lymphocytes via liposomes, and then the lymphocytes were returned to the patient's body.
   c. Lymphocytes were removed from a SCID patient, the normal *ADA* gene was transferred into the lymphocytes via a retrovirus, and then the lymphocytes were returned to the patient's body.
   d. None of the above describe the approach used in the trial.

24.6 PERSONALIZED MEDICINE

Learning Outcomes:
1. Define *personalized medicine* and *molecular profiling*.
2. Describe specific ways in which personalized medicine affects patient care.

As mentioned, personalized medicine, also known as **precision medicine,** is the use of information about a patient's genotype and other clinical data in order to select a medication, therapy, or preventative measure that is specifically suited to that patient. As we gain a better understanding of human genes and disease states, researchers expect that personalized medicine will become an increasingly important aspect of health care. In this section, we will begin by examining how personalized medicine can affect treatment options for cancer patients. We will then consider how personalized medicine may play a role in determining the correct dosage for certain types of drugs.

Molecular Profiling Is Increasingly Used to Classify Tumors and Improve Treatment Options

Traditionally, different types of tumors have been classified according to their appearance under a microscope. Although this approach is useful, a major drawback is that two tumors may have a very similar microscopic appearance but yet have very different underlying genetic changes and clinical outcomes. For this reason, researchers and clinicians are turning to methods that enable them to understand the molecular changes that occur in diseases such as cancer. This general approach is called **molecular profiling.**

In cancer research and treatment, which is discussed in Chapter 25, molecular profiling involves the identification of the genes that play a role in the development of cancer. Why is this useful?

- First, molecular profiling can distinguish between tumors that look very similar under the microscope.
- Second, researchers are optimistic that molecular profiling may lead to improved treatment options. As we gain a better understanding of the genetic changes associated with particular types of cancers, researchers may be able to develop drugs that specifically target the proteins that are coded by cancer-causing gene mutations. As discussed Chapter 25, the drug imatinib mesylate, which is used to treat chronic myelogenous leukemia, was developed in this way.
- As another example, about 70% of all breast cancers exhibit an overexpression of the estrogen receptor. These types of breast cancer are better treated with drugs that either block the estrogen receptor or block the synthesis of estrogen. Therefore, the drug tamoxifen, which is an antagonist of the estrogen receptor, is used to treat tumors in which the estrogen receptor is overexpressed.

DNA Microarrays Are Used in the Molecular Profiling of Tumors

DNA microarrays, described in Chapter 23, are often used as a tool in the molecular profiling of tumors. The goal is to identify genes whose patterns of expression correlate with each other. In the study of cancer, researchers can compare cancer cells with normal cells and identify groups of genes that are turned on in the cancer cells and off in the normal cells, and other groups of genes that are turned off in the cancer cells and on in the normal cells. Likewise, researchers can compare two different types of tumors and identify groups of genes that show different patterns of expression.

As an example, **Figure 24.12a** shows a computer-generated image that presents the results of a microarray analysis of 47 samples, most of which came from the tumors of patients with a type of cancer called diffuse large B-cell lymphoma (DLBCL). Each column represents the expression pattern of a set of genes from a particular sample. Genes that are expressed are shown in red; those that are not expressed are shown in green.

During the course of these studies, the researchers identified two different patterns of gene expression.

- The tumor samples on the left side showed a set of genes (next to the orange bar) that tended to be turned on in the tumor and another set of genes (next to the blue bar) that tended to be turned off in the tumor. This pattern of gene expression was similar to the pattern found in a type of B cell called a germinal center B cell.
- In contrast, the tumor samples on the right side showed the opposite pattern. The upper genes tended to be turned off in these patients, and the lower genes were turned on. These samples showed a gene expression pattern found in activated B cells.

These results suggest that the two groups of tumors may have originated in B cells at different stages of development—those on

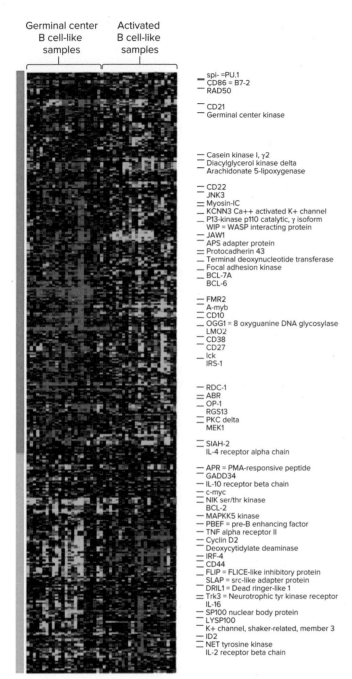

Germinal center B cell-like samples   Activated B cell-like samples

spi- =PU.1
CD86 = B7-2
RAD50

CD21
Germinal center kinase

Casein kinase I, γ2
Diacylglycerol kinase delta
Arachidonate 5-lipoxygenase

CD22
JNK3
Myosin-IC
KCNN3 Ca++ activated K+ channel
P13-kinase p110 catalytic, γ isoform
WIP = WASP interacting protein
JAW1
APS adapter protein
Protocadherin 43
Terminal deoxynucleotide transferase
Focal adhesion kinase
BCL-7A
BCL-6

FMR2
A-myb
CD10
OGG1 = 8 oxyguanine DNA glycosylase
LMO2
CD38
CD27
lck
IRS-1

RDC-1
ABR
OP-1
RGS13
PKC delta
MEK1

SIAH-2
IL-4 receptor alpha chain

APR = PMA-responsive peptide
GADD34
IL-10 receptor beta chain
c-myc
NIK ser/thr kinase
BCL-2
MAPKK5 kinase
PBEF = pre-B enhancing factor
TNF alpha receptor II
Cyclin D2
Deoxycytidylate deaminase
IRF-4
CD44
FLIP = FLICE-like inhibitory protein
SLAP = src-like adapter protein
DRIL1 = Dead ringer-like 1
Trk3 = Neurotrophic tyr kinase receptor
IL-16
SP100 nuclear body protein
LYSP100
K+ channel, shaker-related, member 3
ID2
NET tyrosine kinase
IL-2 receptor beta chain

(a) Cluster analysis

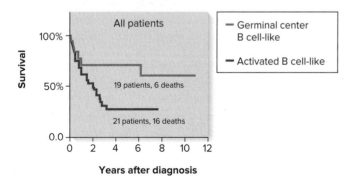

(b) Patient outcomes

FIGURE 24.12 **The use of DNA microarrays to classify types of tumors.** **(a)** Forty-seven samples, mostly from patients with diffuse large B-cell lymphoma (DLBCL), were subjected to a DNA microarray analysis. The DNA microarray data were then analyzed to identify genes that are coordinately expressed. The figure shown here is a graphical compilation of the microarray analyses. Each column represents one sample; each row represents the expression of a particular gene. The names of some of the genes are shown along the right side. (Note: The rows and columns are not easily resolved in this illustration.) Genes highly expressed are shown in red; those not expressed are shown in green. One group of samples had an expression pattern similar to that found in germinal center B cells; the other group had an expression pattern typical of activated B cells. **(b)** Survival rates of the patients with DLBCL.

(a): Dr. Pat Brown

the left originated in germinal center B cells, whereas those on the right originated in activated B cells.

Furthermore, the patients from whom these tumors were derived also appeared to have very different clinical outcomes (**Figure 24.12b**). The patients whose tumors had a pattern of gene expression similar to that in activated B cells had a significantly lower overall survival rate than did the other patients.

A Patient's Genotype Is Important in Determining the Proper Dosage of Certain Drugs

Pharmacogenetics is the study or clinical testing of genetic variation that causes differing responses to drugs. The proper dosage for any given drug depends on a variety of factors. Some key factors include the following:

- For drugs that are taken orally, the rate of transport of the drug from the digestive tract into the bloodstream
- The rate of transport of the drug into the appropriate cells where the drug exerts its effect
- The ability of the drug to affect the function of a specific target protein
- The ability of the drug to be taken up and metabolized by the liver
- The rate of excretion of the drug from the body

These five factors are affected by genetics because proteins, which are coded by genes, are directly involved. For example, transport proteins are often required for the uptake of drugs into the bloodstream, into specific cell types, and into liver cells. Enzymes, which are proteins, are involved in metabolizing drugs into inactive products. Finally, nearly all drugs exert their effects by binding to specific target proteins and altering their function. For example, acetylsalicylic acid, more commonly known as aspirin, binds to a protein called cyclooxygenase, thereby inhibiting the protein's enzymatic function and reducing inflammation.

Why is genetics important in the proper dosage of drugs? The answer is that genetic variation in the human population often affects the function of proteins that are involved with drug transport, drug metabolism, and the ability of drugs to affect their target proteins.

To understand how genetic variation can affect the proper dosage of a specific drug, let's consider drug metabolism by the liver. Many drugs are broken down in the liver by a family of related enzymes called cytochrome P450. An example is warfarin (Coumadin), which is used clinically as an anticoagulant, but requires periodic monitoring and is associated with adverse side effects. This drug is metabolized by a cytochrome P450 enzyme designated CYP2C9. Researchers have identified over 80 variants (SNPs) in the gene that codes CYP2C9 in human populations. Some of these variants affect CYP2C9 function. Clinically, the variable activity of CYP2C9 can result in four different levels of warfarin metabolism, which are described as ultrarapid, extensive, intermediate, and poor.

Currently, the dose of warfarin given to a patient is either determined by a "one size fits all" approach or based on consideration of characteristics such as sex, age, and liver function. Adjustments to the dosage are made based on periodic blood tests that measure the level of blood coagulation. Even so, overcoagulation and undercoagulation remain a problem in a significant number of patients. Recently, genetic tests are available that help doctors determine the proper warfarin dosage for their patients. For example, a person with an ultrarapid metabolizer genotype requires a higher dosage because the drug tends to be rapidly broken down in that person's body. By comparison, someone with a poor metabolizer genotype requires a lower dosage. In the future, such genetic tests may be routinely used by doctors to determine the proper drug dosage.

As discussed in Chapter 22, DNA-sequencing technologies have advanced to the point where the sequencing of a person's entire genome is inexpensive enough to be used as a routine diagnostic procedure. Therefore, many clinicians are predicting that it will become common practice for patients' genome sequences to be determined and analyzed to improve care. As we gain a better understanding of how genetic variation affects drug action, transport, and metabolism, pharmacogenetics is expected to play an increasingly important role in personalizing health care.

24.6 COMPREHENSION QUESTION

1. Personalized medicine may be used
   a. to characterize types of tumors.
   b. to predict the outcome of certain types of cancers.
   c. to determine the proper dosage of drugs.
   d. in all of the above.

KEY TERMS

Introduction: personalized medicine (precision medicine)
24.1: monozygotic (MZ) twins, dizygotic (DZ) twins, concordance, incomplete penetrance, age of onset, simple Mendelian inheritance, autosomal, recessive, haploinsufficiency, gain-of-function mutation, dominant-negative mutation, locus heterogeneity
24.2: haplotype, founder, linkage disequilibrium, International HapMap Project, HapMap, genome-wide association study (GWAS)

24.3: genetic testing, genetic screening, noninvasive prenatal testing (NIPT), amniocentesis, chorionic villus sampling, preimplantation genetic diagnosis (PGD), in vitro fertilization (IVF)
24.4: prion
24.5: gene therapy, ex vivo approach
24.6: precision medicine, molecular profiling, pharmacogenetics

CHAPTER SUMMARY

- Thousands of genetic diseases are known to afflict people.

24.1 Inheritance Patterns of Genetic Diseases

- A genetic basis for a human disease may be suggested by a variety of observations (see Figure 24.1).
- Thousands of human genetic diseases follow simple Mendelian patterns of inheritance that can be determined by pedigree analysis. These patterns include autosomal recessive, autosomal dominant, X-linked recessive, and X-linked dominant inheritance (see Figures 24.2–24.4, Tables 24.1–24.4).
- Autosomal recessive diseases are usually caused by loss-of-function mutations, whereas autosomal dominant diseases may be caused by haploinsufficiency, gain-of-function mutations, or dominant-negative mutations.
- Many genetic diseases exhibit locus heterogeneity, which means that they may be caused by mutations in more than one gene.

24.2 Detection of Disease-Causing Alleles via Haplotypes and via Genome-Wide Association Studies

- A haplotype is a linkage of alleles or molecular markers on a single chromosome (see Figure 24.5).
- Disease-causing mutations may originate in a founder with a specific haplotype (see Figure 24.6).
- Researchers may identify a disease-causing allele by determining its location within a chromosome by a haplotype association study or directly identify the allele by a genome-wide association study (GWAS) (see Figures 24.7, 24.8).

24.3 Genetic Testing and Screening

- Genetic testing for abnormalities can be performed in a variety of ways (see Table 24.5).
- Genetic screening is population-wide genetic testing.

• Genetic testing by means of noninvasive prenatal testing (NIPT), amniocentesis, chorionic villus sampling, or preimplantation genetic diagnosis (PGD) can be done prior to birth (see Figure 24.9).

24.4 Prions

• Prions are disease-causing agents composed entirely of protein. Prion-related diseases can be acquired by eating products from an infected animal or may be due to carrying a rare allele of the *PrP* gene (see Table 24.6).

• Prions cause disease because the prion protein in the abnormal conformation (PrP^{Sc}) is able to convert a protein in the normal form (PrP^C) to the abnormal conformation (see Figure 24.10).

24.5 Human Gene Therapy

• Human gene therapy is the introduction of cloned genes into somatic cells or the modification of existing genes in an attempt to treat or cure a disease (see Table 24.7).

• The first human gene therapy trial was aimed at treating adenosine deaminase deficiency (see Figure 24.11).

24.6 Personalized Medicine

• Personalized medicine is the use of information about a patient's genotype and other clinical data in order to select a medication, therapy, or preventative measure that is specifically suited to that patient.

• Molecular profiling is used to classify tumors, which may affect treatment options. Methods used in such profiling include DNA microarrays (see Figure 24.12).

• Pharmacogenetics, which is the study or clinical testing of genetic variation that causes differing responses to drugs, is likely to play an increasing role in determining the proper dosages of drugs given to patients.

PROBLEM SETS & INSIGHTS

MORE GENETIC TIPS

1. The pedigree presented here shows the incidence of a human disease known as familial hypercholesterolemia in a certain family.

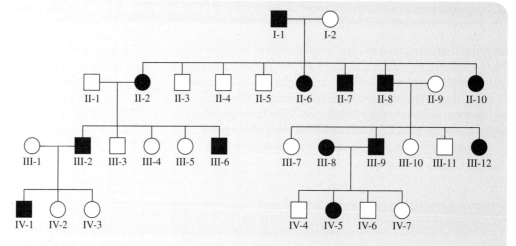

This disorder is characterized by an elevated level of serum cholesterol in the blood. Though relatively rare, this genetic abnormality can be a contributing factor to heart attacks. At the molecular level, this disease is caused by a defective gene that codes a protein called low-density lipoprotein receptor (LDLR). In the bloodstream, serum cholesterol is bound to a carrier protein known as low-density lipoprotein (LDL). LDL binds to LDLR, which enables cells to absorb cholesterol. When LDLR is defective, it becomes more difficult for the cells to absorb cholesterol. This explains why the blood level of cholesterol remains high. Based on the pedigree, what is the most likely pattern of inheritance of this disorder?

T OPIC: *What topic in genetics does this question address?* The topic is inheritance patterns in humans. More specifically, the question is about identifying the inheritance pattern of familial hypercholesterolemia from the pedigree of a certain family.

I NFORMATION: *What information do you know based on the question and your understanding of the topic?* In the question, you are given a pedigree to analyze. From your understanding of the topic, you may remember that single-gene disorders can follow an autosomal recessive, autosomal dominant, X-linked recessive, or X-linked dominant pattern of inheritance.

P ROBLEM-SOLVING **S** TRATEGIES: *Predict the outcome. Compare and contrast.* One strategy to solve this problem is to consider what the possible patterns of inheritance predict with regard to the traits found in parents and offspring. You can compare and contrast the four patterns described in Section 24.1, and see if you can rule any of them out.

ANSWER: The pedigree is consistent with an autosomal dominant pattern of inheritance. An affected individual always has an affected parent. We can rule out the other possible patterns as follows:

The pattern can't be autosomal recessive, because two affected parents (III-8 and III-9) produced some unaffected offspring.

It can't be X-linked recessive, because an affected mother (III-8, who would have to be homozygous) produced unaffected sons.

It can't be X-linked dominant, because an affected father (III-2) produced unaffected daughters.

---

2. Some prion-related diseases, such as familial fatal insomnia, are inherited. How would you expect the mutation has altered the *PrP* gene in this case? Would it have affected the promoter, the regulatory sequences, or the coding sequence of the *PrP* gene?

T OPIC: *What topic in genetics does this question address?* The topic is prion-related diseases. More specifically, the question is

about predicting the effects of mutations that are involved with inherited prion-related diseases.

I NFORMATION: *What information do you know based on the question and your understanding of the topic?* From the question, you know that some prion-related diseases are inherited. From your understanding of the topic, you may remember that these diseases are characterized by a prion protein that is more likely to convert to the abnormal form, PrP^{Sc}.

P ROBLEM-SOLVING **S** TRATEGY: *Relate structure and function.* One strategy to solve this problem is to consider that the inherited form of the disease affects the ability of the prion protein to make conformational changes. The mutation has an effect on protein structure that alters protein function.

ANSWER: The mutation is likely to be in the coding sequence. It alters the amino acid sequence of the prion protein in such a way that it becomes more likely to convert to the abnormal form, PrP^{Sc}.

Conceptual Questions

C1. With regard to pedigree analysis, make a list of observations that distinguish recessive, dominant, and X-linked patterns of inheritance.

C2. Explain, at the molecular level, why human genetic diseases often follow simple Mendelian patterns of inheritance, whereas most normal (non-disease-related) traits, such as the shape of your nose or the size of your head, are governed by multiple gene interactions.

C3. Many genetic disorders exhibit locus heterogeneity. Define *locus heterogeneity*, and give two examples of disorders that exhibit it. How does locus heterogeneity confound a pedigree analysis?

C4. In general, why do changes in chromosome structure or number tend to affect an individual's phenotype? Explain why some changes in chromosome structure, such as reciprocal translocations, do not.

C5. We often speak of diseases such as phenylketonuria (PKU) and achondroplasia as having a genetic basis. Explain whether the following statements are accurate with regard to the genetic basis of any human disease (not just PKU and achondroplasia).

  A. An individual must inherit two copies of a mutant allele to have disease symptoms.

  B. A genetic predisposition means that an individual has inherited one or more alleles that make it more likely that this individual will develop disease symptoms than other individuals in a population will.

  C. A genetic predisposition to develop a disease may be passed from parents to offspring.

  D. The genetic basis for a disease is always more important than the environment.

C6. Figure 24.1 illustrates albinism in two different species. Describe two other genetic disorders found in both humans and other animals.

C7. Discuss why a genetic disease might have a particular age of onset. Would an infectious disease have an age of onset? Explain why or why not.

C8. Gaucher disease (type I) is due to a defect in a gene that codes a protein called acid β-glucosidase. This enzyme plays a role in carbohydrate metabolism within lysosomes. The gene is located on the long arm of chromosome 1. People who inherit two defective copies of this gene exhibit Gaucher disease, the major symptoms of which include an enlarged spleen, bone lesions, and changes in skin pigmentation. Let's suppose a phenotypically unaffected female, whose male parent had Gaucher disease, has a child with a phenotypically unaffected male, whose female parent had Gaucher disease.

  A. What is the probability that this child will have the disease?

  B. What is the probability that this child will have two normal copies of this gene?

  C. If this couple has five children, what is the probability that one of them will have Gaucher disease and four will be phenotypically unaffected?

C9. Ehler-Danlos syndrome is a rare disorder caused by a mutation in a gene that codes a protein called collagen (type 3 A1). Collagen is found in the extracellular matrix that plays an important role in the formation of skin, joints, and other connective tissues. People with Ehler-Danlos syndrome have extraordinarily flexible skin and very loose joints. The pedigree shown next contains several individuals who have this syndrome, represented with black symbols. Based on this pedigree, does the syndrome follow autosomal recessive, autosomal dominant, X-linked recessive, or X-linked dominant inheritance? Explain your reasoning.

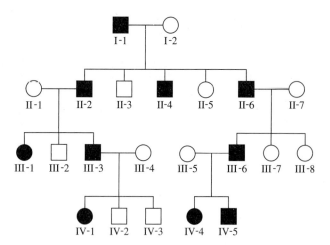

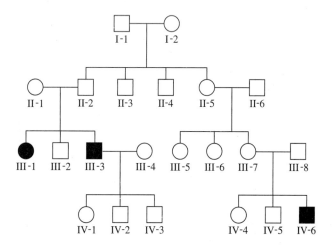

C10. Hurler syndrome is due to a mutation in a gene that codes a protein called α-l-iduronidase. This protein functions within lysosomes as an enzyme that breaks down mucopolysaccharides (a type of polysaccharide that has many acidic groups attached). When this enzyme is defective, excessive amounts of the mucopolysaccharides dermatan sulfate and heparin sulfate accumulate within the lysosomes, especially in liver cells and connective tissue cells. This accumulation leads to symptoms such as an enlarged liver and spleen, bone abnormalities, corneal clouding, heart problems, and severe neurological problems. The following pedigree contains three individuals with Hurler syndrome, represented with black symbols. Based on this pedigree, does this syndrome appear to follow autosomal recessive, autosomal dominant, X-linked recessive, or X-linked dominant inheritance? Explain your reasoning.

C11. Like Hurler syndrome, Fabry disease involves an abnormal accumulation of substances within lysosomes. However, the lysosomes of individuals with Fabry disease show an abnormal accumulation of lipids. The defective enzyme is α-galactosidase A, which is a lysosomal enzyme that functions in lipid metabolism. The enzymatic defect causes cell damage, especially to the kidneys, heart, and eyes. The gene that codes α-galactosidase A is found on the X chromosome. Let's suppose a phenotypically unaffected couple produces two male offspring with Fabry disease and one phenotypically unaffected female offspring. What is the probability that the female offspring will have an affected male offspring?

C12. Achondroplasia, which results in short stature, is caused by an autosomal dominant mutation that affects the gene that codes a fibroblast growth factor receptor. Among 1,422,000 live births, the number of babies born with achondroplasia was 31. Among those 31 babies, 18 of them had at least one parent with achondroplasia. The other 13 babies had two unaffected parents. How do you explain those 13 babies, assuming that the mutant allele has 100% penetrance? What are the odds that these 13 individuals will pass this mutant gene to their offspring?

C13. Lesch-Nyhan syndrome is due to a mutation in a gene that codes a protein called hypoxanthine-guanine phosphoribosyltransferase (HPRT). HPRT is an enzyme that functions in purine metabolism. People with this syndrome have severe neurodegeneration and loss of motor control. The following pedigree contains several individuals with Lesch-Nyhan syndrome, shown with black symbols. Based on this pedigree, does this syndrome appear to be inherited via an autosomal recessive, autosomal dominant, X-linked recessive, or X-linked dominant pattern? Explain your reasoning.

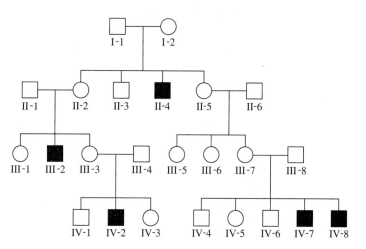

C14. Marfan syndrome is due to a mutation in a gene that codes a protein called fibrillin-1. The syndrome is inherited as a dominant trait. The fibrillin-1 protein is the main constituent of extracellular microfibrils. These microfibrils can exist as individual fibers or associate with a protein called elastin to form elastic fibers. People with Marfan syndrome tend to be unusually tall with long limbs, and they may have defects in their heart valves and aorta. Let's suppose a phenotypically unaffected female has a child with a male who has Marfan syndrome.

A. What is the probability this child will have the disease?

B. If this couple has three children, what is the probability that none of them will have Marfan syndrome?

C15. Sandhoff disease is due to a mutation in a gene that codes a protein called hexosaminidase B. This disease has symptoms that are similar to those of Tay-Sachs disease. Weakness begins in the first 6 months of life. Individuals exhibit early blindness and progressive mental and motor deterioration. The family in the following pedigree has three individuals with Sandhoff disease, indicated with black symbols.

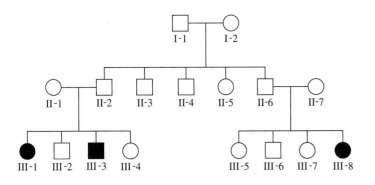

A. Based on this pedigree, does this syndrome appear to follow autosomal recessive, autosomal dominant, X-linked recessive, or X-linked dominant inheritance? Explain your reasoning.

B. What is the likelihood that individuals II-1, II-2, II-3, II-4, II-5, II-6, and II-7 carry a mutant allele of the gene coding hexosaminidase B?

C16. What are the two assumptions that underlie the identification of disease-causing alleles via haplotypes?

C17. What is the purpose of the International HapMap Project? How will it help researchers who study disease-causing alleles?

C18. What is a prion? Explain how a prion relies on normal cellular proteins to cause a disease such as mad cow disease.

C19. Some people have a genetic predisposition for developing prion-related diseases. Examples of these diseases are described in Table 24.6. In the case of Gerstmann-Sträussler-Scheinker disease, the age of onset is typically 30–50 years, and the duration of the disease (which leads to death) is about 5 years. Suggest a possible explanation why someone can live for a relatively long time without symptoms and then succumb to the disease in a relatively short time.

Experimental Questions

E1. Which of the following observations suggest that a disease has a genetic basis?

A. The disease is less likely to be exhibited by relatives that live apart than by relatives that live together.

B. The frequency of the disease is unusually high in a small group of genetically related individuals who live in southern Spain.

C. The disease symptoms usually begin around the age of 40.

D. It is more likely that monozygotic twins will both be affected by the disease than will dizygotic twins.

E2. Section 24.1 discussed the types of observations that suggest a disease is inherited. Which of these observations do you find the least convincing? Which do you find the most convincing? Explain your answer.

E3. What is meant by the term *genetic testing*? How do testing at the protein level and testing at the DNA level differ? Describe five different techniques used in genetic testing.

E4. A particular disease occurs in a group of South Amerindians. During the 1920s, many of them migrated to Central America. In the Central American group, the disease is never observed. Discuss whether or not you think the disease has a genetic component. What types of further observations would you make?

E5. Chapter 20 describes a method known as Western blotting that can be used to detect a polypeptide that is translated from a particular mRNA. In this method, a particular polypeptide or protein is detected by an antibody that specifically recognizes a segment of its amino acid sequence. After the antibody binds to the polypeptide within a gel, a secondary antibody (which is labeled) is used to visualize the polypeptide as a dark band. For example, an antibody that recognizes α-galactosidase A could be used to specifically detect that protein on a gel. The enzyme α-galactosidase A is defective in individuals with Fabry disease, which shows an X-linked recessive pattern of inheritance. Amy, Nan, and Pete are siblings, and Pete has Fabry disease.

Aileen, Jason, and Jerry are siblings, and Jerry has Fabry disease. Amy, Nan, and Pete are not related to Aileen, Jason, and Jerry. Amy, Nan, and Aileen are concerned that they could be carriers of a defective α-galactosidase A gene. A sample of cells was obtained from each of these six individuals and subjected to Western blotting, using an antibody against α-galactosidase A. Samples were also obtained from two unrelated normal individuals (lanes 7 and 8). The results are shown here.

Samples from:

Lane 1. Amy
Lane 2. Nan
Lane 3. Pete
Lane 4. Aileen
Lane 5. Jason
Lane 6. Jerry
Lane 7. Normal male
Lane 8. Normal female

(Note: Due to X-chromosome inactivation in females, the amount of expression of genes on the single X chromosome in males is equal to the amount of expression from genes on both X chromosomes in females.)

A. Explain the type of mutation (e.g., missense, nonsense, promoter, etc.) that caused Fabry disease in Pete and Jerry.

B. What would you tell Amy, Nan, and Aileen regarding the likelihood that they are carriers of the mutant allele and the probability of having affected offspring?

E6. An experimental assay for the blood-clotting protein called clotting factor IX is available. A blood sample was obtained from each individual in the following pedigree. The amount of factor IX protein, shown within each symbol on the pedigree, is expressed as a percentage of the average amount observed in individuals who do not carry a mutant copy of the gene.

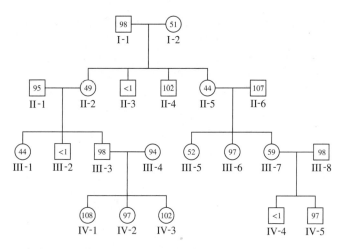

What is the likely genotype of each person in this pedigree?

E7. Treatment of adenosine deaminase (ADA) deficiency is an example of ex vivo gene therapy. Why is this therapy called ex vivo? Can ex vivo gene therapy be used to treat all inherited diseases? Explain.

E8. Several research studies are under way that involve the use of gene therapies to inhibit the growth of cancer cells. As discussed in Chapter 25, oncogenes are mutant genes that are overexpressed and cause cancer. New gene therapies are aimed at silencing oncogenes by producing antisense RNA that recognizes the mRNA transcribed from oncogenes. Based on your understanding of antisense RNA (discussed in Chapter 14), explain how this strategy would prevent the growth of cancer cells.

E9. Explain how DNA microarrays are used in molecular profiling of cancerous tumors.

Questions for Student Discussion/Collaboration

1. Make a list of the benefits that may arise from genetic testing as well as possible negative consequences. Discuss the items on your list.

2. Discuss the advantages and disadvantages of gene therapy. A limited amount of funding is available for gene therapy research. Make a priority list of the top three diseases for which you would fund research. Discuss your choices.

Note: All answers are available for the instructor in Connect; the answers to the even-numbered questions and all of the Concept Check and Comprehension Questions are in Appendix B.

CHAPTER OUTLINE

Cigarette smoking and lung cancer. *Cigarette smoke contains chemicals that are known to mutate genes in the cells of a person's lungs, thereby leading to lung cancer. Lung cancer remains the top cause of deaths from cancer in the United States, and 87% of deaths due to lung cancer are linked to smoking.*

Grantly Lynch /UK Stock Images Ltd./Alamy Stock Photo

25

GENETIC BASIS OF CANCER

Cancer is a disease that affects multicellular organisms and is characterized by uncontrolled cell division. Worldwide, cancer is the second leading cause of death in humans, exceeded only by heart disease. In the United States, approximately 1.5 million people are diagnosed with cancer each year; over 0.5 million will die from the disease. Overall, about one in four Americans will die from cancer.

For about 10% of cancers, a higher predisposition to develop the disease is an inherited trait. Most cancers, though, perhaps 90%, do not involve genetic changes that are passed from parent to offspring. Rather, cancer is usually an acquired condition that typically occurs later in life. At least 80% of all cancers in humans are related to exposure to **carcinogens,** environmental agents that increase the likelihood of developing cancer.

Most carcinogens, such as UV light and certain chemicals in cigarette smoke, are mutagens that promote genetic changes in somatic cells. These genetic changes can alter gene expression in a way that ultimately affects cell division, leading to cancer. In addition, researchers have determined that epigenetic changes also contribute to many forms of cancer. In this chapter, we will begin with an overview of the basic features of cancer. We will then consider how genetic changes—that is, cancer-causing mutations—can alter

gene expression and promote uncontrolled cell growth. We then explore how epigenetic changes can alter gene expression and contribute to the development of cancer. Finally, we will examine various treatment options in the battle against human cancers.

25.1   OVERVIEW OF CANCER

Learning Outcomes:

1. Describe the key characteristics of cancer.
2. Compare and contrast oncogenes and tumor-suppressor genes.

As mentioned, cancer is a disease characterized by uncontrolled cell division. It is a genetic disease at the cellular level. More than 100 kinds of human cancers have been identified, and they are classified according to the type of cell that has become cancerous. Though cancer is a diverse collection of many diseases, some characteristics are common to all cancers:

- Most cancers originate in a single cell. This single cell, as well as its line of daughter cells, usually undergoes a series of genetic and epigenetic changes that accumulate during cell division. As a result, a cancerous growth can be considered to be **clonal** in origin. A hallmark of a cancer cell is that it divides to produce two daughter cancer cells.

- At the cellular and genetic levels, cancer is usually a multistep process that begins with a precancerous genetic change—a **benign** growth—that is followed by additional genetic and epigenetic changes that lead to cancerous cell growth (**Figure 25.1**).

- When cells have become cancerous, the growth is described as **malignant.** Cancer cells are **invasive**—they can invade healthy tissues—and **metastatic**—they can migrate to other parts of the body and cause secondary tumors.

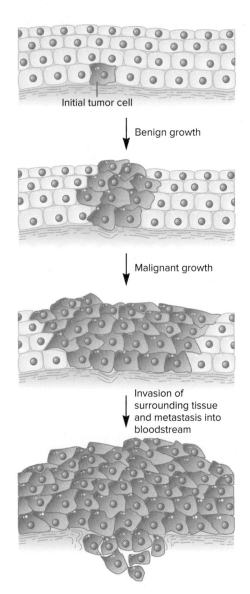

Initial tumor cell

Benign growth

Malignant growth

Invasion of surrounding tissue and metastasis into bloodstream

FIGURE 25.1 Progression of cellular growth leading to cancer.

Genes→Traits In a healthy individual, one or more gene mutations convert a normal cell into a tumor cell. This tumor cell divides to produce a benign tumor. Additional genetic changes in the tumor cells may occur, leading to malignant growth. At this malignant stage, the tumor cells invade surrounding tissues, and some malignant cells may metastasize by traveling through the bloodstream to other parts of the body where they can grow and cause secondary tumors. As a trait, cancer can be viewed as a series of genetic and epigenetic changes that eventually lead to uncontrolled cell growth.

Both genetic and epigenetic changes can promote the proliferation of cancer cells. Most genetic changes that cause cancer involve mutations in particular genes. The effects of these mutations are placed into two broad categories:

- An **oncogene** is produced when a mutation alters a gene in a way that results in overactivity—the amount and/or function of the gene product is abnormally high and that higher activity promotes cancerous growth. For example, a protein may be overproduced that promotes cell division.
- A **tumor-suppressor gene** codes a gene product that prevents cancer. A loss-of-function mutation in a tumor-suppressor gene may allow cancerous growth to occur.

Therefore, oncogenes and tumor-suppressor genes are distinguished by whether a cancer-causing mutation in such a gene increases or decreases the activity of gene products.

As discussed in Chapter 16, genes are also regulated by epigenetic mechanisms, such as DNA methylation and histone modifications. These epigenetic changes can promote cancer by increasing the expression of oncogenes or inhibiting the expression of tumor-suppressor genes. In Sections 25.2 and 25.3, we will explore how mutations promote cancer by affecting oncogenes and tumor-suppressor genes, respectively, and in Section 25.4, we will consider epigenetic effects.

25.1 COMPREHENSION QUESTION

1. A mutant gene that promotes cancer when it is overexpressed is called
   a. a tumor-suppressor gene.
   b. an oncogene.
   c. both a and b.
   d. neither a nor b.

25.2 ONCOGENES

Learning Outcomes:

1. Explain how gain-of-function mutations in genes involved in cell-signaling pathways can convert proto-oncogenes into oncogenes.
2. Describe the different types of genetic changes that can convert a proto-oncogene into an oncogene.
3. Give examples of viruses that carry oncogenes.

As noted in Section 25.1, an oncogene is produced when a mutation alters a gene in way that results in overactivity—the amount and/or function of the gene product is abnormally high and that higher activity promotes cancerous growth. A normal, nonmutated gene that has the potential to mutate into an oncogene is termed a **proto-oncogene.** To become an oncogene, a proto-oncogene incurs a gain-of-function mutation, which we discussed in Chapter 24 in the context of autosomal dominant disorders.

For a protein-coding gene, a gain-of-function mutation that produces an oncogene typically has one of three possible effects:

- The amount of the coded protein is greatly increased.
- A change occurs in the structure of the coded protein that causes it to be overly active.
- The coded protein is expressed in a cell type where it is not normally expressed.

However, other possible changes can cause overactivity and contribute to cancer:

- As discussed in Chapter 17 (see Section 17.7), mutations that promote cancer can occur in genes for non-coding RNAs (ncRNAs). For example, a high level of expression of an ncRNA called HOTAIR is sometimes found in primary breast tumors and is a significant predictor of metastasis and death.
- Later, in Section 25.4, we will explore how epigenetic changes can also increase the expression of oncogenes.

In this section, we will explore the possible ways that oncogenes affect cell growth and examine the types of genetic changes that convert proto-oncogenes into oncogenes. We will also consider how viruses can introduce oncogenes into cells.

Oncogenes Have Gain-of-Function Mutations That May Affect Proteins Involved in Cell Division Pathways

As described in Chapter 2, eukaryotic cells destined to divide advance through a series of stages known as the cell cycle (refer back to Figure 2.5). The phases consist of G_1 (first gap), S (synthesis of DNA, the genetic material), G_2 (second gap), and M phase (mitosis and cytokinesis). The G_1 phase is a period in the cell cycle when a cell may become committed to divide.

Depending on the conditions, a cell in the G_1 phase may accumulate molecular changes that cause it to advance through the rest of the cell cycle. When this occurs, cell biologists say that a cell has reached a special control point called the restriction point. The commitment to divide is based on a variety of factors. For example, environmental conditions, such as the presence of sufficient nutrients, are important for cell division. In addition, multicellular organisms rely on signaling molecules to coordinate cell division throughout the body. These signaling molecules are often called **growth factors** because they promote cell division.

As researchers began to investigate cancer at the molecular level, they wanted to understand how mutant genes promote abnormal cell division. In parallel with cancer research, cell biologists have studied the roles that normal cellular proteins play in cell division. As mentioned, the cell cycle is regulated in part by growth factors that bind to cell surface receptors and initiate a cascade of cellular events that lead eventually to cell division.

Figure 25.2 considers a growth factor called epidermal growth factor (EGF), which is secreted from endocrine cells and stimulates epidermal cells, such as skin cells, to divide. This process occurs as follows:

1. EGF binds to its receptor.
2. This binding results in the activation of a cell-signaling pathway within the epidermal cell.

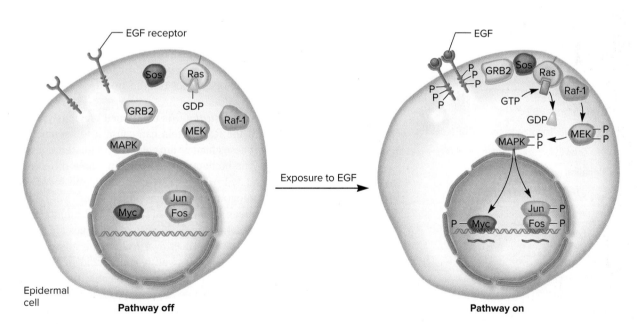

FIGURE 25.2 **The activation of a cell-signaling pathway by a growth factor.** In this example, epidermal growth factor (EGF) binds to two EGF receptors, causing them to dimerize and phosphorylate each other. An intracellular protein called GRB2 binds to the phosphorylated EGF receptor, and it is subsequently bound by another protein called Sos. The binding of Sos to GRB2 enables Sos to activate a protein called Ras. This activation involves the release of GDP and the binding of GTP. The activated Ras/GTP complex then activates Raf-1, which is a protein kinase. Raf-1 phosphorylates MEK, and then MEK phosphorylates MAPK. More than one MAPK may be involved. Finally, the phosphorylated form of MAPK activates transcription factors, such as Myc, Jun, and Fos. This activation leads to the transcription of genes, which code proteins that promote cell division.

3. This pathway causes a change in gene transcription; the transcription of specific genes is activated in response to the growth factor.

4. After these genes are transcribed and the mRNAs are translated into proteins, the proteins promote the advancement through the cell cycle. In other words, the cell is stimulated to divide.

Figure 25.2 shows just one example of a pathway involving a growth factor and gene activation. Eukaryotic species produce many different growth factors.

The mutations that convert proto-oncogenes into oncogenes have been analyzed in many types of cancers. Oncogenes commonly code proteins that function in cell-signaling pathways related to cell division (**Table 25.1**). The examples shown here are all protein-coding genes. The proteins they code include growth factors, growth factor receptors, intracellular signaling proteins, and transcription factors. The overexpression of oncogenes causes cell division to occur too often.

As an example, mutations that alter the amino acid sequence of the Ras protein have been shown to cause functional abnormalities. The Ras protein is a GTPase, which hydrolyzes GTP to GDP + P_i (**Figure 25.3**). Therefore, after it has been activated, the Ras protein returns to its inactive state by hydrolyzing GTP. Certain mutations that convert the normal *ras* gene into an oncogene decrease the ability of the Ras protein to hydrolyze GTP. This results in a greater amount of the active

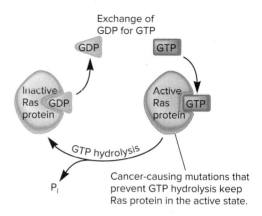

FIGURE 25.3 **Functional cycle of the Ras protein.** The binding of GTP to Ras activates the function of Ras and promotes cell division. The hydrolysis of GTP to GDP and P_i converts the active form of Ras to an inactive form.

CONCEPT CHECK: How would a mutation that prevents the Ras protein from hydrolyzing GTP affect the cell-signaling pathway in Figure 25.2?

GTP-bound form of the Ras protein. In this way, such mutations keep the signaling pathway turned on, thereby stimulating the cell to divide.

Genetic Changes in Proto-Oncogenes Convert Them to Oncogenes

How do specific genetic alterations convert proto-oncogenes into oncogenes? By isolating and studying oncogenes at the molecular level, researchers have discovered four main ways this occurs. These genetic alterations can be categorized as missense mutation, gene amplification (i.e., an increase in copy number), chromosomal translocation, or viral integration. In addition to these four types of changes, epigenetic modifications can also convert proto-oncogenes into oncogenes (see Section 25.4).

Missense Mutation.   As mentioned previously, changes in the structure of the Ras protein can cause it to become permanently activated. These changes are caused by a missense mutation in the *ras* gene. The human genome contains three different but evolutionarily related *ras* genes: H-*ras*, N-*ras*, and K-*ras*. All three of these homologous genes code proteins with very similar amino acid sequences consisting of 188 or 189 amino acids in total. Missense mutations in these normal *ras* genes are associated with particular forms of cancer. For example, a missense mutation in H-*ras* that changes the twelfth amino acid in the coded protein from a glycine to a valine is responsible for the conversion of H-*ras* into an oncogene. Experimentally, chemical carcinogens have been shown to cause these missense mutations and thereby lead to cancer.

Gene Amplification.   Another genetic event that may convert proto-oncogenes to oncogenes is gene amplification, or an abnormal increase in the copy number of a proto-oncogene.

TABLE 25.1
Examples of Proto-Oncogenes That Can Mutate into Oncogenes*

Gene	Cellular Function of Coded Protein
Growth factors	
sis	Platelet-derived growth factor
int-2	Fibroblast growth factor
Growth factor receptors	
erbB	Growth factor receptor for EGF (epidermal growth factor)
fms	Growth factor receptor for NGF (nerve growth factor)
Intracellular signaling proteins	
ras	GTP/GDP-binding protein
raf	Serine/threonine kinase
src	Tyrosine kinase
abl	Tyrosine kinase
Transcription factors	
myc	Transcription factor
jun	Transcription factor
fos	Transcription factor

*The genes included in this table are found in humans as well as other vertebrate species. Many of these genes were initially identified in retroviruses. Most of the genes have been given three-letter names that are abbreviations for the type of cancer the oncogene causes or the type of virus in which the gene was first identified.

An increase in gene copy number is expected to increase the amount of the coded protein, thereby contributing to malignancy. Gene amplification does not normally happen in mammalian cells, but it is a common occurrence in cancer cells. For example, Robert Gallo and Mark Groudine discovered that a gene called c-*myc* was amplified in a human leukemia cell line. Many human cancers are associated with the amplification of particular oncogenes. In such cases, the extent of oncogene amplification may be correlated with the progression of malignancy in tumors. These include the amplification of N-*myc* in neuroblastomas and of *erbB-2* in breast carcinomas.

Chromosomal Translocation.   A third type of genetic alteration that can lead to cancer is a chromosomal translocation. Abnormalities in chromosome structure are common in cancer cells, and specific types of chromosomal translocations have been identified in certain types of tumors. In 1960, Peter Nowell and David Hungerford discovered that chronic myelogenous leukemia (CML) is correlated with the presence of a shortened version of chromosome 22, which they called the Philadelphia chromosome after the city where it was discovered. Rather than being due to a deletion, this shortened chromosome is the result of a reciprocal translocation between chromosomes 9 and 22.

Later studies revealed that this translocation activates a proto-oncogene, *abl*, in an unusual way (**Figure 25.4**). The reciprocal translocation involves breakpoints within the *abl* and *bcr* genes. Following the reciprocal translocation, the coding sequence of the *abl* gene fuses with the promoter and coding sequence of the *bcr* gene. This yields an oncogene that codes an abnormal fusion protein, which contains the polypeptide sequences coded from both genes. The *abl* gene codes a tyrosine kinase enzyme, which uses ATP to attach phosphate groups onto target proteins. This phosphorylation activates certain proteins involved with cell division. Normally, the *abl* gene is highly regulated. However, in the Philadelphia chromosome, the fusion gene is controlled by the *bcr* promoter, which is active in white blood cells. This explains why this fusion causes a type of cancer called a leukemia, which is characterized by a proliferation of white blood cells.

Viral Integration.   A fourth way that oncogenes can arise is via viral integration. As part of their reproductive cycle, certain viruses integrate their genomes into the chromosomal DNA of their host cell. If the integration occurs next to a proto-oncogene, a viral promoter or enhancer may cause the proto-oncogene to be overexpressed. For example, in certain lymphomas that occur in birds, the genome of the avian leukosis virus has been found to be integrated next to the c-*myc* gene, enhancing its level of transcription.

Certain Viruses Cause Cancer by Carrying Oncogenes into Cells

As mentioned, most types of cancers are caused by mutagens that alter the structure and expression of genes. Some viruses, however, are known to cause cancer in plants, humans, and other animals. Many of these viruses also infect normal laboratory-grown cells and convert them into malignant cells. The first known virus of this type, the Rous sarcoma virus (RSV), was isolated from chicken sarcomas by Peyton Rous in 1911.

During the 1970s, RSV research led to the identification of the first gene known to cause cancer. Researchers investigated RSV by infecting chicken fibroblast cells, which are a type of connective-tissue cell, in the laboratory. This causes the chicken fibroblasts to grow like cancer cells. During the course of their studies, researchers identified mutant RSV strains that infected and proliferated within chicken cells without transforming those cells into malignant cells. These RSV strains were determined to contain a defective viral gene. In contrast, with other strains in which this gene is functional, cancer occurs. This viral gene was designated *src* for <u>sarc</u>oma, the type of cancer it causes. The *src* gene was the first example of an oncogene, a mutant gene that promotes cancer.

Since those early studies of RSV, many other retroviruses carrying oncogenes have been investigated. The characterization of such oncogenes has led to the identification of several genes with cancer-causing potential. In addition to retroviruses, several viruses with DNA genomes cause tumors, and some of these are known to cause cancer in humans (**Table 25.2**). Researchers estimate that up to 15% of all human cancers are associated with viruses.

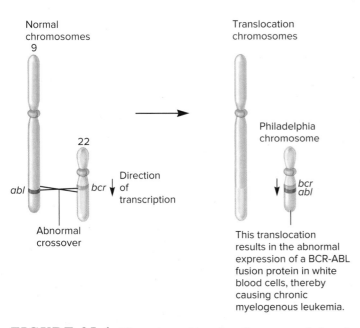

FIGURE 25.4   **The reciprocal translocation commonly found in people with chronic myelogenous leukemia.**

Genes→Traits   In healthy individuals, the *abl* gene is located on chromosome 9, and the *bcr* gene is on chromosome 22. In chronic myelogenous leukemia, a reciprocal translocation causes the *abl* gene to fuse with the *bcr* gene. This combined gene, under the control of the *bcr* promoter, codes an abnormal fusion protein that overexpresses the tyrosine kinase function of the ABL protein in white blood cells and leads to leukemia.

CONCEPT CHECK: Why does this translocation cause leukemia rather than cancer in a different tissue type, such as the lung?

TABLE 25.2

Examples of Viruses That May Cause Cancer

Virus	Description
Retroviruses	
Rous sarcoma virus (RSV)	Causes sarcomas in chickens
Hardy-Zuckerman-4 feline sarcoma virus	Causes sarcomas in cats
DNA viruses	
Hepatitis B	Causes liver cancer in several species, including humans
Papillomavirus	Causes benign tumors and malignant carcinomas in several species, including humans; causes cervical cancer in humans
Herpes virus	Causes carcinoma in frogs and T-cell lymphoma in chickens. A human herpes virus, Epstein-Barr virus, is a causative agent in Burkitt lymphoma, which occurs primarily in immunosuppressed individuals such as AIDS patients.

25.2 COMPREHENSION QUESTION

1. Which of the following is a type of genetic change that could produce an oncogene?
   a. Missense mutation
   b. Gene amplification
   c. Chromosomal translocation
   d. All of the above can produce an oncogene.

25.3 TUMOR-SUPPRESSOR GENES

Learning Outcomes:

1. Using the *rb* and *p53* genes as examples, describe how tumor-suppressor genes prevent cancer, and explain what happens when their function is lost.
2. List the two main categories of tumor-suppressor genes.
3. Summarize the types of genetic changes that can inactivate tumor-suppressor genes.
4. Describe how cancer usually involves multiple genetic changes that occur as a progression.
5. Explain how inheriting a mutant tumor-suppressor gene can result in an increased susceptibility to developing cancer.

In the previous section, we considered how oncogenes promote cancer as a result of gain-of-function mutations. An oncogene is an abnormally activated proto-oncogene that initiates uncontrolled cell growth. We now turn our attention to a second category of genes called tumor-suppressor genes. As the name suggests, the role of a tumor-suppressor gene is to prevent cancerous growth. Therefore, when a tumor-suppressor gene becomes inactivated by mutation, cancer is more

likely to occur. It is a loss-of-function mutation in a tumor-suppressor gene that promotes cancer. In this section, we will explore the possible ways that tumor-suppressor genes affect cell growth and examine the types of genetic changes that may inactivate them.

Tumor-Suppressor Genes, such as the *rb* Gene, Play a Role in Preventing Cancer

The first identification of a human tumor-suppressor gene involved studies of retinoblastoma, a tumor that occurs in the retina of the eye. Some people have inherited a predisposition to develop this disease within the first few years of life. By comparison, the noninherited form of retinoblastoma tends to occur later in life but only rarely.

Based on these differences, in 1971, Alfred Knudson proposed a two-hit model for retinoblastoma. According to this model, retinoblastoma requires two mutations to occur. People with the hereditary form have already received one mutant gene from one of their parents. They need only one additional mutation in the other copy of this tumor-suppressor gene to develop the disease. Because the retina has millions of cells, it is relatively likely that a mutation may occur in one or more of these cells at an early age, leading to the disease. However, people with the noninherited form of the disease must have two mutations in the same retinal cell to cause the disease. Because two rare events are much less likely to occur than a single such event, the noninherited form of this disease is expected to occur much later in life and only rarely. Therefore, this hypothesis explains the different populations typically affected by the inherited and noninherited forms of retinoblastoma.

Since Knudson's original hypothesis, molecular studies have confirmed the two-hit hypothesis for the occurrence of retinoblastoma. The gene in which mutations occur is designated *rb* (for retinoblastoma). This tumor-suppressor gene is found on the long arm of chromosome 13. Most people have two functional copies of the *rb* gene. Persons with hereditary retinoblastoma have inherited one functional and one defective copy. In nontumorous cells throughout the body, they have one functional copy and one defective copy of *rb*. However, in retinal tumor cells, the functional *rb* gene has also suffered the second hit (i.e., a mutation), which renders it defective. Without the tumor-suppressor function, cells are allowed to grow and divide in an unregulated manner, which ultimately leads to retinoblastoma.

More recent studies have revealed how the Rb protein suppresses the proliferation of cancer cells (**Figure 25.5**). The Rb protein regulates a transcription factor called E2F, which activates genes whose expression is required for advancement through the cell cycle. (The eukaryotic cell cycle is described in Chapter 2; see Figure 2.5.) The binding of the Rb protein to E2F inhibits its activity and prevents the cell from advancing through the cell cycle. As discussed later in this section, when a normal cell is supposed to divide, proteins called cyclins bind to cyclin-dependent protein kinases (CDKs). This activates the kinases, which then phosphorylate the Rb protein. The phosphorylated form of the Rb protein is released from E2F, thereby allowing E2F to activate genes needed to advance through the cell cycle. What happens when both copies of the *rb* gene are rendered inactive by mutation? The answer is that the E2F protein is always active, which explains why uncontrolled cell division occurs.

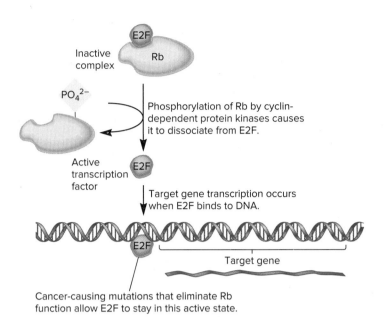

FIGURE 25.5 **Interactions between the Rb and E2F proteins.** The binding of the Rb protein to the transcription factor E2F inhibits the ability of E2F to function. This prevents cell division. For cell division to occur, cyclins bind to cyclin-dependent protein kinases, which then phosphorylate the Rb protein. The phosphorylated Rb protein is released from E2F. The free form of E2F can activate target genes needed to advance through the cell cycle.

CONCEPT CHECK: If a cell cannot make any Rb protein, how will this affect the function of E2F?

The Vertebrate *p53* Gene Is a Master Tumor-Suppressor Gene That Senses DNA Damage

After the *rb* gene, the second tumor-suppressor gene discovered was the *p53* gene. The *p53* gene is one of the most commonly altered genes involved in human cancers. About 50% of all human cancers are associated with defects in *p53*. These include malignant tumors of the lung, breast, esophagus, liver, bladder, and brain, as well as sarcomas, lymphomas, and leukemias. For this reason, an enormous amount of research has been aimed at elucidating the function of the p53 protein.

A primary role of the p53 protein is to determine if a cell has incurred DNA damage. The expression of the *p53* gene is caused by the formation of damaged DNA. The event that induces *p53* gene expression appears to be a double-strand break in the DNA. The p53 protein functions as a transcription factor. If damage is detected, p53 protein can promote three types of cellular pathways aimed at preventing the proliferation of cells with damaged DNA:

- When confronted with DNA damage, the expression of *p53* activates genes involved with DNA repair. This may prevent the accumulation of mutations that activate oncogenes or inactivate tumor-suppressor genes.
- If a cell is in the process of dividing, it can arrest itself in the cell cycle. By stopping the cell cycle, a cell gains time to repair its DNA and avoids producing two mutant daughter cells. To stop the cell cycle, the p53 protein stimulates

the expression of another gene termed *p21*. The p21 protein inhibits the formation of cyclin/CDK complexes that are needed to advance from the G_1 phase of the cell cycle to the S phase.

- A drastic pathway occurs when the expression of p53 initiates a series of events called **apoptosis,** or programmed cell death. In response to DNA-damaging agents, a cell may self-destruct if adequate DNA repair is not possible. Apoptosis is an active process that involves cell shrinkage, chromatin condensation, and DNA degradation. This process is facilitated by proteases known as **caspases,** which are sometimes called the "executioners" of the cell. Caspases digest selected cellular proteins, such as components of the intracellular cytoskeleton. This causes the cell to break down into small vesicles that are eventually phagocytized by cells of the immune system. In this way, an organism can eliminate cells with cancer-causing potential.

Tumor-Suppressor Genes Usually Code Proteins That Regulate Cell Division or Maintain Genome Integrity

Since the early discovery of the *rb* and *p53* genes, researchers have identified many tumor-suppressor genes that, when defective, contribute to the development and progression of cancer. The ones listed in **Table 25.3** are protein-coding genes. What are the general

TABLE 25.3	
Functions of Selected Tumor-Suppressor Genes	
Gene	**Function**
Genes that negatively regulate cell division	
rb	The Rb protein is a negative regulator of E2F (see Figure 25.5). The inhibition of E2F prevents the transcription of certain genes required for DNA replication and cell division.
p16	The protein kinase p16 negatively regulates cyclin-dependent kinases and thereby controls the transition from the G_1 phase of the cell cycle to the S phase.
NF1	The NF1 protein stimulates Ras to hydrolyze its GTP to GDP. Loss of NF1 function causes the Ras protein to be overactive, which promotes cell division.
APC	APC is a negative regulator of a cell-signaling pathway that leads to the activation of genes that promote cell division.
Genes that maintain genome integrity	
p53	p53 is a transcription factor that acts as a checkpoint protein and positively regulates a few specific target genes and negatively regulates others in a general manner. It acts as a sensor of DNA damage. It can prevent advancement through the cell cycle and also can promote apoptosis.
BRCA-1, BRCA-2	BRCA1 and BRCA2 proteins are both involved in the cellular defense against DNA damage. These proteins facilitate DNA repair and can promote apoptosis if repair is not achieved.

functions of these genes? They tend to fall into two broad categories: genes that negatively regulate cell division and genes that maintain genome integrity.

Negative Regulators of Cell Division.

Some tumor-suppressor genes code proteins that directly affect the regulation of cell division. The *rb* gene is an example. As mentioned earlier in this section, the Rb protein negatively regulates E2F. If both copies of the *rb* gene are inactivated, the growth of cells is accelerated. Therefore, loss of function of these kinds of negative regulators has a direct effect on the abnormal cell division rates seen in cancer cells. In other words, when the Rb protein is not present, a cell becomes more likely to divide.

Maintenance of Genome Integrity.

Alternatively, other tumor-suppressor genes play a role in the proper maintenance of the integrity of the genome. The term **genome maintenance** refers to cellular mechanisms that prevent mutations from occurring and/or prevent mutant cells from surviving or dividing. The proteins coded by the genes that participate in genome maintenance help to ensure that gene mutations or changes in chromosome structure and number do not occur or, if they do, are not transmitted to daughter cells. Such proteins can be subdivided into two classes: checkpoint proteins and DNA repair proteins.

Figure 25.6 shows a simplified diagram of cell-cycle control. Proteins called cyclins and cyclin-dependent protein kinases (CDKs) are responsible for advancing a cell through the four

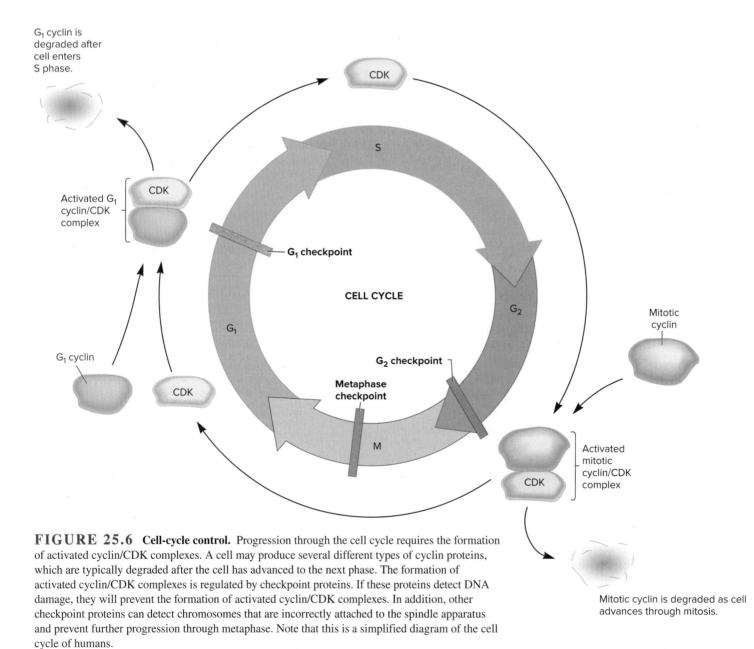

FIGURE 25.6 Cell-cycle control. Progression through the cell cycle requires the formation of activated cyclin/CDK complexes. A cell may produce several different types of cyclin proteins, which are typically degraded after the cell has advanced to the next phase. The formation of activated cyclin/CDK complexes is regulated by checkpoint proteins. If these proteins detect DNA damage, they will prevent the formation of activated cyclin/CDK complexes. In addition, other checkpoint proteins can detect chromosomes that are incorrectly attached to the spindle apparatus and prevent further progression through metaphase. Note that this is a simplified diagram of the cell cycle of humans.

CONCEPT CHECK: What is a checkpoint?

phases of the cell cycle. For example, an activated G_1 cyclin/CDK complex is necessary to advance from the G_1 to the S phase. Human cells produce several types of cyclins and CDKs.

The formation of activated cyclin/CDK complexes is regulated by a variety of factors. One means of regulation is via **checkpoint proteins** that monitor the state of the cell and stop the progression through the cell cycle if abnormalities, such as DNA breaks or improperly segregated chromosomes, are detected. These proteins are called checkpoint proteins because their role is to <u>check</u> the integrity of the genome and prevent cells from advancing past a certain <u>point</u> in the cell cycle if genetic abnormalities are detected. Therefore, these proteins provide a mechanism for stopping the accumulation of genetic abnormalities that could produce cancer cells within the body. When checkpoint proteins are lost, cell division may not be directly accelerated. However, when checkpoint proteins do not function properly, undesirable genetic changes that could cause cancerous growth are more likely to occur.

Several checkpoint proteins regulate the cell cycle of human cells. Figure 25.6 shows three of the major checkpoints where these proteins exert their effects. Both the G_1 and G_2 checkpoints involve the functions of proteins that sense if the DNA has incurred damage. If so, checkpoint proteins, such as p53, prevent the formation of active cyclin/CDK complexes. This stops the advancement of the cell cycle. A checkpoint also exists in metaphase. This checkpoint is monitored by proteins that sense if a chromosome is not correctly attached to the spindle apparatus, making it likely that it will be improperly segregated.

In addition to checkpoint proteins, a second class of proteins involved with genome maintenance consists of DNA repair proteins, which were discussed in Chapter 19 (refer back to Table 19.7). The genes coding such proteins are inactivated in certain forms of cancer. The loss of a DNA repair protein makes it more likely for a cell to accumulate mutations that could create an oncogene or eliminate the function of a tumor-suppressor gene. In Chapter 19, we considered how defects in the genes that code proteins involved in nucleotide excision repair are responsible for the disease called xeroderma pigmentosum, which results in a predisposition to developing skin cancer (refer back to the chapter-opening photo for Chapter 19). As discussed later in this section, defects in DNA mismatch repair proteins can contribute to colorectal cancer. In these cases, the loss of a DNA repair protein contributes to a higher mutation rate, which makes it more likely for other genes to incur cancer-causing mutations.

miRNAs and Cancer.   As discussed in Chapter 17, the levels of specific miRNAs are often altered in cancer cells. A well-studied example of the role of miRNAs in cancer involves a group of several different miRNAs called the miR-200 family. These miRNAs are often involved in cancers that are derived from epithelial cells, such as skin or intestinal cells. Low levels of expression of miR-200 members have been associated with many types of cancer, including bladder cancer, melanoma, stomach cancer, and colorectal cancer. The miR-200 family plays an essential role in tumor suppression by promoting cell adhesion, thereby preventing metastasis. In cancer cells, low levels of miR-200 diminish cell adhesion, which allows the cancer cells to travel to other parts of the body.

Tumor-Suppressor Genes Can Be Silenced by Loss-of-Function Mutations or Chromosome Loss

Thus far, we have considered the functions of proteins and miRNAs that are coded by tumor-suppressor genes. Cancer biologists also want to understand how tumor-suppressor genes are inactivated, because this knowledge may aid in the prevention of cancer. Researchers have identified two different types of genetic changes that diminish the function of tumor-suppressor genes:

- A mutation can occur specifically within a tumor-suppressor gene to inactivate its function. For example, a mutation could inactivate the promoter of a tumor-suppressor gene or introduce an early stop codon in the coding sequence. Either of these would prevent the expression of a functional protein.
- Many types of cancer are associated with aneuploidy. As discussed in Chapter 8, aneuploidy involves the loss or addition of one or more chromosomes, so the total number of chromosomes is not an even multiple of a set. In some cases, chromosome loss may contribute to the progression of cancer because the lost chromosome carries one or more tumor-suppressor genes.

In addition to genetic changes, epigenetic changes can also inactivate tumor-suppressor genes. These are described in Section 25.4.

Most Forms of Cancer Involve Multiple Genetic Changes Leading to Malignancy

The discovery of oncogenes and tumor-suppressor genes, along with the development of molecular techniques that can detect genetic alterations, has enabled researchers to study the progression of certain forms of cancer at the molecular level. Many cancers begin with a benign genetic alteration that, over time and with additional mutations, progresses to malignancy. Furthermore, a malignancy can continue to accumulate genetic changes that make it even more difficult to treat. For example, some tumors may acquire mutations that cause them to be resistant to chemotherapeutic agents.

In 1990, Eric Fearon and Bert Vogelstein proposed a series of genetic changes that lead to colorectal cancer, the second most common cancer in the United States. As shown in **Figure 25.7**, colorectal cancer is derived from cells in the mucosa of the colon. The loss of function of *APC*, a tumor-suppressor gene on chromosome 5, leads to an increased proliferation of mucosal cells and the development of a benign polyp, a noncancerous growth. Additional genetic changes involving the loss of other tumor-suppressor genes and the activation of an oncogene (namely, *ras*) eventually lead to the development of a carcinoma. In Figure 25.7, the genetic changes that lead to colon cancer are portrayed as occurring in an orderly sequence. However, it is the total number of genetic changes, not their exact order, that is important.

By analyzing different types of tumors, researchers have identified a large number of genes that are mutated in cancer cells. Though not all of these mutant genes have been shown to directly affect the growth rate of cells, such mutations are likely to be found in tumors because they provide some type of growth advantage for

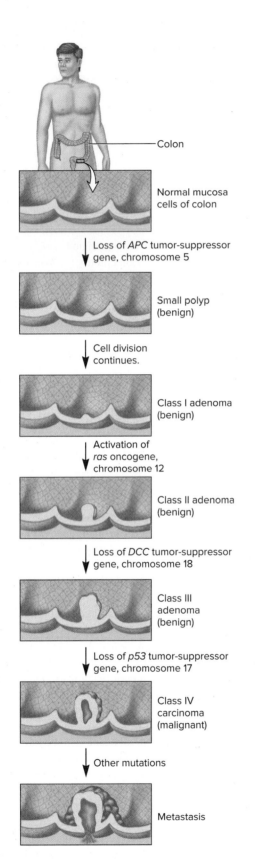

Colon

Normal mucosa
cells of colon

Loss of *APC* tumor-suppressor
gene, chromosome 5

Small polyp
(benign)

Cell division
continues.

Class I adenoma
(benign)

Activation of
ras oncogene,
chromosome 12

Class II adenoma
(benign)

Loss of *DCC* tumor-suppressor
gene, chromosome 18

Class III
adenoma
(benign)

Loss of *p53* tumor-suppressor
gene, chromosome 17

Class IV
carcinoma
(malignant)

Other mutations

Metastasis

FIGURE 25.7 Multiple genetic changes leading to colorectal cancer.

the cell population from which the cancer developed. For example, certain mutations may enable cells to metastasize to neighboring locations. These mutations may not affect growth rate, but they provide an advantage in that the cancer cells are not limited to growing in a particular location and can migrate to new locations.

Researchers have estimated that about 300 different protein-coding genes may play a role in the development of human cancers. Because the approximate genome size is about 20,000 protein-coding genes, this observation indicates that about 1.5% of those genes have the potential to promote cancer if their function is altered by a mutation. More recently, researchers have discovered that mutations in non-coding RNAs are also associated with certain forms of cancer. This topic is discussed in Chapter 17.

In addition to mutations within specific genes, other common genetic changes associated with cancer are alterations in chromosome structure and number. **Figure 25.8** compares the

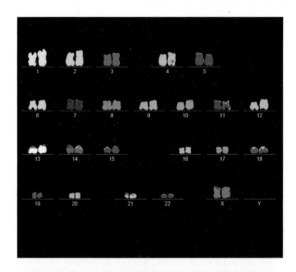

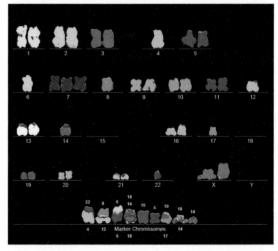

FIGURE 25.8 A comparison of chromosomes found in a normal human cell and those found in a tumor cell from the same person. The bottom set of chromosomes, from a tumor cell, is highly abnormal, with missing copies of some chromosomes and an extra copy of chromosome 7. The chromosomes designated "Marker Chromosomes" in this figure are made of fused pieces of different chromosomes.
Image courtesy of the Duesberg Lab, University of California, Berkeley

chromosome compositions of a normal cell and a tumor cell taken from the same female. The chromosome composition of the tumor cell is quite bizarre.

- Some chromosomes are missing. If tumor-suppressor genes were on these missing chromosomes, their function is lost as well.
- Figure 25.8 also shows that chromosome 7 is present in three copies in the tumor cell. If this chromosome carries proto-oncogenes, the expression of those genes may be overactive.
- Tumor cells often have chromosomes with translocations (designated marker chromosomes in this figure). Such translocations may create fusion genes (as in the case of the Philadelphia chromosome discussed earlier in this chapter), or they may place two genes close enough together so that the regulatory sequences of one gene affect the expression of the other.

Inherited Forms of Cancers Are Usually Caused by Defects in Tumor-Suppressor Genes

Before we end our discussion of the genetic basis of cancer, let's consider which genes are most likely to be affected in inherited forms of the disease. As mentioned previously, about 5–10% of all cases of cancer involve inherited (germ-line) mutations. These familial forms of cancer occur because people have inherited one or more mutant genes that give them an increased susceptibility to developing cancer. Inheriting the mutations does not mean these people will definitely get cancer, but they are more likely to develop the disease than are individuals in the general population.

When individuals have family members who have developed certain forms of cancer, they may be tested to determine if they also carry a mutant gene. For example, von Hippel-Lindau disease and familial adenomatous polyposis are examples of syndromes for which genetic testing to identify at-risk family members is considered the standard of care.

What types of genes are mutant in familial cancers? Most inherited forms of cancer involve a defect in a tumor-suppressor gene (**Table 25.4**). In these cases, the affected individual is heterozygous, with one functional and one inactive allele.

As revealed by analyses of human pedigrees, a predisposition for developing cancer is inherited in a dominant fashion because a heterozygote exhibits this predisposition. **Figure 25.9a** shows a pedigree for familial breast cancer. In this case, individuals with the disorder have inherited a loss-of-function mutation in one *BRCA-1* gene. As seen in the pedigree, the development of breast cancer shows a dominant pattern of inheritance with incomplete penetrance. Most affected individuals have an affected parent.

At the cellular level, however, the actual development of cancer is recessive. A cell must be homozygous for a loss-of-function allele to become cancerous. How does this occur? An individual initially is heterozygous, but then a somatic mutation in the normal *BRCA-1* gene eliminates its function. This somatic mutation makes the cell homozygous for two loss-of-function

TABLE 25.4	
Inherited Mutant Genes That Confer a Predisposition to Developing Cancer	
Gene*	**Type of Cancer**
Tumor-suppressor genes†	
VHL	Causes von Hippel-Lindau disease, which is typically characterized by a clear-cell renal carcinoma
APC	Familial adenomatous polyposis and familial colon cancer
rb	Retinoblastoma
p53	Li-Fraumeni syndrome, which is characterized by a wide spectrum of tumors, including soft-tissue and bone sarcomas, brain tumors, adenocortical tumors, and premenopausal breast cancers
BRCA-1	Familial breast cancer
BRCA-2	Familial breast cancer
NF1	Neurofibromatosis
MSH2	Hereditary malignant melanoma
MLH1	Nonpolyposis colorectal cancer
XP-A to *XP-G*	UV-sensitive forms of cancer such as basal cell carcinoma
Oncogenes	
RET	Multiple endocrine neoplasia type 2

*Many of the genes included in this table are mutated in more than one type of cancer. The cancers listed are those in which it has been firmly established that a predisposition to develop the disease is commonly due to germ-line mutations in the designated gene.
†*MSH2, MLH1,* and *XP-A* to *XP-G* code proteins that are involved in DNA repair.

alleles (**Figure 25.9b**). This phenomenon is called **loss of heterozygosity (LOH)**—the loss of function of a normal allele when the other allele was already inactivated.

25.3 COMPREHENSION QUESTIONS

1. Cancer is more likely to occur when tumor-suppressor genes are
   a. overexpressed.
   b. expressed in the wrong cell type.
   c. inactivated.
   d. expressed at the wrong stage of development.

2. Normal (nonmutant) tumor-suppressor genes often function
   a. as negative regulators of cell division.
   b. in the maintenance of genome integrity.
   c. in the stimulation of cell division.
   d. as both a and b.

3. Most forms of cancer involve
   a. the activation of a single oncogene.
   b. the inactivation of a single tumor-suppressor gene.
   c. the activation of multiple oncogenes.
   d. the activation of multiple oncogenes and the inactivation of multiple tumor-suppressor genes.

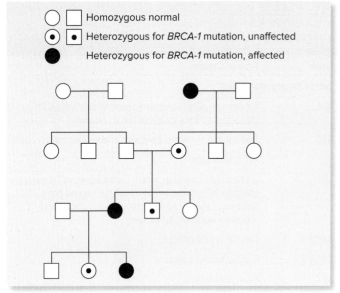

(a) Pedigree for familial breast cancer

(b) Development of breast cancer at the cellular level

FIGURE 25.9 Inheritance pattern of familial breast cancer. (a) A family pedigree that involves a loss-of-function mutation in the *BRCA-1* gene. Samples of cells from individuals in this pedigree were tested to determine if they carry a mutation in the *BRCA-1* gene. Affected individuals are represented with black symbols. The disorder follows a dominant pattern of inheritance. (Note: Males can occasionally develop breast cancer, although it is much more common in females.) **(b)** Breast cancer at the cellular level. Normal cells in an affected individual are heterozygous, whereas cancer cells in the same individual have lost their heterozygosity. Therefore, at the cellular level, cancer is recessive because both alleles must be inactivated for it to occur.

CONCEPT CHECK: Explain why familial breast cancer shows a dominant pattern of inheritance in a pedigree even though it is recessive at the cellular level.

25.4 ROLE OF EPIGENETICS IN CANCER

Learning Outcomes:

1. Explain how epigenetics plays a role in cancer.
2. Describe the underlying causes of epigenetic changes associated with cancer.

Thus far, we have focused on genetic changes that promote cancer. One of the most active fields in genetics involves the study of how epigenetic changes contribute to cancer and other human diseases. We are probably seeing only "the tip of the iceberg" in our understanding of this topic.

Researchers have identified many examples in which epigenetic changes are associated with a particular disease. The role of epigenetics and disease has been most extensively studied with regard to cancer. For some forms of cancer, the evidence suggests a cause-and-effect relationship. In this section, we will consider how epigenetics may play an important role in the development of cancer. We will first explore the types of epigenetic changes that are associated with cancer, and then examine how abnormalities may arise when epigenetic changes occur.

Abnormalities in Chromatin Modification Are Common in Cancer Cells

Three general types of chromatin modifications are often found to be abnormal in cancer cells. These abnormalities in chromatin modification are epigenetic changes, because they are passed from a cancer cell to its daughter cells:

- **DNA methylation.** A particularly common alteration in cancer cells is a change in the level of DNA methylation. For example, hypermethylation—an abnormally high level of methylation, typically at CpG islands, is often observed. This hypermethylation may promote cancer by inhibiting the expression of tumor-suppressor genes.
- **Covalent modification of histones.** As described in Chapter 15, the covalent modification of histones can affect the expression of genes, either activating or inhibiting transcription depending on the pattern of modification. Histone tails are subject to a variety of modifications, including the attachment of acetyl, methyl, and phosphate groups (refer back to Figure 15.5). Variation in the covalent modification of histones has been shown to occur at specific genes in cancer cells. Depending on the specific type of modification, such changes could increase the expression

of oncogenes or inhibit the expression of tumor-suppressor genes.

- **Chromatin remodeling.** Another important chromatin modification is chromatin remodeling, in which the locations of nucleosomes are changed. Abnormalities in the locations of nucleosomes have been frequently found in cancer cells. Depending on how the nucleosomes are rearranged, such changes could increase the expression of oncogenes or inhibit the expression of tumor-suppressor genes.

Epigenetic Changes Associated with Cancer May Arise Because Mutations Occur in Genes That Code Chromatin-Modifying Proteins or Because Environmental Agents Alter the Functions of Chromatin-Modifying Proteins

Thus far, we have considered three general types of epigenetic changes that contribute to cancer. Researchers want to understand how these changes arise. In other words, what causes chromatin modifications to become abnormal and promote cancer? The underlying cause of the abnormality can be placed in two general categories:

1. Mutations may occur in genes that code chromatin-modifying proteins.
2. Environmental agents may alter the functions of chromatin-modifying proteins.

Mutations in Genes That Code Chromatin-Modifying Proteins.
Chromatin modifications are carried out by an array of different proteins. Mutations in the genes that code these proteins are a common occurrence in many or perhaps most types of cancer (**Table 25.5**). The mutation is a genetic event, but it has an epigenetic effect because it alters the function of a chromatin-modifying protein. For example, a mutation may occur in a gene that codes a DNA methyltransferase, thereby inhibiting that enzyme's function. Such inhibition will decrease the level of DNA methylation, and this epigenetic change can be passed from one cancer cell to its daughter cells. As noted in Table 25.5, this type of mutation is often found in cells that give rise to acute myeloid leukemia.

In some cases, the mutations described in Table 25.5 increase the function of the chromatin-modifying proteins, but it is more common for a mutation to inhibit their function. In either case, the mutation may have a widespread effect on gene expression. As discussed in Chapter 15, chromatin modifications can convert chromatin from a closed (transcriptionally inactive) to an open (transcriptionally active) state, and vice versa. When the function of chromatin-modifying proteins is altered by mutation, this may cause other genes to be overexpressed by converting chromatin to an open conformation or it may inhibit gene expression by converting chromatin to a closed conformation. Such changes may increase the expression of oncogenes or inhibit the expression of tumor-suppressor genes, respectively. Both types of change can contribute to cancer.

As a cautionary note, the relationships presented in Table 25.5 are based on observed correlations between epigenetic changes and particular types of cancer. When a statistically significant correlation coefficient is obtained, how do we interpret its meaning? Such a result suggests a true **association**—changes in the two variables follow a pattern. For example, with a positive correlation, when one variable increases, the other variable also increases. However, such an association does not necessarily imply a cause-and-effect relationship. When considering the role of epigenetic changes and cancer, an association can arise in three common ways:

- The epigenetic changes directly contribute to the proliferation of cancer cells. There is a cause-and-effect relationship.
- Conversely, cancerous cells may arise first, and molecular alterations in the cancerous cells may cause subsequent epigenetic changes to happen. This is also a cause-and-effect relationship, but in the opposite direction.
- The association is indirect because a third factor is involved. For example, a toxic agent in the environment may cause a type of cancer, and it may also cause particular types of epigenetic changes even though those epigenetic changes do not contribute to the disease.

In general, correlation coefficients are quite useful in identifying associations between two variables. Caution is necessary, however, because a statistically significant correlation coefficient, by itself, cannot prove that the association is due to cause and effect. Even so, research studies that identify associations are very useful because they provide the motivation to carry out further research to determine if a cause-and-effect relationship exists.

TABLE 25.5

Mutations in Genes That Code Chromatin-Modifying Proteins and Their Occurrence in Different Types of Cancers

Type of Modification	Type of Protein Coded by Mutant Gene	Protein Function	Particular Cancer(s) in Which Mutant Gene Is Observed
DNA methylation	DNA methyltransferase	Methylates DNA	Acute myeloid leukemia
Histone modification	Histone acetyltransferase	Attaches acetyl groups to histones	Colorectal, breast, and pancreatic cancer
Histone modification	Histone methyltransferase	Attaches methyl groups to histones	Renal and breast cancer
Histone modification	Histone demethylase	Removes methyl groups from histones	Multiple myeloma and esophageal cancer
Histone modification	Histone kinase	Attaches phosphate groups to histones	Medulloblastoma, glioma
Chromatin remodeling	SWI/SNF complex	Alters the positions of histones	Lung, breast, prostate, and pancreatic cancer

TABLE 25.6

Environmental Agents That Are Associated with Cancer and Are Known to Cause Epigenetic Changes

Environmental Agent	Occurrence	Particular Cancers Associated with Agent
Polycyclic aromatic hydrocarbons	Tobacco smoke, automobile exhaust, charbroiled food	Lung, breast, stomach, and skin cancer
Benzene	Tobacco smoke, automobile exhaust	Leukemia, lymphoma, multiple myeloma
Endocrine disruptors (e.g., diethylstilbestrol)	Insecticides, fungicides, herbicides, and some types of plastic	Breast, prostate, and thyroid cancer
Cadmium	Tobacco products, production of batteries	Lung and breast cancer
Nickel	Occupational exposure in mining, welding, and electroplating, and in the manufacturing of jewelry, stainless steel, and batteries	Lung and nasal cancer
Arsenic	Lead alloy, feed additive in agriculture, insecticides	Skin, bladder, kidney, and liver cancer

Environmental Agents That Alter the Functions of Chromatin-Modifying Proteins. A second way that chromatin modification can be abnormally altered is by exposure to environmental agents. Such agents may directly alter the functions of chromatin-modifying proteins, or they may initiate a series of cellular changes that ultimately affect the functions of chromatin-modifying proteins. **Table 25.6** lists several examples of environmental agents that are known to cause epigenetic changes involving abnormalities in DNA methylation, histone modification, and/or chromatin remodeling. These agents are also associated with particular forms of cancer.

For some of the examples listed in Table 25.6, scientific evidence indicates that the association is causative. For example, certain agents in tobacco smoke have been shown to cause cellular changes that underlie lung cancer. Furthermore, some of the chemicals in tobacco smoke, such as polycyclic aromatic hydrocarbons, are both mutagenic and epimutagenic, which makes them particularly potent at promoting changes that may lead to cancer. Alternatively, some of the agents listed in Table 25.6 show an association with particular cancers, but researchers are still trying to determine if the epigenetic changes caused by these agents actually promote changes that result in cancer.

25.4 COMPREHENSION QUESTIONS

1. Which of the following types of epigenetic changes may promote cancer?
   a. DNA methylation
   b. Covalent modification of histones
   c. Chromatin remodeling
   d. All of the above may promote cancer.
2. The underlying cause(s) of epigenetic changes associated with cancer may be
   a. mutations in genes that code chromatin-modifying proteins.
   b. environmental agents that alter the function of chromatin-modifying proteins.
   c. mutations in genes that code proteins that directly accelerate cell growth.
   d. all of the above.
   e. both a and b.

25.5 CANCER THERAPEUTICS

Learning Outcomes:

1. List and briefly describe five different options for the treatment of cancer.
2. Explain why some cancer drugs are aimed at actively dividing cells, and discuss a few examples.
3. Describe examples in which targeted drug therapy is used to kill certain types of cancer cells.
4. Define *immunotherapy*, and briefly discuss different types.
5. Compare and contrast potential cancer treatments that are aimed at non-coding RNAs and epigenetic changes.

In the United States, cancer is the second leading cause of death, just behind heart disease. For this reason, an enormous amount of research has been directed at developing treatment options for cancer. Many of these treatments have resulted from studies that revealed the underlying molecular causes of cancer. We have already discussed many examples of such causes in Sections 25.2 through 25.4. In this section, we will explore various types of current and potential treatment options, and relate most of them to our molecular and cellular understanding of cancer.

Many Different Treatment Options Are Available to Combat Cancer

During the past century or so, many different types of cancer treatments have become available. The type of treatment that a patient receives depends on the type of cancer that has been diagnosed and how advanced it is. Some people with cancer will undergo only one type of treatment, but most undergo a combination of two or more. Most cancer treatments fall into a few main categories (**Table 25.7**):

Physical Methods. For solid tumors such as those found in breast and lung cancer, a common treatment option is surgery. The goal of surgery is to remove the cancer or as much of it as possible. However, in some cases, removal of a tumor may not be possible or may not be the best treatment option. Instead, tumor cells can be killed while still in the body, a process called ablation. In cryoablation, cancer cells are killed with a repeated freezing and

TABLE 25.7

Treatment Options for Cancer

Treatment	Description
Physical methods	Removes or physically kills cancer cells; includes surgery and ablation.
Killing actively dividing cells	Radiation and chemotherapy are usually used to kill cells that are actively dividing.
Targeted drug therapy	A specific protein is targeted that has an abnormal structure and/or is overactive in cancer cells.
Hormone therapy	Decreases the level of hormones or blocks their effects to stop or slow the growth of cancer cells.
Immunotherapy	Uses the immune system or components from the immune system to fight cancer.

thawing process. Radiofrequency ablation uses electrical energy to heat cancer cells, causing them to die.

Killing Actively Dividing Cells. Because cancer cells exhibit uncontrolled cell division, a common approach in cancer treatment is to use therapies that exert their effects on actively dividing cells. Radiation therapy uses high-energy beams, such as X-rays or proton beams, whereas chemotherapy uses drugs to kill cancer cells. These types of treatments can kill cells in a variety of ways, such as by damaging DNA, interfering with DNA replication, and/or inhibiting specific steps in cell division. We will consider the mechanism of action of a few chemotherapy drugs later in this section.

Targeted Drug Therapy. **Targeted drug therapy** focuses on abnormalities within cancer cells that allow them to survive and proliferate. Typically, the drug targets a specific protein that has an abnormal structure and/or is overactive in cancer cells. We will examine specific examples later in this section.

Hormone Therapy. Some types of cancer cells are stimulated to divide by natural hormones. For example, the hormone estrogen can promote the proliferation of certain types of breast cancer. Decreasing the level of those hormones in the body or blocking their effects may cause the cancer cells to stop growing or grow more slowly.

Immunotherapy. **Immunotherapy** uses the immune system or components from the immune system to fight cancer. As discussed later, some forms of immunotherapy help a patient's immune system to detect and destroy cancer cells. In other forms of immunotherapy, components from the immune system are synthesized in large amounts in a laboratory and then injected into patients to directly attack cancer cells.

Thus far, surgery and/or agents that kill actively dividing cells (radiation therapy and chemotherapy) remain the most common methods for cancer treatment. However, as we gain a greater understanding of the genetic and epigenetic changes that occur in various types of cancer, the newer approaches of targeted drug therapy, immunotherapy, and hormone therapy are gaining wider use. In the rest of this section, we'll take a closer look at drugs that kill actively dividing cells, targeted drug therapy, and immunotherapy.

A Wide Variety of Cancer Drugs Are Aimed at Killing Actively Dividing Cells

As mentioned, the mechanism of action of many chemotherapy drugs is to kill cells that are actively dividing, which includes cancer cells. However, some noncancerous cells in the body are also actively dividing. The normal cells most likely to be damaged by this type of chemotherapy include blood-forming cells in the bone marrow, hair follicles, and cells that line the digestive tract. Common side effects caused by chemotherapy are fatigue, hair loss, anemia, nausea, and loss of appetite.

Let's consider a few examples in which a chemotherapy drug kills cancer cells because they are actively dividing:

DNA-Damaging Agents. A category of drugs called alkylating agents directly damage DNA to prevent cancer cells from undergoing cell division. They have a chemical structure that contains a bifunctional nitrogen mustard, which includes two reactive alkyl groups. (The action of nitrogen mustard is described in Chapter 19.) The bifunctional group can produce a crosslink in DNA when both alkyl groups form covalent bonds with two guanine bases in DNA. During cell division, such crosslinking may result in single- and double-strand DNA breaks, and ultimately cause cell death. Alkylating agents are used to treat many different cancers, including those of the lung, breast, and ovary as well as leukemia, lymphoma, Hodgkin disease, multiple myeloma, and sarcoma. Examples of alkylating agents are busulfan and cyclophosphamide.

Disruptors of DNA Replication. Some chemotherapy drugs interfere with DNA replication, which only occurs when cells are dividing. For example, the drug irinotecan inhibits topoisomerase I, which is a protein required for DNA replication. Inhibition of topoisomerase I blocks DNA replication and leads to double-strand breaks. The double-strand breaks, in turn, cause cell death via apoptosis. The sensitivity of cells to irinotecan and other topoisomerase I inhibitors largely depends on the amount of topoisomerase I inside the cell. Cancer cells typically express higher levels of this protein and therefore are more likely to be killed. Certain types of cancer cells, such as those found in colon cancer and lung cancer, are especially sensitive to topoisomerase I inhibitors.

Cell Cycle Inhibitors. Cell cycle inhibitors, also called mitotic inhibitors, stop cells from dividing to form new cells. Depending on the drug, they can halt cell division in one or more phases of the cell cycle. (Note: The cell cycle is discussed in Chapter 2; see Figure 2.5.) For example, vinblastine and vincristine are chemotherapy drugs isolated from the leaves of the

Madagascar periwinkle plant, *Catharanthus roseus.* They inhibit the cell cycle by interfering with the mitotic spindle. They are used to treat many different types of cancer, including breast, lung, myelomas, lymphomas, and leukemias.

Targeted Drug Therapies Are Usually Aimed at Specific Proteins That Are Produced by Cancer Cells

As discussed in Section 25.2, cancer cells may produce specific proteins in overabundance and/or produce proteins that have altered structures. These changes contribute to cancerous growth. Targeted drug therapy usually requires **biomarker testing,** which is a way to identify genes, proteins, and other substances that have become abnormal in the cancer cells of a given patient. To carry out biomarker testing, a sample of cancer cells is removed from a patient and tested for the types of genetic and biochemical abnormalities they may contain. Biomarker testing provides the information that is needed to choose appropriate types of targeted drug therapy.

Several types of cancers are currently treated with targeted drug therapy; these include breast cancer, chronic myeloid leukemia, colorectal cancer, lung cancer, lymphoma, and melanoma. Let's consider a few examples of targeted drug therapy.

Targeting the Estrogen Receptor.   In most types of breast cancer, the cancer cells produce too much of a protein called the estrogen receptor (ER). Estrogen binds to this overly expressed receptor and thereby promotes tumor growth. If biomarker testing reveals that a tumor overexpresses ER—that is, it is ER positive—targeted drug therapy options are available. For example, tamoxifen is a drug that acts by binding to the estrogen receptor. When tamoxifen binds to this receptor, it prevents the binding of estrogen, and thereby stops cell division (**Figure 25.10a**). The effects of tamoxifen depend on the cell type. In breast cancer cells, tamoxifen has mostly an inhibitory effect on the estrogen receptor; as a result, the cancer cells do not divide and eventually die due to that effect.

Targeting the Epidermal Growth Factor Receptor.   Many types of cancer cells, including those found in certain forms of breast cancer and colorectal cancer overexpress a protein called human epidermal growth factor receptor 2 (HER2). The binding of the epidermal growth factor to this receptor promotes cell division; cell division becomes cancerous in cells that overexpress HER2. Tumors that are HER2 positive may be treated with drugs that block this receptor and thereby help to stop or slow cancer growth. Certain types of tyrosine kinase inhibitors bind to the tyrosine kinase domain in HER2 and prevent it from autophosphorylating itself. This stops the activation of the signaling pathway (**Figure 25.10b**). An example is dacomitinib. Another option is the use of monoclonal antibodies, which is a form of immunotherapy that is discussed later in this section.

Targeting the ABL Protein.   As discussed in Section 25.2, chronic myeloid leukemia (CML) may be caused by a fusion between the *bcr* and *abl* genes (refer back to Figure 25.4). Interestingly, the study of the *abl* gene has led to an effective treatment

for CML. The drug imatinib fits into the active site of the BCR-ABL fusion protein, preventing ATP from binding there. Without ATP, the BCR-ABL fusion protein cannot phosphorylate certain signaling proteins (**Figure 25.10c**). This prevents the fusion protein from stimulating cell division, and thereby provides an effective treatment option. In CML, imatinib targets the BCR-ABL fusion protein, but it can also target ABL alone, and is used to treat some other forms of leukemia and certain types of stomach cancer.

Immunotherapy Uses the Immune System or Components from the Immune System to Kill Cancer Cells

In addition to fighting pathogens such as viruses and bacteria, the immune system in mammals plays a key role in preventing cancer. It is able to recognize abnormal proteins that are produced from cancer cells and elicit a series of events that ultimately kills those cancer cells. This immune surveillance is happening all the time in the bodies of healthy people. Unfortunately, cancer may occur if cells incur a series of changes that either outcompete or evade the immune system. Also, as people age, the immune system becomes less robust, which is one reason why cancer is more common in older people.

Several types of immunotherapy are used to treat cancer, including the following.

Monoclonal Antibodies.   Antibodies are proteins that are made by the immune system and that recognize antigens and mount a response against them or the cells that are producing them. **Monoclonal antibodies (mAbs)** are a homogenous population of antibodies that are made in a laboratory and recognize one specific antigen. To combat cancer, it is possible to produce mAbs that recognize a protein found in cancer cells. If the mAbs are viewed as a drug, this is another example of a targeted drug treatment. How do mAbs kill cancer cells? Most of them work in one of three ways:

- Cancer cells that are coated in mAbs may be more easily detected by a cancer patient's immune system and targeted for destruction.
- Researchers can design mAbs to directly attack and kill cancer cells. For this to happen, some toxic agent is attached to the mAbs, such as a radioactive particle or a chemotherapy drug.
- For some cancers, mAbs may inhibit the function of a protein that is overexpressed in cancer cells. As mentioned earlier, certain types of cancer cells overexpress HER2. Tumors that are HER2 positive may be treated with mAbs that can bind to the extracellular component of the HER2 protein and prevent natural growth factors from binding to the receptor. This prevents receptor activation. Cetuximab and panitumumab are mAbs that are effective agents against HER2-positive breast and colorectal cancers.

T-cell Transfer Therapy.   T cells are a type of immune system cell involved in killing cancer cells. In T-cell transfer therapy, a patient's own immune cells are taken from a tumor, and those T cells that are most active against the cancer cells are selected or

Without drug

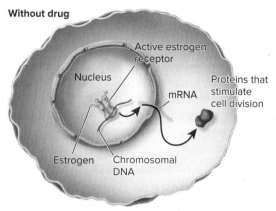

Estrogen binds to the estrogen receptor. The receptor stimulates genes to make mRNA. The mRNAs are translated into proteins that stimulate cell division.

(a) Targeting the estrogen receptor with tamoxifen

With drug

Tamoxifen binds to the estrogen receptor. The receptor does not stimulate genes to make mRNA. Cell division is stopped and eventually cell death occurs.

Without drug

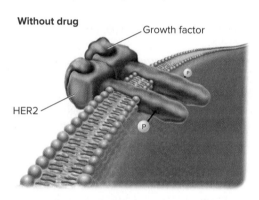

When growth factor binds, HER2 is autophosphorylated and stimulates the cell to divide.

(b) Targeting HER2 with dacomitinib

With drug

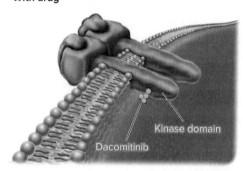

Dacomitinib binds to kinase domain and prevents autophosphorylation of HER2. This stops cell division.

Without drug

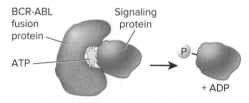

BCR-ABL fusion protein binds ATP and phosphorylates a signaling protein. The phosphorylated signaling protein stimulates cell division.

(c) Targeting the BCR-ABL fusion protein with imatinib

With drug

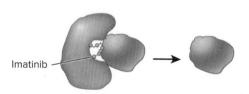

The binding of imatinib to the BCR-ABL fusion protein prevents the binding of ATP. The signaling protein is not phosphorylated and cell division is stopped.

FIGURE 25.10 **Examples of targeted drug therapy for cancer.** See text for further details.

CONCEPT CHECK: Which of these three drugs blocks a hormone from binding to its receptor?

they are changed in the laboratory to better attack the cancer cells. The T cells are grown in large batches, and then put back into a patient's body to destroy the cancer cells.

Checkpoint Inhibitors. Immune checkpoints are a normal part of the immune system; they keep immune responses from becoming too strong. Immune checkpoint inhibitors are drugs that inhibit immune checkpoints. Therefore, these drugs allow immune cells to respond more strongly to cancer.

Immune System Modulators. Various drugs and proteins are known to enhance the body's immune response against cancer. Some of these agents affect specific parts of the immune system, whereas others affect the immune system in a more general way.

Vaccines. Vaccines are discussed in Chapter 21 (refer back to Section 21.2). Vaccines used to treat cancer work by boosting a patient's immune system's response to cancer cells. They are used in people who already have cancer. Such vaccines can be made in different ways. For example, a cancer treatment vaccine can be made from a sample of tumor cells from a patient's body, or it may be made from protein antigens that are usually found on cancer cells of many people with a specific type of cancer.

Drug Therapy May Be Aimed at Epigenetic Changes

A potentially exciting application of the research on epigenetic changes associated with cancer is the development of new drugs that may reverse these changes, thereby providing another treatment option for cancer patients. Researchers are actively investigating drugs that may inhibit cancer cells by affecting either DNA methylation or covalent histone modifications.

For example, inhibitors of DNA methyltransferase, the enzyme that attaches methyl groups to DNA, are being developed to treat certain forms of cancer, including leukemia—a cancer of white blood cells. Drugs such as 5-azacytidine and decitabine, which inhibit DNA methyltransferase, have shown some promising results when used in conjunction with other anticancer drugs. In some cases, clinical improvement in patients with leukemia has been associated with a decrease in DNA methylation. Although the specific mechanisms for patient improvement are not well understood at the molecular level, one possibility is that the lower level of DNA methylation has reversed the inhibition of tumor-suppressor genes.

Drug Therapy May Be Used to Alter Non-coding RNAs

Historically, most drugs that are used to treat human diseases bind to cellular proteins. This bias developed because we have a much better understanding of the effects of proteins on cell structure and function than we do about the effects of non-coding RNAs (ncRNAs). (Note: ncRNAs are discussed in Chapter 17.) However, as we learn more about the functions of ncRNAs and their roles in human diseases, researchers are beginning to conduct studies to determine if ncRNAs and components needed for the processing of microRNAs (miRNAs) may be effective targets for novel treatment therapies. So far, much of this work has centered on miRNAs and cancer. Although the development of this methodology is still in its infancy, researchers are optimistic that the approach may be applicable to other types of diseases and other categories of ncRNAs.

Therapies That Inhibit miRNA Function. One novel strategy for treating certain forms of cancer is the use of **anti-miRNA oligonucleotides (AMOs)** to inhibit miRNA function. An AMO is complementary to a specific miRNA and can base-pair with it. This action can block the ability of the miRNA to function and/or cause it to be degraded. Some AMOs contain chemical modifications that cause them to bind more tightly to their complementary miRNAs. Examples of these AMOs include the following:

- **Locked nucleic acids (LNAs)** contain a ribose sugar that has an extra bridge connecting the 2′ oxygen and 4′ carbon. The bridge locks the ribose in a conformation that causes it to bind more tightly to the complementary miRNA.
- **Antagomirs** have one or more base modifications that may promote a stronger binding to their complementary miRNAs.

Studies involving mice have shown that AMOs may be useful in the treatment of certain types of cancer. For example, in a mouse strain that develops mammary tumors, the intravenous injection of an AMO that binds to a particular miRNA called miR-10b inhibited the onset of metastasis and suppressed the dissemination of cancer cells to the lungs. However, the silencing of a single miRNA might not be sufficient to inhibit the growth of most types of cancer cells. Recent research suggests that several miRNAs may need to be simultaneously inhibited to slow or prevent the progression of the cancer.

Therapies That Restore miRNA Function. Some miRNAs behave like tumor suppressors. When their function is decreased, cancer may occur. For this reason, researchers are developing strategies to restore the function of miRNAs that have been down-regulated in cancer cells. This approach has been called miRNA replacement therapy. Two examples of this therapy are as follows:

- A viral vector was used to restore a particular miRNA called miR-26a in mice with a type of liver cancer called hepatocellular carcinoma. This therapy inhibited the progression of cancer.
- Another approach for increasing the levels of miRNAs is to target the miRNA-processing machinery. The drug enoxacin enhances the miRNA-processing machinery in certain cancer cells and thereby results in an increase of many miRNAs. Recent research has shown that this drug may inhibit the growth of cancer. In mice, the drug did not affect miRNA levels in normal cells and was not associated with harmful side effects.

25.5 COMPREHENSION QUESTIONS

1. Which of the following is *not* a treatment option for cancer?
   a. Surgery
   b. Radiation therapy and/or chemotherapy
   c. Targeted drug therapy
   d. All of the above are treatment options for cancer.
2. For targeted drug therapy, biomarker testing is needed to
   a. identify proteins that are abnormal or overexpressed in cancer cells.
   b. put a mark on cancer cells so the drug can specifically enter them.
   c. introduce a new gene into cancer cells that will inhibit cell division.
   d. determine if DNA replication can be inhibited.

KEY TERMS

Introduction: cancer, carcinogen
25.1: clonal, benign, malignant, invasive, metastatic, oncogene, tumor-suppressor gene
25.2: proto-oncogene, growth factors
25.3: apoptosis, caspases, genome maintenance, checkpoint protein, loss of heterozygosity (LOH)

25.4: association
25.5: targeted drug therapy, immunotherapy, biomarker testing, monoclonal antibodies (mAbs), anti-miRNA oligonucleotides (AMOs), locked nucleic acids (LNAs), antagomirs

CHAPTER SUMMARY

- Cancer is a disease that affects multicellular organisms and is characterized by uncontrolled cell division.

25.1 Overview of Cancer

- Cancer is clonal in origin. It usually develops via a multistep process that involves several mutations, begins with benign growth, and advances to growth that is invasive and metastatic (see Figure 25.1).
- Cancer can be caused by mutations that overexpress oncogenes or inhibit the expression of tumor-suppressor genes. Epigenetic changes can also promote cancer.

25.2 Oncogenes

- Oncogenes often exert their effects via intracellular signaling pathways that control the cell cycle (see Figure 25.2).
- Oncogenes arise due to gain-of-function mutations in proto-oncogenes. An example is a mutation in the *ras* gene that decreases the ability of the Ras protein to hydrolyze GTP (see Table 25.1, Figure 25.3).
- Types of genetic changes that can convert proto-oncogenes to oncogenes include missense mutation, gene amplification, chromosomal translocation, and viral integration (see Figure 25.4).
- Certain viruses cause cancer by carrying oncogenes into cells (see Table 25.2).

25.3 Tumor-Suppressor Genes

- Knudson proposed a two-hit model to explain the occurrence of retinoblastoma (see Figure 25.5).
- The vertebrate *p53* gene senses DNA damage and prevents damaged cells from dividing. Its expression may result in apoptosis.
- Most tumor-suppressor genes code proteins that are negative regulators of cell division or play a role in genome maintenance (see Table 25.3).

- Checkpoint proteins prevent cells that may harbor genetic abnormalities from advancing through the cell cycle (see Figure 25.6).
- Tumor-suppressor genes can be silenced by loss-of-function mutations or chromosome loss.
- Most forms of cancer involve multiple genetic changes (see Figure 25.7).
- Alterations in chromosome structure and number are common in cancer cells (see Figure 25.8).
- An inherited predisposition to develop cancer is usually caused by a mutation in a tumor-suppressor gene (see Table 25.4).
- Familial breast cancer exhibits a dominant pattern of inheritance. At the cellular level, loss of heterozygosity (LOH) promotes cancer (see Figure 25.9).

25.4 Role of Epigenetics in Cancer

- Epigenetic changes associated with cancer may arise because mutations occur in genes that code chromatin-modifying proteins or because environmental agents alter the functions of chromatin-modifying proteins. Either of these events can activate oncogenes or inactivate tumor-suppressor genes (see Tables 25.5, 25.6).

25.5 Cancer Therapeutics

- Many different treatment options are available to treat cancer (see Table 25.7).
- Radiation therapy and many chemotherapy drugs are aimed at killing cancer cells that are actively dividing.
- Targeted drug therapy uses drugs that bind to proteins that promote cancerous growth and are overproduced or abnormal in cancer cells (see Figure 25.10).
- Immunotherapy uses the immune system or components from the immune system to kill cancer cells.
- Drug therapy may be aimed at epigenetic changes or noncoding RNAs.

PROBLEM SETS & INSIGHTS

MORE GENETIC TIPS   **1.** In lung tumors in cigarette smokers, a mutation that is commonly found is a mutation in the *ras* gene that prevents the Ras protein from hydrolyzing GTP. Look back at the intracellular signaling pathway involving EGF shown in Figure 25.2. How would such a mutation affect cell growth?

TOPIC: *What topic in genetics does this question address?* The topic is cancer. More specifically, the question asks you to explain how a mutation in the *ras* gene could promote cancer.

INFORMATION: *What information do you know based on the question and your understanding of the topic?* In the question, you are given the effects of a mutation on the function of the Ras protein. From your understanding of the topic, you may remember that the Ras protein is active in its GTP-bound form and inactive in its GDP-bound form.

PROBLEM-SOLVING **S** TRATEGY: *Compare and contrast.* One strategy to solve this problem is to compare and contrast the functions of the normal and mutant Ras proteins. If the mutant protein is more active, that will promote cell division, whereas, if it is less active, that will inhibit cell division.

ANSWER: In this case, the mutation would cause the Ras protein to remain active for a longer period of time and thereby keep the intracellular signaling pathway activated. This would cause cell growth (division) to occur when it is not supposed to, and thereby contribute to cancerous cell growth.

2. Oncogenes sometimes result from genetic rearrangements (e.g., translocations) that produce gene fusions. An example is the formation of a Philadelphia chromosome, in which a reciprocal translocation between chromosomes 9 and 22 leads to fusion of the first part of the *bcr* gene with the *abl* gene. Suggest two different reasons why a gene fusion can create an oncogene.

TOPIC: *What topic in genetics does this question address?* The topic is chromosomal rearrangements and how they may affect gene expression. More specifically, the question is about the translocation that produces a Philadelphia chromosome.

INFORMATION: *What information do you know based on the question and your understanding of the topic?* From the question, you know that the Philadelphia chromosome has a fusion of the *bcr* and *abl* genes due to a chromosomal translocation. From your understanding of the topic, you may remember that the promoter controls the transcription of a gene and that the coding sequence determines the structure and function of the protein.

PROBLEM-SOLVING **S** TRATEGIES: *Relate structure and function. Predict the outcome.* One strategy to solve this problem is to relate the structural change in the gene to possible changes in gene expression and protein function.

ANSWER: An oncogene is derived from a genetic change that abnormally activates the expression of a gene that plays a role in cell division. A genetic change that creates a gene fusion can abnormally activate the expression of a gene in two ways.

The first way is at the level of transcription. The promoter and part of the coding sequence of one gene may become fused with the coding sequence of the other gene. For example, the promoter and part of the coding sequence of the *bcr* gene may fuse with the coding sequence of the *abl* gene. After this has occurred, the *abl* gene is under the control of the *bcr* promoter, rather than its normal promoter. Because the *bcr* promoter is turned on in different cells than the *abl* promoter is, overexpression of *abl* occurs in certain cell types where its product is not normally found.

A second way that a gene fusion can cause abnormal activation is at the level of protein structure. A fusion protein has parts of two different polypeptides. One portion of a fusion protein may affect the structure of the second portion in such a way that the second portion becomes abnormally active, or vice versa.

Conceptual Questions

C1. Identify the key properties of cancer cells that distinguish them from noncancerous (normal) cells.

C2. What is the difference between an oncogene and a tumor-suppressor gene? Give two examples of each type of gene.

C3. What is a proto-oncogene? What are the typical functions of proteins coded by proto-oncogenes? At the level of protein function, what are the general ways that proto-oncogenes can be converted to oncogenes?

C4. What is a retroviral oncogene? Is it necessary for viral infection and proliferation? How have retroviruses acquired oncogenes?

C5. A genetic predisposition to developing cancer is usually inherited as a dominant trait. At the level of cellular function, are the alleles involved actually dominant? Explain why some individuals who

have inherited these dominant alleles do not develop cancer during their lifetime.

C6. List four types of genetic changes that commonly convert a proto-oncogene to an oncogene. Explain how the genetic changes are expected to alter the activity of the gene product.

C7. Which of the following mutations affecting proteins involved in the intracellular signaling pathway that is activated by epidermal growth factor (EGF) would you expect to promote cancer?

   A. A mutation that prevents GTP from binding to Ras

   B. A mutation that keeps MAPK active for a longer period of time than normal

   C. A mutation that prevents the dephosphorylation of the EGF receptor

C8. The effects of point mutations within the coding sequence of a gene are discussed in Chapter 19 (refer back to Table 19.1). For each of the following types of point mutations to promote cancer, would you expect the mutation to occur in a proto-oncogene, a tumor-suppressor gene, or either kind of gene?

A. Nonsense mutation

B. Missense mutation

C. Frameshift mutation

C9. What are the two main functions of tumor-suppressor genes?

C10. A random change in the sequence of a gene is more likely to result in a loss-of-function mutation rather than a gain-of-function mutation. Also, cancer usually results from multiple genetic changes, as illustrated in Figure 25.7. Based on these ideas, which kinds of cancer-causing mutation would you expect to happen first, the inactivation of tumor-suppressor genes or the activation of oncogenes? After such an early mutation has happened, explain why further mutations that promote uncontrolled cell division become more likely to occur.

C11. Describe how changes in chromosome structure and number may promote cancer.

C12. Certain inherited forms of cancer, such as breast cancer involving the *BRCA-1* gene, show a dominant pattern of inheritance in a family pedigree. However, at the cellular level, the mutant allele is actually recessive. Explain these two (seemingly contradictory) observations.

C13. Relatively few inherited forms of cancer involve the inheritance of mutant oncogenes. Instead, most inherited forms of cancer are due to defects in tumor-suppressor genes. Give two or more reasons why inherited forms of cancer seldom involve activated oncogenes.

C14. The *rb* gene codes a protein that inhibits E2F, a transcription factor that activates several genes involved in cell division. Mutations in *rb* are associated with certain forms of cancer, such as retinoblastoma. Under each of the following conditions, would you expect cancer to occur?

A. One copy of the *rb* gene is defective; both copies of the *E2F* gene are functional.

B. Both copies of *rb* are defective; both copies of *E2F* are functional.

C. Both copies of *rb* are defective; one copy of *E2F* is defective.

D. Both copies of both *rb* and *E2F* are defective.

C15. A *p53* knockout mouse in which both copies of *p53* are defective has been produced by researchers. This type of mouse appears normal at birth. However, it is highly sensitive to UV light. Based on your knowledge of *p53*, explain the normal appearance at birth and the high sensitivity to UV light.

C16. With regard to cancer cells, which of the following statements are *true*?

A. Cancer cells are clonal, which means that they are derived from a single mutant cell.

B. To become cancerous, cells usually accumulate multiple genetic changes that eventually result in uncontrolled growth.

C. Most cancers are caused by oncogenic viruses.

D. Cancer cells have lost the ability to properly regulate cell division.

C17. When the DNA of a human cell becomes damaged, the *p53* gene is activated. What is the general function of the p53 protein? Is it an enzyme, transcription factor, cell-cycle protein, or something else? Describe three ways in which the synthesis of the p53 protein affects cellular function. Why is it beneficial for these three things to happen when a cell's DNA has been damaged?

C18. How can environmental agents that do not cause gene mutations contribute to cancer?

C19. Would epigenetic changes that promote cancer be passed to offspring?

C20. What are the more established methods of cancer treatment, and which ones are more recent and rely on a greater understanding of cancer at the molecular and cellular levels?

C21. For each of the following cancer treatments, briefly explain how the targeted drug therapy works.

A. Targeting the estrogen receptor with tamoxifen

B. Targeting HER2 with daconitinib

C. Targeting the BCR-ABL protein with imatinib

C22. Compare and contrast anti-miRNA oligonucleotides (AMOs), locked nucleic acids (LNAs), and antagomirs, which may eventually be used to treat certain forms of cancer in humans.

C23. What is miRNA replacement therapy? Describe three examples of this treatment approach.

Experimental Questions

E1. Discuss ways to distinguish whether a particular form of cancer involves an inherited predisposition or is due strictly to (postzygotic) somatic mutations. In your answer, consider that only one mutation may be inherited, but the cancer might develop only after several somatic mutations.

E2. The codon change (Gly-12 to Val-12) in human H-*ras* that converts it to oncogenic H-*ras* has been associated with many types of cancers. For this reason, researchers would like to develop drugs to inhibit oncogenic H-*ras*. Based on your understanding of the Ras protein, what types of drugs might you develop? In other words, what would be the structure of the drugs, and how would they inhibit Ras protein? How would you test the efficacy of the drugs? What might be some side effects?

E3. A research study indicated that an agent in cigarette smoke caused the silencing of the tumor-suppressor gene called *p53*. However, using sequencing, no mutation was found in the DNA sequence for this gene. Give two possible explanations for these results.

E4. Let's suppose you were interested in developing drugs to prevent epigenetic changes that may contribute to cancer. What cellular proteins would be the target of your drugs? What possible side effects might your drugs cause?

Questions for Student Discussion/Collaboration

1. Go to the PubMed website and search using the words *epigenetic* and *cancer*. Scan through the journal articles you retrieve, and make a list of environmental agents that may cause epigenetic changes that contribute to cancer.

2. The U.S. government has finite funds to devote to cancer research. Discuss which of the following areas of research you think should receive the most funding.

   A. Identifying and characterizing oncogenes and tumor-suppressor genes

   B. Identifying agents in our environment that cause cancer

   C. Identifying viruses that cause cancer

   D. Devising methods aimed at killing cancer cells in the body

   E. Informing the public of the risks involved in exposure to carcinogens

   In the long run, in which of these areas would you expect successful research to be the most effective in decreasing human mortality due to cancer?

Note: All answers are available for the instructor in Connect; the answers to the even-numbered questions and all of the Concept Check and Comprehension Questions are in Appendix B.

A fruit fly (Drosophila melanogaster). The fruit fly is a model organism that has been widely used in the study of developmental genetics.
N A Callow/Avalon

26

DEVELOPMENTAL GENETICS

Multicellular organisms, such as animals and plants, begin their lives with a fairly simple organization, namely, a fertilized egg, and then proceed step by step to achieve a much more complex arrangement. As this process occurs, cells divide and change their characteristics to become highly specialized units within a multicellular individual. Each cell in an organism has its own particular role. In animals, for example, muscle cells produce movement, and intestinal cells facilitate the absorption of nutrients. This division of labor among the various cells and organs of the body promotes the survival of the individual.

Developmental genetics is concerned with the roles that genes play in orchestrating the changes that occur during development. In this chapter, we will examine how the sequential actions of genes provide a program for the development of an organism from a fertilized egg to an adult. The last couple of decades have seen staggering advances in our understanding of developmental genetics at the molecular level.

Scientists have chosen a few experimental organisms, such as the fruit fly (*Drosophila melanogaster*), nematode (*Caenorhabditis elegans*), zebrafish (*Danio rerio*), mouse (*Mus musculus*), and a flowering plant (*Arabidopsis thaliana*), to use in research into the identification and characterization of the genes required for development to proceed. In certain organisms, notably *Drosophila*, many

of the genes that play a critical role in the early stages of development have been identified. Researchers are now studying how the proteins coded by these genes control the course of development. In this chapter, we will explore the body plans of invertebrates, vertebrates, and plants and consider many genes whose actions in governing the developmental process are well understood.

26.1 OVERVIEW OF ANIMAL DEVELOPMENT

Learning Outcomes:

1. List the four types of events that give rise to a body pattern.
2. Define *positional information,* and describe three ways in which it is conveyed.
3. Describe the Notch signaling pathway, and explain how it may affect embryonic development in animals.
4. Outline the four overlapping phases of animal development.

Although the details differ widely among various species, the general features and steps in development are largely conserved within the animal kingdom. In this section, we will examine the types of molecules that promote developmental changes and consider how genes play the underlying role in orchestrating a plan of development that advances from a fertilized egg to an adult.

679

The Generation of a Body Pattern Depends on Four Types of Cellular Events

Multicellular development in animals (and plants) follows a body plan, or pattern. The term **body pattern** refers to the spatial arrangement of different regions of the body. At the cellular level, the body pattern is due to the arrangement of cells and their specialization.

The progressive growth of a fertilized egg into an adult organism involves four types of cellular events: cell division, cell migration, cell differentiation, and cell death (apoptosis) (**Figure 26.1**). The coordination of these four events leads to the formation of a body with a particular pattern. As we will see, the timing of gene expression and the localization of gene products at precise regions in the fertilized egg and early embryo are the critical phenomena that underlie this coordination.

Positional Information Controls How Cells Behave During Development

Before we consider how genes affect development, let's examine a central concept in developmental biology known as **positional information.** For an organism to develop a body pattern with unique morphological and cellular features, each cell of the body

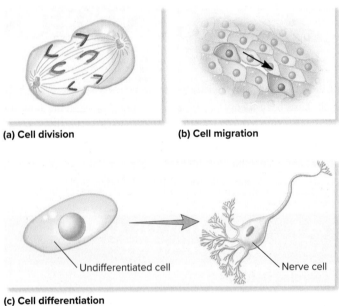

(a) Cell division   (b) Cell migration

(c) Cell differentiation

Undifferentiated cell   Nerve cell

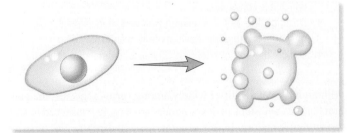

(d) Cell death (apoptosis)

FIGURE 26.1 **Four types of cellular events that occur during the course of development.**

must receive signals—positional information—that affect its course of development.

A **morphogen** is a molecule that conveys positional information and promotes developmental changes over time. It is often a diffusible molecule that acts in a concentration-dependent manner to influence the developmental fate of a cell. **Cell fate** is determined by developmental events that will ultimately cause a cell or group of cells to have a particular destiny. The fate of a cell describes its future identity, or the identity of its daughter cells, before it is actually phenotypically detectable. For example, an embryonic cell may have the fate of becoming a neuron long before it has the molecular and morphological features of a neuron.

Morphogens provide the positional information that stimulates a cell to divide, migrate, differentiate, or die. For example, when a region of a *Drosophila* embryo is exposed to a high concentration of the morphogen known as Bicoid, it develops structures that are characteristic of the anterior region of the body. Many morphogens function as transcription factors that regulate the expression of many genes. This topic is discussed in greater detail later in this chapter.

How do morphogens control pattern development? Within an oocyte and during embryonic development, morphogens typically are distributed along a concentration gradient. In other words, the concentration of a morphogen varies from low to high in different regions of the developing organism. A key feature of morphogens is that they act in a concentration-dependent manner. In a particular concentration range, a morphogen or a combination of two or more different morphogens restricts a cell to a specific developmental pathway.

During the earliest stages of development of certain species, morphogenic gradients are preestablished within the oocyte (**Figure 26.2a**). Following fertilization, the zygote subdivides into many smaller cells. Due to the preestablished gradients of morphogens within the oocyte, these smaller cells have higher or lower concentrations of morphogens, depending on their location in the embryo. In this way, morphogen gradients in the oocyte provide positional information that is important in establishing two main axes in the embryo: the anteroposterior axis and the dorsoventral axis, which are described in Section 26.2.

Morphogen gradients are also established in the embryo by secretion and transport to neighboring cells (**Figure 26.2b**). A certain cell or group of cells may synthesize and secrete a morphogen at a specific stage of development. After secretion, the morphogen is transported to neighboring cells. The concentration of the morphogen is usually highest near the cells that secrete it. The morphogen then influences the developmental fate of the cells exposed to it. The process by which a cell or group of cells governs the developmental fate of neighboring cells is known as **induction.**

In addition to the effects of morphogens, positional information is conveyed by **cell adhesion** (**Figure 26.2c**). Each cell makes its own collection of surface receptors that enable it to adhere to other cells and/or to the extracellular matrix (ECM), which consists primarily of carbohydrates and fibrous proteins. Such receptors are known as **cell adhesion molecules (CAMs).** A cell may acquire positional information via the combination of contacts it makes with other cells or with the ECM. Different cell types

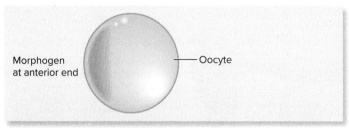

(a) Preestablished morphogenic gradient in an oocyte

Morphogen
at anterior end

Oocyte

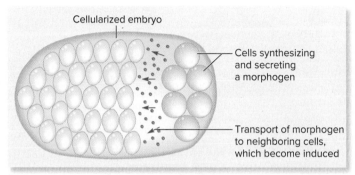

Cellularized embryo

Cells synthesizing
and secreting
a morphogen

Transport of morphogen
to neighboring cells,
which become induced

**(b) Asymmetric secretion and extracellular transport of a morphogen
and induction of neighboring cells in an embryo**

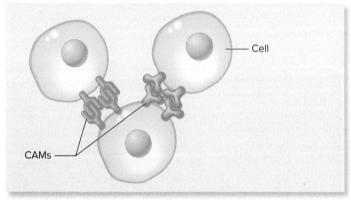

Cell

CAMs

(c) Cell adhesion conveying positional information

FIGURE 26.2 Three molecular mechanisms for conveying positional information.

CONCEPT CHECK: Which of these mechanisms involve(s) diffusible morphogens?

express unique combinations of CAMs in order to self-sort during embryonic development. Such CAMs are thought to determine which cells tend to stay connected, even as widespread rearrangements occur in a developing embryo.

The phenomenon of cell adhesion, and its role in multicellular development, was first recognized by Henry Wilson in 1907. Wilson passed multicellular sponges through a sieve, which disaggregated them into individual cells. Remarkably, the cells actively migrated until they adhered to one another to form a new sponge, complete with the chambers and canals that characterize a sponge's internal structure! When sponge cells from different species were mixed, they sorted themselves properly, adhering only to cells of the same species. Overall, these results indicate that cell adhesion plays an important role in governing the position a cell adopts during development.

Cell-to-Cell Contacts May Activate Signaling Pathways That Lead to Developmental Changes

As we have seen in Figure 26.2c, CAMs provide positional information and allow neighboring cells to adhere to one another. As embryonic development advances, adjacent cells also influence each other's fate via cell signaling pathways. In animals, one example is cell signaling via the **Notch receptor.** This receptor is named after the notched-wing phenotype of fruit flies that are heterozygous for a loss-of-function mutation in the *Notch* gene, which codes the receptor. *Drosophila* has one *Notch* gene, whereas mammals have four *Notch* genes. During embryonic development, Notch signaling plays a key role in cell differentiation, cell proliferation, and apoptosis. In adults, Notch signaling affects wound healing and the growth of cancer cells.

With regard to cell differentiation, Notch signaling can influence the fate of adjacent cells in two common ways. In some cases, Notch signaling promotes different cell fates for two neighboring cells, a process known as **lateral inhibition.** For example, in one embryonic region, one cell might become a neuron and the adjacent cell could become a muscle cell. Alternatively, in another region of the embryo, the opposite process, called **lateral induction,** may occur. During lateral induction, adjacent cells adopt identical or similar cell fates.

Even though Notch signaling may have diverse effects during development, the core components of Notch signaling are highly conserved among all animals. The general steps in Notch signaling are shown in **Figure 26.3**.

1. A membrane-bound Notch ligand in a signal-sending cell binds to the Notch receptor, which is found in the signal-receiving cell. In particular, two classes of ligands, referred to as Delta-like and Jagged-like, can bind to Notch receptors.
2. The signal-sending cell attempts to endocytose the Notch receptor. This exerts a force on the Notch receptor that changes the Notch negative regulatory region (NRR) from a closed to an open conformation. This conformational change allows an enzyme called metalloprotease to cleave the Notch receptor at the extracellular surface, releasing the Notch ligand-binding domain. Another enzyme, called secretase, is then able to cleave near the intracellular surface, thereby releasing the Notch intracellular domain (NICD) into the cytosol.
3. The NICD moves into the nucleus, associates with transcription factors such as CoA and CSL, and influences the transcription of specific target genes.

Even though the core components of Notch signaling are highly conserved among different embryonic cells in all animal species, the effects of Notch signaling can be quite diverse. For example, it can cause lateral inhibition, lateral induction, cell proliferation, or apoptosis. How can the same signaling components produce such diverse effects? Though the answer is not completely understood, specific variables are known to affect Notch signaling. These include the following.

Signaling Duration. The Notch receptor is used only once. The number of Notch receptors and the Notch ligands in adjacent cells

FIGURE 26.3 **Notch signaling in animals.** See the text for added details regarding the steps in this signaling pathway.

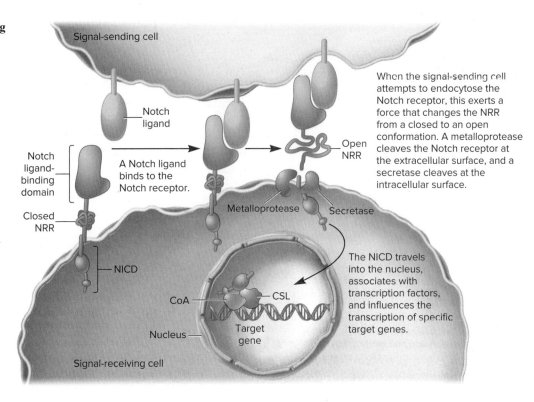

Signal-sending cell

Notch ligand

Notch ligand-binding domain

A Notch ligand binds to the Notch receptor.

Closed NRR

Open NRR

When the signal-sending cell attempts to endocytose the Notch receptor, this exerts a force that changes the NRR from a closed to an open conformation. A metalloprotease cleaves the Notch receptor at the extracellular surface, and a secretase cleaves at the intracellular surface.

Metalloprotease   Secretase

NICD

The NICD travels into the nucleus, associates with transcription factors, and influences the transcription of specific target genes.

CoA   CSL

Nucleus   Target gene

Signal-receiving cell

affect the duration of Notch signaling. A longer duration can promote the transcription of more genes in the signal-receiving cell.

Signaling Topology.   In some cases, Notch ligands may be preferentially located on one side of a cell and thereby affect which adjacent cells within an embryo receive the signal.

Nuclear Factors.   The cells throughout an embryo differ with regard to transcription factors that are found within their cell nucleus. Also, changes in chromatin structure, such as covalent histone modifications, vary throughout the embryo. Therefore, when the NICD enters the nucleus, the effects on gene transcription may be different.

Covalent Modifications of the Notch Receptor.   The formation and function of the Notch receptor are regulated by a variety of covalent modifications including proteolysis, glycosylation, phosphorylation, hydroxylation, and ubiquitination.

Crosstalk Among Different Signaling Pathways.   In addition to Notch signaling, embryonic cells express several other signaling pathways that can influence cell fate. These signaling pathways can activate or inhibit each other, a phenomenon called crosstalk.

The Study of Mutants with Disrupted Developmental Patterns Has Identified Genes That Control Development

Mutations that alter the course of development in experimental organisms, such as *Drosophila*, have greatly contributed to our understanding of the normal process of development. For example, **Figure 26.4**

FIGURE 26.4 **The *bithorax* mutation in *Drosophila*.**

Genes→Traits A normal fly has two wings on the second thoracic segment and two halteres on the third thoracic segment. However, this mutant fly contains multiple mutations in a cluster of genes called the *bithorax* complex. In this fly, the third thoracic segment has the same characteristics as the second thoracic segment, thereby producing a fly with four wings instead of the normal two.

David Scharf/Science Source

is a photograph of a fruit fly that carries multiple mutations in a cluster of genes called *bithorax*. This mutant fly has four wings instead of two; the halteres (balancing organs that resemble miniature wings), which are normally found on the third thoracic segment, have been changed into wings, normally found only on the second thoracic segment. (The term *bithorax* refers to the observation that the characteristics of the second thoracic segment are duplicated.)

Edward Lewis, a pioneer in the genetic study of development, became interested in the bithorax phenotype and began investigating it in 1946. Researchers later discovered that the mutant chromosomal region contains a complex of three genes involved in specifying developmental pathways in the fly. A gene that plays a central role in specifying the final identity of a body region is called a **homeotic gene.** We will discuss particular examples of homeotic genes later in this chapter.

During the 1960s and 1970s, interest in the relationship between genetics and embryology blossomed as biologists began to appreciate the role of genetics at the molecular and cellular levels. It became clear that the genomes of multicellular organisms contain groups of genes that initiate a program of development involving networks of gene regulation. By identifying mutant alleles that disrupt development, geneticists have begun to unravel the pattern of gene expression that underlies the normal pattern of multicellular development.

Animal Development Occurs in Four Overlapping Phases

With the exception of sponges, all animal bodies are organized along axes. Some simple animals, such as jellyfish, exhibit radial symmetry in which the body is circularly symmetrical around an axis. However, most animals, including humans, are **bilaterians,** meaning that they have an anteroposterior axis with left-right symmetry. Our discussion of animal development in this chapter will focus on the bilaterians. For these species, the body is typically organized along two major axes: an anteroposterior axis and a dorsoventral axis. We will consider these axes in Section 26.2 when we discuss invertebrate development. Left-right symmetry occurs relative to the anteroposterior axis.

By comparing a variety of bilaterians, researchers have discovered that development generally proceeds in four overlapping phases (**Figure 26.5**):

1. *Formation of body axes.* The first phase of pattern development is the establishment of the body axes.
2. *Segmentation of the body.* In the second phase, the body is subdivided into segments. In invertebrates, the segments may remain morphologically distinct, whereas in vertebrates, distinct segments are obvious only in the early stages of development.
3. *Determination of structures within the segments.* As the segments form, groups of cells become destined to develop into particular structures and cell types. This process, called **determination,** occurs before the structures and cell types have changed their morphologies.
4. *Cell differentiation.* Toward the end of development, cells differentiate into particular cell types. This results in tissues and organs with specific morphologies.

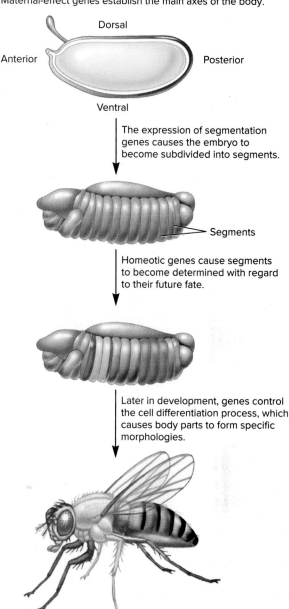

Maternal-effect genes establish the main axes of the body.

Dorsal

Anterior   Posterior

Ventral

The expression of segmentation genes causes the embryo to become subdivided into segments.

Segments

Homeotic genes cause segments to become determined with regard to their future fate.

Later in development, genes control the cell differentiation process, which causes body parts to form specific morphologies.

FIGURE 26.5   **The four overlapping phases of animal development.**

CONCEPT CHECK: How does genetics play a role in development?

In Sections 26.2 and 26.3, we will explore the phases of animal development in greater detail and discuss examples of genes that play a key role in these four phases.

26.1 COMPREHENSION QUESTIONS

1. Positional information may provide a cue for a cell to
   a. divide.
   b. migrate.
   c. differentiate.
   d. undergo apoptosis.
   e. do any of these.

2. Molecules that convey positional information include
   a. diffusible morphogens.
   b. cell adhesion molecules.
   c. ATP.
   d. both a and b.

3. Which of the following is the correct order for the four developmental phases in animals?
   A. Segmentation of the body
   B. Determination
   C. Cell differentiation
   D. Formation of body axes
   a. A, B, C, D
   b. A, D, C, B
   c. D, A, B, C
   d. D, A, C, B

26.2 INVERTEBRATE DEVELOPMENT

Learning Outcomes:

1. List the stages of *Drosophila* development.
2. Compare and contrast the effects of maternal effect genes, gap genes, and homeotic genes on *Drosophila* development.
3. Explain how an understanding of cell lineages in *Caenorhabditis elegans* aids in the identification of mutations that affect the timing of developmental changes.

This discussion of invertebrate development focuses on two model organisms, *Drosophila melanogaster* and *Caenorhabditis elegans*, that have been pivotal to our understanding of developmental genetics. *Drosophila* has been studied for a variety of reasons. First, the techniques for generating and analyzing mutants in this organism are highly advanced, and researchers have identified many mutant strains with altered developmental pathways. Second, at the embryonic and larval stages, *Drosophila* is large enough to conduct experiments in which portions of the body are transplanted to different sites. It is also small enough to examine under a microscope to determine where particular genes are expressed at critical stages of development. By comparison, *C. elegans* is studied by developmental geneticists because of its simplicity. The adult organism is a small transparent worm composed of only about 1000 somatic cells. From the fertilized egg onward, the pattern of cell division and the developmental fate of each cell are completely known.

In this section, we will begin by examining the stages of *Drosophila* development. We will then turn to embryonic development, because it is during this stage that the overall body plan is determined. We will see how the timing of the expression of particular genes and the localization of gene products within the embryo influence the developmental process. We will then briefly consider development in *C. elegans* and examine how the timing of gene expression plays a key role in determining the developmental fate of particular cells in this organism.

Drosophila Advances Through Several Developmental Stages to Become an Adult

Figure 26.6 illustrates the general sequence of events in *Drosophila* development. The oocyte is the cell most critical in determining the pattern of development in the adult organism. It is an elongated cell with preestablished axes (Figure 26.6a). After fertilization takes place, the zygote goes through a series of nuclear divisions that are not accompanied by cytoplasmic division. Initially, the resulting nuclei are scattered throughout the yolk, but eventually they migrate to the periphery of the cytoplasm. This is the syncytial blastoderm stage (Figure 26.6b).

After the nuclei line up along the cell membrane, individual cells are formed as portions of the cell membrane surround each nucleus, creating a cellular blastoderm (Figure 26.6c). This structure is composed of a sheet of cells surrounding the yolk in the center. In this arrangement, the cells are distributed asymmetrically. At the posterior end are a group of cells called the pole cells—the primordial germ cells that eventually give rise to gametes in the adult organism. After blastoderm formation is complete, dramatic changes occur during gastrulation (Figure 26.6d). This stage produces three cell layers known as the ectoderm, mesoderm, and endoderm.

As this process occurs, the embryo subdivides into detectable units. Initially, shallow grooves partition the embryo into 14 **parasegments,** which are transient subdivisions. A short time later, these grooves disappear, and new boundaries are formed that divide the embryo into morphologically discrete **segments.** Figure 26.6e shows the segmented pattern of a *Drosophila* embryo about 10 hours after fertilization.

At the end of **embryogenesis,** which lasts about 18–22 hours, a larva hatches from the eggshell (Figure 26.6f) and begins feeding. *Drosophila* has three larval stages, separated by molts. During molting, the larva sheds its cuticle, a hardened extracellular shell that is secreted by the epidermis. Each larval stage between two molts is called an instar.

After the third larval stage, a pupa is formed and undergoes metamorphosis. During metamorphosis, groups of cells called imaginal disks, which were produced earlier in development, differentiate into structures found in the adult fly, such as the head, wings, legs, and abdomen. At 10–14 days after fertilization, an adult fly emerges from the pupal case. In *Drosophila*, as in all bilateral animals, an adult body is organized along two major axes: the **anteroposterior axis** and the **dorsoventral axis** (Figure 26.6g). The **left-right axis** is oriented relative to the anteroposterior axis. An additional axis, used mostly for designating limb parts, is the **proximodistal axis.** *Proximal* refers to a location that is closer to the center of the body, whereas *distal* is a location farther away.

The Early Stages of Embryonic Development Determine the Pattern of Structures in the Adult Organism

Although many interesting developmental events occur during the three larval stages and the pupal stage of *Drosophila*, we will focus most of our attention on the events that occur in the oocyte and during embryonic development. Even before hatching, the embryo has

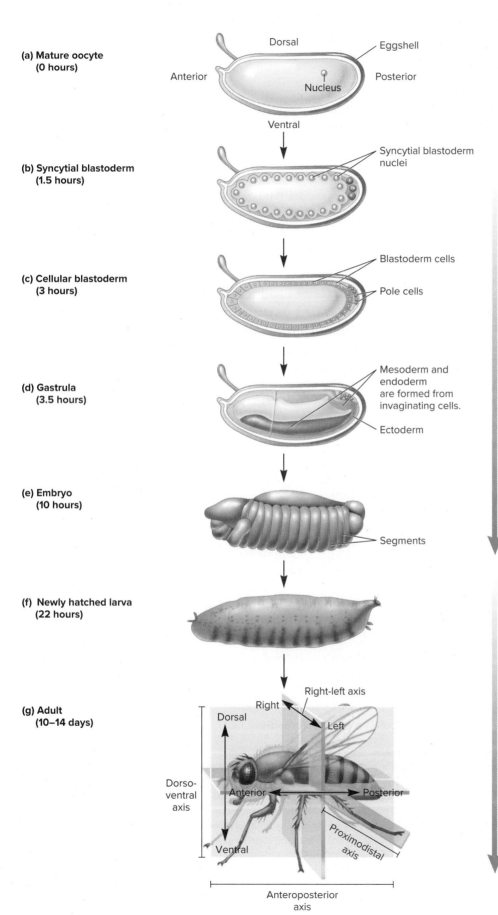

(a) Mature oocyte (0 hours)

(b) Syncytial blastoderm (1.5 hours)

(c) Cellular blastoderm (3 hours)

(d) Gastrula (3.5 hours)

(e) Embryo (10 hours)

(f) Newly hatched larva (22 hours)

(g) Adult (10–14 days)

Progression of Developmental Events

Preestablishment of axes in oocyte, by nurse cells in the ovary. Nurse cells are not shown.

Formation of body segments

Formation of structures within each segment

FIGURE 26.6 Developmental stages of *Drosophila*. (**a**) Oocyte. (Note: The outer gray layer is an eggshell.) (**b**) Syncytial blastoderm. (**c**) Cellular blastoderm. (**d**) Gastrula. (**e**) Embryo at 10 hours. (Note: The eggshell surrounding the embryo is not shown.) (**f**) Newly hatched larva. (**g**) Adult. (Note: Prior to the adult stage, metamorphosis occurs during a pupal stage. The adult emerges from the pupal case.)

TABLE 26.1

Examples of *Drosophila* Genes That Play a Role in Pattern Development

Description	Examples
Maternal effect genes play a role in determining the axes of development. Also, certain genes govern the formation of the extreme terminal (anterior and posterior) regions.	Anterior: *bicoid, exuperantia, swallow, staufen.* Posterior: *nanos, cappuccino, oskar pumilio, spire, staufen, tudor, vasa.* Terminal: *torso, torsolike, Trunk, NTF-1.* Dorsoventral: *Toll, cactus, dorsal, easter, gurken, nudel, pelle, pipe, snake, spatzle.*
Segmentation genes play a role in promoting the subdivision of the embryo into segments. The three types are gap genes, pair-rule genes, and segment-polarity genes.	Gap genes: *empty spiracles, giant, huckebein, hunchback, knirps, Krüppel, tailless, orthodenticle.* Pair-rule genes: *even-skipped, hairy, runt, fushi tarazu, paired.* Segment-polarity genes: *frizzled, frizzled-2, engrailed, patched, smoothened, hedgehog, wingless, gooseberry.*
Homeotic genes play a role in determining the fate of particular segments. *Drosophila* has two clusters of homeotic genes, known as the *Antennapedia* complex and the *bithorax* complex.	*Antennapedia* complex: *labial, proboscipedia, Deformed, Sex combs reduced.* Bithorax complex: *Ultrabithorax, abdominal A, Abdominal B.*

developed the basic body plan that is eventually found in the adult (see Figure 26.6e). In other words, during the early stages of development, the embryo is divided into segments that correspond to the segments of the larva and adult. Therefore, an understanding of how these segments form is critical to our understanding of pattern formation.

In *Drosophila*, the establishment of the body axes and division of the body into segments involves the participation of a few dozen genes. **Table 26.1** lists many of the important genes governing pattern formation during embryonic development. These genes were identified by characterizing mutants that had altered development patterns, and their names are often based on the phenotypes observed in the mutants. It is beyond the scope of this text to describe how all of these genes exert their effects during embryonic development. Instead, we will consider a few examples that illustrate how the expression of a particular gene and the localization of its gene product have a specific effect on the pattern of development.

The Gene Products of Maternal Effect Genes Are Deposited Asymmetrically into the Oocyte and Establish the Anteroposterior and Dorsoventral Axes at a Very Early Stage of Development

The first stage in *Drosophila* embryonic pattern development is the establishment of the body axes. This occurs before the embryo becomes segmented. The morphogens necessary to establish these axes are distributed prior to fertilization. During oogenesis, gene products such as mRNAs, which are important in early developmental stages, are deposited asymmetrically within the egg. Later, after the egg has been fertilized and development begins, these gene products establish developmental programs that govern the formation of the body axes of the embryo.

As shown in **Figure 26.7**, a few products of maternal effect genes act as key morphogens, or receptors for morphogens, that initiate changes in embryonic development. As shown in the figure, these gene products are deposited asymmetrically in the egg. For example, the product of the *bicoid* gene is necessary to initiate development of the anterior structures of the organism. During oogenesis, the mRNA from *bicoid* accumulates in the anterior region of the oocyte. In contrast, the mRNA from the *nanos* gene accumulates in the posterior end. Later in development, the *nanos* mRNA is translated into a protein, which functions to promote posterior development. The *nanos* gene is required for the formation of the abdomen.

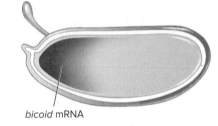

bicoid mRNA

(a) Anterior distribution of *bicoid* mRNA

nanos mRNA

(b) Posterior distribution of *nanos* mRNA

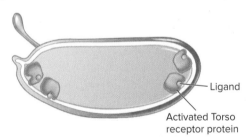

Ligand

Activated Torso receptor protein

(c) Terminal distribution of Torso receptor protein

Ligand

Activated Toll receptor protein

(d) Ventral distribution of Toll receptor protein

FIGURE 26.7   Establishment of the axes of polarity in *Drosophila*. This figure shows some of the maternal effect gene products deposited in the oocyte that are critical in the establishment of the anteroposterior and dorsoventral axes. **(a)** The *bicoid* mRNA is distributed in the anterior end of the oocyte and promotes the formation of anterior structures. **(b)** The *nanos* mRNA is localized to the posterior end and promotes the formation of posterior structures. **(c)** The Torso receptor protein is found in the plasma membrane and is activated by ligand binding at either end of the oocyte. It causes the formation of structures that are found only at the ends of the organism. **(d)** The Toll receptor protein is activated by ligand binding at the ventral side of the embryo and establishes the dorsoventral axis. (Note: The Torso and Toll receptor proteins are distributed throughout the plasma membranes of the oocyte, but they are activated by ligand binding only in the regions shown in this figure. The gray region surrounding the oocyte and embryo is the eggshell; the oocyte is inside the eggshell. Torso and Toll proteins are embedded in the plasma membrane of the oocyte and embryonic cells.)

CONCEPT CHECK: Describe the orientations of the anteroposterior and dorsoventral axes.

In addition to being affected by morphogens such as Bicoid and Nanos, the development of the structures at the extreme anterior and posterior ends of the embryo is regulated, in part, by a receptor protein called Torso. This protein is activated by the binding of a signaling molecule, called a ligand, which occurs only at the anterior and posterior ends of the egg. Such activation is necessary for the formation of the terminal ends of the embryo. The dorsoventral axis is established by the actions of several proteins, including a receptor protein known as Toll. This receptor is found in plasma membranes of cells throughout the embryo. However, ligand binding is needed to activate Toll, and this binding occurs only along the ventral midline of the embryo.

The Morphogen Bicoid Is a Transcription Factor That Controls the Development of Anterior Structures

Let's now take a closer look at the molecular mechanism of one morphogen, Bicoid. The *bicoid* gene got its name because a larva whose female parent is defective in this gene develops with two posterior ends (**Figure 26.8**). This allele exhibits a maternal effect pattern of inheritance, in which the genotype of the female parent determines the phenotype of the offspring (see Chapter 5). A female with one or two copies of a functional *bicoid* gene (*bicoid*⁺) produces larva that are morphologically normal. In contrast, a female fly that is homozygous for an inactive *bicoid* allele produces 100% abnormal offspring (see Figure 26.8b). Such abnormal offspring are produced even when the female is mated to a male that is homozygous for the normal *bicoid*⁺ allele. In other words, the genotype of the female parent determines the phenotype of the offspring. This occurs because the *bicoid* gene product is provided to the oocyte via maternal nurse cells.

In the ovaries of female flies, the nurse cells are localized asymmetrically toward the anterior end of the oocyte. During oogenesis, gene products are transferred from nurse cells into the oocyte via cell-to-cell connections called cytoplasmic bridges. Maternally coded gene products enter one side of the oocyte, which becomes the anterior end of the embryo (**Figure 26.9a**). The *bicoid* gene is transcribed in the nurse cells, and *bicoid* mRNA is transported into the anterior end of the oocyte. The 3′ end of *bicoid* mRNA contains a

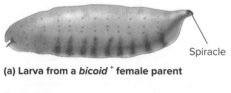

(a) Larva from a *bicoid*⁺ female parent

(b) Larva from a *bicoid*⁻ female parent

FIGURE 26.8 The *bicoid* mutation in *Drosophila.* (a) A larva from a normal *bicoid*⁺ female parent. (b) A larva from a homozygous *bicoid*⁻ female parent in which both ends of the larva develop posterior structures. For example, both ends develop a spiracle, which normally is found only at the posterior end. This is a lethal condition.

CONCEPT CHECK: What is the normal function of the Bicoid protein?

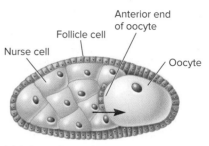

(a) Schematic drawing of transport of maternal effect gene products into a developing oocyte

(b) Schematic drawing of in situ hybridization of *bicoid* mRNA

(c) Schematic drawing of immunostaining of Bicoid protein in a fertilized egg

FIGURE 26.9 **Asymmetrical localization of gene products during oogenesis in *Drosophila*.** (a) The nurse cells transport gene products into the anterior (left) end of the developing oocyte. (b) Result of an in situ hybridization experiment showing that the *bicoid* mRNA is trapped near the anterior end of the oocyte. (c) The *bicoid* mRNA is translated into protein soon after fertilization. The location of the Bicoid protein at the anterior end is revealed by immunostaining using an antibody that specifically recognizes this protein and labels it a dark brown.

CONCEPT CHECK: Where are maternal effect gene products made? Where do they go?

signal that is recognized by RNA-binding proteins that are necessary for the transport of this mRNA into the oocyte. After it enters the oocyte, the *bicoid* mRNA is trapped at the anterior end.

How can researchers determine the location of *bicoid* mRNA in the oocyte and resulting zygote? **Figure 26.9b** shows an in situ hybridization experiment in which a *Drosophila* egg was examined via a labeled probe complementary to the *bicoid* mRNA. (The technique of in situ hybridization is described in Chapter 22; see Figure 22.2.) As seen here, the *bicoid* mRNA is highly concentrated near the anterior end of the egg cell. Following fertilization, the *bicoid* mRNA is translated, and a gradient of Bicoid protein is established, as shown in **Figure 26.9c**.

After fertilization occurs, Bicoid protein functions as a transcription factor. Depending on the distribution of the Bicoid protein, this transcription factor activates genes only in certain regions of the embryo. For example, Bicoid stimulates the transcription of a gene called *hunchback* in the anterior half of the embryo, but its concentration is too low in the posterior half to activate the *hunchback* gene there.

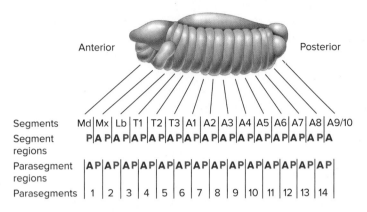

Segments | Md | Mx | Lb | T1 | T2 | T3 | A1 | A2 | A3 | A4 | A5 | A6 | A7 | A8 | A9/10

FIGURE 26.10 **A comparison of segments and parasegments in the *Drosophila* embryo.** Note that the parasegments and segments are out of register. The posterior (P) and anterior (A) regions are shown for each segment. The head segments are Md (mandibular), Mx (maxillary), and Lb (labial). The thoracic and abdominal segments are designated with the letters T and A, respectively.

Gap, Pair-Rule, and Segment-Polarity Genes Act Sequentially to Divide the *Drosophila* Embryo into Segments

After the anteroposterior and dorsoventral regions of the embryo have been established by maternal effect genes, the next stage of the developmental process organizes the embryo transiently into parasegments and then permanently into segments. The segmentation pattern of the embryo is shown in **Figure 26.10**. This pattern is maintained, or "remembered," throughout the rest of development. In other words, each segment of the embryo gives rise to unique morphological features in the adult. For example, segment T2 becomes a thoracic segment with a pair of legs and a pair of wings, and A8 becomes a segment of the abdomen.

Figure 26.10 shows the overlapping relationship between parasegments and segments. As you can see, the boundaries of the segments are out of register with the boundaries of the parasegments.

The parasegments are the locations where gene expression is controlled spatially. The anterior region of each segment coincides with the posterior region of a parasegment; the posterior region of a segment coincides with the anterior region of the next parasegment. The pattern of gene expression that occurs in the posterior region of one parasegment and the anterior region of an adjacent parasegment results in the formation of a segment.

Now that you have a general understanding of the way the *Drosophila* embryo is subdivided, we can examine how particular genes cause it to become segmented in this pattern. Three classes of genes, collectively called **segmentation genes,** play a role in the formation of body segments: gap genes, pair-rule genes, and segment-polarity genes. The expression and activation patterns of these genes in specific regions of the embryo cause it to become segmented.

How were the three classes of segmentation genes discovered? In the 1970s, segmentation genes were identified by Christiane Nüsslein-Volhard and Eric Wieschaus, who undertook a systematic search for mutations affecting embryonic development in *Drosophila*. Their pioneering effort identified most of the genes required for the embryo to develop a segmented pattern.

Figure 26.11 shows a few of the phenotypic effects that occur when a particular segmentation gene is defective. The gray boxes indicate the regions missing in the resulting larvae.

- A mutation in a gap gene known as *Krüppel* causes a contiguous section of the larva to be missing (Figure 26.11a). In other words, a gap of several segments has occurred.
- By comparison, a defect in a pair-rule gene causes alternating regions to be deleted (Figure 26.11b). For example, when the *even-skipped* gene is defective, portions of alternating segments in the resulting larva are missing.
- Segment-polarity mutations cause individual segments to be missing either an anterior or a posterior region. Figure 26.11c shows a mutation in a segment-polarity gene known as *gooseberry*. When this gene is defective, the anterior portion of each segment is missing from the larva. In the case of

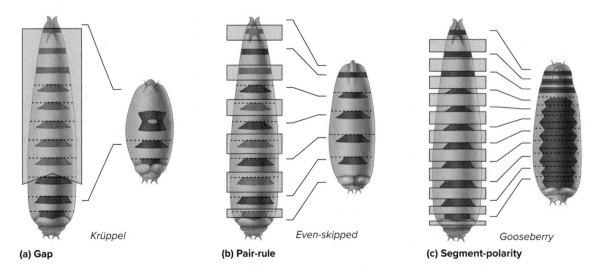

(a) Gap   *Krüppel*   (b) Pair-rule   *Even-skipped*   (c) Segment-polarity   *Gooseberry*

FIGURE 26.11 **Phenotypic effects in *Drosophila* larvae of mutations in segmentation genes.** The effects shown here are caused by mutations in **(a)** a gap gene, **(b)** a pair-rule gene, and **(c)** a segment-polarity gene. The dashed lines indicate the boundaries between segments.

CONCEPT CHECK: Describe the difference between the effects of a mutation in a gap gene and a mutation in a pair-rule gene.

segment-polarity mutants, the segments adjacent to the deleted regions exhibit a mirror-image duplication.

Overall, the phenotypic effects of mutant segmentation genes provided geneticists with important clues regarding the roles of these genes in the developmental process of segmentation.

Figure 26.12 presents a partial, simplified scheme of the genetic hierarchy that leads to a segmented pattern in the *Drosophila* embryo. Keep in mind that although this figure presents the general sequence of events that occur during the early stages of embryonic development, many more genes are actually involved in this process (refer back to Table 26.1). As presented in this figure, the following steps occur:

1. Maternal effect gene products, such as *bicoid* mRNA, are deposited asymmetrically into the oocyte. These gene products form a gradient that later influences the formation of axes, such as the anteroposterior axis.

1. Maternal effect genes establish the anteroposterior and dorsoventral axes. The products of maternal effect genes activate gap genes.

Asymmetrical localization of maternal effect gene products

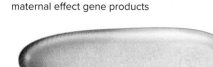

Examples

Asymmetrical localization of Bicoid protein. Other maternal effect gene products (not shown) are also asymmetrically localized.

2. Gap gene products act as genetic regulators of pair-rule genes. They bind to stripe-specific enhancers that are located adjacent to pair-rule genes.

Gap gene expression

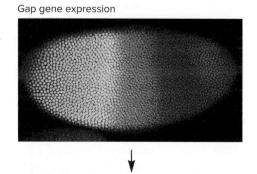

Gap gene expression occurs as broad bands in the embryo. In this photo, Hunchback protein is shown in green at the anterior end, and Krüppel protein is shown in red in the middle. Their region of overlap is yellow. Other gap genes (not shown) are also expressed.

3. The expression of a pair-rule gene in a stripe defines the boundary of a parasegment. Pair-rule gene products regulate the expression of segment-polarity genes.

Pair-rule gene expression

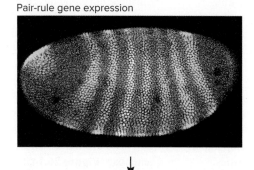

Pair-rule genes are expressed in alternating stripes. Each stripe corresponds to a parasegment. In this photo, the *even-skipped* gene product is expressed in the light bands that correspond to odd-numbered parasegments.

4. Segment-polarity genes define the anterior or posterior region of each parasegment.

Segment-polarity gene expression

Segment-polarity genes are expressed in either an anterior or posterior region. This photo shows the product of the *engrailed* gene in the anterior region of each parasegment.

FIGURE 26.12 **Overview of the genetic hierarchy leading to segmentation in *Drosophila*.** (Note: Between steps 3 and 4, the elongated embryo bends in the middle and folds back on itself.)

Jim Langeland, Steve Paddock and Sean Carroll/University of Wisconsin-Madison

2. In contrast to maternal effect genes, which are expressed during oogenesis, **zygotic genes** are expressed after fertilization. The first zygotic genes to be activated are the gap genes. Maternal effect gene products are responsible for activating the gap genes. As shown in step 2 of Figure 26.12, gap gene products divide the embryo into a series of broad bands or regions. These bands do not correspond to parasegments or segments within the embryo.

3. Products of the gap genes and maternal effect genes then activate the pair-rule genes. The photograph in step 3 of Figure 26.12 illustrates the alternating pattern of expression of the *even-skipped* gene. The expression of pair-rule genes in stripes defines the boundaries of parasegments. *Even-skipped* is expressed in the odd-numbered parasegments. The pair-rule genes divide the broad regions established by gap genes into seven bands, or stripes.

How does this alternating pattern of gene expression arise? The answer is that the *even-skipped* gene is regulated in a complex way that involves the binding of activators and repressors to an enhancer that controls transcription. In stripes where *even-skipped* is expressed, certain activators have a stronger effect, whereas in regions where *even-skipped* is not expressed, repressors exert a stronger effect. Several different activators and repressors are expressed in a gradient from the anterior to posterior end of the embryo. In some regions along the anteroposterior axis, the concentration of activators is higher, and in other regions the concentration of repressors is higher; this variation gives rise to a series of stripes. Question 3 in More Genetic TIPS at the end of this chapter provides a more detailed explanation of how the *even-skipped* gene can be expressed in this pattern of alternating stripes.

4. Once the pair-rule genes are activated in an alternating banding arrangement, their gene products then regulate the expression of segment-polarity genes. The segment-polarity genes divide the embryo into 14 stripes, one within each parasegment. As shown in step 4 of Figure 26.12, the segment-polarity gene *engrailed* is expressed in the anterior region of each parasegment. Another segment-polarity gene, *wingless*, is expressed in the posterior region. Later in development, the anterior region of one parasegment and the posterior region of an adjacent parasegment develop into a segment with particular morphological characteristics.

Homeotic Genes Control the Phenotypic Characteristics of Each Segment

Thus far, we have considered how the *Drosophila* embryo becomes organized along axes and then into a segmented body pattern. Now let's examine how each segment develops its unique morphological features. As mentioned earlier, geneticists often use the term *cell fate* to describe the morphological features that a cell or group of cells will ultimately adopt. For example, the fate of the cells in segment T2 in the *Drosophila* embryo is to develop into a thoracic segment containing two legs and two wings. In *Drosophila*, the fate of the cells in each segment of the body is determined at a very early stage of embryonic development, long before the morphological features become apparent.

Our understanding of developmental fate has been greatly aided by the identification of mutant genes that alter cell fates. In animals, the first mutant of this type was described by Ernst Gustav Kraatz in 1876, who observed a sawfly (*Cimbex axillaris*) in which part of an antenna was replaced with a leg. During the late nineteenth century, William Bateson collected many of these types of observations and published them in 1894 in a book entitled *Materials for the Study of Variation Treated with Especial Regard to Discontinuity in the Origin of Species*. In this book, Bateson coined the term **homeotic** to describe mutants in which one body part is replaced by another, such as the transformation of the antennae of insects into legs.

As was mentioned in Section 26.1, Edward Lewis began to study strains of *Drosophila* having homeotic mutations. This work, which began in 1946, was the first systematic study of homeotic genes. Each homeotic gene controls the fate of a particular region of the body. *Drosophila* contains two clusters of homeotic genes called the *Antennapedia* complex and the *bithorax* complex. **Figure 26.13** shows the organization of genes within these complexes. As discussed later in this chapter, these types of genes, which are called *Hox* genes, are found in all animals except early diverging groups such as ctenophores and sponges. The *Antennapedia* complex contains five *Hox* genes, designated *lab*, *pb*, *Dfd*, *Scr*, and *Antp*. The *bithorax* complex has three genes, *Ubx*, *abd-A*, and *Abd-B*. Both of these complexes are located on chromosome 3 in *Drosophila*, but a large segment of DNA separates them.

As noted in Figure 26.13, the order of these genes on chromosome 3 correlates with their pattern of expression along the anteroposterior axis of the body. For example, *lab* is expressed in an anterior segment and governs the formation of mouth structures. The *Antp* gene is expressed strongly in the thoracic region during embryonic development and controls the formation of thoracic structures. Transcription of the *Abd-B* gene occurs in the posterior region of the embryo. This gene controls the formation of the abdominal segments at the posterior end of the body.

The role of homeotic genes in determining the identity of particular segments has been revealed by mutations that alter the functions of these genes. For example, the *Antp* gene is normally expressed in the thoracic region. The *Antennapedia* mutation causes the *Antp* gene to also be expressed in the region where the antennae are made. These mutant flies have legs in the place of antennae (**Figure 26.14**)! This is an example of a **gain-of-function mutation**. In this case, the *Antp* gene is expressed normally in the thoracic region and also expressed abnormally in the anterior segment that normally gives rise to the antennae.

Investigators have also studied many **loss-of-function mutations** in homeotic genes. In this case, when a particular homeotic gene is defective, the region it normally governs is usually controlled by the homeotic gene that acts in the adjacent anterior region. For example, the *Ubx* gene functions within parasegments 5 and 6. If this gene is missing, this section of the fly becomes converted to the morphological features that would normally be produced from parasegment 4.

Homeotic genes are part of the genetic hierarchy that produces the morphological characteristics of the fruit fly. How are

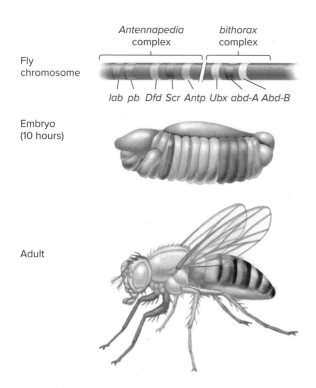

FIGURE 26.13   **Expression pattern of homeotic genes in** **Drosophila.** The homeotic genes are found in two complexes termed *Antennapedia* and *bithorax*. The order of homeotic genes, *labial* (*lab*), *proboscipedia* (*pb*), *Deformed* (*Dfd*), *Sex combs reduced* (*Scr*), *Antennapedia* (*Antp*), *Ultrabithorax* (*Ubx*), *abdominal A* (*abd-A*), and *Abdominal B* (*Abd-B*), correlates with the order of expression in the embryo. The expression pattern of seven of these genes is shown. The *lab* gene (purple) is expressed in the region that eventually gives rise to mouth structures. *Dfd* (light green) is expressed in the region that will form much of the head. *Scr* (forest green) and *Antp* (light blue) are expressed in embryonic segments that give rise to thoracic segments. *Ubx* (gray), *abd-A* (red), and *Abd-B* (yellow) are expressed in posterior segments that will form the abdomen. The order of gene expression, from anterior to posterior, parallels the order of genes along the chromosome. (Note: Some of the homeotic gene names are capitalized because the first mutation found in the gene was dominant, whereas others are not capitalized because the first mutation found in the gene was recessive.)

CONCEPT CHECK: Explain how the physical arrangement of the homeotic genes on the *Drosophila* chromosome correlates with their effects on phenotype.

they regulated? Homeotic genes are controlled by gap genes and pair-rule genes, and they are also regulated by interactions among themselves. Epigenetic modifications maintain the patterns of homeotic gene expression after the initial patterns are established by segmentation genes.

As described in Chapter 16, trithorax and polycomb group complexes promote epigenetic modifications that result in gene activation or repression, respectively. In regions of the embryo where the homeotic genes are active, the trithorax group complexes may maintain an open conformation of chromatin, which promotes the transcription of those genes. Alternatively, polycomb group complexes repress the expression of homeotic genes in regions where they should not act. One way this is accomplished is by changing the structure of chromatin to convert it to a closed

(a) Normal fly   **(b) *Antennapedia* mutant**

FIGURE 26.14   The *Antennapedia* mutation in *Drosophila.*
Genes→Traits **(a)** A normal fly with antennae. **(b)** This mutant fly has a gain-of-function mutation in which the *Antp* gene is expressed in the embryonic segment that normally gives rise to antennae. The expression of *Antp* causes this region to have legs rather than antennae.
(a): Juergen Berger/Science Source; (b): Eye of Science/Science Source

conformation in which transcription is inhibited (refer back to Figure 16.13). Overall, the concerted actions of many gene products cause the homeotic genes to be expressed only in the appropriate region of the embryo, as shown in Figure 26.13.

Because they are part of a genetic hierarchy in which genes activate other genes, perhaps it is not surprising that homeotic genes code transcription factors. The coding sequence of homeotic genes contains a 180-bp consensus sequence known as a **homeobox** (**Figure 26.15a**). This sequence was first discovered in the *Antp* and *Ubx* genes, and it has since been found in all *Drosophila* homeotic genes and in some other genes affecting pattern development, such as *bicoid*. The protein domain coded by a homeobox is called a **homeodomain**. The arrangement of α helices within the homeodomain promotes the binding of the protein to the major groove of DNA (**Figure 26.15b**). In this way, homeotic proteins can bind to DNA in a sequence-specific manner. In addition to DNA-binding ability, homeotic proteins also contain a transcriptional activation domain that functions to activate the genes to which the homeodomain can bind.

The transcription factors coded by homeotic genes activate other genes that code proteins that produce the morphological characteristics of each segment. Much current research is attempting to identify these genes and determine how their expression in particular regions of the embryo leads to morphological changes in the embryo, larva, and adult.

The Developmental Fate of Each Cell in the Nematode *Caenorhabditis elegans* Is Known

We now turn our attention to another invertebrate, *C. elegans*, a nematode that has been the subject of numerous studies in developmental genetics. The embryo develops within an eggshell and hatches when it reaches a size of 550 cells. After hatching, it continues to grow and mature as it passes through four successive larval stages. It takes about 3 days for a fertilized egg to develop into an adult worm that is 1 mm in length. With regard to sex, *C. elegans* can be a male (and produce only sperm) or a hermaphrodite (capable of producing sperm and egg cells). An adult male is composed of 1031 somatic cells and

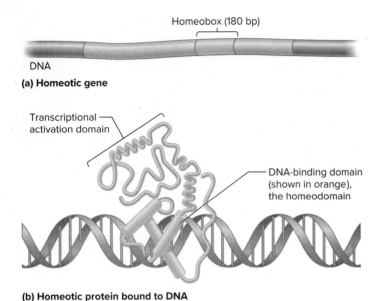

Homeobox (180 bp)

DNA

(a) Homeotic gene

Transcriptional activation domain

DNA-binding domain (shown in orange), the homeodomain

(b) Homeotic protein bound to DNA

FIGURE 26.15 Molecular features of homeotic proteins.
(a) A homeotic gene (shown in tan and orange) contains a 180-bp sequence called the homeobox (shown in orange). **(b)** When a homeotic gene is expressed, it produces a protein that functions as a transcription factor. The homeobox codes a region of the protein called a homeodomain, which binds to the major groove of DNA. These DNA-binding sites are found within genetic regulatory elements (i.e., enhancers). The enhancers are found in the vicinity of promoters that are turned on by homeotic proteins. For this to occur, the homeotic protein also contains a transcriptional activation domain, which activates the transcription of a gene after the homeodomain has bound to the DNA.

produces about 1000 sperm. A hermaphrodite consists of 959 somatic cells and produces about 2000 gametes (both sperm and eggs).

A remarkable feature of this organism is that the pattern of cellular development remains constant from worm to worm. In the early 1960s, Sydney Brenner pioneered the effort to study the pattern of cell division in *C. elegans* and establish it as a model organism. Because *C. elegans* is transparent and composed of relatively few cells, researchers can follow cell division step by step under the microscope, beginning with a fertilized egg and ending with an adult worm. Researchers can identify a particular cell at an embryonic stage, follow that cell as it divides, and observe where its descendant cells are located in the adult. An illustration that depicts how cell division proceeds is called a **cell lineage diagram.** It depicts the cell division patterns and the fates of any cell's descendants.

Figure 26.16 shows a partial cell lineage diagram for a *C. elegans* hermaphrodite. At the first cell division, the egg divides to produce two cells, called AB and P_1. AB then divides into two cells, ABa and ABp; and P_1 divides into two cells, EMS and P_2. The EMS cell then divides into two cells, called MS and E. The cellular descendants of the E cell give rise to the worm's intestine. In other words, the fate of the E cell's descendants is to develop into intestinal cells. This diagram also illustrates the concept of a **cell lineage,** a series of cells that are derived from a particular cell by cell division. For example, the EMS cell, E cell, and intestinal cells are all part of the same cell lineage.

Why does a cell lineage diagram for an organism provide an important experimental tool? It allows researchers to investigate how gene expression in any cell, at any stage of development, may affect the outcome of a cell's fate. The experiment described next demostrated that the timing of gene expression is an important factor in the fate of a cell's descendants.

FIGURE 26.16 A cell lineage diagram for the nematode *Caenorhabditis elegans*. This partial cell lineage diagram illustrates how the cells divide to produce different regions of the adult worm. The fate of the intestinal cell lineage is shown in greater detail than the other cell lineages. A complete cell lineage diagram is known for this organism, although its level of detail is beyond the scope of this text. (Note: The lowercase letters a and p stand for *anterior* and *posterior*, respectively, which indicate the relative positions of the daughter cells.)

CONCEPT CHECK: What is a cell lineage?

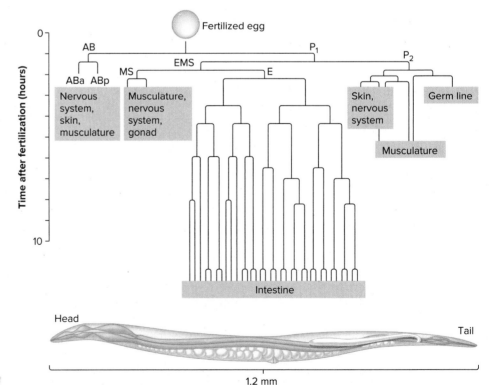

Fertilized egg

AB P_1 P_2

EMS

ABa ABp   MS   E

Nervous system, skin, musculature

Musculature, nervous system, gonad

Skin, nervous system

Germ line

Musculature

Intestine

Time after fertilization (hours)

0

10

Head

Tail

1.2 mm

Mutations Can Disrupt the Timing of Developmental Changes in *C. elegans*.

Our discussion of *Drosophila* development focused on how the spatial expression and localization of gene products can lead to a particular pattern of embryonic development. An important issue in development is the timing of developmental changes at the cellular level. The cells of a multicellular organism must divide at the proper time and differentiate into the correct cell types. If the timing of these processes is not coordinated, certain tissues will develop too early or too late, disrupting the developmental process.

In *C. elegans*, the timing of developmental events can be examined at the cellular level. As mentioned, the fate of each cell in *C. elegans* has been determined. Using a microscope, a researcher can focus on a particular cell within an embryo of this transparent worm and watch it divide into two cells, then four cells, and so forth. Therefore, the researcher can judge whether a cell is behaving as it should during the developmental process.

To identify genes that play a role in the timing of cell fates, researchers have searched for mutant alleles that disrupt the normal timing process. In a collaboration in the late 1970s, H. Robert Horvitz and John Sulston set out to identify mutant alleles in *C. elegans* that disrupt cell fates or the timing of cell fates. Using a microscope, they screened thousands of worms for altered morphologies that might indicate an abnormality in development. During this screening process, one of the phenotypic abnormalities they found was a defective egg-laying phenotype. They reasoned that because the egg-laying system depends on multiple cell types (vulval cells, muscle cells, and nerve cells), an abnormality

in any of the cell lineages leading to these cell types might cause an inability to lay eggs.

In *C. elegans*, a defective egg-laying phenotype is easy to identify, because the hermaphrodite is able to fertilize its own eggs but unable to lay them. When this occurs, the eggs actually hatch within the hermaphrodite's body. This leads to the death of the hermaphrodite as it becomes filled with hatching worms. This defective egg-laying phenotype, in which the hermaphrodite becomes filled with its own offspring, is called a "bag of worms." Eventually, the newly hatched larvae eat their way out and can be saved for further study.

In Horvitz and Sulston's initial study, published in 1980, the defective egg-laying phenotype produced several mutant strains that were defective in particular cell lineages. A few years later, in the experiment described in **Figure 26.17**, Victor Ambros and H. Robert Horvitz took this same approach and were able to identify genes that play a key role in the timing of cell fates. They began with wild-type *C. elegans* and three mutant lines designated n536, n355, and n540. All three mutant lines showed an egg-laying defect. After the larvae hatched, the researchers observed the fates of particular cells via microscopy. This involved spending hours looking at the nuclei of specific cells (which are relatively easy to see in this transparent worm) and timing the patterns of cell division. The patterns in the mutant and wild-type strains were then used to create cell lineage diagrams for particular cells.

THE HYPOTHESIS

Mutations that cause a defective egg-laying phenotype may affect the timing of cell fates within lineages.

TESTING THE HYPOTHESIS

FIGURE 26.17 Identification of mutations that affect the timing of development in *C. elegans*.

Starting material: Prior to this work, many laboratories had screened thousands of *C. elegans* worms and identified particular mutant strains that were defective in egg-laying. (Note: When mutated, many different genes may cause a defective egg-laying phenotype. Only some of them are expected to be genes that alter the timing of cell fate within a particular cell lineage.)

	Experimental level	Conceptual level

1. Obtain a large number of *C. elegans* strains that have a defective egg-laying phenotype. This experiment focuses on three mutant strains designated n536, n355, and n540. The wild-type strain was also studied as a control.

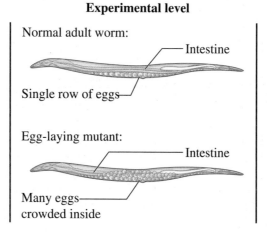

Normal adult worm:
— Intestine
Single row of eggs

Egg-laying mutant:
— Intestine
Many eggs crowded inside

2. Right after hatching, observe the fate of particular cells via microscopy. In this example, a researcher began watching a cell called the T cell and monitored its division pattern, and the pattern of subsequent daughter cells, during the larval stages. These patterns were examined in both wild-type and defective egg-laying worms.

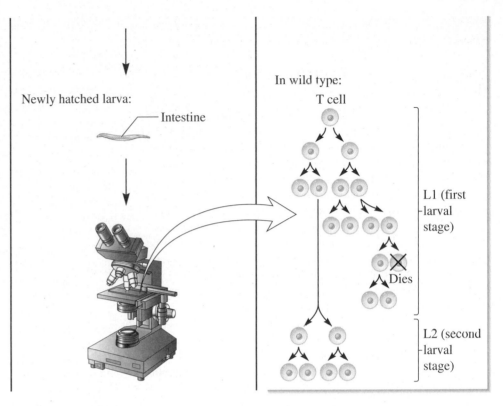

Newly hatched larva:
— Intestine

In wild type:
T cell

L1 (first -larval stage)

Dies

L2 (second -larval stage)

THE DATA

T-cell lineages in different strains of *C. elegans*

Larval stage/hour

Wild type	n536	n355	n540

L1
— 10
L2 ⊢ 20
L3 ⊢ 30
L4 ⊢ 40

● Neurons
● Epidermal cells
Ⓧ Programmed to die

Source: Ambros, Victor, and Horvitz, H. Robert, "Heterochronic mutants of the nematode *Caenorhabditis elegans*," *Science*, vol. 226, no. 4673, October 26, 1984, 409–416.

INTERPRETING THE DATA

Cell lineage diagrams involving cells derived from a cell called a T cell are shown in the data. As seen here, the wild-type strain follows a particular pattern of cell division for the T-cell lineage. Each division event occurs at specific times during the L1 and L2 larval stages. In the normal strain, the T cell divides during the L1 larval stage to produce

a T.a and a T.p cell. The T.a cell also divides during L1 to produce a T.aa and a T.ap cell. The T.p cell divides during L1 to produce T.pa and T.pp. These cells also divide during L1, eventually producing five neurons (labeled in blue) and one cell that is programmed to die (designated with an X). During the L2 larval stage, the T.ap cell resumes division to produce four cells: three epidermal cells (labeled in red) and one neuron.

The other T-cell lineages shown in the data are from worms that carry mutations causing an egg-laying defect. Later research revealed that these three mutations are located in a gene called *lin-14*. The allele designated n536 caused the reiteration of the normal events of L1 during the L2 larval stage. In L2, the only cell of this lineage that is supposed to divide is T.ap. In worms carrying the n536 allele, however, this cell behaves as if it were a T cell, rather than a T.ap cell, and produces a group of cells identical to what a T cell normally produces during the L1 stage. In the L3 stage, the cells in the n536 strain behave as if they were in L2. In addition to the egg-laying defect, the phenotypic outcome of this irregularity in the timing of cell fates is a worm that has several additional cells derived from the T-cell lineage.

The n355 allele causes multiple reiterations. This strain continues to reiterate the normal events of L1 during the L2, L3, and L4 stages. In contrast, the n540 allele has an opposite effect on the T-cell lineage. During the L1 larval stage, the T cell behaves as if it were a T.ap cell in the L2 stage. In this case, it skips the divisions and cell fates of the L1 and proceeds directly to cell fates that occur during the L2 stage.

The types of mutations described here are called heterochronic mutations. The term *heterochrony* refers to a change in the relative timing of developmental events. In **heterochronic mutations,** the timing of the fates of particular cell lineages is not synchronized with the development of the rest of the organism. More recent molecular data have shown that this is due to an irregular pattern of gene expression.

- In wild-type worms, the lin-14 protein accumulates during the L1 stage and promotes the T-cell division pattern shown for the wild type. During L2, the lin-14 protein diminishes to negligible levels.
- The n536 and n355 alleles are examples of gain-of-function mutations. In these strains, the lin-14 protein persists during later larval stages. The n536 allele produces this protein

for one additional cell division, whereas the n355 allele continues to express *lin-14* for several cell divisions.

- By comparison, the n540 allele is the result of a loss-of-function mutation. This allele causes lin-14 to be inactive during L1, so it cannot promote the normal L1 pattern of cell division and cell fate.

Overall, the results described in this experiment are consistent with the idea that the precise timing of *lin-14* expression during development is necessary to correctly control the fates of particular cells in *C. elegans*. Mutations that alter the expression of *lin-14* lead to phenotypic abnormality, including the inability to lay eggs. This detrimental phenotypic consequence illustrates the importance of the correct timing for cell division and differentiation during development.

26.2 COMPREHENSION QUESTIONS

1. The expression of maternal effect genes directly leads to
   a. the establishment of body axes.
   b. segmentation.
   c. determination.
   d. cell differentiation.

2. Which of the following are types of segmentation genes?
   a. Gap genes
   b. Pair-rule genes
   c. Segment-polarity genes
   d. All of the above are types of segmentation genes.

3. The expression of homeotic genes leads to
   a. the establishment of body axes.
   b. the formation of segments in the embryo.
   c. the determination of structures within segments.
   d. cell differentiation.

4. Homeotic genes code proteins that function as
   a. cell-signaling proteins.
   b. transcription factors.
   c. hormones.
   d. all of the above.

26.3 VERTEBRATE DEVELOPMENT

Learning Outcomes:

1. Describe the relationship between homeotic genes in *Drosophila* and those in mice.
2. Explain how the expression of *Hox* genes affects vertebrate development.
3. Describe how cell differentiation is controlled in vertebrates.

Biologists have studied the morphological features of development in many vertebrate species. Historically, amphibians and

birds have been studied extensively, because their eggs are rather large and easy to manipulate. For example, certain developmental stages of the frog and the chicken have been described in great detail. In more recent times, the successes obtained with *Drosophila* have shown the great power of genetic analyses in elucidating the underlying molecular mechanisms that govern biological development. With this knowledge, many researchers are attempting to understand the genetic pathways that govern the development of the more complex body structures found in vertebrate organisms.

Several vertebrate species have been the subject of developmental studies. These include the mouse (*Mus musculus*), the frog (*Xenopus laevis*), and the zebrafish (*Danio rerio*). In this section, we will primarily discuss the genes that are important in mammalian development, particularly those that have been characterized in the mouse, one of the best-studied mammals. As we will see, many genes affecting its developmental pathways have been cloned and characterized. We will also examine how these genes affect the course of vertebrate development.

Homeotic Genes Are Found in Vertebrates

Vertebrates typically have long generation times and produce relatively few offspring. Therefore, it is usually not practical to screen large numbers of embryos or offspring in search of mutant phenotypes with developmental defects. As an alternative, a successful way of identifying genes that affect vertebrate development has been the use of molecular techniques to identify vertebrate genes similar to those that control development in simpler organisms such as *Drosophila*.

As discussed in Chapters 23 and 29, species that are evolutionarily related to each other carry genes with similar DNA sequences. When two or more genes have similar sequences because they are derived from the same ancestral gene, they are called **homologous genes.** Homologous genes found in different species are termed **orthologs.**

Researchers have found groups of homeotic genes in vertebrate species that are homologous to those in the fruit fly. These groups of homeotic genes are called ***Hox* complexes.** *Hox* is a contraction of <u>ho</u>meo<u>box</u>, which is a domain found in all *Hox*

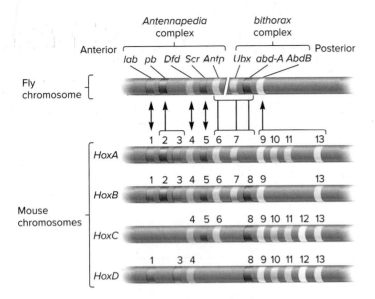

FIGURE 26.18 A comparison of homeotic genes in *Drosophila* and the mouse. The mouse genome contains four gene complexes, *HoxA* through *HoxD*, that correspond to certain homeotic genes found in *Drosophila*. Thirteen different types of homeotic genes are found in the mouse, although each *Hox* complex does not contain all 13 genes. In this drawing, orthologous genes are aligned in columns. The arrows and brackets indicate the evolutionary relationships between the *Drosophila* and mouse genes. For example, *lab* is the ortholog of *HoxA-1*, *HoxB-1*, and *HoxD-1*; *pb* is an ortholog of *HoxA-2*, *HoxA-3*, *HoxB-2*, *HoxB-3*, and *HoxD-3*. In *Drosophila*, the homeotic genes are located on chromosome 3. In the mouse, the chromosomal locations are as follows: *HoxA* (chromosome 6), *HoxB* (chromosome 11), *HoxC* (chromosome 15), and *HoxD* (chromosome 2).

CONCEPT CHECK: What is an ortholog?

genes (see Figure 26.15). As shown in **Figure 26.18**, the mouse has four *Hox* complexes, designated *HoxA* (on chromosome 6), *HoxB* (on chromosome 11), *HoxC* (on chromosome 15), and *HoxD* (on chromosome 2). A total of 39 genes are found in the four complexes. Thirteen different types of homeotic genes occur within the four *Hox* complexes, although none of the four complexes contains representatives of all 13 types of genes. The addition of *Hox* genes into the genomes of animals has allowed certain animal species to develop more complex body plans.

The homeotic genes in fruit flies and mammals are homologous to each other, as shown by the arrows and brackets in Figure 26.17. Because of the known roles of homeotic genes, this observation indicates that fundamental similarities occur in the ways that animals as different as fruit flies and mammals undergo embryonic development. These similarities suggest that a common plan of body development is found in all animals with bilateral symmetry.

As with the *Antennapedia* and *bithorax* complexes in *Drosophila*, the arrangement of *Hox* genes on the mouse chromosomes reflects their pattern of expression from the anterior to the posterior end of the animal (**Figure 26.19a**). This phenomenon is seen in more detail in **Figure 26.19b**, which shows the expression pattern for a group of *HoxB* genes in a mouse embryo. Overall, these results are consistent with the idea that the *Hox* genes play a role in determining the fates of segments along the anteroposterior axis.

Researchers Use Reverse Genetics to Understand *Hox* Gene Function

Currently, researchers are trying to understand the functional roles that the genes in the *Hox* complexes play in vertebrate development. Great advances in developmental genetics have been made by studying mutant alleles in genes that control development in *Drosophila*. In mice, however, few natural mutations have been identified that affect development. This has made it more difficult to understand the role that genetics plays in the development of the mouse and other vertebrate organisms.

How have researchers overcome this problem? Geneticists are taking an approach known as **reverse genetics.** In this strategy, researchers first identify the wild-type gene using molecular methods. The next step is to create a mutant version of a *Hox* gene using gene-editing techniques, such as those described in Chapter 20. Eliminating the function of the wild-type gene produces a **gene knockout.** In this way, researchers can determine how the mutant allele affects the phenotype of the mouse.

The strategy of reverse genetics means that the experimental steps occur in an order opposite to that of the conventional approach used in *Drosophila* and other organisms. In the fly, mutant alleles were identified by their phenotype first, and then they were identified at the molecular level. This approach is called **forward genetics.** In contrast, the mouse genes were first identified at the molecular level, mutations were made to these genes, and then the mice carrying the mutated genes were analyzed with regard to their phenotypic effects.

In recent years, many laboratories have used a reverse genetic approach to understand how many different genes, including *Hox* genes, affect vertebrate development.

- In *Drosophila*, loss-of-function alleles for homeotic genes usually produce an anterior transformation. This means that the segment where the defective homeotic gene is normally expressed instead exhibits characteristics that resemble the adjacent anterior segment. Similarly, certain gene knockouts (e.g., *HoxA-2*, *HoxB-4*, and *HoxC-8*) also produce anterior transformations within particular regions of the mouse.
- However, knockouts of other *Hox* genes (e.g., *HoxA-11*) result in posterior transformations, and knockouts of *HoxA-3* and *HoxA-1* exhibit abnormalities in morphology but no clear homeotic transformations.
- Interestingly, a *HoxA-5* knockout in mice shows evidence of both anterior and posterior transformations, which are also seen in *Drosophila* when the *HoxA-5* ortholog, *Scr*, is knocked out.

Overall, the current evidence indicates that the *Hox* genes in vertebrates play a key role in patterning the anteroposterior axis. During the evolution of animals, increases in the number of *Hox* genes and changes in their patterns of expression have had an important effect on morphologies. As an example, let's consider the expression of the *HoxC-6* gene. The *HoxC-6* gene is expressed during embryonic development prior to vertebrae formation. Differences in the

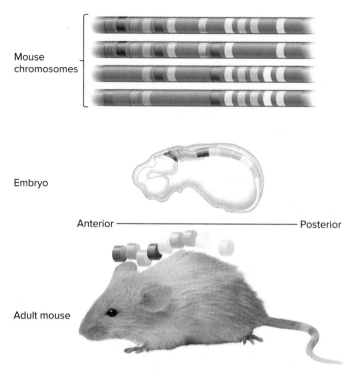

(a) Correlation between the arrangement of *Hox* genes and their expression

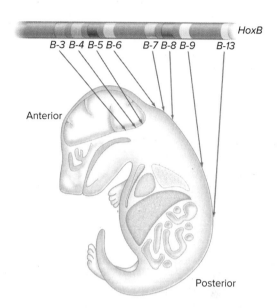

(b) Anterior expression boundaries for a series of *HoxB* genes in a mouse embryo

FIGURE 26.19 **Expression pattern of *Hox* genes in the mouse.** **(a)** A schematic illustration of *Hox* gene expression in the embryo and the corresponding regions in the adult. **(b)** A more detailed description of the expression of *HoxB* genes in a mouse embryo. The arrows indicate the anterior expression boundaries for the genes *HoxB-3* to *HoxB-13*. The order of *Hox* gene expression, from the anterior end to the posterior end of the embryo, is the same as the order of the genes on the chromosome.

(a): G.K. & Vikki Hart/Photodisc/Getty Images

relative position of its expression correlate with the number of neck vertebrae produced (**Figure 26.20**). In the mouse, which has a relatively short neck, *HoxC-6* expression occurs posterior to the region of the early embryo that later develops into vertebrae 7. In contrast, *HoxC-6* expression in the chicken and the goose occurs much farther back, posterior to vertebrae 14 and 17, respectively. The forelimbs also arise at this boundary in all vertebrates. However, snakes, which have no neck or forelimbs, do not have such a boundary because *HoxC-6* expression begins toward their heads.

Genes That Code Transcription Factors Also Play a Key Role in Cell Differentiation

Thus far, we have focused our attention on patterns of gene expression that occur during the very early stages of development. These genes control the basic body plan of the organism. As this process occurs, cells become **determined.** As mentioned earlier, determination is the process by which a cell is destined to become a particular cell type. In other words, its fate is predetermined, and it will eventually become a particular type of cell, such as a neuron. Determination occurs long before a cell becomes **differentiated.** This term means that a cell's morphology and function have changed, usually permanently, transforming it into a highly specialized cell type. For example, an undifferentiated mesodermal cell may differentiate into a

specialized muscle cell, or an undifferentiated ectodermal cell may differentiate into a neuron.

At the molecular level, the profound morphological differences between muscle cells and neurons arise from gene regulation. Though muscle cells and neurons contain the same set of genes, they regulate the expression of those genes in very different ways. Certain genes that are transcriptionally active in muscle cells are inactive in neurons, and vice versa. Therefore, muscle cells and neurons express different proteins, which affect the characteristics of the respective cells in distinct ways. In this manner, differential gene regulation underlies cell differentiation.

Researchers have identified specific genes that cause cells to differentiate into particular cell types. For example, in 1987, Harold Weintraub and colleagues identified a gene, which they called *MyoD.* This gene plays a key role in skeletal muscle-cell differentiation. Experimentally, when the cloned *MyoD* gene was introduced into fibroblast cells in a laboratory, the fibroblasts differentiated into skeletal muscle cells. This result was particularly remarkable because fibroblasts normally differentiate in vivo into osteoblasts (bone cells), chondrocytes (cartilage cells), adipocytes (fat cells), and smooth muscle cells, but they never differentiate into skeletal muscle cells.

Since this initial discovery, researchers have found that *MyoD* belongs to a small group of genes that initiate muscle development. Besides *MyoD,* these genes include *Myogenin, Myf5,*

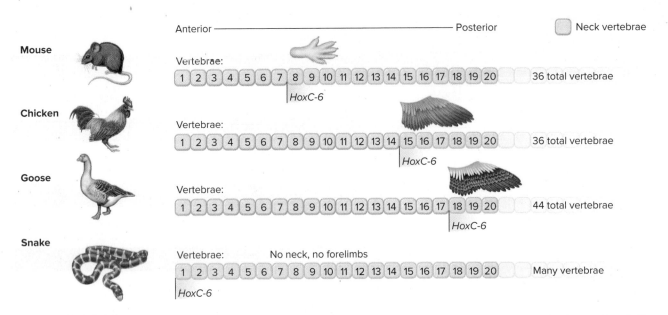

FIGURE 26.20 **Expression of the *HoxC-6* gene in different species of vertebrates.** The pink region under the vertebrae indicates the general area of *HoxC-6* gene expression in the mouse, chicken, goose, and snake. The vertical black line shows the anterior boundary of *HoxC-6* gene expression, which defines the posterior boundary of neck vertebrae.

CONCEPT CHECK: Briefly describe how the *HoxC-6* gene affects vertebrate development.

and *Mrf4*. All four of these genes code transcription factors that contain a **basic helix-loop-helix (bHLH) domain,** in which one α-helix is connected to another α-helix by a loop. One helix is smaller and promotes dimer formation with another protein also having a bHLH domain. The larger α-helix contains basic amino acids (arginines and lysines) that are responsible for DNA binding and the activation of genes specific to functions of skeletal muscle cells. Because of their common structural features and their role in muscle-cell differentiation, the transcription factors coded by *MyoD*, *Myogenin*, *Myf5*, and *Mrf4* constitute a family called **myogenic bHLH proteins.** They are found in all vertebrates and have been identified in several invertebrates, such as *Drosophila* and *C. elegans*. In all cases, the myogenic bHLH genes are activated during skeletal muscle-cell development.

Two key features enable myogenic bHLH proteins to promote skeletal muscle-cell differentiation. First, the basic domain binds to a muscle-cell-specific enhancer sequence; this sequence is adjacent to genes that are expressed only in muscle cells (**Figure 26.21**). Therefore, when myogenic bHLH proteins are activated, they can bind to these enhancers and activate the expression of many different muscle-cell-specific genes. In this way, myogenic bHLH proteins function as master switches that activate the expression of muscle-cell-specific genes. When the coded proteins are synthesized, they change the characteristics of an undifferentiated cell into those of a highly specialized skeletal muscle cell.

A second key feature of myogenic bHLH proteins is that their activity is regulated by dimerization. As shown in Figure 26.21, the resulting heterodimers—dimers formed from two different proteins—may be activating or inhibitory.

- When a heterodimer forms between a myogenic bHLH protein and an E protein, which also contains a bHLH domain, the heterodimer binds to the DNA and activates gene expression (Figure 26.21a).
- However, when a heterodimer forms between a myogenic bHLH protein and a protein called Id (for inhibitor of differentiation), the heterodimer cannot bind to the DNA, because the Id protein lacks the basic amino acids that are needed for DNA binding (Figure 26.21b).

The Id protein is produced during early stages of embryonic development and prevents myogenic bHLH proteins from promoting muscle-cell differentiation too soon. At later stages of development, the amount of Id protein decreases, and myogenic bHLH proteins can then combine with E proteins to induce muscle-cell differentiation.

26.3 COMPREHENSION QUESTIONS

1. *Hox* genes code transcription factors that
   a. control segmentation.
   b. promote determination.
   c. cause cell differentiation.
   d. do all of the above.
2. A cell that is _____ has a particular morphology and function.
   a. determined
   b. differentiated
   c. undergoing apoptosis
   d. dividing

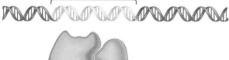

(a) Binding of a myogenic bHLH protein–E heterodimer to a muscle-cell-specific enhancer

(b) Inability of a myogenic bHLH protein–Id heterodimer to bind to the DNA

FIGURE 26.21 Regulation of muscle-cell-specific genes by myogenic bHLH proteins.

CONCEPT CHECK: At which stage of development does the Id protein function? Why is its function important?

3. Myogenic bHLH proteins are _____ that promote _____.
   a. cell-signaling proteins, muscle-cell differentiation
   b. cell-signaling proteins, muscle-cell determination
   c. transcription factors, muscle-cell differentiation
   d. transcription factors, muscle-cell determination

26.4 PLANT DEVELOPMENT

Learning Outcomes:

1. Describe how plant growth occurs from meristems.
2. Explain how homeotic genes control flower development.

In developmental plant biology, the model organism for genetic analysis is *Arabidopsis thaliana* (**Figure 26.22**). Unlike most flowering plants, which have long generation times and large genomes, *Arabidopsis* has a generation time of about 2 months and a genome size of 1.4×10^8 bp, which is similar to the genome sizes of *Drosophila* and *C. elegans*. A flowering *Arabidopsis* plant produces a large number of seeds and is small enough to be grown in the laboratory. Like *Drosophila*, *Arabidopsis* can be subjected to mutagens to generate mutations that alter developmental processes. The small genome size of this organism makes it relatively easy to map these mutant alleles and eventually identify the relevant genes (as described in Chapter 22).

FIGURE 26.22 The model organism *Arabidopsis*. The plant is relatively small, making it easy to grow many specimens in a laboratory.
Dr. Jeremy Burgess/Science Source

The morphological patterns of growth are markedly different between animals and plants. As described previously, animal embryos become organized along anteroposterior, dorsoventral, and left-right axes, and then they subdivide into segments. By comparison, the form of plants has two key features:

- The first is the root-shoot axis. Most plant growth occurs via cell division near the tips of the shoots and roots.
- Second, plant growth occurs in a well-defined radial or circular pattern. For example, rings of dividing cells are found in the stems of plants. Growing stems also produce buds that give rise to additional branches, leaves, and flowers. Overall, the radial pattern in which a plant shoot gives off the buds that produce branches, leaves, and flowers is an important mechanism that determines much of the plant's general morphology.

At the cellular level, too, plant development differs markedly from animal development. For example, cell migration does not

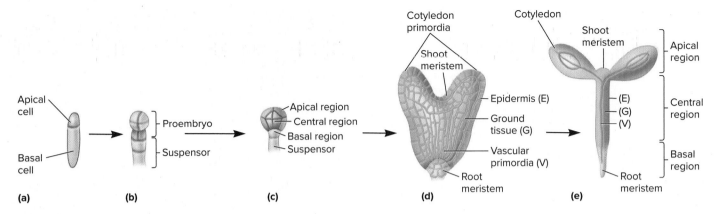

FIGURE 26.23 Developmental steps in the formation of a plant embryo.

occur during plant development. In addition, the development of a plant does not rely on morphogens that are deposited asymmetrically in the oocyte. In plants, an entirely new individual can be regenerated from many types of somatic cells. In other words, many plant cells are **totipotent,** meaning that they have the ability to differentiate into every cell type and to produce an entire new individual. By comparison, animal development typically relies on the organization within an oocyte as a starting point for development.

In spite of these apparent differences, the underlying molecular mechanisms of pattern development in plants still share some similarities with those in animals. In this section, we will consider a few examples in which genes that code transcription factors play a key role in plant development.

Plant Growth Occurs from Meristems Formed During Embryonic Development

Figure 26.23 illustrates a common sequence of events that takes place in the development of seed plants such as *Arabidopsis*.

After fertilization, the first cellular division is asymmetrical and produces a smaller cell, called the apical cell, and a larger cell, called a basal cell (Figure 26.23a). The apical cell gives rise to most of the embryo, and it later develops into the shoot of the plant. The basal cell gives rise to the root, along with the suspensor that produces extraembryonic tissue required for seed formation. At the heart stage, which is composed of only about 100 cells, the basic organization of the plant has been established (Figure 26.23d). The cotyledon primordia develop into **cotyledons,** which are structures that will become the first leaves of the seedling. The **shoot meristem** arises from a group of cells located between the cotyledons. These cells are the precursors that will produce the shoot of the plant, along with lateral structures such as leaves and flowers. The **root meristem** is located on the opposite side from the shoot meristem and creates the root.

A meristem contains an organized group of actively dividing stem cells. As discussed in Chapter 21, stem cells retain the ability to divide and differentiate into multiple cell types. As they grow, meristems produce offshoots of proliferating cells. On a shoot meristem, for example, these offshoots or buds give rise to structures such as leaves and flowers. The organization of a shoot

meristem is shown in **Figure 26.24**. The meristem has three areas called the **organizing center,** the **central zone,** and the **peripheral zone.** The role of the organizing center is to ensure the proper organization of the meristem and preserve the correct number of actively dividing stem cells. The central zone is an area where undifferentiated stem cells are always maintained. The peripheral zone contains dividing cells that eventually differentiate into plant structures. For example, the peripheral zone may form a bud, which will produce a leaf or flower.

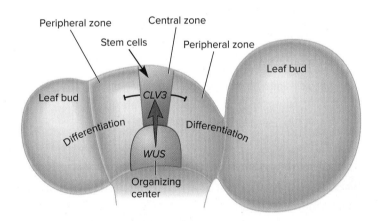

FIGURE 26.24 Organization of a shoot meristem. The organization of a shoot meristem is controlled by the *WUS* and *CLV3* genes, whose full names are *Wuschel* and *clavata,* respectively. The *WUS* gene is expressed in the organizing center and induces the cells in the central zone to become undifferentiated stem cells. The red arrow indicates that the WUS protein induces these central zone stem cells to turn on the *CLV3* gene, which codes a protein that is secreted and binds to receptors in the cells of the peripheral zone. The black lines with short vertical segments at their ends indicate that the CLV3 protein prevents the cells in the peripheral zone from expressing the *WUS* gene. This inhibition limits the area of *WUS* gene expression to the underlying organizing center, thereby maintaining a small population of stem cells at the growing tip. The cells in the peripheral zone are allowed to divide and eventually differentiate into lateral structures such as leaves.

CONCEPT CHECK: Why is it important to maintain the correct number of stem cells in the growing tip?

In *Arabidopsis*, the organization of a shoot meristem is controlled by two critical genes called *WUS* and *CLV3*. The *WUS* gene codes a transcription factor that is expressed in the organizing center (Figure 26.24). The expression of the *WUS* gene induces the adjacent cells in the central zone to become undifferentiated stem cells. These stem cells then turn on the *CLV3* gene, which codes a protein. The CLV3 protein is secreted and binds to receptors in the cells of the central and peripheral zones, preventing them from expressing the *WUS* gene. This inhibition limits the area of *WUS* gene expression to the underlying organizing center, thereby maintaining a small population of stem cells at the growing tip. A shoot meristem in *Arabidopsis* contains only about 100 cells. The inhibition of *WUS* expression in the peripheral cells allows them to begin the process of cell differentiation so that they can produce structures such as leaves and flowers.

Plant Homeotic Genes Control Flower Development

Although the term *homeotic* was coined by William Bateson to describe homeotic mutations in animals, the first known mutations in homeotic genes were observed in plants. In ancient Greece and Rome, for example, double flowers in which stamens were replaced by petals were noted. In current research, geneticists have been studying these types of mutations to better understand developmental pathways in plants. Many homeotic mutations affecting flower development have been identified in *Arabidopsis* and in the snapdragon (*Antirrhinum majus*).

A normal *Arabidopsis* flower is composed of four concentric whorls of structures (**Figure 26.25a**). The outer whorl contains four sepals, which protect the flower bud before it opens. The second whorl is composed of four petals, and the third whorl contains six stamens, the structures that make the male gametophyte,

pollen. Finally, the innermost whorl contains two carpels, which are fused together. The fused carpel produces the female gametophyte. An example of a homeotic mutant in *Arabidopsis* is shown in **Figure 26.25b**. The sepals have been transformed into carpels, and the petals into stamens.

After analyzing the effects of many different homeotic mutations in *Arabidopsis*, Elliot Meyerowitz and colleagues proposed the **ABC model** for flower development. In this model, three classes of genes, called *A*, *B*, and *C*, govern the formation of sepals, petals, stamens, and carpels. More recently, a fourth category of genes called the *Sepallata* genes (*SEP* genes) has been found to be required for this process. **Figure 26.26** illustrates how these genes affect normal flower development in *Arabidopsis*. Gene *A* and SEP gene products are made in tissues that will become the outermost whorl (whorl 1) and promote sepal formation. In tissues that will form whorl 2, gene products of gene *A*, gene *B*, and *SEP* genes are made, which promotes petal formation. The expression of gene *B*, gene *C*, and *SEP* genes causes stamens to be made in whorl 3. Finally, gene *C* and *SEP* genes promote carpel formation in whorl 4.

What is the molecular explanation for how the homeotic mutants of the *A*, *B*, *C*, or *SEP* genes cause their phenotypic effects? In the original ABC model, it was proposed that genes *A* and *C* repress each other's expression, and gene *B* functions independently. In a mutant plant in which gene *A* is not expressed, gene *C* is also expressed in tissues that give rise to whorls 1 and 2. This produces a carpel-stamen-stamen-carpel arrangement of those whorls. When gene *B* is defective, a flower cannot make petals or stamens. Therefore, a gene *B* defect yields a flower with a sepal-sepal-carpel-carpel arrangement. When gene *C* is defective, gene *A* is expressed in tissues that give rise to all four whorls. This results in a sepal-petal-petal-sepal pattern. If the expression of *SEP* genes is defective, the flower consists entirely of sepals, which is the origin of the gene's name.

(a) Normal flower   **(b) Single homeotic mutant**   **(c) Triple mutant**

FIGURE 26.25 **Examples of homeotic mutations in *Arabidopsis*.** (**a**) A normal flower. It is composed of four concentric whorls of structures: sepals, petals, stamens, and carpel. (**b**) A homeotic mutant in which the sepals have been transformed into carpels and the petals have been transformed into stamens. (**c**) A triple mutant in which all of the whorls have been changed into leaves.

(a) Iva Vagnerova/Shutterstock; (b-c) John Bowman

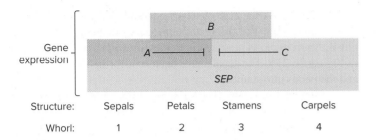

FIGURE 26.26 **The ABC model of homeotic gene action in** *Arabidopsis.* (Note: This is a revised model based on the identification of *SEP* genes. The black lines with a vertical segment at one end indicate that the gene *A* product represses the gene *C* product, and vice versa.)

CONCEPT CHECK: What would be the expected result if gene *A* was inactivated?

What happens if genes *A*, *B*, and *C* are all defective? As shown in **Figure 26.25c**, this triple mutant produces a flower that is composed entirely of leaves! These results indicate that the leaf structure is the default pathway and that the *A*, *B*, *C*, and *SEP* genes cause development to deviate from a leaf structure and produce something else. Thus, the sepals, petals, stamens, and carpels can be viewed as modified leaves. Interestingly, philosopher and poet Johann Goethe originally proposed this idea—that flower formation comes from modifications of the leaf—over 200 years ago.

Arabidopsis has two types of gene *A* (*apetala1* and *apetala2*), two types of gene *B* (*apetala3* and *pistillata*), one type of gene *C* (*agamous*), and three *SEP* genes (*SEP1*, *SEP2*, and *SEP3*). All of these plant homeotic genes code transcription factor proteins that contain a DNA-binding domain and a dimerization domain. However, the *Arabidopsis* homeotic genes do not contain a sequence similar to the homeobox found in animal homeotic genes.

Like the homeotic genes in *Drosophila*, plant homeotic genes are part of a hierarchy of gene regulation. Genes that are expressed within a flower bud produce proteins that activate the expression of these homeotic genes. Once they are transcriptionally activated, the homeotic genes then regulate the expression of other genes, the products of which promote the formation of sepals, petals, stamens, or carpels.

26.4 COMPREHENSION QUESTIONS

1. The growth of plants is due to the division of _____, which are found in _____.
   a. stem cells, the shoots
   b. stem cells, apical and basal meristems
   c. somatic cells, the shoots
   d. somatic cells, apical and basal meristems

2. Flower development occurs in _____ according to _____.
   a. 3 whorls, maternal-effect genes
   b. 3 whorls, the ABC model
   c. 4 whorls, maternal-effect genes
   d. 4 whorls, the ABC model

26.5 SEX DETERMINATION IN ANIMALS

Learning Outcome:

1. Outline the molecular mechanism of sex determination in *Drosophila* and mammals.

To end our discussion of development, let's consider how genetics plays a role in **sex determination**—the process that governs the development of male and female individuals. In the animal kingdom, the existence of two sexes is nearly universal. As discussed in Chapter 2 (see Figure 2.15), the underlying factors that determine female versus male development vary widely. In animals, sex determination is often caused by differences in chromosomal composition. In flies, the ratio of the number of X chromosomes to the number of sets of autosomes determines sex. By comparison, in mammals, the presence of the *SRY* gene on the Y chromosome causes maleness.

The adoption of one of two sexual fates is an event that has been studied in great detail in several species. Researchers have discovered that sex determination is a process controlled genetically by a hierarchy of genes that exert their effects in early embryonic development. In this section, we will consider features of these hierarchies in *Drosophila* and mammals.

In *Drosophila*, Sex Determination Involves a Regulatory Cascade That Includes Alternative Splicing

In diploid fruit flies, XX flies develop into females, and X0 or XY flies become males. The ratio of the number of X chromosomes to the number of sets of autosomes (A) is the determining factor. In diploid flies that carry two sets of chromosomes, the X:A ratio is 1.0 in females versus 0.5 in males. Although male fruit flies usually carry a Y chromosome, it is not necessary for male development. The mechanism of sex determination in *Drosophila* begins in early embryonic development and involves a regulatory cascade composed of several genes. Females follow one of two alternative pathways, and males follow the other. Simplified versions of these pathways are depicted in **Figure 26.27**.

Let's begin with the pathway that produces female flies. In females, the higher X:A ratio results in the embryonic expression of a gene designated *Sxl* (Figure 26.27a). The *Sxl* gene product is a protein that functions in the splicing of pre-mRNAs. In female embryos, the Sxl protein enhances its own expression by splicing its own pre-mRNA, in a process termed an **autoregulatory loop.** In addition, it splices the pre-mRNA from two other genes called *msl-2* and *tra*. The Sxl protein promotes the splicing of the *msl-2* pre-mRNA in a way that introduces an early stop codon in the coding sequence, thereby producing a shortened version of the msl-2 protein that is functionally inactive. The Sxl protein also promotes the splicing of the *tra* pre-mRNA to produce an mRNA that is translated into a functional protein. Therefore, *Sxl* activates *tra*.

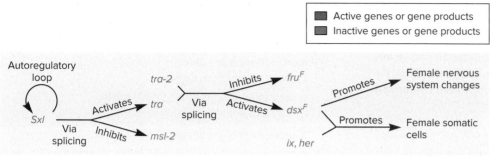

(a) Female pathway (X:A = 1.0)

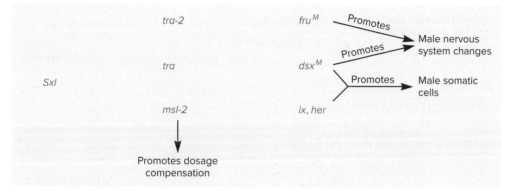

(b) Male pathway (X:A = 0.5)

FIGURE 26.27 **Sex determination pathways for *Drosophila melanogaster*.** Genes or gene products that are functionally expressed are shown in light orange; those that are not expressed are shown in blue. The gene names are abbreviations for the phenotypes that result from mutations that cause loss of function or aberrant expression. These are as follows: *Sxl* (s̲e̲x̲ l̲ethal), *msl* (m̲ale s̲e̲x̲ l̲ethal), *tra* (t̲r̲a̲nsformer), *dsx* (d̲ouble s̲e̲x̲), *fru* (f̲r̲u̲itless), *ix* (i̲nterse̲x̲), and *her* (h̲e̲r̲maphrodite). (Note: These are simplified pathways. More gene products are involved than are shown here.)

The product of the *tra* gene and a constitutively expressed product from a gene called *tra-2* are also splicing factors. In the female, they cause the alternative splicing of the pre-mRNAs that are expressed from the *fru* and *dsx* genes. The tra and tra-2 proteins cause these pre-mRNAs to be spliced into mRNAs designated *fru^F* and *dsx^F*, respectively. The female-specific *fru^F* mRNA is not translated into a sex-specific protein. However, the *dsx^F* mRNA, together with two other proteins coded by the *ix* and *her* genes, promotes female sexual development and controls some aspects of female-specific behavior via the central nervous system. The dsx^F protein is known to be a transcription factor that regulates certain genes that promote these changes.

How are males produced? In X0 or XY flies, the *Sxl* gene is transcriptionally activated, but it is spliced in a way that places an early stop codon in the coding sequence. Therefore, a functional Sxl protein is not made (Figure 26.27b). The absence of *Sxl* expression permits the expression of *msl-2*, which promotes dosage compensation. In fruit flies, dosage compensation is accomplished by increasing the expression of X-linked genes in the male to a level that is twofold higher than it is in the female. Therefore, even though the male has only one X chromosome, the expression of X-linked genes is approximately equal in males and females. The absence of *Sxl* expression in male embryos also promotes the development of maleness. Without *Sxl* expression, the *tra* mRNA is not properly spliced, so *tra* is not expressed. Without the tra protein, the *fru* and *dsx* mRNAs are spliced in a different way to produce mRNAs designated *fru^M* and *dsx^M*. The dsx^M protein is a transcription factor that regulates several different genes, thereby promoting male development. In addition, the *fru^M* gene product is necessary for the regulation of genes involved in male-specific behaviors.

In Mammals, the *SRY* Gene on the Y Chromosome Determines Maleness

In most mammals, such as humans, mice, and marsupials, the presence of the *SRY* gene on the Y chromosome plays a key role in determining maleness. In cases of abnormal sex chromosome composition, such as XXY, an individual develops into a male. The *SRY* gene, which is located on the Y chromosome, causes the sex determination pathway to follow a male developmental scheme. The *SRY* gene codes a protein named testis-determining factor (TDF) that contains a DNA-binding domain called an HMG box, which is found in a broad category of DNA-binding proteins known as the h̲igh-m̲obility g̲roup. TDF is a member of the SOX (S̲RY-like b̲o̲x̲) family of DNA-binding proteins.

Sex determination in mammals, like that in fruit flies, involves a regulatory cascade that is initiated in early embryonic development. However, the pathway is complex, and many details are not completely understood. Several genes that have been identified in mammals are expressed very early in embryonic development and may be directly or indirectly involved in turning on the *SRY* gene. For example, the *WT1* gene is expressed in the early embryo prior to sexual differentiation and may activate *SRY* expression.

Once the *SRY* gene is activated, the coded TDF, along with a protein called SFI, upregulates genes such as *SOX9* that code other transcription factors. The SOX9 protein turns on genes that promote the development of the primary sex cords, which later develop into seminiferous tubules, turning the gonad into a testis rather than an ovary. Cells within the testis secrete testosterone and anti-Müllerian hormone, which contribute to male development.

Researchers postulate that *SRY* expression also causes the expression of other genes that promote male development, such as *DMRT1*. The *DMRT1* gene in mammals is evolutionarily related to the *dsx* gene in *Drosophila*. Both genes code transcription factors involved in the differentiation of the testes. However, all of the target genes that are turned on by TDF have not been definitively identified.

In female mammals, the *DAX1* gene, which is X-linked, is thought to prevent male development. *DAX1* codes a hormone-receptor protein. In XX mammals, *DAX1* expression remains high, in contrast to XY individuals. XY mammals with two copies of the *DAX1* gene develop into females when *SRY* gene expression is low due to a mutation in that gene. This outcome indicates that *DAX1* can inhibit the effects of the *SRY* gene. However, the *DAX1* gene is not needed for female development because XX mice lacking the *DAX1* gene develop as normal females.

26.5 COMPREHENSION QUESTIONS

1. A key event that initially determines female or male development in *Drosophila* is the
   a. transcription of the *Sxl* gene.
   b. alternative splicing of the *Sxl* pre-mRNA.
   c. expression of the *ix* gene.
   d. expression of the *her* gene.
2. A mammalian embryo that is XY but is missing the *SRY* gene would be expected to develop into
   a. a male.
   b. a female.
   c. a hermaphrodite.
   d. none of the above because sex differentiation would not occur.

KEY TERMS

Introduction: developmental genetics
26.1: body pattern, positional information, morphogen, cell fate, induction, cell adhesion, cell adhesion molecule (CAM), Notch receptor, lateral inhibition, lateral induction, homeotic gene, bilaterian, determination
26.2: parasegments, segments, embryogenesis, anteroposterior axis, dorsoventral axis, left-right axis, proximodistal axis, segmentation gene, zygotic gene, homeotic, gain-of-function mutation, loss-of-function mutation, homeobox, homeodomain, cell lineage diagram, cell lineage, heterochronic mutation

26.3: homologous genes, orthologs, *Hox* complex, reverse genetics, gene knockout, forward genetics, determined cell, differentiated cell, basic helix-loop-helix (bHLH) domain, myogenic bHLH protein
26.4: totipotent, cotyledons, shoot meristem, root meristem, organizing center, central zone, peripheral zone, ABC model
26.5: sex determination, autoregulatory loop

CHAPTER SUMMARY

- Developmental genetics is concerned with the roles genes play in orchestrating the changes that occur during development.

26.1 Overview of Animal Development

- During development of multicellular organisms, cells divide, migrate, differentiate, or undergo apoptosis (see Figure 26.1).
- A morphogen is a molecule that conveys positional information to cells and promotes their developmental changes.
- Three molecular mechanisms that convey positional information are (1) a preestablished morphogenic gradient in an oocyte, (2) asymmetrical secretion of a morphogen and induction of neighboring cells in an embryo, and (3) cell adhesion (see Figure 26.2).
- The Notch signaling pathway influences development in adjacent embryonic cells (Figure 26.3).
- The study of mutants with disrupted developmental patterns has identified genes that control development (see Figure 26.4).

- Development in animals involves four overlapping phases: formation of body axes, segmentation, determination, and cell differentiation (see Figure 26.5).

26.2 Invertebrate Development

- *Drosophila* proceeds through several developmental stages from fertilized egg to adult. Various sets of genes are responsible for developmental changes (see Figure 26.6, Table 26.1).
- Maternal effect genes establish the anteroposterior and dorsoventral axes as a result of their asymmetrical distribution. An example of a maternal effect gene is *bicoid*, which controls the formation of anterior structures (see Figures 26.7, 26.8, 26.9).
- The *Drosophila* embryo is divided into segments (see Figure 26.10).
- Three categories of segmentation genes, called gap, pair-rule, and segment-polarity genes, have been identified based on the

- effects they have on development when they are mutant (see Figure 26.11).
- A hierarchy of gene expression, which includes maternal effect genes, gap genes, pair-rule genes, and segment-polarity genes, gives rise to a segmented *Drosophila* embryo (see Figure 26.12).
- Homeotic genes control the developmental fate of particular segments in *Drosophila* (see Figures 26.13, 26.14).
- Homeotic proteins contain a DNA-binding domain and a transcriptional activation domain (see Figure 26.15).
- For the nematode *C. elegans*, a cell lineage diagram depicts the cell division patterns and the fates of each cell's descendants (see Figure 26.16).
- Heterochronic mutations in *C. elegans* disrupt the timing of developmental changes (see Figure 26.17).

26.3 Vertebrate Development

- Homeotic genes in vertebrates are found in groups called *Hox* complexes (see Figure 26.18).
- *Hox* genes control the fates of regions along the anteroposterior axis of a vertebrate (see Figures 26.19, 26.20).

- Transcription factors also control cell differentiation. An example is the family of myogenic bHLH proteins, which causes cells to differentiate into skeletal muscle cells (see Figure 26.21).

26.4 Plant Development

- *Arabidopsis thaliana* is a model organism for studying plant development (see Figure 26.22).
- Plant growth occurs from shoot and root meristems (see Figure 26.23).
- The expression of the *WUS* and *CLV3* genes maintains the correct number of stem cells in the central zone of a shoot meristem (see Figure 26.24).
- The ABC model describes how homeotic genes control flower development in plants (see Figures 26.25, 26.26).

26.5 Sex Determination in Animals

- At the molecular level, sex determination is controlled by pathways that activate specific genes or proteins and inactivate others (see Figure 26.27).

PROBLEM SETS & INSIGHTS

MORE GENETIC TIPS

1. Explain the functional roles of maternal effect genes, gap genes, pair-rule genes, and segment-polarity genes in *Drosophila* development.

T **OPIC:** *What topic in genetics does this question address?* The topic is *Drosophila* development. More specifically, the question is about the functions of certain categories of genes in this process.

I **NFORMATION:** *What information do you know based on the question and your understanding of the topic?* In the question, you are reminded of a few categories of genes that are involved in *Drosophila* development. From your understanding of the topic, you may recall how these genes are expressed in the embryo and how they affect the establishment of body axes and the process of segmentation.

P **ROBLEM-SOLVING** **S** **TRATEGY:** *Describe the steps.* To solve this problem, it is helpful to sort out the steps of embryonic development in *Drosophila* and focus on the roles of these four categories of genes (see Figure 26.12).

ANSWER: These genes are involved in establishing body axes and promoting segmentation of the *Drosophila* embryo. The asymmetrical distribution of maternal effect gene products in the oocyte establishes the anteroposterior and dorsoventral axes. These gene products also control the expression of the gap genes, which are expressed as broad bands in certain regions of the embryo. The overlapping expression of maternal effect genes and gap genes controls the pair-rule genes, which are expressed in alternating stripes. A stripe corresponds to a parasegment. Within each parasegment, the expression of segment-polarity genes defines an anterior and posterior region. With regard to morphology, an anterior region of one parasegment and the posterior region of an adjacent parasegment form a segment of the embryo.

2. Mutations in genes that control the early stages of development are often lethal (e.g., see Figure 26.8b). To circumvent this problem in studies of early development, developmental geneticists may try to isolate temperature-sensitive developmental mutants, or *ts* alleles. If an embryo carries a *ts* allele, it will develop correctly at the permissive temperature (e.g., 25°C) but will fail to develop if incubated at the nonpermissive temperature (e.g., 30°C). In most cases, *ts* alleles result from missense mutations that slightly alter the amino acid sequence of a protein, causing a change in its structure that prevents it from working properly at the nonpermissive temperature. In research, *ts* alleles are particularly useful because they can provide insight regarding the stage of development when the protein coded by the gene is necessary. Researchers can take groups of embryos that carry a *ts* allele and expose them to the permissive or nonpermissive temperature at different stages of development. In the experiment whose results are presented in the following table, embryos were divided into five groups and exposed to the permissive or nonpermissive temperature at different times after fertilization.

Time After Fertilization (hours):	Group				
	1	2	3	4	5
0–1	25°C	25°C	25°C	25°C	25°C
1–2	25°C	30°C	25°C	25°C	25°C
2–3	25°C	25°C	30°C	25°C	25°C
3–4	25°C	25°C	25°C	30°C	25°C
4–5	25°C	25°C	25°C	25°C	30°C
5–6	25°C	25°C	25°C	25°C	25°C
SURVIVAL:	Yes	Yes	Yes	No	Yes

Explain these results.

TOPIC: *What topic in genetics does this question address?* The topic is development. More specifically, the question asks you to determine when a protein that is coded by a *ts* allele is needed during early development.

INFORMATION: *What information do you know based on the question and your understanding of the topic?* In the question, you are told that *ts* alleles are useful, because many other kinds of alleles that affect development may be lethal. You are also given data regarding the survival of embryos that have been subjected to different patterns of permissive and nonpermissive temperatures from 0 to 6 hours after fertilization. From your understanding of the topic, you may realize that a protein coded by a *ts* allele may prevent survival if the embryo is exposed to the nonpermissive temperature when that protein is needed.

PROBLEM-SOLVING **S**TRATEGIES: *Analyze data. Compare and contrast.* To solve this problem, you need to analyze the data by comparing the timing of different incubation temperatures and determining which ones allow survival. If the protein is critical at a particular stage of development, exposure to the nonpermissive temperature during that stage will prevent survival.

ANSWER: The embryos fail to survive if they are subjected to the nonpermissive temperature between 3 and 4 hours after fertilization, but they do survive if subjected to the nonpermissive temperature at other times of development. These results indicate that this protein plays a crucial role at the stage of development that occurs 3–4 hours after fertilization.

---

3. An intriguing question in developmental genetics is this one: How can a particular gene, such as *even-skipped*, be expressed in the alternating banding pattern seen in step 3 in Figure 26.12? Here is another way of asking this question: How is the positional information within the broad bands due to the gap genes able to be deciphered in a way that causes the pair-rule genes to be expressed in this alternating banding pattern? The answer lies in a complex mechanism of genetic regulation.

Certain pair-rule genes have several stripe-specific enhancers that are controlled by multiple transcription factors. A stripe-specific enhancer is typically a short segment of DNA, 300–500 bp in length, that contains regulatory elements recognized by several different transcription factors.

In 1992, Michael Levine and colleagues investigated stripe-specific enhancers located near the promoter of the *even-skipped* gene. A segment of DNA, termed the stripe 2 enhancer, controls the expression of the *even-skipped* gene; this enhancer is responsible for the expression of the *even-skipped* gene in stripe 2, which corresponds to parasegment 3 of the embryo. The stripe 2 enhancer is a segment of DNA that contains binding sites for four transcription factors that are the products of the *Krüppel*, *bicoid*, *hunchback*, and *giant* genes. The Hunchback and Bicoid transcription factors bind to this enhancer and activate the transcription of the *even-skipped* gene. In contrast, the transcription factors coded by the *Krüppel* and *giant* genes bind to the stripe 2 enhancer and repress transcription of the *even-skipped* gene. The figure below describes the concentrations of

these four transcription factor proteins in the region of parasegments 2, 3, and 4. Parasegment 2 corresponds to stripe 2 in the *Drosophila* embryo.

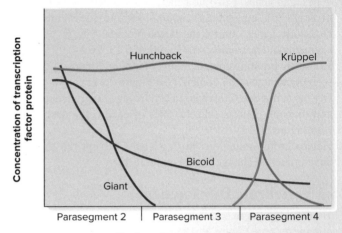

Region of *Drosophila* embryo

To study stripe-specific enhancers, researchers have constructed artificial genes in which the enhancer is linked to a reporter gene, whose expression is easy to detect. The following photo shows the results of an experiment in which an artificial gene was made by putting the stripe 2 enhancer next to the gene that codes β-galactosidase. This artificial gene was introduced into *Drosophila*, and then embryos containing this gene were analyzed for β-galactosidase activity. If a region of the embryo is expressing β-galactosidase, the region will stain darkly because β-galactosidase converts a colorless compound into a dark blue compound.

Courtesy of Stephen Small/New York University

Explain these results.

TOPIC: *What topic in genetics does this question address?* The topic is development. More specifically, the question asks about the mechanism of regulation of a pair-rule gene.

INFORMATION: *What information do you know based on the question and your understanding of the topic?* From the question, you know that pair-rule genes have stripe-specific enhancers that are regulated by multiple transcription factors. You are reminded that particular transcription factors inhibit transcription and others enhance transcription, and you are given data regarding the expression of these transcription factors in parasegments 2, 3, and 4. You have also learned that researchers can place the stripe-specific enhancer next to a reporter gene that codes a protein that can produce a dark blue band in the embryo.

P ROBLEM-SOLVING S TRATEGIES: *Analyze data. Predict the outcome.* To solve this problem, you need to look at the expression of the transcription factors in parasegments 2, 3, and 4, and predict if the reporter gene would be expressed.

ANSWER: As shown in the graph, the concentrations of Hunchback and Bicoid are relatively high in the region of the embryo corresponding to stripe 2 (which is parasegment 3). The levels of Krüppel and

Giant are very low in this region. Therefore, the high levels of activators and low levels of repressors cause the *even-skipped* gene to be transcribed. In this experiment, β-galactosidase was made only in stripe 2 (i.e., parasegment 3). These results show that the stripe-2-specific enhancer controls gene expression only in parasegment 3. Because we know that the *even-skipped* gene is expressed as several alternating stripes (as seen in Figure 26.12), the *even-skipped* gene must contain other stripe-specific enhancers that allow it to be expressed in the other parasegments.

Conceptual Questions

C1. What four types of cellular processes must occur to enable a fertilized egg to develop into an adult multicellular animal? Briefly discuss the role of each process.

C2. Explain how cell adhesion molecules influence embryonic development in animals.

C3. Describe the steps in the Notch signaling pathway, and explain how this pathway may affect the fate of adjacent embryonic cells in animals.

C4. The body axes of the fruit fly are shown in Figure 26.6g. Are the following statements true or false with regard to body axes in the mouse?

    A. Along the anteroposterior axis, the head is posterior to the tail.

    B. Along the dorsoventral axis, the vertebrae of the back are dorsal to the stomach.

    C. Along the dorsoventral axis, the feet are dorsal to the hips.

    D. Along the proximodistal axis, the feet on the hind legs are distal to the upper parts of the hind legs.

C5. If you observed a fruit fly with the following developmental abnormalities, would you guess that a mutation has occurred in a segmentation gene or a homeotic gene? Explain your guess.

    A. Three abdominal segments were missing.

    B. One abdominal segment had legs.

    C. A fly with the correct number of segments had two additional thoracic segments and two fewer abdominal segments.

C6. Which of the following statement(s) is/are true with regard to positional information in *Drosophila*?

    A. Morphogens are a type of molecule that conveys positional information.

    B. Gradients of morphogen distribution are established only in the oocyte, prior to fertilization.

    C. Cell adhesion molecules also provide positional information to a cell.

C7. Discuss the morphological differences between the parasegments and segments of *Drosophila*. Discuss the evidence, providing specific examples, that suggests that the parasegments of the embryo are the subdivisions for the organization of gene expression.

C8. The following are schematic diagrams of mutant *Drosophila* larvae.

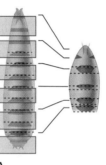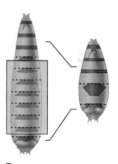

A.   **B.**

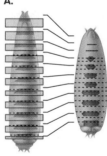

C.

The left side of each pair shows a wild-type larva, with gray boxes indicating the sections that are missing in the mutant larva. Which type of gene is defective in each larva: a gap gene, a pair-rule gene, or a segment-polarity gene?

C9. Explain what a morphogen is, and describe how it exerts its effects. What do you expect will happen when a morphogen is expressed in the wrong place in an embryo? List five examples of morphogens that function in *Drosophila*.

C10. What is positional information? Discuss three different ways that cells obtain positional information. Which of these three ways do you think is the most important for the formation of a segmented body pattern in *Drosophila*?

C11. Gradients of morphogens can be preestablished in the oocyte. Also, later in development, morphogens can be secreted from cells. How are these two processes similar and different?

C12. Discuss how the anterior portion of the anteroposterior axis is established in *Drosophila*. What aspects of oogenesis are critical in establishing this axis? What do you think would happen if the *bicoid* mRNA was not trapped at the anterior end but instead diffused freely throughout the oocyte?

C13. Describe the function of the Bicoid protein. Explain how its ability to exert its effects in a concentration-dependent manner is a critical feature of its function.

C14. With regard to development, what are the roles of the maternal effect genes and the zygotic genes? Which types of genes are needed earlier in the development process?

C15. Discuss the role of homeotic genes in development. Explain what happens to the phenotype of a fruit fly when a gain-of-function mutation in a homeotic gene causes the coded protein to be expressed in an abnormal region of the embryo. What are the consequences of a loss-of-function mutation in such a gene?

C16. Describe the molecular features of the homeobox and homeodomain. Explain how these features are important in the function of homeotic genes.

C17. What would you predict to be the phenotype of a *Drosophila* larva whose female parent was homozygous for a loss-of-function allele in the *nanos* gene?

C18. Based on the photographs in Figure 26.14, in which segment(s) is the *Antp* gene normally expressed?

C19. If a mutation in a homeotic gene produced each of the following phenotypes, would you expect it to be a loss-of-function or a gain-of-function mutation? Explain your answer in each case.

 A. An abdominal segment has antennae attached to it.

 B. The most anterior abdominal segment resembles the most posterior thoracic segment.

 C. The most anterior thoracic segment resembles the most posterior abdominal segment.

C20. Explain how loss-of-function mutations in each of the following categories of genes would affect the morphologies of *Drosophila* larvae.

 A. Gap genes

 B. Pair-rule genes

 C. Segment-polarity genes

C21. What is the difference between a maternal effect gene and a zygotic gene? Of the following genes that play a role in *Drosophila* development, which are maternal effect genes and which are zygotic? Explain your answers.

 A. *nanos*

 B. *Antp*

 C. *bicoid*

 D. *lab*

C22. Cloning of mammals (such as Dolly the sheep) is described in Chapter 21. Based on your understanding of animal development, explain why an enucleated egg is needed to clone a mammal. In other words, what features of the oocyte are essential for animal development?

C23. A hypothetical cell lineage is shown next.

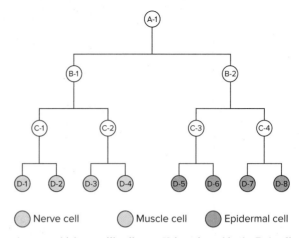

○ Nerve cell   ◐ Muscle cell   ● Epidermal cell

A gene, which we will call gene *X*, is activated in the B-1 cell, so the B-1 cell will advance through the proper developmental stages to produce three nerve cells (D-1, D-2, and D-3) and one muscle cell (D-4). Gene *X* is normally inactivated in the A-1, C-1, and C-2 cells, as well as the four D cells, D-1 through D-4. Draw the expected cell lineage if a heterochronic mutation had each of the following effects:

 A. Gene *X* is turned on one cell division too early.

 B. Gene *X* is turned on one cell division too late.

C24. What is a heterochronic mutation? How does it affect the phenotypic outcome of an organism? What phenotypic effects would you expect if a heterochronic mutation affected the cell lineage that determines the fates of intestinal cells?

C25. Discuss the similarities and differences between the *bithorax* and *Antennapedia* complexes in *Drosophila* and the *Hox* gene complexes in mice.

C26. What is cell differentiation? Discuss the role of myogenic bHLH proteins in the differentiation of muscle cells. Explain how these proteins work at the molecular level. In your answer, explain how protein dimerization is key to gene regulation.

C27. The gene family that codes the myogenic bHLH proteins in mammals plays a key role in skeletal muscle-cell differentiation, whereas the *Hox* genes are homeotic genes that play a role in the differentiation of particular regions of the body. Explain how the functions of these genes are similar and different.

C28. What is a totipotent cell? In each of the following types of organisms, which cells are totipotent?

 A. Humans   C. Yeast

 B. Corn   D. Bacteria

C29. What is a meristem? Explain the role of the meristem in plant development.

C30. Discuss the morphological differences between animal and plant development. How are the developmental processes different at the cellular level? How are they similar at the genetic level?

C31. Predict the phenotypic consequences of each of the following mutations:

 A. *apetala1* defective

 B. *pistillata* defective

 C. *apetala1* and *pistillata* defective

C32. Explain how alternative splicing affects sex determination in *Drosophila*.

Experimental Questions

E1. Researchers have used the gene cloning methods described in Chapter 20 to clone the *bicoid* gene and express large amounts of the Bicoid protein. The Bicoid protein was then injected into the posterior end of a zygote immediately after fertilization. What phenotypic results would you expect? What do you think would happen if the Bicoid protein was injected into a segment of a larva?

E2. Compare and contrast the experimental advantages of *Drosophila* and *C. elegans* for studying developmental genetics.

E3. What is meant by the term *cell fate*? What is a cell lineage diagram? Discuss the experimental advantage of having a cell lineage diagram. What is a cell lineage?

E4. Explain why a cell lineage diagram is necessary to determine if a mutation is heterochronic.

E5. Explain the rationale behind the use of the "bag of worms" phenotype as a way to identify heterochronic mutations.

E6. Shown below are cell lineages determined from analyses of hypodermal cells in wild-type and mutant strains of *C. elegans*.

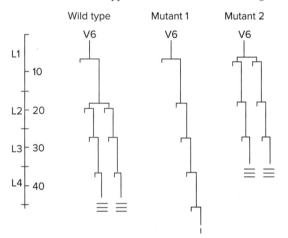

Explain the nature of the mutations in the altered strains.

E7. Look back at question 2 in More Genetic TIPS before answering this question. *Drosophila* embryos carrying a *ts* mutation were exposed to the permissive (25°C) or nonpermissive (30°C) temperature at different stages of development. Explain these results.

Time After	Group				
Fertilization (hours)	1	2	3	4	5
0–1	25°C	25°C	25°C	25°C	25°C
1–2	25°C	30°C	25°C	25°C	25°C
2–3	25°C	25°C	30°C	25°C	25°C
3–4	25°C	25°C	25°C	30°C	25°C
4–5	25°C	25°C	25°C	25°C	30°C
5–6	25°C	25°C	25°C	25°C	25°C
SURVIVAL:	Yes	No	No	Yes	Yes

E8. All of the homeotic genes in *Drosophila* have been cloned. As discussed in Chapter 20, cloned genes can be manipulated in vitro. They can be subjected to cutting and pasting, gene editing, and other alterations. After *Drosophila* genes have been altered in vitro, they can be inserted into a transposon vector (i.e., a P element vector), which can be injected into *Drosophila* embryos. The P

element then transposes into the chromosomes, thereby introducing one or more copies of the altered gene into the *Drosophila* genome. This method is termed P element transformation.

With this information in mind, how would you create a gain-of-function mutation that caused the *Antp* gene to be expressed where the *abd-A* gene is normally expressed? What phenotype would you expect for flies that carried this altered gene?

E9. You need to refer to the answer to question 3 in More Genetic TIPS to answer this question. If the artificial gene containing the stripe 2 enhancer and the gene encoding β-galactosidase were found within an embryo that also had a loss-of-function mutation in each of the following genes, what results would you expect? In other words, would there be a stripe or not? Explain why.

A. *Krüppel*

B. *bicoid*

C. *hunchback*

D. *giant*

E10. Two techniques commonly used to study the expression patterns of genes that play a role in development are Northern blotting and in situ hybridization. As described in Chapter 20, Northern blotting is used to detect RNA that is transcribed from a particular gene. In this method, a specific RNA is detected by using a short segment of cloned DNA as a probe. The DNA probe, which is labeled, is complementary to the RNA that the researcher wishes to detect. After the DNA probe binds to the RNA within a blot of a gel, the RNA is visualized as a labeled band on a nylon membrane. For example, a DNA probe that is complementary to the *bicoid* mRNA could be used to specifically detect the amount of that mRNA in a blot.

A second technique, termed fluorescence in situ hybridization (FISH), is used to identify the locations of genes on chromosomes. This technique is also used to locate gene products within oocytes, embryos, and larvae. Thus, it has been commonly used by developmental geneticists to understand the expression patterns of genes during development. The schematic drawing in Figure 26.9b is the result of the application of the FISH technique. In this case, the probe was complementary to *bicoid* mRNA.

Here is the question: Suppose a researcher has three different *Drosophila* strains that have loss-of-function mutations in the *bicoid* gene. We will call the mutant genes *bicoid-A*, *bicoid-B*, and *bicoid-C*; the wild type is designated *bicoid⁺*. To study these mutations, phenotypically normal female flies that are homozygous for each *bicoid* mutation were obtained, and their oocytes were analyzed using Northern blotting and FISH. A wild-type strain was also analyzed as a control. In other words, RNA was isolated from some oocytes and analyzed by Northern blotting, and some oocytes were subjected to in situ hybridization. In both cases, the probe was complementary to the *bicoid* mRNA. The results are shown next.

Northern blot

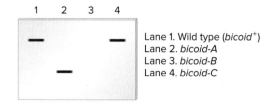

Lane 1. Wild type (*bicoid⁺*)
Lane 2. *bicoid-A*
Lane 3. *bicoid-B*
Lane 4. *bicoid-C*

In situ hybridization

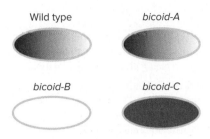

Wild type   bicoid-A

bicoid-B   bicoid-C

A. How can phenotypically normal female flies be homozygous for a loss-of-function allele in the *bicoid* gene?

B. Explain the type of mutation (e.g., deletion, point mutation, etc.) in each of the three strains. Explain how the mutation may cause a loss of normal function for the *bicoid* gene product.

C. Discuss how the use of both techniques provides more definitive information than using just one of the techniques.

E11. Explain one experimental strategy for determining the functional role of the mouse *HoxD-3* gene.

E12. For the experiment shown in Figure 26.17, suggest reasons why the n536, n355, and n540 strains have an egg-laying defect.

E13. Another way to study the role of proteins (e.g., transcription factors) that function in development is to microinject the mRNA that codes a protein, or the purified protein itself, into an oocyte or embryo, and then determine how this affects the subsequent development of the embryo, larva, and adult. For example, if Bicoid protein is injected into the posterior region of an oocyte, the resulting embryo will develop into a larva that has anterior structures at both ends. Based on your understanding of the function of each developmental gene, what would be the predicted phenotype following microinjection of each of the following proteins or mRNAs?

A. *Nanos* mRNA injected into the anterior end of an oocyte

B. *Antp* protein injected into the posterior end of an embryo

C. *Toll* mRNA injected into the dorsal side of an early embryo

E14. Why have geneticists used reverse genetics to study the genes involved in vertebrate development? Explain how this strategy differs from traditional genetic analyses like those done by Mendel.

Questions for Student Discussion/Collaboration

1. Compare and contrast the experimental advantages and disadvantages of *Drosophila*, *C. elegans*, mammals, and *Arabidopsis*.

2. It seems that developmental genetics boils down to a complex network of gene regulation. Try to draw a structure of this network for *Drosophila*. How many genes do you think are necessary to complete the developmental network for the fruit fly? How many genes do you think are needed for a network to specify one segment? Do you think it is more difficult to identify genes that are involved in the beginning, middle, or end of this network? Suppose you were trying to identify all of the genes needed for development in a chicken. Knowing what you know about *Drosophila* development, would you first try to identify genes necessary for early development, or would you begin by identifying genes involved in cell differentiation?

3. At the molecular level, how do you think a gain-of-function mutation in a developmental gene might cause it to be expressed in the wrong place or at the wrong time? Explain what type of DNA sequence would be altered.

Note: All answers are available for the instructor in Connect; the answers to the even-numbered questions and all of the Concept Check and Comprehension Questions are in Appendix B.

The African cheetah. *This species has a relatively low level of genetic variation because the population was reduced to a small size approximately 10,000–12,000 years ago.*

Riccardo Vallini Pics/Getty Images

27

POPULATION GENETICS

Until now, we have primarily focused our attention on genes within individuals and their related family members. In this chapter and Chapters 28 and 29, we turn to the study of genes in a population or species. The field of **population genetics** is concerned with the extent of genetic variation within a group of individuals and changes in that variation over time. The field of population genetics emerged as a branch of genetics in the 1920s and 1930s. Its mathematical foundations were developed by theoreticians who extended the principles of Gregor Mendel and Charles Darwin by deriving formulas to explain the occurrence of genotypes within populations. Development of these foundations can be largely attributed to three scientists: Ronald Fisher, Sewall Wright, and J. B. S. Haldane. As we will see, support for their mathematical theories was provided by several researchers who analyzed the genetic composition of natural and experimental populations.

More recently, population geneticists have used techniques to probe genetic variation at the molecular level. In addition, staggering advances in computer technology have aided population geneticists in the analysis of their genetic theories and data. In this chapter, we will explore the genetic variation that occurs in populations and consider the reasons why the genetic composition of a population may change over the course of several generations.

Much of this chapter explores how equations are used to predict changes in genetic variation over time. A better appreciation for applying these equations can be obtained through the use of computer simulations in which you can manipulate variables and see how such manipulations affect the outcome. Many such simulations are freely available online.

27.1 GENES IN POPULATIONS AND THE HARDY-WEINBERG EQUATION

Learning Outcomes:

1. Define *gene pool* and *population*.
2. Describe the extent of polymorphism in natural populations.
3. Use the Hardy-Weinberg equation to calculate allele and genotype frequencies.

Population genetics may seem like a significant departure from other topics in this text, but it is a direct extension of our understanding of Mendel's laws of inheritance, molecular genetics, and the ideas of Darwin, which are described in Chapter 29. In the field of population genetics, the focus shifts away from the individual and onto the population of which the individual is a member. A common focus in population genetics is variation among alleles. Conceptually, all of the alleles of every gene in a population make up the **gene pool.**

One way to view gene pools is to consider them on a generation-to-generation basis. From this viewpoint, individuals of one generation constitute a gene pool. In turn, individuals that reproduce contribute to the gene pool of the next generation. Population geneticists study the genetic variation within gene pools and how such variation changes from one generation to the next. In this section, we will examine some of the general features of populations and gene pools.

A Population Is a Group of Interbreeding Individuals That Share a Gene Pool

In genetics, the term *population* has a very specific meaning. With regard to sexually reproducing species, a **population** is a group of individuals of the same species that occupy the same region and can interbreed with one another. Many species occupy a wide geographic range and are divided into discrete populations. For example, distinct populations of a given species may be located on different continents, or populations on the same continent may be divided by a geographical barrier such as a large mountain range.

A large population usually is composed of smaller groups called **local populations.** The members of a local population are far more likely to breed among themselves than with individuals of the same species from a more distant population. Local populations are often separated from each other by moderate geographic barriers.

As shown in **Figure 27.1,** the large ground finch (*Geospiza magnirostris*) is found on a small volcanic island called Daphne Major, which is one of the Galápagos Islands. Daphne Major is located north of the much larger Santa Cruz Island. The population of large ground finches on Daphne Major constitutes a local population of this species. Breeding is much more likely to occur

among members of this local population than between members of neighboring populations of the large ground finch. On relatively rare occasions, however, a bird may fly from Daphne Major to Santa Cruz Island, which means that breeding between the two different local populations is possible.

Populations typically are dynamic units that change from one generation to the next. A population may change its size, geographic location, and/or genetic composition. With regard to size, natural populations commonly go through cycles of "feast or famine," during which environmental factors cause their numbers to swell or shrink. In addition, natural predators or disease may periodically decrease the size of a population to significantly lower levels; the population later may rebound to its original size. Populations or individuals within populations may migrate to a new site and establish a distinct population in this location. The environment at this new geographic location may differ from the original site. What are the consequences of such changes? As populations change their sizes and locations, their genetic composition generally changes as well. As described later in this chapter, population geneticists have developed mathematical models that predict how the gene pool will change in response to fluctuations in size, migration, and natural selection.

At the Population Level, Some Genes May Be Monomorphic, but Most Are Polymorphic

In population genetics, the term **polymorphism** (meaning "many forms") refers to the observation that many traits display variation within a population. Historically, polymorphism first referred to variation in traits that are observable with the unaided eye. Polymorphisms in color and pattern have long attracted the attention of population geneticists. Some of the well-studied variations include yellow and red varieties of the elderflower orchid and brown, pink, and yellow shells in land snails, which are discussed later in this chapter. **Figure 27.2** illustrates a striking example of polymorphism in the Hawaiian happy-face spider (*Theridion grallator*). The three individuals shown in this figure are from the same species, but they differ in alleles that affect color and pattern.

What is the underlying cause of polymorphism? At the DNA level, polymorphism may be due to two or more alleles that influence the phenotype of the individual that inherits them. In other words, it is due to genetic variation. Geneticists also use the term **polymorphic** to describe a gene that commonly exists as two or more alleles in a population. By comparison, a **monomorphic** gene exists predominantly as a single allele in a population. By convention, when a single allele is found in at least 99% of all cases, the gene is considered monomorphic.

At the level of a particular gene, polymorphism may involve various types of changes such as a deletion of a significant region of the gene, a duplication of a region, or a change in a single nucleotide. This last phenomenon is called a **single-nucleotide polymorphism (SNP).** SNPs are the smallest type of genetic change that can occur within a given gene and are also the most common. In humans, for example, SNPs represent 90% of all the variation in DNA sequences that occurs among different people. SNPs are found very frequently in genes. In the human population, a gene that is 2000–3000 bp in length contains, on average, 10

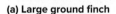

(a) Large ground finch   **(b) A view of Daphne Major (the island in the distance) from Santa Cruz Island**

FIGURE 27.1   A local population of the large ground finch.
(a) The large ground finch (*Geospiza magnirostris*) on Daphne Major.
(b) A view of Daphne Major, one of the Galápagos Islands, from Santa Cruz Island.

(a): LABETAA Andre/Shutterstock; (b): Deborah Freund

CONCEPT CHECK: What does the term *local population* mean?

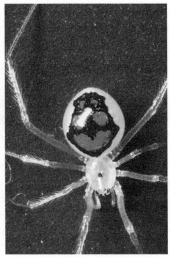

FIGURE 27.2 **Polymorphism in the Hawaiian happy-face spider (*Theridion grallator*).**

Genes→Traits These three spiders are members of the same species and carry the same genes. However, several genes that affect pigmentation patterns are polymorphic, meaning that each of these genes exists in multiple alleles within the population. This polymorphism within the Hawaiian happy-face spider population produces members that look quite different from each other.

(All): Geoff Oxford

CONCEPT CHECK: Are polymorphisms common or rare in natural populations?

different sites that are polymorphic. The high frequency of SNPs indicates that polymorphism is the norm for most human genes. Likewise, relatively large, healthy populations of nearly all species exhibit a high level of genetic variation due to possessing SNPs within most genes.

Within a population, the alleles of a given gene may arise through different types of genetic changes. **Figure 27.3** considers a gene that exists in multiple forms in humans. This example is a short segment of DNA found within the human β-globin gene. The top sequence is an allele designated Hb^A, whereas the middle sequence is an allele called Hb^S. These alleles differ from each other by a single nucleotide, so they provide an example of a SNP. As discussed in Chapter 4, the Hb^S allele causes sickle cell disease in a homozygote. The bottom sequence has a short, 5-bp deletion compared with the other two alleles. This deletion results in a nonfunctional β-globin polypeptide. Therefore, the bottom sequence is an example of a loss-of-function allele.

Population Genetics Is Concerned with Allele, Genotype, and Phenotype Frequencies

As we have seen, population geneticists want to understand the prevalence of polymorphic genes within populations. Their goal is to identify the causative factors that govern changes in genetic variation. Much of their work evaluates the frequency of alleles in a quantitative way. Calculations of two fundamental values are central to population genetics: **allele frequencies** and **genotype frequencies.** The allele and genotype frequencies are defined as follows:

$$\text{Allele frequency} = \frac{\text{Number of copies of a specific allele in a population}}{\text{Total number of all types of alleles for that gene in a population}}$$

$$\text{Genotype frequency} = \frac{\text{Number of individuals with a particular genotype in a population}}{\text{Total number of individuals in a population}}$$

Though these two frequencies are related, a clear distinction between them must be kept in mind. As an example, let's consider a hypothetical population of 100 frogs with regard to two alleles that are incompletely dominant. The G^D allele confers dark green color whereas the G^L allele confers light green color. The population has the following phenotypes and genotypes:

64 dark green frogs with genotype $G^D G^D$
32 medium green frogs with genotype $G^D G^L$
4 light green frogs with genotype $G^L G^L$

FIGURE 27.3 **The relationship between alleles and various types of mutations.** The DNA sequence shown here is a small portion of the β-globin gene in humans. Mutations have altered the gene to create the three different alleles in this figure. The top two alleles differ by a single base pair and thus exhibit what is known as a single-nucleotide polymorphism (SNP). The bottom allele has a 5-bp deletion that begins right after the arrowhead. The deletion results in a nonfunctional polypeptide, so the allele is a loss-of-function allele.

Region of the human β-globin gene

These are three different alleles of the human β-globin gene. Many more have been identified.

```
A C T C C T   G A G G A A
T G A G G A   C T C C T T
```
Hb^A allele

These two alleles are an example of a single-nucleotide polymorphism in the human population.

```
A C T C C T   G T A G G A A
T G A G G A   C A T C C T T
```
Hb^S allele

```
A C T C C     A A
T G A G G     T T
```
Loss-of-function allele

Site of a 5-bp deletion

For calculating an allele frequency, homozygous individuals have two copies of an allele, whereas heterozygotes have only one. For example, in tallying the G^L allele, we see that each of the 32 heterozygotes has one copy of the G^L allele, and each of the 4 (light green) homozygotes has two copies. The allele frequency for G^L equals

$$G^L = \frac{32 + 2(4)}{2(64) + 2(32) + 2(4)}$$

$$= \frac{40}{200} = 0.2, \text{ or } 20\%$$

This result tells us that the allele frequency of G^L is 20%. In other words, 20% of the alleles for this gene in the population are the G^L allele.

Let's now calculate the genotype frequency of $G^L G^L$ (light green) frogs:

$$G^L G^L = \frac{4}{64 + 32 + 4}$$

$$= \frac{4}{100} = 0.04, \text{ or } 4\%$$

We see that 4% of the individuals in this population are light green frogs.

Allele and genotype frequencies are always less than or equal to 1 (i.e., ≤100%). If a gene is monomorphic, the allele frequency for the single allele will be equal to or slightly less than a value of 1.0. For polymorphic genes, if we add up the frequencies for all of the alleles in the population, we should obtain a value of 1.0. In our frog example, the allele frequency of G^L equals 0.2. The frequency of the other allele, G^D, equals 0.8. If we add the two together, we obtain a value of 0.2 + 0.8 = 1.0.

In addition to allele and genotype frequencies, a third key concept in population genetics is **phenotype frequency,** which is the frequency of a particular phenotype in a population. Phenotype frequencies are particularly important in population genetics because natural selection acts on phenotypes, which are determined by genotypes. This topic is discussed in Section 27.3. In medicine, clinicians may be interested in the phenotype frequency of individuals harboring a genetic disorder. In that case, they may group phenotypes into two categories—those individuals that are affected with a disorder and unaffected individuals.

How do we compute phenotype frequencies? The answer is that it depends on the pattern of inheritance. Two examples are described next:

- When a trait follows incomplete dominance, as in our frog example, each genotype frequency correlates with a phenotype frequency. For example, the genotype frequency of $G^L G^L$ frogs is 4%, which is the same as the phenotype frequency of light green frogs.
- However, for alleles that are dominant, the frequency of the homozygous dominant individuals is added to the frequency of heterozygotes to compute the phenotype frequency. As an example, let's consider cystic fibrosis, which is a recessive disorder in humans. The genotype frequency of individuals homozygous for the dominant (non-disease-causing) allele is 0.960 and that of heterozygotes is 0.039. Both genotypes are unaffected with the disease. Therefore, the phenotype frequency of unaffected individuals is 0.960 + 0.039 = 0.999, or 99.9%.

The Hardy-Weinberg Equation Can Be Used to Calculate Genotype Frequencies Based on Allele Frequencies

Now that you have a general understanding of genes in populations, we can begin to relate these concepts to mathematical expressions as a way to examine whether allele and genotype frequencies are changing over the course of many generations. In 1908, mathematician Godfrey Hardy and physician Wilhelm Weinberg independently derived a simple mathematical expression that predicts stability of allele and genotype frequencies from one generation to the next. The maintenance of stability of these frequencies is called **Hardy-Weinberg equilibrium,** because (under a given set of conditions, described later in this section) allele and genotype frequencies do not change over the course of many generations.

Why is Hardy-Weinberg equilibrium a useful concept? An equilibrium is a null hypothesis, which suggests that evolutionary change is not occurring. In reality, however, populations rarely achieve such an equilibrium. Therefore, the main usefulness of Hardy-Weinberg equilibrium is that it provides a framework that can be used to understand changes in allele frequencies within a population when such an equilibrium is violated.

To appreciate Hardy-Weinberg equilibrium, let's return to our hypothetical frog example in which a gene is polymorphic and exists as two different alleles: G^D and G^L. If the allele frequency of G^D is denoted by the variable p, and the allele frequency of G^L by q, then

$$p + q = 1$$

For example, if $p = 0.8$, then q must be 0.2. In other words, if the allele frequency of G^D equals 80%, the remaining 20% of alleles must be G^L, because the allele frequencies must add up to 100%.

Hardy-Weinberg equilibrium relates allele frequencies and genotype frequencies. For a diploid species, each individual inherits two copies of most genes. Hardy-Weinberg equilibrium assumes that two gametes are chosen at random from the gene pool. Therefore, because gametes are chosen independently, we can use the product rule (discussed in Chapter 3), and multiply the sum $(p + q)$ by itself to represent genotype frequencies. Because $p + q = 1$, we know that $p + q$ multiplied by $p + q$ also equals 1, because $1 \times 1 = 1$:

$$(p + q)(p + q) = 1$$

$$p^2 + 2pq + q^2 = 1 \text{ (\textbf{Hardy-Weinberg equation})}$$

The Hardy-Weinberg equation applies to a gene in a diploid species that is found in only two alleles. If this equation is applied to our hypothetical frog population, in which a gene exists in alleles designated G^D and G^L, then

p^2 equals the genotype frequency of $G^D G^D$

$2pq$ equals the genotype frequency of $G^D G^L$

q^2 equals the genotype frequency of $G^L G^L$

If $p = 0.8$ and $q = 0.2$ and if the population is in Hardy-Weinberg equilibrium, then

$$G^D G^D = p^2 = (0.8)^2 = 0.64$$

$$G^D G^L = 2pq = 2(0.8)(0.2) = 0.32$$

$$G^L G^L = q^2 = (0.2)^2 = 0.04$$

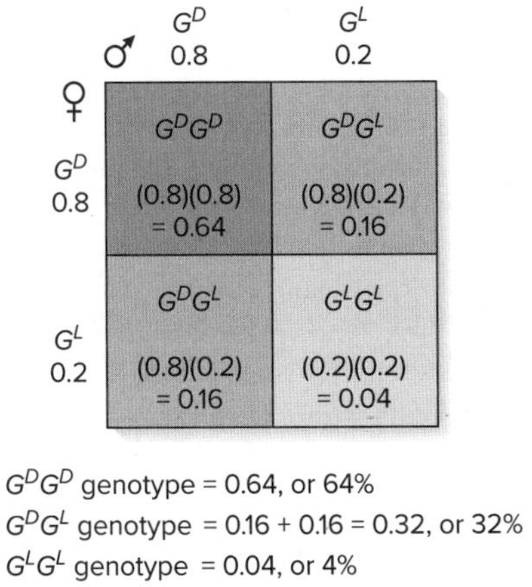

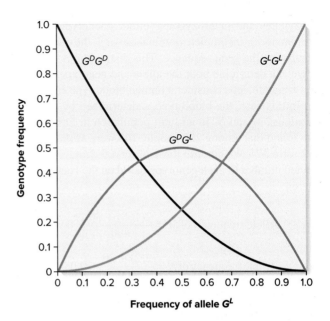

G^DG^D genotype = 0.64, or 64%
G^DG^L genotype = 0.16 + 0.16 = 0.32, or 32%
G^LG^L genotype = 0.04, or 4%

FIGURE 27.4 **A comparison between allele frequencies and genotype frequencies in a Punnett square.** In a population in Hardy-Weinberg equilibrium, the frequency of gametes carrying a particular allele is equal to the allele frequency in the population.

FIGURE 27.5 **The relationship between allele frequencies and genotype frequencies according to the Hardy-Weinberg equation.** This graph assumes that G^D and G^L are the only two alleles for this gene.

In other words, if the allele frequency of G^D is 80% and the allele frequency of G^L is 20%, the genotype frequency of G^DG^D is 64%, G^DG^L is 32%, and G^LG^L is 4%.

To illustrate the relationship between allele frequencies and genotypes, **Figure 27.4** compares the values obtained with the Hardy-Weinberg equation to the way that gametes combine randomly with each other to produce offspring. In a population, the frequency of a gamete carrying a particular allele is equal to the allele frequency in that population. In this example, the frequency of a gamete carrying the G^D allele equals 0.8.

We can use the product rule to determine the frequency of genotypes. For example, the frequency of a G^DG^D homozygote is (0.8)(0.8) = 0.64, or 64%. Also, the probability of inheriting both G^L alleles is (0.2)(0.2) = 0.04, or 4%. As Figure 27.4 indicates, heterozygotes can be produced in two different ways. An offspring could inherit the G^D allele from its male parent and G^L from its female parent, or G^D from its female parent and G^L from its male parent. Therefore, the frequency of heterozygotes is $pq + pq$, which equals $2pq$; in our example, this is 2(0.8)(0.2) = 0.32, or 32%.

In the absence of evolutionary changes, the Hardy-Weinberg equation predicts an equilibrium—unchanging allele and genotype frequencies from generation to generation—if certain conditions are met in a population. With regard to a particular gene of interest, these conditions are as follows:

1. *No new mutations.* The gene of interest does not incur any new mutations.
2. *No genetic drift.* The population is so large that allele frequencies do not change due to random sampling effects.
3. *No migration.* Individuals or their gametes do not travel between different populations. (Note: For some species, gametes can be dispersed via wind or water.)
4. *No natural selection.* All of the genotypes result in phenotypes that have the same reproductive success.
5. *Random mating or breeding.* With respect to the gene of interest, the members of the population reproduce with

each other without regard to their phenotypes and genotypes. (Note: The term *mating* is used to describe sexual reproduction among animals. In the case of plants, the term *breeding* is commonly used.)

The Hardy-Weinberg equation provides a quantitative relationship between allele and genotype frequencies in a population. If the preceding five conditions are met, a Hardy-Weinberg equilibrium will be achieved after one generation of random mating. **Figure 27.5** shows this relationship for different allele frequencies of G^D and G^L. As expected, when the allele frequency of G^L is very low, the G^DG^D genotype predominates; when the G^L allele frequency is high, the G^LG^L homozygote is most prevalent in the population. When the allele frequencies of G^D and G^L are intermediate in value, the heterozygote predominates.

In reality, no population satisfies Hardy-Weinberg equilibrium completely. Nevertheless, in large natural populations with little migration and negligible natural selection, Hardy-Weinberg equilibrium may be nearly approximated for certain genes. In addition, Hardy-Weinberg equilibrium can be extended to situations in which a single gene exists in three or more alleles, as described in question 2 of More Genetic TIPS at the end of the chapter.

The Hardy-Weinberg Equation Can Be Used to Detect Evolutionary Change

As discussed in Chapter 3, the chi square test is used to evaluate the agreement between observed and expected data. We can use a chi square test to determine whether a population exhibits Hardy-Weinberg equilibrium for a particular gene. To carry out this calculation, it is necessary to distinguish between homozygotes and heterozygotes, either phenotypically (due to codominance or incomplete dominance) or at the molecular level by analyzing the gene sequence. This distinction is necessary so that we can determine both the allele and genotype frequencies. As an example, let's consider a human blood type called the MN type. In this case, the blood type is determined by two codominant alleles, *M* and *N*. In an Inuit population in East Greenland, it was found that among 200 people, 168 were *MM*, 30 were *MN*, and 2 were *NN*. We can use these observed data to calculate the expected number of each genotype based on the Hardy-Weinberg equation:

$$\text{Allele frequency of } M = \frac{2(168) + 30}{400} = 0.915$$

$$\text{Allele frequency of } N = \frac{2(2) + 30}{400} = 0.085$$

Expected frequency of $MM = p^2 = (0.915)^2 = 0.837$

Expected number of *MM* individuals $= 0.837 \times 200 = 167.4$
(or 167 if rounded to the nearest individual)

Expected frequency of $NN = q^2 = (0.085)^2 = 0.007$

Expected number of *NN* individuals $= 0.007 \times 200 = 1.4$ (or 1 if rounded to the nearest individual)

Expected frequency of $MN = 2pq = 2 \times 0.915 \times 0.085 = 0.156$

Expected number of *MN* individuals $= 0.156 \times 200 = 31.2$ (or 31 if rounded to the nearest individual)

$$\chi^2 = \frac{(O_1 - E_1)^2}{E_1} + \frac{(O_2 - E_2)^2}{E_2} + \frac{(O_3 - E_3)^2}{E_3}$$

$$\chi^2 = \frac{(168 - 167)^2}{167} + \frac{(30 - 31)^2}{31} + \frac{(2 - 1)^2}{1}$$

$$\chi^2 = 1.04$$

We need to refer back to Table 3.1 in Chapter 3 to evaluate the calculated chi square value. For a chi square value of 1.04 with 1 degree of freedom (*df*), the *P* value is between 0.5 and 0.2, which is well within the acceptable range. Therefore, we fail to reject the null hypothesis. In this case, the alleles for this gene appear to be in Hardy-Weinberg equilibrium.

When researchers have investigated other genes in various populations, a high chi square value is sometimes obtained, and the hypothesis that the allele and genotype frequencies are in Hardy-Weinberg equilibrium is rejected. In these cases, we would say that the population is in **disequilibrium.** Deviation from Hardy-Weinberg equilibrium indicates evolutionary change. As discussed in later sections of this chapter, factors such as natural selection, genetic drift, migration, and inbreeding may disrupt Hardy-Weinberg equilibrium. Therefore, when population geneticists discover that a population is not in equilibrium, they try to determine which factors are at work.

GENETIC TIPS   **THE QUESTION:** One particularly useful feature of the Hardy-Weinberg equation is that it allows us to estimate the frequency of heterozygotes for recessive genetic diseases, assuming that Hardy-Weinberg equilibrium exists. As an example, let's consider cystic fibrosis, which is a genetic disease in humans involving a gene that codes a chloride transporter. Persons with this disorder have an irregularity in salt and water balance. One of the symptoms is thick mucus in the lungs that can contribute to repeated lung infections. In populations of Northern European descent, the frequency of affected individuals is approximately 1 in 2500. Because cystic fibrosis is a recessive disorder, affected individuals are homozygotes. Assuming that the population of people of Northern European descent is in Hardy-Weinberg equilibrium, what is the frequency of individuals who are heterozygous carriers?

T OPIC: *What topic in genetics does this question address?* The topic is predicting the frequency of heterozygotes in a population. More specifically, the question is about predicting the frequency of heterozygotes carrying the recessive allele that causes cystic fibrosis.

I NFORMATION: *What information do you know based on the question and your understanding of the topic?* From the question, you know the frequency of homozygotes, who have the disease. From your understanding of the topic, you may realize that you can use the Hardy-Weinberg equation to determine allele and genotype frequencies.

P ROBLEM-SOLVING **S** TRATEGY: *Make a calculation.* One strategy to solve this problem is to use the Hardy-Weinberg equation to first determine the allele frequencies for the disease-causing allele and the non-disease-causing allele, and then use these allele frequencies to calculate the genotype frequency of heterozygotes.

If *q* represents the allele frequency of the disease-causing allele, then

$$q^2 = 1/2500$$
$$q^2 = 0.0004$$

We take the square root to determine *q*:

$$q = \sqrt{0.0004}$$
$$q = 0.02$$

If *p* represents the non-disease-causing allele,

$$p = 1 - q$$
$$p = 1 - 0.02 = 0.98$$

ANSWER: The frequency of heterozygous carriers is
$$2pq = 2(0.98)(0.02) = 0.0392, \text{ or } 3.92\%$$

27.1 COMPREHENSION QUESTIONS

1. A gene pool is
   a. all alleles of the genes in a single individual.
   b. all of the genes in the gametes produced by a single individual.
   c. all alleles of all genes in a population of individuals.
   d. the random mixing of alleles during sexual reproduction.

2. In natural populations, most genes are
   a. polymorphic.   c. recessive.
   b. monomorphic.   d. both a and c.

3. A gene exists in two alleles designated *D* and *d*. If 48 copies of this gene are the *D* allele and 152 are the *d* allele, what is the allele frequency of *D*?
   a. 0.24   c. 0.38
   b. 0.32   d. 0.76

4. The allele frequency of *C* is 0.4 and that of *c* is 0.6. If the population is in Hardy-Weinberg equilibrium, what is the frequency of heterozygotes?
   a. 0.16   c. 0.26
   b. 0.24   d. 0.48

27.2 AN INTRODUCTION TO MICROEVOLUTION

Learning Outcomes:

1. Define *microevolution*.
2. Explain the role of mutation in microevolution.
3. List the mechanisms that may cause allele and genotype frequencies to change significantly from one generation to the next.

Biological evolution, or simply **evolution,** involves a change in genetic variation in a population from one generation to the next. The term **microevolution** refers to changes in a population's gene pool with regard to particular alleles over measurable periods of time. The study of microevolution focuses on changes in allele or genotype frequencies for one or more genes.

Microevolution is rooted in two related phenomena (**Table 27.1**). First, the introduction of new genetic variation into a population is one essential factor underlying microevolution. As discussed in Section 27.8, gene variation can originate by a variety of mechanisms. For example, new alleles of preexisting genes can arise by random mutations. Such events provide a continuous source of new variation to populations. However, new mutations are relatively rare. For example, a common rate of mutation in a given gene may be on the order of one new mutation per one million copies of the gene per generation. Therefore, even though new mutations are a vital source of genetic variation, they do not, by themselves, act as a major factor in promoting widespread changes in a population.

Microevolution also involves the action of evolutionary mechanisms that alter the prevalence of a given allele or genotype in a population. These mechanisms are natural selection, genetic drift, migration, and nonrandom mating (see Table 27.1). The collective contributions of these evolutionary mechanisms over the course of many generations have the potential to promote widespread genetic changes in a population. In the following sections, we will examine how these mechanisms can affect the type of genetic variation that occurs when a gene exists in two or more alleles in a population. As you will learn, these mechanisms may

TABLE 27.1

Factors That Govern Microevolution

Source of New Genetic Variation*

Mutation	Throughout most of this chapter, we consider allelic variation. Random mutations within preexisting genes introduce new alleles into populations, but at a very low rate. New mutations may be beneficial, neutral, or deleterious. For new alleles to rise to a significant percentage in a population, other evolutionary mechanisms (i.e., natural selection, genetic drift, and/or migration) must operate on them.

Mechanisms That Alter Existing Genetic Variation

Natural selection	This is the phenomenon in which certain phenotypes have greater reproductive success than do other phenotypes. For example, natural selection may be related to the survival of members to reproductive age.
Genetic drift	This is a change in genetic variation from generation to generation due to random sampling error. Allele frequencies may change as a matter of chance from one generation to the next. This tends to have a greater effect in a small population.
Migration	Migration can occur between two different populations. The introduction of migrants into a recipient population may change the allele frequencies of that population.
Nonrandom mating	This is the phenomenon in which individuals select mates based on their phenotypes or genetic lineage. This can alter the relative proportion of homozygotes and heterozygotes predicted by the Hardy-Weinberg equation, but does not change allele frequencies.

*Allelic variation is just one source of new genetic variation. Section 27.8 considers a variety of mechanisms through which new genetic variation can occur.

cause a particular allele to be favored, or they may create a balance where two or more alleles are maintained in a population.

27.2 COMPREHENSION QUESTION

1. Which of the following is a factor that, by itself, does *not* promote widespread changes in allele or genotype frequencies?
   a. New mutation   d. Migration
   b. Natural selection   e. Nonrandom mating
   c. Genetic drift

27.3 OVERVIEW OF NATURAL SELECTION

Learning Outcomes:

1. List four key principles of natural selection.
2. Define *Darwinian fitness*.
3. Explain how fitness relates to the Hardy-Weinberg equation according to the general selection model.
4. Give an example of how genome-wide selection scans can reveal the effects of natural selection.

In the 1850s, Charles Darwin and Alfred Russel Wallace independently proposed the theory of evolution by **natural selection.** According to this theory, phenotypes may vary with regard to reproductive success. Because the phenotypes of individuals are largely rooted in the alleles they carry, those individuals with higher reproductive success are more likely to produce offspring, and thereby pass certain alleles to the next generation. Natural selection acts on phenotypes, which are governed by individuals' genotypes.

In this section, we will begin by considering the key principles of natural selection. We will then explore the concept of fitness, which is a measure of reproductive success. We will next examine a general fitness model that relates fitness to the Hardy-Weinberg equation. Finally, we will take a look at how genome-wide selection scans (GWSS), which are similar to genome-wide association studies (GWAS), discussed in Chapter 24, can provide insight about how natural selection can alter allele frequencies in populations that occupy different environments.

Natural Selection Is a Process That Depends on Differences in Reproductive Success

Reproductive success is the likelihood that an individual will contribute fertile offspring to the next generation. What factors contribute to reproductive success? Some individuals may have characteristics that make them better adapted to their environment. These individuals are more likely to survive to reproductive age and contribute offspring to the next generation. In sexually reproducing species, the ability to find a mate and an individual's fertility are also key factors that contribute to reproductive success.

A modern restatement of the principles of natural selection can relate our knowledge of molecular genetics to the phenotypes of individuals.

1. Within a population, allelic variation arises in various ways, such as through random mutations that cause differences in DNA sequences. A mutation that creates a new allele may alter the amino acid sequence of the coded protein, which, in turn, may alter the function of the protein.
2. Some alleles may code proteins that enhance an individual's survival or reproductive capability compared with other members of the population. For example, an allele may produce a protein that functions more efficiently at a higher temperature, conferring on the individual a greater probability of survival in a hot climate.
3. Individuals with beneficial alleles have phenotypes that make them more likely to reproduce and contribute to the gene pool of the next generation.
4. Over the course of many generations, allele frequencies of many different genes may change due to natural selection, thereby significantly altering the characteristics of a population or species. The net result of natural selection is a population that is better adapted to its environment and more successful at reproduction. This form of evolution is sometimes called **adaptive evolution.** Even so, it should be emphasized that species are not perfectly adapted to their

environments, because mutations are random events and because the environment tends to change from generation to generation.

Darwinian Fitness Is a Measure of Reproductive Success

The relationship between genotypes and reproductive success lies at the heart of natural selection. The concept called **Darwinian fitness** (or simply, **fitness**) is the relative likelihood that one genotype will contribute to the gene pool of the next generation compared to other genotypes. As mentioned, natural selection acts on phenotypes that are derived from individuals' genotypes. Although Darwinian fitness often correlates with physical fitness, the two concepts are not identical. Darwinian fitness is a measure of reproductive success. An extremely fertile genotype may have a higher Darwinian fitness than a less fertile genotype that appears more physically fit.

To consider Darwinian fitness, let's use the example of a gene existing in A and a alleles. If the three genotypes have the same level of mating success and fertility, we can assign a fitness value to each genotype class based on the likelihood of its individuals surviving to reproductive age. For example, let's suppose that the relative survival to adulthood of each of the three genotype classes is as follows: For every five AA individuals that survive, four Aa individuals survive, and one aa individual survives. **Relative fitness** (w) is the reproductive success of a genotype relative to other genotypes in the population. By convention, the genotype with the highest reproductive ability is given a fitness value of 1.0. The other genotypes are assigned fitness values relative to this 1.0 value:

$$\text{Fitness of } AA: w_{AA} = 1.0$$
$$\text{Fitness of } Aa: w_{Aa} = 4/5 = 0.8$$
$$\text{Fitness of } aa: w_{aa} = 1/5 = 0.2$$

Differences in reproductive success among genotypes may be due to various factors.

- In this example, the fittest genotype is more likely to survive to reproductive age.
- In other situations, the fittest genotype is more likely to mate. For example, a bird with brightly colored feathers may have an easier time attracting a mate than a bird with duller plumage.
- Finally, a third possibility is that the fittest genotype may be more fertile. The individual with that genotype may produce a higher number of gametes or gametes that are more successful at fertilization.

Also keep in mind that the above discussion of relative fitness presents the simplified case in which a single gene affects reproductive success. However, most traits are affected by allelic variation involving multiple genes. In Chapter 28, we will examine quantitative traits, such as weight and height, which are determined by alleles of many different genes. When natural selection acts on quantitative traits, relative fitness values are determined by allelic variation of multiple genes, not just one.

TABLE 27.2

General Selection Model

				Total
Genotype	AA	Aa	aa	–
Relative fitness	w_{AA}	w_{Aa}	w_{aa}	–
Starting frequency based on Hardy-Weinberg equilibrium	p^2	$2pq$	q^2	1.0
Relative contribution to the next generation	$p^2 w_{AA}$	$2pq\,w_{Aa}$	$q^2 w_{aa}$	$\overline{w}$
Genotype frequency after one generation of selection	$\dfrac{p^2 w_{AA}}{\overline{w}}$	$\dfrac{2pq\,w_{Aa}}{\overline{w}}$	$\dfrac{q^2 w_{aa}}{\overline{w}}$	1.0

The General Selection Model Relates Fitness to the Hardy-Weinberg Equation

Population geneticists want to predict changes in genotype frequencies based on relative fitness values. In other words, they want to predict the effects of natural selection. **Table 27.2** shows a **general selection model** that relates genotypes, relative fitness values, and changes in genotype frequencies based on natural selection. In this example, a gene exists in two alleles, which are designated A and a. The relative fitness values for the three possible genotypes are w_{AA}, w_{Aa}, and w_{aa}. In this model, the starting genotype frequencies are based on Hardy-Weinberg equilibrium.

Each of the three genotypes makes a relative contribution to the next generation based on its initial frequency times its fitness. As seen in Table 27.2, when the genotype frequencies are weighted by the relative fitness values, they add up to a value known as the **mean fitness of the population** ($\overline{w}$). The model allows us to predict the genotype frequencies after one generation of selection by dividing their relative contribution by $\overline{w}$ (see the bottom row of Table 27.2). In Section 27.4, we will apply the general selection model to a pattern of natural section called directional selection.

Genome-Wide Selection Scans Can Provide Insight Regarding the Effects of Natural Selection

As discussed in Chapter 24, a **genome-wide association study (GWAS)** is an examination of a genome-wide set of genetic variants among many different individuals to see if any variant is associated with a disease or other type of human trait (refer back to Figure 24.8). A similar approach, called a **genome-wide selection scan (GWSS),** attempts to identify genetic differences between populations of the same or closely related species that are found in different natural environments and/or exhibit different adaptive traits. In other words, a GWSS is aimed at identifying genetic differences that are due to natural selection.

How is a GWSS carried out? First, the DNA sequences of entire genomes are obtained from many individuals in two or more populations. These sequences are then compared to each other using computer software. One common approach is to identify single nucleotide polymorphisms (SNPs) that differ between populations. (SNPs are single base-pair differences within a population; an example of an SNP was presented earlier in this chapter in

Figure 27.3.) The alleles of a particular gene usually differ in the SNPs that they contain.

Although a description of the computer software used in a GWSS is beyond the scope of this textbook, such studies are often aimed at detecting **positive selection** in a particular population. Positive selection is an increase in the frequency of a particular allele that occurs because the allele is favored by natural selection. For example, researchers could compare genome sequences from two populations of a species of snakes, one found along the coast of California and another found in the Rocky Mountains. Such a comparison may reveal genetic variation (e.g., allelic variation among particular genes) that favors coastal life versus those that favor mountain life. Put another way, the GWSS may identify alleles that were selected because they provide an advantage in a coastal environment versus other alleles that provide an advantage in the mountains. **Table 27.3** describes examples of genes in which particular alleles were favored by positive selection.

Genome-wide selection scans typically reveal chromosomal regions, not just single genes, which differ between populations that are found in different environments. Such regions contain alleles that have attained a high frequency or are fixed in a population due to positive selection. However, not all of the alleles in such regions are due to positive selection; some of them may be due to **genetic hitchhiking.** As one or more positively selected alleles increase in frequency, nearby linked alleles on the

TABLE 27.3

Examples of Positive Selection Revealed by Genome-Wide Selection Scans

Populations	Differences in Allelic Variation
Marine and freshwater populations of three-spined stickleback fish (*Gasterosteus aculeatus*)	Allelic differences have been identified in many genes that play a role in adaptation to a marine or a freshwater environment. For example, allelic variation was noted in genes that code proteins that control collecting-tubule length and diameter in kidneys. Fish living in freshwater produce copious amounts of hypotonic urine compared to marine fish. They require longer and wider collecting tubules to do so.
Chicken (*Gallus gallus domesticus*) populations obtained from different environments	Chickens have been domesticated in many parts of the world. Genome-wide comparisons have revealed adaptation to their local climates. For example, Egyptian chickens exhibit allelic variation that is thought to convey adaptation to heat, solar radiation, and immunity. High-altitude-adapted East-African populations show allelic variation associated with genes that play a role in angiogenesis (formation of blood vessels), oxygen-heme binding, and oxygen-heme transport. This variation facilitates breathing at high altitudes.
Crow populations (genus *Corvus*)	In Africa, some crow species are all black, whereas closely related species are pied and contain white undersides. This color variation is thought to prevent interspecies breeding. Allelic variation was identified in genes that code proteins that are components of the Wnt signaling pathway that controls the deposition of pigment (i.e., melanogenesis).

chromosome will "hitchhike" along with the selected allele(s) and thereby increase in frequency, even though those hitchhiked alleles do not confer a selective advantage.

27.3 COMPREHENSION QUESTIONS

1. Darwinian fitness is a measure of
   a. survival.
   b. reproductive success.
   c. heterozygosity of the gene pool.
   d. polymorphisms in a population.

2. According to the general selection model, each genotype makes a relative contribution to the next generation based on
   a. its initial frequency.
   b. its fitness.
   c. its initial frequency times its fitness.
   d. its initial frequency divided by its fitness.

27.4 PATTERNS OF NATURAL SELECTION

Learning Outcomes:

1. Compare and contrast directional, balancing, disruptive, and stabilizing selection.
2. Use equations to predict the effects of directional and balancing selection.

In Section 27.3, we examined some of the general features of natural selection. By studying species in their native environments, population geneticists have determined that natural selection can follow different types of patterns. These patterns are based on the relative fitness values of the members of one or more populations. In this section, we will explore four patterns of natural selection called directional, stabilizing, disruptive, and balancing selection. In most of the examples described in this section, natural selection leads to adaptation that makes members of a population more likely to survive to reproductive age.

Directional Selection Favors the Extreme Phenotype

Directional selection favors individuals that are at one extreme of a phenotypic distribution and are more likely to survive and reproduce in a particular environment. In some cases, directional selection may act on phenotypes that are largely determined by the alleles of a single gene. For example, the level of resistance to an insecticide in a mosquito population may be determined by alleles of a single gene. In other cases, directional selection may act on phenotypes that are determined by multiple genes. For example, body weight in mammals is influenced by alleles of many different genes. If directional selection favored higher body weight, this would affect the allele frequencies of many different genes.

Different phenomena may initiate the process of directional selection.

- One way that directional selection may arise is that a new allele may be introduced into a population by mutation, and the new allele may promote higher fitness in individuals that carry it (**Figure 27.6**). If the homozygote carrying the favored allele has the highest fitness value, directional selection may cause this favored allele to eventually become the predominant allele in the population, perhaps even giving rise to a monomorphic gene.

- Another possibility is that a population may be exposed to a prolonged change in the environment in which it lives. Under the new environmental conditions, the relative fitness values may change to favor one or more genotypes, which will promote the elimination of other genotypes. As an example, let's suppose a population of finches on the mainland already has genetic variation in beak size. A small number of birds migrate to an island where the seeds are generally larger than they are on the mainland. In this new environment, birds with larger beaks have higher fitness because they are better able to crack open the larger seeds and thereby survive to reproductive age. Over the course of many generations, directional selection produces a population of birds carrying alleles that promote larger beak size.

The General Selection Model Can Be Applied to Directional Selection

As discussed in Section 27.3, the general selection model relates the Hardy-Weinberg equation to fitness values. For directional selection to occur, the fitness values must follow one of two patterns:

- One type of homozygote and the heterozygote have the same fitness value, and the other homozygote has a lower fitness. For example, $w_{AA} = w_{Aa} > w_{aa}$.
- One type of homozygote has the highest fitness, the heterozygote has intermediate fitness, and the other homozygote has the lowest fitness. For example, $w_{AA} > w_{Aa} > w_{aa}$.

If either of these fitness scenarios is observed, we can calculate how directional selection will change genotype frequencies using the general selection model shown in Table 27.2. Let's suppose the starting allele frequencies are $A = 0.5$ and $a = 0.5$, and the three fitness values, which are based on relative survival to reproductive age, are

$$w_{AA} = 1.0$$
$$w_{Aa} = 0.8$$
$$w_{aa} = 0.2$$

We begin by calculating the mean fitness of the population:

$$p^2 w_{AA} + 2pq\, w_{Aa} + q^2 w_{aa} = \overline{w}$$

$$\overline{w} = (0.5)^2(1) + 2(0.5)(0.5)(0.8) + (0.5)^2(0.2)$$

$$\overline{w} = 0.25 + 0.4 + 0.05 = 0.7$$

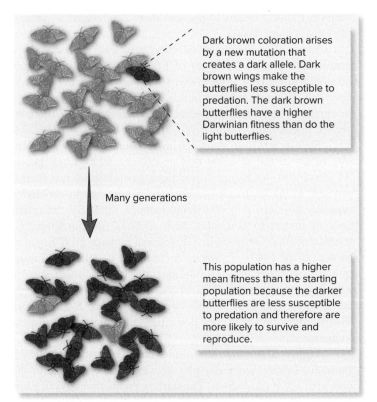

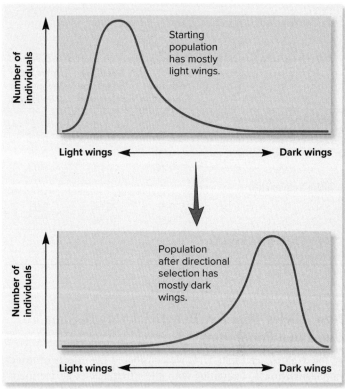

(a) An example of directional selection

(b) Graphical representation of directional selection

FIGURE 27.6   Directional selection. (a) A new mutation arises in a population that confers higher Darwinian fitness. In this example, butterflies with dark wings are more likely to survive and reproduce. Over many generations, directional selection favors the prevalence of darker individuals. (b) A graphical representation of directional selection.

CONCEPT CHECK: With respect to this pattern of natural selection, explain the meaning of the word *directional*.

After one generation of directional selection,

Frequency of AA genotype: $\dfrac{p^2 w_{AA}}{\overline{w}} = \dfrac{(0.5)^2(1)}{0.7} = 0.36$

Frequency of Aa genotype: $\dfrac{2pq\, w_{Aa}}{\overline{w}} = \dfrac{2(0.5)(0.5)(0.8)}{0.7} = 0.57$

Frequency of aa genotype: $\dfrac{q^2 w_{aa}}{\overline{w}} = \dfrac{(0.5)^2(0.2)}{0.7} = 0.07$

We can also calculate the changes in allele frequencies in the following way:

Allele frequency of A: $p_A = \dfrac{p^2 w_{AA}}{\overline{w}} + \dfrac{pq\, w_{Aa}}{\overline{w}}$

$\qquad\qquad = \dfrac{(0.5)^2(1)}{0.7} + \dfrac{(0.5)(0.5)(0.8)}{0.7} = 0.64$

Allele frequency of a: $q_a = \dfrac{q^2 w_{aa}}{\overline{w}} + \dfrac{pq\, w_{Aa}}{\overline{w}}$

$\qquad\qquad = \dfrac{(0.5)^2(0.2)}{0.7} + \dfrac{(0.5)(0.5)(0.8)}{0.7} = 0.36$

After one generation, the allele frequency of A has increased from 0.5 to 0.64, and that of a has decreased from 0.5 to 0.36. These changes have occurred because the AA genotype has the highest fitness, whereas the Aa and aa genotypes have progressively lower fitness values.

Another interesting result of natural selection is that it raises the mean fitness of the population. If we assume that the individual fitness values are constant, the mean fitness of this next generation is

$$\begin{aligned} \overline{w} &= p^2 w_{AA} + 2pq\, w_{Aa} + q^2 w_{aa} \\ &= (0.64)^2(1) + 2(0.64)(0.36)(0.8) + (0.36)^2(0.2) \\ &= 0.8 \end{aligned}$$

The mean fitness of the population has increased from 0.7 to 0.8.

What are the consequences of natural selection at the population level? After one or more generations, the population is better adapted to its environment than the population(s) in the preceding generation(s). Another way of viewing this calculation is that the subsequent population has a greater reproductive potential than the previous one.

Table 27.4 shows the effects of directional selection for two generations, assuming that the individual fitness values remain constant. As we can see, the general trend is to increase the frequency of AA, decrease that of aa, and increase the mean fitness of the population. The allele frequency of A steadily increases and that of a decreases. If this process were to occur for many generations, the Aa and aa genotypes and the a allele may be eliminated from the population.

TABLE 27.4

Directional Selection for Two Generations Based on Constant Fitness Values of w_{AA} (1.0), w_{Aa} (0.8), and w_{aa} (0.2)

	Starting Generation	After One Generation of Selection	After Two Generations of Selection
Genotype frequencies			
AA	0.25	0.36	0.51
Aa	0.50	0.57	0.46
aa	0.25	0.07	0.03
Allele frequencies			
A	0.50	0.64	0.85
a	0.50	0.36	0.15
Mean fitness of the population			
	0.70	0.80	0.93

New Alleles That Are Beneficial May Become Fixed in a Population

In the example in Table 27.4, we considered the effects of directional selection by beginning with allele frequencies at intermediate levels (namely, $A = 0.5$ and $a = 0.5$). **Figure 27.7** illustrates what would happen if a new mutation introduced the A allele into a population that was originally monomorphic for the a allele. As before, the AA homozygote has a fitness value of 1.0, the Aa heterozygote has a value of 0.8, and the recessive aa homozygote has a value of 0.2. Initially, the A allele is at a very low frequency in the population. If it is not lost initially due to genetic drift, its

frequency slowly begins to rise and then, at intermediate values, rises much more rapidly.

Eventually, directional selection may lead to fixation of a beneficial allele at a frequency of 1.0, or 100%. However, a new beneficial allele is in a precarious situation when its frequency is very low. As discussed later in this chapter, genetic drift is likely to eliminate new mutations, even beneficial ones, as a result of chance fluctuations.

Researchers have identified many naturally occurring examples of directional selection that are determined by single genes. As mentioned in Chapter 7, resistance to antibiotics is a critical problem in the treatment of infections. Resistance to antibiotics typically develops in a directional manner. Similarly, the resistance of insects to pesticides, such as DDT (dichlorodiphenyltrichloroethane), develops in a directional manner. DDT began to be used in the 1940s as a way to decrease the populations of mosquitoes and other insects. Subsequently, certain insect species have become resistant to DDT by a dominant mutation in a single gene, which codes an enzyme called glutathione S-transferase. The resulting mutant enzyme detoxifies DDT, making it harmless to the insect.

Figure 27.8 shows the results of an experiment in which mosquito larvae (*Aedes aegypti*) were exposed to DDT over the course of seven generations. The starting population showed a low level of DDT resistance, as evidenced by the low percentage of survivors after exposure to DDT. By comparison, after seven generations, nearly 100% of the population was DDT-resistant. These results illustrate the power of directional selection in promoting change in a population. Since the 1950s, resistance to nearly every known insecticide has evolved within 10 years of its commercial introduction!

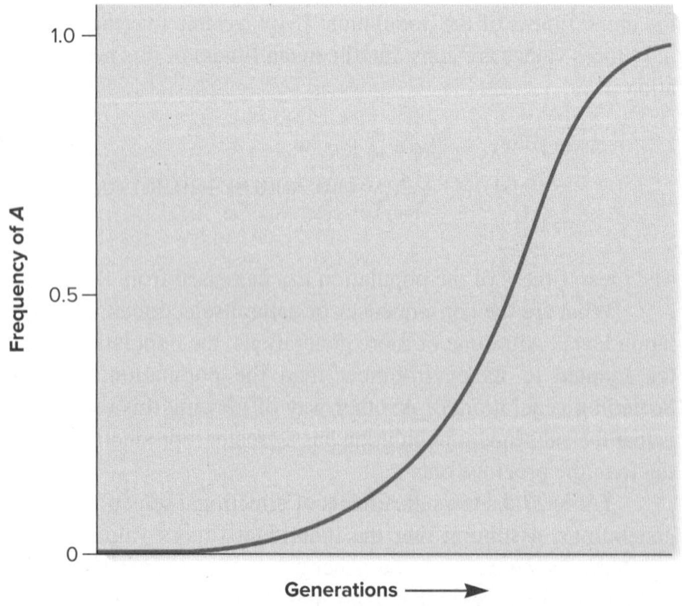

FIGURE 27.7 The fate of a beneficial allele that is introduced into a population as a new mutation. A new allele (A) is beneficial in the homozygous condition: $w_{AA} = 1.0$. The heterozygote, Aa ($w_{Aa} = 0.8$), and the homozygote, aa ($w_{aa} = 0.2$), have lower fitness values.

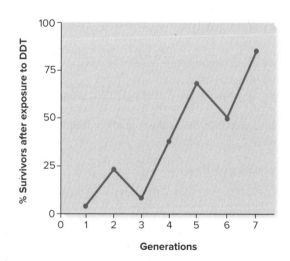

FIGURE 27.8 **Directional selection for DDT resistance in a mosquito population.** In this experiment, mosquito larvae (*Aedes aegypti*) were exposed to DDT at a concentration of 10 mg/L. The percentage of survivors was recorded, and then the survivors of each generation were used as parents for the next generation.

CONCEPT CHECK: In this example, is directional selection promoting genetic diversity? Explain.

EXPERIMENT 27A

The Grants Have Observed Directional Selection in Galápagos Finches

Let's now turn to a study that appears to demonstrate directional selection in action. Beginning in 1973, Peter Grant, Rosemary Grant, and their colleagues studied the process of natural selection in finches found on the Galápagos Islands. They focused much of their research on one of the Galápagos Islands known as Daphne Major. This small island (0.34 km^2) has a moderate degree of isolation (8 km from the nearest island), an undisturbed habitat, and a resident population of finches, including the medium ground finch, *Geospiza fortis* (**Figure 27.9**).

FIGURE 27.9   The medium ground finch (*Geospiza fortis*), which is found on Daphne Major.
Ralph Lee Hopkins/Getty Images

To study natural selection, the Grants observed various traits in the medium ground finch, including beak size, over the course of many years. The medium ground finch has a relatively small beak, suitable for crushing and breaking open small, tender seeds. Beak size is an example of a quantitative trait, which is controlled by multiple genes. The Grants quantified beak size among the medium ground finches of Daphne Major by carefully measuring beak depth (a measurement of the beak from top to bottom, at its base) on individual birds and then releasing them. During the course of their studies, they compared the beak sizes of parents and offspring by examining many broods over several years. The depth of the beak was transmitted from parents to offspring, regardless of environmental conditions, indicating that differences in beak sizes are due to genetic differences in the population. In other words, they found that beak depth is a heritable trait.

By measuring many birds every year, the Grants were able to assemble a detailed portrait of natural selection from generation to generation. In the study shown in **Figure 27.10**, they measured beak depth in 1976 and 1978.

THE HYPOTHESIS

Beak size will be influenced by natural selection. Environments that produce larger seeds will select for birds with larger beaks.

TESTING THE HYPOTHESIS

FIGURE 27.10   Natural selection in medium ground finches of Daphne Major.

	Experimental level	**Conceptual level**
1. In 1976, measure beak depth in parents and offspring of the species *G. fortis*.	Capture birds and measure beak depth. 	This is a way to measure a trait that may be subject to natural selection.
2. Repeat the procedure on offspring that were born in 1978 and had reached mature size. A drought had occurred in 1977 that caused plants on the island to produce mostly larger seeds and relatively few small seeds.	Capture birds and measure beak depth. 	This is a way to measure a trait that may be subject to natural selection.

THE DATA

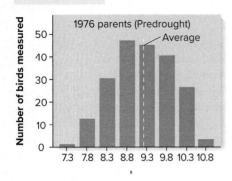
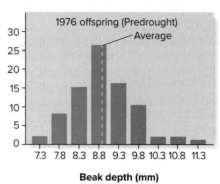
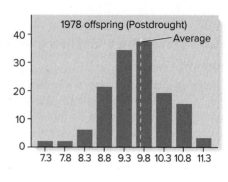

Beak depth (mm)

Source: Grant, Rosemary B., and Grant, Peter R., "What Darwin's finches can teach us about the evolutionary origin and regulation of biodiversity," *Bioscience, vol.* 53, no. 10, October 1, 2003, 965–997.

INTERPRETING THE DATA

In the wet year of 1976, the plants of Daphne Major produced the small seeds that these finches were able to eat in abundance. However, a drought occurred in 1977. During this year, the plants on Daphne Major tended to produce few of the smaller seeds, which the finches rapidly consumed. To survive, the finches resorted to eating larger, drier seeds, which were harder to crush. As a result, the birds that survived tended to have larger beaks, because they were better able to break open these large seeds. In the year after the drought, the average beak depth of offspring in the population increased to approximately 9.8 mm because the surviving birds with larger beaks passed this trait on to their offspring. This change is likely to be due to directional selection (see Figure 27.6), although genetic drift could also contribute to the observed difference. Overall, the results of the study illustrate the power of natural selection to alter the nature of a trait, in this case, beak depth, in a given population.

Balanced Polymorphisms May Arise as a Result of Heterozygote Advantage or Negative Frequency-Dependent Selection

Thus far, we have focused on directional selection, which tends to eliminate alleles that confer lower fitness values on individuals who carry them. We will now turn to other patterns of natural selection. A common misperception is that natural selection always eliminates "weaker alleles" from a population, as suggested by Table 27.4. Not so. Researchers have discovered certain patterns of natural selection that actually favor the maintenance of two or more alleles in a population. One example is called **balancing selection.** Two types of balancing selection are heterozygote advantage and negative frequency-dependent selection.

Heterozygote Advantage. For genetic variation involving a single gene, balancing selection may arise when the heterozygote has a higher fitness than either of the homozygotes for that gene, a situation called **heterozygote advantage.** In this case, an equilibrium is reached in which both alleles are maintained in the population. If the relative fitness values are known for each of the genotypes, the allele frequencies at equilibrium can be calculated. To do so, we must consider the **selection coefficient (s),** which measures the degree to which a genotype is selected against:

$$s = 1 - w$$

By convention, the genotype with the highest fitness has an s value of zero. Genotypes at a selective disadvantage have s values that are greater than 0 but less than or equal to 1.0. An extreme case is a recessive lethal allele, which would have an s value of 1.0 in the homozygote, whereas the s value in the heterozygote could be 0.

Let's consider genotypes with the following relative fitness values:

$$w_{AA} = 0.7$$
$$w_{Aa} = 1.0$$
$$w_{aa} = 0.4$$

The selection coefficients are

$$s_{AA} = 1 - 0.7 = 0.3$$
$$s_{Aa} = 1 - 1.0 = 0$$
$$s_{aa} = 1 - 0.4 = 0.6$$

The population reaches an equilibrium when

$$s_{AA}p = s_{aa}q$$

If we take this equation, let $q = 1 - p$, and then solve for p, we obtain:

$$p = \text{Allele frequency of } A = \frac{s_{aa}}{s_{AA} + s_{aa}}$$

$$= \frac{0.6}{0.3 + 0.6} = 0.67$$

If we let $p = 1 - q$ and then solve for q, we get:

$$q = \text{Allele frequency of } a = \frac{s_{AA}}{s_{AA} + s_{aa}}$$

$$= \frac{0.3}{0.3 + 0.6} = 0.33$$

In this example, balancing selection maintains the two alleles in the population at frequencies of 0.67 for A and 0.33 for a.

Heterozygote advantage can sometimes explain the high frequency of alleles that are harmful in a homozygous condition. A classic example is the Hb^S allele of the human β-globin gene. A homozygous Hb^SHb^S individual exhibits sickle cell disease, a disorder characterized by the sickling of the red blood cells. The Hb^SHb^S homozygote has a lower fitness than a homozygote with two copies of the more common β-globin allele, Hb^AHb^A. However, the heterozygote, Hb^AHb^S, has a higher level of fitness than either homozygote in areas where malaria is endemic (**Figure 27.11**). Compared with Hb^AHb^A homozygotes, heterozygotes have a 10–15% higher chance of survival if infected by the malarial parasite, *Plasmodium falciparum*. Therefore, the Hb^S allele is maintained in populations in areas where malaria is prevalent, even though the allele is detrimental in the homozygous state.

In addition to sickle cell disease, other gene mutations that cause genetic diseases in humans in the homozygous state are thought to be prevalent because of heterozygote advantage. For example, the high prevalence of the allele causing cystic fibrosis may be related to this phenomenon, but the advantage that a heterozygote may possess has not been determined.

Negative Frequency-Dependent Selection. A second mechanism of balancing selection is **negative frequency-dependent selection.** In this pattern of natural selection, the fitness of a genotype decreases when its frequency becomes higher. In other words, rare individuals have a higher fitness than more common individuals. Therefore, rare individuals are more likely to reproduce, whereas common individuals are less likely, thereby producing a balanced polymorphism in which no genotype becomes too rare or too common.

An interesting example of negative frequency-dependent selection involves the elderflower orchid, *Dactylorhiza sambucina* (**Figure 27.12**). Throughout the range of this plant in central and southern Europe, both yellow- and red-flowered individuals are prevalent. A proposed explanation for this polymorphism is related to the orchid's pollinators, which are mainly bumblebees such as *Bombus lapidarius* and *Bombus terrestris*. The pollinators increase their visits to the flowers of one color of *D. sambucina* as that color becomes less common in a given area. One reason this may occur is that *D. sambucina* is a rewardless flower; that is, it does not provide its pollinators with any reward for visiting, such as sweet nectar. Pollinators learn that the flowers of the more common color in a given area do not offer a reward, and they increase their visits to the flowers of the less common color. Thus, the relative fitness of the less common flower increases.

Disruptive Selection Favors Multiple Phenotypes in Heterogeneous Environments

As we have just seen, polymorphisms may arise as a result of balancing selection. Researchers have discovered another pattern of natural selection that favors the maintenance of two or more alleles in heterogeneous environments. This pattern, called **disruptive selection,** also known as diversifying selection, favors the

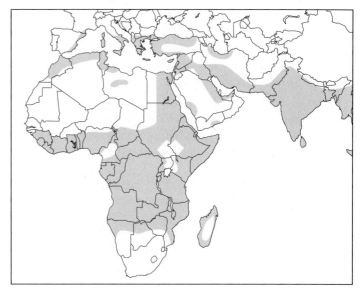

(a) Malaria prevalence

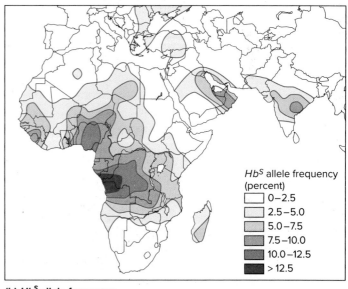

Hb^S allele frequency (percent)

☐	0–2.5
☐	2.5–5.0
☐	5.0–7.5
☐	7.5–10.0
☐	10.0–12.5
■	> 12.5

(b) Hb^S allele frequency

FIGURE 27.11 **The geographic relationship between malaria and the frequency of the sickle cell allele in human populations.** (a) The geographic prevalence of malaria in Africa and surrounding areas. (b) The frequency of the Hb^S allele in the same areas.

Genes→Traits The sickle cell allele of the β-globin gene is maintained in human populations by balancing selection. In areas where malaria is endemic, the heterozygote carrying one copy of the Hb^S allele has a greater fitness than either of the corresponding homozygotes (Hb^AHb^A and Hb^SHb^S). Therefore, even though the Hb^SHb^S homozygotes suffer the detrimental consequences of sickle cell disease, this negative outcome is balanced by the beneficial effects of malarial resistance in the heterozygotes.

CONCEPT CHECK: Explain why the Hb^S allele is prevalent in certain regions even though it is detrimental in the homozygous condition.

survival of two or more different phenotypes (**Figure 27.13**). This pattern of selection typically acts on traits that are determined by multiple genes. In disruptive selection, the fitness values of particular genotypes are higher in one environment and lower in a different one. Disruptive selection is likely to occur in populations

FIGURE 27.12 **The two color variations found in the elder-flower orchid, *Dactylorhiza sambucina*.** The two colors are maintained in the population due to negative frequency-dependent selection.

Paul Harcourt Davies/SPL/Science Source

CONCEPT CHECK: Explain how negative frequency-dependent selection works.

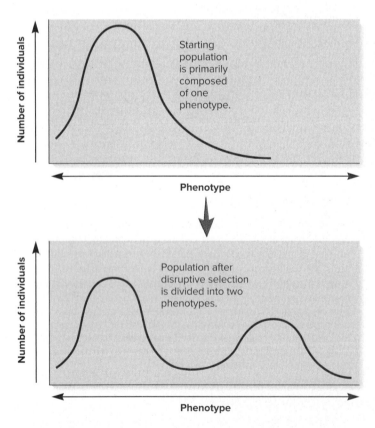

FIGURE 27.13 **Disruptive selection.** Over time, this pattern of selection favors two or more phenotypes that occupy heterogeneous environments.

CONCEPT CHECK: Does disruptive selection favor polymorphism? Explain why or why not.

that occupy diverse environments; some members of the species are more likely to survive and reproduce in each type of environmental condition.

As an example, **Figure 27.14a** shows a species of land snails, *Cepaea nemoralis*, that live in woods and open fields. This snail is polymorphic with respect to the color and banding pattern of the shell. In 1954, Arthur Cain and Philip Sheppard found that shell color was correlated with the environment.

- As shown in **Figure 27.14b**, the highest frequency of brown shell color was found among snails in beech woods, where there are wide expanses of dark soil. The frequency of brown shells was substantially lower in other environments.
- By comparison, pink-shelled snails were most common in the leaf litter of both beech woods and deciduous woods.
- The yellow-shelled snails were most abundant in the sunny, grassy areas of hedgerows and rough herbage.

(a) Land snails

Habitat	Brown	Pink	Yellow
Beech woods	0.23	0.61	0.16
Deciduous woods	0.05	0.68	0.27
Hedgerows	0.05	0.31	0.64
Rough herbage	0.004	0.22	0.78

(b) Frequency of shell color

FIGURE 27.14 **Polymorphism in the land snail, *Cepaea nemoralis*.** **(a)** This species of snail can have shells with several different colors and banding patterns. **(b)** Coloration of the shells is correlated with the specific environments where the snails are located.

Genes→Traits Shell coloration is an example of genetic polymorphism in heterogeneous environments; the genes governing shell coloration are polymorphic. The predation of snails is correlated with their ability to be camouflaged in their natural environment. Snails with brown shells are most prevalent in beech woods, where the soil is dark. Pink-shelled snails are most abundant in the leaf litter of beech woods and deciduous woods. Yellow-shelled snails are most prevalent in sunnier locations, such as hedgerows and rough herbage.

R. Koenig/BLICKWINKEL/age fotostock

Researchers have suggested that, in this case, the disruptive selection can be explained by different levels of predation by thrushes. Depending on the environment, snails with a certain phenotype may be more easily seen by the predators than snails with the other phenotypes. Migration can occasionally occur among the snail populations, which keeps the polymorphism in balance among these different environments.

Stabilizing Selection Favors Individuals with Intermediate Phenotypes

In **stabilizing selection,** the extreme phenotypes for a trait are selected against, and those individuals with intermediate phenotypes have the highest fitness values. Stabilizing selection is typically directed at quantitative traits, such as body weight and offspring number, which are determined by multiple genes. Stabilizing selection tends to decrease genetic diversity for genes affecting such traits because it eliminates alleles that cause a greater variation in phenotypes.

In 1947, David Lack proposed that stabilizing selection may apply to clutch size in birds. Under stabilizing selection, birds that lay too many or too few eggs have lower fitness values than those that lay an intermediate number (**Figure 27.15**). Laying too many eggs may cause many offspring to die due to inadequate parental care and food. In addition, the strain on the parents themselves may decrease their likelihood of survival and therefore their ability to produce more offspring. Having too few offspring, on the other hand, does not contribute many individuals to the next generation. Therefore, the most successful parents are those that produce an intermediate clutch size.

In the 1980s, Lars Gustafsson and colleagues examined the phenomenon of stabilizing selection in the collared flycatcher, *Ficedula albicollis*, on the island of Gotland, which is southeast of the mainland of Sweden. They discovered that Lack's hypothesis that clutch size is subject to the action of stabilizing selection appears to be true for this species.

27.4 COMPREHENSION QUESTIONS

1. Let's suppose that three genotypes—*BB*, *Bb*, and *bb*—have fitness values of 0.47, 0.77, and 1.0, respectively. This situation would promote _____ selection and lead eventually to the fixation of the _____ genotype(s).
   a. directional, *BB*
   b. directional, *bb*
   c. balancing, *Bb*
   d. diversifying, *BB* and *bb*

2. Within a particular population, dark-colored rats are more likely to survive than more light-colored individuals. This situation is likely to result in
   a. directional selection.
   b. stabilizing selection.
   c. disruptive selection.
   d. balancing selection.

3. A population occupies heterogeneous environments in which the fitness of some genotypes is higher in one environment and the fitness of other individuals is higher in another environment. This situation is likely to result in
   a. directional selection.
   b. stabilizing selection.
   c. disruptive selection.
   d. balancing selection.

4. A gene exists in two alleles, and the heterozygote has the highest fitness. This situation is likely to result in
   a. directional selection.
   b. stabilizing selection.
   c. disruptive selection.
   d. balancing selection.

27.5 GENETIC DRIFT

Learning Outcomes:
1. Define *genetic drift*.
2. Explain how population size affects genetic drift, and calculate the probabilities of the outcomes of this process.
3. Compare and contrast the bottleneck effect and the founder effect.

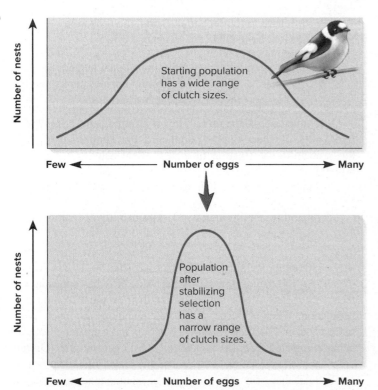

FIGURE 27.15 Stabilizing selection. In this pattern of natural selection, the extremes of a phenotypic distribution are selected against. Those individuals with intermediate traits have the highest fitness. This results in a population with less diversity and more uniform traits.

CONCEPT CHECK: In general, why does stabilizing selection decrease genetic diversity?

In the 1930s, geneticist Sewall Wright played a key role in developing the concept of **genetic drift,** which refers to changes in allele frequencies in a population due to random fluctuations. As a matter of chance, the frequencies of alleles found in gametes that unite to form zygotes vary from generation to generation. Over the long run, genetic drift usually results in either the loss of an allele or its fixation at 100% in the population. The process is random

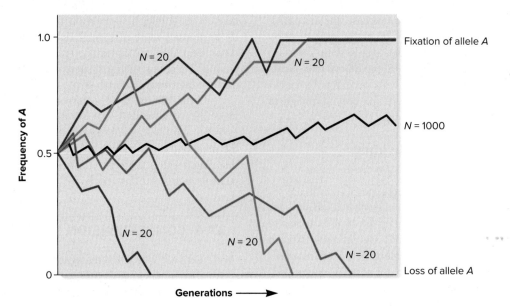

FIGURE 27.16 **A hypothetical simulation of genetic drift.** In all cases, the starting allele frequencies are $A = 0.5$ and $a = 0.5$. The colored lines illustrate five populations for which $N = 20$; the black line shows a population for which $N = 1000$.

CONCEPT CHECK: How does population size affect genetic drift?

with regard to particular alleles. Genetic drift can lead to the loss or fixation of deleterious, neutral, or beneficial alleles. The rate at which this occurs depends on the population size and on the initial allele frequencies.

Figure 27.16 illustrates the potential consequences of genetic drift in one large ($N = 1000$) and five small ($N = 20$) populations. At the beginning of this hypothetical simulation, all of these populations have identical allele frequencies: $A = 0.5$ and $a = 0.5$. In the five small populations, this allele frequency fluctuates substantially from generation to generation. Eventually, one of the alleles is eliminated and the other is fixed at 100%. At this point, the gene has become monomorphic and does not fluctuate any further. By comparison, the allele frequencies in the large population fluctuate much less, because random sampling error is expected to have a smaller effect. Nevertheless, genetic drift leads to homozygosity even in large populations, but it takes many more generations to reach that result.

Now let's ask two questions:

1. How many new mutations do we expect in a natural population?
2. How likely is it that any new mutation will be either fixed in or eliminated from a population due to genetic drift?

With regard to the first question, the average number of new mutations depends on the mutation rate (μ) and the number of individuals in a population (N). If each individual has two copies of the gene of interest, the expected number of new mutations in this gene is

$$\text{Expected number of new mutations} = 2N\mu$$

From this equation, we see that a new mutation is more likely to occur in a large population than in a small one. This makes sense,

because the larger population has more copies of the gene that could be mutated.

With regard to the second question, the probability of fixation of a newly arising allele due to genetic drift is

$$\text{Probability of fixation} = \frac{1}{2N} \quad \begin{array}{l} \text{(assuming equal numbers of} \\ \text{males and females contribute} \\ \text{to the next generation)} \end{array}$$

In other words, the probability of fixation is the same as the initial allele frequency in the population. For example, if $N = 20$, the probability of fixation of a new allele equals $1/(2 \times 20)$, or 2.5%.

Conversely, a new allele may be lost from the population:

$$\text{Probability of elimination} = 1 - \text{probability of fixation}$$
$$= 1 - \frac{1}{2N}$$

If $N = 20$, the probability of elimination equals $1 - 1/(2 \times 20)$, or 97.5%. As you may have noticed, the value of N has opposing effects with regard to new mutations and their eventual fixation in a population. When N is very large, new mutations are much more likely to occur. Each new mutation, however, has a greater chance of being eliminated from the population due to genetic drift. On the other hand, when N is small, the probability of new mutations is also small, but if they occur, the likelihood of fixation is relatively large.

Now that we appreciate the phenomenon of genetic drift, we can ask a third question:

3. If fixation of a new allele does occur, how many generations is it likely to take?

The formula for calculating this also depends on the number of individuals in the population:

$$\bar{t} = 4N$$

where

 $\bar{t}$ equals the average number of generations needed to achieve fixation

 N equals the number of individuals in the population, assuming that males and females contribute equally to each succeeding generation

As you may have expected, allele fixation takes much longer in large populations. If a population has 1 million breeding members, it takes, on average, 4 million generations, perhaps an insurmountable period of time, to reach fixation. In a small group of 100 individuals, however, fixation takes only 400 generations, on average.

In nature, allele frequencies in small populations are more susceptible to genetic drift. This susceptibility commonly arises as a result of the bottleneck effect or the founder effect.

Bottleneck Effect. Changes in population size may influence genetic drift via the **bottleneck effect** (**Figure 27.17**). In nature, a population can be reduced dramatically in size by events such as earthquakes, floods, droughts, or human destruction of habitat. Such events may randomly eliminate most members of the population without regard to their genetic composition. The bottleneck may initiate genetic drift because the population of survivors may have allele frequencies that differ from those of the original population. In addition, allele frequencies are expected to drift substantially during the generations when the population size is small. In extreme cases, alleles may even be eliminated. Eventually, the population that experienced the bottleneck may regain its original size. However, the new population will have less genetic variation than the original large population.

As an example, the African cheetah population lost a substantial amount of its genetic variation due to a bottleneck effect (Figure 27.17b). DNA analysis by population geneticists has suggested that a severe bottleneck occurred approximately 10,000–12,000 years ago, when the population size of cheetahs was dramatically reduced. The population eventually rebounded, but the bottleneck effect significantly decreased the genetic variation.

Founder Effect. Geography and population size may also influence genetic drift via the **founder effect.** The key difference between the bottleneck effect and the founder effect is that the founder effect involves migration; a small group of individuals separates from a larger population and establishes a colony in a new location. For example, a few individuals from a large continental population may move to an island and become the founders of a population in that location. The founder effect has two important consequences:

- The founding population is expected to have less genetic variation than the original population from which it was derived.
- As a matter of chance, the allele frequencies in the founding population may differ markedly from those of the original population.

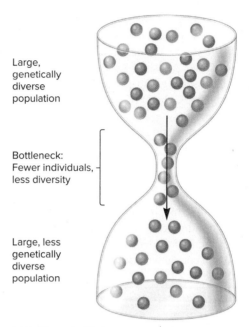

(a) Bottleneck effect

(b) An African cheetah

FIGURE 27.17 **The bottleneck effect, an example of a mechanism leading to genetic drift.** (**a**) A representation of the bottleneck effect. Note that the genetic variation denoted by the green balls has been lost. (**b**) The African cheetah. The modern species has low genetic variation due to a bottleneck that is thought to have occurred about 10,000–12,000 years ago.

nwdph/Shutterstock

CONCEPT CHECK: What is happening at the bottleneck? Describe the effect of genetic drift following the bottleneck.

Population geneticists have studied many examples of isolated populations that were started from a few members of another population. In the 1960s, Victor McKusick studied allele frequencies in the Old Order Amish of Lancaster County, Pennsylvania. At that time, this was a group of about 8000 people, descended from just three couples who immigrated to the United States in the eighteenth century.

Among this population of 8000, a genetic disease known as the Ellis-van Creveld syndrome (a recessive disorder resulting in short stature) was found at a frequency of 0.07, or 7%. By comparison, this disorder is extremely rare in other human populations, even the population in which the founding members of the Lancaster County Amish population had originated. The high frequency of this syndrome in the Amish population is a chance occurrence due to the founder effect. The recessive allele can be traced back to one couple who came to the area in 1744.

27.5 COMPREHENSION QUESTIONS

1. Genetic drift is
   a. a change in allele frequencies due to random fluctuations.
   b. likely to result in allele loss or fixation over the long run.
   c. more pronounced in smaller populations.
   d. all of the above.
2. Which of the following influences on genetic drift involve(s) the migration of a population from one location to another?
   a. The bottleneck effect
   b. The founder effect
   c. Both a and b
   d. None of the above

27.6 MIGRATION

Learning Outcome:

1. Explain how migration affects allele frequencies between neighboring populations, and calculate the magnitude of such a change.

We have just seen how migration to a new location by a relatively small group can lead to genetic drift, resulting in a population with altered allele frequencies. In addition, migration between two different established populations can alter allele frequencies. **Gene flow** refers to the transfer of genes from one population (a donor population) to another, thereby changing the gene pool of the recipient population. One way this occurs is by the migration of fertile individuals from one population to another population and the successful breeding of such migrants with the members of the recipient population. Gene flow depends not only on migration, but also on the ability of the migrants' alleles to be passed to subsequent generations.

To determine the effects of migration, we need to consider three populations: the original donor population, the original recipient population, and the population that has new members due to migration. We call this third population a **conglomerate** because it is composed of members of both the donor and recipient populations. To calculate changes in allele frequencies in the conglomerate population, we must know the original allele frequencies in the donor and recipient populations prior to migration. In addition, we must know what proportion of the conglomerate population the migrants represent. With these data, we can calculate the change in allele frequency in the conglomerate population using the following equation:

$$\Delta p_C = m(p_D - p_R)$$

where

Δp_C is the change in allele frequency in the conglomerate population

p_D is the allele frequency in the donor population

p_R is the allele frequency in the original recipient population

m is the proportion of migrants in the conglomerate population, that is

$$m = \frac{\text{Number of migrants in the conglomerate population}}{\text{Total number of individuals in the conglomerate population}}$$

As an example, let's suppose the allele frequency of A is 0.7 in the donor population and 0.3 in the recipient population. A group of 20 individuals migrates and joins the recipient population, which originally had 80 members. Thus,

$$m = \frac{20}{20 + 80}$$
$$= 0.2$$
$$\Delta p_C = m(p_D - p_R)$$
$$= 0.2(0.7 - 0.3)$$
$$= 0.08$$

We can now calculate the allele frequency in the conglomerate population:

$$p_C = p_R + \Delta p_C$$
$$= 0.3 + 0.08 = 0.38$$

Therefore, in the conglomerate population, the allele frequency of A has changed from 0.3 (its value in the recipient population before migration occurred) to 0.38. This increase in allele frequency arises from the higher allele frequency of A in the donor population.

In our example, we considered the consequences of a unidirectional migration from a donor to a recipient population. In nature, it is common for individuals to migrate in both directions. What are the main consequences of bidirectional migration? Depending on its rate, such migration tends to reduce differences in allele frequencies between neighboring populations. Populations that frequently mix their gene pools via migration tend to have similar allele frequencies, whereas isolated populations are expected to have more dissimilar allele frequencies.

Migration can also enhance genetic diversity within a population. As mentioned, new mutations are relatively rare events. Therefore, a particular mutation may arise in only one population. Migration may then introduce this new allele into neighboring populations.

27.6 COMPREHENSION QUESTION

1. Gene flow depends on
   a. migration.
   b. the ability of migrant alleles to be passed to subsequent generations.
   c. genetic drift.
   d. both a and b.

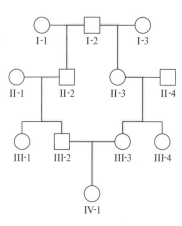

FIGURE 27.18 **A human pedigree containing inbreeding.**
Individual IV-1 is the result of inbreeding because IV-1's parents are related. The inbreeding path is highlighted in yellow.

CONCEPT CHECK: How does inbreeding affect the likelihood that recessive traits will be expressed? Explain.

27.7 NONRANDOM MATING

Learning Outcomes:

1. Define *assortative mating, inbreeding,* and *outbreeding.*
2. Calculate the inbreeding coefficient for an individual in a given pedigree.
3. Explain how inbreeding affects Hardy-Weinberg equilibrium.

As mentioned earlier, one of the conditions required to establish Hardy-Weinberg equilibrium is random mating (or random breeding), which means that individuals reproduce with each other irrespective of their genotypes and phenotypes. In many cases, particularly in human populations, this condition is sometimes violated.

When sexual reproduction is nonrandom in a population, the process is called **assortative mating.**

- Positive assortative mating occurs when individuals with similar phenotypes reproduce with each other.
- The opposite situation, where dissimilar phenotypes preferentially reproduce, is called negative assortative mating.
- Individuals may reproduce with other individuals that are part of the same genetic lineage. Reproduction between two genetically related individuals, such as cousins, is called **inbreeding.** It sometimes occurs in human societies and is more likely to take place in nature when population size becomes very limited.
- **Outbreeding,** which is reproduction between unrelated individuals, can create hybrids that are heterozygous for many genes.

In the absence of other evolutionary processes, inbreeding and outbreeding do not affect allele frequencies in a population. However, they do alter the relative proportions of homozygotes and heterozygotes that are predicted by the Hardy-Weinberg equation.

Inbreeding involves a smaller gene pool because the reproducing individuals are related genetically. In the 1940s, Gustave Malécot developed methods to quantify the degree of inbreeding. The **inbreeding coefficient (F)** is the probability that two alleles for a given gene in a particular individual will be identical because both copies are due to descent from a common ancestor. An inbreeding coefficient can be computed by analyzing the degree of relatedness within a pedigree.

Inbreeding in a Pedigree. **Figure 27.18** shows a human pedigree involving a mating between cousins. Individuals III-2 and III-3 are cousins and have produced IV-1. This female offspring is said to be inbred, because IV-1's parents are genetically related to each other.

Let's determine the inbreeding coefficient for IV-1. To begin this calculation, we must first identify all of this individual's common ancestors. A common ancestor is anyone who is an ancestor to both of an individual's parents. In Figure 27.18, IV-1 has one common ancestor, I-2, who is a grandparent of both III-2 and III-3.

Our next step is to determine the inbreeding path(s). An inbreeding path for an individual is the shortest path through the pedigree that includes both parents and the common ancestor. In a pedigree containing an inbred individual, there is an inbreeding path for each common ancestor. The length of each inbreeding path is calculated by adding together all of the individuals in the path except the individual of interest. In this case, there is only one path because IV-1 has only one common ancestor.

To add the members of the path, we begin with individual IV-1, but we do not count IV-1. We then move to the male parent (III-2); to a paternal grandparent (II-2); and to I-2, the common ancestor. From the common ancestor, we move down to a maternal grandparent (II-3); and finally to the female parent (III-3). This path has five members. Finally, to calculate the inbreeding coefficient (F), we use the following formula:

$$F = \sum (1/2)^n (1 + F_A)$$

where

F is the inbreeding coefficient of the individual of interest

n is the number of individuals in the inbreeding path, excluding the inbred offspring

F_A is the inbreeding coefficient of the common ancestor

$\sum$ indicates that we add together $(1/2)^n (1 + F_A)$ for each inbreeding path

In this case, there is only one common ancestor and, therefore, only one inbreeding path. Also, we do not know anything about the heritage of the common ancestor, so we assume that F_A is zero. Thus, for the example of Figure 27.18,

$$F = \sum(1/2)^n(1 + 0)$$

$$= (1/2)^5 = 1/32 = 0.03125, \text{ or } 3.125\%$$

What does this value mean? The inbreeding coefficient, 3.125%, tells us the probability that a gene in the inbred individual (IV-1) is homozygous due to inheritance of alleles from a common ancestor (I-2). In this case, each gene in individual IV-1 has a 3.125% chance of being homozygous because the individual has inherited the same allele twice from the great-grandparent (I-2), once through each parent.

As an example, let's suppose that the common ancestor (I-2) is heterozygous for the gene involved in cystic fibrosis (CF). I-2's genotype is Cc, where c is the recessive allele that causes CF. Since F is 3.125%, there is a 3.125% probability that the inbred individual (IV-1) is homozygous (CC or cc) for this gene because IV-1 has inherited both copies from a great-grandparent (I-2). There is a 1.56% probability of inheriting both normal alleles (CC) and a 1.56% probability of inheriting both mutant alleles (cc). The inbreeding coefficient is denoted by the letter F (for fixation) because it is the probability that an allele will be fixed in the homozygous condition. The term *fixation* signifies that a homozygous individual can pass only one type of allele to offspring.

In other pedigrees, an individual may have two or more common ancestors. In this case, the inbreeding coefficient F is calculated using the sum of the numbers of individuals in the inbreeding paths. Such an example is described in question 4 of More Genetic TIPS at the end of this chapter.

Nonrandom Mating in a Population.   The effects of inbreeding and outbreeding can also be examined at the population level. To illustrate such an analysis, let's consider a situation in which the frequency of allele A is p and the frequency of allele a is q. In a given population, the genotype frequencies are determined in the following way:

$p^2 + fpq$ equals the frequency of AA

$2pq(1 - f)$ equals the frequency of Aa

$q^2 + fpq$ equals the frequency of aa

where f is a measure of how much the genotype frequencies deviate from Hardy-Weinberg equilibrium due to nonrandom mating. The value of f ranges from -1 to $+1$. When inbreeding occurs, the value is greater than zero. When outbreeding occurs, the value is less than zero.

As an example, let's suppose that $p = 0.8$, $q = 0.2$, and $f = 0.25$. We can calculate the frequencies of the AA, Aa, and aa genotypes under these conditions as follows:

$AA = p2 + fpq = (0.8)2 + (0.25)(0.8)(0.2) = 0.68$

$Aa = 2pq(1 - f) = 2(0.8)(0.2)(1 - 0.25) = 0.24$

$aa = q2 + fpq = (0.2)2 + (0.25)(0.8)(0.2) = 0.08$

There will be 68% AA homozygotes, 24% heterozygotes, and 8% aa homozygotes. If mating had been random (i.e., $f = 0$), the genotype frequency of AA would be p^2, which equals 64%, and aa would be q^2, which equals 4%. The frequency of heterozygotes would be $2pq$, which equals 32%. Comparing these sets of numbers, we see that inbreeding raises the proportions of homozygotes and decreases the proportion of heterozygotes. In natural populations, the value of f tends to become larger as a population becomes smaller, because each individual is more limited as to the choice of another individual for sexual reproduction.

What are the consequences of inbreeding in a population? From an agricultural viewpoint, it results in a higher proportion of homozygotes, which may exhibit a desirable trait. For example, an animal breeder may use inbreeding to produce animals that are larger because they have become homozygous for alleles promoting larger size. On the negative side, many genetic diseases are inherited in a recessive manner (see Chapter 24). For these disorders, inbreeding increases the likelihood that an individual will be homozygous for the recessive allele and therefore afflicted with the disease.

Also, in natural populations, inbreeding lowers the mean fitness of the population if homozygous offspring have lower fitness values. The lowering of mean fitness can be a serious problem as natural populations become smaller due to human destruction of their habitats. As a population shrinks, inbreeding becomes more likely because individuals have fewer potential mates from which to choose. The inbreeding, in turn, produces homozygotes that are less fit, thereby decreasing the reproductive success of the population. This phenomenon is called **inbreeding depression.**

Conservation biologists sometimes try to circumvent this problem by introducing individuals from one population into another. For example, the endangered Florida panther (*Felis concolor coryi*) suffers from inbreeding-related defects, which include poor sperm quality and quantity and morphological abnormalities. To help alleviate these effects, panthers of the same species from Texas have been introduced into the Florida population.

27.7 COMPREHENSION QUESTION

1. Inbreeding is sexual reproduction between individuals that are
   a. homozygous.
   b. heterozygous.
   c. part of the same genetic lineage.
   d. both a and c.

27.8   SOURCES OF NEW GENETIC VARIATION

Learning Outcomes:

1. List different sources of genetic variation.
2. Define *mutation rate,* and calculate how it affects allele frequencies in populations.
3. Outline the processes of exon shuffling and horizontal gene transfer, and explain how they produce genetic variation.
4. Identify common types of repetitive sequences, and explain how they are used in DNA fingerprinting.

In the previous sections, we primarily focused on genetic variation in cases when a single gene exists in two or more alleles. These simplified scenarios allow us to appreciate the general principles behind evolutionary mechanisms. However, as researchers have analyzed genetic variation at the molecular, cellular, and population levels, they have come to understand that new genetic variation occurs in many ways (**Table 27.5**).

Among eukaryotic species, sexual reproduction is an important way that new genetic variation arises among offspring. In Chapters 2 and 6, we considered how independent assortment and crossing over during sexual reproduction may produce new combinations of alleles among various genes, thereby producing new genetic variation in the resulting offspring. Similarly, in Chapter 29, we will consider how breeding between members of different species may produce hybrid offspring that harbor new combinations of genetic material. Such hybridization events have been important in the evolution of new species, particularly those in the plant kingdom.

Though prokaryotic species reproduce asexually, they also have mechanisms for gene transfer, such as conjugation, transduction, and transformation (see Chapter 7). These mechanisms are important for fostering genetic variation among bacterial and archaeal populations.

Rare mutations in DNA may also give rise to new types of variation (see Table 27.1). As discussed earlier in this chapter (see Figure 27.3) and in Chapter 19, mutations may occur within a particular gene to create new alleles of that gene. Such allelic variation is common in natural populations. Also, as described in Chapter 8, gene duplications may create a gene family; each family member acquires independent mutations and often evolves more specialized functions. An example is the globin gene family (refer back to Figure 8.7). Changes in chromosome structure and number are also important in the evolution of new species (see Chapter 29). In this section, we will examine some additional mechanisms through which an organism can acquire new genetic variation. These include exon shuffling, horizontal gene transfer, and changes in repetitive sequences. The diversity of mechanisms for fostering genetic variation underscores its profound importance in the evolution of species that are both well adapted to their native environments and successful at reproduction.

TABLE 27.5
Sources of New Genetic Variation That Occurs in Populations

Type	Description
Independent assortment	The independent segregation of different chromosomes may give rise to new combinations of alleles in offspring (see Chapter 3).
Crossing over	Recombination (crossing over) between homologous chromosomes can also produce new combinations of alleles that are located on the same chromosome (see Chapter 6).
Interspecies crosses	On occasion, members of different species may breed with each other to produce hybrid offspring. This topic is discussed in Chapter 29.
Prokaryotic gene transfer	Prokaryotic species have mechanisms of genetic transfer such as conjugation, transduction, and transformation (see Chapter 7).
New alleles	Point mutations can occur within a gene to create single-nucleotide polymorphisms (SNPs). In addition, genes can be altered by small deletions and additions. Gene mutations are discussed in Chapter 19.
Gene duplications	Events, such as misaligned crossovers, can add additional copies of a gene to a genome and lead to the formation of gene families (see Chapter 8).
Chromosome structure and number	Chromosome structure may be changed by deletions, duplications, inversions, and translocations. Changes in chromosome number result in aneuploid, polyploid, and alloploid offspring. These mechanisms are discussed in Chapters 8 and 29.
Exon shuffling	New genes can be created when exons of a preexisting gene are rearranged to make a gene that codes a protein with a new combination of domains.
Horizontal gene transfer	Genes from one species can be introduced into a different strain of the same species or into another species and become incorporated into that species' genome.
Changes in repetitive sequences	Short repetitive sequences are common in genomes due to the occurrence of transposable elements and tandem arrays. The numbers and lengths of repetitive sequences tend to show considerable variation in natural populations.

Mutations Can Be Deleterious, Neutral, or Beneficial

As discussed in Chapters 8 and 19, mutations involve changes in gene sequences, chromosome structure, and/or chromosome number. Mutations are random events that occur spontaneously at a low rate or are caused by mutagens at a higher rate. In 1926, geneticist Sergei Chetverikov was the first to suggest that mutational variability provides the raw material for evolution but does not constitute a significant evolutionary change. In other words, mutation can produce new alleles in a population but does not substantially alter allele frequencies. Chetverikov proposed that populations in nature absorb mutations like a sponge and retain them in a heterozygous condition, thereby providing a source of variability that may lead to future change.

Population geneticists often consider how new mutations affect the survival and reproductive potential of the individual that inherits them. A new mutation may be deleterious, neutral, or beneficial, depending on its effect. For genes that code proteins, the effects of new mutations depend on their influence on protein function. Deleterious and neutral mutations are far more likely to occur than beneficial ones. For example, alleles can be altered in many different ways that render a coded protein defective. As discussed in Chapter 19, deletions and point mutations such as missense mutations, nonsense mutations, and frameshift mutations all may cause a gene to express a protein that is nonfunctional or less functional than the wild-type protein. Also, mutations in non-coding sequences can alter gene expression (refer back to Table 19.2).

Neutral mutations, which are not acted upon by natural selection, occur in several different ways. For example, a neutral mutation can change the base in the wobble position without affecting the amino acid sequence of the coded protein (refer back to Section 13.4 in Chapter 13), or a neutral mutation can be a missense mutation that has no effect on protein function. Such point mutations happen at specific sites within the coding sequence. Neutral mutations can also occur within introns, the non-coding sequences of genes.

By comparison, beneficial mutations are relatively uncommon. To be advantageous, a new mutation might alter the amino acid sequence of a protein to yield a better-functioning product. Such mutations occur less frequently than deleterious or neutral mutations.

The Mutation Rate Is the Likelihood of a New Mutation

The **mutation rate** is defined as the likelihood that a gene will be altered by a new mutation. The rate is typically expressed as the number of new mutations in a given gene per generation. A common value for the mutation rate is in the range of 1 in 100,000 to 1 in 1,000,000, or 10^{-5} to 10^{-6} per generation. However, mutation rates vary depending on species, cell type, chromosomal location, and gene size. Furthermore, in experimental studies, the mutation rate is usually measured by following the change of a normal (functional) gene to a deleterious (nonfunctional) allele. The mutation rate producing beneficial alleles is expected to be substantially lower.

It is clear that new mutations provide genetic variability, but population geneticists also want to know how much the mutation rate affects the allele frequencies in a population. Can random mutations have a large effect on allele frequencies over time? To answer this question, let's take this simple case: A gene exists as an allele, A; the allele frequency of A is denoted by the variable p. A mutation can convert the A allele into a different allele called a. The allele frequency of a is designated by q. The conversion of the A allele into the a allele by mutation occurs at a rate that is designated μ. If we assume that the rate of the reverse mutation (a to A) is negligible, the change in allele frequency of the a allele (Δq) after one generation is

$$\Delta q = \mu p$$

As an example, let's consider the following conditions:

$p = 0.8$ (i.e., frequency of A is 80%)

$q = 0.2$ (i.e., frequency of a is 20%)

$\mu = 10^{-5}$ (hypothetical mutation rate for the conversion of A to a)

$\Delta q = (10^{-5})(0.8) = (0.00001)(0.8) = 0.000008$

Therefore, in the next generation (designated $n + 1$),

$$q_{n+1} = 0.2 + 0.000008 = 0.200008$$
$$p_{n+1} = 0.8 - 0.000008 = 0.799992$$

As we can see from this calculation, new mutations do not significantly alter the allele frequencies in a single generation.

We can use the following equation to calculate the change in allele frequency after any number of generations:

$$(1 - \mu)^t = \frac{p_t}{p_0}$$

where

μ is the mutation rate for the conversion of A to a

t is the number of generations

p_0 is the allele frequency of A in the starting generation

p_t is the allele frequency of A after t generations

As an example, let's suppose that the allele frequency of A is 0.8 and $\mu = 10^{-5}$, and we want to know what the allele frequency will be after 1000 generations ($t = 1000$). Plugging these values into the preceding equation and solving for p_t gives

$$(1 - 0.00001)^{1000} = \frac{p_t}{0.8}$$
$$p_t = 0.792$$

Therefore, after 1000 generations, the frequency of A has dropped only from 0.8 to 0.792. Again, these results point to how slowly the occurrence of new mutations changes allele frequencies. In natural populations, the rate of new mutation is rarely a significant catalyst in shaping allele frequencies. Instead, other processes, such as natural selection and genetic drift, have far greater effects on allele frequencies.

New Genes Are Produced in Eukaryotes via Exon Shuffling

Sources of new genetic variation are revealed when the parts of genes that code protein domains are compared within a single species. Many proteins, particularly those found in eukaryotic species, have a modular structure composed of two or more domains with different functions. For example, certain transcription factors have discrete domains involved with hormone binding, dimerization, and DNA binding. As described in Chapter 15, the glucocorticoid receptor has a domain that binds the hormone, a second domain that facilitates protein dimerization, and a third that allows the glucocorticoid receptor to bind to a glucocorticoid response element (GRE) next to particular genes (refer back to Figure 15.12). By comparing the modular structure of eukaryotic proteins with the genes that code them, geneticists have discovered that each domain tends to be coded by one coding sequence, or exon, or by a series of two or more adjacent exons.

During the evolution of eukaryotic species, many new genes have been created by a process known as **exon shuffling,** in which an exon and its flanking introns from one gene are inserted into another gene, thereby producing a new gene that codes a protein with an additional domain (**Figure 27.19**). This process may also involve the duplication and rearrangement of exons. Exon shuffling results in novel genes that express proteins with diverse functional domains. Such proteins can then alter traits in the organism and may be acted on by evolutionary processes, such as natural selection and genetic drift.

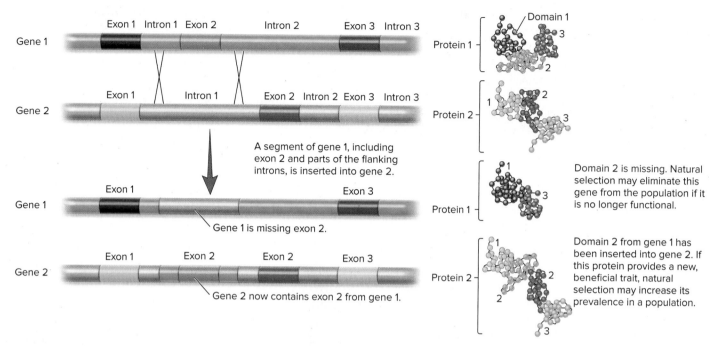

FIGURE 27.19 **The process of exon shuffling.** In this example, a segment of one gene containing an exon and its flanking introns has been inserted into another gene. A rare, abnormal crossover event called nonhomologous recombination may cause this to happen. This results in proteins that have new combinations of domains and possibly new functions.

Exon shuffling may occur by more than one mechanism. Transposable elements, which are described in Chapter 10, may promote the insertion of exons into the coding sequences of genes. Alternatively, an abnormal crossover event could promote the insertion of an exon into another gene (this is the case in Figure 27.19). Such an event is called nonhomologous recombination because the two regions involved in the crossover are not homologous to each other.

New Genes Are Acquired via Horizontal Gene Transfer

Species also accumulate genetic changes by a process called **horizontal gene transfer,** in which an organism incorporates genetic material from another organism without being the offspring of that organism. This process often involves the exchange of genetic material between different species. **Figure 27.20** illustrates one possible mechanism for horizontal gene transfer. In this example, a eukaryotic cell has engulfed a bacterium by endocytosis. During the degradation of the bacterium, a bacterial gene escapes to the nucleus of the cell, where it is inserted into one of the chromosomes. In this way, a gene has been transferred from a bacterial species to a eukaryotic species.

By analyzing gene sequences among many different species, researchers have discovered that horizontal gene transfer is a common phenomenon. This process can occur from prokaryotes to eukaryotes, from eukaryotes to prokaryotes, between different species of prokaryotes, and between different species of eukaryotes.

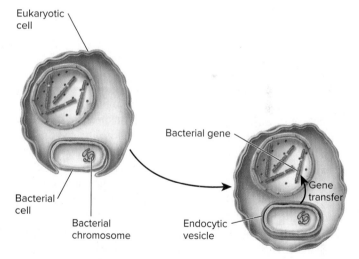

FIGURE 27.20 **Horizontal gene transfer from a bacterium to a eukaryote.** In this example, a bacterium is engulfed by a eukaryotic cell, and a bacterial gene is transferred to one of the eukaryotic chromosomes.

Gene transfer among bacterial species is relatively widespread. As discussed in Chapter 7, bacterial species can carry out three natural mechanisms of gene transfer, known as conjugation, transduction, and transformation. By analyzing the genomes of bacterial species, scientists have determined that many genes within a given bacterial genome are derived from genes acquired from other species via horizontal gene transfer. Genome studies have suggested that as much as 20–30% of the variation in the genetic composition of modern prokaryotic species can be attributed to this process. For example, roughly 17% of the genes of *E. coli* and of *Salmonella typhimurium* have been acquired from other species via horizontal gene transfer during the past 100 million years. The roles of these acquired genes are quite varied, though they commonly involve functions that are readily acted on by natural selection. These functions include antibiotic resistance, the ability to degrade toxic compounds, and pathogenicity (the ability to cause disease).

Genetic Variation Is Produced via Changes in Repetitive Sequences

Another source of genetic variation involves changes in **repetitive sequences**—short sequences, typically a few base pairs to a few thousand base pairs long, that are repeated many times within a species' genome. Repetitive sequences are usually of two types. First, transposable elements (TEs) are DNA sequences that can move from place to place in a species' genome (see Chapter 10). The prevalence and movement of TEs provide a great deal of genetic variation between species and within a single species. In certain eukaryotic species, TEs have become fairly abundant (see Table 10.2).

A second type of repetitive sequence is nonmobile and consists of short sequences that are tandemly repeated. In a **microsatellite** (also called a short tandem repeat, STR), the repeat unit is usually 1–6 bp long, and the whole tandem repeat is less than a couple hundred base pairs in length. For example, the most common microsatellite encountered in humans is the sequence $(CA)_N$, where N may range from 5 to more than 50. In other words, this dinucleotide sequence can be tandemly repeated 5–50 or more times. The $(CA)_N$ microsatellite is found, on average, about every 10,000 bases in the human genome.

In a **minisatellite,** the repeat unit is typically 6–80 bp in length, and the size of the minisatellite ranges from 1 kbp to 20 kbp. An example of a minisatellite in humans is telomeric DNA. In a human sperm cell, for example, the repeat unit is 6 bp and the size of a telomere is about 15 kbp. (Note: Tandem repeat sequences are called satellites because they sediment away from the rest of the chromosomal DNA during equilibrium density centrifugation.)

Alterations in the sequences of microsatellites often escape the proofreading function of DNA polymerase. Therefore, microsatellites tend to be hotspots for a type of mutation in which the number of tandem repeats changes. For example, a microsatellite with a 4-bp repeat unit and a length of 64 bp may undergo a mutation that adds three more repeat units and becomes 76 bp long. How does this occur? One common mechanism to explain this phenomenon is the slippage of DNA polymerase off the template strand during DNA replication due to the formation of a hairpin structure. As discussed in Chapter 19, this mechanism can explain trinucleotide repeat expansion (TNRE) (refer back to Figure 19.11).

Because repetitive sequences tend to vary within a population, they have become a common tool that geneticists use in a variety of ways. For example, as described in Chapter 22, microsatellites are used as molecular markers to map the locations of genes. Also, population geneticists analyze microsatellites or minisatellites to study variation at the population level and to determine the relationships among individuals and between neighboring populations. The sizes of microsatellites and minisatellites found in closely related individuals tend to be more similar than the sizes of those in unrelated individuals. As described next, this phenomenon is the basis for DNA fingerprinting.

DNA Fingerprinting Is Used for Identification and Relationship Testing

The technique of **DNA fingerprinting,** also known as **DNA profiling,** analyzes individuals based on the occurrence of repetitive sequences in their genome. When subjected to traditional DNA fingerprinting, chromosomal DNA gives rise to a series of bands on a gel (**Figure 27.21**). The sizes and order of bands constitute an

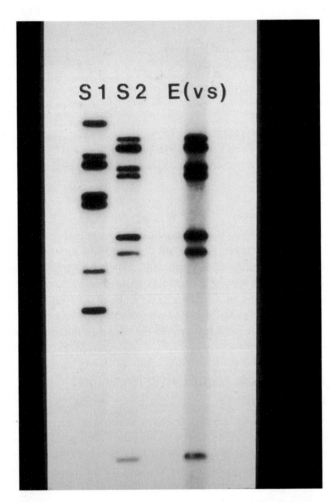

FIGURE 27.21 A comparison of two DNA fingerprints. The chromosomal DNA from two different individuals (suspect 1 is S1, and suspect 2 is S2) was subjected to DNA fingerprinting. The DNA evidence from a crime scene, E(vs), was also subjected to DNA fingerprinting. Following the hybridization of a labeled probe, the DNA appears as a series of bands on a gel. The dissimilarity in the pattern of these bands distinguishes different individuals, much as the differences in physical fingerprint patterns can be used for identification. As seen here, the DNA from S2 matches the DNA found at the crime scene.

Leonard Lessin

CONCEPT CHECK: What are two common applications of DNA fingerprinting?

individual's DNA fingerprint. Like the human fingerprint, the DNA of each individual has a distinctive pattern. It is the unique patterns of these bands that make it possible to distinguish individuals.

A comparison of the DNA fingerprints among different individuals has found two applications:

- DNA fingerprinting is used as a method of identification. For example, in forensics, DNA fingerprinting can identify a crime suspect.
- DNA fingerprinting is also used for relationship testing. Closely related individuals have more similar DNA fingerprints than do distantly related ones (see question 5 in More Genetic TIPS at the end of this chapter). In humans, such similarity is useful for paternity testing. In population genetics, DNA fingerprinting can provide evidence regarding the degree of relatedness among members of a population. Such information may help geneticists determine if a population is likely to be suffering from inbreeding depression.

The development of DNA fingerprinting relied on the identification of DNA sequences that vary in length among members of a population. This naturally occurring variation causes each individual to have a unique DNA fingerprint. In the 1980s, Alec Jeffreys and colleagues found that certain minisatellites within human chromosomes are particularly variable in their lengths. As discussed earlier, minisatellites tend to vary within populations due to changes in the number of tandem repeats at each site.

DNA fingerprinting is now done using the technique of polymerase chain reaction (PCR), which amplifies microsatellites. Like minisatellites, microsatellites are found in multiple sites in the genome of humans and other species and vary in length among different individuals.

1. In this procedure, the microsatellites from a sample of DNA are amplified by PCR using primers that flank the repetitive region.
2. The amplified microsatellites are fluorescently labeled.
3. They are then separated by gel electrophoresis according to their molecular masses.
4. As in automated DNA sequencing, described in Chapter 20, a laser excites the fluorescent molecules within a microsatellite, and a detector records the amount of fluorescence emission for each microsatellite.
5. As shown in **Figure 27.22**, this type of DNA fingerprint yields a series of peaks, each peak corresponding to a characteristic molecular mass. In this automated approach, the pattern of peaks rather than bands constitutes an individual's DNA fingerprint.

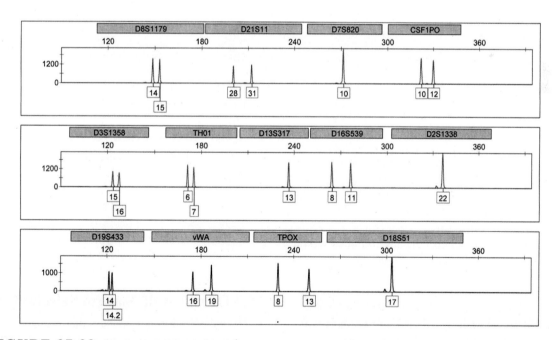

ONLINE ANIMATION

FIGURE 27.22 Automated DNA fingerprinting. In automated DNA fingerprinting, microsatellites in a sample of DNA are amplified, using primers that recognize the ends of microsatellites. The microsatellite fragments are fluorescently labeled and then separated by gel electrophoresis or liquid chromatography. The fluorescent molecules within each microsatellite are excited with a laser, and the amount of fluorescence is measured by a fluorescence detector. A printout from the detector is shown here. The gray boxes contain the names of 13 different microsatellites. The peaks show the relative amounts of each microsatellite. The boxes beneath each peak indicate the number of tandem repeats in a given microsatellite. In this example, the individual is heterozygous for certain microsatellites (e.g., D8S1179) and homozygous for others (e.g., D7S820).

27.8 COMPREHENSION QUESTIONS

1. The mutation rate is
   a. the likelihood that a new mutation will occur in a given gene.
   b. too low to substantially change allele frequencies in a population.
   c. lower for mutations that create beneficial alleles.
   d. All of the above are true of the mutation rate.

2. The transfer of an antibiotic resistance gene from one bacterial species to a different species is an example of
   a. exon shuffling.   c. genetic drift.
   b. horizontal gene transfer.   d. migration.

3. DNA fingerprinting analyzes the DNA from individuals on the basis of the occurrence of _____ in their genomes.
   a. repetitive sequences
   b. abnormalities in chromosome structure
   c. specific genes
   d. viral insertions

KEY TERMS

Introduction: population genetics

27.1 gene pool, population, local population, polymorphism, polymorphic, monomorphic, single-nucleotide polymorphism (SNP), allele frequency, genotype frequency, phenotype frequency, Hardy-Weinberg equilibrium, Hardy-Weinberg equation, disequilibrium

27.2: biological evolution (evolution), microevolution

27.3: natural selection, reproductive success, adaptive evolution, Darwinian fitness (fitness), relative fitness (*w*), general selection model, mean fitness of the population ($\overline{w}$), genome-wide association study (GWAS), genome-wide selection scan (GWSS), positive selection, genetic hitchhiking

27.4: directional selection, balancing selection, heterozygote advantage, selection coefficient (*s*) negative frequency-dependent selection, disruptive selection, stabilizing selection

27.5: genetic drift, bottleneck effect, founder effect

27.6: gene flow, conglomerate

27.7: assortative mating, inbreeding, outbreeding, inbreeding coefficient (*F*), inbreeding depression

27.8: mutation rate, exon shuffling, horizontal gene transfer, repetitive sequences, microsatellite, minisatellite, DNA fingerprinting (DNA profiling)

CHAPTER SUMMARY

- Population genetics is concerned with the extent of genetic variation within a group of individuals and changes in that variation over time.

27.1 Genes in Populations and the Hardy-Weinberg Equation

- All of the alleles of every gene in a population constitute the population's gene pool.
- For sexually reproducing organisms, a population is a group of individuals of the same species that occupy the same region and can interbreed with one another (see Figure 27.1).
- In population genetics, polymorphism refers to inherited traits or genes that exhibit variation in a population (see Figure 27.2).
- Single-nucleotide polymorphisms (SNPs) are the most common type of variation among genes (see Figure 27.3).
- Geneticists analyze genetic variation by determining allele and genotype frequencies.
- The Hardy-Weinberg equation can be used to calculate genotype frequencies based on allele frequencies (see Figures 27.4, 27.5).
- Deviation from Hardy-Weinberg equilibrium indicates that evolutionary change is occurring.

- The Hardy-Weinberg equation can be used to estimate the frequency of heterozygous carriers.

27.2 An Introduction to Microevolution

- Microevolution refers to changes in a population's gene pool from generation to generation.
- Mutations are the source of new genetic variation. However, the occurrence of new mutations does not greatly change allele frequencies because it happens at a very low rate. Other factors, such as natural selection, genetic drift, migration, and nonrandom mating, may alter allele and/or genotype frequencies (see Table 27.1).

27.3 Overview of Natural Selection

- Natural selection is a process that changes allele frequencies from one generation to the next based on fitness, which is a measure of the relative reproductive success of different genotypes.
- The general selection model relates fitness to the Hardy-Weinberg equation (Table 27.2).
- Genome-wide selection scans can provide insight regarding the effects of natural selection (Table 27.3).

27.4 Patterns of Natural Selection

- Directional selection favors an extreme phenotype (see Figures 27.6, 27.7, 27.8).
- The general selection model can be used to predict changes in genotype and allele frequencies from one generation to the next due to directional selection (Table 27.4).
- The Grants observed natural selection in a finch population. The selection involved a change in beak size due to drought conditions that produced larger seeds (see Figures 27.9, 27.10).
- Balancing selection results in stable polymorphism. It may occur as a result of heterozygote advantage or negative frequency-dependent selection (see Figures 27.11, 27.12).
- Disruptive selection favors multiple phenotypes in heterogeneous environments (see Figures 27.13, 27.14).
- Stabilizing selection favors individuals with intermediate phenotypes (see Figure 27.15).

27.5 Genetic Drift

- Genetic drift involves changes in allele frequencies in a population due to random fluctuations. Over the long run, it often results in allele fixation or loss. The effect of genetic drift occurs more rapidly in small populations (see Figure 27.16).
- Two mechanisms that can influence genetic drift are the bottleneck effect and the founder effect (see Figure 27.17).

27.6 Migration

- Migration can alter allele frequencies. It tends to reduce differences in allele frequencies between neighboring populations and increase genetic diversity within a population.

27.7 Nonrandom Mating

- Nonrandom mating may alter the genotype frequencies that are predicted by the Hardy-Weinberg equation. Inbreeding results in a higher proportion of homozygotes in a population (see Figure 27.18).

27.8 Sources of New Genetic Variation

- A variety of different mechanisms can bring about genetic variation (see Table 27.5).
- The mutation rate is the likelihood that a gene will be altered by a new mutation.
- New genes can be produced in eukaryotes by exon shuffling (see Figure 27.19).
- A species may acquire a new gene from another species via horizontal gene transfer (see Figure 27.20).
- A common source of genetic variation in populations involves changes in repetitive sequences, such as microsatellites.
- DNA fingerprinting is a technique that relies on variation in repetitive sequences within a population. It is used as a means of identification and in relationship testing (see Figures 27.21, 27.22).

PROBLEM SETS & INSIGHTS

MORE GENETIC TIPS

1. The phenotype frequency of the inability to taste phenylthiocarbamide (PTC) in a particular human population is approximately 4%, or 0.04. The inability to taste this bitter substance is due to homozygosity of a recessive allele. If you assume there are only two alleles in the population, namely, T (for tasters) and t (for nontasters), and that the population is in Hardy-Weinberg equilibrium, calculate the frequencies of these two alleles.

T OPIC: *What topic in genetics does this question address?* The topic is predicting allele frequencies in a population. More specifically, the question asks you to predict the frequencies of alleles that affect people's ability to taste phenylthiocarbamide (PTC).

I NFORMATION: *What information do you know based on the question and your understanding of the topic?* From the question, you know the frequency of homozygotes who are nontasters in a certain population. From your understanding of the topic, you may realize that you can use the Hardy-Weinberg equation to determine allele frequencies if you know the recessive genotype frequency.

P ROBLEM-SOLVING S TRATEGY: *Make a calculation.* Let p = the allele frequency of the taster allele, and q = the allele frequency of the nontaster allele. The genotype frequency of

nontasters is 0.04. This is the frequency of the genotype tt, which in this case is equal to q^2:

$$q^2 = 0.04$$

To determine the frequency q of the nontaster allele, you take the square root of both sides of this equation:

$$q = 0.2$$

With this value, you can calculate the frequency p of the taster allele:

$$p = 1 - q$$
$$= 1 - 0.2 = 0.8$$

ANSWER: The frequency of the nontaster allele is 0.2, or 20%, and that of the taster allele is 0.8, or 80%.

---

2. The Hardy-Weinberg equation can be expanded to include situations involving three or more alleles. In its standard (two-allele) form, the Hardy-Weinberg equation indicates that each individual inherits two copies of each gene, one from each parent. For a two-allele situation, the equation can also be written as $(p + q)^2 = 1$. (Note: The number 2 in this equation reflects the fact that the genotype is due to the inheritance of two alleles, one from each parent.)

This equation can be expanded to include three or more alleles. For example, let's consider a situation in which a gene exists as three alleles: *A1*, *A2*, and *A3*. The allele frequency of *A1* is designated by the letter *p*, that of *A2* by the letter *q*, and that of *A3* by the letter *r*. Under these circumstances, the Hardy-Weinberg equation becomes

$$(p + q + r)^2 = 1$$
$$p^2 + q^2 + r^2 + 2pq + 2pr + 2qr = 1$$

where

p^2 is the genotype frequency of *A1A1*
q^2 is the genotype frequency of *A2A2*
r^2 is the genotype frequency of *A3A3*
$2pq$ is the genotype frequency of *A1A2*
$2pr$ is the genotype frequency of *A1A3*
$2qr$ is the genotype frequency of *A2A3*

Here is the question: As discussed in Chapter 4, the gene that determines human blood types exists in three alleles. In a Japanese population, the allele frequencies are

$I^A = 0.28$
$I^B = 0.17$
$i = 0.55$

Based on these allele frequencies and assuming Hardy-Weinberg equilibrium, calculate the genotype frequencies and blood type frequencies in this Japanese population.

TOPIC: *What topic in genetics does this question address?* The topic is Hardy-Weinberg equilibrium. More specifically, the question is about using the Hardy-Weinberg equation in a situation in which there are three alleles for a given gene.

INFORMATION: *What information do you know based on the question and your understanding of the topic?* In the question, you are given an expanded version of the Hardy-Weinberg equation. From your understanding of the topic, you may realize that you can use the Hardy-Weinberg equation to determine genotype frequencies if you know the allele frequencies. The genotype frequencies can then be used to find the blood type frequencies.

PROBLEM-SOLVING **S**TRATEGY: *Make a calculation.* Use the equation given in the question, and let *p* represent the allele frequency of I^A, *q* represent that of I^B, and *r* represent that of *i*.

ANSWER:
p^2 (the genotype frequency of $I^A I^A$, which produces type A blood) = $(0.28)^2 = 0.08$

q^2 (the genotype frequency of $I^B I^B$, which produces type B blood) = $(0.17)^2 = 0.03$

r^2 (the genotype frequency of *ii*, which produces type O blood) = $(0.55)^2 = 0.30$

$2pq$ (the genotype frequency of $I^A I^B$, which produces type AB blood) = $2(0.28)(0.17) = 0.09$

$2pr$ (the genotype frequency of $I^A i$, which produces type A blood) = $2(0.28)(0.55) = 0.31$

$2qr$ (the genotype frequency of $I^B i$, which produces type B blood) = $2(0.17)(0.55) = 0.19$

Type A = 0.08 + 0.31 = 0.39, or 39%

Type B = 0.03 + 0.19 = 0.22, or 22%

Type O = 0.30, or 30%

Type AB = 0.09, or 9%

3. Let's suppose that pigmentation in a species of insect is controlled by a single gene that exists in two alleles, *D* for dark and *d* for light. The heterozygote, *Dd*, is intermediate in color. In a heterogeneous environment, the allele frequencies are $D = 0.7$ and $d = 0.3$. This polymorphism is maintained because the environment has some dimly lit forested areas and some sunny fields. During a hurricane, 1000 of these insects are blown to a completely sunny area. In this environment, the relative fitness values are $DD = 0.3$, $Dd = 0.7$, and $dd = 1.0$. Calculate the allele frequencies in the next generation.

TOPIC: *What topic in genetics does this question address?* The topic is natural selection. More specifically, the question is about directional selection in an insect population that has moved to a new location.

INFORMATION: *What information do you know based on the question and your understanding of the topic?* In the question, you are given the relative fitness values for genotypes in a population of insects. From your understanding of the topic, you may realize that directional selection favors the extreme (light) phenotype in the new environment.

PROBLEM-SOLVING **S**TRATEGY: *Make a calculation.* To solve this problem, you first need to calculate the mean fitness of the population and then calculate the allele frequencies of the next generation. To calculate the mean fitness of the population:

$$p^2 w_{DD} + 2pq\, w_{Dd} + q^2 w_{dd} = \overline{w}$$
$$\overline{w} = (0.7)^2 (0.3) + 2(0.7)(0.3)(0.7) + (0.3)^2 (1.0)$$
$$= 0.15 + 0.29 + 0.09 = 0.53$$

After one generation of seleciton, the allele frequencies will be

Allele frequency of D: $p_D = \dfrac{p^2 w_{DD}}{\overline{w}} + \dfrac{pq w_{Dd}}{\overline{w}}$
$$= \dfrac{(0.7)^2 (0.3)}{0.53} + \dfrac{(0.7)(0.3)(0.7)}{0.53}$$
$$= 0.55$$

Allele frequency of d: $q_d = \dfrac{q^2 w_{dd}}{\overline{w}} + \dfrac{pq w_{Dd}}{\overline{w}}$
$$= \dfrac{(0.3)^2 (1.0)}{0.53} + \dfrac{(0.7)(0.3)(0.7)}{0.53}$$
$$= 0.45$$

ANSWER: After one generation, the allele frequency of *D* has decreased from 0.7 to 0.55, and that of *d* has increased from 0.3 to 0.45.

4. Using the pedigree shown next, answer the following questions with regard to individual VII-1.

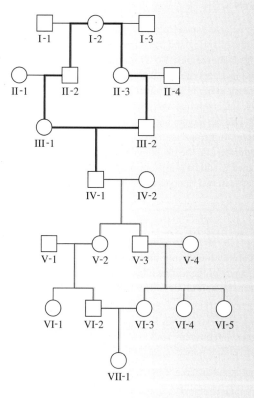

A. Who are the common ancestors of VII-1's parents?
B. What is the inbreeding coefficient for VII-1?

TOPIC: *What topic in genetics does this question address?* The topic is inbreeding. More specifically, the questions ask you to identify the common ancestors in a pedigree and calculate the inbreeding coefficient for an individual.

INFORMATION: *What information do you know based on the question and your understanding of the topic?* In the question, you are given a family pedigree. From your understanding of the topic, you may remember that a common ancestor is someone who is an ancestor of both of a person's parents. You may also remember that geneticists have derived an equation to calculate the inbreeding coefficient.

PROBLEM-SOLVING **S**TRATEGIES: *Compare and contrast. Make a calculation.* To solve this problem, you first need to compare the members of the pedigree and identify which of them are common ancestors of both of the individual's parents. You can then use the inbreeding equation to calculate the inbreeding coefficient.

ANSWER:

A. The common ancestors are IV-1 and IV-2. They are the grandparents of VI-2 and VI-3, who are the parents of VII-1. (Follow the yellow-highlighted lines.)

B. The inbreeding coefficient is calculated using the formula

$$F = \sum (1/2)^n (1 + F_A)$$

In this case, there are two common ancestors, IV-1 and IV-2. Also, IV-1 is inbred, because I-2 is a common ancestor to both of IV-1's parents. The first step is to calculate F_A, the inbreeding coefficient for IV-1, the common ancestor. The inbreeding path for IV-1, which is highlighted in red, contains five people: III-1, II-2, I-2, II-3, and III-2. Therefore,

$$n = 5$$
$$F_A = (1/2)^5 = 0.03$$

Now you can calculate the inbreeding coefficient for VII-1. Each inbreeding path for VII-1, highlighted in yellow, contains five people: VI-2, V-2, IV-1, V-3, and VI-3; and VI-2, V-2, IV-2, V-3, and VI-3. Thus,

$$F = (1/2)^5 (1 + 0.03) + (1/2)^5 (1 + 0)$$
$$= 0.032 + 0.031 = 0.063$$

5. An important application of DNA fingerprinting is relationship testing. Persons who are related genetically have some bands or peaks in common. The number they share depends on the closeness of their genetic relationship. For example, offspring are expected to receive half of their minisatellites from one parent and the rest from the other. The diagram presented next schematically shows traditional DNA fingerprints of an offspring, the female parent, and two potential male parents.

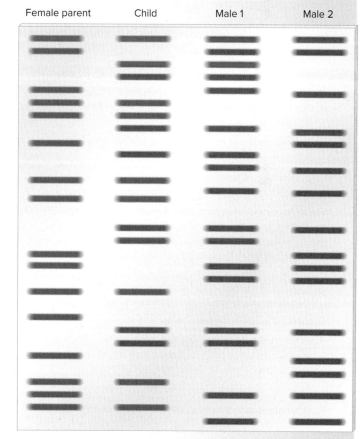

In paternity testing, the offspring's DNA fingerprint is first compared with that of the female parent. The bands that the offspring has in common with the female parent are shown in purple. The bands that are not similar between the offspring and the female parent must have been inherited from the male parent. These bands are shown in red in the offspring's DNA fingerprint. Is male 1 or male 2 the offspring's parent?

TOPIC: *What topic in genetics does this question address?* The topic is DNA fingerprinting. More specifically, the question is about using DNA fingerprinting to determine paternity.

INFORMATION: *What information do you know based on the question and your understanding of the topic?* From the question, you know the pattern of bands in the DNA fingerprints of a female parent, an offspring, and two potential male parents. From your understanding of the topic, you may remember that an offspring shares 50% of its bands with each parent.

PROBLEM-SOLVING **S**TRATEGIES: *Analyze data. Compare and contrast.* An approach to solve this problem is to compare the bands of the offspring with those of each of the potential male parents.

ANSWER: Male 2 does not have many of the red (paternal) bands seen in the offspring's fingerprint. Therefore, male 2 can be excluded as being the male parent of this offspring. However, male 1 has all of the paternal bands. He is very likely to be the male parent.

Note: Geneticists can calculate the likelihood that the matching bands between the offspring and a prospective male parent could occur as a matter of random chance. To do so, they analyze the frequency of each band in a reference population (e.g., people of Northern European descent living in the United States). For example, let's suppose that DNA fingerprinting analyzed 40 bands. Of these, 20 bands matched with the mother and 20 bands matched with a prospective male parent. If the probability of each of these bands in a reference population was 1/4, the likelihood of such a match occurring by random chance would be $(1/4)^{20}$, or roughly 1 in 1 trillion. Therefore, a match between two samples is rarely a matter of random chance.

Conceptual Questions

C1. What is a gene pool? How is a gene pool described in a quantitative way?

C2. In genetics, what does the term *population* mean? Pick any species you like and describe how its population might change over the course of many generations.

C3. What is genetic polymorphism? What is the source of genetic variation?

C4. Identify each of the following as an example of allele, genotype, and/or phenotype frequency:

A. Approximately 1 in 2500 people of Northern European descent is born with cystic fibrosis.

B. The percentage of carriers of the sickle cell allele in West Africa is approximately 13%.

C. The number of new mutations per generation resulting in achondroplasia, a genetic disorder, is approximately 5×10^{-5}.

C5. The term *polymorphism* can refer to both genes and traits. Explain what is meant by a polymorphic gene and a polymorphic trait. If a gene is polymorphic, does the trait that the gene affects also have to be polymorphic? Explain why or why not.

C6. Cystic fibrosis (CF) is a recessive autosomal disorder. In certain populations of Northern European descent, the number of people born with this disorder is about 1 in 2500. Assuming Hardy-Weinberg equilibrium for this trait:

A. What are the frequencies for the common (non-disease-causing) allele and the mutant (disease-causing) allele?

B. What are the genotype frequencies for homozygous unaffected, heterozygous, and homozygous affected individuals?

C. Assuming random mating, what is the probability that two phenotypically unaffected heterozygous carriers will choose each other as mates?

C7. For a gene existing in two alleles, what are the allele frequencies when the heterozygote frequency is at its maximum value, assuming Hardy-Weinberg equilibrium? What if there are three alleles?

C8. In a population, the frequencies of two alleles are $B = 0.67$ and $b = 0.33$. The genotype frequencies are $BB = 0.50$, $Bb = 0.37$, and $bb = 0.13$. Do these numbers suggest inbreeding? Explain why or why not.

C9. The ability to roll your tongue is inherited as a recessive trait. The frequency of the rolling allele is approximately 0.6, and that of the dominant (nonrolling) allele is 0.4. What is the frequency of individuals who can roll their tongues?

C10. What evolutionary mechanisms can cause allele frequencies to change and possibly lead to a genetic polymorphism? Discuss the relative importance of each type of process.

C11. What is the difference between genetic drift and natural selection? Describe an example of each. At the molecular level, explain how mutations can be neutral (and change in frequency via genetic drift) or adaptive (and change in frequency via natural selection).

C12. What is Darwinian fitness? What types of characteristics can promote high fitness values? Give several examples.

C13. What is the mean fitness of a population? How does its value change in response to natural selection?

C14. Describe the similarities and differences among directional, balancing, disruptive, and stabilizing selection.

C15. Is each of the following examples due to directional, disruptive, balancing, or stabilizing selection?

A. Polymorphisms in color and banding pattern of the shells of land snails, as shown in Figure 27.14

B. Thick fur among mammals living in cold climates

C. Birth weight in humans

D. Sturdy stems and leaves among plants exposed to windy climates

C16. With regard to the term *genetic drift*, what is drifting? Why is this an appropriate term to describe this phenomenon?

C17. Why is genetic drift more significant in small populations? Why does it take longer for genetic drift to cause allele fixation in large populations than in small ones?

C18. A group of four birds (two males and two females) flies to a new location and starts a new breeding colony. Three of the birds are homozygous AA, and one bird is heterozygous Aa.

A. What is the probability that the a allele will become fixed in the population via genetic drift?

B. If fixation of the a allele occurs, how long will it take?

C. How will the growth of the population, from generation to generation, affect the answers to parts A and B? Explain.

C19. Describe what happens to allele frequencies as a result of the bottleneck effect. Discuss the relevance of this effect with regard to species that are approaching extinction.

C20. With regard to genetic drift, is each of the following statements *true* or *false*? If a statement is false, explain why.

A. Over the long run, genetic drift leads to allele fixation or loss.

B. When a new mutation occurs within a population, genetic drift is more likely to cause the loss of the new allele rather than the fixation of the new allele.

C. Genetic drift promotes genetic diversity in large populations.

D. Genetic drift is more significant in small populations.

C21. When two populations frequently intermix due to migration, what are the long-term consequences with regard to allele frequencies and genetic variation?

C22. Two populations of antelope are separated by a mountain range. Some antelope occasionally migrate from one population to the other. Migration can occur in either direction. Explain how migration affects the following phenomena:

A. Genetic diversity in the two populations

B. Allele frequencies in the two populations

C. Genetic drift in the two populations

C23. Does inbreeding affect allele frequencies? Why or why not? How does it affect genotype frequencies? With regard to rare recessive diseases, what are the consequences of inbreeding in human populations?

C24. Using the pedigree shown here, answer the following questions for individual VI-1.

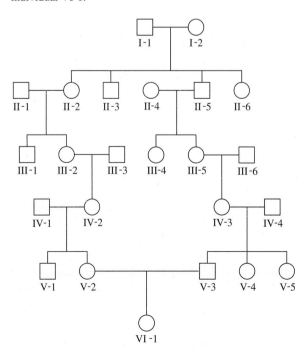

A. Is this individual inbred?

B. If so, who is/are the common ancestor(s) of the parents of VI-1?

C. Calculate the inbreeding coefficient for VI-1.

D. Are the parents of VI-1 inbred?

C25. A family pedigree is shown here.

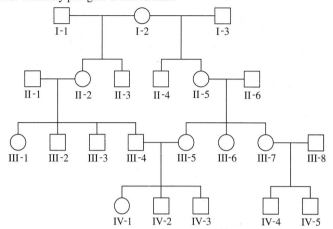

A. What is the inbreeding coefficient for individual IV-3?

B. Based on the data shown in this pedigree, is individual IV-4 inbred?

C26. A family pedigree is shown here.

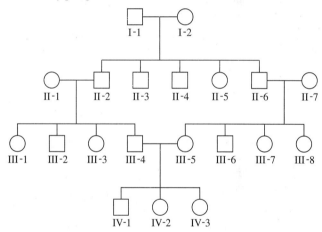

A. What is the inbreeding coefficient for individual IV-2? Who is/are common ancestor(s) of the parents of IV-2?

B. Based on the data shown in this pedigree, is individual III-4 inbred?

C27. Antibiotics are commonly used to combat bacterial and fungal infections. During the past several decades, however, antibiotic-resistant strains of microorganisms have become alarmingly prevalent. This resistance has undermined the effectiveness of antibiotics in treating many types of infectious disease. Discuss how the following processes that alter allele frequencies may have contributed to the emergence of antibiotic-resistant strains:

A. Random mutation

B. Genetic drift

C. Natural selection

C28. Let's suppose the mutation rate for converting allele B into allele b is 10^{-4} per generation. The current allele frequencies are $B = 0.6$ and $b = 0.4$. How long will it take for the allele frequencies to equal each other, assuming that no genetic drift takes place?

Experimental Questions

E1. You will need to be familiar with the techniques described in Chapter 20 to answer this question. Genetic polymorphisms can be detected using a variety of cellular and molecular techniques. Which techniques would you use to detect genetic polymorphisms at the following levels?

A. DNA level

B. RNA level

C. Polypeptide level

E2. You will need to refer to question 2 in More Genetic TIPS to answer this question. The gene for coat color in rabbits can exist in four alleles designated C (full coat color), c^{ch} (chinchilla), c^h (Himalayan), and c (albino). In a population of rabbits in Hardy-Weinberg equilibrium, the allele frequencies are

$$C = 0.34$$
$$c^{ch} = 0.17$$
$$c^h = 0.44$$
$$c = 0.05$$

Assume that C is dominant to the other three alleles, c^{ch} is dominant to c^h and c, and c^h is dominant to c.

A. What is the genotype frequency of rabbits with an albino phenotype?

B. Among 1000 rabbits, how many would you expect to have a Himalayan coat color?

C. Among 1000 rabbits, how many would be heterozygotes with a chinchilla coat color?

E3. In a large herd of 5468 sheep, 76 animals have yellow fat, and the rest have white fat. Yellow fat is inherited as a recessive trait. This herd is assumed to be in Hardy-Weinberg equilibrium.

A. What are the frequencies of the white and yellow fat alleles in this population?

B. Approximately how many sheep with white fat are heterozygous carriers of the yellow allele?

E4. The human MN blood group is determined by two codominant alleles, M and N. The following data were obtained from five human populations:

Genotype frequencies (%)

Population	Place	MM	MN	NN
Inuit	East Greenland	83.5	15.6	0.9
Navajo	New Mexico	84.5	14.4	1.1
Finns	Karajala	45.7	43.1	11.2
Russians	Moscow	39.9	44.0	16.1
Aboriginal Australians	Queensland	2.4	30.4	67.2

Source: Spiess, Eliot B., *Genes in Populations.* John Wiley & Sons, 1989, 774.

A. Calculate the allele frequencies in these five populations.

B. Which populations appear to be in Hardy-Weinberg equilibrium?

C. Which populations do you think have experienced significant intermixing due to migration?

E5. You will need to refer to question 2 in More Genetic TIPS before answering this question. In an island population, the following data were obtained for the numbers of people with each of the four blood types:

Type O	721
Type A	932
Type B	235
Type AB	112

Is this population in Hardy-Weinberg equilibrium? Explain your answer.

E6. Resistance to the poison warfarin is a genetically determined trait in rats. Homozygotes carrying the resistance allele ($W^R W^R$) have a lower fitness because they suffer from vitamin K deficiency, but heterozygotes ($W^R W^S$) do not have this deficiency. However, the heterozygotes are still resistant to warfarin. In an area where warfarin is applied, a heterozygote has a survival advantage. Due to warfarin resistance, a heterozygote is also more fit than a homozygote ($W^S W^S$) that is sensitive to warfarin. If the relative fitness values for $W^R W^S$, $W^R W^R$, and $W^S W^S$ individuals are 1.0, 0.37, and 0.19, respectively, in areas where warfarin is applied, calculate the allele frequencies at equilibrium. How would this equilibrium be affected if the rats were no longer exposed to warfarin?

E7. A gene exists in two alleles, B and b. In a population in Hardy-Weinberg equilibrium, the allele frequencies are 0.7 and 0.3, respectively. A new predator infiltrates the region occupied by this population and the fitness values become 1.0 (for BB), 0.8 (for Bb), and 0.4 (for bb). According to the general selection model, what would be the genotype and allele frequencies after one generation of directional selection?

E8. What is the goal of a genome-wide selection scan (GWSS)? What does it mean when a GWSS is used to detect positive selection? How might the conclusions of a GWSS be affected by genetic hitchhiking?

E9. Describe, in as much experimental detail as possible, how you would test the hypothesis that the distribution of shell color among land snails is due to predation.

E10. In the Grants' study of the medium ground finch, do you think the pattern of natural selection was directional, stabilizing, disruptive, or balancing? Explain your answer. If the environment remained dry indefinitely (for many years), what do you think would be the long-term outcome?

E11. A recessive lethal allele has achieved a frequency of 0.22 due to genetic drift in a very small population. Based on natural selection, how would you expect the allele frequencies to change in the next three generations? (Note: Your calculation can assume that genetic drift is not altering allele frequencies in either direction.)

E12. Among a large population of 2 million gray mosquitoes, one mosquito is heterozygous for a body color gene; this mosquito has one gray allele and one blue allele. There is no selective advantage or disadvantage for either gray or blue body color. All the other mosquitoes carry two copies of the gray allele.

A. What is the probability of fixation of the blue allele?

B. If fixation does occur, how many generations is it likely to take?

C. Qualitatively, how would the answers to parts A and B be affected if the blue allele conferred a slight survival advantage?

E13. In a donor population, the allele frequencies for the common (Hb^A) and sickle cell (Hb^S) alleles are 0.9 and 0.1, respectively. A group of 550 individuals from this donor population migrates to join a recipient population containing 10,000 individuals; in the recipient population, the allele frequencies are $Hb^A = 0.99$ and $Hb^S = 0.01$.

A. Calculate the allele frequencies in the conglomerate population.

B. Assuming that the donor and recipient populations are each in Hardy-Weinberg equilibrium, calculate the genotype frequencies in the conglomerate population prior to any mating between the donor and recipient populations.

C. What will be the genotype frequencies of the conglomerate population in the next generation, assuming that it achieves Hardy-Weinberg equilibrium in one generation?

E14. Look back at question 5 in More Genetic TIPS before answering this question. Shown next are traditional DNA fingerprints of five people: a child, the female parent, and three potential male parents:

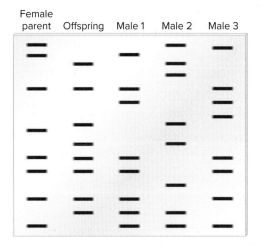

Female parent Offspring Male 1 Male 2 Male 3

Which males can be ruled out as being a parent? Explain your answer. If one of the males could be the child's parent, explain the general strategy for calculating the likelihood that the DNA fingerprint of the potential male parent could match the child's by chance alone.

E15. What is DNA fingerprinting? How can it be used in human identification?

E16. When analyzing the DNA fingerprints of a male named Jack and Jack's biological child, a technician examined 50 peaks and found that 30 of them were a perfect match. In other words, 30 out of 50 peaks, or 60%, were a perfect match. Is this percentage too high; that is, would you expect a value of only 50%? Explain why or why not.

E17. What would you expect to be the minimum percentage of matching peaks in DNA fingerprints for each of the following pairs of individuals?

A. Child and parent

B. A male and female sibling

C. Child and grandparent

Questions for Student Discussion/Collaboration

1. Discuss examples of positive and negative assortative mating in natural populations, human populations, and agriculturally important species.

2. Discuss the role of mutation in the origin of genetic polymorphisms. Suppose that a genetic polymorphism involves two alleles that have frequencies of 0.45 and 0.55. Describe three different scenarios to explain these observed allele frequencies. You can propose that the mutations that produced the polymorphism are neutral, beneficial, or deleterious.

3. Most new mutations are detrimental, yet rare beneficial mutations can be adaptive. With regard to the fate of new mutations, discuss whether you think it is more important for natural selection to select against detrimental alleles or to select in favor of beneficial ones. Which do you think is more significant in human populations?

Note: All answers are available for the instructor in Connect; the answers to the even-numbered questions and all of the Concept Check and Comprehension Questions are in Appendix B.

Domesticated wheat. *The color of the hulls of wheat seeds ranges from a dark red to white, making hull color an example of a quantitative trait.*
Robert Glusie/Photodisc/Getty Images

28

COMPLEX AND QUANTITATIVE TRAITS

In this chapter, we will examine **complex traits**—characteristics that are determined by several genes and are significantly influenced by environmental factors. Most of the complex traits that we will consider are also called **quantitative traits** because they can be described numerically. In humans, quantitative traits include height, the shape of the nose, and the rate of food metabolism, to name a few examples. The field of genetics that studies the mode of inheritance of complex and quantitative traits is called **quantitative genetics.**

Quantitative genetics is an important branch of genetics for several reasons. In agriculture, most of the key characteristics of interest to plant and animal breeders are quantitative traits. These include traits such as weight, fruit size, resistance to disease, and the ability to withstand harsh environmental conditions. As we will see, genetic techniques have improved our ability to develop strains of agriculturally important species with desirable quantitative traits. In addition, many human diseases, such as asthma and diabetes, are viewed as complex traits because they are influenced by several genes. Quantitative genetics is also important in the study of evolution. Many of the traits that allow a species to adapt to its environment are quantitative. Examples include the swift speed of the cheetah and the sturdiness of tree branches in windy climates. The importance of quantitative traits in the evolution of species will be discussed in Chapter 29.

In this chapter, we will examine how genes and the environment contribute to the phenotypic expression of quantitative traits.

We will begin with an overview of quantitative traits and the use of statistical methods to analyze them. We will then look at experimental ways to identify quantitative trait loci (QTLs)—locations on chromosomes containing genes that affect the outcome of quantitative traits. Advances in genetic mapping strategies have enabled researchers to identify these genes. Last, we will discuss heritability, which is a measure of the amount of phenotypic variation in a population due to genetic variation, and consider various ways of analyzing and modifying the genetic variation that affects phenotype.

28.1 OVERVIEW OF COMPLEX AND QUANTITATIVE TRAITS

Learning Outcomes:

1. List examples of complex and quantitative traits.
2. Explain how phenotypes for quantitative traits may be described with a frequency distribution.

When we compare characteristics among members of the same species, the differences may be complex or quantitative rather than qualitative. Humans, for example, have the same basic anatomical features (two eyes, two ears, and so on), but they differ in quantitative ways. People vary with regard to height, weight, the shape of facial features, skin and hair pigmentation, and many other

TABLE 28.1
Examples of Complex and Quantitative Traits

Type of Trait	Examples
Anatomical traits	Height, weight, number of bristles in *Drosophila*, ear length in corn, and degree of pigmentation in flowers and skin
Physiological traits	Metabolic traits, speed of running and flight, ability to withstand harsh temperatures, and milk production in mammals
Behavioral traits	Mating calls, courtship rituals, ability to learn a maze, and ability to grow or move toward light
Diseases	Predisposition toward heart disease, hypertension, cancer, diabetes, asthma, and arthritis

characteristics. As shown in **Table 28.1**, quantitative traits can be categorized as anatomical, physiological, or behavioral. In addition, many human diseases are viewed as complex traits because they are influenced by many genes and environmental factors. Three of the leading causes of death worldwide—heart disease, cancer, and diabetes—are considered complex traits.

Quantitative traits are described numerically. Most traits are either quantitative or have a quantitative aspect to them. Height and weight are quantitative traits that can be measured in centimeters (or inches) and kilograms (or pounds), respectively. The number of bristles on a fruit fly's body can be counted, and metabolic rate can be assessed as the amount of glucose burned per minute. Behavioral traits can also be quantified. A mating call can be evaluated with regard to its duration, sound level, and pattern. The ability to learn a maze can be described as the amount of time and/or the number of repetitions it takes to master the skill. Some traits, like hair color in humans, may seem discrete yet have a quantitative aspect to them. For example, even though one type of hair color can be described as brown, the amount of brown pigment may vary from light brown to dark brown.

The manner in which complex and quantitative traits are numerically described can vary.

- Most commonly quantitative traits, such as height and weight, are viewed as **continuous traits**—traits that do not fall into discrete categories.
- Some, such as bristle number in *Drosophila*, are **meristic traits**—traits that can be counted and expressed in whole numbers.
- Some diseases are referred to as **threshold traits**—traits that are inherited through the contributions of many genes. For diseases, such as diabetes, heart disease, and cancer, the alleles of several different genes contribute to the likelihood that an individual will develop the disease; a certain threshold must be reached in which the number of disease-causing alleles results in the development of the disease.

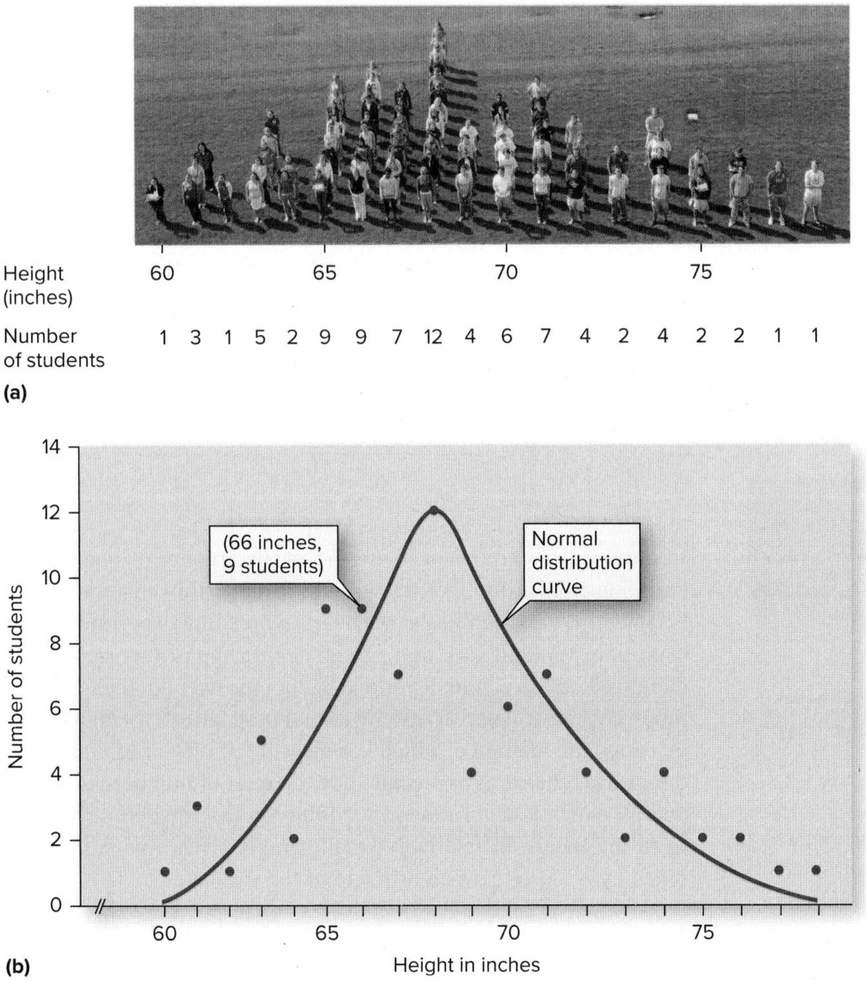

Many Quantitative Traits Exhibit a Continuum of Phenotypic Variation That Follows a Normal Distribution

As mentioned, most quantitative traits show a continuum of phenotypic variation within a group of individuals. For such traits, it may be impossible to place organisms into a discrete phenotypic category. For example, **Figure 28.1a** is a photograph showing the range of heights of 82 college students. Though height is found at minimum and maximum values, the range of heights between these values is fairly continuous.

How do geneticists describe traits that show a continuum of phenotypes? Because most quantitative traits do not naturally fall into a small number of discrete categories, an alternative way to describe them is to use a **frequency distribution.** To construct a frequency distribution, the trait is divided arbitrarily into a number of convenient, discrete phenotypic categories. For example, in Figure 28.1, the range of heights is partitioned into

FIGURE 28.1 Normal distribution of a quantitative trait. (a) The distribution of heights of 82 college students. **(b)** A frequency distribution for the heights of students shown in part (a).
David Hyde/Wayne Falda/McGraw Hill

CONCEPT CHECK: Is height a discontinuous (discrete) trait, or does it exhibit a continuum?

1-inch intervals. Then a graph is made that shows the number of individuals found in each of the categories.

Figure 28.1b shows a frequency distribution for the heights of the students pictured in Figure 28.1a. The height measurement is plotted along the x-axis, and the number of individuals who exhibit that phenotype is plotted on the y-axis. The values along the x-axis are divided into the discrete 1-inch intervals that define the phenotypic categories, even though height is essentially continuous within a group of individuals. For example, in Figure 28.1a, 9 students were between 65.5 and 66.5 inches in height, which is plotted as the point (66 inches, 9 students) on the graph in Figure 28.1b. This type of analysis can be conducted with any group of individuals who vary with regard to a quantitative trait.

The line in the frequency distribution depicts a **normal distribution,** a distribution for a large sample in which the trait of interest varies in a symmetrical way around an average value. The distribution of measurements of many biological characteristics is approximated by a symmetrical bell-shaped curve like that in Figure 28.1b. Normal distributions are common when the phenotype is determined by the cumulative effect of many small independent factors.

28.1 COMPREHENSION QUESTIONS

1. Which of the following is an example of a quantitative trait?
   a. Height
   b. Rate of glucose metabolism
   c. Ability to learn a maze
   d. All of the above are quantitative traits.
2. Saying that a quantitative trait exhibits a continuum means that
   a. the numerical value for the trait increases with the age of the individual.
   b. environmental effects are additive.
   c. the phenotypes for the trait are continuous and do not fall into discrete categories.
   d. the trait continuously changes during the life of an individual.

28.2 STATISTICAL METHODS FOR EVALUATING QUANTITATIVE TRAITS

Learning Outcome:

1. Calculate the mean, variance, standard deviation, and correlation coefficient for a quantitative trait, and explain the meanings of these statistics.

In the nineteenth century, Francis Galton and Karl Pearson showed that many traits in humans and domesticated animals are quantitative in nature. To understand the underlying genetic basis of these traits, they founded what became known as the **biometric field** of genetics, which involves the statistical study

of biological traits. During this period, Galton and Pearson developed various statistical tools for studying the variation of quantitative traits within groups of individuals. Many of these tools are still in use today. In this section, we will examine how statistical tools are used to analyze the variation of quantitative traits within groups.

Statistical Methods Are Used to Evaluate a Frequency Distribution Quantitatively

Statistical tools can be used to analyze a frequency distribution in a number of ways. One statistical measure you are probably familiar with is called the **mean,** which is the sum of all the values in the group divided by the number of individuals in the group. The mean is computed using the following formula:

$$\overline{X} = \frac{\Sigma X}{N}$$

where

$\overline{X}$ is the mean
ΣX is the sum of all the values in the group
N is the number of individuals in the group

A more generalized form of this equation can be used:

$$\overline{X} = \frac{\Sigma f_i X_i}{N}$$

where

$\overline{X}$ is the mean
$\Sigma f_i X_i$ is the sum of all the values in the group; each value (X_i) in the group is multiplied by its frequency (f_i) in the group
N is the number of individuals in the group

For example, let's suppose a group of corn ears have the following lengths (rounded to the nearest centimeter): 15, 14, 13, 14, 15, 16, 16, 17, 15, and 15. Then

$$\overline{X} = \frac{4(15) + 2(14) + 13 + 2(16) + 17}{10}$$

$$\overline{X} = 15\,cm$$

In genetics, we are often interested in the amount of phenotypic variation in a group. As we will see later in this chapter and in Chapter 29, variation lies at the heart of breeding experiments and evolution. Without variation, selective breeding is impossible, and natural selection cannot favor one phenotype over another. A common way to evaluate variation within a population is with a statistic called the **variance,** which is a measure of the variation around the mean. It helps us appreciate how far a set of numbers is spread out. Variance is determined by adding up the squared deviations from the mean. In relatively small sample sizes, this sum is divided by $N - 1$ to give an estimate of the variance.

$$V_X = \frac{\Sigma f_i (X_i - \overline{X})^2}{N - 1}$$

where

V_X	is the variance
$X_i - \overline{X}$	is the difference between each value and the mean
N	is the number of observations

For example, if we use the values given previously for the lengths of ears of corn, the variance in length is calculated as follows:

$$\Sigma f_i (X_i - \overline{X})^2 = 4(15 - 15)^2 + 2(14 - 15)^2$$
$$+ (13 - 15)^2 + 2(16 - 15)^2 + (17 - 15)^2$$

$$\Sigma f_i (X_i - \overline{X})^2 = 0 + 2 + 4 + 2 + 4$$

$$\Sigma f_i (X_i - \overline{X})^2 = 12 \text{ cm}^2$$

$$V_X = \frac{\Sigma f_i (X_i - \overline{X})^2}{N - 1}$$

$$V_X = \frac{12 \text{ cm}^2}{9}$$

$$V_X = 1.33 \text{ cm}^2$$

Although variance is a measure of the variation around the mean, this statistic may be difficult to understand intuitively because its value is computed from squared deviations. For example, weight can be measured in grams; the corresponding variance is measured in grams squared. Even so, variances are centrally important to the analysis of quantitative traits because they are additive under certain conditions. This means that the variances for different factors that contribute to a quantitative trait, such as genetic and environmental factors, can be added together to predict the total variance for that trait. Later, we will examine how this property is useful in predicting the outcomes of genetic crosses.

To gain a more intuitive grasp of variation, we can take the square root of the variance. This statistic is called the **standard deviation** (*SD*). Using the example of the lengths of the group of corn ears, the standard deviation is

$$SD = \sqrt{V_X} = \sqrt{1.33}$$
$$SD = 1.15 \text{ cm}$$

If the values in a population follow a normal distribution, it is easier to appreciate the amount of variation by considering the standard deviation. **Figure 28.2** illustrates the relationship between the standard deviation and the percentages of individuals that deviate from the mean. Approximately 68% of all individuals in a population have values within one standard deviation from the mean, in either the positive or the negative direction. About 95% are within two standard deviations, and 99.7% are within three standard deviations, in either the positive or negative direction. When a quantitative characteristic follows a normal distribution, less than 0.3% of the individuals have values that are more or less than three standard deviations from the mean of the population. In our ears of corn example, three standard deviations equals 3.45 cm. Therefore, we expect approximately 0.3% of the ears of corn have lengths less than 11.55 cm or greater than 18.45 cm, assuming that corn ear length follows a normal distribution.

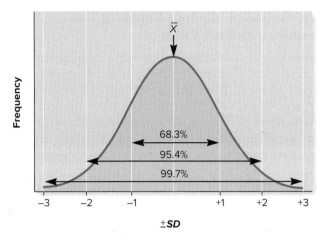

FIGURE 28.2   **The relationship between the standard deviation and the proportions of individuals in a normal distribution.** For example, approximately 68% of the individuals in a population are between the mean and one standard deviation (1 *SD*) above or below the mean.

CONCEPT CHECK: What percentage of individuals fall more than 2 *SD*s above the mean?

GENETIC TIPS   **THE QUESTION:** In a population of 100 male fruit flies, the mean abdomen length is 2.0 mm and the standard deviation for that trait is 0.3 mm. If you assume that abdomen length follows a normal distribution, what percentage of male flies will have an abdomen length equal to or greater than 2.6 mm?

T **OPIC:** *What topic in genetics does this question address?* The topic is the use of values of the mean and standard deviation to determine the percentage of individuals that deviate a certain amount from the mean for a population.

I **NFORMATION:** *What information do you know based on the question and your understanding of the topic?* From the question, you know the mean and standard deviation for abdomen length in a population of fruit flies. From your understanding of the topic, you may remember that a relationship exists between the standard deviation and the proportions of individuals in a normal distribution (see Figure 28.2).

P **ROBLEM-SOLVING** **S** **TRATEGY:** *Make a calculation.* To begin to solve this problem, you first need to consider how far an abdomen length of 2.6 mm deviates from the mean. The mean for the population is 2.0 mm, so this abdomen length deviates 0.6 mm from the mean. If you divide 0.6 mm by the standard deviation, which is 0.3 mm, you find that this abdomen length deviates 2 *SD*s from the mean.

According to Figure 28.2, about 95.4% of all male flies will have abdomen lengths that are within 2 *SD*s from the mean, which means that 4.6% fall outside of 2 *SD*s. Half of these will be 2 *SD*s or more below the mean, and half of them, or 2.3% of them, will be 2 *SD*s or more above it.

ANSWER: Only 2.3% of the male flies in this population will have an abdomen length equal to or greater than 2.6 mm.

Some Statistical Methods Compare Two Variables with Each Other

In many genetics problems, it is useful to compare two different variables. For example, we may wish to consider the occurrence of two different phenotypic traits. Do obese animals have larger hearts? Are blue eyes more likely to occur in people with blond hair? A second type of comparison is between traits and environmental factors. Does insecticide resistance occur more frequently in areas that have been exposed to insecticides? Is heavy body weight more prevalent in colder climates? Finally, a third type of comparison is between traits and genetic relationships. Do tall parents tend to produce tall offspring? Do females with diabetes tend to have male siblings with diabetes?

To gain insight into such questions, a statistic known as the correlation coefficient is often applied. To calculate this statistic, we first need to determine the **covariance,** which describes the degree of variation between two variables within a group of individuals. The covariance is similar to the variance, except that it is calculated by multiplying together the deviations of two different variables rather than squaring the deviations from a single factor.

$$CoV_{(X,Y)} = \frac{\Sigma f_i[(X_i - \overline{X})(Y_i - \overline{Y})]}{N-1}$$

where

$CoV_{(X,Y)}$	is the covariance between X and Y values
X_i	represents the values for one variable, and $\overline{X}$ is the mean value in the group
Y_i	represents the values for another variable, and $\overline{Y}$ is the mean value in the group
N	is the total number of pairs of observations

As an example, let's compare the weights of female cattle (i.e., cows) and those of their adult female offspring. A farmer might be interested in this relationship to determine if genetic variation plays a role in the weights of cattle. The following data give the weights at 5 years of age for 10 different cows and their female offspring.

Cow's Weight (kg)	Offspring's Weight (kg)	$X_i - \overline{X}$	$Y_i - \overline{Y}$	$(X_i - \overline{X})(Y_i - \overline{Y})$
570	568	−26	−30	780
572	560	−24	−38	912
599	642	3	44	132
602	580	6	−18	−108
631	586	35	−12	−420
603	642	7	44	308
599	632	3	34	102
625	580	29	−18	−522
584	605	−12	7	−84
575	585	−21	−13	273
$\overline{X} = 596$	$\overline{Y} = 598$			$\Sigma = 1373$
$SD_X = 21.1$	$SD_Y = 30.5$			

$$CoV_{(X,Y)} = \frac{\Sigma f_i[(X_i - \overline{X})(Y_i - \overline{Y})]}{N-1}$$

$$CoV_{(X,Y)} = \frac{1373}{10-1}$$

$$CoV_{(X,Y)} = 152.6$$

After we calculate the covariance, we can evaluate the strength of the association between the two variables by calculating a **correlation coefficient** (*r*). This value, which ranges between −1 and +1, indicates how two factors vary in relation to each other. The correlation coefficient is calculated using the following formula:

$$r_{(X,Y)} = \frac{CoV_{(X,Y)}}{SD_X \, SD_Y}$$

A positive *r* value means that the two factors tend to vary in the same way relative to each other; as one factor increases, the other increases with it. A value of zero indicates that the two factors do not vary in a consistent way relative to each other; the values of the two factors are not related. Finally, a negative correlation, in which the correlation coefficient is negative, indicates that the two factors tend to vary in opposite ways to each other; as one factor increases, the other decreases.

Let's use the data on 5-year weights for cows and their female offspring to calculate a correlation coefficient:

$$r_{(X,Y)} = \frac{152.6}{(21.1)(30.5)}$$

$$r_{(X,Y)} = 0.237$$

The result is a positive correlation between the 5-year weights of cows and offspring. In other words, the positive correlation coefficient suggests that heavy cows tend to have heavy offspring and lighter cows have lighter offspring.

How do we interpret the value of *r*? After a correlation coefficient has been calculated, we need to consider whether the *r* value represents a true association between the two variables or if it could simply be due to chance. To accomplish this, we can test the hypothesis that there is no real correlation (i.e., the null hypothesis, *r* = 0). The null hypothesis is that the observed *r* value differs from zero only as a result of random sampling error. We followed a similar approach in the chi square analysis described in Chapter 3.

Like the chi square value, the significance of the correlation coefficient is directly related to sample size and the degrees of freedom (*df*). In testing the significance of correlation coefficients, *df* equals $N - 2$, because two variables are involved; N equals the number of paired observations. We reject the null hypothesis if the correlation coefficient has a significance level that is less than 0.05 (less than 5%) or that is less than 0.01 (less than 1%). These are called the 5% and 1% significance levels, respectively. **Table 28.2** shows the relationship between the *r* values and degrees of freedom at the 5% and 1% significance levels.

Use of Table 28.2 is valid only if certain assumptions are met. First, the values of X and Y in the study must have been obtained by an unbiased sampling of the entire population. In addition, this approach assumes that the values of X and Y follow a

TABLE 28.2

Values of *r* at the 5% and 1% Significance Levels

Degrees of Freedom (*df*)*	5%	1%
1	0.997	1.000
2	0.950	0.990
3	0.878	0.959
5	0.754	0.874
10	0.576	0.708
15	0.482	0.606
20	0.423	0.537
25	0.381	0.487
30	0.349	0.449
40	0.304	0.393
50	0.273	0.354
60	0.250	0.325
80	0.217	0.283
100	0.195	0.254
150	0.159	0.208
200	0.138	0.181
300	0.113	0.148
400	0.098	0.128
500	0.088	0.115
1000	0.062	0.081

*Note: *df* equals *N* − 2.
Source: Spence, Janet T., Underwood, Benton J., *Elementary Statistics.* Prentice-Hall, 1976, 282.

normal distribution, like that in Figure 28.1, and that the relationship between *X* and *Y* is linear.

To illustrate the use of Table 28.2, let's consider the correlation coefficient we have just calculated for 5-year weights of cows and their female offspring. In this case, we obtained a value of 0.237 for *r*, and the value of *N* was 10. Under these conditions, *df* equals 10 (the number of paired observations) minus 2, which equals 8. To be valid at a 5% significance level, the value of *r* would have to be 0.632 or higher. Because the value we obtained is much less than this, it is fairly likely that this value could have occurred because of random sampling error. In this case, we cannot reject the null hypothesis, and, therefore, we cannot conclude the positive correlation is due to a true association between the weights of cows and their female offspring.

In an actual experiment, however, a researcher would examine many more pairs of cows and offspring, perhaps 500–1000. If a correlation coefficient of 0.237 was observed for *N* = 1000, the value would be significant at the 1% level. In this case, we would reject the null hypothesis that weights are not associated with each other. Instead, we would conclude that a real association occurs between the weights of cows and their offspring. In research studies, these kinds of experiments have been done for cattle weights,

and the correlation coefficients between cows and offspring have often been found to be significant.

If a statistically significant correlation coefficient is obtained, how do we interpret its meaning? An *r* value that is statistically significant suggests a true association, but it does not necessarily imply a cause-and-effect relationship. When parents and offspring display a significant correlation for a trait, we should not jump to the conclusion that genetics is the underlying cause of the positive association. In many cases, parents and offspring share similar environments, so the positive association might be rooted in environmental factors. In general, a correlation coefficient is quite useful in identifying a positive or negative association between two variables. We should use caution, however, because this statistic, by itself, cannot prove that the association is due to cause and effect.

28.2 COMPREHENSION QUESTIONS

1. The variance is
   a. a measure of the variation around the mean.
   b. computed as a squared deviation.
   c. higher when there is less phenotypic variation.
   d. Both a and b are correct.
2. Which of the following statistics is used to compare two variables?
   a. Mean
   b. Correlation coefficient
   c. Variance
   d. Standard deviation

28.3 POLYGENIC INHERITANCE

Learning Outcome:

1. Describe how polygenic inheritance may result in a continuum of phenotypes.

Thus far, we have seen that quantitative traits tend to show a continuum of variation and can be analyzed with various statistical tools. In this section, we begin to focus on the genetic basis of such traits. Quantitative traits are usually **polygenic,** which means they are controlled by multiple genes. The term **polygenic inheritance** refers to the transmission of any trait that is governed by two or more different genes. In this section, we will consider how the number of genes involved and environmental factors influence the continuum of variation that is exhibited by quantitative traits.

For Quantitative Traits That Are Polygenic, Each Gene May Contribute to the Trait in an Additive Way

The first experiment demonstrating polygenic inheritance was conducted by Herman Nilsson-Ehle in 1909 who studied the inheritance of red pigment in the hulls of bread wheat, *Triticum aestivum* (**Figure 28.3a**). When true-breeding plants with white hulls were crossed to true-breeding plants with red hulls, the F_1 generation had an intermediate color. When the F_1 generation was allowed to self-fertilize, a variation in redness was observed in the F_2 generation, including dark red, medium red, intermediate red, light red, and white. An undiscriminating observer might conclude that this F_2 generation displayed a continuous variation in hull color. However, as shown in **Figure 28.3b**, Nilsson-Ehle carefully categorized the colors of the hulls and discovered that they exhibited a 1:4:6:4:1 ratio. This ratio can be explained by making two conclusions:

- This species has two different genes that control hull color, with each gene existing in a red or white allele.
- The two genes contribute additively to the color of the hull. In this case, each red allele carried by a given plant contributed to the red color of the hull. Plants carrying more red alleles had hulls with a deeper red color.

Polygenic Inheritance and Environmental Factors May Produce a Continuum of Phenotypes

As we have just seen, Nilsson-Ehle categorized the genotypes for wheat hull color into five discrete phenotypic categories, ranging from dark red to white. However, for many quantitative traits, it is difficult or impossible to categorize individuals into discrete phenotypes. In general, as the number of genes controlling a trait increases and the influence of environment variation becomes greater, the categorization of genotypes into discrete categories of phenotypes becomes increasingly problematic. Therefore, a Punnett square cannot be used to analyze most quantitative traits. Instead, statistical methods, which are described in Section 28.2, must be employed.

Figure 28.4 illustrates how gene number and the environment affect the ability of genotypes to produce discrete categories of phenotypes for a quantitative trait, in this case, seed weight.

(a) Wheat with red hulls (top) or white hulls (bottom)

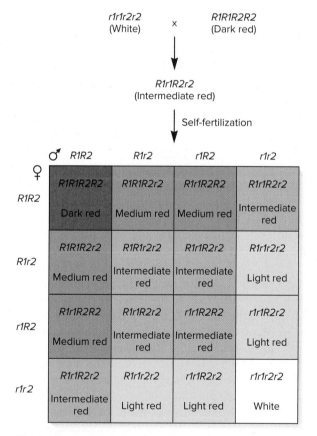

(b) Inheritance of hull color in wheat

FIGURE 28.3 **The Nilsson-Ehle experiment on the relationship of continuous variation to polygenic inheritance in wheat. (a)** Red (top) and white (bottom) varieties of wheat, *Triticum aestivum*. **(b)** Nilsson-Ehle carefully categorized the colors of the hulls in the F_2 generation and discovered that they exhibited a 1:4:6:4:1 ratio. This pattern occurs because the contributions of the red alleles are additive.

Genes→Traits In this example, two genes, each with two alleles (red and white), govern hull color. Offspring can display a range of colors, depending on how many copies of the red allele they inherit. If an offspring is homozygous for the red allele of both genes, it will have dark red hulls. By comparison, if it carries three red alleles and one white allele, it will be medium red (which is not quite as deep in color). Thus, this quantitative trait can exhibit a range of phenotypes from dark red to white.

(a): (top) Nigel Cattlin/Science Source; (bottom) irin-k/age fotostock

CONCEPT CHECK: When we say that alleles are additive, what does it mean?

Low environmental effect

High environmental effect

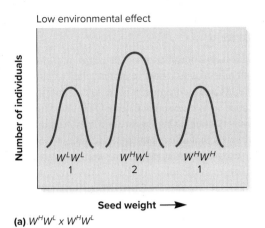

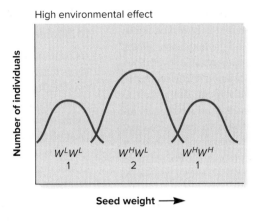

(a) $W^H W^L \times W^H W^L$

Low environmental effect

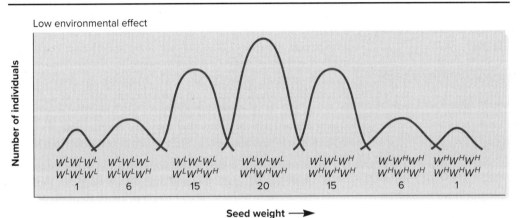

High environmental effect

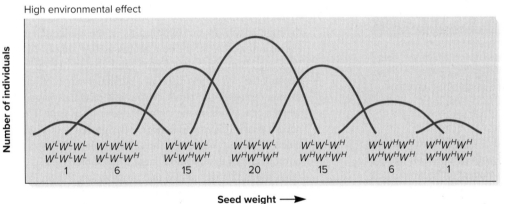

(b) $W^H W^L \; W^H W^L \; W^H W^L \times W^H W^L \; W^H W^L \; W^H W^L$

FIGURE 28.4 **How genotypes and phenotypes may overlap for quantitative traits.** **(a)** Situations in which seed weight is controlled by one gene, existing in light (W^L) and heavy (W^H) alleles. **(b)** Situations in which seed weight is governed by three genes instead of one, each existing in light and heavy alleles. (Note: The 1:2:1 and 1:6:15:20:15:6:1 ratios were derived by using a Punnett square and assuming a cross between individuals that are both heterozygous for one or three different genes, respectively.)

Genes→Traits The ability of geneticists to correlate genotype and phenotype depends on how many genes are involved and how much the environment causes the phenotype to vary. In part (a), a single gene influences seed weight. In the graph on the left side, environmental variation does not cause much variation in seed weight. No overlap in seed weight is observed for the $W^L W^L$, $W^H W^L$, and $W^H W^H$ genotypes. In the graph on the right side, environmental variation has a greater effect on seed weight. In this case, a few individuals with the $W^L W^L$ genotype produce seeds having the same weight as seeds of a few individuals with the $W^H W^L$ genotype; and a few individuals with the $W^H W^L$ genotype produce seeds having the same weight as seeds from individuals with the $W^H W^H$ genotype. As shown in part (b), it becomes even more difficult to distinguish genotype based on phenotype when three genes are involved. The overlaps are minor when environmental variation does not cause much seed weight variation. However, when environmental variation has a greater effect on phenotype, the overlaps between phenotypes due to different genotypes are very pronounced, and the trait appears to follow a continuum of variation.

CONCEPT CHECK: Explain how gene number and environmental variation affect the overlaps between phenotypes due to different genotypes.

Before we consider these graphs, let's discuss the possible effects of the environment:

- First, the environment may or may not have a large amount of variation. Some plants may be exposed to much more sunlight than others, or they may be planted in better soil or receive more rain.
- Second, such environmental variation may or may not have a great effect on seed weight. For example, in one species of plant, receiving low or high amounts of rain may have a great effect on seed weight, whereas in another species, such variation in rainfall may have a minimal effect.

One Gene Affecting Seed Weight. Figure 28.4a shows a situation in which seed weight is controlled by one gene with light (W^L) and heavy (W^H) alleles. A heterozygous plant ($W^H W^L$) is allowed to self-fertilize. When seed weight is only slightly influenced by environmental variation, as seen on the left, the $W^L W^L$, $W^H W^L$, and $W^H W^H$ genotypes result in well-defined phenotypic categories. When environmental variation has a greater effect on seed weight, as shown on the right, more phenotypic variation is found in seed weight for each genotype. The variation in the frequency distribution on the right is much higher. Even so, most genotypes can still be categorized into the three main classes.

Three Genes Affecting Seed Weight.   By comparison, Figure 28.4b illustrates a situation in which seed weight is governed not by one gene, but by three genes, each existing in light and heavy alleles. A cross between two heterozygotes is expected to produce seven genotypes in a 1:6:15:20:15:6:1 ratio. When the variation in environmental factors is low and/or plays a minor role in the outcome of this trait, as shown in the upper graph in Figure 28.4b, nearly all individuals fall within a phenotypic category that corresponds to their genotype. When the environment has a greater effect on phenotype, as shown in the lower graph, the situation becomes more ambiguous. For example, individuals with one W^L allele and five W^H alleles have a phenotype that overlaps with that of individuals having six W^H alleles or two W^L alleles and four W^H alleles. Therefore, it becomes difficult to categorize each genotype into a unique phenotypic group. Instead, the trait displays a continuum ranging from light to heavy seed weight.

28.3 COMPREHENSION QUESTION

1. For many quantitative traits, phenotypes due to different genotypes tend to overlap because
   a. the trait changes over time.
   b. the trait is polygenic.
   c. environmental variation affects the trait.
   d. both b and c are true.

28.4   IDENTIFICATION OF GENES THAT CONTROL QUANTITATIVE TRAITS

Learning Outcomes:

1. Define *quantitative trait locus.*
2. Explain how quantitative trait loci are mapped on chromosomes using molecular markers.

A goal of researchers working in the field of quantitative genetics is to identify the genes that are associated with complex and quantitative traits. This can be a challenging endeavor because such traits are usually polygenic. In the past few decades, the development of many molecular techniques has made it easier for researchers to identify these genes. These include the following:

- *Genome-wide association studies (GWAS).* In Chapter 24, we considered how a GWAS may be used to identify genes that contribute to human diseases (refer back to Figure 24.8). This same approach can be used to identify genes that contribute to quantitative traits.
- *QTL mapping.* A second approach, called QTL mapping, identifies regions within chromosomes that contain **quantitative trait loci** (singular: **locus**) (**QTLs**); a QTL is a site (i.e., a locus) that contains one or more genes that contribute to a quantitative trait. QTL mapping determines the locations of QTLs by identifying their linkage to molecular markers.

In this section, we will explore how QTL mapping is carried out.

QTL Mapping Relies on Linkage to Molecular Markers

As mentioned, the location on a chromosome that carries one or more genes that affect the outcome of a quantitative trait is called a quantitative trait locus (QTL). QTLs are chromosomal regions identified by genetic mapping. Because such mapping locates a QTL to a chromosomal region, a QTL may contain a single gene, or it may contain two or more closely linked genes that affect a quantitative trait.

To map genes, researchers may determine their locations by identifying their linkage to molecular markers. This approach is described in Chapter 24 (see Figures 24.5 and 24.6). Similarly, the basis of **QTL mapping** is the association between genetically determined phenotypes for quantitative traits and molecular markers such as restriction fragment length polymorphisms (RFLPs), microsatellites, and single-nucleotide polymorphisms (SNPs), which are described in Chapter 22 (refer back to Table 22.1). In this approach, a researcher identifies QTLs that are close to particular molecular markers whose locations on a chromosome are already known.

A general strategy for QTL mapping is presented in **Figure 28.5**, which focuses on two different strains of a diploid plant species with four chromosomes per set. The strains are highly inbred, which means that they are homozygous for most molecular markers and genes. They differ in two important ways:

- First, the two strains differ with regard to many molecular markers; the sites of many markers on each chromosome are already known. These markers are designated 1A and 1B, 2A and 2B, and so forth. The markers 1A and 1B mark the same chromosomal location in this species, namely, the upper tip of chromosome 1. However, the two markers are distinguishable in the two strains at the molecular level. For example, 1A might be a microsatellite that is 148 bp, whereas 1B might be a microsatellite that is 212 bp.
- Second, the two strains differ in a quantitative trait of interest. In this example, the strain on the left produces large fruit, whereas the strain on the right produces small fruit. The unknown genes affecting this trait are designated with a black or blue QTL label. A black QTL indicates a site that harbors one or more alleles that promote large fruit. A blue QTL is at the same site but carries alleles that promote small fruit. Prior to conducting their crosses, researchers would not know the chromosomal locations of the QTLs shown in this figure. The purpose of the experiment is to determine their locations.

With these ideas in mind, the QTL mapping strategy shown in Figure 28.5 begins by crossing the two inbred strains to each other and then backcrossing the F_1 offspring to both parental strains. This produces a second generation with a great degree of variation. The offspring from these backcrosses are then characterized in two ways. First, they are examined for their fruit size, and second, a cell sample from each individual is analyzed to determine which molecular markers are found on its chromosomes. The goal is to find an association between particular molecular markers and fruit size. Let's consider some examples:

- 2A is strongly associated with large size, whereas 2B is strongly associated with small size.

Note: The locations of QTLs are not known at the start of this experiment.

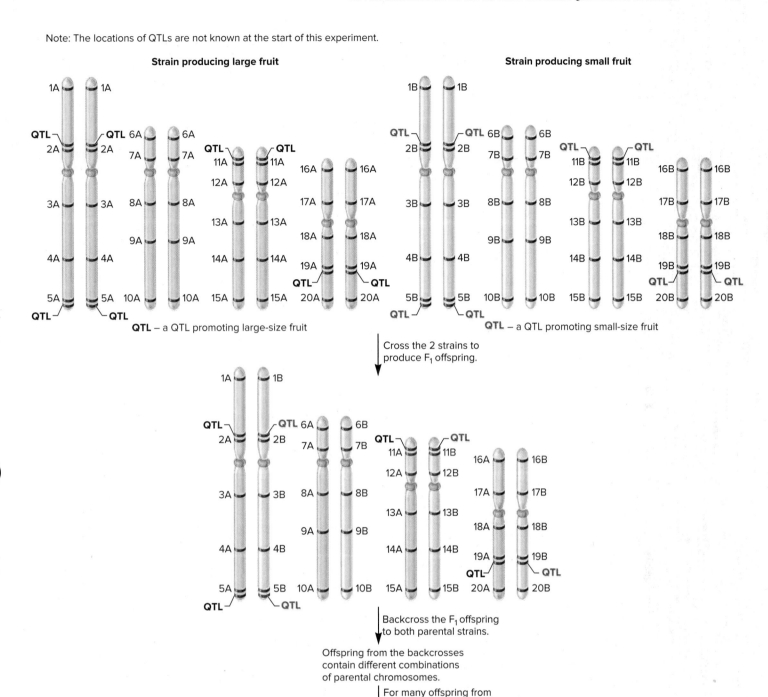

Cross the 2 strains to produce F₁ offspring.

Backcross the F₁ offspring to both parental strains.

Offspring from the backcrosses contain different combinations of parental chromosomes.

For many offspring from the backcrosses, determine fruit size and molecular marker composition.

2A, 5A, 11A, and 19A are strongly associated with large fruit size. This suggests that 4 QTLs lie close to these markers.

FIGURE 28.5 **A general strategy for QTL mapping using molecular markers.** Two different inbred strains have four chromosomes per set. The strain on the left produces large fruit, and the strain on the right produces small fruit. The goal of this mapping strategy is to locate the unknown genes affecting this trait, which are designated with a QTL label. A black QTL indicates a site carrying one or more alleles that promote large fruit, and a blue QTL is a site that carries alleles that promote small fruit. The two strains differ with regard to many molecular markers designated 1A and 1B, 2A and 2B, and so forth. The two strains are crossed, and then the F₁ offspring are backcrossed to both parental strains. Many offspring from the backcrosses are then examined to assess their fruit size and to determine which molecular markers are found on their chromosomes. The data are analyzed by computer programs that can statistically associate the phenotype (i.e., fruit size) with particular markers. Markers found throughout the genome of this species provide a way to locate several different genes that may affect the outcome of a single quantitative trait. In this case, the analysis predicts that four QTLs promoting heavier fruit weight are linked to regions of the chromosomes carrying the following markers: 2A, 5A, 11A, and 19A.

CONCEPT CHECK: What are the two ways that the two strains in this figure differ?

- 9A and 9B are not associated with large or small size, because a QTL affecting this trait is not found on this chromosome.
- 14A and 14B, which are fairly far away from a QTL, are not strongly associated with any particular QTL. Markers that are on the same chromosome but far away from a QTL are often separated from the QTL by crossing over that occurs during meiosis in the F_1 heterozygote. Only closely linked markers are strongly associated with a particular QTL.

Overall, QTL mapping involves the analysis of a large number of markers and offspring. The data are analyzed by computer programs that can statistically associate the phenotype (e.g., fruit size) with particular markers. Markers found throughout the genome of a species provide a way to identify the locations of several different genes that possess allelic differences that may affect the outcome of a quantitative trait.

In 1988, an investigation using QTL mapping by Andrew Paterson and colleagues examined the inheritance of quantitative traits in tomato plants. They studied a domesticated strain of tomato plant and a wild South American green-fruited variety. These two strains differed in their RFLPs, and they also exhibited dramatic differences in three agriculturally important characteristics: fruit mass, soluble solids content, and fruit pH.

The researchers crossed the two strains and then backcrossed the offspring to the domesticated tomato. The researchers then examined 237 plants with regard to 70 known RFLP markers. In addition, 5–20 tomatoes from each plant were analyzed for fruit mass, soluble solids content, and fruit pH. Using this approach, the researchers were able to map genes contributing much of the variation in these traits to particular sites on the tomato chromosomes. They identified six QTLs causing variation in fruit mass, four affecting soluble solids content, and five with effects on fruit pH.

More recently, the DNA sequence of the entire genome of many species has been determined. In such cases, the mapping of QTLs to a defined chromosomal region may allow researchers to analyze the DNA sequence in that region and to identify one or more genes that influence a quantitative trait of interest.

28.4 COMPREHENSION QUESTIONS

1. A QTL is a _____ where one or more genes affecting a quantitative trait are _____.
   a. site in a cell, located
   b. site on a chromosome, located
   c. site in a cell, expressed
   d. site on a chromosome, expressed
2. To map QTLs, researchers cross strains that differ with regard to
   a. a quantitative trait.
   b. molecular markers.
   c. a quantitative trait and molecular markers.
   d. a quantitative trait and a discontinuous trait.

28.5 HERITABILITY

Learning Outcomes:

1. Use an equation that describes the relationship among phenotypic variance, genetic variance, and environmental variance.
2. Describe how interactions and associations between genotype and environmental factors may affect phenotypic variance.
3. Define *heritability*, and distinguish between broad-sense heritability and narrow-sense heritability.
4. Calculate narrow-sense heritability using correlation coefficients.

As we have just seen, recently developed techniques in molecular mapping have enabled researchers to identify the genes that contribute to a quantitative trait. The other key factor that affects the phenotypic outcome of quantitative traits is the environment. All traits of organisms are influenced by both genetics and the environment, and this kind of interaction is particularly pertinent in the study of quantitative traits. Researchers want to understand how variation, both genetic and environmental, affects phenotypes. In this section, we will examine how geneticists analyze the genetic and environmental components that affect quantitative traits.

Both Genetic Variance and Environmental Variance May Contribute to Phenotypic Variance

Earlier, we examined the amount of phenotypic variation within a group by calculating the variance. Geneticists partition quantitative trait variation into components that are attributable to the following different causes:

Genetic variance (V_G)

Environmental variance (V_E)

Variance due to interactions between genetic and environmental factors ($V_{G \times E}$)

Variance due to associations between genetic and environmental factors ($V_{G \leftrightarrow E}$)

Let's begin by considering a simple situation in which V_G and V_E are the only factors that determine phenotypic variance, and the genetic and environmental factors are independent of each other. If so, then the total variance for a trait in a group of individuals is

$$V_P = V_G + V_E$$

where

V_P is the total phenotypic variance

V_G is the relative amount of variance due to genetic variation

V_E is the relative amount of variance due to environmental variation

Why is this equation useful? The partitioning of variance into genetic and environmental components allows us to estimate their relative importance in influencing the phenotypic variance within a group. If V_G is very high and V_E is very low, genetics plays the greater role in promoting phenotypic variation within a group. Alternatively, if V_G is low and V_E is high, environmental factors underlie much of the phenotypic variation. As described later in this chapter, a livestock breeder might want to apply selective breeding if V_G for an important quantitative trait is high. In this way, the characteristics of the herd may be improved. Alternatively, if V_G is negligible, it would make more sense to investigate and manipulate the environmental causes of phenotypic variation.

With experimental and domesticated species, one possible way to determine V_G and V_E is by comparing the variation in traits between genetically identical and genetically disparate groups. For example, researchers have used the practice of **inbreeding** to develop genetically homogeneous strains of mice. Inbreeding in mice involves many generations of matings between male and female siblings, which eventually produces strains that are **monomorphic**—carry the same allele—for all of their genes. Within such an inbred strain of mice, V_G equals zero. Therefore, all phenotypic variance is due to V_E. When studying quantitative traits such as weight, an experimenter might want to know the genetic and environmental variance for a different, genetically heterogeneous group of mice. To do so, genetically homogeneous and genetically heterogeneous mice could be raised under the same environmental conditions and their weights measured (in grams, g). The phenotypic variance for weight could then be calculated as described earlier. Let's suppose we obtained the following results:

$V_P = 15$ g^2 for the group of genetically homogeneous mice

$V_P = 22$ g^2 for the group of genetically heterogeneous mice

In the case of the homogeneous mice, $V_P = V_E$, because V_G equals 0. Therefore, V_E equals 15 g^2. To estimate V_G for the heterogeneous group of mice, we could assume that V_E is the same for them as it is for the homogeneous mice, because the two groups were raised in the same type of environment. This assumption allows us to calculate the genetic variance in weight for the heterogeneous, mice:

$$V_p = V_G + V_E$$
$$22 \text{ g}^2 = V_G + 15 \text{ g}^2$$
$$V_G = 7 \text{ g}^2$$

This result tells us that some of the phenotypic variance in the genetically heterogeneous group is due to the environment (namely, 15 g^2), and some (7 g^2) is due to genetic variation in alleles that affect weight.

Phenotypic Variance May Also Be Influenced by Interactions and Associations Between Genotype and the Environment

Thus far, we have considered the simple situation in which genetic variance and environmental variance are independent of each other and affect the phenotypic variance in an additive way. As another example, let's suppose that three genotypes, $U^N U^N$, $U^N U^D$, and $U^D U^D$, affect height, producing tall, intermediate, and short plants, respectively. Greater sunlight makes the plants grow taller regardless of their genotypes. In this case, our assumption that $V_P = V_G + V_E$ is reasonably valid.

Genotype-Environment Interaction. For the $U^N U^N$, $U^N U^D$, and $U^D U^D$ plants, let's consider a different environmental factor,

namely, minerals in the soil. As a hypothetical example, let's suppose the U^N allele codes a fully functional (normal) protein involved with mineral uptake from the soil; the designation U^N, refers to uptake that is normal. However, the U^D allele carries a mutation that causes a decrease in the affinity of this protein for certain minerals; the designation U^D, refers to uptake that is defective. When grown in standard soil, the $U^N U^D$ and $U^D U^D$ plants are shorter because they do not take up enough minerals to support maximal growth, whereas the $U^N U^N$ plants are not limited by mineral uptake.

According to this hypothetical scenario, adding minerals to the soil enables the defective protein to transport more minerals into the roots, thereby enhancing the growth rate of $U^D U^D$ plants by a large amount and the $U^N U^D$ plants by a smaller amount, because the minerals are more easily taken up by the plants (**Figure 28.6**). The height of $U^N U^N$ plants is not affected by mineral supplementation. When the environmental effects on phenotype differ according to genotype, this phenomenon is called a **genotype-environment interaction.** Variation due to interactions between genetic and environmental factors is termed $V_{G \times E}$, as noted earlier.

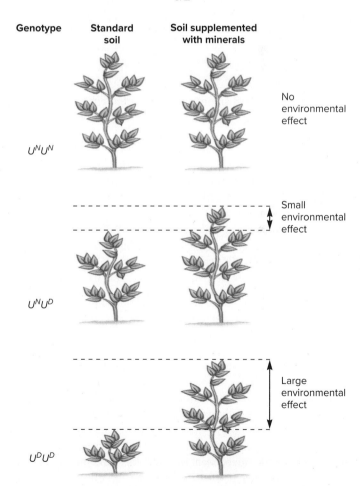

FIGURE 28.6 A schematic illustration of a genotype-environment interaction. When grown in standard soil, plants with the three genotypes $U^N U^N$, $U^N U^D$, and $U^D U^D$ show tall, intermediate, and short heights, respectively. When the soil is supplemented with minerals, a large effect on height is seen in the plants with the $U^D U^D$ genotype and a smaller effect in those with the $U^N U^D$ genotype. The plants with the $U^N U^N$ genotype are unaffected by the environmental change.

TABLE 28.3

Longevity of Two Strains of *Drosophila melanogaster**

	Strain *A*		Strain *B*	
Temperature	Male	Female	Male	Female
Standard	33.6	39.5	37.5	28.9
High	36.3	33.9	23.2	28.6
Low	77.5	48.3	45.8	77.0

*Longevity was measured in the mean number of days of survival. Strains *A* and *B* were inbred strains of *D. melanogaster* called Oregon and 2b, respectively. The standard-, high-, and low-temperature conditions were 25°C, 29°C, and 14°C, respectively.

Interactions between genetic and environmental factors are common. As an example, **Table 28.3** shows results from a study conducted in 2000 by Cristina Vieira, Trudy Mackay, and colleagues in which they investigated the genotype-environment interaction for QTLs affecting life spans in *Drosophila melanogaster*. The data in the table compare the life spans in days of male and female flies from two different strains of *D. melanogaster* raised at different temperatures. Because males and females differ in their sex chromosomes and gene expression patterns, they can be viewed as having different genotypes. The effects of environmental changes depended greatly on the strain and the sex of the flies.

- Under standard culture conditions, the females of strain *A* had the longest life span, whereas females of strain *B* had the shortest.
- In strain *A*, high temperature increased the longevity of males and decreased the longevity of females. In contrast, under hotter conditions, the longevity of males of strain *B* was dramatically reduced, whereas females of this same strain were not significantly affected.
- Lower growth temperature also had different effects in these two strains. Although low temperature increased the longevity of both strains, the effects were most dramatic in the males of strain *A* and the females of strain *B*.

Taken together, these results illustrate the potential complexity of the effects of genotype-environment interaction on a quantitative trait such as life span.

Genotype-Environment Association.
Another complication confronting geneticists is that genotypes may not be randomly distributed in all possible environments. When certain genotypes are preferentially found in particular environments, this phenomenon is called a **genotype-environment association** ($V_{G \leftrightarrow E}$). When such an association occurs, the effects of genotype and environment are not independent of each other, and the association needs to be considered when determining the effects of genetic and environmental variance on the total phenotypic variance.

Genotype-environment associations are very common in human genetics, because members of large families tend to have more similar environments than do members of the population as a whole. One way to evaluate this type of effect is to compare individuals who have different genetic relationships, such as identical versus fraternal twins. We will examine this approach later in this section. Another strategy that geneticists might follow is to analyze siblings

that have been adopted by different parents at birth. Their environmental conditions tend to be more disparate, and this may help to minimize the effects of genotype-environment associations.

Heritability Is the Relative Amount of Phenotypic Variance That Is Due to Genetic Variance

Another way to view variance is to focus our attention on the genetic contribution to phenotypic variance. The term **heritability** refers to the amount of phenotypic variance due to genetic variance within a specific group of individuals raised in a particular environment. Both genes and the environment are essential to produce the traits of an organism. However, variation of a trait in a population may be due entirely to environmental variation, entirely to genetic variation, or, more commonly, to a combination of the two.

If all of the phenotypic variance in a group is due to genetic variance, the heritability will have a value of 1. If all of the phenotypic variation is due to environmental effects, the heritability will equal 0. For most groups of organisms, the heritability for a given trait lies between these two extremes. For example, both genes and diet affect the size an individual will attain. In a given population, some individuals inherit alleles that tend to make them larger, and a proper diet also promotes larger size. Other individuals inherit alleles that tend to make them smaller, and an inadequate diet may contribute to small size. Taken together, both genetics and the environment affect the amount of phenotypic variation for a trait such as size.

If we assume that environment and genetics are independent and are the only two factors affecting phenotype, then

$$h_B^2 = V_G/V_P$$

h_B^2 is the heritability in the broad sense

V_G is the variance due to genetics

V_P is the total phenotypic variance, which equals $V_G + V_E$

The heritability defined here, h_B^2, called **broad-sense heritability**, takes into account different types of genetic variation that may affect the phenotype. As we have seen throughout this text, genes can affect phenotypes in various ways.

- The Nilsson-Ehle experiment described earlier in this chapter showed that the alleles determining hull color in wheat affect the phenotype in an additive way. A heterozygote exhibits a phenotype that is intermediate between the respective homozygotes.
- Alternatively, alleles affecting other traits may have a dominant/recessive relationship. In this case, the alleles are not strictly additive, because the heterozygote has a phenotype closer to, or perhaps the same as, the homozygote containing two copies of the dominant allele. For example, Mendel discovered that both *PP* and *Pp* pea plants have purple flowers.
- In addition, another complicating factor is epistasis (discussed in Chapter 4), in which the alleles for one gene can mask the phenotypic expression of the alleles of another gene.

To account for these differences, geneticists usually subdivide V_G into three different genetic categories:

$$V_G = V_A + V_D + V_I$$

where

V_A is the variance due to the additive effects of alleles

V_D is the variance due to the effects of alleles that follow a dominant/recessive pattern of inheritance

V_I is the variance due to the effects of alleles that interact in an epistatic manner

When analyzing quantitative traits, geneticists may focus on V_A and ignore the contributions of V_D and V_I. They do this for scientific as well as practical reasons. For some quantitative traits, the additive effects of alleles may play a primary role in the phenotypic outcome. In addition, when the alleles behave additively, we can predict the outcomes of crosses based on the quantitative characteristics of the parents. The heritability of a trait due to the additive effects of alleles is called **narrow-sense heritability:**

$$h_N{}^2 = V_A/V_P$$

For many quantitative traits, the value of V_A may be relatively large compared with the values of V_D and V_I. In such cases, the determination of the narrow-sense heritability provides an estimate of the broad-sense heritability.

How can the narrow-sense heritability be determined? In this chapter, we will consider two common ways. As discussed later in Section 28.6, one way to calculate the narrow-sense heritability involves selective breeding practices, as is done with agricultural species. A second common strategy for determining narrow-sense heritability involves measurement of a quantitative trait among groups of genetically related individuals. For example, agriculturally important traits, such as egg weight in poultry, can be analyzed in this way. To calculate the heritability, a researcher determines the observed egg weights between individuals whose genetic relationships are known, such as female parents and their female offspring. These data can then be used to compute a correlation coefficient between the phenotypes of parent and offspring, using the methods described earlier. The narrow-sense heritability is then calculated using this formula:

$$h_N{}^2 = r_{obs}/r_{exp}$$

where

r_{obs} is the observed phenotypic correlation coefficient between related individuals

r_{exp} is the expected correlation coefficient based on the known genetic relationship

In our example, r_{obs} is the observed phenotypic correlation coefficient between parent and offspring. In one research study, the observed phenotypic correlation coefficient for egg weights between female parents and offspring was found to be about 0.3 (although this varies among chicken breeds). The expected correlation coefficient, r_{exp}, is based on the known genetic relationship. A parent and offspring share 50% of their genetic material, so r_{exp} equals 0.50, which gives

$$h_N{}^2 = r_{obs}/r_{exp}$$
$$= 0.3/0.50$$
$$= 0.60$$

(Note: For siblings, $r_{exp} = 0.50$; for identical twins, $r_{exp} = 1.0$; and for a relationship between an aunt or uncle and a niece or nephew, $r_{exp} = 0.25$.) According to this calculation, about 60% of the phenotypic variance in egg weight is due to additive genetic variance; the other 40% is due to the environment. This calculation assumes that V_D and V_I are negligible.

When calculating heritabilities from correlation coefficients, keep in mind that such a computation also assumes that genetics and the environment are independent variables. However, this is not always the case. The environments of parents and offspring are often more similar to each other than are the environments of unrelated individuals. As mentioned earlier, there are several ways to minimize this confounding factor. First, in human studies, researchers may analyze the heritabilities from correlation coefficients between adopted children and their biological parents. Alternatively, they can examine a variety of relationships (aunt and niece, identical twins versus fraternal twins, and so on) and see if the heritability values are roughly the same in all cases. This approach was applied in the study that is described next.

EXPERIMENT 28A

The Heritability of Dermal Ridge Count in Human Fingerprints Is Very High

Fingerprints are inherited as a quantitative trait. It has long been known that identical twins have fingerprints that are very similar, whereas fraternal twins show considerably less similarity. Galton was the first researcher to study fingerprint patterns, but this trait became more amenable to genetic studies in the 1920s, when Kristine Bonnevie, a Norwegian geneticist, developed a method for counting the number of ridges within a human fingerprint.

As shown in **Figure 28.7**, human fingerprints can be categorized as having an arch, a loop, or a whorl, or a combination of these patterns. The primary difference among these patterns is the number of triple junctions, each known as a triradius (Figure 28.7b and c). At a triradius, a ridge emanates in three different directions. An arch has zero triradii, a loop has one, and a whorl has two. In Bonnevie's method of counting, a line is drawn from a triradius to the center of the fingerprint. The ridges that touch this line are then counted. (Note: The triradius ridge itself is not counted, and the last ridge is not counted if it forms the center of the fingerprint.) With this method, one can obtain a ridge count for all 10 fingers. Bonnevie conducted a study on a small population and found that correlation coefficients for ridge counts were relatively high in genetically related individuals.

Sarah Holt, who was also interested in the inheritance of this quantitative trait, carried out a more extensive analysis of ridge counts by examining the fingerprint patterns of a large group of people and their close relatives. In the experiment shown in **Figure 28.8**, the ridge counts for pairs of related individuals were determined by the method described in Figure 28.7. The

(a) Arch (no triradius)

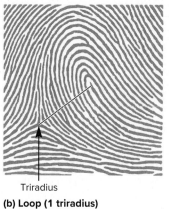

Triradius

(b) Loop (1 triradius)

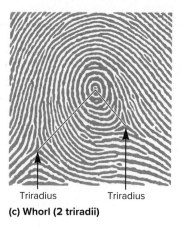

Triradius   Triradius

(c) Whorl (2 triradii)

FIGURE 28.7 Human finger-prints and the ridge count method of Bonnevie. **(a)** This print has an arch rather than a triradius. The ridge count is zero. **(b)** This print has one triradius. A straight line is drawn from the triradius to the center of the print. The number of ridges touching this straight line is 13. **(c)** This print has two triradii. Straight lines are drawn from both triradii to the center. There are 16 ridges touching the left line and 7 touching the right line, giving a total ridge count of 23.

correlation coefficients for ridge counts were then calculated for pairs of related or unrelated individuals. To estimate the narrow-sense heritability, the observed correlation coefficients were then divided by the expected correlation coefficients based on the known genetic relationships.

THE HYPOTHESIS

Dermal ridge count has a genetic component. The goal of this experiment was to determine the contribution of genetics to the variation in dermal ridge counts.

TESTING THE HYPOTHESIS   FIGURE 28.8 Heritability of human fingerprint patterns.

Starting material: A group of human subjects from Great Britain.

	Experimental level	Conceptual level
1. Take a person's finger and blot it onto an ink pad.		
2. Roll the person's finger onto a recording surface to obtain a print.	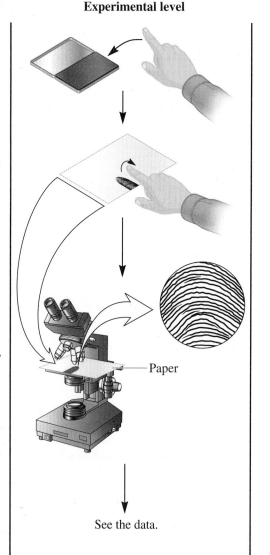	This is a method to measure a quantitative trait.
3. With a low-power binocular microscope, count the number of ridges, using the method described in Figure 28.7.	Paper	
4. Calculate the correlation coefficients between different pairs of individuals as described earlier in this chapter.	See the data.	The correlation coefficient provides a way to determine the heritability for the quantitative trait.

THE DATA

Type of Relationship	Number of Pairs Examined	Correlation Coefficient (r_{obs})	Heritability (r_{obs}/r_{exp})
Parent-child	810	0.48 ± 0.04*	0.96
Parent-parent	200	0.05 ± 0.07	–[†]
Sibling-sibling	642	0.50 ± 0.04	1.00
Identical twins	80	0.95 ± 0.01	0.95
Fraternal twins	92	0.49 ± 0.08	0.98
		Average heritability =	0.97

*The value following $\pm$ is the standard error of the mean.
[†]We cannot calculate a heritability value in this case because the value for r_{exp} is not known. Nevertheless, the value for r_{obs} is very low, suggesting that there is a negligible correlation between unrelated individuals.
S. B. Holt (1961). Quantitative genetics of fingerprint patterns. *Br Med Bull 17*, 247–250.

INTERPRETING THE DATA

As seen in the data, the results indicate that genetics plays the major role in explaining the variation in this trait. Genetically unrelated individuals (namely, those in parent-parent relationships) have a negligible correlation for this trait. By comparison, individuals who are genetically related have a substantially higher correlation coefficient. When the observed correlation coefficient is divided by the expected correlation coefficient based on the known genetic relationships, the average heritability value is 0.97, which is very close to 1.0.

What do these high heritability values mean? They indicate that nearly all of the phenotypic variance in fingerprint pattern is due to genetic variance. Significantly, fraternal and identical twins have substantially different observed correlation coefficients, even though we expect they have been raised in very similar environments. These results support the idea that genetics is playing the major role in promoting variation and the results are not biased heavily by environmental similarities that may be associated with genetically related individuals. From an experimental viewpoint, the results show us how the determination of correlation coefficients between related individuals can provide insight into the relative contributions of genetics and environment to the variation of a quantitative trait.

28.5 COMPREHENSION QUESTIONS

1. In a population of squirrels in North Carolina, the heritability for body weight is high. This means that
   a. body weight is primarily controlled by genes.
   b. the environment has little influence on body weight.
   c. the variance in body weight is mostly due to genetic variation.
   d. Both a and b are correct.
2. If two or more different genotypes are not affected by environmental variation in the same way, this outcome is due to
   a. a genotype-environment association.
   b. a genotype-environment interaction.
   c. the additive effects of alleles.
   d. both a and b.
3. One way to estimate narrow-sense heritability for a given trait is to compare _____ for _____.
   a. variances, related pairs of individuals
   b. correlation coefficients, related pairs of individuals
   c. variances, unrelated pairs of individuals
   d. correlation coefficients, unrelated pairs of individuals

28.6 SELECTIVE BREEDING

Learning Outcomes:
1. Describe the effects of selective breeding.
2. Calculate heritability from the results of a selective breeding experiment.
3. Explain how dominance and heterozygote advantage may contribute to the beneficial characteristics of hybrids.

The term **selective breeding** refers to programs and procedures designed to modify phenotypes in economically important species of plants and animals. This approach, also called **artificial selection,** is related to natural selection, discussed in Chapter 27. In forming the theory of natural selection, Charles Darwin was influenced by observations of selective breeding by pigeon fanciers and other breeders. The primary difference between artificial and natural selection is how the parents are chosen. Natural selection is due to natural variation in reproductive success. In artificial selection, the breeder chooses individuals that possess traits that are desirable from a human perspective. In this section, we will examine the effects of selective breeding and consider its relationship to heritability.

Selective Breeding of Species Can Alter Quantitative Traits Dramatically

For centuries, humans have been practicing selective breeding to obtain domesticated species with interesting or agriculturally useful characteristics. A good example is the common house pet, the dog. All domesticated dogs are derived from the gray wolf (*Canis lupus*). The various breeds of dogs have been obtained by selective breeding strategies that typically focus on morphological traits (size, fur color, etc.) and behavioral traits (ability to hunt, friendliness toward humans, etc.). As shown in **Figure 28.9**, the modification of quantitative traits in a species by selective breeding can have striking results. When comparing a greyhound with a bulldog, the magnitude of the differences is amazing. They hardly look like members of the same species.

A QTL study in 2007 by Nathan Sutter and colleagues indicated that the size of dogs is determined, in part, by alleles of the *Igf1* gene, which codes a growth hormone called insulin-like growth factor 1 (Igf1). A particular allele of this gene was found to be common to all small breeds of dogs and nearly absent from

Greyhound

German shepherd

Bulldog

Cocker spaniel

FIGURE 28.9 **Some common breeds of dogs that have been obtained by selective breeding.**

Genes→Traits By selecting parents carrying the alleles that have a desired effect on certain quantitative traits, dog breeders have produced breeds with distinctive sets of traits. For example, the bulldog has alleles that produce short legs and a flat face. By comparison, the corresponding genes in a German shepherd have alleles that produce longer legs and a more pointy snout. All of the dogs shown in this figure carry the same kinds of genes (e.g., many genes that affect their sizes, shapes, and fur color). However, the alleles for many of these genes are different among these dogs, thereby producing breeds with strikingly different phenotypes.

(greyhound): vizland/123RF; (German shepherd): "Pavel Shlykov/Shutterstock"; (bulldog): WilleeCole/iStockphoto/Getty Images; (cocker spaniel): "Labrador Photo Video/Shutterstock"

CONCEPT CHECK: What are the similarities and differences between natural selection and selective breeding?

very large breeds, suggesting that this allele is one of several genes that influences body size in small breeds of dogs.

Most of the food we eat is obtained from species that have been modified profoundly by selective breeding strategies. These food products include grains, fruits, vegetables, meat, milk, and juices. **Figure 28.10** illustrates how certain characteristics in the wild mustard plant (*Brassica oleracea*) have been modified by selective breeding to create several varieties of important domesticated crops. The wild plant is native to Europe and Asia, and plant breeders began to modify its traits approximately 4000 years ago. As the figure shows, certain quantitative traits in the domesticated strains, such as stems and lateral buds, differ considerably from those of the original wild species.

The phenomenon that underlies selective breeding is genetic variation. Within a group of individuals, allelic variation may affect the outcome of quantitative traits. The fundamental strategy of the selective breeder is to choose parents that will pass on to their offspring alleles that produce desirable phenotypic characteristics. For example, if a breeder wants large cattle, the largest members of the herd are chosen as parents for the next generation. These large cattle will transmit an array of alleles to their offspring that confer large size. The breeder often chooses genetically related individuals (e.g., siblings) as the parental stock. As mentioned previously, reproduction between genetically related individuals is known as inbreeding. Some of the consequences of inbreeding are also described in Chapter 27.

What is the outcome when selective breeding is conducted to modify a quantitative trait? **Figure 28.11** shows the results of a program begun at the Illinois Agricultural Experiment Station in 1896, before the rediscovery of Mendel's laws. This experiment began with 163 ears of corn with an oil content ranging from 4% to 6%. In each of 80 succeeding generations, corn plants were divided into two separate groups. In one group, several members with the highest oil content were chosen as parents of the next generation. In the other group, several members with the lowest oil content were chosen. After 77 generations, the oil content in the first group rose to over 18%; in the other group, it dropped to less than 1%. These results show that selective breeding can modify quantitative traits in a very directed manner.

When comparing the curves in Figure 28.11, keep in mind that quantitative traits are often at an intermediate value in unselected populations. Therefore, artificial selection can increase or decrease the magnitude of the trait. In this case, oil content can go up or down. Artificial selection tends to be the most rapid and effective in changing the frequency of alleles that are within an intermediate range in a starting population, such as 0.2 to 0.8.

Figure 28.11 also shows the phenomenon known as a **selection limit**—after many generations a plateau is reached where artificial selection is no longer effective. A selection limit may occur for two reasons:

- Presumably, the starting population possessed a large amount of genetic variation, which contributed to the

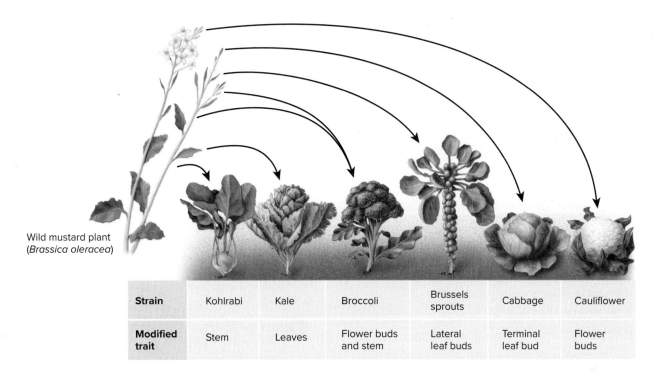

FIGURE 28.10   Crop plants developed by selective breeding of the wild mustard plant (*Brassica oleracea*).

Genes→Traits   The wild mustard plant carries a large amount of genetic (i.e., allelic) variation, which was used by plant breeders to produce modern strains of plants that are agriculturally desirable and economically important. For example, by selecting for alleles that promote the formation of large lateral leaf buds, the strain Brussels sprouts was created. By selecting for alleles that alter leaf morphology, kale was developed. Although these six agricultural plants look quite different from each other, they carry many of the same alleles as the wild mustard. However, they differ in alleles affecting the formation of stems, leaves, flower buds, and leaf buds.

CONCEPT CHECK: Identify the types of traits that have been subjected to selective breeding to develop these crop plants.

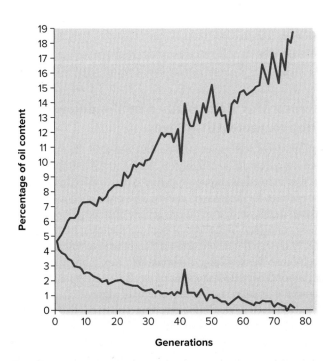

FIGURE 28.11   Results of selective breeding for a high and low oil content, a quantitative trait in corn.

CONCEPT CHECK: What are two reasons why a selection limit is reached at which artificial selection no longer has an effect?

diversity in phenotypes. Because the parents were carefully chosen, each succeeding generation has a higher proportion of the desirable alleles. However, after many generations, the population may be nearly monomorphic for all or most of the desirable alleles that affect the trait of interest. At this point, additional selective breeding will have no effect. When this occurs, the heritability for the trait is near zero, because nearly all genetic variation for the trait of interest has been eliminated from the population. Without the introduction of new mutations into the population, further selection is not possible.

- A second reason for a selection limit is related to fitness. Some alleles that accumulate in a population due to artificial selection may have a negative influence on the population's mean fitness. A selection limit is reached in which the desired effects of artificial selection are balanced by the negative effects on fitness.

Selective Breeding Provides a Way of Estimating Heritability

In artificial selection experiments, the response to selection is a common way to estimate the narrow-sense heritability in a starting population. The narrow-sense heritability measured in this way is also called the **realized heritability.** It is calculated using this formula:

$$h_N^2 = \frac{R}{S}$$

where

R   is the response in the offspring to selection. It is the difference between the mean of the offspring and the mean of the starting population.

S   is the selection differential in the parents. It is the difference between the mean of the parents and the mean of the starting population.

Here,

$$R = \overline{X}_o - \overline{X}$$
$$S = \overline{X}_p - \overline{X}$$

where

$\overline{X}$ is the mean of the starting population
$\overline{X}_o$ is the mean of the offspring
$\overline{X}_p$ is the mean of the parents

So,

$$h_N^2 = \frac{\overline{X}_o - \overline{X}}{\overline{X}_p - \overline{X}}$$

The narrow-sense heritability is the proportion of the variance in phenotype that can be used to predict changes in the population mean when selection is practiced.

As an example, we can consider the trait of bristles in fruit flies, which are hairlike structures that protrude from the body. Let's suppose we began with a population of fruit flies in which the average bristle number was 37.5. The parents chosen from this population had an average bristle number of 40. The offspring of the next generation had an average bristle number of 38.7. With these values, the realized heritability is

$$h_N^2 = \frac{38.7 - 37.5}{40 - 37.5}$$

$$h_N^2 = \frac{1.2}{2.5}$$

$$h_N^2 = 0.48$$

This result tells us that about 48% of the phenotypic variance is due to the additive effects of the alleles that affect bristle number.

As we have just seen, selective breeding can be used to predict narrow-sense heritability. Alternatively, if we already know the narrow-sense heritability, we can predict the outcome of selective breeding. In this case, the goal is to predict the mean phenotypes of offspring. If we rearrange the realized heritability equation, we get

$$R = h_N^2 S$$
$$\overline{X}_o - \overline{X} = h_N^2 \left(\overline{X}_p - \overline{X} \right)$$

The last equation is referred to as the breeder's equation, because it is used to calculate the mean phenotype of offspring based on the mean phenotype of the parents, the mean phenotype of the starting population, and the narrow-sense heritability. An example of the use of this equation is described next.

T OPIC: *What topic in genetics does this question address?* The topic is heritability. More specifically, the question is about using heritability to predict the phenotypes of offspring.

I NFORMATION: *What information do you know based on the question and your understanding of the topic?* From the question, you know the narrow-sense heritability for potato weight in a population, and you know the mean potato weight for the population and the weights of potatoes from selected parents. From your understanding of the topic, you may remember that $R = h_N^2 S$.

P ROBLEM-SOLVING **S** TRATEGIES: *Make a calculation. Predict the outcome.* To solve this problem, you first need to calculate the mean weight of potatoes produced from the parents and then use the equation described above. The mean potato weight for the parental plants is 2.0 pounds. Solving for the mean weight for the offspring gives

$$R = h_N^2 S$$
$$\overline{X}_o - \overline{X} = h_N^2 (\overline{X}_p - \overline{X})$$
$$\overline{X}_o - 1.4 = 0.42(2.0 - 1.4)$$
$$\overline{X}_o = 1.65 \text{ pounds}$$

ANSWER: The predicted mean weight of potatoes from the offspring is 1.65 pounds.

Heterosis May Be Explained by Dominance or Heterozygote Advantage

As we have just seen, selective breeding can alter the phenotypes of domesticated species in a highly directed way. An unfortunate consequence of inbreeding, however, is that it may inadvertently promote homozygosity for deleterious alleles. This phenomenon is called **inbreeding depression.** In addition, genetic drift, described in Chapter 27, may contribute to the loss of beneficial alleles. In agriculture, it is widely observed that when two different inbred strains are crossed to each other, the resulting offspring are often more vigorous (e.g., larger or longer-lived) than either of the inbred parental strains. This phenomenon is called **heterosis,** or **hybrid vigor.**

In modern agricultural breeding practices, many strains of plants and animals are hybrids produced by crossing two different inbred lines. Much of the success of agricultural breeding programs is founded in heterosis. In rice, for example, hybrid strains have a 15–20% yield advantage over the best conventional inbred varieties under similar cultivation conditions.

Two different phenomena may contribute to heterosis. In 1908, Charles Davenport proposed the dominance hypothesis, in

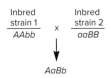

The recessive alleles (*a* and *b*) are slightly harmful in the homozygous condition.

The hybrid offspring is more vigorous, because the harmful effects of the recessive alleles are masked by the dominant alleles.

Neither the *A1* nor *A2* allele is recessive.

The hybrid offspring is more vigorous, because the heterozygous combination of alleles exhibits heterozygote advantage. This means that the *A1A2* heterozygote is more vigorous than either the *A1A1* or *A2A2* homozygote.

(a) The dominance hypothesis   **(b) The heterozygote advantage hypothesis**

FIGURE 28.12 **Mechanisms to explain heterosis.** The two common explanations are **(a)** the dominance hypothesis and **(b)** the heterozygote advantage hypothesis.

which the effects of dominant alleles explain the favorable outcome in a hybrid (**Figure 28.12a**). He suggested that highly inbred strains have become homozygous for recessive alleles of one or more genes. In the homozygous state, the recessive alleles are somewhat deleterious (but not lethal). Because the homozygosity occurs by chance, two different inbred strains are likely to be homozygous for recessive alleles in different genes. Therefore, when they are crossed to each other, the resulting hybrids are heterozygous and do not suffer the consequences of homozygosity for deleterious recessive alleles. In other words, the benefit of the dominant alleles explains the observed heterosis. Steven Tanksley, working with colleagues in China, found that heterosis in rice

seems to be due to the phenomenon of dominance. This is a common explanation for heterosis.

In 1908, George Shull and Edward East proposed a second hypothesis, called the overdominance hypothesis and renamed the heterozygote advantage hypothesis here (**Figure 28.12b**). As discussed in Chapter 4, heterozygote advantage, also known as overdominance, occurs when a heterozygote has greater reproductive success than either corresponding homozygote. According to this idea, heterosis can occur because the resulting hybrids are heterozygous for one or more genes that display heterozygote advantage. The heterozygote is more vigorous (and therefore more likely to reproduce) than either homozygote. Charles Stuber and his colleagues have found that several QTLs for grain yield in corn support the heterozygote advantage hypothesis.

28.6 COMPREHENSION QUESTIONS

1. For selective breeding to be successful, the starting population must
   a. have genetic variation that affects the trait of interest.
   b. be very large.
   c. have the potential for phenotypic variation caused by environmental effects.
   d. have very little phenotypic variation.

2. The mean weight of cows in a population is 520 kg. Animals with a mean weight of 540 kg are used as parents and produce offspring that have a mean weight of 535 kg. What is the narrow-sense heritability (h_N^2) for body weight in this population of cows?
   a. 0.25   c. 0.75
   b. 0.5   d. 1.0

KEY TERMS

Introduction: complex traits, quantitative traits, quantitative genetics
28.1: continuous traits, meristic traits, threshold traits, frequency distribution, normal distribution
28.2: biometric field, mean, variance, standard deviation (*SD*), covariance, correlation coefficient (*r*)
28.3: polygenic, polygenic inheritance

28.4: quantitative trait locus (QTL), QTL mapping
28.5: inbreeding, monomorphic, genotype-environment interaction, genotype-environment association, heritability, broad-sense heritability, narrow-sense heritability
28.6: selective breeding (artificial selection), selection limit, realized heritability, inbreeding depression, heterosis (hybrid vigor)

CHAPTER SUMMARY

- Quantitative genetics is the field of genetics concerned with complex and quantitative traits.

28.1 Overview of Complex and Quantitative Traits

- Quantitative traits can be categorized as anatomical, physiological, or behavioral; many human diseases are viewed as complex traits (see Table 28.1).

- Quantitative traits often exhibit a continuum of phenotypic variation that follows a normal distribution (see Figure 28.1).

28.2 Statistical Methods for Evaluating Quantitative Traits

- Statistical methods, including calculations of the mean, variance, standard deviation, covariance, and correlation coefficient, are used to analyze quantitative traits (see Figure 28.2, Table 28.2).

28.3 Polygenic Inheritance

- Polygenic inheritance refers to the transmission of any trait that is governed by two or more different genes.
- Polygenic inheritance and environmental factors may produce a continuum of phenotypes for a quantitative trait (see Figures 28.3, 28.4).

28.4 Identification of Genes That Control Quantitative Traits

- A location on a chromosome that carries one or more genes that affect the outcome of a quantitative trait is called a quantitative trait locus (QTL).
- QTLs are identified by their proximity to known molecular markers (see Figure 28.5).

28.5 Heritability

- Genetic variance and environmental variance may contribute additively to the total phenotypic variance.
- Genetic variance and environmental variance may exhibit interaction or association (see Figure 28.6, Table 28.3).
- Heritability is the amount of phenotypic variance due to genetic variance within a specific group of individuals raised in a particular environment.

- Broad-sense heritability takes into account different types of genetic variation that may affect the phenotype, including the additive effects of alleles, effects due to dominant/recessive relationships, and effects due to epistatic interactions.
- Narrow-sense heritability is heritability that is due to the additive effects of alleles.
- Holt determined that dermal ridge count has a very high heritability value in humans (see Figures 28.7, 28.8).

28.6 Selective Breeding

- Selective breeding refers to programs and procedures designed to modify phenotypes in economically important species of plants and animals (see Figures 28.9, 28.10).
- Starting with a genetically diverse population, selective breeding can usually modify a trait in a desired direction until a selection limit is reached (see Figure 28.11).
- The outcome of selective breeding can be used to estimate heritability. The estimated value is called realized heritability.
- Heterosis is the phenomenon in which crossing of different inbred strains produces hybrids that are more vigorous than the inbred strains. This outcome may be due to dominance or heterozygote advantage (see Figure 28.12).

PROBLEM SETS & INSIGHTS

MORE GENETIC TIPS   **1.** The following data describe the 6-week weights (in grams) of mice and their offspring of the same sex:

Parent's weight (g)	Offspring's weight (g)
24	26
21	24
24	22
27	25
23	21
25	26
22	24
25	24
22	24
27	24

Calculate the correlation coefficient.

T OPIC: *What topic in genetics does this question address?* The topic is calculating a correlation coefficient.

I NFORMATION: *What information do you know based on the question and your understanding of the topic?* From the

question, you know the 6-week weights of mice and their same-sex offspring. From your understanding of the topic, you may remember that you first need to calculate the mean weights and standard deviations of the parents and offspring, and then calculate the covariance. The correlation coefficient is computed as the covariance divided by the product of the standard deviations.

P ROBLEM-SOLVING S TRATEGY: *Make a calculation.* To begin to solve this problem, you first need to calculate the means and standard deviations for each group:

$$\overline{X}_{parents} = \frac{24 + 21 + 24 + 27 + 23 + 25 + 22 + 25 + 22 + 27}{10} = 24$$

$$\overline{X}_{offspring} = \frac{26 + 24 + 22 + 25 + 21 + 26 + 24 + 24 + 24 + 24}{10} = 24$$

$$SD_{parents} = \frac{\sqrt{0 + 9 + 0 + 9 + 1 + 1 + 4 + 1 + 4 + 9}}{9} = 2.1$$

$$SD_{offspring} = \frac{\sqrt{4 + 0 + 4 + 1 + 9 + 4 + 0 + 0 + 0 + 0}}{9} = 1.6$$

Next, you calculate the covariance.

$$CoV_{(p, o)} = \frac{\Sigma\,[(X_p - \overline{X}_p)(X_o - \overline{X}_o)]}{N - 1}$$

$$= \frac{0 + 0 + 0 + 3 + 3 + 2 + 0 + 0 + 0 + 0}{9}$$

$$= 0.9$$

Finally, you calculate the correlation coefficient:

$$r_{(p, o)} = \frac{Co\,V_{(p,o)}}{SD_p\,SD_o}$$

$$r_{(p, o)} = \frac{0.9}{(2.1)(1.6)}$$

$$r_{(p, o)} = 0.27$$

ANSWER: The correlation coefficient is 0.27.

---

2. A family of farmers wants to increase the average body weight in a herd of cattle. They begin with a herd having a mean weight of 595 kg and choose as the breeding stock individual cattle that have a mean weight of 625 kg. Twenty offspring are obtained, having the following weights in kilograms: 612, 587, 604, 589, 615, 641, 575, 611, 610, 598, 589, 620, 617, 577, 609, 633, 588, 599, 601, and 611. Calculate the realized heritability for body weight in this herd.

TOPIC: *What topic in genetics does this question address?* The topic is heritability. More specifically, the question is about calculating the realized heritability based on the outcomes of crosses.

INFORMATION: *What information do you know based on the question and your understanding of the topic?* From the question, you know the mean weight of a herd of cattle and the mean weight of selected parents. You also know the weights of 20 offspring. From your understanding of the topic, you may remember that

$$h_N^2 = \frac{R}{S}$$

PROBLEM-SOLVING **S**TRATEGY: *Make a calculation.* To solve this problem, you first need to calculate the mean weight of the offspring and then use the equation above.

$$h_N^2 = \frac{R}{S}$$
$$= \frac{\overline{X}_o - \overline{X}}{\overline{X}_p - \overline{X}}$$

You already know the mean weight of the starting population (595 kg) and the mean weight of the selected parents (625 kg). The only value missing is the mean weight of the offspring, $\overline{X}_o$.

$$\overline{X}_o = \frac{\text{Sum of the offspring's weights}}{\text{Number of offspring}}$$

$$\overline{X}_o = 604\,\text{kg}$$

$$h_N^2 = \frac{604 - 595}{625 - 595}$$
$$= 0.3$$

ANSWER: The realized heritability is 0.3.

---

3. Are the following statements regarding heritability *true* or *false*?
A. Heritability applies to a specific population raised in a particular environment.
B. Heritability in the narrow sense takes into account all types of genetic variance.
C. Heritability is a measure of the amount that genetics contributes to the outcome of a trait.

TOPIC: *What topic in genetics does this question address?* The topic is heritability.

INFORMATION: *What information do you know based on the question and your understanding of the topic?* In the question, you are given statements regarding heritability. From your understanding of the topic, you may remember the definitions of *heritability* and *narrow-sense heritability*.

PROBLEM-SOLVING **S**TRATEGY: *Define key terms.* One strategy to solve this problem is to recall the definitions of *heritability* and *narrow-sense heritability*. See Section 28.5.

ANSWER:
A. True
B. False. Narrow-sense heritability considers only the effects of additive alleles.
C. False. Heritability is a measure of the amount of phenotypic variance that is due to genetic variance; it applies to the phenotypic variance of a specific population raised in a particular environment.

---

4. For each of the following relationships, correlation coefficients for height were determined for 15 pairs of individuals:

Parent-offspring: 0.41
Grandparent-grandchild: 0.18
Siblings (nontwins): 0.40
Fraternal twins: 0.41
Identical twins: 0.83

What is the average heritability for height in this group?

TOPIC: *What topic in genetics does this question address?* The topic is heritability. More specifically it is about using correlation coefficients to determine narrow-sense heritability.

INFORMATION: *What information do you know based on the question and your understanding of the topic?* In the question, you are given data regarding the correlation coefficients for height for five different types of related pairs of individuals. From your understanding of the topic, you may recall that you can use these correlation coefficients to calculate narrow-sense heritability.

PROBLEM-SOLVING **S**TRATEGY: *Make a calculation.* To solve this problem, you need to use the following equation:

$$h_N^2 = r_{obs}/r_{exp}$$

The value for r_{exp} comes from the known genetic relationships:

Parent-child	$r_{obs} = 0.41$	$r_{exp} = 0.5$	$h_N^2 = 0.82$
Grandparent-grandchild	$r_{obs} = 0.18$	$r_{exp} = 0.25$	$h_N^2 = 0.72$
Siblings (nontwins)	$r_{obs} = 0.40$	$r_{exp} = 0.5$	$h_N^2 = 0.80$
Fraternal twins	$r_{obs} = 0.41$	$r_{exp} = 0.5$	$h_N^2 = 0.82$
Identical twins	$r_{obs} = 0.83$	$r_{exp} = 1.0$	$h_N^2 = 0.83$

ANSWER: Taking the average of the five h_N^2 values gives an average heritability of 0.80.

Conceptual Questions

C1. Give several examples of quantitative traits. How are these traits described within groups of individuals?

C2. At the molecular level, explain why quantitative traits often exhibit a continuum of phenotypes within a population. How does the environment help produce this continuum?

C3. What is a normal distribution? Discuss the normal curve with regard to quantitative traits within a population. What is the relationship between the standard deviation and the normal distribution?

C4. Explain the difference between a continuous trait and a discontinuous (discrete) trait. Give two examples of each. Are quantitative traits likely to be continuous or discontinuous? Explain why.

C5. What is a frequency distribution? Explain how the graph of such a distribution is plotted for a quantitative trait that is continuous.

C6. The variance for weight in a particular herd of cattle is 484 pounds^2. The mean weight is 562 pounds. How heavy would an animal have to be to rank in the top 2.5% of the herd? The bottom 0.13%?

C7. Two different varieties of potato plants produce potatoes with the same mean weight of 1.5 pounds. One variety has a very low variance for potato weight, and the other has a much higher variance.

A. Discuss the possible reasons for the differences in variance.

B. If you were a potato farmer, would you rather raise a variety with a low or a high variance? Explain your answer from a practical point of view.

C. If you were a potato breeder and you wanted to develop plants that produced heavier potatoes, would you choose the variety with a low or high variance? Explain your answer.

C8. If $r = 0.5$ and $N = 4$, would you conclude that a positive correlation exists between the two variables? Explain your answer. What if $N = 500$?

C9. What does it mean when a correlation coefficient is negative? Can you think of examples?

C10. When a correlation coefficient is statistically significant, what do you conclude about the two variables? What do the results mean with regard to cause and effect?

C11. What is polygenic inheritance? Discuss the issues that make polygenic inheritance difficult to study.

C12. What is a quantitative trait locus (QTL)? Does a QTL contain one gene or multiple genes? What technique is commonly used to identify QTLs?

C13. Suppose that weight in a species of mammal is a polygenic trait and each gene exists as a heavy and light allele. If the allele frequencies in the population are equal for both types of alleles (i.e., 50% heavy alleles and 50% light alleles), what percentage of individuals will be homozygous for the light alleles in all of the genes affecting this trait, if the trait was determined by the following number of genes?

A. Two

B. Three

C. Four

C14. The broad-sense heritability for a trait equals 1.0. In your own words, explain what this value means. Would you conclude that the environment is unimportant in the outcome of this trait? Explain your answer.

C15. From an agricultural point of view, discuss the advantages and disadvantages of selective breeding. It is common for plant breeders to take two different, highly inbred strains, which are the product of many generations of selective breeding, and cross them to make hybrids. How does this approach overcome some of the disadvantages of selective breeding?

C16. Many beautiful varieties of roses have been produced, particularly in the last few decades. These newer varieties often have very striking and showy flowers, making them desirable as horticultural specimens. However, breeders and novices alike have noticed that some of these newer varieties are not very fragrant compared with the older, more traditional varieties. From a genetic point of view, suggest an explanation why some of these newer varieties with superb flowers are not as fragrant.

C17. In your own words, explain the meaning of the term *heritability*. Why is a heritability value valid only for a particular population of individuals raised in a particular environment?

C18. What is the difference between broad-sense heritability and narrow-sense heritability? Why is narrow-sense heritability such a useful concept in the field of agricultural genetics?

C19. The heritability for egg weight in a group of chickens on a farm in Maine is 0.95. Are the following statements regarding this heritability *true* or *false*? If a statement is false, explain why.

A. The environment in Maine has very little effect on the outcome of this trait.

B. Nearly all of the phenotypic variance for this trait in this group of chickens is due to genetic variance.

C. The trait is polygenic and likely to involve a large number of genes.

D. Based on the observation of the heritability in the Maine chickens, it is reasonable to conclude that the heritability for egg weight in a group of chickens on a farm in Montana is also very high.

C20. In a fairly large population of people living in a commune in the southern United States, everyone cares about good nutrition. All members of this population eat very nutritious foods, and their diets are very similar. How do you think the heights of individuals in this commune population would compare with those of the general population in the following categories?

A. Mean height

B. Heritability for height

C. Genetic variation for alleles that affect height

C21. When artificial selection is practiced over many generations, eventually a plateau is reached in which further selection has little effect on the outcome of the trait. This phenomenon is illustrated in Figure 28.11. Explain why it occurs.

C22. Discuss whether a natural population of wolves or a domesticated population of German shepherds is more likely to have a higher heritability for the trait of size.

C23. With regard to heterosis, is each of the following statements consistent with the dominance hypothesis, the heterozygote advantage hypothesis, or both?

A. Strains that have been highly inbred have become monomorphic for one or more recessive alleles that are somewhat detrimental to the organism.

B. Hybrid vigor occurs because highly inbred strains are monomorphic for many genes, whereas hybrids are more likely to be heterozygous for those same genes.

C. If a gene exists in two alleles, hybrids are more vigorous because heterozygosity for the gene is more beneficial than homozygosity of either allele.

Experimental Questions

E1. Here are data consisting of heights and weights of 10 male college students.

Height (cm)	Weight (kg)
159	48
162	50
161	52
175	60
174	64
198	81
172	58
180	74
161	50
173	54

A. Calculate the correlation coefficient for height and weight for this group.

B. Is the correlation coefficient statistically significant? Explain.

E2. The abdomen length (in millimeters) was measured in 15 male *Drosophila*, and the following data were obtained: 1.9, 2.4, 2.1, 2.0, 2.2, 2.4, 1.7, 1.8, 2.0, 2.0, 2.3, 2.1, 1.6, 2.3, and 2.2. Calculate the mean, variance, and standard deviation for this population of male fruit flies.

E3. Restriction fragment length polymorphisms (RFLPs), which are described in Chapter 22, are a type of molecular marker that varies in length and can be used in mapping studies (refer back to Table 22.1). Using RFLPs in an analysis of head weight in one strain of cabbage, you determine that seven QTLs affect this trait. In another strain of cabbage, you find that only four QTLs affect this trait. Both strains of cabbage are the same species, although they may have different degrees of inbreeding. Explain how one strain can have seven QTLs and another strain four QTLs for exactly the same trait. Is the second strain missing three genes?

E4. As mentioned in question E3, RFLPs are a type of molecular marker that varies in length. From an experimental viewpoint, what does it mean to say that an RFLP is associated with a trait? Let's suppose that two strains of pea plants differ in two RFLPs that are linked to two genes governing pea size. RFLP-1 is found in 2000-bp and 2700-bp bands, and RFLP-2 is found in 3000-bp and 4000-bp bands. The plants producing large peas have RFLP-1 (2000 bp) and RFLP-2 (3000 bp); those producing small peas have RFLP-1 (2700 bp) and RFLP-2 (4000 bp). A cross is made between these two strains, and the F_1 offspring are allowed to self-fertilize. Five phenotypes are observed: small peas, small-medium peas, medium peas, medium-large peas, and large peas. Assume that each of the two genes makes an equal contribution to pea size and that the genetic variance is additive. Identify the bands that you would expect to obtain on a gel, and explain what RFLP banding patterns you would expect to observe for these five phenotypic categories. (Note: Certain phenotypic categories may have more than one possible banding pattern.)

E5. As mentioned in question E3, RFLPs are a type of molecular marker that varies in length. Let's suppose that two strains of pigs differ in 500 RFLPs. One strain is much larger, on average, than the other. The pigs are crossed to each other, and the members of the F_1 generation are also crossed among themselves to produce an F_2 generation. Three distinct RFLPs, which are not close to each other, are associated with F_2 pigs that are larger. How would you interpret these results?

E6. Outline the steps you would follow to determine the number of genes that influence the yield of rice. Describe the results you might get if rice yield is governed by variation in six different genes.

E7. In a wild strain of tomato plants, the phenotypic variance for tomato weight is 3.2 g^2. In another strain of highly inbred tomato plants raised under the same environmental conditions, the phenotypic variance for tomato weight is 2.2 g^2. With regard to the wild strain,

A. Estimate V_G.

B. What is h_B^2?

C. Assuming that all of the genetic variance is additive, what is h_N^2?

E8. The average thorax length in a *Drosophila* population is 1.01 mm. You want to practice selective breeding to obtain larger flies. To do so, you choose 10 parents (5 males and 5 females) that have thoraxes of the following lengths (in mm): 0.97, 0.99, 1.05, 1.06, 1.03, 1.21, 1.22, 1.17, 1.19, 1.20. You mate them and then determine the thorax lengths of 30 offspring (half male and half female): 0.99, 1.15, 1.20, 1.33, 1.07, 1.11, 1.21, 0.94, 1.07, 1.11, 1.20, 1.01, 1.02, 1.05, 1.21, 1.22, 1.03, 0.99, 1.20, 1.10, 0.91, 0.94, 1.13, 1.14, 1.20, 0.89, 1.10, 1.04, 1.01, 1.26. Calculate the realized heritability of thorax length in this group of flies.

E9. In a strain of mice, the average 6-week body weight is 25 g, and the narrow-sense heritability for this trait is 0.21.

A. What would be the average weight of the offspring if parents with a mean weight of 27 g were chosen?

B. What mean parental weight would you have to choose to obtain offspring with an average weight of 26.5 g?

E10. A danger in computing heritability values from studies involving genetically related individuals is the possibility that these individuals share more similar environments than do unrelated individuals. In the experiment shown in Figure 28.8, which data are the most compelling evidence that ridge count is not caused by genetically related individuals sharing common environments? Explain.

E11. A large, genetically heterogeneous group of tomato plants was used as the original breeding stock by two different breeders, named Mary and Hector. Each breeder was given 50 seeds and began a selective breeding strategy, much like that whose results are shown in Figure 28.11. The seeds were planted, and the breeders selected the 10 plants with the highest mean tomato weights as the breeding stock for the next generation. This process was repeated over the course of 12 growing seasons, and the following data were obtained:

Mean Weights of Tomatoes (pounds)

Year	Mary's Tomatoes	Hector's Tomatoes
1	0.7	0.8
2	0.9	0.9
3	1.1	1.2
4	1.2	1.3
5	1.3	1.3
6	1.4	1.4
7	1.4	1.5
8	1.5	1.5
9	1.5	1.5
10	1.5	1.5
11	1.5	1.5
12	1.5	1.5

A. Explain these results.

B. Another tomato breeder, named Martin, got some seeds from Mary's and Hector's tomato strains after 12 generations, grew the plants, and then crossed them to each other. The mean weight of the tomatoes in these hybrids was 1.7 pounds. For a period of 5 years, Martin subjected these hybrids to the same selective breeding strategy that Mary and Hector had followed, and obtained the following results:

Mean Weight of Tomatoes (pounds)

Year	Martin's Tomatoes
1	1.7
2	1.8
3	1.9
4	2.0
5	2.0

Explain Martin's data. Why was Martin able to obtain tomatoes heavier than 1.5 pounds, whereas Mary's and Hector's strains appeared to plateau at this weight?

E12. For 15 pairs of individuals in each of the following relationship categories, correlation coefficients for height were determined:

  Parent-offspring: 0.36

  Grandparent-grandchild: 0.17

  Siblings (nontwins): 0.39

  Fraternal twins: 0.40

  Identical twins: 0.77

What is the average heritability for height in this group of relatives?

E13. An animal breeder had a herd of sheep with a mean weight of 254 pounds at 3 years of age. Animals with a mean weight of 281 pounds were chosen as parents for the next generation. When those offspring reached 3 years of age, their mean weight was 269 pounds.

A. Calculate the narrow-sense heritability for weight in this herd.

B. Using the heritability value that you calculated in part A, what mean parental weight would you have to choose to get offspring that weigh 275 pounds on average (at 3 years of age)?

E14. The trait of blood pressure in humans has a frequency distribution that is similar to a normal distribution. The following graph shows the ranges of blood pressures for a selected population of people. The red line depicts the frequency distribution of the systolic pressures for the entire population. Several individuals with high blood pressure were identified, and the blood pressures of their relatives were determined. This frequency distribution is depicted with a blue line. (Note: The blue line does not include the people who were identified with high blood pressure; it includes only their relatives.)

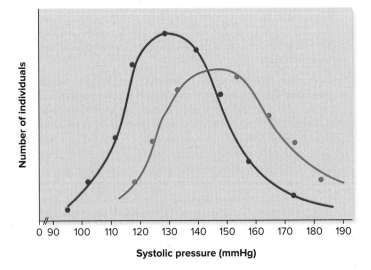

What do these data suggest with regard to a genetic basis for high blood pressure? What statistical approach could you use to determine the heritability for this trait?

Questions for Student Discussion/Collaboration

1. Explain why heritability is an important consideration in agriculture.

2. From a biological viewpoint, speculate as to why many traits seem to fit a normal distribution. Students with a background in math and statistics may want to explain how a normal distribution is generated and what it means. Can you think of biological examples that do not fit a normal distribution?

3. What is heterosis? Discuss whether it is caused by a single gene or several genes. Discuss the two major hypotheses proposed to explain heterosis. Which do you think is more likely to be correct?

Note: All answers are available for the instructor in Connect; the answers to the even-numbered questions and all of the Concept Check and Comprehension Questions are in Appendix B.

CHAPTER OUTLINE

The evolution of eyes. *Developmental biologists have recently discovered that the eyes of many diverse species, including fruit flies, frogs, mice, and people, are under the control of the homologous gene called Pax6, suggesting that eyes may have originated once during the evolution of animals.*
Stephen Welstead/SuperStock

29

EVOLUTIONARY GENETICS

As discussed in Chapter 17 (see Section 17.1), the emergence of life on Earth probably began with an **RNA world** in which RNA molecules had three key properties: storing information, self replication, and catalytic functions. Over time, the RNA world evolved into the DNA/RNA/protein world that exists today. This chapter will focus on how evolution has occurred in the DNA/RNA/protein world.

Biological evolution, or simply **evolution,** is the accumulation of heritable changes in one or more characteristics of a population or species from one generation to the next. Evolution is a process. It can be viewed on a small scale as it relates to a single gene, or it can be viewed on a larger scale as it relates to the formation of new species. In Chapter 27, we examined several factors that cause allele frequencies to change in populations. This process, also known as **microevolution,** accounts for the changing composition of gene pools with regard to particular alleles over measurable periods of time. As we have seen, several evolutionary mechanisms, such as mutation, genetic drift, migration, natural selection, and inbreeding, affect the allele and genotype frequencies within natural populations. On a microevolutionary scale, evolution can be viewed as a change in allele frequency over time.

A goal of this chapter is to relate phenotypic changes that occur during evolution to the underlying genetic changes that cause them to happen. In the first part of the chapter, we will consider evolution on a large scale, which leads to the origin of new species. The question of how species form has been central to the development of evolutionary theory. The term **macroevolution** refers to large-scale evolutionary changes that create new species and higher taxa. It concerns the establishment of the diversity of organisms over long periods of time through the accumulated evolution and extinction of many species.

In the last section of this chapter, we will link molecular genetics to the evolution of species. Techniques for analyzing chromosomes and DNA sequences have greatly enhanced our understanding of evolutionary processes at the molecular level. The term **molecular evolution** refers to molecular-level changes in the genetic material that underlie the process of evolution. Such changes may be phenotypically neutral, or they may promote the phenotypic changes associated with evolution. In this chapter, we will examine how molecular data can provide information about the phylogenetic relationships among different organisms.

The topic of molecular evolution is a fitting way to end our discussion of genetics because it integrates the ongoing theme of this text—the relationship between molecular genetics and traits—in the broadest and most profound ways. Theodosius Dobzhansky, an influential evolutionary scientist, once said, "Nothing in biology makes sense except in the light of evolution." The extraordinarily diverse and seemingly bizarre array of species on our planet can be explained naturally within the context of evolution. An examination of molecular evolution allows us to make sense of the existence of these species at both the population and the molecular levels.

29.1 OVERVIEW OF EVOLUTION

Learning Outcomes:

1. Describe the observations that led to Darwin's theory of evolution.
2. Explain two factors that lead to adaptive evolution.

A key topic in evolutionary genetics is the origin of species. How do new species come into existence? In this section, we will begin by considering the experiences that allowed Darwin to explain how evolutionary change leads to the formation of new species. We will then discuss the two key factors that allow evolutionary change to happen.

Darwin Formulated the Theory of Evolution by Comparing Different Disciplines

Charles Darwin, a naturalist born in 1809, proposed the theory of evolution and provided evidence that existing species have evolved from preexisting ones. Like many great scientists, Darwin had a broad background in science and was often able to see connections among different disciplines. Three areas proved particularly important regarding the theory of evolution:

- Darwin's thinking was influenced by the field of geology. According to geologist Charles Lyell, the processes that alter Earth are uniform through time. This view, which was known as uniformitarianism, suggested that Earth is very old and that slow geological processes can lead eventually to substantial changes in Earth's characteristics.
- Darwin's own experimental observations also has a major impact. The famous voyage on *HMS Beagle*, which lasted from 1832 to 1836, involved a careful examination of many different species. Darwin observed the similarities among many discrete species, yet noted the differences that enabled them to be adapted to their environmental conditions. Some of this work included observations of the distinctive adaptations of island species. For example, the finches found on the Galápagos Islands had unique phenotypic characteristics compared with those of similar finches found on the mainland.
- A third important influence on Darwin was a paper published in 1798, "Essay on the Principle of Population," by Thomas Malthus, an economist. Malthus asserted that the population size of humans has the potential to increase exponentially. However, such potential increases are not realized because of factors that limit population growth, such as famine, war, and disease. Therefore, not all offspring are able to survive and reproduce.

With these three perspectives in mind, Darwin had largely formulated the theory of evolution by natural selection by the mid-1840s. However, instead of publishing this work right away, Darwin spent several years studying barnacles. Charles Lyell strongly encouraged Darwin to publish this work. In 1856, Darwin began to write a long book to explain the theory of evolution. In 1858, however, Alfred Russel Wallace, a naturalist working in the East Indies, sent Darwin an unpublished manuscript to read prior to its publication. In it, Wallace proposed the same ideas concerning evolution. Darwin therefore quickly excerpted some writings on this subject, and two papers, describing each person's work, were published in the *Proceedings of the Linnaean Society of London*. These papers were not widely recognized.

A short time later, however, Darwin finished the book, *On the Origin of Species by Means of Natural Selection*, which explained evolution in greater detail and, importantly, with experimental support. This book, which received high praise from many scientists and scorn from others, started a great debate concerning natural selection. Although some of Darwin's ideas were incomplete because the genetic basis of traits was not understood at that time, this work represents one of the most important contributions to our understanding of biology.

Genetic Variation and Natural Selection Underlie Adaptive Evolution

Darwin called evolution "the theory of descent with modification through variation and natural selection." This form of evolution, which is sometimes called **adaptive evolution,** is the result of two key factors: genetic variation and natural selection. A modern interpretation of evolution can view these two factors at the species level (macroevolution) and at the level of genes in populations (microevolution).

1. *Genetic variation at the species level.* As we saw in Chapter 27, genetic variation is a consistent feature of natural populations. Darwin observed that many species exhibit a great amount of phenotypic variation. Although the theory of evolution preceded Mendel's pioneering work in genetics, Darwin (as well as many other people of this time period) observed that offspring resemble their parents more than they do unrelated individuals. Therefore, Darwin assumed that traits are passed from parent to offspring. However, the genetic basis for the inheritance of traits was not understood at that time.

   At the gene level. Genetic variation can involve allelic differences in genes. These differences are caused by random mutations. Different alleles may affect the functions of the proteins they code, thereby affecting the phenotype of the organism. Likewise, changes in chromosome structure and number may affect gene expression, thereby influencing the phenotype of the individual.

2. *Natural selection at the species level.* Darwin agreed with Malthus that most species produce many more offspring than will survive and reproduce, resulting in an ongoing struggle for existence. Over the course of many generations, those individuals that happen to possess the most favorable traits will dominate the composition of the population. The result of natural selection is to make a species better adapted to its environment and more successful at reproduction.

At the gene level. Some alleles code proteins that provide the individual with a selective advantage. Over time, natural selection may change the allele frequencies of genes, thereby leading to the fixation of beneficial alleles and the elimination of detrimental alleles.

29.1 COMPREHENSION QUESTION

1. Evolution that results because certain genotypes have a higher reproductive success is based on
   a. genetic variation.
   b. natural selection.
   c. genetic drift.
   d. both a and b.

29.2 IDENTIFICATION OF SPECIES AND MECHANISMS OF REPRODUCTIVE ISOLATION

Learning Outcomes:

1. Outline the characteristics used to distinguish species.
2. Define *reproductive isolation*, and give examples of prezygotic and postzygotic isolating mechanisms.
3. Define *species concept*.

To consider how biologists study the evolution of new species, we need to begin with a definition. Although many definitions are possible, a common one is that a **species** is a group of organisms that maintains a distinctive set of attributes in nature.

How many different species are on Earth? The number is astounding. Currently, about 2 million species have been identified and cataloged. However, this number does not include a vast number of species that have yet to be classified. The existence of unclassified species is particularly common among bacteria and archaea, which are difficult to categorize into distinct species. Also, new invertebrate and even vertebrate species are still being found in the far reaches of pristine habitats. Common estimates of the total number of species range from 5 million to 50 million! In this section, we will examine how species are identified and how

they are prevented from breeding with other species via reproductive isolating mechanisms.

Each Species Is Identified Using Characteristics and Histories That Distinguish It from Other Species

When studying natural populations, evolutionary biologists are often confronted with situations in which some differences between two populations are apparent, but it is difficult to decide whether the two populations truly represent separate species. When two or more geographically restricted populations of the same species display one or more traits that are somewhat different but not enough to warrant their placement into different species, biologists sometimes classify such groups as **subspecies.** Similarly, many bacterial species are subdivided into **ecotypes.** Each ecotype is a genetically distinct population adapted to its local environment.

Members of the same species share an evolutionary history that is distinct from other species. Although this may seem like a reasonable way to characterize a given species, evolutionary biologists would agree that the identification of many species is a difficult undertaking. What criteria do we use to distinguish species? How many differences must exist between two populations to classify them as distinct species? Such questions are often difficult to answer.

The characteristics that a biologist uses to identify a species depend, in large part, on the species in question. For example, the traits used to distinguish insect species are quite different from those used to identify bacterial species. The relatively high level of horizontal gene transfer among bacteria presents special challenges in grouping these organisms into species. The division of bacteria into separate species is usually very difficult and, at times, seemingly arbitrary. The most commonly used characteristics to identify species are morphological traits, their inability to interbreed with members of other species, molecular features, ecological factors, and evolutionary relationships. A comparison of these characteristics will help you to appreciate the various approaches that biologists use to identify the bewildering array of species on our planet.

Morphological Traits. One way to establish that a population constitutes a unique species is based on the members' physical characteristics. Organisms are classified as the same species if their anatomical traits appear to be very similar. Likewise, microorganisms can be classified according to morphological traits at the cellular level. By comparing many different morphological traits, biologists may decide that certain populations constitute a unique species.

Although an analysis of morphological traits is a common way for biologists to establish that a particular group constitutes a species, this approach has a few drawbacks. First, researchers may have trouble deciding how many traits to consider. In addition, quantitative traits such as size and weight, which vary in a continuous way among members of the same species, are difficult to analyze. Another drawback is that the degree of dissimilarity that distinguishes different species may not show a simple relationship; the members of the same species sometimes look very different, and conversely, members of different species sometimes look remarkably similar to each other. For example, **Figure 29.1a**

(a) Frogs of the same species

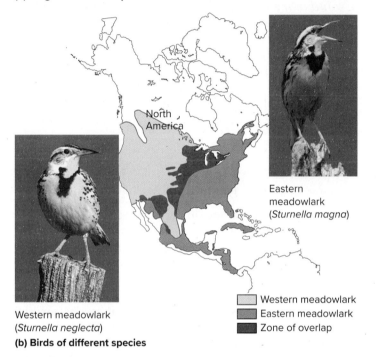

Eastern
meadowlark
(*Sturnella magna*)

Western meadowlark
(*Sturnella neglecta*)

☐ Western meadowlark
☐ Eastern meadowlark
☐ Zone of overlap

(b) Birds of different species

FIGURE 29.1 **Morphological dissimilarities and similarities in species.** **(a)** In some cases, two members of the same species can look quite different. These photographs show two members of the same species, the dyeing poison frog (*Dendrobates tinctorius*). **(b)** In comparison, members of different species can look quite similar, as illustrated by the western meadowlark (*Sturnella neglecta*) and eastern meadowlark (*Sturnella magna*).

(a): (left) Mark Smith/Science Source; (right) Pascal Goetgheluck/Ardea Picture;
(b): (left) Rod Planck/Science Source; (right) Ron Austing/Science Source

CONCEPT CHECK: Does this figure illustrate a strength or a drawback of using morphological traits to establish species?

shows two different frogs of the species *Dendrobates tinctorius*, commonly called the dyeing poison frog. This species, which is found in South America, exists in many different-colored morphs, which are individuals of the same species that have noticeably dissimilar appearances. In contrast, **Figure 29.1b** shows two different species of meadowlarks, named the western meadowlark (*Sturnella neglecta*) and the eastern meadowlark (*Sturnella magna*). These two species are nearly identical in shape, coloration, and habitat, and their ranges overlap in the central United States.

Inability to Interbreed (Reproductive Isolation). Why do biologists describe two species, such as the western and eastern meadowlarks, as being different if they are morphologically similar? A key reason is that biologists have discovered that they are unable to breed with each other in nature. In the zone of overlap, very little interspecies mating takes place between western and eastern meadowlarks, largely due to differences in their songs. The song of the western meadowlark is a long series of flutelike gurgling notes that go down the scale. By comparison, the eastern meadowlark's song is a simple series of whistles, typically about four or five notes. These differences in songs enable meadowlarks to recognize potential mates as members of their own species.

Therefore, a second way to identify a species is by the inability of its members to interbreed with individuals of other species. In the late 1920s, geneticist Theodosius Dobzhansky proposed that each species is unable to successfully interbreed with other species—a phenomenon called **reproductive isolation.** In 1942, evolutionary biologist Ernst Mayr expanded on the ideas of Dobzhansky to provide a definition of a species. According to Mayr, a key feature of sexually reproducing species is that, in nature, the members of one species have the potential to interbreed with one another to produce viable, fertile offspring but cannot successfully interbreed with members of other species.

Reproductive isolation has been used to distinguish many plant and animal species, especially those that look alike but do not interbreed. How does reproductive isolation occur? **Table 29.1** describes several ways. These are classified as **prezygotic isolating mechanisms,** which prevent the formation of a zygote, and **postzygotic isolating mechanisms,** which prevent the development of a viable and fertile individual after fertilization has taken place. Species that are reproductively isolated in nature may interbreed when kept in captivity. For example, different species of the genus *Drosophila* rarely mate with each other in nature. In the laboratory, however, it is fairly easy to produce interspecies hybrids.

Although reproductive isolation has been commonly used to classify species, it suffers from four main drawbacks.

- In nature, it may be difficult to determine if two populations are reproductively isolated, particularly if they are populations with nonoverlapping geographical ranges.
- Biologists have noted many cases in which two different species can interbreed in nature yet consistently maintain themselves as separate species. For example, different species of yucca plants, such as *Yucca pallida* and *Yucca constricta*, do interbreed in nature yet typically maintain populations with distinct characteristics. For this reason, they are viewed as distinct species.
- Reproductive isolation does not apply to asexual species such as bacteria. Likewise, some species of plants and fungi only reproduce asexually.
- The criterion of reproductive isolation cannot be applied to extinct species.

For these reasons, reproductive isolation has been primarily used to distinguish closely related species of modern animals and plants that reproduce sexually.

TABLE 29.1

Mechanisms of Reproductive Isolation Among Different Species

Prezygotic Isolating Mechanisms

Habitat isolation	Species occupy different habitats, so they never come in contact with each other.
Temporal isolation	Species have different mating or flowering seasons, mate at different times of day, or become sexually active at different times of the year.
Sexual isolation	Sexual attraction between males and females of different animal species is limited due to differences in behavior, physiology, or morphology.
Mechanical isolation	The anatomical structures of genitalia prevent mating between different species.
Gametic isolation	Gametic transfer takes place, but the gametes fail to unite with each other. This can occur because the male and female gametes fail to attract, because they are unable to fuse, or because the male gametes are inviable in the female reproductive tract of another species.

Postzygotic Isolating Mechanisms

Hybrid inviability	The egg of one species is fertilized by the sperm from another species, but the fertilized egg fails to develop past early embryonic stages.
Hybrid sterility	The interspecies hybrid survives, but it is sterile. For example, a mule, which is sterile, is a cross between a female horse (*Equus caballus*) and a male donkey (*Equus asinus*).
Hybrid breakdown	The F_1 interspecies hybrid is viable and fertile, but succeeding generations (i.e., F_2, etc.) become increasingly inviable. This is usually due to the formation of less fit genotypes by genetic recombination.

Molecular Features. Molecular features are now commonly used to determine if two populations are different species. Evolutionary biologists may compare DNA sequences within genes, gene order along chromosomes, chromosome structure, and chromosome number to identify similarities and differences among different populations. DNA sequence differences are often used to compare populations. For example, researchers may compare the DNA sequences of the gene that codes 16S rRNA in two bacterial populations as a way to decide if the populations represent different species. When the sequences are very similar, such populations are usually judged to be the same species. However, it may be difficult to draw the line when separating groups into different species. Is a 2% difference in genome sequences sufficient to warrant placement into two different species, or do we need a 5% difference?

Ecological Factors. A variety of factors related to the habitats in which organisms exist can be used to distinguish one species from another. For example, certain species of warblers can be distinguished by the habitat in which they forage for food. Some species search the ground for food, others forage in bushes or small

trees, and some species primarily forage in tall trees. Such habitat differences are used to distinguish different species that look morphologically similar.

Many bacterial species have been categorized as distinct species based on ecological factors. Bacterial cells of the same species are likely to use the same types of resources (e.g., sugars and vitamins) and grow under the same types of conditions (e.g., temperature and pH). However, a drawback of this approach is that different groups of bacteria sometimes display very similar growth characteristics, and members of the same species may show great variation in the growth conditions they will tolerate.

Evolutionary Relationships. In Section 29.4, we will examine the methods used to produce tree diagrams that describe the evolutionary relationships among different species. In some cases, such relationships are based on an analysis of the fossil record. Alternatively, another way to establish evolutionary relationships is by the analysis of DNA sequences. Researchers can obtain samples of cells from different individuals and compare the genes within those cells to see how similar or different they are.

A Species Concept Is a Way to Define What a Species Is and/or Distinguish Different Species

A **species concept** is a way to define what a species is and/or provide an approach for distinguishing one species from another. Several different species concepts have been proposed; we will consider a few examples.

Biological Species Concept. In 1942, Ernst Mayr proposed an early species concept called the **biological species concept.** According to this idea, a species is a group of individuals whose members have the potential to interbreed in nature to produce viable, fertile offspring but cannot successfully interbreed with members of other species. The biological species concept emphasizes reproductive isolation as the most important criterion for delimiting species.

Evolutionary Species Concept. Another example of a species concept is the **evolutionary species concept** proposed by American paleontologist George Gaylord Simpson in 1961. According to this idea, species should be defined based on the separate evolution of their lineages.

Ecological Species Concept. A third example is the **ecological species concept,** described by American evolutionary biologist Leigh Van Valen in 1976. According to this viewpoint, each species occupies an ecological niche, which is the unique set of habitat resources that the species requires, as well as its influence on the environment and other species.

General Lineage Concept. Most evolutionary biologists agree that different methods are needed to distinguish the vast array of species on Earth. Even so, some evolutionary biologists have questioned whether it is valid to have many different species concepts. In 1998, Kevin de Queiroz suggested that there is only a single general species concept, which concurs with Simpson's evolutionary species concept and includes all previous concepts. According to de Queiroz's **general lineage concept,** each species is a population of an independently evolving lineage. Each species has evolved from a specific series of ancestors and, as a consequence, forms a group of organisms with a particular set of characteristics. Multiple criteria are used to determine if a population is part of an independent evolutionary lineage, and thus a species, distinct from others. Typically, researchers use analyses of morphology, reproductive isolation, DNA sequences, and ecology to determine if a population or group of populations is distinct from others. Because of its generality, the general lineage concept has received significant support.

29.2 COMPREHENSION QUESTIONS

1. Characteristics that are used to identify species include
   a. morphological traits.
   b. reproductive isolation.
   c. molecular features.
   d. ecological factors.
   e. evolutionary relationships.
   f. all of the above.

2. Which of the following is an example of a postzygotic isolating mechanism?
   a. habitat isolation
   b. temporal isolation
   c. mechanical isolation
   d. hybrid sterility
   e. All of the above are examples of postzygotic isolating mechanisms.

29.3   SPECIATION

Learning Outcomes:

1. Compare and contrast anagenesis and cladogenesis.
2. Describe the key features of allopatric and sympatric speciation.
3. Define *hydrid zone*, and give an example.

Speciation is the evolutionary process by which populations evolve to become distinct species. As mentioned in Section 29.2, somewhere between 5 and 50 million species are estimated to exist on Earth. These numbers tell us that speciation is a common process. In this section, we will explore different ways that it can occur.

Speciation Usually Occurs via a Branching Process Called Cladogenesis

Speciation can occur by **anagenesis** (from the Greek *ana*, "up," and *genesis*, "origin") in which a single species evolves into a different species over the course of many generations (**Figure 29.2a**). However, most evolutionary biologists would argue that anagenesis rarely occurs, though it could happen if an entire species was confined to a single environment for a long period of time.

Speciation primarily occurs via **cladogenesis** (from the Greek *clados*, "branch"), which involves the division of a single species into two or more species (**Figure 29.2b**). This process of speciation increases species diversity. Although cladogenesis is usually thought of as a splitting process, it commonly occurs as a budding process, which results in the retention of the original species with the addition of one or more new species with different characteristics. If we view evolution as a tree, the new species bud from the original species and develop characteristics that prevent them from breeding with the original one.

Depending on the geographic locations of the evolving population(s) and the environment that a species occupies, cladogenesis is categorized as allopatric or sympatric

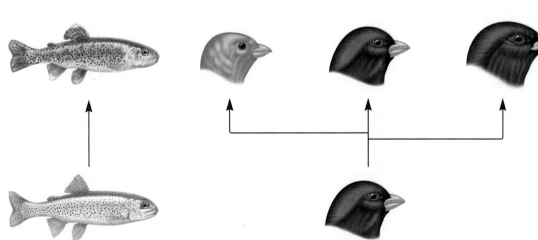

(a) Anagenesis

(b) Cladogenesis—prevailing mechanism of speciation

FIGURE 29.2
A comparison between anagenesis and cladogenesis, two patterns of speciation. (a) Anagenesis is the change of one species into another. (b) Cladogenesis involves a process in which one original species is separated into two or more species.

CONCEPT CHECK: Which is more common: anagenesis or cladogenesis?

TABLE 29.2

Common Genetic Mechanisms That Underlie Allopatric and Sympatric Speciation

Type of Speciation	Common Genetic Mechanisms Responsible for Speciation
Allopatric—two large populations are separated by geographic barriers.	Many small genetic differences may accumulate over a long period, leading to reproductive isolation. Some of these genetic differences may be adaptive, whereas others are neutral.
Allopatric—a small founding population separates from the main population.	Genetic drift may lead to the rapid formation of a new species. If a group has moved to an environment that is different from its previous one, natural selection is expected to favor beneficial alleles and eliminate harmful alleles.
Sympatric—within a population occupying a single habitat in a range, a small group evolves into a reproductively isolated species.	One mechanism involves an abrupt genetic change that leads to reproductive isolation. For example, a mutation may affect gamete recognition. In plants, the formation of a tetraploid may lead to the formation of a new species because the interspecies hybrid (i.e., offspring of a cross between diploid and tetraploid plants) is triploid and sterile. Alternatively, members of a population may occupy different local environments that are continuous with each other. For example, they may feed on different sources. Over long periods of time, genetic changes may gradually accumulate and lead to speciation.

(**Table 29.2**). **Allopatric speciation** (from the Greek *allos*, "other," and Latin *patria*, "homeland") is thought to be the most prevalent way for a species to diverge. It happens when some members of a species become geographically separated from the other members. **Sympatric speciation** (from the Greek *sym*, "together") occurs when a new species arises in the same geographic area as the species from which it was derived. As described in Table 29.2, various genetic mechanisms underlie these forms of speciation. Next, we will examine both allopatric and sympatric speciation in greater detail.

Allopatric Speciation Occurs When Populations Become Separated from Each Other

Allopatric speciation occurs in two common ways, both of which involve geographic separation.

Geological Processes. Allopatric speciation can occur by the geographic subdivision of large populations via geological processes. For example, a mountain range may emerge and split a species that occupies the lowland regions, or a creeping glacier may divide a population. **Figure 29.3** shows an interesting example in which geological separation promoted speciation. Two species of antelope squirrels occupy opposite rims of the Grand Canyon. On the south rim is Harris's antelope squirrel (*Ammospermophilus harrisi*), whereas a closely related white-tailed antelope squirrel (*Ammospermophilus leucurus*) is found on the north rim. Presumably, these two species evolved from a common species that existed before the canyon was formed. Over time, the accumulation of genetic changes in the two separated populations led to the formation of two morphologically distinct species. Interestingly, birds that can easily fly across the canyon have not diverged into different species on the opposite rims.

Founder Effect. Allopatric speciation can also occur via a second mechanism, known as the **founder effect,** which is thought to be more rapid and frequent than allopatric speciation caused by geological events. The founder effect, which was discussed in

A. harrisi   *A. leucurus*

FIGURE 29.3 **An example of allopatric speciation: two closely related species of antelope squirrels that occupy opposite rims of the Grand Canyon.**

Genes→Traits   Harris's antelope squirrel (*Ammospermophilus harrisi*) is found on the south rim of the Grand Canyon, whereas the white-tailed antelope squirrel (*Ammospermophilus leucurus*) is found on the north rim. These two species evolved from a common species that existed before the canyon was formed. After the canyon was formed, the two separated populations accumulated genetic changes due to mutation, genetic drift, and natural selection that eventually led to the formation of two distinct species.

(Left): MichaelStubblefield/Getty Images; (right): B Christopher/Alamy Stock Photo

Chapter 27, occurs when a small group migrates to a new location that is geographically separated from the main population. For example, a storm may force a small group of birds from the mainland to a distant island. In this case, the migration of individuals between the island and the mainland is a very infrequent event. In a relatively short time, the founding population on the island may evolve into a new species.

Two evolutionary mechanisms may contribute to this rapid evolution:

- First, genetic drift may quickly lead to the random fixation of certain alleles and the elimination of other alleles from the population.
- Another factor is natural selection. The environment on the island may differ significantly from the mainland environment. For this reason, natural selection on the island may favor different types of alleles.

Hybrid Zones Are Areas Where Separated Populations Can Interbreed

Before ending our discussion of allopatric speciation, let's consider a common situation in which geographic separation is not complete. As a population is separating into two (or more) distinct populations due to geographic barriers or at some time after that separation has happened, the two populations may come in contact with each other in **hybrid zones.** A hybrid zone is an area where interbreeding between two populations may occur, producing hybrids. **Figure 29.4** shows a hybrid zone along a mountain pass that connects two hypothetical deer populations. These could be closely related deer species, or they could be the same species in which each population has some unique characteristics. Whether or not these populations remain genetically distinct depends on the characteristics of the hybrids that are produced within the hybrid zone and on the level at which interbreeding occurs.

By studying hybrid zones over time, researchers have determined that three different outcomes are common: reinforcement, fusion, and stability. **Figure 29.5** illustrates these outcomes when a hybrid zone has formed at some time after two populations have diverged from each other. These outcomes either increase, decrease, or maintain reproductive isolation between the two populations.

Reinforcement: Increasing Reproductive Barriers Between Two Populations.
The term **reinforcement** indicates that gene flow through the hybrid zone from one population to the other is very limited and thereby reinforces the reproductive isolation between the two populations (Figure 29.5a). How does this happen? A common reason is that the hybrids are less fit than the members of the two original populations. Many such hybrid zones

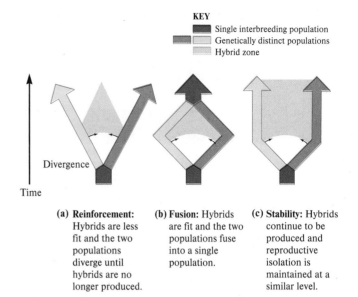

KEY
- Single interbreeding population
- Genetically distinct populations
- Hybrid zone

(a) Reinforcement: Hybrids are less fit and the two populations diverge until hybrids are no longer produced.

(b) Fusion: Hybrids are fit and the two populations fuse into a single population.

(c) Stability: Hybrids continue to be produced and reproductive isolation is maintained at a similar level.

FIGURE 29.5 Three different outcomes of hybrid zones over time. In this example, a hybrid zone is formed after two populations have diverged from each other. In part **(a)**, reproductive isolation increases between the two populations, whereas in part **(b)**, it decreases. In part **(c)**, reproductive isolation remains relatively constant, and the hybrid zone is maintained.

CONCEPT CHECK: Which of these three scenarios will allow speciation to occur?

are also **tension zones,** hybrid zones in which the hybrids are selected against and the two populations on either side of the zone are adapted to different environments.

Because the lower fitness of hybrids minimizes gene flow, the hybrid zone reinforces reproductive isolation between the two populations. As the two populations continue to diverge, the formation of hybrids becomes less likely and the hybrid zone may no longer exist. If the two populations were already distinct species, reinforcement will maintain them as such. If the starting populations were the same species with unique characteristics, each population may eventually become a distinct species as both populations accumulate more genetic changes that affect their characteristics and promote reproductive isolation.

Fusion: Decreasing Reproductive Barriers Between Two Populations.
A hybrid zone that promotes fusion has an effect that is opposite to that of reinforcement. When fusion occurs, the hybrids are fit, and they enable gene flow between the two populations. This gene flow will make the two populations more similar genetically and reduce reproductive isolating mechanisms. If the extent of the gene flow is substantial, the two populations may eventually fuse back into a single interbreeding population (Figure 29.5b).

Stability: Maintaining Reproductive Barriers and the Hybrid Zone.
In some cases, a hybrid zone may be maintained for long periods of time, a condition called stability (Figure 29.5c).

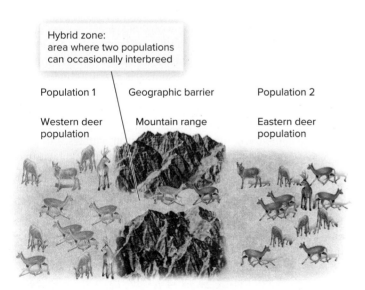

Hybrid zone: area where two populations can occasionally interbreed

| Population 1 | Geographic barrier | Population 2 |
| Western deer population | Mountain range | Eastern deer population |

FIGURE 29.4 An example of a hybrid zone. Two populations of deer are separated by a mountain range. A hybrid zone exists in a mountain pass, where occasional interbreeding may occur.

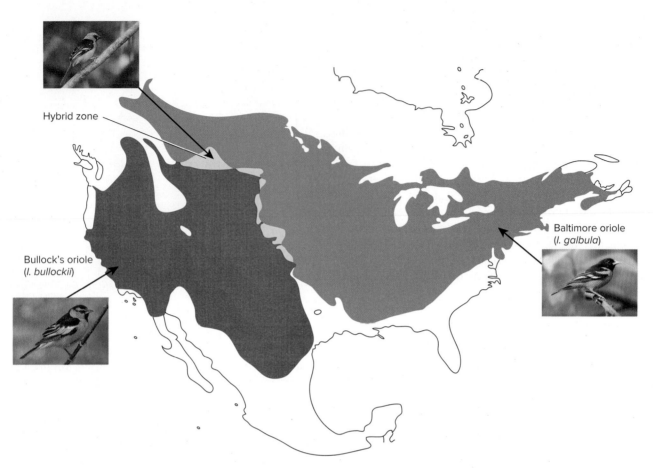

FIGURE 29.6 **A hybrid zone between Bullock's orioles and Baltimore orioles.** Bullock's oriole has a western range shown in red, and the Baltimore oriole has an eastern range shown in blue. The hybrid zone is yellow.

(top left) vagabond54/Shutterstock; (bottom left) J. Omar Hansen/Shutterstock; (right) Wang LiQiang/Shutterstock

An example of a stable hybrid zone involves Bullock's orioles (*Icterus bullockii*) and Baltimore orioles (*Icterus galbula*). Populations of Bullock's and Baltimore orioles largely came into contact following the conversion of the Great Plains to agriculture and the consequent planting of trees, which allowed Baltimore orioles to spread west and Bullock's orioles to spread east. As shown in the map of **Figure 29.6**, these two species form a hybrid zone in the central United States.

A key factor that prevents the fusion of these two bird species is thought to be the lowered fitness of hybrids. Researchers have speculated that one reason for this lowered fitness is related to molting (the shedding of feathers). Baltimore orioles typically molt in the summer after breeding but before migrating. Bullock's orioles molt during migration in the fall and on their wintering grounds. However, the hybrids molt twice, which is a significant energy drain. Even so, the hybrids are viable and fertile to the extent that the hybrid zone shown in Figure 29.6 has been maintained for many decades.

Sympatric Speciation Occurs Within the Same Geographic Area

As mentioned, sympatric speciation is a process in which a new species arises in the same geographic area as the species from which it

was derived. Two common mechanisms that result in sympatric species are polyploidy and adaptation to local environments.

Polyploidy. In plants, a common way for sympatric speciation to occur is the formation of polyploids. As discussed in Chapter 8, complete nondisjunction of chromosomes during gamete formation can increase the number of chromosome sets within a single species (autopolyploidy) or between different species (allopolyploidy). Polyploidy is a major form of speciation in plants. In ferns and flowering plants, at least 30% of the species are polyploid. By comparison, polyploidy is much less common in animals, but it can occur. For example, some species of reptiles and amphibians have been identified that are polyploids derived from diploid relatives.

The formation of polyploids can lead abruptly to reproductive isolation. As an example, let's consider the probable events that led to the formation of a natural species of common hemp nettle known as *Galeopsis tetrahit*. This species is thought to be an allotetraploid derived from two diploid species: *Galeopsis pubescens* and *Galeopsis speciosa*.

As shown in **Figure 29.7a**, *G. tetrahit* has 32 chromosomes, whereas the two diploid species have 16 chromosomes each ($2n = 16$). **Figure 29.7b** illustrates the chromosomal composition of offspring derived from crosses between the allotetraploid and

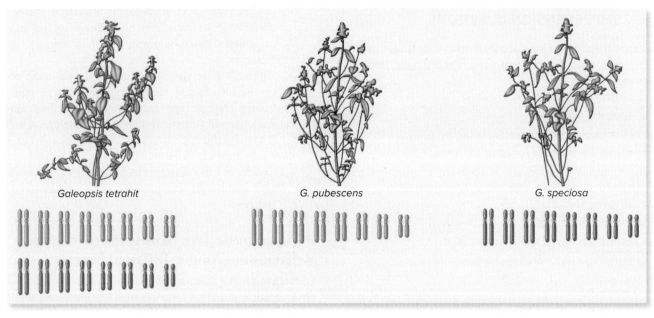

(a) Chromosomal composition of three *Galeopsis* species

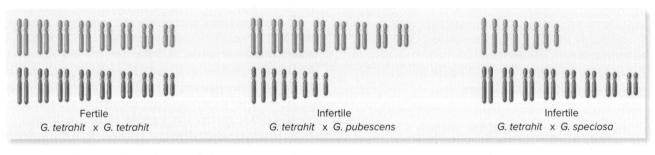

(b) Outcome of intraspecies and interspecies crosses

FIGURE 29.7 **A comparison of crosses between three natural species of hemp nettle that vary in the number of chromosome sets.**
(a) *Galeopsis tetrahit* is an allotetraploid that is thought to be derived from *Galeopsis pubescens* and *Galeopsis speciosa*. **(b)** If *G. tetrahit* is crossed with either of the other two species, the F$_1$ hybrid offspring will be monoploid for one chromosome set and diploid for the other. The F$_1$ offspring are likely to be sterile, because they will produce highly aneuploid gametes.

CONCEPT CHECK: Why is *G. tetrahit* reproductively isolated from the other two species?

the diploid species. The allotetraploid crossed to another allotetraploid produces an allotetraploid. The allotetraploid is fertile, because all of its chromosomes occur in homologous pairs that can segregate evenly during meiosis. However, a cross between an allotetraploid and a diploid produces an offspring that is monoploid for one chromosome set and diploid for the other. These offspring are expected to be sterile, because they produce highly aneuploid gametes that have incomplete sets of chromosomes. This hybrid sterility renders the allotetraploid reproductively isolated from the diploid species.

Adaptation to Local Environments. Sympatric speciation may also occur when members of a population occupy different local environments that are continuous with each other. An example of this type of sympatric speciation was described by Jeffrey Feder, Guy Bush, and colleagues.

They studied the North American apple maggot fly (*Rhagoletis pomonella*). Mating in these flies occurs on the host plant, primarily on the host fruit in which eggs are deposited and larvae develop. The larva originally fed on the fruit of native hawthorn trees. However, the introduction of apple trees into North America approximately 200 years ago provided a new food source for this species. Apple trees and hawthorn trees can be viewed as different local environments for the flies. The apple-feeding populations of this species develop more rapidly because apples mature more quickly than hawthorn fruit. The result is partial temporal isolation (see Table 29.1). Although the two populations—those that feed on apple trees and those that feed on hawthorn trees—are considered subspecies, evolutionary biologists speculate that they may eventually become distinct species due to reproductive isolation and the accumulation of independent mutations.

29.3 COMPREHENSION QUESTIONS

1. A pair of birds flies to a deserted island and establishes a colony. Over time, this population evolves into a new species. This is an example of
   a. allopatric speciation.
   b. sympatric speciation.
   c. both allopatric and sympatric speciation.
   d. neither allopatric nor sympatric speciation.
2. The formation of polyploids is common in plants and can abruptly produce a new species. This is an example of
   a. allopatric speciation.
   b. sympatric speciation.
   c. both allopatric and sympatric speciation.
   d. neither allopatric nor sympatric speciation.

29.4 PHYLOGENETIC TREES

Learning Outcomes:
1. Describe the key features of phylogenetic trees.
2. Compare and contrast the cladistic and phenetic approaches for constructing phylogenetic trees.
3. Define *horizontal gene transfer*, and explain how it affects the relationships in phylogenetic trees.

Thus far, we have considered the factors that play a role in the formation of new species. In this section, we will examine **phylogeny**—the sequence of events involved in the evolutionary development of a species or group of species. A **phylogenetic tree** is a diagram that describes a phylogeny. Such a tree is a hypothesis concerning the evolutionary relationships among different species. In this section, we will examine the general features of phylogenetic trees, how they can be constructed, and the types of information they reveal.

Phylogenetic Trees Are Based on Homology

Homology refers to similarities among various species that occur because the species are derived from a common ancestor. Attributes that are the result of homology are said to be **homologous.** For example, the wing of a bat, the arm of a human, and the front leg of a cat are homologous structures. By comparison, a bat wing and an insect wing are not homologous; they arose independently of each other. When constructing phylogenetic trees, researchers identify homologous features that are shared by some species but not by others. Making such distinctions allows them to group species based on their shared characteristics. Researchers typically study homology at the level of morphological traits or at the level of genes.

- Historically, comparisons of morphological similarities and differences have been used to construct evolutionary trees. In this approach, species that share certain

characteristics (i.e., homologous traits) tend to be placed closer together on a tree. In addition, species have been categorized based on physiology, biochemistry, and even behavior.
- Although these approaches continue to be used, researchers are increasingly using molecular data to infer evolutionary relationships. In 1963, Linus Pauling and Emile Zuckerkandl were the first to suggest the use of molecular data to establish evolutionary relationships. When homologous genes in different species are compared, the DNA sequences from closely related species are more similar to each other than are the sequences from distantly related species.

A Phylogenetic Tree Depicts the Evolutionary Relationships Among Different Species

Let's first take a look at what information is found within a phylogenetic tree and the form in which it is presented. **Figure 29.8** shows a hypothetical phylogenetic tree of the relationships between various butterfly species labeled A through J. The vertical axis represents time, with the oldest species at the bottom.

The prevailing mechanism of speciation is cladogenesis, in which a species diverges into two or more species. The nodes, or branch points, in a phylogenetic tree represent times when cladogenesis occurred. For example, approximately 12 million years ago (mya), species A diverged into species A and species B by cladogenesis. The tips of branches are occupied by either species that became extinct in the past (species A, B, D, and C) or modern species that are at the top of the tree (E, F, G, I, H, and J).

By studying the branch points of a phylogenetic tree, researchers can group species according to common ancestry. A **monophyletic group,** also known as a **clade,** is a group of species consisting of all descendants of the group's most common ancestor. For example, the group highlighted in light green in Figure 29.8 is a clade derived from the common ancestor labeled C. Likewise, the entire tree shown in Figure 29.8 forms a clade, with species A as the common ancestor. As we see in this figure, smaller and more recent clades are subsets of larger ones.

The phylogenetic tree in Figure 29.8 includes ancestral species. This type of tree could be obtained by examining the fossil record. However, many phylogenetic trees do not include ancestral species, but instead focus on the relationships among modern species.

A Phylogenetic Tree Can Be Constructed Using a Cladistic or Phenetic Approach

Now that you appreciate the concept of a phylogenetic tree, let's turn our attention to how biologists actually construct them. The most popular methods of building phylogenetic trees can be classified into two broad categories called phenetics and cladistics. A **phenetic approach** constructs a phylogenetic tree

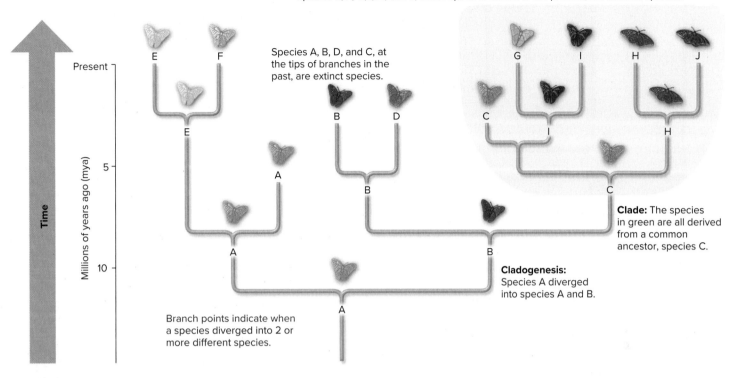

Species E, F, G, I, H, and J, at the tips of branches in the present, are modern species.

Species A, B, D, and C, at the tips of branches in the past, are extinct species.

Clade: The species in green are all derived from a common ancestor, species C.

Cladogenesis: Species A diverged into species A and B.

Branch points indicate when a species diverged into 2 or more different species.

FIGURE 29.8   How to read a phylogenetic tree. This hypothetical tree shows the proposed relationships among various butterfly species.

CONCEPT CHECK: What is a clade?

by considering the overall similarities among a group of species without trying to understand their evolutionary history. Such trees are called **phenograms.** By comparison, in a **cladistic approach,** a phylogenetic tree is constructed by considering the various possible pathways of evolution and then choosing the most plausible tree. In this approach, a goal is to group species based on knowledge concerning traits that arose earlier in evolution versus traits that appeared later. Such trees are called **cladograms.**

Which approach is better? For data involving morphological traits and for the construction of complex evolutionary trees that include many taxonomic levels, the cladistic approach is generally superior. Even so, both approaches are used, and many evolutionary biologists construct their evolutionary trees based on a combination of cladistics and phenetics. We will briefly consider both types of methods next.

Cladistic Approach.   In the 1950s, Willi Hennig proposed that evolutionary relationships should be inferred from new features shared by descendants of a common ancestor. A cladistic approach compares features, also called **characters,** that are either shared or not shared by different species. These can be morphological features, such as the shapes of the front limbs, or molecular characteristics, such as sequences of homologous

genes. Such characters may come in different versions called **character states.**

Those characters that are shared with a distant ancestor are called **ancestral characters** (also called **primitive characters**). Such characters are viewed as being older—ones that arose earlier in evolution. In contrast, a **shared derived character,** or **synapomorphy,** is a character state that is shared by a group of organisms but not by a distant common ancestor. Compared with ancestral characters, shared derived characters have arisen more recently on an evolutionary time scale. For example, among mammals, only some species, such as whales and dolphins, have flippers. In this case, flippers are derived from the two front limbs of an ancestral species.

The word *derived* refers to the observation that evolution involves the modification of traits in preexisting species. In other words, the features of newer populations of organisms are derived from changes in preexisting populations. The basis of the cladistic approach is to analyze many shared derived characters among groups of species to deduce the pathway that gave rise to the species.

To illustrate the distinction between ancestral and shared derived characters, let's consider how a cladogram can be constructed based on molecular data such as a sequence of a gene. Our example uses molecular data obtained from seven different

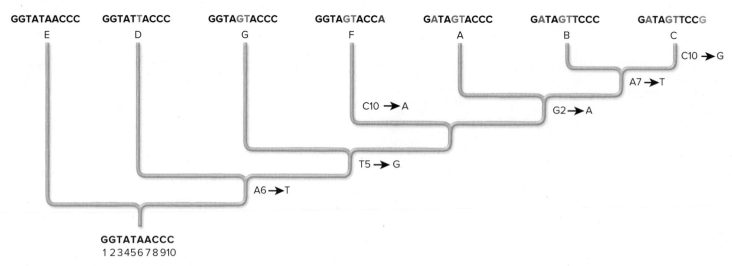

FIGURE 29.9 **Shared derived characters involving a molecular trait.** This phylogenetic tree illustrates a cladogram of relationships involving homologous gene sequences found in seven hypothetical species. Mutations that alter an ancestral DNA sequence are shared among certain species, thereby allowing the construction of a cladogram.

CONCEPT CHECK: What is the difference between an ancestral character and a shared derived character?

hypothetical species called A through G. The same gene was sequenced from the seven species; a portion of the gene sequences is shown here:

A: GATAGTACCC   E: GGTATAACCC
B: GATAGTTCCC   F: GGTAGTACCA
C: GATAGTTCCG   G: GGTAGTACCC
D: GGTATTACCC

In a cladogram, an **ingroup** is a species or group of species in which a researcher is interested and which is hypothesized to be monophyletic. By comparison, an **outgroup** is a species or group of species that possess characteristics that set it apart as being more distantly related to the ingroup. For example, the outgroup could be a reptile, whereas the ingroup could be a group of mammalian species. The root of a cladogram is placed between the outgroup and the ingroup. In the cladogram of **Figure 29.9**, the outgroup is species E. This may have been inferred because the other species share traits that are not found in species E. The other species (A, B, C, D, F, and G) form the ingroup. For these data, a mutation that changes the DNA sequence is analogous to a modification of a characteristic. In other words, differences in DNA sequence represent different character states. Species that share such genetic changes possess shared derived characters because the new genetic sequence was derived from a more ancestral sequence.

Now that you understand some of the general principles of cladistics, let's consider the steps a researcher follows to construct a cladogram using this approach.

1. ***Choose the species in whose evolutionary relationships you are interested.*** In a simple cladogram, individual species are compared with each other. In more complex cladograms, species may be grouped into larger taxa (e.g., families) and compared with each other. If such grouping is done, the groups must be clades for the results to be reliable.

2. ***Choose characters for comparing different species.*** As mentioned, a character is a general feature of an organism. Characters may come in different versions called character states. For example, a base at a particular location in a gene can be considered a character, and this character could exist in different character states, such as A, T, G, or C, due to mutations.

3. ***Determine the polarity of character states.*** In other words, determine if a character state is ancestral or derived. Comparisons are made within the ingroup and between the ingroup and the outgroup.

4. ***Group species (or higher taxa) based on shared derived characters.***

5. ***Build a cladogram based on the following principles:***
   - All species (or higher taxa) are placed on tips in the phylogenetic tree, not at branch points. A cladogram does not include ancestral species at branch points.
   - Each cladogram branch point should have a list of one or more shared derived characters that are common to all species above the branch point unless the character is later modified.
   - All shared derived characters appear together only once in a cladogram unless they independently arose during evolution more than once in the ancestors of different clades.

6. ***Choose the best cladogram among possible options.*** When grouping species (or higher taxa), more than one cladogram is possible. Therefore, analyzing the data and

producing the most likely cladogram is a key aspect of this process. As described next, different approaches can be followed to achieve this goal.

The greatest challenge in a cladistic approach is to determine the correct order of events. It may not always be obvious which characters are ancestral and came earlier and which are derived and came later in evolution. Different approaches can be used to deduce the correct order.

- First, for morphological traits, a common way to deduce the order of events is to analyze fossils and determine the relative dates that certain traits arose.
- A second strategy assumes that the best hypothesis is the one that requires the fewest number of evolutionary changes. This concept, called the **principle of parsimony,** states that the preferred hypothesis is the one that is the simplest. For example, if two species possess a tail, we would initially assume that a tail arose once during evolution and that both species have descended from a common ancestor with a tail. Such a hypothesis is simpler than assuming that tails arose twice during evolution.

Let's consider a simple example to illustrate how the principle of parsimony can be used. This example involves molecular data obtained from four different hypothetical species called A through D. In these species, a homologous region of DNA was sequenced as shown here:

         1 2 3 4 5
    A: GTACA (outgroup)
    B: GACAG
    C: GTCAA
    D: GACCG

Many different trees could be constructed from this information. Three of them are shown in **Figure 29.10**. These three trees hypothesize that species A is the outgroup. This assumption may be based on the sources of the four sequences. For example, species A might be a reptile, whereas species B, C, and D are mammals. In these examples, tree 1 requires seven mutations, tree 2 requires six, and tree 3 requires only five. Because tree 3 requires the fewest number of mutations, it is considered the most parsimonious. Based on the principle of parsimony, tree 3 is the most likely option.

Phenetic Approach. As mentioned earlier, a phenetic approach constructs a phylogenetic tree based on overall similarities among a group of species. A relatively simple method used in phenetics is abbreviated UPGMA (<u>u</u>nweighted <u>p</u>air <u>g</u>roup <u>m</u>ethod with <u>a</u>rithmetic mean). When comparing gene sequences, this method assumes that after two species have diverged from each other, the rate of accumulation of neutral mutations in a homologous region of DNA is approximately the same in the two species. As discussed in Section 29.5, this means that the molecular clock for neutral mutations is linear. If this assumption is met, closely related species (i.e., those that have diverged more recently) have

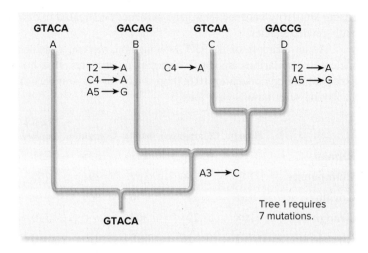

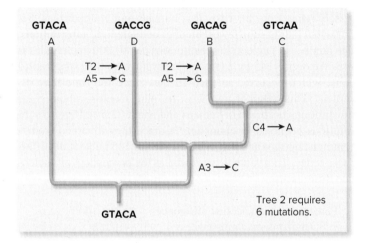

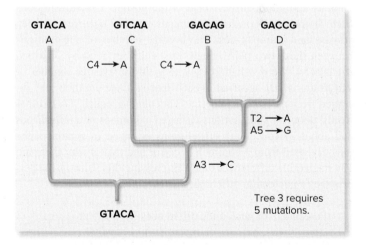

FIGURE 29.10 **The cladistic approach: choosing a cladogram from molecular genetic data.** This figure shows three different phylogenetic trees for the evolution of a short DNA sequence, but many more are possible. According to the principle of parsimony, the cladogram shown in tree 3 is the more plausible choice because it requires only five mutations. When constructing cladograms based on long genetic sequences, researchers use computers to generate trees with the fewest possible genetic changes.

CONCEPT CHECK: What is the principle of parsimony?

greater similarity between their gene sequences compared to distantly related species.

As an example of the UPGMA method, we can consider nucleotide similarities among five species of primates. In a homologous region containing 10,000 bp, the following numbers of nucleotide substitutions were found:

	Human	Chimpanzee	Gorilla	Orangutan	Rhesus monkey
Human	0	145	151	298	751
Chimpanzee	145	0	157	294	755
Gorilla	151	157	0	304	739
Orangutan	298	294	304	0	710
Rhesus monkey	751	755	739	710	0

To construct a tree, we begin with the pair that is the most closely related. In this case, it is humans and chimpanzees, because that pair has the fewest number of nucleotide substitutions. The number of nucleotide substitutions per 10,000 nucleotides is 145. We divide 145 by 2 to yield the average number of substitutions, which occurred in each species since the time they diverged. This number is 72.5. In other words, about 72.5 substitutions occurred in humans and about 72.5 (different) substitutions occurred in chimpanzees since they diverged from a common ancestor and that is why there are 145 nucleotide differences today. This value is expressed as the percentage of nucleotide differences:

$$\text{Percentage of nucleotide differences (humans vs. chimpanzees)} = \frac{72.5 \times 100}{10,000} = 0.725$$

Next, we consider the species that is the most closely related to humans and chimpanzees. That is the gorilla: 151 substitutions occurred between gorillas and humans, and 157 occurred between gorillas and chimpanzees. The average number of substitutions between these two pairs equals 151 + 157 divided by 2, which equals 154. We divide 154 by 2 to yield the average number of substitutions that occurred in each species since the time they diverged from a common ancestor. This number equals 77. In other words, about 77 substitutions occurred in humans or chimpanzees and about 77 (different) substitutions occurred in gorillas since they diverged from a common ancestor and that is why there are 154 nucleotide differences today. Expressing this value as the percentage of nucleotide differences gives:

$$\text{Percentage of nucleotide differences (gorillas vs. humans/chimpanzees)} = \frac{77 \times 100}{10,000} = 0.77$$

Next, we consider the species that is the most closely related to humans, chimpanzees, and gorillas. That is the orangutan: 298 substitutions occurred between orangutans and humans, 294 between orangutans and chimpanzees, and 304 between orangutans and gorillas. The average number of substitutions between these three pairs equals 298 + 294 + 304 divided by 3, which equals 299. We divide 299 by 2 to yield the average number of substitutions that occurred in each species since the time the

species diverged from a common ancestor. This number is approximately equal to 149. We express this value as the percentage of nucleotide differences:

$$\text{Percentage of nucleotide differences (orangutans vs. humans/chimpanzees/gorillas)} = \frac{149 \times 100}{10,000} = 1.49$$

Finally, we consider the most distantly related species, which is the rhesus monkey: 751 substitutions occurred between rhesus monkeys and humans, 755 between rhesus monkeys and chimpanzees, 739 between rhesus monkeys and gorillas, and 710 between rhesus monkeys and orangutans. The average number of substitutions between these four pairs equals 751 + 755 + 739 + 710 divided by 4, which equals 739. We divide 739 by 2 to yield the average number of substitutions that occurred in each species since the time the species diverged from a common ancestor. This number approximately equals 369. We express this value as the percentage of nucleotide differences:

$$\text{Percentage of nucleotide differences (rhesus monkeys vs. humans/chimpanzees/gorillas/orangutans)} = \frac{369 \times 100}{10,000} = 3.69$$

With these data, we can construct the phylogenetic tree shown in **Figure 29.11**. This illustration shows the relative evolutionary relationships among these primate species.

The Maximum Likelihood Approach and Bayesian Methods Are Also Used to Discriminate Among Possible Phylogenetic Trees

Evolutionary biologists also apply other approaches, such as maximum likelihood and Bayesian methods, when proposing and evaluating phylogenetic trees. These methods involve the use of an

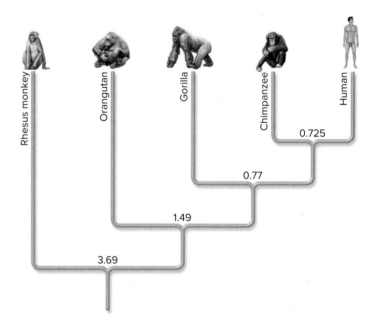

FIGURE 29.11 Construction of a phylogenetic tree using the UPGMA method.

evolutionary model—a set of assumptions about how evolution is likely to happen. For example, mutations affecting the third base in a codon are often neutral because they don't affect the amino acid sequence of the coded protein and therefore don't affect the fitness of an organism. As discussed later in this chapter, such neutral mutations are more likely to become prevalent in a population than are mutations in the first or second base. Therefore, one possible assumption of an evolutionary model is that neutral mutations are more likely than nonneutral ones.

When using an approach called **maximum likelihood,** researchers ask this question: What is the probability that an evolutionary model and a proposed phylogenetic tree would give rise to the observed data? The rationale is that a phylogenetic tree that gives a higher probability of producing the observed data is preferred to any trees that give a lower probability. By comparison, researchers using **Bayesian methods** ask this question: What is the probability that a particular phylogenetic tree is correct, given the observed data and a particular evolutionary model? Though the computational strategies of maximum likelihood and Bayesian methods are different (and beyond the scope of this text), the goal of both approaches is to identify one or more trees that are most likely to be correct based on an evolutionary model and the available data.

Phylogenetic Trees Refine Our Understanding of Evolutionary Relationships

In molecular evolutionary studies, the DNA sequences of many genes have been obtained from a wide range of sources. Many different gene sequences have been used to construct phylogenetic trees. The gene coding 16S rRNA (bacterial) or 18S rRNA (eukaryotic), an rRNA found in the small ribosomal subunit, is commonly analyzed. This gene has been sequenced from thousands of different species. Because rRNA is universal in all living organisms, its function was established at an early stage in the evolution of life on this planet, and its sequence has changed fairly slowly. Presumably, most mutations in this gene are deleterious, so the occurrence of new neutral or beneficial alleles is relatively rare. This limitation causes this gene sequence to change very slowly during evolution. Furthermore, 16S and 18S rRNAs are rather large molecules and therefore contain a large amount of sequence information.

In 1977, Carl Woese analyzed 16S and 18S rRNA sequences and proposed three main evolutionary branches called **domains: the Bacteria,** the **Archaea,** and the **Eukaryotes** (also called Eukarya). From these types of genetic analyses, it has become apparent that all living organisms are connected through a complex evolutionary tree.

Although the work of Woese was a breakthrough in our appreciation of evolution, more recent molecular genetic data have shed new light on the classification of species. Specifically, biologists once categorized eukaryotic species into four kingdoms: fungi (Fungi), animals (Animalia), plants (Plantae), and protists (Protista). However, recent models propose several major groups, called **supergroups,** as a way to organize eukaryotes into monophyletic groups.

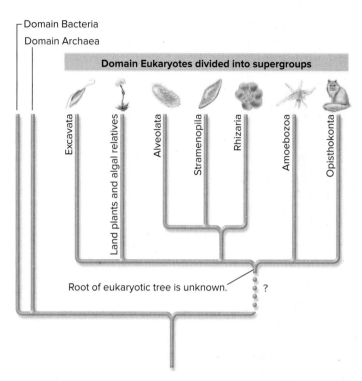

FIGURE 29.12 **A modern cladogram for eukaryotes.** Each supergroup shown here is hypothesized to be monophyletic. This tree should be considered a working hypothesis. The arrangement of these supergroups relative to each other is not entirely certain. Also, there are some eukaryotic species that cannot be placed in any of these supergroups.

CONCEPT CHECK: Discuss where protists are found in this newer organization of eukaryotic species.

Figure 29.12 shows a diagram that hypothesizes seven eukaryotic supergroups. Each supergroup contains distinctive types of protists. In addition, the kingdom Plantae is found within the supergroup Land plants and algal relatives, and the kingdoms Fungi and Animalia are within the supergroup Opisthokonta. As seen in this figure, molecular data and newer ways of building evolutionary trees reveal that protists played a key role in the evolution of many diverse groups of eukaryotic species, producing several large monophyletic supergroups.

Horizontal Gene Transfer Also Contributes to the Evolution of Species

The types of phylogenetic trees considered thus far are examples that show **vertical evolution,** in which species evolve from preexisting species by the accumulation of gene mutations and by changes in chromosome structure and number. Vertical evolution involves genetic changes in a series of ancestors that form a lineage. In addition to vertical evolution, however, species accumulate genetic changes by another mechanism called **horizontal gene transfer.** This refers to a process in which an organism incorporates genetic material from another organism without being the offspring of that organism. It often involves the exchange of genetic material between different species.

The traditional viewpoint is that the three domains of life arose from a single type of unicellular species called the **last common universal ancestor** (**Figure 29.13**). Though the precise timeline is difficult to pinpoint, bacteria diverged from archaea at least 2.5 billion years ago. At a later time, eukaryotes emerged. A scenario to explain how they emerged involved an archaeal host cell that established an endosymbiotic relationship with bacteria. With regard to the archaeal host cell, genome studies indicate that eukaryotes are most closely related to a superphylum called Asgard archaea, which have many eukaryotic cellular features. Even so, major contributions to the eukaryotic genome have come from bacteria. For example, molecular analyses indicate that mitochondria and chloroplasts are derived from bacterial endosymbiotic relationships.

How has horizontal gene transfer affected evolution? Analyses of many genomes suggests that horizontal gene transfer was prevalent during the early stages of evolution, when all organisms were unicellular, but continued even after the divergence of the three major domains of life (see Figure 29.13). With regard to modern organisms, horizontal gene transfer remains prevalent among prokaryotic species. By comparison, this process is less common among eukaryotes, though it does occur. Researchers have speculated that multicellularity and sexual reproduction have presented barriers to horizontal gene transfer among eukaryotes. For a gene to be transmitted to eukaryotic offspring, it must be transferred into a eukaryotic gamete or a cell that gives rise to gametes.

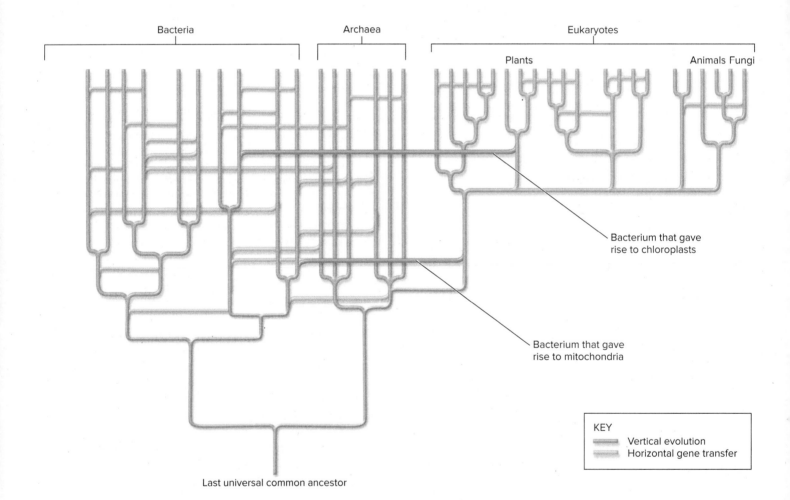

FIGURE 29.13   A revised scenario for the evolution of life, incorporating the concept of horizontal gene transfer. This phylogenetic tree shows a classification of life on Earth that includes the contribution of horizontal gene transfer in the evolution of species on our planet. This phenomenon was prevalent during the early stages of evolution when all organisms were unicellular. Horizontal gene transfer continues to be a prominent factor in the speciation of bacteria and archaea. (Note: This tree is meant to be schematic. Figure 29.12 is a more realistic representation of the evolutionary relationships among modern species. In this diagram, the endosymbiosis that gave rise to chloroplasts is shown as a single event, though multiple transfers of plastids have occurred among different supergroups during the evolution of life on Earth.)

CONCEPT CHECK: What is horizontal gene transfer?

29.4 COMPREHENSION QUESTIONS

1. Phylogenetic trees are based on
   a. natural selection.
   b. genetic drift.
   c. homology.
   d. none of the above.

2. A shared derived character is
   a. derived from an ancestral character.
   b. more recently developed than an ancestral character.
   c. found in all or most of the members of an ingroup.
   d. all of the above.

3. An approach that is used to construct and/or choose the most likely phylogenetic tree is
   a. cladistics and the principle of parsimony.
   b. phenetics.
   c. maximum likelihood or Bayesian methods.
   d. all of the above.

4. Horizontal gene transfer is a process in which genetic material from an organism is
   a. transferred from cell to cell.
   b. transferred to its offspring.
   c. transferred to another organism that is not the offspring of the first organism.
   d. released into the environment.

29.5 MOLECULAR EVOLUTION

Learning Outcomes:

1. Compare and contrast orthologs and paralogs.
2. Describe how researchers analyze orthologs to construct phylogenetic trees.
3. Outline the neutral theory of evolution.
4. Explain the concept of a molecular clock, and describe how it is used in dating a phylogenetic tree.
5. Describe how changes in chromosome structure and number are associated with molecular evolution.

Recall from the chapter introduction that molecular evolution consists of molecular-level changes in the genetic material that underlie the process of evolution. Such changes may be phenotypically neutral, or they may underlie the phenotypic changes associated with adaptive evolution.

Differences in nucleotide sequences are quantitative and can be analyzed using mathematical principles in conjunction with computer programs. Evolutionary changes at the DNA level can be objectively compared among different species to establish evolutionary relationships. Furthermore, this approach can be used to compare any two existing organisms, no matter how greatly they differ in their morphological traits. For example, we can compare DNA sequences between humans and bacteria, or between plants and fruit flies. Such comparisons would be very difficult at a morphological level. In this section, we will examine how evolution occurs at the molecular level.

Homologous Genes Are Derived from a Common Ancestral Gene

As we have seen, a phylogenetic tree is based on homology—that is, similarities among various species that occur because the species are derived from a common ancestor. In this section, we will focus on genetic homology. Two genes are said to be homologous if they are derived from the same ancestral gene. During evolution, a single species may diverge into two or more different ones. When two homologous genes are found in different species, these genes are termed **orthologs.** In Chapter 26, we considered orthologs of *Hox* genes, which have been identified in the fruit fly and the mouse.

In addition, two or more homologous genes can be found within a single species. These are termed paralogous genes, or **paralogs.** As discussed in Chapter 8, rare gene duplication events can produce multiple copies of a gene and ultimately lead to the formation of a **gene family.** A gene family consists of two or more paralogs within the genome of a particular species.

Figure 29.14a shows examples of both orthologs and paralogs in the globin gene family. Hemoglobin is an oxygen-carrying protein found in all vertebrate species. It is composed of two different subunits, coded by the α-globin and β-globin genes. Figure 29.14a shows an alignment of the deduced amino acid sequences coded by these genes. The sequences are homologous between humans and horses because of the evolutionary relationship between these two species. We say that the human and horse α-globin genes are orthologs of each other, as are the human and horse β-globin genes. The α-globin and β-globin genes in humans are paralogs of each other.

As shown in **Figure 29.14b**, the sequences of the orthologs are more similar to each other than they are to the paralogs. For example, the sequences of human and horse β globins show 25 differences, whereas human β globin and human α globin show 84 differences. What do these results mean? They indicate that the gene duplications that created the α-globin and β-globin genes occurred long before the evolutionary divergence that produced different species of mammals. For this reason, a greater amount of time has elapsed for the α- and β-globin genes to accumulate changes compared to the amount of time since the evolutionary divergence of mammalian species. This idea is schematically shown in **Figure 29.15**.

Based on the analysis of genetic sequences, evolutionary biologists have estimated that the gene duplication that produced the α-globin and β-globin gene lineages occurred approximately 400 mya, whereas the speciation events that resulted in different species of mammals occurred less than 200 mya. Therefore, the α-globin and β-globin genes have had much more time to accumulate changes relative to each other.

Hemoglobin

α	1	—	2	3	4	5	6	7	8	9	10	11	12	13	14	15	16	17	18	19	20	21	22	23	24	25	26	27	28	29	30
Human	Val	—	Leu	Ser	Pro	Ala	Asp	Lys	Thr	Asn	Val	Lys	Ala	Ala	Trp	Gly	Lys	Val	Gly	Ala	His	Ala	Gly	Glu	Tyr	Gly	Ala	Glu	Ala	Leu	Glu
Horse	Val	—	Leu	Ser	Ala	Ala	Asp	Lys	Thr	Asn	Val	Lys	Ala	Ala	Trp	Ser	Lys	Val	Gly	Gly	His	Ala	Gly	Glu	Val	Gly	Ala	Glu	Ala	Leu	Glu

β	1	2	3	4	5	6	7	8	9	10	11	12	13	14	15	16	17	18	—	—	19	20	21	22	23	24	25	26	27	28	29
Human	Val	His	Leu	Thr	Pro	Glu	Glu	Lys	Ser	Ala	Val	Thr	Ala	Leu	Trp	Gly	Lys	Val	—	—	Asn	Val	Asp	Glu	Val	Gly	Gly	Glu	Ala	Leu	Gly
Horse	Val	Gln	Leu	Ser	Gly	Glu	Glu	Lys	Ala	Ala	Val	Leu	Ala	Leu	Trp	Asp	Lys	Val	—	—	Asn	Glu	Glu	Glu	Val	Gly	Gly	Glu	Ala	Leu	Gly

α	31	32	33	34	35	36	37	38	39	40	41	42	43	44	45	46	—	47	48	49	50	—	—	—	—	—	51	52	53	54	55
Human	Arg	Met	Phe	Leu	Ser	Phe	Pro	Thr	Thr	Lys	Thr	Tyr	Phe	Pro	His	Phe	—	Asp	Leu	Ser	His	—	—	—	—	—	Gly	Ser	Ala	Gln	Val
Horse	Arg	Met	Phe	Leu	Gly	Phe	Pro	Thr	Thr	Lys	Thr	Tyr	Phe	Pro	His	Phe	—	Asp	Leu	Ser	His	—	—	—	—	—	Gly	Ser	Ala	Gln	Val

β	30	31	32	33	34	35	36	37	38	39	40	41	42	43	44	45	46	47	48	49	50	51	52	53	54	55	56	57	58	59	60
Human	Arg	Leu	Leu	Val	Val	Tyr	Pro	Trp	Thr	Gln	Arg	Phe	Phe	Glu	Ser	Phe	Gly	Asp	Leu	Ser	Thr	Pro	Asp	Ala	Val	Met	Gly	Asn	Pro	Lys	Val
Horse	Arg	Leu	Leu	Val	Val	Tyr	Pro	Trp	Thr	Gln	Arg	Phe	Phe	Asp	Ser	Phe	Gly	Asp	Leu	Ser	Asn	Pro	Gly	Ala	Val	Met	Gly	Asn	Pro	Lys	Val

α	56	57	58	59	60	61	62	63	64	65	66	67	68	69	70	71	72	73	74	75	76	77	78	79	80	81	82	83	84	85	86
Human	Lys	Gly	His	Gly	Lys	Lys	Val	Ala	Asp	Ala	Leu	Thr	Asn	Ala	Val	Ala	His	Val	Asp	Asp	Met	Pro	Asn	Ala	Leu	Ser	Ala	Leu	Ser	Asp	Leu
Horse	Lys	Ala	His	Gly	Lys	Lys	Val	Gly	Asp	Ala	Leu	Thr	Leu	Ala	Val	Gly	His	Leu	Asp	Asp	Leu	Pro	Gly	Ala	Leu	Ser	Asp	Leu	Ser	Asn	Leu

β	61	62	63	64	65	66	67	68	69	70	71	72	73	74	75	76	77	78	79	80	81	82	83	84	85	86	87	88	89	90	91
Human	Lys	Ala	His	Gly	Lys	Lys	Val	Leu	Gly	Ala	Phe	Ser	Asp	Gly	Leu	Ala	His	Leu	Asp	Asn	Leu	Lys	Gly	Thr	Phe	Ala	Thr	Leu	Ser	Glu	Leu
Horse	Lys	Ala	His	Gly	Lys	Lys	Val	Leu	His	Ser	Phe	Gly	Glu	Gly	Val	His	His	Leu	Asp	Asn	Leu	Lys	Gly	Thr	Phe	Ala	Ala	Leu	Ser	Glu	Leu

α	87	88	89	90	91	92	93	94	95	96	97	98	99	100	101	102	103	104	105	106	107	108	109	110	111	112	113	114	115	116	117
Human	His	Ala	His	Lys	Leu	Arg	Val	Asp	Pro	Val	Asn	Phe	Lys	Leu	Leu	Ser	His	Cys	Leu	Leu	Val	Thr	Leu	Ala	Ala	His	Leu	Pro	Ala	Glu	Phe
Horse	His	Ala	His	Lys	Leu	Arg	Val	Asp	Pro	Val	Asn	Phe	Lys	Leu	Leu	Ser	His	Cys	Leu	Leu	Ser	Thr	Leu	Ala	Val	His	Leu	Pro	Asn	Asp	Phe

β	92	93	94	95	96	97	98	99	100	101	102	103	104	105	106	107	108	109	110	111	112	113	114	115	116	117	118	119	120	121	122
Human	His	Cys	Asp	Lys	Leu	His	Val	Asp	Pro	Glu	Asn	Phe	Arg	Leu	Leu	Gly	Asn	Val	Leu	Val	Cys	Val	Leu	Ala	His	His	Phe	Gly	Lys	Glu	Phe
Horse	His	Cys	Asp	Lys	Leu	His	Val	Asp	Pro	Glu	Asn	Phe	Arg	Leu	Leu	Gly	Asn	Val	Leu	Ala	Val	Val	Leu	Ala	Arg	His	Phe	Gly	Lys	Asp	Phe

α	118	119	120	121	122	123	124	125	126	127	128	129	130	131	132	133	134	135	136	137	138	139	140	141
Human	Thr	Pro	Ala	Val	His	Ala	Ser	Leu	Asp	Lys	Phe	Leu	Ala	Ser	Val	Ser	Thr	Val	Leu	Thr	Ser	Lys	Tyr	Arg
Horse	Thr	Pro	Ala	Val	His	Ala	Ser	Leu	Asp	Lys	Phe	Leu	Ser	Ser	Val	Ser	Thr	Val	Leu	Thr	Ser	Lys	Tyr	Arg

β	123	124	125	126	127	128	129	130	131	132	133	134	135	136	137	138	139	140	141	142	143	144	145	146
Human	Thr	Pro	Pro	Val	Gln	Ala	Ala	Tyr	Gln	Lys	Val	Val	Ala	Gly	Val	Ala	Asn	Ala	Leu	Ala	His	Lys	Tyr	His
Horse	Thr	Pro	Glu	Leu	Gln	Ala	Ser	Tyr	Gln	Lys	Val	Val	Ala	Gly	Val	Ala	Asn	Ala	Leu	Ala	His	Lys	Tyr	His

(a) Alignment of human and horse globin polypeptides

Orthologs	Number of Amino Acid Differences
Human α globin vs. horse α globin	18 out of 141
Human β globin vs. horse β globin	25 out of 146
Paralogs	
Human α globin vs. human β globin	84 out of 146
Horse α globin vs. horse β globin	81 out of 146

(b) A comparison of amino acid differences between orthologs and paralogs

FIGURE 29.14 **A comparison of the α- and β-globin polypeptides from humans and horses.** (a) An alignment of the deduced amino acid sequences obtained by sequencing the exon portions of the corresponding genes. The gaps indicate where additional amino acids are found in the sequence of myoglobin, another polypeptide coded by a member of this gene family. (b) A comparison of amino acid differences between orthologs and paralogs of human and horse globin genes.

CONCEPT CHECK: What is the difference between a paralog and an ortholog?

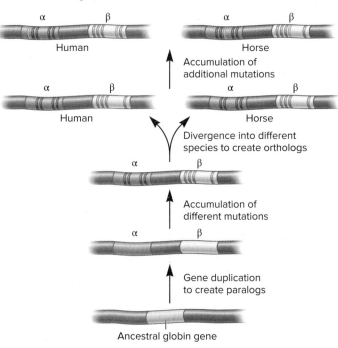

FIGURE 29.15 **Evolution of paralogs and orthologs.** In this schematic illustration, the ancestral globin gene duplicated to create the α- and β-globin genes, which are paralogs. Over time, these two paralogs accumulated different mutations, designated with red and blue lines. At a later point in evolution, a species divergence occurred to create different mammalian species, including humans and horses. Over time, the orthologs also accumulated different mutations, designated with green lines. As noted here, the orthologs have fewer differences than the paralogs because the gene duplication occurred prior to the species divergence. (Note: Actually, gene duplications occurred several times to produce a large globin gene family, but this is a simplified example that shows only one gene duplication event. Also, chromosomal rearrangements have placed the α- and β-globin genes on different chromosomes.)

Scientists Can Compare Orthologs Among Living and Extinct Flightless Birds to Establish Evolutionary Relationships

A common way to construct phylogenetic trees is by comparing orthologs. The majority of such trees have been constructed from molecular data using DNA samples collected from living species. With this approach, we can infer the prehistoric changes that gave rise to present-day DNA sequences. In addition, scientists have discovered that it is occasionally possible to obtain DNA sequence information from species that lived in the past. In 1984, the first successful attempt at determining DNA sequences from an extinct species was accomplished by groups of researchers at the University of California at Berkeley and the San Diego Zoo, including Russell Higuchi, Barbara Bowman, Mary Freiberger, Oliver Ryder, and Allan Wilson. They obtained a sample of dried muscle from a museum specimen of the quagga (*Equus quagga*), a zebra-like species that became extinct in 1883. This piece of muscle tissue was obtained from an animal that had died 140 years ago. A sample of its skin and muscle was preserved in salt in the Museum of Natural History at Mainz, Germany. The researchers extracted DNA from the sample, cloned pieces of it into vectors, and then sequenced the quagga DNA. This pioneering study opened the field of **ancient DNA analysis,** also known as **molecular paleontology.**

Since the mid-1980s, many researchers have become excited about the information that might be derived from sequencing DNA obtained from specimens of extinct species. Currently there is debate about how long DNA can remain significantly intact after an organism has died. Over time, the structure of DNA is degraded by hydrolysis and the loss of purines. Nevertheless, under certain conditions (e.g., cold temperature, low oxygen), DNA samples may remain stable for as long as 50,000–100,000 years, and perhaps longer.

In most studies involving prehistoric specimens (in particular, those that are much older than the salt-preserved quagga sample), the ancient DNA is extracted from bone, dried muscle, or preserved skin. These samples are often obtained from museum specimens that have been gathered by archaeologists. PCR technology, described in Chapter 20, is conducted to amplify very small amounts of DNA using PCR primers that typically flank a region within the 12S rRNA gene, a mitochondrial gene. By comparing orthologs of the 12S rRNA among different species, researchers have achieved some success at elucidating the phylogenetic relationships between modern and extinct species.

In the experiment presented in **Figure 29.16,** Alan Cooper, Cécile Mourer-Chauviré, Geoffrey Chambers, Arndt von Haeseler, Allan Wilson, and Svante Pääbo investigated the evolutionary relationships among some extinct and modern species of flightless birds. Two groups of flightless birds, the moas and the kiwis, existed in New Zealand during the Pleistocene era. The moas are now extinct, although 11 species were formerly present. In this study, the researchers investigated the phylogenetic relationships among four extinct species of moas that were available as museum samples, three kiwis of New Zealand, and several other (nonextinct) species of flightless birds. These included the emu and the cassowary (found in Australia and New Guinea), the ostrich (found in Africa and formerly Asia), and two rheas (found in South America).

In this work, the researchers wanted to compare orthologs of the 12S rRNA gene. To do so, they collected samples from the various species and used PCR to amplify the 12S rRNA gene from each sample. This provided enough DNA to subject the gene to DNA sequencing. The sequences of the 12S rRNA gene orthologs were aligned using computer programs as described in Chapter 23.

THE GOAL (DISCOVERY-BASED SCIENCE)

Because DNA is a relatively stable molecule, it can be isolated from a preserved sample of a deceased organism and then subjected to PCR and DNA sequencing. A comparison of DNA sequences of extinct and modern species may help elucidate the phylogenetic relationships among the species.

ACHIEVING THE GOAL   **FIGURE 29.16** **DNA analysis of orthologs reveals phylogenetic relationships among extinct and modern flightless birds.**

Starting material: Tissue samples from four extinct species of moas were obtained from museum specimens. Tissue samples were also obtained from three species of kiwis, one emu, one cassowary, one ostrich, and two species of rhea.

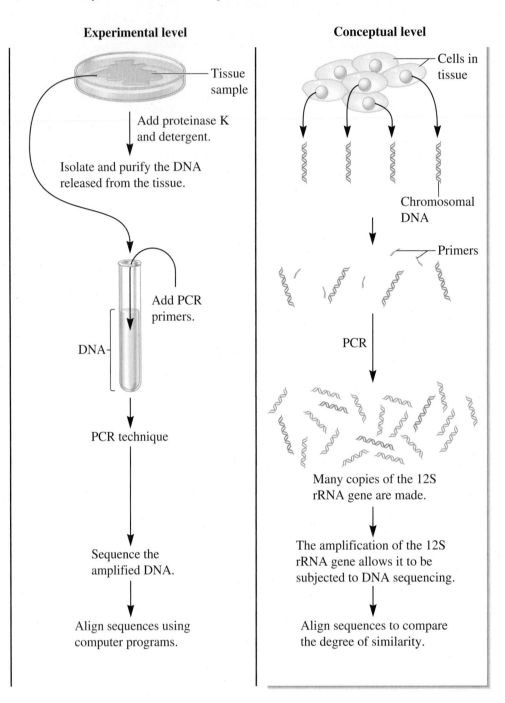

Experimental level

Conceptual level

1. For cellular samples, treat with proteinase K (which digests protein) and a detergent that dissolves cell membranes. This releases the DNA from the cells.

Tissue sample

Add proteinase K and detergent.

Isolate and purify the DNA released from the tissue.

Cells in tissue

Chromosomal DNA

2. Individually, mix the DNA samples with a pair of PCR primers that are complementary to the 12S rRNA gene.

DNA

Add PCR primers.

Primers

3. Subject the samples to PCR. See Chapter 20 (Figure 20.6) for a description of PCR.

PCR technique

PCR

Many copies of the 12S rRNA gene are made.

4. Subject the amplified DNA fragments to DNA sequencing. See Chapter 20 for a description of DNA sequencing.

Sequence the amplified DNA.

The amplification of the 12S rRNA gene allows it to be subjected to DNA sequencing.

5. Align the DNA sequences to each other. Methods of DNA sequence alignment are described in Chapter 23.

Align sequences using computer programs.

Align sequences to compare the degree of similarity.

THE DATA

```
MOA 1      GCTTAGCCCTAAATCCAGATACTTACCCTACACAAGTATCCGCCCGAGAACTACGAGCACAAACGCTTAAAACTCTAAGGACTTGGCGGTGCCCCAAACCCACCTAGAGGAGCCTGTTCTATAATCGATAATCCACGATA
MOA 2      .........................................................................................................................................
MOA 3      ...G.........T...........................T...............................................T.......................................
MOA 4      ............................C..........................................................................C.....T...
KIWI 1     ..........T.G....GT..CT...C.........................................T.......................C....
KIWI 2     ..........T.G.G...AT..CT...C.........................................T.......................C....
KIWI 3     ..........T.G.G..G.AT..C...C.........................................T.......................C....
EMU        ........TT....C..T..CAG..C..T.......................................T.......................C....
CASSOWARY  ........TT....CG.TA..CTG..........................................T.......................C....
OSTRICH    .......T..AT....C.CT.............................................T.......................T
RHEA 1     .......T........C.CT.............................................T.......................C....
RHEA 2     ............C.......C.C....................................T.......................C....
```

```
MOA 1      CACCCGACCATCCCTCGCCCGT-GCAGCCTACATACCGCGTCCCCAGCCCGCCT--AATGAAAG-AACAATAGCGAGCACAACAGCCCTCCCCCGCTAACAAGACAGGTCAAGGTATAGCATATGAGATGGAAGAAATG
MOA 2      ......................A--.......................................TCA-.....................................
MOA 3      .......T.T..A-.........................-...............TA---.T..............................
MOA 4      .......T.T..A-.........................T..AC--.............................
KIWI 1     ....A..T.T..AAC-A.......T.........G...T...AA...G.....C...A.....TA-..A.........................C.
KIWI 2     ....A..T.T..AAC-A.......T.........G...T...AA...G.....C...A.....TA-..A.........................C.
KIWI 3     ....A..T.T..AAC-A.........G.......AA.......GC.........TACA-..A.........................CC.C.......G..
EMU        ...AG....T.T..AA.-A.........G..............--.......T...AC--TT.........................G..
CASSOWARY  ....A..T.T..AA.TA.........G.....---.G..G.........T...AC--T.........................G..
OSTRICH    ....A..C..T..A--T.........G.........C---..G..........T...A.........................GAG.....
RHEA 1     ....T.T..A-.........................TA.G....C..AG...T..T..TA---.........................G..
RHEA 2     ....T.T..A-.........................TA...G.....C..A..T.T..TA---.........G.........................
```

```
MOA 1      GGCTACATTTTCTAACATAGAACACCC-------------ACGAAAGAGAAGGTGAAACCCTCCTCAAAAGGCGGATTTAGCAGTAAAATAGAACAAGAATGCCTATTTTAAGCCCGGCCCTGGGGC
MOA 2      .........................-------------...........A...T....G.......................T........
MOA 3      ...............T-------------.............G.......G.....C...C...C.......T......
MOA 4      .........................-------------.A......G.....G...C..C.......T......
KIWI 1     .......A.....T.T-------------A.GGT...T.-C..T.G.......C...T..GA.T...--.T....A.....
KIWI 2     .......A.....T.T-------------A.GGT...T.C..T.G.......C...T..GA.T...--.T....A.....
KIWI 3     .......A.....T.T-------------A.GGTA...T.C..T.G.......C...T..A.T.........A.....
EMU        .......T.T-------------AG.T...T.AC.T...G.......C...T..GA.T....A-.T...T.A...
CASSOWARY  .......T-------------A..G.T...T.A..T.G.......C...T..GA.T....A--......A...
OSTRICH    .......T.A-------------.G.TA...T.A.........G.......C...T..GA.T....-T...T.A...
RHEA 1     ...TC.....A-------------.G....GGCA....AC....CG.......G...G.TC...A..C.C.........A...
RHEA 2     ...GTC....G-------------.GGCA....AC....CG.......G.G.G.TC...A..C.C.........A...
```

Source: Cooper, Alan, Mourer-Chauvire, Cecile, Chambers, Geoffrey K., von Haeseler, Arndt, Wilson, Allan C., and Paabo, Svante, "Independent origins of New Zealand moas and kiwis," *Proceedings of the National Academy of Science*, vol. 89, no.18, September 15, 1992, 8741–8744.

INTERPRETING THE DATA

The data present a multiple sequence alignment of the amplified DNA sequences. The first line shows the DNA sequence of one extinct moa species, and underneath it are the sequences of the other species. When the other sequences are identical to the first sequence, a dot is placed in the corresponding position. When the sequences are different, the changed base (A, T, G, or C) is placed there. In a few regions, the genes are different lengths. In these cases, one or more dashes indicate where the sequences differ in length.

As you can see from the large number of dots, the sequences among these flightless birds are very similar. To establish evolutionary relationships, researchers focus on sites where the gene sequences are not identical. At these sites, base changes have occurred, thereby identifying species with shared derived characters. Some surprising results were obtained.

- Certain sites in the DNA sequences from the kiwis (a New Zealand species) are the same as the sequence from the ostrich (an African species), but different from those of the moas, which were once found in New Zealand.
- Several sites in the DNA sequences of the kiwis are the same as the emu and cassowary (found in Australia and New Guinea), but different from the moas.

Contrary to their original expectations, the researchers concluded that the kiwis are more closely related to Australian and African flightless birds than they are to the moas. They proposed that New Zealand was colonized twice by ancestors of flightless birds. As shown in **Figure 29.17**, the researchers constructed a new phylogenetic tree to illustrate the relationships among these modern and extinct species.

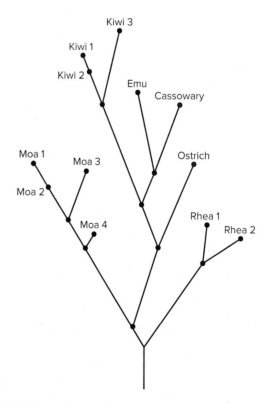

FIGURE 29.17 A revised phylogenetic tree of moas, kiwis, emus, cassowaries, ostriches, and rheas based on a comparison of orthologs.

Genetic Variation at the Molecular Level Is Associated with Neutral Changes in Gene Sequences

As we have seen, the globin genes in mammals and the 12S rRNA genes in flightless birds exhibit variation in their sequences. Is such variation due primarily to mutations that have been favored by natural selection or to genetic drift?

A **nonneutral mutation** is one that affects the phenotype of the organism and can be acted on by natural selection. Such a mutation may only subtly alter the phenotype of an organism, or it may have a major effect. According to Darwin, natural selection is the agent that leads to evolutionary change in populations. It selects for individuals with the highest Darwinian fitness and often promotes the establishment of beneficial alleles and the elimination of deleterious ones. Therefore, many geneticists have assumed that natural selection is the dominant factor in changing the genetic composition of natural populations, thereby leading to variation.

In opposition to this viewpoint, in 1968, Motoo Kimura proposed the **neutral theory of evolution.** According to this theory, most genetic variation observed in natural populations is due to the accumulation of neutral mutations that do not affect the phenotype of the organism and are not acted on by natural selection. For example, a mutation within a protein-coding gene that changes a glycine codon from GGG to GGC does not affect the amino acid sequence of the coded protein. Because neutral mutations do not affect phenotype, they spread throughout a population according to their frequency of appearance and to genetic drift. This theory has been called "survival of the luckiest" and also **non-Darwinian evolution** to contrast it with Darwin's theory, which focuses on fitness. Kimura agreed with Darwin that natural selection is responsible for adaptive changes in a species during evolution. The main distinction is that most variation in gene sequences is neutral with respect to natural selection.

In support of the neutral theory of evolution, Kimura and colleague Tomoko Ohta outlined five principles that govern the evolution of genes at the molecular level:

1. For each protein, the rate of evolution, in terms of amino acid substitutions, is approximately constant with regard to neutral substitutions that do not affect protein structure or function.

   Evidence: As an example, the amount of genetic variation between the coding sequences of the human α-globin and β-globin genes is approximately the same as the difference between the α-globin and β-globin genes in the horse (shown earlier in Figure 29.14b). This type of comparison holds true for many different genes compared among many different species.

2. Proteins that are functionally less important for the survival of an organism or parts of a protein that are less important for its function tend to evolve faster than more important proteins or regions of a protein. In other words, during evolution, less important proteins or protein domains accumulate amino acid substitutions more rapidly than important ones do.

   Evidence: Certain proteins are critical for survival, and their structure is precisely suited to their function. Examples are the histone proteins necessary for nucleosome formation in eukaryotes. Histone genes tolerate very few mutations and have evolved extremely slowly. By comparison, genes that code fibrinopeptides, which bind to fibrinogen to form a blood clot, evolve very rapidly. Presumably, the sequence of amino acids in these polypeptides is not very important for allowing them to aggregate and form a clot.

3. Amino acid substitutions that do not significantly alter the existing structure and function of a protein are found more commonly than disruptive amino acid changes.

   Evidence: When comparing the coding sequences within homologous genes of modern species, nucleotide differences are more likely to be observed in the wobble base than in the first or second base within a codon. Mutations in the wobble base are often silent because they do not change the amino acid sequence of the protein. In addition, conservative substitutions (i.e., a substitution with a similar amino acid, such as a nonpolar amino acid for another nonpolar amino acid) are fairly common. By comparison, nonconservative substitutions—those that significantly alter the structure and function of a protein—are less frequent. Nonsense and frameshift mutations are very rare within the coding sequences of genes. Also, intron sequences evolve more rapidly than exon sequences.

4. Gene duplication often precedes the emergence of a gene having a new function.

   Evidence: When a single copy of a gene exists in a species, it usually plays a functional role similar to that of the homologous gene found in another species. Gene duplications have produced gene families in which each family member has evolved to have a somewhat different functional role. An example is the globin gene family described in Chapter 8.

5. Selective elimination of definitely deleterious mutations and the random fixation of selectively neutral or very slightly deleterious alleles occur far more frequently during evolution than selection of advantageous mutants.

   Evidence: As mentioned in principle 3, silent and conservative mutations are much more common than nonconservative substitutions. Presumably nonconservative mutations usually have a negative effect on the phenotype of the organism, so they are effectively eliminated from the population by natural selection. On rare occasions, however, an amino acid substitution due to a mutation may have a beneficial effect on the phenotype. For example, a nonconservative mutation in the β-globin gene produced the Hb^S allele, which gives an individual resistance to malaria in the heterozygous condition.

In general, the DNA sequencing of hundreds of thousands of different genes from thousands of species has provided compelling support for these five principles of gene evolution at the molecular level. When it was first proposed, the neutral theory of evolution sparked a great debate. Some geneticists, called selectionists, strongly opposed this theory. However, the debate largely cooled after Tomoko Ohta incorporated the concept of nearly neutral mutations into the theory. Nearly neutral mutations have a minimal effect on phenotype—they may be slightly beneficial or slightly detrimental. Ohta suggested that the prevalence of such alleles can depend mostly on natural selection or mostly on genetic drift, depending on the population size.

Why do evolutionary biologists care about neutral or nearly neutral mutations? One reason is that their prevalence is used as a tool to add a time scale to phylogenetic trees. This topic is discussed next.

Molecular Clocks Can Be Used to Date the Divergence of Species

According to the neutral theory of evolution, most of the genetic variation observed in natural populations is due to neutral mutations. In a sense, the relatively constant rate of neutral or nearly neutral mutations acts as a **molecular clock** with which to measure evolutionary time. According to this idea, neutral mutations become fixed in a population at a rate that is proportional to the rate of mutation per generation. On this basis, the genetic divergence between species that is due to neutral mutations reflects the time elapsed since they diverged from a common ancestor.

Figure 29.18a shows an example of a molecular clock derived from a study by Francisco Ayala involving superoxide dismutase found in 27 different fruit fly species. (Superoxide dismutase is an enzyme that protects cells against harmful free radicals.) Twenty-three species were in the genus *Drosophila*, two in the genus *Chymomyza*, one in the genus *Scaptodrosophila*, and one in the genus *Ceratitis*. The genus *Ceratitis* is in the family Tephritidae and is more distantly related to the other 26 species, which are in the family Drosophilidae. In this figure, the y-axis is a measure of the average number of amino acid differences in superoxide dismutase between pairs of species or between groups of species. The x-axis plots the amount of time that has elapsed since a pair of species or two different groups diverged from a common ancestor. For example, the yellow dot represents the average number of amino acid differences between *Ceratitis* (in family Tephritidae) versus the other species (in family Drosophilidae). Approximately 30 amino acid differences were observed. By comparison, each of the two red dots compares one of the *Chymomyza* species with all of the *Drosophila* species. The blue dots are pairwise comparisons of some *Drosophila* species with each other or with the *Scaptodrosophila* species.

 FIGURE 29.18 **A molecular clock.** According to the concept of a molecular clock, neutral or nearly neutral mutations accumulate over evolutionary time at a fairly constant rate. **(a)** The clock shown here is based on an analysis of superoxide dismutase found in 27 different fruit fly species. Twenty-three species were in the genus *Drosophila*, two in the genus *Chymomyza*, one in the genus *Scaptodrosophila*, and one in the genus *Ceratitis*. Comparisons of the amino acid sequence of this enzyme between species or groups of species showed that those that diverged more recently tend to have fewer differences than those whose common ancestor lived in the very distant past. **(b)** A phylogenetic tree for the species shown in part (a).

Source: Ayala, Francisco J., "Vagaries of the molecular clock," *Proceedings of the National Academy of Sciences*, vol. 94, no. 15, July 22, 1997, 7776–7783.

CONCEPT CHECK: With regard to phylogenetic trees, how is a molecular clock useful?

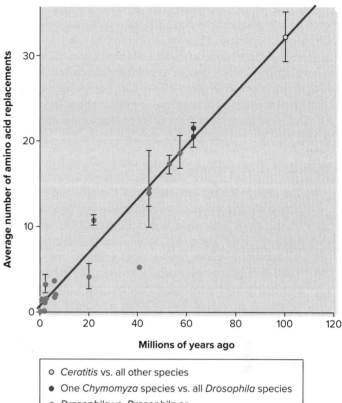

(a) An example of a molecular clock

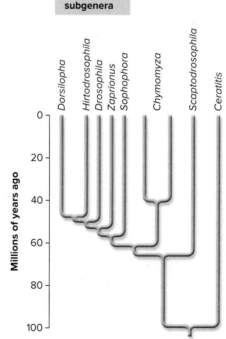

(b) A phylogenetic tree with a time scale

As a general trend, species that diverged a longer time ago show a greater number of amino acid differences than species that diverged more recently. The explanation for this phenomenon is that species accumulate independent mutations (e.g., nearly neutral mutations) after they have diverged from each other. A longer period of time since divergence allows for a greater accumulation of mutations that makes the sequences differ between species.

To further understand the concept of a molecular clock, let's consider how the molecular clock data are related to a phylogenetic tree (**Figure 29.18b**). In this diagram, the *Drosophila* genus is divided into five subgenera.

- The divergence between the genus *Ceratitis* and the other genera occurred a long time ago, nearly 100 mya. The molecular clock data described in Figure 29.18a showed a relatively large number of amino acid differences (about 30) between *Ceratitis* and the other genera.
- The divergence between *Chymomyza* species and *Drosophila* species occurred more recently—about 65 mya as shown in Figure 29.18b. The molecular clock data revealed fewer amino acid differences (about 20) between *Chymomyza* and *Drosophila* species (see Figure 29.18a).
- The five subgenera of *Drosophila* diverged the most recently and had the fewest number of amino acid differences.

Figure 29.18a suggests a linear relationship between the number of sequence changes and the time of divergence. Such a relationship indicates that the observed rate of neutral or nearly neutral mutations remains constant over millions of years. For example, the linear relationship shown in Figure 29.18a indicates that a pair of species showing 15 amino acid differences diverged from a common ancestor that existed about 50 mya, whereas a pair of species showing 30 amino acid differences diverged from a common ancestor that existed approximately 100 mya. In other words, twice as many amino acid differences (30 versus 15) correlate with a divergence that occurred twice as many years ago (100 mya versus 50 mya). Although actual data sometimes show a relatively linear relationship over a defined period, evolutionary biologists have discovered that molecular clocks are often not linear. This is particularly true over very long periods or when comparing distantly related taxa. When comparing different species or groups of species, several factors can contribute to the nonlinearity of molecular clocks. These include the following:

- Differences in population sizes that may affect the relative effects of genetic drift and natural selection
- Differences in mutation rates among different species
- Differences in the generation times of the species being analyzed
- Differences in the relative number of sites in a gene or protein that are susceptible to neutral mutations

To produce a time scale, researchers need to calibrate their molecular clocks. How much time does it take to accumulate a certain percentage of nucleotide changes? To perform such a calibration, researchers must have information regarding the date when two species diverged from a common ancestor. Such information could come from the fossil record, for example.

The genetic differences between those species are then divided by the time elapsed since their divergence from a common ancestor to calculate a rate of change. For example, fossil evidence suggests that humans and chimpanzees diverged from a common ancestor approximately 6 mya. The percentage of nucleotide differences in mitochondrial DNA between humans and chimpanzees is 12%. From these data, the molecular clock for nucleotide changes in the mitochondrial DNA sequences of primates is calibrated at roughly 2% of nucleotides changing per million years.

Evolution Is Associated with Changes in Chromosome Structure and Number

Thus far, we have focused on mutations that alter the DNA sequences within genes. In addition to gene mutations, other types of changes, such as gene duplications, transpositions, inversions, translocations, and changes in chromosome number, are important factors in evolution.

As discussed earlier, changes in chromosome structure and/or number may lead to reproductive isolation and the origin of new species. As an example of variation in chromosome structure among closely related species, **Figure 29.19** compares the banding patterns of the three largest chromosomes in humans and the corresponding chromosomes in chimpanzees, gorillas, and orangutans. The banding patterns are strikingly similar because these species are closely related evolutionarily. However, some interesting differences are observed.

- Humans have one large chromosome 2, but this chromosome is divided into two separate chromosomes in the other three species. This explains why humans have 23 pairs of chromosomes and the other species have 24. This difference may have arisen through a fusion of the two smaller chromosomes during the evolution of the human lineage, which is an example of a Robertsonian translocation, described in Chapter 8.
- Another interesting difference in chromosome structure is seen in chromosome 3. The banding patterns among humans, chimpanzees, and gorillas are very similar, but the orangutan has a large inversion that flips the arrangement of bands in the centromere region.

Synteny Groups Contain the Same Groups of Linked Genes

With the advent of molecular techniques, researchers can analyze the chromosomes of two or more different species and identify regions that contain the same groups of linked genes, which are called **synteny groups.** Within a particular synteny group, the same types of genes are found in the same order. In 1995, Graham Moore and colleagues analyzed the locations of molecular markers on the chromosomes of several cereal grasses, including rice (*Oryza sativa*), wheat (*Triticum aestivum*), maize (*Zea mays*), foxtail millet (*Setaria italica*), sugarcane (*Saccharum officinarum*), and sorghum (*Sorghum vulgare*). From this analysis, they were able to identify several large synteny groups that are common to most of these species (**Figure 29.20**).

FIGURE 29.19 **A comparison of banding patterns among the three largest human chromosomes and the corresponding chromosomes in the chimpanzee, gorilla, and orangutan.** This is a schematic drawing of late prophase chromosomes. The conventional numbering system of the banding patterns is shown next to the human chromosomes.

CONCEPT CHECK: Describe two differences among these chromosomes.

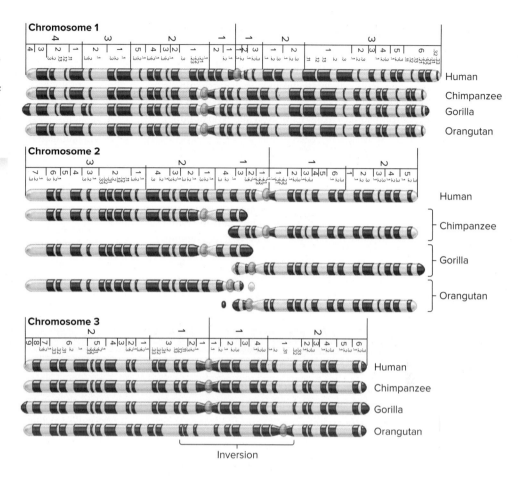

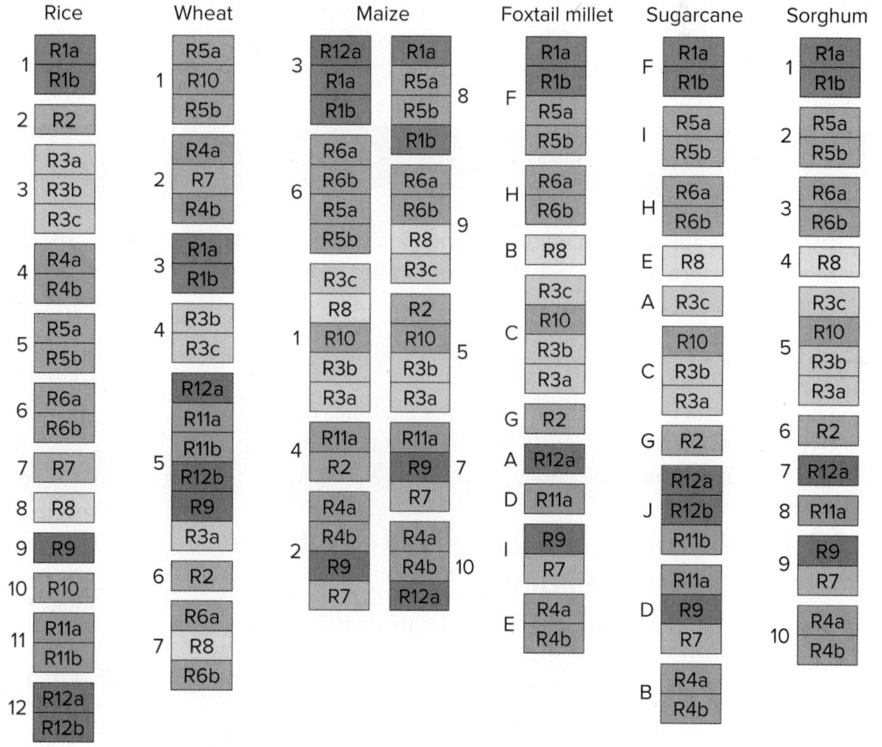

FIGURE 29.20 **Synteny groups in cereal grasses.** The synteny groups are named according to their locations on the set of 12 chromosomes from rice, which are shown on the left. In some cases, a synteny group or portion of a group may have incurred an inversion, but these are not shown in this figure. Also, the genome of maize contains two copies of most synteny groups, suggesting that it is derived from an ancestral polyploid species.

As an example, let's compare rice (12 chromosomes per set) with wheat (7 chromosomes per set).

- In rice, chromosome 6 contains two synteny groups designated R6a and R6b, whereas chromosome 8 contains a single synteny group, R8.
- In wheat, chromosome 7 consists of R6a and R6b at either end, and R8 is sandwiched in the middle.

One possible explanation for these differences could be a chromosomal rearrangement during the evolution of wheat in which R8 became inserted into the middle of a chromosome containing R6a and R6b. Overall, the evolution of cereal grass species has maintained most of the same types of genes, but many chromosomal rearrangements have occurred. These types of rearrangements promote reproductive isolation.

29.5 COMPREHENSION QUESTIONS

1. Homologous genes found in different species are called
   a. orthologs.
   c. analogs.
   b. paralogs.
   d. none of these.

2. A molecular clock may not be linear because
   a. mutation rates may differ among different species.
   b. differences in population sizes may affect the relative effects of natural selection and genetic drift.
   c. different species may differ in their generation times.
   d. all of the above may occur.

3. When the chromosomes of closely related species are compared,
   a. the banding patterns are often similar.
   b. a few structural alterations may be seen.
   c. a change in chromosome number may be seen.
   d. all of the above are commonly observed.

KEY TERMS

Introduction: RNA world, biological evolution (evolution), microevolution, macroevolution, molecular evolution

29.1: adaptive evolution

29.2: species, subspecies, ecotypes, reproductive isolation, prezygotic isolating mechanism, postzygotic isolating mechanism, species concept, biological species concept, evolutionary species concept, ecological species concept, general lineage concept

29.3: speciation, anagenesis, cladogenesis, allopatric speciation, sympatric speciation, founder effect, hybrid zones, reinforcement, tension zones

29.4: phylogeny, phylogenetic tree, homology, homologous, monophyletic group (clade), phenetic approach, phenogram, cladistic approach, cladogram, character, character states, ancestral character (primitive character), shared derived character (synapomorphy), ingroup, outgroup, principle of parsimony, maximum likelihood, Bayesian methods, domain, Bacteria, Archaea, Eukaryotes, supergroups, vertical evolution, horizontal gene transfer, last common universal ancestor

29.5: orthologs, paralogs, gene family, ancient DNA analysis (molecular paleontology), nonneutral mutation, neutral theory of evolution (non-Darwinian evolution), molecular clock, synteny groups

---

CHAPTER SUMMARY

- Evolution is the accumulation of heritable changes in one or more characteristics of a population or species from one generation to the next.

29.1 Overview of Evolution

- Darwin's theory of evolution, which is sometimes called adaptive evolution, is based on two fundamental factors: genetic variation and natural selection.

29.2 Identification of Species and Mechanisms of Reproductive Isolation

- The most commonly used characteristics to identify species are morphological traits, inability to interbreed, molecular features, ecological factors, and evolutionary relationships. However, each of these has its drawbacks (see Figure 29.1).
- Mechanisms of reproductive isolation can be prezygotic or postzygotic (see Table 29.1).

- A species concept is a way to define what a species is and/or to distinguish one species from another. The general lineage concept is receiving wide support.

29.3 Speciation

- Speciation is an evolutionary process by which populations evolve to become distinct species. It usually occurs via cladogenesis, in which a single species diverges into two or more species (see Figure 29.2).
- Cladogenesis can occur via allopatric or sympatric speciation. Allopatric speciation may involve the formation of hybrid zones (see Table 29.2, Figures 29.3, 29.4, 29.5, 29.6, 29.7).

29.4 Phylogenetic Trees

- A phylogenetic tree is a diagram that describes a phylogeny—the sequence of events involved in the evolutionary development of a species or group of species (see Figure 29.8).

- Different methods, including cladistics and phenetics, are used to construct phylogenetic trees. One way to evaluate the validity of possible trees is to apply the principle of parsimony (see Figures 29.9, 29.10, 29.11).
- The maximum likelihood approach and Bayesian methods are used to discriminate among possible phylogenetic trees.
- Phylogenetic trees can depict the relationships among groups of species such as domains and supergroups (see Figure 29.12).
- In addition to vertical evolution, horizontal gene transfer has played an important role in the evolution of life on Earth (see Figure 29.13).

29.5 Molecular Evolution

- Molecular evolution consists of molecular-level changes in the genetic material that underlie the process of evolution.

- Homologous genes are derived from the same ancestral gene. They can be orthologs or paralogs (see Figures 29.14, 29.15).
- Phylogenetic trees can be constructed by comparing orthologs from extinct and modern species (see Figures 29.16, 29.17).
- The neutral theory of evolution proposes that most variation in gene sequences is due to mutations that are neutral or nearly neutral with regard to phenotype. Such variation accumulates in populations largely due to genetic drift.
- The relatively constant rate of neutral or nearly neutral mutations acts as a molecular clock with which to measure evolutionary time (see Figure 29.18).
- Evolution is also associated with changes in chromosome structure and number. Synteny groups are regions on the chromosomes of two or more species that contain the same groups of linked genes (see Figures 29.19, 29.20).

PROBLEM SETS & INSIGHTS

MORE GENETIC TIPS
1. Explain why orthologs have sequences that are similar but not identical.

T OPIC: What topic in genetics does this question address? The topic is homology. More specifically, the question asks you to explain the similarities in the sequences of homologous genes.

I NFORMATION: What information do you know based on the question and your understanding of the topic? In the question, you are reminded that homologous genes have sequences that are similar but not identical. From your understanding of the topic, you may recall that homologous genes are derived from the same ancestral gene.

P ROBLEM-SOLVING S TRATEGY: Describe the steps. One strategy to solve this problem is to describe the steps that create orthologs. Consider how those steps affect the gene sequences.

ANSWER: Orthologs are homologous genes that are derived from the same ancestral gene. Therefore, as a starting point, they had identical sequences. However, after a species diverges into two or more different species, each ortholog accumulates random mutations that other orthologs may not acquire. These random mutations change the original gene sequence. Therefore, much of the sequence remains identical among orthologs, but some of it is altered due to the accumulation of independent random mutations.

2. A codon for leucine is UUA. A mutation causing a single-base substitution in a gene can change this codon in the transcribed mRNA into GUA (valine), AUA (isoleucine), CUA (leucine), UGA (stop), UAA (stop), UCA (serine), UUG (leucine), UUC (phenylalanine), or UUU (phenylalanine). According to the neutral theory of evolution, which of these mutations would you expect to be the most likely to be found within a natural population? Explain.

T OPIC: What topic in genetics does this question address? The topic is the neutral theory of evolution. More specifically, you are given a codon and asked to predict which changes to that codon would be most likely to persist in a natural population according to the neutral theory of evolution.

I NFORMATION: What information do you know based on the question and your understanding of the topic? In the question, you are given a codon and the possible codons that could be produced in transcribed mRNA by a single-base substitution. From your understanding of the neutral theory of evolution, you may recall that mutations that disrupt protein structure are more likely to be eliminated from a population compared to conservative substitutions that do not affect or have a minimal effect on protein structure.

P ROBLEM-SOLVING S TRATEGIES: Compare and contrast. Relate structure and function. The neutral theory proposes that neutral mutations accumulate to the greatest extent in a population. One strategy to solve this problem is to compare and contrast the original codon (which codes leucine) with each of the possible mutant codons and evaluate whether each mutation is likely to be neutral or likely to inhibit protein function.

ANSWER: Leucine is a nonpolar amino acid. For a UUA codon, single-base changes that result in CUA and UUG are silent, so these mutations would be the most likely to persist in a natural population. Likewise, conservative substitutions yielding other nonpolar amino acids such as isoleucine (AUA), valine (GUA), and phenylalanine (UUC and UUU) may not affect protein structure and function, so they may also occur and not be eliminated rapidly by natural selection. The polar amino acid serine (UCA) results from a nonconservative substitution; you would predict that this mutation is more likely to inhibit protein function. Therefore, it may be less likely to be found. Finally, the stop codons, UGA and UAA, would be expected to diminish or eliminate protein function. These types of mutations are selected against and, therefore, are not usually found in natural populations.

3. Evolution is associated with changes in chromosome structure and number. As described in Figure 29.19, chromosome 2 in humans is divided into two distinct chromosomes in chimpanzees, gorillas, and orangutans. In addition, chromosome 3 in the orangutan has a large inversion not found in chromosome 3 in the other three primates. Discuss the potential role of these types of changes in the evolution of these primate species. (Note: You may want to refer back to Chapter 8 before answering this question.)

T OPIC: *What topic in genetics does this question address?*
The topic is changes in chromosome structure and number. More specifically, the question is about how certain changes in chromosomes 2 and 3 may have affected primate evolution.

I NFORMATION: *What information do you know based on the question and your understanding of the topic?* In the question, you are referred to Figure 29.19, which shows differences in certain primate chromosomes. From your understanding of the topics discussed in Chapter 8, you may recall that crossing over and certain segregation patterns may result in imbalances in genetic material that may be transmitted to offspring.

P ROBLEM-SOLVING **S** TRATEGIES: *Make a drawing. Predict the outcome.* One approach to solve this problem is to make one or more drawings (e.g., see Figure 8.11) and then predict the outcome. If an offspring inherits too much or too little genetic material, such an event is likely to be detrimental.

ANSWER: As discussed in Chapter 8, changes in chromosome structure, such as inversions and balanced translocations, may not have any phenotypic effects. Likewise, the division of a single chromosome into two distinct chromosomes may not have any phenotypic

effect as long as the total amount of genetic material remains the same. Overall, the types of changes in chromosome structure and number shown in Figure 29.19 may not have caused any changes in the phenotypes of primates. However, the changes would be expected to promote reproductive isolation. For example, if a gorilla mated with an orangutan, the offspring would be an inversion heterozygote for chromosome 3. As shown in Figure 8.11, crossing over during gamete formation in an inversion heterozygote may produce chromosomes that have too much or too little genetic material. This is particularly likely if the inversion is fairly large (like the one shown in Figure 29.19). The inheritance of too much or too little genetic material is likely to be detrimental or even lethal. For this reason, the hybrid offspring of a gorilla and orangutan would probably be infertile. (Note: In reality, there are several other reasons why interspecies matings between gorillas and orangutans do not produce viable offspring.)

Overall, the primary effect of changes in chromosome structure and number, like the ones shown in Figure 29.19, is to promote reproductive isolation. Once two populations become reproductively isolated, they accumulate different mutations, and over the course of many generations, the accumulation leads to two different species with distinct characteristics.

Conceptual Questions

C1. Discuss the two factors on which adaptive evolution is based.

C2. Evolution, which involves genetic changes in a population of organisms over time, is often described as the unifying theme in biology. Discuss how evolution is unifying at the molecular and cellular levels.

C3. What is a species? What types of observations do researchers analyze when trying to identify species?

C4. What is meant by the term *reproductive isolation*? Give several examples.

C5. Would each of the following examples of reproductive isolation be considered a result of a prezygotic or postzygotic mechanism?

   A. Horses and donkeys can interbreed to produce mules, but the mules are infertile.

   B. Three species of the orchid genus *Dendrobium* produce flowers 8 days, 9 days, and 11 days after a rainstorm. The flowers remain open for 1 day.

   C. Two species of fish release sperm and eggs into seawater at the same time, but the sperm of one species do not fertilize the eggs of the other species.

   D. Two tree frogs, *Hyla chrysoscelis* (diploid) and *Hyla versicolor* (tetraploid), can produce viable offspring, but the offspring are sterile.

C6. Distinguish between anagenesis and cladogenesis. Which mechanism of speciation is more prevalent? Why?

C7. Describe three or more genetic mechanisms that may lead to the rapid evolution of a new species. Which of these genetic mechanisms are influenced by natural selection, and which are not?

C8. Identify the mechanism of speciation (allopatric, allopatric with hybrid zones, or sympatric) most likely to occur under each of the following conditions:

   A. A pregnant female rat is transported by an ocean liner to a new continent.

   B. A meadow containing several species of grasses is exposed to a pesticide that promotes nondisjunction.

   C. In a very large lake containing several species of fish, the water level gradually falls over the course of several years. Eventually, the large lake becomes subdivided into smaller lakes, some of which are connected by narrow streams.

C9. Alloploids are produced by crosses involving two different species. Explain why alloploids may be reproductively isolated from the two original species from which they were derived. Explain why alloploids are usually sterile, whereas allotetraploids (containing a diploid set from each species) are commonly fertile.

C10. Discuss whether the phenomenon of reproductive isolation applies to bacteria, which reproduce asexually. How would a geneticist divide bacteria into separate species?

C11. Discuss the major differences among allopatric speciation, allopatric speciation with hybrid zones, and sympatric speciation.

C12. The following are DNA sequences from two homologous genes:

TTGCATAGGCATACCGTATGATATCGAAAACTAGAAAAATAGGGCGATAGCTA
GTATGTTATCGAAAAGTAGCAAAATAGGGCGATAGCTACCCAGACTACCGGAT

   The two sequences, however, do not begin and end at the same location. Try to line them up according to their homologous regions.

C13. What is meant by the term *molecular clock*? How is this concept related to the neutral theory of evolution?

C14. Would the rate of deleterious or beneficial mutations be a good basis for a molecular clock? Why or why not?

C15. Which would you expect to exhibit a faster rate of evolutionary change, the nucleotide sequence of a gene or the amino acid sequence of the coded polypeptide of the same gene? Explain your answer.

C16. When the coding sequences of a protein-coding gene are compared among closely related species, certain regions are commonly found to have evolved more rapidly (i.e., have tolerated more changes in sequence) than other regions. Explain why different regions of a protein-coding gene evolve at different rates.

C17. Plant seeds contain storage proteins that are coded by the plant's genes. When a seed germinates, these proteins are rapidly hydrolyzed (i.e., the covalent bonds between amino acids within the polypeptides are broken), which releases amino acids for the developing seedling. Would you expect the genes that code plant storage proteins to evolve more slowly or more rapidly than genes that code enzymes? Explain your answer.

C18. Take a look at the α-globin and β-globin amino acid sequences in Figure 29.14. Which sequences are more similar, the α globin in humans and the α globin in horses, or the α globin in humans and the β globin in humans? Based on your answer, would you conclude that the gene duplication that gave rise to the α-globin and β-globin genes occurred before or after the divergence of humans and horses? Explain your reasoning.

C19. Compare and contrast the neutral theory of evolution and the Darwinian theory of evolution. Explain why the neutral theory of evolution is sometimes called non-Darwinian evolution.

C20. For each of the following examples, discuss whether the observed result is due to neutral mutation, mutation(s) that have been acted on by natural selection, or both:

A. When sequences of homologous genes are compared, differences in the coding sequence are most common at the base in the wobble position (i.e., the third base in each codon).

B. For a protein-coding gene, the regions that code portions of the polypeptide that are vital for structure and function are less likely to display mutations than other regions of the gene.

C. When the sequences of homologous genes are compared, introns usually have more sequence differences than exons.

C21. As discussed in Chapter 27, genetic variation is prevalent in natural populations. This variation is revealed via DNA sequencing of genes. In regard to the neutral theory of evolution, discuss the relative importance of natural selection acting against detrimental mutations, natural selection acting in favor of beneficial mutations, and neutral mutations in accounting for the genetic variation we see in natural populations.

C22. If you were comparing the karyotypes of species that are closely related evolutionarily, what types of similarities and differences would you expect to find?

Experimental Questions

E1. Two populations of snakes are separated by a river. The snakes cross the river only on rare occasions. The snakes in the two populations look very similar to each other, except that the members of the population on the eastern bank of the river have a yellow spot on the top of their head, whereas the members of the western population have an orange spot on the top of their head. Discuss two experimental methods that you might use to determine whether the two populations are members of the same species or members of different ones.

E2. Sympatric speciation via allotetraploidy has been proposed as a common mechanism for speciation. Let's suppose you were interested in the origin of certain grass species in southern California. Experimentally, how would you go about determining if some of the grass species are the result of allotetraploidy?

E3. Two diploid species of closely related frogs, which we will call species A and species B, were analyzed with regard to the genes that code an enzyme called hexokinase. Species A has two distinct copies of this gene: *A1* and *A2*. In other words, the genotype of this diploid species is *A1A1 A2A2*. Species B has three copies of the hexokinase gene, which we will call *B1*, *B2*, and *B3*. A diploid individual of species B has the genotype *B1B1 B2B2 B3B3*. These hexokinase genes from the two species were subjected to DNA sequencing, and the percentages of identical sequences were compared between pairs of genes. The results are shown next.

Percentages of Identical DNA Sequences

	A1	*A2*	*B1*	*B2*	*B3*
A1	100	62	54	94	53
A2	62	100	91	49	92
B1	54	91	100	67	90
B2	94	49	67	100	64
B3	53	92	90	64	100

If you assume that hexokinase genes were never lost in the evolution of these frog species, how many distinct hexokinase genes do you think there were in the most recent ancestor that preceded the divergence of these two species? Explain your answer. Also, explain why species B has three distinct copies of this gene, whereas species A has only two.

E4. A researcher sequenced a portion of a bacterial gene and obtained the following base sequence, beginning with the start codon, which is underlined:

<u>ATG</u> CCG GAT TAC CCG GTC CCA AAC AAA ATG ATC GGC CGC CGA ATC TAT CCC

The bacterial strain that contained this gene has been maintained in a laboratory and grown serially for many generations. Recently, another researcher working in the laboratory isolated DNA from the bacterial strain and sequenced the same region. The following results were obtained:

<u>ATG</u> CCG GAT TAT CCG GTC CCA AAT AAA ATG ATC GGC CGC CGA ATC TAC CCC

Explain why the differences in the sequence may have occurred.

E5. F_1 hybrids between two species of cotton, *Gossypium barbadense* and *Gossypium hirsutum*, are very vigorous plants. However, crosses of F_1 hybrids produce many seeds that do not germinate and a high percentage of very weak F_2 offspring. Suggest two reasons for these observations.

E6. A species of antelope has 20 chromosomes per set. The species is divided by a mountain range into two separate populations, which we will call the eastern and western populations. In a comparison of the karyotypes of these two populations, it was discovered that the members of the eastern population are homozygous for a large inversion within chromosome 14. How would this inversion affect the interbreeding between the two populations? Could such an inversion play an important role in speciation?

E7. Explain why molecular techniques were needed to provide evidence for the neutral theory of evolution.

E8. Prehistoric specimens often contain minute amounts of ancient DNA. What technique can be used to increase the amount of DNA in an older sample? Explain how this technique is carried out and how it increases the amount of a specific region of DNA.

E9. From the results of the experiment of Figure 29.16, explain how we know that the kiwis are more closely related to the emu and the cassowary than to the moas. Cite particular regions in the sequences that support your answer.

E10. In Chapter 22, a technique called fluorescence in situ hybridization (FISH) is described. In this method, a labeled piece of DNA is hybridized to a set of chromosomes. Suppose that you cloned a piece of DNA from *Galeopsis pubescens* (see Figure 29.7) and used it as a labeled probe for fluorescence in situ hybridization. What would you expect to happen if this DNA probe were hybridized to the *G. speciosa* or *G. tetrahit* chromosomes? Describe the expected results.

E11. A team of researchers has obtained a dinosaur bone (*Tyrannosaurus rex*) and has attempted to extract ancient DNA from it. Using primers for the 12S rRNA mitochondrial gene, they carried out PCR and obtained a DNA segment that had a sequence homologous to a sequence in crocodile DNA. Other scientists are skeptical that this sequence is really from the dinosaur. Instead, they believe that it may have come from contamination by more recent DNA, such as the remains of a reptile that lived much more recently. What criteria might you use to establish the credibility of the dinosaur sequence?

E12. Discuss how the principle of parsimony can be used in a cladistic approach to constructing a phylogenetic tree.

E13. A homologous DNA region, which was 20,000 bp in length, was sequenced from four different species. The following numbers of nucleotide differences were obtained:

	Species A	Species B	Species C	Species D
Species A	0	443	719	465
Species B	443	0	744	423
Species C	719	744	0	723
Species D	465	423	723	0

Construct a phylogenetic tree that describes the evolutionary relationships among these four species using the UPGMA method. Your tree should include values that show the percentages of nucleotide substitutions.

E14. As discussed in this chapter and Chapter 27, genes are sometimes transferred between different species via horizontal gene transfer. Discuss how horizontal gene transfer might lead to misleading results when constructing a phylogenetic tree. How could you overcome this problem?

Questions for Student Discussion/Collaboration

1. The raw material for evolution is random mutation. Discuss whether or not you view evolution as a random process.

2. Compare the mechanisms of speciation that proceed slowly to those that proceed more rapidly. Make a list of the slow and fast mechanisms. With regard to genetic change, what features do slow and rapid speciation have in common? What features are different?

3. Do you think that Darwin would object to the neutral theory of evolution?

Note: All answers are available for the instructor in Connect; the answers to the even-numbered questions and all of the Concept Check and Comprehension Questions are in Appendix B.

APPENDIX
EXPERIMENTAL TECHNIQUES

A

OUTLINE

A.1 METHODS FOR GROWING CELLS

Researchers often grow cells in a laboratory as a way to study their properties. A population of cells grown in a laboratory is known as a **cell culture.** Cell culturing offers several technical advantages. The primary advantage is that the growth medium is defined and can be controlled. Minimal growth medium contains the bare essentials for cell growth: salts, a carbon source, an energy source, essential vitamins, amino acids, and trace elements. In their experiments, geneticists often compare strains that can grow in minimal media and mutant strains that cannot grow unless the medium is supplemented with additional components. A rich growth medium contains many more components than are required for growth.

Researchers also add substances to the culture medium for other experimental reasons. For example, radioisotopes can be added to the culture medium to radiolabel cellular macromolecules. In addition, an experimenter might add a hormone to the growth medium and then monitor the cells' response to the hormone. In all of these cases, cell culturing is advantageous because the experimenter can control and vary the composition of the growth medium.

The first step in creating a cell culture is the isolation of a cell population that the researchers wish to study. For bacteria, such as *Escherichia coli*, and eukaryotic microorganisms, such as yeast and *Neurospora*, researchers simply obtain a sample of cells from a colleague or a stock center. For animal or plant tissues, the procedure is a bit more complicated. When cells are contained within a complex tissue, they must first be dispersed by treating the tissue with agents that separate it into individual cells to create a cell suspension.

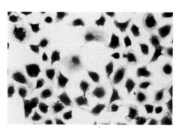

(a) Fibroblast (animal cell) culture   **(b) Bacterial colonies**

FIGURE A.1 Growth of cells on solid growth media. (a) This micrograph shows fibroblasts growing as a monolayer on a solid growth medium. **(b)** Bacterial cells form colonies that are a clonal population of cells derived from a single cell.

(a) De Agostini Picture Library/age fotostock; (b) Science Photo Library/CNRI/Getty Images

Once a desired population of cells has been obtained, researchers can grow them in a laboratory (i.e., in vitro) either suspended in a liquid growth medium or attached to a solid surface such as agar. Both methods have been commonly used in the experiments considered throughout this text. Liquid culture is often used when researchers want to obtain a large quantity of cells and isolate individual cellular components, such as nuclei or DNA. **Figure A.1** shows animal cells and bacteria cells that are being grown on solid growth media. Solid media are used to study cancer cells, because such cells can be distinguished by the formation of foci in which malignant cells pile up on top of each other. In gene-cloning experiments with bacteria and yeast, solid media are also used. Each colony of cells is a clone of cells that is derived from a single cell that divided to produce many cells (Figure A.1b). As discussed in Chapter 20, a solid medium is used in the isolation of individual clones that contain a desired gene.

A.2 MICROSCOPY

Microscopy is a technique for observing things that cannot be seen (or can hardly be seen) with the unaided eye. A key aspect of microscopy is **resolution,** which is the minimum distance between two objects that enables them to be seen as separate from each other. The ability to resolve two points as being separate depends

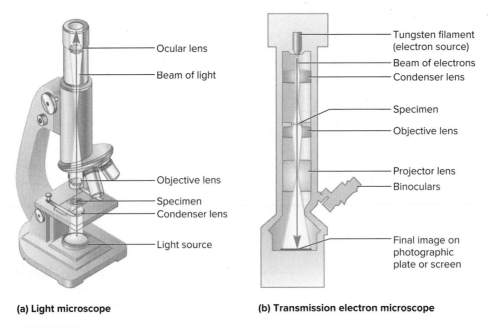

FIGURE A.2 Design of (a) optical (light) and (b) transmission electron microscopes.

on several factors, including the wavelength of the illumination source (light or electron beam), the medium in which the sample is immersed, and the structural features of the microscope (which are beyond the scope of this text).

As shown in **Figure A.2**, two widely used kinds of microscopes are the optical (light) microscope and the transmission electron microscope (TEM). The light microscope is used to resolve cellular structures to a limit of approximately 0.3 μm. (For comparison, a typical bacterium is about 1 μm long.) At this resolution, the individual cell organelles in eukaryotic cells can be discerned, and chromosomes are also visible. Karyotyping is accomplished via light microscopy after the chromosomes have been treated with stains. A variation of light microscopy known as fluorescence microscopy is often used to highlight a particular feature of a chromosome or cellular structure. The technique of fluorescence in situ hybridization (FISH; see Chapter 22) makes use of this type of microscope. Also, optical modifications in certain light microscopes (e.g., phase contrast and differential interference) can be used to exaggerate the differences in densities between neighboring cells or cell structures. These kinds of light microscopes are useful in monitoring cell division in living (unstained) cells or in transparent worms (as in Figure 26.17).

The structural details of large macromolecules such as DNA and ribosomes are not observable by light microscopy. The coarse topology of these macromolecules can be determined by using electron microscopy. Electron microscopes have a limit of resolution of about 2 nm, which is about 100 times finer than the best light microscopes. The primary advantage of electron microscopy over light microscopy is its better resolution. Disadvantages include a much higher expense and more extensive sample preparation. In transmission electron microscopy, the

sample is bombarded with an electron beam. This requires that the sample be dried, fixed, and usually coated with a heavy metal that absorbs electrons.

A.3 SEPARATION METHODS

Biologists often wish to take complex systems and separate them into less complex components. For example, the cells within a complex tissue can be separated into individual cells, or the macromolecules within cells can be separated from the other cellular components. In this section, we will focus primarily on methods aimed at separating and purifying macromolecules.

Disruption of Cellular Components

In many experiments described in this text, researchers have obtained a sample of cells and wish to isolate particular components from those cells. For example, researchers may want to purify a protein that functions as a transcription factor. To do so, they would begin with a sample of cells that synthesize this protein and then break open the cells using one of the methods described in **Table A.1**. In eukaryotes, the breakage of cells releases the soluble proteins from the cells; it also dissociates the cell organelles that are bounded by membranes. This mixture of proteins and cell organelles can then be isolated and purified by centrifugation and chromatographic methods, which are described next.

Centrifugation

Centrifugation is a method commonly used to separate cell organelles and macromolecules. A **centrifuge** contains a motor that causes a rotor holding centrifuge tubes to spin very rapidly.

TABLE A.1
Common Methods of Cell Disruption

Method	Description
Sonication	The exposure of cells to intense sound waves, which breaks the cell membranes.
French press	The passage of cells through a small aperture under high pressure, which breaks the cell membranes and cell wall.
Homogenization	Cells are placed in a tube that contains a pestle. When the pestle is spun, the cells are squeezed through the small space between the pestle and the glass wall of the tube, thereby breaking them.
Osmotic shock	The transfer of cells into a hypo-osmotic medium. The cells take up water and eventually burst.

As the rotor spins, particles move toward the bottom of a centrifuge tube; the rate at which they move depends on several factors, including their densities, sizes, and shapes and the viscosity of the medium. The rate at which macromolecules or cell organelles sediment to the bottom of a centrifuge tube is called the **sedimentation coefficient,** which is normally expressed in Svedberg units (S): $1\ S = 1 \times 10^{-13}$ second.

When a sample contains a mixture of macromolecules or cell organelles, it is likely that different components will sediment at different rates. This phenomenon is the basis for the method of **differential centrifugation,** shown in **Figure A.3**. As seen here, particles with large sedimentation coefficients reach the bottom of the tube more quickly than those with smaller coefficients. Researchers can follow two different strategies that use differential centrifugation as a separation technique. One way is to separate

the **supernatant** from the **pellet** following centrifugation. The pellet is a collection of particles found at the bottom of the tube, and the supernatant is the liquid found above the pellet.

In Figure A.3, when the experimenter subjected the sample to a low-speed spin, most of the particles with large sedimentation coefficients were found in the pellet, whereas most of the particles with small and intermediate coefficients were found in the supernatant. A high-speed spin of the supernatant then separated the small and intermediate particles. Thus, differential centrifugation provides a way of segregating these three types of particles.

A second way to separate particles using centrifugation is to collect fractions. A **fraction** is a portion of the liquid contained within a centrifuge tube. The collection of fractions is done when the solution within the centrifuge tube is composed of a gradient. For example, as shown in **Figure A.4**, the solution at the top of the tube has a lower concentration of cesium chloride (CsCl) than that at the bottom. In this experiment, a sample is layered on the top of the gradient and then centrifuged. In this example, the DNA

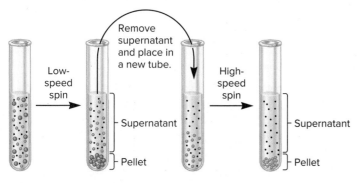

- ● Particles with large sedimentation coefficients
- ● Particles with intermediate sedimentation coefficients
- · Particles with small sedimentation coefficients

FIGURE A.3 The method of differential centrifugation. A sample containing a mixture of particles with different sedimentation coefficients is placed in a centrifuge tube. The tube is subjected to a low-speed spin that causes the particles with large sedimentation coefficients to form a pellet at the bottom of the tube. After a high-speed spin of the supernatant, the particles with an intermediate sedimentation coefficient are found in the pellet, and those with a small sedimentation coefficient are in the liquid supernatant.

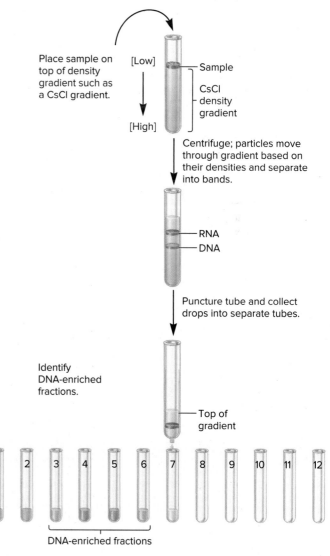

FIGURE A.4 Gradient centrifugation and the collection of fractions.

fragments and RNA molecules separate from each other because they have different sedimentation coefficients. The experimenter then punctures the bottom of the tube and collects fractions. The DNA fragments, which are heavier, come out of the tube in the earlier fractions; the RNA molecules are collected in later fractions.

A type of gradient centrifugation that may also be used to separate macromolecules and organelles is **equilibrium density centrifugation.** In this method, the particles sediment through the gradient, reaching positions where the density of each particle matches the density of the solution. At this point, the particle is at equilibrium and does not move any farther toward the bottom of the tube.

Chromatography and Gel Electrophoresis

Chromatography is a method of separating different macromolecules or smaller molecules based on their chemical and physical properties. In this method, a sample is dissolved in a liquid solvent and exposed to some type of matrix, such as a gel, a column containing beads, or a thin strip of paper. The degree to which the molecules interact with the matrix depends on their chemical and physical characteristics. For example, a positively charged molecule binds tightly to a negatively charged matrix, but a neutral molecule does not.

Figure A.5 illustrates how column chromatography can be used to separate molecules that differ with respect to charge. Prior to this experiment, a column is packed with beads that are positively charged. There is plenty of space between the beads for molecules to flow from the top of the column to the bottom. However, if the molecules are negatively charged, they will spend some of their time binding to the positive charges on the surface of the beads. In the example shown in Figure A.5, the red proteins are positively charged and, therefore, flow rapidly from the top of the column to the bottom. They emerge in the fractions that are collected early in this experiment. The blue proteins, however, are

negatively charged and tend to bind to the beads. The binding of the blue proteins to the beads can be disrupted by changing the ionic strength or pH of the solution that is added to the column. Eventually, the blue proteins will be eluted (i.e., will leave the column) in later fractions.

Researchers use many variations of column chromatography to separate molecules and macromolecules.

- The type of chromatography shown in Figure A.5 is called ion-exchange chromatography, because its basis for separation depends on the charge of the molecules.
- In another type of column chromatography, known as gel filtration chromatography, the beads are porous. Smaller molecules are temporarily trapped within the beads, whereas larger molecules flow between the beads. In this way, gel filtration separates molecules on the basis of size.
- To separate different types of macromolecules, such as proteins, researchers may use a bead with a preattached molecule that binds specifically to the protein they want to purify. For example, if a transcription factor binds a particular DNA sequence as part of its function, the beads within a column can have this DNA sequence preattached to them. The transcription factor binds tightly to the DNA attached to these beads, whereas all other proteins are eluted rapidly from the column. This form of chromatography is called affinity chromatography, because the beads have a special affinity for the macromolecule of interest.
- In high-performance liquid chromatography, a pump is used to pass a pressurized liquid solvent containing the sample mixture through a column. The pump allows researchers to control the flow rate through the column, which often facilitates greater success in the purification of particular molecules or macromolecules.

Column chromatography, which uses beads packed into a column, is typically used to separate macromolecules, such as mixtures of DNA fragments or proteins. Other types of chromatography are more commonly used to separate small molecules.

- In paper chromatography, molecules pass through a matrix composed of paper. The rate of movement of molecules through the paper depends on their degree of interaction with the solvent and the paper.
- In thin-layer chromatography, a matrix is spread out as a very thin layer on a rigid support such as a glass plate.

Gel electrophoresis combines chromatography and electrophoresis to separate molecules and macromolecules. As its name suggests, the matrix used in gel electrophoresis is composed of a gel.

1. As shown in **Figure A.6**, samples are loaded into wells at one end of the gel.
2. An electric field is applied across the gel. This electric field causes charged molecules to migrate from one side of the gel to the other. The migration of ions or charged molecules in response to an electric field is called **electrophoresis.** In the examples of gel electrophoresis found in this

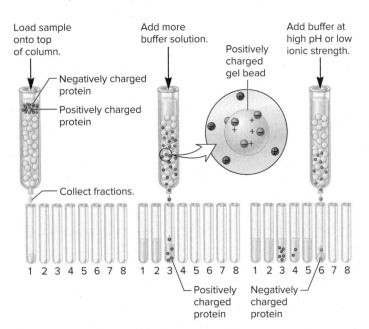

Load sample onto top of column.

Add more buffer solution.

Add buffer at high pH or low ionic strength.

Positively charged gel bead

Negatively charged protein

Positively charged protein

Collect fractions.

1 2 3 4 5 6 7 8   1 2 3 4 5 6 7 8   1 2 3 4 5 6 7 8

Positively charged protein

Negatively charged protein

FIGURE A.5 **Ion-exchange chromatography.**

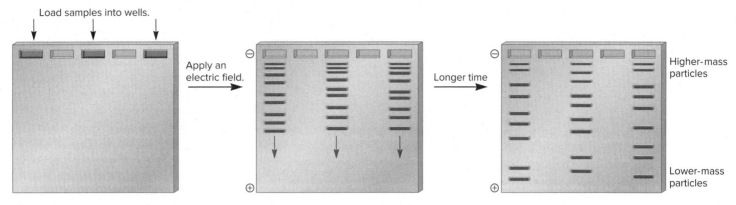

(a) Separation of a mixture of particles by gel electrophoresis

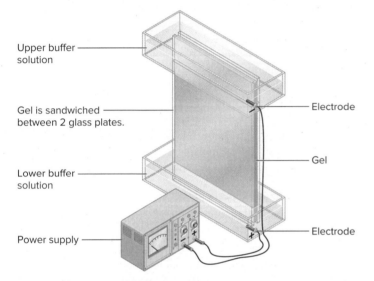

(b) Apparatus used in gel electrophoresis

FIGURE A.6 **Acrylamide gel electrophoresis of DNA fragments.**

text, the macromolecules within the sample migrate toward the positive end of the gel.

3. In most forms of gel electrophoresis, a mixture of macromolecules is separated according to their molecular masses. Small proteins or DNA fragments move to the bottom of the gel more quickly than larger ones. Because the samples are loaded in rectangular wells at the top of the gel, the molecules within the sample are separated into bands within the gel.

4. These bands of separated macromolecules can be visualized with stains. For example, ethidium bromide is a stain that binds to DNA and RNA and can be seen under ultraviolet light.

The two most commonly used gels are polymers made from acrylamide or agarose. Proteins typically are separated on polyacrylamide gels, whereas DNA fragments are separated on agarose gels. Occasionally, researchers use polyacrylamide gels to separate DNA fragments that are relatively small (less than 1000 bp in length).

A.4 METHODS FOR MEASURING CONCENTRATIONS OF MOLECULES AND DETECTING RADIOISOTOPES AND ANTIGENS

To understand the structure and function of cells, researchers often need to detect the presence of molecules and macromolecules and to measure their concentrations. In this section, we consider a variety of methods for detecting and measuring the concentrations of biological molecules and macromolecules.

Spectroscopy

Macromolecules found in living cells, such as proteins, DNA, and RNA, are fairly complex molecules that can absorb radiation (e.g., light). Likewise, small molecules such as amino acids and nucleotides can also absorb light. A device known as a **spectrophotometer** is used by researchers to determine how much radiation at various wavelengths a sample absorbs. The amount of absorption can be used to determine the concentration of particular molecules within a sample, because each type of molecule or macromolecule has its own characteristic wavelength(s) of absorption, called its absorption spectrum.

A spectrophotometer typically has two light sources, which can emit ultraviolet or visible light. As shown in **Figure A.7**, the light is passed through a monochromator, which allows the passage of a desired range of wavelengths and filters out undesired wavelengths. The light of desired wavelengths then strikes a sample contained within a cuvette. Some of the incident light is absorbed, and some is not. The amount and wavelengths of light that are absorbed depend on the concentration and structures of the molecules and macromolecules in the cuvette. The unabsorbed light passes through the sample and is detected by the spectrophotometer. The amount of light that strikes the detector is subtracted from the amount of incident light, yielding the measure of absorption. In this way, the spectrophotometer provides an absorption reading for the sample. This reading can be

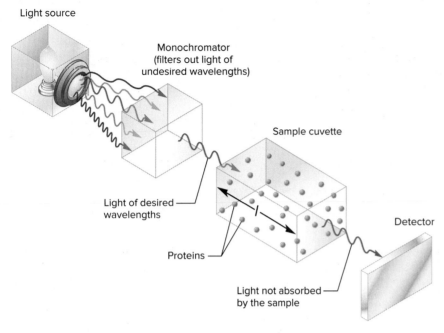

FIGURE A.7 Design of a spectrophotometer.

used to calculate the concentration of particular molecules or macromolecules in a sample.

Detection of Radioisotopes

A radioisotope is an unstable form of an atom that decays to a more stable form by emitting α-, β-, or γ-rays, which are types of ionizing radiation. In research, radioisotopes that are β and/or γ emitters are commonly used. A β-ray consists of emitted electrons, and a γ-ray consists of emitted photons. Some radioisotopes commonly used in biological experiments are shown in **Table A.2**.

Experimentally, radioisotopes are used because they are easy to detect. Therefore, if a particular compound is radiolabeled, its presence can be detected specifically throughout the course of the experiment. For example, if a nucleotide is radiolabeled with ^{32}P, a researcher can determine whether the nucleotide containing the radioisotope becomes incorporated into newly made DNA or whether it remains as the free nucleotide. Researchers commonly use two different methods of detecting radioisotopes: scintillation counting and autoradiography.

The technique of **scintillation counting** permits a researcher to count the number of radioactive emissions from a sample containing a population of radioisotopes. In this approach, the sample is dissolved in a solution (called the scintillant) that contains organic solvents and one or more compounds known as fluors. When radioisotopes emit ionizing radiation, the energy is absorbed by the fluors in the solvent. This excites the fluor molecules, causing their electrons to be boosted to higher energy levels. The excited electrons return to lower, more stable energy levels by releasing photons of light. Thus, when a fluor is struck by ionizing radiation, it also absorbs the energy and then releases a photon of light within a particular wavelength range.

The role of the device known as a scintillation counter is to count the photons of light emitted by the fluor. **Figure A.8** shows a scintillation counter. To use this device, a researcher dissolves a sample in a scintillant and then places the sample in a scintillation vial. The vial is then placed in the scintillation counter, which detects the amount of radioactivity. The scintillation counter contains several rows for the loading and analysis of many scintillation vials. After they have been loaded, the scintillation counter counts

TABLE **A.2**			
Some Useful Isotopes in Genetics			
Isotope	**Stable or Radioactive**	**Emission**	**Half-life**
^{2}H	Stable		
^{3}H	Radioactive	β	12.3 years
^{13}C	Stable		
^{14}C	Radioactive	β	5730 years
^{15}N	Stable		
^{18}O	Stable		
^{24}Na	Radioactive	β (and γ)	15 hours
^{32}P	Radioactive	β	14.3 days
^{35}S	Radioactive	β	87.4 days
^{45}Ca	Radioactive	β	164 days
^{59}Fe	Radioactive	β (and γ)	45 days
^{131}I	Radioactive	β (and γ)	8.1 days

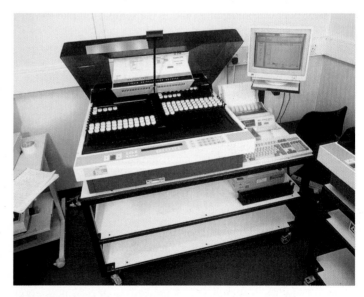

FIGURE A.8 A scintillation counter.
Nigel Cattlin/Science Source

the amount of radioactivity in each vial and provides the researcher with a printout of the amount of radioactivity (in counts per minute) in each vial.

A second way of detecting radioisotopes is via **autoradiography.** This technique is not as quantitative as scintillation counting, because it does not provide the experimenter with a precise measure of the amount of radioactivity in counts per minute. However, autoradiography has the great advantage that it can detect the location of radioisotopes as they are found in macromolecules or cells. For example, autoradiography is used to identify a particular band on a gel or to map the location of a gene within an intact chromosome.

To conduct autoradiography, a sample containing a radioisotope is fixed and usually dried. If it is a cellular sample, it also may be thin-sectioned. The sample is then pressed next to X-ray film (in the dark) and placed in a lightproof cassette. When a radioisotope decays, it emits a β- or γ-ray, which may strike a thin layer of photoemulsion next to the film. The photoemulsion contains a silver salt such as AgBr. When a radioactive particle is emitted and strikes the photoemulsion, a silver grain is deposited on the film. This produces a dark spot on the film, which correlates with the original location of the radioisotope in the sample. Thus, the dark image on the film reveals the location(s) of the radioisotopes in the sample.

Detection of Antigens by Radioimmunoassay

Antibodies, also known as **immunoglobulins,** are proteins that are used to ward off infection by foreign substances; they are produced by cells of the immune system of vertebrates. Antibodies bind to structures on the surface of foreign substances known as **epitopes;** the foreign substance is called an **antigen.** A particular antibody binds to a particular antigen with a very high degree of specificity. For this reason, antibodies have been used extensively by researchers to detect particular antigens. For example, a human protein such as hemoglobin can be injected into a rabbit. Human hemoglobin is a foreign substance in the rabbit's bloodstream. Therefore, the rabbit makes antibodies that specifically recognize human hemoglobin and are able to destroy it. Researchers can isolate and purify these antibodies from a sample of the rabbit's blood and then use them to detect human hemoglobin in their experiments.

A **radioimmunoassay** is a method for measuring the amount of an antigen in a biological sample. The steps in this method are shown in **Figure A.9a.** The researcher begins with two tubes that have a known amount of radiolabeled antigen (shown in blue). An unknown amount of the same antigen, which is not radiolabeled (shown in orange), is added to the tube on the right. The nonradiolabeled antigen comes from a biological sample; the goal of this experiment is to determine how much of this antigen is contained within the sample. Next, a known amount of antibody is added to each of the two tubes. The amount of the antibody is less than the amount of the antigen, so the nonlabeled and radiolabeled antigens compete with each other for binding to the antibody. After binding, a precipitating agent such as an anti-immunoglobulin antibody is added, and the precipitate is centrifuged to the bottom

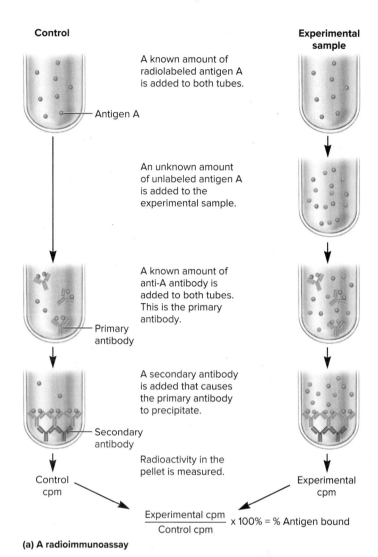

A known amount of radiolabeled antigen A is added to both tubes.

Antigen A

An unknown amount of unlabeled antigen A is added to the experimental sample.

A known amount of anti-A antibody is added to both tubes. This is the primary antibody.

Primary antibody

A secondary antibody is added that causes the primary antibody to precipitate.

Secondary antibody

Radioactivity in the pellet is measured.

Control cpm

Experimental cpm

$$\frac{\text{Experimental cpm}}{\text{Control cpm}} \times 100\% = \% \text{ Antigen bound}$$

(a) A radioimmunoassay

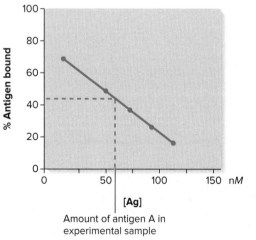

(b) Standard curve

FIGURE A.9 The method of radioimmunoassay (a) and the construction of a standard curve (b). In the standard curve, the dashed line corresponds to the amount of antigen (Ag) bound by an unknown sample. This amounts to an antigen concentration between 50 and 75 nanomolar (nM).

of the tube. The radioactivity in the precipitate is then determined by scintillation counting.

To calculate the amount of antigen in the sample being assayed, the researcher must determine the percentage of antibody that has bound to nonlabeled antigen. To do so, a second component of the experiment is to develop a standard curve in which a fixed amount of radiolabeled antigen is mixed with varying amounts of unlabeled antigen (**Figure A.9b**). Using this standard curve, a researcher can determine how much antigen is found in the unknown sample. For example, as indicated by the dashed lines, if the unknown sample had about 45% of the antibody bound, then the concentration of antigen in the sample is between 50 and 75 nanomolar (nM).

Radioimmunoassays are used to determine the concentrations of many different kinds of antigens. This includes small molecules such as hormones or macromolecules such as proteins.

APPENDIX
SOLUTIONS TO EVEN-NUMBERED
PROBLEMS AND ALL
COMPREHENSION AND CONCEPT
CHECK QUESTIONS

CHAPTER 1

Answers to Comprehension Questions

1.1: d, c, b, b

1.2: d, d, a, c

1.3: d, a

1.4: d

1.5: c, d

1.6: b, a

Concept Check Questions (follow figure legends)

FIGURE 1.1 Understanding our genes may help to diagnose inherited diseases. It may also lead to the development of drugs to combat diseases. Other answers are possible.

FIGURE 1.2 Many ethical issues are associated with human cloning. Is it the wrong thing to do? Does it conflict with an individual's religious views? And so on.

FIGURE 1.3 Because females mate only once, sorting out the male mosquitoes and releasing sterile males into the environment can limit mosquito reproduction.

FIGURE 1.4 DNA is a macromolecule.

FIGURE 1.5 DNA and proteins are found in chromosomes. A small amount of RNA may also be associated with chromosomes when transcription is occurring, and, as discussed in Chapter 17, some non-coding RNAs may bind to chromosomes.

FIGURE 1.6 The information to make a polypeptide is stored in DNA.

FIGURE 1.7 The dark-colored butterfly has a more active pigment-synthesizing enzyme.

FIGURE 1.8 Genetic variation is the reason these frogs look different.

FIGURE 1.9 These are examples of variation in chromosome number.

FIGURE 1.10 A corn gamete contains 10 chromosomes. (The leaf cells are diploid.)

FIGURE 1.11 The horse populations adapted to their environment, which has gradually changed over the course of many years.

FIGURE 1.12 There are several possible examples of other model organisms, including rats and frogs.

End-of-Chapter Questions:

Conceptual Questions

C2. A chromosome is a very long polymer of DNA. A gene is a specific sequence of bases within that polymer; the sequence of bases distinguishes a gene from other genes. Genes are located in chromosomes, which are found within living cells.

C4. At the molecular level, a gene (a sequence of bases in DNA) is first transcribed into RNA. Most genes are transcribed into an mRNA, which codes a polypeptide. The genetic code within the mRNA is used to synthesize a polypeptide with a particular amino acid sequence. This second process is called translation.

C6. Genetic variation is the occurrence of genetic differences in members of the same species or among different species. Within any population, variation may occur in the genetic material. Variation may occur in particular genes, so some individuals carry one allele and other individuals carry a different allele. An example would be alleles that cause differences in coat color among mammals. Variation may also occur in chromosome structure and number. In plants, differences in chromosome number can affect disease resistance.

C8. You can pick almost any trait. For example, flower color in petunias would be an interesting choice. Some petunias are red and others are purple. There must be different alleles of a flower color gene that affect this trait in petunias. In addition, the amounts of sunlight, fertilizer, and water also affect the intensity of flower color.

C10. A DNA sequence is a sequence of nucleotides. Each nucleotide may have one of four different bases (i.e., A, T, G, or C). When we speak of a DNA sequence, we focus on the sequence of bases.

C12. A. A gene is a segment of DNA. For most genes, the expression of the gene results in the production of a functional protein. The functioning of proteins within living cells affects the traits of an organism.

B. A gene is a segment of DNA that usually codes the information for the production of a specific protein. Genes are found within chromosomes. Many genes are found within a single chromosome.

C. An allele is an alternative version of a particular gene. For example, suppose that a plant has a flower color gene. One allele of this gene may produce a white flower, and a different allele may produce an orange flower. The white allele and the orange allele are two versions of the flower color gene.

D. A DNA sequence is a sequence of nucleotides. The information within a DNA sequence (which is transcribed into an RNA sequence) specifies the amino acid sequence within a polypeptide.

C14. A. How genes and traits are transmitted from parents to offspring

B. How the genetic material functions at the molecular and cellular levels

C. Why genetic variation exists in populations and how it changes over the course of many generations

Experimental Questions

E2. DNA sequencing is used primarily by molecular geneticists. The sequence of DNA is a molecular characteristic of DNA. However, as you will learn throughout this textbook, the sequence of DNA is also of interest to transmission and population geneticists.

E4. A. Transmission geneticists. Dog breeders are interested in how genetic crosses affect the traits of dogs.

B. Molecular geneticists. *E. coli* is a good model organism for studying genetics at the molecular level.

C. Both transmission geneticists and molecular geneticists. Fruit flies are easy to cross for studying the transmission of genes and traits from parents to offspring. Molecular geneticists have also studied many genes in fruit flies to see how they function at the molecular level.

D. Population geneticists. Most wild animals and plants would be of interest to population geneticists. In the wild, you cannot make controlled crosses. But you can study genetic variation within a population and try to understand its relationship to the environment.

E. Transmission geneticists. Agricultural breeders are interested in how genetic crosses affect the outcome of traits.

CHAPTER 2

Answers to Comprehension Questions

2.1: c, d, a
2.2: d, b, b
2.3: b, d

2.4: a, c
2.5: c, d
2.6: d, a, b

Concept Check Questions (follow figure legends)

FIGURE 2.1   *Compartmentalization* means that the cells have membrane-bound compartments.

FIGURE 2.2   The chromosomes would not be spread out very well and would probably be overlapping. It would be difficult to see individual chromosomes.

FIGURE 2.3   Homologs are similar in size and banding pattern, and they carry the same types of genes. However, the alleles of a given gene may be different on the two homologs.

FIGURE 2.4   Filaments of FtsZ assemble into a ring at the future site of the septum and recruit other proteins to this site to produce a cell wall between the two daughter cells.

FIGURE 2.5   In the G_1 phase of the cell cycle, a cell may be preparing to divide. By comparison, the G_0 phase is the phase in which a cell is either temporarily not advancing through the cell cycle or never dividing again.

FIGURE 2.6   Homologs are genetically similar; one is inherited from the female parent and the other from the male parent. By comparison, chromatids are the product of DNA replication. The chromatids within a pair of sister chromatids are genetically identical.

FIGURE 2.7   One end of a kinetochore microtubule is attached to a kinetochore on a chromosome. The other end is within the centrosome.

FIGURE 2.8   Anaphase.

FIGURE 2.9   Ingression occurs because myosin shortens the contractile ring, which is formed from myosin proteins and actin filaments.

FIGURE 2.10   The end result of crossing over is that homologous chromosomes have exchanged pieces.

FIGURE 2.11   The cells at the end of meiosis are haploid, whereas the mother cell is diploid.

FIGURE 2.12   In metaphase of mitosis, each pair of sister chromatids is attached to both poles, whereas in metaphase of meiosis I, each pair of sister chromatids is attached to just one pole.

FIGURE 2.13   Polar bodies are small cells that are produced during oogenesis and then degenerate.

FIGURE 2.14   All cell nuclei in the embryo sac are haploid. The central cell has two haploid nuclei and the other cells, including the egg, have just one.

FIGURE 2.15   In the X-Y system, the presence of the Y chromosome causes maleness, whereas in the X-0 system, it is the ratio between the number of X chromosomes and number of sets of autosomes that determines sex. A ratio of 0.5 results in a male and a ratio of 1.0 produces a female.

FIGURE 2.16   A higher average temperature would favor a high percentage of male alligators. This might limit population size by decreasing the number of females, which are the egg producers.

End-of-Chapter Questions:

Conceptual Questions

C2.   A homolog is one of the members of a chromosome pair. Homologs are usually the same size and carry the same types of genes in the same order. They may differ in that the genes they carry may be different alleles.

C4.   Metaphase is the organization phase, and anaphase is the separation phase.

C6.   In metaphase of meiosis I, each pair of chromatids is attached to only one pole via the kinetochore microtubules. In metaphase of mitosis, there are two attachments (i.e., to both poles). If a chromatid did not attach to a kinetochore microtubule, that chromosome would not migrate to a pole and might not become enclosed in a nuclear membrane after telophase. If left out in the cytosol, it would eventually be degraded.

C8.   The reduction occurs because there is a single DNA replication event but two cell divisions. Because of the nature of separation during anaphase of meiosis I, each cell receives one copy of each type of chromosome.

C10.   It means that the maternally derived and paternally derived chromosomes are randomly aligned along the metaphase plate during metaphase of meiosis I. Refer to Figures 2.11 and 3.13.

C12.   The number of different, random arrangements equals 2^n, where n is the number of chromosomes per set. In this case there are three per set, so the possible number of arrangements equals 2^3, which is 8.

C14.   The probability of inheriting only paternal chromosomes would be much lower because pieces of maternal chromosomes would be mixed with the paternal chromosomes. Therefore, inheriting a chromosome that was completely paternally derived would be unlikely.

C16.   During interphase, the chromosomes are greatly extended. In this conformation, they might get tangled up with each other and not sort properly during meiosis and mitosis. The condensation process probably occurs so that the chromosomes can align along the equatorial plate during metaphase without getting tangled up.

C18.   During prophase of meiosis II, your drawing should show four replicated chromosomes (i.e., four structures that look like X's). Each chromosome is one replicated homolog. During prophase of mitosis, there should be eight replicated chromosomes (i.e., eight X's). During prophase of mitosis, there are pairs of homologs. The main difference is that prophase of meiosis II has a single copy of each of the four chromosomes, whereas prophase of mitosis has four pairs of homologs. At the end of meiosis I, each daughter cell has received only one copy of a homologous pair, not both. This is due to the alignment of homologs during metaphase I and their separation during anaphase I.

C20.   DNA replication does not take place during interphase II. The chromosomes at the end of telophase of meiosis I have already replicated (i.e., they are found in pairs of sister chromatids). During meiosis II, the sister chromatids separate from each other, yielding individual chromosomes.

C22.   A. 20   C. 30
        B. 10   D. 20

C24.   A. Dark males and light females; reciprocal: all dark offspring

        B. All dark offspring; reciprocal: dark females and light males

        C. All dark offspring; reciprocal: dark females and light males

        D. All dark offspring; reciprocal: dark females and light males

C26.   To produce sperm, a spermatogonial cell first goes through mitosis to produce two cells. One of these remains a spermatogonial cell and the other progresses through meiosis. In this way, the testes continue to maintain a population of spermatogonial cells.

C28.   A. *ABC, ABc, AbC, Abc, aBC, abC, aBc, abc*

        B. *ABC, AbC*

        C. *ABC, ABc, aBC, aBc*

        D. *Abc, abc*

C30.   A. The mosquito is a male because the ratio of X chromosomes to sets of autosomes is ½, or 0.5.

B. The mosquito is a female because the ratio of X chromosomes to sets of autosomes is 1.0.

C. The mosquito is a male because the ratio of X chromosomes to sets of autosomes is 0.5.

D. The mosquito is a female because the ratio of X chromosomes to sets of autosomes is 1.0.

Experimental Questions

E2. The basic strategy is to set up a pair of reciprocal crosses. The phenotype of male offspring is usually the easiest way to discern the two patterns. If the gene is on the Y chromosome, the trait will be passed only from a male parent to a male offspring. If it is on the X chromosome, the trait will be passed from a female parent to a male offspring.

E4 You could karyotype other members of the family to see if affected members always carry the abnormal chromosome.

E6. A. 18 pg   C. 9 pg

  B. 18 pg   D. 4.5 pg

Questions for Student Discussion/Collaboration

2. It's not possible to give a direct answer, but the point is to be able to draw chromosomes in different configurations and understand the various phases. The chromosomes may or may not be the following:

  1. In homologous pairs

  2. Connected as sister chromatids

  3. Associated in bivalents

  4. Lined up in metaphase

  5. Moving toward the poles

  and so on.

CHAPTER 3

Answers to Comprehension Questions

3.1:   d, a, b

3.2:   c, b

3.3:   b, c, a

3.4:   c, a

3.5:   b, c (recessive is possible though dominant seems more likely)

3.6:   a, b (using the binomial expansion equation), c

Concept Check Questions (follow figure legends)

FIGURE 3.2   The male gametes are found within pollen grains.

FIGURE 3.3   The white-flowered plant is providing the sperm, and the purple-flowered plant is providing the eggs.

FIGURE 3.4   A true-breeding strain maintains the same trait over the course of several generations of self-fertilization.

FIGURE 3.6   Segregation means that the *T* and *t* alleles separate from each other, and so a haploid cell receives one of them, but not both.

FIGURE 3.7   According to the hypothesis of linked assortment, two different genes are linked. The alleles of the same gene are not linked.

FIGURE 3.9   Independent assortment allows new combinations of alleles of different genes to be found in future generations of offspring.

FIGURE 3.10   Such a parent could make two types of gametes, *Ty* and *ty*, in equal proportions.

FIGURE 3.12   Homologous chromosomes separate at anaphase of meiosis I.

FIGURE 3.13   If you view the left and right sides as being distinctly different, these chromosomes could line up in eight different ways.

FIGURE 3.15   Horizontal lines connect two individuals that have offspring together, and they connect all offspring produced by the same two parents.

End-of-Chapter Questions:

Conceptual Questions

C2. In plants, cross-fertilization occurs when the pollen and eggs come from different plants, whereas in self-fertilization, they come from the same plant.

C4. A true-breeding organism is a homozygote that has two copies of the same allele.

C6. Diploid organisms contain two copies of each type of gene. When they make gametes, only one copy of each gene is found in a gamete. Two alleles cannot stay together within the same gamete.

C8. Genotypes: 1 *Tt* : 1 *tt*

  Phenotypes: 1 tall : 1 short

C10. Here, *c* is the recessive allele for constricted pods; *Y* is the dominant allele for yellow color. The cross is *ccYy* × *CcYy*. Follow the directions for setting up a Punnett square, as described in Section 3.3. The genotypic ratio is 2 *CcYY* : 4 *CcYy* : 2 *Ccyy* : 2 *ccYY* : 4 *ccYy* : 2 *ccyy*. This 2:4:2:2:4:2 ratio can be reduced to a 1:2:1:1:2:1 ratio. The phenotypic ratio is 6 smooth pods, yellow seeds : 2 smooth pods, green seeds : 6 constricted pods, yellow seeds : 2 constricted pods, green seeds. This 6:2:6:2 ratio can be reduced to a 3:1:3:1 ratio.

C12. Offspring with a nonparental phenotype are consistent with the law of independent assortment. If two different traits were always transmitted together as a unit, it would not be possible to get nonparental combinations of traits. For example, if a true-breeding parent had two dominant traits and was crossed to a true-breeding parent having the two recessive traits, the F_2 offspring could not have one recessive and one dominant trait. However, because independent assortment can occur, it is possible for F_2 offspring to have one dominant and one recessive trait.

C14. A. Barring a new mutation during gamete formation, the probability is 100%, because the parents must be heterozygotes in order to produce a child with a recessive disorder.

  B. Construct a Punnett square. There is a 50% chance of a heterozygous offspring.

  C. Use the product rule. The chance of being phenotypically unaffected is 0.75 (i.e., 75%), so the answer is $0.75 \times 0.75 \times 0.75 = 0.422$, which is 42.2%.

  D. Use the binomial expansion equation, where $n = 3$, $x = 2$, $p = 0.75$, $q = 0.25$. The answer is 0.422, or 42.2%.

C16. First construct a Punnett square. The probabilities are 75% for solid coat and 25% for spotted coat.

  A. Use the binomial expansion equation, where $n = 5$, $x = 4$, $p = 0.75$, $q = 0.25$. The answer is 0.396, or 39.6% of the time.

  B. You can use the binomial expansion equation for each litter. For the first litter, $n = 6$, $x = 4$, $p = 0.75$, $q = 0.25$; for the second litter, $n = 5$, $x = 5$, $p = 0.75$, $q = 0.25$. Because the litters are in a specified order, you use the product rule and multiply the probability of the first litter times the probability of the second litter. The answer is 0.070, or 7.0%.

  C. To calculate the probability of the first litter, you use the product rule and multiply the probability of the first pup (0.75) times the probability of the remaining four. You use the binomial expansion equation to calculate the probability of the remaining four, where $n = 4$, $x = 3$, $p = 0.75$, $q = 0.25$. The probability of the first litter is 0.316. To calculate the probability of the second litter, you use the product rule and multiply the probability of the first pup (0.25) times the probability of the second pup (0.25) times the probability of the remaining five. To calculate the probability of the remaining five, you use the binomial expansion equation, where $n = 5$, $x = 4$, $p = 0.75$, $q = 0.25$. The probability of the second litter is 0.025. To get the probability of these two litters occurring in this order, you use the product rule and multiply the probability of the first litter (0.316) times the probability of the second litter (0.025). The answer is 0.008, or 0.8%.

D. Because the order of the first two is specified, you use the product rule and multiply the probability of the firstborn (0.75) times the probability of the second born (0.25) times the probability of the remaining four. You use the binomial expansion equation to calculate the probability of the remaining four pups, where $n = 4$, $x = 2$, $p = 0.75$, $q = 0.25$. The answer is 0.040, or 4.0%.

C18. A. Use the product rule:

$(\frac{1}{4})(\frac{1}{4}) = \frac{1}{16}$

B. Use the binomial expansion equation, where $n = 4$, $p = \frac{1}{4}$, $q = \frac{3}{4}$, $x = 2$:

$P = 0.21$, or 21%

C. Use the product rule:

$(\frac{1}{4})(\frac{3}{4})(\frac{3}{4}) = 0.14$, or 14%

C20. A. $\frac{1}{4}$

B. 1, or 100%

C. $(\frac{3}{4})(\frac{3}{4})(\frac{3}{4}) = \frac{27}{64} = 0.42$, or 42%

D. Use the binomial expansion equation, where $n = 7$, $p = \frac{3}{4}$, $q = \frac{1}{4}$, $x = 3$:

$P = 0.058$, or 5.8%

E. The probability that the first plant is tall is $\frac{3}{4}$. To calculate the probability that among the next four, any two will be tall, use the binomial expansion equation, where $n = 4$, $p = \frac{3}{4}$, $q = \frac{1}{4}$, and $x = 2$:

$P = 0.21$

Then calculate the overall probability of these two events:

$(\frac{3}{4})(0.21) = 0.16$, or 16%

C22. The gamete violates the law of segregation because it has two copies of one gene. The two alleles for the *A* gene did not segregate from each other.

C24. Based on the pedigree provided, the disease is likely to be a dominant trait because a child with Marfan syndrome always has an affected parent. In fact, it is a dominant disorder.

C26. It is impossible for F$_1$ individuals to be true-breeding because they are all heterozygotes.

C28. 2 *TY*, *tY*, 2 *Ty*, *ty*, *TTY*, *TTy*, 2 *TtY*, 2 *Tty*

This question is a bit tricky, but 2 *TY* and 2 *Ty* gametes are made because either of the two *T* alleles can combine with *Y* or *y*. Also, there are 2 *TtY* and 2 *Tty* because either of the two *T* alleles can combine with *t* and then combine with *Y* or *y*.

C30. The genotype of the F$_1$ plants is *Tt Yy Rr*. According to the laws of segregation and independent assortment, the alleles of each gene will segregate from each other, and the alleles of different genes will randomly assort into gametes. A *Tt Yy Rr* individual could make eight types of gametes: *TYR*, *TyR*, *Tyr*, *TYr*, *tYR*, *tyR*, *tYr*, and *tyr*, in equal proportions (i.e., $\frac{1}{8}$ of each type of gamete). To determine genotypes and phenotypes, you could make a large Punnett square that would contain 64 boxes. You would need to line up the eight possible gametes across the top and along the side and then fill in the 64 boxes. Alternatively, you could use either the multiplication method or the forked-line method described in Figure 3.11. The genotypes and phenotypes are as follows:

1 *TT YY RR*

2 *TT Yy RR*

2 *TT YY Rr*

2 *Tt YY RR*

4 *TT Yy Rr*

4 *Tt Yy RR*

4 *Tt YY Rr*

8 *Tt Yy Rr* = 27 tall, yellow, round

1 *TT yy RR*

2 *Tt yy RR*

2 *TT yy Rr*

4 *Tt yy Rr* = 9 tall, green, round

1 *TT YY rr*

2 *TT Yy rr*

2 *Tt YY rr*

4 *Tt Yy rr* = 9 tall, yellow, wrinkled

1 *tt YY RR*

2 *tt Yy RR*

2 *tt YY Rr*

4 *tt Yy Rr* = 9 short, yellow, round

1 *TT yy rr*

2 *Tt yy rr* = 3 tall, green, wrinkled

1 *tt yy RR*

2 *tt yy Rr* = 3 short, green, round

1 *tt YY rr*

2 *tt Yy rr* = 3 short, yellow, wrinkled

1 *tt yy rr* = 1 short, green, wrinkled

C32. Because this is a dominant trait, the wooly-haired male must carry at least one copy of the wooly hair allele. Because the female parent of the wooly-haired male does not have wooly hair, the wooly-haired male must also have inherited one copy of the normal allele. Therefore, the wooly-haired male is a heterozygote. This heterozygous wooly-haired male has a 50% chance of passing the wooly hair allele to each offspring; each offspring has a 50% chance of passing the allele to their offspring; and those grandchildren have a 50% chance of passing the allele to their offspring (the wooly-haired male's great-grandchildren). Because this is an ordered sequence of independent events, you use the product rule: $0.5 \times 0.5 \times 0.5 = 0.125$, or 12.5%. If no other Scandinavians are on the island, the probability that the offspring do not have wooly hair is $100\% - 12.5\% = 87.5\%$. You use the binomial expansion equation to determine the probability that one out of eight great-grandchildren will have wooly hair, where $n = 8$, $x = 1$, $p = 0.125$, $q = 0.875$. The answer is 0.393, or 39.3%.

C34. A. Use the product rule. If the brown-eyed parent is a heterozygote, there is a 50% chance of having an offspring with brown eyes at each birth: $(0.5)^7 = 0.0078$, or 0.78%. This is a pretty small probability.

B. However, the brown-eyed parent must be a heterozygote if an eighth child has blue eyes because brown eyes is a dominant trait. To have a child with blue eyes, the brown-eyed parent would have to pass a blue-eye allele to the offspring. In this case, the probability of being heterozygous is 100%.

Experimental Questions

E2. The difference lies in where the pollen comes from. In self-fertilization, the pollen and eggs come from the same plant. In cross-fertilization, they come from different plants.

E4. According to Mendel's law of segregation, the genotypic ratio should be 1 homozygote dominant : 2 heterozygotes : 1 homozygote recessive. The data table considers only the plants with a dominant phenotype. The genotypic ratio should be 1 homozygote dominant : 2 heterozygotes. The homozygote dominants would be true-breeding, but the heterozygotes would not be true-breeding. This 1:2 ratio is very close to what Mendel observed.

E6. All three offspring had black fur. The ovaries from the albino female could produce eggs with only the dominant black allele (because they were obtained from a true-breeding black female). The actual phenotype of the albino female parent does not matter. Therefore, all offspring are heterozygotes (*Bb*) and have black fur.

E8. If you construct a Punnett square according to Mendel's laws, you expect a 9:3:3:1 ratio. Because a total of 556 offspring were observed, the expected numbers of offspring with the different phenotypes are

$556 \times \frac{9}{16} = 313$ round, yellow

556 × ³⁄₁₆ = 104 wrinkled, yellow

556 × ³⁄₁₆ = 104 round, green

556 × ¹⁄₁₆ = 35 wrinkled, green

If you substitute the observed and expected values into the chi square equation, you get a value of 0.51. With four categories, the degrees of freedom equal $n - 1$, or 3. If you look up the value of 0.51 in the chi square table (see Table 3.1), you see that it falls between P values of 0.80 and 0.95. This means that the probability is between 80% and 95% that any deviation between observed results and expected results was caused by random sampling error. Therefore, you accept the hypothesis. In other words, the results are consistent with the law of independent assortment.

E10. A. Let c^+ represent normal wings and c represent curved wings and e^+ represent gray body and e represent ebony body.

Parental cross: $cce^+e^+ \times c^+c^+ee$

F_1 generation is heterozygous: c^+ce^+e

A cross of an F_1 offspring with a fly with curved wings and ebony body is represented as follows:

$c^+ce^+e \times ccee$

The F_2 offspring will have this 1:1:1:1 ratio of genotypes:

$c^+ce^+e : c^+cee : cce^+e : ccee$

B. The phenotypic ratio of the F_2 flies will be 1:1:1:1:

normal wings, gray body : normal wings, ebony bodies : curved wings, gray bodies : curved wings, ebony bodies

C. From part B, the F_2 offspring consists of ¼ of each phenotypic category. There are a total of 444 offspring. The expected number of each category is ¼ × 444, which equals 111.

$$\chi^2 = \frac{(114-111)^2}{111} + \frac{(105-111)^2}{111} + \frac{(111-111)^2}{111} + \frac{(114-111)^2}{111}$$

$$\chi^2 = 0.49$$

With 3 degrees of freedom, a value of 0.49 or greater is likely to occur between 80% and 95% of the time. Therefore, the experimental data are consistent with the expected outcome.

E12. Use the basic chi square analysis. You expect a 3:1 ratio, or ¾ of the dominant phenotype and ¼ of the recessive phenotype. The observed and expected values are as follows (rounded to the nearest whole number):

Observed*	Expected	$\frac{(O-E)^2}{E}$
5474	5493	0.066
1850	1831	0.197
6022	6017	0.004
2001	2006	0.012
705	697	0.092
224	232	0.276
882	886	0.018
299	295	0.054
428	435	0.113
152	145	0.338
651	644	0.076
207	215	0.298
787	798	0.152
277	266	0.455
		$\chi^2 = 2.15$

*Due to rounding, the observed and expected values may not add up to precisely the same number.

Because $n = 14$, there are 13 degrees of freedom. To find the value 2.15 in the chi square table, you have to look between 10 and 15 degrees of freedom. In either case, the value 2.15 or greater is expected to occur more than 99% of the time. Therefore, the data are consistent with the law of segregation.

E14. Perhaps the most convincing evidence was that all white-eyed flies of the F_2 generation were males. This suggests a link between sex determination and the inheritance of this trait. Because sex determination in fruit flies is determined by the number of X chromosomes, this outcome suggests a relationship between the inheritance of the X chromosome and the inheritance of this trait.

E16. If you use the data from the F_1 mating (i.e., the F_2 results), there were 3470 red-eyed flies. You expect a 3:1 ratio between red- and white-eyed flies. Therefore, assuming that all red-eyed offspring survived, there should have been about 1157 (i.e., 3470/3) white-eyed flies. However, there were only 782. If you divide 782 by 1157, you get a value of 0.676, or a 67.6% survival rate.

Questions for Student Discussion/Collaboration

2. If you construct a Punnett square, the following probabilities will be obtained:

tall with axial flowers, ⅜

short with terminal flowers, ⅛

The probability of being tall with axial flowers or short with terminal flowers is then:

⅜ + ⅛ = 4⁄8 = ½

You use the product rule to calculate the probability of the ordered events of the first three offspring being tall and axial or short and terminal and the fourth offspring being tall and axial:

(½)(½)(½)(⅜) = ³⁄₆₄ = 0.047 = 4.7%

CHAPTER 4

Answers to Comprehension Questions

4.1: d	4.6: b
4.2: d, d, c	4.7: c
4.3: c	4.8: d
4.4: d, a	4.9: b, d
4.5: c, a	

Concept Check Questions (follow figure legends)

FIGURE 4.1 Both of these colors are considered wild type because both of them are common in natural populations.

FIGURE 4.2 Yes. The *PP* homozygote probably makes twice as much of the protein than is needed for the purple color.

FIGURE 4.3 Individual III-2 shows the effect of incomplete penetrance.

FIGURE 4.4 Genes and the environment determine an organism's traits.

FIGURE 4.5 In a heterozygous four-o'clock plant, 50% of the functional protein is not enough to give a red color.

FIGURE 4.6 It is more likely to observe incomplete dominance at the molecular or cellular level.

FIGURE 4.7 In this case, the heterozygote is resistant to malaria.

FIGURE 4.8 The scenario in part (a) explains the heterozygote advantage for the sickle cell allele.

FIGURE 4.9 The *i* allele is a loss-of-function allele.

FIGURE 4.10 The key feature of the pedigree that points to X-linked inheritance is that only males are affected with the disorder. Also, carrier females often have male siblings that are affected with the disorder.

FIGURE 4.11 The reciprocal cross yields a different result because females carry two copies of an X-linked gene, whereas males have only one.

FIGURE 4.12 Homologous regions of the X and Y chromosomes are important for chromosome pairing (synapsis) during meiosis.

FIGURE 4.13 A heterozygous female cow would not have scurs.

FIGURE 4.14 When a trait is expressed only in males or females, this is possibly due to differences in the levels of sex hormones or to developmental factors that differ between males and females.

FIGURE 4.15 The heterozygote has one copy of the wild-type allele (m), which allows for development to proceed in a way that is not too far from normal. Having two mutant copies of the gene (MM) probably adversely affects development to a degree that is incompatible with survival.

FIGURE 4.17 *Epistasis* means that the alleles of one gene mask the phenotypic effects of the alleles of a different gene. *Complementation* means that two strains exhibiting the same recessive trait will produce offspring that show the dominant (wild-type) trait. This usually indicates that the alleles for the recessive trait are in two different genes.

FIGURE 4.18 In some cases, a single gene knockout does not have an effect due to gene redundancy. Other explanations are also possible.

FIGURE 4.19 The two genes determining seed capsule shape are redundant. Having one functional allele of either gene produces a triangular capsule. If both genes are inactive, an ovate capsule is produced.

End-of-Chapter Questions:

Conceptual Questions

C2.   Sex-influenced traits are affected by the sex of the individual even though the gene that governs the trait is autosomally inherited. Scurs in certain breeds of cattle is an example. A sex-influenced trait is dominant in one sex and recessive in the other. The expression of a sex-limited trait is limited to one sex. For example, colorful plumage in certain species of birds is limited to the male sex. Sex-linked traits are traits whose genes are found on the sex chromosomes. Examples in humans include hemophilia and color blindness.

C4.   If the functional allele is dominant, it tells you that one copy of the gene produces a sufficient amount of the protein coded by the gene. Having twice as much of this protein, as in the dominant homozygote, does not alter the phenotype. If the functional allele is incompletely dominant, one copy of that allele does not produce the same trait as is observed in the homozygote with two functional alleles.

C6.   The ratio would be 1 normal : 2 star-eyed.

C8.   If individual 1 is *ii*, individual 2 could be $I^A i$, $I^A I^A$, $I^B i$, $I^B I^B$, or $I^A I^B$.

   If individual 1 is $I^A i$ or $I^A I^A$, individual 2 could be $I^B i$, $I^B I^B$, or $I^A I^B$.

   If individual 1 is $I^B i$ or $I^B I^B$, individual 2 could be $I^A i$, $I^A I^A$, or $I^A I^B$.

   Assuming that individual 1 is the parent of individual 2:

   If individual 1 is *ii*, individual 2 could be $I^A i$ or $I^B i$.

   If individual 1 is $I^A i$, individual 2 could be $I^B i$ or $I^A I^B$.

   If individual 1 is $I^A I^A$, individual 2 could be $I^A I^B$.

   If individual 1 is $I^B i$, individual 2 could be $I^A i$ or $I^A I^B$.

   If individual 1 is $I^B I^B$, individual 2 could be $I^A I^B$.

C10.   A child with type O blood (genotype *ii*) received the *i* allele from both parents. Therefore, the male parent could not be $I^A I^B$, $I^B I^B$, or $I^A I^A$. because these three genotypes do not have an *i* allele. Genotypically, the male parent could be $I^A i$, $I^B i$, or *ii* and have type A, B, or O blood, respectively.

C12.   It might be called codominance at the "hair level" because one or the other allele is dominant with regard to each individual hair. However, this is not the same as codominance in blood types, in which every cell can express both alleles.

C14.   A. X-linked recessive (unaffected female parents transmit the trait to male offspring)

   B. Autosomal recessive (affected offspring have two unaffected parents)

C16.   First set up the following Punnett square:

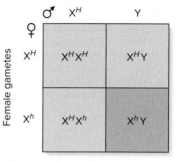

Male gametes

   A. ¼

   B. $(\frac{3}{4})(\frac{3}{4})(\frac{3}{4})(\frac{3}{4}) = \frac{81}{256}$

   C. ¾

   D. The probability of an affected offspring is ¼, and the probability of an unaffected offspring is ¾. For this problem, you use the binomial expansion equation with $x = 2$, $n = 5$, $p = \frac{1}{4}$, and $q = \frac{13}{4}$. The answer is 0.26, or 26% of the time.

C18.   You know that the parents must be heterozygotes for both genes.

   The genotypic ratio of their offspring is 1 $S^P S^P$: 2 $S^P S^A$: 1 $S^A S^A$.

   The phenotypic ratio depends on sex: 1 $S^P S^P$ male with scurs : 1 $S^P S^P$ female with scurs : 2 $S^P S^A$ males with scurs : 2 $S^P S^A$ females without scurs : 1 $S^A S^A$. male without scurs : 1 $S^A S^A$. female without scurs.

   A. 50%

   B. ⅛, or 12.5%

   C. $(\frac{3}{8})(\frac{3}{8})(\frac{3}{8}) = \frac{27}{512} = 0.05$, or 5%

C20.   The injury probably occurred in the summer. In a Siamese cat, dark fur occurs in cooler regions of the body. If the fur grows during the summer, these regions are likely to be somewhat warmer, and therefore the fur will be lighter.

C22.   A. Could be possible.

   B. No, because an unaffected male has an affected female offspring.

   C. No, because two unaffected parents have affected children.

   D. No, because an unaffected male parent has an affected female offspring.

   E. No, because both sexes exhibit the trait.

   F. Could be possible.

C24.   To decide, you would look at the pattern within families over the course of many generations. For a recessive trait, 25% of the offspring within a family are expected to be affected if both parents are unaffected carriers, and 50% of the offspring are expected to be affected if one parent is affected. You need to look at many families and see if these 25% and 50% values are approximately true. Incomplete penetrance would not necessarily produce such numbers. Also, for very rare alleles, incomplete penetrance would probably yield a much higher frequency of affected parents producing affected offspring. For rare recessive disorders, it is most likely that both parents are heterozygous carriers. Finally, the most informative pedigrees would include situations in which two affected parents produce children. If they produced an unaffected offspring, this would indicate incomplete penetrance. If all their offspring are affected, this would be consistent with recessive inheritance.

C26.   The probability of a heterozygote passing the allele to offspring is 50%. The probability of an affected offspring expressing the trait is 80%. You use the product rule to determine the likelihood of these two independent events: The probability is $(0.5)(0.8) = 0.4$, or 40%.

C28.   This pattern of inheritance is an example of incomplete dominance. The heterozygous horses are palominos. For example, if C^{Ch} represents chestnut and C^{Cr} represents cremello, the chestnut horses are $C^{Ch} C^{Ch}$, the cremello horses are $C^{Cr} C^{Cr}$, and the palominos are $C^{Ch} C^{Cr}$.

Experimental Questions

E2. Two redundant genes are involved in the feathering of shanks in chickens. The unfeathered Buff Rocks are homozygous for recessive alleles of both genes. The Black Langhans are homozygous for dominant alleles of both genes. In a Punnett square depicting the F_2 generation (which results from a cross of a double heterozygote to another double heterozygote), 1 out of 16 offspring are homozygous for the recessive alleles of both genes and will have unfeathered shanks. The other 15 will have feathered shanks because they have at least one dominant allele for one of the two (redundant) genes.

E4. The reason all the puppies had black hair is because albino alleles are found in two different genes. If you let the letters *A* and *B* represent the two different pigmentation genes, then one of the albino parents is *AAbb*, and the other is *aaBB*. Their offspring are *AaBb* and therefore are not albinos because they have one dominant copy of each gene.

E6. In general, you cannot distinguish between autosomal and pseudoautosomal inheritance from a pedigree analysis. Females and males have an equal probability of passing the alleles of interest to their female and male offspring. However, if an offspring had a chromosomal abnormality, you might be able to tell. For example, an offspring that was X0 would produce less surface antigen, and an offspring that was XXX or XYY or XXY would produce an extra amount. This would lead you to suspect that the gene is located on the sex chromosomes.

E8. The ratios would be 3 yellow head, green underside : 3 white head, blue underside : 1 all yellow : 1 all white.

E10. In this case, you expect a 9:7 ratio between red and white flowers. In other words, $9/16$ of the offspring will have red flowers and $7/16$ will have white. Because there are a total of 345 plants, the expected values are

$$9/16 \times 345 = 194 \text{ red}$$

$$7/16 \times 345 = 151 \text{ white}$$

Substituting these values into the chi square equation gives a chi square value of 0.58. With 1 degree of freedom, this chi square value is too small to reject the hypothesis. Therefore, it can be accepted.

E12. The results obtained when crossing two F_1 offspring appear to yield a 9:3:3:1 ratio, which would be expected if eye color is affected by two different genes that exist in dominant and recessive alleles. Neither gene is X-linked. Let pr^+ represent the red allele of the first gene and pr the purple allele. Let sep^+ represent the red allele of the second gene and sep the sepia allele.

The first cross is $prpr\ sep^+sep^+ \times pr^+pr^+ sep\ sep$.

All the F_1 offspring will be $pr^+pr\ sep^+sep$. They have red eyes because they have a dominant red allele for each gene. When the F_1 offspring are crossed to each other, the following results will be obtained:

♀ \ ♂	pr^+sep^+	pr^+sep	$pr\ sep^+$	$pr\ sep$
pr^+sep^+	pr^+pr^+ sep^+sep^+ Red	pr^+pr^+ sep^+sep Red	pr^+pr sep^+sep^+ Red	pr^+pr sep^+sep Red
pr^+sep	pr^+pr^+ sep^+sep Red	pr^+pr^+ $sep\ sep$ Sepia	pr^+pr sep^+sep Red	pr^+pr $sep\ sep$ Sepia
$pr\ sep^+$	pr^+pr sep^+sep^+ Red	pr^+pr sep^+sep Red	$pr\ pr$ sep^+sep^+ Purple	$pr\ pr$ sep^+sep Purple
$pr\ sep$	pr^+pr sep^+sep^+ Red	pr^+pr $sep\ sep$ Sepia	$pr\ pr$ sep^+sep Purple	$pr\ pr$ $sep\ sep$ Purplish sepia

In this case, one gene exists as the red (dominant) or purple (recessive) allele, and the second gene exists as the red (dominant) or sepia (recessive) allele. If an offspring is homozygous for the purple allele, it will have purple eyes. Similarly, if an offspring is homozygous for the sepia allele, it will have sepia eyes. An offspring homozygous for both recessive alleles has purplish sepia eyes. To have red eyes, an offspring must have at least one copy of the dominant red allele for both genes. Based on an expected 9 red : 3 purple : 3 sepia : 1 purplish sepia, the observed and expected numbers of offspring are as follows:

Observed	Expected
146 purple eyes	148 purple eyes ($791 \times 3/16$)
151 sepia eyes	148 sepia eyes ($791 \times 3/16$)
50 purplish sepia eyes	49 purplish sepia eyes ($791 \times 1/16$)
444 red eyes	445 red eyes ($791 \times 9/16$)
791 total offspring	

Substituting the observed and expected values into the chi square equation yields a chi square value of about 0.11. With 3 degrees of freedom, this is well within the expected range of values, so you cannot reject the hypothesis that purple and sepia alleles are in two different genes and that these recessive alleles are epistatic to each other.

E14. With X-linked recessive inheritance, it is much more common for males to be affected. With autosomal recessive inheritance, there is an equal chance of males and females being affected (unless there is a sex influence, in which an allele is dominant in one sex but recessive in the opposite sex). For X-linked dominant inheritance, affected males will produce 100% affected female offspring and will not transmit the trait to male offspring. This is not true for autosomal dominant traits, where there is an equal chance of males and females being affected.

Questions for Student Discussion/Collaboration

2. Let's refer to the alleles as *B* dominant, *b* recessive and *G* dominant, *g* recessive.

   The parental cross is *BBGG × bbgg*.

   All of the F_1 offspring are *BbGg*.

   If you make a Punnett square, the genotypes that are homozygous for the *b* allele and have at least one copy of the dominant *G* allele are gray. To explain this phenotype, you could hypothesize that the *B* allele codes an enzyme that can make lots of pigment, whether or not the *G* allele is present. Therefore, you get a black chaff phenotype when one *B* allele is inherited. The *G* allele codes a somewhat redundant enzyme, but maybe it does not function quite as well or its pigment product may not be as dark. Therefore, in the absence of a *B* allele, the *G* allele will give a gray chaff phenotype.

CHAPTER 5

Answers to Comprehension Questions

5.1: d, d
5.2: b, a, d
5.3: d, c, a, b
5.4: b, a, c, b

Concept Check Questions (follow figure legends)

FIGURE 5.1   The F_2 offspring are all dextral because all F_1 females are *Dd*, and the genotype of the female parent determines the phenotype of the offspring.

FIGURE 5.2   The oocyte will receive both *D* and *d* gene products.

FIGURE 5.3   The Barr body is more brightly stained because it is very compact.

FIGURE 5.4   X-chromosome inactivation first occurs during embryonic development. It is then maintained into adulthood.

FIGURE 5.5   They migrate differently because their amino acid sequences are slightly different.

FIGURE 5.7   Only the maintenance phase occurs in an adult female.

FIGURE 5.8   All offspring would be normal size because they would inherit a functional copy of the gene from their male parent.

FIGURE 5.9   Erasure allows eggs to transmit unmethylated copies of the gene to the offspring.

FIGURE 5.10   Maintenance methylation is methylation that occurs when a methylated gene replicates and each daughter strand is also methylated. It occurs in somatic cells. De novo methylation is the methylation of a gene that is not already methylated. It occurs in germ-line cells.

FIGURE 5.11   The offspring on the left side did not receive a copy of either gene from the female parent. The *AS* gene is silenced during sperm formation in the male parent, so the offspring does not have an active copy of the *AS* gene and therefore has Angelman syndrome. By comparison, the offspring received an active *PWS* gene from the male parent and therefore does not have Prader-Willi syndrome.

FIGURE 5.12   A nucleoid is not surrounded by a membrane, whereas a cell nucleus is surrounded by a double membrane.

FIGURE 5.13   Mitochondria need genes that code rRNAs and tRNAs in order to synthesize polypeptides involved in the production of ATP within the mitochondrial matrix.

FIGURE 5.14   A reciprocal cross is a cross in which the sexes and phenotypes of the parents are reversed compared to a first cross.

FIGURE 5.15   No. Once a patch of tissue is white, it has lost all of the normal chloroplasts, so it could not produce a patch of green tissue unless a rare mutation occurred.

FIGURE 5.17   Chloroplasts and mitochondria have lost most of their genes during evolution. Many of these genes have been transferred to the cell nucleus.

End-of-Chapter Questions:

Conceptual Questions

C2.   A maternal effect gene is one for which the genotype of the female parent determines the phenotype of the offspring. At the cellular level, this happens because maternal effect genes are expressed in diploid nurse cells and then the gene products are transported into the egg. These gene products are mRNAs and/or proteins that play key roles in the early steps of embryonic development.

C4.   The genotype of female P must be bic^-bic^-. That is why female P produces abnormal offspring. Because female P is alive and able to produce offspring, female GP must have been bic^+bic^- and passed the bic^- allele to female P. Male GP also must have passed the bic^- allele to female P because female P is bic^-bic^-. Male GP could be either bic^+bic^- or bic^-bic^-.

C6.   For clarity, let's call the female parent with the disorder female P and the maternal grandparents female GP and male GP, where P stands for parent and GP stands for grandparent. Female P must be heterozygous and is phenotypically abnormal because female GP must have been homozygous for the nonfunctional recessive allele. However, because female P produces all normal offspring, female P must have inherited the functional dominant allele from male GP. Female P produces all normal offspring because the gene causing the disorder is a maternal effect gene, and the gene product of the functional dominant allele is transferred to the egg.

C8.   Maternal effect genes exert their effects early in development because the gene products are transferred from nurse cells to eggs. The gene products, mRNAs and proteins, do not last very long before they are degraded. Therefore, they can exert their effects only during early stages of embryonic development.

C10.   Dosage compensation refers to the phenomenon that the level of expression of genes on the sex chromosomes is the same in males and females, even though their numbers of sex chromosomes differ. In many species, dosage compensation seems necessary to keep the balance of gene expression between the autosomes and sex chromosomes similar between the two sexes.

C12.   In mammals, such as humans, one of the X chromosomes is inactivated in females; in *Drosophila*, the level of transcription on the X chromosome in males is doubled; in *C. elegans*, the level of transcription of the X chromosome in hermaphrodites is decreased by 50% relative to that of males.

C14.   X-chromosome inactivation begins with the counting of Xics. If there are two X chromosomes, one is targeted for inactivation during initiation. During embryogenesis, this inactivation begins at the Xic locus and spreads to both ends of the X chromosome until the chromosome becomes a highly condensed Barr body. The *Tsix* gene plays a role in the choice of the X chromosome that remains active. The *Xist* gene, which is located in the Xic region, remains transcriptionally active on the inactivated X chromosome. It is thought to play an important role in X-chromosome inactivation by coating the inactive X chromosome with *Xist* RNA. After X-chromosome inactivation has been completed, it is maintained for the same X chromosome in somatic cells during subsequent cell divisions. In germ cells, however, the X chromosomes are not inactivated, so an egg can transmit either copy of an active (noncondensed) X chromosome.

C16.   A. One   C. Two

      B. Zero   D. Zero

C18.   The offspring inherited X^B from its female parent and X^O and Y from its male parent. It is an XXY animal, which is male (but somewhat feminized).

C20.   The erasure and reestablishment phase occurs during gametogenesis. It is necessary to erase the imprint because each sex will transmit either inactive or active alleles of a gene. In somatic cells, the two alleles for a gene are imprinted according to the sex of the parent from which the allele was inherited.

C22.   A person born with paternal uniparental disomy 15 would have Angelman syndrome, because this individual would not have an active copy of the *AS* gene; the paternally inherited copies of the *AS* gene are silenced. This female does not have a deletion in either copy of chromosome 15 and therefore would produce unaffected offspring.

C24.   In some species, such as marsupials, X-chromosome inactivation depends on the sex. This is similar to genomic imprinting. Also, once X-chromosome inactivation occurs during embryonic development, it is maintained throughout the rest of the life of the organism, which is also similar to imprinting. X-chromosome inactivation in some placental mammals is different from genomic imprinting in that it is not sex-dependent. The X chromosome that is inactivated in each embryonic cell can be inherited from the female parent or male parent. No marking process of the X chromosome occurs during gametogenesis. In contrast, genomic imprinting always involves a marking process during gametogenesis.

C26.   A reciprocal cross is the second of two crosses that involve the same genotypes of two parents, but the sexes of the parents are opposite in the two crosses. For autosomal inheritance, the cross in this case is female *FF* × male *ff*, and the reciprocal cross is female *ff* × male *FF*. Autosomal inheritance gives the same result for both crosses because the autosomes are transmitted from parent to offspring in the same way for both sexes. For both crosses, the offspring would be *Ff* and be able to fly. However, for extranuclear inheritance, the mitochondria are not transmitted via the gametes in the same way for both sexes. For maternal inheritance, the reciprocal crosses would show that the gene is always inherited from the female parent. If the female carried the *F* allele and the male carried the *f* allele, the offspring would be able to fly. In the reciprocal cross, the female would carry the *f* allele and the male would carry the *F* allele, and the offspring would not be able to fly.

C28. The three general cellular mechanisms that can promote maternal inheritance are:

(1) Lack of entry of sperm mitochondria into the oocyte. An example is the sperm of Chinese hamsters.

(2) Destruction of sperm mitochondrial DNA prior to fertilization. In *Drosophila*, sperm mitochondrial DNA is destroyed by an endonuclease.

(3) Destruction of sperm mitochondria after fertilization. In most mammals, ubiquitinylation targets the sperm mitochondria for destruction via proteasomes or lysosomes.

C30. Mitochondria and chloroplasts evolved from an endosymbiotic relationship in which bacteria took up residence within primordial eukaryotic cells. Throughout evolution, there has been a movement of genes out of the organellar genomes and into the nuclear genome. The genomes of modern mitochondria and chloroplasts contain only a fraction of the genes necessary for organellar structure and function. Nuclear genes code most of the proteins that function within chloroplasts and mitochondria. Long ago, these genes were originally in the mitochondrial and chloroplast genomes but have been subsequently transferred to the nuclear genome.

C32. The tendency to develop this form of leukemia appears to be inherited from the female parent, much like the inheritance of mitochondria. To show that it is not, you could separate newborn mice from the female parent with AMLV and place them with lactating females that do not carry AMLV. These offspring would not be expected to develop leukemia, even though their female parent would.

Experimental Questions

E2. The first type of observation was based on cytological studies. The presence of the Barr body in female cells was consistent with the idea that one of the X chromosomes was highly condensed. The second type of observation was based on genetic mutations. A variegated phenotype that is found only in females is consistent with the idea that certain patches express one allele and other patches express the other allele. This variegated phenotype would occur only if the inactivation of one X chromosome happened at an early stage of embryonic development and was inherited permanently thereafter.

E4. The pattern of inheritance is consistent with imprinting. In every cross, the allele that is inherited from the male parent is expressed in the offspring, but the allele inherited from the female parent is not.

E6. You assume that the snails in the large colony on the second island are true-breeding with the genotype *DD*. Let the male snail from the deserted island mate with a female snail from the large colony. Then let the F_1 snails mate with each other to produce an F_2 generation, and let the F_2 snails mate with each other to produce an F_3 generation. Here are the expected results:

Female *DD* × Male *DD*

All F_1 snails coil to the right.

All F_2 snails coil to the right.

All F_3 snails coil to the right.

Female *DD* × Male *Dd*

All F_1 snails coil to the right.

All F_2 snails coil to the right because all of the F_1 females are *DD* or *Dd*.

$15/16$ of F_3 snails coil to the right; $1/16$ of F_3 snails coil to the left (because $15/16$ of the F_2 females are *dd*).

Female *DD* × Male *dd*

All F_1 snails coil to the right.

All F_2 snails coil to the right because all of the F_1 females are *Dd*.

$3/4$ of F_3 snails coil to the right; $1/4$ of F_3 snails coil to the left (because $1/4$ of the F_2 females are *dd*).

E8. Let's first consider the genotypes of male A and male B. Male A must have two copies of the normal-size allele of the *Igf2* gene. You know

this because male A's female parent was *Igf2 Igf2*; the male parent of male A must have been a heterozygote, *Igf2 Igf2⁻*, because half of the litter that contained male A were small-size offspring. But because male A was not small size, it must have inherited the normal-size allele from its male parent. Therefore, male A must be *Igf2 Igf2*. You cannot be completely sure of the genotype of male B. It must have inherited the normal-size *Igf2* allele from its male parent because male B is normal size. You do not know the genotype of male B's female parent, but it could be either *Igf2⁻ Igf2⁻* or *Igf2 Igf2⁻*. In either case, the female parent of male B could pass the *Igf2⁻* allele to an offspring, but you do not know for sure if that happened. So, male B could be either *Igf2 Igf2⁻* or *Igf2 Igf2*.

For the *Igf2* gene, you know that the maternal allele is silenced. Therefore, the genotypes and phenotypes of females A and B are irrelevant. The phenotype of the offspring is determined only by the allele that is inherited from the male parent. Because you know that male A has to be *Igf2 Igf2*, you know that it can produce only normal-size offspring. Because both females A and B produced small-size offspring, male A cannot be their male parent. In contrast, male B could be either *Igf2 Igf2* or *Igf2 Igf2⁻*. Because both females gave birth to some small-size babies (and because male A and male B were the only two male mice in the cage), you conclude that male B must be *Igf2 Igf2⁻* and is the male parent of both litters.

E10. In fruit flies, the expression of a male's X-linked genes is doubled. In mice, one of the two X chromosomes is inactivated; that is why females and males produce the same total amount of mRNA for most X-linked genes. In *C. elegans*, the expression of a hermaphrodite's X-linked genes is halved. Overall, the total amount of expression of X-linked genes is the same in males and females (or hermaphrodites) of these three species. In fruit flies and *C. elegans*, heterozygous females and hermaphrodites express 50% of each allele compared with a homozygous male, so heterozygous females and hermaphrodites produce the same total amount of mRNA from X-linked genes as males do. Note: In heterozygous females of fruit flies, mice, and *C. elegans*, there is 50% of each gene product (compared to hemizygous males and homozygous females).

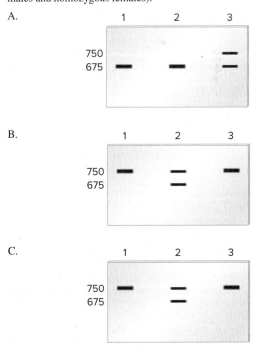

Questions for Student Discussion/Collaboration

2. Most of the genes originally within mitochondria and chloroplasts have either been lost or transferred to the nucleus. Therefore, mitochondria and chloroplasts no longer have most of the genes that would be needed for them to survive as independent organisms.

CHAPTER 6

Answers to Comprehension Questions

6.1: b, d

6.2: a, a, b

6.3: d, d, b

6.4: a, b

6.5: b

Concept Check Questions (follow figure legends)

FIGURE 6.1   The types of offspring found in excess are those with purple flowers and long pollen and those with red flowers and round pollen.

FIGURE 6.2   No, such a crossover would not change the arrangement of the *B* and *A* alleles.

FIGURE 6.3   A single crossover can produce offspring with the following phenotypes: gray body, red eyes, miniature wings; gray body, white eyes, miniature wings; yellow body, red eyes, long wings; and yellow body, white eyes, long wings.

FIGURE 6.4   When genes are relatively close together, as is the case for the body color and eye color genes considered in this figure, a crossover is relatively unlikely to occur between them. Therefore, the nonrecombinant offspring are more common.

FIGURE 6.5   The reason is that the eye color and wing length genes are farther apart from each other on the X chromosome than are the body color and eye color genes.

FIGURE 6.7   Genetic maps are useful: (1) for understanding the complexity and organization of the genome of a species and the underlying basis of inherited traits; (2) in cloning genes; (3) in understanding evolution and the evolutionary relationships among different species; (4) in diagnosing and treating inherited diseases in humans and predicting the likelihood of a couple having offspring with certain inherited diseases; (5) in helping breeders of livestock and crops to improve strains.

FIGURE 6.8   Crossing over occurred during oogenesis in the female parent of the recombinant offspring.

FIGURE 6.9   The possible outcomes of multiple crossovers prevent the maximum percentage of recombinant offspring from exceeding 50%.

FIGURE 6.13   Mitotic recombination occurs in a somatic cell.

End-of-Chapter Questions:

Conceptual Questions

C2.   An independent assortment hypothesis is proposed because it allows you to calculate the expected values based on Mendel's ratios. Using the observed and expected values, you can determine whether or not the deviations between the observed and expected values are too large to occur as a matter of chance. If the deviations are very large, you reject the hypothesis of independent assortment.

C4.   A single crossover produces *ABC*, *Abc*, *aBC*, and *abc*.

A. Chromatids 2 and 3, between genes *B* and *C*

B. Chromatids 1 and 4, between genes *A* and *B*

C. Chromatids 1 and 4, between genes *B* and *C*

D. Chromatids 2 and 3, between genes *A* and *B*

C6.   The likelihood of scoring a basket will be greater if the basket is larger. Similarly, the chances of a crossover initiating in a region between two genes is proportional to the size of the region between the two genes. A finite number of crossovers (usually only a few) occur between homologous chromosomes during meiosis, and the likelihood that a crossover will occur in a region between two genes depends on how big that region is.

C8.   The pedigree suggests a linkage between the dominant allele causing nail-patella syndrome and the *I^B* allele of the gene for ABO blood type. In every case, the individual who inherits the *I^B* allele also inherits this disorder.

C10.   *Ass-1* **43** *Sdh-1* **5** *Hdc* **9** *Hao-1* **6** *Odc-2* **8** *Ada*

The numbers in boldface indicate the distances between the genes.

C12.   The inability to detect double crossovers causes the map distance to be underestimated. In other words, more crossovers occur in the region than you realize. When there is a double crossover, there are no recombinant offspring (in a two-factor cross). Therefore, the second crossover cancels out the effects of the first crossover.

C14.   The key feature is that all the products of a single meiosis are contained within a single sac. The spores in this sac can be dissected, and then their genetic traits can be analyzed individually.

C16.   If the chromosomes on the right side (shown below), labeled 2 and 4, move into one daughter cell, that will lead to a patch that is albino and has long fur. The other cell will receive chromosomes 1 and 3, which will produce a patch that has dark, short fur.

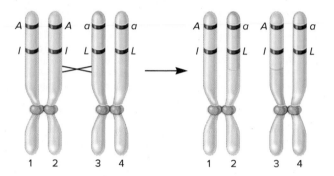

Experimental Questions (Including Many Mapping Questions)

E2.   Stern could have used a strain with a single abnormal X chromosome. In this case, the recombinant chromosomes would either have a deletion at one end or an extra piece of the Y chromosome at the other end, but not both.

E4.   A testcross cannot produce more than 50% recombinant offspring because multiple crossovers can yield an average maximum value of only 50%. When a testcross does produce 50% recombinant offspring, it can mean two different things. Either the two genes are on different chromosomes, or the two genes are on the same chromosome but at least 50 mu apart.

E6.   If two genes are more than 50 mu apart, you would need to map genes between them to show that the two genes are actually in the same linkage group. For example, if gene *A* was 55 mu from gene *B*, there might be a third gene (e.g., gene *C*) that was 20 mu from *A* and 35 mu from *B*. These results would indicate that *A* and *B* are 55 mu apart, assuming that two-factor testcrosses between genes *A* and *B* yielded 50% recombinant offspring.

E8.   A. Because the two genes are 12 mu apart, you expect 12% (or 120) recombinant offspring. The predicted numbers of offspring with each genotype would be approximately 60 *Aabb*, 60 *aaBb*, 440 *AaBb*, and 440 *aabb*.

B. In this case, the predicted numbers would be 60 *AaBb*, 60 *aabb*, 440 *Aabb*, and 440 *aaBb*.

E10.   Due to the large distance between the two genes, they will assort independently even though they are actually on the same chromosome. With independent assortment, you expect 50% parental and 50% recombinant offspring. Therefore, this cross is expected to produce 150 offspring in each of the four phenotypic categories.

E12.   A. If the two genes independently assort, then the predicted ratio is 1:1:1:1. There are a total of 390 offspring. The expected number of offspring in each category is about 98. Substituting into the chi square equation gives

$$\chi^2 = \frac{(117-98)^2}{98} + \frac{(115-98)^2}{98} + \frac{(78-98)^2}{98} + \frac{(80-98)^2}{98}$$

$$\chi^2 = 3.68 + 2.95 + 4.08 + 3.31$$

$$\chi^2 = 14.02$$

Looking up this value in the chi square table under 1 degree of freedom, you reject the hypothesis that the genes independently assort, because the chi square value is above 3.841.

B.
$$\text{Map distance} = \frac{78 + 80}{117 + 115 + 78 + 80}$$
$$= 40.5 \, \text{mu}$$

Because the value is relatively close to 50 mu, it is probably a significant underestimate of the true distance between these two genes.

E14.   The percentage of recombinants for green, yellow and wide, narrow is 7%, or 0.07; the offspring will include 3.5% of the green, narrow phenotype and 3.5% of the yellow, wide phenotype. The remaining 93% of the offspring are nonrecombinants, and their phenotypes will be 46.5% green, wide and 46.5% yellow, narrow. The third gene assorts independently. The phenotypes will be 50% long and 50% short with respect to each of the other two genes. To calculate the number of offspring in each category out of a total of 800, you multiply 800 by the appropriate percentages.

(0.465 green, wide)(0.5 long)(800) = 186 green, wide, long

(0.465 yellow, narrow)(0.5 long)(800) = 186 yellow, narrow, long

(0.465 green, wide)(0.5 short)(800) = 186 green, wide, short

(0.465 yellow, narrow)(0.5 short)(800) = 186 yellow, narrow, short

(0.035 green, narrow)(0.5 long)(800) = 14 green, narrow, long

(0.035 yellow, wide)(0.5 long)(800) = 14 yellow, wide, long

(0.035 green, narrow)(0.5 short)(800) = 14 green, narrow, short

(0.035 yellow, wide)(0.5 short)(800) = 14 yellow, wide, short

E16.   Let's use the following symbols: G for green pods, g for yellow pods, S for green seedlings, s for bluish green seedlings, C for normal plants, c for creepers. The parental cross is $GG\ SS\ CC$ crossed to $gg\ ss\ cc$.

The F_1 plants are all $Gg\ Ss\ Cc$. If the genes are linked, the alleles G, S, and C will be linked on one chromosome, and the alleles g, s, and c will be linked on the homologous chromosome.

The testcross is F_1 plants, which are $Gg\ Ss\ Cc$, crossed to $gg\ ss\ cc$.

To measure the distances between the genes, you can separate the data into gene pairs.

Pod color, seedling color

2210 green pods, green seedlings—nonrecombinant

296 green pods, bluish green seedlings—recombinant

2198 yellow pods, bluish green seedlings—nonrecombinant

293 yellow pods, green seedlings—recombinant

$$\text{Map distance} = \frac{296 + 293}{2210 + 296 + 2198 + 293} \times 100 = 11.8 \, \text{mu}$$

Pod color, plant stature

2340 green pods, normal—nonrecombinant

166 green pods, creeper—recombinant

2323 yellow pods, creeper—nonrecombinant

168 yellow pods, normal—recombinant

$$\text{Map distance} = \frac{166 + 168}{2340 + 166 + 2323 + 168} \times 100 = 6.7 \, \text{mu}$$

Seedling color, plant stature

2070 green seedlings, normal—nonrecombinant

433 green seedlings, creeper—recombinant

2056 bluish green seedlings, creeper—nonrecombinant

438 bluish green seedlings, normal—recombinant

$$\text{Map distance} = \frac{433 + 438}{2070 + 433 + 2056 + 438} \times 100 = 17.4 \, \text{mu}$$

The order of the genes is seedling color, pod color, and plant stature (or the opposite order). Pod color is in the middle. If you use the two shortest distances to construct our map, you obtain:

S **11.8** *G* **6.7** *C*

The numbers in boldface indicate the distances between the genes.

E18.   To answer this question, you can consider genes in pairs. Let's consider the two gene pairs that are closest together. The distance between the wing length and eye color genes is 12.5 mu. For the cross described in the question, you expect 87.5% of offspring to have long wings and red eyes or short wings and purple eyes, and 12.5% to have long wings and purple eyes or short wings and red eyes. Therefore, you expect 43.75% to have long wings and red eyes, 43.75% to have short wings and purple eyes, 6.25% to have long wings and purple eyes, and 6.25% to have short wings and red eyes. With 1000 offspring, you expect 438 to have long wings and red eyes, 438 to have short wings and purple eyes, 62 to have long wings and purple eyes, and 62 to have short wings and red eyes (rounding to the nearest whole numbers).

The distance between the eye color and body color genes is 6 mu. For the given cross, you expect 94% of offspring to be nonrecombinant (red eyes and gray body or purple eyes and black body) and 6% to be recombinant (red eyes and black body or purple eyes and gray body). Therefore, of the 438 offspring with long wings and red eyes, you expect 94% of them (or about 412) to have long wings, red eyes, and gray body and 6% of them (or about 26) to have long wings, red eyes, and black bodies. Of the 438 offspring with short wings and purple eyes, you expect about 412 to have short wings, purple eyes, and black bodies and 26 to have short wings, purple eyes, and gray bodies.

Of the 62 offspring with long wings and purple eyes, you expect 94% of them (or about 58) to have long wings, purple eyes, and black bodies and 6% of them (or about 4) to have long wings, purple eyes, and gray bodies. Of the 62 offspring with short wings and red eyes, you expect 94% (or about 58) to have short wings, red eyes, and gray bodies and 6% (or about 4) to have short wings, red eyes, and black bodies.

In summary,

Long wings, red eyes, gray body	412
Long wings, purple eyes, gray body	4
Long wings, red eyes, black body	26
Long wings, purple eyes, black body	58
Short wings, red eyes, gray body	58
Short wings, purple eyes, gray body	26
Short wings, red eyes, black body	4
Short wings, purple eyes, black body	412

The flies with long wings, purple eyes, and gray bodies and those with short wings, red eyes, and black bodies are produced by a double crossover event.

E20.   Yes, it would be possible. You would begin with females that have one X chromosome that is X^{Nl} and the other X chromosome that is X^{nL}. These females have to be mated to $X^{NL}Y$ males because a living male cannot carry the n or l allele. In the absence of crossing over, a mating between $X^{Nl}X^{nL}$ females to $X^{NL}Y$ males should not produce any surviving male offspring. However, during oogenesis in these heterozygous female mice, there could be a crossover in the region between the two genes, which would produce an X^{NL} chromosome and an X^{nl} chromosome. Male offspring inheriting these recombinant chromosomes will be either $X^{NL}Y$ or $X^{nl}Y$ (whereas nonrecombinant males will be $X^{nL}Y$ or $X^{Nl}Y$). Only the male mice that inherit $X^{NL}Y$ will live. The living males represent only half of the recombinant offspring. (The other half are $X^{nl}Y$, which are born dead.)

To compute the map distance, you would use this formula:

$$\text{Map distance} = \frac{2(\text{number of male living offspring})}{\text{number of males born dead} + \text{number of males born alive}}$$

E22. A. The first one is a parental ditype, the second one is a nonparental ditype, and the third one is a tetratype.

B. The second and third types contain spores that are due to crossing over.

Questions for Student Discussion/Collaboration

2. The X and Y chromosomes are not completely distinct linkage groups. You might describe them as overlapping linkage groups with most of their genes not common to both.

CHAPTER 7

Answers to Comprehension Questions

7.1: c	7.4: a
7.2: a, d	7.5: b, c
7.3: c, b	7.6: d

Concept Check Questions (in figure legends)

FIGURE 7.1 To grow, the colonies must have functional copies of all five genes. This could occur by the transfer of the *met*⁺ and *bio*⁺ genes to the *met*⁻ *bio*⁻ *thr*⁺ *leu*⁺ *thi*⁺ strain or the transfer of the *thr*⁺, *leu*⁺, and *thi*⁺ genes to the *met*⁺ *bio*⁺ *thr*⁻ *leu*⁻ *thi*⁻ strain.

FIGURE 7.2 Because bacteria are too large to pass through the filter, the U-tube apparatus can allow the experimenter to determine if direct cell-to-cell contact is necessary for gene transfer to occur.

FIGURE 7.3 It would be found in an F⁺ strain.

FIGURE 7.4 Relaxase is a part of the relaxosome, which cuts the F factor and thus initiates its transfer to the recipient cell. The coupling factor guides the DNA strand to the exporter, which transports it to the recipient cell.

FIGURE 7.5 An F′ factor carries a portion of the bacterial chromosome, whereas an F factor does not.

FIGURE 7.6 Because conjugation occurred for a longer period of time, *pro*⁺ was transferred to the recipient cell at the bottom right.

FIGURE 7.8 This type of map is based on the timing of gene transfer in conjugation experiments, which is measured in minutes.

FIGURE 7.9 The *lacZ* gene is closer to the origin of transfer; its time of entry occurred at 16 minutes.

FIGURE 7.10 The normal process is for bacteriophage DNA to be incorporated into new phages. In transduction, a segment of the bacterial DNA is incorporated into at least one new phage.

FIGURE 7.11 If the recipient cell did not have a *lys*⁻ gene, the *lys*⁺ DNA could be incorporated into the bacterial chromosome by nonhomologous recombination.

End-of-Chapter Questions:

Conceptual Questions

C2. Conjugation is not a form of sexual reproduction, in which two distinct parents produce gametes that unite to form a new individual. However, conjugation is similar to sexual reproduction in the sense that the genetic material from two cells is mixed. Conjugation does not involve a mixing of two genomes, one from each gamete. Instead, genetic material from one cell is transferred to another. This transfer can alter the combination of genetic traits in the recipient cell.

C4. An F⁺ strain contains a separate, circular piece of DNA that has its own origin of transfer. An Hfr strain has its origin of transfer integrated into the bacterial chromosome. An F⁺ strain can transfer only the DNA contained on the F factor. Given enough time, an Hfr strain can actually transfer the entire bacterial chromosome to the recipient cell.

C6. Sex pili promote the binding of donor and recipient cells.

C8. Though exceptions are common, interspecies genetic transfer via conjugation is not likely because the cell surfaces do not interact correctly. Interspecies genetic transfer via transduction is also not very likely because each species of bacteria is sensitive to particular bacteriophages. The most likely form of genetic transfer between bacterial species is transformation. A consequence of interspecies genetic transfer is that new genes can be introduced into a bacterial species from another species. For example, interspecies genetic transfer could provide the recipient bacterium with a new trait, such as resistance to an antibiotic.

C10. The transfer of fertility plasmids such as F factors does not require recombination.

C12. A. If it occurs in a single step, transformation is the most likely mechanism because conjugation does not usually occur between different species, particularly distantly related species, and transduction is not likely because different species are not usually infected by the same bacteriophages.

B. The genetic transfer could have occurred directly, but it is more likely to have involved multiple steps.

C. The use of antibiotics selects for the survival of bacteria that have genes conferring resistance to the drugs. If a population of bacteria is exposed to an antibiotic, those carrying the resistance genes will survive, and their relative numbers will increase in subsequent generations.

Experimental Questions

E2. Mix the two strains together and then put some of the mixture on plates containing streptomycin and some on plates without streptomycin. If colonies grow on both types of plates, then the *thr*⁺, *leu*⁺, and *thi*⁺ genes are being transferred to the *met*⁺ *bio*⁺ *thr*⁻ *leu*⁻ *thi*⁻ strain. If colonies are found only on the plates that lack streptomycin, then the *met*⁺ and *bio*⁺ genes are being transferred to the *met*⁻ *bio*⁻ *thr*⁺ *leu*⁺ *thi*⁺ strain. This answer assumes a one-way transfer of genes from a donor to a recipient strain.

E4. An interrupted mating experiment is a procedure in which two bacterial strains are allowed to conjugate, and then conjugation is interrupted at various time points. The interruption occurs by agitation of the solution in which the bacteria are found. This type of study is used to map the locations of genes on the bacterial chromosome. It is necessary to interrupt conjugation so that the time elapsed varies, thus providing information about the order of transfer: which gene transferred first, second, and so on.

E6. Mate unknown strains *A* and *B* to the F⁻ strain in your lab that is resistant to streptomycin and cannot metabolize lactose. Use two separate tubes (i.e., strain *A* plus the F⁻ strain in one tube, and strain *B* plus the F⁻ strain in the other tube). Plate the mated cells on a growth medium containing lactose plus streptomycin. If colonies grow on this medium, the unknown strain has to be strain *A*, the F⁺ strain that has lactose utilization genes on its F factor.

E8. A. The curve for *hisE*⁺ intersects the *x*-axis at about 3 minutes, and that for *pheA*⁺ intersects it at about 24 minutes. These values are the times at which the genes entered the recipient cells. Therefore, the distance between these two genes is 21 minutes (i.e., 24 − 3).

B. _____

  ↑   4 ↑   17   ↑

  hisE   *pabB*   *pheA*

Questions for Student Discussion/Collaboration

2. Conjugation requires direct cell contact, which is mediated by proteins that are found in the same species. So, it is less likely to occur between different species unless they are very closely related evolutionarily. Similarly, gene transfer via transduction involves bacteriophages that are usually species-specific. Gene transfer via transformation is most likely to occur between different species. Interspecies gene transfer has several potential consequences, including antibiotic resistance and the ability to survive under new growth conditions.

CHAPTER 8

Answers to Comprehension Questions

8.1: c, d   8.5: b

8.2: c   8.6: a, d

8.3: a, d   8.7: c, b

8.4: b, d, d   8.8: c, b

Concept Check Questions (follow figure legends)

FIGURE 8.1   The staining of chromosomes results in banding patterns that make it easier to distinguish chromosomes that are similar in size and have similar centromere locations.

FIGURE 8.2   Deletions and duplications alter the total amount of genetic material.

FIGURE 8.3   If a chromosomal fragment does not contain a centromere, it will not segregate properly during cell division. If it remains outside the nucleus, it will be degraded.

FIGURE 8.5   Nonallelic homologous recombination occurs when the pairing of sites within two homologous chromosomes involves sites that are not alleles of the same gene. The chromosomes are misaligned and crossing over results in one chromosome with a duplication and another chromosome with a deletion.

FIGURE 8.11   These chromosomes need an inversion loop to allow their homologous genes to align with each other. For the inverted and noninverted regions to pair correctly, a loop must form.

FIGURE 8.12   The mechanism shown in part (b) may be promoted if the same transposable elements are found in different chromosomes. These elements could lead to pairing between nonhomologous chromosomes, and a crossover could subsequently occur.

FIGURE 8.13   Two out of six gametes (the two on the left) will produce a viable offspring with a normal phenotype. Therefore, the probability is $2/6$, or $1/3$.

FIGURE 8.14   These chromosomes form a translocation cross because their homologous regions are pairing with each other.

FIGURE 8.15   Aneuploid, monosomic (having monosomy 3).

FIGURE 8.16   The genes on chromosome 2 will be present in single copies, whereas the genes on the other chromosomes will be present in two copies. The expression of genes on chromosome 2 will be less than (perhaps only 50% of) their expression in a normal individual. The result is an imbalance between the products of genes on chromosome 2 and the products of genes on the other chromosomes.

FIGURE 8.19   About 512.

FIGURE 8.20   Polyploid plants are often more vigorous and disease-resistant. They may have larger flowers and produce more fruit.

FIGURE 8.21   During meiosis in a triploid individual, the homologs cannot pair properly. This results in highly aneuploid gametes, which are usually inviable.

FIGURE 8.22   *Nondisjunction* means that pairs of chromosomes are not separating from each other properly during meiosis.

FIGURE 8.24   In autopolyploidy, multiple sets of chromosomes come from the same species. In allopolyploidy, multiple sets of chromosomes come from at least two different species.

End-of-Chapter Questions:

Conceptual Questions

C2.   Small deletions and duplications are less likely to affect phenotype simply because they usually involve fewer genes. If a small deletion did have a phenotypic effect, you would conclude that at least one gene in the deleted region is required for a normal phenotype.

C4.   A gene family is a group of genes that are derived from the process of gene duplication. The genes have similar sequences, but

the sequences have some differences due to the accumulation of mutations over many generations. The members of a gene family usually code proteins with similar but specialized functions. The specialization may occur in different cells or at different stages of development.

C6.   The abnormal chromosome has a pericentric inversion.

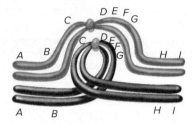

C8.   There would be four products after meiosis. One would be a normal chromosome, and one would contain the inversion shown in the diagram in question C8. The other two chromosomes would be dicentric or acentric with the following orders of genes:

centromere   centromere

$\underline{A \downarrow \; B C D E F G H I J D C B \downarrow \; A}$ Dicentric

$\underline{M L K J I H G F E K L M}$ Acentric

C10. In the absence of crossing over, alternate segregation would yield half of the cells with two normal chromosomes and half with chromosomes with a balanced translocation. With adjacent-1 segregation, all cells would have chromosomes with unbalanced translocations. Two cells would have

$$A B C D E + A I J K L M$$

And the other two cells would have

$$H B C D E + H I J K L M$$

C12. One of the parents may carry a balanced translocation between chromosomes 5 and 7. The phenotypically abnormal offspring has inherited an unbalanced translocation due to the segregation of translocated chromosomes during meiosis.

C14. A deletion and an unbalanced translocation are more likely to have phenotypic effects because they create genetic imbalances. With a deletion, there are too few copies of several genes, and with an unbalanced translocation, there are too many. Pericentric inversions and reciprocal translocations are not likely to affect phenotype unless a breakpoint occurs in a critical gene or they cause a position effect, but they may affect fertility and cause problems in future offspring.

C16. The order of genes on chromosome 7 in the phenotypically affected offspring is due to a crossover within the inverted region. You can draw the inversion loop as shown in Figure 8.11a. The crossover occurred between *P* and *U*.

C18. This person has a total of 46 chromosomes. However, this person is considered to be aneuploid rather than euploid, because one of the sets is missing a sex chromosome and one set has an extra copy of chromosome 21.

C20. The potential benefit may be related to genetic balance. In aneuploidy, there is an imbalance in gene expression between the chromosomes found in their normal number versus those that have either too many or too few copies. In polyploidy, the balance in gene expression is maintained.

C22. The red-eyed male offspring is the result of nondisjunction during oogenesis. The female parent produced an egg without any sex chromosomes. The male parent transmitted a single X chromosome carrying the red allele. This produces an X0 male offspring with red eyes.

C24. Humans with trisomy 13, 18, or 21 can survive because these chromosomes are small and probably contain fewer genes compared to the larger chromosomes. Individuals with abnormal numbers of X chromosomes can survive because the extra copies are converted to transcriptionally inactive Barr bodies. The other trisomies are lethal because the trisomic chromosomes cause a great imbalance in the level of gene expression relative to that of normal diploid chromosomes.

C26. Endopolyploidy means that a particular area of somatic tissue is polyploid even though the rest of the organism is not. The biological significance is not entirely understood, although it has been speculated that an increase in chromosome number in certain cells may enhance their ability to produce specific gene products that are needed in great abundance.

C28. In certain types of cells in *Drosophila*, such as salivary gland cells, the homologous chromosomes pair with each other and then replicate approximately nine times to produce a polytene chromosome. The centromeres from each chromosome aggregate at the chromocenter. This forms a structure with six arms: a single arm from each of the two telocentric chromosomes (X and 4) and two arms each from the metacentric chromosomes (2 and 3).

C30. The turtles are two distinct species that appear phenotypically identical. The turtles with 48 chromosomes are polyploids (i.e., tetraploids) relative to the species with 24 chromosomes. In animals, it is somewhat hard to imagine how this polyploidy could occur because most animals cannot self-fertilize; thus, there had to be two animals (i.e., one male and one female) that became tetraploids. It is easy to imagine how one animal could become a tetraploid; complete nondisjunction could occur during the first cell division of a fertilized egg, thereby creating a tetraploid cell that continued to develop into a tetraploid animal. This nondisjunction would have to happen independently (i.e., in one female and one male) to create a tetraploid species. If you mated a tetraploid turtle with a diploid turtle, the offspring would be triploid and probably phenotypically normal. However, the triploid offspring would be sterile because they would make highly aneuploid gametes.

C32. Polyploid, triploid, and euploid should not be used.

C34. The older sibling carries a translocation involving chromosome 21: probably a translocation in which nearly all of chromosome 21 is translocated to chromosome 14. This sibling would have one normal copy of chromosome 14, one normal copy of chromosome 21, and the translocated chromosome that contains both chromosome 14 and chromosome 21. The sibling is phenotypically normal because the total amount of genetic material is normal, although the total number of chromosomes is 45 (because chromosome 14 and chromosome 21 are fused into a single chromosome). The young child with Down syndrome has inherited the translocated chromosome, but also must have one copy of chromosome 14 and two copies of chromosome 21. This child has the equivalent of three copies of chromosome 21 (i.e., two normal copies and one copy fused with chromosome 14) and therefore has familial Down syndrome. One of the parents of these two children is probably normal with regard to karyotype (i.e., the parent has 46 normal chromosomes). The other parent likely has a karyotype like that of the phenotypically normal sibling.

C36. Nondisjunction is a mechanism whereby the chromosomes do not segregate equally into the two daughter cells. This can occur during meiosis to produce cells with altered numbers of chromosomes, or it can occur during mitosis to produce an individual with genetic mosaicism. A third way to alter chromosome number is by an interspecies cross that produces an alloploid.

C38. A mutation occurred during early embryonic development to create the blue patch of tissue. One possibility is mitotic nondisjunction in which the two chromosomes carrying the *b* allele went to one cell and the two chromosomes carrying the *B* allele went to the other daughter cell. A second possibility is that the chromosome carrying the *B* allele was lost. A third possibility is that the *B* allele was deleted.

C40. In meiotic nondisjunction, the bivalents are not separating correctly during meiosis I. During mitotic nondisjunction, the sister chromatids are not separating properly.

C42. Complete nondisjunction occurs during meiosis I, so one nucleus receives all the chromosomes and the other nucleus does not get any. The nucleus with all the chromosomes then proceeds through normal meiosis II to produce two haploid sperm cells.

Experimental Questions

E2. Due to the mistake, the ratio of green-to-red fluorescence would be 0.5 in regions where the cancer cells had a normal amount of DNA. If a duplication occurred on both chromosomes of cancer cells, the ratio would be 1.0. If a deletion occurred on a single chromosome, the ratio would be 0.25.

E4. Colchicine interferes with the spindle apparatus and thereby causes nondisjunction. At high concentrations, it can cause complete nondisjunction and produce polyploid cells.

E6. First, you would cross the two strains of tomatoes. It is difficult to predict the phenotype of the offspring. Nevertheless, you would keep crossing offspring to each other and backcrossing them to the parental strains until you obtained a strain that was resistant to heat and the viral pathogen and produced great-tasting tomatoes. You could then make this strain tetraploid by treatment with colchicine. If you crossed the tetraploid strain with your great-tasting diploid strain that was resistant to heat and the viral pathogen, you may get a triploid that had these characteristics. This triploid would probably be seedless.

E8. A polytene chromosome is formed when a chromosome replicates many times, and the chromatids lie side by side. The homologous chromosomes also lie side by side. Therefore, if there is a deletion, there will be a loop. The loop is the segment that is not deleted from one of the two homologs.

Questions for Student Discussion/Collaboration

2. There are many possibilities. The fields of agriculture and botany offer many examples. In the insect world, there are interesting examples of variation in euploidy affecting sex determination. Among amphibians and reptiles, there are also several examples of closely related species that have variation in euploidy.

4. 1. Polyploids are often more robust and disease resistant.

   2. Allopolyploids may have useful combinations of traits.

   3. Hybrids are often more vigorous.

   4. Strains with an odd number of chromosome sets (e.g., triploids) are usually seedless.

CHAPTER 9

Answers to Comprehension Questions

9.1: d	9.5: d, c
9.2: c	9.6: d, a, a
9.3: c, d	9.7: d
9.4: a	

Concept Check Questions (follow figure legends)

FIGURE 9.1 In this experiment, the type R bacteria had taken up genetic material from the heat-killed type S bacteria, which converted the type R bacteria into type S. This enabled the bacteria to proliferate within the mouse and kill it.

FIGURE 9.2   RNase or protease was added to a DNA extract to rule out the possibility that small amounts of contaminating RNA or protein were responsible for converting the type R bacteria into type S.

FIGURE 9.5   Ribose and uracil are not found in DNA.

FIGURE 9.8   Deoxyribose and phosphate form the backbone of a DNA strand.

FIGURE 9.13   Hydrogen bonding between bases in the base pairs and base stacking hold the DNA strands together.

FIGURE 9.14   The major and minor grooves are indentations in the outer surface of the DNA double helix where the bases make contact with water in the surroundings. The major groove is wider than the minor groove.

FIGURE 9.15   B DNA is a right-handed helix and the backbone is helical, whereas Z DNA is a left-handed helix and the backbone appears to zigzag slightly. Z DNA has the bases tilted relative to the central axis, whereas they are perpendicular to that axis in B DNA. There are also minor differences in the number of base pairs per turn.

FIGURE 9.16   Covalent bonds within phosphodiester linkages hold nucleotides together in an RNA strand.

FIGURE 9.17   A bonds with U and G bonds with C.

End-of-Chapter Questions:

Conceptual Questions

C2.   The transformation process is described in Chapter 7.

1. A fragment of DNA binds to the cell surface.
2. It penetrates the cell membrane.
3. It enters the cytoplasm.
4. It recombines with the chromosome.
5. The genes within the DNA are expressed (i.e., transcription and translation).
6. The gene products create a capsule. That is, they are enzymes that synthesize a capsule using cellular molecules as building blocks.

C4.   The building blocks of a nucleotide are a sugar (ribose or deoxyribose), a nitrogenous base, and a phosphate group. In a nucleotide, the phosphate is already linked to the $5'$ position on the sugar. When two nucleotides link together, a phosphate on one nucleotide forms a covalent bond with the $3'$ hydroxyl group on another nucleotide.

C6.   The structure is a phosphate group connecting two sugars at the $3'$ and $5'$ positions, as shown in Figure 9.8.

C8. $3'$–CCGTAATGTGATCCGGA–$5'$

C10.   A drawing of a DNA helix with 10 bp per turn will look like the drawing on the left in Figure 9.13. To have 15 bp per turn, you need to add 5 more base pairs, but the helix should still make only one complete turn.

C12.   The nucleotide bases are accessible in the major and minor grooves. Phosphates and sugars are found in the backbone. If a DNA-binding protein does not recognize a nucleotide sequence, it probably is not binding in the grooves but instead is binding to the DNA backbone (i.e., composed of alternating sugars and phosphates). DNA-binding proteins that recognize a base sequence must bind into the major or minor groove of DNA, which is where the bases are accessible to such proteins. Most DNA-binding proteins that recognize a base sequence fit into the major groove. By comparison, other DNA-binding proteins, such as histones, which do not recognize a base sequence, bind to the DNA backbone.

C14.   The structure of deoxyribose is shown in Figure 9.5. The numbering of the carbon atoms begins at the carbon that is to the right of the ring's oxygen and proceeds in a clockwise direction. *Antiparallel* means that the backbones run in the opposite directions. In one strand, the sugar carbons are oriented in the $3'$ to $5'$ direction, and in the other strand, they are oriented in the $5'$ to $3'$ direction.

C16.   Double-stranded RNA and DNA both form a helical structure due to base pairing. The structures differ in that the number of base pairs per turn is slightly different and RNA follows an AU/GC rule for base pairing, whereas DNA follows the AT/GC rule.

C18.   The nucleotide base sequence is the structural feature that allows DNA to store information.

C20.   G = 32%, C = 32%, A = 18%, T = 18%

C22.   Lysine and arginine, which are positively charged, and the polar amino acids might be interacting with the DNA.

C24.   This DNA molecule contains 280 bp. There are 10 bp per turn, so there are 28 complete turns.

C26.   A hydroxyl group is at the $3'$ end, and a phosphate group is at the $5'$ end.

C28.   Not necessarily. The AT/GC rule applies only to double-stranded DNA molecules.

C30.   The first thing you need to do is to determine how many base pairs are in this DNA molecule. The linear length of 1 bp is 0.34 nm, which equals 0.34×10^{-9} m, and 1 centimeter (cm) equals 10^{-2} meters. Thus, you have

$$\frac{10^{-2}}{0.34 \times 10^{-9}} = 2.9 \times 10^7 \,\text{bp}$$

There are approximately 2.9×10^7 bp in this DNA molecule, which equals 5.8×10^7 nucleotides. If 15% are adenine, then 15% must also be thymine. This leaves 70% as the total percentage for cytosine and guanine. Because cytosine and guanine bind to each other, there must be 35% cytosine and 35% guanine. Multiply 5.8×10^7 times 0.35:

$$(5.8 \times 10^7)(0.35) = 2.0 \times 10^7 \,\text{cytosines (about 20 million cytosines)}$$

C32.   The methyl group is not attached to one of the atoms that hydrogen bonds with guanine, so methylation would not directly affect hydrogen bonding. It could indirectly affect hydrogen bonding if it perturbed the structure of DNA. Methylation may affect gene expression because it can alter the ability of proteins to recognize DNA sequences. For example, a protein might bind into the major groove by interacting with a sequence of bases that includes one or more cytosines. If the cytosines are methylated, this may prevent a protein from binding into the major groove properly. Alternatively, methylation could enhance protein binding.

Experimental Questions

E2.   A. There are several possible reasons why most of the type R bacteria were not transformed.

1. Most of the cells did not take up any of the type S DNA.
2. The type S DNA was usually degraded after it entered the type R bacteria.
3. The type S DNA was usually not expressed in the type R bacteria.

B. The antibody and centrifugation steps were used to remove the bacteria that had not been transformed. It enabled the researchers to determine the phenotype of the bacteria that had been transformed. If this step had been omitted, there would have been so many colonies on the plate that it would have been difficult to identify any transformed bacterial colonies, because they would have represented a very small proportion of the total number of bacterial colonies.

C. The researchers were trying to establish that it was really the DNA in the DNA extract that was the genetic material. It was possible that the extract was not entirely pure and contained contaminating RNA or protein. However, treatment with RNase or protease did not prevent transformation, indicating that RNA or proteins were not the genetic material. In contrast, treatment with DNase blocked transformation, confirming that DNA is the genetic material.

E4.   1. You can make many different shapes.

2. You can move things around very quickly with a mouse.

3. You can use mathematical formulas to fit things together in a systematic way.

4. Computers are very fast.

5. You can store the information you have obtained from model building in a computer file.

E6.   If the RNA was treated with RNase, the plants would not be expected to develop lesions. If it was treated with DNase or protease, lesions would still develop because RNA is the genetic material, and DNase and protease do not destroy RNA.

Questions for Student Discussion/Collaboration

2.   There are many possibilities. You could treat with a DNA-specific chemical and show that it causes heritable mutations. Perhaps you could inject an oocyte with a piece of DNA and produce a mouse with a new trait.

CHAPTER 10

Answers to Comprehension Questions

10.1: d   10.5: c, d, b

10.2: d, b, d   10.6: b, b, c

10.3: e, b   10.7: d, c

10.4: a

Concept Check Questions (follow figure legends)

FIGURE 10.1   The sequences that comprise genes constitute most of a prokaryotic genome.

FIGURE 10.2   Two.

FIGURE 10.6   Strand separation is beneficial because it allows certain processes such as DNA replication and RNA transcription to proceed.

FIGURE 10.7   The DNA held in the lower jaws is cut, and then the DNA held in the upper jaws passes through the break in that DNA. ATP molecules are hydrolyzed to provide the energy for this step.

FIGURE 10.8   Eukaryotic chromosomes have centromeres and telomeres, which prokaryotic chromosomes do not have. Also, eukaryotic chromosomes typically have many more repetitive sequences.

FIGURE 10.9   One reason for the variation in genome size is that the number of genes varies among different eukaryotes. A second reason is that the number of repetitive sequences varies greatly.

FIGURE 10.11   Retrotransposition always causes the TEs to increase in number. Simple transposition by DNA transpososons, by itself, does not increase the number of copies of a TE. However, simple transposition can increase the number if it occurs around the time of DNA replication (see Figure 10.14).

FIGURE 10.15   Reverse transcriptase uses RNA as a template to make a strand of DNA.

FIGURE 10.17   A nucleosome has a diameter of about 11 nm at its widest point.

FIGURE 10.21   The SMC protein is responsible for forming the loop. The CTCFs bind to each other to prevent further loop growth and stabilize the bottom of the loop.

FIGURE 10.23   One way is the formation of a loop domain via SMC protein and CTCFs. Within the loop domain, addition loops can form due to protein-protein interactions, such as between regulatory and general transcription factors.

FIGURE 10.24   Active genes are found in the more loosely packed regions of euchromatin.

FIGURE 10.25   A chromosome territory is a discrete region in the cell nucleus that is occupied by a single chromosome.

FIGURE 10.29   Condensin II forms a protein scaffold, and it promotes loop formation. The loops that are formed by condensin II are anchored to a scaffold of condensin II proteins.

FIGURE 10.30   At the beginning of prophase, cohesins line the region between sister chromatids. Cohesin is first released from the arms of the sister chromatids but remains in the centromere region. At anaphase, the cohesin at the centromere region is degraded, which allows the sister chromatids to separate.

End-of-Chapter Questions:

Conceptual Questions

C2.   A bacterium with two nucleoids is similar to a diploid eukaryotic cell in that it has two copies of each gene. The bacterium is different, however, with regard to alleles. A eukaryotic cell can have two different alleles for the same gene. For example, a cell from a pea plant can be heterozygous, *Tt*, for the gene that affects height. By comparison, a bacterium with two nucleoids has two identical chromosomes. Therefore, a bacterium with two nucleoids is homozygous for its chromosomal genes. (Note: As discussed in Chapter 7, a bacterium can contain another piece of DNA, called an F′ factor, that can carry a few genes. The alleles on an F′ factor can be different from the alleles on the bacterial chromosome.)

C4.   DNA is a double helix. A helix is a coiled structure. Supercoiling involves adding coils to a structure that is already a coil. In B DNA, positive supercoiling is called overwinding because it occurs in the same direction (i.e., right handed) as the coiling of the double helix, whereas negative supercoiling is called underwinding because it occurs in the opposite direction. Both positive and negative supercoiling make the DNA more compact. Z DNA is a left-handed helix. Positive supercoiling in Z DNA is in a left-handed direction, whereas negative supercoiling is in the right-handed direction. These are opposite to the directions of positive and negative supercoiling in B DNA.

C6.   A. The three twists create either three fewer or three more turns, for a total of 7 or 13 turns, respectively.

B. If the helix now has seven turns, it is left-handed. The three right-handed twists you made would result in three fewer turns in a left-handed helix. If the helix now has 13 turns, it is right-handed. The three right-handed twists you made would add three more turns to a right-handed helix [compare parts (a) and (d) of Figure 10.4].

C. The twisting motion probably did not create supercoils because the two strings are not tightly interacting with each other. It's easy for the two strings to change the number of coils.

D. If you glued the two strings together with rubber cement, the three additional twists would probably create supercoils. A glued pair of strings is more like the DNA double helix. In DNA, the two strands are hydrogen-bonded to each other. The hydrogen bonding is analogous to glue. Additional twisting motions tend to create supercoils rather than alter the number of coils.

C8.   Topoisomers differ with regard to the number of supercoils they contain. They are identical with regard to the number of base pairs in the double helix.

C10.   Centromeres are structures found in eukaryotic chromosomes that provide an attachment site for kinetochore proteins, allowing the chromosomes to be sorted (i.e., segregated) during mitosis and meiosis. They are most important during M phase.

C12.   The ends of a short region of the sequence would be flanked by direct repeats. This is a universal characteristic of all transposable elements. In addition, many elements would contain IRs or LTRs that are involved in the transposition process. You might also look for the presence of a transposase or reverse transcriptase gene, although this is not an absolute requirement, because nonautonomous elements typically lack such a gene.

C14. Direct repeats are produced because transposase or integrase produces staggered cuts in the two strands of chromosomal DNA. The transposable element is then inserted into this site, which temporarily leaves two gaps. The gaps are filled in by DNA polymerase. Because this gap filling occurs at sites that are complementary, the two gaps end up with the exact same sequence, thereby creating direct repeats.

C16. Transposable elements are mutagens, because they alter (disrupt) the sequences of chromosomes and of genes within chromosomes. They do this by inserting themselves into genes.

C18. A. LTR retrotransposons and non-LTR retrotransposons

B. Insertion elements and simple transposons

C. All four types are flanked by direct repeats.

D. Insertion elements and simple transposons

C20. A deletion may occur when there is recombination between two different transposable elements, such as two transposons, that align in the same direction. The region between the transposons is deleted. An inversion can happen when recombination occurs between two different transposons that are oriented in opposite directions. A translocation can result when recombination occurs between transposons that are located on different (nonhomologous) chromosomes. In other words, the transposons on different chromosomes align themselves and a crossover occurs. This will produce a translocation, as shown here:

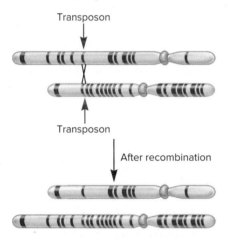

Transposon

Transposon

After recombination

C22. A nucleosome is composed of double-stranded DNA wrapped 1.65 times around an octamer of histones. The zigzag model is a three-dimensional model that describes how nucleosomes are arranged. Nucleosomes zigzag back and forth with a straight linker region.

C24. In some archaea, loop domains are formed due to the binding of an Alba protein to two different sites in the DNA. In eukaryotes, loop domains form due to the actions of SMC protein and CTCFs via the loop exclusion model. See Figure 10.21.

C26. Heterochromatin consists of the tightly compacted regions of chromosomes in nondividing cells. These regions of the chromosome are usually transcriptionally inactive. Less compacted regions, known as euchromatin, are capable of gene transcription. Both constitutive and facultative heterochromatin are highly compacted. Constitutive heterochromatin is found around the centromere and at the telomeres, which contain repetitive sequences and relatively few genes. Facultative heterochromatin is found in short regions located between the centromere and the telomeres. These regions contain many genes. Regions of facultative heterochromatin may vary among different cell types.

C28. Metaphase chromosomes are uniformly compact due to the formation of radial loop arrays. By comparison, interphase chromosomes have compact regions (heterochromatin) and less compact regions (euchromatin). Gene transcription and DNA replication occur during interphase because the DNA is less compact than M phase.

C30. The function of the protein scaffold is to hold the bottoms of the radial loops in place. Also, the protein complex that forms the scaffold, condensin II, functions in the formation of the radial loops.

C32. Both B and C contain SMC proteins.

C34. The function of cohesin is to promote binding (i.e., cohesion) between sister chromatids.

Experimental Questions

E2. Supercoiled DNA has a relatively compact structure. You could add different purified topoisomerases to the DNA sample and observe, via microscopy, how they affect the structure. For example, DNA gyrase relaxes positive supercoils, and topoisomerase I relaxes negative supercoils. If you added topoisomerase I to a DNA preparation and the DNA became less compacted, then the DNA was negatively supercoiled.

E4. DNase I cuts the DNA in the linker region. A single nucleosome plus one half of a linker region on each side of it contains about 200 bp of DNA. At lower DNase I concentrations, a linker region was sometimes uncut. Therefore, a segment of DNA could be derived from two or more nucleosomes, and its length would be a multiple of 200 bp.

E6. You would get DNA fragments about 446 to 496 bp long (i.e., 146 bp plus 300 to 350 bp).

E8. Histones are positively charged, and DNA is negatively charged. Thus, they bind to each other by ionic interactions. Salt is composed of positively charged ions and negatively charged ions. For example, when dissolved in water, NaCl separates into individual ions, Na$^+$ and Cl$^-$. When chromatin is exposed to a salt such as NaCl, the positively charged Na$^+$ can bind to the DNA, and the negatively charged Cl$^-$ can bind to the histones. This formation of new bonds would separate the histones from the DNA.

E10. A. Because the *Alu* sequence is interspersed throughout all chromosomes, there would be many brightly colored spots along all chromosomes.

B. Only the region near the centromere of the X chromosome would be brightly colored.

Questions for Student Discussion/Collaboration

2. This is a matter of opinion. It seems strange to have so much DNA that has no obvious function. It's a waste of energy. Perhaps the highly repetitive sequences have a function that we don't know about yet. On the other hand, evolution does allow bad things to accumulate within genomes, such as genes that cause diseases, etc. Perhaps the prevalence of repetitive sequences is just another example of the negative consequences of evolution.

CHAPTER 11

Answers to Comprehension Questions

11.1: c, d, b	11.4: a, d
11.2: b, d	11.5: b, d, c, b
11.3: a, d	

Concept Check Questions (follow figure legends)

FIGURE 11.1 The two features that allow DNA to be replicated are its double-stranded structure and the base pairing between A and T and between G and C.

FIGURE 11.5 The DnaA boxes are recognized by DnaA proteins, which bind to them and cause the DNA strands to separate at the AT-rich region.

FIGURE 11.6 Two replication forks are formed at the origin.

FIGURE 11.7 Primase is needed for DNA replication because DNA polymerase cannot initiate DNA replication on a bare template strand.

FIGURE 11.8   The template strand is read in the 3′ to 5′ direction.

FIGURE 11.10   The leading strand is synthesized as one, long continuous strand in the same direction in which the replication fork is moving. The lagging strand is formed from Okazaki fragments produced in the direction away from the replication fork.

FIGURE 11.12   Being associated with one another in a complex allows coordination between the various replication enzymes. For example, helicase and primase can work together to make multiple primers for the lagging strand.

FIGURE 11.14   Yes, it is necessary so that the strands can move relative to each other and the catenanes can separate.

FIGURE 11.16   The oxygen in the newly made ester bond comes from the sugar.

FIGURE 11.19   Because eukaryotic chromosomes are so large, multiple origins of replication are needed so that the DNA can be replicated in a reasonable length of time.

FIGURE 11.20   A G-quadruplex is stabilized by the formation of a few G-quartets. Each G-quartet is composed of four guanines that hydrogen bond to each other.

FIGURE 11.25   Six times (36 divided by 6).

End-of-Chapter Questions:

Conceptual Questions

C2. Bidirectional replication refers to the fact that DNA replication proceeds in both directions starting from one origin.

C4. A.
```
                    TTGGHTGUTGG
                    HHUUTHUGHUU
```
B.
```
                    TTGGHTGUTGG
                    HHUUTHUGHUU
                         ↓
           TTGGHTGUTGG CCAAACACCAA
           AACCCACAACC HHUUTHUGHUU
                         ↓
   TTGGHTGUTGG TTGGGTGTTGG CCAAACACCAA CCAAACACCAA
   AACCCACAACC AACCCACAACC GGTTTGTGGTT HHUUTHUGHUU
```

C6. Given 4,600,000 bp of DNA and assuming that DNA replication occurs at a rate of 750 nucleotides per second, the time required if there was a single replication fork would be

$$4{,}600{,}000/750 = 6133 \text{ seconds, or } 102.2 \text{ minutes}$$

Because replication is bidirectional, the time required is

$$\frac{102.2}{2} = 51.1 \text{ minutes}$$

Actually, this is an average value based on a variety of growth conditions. Under optimal growth conditions, replication can occur substantially faster.

With regard to base pair mistakes, if you assume an error rate of 1 mistake per 100,000,000 nucleotides:

$4{,}600{,}000 \times 1000$ bacteria $= 4{,}600{,}000{,}000$ nucleotides of replicated DNA

$4{,}600{,}000{,}000/100{,}000{,}000 = 46$ mistakes

When you think about it, this is pretty amazing. In this bacterial population, only 46 base pair mistakes would be made in 1000 bacteria, each containing 4.6 million bp of DNA.

C8. DNA polymerase would slide from right to left. The sequence of the new strand would be

3′–CTAGGGCTAGGCGTATGTAAATGGTCTAGTGGTGG–5′

C10. A. In Figure 11.5, the first, second, and fourth DnaA boxes are running in the same direction, and the third and fifth are running in the opposite direction. Once you realize that, you can see that the sequences are very similar to each other.

B. After flipping the third and fifth boxes and then lining all 5 of them up, the consensus sequence is

```
TGTGGATAA
ACACCTATT
```

C. This consensus sequence is nine base pairs long. Because there are four kinds of nucleotides (i.e., A, T, G, and C), the chance of this sequence occurring by random chance is 4^{-9}, or once every 262,144 nucleotides. Because the *E. coli* chromosome is more than 10 times longer than this, it is fairly likely that this consensus sequence occurs elsewhere. The reason why there are not multiple origins of replication, however, is because the origin has five copies of the consensus sequence very close together. The chance of having five copies of this consensus sequence occurring close together (as a matter of random chance) is very small.

C12. 1. According to the AT/GC rule, a pyrimidine always hydrogen bonds with a purine. A transition mutation does lead to a pyrimidine hydrogen bonding with a purine, but a transversion mutation causes a purine to hydrogen bond with a purine or a pyrimidine to hydrogen bond with a pyrimidine. The structure of the double helix makes it much more difficult for this latter type of hydrogen bonding to occur.

2. The induced-fit phenomenon at the active site of DNA polymerase makes it unlikely for DNA polymerase to catalyze covalent bond formation if the wrong nucleotide is bound to the template strand. A transition mutation has a detrimental effect on the interaction between the bases in opposite strands, but it is not as bad as the fit caused by a transversion mutation. In a transversion, a purine is opposite another purine, or a pyrimidine is opposite a pyrimidine. Either of these pairings is a very bad fit.

3. The proofreading function of DNA polymerase is able to detect and remove an incorrect nucleotide that has been incorporated into the growing strand. A transversion mutation will cause a larger distortion in the structure of the double helix and make it more likely to be detected by the proofreading function of DNA polymerase.

C14. Primase and DNA polymerase are able to knock the single-strand binding proteins off the template DNA.

C16. 1. DnaA proteins recognize the origin of replication.

2. They promote the separation of DNA strands.

3. They recruit DNA helicase to the region.

C18. An Okazaki fragment is a short segment of newly made DNA that becomes part of the lagging strand. It is necessary to make these short fragments because in the lagging strand, the replication fork is exposing nucleotides in a 5′ to 3′ direction, but DNA polymerase is sliding along the template strand in a 3′ to 5′ direction, away from the replication fork. Therefore, the newly made lagging strand is synthesized in short pieces that are eventually attached to each other.

C20. DNA polymerase has the ability to recognize a mismatched nucleotide in the newly made strand and remove it by cleavage via its exonuclease activity. Proofreading occurs in a 3′ to 5′ direction. After the mistake is corrected, DNA polymerase resumes DNA synthesis in the 5′ to 3′ direction.

C22. The inability of DNA polymerase to synthesize DNA in the 3′ to 5′ direction and its need for a primer prevent replication at the 3′ end of the DNA strands. Telomerase differs from DNA polymerase in that it uses a short RNA sequence, which is part of its structure, as a template for DNA synthesis. Because it uses this sequence many times in a row, it produces a tandemly repeated sequence in the telomeres at the 3′ ends of linear chromosomes.

C24. 50

C26. The ends labeled B and C could not be replicated by DNA polymerase. DNA polymerase makes a strand in the 5′ to 3′ direction using a template strand that is running in the 3′ to 5′ direction. Also, DNA polymerase requires a primer. At the ends labeled B and C, there is no place (upstream) for a primer to be made.

C28. As shown in Figure 11.25, the first step in replicating a telomere involves binding of telomerase to the telomere. The 3′ overhang binds to the complementary RNA in telomerase. For this reason, a 3′ overhang is necessary for telomerase to replicate the telomere region.

Experimental Questions

E2. A. In this case, the researcher would probably still see a band of DNA, but only a heavy band.

   B. The researcher would probably not see a band because the DNA would not be released from the bacteria. The bacteria would sediment to the bottom of the tube.

   C. The researcher would not see a band. UV light is needed to see the DNA, which absorbs light in the UV region.

E4. Rapid-stop mutants immediately stop DNA replication when the cells are shifted to a nonpermissive temperature. Slow-stop mutants allow the round of DNA replication that is occurring when the temperature is shifted to finish, but they cannot initiate a new round of DNA replication. Rapid-stop mutants affect proteins that are needed to make DNA at a replication fork, whereas slow-stop mutants affect proteins that are needed to initiate DNA replication at an origin of replication.

E6. A. Heat is used to separate the DNA strands, so DNA helicase is not needed.

   B. Each primer must have a sequence that is complementary to one of the DNA strands. There are two types of primers, and each type binds to one of the two complementary strands.

   C. A thermophilic DNA polymerase is used because DNA polymerases isolated from nonthermophilic species would be permanently inactivated during the heating phase of the PCR cycle. Remember that DNA polymerase is a protein, and most proteins are denatured by heating. However, proteins from thermophilic organisms have evolved to withstand heat, which is why these organisms survive at high temperatures.

   D. With each cycle, the amount of DNA is doubled. Because there are initially 10 copies of the DNA, there will be 10×2^{27} copies after 27 cycles; $10 \times 2^{27} = 1.34 \times 10^9 = 1.34$ billion copies of DNA. As you can see, PCR can amplify the amount of DNA by a staggering amount!

Questions for Student Discussion/Collaboration

2. DNA synthesis at a replication fork is very similar in bacteria and eukaryotes, involving DNA helicase, topoisomerase II, single-strand binding proteins, primase, DNA polymerase, and ligase. Eukaryotes have more types of DNA polymerases, and different ones are involved in synthesizing the leading and lagging strands. In bacteria, the RNA primers are removed by DNA polymerase I, whereas in eukaryotes, they are removed by flap endonuclease. Bacteria have a single origin of replication that is recognized by DnaA proteins. Eukaryotes have multiple origins of replications that are recognized by ORC. After ORC binds, other proteins including MCM helicase complete a process called DNA replication licensing. Another difference is that the ends of eukaryotic chromosomes are replicated by telomerase.

CHAPTER 12

Answers to Comprehension Questions

12.1: a, c   12.4: d, c, b, d

12.2: d, b, a, b   12.5: d

12.3: b, c, d, b

Concept Check Questions (follow figure legends)

FIGURE 12.2 The mutation would not affect the length of the RNA, because it would not terminate transcription. However, the coded polypeptide would be shorter.

FIGURE 12.5 For a specific type of sequence in DNA, the consensus sequence consists of the most common base found at each location within a group of sequences.

FIGURE 12.6 Parts of σ-factor protein must fit into the major groove so that the protein can recognize a promoter and form hydrogen bonds with its bases.

FIGURE 12.7 The −10 sequence is AT-rich, so it has fewer hydrogen bonds between strands compared to a region with a lot of G-C base pairs.

FIGURE 12.10 The mutation would prevent ρ-dependent termination of transcription.

FIGURE 12.11 NusA helps RNA polymerase to pause, which facilitates transcriptional termination.

FIGURE 12.13 The location of the TATA box determines the precise starting point for the transcription of eukaryotic protein-coding genes.

FIGURE 12.14 The phosphorylation of the CTD allows RNA polymerase to proceed to the elongation phase of transcription.

FIGURE 12.17 It has been cut in four places: in one place by RNase P, one by RNase Z, and two by the splicing endonuclease.

FIGURE 12.18 Splicing via a spliceosome is very common in eukaryotes.

FIGURE 12.20 snRNPs are involved in recognizing the intron boundaries, cutting out the intron, and connecting the two adjacent exons.

FIGURE 12.22 Exon 4 would be spliced out. Exon 3 would be included in the mRNA.

FIGURE 12.23 The 7-methylguanosine cap plays a role in the exit of mRNAs from the nucleus, the binding of mRNA to a ribosome during initiation of translation, and the efficient splicing of introns.

FIGURE 12.25 A functional consequence of RNA editing is that the amino acid sequence of a coded polypeptide may be changed, which can affect the polypeptide's function.

End-of-Chapter Questions:

Conceptual Questions

C2. The release of σ factor marks the transition to the elongation stage of transcription.

C4. GGCATTGTCA

C6. The most highly conserved positions are the first, second, and sixth. In general, when promoter sequences are conserved, they are more likely to be important for binding. That explains why changes are not found at these positions. If a mutation altered such a position, the promoter would probably not work very well; such an unfavorable mutation would likely be eliminated by natural selection. By comparison, changes are tolerated occasionally at the fourth position and frequently at the third and fifth positions. The positions that tolerate changes are less important for binding of σ factor.

C8. This mutation will not affect transcription. However, it will affect translation by preventing the initiation of polypeptide synthesis.

C10. Sigma factor can slide along the major groove of the DNA. This protein contains α-helices that fit into the major groove of the DNA. This ability to fit into the groove allows it to recognize base sequences that are exposed there. When σ factor encounters a promoter sequence, hydrogen bonding between it and the bases in that sequence can promote a tight and specific interaction.

C12. The complementarity rule for RNA synthesis is as follows: G in DNA binds to C in RNA; C in DNA binds to G in RNA; A in DNA binds to U in RNA; and T in DNA binds to A in RNA. The template strand is

    3′–CCGTACGTAATGCCGTAGTGTGATCCCTAG–5′

and the coding strand is

    5′–GGCATGCATTACGGCATCACACTAGGGATC–3′

The promoter is located toward the left end (the 3′ end) of the template strand.

C14. When transcriptional termination occurs, the hydrogen bonds between the DNA and the part of the newly made RNA transcript that is located in the open complex are broken.

C16. DNA helicase and ρ-protein bind to a nucleic acid strand and travel in the 5′ to 3′ direction. When they encounter a double-stranded region, they break the hydrogen bonds between complementary strands. The ρ-protein differs from DNA helicase in that it moves along an RNA strand, whereas DNA helicase moves along a DNA strand. The function of DNA helicase is to promote DNA replication; the function of ρ-protein is to promote transcriptional termination.

C18. A. Mutations that alter the uracil-rich region by introducing guanines and cytosines and mutations that prevent the formation of the stem-loop structure

   B. Mutations that alter the termination sequence and mutations that alter the ρ-protein recognition site

   C. Eventually, somewhere downstream from the gene, another transcriptional termination sequence would be found, and transcription would terminate there. This second termination sequence might be in a random location, or it might be at the end of an adjacent gene.

C20. Core promoters in eukaryotic protein-coding genes are somewhat variable with regard to the pattern of sequence elements. In protein-coding genes that are transcribed by RNA polymerase II, it is common to have a TATA box, which is about 25 bp upstream from a transcriptional start site. The TATA box is important in the identification of the transcriptional start site and the assembly of RNA polymerase and various transcription factors. The transcriptional start site defines where transcription actually begins. In addition, downstream promoter elements (DPEs) are typically found after the transcriptional start site. DPEs provide a binding site for TFIID.

C22. The two models are presented in Figure 12.15. The allosteric model is more like ρ-independent termination, whereas the torpedo model is more like ρ-dependent termination. In the torpedo model, a protein knocks RNA polymerase off the DNA, much like the effect of ρ-protein.

C24. Hydrogen bonding is usually the predominant type of interaction when proteins and DNA are involved in an assembly-and-disassembly process. In addition, ionic bonding and hydrophobic interactions can occur. Covalent interactions will not occur. High temperature and low salt concentration tend to break hydrogen bonds. Therefore, high temperature and low salt concentration will inhibit assembly and stimulate disassembly.

C26. Near the 5′ end, the precursor tRNA is recognized by RNaseP, which is an endonuclease that cuts the precursor tRNA. The action of RNaseP produces the correct 5′ end of the mature tRNA in all species. RNase Z, also called 3′-tRNase, cleaves next to a nucleotide that contains the discriminator base. All tRNAs have a CCA sequence at their 3′ end. In eukaryotes, the CCA sequence is added next to the nucleotide with the discriminator base by an enzyme called tRNA nucleotidyltransferase. Some pre-tRNAs contain introns. An intron is removed by a splicing endonuclease that makes a cut on each side of the intron. The intron is released, and then the remaining ends in the tRNA are ligated together by an enzyme called tRNA ligase. Certain bases in tRNA molecules are covalently modified to alter their structure.

C28. A ribozyme is an RNA molecule with catalytic activity. Examples are RNaseP and self-splicing group I and II introns. The spliceosome contains a catalytic RNA as well.

C30. Self-splicing means that an RNA molecule can splice itself without the aid of a protein. Group I and II introns can be self-splicing, although proteins can also enhance the rate of that splicing.

C32. In alternative splicing, variation occurs in the pattern of splicing, and so the resulting mRNAs contain alternative combinations of exons. The biological significance of this variation is that two or more different proteins can be produced from a single gene. This is a more efficient use of the genetic material. In multicellular organisms, alternative splicing is often used in a cell-specific manner.

C34. As shown in Figure 12.21, the unique feature of the smooth muscle mRNA for α-tropomyosin is that it contains exon 2. Splicing factors that are found only in smooth muscle cells may recognize the splice junction at the 3′ end of intron 1 and the 5′ end of intron 2 and promote splicing at these sites. This would cause exon 2 to be included in the mRNA. Furthermore, because smooth muscle mRNA does not contain exon 3, a splicing repressor may bind to the 3′ end of intron 2. This would promote exon skipping, so exon 3 would not be contained in the mRNA.

C36. As shown in Figure 12.18a, guanosine, which binds to the guanosine-binding site, does not have a phosphate group attached to it. This guanosine is the nucleoside that winds up at the 5′ end of the intron. Therefore, the intron does not have a phosphate group at its 5′ end.

C38. U5

Experimental Questions

E2. The 1100-nucleotide band would be observed in the sample from a normal individual (lane 1). A deletion that removed the −50 to −100 region would greatly diminish transcription, so the homozygote would produce hardly any of the transcript (just a faint band, as shown in lane 2), and the heterozygote would produce roughly half as much of the 1100-nucleotide transcript (lane 3) compared with a normal individual. A nonsense codon would not have an effect on transcription; it affects only translation. So the individual with this mutation would produce a normal amount of the 1100-nucleotide transcript (lane 4). A mutation that removed the AG sequence at the 3′ splice site would prevent splicing. Therefore, this individual would produce a 1550-nucleotide transcript (actually, 1547 to be precise, 1550 minus 3). The Northern blot is shown here:

E4. A. The movement would not be slowed down because ρ protein would not bind to the mRNA coded by a gene that is terminated in a ρ-independent manner. The mRNA from such genes does not contain the sequence near the 3′ end that acts as a recognition site for the binding of ρ protein.

   B. The movement would be slowed down because ρ protein would bind to the mRNA.

   C. The movement would be slowed down because U1 would bind to the pre-mRNA.

   D. The movement would not be slowed down because U1 would not bind to mRNA that has already had its introns removed. U1 binds only to pre-mRNA.

E6. A. mRNA molecules bind to the poly-dT column because they have a polyA tail. The string of adenines in the polyA tail is complementary to a stretch of thymines in the poly-dT column, so the mRNAs will hydrogen bond to to the poly-T column. To purify mRNAs, you begin with a sample of cells; the cells need to be broken open by some technique such as homogenization or sonication. This will release the RNAs and other cellular macromolecules. The large cellular structures (organelles, membranes, etc.) can be removed from the cell extract by a centrifugation step. The large cellular structures will be found in the pellet, whereas soluble molecules such as RNA and proteins will stay in the supernatant. The supernatant is then poured over the poly-dT column. The mRNAs will bind to the poly-dT column and other molecules (i.e., other types of RNAs and proteins) will flow through the column. The mRNAs bind to the poly-dT column via hydrogen bonds, and to break those bonds, you can add a solution that contains a very low salt concentration and/or has a high pH. This will release the mRNAs, which can then be collected in a low salt/high pH solution as it dripped from the column.

B. The basic strategy is to attach a short stretch of nucleotides to the column matrix that is complementary to the type of RNA that you want to purify. For example, if an rRNA contained a sequence 5′–AUUCCUCCA–3′, a researcher could chemically synthesize an oligonucleotide with the sequence 3′–TAAGGAGGT–5′ and attach it to the column matrix. To purify rRNA, the researcher would use this 3′–TAAGGAGGT–5′ column and follow the general strategy described in part A.

Questions for Student Discussion/Collaboration

2. RNA transcripts come in two basic types: those that function as RNAs (e.g., tRNA, rRNA, etc.) and those that are translated (i.e., mRNA). As described in this chapter, these transcripts play many functional roles. RNAs that form complexes with proteins carry out some interesting roles. In some cases, the role is to bind other types of RNA molecules. For example, rRNA in bacteria plays a role in binding mRNA. In other cases, the RNA plays a catalytic role. An example is RNaseP. The structure and function of an RNA molecule may be enhanced by forming a complex with one or more proteins.

CHAPTER 13

Answers to Comprehension Questions

13.1: a, c
13.2: a, d, b, c
13.3: c, b
13.4: c, c, b
13.5: d, d
13.6: d, a, c, c
13.7: a
13.8: b

Concept Check Questions (follow figure legends)

FIGURE 13.1 Alkaptonuria occurs when homogentisic acid oxidase is defective.

FIGURE 13.2 The strain 2 mutants are unable to convert O-acetylhomoserine into cystathionine.

FIGURE 13.3 The role of DNA in polypeptide synthesis is storage. It stores the information that specifies the amino acid sequence of a polypeptide.

FIGURE 13.5 Tryptophan and phenylalanine are the least soluble in water.

FIGURE 13.6 Hydrogen bonding is responsible for the formation of secondary structures of polypeptides.

FIGURE 13.8 Only one radiolabeled amino acid was found in each sample. If radioactivity was trapped on the filter, this meant that the codon triplet specified that particular amino acid.

FIGURE 13.10 A tRNA molecule has an anticodon that recognizes a codon in the mRNA. It also has an acceptor stem where the correct amino acid binds.

FIGURE 13.11 A charged tRNA has an amino acid attached to it.

FIGURE 13.12 The wobble rules allow a smaller number of tRNAs to recognize all of the possible mRNA codons.

FIGURE 13.14 A site in an mRNA promotes the binding of the mRNA to the ribosome during initiation. The codons in an mRNA specify the polypeptide sequence during elongation. The stop codon terminates translation.

FIGURE 13.16 The Shine-Dalgarno sequence in bacterial mRNA is complementary to a region in the 16S rRNA within the small ribosomal subunit. These complementary sequences hydrogen bond with each other.

FIGURE 13.17 Peptidyl transferase catalyzes the formation of a peptide bond between the amino acid at the A site and the last amino acid in the growing polypeptide.

FIGURE 13.18 The three-dimensional structures of release factors, which are proteins, resemble those of tRNAs.

FIGURE 13.20 The mutation would cause the overproduction of ferritin, because ferritin synthesis would occur even if iron levels were low.

End-of-Chapter Questions:

Conceptual Questions

C2. When we say the genetic code possesses degeneracy, it means that more than one codon can specify the same amino acid. For example, GGG, GGC, GGA, and GGU all specify glycine. In general, the genetic code is nearly universal, because it is used in the same way by viruses, prokaryotes, fungi, plants, and animals. As shown in Table 13.2, there are a few exceptions, which occur primarily in protists and also in yeast and mammalian mitochondria.

C4. A. The mutant tRNA would recognize glycine codons in the mRNA but would place the amino acid tryptophan where glycine was supposed to be in the polypeptide.

B. This mutation tells us that the aminoacyl-tRNA synthetase is primarily recognizing regions of the tRNA molecule other than the anticodon region. In other words, tryptophanyl-tRNA synthetase (the aminoacyl-tRNA synthetase that attaches tryptophan) primarily recognizes other regions of the tRNA^{Trp} sequence (that is, other than the anticodon region), such as the T- and D-loops. If aminoacyl-tRNA synthetases recognized only the anticodon region, you would expect glycyl-tRNA synthetase to recognize this mutant tRNA and attach glycine. That is not what happens.

C6. A. Three tRNAs are needed. There are six leucine codons: UUA, UUG, CUU, CUC, CUA, and CUG. The anticodon AAU would recognize UUA and UUG. Two other tRNAs are needed to efficiently recognize the other four leucine codons. These could be GAG and GAU or GAA and GAU.

B. One tRNA is needed. There is only one methionine codon, AUG, so only one tRNA with the anticodon UAC is needed.

C. Three tRNAs are needed. There are six serine codons: AGU, AGC, UCU, UCC, UCA, and UCG. Only one tRNA is needed to recognize AGU and AGC. This tRNA could have the anticodon UCG or UCA. Two tRNAs are needed to efficiently recognize the other four tRNAs. These could be AGG and AGU or AGA and AGU.

C8. 3′–CUU–5′ or 3′–CUC–5′

C10. It can recognize 5′–GGU–3′, 5′–GGC–3′, and 5′–GGA–3′. All of these specify glycine.

C12. All tRNA molecules have some basic features in common. They all have a cloverleaf structure with three stem-loops. The second stem-loop contains the anticodon sequence that recognizes the codon sequence in mRNA. At the 3′ end of each tRNA, there is an acceptor stem, with the sequence CCA, that serves as an attachment site for an amino acid. Most tRNAs also have base modifications within their nucleotide sequences.

C14. The role of aminoacyl-tRNA synthetase is to specifically recognize tRNA molecules and attach the correct amino acid to them. This ability is sometimes referred to as the "second genetic code" because the specificity of the attachment is a critical step in deciphering the genetic code. For example, if a tRNA has a 3′–GGG–5′ anticodon, it will recognize a 5′–CCC–3′ codon, which should specify proline. It is essential that the aminoacyl-tRNA synthetase known as prolyl-tRNA-synthetase recognize this tRNA and attach proline to the 3′ end. The other aminoacyl-tRNA synthetases should not recognize this tRNA.

C16. Bases that have been chemically modified can occur at various locations within a tRNA molecule. The significance of all of these modifications is not entirely known. However, within the anticodon region, base modification may alter base pairing to allow the anticodon to recognize two or more different bases within the codon.

C18. No, it is not. According to the wobble rules, the 5′ base in the anticodon of a tRNA can recognize two or more bases in the third (3′) position of the mRNA. Therefore, any given cell type synthesizes far fewer than 61 types of tRNAs.

C20. The assembly process for the ribosomal subunits is very complex at the molecular level. In eukaryotes, 33 proteins and one rRNA assemble to form a 40S subunit, and 49 proteins and three rRNAs assemble to form a 60S subunit. This assembly occurs within the nucleolus.

C22. A. On the surface of the 30S subunit and at the interface between the two subunits

B. Within the 50S subunit

C. Within the 50S subunit

D. Within the 30S subunit

C24. Most bacterial mRNAs contain a Shine-Dalgarno sequence, which is necessary for the binding of the mRNA to the small ribosomal subunit. This sequence, UUAGGAGGU, is complementary to a sequence in the 16S rRNA. Due to this complementarity, these sequences hydrogen bond to each other during the initiation stage of translation.

C26. The ribosome binds at the 5′ end of the mRNA and then scans in the 3′ direction in search of an AUG start codon. If it finds one that satisfies Kozak's rules, it will begin translation at that site. Aside from an AUG start codon, two other important features are a guanosine at the +4 position and a purine at the −3 position.

C28. The A (aminoacyl) site is the location where a tRNA carrying a single amino acid initially binds. The only exception is the initiator tRNA, which binds to the P (peptidyl) site. The growing polypeptide chain is removed from the tRNA in the P site and transferred to the amino acid attached to the tRNA in the A site. The ribosome translocates in the 3′ direction, with the result that the two tRNAs in the P and A sites are moved to the E (exit) and P sites, and the uncharged tRNA in the E site is released.

C30. The initiation phase involves the binding of the Shine-Dalgarno sequence to the rRNA in the 30S subunit. The elongation phase involves the binding of anticodons in tRNA to codons in mRNA.

C32. A. The E site and the P site. (Note: A tRNA without an amino acid attached is found only briefly in the P site, just before translocation occurs.)

B. The P site and the A site. (Note: A tRNA with a polypeptide chain attached is found only briefly in the A site, just before translocation occurs.)

C. Usually the A site, except for the initiator tRNA, which can be found in the P site.

C34. The tRNAs bind to the mRNA because their anticodon sequences are complementary to the mRNA's codon sequences. When the ribosome translocates in the 5′ to 3′ direction, the tRNAs remain bound to their complementary codons, and the two tRNAs shift from the A site and P site to the P site and E site. If the ribosome moved in the 3′ direction, it would have to dislodge the tRNAs and drag them to a new position where they would not (necessarily) be complementary to the mRNA.

C36. 52

C38. The binding of IRP to the IRE inhibits the translation of ferritin mRNA and enhances the stability of the transferrin receptor mRNA. The increase in the stability of transferrin receptor mRNA increases the concentration of this mRNA and ultimately leads to more transferrin receptor proteins. Conditions of low iron promote the binding of IRP to the IRE, leading to a decrease in ferritin proteins and an increase in transferrin receptor proteins. When the iron concentration is high, iron binds to IRP, causing it to be released from the IRE. This allows the ferritin mRNA to be translated and also causes a decrease in the stability of transferrin receptor mRNA. Under these conditions, more ferritin proteins are made, and less transferrin receptors are made.

Experimental Questions

E2. The initiation phase of translation is very different in bacteria and in eukaryotes, so mRNAs would not be translated very efficiently in either case. A bacterial mRNA would not be translated very efficiently

in a eukaryotic translation system, because it lacks a cap structure attached to its 5′ end. A eukaryotic mRNA would not have a Shine-Dalgarno sequence near its 5′ end, so it would not be translated very efficiently in a bacterial translation system.

E4. A polysome is a structure in which a single mRNA is being translated into polypeptides by many ribosomes.

E6 To estimate the mass of the shorter protein with 421 amino acids, divide 421 by 572 and multiply by 68.6 kDa, which equals 50.5 kDa.

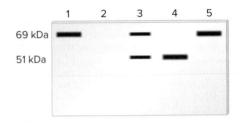

E8. A. If codon usage was significantly different between kangaroo and yeast cells, this would inhibit the translation process. For example, if the preferred leucine codon in kangaroos was CUU, translation of kangaroo mRNA would probably be slow in a yeast translation system. The cell-free translation system from yeast cells would primarily contain leucine tRNAs with AAC as the anticodon sequence, because this tRNA^{Leu} would match the preferred yeast leucine codon, which is UUG. In a yeast translation system, there probably would not be a large amount of tRNA with an anticodon of GAA, which would match the preferred leucine codon, CUU, of kangaroos. For this reason, kangaroo mRNA would not be translated very well in a yeast translation system, but it probably would be translated to some degree.

B. The advantage of codon bias is that a cell can rely on a smaller population of tRNA molecules to efficiently translate its proteins. A disadvantage is that mutations, which do not change the amino acid sequence but do change a codon (e.g., UUG to UUA), may inhibit the production of a polypeptide if a preferred codon is changed to a nonpreferred codon.

Questions for Student Discussion/Collaboration

2. These could be long lists. There are similarities along several lines:

1. Both processes involve a great deal of molecular recognition, either between two nucleic acid molecules or between proteins and nucleic acid molecules. (You may see these as similarities or differences, depending on your point of view.)

2. Biosynthesis occurs in both processes. Small building blocks are being connected together. This requires an input of energy.

3. There are genetic signals that determine the beginning and ending of these processes.

There are also many differences:

1. Transcription produces an RNA molecule with a similar structure to the DNA, whereas translation produces a polypeptide with a structure that is very different from RNA.

2. Depending on your point of view, it seems that translation is more biochemically complex, requiring more proteins and RNA molecules to accomplish the task.

CHAPTER 14

Answers to Comprehension Questions

14.1: c, a

14.2: b, d, b, d

14.3: a, b, a

14.4: d, d

14.5: a

Concept Check Questions (follow figure legends)

FIGURE 14.1   Gene regulation enhances the efficiency of energy use. A cell does not waste energy making RNAs and proteins it does not need.

FIGURE 14.3   The *lacZ*, *lacY*, and *lacA* genes are under the control of the *lac* promoter.

FIGURE 14.5   This repressor protein is bound to the *lac* operon when allolactose is not bound to the repressor, which occurs when lactose is not available.

FIGURE 14.9   Regulation via the repressor protein allows the cell to avoid turning on the operon in the absence of lactose. Regulation by the activator protein allows the cell to choose between glucose and lactose.

FIGURE 14.10   The data for the cases in which O_2 and O_3 are deleted indicate that O_1 is not the only *lac* operator site.

FIGURE 14.12   Tryptophan acts as a corepressor that causes trp repressor to bind to the *trp* operon and repress transcription.

FIGURE 14.13   Hydrogen bonding between complementary regions of the mRNA transcribed from the beginning of the *trp* operon causes stem-loops to form.

FIGURE 14.14   The presence of tryptophan causes the *trpL* gene to be translated up to its stop codon. This blocks region 2 from hydrogen bonding with any other region, which allows a 3–4 stem-loop to form. The 3–4 stem-loop causes transcriptional termination.

FIGURE 14.15   The *micF* antisense RNA binds to the *ompF* mRNA and inhibits its translation.

FIGURE 14.16   Feedback inhibition prevents a bacterium from overproducing the product of a metabolic pathway.

FIGURE 14.17   The RNA conformation with the antiterminator stem-loop favors transcription.

FIGURE 14.18   The RNA conformation with the Shine-Dalgarno antisequestor favors translation.

End-of-Chapter Questions:

Conceptual Questions

C2.   In bacteria, gene regulation greatly enhances the efficiency of cell growth. It takes a lot of energy to transcribe and translate genes. Therefore, a cell is much more efficient and better at competing in its environment if it expresses a gene only when the gene product is needed. For example, a bacterium will express the genes that are necessary for lactose metabolism only when the bacterium is exposed to lactose. When the environment is missing lactose, these genes are turned off. Similarly, when tryptophan levels are high within the cytoplasm, the genes required for tryptophan biosynthesis are repressed.

C4.   A. RTF

  B. Small effector molecule

  C. DNA segment

  D. Small effector molecule

  E. RTF

  F. DNA segment

  G. Small effector molecule

C6.   A mutation that has a *cis*-effect occurs within a genetic regulatory sequence, such as an operator site, that affects the binding of a genetic regulatory protein. A *cis*-effect mutation affects only the adjacent genes that the genetic regulatory sequence controls. A mutation having a *trans*-effect is usually in a gene that codes a genetic regulatory protein. A *trans*-effect mutation can be complemented in a merozygote experiment by the introduction of a normal gene that codes the regulatory protein.

C8.   A. No transcription would take place. The *lac* operon could not be expressed.

  B. No regulation would take place. The operon would be continuously turned on.

  C. The rest of the operon would function normally, but no galactoside transacetylase would be produced.

C10.   Diauxic growth refers to the phenomenon in which a cell first uses up one type of sugar (e.g., glucose) before it begins to metabolize a second type (e.g., lactose). This process is governed by gene regulation. When a bacterial cell is exposed to both glucose and lactose, the uptake of glucose causes the cAMP levels in the cell to fall. When this occurs, the catabolite activator protein (CAP) is removed from the *lac* operon, so it is not able to be activated by CAP.

C12.   A mutation that prevented lac repressor from binding to the operator would make the *lac* operon constitutive only in the absence of glucose. However, this mutation would not be entirely constitutive because transcription would be inhibited in the presence of glucose. The disadvantage of constitutive expression of the *lac* operon is that the bacterial cell would waste a lot of energy transcribing the genes and translating the mRNAs when lactose was not present.

C14.   A. Attenuation will not occur because a 2–3 stem-loop will form.

  B. Attenuation will occur because a 2–3 stem-loop cannot form, and so a 3–4 stem-loop will form.

  C. Attenuation will not occur because a 3–4 stem-loop cannot form.

  D. Attenuation will not occur because a 3–4 stem-loop cannot form.

C16.   The addition of G and C into the U-rich sequence would prevent attenuation. The U-rich sequence promotes the dissociation of the mRNA from the DNA, when the terminator stem-loop forms. This causes RNA polymerase to dissociate from the DNA and thereby causes transcriptional termination. The UGGUUGUC sequence would probably not dissociate because of the presence of G and C. Remember that G-C base pairs have three hydrogen bonds and are more stable than A-U base pairs, which have only two hydrogen bonds.

C18.   It takes a lot of cellular energy to translate mRNAs into proteins. A cell wastes less energy if it prevents the initiation of translation rather than stopping the process at a later stage, such as elongation or termination.

C20.   One possible mechanism is that histidine acts as corepressor that shuts down the transcription of the histidine synthetase gene. A second mechanism might be that histidine acts as an inhibitor via feedback inhibition. A third possibility is that histidine inhibits the ability of the mRNA that codes histidine synthetase to be translated. Perhaps histidine induces a gene that codes an antisense RNA. If the amount of histidine synthetase protein was identical in the presence and absence of extracellular histidine, a feedback inhibition mechanism is favored, because this would affect only the activity of histidine synthetase, not the amount of the enzyme. The other two mechanisms would diminish the amount of this enzyme.

C22.   The two proteins are similar in that both bind to a segment of DNA and repress transcription. They are different in three ways: (1) They recognize different effector molecules (i.e., lac repressor recognizes allolactose, and trp repressor recognizes tryptophan). (2) Allolactose causes lac repressor to be released from its operator, while tryptophan causes trp repressor to bind to its operator. (3) The sequences of the operator sites that these two proteins recognize are different. Otherwise, lac repressor could bind to the *trp* operator, and trp repressor could bind to the *lac* operator.

Experimental Questions

E2.   In the samples loaded into lanes 1 and 4, the repressor should bind to the operator because no lactose is present. In the sample loaded into lane 4, the CAP protein could still bind cAMP because there is no glucose. However, there is no difference between lanes 1 and 4, so it does not look like CAP can activate transcription when lac repressor is bound. If you compare the samples loaded into lanes 2 and 3, lac repressor would not be bound in either case, and CAP would not be bound in the sample loaded into lane 3. There is less transcription in lane 3 compared to lane 2, but because there is some transcription in lane 3, you can conclude that the removal of CAP (because cAMP levels are low) is not entirely effective at preventing transcription. Overall, the results indicate that the binding of lac repressor is more effective at preventing transcription of the *lac* operon than is the removal of CAP.

E4. A. Yes, if the researcher did not sonicate the cells, then β-galactosidase would not be released from them, and not much yellow color would be observed. (Note: A little yellow may be observed because some β-ONPG may be taken into the cells.)

B. No, yellow color should still be observed in the first two tubes even if the researcher forgot to add lactose because the unmated strain does not have a functional lac repressor.

C. Yes, if the researcher forgot to add β-ONPG, no yellow color would be observed because the cleavage of β-ONPG by β-galactosidase is what produces the yellow color.

E6. You could conjugate a strain that has an F′ factor carrying a normal *lac* operon and a normal *lacI* gene to this mutant strain. Because the mutation is in the operator site, you would still continue to get expression of β-galactosidase, even in the absence of lactose.

E8. Diauxic growth refers to the sequential use of two sugars by a population of bacterial cells. For example, *E. coli* will use glucose first and then use lactose. In the first phase of growth, the presence of glucose inhibits the synthesis of cAMP. When cAMP levels are low, it can't bind to CAP and therefore the *lac* operon cannot be activated by CAP and transcription of the *lac* operon is low. Once the glucose is used up, cAMP levels will rise (during the lag phase) and eventually cAMP will bind to CAP and the *lac* operon will be expressed at high levels. In the second growth phase, this allows the population of *E. coli* cells to then use lactose efficiently and further increase in numbers.

Questions for Student Discussion/Collaboration

2. A DNA loop may inhibit transcription by preventing RNA polymerase from recognizing the promoter. Or, it may inhibit transcription by preventing the formation of the open complex. Alternatively, a bend in the DNA may enhance transcription by exposing the base sequence that σ factor of RNA polymerase recognizes. The bend may expose the major groove in such a way that this base sequence is more accessible to binding by σ factor and RNA polymerase.

CHAPTER 15

Answers to Comprehension Questions

15.1: b, b, b, d   15.4: d, a, b, d, b
15.2: d, a   15.5: d
15.3: d, b

Concept Check Questions (follow figure legends)

FIGURE 15.1 An α helix can bind into the major groove of DNA and recognize a specific sequence of bases.

FIGURE 15.4 Eviction of histone octamers may allow certain proteins to access particular sites in the DNA and bind there, which could affect transcription.

FIGURE 15.5 Histone modifications may directly affect the interaction between histones and the DNA, or they may affect the binding of other proteins to the chromatin.

FIGURE 15.7 In part (a), DNA methylation blocks an activator protein from binding to the DNA. This prevents transcriptional activation.

FIGURE 15.8 De novo methylation occurs on unmethylated DNA, whereas maintenance methylation occurs on hemimethylated DNA.

FIGURE 15.9 An NFR is needed at the core promoter so that the preinitiation complex can assemble there.

FIGURE 15.10 Chromatin-remodeling complexes evict histones and thereby provide space for the assembly of the preinitiation complex on the DNA in the core promoter region.

FIGURE 15.11 When an activator interacts with mediator, mediator phosphorylates CTD, which causes RNA polymerase II to proceed to the elongation phase of transcription. Promoter escape occurs.

FIGURE 15.12 The glucocorticoid receptor binds next to the core promoter only when there is a GRE in an enhancer in the vicinity of the gene.

FIGURE 15.14 The antibody recognizes and binds to the protein of interest that is covalently crosslinked to DNA. The antibody is also attached to a heavy bead, which makes it easy to separate it and the DNA-protein complex from the rest of the cellular components by centrifugation. This provides an easy way to purify the protein of interest along with its attached DNA.

End-of-Chapter Questions:

Conceptual Questions

C2. Combinatorial control means that most eukaryotic genes, particularly those found in multicellular species, are transcriptionally regulated by many factors. These include: (1) one or more activator proteins may stimulate transcription; (2) one or more repressor proteins may inhibit transcription; (3) the function of activators and repressors may be modulated in a variety of ways, including the binding of small effector molecules, protein-protein interactions, and covalent modifications; (4) regulatory proteins may alter the composition or arrangement of nucleosomes in the vicinity of a promoter, thereby affecting transcription, or histone proteins may be covalently modified; (5) DNA methylation may inhibit transcription, either by preventing the binding of an activator protein or by recruiting proteins that change the structure of chromatin in a way that inhibits transcription; and (6) the formation of heterochromatin may inhibit gene expression in localized regions of a chromosome.

C4. Regulatory elements are relatively short DNA sequences that are recognized by regulatory transcription factors. After a regulatory transcription factor binds to a regulatory element, it will affect the rate of transcription, either activating it or repressing it, depending on the action of the regulatory protein. Regulatory elements may be located in the upstream region near the core promoter, but they can be located almost anywhere (i.e., upstream or downstream) and even quite far from the promoter. Typically, they are not close to the core promoter.

C6. Transcriptional activation occurs when a regulatory transcription factor binds to a regulatory element and activates transcription. Such proteins, called activators, may interact with coactivators, such as chromatin-remodeling complexes and histone-modifying enzymes, or they may interact with TFIID and/or mediator to promote the assembly of RNA polymerase II and general transcription factors at the promoter region. They can also alter the structure of chromatin so that RNA polymerase II and transcription factors are able to gain access to the promoter. Transcriptional inhibition occurs when a regulatory transcription factor inhibits transcription. Such a repressor may interact with coactivators, such as chromatin-remodeling complexes and histone-modifying enzymes, or it may interact with TFIID and/or mediator to inhibit RNA polymerase II.

C8. A. DNA binding
B. DNA binding
C. Protein dimerization

C10. ATP-dependent chromatin-remodeling complexes may change the positions of nucleosomes, evict histones, and/or replace histones with histone variants.

C12. The attraction between DNA and histones occurs because the histones are positively charged and the DNA is negatively charged. The covalent attachment of acetyl groups decreases the amount of positive charge on the histone proteins and thereby may loosen the binding of the DNA. In addition, histone acetylation may attract to the region proteins that reduce chromatin compaction.

C14. DNA methylation is the attachment of a methyl group to a base within the DNA. In many eukaryotic species, the attachment occurs primarily on cytosines at a CG sequence. After de novo methylation has occurred, the alteration is passed from mother cell to daughter cell. Because DNA replication is semiconservative, the newly made DNA contains one strand that is methylated and one that is not. DNA methyltransferase recognizes this hemimethylated DNA and methylates the cytosine in the unmethylated DNA strand; this event is called maintenance methylation.

C16. A CpG island is a stretch of 1000–2000 base pairs that contains a large number of CpG sites. CpG islands are often located near promoters. Methylation of these islands inhibits transcription. This inhibition may be the result of the inability of the transcriptional activators to recognize the methylated promoter and the effects of methyl-CpG-binding proteins, which may promote a closed chromatin conformation.

C18. See Figure 15.10.

C20. In the preinitiation complex, RNA polymerase, general transcription factors, and mediator are wrapped around DNA in which the strands have not yet separated. The separation of DNA strands marks the formation of the open complex.

C22. A potentially harmful consequence is that transcription may be initiated at multiple points within a gene, thereby producing many nonfunctional transcripts. This result would be a waste of energy.

C24. For the glucocorticoid receptor to bind to a GRE, a steroid hormone must first enter the cell. The hormone then binds to the glucocorticoid receptor, which releases HSP90. The release of HSP90 exposes a nuclear localization signal (NLS) within the receptor, which enables it to dimerize and then enter the nucleus. Once inside the nucleus, the dimer binds to a GRE, which activates transcription of the adjacent gene.

C26. Phosphorylation of the CREB protein causes it to act as a transcriptional activator. Unphosphorylated CREB protein can still bind to a CRE, but it does not stimulate transcription via CBP.

C28. A. The glucocorticoid receptor binds the glucocorticoid hormone with a certain affinity. The binding is a reversible process. Eventually, the glucocorticoid hormone will be degraded by the cell. Once the concentration of the hormone falls below the affinity of the hormone for the receptor, the receptor will no longer have the hormone bound to it. When the hormone is released, the glucocorticoid receptor will change its conformation, and it will no longer bind to the DNA.

B. An enzyme that can remove phosphate groups from the CREB protein is needed (such an enzyme is called a phosphatase). When the phosphates are removed, the CREB protein will stop activating transcription.

C30. The sequence shown in part A could work as a regulatory element; but the ones in parts B and C could not. The sequence that is recognized by the transcriptional activator is 5′–GTAG–3′ in one strand and 3′–CATC–5′ in the opposite strand. This is the same as the sequence shown in part A. In B and C, however, the arrangement is 5′–GATG–3′ and 3′–CATC–5′. In the sequences in B and C, the two middle bases (i.e., A and T) are not in the correct order.

C32. The mutation might have resulted in a defect in any of the following:

1. Adrenaline receptor
2. G protein
3. Adenylyl cyclase
4. Protein kinase A
5. CREB protein
6. CREs of the tyrosine hydroxylase gene

If other genes were properly up-regulated by the CREB protein, you could conclude that the mutation is probably within the tyrosine hydroxylase gene itself. Perhaps a CRE has been mutated and no longer recognizes the CREB protein.

Experimental Questions

E2. The results described in the question indicate that the fibroblasts perform maintenance methylation because they can replicate and methylate DNA if it has already been methylated. However, the fibroblasts do not perform de novo methylation, because if the donor DNA was unmethylated, the DNA in the daughter cells remains unmethylated.

E4. Based on these results, enhancers that are recognized by one or more activators are located in regions A, D, and E. When these enhancers are deleted, the level of transcription is decreased. There also appears to be an enhancer in region B that is recognized by one or more repressors, because a deletion of this region increases the rate of transcription. There do not seem to be any response elements in region C, or at least not any that function in muscle cells. Region F contains the core promoter, so the deletion of this region inhibits transcription.

E6. The results of the EMSA indicate that protein X binds to the DNA fragment and retards its mobility (lanes 3 and 4). However, hormone X is not required for DNA binding. Because the hormone does produce transcriptional activation, it must play some other role. Perhaps the hormone activates a signaling pathway that leads to the phosphorylation of the transcription factor, and phosphorylation is necessary for transcriptional activation. This role is similar to that of the CREB protein, which is activated by phosphorylation. The CREB protein can bind to the DNA whether or not it is phosphorylated.

Questions for Student Discussion/Collaboration

2. Probably the most efficient method would be to systematically make deletions of progressively smaller sizes. For example, you could begin by deleting 20,000 bp on either side of the gene and see if that affects transcription. If you found that only the deletion on the 5′ end of the gene had an effect, you could then start making deletions from the 5′ end, perhaps in 10,000-bp or 5000-bp increments until you localized response elements. You would then make smaller deletions in that localized region until the trimmed piece was down to a hundred or a few dozen nucleotides. At this point, you might conduct gene editing, as described in Chapter 20, to specifically identify the regulatory element sequence.

CHAPTER 16

Answers to Comprehension Questions

16.1: f, c, b, d	16.4: c, d
16.2: d, c, c, d	16.5: a, c
16.3: c, b, d	

Concept Check Questions (follow figure legends)

FIGURE 16.2 Following DNA replication, each DNA strand is hemimethylated. As discussed in Chapter 15, hemimethylated DNA becomes fully methylated via maintenance methylation, which is a routine event in mammalian and plant cells. If another copy of the same gene in the cell was not methylated to begin with, it would need to undergo de novo methylation, which is not a routine event. De novo methylation is highly regulated and typically only occurs during gamete formation or embryonic development.

FIGURE 16.4 A reader domain recognizes a specific type of structure, such as an H3K9me3 modification on a nucleosome, and binds to it.

FIGURE 16.5 Each nucleosome contains two copies of histone H3, so there can be a maximum of two H3K9me3 modifications in a single nucleosome.

FIGURE 16.6 Liquid-like compartments are formed when macromolecules become concentrated in a given location and come out of solution.

FIGURE 16.9 The histone octamers are randomly distributed so that some nucleosomes contain the original octamers and other nearby nucleosomes contain newly synthesized octamers.

FIGURE 16.11 The choice is made during embryonic development.

FIGURE 16.13 A knotlike structure would prevent the binding of proteins, such as TFIID and RNA polymerase, to the gene, thereby inhibiting transcription.

FIGURE 16.14 It was converted to a *B'* allele by a paramutation.

FIGURE 16.16 At the level of DNA sequences, queens and worker bees are not different from each other. However, epigenetic changes cause differences in gene expression, which explains the morphological differences between a queen and the worker bees.

End-of-Chapter Questions:

Conceptual Questions

C2. 1. DNA methylation—the attachment of methyl groups to cytosines in DNA. This often silences transcription.

2. Covalent histone modification—the covalent attachment of groups to the amino terminal tails of histones. This may silence or activate genes.

3. Chromatin remodeling—changes in the positions of nucleosomes. This may lead to a closed or open conformation for transcription.

4. Histone variants—replacement of standard histones with histone variants. This may silence or activate genes.

5. Feedforward loop—the activation of a gene coding a transcription factor. After the transcription factor is made, it continues to activate its own expression, as well as the expression of other genes.

C4. In a *cis*-epigenetic mechanism, the pattern of gene modification is maintained when two or more copies of a gene are found in the same cell. One copy may be silenced and the other is active. In a *trans*-epigenetic mechanism, soluble proteins, such as transcription factors, are responsible for maintaining gene activation. In this case, all copies of the gene will be active. The expression of the copy of the *Igf2* gene inherited from the male parent is due to a *cis*-epigenetic mechanism.

C6. Constitutive heterochromatin is located close to a centromere, in what is called the pericentric region, and also at telomeres. Facultative heterochromatin may be found at multiple discrete sites that are located between the centromere and telomeres.

C8. HP1 forms crossbridges that bring nucleosomes closer together. LADs bind to the nuclear lamina. Heterochromatin droplets form via liquid-liquid phase separation.

C10. Liquid-liquid phase separation (LLPS) is a phenomenon in which liquid-like compartments are formed by macromolecules that become concentrated in a given location and come out of solution and form droplets. The environments within phase-separated droplets may favor the importing of certain substances and the exporting of others. The environment within a heterochromatic droplet may favor the exporting of components that are involved with gene transcription and the importing of components that are needed for heterochromatin formation. Therefore, the chromatin within the droplet would be transcriptionally inactive and maintain a heterochromatic structure.

C12. Heterochromatin formation is maintained by: (1) the methylation of hemimethylated DNA; (2) the action of chromatin-modifying enzymes that were associated with the original histones, which, in turn, promote the binding of new HP1 proteins to form dimers that link nucleosomes together; and (3) recruitment of chromatin-modifying enzymes and chromatin-remodeling complexes to the newly replicated region via DNA polymerase.

C14. The *Igf2* gene inherited from the female parent will be expressed. For silencing to occur, a loop must form. The loop could not form if the ICR was missing.

C16. After the *Xist* RNA coats the X chromosome, it recruits proteins that silence genes, such as the gene that codes for DNA methyltransferase, and proteins that make the chromosome more compact to form a Barr body. One gene on the inactivated X chromosome that is not inactivated is the *Xist* gene.

C18. Pioneer factors have the unique ability to recognize and bind to specific DNA sequences that are exposed on the surface of a nucleosome. After binding to a nucleosome, pioneer factors subsequently recruit chromatin-remodeling complexes and histone-modifying enzymes that carry out epigenetic changes such as histone eviction and covalent modifications. These events may influence the ability of other transcription factors to bind to enhancer sequences. Pioneer factors can also decrease the level of DNA methylation by binding to CpG islands, thereby blocking access to DNA methyltransferases.

C20. The trithorax and polycomb group complexes are involved with gene activation and gene repression during development, respectively. They cause epigenetic changes that allow genes to either remain active or be permanently repressed. Such changes occur during embryonic development and are maintained during subsequent stages.

C22. The consequences would be that many genes would be expressed in cell types where they should not be expressed. This would cause abnormalities in development and would likely be lethal.

C24. Though they don't change the DNA sequence, epigenetic modifications can affect gene expression. (As discussed in Chapter 25, such changes could increase gene expression and thereby result in oncogenes, or they could inhibit the expression of tumor suppressor genes.) Either type of change could contribute to cancer. For example, DNA methylation of a tumor suppressor gene could promote cancer.

C26. The expression of the *FLC* gene inhibits flowering. Cold temperatures inhibit the expression of the *FLC* gene, which allows flowering to occur.

Experimental Questions

E2. The coat color of the offspring depends on the expression of the *Agouti* gene. In this case, the *Agouti* gene is under the control of a promoter that is found in a transposable element, causing the gene to be overexpressed. When the promoter is not methylated and the gene is overexpressed, the coat color is yellow because the *Agouti* gene product promotes the synthesis of yellow pigment. When the promoter is methylated, the *Agouti* gene is inhibited, and coat color is darker because less yellow pigment is made.

E4. The F$_3$ offspring would be *B'B'*, and their phenotype would be green. This outcome occurs because the *B'* allele converts the *B-I* allele to *B'*.

Questions for Student Discussion/Collaboration

2. Epigenetic gene regulation and variation in DNA sequences are similar in that both can affect gene expression, thereby affecting phenotype. One key difference is reversibility. Barring rare mutations, changes in DNA sequences are not usually reversible. In contrast, epigenetic modifications are reversible, though they may be permanent during the life of a particular individual.

CHAPTER 17

Answers to Comprehension Questions

17.1: e, b, b	17.5: d
17.2: d	17.6: d, c, d
17.3: d, d	17.7: d, d
17.4: d	

Concept Check Questions (follow figure legends)

FIGURE 17.1 DNA, RNAs, proteins, and small molecules can bind to an ncRNA.

FIGURE 17.4 HOTAIR binds next to the target gene because a segment of HOTAIR is complementary to the GA-rich region.

FIGURE 17.5 The researchers inserted cloned genes that coded chalcone synthase or dihydroflavono-l4-reductase, which are enzymes involved in the synthesis of flower pigments. Having more copies of these genes was expected to cause a greater level of pigment synthesis, so a white flower was unexpected.

FIGURE 17.7 RISC binds to a specific mRNA because the miRNA or siRNA within RISC is complementary to that mRNA. The bonding is hydrogen bonding between complementary bases.

FIGURE 17.8 The molecule called miR-156 prevents the transition to adult growth.

FIGURE 17.9 A C/D box snoRNA causes methylation of an rRNA.

FIGURE 17.10 The hydrolysis of GTP promotes the release of SRP, which allows translation to resume.

FIGURE 17.11 The crRNA binds directly to one of the strands of the bacteriophage DNA via hydrogen bonding between complementary bases.

FIGURE 17.12 They cause TE RNA to be degraded or prevent the TE from being transcribed.

End-of-Chapter Questions:

Conceptual Questions

C2.   A. Scaffold and guide

     B. Ribozyme

     C. Guide

     D. Guide

C4.   First, HOTAIR acts as a scaffold for the binding of PRC2 and LSD1. Second, HOTAIR acts as a guide and binds to GA-rich sequences next to particular target genes. The histones associated with those genes are then covalently modified via the actions of PRC2 and LSD1.

C6.   First, double-stranded RNA could be produced by transcription of a gene, as is the case with pri-miRNA. Second, it could come from a virus. Third, double-stranded RNA could be made in a laboratory.

C8.   A snoRNA functions as a scaffold for the binding of a specific set of proteins to form a small nucleolar ribonucleoprotein (snoRNP). In addition, antisense sequences in a snoRNA allow it to act as a guide to direct the snoRNP to an appropriate rRNA.

C10. If the SRP RNA did not stimulate the GTPase activities, SRP would remain bound to the polypeptide and translation would not resume.

C12. It is a guide, directing crRNA to Cas9.

C14. See Figure 17.12. The role of piRNAs is to guide the PIWI proteins to the TE RNA, where they either cause transcription to be inhibited or cause the TE RNA to be degraded.

C16. The miR-200 family plays an essential role in tumor suppression by inhibiting an event called the epithelial-mesenchymal transition (EMT), which is the initiating step of metastasis. During the EMT, cells lose their adhesion to neighboring cells. This loss of adhesion is associated with a decrease in expression of E-cadherin, which is a membrane protein that adheres adjacent cells to each other. When E-cadherin levels are low, cells can more easily move to new sites in the body. In an adult, when the miR-200 family of miRNAs is expressed at normal levels, these miRNAs inhibit EMT. This inhibition maintains a normal level of E-cadherin and thereby prevents metastasis. However, if the miR-200 miRNAs are expressed at low levels, EMT is not inhibited, and metastasis can occur.

Experimental Questions

E2.   The researchers were injecting pre-siRNA. In this case, it was a perfect match to the *mex-3* RNA and caused its degradation.

E4.   The CRISPR-Cas system can be used to edit genes by causing a small deletion within a gene or producing a change in the sequence of a gene. In this system, tracrRNA and crRNA form one RNA called a single guide RNA (sgRNA). The sgRNA recognizes both the Cas9 protein and a target gene that a researcher wants to mutate. After the sgRNA-Cas9 complex binds to the target gene, Cas9 cleaves the gene. Following nonhomologous end joining, this can result in a small segment of the gene being deleted, or if donor DNA is added, a double crossover can introduce a point mutation into the target gene.

Questions for Student Discussion/Collaboration

2.   You will find many examples of roles played by ncRNAs that are not discussed in this text. It should be easy and interesting to make a list.

CHAPTER 18

Answers to Comprehension Questions

18.1: c, d   18.3: b, b

18.2: c, d   18.4: b, c, c

Concept Check Questions (follow figure legends)

FIGURE 18.1 Viruses vary in their genomes. Viral genomes differ with regard to size, RNA versus DNA, and single-stranded versus double-stranded. The structures of viruses also vary with regard to the complexity of their capsids and whether or not they have an envelope.

FIGURE 18.3 A virus can remain latent following the integration of its genome into the genome of the host cell.

FIGURE 18.4 The lytic cycle produces new phage particles.

FIGURE 18.5 Emerging viruses are viruses that have arisen recently and are more likely than previous strains to cause infection.

FIGURE 18.9 Neither cycle could occur. The N protein is needed to extend transcription from P_L and begin the lysogenic cycle and to extend transcription from P_R and begin the lytic cycle.

FIGURE 18.10 This promoter is used to express the λ repressor that ensures maintenance of the lysogenic cycle.

FIGURE 18.11 An adequate supply of nutrients or exposure to UV light will favor a switch to the lytic cycle.

FIGURE 18.13 The two enzymatic functions are the synthesis of DNA using RNA or DNA as a template and the digestion of RNA via the RNase H component.

FIGURE 18.15 Fully spliced HIV RNA is needed during the early stage of the synthesis of HIV components.

End-of-Chapter Questions:

Conceptual Questions

C2.   All viruses have a nucleic acid genome and a capsid composed of proteins. Some eukaryotic viruses are surrounded by an envelope that consists of a membrane with embedded proteins.

C4.   A viral envelope is a membrane with embedded proteins. It is made when the virus buds from the host cell, taking with it a portion of the host cell's plasma membrane.

C6.   The attachment step usually involves the binding of the virus to a specific protein on the surface of the host cell. Only certain cell types will make that specific protein.

C8.   Reverse transcriptase is used to copy the viral RNA into DNA so that it can be integrated into a chromosome of the host cell.

C10. A temperate phage can follow either the lytic or lysogenic cycle, whereas a virulent phage can follow only the lytic cycle.

C12. In the lytic cycle, the virus directs the bacterial cell to make more virus particles until eventually the cell lyses and releases them. In the lysogenic cycle, the viral genome is incorporated into the host cell's genome as a prophage. It remains there in a dormant state until some stimulus causes it to excise itself from the bacterial chromosome and enter the lytic cycle.

C14. A. The lysogenic cycle would occur because cro protein is necessary to initiate the lytic cycle.

     B. The lytic cycle would occur because *cI* codes the λ repressor, which prevents the lytic cycle.

     C. The lytic cycle would occur because cII protein is necessary to initiate the lysogenic cycle.

D. Both cycles might try to initiate, but the lysogenic cycle would fail because the viral genome would be unable to integrate into the host chromosome.

E. Neither cycle could occur.

C16. P_{RE} is activated by the cII-cIII complex. However, later in the lysogenic cycle, the amount of the cII-cIII complex falls. This decrease prevents further synthesis of the λ repressor. However, the λ repressor can activate its own transcription from P_{RM}. This activation will maintain the lysogenic cycle.

C18. The process is summarized in Figure 18.13.

C20. RNase H is needed to remove the HIV RNA as the HIV DNA is being made.

C22. The Vpr protein facilitates the transport of the preintegration complex into the host-cell nucleus.

C24. Fully spliced HIV RNA is used in the translation of the Nef, Tat, and Rev proteins. It is needed during the early stage of HIV proliferation.

Incompletely spliced HIV RNA is involved in the translation of the Vif, Env, Vpu, and Vpr proteins. It is needed in a later stage of proliferation.

Unspliced HIV RNA is packaged into HIV particles. It also plays a role in the translation of the Gag and Gag-Pol polyproteins. It is needed in a later stage.

C26. An HIV particle acquires its envelope in a budding process. The gp41 and gp120 proteins reach the plasma membrane by traveling through the ER and the Golgi. The Gag polyproteins associate with each other and cause the plasma membrane to bud from the cell surface. The MA portion of Gag polyprotein binds to gp41, which ensures that gp41 and gp120 are contained within the envelope.

Experimental Questions

E2. Electron microscopy must be used to visualize a virus.

E4. Following infection, the features observed on the infected leaves correlated with the RNA that was initially found in the reconstituted virus. Also, the biochemical composition of the capsid protein in newly made viruses was consistent with the RNA that was found in the reconstituted virus.

E6. Drugs that bind to and inhibit HIV protease prevent the cleavage of the Gag and Gag-pol polyproteins. This cleavage is needed for the maturation of HIV.

E8. A cell that has a λ prophage is making a significant amount of the λ repressor. If another phage infects that cell, the λ repressor inhibits transcription from P_R and P_L, thereby inhibiting the early steps that are required for either the lytic or lysogenic cycle.

E10. If the F⁻ strain is lysogenic for phage λ, the λ repressor is already being made in that cell. If the F⁻ strain receives genetic material from an Hfr strain, you would not expect the material to have an effect on the lysogenic cycle, which is already established in the F⁻ cell. However, if the Hfr strain is lysogenic for λ and the F⁻ strain is not, the Hfr strain could transfer the integrated λ DNA (i.e., the prophage) to the F⁻ strain. The cytoplasm of the F⁻ strain would not contain any λ repressor. Therefore, this λ DNA could choose between the lytic and lysogenic cycle. If it follows the lytic cycle, the F⁻ recipient bacterium will lyse.

Questions for Student Discussion/Collaboration

2. The diagram should have three steps:

Step 1. In the absence of the λ repressor, P_R will be turned on, which will allow the synthesis of the cII and cro proteins.

Step 2. cII will activate P_L, which will lead to the synthesis of integrase and exisionase, so the λ genome will be excised from the E. coli genome.

Step 3. In the absence of the λ repressor, cro will be able to activate P_R; this activation will eventually result in synthesis of the proteins required for λ DNA replication and coat protein synthesis.

CHAPTER 19

Answers to Comprehension Questions

19.1: b, c, a, d   19.4: b, b, a
19.2: b   19.5: c, d, a, d, b
19.3: c, d, c   19.6: a, d, a, d

Concept Check Questions (follow figure legends)

FIGURE 19.2 A position effect occurs when a change in the position of a gene along a chromosome alters the expression of the gene.

FIGURE 19.3 In part (b), the DNA sequence of the eye color gene has not been changed. The change in its expression compared with part (a) is due to a position effect.

FIGURE 19.5 No. A trait due to a somatic mutation is not passed from parent to offspring.

FIGURE 19.7 The probability is 75% because three of the four nucleotides that may be added to the newly made strand opposite the apurinic site are not the correct one.

FIGURE 19.8 The deamination of 5-methylcytosine is more difficult to repair because thymine is a base normally found in DNA. This makes it difficult for DNA repair enzymes to distinguish between the correct and altered strand.

FIGURE 19.10 A reactive oxygen species is an oxygen-containing chemical that may react with cellular molecules in a way that may cause harm.

FIGURE 19.13 5-Bromouracil causes a transition.

FIGURE 19.14 Thymine dimer formation is often the result of exposure to UV light. It most commonly occurs in skin cells.

FIGURE 19.15 Enzymes within the rat liver extract may convert nonmutagenic molecules into mutagenic forms. Adding the extract allows researchers to identify molecules that may be mutagenic in people.

FIGURE 19.16 Photoreactivation via photolyase is particularly valuable to plants, which are exposed to sunlight throughout the day and therefore susceptible to thymine dimer formation.

FIGURE 19.18 Cuts are made on both sides of the damaged region of the DNA so that the region can be removed by UvrD.

FIGURE 19.19 MutH distinguishes between the parental strand and the newly made daughter strand, which ensures that the daughter strand is repaired.

FIGURE 19.21 An advantage of nonhomologous end joining is that it can occur at any stage of the cell cycle. A disadvantage is that it may be imprecise and result in a short deletion in the DNA.

FIGURE 19.22 An advantage of genetic recombination is that it may foster genetic diversity, which may produce organisms that have reproductive advantages.

FIGURE 19.23 A heteroduplex region may be produced after branch migration because the two strands are not perfectly complementary.

FIGURE 19.24 A D-loop is a structure formed during the recombination process when a DNA strand from one homologous chromatid invades the same region of the other homologous chromatid.

FIGURE 19.26 The region of the upper chromosome that contained the genetic sequence that comprised the *b* allele, which may have included one or more base sequence differences, was digested away following the double-strand break in that chromosome. The same region in the homologous chromosome, which carried the *B* allele, was then used as a template to make another copy of the *B* allele.

End-of-Chapter Questions:

Conceptual Questions

C2. It is a gene mutation, a point mutation, a base substitution, a transition mutation, a deleterious mutation, a mutant allele, a nonsense mutation, a conditional mutation, a temperature-sensitive mutation, and a lethal allele.

C4. A. A nonsense mutation would probably inhibit protein function, particularly if it did not occur near the end of the coding sequence.

B. A missense mutation may or may not affect protein function, depending on the nature of the amino acid substitution and on whether the substitution is in a critical region of the protein.

C. An up promoter mutation would increase the amount of functional protein.

D. This mutation may affect protein function if the alteration in splicing changes an exon in the mRNA in a way that results in a protein with an altered structure.

C6. A. Not appropriate, because the second mutation occurred at a different codon

B. Appropriate

C. Not appropriate, because the second mutation is in the same gene as the first mutation

D. Appropriate

C8. Nonsense and frameshift mutations would be most likely to disrupt protein function. A nonsense mutation would cause the protein to be much shorter, and a frameshift mutation would alter the amino acid sequence downstream from the mutation. A missense mutation only affects a single amino acid, so it is less likely to disrupt protein function, but it could do so if it altered an important region of the protein. A silent mutation would not alter protein function.

C10. One possibility is that a translocation may move a gene next to a heterochromatic region of another chromosome, thereby diminishing the gene's expression, or it may move the gene next to a euchromatic region and increase its expression. Another possibility is that the translocation may move the gene to a position next to a different promoter or regulatory sequence that may then influence the gene's expression.

C12. A. The point mutation does not have a position effect; the position (i.e., chromosomal location) of a gene has not been altered.

B. The translocation does have a position effect; the expression of the gene will be altered because it has been moved to a new chromosomal location.

C. The inversion does have a position effect; the expression of the gene will be altered because it has been moved to a new chromosomal location.

C14. If a mutation within a germ-line cell is passed to an offspring, all cells of the offspring's body will carry the mutation. A somatic mutation affects only the somatic cell in which it originated and all of the daughter cells that the somatic cell produces. If a somatic mutation occurs early during embryonic development, it may affect a fairly large region of the organism. Because germ-line mutations affect the entire organism, they are potentially more harmful (or beneficial), but this is not always the case. Somatic mutations can cause quite harmful effects, such as cancer.

C16. UV light can cause thymine dimers, which can interfere with DNA replication because DNA polymerase cannot slide past such a dimer and add bases to the growing strand. Alkylating mutagens such as nitrous acid will cause mistakes in base pairing during DNA replication. For example, an alkylated cytosine will base-pair with adenine, thereby creating a mutation in the newly made strand. A third example of a mutagen that can interfere with DNA replication is 5-bromouracil, which is a thymine analog. It may base-pair with guanine instead of adenine during DNA replication.

C18. During TNRE, a trinucleotide repeat sequence gets longer. If someone was mildly affected with a TNRE disorder, they might be concerned that expansion of the repeats might occur during gamete formation, yielding offspring more severely affected with the disorder, a phenomenon called anticipation, whose occurrence depends on the sex of the parent with the TNRE disorder.

C20. According to the random mutation theory, spontaneous mutations can occur in any gene and do not require exposure of the organism to an environmental condition that causes specific types of mutation. However, the structure of chromatin may cause certain regions of the

DNA to be more susceptible to random mutations. For example, DNA in an open conformation may be more accessible to mutagens and more likely to incur mutations. Similarly, hot spots—certain regions of a gene that are more likely to mutate than other regions—can occur within a gene. Also, some genes mutate at a higher rate because they are larger than others, which means a greater chance of mutation.

C22. Nucleotide excision repair or homologous recombination repair could fix this damage.

C24. *Anticipation* means that a repeat sequence expands even further in subsequent generations. Anticipation seems to depend on the sex of the parent with the TNRE disorder.

C26. The mutation frequency is the total number of mutant alleles divided by the total number of alleles of that gene in the population. If there are 1,422,000 babies, there are 2,844,000 copies of this gene (because each baby has two copies). The mutation frequency is 31/2,844,000, which equals 1.09×10^{-5}. The mutation rate is the number of new mutations in a gene per generation. There are 13 babies who did not have a parent with achondroplasia; thus, 13 is the number of new mutations. Because the mutation rate is calculated as the number of new mutations in a given gene per generation, you divide 13 by 2,844,000. In this case, the mutation rate is 4.6×10^{-6}.

C28. The effects of mutations are cumulative. If one mutation occurs in a cell, this mutation will be passed to the daughter cells. If a mutation then occurs in a daughter cell, that cell will have two mutations. These two mutations will be passed to the next generation of daughter cells, and so forth. The accumulation of many mutations eventually kills the cells. That is why mutagens are more effective at killing dividing cells than nondividing cells. It is because the number of mutations accumulates to a lethal level.

There are two main side effects to these types of treatment. First, some normal (noncancerous) cells of the body, particularly skin cells and intestinal cells, are actively dividing. These cells are killed by chemotherapy and radiation therapy. Secondly, it is possible that the therapy may produce mutations that will cause noncancerous cells to become cancerous.

C30. A. Likely

B. Not likely; the albino trait affects the entire individual.

C. Not likely; the trait of early fruit production affects the entire apple tree.

D. Likely

C32. Mismatch repair is aimed at eliminating base pair mismatches that may have occurred during DNA replication. In such a case, the wrong base is in the newly made strand. The binding of MutH, which occurs on a hemimethylated sequence, provides a mechanism to distinguish between the unmethylated and methylated strands. In other words, MutH binds to the hemimethylated DNA in a way that allows the mismatch repair system to distinguish which strand is methylated and which is not.

C34. Because sister chromatids are genetically identical, an advantage of homologous recombination repair is that it can be an error-free mechanism to repair a double-strand break. A disadvantage, however, is that HRR occurs only during the S and G_2 phases of the cell cycle in eukaryotes or following DNA replication in bacteria. An advantage of nonhomologous end joining is that it doesn't involve a sister chromatid, so it can occur at any stage of the cell cycle. However, a disadvantage is that NHEJ can result in small deletions in the region that has been repaired. Overall, NHEJ is a quick but error-prone repair mechanism, while HRR is a more accurate method of repair that is limited to certain stages of the cell cycle.

C36. The underlying genetic defect that causes xeroderma pigmentosum is a defect in one of the genes that code a polypeptide involved with nucleotide excision repair. Individuals with XP thus have a deficiency in repairing DNA abnormalities such as thymine dimers and chemically modified or missing bases. Therefore, they are very sensitive to environmental agents such as UV light, which is more likely to cause mutations in these people than in unaffected individuals. For this reason, people with XP have pigmentation abnormalities and premalignant lesions and a high predisposition for developing skin cancer.

C38. Both types of repair systems recognize an abnormality in the DNA and excise a segment of the strand containing the abnormality. The normal strand is then used as a template to synthesize a complementary strand of DNA. The systems differ in the types of abnormalities they detect. The mismatch repair system detects base pair mismatches, while the excision repair system recognizes thymine dimers, chemically modified bases, missing bases, and certain types of crosslinks. The mismatch repair system operates immediately after DNA replication, allowing it to distinguish between the daughter strand (which contains the wrong base) and the parental strand. The excision repair system can operate at any time in the cell cycle.

C40. At the molecular level, SCE and recombination involving homologs are very similar. Identical (SCE) or similar (homologous recombination) segments of DNA line up and then cross over. Due to the molecular similarities of the two processes, you would expect that the same types of proteins would catalyze both events. However, at the genetic level, the events are different. SCE does not result in the recombination of alleles because the chromatids are genetically identical. Homologous recombination usually results in a new combination of alleles after a crossover has taken place.

C42. The steps are shown in Figure 19.24.

A. The ends of the broken strands would not be recognized and degraded.

B. The single-stranded ends would not be recognized, and strand invasion of the homologous strand would not occur.

C. A Holliday junction would not form.

D. Branch migration would not occur, and resolution to separate the chromosomes would not occur.

C44. Usually, homologous recombination does not create any new mutations in particular genes, unless gene conversion has occurred. However, homologous recombination does rearrange the combinations of alleles along particular chromosomes. This can be viewed as a mutation, because the sequence of a chromosome has been altered in a heritable fashion.

C46. It depends on where the breaks occur in the DNA strands during resolution. If the two breaks occur in the crossed DNA strands, nonrecombinant chromosomes result. If the two breaks occur in the uncrossed strands, the result is a pair of recombinant chromosomes.

C48. The Holliday model proposes two breaks, one in each chromatid, after which the two chromatids exchange a single strand of DNA. The double-strand break model proposes two breaks, both in the same chromatid. As in the Holliday model, single-strand migration occurs between both homologs. The double-strand break model also includes strand degradation and gap repair synthesis. Finally, in the double-strand break model, two Holliday junctions are produced, not just one.

C50. The RecA protein binds to single-stranded DNA and promotes strand invasion. It also promotes the displacement of the complementary strand to form a D-loop.

C52. RecG and RuvABC bind to Holliday junctions. They do not necessarily recognize a DNA sequence, but instead recognize a region of crossing over (i.e., a four-stranded structure).

Experimental Questions

E2. To show that antibiotic resistance is due to random mutations, you could follow the same basic strategy shown in Figure 19.6, except the secondary plates would contain the antibiotic instead of T1 phage. If the antibiotic resistance arose as a result of random mutations in the bacteria on the master plate, you would expect the antibiotic-resistant colonies to appear at the same locations on two different secondary plates.

E4. You would expose the bacteria to the physical agent. Use of the rat liver extract is probably not necessary for two reasons. First, a physical mutagen is not something that a person would eat. Therefore, the actions of digestion via the liver are probably irrelevant with regard to a physical mutagen. Second, the rat liver extract would not be expected to alter the properties of a physical mutagen.

E6. The results suggest that the mutant strain is defective with regard to an excision repair system. If you compare the results for the normal and mutant strains that have been incubated for 2 hours at 37°C, much of the radioactivity in the normal strain has been transferred to the soluble fraction because it has been excised. For the mutant strain, however, less of the radioactivity has been transferred to the soluble fraction, suggesting that this strain is not as efficient at removing thymine dimers.

Questions for Student Discussion/Collaboration

2. The worst time to be exposed to mutagens is at very early stages of embryonic development. An early embryo is most sensitive to a mutation because the mutation will affect a large region of the body. Adults must also worry about mutagens for several reasons. Mutations in somatic cells can cause cancer, a topic discussed in Chapter 25. Also, adults should be careful to avoid mutagens that may affect the ovaries or testes because these mutations could be passed to offspring.

CHAPTER 20

Answers to Comprehension Questions

20.1: d, b, c, a, a   20.4: c

20.2: c, b, d   20.5: b, b, b

20.3: b   20.6: b, c

Concept Check Questions (follow figure legends)

FIGURE 20.1   Eight hydrogen bonds hold the DNA fragments together.

FIGURE 20.2   In this experiment, the selectable marker selects for the growth of bacteria that have taken up a plasmid; they can grow in the presence of ampicillin.

FIGURE 20.3   The enzyme is called reverse transcriptase because it catalyzes the opposite of transcription. It uses an RNA template to make DNA, whereas during transcription a DNA template is used to make RNA.

FIGURE 20.4   The advantage of making a cDNA library is that the inserted DNA lacks introns. This feature is useful if a researcher wants to focus attention on the coding sequence of a gene or wants to express the gene in cells that do not splice out introns.

FIGURE 20.5   The exonuclease digests DNA strands in the 5′ to 3′ direction and generates DNA molecules with 3′ single-stranded regions. The single-stranded regions are able to hydrogen bond with complementary sites on other DNA molecules.

FIGURE 20.7   The product that predominates after four cycles is the type of DNA fragment containing only the region flanked by the primers. This occurs because this type of fragment is used in each cycle as a template to make another copy of itself.

FIGURE 20.9   The probe must be cleaved to separate the quencher from the reporter.

FIGURE 20.12   They differ in their lengths (shorter DNA segments are closer to the bottom of the gel) and in the colors of their bands on the gel. Each color corresponds to the last dideoxyribonucleotide added to the DNA strand.

FIGURE 20.13   In the bacterial defense system, tracrRNA and crRNA are separate molecules. In gene editing using CRISPR-Cas technology, tracrRNA and crRNA are covalently linked to form a single guide RNA molecule (sgRNA).

FIGURE 20.16   The secondary antibody is labeled or attached to an enzyme, which provides a way to detect the protein of interest.

FIGURE 20.18   The binding of a protein to DNA prevents DNase I from cleaving the DNA in the region where the protein is bound.

End-of-Chapter Questions:

Conceptual Questions

C2. A restriction enzyme recognizes and binds to a specific DNA sequence and then cleaves a (covalent) ester bond in each of two DNA strands.

C4. cDNA is DNA that is made using RNA as the starting material. Unlike eukaryotic genomic DNA, it lacks introns.

Experimental Questions

E2. Remember that A-T base pairs form two hydrogen bonds, whereas G-C base pairs form three hydrogen bonds. The order (from stickiest to least sticky) is as follows:

*Bam*HI = *Sau*3AI = *Pst*I > *Eco*RI > *Nae*I

E4. In conventional gene cloning, many copies are made because the vector replicates to a high copy number within the cell, and the cells divide to produce many more cells. In PCR, the replication of the DNA to produce many copies is facilitated by primers, deoxyribonucleoside triphosphates (dNTPs), and *Taq* polymerase.

E6. A recombinant vector is a vector that has a piece of foreign DNA inserted into it. The foreign DNA came from somewhere else, such as the chromosomal DNA of some organism. To construct a recombinant vector, the vector and the source of foreign DNA are digested with the same restriction enzyme. The complementary ends of the fragments are allowed to hydrogen bond to each other (i.e., sticky ends are allowed to bind), and then DNA ligase is added to create covalent bonds. In some cases, a piece of the foreign DNA will become ligated to the vector, thereby creating a recombinant vector. In other cases, the two ends of the vector ligate back together, restoring the vector to its original structure.

As shown in Figure 20.2, the insertion of foreign DNA can be detected using X-Gal. The insertion of the foreign DNA causes the inactivation of the *lacZ* gene. The *lacZ* gene codes the enzyme β-galactosidase, which converts the colorless compound X-Gal to a blue compound. If the *lacZ* gene is inactivated by the insertion of foreign DNA, the enzyme will not be produced, and the bacterial colonies will be white. If the vector has simply recircularized and the *lacZ* gene remains intact, the enzyme will be produced, and the colonies will be blue.

E8. It is necessary to use cDNA so that the gene will not carry any introns. Bacterial cells do not contain spliceosomes (which are described in Chapter 12). To express a eukaryotic protein in bacteria, a researcher clones cDNA into the bacteria, because the cDNA does not contain introns.

E10. (1) You can control the directionality of the hydrogen-bonding process. (2) You can control the order in which DNA fragments will be connected to each other. (3) You do not need to isolate and purify chromosomal DNA fragments, because the DNA fragments are generated via PCR.

E12. Initially, the mRNA would be mixed with reverse transcriptase and nucleotides to create a complementary strand of DNA. Reverse transcriptase also needs a primer, which could be a primer that is known to be complementary to the β-globin mRNA. Alternatively, since mature mRNAs have a polyA tail, the primer could consist of many Ts, called a poly-dT primer. After the complementary DNA strand had been made, the sample is then mixed with primers, *Taq* polymerase, and nucleotides and subjected to the standard PCR protocol. (Note: The PCR reaction would have two kinds of primers. One primer would be complementary to the 5′ end of the mRNA and would be unique to the β-globin sequence. The other primer would be complementary to the 3′ end. This second primer could be a poly-dT primer, or it could be a unique primer that would bind slightly upstream from the polyA-tail region.)

E14. The exponential phase of qPCR is analyzed because it is during this phase that the amount of PCR product is proportional to the amount of the original DNA in the sample.

E16. AGGTCGGTTGCCATCGCAATAATTTCTGCCTGAACCCAATA

E18. There are many different mutations you could make. For example, you could mutate every other base and see what happens. It would be best to introduce very nonconservative mutations such as switching a purine for a pyrimidine or a pyrimidine for a purine. If the mutation prevents protein binding in an electrophoretic mobility shift assay, then the mutation is probably within the response element. If the mutation has no effect on protein binding, it probably is outside the response element.

E20. Binding occurs due to the hydrogen bonding of complementary sequences. Due to the chemical properties of DNA and RNA strands, they form double-stranded regions when the base sequences are complementary.

E22. The purpose of a Northern blotting experiment is to identify a specific RNA within a mixture of many RNA molecules, using a fragment of cloned DNA as a probe. Such an experiment can tell you if a gene is transcribed in a particular cell or at a particular stage of development. It can also tell you if a pre-mRNA is alternatively spliced into two or more mRNAs of different sizes.

E24. It appears that this mRNA is alternatively spliced to create one product with a higher molecular mass and another with a lower molecular mass. Nerve cells produce a large amount of the larger mRNA, whereas spleen cells produce a moderate amount of the smaller mRNA. Both types are produced in small amounts by the muscle cells. It appears that kidney cells do not transcribe this gene.

E26. Western blotting

E28. The products of protein-coding genes are polypeptides with particular amino acid sequences. Antibodies can specifically recognize polypeptides due to their amino acid sequence. Therefore, an antibody can detect whether or not a cell is making a particular type of polypeptide, which is a component of a protein.

E30. The rationale underlying the electrophoretic mobility shift assay is that a segment of DNA with a protein bound to it will migrate more slowly through a gel than will the same DNA without any bound protein. A shift in a DNA band to a higher molecular mass provides a way to identify DNA-binding proteins.

E32. TFIIA/TFIID can bind to this DNA fragment by themselves, as shown by the band in lane 2. However, TFIIB and RNA polymerase II cannot bind to the DNA by themselves (see lanes 3 and 4). As seen in lane 5, TFIIB can bind if TFIIA/TFIID are also present, because the band is higher than TFIIA/TFIID alone (compare lanes 2 and 5). In contrast, RNA polymerase II cannot bind to the DNA when only TFIIA/TFIID are present. The band in lane 6 is at the same location in the gel as that in lane 2, indicating that only TFIIA/TFIID are bound. Finally, in lane 7, when all components are present, the band is higher than when both TFIIA/TFIID and TFIIB are present (compare lanes 5 and 7). This result means that all four protein complexes are bound to the DNA. Taken together, the results indicate that TFIIA/TFIID can bind by themselves, TFIIB needs TFIIA/TFIID to bind, and RNA polymerase II needs all three proteins to bind to the DNA.

E34. In lane 2, the gel does not contain any bands for fragments with lengths from about 350 bp to 175 bp. Thus, that region of the 400-bp-long DNA segment is being covered up; the "footprint" is about 175 bp long.

Questions for Student Discussion/Collaboration

2. 1. Does a particular amino acid within a protein sequence play a critical role in the protein's structure or function?

2. Does a DNA sequence function as a promoter?

3. Does a DNA sequence function as a regulatory site?

4. Does a DNA sequence function as a splicing junction?

5. Is a sequence important for correct translation?

6. Is a sequence important for RNA stability? And many others . . .

CHAPTER 21

Answers to Comprehension Questions

21.1: d, b, c   21.4: c, d

21.2: c, c, d   21.5: b, d

21.3: a, b

Concept Check Questions (follow figure legends)

FIGURE 21.1   Cyanogen bromide (CNBr) separates β-galactosidase from the A or B chain of human insulin.

FIGURE 21.4   In a gene knockout, the function of a gene is eliminated. For diploid organisms, both copies are inactivated. In a gene knockin (in mice), a gene is added to a noncritical site in the genome.

FIGURE 21.6   The β-lactoglobulin promoter is used because it is expressed in mammary cells.

FIGURE 21.7   The nucleus of the oocyte is removed so that the resulting organism contains (nuclear) genetic material only from the somatic cell.

FIGURE 21.8   No. Carbon Copy did not receive any genetic material from a different species.

FIGURE 21.9   When stem cells divide, they produce one cell that remains a stem cell and another cell that differentiates. This pattern maintains a population of stem cells.

FIGURE 21.11   Hematopoietic stem cells are multipotent.

FIGURE 21.15   Only the T DNA within the T-DNA vector is transferred to a plant.

End-of-Chapter Questions:

Conceptual Questions

C2.   *Agrobacterium radiobacter* synthesizes an antibiotic that kills *Agrobacterium tumefaciens*. The genes that are necessary for antibiotic biosynthesis and resistance are plasmid-coded and can be transferred during interspecies conjugation. If *A. tumefaciens* received this plasmid during conjugation, it would be resistant to the antibiotic. Therefore, use of a conjugation-deficient strain of *A. radiobacter* prevents the occurrence of resistant *A. tumefaciens*.

C4.   A biological control agent is an organism that prevents the harmful effects of some other agent in the environment. Examples include *Bacillus thuringiensis*, a bacterium that synthesizes compounds that act as toxins to kill insects, and *A. radiobacter,* which is used to prevent crown gall disease caused by *A. tumefaciens.*

C6.   1. Whole-pathogen vaccines consist of entire pathogens that have been completely inactivated or weakened so that they cannot cause serious disease symptoms. These include inactivated vaccines to prevent influenza, hepatitis A, and rabies and attenuated vaccines to prevent measles, mumps, and rubella (MMR combined vaccine), chickenpox, and yellow fever.

2. Viral vector vaccines use a modified version of a virus that is different from the virus that the vaccine is directed against. They are used to prevent Ebola and COVID-19.

3. Subunit vaccines contain only certain components, or antigens, that best stimulate the immune system. They are used to prevent COVID-19 and bacterial infections, such as those caused by *Streptococcus pneumoniae.*

4. Nucleic acid vaccines introduce genetic material coding the protein antigen or antigens against which an immune response is sought. The body's own cells then use this genetic material to produce the antigen(s). They are used to prevent West Nile virus infection in horses and COVID-19 in people.

C8.   A mouse model is a strain of mice that carries a mutation in a gene that is analogous to a mutation in a human gene that causes a disease or it carries a human gene that is mutant. These mice can be used to study the disease and to test potential therapeutic agents.

C10.   The T DNA gets transferred to the plant cell; it is then incorporated into the plant cell's genome.

C12.   A. With regard to maternal effect genes, the phenotype is determined by the animal that donated the egg. It is the cytoplasm of the egg that accumulates the gene products of maternal effect genes.

B. The extranuclear traits depend on the mitochondrial genome. Mitochondria are found in the egg and in the somatic cell. So, theoretically, both animals could contribute extranuclear traits.

C. The cloned animal would be genetically identical to the animal that donated the nucleus with regard to traits that are determined by nuclear genes, which are expressed during the lifetime of the organism. The cloned animal could differ from the animal that donated the nucleus with regard to traits that are determined by maternal effect genes and mitochondrial genes. Thus, the animal is not a true clone, but it is likely that it would greatly resemble the animal that donated the nucleus, because the vast majority of genes are found in the cell nucleus.

C14.   Some people are concerned about the release of genetically engineered microorganisms into the environment. The fear is that such organisms may continue to proliferate and it may not be possible to stop them. A second concern involves the use of genetically engineered organisms as food. Some people are worried that genetically engineered organisms may pose an unknown health risk. A third issue is ethics. Some people feel that it is morally wrong to tamper with the genetics of organisms. This objection may also apply to genetic techniques such as cloning, stem cell research, and gene therapy.

Experimental Questions

E2.   One possibility is to clone the toxin-producing genes from *B. thuringiensis* and introduce them into *P. syringae*. This altered bacterial strain would have the advantage of not needing repeated applications. However, it would be a recombinant strain and might be viewed in a negative light by people who are hesitant to use recombinant organisms in the field. By comparison, *B. thuringiensis* is a naturally occurring species.

E4.   A kanamycin-resistance gene is contained within the T DNA. Exposure to kanamycin selects for the growth of plant cells that have incorporated the T DNA into their genome. The carbenicillin kills *A. tumefaciens.* The plant growth hormones promote the regeneration of an entire plant from somatic cells. If kanamycin was left out, it would not be possible to select for the growth of cells that had taken up the T DNA.

E6.   A gene knockout is the elimination of the function of a particular gene. For autosomal genes in animals and plants, a gene knockout produces a homozygote with a defect in both copies of the gene. If a gene knockout has no phenotypic effect, the gene may be redundant. In other words, there may be multiple genes within the genome that can carry out the same function. Another reason why a gene knockout may not have a phenotypic effect is because of the environment. As an example, let's say a mouse gene is required for the synthesis of a vitamin. If the researchers were providing food that contained the vitamin, the knockout mouse that was lacking this gene would have a normal phenotype; it would survive just fine. Sometimes, researchers have trouble identifying the effects of a gene knockout until they modify the environmental conditions in which the animals are raised.

E8.   A. Dolly's chromosomes may have seemed older because they were already old when they were in the nucleus that was incorporated into the enucleated egg. They had already become significantly shortened in the mammary cells. This shortening was not repaired by the egg.

B. The age of Molly's female parent does not matter. Remember that shortening does not occur in germ cells. However, the eggs from Molly's female parent are older than they should be by about 6 or 7 years, because the egg that gave rise to Molly's female parent received its diploid set of chromosomes from a sheep that was 6 years old, and the cells were grown in culture for a few doublings before a mammary cell was fused with an enucleated egg. Therefore, the calculation for Molly's chromosomes is as follows: 6 or 7 years (the age of the mammary cells that produced the female parent's germ-line cells) plus 8 years (the age of Molly), which equals 14 or 15 years. However, only half of Molly's chromosomes would appear to be 14 or 15 years old. The other half of Molly's chromosomes, which were inherited from the male parent, would appear to be 8 years old.

C. Chromosome shortening suggests that aging has occurred in the somatic cell, and this aging has been passed to the cloned organism. If cloning was done over the course of many generations, this might eventually have a major impact on the life span of the cloned organism. It may die much earlier than if it was not a clone. However, chromosome shortening may not always occur. It did not seem to occur in mice that were cloned for six consecutive generations.

E10. Reproductive cloning means the cloning of entire multicellular organisms. In plants, this is easy. Most species of plants can be cloned by asexual cuttings. In animals, cloning occurs naturally, as in identical twins. Identical twins are genetic replicas of each other because they begin from the same fertilized egg. (Note: There could be some somatic mutations that occur in identical twins that would make them slightly different.) Recently, as in the case of Dolly, reproductive cloning has become possible by fusing somatic cells with enucleated eggs. The advantage, from an agricultural point of view, is that reproductive cloning could allow farmers to choose the best animal in a herd and make many clones from it. Breeding would no longer be necessary. Also, breeding may be less reliable because the offspring inherit traits from both parents.

Questions for Student Discussion/Collaboration

2. This is a matter of opinion. Examples include general traits such as size, productivity, flavor, disease resistance, pesticide resistance, and many more. Several techniques, such as gene editing and gene addition, which are discussed in this chapter and Chapter 20, could be used to produce transgenic plants and animals.

CHAPTER 22

Answers to Comprehension Questions

22.1: a   22.5: b, d, b
22.2: d   22.6: a, b, b
22.3: c, d, a 22.7: c
22.4: b

Concept Check Questions (follow figure legends)

FIGURE 22.1 A genetic map is a diagram that shows the order and relative distances between genes or other DNA segments on a chromosome.

FIGURE 22.2 The probe binds to a specific site because that site has a DNA sequence that is complementary to the sequence of the probe.

FIGURE 22.5 Microsatellites for which the number of the repeat sequence varies are polymorphic.

FIGURE 22.7 Chromosome walking is used to identify a gene that may exist as a disease-causing allele. The disease-causing allele is already known to be close to another gene or a molecular marker.

FIGURE 22.9 The sequence is determined in real time as the DNA is being synthesized.

FIGURE 22.11 A metagenomics study produces a collection of DNA sequences from an environmental sample.

End-of-Chapter Questions:

Conceptual Questions

C2. A. Yes

B. No; this is only one chromosome in the human genome.

C. Yes

D. Yes

Experimental Questions

E2. They are complementary to each other.

E4. Because normal cells contain two copies of chromosome 14, you would expect the probe to bind to complementary DNA sequences on both of these chromosomes. If the probe recognizes only one of two chromosomes, this means that one of the copies of chromosome 14 has been lost, or it has suffered a deletion in the region where the probe binds. The loss of this genetic material may be related to the uncontrollable cell growth of the malignant tumor.

E6. After the chromosomes have been fixed to a slide, it is possible to add two or more different probes that recognize different sequences (i.e., different sites) within the genome. Each probe has a different fluorescent molecule bound to it. Usually, a researcher will use computer imagery that recognizes the wavelength of the fluorescence emitted by each probe and then assigns that wavelength a bright color. The color seen by the researcher is not the actual color emitted by the probe; it is a secondary color assigned by the computer. In a sense, the probes, with the aid of a computer, are "painting" the regions of the chromosomes that they recognize. An example of chromosome painting is shown in Figure 22.3. In this example, human chromosome 5 is painted with six different colors.

E8. The first step involves the determination of the base sequences of many DNA segments that are derived from chromosomal (genomic) DNA. This can be achieved via short-read sequencing (SRS) or long-read sequencing (LRS). The second step is de novo genome assembly in which overlapping regions allow researchers to align the DNA sequences into a complete DNA sequence as it would occur along a chromosome.

E10. The aim of both techniques is to identify disease-causing alleles in humans by starting at a nearby molecular marker and walking toward the gene of interest. In chromosome walking, the steps involve the insertion of DNA segments into cloning vectors, the subcloning of smaller segments, and use of the subclones to screen a genomic library. By contrast, primer walking does not involve any cloning or library screening steps. Primer walking only involves DNA sequencing. The sequencing begins at a primer that binds to the site of the molecular marker. Each sequencing run is used to generate the next primer, which is closer to the gene of interest. Note: Both methods involve the comparison of DNA sequences between affected and unaffected individuals.

E12. Shotgun sequencing is a genome-sequencing strategy that involves breaking the genome into a collection of many DNA fragments that are sequenced individually.

E14. A. One homolog contains the STS-1 that is 289 bp and the STS-2 that is 422 bp, whereas the other homolog contains the STS-1 that is 211 bp and the STS-2 that is 115 bp. This conclusion is based on the observation that 28 of the sperm have either the 289-bp and 422-bp bands or the 211-bp and 115-bp bands.

B. There are two recombinant sperm; see lanes 12 and 18. Because there are two recombinant sperm out of a total of 30,

$$\text{Map distance} = \frac{2}{30} \times 100$$
$$= 6.7\,\text{mu}$$

C. In theory, this approach could be used. However, there is not enough DNA in one sperm to carry out an RFLP analysis unless the DNA is amplified by PCR.

E16. A. The general strategy is shown in Figure 22.7a. You begin at a certain location on a chromosome and then walk toward the gene of interest. You begin with a clone that has a marker that is known to map relatively close to the gene of interest. A piece of DNA at the end of the insert is subcloned and then hybridized to an adjacent clone in a genomic library. This is the first step. The end of this clone is subcloned to make the next step, and so on. Eventually, after many steps, you will arrive at the gene of interest.

B. In this example, you would begin at STS-3. If you walked a few steps and arrived at STS-2, you would know that you were walking in the wrong direction.

C. This is a challenging aspect of chromosome walking. Basically, you would walk toward gene *X* using DNA from a normal individual and DNA from an individual with a mutant gene *X*. When you found a site where the sequences differed between the normal and mutant individual, you might have found gene *X*. You would eventually have to confirm this by analyzing the DNA sequence of this region and confirming that it codes a functional gene.

E18. You can use the following equation to calculate the probability that a base will not be sequenced when shotgun sequencing is used:

$$P = e^{-m}$$

where

P is the probability that a base will be left unsequenced;

e is the base of the natural logarithm, $e = 2.72$;

m is the number of bases sequenced, divided by the total genome size.

In this case, *m* is equal to 19 divided by 4.4, which equals 4.3. Thus,

$$P = e^{-m} = e^{-4.3} = 0.0136 = 1.36\%$$

This means that if a researcher randomly sequences 19 Mb, the researcher is likely to miss only 1.36% of the genome. With a genome size of 4.4 Mb, about 59,840 base pairs out of approximately 4,400,000 will be unsequenced.

E20. Sequence by synthesis is a next-generation sequencing technology in which the base sequence is determined as the DNA strand is being made. Illumina sequencing is an example.

Questions for Student Discussion/Collaboration

2. This is a matter of opinion. Many people would say that the ability to identify many human genetic diseases is the most important goal. In addition, the Human Genome Project will provide a better understanding of how humans are constructed at the molecular level. As we gain a greater understanding of our genetic makeup, some people are worried that this may lead to greater discrimination. Insurance companies or health care providers could refuse to cover people who are known to carry genetic abnormalities. Similarly, employers could make their employment decisions based on the genetic makeup of potential employees rather than their past accomplishments. At the family level, genetic information may affect how people choose mates, and whether or not they decide to have children.

CHAPTER 23

Answers to Comprehension Questions

23.1: c, c, a   23.4: d

23.2: d, d, a, d   23.5: d, c

23.3: a, b

Concept Check Questions (follow figure legends)

FIGURE 23.1   A key point is that mRNA is made only when a gene is expressed. In this experiment, mRNA is first isolated and then used to make cDNA, which is fluorescently labeled. The fluorescent spots indicate which genes have been transcribed to produce mRNA.

FIGURE 23.3   Each of these mechanisms alters the structure and possibly the function of a protein, thereby increasing protein diversity.

FIGURE 23.5   The purpose of tandem mass spectrometry is to determine the amino acid sequences of peptides and identify a protein.

FIGURE 23.7   The first two reading frames contain many stop codons, which would prevent the translation of a full-length protein. Only the third reading frame has a long stretch without any stop codons.

FIGURE 23.8   These homologous genes have similar sequences because they were derived from the same ancestral gene. They are not identical because they have accumulated random mutations after the two species of bacteria diverged from each other.

FIGURE 23.9   If functionally important sites were changed due to mutations, such mutations would be likely to inhibit the function of the protein. This could have a detrimental effect on survival or reproduction. If so, the mutation would be eliminated via natural selection. Therefore, the original (functional) site remains the same—that is, conserved—during evolution.

End-of-Chapter Questions:

Conceptual Questions

C2. There are two main reasons why the proteome is larger than the genome. The first reason involves the processing of pre-mRNAs, a phenomenon that occurs primarily in eukaryotic species. RNA splicing and editing can alter the codon sequence of the mRNAs that are made and thereby produce alternative forms of proteins that have different amino acid sequences. The second reason for the larger size of the proteome is posttranslational modification. There are many ways in which a given protein's structure can be covalently modified by cellular enzymes, including proteolytic processing, disulfide bond formation, glycosylation, attachment of a lipid, phosphorylation, methylation, and acetylation, to name a few.

C4. Other types of short sequence elements include centromere sequences, origins of replication, telomeric sequences, repetitive sequences, and enhancers. (There are other examples.)

C6. There are a few interesting features. Sequences 1 and 2 are similar to each other, as are sequences 3 and 4. There are a few places where individual amino acids are conserved among all five sequences. These amino acids may be particularly important with regard to function.

C8. A gap is needed when two homologous sequences are not the same length. Because homologous sequences are derived from the same ancestral gene, they were originally the same length. However, during evolution, the sequences can incur deletions and/or additions that make them shorter or longer than the sequence in the original ancestral gene. If one gene incurs a deletion, a gap will need to be inserted in this gene's sequence in order to align it with a homologous gene. If an addition occurs in a gene's sequence, a gap will be needed in the homologous gene sequence in order to align the two sequences.

Experimental Questions

E2. The type of molecule that is sequenced is cDNA.

E4. In tandem mass spectroscopy, the first spectrometry step determines the mass of a peptide fragment from a protein of interest. The second spectrometry step determines the masses of progressively smaller fragments derived from that peptide. Because the masses of all the amino acids are known, the molecular masses of these smaller fragments reveal the amino acid sequence of the peptide. With peptide sequence information, it is possible to use the genetic code and produce DNA sequences that could code such a peptide. More than one sequence is possible, due to the degeneracy of the genetic code. These sequences are used as query sequences to search a genomic database, which will (hopefully) locate a match. The genomic sequence can then be analyzed to determine the entire coding sequence for the protein of interest.

E6. One strategy is search by signal, which relies on known sequences such as promoters, start and stop codons, and splice sites to help predict whether or not a DNA sequence contains a protein-coding gene. This approach attempts to identify a region that contains a promoter sequence, then a start codon, a coding sequence, and a stop codon. A second strategy is search by content. The goal is to identify sequences whose nucleotide content differs significantly from a random distribution, because such a difference is usually due to codon bias. This approach attempts to locate coding regions by identifying regions where the nucleotide content displays a bias. A third approach for locating protein-coding genes is to search for long open reading frames within a DNA sequence. An open reading frame is a sequence that does not contain any stop codons.

E8. By searching a database, you can identify genetic sequences that are homologous to a newly determined sequence. In most cases, homologous sequences carry out identical or very similar functions. Therefore, if a database search locates a homologous sequence whose function is already understood, this provides an important clue regarding the function of the newly determined sequence.

E10. The advantages of running the computer program are speed and accuracy. Once the program has been created, and a file has been entered into a computer, the program can analyze long genetic sequences quickly and accurately.

E12. A sequence element is a specialized base sequence or amino acid sequence with a particular meaning or function. Two examples are a stop codon (e.g., UAA), which is a base sequence element, and an amino acid sequence that is a site for protein glycosylation (e.g., asparagine–any amino acid–serine [or threonine]), which is an amino acid sequence element, or motif. The computer program does not create these sequence elements. The program is given information about sequence elements, which comes from genetic research. Scientists have conducted experiments to identify the sequence of bases that constitute a stop codon and the sequence of amino acids where proteins are glycosylated. Once this information is known from research, it can be incorporated into computer programs, and then the programs can analyze new genetic sequences and identify the occurrence of stop codons and glycosylation sites.

E14. A. Because most of the globins contain a histidine at position 119, that is likely to be the amino acid coded by the ancestral codon. The histidine codon mutated into an arginine codon after the occurrence of the gene duplication that produced the ζ-globin gene. This would be after the emergence of primates or within the last 10 or 20 million years.

B. You do not know if the ancestral globin gene coded a glycine or a proline at codon 121. The mutation probably occurred after the duplication that produced the α-globin and β-globin families, but before the gene duplications that gave rise to the multiple copies of α- and β-globin genes on chromosome 16 and chromosome 11, respectively. Therefore, it occurred between 300 mya and 200 mya.

C. All of the β globins contain glutamic acid at position 103, and all of the α globins contain valine at that position, except for θ globin. You do not know if the ancestral globin gene coded valine or glutamic acid at codon 121. Nevertheless, a mutation, converting one to the other, probably occurred after the duplication that produced the α-globin and β-globin gene families, but before the gene duplications that gave rise to the multiple copies of α- and β-globin genes on chromosome 16 and chromosome 11, respectively. Therefore, it occurred between 300 mya and 200 mya. The mutation that produced the alanine codon in the θ-globin gene probably occurred after the gene duplication that produced this gene. This would be after the emergence of mammals (i.e., sometime within the last 200 million years).

E16. As noted in the answer to question 3 in More Genetic TIPS, a serine codon was likely to be the ancestral codon. From Table 13.1, you can see that an AGU or AGC codon for serine (Ser) could change into a codon for asparagine (Asn), threonine (Thr), or isoleucine (Ile) by a single base change. In contrast, UCU, UCC, UCA, and UCG codons, which also code for serine, could not change into codons for asparagine or isoleucine by a single base change. Therefore, the two likely scenarios are as shown next. The

mutated base is underlined. (The mutations would actually occur in the DNA, although the sequences of the RNA codons are shown here.)

Ancestral codon

A<u>C</u>U (Thr) ← AGU (Ser) → A<u>A</u>U (Asn)

↓

A<u>U</u>U (Ile)

Ancestral codon

A<u>C</u>C (Thr) ← AGC (Ser) → A<u>A</u>C (Asn)

↓

A<u>U</u>C (Ile)

Questions for Student Discussion/Collaboration

2. You need to use a computer to answer this question. You will need to be able to access programs that can translate DNA sequences and search databases.

CHAPTER 24

Answers to Comprehension Questions

24.1: b, a, b, b   24.4: c

24.2: c, c   24.5: d, c

24.3: a, c   24.6: d

Concept Check Questions (follow figure legends)

FIGURE 24.2 The key feature is that both parents of affected offspring are unaffected by the disease. The parents are heterozygous carriers.

FIGURE 24.3 The key feature is that affected offspring have an affected parent.

FIGURE 24.4 The key feature is that all affected individuals are males. Furthermore, these males all have female parents who were descendants of Queen Victoria.

FIGURE 24.5 A haplotype is a linkage of alleles or molecular markers on a single chromosome.

FIGURE 24.7 The original mutation in the *HTT* gene that resulted in the allele that causes Huntington disease occurred in a germ-line cell or gamete of the founder and that allele was closely linked to the G8-C marker. Therefore, individuals in this pedigree that carry the G8-C marker are also likely to carry the disease-causing allele.

FIGURE 24.10 The PrP^{Sc} protein may come from eating infected animal products or may be due to a genetic predisposition that occasionally causes the PrP^{C} protein to convert to the PrP^{Sc} protein.

End-of-Chapter Questions:

Conceptual Questions

C2. When a disease-causing allele affects a trait, it is causing a deviation from normality, but the gene involved is not usually the only gene that governs the trait. For example, an allele causing hemophilia prevents the normal blood-clotting pathway from operating correctly. Inheritance of hemophilia follows a simple Mendelian pattern because a single gene affects the phenotype. Even so, it is known that normal blood clotting is due to the actions of many genes.

C4. Changes in chromosome number and unbalanced changes in chromosome structure tend to affect phenotype because they create an imbalance of gene expression. For example, individuals with Down syndrome have three copies of chromosome 21 and, therefore, three copies of all the genes on chromosome 21. This leads to a relative overexpression of genes that are located on chromosome 21 compared with the genes on other chromosomes. Reciprocal translocations and inversions often lack phenotypic consequences because the total amount of genetic material is not altered, and the level of gene expression is not significantly changed.

C6. There are many possible answers; here are a few. There are diabetic people and mice. There are forms of inherited obesity in people and mice. Hip dysplasia is found in people and dogs.

C8. A. Because a person must inherit two defective copies of the gene that codes β-glucosidase and the gene is known to be on chromosome 1, inheritance of Gaucher disease follows an autosomal recessive pattern. Both members of the couple must be heterozygous, because they each have one affected parent (who had to transmit the mutant allele to them) and their phenotypes are unaffected (so they must have received the normal allele from their other parent). Because both members of the couple are heterozygotes, there is a ¼ chance that they will produce an affected child (a homozygote) with Gaucher disease. If you let *G* represent the nonmutant allele and *g* the mutant allele:

	♂ *G*	*g*
♀ *G*	*GG* Normal	*Gg* Normal
g	*Gg* Normal	*gg* Gaucher

B. From the Punnett square in part A, you can see that there is also a ¼ chance that the child will be a homozygote with two normal copies of the gene.

C. You need to apply the binomial expansion equation to answer this question (see Chapter 3 for an explanation of the equation). In this case, $n = 5$, $x = 1$, $p = 0.25$, $q = 0.75$. The answer is 0.396, or 39.6%.

C10. The pattern of inheritance is autosomal recessive. All of the affected individuals do not have affected parents. Also, the disorder is found in both males and females. If the pattern of inheritance was X-linked recessive, individual III-1 would have to have an affected male parent, but III-1's male parent (II-2) is unaffected.

C12. The 13 babies with unaffected parents have acquired a new mutation. In other words, during spermatogenesis or oogenesis or after the egg was fertilized, a new mutation occurred in the fibroblast growth factor gene. These 13 individuals have the same chance of passing the mutant allele to their offspring as the 18 individuals who inherited the mutant allele from a parent. The probability is 50%.

C14. Because this disease is inherited as a dominant trait, the female parent must have two normal copies of the gene, and the male parent (who is affected) is most likely to be a heterozygote. (Note: The male parent could be a homozygote, but this is extremely unlikely because the dominant allele is very rare.) If you let *M* represent the mutant Marfan allele and *m* the normal allele, you can construct the following Punnett square:

	♂ *M*	*m*
♀ *m*	*Mm* Marfan	*mm* Normal
m	*Mm* Marfan	*mm* Normal

B. There is a 50% chance that the child will have the disease.

C. You need to use the product rule. The odds of having an unaffected child are 50%. If you multiply $0.5 \times 0.5 \times 0.5$, this equals 0.125, or a 12.5% chance of having three unaffected offspring.

C16. 1. The disease-causing allele had its origin in a single individual known as the founder, who lived many generations ago. Since that time, the allele has spread throughout portions of the human population.

2. When the disease-causing allele originated in the founder, it occurred in a region on a chromosome with a particular haplotype. The haplotype is not likely to have changed over the course of several generations if the disease-causing allele and markers in this region are very close together.

C18. A prion is a protein that behaves like an infectious agent. The infectious form of the prion protein has an abnormal conformation. This abnormal conformation is represented as PrP^{Sc}, and the normal conformation as PrP^{C}. An individual can be "infected" with the abnormal conformation of the protein by eating products from an infected animal, or the prion protein may convert spontaneously to the abnormal conformation. A prion protein in the PrP^{Sc} conformation can bind to a prion protein in the PrP^{C} conformation and convert it to the PrP^{Sc} form. An accumulation of prions in the PrP^{Sc} form is what causes the disease symptoms.

Experimental Questions

E2. Perhaps the least convincing is the higher incidence of the disease in particular populations. Because populations living in specific geographic locations are exposed to unique environments, it is difficult to distinguish genetic from environmental causes for a particular disease. The most convincing evidence might be the higher incidence of the disease in related individuals or the ability to correlate the disease with the presence of a mutant gene. Overall, however, the conclusion that a disease has a genetic component should be based on as many observations as possible.

E4. You would probably conclude that the disease is unlikely to have a genetic component. If it was rooted primarily in genetics, it would be likely to be observed in the Central American population. Of course, there is a chance that very few or none of the people who migrated to Central America were carriers of the mutant gene, which is somewhat unlikely for a large migrating population. By comparison, you might suspect that an environmental agent present in South America but not present in Central America may underlie the disease. Researchers could try to search for this environmental agent (e.g., a pathogenic organism).

E6. Males I-1, II-1, II-4, II-6, III-3, III-8, and IV-5 have a normal copy of the gene. Males II-3, III-2, and IV-4 are hemizygous for an inactive mutant allele. Females III-4, III-6, IV-1, IV-2, and IV-3 have two normal copies of the gene, whereas females I-2, II-2, II-5, III-1, III-5, and III-7 are heterozygous carriers of a mutant allele.

E8. It is the gene product (i.e., the polypeptide) of an oncogene that causes cancerous cell growth. The antisense RNA from the gene introduced via gene therapy would bind to the mRNA from an oncogene. This would prevent the translation of the mRNA into the polypeptide and thereby prevent cancerous cell growth. Furthermore, the formation of double-stranded RNA would result in RNA interference.

Questions for Student Discussion/Collaboration

2. One advantage of gene therapy is that the introduction of a normal gene into cells harboring a defective gene has the potential to cure symptoms caused by the defective gene. For persons with certain inherited genetic abnormalities, this may be the only option for treating the disease. Disadvantages include the inability to target the appropriate cells and the potential side effects, such as life-threatening immune reactions. The choice of the three top diseases for which to fund research is a matter of opinion, but the decision might be based on factors such as the number of people affected with the disease, the severity of the disease, current treatment options to treat the disease (if any), and the potential of gene therapy to successfully treat the disease.

CHAPTER 25

Answers to Comprehension Questions

25.1: b   25.4: d, e
25.2: d   25.5: d, a
25.3: c, d, d

Concept Check Questions (follow figure legends)

FIGURE 25.3 Such a mutation would keep the Ras protein in an active state and thereby promote cancerous growth. The cell-signaling pathway would be turned on.

FIGURE 25.4 The *bcr* gene is expressed in white blood cells. This translocation causes the *abl* gene to be under the control of the *bcr* promoter. The abnormal expression of *abl* in white blood cells causes leukemia.

FIGURE 25.5 E2F will be active all of the time, which will lead to uncontrolled cell division.

FIGURE 25.6 A checkpoint is a point in the cell cycle at which checkpoint proteins determine if the cell is in the proper condition to divide. If an abnormality such as DNA damage is detected, these proteins will halt the cell cycle.

FIGURE 25.9 An individual with a predisposition for developing familial breast cancer inherits only one copy of the mutant allele. Therefore, the disease shows a dominant pattern of inheritance. However, for the individual to actually get breast cancer, the other allele must become mutant in somatic cells.

FIGURE 25.10 Only tamoxifen blocks the binding of a hormone (estrogen) to a receptor. Dacomitinib does not block hormone binding; it prevents autophosphorylation. Imatinib blocks the binding of ATP, which is not a hormone.

End-of-Chapter Questions:

Conceptual Questions

C2. An oncogene can cause cancer if it is abnormally activated, whereas a tumor-suppressor gene needs to be inactivated to cause cancer. The *ras* and *src* genes are examples of oncogenes, and *rb* and *p53* are tumor-suppressor genes.

C4. A retroviral oncogene is a cancer-causing gene found within the genome of a retrovirus. It is not necessary for viral infection and proliferation. Oncogene-defective viral strains are able to infect cells and multiply normally. It is thought that retroviruses acquire oncogenes via their reproductive cycle. A retrovirus may integrate into cellular DNA next to a proto-oncogene. Later in its reproductive cycle, it may transcribe this proto-oncogene and thereby incorporate it into its viral genome. A high copy number of the virus or additional mutations

may lead to the overexpression of the proto-oncogene and thereby cause cancer.

C6. Conversion of a proto-oncogene to an oncogene can occur via missense mutation, gene amplification, chromosomal translocation, or viral integration. The genetic changes are expected to increase the amount of the coded protein or alter its function in a way that makes it more active.

C8. A. A nonsense mutation would cause gene inactivation, so it would occur in a tumor-suppressor gene.

B. A missense mutation could either enhance or inhibit protein function, so it could occur in either a tumor-suppressor gene or a proto-oncogene.

C. A frameshift mutation would most likely cause gene inactivation, so it would occur in a tumor-suppressor gene.

C10. Loss-of-function mutations in tumor-suppressor genes are more likely to happen first. If the mutations are in tumor-suppressor genes that negatively regulate cell division, they could lead directly to an increase in cell division. Alternatively, such loss-of-function mutations could occur in tumor-suppressor genes that maintain genome integrity. When the functions of the proteins these genes code are lost, it becomes more likely for mutations to occur. These additional mutations could inactivate more tumor-suppressor genes or activate oncogenes. Usually, the activation of oncogenes occurs later in a tumor's progression.

C12. Most inherited forms of cancer are due to loss-of-function alleles in tumor-suppressor genes. An individual who has a higher predisposition to develop such a cancer is a heterozygote. The inheritance of a single (loss-of-function) allele by an offspring causes this higher predisposition. Therefore, the cancer behaves like a dominant disorder at the level of a family pedigree. At the cellular level, the other normal allele must incur a loss-of-function mutation for cell division to be affected. In other words, at the cellular level, the gene is homozygous for a loss-of-function allele when it promotes cancer.

C14. A. No, because E2F is inhibited

B. Yes, because E2F is not inhibited

C. Yes, because E2F is not inhibited

D. No, because no E2F is produced

C16. A. True

B. True

C. False, most cancer cells are caused by mutations that result from environmental mutagens.

D. True

C18. Though they don't change the DNA sequence, epigenetic modifications can affect gene expression. Such changes could increase gene expression and thereby result in oncogenes or they could inhibit the expression of tumor-suppressor genes. Either type of change could contribute to cancer. For example, DNA methylation of a tumor-suppressor gene could promote cancer.

C20. The more established methods for cancer treatment are surgery and the use of radiation therapy and chemotherapy to kill actively dividing cells. Newer methods, including targeted drug therapy, immunotherapy, and hormone therapy, usually rely on a greater understanding of cancer cells at the molecular and cellular levels.

C22. The term *anti-miRNA oligonucleotide* is a broad term that refers to an oligonucleotide that inhibits miRNA function. Two types are locked nucleic acids and antagomirs. Locked nucleic acids contain a ribose sugar that has an extra bridge connecting the 2′ oxygen and the 4′ carbon. The bridge locks the ribose in a conformation that causes it to bind more tightly to the complementary miRNAs. Antagomirs have one or more base modifications that may promote stronger binding to complementary miRNAs.

Experimental Questions

E2. One possible category of drugs would be GDP analogues (i.e., compounds whose structures resemble that of GDP). Perhaps you could find a GDP analogue that binds to the Ras protein and locks it in the inactive conformation.

One way to test the efficacy of such a drug would be to incubate the drug with a type of cancer cell that is known to have an overactive Ras protein, and then plate the cells on a solid medium. If the drug locked the Ras protein in the inactive conformation, it should inhibit malignant growth of the cells.

There are possible side effects of such drugs. First, they might block the growth of normal cells, because Ras protein plays a role in normal cell proliferation. Second (as you know if you have taken a cell biology course), there are many GTP/GDP-binding proteins in cells, and the drugs could somehow inhibit cell growth and function by interacting with these proteins.

E4. You would want to target proteins that carry out epigenetic changes, such as DNA methyltransferase, enzymes that covalently modify histone proteins, chromatin remodeling complexes, and perhaps trithorax and polycomb group complexes. The possible side effects are somewhat difficult to predict but the general problem is that the drugs may cause genes in normal cells to be turned on or turned off when they should not be. Such effects could result in many different types of cellular dysfunction.

Questions for Student Discussion/Collaboration

2. There isn't a clearly correct answer to this question, but it should stimulate a great deal of discussion.

CHAPTER 26

Answers to Comprehension Questions

26.1: e, d, c   26.4: b, d

26.2: a, d, c, b   26.5: b, b

26.3: b, b, c

Concept Check Questions (follow figure legends)

FIGURE 26.2 The mechanisms in parts (a) and (b) involve diffusible morphogens.

FIGURE 26.5 Genes code proteins that control the changes that cause development to happen. The process is largely a hierarchy in which certain sets of genes control other sets of genes.

FIGURE 26.7 The anteroposterior axis runs from head to tail (abdomen in fruit flies). The dorsoventral axis is oriented from back (spine in vertebrates) to front.

FIGURE 26.8 The Bicoid protein controls the development of anterior structures in the fruit fly.

FIGURE 26.9 Maternal effect gene products are made in nurse cells and then transported into the oocyte.

FIGURE 26.11 A mutation in a gap gene causes several adjacent segments to be missing, whereas a mutation in a pair-rule gene causes regions in alternating segments to be missing.

FIGURE 26.13 The arrangement of homeotic genes on the chromosome correlates with their pattern of expression along the anteroposterior axis of the body.

FIGURE 26.16 A cell lineage is a series of cells that are derived from a particular cell via cell division.

FIGURE 26.18 An ortholog is a gene that is homologous to another gene found in a different species.

FIGURE 26.20 The expression of the *HoxC-6* gene appears to determine the boundary between the neck and thoracic region in vertebrates.

FIGURE 26.21 The Id protein functions during early stages of embryonic development. It prevents myogenic bHLH proteins from promoting muscle cell differentiation too soon.

FIGURE 26.24 Maintaining the correct number of stem cells in the growing tip is necessary to promote proper growth and development.

FIGURE 26.26 The arrangement of the whorls would be carpel-stamen-stamen-carpel.

End-of-Chapter Questions:

Conceptual Questions

C2. Each embryonic cell expresses a distinct set of cell adhesion molecules (CAMs), which provide positional information to neighboring cells. The unique combinations of CAMs allow cells to self-sort during embryonic development. The CAMs are thought to determine which cells prefer to stay connected.

C4. A. False; the head is anterior to the tail.

B. True

C. False; the feet are ventral to the hips.

D. True

C6. A. True

B. False; such gradients are also established after fertilization during embryonic development.

C. True

C8. A. The mutant larva has a mutation in a pair-rule gene (*runt*).

B. The mutant larva has a mutation in a gap gene (*knirps*).

C. The mutant larva has a mutation in a segment-polarity gene (*patched*).

C10. Positional information consists of signals in the form of morphogens or cell adhesion molecules that provide a cell with information regarding its position relative to other cells. In *Drosophila*, the formation of a segmented body pattern relies initially on the spatial location of maternal effect gene products. These gene products lead to the sequential activation of the segmentation genes. Positional information can be conveyed to cells via morphogens that are found within the oocyte, morphogens secreted from cells during development, or cell-to-cell contact. Although all three of these are important, morphogens in the oocyte have the greatest influence on the overall body structure.

C12. The anterior portion of the anteroposterior axis is established by the action of the Bicoid protein. During oogenesis, the mRNA for Bicoid enters the anterior end of the oocyte and is sequestered there to establish an anterior (high) to posterior (low) gradient. Later, when the mRNA is translated, the Bicoid protein in the anterior region establishes a genetic hierarchy that leads to the formation of anterior structures. If Bicoid was not trapped in the anterior end, it is likely that anterior structures would not form.

C14. Maternal effect gene products influence the establishment of the main body axes, anteroposterior and dorsoventral, and also the formation of structures in the terminal regions. They are expressed during oogenesis and are necessary very early in development. Zygotic genes, particularly the three classes of segmentation genes, are necessary after the axes have been established. The segmentation genes are expressed after fertilization.

C16. The coding sequence of homeotic genes contains a 180-bp consensus sequence known as a homeobox. The protein domain coded by the homeobox is called a homeodomain. The homeodomain contains three conserved sequences that are folded into α-helical conformations. The arrangement of these α helices promotes the binding of the protein into the major groove of the DNA. In this way, homeotic proteins are able to bind to DNA in a sequence-specific manner and thereby activate particular genes.

C18. It is normally expressed in the three thoracic segments that have legs (T1, T2, and T3).

C20. A. When a mutation inactivates a gap gene, a contiguous section of the larva is missing.

B. When a mutation inactivates a pair-rule gene, some regions that are derived from alternating parasegments are missing.

C. When a mutation inactivates a segment-polarity gene, portions are missing at either the anterior or posterior end of the segment.

C22. Proper development in mammals is likely to require the products of maternal effect genes that play a key role in initiating embryonic development. The adult body plan is merely an expansion of the embryonic body plan, which is established in the oocyte. Because the starting point for the development of an embryo is the oocyte, an enucleated egg is needed to clone a mammal.

C24. A heterochronic mutation is one that alters the normal timing of expression for a gene involved in development. The gene may be expressed too early or too late, which would cause a cell lineage to be out of sync with development of the rest of the body. If a heterochronic mutation affected the intestine, the animal might end up with too many intestinal cells if it was a gain-of-function mutation or too few if it was a loss-of-function mutation. In either case, the effects might be detrimental because the growth of the intestine must be coordinated with the growth of the rest of the animal.

C26. Cell differentiation is the specialization of a cell into a particular cell type. In the case of skeletal muscle cells, the bHLH proteins play a key role in the initiation of cell differentiation. When bHLH proteins are activated, they are able to bind to enhancers and activate the expression of many different muscle-cell-specific genes. In this way, myogenic bHLH proteins initiate the synthesis of many muscle-cell-specific proteins. When these proteins are synthesized, they cause the characteristics of a cell to become those of a muscle cell. The activity of myogenic bHLH proteins is regulated by dimerization. When a heterodimer forms between a myogenic bHLH protein and an E protein, it activates gene expression. However, when a heterodimer forms between a myogenic bHLH protein and a protein called Id, the heterodimer is unable to bind to DNA. The Id protein is produced during early stages of development and prevents myogenic bHLH proteins from promoting muscle cell differentiation too soon. At later stages of development, the amount of Id protein falls, and myogenic bHLH proteins can combine with E proteins to induce muscle cell differentiation.

C28. A totipotent cell is a cell that has the potential to create a complete organism.

A. In humans, a fertilized egg is totipotent, and the cells that exist during the first few embryonic divisions are totipotent. However, after several divisions, embryonic cells lose their totipotency and, instead, are destined to become particular tissues within the body.

B. In plants, many cells are totipotent.

C. Because yeast are unicellular, one cell is a complete individual. Therefore, yeast cells are totipotent; they can produce new individuals by cell division.

D. Because bacteria are unicellular, one cell is a complete individual. Therefore, bacterial cells are totipotent; they can produce new individuals by cell division.

C30. Animals begin their development from an egg, and then the body is organized along anteroposterior and dorsoventral axes. The formation of an adult organism is an expansion of the embryonic body plan. Plants grow primarily from two meristems: shoot and root meristems. At the cellular level, plant development is different in that it does not involve cell migration, and many plant cells are totipotent. Animals require organization within an oocyte to begin development. At the genetic level, however, animal development and plant development are similar in that a gene hierarchy produces transcription factors that govern pattern formation and cell specialization.

C32. The *tra* and *tra-2* gene products are splicing factors. In the female, they cause the alternative splicing of the pre-mRNAs that are expressed from the *fru* and *dsx* genes. The tra and tra-2 proteins cause these pre-mRNAs to be spliced into mRNAs designated *fru^F* and *dsx^F*. The absence of *Sxl* expression in the male prevents *tra* from being expressed. Without the tra protein, the *fru* and *dsx* mRNAs are spliced in a different way to produce mRNAs designated *fru^M* and *dsx^M*.

Experimental Questions

E2. *Drosophila* has an advantage in that researchers have identified many mutant alleles that alter development in specific ways. The hierarchy of gene regulation is particularly well understood in the fruit fly. *C. elegans* has the advantage of simplicity, and having complete knowledge of cell fate in this organism enables researchers to explore how the timing of gene expression is critical to the developmental process.

E4. To investigate whether a mutation is affecting the timing of developmental decisions, a researcher needs to know the normal time or stage of development when cells are supposed to divide and what types of cells will be produced. A cell lineage diagram provides this. Having this information, the researcher can then determine if particular mutations alter the timing when cell division occurs.

E6. Mutant 1 has a gain-of-function mutation; it keeps reiterating the L1 pattern of division. Mutant 2 has a loss-of-function mutation; it skips the L1 pattern and immediately follows an L2 pattern.

E8. As discussed in Chapter 15, most eukaryotic genes have a core promoter that is adjacent to the coding sequence; regulatory elements that control the transcription rate at the core promoter are typically upstream from that site. Therefore, to get the *Antp* gene product expressed where the *abd-A* gene product is normally expressed, you would link the upstream genetic regulatory region of the *abd-A* gene to the coding sequence of the *Antp* gene. This segment would be inserted into the middle of a P element (see below). The DNA would then be introduced into an embryo by P element transformation.

P element	*abd-A* regulatory region	*Antp* coding sequence	P element

The *Antp* gene product is normally expressed in the thoracic region and produces segments with legs, as illustrated in Figure 26.12. Therefore, because the *abd-A* gene product is normally expressed in the anterior abdominal segments, you might predict that the DNA shown above would produce a fly with legs attached to the segments that are supposed to be the anterior abdominal segments. In other words, the anterior abdominal segments might resemble thoracic segments with legs.

E10. A. The female flies must have had female parents that were heterozygous, with a (dominant) normal allele and the mutant allele. Their male parents were either homozygous for the mutant allele or heterozygous like the female parents. The female flies inherited a mutant allele from both parents. Nevertheless, because their female parent was heterozygous, having the normal (dominant) allele and the mutant allele, and because the *bicoid* gene is a maternal effect gene, their phenotype is based on the genotype of their female parent. The normal allele is dominant, so they have a normal phenotype.

B. *Bicoid-A* appears to have a deletion that removes part of the sequence of the *bicoid* gene, resulting in a shorter mRNA. *Bicoid-B* could also have a deletion that removes all of the sequence of the *bicoid* gene, or it could have a promoter mutation that prevents the expression of the gene. *Bicoid-C* seems to have a point mutation that does not affect the amount of the *bicoid* mRNA.

C. With regard to function, all three mutations are known to be loss-of-function mutations. The mutation in *bicoid-A* probably eliminates function by truncating the Bicoid protein, which is a transcription factor. The mutation in *bicoid-A* probably shortens this protein, thereby inhibiting its function. The mutation in *bicoid-B* prevents expression of the *bicoid* mRNA. Therefore, none of the Bicoid protein is made, and this explains the loss of function. The mutation in *bicoid-C* seems to prevent the proper localization of the *bicoid* mRNA in the oocyte. There must be proteins within the oocyte that recognize specific sequences in the *bicoid* mRNA and trap it in the anterior end of the oocyte. This mutation must change these sequences and prevent these proteins from recognizing the bicoid mRNA.

E12. An egg-laying defect is somehow related to an abnormal anatomy. The n540 strain has fewer neurons compared to a normal worm. Perhaps the n540 strain is unable to lay eggs because it is missing neurons that are needed for egg laying. The n536 and n355 strains have an abnormal abundance of neurons. Perhaps this overabundance also interferes with the proper neural signals needed for egg laying.

E14. Geneticists interested in vertebrate development have used reverse genetics because it has been difficult for them to identify mutations in developmental genes based on phenotypic effects in the embryo. The problem lies in the difficulty associated with screening a large number of vertebrate embryos in search of abnormal ones that carry mutant genes. It is easy to have thousands of flies in a laboratory, but it is not easy to have thousands of mice. Instead, it is easier to clone the normal gene based on its homology to invertebrate genes and then make mutations in vitro. These mutations can be introduced into a mouse to create a gene knockout. This strategy is the opposite of that used by Mendel, who characterized genes by first identifying phenotypic variants (e.g., tall versus short, green seeds versus yellow seeds).

Questions for Student Discussion/Collaboration

2. You should try to make a flow diagram that begins with maternal effect genes, then gap genes, pair-rule genes, and segment-polarity genes. The gene regulation pathways then lead to homeotic genes and finally to genes that code proteins that affect cell structure and function. It's almost impossible to make an accurate flow diagram because there are so many gene interactions, but it is instructive to think about developmental genetics in this way. It is probably easier to identify mutant phenotypes that affect later stages of development because they are less likely to be lethal. However, modern methods can screen for mutations that affect early stages of development, as described in question 2 in More Genetic TIPS. To identify all of the genes necessary for chicken development, you might begin with genes involved in early development, but this plan assumes that you have some way to identify them. If they have already been identified, you would try to identify the genes that they stimulate or repress. This could be done using molecular methods described in Chapters 14, 15, and 20.

CHAPTER 27

Answers to Comprehension Questions

27.1: c, a, a, d	27.5: d, b
27.2: a	27.6: d
27.3: b, c	27.7: c
27.4: b, a, c, d	27.8: d, b, a

Concept Check Questions (follow figure legends)

FIGURE 27.1   A local population is a group of individuals that are more likely to interbreed with each other than with members of a more distant population.

FIGURE 27.2   Polymorphisms are very common in nearly all natural populations.

FIGURE 27.6   The word *directional* means that selection is favoring a particular phenotype, moving the population toward prevalence of that phenotype.

FIGURE 27.8   No, directional selection is not promoting genetic diversity because it is eliminating individuals that are sensitive to DDT.

FIGURE 27.11   The Hb^S allele is an advantage in the heterozygous condition because it confers resistance to malaria. This heterozygote advantage outweighs the homozygote disadvantage.

FIGURE 27.12   With negative frequency-dependent selection, the rarer phenotype has a higher fitness, which improves its reproductive success.

FIGURE 27.13   Yes, disruptive selection fosters polymorphism. The fitness values of the resulting phenotypes depend on the environment. Some phenotypes are the fittest in one environment, whereas other phenotypes are the fittest in another environment.

FIGURE 27.15   Stabilizing selection decreases genetic diversity because it eliminates individuals that carry alleles that promote more extreme phenotypes.

FIGURE 27.16   Genetic drift tends to have a greater effect in small populations. It can lead to the rapid loss or fixation of an allele.

FIGURE 27.17   At the bottleneck, genetic diversity may be reduced because there are fewer individuals in the population. During the bottleneck, genetic drift may promote the loss of certain alleles and the fixation of other alleles, thereby lowering the genetic variation even more.

FIGURE 27.18   Inbreeding increases the likelihood of homozygosity, and therefore tends to increase the likelihood that an individual will exhibit a recessive trait. This occurs because an individual can inherit both copies of the same allele from a common ancestor.

FIGURE 27.21   DNA fingerprinting is used for identification and for relationship testing.

End-of-Chapter Questions:

Conceptual Questions

C2. A population is a group of interbreeding individuals. Let's consider a squirrel population in a forested area. Over the course of many generations, several things could happen to this population. A forest fire, for example, could dramatically decrease the number of individuals and thereby create a bottleneck. This would decrease the genetic diversity of the population. A new predator may enter the region, and natural selection may then select for the survival of squirrels that are best able to evade the predator. Another possibility is that a group of squirrels within the population may migrate to a new region and found a new squirrel population.

C4. A. Phenotype frequency and genotype frequency

   B. Genotype frequency

   C. Allele frequency

C6. A. The genotype frequency of homozygotes for the mutant allele is $\frac{1}{2500}$, or 0.0004. This is equal to q^2. The allele frequency is the square root of this value, which equals 0.02. The frequency of the corresponding dominant allele is $1 - 0.02 = 0.98$.

   B. The genotype frequency for homozygous affected individuals is 0.0004; for unaffected homozygous individuals, it is $(0.98)^2 = 0.96$; and for heterozygous individuals, it is $2(0.98)(0.02) = 0.039$.

   C. If a person is known to be a heterozygous carrier, the chances that this person will happen to choose another heterozygous carrier as a mate is equal to the frequency of heterozygous carriers in the population, which is 0.039, or 3.9%. The chances that two heterozygous carriers will choose each other as mates equals $0.039 \times 0.039 = 0.0015$, or 0.15%.

C8. Applying the Hardy-Weinberg equation gives these genotype frequencies:

   $BB = (0.67)^2 = 0.45$, or 45%
   $Bb = 2(0.67)(0.33) = 0.44$, or 44%
   $bb = (0.33)^2 = 0.11$, or 11%

   The actual data show a higher percentage of homozygotes (compare 45% with 50% and 11% with 13%) and a lower percentage of heterozygotes (compare 44% with 37%) than expected. Therefore, these data are consistent with inbreeding, which increases the percentage of homozygotes and decreases the percentage of heterozygotes.

C10. Migration, genetic drift, and natural selection are the main mechanisms that alter allele frequencies within a population. Natural selection acts to eliminate harmful alleles and promote beneficial alleles. Genetic drift involves random changes in allele frequencies that may eventually lead to elimination or fixation of alleles. It is thought to be important in the establishment of neutral alleles in a population, and it is largely responsible for the variation seen in natural populations. Migration is important because it introduces new alleles into neighboring populations.

C12. Darwinian fitness is the relative likelihood that an individual with a particular genotype will survive and contribute to the gene pool of the next generation, as compared to other genotypes. The genotype with the highest reproductive success is given a fitness value of 1.0. Characteristics that promote survival, ability to attract a mate, or an enhanced fertility are expected to promote Darwinian fitness. Examples are the thick fur of a polar bear, which helps it to survive in a cold climate; the bright plumage of male birds, which helps them to attract a mate; and the high number of gametes released by certain species of fish, which enhances their fertility.

C14. All of these patterns of natural selection favor one or more phenotypes because such phenotypes have a reproductive advantage. However, the patterns differ with regard to whether a single phenotype or multiple phenotypes are favored and whether the phenotype that is favored is in the middle of the phenotypic range or at one or both extremes. Directional selection favors an extreme phenotype. Over time, directional selection is expected to favor the fixation of alleles that result in such a phenotype. Disruptive selection favors two or more phenotypes. It will lead to a polymorphism for the trait. Mechanisms of balancing selection are heterozygote advantage and negative-frequency dependent selection. These promote a stable polymorphism in a population. Stabilizing selection favors individuals with intermediate phenotypes. It tends to decrease genetic diversity because alleles for extreme phenotypes are eliminated.

C16. In genetic drift, allele frequencies are drifting. *Genetic drift* is an appropriate term because the word *drift* implies a random process. Nevertheless, drift can be directional. A boat may drift from one side of a lake to another. It would not drift in a straight path, but the drifting process will alter its location. Similarly, allele frequencies can "drift" up or down and eventually lead to the elimination or fixation of particular alleles within a population.

C18. A. Probability of fixation for allele $a = 1/2N = 1/2(4) = 1/8$, or 0.125

   B. $\bar{t} = 4N = 4(4) = 16$ generations

   C. The calculations for parts A and B assume a constant population size. If the population grows after it was founded by the four individuals, the probability of fixation will be lower and the time it takes for fixation to occur will be longer.

C20. A. True

   B. True

   C. False; it causes allele loss or fixation, which results in less genetic diversity.

   D. True

C22. A. Migration will increase the genetic diversity in both populations. A random mutation could occur in one population to create a new allele. This new allele could be introduced into the other population via migration.

   B. The allele frequencies in the two populations will tend to be similar to each other, due to the intermixing of the alleles.

   C. The rate of genetic drift depends on population size. When two populations intermix, this has the effect of increasing the overall population size. In a sense, the two smaller populations behave somewhat like one big population. Therefore, the effects of genetic drift are lessened when the individuals in two populations can migrate. The net effect is that allele loss and allele fixation are less likely to occur due to genetic drift.

C24. A. Yes

   B. The common ancestors are I-1 and I-2.

C. $F = \sum(1/2)^n(1 + F_A)$

   $F = (1/2)^9 + (1/2)^9$

   $F = 1/512 + 1/512 = 2/512 = 0.0039$

   D. You cannot conclude based on this pedigree that the parents of VI-1 are inbred.

C26. A. The inbreeding coefficient is calculated using this formula:

   $$F = \sum(1/2)^n(1 + F_A)$$

   In this case, there are two common ancestors, I-1 and I-2. Because you have no prior history for I-1 or I-2, you assume they are not inbred, which makes $F_A = 0$. The two inbreeding paths for IV-2 each contain five people: III-4, II-2, I-1, II-6, and III-5, and III-4, II-2, I-2, II-6, and III-5. Therefore, $n = 5$ for both paths.

   $$F = (1/2)^5 (1 + 0) + (1/2)^5 (1 + 0)$$

   $$F = 0.031 + 0.031 = 0.062$$

   B. Based on the data shown in this pedigree, individual III-4 is not inbred.

C28. You can use the following equation to calculate the number of generations:

   $$(1 - \mu)^t = \frac{p_t}{p_0}$$

   $$(1 - 10^{-4})^t = 0.5/0.6 = 0.833$$

   $$(0.9999)^t = 0.833$$

   $$t = 1827 \text{ generations}$$

Experimental Questions

E2. Question 2 in More Genetic TIPS shows how the Hardy-Weinberg equation can be modified to include three or more alleles. In this case:

   $$(p + q + r + s)^2 = 1$$

   $$p^2 + q^2 + r^2 + s^2 + 2pq + 2qr + 2qs + 2rp + 2rs + 2sp = 1$$

   Let $p = C$, $q = c^{ch}$, $r = c^h$, and $s = c$.

   A. The genotype frequency of albino rabbits is s^2:

   $$s^2 = (0.05)^2 = 0.0025 = 0.25\%$$

   B. Himalayan is dominant to albino but recessive to full and chinchilla. Therefore, genotypes for Himalayan rabbits are represented by r^2 and by $2rs$:

   $$r^2 + 2rs = (0.44)^2 + 2(0.44)(0.05) = 0.24 = 24\%$$

   Among 1000 rabbits, about 240 would have a Himalayan coat color.

   C. Chinchilla is dominant to Himalayan and albino but recessive to full coat color. Therefore, heterozygotes with chinchilla coat color are represented by $2qr$ and by $2qs$:

   $$2qr + 2qs = 2(0.17)(0.44) + 2(0.17)(0.05) = 0.17, \text{ or } 17\%$$

   Among 1000 rabbits, about 170 would be heterozygotes with chinchilla fur.

E4. A.

Inuit	$M = 0.913$	$N = 0.087$
Navajo	$M = 0.917$	$N = 0.083$
Finns	$M = 0.673$	$N = 0.327$
Russians	$M = 0.619$	$N = 0.381$
Aboriginal Australians	$M = 0.176$	$N = 0.824$

   B. To determine if any of these populations are in equilibrium, you can use the Hardy-Weinberg equation and calculate the expected number of individuals with each genotype. For example:

   Inuit $MM = (0.913)^2 = 0.833 = 83.3\%$

   $MN = 2(0.913)(0.087) = 0.159 = 15.9\%$

   $NN = (0.087)^2 = 0.0076 = 0.76\%$

In general, these values agree pretty well with an equilibrium. The same is true for the other four populations.

C. Based on their similar allele frequencies, the Inuit and Navajo seem to have interbred, as have the Finns and Russians.

E6. The selection coefficients are as follows.

For $W^S W^S$, $s = 1 - 0.19 = 0.81$

For $W^R W^R$, $s = 1 - 0.37 = 0.63$

At equilibrium, the allele frequency of W^S is $0.63/(0.63 + 0.81) = 0.44$, or 44%

At equilibrium, the allele frequency of W^R is $0.81/(0.63 + 0.81) = 0.56$, or 56%

If the rats were not exposed to warfarin, the equilibrium would no longer exist, and natural selection would tend to eliminate the warfarin-resistance allele because the homozygotes are deficient in vitamin K.

E8. A GWSS is aimed at identifying genetic differences that are due to natural selection among different populations. If a GWSS is used to detect positive selection, this means it is attempting to detect an increase in the frequency of one or more alleles because such allele(s) is/are favored by natural selection. However, not all the increases in allele frequency that are identified in a GWSS may be due to positive selection. Some of them may be due to genetic hitchhiking. As one or more positively selected alleles increase in frequency, nearby linked alleles on the chromosome will "hitchhike" along with it, and thereby increase in frequency, even though such alleles do not confer a selective advantage.

E10. The pattern of natural selection was directional. Over the long run, directional selection may lead to the loss of certain alleles and the fixation of others. In this case, alleles promoting smaller beak size might be lost from the population, while alleles promoting larger beak size could become fixed.

E12. A. Probability of fixation = $\frac{1}{2}N$ (assuming equal numbers of males and females contributing to the next generation)
$$= \frac{1}{2}(2,000,000)$$
$$= 1 \text{ in } 4,000,000 \text{ chance}$$

B. $\bar{t} = 4N$

where

$\bar{t}$ = the average number of generations to achieve fixation

N = the number of individuals in population, assuming that males and females contribute equally to each succeeding generation

In this case,

$\bar{t} = 4(2 \text{ million}) = 8 \text{ million generations}$

C. If the blue allele conferred a selective advantage, the value calculated in part A would be slightly larger; there would be a higher chance of allele fixation. The value calculated in part B would be smaller; it would take a shorter period of time to reach fixation.

E14. Male 2 is the potential male parent, because male 2 has the bands found in the offspring but not found in the female parent. To calculate the likelihood of the match being due to chance alone, you would have to know the probability of having each of the types of bands that match. In this case, for example, male 2 and the offspring have four bands in common. As a simple calculation, you could eliminate the four bands the offspring shares with the female parent. If the probability of having each paternal band is $\frac{1}{4}$, the chances that this person is not the male parent are $(\frac{1}{4})^4$, which equals 1/256.

E16. This percentage is not too high. Based on their genetic relationship, you expect Jack and Jack's offspring to share at least 50% of the same bands in a DNA fingerprint. However, the value can be higher than that because the female parent and Jack may have some bands in common, even though they are not genetically related. For example, at one site in the genome, Jack may be heterozygous for

4100-bp and 5200-bp minisatellites, and the female parent may also be heterozygous in this same region and have 4100-bp and 4700-bp minisatellites. Jack could pass the 5200-bp minisatellite to the offspring, and the female parent could pass the 4100-bp minisatellite. Thus, the offspring's fingerprint would show the 4100-bp and 5200-bp bands. This would be a perfect match to both of the Jack's bands, even though Jack transmitted only the 5200-bp minisatellite to the offspring. The 4100-bp band would match because Jack and the female parent happened to have a minisatellite in common. Therefore, the 50% estimate of matching bands in a DNA fingerprint based on genetic relationships is a minimum estimate. The value can be higher than that.

Questions for Student Discussion/Collaboration

2. Mutation is responsible for creating new alleles, but the rate of new mutations is so low that it cannot explain allele frequencies in this range. Let's call the two alleles B and b and assume that B was the original allele and b is a more recent allele that arose as a result of mutation. Three scenarios can explain the allele frequencies:

A. The b allele arose via a neutral mutation and reached its present frequency by genetic drift. It hasn't yet been eliminated or fixed.

B. The b allele arose via a beneficial mutation, and its frequency is increasing due to natural selection. However, there hasn't been enough time to reach fixation.

C. The Bb heterozygote has a selective advantage, leading to a balanced polymorphism.

CHAPTER 28

Answers to Comprehension Questions

28.1: d, c

28.2: d, b

28.3: d

28.4: b, c

28.5: c, b, b

28.6: a, c

Concept Check Questions (follow figure legends)

FIGURE 28.1 In most populations (like the one shown), height exhibits a continuum.

FIGURE 28.2 About 95.4% are within two standard deviations of the mean, which means that 4.6% are outside of this range. Half of them, or 2.3%, fall more than two standard deviations above the mean.

FIGURE 28.3 When alleles are additive, they contribute in an incremental way to the outcome of a trait. Having three red alleles, for example, will make the hull color of a wheat plant darker red than it will be with two red alleles.

FIGURE 28.4 Increases in gene number and more environmental variation tend to cause greater overlaps between phenotypes due to different genotypes.

FIGURE 28.5 The two strains differ with regard to a quantitative trait, and they differ in their molecular markers.

FIGURE 28.9 Both natural selection and selective breeding affect allele frequencies as a result of differences in reproductive success. In the case of natural selection, reproductive success is determined by environmental conditions. For selective breeding, reproductive success is determined by the people who choose the parents for breeding.

FIGURE 28.10 The strains differ with regard to the relative sizes of different parts of the plants. These include the stems, leaves, leaf buds, and flower buds.

FIGURE 28.11 A selection limit may be reached (1) because a population has become monomorphic for all of the desirable alleles or (2) because the desired effects of artificial selection are balanced by the negative effects on fitness.

End-of-Chapter Questions:

Conceptual Questions

C2. Quantitative traits often exhibit a continuum of phenotype because they are usually influenced by multiple genes that exist as multiple alleles. A large amount of environmental variation will also increase the overlap between genotypes and phenotypes for polygenic traits.

C4. A discontinuous trait is one that falls into discrete categories. Examples include brown eyes versus blue eyes in humans and purple versus white flowers in pea plants. A continuous trait is one that does not fall into discrete categories. Examples include height in humans and fruit weight in tomatoes. Most quantitative traits are continuous; the trait falls within a range of values. The reason why quantitative traits are continuous is because they are usually polygenic and greatly influenced by the environment. As shown in Figure 28.4b, this tends to create overlaps between phenotypes due to different genotypes.

C6. To rank in the top 2.5% requires a weight about 2 standard deviations above the mean. If you take the square root of the variance, the standard deviation would be 22 pounds. To rank in the top 2.5%, an animal would have to weigh at least 44 pounds more than the mean weight, or 562 + 44 = 606 pounds. To rank in the bottom 0.13%, an animal would have to be 3 standard deviations lighter than the mean weight, or at least 66 pounds lighter, which gives a weight of 496 pounds.

C8. This is a positive correlation, but it could have occurred as a matter of chance alone. According to Table 28.2, this value could have occurred due to random sampling error. Further investigation would be needed to determine if there is a significant correlation, such as examining a greater number of pairs of individuals. If $N = 500$, the correlation would be statistically significant, and you would conclude that the correlation did not occur as a matter of random chance. However, you could not conclude that it occurred due to cause and effect.

C10. When a correlation coefficient is statistically significant, it means that the association is likely to have occurred for reasons other than random sampling error. It may indicate cause and effect but not necessarily. For example, large parents may have large offspring due to genetics (cause and effect). However, such a correlation may be related to the sharing of similar environments rather than being due to cause and effect.

C12. A quantitative trait locus is a site on a chromosome that contains one or more genes that affect a quantitative trait. It is possible for a QTL to contain one gene, or it may contain two or more closely linked genes. QTL mapping, which analyzes linkage to known molecular markers, is commonly used to determine the locations of QTLs.

C14. If the broad-sense heritability equals 1.0, it means that all the variation in the population is due to genetic variation rather than environmental variation. It does not mean that the environment is unimportant in the outcome of the trait. Under another set of environmental conditions, the trait may have turned out quite differently.

C16. When a species is subjected to selective breeding, the breeder is focusing on improving particular traits. In this case, rose breeders have focused on the size and quality of the flowers. Because the breeder usually selects a small number of individuals (e.g., the ones with best flowers) as the breeding stock for the next generation, this selection may lead to a decrease in the allelic diversity of other genes. For example, several genes affect flower fragrance. In an unselected population, these genes may exist as "fragrant" alleles and "nonfragrant" alleles. After many generations of breeding for large flowers, the fragrant alleles may be lost from the population, just as a matter of random chance. This is a common problem of selective breeding. As you select for an improvement in one trait, you may inadvertently diminish the quality of an unselected trait. Others have suggested that the lack of fragrance may be related to flower structure and function. Perhaps the amount of energy that a flower uses to make beautiful petals somehow diminishes its capacity to make fragrance.

C18. Broad-sense heritability takes into account all genetic factors that affect the phenotypic variation in a trait. Narrow-sense heritability considers only alleles that behave in an additive fashion. In many cases, the alleles affecting quantitative traits appear to behave additively. More importantly, if a breeder assumes that the heritability of a trait is due to the additive effects of alleles, it is possible to predict the outcome of selective breeding. This estimated narrow-sense heritability is termed the realized heritability.

C20. A. Due to good nutrition, the individuals in the commune would tend to grow taller, increasing the mean height of that population. If the genetic variation affecting height was similar to the general population, the average member of the commune population would be taller than the average member of the general population.

B. If the environment is fairly homogeneous, then heritability values tend to be higher because the environment contributes less to the amount of variation in the trait. Therefore, in the commune, the heritability might be higher, because the members uniformly practice good nutrition. On the other hand, because the commune population is smaller than the general population, the amount of genetic variation might be less, which would make the heritability lower. However, because the problem states that the commune population is fairly large, you would probably assume that the amount of genetic variation is similar to that in the general population. Overall, the best guess would be that the heritability in the commune population is higher because of the uniform nutrition standards.

C. For the same reasons as stated in part B, the amount of variation would probably be similar to that in the general population, because the commune population is fairly large. As a general answer, larger populations tend to have more genetic variation. Therefore, the general population probably has a bit more variation.

C22. A natural population of animals is more likely to have a higher genetic diversity than a domesticated population. The reason for this is that domesticated populations have been subjected to many generations of selective breeding, which decreases the genetic diversity. Therefore, V_G is likely to be higher for the natural population. The other issue is the environment. It is difficult to say which group would have a more homogeneous environment. In general, natural populations tend to have a more heterogeneous environment, but not always. If the environment is more heterogeneous, this tends to cause more phenotypic variation, which makes V_E higher. With regard to heritability in the broad sense:

$$\text{Heritability} = V_G/V_T$$
$$= V_G/(V_G + V_E)$$

When V_G is high, heritability increases. When V_E is high, heritability decreases. In the natural wolf population, you expect V_G to be high. In addition, you would guess that V_E might be high as well (but that is less certain). Nevertheless, if this were the case, the heritability of size for the wolf population might be similar to that for the domestic population. This would occur because the high V_G of the wolf population would be balanced by its high V_E. On the other hand, if V_E is not that high in the wolf population, or if it is fairly high in the domestic population, then the wolf population would have a higher heritability for size.

Experimental Questions

E2. To calculate the mean, you add the values together and divide by the total number of flies:

$$\text{Mean} = \frac{1.9 + 2(2.4) + 2(2.1) + 3(2.0) + 2(2.2) + 1.7 + 1.8 + 2(2.3) + 1.6}{15}$$
$$= 2.1$$

The variance is the sum of the squared deviations from the mean divided by $N - 1$. The mean value of 2.1 must be subtracted from each value, and then the square of each resulting value is taken. These 15 values are added together and then divided by 14 (which is $N - 1$):

$$\text{Variance} = \frac{0.85}{14}$$
$$= 0.061$$

The standard deviation is the square root of the variance:

Standard deviation = 0.25

E4. When we say an RFLP is associated with a trait, we mean that a gene that influences the trait is closely linked to that RFLP. At the chromosomal level, the gene of interest is so closely linked to the RFLP that a crossover almost never occurs in between them.

(Note: Each plant inherits four RFLPs, but it may be homozygous for one or two of them.) The bands on the gel would be as follows:

Small: 2700 bp and 4000 bp (homozygous for both)

Small–medium: 2700 (homozygous), 3000, and 4000 bp; or 2000, 2700, and 4000 bp (homozygous)

Medium: 2000 bp and 4000 bp (homozygous for both); or 2700 bp and 3000 bp (homozygous for both); or 2000, 2700, 3000, and 4000 bp

Medium–large: 2000 (homozygous), 3000, and 4000 bp; or 2000, 2700, and 3000 bp (homozygous)

Large: 2000 bp and 3000 bp (homozygous for both)

E6. Let's assume there is an extensive map of molecular markers for the rice genome. You would begin with two strains of rice, one with a high yield and one with a low yield, that differ greatly with regard to the molecular markers they carry. You would make crosses between these two strains to get F_1 hybrids. You would then backcross the F_1 hybrids to either of the parental strains and then examine hundreds of offspring with regard to yields and molecular markers. In this case, the expected results would be that six different markers in the high-producing strain would be correlated with offspring that produce higher yields. You might get fewer than six bands if some of these genes are closely linked and associated with the same marker. You also might get fewer than six if the two parental strains have the same marker that is associated with one or more of the genes that affect yield.

E8. $h_N^2 = \dfrac{R}{S}$

$R = \bar{X}_o - \bar{X}$

$S = \bar{X}_p - \bar{X}$

In this problem:

$\bar{X}$ equals 1.01 (as given in the problem)

$\bar{X}_o$ equals 1.09 (by calculating the mean for the offspring)

$\bar{X}_p$ equals 1.11 (by calculating the mean for the parents)

$R = 1.09 - 1.01 = 0.08$

$S = 1.11 - 1.01 = 0.10$

$h_N^2 = 0.08/0.10 = 0.8$ (which is a pretty high heritability value)

E10. The results for identical and fraternal twins, who probably share very similar environments but who differ in the amount of genetic material they share, are a strong argument against an environmental bias. The differences in the observed correlations (0.49 versus 0.95) are consistent with the differences in the expected correlations (0.5 versus 1.0).

E12. $h_N^2 = r_{obs}/r_{exp}$

The value for r_{exp} comes from the known genetic relationships:

Parent-offspring $r_{obs} = 0.36$, $r_{exp} = 0.5$, $h_N^2 = 0.72$

Grandparent-grandchild $r_{obs} = 0.17$, $r_{exp} = 0.25$, $h_N^2 = 0.68$

Siblings (nontwins) $r_{obs} = 0.39$, $r_{exp} = 0.5$, $h_N^2 = 0.78$

Fraternal twins $r_{obs} = 0.40$, $r_{exp} = 0.5$, $h_N^2 = 0.80$

Identical twins $r_{obs} = 0.77$, $r_{exp} = 1.0$, $h_N^2 = 0.77$

The average heritability is 0.75.

E14. These data suggest that there might be a genetic component to blood pressure, because the relatives of people with high blood pressure also seem to have high blood pressure. Of course, more extensive studies would need to be conducted to determine the role of the environment. To calculate heritability, the first thing to do is to calculate the correlation coefficient between relatives to see if it is statistically significant. If it is, then you could follow the approach described in the experiment featured in Figure 28.8. You would determine the correlation coefficients between genetically related individuals as a way to determine the heritability for the trait. In this approach, heritability equals r_{obs}/r_{exp}. It would be important to include genetically related pairs that were raised apart (e.g., aunts or uncles and nieces or nephews) to see if they had a similar heritability value compared to genetically related pairs raised in the same environment (e.g., siblings). If their values were similar, this would give you some confidence that the heritability value is due to genetics and not due to fact that relatives often share similar environments.

Questions for Student Discussion/Collaboration

2. Most traits depend on the influence of many genes. Also, genetic variation is a common phenomenon in most populations. Therefore, most individuals have a variety of alleles that contribute to a given trait. For quantitative traits that involve the size of the whole body or the size of body parts, some alleles may promote a bigger size, and other alleles may promote a smaller size. If a population contains many different genes and alleles that govern a quantitative trait such as size, most individuals will have an intermediate phenotype because they will have inherited some alleles for larger size and some alleles for smaller size. Fewer individuals will inherit a predominance of either type of allele. An example of a quantitative trait that does not fit a normal distribution is snail shell pigmentation. The dark-shelled snails and light-shelled snails are favored rather than the intermediate colors because they are less susceptible to predation in certain environments.

CHAPTER 29

Answers to Comprehension Questions

29.1: b

29.2: f, d

29.3: a, b

29.4: c, d, d, c

29.5: a, d, d

Concept Check Questions (follow figure legends)

FIGURE 29.1   It illustrates a potential drawback of using morphological traits to establish species.

FIGURE 29.2   Cladogenesis is much more common.

FIGURE 29.5   The scenarios in parts (a) and (c) will allow speciation to occur.

FIGURE 29.7   The offspring that would be produced from a cross between *G. tetrahit* and either of the other two species would have two sets of chromosomes from one species plus one set from the other. The chromosomes that exist as a single set cannot segregate evenly during meiosis, thereby causing sterility. This sterility is why *G. tetrahit* is reproductively isolated from the other two species.

FIGURE 29.8   A clade is a group of species consisting of all descendants of the group's most common ancestor.

FIGURE 29.9   A shared derived character arose more recently (on an evolutionary time scale) compared to an ancestral character. A shared derived character is shared by a group of organisms but not by a distant common ancestor.

FIGURE 29.10   The principle of parsimony states that the preferred hypothesis is the one that is the simplest.

FIGURE 29.12   The protists are distributed within the seven supergroups of Eukaryotes.

FIGURE 29.13   Horizontal gene transfer is the transfer of genetic material from one individual organism to another that is not an offspring of the first organism. It can occur between different species.

FIGURE 29.14   Orthologs are homologous genes found in different species, whereas paralogs are homologous genes found in the same species.

FIGURE 29.18   A molecular clock allows researchers to put a time scale on a phylogenetic tree.

FIGURE 29.19   In humans, chromosome 2 is a single chromosome, but it is divided into two chromosomes in the other species. On chromosome 3, the orangutan has a large inversion that the other species do not have.

End-of-Chapter Questions:

Conceptual Questions

C2.   Evolution is unifying because all living organisms on this planet evolved from an interrelated group of common ancestors. At the molecular level, all organisms have a great deal in common. With the exception of some viruses, they all use DNA as their genetic material. This DNA is found within chromosomes, and the sequence of the DNA is organized into units called genes. Most genes are protein-coding genes that code the amino acid sequences of polypeptides. Polypeptides fold to form functional units called proteins. At the cellular level, all living organisms also share many similarities. For example, living cells share many of the same basic features including a plasma membrane, ribosomes, enzymatic pathways, and so on. In addition, as discussed in Chapter 5, the mitochondria and chloroplasts of eukaryotic cells are evolutionarily derived from bacterial cells.

C4.   Reproductive isolation occurs when two species are unable to mate and produce viable offspring. As summarized in Table 29.1, several prezygotic and postzygotic isolating mechanisms can prevent interspecies matings.

C6.   Anagenesis is the evolution of one species into another, whereas cladogenesis is the divergence of one species into two or more species. Of the two, cladogenesis is more common. There may be many reasons why. An abrupt genetic change such as the development of alloploidy may produce a new species from a preexisting one. Also, migrations of a few members of species into a new region may lead to the formation of a new species in the region (i.e., allopatric speciation).

C8.   A. Allopatric

  B. Sympatric

  C. At first, allopatric speciation with hybrid zones will occur. Eventually, when smaller lakes are formed, allopatric speciation will occur.

C10.   Reproductive isolation does not really apply to bacteria. Two different bacteria of the same species do not produce gametes that have to fuse to produce an offspring, although bacteria can exchange genetic material (as described in Chapter 7). For this reason, it is difficult to distinguish different species of bacteria. A geneticist would probably divide bacteria into different species based on their cellular traits and the sequences of their DNA. Historically, bacteria were first categorized into different species based on morphological and physiological differences. Later, when genetic tools such as DNA sequencing became available, the previously identified species could be categorized based on genetic sequences. One issue that makes categorization rather difficult is that a species of bacteria can exist as closely related strains that may have a small number of genetic differences.

C12.   The sequences line up where the two Gs are underlined:

TTGCATAGGCATACC**G**TATGATATCGAAAACTAGAAAAATAGGGCGATAGCTA
GTATGTTATCGAAAAGTAGCAAAATAGGGCGATAGCTACCCAGACTACCGGAT

C14.   The rate of deleterious or beneficial mutations would probably not be a good molecular clock. The rate at which such mutations arise might be relatively constant, but the rate of elimination or fixation would probably be quite variable. Alleles are acted upon by natural selection. As environmental conditions change, the degree to which natural selection will favor beneficial alleles and eliminate deleterious alleles also changes. For example, natural selection favors the sickle cell allele in regions where malaria is prevalent but not in other regions. Therefore, the prevalence of this allele does not depend solely on its rate of formation and genetic drift.

C16.   Some regions of a polypeptide are particularly important for the structure or function of a protein. For example, a region of a polypeptide may be the active site of an enzyme. The amino acids that are found within the active site are likely to be precisely located for the binding of the enzyme's substrate and/or for catalysis. Changes in the amino acid sequence of the active site usually have a detrimental effect on the enzyme's function. Therefore, polypeptide sequences that are critical to structure or function (like those found in active sites) are not likely to change. If they did change, natural selection would tend to prevent the change from being transmitted to future generations. In contrast, other regions of a polypeptide are less important. These other regions would be more tolerant of changes in amino acid sequence and therefore would evolve more rapidly. When related protein sequences are compared, regions that are important for function can often be identified because they show less sequence variation.

C18.   The α-globin sequences in humans and horses are more similar to each other than are the α-globin and β-globin sequences in humans. This evidence suggests that the gene duplication that produced the α-globin and β-globin genes occurred first. After this gene duplication occurred, each gene accumulated several different mutations that caused the sequences of the two genes to diverge. At a much later time during the evolution of mammals, a split occurred that produced different branches in the evolutionary tree of mammals. One branch eventually led to horses and a different branch led to humans. After these mammalian branches were produced, some additional mutations occurred in the α- and β-globin genes. This explains why the α-globin gene is not exactly the same in humans and horses. However, it is more similar in those two species than the α- and β-globin genes within humans because the divergence of humans and horses occurred much more recently than the gene duplication that produced the α- and β-globin genes. In other words, there has been much less time for the α-globin gene in humans to diverge from the α-globin gene in horses.

C20.   A. This difference is due to neutral mutation. Mutations that change the base in the wobble position are neutral when they do not affect the amino acid sequence.

  B. This is the result of the action of natural selection. Random mutations that occur in vital regions of a polypeptide sequence are likely to inhibit function. Therefore, these types of mutations are eliminated by natural selection. That is why they are relatively rare.

  C. This is the result of a combination of neutral mutation and the effect of natural selection. The prevalence of mutations in introns is due to the accumulation of neutral mutations. Most mutations within introns do not have any effect on the expression of the exons, which contain the polypeptide-coding sequences. In contrast, mutations within exons are more likely to be subject to natural selection, and those that affect vital regions of the coded polypeptides are likely to inhibit function. Natural selection tends to eliminate these mutations. Therefore, mutations within exons are less likely than mutations within introns.

C22.   Generally, you would expect the karyotypes to show a similar number of chromosomes with very similar banding patterns. However, there may be a few notable differences. An occasional translocation could change the size of a chromosome in one or more species or the number of chromosomes between two different species. Also, an occasional inversion may alter the banding patterns in two closely related species.

Experimental Questions

E2.   Perhaps the easiest way to determine whether allotetraploidy is present among related species is by chromosomal examination. A researcher could karyotype the chromosomes from many different grass species and look for homologous chromosomes that have similar banding patterns. This comparison may identify allotetraploids that contain a diploid set of chromosomes from two different species.

E4. The differences that have occurred in this sequence are due to neutral mutations. In all cases, the base in the wobble position has changed, and the change will not affect the amino acid sequence of the coded polypeptide. Therefore, a reasonable explanation is that the gene has accumulated random neutral mutations over the course of many generations. This observation is consistent with the neutral theory of evolution. A second explanation is that one of these two researchers made a few experimental mistakes when determining the base sequence.

E6. Inversions do not affect the total amount of genetic material. Usually, inversions do not affect the phenotype of the organism. Therefore, if members of the two antelope populations were to interbreed, the offspring would probably be viable because they would have inherited a normal amount of genetic material from each parent. However, such offspring would be inversion heterozygotes. As illustrated in Chapter 8 (see Figure 8.11), crossing over during meiosis may create chromosomes that have too much or too little genetic material. If these unbalanced chromosomes are passed to the next generation of offspring, the offspring may not survive. For this reason, inversion heterozygotes (that are phenotypically normal) may not be very fertile because many of their offspring will die. Because inversion heterozygotes are less fertile, the eastern and western populations will tend to be reproductively isolated. Over time, this will aid in the independent evolution of the two populations and ultimately promote the evolution of the two populations into separate species.

E8. The technique of PCR is used to amplify the amount of DNA in a sample. To carry out PCR, you use oligonucleotide primers that are complementary to the region of the DNA that is to be amplified. For example, as demonstrated in the experiment in Figure 29.16, PCR primers that are complementary to and flank the 12S rRNA gene can be used to amplify that gene. The technique of PCR is described in Chapter 20.

E10. Because *G. speciosa* and *G. pubescens* are closely related, the probe is likely to bind to homologous sites in both genomes. You would expect two bright spots in the results of the in situ experiment for the *G. speciosa* chromosomes. The *G. tetrahit* chromosomes would show four spots because that species carries two copies of genomes from both *G. speciosa* and *G. pubescens*.

E12. Applying the principle of parsimony leads to a phylogenetic tree that requires the fewest number of evolutionary changes. When working with molecular data, researchers can run computer programs that compare DNA sequences from homologous genes of different species and construct a tree that requires the fewest numbers of mutations. Such a tree is the most likely pathway for the evolution of such species.

E14. If you constructed a phylogenetic tree using a gene sequence that was the result of horizontal gene transfer, the results might be misleading. For example, if a bacterial gene was transferred to a protist, and you used that gene to construct your tree, the results would suggest that the bacterium and protist are very closely related, which they are not. Evolutionary biologists can overcome this problem by choosing several different genes to use in the construction of a phylogenetic tree, assuming that most of them are not transferred horizontally.

Questions for Student Discussion/Collaboration

2. The founder effect and the formation of an allotetraploid are examples of rapid mechanisms of speciation. In addition, some single gene mutations may have a great impact on phenotype and lead to the rapid evolution of a new species by cladogenesis. Geological processes may promote the slow accumulation of alleles and alter a species' characteristics more gradually. In this case, it is the accumulation of many phenotypically minor genetic changes that ultimately leads to reproductive isolation. Slow and fast mechanisms of speciation both result in reproductive isolation, which is a prerequisite for the evolution of new species. Fast mechanisms tend to involve small populations and a fewer number of genetic changes. Slower mechanisms may involve larger populations and involve the accumulation of a large number of genetic changes that each contributes in a small way.

GLOSSARY

2-aminopurine a base analog that acts as a chemical mutagen.

3′-untranslated region in an mRNA, the region at the 3′ end that follows the stop codon and does not code a polypeptide.

5′-untranslated region in an mRNA, the region at the 5′ end that precedes the stop codon and does not code a polypeptide.

5-bromouracil a base analog that acts as a chemical mutagen.

A

ABC model a model for flower development.

abnormal displaying one or more characteristics that are not very common among members of a given species.

acentric fragment a fragment of a chromosome that lacks a centromere.

acquired antibiotic resistance the acquisition of antibiotic resistance by a bacterial strain, which may result from genetic alterations in the bacterial genome, but often occurs because a bacterium has taken up a gene or plasmid from another bacterial strain.

acridine dye a type of chemical mutagen that intercalates between adjacent base pairs in DNA and causes frameshift mutations.

acrocentric describes a chromosome with the centromere significantly off center, but not at the end.

activator a regulatory protein that binds to DNA and increases the rate of transcription.

adaptive evolution evolution that occurs due to natural selection acting on genetic variation.

adaptor hypothesis a hypothesis that proposes a tRNA has two functions: recognizing a three-base codon sequence in mRNA and carrying an amino acid that is specific for that codon.

adenine (A) a purine base found in DNA and RNA. It base-pairs with thymine in DNA.

age of onset the age at which symptoms of a disease first appear.

alkaptonuria a human genetic disorder involving the accumulation of homogentisic acid in the body due to a lack of the enzyme homogentisic acid oxidase.

alkyltransferase an enzyme that can remove methyl or ethyl groups from guanine bases.

allele an alternative form of a specific gene.

allele frequency the number of copies of a particular allele in a population divided by the total number of all alleles for that gene in the population.

allelic variation genetic variation in a population that involves the occurrence of two or more different alleles for a particular gene.

allodiploid describes an organism that contains one set of chromosomes from two different species.

allopatric speciation an evolutionary phenomenon in which speciation occurs when some members of a species become geographically separated from the other members.

alloploid describes an organism that contains sets of chromosomes from two or more different species.

allopolyploid describes an organism that contains two (or more) sets of chromosomes from two (or more) species.

allosteric enzyme an enzyme that contains two binding sites: a catalytic site and a regulatory site.

allosteric regulation the phenomenon in which a small effector molecule binds to a noncatalytic site on a protein and causes a conformational change that regulates the protein's function.

allosteric site the site on a protein where a small effector molecule binds to regulate the function of the protein.

allotetraploid describes an organism that contains two sets of chromosomes from two different species, for a total of four sets.

α helix a type of secondary structure found in proteins.

alternative exon an exon that is not always found in mature mRNAs. It is only found in certain types of alternatively spliced mRNAs.

alternative splicing the phenomenon that a pre-mRNA can be spliced in more than one way.

Ames test a test using strains of a bacterium, *Salmonella typhimurium,* to determine if a substance is a mutagen.

amino acid a building block of polypeptides and proteins. It contains an amino group, a carboxyl group, and a side chain.

aminoacyl site (A site) a site on a ribosome where a charged tRNA initially binds.

aminoacyl-tRNA see *charged tRNA.*

aminoacyl-tRNA synthetase an enzyme that catalyzes the attachment of a specific amino acid to the correct tRNA.

amino-terminus the location of the first amino acid in a polypeptide. The amino acid at the amino-terminus retains a free amino group that is not covalently attached to the second amino acid.

amniocentesis a method of obtaining cellular material from a fetus for the purpose of genetic testing.

anagenesis a mechanism for speciation in which a single species is transformed into a different species over the course of many generations.

anaphase the fourth phase of mitosis. As anaphase proceeds, half of the chromosomes move to one pole, and the other half move to the other pole.

ancestral character a characteristic that is shared with a distant ancestor. Also called a *primitive character.*

ancient DNA analysis analysis of DNA that is extracted from the remains of extinct species. Also referred to as *molecular paleontology.*

aneuploid not euploid. Refers to a variation in chromosome number such that the total number of chromosomes is not an exact multiple of a set (or of the number *n*).

annotation in computer files containing genetic sequences, additional information such as a concise description of a sequence, the name of the organism from which it was obtained, and the function of the coded protein, if known. Also referred to as *metadata.*

antagomir an AMO that has one or more base modifications that may promote a stronger binding to the complementary miRNA.

anteroposterior axis in animals, the axis that runs from the head (anterior) to the tail or base of the spine or abdomen (posterior).

anther the structure in flowering plants in which pollen grains form.

antibiotic any substance produced by a microorganism that inhibits the growth of other microorganisms, such as pathogenic bacteria.

antibodies proteins that are produced by the immune systems of vertebrates, which recognize foreign material (namely, viruses, bacteria, and so forth) and target it for destruction. Also known as *immunoglobins.*

antibody microarray a small silica, glass, or plastic slide that is dotted with many different antibodies, which recognize particular amino acid sequences within proteins.

anticipation the phenomenon in which the severity of an inherited disease tends to get worse in subsequent generations.

anticodon a three-nucleotide sequence in tRNA that is complementary to a codon in mRNA.

antigens foreign substances that elicit an immune response because they are recognized by antibodies.

anti-miRNA oligonucleotide (AMO) a short ncRNA that is complementary to a particular miRNA.

antiparallel refers to an arrangement in a double helix in which one strand is running in the 5′ to 3′ direction, while the other strand runs in the 3′ to 5′ direction.

antisense RNA an RNA strand that is complementary to a strand of mRNA.

antitermination a function of certain proteins, such as the N protein in bacteria, which is to prevent transcriptional termination.

antiterminator a secondary structure in RNA that prevents early transcriptional termination.

AP endonuclease a DNA repair enzyme that recognizes a DNA region that is missing a base and makes a cut in the DNA backbone near that site.

apoptosis programmed cell death.

apurinic site a site in DNA that is missing a purine base.

Archaea one of the three domains of life. Archaea tend to live in extreme environments and are less common than bacteria.

ARS elements DNA sequences found in yeast that function as origins of replication.

artificial selection see *selective breeding.*

artificial transformation transformation of bacteria that occurs via experimental treatments.

ascus (pl. asci) a sac that contains haploid spores of fungi (i.e., yeast or molds).

asexual reproduction a form of reproduction that does not involve the union of gametes; at the cellular level, a preexisting cell divides to produce two new cells.

association in statistics, when two or more variables vary according to some pattern.

assortative mating sexual reproduction in which individuals preferentially breed with each other based on their phenotypes.

astral microtubules the microtubules that emanate outward from the centrosome toward the plasma membrane.

AT/GC rule in DNA, the phenomenon in which an adenine base in one strand always hydrogen bonds with a thymine base in the opposite strand, and a guanine always hydrogen bonds with a cytosine.

ATP-dependent chromatin remodeling changes in chromatin structure that are due to the action of chromatin-remodeling complexes and that affect the positions and/or compositions of nucleosomes.

AT-rich region a region in a bacterial origin of replication that contains high amounts of adenine (A) and thymine (T) and is the site where the DNA strands initially separate.

attachment sites sites in a host cell chromosome and in phage DNA that allow the viral genome to become a provirus or prophage.

attenuation a mechanism of genetic regulation, observed in the *trp* operon, in which a short RNA is made but its synthesis is terminated before RNA polymerase can transcribe the rest of the operon.

attenuator sequence a sequence found in certain operons (e.g., the *trp* operon) in bacteria that stops transcription soon after it has begun.

automated DNA sequencing the use of fluorescently labeled dideoxyribonucleotides and a fluorescence detector to sequence DNA.

autonomous element a transposable element that contains all the information necessary for transposition or retrotransposition to occur.

autopolyploid a polyploid produced within a single species due to nondisjunction.

autoradiography a technique that involves the use of X-ray film to detect the location of radioisotopes as they are found in macromolecules or cells. It is used to identify a particular band on a gel or to map the location of a gene within an intact chromosome.

autoregulatory loop a form of gene regulation in which a protein, such as a splicing factor or a transcription factor, regulates its own expression.

autosomal refers to a gene that is on an autosome, not on a sex chromosome.

autosomes chromosomes that are not sex chromosomes.

auxotroph a strain that cannot synthesize a particular nutrient and needs that nutrient to be added to its growth medium.

B

B DNA the predominant form of DNA in living cells. It is a right-handed DNA helix with 10 bp per turn.

backbone the portion of a DNA or RNA strand that is composed of covalently linked phosphates and sugar molecules.

Bacteria one of the three domains of life. Bacteria are prokaryotic species.

bacterial artificial chromosome (BAC) a type of cloning vector that propagates in bacteria and is used to clone large fragments of DNA.

bacteriophage a virus that infects bacteria. Also referred to as a *phage*.

balanced translocation a translocation, such as a reciprocal translocation, in which the total amount of genetic material is not altered.

balancing selection a pattern of natural selection that favors the maintenance of two or more

alleles in a population; it may be due to heterozygote advantage or negative frequency-dependent selection.

Barr body a structure in the interphase nuclei of somatic cells of female mammals that is a highly condensed X chromosome.

basal transcription in eukaryotes, a low level of transcription produced by the core promoter. The binding of transcription factors to enhancer elements may increase transcription above the basal level.

basal transcription apparatus the minimum components that are needed to transcribe a eukaryotic gene; these include TFIID, TFIIB, TFIIF, TFIIE, TFIIH, RNA polymerase II, and a DNA sequence containing a TATA box and transcriptional start site.

base excision repair a type of DNA repair in which a modified base is removed from a DNA strand. Following base removal, a short region of the DNA strand is removed and then resynthesized using the complementary strand as a template.

base pair (bp) the structure in which two nucleotides in opposite strands of DNA hydrogen bond with each other. For example, an A-T base pair is a structure in which an adenine-containing nucleotide in one DNA strand hydrogen bonds with a thymine-containing nucleotide in the complementary strand.

base-pair mismatch a DNA abnormality in which two bases opposite each other in a double helix do not conform to the AT/GC rule. For example, if A were opposite C, that would be a base-pair mismatch.

base stacking In DNA, the orientation of base pairs in which the flat sides of the bases are facing each other.

base substitution a point mutation in which one base is substituted for another.

basic helix-loop-helix (bHLH) domain a domain found in transcription factors that enables them to dimerize and bind to DNA.

Bayesian methods approaches to evaluating phylogenetic trees based on the probability that a particular phylogenetic tree is correct, given the observed data and a particular evolutionary model.

behavioral trait a trait that affects the way an organism responds to some aspect of its environment. An example is the ability to learn a maze.

beneficial mutation a mutation that enhances the survival or reproductive success of an organism.

benign refers to a noncancerous tumor that is not invasive and cannot metastasize.

β sheet a type of secondary structure found in proteins.

bidirectional (1) the manner in which two replication forks move, in opposite directions outward from the origin; (2) refers to a regulatory element that can function in either the forward or the reverse direction.

bidirectional replication the phenomenon in which two DNA replication forks move in opposite directions from an origin of replication.

bilaterian an animal that has an anteroposterior axis with left-right symmetry.

binary fission the physical process whereby a bacterial cell divides into two daughter cells. During this event, the two daughter cells become separated by the formation of a septum.

binomial expansion equation an equation used to solve genetic problems involving a given set of two unordered outcomes.

biodegradation the breakdown of a larger molecule, such as a pollutant in the

environment, into smaller molecules via cellular enzymes.

bioinformatics the use of computers, mathematical tools, and statistical techniques to record, store, and analyze biological information.

biolistic gene transfer the use of microprojectiles to introduce DNA into plant cells to produce transgenic plants.

biological control the control of a pest by the introduction of a natural enemy, predator, or a biological product.

biological evolution the accumulation of heritable changes in one or more characteristics of a population or species from one generation to the next.

biological species concept definition of a species as a group of individuals whose members have the potential to interbreed in nature to produce viable, fertile offspring, but that cannot interbreed successfully with members of other species.

biomarker testing a way to identify genes, proteins, and other substances that have become abnormal in the cancer cells of a given patient.

biometric field a field of genetics that involves the statistical study of biological traits.

bioremediation the use of living organisms or their products to decrease pollutants in the environment.

biotechnology the use of living organisms or materials they produce in the development of products or processes that are beneficial to humans.

biotransformation the conversion of one molecule to another via cellular enzymes. This term is often used to describe the conversion of a toxic molecule into a nontoxic one.

bivalent a structure in which two pairs of homologous sister chromatids have synapsed (i.e., aligned) with each other. Also called a *tetrad*.

BLAST (basic local alignment search tool) a computer program that can start with a particular genetic sequence and then locate similar sequences within a large database.

body pattern the spatial arrangement of different regions of the body. At the cellular level, the body pattern is due to the arrangement of cells and their specialization.

bottleneck effect a mechanism that can give rise to genetic drift; occurs when most members of a population are eliminated without any regard to their genetic composition.

branch migration the lateral movement of a Holliday junction.

breakpoint a region where two chromosome pieces break apart and rejoin with other chromosome pieces.

broad-sense heritability heritability that takes into account different types of genetic variation that may affect phenotype.

C

cAMP response element-binding protein see *CREB protein.*

cancer a disease characterized by uncontrolled cell division.

capping the covalent attachment of a 7-methylguanosine to the 5′ end of mRNAs in eukaryotes.

capsid the protein coat of a virus, which encloses its genome.

CAP site a DNA sequence that is recognized by the activator protein CAP.

carbohydrates organic molecules with the general formula $C_n(H_2O)_n$. An example of a simple carbohydrate is the sugar glucose. Large carbohydrates are composed of multiple sugar units.

carboxyl-terminus the location of the last amino acid in a polypeptide chain. The amino acid at the carboxyl-terminus retains a free carboxyl group that is not covalently attached to another amino acid.

carcinogen an environmental agent that increases the likelihood of developing cancer.

caspases proteolytic enzymes that play a role in apoptosis.

catabolite activator protein (CAP) a genetic regulatory protein found in bacteria.

catabolite repression the phenomenon in which a catabolite (e.g., glucose) represses the expression of certain genes (e.g., the *lac* operon).

catenanes intertwined circular molecules.

CCCTC binding factor (CTCF) a protein that recognizes CCCTC base sequences and binds to the DNA and to another CTCF molecule, thereby promoting the stabilization of a loop domain.

cDNA library a DNA library whose recombinant vectors carry cDNA inserts.

cell adhesion the binding of the surfaces of cells to each other or of the surface of a cell to the extracellular matrix.

cell adhesion molecule (CAM) a molecule (e.g., protein or carbohydrate) that acts as a surface receptor and plays a role in cell adhesion.

cell culture a population of cells grown in a laboratory for research purposes.

cell cycle in eukaryotic cells, a series of stages through which a cell progresses in order to divide. The phases are G for gap, S for synthesis (of the genetic material), and M for mitosis (which includes cytokinesis). There are two G phases, G_1 and G_2.

cell fate the future identity of a cell, or the identity of its daughter cells, before it is actually phenotypically detectable.

cell-free translation system an experimental mixture that can synthesize polypeptides.

cell lineage a series of cells that are derived from a particular cell by cell division.

cell lineage diagram an illustration of the cell division patterns and the fates of any cell's descendants.

cell plate the structure that forms between two daughter plant cells and leads to the separation of the cells by the formation of an intervening cell wall.

cellular level refers to an observation or experimentation at the level of individual cells.

centiMorgan (cM) a unit of map distance obtained from genetic crosses; equivalent to a map unit (mu). Named in honor of Thomas Hunt Morgan.

central dogma of genetics the idea that the usual flow of genetic information is from DNA to RNA to polypeptide (protein). In addition, DNA replication serves to copy the genetic information so that it can be transmitted from cell to cell and from parent to offspring.

central zone in plants, an area in the meristem where undifferentiated stem cells are always maintained.

centrifugation a method to separate cell organelles and macromolecules in which samples are placed in tubes and spun very rapidly. The rate at which particles move toward the bottom of the tube depends on their densities, sizes, shapes, and the viscosity of the medium.

centrifuge a machine that contains a motor, which causes a rotor holding centrifuge tubes to spin very rapidly.

centrioles a pair of structures within each centrosome of animal cells.

centromere a segment of a eukaryotic chromosome that provides an attachment site for the kinetochore.

centrosome a cellular structure from which microtubules emanate.

chain termination an event that stops the growth of a DNA strand, RNA strand, or polypeptide.

chaperone a protein that facilitates the folding of some polypeptides.

character in genetics, a general characteristic such as eye color.

character states different versions of a character.

Chargaff's rule the observation that in DNA the amounts of A and T are equal, as are the amounts of G and C.

charged tRNA a tRNA that has an amino acid attached to its 3′ end by a covalent bond.

checkpoint protein a protein that monitors the conditions of DNA and chromosomes and may prevent a cell from progressing through the cell cycle if an abnormality is detected.

chi square (χ^2) test a commonly used statistical method for determining the goodness of fit. This method can be used to analyze population data in which the members of the population fall into different categories.

chiasma (pl. chiasmata) the site where crossing over occurs between two chromosomes. It resembles the Greek letter chi, χ.

chiasma formation a physical exchange of pieces between homologous chromosomes that most commonly occurs during prophase of meiosis I.

chloroplast DNA (cpDNA) the genetic material found within a chloroplast.

chorionic villus sampling a method for obtaining cellular material from a fetus for the purpose of genetic testing.

chromatids following chromosomal replication in eukaryotes, the two copies that remain attached to each other in the form of sister chromatids.

chromatin the complex of DNA and proteins that is found within eukaryotic chromosomes.

chromatin immunoprecipitation sequencing (ChIP-Seq) a technique that is used to determine where in a genome a particular protein binds to the DNA.

chromatin remodeling see *ATP-dependent chromatin remodeling.*

chromatography a method of separating different macromolecules and small molecules based on their chemical and/or physical properties. A sample is dissolved in a liquid solvent and exposed to some type of matrix, such as a gel, a column containing beads, or a thin strip of paper.

chromocenter the central point where the chromosomes of a polytene chromosome aggregate.

chromosome the structures within living cells that contain the genetic material. Genes are physically located within the chromosomes. Biochemically, a chromosome contains a very long segment of DNA, which is the genetic material, and proteins, which are bound to the DNA and provide it with an organized structure.

chromosome conformation capture methods a set of related techniques used to analyze the spatial organization of chromatin in a living cell; also referred to as *3C methods.*

chromosome map see *genetic map.*

chromosome painting the use of fluorescently labeled probes to identify multiple sites along one or more chromosomes. The probes are usually assigned different computer-generated colors.

chromosome territory a region in the cell nucleus occupied by a chromosome. The chromosome territories are nonoverlapping.

chromosome theory of inheritance a theory of Sutton and Boveri that the inheritance patterns

of traits can be explained by the transmission patterns of chromosomes during meiosis and fertilization.

chromosome walking a method in which a mapped gene or molecular marker provides a starting point from which to molecularly "walk" toward a gene of interest via overlapping clones.

***cis*-acting element** a sequence of DNA, such as a regulatory element, that exerts a *cis*-effect.

***cis*-effect** in bacteria and archaea, an effect on gene expression due to a genetic sequence that is adjacent to the gene(s) it regulates; in eukaryotes, an effect on gene expression due to a genetic sequence within the same chromosome and often immediately adjacent to the gene of interest.

***cis*-epigenetic mechanism** an epigenetic mechanism that may affect only one copy of a gene in a cell that has two copies of the gene. It may be caused by DNA methylation or changes in chromatin structure.

clade see *monophyletic group.*

cladistic approach a way to construct a phylogenetic tree, called a cladogram, by considering the various possible pathways of evolution and then choosing the most plausible tree.

cladogenesis a mechanism of speciation that involves the division of a single species into two or more species.

cladogram a phylogenetic tree that has been constructed using a cladistic approach.

cleavage furrow a constriction that causes the division of an animal cell into two cells during cytokinesis.

clonal relating to a clone or the process of cloning. For example, a clonal population of cells is a group of cells that are derived from the same cell.

clone (1) A group of genetically identical cells that were derived from a single cell; (2) an individual that has been produced from a somatic cell of another individual, such as the sheep Dolly; (3) many copies of a DNA fragment that are propagated within a vector or produced by PCR.

closed complex a complex of RNA polymerase, transcription factors, and a promoter prior to the separation of DNA strands and the transcription of RNA.

closed conformation a conformation of chromatin that cannot be transcribed and may be tightly packed.

coactivator an activator protein that increases the rate of transcription but does not directly bind to the base sequence of the enhancer.

coding strand the strand of DNA in protein-coding genes that is not used as a template for mRNA synthesis.

codominance a pattern of inheritance in which two alleles are both expressed in the heterozygous condition. For example, a person with the genotype $I^A I^B$ has the blood type AB and expresses both surface antigens A and B.

codon a sequence of three nucleotides in mRNA that functions in translation. A start codon, which usually specifies methionine, initiates translation, and a stop codon terminates translation. The other codons specify the amino acids within a polypeptide according to the genetic code.

codon bias in a given species, the phenomenon in which certain codons are used more frequently than others.

cohesin a multiprotein complex that facilitates the alignment of sister chromatids.

colinearity the correspondence between the sequence of codons in the DNA coding strand and the amino acid sequence of a polypeptide.

combinatorial control the phenomenon widely observed in eukaryotes in which the combination of many factors determines the expression of any given gene.

comparative genomic hybridization (CGH) a hybridization technique to determine if cells (e.g., cancer cells) have changes in chromosome structure, such as deletions or duplications.

comparative genomics using information from genome-sequencing projects to understand the genetic variation between different populations and evolutionary relationships among different species.

competence factors proteins that are needed for bacterial cells to become naturally transformed by extracellular DNA.

competence-stimulating peptide (CSP) a peptide secreted by certain species of bacteria that allows them to become competent for transformation.

competent cells cells that can take up DNA from the environment or an extracellular medium.

complementary describes sequences in two DNA strands that match each other according to the AT/GC rule. For example, if one strand has the sequence 5′-ATGGCGGATTT-3′, then the complementary strand must be 3′-TACCGCCTAAA-5′; in DNA, complementary sequences are also antiparallel.

complementary DNA (cDNA) DNA that is made from an RNA template by the action of reverse transcriptase.

complementation a phenomenon in which the presence of two different mutant alleles in the same organism produces a wild-type phenotype. It usually happens because the two mutations are in different genes, so the organism carries one copy of each mutant allele and one copy of each wild-type allele.

complete nondisjunction event in which all of the chromosomes fail to segregate properly during meiosis or mitosis and remain in one of the two daughter cells.

complex traits characteristics that are determined by several genes and are significantly influenced by environmental factors.

computer data file a collection of information in a form suitable for storage and manipulation on a computer.

computer program a series of operations that can manipulate and analyze data in a desired way.

concordance in genetics, the degree to which a trait or disorder is inherited, determined by how many pairs of twins both exhibit it.

condensed forming a more compact structure; used to describe chromatids in prophase.

condensin multiprotein complexes that play a role in the condensation of interphase chromosomes to produce metaphase chromosomes; the two types are condensin I and condensin II, which play distinct roles in the condensation process.

conditional lethal allele an allele that is lethal only under certain environmental conditions.

conditional mutant a mutant whose phenotype depends on the environmental conditions, such as a temperature-sensitive (ts) mutant.

conglomerate population a population composed of members of an original population plus new members that have migrated from another population.

conjugation a form of genetic transfer between bacteria that involves direct physical interaction between two bacterial cells. One bacterium acts as donor and transfers genetic material to a recipient cell.

conjugation bridge a connection between two bacterial cells that provides a passageway for DNA transfer during conjugation.

consensus sequence the set of the most commonly occurring bases within a specific type of sequence in DNA.

conservative model an incorrect model of DNA replication in which both parental strands of DNA remain together following DNA replication.

conserved site a particular base within a gene or a particular amino acid within a polypeptide that is identical or similar across multiple species.

constitutive exon an exon that is always found in mature mRNAs following splicing.

constitutive gene a gene that is not regulated and has essentially constant levels of expression over time.

constitutive heterochromatin regions of chromosomes that are heterochromatic in all cell types of a multicellular organism and are permanently transcriptionally inactive.

continuous traits quantitative traits that fall along a continuum rather than occurring in discrete categories.

copy number variation (CNV) a type of structural variation in which a segment of DNA that is 1000 bp or more in length commonly exhibits copy number differences among members of the same species.

core enzyme an enzyme composed of subunits that are needed for catalytic activity, as in the core enzyme of RNA polymerase.

core promoter a relatively short DNA sequence that is necessary for transcription to take place. It provides the binding site for general transcription factors and RNA polymerase.

corepressor in bacteria, a small effector molecule that binds to a repressor protein, thereby causing the repressor protein to bind to DNA and inhibit transcription; in eukaryotes, a protein or protein complex that interacts with a repressor protein and inhibits transcription.

correlation coefficient (r) a statistic with a value that ranges between −1 and +1. It indicates how two factors vary in relation to each other.

cotyledons food-storing structures that will become the first leaves of a plant seedling.

covalent histone modification the covalent attachment of functional groups to histone proteins. These modifications may affect chromatin structure and constitute a histone code.

covariance a statistic that describes the degree of variation between two variables within a group of individuals.

CpG island a group of CG base sequences that may be clustered near a promoter region of a gene. The methylation of the cytosine bases usually inhibits transcription.

CREB protein (cAMP response element-binding protein) a regulatory transcription factor that becomes activated in response to specific cell-signaling molecules that cause an increase in the cytoplasmic concentration of cAMP.

CRISPR-Cas system a system found in bacteria and archaea that provides defense against viruses, plasmids, and transposable elements.

CRISPR-Cas technology a method that uses the components of the CRISPR-Cas system of genome defense found in prokaryotes to introduce mutations into genes.

cross immunity the phenomenon in which immunity to one infectious agent also confers immunity to a closely related agent.

cross a breeding between two distinct individuals. An analysis of their offspring may be conducted to understand how traits are passed from parent to offspring.

cross-fertilization a cross in which the male and female gametes come from separate individuals.

crossing over see *chiasma formation.*

C-terminus see *carboxyl-terminus.*

cycle threshold method (Cₜ method) in quantitative PCR, a method of determining the starting amount of DNA based on a threshold level at which the accumulation of fluorescence is significantly greater than the background level.

cyclic-AMP (cAMP) in bacteria, a small effector molecule that binds to CAP (catabolite activator protein). In eukaryotes, cAMP functions as a second messenger in a variety of intracellular signaling pathways.

cytogenetic mapping determining the locations of specific genetic sequences within chromosomes using microscopy. Also called *cytological mapping.*

cytogeneticist a scientist who studies chromosomes under the microscope.

cytogenetics the field of genetics that involves the microscopic examination of chromosomes.

cytokinesis the division of a single cell into two cells. The two nuclei produced in mitosis are segregated into separate daughter cells during cytokinesis.

cytological mapping see *cytogenetic mapping.*

cytoplasmic inheritance see *extranuclear inheritance.*

cytosine (C) a pyrimidine base found in DNA and RNA. It base-pairs with guanine in DNA.

D

Darwinian fitness the relative likelihood that a genotype will contribute to the gene pool of the next generation compared to other genotypes.

database a large number of computer data files, such as those containing genetic sequences, collected and stored in a single location.

daughter strands in DNA replication, the two newly made strands of DNA.

de novo genome assembly a computer-based method to determine the order of the DNA fragments as they would occur along the chromosome(s) in a given species' genome; the order is deduced from overlaps of the DNA sequences among different DNA fragments.

de novo methylation the methylation of DNA that has not been previously methylated. This is usually a highly regulated event.

deamination the removal of an amino group from a molecule. For example, the removal of an amino group from cytosine produces uracil.

decoding function the ability of 16S rRNA to detect when an incorrect tRNA is bound at the A site and prevent elongation until the mispaired tRNA is released from the A site.

decondensed less tightly compacted; used to describe chromosomes during interphase.

deficiency the condition in which a segment of chromosomal material is missing.

degeneracy in genetics, this term means that more that one codon specifies the same amino acid. For example, the codons GGU, GGC, GGA, and GGG all specify the amino acid glycine.

degrees of freedom in a statistical analysis, the number of categories that are independent of each other.

deleterious mutation a mutation that is detrimental with regard to its effect on phenotype; it decreases the chances of survival and reproduction.

deletion a type of change affecting a chromosome in which a segment of DNA is removed.

deoxyribonucleic acid (DNA) the genetic material. It is a double-stranded structure, with each strand composed of repeating units of deoxyribonucleotides.

deoxyribose the sugar found in DNA.

depurination the removal of a purine base from DNA.

determination the process in which a cell or region of an organism adopts a particular fate that will be realized later in development.

determined describes a cell that is destined to differentiate into a specific cell type.

development in multicellular species, the series of genetically programmed stages in which a fertilized egg becomes an embryo and eventually develops into an adult.

developmental genetics the area of genetics concerned with the roles genes play in orchestrating the changes that occur during development.

diauxic growth the sequential use of two sugars by a growing population of bacterial cells.

dicentric describes a chromosome with two centromeres.

dicentric bridge the region between the two centromeres in a dicentric chromosome.

dideoxy sequencing a method of DNA sequencing that uses dideoxyribonucleotides to terminate the growth of DNA strands.

dideoxyribonucleotide (ddNTP) a nucleotide used in DNA sequencing that is missing the 3′—OH group. If a dideoxyribonucleotide is incorporated into a DNA strand, it stops any further growth of the strand.

differential centrifugation a form of centrifugation involving a series of centrifugation steps in which the supernatant or pellet is used in each subsequent centrifugation step.

differentially methylated region (DMR) a site that is methylated during spermatogenesis or oogenesis, but not both, and plays a role in imprinting.

differentiated describes a cell that has become a specialized type of cell within a multicellular organism.

dimeric DNA polymerase a complex of two DNA polymerase holoenzymes that move as a unit during DNA replication.

dioecious in plants, a species in which sporophytes are divided into those that produce only male gametophytes and those that produce only female gametophytes.

diploid an organism or cell that contains two sets of chromosomes.

direct repeats (DRs) short DNA sequences that flank transposable elements in which the DNA sequence is repeated in the same direction.

directional selection natural selection that favors an extreme phenotype. This usually leads to the fixation of the favored allele.

directionality in DNA and RNA, refers to the 5′ to 3′ orientation of nucleotides in a strand; in proteins, refers to the linear arrangement of amino acids from the N-terminus to the C-terminus.

discovery-based science experimentation that does not require a preconceived hypothesis. In some cases, the goal is to collect data to be able to formulate a hypothesis.

disequilibrium in population genetics, refers to the condition of a population that is not in Hardy-Weinberg equilibrium.

dispersive model an incorrect model for DNA replication in which segments of parental DNA and newly made DNA are interspersed in both strands following the replication process.

disruptive selection natural selection that favors the maintenance of two or more alleles in heterogeneous environments, resulting in two or more phenotypes. Also referred to as *diversifying selection.*

dizygotic (DZ) twins also known as fraternal twins; twins formed from separate pairs of sperm and egg cells.

DNA see *deoxyribonucleic acid.*

DNA fingerprinting a technology for identifying a particular individual based on the properties of that individual's DNA. Also known as *DNA profiling.*

DNA gap repair synthesis the synthesis of DNA in a region where part of a DNA strand has been previously removed, usually by a DNA repair enzyme or by an enzyme involved in homologous recombination.

DNA gyrase an enzyme that introduces negative supercoils into DNA using energy from ATP and that can also relax positive supercoils when they occur. Also known as *topoisomerase II.*

DNA N-glycosylase an enzyme that can recognize an abnormal base and cleave the bond between it and the sugar in the DNA backbone.

DNA helicase an enzyme that separates the two strands of DNA.

DNA library a collection of many recombinant vectors, each carrying a particular fragment of DNA from a larger source. For example, each recombinant vector in a DNA library might carry a small segment of chromosomal DNA from a particular species.

DNA ligase an enzyme that catalyzes the formation of covalent bonds within the sugar-phosphate backbones of two DNA strands.

DNA methylation a regulatory mechanism in which an enzyme covalently attaches a methyl group (—CH_3) to a base in DNA. In eukaryotes, the base is cytosine. In prokaryotes, both adenine and cytosine can be methylated.

DNA methyltransferase the enzyme that attaches a methyl group to cytosine in eukaryotes and prokaryotes or to adenine in prokaryotes.

DNA microarray a small silica, glass, or plastic slide that is dotted with many different DNA sequences, each corresponding to a short sequence within a known gene. Also referred to as a *gene chip.*

DNA polymerase an enzyme that catalyzes the formation of covalent bonds between nucleotides to form a strand of DNA.

DNA profiling see *DNA fingerprinting.*

DNA replication licensing in eukaryotes, occurs when MCM helicase is bound at an origin of replication, enabling the formation of two replication forks.

DNA replication the process in which original DNA strands are used as templates for the synthesis of new DNA strands.

DNA sequencing a method for determining the base sequence in a segment of DNA.

DNA supercoiling the formation of additional coils in DNA due to twisting forces.

DNA translocase a catalytic ATPase subunit that moves along the DNA. It is a component of chromatin-remodeling complexes.

DNA uptake signal sequences DNA sequences found in certain species of bacteria that are needed for a DNA fragment to be taken up during transformation.

DnaA box a DNA sequence that serves as a recognition site for the binding of a DnaA protein, which is involved in the formation of a replication fork.

DnaA protein a protein that binds to a DnaA box sequence at the origin of replication in bacteria and initiates DNA replication.

DNase an enzyme that cuts the sugar-phosphate backbone in DNA.

DNase I footprinting a method for studying protein-DNA interactions in which the binding of a protein to DNA protects the DNA from digestion by DNase I.

domain (1) a region of a protein that has a specific function; (2) one of the major evolutionary branches of life, which are Bacteria, Archaea, and Eukaryotes.

dominant describes an allele that determines the phenotype in the heterozygous condition. For example, if a plant is *Tt* and has a tall phenotype, the *T* (tall) allele is dominant over the *t* (short) allele.

dominant-negative mutation a mutation that produces an altered gene product that acts antagonistically to the normal gene product. Shows a dominant pattern of inheritance.

dorsoventral axis in animals, the axis from the upper side of the back (e.g., the spine in humans) to the opposite side (e.g., the stomach in humans).

dosage compensation the phenomenon observed in species with sex chromosomes, in which one of the sex chromosomes is altered so that males and females have similar levels of gene expression, even though they do not have the same complement of sex chromosomes.

double helix the structure that results when two strands of DNA (or sometimes strands of RNA) interact with each other to form a double-stranded helical structure.

double-strand break model a model for homologous recombination in which the event that initiates recombination is a double-strand break in one of two homologous chromatids.

down promoter mutation a mutation in a promoter that decreases the rate of transcription.

down regulation genetic regulation that leads to a decrease in gene expression.

droplet organelle an organelle in which aggregated solutes, such as proteins and RNA molecules, separate from the bulk solvent and form a droplet.

duplication the repetition of a segment of DNA more than once within a chromosome and/or within a genome.

dyad see *sister chromatids.*

E

ecological species concept a species concept in which each species occupies an ecological niche, which is the unique set of habitat resources that the species requires, as well as the species' influence on the environment and other species.

ecotypes genetically distinct populations of bacterial species that are adapted to their local environments.

egg cell a female gamete that is usually very large and nonmotile; also known as an ovum or egg.

electrophoresis the migration of ions or charged molecules in response to an electric field.

electrophoretic mobility shift assay a technique for studying protein-DNA interactions (or protein-RNA interactions) in which the binding of protein to a DNA fragment (or the binding of protein to an RNA molecule) slows down its mobility during gel electrophoresis.

electroporation the use of electric current to create temporary pores in the plasma membrane of a cell to allow entry of DNA.

elongation (1) in transcription, the synthesis of an RNA transcript using DNA as a template; (2) in translation, the synthesis of a polypeptide using the information within mRNA.

embryo sac in flowering plants, the female gametophyte that contains an egg cell.

embryogenesis an early stage of animal and plant development resulting in an embryo with organized tissue layers and a body plan.

embryonic germ cell (EG cell) a type of pluripotent stem cell found in the gonads of a fetus.

embryonic stem cell (ES cell) a type of pluripotent stem cell found in the inner cell mass of the blastocyst.

emerging virus a virus that has arisen recently and is more likely to cause infection than previous strains.

empirical approach a strategy in which experiments are designed to determine quantitative relationships as a way to derive laws that govern biological, chemical, or physical phenomena.

endonuclease an enzyme that can cleave a bond between adjacent nucleotides within a DNA or RNA strand.

endopolyploidy in a diploid individual, the phenomenon in which certain cells of the body are polyploid.

endosperm in flowering plants, the material in the seed, which is $3n$ and nourishes the developing embryo.

endosymbiosis a symbiotic relationship in which the symbiont actually lives inside (*endo-*) the host.

endosymbiosis theory the theory that the ancient origin of chloroplasts and mitochondria was the result of certain species of bacteria taking up residence within primordial eukaryotic cells.

enhancement changing traits that are not involved with causing a disease.

enhancer a DNA region in eukaryotes, usually 50–1000 bp in length, that contains one or more regulatory elements. The binding of an activator to an enhancer increases the rate of transcription, whereas the binding of a repressor to an enhancer decreases the rate of transcription.

environment the surroundings in which an organism exists.

enzyme a protein that functions to accelerate chemical reactions within the cell.

enzyme adaptation the phenomenon in which a particular enzyme appears within a living cell only after the cell has been exposed to the substrate for that enzyme.

epigenetic inheritance an inheritance pattern in which a modification occurs to a nuclear gene or chromosome that alters gene expression, but is not permanent over the course of many generations.

epigenetics the study of mechanisms that lead to changes in gene expression that can be passed from cell to cell and are reversible, but do not involve a change in the DNA sequence.

epimutation a heritable change in gene expression that does not alter the sequence of DNA.

episome a genetic element that can replicate independently of the chromosomal DNA but can also occasionally integrate into the chromosomal DNA.

epistasis an inheritance pattern where one gene can mask the phenotypic effects of a different gene.

epitope the structure on the surface of an antigen that is recognized by an antibody.

equilibrium density centrifugation a form of centrifugation in which the particles sediment through the gradient, reaching a position where the density of the particle matches the density of the solution.

ER signal sequence a sequence of about 6–12 amino acids that are predominantly hydrophobic and usually located near the N-terminus; this sequence is recognized by a protein in signal recognition particle (SRP).

eraser domain a domain in a protein that removes posttranslational modifications from proteins in chromatin.

error-prone replication a form of DNA replication carried out by translesion-replicating polymerases that results in a high rate of mutation.

essential gene a gene that is essential for survival.

ethyl methanesulfonate (EMS) a type of chemical mutagen that alkylates bases (i.e., attaches methyl or ethyl groups).

euchromatin less compacted regions of chromosomes, where DNA may be transcriptionally active.

Eukaryotes one of the three domains of life. A defining feature of these organisms is that their cells contain a membrane-bound nucleus. Some simple eukaryotic species are single-celled protists and yeast; more complex multicellular species include fungi, plants, and animals.

euploid describes an organism in which the chromosome number is an exact multiple of a chromosome set.

E-value with regard to results from the BLAST program, the number of times that the match or a better one would be expected to occur purely by random chance in a search of an entire database.

evolution see *biological evolution.*

evolutionary species concept a species concept in which each species is defined based on the separate evolution of its lineage.

ex vivo approach in the case of gene therapy, refers to genetic manipulations that occur outside the body.

exit site (E site) a site on a ribosome where an uncharged tRNA binds and then is released.

exon a region of an RNA molecule that remains after splicing has removed the introns. In mRNA, the coding sequence of a polypeptide is contained within the exons.

exon shuffling the phenomenon in which an exon and its flanking introns from one gene are inserted into another gene.

exon skipping the splicing out of an exon from a pre-mRNA so that it is not included in the mature mRNA.

exonuclease an enzyme that digests an RNA or DNA strand from one end by cleaving the bond between the two nucleotides at that end.

expression vector a cloning vector containing a promoter that causes the gene of interest to be transcribed into RNA when the vector is introduced into a host cell.

expressivity the degree to which a trait is expressed. For example, flowers with deep red color have a high expressivity of the red allele.

extranuclear inheritance the inheritance of genetic material that is not found within the cell nucleus Also called *cytoplasmic inheritance.*

F

F factor a fertility factor that is found in certain strains of bacteria and carries genes that allow the bacteria to conjugate. Strains of bacteria that contain an F factor are designated F^+; strains without any F factor are designated F^-.

F′ factors an F factor that also carries genes derived from the bacterial chromosome.

F_1 generation the offspring produced from a cross of a parental generation.

F_2 generation the offspring produced from a cross or self-fertilization of individuals in the F_1 generation.

facultative heterochromatin heterochromatin that differs among different cells of the body.

feedback inhibition the phenomenon in which the final product of a metabolic pathway inhibits an enzyme that acts early in the pathway.

feedforward loop a mechanism in which a gene that codes a transcription factor is regulated by that same transcription factor.

female In the case of humans, females usually have ovaries at birth, have the potential to produce oocytes (eggs) if fertile, and are usually XX.

fidelity a term used to describe the accuracy of a process. If there are few mistakes, a process has a high fidelity.

fitness see *Darwinian fitness.*

flap endonuclease an enzyme that removes small RNA flaps that are generated by the action of DNA polymerase δ on an RNA primer during DNA replication.

fluorescence in situ hybridization (FISH) a form of in situ hybridization in which the DNA probe is fluorescently labeled.

forked-line method a method to solve independent assortment problems in which lines are drawn to connect particular genotypes.

forward genetics the traditional method of genetics research, in which mutant alleles are first identified by their effect on phenotype and then identified at the molecular level.

founder with regard to genetic diseases, an individual who lived many generations ago and was the person in which the disease-causing allele originated.

founder effect a change in allele frequency that occurs when a small group of individuals separates from a larger population and establishes a colony in a new location.

fraction (1) following centrifugation, a portion of the liquid contained within a centrifuge tube; (2) following column chromatography, a portion of the liquid that has been eluted from a column.

frameshift mutation a mutation that involves the addition or deletion of a number of nucleotides not divisible by 3, which shifts the reading frame of the codons downstream from the mutation.

frequency distribution a graph that displays the numbers of individuals that are found in each of several phenotypic categories.

functional genomics the study of gene function at the genome level. It involves the study of many genes simultaneously.

functional protein microarray a type of protein microarray that monitors a particular kind of protein function, such as the ability to bind a specific drug.

G

G bands the chromosomal banding pattern that is observed when the chromosomes have been treated with the chemical called Giemsa stain.

G_1 phase a phase of the eukaryotic cell cycle in which a cell may prepare to divide, depending on the cell type and conditions.

G_2 phase A phase of the eukaryotic cell cycle between S phase and M phase in which a cell accumulates the materials necessary for nuclear and cell division.

gain-of-function mutation a mutation that changes a gene product so that it gains a new or abnormal function.

gamete a reproductive cell (usually haploid) that can unite with another reproductive cell to create a zygote. Sperm and egg cells are types of gametes.

gametogenesis the production of gametes (e.g., sperm or egg cells).

gametophyte the haploid generation of plants.

gel electrophoresis a method that combines chromatography and electrophoresis to separate molecules and macromolecules. Samples are loaded into wells at one end of the gel, and an electric field is applied across the gel that causes charged molecules to migrate from one side of the gel to the other.

gel retardation assay see *electrophoretic mobility shift assay*.

gender society's idea of what it means to be a woman or a man.

gender identity a person's inner feelings of whether they are a woman or a man, both, or neither.

gene a unit of heredity that may influence the outcome of a trait in an organism. At the molecular level, a gene is a segment of DNA that contains the information to make a functional product, either an RNA molecule or a polypeptide.

gene activation a series of events that enable a gene to be transcribed to produce an RNA molecule.

gene addition the insertion of a cloned gene into a site in a chromosome of a living cell.

gene chip see *DNA microarray*.

gene cloning the production of many copies of a gene using molecular methods, such as PCR or the introduction of the gene into a vector that replicates in a host cell.

gene conversion the phenomenon in which one allele is converted to the allele on the homologous chromosome due to recombination or DNA repair.

gene duplication an increase in the copy number of a gene. Can lead to the evolution of a gene family.

gene editing experimentally altering the sequence of a gene.

gene expression the process by which the information within a gene is accessed, first to synthesize RNA and polypeptides, and eventually to affect the properties of cells and the phenotype of multicellular organisms.

gene family two or more genes within a single species that are similar to each other because they were derived from the same ancestral gene.

gene flow transfer of genes from one population (a donor population) to another via the transfer of fertile individuals, thereby changing the recipient population's gene pool.

gene interaction the phenomenon in which two or more different genes influence the outcome of a single trait.

gene knockin a genetic modification in which a gene of interest has been inserted into a particular site in a genome.

gene knockout in the case of diploid species, the condition in which both copies of a gene have been altered to an inactive form.

gene modification the phenomenon in which an allele of one gene modifies the phenotypic outcome of the alleles of a different gene.

gene mutation a mutation that affects only a single gene.

gene pool all of the alleles of every gene within a particular population.

gene prediction the process of identifying regions of genomic DNA that code genes.

gene redundancy the phenomenon in which one gene compensates for the loss of function of another gene.

gene regulation the phenomenon in which the level of gene expression can vary under different conditions.

gene repression a mechanism that inhibits the transcription of a gene and thereby results in a lower level of RNA synthesis from that gene.

gene silencing a phenomenon in which a gene is subjected to changes in chromatin structure that are more permanent in nature; this usually involves a change of euchromatin into heterochromatin.

gene therapy the introduction of cloned genes into somatic cells or the modification of existing genes in order to treat or cure a disease.

general lineage concept a species concept which states that each species is a population of an independently evolving lineage.

general selection model a model that relates genotypes, relative fitness values, and changes in genotype frequencies based on natural selection.

general transcription factor (GTF) one of several proteins that are necessary to initiate basal transcription at the core promoter.

genetic approach in research, the study of mutant genes that have abnormal function. By studying mutant genes, researchers may better understand normal genes and normal biological processes.

genetic code the correspondence between a codon (i.e., a sequence of three bases in an mRNA molecule) and the functional role that the codon plays during translation. Each codon specifies a particular amino acid or the end of translation.

genetic cross the breeding of two selected individuals and the analysis of their offspring in an attempt to understand how traits are passed from parent to offspring.

genetic drift changes in allele frequencies in a population due to random fluctuations.

genetic hitchhiking The phenomenon in which one or more alleles increase in frequency because they are linked to alleles that are subjected to positive selection; the hitchhiked alleles do not provide a selective advantage.

genetic linkage the phenomenon in which genes that are close together on the same chromosome tend to be transmitted as a unit.

genetic linkage map see *genetic map*.

genetic map a diagram that shows the relative locations of genes or other DNA segments on a chromosome.

genetic mapping any method used to determine the linear order and distance of separation among genes that are linked to each other along the same chromosome. This term is also used to describe the use of genetic crosses to determine the linear order of genes. See also *linkage mapping*.

genetic mosaic an individual that has somatic regions that differ genotypically from each other.

genetic polymorphism when two or more wild-type alleles occur in a population; each wild-type allele is found at a frequency of 1% or higher.

genetic recombination (1) the process in which chromosomes are broken and then rejoined to form a novel genetic combination; (2) the process in which alleles are assorted and passed to offspring in combinations that are different from those found in the parents.

genetic screening the use of testing methods at the population level to determine if individuals are heterozygous carriers for or have a genetic disease.

genetic testing the use of testing methods to analyze an individual's genes or gene products. In many cases, the goal is to determine if the individual carries a genetic abnormality.

genetic transfer the transfer of genetic material from one bacterial cell to another.

genetic variation differences in inherited traits among members of the same species or among different species.

genetically modified organism (GMO) an organism that has received genetic material via recombinant DNA technology.

genetics the study of heredity; the branch of biology that deals with heredity and variation.

genome all of the chromosomes and DNA sequences that a cell, an organism, or a species can possess.

genome maintenance cellular mechanisms that prevent mutations from occurring and/or prevent mutant cells from surviving or dividing.

genome-sequencing projects research endeavors that have the ultimate goal of determining the sequence of DNA bases of the entire genome of a given species.

genome-wide association study (GWAS) an examination of a genome-wide set of genetic variants among many different individuals to see if any variant is associated with a disease or other type of trait.

genome-wide selection scan (GWSS) a study that is aimed at identifying genetic differences between populations of the same or closely related species that are found in different natural environments and/or exhibit different adaptive traits.

genomic imprinting a pattern of inheritance that involves a change in a single gene or chromosome during gamete formation. Depending on whether the modification occurs during spermatogenesis or oogenesis, imprinting governs whether an offspring will express a gene that has been inherited from its female parent or male parent.

genomic library a DNA library of recombinant vectors that carry chromosomal DNA fragments.

genomics the molecular analysis of the entire genome of a species.

genotype the genetic composition of an individual, especially in terms of the alleles for particular genes.

genotype-environment association the phenomenon that occurs when certain genotypes are preferentially found in particular environments.

genotype-environment interaction the phenomenon that occurs when the environmental effects on phenotype differ according to genotype.

genotype frequency the number of individuals with a particular genotype in a population divided by the total number of individuals in the population.

germ cells the gametes (i.e., sperm and egg cells).

germ line refers to cells that give rise to gametes.

germ-line mutation a mutation in a cell of the germ line.

Gibson assembly A method of cloning that can link multiple DNA fragments to each other in a specific order.

glucocorticoid receptor a type of steroid receptor that functions as a regulatory transcription factor.

goodness of fit the degree to which the observed data and the data predicted from a hypothesis are similar to each other. If the observed and predicted data are very similar, the goodness of fit is high.

G-quadraplex a four-stranded structure in DNA that is stabilized by hydrogen bonding between four guanines.

grooves in DNA, the indentations where the atoms of the bases are in contact with the water in the surrounding cellular fluid. In B DNA, there is a smaller minor groove and a larger major groove.

group I intron a type of intron found in self-splicing RNA that uses free guanosine in its splicing mechanism.

group II intron a type of intron found in self-splicing RNA that uses an adenine nucleotide within the intron itself in its splicing mechanism.

growth factors signaling molecules that bind to cell surface receptors and influence cell division.

guanine (G) a purine base found in DNA and RNA. It base-pairs with cytosine in DNA.

GWAS see *genome-wide association study*.

H

haplodiploid describes a species, such as certain bees, in which one sex is haploid (e.g., male) and the other sex is diploid (e.g., female).

haploid refers to a cell that contains half the genetic material found in somatic cells. For a species that is diploid, a haploid gamete contains a single set of chromosomes.

haploinsufficiency the phenomenon in which an individual has only a single functional copy of a gene and that single functional copy does not produce a normal phenotype. Shows a dominant pattern of inheritance.

haplotype the linkage of particular alleles or molecular markers on a single chromosome.

HapMap an extensive catalog of common human genetic variants being produced by the International HapMap Project.

Hardy-Weinberg equation $p^2 + 2pq + q^2 = 1$.

Hardy-Weinberg equilibrium the phenomenon by which, under certain conditions, allele frequencies are maintained in a stable condition and genotypes can be predicted according to the Hardy-Weinberg equation.

helix-turn-helix motif a structure found in transcription factor proteins that promotes binding to the major groove of DNA.

hemizygous indicates that a male has a single copy of an X-linked gene. A male mammal is said to be hemizygous for X-linked genes.

heritability the amount of phenotypic variation due to genetic variance within a specific group of individuals raised in a particular environment.

heterochromatin during interphase, highly compacted regions of chromosomes in which the DNA is usually transcriptionally inactive.

heterochronic mutation a mutation that alters the timing of expression of a gene and thereby alters the outcomes of cell fates of particular cell lineages.

heterodimer a combined structure (a dimer) formed when two polypeptides coded by different genes bind to each other.

heteroduplex a region of double-stranded DNA that contains one or more base-pair mismatches.

heterogametic sex in species with two types of sex chromosomes, the heterogametic sex produces two types of gametes. For example, in mammals, the male is the heterogametic sex, because a sperm can contain either an X or a Y chromosome.

heterogamous describes a species that produces two morphologically different types of gametes (e.g., sperm and eggs).

heteroplasmy a condition in which a cell has variation in a particular type of organelle. For example, a plant cell could contain some chloroplasts that make chlorophyll and other chloroplasts that do not.

heterosis the phenomenon in which hybrids display traits superior to those of either parental strain; also referred to as *hybrid vigor*. Heterosis is usually different from heterozygote advantage, because the hybrid may be heterozygous for many genes, not just a single gene, and because the superior phenotype may be due to the masking of deleterious recessive alleles.

heterozygote advantage an inheritance pattern in which a heterozygote has greater reproductive success than either of the corresponding homozygotes. Also called *overdominance*.

heterozygous describes a diploid individual that has different versions (i.e., two different alleles) of the same gene.

Hfr strain a bacterial strain in which an F factor has become integrated into the bacterial chromosome. During conjugation, an Hfr strain can transfer segments of the bacterial chromosome (Hfr stands for "high frequency of recombination").

higher-order structure with regard to chromatin, any assemblage of nucleosomes that assumes a reproducible conformation in three-dimensional space.

highly repetitive sequences base sequences that are found tens of thousands or even millions of times throughout a genome.

high-throughput sequencing the ability to sequence large amounts of DNA in a short period of time. It usually involves the simultaneous sequencing of many samples.

histone acetyltransferase (HAT) an enzyme that attaches an acetyl group to the amino-terminal tail of a histone protein.

histone code hypothesis the hypothesis that the pattern of histone modification acts much like a language or code in specifying alterations in chromatin structure.

histone deacetylase (HDAC) an enzyme that removes acetyl groups from the amino-terminal tails of histone proteins.

histone proteins (histones) a group of proteins involved in forming the nucleosome, the repeating structural unit within eukaryotic chromatin.

histone variants histone proteins whose amino acid sequences are slightly different from those of the standard histones. They often play specialized roles in chromatin structure and function.

holandric genes see *Y-linked genes*.

Holliday junction a site where an unresolved crossover has occurred between two homologous chromosomes.

Holliday model a model to explain the molecular mechanism of homologous recombination.

holoenzyme an enzyme containing all of the functional subunits, such as the RNA polymerase holoenzyme, which has σ factor associated with the core enzyme.

homeobox a 180-bp consensus sequence found in homeotic genes.

homeodomain the protein domain coded by the homeobox. The homeodomain promotes the binding of the protein to the major groove of DNA.

homeotic term that was originally coined to describe mutants in which one body part is replaced by another.

homeotic gene a gene that functions in governing the developmental fate of a particular region of the body.

homodimer a combined structure (a dimer) formed when two polypeptides coded by the same gene bind to each other.

homogametic sex in species with two types of sex chromosomes, the homogametic sex produces only one type of gamete. For example, in mammals, the female is the homogametic sex, because an egg can only contain an X chromosome.

homologous describes attributes that are the result of homology. In the case of genes, this term describes two genes that are derived from the same ancestral gene. Homologous genes have similar DNA sequences. In the case of chromosomes, the two homologs of a chromosome pair are said to be homologous to each other.

homologous recombination the exchange of identical or similar DNA segments between homologous chromosomes.

homologous recombination repair (HRR) a mechanism for repairing double-strand breaks in which the DNA strands from a sister chromatid are used to repair a lesion in the other sister chromatid. Also called *homology-directed repair*.

homologs structures that are similar to each other due to descent from a common ancestor. For example, a homolog can be one of the chromosomes in a pair of chromosomes. Genes can also be homologs when they are descended from a common gene.

homology similarities among various species that occur because the species are derived from a common ancestor.

homology-directed repair see *homologous recombination repair (HRR)*.

homozygous describes a diploid individual that has two identical alleles of a particular gene.

horizontal gene transfer the transfer of genes from one individual to another individual that is not its offspring.

host cell a cell that is infected with a virus or bacterium, or one that harbors a vector.

host range the number of species that a virus or other pathogen can infect.

hot spots regions within a gene that are more likely to mutate than others.

housekeeping gene a gene that codes a protein required in most cells of a multicellular organism.

***Hox* complex** a group of several homeotic genes located in a particular chromosomal region in vertebrates.

Human Genome Project a worldwide collaborative project that provided a detailed map of the human genome and obtained its complete DNA sequence.

human immunodeficiency virus (HIV) an enveloped human virus that contains two copies of single-stranded RNA and is responsible for acquired immunodeficiency syndrome (AIDS).

hybrid (1) an offspring obtained from a hybridization experiment; (2) a cell produced from a cell fusion experiment in which the two separate nuclei have fused to make a single nucleus.

hybrid dysgenesis the production of defective hybrid offspring of *Drosophila*, due to the phenomenon that P elements can transpose freely.

hybrid vigor see *heterosis*.

hybrid zone an area where interbreeding between two populations may occur, producing hybrids.

hybridization (1) the breeding of two organisms of the same species with different characteristics; (2) the phenomenon in which two single-stranded DNA molecules from different sources bind with one another to form a hybrid molecule.

hypothesis testing one experimental approach for conducting science. It involves the formation of a hypothesis, which is followed by experimentation, so that scientists may reach verifiable conclusions about the world in which they live.

I

immunoglobulins (Igs) see *antibodies*.

immunoprecipitation the use of antibodies to cause other molecules, such as proteins, to precipitate, which allows them to be collected by centrifugation.

immunotherapy use of the immune system or components from the immune system to fight cancer.

imprinting see *genomic imprinting*.

imprinting control region (ICR) a DNA region that is differentially methylated and plays a role in genomic imprinting.

in situ hybridization a technique used to cytogenetically map the locations of genes or other DNA sequences within large eukaryotic chromosomes. In this method, a complementary probe is used to detect the location of a gene within a set of chromosomes.

in vitro fertilization (IVF) in the case of humans, the fertilization of an egg outside of a female's body.

inborn error of metabolism a genetic disorder that involves a defect in a metabolic enzyme.

inbreeding sexual reproduction between two genetically related individuals.

inbreeding coefficient (*F*) the probability that two alleles for a given gene in a particular individual will be identical because both copies are due to descent from a common ancestor.

inbreeding depression the phenomenon in which inbreeding produces homozygotes that are less fit, thereby decreasing the reproductive success of a population.

inclusion the practice that strives to include all types of people, regardless of their differences, in many areas such as education and the workplace; it involves including and embracing people from various backgrounds and those who might otherwise be excluded or marginalized.

incomplete dominance a pattern of inheritance in which a heterozygote that carries two different alleles exhibits a phenotype that is intermediate to those of the corresponding homozygous individuals. For example, a heterozygote may have pink flowers, whereas the homozygotes have red or white flowers.

incomplete penetrance the phenomenon in which an allele that is expected to cause a particular phenotype does not.

indel the insertion or deletion of a DNA segment in a chromosome.

induced refers to a gene that has been transcriptionally activated by an inducer.

induced mutation a change in DNA structure caused by an environmental agent.

inducer a small effector molecule that binds to a repressor or activator and causes the rate of transcription to increase.

inducible gene a gene that is regulated by an inducer, which is a small effector molecule that causes the rate of transcription to increase.

induction (1) the effects of an inducer in increasing the transcription of a gene; (2) the process by which a cell or group of cells governs the developmental fate of neighboring cells.

ingroup in cladistics, a species or group of species in which a researcher is interested and which is hypothesized to be monophyletic.

inhibitor a small effector molecule that binds to an activator protein, preventing the protein from binding to the DNA and thereby inhibiting transcription.

initiation (1) in transcription, the stage that involves the initial binding of RNA polymerase to the promoter in order to begin RNA synthesis; (2) in translation, the formation of a complex between mRNA, the initiator tRNA, and the ribosomal subunits.

initiator tRNA during translation, the tRNA that recognizes the start codon in the mRNA.

insertion see *duplication*.

insertion element (IS element) the simplest transposable element; commonly found in bacteria.

insulator a structure that limits genetic interactions, such as the effects of regulatory elements, to a specific region of a chromosome.

integrase an enzyme that functions in the integration of viral DNA or a retrotransposon into a chromosome.

intergenic region in a chromosome, a nontranscribed region of DNA that lies between two adjacent genes.

intergenic suppressor a suppressor mutation that occurs in a different gene than the gene that contains the first mutation.

International HapMap Project a worldwide effort to identify common human genetic variations.

interphase the series of phases G_1, S, and G_2, during which a eukaryotic cell spends most of its life.

interrupted mating a method used in conjugation experiments in which the length of time that the bacteria spend conjugating is stopped by a blender treatment or other type of harsh agitation.

intersex structures anatomical features that appear differently than those typically associated with males or females.

interstitial deletion deletion in which an internal segment is lost from a linear chromosome.

intervening sequence A segment of RNA that is removed during RNA splicing. Also known as an *intron*.

intragenic suppressor a suppressor mutation that is within the same gene as the first mutation that it suppresses.

intrinsic termination see ρ-*independent termination*.

intron see *intervening sequence*.

invasive refers to the ability of cancer cells to invade healthy tissue.

inversion heterozygote a diploid individual that carries one normal chromosome and a homologous chromosome with an inversion.

inversion a change in the orientation of genetic material along a chromosome such that a segment is flipped or reversed from the normal order.

inversion loop the loop structure that is formed when the homologous chromosomes of an inversion heterozygote attempt to align themselves (i.e., synapse) during meiosis.

inverted repeats (IRs) DNA sequences found in transposable elements that are identical (or very similar) but run in opposite directions.

iron regulatory protein (IRP) a translational regulatory protein that recognizes iron response elements that are found in specific mRNAs. It may inhibit translation or stabilize the mRNA.

iron response element (IRE) a sequence in mRNAs that is recognized by the iron regulatory protein.

isoacceptor tRNAs two or more tRNAs that differ at the wobble position but can recognize the same codon.

isoelectric focusing a form of gel electrophoresis in which a protein migrates to the point in a gel where its net charge is zero.

isogamous describes a species that produces morphologically similar gametes.

K

karyotype a photographic representation of all the chromosomes within a cell. It reveals how many chromosomes are found within an actively dividing somatic cell.

kinetochore a protein complex bound to a centromere of a chromosome during meiosis and mitosis.

kinetochore microtubules the microtubules that are connected to kinetochores on chromosomes.

Kozak's rules a set of rules that identify the most favorable types of bases to flank a eukaryotic start codon in an mRNA molecule.

L

lac repressor a protein that binds to the operator site of the *lac* operon and inhibits transcription.

lagging strand a strand during DNA replication that is synthesized as short Okazaki fragments in the direction away from the replication fork.

lamina-associated domains (LADs) heterochromatic chromosomal regions that are attached to the nuclear lamina that lines the inner nuclear membrane.

last common universal ancestor the ancestor that gave rise to all living species on Earth.

latent used to describe a virus that is inactive, that is, is within a cell but not making new viruses. A virus may remain latent for a long time.

lateral induction an effect of Notch signaling during embryonic development in which adjacent cells acquire identical or similar cell fates.

lateral inhibition an effect of Notch signaling during embryonic development in which adjacent cells acquire different cell fates.

leading strand a strand during DNA replication that is synthesized continuously toward the replication fork.

left-right axis in bilaterians, the axis that determines left-right symmetry.

lethal allele an allele that may cause the death of an organism.

lethal mutation a mutation that produces an allele that results in the death of a cell or an organism.

LINEs in mammals, long interspersed elements that are usually 1 to 10 kbp in length and occur in 20,000 to 1,000,000 copies per genome.

linkage see *genetic linkage*.

linkage disequilibrium the phenomenon that exists when alleles and molecular markers are associated with each other at a frequency that is significantly higher than expected by random chance.

linkage group a single chromosome, or all of the genes found on a single chromosome, which are linked together.

linkage mapping determining the relative spacing and order of genes along a chromosome by analyzing the outcomes of crosses.

lipid a general name given to an organic molecule that is insoluble in water. Cell membranes contain large amounts of lipids.

liquid-liquid phase separation (LLPS) a phenomenon in which liquid-like compartments are formed by macromolecules that become concentrated in a given location and come out of solution.

local population a segment of a population that is somewhat isolated. Members of a local population are more likely to breed with each other than with individuals of the same species from a more distant population.

locked nucleic acid (LNA) an AMO that contains a ribose sugar that has an extra bridge connecting the 2′ oxygen and 4′ carbon. The bridge locks the ribose in a conformation that causes it to bind more tightly to the complementary miRNA.

locus (pl. loci) the site within a genetic map where a specific gene or other DNA segment is found

locus heterogeneity the phenomenon in which a particular type of disease or trait may be caused by mutations in two or more different genes.

long non-coding RNA (lncRNA) an ncRNA that is longer than 200 nucleotides.

long-read sequencing (LRS) methods of DNA sequencing that produce base sequences with lengths in the range from 10,000 to 30,000 bases, and sometimes longer.

long terminal repeats (LTRs) sequences containing many short segments that are tandemly repeated. They are found in retroviruses and viral-like retrotransposons.

loop domain a segment of chromatin that is organized into a loop.

loop extrusion model a mechanism for loop formation in eukaryotes in which SMC proteins promote the formation of a loop, and the bottom of the loop is held in place by SMC proteins.

loss-of-function allele an allele of a gene that codes an RNA or protein that is nonfunctional or compromised in function.

loss-of-function mutation a change in a genetic sequence that creates a loss-of-function allele.

loss of heterozygosity (LOH) the phenomenon in which a heterozygous somatic cell incurs a genetic change that inactivates the single functional allele.

LTR retrotransposon a type of retrotransposon that is evolutionarily related to retroviruses and has long terminal repeats.

Lyon hypothesis a hypothesis to explain the pattern of X-chromosome inactivation seen in mammals. Initially, both X chromosomes are active. However, at an early stage of embryonic development, one of the two X chromosomes is randomly inactivated in each somatic cell.

lysogenic cycle a type of growth cycle for a phage in which the phage integrates its genetic material into the chromosome of the bacterium. This integrated phage DNA can exist in a dormant state for a long time, during which no new bacteriophages are made.

lysogeny condition of latency in bacteriophages.

lytic cycle a type of growth cycle for a bacteriophage in which the phage directs the synthesis of many copies of its genetic material and coat proteins. These components then assemble to make new phages. When synthesis and assembly are completed, the bacterial host cell is lysed, and the newly made phages are released into the environment.

M

M phase the phase of the eukaryotic cell cycle in which nuclear division, either mitosis or meiosis, occurs. It is divided into prophase, prometaphase, metaphase, anaphase, and telophase.

macrodomain a unit of organization in a bacterial chromosome that consists of many adjacent microdomains.

macroevolution evolutionary changes that occur at or above the species level and produce relatively large changes in form and function that are sufficient to produce new species and higher taxa.

macromolecule a large organic molecule composed of smaller building blocks. Examples include DNA, RNA, proteins, and large carbohydrates.

maintenance methylation the methylation of hemimethylated DNA following DNA replication.

major groove a wide indentation in the DNA double helix in which the bases have access to water.

male In the case of humans, males usually have testes at birth, have the potential to produce sperm if fertile, and are usually XY.

malignant describes a tumor composed of cancerous cells.

map distance the relative distance between sites (e.g., genes) along a single chromosome. In a testcross, it is defined as the number of recombinant offspring divided by the total number of offspring, multiplied by 100.

map unit (mu) a unit of map distance obtained from genetic crosses. One map unit is equivalent to 1% recombinant offspring in a testcross.

mapping the experimental process of determining the relative locations of genes or other segments of DNA on individual chromosomes.

mass spectrometry a technique to accurately measure the mass of a molecule, such as a peptide that is a fragment of a protein.

maternal effect an inheritance pattern for certain nuclear genes in which the genotype of the female parent directly determines the phenotypic traits of the offspring.

maternal inheritance a type of extranuclear inheritance of DNA that occurs through the cytoplasm of the egg.

maturase a protein that enhances the rate of self-splicing of group I and II introns.

maximum likelihood a strategy for choosing the best phylogenetic tree based on the probability that an evolutionary model and a proposed phylogenetic tree would give rise to the observed data.

MCM helicase a group of six proteins needed to complete a process called DNA replication licensing, which is necessary for the formation of two replication forks at an origin of replication in eukaryotes.

mean the sum of all the values in a group divided by the number of individuals in the group.

mean fitness of the population ($\bar{w}$) the average fitness of a population, calculated by considering the frequencies and fitness values for all genotypes.

mediator a large protein complex that interacts with RNA polymerase II and various regulatory transcription factors. Depending on its interactions with regulatory transcription factors, mediator may stimulate or inhibit RNA polymerase II.

meiosis a form of nuclear division in which the sorting process results in the production of haploid cells from a diploid cell.

meiotic nondisjunction the event in which chromosomes do not segregate properly during meiosis.

Mendel's law of independent assortment two different genes will randomly assort their alleles during gamete formation (if they are not linked).

Mendel's law of segregation the two copies of a gene segregate (or separate) from each other during the process that gives rise to gametes.

Mendelian inheritance a pattern of inheritance that follows Mendel's laws; this pattern involves the transmission of eukaryotic genes that are located on the chromosomes found within the cell nucleus.

meristic traits traits that can be counted and expressed in whole numbers.

merozygote a partial diploid strain of bacteria containing F′ factor genes.

messenger RNA (mRNA) a type of RNA that is transcribed from a protein-coding gene and contains the information for the synthesis of a polypeptide.

metacentric describes a chromosome with the centromere near the middle.

metadata see *annotations*

metagenome a collection of genes from an environmental sample.

metagenomics the study of a complex mixture of genetic material obtained from an environmental sample.

metalloribozyme a catalytically active RNA that uses one or more metal ions in its catalytic action.

metaphase the third phase of mitosis. The chromosomes align along the central plane of the spindle apparatus, and the formation of the spindle is completed.

metaphase chromosome the highly compacted form of a eukaryotic chromosome that is observed during metaphase of mitosis.

metaphase plate the plane along which pairs of sister chromatids align during metaphase.

metastatic describes cancer cells that have migrated to other parts of the body.

methylation see *DNA methylation*.

methyl-CpG-binding protein a protein that binds to a CpG island when it is methylated.

microdomain a loop of DNA that is found in a bacterial chromosome and is typically 10 kbp in length.

microevolution changes in a population's gene pool with regard to particular alleles over measurable periods of time.

microinjection the use of microscopic-sized needles to inject a substance, such as DNA, into cells.

microRNAs (miRNAs) ncRNAs that are transcribed from eukaryotic genes that are endogenous—they are normally found in the genome. They silence the expression of mRNAs via RNA interference.

microsatellite a molecular marker composed of many repeated copies of a short sequence. Microsatellites are interspersed throughout a genome and are quite variable in length among different individuals. They can be amplified by PCR.

microscopy the use of a microscope to view cells or subcellular structures.

microtubule-organizing center (MTOC) a structure in a eukaryotic cell from which microtubules grow.

minimal medium a type of growth medium for microorganisms that contains a mixture of nutrients that are required for growth; nothing additional has been added.

minisatellite a repetitive sequence that was formerly used in DNA fingerprinting. Its use has been largely superseded by smaller repetitive sequences called microsatellites.

minor groove a narrow indentation in the DNA double helix in which the bases have access to water.

minute a unit of measure used in bacterial conjugation experiments; refers to the relative time it takes for genes to first enter an F⁻ recipient strain during conjugation.

mismatch repair system a DNA repair system that recognizes base pair mismatches and repairs the newly made daughter strand that contains the incorrect base.

missense mutation a base substitution that leads to a change in the amino acid sequence of the polypeptide that is coded by a gene.

mitochondrial DNA (mtDNA) the DNA found within mitochondria.

mitosis a type of nuclear division into two nuclei, during which each daughter cell receives the same complement of chromosomes.

mitotic nondisjunction an event in which chromosomes do not segregate properly during mitosis.

mitotic recombination crossing over that occurs during mitosis and produces a pair of recombinant chromosomes.

mitotic spindle apparatus during cell division in eukaryotic cells, a structure composed of microtubules that sorts the chromosomes.

mitotic spindle see *mitotic spindle apparatus*.

model organism an organism studied by many researchers so that they can more easily compare their results and begin to unravel the properties of a given species.

moderately repetitive sequences base sequences that are found a few hundred to several thousand times in a genome.

molecular clock the use of the rate of neutral or nearly neutral mutations as a tool to measure evolutionary time.

molecular evolution molecular-level changes in the genetic material that underlie the process of evolution.

molecular genetics the study of DNA structure and function at the molecular level.

molecular level the observation of living things at the level of individual molecules.

molecular marker a segment of DNA that is found at a specific site along a chromosome and has properties that enable it to be uniquely recognized using molecular tools, such as PCR and gel electrophoresis.

molecular paleontology see *ancient DNA analysis.*

molecular pharming the use of biotechnology to produce medically important proteins in the mammary glands of livestock and in agricultural plants.

molecular profiling the use of methods that enable researchers to understand the molecular changes that occur in diseases, such as cancer.

monad a single chromatid within a dyad.

monoallelic expression in the case of genomic imprinting, refers to the phenomenon that only one of the two alleles of a given gene is transcriptionally expressed.

monoclonal antibodies (mAbs) a homogenous population of antibodies that are made in a laboratory and recognize one specific antigen.

monoecious in plants, a species in which male and female gametophytes are produced on a single (sporophyte) individual.

monohybrid an individual produced from a single-factor cross in which the parents had different variants for a single character.

monomorphic describes a trait that is found in only one form in a population or a gene that is found as only one allele in a population.

monophyletic group a group of species consisting of all descendents of the group's most common ancestor. Also referred to as a *clade.*

monosomic describes a diploid cell or organism that is missing a chromosome (i.e., $2n - 1$).

monozygotic (MZ) twins twins that are genetically identical because they were formed from the same sperm and egg.

morph a form or phenotype within a species. For example, red eyes and white eyes are different eye color morphs of fruit flies.

morphogen a molecule that conveys positional information and promotes developmental changes.

morphological trait a trait that affects the morphology (physical form) of an organism. An example is eye color.

mosaicism condition in which the cells of part of an organism differ genetically from those of the rest of the organism.

motif (1) in proteins, the name given to an amino acid sequence or domain that has a very similar structure and function in many proteins; (2) in DNA or RNA, the name given to a particular nucleotide base sequence that has a specific function; (3) in genetics, a specialized sequence with a particular meaning or function.

mouse model a strain of mice that carries a mutation that is analogous to a disease-causing mutation of a human gene; such mice often have disease symptoms that are similar to the human symptoms.

mRNA see *messenger RNA.*

multicellularity consisting of more than one cell.

multinomial expansion equation an equation used to solve genetic problems involving three or more types of unordered outcomes.

multiple alleles two or more alleles of the same gene found within a population.

multiple-sequence alignment the aligning by a computer program of two or more DNA or amino acid sequences based on their homology to each other.

multiplication method a method for solving independent assortment problems in which the probabilities of the outcome for each gene are multiplied together.

multipotent a type of stem cell that can differentiate into any of several different types of cells.

mutable site a site in a chromosome that tends to break at a fairly high rate due to the presence of a transposable element. Also called a *mutable locus.*

mutagen an agent that can alter the structure of DNA, causing a mutation.

mutant allele an allele that has been created because a wild-type allele has been altered by mutation.

mutation a permanent change in the genetic material that can be passed from cell to cell or, if it occurs in reproductive cells, from parent to offspring.

mutation frequency the number of mutant genes divided by the total number of copies of that gene within a population.

mutation rate the likelihood that a gene will be altered by a new mutation.

myogenic bHLH protein a type of transcription factor involved in muscle-cell differentiation.

N

narrow-sense heritability heritability that takes into account only the additive effects of alleles.

natural selection the process whereby differential fitness acts on the gene pool. When a mutation creates a new allele that is beneficial, the allele may become prevalent within future generations because the individuals possessing the allele are more likely to reproduce and pass it to their offspring.

natural transformation a natural process of transformation that occurs in certain species of bacteria.

negative control transcriptional regulation by a repressor protein.

negative frequency-dependent selection a pattern of natural selection in which the fitness of a genotype decreases when its frequency becomes higher.

neutral mutation a mutation that has no detectable effect on protein function or no detectable effect on the survival of the organism.

neutral theory of evolution the theory that most genetic variation observed in natural populations is due to the accumulation of neutral mutations. Also known as *non-Darwinian evolution.*

next-generation sequencing technologies newer DNA-sequencing technologies that are more rapid and inexpensive than the dideoxy method.

nitrogen mustard an alkylating agent that can cause mutations in DNA.

nitrous acid a type of chemical mutagen that deaminates bases, replacing amino groups with keto groups.

nonallelic homologous recombination recombination that occurs at homologous sites within chromosomes, where the sites are not alleles of the same gene. Such misaligned crossovers are often due to the occurrence of repetitive sequences.

nonautonomous element a transposable element that lacks a gene such as the one that codes transposase or reverse transcriptase, which is necessary for transposition to occur.

non-coding RNAs (ncRNAs) RNAs that do not code polypeptides.

non-Darwinian evolution see *neutral theory of evolution.*

nondisjunction event in which chromosomes do not segregate properly during mitosis or meiosis.

nonessential genes genes that are not absolutely required for survival, although they are likely to be beneficial to the organism.

nonhomologous end joining (NHEJ) a repair mechanism for double-strand breaks in which the ends of the DNA are pieced back together.

nonhomologous recombination the exchange of DNA between nonhomologous segments of chromosomes or plasmids.

noninvasive prenatal testing (NIPT) a type of genetic test on a fetus that is done by collecting a blood sample from the female parent and analyzing cell-free DNA (cfDNA); it is usually used to test for abnormalities in chromosome number.

non-LTR retrotransposon a type of retrotransposon that does not have long terminal repeats.

nonneutral mutation a mutation that affects the phenotype of the organism and can be acted on by natural selection.

nonparental refers to a combination of traits not found in the true-breeding parental generation. Also called *recombinant.*

nonparental ditype (NPD) an ascus that contains four spores that all have a nonparental combination of alleles.

nonrecombinant refers to a cell or offspring that carries the same combinations of alleles found in the chromosomes of their parents or in the chromosomes of the parental generation in linkage experiments.

nonsense codon see *stop codon.*

nonsense mutation a mutation that involves a change from a codon that specifies an amino acid to a stop codon.

nontemplate strand a strand of DNA that is not used as a template during transcription.

norm of reaction the effects of environmental variation on an individual's phenotype.

normal displaying one or more characteristics that are common among members of a given species.

normal distribution a distribution for a large sample in which the trait of interest varies in a symmetrical way around an average value.

Northern blotting a technique used to detect a specific RNA within a mixture of many RNA molecules.

Notch receptor a cell receptor that is involved in cell signaling between adjacent cells in embryonic development, which plays a key role in cell differentiation, cell proliferation, and apoptosis.

N-terminus see *amino-terminus.*

nuclear genes genes that are located on chromosomes found in the cell nucleus of eukaryotic cells.

nuclear lamina (NL) a fibrous layer of proteins that lines the inner nuclear membrane of the cell nucleus in most eukaryotic cells.

nucleic acid RNA or DNA. A macromolecule that is composed of repeating nucleotide units.

nucleoid a darkly staining region that contains the genetic material of mitochondria, chloroplasts, or bacteria.

nucleoid-associated proteins (NAPs) a set of DNA-binding proteins found in bacteria that facilitate chromosome compaction and organization.

nucleolus an organelle within the nucleus of eukaryotic cells where the assembly of ribosomal subunits occurs.

nucleoprotein a complex containing DNA (or RNA) and a protein.

nucleoside structure in which a base is attached to a sugar, but no phosphate is attached to the sugar.

nucleosome the repeating structural unit within eukaryotic chromatin and also within some archaeal chromosomes. It is composed of double-stranded DNA wrapped around histone proteins. In eukaryotes, it is an octamer of histone proteins.

nucleosome-free region (NFR) a region within the DNA of a eukaryote where nucleosomes are not found.

nucleotide the repeating structural unit of nucleic acids, composed of a sugar, one to three phosphates, and a nitrogen-containing base.

nucleotide excision repair (NER) a DNA repair mechanism in which several nucleotides in the damaged strand are removed from the DNA and the undamaged strand is used as a template to resynthesize a normal strand.

nucleus a membrane-bound organelle in eukaryotic cells where the chromosomes are found.

null hypothesis a hypothesis that assumes that there is no real difference between the observed and expected values.

O

Okazaki fragments short segments of DNA that are synthesized to produce the lagging strand during DNA replication.

oncogene a mutant gene that is overexpressed and thus promotes cancerous growth.

one-gene/one-enzyme hypothesis the idea, which later needed to be modified, that one gene codes one enzyme.

oogenesis the production of egg cells.

open complex the region of separation of two DNA strands where RNA polymerase can use one of the strands as a template for transcription.

open conformation a loosely packed structure of chromatin that can be transcribed.

open reading frame (ORF) a region in a genetic sequence that does not contain stop codons.

operator see *operator site*.

operator site a sequence of nucleotides in bacterial DNA that provides a binding site for a genetic regulatory protein. Often referred to simply as the *operator*.

operon an arrangement in DNA in which two or more genes are found within a regulatory unit that is under the transcriptional control of a single promoter.

organelle a specialized structure within a cell that is surrounded by a single or double membrane or caused by liquid-liquid phase separation.

organism level the level of observation or experimentation that involves a whole organism.

organizing center in plants, a region of the meristem that ensures the proper organization of the meristem and preserves the correct number of actively dividing stem cells.

orientation-independent refers to certain types of genetic regulatory elements that can function in either the forward or the reverse direction. Certain enhancers are orientation-independent.

origin of replication a site on a chromosome that functions as an initiation site for the assembly of several proteins that begin the process of DNA replication.

origin of transfer the location on an F factor or within the chromosome of an *Hfr* strain that is the initiation site for the transfer of DNA from one bacterium to another during conjugation.

origin recognition complex (ORC) a group of proteins found in eukaryotes that acts as the first initiator of preRC assembly to begin DNA replication.

orthologs homologous genes (genes derived from the same ancestral gene) that are found in different species.

outbreeding sexual reproduction between genetically unrelated individuals.

outgroup in cladistics, a species or group of species that is more distantly related to the ingroup.

ovary (1) in plants, the structure in which the ovules develop; (2) in animals, the structure that produces egg cells and female hormones.

ovule the structure in flowering plants where the female gametophyte (i.e., embryo sac) is produced.

ovum a female gamete; also known as an egg cell or egg.

oxidative DNA damage changes in DNA structure that are caused by reactive oxygen species (ROS).

oxidative stress an imbalance between the production of reactive oxygen species (ROS) and an organism's ability to break them down.

P

P generation the parental generation in a genetic cross.

P value the probability, found in a chi square table, that the deviations between observed and expected values are due to random chance.

palindromic describes sequences in the two strands of a DNA molecule that are identical when read in opposite directions.

pandemic a disease that occurs over a wide geographic area and usually affects a high proportion of the population.

paracentric inversion an inversion in which the centromere is found outside the inverted region.

paralogs two or more homologous genes that are found within a single species and constitute a gene family.

paramutable describes an allele whose expression can be changed by paramutation.

paramutagenic describes an allele that can cause a paramutation, which changes the expression of another allele (the paramutable allele).

paramutation an interaction between two alleles of a given gene in which one allele induces a heritable change in the other allele without changing that allele's DNA sequence.

parasegments transient subdivisions that occur in the *Drosophila* embryo prior to the formation of segments.

parental ditype (PD) an ascus that contains four spores with the parental combinations of alleles.

parental generation in a genetic cross, the first generation in the experiment. In Mendel's studies, the parental generation was true-breeding with regard to particular traits.

parental strand see *template strand*.

partial monosomy the situation in which an individual carries a deletion of a segment of a chromosome.

partial trisomy the situation in which an individual carries a duplication of a segment of a chromosome.

particulate theory of inheritance a theory proposed by Mendel. It states that the factors that govern traits are inherited as discrete units that remain unchanged as they are passed from parent to offspring.

pattern recognition in bioinformatics, this term refers to the ability of a computer program to recognize a pattern of symbols.

pedigree analysis genetic analysis using information contained within family trees. In this approach, the aim is to determine the type of inheritance pattern that a gene follows.

pedigrees charts representing family relationships.

pellet a collection of particles found at the bottom of a centrifuge tube.

peptide bond a covalent bond formed between the carboxyl group in one amino acid in a polypeptide and the amino group in the next amino acid.

peptidyl site (P site) a site on a ribosome that carries a tRNA along with a polypeptide.

peptidyl transfer the step during the elongation stage of translation in which the polypeptide is removed from the tRNA in the P site and transferred to the amino acid at the A site.

peptidyl transferase a complex that functions during translation to catalyze the formation of a peptide bond between the amino acid in the A site of the ribosome and the growing polypeptide.

pericentric inversion an inversion in which the centromere is located within the inverted region of the chromosome.

peripheral zone in plants, an area in the meristem that contains dividing cells that eventually differentiate into plant structures.

personalized medicine the use of information about a patient's genotype and other clinical data in order to select a medication, therapy, or preventative measure that is specifically suited to that patient. Also known as *precision medicine*.

phage see *bacteriophage*.

phage λ a bacteriophage that infects *E. coli* and has double-stranded DNA as its genome.

pharmacogenetics the study or clinical testing of genetic variation that causes differing responses to drugs.

phenetic approach the construction of a phylogenetic tree, known as a phenogram, based on the overall similarities among a group of species without understanding their evolutionary history.

phenogram a phylogenetic tree that has been constructed using a phenetic approach.

phenotype the observable traits of an organism.

phenotype frequency the frequency of a particular phenotype in a population.

phenylketonuria (PKU) a human genetic disorder arising from a defect in phenylalanine hydroxylase.

phosphodiester linkage in a DNA or RNA strand, a linkage in which a phosphate group connects two sugar molecules.

photolyase an enzyme found in bacteria, fungi, most plants, and some animals that can recognize and split thymine dimers, which returns the DNA to its original condition.

photoreactivation a type of DNA repair mechanism in which photolyase recognizes and splits thymine dimers; the mechanism requires light.

phylogenetic tree a diagram that describes a phylogeny and constitutes a hypothesis concerning the evolutionary relationships among different species.

phylogeny the sequence of events involved in the evolutionary development of a species or group of species.

physical mapping determining the locations of and distances between genes and other genetic sequences on a chromosome using DNA-cloning and/or DNA-sequencing techniques.

physiological trait a trait that affects a cellular or body function. An example is the rate of glucose metabolism.

pioneer factors transcription factors that have the ability to recognize and bind to specific DNA sequences that are exposed on the surface of a nucleosome and then recruit chromatin-remodeling complexes and histone-modifying enzymes that carry out epigenetic changes such as histone eviction and covalent modifications.

piRNA-induced silencing complex (piRISC) a complex found in animal cells that is composed of piRNA and PIWI proteins and that silences transposable elements.

piRNA-induced transcriptional silencing complex (piRITS) a complex found in animal cells that is composed of piRNA and PIWI proteins; it silences transposable elements by preventing their transcription.

PIWI-interacting RNA (piRNA) a type of ncRNA that interacts with PIWI proteins and provides defense against the movement and insertion of transposable elements.

plasmid a general term for a DNA molecule (most often circular) that exists independently of the chromosomal DNA. Some plasmids are used as vectors in cloning experiments.

pleiotropy the multiple effects of a single gene on the phenotype of an organism.

pluripotent describes a stem cell that can differentiate into any or almost any type of cell found in the adult organism.

point mutation a change in a single base pair within DNA.

polar microtubules the microtubules that project toward the region where the chromosomes will be found during mitosis; those that overlap with each other play a role in pushing the spindle poles apart.

pollen grain the male gametophyte of flowering plants; also called *pollen.*

polyA tail the string of adenine nucleotides at the 3′ end of eukaryotic mRNAs.

polyadenylation the process of attaching a string of adenine nucleotides to the 3′ end of eukaryotic pre-mRNAs.

polycistronic mRNA an mRNA that is transcribed from an operon and codes two or more proteins.

polycomb group (PcG) a set of protein complexes that are key regulators of epigenetic changes that are programmed during development. They cause gene repression.

polycomb response element (PRE) a DNA element in *Drosophila* that is initially recognized by a PRE-binding protein, which then recruits PRC2 to the site.

polygenic refers to a trait that is controlled by multiple genes.

polygenic inheritance the transmission of any trait that is governed by two or more different genes.

polymerase chain reaction (PCR) the method for amplifying a DNA region involving the sequential use of oligonucleotide primers and *Taq* polymerase.

polymerase switch an exchange of one type of DNA polymerase for another type during DNA replication.

polymorphic describes a trait or gene (or other segment of DNA) that is found in two or more forms in a population.

polymorphism (1) the prevalence of two or more phenotypic forms in a population; (2) the phenomenon in which a gene exists in two or more alleles within a population.

polypeptide a linear sequence of amino acids that is the product of mRNA translation. One or more polypeptides fold and associate with each other to form a functional protein.

polyploid describes an organism or cell with three or more sets of chromosomes.

polyribosome an mRNA transcript that has many bound ribosomes in the act of translation.

polysome see *polyribosome.*

polytene chromosome aggregation of chromosomes found in certain cells, such as *Drosophila* salivary cells, in which homologous chromosomes have synapsed and replicated many times and the copies lie side by side.

population a group of individuals of the same species that occupy the same region and (for sexually reproducing species) can interbreed with one another.

population genetics the field of genetics that is primarily concerned with the extent of genetic variation within a group of individuals and changes in that variation over time.

population level the level of observation or experimentation that involves a population of organisms.

position effect a change in phenotype that occurs when the location of a gene changes from one chromosomal site to a different one.

positional information signals in the form of chemical substances and other environmental cues that enable a cell to deduce its position relative to other cells.

positive control transcriptional regulation by an activator protein.

positive interference the phenomenon in which a crossover that occurs in one region of a chromosome decreases the probability that another crossover will occur nearby.

positive selection an increase in the frequency of a particular allele that occurs because the allele is favored by natural selection.

posttranslational describes events that occur after translation is completed.

posttranslational covalent modification the chemical alteration of a protein or the covalent attachment of a molecule or functional group to a protein after it has been synthesized via translation.

posttranslational regulation the functional control of proteins that are already present in the cell.

postzygotic isolating mechanism a mechanism of reproductive isolation that prevents the development of a viable or fertile offspring after fertilization has occurred.

precision medicine the use of information about a patient's genotype and other clinical data in order to select a medication, therapy, or preventative measure that is specifically suited to that patient. Also known as *personalized medicine.*

preimplantation genetic diagnosis (PGD) a form of genetic testing in which an embryo obtained via in vitro fertilization is tested for genetic abnormalities prior to implantation within the uterus.

preinitiation complex a closed complex formed by the assembly of general transcription factors, mediator, and RNA polymerase II at the core promoter, which initiates transcription.

preintegration complex a complex of integrase, other proteins, and double-stranded HIV DNA that will be integrated into the host chromosomal DNA.

pre-mRNA in eukaryotes, a long transcript corresponding to the entire sequence of a protein-coding gene, which is produced within the nucleus during transcription. This pre-mRNA is usually altered by splicing and other modifications before it exits the nucleus.

prereplication complex (preRC) in eukaryotes, an assembly of at least 14 different proteins, including a group of proteins called the origin recognition complex (ORC) that are required to initiate DNA replication.

prezygotic isolating mechanism a mechanism of reproductive isolation that prevents the formation of a zygote.

Pribnow box the TATAAT sequence that is often found at the −10 site of a bacterial promoter.

primary structure with regard to proteins, the linear sequence of amino acids that forms a polypeptide.

primase an enzyme that synthesizes short RNA primers during DNA replication.

primer a short segment of DNA or RNA that initiates DNA replication.

primer annealing in PCR, the process in which primers bind to a strand of template DNA.

primer extension in PCR, the step during which complementary strands of DNA are synthesized from the denatured template DNA, starting at the primers.

primer walking a method to identify disease-causing alleles in which a primer that recognizes a known region of DNA, such as a molecular marker, guides the DNA sequencing to that gene of interest.

primitive character see *ancestral character.*

primosome a protein complex that includes DNA helicase and primase.

principle of parsimony the concept that the preferred hypothesis is the one that is the simplest.

prion an infectious particle that causes any of several types of neurodegenerative diseases affecting humans or livestock. It is composed entirely of protein.

probability the chance that an outcome will occur in the future.

processing body (P-body) the cellular structure where a RISC-mRNA complex is stored until it is reused or degraded.

processive enzyme an enzyme, such as RNA or DNA polymerase, which glides along the template strand and does not dissociate from it while catalyzing the covalent attachment of nucleotides.

product rule the probability that two or more independent outcomes will occur is equal to the product of their individual probabilities.

proflavin an acridine dye, which is a chemical mutagen that causes frameshift mutations.

prokaryotes another name for bacteria and archaea. The term refers to the observation that their chromosomes are not contained within a membrane-bound nucleus in the cell.

prometaphase the second phase of mitosis. During this phase, the nuclear membrane vesiculates, and the mitotic spindle is completely formed.

promoter a sequence within a gene that initiates (i.e., promotes) transcription.

promoter escape occurs at the core promoter in eukaryotes; the GTFs and mediator release their binding to RNA polymerase II, thereby allowing elongation to occur.

proofreading function the ability of DNA polymerase to remove mismatched bases from a newly made strand.

prophage phage DNA that has been integrated into the bacterial chromosome.

prophase the first phase of mitosis. The chromosomes have already replicated and begin to condense. The mitotic spindle apparatus starts to form.

protandrous hermaphrodite an organism that first becomes a male and can later transform into a female.

protease an enzyme that digests the polypeptide backbone in a protein.

protein a functional unit composed of one or more polypeptides.

protein-coding genes genes that carry the information to produce mRNA and that code the amino acid sequence of polypeptides.

protein microarray a small silica, glass, or plastic slide that is dotted with many different proteins.

proteome the collection of all the proteins that a given cell or organism makes.

proteomics the study of protein function at the genome level. It involves the study of many proteins simultaneously.

protobiont a structure that preceded the emergence of living cells; it was an aggregate of molecules and macromolecules that acquired a boundary, such as a lipid bilayer, that allowed it to maintain an internal chemical environment distinct from that of its surroundings.

proto-oncogene a normal cellular gene that does not cause cancer but which may incur a gain-of-function mutation that causes abnormally high expression.

prototroph a strain that does not need a particular nutrient included in its growth medium.

provirus viral DNA that has been integrated into the chromosome of a eukaryotic host cell.

proximal promoter pausing a phenomenon in eukaryotes in which RNA polymerase II pauses in RNA synthesis while it is still fairly close to the transcriptional start site, usually within 60 bp.

proximodistal axis in animals, an axis for designating positions on limbs in which the part of the limb attached to the trunk is proximal, whereas the end of the limb is distal.

pseudoautosomal inheritance the inheritance pattern of genes that are found on both the X and Y chromosomes. Even though such genes are located physically on the sex chromosomes, their pattern of inheritance is identical to that of autosomal genes.

Punnett square a diagrammatic method in which the gametes that two parents can produce are aligned next to a square grid as a way to predict the types of offspring the parents are expected to produce and in what proportions.

purine a type of nitrogenous base that has a double-ring structure. Examples are adenine and guanine.

pyrimidine a type of nitrogenous base that has a single-ring structure. Examples are cytosine, thymine, and uracil.

pyrrolysine a nonstandard amino acid that may be incorporated into polypeptides during translation.

Q

QTL mapping the determination of the locations of QTLs using mapping methods, such as genetic crosses coupled with analysis of molecular markers.

quantitative genetics the field of genetics that studies complex and quantitative traits.

quantitative PCR (qPCR) a method of PCR in which the synthesis of DNA is monitored in real time. This method can quantitate the starting amount of DNA in a sample.

quantitative trait locus (QTL) the location on a chromosome of one or more genes that affect the outcome of a quantitative trait.

quantitative traits traits, usually polygenic in nature, that can be described using numbers.

quaternary structure the structure of a functional protein formed when two or more polypeptides associate with each other.

R

R group refers to a grouping of atoms, such as the side chain of an amino acid, which has particular chemical properties.

radial loop arrays in metaphase chromosomes, the arrangement of loops in a circular and spiral manner around a condensin II scaffold.

radioimmunoassay a method for measuring the amount of an antigen in a biological sample.

random mutation theory according to this theory, mutations are random events—they can occur in any gene and do not require exposure of an organism to an environmental condition that causes specific types of mutations.

random sampling error the deviation between the observed and expected outcomes.

reactive oxygen species (ROS) products of oxygen metabolism that are produced in all aerobic organisms and that can, if they accumulate, damage cellular molecules, including DNA, proteins, and lipids.

reader domain a domain in a protein that controls the binding of that protein to a nucleosome.

reading frame a series of codons determined by reading bases in groups of three beginning with the start codon as a frame of reference.

realized heritability a form of narrow-sense heritability that is observed when selective breeding is practiced.

recessive a trait or allele that is masked by the presence of a dominant trait or allele.

recessive epistasis a form of epistasis in which an individual must be homozygous for either of two recessive alleles to mask a particular phenotype.

reciprocal crosses a pair of crosses in which the traits of the two parents differ with regard to sex. For example, one cross could be a red-eyed female fly and a white-eyed male fly, and the reciprocal cross would be a red-eyed male fly and a white-eyed female fly.

reciprocal translocation rearrangment in which two different chromosomes exchange pieces.

recombinant (1) refers to a cell or offspring that carries a new combination of alleles or traits due to crossing over or due to the independent assortment of chromosomes; (2) describes DNA molecules that are produced by molecular techniques in which segments of DNA are joined to each other in ways that differ from their original arrangement in their native chromosomal sites. The cloning of DNA into vectors is an example.

recombinant DNA molecules molecules that are produced in a test tube by covalently linking DNA fragments from two different sources.

recombinant DNA technology the use of in vitro molecular techniques to manipulate fragments of DNA to produce new arrangements.

recombinant offspring an offspring that has inherited a combination of alleles or a combination of traits that is different from either of their parents. For unlinked genes, recombinant offspring are produced via independent assortment. For linked genes, recombinant offspring are produced by crossing over.

recombinant vector a vector that contains an inserted fragment of DNA, such as a gene from a chromosome.

recombination frequency see *map distance.*

regulatory element (1) a short segment of DNA that is recognized by a regulatory transcription factor. The binding of the transcription factor affects the rate of transcription. Also referred to as a *regulatory sequence* or a *response element;*(2) a short segment of RNA that is recognized by a translational regulatory protein. The binding of the protein affects the rate of translation or the rate of RNA degradation.

regulatory sequence see *regulatory element.*

regulatory transcription factor (RTF) a protein or protein complex that binds to a regulatory element and influences the rate of transcription via RNA polymerase.

reinforcement the phenomenon in which gene flow through a hybrid zone from one population to the other is very limited and thereby reinforces the reproductive isolation between the two populations.

relative fitness (*w*) the reproductive success of a genotype relative to the maximum reproductive success of other genotypes in the population.

relaxosome a protein complex that recognizes the origin of transfer in F factors and other conjugative plasmids, cuts one DNA strand, and aids in the transfer of the T DNA.

release factor a protein that recognizes a stop codon and promotes termination of translation and the release of the completed polypeptide.

repetitive sequences short DNA sequences that occur many times within a species' genome.

replica plating a technique in which replicas of bacterial colonies are transferred to new growth plates.

replication fork the region in which two DNA strands have separated and new strands are being synthesized.

replisome a complex that consists of a primosome and dimeric DNA polymerase.

repressible gene a gene that is regulated by a corepressor or an inhibitor, which are small effector molecules that cause the rate of transcription to decrease.

repressor a regulatory protein that binds to DNA and inhibits the rate of transcription.

reproductive cloning biotechnology methods that produce two or more genetically identical individuals and may involve the use of genetic material from somatic cells.

reproductive isolation the inability of a species to successfully interbreed with other species.

reproductive success the likelihood that an individual will contribute fertile offspring to the next generation.

resolution (1) the last two steps of homologous recombination, in which the entangled DNA strands become resolved into two separate chromosomes; (2) in microscopy, the minimum distance between two objects that enables them to be seen as being separate from each other.

response element see *regulatory element.*

restriction endonuclease see *restriction enzyme.*

restriction enzyme an endonuclease that cleaves DNA. The restriction enzymes used in cloning experiments bind to specific base sequences and then cleave the DNA backbone at two defined locations, one in each strand. Also referred to as a *restriction endonuclease.*

restriction point a point in the G_1 phase of the cell cycle at which a cell becomes committed to divide.

retroelement see *retrotransposon.*

retrotransposition a form of transposition in which the transposable element is transcribed into RNA. The RNA is then used as a template by reverse transcriptase to synthesize a DNA molecule that is integrated into a new region of the genome via integrase.

retrotransposon a type of transposable element that moves via an RNA intermediate.

reverse genetics an experimental strategy in which researchers first identify the wild-type gene using cloning methods. The next step is to make a mutant version of the wild-type gene, introduce it into an organism, and see how the mutant gene affects the phenotype of the organism.

reverse transcriptase PCR (RT-PCR) a modification of PCR in which the first round of replication involves the use of RNA and reverse transcriptase to make a complementary strand of DNA.

reverse transcriptase an enzyme that uses an RNA template to make a single-stranded or double-stranded DNA molecule.

reversion a mutation that changes a mutant allele back to a wild-type allele.

ρ-dependent termination transcriptional termination that requires the participation of rho (ρ) protein.

ρ-independent termination transcription termination that does not require the ρ (rho) protein; also known as intrinsic termination.

rho (ρ) protein a protein that is involved in transcriptional termination for certain bacterial genes.

ribonucleic acid (RNA) a nucleic acid that is composed of ribonucleotides. In living cells, RNA is synthesized via the transcription of DNA.

ribose the sugar found in RNA.

ribosomal RNA (rRNA) A type of RNA that is found in ribosomes and is necessary for translation to take place.

ribosome a large macromolecular structure that acts as the catalytic site for polypeptide synthesis. The ribosome allows the mRNAs and tRNAs to be positioned correctly as the polypeptide is made.

ribosome-binding site a short sequence in bacterial mRNA that binds to a ribosome and initiates translation. Also known as the *Shine-Dalgarno sequence*.

riboswitch an element for regulating transcription, translation, RNA stability, or splicing in which an RNA molecule can switch between two secondary conformations based on whether or not a small molecule, such as TPP, binds to the RNA.

ribozyme an RNA molecule with catalytic activity.

RNA editing the process in which a change is made in the base sequence of an RNA molecule that involves additions or deletions of particular nucleotides or conversion of one type of base to a different type.

RNA-induced silencing complex (RISC) the complex of double-stranded RNA and proteins that mediates RNA interference.

RNA interference (RNAi) the phenomenon in which double-stranded RNA targets complementary mRNAs within a cell for silencing or degradation.

RNA polymerase an enzyme that synthesizes a strand of RNA using a DNA strand as a template.

RNA primer a short strand of RNA, made by primase, that is used to elongate a strand of DNA during DNA replication.

RNA sequencing (RNA-Seq) a technology for determining the sequences of RNA molecules isolated from a sample of cells.

RNA splicing the process in which pieces of RNA are removed and the remaining pieces are covalently attached to each other.

RNA world a period on Earth in which RNA molecules, but not DNA or proteins, were found within protobionts; the RNA world preceded the existence of living cells.

RNase an enzyme that cuts the sugar-phosphate backbone in RNA.

Robertsonian translocation the structure produced when two telocentric chromosomes fuse at their short arms.

root meristem an actively dividing group of cells that gives rise to root structures.

S

S phase a phase of the eukaryotic cell cycle, between the G_1 and G_2 phases, during which a cell replicates its chromosomes.

scientific method one experimental approach for conducting science. It involves the formation of a hypothesis, which is followed by experimentation, so that scientists may reach verifiable conclusions about the world in which they live.

scintillation counting a technique that permits a researcher to count the number of radioactive emissions from a sample containing a population of radioisotopes.

search by content in bioinformatics, an approach in which a computer program predicts the location of a gene based on the observation that the nucleotide content of a particular region differs significantly (due to codon bias) from a random distribution.

search by signal in bioinformatics, an approach in which a computer program relies on known sequences such as promoters, start and stop codons, and splice sites to help predict whether or not a DNA sequence contains a protein-coding gene.

secondary structure a regular repeating pattern of molecular structure, such as the DNA double helix or the α helix and β sheet found in proteins.

sedimentation coefficient the rate at which a macromolecule or cell organelle sediments to the bottom of a centrifuge tube, normally expressed in Svedberg units (S): $1 S = 1 \times 10^{-13}$ second.

segmental duplication a duplication in which a small segment of a chromosome acquires more than one copy of the same gene.

segmentation gene in animals, a gene whose coded product is involved in the development of body segments.

segments morphologically discrete body subdivisions that develop in the embryo of a species such as *Drosophila*.

segregate to place two things in separate locations. For example, homologous chromosomes segregate into different gametes.

selectable marker a gene that provides a selectable phenotype in a cloning experiment. Many selectable markers are genes that confer antibiotic resistance.

selection coefficient (s) a measure of the degree to which a genotype is selected against, equal to 1 minus the fitness value: $s = 1 - w$.

selection limit the phenomenon in which several generations of artificial selection results in a plateau where artificial selection is no longer effective.

selective breeding programs and procedures designed to modify phenotypes in economically important species of plants and animals. Also called *artificial selection*.

selenocysteine a nonstandard amino acid that may be incorporated into polypeptides during translation.

self-fertilization fertilization that involves the union of male and female gametes derived from the same parent.

self-splicing RNA splicing mechanism that occurs without the aid of any proteins or other RNAs.

selfish DNA hypothesis the idea that transposable elements exist because they possess characteristics that allow them to multiply within the chromosomal DNA of living host cells even though they do not provide any selective advantage.

semiconservative model the correct model for DNA replication in which the newly made double-stranded DNA contains one parental strand and one daughter strand.

semilethal alleles lethal alleles that kill some individuals but not all.

semisterility condition in which an individual has a lowered fertility.

senescent describes a cell that is no longer capable of dividing.

sense codon a codon that codes a specific amino acid.

sequence by synthesis (SBS) a method of DNA sequencing in which the base sequence is determined in real time as a DNA strand is being made.

sequence complexity the number of times a particular base sequence appears throughout the genome of a given species.

sequence element in genetics, a specialized sequence with a particular meaning or function. Also referred to as a *motif*.

sequence recognition in bioinformatics, the ability of a computer program to recognize particular sequences.

sequence-tagged site (STS) a genetic marker that is produced when a pair of PCR primers copies a single stretch of DNA within a set of chromosomes; an STS can include a microsatellite or a SNP.

sequencing ladder a series of bands on a gel that can be followed in order (e.g., from the bottom of the gel to the top of the gel) to determine the base sequence of a strand of DNA.

sex with regard to humans, refers to physical characteristics that determine whether an individual is categorized as female, male, or intersex. These characteristics include external and internal body parts associated with reproduction, the type of gamete an individual has the potential to produce, and chromosome composition.

sex chromosomes a pair of chromosomes (e.g., X and Y in mammals) that differ between males and females and determine sex in a species.

sex determination the process that governs the development of male and female individuals.

sex-influenced inheritance an inheritance pattern in which an allele is dominant in one sex but recessive in the opposite sex.

sex-limited inheritance an inheritance pattern in which a trait is found in only one of the two sexes. An example of such a trait is beard development in adult human males.

sex-linked gene a gene that is located on only one of the sex chromosomes.

sex pilus (pl. pili) a structure on the surface of bacterial cells that acts as an attachment site to promote the binding of bacteria to each other.

sexual dimorphism phenomenon in which the males and females of a species are morphologically distinct.

sexual orientation an enduring pattern of romantic, emotional, and/or sexual attraction to persons of the opposite sex or gender, the same sex or gender, more than one sex and/or gender, or neither sex nor any gender.

sexual reproduction the process whereby parents make gametes (e.g., sperm and egg) that fuse with each other in the process of fertilization to begin the development of a new organism.

shared derived character a characteristic shared by a group of organisms but not by a distant common ancestor. Also referred to as a *synapomorphy*.

Shine-Dalgarno sequence see *ribosome-binding site*.

shoot meristem an actively dividing group of cells that gives rise to shoot structures.

short-read sequencing (SRS) methods of DNA sequencing that produce base sequences up to a few hundred bases in length.

shotgun sequencing a genome-sequencing strategy that involves breaking the genome into a collection of many DNA fragments that are sequenced individually.

shuttle vector a cloning vector that can propagate in two or more different species, such as *E. coli* and yeast.

side chain in an amino acid, the chemical structure attached to the carbon atom (i.e., the α carbon) that is located between the amino and carboxyl groups.

sigma (σ) factor a transcription factor that recognizes bacterial promoter sequences and facilitates the binding of RNA polymerase to the promoter.

signal recognition particle (SRP) an RNA-protein complex that targets proteins to the plasma membrane in bacteria or archaea or to the endoplasmic reticulum membrane in eukaryotes.

silent mutation a mutation that does not alter the amino acid sequence of the polypeptide coded by a gene even though the base sequence has changed.

simple Mendelian inheritance an inheritance pattern involving a simple, dominant/recessive relationship that produces offspring in which observed ratios of traits clearly obey Mendel's laws.

simple translocation rearrangement in which one piece of a chromosome becomes attached to a different chromosome.

simple transposition a cut-and-paste mechanism for transposition in which a transposable element is removed from one site and then inserted into another.

simple transposon a small transposable element that carries one or more genes that are not required for transposition.

SINEs in the genomes of mammals, short interspersed elements that are less than 500 bp in length.

single-factor cross a cross in which an experimenter is observing the outcome for only a single character.

single-nucleotide polymorphism (SNP) a genetic polymorphism within a population in which two alleles of the gene differ by a single nucleotide.

single-strand binding proteins proteins that bind to both of the single strands of DNA during DNA replication and prevent them from re-forming a double helix.

sister chromatid exchange (SCE) the phenomenon in which crossing over occurs between sister chromatids, which thereby exchange identical genetic material.

sister chromatids pairs of replicated chromosomes that are attached to each other at the centromere. Sister chromatids are genetically identical.

small interfering RNAs (siRNAs) ncRNAs that originate from sources that are exogenous, which means they are not normally made by cells. They silence mRNAs via RNA interference.

small nucleolar ribonucleoprotein (snoRNP) a complex between a snoRNA and several proteins.

small nucleolar RNAs (snoRNAs) ncRNAs that are found in high amounts in the nucleolus and promote the covalent modification of rRNAs.

small regulatory RNA a non-coding RNA that is shorter than 200 nucleotides.

SMC protein any of a category of proteins that use energy from ATP to catalyze the formation of loop domains; SMC stands for structural maintenance of chromosomes.

snRNP a subunit of a spliceosome that consists of small nuclear RNA and a set of proteins.

somatic cell any cell of the body except for gametes and germ-line cells that give rise to gametes.

somatic mutation a mutation in a somatic cell.

speciation an evolutionary process by which populations evolve to become distinct species.

species a group of organisms that maintains a distinctive set of attributes in nature.

species concept a way to define what a species is and/or provide an approach for distinguishing one species from another.

spectrophotometer a device used by researchers to determine how much radiation at various wavelengths a sample absorbs.

sperm cell a male gamete. Sperm are small and usually travel relatively far distances to reach the female gamete.

spermatogenesis the production of sperm cells.

spindle apparatus see *mitotic spindle apparatus*.

spindle pole during cell division in eukaryotes, one of two sites in the cell where microtubules originate.

spliceosome a multisubunit complex that plays a key role in the splicing of eukaryotic pre-mRNAs.

splicing factor a protein that regulates the choice of splice sites during the process of alternative splicing of pre-mRNAs.

spontaneous mutation a change in DNA structure that results from natural biological or chemical processes.

spores haploid cells that are produced by certain species, such as fungi (i.e., yeast and molds).

sporophyte the diploid generation of plants.

SR protein a type of splicing factor.

stabilizing selection natural selection that favors individuals with an intermediate phenotype.

standard deviation a statistic that is computed by taking the square root of the variance.

start codon a three-base sequence in mRNA that initiates translation. It is usually 5′-AUG-3′ and codes methionine.

stem cell a cell that has the capacity to divide and to differentiate into one or more specific cell types.

steroid receptor a category of regulatory transcription factor that responds to a steroid hormone, which binds directly to it. An example is the glucocorticoid receptor.

stigma the structure in flowering plants on which the pollen grain lands, stimulating the growth of the pollen tube that enables sperm cells to reach the egg cells.

stop codon a three-base sequence in mRNA that signals the end of translation of a polypeptide. The three stop codons are 5′–UAA–3′, 5′–UAG–3′, and 5′–UGA–3′.

strain within a given species, a group that displays one or more genetic differences compared to another group.

strand in DNA or RNA, a long linear polymer formed of nucleotides covalently linked together.

subcloning the procedure of making smaller DNA clones from a larger one.

submetacentric describes a chromosome in which the centromere is slightly off center.

subspecies within a species, two or more geographically restricted populations that exhibit some distinct characteristics that differ from other members of the species, but not enough to warrant their placement into different species.

subunit a component of a larger complex. In a protein, each subunit is a single polypeptide.

supergroups monophyletic groups into which evolutionary biologists have recently subdivided eukaryotic species.

supernatant following centrifugation, the fluid that is found above the pellet.

suppressor a mutation at a second site that suppresses the phenotypic effects of another mutation. Also referred to as a suppressor mutation.

suppressor mutation see *suppressor*.

sympatric speciation a form of speciation that occurs when members of a species diverge while occupying the same habitat within the same geographical area.

synapomorphy see *shared derived character*.

synapsis the event in which homologous chromosomes recognize each other and then align themselves along their entire lengths.

synonymous codons two or more different codons that specify the same amino acid.

synteny the situation in which two or more genes are located on the same chromosome.

synteny groups groups of linked genes found in chromosomes of two or more species.

T

T DNA (1) during conjugation, the strand of F-factor DNA that is transferred to a recipient cell; (2) a segment of DNA found within a Ti plasmid that is transferred from a bacterium to an infected plant cell.

T-DNA vectors vectors that carry T DNA and are used to introduce cloned genes into plant cells.

tandem array a region in DNA in which a very short nucleotide sequence is repeated many times in a row. Also known as a *tandem repeat*.

tandem mass spectrometry the sequential use of two mass spectrometers, a procedure that can be used to determine the sequence of amino acids in a polypeptide.

Taq **polymerase** a thermostable form of DNA polymerase used in PCR experiments.

target-site duplications see *direct repeats (DRs)*.

target-site primed reverse transcription (TPRT) a mechanism of non-LTR retrotransposition that involves reverse transcription at the target site where the retrotransposon is inserted.

targeted drug therapy drug therapy in which a drug targets a specific protein that has an abnormal structure and/or is overactive in cancer cells.

TATA box a sequence found within eukaryotic core promoters that determines the starting site for transcription. The TATA box is recognized by a TATA-binding protein, which is a component of TFIID.

tautomeric shift a temporary change in chemical structure, such as an alternation between the keto and enol forms of the bases that are found in DNA.

tautomers chemically similar forms of certain small molecules, such as bases, which can spontaneously interconvert.

telocentric describes a chromosome with its centromere at one end.

telomerase the enzyme that recognizes telomeric sequences at the ends of eukaryotic chromosomes and synthesizes additional repeats of those sequences.

telomerase reverse transcriptase (TERT) the enzyme subunit within telomerase that uses RNA as a template to make DNA.

telomerase RNA component (TERC) the RNA component of telomerase.

telomeres specialized repeated sequences of DNA found at the ends of eukaryotic chromosomes.

telophase the fifth stage of mitosis. The chromosomes have reached their respective poles and decondense.

temperate phage a bacteriophage that can spend some of its time in a lysogenic cycle.

temperature-sensitive (ts) lethal allele an allele that is lethal only in a certain environmental temperature range.

temperature-sensitive (ts) mutant a mutant that has a normal phenotype at a permissive temperature but a different phenotype, such as failure to grow, at a nonpermissive temperature.

temperature-sensitive allele an allele for which the resulting phenotype depends on the environmental temperature.

template DNA a sample of DNA, such as chromosomal DNA, that is used in a PCR experiment.

template strand a strand of DNA that is used to synthesize a complementary strand of DNA or RNA.

tension zone a hybrid zone in which the hybrids are selected against and the two populations on either side of the zone are adapted to different environments.

terminal deletion deletion in which a segment is lost from the end of a linear chromosome.

termination (1) in transcription, the release of the newly made RNA transcript and RNA polymerase from the DNA; (2) in translation, the release of the polypeptide and the last tRNA and the disassembly of the ribosomal subunits and mRNA.

termination codon see *stop codon.*

termination sequences (ter sequences) in *E. coli*, a pair of sequences in the chromosome that bind a protein known as the termination utilization substance (Tus), which stops the movement of the replication forks.

terminator a sequence within a gene that signals the end of transcription.

tertiary structure the three-dimensional structure of a macromolecule, such as the tertiary structure of a polypeptide.

testcross (1) an experimental cross between a recessive individual and an individual whose genotype the experimenter wishes to determine; (2) an experimental cross used for mapping the distance between genes in which an individual that is heterozygous for two or more genes is crossed to an individual that is homozygous recessive for those same genes.

tetrad (1) the structure formed by the association of four sister chromatids during meiosis; (2) a group of four fungal spores contained within an ascus.

tetraploid describes an organism or cell with four sets of chromosomes (i.e., 4*n*).

tetratype (T) an ascus that has two spores with the parental combinations of alleles and two spores with nonparental combinations.

thermocycler a device that automates the timing of temperature changes in each cycle of a PCR experiment.

third-generation sequencing technologies newer methods of DNA sequencing that can provide a DNA sequence from a single DNA molecule.

three parent babies a type of reproductive technology in which a baby is produced from an egg that carries nuclear DNA from a female with heteroplasmy, nuclear DNA from a male partner, and mitochrondrial DNA from a donor female.

three-factor cross a cross in which the experimenter follows the inheritance of three different characters.

threshold traits traits that are inherited through the contributions of many genes.

thymine (T) a pyrimidine base found in DNA. It base-pairs with adenine in DNA.

thymine dimer two adjacent thymine bases in a DNA strand that have become covalently linked.

tissue-specific gene a gene that is highly regulated and is expressed only in a particular cell type.

topoisomerase I an enzyme that alters the degree of supercoiling in DNA by relaxing negative supercoils.

topoisomerase II see *DNA gyrase.*

topoisomers DNA conformations that differ with regard to supercoiling.

topologically associating domains (TADs) a chromatin region in which segments of DNA within each TAD are more likely to interact with each other than they are with segments in other neighboring TADs.

totipotent describes a stem cell that has the genetic potential to produce an entire individual. A somatic plant cell or a fertilized egg is totipotent.

traits characteristics of an organism; also specific properties of a character, such as tall and short heights in pea plants.

transcription the process of synthesizing RNA from a DNA template.

transcription factors a category of proteins that influence the ability of RNA polymerase to transcribe DNA into RNA.

transcriptional start site the site in a gene where transcription begins.

transcriptome the set of all RNA molecules, including mRNAs and non-coding RNAs, that are transcribed in one cell or in a population of cells.

transduction a form of genetic transfer between bacterial cells in which a virus (bacteriophage) transfers bacterial DNA from one bacterium to another.

***trans*-acting factor** a regulatory protein that binds to a regulatory element in DNA and exerts a *trans*-effect.

***trans*-effect** an effect on gene expression that occurs even though two DNA segments are not physically adjacent to each other. *Trans*-effects are mediated by diffusible regulatory proteins.

***trans*-epigenetic mechanism** an epigenetic mechanism that affects all of the genes of a given type within a cell. It may be caused by diffusible transcription factors that are involved in a feedback loop.

transfer RNA (tRNA) a type of RNA used in translation that carries an amino acid. The anticodon in tRNA is complementary to a codon in the mRNA.

transformation (1) when a bacterial cell takes up a plasmid vector or a segment of chromosomal DNA from the environment; (2) when a normal cell is converted into a malignant cell.

transgene a gene from one species that is introduced into another species.

transgenerational epigenetic inheritance transmission of an epigenetic change from parent to offspring.

transgenic organism an organism that has DNA from another species incorporated into its genome via recombinant DNA techniques.

transition a point mutation involving a change of a pyrimidine to another pyrimidine (e.g., C to T) or a purine to another purine (e.g., A to G).

translation the process in which the sequence of codons within mRNA provides the information to synthesize the sequence of amino acids that constitutes a polypeptide.

translational regulatory protein a protein that regulates translation.

translational repressor a protein that binds to an mRNA and inhibits its ability to be translated.

translesion synthesis (TLS) the synthesis of DNA over a template strand that harbors some type of DNA damage. This occurs via translesion-replicating polymerases.

translesion-replicating polymerase a type of DNA polymerase that can replicate over a DNA region that contains an abnormal structure (i.e., a lesion).

translocation (1) rearrangement in which one segment of a chromosome breaks off and becomes attached to a different chromosome or a different part of the same chromosome; (2) event that occurs when a ribosome moves from one codon in an mRNA to the next codon.

translocation cross the structure that is formed when chromosomes that have undergone a reciprocal translocation attempt to synapse during meiosis. This structure contains two normal (nontranslocated) chromosomes and two translocated chromosomes. A total of eight chromatids are found within the cross.

transposable element (TE) a short segment of DNA that can be inserted in multiple locations within chromosomal DNA.

transposase the enzyme that catalyzes the movement of transposable elements that move as DNA elements.

transposition a process in which a short segment of DNA called a transposable element is inserted into a new location in the genome.

transposon a type of transposable element that moves via simple transposition catalyzed by the enzyme transposase.

transversion a point mutation in which a purine is interchanged with a pyrimidine, or vice versa.

trimethylation the attachment of three methyl groups to a single amino acid, such as lysine. It is a function of specific proteins in TrxG and PcG complexes.

trinucleotide repeat expansion (TNRE) a type of mutation that involves an increase in the number of tandemly repeated trinucleotide sequences.

triploid describes an organism or cell with three sets of chromosomes.

trisomic describes a diploid cell or organism with one extra chromosome (i.e., 2*n* + 1).

trithorax group (TrxG) a set of protein complexes that are key regulators of epigenetic changes that are programmed during development. They cause gene activation.

tRNA see *transfer RNA.*

trp repressor a protein that binds to the operator site of the *trp* operon and inhibits transcription.

true-breeding strain a strain that continues to produce the same trait through several generations of self-fertilization.

tumor-suppressor gene a gene that functions to inhibit cancerous growth.

two-dimensional (2D) gel electrophoresis a technique for separating proteins that involves isoelectric focusing followed by SDS-gel electrophoresis.

two-factor cross a cross in which an experimenter follows the inheritance of two different characters.

U

unbalanced translocation a translocation in which a significant portion of genetic material is duplicated and/or deleted.

unipotent decribes a stem cell that can differentiate into only a single type of cell.

up promoter mutation a mutation in a promoter that increases the rate of transcription.

up regulation genetic regulation that leads to an increase in gene expression.

uracil (U) a pyrimidine base found in RNA.

V

vaccination the practice of administering a vaccine.

vaccine a biological preparation that provides active acquired immunity to a particular infectious disease or to a disease such as cancer.

variance a measure of the variation around the mean within a population.

variants specific properties of a character, such as tall and short heights in pea plants.

vector a small DNA molecule that is used as a carrier of a DNA segment in a cloning experiment.

vernalization the phenomenon that certain species of plants must be exposed to the cold before they can flower.

vertical evolution the evolution of species from preexisting species by the accumulation of gene mutations and by changes in chromosome structure and number. Vertical evolution involves genetic changes in a series of ancestors that form a lineage.

viral envelope a lipid bilayer that is derived from the plasma membrane of the host cell and embedded with viral spike glycoproteins. The envelope encloses the capsid.

viral genome the genetic material of a virus.

viral reproductive cycle the series of steps that results in the production of new viruses.

virulent phage a phage that can only follow a lytic cycle.

virus a nonliving, infectious particle that contains nucleic acid (DNA or RNA) as its genetic material, surrounded by a capsid of proteins. Some viruses also have an envelope consisting of a membrane embedded with spike proteins.

W

Western blotting a technique used to detect a specific protein within a mixture of proteins.

wild type a relatively prevalent genotype in a natural population.

wild-type allele an allele that is fairly prevalent in a natural population, generally found in more than 1% of the population. For polymorphic genes, there is more than one wild-type allele.

wobble rules rules that govern the binding specificity between the third base in a codon and the first base in an anticodon.

writer domain a domain in a protein that functions in the posttranslational modification of proteins in chromatin.

X

X-chromosome inactivation (XCI) a process in which mammals equalize the expression of X-linked genes by randomly turning off one X chromosome in the somatic cells of females.

X-inactivation center (Xic) a site on the X chromosome that appears to play a critical role in X-chromosome inactivation.

X-linked alleles see *X-linked genes*.

X-linked genes genes (or alleles of genes) that are physically located on the X chromosome.

X-linked inheritance an inheritance pattern in certain species that involves genes that are located only on the X chromosome.

X-linked recessive an inheritance pattern in which a gene is found on the X chromosome and the disease-causing allele is recessive relative to a corresponding dominant allele.

xenotransplantation the transplantation of cells, tissues, or organs from one animal species to another.

Y

Y-linked genes (alleles) genes (or alleles of genes) that are located only on the Y chromosome.

yeast artificial chromosome (YAC) a cloning vector propagated in yeast that can reliably contain very large inserted fragments of DNA.

Z

Z DNA a left-handed DNA double helix that is found occasionally in living cells.

zigzag model a model for nucleosome interaction in which nucleosomes zigzag back and forth with a straight linker region.

zygotic gene a gene that is expressed after fertilization.

INDEX

T